HANDBOOK OF DISCRETE AND COMBINATORIAL MATHEMATICS

KENNETH H. ROSEN
AT&T Laboratories
Editor-in-Chief

JOHN G. MICHAELS
SUNY Brockport
Project Editor

JONATHAN L. GROSS
Columbia University
Associate Editor

JERROLD W. GROSSMAN
Oakland University
Associate Editor

DOUGLAS R. SHIER
Clemson University
Associate Editor

CRC Press
Boca Raton   London   New York   Washington, D.C.

FIRST INDIAN REPRINT, 2010

Library of Congress Cataloging-in-Publication Data

Handbook of discrete and combinatorial mathematics / Kenneth H. Rosen, editor in chief, John G. Michaels, project editor...[et al.].
    p. cm.
Includes bibliographical references and index.
ISBN 0-8493-0149-1 (alk. paper)
    1. Combinatorial analysis—Handbooks, manuals, etc. 2. Computer science—Mathematics—Handbooks, manuals, etc. I. Rosen, Kenneth H. II. Michaels, John G.

QA164.H36 1999
511.'6—dc21   99-04378

This book contains information obtained from authentic and highly regarded sources. Reprinted material is quoted with permission, and sources are indicated. A wide variety of references are listed. Reasonable efforts have been made to publish reliable data and information, but the author and the publisher cannot assume responsibility for the validity of all materials or for the consequences of their use.

Neither this book nor any part may be reproduced or transmitted in any form or by any means, electronic or mechanical, including photocopying, microfilming, and recording, or by any information storage or retrieval system, without prior permission in writing from the publisher.

All rights reserved. Authorization to photocopy items for internal or personal use, or the personal or internal use of specific clients, may be granted by CRC Press LLC, provided that $.50 per page photocopied is paid directly to Copyright clearance Center, 222 Rosewood Drive, Danvers, MA 01923 USA. The fee code for users of the Transactional Reporting Service is ISBN 0-8493-0149-1/00/$0.00+$.50. The fee is subject to change without notice. For organizations that have been granted a photocopy license by the CCC, a separate system of payment has been arranged.

The consent of CRC Press LLC does not extend to copying for general distribution, for promotion, for creating new works, or for resale. Specific permission must be obtained in writing from CRC Press LLC for such copying.

Direct all inquiries to CRC Press LLC, 2000 N.W. Corporate Blvd., Boca Raton, Florida 33431.

Trademark Notice: Product or corporate names may be trademarks or registered trademarks, and are used only for identification and explanation, without intent to infringe.

Visit the CRC Press Web site at www.crcpress.com

© 2000 by CRC Press LLC

No claim to original U.S. Government works
International Standard Book Number 0-8493-0149-1
Library of Congress Card Number 99-04378

Printed and bound in India by Replika Press Pvt. Ltd.

FOR SALE IN SOUTH ASIA ONLY

CONTENTS

1. FOUNDATIONS 1
- 1.1 Propositional and Predicate Logic — *Jerrold W. Grossman* 12
- 1.2 Set Theory — *Jerrold W. Grossman* 21
- 1.3 Functions — *Jerrold W. Grossman* 31
- 1.4 Relations — *John G. Michaels* 40
- 1.5 Proof Techniques — *Susanna S. Epp* 50
- 1.6 Axiomatic Program Verification — *David Riley* 61
- 1.7 Logic-Based Computer Programming Paradigms — *Mukesh Dalal* 67

2. COUNTING METHODS 81
- 2.1 Summary of Counting Problems — *John G. Michaels* 84
- 2.2 Basic Counting Techniques — *Jay Yellen* 90
- 2.3 Permutations and Combinations — *Edward W. Packel* 96
- 2.4 Inclusion/Exclusion — *Robert G. Rieper* 107
- 2.5 Partitions — *George E. Andrews* 113
- 2.6 Burnside/Pólya Counting Formula — *Alan C. Tucker* 120
- 2.7 Möbius Inversion Counting — *Edward A. Bender* 127
- 2.8 Young Tableaux — *Bruce E. Sagan* 129

3. SEQUENCES 135
- 3.1 Special Sequences — *Thomas A. Dowling and Douglas R. Shier* 138
- 3.2 Generating Functions — *Ralph P. Grimaldi* 171
- 3.3 Recurrence Relations — *Ralph P. Grimaldi* 178
- 3.4 Finite Differences — *Jay Yellen* 189
- 3.5 Finite Sums and Summation — *Victor S. Miller* 195
- 3.6 Asymptotics of Sequences — *Edward A. Bender* 201
- 3.7 Mechanical Summation Procedures — *Kenneth H. Rosen* 204

4. NUMBER THEORY 213
- 4.1 Basic Concepts — *Kenneth H. Rosen* 219
- 4.2 Greatest Common Divisors — *Kenneth H. Rosen* 226
- 4.3 Congruences — *Kenneth H. Rosen* 231
- 4.4 Prime Numbers — *Jon F. Grantham and Carl Pomerance* 236
- 4.5 Factorization — *Jon F. Grantham and Carl Pomerance* 255
- 4.6 Arithmetic Functions — *Kenneth H. Rosen* 259
- 4.7 Primitive Roots and Quadratic Residues — *Kenneth H. Rosen* 268
- 4.8 Diophantine Equations — *Bart E. Goddard* 281
- 4.9 Diophantine Approximation — *Jeff Shalit* 289
- 4.10 Quadratic Fields — *Kenneth H. Rosen* 295

5. ALGEBRAIC STRUCTURES — John G. Michaels 299
  5.1 Algebraic Models 305
  5.2 Groups 307
  5.3 Permutation Groups 319
  5.4 Rings 323
  5.5 Polynomial Rings 329
  5.6 Fields 331
  5.7 Lattices 341
  5.8 Boolean Algebras 344

6. LINEAR ALGEBRA 355
  6.1 Vector Spaces — *Joel V. Brawley* 361
  6.2 Linear Transformations — *Joel V. Brawley* 371
  6.3 Matrix Algebra — *Peter R. Turner* 377
  6.4 Linear Systems — *Barry Peyton and Esmond Ng* 392
  6.5 Eigenanalysis — *R. B. Bapat* 405
  6.6 Combinatorial Matrix Theory — *R. B. Bapat* 417

7. DISCRETE PROBABILITY 427
  7.1 Fundamental Concepts — *Joseph R. Barr* 432
  7.2 Independence and Dependence — *Joseph R. Barr* 435
  7.3 Random Variables — *Joseph R. Barr* 441
  7.4 Discrete Probability Computations — *Peter R. Turner* 448
  7.5 Random Walks — *Patrick Jaillet* 452
  7.6 System Reliability — *Douglas R. Shier* 459
  7.7 Discrete-Time Markov Chains — *Vidyadhar G. Kulkarni* 468
  7.8 Queueing Theory — *Vidyadhar G. Kulkarni* 477
  7.9 Simulation — *Lawrence M. Leemis* 484

8. GRAPH THEORY 495
  8.1 Introduction to Graphs — *Lowell W. Beineke* 509
  8.2 Graph Models — *Jonathan L. Gross* 525
  8.3 Directed Graphs — *Stephen B. Maurer* 526
  8.4 Distance, Connectivity, Traversability — *Edward R. Scheinerman* 539
  8.5 Graph Invariants and Isomorphism Types — *Bennet Manvel* 549
  8.6 Graph and Map Coloring — *Arthur T. White* 557
  8.7 Planar Drawings — *Jonathan L. Gross* 567
  8.8 Topological Graph Theory — *Jonathan L. Gross* 574
  8.9 Enumerating Graphs — *Paul K. Stockmeyer* 580
  8.10 Algebraic Graph Theory — *Michael Doob* 586
  8.11 Analytic Graph Theory — *Stefan A. Burr* 590
  8.12 Hypergraphs — *Andreas Gyarfas* 595

9. TREES 603
  9.1 Characterizations and Types of Trees — *Lisa Carbone* 607
  9.2 Spanning Trees — *Uri Peled* 616
  9.3 Enumerating Trees — *Paul Stockmeyer* 622

10. NETWORKS AND FLOWS 629
10.1 Minimum Spanning Trees — *J. B. Orlin and Ravindra K. Ahuja* 633
10.2 Matchings — *Douglas R. Shier* 641
10.3 Shortest Paths — *J. B. Orlin and Ravindra K. Ahuja* 652
10.4 Maximum Flows — *J. B. Orlin and Ravindra K. Ahuja* 663
10.5 Minimum Cost Flows — *J. B. Orlin and Ravindra K. Ahuja* 673
10.6 Communication Networks — *David Simchi-Levi and Sunil Chopra* 683
10.7 Difficult Routing and Assignment Problems — *Bruce L. Golden and Bharat K. Kaku* .. 692
10.8 Network Representations and Data Structures — *Douglas R. Shier* 706

11. PARTIALLY ORDERED SETS 717
11.1 Basic Poset Concepts — *Graham Brightwell and Douglas B. West* 724
11.2 Poset Properties — *Graham Brightwell and Douglas B. West* 738

12. COMBINATORIAL DESIGNS 753
12.1 Block Designs — *Charles J. Colbourn and Jeffrey H. Dinitz* 759
12.2 Symmetric Designs & Finite Geometries — *Charles J. Colbourn and Jeffrey H. Dinitz* . 770
12.3 Latin Squares and Orthogonal Arrays — *Charles J. Colbourn and Jeffrey H. Dinitz* ... 778
12.4 Matroids — *James G. Oxley* 786

13. DISCRETE AND COMPUTATIONAL GEOMETRY 797
13.1 Arrangements of Geometric Objects — *Ileana Streinu* 805
13.2 Space Filling — *Karoly Bezdek* 824
13.3 Combinatorial Geometry — *János Pach* 830
13.4 Polyhedra — *Tamal K. Dey* 839
13.5 Algorithms and Complexity in Computational Geometry — *Jianer Chen* ... 844
13.6 Geometric Data Structures and Searching — *Dina Kravets* 853
13.7 Computational Techniques — *Nancy M. Amato* 861
13.8 Applications of Geometry — *W. Randolph Franklin* 867

14. CODING THEORY AND CRYPTOLOGY — *Alfred J. Menezes and Paul C. van Oorschot* 889
14.1 Communication Systems and Information Theory 896
14.2 Basics of Coding Theory 900
14.3 Linear Codes 903
14.4 Bounds for Codes 915
14.5 Nonlinear Codes 917
14.6 Convolutional Codes 918
14.7 Basics of Cryptography 923
14.8 Symmetric-Key Systems 927
14.9 Public-Key Systems 935

15. DISCRETE OPTIMIZATION 955
15.1 Linear Programming — *Beth Novick* 959
15.2 Location Theory — *S. Louis Hakimi* 986
15.3 Packing and Covering — *Sunil Chopra and David Simchi-Levi* 996
15.4 Activity Nets — *S. E. Elmaghraby* 1006
15.5 Game Theory — *Michael Mesterton-Gibbons* 1016
15.6 Sperner's Lemma and Fixed Points — *Joseph R. Barr* 1027

16. THEORETICAL COMPUTER SCIENCE ... 1039
- 16.1 Computational Models — *Jonathan L. Gross* ... 1048
- 16.2 Computability — *William Gasarch* ... 1062
- 16.3 Languages and Grammars — *Aarto Salomaa* ... 1066
- 16.4 Algorithmic Complexity — *Thomas Cormen* ... 1077
- 16.5 Complexity Classes — *Lane Hemaspaandra* ... 1085
- 16.6 Randomized Algorithms — *Milena Mihail* ... 1091

17. INFORMATION STRUCTURES ... 1101
- 17.1 Abstract Datatypes — *Charles H. Goldberg* ... 1108
- 17.2 Concrete Data Structures — *Jonathan L. Gross* ... 1117
- 17.3 Sorting and Searching — *Jianer Chen* ... 1125
- 17.4 Hashing — *Viera Krnanova Proulx* ... 1139
- 17.5 Dynamic Graph Algorithms — *Joan Feigenbaum and Sampath Kannan* ... 1142

BIOGRAPHIES — *Victor J. Katz* ... 1153

INDEX ... 1173

PREFACE

The importance of discrete and combinatorial mathematics has increased dramatically within the last few years. The purpose of the *Handbook of Discrete and Combinatorial Mathematics* is to provide a comprehensive reference volume for computer scientists, engineers, mathematicians, and others, such as students, physical and social scientists, and reference librarians, who need information about discrete and combinatorial mathematics.

This book is the first resource that presents such information in a ready-reference form designed for use by all those who use aspects of this subject in their work or studies. The scope of this book includes the many areas generally considered to be parts of discrete mathematics, focusing on the information considered essential to its application in computer science and engineering. Some of the fundamental topic areas covered include:

logic and set theory	graph theory
enumeration	trees
integer sequences	network sequences
recurrence relations	combinatorial designs
generating functions	computational geometry
number theory	coding theory and cryptography
abstract algebra	discrete optimization
linear algebra	automata theory
discrete probability theory	data structures and algorithms.

Format

The material in the *Handbook* is presented so that key information can be located and used quickly and easily. Each chapter includes a glossary that provides succinct definitions of the most important terms from that chapter. Individual topics are covered in sections and subsections within chapters, each of which is organized into clearly identifiable parts: definitions, facts, and examples. The definitions included are carefully crafted to help readers quickly grasp new concepts. Important notation is also highlighted in the definitions. Lists of facts include:

- information about how material is used and why it is important
- historical information
- key theorems
- the latest results
- the status of open questions
- tables of numerical values, generally not easily computed
- summary tables
- key algorithms in an easily understood pseudocode
- information about algorithms, such as their complexity
- major applications
- pointers to additional resources, including websites and printed material.

Facts are presented concisely and are listed so that they can be easily found and understood. Extensive crossreferences linking parts of the handbook are also provided. Readers who want to study a topic further can consult the resources listed.

The material in the *Handbook* has been chosen for inclusion primarily because it is important and useful. Additional material has been added to ensure comprehensiveness so that readers encountering new terminology and concepts from discrete mathematics in their explorations will be able to get help from this book.

Examples are provided to illustrate some of the key definitions, facts, and algorithms. Some curious and entertaining facts and puzzles that some readers may find intriguing are also included.

Each chapter of the book includes a list of references divided into a list of printed resources and a list of relevant websites.

How This Book Was Developed

The organization and structure of the *Handbook* were developed by a team which included the chief editor, three associate editors, the project editor, and the editor from CRC Press. This team put together a proposed table of contents which was then analyzed by members of a group of advisory editors, each an expert in one or more aspects of discrete mathematics. These advisory editors suggested changes, including the coverage of additional important topics. Once the table of contents was fully developed, the individual sections of the book were prepared by a group of more than 70 contributors from industry and academia who understand how this material is used and why it is important. Contributors worked under the direction of the associate editors and chief editor, with these editors ensuring consistency of style and clarity and comprehensiveness in the presentation of material. Material was carefully reviewed by authors and our team of editors to ensure accuracy and consistency of style.

The CRC Press Series on Discrete Mathematics and Its Applications

This *Handbook* is designed to be a ready reference that covers many important distinct topics. People needing information in multiple areas of discrete and combinatorial mathematics need only have this one volume to obtain what they need or for pointers to where they can find out more information. Among the most valuable sources of additional information are the volumes in the CRC Press Series on Discrete Mathematics and Its Applications. This series includes both Handbooks, which are ready references, and advanced Textbooks/Monographs. More detailed and comprehensive coverage in particular topic areas can be found in these individual volumes:

Handbooks

- *The CRC Handbook of Combinatorial Designs*
- *Handbook of Discrete and Computational Geometry*
- *Handbook of Applied Cryptography*

Textbooks/Monographs

- *Graph Theory and its Applications*
- *Algebraic Number Theory*
- *Quadratics*

- *Design Theory*
- *Frames and Resolvable Designs: Uses, Constructions, and Existence*
- *Network Reliability: Experiments with a Symbolic Algebra Environment*
- *Fundamental Number Theory with Applications*
- *Cryptography: Theory and Practice*
- *Introduction to Information Theory and Data Compression*
- *Combinatorial Algorithms: Generation, Enumeration, and Search*

Feedback

To see updates and to provide feedback and errata reports, please consult the Web page for this book. This page can be accessed by first going to the CRC website at
   http://www.crcpress.com
and then following the links to the Web page for this book.

Acknowledgments

First and foremost, we would like to thank the original CRC editor of this project, Wayne Yuhasz, who commissioned this project. We hope we have done justice to his original vision of what this book could be. We would also like to thank Bob Stern, who has served as the editor of this project for his continued support and enthusiasm for this project. We would like to thank Nora Konopka for her assistance with many aspects in the development of this project. Thanks also go to Susan Fox, for her help with production of this book at CRC Press.

We would like to thank the many people who were involved with this project. First, we would like to thank the team of advisory editors who helped make this reference relevant, useful, unique, and up-to-date. We also wish to thank all the people at the various institutions where we work, including the management of AT&T Laboratories for their support of this project and for providing a stimulating and interesting atmosphere.

Project Editor John Michaels would like to thank his wife Lois and daughter Margaret for their support and encouragement in the development of the *Handbook*. Associate Editor Jonathan Gross would like to thank his wife Susan for her patient support, Associate Editor Jerrold Grossman would like to thank Suzanne Zeitman for her help with computer science materials and contacts, and Associate Editor Douglas Shier would like to thank his wife Joan for her support and understanding throughout the project.

ADVISORY EDITORIAL BOARD

Andrew Odlyzko — Chief Advisory Editor
AT&T Laboratories

Stephen F. Altschul
National Institutes of Health

Frank Harary
New Mexico State University

George E. Andrews
Pennsylvania State University

Alan Hoffman
IBM

Francis T. Boesch
Stevens Institute of Technology

Bernard Korte
Rheinische Friedrich-Wilhems-Univ.

Ernie Brickell
Certco

Jeffrey C. Lagarias
AT&T Laboratories

Fan R. K. Chung
Univ. of California at San Diego

Carl Pomerance
University of Georgia

Charles J. Colbourn
University of Vermont

Fred S. Roberts
Rutgers University

Stan Devitt
Waterloo Maple Software

Pierre Rosenstiehl
Centre d'Analyse et de Math. Soc.

Zvi Galil
Columbia University

Francis Sullivan
IDA

Keith Geddes
University of Waterloo

J. H. Van Lint
Eindhoven University of Technology

Ronald L. Graham
Univ. of California at San Diego

Scott Vanstone
University of Waterloo

Ralph P. Grimaldi
Rose-Hulman Inst. of Technology

Peter Winkler
Bell Laboratories

CONTRIBUTORS

Ravindra K. Ahuja
University of Florida

Nancy M. Amato
Texas A&M University

George E. Andrews
Pennsylvania State University

R. B. Bapat
Indian Statistical Institute

Joseph R. Barr
SPSS

Lowell W. Beineke
Purdue University — Fort Wayne

Edward A. Bender
University of California at San Diego

Karoly Bezdek
Cornell University

Joel V. Brawley
Clemson University

Graham Brightwell
London School of Economics

Stefan A. Burr
City College of New York

Lisa Carbone
Harvard University

Jianer Chen
Texas A&M University

Sunil Chopra
Northwestern University

Charles J. Colbourn
University of Vermont

Thomas Cormen
Dartmouth College

Mukesh Dalal
i2 Technologies

Tamal K. Dey
Indian Institute of Technology Kharagpur

Jeffrey H. Dinitz
University of Vermont

Michael Doob
University of Manitoba

Thomas A. Dowling
Ohio State University

S. E. Elmaghraby
North Carolina State University

Susanna S. Epp
DePaul University

Joan Feigenbaum
AT&T Laboratories

W. Randolph Franklin
Rensselaer Polytechnic Institute

William Gasarch
University of Maryland

Bart E. Goddard
Texas A&M University

Charles H. Goldberg
Trenton State College

Bruce L. Golden
University of Maryland

Jon F. Grantham
IDA

Ralph P. Grimaldi
Rose-Hulman Inst. of Technology

Jonathan L. Gross
Columbia University

Jerrold W. Grossman
  Oakland University

Andreas Gyarfas
  Hungarian Academy of Sciences

S. Louis Hakimi
  University of California at Davis

Lane Hemaspaandra
  University of Rochester

Patrick Jaillet
  University of Texas at Austin

Bharat K. Kaku
  American University

Sampath Kannan
  University of Pennsylvania

Victor J. Katz
  Univ. of the District of Columbia

Dina Kravets
  Sarnoff Corporation

Vidyadhar G. Kulkarni
  University of North Carolina

Lawrence M. Leemis
  The College of William and Mary

Bennet Manvel
  Colorado State University

Stephen B. Maurer
  Swarthmore College

Alfred J. Menezes
  University of Waterloo

Michael Mesterton-Gibbons
  Florida State University

John G. Michaels
  SUNY Brockport

Milena Mihail
  Georgia Institute of Technology

Victor S. Miller
  Center for Communications
  Research — IDA

Esmond Ng
  Lawrence Berkeley National Lab.

Beth Novick
  Clemson University

James B. Orlin
  Massachusetts Inst. of Technology

James G. Oxley
  Louisiana State University

János Pach
  City College CUNY, and
  Hungarian Academy of Sciences

Edward W. Packel
  Lake Forest College

Uri Peled
  University of Illinois at Chicago

Barry Peyton
  Oak Ridge National Laboratory

Carl Pomerance
  University of Georgia

Viera Krnanova Proulx
  Northeastern University.

Robert G. Rieper
  William Patterson University

David Riley
  University of Wisconsin

Kenneth H. Rosen
  AT&T Laboratories

Bruce E. Sagan
  Michigan State University

Aarto Salomaa
  University of Turku, Finland

Edward R. Scheinerman
  Johns Hopkins University

Jeff Shalit
  University of Waterloo

Douglas R. Shier
  Clemson University

David Simchi-Levi
  Northwestern University

Paul K. Stockmeyer
  The College of William and Mary

Ileana Streinu
  Smith College

Alan C. Tucker
  SUNY Stony Brook

Peter R. Turner
  United States Naval Academy

Paul C. van Oorschot
  Entrust Technologies

Douglas B. West
  University of Illinois at Champaign-Urbana

Arthur T. White
  Western Michigan University

Jay Yellen
  Florida Institute of Technology

1

FOUNDATIONS

1.1 Propositional and Predicate Logic — Jerrold W. Grossman
    1.1.1 Propositions and Logical Operations
    1.1.2 Equivalences, Identities, and Normal Forms
    1.1.3 Predicate Logic

1.2 Set Theory — Jerrold W. Grossman
    1.2.1 Sets
    1.2.2 Set Operations
    1.2.3 Infinite Sets
    1.2.4 Axioms for Set Theory

1.3 Functions — Jerrold W. Grossman
    1.3.1 Basic Terminology for Functions
    1.3.2 Computational Representation
    1.3.3 Asymptotic Behavior

1.4 Relations — John G. Michaels
    1.4.1 Binary Relations and Their Properties
    1.4.2 Equivalence Relations
    1.4.3 Partially Ordered Sets
    1.4.4 n-ary Relations

1.5 Proof Techniques — Susanna S. Epp
    1.5.1 Rules of Inference
    1.5.2 Proofs
    1.5.3 Disproofs
    1.5.4 Mathematical Induction
    1.5.5 Diagonalization Arguments

1.6 Axiomatic Program Verification — David Riley
    1.6.1 Assertions and Semantic Axioms
    1.6.2 NOP, Assignment, and Sequencing Axioms
    1.6.3 Axioms for Conditional Execution Constructs
    1.6.4 Axioms for Loop Constructs
    1.6.5 Axioms for Subprogram Constructs

1.7 Logic-based Computer Programming Paradigms — Mukesh Dalal
    1.7.1 Logic Programming
    1.7.2 Fuzzy Sets and Logic
    1.7.3 Production Systems
    1.7.4 Automated Reasoning

INTRODUCTION

This chapter covers material usually referred to as the foundations of mathematics, including logic, sets, and functions. In addition to covering these foundational areas, this chapter includes material that shows how these topics are applied to discrete mathematics, computer science, and electrical engineering. For example, this chapter covers methods of proof, program verification, and fuzzy reasoning.

GLOSSARY

action: a literal or a print command in a production system.

aleph-null: the cardinality, $\aleph_0$, of the set $\mathcal{N}$ of natural numbers.

AND: the logical operator for conjunction, also written $\wedge$.

antecedent: in a conditional proposition $p \to q$ ("if p then q") the proposition p ("if-clause") that precedes the arrow.

antichain: a subset of a poset in which no two elements are comparable.

antisymmetric: the property of a binary relation R that if aRb and bRa, then $a = b$.

argument form: a sequence of statement forms each called a *premise* of the argument followed by a statement form called a *conclusion* of the argument.

assertion (or **program assertion**): a program comment specifying some conditions on the values of the computational variables; these conditions are supposed to hold whenever program flow reaches the location of the assertion.

asymmetric: the property of a binary relation R that if aRb, then $b\not{R}a$.

asymptotic: A function f is asymptotic to a function g, written $f(x) \sim g(x)$, if $f(x) \neq 0$ for sufficiently large x and $\lim_{x \to \infty} \frac{g(x)}{f(x)} = 1$.

atom (or **atomic formula**): simplest formula of predicate logic.

atomic formula: See *atom*.

atomic proposition: a proposition that cannot be analyzed into smaller parts and logical operations.

automated reasoning: the process of proving theorems using a computer program that can draw conclusions that follow logically from a set of given facts.

axiom: a statement that is assumed to be true; a postulate.

axiom of choice: the assertion that given any nonempty collection $\mathcal{A}$ of pairwise disjoint sets, there is a set that consists of exactly one element from each of the sets in $\mathcal{A}$.

axiom (or **semantic axiom**): a rule for a programming language construct prescribing the change of values of computational variables when an instruction of that construct-type is executed.

basis step: a proof of the basis premise (first case) in a proof by mathematical induction.

big-oh notation: f is $O(g)$, written $f = O(g)$, if there are constants C and k such that $|f(x)| \leq C|g(x)|$ for all $x > k$.

bijection (or **bijective function**): a function that is one-to-one and onto.

bijective function: See *bijection*.

binary relation from a set A to a set B: any subset of $A \times B$.

binary relation on a set A: a binary relation from A to A; i.e., a subset of $A \times A$.

body of a clause $A_1, \ldots, A_n \leftarrow B_1, \ldots, B_m$ in a logic program: the literals $B_1, \ldots, B_m$ after $\leftarrow$.

cardinal number (or **cardinality**) of a set: for a finite set, the number of elements; for an infinite set, the order of infinity. The cardinal number of S is written $|S|$.

cardinality: See cardinal number.

Cartesian product (of sets A and B): the set $A \times B$ of ordered pairs (a, b) with $a \in A$ and $b \in B$ (more generally, the **iterated Cartesian product** $A_1 \times A_2 \times \cdots \times A_n$ is the set of ordered n-tuples $(a_1, a_2, \ldots, a_n)$, with $a_i \in A_i$ for each i).

ceiling (of x): the smallest integer that is greater than or equal to x, written $\lceil x \rceil$.

chain: a subset of a poset in which every pair of elements are comparable.

characteristic function (of a set S): the function from S to $\{0, 1\}$ whose value at x is 1 if $x \in S$ and 0 if $x \notin S$.

clause (in a logic program): closed formula of the form $\forall x_1 \ldots \forall x_s (A_1 \vee \cdots \vee A_n \leftarrow B_1 \wedge \cdots \wedge B_m)$.

closed formula: for a function value $f(x)$, an algebraic expression in x.

closure (of a relation R with respect to a property $\mathcal{P}$): the relation S, if it exists, that has property $\mathcal{P}$ and contains R, such that S is a subset of every relation that has property $\mathcal{P}$ and contains R.

codomain (of a function): the set in which the function values occur.

comparable: Two elements in a poset are comparable if they are related by the partial order relation.

complement (of a relation): given a relation R, the relation $\overline{R}$ where $a\overline{R}b$ if and only if $a\cancel{R}b$.

complement (of a set): given a set A in a "universal" domain U, the set $\overline{A}$ of objects in U that are not in A.

complement operator: a function $[0, 1] \to [0, 1]$ used for complementing fuzzy sets.

complete: property of a set of axioms that it is possible to prove all true statements.

complex number: a number of the form $a + bi$, where a and b are real numbers, and $i^2 = -1$; the set of all complex numbers is denoted $\mathcal{C}$.

composite key: given an n-ary relation R on $A_1 \times A_2 \times \cdots \times A_n$, a product of domains $A_{i_1} \times A_{i_2} \times \cdots \times A_{i_m}$ such that for each m-tuple $(a_{i_1}, a_{i_2}, \ldots, a_{i_m}) \in A_{i_1} \times A_{i_2} \times \cdots \times A_{i_m}$, there is at most one n-tuple in R that matches $(a_{i_1}, a_{i_2}, \ldots, a_{i_m})$ in coordinates $i_1, i_2, \ldots, i_m$.

composition (of relations): for R a relation from A to B and S a relation from B to C, the relation $S \circ R$ from A to C such that $a(S \circ R)c$ if and only if there exists $b \in B$ such that aRb and bSc.

composition (of functions): the function $f \circ g$ whose value at x is $f(g(x))$.

compound proposition: a proposition built up from atomic propositions and logical connectives.

computer-assisted proof: a proof that relies on checking the validity of a large number of cases using a special purpose computer program.

conclusion (of an argument form): the last statement of an argument form.

conclusion (of a proof): the last proposition of a proof; the objective of the proof is demonstrating that the conclusion follows from the premises.

condition: the disjunction $A_1 \vee \cdots \vee A_n$ of atomic formulas.

conditional statement: the compound proposition $p \rightarrow q$ ("if p then q") that is true except when p is true and q is false.

conjunction: the compound proposition $p \wedge q$ ("p and q") that is true only when p and q are both true.

conjunctive normal form: for a proposition in the variables $p_1, p_2, \ldots, p_n$, an equivalent proposition that is the conjunction of disjunctions, with each disjunction of the form $x_{k_1} \vee x_{k_2} \vee \cdots \vee x_{k_m}$, where x_{k_j} is either p_{k_j} or $\neg p_{k_j}$.

consequent: in a conditional proposition $p \rightarrow q$ ("if p then q") the proposition q ("then-clause") that follows the arrow.

consistent: property of a set of axioms that no contradiction can be deduced from the axioms.

construct (or **program construct**): the general form of a programming instruction such as an assignment, a conditional, or a while-loop.

continuum hypothesis: the assertion that the cardinal number of the real numbers is the smallest cardinal number greater than the cardinal number of the natural numbers.

contradiction: a self-contradictory proposition, one that is always false.

contradiction (in an indirect proof): the negation of a premise.

contrapositive (of the conditional proposition $p \rightarrow q$): the conditional proposition $\neg q \rightarrow \neg p$.

converse (of the conditional proposition $p \rightarrow q$): the conditional proposition $q \rightarrow p$.

converse relation: another name for the inverse relation.

corollary: a theorem that is derived as an easy consequence of another theorem.

correct conclusion: the conclusion of a valid proof, when all the premises are true.

countable set: a set that is finite or denumerable.

counterexample: a case that makes a statement false.

definite clause: clause with at most one atom in its head.

denumerable set: a set that can be placed in one-to-one correspondence with the natural numbers.

diagonalization proof: any proof that involves something analogous to the diagonal of a list of sequences.

difference: a binary relation $R - S$ such that $a(R - S)b$ if and only if aRb is true and aSb is false.

difference (of sets): the set $A - B$ of objects in A that are not in B.

direct proof: a proof of $p \rightarrow q$ that assumes p and shows that q must follow.

disjoint (pair of sets): two sets with no members in common.

disjunction: the statement $p \vee q$ ("p or q") that is true when at least one of the two propositions p and q is true; also called *inclusive or*.

disjunctive normal form: for a proposition in the variables $p_1, p_2, \ldots, p_n$, an equivalent proposition that is the disjunction of conjunctions, with each conjunction of the form $x_{k_1} \wedge x_{k_2} \wedge \cdots \wedge x_{k_m}$, where x_{k_j} is either p_{k_j} or $\neg p_{k_j}$.

disproof: a proof that a statement is false.

divisibility lattice: the lattice consisting of the positive integers under the relation of divisibility.

domain (of a function): the set on which a function acts.

element (of a set): member of the set; the notation $a \in A$ means that a is an element of A.

elementary projection function: the function $\pi_i \colon X_1 \times \cdots \times X_n \to X_i$ such that $\pi(x_1, \ldots, x_n) = x_i$.

empty set: the set with no elements, written $\emptyset$ or $\{\ \}$.

epimorphism: an onto function.

equality (of sets): property that two sets have the same elements.

equivalence class: given an equivalence relation on a set A and $a \in A$, the subset of A consisting of all elements related to a.

equivalence relation: a binary relation that is reflexive, symmetric, and transitive.

equivalent propositions: two compound propositions (on the same simple variables) with the same truth table.

existential quantifier: the quantifier $\exists x$, read "there is an x".

existentially quantified predicate: a statement $(\exists x)P(x)$ that there exists a value of x such that $P(x)$ is true.

exponential function: any function of the form b^x, b a positive constant, $b \neq 1$.

fact set: set of ground atomic formulas.

factorial (function): the function $n!$ whose value on the argument n is the product $1 \cdot 2 \cdot 3 \ldots n$; that is, $n! = 1 \cdot 2 \cdot 3 \ldots n$.

finite: property of a set that it is either empty or else can be put in a one-to-one correspondence with a set $\{1, 2, 3, \ldots, n\}$ for some positive integer n.

first-order logic: See *predicate calculus*.

floor (of x): the greatest integer less than or equal to x, written $\lfloor x \rfloor$.

formula: a logical expression constructed from atoms with conjunctions, disjunctions, and negations, possibly with some logical quantifiers.

full conjunctive normal form: conjunctive normal form where each disjunction is a disjunction of all variables or their negations.

full disjunctive normal form: disjunctive normal form where each conjunction is a conjunction of all variables or their negations.

fully parenthesized proposition: any proposition that can be obtained using the following recursive definition: each variable is fully parenthesized, if P and Q are fully parenthesized, so are $(\neg P)$, $(P \wedge Q)$, $(P \vee Q)$, $(P \to Q)$, and $(P \leftrightarrow Q)$.

function $f \colon A \to B$: a rule that assigns to every object a in the domain set A exactly one object $f(a)$ in the codomain set B.

functionally complete set: a set of logical connectives from which all other connectives can be derived by composition.

fuzzy logic: a system of logic in which each statement has a truth value in the interval $[0, 1]$.

fuzzy set: a set in which each element is associated with a number in the interval $[0, 1]$ that measures its degree of membership.

generalized continuum hypothesis: the assertion that for every infinite set S there is no cardinal number greater than $|S|$ and less than $|\mathcal{P}(S)|$.

goal: a clause with an empty head.

graph (of a function): given a function $f: A \to B$, the set $\{(a, b) \mid b = f(a)\} \subseteq A \times B$.

greatest lower bound (of a subset of a poset): an element of the poset that is a lower bound of the subset and is greater than or equal to every other lower bound of the subset.

ground formula: a formula without any variables.

halting function: the function that maps computer programs to the set $\{0, 1\}$, with value 1 if the program always halts, regardless of input, and 0 otherwise.

Hasse diagram: a directed graph that represents a poset.

head (of a clause $A_1, \ldots, A_n \leftarrow B_1, \ldots, B_m$): the literals $A_1, \ldots, A_n$ before $\leftarrow$.

identity function (on a set): given a set A, the function from A to itself whose value at x is x.

image set (of a function): the set of function values as x ranges over all objects of the domain.

implication: formally, the relation $P \Rightarrow Q$ that a proposition Q is true whenever proposition P is true; informally, a synonym for the conditional statement $p \to q$.

incomparable: two elements in a poset that are not related by the partial order relation.

induced partition (on a set under an equivalence relation): the set of equivalence classes under the relation.

independent: property of a set of axioms that none of the axioms can be deduced from the other axioms.

indirect proof: a proof of $p \to q$ that assumes $\neg q$ is true and proves that $\neg p$ is true.

induction: See *mathematical induction*.

induction hypothesis: in a mathematical induction proof, the statement $P(x_k)$ in the induction step.

induction step: in a mathematical induction proof, a proof of the induction premise "if $P(x_k)$ is true, then $P(x_{k+1})$ is true".

inductive proof: See *mathematical induction*.

infinite (set): a set that is not finite.

injection (or ***injective function***): a one-to-one function.

instance (of a formula): formula obtained using a substitution.

instantiation: substitution of concrete values for the free variables of a statement or sequence of statements; an instance of a production rule.

integer: a whole number, possibly zero or negative; i.e., one of the elements in the set $\mathcal{Z} = \{\ldots, -2, -1, 0, 1, 2, \ldots\}$.

intersection: the set $A \cap B$ of objects common to both sets A and B.

intersection relation: for binary relations R and S on A, the relation $R \cap S$ where $a(R \cap S)b$ if and only if aRb and aSb.

interval (in a poset): given $a \leq b$ in a poset, a subset of the poset consisting of all elements x such that $a \leq x \leq b$.

inverse function: for a one-to-one, onto function $f: X \to Y$, the function $f^{-1}: Y \to X$ whose value at $y \in Y$ is the unique $x \in X$ such that $f(x) = y$.

inverse image (under $f: X \to Y$ of a subset $T \subseteq Y$): the subset $\{x \in X \mid f(x) \in T\}$, written $f^{-1}(T)$.

inverse relation: for a binary relation R from A to B, the relation R^{-1} from B to A where $bR^{-1}a$ if and only if aRb.

invertible (function): a one-to-one and onto function; a function that has an inverse.

irrational number: a real number that is not rational.

irreflexive: property of a binary relation R on A that $a\not{R}a$, for all $a \in A$.

lattice: a poset in which every pair of elements has both a least upper bound and a greatest lower bound.

least upper bound (of a subset of a poset): an element of the poset that is an upper bound of the subset and is less than or equal to every other upper bound of the subset.

lemma: a theorem that is an intermediate step in the proof of a more important theorem.

linearly ordered: the property of a poset that every pair of elements are comparable, also called *totally ordered*.

literal: an atom or its negation.

little-oh notation: f is $o(g)$ if $\lim_{x \to \infty} \left| \frac{f(x)}{g(x)} \right| = 0$.

logarithmic function: a function $\log_b x$ (b a positive constant, $b \neq 1$) defined by the rule $\log_b x = y$ if and only if $b^y = x$.

logic program: a finite sequence of definite clauses.

logically equivalent propositions: compound propositions that involve the same variables and have the same truth table.

logically implies: A compound proposition P logically implies a compound proposition Q if Q is true whenever P is true.

loop invariant: an expression that specifies the circumstance under which the loop body will be executed again.

lower bound (for a subset of a poset): an element of the poset that is less than or equal to every element of the subset.

mathematical induction: a method of proving that every item of a sequence of propositions such as $P(n_0), P(n_0 + 1), P(n_0 + 2), \ldots$ is true by showing: (1) $P(n_0)$ is true, and (2) for all $n \geq n_0$, $P(n) \to P(n+1)$ is true.

maximal element: in a poset an element that has no element greater than it.

maximum element: in a poset an element greater than or equal to every element.

membership function (in fuzzy logic): a function from elements of a set to $[0,1]$.

membership table (for a set expression): a table used to calculate whether an object lies in the set described by the expression, based on its membership in the sets mentioned by the expression.

minimal element: in a poset an element that has no element smaller than it.

minimum element: in a poset an element less than or equal to every element.

monomorphism: a one-to-one function.

multi-valued logic: a logic system with a set of more than two truth values.

multiset: an extension of the set concept, in which each element may occur arbitrarily many times.

mutually disjoint (family of sets): (See *pairwise disjoint*.)

n-ary predicate: a statement involving n variables.

n-ary relation: any subset of $A_1 \times A_2 \times \cdots \times A_n$.

naive set theory: set theory where any collection of objects can be considered to be a valid set, with paradoxes ignored.

NAND: the logical connective "not and".

natural number: a nonnegative integer (or "counting" number); i.e., an element of $\mathcal{N} = \{0, 1, 2, 3, \ldots\}$. *Note*: Sometimes 0 is not regarded as a natural number.

negation: the statement $\neg p$ ("not p") that is true if and only if p is not true.

NOP: pronounced "no-op", a program instruction that does nothing to alter the values of computational variables or the order of execution.

NOR: the logical connective "not or".

NOT: the logical connective meaning "not", used in place of $\neg$.

null set: the set with no elements, written $\emptyset$ or $\{\ \}$.

omega notation: f is $\Omega(g)$ if there are constants C and k such that $|g(x)| \leq C|f(x)|$ for all $x > k$.

one-to-one (function): a function $f: X \to Y$ that assigns distinct elements of the codomain to distinct elements of the domain; thus, if $x_1 \neq x_2$, then $f(x_1) \neq f(x_2)$.

onto (function): a function $f: X \to Y$ whose image equals its codomain; i.e., for every $y \in Y$, there is an $x \in X$ such that $f(x) = y$.

OR: the logical operator for disjunction, also written $\vee$.

pairwise disjoint: property of a family of sets that each two distinct sets in the family have empty intersection; also called *mutually disjoint*.

paradox: a statement that contradicts itself.

partial function: a function $f: X \to Y$ that assigns a well-defined object in Y to some (but not necessarily all) the elements of its domain X.

partial order: a binary relation that is reflexive, antisymmetric, and transitive.

partially ordered set: a set with a partial order relation defined on it.

partition (of a set): given a set S, a pairwise disjoint family $\mathcal{P} = \{A_i\}$ of nonempty subsets of S whose union is S.

Peano definition: a recursive description of the natural numbers that uses the concept of successor.

Polish prefix notation: the style of writing compound propositions in prefix notation where sometime the usual operand symbols are replaced as follows: N for $\neg$, K for $\wedge$, A for $\vee$, C for $\rightarrow$, E for $\leftrightarrow$.

poset: a partially ordered set.

postcondition: an assertion that appears immediately after the executable portion of a program fragment or of a subprogram.

postfix notation: the style of writing compound logical propositions where operators are written to the right of the operands.

power (of a relation): for a relation R on A, the relation R^n on A where $R^0 = I$, $R^1 = R$ and $R^n = R^{n-1} \circ R$ for all $n > 1$.

power set: given a set A, the set $\mathcal{P}(A)$ of all subsets of A.

precondition: an assertion that appears immediately before the executable portion of a program fragment or of a subprogram.

predicate: a statement involving one or more variables that range over various domains.

predicate calculus: the symbolic study of quantified predicate statements.

prefix notation: the style of writing compound logical propositions where operators are written to the left of the operands.

premise: a proposition taken as the foundation of a proof, from which the conclusion is to be derived.

prenex normal form: the form of a well-formed formula in which every quantifier occurs at the beginning and the scope is whatever follows the quantifiers.

preorder: a binary relation that is reflexive and transitive.

primary key: for an n-ary relation on $A_1, A_2, \ldots, A_n$, a coordinate domain A_j such that for each $x \in A_j$ there is at most one n-tuple in the relation whose jth coordinate is x.

production rule: a formula of the form $C_1, \ldots, C_n \rightarrow A_1, \ldots, A_m$ where each C_i is a condition and each A_i is an action.

production system: a set of production rules and a fact set.

program construct: See *construct*.

program fragment: any sequence of program code, from a single instruction to an entire program.

program semantics (or ***semantics***): the meaning of an instruction or of a program fragment; i.e., the effect of its execution on the computational variables.

projection function: a function defined on a set of n-tuples that selects the elements in certain coordinate positions.

proof (of a conclusion from a set of premises): a sequence of statements (called steps) terminating in the conclusion, such that each step is either a premise or follows from previous steps by a valid argument.

proof by contradiction: a proof that assumes the negation of the statement to be proved and shows that this leads to a contradiction.

proof done by hand: a proof done by a human without the use of a computer.

proper subset: given a set S, a subset T of S such that S contains at least one element not in T.

proposition: a declarative sentence or statement that is unambiguously either true or false.

propositional calculus: the symbolic study of propositions.

range (of a function): the image set of a function; sometimes used as synonym for codomain.

rational number: the ratio $\frac{a}{b}$ of two integers such that $b \neq 0$; the set of all rational numbers is denoted $\mathcal{Q}$.

real number: a number expressible as a finite (i.e., terminating) or infinite decimal; the set of all real numbers is denoted $\mathcal{R}$.

recursive definition (of a function with domain $\mathcal{N}$): a set of initial values and a rule for computing $f(n)$ in terms of values $f(k)$ for $k < n$.

recursive definition (of a set S): a form of specification of membership of S, in which some *basis* elements are named individually, and in which a computable rule is given to construct each other element in a finite number of steps.

refinement of a partition: given a partition $\mathcal{P}_1 = \{A_j\}$ on a set S, a partition $\mathcal{P}_2 = \{B_i\}$ on the same set S such that every $B_i \in \mathcal{P}_2$ is a subset of some $A_j \in \mathcal{P}_1$.

reflexive: the property of a binary relation R that aRa.

relation (from set A to set B): a binary relation from A to B.

relation (on a set A): a binary relation from A to A.

restriction (of a function): given $f: X \to Y$ and a subset $S \subseteq X$, the function $f|S$ with domain S and codomain Y whose rule is the same as that of f.

reverse Polish notation: postfix notation.

rule of inference: a valid argument form.

scope (of a quantifier): the predicate to which the quantifier applies.

semantic axiom: See *axiom*.

semantics: See *program semantics*.

sentence: a well-formed formula with no free variables.

sequence (in a set): a list of objects from a set S, with repetitions allowed; that is, a function $f: \mathcal{N} \to S$ (an infinite sequence, often written $a_0, a_1, a_2, \ldots$) or a function $f: \{1, 2, \ldots, n\} \to S$ (a finite sequence, often written $a_1, a_2, \ldots, a_n$).

set: a well-defined collection of objects.

singleton: a set with one element.

specification: in program correctness, a precondition and a postcondition.

statement form: a declarative sentence containing some variables and logical symbols which becomes a proposition if concrete values are substituted for all free variables.

string: a finite sequence in a set S, usually written so that consecutive entries are juxtaposed (i.e., written with no punctuation or extra space between them).

strongly correct code: code whose execution terminates in a computational state satisfying the postcondition, whenever the precondition holds before execution.

subset of a set S: any set T of objects that are also elements of S, written $T \subseteq S$.

substitution: a set of pairs of variables and terms.

surjection (or **surjective function**): an onto function.

symmetric: the property of a binary relation R that if aRb then bRa.

symmetric difference (of relations): for relations R and S on A, the relation $R \oplus S$ where $a(R \oplus S)b$ if and only if exactly one of the following is true: aRb, aSb.

symmetric difference (of sets): for sets A and B, the set $A \oplus B$ containing each object that is an element of A or an element of B, but not an element of both.

system of distinct representatives: given sets $A_1, A_2, \ldots, A_n$ (some of which may be equal), a set $\{a_1, a_2, \ldots, a_n\}$ of n distinct elements with $a_i \in A_i$ for $i = 1, 2, \ldots, n$.

tautology: a compound proposition whose form makes it always true, regardless of the truth values of its atomic parts.

term (in a domain): either a fixed element of a domain S or an S-valued variable.

theorem: a statement derived as the conclusion of a valid proof from axioms and definitions.

theta notation: f is $\Theta(g)$, written $f = \Theta(g)$, if there are positive constants C_1, C_2, and k such that $C_1|g(x)| \leq |f(x)| \leq C_2|g(x)|$ for all $x > k$.

totally ordered: the property of a poset that every pair of elements are comparable; also called *linearly ordered*.

transitive: the property of a binary relation R that if aRb and bRc, then aRc.

transitive closure: for a relation R on A, the smallest transitive relation containing R.

transitive reduction (of a relation): a relation with the same transitive closure as the original relation and with a minimum number of ordered pairs.

truth table: for a compound proposition, a table that gives the truth value of the proposition for each possible combination of truth values of the atomic variables in the proposition.

two-valued logic: a logic system where each statement has exactly one of the two values: true or false.

union: the set $A \cup B$ of objects in one or both of the sets A and B.

union relation: for R and S binary relations on A, the relation $R \cup S$ where $a(R \cup S)b$ if and only if aRb or aSb.

universal domain: the collection of all possible objects in the context of the immediate discussion.

universal quantifier: the quantifier $\forall x$, read "for all x" or "for every x".

universally quantified predicate: a statement $(\forall x)P(x)$ that $P(x)$ is true for every x in its universe of discourse.

universe of discourse: the range of possible values of a variable, within the context of the immediate discussion.

upper bound (for a subset of a poset): an element of the poset that is greater than or equal to every element of the subset.

valid argument form: an argument form such that in any instantiation where all the premises are true, the conclusion is also true.

Venn diagram: a figure composed of possibly overlapping circles or ellipses, used to picture membership in various combinations of the sets.

verification (of a program): a formal argument for the correctness of a program with respect to its specifications.

weakly correct code: code whose execution results in a computational state satisfying the postcondition, whenever the precondition holds before execution and the execution terminates.

well-formed formula (**wff**): a proposition or predicate with quantifiers that bind one or more of its variables.

well-ordered: property of a set that every nonempty subset has a minimum element.

well-ordering principle: the axiom that every nonempty subset of integers, each greater than a fixed integer, contains a smallest element.

XOR: the logical connective "not or".

Zermelo-Fraenkel axioms: a set of axioms for set theory.

zero-order logic: propositional calculus.

1.1 PROPOSITIONAL AND PREDICATE LOGIC

Logic is the basis for distinguishing what may be correctly inferred from a given collection of facts. Propositional logic, where there are no quantifiers (so quantifiers range over nothing) is called zero-order logic. Predicate logic, where quantifiers range over members of a universe, is called first-order logic. Higher-order logic includes second-order logic (where quantifiers can range over relations over the universe), third-order logic (where quantifiers can range over relations over relations), and so on. Logic has many applications in computer science, including circuit design (§5.8.3) and verification of computer program correctness (§1.6). This section defines the meaning of the symbolism and various logical properties that are usually used without explicit mention. [FlPa88], [Me79], [Mo76]

In this section, only two-valued logic is studied; i.e., each statement is either true or false. Multi-valued logic, in which statements have one of more than two values, is discussed in §1.7.2.

1.1.1 PROPOSITIONS AND LOGICAL OPERATIONS

Definitions:

A **truth value** is either true or false, abbreviated T and F, respectively.

A **proposition** (in a natural language such as English) is a declarative sentence that has a well-defined truth value.

A **propositional variable** is a mathematical variable, often denoted by p, q, or r, that represents a proposition.

Propositional logic (or **propositional calculus** or **zero-order logic**) is the study of logical propositions and their combinations using logical connectives.

A **logical connective** is an operation used to build more complicated logical expressions out of simpler propositions, whose truth values depend only on the truth values of the simpler propositions.

A proposition is **atomic** or **simple** if it cannot be syntactically analyzed into smaller parts; it is usually represented by a single logical variable.

A proposition is **compound** if it contains one or more logical connectives.

A **truth table** is a table that prescribes the defining rule for a logical operation. That is, for each combination of truth values of the operands, the table gives the truth value of the expression formed by the operation and operands.

The unary connective **negation** (denoted by $\neg$) is defined by the following truth table:

p	$\neg p$
T	F
F	T

Note: The negation $\neg p$ is also written p', $\bar{p}$, or $\sim p$.

The common binary connectives are:

$p \wedge q$	**conjunction**	p and q
$p \vee q$	**disjunction**	p or q
$p \to q$	**conditional**	if p then q
$p \leftrightarrow q$	**biconditional**	p if and only if q
$p \oplus q$	**exclusive or**	p xor q
$p \downarrow q$	**not or**	p nor q
$p \mid q$ or $p \uparrow q$	**not and**	p nand q

The connective $\mid$ is called the *Sheffer stroke*. The connective $\downarrow$ is called the *Peirce arrow*. The values of the compound propositions obtained by using the binary connectives are given in the following table:

p	q	$p \vee q$	$p \wedge q$	$p \to q$	$p \leftrightarrow q$	$p \oplus q$	$p \downarrow q$	$p \mid q$
T	T	T	T	T	T	F	F	F
T	F	T	F	F	F	T	F	T
F	T	T	F	T	F	T	F	T
F	F	F	F	T	T	F	T	T

In the conditional $p \to q$, p is the **antecedent** and q is the **consequent**. The conditional $p \to q$ is often read informally as "p implies q".

Infix notation is the style of writing compound propositions where binary operators are written between the operands and negation is written to the left of its operand.

Prefix notation is the style of writing compound propositions where operators are written to the left of the operands.

Postfix notation (or **reverse Polish notation**) is the style of writing compound propositions where operators are written to the right of the operands.

Polish notation is the style of writing compound propositions where operators are written using prefix notation and where the usual operand symbols are replaced as follows: N for $\neg$, K for $\wedge$, A for $\vee$, C for $\to$, E for $\leftrightarrow$. (Jan Lukasiewicz, 1878–1956)

A **fully parenthesized proposition** is any proposition that can be obtained using the following recursive definition: each variable is fully parenthesized, if P and Q are fully parenthesized, so are $(\neg P)$, $(P \wedge Q)$, $(P \vee Q)$, $(P \to Q)$, and $(P \leftrightarrow Q)$.

Facts:

1. The conditional connective $p \to q$ represents the following English constructs:
   - if p then q
   - p only if q
   - q follows from p
   - p is a sufficient condition for q
   - q if p
   - p implies q
   - q whenever p
   - q is a necessary condition for p.

2. The biconditional connective $p \leftrightarrow q$ represents the following English constructs:
   - p if and only if q (often written p iff q)
   - p and q imply each other
   - p is a necessary and sufficient condition for q
   - p and q are equivalent.

3. In computer programming and circuit design, the following notation for logical operators is used: p AND q for $p \wedge q$, p OR q for $p \vee q$, NOT p for $\neg p$, p XOR q for $p \oplus q$, p NOR q for $p \downarrow q$, p NAND q for $p \mid q$.

4. *Order of operations*: In an unparenthesized compound proposition using only the five standard operators $\neg, \wedge, \vee, \to$, and $\leftrightarrow$, the following order of precedence is typically used when evaluating a logical expression, at each level of precedence moving from left to right: first $\neg$, then $\wedge$ and $\vee$, then $\to$, finally $\leftrightarrow$. Parenthesized expressions are evaluated procceding from the innermost pair of parentheses outward, analogous to the evaluation of an arithmetic expression.

5. It is often preferable to use parentheses to show precedence, except for negation operators, rather than to rely on precedence rules.

6. No parentheses are needed when a compound proposition is written in either prefix or postfix notation. However, parentheses may be necessary when a compound proposition is written in infix notation.

7. The number of nonequivalent logical statements with two variables is 16, because each of the four lines of the truth table has two possible entries, T or F. Here are examples of compound propositions that yield each possible combination of truth values. (**T** represents a tautology and **F** a contradiction. See §1.1.2.)

$p\ q$	**T**	$p \vee q$	$q \to p$	$p \to q$	$p \mid q$	p	q	$p \leftrightarrow q$
$T\ T$	T	T	T	T	F	T	T	T
$T\ F$	T	T	T	F	T	T	F	F
$F\ T$	T	T	F	T	T	F	T	F
$F\ F$	T	F	T	T	T	F	F	T

$p\ q$	$p \oplus q$	$\neg q$	$\neg p$	$p \wedge q$	$p \wedge \neg q$	$\neg p \wedge q$	$p \downarrow q$	**F**
$T\ T$	F	F	F	T	F	F	F	F
$T\ F$	T	T	F	F	T	F	F	F
$F\ T$	T	F	T	F	F	T	F	F
$F\ F$	F	T	T	F	F	F	T	F

8. The number of different possible logical connectives on n variables is 2^{2^n}, because there are 2^n rows in the truth table.

Examples:

1. "1+1 = 3" and "Romulus and Remus founded New York City" are false propositions.
2. "$1 + 1 = 2$" and "The year 1996 was a leap year" are true propositions.
3. "Go directly to jail" is not a proposition, because it is imperative, not declarative.

4. "$x > 5$" is not a proposition, because its truth value cannot be determined unless the value of x is known.

5. "This sentence is false" is not a proposition, because it cannot be given a truth value without creating a contradiction.

6. In a truth table evaluation of the compound proposition $p \vee (\neg p \wedge q)$ from the innermost parenthetic expression outward, the steps are to evaluate $\neg p$, next $(\neg p \wedge q)$, and then $p \vee (\neg p \wedge q)$:

p	q	$\neg p$	$(\neg p \wedge q)$	$p \vee (\neg p \wedge q)$
T	T	F	F	T
T	F	F	F	T
F	T	T	T	T
F	F	T	F	F

7. The statements in the left column are evaluated using the order of precedence indicated in the fully parenthesized form in the right column:

$$p \vee q \wedge r \qquad ((p \vee q) \wedge r)$$
$$p \leftrightarrow q \rightarrow r \qquad (p \leftrightarrow (q \rightarrow r))$$
$$\neg q \vee \neg r \rightarrow s \wedge t \qquad (((\neg q) \vee (\neg r)) \rightarrow (s \wedge t))$$

8. The infix statement $p \wedge q$ in prefix notation is $\wedge p q$, in postfix notation is $p q \wedge$, and in Polish notation is $K p q$.

9. The infix statement $p \rightarrow \neg(q \vee r)$ in prefix notation is $\rightarrow p \neg \vee q r$, in postfix notation is $p q r \vee \neg \rightarrow$, and in Polish notation is $C p N A q r$.

1.1.2 EQUIVALENCES, IDENTITIES, AND NORMAL FORMS

Definitions:

A **tautology** is a compound proposition that is always true, regardless of the truth values of its underlying atomic propositions.

A **contradiction** (or **self-contradiction**) is a compound proposition that is always false, regardless of the truth values of its underlying atomic propositions. (The term self-contradiction is used for such a proposition when discussing indirect mathematical arguments, because "contradiction" has another meaning in that context. See §1.5.)

A compound proposition P **logically implies** a compound proposition Q, written $P \Rightarrow Q$, if Q is true whenever P is true. In this case, P is **stronger than** Q, and Q is **weaker than** P.

Compound propositions P and Q are **logically equivalent**, written $P \equiv Q$, $P \Leftrightarrow Q$, or P iff Q, if they have the same truth values for all possible truth values of their variables.

A logical equivalence that is frequently used is sometimes called a **logical identity**.

A collection $\mathcal{C}$ of connectives is **functionally complete** if every compound proposition is equivalent to a compound proposition constructed using only connectives in $\mathcal{C}$.

A **disjunctive normal expression** in the propositions $p_1, p_2, \ldots, p_n$ is a disjunction of one or more propositions, each of the form $x_{k_1} \wedge x_{k_2} \wedge \cdots \wedge x_{k_m}$, where x_{k_j} is either p_{k_j} or $\neg p_{k_j}$.

A **disjunctive normal form** (**DNF**) for a proposition P is a disjunctive normal expression that is logically equivalent to P.

A **conjunctive normal expression** in the propositions $p_1, p_2, \ldots, p_n$ is a conjunction of one or more compound propositions, each of the form $x_{k_1} \vee x_{k_2} \vee \cdots \vee x_{k_m}$, where x_{k_j} is either p_{k_j} or $\neg p_{k_j}$.

A **conjunctive normal form (CNF)** for a proposition P is a conjunctive normal expression that is logically equivalent to P.

A compound proposition P using only the connectives $\neg$, $\wedge$, and $\vee$ has a **logical dual** (denoted P' or P^d), obtained by interchanging $\wedge$ and $\vee$ and interchanging the constant **T** (true) and the constant **F** (false).

The **converse** of the conditional proposition $p \to q$ is the proposition $q \to p$.

The **contrapositive** of the conditional proposition $p \to q$ is the proposition $\neg q \to \neg p$.

The **inverse** of the conditional proposition $p \to q$ is the proposition $\neg p \to \neg q$.

Facts:
1. $P \Leftrightarrow Q$ is true if and only if $P \Rightarrow Q$ and $Q \Rightarrow P$.
2. $P \Leftrightarrow Q$ is true if and only if $P \leftrightarrow Q$ is a tautology.
3. Table 1 lists several logical identities.
4. There are different ways to establish logical identities (equivalences):
    - truth tables (showing that both expressions have the same truth values);
    - using known logical identities and equivalence to establish new ones;
    - taking the dual of a known identity (Fact 7).
5. Logical identities are used in circuit design to simplify circuits. See §5.8.4.
6. Each of the following sets of connectives is functionally complete:

$$\{\wedge, \vee, \neg\}, \quad \{\wedge, \neg\}, \quad \{\vee, \neg\}, \quad \{\,|\,\}, \quad \{\downarrow\}.$$

However, these sets of connectives are not functionally complete:

$$\{\wedge\}, \quad \{\vee\}, \quad \{\wedge, \vee\}.$$

7. If $P \Leftrightarrow Q$ is a logical identity, then so is $P' \Leftrightarrow Q'$, where P' and Q' are the duals of P and Q, respectively.
8. Every proposition has a disjunctive normal form and a conjunctive normal form, which can be obtained by Algorithms 1 and 2.

Algorithm 1: Disjunctive normal form of proposition P.

write the truth table for P
for each line of the truth table on which P is true, form a "line term"
$x_1 \wedge x_2 \wedge \cdots \wedge x_n$, where $x_i := p_i$ if p_i is true on that line of the truth table
and $x_i := \neg p_i$ if p_i is false on that line
form the disjunction of all these line terms

Algorithm 2: Conjunctive normal form of proposition $P \triangleright$

write the truth table for P
for each line of the truth table on which P is false, form a "line term"
$x_1 \vee x_2 \vee \cdots \vee x_n$, where $x_i := p_i$ if p_i is false on that line of the truth table
and $x_i := \neg p_i$ if p_i is true on that line
form the conjunction of all these line terms

Section 1.1 PROPOSITIONAL AND PREDICATE LOGIC

Table 1 Logical identities.

name	rule
Commutative laws	$p \wedge q \Leftrightarrow q \wedge p \qquad p \vee q \Leftrightarrow q \vee p$
Associative laws	$p \wedge (q \wedge r) \Leftrightarrow (p \wedge q) \wedge r \qquad p \vee (q \vee r) \Leftrightarrow (p \vee q) \vee r$
Distributive laws	$p \wedge (q \vee r) \Leftrightarrow (p \wedge q) \vee (p \wedge r)$
	$p \vee (q \wedge r) \Leftrightarrow (p \vee q) \wedge (p \vee r)$
DeMorgan's laws	$\neg(p \wedge q) \Leftrightarrow (\neg p) \vee (\neg q) \qquad \neg(p \vee q) \Leftrightarrow (\neg p) \wedge (\neg q)$
Excluded middle	$p \vee \neg p \Leftrightarrow \mathbf{T}$
Contradiction	$p \wedge \neg p \Leftrightarrow \mathbf{F}$
Double negation law	$\neg(\neg p) \Leftrightarrow p$
Contrapositive law	$p \rightarrow q \Leftrightarrow \neg q \rightarrow \neg p$
Conditional as disjunction	$p \rightarrow q \Leftrightarrow \neg p \vee q$
Negation of conditional	$\neg(p \rightarrow q) \Leftrightarrow p \wedge \neg q$
Biconditional as implication	$(p \leftrightarrow q) \Leftrightarrow (p \rightarrow q) \wedge (q \rightarrow p)$
Idempotent laws	$p \wedge p \Leftrightarrow p \qquad p \vee p \Leftrightarrow p$
Absorption laws	$p \wedge (p \vee q) \Leftrightarrow p \qquad p \vee (p \wedge q) \Leftrightarrow p$
Dominance laws	$p \vee \mathbf{T} \Leftrightarrow \mathbf{T} \qquad p \wedge \mathbf{F} \Leftrightarrow \mathbf{F}$
Exportation law	$p \rightarrow (q \rightarrow r) \Leftrightarrow (p \wedge q) \rightarrow r$
Identity laws	$p \wedge \mathbf{T} \Leftrightarrow p \qquad p \vee \mathbf{F} \Leftrightarrow p$

Examples:

1. The proposition $p \vee \neg p$ is a tautology (the *law of the excluded middle*).
2. The proposition $p \wedge \neg p$ is a self-contradiction.
3. A proof that $p \leftrightarrow q$ is logically equivalent to $(p \wedge q) \vee (\neg p \wedge \neg q)$ can be carried out using a truth table:

p	q	$p \leftrightarrow q$	$\neg p$	$\neg q$	$p \wedge q$	$\neg p \wedge \neg q$	$(p \wedge q) \vee (\neg p \wedge \neg q)$
T	T	T	F	F	T	F	T
T	F	F	F	T	F	F	F
F	T	F	T	F	F	F	F
F	F	T	T	T	F	T	T

Since the third and eighth columns of the truth table are identical, the two statements are equivalent.

4. A proof that $p \leftrightarrow q$ is logically equivalent to $(p \wedge q) \vee (\neg p \wedge \neg q)$ can be given by a series of logical equivalences. Reasons are given at the right.

$p \leftrightarrow q \Leftrightarrow (p \rightarrow q) \wedge (q \rightarrow p)$ biconditional as implication
$\Leftrightarrow (\neg p \vee q) \wedge (\neg q \vee p)$ conditional as disjunction
$\Leftrightarrow [(\neg p \vee q) \wedge \neg q] \vee [(\neg p \vee q) \wedge p]$ distributive law
$\Leftrightarrow [(\neg p \wedge \neg q) \vee (q \wedge \neg q)] \vee [(\neg p \wedge p) \vee (q \wedge p)]$ distributive law
$\Leftrightarrow [(\neg p \wedge \neg q) \vee \mathbf{F}] \vee [\mathbf{F} \vee (q \wedge p)]$ contradiction
$\Leftrightarrow [(\neg p \wedge \neg q) \vee \mathbf{F}] \vee [(q \wedge p) \vee \mathbf{F}]$ commutative law
$\Leftrightarrow (\neg p \wedge \neg q) \vee (q \wedge p)$ identity law
$\Leftrightarrow (\neg p \wedge \neg q) \vee (p \wedge q)$ commutative law
$\Leftrightarrow (p \wedge q) \vee (\neg p \wedge \neg q)$ commutative law

5. The proposition $p \downarrow q$ is logically equivalent to $\neg(p \vee q)$. Its DNF is $\neg p \wedge \neg q$, and its CNF is $(\neg p \vee \neg q) \wedge (\neg p \vee q) \wedge (p \vee \neg q)$.

6. The proposition $p|q$ is logically equivalent to $\neg(p \wedge q)$. Its DNF is $(p \wedge \neg q) \vee (\neg p \wedge q) \vee (\neg p \wedge \neg q)$, and its CNF is $\neg p \vee \neg q$.

7. The DNF and CNF for Examples 5 and 6 were obtained by using Algorithm 1 and Algorithm 2 to construct the following table of terms:

p	q	$p \downarrow q$	DNF terms	CNF terms	p	q	$p \mid q$	DNF terms	CNF terms
T	T	F		$\neg p \vee \neg q$	T	T	F		$\neg p \vee \neg q$
T	F	F		$\neg p \vee q$	T	F	T	$p \wedge \neg q$	
F	T	F		$p \vee \neg q$	F	T	T	$\neg p \wedge q$	
F	F	T	$\neg p \wedge \neg q$		F	F	T	$\neg p \wedge \neg q$	

8. The dual of $p \wedge (q \vee \neg r)$ is $p \vee (q \wedge \neg r)$.

9. Let S be the proposition in three propositional variables p, q, and r that is true when precisely two of the variables are true. Then the disjunctive normal form for S is

$$(p \wedge q \wedge \neg r) \vee (p \wedge \neg q \wedge r) \vee (\neg p \wedge q \wedge r)$$

and the conjunctive normal form for S is

$$(\neg p \vee \neg q \vee \neg r) \wedge (\neg p \vee q \vee r) \wedge (p \vee \neg q \vee r) \wedge (p \vee q \vee \neg r) \wedge (p \vee q \vee r).$$

1.1.3 PREDICATE LOGIC

Definitions:

A **predicate** is a declarative statement with the symbolic form $P(x)$ or $P(x_1, \ldots, x_n)$ about one or more variables x or $x_1, \ldots, x_n$ whose values are unspecified.

Predicate logic (or **predicate calculus** or **first-order logic**) is the study of statements whose variables have quantifiers.

The **universe of discourse** (or **universe** or **domain**) of a variable is the set of possible values of the variable in a predicate.

An **instantiation** of the predicate $P(x)$ is the result of substituting a fixed constant value c from the domain of x for each free occurrence of x in $P(x)$. This is denoted by $P(c)$.

The **existential quantification** of a predicate $P(x)$ whose variable ranges over a domain set D is the proposition $(\exists x \in D)P(x)$ or $(\exists x)P(x)$ that is true if there is at least one c in D such that $P(c)$ is true. The *existential quantifier symbol*, $\exists$, is read "there exists" or "there is".

The **universal quantification** of a predicate $P(x)$ whose variable ranges over a domain set D is the proposition $(\forall x \in D)P(x)$ or $(\forall x)P(x)$, which is true if $P(c)$ is true for every element c in D. The *universal quantifier symbol*, $\forall$, is read "for all", "for each", or "for every".

The **unique existential quantification** of a predicate $P(x)$ whose variable ranges over a domain set D is the proposition $(\exists! x)P(x)$ that is true if $P(c)$ is true for exactly one c in D. The *unique existential quantifier symbol*, $\exists!$, is read "there is exactly one".

The **scope** of a quantifier is the predicate to which it applies.

A variable x in a predicate $P(x)$ is a **bound variable** if it lies inside the scope of an x-quantifier. Otherwise it is a **free variable**.

A **well-formed formula** (**wff**) (or **statement**) is either a proposition or a predicate with quantifiers that bind one or more of its variables.

A **sentence** (**closed wff**) is a well-formed formula with no free variables.

A well-formed formula is in **prenex normal form** if all the quantifiers occur at the beginning and the scope is whatever follows the quantifiers.

A well-formed formula is **atomic** if it does not contain any logical connectives; otherwise the well-formed formula is **compound**.

Higher-order logic is the study of statements that allow quantifiers to range over relations over a universe (second-order logic), relations over relations over a universe (third-order logic), etc.

Facts:

1. If a predicate $P(x)$ is atomic, then the scope of $(\forall x)$ in $(\forall x)P(x)$ is implicitly the entire predicate $P(x)$.

2. If a predicate is a compound form, such as $P(x) \wedge Q(x)$, then $(\forall x)[P(x) \wedge Q(x)]$ means that the scope is $P(x) \wedge Q(x)$, whereas $(\forall x)P(x) \wedge Q(x)$ means that the scope is only $P(x)$, in which case the free variable x of the predicate $Q(x)$ has no relationship to the variable x of $P(x)$.

3. Universal statements in predicate logic are analogues of conjunctions in propositional logic. If variable x has domain $D = \{x_1, \ldots, x_n\}$, then $(\forall x \in D)P(x)$ is true if and only if $P(x_1) \wedge \cdots \wedge P(x_n)$ is true.

4. Existential statements in predicate logic are analogues of disjunctions in propositional logic. If variable x has domain $D = \{x_1, \ldots, x_n\}$, then $(\exists x \in D)P(x)$ is true if and only if $P(x_1) \vee \cdots \vee P(x_n)$ is true.

5. Adjacent universal quantifiers [existential quantifiers] can be transposed without changing the meaning of a logical statement:

$$(\forall x)(\forall y)P(x,y) \Leftrightarrow (\forall y)(\forall x)P(x,y)$$
$$(\exists x)(\exists y)P(x,y) \Leftrightarrow (\exists y)(\exists x)P(x,y).$$

6. Transposing adjacent logical quantifiers of different types can change the meaning of a statement. (See Example 4.)

7. Rules for negations of quantified statements:

$$\neg(\forall x)P(x) \Leftrightarrow (\exists x)[\neg P(x)]$$
$$\neg(\exists x)P(x) \Leftrightarrow (\forall x)[\neg P(x)]$$
$$\neg(\exists ! x)P(x) \Leftrightarrow \neg(\exists x)P(x) \vee (\exists y)(\exists z)[(y \neq z) \wedge P(y) \wedge P(z)].$$

8. Every quantified statement is logically equivalent to some statement in prenex normal form.

9. Every statement with a unique existential quantifier is equivalent to a statement that uses only existential and universal quantifiers, according to the rule:

$$(\exists ! x)P(x) \Leftrightarrow (\exists x)\big[P(x) \wedge (\forall y)[P(y) \to (x = y)]\big]$$

where $P(y)$ means that y has been substituted for all free occurrences of x in $P(x)$, and where y is a variable that does not occur in $P(x)$.

10. If a statement uses only the connectives $\vee$, $\wedge$, and $\neg$, the following equivalences can be used along with Fact 7 to convert the statement into prenex normal form. The letter A represents a wff without the variable x.

$$(\forall x)P(x) \wedge (\forall x)Q(x) \Leftrightarrow (\forall x)[P(x) \wedge Q(x)]$$
$$(\forall x)P(x) \vee (\forall x)Q(x) \Leftrightarrow (\forall x)(\forall y)[P(x) \vee Q(y)]$$
$$(\exists x)P(x) \wedge (\exists x)Q(x) \Leftrightarrow (\exists x)(\exists y)[P(x) \wedge Q(y)]$$
$$(\exists x)P(x) \vee (\exists x)Q(x) \Leftrightarrow (\exists x)[P(x) \vee Q(x)]$$
$$(\forall x)P(x) \wedge (\exists x)Q(x) \Leftrightarrow (\forall x)(\exists y)[P(x) \wedge Q(y)]$$
$$(\forall x)P(x) \vee (\exists x)Q(x) \Leftrightarrow (\forall x)(\exists y)[P(x) \vee Q(y)]$$
$$A \vee (\forall x)P(x) \Leftrightarrow (\forall x)[A \vee P(x)]$$
$$A \vee (\exists x)P(x) \Leftrightarrow (\exists x)[A \vee P(x)]$$
$$A \wedge (\forall x)P(x) \Leftrightarrow (\forall x)[A \wedge P(x)]$$
$$A \wedge (\exists x)P(x) \Leftrightarrow (\exists x)[A \wedge P(x)].$$

Examples:

1. The statement $(\forall x \in \mathcal{R})(\forall y \in \mathcal{R})[x + y = y + x]$ is syntactically a predicate preceded by two universal quantifiers. It asserts the commutative law for the addition of real numbers.

2. The statement $(\forall x)(\exists y)[xy = 1]$ expresses the existence of multiplicative inverses for all number in whatever domain is under discussion. Thus, it is true for the positive real numbers, but it is false when the domain is the entire set of reals, since zero has no multiplicative inverse.

3. The statement $(\forall x \neq 0)(\exists y)[xy = 1]$ asserts the existence of multiplicative inverses for nonzero numbers.

4. $(\forall x)(\exists y)[x + y = 0]$ expresses the true proposition that every real number has an additive inverse, but $(\exists y)(\forall x)[x+y = 0]$ is the false proposition that there is a "universal additive inverse" that when added to any number always yields the sum 0.

5. In the statement $(\forall x \in \mathcal{R})[x+y = y+x]$, the variable x is bound and the variable y is free.

6. "Not all men are mortal" is equivalent to "there exists at least one man who is not mortal". Also, "there does not exist a cow that is blue" is equivalent to the statement "every cow is a color other than blue".

7. The statement $(\forall x)\,P(x) \to (\forall x)\,Q(x)$ is not in prenex form. An equivalent prenex form is $(\forall x)(\exists y)\,[P(y) \to Q(x)]$.

8. The following table illustrates the differences in meaning among the four different ways to quantify a predicate with two variables:

statement	meaning
$(\exists x)(\exists y)[x+y=0]$	There is a pair of numbers whose sum is zero.
$(\forall x)(\exists y)[x+y=0]$	Every number has an additive inverse.
$(\exists x)(\forall y)[x+y=0]$	There is a universal additive inverse x.
$(\forall x)(\forall y)[x+y=0]$	The sum of every pair of numbers is zero.

9. The statement $(\forall x)(\exists! y)[x+y=0]$ asserts the existence of *unique* additive inverses.

1.2 SET THEORY

Sets are used to group objects and to serve as the basic elements for building more complicated objects and structures. Counting elements in sets is an important part of discrete mathematics.

Some general reference books that cover the material of this section are [FlPa88], [Ha60], [Ka50].

1.2.1 SETS

Definitions:

A *set* is any well-defined collection of objects, each of which is called a **member** or an **element** of the set. The notation $x \in A$ means that the object x is a member of the set A. The notation $x \notin A$ means that x is not a member of A.

A *roster* for a finite set specifies the membership of a set S as a list of its elements within braces, i.e., in the form $S = \{a_1, \ldots, a_n\}$. Order of the list is irrelevant, as is the number of occurrences of an object in the list.

A *defining predicate* specifies a set in the form $S = \{\, x \mid P(x)\,\}$, where $P(x)$ is a predicate containing the free variable x. This means that S is the set of all objects x (in whatever domain is under discussion) such that $P(x)$ is true.

A *recursive description* of a set S gives a roster B of *basic objects* of S and a set of operations for constructing additional objects of S from objects already known to be in S. That is, any object that can be constructed by a finite sequence of applications of the given operations to objects in B is also a member of S. There may also be a list of axioms that specify when two sequences of operations yield the same result.

The set with no elements is called the **null set** or the **empty set**, denoted $\emptyset$ or $\{\ \}$.

A *singleton* is a set with one element.

The set $\mathcal{N}$ of **natural numbers** is the set $\{0, 1, 2, \ldots\}$. (Sometimes 0 is excluded from the set of natural numbers; when the set of natural numbers is encountered, check to see how it is being defined.)

The set $\mathcal{Z}$ of **integers** is the set $\{\ldots, -2, -1, 0, 1, 2, \ldots\}$.

The set $\mathcal{Q}$ of **rational numbers** is the set of all fractions $\frac{a}{b}$ where a is any integer and b is any nonzero integer.

The set $\mathcal{R}$ of **real numbers** is the set of all numbers that can be written as terminating or nonterminating decimals.

The set $\mathcal{C}$ of **complex numbers** is the set of all numbers of the form $a + bi$, where $a, b \in \mathcal{R}$ and $i = \sqrt{-1}$ ($i^2 = -1$).

Sets A and B are **equal**, written $A = B$, if they have exactly the same elements:
$$A = B \Leftrightarrow (\forall x)\left[(x \in A) \leftrightarrow (x \in B)\right].$$

Set B is a **subset** of set A, written $B \subseteq A$ or $A \supseteq B$, if each element of B is an element of A:
$$B \subseteq A \Leftrightarrow (\forall x)\left[(x \in B) \rightarrow (x \in A)\right].$$

Set B is a **proper subset** of A if B is a subset of A and A contains at least one element not in B. (The notation $B \subset A$ is often used to indicate that B is a proper subset of A, but sometimes it is used to mean an arbitrary subset. Sometimes the proper subset relationship is written $B \subsetneq A$, to avoid all possible notational ambiguity.)

A set is **finite** if it is either empty or else can be put in a one-to-one correspondence with the set $\{1, 2, 3, \ldots, n\}$ for some positive integer n.

A set is **infinite** if it is not finite.

The **cardinality** $|S|$ of a finite set S is the number of elements in S.

A **multiset** is an unordered collection in which elements can occur arbitrarily often, not just once. The number of occurrences of an element is called its **multiplicity**.

An **axiom** (**postulate**) is a statement that is assumed to be true.

A set of axioms is **consistent** if no contradiction can be deduced from the axioms.

A set of axioms is **complete** if it is possible to prove all true statements.

A set of axioms is **independent** if none of the axioms can be deduced from the other axioms.

A **set paradox** is a question in the language of set theory that seems to have no unambiguous answer.

Naive set theory is set theory where any collection of objects can be considered to be a valid set, with paradoxes ignored.

Facts:
1. The theory of sets was first developed by Georg Cantor (1845–1918).
2. $A = B$ if and only if $A \subseteq B$ and $B \subseteq A$.
3. $\mathcal{N} \subset \mathcal{Z} \subset \mathcal{Q} \subset \mathcal{R} \subset \mathcal{C}$.
4. Every rational number can be written as a decimal that is either terminating or else repeating (i.e., the same block repeats end-to-end forever).
5. Real numbers can be represented as the points on the number line, and include all rational numbers and all irrational numbers (such as $\sqrt{2}$, π, e, etc.).
6. There is no set of axioms for set theory that is both complete and consistent.
7. Naive set theory ignores paradoxes. To avoid such paradoxes, more axioms are needed.

Examples:
1. The set $\{\, x \in \mathcal{N} \mid 3 \leq x < 10 \,\}$, described by the defining predicate $3 \leq x < 10$ is equal to the set $\{3, 4, 5, 6, 7, 8, 9\}$, which is described by a roster.
2. If A is the set with two objects, one of which is the number 5 and other the set whose elements are the letters x, y, and z, then $A = \{5, \{x, y, z\}\}$. In this example, $5 \in A$, but $x \notin A$, since x is not either member of A.
3. The set E of even natural numbers can be described recursively as follows:
    Basic objects: $0 \in E$,
    Recursion rule: if $n \in E$, then $n + 2 \in E$.
4. *The liar's paradox*: A person says "I am lying". Is the person lying or is the person telling the truth? If the person is lying, then "I am lying" is false, and hence the person is telling the truth. If the person is telling the truth, then "I am lying" is true, and the person is lying. This is also called the *paradox of Epimenides*. This paradox also results from considering the statement "This statement is false".

5. *The barber paradox*: In a small village populated only by men there is exactly one barber. The villagers follow the following rule: the barber shaves a man if and only if the man does not shave himself. Question: does the barber shave himself? If "yes" (i.e., the barber shaves himself), then according to the rule he does not shave himself. If "no" (i.e., the barber does not shave himself), then according to the rule he does shave himself. This paradox illustrates a danger in describing sets by defining predicates.

6. *Russell's paradox*: This paradox, named for the British logician Bertrand Russell (1872–1970), shows that the "set of all sets" is an ill-defined concept. If it really were a set, then it would be an example of a set that is a member of itself. Thus, some "sets" would contain themselves as elements and others would not. Let S be the "set" of "sets that are not elements of themselves"; i.e., $S = \{ A \mid A \notin A \}$. Question: is S a member of itself? If "yes", then S is not a member of itself, because of the defining membership criterion. If "no", then S is a member of itself, due to the defining membership criterion. One resolution is that the collection of all sets is not a set. (See Chapter 4 of [MiRo91].)

7. Paradoxes such as those in Example 6 led Alfred North Whitehead (1861–1947) and Bertrand Russell to develop a version of set theory by categorizing sets based on *set types*: $T_0, T_1, \ldots$. The lowest type, T_0, consists only of individual elements. For $i > 0$, type T_i consists of sets whose elements come from type T_{i-1}. This forces sets to belong to exactly one type. The expression $A \in A$ is always false. In this situation Russell's paradox cannot happen.

1.2.2 SET OPERATIONS

Definitions:

The *intersection* of sets A and B is the set $A \cap B = \{ x \mid (x \in A) \wedge (x \in B) \}$. More generally, the intersection of any family of sets is the set of objects that are members of every set in the family. The notation

$$\bigcap_{i \in I} A_i = \{ x \mid x \in A_i \text{ for all } i \in I \}$$

is used for the intersection of the family of sets A_i indexed by the set I.

Two sets A and B are *disjoint* if $A \cap B = \emptyset$.

A collection of sets $\{ a_i \mid i \in I \}$ is *disjoint* if $\bigcap_{i \in I} A_i = \emptyset$.

A collection of sets is *pairwise disjoint* (or *mutually disjoint*) if every pair of sets in the collection are disjoint.

The *union* of sets A and B is the set $A \cup B = \{ x \mid (x \in A) \vee (x \in B) \}$. More generally, the union of a family of sets is the set of objects that are members of at least one set in the family. The notation

$$\bigcup_{i \in I} A_i = \{ x \mid x \in A_i \text{ for some } i \in I \}$$

is used for the union of the family of sets A_i indexed by the set I.

A *partition* of a set S is a pairwise disjoint family $\mathcal{P} = \{A_i\}$ of nonempty subsets whose union is S.

The partition $\mathcal{P}_2 = \{B_i\}$ of a set S is a *refinement* of the partition $\mathcal{P}_1 = \{A_j\}$ of the same set if for every subset $B_i \in \mathcal{P}_2$ there is a subset $A_j \in \mathcal{P}_1$ such that $B_i \subseteq A_j$.

The *complement* of the set A is the set $\overline{A} = U - A = \{ x \mid x \notin A \}$ containing every object not in A, where the context provides that the objects range over some specific universal domain U. (The notation A' or A^c is sometimes used instead of $\overline{A}$.)

The **set difference** is the set $A - B = A \cap \overline{B} = \{x \mid (x \in A) \wedge (x \notin B)\}$. The set difference is sometimes written $A \setminus B$.

The **symmetric difference** of A and B is the set $A \oplus B = \{x \mid (x \in A - B) \vee (x \in B - A)\}$. This is sometimes written $A \triangle B$.

The **Cartesian product** $A \times B$ of two sets A and B is the set $\{(a,b) \mid (a \in A) \wedge (b \in B)\}$, which contains all ordered pairs whose first coordinate is from A and whose second coordinate is from B. The Cartesian product of $A_1, \ldots, A_n$ is the set $A_1 \times A_2 \times \cdots \times A_n = \prod_{i=1}^{n} A_i = \{(a_1, a_2, \ldots, a_n) \mid (\forall i)(a_i \in A_i)\}$, which contains all ordered n-tuples whose ith coordinate is from A_i. The Cartesian product $A \times A \times \cdots \times A$ is also written A^n. If S is any set, the Cartesian product of the collection of sets A_s, where $s \in S$, is the set $\prod_{s \in S} A_s$ of all functions $f: S \to \bigcup_{s \in S} A_s$ such that $f(s) \in A_s$ for all $s \in S$.

The **power set** of A is the set $\mathcal{P}(A)$ of all subsets of A. The alternative notation 2^A for $\mathcal{P}(A)$ emphasizes the fact that the power set has 2^n elements if A has n elements.

A **set expression** is any expression built up from sets and set operations.

A **set equation** (or **set identity**) is an equation whose left side and right side are both set expressions.

A **system of distinct representatives** (**SDR**) for a collection of sets $A_1, A_2, \ldots, A_n$ (some of which may be equal) is a set $\{a_1, a_2, \ldots, a_n\}$ of n distinct elements such that $a_i \in A_i$ for $i = 1, 2, \ldots, n$.

A **Venn diagram** is a family of n simple closed curves (typically circles or ellipses) arranged in the plane so that all possible intersections of the interiors are nonempty and connected. (John Venn, 1834–1923)

A Venn diagram is **simple** if at most two curves intersect at any point of the plane.

A Venn diagram is **reducible** if there is a sequence of curves whose iterative removal leaves a Venn diagram at each step.

A **membership table** is a table used to calculate whether an object lies in the set described by a set expression, based on its membership in the sets mentioned by the expression.

Facts:

1. If a collection of sets is pairwise disjoint, then the collection is disjoint. The converse is false.

2. The following figure illustrates Venn diagrams for two and three sets.

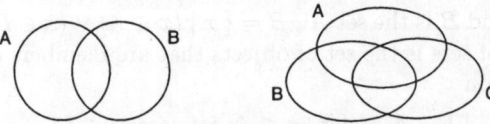

3. The following figure gives the Venn diagrams for sets constructed using various set operations.

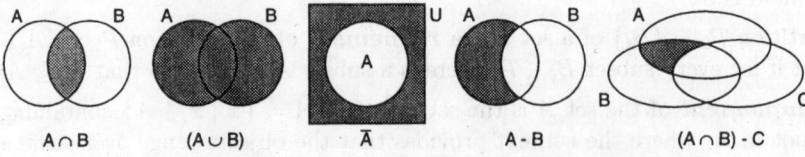

4. Intuition regarding set identities can be gleaned from Venn diagrams, but it can be misleading to use Venn diagrams when proving theorems unless great care is taken to make sure that the diagrams are sufficiently general to illustrate all possible cases.

5. Venn diagrams are often used as an aid to inclusion/exclusion counting. (See §2.4.)

6. Venn gave examples of Venn diagrams with four ellipses and asserted that no Venn diagram could be constructed with five ellipses.

7. Peter Hamburger and Raymond Pippert (1996) constructed a simple, reducible Venn diagram with five congruent ellipses. (Two ellipses are *congruent* if they are the exact same size and shape, and differ only by their placement in the plane.)

8. Many of the logical identities given in §1.1.2 correspond to set identities, given in the following table.

name	rule
Commutative laws	$A \cap B = B \cap A \quad A \cup B = B \cup A$
Associative laws	$A \cap (B \cap C) = (A \cap B) \cap C$ $A \cup (B \cup C) = (A \cup B) \cup C$
Distributive laws	$A \cap (B \cup C) = (A \cap B) \cup (A \cap C)$ $A \cup (B \cap C) = (A \cup B) \cap (A \cup C)$
DeMorgan's laws	$\overline{A \cap B} = \overline{A} \cup \overline{B} \quad \overline{A \cup B} = \overline{A} \cap \overline{B}$
Complement laws	$A \cap \overline{A} = \emptyset \quad A \cup \overline{A} = U$
Double complement law	$\overline{\overline{A}} = A$
Idempotent laws	$A \cap A = A \quad A \cup A = A$
Absorption laws	$A \cap (A \cup B) = A \quad A \cup (A \cap B) = A$
Dominance laws	$A \cap \emptyset = \emptyset \quad A \cup U = U$
Identity laws	$A \cup \emptyset = A \quad A \cap U = A$

9. In a computer, a subset of a relatively small universal domain can be represented by a bit string. Each bit location corresponds to a specific object of the universal domain, and the bit value indicates the presence (1) or absence (0) of that object in the subset.

10. In a computer, a subset of a relatively large ordered datatype or universal domain can be represented by a *binary search tree*.

11. For any two finite sets A and B, $|A \cup B| = |A| + |B| - |A \cap B|$ (inclusion/exclusion principle). (See §2.3.)

12. Set identities can be proved by any of the following:
    - a containment proof: show that the left side is a subset of the right side and the right side is a subset of the left side;
    - a membership table: construct the analogue of the truth table for each side of the equation;
    - using other set identities.

13. For all sets A, $|A| < |\mathcal{P}(A)|$.

14. **Hall's theorem**: A collection of sets $A_1, A_2, \ldots, A_n$ has a system of distinct representatives if and only if for all $k = 1, \ldots, n$ every collection of k subsets $A_{i_1}, A_{i_2}, \ldots, A_{i_k}$ satisfies $|A_{i_1} \cup A_{i_2} \cup \cdots \cup A_{i_k}| \geq k$.

15. If a collection of sets $A_1, A_2, \ldots, A_n$ has a system of distinct representatives and if an integer m has the property that $|A_i| \geq m$ for each i, then:
   - if $m \geq n$ there are at least $\frac{m!}{(m-n)!}$ systems of distinct representatives;
   - if $m < n$ there are at least $m!$ systems of distinct representatives.

16. Systems of distinct representatives can be phrased in terms of 0-1 matrices and graphs. See §6.6.1, §8.12, and §10.4.3.

Examples:

1. $\{1,2\} \cap \{2,3\} = \{2\}$.
2. The collection of sets $\{1,2\}, \{4,5\}, \{6,7,8\}$ is pairwise disjoint, and hence disjoint.
3. The collection of sets $\{1,2\}, \{2,3\}, \{1,3\}$ is disjoint, but not pairwise disjoint.
4. $\{1,2\} \cup \{2,3\} = \{1,2,3\}$.
5. Suppose that for every positive integer n, $[j \bmod n] = \{k \in \mathcal{Z} \mid k \bmod n = j\}$, for $j = 0, 1, \ldots, n-1$. (See §1.3.1.) Then $\{[0 \bmod 3], [1 \bmod 3], [2 \bmod 3]\}$ is a partition of the integers. Moreover, $\{[0 \bmod 6], [1 \bmod 6], \ldots, [5 \bmod 6]\}$ is a refinement of this partition.
6. Within the context of $\mathcal{Z}$ as universal domain, the complement of the set of positive integers is the set consisting of the negative integers and 0.
7. $\{1,2\} - \{2,3\} = \{1\}$.
8. $\{1,2\} \times \{2,3\} = \{(1,2), (1,3), (2,2), (2,3)\}$.
9. $\mathcal{P}(\{1,2\}) = \{\emptyset, \{1\}, \{2\}, \{1,2\}\}$.
10. If L is a line in the plane, and if for each $x \in L$, C_x is the circle of radius 1 centered at point x, then $\bigcup_{x \in L} C_x$ is an infinite strip of width 2, and $\bigcap_{x \in L} C_x = \emptyset$.
11. The five-fold Cartesian product $\{0,1\}^5$ contains 32 different 5-tuples, including, for instance, $(0,0,1,0,1)$.
12. The set identity $\overline{A \cap B} = \overline{A} \cup \overline{B}$ is verified by the following membership table. Begin by listing the possibilities for elements being in or not being in the sets A and B, using 1 to mean "is an element of" and 0 to mean "is not an element of". Proceed to find the element values for each combination of sets. The two sides of the equation are the same since the columns for $\overline{A \cap B}$ and $\overline{A} \cup \overline{B}$ are identical:

A	B	$A \cap B$	$\overline{A \cap B}$	$\overline{A}$	$\overline{B}$	$\overline{A} \cup \overline{B}$
1	1	1	0	0	0	0
1	0	0	1	0	1	1
0	1	0	1	1	0	1
0	0	0	1	1	1	1

13. The collection of sets $A_1 = \{1,2\}$, $A_2 = \{2,3\}$, $A_3 = \{1,3,4\}$ has systems of distinct representatives, for example $\{1,2,3\}$ and $\{2,3,4\}$.

14. The collection of sets $A_1 = \{1,2\}$, $A_2 = \{1,3\}$, $A_3 = \{2,3\}$, $A_4 = \{1,2,3\}$, $A_5 = \{2,3,4\}$ does not have a system of distinct representatives since $|A_1 \cup A_2 \cup A_3 \cup A_4| < 4$.

1.2.3 INFINITE SETS

Definitions:

The **Peano definition** for the natural numbers $\mathcal{N}$:
- 0 is a natural number;
- every natural number n has a successor $s(n)$;
- axioms:
  ⋄ 0 is not the successor of any natural number;
  ⋄ two different natural numbers cannot have the same successor;
  ⋄ if $0 \in T$ and if $(\forall n \in \mathcal{N}) \left[(n \in T) \to (s(n) \in T) \right]$, then $T = \mathcal{N}$.

(This axiomatization is named for Giuseppe Peano, 1858–1932.)

A set is **denumerable** (or **countably infinite**) if it can be put in a one-to-one correspondence with the set of natural numbers $\{0, 1, 2, 3, \ldots\}$. (See §1.3.1.)

A **countable** set is a set that is either finite or denumerable. All other sets are **uncountable**.

The **ordinal numbers** (or **ordinals**) are defined recursively as follows:
- the empty set is the ordinal number 0;
- if α is an ordinal number, then so is the *successor* of α, written α^+ or $\alpha + 1$, which is the set $\alpha \cup \{\alpha\}$;
- if β is any set of ordinals closed under the successor operation, then β is an ordinal, called a *limit ordinal*.

The ordinal α is said to be **less than** the ordinal β, written $\alpha < \beta$, if $\alpha \subseteq \beta$ (which is equivalent to $\alpha \in \beta$).

The **sum** of ordinals α and β, written $\alpha + \beta$, is the ordinal corresponding to the well-ordered set given by all the elements of α in order, followed by all the elements of β (viewed as being disjoint from α) in order. (See Fact 26 and §1.4.3.)

The **product** of ordinals α and β, written $\alpha \cdot \beta$, is the ordinal equal to the Cartesian product $\alpha \times \beta$ with ordering $(a_1, b_1) < (a_2, b_2)$ whenever $b_1 < b_2$, or $b_1 = b_2$ and $a_1 < a_2$ (this is reverse lexicographic order).

Two sets have the *same* **cardinality** (or are **equinumerous**) if they can be put into one-to-one correspondence (§1.3.1.). When the equivalence relation "equinumerous" is used on all sets (see §1.4.2.), the sets in each equivalence class have the same **cardinal number**. The cardinal number of a set A is written $|A|$. It can also be regarded as the smallest ordinal number among all those ordinal numbers with the same cardinality.

An **order relation** can be defined on cardinal numbers of sets by the rule $|A| \leq B$ if there is a one-to-one function $f \colon A \to B$. If $|A| \leq |B|$ and $|A| \neq |B|$, write $|A| < |B|$.

The **sum** of cardinal numbers $\mathbf{a}$ and $\mathbf{b}$, written $\mathbf{a} + \mathbf{b}$, is the cardinal number of the union of two disjoint sets A and B such that $|A| = \mathbf{a}$ and $|B| = \mathbf{b}$.

The **product** of cardinal numbers $\mathbf{a}$ and $\mathbf{b}$, written $\mathbf{ab}$, is the cardinal number of the Cartesian product of two sets A and B such that $|A| = \mathbf{a}$ and $|B| = \mathbf{b}$.

Exponentiation of cardinal numbers, written $\mathbf{a}^{\mathbf{b}}$, is the cardinality of the set A^B of all functions from B to A, where $|A| = \mathbf{a}$ and $|B| = \mathbf{b}$.

Facts:

1. Axiom 3 in the Peano definition of the natural numbers is the principle of mathematical induction. (See §1.5.6.)

2. The finite cardinal numbers are written $0, 1, 2, 3, \ldots$.

3. The cardinal number of any finite set with n elements is n.

4. The first infinite cardinal numbers are written $\aleph_0, \aleph_1, \aleph_2, \ldots, \aleph_\omega, \ldots$.

5. For each ordinal α, there is a cardinal number $\aleph_\alpha$.

6. The cardinal number of any denumerable set, such as $\mathcal{N}$, $\mathcal{Z}$, and $\mathcal{Q}$, is $\aleph_0$.

7. The cardinal number of $\mathcal{P}(\mathcal{N})$, $\mathcal{R}$, and $\mathcal{C}$ is denoted **c** (standing for the *continuum*).

8. The set of algebraic numbers (all solutions of polynomials with integer coefficients) is denumerable.

9. The set $\mathcal{R}$ is uncountable (proved by Georg Cantor in late 19th century, using a diagonal argument). (See §1.5.7.)

10. Every subset of a countable set is countable.

11. The countable union of countable sets is countable.

12. Every set containing an uncountable subset is uncountable.

13. The **continuum problem**, posed by Georg Cantor (1845–1918) and restated by David Hilbert (1862–1943) in 1900, is the problem of determining the cardinality, $|\mathcal{R}|$, of the real numbers.

14. The **continuum hypothesis** is the assertion that $|\mathcal{R}| = \aleph_1$, the first cardinal number larger than $\aleph_0$. Equivalently, $2^{\aleph_0} = \aleph_1$. (See Fact 35.) Kurt Gödel (1906–1978) proved in 1938 that the continuum hypothesis is consistent with various other axioms of set theory. Paul Cohen (born 1934) demonstrated in 1963 that the continuum hypothesis cannot be proved from those other axioms; i.e., it is independent of the other axioms of set theory.

15. The **generalized continuum hypothesis** is the assertion that $2^{\aleph_\alpha} = \aleph_{\alpha+1}$ for all ordinals α. That is, for infinite sets there is no cardinal number strictly between $|S|$ and $|\mathcal{P}(S)|$.

16. The generalized continuum hypothesis is consistent with and independent of the usual axioms of set theory.

17. There is no largest cardinal number.

18. $|A| < |\mathcal{P}(A)|$ for all sets A.

19. *Schröder-Bernstein theorem*: If $|A| \leq |B|$ and $|B| \leq |A|$, then $|A| = |B|$. (This is also called the Cantor-Schröder-Bernstein theorem.)

20. The ordinal number $1 = 0^+ = \{\emptyset\} = \{0\}$, the ordinal number $2 = 1^+ = \{0, 1\}$, etc. In general, for finite ordinals, $n + 1 = n^+ = \{0, 1, 2, \ldots, n\}$.

21. The first limit ordinal is $\omega = \{0, 1, 2, \ldots\}$. Then $\omega + 1 = \omega^+ = \omega \cup \{\omega\} = \{0, 1, 2, \ldots, \omega\}$, and so on. The next limit ordinal is $\omega + \omega = \{0, 1, 2, \ldots, \omega, \omega + 1, \omega + 2, \ldots\}$, also denoted $\omega \cdot 2$. The process never stops, because the next limit ordinal can always be formed as the union of the infinite process that has gone before.

22. Limit ordinals have no immediate predecessors.

23. The first ordinal that, viewed as a set, is not countable, is denoted ω_1.

24. For ordinals the following are equivalent: $\alpha < \beta$, $\alpha \in \beta$, $\alpha \subset \beta$.

25. Every set of ordinal numbers has a smallest element; i.e., the ordinals are well-ordered. (See §1.4.3.)

26. Ordinal numbers correspond to well-ordered sets (§1.4.3). Two well-ordered sets represent the same ordinal if they can be put into an order-preserving one-to-one correspondence.

27. Addition and multiplication of ordinals are associative operations.

28. Ordinal addition and multiplication for finite ordinals (those less than ω) are the same as ordinary addition and multiplication on the natural numbers.

29. Addition of infinite ordinals is not commutative. (See Example 2.)

30. Multiplication of infinite ordinals is not commutative. (See Example 3.)

31. The ordinals 0 and 1 are identities for addition and multiplication, respectively.

32. Multiplication of ordinals is distributive over addition on the left: $\alpha(\beta + \gamma) = \alpha\beta + \alpha\gamma$. It is not distributive on the right.

33. In the definition of the cardinal number $\mathbf{a}^{\mathbf{b}}$, when $\mathbf{a} = 2$, the set A can be taken to be $A = \{0, 1\}$ and an element of A^B can be identified with a subset of B (namely, those elements of B sent to 1 by the function). Thus $2^{|B|} = |\mathcal{P}(B)|$, the cardinality of the power set of B.

34. If $\mathbf{a}$ and $\mathbf{b}$ are cardinals, at least one of which is infinite, then $\mathbf{a} + \mathbf{b} = \mathbf{a} \cdot \mathbf{b} =$ the larger of $\mathbf{a}$ and $\mathbf{b}$.

35. $\mathbf{c}^{\aleph_0} = \aleph_0^{\aleph_0} = 2^{\aleph_0}$

36. The usual rules for finite arithmetic continue to hold for infinite cardinal arithmetic (commutativity, associativity, distributivity, and rules for exponents).

Examples:
1. $\omega_1 > \omega \cdot 2$, $\omega_1 > \omega^2$, $\omega_1 > \omega^\omega$.
2. $1 + \omega = \omega$, but $\omega + 1 > \omega$.
3. $2 \cdot \omega = \omega$, but $\omega \cdot 2 > \omega$.
4. $\aleph_0 \cdot \aleph_0 = \aleph_0 + \aleph_0 = \aleph_0$.

1.2.4 AXIOMS FOR SET THEORY

Set theory can be viewed as an axiomatic system, with undefined terms "set" (the universe of discourse) and "is an element of" (a binary relation denoted $\in$).

Definitions:

The **Axiom of choice** (AC) states: If $\mathcal{A}$ is any set whose elements are pairwise disjoint nonempty sets, then there exists a set X that has as its elements exactly one element from each set in $\mathcal{A}$.

The **Zermelo-Fraenkel (ZF) axioms** for set theory: (The axioms are stated informally.)

- *Extensionality* (equality): Two sets with the same elements are equal.
- *Pairing*: For every a and b, the set $\{a, b\}$ exists.
- *Specification* (subset): If A is a set and $P(x)$ is a predicate with free variable x, then the subset of A exists that consists of those elements $c \in A$ such that $P(c)$ is true. (The specification axiom guarantees that the intersection of two sets exists.)
- *Union*: The union of a set (i.e., the set of all the elements of its elements) exists. (The union axiom together with the pairing axiom implies the existence of the union of two sets.)
- *Power set*: The power set (set of all subsets) of a set exists.
- *Empty set*: The empty set exists.
- *Regularity* (foundation): Every nonempty set contains a "foundational" element; that is, every nonempty set contains an element that is not an element of any other element in the set. (The regularity axiom prevents anomalies such as a set being an element of itself.)
- *Replacement*: If f is a function defined on a set A, then the collection of images $\{f(a) \mid a \in A\}$ is a set. The replacement axiom (together with the union axiom) allows the formation of large sets by expanding each element of a set into a set.
- *Infinity*: An infinite set, such as ω (§1.2.3), exists.

Facts:

1. The axiom of choice is consistent with and independent of the other axioms of set theory; it can be neither proved nor disproved from the other axioms of set theory.

2. The axioms of ZF together with the axiom of choice are denoted ZFC.

3. The following propositions are equivalent to the axiom of choice:
   - *The well-ordering principle*: Every set can be well-ordered; i.e., for every set A there exists a total ordering on A such that every subset of A contains a smallest element under this ordering.
   - *Generalized axiom of choice* (functional version): If $\mathcal{A}$ is any collection of nonempty sets, then there is a function f whose domain is $\mathcal{A}$, such that $f(X) \in X$ for all $X \in \mathcal{A}$.
   - *Zorn's lemma*: Every nonempty partially ordered set in which every chain (totally ordered subset) contains an upper bound (an element greater than all the other elements in the chain) has a maximal element (an element that is less than no other element). (§1.4.3.)
   - *The Hausdorff maximal principle*: Every chain in a partially ordered set is contained in a maximal chain (a chain that is not strictly contained in another chain). (§1.4.3.)
   - *Trichotomy*: Given any two sets A and B, either there is a one-to-one function from A to B, or there is a one-to-one function from B to A; i.e., either $|A| \leq |B|$ or $|B| \leq |A|$.

1.3 FUNCTIONS

A function is a rule that associates to each object in one set an object in a second set (these sets are often sets of numbers). For instance, the expected population in future years, based on demographic models, is a function from calendar years to numbers. Encryption is a function from confidential information to apparent nonsense messages, and decryption is a function from apparent nonsense back to confidential information. Computer scientists and mathematicians are often concerned with developing methods to calculate particular functions quickly.

1.3.1 BASIC TERMINOLOGY FOR FUNCTIONS

Definitions:

A **function** f from a set A to a set B, written $f: A \to B$, is a rule that assigns to every object $a \in A$ exactly one element $f(a) \in B$. The set A is the **domain** of f; the set B is the **codomain** of f; the element $f(a)$ is the **image** of a or the **value** of f at a. A function f is often identified with its **graph** $\{(a,b) \mid a \in A \text{ and } b = f(a)\} \subseteq A \times B$.

Note: The function $f: A \to B$ is sometimes represented by the "maps to" notation $x \mapsto f(x)$ or by the variation $x \mapsto expr(x)$, where $expr(x)$ is an expression in x. The notation $f(x) = expr(x)$ is a form of the "maps to" notation without the symbol $\mapsto$.

The rule defining a function $f: A \to B$ is called **well-defined** since to each $a \in A$ there is associated exactly one element of B.

If $f: A \to B$ and $S \subseteq A$, the **image** of the subset S under f is the set $f(S) = \{f(x) \mid x \in S\}$.

If $f: A \to B$ and $T \subseteq B$, the **pre-image** or **inverse image** of the subset T under f is the set $f^{-1}(T) = \{x \mid f(x) \in T\}$.

The **image** of a function $f: A \to B$ is the set $f(A) = \{f(x) \mid x \in A\}$.

The **range** of a function $f: A \to B$ is the image set $f(A)$. (Some authors use "range" as a synonym for "codomain".)

A function $f: A \to B$ is **one-to-one** (**1–1**, **injective**, or a **monomorphism**) if distinct elements of the domain are mapped to distinct images; i.e., $f(a_1) \neq f(a_2)$ whenever $a_1 \neq a_2$. An **injection** is an injective function.

A function $f: A \to B$ is **onto** (**surjective**, or an **epimorphism**) if every element of the codomain B is the image of at least one element of A; i.e., if $(\forall b \in B)(\exists a \in A)$ $[f(a) = b]$ is true. A **surjection** is a surjective function.

A function $f: A \to B$ is **bijective** (or a **one-to-one correspondence**) if it is both injective and surjective; i.e., it is 1–1 and onto. A **bijection** is a bijective function.

If $f: A \to B$ and $S \subseteq A$, the **restriction** of f to S is the function $f_S: S \to B$ where $f_S(x) = f(x)$ for all $x \in S$. The function f is an **extension** of f_S. The restriction of f to S is also written $f|_S$.

A **partial function** on a set A is a rule f that assigns to each element in a subset of A exactly one element of B. The subset of A on which f is defined is the **domain of definition** of f. In a context that includes partial functions, a rule that applies to all of A is called a **total function**.

Given a 1–1 onto function $f: A \to B$, the **inverse function** $f^{-1}: B \to A$ has the rule that for each $y \in B$, $f^{-1}(y)$ is the object $x \in A$ such that $f(x) = y$.

If $f: A \to B$ and $g: B \to C$, then the **composition** is the function $g \circ f: A \to C$ defined by the rule $(g \circ f)(x) = g(f(x))$ for all $x \in A$. The function to the right of the raised circle is applied first.

Note: Care must be taken since some sources define the composition $(g \circ f)(x) = f(g(x))$ so that the order of application reads left to right.

If $f: A \to A$, the **iterated functions** $f^n: A \to A$ ($n \geq 2$) are defined recursively by the rule $f^n(x) = f \circ f^{n-1}(x)$.

A function $f: A \to A$ is **idempotent** if $f \circ f = f$.

A function $f: A \to A$ is an **involution** if $f \circ f = i_A$. (See Example 1.)

A function whose domain is a Cartesian product $A_1 \times \cdots \times A_n$ is often regarded as a function of n variables (also called a **multivariate** function), and the value of f at $(a_1, \ldots, a_n)$ is usually written $f(a_1, \ldots, a_n)$.

An (**n-ary**) **operation** on a set A is a function $f: A^n \to A$, where $A^n = A \times \cdots \times A$ (with n factors in the product). A 1-ary operation is called **monadic** or **unary**, and a 2-ary operation is called **binary**.

Facts:

1. The graph of a function $f: A \to B$ is a binary relation on $A \times B$. (§1.4.1.)

2. The graph of a function $f: A \to B$ is a subset S of $A \times B$ such that for each $a \in A$ there is exactly one $b \in B$ such that $(a, b) \in S$.

3. In general, two or more different objects in the domain of a function might be assigned the same value in the codomain. If this occurs, the function is not 1–1.

4. If $f: A \to B$ is bijective, then: $f \circ f^{-1} = i_B$ (Example 1), $f^{-1} \circ f = i_A$, f^{-1} is bijective, and $(f^{-1})^{-1} = f$.

5. Function composition is associative: $(f \circ g) \circ h = f \circ (g \circ h)$, whenever $h: A \to B$, $g: B \to C$, and $f: C \to D$.

6. Function composition is not commutative; that is, $f \circ g \neq g \circ f$ in general. (See Example 12.)

7. *Set operations with functions:* If $f: A \to B$ with $S_1, S_2 \subseteq A$ and $T_1, T_2 \subseteq B$, then:
   - $f(S_1 \cup S_2) = f(S_1) \cup f(S_2)$;
   - $f(S_1 \cap S_2) \subseteq f(S_1) \cap f(S_2)$, with equality if f is injective;
   - $f(\overline{S_1}) \supseteq \overline{f(S_1)}$ (i.e., $f(A - S_1) \supseteq B - f(S_1)$), with equality if f is injective;
   - $f^{-1}(T_1 \cup T_2) = f^{-1}(T_1) \cup f^{-1}(T_2)$;
   - $f^{-1}(T_1 \cap T_2) = f^{-1}(T_1) \cap f^{-1}(T_2)$;
   - $f^{-1}(\overline{T_1}) = \overline{f^{-1}(T_1)}$ (i.e., $f^{-1}(B - T_1) = A - f^{-1}(T_1)$);
   - $f^{-1}(f(S_1)) \supseteq S_1$, with equality if f is injective;
   - $f(f^{-1}(T_1)) \subseteq T_1$, with equality if f is surjective.

8. If $f: A \to B$ and $g: B \to C$ are both bijective, then $(g \circ f)^{-1} = f^{-1} \circ g^{-1}$.

9. If an operation $*$ (such as addition) is defined on a set B, then that operation can be extended to the set of all functions from a set A to B, by setting $(f * g)(x) = f(x) * g(x)$.

10. *Numbers of functions*: If $|A| = m$ and $|B| = n$, the numbers of different types of functions $f: A \to B$ are given in the following list:
- all: n^m (§2.2.1)
- one-to-one: $P(n, m) = n(n-1)(n-2)\ldots(n-m+1)$ if $n \geq m$ (§2.2.1)
- onto: $\sum_{j=0}^{n}(-1)^j \binom{n}{j}(n-j)^m$ if $m \geq n$ (§2.4.2)
- partial: $(n+1)^m$ (§2.3.2)

Examples:
1. The following are some common functions:
   - **exponential function to base b** (for $b > 0$, $b \neq 1$): the function $f: \mathcal{R} \to \mathcal{R}^+$ where $f(x) = b^x$. (See the following figure.) ($\mathcal{R}^+$ is the set of positive real numbers.)
   - **logarithm function with base b** (for $b > 0$, $b \neq 1$): the function $\log_b: \mathcal{R}^+ \to \mathcal{R}$ that is the inverse of the exponential function to base b; that is,
     $$\log_b x = y \text{ if and only if } b^y = x.$$
   - **common logarithm function**: the function $\log_{10}: \mathcal{R}^+ \to \mathcal{R}$ (also written log) that is the inverse of the exponential function to base 10; i.e., $\log_{10} x = y$ when $10^y = x$. (See the following figure.)
   - **binary logarithm function**: the function $\log_2: \mathcal{R}^+ \to \mathcal{R}$ (also denoted log or lg) that is the inverse of exponential function to base 2; i.e., $\log_2 x = y$ when $2^y = x$. (See the following figure.)
   - **natural logarithm function**: the function $\ln: \mathcal{R}^+ \to \mathcal{R}$ is the inverse of the exponential function to base e; i.e., $\ln(x) = y$ when $e^y = x$, where $e = \lim_{n \to \infty}(1 + \frac{1}{n})^n \approx 2.718281828459$. (See the following figure.)

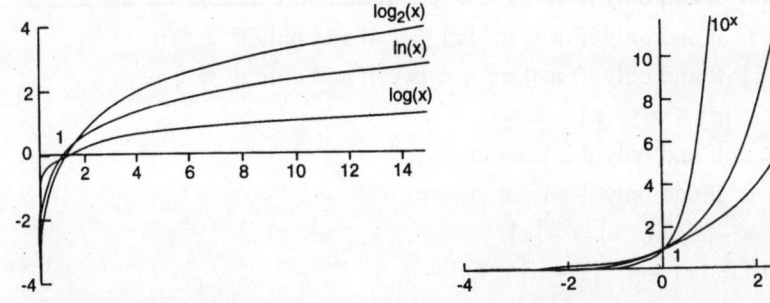

   - **iterated logarithm**: the function $\log^*: \mathcal{R}^+ \to \{0, 1, 2, \ldots\}$ where $\log^* x$ is the smallest nonnegative integer k such that $\log^{(k)} x \leq 1$; the function $\log^{(k)}$ is defined recursively by
     $$\log^{(k)} x = \begin{cases} x & \text{if } k = 0 \\ \log(\log^{(k-1)} x) & \text{if } \log^{(k-1)} x \text{ is defined and positive} \\ \text{undefined} & \text{otherwise.} \end{cases}$$
   - **mod function**: for a given positive integer n, the function $f: \mathcal{Z} \to \mathcal{N}$ defined by the rule $f(k) = k \bmod n$, where $k \bmod n$ is the remainder when the division algorithm is used to divide k by n. (See §4.1.2.)
   - **identity function** on a set A: the function $i_A: A \to A$ such that $i_A(x) = x$ for all $x \in A$.

- **characteristic function** of S: for $S \subseteq A$, the function $\chi_S: A \to \{0,1\}$ given by $\chi_S(x) = 1$ if $x \in S$ and $\chi_S(x) = 0$ if $x \notin S$.
- **projection function**: the function $\pi_j: A_1 \times \cdots \times A_n \to A_j$ $(j = 1, 2, \ldots, n)$ such that $\pi_j(a_1, \ldots, a_n) = a_j$.
- **permutation**: a function $f: A \to A$ that is 1-1 and onto.
- **floor function** (sometimes referred to, especially in number theory, as the **greatest integer function**): the function $\lfloor \; \rfloor: \mathcal{R} \to \mathcal{Z}$ where $\lfloor x \rfloor =$ the greatest integer less than or equal to x. The floor of x is also written $[x]$. (See the following figure.) Thus $\lfloor \pi \rfloor = 3$, $\lfloor 6 \rfloor = 6$, and $\lfloor -0.2 \rfloor = -1$.
- **ceiling function**: the function $\lceil \; \rceil: \mathcal{R} \to \mathcal{Z}$ where $\lceil x \rceil =$ the smallest integer greater than or equal to x. (See the following figure.) Thus $\lceil \pi \rceil = 4$, $\lceil 6 \rceil = 6$, and $\lceil -0.2 \rceil = 0$.

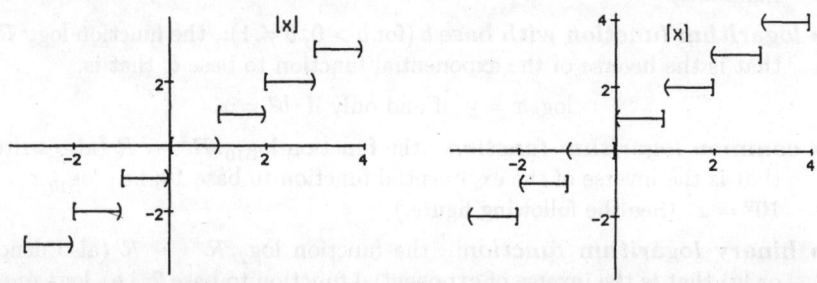

2. The floor and ceiling functions are total functions from the reals $\mathcal{R}$ to the integers $\mathcal{Z}$. They are onto, but not one-to-one.
3. *Properties of the floor and ceiling functions (m and n represent arbitrary integers)*:
    - $\lfloor x \rfloor = n$ if and only if $n \le x < n+1$ if and only if $x - 1 < n \le x$;
    - $\lceil x \rceil = n$ if and only if $n - 1 < x \le n$ if and only if $x \le n < x + 1$;
    - $\lfloor x \rfloor < n$ if and only if $x < n$; $\lceil x \rceil \le n$ if and only if $x \le n$;
    - $n \le \lfloor x \rfloor$ if and only if $n \le x$; $n < \lceil x \rceil$ if and only if $n < x$;
    - $x - 1 < \lfloor x \rfloor \le x \le \lceil x \rceil < x + 1$;
    - $\lfloor x \rfloor = x$ if and only if x is an integer;
    - $\lceil x \rceil = x$ if and only if x is an integer;
    - $\lfloor -x \rfloor = -\lceil x \rceil$; $\lceil -x \rceil = -\lfloor x \rfloor$;
    - $\lfloor x + n \rfloor = \lfloor x \rfloor + n$; $\lceil x + n \rceil = \lceil x \rceil + n$;
    - the interval $[x_1, x_2]$ contains $\lfloor x_2 \rfloor - \lceil x_1 \rceil + 1$ integers;
    - the interval $[x_1, x_2)$ contains $\lceil x_2 \rceil - \lceil x_1 \rceil$ integers;
    - the interval $(x_1, x_2]$ contains $\lfloor x_2 \rfloor - \lfloor x_1 \rfloor$ integers;
    - the interval (x_1, x_2) contains $\lceil x_2 \rceil - \lfloor x_1 \rfloor - 1$ integers;
    - if $f(x)$ is a continuous, monotonically increasing function, and whenever $f(x)$ is an integer, x is also an integer, then $\lfloor f(x) \rfloor = \lfloor f(\lfloor x \rfloor) \rfloor$ and $\lceil f(x) \rceil = \lceil f(\lceil x \rceil) \rceil$;
    - if $n > 0$, then $\lfloor \frac{x+m}{n} \rfloor = \lfloor \frac{\lfloor x \rfloor + m}{n} \rfloor$ and $\lceil \frac{x+m}{n} \rceil = \lceil \frac{\lceil x \rceil + m}{n} \rceil$ (a special case of the preceding fact);
    - if $m > 0$, then $\lfloor mx \rfloor = \lfloor x \rfloor + \lfloor x + \frac{1}{m} \rfloor + \cdots + \lfloor x + \frac{m-1}{m} \rfloor$.
4. The logarithm function $\log_b x$ is bijective from the positive reals $\mathcal{R}^+$ to the reals $\mathcal{R}$.

5. The logarithm function $x \mapsto \log_b x$ is the inverse of the function $x \mapsto b^x$, if the codomain of $x \mapsto b^x$ is the set of positive real numbers. If the domain and codomain are considered to be $\mathcal{R}$, then $x \mapsto \log_b x$ is only a partial function, because the logarithm of a nonpositive number is not defined.

6. All logarithm functions are related according to the following change of base formula: $\log_b x = \frac{\log_a x}{\log_a b}$.

7. $\log^* 2 = 1$, $\log^* 4 = 2$, $\log^* 16 = 3$, $\log^* 65536 = 4$, $\log^* 2^{65536} = 5$.

8. The diagrams in the following figure illustrate a function that is onto but not 1-1 and a function that is 1-1 but not onto.

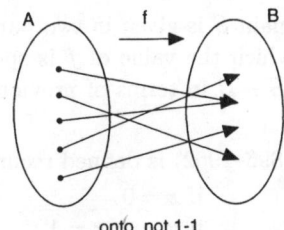
onto, not 1-1

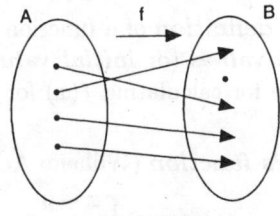
1-1, not onto

9. If the domain and codomain are considered to be the nonnegative reals, then the function $x \mapsto x^2$ is a bijection, and $x \mapsto \sqrt{x}$ is its inverse.

10. If the codomain is considered to be the subset of complex numbers with polar coordinate $0 \leq \theta < \pi$, then $x \mapsto \sqrt{x}$ can be regarded as a total function.

11. Division of real numbers is a multivariate function from $\mathcal{R} \times (\mathcal{R} - \{0\})$ to $\mathcal{R}$, given by the rule $f(x, y) = \frac{x}{y}$. Similarly, addition, subtraction, and multiplication are functions from $\mathcal{R} \times \mathcal{R}$ to $\mathcal{R}$.

12. If $f(x) = x^2$ and $g(x) = x + 1$, then $(f \circ g)(x) = (x+1)^2$ and $(g \circ f)(x) = x^2 + 1$. (Therefore, composition of functions is not commutative.)

13. *Collatz conjecture:* If $f \colon \{1, 2, 3, \ldots\} \to \{1, 2, 3, \ldots\}$ is defined by the rule $f(n) = \frac{n}{2}$ if n is even and $f(n) = 3n + 1$ if n is odd, then for each positive integer m there is a positive integer k such that the iterated function $f^k(m) = 1$. It is not known whether this conjecture is true.

1.3.2 COMPUTATIONAL REPRESENTATION

A given function may be described by several different rules. These rules can then be used to evaluate specific values of the function. There is often a large difference in the time required to compute the value of a function using different computational rules. The speed usually depends on the representation of the data as well as on the computational process.

Definitions:

A (*computational*) *representation* of a function is a way to calculate its values.

A *closed formula* for a function value $f(x)$ is an algebraic expression in the argument x.

A *table of values* for a function $f \colon A \to B$ with finite domain A is any explicit representation of the set $\{(a, f(a)) \in A \times B \mid a \in A\}$.

An **infinite sequence** in a set S is a function from the natural numbers $\{0, 1, 2, \ldots\}$ to the set S. It is commonly represented as a list $x_0, x_1, x_2, \ldots$ such that each $x_j \in S$. Sequences are often permitted to start at the index 1 or elsewhere, rather than 0.

A **finite sequence** in a set S is a function from $\{1, 2, \ldots, n\}$ to the set S. It is commonly represented as a list $x_1, x_2, \ldots, x_n$ such that each $x_j \in S$. Finite sequences are often permitted to start at the index 0 (or at some other value of the index), rather than at the index 1.

A value of a sequence is also called an **entry**, an **item**, or a **term**.

A **string** is a representation of a sequence as a list in which the successive entries are juxtaposed without intervening punctuation or extra spacing.

A **recursive definition** of a function f with domain S is given in two parts: there is a set of **base values** (or **initial values**) B on which the value of f is specified, and there is a rule for calculating $f(x)$ for every $x \in S - B$ in terms of previously defined values of f.

Ackermann's function (Wilhelm Ackermann, 1896–1962) is defined recursively by

$$A(x, y, z) = \begin{cases} x + y & \text{if } z = 0 \\ 0 & \text{if } y = 0, z = 1 \\ 1 & \text{if } y = 0, z = 2 \\ x & \text{if } y = 0, z > 2 \\ A(x, A(x, y-1, z), z-1) & \text{if } y, z > 0. \end{cases}$$

An alternative version of Ackermann's function, with two variables, is defined recursively by

$$A(m, n) = \begin{cases} n + 1 & \text{if } m = 0 \\ A(m-1, 1) & \text{if } m > 0, n = 0 \\ A(m-1, A(m, n-1)) & \text{if } m, n > 0. \end{cases}$$

Another alternative version of Ackermann's function is defined recursively by the rule $A(n) = A_n(n)$, where $A_1(n) = 2n$ and $A_m(n) = A_{m-1}^{(n)}(1)$ if $m \geq 2$.

The (input-independent) **halting function** maps computer programs to the set $\{0, 1\}$, with value 1 if the program always halts, regardless of input, and 0 otherwise.

Facts:

1. If $f: \mathcal{N} \to \mathcal{R}$ is recursively defined, the set of base values is frequently the set $\{f(0), f(1), \ldots, f(j)\}$ and there is a rule for calculating $f(n)$ for every $n > j$ in terms of $f(i)$ for one or more $i < n$.

2. There are functions whose values cannot be computed. (See Example 5.)

3. There are recursively defined functions that cannot be represented by a closed formula.

4. It is possible to find closed formulas for the values of some functions defined recursively. See Chapter 3 for more information.

5. Computer software developers often represent a table as a *binary search tree* (§17.2).

6. In Ackermann's function of three variables $A(x, y, z)$, as the variable z ranges from 0 to 3, $A(x, y, z)$ is the sum of x and y, the product of x and y, x raised to the exponent y, and the iterated exponentiation of x y times. That is, $A(x, y, 0) = x + y$, $A(x, y, 1) = xy$, $A(x, y, 2) = x^y$, $A(x, y, 3) = x^{x^{\cdot^{\cdot^{\cdot^x}}}}$ (y xs in the exponent).

7. The version of Ackermann's function with two variables, $A(x, y)$, has the following properties: $A(1, n) = n + 2$, $A(2, n) = 2n + 3$, $A(3, n) = 2^{n+3} - 3$.

8. $A(m, n)$ is an example of a well-defined total function that is computable, but not primitive recursive. (See §16.)

Examples:

1. The function that maps each month to its ordinal position is represented by the table
$$\{(Jan, 1), (Feb, 2), \ldots, (Dec, 12)\}.$$

2. The function defined by the recurrence relation
$$f(0) = 0; \quad f(n) = f(n-1) + 2n - 1 \text{ for } n \geq 1$$
has the closed form $f(x) = x^2$.

3. The function defined by the recurrence relation
$$f(0) = 0, f(1) = 1; \quad f(n) = f(n-1) + f(n-2) \text{ for } n \geq 2$$
generates the Fibonacci sequence $0, 1, 1, 2, 3, 5, 8, \ldots$ (see §3.1.2) and has the closed form
$$f(n) = \frac{(1+\sqrt{5})^n - (1-\sqrt{5})^n}{2^n \sqrt{5}}.$$

4. The factorial function $n!$ is recursively defined by the rules
$$0! = 1; \quad n! = n \cdot (n-1)! \text{ for } n \geq 1.$$
It has no known closed formula in terms of elementary functions.

5. It is impossible to construct an algorithm to compute the halting function.

6. The halting function from the Cartesian product of the set of computer programs and the set of strings to $\{0, 1\}$ whose value is 1 if the program halts when given that string as input and 0 if the program does not halt when given that string as input is noncomputable.

7. The following is not a well-defined function $f: \{1, 2, 3, \ldots\} \to \{1, 2, 3, \ldots\}$
$$f(n) = \begin{cases} 1 & \text{if } n = 1 \\ 1 + f(\frac{n}{2}) & \text{if } n \text{ is even} \\ f(3n - 1) & \text{if } n \text{ is odd}, n > 1 \end{cases}$$
since evaluating $f(5)$ leads to the contradiction $f(5) = f(5) + 3$.

8. It is not known whether the following is a well-defined function $f: \{1, 2, 3, \ldots\} \to \{1, 2, 3, \ldots\}$
$$f(n) = \begin{cases} 1 & n = 1 \\ 1 + f(\frac{n}{2}) & n \text{ even} \\ f(3n + 1) & n \text{ odd}, n > 1. \end{cases}$$

(See §1.3.1, Example 13.)

1.3.3 ASYMPTOTIC BEHAVIOR

The asymptotic growth of functions is commonly described with various special pieces of notation and is regularly used in the analysis of computer algorithms to estimate the length of time the algorithms take to run and the amount of computer memory they require.

Definitions:

A function $f: \mathcal{R} \to \mathcal{R}$ or $f: \mathcal{N} \to \mathcal{R}$ is **bounded** if there is a constant k such that $|f(x)| \leq k$ for all x in the domain of f.

For functions $f, g: \mathcal{R} \to \mathcal{R}$ or $f, g: \mathcal{N} \to \mathcal{R}$ (sequences of real numbers) the following are used to compare their growth rates:

- f is **big-oh** of g (g **dominates** f) if there exist constants C and k such that $|f(x)| \leq C|g(x)|$ for all $x > k$.
  Notation: f is $O(g)$, $f(x) \in O(g(x))$, $f \in O(g)$, $f = O(g)$.

- f is **little-oh** of g if $\lim_{x \to \infty} \left| \frac{f(x)}{g(x)} \right| = 0$; i.e., for every $C > 0$ there is a constant k such that $|f(x)| \leq C|g(x)|$ for all $x > k$.
  Notation: f is $o(g)$, $f(x) \in o(g(x))$, $f \in o(g)$, $f = o(g)$.

- f is **big omega of** g if there are constants C and k such that $|g(x)| \leq C|f(x)|$ for all $x > k$.
  Notation: f is $\Omega(g)$, $f(x) \in \Omega(g(x))$, $f \in \Omega(g)$, $f = \Omega(g)$.

- f is **little omega of** g if $\lim_{x \to \infty} \left| \frac{g(x)}{f(x)} \right| = 0$.
  Notation: f is $\omega(g)$, $f(x) \in \omega(g(x))$, $f \in \omega(g)$, $f = \omega(g)$.

- f is **theta of** g if there are positive constants C_1, C_2, and k such that $C_1|g(x)| \leq |f(x)| \leq C_2|g(x)|$ for all $x > k$.
  Notation: f is $\Theta(g)$, $f(x) \in \Theta(g(x))$, $f \in \Theta(g)$, $f = \Theta(g)$, $f \approx g$.

- f is **asymptotic** to g if $\lim_{x \to \infty} \frac{g(x)}{f(x)} = 1$. This relation is sometimes called **asymptotic equality**.
  Notation: $f \sim g$, $f(x) \sim g(x)$.

Facts:

1. The notations $O(\ )$, $o(\ )$, $\Omega(\ )$, $\omega(\ )$, and $\Theta(\ )$ all stand for *collections* of functions. Hence the equality sign, as in $f = O(g)$, does not mean equality of functions.

2. The symbols $O(g)$, $o(g)$, $\Omega(g)$, $\omega(g)$, and $\Theta(g)$ are frequently used to represent a typical element of the class of functions it represents, as in an expression such as $f(n) = n \log n + o(n)$.

3. *Growth rates*:
   - $O(g)$: the set of functions that grow no more rapidly than a positive multiple of g;
   - $o(g)$: the set of functions that grow less rapidly than a positive multiple of g;
   - $\Omega(g)$: the set of functions that grow at least as rapidly as a positive multiple of g;
   - $\omega(g)$: the set of functions that grow more rapidly than a positive multiple of g;
   - $\Theta(g)$: the set of functions that grow at the same rate as a positive multiple of g.

4. Asymptotic notation can be used to describe the growth of infinite sequences, since infinite sequences are functions from $\{0, 1, 2, \ldots\}$ or $\{1, 2, 3, \ldots\}$ to $\mathcal{R}$ (by considering the term a_n as $a(n)$, the value of the function $a(n)$ at the integer n).

5. The big-oh notation was introduced in 1892 by Paul Bachmann (1837–1920) in the study of the rates of growth of various functions in number theory.

6. The big-oh symbol is often called a *Landau symbol*, after Edmund Landau (1877–1938), who popularized this notation.

7. Properties of big-oh:
   - if $f \in O(g)$ and c is a constant, then $cf \in O(g)$;
   - if $f_1, f_2 \in O(g)$, then $f_1 + f_2 \in O(g)$;
   - if $f_1 \in O(g_1)$ and $f_2 \in O(g_2)$, then
      ◦ $(f_1 + f_2) \in O(g_1 + g_2)$
      ◦ $(f_1 + f_2) \in O(\max(|g_1|, |g_2|))$
      ◦ $(f_1 f_2) \in O(g_1 g_2)$;
   - if f is a polynomial of degree n, then $f \in O(x^n)$;
   - if f is a polynomial of degree m and g a polynomial of degree n, with $m \geq n$, then $\frac{f}{g} \in O(x^{m-n})$;
   - if f is a bounded function, then $f \in O(1)$;
   - for all $a, b > 1$, $O(\log_a x) = O(\log_b x)$;
   - if $f \in O(g)$ and $|h(x)| \geq |g(x)|$ for all $x > k$, then $f \in O(h)$;
   - if $f \in O(x^m)$, then $f \in O(x^n)$ for all $n > m$.

8. Some of the most commonly used benchmark big-oh classes are: $O(1)$, $O(\log x)$, $O(x)$, $O(x \log x)$, $O(x^2)$, $O(2^x)$, $O(x!)$, and $O(x^x)$. If f is big-oh of any function in this list, then f is also big-oh of each of the following functions in the list:

$$O(1) \subset O(\log x) \subset O(x) \subset O(x \log x) \subset O(x^2) \subset O(2^x) \subset O(x!) \subset O(x^x).$$

The benchmark functions are drawn in the following figure.

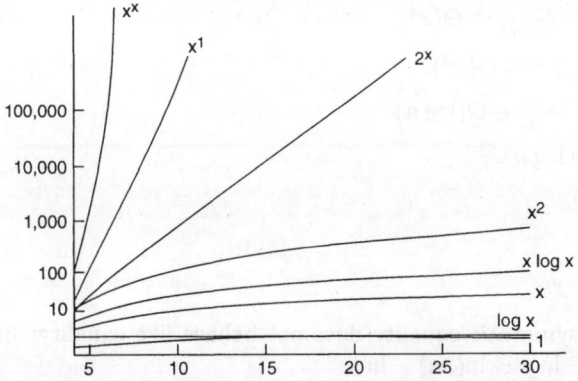

9. Properties of little-oh:
   - if $f \in o(g)$, then $cf \in o(g)$ for all nonzero constants c;
   - if $f_1 \in o(g)$ and $f_2 \in o(g)$, then $f_1 + f_2 \in o(g)$;
   - if $f_1 \in o(g_1)$ and $f_2 \in o(g_2)$, then
      ◦ $(f_1 + f_2) \in o(g_1 + g_2)$
      ◦ $(f_1 + f_2) \in o(\max(|g_1|, |g_2|))$
      ◦ $(f_1 f_2) \in o(g_1 g_2)$;
   - if f is a polynomial of degree m and g a polynomial of degree n with $m < n$, then $\frac{f}{g} \in o(1)$;
   - the set membership $f(x) \in L + o(1)$ is equivalent to $f(x) \to L$ as $x \to \infty$, where L is a constant.

10. If $f \in o(g)$, then $f \in O(g)$; the converse is not true.
11. If $f \in O(g)$ and $h \in o(f)$, then $h \in o(g)$.
12. If $f \in o(g)$ and $h \in O(f)$, then $h \in O(g)$.
13. If $f \in O(g)$ and $h \in O(f)$, then $h \in O(g)$.
14. If $f_1 \in o(g_1)$ and $f_2 \in O(g_2)$, then $f_1 f_2 \in o(g_1 g_2)$.
15. $f \in O(g)$ if and only if $g \in \Omega(f)$.
16. $f \in \Theta(g)$ if and only if $f \in O(g)$ and $g \in O(f)$.
17. $f \in \Theta(g)$ if and only if $f \in O(g)$ and $f \in \Omega(g)$.
18. If $f(x) = a_n x^n + \cdots + a_1 x + a_0$ $(a_n \neq 0)$, then $f \sim a_n x^n$.
19. $f \sim g$ if and only if $\left(\frac{f}{g} - 1\right) \in o(1)$ (provided $g(x) = 0$ only finitely often).

Examples:
1. $5x^8 + 10^{200} x^5 + 3x + 1 \in O(x^8)$.
2. $x^3 \in O(x^4)$, $x^4 \notin O(x^3)$.
3. $x^3 \in o(x^4)$, $x^4 \notin o(x^3)$.
4. $x^3 \notin o(x^3)$.
5. $x^2 \in O(5x^2)$; $x^2 \notin o(5x^2)$.
6. $\sin(x) \in O(1)$.
7. $\frac{x^7 - 3x}{8x^3 + 5} \in O(x^4)$; $\frac{x^7 - 3x}{8x^3 + 5} \in \Theta(x^4)$
8. $1 + 2 + 3 + \cdots + n \in O(n^2)$.
9. $1 + \frac{1}{2} + \frac{1}{3} + \cdots + \frac{1}{n} \in O(\log n)$.
10. $\log(n!) \in O(n \log n)$.
11. $8x^5 \in \Theta(3x^5)$.
12. $x^3 \in \Omega(x^2)$.
13. $2^n + o(n^2) \sim 2^n$.
14. Sometimes asymptotic equality does not behave like equality: $\ln n \sim \ln(2n)$, but $n \not\sim 2n$ and $\ln n - \ln n \not\sim \ln(2n) - \ln n$.
15. $\pi(n) \sim \frac{n}{\ln n}$ where $\pi(n)$ is the number of primes less than or equal to n.
16. If p_n is the nth prime, then $p_n \sim n \ln n$.
17. Stirling's formula: $n! \sim \sqrt{2\pi n} \left(\frac{n}{e}\right)^n$.

1.4 RELATIONS

Relationships between two sets (or among more that two sets) occur frequently throughout mathematics and its applications. Examples of such relationships include integers and their divisors, real numbers and their logarithms, corporations and their customers, cities and airlines that serve them, people and their relatives. These relationships can be described as subsets of product sets.

Functions are a special type of relation. Equivalence relations can be used to describe similarity among elements of sets and partial order relations describe the relative size of elements of sets.

1.4.1 BINARY RELATIONS AND THEIR PROPERTIES

Definitions:

A *binary relation* from set A to set B is any subset R of $A \times B$.

An element $a \in A$ *is related to* $b \in B$ in the relation R if $(a, b) \in R$, often written aRb. If $(a,b) \notin R$, write $a\not{R}b$.

A *binary relation* (*relation*) on a set A is a binary relation from A to A; i.e., a subset of $A \times A$.

A binary relation R on A can have the following properties (to have the property, the relation must satisfy the property for all $a, b, c \in A$):

- *reflexivity*: aRa
- *irreflexivity*: $a\not{R}a$
- *symmetry*:   if aRb, then bRa
- *asymmetry*:   if aRb, then $b\not{R}a$
- *antisymmetry*:   if aRb and bRa, then $a = b$
- *transitivity*:   if aRb and bRc, then aRc
- *intransitivity*:   if aRb and bRc, then $a\not{R}c$

Binary relations R and S from A to B can be combined in the following ways to yield other relations:

- *complement* of R: the relation $\overline{R}$ from A to B where $a\overline{R}b$ if and only if $a\not{R}b$ (i.e., $\neg(aRb)$)
- *difference*:   the binary relation $R - S$ from A to B such that $a(R - S)b$ if and only if aRb and $\neg(aSb)$
- *intersection*:   the relation $R \cap S$ from A to B where $a(R \cap S)b$ if and only if aRb and aSb
- *inverse* (*converse*):   the relation R^{-1} from B to A where $bR^{-1}a$ if and only if aRb
- *symmetric difference*:   the relation $R \oplus S$ from A to B where $a(R \oplus S)b$ if and only if exactly one of the following is true: aRb, aSb
- *union*:   the relation $R \cup S$ from A to B where $a(R \cup S)b$ if and only if aRb or aSb.

The *closure* of a relation R with respect to a property $\mathcal{P}$ is the relation S, if it exists, that has property $\mathcal{P}$ and contains R, such that S is a subset of every relation that has property $\mathcal{P}$ and contains R.

A relation R on A is *connected* if for all $a, b \in A$ with $a \neq b$, either aRb or there are $c_1, c_2, \ldots, c_k \in A$ such that $aRc_1, c_1Rc_2, \ldots, c_{k-1}Rc_k, c_kRb$.

If R is a relation on A, the **connectivity relation associated with** R is the relation R' where $aR'b$ if and only if aRb or there are $c_1, c_2, \ldots, c_k \in A$ such that $aRc_1, c_1Rc_2, \ldots, c_{k-1}Rc_k, c_kRb$.

If R is a binary relation from A to B and if S is a binary relation from B to C, then the **composition** of R and S is the binary relation $S \circ R$ from A to C where $a(S \circ R)c$ if and only if there is an element $b \in B$ such that aRb and bSc.

The **nth power** (n a nonnegative integer) of a relation R on a set A, is the relation R^n, where $R^0 = \{(a, a) \mid a \in A\} = I_A$ (see Example 4), $R^1 = R$ and $R^n = R^{n-1} \circ R$ for all integers $n > 1$.

A **transitive reduction** of a relation, if it exists, is a relation with the same transitive closure as the original relation and with a minimal superset of ordered pairs.

Notation:
1. If a relation R is symmetric, aRb is often written $a \sim b$, $a \approx b$, or $a \equiv b$.
2. If a relation R is antisymmetric, aRb is often written $a \leq b$, $a < b$, $a \subset b$, $a \subseteq b$, $a \preceq b$, $a \prec b$, or $a \sqsubseteq b$.

Facts:
1. A binary relation R from A to B can be viewed as a function from the Cartesian product $A \times B$ to the boolean domain {TRUE, FALSE} (often written $\{T, F\}$). The truth value of the pair (a, b) determines whether a is related to b.
2. Under the *infix convention* for a binary relation, aRb (*a is related to b*) means $R(a, b) = $ TRUE; $a\cancel{R}b$ (*a is not related to b*) means $R(a, b) = $ FALSE.
3. A binary relation R from A to B can be represented in any of the following ways:
    - a set $R \subseteq A \times B$, where $(a, b) \in R$ if and only if aRb (this is the definition of R);
    - a directed graph D_R whose vertices are the elements of $A \cup B$, with an edge from vertex a to vertex b if aRb (§8.3.1);
    - a matrix (the adjacency matrix for the directed graph D_R): if $A = \{a_1, \ldots, a_m\}$ and $B = \{b_1, \ldots, b_n\}$, the matrix for the relation R is the $m \times n$ matrix M_R with entries m_{ij} where $m_{ij} = 1$ if a_iRb_j and $m_{ij} = 0$ otherwise.
4. R is a reflexive relation on A if and only if $\{(a, a) \mid a \in A\} \subseteq R$; i.e., R is a reflexive relation on A if and only if $I_A \subseteq R$.
5. R is symmetric if and only if $R = R^{-1}$.
6. R is an antisymmetric relation on A if and only if $R \cap R^{-1} \subseteq \{(a, a) \mid a \in A\}$.
7. R is transitive if and only if $R \circ R \subseteq R$.
8. A relation R can be both symmetric and antisymmetric. See the first example in Table 2.
9. For a relation R that is both symmetric and antisymmetric: R is reflexive if and only if R is the equality relation on some set; R is irreflexive if and only if $R = \emptyset$.
10. The closure of a relation R with respect to a property $\mathcal{P}$ is the intersection of all relations Q with property $\mathcal{P}$ such that $R \subseteq Q$, if there is at least one such relation Q.
11. The transitive closure of a relation R is the connectivity relation R' associated with R, which is equal to the union $\bigcup_{i=1}^{\infty} R^i$ of all the positive powers of the relation.
12. A transitive reduction of a relation may contain pairs not in the original relation (Example 8).

13. Transitive reductions are not necessarily unique (Example 9).
14. If R is a relation on A and $x, y \in A$ with $x \neq y$, then x is related to y in the transitive closure of R if and only if there is a nontrivial directed path from x to y in the directed graph D_R of the relation.
15. The following table shows how to obtain various closures of a relation and gives the matrices for the various closures of a relation R with matrix M_R on a set A where $|A| = n$.

relation	set	matrix
reflexive closure	$R \cup \{(a,a) \mid a \in A\}$	$M_R \vee I_n$
symmetric closure	$R \cup R^{-1}$	$M_R \vee M_{R^{-1}}$
transitive closure	$\bigcup_{i=1}^{n} R^i$	$M_R \vee M_R^{[2]} \vee \cdots \vee M_R^{[n]}$

The matrix I_n is the $n \times n$ identity matrix, $M_R^{[i]}$ is the ith boolean power of the matrix M_R for the relation R, and $\vee$ is the join operator (defined by $0 \vee 0 = 0$ and $0 \vee 1 = 1 \vee 0 = 1 \vee 1 = 1$).

16. The following table provides formulas for the number of binary relations with various properties on a set with n elements.

type of relation	number of relations
all relations	2^{n^2}
reflexive	$2^{n(n-1)}$
symmetric	$2^{n(n+1)/2}$
transitive	no known simple closed formula (§3.1.7)
antisymmetric	$2^n \cdot 3^{n(n-1)/2}$
asymmetric	$3^{n(n-1)/2}$
irreflexive	$2^{n(n-1)}$
equivalence (§1.4.2)	$B_n =$ Bell number $= \sum_{k=1}^{n} \left\{ {n \atop k} \right\}$ where $\left\{ {n \atop k} \right\}$ is a Stirling subset number (§2.4.2)
partial order (§1.4.3)	no known simple closed formula (§3.1.7)

Algorithm:
1. Warshall's algorithm, also called the Roy-Warshall algorithm (B. Roy and S. Warshall described the algorithm in 1959 and 1960, respectively), Algorithm 1, is an algorithm of order n^3 for finding the transitive closure of a relation on a set with n elements. (Stephen Warshall, born 1935)

Algorithm 1: Warshall's algorithm.

input: $M = [m_{ij}]_{n \times n} =$ the matrix representing the binary relation R
output: $M =$ the transitive closure of relation R
for $k := 1$ to n
  for $i := 1$ to n
   for $j := 1$ to n
    $m_{ij} := m_{ij} \vee (m_{ik} \wedge m_{kj})$

Chapter 1 FOUNDATIONS

Examples:

1. Some common relations and whether they have certain properties are given in the following table:

set	relation	reflexive	symmetric	antisymmetric	transitive
any nonempty set	$=$	yes	yes	yes	yes
any nonempty set	$\neq$	no	yes	no	no
$\mathcal{R}$	$\leq$ (or $\geq$)	yes	no	yes	yes
$\mathcal{R}$	$<$ (or $>$)	no	no	yes	yes
positive integers	is a divisor of	yes	no	yes	yes
nonzero integers	is a divisor of	yes	no	no	yes
integers	congruence mod n	yes	yes	no	yes
any set of sets	$\subseteq$ (or $\supseteq$)	yes	no	yes	yes
any set of sets	$\subset$ (or $\supset$)	no	no	yes	yes

2. If A is any set, the *universal relation* is the relation R on $A \times A$ such that aRb for all $a, b \in A$; i.e., $R = A \times A$

3. If A is any set, the *empty relation* is the relation R on $A \times A$ where aRb is never true; i.e., $R = \emptyset$.

4. If A is any set, the relation R on A where aRb if any only if $a = b$ is the *identity* (or *diagonal*) *relation* $I = I_A = \{(a, a) \mid a \in A\}$, which is also written Δ or Δ_A.

5. Every function $f: A \to B$ induces a binary relation R_f from A to B under the rule $aR_f b$ if and only if $f(a) = b$.

6. For $A = \{2, 3, 4, 6, 12\}$, suppose that aRb means that a is a divisor of b. Then R can be represented by the set
$\{(2, 2), (2, 4), (2, 6), (2, 12), (3, 3), (3, 6), (3, 12), (4, 4), (4, 12), (6, 6), (6, 12), (12, 12)\}$.
The relation R can also be represented by the digraph with the following adjacency matrix

$$\begin{pmatrix} 1 & 0 & 1 & 1 & 1 \\ 0 & 1 & 0 & 1 & 1 \\ 0 & 0 & 1 & 0 & 1 \\ 0 & 0 & 0 & 1 & 1 \\ 0 & 0 & 0 & 0 & 1 \end{pmatrix}$$

7. The transitive closure of the relation $\{(1, 3), (2, 3), (3, 2)\}$ on $\{1, 2, 3\}$ is the relation $\{(1, 2), (1, 3), (2, 2), (2, 3), (3, 2), (3, 3)\}$.

8. The transitive closure of the relation $R = \{(1, 2), (2, 3), (3, 1)\}$ on $\{1, 2, 3\}$ is the universal relation $\{1, 2, 3\} \times \{1, 2, 3\}$. A transitive reduction of R is the relation given by $\{(1, 3), (3, 2), (2, 1)\}$. This shows that a transitive reduction may contain pairs that are not in the original relation.

9. If $R = \{(a, b) \mid aRb$ for all $a, b \in \{1, 2, 3\}\}$, then the relations $\{(1, 2), (2, 3), (3, 1)\}$ and $\{(1, 3), (3, 2), (2, 1)\}$ are both transitive reductions for R. Thus, transitive reductions are not unique.

1.4.2 EQUIVALENCE RELATIONS

Equivalence relations are binary relations that describe various types of similarity or "equality" among elements in a set. The elements that look alike or behave in a similar way are grouped together in equivalence classes, resulting in a partition of the set. Any element chosen from an equivalence class essentially "mirrors" the behavior of all elements in that class.

Definitions:

An *equivalence relation* on A is a binary relation on A that is reflexive, symmetric, and transitive.

If R is an equivalence relation on A, the *equivalence class* of $a \in A$ is the set $R[a] = \{b \in A \mid aRb\}$. When it is clear from context which equivalence relation is intended, the notation for the induced equivalence class can be abbreviated $[a]$.

The *induced partition* on a set A under an equivalence relation R is the set of equivalence classes.

Facts:

1. A nonempty relation R is an equivalence relation if and only if $R \circ R^{-1} = R$.
2. The induced partition on a set A actually is a partition of A; i.e., the equivalence classes are all nonempty, every element of A lies in some equivalence class, and two classes $[a]$ and $[b]$ are either disjoint or equal.
3. There is a one-to-one correspondence between the set of all possible equivalence relations on a set A and the set of all possible partitions of A. (Fact 2 shows how to obtain a partition from an equivalence relation. To obtain an equivalence relation from a partition of A, define R by the rule aRb if and only if a and b lie in the same element of the partition.)
4. For any set A, the coarsest partition (with only one set in the partition) of A is induced by the equivalence relation in which every pair of elements are related. The finest partition (with each set in the partition having cardinality 1) of A is induced by the equivalence relation in which no two different elements are related.
5. The set of all partitions of a set A is partially ordered under refinement (§1.2.2 and §1.4.3). This partial ordering is a lattice (§5.7).
6. To find the smallest equivalence relation containing a given relation, first take the transitive closure of the relation, then take the reflexive closure of that relation, and finally take the symmetric closure.

Examples:

1. For any function $f: A \to B$, define the relation $a_1 R a_2$ to mean that $f(a_1) = f(a_2)$. Then R is an equivalence relation. Each induced equivalence class is the inverse image $f^{-1}(b)$ of some $b \in B$.
2. Write $a \equiv b \pmod{n}$ ("a is congruent to b modulo n") when a, b and $n > 0$ are integers such that $n \mid b - a$ (n divides $b - a$). Congruence mod n is an equivalence relation on the integers.
3. The equivalence relation of congruence modulo n on the integers $\mathcal{Z}$ yields a partition with n equivalence classes: $[0] = \{kn \mid k \in \mathcal{Z}\}, [1] = \{1 + kn \mid k \in \mathcal{Z}\}, [2] = \{2 + kn \mid k \in \mathcal{Z}\}, \ldots, [n-1] = \{(n-1) + kn \mid k \in \mathcal{Z}\}$.

4. The isomorphism relation on any set of groups is an equivalence relation. (The same result holds for rings, fields, etc.) (See Chapter 5.)

5. The congruence relation for geometric objects in the plane is an equivalence relation.

6. The similarity relation for geometric objects in the plane is an equivalence relation.

1.4.3 PARTIALLY ORDERED SETS

Partial orderings extend the relationship of $\leq$ on real numbers and allow a comparison of the relative "size" of elements in various sets. They are developed in greater detail in Chapter 11.

Definitions:

A **preorder** on a set S is a binary relation $\leq$ on S that has the following properties for all $a, b, c \in S$:
- reflexive: $a \leq a$
- transitive: if $a \leq b$ and $b \leq c$, then $a \leq c$.

A **partial ordering** (or **partial order**) on a set S is a binary relation $\leq$ on S that has the following properties for all $a, b, c \in S$:
- reflexive: $a \leq a$
- antisymmetric: if $a \leq b$ and $b \leq a$, then $a = b$
- transitive: if $a \leq b$ and $b \leq c$, then $a \leq c$.

Notes: The expression $c \geq b$ means that $b \leq c$. The symbols $\preceq$ and $\succeq$ are often used in place of $\leq$ and $\geq$. The expression $a < b$ (or $b > a$) means that $a \leq b$ and $a \neq b$.

A **partially ordered set** (or **poset**) is a set with a partial ordering defined on it.

A **directed ordering** on a set S is a partial ordering that also satisfies the following property: if $a, b \in S$, then there is a $c \in S$ such that $a \leq c$ and $b \leq c$.

Note: Some authors do not require that antisymmetry hold in the definition of directed ordering.

Two elements a and b in a poset are **comparable** if either $a \leq b$ or $b \leq a$. Otherwise, they are **incomparable**.

A **totally ordered** (or **linearly ordered**) set is a poset in which every pair of elements are comparable.

A **chain** is a subset of a poset in which every pair of elements are comparable.

An **antichain** is a subset of a poset in which no two distinct elements are comparable.

An **interval** in a poset $(S, \leq)$ is a subset $[a, b] = \{\, x \mid x \in S, a \leq x \leq b \,\}$.

An element b in a poset is **minimal** if there exists no element c such that $c < b$.

An element b in a poset is **maximal** if there exists no element c such that $c > b$.

An element b in a poset S is a **maximum element** (or **greatest element**) if every element c satisfies the relation $c \leq b$.

An element b in a poset S is a **minimum element** (or **least element**) if every element c satisfies the relation $c \geq b$.

A **well-ordered** set is a poset $(S, \leq)$ in which every nonempty subset contains a minimum element.

An element b in a poset S is an **upper bound** for a subset $U \subseteq S$ if every element c of U satisfies the relation $c \leq b$.

An element b in a poset S is a **lower bound** for a subset $U \subseteq S$ if every element c of U satisfies the relation $c \geq b$.

A **least upper bound** for a subset U of a poset S is an upper bound b such that if c is any other upper bound for U then $c \geq b$.

A **greatest lower bound** for a subset U of a poset S is a lower bound b such that if c is any other lower bound for U then $c \leq b$.

A **lattice** is a poset in which every pair of elements, x and y, have both a least upper bound $\mathrm{lub}(x, y)$ and a greatest lower bound $\mathrm{glb}(x, y)$ (§5.7).

The **Cartesian product** of two posets $(S_1, \leq_1)$ and $(S_2, \leq_2)$ is the poset with domain $S_1 \times S_2$ and relation $\leq_1 \times \leq_2$ given by the rule $(a_1, a_2) \leq_1 \times \leq_2 (b_1, b_2)$ if and only if $a_1 \leq_1 b_1$ and $a_2 \leq_2 b_2$.

The element c **covers** another element b in a poset if $b < c$ and there is no element d such that $b < d < c$.

A **Hasse diagram** (**cover diagram**) for a poset $(S, \leq)$ is a directed graph (§11.8) whose vertices are the elements of S such that there is an arc from b to c if c covers b, all arcs are directed upward on the page when drawing the diagram, and arrows on the arcs are omitted.

Facts:

1. R is a partial order on a set S if and only if R^{-1} is a partial order on S.
2. The only partial order that is also an equivalence relation is the relation of equality.
3. The Cartesian product of two posets, each with at least two elements, is not totally ordered.
4. In the Hasse diagram for a poset, there is a path from vertex b to vertex c if and only if $b \leq c$. (When $b = c$, it is the path of length 0.)
5. Least upper bounds and greatest lower bounds are unique, if they exist.

Examples:

1. The positive integers are partially ordered under the relation of divisibility, in which $b \leq c$ means that b divides c. In fact, they form a lattice (§5.7.1), called the *divisibility lattice*. The least upper bound of two numbers is their least common multiple, and the greatest lower bound is their greatest common divisor.
2. The set of all powers of two (or of any other positive integer) forms a chain in the divisibility lattice.
3. The set of all primes forms an antichain in the divisibility lattice.
4. The set $\mathcal{R}$ of real numbers with the usual definition of $\leq$ is a totally ordered set.
5. The set of all logical propositions on a fixed set of logical variables $p, q, r, \ldots$ is partially ordered under inverse implication, so that $B \leq A$ means that $A \to B$ is a tautology.
6. The complex numbers, ordered under magnitude, do *not* form a poset, because they do not satisfy the axiom of antisymmetry.

7. The set of all subsets of any set forms a lattice under the relation of subset inclusion. The least upper bound of two subsets is their union, and the greatest lower bound is their intersection. Part (a) in the following figure gives the Hasse diagram for the lattice of all subsets of $\{a, b, c\}$.

8. Part (b) of the following figure shows the Hasse diagram for the lattice of all positive integer divisors of 12.

9. Part (c) of the following figure shows the Hasse diagram for the set $\{1, 2, 3, 4, 5, 6\}$ under divisibility.

10. Part (d) of the following figure shows the Hasse diagram for the set $\{1, 2, 3, 4\}$ with the usual definition of $\leq$.

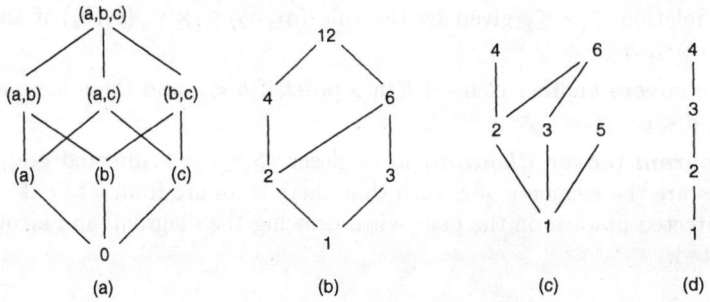

11. *Multilevel security policy*: The flow of information is often restricted by using security clearances. Documents are put into security classes, (L, C), where L is an element of a totally ordered set of authority levels (such as "unclassified", "confidential", "secret", "top secret") and C is a subset (called a "compartment") of a set of subject areas. The subject areas might consist of topics such as agriculture, Eastern Europe, economy, crime, and trade. A document on how trade affects the economic structure of Eastern Europe might be assigned to the compartment {trade, economy, Eastern Europe}. The set of security classes is made into a lattice by the rule: $(L_1, C_1) \leq (L_2, C_2)$ if and only if $L_1 \leq L_2$ and $C_1 \subseteq C_2$. Information is allowed to flow from class (L_1, C_1) to class (L_2, C_2) if and only if $(L_1, C_1) \leq (L_2, C_2)$. For example, a document with security class (secret, {trade, economy}) flows to both (top secret, {trade, economy}) and (secret, {trade, economy, Eastern Europe}), but not vice versa. This set of security classes forms a lattice (§5.7.1).

1.4.4 *n*-ARY RELATIONS

Definitions:

An **n-ary relation** on sets $A_1, A_2, \ldots, A_n$ is any subset R of $A_1 \times A_2 \times \cdots \times A_n$.

The sets A_i are called the **domains** of the relation and the number n is called the **degree** of the relation.

A **primary key** of an n-ary relation R on $A_1 \times A_2 \times \cdots \times A_n$ is a domain A_i such that each $a_i \in A_i$ is the ith coordinate of at most one n-tuple in R.

A *composite key* of an n-ary relation R on $A_1 \times A_2 \times \cdots \times A_n$ is a product of domains $A_{i_1} \times A_{i_2} \times \cdots \times A_{i_m}$ such that for each m-tuple $(a_{i_1}, a_{i_2}, \ldots, a_{i_m}) \in A_{i_1} \times A_{i_2} \times \cdots \times A_{i_m}$, there is at most one n-tuple in R that matches $(a_{i_1}, a_{i_2}, \ldots, a_{i_m})$ in coordinates $i_1, i_2, \ldots, i_m$.

The *projection function* $P_{i_1,i_2,\ldots,i_k} : A_1 \times A_2 \times \cdots \times A_n \to A_{i_1} \times A_{i_2} \times \cdots \times A_{i_k}$ is given by the rule

$$P_{i_1,i_2,\ldots,i_k}(a_1, a_2, \ldots, a_n) = (a_{i_1}, a_{i_2}, \ldots, a_{i_k}).$$

That is, $P_{i_1,i_2,\ldots,i_k}$ selects the elements in coordinate positions $i_1, i_2, \ldots, i_k$ from the n-tuple $(a_1, a_2, \ldots, a_n)$.

The *join* $J_k(R, S)$ of an m-ary relation R and an n-ary relation S, where $k \leq m$ and $k \leq n$, is a relation of degree $m + n - k$ such that

$$(a_1, \ldots, a_{m-k}, c_1, \ldots, c_k, b_1, \ldots, b_{n-k}) \in J_k(R, S)$$

if and only if

$$(a_1, \ldots, a_{m-k}, c_1, \ldots, c_k) \in R \text{ and } (c_1, \ldots, c_k, b_1, \ldots, b_{n-k}) \in S.$$

Facts:

1. An n-ary relation on sets $A_1, A_2, \ldots, A_n$ can be regarded as a function R from $A_1 \times A_2 \times \cdots \times A_n$ to the Boolean domain {TRUE, FALSE}, where $(a_1, a_2, \ldots, a_n) \in R$ if and only if $R(a_1, a_2, \ldots, a_n) = $ TRUE.

2. n-ary relations are essential models in the construction of database systems.

Examples:

1. Let A_1 be the set of all men and A_2 the set of all women, in a nonpolygamous society. Let mRw mean that m and w are presently married. Then each of A_1 and A_2 is a primary key.

2. Let A_1 be the set of all telephone numbers and A_2 the set of all persons. Let nRp mean that telephone number n belongs to person p. Then A_1 is a primary key if each number is assigned to at most one person, and A_2 is a primary key if each person has at most one phone number.

3. In a conventional telephone directory, the name and address domains can form a composite key, unless there are two persons with the same name (no distinguishing middle initial or suffix such as "Jr.") at the same address.

4. Let $A = B = C = \mathcal{Z}$, and let R be the relation on $A \times B \times C$ such that $(a, b, c) \in R$ if and only if $a + b = c$. The set $A \times B$ is a composite key. There is no primary key.

5. Let $A = $ all students at a certain college, $B = $ all student ID numbers being used at the college, $C = $ all major programs at the college. Suppose a relation R is defined on $A \times B \times C$ by the rule $(a, b, c) \in R$ means student a with ID number b has major c. If each student has exactly one major and if there is a one-to-one correspondence between students and ID numbers, then A and B are each primary keys.

6. Let $A = $ all employee names at a certain corporation, $B = $ all Social Security numbers, $C = $ all departments, $D = $ all job titles, $E = $ all salary amounts, and $F = $ all calendar dates. On $A \times B \times C \times D \times E \times F \times F$ let R be the relation such that $(a, b, c, d, e, f, g) \in R$ means employee named a with Social Security number b works in department c, has job title d, earns an annual salary e, was hired on date f, and had the most recent performance review on date g. The projection $P_{1,5}$ (projection onto $A \times E$) gives a list of employees and their salaries.

1.5 PROOF TECHNIQUES

A proof is a derivation of new facts from old ones. A proof makes possible the derivation of properties of a mathematical model from its definition, or the drawing of scientific inferences based on data that have been gathered. Axioms and postulates capture all basic truths used to develop a theory. Constructing proofs is one of the principal activities of mathematicians.

Furthermore, proofs play an important role in computer science — in such areas as verification of the correctness of computer programs, verification of communications protocols, automatic reasoning systems, and logic programming.

1.5.1 RULES OF INFERENCE

Definitions:

A *proposition* is a declarative sentence that is unambiguously either true or false. (See §1.1.1.)

A *theorem* is a proposition derived as the conclusion of a valid proof from axioms and definitions.

A *lemma* is a theorem that is an intermediate step in the proof of a more important theorem.

A *corollary* is a theorem that is derived as an easy consequence of another theorem.

A *statement form* is a declarative sentence containing some variables and logical symbols, such that the sentence becomes a proposition if concrete values are substituted for all the free variables.

An *argument form* is a sequence of statement forms.

The final statement form in an argument form is called the ***conclusion*** (of the argument). The conclusion is often preceded by the word "therefore" (symbolized $\therefore$).

The statement forms preceding the conclusion in an argument form are called ***premises*** (of the argument).

If concrete values are substituted for the free variables of an argument form, an ***argument of that form*** is obtained.

An *instantiation* of an argument is the substitution of concrete values into all free variables of the premises and conclusion.

A *valid argument form* is an argument form such that in every instantiation in which all the premises are true, the conclusion is also true.

A *rule of inference* is an alternative name for a valid argument form, which is used when the form is frequently applied.

Facts:

1. *Substitution rule*: Any variable occurring in an argument may be replaced by an expression of the same type without affecting the validity of the argument, as long as the replacement is made everywhere the variable occurs.

2. The following table gives rules of inference for arguments with compound statements.

name	argument form	name	argument form
Modus ponens (method of affirming)	$p \to q$ p $\therefore q$	Modus tollens (method of denying)	$p \to q$ $\neg q$ $\therefore \neg p$
Hypothetical syllogism	$p \to q$ $q \to r$ $\therefore p \to r$	Disjunctive syllogism	$p \vee q$ $\neg p$ $\therefore q$
Disjunctive addition	p $\therefore p \vee q$	Dilemma by cases	$p \vee q$ $p \to r$ $q \to r$ $\therefore r$
Constructive dilemma	$p \vee r$ $p \to q$ $r \to s$ $\therefore q \vee s$	Destructive dilemma	$\neg q \vee \neg s$ $p \to q$ $r \to s$ $\therefore \neg p \vee \neg r$
Conjunctive addition	p q $\therefore p \wedge q$	Conditional proof	p $p \wedge q \to r$ $\therefore q \to r$
Conjunctive simplification	$p \wedge q$ $\therefore p$	Rule of contradiction	given contradiction c $\neg p \to c$ $\therefore p$

3. The following table gives rules of inference for arguments with quantifiers.

name	argument form
Universal instantiation	$(\forall x \in D)\, Q(x)$ $\therefore Q(a)$ (a any particular element of D)
Generalizing from the generic particular	$Q(a)$ (a an arbitrarily chosen element of D) $\therefore (\forall x \in D)\, Q(x)$
Existential specification	$(\exists x \in D)\, Q(x)$ $\therefore Q(a)$ (for at least one $a \in D$)
Existential generalization	$Q(a)$ (for at least one element $a \in D$) $\therefore (\exists x \in D)\, Q(x)$

4. Substituting $R(x) \to S(x)$ in place of $Q(x)$ and z in place of x in generalizing from the generic particular gives the following inferential rule:

Universal modus ponens: $R(a) \to S(a)$ for any particular but arbitrarily chosen $a \in D$
$\therefore (\forall z \in D)\, [R(z) \to S(z)]$.

5. The rule of generalizing from the generic particular determines the outline of most mathematical proofs.

6. The rule of existential specification is used in deductive reasoning to give names to quantities that are known to exist but whose exact values are unknown.

7. A useful strategy for determining whether a statement is true is to first try to prove it using a variety of approaches and proof methods. If this is unsuccessful, the next step may be to try to disprove the statement, such as by trying to construct or prove the existence of a counterexample. If this does not work, the next step is to try to prove the statement again, and so on. This is one of the many ways in which many mathematicians attempt to develop new results.

Examples:

1. Suppose that D is the set of all objects in the physical universe, $P(x)$ is "x is a human being", $Q(x)$ is "x is mortal", and a is the Greek philosopher Socrates.

argument form	an argument of that form
$(\forall x \in D)\,[P(x) \to Q(x)]$	$\forall$ objects x, (x is a human being) $\to$ (x is mortal).
	(*informally*: All human beings are mortal.)
$P(a)$ (for particular $a \in D$)	Socrates is a human being.
$\therefore Q(a)$	$\therefore$ Socrates is mortal.

2. The argument form shown below is invalid: there is an argument of this form (shown next to it) that has true premises and a false conclusion.

argument form	an argument of that form
$(\forall x \in D)\,[P(x) \to Q(x)]$	$\forall$ objects x, (x is a human being) $\to$ (x is mortal).
	(*informally*: All human beings are mortal.)
$Q(a)$ (for particular $a \in D$)	My cat Bunbury is mortal.
$\therefore P(a)$	$\therefore$ My cat Bunbury is a human being.

In this example, D is the set of all objects in the physical universe, $P(x)$ is "x is a human being", $Q(x)$ is "x is mortal", and a is my cat Bunbury.

3. The distributive law for real numbers, $(\forall a, b, c \in \mathcal{R})[ac + bc = (a+b)c]$, implies that $2\sqrt{2} + 3\sqrt{2} = (2+3)\sqrt{2}$ (because 2, 3, and $\sqrt{2}$ are particular real numbers).

4. Since 2 is a prime number that is not odd, the rule of existential generalization implies the truth of the statement "$\exists$ a prime number n such that n is not odd".

5. To prove that the square of every even integer is even, by the rule of generalizing from the generic particular, begin by supposing that n is any particular but arbitrarily chosen even integer. The job of the proof is to deduce that n^2 is even.

6. By definition, every even integer equals twice some integer. So if at some stage of a reasoning process there is a particular even integer n, it follows from the rule of existential specification that $n = 2k$ for *some* integer k (even though the numerical values of n and k may be unknown).

1.5.2 PROOFS

Definitions:

A (**logical**) **proof** of a statement is a finite sequence of statements (called the **steps** of the proof) leading from a set of premises to the given statement. Each step of the proof must either be a premise or follow from some previous steps by a valid rule of inference.

In a **mathematical proof**, the set of premises may contain any item of previously proved or agreed upon mathematical knowledge (definitions, axioms, theorems, etc.) as well as the specific hypotheses of the statement to be proved.

A **direct proof** of a statement of the form $p \to q$ is a proof that assumes p to be true and then shows that q is true.

An **indirect proof** of a statement of the form $p \to q$ is a proof that assumes that $\neg q$ is true and then shows that $\neg p$ is true. That is, a proof of this form is a direct proof of the contrapositive $\neg q \to \neg p$.

A **proof by contradiction** assumes the negation of the statement to be proved and shows that this leads to a contradiction.

Facts:

1. A useful strategy to determine if a statement of the form $(\forall x \in D)[P(x) \to Q(x)]$ is true or false is to imagine an element $x \in D$ that satisfies $P(x)$ and, using this assumption (and other facts), investigate whether x must also satisfy $Q(x)$. If the answer for all such x is "yes", the given statement is true and the result of the investigation is a direct proof. If it is possible to find an $x \in D$ for which $Q(x)$ is false, the statement is false and this value of x is a counterexample. If the investigation shows that is not possible to find an $x \in D$ for which $Q(x)$ is false, the given statement is true and the result of the investigation is a proof by contradiction.

2. There are many types of techniques that can be used to prove theorems. Table 2 describes how to approach proofs of various types of statements.

Examples:

1. In the following direct proof (see Table 1, item 2), the domain D is the set of all pairs of integers, x is (m, n), and the predicate $P(m, n)$ is "if m and n are even, then $m + n$ is even".

> **Theorem**: For all integers m and n, if m and n are even, then $m + n$ is even.
> **Proof**: Suppose m and n are arbitrarily chosen even integers. [$m + n$ must be shown to be even.]
> 1. $\therefore$ $m = 2r$, $n = 2s$ for some integers r and s (by definition of even)
> 2. $\therefore$ $m + n = 2r + 2s$ (by substitution)
> 3. $\therefore$ $m + n = 2(r + s)$ (by factoring out the 2)
> 4. $r + s$ is an integer (it is a sum of two integers)
> 5. $\therefore$ $m + n$ is even (by definition of even)

The following partial expansion of the proof shows how some of the steps are justified by rules of inference combined with previous mathematical knowledge:

1. Every even integer equals twice some integer:
   $[\forall$ even $x \in \mathcal{Z}$ $(x = 2y$ for some $y \in \mathcal{Z})]$
   m is a particular even integer.
   $\therefore m = 2r$ for some integer r.

3. Every integer is a real number: $[\forall n \in \mathcal{Z}$ $(n \in \mathcal{R})]$
   ($\forall$ integer n, n is a real number.)
   r and s are particular integers.
   $\therefore$ r and s are real numbers.
   The distributive law holds for real numbers: $[\forall a, b, c \in \mathcal{R}$ $(ab + ac = a(b + c))]$
   $2, r$, and s are particular real numbers.
   $\therefore 2r + 2s = 2(r + s)$.

4. Any sum of two integers is an integer: $[\forall m, n \in \mathcal{Z}$ $(m + n \in \mathcal{Z})]$
   r and s are particular integers.
   $\therefore r + s$ is an integer.

Table 1 Techniques of proof.

statement	technique of proof
$p \to q$	*Direct proof*: Assume that p is true. Use rules of inference and previously accepted axioms, definitions, theorems, and facts to deduce that q is true.
$(\forall x \in D)P(x)$	*Direct proof*: Suppose that x is an arbitrary element of D. Use rules of inference and previously accepted axioms, definitions, and facts to deduce that $P(x)$ is true.
$(\exists x \in D)P(x)$	*Constructive direct proof*: Use rules of inference and previously accepted axioms, definitions, and facts to actually find an $x \in D$ for which $P(x)$ is true.
	Nonconstructive direct proof: Deduce the existence of x from other mathematical facts without a description of how to compute it.
$(\forall x \in D)(\exists y \in E)P(x,y)$	*Constructive direct proof*: Assume that x is an arbitrary element of D. Use rules of inference and previously accepted axioms, definitions, and facts to show the existence of a $y \in E$ for which $P(x,y)$ is true, in such a way that y can be computed as a function of x.
	Nonconstructive direct proof: Assume x is an arbitrary element of D. Deduce the existence of y from other mathematical facts without a description of how to compute it.
$p \to q$	*Proof by cases*: Suppose $p \equiv p_1 \vee \cdots \vee p_k$. Prove that each conditional $p_i \to q$ is true. The basis for division into cases is the logical equivalence $[(p_1 \vee \cdots \vee p_k) \to q] \equiv [(p_1 \to q) \wedge \cdots \wedge (p_k \to q)]$.
$p \to q$	*Indirect proof* or *Proof by contraposition*: Assume that $\neg q$ is true (that is, assume that q is false). Use rules of inference and previously accepted axioms, definitions, and facts to show that $\neg p$ is true (that is, p is false).
$p \to q$	*Proof by contradiction*: Assume that $p \to q$ is false (that is, assume that p is true and q is false). Use rules of inference and previously accepted axioms, definitions, and facts to show that a contradiction results. This means that $p \to q$ cannot be false, and hence must be true.
$(\exists x \in D)P(x)$	*Proof by contradiction*: Assume that there is no $x \in D$ for which $P(x)$ is true. Show that a contradiction results.
$(\forall x \in D)P(x)$	*Proof by contradiction*: Assume that there is some $x \in D$ for which $P(x)$ is false. Show that a contradiction results.
$p \to (q \vee r)$	*Proof of a disjunction*: Prove that one of its logical equivalences $(p \wedge \neg q) \to r$ or $(p \wedge \neg r) \to q$ is true.
$p_1, \ldots, p_k$ are equivalent	*Proof by cycle of implications*: Prove $p_1 \to p_2$, $p_2 \to p_3$, $\ldots$, $p_{k-1} \to p_k$, $p_k \to p_1$. This is equivalent to proving $(p_1 \to p_2) \wedge (p_2 \to p_3) \wedge \cdots \wedge (p_{k-1} \to p_k) \wedge (p_k \to p_1)$.

5. Any integer that equals twice some integer is even: $[\forall x \in \mathcal{Z}$ (if $x = 2y$ for some $y \in \mathcal{Z}$, then x is even.)]
   $2(r + s)$ equals twice the integer $r + s$.
   $\therefore 2(r + s)$ is even.

2. *A constructive existence proof*:

   Theorem: Given any integer n, there is an integer m with $m > n$.

   Proof: Suppose that n is an integer. Let $m = n + 1$. Then m is an integer and $m > n$.

The proof is constructive because it established the existence of the desired integer m by showing that its value can be computed by adding 1 to the value of n.

3. *A Nonconstructive existence proof*:

   Theorem: Given a nonnegative integer n, there is always a prime number p that is greater than n.

   Proof: Suppose that n is a nonnegative integer. Consider $n! + 1$. Then $n! + 1$ is divisible by some prime number p because every integer greater than 1 is divisible by a prime number, and $n! + 1 > 1$. Also, $p > n$ because when $n! + 1$ is divided by any positive integer less than or equal to n, the remainder is 1 (since any such number is a factor of $n!$).

The proof is a nonconstructive existence proof because it demonstrated the existence of the number p, but it offered no computational rule for finding it.

4. *A proof by cases*:

   Theorem: For all odd integers n, the number $n^2 - 1$ is divisible by 8.

   Proof: Suppose n is an odd integer. When n is divided by 4, the remainder is 0, 1, 2, or 3. Hence n has one of the four forms $4k$, $4k + 1$, $4k + 2$, or $4k + 3$ for some integer k. But n is odd. So $n \neq 4k$ and $n \neq 4k + 2$. Thus either $n = 4k + 1$ or $n = 4k + 3$ for some integer k.

   Case 1 $[n = 4k + 1$ for some integer $k]$: In this case $n^2 - 1 = (4k + 1)^2 - 1 = 16k^2 + 8k + 1 - 1 = 16k^2 + 8k = 8(2k^2 + k)$, which is divisible by 8 because $2k^2 + k$ is an integer.

   Case 2 $[n = 4k + 3$ for some integer $k]$: In this case $n^2 - 1 = (4k + 3)^2 - 1 = 16k^2 + 24k + 9 - 1 = 16k^2 + 24k + 8 = 8(2k^2 + 3k + 1)$, which is divisible by 8 because $2k^2 + 3k + 1$ is an integer.

   So in either case $n^2 - 1$ is divisible by 8, and thus the given statement is proved.

5. *A proof by contraposition*:

   Theorem: For all integers n, if n^2 is even, then n is even.

   Proof: Suppose that n is an integer that is not even. Then when n is divided by 2 the remainder is 1, or, equivalently, $n = 2k + 1$ for some integer k. By substitution, $n^2 = (2k + 1)^2 = 4k^2 + 4k + 1 = 2(2k^2 + 2k) + 1$. It follows that when n^2 is divided by 2 the remainder is 1 (because $2k^2 + 2k$ is an integer). Thus, n^2 is not even.

In this proof by contraposition, a direct proof of the contrapositive "if n is not even, then n^2 is not even" was given.

6. *A proof by contradiction*:

   Theorem: $\sqrt{2}$ is irrational.

   Proof: Suppose not; that is, suppose that $\sqrt{2}$ were a rational number. By definition of rational, there would exist integers a and b such that $\sqrt{2} = \frac{a}{b}$, or, equivalently, $2b^2 = a^2$. Now the prime factorization of the left-hand side of this equation contains an odd number of factors and that of the right-hand side contains an even number of factors (because every prime factor in an integer occurs twice in the prime factorization of the square of that integer). But this is impossible because the prime factorization of every integer is unique. This yields a contradiction, which shows that the original supposition was false. Hence $\sqrt{2}$ is irrational.

7. *A proof by cycle of implications*:

   Theorem: For all positive integers a and b, the following statements are equivalent:

   (1) a is a divisor of b;
   (2) the greatest common divisor of a and b is a;
   (3) $\lfloor \frac{b}{a} \rfloor = \frac{b}{a}$.

   Proof: Let a and b be positive integers.

   (1) $\rightarrow$ (2): Suppose that a is a divisor of b. Since a is also a divisor of a, a is a common divisor of a and b. But no integer greater than a is a divisor of a. So the greatest common divisor of a and b is a.

   (2) $\rightarrow$ (3): Suppose that the greatest common divisor of a and b is a. Then a is a divisor of both a and b, so $b = ak$ for some integer k. Then $\frac{b}{a} = k$, an integer, and so by definition of floor, $\lfloor \frac{b}{a} \rfloor = k = \frac{b}{a}$.

   (3) $\rightarrow$ (1): Suppose that $\lfloor \frac{b}{a} \rfloor = \frac{b}{a}$. Let $k = \lfloor \frac{b}{a} \rfloor$. Then $k = \lfloor \frac{b}{a} \rfloor = \frac{b}{a}$, and k is an integer by definition of floor. Multiplying the outer parts of the equality by a gives $b = ak$, so by definition of divisibility, a is a divisor of b.

8. *A proof of a disjunction*:

   Theorem: For all integers a and p, if p is prime, then either p is a divisor of a, or a and p have no common factor greater than 1.

   Proof: Suppose a and p are integers and p is prime, but p is not a divisor of a. Since p is prime, its only positive divisors are 1 and p. So, since p is not a divisor of a, the only possible positive common divisor of a and p is 1. Hence a and p have no common divisor greater than 1.

1.5.3 DISPROOFS

Definitions:

A *disproof* of a statement is a proof that the statement is false.

A *counterexample* to a statement of the form $(\forall x \in D) P(x)$ is an element $b \in D$ for which $P(b)$ is false.

Facts:

1. The method of disproof by counterexample is based on the following fact:
$$\neg[(\forall x \in D)\, P(x)] \Leftrightarrow (\exists x \in D)\,[\neg P(x)].$$

2. The following table describes how to give various types of disproofs:

statement	technique of disproof
$(\forall x \in D) P(x)$	*Constructive disproof by counterexample*: Exhibit a specific $a \in D$ for which $P(a)$ is false.
$(\forall x \in D) P(x)$	*Existence disproof*: Prove the existence of some $a \in D$ for which $P(a)$ is false.
$(\exists x \in D) P(x)$	Prove that there is no $a \in D$ for which $P(a)$ is true.
$(\forall x \in D)\,[P(x) \to Q(x)]$	Find an element $a \in D$ with $P(a)$ true and $Q(a)$ false.
$(\forall x \in D)(\exists y \in E)\, P(x,y)$	Find an element $a \in D$ with $P(a,y)$ false for every $y \in E$.
$(\exists x \in D)(\forall y \in E)\, P(x,y)$	Prove that there is no $a \in D$ for which $P(a,y)$ is true for every possible $a \in E$.

Examples:

1. The statement $(\forall a, b \in \mathcal{R})\,[a^2 < b^2 \to a < b]$ is disproved by the following counterexample: $a = 2$, $b = -3$. Then $a^2 < b^2$ (because $4 < 9$) but $a \not< b$ (because $2 \not< -3$).
2. The statement "every prime number is odd" is disproved by the following counterexample: $n = 2$, since n is prime and not odd.

1.5.4 MATHEMATICAL INDUCTION

Definitions:

The **principle of mathematical induction (weak form)** is the following rule of inference for proving that all the items in a list $x_0, x_1, x_2, \ldots$ have some property $P(x)$:

$P(x_0)$ is true	basis premise
$(\forall k \geq 0)\,$ [if $P(x_k)$ is true, then $P(x_{k+1})$ is true]	induction premise
$\therefore (\forall n \geq 0)\,[P(x_n)$ is true].	conclusion

The antecedent $P(x_k)$ in the induction premise "if $P(x_k)$ is true, then $P(x_{k+1})$ is true" is called the **induction hypothesis**.

The **basis step** of a proof by mathematical induction is a proof of the basis premise.

The **induction step** of a proof by mathematical induction is a proof of the induction premise.

The **principle of mathematical induction (strong form)** is the following rule of inference for proving that all the items in a list $x_0, x_1, x_2, \ldots$ have some property $P(x)$:

$P(x_0)$ is true	basis premise
$(\forall k \geq 0)\,$ [if $P(x_0), P(x_1), \ldots, P(x_k)$ are all true, then $P(x_{k+1})$ is true]	(strong) induction premise
$\therefore (\forall n \geq 0)\,[P(x_n)$ is true].	conclusion

The **well-ordering principle for the integers** is the following axiom: If S is a nonempty set of integers such that every element of S is greater than some fixed integer, then S contains a least element.

Facts:

1. Typically, the principle of mathematical induction is used to prove that one of the following sequences of statements is true: $P(0), P(1), P(2), \ldots$ or $P(1), P(2), P(3), \ldots$. In these cases the principle of mathematical induction has the form: if $P(0)$ is true and $P(n) \to P(n+1)$ is true for all $n \geq 0$, then $P(n)$ is true for all $n \geq 0$; or if $P(1)$ is true and $P(n) \to P(n+1)$ is true for all $n \geq 1$, then $P(n)$ is true for all $n \geq 1$

2. If the truth of $P(n+1)$ can be obtained from the previous statement $P(n)$, the weak form of the principle of mathematical induction can be used. If the truth of $P(n+1)$ requires the use of one or more statements $P(k)$ for $k \leq n$, then the strong form should be used.

3. Mathematical induction can also be used to prove statements that can be phrased in the form "For all integers $n \geq k$, $P(n)$ is true".

4. Mathematical induction can often be used to prove summation formulas and inequalities.

5. There are alternative forms of mathematical induction, such as the following:
   - if $P(0)$ and $P(1)$ are true, and if $P(n) \to P(n+2)$ is true for all $n \geq 0$, then $P(n)$ is true for all $n \geq 0$;
   - if $P(0)$ and $P(1)$ are true, and if $[P(n) \wedge P(n+1)] \to P(n+2)$ is true for all $n \geq 0$, then $P(n)$ is true for all $n \geq 0$.

6. The weak form of the principle of mathematical induction, the strong form of the principle of mathematical induction, and the well-ordering principle for the integers are all regarded as axioms for the integers. This is because they cannot be derived from the usual simpler axioms used in the definition of the integers. (See the Peano definition of the natural numbers in §1.2.3.)

7. The weak form of the principle of mathematical induction, the strong form of the principle of mathematical induction, and the well-ordering principle for the integers are all equivalent. In other words, each of them can be proved from each of the others.

8. The earliest recorded use of mathematical induction occurs in 1575 in the book *Arithmeticorum Libri Duo* by Francesco Maurolico, who used the principle to prove that the sum of the first n odd positive integers is n^2.

Examples:

1. *A proof using the weak form of mathematical induction*: (In this proof, $x_0, x_1, x_2, \ldots$ is the sequence $1, 2, 3, \ldots$, and the property $P(x_n)$ is the equation $1 + 2 + \cdots + n = \frac{n(n+1)}{2}$.)

   Theorem: For all integers $n \geq 1$, $1 + 2 + \cdots + n = \frac{n(n+1)}{2}$.

   Proof:
   Basis Step: For $n = 1$ the left-hand side of the formula is 1, and the right-hand side is $\frac{1(1+1)}{2}$, which is also equal to 1. Hence $P(1)$ is true.
   Induction Step: Let k be an integer, $k \geq 1$, and suppose that $P(k)$ is true. That is, suppose that $1 + 2 + \cdots + k = \frac{k(k+1)}{2}$ (the induction hypothesis) is true. It must be shown that $P(k+1)$ is true: $1 + 2 + \cdots + (k+1) = \frac{(k+1)((k+1)+1)}{2}$,

or, equivalently, that $1 + 2 + \cdots + (k+1) = \frac{(k+1)(k+2)}{2}$. But, by substitution from the induction hypothesis,

$$1 + 2 + \cdots + (k+1) = (1 + 2 + \cdots + k) + (k+1)$$
$$= \frac{k(k+1)}{2} + (k+1)$$
$$= \frac{(k+1)(k+2)}{2}.$$

Thus, $1 + 2 + \cdots + (k+1) = \frac{(k+1)(k+2)}{2}$ is true.

2. *A proof using the weak form of mathematical induction:*

   Theorem: For all integers $n \geq 4$, $2^n < n!$.

   Proof:
   Basis Step: For $n = 4$, $2^4 < 4!$ is true since $16 < 24$.
   Induction Step: Let k be an integer, $k \geq 4$, and suppose that $2^k < k!$ is true. The following shows that $2^{k+1} < (k+1)!$ must also be true:
   $$2^{k+1} = 2 \cdot 2^k < 2 \cdot k! < (k+1)k! = (k+1)!.$$

3. *A proof using the weak form of mathematical induction:*

   Theorem: For all integers $n \geq 8$, n cents in postage can be made using only 3-cent and 5-cent stamps.

   Proof: Let $P(n)$ be the predicate "n cents postage can be made using only 3-cent and 5-cent stamps".
   Basis Step: $P(8)$ is true since 8 cents in postage can be made using one 3-cent stamp and one 5-cent stamp.
   Induction Step: Let k be an integer, $k \geq 8$, and suppose that $P(k)$ is true. The following shows that $P(k+1)$ must also be true. If the pile of stamps for k cents postage has in it any 5-cent stamps, then remove one 5-cent stamp and replace it with two 3-cent stamps. If the pile for k cents postage has only 3-cent stamps, there must be at least three 3-cent stamps in the pile (since $k \neq 3$ or 6). Remove three 3-cent stamps and replace them with two 5-cent stamps. In either case, a pile of stamps for $k+1$ cents postage results.

4. *A proof using an alternative form of mathematical induction (Fact 5):*

   Theorem: For all integers $n \geq 0$, $F_n < 2^n$. (F_k are Fibonacci numbers. See §3.1.2.)

   Proof: Let $P(n)$ be the predicate "$F_n < 2^n$".
   Basis Step: $P(0)$ and $P(1)$ are both true since $F_0 = 0 < 1 = 2^0$ and $F_1 = 1 < 2 = 2^1$.
   Induction Step: Let k be an integer, $k \geq 0$, and suppose that $P(k)$ and $P(k+1)$ are true. Then $P(k+2)$ is also true: $F_{k+2} = F_k + F_{k+1} < 2^k + 2^{k+1} < 2^{k+1} + 2^{k+1} = 2 \cdot 2^{k+1} = 2^{k+2}$.

5. *A proof using the strong form of mathematical induction:*

   Theorem: Every integer $n \geq 2$ is divisible by some prime number.

   Proof: Let $P(n)$ be the sentence "n is divisible by some prime number".
   Basis Step: Since 2 is divisible by 2 and 2 is a prime number, $P(2)$ is true.
   Induction Step: Let k be an integer with $k > 2$, and suppose that $P(i)$ (the induction hypothesis) is true for all integers i with $2 \leq i < k$. That is, suppose for all integers i with $2 \leq i < k$ that i is divisible by a prime number. (*It must now be shown that k is divisible by a prime number.*)

Now either the number k is prime or k is not prime. If k is prime, then k is divisible by a prime number, namely itself. If k is not prime, then $k = a \cdot b$ where a and b are integers, with $2 \leq a < k$ and $2 \leq b < k$. By the induction hypothesis, the number a is divisible by a prime number p, and so $k = ab$ is also divisible by that prime p. Hence, regardless of whether k is prime or not, k is divisible by a prime number.

6. *A proof using the well-ordering principle*:

   Theorem: Every integer $n \geq 2$ is divisible by some prime number.

   Proof: Suppose, to the contrary, that there exists an integer $n \geq 2$ that is divisible by no prime number. Thus, the set S of all integers ≥ 2 that are divisible by no prime number is nonempty. Of course, no number in S is prime, since every number is divisible by itself.

   By the well-ordering principle for the integers, the set S contains a least element k. Since k is not prime, there must exist integers a and b with $2 \leq a < k$ and $2 \leq b < k$, such that $k = a \cdot b$. Moreover, since k is the least element of the set S and since both a and b are smaller than k, it follows that neither a nor b is in S. Hence, the number a (in particular) must be divisible by some prime number p. But then, since a is a factor of k, the number k is also divisible by p, which contradicts the fact that k is in S. This contradiction shows that the original supposition is false, or, in other words, that the theorem is true.

7. *A proof using the well-ordering principle*:

   Theorem: Every decreasing sequence of nonnegative integers is finite.

   Proof: Suppose $a_1, a_2, \ldots$ is a decreasing sequence of nonnegative integers: $a_1 > a_2 > \cdots$. By the well-ordering principle, the set $\{a_1, a_2, \ldots\}$ contains a least element, a_n. This number must be the last in the sequence (and hence the sequence is finite). If a_n is not the last term, then $a_{n+1} < a_n$, which contradicts the fact that a_n is the smallest element.

1.5.5 DIAGONALIZATION ARGUMENTS

Definition:

The **diagonal** of an infinite list of sequences $s_1, s_2, s_3, \ldots$ is the infinite sequence whose jth element is the jth entry of sequence s_j.

A **diagonalization proof** is any proof that involves the diagonal of a list of sequences, or something analogous to this.

Facts:

1. A diagonalization argument can be used to prove the existence of nonrecursive functions.

2. A diagonalization argument can be used to prove that no computer algorithm can ever be developed to determine whether an arbitrary computer program given as input with a given set of data will terminate (*the Turing Halting Problem*).

3. A diagonalization argument can be used to prove that every mathematical theory (under certain reasonable hypotheses) will contain statements whose truth or falsity is impossible to determine within the theory (*Gödel's Incompleteness Theorem*).

Example:

1. *A diagonalization proof*:

    Theorem: The set of real numbers between 0 and 1 is uncountable. (Georg Cantor, 1845–1918)

    Proof: Suppose, to the contrary, that the set of real numbers between 0 and 1 is countable. The decimal representations of these numbers can be written in a list as follows:

    $$0.a_{11}a_{12}a_{13}\ldots a_{1n}\ldots$$
    $$0.a_{21}a_{22}a_{23}\ldots a_{2n}\ldots$$
    $$0.a_{31}a_{32}a_{33}\ldots a_{3n}\ldots$$
    $$\vdots$$
    $$0.a_{n1}a_{n2}a_{n3}\ldots a_{nn}\ldots$$
    $$\vdots$$

    From this list, construct a new decimal number $0.b_1 b_2 b_3 \ldots b_n \ldots$ by specifying that

    $$b_i = \begin{cases} 5 & \text{if } a_{ii} \neq 5 \\ 6 & \text{if } a_{ii} = 5. \end{cases}$$

    For each integer $i \geq 1$, $0.b_1 b_2 b_3 \ldots b_n \ldots$ differs from the ith number in the list in the ith decimal place, and hence $0.b_1 b_2 b_3 \ldots b_n \ldots$ is not in the list. Consequently, no such listing of all real numbers between 0 and 1 is possible, and hence, the set of real numbers between 0 and 1 is uncountable.

1.6 AXIOMATIC PROGRAM VERIFICATION

Axiomatic program verification is used to prove that a sequence of programming instructions achieves its specified objective. Semantic axioms for the programming language constructs are used in a formal logic argument as rules of inference. Comments called *assertions*, within the sequence of instructions, provide the main details of the argument. The presently high expense of creating verified software can be justified for code that is frequently reused, where the financial benefit is otherwise adequately large, or where human life is concerned, for instance, in airline traffic control. This section presents a representative sample of axioms for typical programming language constructs.

1.6.1 ASSERTIONS AND SEMANTIC AXIOMS

The correctness of a program can be argued formally based on a set of semantic axioms that define the behavior of individual programming language constructs [Fl67], [Ho69], [Ap81]. (Some alternative proofs of correctness use denotational semantics [St77], [Sc86] or operational semantics [We72].) In addition, it is possible to synthesize code, using techniques that permit the axioms to guide the selection of appropriate instructions [Di76], [Gr81]. Code specifications and intermediate conditions are expressed in the form of program assertions.

Definitions:

An **assertion** is a program comment containing a logical statement that constrains the values of the computational variables. These constraints are expected to hold when execution flow reaches the location of the assertion.

A **semantic axiom** for a type of programming instruction is a rule of inference that prescribes the change of value of the variables of computation caused by the execution of that type of instruction.

The assertion **false** represents an inconsistent set of logical conditions. A computer program cannot meet such a specification.

Given two constraints A and B on computational variables, a statement that B follows from A purely for reasons of logic and/or mathematics is called a **logical implication**.

The **postcondition** for an instruction or program fragment is the assertion that immediately follows it in the program.

The **precondition** for an instruction or program fragment is the assertion that immediately precedes it in the program.

The assertion **true** represents the empty set of logical conditions.

Notation:

1. To say that whenever the precondition {Apre} holds, the execution of a program fragment called "Code" will cause the postcondition {Apost} to hold, the following notation styles can be used:

   - Horizontal notation: {Apre} Code {Apost}
   - Vertical notation:
                                    {Apre}
                                    Code
                                    {Apost}.
   - Flowgraph notation:

   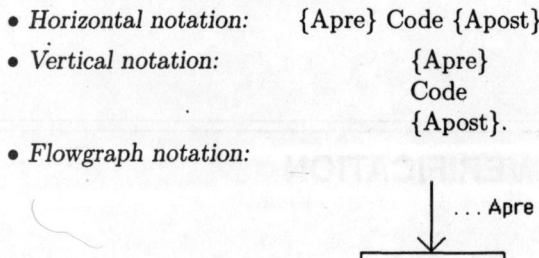

2. Curly braces { ... } enclose assertions in generic program code. They do not denote a set.

3. Semantic axioms have a finite list of premises and a conclusion. They are represented in the following format:

   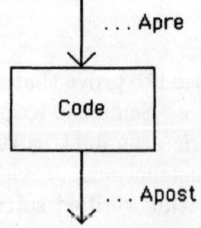

                              {Premise 1}
                                   ⋮
                              {Premise n}
                              - - - - - - - - -
                              {Conclusion}

4. The circumstance that A logically implies B is denoted $A \Rightarrow B$.

1.6.2 NOP, ASSIGNMENT, AND SEQUENCING AXIOMS

Formal axioms of pure mathematical consequence (no operation, from a computational perspective) and of straight-line sequential flow are used as auxiliaries to verify correctness, even of sequences of simple assignment statements.

Definitions:

A **NOP** ("no-op") is a (possibly empty) program fragment whose execution does not alter the state of any computational variables or the sequence of flow.

The **Axiom of NOP** states:

$\{Apre\} \Rightarrow \{Apost\}$ Premise 1

$\{Apre\}$ NOP $\{Apost\}$ Conclusion

Note: The Axiom of NOP is frequently applied to empty program fragments in order to facilitate a clear logical argument.

An **assignment instruction** $X := E;$ means that the variable X is to be assigned the value of the expression E.

In a logical assertion $A(X)$ with possible instances of the program variable X, the **result of replacing each instance** of X in A by the program expression E is denoted $A(X \leftarrow E)$.

The **Axiom of Assignment** states:

$\{true\}$ No premises

$\{A(X \leftarrow E)\} X := E; \{A(X)\}$ Conclusion

The following **Axiom of Sequence** provides that two consecutive instructions in the program code are executed one immediately after the other:

$\{Apre\}$ Code1 $\{Amid\}$ Premise 1
$\{Amid\}$ Code2 $\{Apost\}$ Premise 2

$\{Apre\}$ Code1, Code2 $\{Apost\}$ Conclusion

(Commas are used as separators in program code.)

Examples:

1. *Example of NOP*: Suppose that X is a numeric program variable.

   $\{X = 3\} \Rightarrow \{X > 0\}$ mathematical fact

   $\{X = 3\}$ NOP $\{X > 0\}$ by Axiom of NOP

2. Suppose that X and Y are integer-type program variables. The Axiom of Assignment alone implies correctness of all the following examples:
   (a) $\{X = 4\}$ $X := X * 2;$ $\{X = 8\}$
   $A(X)$ is $\{X = 8\}$; E is $X * 2$; $A(X \leftarrow E)$ is $\{X * 2 = 8\}$, which is equivalent to $\{X = 4\}$.
   (b) $\{true\}$ $X := 2;$ $\{X = 2\}$
   $A(X)$ is $\{X = 2\}$; E is 2; $A(X \leftarrow E)$ is $\{2 = 2\}$, which is equivalent to $\{true\}$.
   (c) $\{(-9 < X) \wedge (X < 0)\}$ $Y := X;$ $\{(-9 < Y) \wedge (Y < 0)\}$
   $A(Y)$ is $\{(-9 < Y) \wedge (Y < 0)\}$; E is X; $A(Y \leftarrow E)$ is $\{(-9 < X) \wedge (X < 0)\}$.

(d) $\{Y = 1\}$ $X := 0$; $\{Y = 1\}$
$A(X)$ is $\{Y = 1\}$; E is 0; $A(X \leftarrow E)$ is $\{Y = 1\}$.

(e) $\{\text{false}\}$ $X := 8$; $\{X = 2\}$
$A(X)$ is $\{X = 2\}$; E is 8; $A(X \leftarrow E)$ is $\{8 = 2\}$, which is equivalent to $\{\text{false}\}$.

3. *Examples of sequence*:
(a) $\{X = 1\}$ $X := X + 1$; $\{X > 0\}$

i. $\{X = 1\} \Rightarrow \{X > -1\}$	mathematics
ii. $\{X = 1\}$ NOP $\{X > -1\}$	Axiom of NOP
iii. $\{X > -1\}$ $X := X + 1$; $\{X > 0\}$	Axiom of Assignment
iv. $\{X = 1\}$ NOP, $X := X + 1$; $\{X > 0\}$	Axiom of Sequence on ii, iii
v. $\{X = 1\}$ $X := X + 1$; $\{X > 0\}$	definition of NOP.

(b) $\{Y = a \wedge X = b\}$ $Z := Y$; $Y := X$; $X := Z$; $\{X = a \wedge Y = b\}$

i. $\{Y = a \wedge X = b\}$ $Z := Y$; $\{Z = a \wedge X = b\}$	Axiom of Assignment
ii. $\{Z = a \wedge X = b\}$ $Y := X$; $\{Z = a \wedge Y = b\}$	Axiom of Assignment
iii. $\{Y = a \wedge X = b\}$ $Z := Y, Y := X$, $\{Z = a \wedge Y = b\}$	Axiom of Sequence on i, ii
iv. $\{Z = a \wedge Y = b\}$ $X := Z$; $\{X = a \wedge Y = b\}$	Axiom of Assignment
v. $\{Y = a \wedge X = b\}$ $Z := Y, Y := X, X := Z$, $\{X = a \wedge Y = b\}$	Axiom of Sequence on iii, iv.

1.6.3 AXIOMS FOR CONDITIONAL EXECUTION CONSTRUCTS

Definitions:

A *conditional assignment construct* is any type of program instruction containing a logical condition and an imperative clause such that the imperative clause is to be executed if and only if the logical condition is true. Some types of conditional assignment contain more than one logical condition and more than one imperative clause.

An *if-then* instruction **if IfCond then ThenCode** has one logical condition (which follows the keyword if) and one imperative clause (which follows the keyword then).

The **Axiom of If-then** states:

$\{Apre \wedge IfCond\}$ ThenCode $\{Apost\}$	Premise 1
$\{Apre \wedge \neg IfCond\} \Rightarrow \{Apost\}$	Premise 2
- -	
$\{Apre\}$ **if** IfCond **then** ThenCode $\{Apost\}$	Conclusion

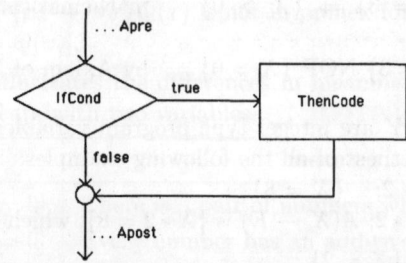

An *if-then-else* instruction **if IfCond then ThenCode else ElseCode** has one logical condition, which follows the keyword if, and two imperative clauses, one after the keyword then, and the other after the keyword else.

The **Axiom of If-then-else** states:

{Apre ∧ IfCond} ThenCode {Apost}   Premise 1
{Apre ∧ ¬IfCond} ElseCode {Apost}   Premise 2
- -
{Apre} if IfCond then ThenCode else ElseCode {Apost}   Conclusion

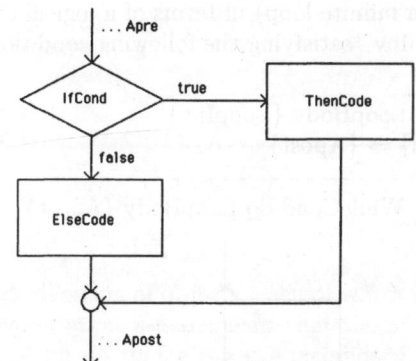

Examples:

1. *If-then*:
{true} if $X = 3$ then $Y := X$; $\{X = 3 \rightarrow Y = 3\}$

  i. $\{X = 3\}\ Y := X;\ \{X = 3\ \wedge\ Y = 3\}$ Axiom of Assignment
  ii. $\{X = 3\ \wedge\ Y = 3\}\ NOP\ \{(X = 3) \rightarrow (Y = 3)\}$ Axiom of NOP
    (Step ii uses a logic fact: $p \wedge q \Rightarrow p \rightarrow q$)
  iii. $\{X = 3\}\ Y := X; \{X = 3 \rightarrow Y = 3\}$ Axiom of Sequence on i, ii
    (Step iii establishes Premise 1 for Ax. of If-then)
  iv. $\{\neg(X = 3)\} \Rightarrow \{X = 3 \rightarrow Y = 3\}$ Logic fact
    (Step iv establishes Premise 2 for Ax. of If-then)
  v. {true} if $X = 3$ then $Y := X$; $\{X = 3 \rightarrow Y = 3\}$ Axiom of If-then on iii, iv.

2. *If-then-else*:
$\{X > 0\}$
if $(X > Y)$ then $M := X$; else $M := Y$;
$\{(X > 0) \wedge (X > Y \rightarrow M = X) \wedge (X \leq Y \rightarrow M = Y)\}$

  i. $\{X > 0\ \wedge\ X > Y\}\ M := X;\ \{X > 0\ \wedge\ (X > Y \rightarrow M = X)\ \wedge\ (X \leq Y \rightarrow M = Y)\}$
    by Axiom of Assignment and Axiom of NOP (establishes Premise 1)
  ii. $\{X > 0\ \wedge\ \neg(X > Y)\}\ M := Y;\ \{X > 0\ \wedge\ (X > Y \rightarrow M = X)\ \wedge\ (X \leq Y \rightarrow M = Y)\}$
    by Axiom of Assignment and Axiom of NOP (establishes Premise 2)
  iii. Conclusion now follows from Axiom of If-then-else.

1.6.4 AXIOMS FOR LOOP CONSTRUCTS

Definitions:

A **while-loop** instruction **while WhileCond do LoopBody** has one logical condition called the **while-condition**, which follows the keyword while, and a sequence of instructions called the **loop-body**. At the outset of execution, the while condition is tested for its truth value. If it is true, then the loop body is executed. This two-step process of test and execute continues until the while condition becomes false, after which the flow of execution passes to whatever program instruction follows the while-loop.

A loop is **weakly correct** if whenever the precondition is satisfied at the outset of execution and the loop is executed to termination, the resulting computational state satisfies the postcondition.

A loop is **strongly correct** if it is weakly correct and if whenever the precondition is satisfied at the outset of execution, the computation terminates.

The **Axiom of While** defines weak correctness of a while-loop (i.e., the axiom ignores the possibility of an infinite loop) in terms of a logical condition called the **loop invariant** denoted "LoopInv" satisfying the following condition:

$\{Apre\} \Rightarrow \{LoopInv\}$ "Initialization" Premise
$\{LoopInv \land WhileCond\}\ LoopBody\ \{LoopInv\}$ "Preservation" Premise
$\{LoopInv \land \neg WhileCond\} \Rightarrow \{Apost\}$ "Finalization" Premise

---

$\{Apre\}$ <u>while</u> $\{LoopInv\}$ WhileCond <u>do</u> LoopBody $\{Apost\}$ Conclusion

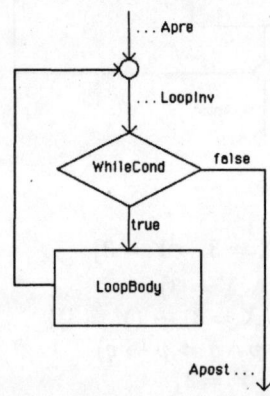

Example:
1. Suppose that J, N, and P are integer-type program variables.
$\{Apre : J = 0 \land P = 1 \land N \geq 0\}$
<u>while</u> $\{LoopInv : P = 2^J \land J \leq N\}$ $(J < N)$ <u>do</u>
    $P := P * 2;$
    $J := J + 1;$
<u>endwhile</u>
$\{Apost : P = 2^N\}$
  i. $\{Apre : J = 0 \land P = 1 \land N \geq 0\} \Rightarrow \{LoopInv : P = 2^J \land J \leq N\}$
        Initialization Premise trivially true by mathematics
 ii. $\{LoopInv \land WhileCond : (P = 2^J \land J \leq N) \land (J < N)\}$
     $P := P * 2;$
     $J := J + 1;$
     $\{LoopInv :\ P = 2^J \land J \leq N\}$
        Preservation Premise proved using by Axiom of Assignment twice
        and Axiom of Sequence
 iii. $\{LoopInv \land \neg WhileCond : (P = 2^J \land J \leq N) \land \neg(J < N)\} \Rightarrow \{Apost : P = 2^N\}$
        Finalization Premise provable by mathematics
 iv. Conclusion now follows from Axiom of While.

Fact:
1. Proof of termination of a loop is usually achieved by mathematical induction.

1.6.5 AXIOMS FOR SUBPROGRAM CONSTRUCTS

The parameterless procedure is the simplest subprogram construct. Procedures with parameters and functional subprograms have somewhat more complicated semantic axioms.

Definitions:

A *procedure* is a sequence of instructions that lies outside the main sequence of instructions in a program. It consists of a *procedure name*, followed by a *procedure body*.

A *call* instruction <u>call</u> ProcName is executed by transferring control to the first executable instruction of the procedure ProcName.

A *return* instruction causes a procedure to transfer control to the executable instruction immediately following the most recently executed call to that procedure. An implicit return is executed after the last instruction in the procedure body is executed. It is good programming style to put a *return* there.

In the following **Axiom of Procedure (*parameterless*)**, *Apre* and *Apost* are the precondition and postcondition of the instruction <u>call</u> ProcName; *ProcPre* and *ProcPost* are the precondition and postcondition of the procedure whose name is *ProcName*.

{Apre} ⇒ {ProcPre}	"Call" Premise
{ProcPre} ProcBody {ProcPost}	"Body" Premise
{ProcPost} ⇒ {Apost}	"Return" Premise

{Apre} <u>call</u> ProcName; {Apost}	Conclusion

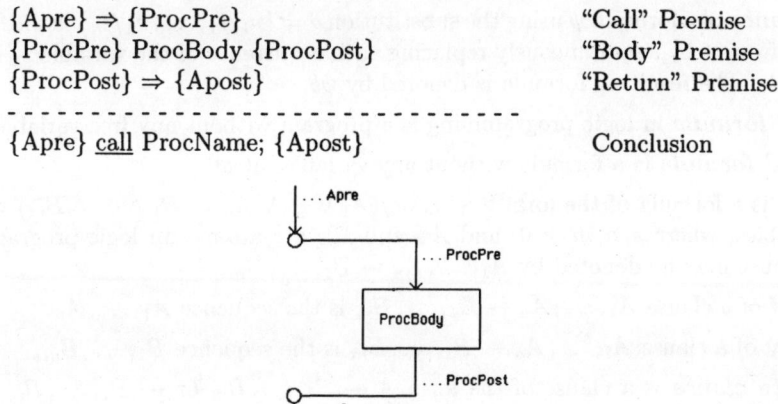

1.7 LOGIC-BASED COMPUTER PROGRAMMING PARADIGMS

Mathematical logic is the basis for several different computer software paradigms. These include logic programming, fuzzy reasoning, production systems, artificial intelligence, and expert systems.

1.7.1 LOGIC PROGRAMMING

A computer program in the imperative paradigm (familiar in languages like C, BASIC, FORTRAN, and ALGOL) is a list of instructions that describes a precise sequence of actions that a computer should perform. To initiate a computation, one supplies the iterative program plus specific input data to the computer. *Logic programming* provides an alternative paradigm in which a program is a list of "clauses", written in predicate logic, that describe an allowed range of behavior for the computer. To initiate a computation, the computer is supplied with the logic program plus another clause called a "goal". The aim of the computation is to establish that the goal is a logical consequence of the clauses constituting the logic program. The computer simplifies the goal by executing the program repeatedly until the goal becomes empty, or until it cannot be further simplified.

Definitions:

A **term** in a domain S is either a fixed element of S or an S-valued variable.

An **n-ary predicate** on a set S is a function $P: S^n \to \{T, F\}$.

An **atomic formula** (or **atom**) is an expression of the form $P(t_1, \ldots, t_n)$, where $n \geq 0$, P is an n-ary predicate, and $t_1, \ldots, t_n$ are terms.

A **formula** is a logical expression constructed from atoms with conjunctions, disjunctions, and negations, possibly with some logical quantifiers.

A **substitution** for a formula is a finite set of the form $\{v_1/t_1, \ldots, v_n/t_n\}$, where each v_i is a distinct variable, and each t_i is a term distinct from v_i.

The **instance** of a formula ψ using the substitution $\theta = \{v_1/t_1, \ldots, v_n/t_n\}$ is the formula obtained from ψ by simultaneously replacing each occurrence of the variable v_i in ψ by the term t_i. The resulting formula is denoted by $\psi\theta$.

A **closed formula** in logic programming is a program without any free variables.

A **ground formula** is a formula without any variables at all.

A **clause** is a formula of the form $\forall x_1 \ldots \forall x_s (A_1 \vee \cdots \vee A_n \leftarrow B_1 \wedge \cdots \wedge B_m)$ with no free variables, where $s, n, m \geq 0$, and A's and B's are atoms. In logic programming, such a clause may be denoted by $A_1, \ldots, A_n \leftarrow B_1, \ldots, B_m$.

The **head** of a clause $A_1, \ldots, A_n \leftarrow B_1, \ldots, B_m$ is the sequence $A_1, \ldots, A_n$.

The **body** of a clause $A_1, \ldots, A_n \leftarrow B_1, \ldots, B_m$ is the sequence $B_1, \ldots, B_m$.

A **definite clause** is a clause of the form $A \leftarrow B_1, \ldots, B_m$ or $\leftarrow B_1, \ldots, B_m$, which contains at most one atom in its head.

An **indefinite clause** is a clause that is not definite.

A **logic program** is a finite sequence of definite clauses.

A **goal** is a definite clause $\leftarrow B_1, \ldots, B_m$ whose head is empty. (Prescribing a goal for a logic program P tells the computer to derive an instance of that goal by manipulating the logical clauses in P.)

An **answer** to a goal G for a logic program P is a substitution θ such that $G\theta$ is a logical consequence of P.

A **definite answer** to a goal G for a logic program P is an answer in which every variable is substituted by a constant.

Facts:

1. A definite clause $A \leftarrow B_1, \ldots, B_m$ represents the following logical constructs:

    If every B_i is true, then A is also true;
    Statement A can be proved by proving every B_i.

2. *Definite answer property*: If a goal G for a logic program P has an answer, then it has a definite answer.

3. The definite answer property does not hold for indefinite clauses. For example, although $G = \exists x Q(x)$ is a logical consequence of $P = \{Q(a), Q(b) \leftarrow\}$, no ground instance of G is a logical consequence of P.

4. Logic programming is Turing-complete (§16.3); i.e., any computable function can be represented using a logic program.

5. Building on the work of logician J. Alan Robinson in 1965, computer scientists Robert Kowalski and Alain Colmerauer of Imperial College and the University of Marseille-Aix, respectively, in 1972 independently developed the programming language PROLOG (PROgramming in LOGic) based on a special subset of predicate logic.

6. The first PROLOG interpreter was implemented in ALGOL-W in 1972 at the University of Marseille-Aix. Since then, several variants of PROLOG have been introduced, implemented, and used in practical applications. The basic paradigm behind all these languages is called Logic Programming.

7. In PROLOG, the relation "is" means equality.

Examples:

1. The following three clauses are definite:
$$P \leftarrow Q, R \qquad P \leftarrow \qquad \leftarrow Q, R.$$

2. The clause $P, S \leftarrow Q, R$ is indefinite.

3. The substitution $\{X/a, Y/b\}$ for the atom $P(X, Y, Z)$ yields the instance $P(a, b, Z)$.

4. The goal $\leftarrow P$ to the program $\{P \leftarrow\}$ has a single answer, given by the empty substitution. This means the goal can be achieved.

5. The goal $\leftarrow P$ to the program $\{Q \leftarrow\}$ has no answer. This means it cannot be derived from that program.

6. The logic program consisting of the following two definite clauses P1 and P2 computes a complete list of the pairs of vertices in an arbitrary graph that have a path joining them:

    P1. $\text{path}(V, V) \leftarrow$
    P2. $\text{path}(U, V) \leftarrow \text{path}(U, W), \text{edge}(W, V)$

Definite clauses P3 and P4 comprise a representation of a graph with nodes 1, 2, and 3, and edges (1,2) and (2,3):

    P3. $\text{edge}(1, 2) \leftarrow$
    P4. $\text{edge}(2, 3) \leftarrow$

The goal G represents a query asking for a complete list of the pairs of vertices in an arbitrary graph that have a path joining them:

    G. $\leftarrow \text{path}(Y, Z)$

There are three distinct answers of the goal G to the logic program consisting of definite clauses P1 to P4, corresponding to the paths (1,2), (1,2,3), and (2,3), respectively:

    A1. $\{Y/1, Z/2\}$
    A2. $\{Y/1, Z/3\}$
    A3. $\{Y/2, Z/3\}$

7. The following logic program computes the Fibonacci sequence $0, 1, 1, 2, 3, 5, 8, 13, \ldots$, where the predicate $fib(N, X)$ is true if X is the Nth number in the Fibonacci sequence:

$fib(0, 0) \leftarrow$
$fib(1, 1) \leftarrow$
$fib(N, X + Y) \leftarrow N > 1, fib(N - 1, X), fib(N - 2, Y)$

The goal "$\leftarrow fib(6, X)$" is answered $\{X/8\}$, the goal "$\leftarrow fib(X, 8)$" is answered $\{X/6\}$, and the goal "$\leftarrow fib(N, X)$" has the following infinite sequence of answers:

$\{N/0, X/0\}$
$\{N/1, X/1\}$
$\{N/2, X/1\}$
$\vdots$

8. Consider the problem of finding an assignment of digits (integers $0, 1, \ldots, 9$) to letters such that adding two given words produces the third given word, as in this example:

$$\begin{array}{r} S\,E\,N\,D \\ +\,M\,O\,R\,E \\ \hline M\,O\,N\,E\,Y \end{array}$$

One solution to this particular puzzle is given by the following assignment:

$D = 0, \quad E = 0, \quad M = 1, \quad N = 0, \quad O = 0, \quad R = 0, \quad S = 9, \quad Y = 0.$

The following PROLOG program solves all such puzzles:

$between(X, X, Z) \leftarrow X < Z.$
$between(X, Y, Z) \leftarrow between(K, Y, Z), X \text{ is } K - 1.$
$val([\,], 0) \leftarrow.$
$val([X|Y], A) \leftarrow val(Y, B), between(0, X, 9), A \text{ is } 10 * B + X.$
$solve(X, Y, Z) \leftarrow val(X, A), val(Y, B), val(Z, C), C \text{ is } A + B.$

The specific example given above is captured by the following goal:

$\leftarrow solve([D, N, E, S], [E, R, O, M], [Y, E, N, O, M]).$

The predicate $between(X, Y, Z)$ means $X \leq Y \leq Z$. The predicate $val(L, N)$ means that the number N is the value of L, where L is the kind of list of letters that occurs on a line of these puzzles. The notation $[X|L]$ means the list obtained by writing list L after item X. The predicate $solve(X, Y, Z)$ means that the value of list Z equals the sum of the values of list X and list Y.

This example illustrates the ease of writing logic programs for some problems where conventional imperative programs are more difficult to write.

1.7.2 FUZZY SETS AND LOGIC

Fuzzy set theory and fuzzy logic are used to model imprecise meanings, such as "tall", that are not easily represented by predicate logic. In particular, instead of assigning either "true" or "false" to the statement "John is tall", fuzzy logic assigns a real number between 0 and 1 that indicates the degree of "tallness" of John. Fuzzy set theory assigns a real number between 0 and 1 to John that indicates the extent to which he is a member of the set of tall people. See [Ka86], [Ka92], [KaLa94], [YaFi94], [YaZa94], [Za65], [Zi91], [Zi93].

Definitions:

A *fuzzy set* $F = (X, \mu)$ consists of a set X (the **domain**) and a **membership function** $\mu\colon X \to [0,1]$. Sometimes the set is written $\{\,(x, \mu(x)) \mid x \in X\,\}$ or $\{\,\mu(x)\,x \mid x \in X\,\}$.

The *fuzzy intersection* of fuzzy sets (A, μ_A) and (B, μ_B) is the fuzzy set $A \cap B$ with domain $A \cap B$ and membership function $\mu_{A \cap B}(x) = \min(\mu_A(x), \mu_B(x))$.

The *fuzzy union* of fuzzy sets (A, μ_A) and (B, μ_B) is the fuzzy set $A \cup B$ with domain $A \cup B$ and membership function $\mu_{A \cup B}(x) = \max(\mu_A(x), \mu_B(x))$.

The *fuzzy complement* of the fuzzy set (A, μ) is the fuzzy set $\neg A$ or $\overline{A}$ with domain A and membership function $\mu_{\overline{A}}(x) = 1 - \mu(x)$.

The *nth constructor* $con(\mu, n)$ of a membership function μ is the function μ^n. That is, $con(\mu, n)(x) = (\mu(x))^n$.

The *nth dilutor* $dil(\mu, n)$ of a membership function μ is the function $\mu^{1/n}$. That is, $dil(\mu, n)(x) = (\mu(x))^{1/n}$.

A *T-norm operator* is a function $f\colon [0,1] \times [0,1] \to [0,1]$ with the following properties:

- $f(x, y) = f(y, x)$ commutativity
- $f(f(x, y), z) = f(x, f(y, z))$ associativity
- if $x \le v$ and $y \le w$, then $f(x, y) \le f(v, w)$ monotonicity
- $f(a, 1) = a$.   1 is a unit element

The *fuzzy intersection* $A \cap_f B$ of fuzzy sets (A, μ_A) and (B, μ_B) **relative to the T-norm operator** f is the fuzzy set with domain $A \cap B$ and membership function $\mu_{A \cap_f B}(x) = f(\mu_A(x), \mu_B(x))$.

An *S-norm operator* is a function $f\colon [0,1] \times [0,1] \to [0,1]$ with the following properties:

- $f(x, y) = f(y, x)$ commutativity
- $f(f(x, y), z) = f(x, f(y, z))$ associativity
- if $x \le v$ and $y \le w$, then $f(x, y) \le f(v, w)$ monotonicity
- $f(a, 1) = 1$.

The *fuzzy union* $A \cup_f B$ of fuzzy sets (A, μ_A) and (B, μ_B) **relative to the S-norm operator** f is the fuzzy set with domain $A \cup B$ and membership function $\mu_{A \cup_f B}(x) = f(\mu_A(x), \mu_B(x))$.

A *complement operator* is a function $f\colon [0,1] \to [0,1]$ with the following properties:

- $f(0) = 1$
- if $x < y$ then $f(x) > f(y)$
- $f(f(x)) = x$.

The *fuzzy complement* $\neg_f A$ of the fuzzy set (A, μ) **relative to the complement operator** f is the fuzzy set with domain A and membership function $\mu_{\neg_f}(x) = f(\mu(x))$.

A *fuzzy system* consists of a base collection of fuzzy sets, intersections, unions, complements, and implications.

A *hedge* is a monadic operator corresponding to linguistic adjectives such as "very", "about", "somewhat", or "quite" that modify membership functions.

A *two-valued logic* is a logic where each statement has exactly one of the two values: true or false.

A ***multi-valued logic*** (***n-valued logic***) is a logic with a set of n (≥ 2) truth values; i.e., there is a set of n numbers $v_1, v_2, \ldots, v_n \in [0,1]$ such that every statement has exactly one truth value v_i.

Fuzzy logic is the study of statements where each statement has assigned to it a truth value in the interval $[0,1]$ that indicates the extent to which the statement is true.

If statements p and q have truth values v_1 and v_2 respectively, the ***truth value*** of $p \vee q$ is $\max(v_1, v_2)$, the truth value of $p \wedge q$ is $\min(v_1, v_2)$, and the truth value of $\neg p$ is $1 - v_1$.

Facts:

1. Fuzzy set theory and fuzzy logic were developed by Lofti Zadeh in 1965.

2. Fuzzy set theory and fuzzy logic are parallel concepts: given a predicate $P(x)$, the fuzzy truth value of the statement $P(a)$ is the fuzzy set value assigned to a as an element of $\{\, x \mid P(x)\,\}$.

3. The usual minimum function $\min(x, y)$ is a T-norm. The usual real maximum function $\max(x, y)$ is an S-norm. The function $c(x) = 1 - x$ is a complement operator.

4. Several other kinds of T-norms, S-norms, and complement operators have been defined.

5. The words "T-norm" and "S-norm" come from multi-valued logics.

6. The only difference between T-norms and S-norms is that the T-norm specifies $f(a, 1) = a$, whereas the S-norm specifies $f(a, 1) = 1$.

7. Several standard classes of membership functions have been defined, including step, sigmoid, and bell functions.

8. Constructors and dilutors of membership functions are also membership functions.

9. The large number of practical applications of fuzzy set theory can generally be divided into three types: machine systems, human-based systems, human-machine systems. Some of these applications are based on fuzzy set theory alone and some on a variety of hybrid configurations involving neurofuzzy approaches, or in combination with neural networks, genetic algorithms, or case-based reasoning.

10. The first fuzzy expert system that set a trend in practical fuzzy thinking was the design of a cement kiln called Linkman, produced by Blue Circle Cement and SIRA in Denmark in the early 1980s. The system incorporates the experience of a human operator in a cement production facility.

11. The Sendai Subway Automatic Train Operations Controller was designed by Hitachi in Japan. In that system, speed control during cruising, braking control near station zones, and switching of control are determined by fuzzy IF-THEN rules that process sensor measurements and consider factors related to travelers' comfort and safety. In operation since 1986, this most celebrated application encouraged many applications based on fuzzy set controllers in the areas of home appliances (refrigerators, vacuum cleaners, washers, dryers, rice cookers, air conditioners, shavers, blood-pressure measuring devices), video cameras (including fuzzy automatic focusing, automatic exposure, automatic white balancing, image stabilization), automotive (fuzzy cruise control, fuel injection, transmission and brake systems), robotics, and aerospace.

12. Applications to finance started with the Yamaichi Fuzzy Fund, which is a fuzzy trading system. This was soon followed by a variety of financial applications world-wide.

13. Research activities will soon result in commercial products related to the use of fuzzy set theory in the areas of audio and video data compression (such as HDTV), robotic arm movement control, computer vision, coordination of visual sensors with mechanical motion, aviation (such as unmanned platforms), and telecommunication.

14. *Current status*: Most applications of fuzzy sets and logic are directly related to structured numerical model-free estimators. Presently, most applications are designed with linguistic variables, where proper levels of granularity are being used in the evaluations of those variables, expressing the ambiguity and subjectivity in human thinking. Fuzzy systems capture expert knowledge and through the processing of fuzzy IF-THEN rules are capable of processing knowledge combining the antecedents of each fuzzy rule, calculating the conclusions, and aggregating them to the final decision.

15. One way to model *fuzzy implication* $A \to B$ is to define $A \to B$ as $\neg_c A \cup_f B$ relative to some complement operator c and to some S-norm operator f. Several other ways have also been considered.

16. A fuzzy system is used computationally to control the behavior of an external system.

17. Large fuzzy systems have been used in specifying complex real-world control systems. The success of such systems depends crucially on the specific engineering parameters. The correct values of these parameters are usually obtained by trial-and-readjustment.

18. A two-valued logic is a logic that assumes the law of the excluded middle: $p \lor \neg p$ is a tautology.

19. Every n-valued logic is a fuzzy logic.

Examples:

1. A committee consisting of five people met ten times during the past year. Person A attended 7 meetings, B attended all 10 meetings, C attended 6 meetings, D attended no meetings, and E attended 9 meetings. The set of committee members can be described by the following fuzzy set that reflects the degree to which each the members attended meetings, using the function $\mu: \{A, B, C, D, E\} \to [0, 1]$ with the rule $\mu(x) = \frac{1}{10}$(number of meetings attended):

$$\{(A, 0.7), (B, 1.0), (C, 0.6), (D, 0.0), (E, 0.9)\},$$

which can also be written as

$$\{0.7A, 1.0B, 0.6C, 0.0D, 0.9E\}.$$

Person B would be considered a "full" member and person D a "nonmember".

2. Four people are rated on amount of activity in a political party, yielding the fuzzy set

$$P_1 = \{0.8A, 0.45B, 0.1C, 0.75D\},$$

and based on their degree of conservatism in their political beliefs, as

$$P_2 = \{0.6A, 0.85B, 0.7C, 0.35D\}.$$

The fuzzy union of the sets is

$$P_1 \cup P_2 = \{0.8A, 0.85B, 0.7C, 0.75D\},$$

the fuzzy intersection is

$$P_1 \cap P_2 = \{0.6A, 0.45B, 0.1C, 0.35D\}$$

and the fuzzy complement of P_1 (measurement of political inactivity) is
$$\overline{P_1} = \{0.2A, 0.55B, 0.9C, 0.25D\}.$$

3. In the fuzzy set with domain T and membership function
$$\mu_T(h) = \begin{cases} 0 & \text{if } h \leq 170 \\ \frac{h-170}{20} & \text{if } 170 < h < 190 \\ 1 & \text{otherwise} \end{cases}$$
the number 160 is not a member, the number 195 is a member, and the membership of 182 is 0.6. The graph of μ_T is given in the following figure.

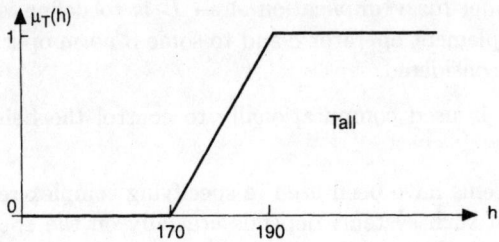

4. The fuzzy set (T, μ_T) of Example 3 can be used to define the fuzzy set "Tall" $= (H, \mu_H)$ of tall people, by the rule $\mu_H(x) = \mu_T(height(x))$ where $height(x)$ is the height of person x calibrated in centimeters.

5. The second constructor $con(\mu_H, 2)$ of the fuzzy set "Tall" can be used to define a fuzzy set "Quite tall", whose graph is given in the following figure.

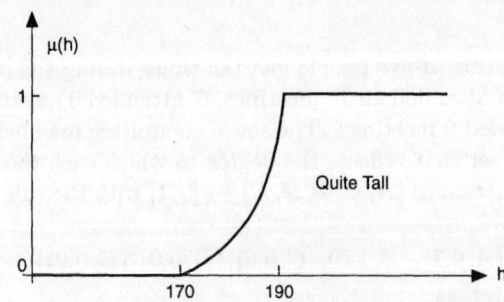

6. The second dilutor $dil(\mu_H, 2)$ of the fuzzy set "Tall" defines the fuzzy set "Somewhat tall", whose graph is given in the following figure.

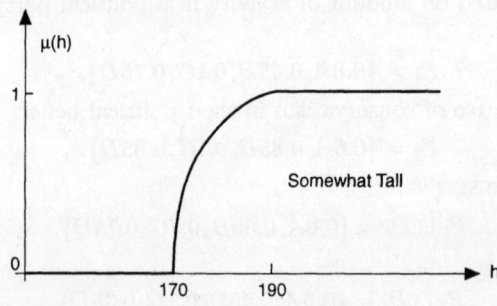

7. The concept of "being healthy" can be modeled using fuzzy logic. The truth value 0.95 could be assigned to "Fran is healthy" if Fran is almost always healthy. The truth value 0.4 could be assigned to "Leslie is healthy" if Leslie is healthy somewhat less than half the time. The truth of the statements "Fran and Leslie are healthy" would be 0.4 and "Fran is not healthy" would be 0.05.

8. *Behavior closed-loop control systems*: The behavior of some closed-loop control systems can be specified using fuzzy logic. For example, consider an automated heater whose output setting is to be based on the readings of a temperature sensor. A fuzzy set "cold" and the implication "very cold $\rightarrow$ high" could be used to relate the temperature to the heater settings. The exact behavior of this system is determined by the degree of the constructor used for "very" and by the specific choices of S-norm and complement operators used to define the fuzzy implication — the "engineering parameters" of the system.

1.7.3 PRODUCTION SYSTEMS

Production systems are a logic-based computer programming paradigm introduced by Allen Newell and Herbert Simon in 1975. They are commonly used in intelligent systems for representing an expert's knowledge used in solving some real-world task, such as a physician's knowledge of making medical diagnoses.

Definitions:

A *fact set* is a set of ground atomic formulas. These formulas represent the information relevant to the system.

A *condition* is a disjunction $A_1 \vee \cdots \vee A_n$, where $n \geq 0$ and each A_i is a literal.

A condition C is **true** in a fact set S if:
- C is empty, or
- C is a positive literal and $C \in S$, or
- C is a negative literal $\neg A$, and $B \notin S$ for each ground instance B of A, or
- $C = A_1 \vee \cdots \vee A_n$, and some condition A_i is true in S.

A *print command* "print(x)", means that the value of the term x is to be printed.

An *action* is either a literal or a print command.

A *production rule* is of the form $C_1, \ldots, C_n \rightarrow A_1, \ldots, A_m$, where $n, m \geq 1$, each C_i is a condition, each A_i is an action, and each variable in each action appears in some positive literal in some condition.

The *antecedent* of the rule $C_1, \ldots, C_n \rightarrow A_1, \ldots, A_m$ is $C_1, \ldots, C_n$.

The *consequent* of the rule $C_1, \ldots, C_n \rightarrow A_1, \ldots, A_m$ is $A_1, \ldots, A_m$.

An *instantiation* of a production rule is the rule obtained by replacing each variable in each positive literal in each condition of the rule by a constant.

A *production system* consists of a fact set and a set of production rules.

Facts:

1. Given a fact set S, an instantiation $C_1, \ldots, C_n \to A_1, \ldots, A_m$ of a production rule denotes the following operation:

if each condition C_i is true in S then
  for each A_i:
    if A_i is an atom, add it to S
    if A_i is a negative literal $\neg B$, then remove B from S
    if A_i is "print(c)", then print c.

2. In addition to "print", production systems allow several other system-level commands.

3. OPS5 and CLIPS are currently the most popular languages for writing production systems. They are available for most operating systems, including UNIX and DOS.

4. To initialize a computation prescribed by a production system, the initial fact set and all the production rules are supplied as input. The command "run1" non-deterministically selects an instantiation of a production rule such that all conditions in the antecedent hold in the fact set, and it "fires" the rule by carrying out the actions in the consequent. The command "run" keeps on selecting and firing rules until no more rule instantiations can be selected.

5. Production systems are Turing complete.

Examples:

1. The fact set $S = \{N(3), 3 > 2, 2 > 1\}$ may represent that "3 is a natural number", that "3 is greater than 2", and that "2 is greater than 1".

2. If the fact set S of Example 1 and the production $N(x) \to \text{print}(x)$ are supplied as input, the command "run" will yield the instantiation $N(3) \to \text{print}(3)$ and fire it to print 3.

3. The production rule $N(x), x > y \to \neg N(x), N(y)$ has $N(3), 3 > 2 \to \neg N(3), N(2)$ as an instantiation. If operated on fact set S of Example 1, this rule will change S to $\{3 > 2, 2 > 1, N(2)\}$.

4. The production system consisting of the following two production rules can be used to add a set of numbers in a fact set:

$\neg S(x) \to S(0)$
$S(x), N(y) \to \neg S(x), \neg N(y), S(x+y)$.

For example, starting with the fact set $\{N(1), N(2), N(3), N(4)\}$, this production system will produce the fact set $\{S(10)\}$.

1.7.4 AUTOMATED REASONING

Computers have been used to help prove theorems by verifying special cases. But even more, they have been used to carry out reasoning without external intervention. Developing computer programs that can draw conclusions from a given set of facts is the goal of automated reasoning. There are now automated reasoning programs

Section 1.7 LOGIC-BASED COMPUTER PROGRAMMING PARADIGMS

that can prove results that people have not been able to prove. Automated reasoning can help in verifying the correctness of computer programs, verifying protocol design, verifying hardware design, creating software using logic programming, solving puzzles, and proving new theorems.

Definitions:

Automated reasoning is the process of proving theorems using a computer program that can draw conclusions which follow logically from a set of given facts.

A *computer-assisted proof* is a proof that relies on checking the validity of a large number of cases using a special purpose computer program.

A *proof done by hand* is a proof done by a human without the use of a computer.

Facts:

1. Computer-assisted proofs have been used to settle several well-known conjectures, including the Four Color Theorem (§8.6.4) and the nonexistence of a finite projective plane of order 10 (§12.2.3).

2. The computer-assisted proofs of both the Four Color Theorem and the nonexistence of a finite projective plane of order 10 rely on having a computer verify certain facts about a large number of cases using special purpose software.

3. Hardware, system software, and special purpose program errors can invalidate a computer-assisted proof. This makes the verification of computer-assisted proofs important. However, such verification may be impractical.

4. Automated reasoning software has been developed for both first-order and higher-order logics. A database of automated reasoning systems can be found at
   http://www-formal.stanford.edu:80/clt/ARS/systems.html

5. Automated reasoning software has been used to prove new results in many areas, including settling long-standing, well-known, open conjectures (such as the Robbins problem described in Example 2).

6. Proofs generated by automated reasoning software can usually be checked without using computers or by using software programs that check the validity of proofs.

7. Proofs done by humans often use techniques ill-suited for implementation in automated proof software.

8. Automatic proof systems rely on proof procedures suitable for computer implementation, such as resolution and the semantic tableaux procedure. (See [Fi96] or [Wo96] for details.)

9. The effectiveness of automatic proof systems depends on following strategies that help programs prove results efficiently.

10. Restriction strategies are used to block paths of reasoning that are considered to be unpromising.

11. Direction strategies are used to help programs select the approaches to take next.

12. Look-ahead strategies let programs draw conclusions before they would ordinarily be drawn following the basic rules of the program.

13. Redundancy-control strategies are used to eliminate some of the redundancy in retained information.

14. There are efforts underway to capture all mathematical knowledge into a database that can be used in automated reasoning systems (see the information about the QED system in Example 3).

Examples:

1. The OTTER system is an automated reasoning system for first order logic developed at Argonne National Laboratory [Wo96]. OTTER has been used to establish many previously unknown results in a wide variety of areas, including algebraic curves, lattices, Boolean algebra, groups, semigroups, and logic. A summary of these results can be found at

   http://www.mcs.anl.gov/home/mccune/ar/new_results

2. The automated reasoning system EQP, developed at Argonne National Laboratory, settled the Robbins problem in 1996. This problem was first proposed in the 1930s by Herbert Robbins, and was actively worked on by many mathematicians. The Robbins problem can be stated as follows. Can the equivalence
$$\neg(\neg p) \Leftrightarrow p$$
be derived from the commutative and associative laws for the "or" operator $\vee$ and the identity
$$\neg(\neg(p \vee q) \vee \neg(p \vee \neg q)) \Leftrightarrow p?$$
The EQP system, using some earlier work that established a sufficient condition for the truth of Robbins' problem, found a 15-step proof of the theorem after approximately 8 days of searching on a UNIX workstation when provided with one of several different search strategies.

3. The goal of the QED Project is to build a repository that represents all important, established mathematical knowledge. It is designed to help mathematicians cope with the explosion of mathematical knowledge and help in developing and verifying computer systems.

REFERENCES

Printed Resources:

[Ap81] K. R. Apt, "Ten years of Hoare's logic: a survey—Part 1", *ACM Transactions of Programming Languages and Systems* 3 (1981), 431–483.

[Di76] E. W. Dijkstra, *A Discipline of Programming*, Prentice-Hall, 1976.

[DuPr80] D. Dubois and H. Prade, *Fuzzy Sets and Systems—Theory and Applications*, Academic Press, 1980.

[Ep95] S. S. Epp, *Discrete Mathematics with Applications*, 2nd ed., PWS, 1995.

[Fi96] M. Fitting, *First Order Logic and Automated Theorem Proving*, 2nd ed., Springer-Verlag, 1996.

[Fl67] R. W. Floyd, "Assigning meanings to programs", *Proceedings of the American Mathematical Society Symposium in Applied Mathematics* 19 (1967), 19–32.

[FlPa88] P. Fletcher and C. W. Patty, *Foundations of Higher Mathematics*, PWS, 1988.

[Gr81] D. Gries, *The Science of Programming*, Springer-Verlag, 1981.

[Ha60] P. Halmos, *Naive Set Theory*, Van Nostrand, 1960.

[Ho69] C. A. R. Hoare, "An axiomatic basis for computer programming", *Communications of the ACM* 12 (1969).

[Ka50] E. Kamke, *Theory of Sets*, translated by F. Bagemihl, Dover, 1950.

[Ka86] A. Kandel, *Fuzzy Mathematical Techniques with Applications*, Addison-Wesley, 1986.

[Ka92] A. Kandel, ed., *Fuzzy Expert Systems*, CRC Press, 1992.

[KaLa94] A. Kandel and G. Langholz, eds., *Fuzzy Control Systems*, CRC Press, 1994.

[Kr95] S. G. Krantz, *The Elements of Advanced Mathematics*, CRC Press, 1995.

[Ll84] J. W. Lloyd, *Foundations of Logic Programming*, 2nd ed., Springer-Verlag, 1987.

[Me79] E. Mendelson, *Introduction to Mathematical Logic*, 2nd ed., Van Nostrand, 1979.

[MiRo91] J. G. Michaels and K. H. Rosen, eds., *Applications of Discrete Mathematics*, McGraw-Hill, 1991.

[Mo76] J. D. Monk, *Mathematical Logic*, Springer-Verlag, 1976.

[ReCl90] S. Reeves and M. Clark, *Logic for Computer Science*, Addison-Wesley, 1990.

[Ro95] K. H. Rosen, *Discrete Mathematics and Its Applications*, 4th ed., McGraw-Hill, 1999.

[Sc86] D. A. Schmidt, *Denotational Semantics—A Methodology for Language Development*, Allyn & Bacon, 1986.

[St77] J. E. Stoy, *Denotational Semantics: The Scott-Strachey Approach to Programming Language Theory*, MIT Press, 1977.

[WaHa78] D. A. Waterman and F. Hayes-Roth, *Pattern-Directed Inference Systems*, Academic Press, 1978.

[We72] P. Wegner, "The Vienna definition language", *ACM Computing Surveys* 4 (1972), 5–63.

[Wo96] L. Wos, *The Automation of Reasoning: An Experiment's Notebook with OTTER Tutorial*, Academic Press, 1996.

[YaFi94] R. R. Yager and D. P. Filev, *Essentials of Fuzzy Modeling and Control*, Wiley, 1994.

[YaZa94] R. R. Yager and L. A. Zadeh, eds., *Fuzzy Sets, Neural Networks and Soft Computing*, Van Nostrand Reinhold, 1994.

[Za65] L. A. Zadeh, "Fuzzy Sets," *Information and Control* 8 (1965), 338–353.

[Zi92] H.-J. Zimmermann, *Fuzzy Set Theory and Its Applications*, 2nd ed., Kluwer, 1992.

Web Resources:

http://plato.stanford.edu/archives/win1997/entries/russell-paradox/ (The on-line Stanford Encyclopedia of Philosophy's discussion of Russell's paradox.)

http://www.austinlinks.com/Fuzzy/ (Quadralay's Fuzzy Logic Archive: a tutorial on fuzzy logic and fuzzy systems, and examples of how fuzzy logic is applied.)

http://www-cad.eecs.berkeley.edu/~fmang/paradox.html (Paradoxes.)

http://www.cut-the-knot.com/selfreference/russell.html (Russell's paradox.)

http://www-formal.stanford.edu:80/clt/ARS/systems.html (A database of existing mechanized reasoning systems.)

http://www-history.mcs.st-and.ac.uk/history/HistTopics/Beginnings_of_set_theory.html#61 (The beginnings of set theory.)

http://www.mcs.anl.gov/home/mccune/ar/new_results (A summary of new results in mathematics obtained with Argonne's Automated Deduction Software.)

http://www.philosophers.co.uk/current/paradox2.htm (Discussion of Russell's paradox written by F. Moorhead for *Philosopher's Magazine*.)

http://www.rbjones.com/rbjpub/logic/log025.htm (Information on logic.)

2

COUNTING METHODS

2.1 Summary of Counting Problems — John G. Michaels
2.2 Basic Counting Techniques — Jay Yellen
    2.2.1 Rules of Sum, Product, and Quotient
    2.2.2 Tree Diagrams
    2.2.3 Pigeonhole Principle
    2.2.4 Solving Counting Problems Using Recurrence Relations
    2.2.5 Solving Counting Problems Using Generating Functions

2.3 Permutations and Combinations — Edward W. Packel
    2.3.1 Ordered Selection: Falling Powers
    2.3.2 Unordered Selection: Binomial Coefficients
    2.3.3 Selection with Repetition
    2.3.4 Binomial Coefficient Identities
    2.3.5 Generating Permutations and Combinations

2.4 Inclusion/Exclusion — Robert G. Rieper
    2.4.1 Principle of Inclusion/Exclusion
    2.4.2 Applying Inclusion/Exclusion to Counting Problems

2.5 Partitions — George E. Andrews
    2.5.1 Partitions of Integers
    2.5.2 Stirling Coefficients

2.6 Burnside/Pólya Counting Formula — Alan C. Tucker
    2.6.1 Permutation Groups and Cycle Index Polynomials
    2.6.2 Orbits and Symmetries
    2.6.3 Color Patterns and Induced Permutations
    2.6.4 Fixed Points and Burnside's Lemma
    2.6.5 Pólya's Enumeration Formula

2.7 Möbius Inversion Counting — Edward A. Bender
    2.7.1 Möbius Inversion

2.8 Young Tableaux — Bruce E. Sagan
    2.8.1 Tableaux Counting Formulas
    2.8.2 Tableaux Algorithms

INTRODUCTION

Many problems in mathematics, computer science, and engineering involve counting objects with particular properties. Although there are no absolute rules that can be used to solve all counting problems, many counting problems that occur frequently can be solved using a few basic rules together with a few important counting techniques. This chapter provides information on how many standard counting problems are solved.

GLOSSARY

binomial coefficient: the coefficient $\binom{n}{k}$ of $x^k y^{n-k}$ in the expansion of $(x+y)^n$.

coloring pattern (with respect to a set of symmetries of a figure): a set of mutually equivalent colorings.

combination (from a set S): a subset of S; any unordered selection from S. A k-combination from a set is a subset of k elements of the set.

combination coefficient: the number $C(n, k)$ (equal to $\binom{n}{k}$) of ways to make an unordered choice of k items from a set of n items.

combination-with-replacement (from a set S): any unordered selection with replacement; a multiset of objects from S.

combination-with-replacement coefficient: the number of ways to choose a multiset of k items from a set of n items, written $C^R(n, k)$.

cycle index: for a permutation group G, the multivariate polynomial P_G obtained by dividing the sum of the cycle structure representations of all the permutations in G by the number of elements of G.

cycle structure (of a permutation): a multivariate monomial whose exponents record the number of cycles of each size.

derangement: a permutation on a set that leaves no element fixed.

exponential generating function (for $\{a_k\}_0^\infty$): the formal sum $\sum_{k=0}^{\infty} a_k \frac{x^k}{k!}$, or any equivalent closed-form expression.

falling power: the product $x^{\underline{k}} = x(x-1)\ldots(x-k+1)$ of k consecutive factors starting with x, each factor decreasing by 1.

Ferrers diagram: a geometric, left-justified, and top-justified array of cells, boxes, dots or nodes representing a partition of an integer, in which each row of dots corresponds to a part of the partition.

Gaussian binomial coefficient: the algebraic expression $\begin{bmatrix} n \\ k \end{bmatrix}$ in the variable q defined for nonnegative integers n and k by $\begin{bmatrix} n \\ k \end{bmatrix} = \frac{q^n - 1}{q-1} \cdot \frac{q^{n-1}-1}{q^2 - 1} \ldots \frac{q^{n+1-k}-1}{q^k - 1}$ for $0 < k \le n$ and $\begin{bmatrix} n \\ 0 \end{bmatrix} = 1$.

generating function (or **ordinary generating function**) for $\{a_k\}_0^\infty$: the formal sum $\sum_{k=0}^{\infty} a_k x^k$, or any equivalent closed-form expression.

hook (of a cell in a Ferrers diagram): the set of cells directly to the right or directly below a given cell, together with the cell itself.

hooklength (of a cell in a Ferrers diagram): the number of cells in the hook of that cell.

Kronecker delta function: the function $\delta(x, y)$ defined by the rule $\delta(x, y) = 1$ if $x = y$ and 0 otherwise.

lexicographic order: the order in which a list of strings would appear in a dictionary.

Möbius function: the function $\mu(m)$ where
$$\mu(m) = \begin{cases} 1 & \text{if } m = 1 \\ (-1)^k & \text{if } m \text{ is a product of } k \text{ distinct primes} \\ 0 & \text{if } m \text{ is divisible by the square of a prime,} \end{cases}$$
or a generalization of this function to partially ordered sets.

multinomial coefficient: the coefficient $\binom{n}{k_1\ k_2\ \ldots\ k_m}$ of $x_1^{k_1} x_2^{k_2} \ldots x_m^{k_m}$ in the expansion of $(x_1 + x_2 + \cdots + x_m)^n$.

ordered selection (of k items from a set S): a nonrepeating list of k items from S.

ordered selection with replacement (of k items from a set S): a possibly-repeating list of k items from S.

ordinary generating function (for the sequence $\{a_k\}_0^\infty$): See *generating function*.

partially ordered set (or *poset*): a set S together with a binary relation $\leq$ that is reflexive, antisymmetric, and transitive, written $(S, \leq)$.

partition: an unordered decomposition of an integer into a sum of positive integers.

Pascal's triangle: a triangular table with the binomial coefficient $\binom{n}{k}$ appearing in row n, column k.

pattern inventory: a generating function that enumerates the number of coloring patterns.

permutation: a one-to-one mapping of a set of elements onto itself, or an arrangement of the set into a list. A k-permutation of a set is an ordered nonrepeating sequence of k elements of the set.

permutation coefficient: the number of ways to choose a nonrepeating list of k items from a set of n items, written $P(n, k)$.

permutation group: a nonempty set P of permutations on a set S, such that P is closed under composition and under inversion.

permutation-with-replacement coefficient: the number of ways to choose a possibly repeating list of k items from a set of n items, written $P^R(n, k)$.

poset: See *partially ordered set*.

problème des ménages: the problem of finding the number of ways that married couples can be seated around a circular table so that no men are adjacent, no women are adjacent, and no husband and wife are adjacent.

problème des rencontres: given balls 1 through n drawn out of an urn one at a time, the problem of finding the probability that ball i is never the ith one drawn.

Stirling cycle number: the number $\left[{n \atop k}\right]$ of ways to partition n objects into k nonempty cycles.

Stirling number of the first kind: the coefficient $s(n, k)$ of x^k in the polynomial $x(x-1)(x-2)\ldots(x-n+1)$.

Stirling number of the second kind: the coefficient $S(n, k)$ of $x^{\underline{k}}$ in the representation $x^n = \sum_k S(n, k) x^{\underline{k}}$ of x^n as a linear combination of falling powers.

Stirling subset number: the number $\left\{{n \atop k}\right\}$ of ways to partition n objects into k nonempty subsets.

symmetry (of a figure): a spatial motion that maps the figure onto itself.

tree diagram: a tree that displays the different alternatives in some counting process.

unordered selection (of k items from a set S): a subset of k items from S.

unordered selection (of k items from a set S with replacement): a selection of k objects in which each object in the selection set S can be chosen arbitrarily often and such that the order in which the objects are selected does not matter.

Young tableau: an array obtained by replacing each cell of a Ferrers diagram by a positive integer.

2.1 SUMMARY OF COUNTING PROBLEMS

Table 1 lists many important counting problems, gives the number of objects being counted, together with a reference to the section of this *Handbook* where details can be found. Table 2 lists several important counting rules and methods, and gives the types of counting problems that can be solved using these rules and methods.

Table 1 Counting problems.

The notation used in this table is given at the end of the table.

objects	number of objects	reference
Arranging objects in a row:		
n distinct objects	$n! = P(n,n) = n(n-1)\ldots 2 \cdot 1$	§2.3.1
k out of n distinct objects	$n^{\underline{k}} = P(n,k) = n(n-1)\ldots(n-k+1)$	§2.3.1
some of the n objects are identical: k_1 of a first kind, k_2 of a second kind, ..., k_j of a jth kind, and where $k_1 + k_2 + \cdots + k_j = n$	$\binom{n}{k_1\, k_2\, \ldots\, k_j} = \frac{n!}{k_1!\,k_2!\ldots k_j!}$	§2.3.2
none of the n objects remains in its original place (derangements)	$D_n = n!\left(1 - \frac{1}{1!} + \cdots + (-1)^n \frac{1}{n!}\right)$	§2.4.2
Arranging objects in a circle (where rotations, but not reflections, are equivalent):		
n distinct objects	$(n-1)!$	§2.2.1
k out of n distinct objects	$\frac{P(n,k)}{k}$	§2.2.1
Choosing k objects from n distinct objects:		
order matters, no repetitions	$P(n,k) = \frac{n!}{(n-k)!} = n^{\underline{k}}$	§2.3.1
order matters, repetitions allowed	$P^R(n,k) = n^k$	§2.3.3
order does not matter, no repetitions	$C(n,k) = \binom{n}{k} = \frac{n!}{k!(n-k)!}$	§2.3.2
order does not matter, repetitions allowed	$C^R(n,k) = \binom{k+n-1}{k}$	§2.3.3

Section 2.1 SUMMARY OF COUNTING PROBLEMS

objects	number of objects	reference
Subsets:		
of size k from a set of size n	$\binom{n}{k}$	§2.3.2
of all sizes from a set of size n	2^n	§2.3.4
of $\{1,\ldots,n\}$, without consecutive elements	F_{n+2}	§3.1.2
Placing n objects into k cells:		
distinct objects into distinct cells	k^n	§2.2.1
distinct objects into distinct cells, no cell empty	$\left\{{n \atop k}\right\}k!$	§2.5.2
distinct objects into identical cells	$\left\{{n \atop 1}\right\}+\left\{{n \atop 2}\right\}+\cdots+\left\{{n \atop k}\right\}=B_n$	§2.5.2
distinct objects into identical cells, no cell empty	$\left\{{n \atop k}\right\}$	§2.5.2
distinct objects into distinct cells, with k_i in cell i ($i=1,\ldots,n$), and where $k_1+k_2+\cdots+k_j=n$	$\binom{n}{k_1\,k_2\,\ldots\,k_j}$	§2.3.2
identical objects into distinct cells	$\binom{n+k-1}{n}$	§2.3.3
identical objects into distinct cells, no cell empty	$\binom{n-1}{k-1}$	§2.3.3
identical objects into identical cells	$p_k(n)$	§2.5.1
identical objects into identical cells, no cell empty	$p_k(n)-p_{k-1}(n)$	§2.5.1
Placing n distinct objects into k nonempty cycles	$\left[{n \atop k}\right]$	§2.5.2
Solutions to $x_1+\cdots+x_n=k$:		
nonnegative integers	$\binom{k+n-1}{k}=\binom{k+n-1}{n-1}$	§2.3.3
positive integers	$\binom{k-1}{n-1}$	§2.3.3
integers where $0\leq a_i\leq x_i$ for all i	$\binom{k-(a_1+\cdots+a_n)+n-1}{n-1}$	§2.3.3
integers where $0\leq x_i\leq a_i$ for one or more i	inclusion/exclusion principle	§2.4.2
integers where $x_1\geq\cdots\geq x_n\geq 1$	$p_n(k)-p_{n-1}(k)$	§2.5.1
integers where $x_1\geq\cdots\geq x_n\geq 0$	$p_n(k)$	§2.5.1
Solutions to $x_1+x_2+\cdots+x_n=n$ in nonnegative integers where $x_1\geq x_2\geq\cdots\geq x_n\geq 0$	$p(n)$	§2.5.1
Solutions to $x_1+2x_2+3x_3+\cdots+nx_n=n$ in nonnegative integers	$p(n)$	§2.5.1

objects	number of objects	reference
Functions from a k-element set to an n-element set:		
all functions	n^k	§2.2.1
one-to-one functions ($n \geq k$)	$n^{\underline{k}} = \frac{n!}{(n-k)!} = P(n,k)$	§2.2.1
onto functions ($n \leq k$)	inclusion/exclusion	§4.2
partial functions	$\binom{k}{0}+\binom{k}{1}n+\binom{k}{2}n^2+\cdots+\binom{k}{k}n^k$ $= (n+1)^k$	§2.3.2
Bit strings of length n:		
all strings	2^n	§2.2.1
with given entries in k positions	2^{n-k}	§2.2.1
with exactly k 0s	$\binom{n}{k}$	§2.3.2
with at least k 0s	$\binom{n}{k}+\binom{n}{k+1}+\cdots+\binom{n}{n}$	§2.3.2
with equal numbers of 0s and 1s	$\binom{n}{n/2}$	§2.3.2
palindromes	$2^{\lceil n/2 \rceil}$	§2.2.1
with an even number of 0s	2^{n-1}	§2.3.4
without consecutive 0s	F_{n+2}	§3.1.2
Partitions of a positive integer n into positive summands:		§2.5.1
total number	$p(n)$	
into at most k parts	$p_k(n)$	
into exactly k parts	$p_k(n) - p_{k-1}(n)$	
into parts each of size $\leq k$	$p_k(n)$	
Partitions of a set of size n:		
all partitions	$B(n)$	§2.5.2
into k parts	$\left\{ {n \atop k} \right\}$	§2.5.2
into k parts, each part having at least 2 elements	$b(n,k)$	§3.1.8
Paths:		
from $(0,0)$ to $(2n,0)$ made up of line segments from (i, y_i) to $(i+1, y_{i+1})$, where integer $y_i \geq 0$, $y_{i+1} = y_i \pm 1$	C_n	§3.1.3
from $(0,0)$ to $(2n,0)$ made up of line segments from (i, y_i) to $(i+1, y_{i+1})$, where integer $y_i > 0$ (for $0 < i < 2n$), $y_{i+1} = y_i \pm 1$	C_{n-1}	§3.1.3
from $(0,0)$ to (m,n) that move 1 unit up or right at each step	$\binom{m+n}{n}$	§2.3.2

objects	number of objects	reference
Permutations of $\{1,\ldots,n\}$:		
all permutations	$n!$	§2.3.1
with k cycles, all cycles of length ≥ 2	$d(n,k)$	§3.1.8
with k descents	$E(n,k)$	§3.1.5
with k excedances	$E(n,k)$	§3.1.5
alternating, n even	$(-1)^{n/2}E_n$	§3.1.7
alternating, n odd	T_n	§3.1.7
Symmetries of regular figures:		§2.6
n-gon	$2n$	
tetrahedron	12	
cube	24	
octahedron	24	
dodecahedron	60	
icosahedron	60	
Coloring regular 2-dimensional & 3-dimensional figures with $\leq k$ colors:		§2.6
corners of an n-gon, allowing rotations and reflections	$\frac{1}{2n}\sum_{d\mid n}\varphi(d)k^{\frac{n}{d}} + \frac{1}{2}k^{\frac{(n+1)}{2}}$, n odd; $\frac{1}{2n}\sum_{d\mid n}\varphi(d)k^{\frac{n}{d}} + \frac{1}{4}(k^{\frac{n}{2}} + k^{\frac{(n+2)}{2}})$, n even	
corners of an n-gon, allowing only rotations	$\frac{1}{n}\sum_{d\mid n}\varphi(d)k^{\frac{n}{d}}$	
corners of a triangle, allowing rotations and reflections	$\frac{1}{6}[k^3 + 3k^2 + 2k]$	
corners of a triangle, allowing only rotations	$\frac{1}{3}[k^3 + 2k]$	
corners of a square, allowing rotations and reflections	$\frac{1}{8}[k^4 + 2k^3 + 3k^2 + 2k]$	
corners of a square, allowing only rotations	$\frac{1}{4}[k^4 + k^2 + 2k]$	
corners of a pentagon, allowing rotations and reflections	$\frac{1}{10}[k^5 + 5k^3 + 4k]$	
corners of a pentagon, allowing only rotations	$\frac{1}{5}[k^5 + 4k]$	

objects	number of objects	reference
corners of a hexagon, allowing rotations and reflections	$\frac{1}{12}[k^6 + 3k^4 + 4k^3 + 2k^2 + 2k]$	
corners of a hexagon, allowing only rotations	$\frac{1}{6}[k^6 + k^3 + 2k^2 + 2k]$	
corners of a tetrahedron	$\frac{1}{12}[k^4 + 11k^2]$	
edges of a tetrahedron	$\frac{1}{12}[k^6 + 3k^4 + 8k^2]$	
faces of a tetrahedron	$\frac{1}{12}[k^4 + 11k^2]$	
corners of a cube	$\frac{1}{24}[k^8 + 17k^4 + 6k^2]$	
edges of a cube	$\frac{1}{24}[k^{12} + 6k^7 + 3k^6 + 8k^4 + 6k^3]$	
faces of a cube	$\frac{1}{24}[k^6 + 3k^4 + 12k^3 + 8k^2]$	
Number of sequences of wins/losses in a $\frac{n+1}{2}$-out-of-n playoff series (n odd)	$2C(n, \frac{n+1}{2})$	§2.3.2
Sequences $a_1, \ldots, a_{2n}$ with n 1s and n -1s, and each partial sum $a_1 + \cdots + a_k \geq 0$	C_n	§3.1.3
Well-formed sequences of parentheses of length $2n$	C_n	§3.1.3
Well-parenthesized products of $n+1$ variables	C_n	§3.1.3
Triangulations of a convex $(n+2)$-gon	C_n	§3.1.3

Notation:

$B(n)$ or B_n: Bell number

$b(n,k)$: associated Stirling number of the second kind

$C_n = \frac{1}{n+1}\binom{2n}{n}$: Catalan number

$C(n,k) = \binom{n}{k} = \frac{n!}{k!(n-k)!}$: binomial coefficient

$d(n,k)$: associated Stirling number of the first kind

E_n: Euler number

φ: Euler phi-function

$E(n,k)$: Eulerian number

F_n: Fibonacci number

$n^{\underline{k}} = n(n-1)\ldots(n-k+1) = P(n,k)$: falling power

$P(n,k) = \frac{n!}{(n-k)!}$: k-permutation

$p(n)$: number of partitions of n

$p_k(n)$: number of partitions of n into at most k summands

$p_k^*(n)$: number of partitions of n into exactly k summands

$\begin{bmatrix} n \\ k \end{bmatrix}$: Stirling cycle number

$\begin{Bmatrix} n \\ k \end{Bmatrix}$: Stirling subset number

T_n: tangent number

Table 2 Methods of counting and the problems they solve.

statement	technique of proof
rule of sum (§2.2.1)	problems that can be broken into disjoint cases, each of which can be handled separately
rule of product (§2.2.1)	problems that can be broken into sequences of independent counting problems, each of which can be solved separately
rule of quotient (§2.2.1)	problems of counting arrangements, where the arrangements can be divided into collections that are all of the same size
pigeonhole principle (§2.2.3)	problems with two sets of objects, where one set of objects needs to be matched with the other
inclusion/exclusion principle (§2.4)	problems that involve finding the size of a union of sets, where some or all the sets in the union may have common elements
permutations (§2.2.1, 2.3.1, 2.3.3)	problems that require counting the number of selections or arrangements, where order within the selection or arrangement matters
combinations (§2.3.2, 2.3.3)	problems that require counting the number of selections or sets of choices, where order within the selection does not matter
recurrence relations (§2.3.6)	problems that require an answer depending on the integer n, where the solution to the problem for a given size n can be related to one or more cases of the problem for smaller sizes
generating functions (§2.3.7)	problems that can be solved by finding a closed form for a function that represents the problem and then manipulating the closed form to find a formula for the coefficients
Pólya counting (§2.6.5)	problems that require a listing or number of patterns, where the patterns are not to be regarded as different under certain types of motions (such as rotations and reflections)
Möbius inversion (§2.7.1)	problems that involve counting certain types of circular permutations

2.2 BASIC COUNTING TECHNIQUES

Most counting methods are based directly or indirectly on the fundamental principles and techniques presented in this section. The rules of sum, product, and quotient are the most basic and are applied more often than any other. The section also includes some applications of the pigeonhole principle, a brief introduction to generating functions, and several examples illustrating the use of tree diagrams and Venn diagrams.

2.2.1 RULES OF SUM, PRODUCT, AND QUOTIENT

Definitions:

The **rule of sum** states that when there are m cases such that the ith case has n_i options, for $i = 1, \ldots, m$, and no two of the cases have any options in common, the total number of options is $n_1 + n_2 + \cdots + n_m$.

The **rule of product** states that when a procedure can be broken down into m steps, such that there are n_1 options for step 1, and such that after the completion of step $i-1$ ($i = 2, \ldots, m$) there are n_i options for step i, the number of ways of performing the procedure is $n_1 n_2 \ldots n_m$.

The **rule of quotient** states that when a set S is partitioned into equal-sized subsets of m elements each, there are $\frac{|S|}{m}$ subsets.

An **m-permutation** of a set S with n elements is a nonrepeating ordered selection of m elements of S, that is, a sequence of m distinct elements of S. An n-permutation is simply called a **permutation** of S.

Facts:
1. The rule of sum can be stated in set-theoretic terms: if sets $S_1, \ldots, S_m$ are finite and pairwise disjoint, then $|S_1 \cup S_2 \cup \cdots \cup S_m| = |S_i| + |S_2| + \cdots + |S_m|$.
2. The rule of product can be stated in set-theoretic terms: if sets $S_1, \ldots, S_m$ are finite, then $|S_1 \times S_2 \times \cdots \times S_m| = |S_1| \cdot |S_2| \cdot \cdots \cdot |S_m|$.
3. The rule of quotient can be stated in terms of the equivalence classes of an equivalence relation on a finite set S: if every class has m elements, then there are $|S|/m$ equivalence classes.
4. Venn diagrams (§1.2.2) are often used as an aid in counting the elements of a subset, as an auxiliary to the rule of sum. This generalizes to the principle of inclusion/exclusion (§2.3).
5. Counting problems can often be solved by using a combination of counting methods, such as the rule of sum and the rule of product.

Examples:
1. *Counting bit strings*: There are 2^n bit strings of length n, since such a bit string consists of n bits, each of which is either 0 or 1.
2. *Counting bit strings with restrictions*: There are 2^{n-2} bit strings of length n ($n \geq 2$) that begin with two 1s, since forming such a bit string consists of filling in $n-2$ positions with 0s or 1s.

3. *Counting palindromes*: A palindrome is a string of symbols that is unchanged if the symbols are written in reverse order, such as *rpnbnpr* or 10011001. There are $k^{\lceil n/2 \rceil}$ palindromes of length n where the symbols are chosen from a set of k symbols.

4. *Counting the number of variable names*: Determine the number of variable names, subject to the following rules: a variable name has four or fewer characters, the first character is a letter, the second and third are letters or digits, and the fourth must be X or Y or Z. Partition the names into four sets, S_1, S_2, S_3, S_4, containing names of length 1, 2, 3, and 4 respectively. Then $|S_1| = 26$, $|S_2| = 26 \times 36$, $|S_3| = 26 \times 36^2$, and $|S_4| = 26 \times 36^2 \times 3$. Therefore the total number of names equals $|S_1|+|S_2|+|S_3|+|S_4| = 135{,}746$.

5. *Counting functions*: There are n^m functions from a set $A = \{a_1, \ldots, a_m\}$ to a set $B = \{b_1, \ldots, b_n\}$. (Construct each function $f : A \to B$ by an m-step process, where step i is to select the value $f(a_i)$.)

6. *Counting one-to-one functions*: There are $n(n-1)\ldots(n-m+1)$ one-to-one functions from $A = \{a_1, \ldots, a_m\}$ to $B = \{b_1, \ldots, b_n\}$. If values $f(a_1), \ldots, f(a_{i-1})$ have already been selected in set B during the first $i-1$ steps, then there are $n-i+1$ possible values remaining for $f(a_i)$.

7. *Counting permutations*: There are $n(n-1)\ldots(n-m+1) = \frac{n!}{(n-m)!}$ m-permutations of an n-element set. (Each one-to-one function in Example 6 may be viewed as an m-permutation of B.) (Permutations are discussed in §2.3.)

8. *Counting circular permutations*: There are $(n-1)!$ ways to seat n people around a round table (where rotations are regarded as equivalent, but the clockwise/counterclockwise distinction is maintained). The total number of arrangements is $n!$ and each equivalence class contains n configurations. By the rule of quotient, the number of arrangements is $\frac{n!}{n} = (n-1)!$.

9. *Counting restricted circular permutations*: If n women and n men are to be seated around a circular table, with no two of the same sex seated next to each other, the number of possible arrangements is $n(n-1)!^2$.

2.2.2 TREE DIAGRAMS

When a counting problem breaks into cases, a tree can be used to make sure that every case is counted, and that no case is counted twice.

Definitions:

A **tree diagram** is a line-drawing of a tree, often with its branches and/or nodes labeled. The **root** represents the start of a procedure and the **branches** at each node represent the options for the next step.

Facts:

1. Tree diagrams are commonly used as an important auxiliary to the rules of sum and product.

2. The objective in a tree-counting approach is often one of the following:
    - the number of leaves (endnodes)
    - the number of nodes
    - the sum of the path products.

Examples:

1. There are 6 possible sequences of wins and losses when the home team (H) plays the visiting team (V) in a best 2-out-of-3 playoff. In the following tree diagram each edge label indicates whether the home team won or lost the corresponding game, and the label at each final node is the outcome of the playoff. The number of different possible sequences equals the number of endnodes — 6.

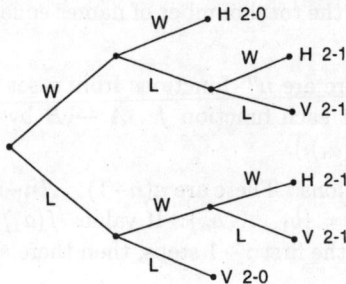

2. Suppose that an experimental process begins by tossing two identical dice. If the dice match, the process continues for a second round; if not, the process stops at one round. Thus, an experimental outcome sequence consist of one or two unordered pairs of numbers from 1 to 6. The three paths in the following tree represent the three different kinds of outcome sequences. The total number of possible outcomes is the sum of the path products $6^2 + 6 \cdot 15 + 15 = 141$.

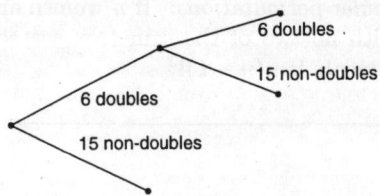

2.2.3 PIGEONHOLE PRINCIPLE

Definitions:

The **pigeonhole principle** (**Dirichlet drawer principle**) states that if $n+1$ objects (pigeons) are placed into n boxes (pigeonholes), then some box contains more than one object. (Peter Gustav Lejeune Dirichlet, 1805–1859)

The **generalized pigeonhole principle** states that if m objects are placed into k boxes, then some box contains at least $\lceil \frac{m}{k} \rceil$ objects.

The **set-theoretic form of the pigeonhole principle** states that if $f: S \to T$ where S and T are finite and any two of the following conditions hold, then so does the third:
- f is one-to-one
- f is onto
- $|S| = |T|$.

Examples:

1. Among any group of eight people, at least two were born on the same day of the week. This follows since there are seven pigeonholes (the seven days of the week) and more than seven pigeons (the eight people).

2. Among any group of 25 people, at least four were born on the same day of the week. This follows from the generalized pigeonhole principle with $m = 25$ and $k = 7$, yielding $\lceil \frac{m}{k} \rceil = \lceil \frac{25}{7} \rceil = 4$.

3. Suppose that a dresser drawer contains many black socks and blue socks. If choosing in total darkness, a person must grab at least three socks to be absolutely certain of having a pair of the same color. The two colors are pigeonholes; the pigeonhole principle says that three socks (the pigeons) are enough.

4. What is the minimum number of points whose placement in the interior of a 2×2 square guarantees that at least two of them are less than $\sqrt{2}$ units apart? Four points are not enough, since they could be placed near the respective corners of the 2×2 square. To see that five is enough, partition the 2×2 square into four 1×1 squares. By the pigeonhole principle, one of these 1×1 squares must contain at least two of the points, and these two must be less than $\sqrt{2}$ units apart.

5. In any set of $n+1$ positive integers, each less than or equal to $2n$, there are at least two such that one is a multiple of the other. To see this, express each of the $n+1$ numbers in the form $2^k \cdot q$, where q is odd. Since there are only n possible odd values for q between 1 and $2n$, at least two of the $n+1$ numbers must have the same q, and the result follows.

6. Let B_1 and B_2 be any two bit strings, each consisting of five ones and five zeros. Then there is a cyclic shift of bit string B_2 so that the resulting string, B_2', matches B_1 in at least five of its positions. For example, if $B_1 = 1010101010$ and $B_2 = 0001110101$, then $B_2' = 1000111010$ satisfies the condition. Observe that there are 10 possible cyclic shifts of bit string B_2. For $i = 1, \ldots, 10$, the ith bit of exactly five of these strings will match the ith bit of B_1. Thus, there is a total of 50 bitmatches over the set of 10 cyclic shifts. The generalized pigeonhole principle implies that there is at least one cyclic shift having $\lceil \frac{50}{10} \rceil = 5$ matching bits.

7. Every sequence of $n^2 + 1$ distinct real numbers must have an increasing or decreasing subsequence of length $n + 1$. Given a sequence $a_1, \ldots, a_{n^2+1}$, for each a_j let d_j and i_j be the lengths of the longest decreasing and increasing subsequences beginning with a_j. This gives a sequence of $n^2 + 1$ ordered pairs (d_j, i_j). If there were no increasing or decreasing subsequence of length $n + 1$, then there are only n^2 possible ordered pairs (d_j, i_j), since $1 \leq d_j \leq n$ and $1 \leq i_j \leq n$. By the pigeonhole principle, at least two ordered pairs must be identical. Hence there are p and q such that $d_p = d_q$ and $i_p = i_q$. If $a_p < a_q$, then the sequence a_p followed by the increasing subsequence starting at a_q gives an increasing subsequence of length greater than i_q — a contradiction. A similar contradiction on the choice of d_p follows if $a_q < a_p$. Hence a decreasing or increasing subsequence of length $n + 1$ must exist.

2.2.4 SOLVING COUNTING PROBLEMS USING RECURRENCE RELATIONS

Certain types of counting problems can be solved by modeling the problem using a recurrence relation (§3.3) and then working with the recurrence relation.

Facts:

1. The following general procedure is used for solving a counting problem using a recurrence relation:
   - let a_n be the solution of the counting problem for the parameter n;
   - determine a recurrence relation for a_n, together with the appropriate number of initial conditions;
   - find the particular value of the sequence that solves the original counting problem by repeated use of the recurrence relation or by finding an explicit formula for a_n and evaluating it at n.

2. There are many techniques for solving recurrence relations which may be useful in the solution of counting problems. Section 3.3 provides general material on recurrence relations and contains many examples illustrating how counting problems are solved using recurrence relations.

Examples:

1. *Tower of Hanoi*: The Tower of Hanoi puzzle consists of three pegs mounted on a board and n disks of different sizes. Initially the disks are on the first peg in order of decreasing size. See the following figure, using four disks. The rules allow disks to be moved one at a time from one peg to another, with no disk ever placed atop a smaller one. The goal of the puzzle is to move the tower of disks to the second peg, with the largest on the bottom. How many moves are needed to solve this puzzle for 64 disks?

   Let a_n be the minimum number of moves to solve the Tower of Hanoi puzzle with n disks. Transferring the $n-1$ smallest disks from peg 1 to peg 3 requires a_{n-1} moves. One move is required to transfer the largest disk to peg 2, and transferring the $n-1$ disks now on peg 3 to peg 2, placing them atop the largest disk requires a_{n-1} moves. Hence, the puzzle with n disks can be solved using $2a_{n-1}+1$ moves. The puzzle for n disks cannot be solved in fewer steps, since then the puzzle with $n-1$ disks could be solved using fewer than a_{n-1} moves. Hence $a_n = 2a_{n-1}+1$. The initial condition is $a_1 = 1$. Iterating shows that $a_n = 2a_{n-1}+1 = 2^2 a_{n-2}+2+1 = \cdots = 2^{n-1}a_1 + 2^{n-2} + \cdots + 2^2 + 2 + 1 = 2^n - 1$. Hence, $2^{64} - 1$ moves are required to solve this problem for 64 disks. (§3.3.3 Example 3 and §3.3.4 Example 1 provide alternative methods for solving this recurrence relation.)

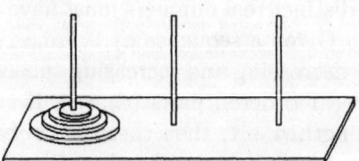

2. *Reve's puzzle*: The Reve's puzzle is the variation of the Tower of Hanoi puzzle that follows the same rules as the Tower of Hanoi puzzle, but uses four pegs.

   The minimum number of moves needed to solve the Reve's puzzle for n disks is not known, but it is conjectured that this number is $R(n) = \sum_{i=1}^{k} i2^{i-1} - \left(\frac{k(k+1)}{2} - n\right)2^{k-1}$ where k is the smallest integer such that $n \leq \frac{k(k+1)}{2}$.

   The following recursive algorithm, the Frame-Stewart algorithm, gives a method for solving the Reve's puzzle by moving n disks from peg 1 to peg 4 in $R(n)$ moves. If $n = 1$, move the single disk from peg 1 to peg 4. If $n > 1$: recursively move the $n - k$ smallest disks from peg 1 to peg 2 using the Frame-Stewart algorithm; then move the k largest disks from peg 1 to peg 4 using the 3-peg algorithm from Example 1 on pegs 1, 3, and 4; and finally recursively move the $n - k$ smallest disks from peg 2 to peg 4 using the Frame-Stewart algorithm.

3. How many strings of 4 decimal digits contain an even number of 0s? Let a_n be the number of strings of n decimal digits that contain an even number of 0s. To obtain such a string: (1) append a nonzero digit to a string of $n-1$ decimal digits that has an even number of 0s, which can be done in $9a_{n-1}$ ways; or (2) append a 0 to a string of $n-1$ decimal digits that has an odd number of 0s, which can be done in $10^{n-1} - a_{n-1}$ ways. Hence $a_n = 9a_{n-1} + (10^{n-1} - a_{n-1}) = 8a_{n-1} + 10^{n-1}$. The initial condition is $a_1 = 9$. It follows that $a_2 = 8a_1 + 10 = 82$, $a_3 = 8a_2 + 100 = 756$, and $a_4 = 8a_3 + 1{,}000 = 7{,}048$.

2.2.5 SOLVING COUNTING PROBLEMS USING GENERATING FUNCTIONS

Some counting problems can be solved by finding a closed form for the function that represents the problem and then manipulating the closed form to find the relevant coefficient.

Facts:

1. Use the following procedure for solving a counting problem by using a generating function:
   - let a_n be the solution of the counting problem for the parameter n;
   - find a closed form for the generating function $f(x)$ that has a_n as the coefficient of x^n in its power series;
   - solve the counting problem by computing a_n by expanding the closed form and examining the coefficient of x^n.

2. Generating functions can be used to solve counting problems that reduce to finding the number of solutions to an equation of the form $x_1 + x_2 + \cdots + x_n = k$, where k is a positive integer and the x_i's are integers subject to constraints.

3. There are many techniques for manipulating generating functions (§3.2, §3.3.5) which may be useful in the solution of counting problems. Section 3.2 contains examples of counting problems solved using generating functions.

Examples:

1. How many ways are there to distribute eight identical cookies to three children if each child receives at least two and no more than four cookies. Let c_n be the number of ways to distribute n identical cookies in this way. Then c_n is the coefficient of x^n in $(x^2 + x^3 + x^4)^3$, since a distribution of n cookies to the three children is equivalent to a solution of $x_1 + x_2 + x_3 = 8$ with $2 \leq x_i \leq 4$ for $i = 1, 2, 3$. Expanding this product shows that c_8, the coefficient of x_8, is 6. Hence there are 6 ways to distribute the cookies.

2. An urn contains colored balls, where each ball is either red, blue, or black, there are at least ten balls of each color, and balls of the same color are indistinguishable. Find the number of ways to select ten balls from the urn, so that an odd number of red balls, an even number of blue balls, and at least five black balls are selected. If x_1, x_2, and x_3 denote the number of red balls, blue balls, and black balls selected, respectively, the answer is provided by the number of nonnegative integer solutions of $x_1 + x_2 + x_3 = 10$ with x_1 odd, x_2 even, $x_3 \geq 5$. This is the coefficient of x^{10} in the generating function $f(x) = (x + x^3 + x^5 + x^7 + x^9 + \cdots)(1 + x^2 + x^4 + x^6 + x^8 + x^{10} + \cdots)(x^5 + x^6 + x^7 + x^8 + x^9 + x^{10} + \cdots)$. Since the coefficient of x^{10} in the expansion is 6, there are six ways to select the balls as specified.

2.3 PERMUTATIONS AND COMBINATIONS

Permutations count the number of arrangements of objects, and combinations count the number of ways to select objects from a set. A permutation coefficient counts the number of ways to arrange a set of objects, whereas a combination coefficient counts the number of ways to select a subset.

2.3.1 ORDERED SELECTION: FALLING POWERS

Falling powers mathematically model the process of selecting k items from a collection of n items in circumstances where the ordering of the selection matters and repetition is not allowed.

Definitions:

An *ordered selection* of k items from a set S is a nonrepeating list of k items from S.

The *falling power* $x^{\underline{k}}$ is the product $x(x-1)\ldots(x-k+1)$ of k decreasing factors starting at the real number x.

The number *n-factorial*, $n!$ (n a nonnegative integer), is defined by the rule $0! = 1$, $n! = n(n-1)\ldots 3\cdot 2\cdot 1$ if $n \geq 1$.

A *permutation of a list* is any rearrangement of the list.

A *permutation of a set* of n items is an arrangement of those items into a list. (Often, such a list and/or the permutation itself is represented by a string whose entries are in the list order.)

A *k-permutation of a set* of n items is an ordered selection of k items from that set. A k-permutation can be written as a sequence or a string.

The *permutation coefficient* $P(n,k)$ is the number of ways to choose an ordered selection of k items from a set of n items; that is, the number of k-permutations.

A *derangement of a list* is a permutation of the entries such that no entry remains in the original position.

Facts:

1. The falling power $x^{\underline{k}}$ is analogous to the ordinary power x^k, which is the product of k constant factors x. The underline in the exponent of the falling power is a reminder that consecutive factors drop.
2. $P(n,k) = n^{\underline{k}} = \frac{n!}{(n-k)!}$.
3. For any integer n, $n^{\underline{n}} = n!$.
4. The numbers $P(n,k) = n^{\underline{k}}$ are given in Table 1.
5. A repetition-free list of length n has approximately $n!/e$ derangements.

Examples:

1. $(4.2)^{\underline{3}} = 4.2 \cdot 3.2 \cdot 2.2 = 29.568$.
2. *Dealing a row of playing cards*: Suppose that five cards are to be dealt from a deck of 52 cards and placed face up in a row. There are $P(52,5) = 52^{\underline{5}} = 52\cdot 51\cdot 50\cdot 49\cdot 48 = 311{,}875{,}200$ ways to do this.

Table 1 Permutation coefficients $P(n,k) = n^{\underline{k}}$.

$n \backslash k$	0	1	2	3	4	5	6	7	8	9	10
0	1										
1	1	1									
2	1	2	2								
3	1	3	6	6							
4	1	4	12	24	24						
5	1	5	20	60	120	120					
6	1	6	30	120	360	720	720				
7	1	7	42	210	840	2,520	5,040	5,040			
8	1	8	56	336	1,680	6,720	20,160	40,320	40,320		
9	1	9	72	504	3,024	15,120	60,480	181,440	362,880	362,880	
10	1	10	90	720	5,040	30,240	151,200	604,800	1,814,400	3,628,800	3,628,800

3. *Placing distinct balls into distinct bins*: k differently-colored balls are to be placed into n bins ($n \geq k$), with at most one ball to a bin. The number of different ways to arrange the balls is $P(n,k) = n^{\underline{k}}$. (Think of the balls as if they were numbered 1 to k, so that placing ball j into a bin corresponds to placing that bin into the jth position of the list.)

4. *Counting ballots*: Each voter is asked to identify 3 top choices from 11 candidates running for office. A first choice vote is worth 3 points, second choice 2 points, and third choice 1 point. Since a completed ballot is an ordered selection in this situation, each voter has $P(11,3) = 11^{\underline{3}} = 11 \cdot 10 \cdot 9 = 990$ distinct ways to cast a vote.

5. *License plate combinations*: The license plates in a state have three letters (from the upper-case Roman alphabet of 26 letters) followed by four digits. There are $P(26,3) = 15{,}600$ ways to select the letters and $P(10,4) = 5{,}040$ ways to select the digits. By the rule of product there are $P(26,3)P(10,4) = 15{,}600 \cdot 5{,}040 = 78{,}624{,}000$ acceptable strings.

6. *Circular permutations of distinct objects*: See Example 8 of §2.2.1. Also see Example 3 of §2.7.1 for problems that allow identical objects.

7. *Increasing and decreasing subsequences of permutations*: Young tableaux (§2.8) can be used to find the number of permutations of $\{1, 2, \ldots, n\}$ with specified lengths of their longest increasing subsequences and longest decreasing subsequences.

2.3.2 UNORDERED SELECTION: BINOMIAL COEFFICIENTS

Binomial coefficients mathematically model the process of selecting k items from a collection of n items in circumstances where the ordering of the selection does not matter, and repetitions are not allowed.

Definitions:

An **unordered selection** of k items from a set S is a subset of k items from S.

A **k-combination** from a set S is an unordered selection of k items.

The **combination coefficient** $C(n,k)$ is the number of k-combinations of n objects.

The **binomial coefficient** $\binom{n}{k}$ is the coefficient of $x^k y^{n-k}$ in the expansion of $(x+y)^n$.

The **extended binomial coefficient (generalized binomial coefficient)** $\binom{n}{k}$ is zero whenever k is negative. When n is a negative integer and k a nonnegative integer, its value is $(-1)^k \binom{k-n-1}{k}$.

The **multicombination coefficient** $C(n: k_1, k_2, \ldots, k_m)$, where $n = k_1 + k_2 + \cdots + k_m$ denotes the number of ways to partition n items into subsets of sizes $k_1, k_2, \ldots, k_m$.

The **multinomial coefficient** $\binom{n}{k_1\ k_2\ \ldots\ k_m}$ is the coefficient of $x_1^{k_1} x_2^{k_2} \ldots x_m^{k_m}$ in the expansion of $(x_1 + x_2 + \cdots + x_m)^n$.

The **Gaussian binomial coefficient** is defined for nonnegative integers n and k by

$$\begin{bmatrix} n \\ k \end{bmatrix} = \frac{q^n-1}{q-1} \cdot \frac{q^{n-1}-1}{q^2-1} \cdot \frac{q^{n-2}-1}{q^3-1} \cdots \frac{q^{n+1-k}-1}{q^k-1} \quad \text{for } 0 < k \leq n$$

and $\begin{bmatrix} n \\ 0 \end{bmatrix} = 1$, where q is a variable. (See also §2.5.1.)

Facts:
1. $C(n,k) = \frac{P(n,k)}{k!} = \frac{n^{\underline{k}}}{k!} = \frac{n!}{k!(n-k)!} = \binom{n}{k}$.
2. *Pascal's recursion*: $\binom{n}{k} = \binom{n-1}{k-1} + \binom{n-1}{k}$, where $n > 0$ and $k > 0$.
3. *Subsets*: There are $C(n,k)$ subsets of size k that can be chosen from a set of size n.
4. The numbers $C(n,k) = \binom{n}{k}$ are given in Table 2. Sometimes the entries in Table 2 are arranged into the form called *Pascal's triangle* (Table 3), in which each entry is the sum of the two numbers diagonally above the number (Pascal's recursion, Fact 2).

Table 2 Combination coefficients (binomial coefficients) $C(n,k) = \binom{n}{k}$.

$n \backslash k$	0	1	2	3	4	5	6	7	8	9	10	11	12
0	1												
1	1	1											
2	1	2	1										
3	1	3	3	1									
4	1	4	6	4	1								
5	1	5	10	10	5	1							
6	1	6	15	20	15	6	1						
7	1	7	21	35	35	21	7	1					
8	1	8	28	56	70	56	28	8	1				
9	1	9	36	84	126	126	84	36	9	1			
10	1	10	45	120	210	252	210	120	45	10	1		
11	1	11	55	165	330	462	462	330	165	55	11	1	
12	1	12	66	220	495	792	924	792	495	220	66	12	1

5. The extended binomial coefficients satisfy Pascal's recursion. Their definition is constructed precisely to achieve this purpose.
6. $C(n: k_1, k_2, \ldots, k_m) = \frac{n!}{k_1! k_2! \ldots k_m!} = \binom{n}{k_1\ k_2\ \ldots\ k_m}$. The number of strings of length n with k_i objects of type i ($i = 1, 2, \ldots, m$) is $\frac{n!}{k_1! k_2! \ldots k_m!}$.
7. $C(n,k) = C(n: k, n-k) = C(n, n-k)$. That is, the number of unordered selections of k objects chosen from n objects is equal to the number of unordered selections of $n-k$ objects chosen from n objects.
8. *Gaussian binomial coefficient identities*:
   - $\begin{bmatrix} n \\ k \end{bmatrix} = \begin{bmatrix} n \\ n-k \end{bmatrix}$;
   - $\begin{bmatrix} n \\ k \end{bmatrix} + \begin{bmatrix} n \\ k-1 \end{bmatrix} q^{n+1-k} = \begin{bmatrix} n+1 \\ k \end{bmatrix}$.

Table 3 Pascal's triangle.

$$
\begin{array}{ccccccccccccccccccccc}
 & & & & & & & & & & 1 & & & & & & & & & & \\
 & & & & & & & & & 1 & & 1 & & & & & & & & & \\
 & & & & & & & & 1 & & 2 & & 1 & & & & & & & & \\
 & & & & & & & 1 & & 3 & & 3 & & 1 & & & & & & & \\
 & & & & & & 1 & & 4 & & 6 & & 4 & & 1 & & & & & & \\
 & & & & & 1 & & 5 & & 10 & & 10 & & 5 & & 1 & & & & & \\
 & & & & 1 & & 6 & & 15 & & 20 & & 15 & & 6 & & 1 & & & & \\
 & & & 1 & & 7 & & 21 & & 35 & & 35 & & 21 & & 7 & & 1 & & & \\
 & & 1 & & 8 & & 28 & & 56 & & 70 & & 56 & & 28 & & 8 & & 1 & & \\
 & 1 & & 9 & & 36 & & 84 & & 126 & & 126 & & 84 & & 36 & & 9 & & 1 & \\
1 & & 10 & & 45 & & 120 & & 210 & & 252 & & 210 & & 120 & & 45 & & 10 & & 1 \\
\end{array}
$$

9. $(1+x)(1+qx)(1+q^2x)\ldots(1+q^{n-1}x) = \sum_{k=0}^{n} \begin{bmatrix} n \\ k \end{bmatrix} q^{k(k-1)/2} x^k$.

10. $\lim_{q \to 1} \begin{bmatrix} n \\ k \end{bmatrix} = \binom{n}{k}$.

11. $\begin{bmatrix} n \\ k \end{bmatrix} = a_0 + a_1 q + a_2 q^2 + \cdots + a_{k(n-k)} q^{k(n-k)}$ where each a_i is an integer and $\sum_{i=0}^{k(n-k)} a_i = \binom{n}{k}$.

Examples:

1. *Subsets*: A set with 20 elements has $C(20, 4)$ subsets with four elements. The total number of subsets of a set with 20 elements is equal to $C(20, 0) + C(20, 1) + \cdots + C(20, 20)$, which is equal to 2^{20}. (See §2.3.4.)

2. *Nondistinct balls into distinct bins*: k identically colored balls are to be placed into n bins ($n \geq k$), at most one ball to a bin. The number of different ways to do this is $C(n, k) = \frac{n^{\underline{k}}}{k!}$. (This amounts to selecting from the n bins the k bins into which the balls are placed.)

3. *Counting ballots*: Each voter is asked to identify 3 choices for trustee from 11 candidates nominated for the position, without specifying any order of preference. Since a completed ballot is an unordered selection in this situation, each voter has $C(11, 3) = \frac{11 \cdot 10 \cdot 9}{3!} = 165$ distinct ways to cast a vote.

4. *Counting bit strings with exactly k 0s*: There are $\binom{n}{k}$ bit strings of length n with exactly k 0s, since each such bit string is determined by choosing a subset of size k from the n positions; 0s are placed in these k positions, and 1s in the remaining positions.

5. *Counting bit strings with at least k 0s*: There are $\binom{n}{k} + \binom{n}{k+1} + \cdots + \binom{n}{n}$ bit strings of length n with at least k 0s, since each such bit string is determined by choosing a subset of size $k, k+1, \ldots,$ or n from the n positions; 0s are placed in these positions, and 1s in the remaining positions.

6. *Counting bit strings with equal numbers of 0s and 1s*: For n even, there are $\binom{n}{n/2}$ bit strings of length n with equal numbers of 0s and 1s, since each such bit string is determined by choosing a subset of size $\frac{n}{2}$ from the n positions; 0s are placed in these positions, and 1s in the remaining positions.

7. *Counting strings with repeated letters*: The word "MISSISSIPPI" has eleven letters, with "I" and "S" appearing four times each, "P" appearing twice, and "M" once. There are $C(11 : 4, 4, 2, 1) = \frac{11!}{4!4!2!1!} = 34{,}650$ possible different strings obtainable by permuting the letters. This counting problem is equivalent to partitioning 11 items into subsets of sizes 4, 4, 2, 1.

8. *Counting circular strings with repeated letters*: See §2.7.1.

9. *Counting paths*: The number of paths in the plane from $(0,0)$ to a point (m,n) $(m, n \geq 0)$ that move one unit upward or one unit to the right at each step is $\binom{m+n}{n}$. Using U for "up" and R for "right", each path can be described by a string of m Rs and n Us.

10. *Playoff series*: In a series of playoff games, such as the World Series or Stanley Cup finals, the winner is the first team to win more that half the maximum number of games possible, n (odd). The winner must win $\frac{n+1}{2}$ games. The number of possible win-loss sequences of such a series is $2C(n, \frac{n+1}{2})$. For example, in the World Series between teams A and B, any string of length 7 with exactly 4 As represents a winning sequence for A. (The string $AABABBA$ means that A won a seven-game series by winning the first, second, fourth, and seventh games; the string $AAAABBB$ means that A won the series by winning the first four games.) There are $C(7,4)$ ways for A to win the World Series, and $C(7,4)$ ways for B to win the World Series.

11. *Dealing a hand of playing cards*: A hand of five cards (where order does not matter) can be dealt from a deck of 52 cards in $C(52,5) = \frac{52^{\underline{5}}}{5!} = 2{,}598{,}960$ ways.

12. *Poker hands*: Table 4 contains the number of combinations of five cards that form various poker hands (where an ace can be high or low):

13. *Counting partial functions*: There are $\binom{k}{0} + \binom{k}{1}n + \binom{k}{2}n^2 + \cdots + \binom{k}{k}n^k$ partial functions $f: A \to B$ where $|A| = k$ and $|B| = n$. Each partial function is determined by choosing a domain of definition for the function, which can be done, for each $j = 0, \ldots, n$, in $\binom{k}{j}$ ways. Once a domain of definition is determined, there are n^j ways to define a function on that set. (The sum can be simplified to $(n+1)^k$.)

14. $\begin{bmatrix}3\\1\end{bmatrix} = \frac{q^3-1}{q-1} = 1 + q + q^2$.

15. $\begin{bmatrix}6\\2\end{bmatrix} = \frac{q^6-1}{q-1} \cdot \frac{q^5-1}{q^2-1} = \frac{q^6-1}{q^2-1} \cdot \frac{q^5-1}{q-1} = (q^4+q^2+1)(q^4+q^3+q^2+q+1) = 1+q+2q^2+2q^3+3q^4+2q^5+2q^6+q^7+q^8$. The sum of these coefficients is $15 = \binom{6}{2}$, as Fact 11 predicts.

16. A particle moves in the plane from $(0,0)$ to $(n-k, k)$ by moving one unit at a time in either the positive x or positive y direction. The number of such paths where the area bounded by the path, the x-axis, and the vertical line $x = n-k$ is i units is equal to a_i, where a_i is the coefficient of q^i in the expansion of the Gaussian binomial coefficient $\begin{bmatrix}n\\k\end{bmatrix}$ in Fact 11.

2.3.3 SELECTION WITH REPETITION

Some problems concerning counting the number of ways to select k objects from a set of n objects permit choices of objects to be repeated. Some of these situations are also modeled by binomial coefficients.

Definitions:

An **ordered selection with replacement** is an ordered selection in which each object in the selection set can be chosen arbitrarily often.

An **ordered selection with specified replacement** fixes the number of times each object is to be chosen.

An **unordered selection with replacement** is a selection in which each object in the selection set can be chosen arbitrarily often.

Table 4 Number of poker hands.

type of hand	formula	explanation
royal flush (ace, king, queen, jack, 10 in same suit)	4	4 choices for a suit, and 1 royal flush in each suit
straight flush (5 cards of 5 consecutive ranks, all in 1 suit, but not a royal flush)	$\binom{4}{1}9$	4 choices for a suit, and in each suit there are 9 ways to get 5 cards in a row
four of a kind (4 cards in 1 rank and a fifth card)	$\binom{13}{1}\binom{48}{1}$	13 choices for a rank, only 1 way to select the 4 cards in that rank, and 48 ways to select a fifth card
full house (3 cards of 1 rank, 2 of another rank)	$13\binom{4}{3}12\binom{4}{2}$	13 ways to select a rank for the 3-of-a-kind and $\binom{4}{3}$ ways to choose 3 of this rank; 12 ways to select a rank for the pair and $\binom{4}{2}$ ways to get a pair of this rank
flush (5 cards in 1 suit, but neither royal nor straight flush)	$4\binom{13}{5}-4\cdot 10$	4 ways to select suit, $\binom{13}{5}$ ways to choose 5 cards in that suit; subtract royal and straight flushes
straight (5 cards in 5 consecutive ranks, but not all of the same suit)	$10\cdot 4^5 - 4\cdot 10$	10 ways to choose 5 ranks in a row and 4 ways to choose a card from each rank; then subtract royal and straight flushes
three of a kind (3 cards of 1 rank, and 2 cards of 2 different ranks)	$13\binom{4}{3}\binom{12}{2}4^2$	13 ways to select 1 rank, $\binom{4}{3}$ ways to choose 3 cards of that rank; $\binom{12}{2}$ ways to pick 2 other ranks and 4^2 ways to pick a card of each of those 2 ranks
two pairs (2 cards in each of 2 different ranks, and a fifth card of a third rank)	$\binom{13}{2}\binom{4}{2}\binom{4}{2}44$	$\binom{13}{2}$ ways to select 2 ranks and $\binom{4}{2}$ ways to choose 2 cards in each of these ranks, and $\binom{44}{1}$ way to pick a nonmatching fifth card
one pair (2 cards in 1 rank, plus 3 cards from 3 other ranks)	$13\binom{4}{2}\binom{12}{3}4^3$	13 ways to select a rank, $\binom{4}{2}$ ways to choose 2 cards in that rank; $\binom{12}{3}$ ways to pick 3 other ranks, and 4^3 ways to pick 1 card from each of those ranks

The **permutation-with-replacement coefficient** $P^R(n,k)$ is the number of ways to choose a possibly repeating list of k items from a set of n items.

The **combination-with-replacement coefficient** $C^R(n,k)$ is the number of ways to choose a multiset of k items from a set of n items.

Facts:

1. An ordered selection with replacement can be thought of as obtaining an ordered list of names, obtained by selecting an object from a set, writing its name, placing it back in the set, and repeating the process.

2. The number of ways to make an ordered selection with replacement of k items from n distinct items (with arbitrary repetition) is n^k. Thus $P^R(n,k) = n^k$.

3. The number of ways to make an ordered selection of n items from a set of q distinct items, with exactly k_i selections of object i, is $\frac{n!}{k_1!k_2!\ldots k_q!}$.

4. An unordered selection with replacement can be thought of as obtaining a collection of names, obtained by selecting an object from a set, writing its name, placing it back in the set, and repeating the process. The resulting collection is a multiset (§1.2.1).

5. The number of ways to make an unordered selection with replacement of k items from a set of n items is $C(n+k-1,k)$. Thus $C^R(n,k) = C(n+k-1,k)$.

Combinatorial interpretation: It is sufficient to show that the k-multisets that can be chosen from a set of n items are in one-to-one correspondence with the bit strings of length $(n+k-1)$ with k ones. To indicate that k_j copies of item j are selected, for $j = 1, \ldots, n$, write a string of k_1 ones, then a "0", then a string of k_2 ones, then another "0", then a string of k_3 ones, then another "0", and so on, until after the string of k_{n-1} ones and the last "0", there appears the final string of k_n ones. The resulting bit string has length $n+k-1$ (since it has k ones and $n-1$ zeros). Every such bit string describes a possible selection. Thus the number of possible selections is $C(n+k-1,k) = C(n+k-1,n-1)$.

6. *Integer solutions to the equation* $x_1 + x_2 + \cdots + x_n = k$:

   - The number of nonnegative integer solutions is $C(n+k-1,k) = C(n+k-1,n-1)$. [In the combinatorial argument of Fact 5, there are n strings of ones. The first string of ones can be regarded as the value for x_1, the second string of ones as the value for x_2, etc.]

   - The number of positive integer solutions is $C(k-1, n-1)$.

   - The number of nonnegative integer solutions where $x_i \geq a_i$ for $i = 1, \ldots, n$ is $C(n+k-1-(a_1+\cdots+a_n), n-1)$ (if $a_1+\cdots+a_n \leq k$). [Let $x_i = y_i + a_i$ for each i, yielding the equation $y_1+y_2+\cdots+y_n = k-(a_1+\cdots+a_n)$ to be solved in nonnegative integers.]

   - The number of nonnegative integer solutions where $x_i \leq a_i$ for $i = 1, \ldots, n$ can be obtained using the inclusion/exclusion principle. See §2.4.2.

Examples:

1. *Distinct balls into distinct bins*: k differently colored balls are to be placed into n bins, with arbitrarily many balls to a bin. The number of different ways to do this is n^k. (Apply the rule of product to the number of possible bin choices for each ball.)

2. *Binary strings*: The number of sequences (bit strings) of length n that can be constructed from the symbol set $\{0,1\}$ is 2^n.

3. *Colored balls into distinct bins with colors repeated*: k balls are colored so that k_1 balls have color 1, k_2 have color 2, ..., and k_q have color q. The number of ways these k balls can be placed into n distinct bins ($n \geq k$), at most one per bin, is $\frac{P(n,k)}{k_1!k_2!\ldots k_q!}$.

Note: This is more general than Fact 2, since n can exceed the sum of all the k_is. If n equals this sum, then $P(n,n) = n!$ and the two formulas agree.

4. When three dice are rolled, the "outcome" is the number of times each of the numbers 1 to 6 appears. For instance, two 3s and a 5 is an outcome. The number of different possible outcomes is $C(6+3-1,3) = \binom{8}{3} = 56$.

5. *Nondistinct balls into distinct bins with multiple balls per bin allowed*: The number of ways that k identical balls can be placed into n distinct bins, with any number of balls allowed in each bin, is $C(n+k-1,k)$.

6. *Nondistinct balls into distinct bins with no bin allowed to be empty*: The number of ways that k identical balls can be placed into n distinct bins, with any number of balls allowed in each bin and no bin allowed to remain empty, is $C(k-1, n-1)$.

7. How many ways are there to choose one dozen donuts when there are 7 different kinds of donuts, with at least 12 of each type available? Order is not important, so a multiset of size 12 is being constructed from 7 distinct types. Accordingly, there are $C(7+12-1, 12) = 18{,}564$ ways to choose the dozen donuts.

8. The number of nonnegative integer solutions to the equation $x_1 + x_2 + \cdots + x_7 = 12$ is $C(7+12-1, 12)$, since this is a rephrasing of Example 7.

9. The number of nonnegative integer solutions to $x_1 + x_2 + \cdots + x_5 = 36$, where $x_1 \geq 4$, $x_3 = 11$ and $x_4 \geq 7$ is $C(17,3)$. [It is easiest to think of purchasing 36 donuts, where at least 4 of type 1, exactly 11 of type 3, and at least 7 of type 4 must be purchased. Begin with an empty bag, and put in 4 of type 1, 11 of type 3, and 7 of type 4. This leaves 14 donuts to be chosen, and they must be of types 1, 2, 4, or 5, which is equivalent to finding the number of nonnegative integer solutions to $x_1 + x_2 + x_4 + x_5 = 14$.]

2.3.4 BINOMIAL COEFFICIENT IDENTITIES

Facts:

1. Table 5 lists some identities involving binomial coefficients.

2. Combinatorial identities, such as those in Table 5, can be proved using either algebraic proofs using techniques such as substitution, differentiation, or the principle of mathematical induction (see Facts 4 and 5); they can also be proved by using combinatorial proofs. (See Fact 3.)

3. The following give combinatorial interpretations of some of the identities involving binomial coefficients in Table 5.
   - *Symmetry*: In choosing a subset of k items from a set of n items, the number of ways to select which k items to include must equal the number of ways to select which $n-k$ items to exclude.
   - *Pascal's recursion*: In choosing k objects from a list of n distinct objects, the number of ways that include the last object is $\binom{n-1}{k-1}$, and the number of ways that exclude the last object is $\binom{n-1}{k}$. Their sum is then the total number of ways to choose k objects from a set of n, namely $\binom{n}{k}$.
   - *Binomial theorem*: The coefficient of $x^k y^{n-k}$ in the expansion $(x+y)^n = (x+y)(x+y)\ldots(x+y)$ equals the number of ways to choose k factors from among the n factors $(x+y)$ in which x contributes to the resultant term.
   - *Counting all subsets*: Summing the numbers of subsets of all possible sizes yields the total number of different possible subsets.

Table 5 Binomial coefficient identities.

Factorial expansion	$\binom{n}{k} = \frac{n!}{k!(n-k)!}, \ k = 0, 1, 2, \ldots, n$
Symmetry	$\binom{n}{k} = \binom{n}{n-k}, \ k = 0, 1, 2, \ldots, n$
Monotonicity	$\binom{n}{0} < \binom{n}{1} < \cdots < \binom{n}{\lfloor n/2 \rfloor}, \ n \geq 0$
Pascal's identity	$\binom{n}{k} = \binom{n-1}{k-1} + \binom{n-1}{k}, \ k = 0, 1, 2, \ldots, n$
Binomial theorem	$(x+y)^n = \sum_{k=0}^{n} \binom{n}{k} x^k y^{n-k}, \ n \geq 0$
Counting all subsets	$\sum_{k=0}^{n} \binom{n}{k} = 2^n, \ n \geq 0$
Even and odd subsets	$\sum_{k=0}^{n} (-1)^k \binom{n}{k} = 0, \ n \geq 0$
Sum of squares	$\sum_{k=0}^{n} \binom{n}{k}^2 = \binom{2n}{n}, \ n \geq 0$
Square of row sums	$\left[\sum_{k=0}^{n} \binom{n}{k}\right]^2 = \sum_{k=0}^{2n} \binom{2n}{k}, \ n \geq 0$
Absorption/extraction	$\binom{n}{k} = \frac{n}{k}\binom{n-1}{k-1}, \ k \neq 0$
Trinomial revision	$\binom{n}{m}\binom{m}{k} = \binom{n}{k}\binom{n-k}{m-k}, \ 0 \leq k \leq m \leq n$
Parallel summation	$\sum_{k=0}^{m} \binom{n+k}{k} = \binom{n+m+1}{m}, \ m, n \geq 0$
Diagonal summation	$\sum_{k=0}^{n-m} \binom{m+k}{m} = \binom{n+1}{m+1}, \ n \geq m \geq 0$
Vandermonde convolution	$\sum_{k=0}^{r} \binom{m}{k}\binom{n}{r-k} = \binom{m+n}{r}, \ m, n, r \geq 0$
Diagonal sums in Pascal's triangle (§2.3.2)	$\sum_{k=0}^{\lfloor n/2 \rfloor} \binom{n-k}{k} = F_{n+1}$ (Fibonacci numbers), $n \geq 0$
Other Common Identities	$\sum_{k=0}^{n} k\binom{n}{k} = n2^{n-1}, \ n \geq 0$
	$\sum_{k=0}^{n} k^2 \binom{n}{k} = n(n+1)2^{n-2}, \ n \geq 0$
	$\sum_{k=0}^{n} (-1)^k k \binom{n}{k} = 0, \ n \geq 0$
	$\sum_{k=0}^{n} \frac{\binom{n}{k}}{k+1} = \frac{2^{n+1}-1}{n+1}, \ n \geq 0$
	$\sum_{k=0}^{n} (-1)^k \frac{\binom{n}{k}}{k+1} = \frac{1}{n+1}, \ n \geq 0$
	$\sum_{k=1}^{n} (-1)^{k-1} \frac{\binom{n}{k}}{k} = 1 + \frac{1}{2} + \frac{1}{3} + \cdots + \frac{1}{n}, \ n > 0$
	$\sum_{k=0}^{n-1} \binom{n}{k}\binom{n}{k+1} = \binom{2n}{n-1}, \ n > 0$
	$\sum_{k=0}^{m} \binom{m}{k}\binom{n}{p+k} = \binom{m+n}{m+p}, \ m, n, p \geq 0, \ n \geq p+m$

- **Sum of squares:** Choose a committee of size n from a group of n men and n women. The left side, rewritten as $\binom{n}{k}\binom{n}{n-k}$, describes the process of selecting committees according to the number of men, k, and the number of women, $n-k$, on the committee. The right side gives the total number of committees possible.

- **Absorption/extraction:** From a group of n people, choose a committee of size k and a person on the committee to be its chairperson. Equivalently, first select a chairperson from the entire group, and then select the remaining $k-1$ committee members from the remaining $n-1$ people.

- *Trinomial revision*: The left side describes the process of choosing a committee of size m from n people and then a subcommittee of size k. The right side describes the process where the subcommittee of size k is first chosen from the n people and then the remaining $m - k$ members of the committee are selected from the remaining $n - k$ people.
- *Vandermonde convolution*: Given m men and n women, form committees of size r. The summands give the numbers of committees broken down by number of men, k, and number of women, $r - k$, on the committee; the right side gives the total number of committees.

4. The formula for counting all subsets can be obtained from the binomial theorem by substituting 1 for x and 1 for y.

5. The formula for even and odd subsets can be obtained from the binomial theorem by substituting 1 for x and -1 for y.

6. A set A of size n has 2^{n-1} subsets with an even number of elements and 2^{n-1} subsets with an odd number of elements. (The *even and odd subsets* identity in Table 5 shows that $\sum \binom{n}{k}$ for k even is equal to $\sum \binom{n}{k}$ for k odd. Since the total number of subsets is 2^n, each side must equal 2^{n-1}.)

2.3.5 GENERATING PERMUTATIONS AND COMBINATIONS

There are various systematic ways to generate permutations and combinations of the set $\{1,\ldots,n\}$.

Definitions:

A list of strings from an ordered set is in **lexicographic order** if the strings are sorted as they would appear in a dictionary.

If the elements in the strings are ordered by a relation $<$, string $a_1 a_2 \ldots a_m$ **precedes** $b_1 b_2 \ldots b_n$ if any of the following happens: $a_1 < b_1$; there is a positive integer k such that $a_1 = b_1, \ldots, a_k = b_k$ and $a_{k+1} < b_{k+1}$; or $m < n$ and $a_1 = b_1, \ldots, a_m = b_m$.

Algorithms:

Algorithms 1, 2, and 5 give ways to generate all permutations, k-permutations, and k-combinations of $\{1, 2, \ldots, n\}$ in lexicographic order. Algorithms 3, 4, and 6 give ways to randomly generate a permutation, k-permutation, and k-combination of $\{1, 2, \ldots, n\}$.

Algorithm 1: Generate the permutations of $\{1, \ldots, n\}$ in lexicographic order.

$a_1 a_2 \ldots a_n := 1 2 \ldots n$
while $a_1 a_2 \ldots a_n \neq n\ n{-}1 \ldots 1$
  $m :=$ the rightmost location such that a_m is followed by a larger number
  $a'_1 a'_2 \ldots a'_{m-1} = a_1 a_2 \ldots a_{m-1}$ {retain everything to the left of a_m}
  $a'_m :=$ the smallest number larger than a_m to the right of a_m
  $a'_{m+1} a'_{m+2} \ldots a'_n :=$ everything else, in ascending order
  $a_1 a_2 \ldots a_n := a'_1 a'_2 \ldots a'_n$
  output $a_1 a_2 \ldots a_n$

Algorithm 2: Generate the k-permutations of $\{1,\ldots,n\}$ in lexicographic order.

$a_1 a_2 \ldots a_k := 1\, 2 \ldots k$ {k a given positive integer less than or equal to n}
while $a_1 a_2 \ldots a_k \neq n\, n{-}1 \ldots n-(k-1)$
   $m :=$ the rightmost location such that a_m is followed by a larger number
   $a'_1 a'_2 \ldots a'_{m-1} := a_1 a_2 \ldots a_{m-1}$ {retain everything to the left of a_m}
   $a'_m :=$ the smallest number larger than a_m to the right of a_m
   $a'_{m+1} a'_{m+2} \ldots a'_k :=$ everything else, in ascending order
   $a_1 a_2 \ldots a_k := a'_1 a'_2 \ldots a'_k$
   output $a_1 a_2 \ldots a_k$

Algorithm 3: Generate a random permutation of $\{1,\ldots,n\}$.

$a_1 a_2 \ldots a_n := 1\, 2 \ldots n$
for $i := 0$ to $n-2$
   interchange a_{n-i} and $a_{r(n-i)}$ {$r(k)$ a randomly chosen integer in $\{1,\ldots,k\}$}
output $a_1 \ldots a_n$ {a randomly chosen permutation of $\{1,\ldots,n\}$}

Algorithm 4: Generate a random k-permutation of $\{1,\ldots,n\}$.

$a_1 a_2 \ldots a_n :=$ a random permutation of $\{1,\ldots,n\}$ {obtained from Algorithm 3}
output $a_1 \ldots a_k$ {a randomly chosen k-permutation of $\{1,\ldots,n\}$}

Algorithm 5: Generate k-combinations of $\{1,\ldots,n\}$ in lexicographic order.

$a_1 a_2 \ldots a_k := 1\, 2 \ldots k$ {first combination in lexicographic order}
while $a_1 a_2 \ldots a_k \neq n{-}k{+}1\, n{-}k{+}2 \ldots n$
   $m :=$ the rightmost location among $1,\ldots,k$ such that a number larger than a_m but smaller than n is not in the combination
   $a'_1 a'_2 \ldots a'_{m-1} := a_1 a_2 \ldots a_{m-1}$ {retain everything to the left of a_m}
   $a'_m := a_m + 1$ {increase a_m by 1}
   $a'_{m+1} a'_{m+2} \ldots a'_k := a_m{+}2\, a_m{+}3 \ldots a_m{+}k{-}m{+}1$ {continue consecutively}
   $a_1 a_2 \ldots a_k := a'_1 a'_2 \ldots a'_k$
output $a_1 a_2 \ldots a_k$ {the members of each k-combination are given in ascending order}

Algorithm 6: Generate random k-combinations of $\{1,\ldots,n\}$.

$a_1 a_2 \ldots a_k :=$ any k-permutation of $\{1,\ldots,n\}$ generated by Algorithm 4
output $a_1 a_2 \ldots a_k$ {ignoring the order in which elements are written, this is a random k-combination}

Examples:

1. The lexicographic order for the 3-permutations of $\{1,2,3\}$ is 123, 132, 213, 231, 312, 321.

2. The lexicographic order of the $C(5,3) = 10$ 3-combinations of $\{1,2,3,4,5\}$ is 123, 124, 125, 134, 135, 145, 234, 235, 245, 345.

3. *Generating permutations*: What permutation follows 3142765 in the lexicographic ordering of the permutations of $\{1,\ldots,7\}$? Step 1 of the while-loop of Algorithm 1 leads to the fourth digit, namely the digit 2, as the first digit from the right that has larger digits following it. Steps 2 and 3 show that the next permutation starts with 3145 since 5 is the smallest digit greater than 2 and following it. Finally, step 4 yields 2, 6, and 7 (in numerical order) as the digits that follow. Thus, the permutation immediately following 3142765 is 3145267.

4. *Generating combinations*: What 5-combination follows 12478 in the lexicographic ordering of 5-combinations of $\{1,\ldots,8\}$? Step 1 of the while-loop of Algorithm 2 leads to the third digit, namely the digit 4, as the first digit from the right that can be safely increased by 1. Step 2 shows that the next permutation starts with 125 since the 3rd digit is increased by 1. Finally, step 3 yields 6 and 7 as the following digits (add 1 to the newly-listed previous digit until the new selection of k digits is complete). Thus, the combination after 12478 is 12567.

2.4 INCLUSION/EXCLUSION

The principle of inclusion/exclusion is used to count the elements in a non-disjoint union of finite sets. Many counting problems can be solved by applying this principle to a well-chosen collection of sets. The techniques involved in this process are best illustrated with examples.

2.4.1 PRINCIPLE OF INCLUSION/EXCLUSION

The number of elements in the union of two finite sets A and B is $|A| + |B|$, provided that the sets have no element in common. In the general case, however, some elements in common to both sets have been included in the sum twice. The sum is adjusted to exclude the double-counting of these common elements by subtracting their number:

$$|A \cup B| = |A| + |B| - |A \cap B|.$$

A Venn diagram (§1.2.2) for these sets is given in the following figure.

Similarly, the number of elements in the union of three finite sets is

$$|A \cup B \cup C| = |A| + |B| + |C| - |A \cap B| - |A \cap C| - |B \cap C| + |A \cap B \cap C|.$$

See the following figure. These simple equations generalize to the case of n sets.

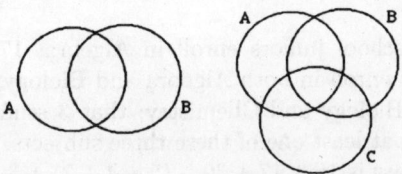

Facts:

1. *Inclusion/exclusion principle*: the number of elements in the union of n finite sets $A_1, A_2, \ldots, A_n$ is:

$$|A_1 \cup A_2 \cup \cdots \cup A_n| = \sum_{1 \leq i \leq n} |A_i| - \sum_{1 \leq i < j \leq n} |A_i \cap A_j| + \sum_{1 \leq i < j < k \leq n} |A_i \cap A_j \cap A_k|$$
$$- \cdots + (-1)^{n+1} |A_1 \cap A_2 \cap \cdots \cap A_n|$$

or, alternatively,

$$|A_1 \cup A_2 \cup \cdots \cup A_n| = \sum_{k=1}^{n} (-1)^{k+1} \sum_{1 \leq i_1 < \cdots < i_k \leq n} |A_{i_1} \cap A_{i_2} \cap \cdots \cap A_{i_k}|$$

Sometimes the inner sum of the alternative formula is denoted S_k.

2. The inclusion/exclusion formula for n sets has $2^n - 1$ terms, one for each possible nonempty intersection. The coefficient of a term is -1 if the term corresponds to intersections of an even number of sets, and $+1$ otherwise.

3. The principle is often applied to the complement of a set. Let A_i be the subset of elements in a universal set U that have property P_i. The number of elements that have properties $P_{i_1}, P_{i_2}, \ldots, P_{i_k}$ is often written $N(P_{i_1} P_{i_2} \ldots P_{i_k})$ and the number of elements that have none of these properties is often written $N(P'_{i_1} P'_{i_2} \ldots P'_{i_k})$. The number of element in U that have none of the properties is:

$$N(P'_1 P'_2 \ldots P'_n) = |U| - \sum_{1 \leq i \leq n} N(P_i) + \sum_{1 \leq i < j \leq n} N(P_i P_j) - \cdots + (-1)^n N(P_1 P_2 \ldots P_n).$$

Examples:

1. Of 70 people surveyed, 37 drink coffee, 23 drink tea, and 25 drink neither. Find the number who drink both coffee and tea. Using C to represent the set of coffee drinkers and T to represent the set of tea drinkers, the size of $C \cap T$ must be found. Since $|\overline{T \cup C}| = 25$, the Venn diagram in part (a) of the following figure shows that $|C \cup T| = 45$. According to the inclusion/exclusion principle,

$$|C \cap T| = |C| + |T| - |C \cup T| = 37 + 23 - 45 = 15,$$

illustrated in part (b) of the figure.

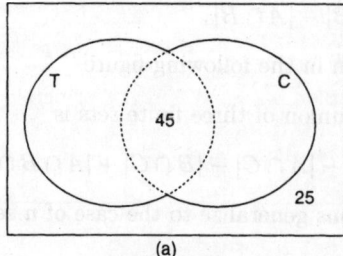
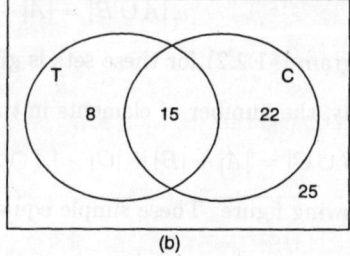

2. Suppose that 16 high-school juniors enroll in Algebra, 17 in Biology, and 30 in Chemistry; that 5 students enroll in both Algebra and Biology, 4 in both Algebra and Chemistry, and 7 in both Biology and Chemistry; that 3 students enroll in all three; and that every junior takes at least one of these three subjects. Then the total number of students in the junior class is $16 + 17 + 30 - (5 + 4 + 7) + 3 = 50$.

3. Each of 11 linguists translates at least one of the languages Amharic and Burmese into English. The numbers who translate only Amharic or Burmese are both odd primes. More linguists translate Burmese than Amharic. How many can translate Amharic?

Based on experimentation or on an analytic approach, the only possible assignment of numbers to regions that fits all these facts leads to 6, as shown in the following figure.

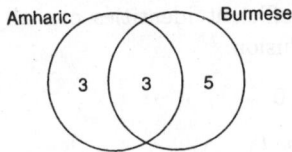

4. At a party for 28 people, three kinds of pizza were served: anchovy, broccoli, and cheese. Everyone ate at least one kind. No two of the seven different possible selections of one or more kinds of pizza were eaten by the same number of partygoers. Each of the three possible exclusive selections (one kind of pizza only) was eaten by an odd number of partygoers, and each of the three possible combinations of two kinds of pizza was eaten by an even number of partygoers. If a total of 18 partygoers ate cheese pizza, how many ate both anchovy and broccoli?

The answer is 2. Experimentation or an analytic approach leads to the possible assignments of numbers to regions that fit all these facts, shown in the following figure.

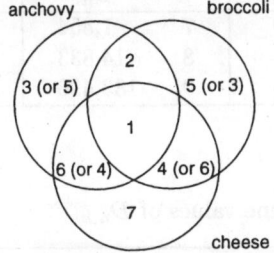

5. To count the number of ways to select a 5-card hand from a standard 52-card deck so that the hand contains at least one card from each of the four suits, let A_1, A_2, A_3, and A_4 be the subsets of 5-card hands that do not contain a club, diamond, heart, or spade, respectively. Then

$$|A_i| = \binom{52-13}{5} = \binom{39}{5} \text{ with } \binom{4}{1} \text{ choices for } i$$

$$|A_i \cap A_j| = \binom{52-26}{5} = \binom{26}{5} \text{ with } \binom{4}{2} \text{ choices for } i \text{ and } j$$

$$|A_i \cap A_j \cap A_k| = \binom{52-39}{5} = \binom{13}{5} \text{ with } \binom{4}{3} \text{ choices for } i, j, \text{ and } k.$$

There are $\binom{52}{5}$ possible 5-card hands, so by complementation and the principle of inclusion/exclusion, those that contain at least one card from each suit is

$$\binom{52}{5} - \binom{4}{1}\binom{39}{5} + \binom{4}{2}\binom{26}{5} - \binom{4}{3}\binom{13}{5} = 685{,}464.$$

2.4.2 APPLYING INCLUSION/EXCLUSION TO COUNTING PROBLEMS

Definitions:

A **derangement** on a set is a permutation that leaves no element fixed. The number of derangements on a set of cardinality n is denoted D_n.

A **rencontre number** $D_{n,k}$ is the number of permutations on a set of n elements that leave exactly k elements fixed.

Facts:
1. The number of onto functions from an n-element set to a k-element set $(n \geq k)$ is
$$\sum_{j=0}^{k}(-1)^j \binom{k}{j}(k-j)^n.$$
(See Example 3.)
2. The following binomial coefficient identities can all be derived by combinatorial arguments using inclusion/exclusion:
   - $\sum_{k=0}^{m}(-1)^k \binom{n}{k}\binom{n-k}{m-k} = 0$
   - $\sum_{k=m}^{n}(-1)^{k-m}\binom{n}{k} = \binom{n-1}{m-1}$
   - $\sum_{k=0}^{n}(-1)^k \binom{n}{k}\left(\frac{n-k+r-1}{r}\right) = \binom{r-1}{n-1}$.
3. $D_n = n!(1 - \frac{1}{1!} + \frac{1}{2!} - \cdots + (-1)^n \frac{1}{n!})$. (See Example 8.)
4. $\frac{D_n}{n!} \to e^{-1} \approx 0.368$ as $n \to \infty$.
5. $D_n = nD_{n-1} + (-1)^n$ for $n \geq 1$.
6. $D_n = (n-1)(D_{n-1} + D_{n-2})$ for $n \geq 2$.
7. The following table gives some values of D_n:

n	D_n	n	D_n	n	D_n	n	D_n
1	0	4	9	7	1,854	10	1,334,961
2	1	5	44	8	14,833	11	14,684,570
3	2	6	265	9	133,496	12	176,214,841

8. $D_{n,0} = D_n$.
9. $D_{n,k} = \binom{n}{k} D_{n-k}$
10. The following table gives some values of $D_{n,k}$:

$n \backslash k$	0	1	2	3	4	5	6	7	8	9	10
0	1										
1	0	1									
2	1	0	1								
3	2	3	0	1							
4	9	8	6	0	1						
5	44	45	20	10	0	1					
6	265	264	135	40	15	0	1				
7	1,854	1,855	924	315	70	21	0	1			
8	14,833	14,832	7,420	2,464	630	112	28	0	1		
9	133,496	133,497	66,744	22,260	5,544	1,134	168	36	0	1	
10	1,334,961	1,334,960	667,485	222,480	55,650	11,088	1,890	240	45	0	1

Examples:
1. The inclusion/exclusion principle can be used to establish the binomial coefficient identity
$$\binom{n}{m} = \sum_{k=1}^{m}(-1)^{k+1}\binom{n-k}{m-k}\binom{n}{k}.$$
Let A_i denote the subset of m-combinations that contain object i. Thus, the k-fold intersection $A_{i_1} \cap A_{i_2} \cap \cdots \cap A_{i_k}$ consists of all the m-combinations that contain all

the objects $i_1, i_2, \ldots, i_k$. Since there are $\binom{n-k}{m-k}$ ways to complete an m-combination in this intersection, it follows that $|A_{i_1} \cap A_{i_2} \cap \cdots \cap A_{i_k}| = \binom{n-k}{m-k}$. Since the k objects themselves can be specified in $\binom{n}{k}$ ways, it follows that

$$\sum_{1 \leq i_1 < i_2 < \cdots < i_k \leq n} |A_{i_1} \cap A_{i_2} \cap \cdots \cap A_{i_k}| = \binom{n-k}{m-k}\binom{n}{k}, \quad k \leq m.$$

Since $A_1 \cup A_2 \cup \cdots \cup A_n$ is the set of all m-combinations selected from $1, 2, \ldots, n$ that contain at least one of the objects $1, 2, \ldots, n$, it must be the set of all m-combinations.

2. *Sieve of Eratosthenes*: The sieve of Eratosthenes (276–194 BCE) is a method for finding all primes less than or equal to a given positive integer n. Begin with the list of integers 2 through n, and delete all multiples of the first number in the list, 2, but not including 2. The first integer remaining after 2 is 3; delete all multiples of 3, not including 3. The first integer remaining after 3 is 5; delete all multiples of 5, not including 5. Continue the process. The remaining integers are the primes less than or equal to n. (See §4.4.2.)

The inclusion/exclusion principle can be used to obtain the number of primes less than or equal to n. (A number $x \leq n$ is prime if and only if x has a prime factor less than or equal to $\lfloor \sqrt{n} \rfloor$.) Let P_i be the property: a number is greater than the ith prime and divisible by the ith prime. Then the number of primes less than or equal to n is $N(P_1' P_2' \ldots P_k')$, where there are k primes less than or equal to $\lfloor \sqrt{n} \rfloor$. (§2.3.1, Fact 3.)

For example, the number of primes less than or equal to 100 is $N(P_1' P_2' P_3' P_4') = 99 - \lfloor \frac{100}{2} \rfloor - \lfloor \frac{100}{3} \rfloor - \lfloor \frac{100}{5} \rfloor - \lfloor \frac{100}{7} \rfloor + \lfloor \frac{100}{2 \cdot 3} \rfloor + \lfloor \frac{100}{2 \cdot 5} \rfloor + \lfloor \frac{100}{2 \cdot 7} \rfloor + \lfloor \frac{100}{3 \cdot 5} \rfloor + \lfloor \frac{100}{3 \cdot 7} \rfloor + \lfloor \frac{100}{5 \cdot 7} \rfloor - \lfloor \frac{100}{2 \cdot 3 \cdot 5} \rfloor - \lfloor \frac{100}{2 \cdot 3 \cdot 7} \rfloor - \lfloor \frac{100}{2 \cdot 5 \cdot 7} \rfloor - \lfloor \frac{100}{3 \cdot 5 \cdot 7} \rfloor + \lfloor \frac{100}{2 \cdot 3 \cdot 5 \cdot 7} \rfloor = 99 - 50 - 33 - 20 - 14 + 16 + 10 + 7 + 6 + 4 + 2 - 3 - 2 - 1 - 0 + 0 = 21$.

3. *Number of onto functions*: The number of onto functions from an n-element set to a k-element set $(n \geq k)$ is $\sum_{j=0}^{k}(-1)^j \binom{k}{j}(k-j)^n$. The number of onto functions from an n-element set to a k-element set equals the number of ways that n different objects can be distributed among k different boxes with none left empty. Let A_i be the subset of distributions with box i empty. Then

$$|A_i| = (k-1)^n \quad \text{with } \binom{k}{1} \text{ choices for } i$$
$$|A_i \cap A_j| = (k-2)^n \quad \text{with } \binom{k}{2} \text{ choices for } i \text{ and } j$$
$$\vdots$$
$$|A_{i_1} \cap A_{i_2} \cap \cdots \cap A_{i_k}| = (k-k)^n \quad \text{with } \binom{k}{k} \text{ choices for } i_1, i_2, \ldots, i_k.$$

The number of distributions that leave no box empty is then $\sum_{j=0}^{k}(-1)^j \binom{k}{j}(k-j)^n$.

The number of onto functions from an n-element set to a k-element set for some values of n and k $(n \geq k)$ are given in the following table.

$n \backslash k$	1	2	3	4	5	6	7	8	9
1	1								
2	1	2							
3	1	6	6						
4	1	14	36	24					
5	1	30	150	240	120				
6	1	62	540	1560	1800	720			
7	1	126	1806	8400	16,800	15,120	5,040		
8	1	254	5796	40,824	126,000	191,520	141,120	40,320	
9	1	510	18,150	186,480	834,120	1,905,120	2,328,480	1,451,520	362,880

4. There are 584 nonnegative integer solutions to $x_1 + x_2 + x_3 + x_4 = 20$ where $x_1 \leq 8$, $x_2 \leq 10$, and $x_3 \leq 5$. [Let A_1 be the set of solutions where $x_1 \geq 9$, A_2 the set of solutions where $x_2 \geq 11$, and A_3 the solutions where $x_3 \geq 6$. The final answer, obtained using the inclusion/exclusion principle and techniques of the Examples of §2.3.3, is equal to $C(23,3) - |A_1 \cup A_2 \cup A_3| = C(23,3) - (C(14,3) + C(12,3) + C(17,3) - C(3,3) - C(8,3) - C(6,3) + 0) = 584.$]

5. The permutations $\begin{pmatrix} 1 & 2 & 3 \\ 2 & 3 & 1 \end{pmatrix}$ and $\begin{pmatrix} 1 & 2 & 3 \\ 3 & 1 & 2 \end{pmatrix}$ are derangements of $1, 2, 3$, but the permutations $\begin{pmatrix} 1 & 2 & 3 \\ 1 & 2 & 3 \end{pmatrix}$, $\begin{pmatrix} 1 & 2 & 3 \\ 1 & 3 & 2 \end{pmatrix}$, $\begin{pmatrix} 1 & 2 & 3 \\ 3 & 2 & 1 \end{pmatrix}$, and $\begin{pmatrix} 1 & 2 & 3 \\ 2 & 1 & 3 \end{pmatrix}$ are not.

6. *Problème des rencontres:* In the *problème des rencontres* (*matching problem*) an urn contains balls numbered 1 through n, and they are drawn out one at a time. A match occurs if ball i is the ith ball drawn. The probability that no matches occur when all the balls are drawn is $\frac{D_n}{n!}$. The problem was studied by Pierre-Rémond de Montmort (1678–1719) who studied the card game treize, in which matchings of pairs of cards were counted when two decks of cards were laid out face-up.

7. *Problème des ménages:* The *problème des ménages*, first raised by François Lucas (1842–1891), requires that n married couples be seated around a circular table so that no men are adjacent, no women are adjacent, and no husband and wife are adjacent. There are $2n! \sum_{i=0}^{n} (-1)^i (n-i)! \binom{2n-i}{i} \frac{2n}{2n-i}$ ways to seat the people. (There are $2n!$ ways to seat the n women. Regardless of how this is done, by the inclusion/exclusion principle there are $\sum_{i=0}^{n} (-1)^i (n-i)! \binom{2n-i}{i} \frac{2n}{2n-i}$ ways to seat the n men.)

8. *Determining the number D_n of derangements of $\{1, \ldots, n\}$:* Let A_i be the subset of permutations that fix object i. The permutations in the subset $A_1 \cup A_2 \cup \cdots \cup A_n$ are those that fix at least one object. Then

$$|A_i| = (n-1)! \quad \text{with } \binom{n}{1} \text{ choices for } i$$
$$|A_i \cap A_j| = (n-2)! \quad \text{with } \binom{n}{2} \text{ choices for } i \text{ and } j$$
$$\vdots$$
$$|A_{i_1} \cap A_{i_2} \cap \cdots \cap A_{i_k}| = (n-k)! \quad \text{with } \binom{n}{k} \text{ choices for } i_1, i_2, \ldots, i_k$$

Complementation and inclusion/exclusion now yield the formula in Fact 3:

$$D_n = n! - \sum_{k=1}^{n} (-1)^{k+1} \binom{n}{k} (n-k)! = n! \sum_{k=0}^{n} (-1)^k \frac{1}{k!}.$$

As n becomes large, $\frac{D_n}{n!}$ approaches $e^{-1} \approx 0.368$ very rapidly.

9. *Hatcheck problem:* The hatchecker at a restaurant neglects to place claim checks on n hats. Each of the n customers is given a randomly selected hat upon exiting. What is the probability that no one receives the correct hat?

There are $n!$ possible permutations of the n hats, and there are D_n cases in which no one gets the correct hat. Thus, by Example 8, the probability is approximately e^{-1}, regardless of the number of diners.

10. *Rook polynomials/Arrangements of objects where there are restrictions on positions in which the objects can be placed:* This describes a family of assignment or matching problems, such as matching applicants to jobs where some applicants cannot be assigned to certain jobs, the problème des ménages, and the problème des rencontres. In terms of matching n applicants to n jobs, set up an $n \times n$ "board of possibilities" where the rows are labeled by the people and the columns are labeled by the jobs. Square (i, j) is a *forbidden square* if applicant i cannot perform job j; the remaining squares are *allowable squares*. An allowable arrangement is an arrangement where only allowable squares are chosen, with exactly one square chosen in each row and column.

These problems can be rephrased in terms of placing rooks on a chessboard: given a chessboard with some squares forbidden, find the number of ways of placing rooks on the allowable squares of the chessboard so that no rook can capture any other rook. (In chess a rook can move any number of squares vertically or horizontally.) For a given $n \times n$ board B, let A_i = the number of ways to place n nontaking rooks on B so that the rook in row i is on a forbidden square. The total number of ways to place n nontaking rooks on allowable squares is

$$n! - |A_1 \cup \cdots \cup A_n| = n! - r_1(B)(n-1)! + r_2(B)(n-2)! - \cdots + (-1)^n r_n(B)0!$$

where the coefficients $r_i(B)$ are the number of ways to place i nontaking rooks on forbidden squares of B.

A *rook polynomial* for an $n \times n$ board B is a polynomial of the form

$$R(x, B) = r_0(B) + r_1(B)x + r_2(B)x^2 + \cdots + r_n(B)x^n,$$

where $r_0(B)$ is defined to be 1.

The numbers $r_i(B)$ can sometimes be found more easily by using a combination of the following two reduction techniques:

- $R(x, B) = R(x, B_1) \cdot R(x, B_2)$, if all forbidden squares of B appear in two disjoint sub-boards B_1 and B_2 (the sub-boards B_1 and B_2 are disjoint if the row labels of B are partitioned into two parts S_1 and S_2, the column labels of B are partitioned into two parts T_1 and T_2, and B_1 is obtained from $S_1 \times T_1$ and B_2 is obtained from $S_2 \times T_2$).

- $R(x, B) = xR(x, B_1) + R(x, B_2)$, where there is a square (i, j) of B, B_1 is obtained from B by removing all squares in row i and all squares in column j, and B_2 is obtained from B by making square (i, j) allowable.

It may be necessary to use these techniques repeatedly to obtain boards that are simple enough that the rook polynomial coefficients can be easily found.

11. Rook polynomials can be used to find the number of derangements of n objects. The forbidden squares of the board B are the squares (i, i). The first reduction technique of Example 10 used repeatedly breaks B into $B_1, \ldots, B_n$ where B_i consists only of square (i, i). Then

$$R(x, B) = R(x, B_1) R(x, B_2) \ldots R(x, B_n) = (1+x) \ldots (1+x) = (1+x)^n = \sum_{i=0}^{n} \binom{n}{i} x^i.$$

Therefore, the number of derangements is

$$n! - \left[\binom{n}{1}(n-1)! - \binom{n}{2}(n-2)! + \cdots + (-1)^{n+1}\binom{n}{n}0!\right] = n!\sum_{k=0}^{n}(-1)^k \frac{1}{k!}.$$

2.5 PARTITIONS

Each way to write a positive integer n as a sum of positive integers is called a partition of n. Similarly, each way to decompose a set S into a family of mutually disjoint nonempty subsets is called a partition of S. In a cyclic partition of a set, the elements of each subset are arranged into cycles, and two cyclic partitions in the same family of subsets are distinct if any of the cycle arrangements are different. The main concerns are counting the number of essentially different partitions of integers and sets, and with counting cyclic partitions of sets.

2.5.1 PARTITIONS OF INTEGERS

A positive integer can be decomposed into a sum of positive integers in various ways, taking into account restrictions on the number of parts or on the properties of the parts.

Definitions:

A *partition* of a positive integer n is a representation of n as the sum of positive integers. The parts are usually written in nonascending order, but order is ignored.

A *Ferrers diagram* of a partition is an array of boxes, nodes, or dots into rows of nonincreasing size so that each row represents one part of the partition.

The *conjugate* of a partition is the partition obtained by transposing the rows and columns of its Ferrers diagram.

A *composition* is a partition in which the order of the parts is taken into account.

A *vector partition* is a decomposition of an n-tuple of nonnegative integers into a sum of nonzero n-tuples of nonnegative integers, where order is ignored

A *vector composition* is the same as a vector partition, except that order is taken into account.

Facts:

1. The following table gives the notation for various functions that count partitions:

function	type of partitions counted
$p(n)$	number of partitions of n
$Q(n)$	number of partitions of n into distinct parts
$\mathcal{O}(n)$	number of partitions of n into odd parts
$p_m(n)$	number of partitions of n with at most m parts
$q_m(n)$	number of partitions of n with no part larger than m
$p(N, M, n)$	number of partitions of n into at most M parts, with each part no larger than N

2. $p(m, n, n) = q_m(n)$.
3. $p(n, m, n) = p_m(n)$.
4. $p_m(n) = q_m(n)$.
5. $\mathcal{O}(n) = \mathcal{Q}(n)$.
6. The number of compositions of n into k parts is $\binom{n-1}{n-k} = \binom{n-1}{k-1}$.
7. The number of compositions of n is 2^{n-1}.
8. The number of compositions of n using no 1s is F_{n-1} (Fibonacci numbers $F_0 = 0$, $F_1 = 1$, $F_2 = 1$, $F_3 = 2, \ldots$).
9. The partition function $p(n)$ satisfies these congruences (see [Kn93] for details):

$$p(5n + 4) \equiv 0 \pmod{5}$$
$$p(7n + 5) \equiv 0 \pmod{7}$$
$$p(11n + 6) \equiv 0 \pmod{11}.$$

10. The partition functions $p(n)$ and $p_m(n)$ satisfy these recurrences:

$$p(n) - p(n-1) - p(n-2) + p(n-5) + p(n-7) + \cdots$$
$$+ (-1)^k p(n - \tfrac{k}{2}(3k-1)) + (-1)^k p(n - \tfrac{k}{2}(3k+1)) + \cdots = 0, \quad n > 0$$

$$p_m(n) = p_m(n-m) + p_{m-1}(n).$$

11. The asymptotic behavior of $p(n)$, $Q(n)$, and $p_m(n)$ is as follows (see [An84] Chapters 5 and 6, [HaRa18], or [Kn93] for details):

$$p(n) \sim \frac{1}{4n\sqrt{3}} e^{\pi\sqrt{2n/3}} \quad \text{as } n \to \infty,$$

$$Q(n) \sim \frac{1}{4 \cdot 3^{1/4}} n^{-3/4} e^{\pi\sqrt{n/3}} \quad \text{as } n \to \infty,$$

$$p_m(n) \sim \frac{n^{m-1}}{m!(m-1)!} \quad \text{as } n \to \infty, \text{ with } m \text{ fixed}.$$

12. The following are generating functions for partition functions:

$$\sum_{n \geq 0} p(n) q^n = \prod_{i=1}^{\infty} (1 + q^i + q^{i+i} + \cdots) = \prod_{i=1}^{\infty} \left(\sum_{m=0}^{\infty} q^{mi} \right) = \prod_{i=1}^{\infty} \frac{1}{1-q^i}$$

$$\sum_{n \geq 0} Q(n) q^n = \prod_{i=1}^{\infty} (1 + q^i)$$

$$\sum_{n \geq 0} p_m(n) q^n = \prod_{i=1}^{m} (1 + q^i + q^{i+i} + \cdots) = \prod_{i=1}^{m} \left(\sum_{m=0}^{\infty} q^{mi} \right) = \prod_{i=1}^{m} \frac{1}{1-q^i}$$

$$\sum_{n \geq 0} p(N, M, n) q^n = \prod_{j=1}^{N} \frac{(1-q^{N+M+1-j})}{(1-q^j)} = \frac{\prod_{j=1}^{N+M}(1-q^j)}{\prod_{j=1}^{N}(1-q^j) \prod_{j=1}^{M}(1-q^j)}.$$

Note: Even though these expressions for $p(N, M, n)$ look like quotients of polynomials they are actually just polynomials of degree NM. They are called *Gaussian polynomials* or *q-binomial coefficients*. (See Chapters 1 and 2 of [An76], Chapter 19 of [HaWr60], or [Ma16] for details. Also see §2.3.2.)

13. The following are additional generating functions for partition functions (see Chapter 2 of [An76] or Section 8.10 of [GaRa90] for details):

$$\sum_{n=1}^{\infty} p(n) q^n = 1 + \sum_{n=1}^{\infty} \frac{q^n}{(1-q)(1-q^2)\cdots(1-q^n)}$$

$$= 1 + \sum_{n=1}^{\infty} \frac{q^{n^2}}{(1-q)^2(1-q^2)^2\cdots(1-q^n)^2}$$

$$\sum_{n=1}^{\infty} Q(n) q^n = 1 + \sum_{n=1}^{\infty} \frac{q^{n(n+1)/2}}{(1-q)(1-q^2)\cdots(1-q^n)}$$

$$= 1 + q + \sum_{n=2}^{\infty} q^n (1+q)(1+q^2) \cdots (1+q^{n-1})$$

$$\sum_{n=1}^{\infty} p_m(n) q^n = 1 + \sum_{n=1}^{\infty} \frac{(1-q^m)(1-q^{m+1})\cdots(1-q^{m+n-1})}{(1-q)(1-q^2)\cdots(1-q^n)} q^n.$$

14. See [GrKnPa94] for an algorithm for generating partitions.

15. The following table gives some values of $p_m(n)$. More extensive tables appear in [GuGwMi58].

$n \backslash m$	0	1	2	3	4	5	6	7	8	9	10
0	1	1	1	1	1	1	1	1	1	1	1
1	0	1	1	1	1	1	1	1	1	1	1
2	0	1	2	2	2	2	2	2	2	2	2
3	0	1	2	3	3	3	3	3	3	3	3
4	0	1	3	4	5	5	5	5	5	5	5
5	0	1	3	5	6	7	7	7	7	7	7
6	0	1	4	7	9	10	11	11	11	11	11
7	0	1	4	8	11	13	14	15	15	15	15
8	0	1	5	10	15	18	20	21	22	22	22
9	0	1	5	12	18	23	26	28	29	30	30
10	0	1	6	14	23	30	35	38	40	41	42

16. The following table gives values for $p(n)$ and $Q(n)$.

n	$p(n)$	$Q(n)$	n	$p(n)$	$Q(n)$	n	$p(n)$	$Q(n)$
0	1	1	17	297	38	34	12,310	512
1	1	1	18	385	46	35	14,883	585
2	2	1	19	490	54	36	17,977	668
3	3	2	20	627	64	37	21,637	760
4	5	2	21	792	76	38	26,015	864
5	7	3	22	1,002	89	39	31,185	982
6	11	4	23	1,255	104	40	37,338	1,113
7	15	5	24	1,575	122	41	44,583	1,260
8	22	6	25	1,958	142	42	53,174	1,426
9	30	8	26	2,436	165	43	63,261	1,610
10	42	10	27	3,010	192	44	75,175	1,816
11	56	12	28	3,718	222	45	89,134	2,048
12	77	15	29	4,565	256	46	105,558	2,304
13	101	18	30	5,604	296	47	124,754	2,590
14	135	22	31	6,842	340	48	147,273	2,910
15	176	27	32	8,349	390	49	173,525	3,264
16	231	32	33	10,143	448	50	204,226	3,658

Examples:

1. The number 4 has five partitions:
$$4 \quad 3+1 \quad 2+2 \quad 2+1+1 \quad 1+1+1+1.$$

2. The number 4 has eight compositions:
$$4 \quad 1+3 \quad 3+1 \quad 2+2 \quad 2+1+1 \quad 1+2+1 \quad 1+1+2 \quad 1+1+1+1.$$

3. The vector partitions of $(2,1)$ are:
$$(2,1) \quad (2,0)+(0,1) \quad (1,0)+(1,0)+(0,1) \quad (1,0)+(1,1)$$

4. The partition $18 = 5 + 4 + 4 + 2 + 1 + 1 + 1$ has the Ferrers diagram in part (a) of the following figure. Its conjugate is the partition $18 = 7 + 4 + 3 + 3 + 1$, with the Ferrers diagram in part (b) of the figure.

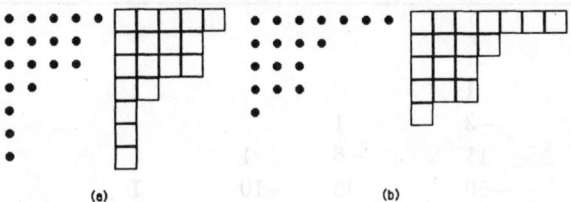

(a)   (b)

5. *Identical balls into identical bins*: The number of ways that n identical balls can be placed into k identical bins, with any number of balls allowed in each bin, is given by $p_k(n)$.

6. *Identical balls into identical bins with no bin allowed to be empty*: The number of ways that n identical balls can be placed into k identical bins $(n \geq k)$, with any number of balls allowed in each bin and no bin allowed to remain empty, is given by $p_k(n) - p_{k-1}(n)$.

2.5.2 STIRLING COEFFICIENTS

Definitions:

A *cyclic partition* of a set is a partition of the set (into disjoint subsets whose union is the entire set) where the elements of each subset are arranged into cycles. Two cyclic partitions using the same family of subsets distinct if any of the cycle arrangements are different.

The *Stirling cycle number* $\left[{n \atop k} \right]$ is the number of ways to partition n objects into k nonempty cycles.

The *Stirling number of the first kind* $s(n,k)$ is the coefficient of x^k in the polynomial $x(x-1)(x-2)\ldots(x-n+1)$. Thus,

$$\sum_{k=0}^{n} s(n,k)x^k = x(x-1)(x-2)\ldots(x-n+1).$$

The *Stirling subset number* $\left\{ {n \atop k} \right\}$ is the number of ways to partition a set of n objects into k nonempty subsets.

The *Stirling numbers of the second kind* $S(n,k)$ are defined implicitly by the equation

$$x^n = \sum_{k=0}^{n} S(n,k)x(x-1)(x-2)\ldots(x-k+1).$$

The *Bell number* B_n is the number of partitions of a set of n objects. (Eric Temple Bell, 1883–1960)

Facts:

1. $s(n,k)(-1)^{n-k} = \left[{n \atop k} \right]$.
2. $S(n,k) = \left\{ {n \atop k} \right\}$.

3. The following table gives Stirling numbers of the first kind, $s(n,k)$:

n\k	0	1	2	3	4	5	6	7	8	9	10
0	1										
1	0	1									
2	0	−1	1								
3	0	2	−3	1							
4	0	−6	11	−6	1						
5	0	24	−50	35	−10	1					
6	0	−120	274	−225	85	−15	1				
7	0	720	−1,764	1,624	−735	175	−21	1			
8	0	−5,040	13,068	−13,132	6,769	−1,960	322	−28	1		
9	0	40,320	−109,584	118,124	−67,284	22,449	−4,536	546	−36	1	
10	0	−362,880	1,026,576	−1,172,700	723,680	−269,325	63,273	−9,450	870	−45	1

4. The following table gives Stirling subset numbers of the second kind, $S(n,k) = \left\{{n \atop k}\right\}$:

n\k	0	1	2	3	4	5	6	7	8	9	10
0	1										
1	0	1									
2	0	1	1								
3	0	1	3	1							
4	0	1	7	6	1						
5	0	1	15	25	10	1					
6	0	1	31	90	65	15	1				
7	0	1	63	301	350	140	21	1			
8	0	1	127	966	1,701	1,050	266	28	1		
9	0	1	255	3,035	7,770	6,951	2,646	462	36	1	
10	0	1	511	9,330	34,501	42,525	22,827	5,880	750	45	1

5. $B_n = \sum_{k=1}^{n} \left\{{n \atop k}\right\}$.

6. The first fifteen Bell numbers are:

$B_1 = 1$ $B_2 = 2$ $B_3 = 5$ $B_4 = 15$
$B_5 = 52$ $B_6 = 203$ $B_7 = 877$ $B_8 = 4,140$
$B_9 = 21,147$ $B_{10} = 115,975$ $B_{11} = 678,570$ $B_{12} = 4,213,597$
$B_{13} = 27,644,437$ $B_{14} = 190,899,322$ $B_{15} = 1,382,958,545$.

7. Table 1 lists some identities involving Stirling numbers.

8. The following give combinatorial interpretations of some of the identities involving Stirling numbers:

- *Stirling cycle number recursion*: When partitioning n objects into k cycles, there are $\left[{n-1 \atop k-1}\right]$ ways in which the last object has a cycle to itself. Otherwise, there are $\left[{n-1 \atop k}\right]$ ways to partition the other $n-1$ objects into k cycles, and then $n-1$ choices of a location into which the last object can be inserted.

Table 1 Stirling number identities.

$$\left[{n \atop k} \right] = (n-1)\left[{n-1 \atop k} \right] + \left[{n-1 \atop k-1} \right], \ (k > 0) \qquad \text{Stirling cycle number recursion}$$

$$\left[{n \atop 0} \right] = \begin{cases} 0, & \text{if } n \neq 0 \\ 1, & \text{if } n = 0 \end{cases}$$

$$\left\{ {n \atop k} \right\} = k\left\{ {n-1 \atop k} \right\} + \left\{ {n-1 \atop k-1} \right\}, \ (k > 0) \qquad \text{Stirling subset number recursion}$$

$$\left\{ {n \atop 0} \right\} = \begin{cases} 0, & \text{if } n \neq 0 \\ 1, & \text{if } n = 0 \end{cases}$$

$$\sum_k \left[{n \atop k} \right]\left\{ {k \atop m} \right\}(-1)^{n-k} = \begin{cases} 0, & \text{if } n \neq m \\ 1, & \text{if } n = m \end{cases}$$

$$\sum_k \left\{ {n \atop k} \right\}\left[{k \atop m} \right](-1)^{n-k} = \begin{cases} 0, & \text{if } n \neq m \\ 1, & \text{if } n = m \end{cases} \qquad \text{Inversion formulas}$$

$$\left\{ {n \atop 1} \right\} = \left\{ {n \atop n} \right\} = 1$$

$$\left\{ {n \atop 2} \right\} = 2^{n-1} - 1$$

$$\left\{ {n \atop k} \right\} k! = \text{the number of onto functions from an } n\text{-set to a } k\text{-set}$$

$$\sum_{k=0}^n \left[{n \atop k} \right] = n!$$

$$\sum_{n=0}^\infty S(n+k, k) x^n = \frac{1}{(1-x)(1-2x)\ldots(1-kx)}$$

$$\sum_{n=0}^\infty \frac{s(n,k) x^n}{n!} = \frac{(\log(1+x))^k}{k!}$$

$$\sum_{n=0}^\infty \frac{S(n,k) x^n}{n!} = \frac{1}{k!}(e^x - 1)^k$$

- *Stirling subset number recursion*: When partitioning n objects into k nonempty subsets, there are $\left\{ {n-1 \atop k-1} \right\}$ ways in which the last object has a subset to itself. Otherwise, there are $\left\{ {n-1 \atop k} \right\}$ ways to partition the other $n-1$ objects into k subsets, and then k choices of a subset into which the last object can be inserted.

- $\sum_{k=0}^n \left[{n \atop k} \right] = n!$: The partitions into cycles are in a one-to-one correspondence with the permutations of n objects, since each permutation can be represented as a composition of disjoint cycles.

Examples:

1. $x(x-1)(x-2)(x-3) = x^4 - 6x^3 + 11x^2 - 6x$, and hence there are $\left[{4 \atop 2} \right] = 11$ permutations of $\{1, 2, 3, 4\}$ with 2 cycles: (12)(34), (13)(24), (14)(23), (1)(234), (1)(324), (2)(134), (2)(314), (3)(124), (3)(214), (4)(123), (4)(213). Also, $s(4,2) = (-1)^{4-2} \cdot 11$.

2. $x^4 = x(x-1)(x-2)(x-3)+6x(x-1)(x-2)+7x(x-1)+x$, and hence there are exactly $\{{4 \atop 2}\} = 7$ set-partitions of $\{1,2,3,4\}$ into two blocks: $\{1\}$ & $\{2,3,4\}$, $\{2\}$ & $\{1,3,4\}$, $\{3\}$ & $\{1,2,4\}$, $\{4\}$ & $\{1,2,3\}$, $\{1,2\}$ & $\{3,4\}$, $\{1,3\}$ & $\{2,4\}$, $\{1,4\}$ & $\{2,3\}$.

2.6 BURNSIDE/PÓLYA COUNTING FORMULA

Burnside's Lemma and Pólya's formula are used to count the number of "really different" configurations, such as tic-tac-toe patterns and placement of beads on a bracelet, in which various symmetries play a role. One of the scientific applications of Pólya's formula is the enumeration of isomers of a chemical compound. From a mathematical perspective, Burnside/Pólya methods count orbits under a permutation group action. (See §5.3.1.)

2.6.1 PERMUTATION GROUPS AND CYCLE INDEX POLYNOMIALS

Definitions:

A **permutation on a set** S is a one-to-one mapping of S onto itself. In this context, the elements of S are called **objects**.

A permutation π of a finite set S is **cyclic** if there is a subcollection of objects that can be arranged in a cycle $(a_1 a_2 \ldots a_n)$ so that each object a_j is mapped by π onto the next object in the cycle and every object of S not in this cycle is fixed by π, that is, mapped to itself.

The **tabular form** of a permutation π on a finite set S is a matrix with two rows. In the first row, each object from S is listed once. Below the object a is its image $\pi(a)$, in this form:
$$\begin{pmatrix} a_1 & a_2 & \cdots & a_n \\ \pi(a_1) & \pi(a_2) & \cdots & \pi(a_n) \end{pmatrix}.$$

The **cycle decomposition (form)** of a permutation π is a concatenation of cyclic permutations whose object subcollections are disjoint and whose product is π. (Sometimes the 1-cycles are explicitly written and sometimes they are omitted.)

A set P of permutations of a set S is **closed under composition** if the composition of each pair of permutations in P is also in P.

A set P of permutations of a set S is **closed under inversion** if for every permutation $\pi \in P$, $\pi^{-1} \in P$.

A **permutation group** $\mathcal{G} = (P, S)$ is a nonempty set P of permutations on a set S such that P is closed under composition and inversion.

The **cycle structure** of a permutation π is an expression (multivariate polynomial) of the form $x_1^{m_1} x_2^{m_2} \ldots x_k^{m_k}$, where m_j is the number of cycles of size j in the cyclic decomposition of π.

The **cycle index** of a permutation group $\mathcal{G}$ is the multivariate polynomial that is the sum of the cycle structures of all the permutations in $\mathcal{G}$, divided by the number of permutations in $\mathcal{G}$. The cycle index polynomial is written $P_{\mathcal{G}}(x_1, x_2, \ldots, x_n)$. (The notation $P_{\mathcal{G}}$ honors George Pólya (1887–1985) who greatly advanced the application of the cycle index polynomial to counting.)

Facts:
1. Every permutation has a tabular form.
2. The tabular form of a permutation is unique up to the order in which the objects of the permuted set are listed in the first row.
3. Every permutation has a cycle decomposition.
4. The cycle decomposition of a permutation into a product of disjoint cyclic permutations is unique up to the order of the factors.
5. The collection of all permutations on a set S forms a permutation group.

Examples:
1. The permutation $\begin{pmatrix} a & b & c & d \\ c & d & a & b \end{pmatrix}$ has the cycle decomposition $(ac)(bd)$.
2. The symmetric group Σ_3 of all 6 possible permutations on $\{a, b, c\}$ has the following elements:
$$(a)(b)(c),\ (ab)(c),\ (ac)(b),\ (a)(bc),\ (abc),\ (acb)$$
with respective cycle structures
$$x_1^3,\ x_1 x_2,\ x_1 x_2,\ x_1 x_2,\ x_3,\ x_3.$$
Thus, the cycle index polynomial is
$$P_{\Sigma_3} = \tfrac{1}{6}\left(x_1^3 + 3x_1 x_2 + 2x_3\right).$$
3. The group Σ_4 of all 24 permutations on $\{a, b, c, d\}$ has the following elements:

$(a)(b)(c)(d)$ $(ab)(c)(d)$ $(ac)(b)(d)$ $(ad)(b)(c)$ $(a)(bc)(d)$ $(a)(bd)(c)$

$(a)(b)(cd)$ $(abc)(d)$ $(acb)(d)$ $(abd)(c)$ $(adb)(c)$ $(acd)(b)$

$(adc)(b)$ $(a)(bcd)$ $(a)(bdc)$ $(ab)(cd)$ $(ac)(bd)$ $(ad)(bc)$

$(abcd)$ $(abdc)$ $(acbd)$ $(acdb)$ $(adbc)$ $(adcb)$

The cycle index polynomial is
$$P_{\Sigma_4} = \tfrac{1}{24}\left[x_1^4 + 6x_1^2 x_2 + 8x_1 x_3 + 3x_2^2 + 6x_4\right].$$

2.6.2 ORBITS AND SYMMETRIES

Definitions:

Given a permutation group $\mathcal{G} = (P, S)$, the **orbit** of $a \in S$ is the set $\{\pi(a) \mid \pi \in P\}$.

A **symmetry of a figure** (or **symmetry motion**) is a spatial motion of the figure onto itself.

Facts:
1. Given a permutation group $\mathcal{G} = (P, S)$, the relation R defined by
$$aRb \iff \text{there exists } \pi \in P \text{ such that } \pi(a) = b$$
is an equivalence relation (§1.4.2), and the equivalence classes under it are precisely the orbits.

2. The set of all symmetries on a figure forms a group.

3. The set of symmetries on a polygon induces a permutation group action on its corner set and a permutation group action on its edge set.

Examples:

1. Acting on the set $\{a, b, c, d, e\}$ is the following permutation group:
$$(a)(b)(c)(d)(e), (ab)(c)(d)(e), (a)(b)(cd)(e), \text{ and } (ab)(cd)(e).$$
The orbits of this group are $\{a, b\}, \{c, d\}, \{e\}$. The cycle index is
$$\tfrac{1}{4}\left[x_1^5 + 2x_1^3 x_2 + x_1 x_2^2\right].$$

2. A square with corners a, b, c, d (in clockwise order) has eight possible symmetries: four rotations in the plane around the center of the square and four reflections (which could also be achieved by 180° spatial rotations out of the plane). See the following figure.

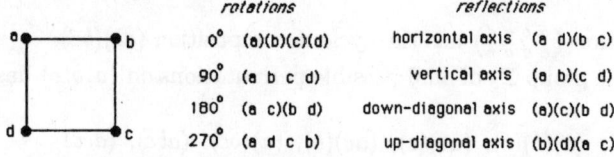

There is only one orbit, $\{a, b, c, d\}$, and the cycle index for the group of symmetries of a square acting on its corner set (the dihedral group D_4) is
$$P_{D_4} = \tfrac{1}{8}\left[x_1^4 + 2x_4 + 3x_2^2 + 2x_1^2 x_2\right].$$

3. A pentagon has 10 different symmetries — five rotations in the plane around the center of the pentagon: $0° = (a)(b)(c)(d)(e), 72° = (abcde), 144° = (acebd), 216° = (adbec)$, and $288° = (aedcb)$, and five reflections (or equivalently, spatial rotations of 180° out of the plane) around axis lines through a corner and the middle of an opposite side: $(a)(be)(cd), (b)(ac)(de), (c)(ae)(bd), (d)(ab)(ce)$, and $(e)(ad)(bc)$. See the following figure. There is only one orbit, $\{a, b, c, d, e\}$, and the associated cycle index is $\tfrac{1}{10}\left[x_1^5 + 4x_5 + 5x_1 x_2^2\right]$.

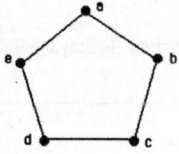

2.6.3 COLOR PATTERNS AND INDUCED PERMUTATIONS

Definitions:

A *coloring* of a set S from a set of n colors is a function from S to the set $\{1, \ldots, n\}$, whose elements are regarded as "colors". The set of all such colorings is denoted $C(S, n)$.

A *corner coloring* of a (polygonal or polyhedral) geometric figure is a coloring of its set of corners.

An *edge coloring* of a geometric figure is a coloring of its set of edges.

Let c_1 and c_2 be colorings of the set S and let π be a permutation of S. Write $\pi(c_1) = c_2$ if $c_1(a) = c_2(\pi(a))$ for every $a \in S$. The correspondence $c_1 \mapsto c_1 \circ \pi^{-1}$ is the **map induced by π on the colorings of S**. (The composition $c_1 \circ \pi^{-1}$ assigns a color to every object $a \in S$, namely the color $c_1(\pi^{-1}(a))$.

Two corner colorings of a figure are **equivalent** if one can be mapped to the other by a symmetry. Similar definitions apply to edge colorings and to face colorings.

Two colorings c_1 and c_2 of a set S are **equivalent under a group** $\mathcal{G} = (P, S)$ if there is a permutation $\pi \in P$ such that $\pi(c_1) = c_2$.

A **corner coloring pattern of a figure with respect to a set of symmetries** is a set of mutually equivalent colorings of the figure.

Facts:

1. Let $\mathcal{G} = (P, S)$ be a permutation group. Then the induced action of P on the set $C(S, n)$ of colorings with n colors is a permutation group action.

2. When P acts on the set $C(S, n)$ of colorings of S, the numbers of permuted objects and orbits, and the cycle index polynomial, are different from when P acts on S itself.

3. In permuting the set S of corners of a figure, a symmetry of a figure simultaneously induces a permutation of the set of all its corner colorings. An analogous fact holds for edge colorings.

Examples:

1. In Example 2 of §2.6.2, a permutation group of 8 elements acts on the four corners of a square. There is only one orbit, and the cycle index is $\frac{1}{8}\left[x_1^4 + 2x_4 + 3x_2^2 + 2x_1^2 x_2\right]$. The following figure shows what happens when the same group acts on the set of black-white colorings. The permuted set has 16 colorings, there are 6 orbits, and the cycle index polynomial is $\frac{1}{8}\left[x_1^{16} + 2x_1^2 x_2 x_4^3 + 3x_1^4 x_2^6 + 2x_1^8 x_2^4\right]$.

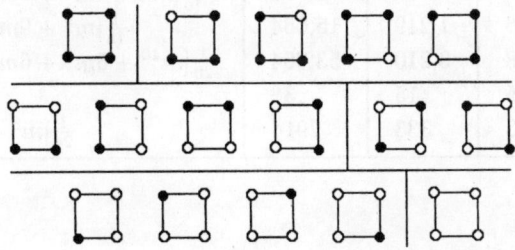

2.6.4 FIXED POINTS AND BURNSIDE'S LEMMA

Definition:

An element $a \in S$ is a **fixed point** of the permutation π if $\pi(a) = a$. The set of all fixed points of π is denoted $\mathrm{fix}(\pi)$.

Facts:

1. The number of fixed points of a permutation π equals the number of 1-cycles in its cycle decomposition.

2. *Burnside's Lemma*: Let $\mathcal{G}$ be a group of permutations acting on a set S. Then the number of orbits induced on S is given by

$$\frac{1}{|\mathcal{G}|} \sum_{\pi \in \mathcal{G}} |\text{fix}(\pi)| \qquad \text{where } \text{fix}(\pi) = \{\, x \in S \mid \pi(x) = x \,\}.$$

Note: The theorem commonly called "Burnside's Lemma" originated with Georg Frobenius (1848–1917). A widely available book by William Burnside (1852–1927) published in 1911 stated and proved the same result, without mentioning its prior discovery.

3. Evaluation of the sum in Burnside's Lemma is simplified by using the cycle index polynomial and Fact 1. For each term in the polynomial, multiply the coefficient by the exponent of x_1, and then sum these products.

4. *Special Burnside's Lemma (for colorings)*: Let $\mathcal{G}$ be a group of permutations acting on a set S. Then the number of orbits induced on $C(S, n)$ (the set of colorings of S from a set of n colors) is given by substituting n for each variable in the cycle index polynomial.

5. The following table gives information on the number of corner coloring patterns of selected figures.

figure	colors			
	2	3	4	m
triangle	4	11	20	$\frac{1}{6}[m^3 + 3m^2 + 2m]$
square	6	21	55	$\frac{1}{8}[m^4 + 2m^3 + 3m^2 + 2m]$
pentagon	8	39	136	$\frac{1}{10}[m^5 + 5m^3 + 4m]$
hexagon	13	92	430	$\frac{1}{12}[m^6 + 3m^4 + 4m^3 + 2m^2 + 2m]$
heptagon	18	198	1,300	$\frac{1}{14}[m^7 + 7m^4 + 6m]$
octagon	30	498	4,183	$\frac{1}{16}[m^8 + 4m^5 + 5m^4 + 2m^2 + 4m]$
nonagon	46	1,219	15,084	$\frac{1}{18}[m^9 + 9m^5 + 2m^3 + 6m]$
decagon	78	3,210	53,764	$\frac{1}{20}[m^{10} + 5m^6 + 6m^5 + 4m^2 + 4m]$
tetrahedron	5	15	36	$\frac{1}{12}[m^4 + 11m^2]$
cube	23	333	2914	$\frac{1}{24}[m^8 + 17m^4 + 6m^2]$

Examples:

1. In Example 1 of §2.6.2, the permutation group is

$$\{(a)(b)(c)(d)(e),\ (ab)(c)(d)(e),\ (a)(b)(cd)(e),\ (ab)(cd)(e)\}.$$

The cycle index is $\frac{1}{4}\left[x_1^5 + 2x_1^3 x_2 + x_1 x_2^2\right]$. By Burnside's Lemma and Fact 3 there are $\frac{1}{4}[1 \cdot 5 + 2 \cdot 3 + 1 \cdot 1] = \frac{12}{4} = 3$ orbits. The orbits are $\{a, b\}, \{c, d\}, \{e\}$.

2. Example 1 of §2.6.3 shows 16 colorings of the corners of the square with colors black or white. There are 6 orbits, and the cycle index for the action on the colorings is $\frac{1}{8}\left[x_1^{16} + 2x_1^2 x_2 x_4^3 + 3x_1^4 x_2^6 + 2x_1^8 x_2^4\right]$. By Burnside's Lemma and Fact 3, there are $\frac{1}{8}[1 \cdot 16 + 2 \cdot 2 + 3 \cdot 4 + 2 \cdot 8] = \frac{48}{8} = 6$ orbits.

It is simpler to apply Special Burnside's Lemma to the cycle index for the action on the square (from Example 2 of §2.6.2), $\frac{1}{8}\left[x_1^4 + 2x_4 + 3x_2^2 + 2x_1^2 x_2\right]$, which yields $\frac{1}{8}\left[1 \cdot 2^4 + 2 \cdot 2 + 3 \cdot 2^2 + 2 \cdot 2^2 \cdot 2\right] = 6$ orbits.

3. (*Continuing Example 3 of §2.6.2*): The cycle index of the group of symmetries of the pentagon is $\frac{1}{10}\left[x_1^5 + 4x_5 + 5x_1x_2^2\right]$. By Special Burnside's Lemma, the number of m-colorings of the corners of an unoriented pentagon is $\frac{1}{10}\left[m^5 + 4m + 5m^3\right]$. For $m = 3$, the formula gives $\frac{1}{10}(243 + 12 + 135) = 39$ 3-coloring patterns of a pentagon.

4. A cube has 24 rotational symmetries, which act on the corners. The identity symmetry has cycle structure x_1^8. There are three additional classes of symmetries, as follows:
(a) Rotations of $90°, 180°$, or $270°$ about an axis line through the middles of opposite faces, for example, through $abcd$ and $efgh$ in part (a) of the following figure. A $90°$ rotation, such as $(abcd)(efgh)$, has cycle structure x_4^2. All $270°$ rotations have that same structure. A $180°$ rotation, such as $(ac)(bd)(eg)(fh)$, has cycle structure x_2^4. There are three pairs of opposite faces, and so the total contribution to the cycle index of opposite-face rotations is $6x_4^2 + 3x_2^4$.

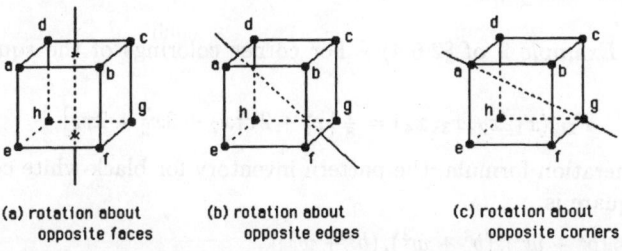

(a) rotation about opposite faces   (b) rotation about opposite edges   (c) rotation about opposite corners

(b) Rotating $180°$ about an axis line through the middles of opposite edges, for example, through edges ad and fg in part (b) of the figure. This rotation, $(ad)(bh)(ce)(fg)$, has cycle structure x_2^4. There are six pairs of opposite edges, and so the total contribution of opposite-edge rotations is $6x_2^4$.

(c) Rotating $120°$ or $240°$ about an axis line through opposite corners, for example, about the line through corners a and g in part (c) of the figure. Any $120°$ rotation, such as $(a)(bde)(chf)(g)$, has cycle structure $x_1^2 x_3^2$. A $240°$ rotation has the same structure. There are four pairs of opposite corners, and so the contribution of opposite-corner rotations is $8x_1^2 x_3^2$.

Collect terms to obtain the cycle index $\frac{1}{24}\left[x_1^8 + 6x_4^2 + 9x_2^4 + 8x_1^2 x_3^2\right]$. Thus, the number of m-colorings of the corners of an unoriented cube is $\frac{1}{24}\left[m^8 + 6m^2 + 9m^4 + 8m^4\right]$. For $m = 2$ and 3, the formula gives 23 2-coloring patterns and 333 3-coloring patterns.

2.6.5 PÓLYA'S ENUMERATION FORMULA

Definition:

A *pattern inventory* is a generating function (§3.2) that enumerates the numbers of coloring patterns of a given figure.

Facts:

1. *Pólya's enumeration formula*: Let $\mathcal{G} = (P, S)$ be a permutation group and let $\{c_1, \ldots, c_n\}$ be a set of names for n colors for the objects of S. Then the pattern inventory with respect to $\mathcal{G}$ for the set of all n-colorings of S is given by substituting $(c_1^j + \cdots + c_n^j)$ for x_j in the cycle index $P_{\mathcal{G}}(x_1, \ldots, x_m)$.

Note: This theorem was published in 1937. Essentially the same result was derived by H. Redfield in 1927.

2. Pólya's enumeration formula has many applications in enumerating various families of graphs. This approach was pioneered by F. Harary. (See [HaPa73].)

3. Pólya's enumeration formula has many applications in which some practical question is modeled as a graph coloring problem.

Examples:

1. The pattern inventory of black-white colorings of the corners of a triangle is
$$1b^3 + 1b^2w + 1bw^2 + 1w^3.$$
This means there is one coloring pattern with all 3 corners black, one with 2 black corners and 1 white corner, etc.

2. (*Continuing Example 2 of §2.6.4*): For corner colorings of the square, the cycle index is
$$P_{D_4}(x_1, x_2, x_3, x_4) = \tfrac{1}{8}\left[x_1^4 + 2x_1^2 x_2 + 3x_2^2 + 2x_4\right].$$
By Pólya's enumeration formula, the pattern inventory for black-white colorings of the corners of the square is
$$P_{D_4}[(b+w), (b^2+w^2), (b^3+w^3), (b^4+w^4)]$$
$$= \tfrac{1}{8}\left[(b+w)^4 + 2(b+w)^2(b^2+w^2) + 3(b^2+w^2)^2 + 2(b^4+w^4)\right]$$
$$= \tfrac{1}{8}\left[8b^4 + 8b^3w + 16b^2w^2 + 8bw^3 + 8w^4\right] = 1b^4 + 1b^3w + 2b^2w^2 + 1bw^3 + 1w^4.$$
This pattern inventory may be confirmed by examining the drawing in Example 1 of §2.6.3.

3. (*Continuing Example 3 of §2.6.4*): For corner colorings of the pentagon, the cycle index is
$$P_{D_5}(x_1, x_2, \ldots, x_5) = \tfrac{1}{10}\left[x_1^5 + 4x_5 + 5x_1 x_2^2\right].$$
By Pólya's enumeration formula, the pattern inventory for black-white colorings of the corners of the pentagon (confirmable by drawing pictures) is
$$P_{D_5}((b+w), (b^2+w^2), (b^3+w^3), (b^4+w^4), (b^5+w^5))$$
$$= \tfrac{1}{10}\left[(b+w)^5 + 4(b^5+w^5) + 5(b+w)(b^2+w^2)^2\right]$$
$$= \tfrac{1}{10}\left[10b^5 + 10b^4 w + 20b^3 w^2 + 20b^2 w^3 + 10bw^4 + 10w^5\right]$$
$$= 1b^5 + 1b^4 w + 2b^3 w^2 + 2b^2 w^3 + 1bw^4 + 1w^5.$$

4. (*Continuing Example 4 of §2.6.4*): For corner colorings of the cube, the cycle index is
$$P_{\mathcal{G}}(x_1, \ldots, x_4) = \tfrac{1}{24}\left[x_1^8 + 6x_4^2 + 9x_2^4 + 8x_1^2 x_3^2\right].$$
By Pólya's enumeration formula, the pattern inventory for black-white colorings of the corners of the cube is
$$P_{\mathcal{G}}\left((b+w), (b^2+w^2), (b^3+w^3), (b^4+w^4)\right)$$
$$= \tfrac{1}{24}\left[(b+w)^8 + 6(b^4+w^4)^2 + 9(b^2+w^2)^4 + 8(b+w)^2(b^3+w^3)^2\right]$$
$$= b^8 + b^7 w + 3b^6 w^2 + 3b^5 w^3 + 7b^4 w^4 + 3b^3 w^5 + 3b^2 w^6 + bw^7 + w^8.$$

5. *Organic chemistry:* Two structurally different compounds with the same chemical formula are called *isomers*. For instance, to two of the six carbons (C) in a ring there might be attached a hydrogen (H), and to each of the four other carbons some other radical (R), thereby yielding the chemical formula $C_6H_2R_4$. The number of different isomers (structurally different arrangements of the radicals) is the same as the number of coloring patterns of a hexagon when two of the corners are "colored" H and four "colored" R. The cycle index for the symmetries of a hexagon, in terms of corner permutations, is

$$P_{D_6}(x_1,\ldots,x_6) = \tfrac{1}{12}\left[x_1^6 + 2x_6 + 2x_3^2 + 4x_2^3 + 3x_1^2 x_2^2\right].$$

Substituting $(H^j + R^j)$ for x^j yields a pattern inventory listing the number of isomers of $C_6 H_i R_{6-i}$:

$$\tfrac{1}{12}\left[(H+R)^6 + 2(H^6+R^6) + 2(H^3+R^3)^2 + 4(H^2+R^2)^3 + 3(H+R)^2(H^2+R^2)^2\right]$$
$$= \tfrac{1}{12}\left[12H^8 + 12H^5 R + 36H^4 R^2 + 36H^3 R^3 + 36H^2 R^4 + 12HR^5 + 12R^6\right]$$
$$= 1H^8 + 1H^5 R + 3H^4 R^2 + 3H^3 R^3 + 3H^2 R^4 + 1HR^5 + 1R^6.$$

The three possible coloring patterns corresponding to $3H^2 R^4$ are shown in the following figure:

2.7 MÖBIUS INVERSION COUNTING

Möbius inversion is an important tool used to solve a variety of counting problems such as counting how many numbers are relatively prime to some given number (without individually checking each smaller number) and counting certain types of circular arrangements. It generalizes the principle of inclusion/exclusion. (Augustus Ferdinand Möbius, 1790–1868)

2.7.1 MÖBIUS INVERSION

Definitions:

The **Kronecker delta function** $\delta(x,y)$ is defined by the rule

$$\delta(x,y) = \begin{cases} 1 & \text{if } x = y \\ 0 & \text{otherwise.} \end{cases}$$

The **Möbius function** is the function μ from the set of positive integers to the set of integers where

$$\mu(m) = \begin{cases} 1 & \text{if } m = 1 \\ (-1)^k & \text{if } m = p_1 p_2 \ldots p_k \text{ (the product of } k \text{ distinct primes)} \\ 0 & \text{if } m \text{ is divisible by the square of a prime.} \end{cases}$$

Note: See Chapter 11 for Möbius functions defined on partially ordered sets.

Facts:

1. For the Möbius function μ defined on the set of positive integers:
   - μ is *multiplicative*: if $\gcd(m,n) = 1$, then $\mu(mn) = \mu(m)\mu(n)$;
   - μ is not *completely multiplicative*: $\mu(mn) = \mu(m)\mu(n)$ is not always true;
   - $\sum_{d|n} \mu(d) = \begin{cases} 1 & \text{if } n = 1 \\ 0 & \text{if } n > 1 \end{cases}$ where the sum is taken over all positive divisors of n.

2. *Möbius inversion formula*: If $f(n)$ and $g(n)$ are defined for all positive integers and $f(n) = \sum_{d|n} g(d)$, then $g(n) = \sum_{d|n} \mu(d) f(n/d)$.

3. For every positive integer n, $n = \sum_{d|n} \phi(d)$. (For example, $6 = \phi(1) + \phi(2) + \phi(3) + \phi(6) = 1 + 1 + 2 + 2$.)

Examples:

1. *Circular permutations with repetitions*: Given an alphabet of m letters, how many circular permutations of length n are possible, if repeated letters are allowed and two permutations are the same if the second can be obtained from the first by rotation? The problem was first solved by Percy A. MacMahon in 1892.

    A circular permutation of length n has a period d, where $d|n$. (The *period* of a circular permutation, viewed as a circular string, is the length of the shortest substring that repeats end-to-end to give the entire string.) Let $g(d)$ be the number of length d circular permutations that have period d. A circular permutation of length n can be constructed from one of length d (where $d|n$) by concatenating it with itself $\frac{n}{d}$ times. For example, the circular permutation $aabbaabb$ (where beginning and end are joined) of period four can be obtained by taking the circular permutation $aabb$ and opening it up at one of four spots between the letters, to obtain any of four linear strings $aabb$, $abba$, $bbaa$, and $baab$. Join one of these to itself, obtaining $aabbaabb$, $abbaabba$, $bbaabbaa$, and $baabbaab$, and then join the beginning and the end to form the circular permutation $aabbaabb$.

    For any positive integer k, there are $dg(d)$ linear strings of length k obtained by taking $\frac{k}{d}$ repetitions of the linear strings of length d that have period d, where $d|k$. Therefore, the total number of linear strings of length k where the objects are chosen from m types is $\sum_{d|k} dg(d) = m^k$. Applying the Möbius inversion formula to m^k and g yields $g(k) = \frac{1}{k} \sum_{d|k} \mu(d) m^{k/d}$. Therefore, the total number of circular permutations of length n where the elements are chosen from an alphabet of size m is $\sum_{d|n} g(d)$, which is equal to

$$\sum_{k|n} \left(\frac{1}{k} \sum_{d|k} \mu(d) m^{k/d} \right).$$

2. *Circular permutations with repetitions with specified numbers of each type of object*: Suppose there are a total of n objects of t types, with a_i of type i $(i = 1, \ldots, t)$, where $a_1 + \cdots + a_t = n$. If $a = \gcd(a_1, \ldots, a_t)$, these circular permutations can be generated as in Example 3 by taking a circular permutation of period d (where $d|a$) with $\frac{a_i d}{n}$ objects of type i $(i = 1, \ldots, t)$, breaking it open, and laying it end-to-end $\frac{n}{d}$ times. Let $g(k)$ be the number of such circular permutations of length k that have period k. Then the total number of linear strings of length n with a_i objects of type i is given by $\sum_{d|a} dg(d) = \frac{n!}{a_1! \ldots a_t!}$.

By the Möbius inversion formula,

$$g(k) = \frac{1}{k} \sum_{d|a} \mu(d) \frac{(k/d)!}{(a_1/d)!(a_2/d)! \ldots (a_t/d)!}.$$

Summing $g(k)$ over all divisors of a gives the desired total number of circular permutations:

$$\sum_{k|a} g(k) = \sum_{k|a} \left(\frac{1}{k} \sum_{d|a} \mu(d) \frac{(k/d)!}{(a_1/d)! \ldots (a_t/d)!} \right).$$

2.8 YOUNG TABLEAUX

Arrays called Young tableaux were introduced by the Reverend Alfred Young (1873–1940). These arrays are used in combinatorics and the theories of symmetric functions, which are the subject of this section. Young tableaux are also used in the analysis of representations of the symmetric group. They make it possible to approach many results about representation theory from a concrete combinatorial viewpoint.

2.8.1 TABLEAUX COUNTING FORMULAS

Definitions:

The **hook** $H_{i,j}$ of cell (i,j) in the Ferrers diagram for a partition λ is the set

$$\{ (k,j) \in \lambda \mid k \geq i \} \cup \{ (i,k) \in \lambda \mid k \geq j \},$$

that is, the set consisting of the cell (i,j), all cells in its row to its right, and all cells in its column below it.

The **hooklength** $h_{i,j}$ of cell (i,j) is the number $|H_{i,j}|$ of cells in its hook.

A **Young tableau** is an array obtained by replacing each cell of the Ferrers diagram by a positive integer.

The **shape** of a Young tableau is the partition corresponding to the underlying Ferrers diagram. The notation $\lambda \vdash n$ indicates that λ partitions the number n.

A Young tableau is **semistandard** (an **SSYT**) if the entries in each row are weakly increasing and the entries in each columns are strictly increasing.

A semistandard Young tableau of shape $\lambda \vdash n$ is **standard** (an **SYT**) if each number $1, \ldots, n$ occurs exactly once as an entry. The number of SYT of shape λ is denoted f_λ.

If G is a group (see §5.2) then an **involution** is an element $g \in G$ such that g^2 is the identity. The number of involutions in the symmetric group S_n (or Σ_n) (the group of all permutations on the set $\{1, 2, \ldots, n\}$) is denoted inv(n).

The following table summarizes notation for Young tableaux:

notation	meaning
$\lambda = (\lambda_1, \ldots, \lambda_l)$	partition (with parts $\lambda_1 \geq \lambda_2 \geq \cdots \geq \lambda_l$)
$\lambda \vdash n$	λ partitions the number n
(i, j)	cell in a Ferrers diagram
$H_{i,j}$	hook of cell (i, j)
$h_{i,j}$	hooklength of hook $H_{i,j}$
f_λ	number of SYT of shape λ
inv(n)	number of involutions in S_n

Facts:

1. *Frame-Robinson-Thrall hook formula* [1954]: The number of SYT of fixed shape λ is
$$f_\lambda = \frac{n!}{\prod_{(i,j) \in \lambda} h_{i,j}}.$$

2. *Frobenius determinantal formula* [1900]: The number of SYT of fixed shape $\lambda = (\lambda_1, \ldots, \lambda_l)$ is the determinant
$$f_\lambda = n! \left| \frac{1}{(\lambda_i + j - i)!} \right|_{1 \leq i, j \leq l}.$$

3. Summations involving the number of SYT:
$$\sum_{\lambda \vdash n} f_\lambda = \text{inv}(n) \qquad \sum_{\lambda \vdash n} f_\lambda^2 = n!.$$

4. Young tableaux can be used to find the number of permutations with specified lengths of their longest increasing subsequences and longest decreasing subsequences. [Be71]

Examples:

1. If $\lambda = (3, 2)$ then a complete list of SYT is:

$$\begin{array}{ccc} 1\ 2\ 3 \\ 4\ 5 \end{array} \quad \begin{array}{ccc} 1\ 2\ 4 \\ 3\ 5 \end{array} \quad \begin{array}{ccc} 1\ 2\ 5 \\ 3\ 4 \end{array} \quad \begin{array}{ccc} 1\ 3\ 4 \\ 2\ 5 \end{array} \quad \begin{array}{ccc} 1\ 3\ 5 \\ 2\ 4 \end{array}$$

2. If $\lambda = (2, 2)$ then a complete list of SSYT with entries at most 3 is:

$$\begin{array}{cc} 1\ 1 \\ 2\ 2 \end{array} \quad \begin{array}{cc} 1\ 1 \\ 3\ 3 \end{array} \quad \begin{array}{cc} 2\ 2 \\ 3\ 3 \end{array} \quad \begin{array}{cc} 1\ 1 \\ 2\ 3 \end{array} \quad \begin{array}{cc} 1\ 2 \\ 2\ 3 \end{array} \quad \begin{array}{cc} 1\ 2 \\ 3\ 3 \end{array}$$

3. For the partition $(3, 2)$, $H_{1,1} = \{(1, 1), (2, 1), (1, 2), (1, 3)\}$. In the following diagram each cell of $(3, 2)$ is replaced with its hooklength.

$$\begin{array}{ccc} 4 & 3 & 1 \\ 2 & 1 \end{array}$$

The hook formula (Fact 1) gives the number of SYT of shape $(3, 2)$: $f_{(3,2)} = \frac{5!}{4 \cdot 3 \cdot 2 \cdot 1^2} = 5$.

The determinantal formula (Fact 2) gives the same result:
$$f_{(3,2)} = 5! \begin{vmatrix} \frac{1}{3!} & \frac{1}{4!} \\ \frac{1}{1!} & \frac{1}{2!} \end{vmatrix} = 5.$$

4. For the partitions of $n = 3$, $f_{(3)} = 1$, $f_{(2,1)} = 2$, $f_{(1,1,1)} = 1$, so the summation formulas become:
$$\sum_{\lambda \vdash 3} f_\lambda = 4 = \text{inv}(3);$$
$$\sum_{\lambda \vdash 3} f_\lambda^2 = 6 = 3!.$$

2.8.2 TABLEAUX ALGORITHMS

Definitions:

An **inner corner** of a partition λ is a cell $(i, j) \in \lambda$ such that $(i+1, j), (i, j+1) \notin \lambda$.

An **outer corner** of a partition λ is a cell $(i, j) \notin \lambda$ such that $(i-1, j), (i, j-1) \in \lambda$.

Algorithm 1: Generate at random a standard tableau of given shape.

input: a shape λ_I such that $\lambda_I \vdash n$ {For a summary of notation, see §2.5.1.}
output: a standard Young tableau of shape λ_I, uniformly at random

$\lambda := \lambda_I$
while λ is nonempty
  {find an inner corner $(i,j) \in \lambda$}
  choose (with probability $\frac{1}{|\lambda|}$) any cell $(i,j) \in \lambda$
  while the current cell (i,j) is not an inner corner
    choose (with probability $\frac{1}{h_{i,j}}$) a pair $(i', j') \in H_{i,j} - \{(i,j)\}$
    $(i,j) := (i', j')$
  assign label n to inner corner (i,j)
  $\lambda := \lambda - \{(i,j)\}$

Examples:

1. The diagrams in the following figure illustrate a plausible sequence of current cells chosen as the first step of the Greene-Nijenhuis-Wilf algorithm [1979] (Algorithm 1) finds an inner corner of a tableau of shape $\lambda = (5, 5, 5, 2)$.

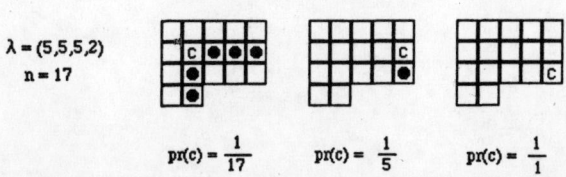

Algorithm 2: Robinson-Schensted.

input: a permutation $\pi \in S_n$ where $\pi = \begin{pmatrix} 1 & 2 & \cdots & n \\ \pi_1 & \pi_2 & \cdots & \pi_n \end{pmatrix}$

output: a pair (P, Q) of standard Young tableaux of the same shape $\lambda \vdash n$

$P_0 := \emptyset; Q_0 := \emptyset$
for $k := 1$ to n
  $r := 1; c := 1; b := \pi_k; P_k := P_{k-1}; exit := \text{FALSE}$
  while $exit = \text{FALSE}$
    {find next insertion row r in tableau P_k}
    while $row_r(P_k) \neq \emptyset$ and $\pi_j > \max\{row_r(P_k)\}$
      $r := r + 1$
    {find next insertion column c in tableau P_k}
    $c := 1$
    while $P_k[r, c] \neq \emptyset$ and $\pi_k < P_k[r, c]$
      $c := c + 1$
    {insert b}
    if $P_k[r, c] = \emptyset$ then
      $P_k[r, c] := b; exit = \text{TRUE}$
    else
      $bb := P_k[r, c]; P_k[r, c] := b; b := bb$
  $Q_k[r, c] := k$
$P := P_n; Q := Q_n$

2. The permutation $\pi = \begin{pmatrix} 1 & 2 & 3 & 4 & 5 & 6 & 7 \\ 6 & 2 & 3 & 1 & 7 & 5 & 4 \end{pmatrix}$ yields this sequence of tableaux pairs (P_k, Q_k) under the Robinson-Schensted algorithm [1938, 1961] (Algorithm 2).

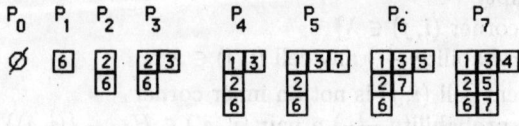

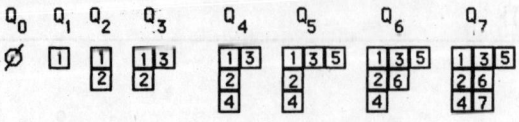

REFERENCES

Printed Resources:

[An76] G. E. Andrews, "The Theory of Partitions", *Encyclopedia of Mathematics and Its Applications*, Vol. 2, Addison-Wesley, 1976. (Reissued: Cambridge University Press, 1984.)

[BeGo75] E. A. Bender and J. R. Goldman, "On the applications of Möbius inversion in combinatorial analysis", *Amer. Math. Monthly* 82 (1975), 789–803. (An expository paper for a general mathematical audience.)

[Be71] C. Berge, *Principles of Combinatorics*, Academic Press, 1971.

[Co78] D. I. A. Cohen, *Basic Techniques of Combinatorial Theory*, Wiley, 1978.

[GaRa90] G. Gasper and M. Rahman, "Basic Hypergeometric Series", *Encyclopedia of Mathematics and Its Applications*, Vol. 35, Cambridge University Press, 1990.

[GrKnPa94] R. L. Graham, D. E. Knuth, and O. Patashnik, *Concrete Mathematics*, 2nd ed., Addison-Wesley, 1994.

[GuGwMi58] H. Gupta, A. E. Gwyther, and J. C. P. Miller, *Tables of Partitions*, Royal Society Math. Tables, Vol. 4, 1958.

[Ha86] M. Hall, Jr., *Combinatorial Theory*, 2nd ed., Wiley-Interscience, 1986. (Section 2.2 of this text contains a brief introduction to Möbius inversion.)

[HaPa73] F. Harary and E. M. Palmer, *Graphical Enumeration*, Academic Press, 1973.

[HaRa18] G. H. Hardy and S. Ramanujan, "Asymptotic formulae in combinatory analysis", Proceedings of the London Mathematical Society, Ser. 2, 17 (1918), 75-115. (Reprinted in *Coll. Papers of S. Ramanujan*, Chelsea, 1962, 276–309.)

[HaWr60] G. H. Hardy and E. M. Wright, *An Introduction to the Theory of Numbers*, 4th ed., Oxford University Press, 1960.

[Ja78] G. D. James, *The Representation Theory of the Symmetric Groups*, Lecture Notes in Mathematics, Vol. 682, Springer-Verlag, 1978.

[JaKe81] G. D. James and A. Kerber, "The Representation Theory of the Symmetric Group", *Encyclopedia of Mathematics and Its Applications*, Vol. 16, Addison-Wesley, 1981.

[Kn69] D. E. Knuth, *The Art of Computer Programming, Vol. I: Fundamental Algorithms*, Addison-Wesley, 1969.

[Kn70] M. I. Knopp, *Modular Functions in Analytic Number Theory*, Markham, 1970.

[Ma95] I. G. Macdonald, *Symmetric Functions and Hall Polynomials*, 2nd ed., Oxford University Press, 1995.

[Ma60] P. A. MacMahon, *Combinatory Analysis*, Vol. 2, Cambridge University Press, London, 1916. (Reissued: Chelsea, 1960.)

[Pa81] E. W. Packel, *The Mathematics of Games and Gambling*, Mathematical Association of America, 1981.

[PóRe87] G. Pólya and R. C. Read, *Combinatorial Enumeration of Groups, Graphs, and Chemical Compounds*, Springer-Verlag, 1987.

[Ro95] K. H. Rosen, *Discrete Mathematics and Its Applications*, 4th ed., McGraw-Hill, 1999.

[Sa91] B. E. Sagan, *The Symmetric Group: Representations, Combinatorial Algorithms, and Symmetric Functions*, Wadsworth, 1991.

[St86] R. P. Stanley, *Enumerative Combinatorics*, Vol. I, Wadsworth, 1986. (Chapter 3 of this text contains an introduction to Möbius functions and incidence algebras.)

[Tu95] A. Tucker, *Applied Combinatorics*, 3rd ed., Wiley, 1995.

[Wi93] H. S. Wilf, *generatingfunctionology*, 2nd ed., Academic Press, 1993.

Web Resources:

http://sue.csc.uvic.ca/~cos/ (Combinatorial object server.)

http://www.cs.sunysb.edu/~algorith/ (The Stony Brook Algorithm Repository; see Section 1.3 on combinatorial problems.)

http://www.schoolnet.ca/vp/ECOS/ (AMOF: the Amazing Mathematical Object Factory.)

3
SEQUENCES

3.1 Special Sequences — Thomas A. Dowling
   3.1.1 Representations of Sequences
   3.1.2 Fibonacci Numbers
   3.1.3 Catalan Numbers
   3.1.4 Bernoulli Numbers and Polynomials
   3.1.5 Eulerian Numbers
   3.1.6 Ramsey Numbers
   3.1.7 Other Sequences
   3.1.8 Miniguide to Sequences

3.2 Generating Functions — Ralph P. Grimaldi
   3.2.1 Ordinary Generating Functions
   3.2.2 Exponential Generating Functions

3.3 Recurrence Relations — Ralph P. Grimaldi
   3.3.1 Basic Concepts
   3.3.2 Linear Homogeneous Recurrence Relations
   3.3.3 Linear Nonhomogeneous Recurrence Relations
   3.3.4 Method of Generating Functions
   3.3.5 Divide-and-conquer Relations

3.4 Finite Differences — Jay Yellen
   3.4.1 The Difference Operator
   3.4.2 Calculus of Differences: Falling and Rising Powers
   3.4.3 Difference Sequences and Difference Tables
   3.4.4 Difference Equations

3.5 Finite Sums and Summation — Victor S. Miller
   3.5.1 Sigma Notation
   3.5.2 Elementary Transformation Rules for Sums
   3.5.3 Antidifferences and Summation Formulas
   3.5.4 Standard Sums

3.6 Asymptotics of Sequences — Edward A. Bender
   3.6.1 Approximate Solutions to Recurrences
   3.6.2 Analytic Methods for Deriving Asymptotic Estimates
   3.6.3 Asymptotic Estimates of Multiply-indexed Sequences

3.7 Mechanical Summation Procedures — Kenneth H. Rosen
   3.7.1 Hypergeometric Series
   3.7.2 Algorithms That Produce Closed Forms for Sums of Hypergeometric Terms
   3.7.3 Certifying the Truth of Combinatorial Identities

INTRODUCTION

Sequences of integers occur regularly in combinatorial applications. For example, the solution to a counting problem that depends on a parameter k can be viewed as the kth term of a sequence. This chapter provides a guide to particular sequences that arise in applied settings. Such (infinite) sequences can often frequently be represented in a finite form. Specifically, sequences can be expressed using generating functions, recurrence relations, or by an explicit formula for the kth term of the sequence.

GLOSSARY

antidifference (of a function f): any function g such that $\Delta g = f$. It is the discrete analogue of antidifferentiation.

ascent (in a permutation π): any index i such that $\pi_i < \pi_{i+1}$.

asymptotic equality (of functions): the function $f(n)$ is asymptotic to $g(n)$, written $f(n) \sim g(n)$, if $f(n) \neq 0$ for sufficiently large n and $\lim_{n \to \infty} \frac{g(n)}{f(n)} = 1$.

Bernoulli numbers: the numbers B_n produced by the recursive definition $B_0 = 1$, $\sum_{j=0}^{n} \binom{n+1}{j} B_j = 0$, $n \geq 1$.

Bernoulli polynomials: the polynomial $B_m(x) = \sum_{k=0}^{m} \binom{m}{k} B_k x^{m-k}$ where B_k is the kth Bernoulli number.

big-oh (of the function f): the set of all functions that do not grow faster than some constant multiple of f, written $O(f(n))$.

big omega (of the function f): the set of all functions that grow at least as fast as some constant multiple of f, written $\Omega(f(n))$.

big theta (of the function f): the set of all functions that grow roughly as fast as some constant multiple of f, written $\Theta(f(n))$.

binomial convolution (of the sequences $\{a_n\}$ and $\{b_n\}$): the sequence whose rth term is formed by summing products of the form $\binom{r}{k} a_k b_{r-k}$.

Catalan number: the number $C_n = \frac{1}{n+1}\binom{2n}{n}$.

characteristic equation: an equation derived from a linear recurrence relation with constant coefficients, whose roots are used to construct solutions to the recurrence relation.

closed form (for a sum): an algebraic expression for the value of a sum with variable limits, which has a fixed number of terms; hence the time needed to calculate it does not grow with the size of the set or interval of summation.

convolution (of the sequences $\{a_n\}$ and $\{b_n\}$): the sequence whose rth term is formed by summing products of the form $a_k b_{r-k}$ where $0 \leq k \leq r$.

de Bruijn sequence: a circular ordering of letters from a fixed alphabet with p letters such that each n consecutive letters (wrapping around from the end of the sequence to the beginning, if necessary) forms a different word.

difference operator: the operator Δ where $\Delta f(x) = f(x+1) - f(x)$ on integer or real-valued functions. It is the discrete analogue of the differentiation operator.

difference sequence (for the sequence $A = \{a_j \mid j = 0, 1, \dots\}$): the sequence $\Delta A = \{a_{j+1} - a_j \mid j = 0, 1, \dots\}$.

difference table (for a function f): a table whose kth row is the kth difference sequence for f.

discordant permutation: a permutation that assigns to every element an image different from those assigned by all other members of a given set of permutations.

dissimilar hypergeometric terms: terms in two hypergeometric series such that their ratio is not a rational function.

divide-and-conquer algorithm: a recursive procedure that solves a given problem by first breaking it into smaller subproblems (of nearly equal size) and then combining their respective solutions.

doubly hypergeometric: property of function $F(n,k)$ that $\frac{F(n+1,k)}{F(n,k)}$ and $\frac{F(n,k+1)}{F(n,k)}$ are rational functions of n and k.

Eulerian number: the number of permutations of $\{1, 2, \ldots, n\}$ with exactly k ascents.

excedance (of a permutation π): any index i such that $\pi_i > i$.

exponential generating function (for the sequence $a_0, a_1, a_2, \ldots$): the function $f(x) = a_0 + a_1 x + a_2 \frac{x^2}{2!} + \cdots$ or any equivalent closed form expression.

falling power (of x): the product $x^{\underline{n}} = x(x-1)(x-2)\ldots(x-n+1)$ of n successive descending factors, starting with x; the discrete analogue of exponentiation.

Fibonacci numbers: the numbers F_n produced by the recursive definition $F_0 = 0$, $F_1 = 1$, $F_n = F_{n-1} + F_{n-2}$ if $n \geq 2$.

figurate number: the number of cells in an array of cells bounded by some regular geometrical figure.

first-order linear recurrence relation with constant coefficients: an equation of the form $C_{n+1}a_{n+1} + C_n a_n = f(n)$, $n \geq 0$, with C_{n+1}, C_n nonzero real constants.

generating function (for the sequence $a_0, a_1, a_2, \ldots$): the function $f(x) = a_0 + a_1 x + a_2 x^2 + \cdots$ or any equivalent closed form expression; sometimes called the *ordinary* generating function for the sequence.

geometric series: an infinite series where the ratio between two consecutive terms is a constant.

Gray code (of size n): a circular ordering of all binary strings of length n in which adjacent strings differ in exactly one bit.

harmonic number: the sum $H_n = \sum_{i=1}^{n} \frac{1}{i}$, which is the discrete analogue of the natural logarithm.

homogeneous recurrence relation: a recurrence relation satisfied by the identically zero sequence.

hypergeometric series: is a series where the ratio of two consecutive terms is a rational function.

indefinite sum (of the function f): the family of all antidifferences of f.

Lah coefficients: the coefficients resulting from expressing the rising factorial in terms of the falling factorials.

linear recurrence relation with constant coefficients: an equation of the form $C_{n+k}a_{n+k} + C_{n+k-1}a_{n+k-1} + \cdots + C_n a_n = f(n)$, $n \geq 0$, where C_{n+i} are real constants with C_{n+k} and C_n nonzero.

little-oh (of the function f): the set of all functions that grow slower than every constant multiple of f, written $o(f(n))$.

little omega (of the function f): the set of all functions that grow faster than every constant multiple of f, written $\omega(f(n))$.

Lucas numbers: the numbers L_n produced by the recursive definition $L_0 = 2$, $L_1 = 1$, $L_n = L_{n-1} + L_{n-2}$ if $n \geq 2$.

nonhomogeneous recurrence relation: a recurrence relation that is not homogeneous.

polyomino: a connected configuration of regular polygons (for example, triangles, squares, or hexagons) in the plane, generalizing a domino.

power sum: the sum of the kth powers of the integers $1, 2, \ldots, n$.

radius of convergence (for the series $\sum a_n x^n$): the number r ($0 \leq r \leq \infty$) such that the series converges for all $|x| < r$ and diverges for all $|x| > r$.

Ramsey number: the number $R(m, n)$ defined as the smallest positive integer k with the following property: if S is a set of size k and the 2-element subsets of S are partitioned into 2 collections, C_1 and C_2, then there is a subset of S of size m such that each of its 2-element subsets belong to C_1 or there is a subset of S of size n such that each of its 2-element sets belong to C_2.

recurrence relation: an equation expressing a term of a sequence as a function of prior terms in the sequence.

rising power (of x): the product $x^{\overline{n}} = x(x+1)(x+2)\ldots(x+n-1)$ of n successive ascending terms, starting with x.

second-order linear recurrence relation with constant coefficients: an equation of the form $C_{n+2}a_{n+2} + C_{n+1}a_{n+1} + C_n a_n = f(n)$, $n \geq 0$, where C_{n+2}, C_{n+1}, C_n are real constants with C_{n+2} and C_n nonzero.

sequence: a function from $\{0, 1, 2, \ldots\}$ to the real numbers (often the integers).

shift operator: the operator E defined by $Ef(x) = f(x+1)$ on integer or real-valued functions.

similar hypergeometric terms: terms in two hypergeometric series such that their ratio is a rational function.

standardized form for a sum: a sum over an integer interval, in which the lower limit of the summation is zero.

Stirling's approximation formula: the asymptotic estimate $\sqrt{2\pi n}(n/e)^n$ for $n!$.

tangent numbers: numbers generated by the exponential generating function $\tan x$.

3.1 SPECIAL SEQUENCES

3.1.1 REPRESENTATIONS OF SEQUENCES

A given infinite sequence $a_0, a_1, a_2, \ldots$ can often be represented in a more useful or more compact form. Namely, there may be a closed form expression for a_n as a function of n, the terms of the sequence may appear as coefficients in a simple generating function, or the sequence may be specified by a recurrence relation. Each representation has advantages, in either defining the sequence or establishing information about its terms.

Definitions:

A *sequence* $\{a_n \mid n \geq 0\}$ is a function from the set of nonnegative integers to the real numbers (often the integers). The *terms* of the sequence $\{a_n \mid n \geq 0\}$ are the values $a_0, a_1, a_2, \ldots$.

A *closed form* for the sequence $\{a_n\}$ is an algebraic expression for a_n as a function of n.

A *recurrence relation* is an equation expressing a term of a sequence as a function of prior terms in the sequence.

A *solution* of a recurrence relation is a sequence whose terms satisfy the relation.

The *generating function* for the sequence $\{a_n\}$ is the function $f(x) = \sum_{i=0}^{\infty} a_i x^i$ or any equivalent closed form expression.

The *exponential generating function* for the sequence $\{a_n\}$ is the function $g(x) = \sum_{i=0}^{\infty} a_i \frac{x^i}{i!}$ or any equivalent closed form expression.

Facts:

1. An important way in which many sequences are represented is by using a recurrence relation (§3.3). Although not all sequences can be represented by useful recurrence relations, many sequences that arise in the solution of counting problems can be so represented.

2. An important way to study a sequence is by using its generating function (§3.2). Information about terms of the sequence can often be obtained by manipulating the generating function.

Examples:

1. The Fibonacci numbers F_n (§3.1.2) arise in many applications and are given by the sequence $0, 1, 1, 2, 3, 5, 8, 13, \ldots$. This infinite sequence can be finitely encoded by means of the recurrence relation

$$F_n = F_{n-1} + F_{n-2}, \quad n \geq 2, \quad \text{with } F_0 = 0 \text{ and } F_1 = 1.$$

Alternatively, a closed form expression for this sequence is given by

$$F_n = \frac{1}{\sqrt{5}}\left[\left(\frac{1+\sqrt{5}}{2}\right)^n - \left(\frac{1-\sqrt{5}}{2}\right)^n\right], \quad n \geq 0.$$

The Fibonacci numbers can be represented in a third way, via the generating function $f(x) = \frac{x}{1-x-x^2}$. Namely, when this rational function is expanded in powers of x, the resulting coefficients generate the sequence values F_n:

$$\frac{x}{1-x-x^2} = 0x^0 + 1x^1 + 1x^2 + 2x^3 + 3x^4 + 5x^5 + 8x^6 + 13x^7 + \cdots.$$

2. Table 1 gives closed form expressions for the generating functions of several combinatorial sequences discussed in this *Handbook*. In this table, r is any real number. Generating functions for other sequences can be found in §3.2.1, Tables 1 and 2.

3. Table 2 gives closed form expressions for the exponential generating functions of several combinatorial sequences discussed in this *Handbook*. Generating functions for other sequences can be found in §3.2.2, Table 3.

4. Table 3 gives recurrence relations defining particular combinatorial sequences discussed in this *Handbook*.

Table 1 Generating functions for particular sequences.

sequence	notation	reference	closed form
$1, 2, 3, 4, 5, \ldots$	$\{n\}$		$\frac{1}{(1-x)^2}$
$1^2, 2^2, 3^2, 4^2, 5^2, \ldots$	$\{n^2\}$		$\frac{1+x}{(1-x)^3}$
$1^3, 2^3, 3^3, 4^3, 5^3, \ldots$	$\{n^3\}$		$\frac{1+4x+x^2}{(1-x)^4}$
$1, r, r^2, r^3, r^4, \ldots$	$\{r^n\}$		$\frac{1}{1-rx}$
Fibonacci	F_n	§3.1.2	$\frac{x}{1-x-x^2}$
Lucas	L_n	§3.1.2	$\frac{2-x}{1-x-x^2}$
Catalan	C_n	§3.1.3	$\frac{1-\sqrt{1-4x}}{2x}$
Harmonic	H_n	§3.1.7	$\frac{1}{1-x}\ln\frac{1}{1-x}$
Binomial	$\binom{m}{n}$	§2.3.2	$(1+x)^m$

Table 2 Exponential generating functions for particular sequences.

sequence	notation	reference	closed form
$1, 1, 1, 1, 1, \ldots$	$\{1\}$		e^x
$1, r, r^2, r^3, r^4, \ldots$	$\{r^n\}$		e^{rx}
Derangements	D_n	§2.4.2	$\frac{e^{-x}}{1-x}$
Bernoulli	B_n	§3.1.4	$\frac{x}{e^x-1}$
Tangent	T_n	§3.1.7	$\tan x$
Euler	E_n	§3.1.7	$\operatorname{sech} x$
Euler	$|E_n|$	§3.1.7	$\sec x$
Stirling cycle number	$\begin{bmatrix}n\\k\end{bmatrix}$	§2.5.2	$\frac{1}{k!}\left[\ln\frac{1}{(1-x)}\right]^k$
Stirling subset number	$\begin{Bmatrix}n\\k\end{Bmatrix}$	§2.5.2	$\frac{1}{k!}\left[e^x-1\right]^k$

3.1.2 FIBONACCI NUMBERS

Fibonacci numbers form an important sequence encountered in biology, physics, number theory, computer science, and combinatorics. [BePhHo88], [PhBeHo86], [Va89]

Definitions:

The **Fibonacci numbers** $F_0, F_1, F_2, \ldots$ are produced by the recursive definition $F_0 = 0$, $F_1 = 1$, $F_n = F_{n-1} + F_{n-2}$, $n \geq 2$.

A **generalized Fibonacci sequence** is any sequence $G_0, G_1, G_2, \ldots$ such that $G_n = G_{n-1} + G_{n-2}$ for $n \geq 2$.

The **Lucas numbers** $L_0, L_1, L_2, \ldots$ are produced by the recursive definition $L_0 = 2$, $L_1 = 1$, $L_n = L_{n-1} + L_{n-2}$, $n \geq 2$. (François Lucas, 1842–1891)

Table 3 Recurrence relations for particular sequences.

sequence	notation	reference	recurrence relation
Derangements	D_n	§2.4.2	$D_n = (n-1)(D_{n-1} + D_{n-2})$, $D_0 = 1, D_1 = 0$
Fibonacci	F_n	§3.1.2	$F_n = F_{n-1} + F_{n-2}$, $F_0 = 0, F_1 = 1$
Lucas	L_n	§3.1.2	$L_n = L_{n-1} + L_{n-2}$, $L_0 = 2, L_1 = 1$
Catalan	C_n	§3.1.3	$C_n = C_0 C_{n-1} + C_1 C_{n-2} + \cdots + C_{n-1} C_0$, $\quad C_0 = 1$
Bernoulli	B_n	§3.1.4	$\sum_{j=0}^{n} \binom{n+1}{j} B_j = 0$, $B_0 = 1$
Eulerian	$E(n, k)$	§3.1.5	$E(n, k) = (k+1)E(n-1, k) + (n-k)E(n-1, k-1)$, $E(n, 0) = 1, n \geq 1$
Binomial	$\binom{n}{k}$	§2.3.2	$\binom{n}{k} = \binom{n-1}{k} + \binom{n-1}{k-1}$, $\binom{n}{0} = 1, n \geq 0$
Stirling cycle number	$\left[\begin{smallmatrix}n\\k\end{smallmatrix}\right]$	§2.5.2	$\left[\begin{smallmatrix}n\\k\end{smallmatrix}\right] = (n-1)\left[\begin{smallmatrix}n-1\\k\end{smallmatrix}\right] + \left[\begin{smallmatrix}n-1\\k-1\end{smallmatrix}\right]$, $\left[\begin{smallmatrix}0\\0\end{smallmatrix}\right] = 1; \left[\begin{smallmatrix}n\\0\end{smallmatrix}\right] = 0, n \geq 1$
Stirling subset number	$\left\{\begin{smallmatrix}n\\k\end{smallmatrix}\right\}$	§2.5.2	$\left\{\begin{smallmatrix}n\\k\end{smallmatrix}\right\} = k\left\{\begin{smallmatrix}n-1\\k\end{smallmatrix}\right\} + \left\{\begin{smallmatrix}n-1\\k-1\end{smallmatrix}\right\}$, $\left\{\begin{smallmatrix}0\\0\end{smallmatrix}\right\} = 1; \left\{\begin{smallmatrix}n\\0\end{smallmatrix}\right\} = 0, n \geq 1$

Facts:

1. The Fibonacci numbers F_n and Lucas numbers L_n for $n = 0, 1, 2, \ldots, 50$ are shown in Table 4.

2. The Fibonacci numbers were initially studied by Leonardo of Pisa (c. 1170–1250), who was the son of Bonaccio; consequently these numbers have been called *Fibonacci* numbers after Leonardo, the son of Bonaccio (Filius Bonaccii).

3. $\lim_{n \to \infty} \frac{F_{n+1}}{F_n} = \lim_{n \to \infty} \frac{L_{n+1}}{L_n} = \frac{1}{2}(1 + \sqrt{5}) \approx 1.61803$, the *golden ratio*.

4. Fibonacci numbers arise in numerous applications in many different areas. For example, they occur in models of population growth of rabbits (Example 3), in modeling plant growth (Example 8), in counting the number of bit strings of length n without consecutive 0s (Example 13), in counting the number of spanning trees of wheel graphs of length n (Example 12), and in a vast number of other contexts. See [Va89] or other books concerning the Fibonacci numbers. There is a journal, the *Fibonacci Quarterly*, devoted to the study of the Fibonacci numbers and related topics. This is a tribute to how widely the Fibonacci numbers arise in mathematics and its applications to other areas. There are also a large number of books, available through the Fibonacci Association, devoted to the Fibonacci numbers and their use. This list can be found on the World Wide Web at

www.sdstate.edu/~wcsc/http/fibbooks.html

Table 4 Fibonacci and Lucas numbers.

n	F_n	L_n	n	F_n	L_n	n	F_n	L_n
0	0	2	17	55	3571	34	5,702,887	12,752,043
1	1	1	18	89	5778	35	9,227,465	20,633,239
2	1	3	19	144	9349	36	14,930,352	33,385,282
3	2	4	20	233	15127	37	24,157,817	54,018,521
4	3	7	21	377	24476	38	39,088,169	87,403,803
5	5	11	22	610	39603	39	63,245,986	141,422,324
6	8	18	23	987	64,079	40	102,334,155	228,826,127
7	13	29	24	1,597	103,682	41	165,580,141	370,248,451
8	21	47	25	2,584	167,761	42	267,914,296	599,074,578
9	34	76	26	4,181	271,443	43	433,494,437	969,323,029
10	55	123	27	6,765	439,204	44	701,408,733	1,568,397,607
11	89	199	28	10,946	710,647	45	1,134,903,170	2,537,720,636
12	144	322	29	17,711	1,149,851	46	1,836,311,903	4,106,118,243
13	233	521	30	28,657	1,860,498	47	2,971,215,073	6,643,838,879
14	377	843	31	46,368	3,010,349	48	4,807,526,976	10,749,957,122
15	610	1,364	32	75,025	4,870,847	49	7,778,742,049	17,393,796,001
16	987	2,207	33	121,393	7,881,196	50	12,586,269,025	28,143,753,123

5. Many properties of the Fibonacci numbers were derived by F. Lucas, who also is responsible for naming them the "Fibonacci" numbers.

6. *Binet form* (Jacques Binet, 1786–1856): If $\alpha = \frac{1}{2}(1+\sqrt{5})$ and $\beta = \frac{1}{2}(1-\sqrt{5})$ then
$$F_n = \frac{\alpha^n - \beta^n}{\sqrt{5}} = \frac{\alpha^n - \beta^n}{\alpha - \beta}, \qquad F_n \sim \frac{\alpha^n}{\sqrt{5}}.$$
Also,
$$L_n = \alpha^n + \beta^n, \qquad L_n \sim \alpha^n.$$

7. $F_n = \frac{1}{2}(F_{n-2} + F_{n+1})$ for all $n \geq 2$. That is, each Fibonacci number is the average of the terms occurring two places before and one place after it in the sequence.

8. $L_n = \frac{1}{2}(L_{n-2} + L_{n+1})$ for all $n \geq 2$. That is, each Lucas number is the average of the terms occurring two places before and one place after it in the sequence.

9. $F_0 + F_1 + F_2 + \cdots + F_n = F_{n+2} - 1$ for all $n \geq 0$.

10. $F_0 - F_1 + F_2 - \cdots + (-1)^n F_n = (-1)^n F_{n-1} - 1$ for all $n \geq 1$.

11. $F_1 + F_3 + F_5 + \cdots + F_{2n-1} = F_{2n}$ for all $n \geq 1$.

12. $F_0 + F_2 + F_4 + \cdots + F_{2n} = F_{2n+1} - 1$ for all $n \geq 0$.

13. $F_0^2 + F_1^2 + F_2^2 + \cdots + F_n^2 = F_n F_{n+1}$ for all $n \geq 0$.

14. $F_1 F_2 + F_2 F_3 + F_3 F_4 + \cdots + F_{2n-1} F_{2n} = F_{2n}^2$ for all $n \geq 1$.

15. $F_1 F_2 + F_2 F_3 + F_3 F_4 + \cdots + F_{2n} F_{2n+1} = F_{2n+1}^2 - 1$ for all $n \geq 1$.

16. If $k \geq 1$ then $F_{n+k} = F_k F_{n+1} + F_{k-1} F_n$ for all $n \geq 0$.

17. *Cassini's Identity:* $F_{n+1} F_{n-1} - F_n^2 = (-1)^n$ for all $n \geq 1$. (Jean Dominique Cassini, 1625–1712)

18. $F_{n+1}^2 + F_n^2 = F_{2n+1}$ for all $n \geq 0$.

19. $F_{n+2}^2 - F_{n+1}^2 = F_n F_{n+3}$ for all $n \geq 0$.

20. $F_{n+2}^2 - F_n^2 = F_{2n+2}$ for all $n \geq 0$.

21. $F_{n+2}^3 + F_{n+1}^3 - F_n^3 = F_{3n+3}$ for all $n \geq 0$.

22. $\gcd(F_n, F_m) = F_{\gcd(n,m)}$. This implies that F_n and F_{n+1} are relatively prime, and that F_k divides F_{nk}.

23. Fibonacci numbers arise as sums of diagonals in Pascal's triangle (§2.3.2):
$$F_{n+1} = \sum_{j=0}^{\lfloor n/2 \rfloor} \binom{n-j}{j} \text{ for all } n \geq 0.$$

24. $F_{3n} = \sum_{j=0}^{n} \binom{n}{j} 2^j F_j$ for all $n \geq 0$.

25. The Fibonacci sequence has the generating function $\frac{x}{1-x-x^2}$. (See §3.1.1.)

26. Fibonacci numbers with negative indices can be defined using the recursive definition $F_{n-2} = F_n - F_{n-1}$. Then $F_{-n} = (-1)^{n-1} F_n$, $n \geq 1$.

27. The units digits of the Fibonacci numbers form a sequence that repeats after 60 terms. (Joseph Lagrange, 1736–1813)

28. The number of binary strings of length n that contain no consecutive 0s is counted by F_{n+2}. (See §3.3.2, Example 12.)

29. $L_0 + L_1 + L_2 + \cdots + L_n = L_{n+2} - 1$ for all $n \geq 0$.

30. $L_0^2 + L_1^2 + L_2^2 + \cdots + L_n^2 = L_n L_{n+1} + 2$ for all $n \geq 0$.

31. $L_n = F_{n-1} + F_{n+1}$, $n \geq 1$. Hence, any formula containing Lucas numbers can be translated into a formula involving Fibonacci numbers.

32. The Lucas sequence $L_0, L_1, L_2, \ldots$ has generating function $\frac{2-x}{1-x-x^2}$. (See §3.1.1.)

33. Lucas numbers with negative indices can be defined by extending the recursive definition. Then $L_{-n} = (-1)^n L_n$ for all $n \geq 1$.

34. $F_n = \frac{L_{n-1} + L_{n+1}}{5}$, $n \geq 1$. Hence, any formula involving Fibonacci numbers can be translated into a formula involving Lucas numbers.

35. If $G_0, G_1, \ldots$ is a sequence of generalized Fibonacci numbers, then $F_n = F_{n-1} G_0 + F_n G_1$ for all $n \geq 1$.

Examples:

1. The Fibonacci number F_8 can be computed, using the initial values $F_0 = 0$ and $F_1 = 1$ and the recurrence relation $F_n = F_{n-1} + F_{n-2}$ repeatedly: $F_2 = F_1 + F_0 = 1 + 0 = 1$, $F_3 = F_2 + F_1 = 1 + 1 = 2$, $F_4 = F_3 + F_2 = 2 + 1 = 3$, $F_5 = F_4 + F_3 = 3 + 2 = 5$, $F_6 = F_5 + F_4 = 5 + 3 = 8$, $F_7 = F_6 + F_5 = 8 + 5 = 13$, $F_8 = F_7 + F_6 = 13 + 8 = 21$.

2. Each male bee (drone) is produced asexually from a female, whereas each female bee is produced from both a male and female. The ancestral tree for a single male bee is shown below. This male has one parent, two grandparents, three great grandparents, and in general F_{k+2} kth-order grandparents, $k \geq 0$.

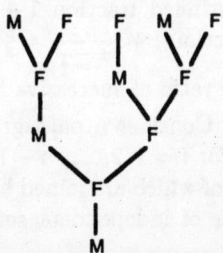

3. *Rabbit breeding*: This problem was originally posed by Fibonacci. A single pair of immature rabbits is introduced into a habitat. It takes two months before a pair of rabbits can breed; each month thereafter each pair of breeding rabbits produces another pair. At the start of months 1 and 2, only the original pair A is present. In the third month, A as well as their newly born pair B are present; in the fourth month, A, B as well as the new pair C (progeny of A) are present; in the fifth month, A, B, C as well as the new pairs D (progeny of A) and E (progeny of B) are present. If P_n is the number of pairs present in month n, then $P_1 = 1$, $P_2 = 1$, $P_3 = 2$, $P_4 = 3$, $P_5 = 5$. In general, P_n equals the number present in the previous month P_{n-1} plus the number of breeding pairs in the previous month (which is P_{n-2}, the number present two months earlier). Thus $P_n = F_n$ for $n \geq 1$.

4. Let S_n denote the number of subsets of $\{1, 2, \ldots, n\}$ that do not contain consecutive elements. For example, when $n = 3$ the allowable subsets are $\emptyset, \{1\}, \{2\}, \{3\}, \{1, 3\}$. Therefore, $S_3 = 5$. In general, $S_n = F_{n+2}$ for $n \geq 1$.

5. Draw n dots in a line. If each domino can cover exactly two such dots, in how many ways can (nonoverlapping) dominoes be placed? The following figure shows the number of possible solutions for $n = 2, 3, 4$. To find a general expression for D_n, the number of possible placements of dominoes with n dots, consider the rightmost dot in any such placement P. If this dot is not covered by a domino, then P minus the last dot determines a solution counted by D_{n-1}. If the last dot is covered by a domino, then the last two dots in P are covered by this domino. Removing this rightmost domino then gives a solution counted by D_{n-2}. Taking into account these two possibilities $D_n = D_{n-1} + D_{n-2}$ for $n \geq 3$ with $D_1 = 1$, $D_2 = 2$. Thus $D_n = F_{n+1}$ for $n \geq 1$.

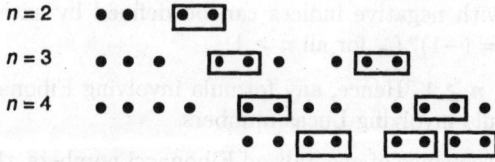

6. *Compositions*: Let T_n be the number of ordered compositions (§2.5.1) of the positive integer n into summands that are odd. For example, $4 = 1 + 3 = 3 + 1 = 1 + 1 + 1 + 1$ and $5 = 5 = 1 + 1 + 3 = 1 + 3 + 1 = 3 + 1 + 1 = 1 + 1 + 1 + 1 + 1$. Therefore, $T_4 = 3$ and $T_5 = 5$. In general, $T_n = F_n$ for $n \geq 1$.

7. *Compositions*: Let B_n be the number of ordered compositions (§2.5.1) of the positive integer n into summands that are either 1 or 2. For example, $3 = 1 + 2 = 2 + 1 = 1 + 1 + 1$ and $4 = 2 + 2 = 1 + 1 + 2 = 1 + 2 + 1 = 2 + 1 + 1 = 1 + 1 + 1 + 1$. Therefore, $B_3 = 3$ and $B_4 = 5$. In general, $B_n = F_{n+1}$ for $n \geq 1$.

8. *Botany*: It has been observed in pine cones (and other botanical structures) that the number of rows of scales winding in one direction is a Fibonacci number while the number of rows of scales winding in the other direction is an adjacent Fibonacci number.

9. *Continued fractions*: The continued fraction $1 + \frac{1}{1} = \frac{2}{1}$, the continued fraction $1 + \frac{1}{1+\frac{1}{1}} = \frac{3}{2}$ and the continued fraction $1 + \frac{1}{1+\frac{1}{1+\frac{1}{1}}} = \frac{5}{3}$. In general, a continued fraction composed entirely of 1s equals the ratio of successive Fibonacci numbers.

10. *Independent sets on a path*: Consider a path graph on vertices $1, 2, \ldots, n$, with edges joining vertices i and $i + 1$ for $i = 1, 2, \ldots, n - 1$. An independent set of vertices (§8.6.3) consists of vertices no two of which are joined by an edge. By an analysis similar to that in Example 5, the number of independent sets in a path graph on n vertices equals F_{n+2}.

11. *Independent sets on a cycle*: Consider a cycle graph on vertices $1, 2, \ldots, n$, with edges joining vertices i and $i+1$ for $i = 1, 2, \ldots, n-1$ as well as vertices n and 1. Then the number of independent sets (§8.6.3) in a cycle graph on n vertices equals L_n.

12. *Spanning trees*: The number of spanning trees of the wheel graph W_n (§8.1.3) equals $L_{2n} - 2$.

13. If A is the 2×2 matrix $\begin{pmatrix} 1 & 1 \\ 1 & 0 \end{pmatrix}$, then $A^n = \begin{pmatrix} F_{n+1} & F_n \\ F_n & F_{n-1} \end{pmatrix}$ for $n \geq 1$.

3.1.3 CATALAN NUMBERS

The sequence of integers called the Catalan numbers arises in counting a variety of combinatorial structures, such as voting sequences, certain types of binary trees, paths in the plane, and triangulations of polygons.

Definitions:

The **Catalan numbers** $C_0, C_1, C_2, \ldots$ satisfy the nonlinear recurrence relation $C_n = C_0 C_{n-1} + C_1 C_{n-2} + \cdots + C_{n-1} C_0$, $n \geq 1$, with $C_0 = 1$. (See §3.3.1, Example 9.) (Eugène Catalan, 1814–1894)

Well-formed (or balanced) sequences of parentheses of length $2n$ are defined recursively as follows: the empty sequence is well-formed; if sequence A is well-formed so is (A); if sequences A and B are well-formed so is AB.

Well-parenthesized products of variables are defined recursively as follows: single variables are well-parenthesized; if A and B are well-parenthesized so is (AB).

Facts:

1. The first 12 Catalan numbers C_n are given in the following table.

n	0	1	2	3	4	5	6	7	8	9	10	11
C_n	1	1	2	5	14	42	132	429	1,430	4,862	16,796	58,786

2. $\lim_{n \to \infty} \frac{C_{n+1}}{C_n} = 4$.

3. Catalan numbers arise in a variety of applications, such as when binary trees on n vertices, triangulations of a convex n-gon, and well-formed sequences of n left and n right parentheses are counted. See the examples below as well as [MiRo91].

4. $C_n = \frac{1}{n+1} \binom{2n}{n}$ for all $n \geq 0$.

5. The Catalan numbers $C_0, C_1, C_2, \ldots$ have the generating function $\frac{1 - \sqrt{1-4x}}{2x}$.

6. $C_n \sim \frac{4^n}{\sqrt{\pi n^3}}$.

7. $C_n = \binom{2n}{n} - \binom{2n}{n-1} = \binom{2n-1}{n} - \binom{2n-1}{n+1}$ for all $n \geq 1$.

8. $C_{n+1} = \frac{2(2n+1)}{n+2} C_n$ for all $n \geq 0$.

Examples:

1. The number of binary trees (§9.1.2) on n vertices is C_n.
2. The number of left-right binary trees (§9.3.3) on $2n + 1$ vertices is C_n.
3. The number of ordered trees (§9.1.2) on n vertices is C_{n-1}.

4. Suppose that a coin is tossed $2n$ times, coming up heads exactly n times and tails exactly n times. The number of sequences of tosses in which the cumulative number of heads is always at least as large as the cumulative number of tails is C_n. For example, when $n = 3$ there are $C_3 = 5$ such sequences of 6 tosses: HTHTHT, HTHHTT, HHTTHT, HHTHTT, HHHTTT.

5. In Example 4, the number of sequences of tosses in which the cumulative number of heads always exceeds the cumulative number of tails (until the very last toss) is C_{n-1}. For example, when $n = 3$ there are $C_2 = 2$ such sequences of 6 tosses: HHTHTT, HHHTTT.

6. *Triangulations*: Let T_n be the number of triangulations of a convex n-gon, using $n-3$ nonintersecting diagonals. For instance, the following figure shows the $T_5 = 5$ triangulations of a pentagon. In general, $T_n = C_{n-2}$ for $n \geq 3$.

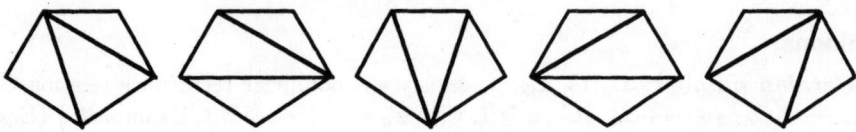

7. Suppose that $2n$ points are placed in fixed positions, evenly distributed on the circumference of a circle. Then there are C_n ways to join n pairs of the points so that the resulting chords do not intersect. The following figure shows the $C_3 = 5$ solutions for $n = 3$.

    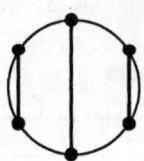

8. *Well-formed sequences of parentheses*: The sequence of parentheses (( ) ( )) involving three left and three right parentheses is well-formed, whereas the sequence ( )) (( ) is not syntactically meaningful. There are five such well-formed sequences in this case:

$$( )( )( ), \ ( )(( )), \ (( ))( ), \ (( )( )), \ ((( ))).$$

Notice that if each left parenthesis is replaced by a H and each right parenthesis by a T, then these five balanced sequences correspond exactly to the five coin tossing sequences listed in Example 4. In general, the number of balanced sequences involving n left and n right parentheses is C_n.

9. Consider the following procedure composed of n nested **for** loops:

    count := 0
    for $i_1 := 1$ **to** 1
      for $i_2 := 1$ **to** $i_1 + 1$
        for $i_3 := 1$ **to** $i_2 + 1$
          ⋮
          for $i_n := 1$ **to** $i_{n-1} + 1$
            count := *count* + 1

Then the value of *count* upon exit from this procedure is C_n.

10. *Well-parenthesized products*: The product $x_1 x_2 x_3$ (relative to some binary "multiplication" operation) can be evaluated as either $(x_1 x_2) x_3$ or $x_1 (x_2 x_3)$. In the former, x_1 and x_2 are first combined and then the result is combined with x_3. In the latter, x_2 and x_3 are first combined and then the result is combined with x_1. Let P_n indicate the number of ways to evaluate the product $x_1 x_2 \ldots x_n$ of n variables, using a binary operation. Note that $P_3 = 2$. In general, $P_n = C_{n-1}$. This was the problem originally studied by Catalan. (See §3.3.1, Example 9.)

11. The numbers $1, 2, \ldots, 2n$ are to be placed in the $2n$ positions of an $2 \times n$ array $A = (a_{ij})$. Such an arrangement is *monotone* if the values increase within each row and within each column. Then there are C_n ways to form a monotone $2 \times n$ array containing the entries $1, 2, \ldots, 2n$. For instance, the following is one of the $C_4 = 14$ monotone 2×4 arrays:

$$A = \begin{pmatrix} 1 & 3 & 5 & 6 \\ 2 & 4 & 7 & 8 \end{pmatrix}.$$

3.1.4 BERNOULLI NUMBERS AND POLYNOMIALS

The Bernoulli numbers are important in obtaining closed form expressions for the sums of powers of integers. These numbers also arise in expansions involving other combinatorial sequences.

Definitions:

The **Bernoulli numbers** B_n satisfy the recurrence relation $\sum_{j=0}^{n} \binom{n+1}{j} B_j = 0$ for all $n \geq 1$, with $B_0 = 1$. (Jakob Bernoulli, 1654–1705)

The **Bernoulli polynomials** $B_m(x)$ are given by $B_m(x) = \sum_{k=0}^{m} \binom{m}{k} B_k x^{m-k}$.

Facts:

1. The first 14 Bernoulli numbers B_n are shown in the following table.

n	0	1	2	3	4	5	6	7	8	9	10	11	12	13
B_n	1	$-\frac{1}{2}$	$\frac{1}{6}$	0	$-\frac{1}{30}$	0	$\frac{1}{42}$	0	$-\frac{1}{30}$	0	$\frac{5}{66}$	0	$-\frac{691}{2730}$	0

2. $B_{2k+1} = 0$ for all $k \geq 1$.
3. The nonzero Bernoulli numbers alternate in sign.
4. $B_n = B_n(0)$.
5. The Bernoulli numbers have the exponential generating function $\sum_{n=0}^{\infty} B_n \frac{x^n}{n!} = \frac{x}{e^x - 1}$.
6. The Bernoulli numbers can be expressed in terms of the Stirling subset numbers (§2.5.2): $B_n = \sum_{j=0}^{n} (-1)^j \{{n \atop j}\} \frac{j!}{j+1}$ for all $n \geq 0$.
7. The Bernoulli numbers appear as coefficients in the Maclaurin expansion of $\tan x$, $\cot x$, $\csc x$, $\tanh x$, $\coth x$, and $\operatorname{csch} x$.
8. The Bernoulli polynomials can be used to obtain closed form expressions for the sum of powers of the first n positive integers. (See §3.5.4.)
9. The first 14 Bernoulli polynomials $B_m(x)$ are shown in Table 5.
10. $\int_0^1 B_m(x)\, dx = 0$ for all $m \geq 1$.

Table 5 Bernoulli polynomials.

n	$B_n(x)$
0	1
1	$x - \frac{1}{2}$
2	$x^2 - x + \frac{1}{6}$
3	$x^3 - \frac{3}{2}x^2 + \frac{1}{2}x$
4	$x^4 - 2x^3 + x^2 - \frac{1}{30}$
5	$x^5 - \frac{5}{2}x^4 + \frac{5}{3}x^3 - \frac{1}{6}x$
6	$x^6 - 3x^5 + \frac{5}{2}x^4 - \frac{1}{2}x^2 + \frac{1}{42}$
7	$x^7 - \frac{7}{2}x^6 + \frac{7}{2}x^5 - \frac{7}{6}x^3 + \frac{1}{6}x$
8	$x^8 - 4x^7 + \frac{14}{3}x^6 - \frac{7}{3}x^4 + \frac{2}{3}x^2 - \frac{1}{30}$
9	$x^9 - \frac{9}{2}x^8 + 6x^7 - \frac{21}{5}x^5 + 2x^3 - \frac{3}{10}x$
10	$x^{10} - 5x^9 + \frac{15}{2}x^8 - 7x^6 + 5x^4 - \frac{3}{2}x^2 + \frac{5}{66}$
11	$x^{11} - \frac{11}{2}x^{10} + \frac{55}{6}x^9 - 11x^7 + 11x^5 - \frac{11}{2}x^3 + \frac{5}{6}x$
12	$x^{12} - 6x^{11} + 11x^{10} - \frac{33}{2}x^8 + 22x^6 - \frac{33}{2}x^4 + 5x^2 - \frac{691}{2730}$
13	$x^{13} - \frac{13}{2}x^{12} + 13x^{11} - \frac{143}{6}x^9 + \frac{286}{7}x^7 - \frac{429}{10}x^5 + \frac{65}{3}x^3 - \frac{691}{210}x$

11. $\frac{dB_m(x)}{dx} = mB_{m-1}(x)$ for all $m \geq 1$.

12. $B_{m+1}(x+1) - B_{m+1}(x) = (m+1)x^m$ for all $m \geq 0$.

13. The Bernoulli polynomials have the following exponential generating function:
$\sum_{m=0}^{\infty} B_m(x)\frac{t^m}{m!} = \frac{te^{xt}}{e^t - 1}$.

3.1.5 EULERIAN NUMBERS

Eulerian numbers are important in counting numbers of permutations with certain numbers of increases and decreases.

Definitions:

Let $\pi = (\pi_1, \pi_2, \ldots, \pi_n)$ be a permutation of $\{1, 2, \ldots, n\}$.

An **ascent** of the permutation π is any index i $(1 \leq i < n)$ such that $\pi_i < \pi_{i+1}$. A **descent** of the permutation π is any index i $(1 \leq i < n)$ such that $\pi_i > \pi_{i+1}$.

An **excedance** of the permutation π is any index i $(1 \leq i \leq n)$ such that $\pi_i > i$. A **weak excedance** of the permutation π is any index i $(1 \leq i \leq n)$ such that $\pi_i \geq i$.

The **Eulerian number** $E(n, k)$ (also written $\langle {n \atop k} \rangle$) is the number of permutations of $\{1, 2, \ldots, n\}$ with exactly k ascents.

Facts:

1. $E(n, k)$ is the number of permutations of $\{1, 2, \ldots, n\}$ with exactly k descents.
2. $E(n, k)$ is the number of permutations of $\{1, 2, \ldots, n\}$ with exactly k excedances.
3. $E(n, k)$ is the number of permutations of $\{1, 2, \ldots, n\}$ with exactly $k + 1$ weak excedances.

4. The Eulerian numbers can be used to obtain closed form expressions for the sum of powers of the first n positive integers (§3.5.4).

5. Eulerian numbers $E(n,k)$ $(1 \leq n \leq 10, 0 \leq k \leq 8)$ are given in the following table.

$n \backslash k$	0	1	2	3	4	5	6	7	8
1	1								
2	1	1							
3	1	4	1						
4	1	11	11	1					
5	1	26	66	26	1				
6	1	57	302	302	57	1			
7	1	120	1,191	2,416	1,191	120	1		
8	1	247	4,293	15,619	15,619	4,293	247	1	
9	1	502	14,608	88,234	156,190	88,234	14,608	502	1
10	1	1,013	47,840	455,192	1,310,354	1,310,354	455,192	47,840	1,013

6. $E(n,0) = E(n, n-1) = 1$ for all $n \geq 1$.

7. Symmetry: $E(n,k) = E(n, n-1-k)$ for all $n \geq 1$.

8. $E(n,k) = (k+1)E(n-1,k) + (n-k)E(n-1, k-1)$ for all $n \geq 2$.

9. $\sum_{k=0}^{n-1} E(n,k) = n!$ for all $n \geq 1$.

10. Worpitzky's identity: $x^n = \sum_{k=0}^{n-1} E(n,k) \binom{x+k}{n}$ for all $n \geq 1$. (Julius Worpitzky, 1835–1895)

11. $E(n,k) = \sum_{j=0}^{k} (-1)^j \binom{n+1}{j}(k+1-j)^n$ for all $n \geq 1$.

12. The Bernoulli numbers (§3.1.4) can be expressed as alternating sums of Eulerian numbers: $B_m = \frac{m}{2^m(2^m-1)} \sum_{k=0}^{m-2} (-1)^k E(m-1, k)$ for $m \geq 2$.

13. The Stirling subset numbers (§2.5.2) can be expressed in terms of the Eulerian numbers: $\left\{ {n \atop m} \right\} = \frac{1}{m!} \sum_{k=0}^{n-1} E(n,k) \binom{k}{n-m}$ for $n \geq m$ and $n \geq 1$.

14. The Eulerian numbers have the following (bivariate) generating function in variables x, t: $\sum_{m=0}^{\infty} \sum_{n=0}^{\infty} E(n,m) x^m \frac{t^n}{n!} = \frac{1-x}{e^{(x-1)t} - x}$.

Examples:

1. The permutation $\pi = (\pi_1, \pi_2, \pi_3, \pi_4) = (1, 2, 3, 4)$ has three ascents since $1 < 2 < 3 < 4$ and it is the only permutation in S_4 with three ascents; note that $E(4,3) = 1$. There are $E(4,1) = 11$ permutations in S_4 with one ascent: $(1,4,3,2)$, $(2,1,4,3)$, $(2,4,3,1)$, $(3,1,4,2)$, $(3,2,1,4)$, $(3,2,4,1)$, $(3,4,2,1)$, $(4,1,3,2)$, $(4,2,1,3)$, $(4,2,3,1)$, and $(4,3,1,2)$.

2. The permutation $\pi = (2, 4, 3, 1)$ has two excedances since $2 > 1$ and $4 > 2$. There are $E(4,2) = 11$ such permutations in S_4.

3. The permutation $\pi = (1,3,2)$ has two weak excedances since $1 \geq 1$ and $3 \geq 2$. There are $E(3,1) = 4$ such permutations in S_3: $(1,3,2)$, $(2,1,3)$, $(2,3,1)$, and $(3,2,1)$.
4. When $n = 3$ Worpitzky's identity (Fact 10) states that
$$x^3 = E(3,0)\binom{x}{3} + E(3,1)\binom{x+1}{3} + E(3,2)\binom{x+2}{3} = \binom{x}{3} + 4\binom{x+1}{3} + \binom{x+2}{3}.$$
This is verified algebraically since $\binom{x}{3} + 4\binom{x+1}{3} + \binom{x+2}{3} = \frac{1}{6}\bigl(x(x-1)(x-2) + 4(x+1)x(x-1) + (x+2)(x+1)x\bigr) = \frac{x}{6}(x^2 - 3x + 2 + 4x^2 - 4 + x^2 + 3x + 2) = \frac{x}{6}(6x^2) = x^3$.

3.1.6 RAMSEY NUMBERS

The Ramsey numbers arise from the work of Frank P. Ramsey (1903–1930), who in 1930 published a paper [Ra30] dealing with set theory that generalized the pigeonhole principle. (Also see §8.11.2.) [GrRoSp80], [MiRo91], [Ro84]

Definitions:

The **Ramsey number** $R(m,n)$ is the smallest positive integer k with the following property: if S is a set of size k and the 2-element subsets of S are partitioned into 2 collections, C_1 and C_2, then there is a subset of S of size m such that each of its 2-element subsets belong to C_1 or there is a subset of S of size n such that each of its 2-element sets belong to C_2.

The **Ramsey number** $R(m_1, \ldots, m_n; r)$ is the smallest positive integer k with the following property: if S is a set of size k and the r-element subsets of S are partitioned into n collections $C_1, \ldots, C_n$, then for some j there is a subset of S of size m_j such that each of its r-element subsets belong to C_j.

The **Schur number** $S(n)$ is the smallest integer k with the following property: if $\{1, \ldots, k\}$ is partitioned into n subsets $A_1, \ldots, A_n$, then there is a subset A_i such that the equation $x + y = z$ has a solution where $x, y, z \in A_i$. (Issai Schur, 1875–1941)

Facts:

1. *Ramsey's theorem*: The Ramsey numbers $R(m,n)$ and $R(m_1, \ldots, m_n; r)$ are well-defined for all $m, n \geq 1$ and for all $m_1, \ldots, m_n \geq 1$, $r \geq 1$.
2. Ramsey numbers $R(m,n)$ can be phrased in terms of coloring edges of the complete graphs K_n: the Ramsey number $R(m,n)$ is the smallest positive integer k such that, if each edge of K_k is colored red or blue, then either the red subgraph contains a copy of K_m or else the blue subgraph contains a copy of K_n. (See §8.11.2.)
3. *Symmetry*: $R(m,n) = R(n,m)$.
4. $R(m,1) = R(1,m) = 1$ for every $m \geq 1$.
5. $R(m,2) = R(2,m) = m$ for every $m \geq 1$.
6. The values of few Ramsey numbers are known. What is currently known about Ramsey numbers $R(m,n)$, for $3 \leq m \leq 10$ and $3 \leq n \leq 10$, and bounds on other Ramsey numbers are displayed in Table 6.
7. If $m_1 \leq m_2$ and $n_1 \leq n_2$, then $R(m_1, n_1) \leq R(m_2, n_2)$.
8. $R(m,n) \leq R(m, n-1) + R(n-1, m)$ for all $m, n \geq 2$.
9. If $m \geq 3$, $n \geq 3$, and if $R(m, n-1)$ and $R(m-1, n)$ are even, then $R(m,n) \leq R(m, n-1) + R(m-1, n) - 1$.
10. $R(m,n) \leq \binom{m+n-2}{m-1}$. (Erdős and Szekeres, 1935)

Table 6 Some classical Ramsey numbers.

The entries in the body of this table are $R(m,n)$ ($m,n \leq 10$) when known, or the best known range $r_1 \leq R(m,n) \leq r_2$ when not known. The Ramsey numbers $R(3,3)$, $R(3,4)$, $R(3,5)$, and $R(4,4)$ were found by A. M. Gleason and R. E. Greenwood in 1955; $R(3,6)$ was found by J. G. Kalbfleisch in 1966; $R(3,7)$ was found by J. E. Graver and J. Yackel in 1968; $R(3,8)$ was found by B. McKay and Z. Ke Min; $R(3,9)$ was found by C. M. Grinstead and S. M. Roberts in 1982; $R(4,5)$ was found by B. McKay and S. Radziszowski in 1993.

$m \backslash n$	3	4	5	6	7	8	9	10
3	6	9	14	18	23	28	36	40-43
4	–	18	25	35-41	49-61	55-84	69-115	80-149
5	–	–	43-49	58-87	80-143	95-216	116-316	141-442
6	–	–	–	102-165	109-298	122-495	153-780	167-1,171
7	–	–	–	–	205-540	216-1,031	227-1,713	238-2,826
8	–	–	–	–	–	282-1,870	295-3,583	308-6,090
9	–	–	–	–	–	–	565-6,625	580-12,715
10	–	–	–	–	–	–	–	798-23,854

Bounds for $R(m,n)$ for $m = 3$ and 4, with $11 \leq n \leq 15$:

$46 \leq R(3,11) \leq 51$ $96 \leq R(4,11) \leq 191$
$52 \leq R(3,12) \leq 60$ $128 \leq R(4,12) \leq 238$
$59 \leq R(3,13) \leq 69$ $131 \leq R(4,13) \leq 291$
$66 \leq R(3,14) \leq 78$ $136 \leq R(4,14) \leq 349$
$73 \leq R(3,15) \leq 89$ $145 \leq R(4,15) \leq 417$

11. The Ramsey numbers $R(m,n)$ satisfy the following asymptotic relationship: $\frac{\sqrt{2}}{e}(1+o(1))m2^{m/2} \leq R(m,m) \leq \binom{2m+2}{m+1} \cdot O((\log m)^{-1})$.
12. There exist constants c_1 and c_2 such that $c_1 m \ln m \leq R(3,m) \leq c_2 m \ln m$.
13. The problem of finding the Ramsey numbers $R(m_1,\ldots,m_n;2)$ can be phrased in terms of coloring edges of the complete graphs K_n. $R(m_1,\ldots,m_n;2)$ is equal to the smallest positive integer k with the following property: no matter how the edges of K_k are colored with the n colors $1, 2, \ldots, n$, there is some j such that K_k has a subgraph K_{m_j} of color j. (The edges of K_k are the 2-element subsets; C_j is the set of edges of color j.)
14. $R(m_1, m_2; 2) = R(m_1, m_2)$.
15. Very little is also known about the numbers $R(m_1, \ldots, m_n; 2)$ if $n \geq 3$.
16. $R(2,\ldots,2;2) = 2$.
17. If each $m_i \geq 3$, the only Ramsey number whose value is known is $R(3,3,3;2) = 17$.
18. $R(m, r, r, \ldots, r; r) = m$ if $m \geq r$.
19. $R(m_1, \ldots m_n; 1) = m_1 + \cdots + m_n - (n-1)$.
20. Ramsey theory is a generalization of the pigeonhole principle. In the terminology of Ramsey numbers, the fact that $R(2,\ldots,2;1) = n+1$ means that $n+1$ is the smallest positive integer with the property that if S has size $n+1$ and the subsets of S are partitioned into n sets $C_1, \ldots, C_n$, then for some j there is a subset of S of size 2 such that each of its elements belong to C_j. Hence, some C_j has at least 2 elements. If S

is a set of $n+1$ pigeons and the subset C_j ($j = 1, \ldots, n$) is the set of pigeons roosting in pigeonhole j, then some pigeonhole must have at least 2 pigeons in it. The Ramsey numbers $R(2, \ldots, 2; 1)$ give the smallest number of pigeons that force at least 2 to roost in the same pigeonhole.

21. *Schur's theorem:* $S(k) \leq R(3, \ldots, 3; 2)$ (where there are k 3s in the notation for the Ramsey number).

22. The following Schur numbers are known: $S(1) = 2$, $S(2) = 5$, $S(3) = 14$.

23. The equation $x + y = z$ in the definition of Schur numbers has been generalized to equations of the form $x_1 + \cdots + x_{n-1} = x_n$, $n \geq 4$. [BeBr82].

24. *Convex sets:* Ramsey numbers play a role in constructing convex polygons. Suppose m is a positive integer and there are n given points, no three of which are collinear. If $n \geq R(m, 5; 4)$, then a convex m-gon can be obtained from m of the n points [ErSz35]. This paper provided the impetus for the study of Ramsey numbers and suggested the possibility of its wide applicability in mathematics.

25. It remains an unsolved problem to find the smallest integer x (which depends on m) such that if $n \geq x$, then a convex m-gon can be obtained from m of the n points.

26. Extensive information on Ramsey number theory, including bounds on Ramsey numbers, can be found at S. Radziszowski's web site:
    http://www.cs.rit.edu/~spr/homepage.html

Examples:

1. If six people are at a party, then either three of these six are mutual friends or three are mutual strangers. If six is replaced by five, the result is not true. These facts follow since $R(3,3) = 6$. (See Fact 2. The six people can be regarded as vertices, with a red edge joining friends and a blue edge joining strangers.)

2. If the set $\{1, \ldots, k\}$ is partitioned into two subsets A_1 and A_2, then the equation $x + y = z$ may or may not have a solution where $x, y, z \in A_1$ or $x, y, z \in A_2$. If $k \geq 5$, a solution is guaranteed since $S(2) = 5$. If $k < 5$, no solution is guaranteed — take $A_1 = \{1, 4\}$ and $A_2 = \{2, 3\}$.

3.1.7 OTHER SEQUENCES

Additional sequences that regularly arise in discrete mathematics are described in this section.

▷ Euler Polynomials

Definition:

The *Euler polynomials* $E_n(x)$ have the exponential generating function $\sum_{n=0}^{\infty} E_n(x) \frac{t^n}{n!} = \frac{2e^{xt}}{e^t + 1}$.

Facts:

1. The first 14 Euler polynomials $E_n(x)$ are shown in Table 7.

2. $E_n(x+1) + E_n(x) = 2x^n$ for all $n \geq 0$.

3. The Euler polynomials can be expressed in terms of the Bernoulli numbers (§3.1.4): $E_{n-1}(x) = \frac{1}{n} \sum_{k=1}^{n} (2 - 2^{k+1}) \binom{n}{k} B_k x^{n-k}$ for all $n \geq 1$.

Table 7 Euler polynomials.

n	$E_n(x)$
0	1
1	$x - \frac{1}{2}$
2	$x^2 - x$
3	$x^3 - \frac{3}{2}x^2 + \frac{1}{4}$
4	$x^4 - 2x^3 + x$
5	$x^5 - \frac{5}{2}x^4 + \frac{5}{2}x^2 - \frac{1}{2}$
6	$x^6 - 3x^5 + 5x^3 - 3x$
7	$x^7 - \frac{7}{2}x^6 + \frac{35}{4}x^4 - \frac{21}{2}x^2 + \frac{17}{8}$
8	$x^8 - 4x^7 + 14x^5 - 28x^3 + 17x$
9	$x^9 - \frac{9}{2}x^8 + 21x^6 - 63x^4 + \frac{153}{2}x^2 - \frac{31}{2}$
10	$x^{10} - 5x^9 + 30x^7 - 126x^5 + 255x^3 - 155x$
11	$x^{11} - \frac{11}{2}x^{10} + \frac{165}{4}x^8 - 231x^6 + \frac{2805}{4}x^4 - \frac{1705}{2}x^2 + \frac{691}{4}$
12	$x^{12} - 6x^{11} + 55x^9 - 396x^7 + 1683x^5 - 3410x^3 + 2073x$
13	$x^{13} - \frac{13}{2}x^{12} + \frac{143}{2}x^{10} - \frac{1287}{2}x^8 + \frac{7293}{2}x^6 - \frac{22165}{2}x^4 + \frac{26949}{2}x^2 - \frac{5461}{2}$

4. The alternating sum of powers of the first n integers can be expressed in terms of the Euler polynomials: $\sum_{j=1}^{n}(-1)^{n-j}j^k = \frac{1}{2}\left[E_k(n+1) + (-1)^n E_k(0)\right]$.

▷ **Euler and Tangent Numbers**

Definitions:

The **Euler numbers** E_n are given by $E_n = 2^n E_n(\frac{1}{2})$, where $E_n(x)$ is an Euler polynomial.

The **tangent numbers** T_n have the exponential generating function $\tan x$: $\sum_{n=0}^{\infty} T_n \frac{x^n}{n!} = \tan x$.

Facts:

1. The first twelve Euler numbers E_n and tangent numbers T_n are shown in the following table.

n	0	1	2	3	4	5	6	7	8	9	10	11
E_n	1	0	-1	0	5	0	-61	0	1,385	0	-50,521	0
T_n	0	1	0	2	0	16	0	272	0	7,936	0	353,792

2. $E_{2k+1} = T_{2k} = 0$ for all $k \geq 0$.
3. The nonzero Euler numbers alternate in sign.

4. The Euler numbers have the exponential generating function $\frac{2}{e^t+e^{-t}} = \operatorname{sech} t$.
5. The exponential generating function for $|E_n|$ is $\sum_{n=0}^{\infty} |E_n| \frac{t^n}{n!} = \sec t$.
6. The tangent numbers can be expressed in terms of the Bernoulli numbers (§3.1.4): $T_{2n-1} = (-1)^{n-1} \frac{4^n(4^n-1)}{2n} B_{2n}$ for all $n \geq 1$.
7. The tangent numbers can be expressed as an alternating sum of Eulerian numbers (§3.1.5): $T_{2n+1} = \sum_{k=0}^{2n} (-1)^{n-k} E(2n+1,k)$ for all $n \geq 0$.
8. $(-1)^n E_{2n}$ counts the number of *alternating* permutations in S_{2n}: that is, the number of permutations $\pi = (\pi_1, \pi_2, \ldots, \pi_{2n})$ on $\{1, 2, \ldots, 2n\}$ with $\pi_1 > \pi_2 < \pi_3 > \pi_4 < \cdots > \pi_{2n}$.
9. T_{2n+1} counts the number of alternating permutations in S_{2n+1}.

Examples:

1. The permutation $\pi = (\pi_1, \pi_2, \pi_3, \pi_4) = (2, 1, 4, 3)$ is alternating since $2 > 1 < 4 > 3$. In all there are $(-1)^2 E_4 = 5$ alternating permutations in S_4: $(2,1,4,3)$, $(3,1,4,2)$, $(3,2,4,1)$, $(4,1,3,2)$, $(4,2,3,1)$.
2. The permutation $\pi = (\pi_1, \pi_2, \pi_3, \pi_4, \pi_5) = (4, 1, 3, 2, 5)$ is alternating since $4 > 1 < 3 > 2 < 5$. In all there are $T_5 = 16$ alternating permutations in S_5.

▷ **Harmonic Numbers**

Definition:

The **harmonic numbers** H_n are given by $H_n = \sum_{i=1}^n \frac{1}{i}$ for $n \geq 0$, with $H_0 = 0$.

Facts:

1. H_n is the discrete analogue of the natural logarithm (§3.4.1).
2. The first twelve harmonic numbers H_n are shown in the following table.

n	0	1	2	3	4	5	6	7	8	9	10	11
H_n	0	1	$\frac{3}{2}$	$\frac{11}{6}$	$\frac{25}{12}$	$\frac{137}{60}$	$\frac{49}{20}$	$\frac{363}{140}$	$\frac{761}{280}$	$\frac{7,129}{2,520}$	$\frac{7,381}{2,520}$	$\frac{83,711}{27,720}$

3. The harmonic numbers can be expressed in terms of the Stirling cycle numbers (§2.5.2): $H_n = \frac{1}{n!} \left[{n+1 \atop 2} \right]$, $n \geq 1$.
4. $\sum_{i=1}^n H_i = (n+1)[H_{n+1} - 1]$ for all $n \geq 1$.
5. $\sum_{i=1}^n i H_i = \binom{n+1}{2}[H_{n+1} - \frac{1}{2}]$ for all $n \geq 1$.
6. $\sum_{i=1}^n \binom{i}{k} H_i = \binom{n+1}{k+1}[H_{n+1} - \frac{1}{k+1}]$ for all $n \geq 1$.
7. $H_n \to \infty$ as $n \to \infty$.
8. $H_n \sim \ln n + \gamma + \frac{1}{2n} - \frac{1}{12n^2} + \frac{1}{120n^4}$, where $\gamma \approx 0.57721\ 56649\ 01533$ denotes Euler's constant.
9. The harmonic numbers have the generating function $\frac{1}{1-x} \ln \frac{1}{1-x}$.

Example:
1. Fact 8 yields the approximation $H_{10} \approx 2.928968257896$. The actual value is $H_{10} = 2.928968253968\ldots$, so the approximation is accurate to 9 significant digits. The approximation $H_{20} \approx 3.597739657206$ is accurate to 10 digits, and the approximation $H_{40} \approx 4.27854303893$ is accurate to 12 digits.

▷ **Gray Codes**

Definition:

A *Gray code* of size n is an ordering $G^n = (g_1, g_2, \ldots, g_{2^n})$ of the 2^n binary strings of length n such that g_k and g_{k+1} differ in exactly one bit, for $1 \leq k < 2^n$. Usually it is required that g_{2^n} and g_1 also differ in exactly one bit.

Facts:
1. Gray codes exist for all $n \geq 1$. Sample Gray codes G^n are shown in this table.

n	G^n							
1	0	1						
2	00	10	11	01				
3	000	100	110	010	011	111	101	001
4	0000	1000	1100	0100	0110	1110	1010	0010	0011
	1011	1111	0111	0101	1101	1001	0001		
5	00000	10000	11000	01000	01100	11100	10100	00100	00110
	10110	11110	01110	01010	11010	10010	00010	00011	10011
	11011	01011	01111	11111	10111	00111	00101	10101	11101
	01101	01001	11001	10001	00001				

2. A Gray code of size $n \geq 2$ corresponds to a Hamilton cycle in the n-cube (§8.4.4).
3. Gray codes correspond to an ordering of all subsets of $\{1, 2, \ldots, n\}$ such that adjacent subsets differ by the insertion or deletion of exactly one element. Each subset A corresponds to a binary string $a_1 a_2 \ldots a_n$ where $a_i = 1$ if $i \in A$, $a_i = 0$ if $i \notin A$.
4. A Gray code G^n can be recursively obtained in the following way:
   - *first half of G^n*: Add a 0 to the end of each string in G^{n-1}.
   - *second half of G^n*: Add a 1 to each string in the reversal of the sequence G^{n-1}.

▷ **de Bruijn Sequences**

Definitions:

A (p, n) *de Bruijn sequence* on the alphabet $\Sigma = \{0, 1, \ldots, p - 1\}$ is a sequence $(s_0, s_1, \ldots, s_{L-1})$ of $L = p^n$ elements $s_i \in \Sigma$ such that each consecutive subsequence $(s_i, s_{i+1}, \ldots, s_{i+n-1})$ of length n is distinct. Here the addition of subscripts is done modulo L so that the sequence is considered as a circular ordering. (Nicolaas G. de Bruijn, born 1918)

The *de Bruijn diagram* $D_{p,n}$ is a directed graph whose vertices correspond to all possible strings $s_1 s_2 \ldots s_{n-1}$ of $n - 1$ symbols from Σ. There are p arcs leaving the vertex $s_1 s_2 \ldots s_{n-1}$, each labeled with a distinct symbol $\alpha \in \Sigma$ and leading to the adjacent node $s_2 s_3 \ldots s_{n-1} \alpha$.

Facts:
1. The de Bruijn diagram $D_{p,n}$ has p^{n-1} vertices and p^n arcs.
2. $D_{p,n}$ is a strongly connected digraph (§11.3.2).
3. $D_{p,n}$ is an Eulerian digraph (§11.3.2).
4. Any Euler circuit in $D_{p,n}$ produces a (p,n) de Bruijn sequence.
5. de Bruijn sequences exist for all p (with $n \geq 1$). Sample de Bruijn sequences are shown in the following table.

(p,n)	a de Bruijn sequence
$(2,1)$	01
$(2,2)$	0110
$(2,3)$	01110100
$(2,4)$	0101001101111000
$(3,2)$	012202110
$(3,3)$	012001110100022212202112102
$(4,2)$	0113102212033230

6. A de Bruijn sequence can be generated from an alphabet $\Sigma = \{0, 1, \ldots, p-1\}$ of p symbols using Algorithm 1.

Algorithm 1:   Generating a (p,n) de Bruijn sequence.
1. Start with the sequence S containing n zeros.
2. Append the largest symbol from Σ to S so that the newly formed sequence S' of n symbols does not already appear as a subsequence of S. Let $S = S'$.
3. Repeat Step 2 as long as possible.
4. When Step 2 cannot be applied, remove the last $n-1$ symbols from S.

Example:
1. The de Bruijn diagram $D_{2,3}$ is shown in the following figure. An Eulerian circuit is obtained by visiting in order the vertices $11, 10, 01, 10, 00, 00, 01, 11, 11$. The de Bruijn sequence 01000111 is obtained by reading off the edge labels α as this circuit is traversed.

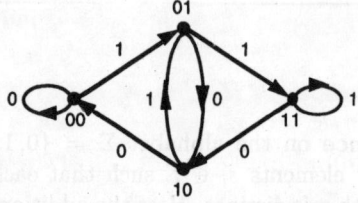

▷ **Self-generating Sequences**

Definition:
Some unusual sequences defined by simple recurrence relations or rules are informally called *self-generating sequences*.

Examples:
1. *Hofstadter G-sequence*: This sequence is defined by $a(n) = n - a(a(n-1))$, with initial condition $a(0) = 0$. The initial terms of this sequence are 0, 1, 1, 2, 3, 3, 4, 4, 5, 6, 7, 8, 8, 9, 9, 10, ... It is easy to show this sequence is well-defined. A formula for the nth term of this sequence is $a(n) = \lfloor (n+1)\mu \rfloor$, where $\mu = (-1 + \sqrt{5})/2$. [Ho79]
2. *Variations of the Hofstader G-sequence about which little is known*: These include the sequence defined by $a(n) = n - a(a(a(n-1)))$ with $a(0) = 1$, whose initial terms are 0, 1, 1, 2, 3, 4, 4, 5, 5, 6, 7, 7, 8, 9, 10, 10, 11, 12, 13, ... and the sequence defined by $a(n) = n - a(a(a(a(n-1))))$ with $a(0) = 1$, whose initial terms are 0, 1, 1, 2, 3, 4, 5, 5, 6, 6, 7, 8, 8, 9, 10, 11, 11, 12, 13, 14, ...
3. The sequence $a(n) = a(n - a(n-1)) + a(n - a(n-2))$, with $a(0) = a(1) = 1$, was also defined by Hofstader. The initial terms of this sequence are 1, 1, 2, 3, 3, 4, 5, 5, 6, 6, 6, 8, 8, 8, 10, 10, 10, 12, ...
4. The intertwined sequence $F(n)$ and $M(n)$ are defined by $F(n) = n - F(M(n-1))$ and $M(n) = n - M(F(n-1))$, with initial conditions $F(0) = 1$ and $M(0) = 0$. The initial terms of the sequence $F(n)$ (sometimes called the "female" sequence of the pair) begins with the terms 1, 1, 2, 2, 3, 3, 4, 5, 5, 6, 6, 7, 8, 8, 9, 9, 10, ... and the initial terms of the sequence $M(n)$ (sometimes called the "male" sequence of the pair) begins with the terms 0, 0, 1, 2, 2, 3, 4, 4, 5, 6, 6, 7, 7, 8, 9, 9, 10, ...
5. *Golomb's self-generating sequence*: This sequence is the unique nondecreasing sequence $a_1, a_2, a_3, \ldots$ with the property that it contains exactly a_k occurrences of the integer k for each integer k. The initial terms of this sequence are 1, 2, 2, 3, 3, 4, 4, 4, 5, 5, 5, 6, 6, 6, 6, ...
6. If $f(n)$ is the largest integer m such that $a_m = n$ where a_k is the kth term of Golomb's self-generating sequence, then $f(n) = \sum_{k=1}^{n} a_k$ and $f(f(n)) = \sum_{k=1}^{n} k a_k$.

3.1.8 MINIGUIDE TO SEQUENCES

This section lists the numerical values of various integer sequences, classified according to the type of combinatorial structure that produces the terms. This listing supplements many of the tables presented in this *Handbook*. A comprehensive tabulation of over 5,400 integer sequences is provided in [SlPl95], arranged in lexicographic order. (See Fact 4.)

Definitions:

The **power sum** $S^k(n) = \sum_{j=1}^{n} j^k$ is the sum of the kth powers of the first n positive integers. The sum of the kth powers of the first n odd integers is denoted $O^k(n) = \sum_{j=1}^{n} (2j-1)^k$.

The **associated Stirling number of the first kind** $d(n,k)$ is the number of k-cycle permutations of an n-element set with all cycles of length ≥ 2.

The **associated Stirling number of the second kind** $b(n,k)$ is the number of k-block partitions of an n-element set with all blocks of size ≥ 2.

The **double factorial** $n!!$ is the product $n(n-2)\ldots 6 \cdot 4 \cdot 2$ if n is an even positive integer and $n(n-2)\ldots 5 \cdot 3 \cdot 1$ if n is an odd positive integer.

The **Lah coefficients** $L(n,k)$ are the coefficients of $x^{\underline{k}}$ (§3.4.2) resulting from the expansion of $x^{\overline{n}}$ (§3.4.2):

$$x^{\overline{n}} = \sum_{k=1}^{n} L(n,k) x^{\underline{k}}.$$

A permutation π is **discordant** from a set A of permutations when $\pi(i) \neq \alpha(i)$ for all i and all $\alpha \in A$. Usually A consists of the identity permutation ι and powers of the n-cycle $\sigma_n = (1\ 2\ \ldots\ n)$ (see §5.3.1).

A **necklace** with n beads in c colors corresponds to an equivalence class of functions from an n-set to a c-set, under cyclic or dihedral equivalence.

A **figurate number** is the number of cells in an array of cells bounded by some regular geometrical figure.

A **polyomino** with p polygons (cells) is a connected configuration of p regular polygons in the plane. The polygons usually considered are either triangles, squares, or hexagons.

Facts:

1. Each entry in the following miniguide lists initial terms of the sequence, provides a brief description, and gives the reference number used in [SlP195].

2. *On-line sequence server*: Sequences can be submitted for identification by e-mail to
    sequences@research.att.com
for lookup on N. J. A. Sloane's *The On-Line Encyclopedia of Integer Sequences*. Sending the word lookup followed by several initial terms of the sequence, each separated by a space but with no commas, will return up to ten matches together with references.

3. A more powerful sequence server is located at superseeker@research.att.com. It tries several algorithms to explain a sequence not found in the table. Requests are limited to one per person per hour.

4. *World Wide Web page*: Sequences can also be accessed and identified using Sloane's web page:
    http://www.research.att.com/~njas/sequences
The entire table of sequences is also accessible from this web page.

Examples:

1. The following initial five terms of an unknown sequence were sent to the e-mail sequence server at sequences@research.att.com
$$\text{lookup 1 2 6 20 70}$$
In this case one matching sequence $M1645$ was identified, corresponding to the central binomial coefficients $\binom{2n}{n}$.

2. After connecting to the web site in Fact 4 and selecting the option "to look up a sequence in the table," a data entry box appears. The initial terms 1 1 2 3 5 8 13 21 were entered into this field and the request was submitted, producing in this case six matching sequences. One of these was the Fibonacci sequence ($M0692$), another was the sequence $a_n = \lceil e^{\frac{n-1}{2}} \rceil$ ($M0693$).

Miniguide to Sequences from Discrete Mathematics

The following miniguide contains a selection of important sequences, grouped by functional problem area (such as graph theory, algebra, number theory). The sequences are listed in a logical, rather than lexicographic, order within each identifiable grouping. This listing supplements existing tables within the *Handbook*. References to appropriate sections of the *Handbook* are also provided. The notation "$Mxxxx$" is the reference number used in [SlP195].

Powers of Integers (§3.1.1, §3.5.4)

1, 2, 4, 8, 16, 32, 64, 128, 256, 512, 1024, 2048, 4096, 8192, 16384, 32768, 65536, 131072
$$2^n \quad \text{[M1129]}$$

1, 3, 9, 27, 81, 243, 729, 2187, 6561, 19683, 59049, 177147, 531441, 1594323, 4782969
$$3^n \quad \text{[M2807]}$$

1, 4, 16, 64, 256, 1024, 4096, 16384, 65536, 262144, 1048576, 4194304, 16777216, 67108864
$$4^n \quad \text{[M3518]}$$

1, 5, 25, 125, 625, 3125, 15625, 78125, 390625, 1953125, 9765625, 48828125, 244140625
$$5^n \quad \text{[M3937]}$$

1, 6, 36, 216, 1296, 7776, 46656, 279936, 1679616, 10077696, 60466176, 362797056
$$6^n \quad \text{[M4224]}$$

1, 7, 49, 343, 2401, 16807, 117649, 823543, 5764801, 40353607, 282475249, 1977326743
$$7^n \quad \text{[M4431]}$$

1, 8, 64, 512, 4096, 32768, 262144, 2097152, 16777216, 134217728, 1073741824, 8589934592
$$8^n \quad \text{[M4555]}$$

1, 9, 81, 729, 6561, 59049, 531441, 4782969, 43046721, 387420489, 3486784401
$$9^n \quad \text{[M4653]}$$

1, 4, 9, 16, 25, 36, 49, 64, 81, 100, 121, 144, 169, 196, 225, 256, 289, 324, 361, 400, 441, 484
$$n^2 \quad \text{[M3356]}$$

1, 8, 27, 64, 125, 216, 343, 512, 729, 1000, 1331, 1728, 2197, 2744, 3375, 4096, 4913, 5832
$$n^3 \quad \text{[M4499]}$$

1, 16, 81, 256, 625, 1296, 2401, 4096, 6561, 1000014641, 20736, 28561, 38416, 50625, 65536
$$n^4 \quad \text{[M5004]}$$

1, 32, 243, 1024, 3125, 7776, 16807, 32768, 59049, 100000, 161051, 248832, 371293, 537824
$$n^5 \quad \text{[M5231]}$$

1, 64, 729, 4096, 15625, 46656, 117649, 262144, 531441, 1000000, 1771561, 2985984
$$n^6 \quad \text{[M5330]}$$

1, 128, 2187, 16384, 78125, 279936, 823543, 2097152, 4782969, 10000000, 19487171
$$n^7 \quad \text{[M5392]}$$

1, 256, 6561, 65536, 390625, 1679616, 5764801, 16777216, 43046721, 100000000, 214358881
$$n^8 \quad \text{[M5426]}$$

1, 512, 19683, 262144, 1953125, 10077696, 40353607, 134217728, 387420489, 1000000000
$$n^9 \quad \text{[M5459]}$$

1, 3, 6, 10, 15, 21, 28, 36, 45, 55, 66, 78, 91, 105, 120, 136, 153, 171, 190, 210, 231, 253, 276
$$S^1(n) \quad \text{[M2535]}$$

1, 5, 14, 30, 55, 91, 140, 204, 285, 385, 506, 650, 819, 1015, 1240, 1496, 1785, 2109, 2470, 2870
$$S^2(n) \text{ [M3844]}$$

1, 9, 36, 100, 225, 441, 784, 1296, 2025, 3025, 4356, 6084, 8281, 11025, 14400, 18496, 23409
$$S^3(n) \text{ [M4619]}$$

1, 17, 98, 354, 979, 2275, 4676, 8772, 15333, 25333, 39974, 60710, 89271, 127687, 178312
$$S^4(n) \text{ [M5043]}$$

1, 33, 276, 1300, 4425, 12201, 29008, 61776, 120825, 220825, 381876, 630708, 1002001
$$S^5(n) \text{ [M5241]}$$

1, 65, 794, 4890, 20515, 67171, 184820, 446964, 978405, 1978405, 3749966, 6735950
$$S^6(n) \text{ [M5335]}$$

1, 129, 2316, 18700, 96825, 376761, 1200304, 3297456, 8080425, 18080425, 37567596
$$S^7(n) \text{ [M5394]}$$

1, 257, 6818, 72354, 462979, 2142595, 7907396, 24684612, 67731333, 167731333, 382090214
$$S^8(n) \text{ [M5427]}$$

1, 512, 19683, 262144, 1953125, 10077696, 40353607, 134217728, 387420489, 1000000000
$$S^9(n) \text{ [M5459]}$$

3, 6, 14, 36, 98, 276, 794, 2316, 6818, 20196, 60074, 179196, 535538, 1602516, 4799354
$$S^n(3) \text{ [M2580]}$$

4, 10, 30, 100, 354, 1300, 4890, 18700, 72354, 282340, 1108650, 4373500, 17312754
$$S^n(4) \text{ [M3397]}$$

5, 15, 55, 225, 979, 4425, 20515, 96825, 462979, 2235465, 10874275, 53201625, 261453379
$$S^n(5) \text{ [M3863]}$$

6, 21, 91, 441, 2275, 12201, 67171, 376761, 2142595, 12313161, 71340451, 415998681
$$S^n(6) \text{ [M4149]}$$

7, 28, 140, 784, 4676, 29008, 184820, 1200304, 7907396, 52666768, 353815700, 2393325424
$$S^n(7) \text{ [M4393]}$$

8, 36, 204, 1296, 8772, 61776, 446964, 3297456, 24684612, 186884496, 1427557524
$$S^n(8) \text{ [M4520]}$$

9, 45, 285, 2025, 15333, 120825, 978405, 8080425, 67731333, 574304985, 4914341925
$$S^n(9) \text{ [M4627]}$$

1, 5, 32, 288, 3413, 50069, 873612, 17650828, 405071317, 10405071317, 295716741928
$$S^n(n) \text{ [M3968]}$$

1, 28, 153, 496, 1225, 2556, 4753, 8128, 13041, 19900, 29161, 41328, 56953, 76636, 101025
$$O^3(n) \text{ [M5199]}$$

1, 82, 707, 3108, 9669, 24310, 52871, 103496, 187017, 317338, 511819, 791660, 1182285
$O^4(n)$ [M5359]

1, 244, 3369, 20176, 79225, 240276, 611569, 1370944, 2790801, 5266900, 9351001, 15787344
$O^5(n)$ [M5421]

Factorial Numbers
1, 1, 2, 6, 24, 120, 720, 5040, 40320, 362880, 3628800, 39916800, 479001600, 6227020800
$n!$ [M1675]

1, 4, 36, 576, 14400, 518400, 25401600, 1625702400, 131681894400, 13168189440000
$(n!)^2$ [M3666]

2, 3, 8, 30, 144, 840, 5760, 45360, 403200, 3991680, 43545600, 518918400, 6706022400
$n! + (n-1)!$ [M0890]

1, 2, 8, 48, 384, 3840, 46080, 645120, 10321920, 185794560, 3715891200, 81749606400
$n!!$, n even [M1878]

1, 1, 3, 15, 105, 945, 10395, 135135, 2027025, 34459425, 654729075, 13749310575
$n!!$, n odd [M3002]

1, 1, 2, 12, 288, 34560, 24883200, 125411328000, 5056584744960000
product of n factorials [M2049]

1, 2, 6, 30, 210, 2310, 30030, 510510, 9699690, 223092870, 6469693230, 200560490130
product of first n primes [M1691]

Binomial Coefficients (§2.3.2)
1, 3, 6, 10, 15, 21, 28, 36, 45, 55, 66, 78, 91, 105, 120, 136, 153, 171, 190, 210, 231, 253, 276
$\binom{n}{2}$ [M2535]

1, 4, 10, 20, 35, 56, 84, 120, 165, 220, 286, 364, 455, 560, 680, 816, 969, 1140, 1330, 1540, 1771
$\binom{n}{3}$ [M3382]

1, 5, 15, 35, 70, 126, 210, 330, 495, 715, 1001, 1365, 1820, 2380, 3060, 3876, 4845, 5985, 7315
$\binom{n}{4}$ [M3853]

1, 6, 21, 56, 126, 252, 462, 792, 1287, 2002, 3003, 4368, 6188, 8568, 11628, 15504, 20349
$\binom{n}{5}$ [M4142]

1, 7, 28, 84, 210, 462, 924, 1716, 3003, 5005, 8008, 12376, 18564, 27132, 38760, 54264, 74613
$\binom{n}{6}$ [M4390]

1, 8, 36, 120, 330, 792, 1716, 3432, 6435, 11440, 19448, 31824, 50388, 77520, 116280, 170544
$\binom{n}{7}$ [M4517]

1, 9, 45, 165, 495, 1287, 3003, 6435, 12870, 24310, 43758, 75582, 125970, 203490, 319770
$\binom{n}{8}$ [M4626]

1, 10, 55, 220, 715, 2002, 5005, 11440, 24310, 48620, 92378, 167960, 293930, 497420, 817190
$$\binom{n}{9} \quad [\text{M4712}]$$

1, 11, 66, 286, 1001, 3003, 8008, 19448, 43758, 92378, 184756, 352716, 646646, 1144066
$$\binom{n}{10} \quad [\text{M4794}]$$

1, 2, 3, 6, 10, 20, 35, 70, 126, 252, 462, 924, 1716, 3432, 6435, 12870, 24310, 48620, 92378
central binomial coefficients $\binom{n}{\lfloor n/2 \rfloor}$ [M0769]

1, 2, 6, 20, 70, 252, 924, 3432, 12870, 48620, 184756, 705432, 2704156, 10400600, 40116600
central binomial coefficients $\binom{2n}{n}$ [M1645]

1, 3, 10, 35, 126, 462, 1716, 6435, 24310, 92378, 352716, 1352078, 5200300, 20058300
$$\binom{2n+1}{n} \quad [\text{M2848}]$$

Stirling Cycle Numbers/Stirling Numbers of the First Kind (§2.5.2)

1, 3, 11, 50, 274, 1764, 13068, 109584, 1026576, 10628640, 120543840, 1486442880
$$\begin{bmatrix} n \\ 2 \end{bmatrix} \quad [\text{M2902}]$$

1, 6, 35, 225, 1624, 13132, 118124, 1172700, 12753576, 150917976, 1931559552
$$\begin{bmatrix} n \\ 3 \end{bmatrix} \quad [\text{M4218}]$$

1, 10, 85, 735, 6769, 67284, 723680, 8409500, 105258076, 1414014888, 20313753096
$$\begin{bmatrix} n \\ 4 \end{bmatrix} \quad [\text{M4730}]$$

1, 15, 175, 1960, 22449, 269325, 3416930, 45995730, 657206836, 9957703756, 159721605680
$$\begin{bmatrix} n \\ 5 \end{bmatrix} \quad [\text{M4983}]$$

1, 21, 322, 4536, 63273, 902055, 13339535, 206070150, 3336118786, 56663366760
$$\begin{bmatrix} n \\ 6 \end{bmatrix} \quad [\text{M5114}]$$

1, 28, 546, 9450, 157773, 2637558, 44990231, 790943153, 14409322928, 272803210680
$$\begin{bmatrix} n \\ 7 \end{bmatrix} \quad [\text{M5202}]$$

2, 11, 35, 85, 175, 322, 546, 870, 1320, 1925, 2717, 3731, 5005, 6580, 8500, 10812, 13566
$$\begin{bmatrix} n \\ n-2 \end{bmatrix} \quad [\text{M1998}]$$

6, 50, 225, 735, 1960, 4536, 9450, 18150, 32670, 55770, 91091, 143325, 218400, 323680
$$\begin{bmatrix} n \\ n-3 \end{bmatrix} \quad [\text{M4258}]$$

24, 274, 1624, 6769, 22449, 63273, 157773, 357423, 749463, 1474473, 2749747, 4899622
$$\begin{bmatrix} n \\ n-4 \end{bmatrix} \quad [\text{M5155}]$$

Stirling Subset Numbers/Stirling Numbers of the Second Kind (§2.5.2)

1, 6, 25, 90, 301, 966, 3025, 9330, 28501, 86526, 261625, 788970, 2375101, 7141686
$$\begin{Bmatrix} n \\ 3 \end{Bmatrix} \quad [\text{M4167}]$$

1, 10, 65, 350, 1701, 7770, 34105, 145750, 611501, 2532530, 10391745, 42355950, 171798901
$$\begin{Bmatrix} n \\ 4 \end{Bmatrix} \quad [\text{M4722}]$$

1, 15, 140, 1050, 6951, 42525, 246730, 1379400, 7508501, 40075035, 210766920, 1096190550
$$\left\{{n \atop 5}\right\}$$ [M4981]

1, 21, 266, 2646, 22827, 179487, 1323652, 9321312, 63436373, 420693273, 2734926558
$$\left\{{n \atop 6}\right\}$$ [M5112]

1, 28, 462, 5880, 63987, 627396, 5715424, 49329280, 408741333, 3281882604, 25708104786
$$\left\{{n \atop 7}\right\}$$ [M5201]

1, 7, 25, 65, 140, 266, 462, 750, 1155, 1705, 2431, 3367, 4550, 6020, 7820, 9996, 12597, 15675
$$\left\{{n \atop n-2}\right\}$$ [M4385]

1, 15, 90, 350, 1050, 2646, 5880, 11880, 22275, 39325, 66066, 106470, 165620, 249900
$$\left\{{n \atop n-3}\right\}$$ [M4974]

1, 31, 301, 1701, 6951, 22827, 63987, 159027, 359502, 752752, 1479478, 2757118, 4910178
$$\left\{{n \atop n-4}\right\}$$ [M5222]

1, 1, 3, 7, 25, 90, 350, 1701, 7770, 42525, 246730, 1379400, 9321312, 63436373, 420693273
$$\max_k \left\{{n \atop k}\right\}$$ [M2690]

Associated Stirling Numbers of the First Kind (§3.1.8)
3, 20, 130, 924, 7308, 64224, 623376, 6636960, 76998240, 967524480, 13096736640
$$d(n, 2)$$ [M3075]

15, 210, 2380, 26432, 303660, 3678840, 47324376, 647536032, 9418945536, 145410580224
$$d(n, 3)$$ [M4988]

2, 20, 210, 2520, 34650, 540540, 9459450, 183783600, 3928374450, 91662070500
$$d(n, n-3)$$ [M2124]

6, 130, 2380, 44100, 866250, 18288270, 416215800, 10199989800, 268438920750
$$d(n, n-4)$$ [M4298]

1, 120, 7308, 303660, 11098780, 389449060, 13642629000, 486591585480, 17856935296200
$$d(2n, n-2)$$ [M5382]

1, 24, 924, 26432, 705320, 18858840, 520059540, 14980405440, 453247114320
$$d(2n+1, n-1)$$ [M5169]

Associated Stirling Numbers of the Second Kind (§3.1.8)
3, 10, 25, 56, 119, 246, 501, 1012, 2035, 4082, 8177, 16368, 32751, 65518, 131053, 262124
$$b(n, 2)$$ [M2836]

15, 105, 490, 1918, 6825, 22935, 74316, 235092, 731731, 2252341, 6879678, 20900922
$$b(n, 3)$$ [M4978]

1, 25, 490, 9450, 190575, 4099095, 94594500, 2343240900
$$b(2n, n-1)$$ [M5186]

1, 56, 1918, 56980, 1636635, 47507460, 1422280860
$$b(2n+1, n-1) \quad [M5315]$$

Lah Coefficients (§3.1.8)
1, 6, 36, 240, 1800, 15120, 141120, 1451520, 16329600, 199584000, 2634508800
$$L(n, 2) \quad [M4225]$$

1, 12, 120, 1200, 12600, 141120, 1693440, 21772800, 299376000, 4390848000, 68497228800
$$L(n, 3) \quad [M4863]$$

1, 20, 300, 4200, 58800, 846720, 12700800, 199584000, 3293136000, 57081024000
$$L(n, 4) \quad [M5096]$$

1, 30, 630, 11760, 211680, 3810240, 69854400, 1317254400, 25686460800, 519437318400
$$L(n, 5) \quad [M5213]$$

1, 42, 1176, 28224, 635040, 13970880, 307359360, 6849722880, 155831195520
$$L(n, 6) \quad [M5279]$$

Eulerian Numbers (§3.1.5)
1, 4, 11, 26, 57, 120, 247, 502, 1013, 2036, 4083, 8178, 16369, 32752, 65519, 131054, 262125
$$E(n, 1) \quad [M3416]$$

1, 11, 66, 302, 1191, 4293, 14608, 47840, 152637, 478271, 1479726, 4537314, 13824739
$$E(n, 2) \quad [M4795]$$

1, 26, 302, 2416, 15619, 88234, 455192, 2203488, 10187685, 45533450, 198410786
$$E(n, 3) \quad [M5188]$$

1, 57, 1191, 15619, 156190, 1310354, 9738114, 66318474, 423281535, 2571742175
$$E(n, 4) \quad [M5317]$$

1, 120, 4293, 88234, 1310354, 15724248, 162512286, 1505621508, 12843262863
$$E(n, 5) \quad [M5379]$$

1, 247, 14608, 455192, 9738114, 162512286, 2275172004, 27971176092, 311387598411
$$E(n, 6) \quad [M5422]$$

1, 502, 47840, 2203488, 66318474, 1505621508, 27971176092, 447538817472
$$E(n, 7) \quad [M5457]$$

Other Special Sequences (§3.1)
1, 1, 2, 3, 5, 8, 13, 21, 34, 55, 89, 144, 233, 377, 610, 987, 1597, 2584, 4181, 6795, 10946, 17711
$$\text{Fibonacci numbers, } n \geq 1 \quad [M0692]$$

1, 3, 4, 7, 11, 18, 29, 47, 76, 123, 199, 322, 521, 843, 1364, 2207, 3571, 5778, 9349, 15127
$$\text{Lucas numbers, } n \geq 1 \quad [M2341]$$

1, 1, 2, 5, 14, 42, 132, 429, 1430, 4862, 16796, 58786, 208012, 742900, 2674440, 9694845
                                          Catalan numbers, $n \geq 0$ [M1459]

1, 3, 11, 25, 137, 49, 363, 761, 7129, 7381, 83711, 86021, 1145993, 1171733, 1195757
                             numerators of harmonic numbers, $n \geq 1$ [M2885]

1, 2, 6, 12, 60, 20, 140, 280, 2520, 2520, 27720, 27720, 360360, 360360, 360360, 720720
                          denominators of harmonic numbers, $n \geq 1$ [M1589]

1, 1, 1, 1, 1, 5, 691, 7, 3617, 43867, 174611, 854513, 236364091, 8553103, 23749461029
                     numerators of Bernoulli numbers $|B_{2n}|$, $n \geq 0$ [M4039]

1, 6, 30, 42, 30, 66, 2730, 6, 510, 798, 330, 138, 2730, 6, 870, 14322, 510, 6, 1919190, 6, 13530
                   denominators of Bernoulli numbers $|B_{2n}|$, $n \geq 0$ [M4189]

1, 1, 5, 61, 1385, 50521, 2702765, 199360981, 19391512145, 2404879675441
                                              Euler numbers $|E_{2n}|$, $n \geq 0$ [M4019]

1, 2, 16, 272, 7936, 353792, 22368256, 1903757312, 209865342976, 29088885112832
                                           tangent numbers T_{2n+1}, $n \geq 0$ [M2096]

1, 1, 2, 5, 15, 52, 203, 877, 4140, 21147, 115975, 678570, 4213597, 27644437, 190899322
                                                      Bell numbers, $n \geq 0$ [M1484]

Numbers of Certain Algebraic Structures (§1.4, §5.2)
1, 1, 1, 2, 1, 1, 1, 3, 2, 1, 1, 2, 1, 1, 1, 5, 1, 2, 1, 2, 1, 1, 1, 3, 2, 1, 3, 2, 1, 1, 1, 7, 1, 1, 1, 4, 1, 1, 1, 3
                                                           abelian groups of order n   [M0064]

1, 1, 1, 2, 1, 2, 1, 5, 2, 2, 1, 5, 1, 2, 1, 14, 1, 5, 1, 5, 2, 2, 1, 15, 2, 2, 5, 4, 1, 4, 1, 51, 1, 2, 1, 14, 1, 2
                                                                   groups of order n   [M0098]

2, 3, 5, 7, 11, 13, 17, 19, 23, 29, 31, 37, 41, 43, 47, 53, 59, 60, 61, 67, 71, 73, 79, 83, 89, 97, 101
                                                             orders of simple groups   [M0651]

60, 168, 360, 504, 660, 1092, 2448, 2520, 3420, 4080, 5616, 6048, 6072, 7800, 7920, 9828
                                                    orders of noncyclic simple groups   [M5318]

1, 1, 2, 5, 16, 63, 318, 2045, 16999, 183231, 2567284, 46749427, 1104891746, 33823827452
                                                  partially ordered sets on n elements   [M1495]

1, 2, 13, 171, 3994, 154303, 9415189, 878222530
                                                   transitive relations on n elements   [M2065]

1, 5, 52, 1522, 145984, 48464496, 56141454464, 229148550030864, 3333310786076963968
                                                 relations on n unlabeled points   [M4010]

1, 2, 1, 2, 3, 6, 9, 18, 30, 56, 99, 186, 335, 630, 1161, 2182, 4080, 7710, 14532, 27594, 52377
                                        binary irreducible polynomials of degree n   [M0116]

Chapter 3 SEQUENCES

Permutations (§5.3.1)
by cycles
1, 1, 1, 3, 15, 75, 435, 3045, 24465, 220185, 2200905, 24209955, 290529855, 3776888115
                                                   no 2-cycles [M2991]

1, 1, 2, 4, 16, 80, 520, 3640, 29120, 259840, 2598400, 28582400, 343235200, 4462057600
                                                   no 3-cycles [M1295]

1, 1, 2, 6, 18, 90, 540, 3780, 31500, 283500, 2835000, 31185000, 372972600, 4848643800
                                                   no 4-cycles [M1635]

0, 1, 1, 3, 9, 45, 225, 1575, 11025, 99225, 893025, 9823275, 108056025, 1404728325
                                            no even length cycles [M2824]

discordant (§2.4.2, §3.1.8)
1, 0, 1, 2, 9, 44, 265, 1854, 14833, 133496, 1334961, 14684570, 176214841, 2290792932
                                     derangements, discordant for ι [M1937]

1, 1, 0, 1, 2, 13, 80, 579, 4738, 43387, 439792, 4890741, 59216642, 775596313, 10927434464
                  menage numbers, discordant for ι and σ_n [M2062]

0, 1, 2, 20, 144, 1265, 12072, 126565, 1445100, 17875140, 238282730, 3407118041
                               discordant for $\iota, \sigma_n, \sigma_n^2$ [M2121]

by order
1, 2, 3, 4, 6, 6, 12, 15, 20, 30, 30, 60, 60, 84, 105, 140, 210, 210, 420, 420, 420, 420, 840, 840
                                                      max order [M0537]

1, 2, 3, 4, 6, 12, 15, 20, 30, 60, 84, 105, 140, 210, 420, 840, 1260, 1540, 2310, 2520, 4620, 5460
                                                      max order [M0577]

1, 2, 4, 16, 56, 256, 1072, 11264, 78976, 672256, 4653056, 49810432, 433429504, 4448608256
                                           order a power of 2 [M1293]

0, 1, 3, 9, 25, 75, 231, 763, 2619, 9495, 35695, 140151, 568503, 2390479, 10349535, 46206735
                                                        order 2 [M2801]

0, 0, 2, 8, 20, 80, 350, 1232, 5768, 31040, 142010, 776600, 4874012, 27027728, 168369110
                                                   order 3 [M1833]

0, 0, 0, 6, 30, 180, 840, 5460, 30996, 209160, 1290960, 9753480, 69618120, 571627056
                                                         order 4 [M4206]

0, 0, 1, 3, 6, 10, 30, 126, 448, 1296, 4140, 17380, 76296, 296088, 1126216, 4940040, 23904000
                                                   odd, order 2 [M2538]

Necklaces (§2.6)
1, 2, 3, 4, 6, 8, 14, 20, 36, 60, 108, 188, 352, 632, 1182, 2192, 4116, 7712, 14602, 27596, 52488
                                                     2 colors, n beads [M0564]

Section 3.1 SPECIAL SEQUENCES   167

1, 3, 6, 11, 24, 51, 130, 315, 834, 2195, 5934, 16107, 44368, 122643, 341802, 956635, 2690844
    3 colors, n beads [M2548]

1, 4, 10, 24, 70, 208, 700, 2344, 8230, 29144, 104968, 381304, 1398500, 5162224, 19175140
    4 colors, n beads [M3390]

1, 5, 15, 45, 165, 629, 2635, 11165, 48915, 217045, 976887, 4438925, 20346485, 93900245
    5 colors, n beads [M3860]

Number Theory (§4.2, §4.3)
2, 3, 5, 7, 11, 13, 17, 19, 23, 29, 31, 37, 41, 43, 47, 53, 59, 61, 67, 71, 73, 79, 83, 89, 97, 101, 103
    primes [M0652]

0, 1, 2, 2, 3, 3, 4, 4, 4, 4, 5, 5, 6, 6, 6, 6, 7, 7, 8, 8, 8, 8, 9, 9, 9, 9, 9, 9, 10, 10, 11, 11, 11, 11, 11, 11
    number of primes $\leq$ n [M0256]

1, 1, 1, 1, 1, 2, 1, 1, 1, 2, 1, 2, 1, 2, 2, 1, 1, 2, 1, 2, 2, 2, 1, 2, 1, 2, 1, 2, 1, 3, 1, 1, 2, 2, 2, 2, 1, 2, 2, 2
    number of distinct primes dividing n [M0056]

2, 3, 5, 7, 13, 17, 19, 31, 61, 89, 107, 127, 521, 607, 1279, 2203, 2281, 3217, 4253, 4423, 9689
    Mersenne primes [M0672]

1, 1, 1, 2, 1, 2, 1, 3, 2, 2, 1, 4, 1, 2, 2, 5, 1, 4, 1, 4, 2, 2, 1, 7, 2, 2, 3, 4, 1, 5, 1, 7, 2, 2, 2, 9, 1, 2, 2, 7
    number of ways of factoring n [M0095]

1, 1, 2, 2, 4, 2, 6, 4, 6, 4, 10, 4, 12, 6, 8, 8, 16, 6, 18, 8, 12, 10, 22, 8, 20, 12, 18, 12, 28, 8, 30, 16
    Euler totient function [M0299]

561, 1105, 1729, 2465, 2821, 6601, 8911, 10585, 15841, 29341, 41041, 46657, 52633, 62745
    Carmichael numbers [M5462]

1, 2, 2, 3, 2, 4, 2, 4, 3, 4, 2, 6, 2, 4, 4, 5, 2, 6, 2, 6, 4, 4, 2, 8, 3, 4, 4, 6, 2, 8, 2, 6, 4, 4, 4, 9, 2, 4, 4, 8
    number of divisors of n [M0246]

1, 3, 4, 7, 6, 12, 8, 15, 13, 18, 12, 28, 14, 24, 24, 31, 18, 39, 20, 42, 32, 36, 24, 60, 31, 42, 40, 56
    sum of divisors of n [M2329]

6, 28, 496, 8128, 33550336, 8589869056, 137438691328, 2305843008139952128
    perfect numbers [M4186]

Partitions (§2.5.1)
1, 1, 2, 3, 5, 7, 11, 15, 22, 30, 42, 56, 77, 101, 135, 176, 231, 297, 385, 490, 627, 792, 1002, 1255
    partitions of n [M0663]

1, 1, 1, 2, 2, 3, 4, 5, 6, 8, 10, 12, 15, 18, 22, 27, 32, 38, 46, 54, 64, 76, 89, 104, 122, 142, 165, 192
    partitions of n into distinct parts [M0281]

1, 3, 6, 13, 24, 48, 86, 160, 282, 500, 859, 1479, 2485, 4167, 6879, 11297, 18334, 29601, 47330
    planar partitions of n [M2566]

Figurate Numbers (§3.1.8)

polygonal

1, 3, 6, 10, 15, 21, 28, 36, 45, 55, 66, 78, 91, 105, 120, 136, 153, 171, 190, 210, 231, 253, 276
                     triangular [M2535]

1, 5, 12, 22, 35, 51, 70, 92, 117, 145, 176, 210, 247, 287, 330, 376, 425, 477, 532, 590, 651, 715
                     pentagonal [M3818]

1, 6, 15, 28, 45, 66, 91, 120, 153, 190, 231, 276, 325, 378, 435, 496, 561, 630, 703, 780, 861, 946
                     hexagonal [M4108]

1, 7, 18, 34, 55, 81, 112, 148, 189, 235, 286, 342, 403, 469, 540, 616, 697, 783, 874, 970, 1071
                     heptagonal [M4358]

1, 8, 21, 40, 65, 96, 133, 176, 225, 280, 341, 408, 481, 560, 645, 736, 833, 936, 1045, 1160, 1281
                     octagonal [M4493]

pyramidal

1, 4, 10, 20, 35, 56, 84, 120, 165, 220, 286, 364, 455, 560, 680, 816, 969, 1140, 1330, 1540, 1771
         3-dimensional triangular, height n [M3382]

1, 5, 14, 30, 55, 91, 140, 204, 285, 385, 506, 650, 819, 1015, 1240, 1496, 1785, 2109, 2470, 2870
         3-dimensional square, height n [M3844]

1, 6, 18, 40, 75, 126, 196, 288, 405, 550, 726, 936, 1183, 1470, 1800, 2176, 2601, 3078, 3610
         3-dimensional pentagonal, height n [M4116]

1, 7, 22, 50, 95, 161, 252, 372, 525, 715, 946, 1222, 1547, 1925, 2360, 2856, 3417, 4047, 4750
         3-dimensional hexagonal, height n [M4374]

1, 8, 26, 60, 115, 196, 308, 456, 645, 880, 1166, 1508, 1911, 2380, 2920, 3536, 4233, 5016, 5890
         3-dimensional heptagonal, height n [M4498]

1, 5, 15, 35, 70, 126, 210, 330, 495, 715, 1001, 1365, 1820, 2380, 3060, 3876, 4845, 5985, 7315
         4-dimensional triangular, height n [M3853]

1, 6, 20, 50, 105, 196, 336, 540, 825, 1210, 1716, 2366, 3185, 4200, 5440, 6936, 8721, 10830
         4-dimensional square, height n [M4135]

1, 7, 25, 65, 140, 266, 462, 750, 1155, 1705, 2431, 3367, 4550, 6020, 7820, 9996, 12597, 15675
         4-dimensional pentagonal, height n [M4385]

1, 8, 30, 80, 175, 336, 588, 960, 1485, 2200, 3146, 4368, 5915, 7840, 10200, 13056, 16473
         4-dimensional hexagonal, height n [M4506]

1, 9, 35, 95, 210, 406, 714, 1170, 1815, 2695, 3861, 5369, 7280, 9660, 12580, 16116, 20349
         4-dimensional heptagonal, height n [M4617]

Polyominoes (§3.1.8)

1, 1, 2, 5, 12, 35, 108, 369, 1285, 4655, 17073, 63600, 238591, 901971, 3426576, 13079255
$\qquad$ **squares, n cells** [M1425]

1, 1, 1, 3, 4, 12, 24, 66, 160, 448, 1186, 3334, 9235, 26166, 73983, 211297
$\qquad$ **triangles, n cells** [M2374]

1, 1, 3, 7, 22, 82, 333, 1448, 6572, 30490, 143552, 683101
$\qquad$ **hexagons, n cells** [M2682]

1, 1, 2, 8, 29, 166, 1023, 6922, 48311, 346543, 2522572, 18598427
$\qquad$ **cubes, n cells** [M1845]

Trees (§9.3)

1, 1, 1, 2, 3, 6, 11, 23, 47, 106, 235, 551, 1301, 3159, 7741, 19320, 48629, 123867, 317955
$\qquad$ **n unlabeled vertices** [M0791]

1, 1, 2, 4, 9, 20, 48, 115, 286, 719, 1842, 4766, 12486, 32973, 87811, 235381, 634847, 1721159
$\qquad$ **rooted, n unlabeled vertices** [M1180]

1, 1, 3, 16, 125, 1296, 16807, 262144, 4782969, 100000000, 2357947691, 61917364224
$\qquad$ **n labeled vertices** [M3027]

1, 2, 9, 64, 625, 7776, 117649, 2097152, 43046721, 1000000000, 25937424601, 743008370688
$\qquad$ **rooted, n labeled vertices** [M1946]

by diameter

1, 2, 5, 8, 14, 21, 32, 45, 65, 88, 121, 161, 215, 280, 367, 471, 607, 771, 980, 1232, 1551, 1933
$\qquad$ **diameter 4, n ≥ 5 vertices** [M1350]

1, 2, 7, 14, 32, 58, 110, 187, 322, 519, 839, 1302, 2015, 3032, 4542, 6668, 9738, 14006, 20036
$\qquad$ **diameter 5, n ≥ 6 vertices** [M1741]

1, 3, 11, 29, 74, 167, 367, 755, 1515, 2931, 5551, 10263, 18677, 33409, 59024, 102984, 177915
$\qquad$ **diameter 6, n ≥ 7 vertices** [M2887]

1, 3, 14, 42, 128, 334, 850, 2010, 4625, 10201, 21990, 46108, 94912, 191562, 380933, 746338
$\qquad$ **diameter 7, n ≥ 8 vertices** [M2969]

1, 4, 19, 66, 219, 645, 1813, 4802, 12265, 30198, 72396, 169231, 387707, 871989, 1930868
$\qquad$ **diameter 8, n ≥ 9 vertices** [M3552]

by height

1, 3, 8, 18, 38, 76, 147, 277, 509, 924, 1648, 2912, 5088, 8823, 15170, 25935, 44042, 74427
$\qquad$ **height 3, n ≥ 4 vertices** [M2732]

1, 4, 13, 36, 93, 225, 528, 1198, 2666, 5815, 12517, 26587, 55933, 116564, 241151, 495417
$\qquad$ **height 4, n ≥ 5 vertices** [M3461]

series-reduced
1, 1, 0, 1, 1, 2, 2, 4, 5, 10, 14, 26, 42, 78, 132, 249, 445, 842, 1561, 2988, 5671, 10981, 21209
n vertices [M0320]

1, 1, 0, 2, 4, 6, 12, 20, 39, 71, 137, 261, 511, 995, 1974, 3915, 7841, 15749, 31835, 64540
rooted, n vertices [M0327]

0, 1, 0, 1, 1, 2, 3, 6, 10, 19, 35, 67, 127, 248, 482, 952, 1885, 3765, 7546, 15221, 30802, 62620
planted, n vertices [M0768]

Graphs (§8.1, §8.3, §8.4, §8.9)
1, 2, 4, 11, 34, 156, 1044, 12346, 274668, 12005168, 1018997864, 165091172592
n vertices [M1253]

chromatic number
4, 6, 7, 7, 8, 9, 9, 10, 10, 10, 11, 11, 12, 12, 12, 13, 13, 13, 13, 14, 14, 14, 15, 15, 15, 15, 16, 16
surface, connectivity n ≥ 1 [M3265]

4, 7, 8, 9, 10, 11, 12, 12, 13, 13, 14, 15, 15, 16, 16, 16, 17, 17, 18, 18, 19, 19, 19, 20, 20, 20, 21
surface, genus n ≥ 0 [M3292]

genus
0, 0, 0, 0, 1, 1, 1, 2, 3, 4, 5, 6, 8, 10, 11, 13, 16, 18, 20, 23, 26, 29, 32, 35, 39, 43, 46, 50, 55, 59, 63
complete graphs, n vertices [M0503]

connected
1, 1, 2, 6, 21, 112, 853, 11117, 261080, 11716571, 1006700565, 164059830476
n vertices [M1657]

1, 1, 0, 2, 5, 32, 234, 3638, 106147, 6039504, 633754161, 120131932774, 41036773627286
series-reduced, n vertices [M1548]

1, 1, 3, 5, 12, 30, 79, 227, 710, 2322, 8071, 29503, 112822, 450141
n edges [M2486]

1, 1, 4, 38, 728, 26704, 1866256, 251548592, 66296291072, 34496488594816
n labeled vertices [M3671]

directed
1, 3, 16, 218, 9608, 1540944, 882033440, 1793359192848, 13027956824399552
n vertices [M3032]

1, 3, 9, 33, 139, 718, 4535
transitive, n vertices [M2817]

1, 1, 2, 4, 12, 56, 456, 6880, 191536, 9733056, 903753248, 154108311168, 48542114686912
tournaments, n vertices [M1262]

1, 4, 29, 355, 6942, 209527, 9535241, 642779354, 63260289423, 8977053873043
$\hspace{6cm}$ transitive, n labeled vertices [M3631]

various
1, 2, 2, 4, 3, 8, 4, 14, 9, 22, 8, 74, 14, 56, 48, 286, 36, 380, 60, 1214, 240, 816, 188, 15506, 464
$\hspace{6cm}$ transitive, n vertices [M0302]

1, 1, 2, 3, 7, 16, 54, 243, 2038, 33120, 1182004, 87723296, 12886193064, 3633057074584
$\hspace{6cm}$ all degrees even, n vertices [M0846]

1, 0, 1, 1, 4, 8, 37, 184, 1782, 31026, 1148626, 86539128, 12798435868, 3620169692289
$\hspace{6cm}$ Eulerian, n vertices [M3344]

1, 0, 1, 3, 8, 48, 383, 6020
$\hspace{6cm}$ Hamiltonian, n vertices [M2764]

1, 2, 2, 4, 3, 8, 6, 22, 26, 176
$\hspace{6cm}$ regular, n vertices [M0303]

0, 1, 1, 3, 10, 56, 468, 7123, 194066, 9743542, 900969091, 153620333545, 48432939150704
$\hspace{6cm}$ nonseparable, n vertices [M2873]

1, 2, 4, 11, 33, 142, 822, 6910
$\hspace{6cm}$ planar, n vertices [M1252]

3.2 GENERATING FUNCTIONS

Generating functions express an infinite sequence as coefficients arising from a power series in an auxiliary variable. The closed form of a generating function is a concise way to represent such an infinite sequence. Properties of the sequence can be explored by analyzing the closed form of an associated generating function. Two types of generating functions are discussed in this section—ordinary generating functions and exponential generating functions. The former arise when counting configurations in which order is not important, while the latter are appropriate when order matters.

3.2.1 ORDINARY GENERATING FUNCTIONS

Definitions:

The (**ordinary**) **generating function** for the sequence $a_0, a_1, a_2, \ldots$ of real numbers is the formal power series $f(x) = a_0 + a_1 x + a_2 x^2 + \cdots = \sum_{i=0}^{\infty} a_i x^i$ or any equivalent closed form expression.

The **convolution** of the sequence $a_0, a_1, a_2, \ldots$ and the sequence $b_0, b_1, b_2, \ldots$ is the sequence $c_0, c_1, c_2, \ldots$ in which $c_t = a_0 b_t + a_1 b_{t-1} + a_2 b_{t-2} + \cdots + a_t b_0 = \sum_{k=0}^{t} a_k b_{t-k}$.

Facts:

1. Generating functions are considered as algebraic forms and can be manipulated as such, without regard to actual convergence of the power series.

2. A rational form (the ratio of two polynomials) is a concise expression for the generating function of the sequence obtained by carrying out long division on the polynomials. (See Example 1.)

3. Generating functions are often useful for constructing and verifying identities involving binomial coefficients and other special sequences. (See Example 10.)

4. Generating functions can be used to derive formulas for the sums of powers of integers. (See Example 17.)

5. Generating functions can be used to solve recurrence relations. (See §3.3.4.)

6. Each sequence $\{a_n\}$ defines a $unique$ generating function $f(x)$, and conversely.

7. *Related generating functions*: Suppose $f(x) = \sum_{k=0}^{\infty} a_k x^k$ and $g(x) = \sum_{k=0}^{\infty} b_k x^k$ are generating functions for the sequences $a_0, a_1, a_2, \ldots$ and $b_0, b_1, b_2, \ldots$, respectively. Table 1 gives some related generating functions.

Table 1 Related generating functions.

generating function	sequence
$x^n f(x)$	$\underbrace{0,0,0,\ldots,0}_{n}, a_0, a_1, a_2, \ldots$
$f(x) - a_n x^n$	$a_0, a_1, \ldots, a_{n-1}, 0, a_{n+1}, \ldots$
$a_0 + a_1 x + \cdots + a_n x^n$	$a_0, a_1, \ldots, a_n, 0, 0, \ldots$
$f(x^2)$	$a_0, 0, a_1, 0, a_2, 0, a_3, \ldots$
$\frac{f(x) - a_0}{x}$	$a_1, a_2, a_3, \ldots$
$f'(x)$	$a_1, 2a_2, 3a_3, \ldots, k a_k, \ldots$
$\int_0^x f(t)\,dt$	$0, a_0, \frac{a_1}{2}, \frac{a_2}{3}, \ldots, \frac{a_k}{k+1}, \ldots$
$\frac{f(x)}{1-x}$	$a_0, a_0 + a_1, a_0 + a_1 + a_2, \ldots$
$rf(x) + sg(x)$	$ra_0 + sb_0, ra_1 + sb_1, ra_2 + sb_2, \ldots$
$f(x)g(x)$	$a_0 b_0, a_0 b_1 + a_1 b_0, a_0 b_2 + a_1 b_1 + a_2 b_0, \ldots$ (convolution of $\{a_n\}$ and $\{b_n\}$)

Examples:

1. The sequence $0, 1, 4, 9, 16, \ldots$ of squares of the nonnegative integers has the generating function $0 + x + 4x^2 + 9x^3 + 16x^4 + \cdots$. However, this generating function has a concise closed form expression, namely $\frac{x + x^2}{1 - 3x + 3x^2 - x^3}$. Verification is obtained by carrying out long division on the indicated polynomials. This concise form can be used to deduce properties involving the sequence, such as an explicit algebraic expression for the sum of squares of the first n positive integers. (See Example 17.)

2. The generating function for the sequence $1, 1, 1, 1, 1, \ldots$ is $1 + x + x^2 + x^3 + x^4 + \cdots = \frac{1}{1-x}$. Differentiating both sides of this expression produces $1 + 2x + 3x^2 + 4x^3 + \cdots = \frac{1}{(1-x)^2}$. Thus, $\frac{1}{(1-x)^2}$ is a closed form expression for the generating function of the sequence $1, 2, 3, 4, \ldots$. (See Table 2.)

Table 2 Generating functions for particular sequences.

sequence	closed form
$1,1,1,1,1,\ldots$	$\frac{1}{1-x}$
$1,1,\ldots,1,0,0,\ldots$ (n 1s)	$\frac{1-x^n}{1-x}$
$1,1,\ldots,1,1,0,1,1,\ldots$ (0 following n 1s)	$\frac{1}{1-x} - x^n$
$1,-1,1,-1,1,-1,\ldots$	$\frac{1}{1+x}$
$1,0,1,0,1,\ldots$	$\frac{1}{1-x^2}$
$1,2,3,4,5,\ldots$	$\frac{1}{(1-x)^2}$
$1,4,9,16,25,\ldots$	$\frac{1+x}{(1-x)^3}$
$1,r,r^2,r^3,r^4,\ldots$	$\frac{1}{1-rx}$
$0,r,2r^2,3r^3,4r^4,\ldots$	$\frac{rx}{(1-rx)^2}$
$0,1,\frac{1}{2},\frac{1}{3},\frac{1}{4},\frac{1}{5},\ldots$	$\ln \frac{1}{1-x}$
$\frac{1}{0!},\frac{1}{1!},\frac{1}{2!},\frac{1}{3!},\frac{1}{4!},\ldots$	e^x
$0,1,-\frac{1}{2},\frac{1}{3},-\frac{1}{4},\frac{1}{5},\ldots$	$\ln(1+x)$
$F_0,F_1,F_2,F_3,F_4,\ldots$	$\frac{x}{1-x-x^2}$
$L_0,L_1,L_2,L_3,L_4,\ldots$	$\frac{2-x}{1-x-x^2}$
$C_0,C_1,C_2,C_3,C_4,\ldots$	$\frac{1-\sqrt{1-4x}}{2x}$
$H_0,H_1,H_2,H_3,H_4,\ldots$	$\frac{1}{1-x} \ln \frac{1}{1-x}$

3. Table 2 gives closed form expressions for the generating functions of particular sequences. In this table, r is an arbitrary real number, F_n is the nth Fibonacci number (§3.1.2), L_n is the nth Lucas number (§3.1.2), C_n is the nth Catalan number (§3.1.3), and H_n is the nth harmonic number (§3.1.7).

4. For every positive integer n, the binomial theorem (§2.3.4) states that
$$(1+x)^n = \binom{n}{0} + \binom{n}{1}x + \binom{n}{2}x^2 + \cdots + \binom{n}{n}x^n = \sum_{k=0}^{n} \binom{n}{k}x^k,$$
so $(1+x)^n$ is a closed form for the generating function of $\binom{n}{0}, \binom{n}{1}, \binom{n}{2}, \ldots, \binom{n}{n}, 0, 0, 0, \ldots$.

5. For every positive integer n, the Maclaurin series expansion for $(1+x)^{-n}$ is
$$(1+x)^{-n} = 1 + (-n)x + \frac{(-n)(-n-1)x^2}{2!} + \cdots$$
$$= 1 + \sum_{k=1}^{\infty} \frac{(-n)(-n-1)(-n-2)\ldots(-n-k+1)}{k!}x^k.$$

Consequently, $(1+x)^{-n}$ is the generating function for the sequence $\binom{-n}{0}, \binom{-n}{1}, \binom{-n}{2}, \ldots$, where $\binom{-n}{k}$ is an extended binomial coefficient (§2.3.2).

Table 3 Examples of binomial-type generating functions.

generating function	expansion
$(1+x)^n$	$\binom{n}{0} + \binom{n}{1}x + \binom{n}{2}x^2 + \cdots + \binom{n}{n}x^n = \sum_{k=0}^{n} \binom{n}{k}x^k$
$(1+rx)^n$	$\binom{n}{0} + \binom{n}{1}rx + \binom{n}{2}r^2x^2 + \cdots + \binom{n}{n}r^nx^n = \sum_{k=0}^{n} \binom{n}{k}r^k x^k$
$(1+x^m)^n$	$\binom{n}{0} + \binom{n}{1}x^m + \binom{n}{2}x^{2m} + \cdots + \binom{n}{n}x^{nm} = \sum_{k=0}^{n} \binom{n}{k}x^{km}$
$(1+x)^{-n}$	$\binom{-n}{0} + \binom{-n}{1}x + \binom{-n}{2}x^2 + \cdots = \sum_{k=0}^{\infty} (-1)^k \binom{n+k-1}{k} x^k$
$(1+rx)^{-n}$	$\binom{-n}{0} + \binom{-n}{1}rx + \binom{-n}{2}r^2x^2 + \cdots = \sum_{k=0}^{\infty} (-1)^k \binom{n+k-1}{k} r^k x^k$
$(1-x)^{-n}$	$\binom{-n}{0} + \binom{-n}{1}(-x) + \binom{-n}{2}(-x)^2 + \cdots = \sum_{k=0}^{\infty} \binom{n+k-1}{k} x^k$
$(1-rx)^{-n}$	$\binom{-n}{0} + \binom{-n}{1}(-rx) + \binom{-n}{2}(-rx)^2 + \cdots = \sum_{k=0}^{\infty} \binom{n+k-1}{k} r^k x^k$
$\dfrac{x^n}{(1-x)^{n+1}}$	$\binom{n}{n}x^n + \binom{n+1}{n}x^{n+1} + \binom{n+2}{n}x^{n+2} + \cdots = \sum_{k=n}^{\infty} \binom{k}{n} x^k$

6. Using Example 5, the expansion of $f(x) = (1-3x)^{-8}$ is

$$(1-3x)^{-8} = (1+y)^{-8} = \sum_{k=0}^{\infty} \binom{-8}{k} y^k = \sum_{k=0}^{\infty} \binom{-8}{k}(-3x)^k.$$

So the coefficient of x^4 in $f(x)$ is $\binom{-8}{4}(-3)^4 = (-1)^4 \binom{8+4-1}{4}(81) = \binom{11}{4}(81) = 26{,}730$.

7. Table 3 gives additional examples of generating functions related to binomial expansions. In this table, m and n are positive integers, and r is any real number.

8. For any real number r, the Maclaurin series expansion for $(1+x)^r$ is

$$(1+x)^r = \binom{r}{0}1 + \binom{r}{1}x + \binom{r}{2}x^2 + \cdots$$

where $\binom{r}{k} = \dfrac{r(r-1)(r-2)\cdots(r-k+1)}{k!}$ if $k > 0$ and $\binom{r}{0} = 1$.

9. Using Example 8, the expansion of $f(x) = \sqrt{1+x}$ is

$$\sqrt{1+x} = (1+x)^{1/2} = \binom{1/2}{0}1 + \binom{1/2}{1}x + \binom{1/2}{2}x^2 + \cdots$$

$$= 1 + \tfrac{1}{2}x + \tfrac{\tfrac{1}{2} \cdot \tfrac{-1}{2}}{2!}x^2 + \tfrac{\tfrac{1}{2} \cdot \tfrac{-1}{2} \cdot \tfrac{-3}{2}}{3!}x^3 + \tfrac{\tfrac{1}{2} \cdot \tfrac{-1}{2} \cdot \tfrac{-3}{2} \cdot \tfrac{-5}{2}}{4!}x^4 + \cdots$$

$$= 1 + \tfrac{1}{2}x - \tfrac{1}{8}x^2 + \tfrac{1}{16}x^3 - \tfrac{5}{128}x^4 + \cdots.$$

Thus $\sqrt{1+x}$ is the generating function for the sequence $1, \tfrac{1}{2}, -\tfrac{1}{8}, \tfrac{1}{16}, -\tfrac{5}{128}, \ldots$.

10. Vandermonde's convolution identity (§2.3.4) can be obtained from the generating functions $f(x) = (1+x)^m$ and $g(x) = (1+x)^n$. First, $(1+x)^m(1+x)^n = (1+x)^{m+n}$. Equating coefficients of x^r on both sides of this equation and using Fact 7 produces

$$\sum_{k=0}^{m} \binom{m}{k}\binom{n}{r-k} = \binom{m+n}{r}.$$

11. Twenty identical computer terminals are to be distributed into five distinct rooms so each room receives at least two terminals. The number of such distributions is the coefficient of x^{20} in the expansion of $f(x) = (x^2 + x^3 + x^4 + \cdots)^5 = x^{10}(1 + x + x^2 + \cdots)^5 = \dfrac{x^{10}}{(1-x)^5}$. Thus the coefficient of x^{20} in $f(x)$ is the coefficient of x^{10} in $(1-x)^{-5}$, which from Table 3 is $\binom{5+10-1}{10} = \binom{14}{10} = 1001$.

12. Suppose in Example 11 that each room can accommodate at most seven terminals. Now the generating function is $g(x) = (x^2 + x^3 + x^4 + x^5 + x^6 + x^7)^5 = x^{10}(1 + x + x^2 + x^3 + x^4 + x^5)^5 = x^{10}\left(\frac{1-x^6}{1-x}\right)^5$. Consequently, the number of allowable distributions is the coefficient of x^{10} in $\left(\frac{1-x^6}{1-x}\right)^5 = (1-x^6)^5(1-x)^{-5} = \left[1 - \binom{5}{1}x^6 + \binom{5}{2}x^{12} - \cdots - x^{30}\right]\left[\binom{-5}{0} + \binom{-5}{1}(-x) + \binom{-5}{2}(-x)^2 + \cdots\right]$. This coefficient is $\left[\binom{-5}{10}(-1)^{10} - \binom{5}{1}\binom{-5}{4}(-1)^4\right] = \binom{14}{10} - \binom{5}{1}\binom{8}{4} = 651$.

13. *Unordered selections with replacement:* k objects are selected from n distinct objects, with repetition allowed. For each of the n distinct objects, the power series $1 + x + x^2 + \cdots$ represents the possible choices (namely none, one, two, ...) for that object. The generating function for all n objects is then
$$f(x) = (1 + x + x^2 + \cdots)^n = \left(\frac{1}{1-x}\right)^n = (1-x)^{-n} = \sum_{k=0}^{\infty} \binom{n+k-1}{k} x^k.$$
The number of selections with replacement is the coefficient of x^k in $f(x)$, namely $\binom{n+k-1}{k}$.

14. Suppose there are p types of objects, with n_i indistinguishable objects of type i. The number of ways to pick a total of k objects (where the number of selected objects of type i is at most n_i) is the coefficient of x^k in the generating function
$$\prod_{i=1}^{p}(1 + x + x^2 + \cdots + x^{n_i}).$$

15. *Partitions:* Generating functions can be found for $p(n)$, the number of partitions of the positive integer n (§2.5.1). The number of 1s that appear as summands in a partition of n is 0 or 1 or 2 or ..., recorded as the terms in the power series $1 + x + x^2 + x^3 + \cdots$. The power series $1 + x^2 + x^4 + x^6 + \cdots$ records the number of 2s that can appear in a partition of n, and so forth. For example, $p(12)$ is the coefficient of x^{12} in
$$(1 + x + x^2 + \cdots)(1 + x^2 + x^4 + \cdots) \cdots (1 + x^{12} + x^{24} + \cdots) = \prod_{i=1}^{12} \frac{1}{1-x^i},$$
or in $(1 + x + x^2 + \cdots + x^{12})(1 + x^2 + x^4 + \cdots + x^{12}) \cdots (1 + x^{12})$. In general, the function $P(x) = \prod_{i=1}^{\infty} \frac{1}{1-x^i}$ is the generating function for the sequence $p(0), p(1), p(2), \ldots$, where $p(0)$ is defined as 1.

16. The function $P_d(x) = (1+x)(1+x^2)(1+x^3)\cdots = \prod_{i=1}^{\infty}(1+x^i)$ generates $Q(n)$, the number of partitions of n into distinct summands (see §2.5.1). The function $P_o(x) = \frac{1}{1-x} \cdot \frac{1}{1-x^3} \cdot \frac{1}{1-x^5} \cdots = \prod_{j=0}^{\infty}(1-x^{2j+1})^{-1}$ is the generating function for $\mathcal{O}(n)$, the number of partitions of n with all summands odd (see §2.5.1). Then
$$P_d(x) = (1+x)(1+x^2)(1+x^3)(1+x^4)\cdots$$
$$= \frac{1-x^2}{1-x} \cdot \frac{1-x^4}{1-x^2} \cdot \frac{1-x^6}{1-x^3} \cdot \frac{1-x^8}{1-x^4} \cdots = \frac{1}{1-x} \cdot \frac{1}{1-x^3} \cdots = P_o(x),$$
so $Q(n) = \mathcal{O}(n)$ for every nonnegative integer n.

17. *Summation formulas:* Generating functions can be used to produce the formula $1^2 + 2^2 + \cdots + n^2 = \frac{1}{6}n(n+1)(2n+1)$. (See §3.5.4 for an extensive tabulation of summation formulas.) Applying Fact 7 to the expansion $(1-x)^{-1} = 1 + x + x^2 + x^3 + \cdots$ produces
$$x\frac{d}{dx}\left[x\frac{d}{dx}(1-x)^{-1}\right] = \frac{x(1+x)}{(1-x)^3} = x + 2^2 x^2 + 3^2 x^3 + \cdots.$$
So $\frac{x(1+x)}{(1-x)^3}$ is the generating function for the sequence $0^2, 1^2, 2^2, 3^2, \ldots$ and, by Fact 7, $\frac{x(1+x)}{(1-x)^4}$ generates the sequence $0^2, 0^2 + 1^2, 0^2 + 1^2 + 2^2, 0^2 + 1^2 + 2^2 + 3^2, \ldots$. Consequently, $\sum_{i=0}^{n} i^2$ is the coefficient of x^n in
$$(x + x^2)(1-x)^{-4} = (x + x^2)\left[\binom{-4}{0} + \binom{-4}{1}(-x) + \binom{-4}{2}(-x)^2 + \cdots\right].$$
The answer is then $\binom{-4}{n-1}(-1)^{n-1} + \binom{-4}{n-2}(-1)^{n-2} = \binom{n+2}{n-1} + \binom{n+1}{n-2} = \frac{1}{6}n(n+1)(2n+1)$.

Table 4 Related exponential generating functions.

generating function	sequence
$xf(x)$	$0, a_0, 2a_1, 3a_2, \ldots, (k+1)a_k, \ldots$
$x^n f(x)$	$\underbrace{0, 0, 0, \ldots, 0}_{n}, P(n,n)a_0, P(n+1,n)a_1, P(n+2,n)a_2, \ldots,$
	$P(n+k,n)a_k, \ldots$
$f'(x)$	$a_1, a_2, a_3, \ldots, a_k, \ldots$
$\int_0^x f(t)dt$	$0, a_0, a_1, a_2, \ldots$
$rf(x) + sg(x)$	$ra_0 + sb_0, ra_1 + sb_1, ra_2 + sb_2, \ldots$
$f(x)g(x)$	$\binom{0}{0}a_0 b_0, \binom{1}{0}a_0 b_1 + \binom{1}{1}a_1 b_0, \binom{2}{0}a_0 b_2 + \binom{2}{1}a_1 b_1 + \binom{2}{2}a_2 b_0, \ldots$
	(binomial convolution of $\{a_k\}$ and $\{b_k\}$)

18. *Catalan numbers*: The Catalan numbers (§3.1.3) $C_0, C_1, C_2, \ldots$ satisfy the recurrence relation $C_n = C_0 C_{n-1} + C_1 C_{n-2} + \cdots + C_{n-1} C_0$, $n \geq 1$, with $C_0 = 1$. (See §3.3.1.) Hence their generating function $f(x) = \sum_{k=0}^{\infty} C_k x^k$ satisfies $xf^2(x) = f(x) - 1$, yielding $f(x) = \frac{1}{2x}(1 - \sqrt{1-4x}) = \frac{1}{2x}(1 - (1-4x)^{1/2})$. (The negative square root is chosen since the numbers C_i cannot be negative.) Applying Example 8 to $(1-4x)^{1/2}$ yields $f(x) = \frac{1}{2x}[1 - \sum_{k=0}^{\infty} \binom{1/2}{k}(-4)^k x^k] = \frac{1}{2x}[1 - \sum_{k=0}^{\infty} \frac{-1}{2k-1}\binom{2k}{k} x^k] = \sum_{k=0}^{\infty} \frac{1}{k+1}\binom{2k}{k} x^k$. Thus $C_n = \frac{1}{n+1}\binom{2n}{n}$.

3.2.2 EXPONENTIAL GENERATING FUNCTIONS

Encoding the terms of a sequence as coefficients of $\frac{x^k}{k!}$ is often helpful in obtaining information about a sequence, such as in counting permutations of objects (where the order of listing objects is important). The functions that result are called exponential generating functions.

Definitions:

The **exponential generating function** for the sequence $a_0, a_1, a_2, \ldots$ of real numbers is the formal power series $f(x) = a_0 + a_1 x + a_2 \frac{x^2}{2!} + \cdots = \sum_{i=0}^{\infty} a_i \frac{x^i}{i!}$ or any equivalent closed form expression.

The **binomial convolution** of the sequence $a_0, a_1, a_2, \ldots$ and the sequence $b_0, b_1, b_2, \ldots$ is the sequence $c_0, c_1, c_2, \ldots$ in which $c_t = \binom{t}{0}a_0 b_t + \binom{t}{1}a_1 b_{t-1} + \binom{t}{2}a_2 b_{t-2} + \cdots + \binom{t}{t}a_t b_0 = \sum_{k=0}^{t} \binom{t}{k} a_k b_{t-k}$.

Facts:

1. Each sequence $\{a_n\}$ defines a unique exponential generating function $f(x)$, and conversely.

2. *Related exponential generating functions*: Suppose $f(x) = \sum_{k=0}^{\infty} a_k \frac{x^k}{k!}$ and $g(x) = \sum_{k=0}^{\infty} b_k \frac{x^k}{k!}$ are exponential generating functions for the sequences $a_0, a_1, a_2, \ldots$ and $b_0, b_1, b_2, \ldots$, respectively. Table 4 gives some related exponential generating functions. $[P(n,k) = \binom{n}{k}k!$ is the number of k-permutations of a set with n distinct objects. (See §2.3.1.)]

Table 5 Exponential generating functions for particular sequences.

sequence	closed form
$1, 1, 1, 1, 1, \ldots$	e^x
$1, -1, 1, -1, 1, \ldots$	e^{-x}
$1, 0, 1, 0, 1, \ldots$	$\frac{1}{2}(e^x + e^{-x})$
$0, 1, 0, 1, 0, \ldots$	$\frac{1}{2}(e^x - e^{-x})$
$0, 1, 2, 3, 4, \ldots$	xe^x
$P(n,0), P(n,1), \ldots, P(n,n), 0, 0, \ldots$	$(1+x)^n$
$\begin{bmatrix} 0 \\ n \end{bmatrix}, \ldots, \begin{bmatrix} n \\ n \end{bmatrix}, \begin{bmatrix} n+1 \\ n \end{bmatrix}, \ldots$	$\frac{1}{n!}\left[\ln \frac{1}{(1-x)}\right]^n$
$\left\{\begin{matrix} 0 \\ n \end{matrix}\right\}, \ldots, \left\{\begin{matrix} n \\ n \end{matrix}\right\}, \left\{\begin{matrix} n+1 \\ n \end{matrix}\right\}, \ldots$	$\frac{1}{n!}[e^x - 1]^n$
$B_0, B_1, B_2, B_3, B_4, \ldots$	$e^{e^x - 1}$
$D_0, D_1, D_2, D_3, D_4, \ldots$	$\frac{e^{-x}}{1-x}$

Examples:

1. The binomial theorem (§2.3.4) gives
$$(1+x)^n = \binom{n}{0} + \binom{n}{1}x + \binom{n}{2}x^2 + \binom{n}{3}x^3 + \cdots + \binom{n}{n}x^n$$
$$= P(n,0) + P(n,1)x + P(n,2)\frac{x^2}{2!} + P(n,3)\frac{x^3}{3!} + \cdots + P(n,n)\frac{x^n}{n!}.$$
Hence $(1+x)^n$ is the exponential generating function for the sequence $P(n,0)$, $P(n,1)$, $P(n,2)$, $P(n,3)$, $\ldots$, $P(n,n), 0, 0, 0, \ldots$.

2. The Maclaurin series expansion for e^x is $e^x = 1 + x + \frac{x^2}{2!} + \frac{x^3}{3!} + \cdots$, so the function e^x is the exponential generating function for the sequence $1, 1, 1, 1, \ldots$. The function $e^{-x} = 1 - x + \frac{x^2}{2!} - \frac{x^3}{3!} + \cdots$ is the exponential generating function for the sequence $1, -1, 1, -1, \ldots$. Consequently,
$$\tfrac{1}{2}(e^x + e^{-x}) = 1 + \frac{x^2}{2!} + \frac{x^4}{4!} + \cdots$$
is the exponential generating function for $1, 0, 1, 0, 1, 0, \ldots$, while
$$\tfrac{1}{2}(e^x - e^{-x}) = x + \frac{x^3}{3!} + \frac{x^5}{5!} + \cdots$$
is the exponential generating function for $0, 1, 0, 1, 0, 1, \ldots$.

3. The function $f(x) = \frac{1}{1-x} = \sum_{i=0}^{\infty} x^i = \sum_{i=0}^{\infty} i! \frac{x^i}{i!}$ is the exponential generating function for the sequence $0!, 1!, 2!, 3!, \ldots$.

4. Table 5 gives closed form expressions for the exponential generating functions of particular sequences. In this table, $\begin{bmatrix} n \\ k \end{bmatrix}$ is a Stirling cycle number, $\left\{\begin{matrix} n \\ k \end{matrix}\right\}$ is a Stirling subset number, B_n is the nth Bell number (§2.5.2), and D_n is the number of derangements of n objects (§2.4.2).

5. The number of ways to permute 5 of the 8 letters in TERMINAL is found using the exponential generating function $f(x) = (1+x)^8$. Here each of the 8 letters in TERMINAL is accounted for by the factor $(1+x)$, where $1(=x^0)$ indicates the letter does not occur in the permutation and $x(=x^1)$ indicates that it does. The coefficient of $\frac{x^5}{5!}$ in $f(x)$ is $\binom{8}{5}5! = P(8,5) = 6{,}720$.

6. The number of ways to permute 5 of the letters in TRANSPORTATION is found as the coefficient of $\frac{x^5}{5!}$ in the exponential generating function $f(x) = (1 + x + \frac{x^2}{2!} + \frac{x^3}{3!})(1 + x + \frac{x^2}{2!})^4(1 + x)^3$. Here the factor $1 + x + \frac{x^2}{2!} + \frac{x^3}{3!}$ accounts for the letter T which can be used 0, 1, 2, or 3 times. The factor $1 + x + \frac{x^2}{2!}$ occurs four times — for each of R, A, N, and O. The letters S, P, and I produce the factor $(1 + x)$. The coefficient of x^5 in $f(x)$ is found to be $\frac{487}{3}$, so the answer is $(\frac{487}{3})5! = 19{,}480$.

7. The number of ternary sequences (made up of 0s, 1s, and 2s) of length 10 with at least one 0 and an odd number of 1s can be found using the exponential generating function

$$f(x) = (x + \frac{x^2}{2!} + \frac{x^3}{3!} + \cdots)(x + \frac{x^3}{3!} + \frac{x^5}{5!} + \cdots)(1 + x + \frac{x^2}{2!} + \frac{x^3}{3!} + \cdots)$$

$$= (e^x - 1)\tfrac{1}{2}(e^x - e^{-x})e^x = \tfrac{1}{2}(e^{3x} - e^{2x} - e^x + 1)$$

$$= \tfrac{1}{2}\left(\sum_{i=0}^{\infty} \frac{(3x)^i}{i!} - \sum_{i=0}^{\infty} \frac{(2x)^i}{i!} - \sum_{i=0}^{\infty} \frac{x^i}{i!} + 1\right).$$

The answer is the coefficient of $\frac{x^{10}}{10!}$ in $f(x)$, which is $\tfrac{1}{2}(3^{10} - 2^{10} - 1^{10}) = 29{,}012$.

8. Suppose in Example 7 that no symbol may occur exactly two times. The exponential generating function is then $f(x) = (1 + x + \frac{x^3}{3!} + \frac{x^4}{4!} + \cdots)^3 = (e^x - \frac{x^2}{2})^3 = e^{3x} - \frac{3}{2}x^2 e^{2x} + \frac{3}{4}x^4 e^x - \frac{1}{8}x^6$. The number of ternary sequences is the coefficient of $\frac{x^{10}}{10!}$ in $f(x)$, namely $3^{10} - \frac{3}{2}(10)(9)2^8 + \frac{3}{4}(10)(9)(8)(7)1^6 = 28{,}269$.

9. Exponential generating functions can be used to count the number of onto functions $\varphi\colon A \to B$ where $|A| = m$ and $|B| = n$. Each such function is specified by the sequence of m values $\varphi(a_1), \varphi(a_2), \ldots, \varphi(a_m)$, where each element $b \in B$ occurs at least once in this sequence. Element b contributes a factor $(x + \frac{x^2}{2!} + \frac{x^3}{3!} + \cdots) = (e^x - 1)$ to the exponential generating function $f(x) = (e^x - 1)^n$. The number of onto functions is the coefficient of $\frac{x^m}{m!}$ in $f(x)$, or $n!$ times the coefficient of $\frac{x^m}{m!}$ in $\frac{(e^x - 1)^n}{n!}$. From Table 5, the answer is then $n!\left\{{m \atop n}\right\}$.

3.3 RECURRENCE RELATIONS

In a number of counting problems, it may be difficult to find the solution directly. However, it is frequently possible to express the solution to a problem of a given size in terms of solutions to problems of smaller size. This interdependence of solutions produces a recurrence relation. Although there is no practical systematic way to solve all recurrence relations, this section contains methods for solving certain types of recurrence relations, thereby providing an explicit formula for the original counting problem. The topic of recurrence relations provides the discrete counterpart to concepts in the study of ordinary differential equations.

3.3.1 BASIC CONCEPTS

Definitions:

A **recurrence relation** for the sequence $a_0, a_1, a_2, \ldots$ is an equation relating the term a_n to certain of the preceding terms a_i, $i < n$, for each $n \geq n_0$.

Section 3.3 RECURRENCE RELATIONS

The recurrence relation is **linear** if it expresses a_n as a linear function of a fixed number of preceding terms. Otherwise the relation is **nonlinear**.

The recurrence relation is **kth-order** if a_n can be expressed in terms of $a_{n-1}, a_{n-2}, \ldots, a_{n-k}$.

The recurrence relation is **homogeneous** if the zero sequence $a_0 = a_1 = \cdots = 0$ satisfies the relation. Otherwise the relation is **nonhomogeneous**.

A kth-order linear homogeneous recurrence relation **with constant coefficients** is an equation of the form $C_n a_n + C_{n-1} a_{n-1} + \cdots + C_{n-k} a_{n-k} = 0$, $n \geq k$, where the C_i are real constants with $C_n \neq 0$, $C_{n-k} \neq 0$. **Initial conditions** for this recurrence relation specify particular values for k of the a_i (typically $a_0, a_1, \ldots, a_{k-1}$).

Facts:

1. A kth-order linear homogeneous recurrence relation with constant coefficients can also be written $C_{n+k} a_{n+k} + C_{n+k-1} a_{n+k-1} + \cdots + C_n a_n = 0$, $n \geq 0$.

2. There are in general an infinite number of solution sequences $\{a_n\}$ to a kth-order linear homogeneous recurrence relation (with constant coefficients).

3. A kth-order linear homogeneous recurrence relation with constant coefficients together with k initial conditions on consecutive terms $a_0, a_1, \ldots, a_{k-1}$ uniquely determines the sequence $\{a_n\}$. This is not necessarily the case for nonlinear relations (see Example 2) or when nonconsecutive initial conditions are specified (see Example 3).

4. The same recurrence relation can be written in different forms by adjusting the subscripts. For example, the recurrence relation $a_n = 3 a_{n-1}$, $n \geq 1$, can be written as $a_{n+1} = 3 a_n$, $n \geq 0$.

Examples:

1. The relation $a_n - a_{n-1}^2 + 2 a_{n-2} = 0$, $n \geq 2$ is a nonlinear homogeneous recurrence relation with constant coefficients. If the initial conditions $a_0 = 0$, $a_1 = 1$ are imposed, this defines a unique sequence $\{a_n\}$ whose first few terms are $0, 1, 1, -1, -1, 3, 11, 115, \ldots$.

2. The first-order (constant coefficient) recurrence relation $a_{n+1}^2 - a_n = 3$, $a_0 = 1$ is nonhomogeneous and nonlinear. Even though one initial condition is specified, this does not uniquely specify a solution sequence. Namely, the two sequences $1, -2, 1, 2, \ldots$ and $1, -2, -1, \sqrt{2}, \ldots$ satisfy the recurrence relation and the given initial condition.

3. The second-order relation $a_{n+2} - a_n = 0$, $n \geq 0$, with nonconsecutive initial conditions $a_1 = a_3 = 0$ does not uniquely specify a solution sequence. Both $a_n = (-1)^n + 1$ and $a_n = 2(-1)^n + 2$ satisfy the recurrence and the given initial conditions.

4. *Compound interest*: If an initial investment of P dollars is made at a rate of r percent compounded annually, then the amount a_n after n years is given by the recurrence relation $a_n = a_{n-1}(1 + \frac{r}{100})$, where $a_0 = P$. [The amount at the end of the nth year is equal to the amount at the end of the $(n-1)$st year, a_{n-1}, plus the interest on a_{n-1}, $\frac{r}{100} a_{n-1}$.]

5. *Fibonacci sequence*: The Fibonacci numbers satisfy the second-order linear homogeneous recurrence relation $a_n - a_{n-1} - a_{n-2} = 0$.

6. *Bit strings*: Let a_n be the number of bit strings of length n. Then $a_0 = 1$ (the empty string) and $a_n = 2 a_{n-1}$ if $n > 0$. [Every bit string of length $n - 1$ gives rise to two bit strings of length n, by placing a 0 or a 1 at the end of the string of length $n - 1$.]

7. *Bit strings with no consecutive 0s*: See §3.3.2 Example 23.

8. *Permutations*: Let a_n denote the number of permutations of $\{1, 2, \ldots, n\}$. Then a_n satisfies the first-order linear homogeneous recurrence relation (with nonconstant coefficients) $a_{n+1} = (n+1)a_n$, $n \geq 1$, $a_1 = 1$. This follows since any n-permutation π can be transformed into an $(n+1)$-permutation by inserting the element $n+1$ into any of the $n+1$ available positions — either at the beginning or end of π, or between two adjacent elements of π. To solve for a_n, repeatedly apply the recurrence relation and its initial condition: $a_n = na_{n-1} = n(n-1)a_{n-2} = n(n-1)(n-2)a_{n-3} = \cdots = n(n-1)(n-2)\ldots 2a_1 = n!$.

9. *Catalan numbers*: The Catalan numbers (§3.1.3, 3.2.1) satisfy the nonlinear homogeneous recurrence relation $C_n - C_0 C_{n-1} - C_1 C_{n-2} - \cdots - C_{n-1} C_0 = 0$, $n \geq 1$, with initial condition $C_0 = 1$. Given the product of $n+1$ variables $x_1 x_2 \ldots x_{n+1}$, let C_n be the number of ways in which the multiplications can be carried out. For example, there are five ways to form the product $x_1 x_2 x_3 x_4$: $((x_1 x_2)x_3)x_4$, $(x_1(x_2 x_3))x_4$, $(x_1 x_2)(x_3 x_4)$, $x_1((x_2 x_3)x_4)$, and $x_1(x_2(x_3 x_4))$. No matter how the multiplications are performed, there will be an outermost product of the form $(x_1 x_2 \ldots x_i)(x_{i+1} \ldots x_{n+1})$. The number of ways in which the product $x_1 x_2 \ldots x_i$ can be formed is C_{i-1} and the number of ways in which the product $x_{i+1} \ldots x_{n+1}$ can be formed is C_{n-i}. Thus, $(x_1 x_2 \ldots x_i)(x_{i+1} \ldots x_{n+1})$ can be obtained in $C_{i-1} C_{n-i}$ ways. Summing these over the values $i = 1, 2, \ldots, n$ yields the recurrence relation.

10. *Tower of Hanoi*: See Example 1 of §2.2.4.

11. *Onto functions*: The number of onto functions $\varphi: A \to B$ can be found by developing a nonhomogeneous linear recurrence relation based on the size of B. Let $|A| = m$ and let a_n be the number of onto functions from A to a set with n elements. Then $a_n = n^m - \binom{n}{1}a_1 - \binom{n}{2}a_2 - \cdots - \binom{n}{n-1}a_{n-1}$, $n \geq 2$, $a_1 = 1$. This follows since the total number of functions from A to B is n^m and the number of functions that map A onto a proper subset of B with exactly j elements is $\binom{n}{j}a_j$.

For example, if $m = 7$ and $n = 4$, applying this recursion gives $a_2 = 2^7 - 2(1) = 126$, $a_3 = 3^7 - 3(1) - 3(126) = 1{,}806$, $a_4 = 4^7 - 4(1) - 6(126) - 4(1{,}806) = 8{,}400$. Thus there are 8,400 onto functions in this case.

3.3.2 HOMOGENEOUS RECURRENCE RELATIONS

It is assumed throughout this subsection that the recurrence relations are linear with constant coefficients.

Definitions:

A **geometric progression** is a sequence $a_0, a_1, a_2, \ldots$ for which $\dfrac{a_1}{a_0} = \dfrac{a_2}{a_1} = \cdots = \dfrac{a_{n+1}}{a_n} = \cdots = r$, the **common ratio**.

The **characteristic equation** for the kth-order recurrence relation $C_n a_n + C_{n-1} a_{n-1} + \cdots + C_{n-k} a_{n-k} = 0$, $n \geq k$, is the equation $C_n r^k + C_{n-1} r^{k-1} + \cdots + C_{n-k} = 0$. The **characteristic roots** are the roots of this equation.

The sequences $\{a_n^{(1)}\}, \{a_n^{(2)}\}, \ldots, \{a_n^{(k)}\}$ are **linearly dependent** if there exist constants $t_1, t_2, \ldots, t_k$, not all zero, such that $\sum_{i=1}^{k} t_i a_n^{(i)} = 0$ for all $n \geq 0$. Otherwise, they are **linearly independent**.

Facts:

1. *General method for solving a linear homogeneous recurrence relation with constant coefficients*: First find the general solution. Then use the initial conditions to find the particular solution.
2. If the k characteristic roots $r_1, r_2, \ldots, r_k$ are distinct, then $r_1^n, r_2^n, \ldots, r_k^n$ are linearly independent solutions of the homogeneous recurrence relation. The general solution is $a_n = c_1 r_1^n + c_2 r_2^n + \cdots + c_k r_k^n$, where $c_1, c_2, \ldots, c_k$ are arbitrary constants.
3. If a characteristic root r has multiplicity m, then $r^n, nr^n, \ldots, n^{m-1}r^n$ are linearly independent solutions of the homogeneous recurrence relation. The linear combination $c_1 r^n + c_2 n r^n + \cdots + c_m n^{m-1} r^n$ is also a solution, where $c_1, c_2, \ldots, c_m$ are arbitrary constants.
4. Facts 2 and 3 can be used together. If there are k characteristic roots $r_1, r_2, \ldots, r_k$, with respective multiplicities $m_1, m_2, \ldots, m_k$ (where some of the m_i can equal 1), the the general solution is a sum of sums, each of the form appearing in Fact 3.
5. *DeMoivre's theorem*: For any positive integer n, $(\cos\theta + i\sin\theta)^n = \cos n\theta + i \sin n\theta$. This result is used to find solutions of recurrence relations when the characteristic roots are complex numbers. (See Example 10.)
6. *Solving first-order recurrence relations*: The solution of the homogeneous recurrence relation $a_{n+1} = d a_n$, $n \geq 0$, with initial condition $a_0 = A$, is $a_n = A d^n$, $n \geq 0$.
7. *Solving second-order recurrence relations*: Let r_1, r_2 be the characteristic roots associated with the second-order homogeneous relation $C_n a_n + C_{n-1} a_{n-1} + C_{n-2} a_{n-2} = 0$. There are three possibilities:

   - r_1, r_2 are distinct real numbers: r_1^n and r_2^n are linearly independent solutions of the recurrence relation. The general solution has the form
   $$a_n = c_1 r_1^n + c_2 r_2^n,$$
   where the constants c_1, c_2 are found from the values of a_n for two distinct values of n (often $n = 0, 1$).

   - r_1, r_2 form a complex conjugate pair $a \pm bi$: The general solution is
   $$a_n = c_1(a+bi)^n + c_2(a-bi)^n = (\sqrt{a^2+b^2})^n (k_1 \cos n\theta + k_2 \sin n\theta),$$
   with $\theta = \arctan(b/a)$. Here $(\sqrt{a^2+b^2})^n \cos n\theta$ and $(\sqrt{a^2+b^2})^n \sin n\theta$ are linearly independent solutions.

   - r_1, r_2 are real and equal: r_1^n and nr_1^n are linearly independent solutions of the recurrence relation. The general solution is
   $$a_n = c_1 r_1^n + c_2 n r_1^n.$$

Examples:

1. The geometric progression $7, 21, 63, 189, \ldots$, with common ratio 3, satisfies the first-order homogeneous recurrence relation $a_{n+1} - 3a_n = 0$ for all $n \geq 0$.
2. The first-order homogeneous recurrence relation $a_{n+1} = 3a_n$, $n \geq 0$, does not determine a *unique* geometric progression. Any geometric sequence with ratio 3 is a solution; for example the geometric progression in Example 1 (with $a_0 = 7$), as well as the geometric progression $5, 15, 45, 135, \ldots$ (with $a_0 = 5$).
3. The first-order recurrence relation $a_{n+1} = 3a_n$, $n \geq 0$, $a_0 = 7$ is easily solved using Fact 6. The general solution is $a_n = 7(3^n)$ for all $n \geq 0$.
4. *Compound interest*: If interest is compounded quarterly, how long does it take for an investment of $500 to double when the annual interest rate is 8%? If a_n denotes the value of the investment after n quarters have passed, then $a_{n+1} = a_n + 0.02 a_n = (1.02)a_n$,

$n \geq 0$, $a_0 = 500$. [Here the quarterly rate is $0.08/4 = 0.02 = 2\%$.] By Fact 6, the solution is $a_n = 500(1.02)^n$, $n \geq 0$. The investment doubles when $1000 = 500(1.02)^n$, so $n = \frac{\log 2}{\log 1.02} \approx 35.003$. Consequently, after 36 quarters (or 9 years) the initial investment of \$500 (more than) doubles.

5. *Population growth*: The number of bacteria in a culture (approximately) triples in size every hour. If there are (approximately) 100,000 bacteria in a culture after six hours, how many were there at the start? Define p_n to be the number of bacteria in the culture after n hours have elapsed. Then $p_{n+1} = 3p_n$ for $n \geq 0$. From Fact 5, $p_n = p_0(3^n)$. So $100{,}000 = p_0(3^6)$ and $p_0 \approx 137$.

6. *Fibonacci sequence*: The Fibonacci sequence $0, 1, 1, 2, 3, 5, 8, 13, \ldots$ arises in varied applications (§3.1.2). Its terms satisfy the second-order homogeneous recurrence relation $F_n = F_{n-1} + F_{n-2}$, $n \geq 2$, with initial conditions $F_0 = 0$, $F_1 = 1$.

An explicit formula can be obtained for F_n using Fact 7. The characteristic equation is $r^2 - r - 1 = 0$, with distinct real roots $\frac{1 \pm \sqrt{5}}{2}$. Thus the general solution is

$$F_n = c_1 \left(\frac{1+\sqrt{5}}{2}\right)^n + c_2 \left(\frac{1-\sqrt{5}}{2}\right)^n$$

Using the initial conditions $F_0 = 0$, $F_1 = 1$ gives $c_1 = \frac{1}{\sqrt{5}}$, $c_2 = -\frac{1}{\sqrt{5}}$ and the explicit formula

$$F_n = \frac{1}{\sqrt{5}} \left[\left(\frac{1+\sqrt{5}}{2}\right)^n - \left(\frac{1-\sqrt{5}}{2}\right)^n \right], \quad n \geq 0.$$

7. *Lucas sequence*: Related to the sequence of Fibonacci numbers is the sequence of Lucas numbers $2, 1, 3, 4, 7, 11, 18, \ldots$ (see §3.1.2). The terms of this sequence satisfy the same second-order homogeneous recurrence relation $L_n = L_{n-1} + L_{n-2}$, $n \geq 2$, but with the different initial conditions $L_0 = 2$, $L_1 = 1$. The formula for L_n is

$$L_n = \left(\frac{1+\sqrt{5}}{2}\right)^n + \left(\frac{1-\sqrt{5}}{2}\right)^n, \quad n \geq 0.$$

8. *Random walk*: A particle undergoes a random walk in one dimension, along the x-axis. Barriers are placed at positions $x = 0$ and $x = T$. At any instant, the particle moves with probability p one unit to the right; with probability $q = 1 - p$ it moves one unit to the left. Let a_n denote the probability that the particle, starting at position $x = n$, reaches the barrier $x = T$ before it reaches the barrier $x = 0$. It can be shown that a_n satisfies the second-order recurrence relation $a_n = pa_{n+1} + qa_{n-1}$ or $pa_{n+1} - a_n + qa_{n-1} = 0$. In this case the two initial conditions are $a_0 = 0$ and $a_T = 1$. The characteristic equation $pr^2 - r + q = (pr - q)(r - 1) = 0$ has roots $1, \frac{q}{p}$. When $p \neq q$, the roots are distinct and the first case of Fact 7 can be used to determine a_n; when $p = q$, the third case of Fact 7 must be used. (Explicit solutions are given in §7.5.2, Fact 10.)

9. The second-order relation $a_n + 4a_{n-1} - 21a_{n-2} = 0$, $n \geq 2$, has the characteristic equation $r^2 + 4r - 21 = 0$, with distinct real roots 3 and -7. The general solution to the recurrence relation is

$$a_n = c_1(3)^n + c_2(-7)^n, \quad n \geq 0,$$

where c_1, c_2 are arbitrary constants.

If the initial conditions specify $a_0 = 1$ and $a_1 = 1$, then solving the equations $1 = a_0 = c_1 + c_2$, $1 = a_1 = 3c_1 - 7c_2$ gives $c_1 = \frac{4}{5}$, $c_2 = \frac{1}{5}$. In this case, the unique solution is

$$a_n = \tfrac{4}{5}3^n + \tfrac{1}{5}(-7)^n, \quad n \geq 0.$$

10. The second-order relation $a_n - 6a_{n-1} + 58a_{n-2} = 0$, $n \geq 2$, has the characteristic equation $r^2 - 6r + 58 = 0$, with complex conjugate roots $r = 3 \pm 7i$. The general solution is
$$a_n = c_1(3+7i)^n + c_2(3-7i)^n, \quad n \geq 0.$$
Using Fact 5, $(3+7i)^n = [\sqrt{3^2 + 7^2}(\cos\theta + i\sin\theta)]^n = (\sqrt{58})^n(\cos n\theta + i\sin n\theta)$, where $\theta = \arctan \frac{7}{3}$. Likewise $(3-7i)^n = (\sqrt{58})^n(\cos n\theta - i\sin n\theta)$. This gives the general solution
$$a_n = (\sqrt{58})^n[(c_1+c_2)\cos n\theta + (c_1-c_2)i\sin n\theta] = (\sqrt{58})^n[k_1 \cos n\theta + k_2 \sin n\theta].$$
If the initial conditions $a_0 = 1$ and $a_1 = 1$ are specified, then $1 = a_0 = k_1$, $1 = a_1 = \sqrt{58}[\cos\theta + k_2 \sin\theta]$, yielding $k_1 = 1$, $k_2 = -\frac{2}{7}$. Thus
$$a_n = (\sqrt{58})^n[\cos n\theta - \tfrac{2}{7}\sin n\theta], \quad n \geq 0.$$

11. The second-order relation $a_{n+2} - 6a_{n+1} + 9a_n = 0$, $n \geq 0$, has the characteristic equation $r^2 - 6r + 9 = (r-3)^2 = 0$, with the repeated roots 3, 3. The general solution to this recurrence is
$$a_n = c_1(3^n) + c_2 n(3^n), \quad n \geq 0.$$
If the initial conditions are $a_0 = 2$ and $a_1 = 4$, then $2 = a_0 = c_1$, $4 = 2(3) + c_2(1)(3)$, giving $c_1 = 2$, $c_2 = -\frac{2}{3}$. Thus
$$a_n = 2(3^n) - \tfrac{2}{3}n(3^n) = 2(3^n - n3^{n-1}), \quad n \geq 0.$$

12. For $n \geq 1$, let a_n count the number of binary strings of length n that contain no consecutive 0s. Here $a_1 = 2$ (for the two strings 0 and 1) and $a_2 = 3$ (for the strings 01, 10, 11). For $n \geq 3$, a string counted in a_n ends in either 1 or 0. If the nth bit is 1, then the preceding $n-1$ bits provide a string counted in a_{n-1}; if the nth bit is 0 then the last two bits are 10, and the preceding $n-2$ bits give a string counted in a_{n-2}. Thus $a_n = a_{n-1} + a_{n-2}$, $n \geq 3$, with $a_1 = 2$ and $a_2 = 3$. The solution to this relation is simply $a_n = F_{n+2}$, the Fibonacci sequence shifted two places. An explicit formula for a_n is obtained using the result in Example 6.

13. The third-order recurrence relation $a_{n+3} - a_{n+2} - 4a_{n+1} + 4a_n = 0$, $n \geq 0$, has the characteristic equation $r^3 - r^2 - 4r + 4 = (r-2)(r+2)(r-1) = 0$, with characteristic roots 2, -2, and 1. The general solution is given by
$$a_n = c_1 2^n + c_2(-2)^n + c_3 1^n = c_1 2^n + c_2(-2)^n + c_3, \quad n \geq 0.$$

14. The general solution of the third-order recurrence relation $a_{n+3} - 3a_{n+2} - 3a_{n+1} + a_n = 0$, $n \geq 0$, is
$$a_n = c_1 1^n + c_2 n 1^n + c_3 n^2 1^n = c_1 + c_2 n + c_3 n^2, \quad n \geq 0.$$
Here the characteristic roots are 1, 1, 1.

15. The fourth-order relation $a_{n+4} + 2a_{n+2} + a_n = 0$, $n \geq 0$, has the characteristic equation $r^4 + 2r^2 + 1 = (r^2 + 1)^2 = 0$. Since the characteristic roots are $\pm i, \pm i$, the general solution is
$$a_n = c_1 i^n + c_2(-i)^n + c_3 n i^n + c_4 n(-i)^n$$
$$= k_1 \cos \tfrac{n\pi}{2} + k_2 \sin \tfrac{n\pi}{2} + k_3 n \cos \tfrac{n\pi}{2} + k_4 n \sin \tfrac{n\pi}{2}, \quad n \geq 0.$$

3.3.3 NONHOMOGENEOUS RECURRENCE RELATIONS

It is assumed throughout this subsection that the recurrence relations are linear with constant coefficients.

Definition:

The kth-order **nonhomogeneous** recurrence relation has the form $C_n a_n + C_{n-1} a_{n-1} + \cdots + C_{n-k} a_{n-k} = f(n)$, $n \geq k$, where $C_n \neq 0$, $C_{n-k} \neq 0$, and $f(n) \neq 0$ for at least one value of n.

Facts:

1. *General solution*: The general solution of the nonhomogeneous kth-order recurrence relation has the form
$$a_n = a_n^{(h)} + a_n^{(p)},$$
where $a_n^{(h)}$ is the general solution of the homogeneous relation $C_n a_n + C_{n-1} a_{n-1} + \cdots + C_{n-k} a_{n-k} = 0$, $n \geq k$, and $a_n^{(p)}$ is a particular solution for the given relation $C_n a_n + C_{n-1} a_{n-1} + \cdots + C_{n-k} a_{n-k} = f(n)$, $n \geq k$.

2. Given a nonhomogeneous first-order relation $C_n a_n + C_{n-1} a_{n-1} = kr^n$, $n \geq 1$, where r and k are nonzero constants,
   - If r^n is *not* a solution of the associated homogeneous relation, then $a_n^{(p)} = Ar^n$ for A a constant.
   - If r^n is a solution of the associated homogeneous relation, then $a_n^{(p)} = Bnr^n$ for B a constant.

3. Given the nonhomogeneous second-order relation $C_n a_n + C_{n-1} a_{n-1} + C_{n-2} a_{n-2} = kr^n$, $n \geq 2$, where r and k are nonzero constants.
   - If r^n is *not* a solution of the associated homogeneous relation, then $a_n^{(p)} = Ar^n$ for A a constant.
   - If $a_n^{(h)} = c_1 r^n + c_2 r_1^n$, for $r \neq r_1$, then $a_n^{(p)} = Bnr^n$ for B a constant.
   - If $a_n^{(h)} = c_1 r^n + c_2 nr^n$, then $a_n^{(p)} = Cn^2 r^n$ for C a constant.

4. Given the kth-order nonhomogeneous recurrence relation $C_n a_n + C_{n-1} a_{n-1} + \cdots + C_{n-k} a_{n-k} = f(n)$. If $f(n)$ is a constant multiple of one of the forms in the first column of Table 1, then the associated *trial solution* $t(n)$ is the corresponding entry in the second column of the table. [Here $A, B, A_0, A_1, \ldots, A_t, r, \alpha$ are real constants.]
   - If no summand of $t(n)$ solves the associated homogeneous relation, then $a_n^{(p)} = t(n)$ is a particular solution.
   - If a summand of $t(n)$ solves the associated homogeneous relation, then multiply $t(n)$ by the *smallest* (positive integer) power of n — say n^s — so that no summand of the adjusted trial solution $n^s t(n)$ solves the associated homogeneous relation. Then $a_n^{(p)} = n^s t(n)$ is a particular solution.
   - If $f(n)$ is a sum of constant multiples of the forms in the first column of Table 1, then (adjusted) trial solutions are formed for each summand using the first two parts of Fact 4. Adding the resulting trial solutions then provides a particular solution of the nonhomogeneous relation.

Table 1 Trial particular solutions for $C_n a_n + \cdots + C_{n-k} a_{n-k} = h(n)$.

$h(n)$	$t(n)$
c, a constant	A
n^t (t a positive integer)	$A_t n^t + A_{t-1} n^{t-1} + \cdots + A_1 n + A_0$
r^n	$A r^n$
$\sin \alpha n$	$A \sin \alpha n + B \cos \alpha n$
$\cos \alpha n$	$A \sin \alpha n + B \cos \alpha n$
$n^t r^n$	$r^n (A_t n^t + A_{t-1} n^{t-1} + \cdots + A_1 n + A_0)$
$r^n \sin \alpha n$	$r^n (A \sin \alpha n + B \cos \alpha n)$
$r^n \cos \alpha n$	$r^n (A \sin \alpha n + B \cos \alpha n)$

Examples:

1. Consider the nonhomogeneous relation $a_n + 4a_{n-1} - 21a_{n-2} = 5(4^n)$, $n \geq 2$. The solution is $a_n = a_n^{(h)} + a_n^{(p)}$, where $a_n^{(h)}$ is the solution of $a_n + 4a_{n-1} - 21a_{n-2} = 0$, $n \geq 2$. So

$$a_n^{(h)} = c_1(3)^n + c_2(-7)^n, \quad n \geq 0.$$

From the third entry in Table 1 $a_n^{(p)} = A(4^n)$ for some constant A. Substituting this into the given nonhomogeneous relation yields $A(4^n) + 4A(4^{n-1}) - 21A(4^{n-2}) = 5(4^n)$. Dividing through by 4^{n-2} gives $16A + 16A - 21A = 80$, or $A = 80/11$. Consequently,

$$a_n = c_1(3)^n + c_2(-7)^n + \tfrac{80}{11} 4^n, \quad n \geq 0.$$

If the initial conditions are $a_0 = 1$ and $a_1 = 2$, then c_1 and c_2 are found using $1 = c_1 + c_2 + 80/11$, $2 = 3c_1 - 7c_2 + 320/11$, yielding

$$a_n = -\tfrac{71}{10}(3^n) + \tfrac{91}{110}(-7)^n + \tfrac{80}{11}(4^n), \quad n \geq 0.$$

2. Suppose the given recurrence relation is $a_n + 4a_{n-1} - 21a_{n-2} = 8(3^n)$, $n \geq 2$. Then it is still true that

$$a_n^{(h)} = c_1(3^n) + c_2(-7)^n, \quad n \geq 0,$$

where c_1 and c_2 are arbitrary constants. By the second part of Fact 3, a particular solution is $a_n^{(p)} = An3^n$. Substituting $a_n^{(p)}$ gives $An3^n + 4A(n-1)3^{n-1} - 21A(n-2)3^{n-2} = 8(3^n)$. Dividing by 3^{n-2} produces $9An + 12A(n-1) - 21A(n-2) = 72$, so $A = 12/5$. Thus

$$a_n = c_1(3^n) + c_2(-7)^n + \tfrac{12}{5} n 3^n, \quad n \geq 0.$$

3. *Tower of Hanoi*: (See Example 1 of §2.2.4.) If a_n is the minimum number of moves needed to transfer the n disks, then a_n satisfies the first-order nonhomogeneous relation

$$a_n = 2a_{n-1} + 1, \quad n \geq 1,$$

where $a_0 = 0$. Here $a_n^{(h)} = c(2^n)$ for an arbitrary constant c, and $a_n^{(p)} = A$, using entry 1 of Table 1. So $A = 2A + 1$ or $A = -1$. Hence $a_n = c(2^n) - 1$ and $0 = a_0 = c(2^0) - 1$ implies $c = 1$, giving

$$a_n = 2^n - 1, \quad n \geq 0.$$

4. How many regions are formed if n lines are drawn in the plane, in general position (no two parallel and no three intersecting at a point)? If a_n denotes the number of regions thus formed, then $a_1 = 2$, $a_2 = 4$, and $a_3 = 7$ are easily determined. A general

formula can be found by developing a recurrence relation for a_n. Namely, if line $n+1$ is added to the diagram with a_n regions formed by n lines, this new line intersects all the other n lines. These intersection points partition line $n+1$ into $n+1$ segments, each of which splits an existing region in two. As a result, $a_{n+1} = a_n + (n+1)$, $n \geq 1$, a first-order nonhomogeneous recurrence relation. Solving this relation with the initial condition $a_1 = 1$ produces $a_n = \frac{1}{2}(n^2 + n + 2)$.

3.3.4 METHOD OF GENERATING FUNCTIONS

Generating functions (see §3.2.1) can be used to solve individual recurrence relations as well as simultaneous systems of recurrence relations. This technique is analogous to the use of Laplace transforms in solving systems of differential equations.

Facts:
1. To solve the kth-order recurrence relation $C_{n+k}a_{n+k} + \cdots + C_n a_n = f(n)$, $n \geq 0$, carry out the following steps:
   - multiply both sides of the recurrence equation by x^{n+k} and sum the result;
   - take this new equation, rewrite it in terms of the generating function $f(x) = \sum_{n=0}^{\infty} a_n x^n$, and solve for $f(x)$;
   - expand the expression found for $f(x)$ in terms of powers of x in order that the coefficient a_n can be identified.
2. To solve a system of kth-order recurrence relations, carry out the following steps:
   - multiply both sides of each recurrence equation by x^{n+k} and sum the results;
   - rewrite the system of equations in terms of the generating functions $f(x)$, $g(x)$, ... for $a_n, b_n, \ldots$, and solve for these generating functions;
   - expand the expressions found for each generating function in terms of powers of x in order that the coefficients $a_n, b_n, \ldots$ can be identified.

Examples:
1. The nonhomogeneous first-order relation $a_{n+1} - 2a_n = 1$, $n \geq 0$, $a_0 = 0$, arises in the *Tower of Hanoi* problem (Example 3 of §3.3.3). Begin by applying the first step of Fact 1:
$$a_{n+1}x^{n+1} - 2a_n x^{n+1} = x^{n+1},$$
$$\sum_{n=0}^{\infty} a_{n+1}x^{n+1} - 2\sum_{n=0}^{\infty} a_n x^{n+1} = \sum_{n=0}^{\infty} x^{n+1}.$$
Then apply the second step of Fact 1:
$$\sum_{n=0}^{\infty} a_{n+1}x^{n+1} - 2x\sum_{n=0}^{\infty} a_n x^n = x\sum_{n=0}^{\infty} x^n,$$
$$(f(x) - a_0) - 2xf(x) = \frac{x}{1-x},$$
$$(f(x) - 0) - 2xf(x) = \frac{x}{1-x}.$$
Solving for $f(x)$ gives
$$f(x) = \frac{x}{(1-x)(1-2x)} = \frac{1}{1-2x} - \frac{1}{1-x} = \sum_{n=0}^{\infty}(2x)^n - \sum_{n=0}^{\infty} x^n = \sum_{n=0}^{\infty}(2^n - 1)x^n.$$
Since a_n is the coefficient of x^n in $f(x)$, $a_n = 2^n - 1$, $n \geq 0$.

2. To solve the nonhomogeneous second-order relation $a_{n+2} - 2a_{n+1} + a_n = 2^n$, $n \geq 0$, $a_0 = 1$, $a_1 = 2$, apply the first step of Fact 1:

$$a_{n+2}x^{n+2} - 2a_{n+1}x^{n+2} + a_n x^{n+2} = 2^n x^{n+2},$$

$$\sum_{n=0}^{\infty} a_{n+2}x^{n+2} - 2\sum_{n=0}^{\infty} a_{n+1}x^{n+2} + \sum_{n=0}^{\infty} a_n x^{n+2} = \sum_{n=0}^{\infty} 2^n x^{n+2}.$$

The second step of Fact 1 produces

$$\sum_{n=0}^{\infty} a_{n+2}x^{n+2} - 2x \sum_{n=0}^{\infty} a_{n+1}x^{n+1} + x^2 \sum_{n=0}^{\infty} a_n x^n = x^2 \sum_{n=0}^{\infty} (2x)^n,$$

$$[f(x) - a_0 - a_1 x] - 2x[f(x) - a_0] + x^2 f(x) = \tfrac{x^2}{1-2x},$$

$$[f(x) - 1 - 2x] - 2x[f(x) - 1] + x^2 f(x) = \tfrac{x^2}{1-2x}.$$

Solving for $f(x)$ gives

$$f(x) = \tfrac{1}{1-2x} = \sum_{n=0}^{\infty} (2x)^n = \sum_{n=0}^{\infty} 2^n x^n.$$

Thus $a_n = 2^n$, $n \geq 0$, is the solution of the given recurrence relation.

3. Fact 2 can be used to solve the system of recurrence relations

$$a_{n+1} = 2a_n - b_n + 2$$
$$b_{n+1} = -a_n + 2b_n - 1$$

for $n \geq 0$, with $a_0 = 0$ and $b_0 = 1$. Multiplying by x^{n+1} and summing yields

$$\sum_{n=0}^{\infty} a_{n+1}x^{n+1} = 2x \sum_{n=0}^{\infty} a_n x^n - x \sum_{n=0}^{\infty} b_n x^n + 2x \sum_{n=0}^{\infty} x^n$$

$$\sum_{n=0}^{\infty} b_{n+1}x^{n+1} = -x \sum_{n=0}^{\infty} a_n x^n + 2x \sum_{n=0}^{\infty} b_n x^n - x \sum_{n=0}^{\infty} x^n.$$

These two equations can be rewritten in terms of the generating functions $f(x) = \sum_{n=0}^{\infty} a_n x^n$ and $g(x) = \sum_{n=0}^{\infty} b_n x^n$ as

$$f(x) - a_0 = 2x f(x) - x g(x) + 2x \tfrac{1}{1-x}$$
$$g(x) - b_0 = -x f(x) + 2x g(x) - x \tfrac{1}{1-x}.$$

Solving this system (with $a_0 = 0$, $b_0 = 1$) produces

$$f(x) = \tfrac{x(1-2x)}{(1-x)^2(1-3x)} = \tfrac{-3/4}{1-x} + \tfrac{1/2}{(1-x)^2} + \tfrac{1/4}{(1-3x)}$$

$$= -\tfrac{3}{4} \sum_{n=0}^{\infty} x^n + \tfrac{1}{2} \sum_{n=0}^{\infty} \binom{-2}{n} x^n + \tfrac{1}{4} \sum_{n=0}^{\infty} (3x)^n$$

$$= -\tfrac{3}{4} \sum_{n=0}^{\infty} x^n + \tfrac{1}{2} \sum_{n=0}^{\infty} \binom{n+1}{n} x^n + \tfrac{1}{4} \sum_{n=0}^{\infty} 3^n x^n$$

and

$$g(x) = \tfrac{1-4x+2x^2}{(1-x)^2(1-3x)} = \tfrac{3/4}{1-x} + \tfrac{1/2}{(1-x)^2} + \tfrac{-1/4}{(1-3x)}$$

$$= \tfrac{3}{4} \sum_{n=0}^{\infty} x^n + \tfrac{1}{2} \sum_{n=0}^{\infty} \binom{n+1}{n} x^n - \tfrac{1}{4} \sum_{n=0}^{\infty} 3^n x^n.$$

It then follows that

$$a_n = -\tfrac{3}{4} + \tfrac{1}{2}(n+1) + \tfrac{1}{4}3^n, \quad n \geq 0$$

$$b_n = \tfrac{3}{4} + \tfrac{1}{2}(n+1) - \tfrac{1}{4}3^n, \quad n \geq 0.$$

3.3.5 DIVIDE-AND-CONQUER RELATIONS

Certain algorithms proceed by breaking up a given problem into subproblems of nearly equal size; solutions to these subproblems are then combined to produce a solution to the original problem. Analysis of such "divide-and-conquer" algorithms results in special types of recurrence relations that can be solved exactly and asymptotically.

Definitions:

The *time-complexity function* $f(n)$ for an algorithm gives the (maximum) number of operations required to solve any instance of size n. The function $f(n)$ is *monotone increasing* if $m < n \Rightarrow f(m) \leq f(n)$ where m and n are positive integers.

A *recursive divide-and-conquer algorithm* splits a given problem of size $n = b^k$ into a subproblems of size $\frac{n}{b}$ each. It requires (at most) $h(n)$ operations to create the subproblems and subsequently combine their solutions.

Let $S = S_b$ be the set of integers $\{1, b, b^2, \ldots\}$ and let $\mathcal{Z}^+$ be the set of positive integers. If $f(n)$ and $g(n)$ are functions on $\mathcal{Z}^+$, then g *dominates* f *on* S, written $f \in O(g)$ on S, if there are positive constants $A \in \mathcal{R}$, $k \in \mathcal{Z}^+$ such that $|f(n)| \leq A|g(n)|$ holds for all $n \in S$ with $n \geq k$.

Facts:

1. The time-complexity function $f(n)$ of a recursive divide-and-conquer algorithm is defined for $n \in S$ and satisfies the recurrence relation
$$f(1) = c,$$
$$f(n) = af(n/b) + h(n), \quad \text{for } n = b^k,\ k \geq 1,$$
where $a, b, c \in \mathcal{Z}^+$ and $b \geq 2$.

2. Solving $f(n) = af(n/b) + c$, $f(1) = c$:
   - If $a = 1$: $f(n) = c(\log_b n + 1)$ for $n \in S$. Thus $f \in O(\log_b n)$ on S. If, in addition, $f(n)$ is monotone increasing, then $f \in O(\log_b n)$ on $\mathcal{Z}^+$.
   - If $a \geq 2$: $f(n) = c(an^{\log_b a} - 1)/(a - 1)$ for $n \in S$. Thus $f \in O(n^{\log_b a})$ on S. If, in addition, $f(n)$ is monotone increasing, then $f \in O(n^{\log_b a})$ on $\mathcal{Z}^+$.

3. Let $f(n)$ be any function satisfying the inequality relations
$$f(1) \leq c,$$
$$f(n) \leq af(n/b) + c, \quad \text{for } n = b^k,\ k \geq 1,$$
where $a, b, c \in \mathcal{Z}^+$ and $b \geq 2$.
   - If $a = 1$: $f \in O(\log_b n)$ on S. If, in addition, $f(n)$ is monotone increasing, then $f \in O(\log_b n)$ on $\mathcal{Z}^+$.
   - If $a \geq 2$: $f \in O(n^{\log_b a})$ on S. If, in addition, $f(n)$ is monotone increasing, then $f \in O(n^{\log_b a})$ on $\mathcal{Z}^+$.

4. Solving for a monotone increasing $f(n)$ where $f(n) = af(n/b) + rn^d$ ($n = b^k$, $k \geq 1$), $f(1) = c$, where $a, b, c, d \in \mathcal{Z}^+$, $b \geq 2$, and r is a positive real number:
   - If $a < b^d$: $f \in O(n^d)$ on $\mathcal{Z}^+$.
   - If $a = b^d$: $f \in O(n^d \log_b n)$ on $\mathcal{Z}^+$.
   - If $a > b^d$: $f \in O(n^{\log_b a})$ on $\mathcal{Z}^+$.

The same asymptotic results hold if inequalities $\leq$ replace equalities in the given recurrence relation.

Examples:

1. If $f(n)$ satisfies the recurrence relation $f(n) = f(\frac{n}{2}) + 3$, $n \in S_2$, $f(1) = 3$, then by Fact 2 $f(n) = 3(\log_2 n + 1)$. Thus $f \in O(\log_2 n)$ on S_2.

2. If $f(n)$ satisfies the recurrence relation $f(n) = 4f(\frac{n}{3}) + 7$, $n \in S_3$, $f(1) = 7$, then by Fact 3 $f(n) = 7(4n^{\log_3 4} - 1)/3$. Thus $f \in O(n^{\log_3 4})$ on S_3.

3. *Binary search*: The binary search algorithm (§17.2.3) is a recursive procedure to search for a specified value in an ordered list of n items. Its complexity function satisfies $f(n) = f(\frac{n}{2}) + 2$, $n \in S_2$, $f(1) = 2$. Since the complexity function $f(n)$ is monotone increasing in the list size n, Fact 2 shows that $f \in O(\log_2 n)$.

4. *Merge sort*: The merge sort algorithm (§17.4) is a recursive procedure for sorting the n elements of a list. It repeatedly divides a given list into two nearly equal sublists, sorts those sublists, and combines the sorted sublists. Its complexity function satisfies $f(n) = 2f(\frac{n}{2}) + (n-1)$, $n \in S_2$, $f(1) = 0$. Since $f(n)$ is monotone increasing and satisfies the inequality relation $f(n) \leq 2f(\frac{n}{2}) + n$, Fact 5 gives $f \in O(n \log_2 n)$.

5. *Matrix multiplication*: The *Strassen* algorithm is a recursive procedure for multiplying two $n \times n$ matrices (see §6.3.3). One version of this algorithm requires seven multiplications of $\frac{n}{2} \times \frac{n}{2}$ matrices and 15 additions of $\frac{n}{2} \times \frac{n}{2}$ matrices. Consequently, its complexity function satisfies $f(n) = 7f(\frac{n}{2}) + 15n^2/4$, $n \in S_2$, $f(1) = 1$. From the third part of Fact 5, $f \in O(n^{\log_2 7})$ on $\mathcal{Z}^+$. This algorithm requires approximately $O(n^{2.81})$ operations to multiply $n \times n$ matrices, compared to $O(n^3)$ for the standard method.

3.4 FINITE DIFFERENCES

The difference and antidifference operators are the discrete analogues of ordinary differentiation and antidifferentiation. Difference methods can be used for curve-fitting and for solving recurrence relations.

3.4.1 THE DIFFERENCE OPERATOR

The difference operator plays a role in combinatorial modeling analogous to that of the derivative operator in continuous analysis.

Definitions:

Let $f: \mathcal{N} \to \mathcal{R}$.

The **difference operator** $\Delta f(x) = f(x+1) - f(x)$ is the discrete analogue of the differentiation operator.

The **kth difference** of f is the operator $\Delta^k f(x) = \Delta^{k-1} f(x+1) - \Delta^{k-1} f(x)$, for $k \geq 1$, with $\Delta^0 f = f$.

The **shift operator** E is defined by $Ef(x) = f(x+1)$.

The **harmonic sum** $H_n = \sum_{i=1}^{n} \frac{1}{i}$ is the discrete analogue of the natural logarithm (§3.1.7).

Note: Most of the results stated in this subsection are also valid for functions on non-discrete domains. The functional notation that is used for most of this subsection, instead of the more usual subscript notation for sequences, makes the results easier to read and helps underscore the parallels between discrete and ordinary calculus.

Facts:
1. Linearity: $\Delta(\alpha f + \beta g) = \alpha \Delta f + \beta \Delta g$, for all constants α and β.
2. Product rule: $\Delta(f(x)g(x)) = (Ef(x))\Delta g(x) + (\Delta f(x))g(x)$. This is analogous to the derivative formula for the product of functions.
3. $\Delta^m x^n = 0$, for $m > n$, and $\Delta^n x^n = n!$.
4. $\Delta^n f(x) = \sum_{k=0}^{n} (-1)^k \binom{n}{k} f(x+n-k)$.
5. $f(x+n) = \sum_{k=0}^{n} \binom{n}{k} \Delta^k f(x)$.
6. Leibniz's theorem: $\Delta^n(f(x)g(x)) = \sum_{k=0}^{n} \binom{n}{k} \Delta^k f(x) \Delta^{n-k} g(x+k)$.
7. Quotient rule: $\Delta\left(\frac{f(x)}{g(x)}\right) = \frac{g(x)\Delta f(x) - f(x)\Delta g(x)}{g(x)g(x+1)}$.
8. The shift operator E satisfies $\Delta f = Ef - f$, written equivalently as $E = 1 + \Delta$.
9. $E^n f(x) = f(x+n)$.
10. The equation $\Delta C(x) = 0$ implies that C is periodic with period 1. Moreover, if the domain is restricted to the integers (e.g., if $C(n)$ is a sequence), then C is constant.

Examples:
1. If $f(x) = x^3$ then $\Delta f(x) = (x+1)^3 - x^3 = 3x^2 + 3x + 1$.
2. The following table gives formulas for the differences of some important functions. In this table, the notation $x^{\underline{n}}$ refers to the nth falling power of x (§3.4.2).

$f(x)$	$\Delta f(x)$
$\binom{x}{n}$	$\binom{x}{n-1}$
$(x+a)^{\underline{n}}$	$n(x+a)^{\underline{n-1}}$
x^n	$\binom{n}{1}x^{n-1} + \binom{n}{2}x^{n-2} + \cdots + 1$
a^x	$(a-1)a^x$
H_x	$x^{\underline{-1}} = \frac{1}{x+1}$
$\sin x$	$2\sin(\frac{1}{2})\cos(x+\frac{1}{2})$
$\cos x$	$-2\sin(\frac{1}{2})\sin(x+\frac{1}{2})$

3. $\Delta^2 f(x) = f(x+2) - 2f(x+1) + f(x)$, from Fact 4.
4. $f(x+3) = f(x) + 3\Delta f(x) + 3\Delta^2 f(x) + \Delta^3 f(x)$, from Fact 5.
5. The shift operator can be used to find the exponential generating function (§3.2.2) for the sequence $\{a_k\}$, where a_k is a polynomial in variable k of degree n.

$$\sum_{k=0}^{\infty} \frac{a_k x^k}{k!} = \sum_{k=0}^{\infty} \frac{E^k(a_0)x^k}{k!} = \left(\sum_{k=0}^{\infty} \frac{x^k E^k}{k!}\right) a_0$$

$$= e^{xE} a_0 = e^{x(1+\Delta)} a_0 = e^x e^{x\Delta} a_0$$

$$= e^x \left(a_0 + \frac{x\Delta a_0}{1!} + \frac{x^2 \Delta^2 a_0}{2!} + \cdots + \frac{x^n \Delta^n a_0}{n!}\right).$$

For example, if $a_k = k^2 + 1$ then $\sum_{k=0}^{\infty} \frac{(k^2+1)x^k}{k!} = e^x(1 + x + x^2)$.

3.4.2 CALCULUS OF DIFFERENCES: FALLING AND RISING POWERS

Falling powers provide a natural analogue between the calculus of finite sums and differences and the calculus of integrals and derivatives. Stirling numbers provide a means of expressing ordinary powers in terms of falling powers and vice versa.

Definitions:

The **nth falling power** of x, written $x^{\underline{n}}$, is the discrete analogue of exponentiation and is defined by
$$x^{\underline{n}} = x(x-1)(x-2)\ldots(x-n+1)$$
$$x^{\underline{-n}} = \frac{1}{(x+1)(x+2)\ldots(x+n)}$$
$$x^{\underline{0}} = 1.$$

The **nth rising power** of x, written $x^{\overline{n}}$, is defined by
$$x^{\overline{n}} = x(x+1)(x+2)\ldots(x+n-1),$$
$$x^{\overline{-n}} = \frac{1}{(x-n)(x-n+1)\ldots(x-1)},$$
$$x^{\overline{0}} = 1.$$

Facts:

1. *Conversion between falling and rising powers*:
$$x^{\underline{n}} = (-1)^n(-x)^{\overline{n}} = (x-n+1)^{\overline{n}} = \frac{1}{(x+1)^{\underline{-n}}},$$
$$x^{\overline{n}} = (-1)^n(-x)^{\underline{n}} = (x+n-1)^{\underline{n}} = \frac{1}{(x-1)^{\overline{-n}}},$$
$$x^{\underline{-n}} = \frac{1}{(x+1)^{\overline{n}}},$$
$$x^{\overline{-n}} = \frac{1}{(x-1)^{\underline{n}}}.$$

2. *Laws of exponents*:
$$x^{\underline{m+n}} = x^{\underline{m}}(x-m)^{\underline{n}},$$
$$x^{\overline{m+n}} = x^{\overline{m}}(x+m)^{\overline{n}}.$$

3. *Binomial theorem*: $(x+y)^{\underline{n}} = \binom{n}{0}x^{\underline{n}} + \binom{n}{1}x^{\underline{n-1}}y^{\underline{1}} + \cdots + \binom{n}{n}y^{\underline{n}}$.

4. The action of the difference operator on falling powers is analogous to the action of the derivative on ordinary powers: $\Delta x^{\underline{n}} = nx^{\underline{n-1}}$.

5. There is no chain rule for differences, but the binomial theorem implies the rule
$$\Delta(x+a)^{\underline{n}} = n(x+a)^{\underline{n-1}}.$$

6. *Newton's theorem*: If $f(x)$ is a polynomial of degree n, then
$$f(x) = \sum_{k=0}^{n} \frac{\Delta^k f(0)}{k!} x^{\underline{k}}.$$
This is an analogue of *Maclaurin's theorem*.

7. If $f(x) = x^n$ then $\Delta^k f(0) = \left\{{n \atop k}\right\} \cdot k!$.

8. Falling powers can be expressed in terms of ordinary powers using Stirling cycle numbers (§2.5.2):
$$x^{\underline{n}} = \sum_{k=1}^{n} \left[{n \atop k}\right](-1)^{n-k}x^k.$$

9. Rising powers can be expressed in terms of ordinary powers using Stirling cycle numbers (§2.5.2):
$$x^{\overline{n}} = \sum_{k=1}^{n} \begin{bmatrix} n \\ k \end{bmatrix} x^k.$$

10. Ordinary powers can be expressed in terms of falling or rising powers using Stirling subset numbers (§2.5.2):
$$x^n = \sum_{k=1}^{n} \begin{Bmatrix} n \\ k \end{Bmatrix} x^{\underline{k}} = \sum_{k=1}^{n} \begin{Bmatrix} n \\ k \end{Bmatrix} (-1)^{n-k} x^{\overline{k}}.$$

Examples:

1. Fact 8 and Table 4 of §2.5.2 give
$$x^{\underline{0}} = x^0,$$
$$x^{\underline{1}} = x^1,$$
$$x^{\underline{2}} = x^2 - x^1,$$
$$x^{\underline{3}} = x^3 - 3x^2 + 2x^1,$$
$$x^{\underline{4}} = x^4 - 6x^3 + 11x^2 - 6x^1.$$

2. Fact 10 and Table 5 of §2.5.2 give
$$x^0 = x^{\underline{0}},$$
$$x^1 = x^{\underline{1}},$$
$$x^2 = x^{\underline{2}} + x^{\underline{1}},$$
$$x^3 = x^{\underline{3}} + 3x^{\underline{2}} + x^{\underline{1}},$$
$$x^4 = x^{\underline{4}} + 6x^{\underline{3}} + 7x^{\underline{2}} + x^{\underline{1}}.$$

3.4.3 DIFFERENCE SEQUENCES AND DIFFERENCE TABLES

New sequences can be obtained from a given sequence by repeatedly applying the difference operator.

Definitions:

The **difference sequence** for the sequence $A = \{ a_j \mid j = 0, 1, \ldots \}$ is the sequence $\Delta A = \{ a_{j+1} - a_j \mid j = 0, 1, \ldots \}$.

The kth **difference sequence** for $f : \mathcal{N} \to \mathcal{R}$ is given by $\Delta^k f(0), \Delta^k f(1), \Delta^k f(2), \ldots$.

The **difference table** for $f : \mathcal{N} \to \mathcal{R}$ is the table T_f whose kth row is the kth difference sequence for f. That is, $T_f[k, l] = \Delta^k f(l) = \Delta^{k-1} f(l+1) - \Delta^{k-1} f(l)$.

Facts:

1. The leftmost column of a difference table completely determines the entire table, via Newton's theorem (Fact 6, §3.4.2).

2. The difference table of an nth degree polynomial consists of $n + 1$ nonzero rows followed by all zero rows.

Examples:

1. If $A = 0, 1, 4, 9, 16, 25, \ldots$ is the sequence of squares of integers, then its difference sequence is $\Delta A = 1, 3, 5, 7, 9, \ldots$. Observe that $\Delta(x^2) = 2x + 1$.

2. The difference table for $x^{\underline{3}}$ is given by

	0	1	2	3	4	5	$\cdots$
$\Delta^0 x^{\underline{3}} = x^{\underline{3}}$	0	0	0	6	24	60	$\cdots$
$\Delta^1 x^{\underline{3}} = 3x^{\underline{2}}$	0	0	6	18	36	$\cdots$	
$\Delta^2 x^{\underline{3}} = 6x^{\underline{1}}$	0	6	12	18	$\cdots$		
$\Delta^3 x^{\underline{3}} = 6$	6	6	6	$\cdots$			
$\Delta^4 x^{\underline{3}} = 0$	0	0	$\cdots$				

3. The difference table for x^3 is given by

	0	1	2	3	4	5	$\cdots$
$\Delta^0 x^3 = x^3$	0	1	8	27	64	125	$\cdots$
$\Delta^1 x^3 = 3x^2 + 3x + 1$	1	7	19	37	61	$\cdots$	
$\Delta^2 x^3 = 6x + 6$	6	12	18	24	$\cdots$		
$\Delta^3 x^3 = 6$	6	6	6	$\cdots$			
$\Delta^4 x^3 = 0$	0	0	$\cdots$				

4. The difference table for 3^x is given by

	0	1	2	3	4	5	$\cdots$
$\Delta^0 3^x = 3^x$	1	3	9	27	81	243	$\cdots$
$\Delta^1 3^x = 2 \cdot 3^x$	2	6	18	54	162	$\cdots$	
$\Delta^2 3^x = 4 \cdot 3^x$	4	12	36	108	$\cdots$		
$\Delta^3 3^x = 8 \cdot 3^x$	8	24	72	$\cdots$			
$\Delta^4 3^x = 16 \cdot 3^x$	16	48	$\cdots$				
$\vdots$	$\vdots$						
$\Delta^k 3^x = 2^k 3^x$							

5. *Application to curve-fitting*: Find the polynomial $p(x)$ of smallest degree that passes through the points: $(0, 5), (1, 5), (2, 3), (3, 5), (4, 17), (5, 45)$. The difference table for the sequence $5, 5, 3, 5, 17, 45$ is

5	5	3	5	17	45	$\cdots$
0	-2	2	12	28	$\cdots$	
-2	4	10	16	$\cdots$		
6	6	6	$\cdots$			
0	0	$\cdots$				

Newton's theorem shows that the polynomial of smallest degree is $p(x) = 5 - x^{\underline{2}} + x^{\underline{3}} = x^3 - 4x^2 + 3x + 5$.

3.4.4 DIFFERENCE EQUATIONS

Difference equations are analogous to differential equations and many of the techniques are as fully developed. Difference equations provide a way to solve recurrence relations.

Definitions:

A *difference equation* is an equation involving the difference operator and/or higher-order differences of an unknown function.

An *antidifference* of the function f is any function g such that $\Delta g = f$. The notation $\Delta^{-1} f$ denotes any such function.

Facts:

1. Any recurrence relation (§3.3) can be expressed as a difference equation, and vice versa, by using Facts 4 and 5 of §3.4.1.

2. The solution to a recurrence relation can sometimes be easily obtained by converting it to a difference equation and applying difference methods.

Examples:

1. To find an antidifference of $10 \cdot 3^x$, use Table 1 (§3.4.1): $\Delta^{-1}(10 \cdot 3^x) = 5\Delta^{-1}(2 \cdot 3^x) = 5 \cdot 3^x + C$. (Also see Table 1 of §3.5.3.)

2. To find an antidifference of $3x$, first express x as $x^{\underline{1}}$ and then use Table 1 (§3.4.1): $\Delta^{-1} 3x = 3\Delta^{-1} x^{\underline{1}} = \frac{3}{2} x^{\underline{2}} + C = \frac{3}{2} x(x-1) + C$.

3. To find an antidifference of x^2, express x^2 as $x^{\underline{2}} + x^{\underline{1}}$ and then use Table 1 (§3.4.1): $\Delta^{-1} x^2 = \Delta^{-1}(x^{\underline{2}} + x^{\underline{1}}) = \Delta^{-1} x^{\underline{2}} + \Delta^{-1} x^{\underline{1}} = \frac{1}{3} x^{\underline{3}} + \frac{1}{2} x^{\underline{2}} + C = \frac{1}{3} x(x-1)(x-2) + \frac{1}{2} x(x-1) + C$.

4. The following are examples of difference equations:
$$\Delta^3 f(x) + x^4 \Delta^2 f(x) - f(x) = 0,$$
$$\Delta^3 f(x) + f(x) = x^2.$$

5. To solve the recurrence relation $a_{n+1} = a_n + 5^n$, $n \geq 0$, $a_0 = 2$, first note that $\Delta a_n = 5^n$. Thus $a_n = \Delta^{-1} 5^n = \frac{1}{4}(5^n) + C$. The initial condition $a_0 = 2$ now implies that $a_n = \frac{1}{4}(5^n + 7)$.

6. To solve the equation $a_{n+1} = (na_n + n)/(n+1)$, $n \geq 1$, the recurrence relation is first rewritten as $(n+1)a_{n+1} - na_n = n$, which is equivalent to $\Delta(na_n) = n$. Thus $na_n = \Delta^{-1} n = \frac{1}{2} n^{\underline{2}} + C$, which implies that $a_n = \frac{1}{2}(n-1) + C(\frac{1}{n})$.

7. To solve $a_n = 2a_{n-1} - a_{n-2} + 2^{n-2} + n - 2$, $n \geq 2$, with $a_0 = 4$, $a_1 = 5$, the recurrence relation is rewritten as $a_{n+2} - 2a_{n+1} + a_n = 2^n + n$, $n \geq 0$. Now, by applying Fact 4 of §3.4.1, the left-hand side may be replaced by $\Delta^2 a_n$. If the antidifference operator is applied twice to the resulting difference equation and the initial conditions are substituted, the solution obtained is
$$a_n = 2^n + \frac{1}{6} n^{\underline{3}} + c_1 n + c_2 = 2^n + \frac{1}{6} n(n-1)(n-2) + 3.$$

3.5 FINITE SUMS AND SUMMATION

Finite sums arise frequently in combinatorial mathematics and in the analysis of running times of algorithms. There are a few basic rules for transforming sums into possibly more tractable equivalent forms, and there is a calculus for evaluating these standard forms.

3.5.1 SIGMA NOTATION

A complex form of symbolic representation of discrete sums using the uppercase Greek letter Σ (sigma) was introduced by Joseph Fourier in 1820 and has evolved into several variations.

Definitions:

The **sigma expression** $\sum_{i=a}^{b} f(i)$ has the value $f(a) + f(a+1) + \cdots + f(b-1) + f(b)$ if $a \leq b$ ($a, b \in \mathcal{Z}$), and 0 otherwise. In this expression, i is the **index of summation** or **summation variable**, which ranges from the **lower limit** a to the **upper limit** b. The interval $[a, b]$ is the **interval of summation**, and $f(i)$ is a **term** or **summand** of the summation.

A sigma expression $S_n = \sum_{i=0}^{n} f(i)$ is in **standardized form** if the lower limit is zero and the upper limit is an integer-valued expression.

A **sigma expression** $\sum_{k \in K} g(k)$ **over the set** K has as its value the sum of all the values $g(k)$, where $k \in K$.

A **closed form** for a sigma expression with an indefinite number of terms is an algebraic expression with a fixed number of terms, whose value equals the sum.

A **partial sum** of the (standardized) sigma expression $S_n = \sum_{i=0}^{n} f(i)$ is the sigma expression $S_k = \sum_{i=0}^{k} f(i)$, where $0 \leq k \leq n$.

An **iterated sum** or **multiple sum** is an expression with two or more sigmas, as exemplified by the double sum $\sum_{i=c}^{d} \sum_{j=a}^{b} f(i, j)$. Evaluation proceeds from the innermost sigma outward.

A lower or upper limit for an inner sum of an iterated sum is **dependent** if it depends on an outer variable. Otherwise, that limit is **independent**.

Examples:

1. The sum $f(1) + f(2) + f(3) + f(4) + f(5)$ may be represented as $\sum_{i=1}^{5} f(i)$.

2. Sometimes the summand is written as an expression, such as $\sum_{n=1}^{50}(n^2 + n)$, which means the same as $\sum_{n=1}^{50} f(n)$, where $f(n) = n^2 + n$. Brackets or parentheses can be used to distinguish what is in the summand of such an "anonymous function" from whatever is written to the immediate right of the sigma expression. They may be omitted when such a summand is very simple.

3. Sometimes the property defining the indexing set is written underneath the Σ, as in the expressions $\sum_{1 \leq k \leq n} a_k$ or $\sum_{k \in K} b_k$.

4. The right side of the equation $\sum_{j=0}^{n} x^j = \frac{x^{n+1} - 1}{x - 1}$ is a closed form for the sigma expression on the left side.

5. The operational meaning of the multiple sum with independent limits $\sum_{i=1}^{3} \sum_{j=2}^{4} \frac{i}{j}$ is first to expand the inner sum, obtaining the single sum $\sum_{i=1}^{3} \left[\frac{i}{2} + \frac{i}{3} + \frac{i}{4}\right]$. Expansion of the outer sum then yields $\left[\frac{1}{2} + \frac{1}{3} + \frac{1}{4}\right] + \left[\frac{2}{2} + \frac{2}{3} + \frac{2}{4}\right] + \left[\frac{3}{2} + \frac{3}{3} + \frac{3}{4}\right] = \frac{13}{2}$.

6. The multiple sum with dependent limits $\sum_{i=1}^{3} \sum_{j=i}^{4} \frac{i}{j}$ is evaluated by first expanding the inner sum, obtaining $\left[\frac{1}{1} + \frac{1}{2} + \frac{1}{3} + \frac{1}{4}\right] + \left[\frac{2}{2} + \frac{2}{3} + \frac{2}{4}\right] + \left[\frac{3}{3} + \frac{3}{4}\right] = 6$.

3.5.2 ELEMENTARY TRANSFORMATION RULES FOR SUMS

Sums can be transformed using a few simple rules. A well-chosen sequence of transformations often simplifies evaluation.

Facts:

1. *Distributivity rule*: $\sum_{k \in K} c a_k = c \sum_{k \in K} a_k$, for c a constant.

2. *Associativity rule*: $\sum_{k \in K} (a_k + b_k) = \sum_{k \in K} a_k + \sum_{k \in K} b_k$.

3. *Rearrangement rule*: $\sum_{k \in K} a_k = \sum_{k \in K} a_{\rho(k)}$, where ρ is a permutation of the integers in K.

4. *Telescoping for sequences*: For any sequence $\{a_j \mid j = 0, 1, \ldots\}$, $\sum_{i=m}^{n} (a_{i+1} - a_i) = a_{n+1} - a_m$.

5. *Telescoping for functions*: For any function $f: \mathcal{N} \to \mathcal{R}$, $\sum_{i=m}^{n} \Delta f(i) = f(n+1) - f(m)$.

6. *Perturbation method*: Given a standardized sum $S_n = \sum_{i=0}^{n} f(i)$, form the equation
$$\sum_{i=0}^{n} f(i) + f(n+1) = f(0) + \sum_{i=1}^{n+1} f(i) = f(0) + \sum_{i=0}^{n} f(i+1).$$
Algebraic manipulation often leads to a closed form for S_n.

7. *Interchanging independent indices of a double sum*: When the lower and upper limits of the inner variable of a double sum are independent of the outer variable, the order of summation can be changed, simply by swapping the inner sigma, limits and all, with the outer sigma. That is,
$$\sum_{i=c}^{d} \sum_{j=a}^{b} f(i,j) = \sum_{j=a}^{b} \sum_{i=c}^{d} f(i,j).$$

8. *Interchanging dependent indices of a double sum*: When either the lower or upper limit of the inner variable j of a double sum of an expression $f(i,j)$ is dependent on the outer variable i, the order of summation can still be changed by swapping the inner sum with the outer sum. However, the limits of the new inner variable i must be written as functions of the new outer variable j so that the entire set of pairs (i,j) over which $f(i,j)$ is summed is the same as before. One particular case of interest is the interchange
$$\sum_{i=1}^{n} \sum_{j=i}^{n} f(i,j) = \sum_{j=1}^{n} \sum_{i=1}^{j} f(i,j).$$

Examples:

1. The following summation can be evaluated using Fact 4 (telescoping for sequences):
$$\sum_{i=1}^{n} \frac{1}{i(i+1)} = -\sum_{i=1}^{n} \left(\frac{1}{i+1} - \frac{1}{i}\right) = 1 - \frac{1}{n+1}.$$

2. Evaluate $S_n = \sum_{i=0}^{n} x^i$, using the perturbation method.

$$\sum_{i=0}^{n} x^i + x^{n+1} = x^0 + \sum_{i=1}^{n+1} x^i = 1 + x \sum_{i=1}^{n+1} x^{i-1},$$

$$S_n + x^{n+1} = 1 + x \sum_{i=0}^{n} x^i = 1 + x S_n,$$

giving $S_n = \frac{x^{n+1}-1}{x-1}$.

3. Evaluate $S_n = \sum_{i=0}^{n} i 2^i$, using the perturbation method.

$$\sum_{i=0}^{n} i 2^i + (n+1) 2^{n+1} = 0 \cdot 2^0 + \sum_{i=1}^{n+1} i 2^i = \sum_{i=0}^{n} (i+1) 2^{i+1},$$

$$S_n + (n+1) 2^{n+1} = 2 \sum_{i=0}^{n} i 2^i + 2 \sum_{i=0}^{n} 2^i = 2 S_n + 2(2^{n+1} - 1),$$

giving $S_n = (n+1) 2^{n+1} - 2(2^{n+1} - 1) = (n-1) 2^{n+1} + 2$.

4. Interchange independent indices of a double sum:

$$\sum_{i=1}^{3} \sum_{j=2}^{4} \tfrac{i}{j} = \sum_{j=2}^{4} \sum_{i=1}^{3} \tfrac{i}{j} = \sum_{j=2}^{4} [\tfrac{1}{j} + \tfrac{2}{j} + \tfrac{3}{j}] = \sum_{j=2}^{4} \tfrac{6}{j} = 6 \sum_{j=2}^{4} \tfrac{1}{j} = 6 [\tfrac{1}{2} + \tfrac{1}{3} + \tfrac{1}{4}] = \tfrac{13}{2}.$$

5. Interchange dependent indices of a double sum:

$$\sum_{i=1}^{3} \sum_{j=i}^{3} \tfrac{i}{j} = \sum_{j=1}^{3} \sum_{i=1}^{j} \tfrac{i}{j} = \sum_{j=1}^{3} \tfrac{1}{j} \sum_{i=1}^{j} i = \tfrac{1}{1} \cdot 1 + \tfrac{1}{2} \cdot 3 + \tfrac{1}{3} \cdot 6 = \tfrac{9}{2}.$$

3.5.3 ANTIDIFFERENCES AND SUMMATION FORMULAS

Some standard combinatorial functions analogous to polynomials and exponential functions facilitate the development of a calculus of finite differences, analogous to the differential calculus of continuous mathematics. The *fundamental theorem of discrete calculus* is useful in deriving a number of summation formulas.

Definitions:

An **antidifference** of the function f is any function g such that $\Delta g = f$, where Δ is the difference operator (§3.4.1). The notation $\Delta^{-1} f$ denotes any such function.

The **indefinite sum** of the function f is the infinite family of all antidifferences of f. The notation $\sum f(x) \delta x + c$ is sometimes used for the indefinite sum to emphasize the analogy with integration.

Facts:

1. *Fundamental theorem of discrete calculus*:

$$\sum_{k=a}^{b} f(k) = \Delta^{-1} f(k) \Big|_{a}^{b+1} = \Delta^{-1} f(b+1) - \Delta^{-1} f(a).$$

Note: The upper evaluation point is one more than the upper limit of the sum.

2. *Linearity*: $\Delta^{-1}(\alpha f + \beta g) = \alpha \Delta^{-1} f + \beta \Delta^{-1} g$, for any constants α and β.

3. *Summation by parts*:

$$\sum_{i=a}^{b} f(i) \Delta g(i) = f(b+1) g(b+1) - f(a) g(a) - \sum_{i=a}^{b} g(i+1) \Delta f(i).$$

This result, which generalizes Fact 5 of §3.5.2, is a direct analogue of integration by parts in continuous analysis.

4. Abel's transformation:
$$\sum_{k=1}^{n} f(k)g(k) = f(n+1)\sum_{k=1}^{n} g(k) - \sum_{k=1}^{n}\left(\Delta f(k)\sum_{r=1}^{k} g(r)\right).$$

5. The following table gives the antidifferences of selected functions. In this table, H_x indicates the harmonic sum (§3.4.1), $x^{\underline{n}}$ is the nth falling power of x (§3.4.2), and $\left\{{n \atop k}\right\}$ is a Stirling subset number (§2.5.2).

$f(x)$	$\Delta^{-1}f(x)$	$f(x)$	$\Delta^{-1}f(x)$
$\binom{x}{n}$	$\binom{x}{n+1}$	$(x+a)^{\underline{n}}$	$\frac{(x+a)^{\underline{n+1}}}{n+1}$, $n \neq -1$
$(x+a)^{\underline{-1}}$	H_{x+a}	a^x	$\frac{a^x}{(a-1)}$, $a \neq 1$
a^x	$\frac{a^x}{(a-1)}$, $a \neq 1$	xa^x	$\frac{a^x}{(a-1)}\left(x - \frac{a}{a-1}\right)$, $a \neq 1$
x^n	$\sum_{k=1}^{n}\frac{\left\{{n \atop k}\right\}}{k+1}x^{\underline{k+1}}$	$(-1)^x$	$\frac{1}{2}(-1)^{x+1}$
$\sin x$	$\frac{-1}{2\sin(\frac{1}{2})}\cos(x-\frac{1}{2})$	$\cos x$	$\frac{1}{2\sin(\frac{1}{2})}\sin(x-\frac{1}{2})$

6. The following table gives finite sums of selected functions.

summation	formula	summation	formula
$\sum_{k=1}^{n} k^{\underline{m}}$	$\frac{(n+1)^{\underline{m+1}}}{m+1}$, $m \neq -1$	$\sum_{k=1}^{n} k^m$	$\sum_{j=1}^{m}\frac{\left\{{m \atop j}\right\}(n+1)^{\underline{j+1}}}{j+1}$
$\sum_{k=0}^{n} a^k$	$\frac{a^{n+1}-1}{a-1}$, $a \neq 1$	$\sum_{k=1}^{n} ka^k$	$\frac{(a-1)(n+1)a^{n+1}-a^{n+2}+a}{(a-1)^2}$, $a \neq 1$
$\sum_{k=1}^{n} \sin k$	$\frac{\sin(\frac{n+1}{2})\sin(\frac{n}{2})}{\sin(\frac{1}{2})}$	$\sum_{k=1}^{n} \cos k$	$\frac{\cos(\frac{n+1}{2})\sin(\frac{n}{2})}{\sin(\frac{1}{2})}$

Examples:

1. $\sum_{k=1}^{n} k^3 = \sum_{k=1}^{n}(k^{\underline{1}} + 3k^{\underline{2}} + k^{\underline{3}}) = \left(\frac{k^{\underline{2}}}{2} + k^{\underline{3}} + \frac{k^{\underline{4}}}{4}\right)\bigg|_{1}^{n+1} = \frac{n^2(n+1)^2}{4}.$

2. To evaluate $\sum_{k=1}^{n} k(k+2)(k+3)$, first rewrite its summand:

$$\sum_{k=1}^{n} k(k+2)(k+3) = \Delta^{-1}[(k+1-1)(k+2)(k+3)]\bigg|_{1}^{n+1}$$
$$= \left[\Delta^{-1}(k+3)^{\underline{3}} - \Delta^{-1}(k+3)^{\underline{2}}\right]\bigg|_{1}^{n+1}$$
$$= \left[\frac{(k+3)^{\underline{4}}}{4} - \frac{(k+3)^{\underline{3}}}{3}\right]\bigg|_{1}^{n+1}$$
$$= \frac{(n+4)^{\underline{4}}}{4} - \frac{(n+4)^{\underline{3}}}{3} + 2$$
$$= \frac{(n+4)(n+3)(n+2)(3n-1)+24}{12}.$$

3. $\sum_{k=1}^{n} k3^k = \Delta^{-1}(k3^k)\bigg|_{1}^{n+1} = 3^k\left[\frac{k}{2} - \frac{3}{4}\right]\bigg|_{1}^{n+1} = \frac{(2n-1)3^{n+1}+3}{4}.$

4. Summation by parts can be used to calculate $\sum_{j=0}^{n} jx^j$, using $f(j) = j$ and $\Delta g(j) = x^j$. Thus $g(j) = x^j/(x-1)$, and Fact 3 yields

$$\sum_{j=0}^{n} jx^j = \frac{(n+1)x^{n+1}}{(x-1)} - 0 - \sum_{j=0}^{n} \frac{x^{j+1}}{(x-1)} = \frac{(n+1)x^{n+1}}{(x-1)} - \frac{x}{x-1}\sum_{j=0}^{n} x^j$$

$$= \frac{(n+1)x^{n+1}}{(x-1)} - \frac{x}{x-1}\frac{x^{n+1}-1}{(x-1)} = \frac{(n+1)(x-1)x^{n+1} - x^{n+2} + x}{(x-1)^2}.$$

5. Summation by parts also yields an antiderivative of $x3^x$:

$$\Delta^{-1}(x3^x) = \Delta^{-1}\left(x\Delta(\tfrac{1}{2} \cdot 3^x)\right) = \tfrac{1}{2}x3^x - \Delta^{-1}(\tfrac{1}{2} \cdot 3^{x+1} \cdot 1) = 3^x\left(\tfrac{x}{2} - \tfrac{3}{4}\right).$$

3.5.4 STANDARD SUMS

Many useful summation formulas are derivable by combinations of elementary manipulation and finite calculus. Such sums can be expressed in various ways, using different combinatorial coefficients. (See §3.1.8.)

Definition:

The **power sum** $S^k(n) = \sum_{j=1}^{n} j^k = 1^k + 2^k + 3^k + \cdots + n^k$ is the sum of the kth powers of the first n positive integers.

Facts:

1. $S^k(n)$ is a polynomial in n of degree $k+1$ with leading coefficient $\frac{1}{k+1}$. The continuous analogue of this fact is the familiar $\int_a^b x^k dx = \frac{1}{k+1}(b^{k+1} - a^{k+1})$.

2. The power sum $S^k(n)$ can be expressed using the Bernoulli polynomials (§3.1.4) as

$$S^k(n) = \tfrac{1}{k+1}[B_{k+1}(n+1) - B_{k+1}(0)].$$

3. When $S^k(n)$ is expressed in terms of binomial coefficients with the second entry fixed at $k+1$, the coefficients are the Eulerian numbers (§3.1.5).

$$S^k(n) = \sum_{i=0}^{k-1} E(k,i)\binom{n+i+1}{k+1}.$$

4. When $S^k(n)$ is expressed in terms of binomial coefficients with the first entry fixed at $n+1$, the coefficients are products of factorials and Stirling subset numbers (§2.5.2).

$$S^k(n) = \sum_{i=1}^{k} i!\left\{{k \atop i}\right\}\binom{n+1}{i+1}.$$

5. Formulas for the power sums described in Facts 1, 3, and 4 are given in Tables 1-3, respectively, for small values of k.

Examples:

1. To find the third power sum $S^3(n) = \sum_{j=1}^{n} j^3$ via Fact 2, use the Bernoulli polynomial $B_4(x) = x^4 - 2x^3 + x^2 - \tfrac{1}{30}$ from Table 5 of §3.1.4. Thus

$$S^3(n) = \tfrac{1}{4}\big[B_4(x)\big]\Big|_0^{n+1} = \frac{(n+1)^4 - 2(n+1)^3 + (n+1)^2}{4} = \frac{n^2(n+1)^2}{4}.$$

Table 1 Sums of powers of integers.

summation	formula
$\sum_{j=1}^{n} j$	$\frac{1}{2}n(n+1)$
$\sum_{j=1}^{n} j^2$	$\frac{1}{6}n(n+1)(2n+1)$
$\sum_{j=1}^{n} j^3$	$\frac{1}{4}n^2(n+1)^2$
$\sum_{j=1}^{n} j^4$	$\frac{1}{30}n(n+1)(2n+1)(3n^2+3n-1)$
$\sum_{j=1}^{n} j^5$	$\frac{1}{12}n^2(n+1)^2(2n^2+2n-1)$
$\sum_{j=1}^{n} j^6$	$\frac{1}{42}n(n+1)(2n+1)(3n^4+6n^3-n^2-3n+1)$
$\sum_{j=1}^{n} j^7$	$\frac{1}{24}n^2(n+1)^2(3n^4+6n^3-n^2-4n+2)$
$\sum_{j=1}^{n} j^8$	$\frac{1}{90}n(n+1)(2n+1)(5n^6+15n^5+5n^4-15n^3-n^2+9n-3)$
$\sum_{j=1}^{n} j^9$	$\frac{1}{20}n^2(n+1)^2(2n^6+6n^5+n^4-8n^3+n^2+6n-3)$

Table 2 Sums of powers and Eulerian numbers.

summation	formula
$\sum_{j=1}^{n} j$	$\binom{n+1}{2}$
$\sum_{j=1}^{n} j^2$	$\binom{n+1}{3} + \binom{n+2}{3}$
$\sum_{j=1}^{n} j^3$	$\binom{n+1}{4} + 4\binom{n+2}{4} + \binom{n+3}{4}$
$\sum_{j=1}^{n} j^4$	$\binom{n+1}{5} + 11\binom{n+2}{5} + 11\binom{n+3}{5} + \binom{n+4}{5}$
$\sum_{j=1}^{n} j^5$	$\binom{n+1}{6} + 26\binom{n+2}{6} + 66\binom{n+3}{6} + 26\binom{n+4}{6} + \binom{n+5}{6}$

Table 3 Sums of powers and Stirling subset numbers.

summation	formula
$\sum_{j=1}^{n} j$	$\binom{n+1}{2}$
$\sum_{j=1}^{n} j^2$	$\binom{n+1}{2} + 2\binom{n+1}{3}$
$\sum_{j=1}^{n} j^3$	$\binom{n+1}{2} + 6\binom{n+1}{3} + 6\binom{n+1}{4}$
$\sum_{j=1}^{n} j^4$	$\binom{n+1}{2} + 14\binom{n+1}{3} + 36\binom{n+1}{4} + 24\binom{n+1}{5}$
$\sum_{j=1}^{n} j^5$	$\binom{n+1}{2} + 30\binom{n+1}{3} + 150\binom{n+1}{4} + 240\binom{n+1}{5} + 120\binom{n+1}{6}$

2. Power sums can be found using antidifferences and Stirling numbers of both types. For example, to find $S^3(n) = \sum_{x=1}^{n} x^3$ first compute
$$\Delta^{-1} x^3 = \Delta^{-1} \left(\left\{ {3 \atop 1} \right\} x^{\underline{1}} + \left\{ {3 \atop 2} \right\} x^{\underline{2}} + \left\{ {3 \atop 3} \right\} x^{\underline{3}} \right) = \frac{x^{\underline{2}}}{2} + x^{\underline{3}} + \frac{x^{\underline{4}}}{4}.$$
Each term $x^{\underline{m}}$ is then expressed in terms of ordinary powers of x
$$x^{\underline{2}} = \left[{2 \atop 2} \right] x^2 - \left[{2 \atop 1} \right] x^1 = x^2 - x,$$
$$x^{\underline{3}} = \left[{3 \atop 3} \right] x^3 - \left[{3 \atop 2} \right] x^2 + \left[{3 \atop 1} \right] x^1 = x^3 - 3x^2 + 2x,$$
$$x^{\underline{4}} = \left[{4 \atop 4} \right] x^4 - \left[{4 \atop 3} \right] x^3 + \left[{4 \atop 2} \right] x^2 - \left[{4 \atop 1} \right] x^1 = x^4 - 6x^3 + 11x^2 - 6x,$$
so $\Delta^{-1} x^3 = \frac{1}{2}(x^2 - x) + (x^3 - 3x^2 + 2x) + \frac{1}{4}(x^4 - 6x^3 + 11x^2 - 6x) = \frac{1}{4}(x^4 - 2x^3 + x^2)$. Evaluating this antidifference between the limits $x = 1$ and $x = n+1$ gives $S^3(n) = \frac{1}{4} n^2 (n+1)^2$. See §3.5.3, Fact 1.

3.6 ASYMPTOTICS OF SEQUENCES

An exact formula for the terms of a sequence may be unwieldy. For example, it is difficult to estimate the magnitude of the central binomial coefficient $\binom{2n}{n} = \frac{(2n)!}{(n!)^2}$ from the definition of the factorial function alone. On the other hand, Stirling's approximation formula (§3.6.2) leads to the asymptotic estimate $\frac{4^n}{\sqrt{\pi n}}$. In applying asymptotic analysis, various "rules of thumb" help bypass tedious derivations. In practice, these rules almost always lead to correct results that can be proved by more rigorous methods. In the following discussions of asymptotic properties, the parameter tending to infinity is denoted by n. Both the subscripted notation a_n and the functional notation $f(n)$ are used to denote a sequence. The notation $f(n) \sim g(n)$ (f is *asymptotic to* g) means that $f(n) \neq 0$ for sufficiently large n and $\lim_{n \to \infty} \frac{g(n)}{f(n)} = 1$.

3.6.1 APPROXIMATE SOLUTIONS TO RECURRENCES

Although recurrences are a natural source of sequences, they often yield only crude asymptotic information. As a general rule, it helps to derive a summation or a generating function from the recurrence before obtaining asymptotic estimates.

Facts:

1. *Rule of thumb*: Suppose that a recurrence for a sequence a_n can be transformed into a recurrence for a related sequence b_n, so that the transformed sequence is approximately homogeneous and linear with constant coefficients (§3.3). Suppose also that ρ is the largest positive root of the characteristic equation for the homogeneous constant coefficient recurrence. Then it is probably true that $\frac{b_{n+1}}{b_n} \sim \rho$; i.e., b_n grows roughly like ρ^n.

2. Nonlinear recurrences are *not* covered by Fact 1.

3. Recurrences without fixed degree such as divide-and-conquer recurrences (§3.3.5), in which the difference between the largest and smallest subscripts is unbounded, are *not* covered by Fact 1. See [GrKn90, Ch. 2] for appropriate techniques.

Examples:

1. Consider the recurrence $D_{n+1} = n(D_n + D_{n-1})$ for $n \geq 1$, and define $d_n = \frac{D_n}{n!}$. Then $d_{n+1} = \frac{n}{n+1}d_n + \frac{1}{n+1}d_{n-1}$, which is quite close to the constant coefficient recurrence $\hat{d}_{n+1} = \hat{d}_n$. Since the characteristic root for this latter approximate recurrence is $\rho = 1$, Fact 1 suggests that $\frac{d_{n+1}}{d_n} \sim 1$, which implies that d_n is close to constant. Thus, we expect the original variable D_n to grow like $n!$. Indeed, if the initial conditions are $D_0 = D_1 = 1$, then $D_n = n!$. With initial conditions $D_0 = 1$, $D_1 = 0$, then D_n is the number of *derangements* of n objects (§2.4.2), in which case D_n is the closest integer to $\frac{n!}{e}$ for $n \geq 1$.

2. The accuracy of Example 1 is unusual. By way of contrast, the number I_n of *involutions* of an n-set (§2.8.1) satisfies the recurrence $I_{n+1} = I_n + nI_{n-1}$ for $n \geq 1$ with $I_0 = I_1 = 1$. By defining $i_n = I_n/(n!)^{1/2}$, then

$$i_{n+1} = \frac{i_n}{(n+1)^{1/2}} + \frac{i_{n-1}}{(1+1/n)^{1/2}},$$

which is nearly the same as the constant coefficient recurrence $\hat{i}_{n+1} = \hat{i}_{n-1}$. The characteristic equation $\rho^2 = 1$ has roots ± 1, so Fact 1 suggests that i_n is nearly constant and hence that I_n grows like $\sqrt{n!}$. The approximation in this case is not so good, because $I_n/\sqrt{n!} \sim e^{\sqrt{n}}/(8\pi e n)^{1/4}$, which is not a constant.

3.6.2 ANALYTIC METHODS FOR DERIVING ASYMPTOTIC ESTIMATES

Concepts and methods from continuous mathematics can be useful in analyzing the asymptotic behavior of sequences.

Definitions:

The **radius of convergence** of the series $\sum a_n x^n$ is the number r such that the series converges for all $|x| < r$ and diverges for all $|x| > r$, where $0 \leq r \leq \infty$.

The **gamma function** is the function $\Gamma(x) = \int_0^\infty t^{x-1} e^{-t}\, dt$.

Facts:

1. *Stirling's approximation*: $n! \sim \sqrt{2\pi n}(\frac{n}{e})^n$.
2. $\Gamma(x+1) = x\Gamma(x)$, $\Gamma(n+1) = n!$, and $\Gamma(\frac{1}{2}) = \sqrt{\pi}$.
3. The radius of convergence of $\sum a_n x^n$ is given by $\frac{1}{r} = \limsup_{n\to\infty} |a_n|^{1/n}$.
4. From Fact 3, it follows that $|a_n|$ tends to behave like r^{-n}. Most analytic methods are refinements of this idea.
5. The behavior of $f(z)$ near singularities on its circle of convergence determines the dominant asymptotic behavior of the coefficients of f. Estimates are often based on Cauchy's integral formula: $a_n = \oint f(z) z^{-n-1}\, dz$.
6. *Rule of thumb*: Consider the set of values of x for which $f(x) = \sum a_n x^n$ is either infinite or undefined, or involves computing a nonintegral power of 0. The absolute value of the least such x is normally the radius of convergence of $f(x)$. If there is no such x, then $r = \infty$.
7. *Rule of thumb*: Suppose that $0 < r < \infty$ is the radius of convergence of $f(x)$, that $g(x)$ has a larger radius of convergence, and that

$$f(x) - g(x) \sim A\bigl(-\ln(1 - \tfrac{x}{r})\bigr)^b \bigl(1 - \tfrac{x}{r}\bigr)^c \quad \text{as } x \to r^-,$$

for some constants A, b, and c, where it is not the case that both $b = 0$ and c is a nonnegative integer. (Often $g(x) = 0$.) Then it is probably true that

$$a_n \sim \begin{cases} A\binom{n-c-1}{n}(\ln n)^b r^{-n}, & \text{if } c \neq 0, \\ Ab(\ln n)^{b-1}/n, & \text{if } c = 0. \end{cases}$$

8. **Rule of thumb**: Let $a(x) = \frac{d \ln f(x)}{d \ln x}$ and $b(x) = \frac{da(x)}{d \ln x}$. Suppose that $a(r_n) = n$ has a solution with $0 < r_n < r$ and that $b(r_n) \in o(n^2)$. Then it is probably true that

$$a_n \sim \frac{f(r_n) r_n^{-n}}{\sqrt{2\pi b(r_n)}}.$$

Examples:

1. The number D_n of derangements has the exponential generating function $f(x) = \sum D_n \frac{x^n}{n!} = \frac{e^{-x}}{1-x}$. Since evaluation for $x = 1$ involves division by 0, it follows that $r = 1$. Since $\frac{e^{-x}}{1-x} \sim \frac{e^{-1}}{1-x}$ as $x \to 1^-$, take $g(x) = 0$, $A = e^{-1}$, $b = 0$, and $c = -1$. Fact 7 suggests that $D_n \sim \frac{n!}{e}$, which is correct.

2. The number b_n of left-right binary n-leaved trees has the generating function $f(x) = \frac{1}{2}(1 - \sqrt{1-4x})$. (See §9.3.3, Facts 1 and 7.) In this case $r = \frac{1}{4}$ since $f(\frac{1}{4})$ requires computing a fractional power of 0. Take $g(x) = \frac{1}{2}$, $A = \frac{1}{2}$, $b = 0$, and $c = \frac{1}{2}$ to suspect from Fact 7 that

$$b_n \sim -\frac{1}{2}\binom{n - \frac{3}{2}}{n} 4^n = \frac{-\Gamma(n - \frac{1}{2}) 4^n}{2\Gamma(n+1)\Gamma(-\frac{1}{2})} \sim \frac{4^{n-1}}{\sqrt{\pi n^3}},$$

which is valid. (Facts 1 and 2 have also been used.) This estimate converges rather rapidly — by the time $n = 40$, the estimate is less than 0.1% below b_{40}.

3. Since $\sum \frac{x^n}{n!} = e^x$, $n!$ can be estimated by taking $a(x) = b(x) = x$ and $r_n = n$ in Fact 8. This gives $\frac{1}{n!} \sim \frac{e^n n^{-n}}{\sqrt{2\pi n}}$, which is Stirling's asymptotic formula.

4. The number B_n of partitions of an n-set (§2.5.2) satisfies $\sum B_n \frac{x^n}{n!} = \exp(e^x - 1)$. In this case, $r = \infty$. Since $a(x) = xe^x$ and $b(x) = x(x+1)e^x$, it follows that r_n is the solution to $r_n \exp(r_n) = n$ and that $b(r_n) = (r_n + 1)n \sim nr_n \in o(n^2)$. Fact 8 suggests

$$B_n \sim \frac{n! \exp(e^{r_n} - 1)}{r_n^n \sqrt{2\pi n r_n}} = \frac{n! \exp(n/r_n - 1)}{r_n^n \sqrt{2\pi n r_n}}.$$

This estimate is correct, though the estimate converges quite slowly, as shown in this table:

n	10	20	100	200
estimate	1.49×10^5	6.33×10^{13}	5.44×10^{115}	7.01×10^{275}
B_n	1.16×10^5	5.17×10^{13}	4.76×10^{115}	6.25×10^{275}
ratio	1.29	1.22	1.14	1.12

Improved asymptotic estimates exist.

5. Analytic methods can sometimes be used to obtain asymptotics when only a functional equation is available. For example, if a_n is the number of n-leaved rooted trees in which each non-leaf node has exactly two children (with left and right not distinguished), the generating function for a_n satisfies $f(x) = x + (f(x)^2 + f(x^2))/2$, from which it can be deduced that $a_n \sim C n^{-3/2} r^{-n}$, where $r = 0.4026975\ldots$ and $C = 0.31877\ldots$ can easily be computed to any desired degree of accuracy. See [BeWi91, p. 394] for more information.

3.6.3 ASYMPTOTIC ESTIMATES OF MULTIPLY-INDEXED SEQUENCES

Asymptotic estimates for multiply-indexed sequences are considerably more difficult to obtain. To begin with, the meaning of a formula such as

$$\binom{n}{k} \sim \frac{2^n \exp(-(n-2k)^2/(2n))}{\sqrt{\pi n/2}}$$

must be carefully stated, because both n and k are tending to ∞, and the formula is valid only when this happens in such a way that $|2n - k| \in o(n^{3/4})$.

Facts:

1. Very little is known about how to obtain asymptotic estimates from multiply-indexed recurrences.

2. Most estimates of multiple summations are based on summing over one index at a time.

3. A few analytic results are available in the research literature. (See [Od95].)

3.7 MECHANICAL SUMMATION PROCEDURES

This section describes mechanical procedures that have been developed to evaluate sums of terms involving binomial coefficients and related factors. These procedures can not only be used to find explicit formulas for many sums, but can also be used to show that no simple closed formulas exist for certain sums. The invention of these mechanical procedures has been a surprising development in combinatorics. The material presented here is mostly adapted from [PeWiZe96], a comprehensive source for material on this topic.

3.7.1 HYPERGEOMETRIC SERIES

Definitions:

A **geometric series** is a series of the form $\sum_{k=0}^{\infty} a_k$ where the ratio between two consecutive terms is a constant, i.e., where the ratio $\frac{a_{k+1}}{a_k}$ is a constant for all $k = 0, 1, 2, \ldots$.

A **hypergeometric series** is a series of the form $\sum_{k=0}^{\infty} t_k$ where $t_0 = 1$ and the ratio of two consecutive terms is a rational function of the summation index k, i.e., the ratio $\frac{t_{k+1}}{t_k} = \frac{P(k)}{Q(k)}$ where $P(k)$ and $Q(k)$ are polynomials in the integer k. The terms of a hypergeometric series are called **hypergeometric terms**.

When the numerator $P(k)$ and denominator $Q(k)$ of this ratio are completely factored to give

$$\frac{P(k)}{Q(k)} = \frac{(k+a_1)(k+a_2)\ldots(k+a_p)}{(k+b_1)(k+b_2)\ldots(k+b_q)(k+1)} x$$

where x is a constant, this hypergeometric series is denoted by

$$_pF_q = \begin{bmatrix} a_1 & a_2 & \ldots & a_p \\ b_1 & b_2 & \ldots & b_q \end{bmatrix}; x \end{bmatrix}.$$

Note: If there is no factor $k+1$ in the denominator $Q(k)$ when it is factored, by convention the factor $k+1$ is added to both the numerator $P(k)$ and denominator $Q(k)$. Also, a horizontal dash is used to indicate the absence of factors in the numerator or in the denominator.

The hypergeometric terms s_n and t_n are **similar**, denoted $s_n \sim t_n$, if their ratio s_n/t_n is a rational function of n. Otherwise, these terms are called **dissimilar**.

Facts:
1. A geometric series is also a hypergeometric series.
2. If s_n is a hypergeometric term, then $\frac{1}{s_n}$ is also a hypergeometric term. (Equivalently, if $\sum_{k=0}^{\infty} s_n$ is a hypergeometric series, then $\sum_{k=0}^{\infty} \frac{1}{s_n}$ also is.)
3. In common usage, instead of stating that the series $\sum_{k=0}^{\infty} s_n$ is a hypergeometric series, it is stated that s_n is a hypergeometric term. This means exactly the same thing.
4. If s_n and t_n are hypergeometric terms, then $s_n \cdot t_n$ is a hypergeometric term. (Equivalently, if $\sum_{k=0}^{\infty} s_n$ and $\sum_{k=0}^{\infty} t_n$ are hypergeometric series, then $\sum_{k=0}^{\infty} s_n t_n$ is a hypergeometric series.)
5. If s_n is a hypergeometric term and s_n is not a constant, then $s_{n+1} - s_n$ is a hypergeometric term similar to s_n.
6. If s_n and t_n are hypergeometric terms and $s_n + t_n \neq 0$ for all n, then $s_n + t_n$ is hypergeometric if and only if s_n and t_n are similar.
7. If $t_n^{(1)}, t_n^{(2)}, \ldots, t_n^{(k)}$ are hypergeometric terms with $\sum_{i=1}^{k} t_n^{(i)} = 0$, then $t_n^{(i)} \sim t_n^{(j)}$ for some i and j with $1 \leq i < j \leq k$.
8. A sum of a fixed number of hypergeometric terms can be expressed as a sum of pairwise dissimilar hypergeometric terms.
9. The terms of a hypergeometric series can be expressed using rising powers $a^{\overline{n}}$ (also known as rising factorials and denoted by $(a)_n$) (see §3.4.2) as follows:

$$_pF_q = \begin{bmatrix} a_1 & a_2 & \cdots & a_p \\ b_1 & b_2 & \cdots & b_q \end{bmatrix} = \sum_{k=0}^{\infty} \frac{(a_1)^{\overline{k}}(a_2)^{\overline{k}} \cdots (a_p)^{\overline{k}}}{(b_1)^{\overline{k}}(b_2)^{\overline{k}} \cdots (b_q)^{\overline{k}}} \frac{x^k}{k!}.$$

10. There are a large number of well-known hypergeometric identities (see Facts 12–17, for example) that can be used as a starting point when a closed form for a sum of hypergeometric terms is sought.
11. There are many rules that transform a hypergeometric series with one parameter set into a different hypergeometric series with a second parameter set. Such transformation rules can be helpful in constructing closed forms for sums of hypergeometric terms.
12. $_1F_1 \begin{bmatrix} 1 \\ 1 ; x \end{bmatrix} = e^x.$
13. $_1F_0 \begin{bmatrix} a \\ - ; x \end{bmatrix} = \frac{1}{(1-x)^a}.$
14. Gauss's $_2F_1$ identity: If b is zero or a negative integer or the real part of $c - a - b$ is positive, then

$$_2F_1 \begin{bmatrix} a & b \\ c ; 1 \end{bmatrix} = \frac{\Gamma(c-a-b)\Gamma(c)}{\Gamma(c-a)\Gamma(c-b)}$$

where Γ is the gamma function (so $\Gamma(n) = (n-1)!$ when n is a positive integer).

15. Kummer's $_2F_1$ identity: If $a - b + c = 1$, then

$$_2F_1 \begin{bmatrix} a & b \\ c ; -1 \end{bmatrix} = \frac{\Gamma(\frac{b}{2}+1)\Gamma(b-a+1)}{\Gamma(b+1)\Gamma(\frac{b}{2}-a+1)}$$

and when b is a negative integer, this can be expressed as

$$_2F_1\begin{bmatrix} a & b \\ c \end{bmatrix};-1\end{bmatrix} = 2\cos(\tfrac{\pi b}{2})\frac{\Gamma(|b|)\Gamma(b-a+1)}{\Gamma(\frac{|b|}{2})\Gamma(\frac{b}{2}-a+1)}.$$

16. **Saalschütz's $_3F_2$ identity**: If $d+e = a+b+c+1$ and c is a negative integer, then

$$_3F_2\begin{bmatrix} a & b & c \\ d & e \end{bmatrix};1\end{bmatrix} = \frac{(d-a)^{\overline{|c|}}(d-b)^{\overline{|c|}}}{d^{\overline{|c|}}(d-a-b)^{\overline{|c|}}}.$$

17. **Dixon's identity**: If $1 + \tfrac{a}{2} - b - c > 0$, $d = a - b + 1$, and $e = a - c + 1$, then

$$_3F_2\begin{bmatrix} a & b & c \\ d & e \end{bmatrix};1\end{bmatrix} = \frac{(\tfrac{a}{2})!(a-b)!(a-c)!(\tfrac{a}{2}-b-c)!}{a!(\tfrac{a}{2}-b)!(\tfrac{a}{2}-c)!(a-b-c)!}.$$

The more familiar form of this identity reads

$$\sum_k (-1)^k \binom{a+b}{a+k}\binom{a+c}{c+k}\binom{b+c}{b+k} = \frac{(a+b+c)!}{a!b!c!}.$$

18. **Clausen's $_4F_3$ identity**: If d is a negative integer or zero and $a+b+c-d = \tfrac{1}{2}$, $e = a+b+\tfrac{1}{2}$, and $a+f = d+1 = b+g$, then

$$_4F_3\begin{bmatrix} a & b & c & d \\ e & f & g \end{bmatrix};1\end{bmatrix} = \frac{(2a)^{\overline{|d|}}(a+b)^{\overline{|d|}}(2b)^{\overline{|d|}}}{(2a+2b)^{\overline{|d|}}(a)^{\overline{|d|}}(b)^{\overline{|d|}}}.$$

Examples:

1. The series $\sum_{k=0}^{\infty} 3 \cdot (-5)^k$ is a geometric series. The series $\sum_{k=0}^{\infty} n2^n$ is not a geometric series.

2. The series $\sum_{k=0}^{\infty} t_k$ is a hypergeometric series when t_k equals 2^k, $(k+1)^2$, $\frac{1}{2k+3}$, or $\frac{1}{(2k+1)(k+3)!}$, but is not hypergeometric when $t_k = 2^k + 1$.

3. The series $\sum_{k=0}^{\infty} \frac{3^k}{k!^4}$ equals $_0F_3\begin{bmatrix} \\ 1 & 1 & 1 \end{bmatrix};3\end{bmatrix}$ since the ratio of the $(k+1)$st and kth terms is $\frac{3}{(k+1)^4}$.

4. A closed form for $S_n = \sum_{k=0}^{\infty}(-1)^k \binom{2n}{k}^2$ can be found by first noting that $S_n = {}_2F_1\begin{bmatrix} -2n & -2n \\ 1 \end{bmatrix};-1\end{bmatrix}$ since the ratio between successive terms of the sum is $\frac{-(k-2n)^2}{(k+1)^2}$. This shows that Kummer's $_2F_1$ identity can be invoked with $a = -2n$, $b = -2n$, and $c = 1$, producing the equality $S_n = \frac{2(-1)^n(2n-1)!}{n!(n-1)!} = (-1)^n \binom{2n}{n}$.

5. An example of a transformation rule for hypergeometric functions is provided by

$$_2F_1\begin{bmatrix} a & b \\ c \end{bmatrix};x\end{bmatrix} = (1-x)^{c-a-b} {}_2F_1\begin{bmatrix} c-a & c-b \\ c \end{bmatrix};x\end{bmatrix}.$$

3.7.2 ALGORITHMS THAT PRODUCE CLOSED FORMS FOR SUMS OF HYPERGEOMETRIC TERMS

Definitions:

A function $F(n,k)$ is called **doubly hypergeometric** if both $\frac{F(n+1,k)}{F(n,k)}$ and $\frac{F(n,k+1)}{F(n,k)}$ are rational functions of n and k.

A function $F(n,k)$ is a **proper hypergeometric term** if it can be expressed as

$$F(n,k) = P(n,k)\frac{\prod_{i=1}^{G}(a_i n + b_i k + c_i)!}{\prod_{i=1}^{H}(u_i n + v_i k + w_i)!} x^k$$

where x is a variable, $P(n,k)$ is a polynomial in n and k, G and H are nonnegative integers, and all the coefficients a_i, b_i, u_i, and v_i are integers.

A function $F(n, k)$ of the form

$$F(n,k) = P(n,k) \frac{\prod_{i=1}^{G}(a_i n + b_i k + c_i)!}{\prod_{i=1}^{H}(u_i n + v_i k + w_i)!} x^k$$

is said to be **well-defined** at (n, k) if none of the terms $(a_i n + b_i k + c_i)$ in the product is a negative integer. The function $F(n,k)$ is defined to have the value 0 if F is well-defined at (n, k) and there is a term $(u_i n + v_i k + w_i)$ in the product that is a negative integer or $P(n, k) = 0$.

Facts:

1. If $F(n, k)$ is a proper hypergeometric term, then there exist positive integers L and M and polynomials $a_{i,j}(n)$ for $i = 0, 1, \ldots, L$ and $j = 0, 1, \ldots, M$, not all zero, such that

$$\sum_{i=0}^{L} \sum_{j=0}^{M} a_{i,j}(n) F(n-j, k-i) = 0$$

for all pairs (n, k) with $F(n, k) \neq 0$ and all the values of $F(n, k)$ in this double sum are well-defined. Moreover, there is such a recurrence with M equal to $M' = \sum_s |b_s| + \sum_t |v_t|$ and L equal to $L' = \deg(P) + 1 + M'(-1 + \sum_s |a_s| + \sum_t |u_t|)$, where the a_i, b_i, u_i, v_i and P come from an expression of $F(n, k)$ as a hypergeometric term as specified in the definition.

2. *Sister Celine's algorithm*: This algorithm, developed in 1945 by Sister Mary Celine Fasenmeyer (1906–1996), can be used to find recurrence relations for sums of the form $f(n) = \sum_k F(n, k)$ where F is a doubly hypergeometric function. The algorithm finds a recurrence of the form $\sum_{i=0}^{L} \sum_{j=0}^{M} a_{i,j}(n) F(n-j, k-i) = 0$ by proceeding as follows:
   - start with trial values of L and M, such as $L = 1$, $M = 1$;
   - assume that a recurrence relation of the type sought exists with these values of L and M, with the coefficients $a_{i,j}(n)$ to be determined, if possible;
   - divide each term in the sum of the recurrence by $F(n, k)$, then reduce each fraction $F(n-j, k-i)/F(n, k)$, simplifying the ratios of factorials so only rational functions of n and k are left;
   - combine the terms in the sum using a common denominator, collecting the numerator into a single polynomial in k;
   - solve the system of linear equations for the $a_{i,j}(n)$ that results when the coefficients of each power of k in the numerator polynomial are equated to zero;
   - if these steps fail, repeat the procedure with larger values of L and M; by Fact 2, this procedure is guaranteed to eventually work.

3. *Gosper's algorithm*: This algorithm, developed by R. W. Gosper, Jr., can be used to determine, given a hypergeometric term t_n, whether there is a hypergeometric term z_n such that $z_{n+1} - z_n = t_n$. When there is such a hypergeometric term z_n, the algorithm also produces such a term.

4. Gosper's algorithm takes a hypergeometric term t_n as input and performs the following general steps (for details see [PeWiZe96]):
   - let $r(n) = t_{n+1}/t_n$; this is a rational function of n since t is hypergeometric;
   - find polynomials $a(n)$, $b(n)$, and $c(n)$ such that $\gcd(a(n), b(n+h)) = 1$ whenever h is a nonnegative integer; this is done using the following steps:

⋄ let $r(n) = K \cdot \frac{f(n)}{g(n)}$ where $f(n)$ and $g(n)$ are monic relatively prime polynomials and K is a constant, let $R(h)$ be the resultant of $f(n)$ and $g(n+h)$ (which is the product of the zeros of $g(n+h)$ at the zeros of $f(n)$), and let $S = \{h_1, h_2, \ldots, h_N\}$ be the set of nonnegative integer zeros of $R(h)$ where $0 \leq h_1 < h_2 < \cdots < h_N$;

⋄ let $p_0(n) = f(n)$ and $q_0(n) = g(n)$; then for $j = 1, 2, \ldots, N$ carry out the following steps:
$$s_j(n) := \gcd(p_{j-1}(n), q_{j-1}(n + h_j))$$
$$p_j(n) := p_{j-1}(n)/s_j(n)$$
$$q_j(n) := q_{j-1}(n)/s_j(n - h_j);$$

- take $a(n) := Kp_N(n)$; $b(n) := q_N(n)$; $c(n) := \prod_{i=1}^{N} \prod_{j=1}^{h_i} s_i(n-j)$;
- find a nonzero polynomial $x(n)$ such that $a(n)x(n+1) - b(n-1)x(n) = c(n)$ if one exists; such a polynomial can be found using the method of undetermined coefficients to find a nonzero polynomial of degree d or less, where the degree d depends on the polynomials $a(n)$, $b(n)$, and $c(n)$. If no such polynomial exists, then the algorithm fails. The degree d is determined by the following rules:

  ⋄ when $\deg a(n) \neq \deg b(n)$ or $\deg a(n) = \deg b(n)$ but the leading coefficients of $a(n)$ and $b(n)$ differ, then $d = \deg c(n) - \max(\deg a(n), \deg b(n))$;

  ⋄ when $\deg a(n) = \deg b(n)$ and the leading coefficients of $a(n)$ and $b(n)$ agree, $d = \max(\deg c(n) - \deg a(n) + 1, (B-A)/L)$ where $a(n) = Ln^k + An^{k-1} + \cdots$ and $b(n-1) = Ln^k + Bn^{k-1} + \cdots$; if this d is negative, then no such polynomial $x(n)$ exists;

- let $z_n = t_n \cdot b(n-1)x(n)/c(n)$; it follows that $z_{n+1} - z_n = t_n$.

5. When Gosper's algorithm fails, this shows that a sum of hypergeometric terms cannot be expressed as a hypergeometric term plus a constant.

6: Programs in both Maple and Mathematica implementing algorithms described in this section can be found at the following sites:
   http://www.cis.upenn.edu/~wilf/AeqB.html
   http://www.math.temple.edu/~zeilberg

Examples:

1. The function $F(n,k) = \frac{1}{5n+2k+2}$ is a proper hypergeometric term since $F(n,k)$ can be expressed as $F(n,k) = \frac{(5n+2k+1)!}{(5n+2k+2)!}$.

2. The function $F(n,k) = \frac{1}{n^2+k^3+5}$ is not a proper hypergeometric term.

3. Sister Celine's algorithm can be used to find a recurrence relation satisfied by the function $f(n) = \sum_k F(n,k)$ where $F(n,k) = k\binom{n}{k}$ for $n = 0, 1, 2, \ldots$. The algorithm proceeds by finding a recurrence relation of the form $a(n)F(n,k) + b(n)F(n+1,k) + c(n)F(n,k+1) + d(n)F(n+1,k+1) = 0$. Since $F(n,k) = k\binom{n}{k}$, this recurrence relation simplifies to $a(n) + b(n) \cdot \frac{n+1}{n+1-k} + c(n) \cdot \frac{n-k}{k} + d(n) \cdot \frac{n+1}{k} = 0$. Putting the left side of this equation over a common denominator and expressing it as a polynomial in k, four equations in the unknowns $a(n)$, $b(n)$, $c(n)$, and $d(n)$ are produced. These equations have the following solutions: $a(n) = t(-1 - \frac{1}{n})$, $b(n) = 0$, $c(n) = t(-1 - \frac{1}{n})$, $d = t$, where t is a constant. This produces the recurrence relation $(-1 - \frac{1}{n})F(n,k) + (-1 - \frac{1}{n})F(n,k+1) + F(n+1,k+1) = 0$, which can be summed over all integers k and simplified to produce the recurrence relation $f(n+1) = 2 \cdot \frac{n+1}{n}f(n)$, with $f(1) = 1$. From this it follows that $f(n) = n2^{n-1}$.

4. As shown in [PeWiZe96], Sister Celine's algorithm can be used to find an identity for $f(n) = \sum_k F(n,k)$ where $F(n,k) = \binom{n}{k}\binom{2n}{k}(-2)^{n-k}$. A recurrence for $F(n,k)$ can be found using her techniques (which can be carried out using either Maple or Mathematica software, for example). An identity that can be found this way is: $-8(n-1)F(n-2,k-1)-2(2n-1)F(n-1,k-1)+4(n-1)F(n-2,k)+2(2n-1)F(n-1,k)+nF(n,k) = 0$. When this is summed over all integers k, the recurrence relation $nf(n) - 4(n-1)f(n-2) = 0$ is obtained. From the definition of f it follows that $f(0) = 1$ and $f(1) = 0$. From the initial conditions and the recurrence relation for $f(n)$, it follows that $f(n) = 0$ when n is odd and $f(n) = \binom{n}{n/2}$ when n is even. (This is known as the *Reed-Dawson identity*.)

5. Gosper's algorithm can be used to find a closed form for $S_n = \sum_{k=1}^n k \cdot k!$. Let $t_n = n \cdot n!$. Following Gosper's algorithm gives $r(n) = \frac{t_{n+1}}{t_n} = \frac{(n+1)^2}{n}$, $a(n) = n+1$, $b(n) = 1$, and $c(n) = n$. The polynomial $x(n)$ must satisfy $(n+1)x(n+1) - x(n) = n$; the polynomial $x(n) = 1$ is such a solution. It follows that $z_n = n!$ satisfies $z_{n+1} - z_n = t_n$. Hence $s_n = z_n - z_1 = n! - 1$ and $S_n = s_{n+1} = (n+1)! - 1$.

6. Gosper's algorithm can be used to show that $S_n = \sum_{k=0}^n k!$ cannot be expressed as a hypergeometric term plus a constant. Let $t_n = n!$. Following Gosper's algorithm gives $r(n) = \frac{t_{n+1}}{t_n} = n+1$, $a(n) = n+1$, $b(n) = 1$, $c(n) = 1$. The polynomial $x(n)$ must satisfy $(n+1)x(n+1) - x(n) = 1$ and must have a degree less than zero. It follows that there is no closed form for $\sum_{k=0}^n k!$ of the type specified.

3.7.3 CERTIFYING THE TRUTH OF COMBINATORIAL IDENTITIES

Definitions:

A pair of functions (F, G) is called a **WZ pair** (after Wilf and Zeilberger) if $F(n+1, k) - F(n, k) = G(n, k+1) - G(n, k)$. If (F, G) is a WZ pair, then F is called the **WZ mate** of G and vice versa.

A **WZ certificate** $R(n, k)$ is a function that can be used to verify the hypergeometric identity $\sum_k f(n, k) = r(n)$ by creating a WZ pair (F, G) with $F(n, k) = \frac{f(n,k)}{r(n)}$ when $r(n) \neq 0$ and $F(n, k) = f(n, k)$ when $r(n) = 0$ and $G(n, k) = R(n, k)F(n, k)$. When a hypergeometric identity is proved using a a WZ certificate, this proof is called a **WZ proof**.

Facts:

1. If (F, G) is a WZ pair such that for each integer $n \geq 0$, $\lim_{k \to \pm\infty} G(n,k) = 0$, then $\sum_k F(n, k)$ is a constant for $n = 0, 1, 2, \ldots$.
2. If (F, G) is a WZ pair such that for each integer k, the limit $f_k = \lim_{n \to \infty} F(n, k)$ exists and is finite, for every nonnegative integer n it is the case that $\lim_{k \to \pm\infty} G(n, k) = 0$, and $\lim_{L \to \infty} \sum_{n \geq 0} G(n, -L) = 0$, then $\sum_{n \geq 0} G(n, k) = \sum_{j \leq k-1}(f_j - F(0, j))$.
3. An identity $\sum_k f(n, k) = r(n)$ can be verified using its WZ certificate $R(n, k)$ as follows:
   - if $r(n) \neq 0$, define $F(n, k)$ by $F(n, k) = \frac{f(n,k)}{r(n)}$, else define $F(n, k) = f(n, k)$; define $G(n, k)$ by $G(n, k) = R(n, k)F(n, k)$;
   - confirm that (F, G) is a WZ pair, i.e., that $F(n+1, k) - F(n, k) = G(n, k+1) - G(n, k)$, by dividing the factorials out and verifying the polynomial identity that results;
   - verify that the original identity holds for a particular value of n.

4. The WZ certificate of an identity $\sum_k f(n,k) = r(n)$ can be found using the following steps:

- if $r(n) \neq 0$, define $F(n,k)$ to be $F(n,k) = \frac{f(n,k)}{r(n)}$, else define $F(n,k)$ to be $F(n,k) = f(n,k)$;
- let $f(k) = F(n+1,k) - F(n,k)$; provide $f(k)$ as input to Gosper's algorithm;
- if Gosper's algorithm produces $G(n,k)$ as output, it is the WZ mate of F and the function $R(n,k) = \frac{G(n,k)}{F(n,k)}$ is the WZ certificate of the identity $\sum_k F(n,k) = C$ where C is a constant.

If Gosper's algorithm fails, this algorithm also fails.

Examples:

1. To prove the identity $f(n) = \sum_k \binom{n}{k}^2 = \binom{2n}{n}$, express it in the form $\sum_k F(n,k) = 1$ where $F(n,k) = \binom{n}{k}^2 / \binom{2n}{n}$. The identity can be proved by taking the function $R(n,k) = \frac{k^2(3n-2k+3)}{2(2n+1)(n-k+1)^2}$ as its WZ certificate. (This certificate can be obtained using Gosper's algorithm.)

2. To prove Gauss's $_2F_1$ identity via a WZ proof, express it in the form $\sum_k F(n,k) = 1$ where $F(n,k) = \frac{(n+k)!(b+k)!(c-n-1)!(c-b-1)!}{(c+k)!(n-1)!(c-n-b-1)!(k+1)!(b-1)!}$. The identity can then be proved by taking the function $R(n,k) = \frac{(k+1)(k+c)}{n(n+1-c)}$ as its WZ certificate. (This certificate can be obtained using Gosper's algorithm.)

REFERENCES

Printed Resources:

[AbSt65] M. Abramowitz and I. A. Stegun, eds., *Handbook of Mathematical Functions*, reprinted by Dover, 1965. (An invaluable and general reference for dealing with functions, with a chapter on sums in closed form.)

[BeWi91] E. A. Bender and S. G. Williamson, *Foundations of Applied Combinatorics*, Addison-Wesley, 1991. (Section 12.4 of this text contains further discussion of rules of thumb for asymptotic estimation.)

[BePhHo88] G. E. Bergum, A. N. Philippou, A. F. Horadam, eds., *Applications of Fibonacci Numbers*, Vols. 2-7, Kluwer Academic Publishers, 1988–1998.

[BeBr82] A. Beutelspacher and W. Brestovansky, "Generalized Schur Numbers" in "Combinatorial Theory," *Lecture Notes in Mathematics*, Vol. 969, Springer-Verlag, 1982, 30–38.

[Br92] R. Brualdi, *Introductory Combinatorics*, 2nd ed., North-Holland, 1992.

[De81] N. G. de Bruijn, *Asymptotic Methods in Analysis*, North-Holland, 1958. Third edition reprinted by Dover Publications, 1981. (This monograph covers a variety of topics in asymptotics from the viewpoint of an analyst.)

[ErSz35] P. Erdős and G. Szekeres, "A Combinatorial Problem in Geometry", *Compositio Mathematica* 2 (1935), 463-470.

[FlSaZi91] P. Flajolet, B. Salvy, and P. Zimmermann, "Automatic average-case analysis of algorithms", *Theoretical Computer Science* 79 (1991), 37–109. (Describes computer software to automate asymptotic average time analysis of algorithms.)

[FlSe98] P. Flajolet and R. Sedgewick, *Analytic Combinatorics*, Addison-Wesley, 1998, to appear. (The first part of this text deals with generating functions and asymptotic methods.)

[GrKnPa94] R. L. Graham, D. E. Knuth, and O. Patashnik, *Concrete Mathematics*, 2nd ed., Addison-Wesley, 1994. (A superb compendium of special sequences, their properties and analytical techniques.)

[GrRoSp80] R. Graham, B. Rothschild, and J. Spencer, *Ramsey Theory*, Wiley, 1980.

[GrKn90] D. H. Greene and D. E. Knuth, *Mathematics for the Analysis of Algorithms*, 3rd ed., Birkhäuser, 1990. (Parts of this text discuss various asymptotic methods.)

[Gr94] R. P. Grimaldi, *Discrete and Combinatorial Mathematics*, 4th ed., Addison-Wesley, 1999.

[Ha75] E. R. Hansen, *A Table of Series and Products*, Prentice-Hall, 1975. (A reference giving many summations and products in closed form.)

[Ho79] D. R. Hofstadter, *Gödel, Escher, Bach*, Basic Books, 1979.

[MiRo91] J. G. Michaels and K. H. Rosen, eds., *Applications of Discrete Mathematics*, McGraw-Hill, 1991. (Chapters 6, 7, and 8 discuss Stirling numbers, Catalan numbers, and Ramsey numbers.)

[Mi87] R. E. Mickens, *Difference Equations*, Van Nostrand Reinhold, 1987.

[NiZuMo91] I. Niven, H. S. Zuckerman, and H. L. Montgomery, *An Introduction to the Theory of Numbers*, 5th ed., Wiley, 1991.

[Od95] A. M. Odlyzko, "Asymptotic Enumeration Methods", in R. L. Graham, M. Grötschel, and L. Lovász (eds.), *Handbook of Combinatorics*, North-Holland, 1995, 1063–1229. (This is an encyclopedic presentation of methods with a 400$^+$ item bibliography.)

[PeWiZe96] M. Petkovšek, H. S. Wilf, and D. Zeilberger, *A=B*, A. K. Peters, 1996.

[PhBeHo86] A. N. Philippou, G. E. Bergum, A. F. Horadam, eds., *Fibonacci Numbers and Their Applications, Mathematics and Its Applications*, Vol. 28, D. Reidel Publishing Company, 1986.

[Ra30] F. Ramsey, "On a Problem of Formal Logic," *Proceedings of the London Mathematical Society* 30 (1930), 264–286.

[Ri58] J. Riordan, *An Introduction to Combinatorial Analysis*, Wiley, 1958.

[Ro84] F. S. Roberts, *Applied Combinatorics*, Prentice-Hall, 1984.

[Ro95] K. H. Rosen, *Discrete Mathematics and Its Applications*, 4th ed., McGraw-Hill, 1999.

[SlPl95] N. J. A. Sloane and S. Plouffe, *The Encyclopedia of Integer Sequences*, 2nd ed., Academic Press, 1995. (The definitive reference work on integer sequences.)

[StMc77] D. F. Stanat and D. F. McAllister, *Discrete Mathematics in Computer Science*, Prentice-Hall, 1977.

[St86] R. P. Stanley, *Enumerative Combinatorics*, vol. IV, Wadsworth, 1986.

[To85] I. Tomescu, *Problems in Combinatorics and Graph Theory*, translated by R. Melter, Wiley, 1985.

[Va89] S. Vajda, *Fibonacci & Lucas Numbers, and the Golden Section*, Halsted Press, 1989.

[Wi93] H. S. Wilf, *generatingfunctionology*, 2nd ed., Academic Press, 1993.

Web Resources:

http://www.cis.upenn.edu/~wilf/AeqB.html   (Contains programs in Maple and Mathematica implementing algorithms described in §3.7.2.)

http://www.cs.rit.edu/~spr/homepage.html   (Contains information on Ramsey numbers.)

http://www.math.temple.edu/~zeilberg (Contains programs in Maple and Mathematica implementing algorithms described in §3.7.2.)

http://www.research.att.com/~njas/sequences   (N. J. A. Sloane's web page; a table of sequences is accessible from this web page.)

http://www.sdstate.edu/~wcsc/http/fobbooks.html   (Contains a list of books that are available through the Fibonacci Association.)

4

NUMBER THEORY

4.1 Basic Concepts — Kenneth H. Rosen
    4.1.1 Numbers
    4.1.2 Divisibility
    4.1.3 Radix Representations

4.2 Greatest Common Divisors — Kenneth H. Rosen
    4.2.1 Introduction
    4.2.2 The Euclidean Algorithm

4.3 Congruences — Kenneth H. Rosen
    4.3.1 Introduction
    4.3.2 Linear and Polynomial Congruences

4.4 Prime Numbers — Jon Grantham and Carl Pomerance
    4.4.1 Basic Concepts
    4.4.2 Counting Primes
    4.4.3 Numbers of Special Form
    4.4.4 Pseudoprimes and Primality Testing

4.5 Factorization — Jon Grantham and Carl Pomerance
    4.5.1 Factorization Algorithms

4.6 Arithmetic Functions — Kenneth H. Rosen
    4.6.1 Multiplicative and Additive Functions
    4.6.2 Euler's Phi-function
    4.6.3 Sum and Number of Divisors Functions
    4.6.4 The Möbius Function and Other Important Arithmetic Functions
    4.6.5 Dirichlet Products

4.7 Primitive Roots and Quadratic Residues — Kenneth H. Rosen
    4.7.1 Primitive Roots
    4.7.2 Index Arithmetic
    4.7.3 Quadratic Residues
    4.7.4 Modular Square Roots

4.8 Diophantine Equations — Bart Goddard
    4.8.1 Linear Diophantine Equations
    4.8.2 Pythagorean Triples
    4.8.3 Fermat's Last Theorem
    4.8.4 Pell's, Bachet's, and Catalan's Equations
    4.8.5 Sums of Squares and Waring's Problem

4.9 Diophantine Approximation Jeff Shallit
    4.9.1 Continued Fractions
    4.9.2 Convergents
    4.9.3 Approximation Theorems
    4.9.4 Irrationality Measures

4.10 Quadratic Fields Kenneth H. Rosen
    4.10.1 Basics
    4.10.2 Primes and Unique Factorization

INTRODUCTION

This chapter covers the basics of number theory. Number theory, a subject with a long and rich history, has become increasingly important because of its applications to computer science and cryptography. The core topics of number theory, such as divisibility, radix representations, greatest common divisors, primes, factorization, congruences, diophantine equations, and continued fractions are covered here. Algorithms for finding greatest common divisors, large primes, and factorizations of integers are described.

There are many famous problems in number theory, including some that have been solved only recently such as Fermat's Last Theorem, and others that have eluded resolution, such as the Goldbach conjecture. The status of such problems is described in this chapter. New discoveries in number theory, such as new large primes, are being made at an increasingly fast pace. This chapter describes the current state of knowledge and provides pointers to Internet sources where the latest facts can be found.

GLOSSARY

algebraic number: a root of a polynomial with integer coefficients.

arithmetic function: a function defined for all positive integers.

Bachet's equation: a diophantine equation of the form $y^2 = x^3 + k$, where k is a given integer.

base: the positive integer b, with $b > 1$, in the expansion $n = a_k b^k + a_{k-1} b^{k-1} + \cdots + a_1 b + a_0$ where $0 \leq a_i \leq b - 1$ for $i = 0, 1, 2, \ldots, k$.

binary coded decimal expansion: the expansion produced by replacing each decimal digit of an integer by the four-bit binary expansion of that digit.

binary representation of an integer: the base two expansion of this integer.

Carmichael number: a positive integer that is a pseudoprime to all bases.

Catalan's equation: the diophantine equation $x^m - y^n = 1$ where solutions in integers greater than 1 are sought for x, y, m, and n.

Chinese remainder theorem: the theorem that states that given a set of congruences $x \equiv a_i \pmod{m_i}$ for $i = 1, 2, \ldots, n$ where the integers m_i, $i = 1, 2, \ldots, n$, are pairwise relatively prime, there is a unique simultaneous solution of these congruences modulo $M = m_1 m_2 \ldots m_n$.

complete system of residues modulo m: a set of integers such that every integer is congruent modulo m to exactly one integer in the set.

composite: a positive integer that has a factor other than 1 and itself.

congruence class of a modulo m: the set of integers congruent to a modulo m.

congruent integers modulo m: two integers with a difference divisible by m.

convergent: a rational fraction obtained by truncating a continued fraction.

continued fraction: a finite or infinite expression of the form $a_0 + 1/(a_1 + 1/(a_2 + \cdots$; usually abbreviated $[a_0, a_1, a_2, \ldots]$.

coprime (integers): integers that have no positive common divisor other than 1; see *relatively prime*.

Dedekind sum: the sum $s(h, k) = \sum_{\mu=1}^{k} \left(\left(\frac{h\mu}{k}\right)\right)\left(\left(\frac{\mu}{k}\right)\right)$ where $((x)) = x - \lfloor x \rfloor - \frac{1}{2}$ if x is not an integer and $((x)) = 0$ if x is an integer.

diophantine approximation: the approximation of a number by numbers belonging to a specified set, often the set of rational numbers.

Diophantine equation: an equation together with the restriction that the only solutions of the equation of interest are those belonging to a specified set, often the set of integers or the set of rational numbers.

Dirichlet's theorem on primes in arithmetic progressions: the theorem that states that there are infinitely many primes in each arithmetic progression of the form $an + b$ where a and b are relatively prime positive integers.

discrete logarithm of a to the base r modulo m: the integer x such that $r^x \equiv a \pmod{m}$, where r is a primitive root of m and $\gcd(a, m) = 1$.

divides: The integer a divides the integer b, written $a \mid b$, if there is an integer c such that $b = ac$.

divisor: (1) an integer d such that d divides a for a given integer a, or (2) the positive integer d that is divided into the integer a to yield $a = dq + r$ where $0 \leq r < d$.

elliptic curve: for prime $p > 3$, the set of solutions (x, y) to the congruence $y^2 \equiv x^3 + ax + b \pmod{p}$, where $4a^3 + 27b^2 \not\equiv 0 \pmod{p}$, together with a special point $\mathcal{O}$, called the point at infinity.

elliptic curve method (ECM): a factoring technique invented by Lenstra that is based on the theory of elliptic curves.

Euler phi-function: the function $\phi(n)$ whose value at the positive integer n is the number of positive integers not exceeding n relatively prime to n.

Euler's theorem: the theorem that states that if n is a positive integer and a is an integer with $\gcd(a, n) = 1$, then $a^{\phi(n)} \equiv 1 \pmod{n}$ where $\phi(n)$ is the value of the Euler phi-function at n.

exactly divides: If p is a prime and n is a positive integer, p^r exactly divides n, written $p^r \| n$, if p^r divides n, but p^{r+1} does not divide n.

factor (of an integer n): an integer that divides n.

factorization algorithm: an algorithm whose input is a positive integer and whose output is the prime factorization of this integer.

Farey series (of order n): the set of fractions $\frac{h}{k}$ where h and k are relatively prime nonnegative integers with $0 \leq h \leq k \leq n$ and $k \neq 0$.

Fermat equation: the diophantine equation $x^n + y^n = z^n$ where n is an integer greater than 2 and x, y, and z are nonzero integers.

Fermat number: a number of the form $2^{2^n} + 1$ where n is a nonnegative integer.

Fermat prime: a prime Fermat number.

Fermat's last theorem: the theorem that states that if n is a positive integer greater than two, then the equation $x^n + y^n = z^n$ has no solutions in integers with $xyz \neq 0$.

Fermat's little theorem: the theorem that states that if p is prime and a is an integer, then $a^p \equiv a \pmod{p}$.

Fibonacci numbers: the sequence of numbers defined by $F_0 = 0$, $F_1 = 1$, and $F_n = F_{n-1} + F_{n-2}$ for $n = 2, 3, 4, \ldots$.

fundamental theorem of arithmetic: the theorem that states that every positive integer has a unique representation as the product of primes written in nondecreasing order.

Gaussian integers: the set of numbers of the form $a + bi$ where a and b are integers and i is $\sqrt{-1}$.

greatest common divisor (gcd) of a set of integers: the largest integer that divides all integers in the set. The greatest common divisor of the integers $a_1, a_2, \ldots, a_n$ is denoted by $\gcd(a_1, a_2, \ldots, a_n)$.

hexadecimal representation (of an integer): the base sixteen representation of this integer.

index of a to the base r modulo m: the smallest nonnegative integer x, denoted $\text{ind}_r a$, such that $r^x \equiv a \pmod{m}$, where r is a primitive root of m and $\gcd(a, m) = 1$.

inverse of an integer a modulo m: an integer $\bar{a}$ such that $a\bar{a} \equiv 1 \pmod{m}$. Here $\gcd(a, m) = 1$.

irrational number: a real number that is not the ratio of two integers.

Jacobi symbol: a generalization of the Legendre symbol. (See §4.7.3.)

Kronecker symbol: a generalization of the Legendre and Jacobi symbols. (See §4.7.3.)

least common multiple (of a set of integers): the smallest positive integer that is divisible by all integers in the set.

least positive residue of a modulo m: the remainder when a is divided by m. It is the smallest positive integer congruent to a modulo m, written $a \bmod m$.

Legendre symbol: the symbol $\left(\frac{a}{p}\right)$ that has the value 1 if a is a square modulo p and -1 if a is not a square modulo p. Here p is a prime and a is an integer not divisible by p.

linear congruential method: a method for generating a sequence of pseudo-random numbers based on a congruence of the form $x_{n+1} \equiv ax_n + c \pmod{m}$.

Mersenne prime: a prime of the form $2^p - 1$ where p is a prime.

Möbius function: the arithmetic function $\mu(n)$ where $\mu(n) = 1$ if $n = 1$, $\mu(n) = 0$ if n has a square factor larger than 1, and $\mu(n) = (-1)^s$ if n is square-free and is the product of s different primes.

modulus: the integer m in a congruence $a \equiv b \pmod{m}$.

multiple of an integer a: an integer b such that a divides b.

multiplicative function: a function f such that $f(mn) = f(m)f(n)$ whenever m and n are relatively prime positive integers.

mutually relatively prime set of integers: integers with no common factor greater than 1.

number field sieve: a factoring algorithm, currently the best one known for large numbers with no small prime factors.

octal representation of an integer: the base eight representation of this integer.

one's complement expansion: an n bit representation of an integer x with $|x| < 2^{n-1}$, where n is a specified positive integer, where the leftmost bit is 0 if $x \geq 0$ and 1 if $x < 0$, and the remaining $n - 1$ bits are those of the binary expansion of x if $x \geq 0$, and the complements of the bits in the expansion of $|x|$ if $x < 0$.

order of an integer a modulo m: the least positive integer t, denoted by $\text{ord}_m a$, such that $a^t \equiv 1 \pmod{m}$. Here $\gcd(a, m) = 1$.

pairwise relatively prime: integers with the property that every two of them are relatively prime.

palindrome: a finite sequence that reads the same forward and backward.

partial quotient: a term a_i of a continued fraction.

Pell's equation: the diophantine equation $x^2 - dy^2 = 1$ where d is a positive integer that is not a perfect square.

perfect number: a positive integer whose sum of positive divisors, other than the integer itself, equals this integer.

periodic base b expansion: a base b expansion where the terms beyond a certain point are repetitions of the same block of integers.

powerful integer: an integer n with the property that p^2 divides n whenever p is a prime that divides n.

primality test: an algorithm that determines whether a positive integer is prime.

prime: a positive integer greater than 1 that has exactly two factors, 1 and itself.

prime factorization: the factorization of an integer into primes.

prime number theorem: the theorem that states that the number of primes not exceeding a positive real number x is asymptotic to $\frac{x}{\log x}$ (where $\log x$ denotes the natural logarithm of x).

prime-power factorization: the factorization of an integer into powers of distinct primes.

primitive root of an integer n: an integer r such that the least positive residues of the powers of r run through all positive integers relatively prime to n and less than n.

probabilistic primality test: an algorithm that determines whether an integer is prime with a small probability of a false positive result.

pseudoprime to the base b: a composite positive integer n such that $b^n \equiv b \pmod{n}$.

pseudo-random number generator: a deterministic method to generate numbers that share many properties with numbers really chosen randomly.

Pythagorean triple: positive integers x, y, and z such that $x^2 + y^2 = z^2$.

quadratic field: the set of number $Q(\sqrt{d}) = \{a + b\sqrt{d} \,|\, a, b \text{ integers}\}$ where d is a square-free integer.

quadratic irrational: an irrational number that is the root of a quadratic polynomial with integer coefficients.

quadratic nonresidue (of m): an integer that is not a perfect square modulo m.

quadratic reciprocity: the law that states that given two odd primes p and q, if at least one of them is of the form $4n + 1$, then p is a quadratic residue of q if and only if q is a quadratic residue of p and if both primes are of the form $4n + 3$, then p is a quadratic residue of q if and only if q is a quadratic nonresidue of p.

quadratic residue (of m): an integer that is a perfect square modulo m.

quadratic sieve: a factoring algorithm invented by Pomerance in 1981.

rational cuboid problem: the unsolved problem of constructing a right parallelepiped with height, width, length, face diagonals, and body diagonal all of integer length.

rational number: a real number that is the ratio of two integers. The set of rational numbers is denoted by Q.

reduced system of residues modulo m: pairwise incongruent integers modulo m such that each integer in the set is relatively prime to m and every integer relatively prime to m is congruent to an integer in the set.

relatively prime (integers): two integers with no common divisor greater than 1; see *coprime*.

remainder (of the integer a when divided by the positive integer d): the integer r in the equation $a = dq + r$ with $0 \leq r < d$, written $r = a \bmod d$.

root (of a function f modulo m): an integer r such that $f(r) \equiv 0 \pmod{m}$.

sieve of Eratosthenes: a procedure for finding all primes less than a specified integer.

smooth number: an integer all of whose prime divisors are small.

square root (of a modulo m): an integer r whose square is congruent to a modulo m.

square-free integer: an integer not divisible by any perfect squares other than 1.

ten most wanted numbers: the large integers on a list, maintained by a group of researchers, whose currently unknown factorizations are actively sought. These integers are somewhat beyond the realm of numbers that can be factored using known techniques.

terminating base-b expansion: a base-b expansion with only a finite number of nonzero coefficients.

totient function: the Euler phi-function.

transcendental number: a complex number that cannot be expressed as the root of an algebraic equation with integer coefficients.

trial division: a factorization technique that proceeds by dividing an integer by successive primes.

twin primes: a pair of primes that differ by two.

two's complement expansion: an n bit representation of an integer x, with $-2^{n-1} \leq x \leq 2^{n-1} - 1$, for a specified positive integer n, where the leftmost bit is 0 if $x \geq 0$ and 1 if $x < 0$, and the remaining $n - 1$ bits are those from the binary expansion of x if $x \geq 0$ and are those of the binary expansion of $2^n - |x|$ if $x < 0$.

ultimately periodic: a sequence (typically a base-k expansion or continued fraction) $(a_i)_{i \geq 0}$ that eventually repeats, that is, there exist k and N such that $a_{n+k} = a_n$ for all $n \geq N$.

unit of a quadratic field: a number ϵ such that $\epsilon | 1$ in the quadratic field.

Waring's problem: the problem of determining the smallest number $g(k)$ such that every integer is the sum of $g(k)$ kth powers of integers.

4.1 BASIC CONCEPTS

The basic concepts of number theory include the classification of numbers into different sets of special importance, the notion of divisibility, and the representation of integers. For more information about these basic concepts, see introductory number theory texts, such as [Ro99].

4.1.1 NUMBERS

Definitions:

The **integers** are the elements of the set $\mathcal{Z} = \{\ldots, -3, -2, -1, 0, 1, 2, 3, \ldots\}$.

The **natural numbers** are the integers in the set $\mathcal{N} = \{0, 1, 2, 3, \ldots\}$.

The **rational numbers** are real numbers that can be written as a/b where a and b are integers with $b \neq 0$. Numbers that are not rational are called **irrational**. The set of rational numbers is denoted by $\mathcal{Q}$.

The **algebraic numbers** are real numbers that are solutions of equations of the form $a_n x^n + \cdots + a_1 x + a_0 = 0$ where a_i is an integer, for $i = 0, 1, \ldots, n$. Real numbers that are not algebraic are called **transcendental**.

Facts:

1. Table 1 summarizes information and notation about some important types of numbers.

2. A real number is rational if and only if its decimal expansion terminates or is periodic. (See §4.1.3).

3. The number $N^{1/m}$ is irrational where N and m are positive integers, unless N is the mth power of an integer n.

4. The number $\log_b a$ is irrational, where a and b are positive integers greater than 1, if there is a prime that divides exactly one of a and b.

Table 1 Types of numbers.

name	definition	examples
natural numbers $\mathcal{N}$	$\{0, 1, 2, \ldots\}$	$0, 43$
integers $\mathcal{Z}$	$\{\ldots, -2, -1, 0, 1, 2, \ldots\}$	$0, 43, -314$
Gaussian integers $\mathcal{Z}[i]$	$\{a + bi \mid a, b \in \mathcal{Z}\}$	$3, 4 + 3i, 7i$
rational numbers $\mathcal{Q}$	$\{\frac{a}{b} \mid a, b \in \mathcal{Z}; b \neq 0\}$	$0, \frac{22}{7}$
quadratic irrationals	irrational root of quadratic equation $a_2 x^2 + a_1 x + a_0 = 0$; all $a_i \in \mathcal{Q}$	$\sqrt{2}, \frac{2+\sqrt{5}}{3}$
irrational numbers	$\mathcal{R} - \mathcal{Q}$	$\sqrt{2}, \pi, e$
algebraic numbers $\overline{\mathcal{Q}}$	root of algebraic equation $a_n x^n + \cdots + a_0 = 0$, $n \geq 1$, $a_0, \ldots, a_n \in \mathcal{Z}$	$i, \sqrt{2}, \sqrt[3]{\frac{3}{2}}$
algebraic integers $\mathcal{A}$	root of monic algebraic equation $x^n + a_{n-1} x^{n-1} + \cdots + a_0 = 0$, $n \geq 1, a_0, a_1, \ldots, a_{n-1} \in \mathcal{Z}$	$i, \sqrt{2}, \frac{1+\sqrt{5}}{2}$
transcendental numbers	$\mathcal{C} - \overline{\mathcal{Q}}$	$\pi, e, i \ln 2$
real numbers $\mathcal{R}$	completion of $\mathcal{Q}$	$0, \frac{1}{3}, \sqrt{2}, \pi$
complex numbers $\mathcal{C}$	$\overline{\mathcal{R}}$ or $\mathcal{R}[i]$	$3 + 2i, e + i\pi$

5. If x is a root of an equation $x^m + a_{m-1} x^{m-1} + \cdots + a_0 = 0$ where the coefficients a_i ($i = 0, 1, \ldots, m-1$) are integers, then x is either an integer or irrational.

6. The set of algebraic numbers is countable (§1.2.3). Hence, almost all real numbers are transcendental. (However, showing a particular number of interest is transcendental is usually difficult.)

7. Both e and π are transcendental. The transcendence of e was proven by Hermite in 1873, and π was proven transcendental by Lindemann in 1882. Proofs of the transcendence of e and π can be found in [IIaWi89].

8. *Gelfond-Schneider theorem:* If α and β are algebraic numbers with α not equal to 0 or 1 and β irrational, then α^β is transcendental. (For a proof see [Ba90].)

9. *Baker's linear forms in logarithms:* If $\alpha_1, \ldots, \alpha_n$ are nonzero algebraic numbers and $\log \alpha_1, \ldots, \log \alpha_n$ are linearly independent over $\mathcal{Q}$, then $1, \log \alpha_1, \ldots, \log \alpha_n$ are linearly independent over $\overline{\mathcal{Q}}$, where $\overline{\mathcal{Q}}$ is the closure of $\mathcal{Q}$. (Consult [Ba90] for a proof and applications of this theorem.)

Examples:

1. The numbers $\frac{11}{17}, -\frac{3345}{7}, -1, \frac{578}{579}$, and 0 are rational.
2. The number $\log_2 10$ is irrational.
3. The numbers $\sqrt{2}, 1 + \sqrt{2}$, and $\frac{1+\sqrt{2}}{5}$ are irrational.

4. The number $x = 0.10100100010000\ldots$, with a decimal expansion consisting of blocks where the nth block is a 1 followed by n 0s, is irrational, since this decimal expansion does not terminate and is not periodic.
5. The decimal expansion of $\frac{22}{7}$ is periodic, since $\frac{22}{7} = 3.\overline{142857}$. However, the decimal expansion of π neither terminates, nor is periodic, with $\pi = 3.141592653589793\ldots$.
6. It is not known whether Euler's constant $\gamma = \lim_{n \to \infty} \left(\sum_{k=1}^{n} \frac{1}{k} - \log n \right)$ (where $\log x$ denotes the natural logarithm of x) is rational or irrational.
7. The numbers 2, $\frac{1}{2}$, $\sqrt{17}$, $\sqrt[3]{5}$, and $1 + \sqrt[6]{2}$ are algebraic.
8. By the Gelfond-Schneider theorem (Fact 8), $\sqrt{2}^{\sqrt{2}}$ is transcendental.
9. By Baker's linear forms in logarithms theorem (Fact 9), since $\log_2 10$ is irrational, it is transcendental.

4.1.2 DIVISIBILITY

The notion of the divisibility of one integer by another is the most basic concept in number theory. Introductory number theory texts, such as [Ro99], [HaWr89], and [NiZuMo91], are good references for this material.

Definitions:

If a and d are integers with $d > 0$, then in the equation $a = dq + r$ where $0 \leq r < d$, a is the **dividend**, d is the **divisor**, q is the **quotient**, and r is the **remainder**.

Let m and n be integers with $m \geq 1$ and $n = dm + r$ with $0 \leq r < m$. Then $n \bmod m$, the value of the **mod** m function at n, is r, the remainder when n is divided by m.

If a and b are integers and $a \neq 0$, then a **divides** b, written $a|b$, if there is an integer c such that $b = ac$. If a divides b, then a is a **factor** or **divisor** of b, and b is a **multiple** of a. If a is a positive divisor of b that does not equal b, then a is a **proper divisor of** b. The notation $a \nmid b$ means that a does not divide b.

A **prime** is a positive integer divisible by exactly two distinct positive integers, 1 and itself. A positive integer, other than 1, that is not prime is called **composite**.

An integer is **square-free** if it is not divisible by any perfect square other than 1.

An integer n is **powerful** if whenever a prime p divides n, p^2 divides n.

If p is prime and n is a positive integer, then p^r **exactly divides** n, written $p^r || n$, if p^r divides n, but p^{r+1} does not divide n.

Facts:
1. If a is a nonzero integer, then $a|0$.
2. If a is an integer, then $1|a$.
3. If a and b are positive integers and $a|b$, then the following statements are true:
   - $a \leq b$;
   - $\frac{b}{a}$ divides b;
   - a^k divides b^k for every positive integer k;
   - a divides bc for every integer c.
4. If a, b, and c are integers such that $a|b$ and $b|c$, then $a|c$.
5. If a, b, and c are integers such that $a|b$ and $a|c$, then $a|bm + cn$ for all integers m and n.

6. If a and b are integers such that $a|b$ and $b|a$, then $a = \pm b$.

7. If a and b are integers and m is a nonzero integer, then $a|b$ if and only if $ma|mb$.

8. *Division algorithm*: If a and d are integers with d positive, then there are unique integers q and r such that $a = dq + r$ with $0 \leq r < d$. (Note: The division algorithm is not an algorithm, in spite of its name.)

9. The quotient q and remainder r when the integer a is divided by the positive integer d are given by $q = \lfloor \frac{a}{d} \rfloor$ and $r = a - d \lfloor \frac{a}{d} \rfloor$, respectively.

10. If a and d are positive integers, then there are unique integers q, r, and e such that $a = dq + er$ where $e = \pm 1$ and $-\frac{d}{2} < r \leq \frac{d}{2}$.

11. There are several divisibility tests that are easily performed using the decimal expansion of an integer. These include:

   - An integer is divisible by 2 if and only if its last digit is even. It is divisible by 4 if and only if the integer made up of its last two digits is divisible by four. More generally, it is divisible by 2^j if and only if the integer made up of the last j decimal digits of n is divisible by 2^j.
   - An integer is divisible by 5 if and only if its last digit is divisible by 5 (which means it is either 0 or 5). It is divisible by 25 if and only if the integer made up of the last two digits is divisible by 25. More generally, it is divisible by 5^j if and only if the integer made up of the last j digits of n is divisible by 5^j.
   - An integer is divisible by 3, or by 9, if and only if the sum of the decimal digits of n is divisible by 3, or by 9, respectively.
   - An integer is divisible by 11 if and only if the integer formed by alternately adding and subtracting the decimal digits of the integer is divisible by 11.
   - An integer is divisible by 7, 11, or 13 if and only if the integer formed by successively adding and subtracting the three-digit integers formed from successive blocks of three decimal digits of the original number, where digits are grouped starting with the rightmost digit, is divisible by 7, 11, or 13, respectively.

12. If $d|b-1$, then $n = (a_k \ldots a_1 a_0)_b$ (this notation is defined in §4.1.3) is divisible by d if and only if the sum of the base b digits of n, $a_k + \cdots + a_1 + a_0$, is divisible by d.

13. If $d|b+1$, then $n = (a_k \ldots a_1 a_0)_b$ is divisible by d if and only if the alternating sum of the base b digits of n, $(-1)^k a_k + \cdots - a_1 + a_0$, is divisible by d.

14. If $p^r \| a$ and $p^s \| b$ where p is a prime and a and b are positive integers, then $p^{r+s} \| ab$.

15. If $p^r \| a$ and $p^s \| b$ where p is a prime and a and b are positive integers, then $p^{\min(r,s)} \| a + b$.

16. There are infinitely many primes. (See §4.4.1.)

17. There are efficient algorithms that can produce large integers that have an extremely high probability of being prime. (See §4.4.4.)

18. *Fundamental theorem of arithmetic*: Every positive integer can be written as the product of primes in exactly one way, where the primes occur in nondecreasing order in the factorization.

19. Many different algorithms have been devised to find the factorization of a positive integer into primes. Using some recently invented algorithms and the powerful computer systems available today, it is feasible to factor integers with over 100 digits. (See §4.5.1.)

20. The relative ease of producing large primes compared with the apparent difficulty of factoring large integers is the basis for an important cryptosystem called RSA. (See Chapter 14.)

Examples:

1. The integers 0, 3, -12, 21, 342, and -1113 are divisible by 3; the integers -1, 7, 29, and -1111 are not divisible by 3.

2. The quotient and remainder when 214 is divided by 6 are 35 and 4, respectively since $214 = 35 \cdot 6 + 4$.

3. The quotient and remainder when -114 is divided by 7 are -17 and 5, respectively since $-114 = -17 \cdot 7 + 5$.

4. With $a = 214$ and $d = 6$, the expansion of Fact 10 is $214 = 36 \cdot 6 - 2$ (so that $e = -1$ and $r = 2$).

5. $11 \bmod 4 = 3$, $100 \bmod 7 = 2$, and $-22 \bmod 5 = 3$.

6. The following are primes: 2, 3, 17, 101, 641. The following are composites: 4, 9, 91, 111, 1001.

7. The integers 15, 105, and 210 are squarefree; the integers 12, 99, and 270 are not.

8. The integers 72 is powerful since 2 and 3 are the only primes that divide 72 and $2^2 = 4$ and $3^2 = 9$ both divide 72, but 180 is not powerful since 5 divides 180, but 5^2 does not.

9. The integer 32,688,048 is divisible by 2,4,8, and 16 since $2|8$, $4|48$, $8|048$, and $16|8,048$, but it is not divisible by 32 since 32 does not divide 88,048.

10. The integer 723,160,823 is divisible by 11 since the alternating sum of its digits, $3 - 2 + 8 - 0 + 6 - 1 + 3 - 2 + 7 = 22$, is divisible by 11.

11. Since $3^3 | 216$, but $3^4 \nmid 216$, it follows that $3^3 || 216$.

4.1.3 RADIX REPRESENTATIONS

The representation of numbers in different bases has been important in the development of mathematics from its earliest days and is extremely important in computer arithmetic. For further details on this topic, see [Kn81], [Ko93], and [Sc85].

Definitions:

The **base b expansion** of a positive integer n, where b is an integer greater than 1, is the unique expansion of n as $n = a_k b^k + a_{k-1} b^{k-1} + \cdots + a_1 b + a_0$ where k is a nonnegative integer, a_j is a nonnegative integer less than b for $j = 0, 1, \ldots, k$ and the initial coefficient $a_k \neq 0$. This expansion is written as $(a_k a_{k-1} \ldots a_1 a_0)_b$.

The integer b in the base b expansion of an integer is called the **base** or **radix** of the expansion.

The coefficients a_j in the base b expansion of an integer are called the base b **digits** of the expansion.

Base 10 expansions are called **decimal** expansions. The digits are called **decimal digits**.

Base 2 expansions are called **binary** expansions. The digits are called **binary digits** or **bits**.

Base 8 expansions are called **octal** expansions.

Base 16 expansions are called **hexadecimal** expansions. The 16 hexadecimal digits are $0, 1, 2, 3, 4, 5, 6, 7, 8, 9, A, B, C, D, E, F$ (where A, B, C, D, E, F correspond to the decimal numbers $10, 11, 12, 13, 14, 15$, respectively).

> **Algorithm 1:** Constructing base b expansions.
>
> **procedure** *base b expansion*(n: positive integer)
> $q := n$
> $k := 0$
> **while** $q \neq 0$
> **begin**
> $\quad a_k := q \bmod b$
> $\quad q := \lfloor \frac{q}{b} \rfloor$
> $\quad k := k + 1$
> **end** {the base b expansion of n is $(a_{k-1} \ldots a_1 a_0)_b$}

The **binary coded decimal** expansion of an integer is the bit string formed by replacing each digit in the decimal expansion of the integer by the four bit binary expansion of that digit.

The **one's complement** expansion of an integer x with $|x| < 2^{n-1}$, for a specified positive integer n, uses n bits, where the leftmost bit is 0 if $x \geq 0$ and 1 if $x < 0$, and the remaining $n-1$ bits are those from the binary expansion of x if $x \geq 0$ and are the complements of the bits in the binary expansion of $|x|$ if $x < 0$. (*Note*: the one's complement representation $11\ldots1$, consisting of n 1s, is usually considered to the negative representation of the number 0.)

The **two's complement** expansion of an integer x with $-2^{n-1} \leq x \leq 2^{n-1}-1$, for a specified positive integer n, uses n bits, where the leftmost bit is 0 if $x \geq 0$ and 1 if $x < 0$, and the remaining $n-1$ bits are those from the binary expansion of x if $x \geq 0$ and are those of the binary expansion of $2^n - |x|$ if $x < 0$.

The **base b expansion** (where b is an integer greater than 1) of a real number x with $0 \leq x < 1$ is the unique expansion of x as $x = \sum_{j=1}^{\infty} \frac{c_j}{b^j}$ where c_j is a nonnegative integer less than b for $j = 1, 2, \ldots$ and for every integer N there is a coefficient $c_n \neq b-1$ for some $n > N$. This expansion is written as $(.c_1 c_2 c_3 \ldots)_b$.

A base b expansion $(.c_1 c_2 c_3 \ldots)_b$ **terminates** if there is a positive integer n such that $c_n = c_{n+1} = c_{n+2} = \cdots = 0$.

A base b expansion $(.c_1 c_2 c_3 \ldots)_b$ is **periodic** if there are positive integers N and k such that $c_{n+k} = c_n$ for all $n \geq N$.

The **periodic base b expansion** $(.c_1 c_2 \ldots c_{N-1} c_N \ldots c_{N+k-1} c_N \ldots c_{N+k-1} c_N \ldots)_b$ is denoted by $(.c_1 c_2 \ldots c_{N-1} \overline{c_N \ldots c_{N+k-1}})_b$. The part of the periodic base b expansion preceding the periodic part is the **pre-period** and the periodic part is the **period**, where the period and pre-period are taken to have minimal possible length.

Facts:

1. If b is a positive integer greater than 1, then every positive integer n has a unique base b expansion.

2. *Converting from base 10 to base b*: Take the positive integer n and divide it by b to obtain $n = bq_0 + a_0$, $0 \leq a_0 < b$. Then divide q_0 by b to obtain $q_0 = bq_1 + a_1$, $0 \leq a_1 < b$. Continue this process, successively dividing the quotients by b, until a quotient of zero is obtained, after k steps. The base b expansion of n is then $(a_{k-1} \ldots a_1 a_0)_b$. (See Algorithm 1.)

3. *Converting from base 2 to base 2^k*: Group the bits in the base 2 expansion into blocks of k bits, starting from the right, and then convert each block of k bits into a base 2^k digit. For example, converting from binary (base 2) to octal (base 8) is done by grouping the bits of the binary expansion into blocks of 3 bits starting from the right and converting each block into an octal digit. Similarly, converting from binary to hexadecimal (base 16) is done by grouping the bits of the binary expansion into blocks of 4 bits starting from the right and converting each block into a hex digit.

4. *Converting from base 2^k to binary (base 2)*: convert each base 2^k digit into a block of k bits and string together these bits in the order the original digits appear. For example, to convert from hexadecimal to binary, convert each hex digit into the block of four bits that represent this hex digit and then string together these blocks of four bits in the correct order.

5. Every positive integer can be expressed uniquely as the sum of distinct powers of two. This follows since every positive integer has a unique base two expansion, with the digits either 0 or 1.

6. There are $\lfloor \log_b n \rfloor + 1$ decimal digits in the base b expansion of the positive integer n.

7. The number x with one's complement representation $(a_{n-1}a_{n-2}\ldots a_1 a_0)$ can be found using the equation
$$x = -a_{n-1}(2^{n-1}-1) + \sum_{i=0}^{n-2} a_i 2^i.$$

8. The number x with two's complement representation $(a_{n-1}a_{n-2}\ldots a_1 a_0)$ can be found using the equation
$$x = -a_{n-1}\cdot 2^{n-1} + \sum_{i=0}^{n-2} a_i 2^i.$$

9. Two's complement representations of integers are often used by computers because addition and subtraction of integers, where these integers may be either positive or negative, can be performed easily using these representations.

10. Define a function $\operatorname{Lg} n$ by the rule
$$\operatorname{Lg} n = \begin{cases} 1 & \text{if } n = 0; \\ 1 + \lfloor \log_2 |n| \rfloor & \text{if } n \neq 0. \end{cases}$$
Then $\operatorname{Lg} n$ is the number of bits in the base 2 expansion of n, not counting the sign bit. (Compare with Fact 6.)

11. The bit operations for the basic operations are given in the following table, adapted from [BaSh96]. This table displays the number of bit operations used by the standard, naive algorithms, doing things bit by bit (addition with carries, subtraction with borrows, standard multiplication by each bit and shifting and adding, and standard division), and a big-oh estimate for the number of bits required to do the operations using the algorithm with the currently best known computational complexity. (The function Lg is defined in Fact 10; the function $\mu(m,n)$ is defined by the rule $\mu(m,n) = m(\operatorname{Lg} n)(\operatorname{Lg}\operatorname{Lg} n)$ if $m \geq n$ and $\mu(m,n) = n(\operatorname{Lg} m)(\operatorname{Lg}\operatorname{Lg} m)$ otherwise.)

operation	number of bits for operation (following naive algorithm)	best known complexity (sophisticated algorithm)
$a \pm b$	$\operatorname{Lg} a + \operatorname{Lg} b$	$O(\operatorname{Lg} a + \operatorname{Lg} b)$
$a \cdot b$	$\operatorname{Lg} a \cdot \operatorname{Lg} b$	$O(\mu(\operatorname{Lg} a, \operatorname{Lg} b))$
$a = qb + r$	$\operatorname{Lg} q \cdot \operatorname{Lg} b$	$O(\mu(\operatorname{Lg} q, \operatorname{Lg} b))$

12. If b is a positive integer greater than 1 and x is a real number with $0 \leq x < 1$, then x can be uniquely written as $x = \sum_{j=1}^{\infty} \frac{c_j}{b^j}$ where c_j is a nonnegative integer less than b for all j, with the restriction that for every positive integer N there is an integer n with $n > N$ and $c_n \neq b - 1$ (in other words, it is not the case that from some point on, all the coefficients are $b - 1$).

13. A periodic or terminating base b expansion, where b is a positive integer, represents a rational number.

14. The base b expansion of a rational number, where b is a positive integer, either terminates or is periodic.

15. If $0 < x < 1$, $x = \frac{r}{s}$ where r and s are relatively prime positive integers, and $s = TU$ where every prime factor of T divides b and $\gcd(U, b) = 1$, then the period length of the base b expansion of x is $\text{ord}_U b$ (defined in §4.7.1) and the pre-period length is the smallest positive integer N such that T divides b^N.

16. The period length of the base b expansion of $\frac{1}{m}$ (b and m positive integers greater than 1) is $m - 1$ if and only if m is prime and b is a primitive root of m. (See §4.7.1.)

Examples:

1. The binary (base 2), octal (base 8), and hexadecimal (base 16) expansions of the integer 2001 are $(11111010001)_2$, $(3721)_8$, and $(7D1)_{16}$, respectively. The octal and hexadecimal expansions can be obtained from the binary expansion by grouping together, from the right, the bits of the binary expansion into groups of 3 bits and 4 bits, respectively.

2. The hexadecimal expansion $2FB3$ can be converted to a binary expansion by replacing each hex digit by a block of four bits to give 10111110110011. (The initial two 0s in the four bit expansion of the initial hex digit 2 are omitted.)

3. The binary coded decimal expansion of 729 is 011100101001.

4. The nine-bit one's complement expansions of 214 and -113 (taking $n = 9$ in the definition) are 011010110 and 110001110.

5. The nine-bit two's complement expansions of 214 and -113 (taking $n = 9$ in the definition) are 011010110 and 110001111.

6. By Fact 7 the integer with a nine-bit one's complement representation of 101110111 equals $-1(256 - 1) + 119 = -136$.

7. By Fact 8 the integer with a nine-bit two's complement representation of 101110111 equals $-256 + 119 = -137$.

8. By Fact 15 the pre-period of the decimal expansion of $\frac{5}{28}$ has length 2 and the period has length 6 since $28 = 4 \cdot 7$ and $\text{ord}_7 10 = 6$. This is verified by noting that $\frac{5}{28} = (.17\overline{857142})_{10}$.

4.2 GREATEST COMMON DIVISORS

The concept of the greatest common divisor of two integers plays an important role in number theory. The Euclidean algorithm, an algorithm for computing greatest common divisors, was known in ancient times and was one of the first algorithms that was studied for what is now called its computational complexity. The Euclidean algorithm and its extensions are used extensively in number theory and its applications, including those to cryptography. For more information about the contents of this section consult [HaWr89], [NiZuMo91], or [Ro99].

4.2.1 INTRODUCTION

Definitions:

The **greatest common divisor** of the integers a and b, not both zero, written $\gcd(a,b)$, is the largest integer that divides both a and b.

The integers a and b are **relatively prime** (or **coprime**) if they have no positive divisors in common other than 1, i.e., if $\gcd(a,b) = 1$.

The **greatest common divisor** of the integers a_i, $i = 1, 2, \ldots, k$, not all zero, written $\gcd(a_1, a_2, \ldots, a_k)$, is the largest integer that divides all the integers a_i.

The integers $a_1, a_2, \ldots, a_k$ are **pairwise relatively prime** if $\gcd(a_i, a_j) = 1$ for $i \neq j$.

The integers $a_1, a_2, \ldots, a_k$ are **mutually relatively prime** if $\gcd(a_1, a_2, \ldots, a_k) = 1$.

The **least common multiple** of nonzero integers a and b, written $\mathrm{lcm}(a,b)$, is the smallest positive integer that is a multiple of both a and b.

The **least common multiple** of nonzero integers $a_1, \ldots, a_k$, written $\mathrm{lcm}(a_1, \ldots, a_k)$, is the smallest positive integer that is a multiple of all the integers a_i, $i = 1, 2, \ldots, k$.

The **Farey series of order n** is the set of fractions $\frac{h}{k}$ where h and k are integers, $0 \leq h \leq k \leq n$, $k \neq 0$, and $\gcd(h,k) = 1$, in ascending order, with 0 and 1 included in the forms $\frac{0}{1}$ and $\frac{1}{1}$, respectively.

Facts:

1. If $d|a$ and $d|b$, then $d|\gcd(a,b)$.
2. If $a|m$ and $b|m$, then $\mathrm{lcm}(a,b)|m$.
3. If a is a positive integer, then $\gcd(0,a) = a$.
4. If a and b are positive integers with $a < b$, then $\gcd(a,b) = \gcd(b \bmod a, a)$.
5. If a and b are integers with $\gcd(a,b) = d$, then $\gcd(\frac{a}{d}, \frac{b}{d}) = 1$.
6. If a, b, and c are integers, then $\gcd(a + cb, b) = \gcd(a,b)$.
7. If a, b, and c are integers with not both a and b zero and $c \neq 0$, then $\gcd(ac, bc) = |c|\gcd(a,b)$.
8. If a and b are integers with $\gcd(a,b) = 1$, then $\gcd(a+b, a-b) = 1$ or 2. (This greatest common divisor is 2 when both a and b are odd.)
9. If a, b, and c are integers with $\gcd(a,b) = \gcd(a,c) = 1$, then $\gcd(a, bc) = 1$.
10. If a, b, and c are mutually relatively prime nonzero integers, then $\gcd(a, bc) = \gcd(a,b) \cdot \gcd(a,c)$.
11. If a and b are integers, not both zero, then $\gcd(a,b)$ is the least positive integer of the form $ma + nb$ where m and n are integers.
12. The probability that two randomly selected integers are relatively prime is $\frac{6}{\pi^2}$. More precisely, if $R(n)$ equals the number of pairs of integers a,b with $1 \leq a \leq n$, $1 \leq b \leq n$, and $\gcd(a,b) = 1$, then $\frac{R(n)}{n^2} = \frac{6}{\pi^2} + O(\frac{\log n}{n})$.
13. If a and b are positive integers, then $\gcd(2^a - 1, 2^b - 1) = 2^{(a,b)} - 1$.
14. If a, b, and c are integers and $a|bc$ and $\gcd(a,b) = 1$, then $a|c$.
15. If a, b, and c are integers, $a|c$, $b|c$ and $\gcd(a,b) = 1$, then $ab|c$.
16. If $a_1, a_2, \ldots, a_k$ are integers, not all zero, then $\gcd(a_1, \ldots, a_k)$ is the least positive integer that is a linear combination with integer coefficients of $a_1, \ldots, a_k$.

17. If $a_1, a_2, \ldots, a_k$ are integers, not all zero, and $d|a_i$ for $i = 1, 2, \ldots, k$, then $d|\gcd(a_1, a_2, \ldots, a_k)$.

18. If $a_1, \ldots, a_n$ are integers, not all zero, then the greatest common divisor of these n integers is the same as the greatest common divisor of the set of $n - 1$ integers made up of the first $n - 2$ integers and the greatest common divisor of the last two. That is, $\gcd(a_1, \ldots, a_n) = \gcd(a_1, \ldots, a_{n-2}, \gcd(a_{n-1}, a_n))$.

19. If a and b are nonzero integers and m is a positive integer, then $\operatorname{lcm}(ma, mb) = m \cdot \operatorname{lcm}(a, b)$.

20. If b is a common multiple of the integers $a_1, a_2, \ldots, a_k$, then b is a multiple of $\operatorname{lcm}(a_1, \ldots, a_k)$.

21. The common multiples of the integers $a_1, \ldots, a_k$ are the integers $0, \operatorname{lcm}(a_1, \ldots, a_k)$, $2 \cdot \operatorname{lcm}(a_1, \ldots, a_k), \ldots$.

22. If $a_1, a_2, \ldots, a_n$ are pairwise relatively prime integers, then $\operatorname{lcm}(a_1, \ldots, a_n) = a_1 a_2 \ldots a_n$.

23. If $a_1, a_2, \ldots, a_n$ are integers, not all zero, then $\operatorname{lcm}(a_1, a_2, \ldots, a_{n-1}, a_n) = \operatorname{lcm}(\operatorname{lcm}(a_1, a_2, \ldots, a_{n-1}), a_n)$.

24. If $a = p_1^{a_1} p_2^{a_2} \cdots p_n^{a_n}$ and $b = p_1^{b_1} p_2^{a_2} \cdots p_n^{b_n}$, where the p_i are distinct primes for $i = 1, \ldots, n$, and each exponent is a nonnegative integer, then

$$\gcd(a, b) = p_1^{\min(a_1, b_1)} p_2^{\min(a_2, b_2)} \ldots p_n^{\min(a_n, b_n)},$$

where $\min(x, y)$ denotes the minimum of x and y, and

$$\operatorname{lcm}(a, b) = p_1^{\max(a_1, b_1)} p_2^{\max(a_2, b_2)} \ldots p_n^{\max(a_n, b_n)},$$

where $\max(x, y)$ denotes the maximum of x and y.

25. If a and b are positive integers, then $ab = \gcd(a, b) \cdot \operatorname{lcm}(a, b)$.

26. If a, b, and c are positive integers, then $\operatorname{lcm}(a, b, c) = \dfrac{abc \cdot \gcd(a, b, c)}{\gcd(a, b) \cdot \gcd(a, c) \cdot \gcd(b, c)}$.

27. If a, b, and c are positive integers, then $\gcd(\operatorname{lcm}(a, b), \operatorname{lcm}(a, c)) = \operatorname{lcm}(a, \gcd(b, c))$ and $\operatorname{lcm}(\gcd(a, b), \gcd(a, c)) = \gcd(a, \operatorname{lcm}(b, c))$.

28. If $\frac{a}{b}$, $\frac{c}{d}$, and $\frac{e}{f}$ are successive terms of a Farey series, then $\frac{c}{d} = \frac{a+e}{b+f}$.

29. If $\frac{a}{b}$ and $\frac{c}{d}$ are successive terms of a Farey series, then $ad - bc = -1$.

30. If $\frac{a}{b}$ and $\frac{c}{d}$ are successive terms of a Farey series of order n, then $b + d > n$.

31. Farey series are named after an English geologist who published a note describing their properties in the *Philosophical Magazine* in 1816. The eminent French mathematician Cauchy supplied proofs of the properties stated, but not proved, by Farey. Also, according to [Di71], these properties had been stated and proved by Haros in 1802.

Examples:

1. $\gcd(12, 15) = 3$, $\gcd(14, 25) = 1$, $\gcd(0, 100) = 100$, and $\gcd(3, 39) = 3$.
2. $\gcd(2^7 3^3 5^4 7^2 11^3 17^3, 2^4 3^5 5^2 7^2 11^2 13^3) = 2^4 3^3 5^2 7^2 11^2$.
3. $\operatorname{lcm}(2^7 3^3 5^4 7^2 11^3 17^3, 2^4 3^5 5^2 7^2 11^2 13^3) = 2^7 3^5 5^4 7^2 11^3 13^3 17^3$.
4. $\gcd(18, 24, 36) = 6$ and $\gcd(10, 25, 35, 245) = 5$.
5. The integers 15, 21, and 35 are mutually relatively prime since $\gcd(15, 21, 35) = 1$. However, they are not pairwise relatively prime since $\gcd(15, 35) = 5$.
6. The integers 6, 35, and 143 are both mutually relatively prime and pairwise relatively prime.
7. The Farey series of order 5 is $\frac{0}{1}, \frac{1}{5}, \frac{1}{4}, \frac{1}{3}, \frac{2}{5}, \frac{1}{2}, \frac{3}{5}, \frac{2}{3}, \frac{3}{4}, \frac{4}{5}, \frac{1}{1}$

4.2.2 THE EUCLIDEAN ALGORITHM

Finding the greatest common divisor of two integers is one of the most common problems in number theory and its applications. An algorithm for this task was known in ancient times by Euclid. His algorithm and its extensions are among the most commonly used algorithms. For more information about these algorithms see [BaSh96] or [Kn81].

Definition:

The *Euclidean algorithm* is an algorithm that computes the greatest common divisor of two integers a and b with $a \leq b$, by replacing them with a and $b \bmod a$, and repeating this step until one of the integers reached is zero.

Facts:

1. *The Euclidean algorithm*: The greatest common divisor of two positive integers can be computed using the recurrence in §4.2.1 Fact 4, together with §4.1.2 Fact 3. The resulting algorithm proceeds by successively replacing a pair of positive integers with a new pair of integers formed from the smaller of the two integers and the remainder when the larger is divided by the smaller, stopping once a zero remainder is reached. The last nonzero remainder is the greatest common divisor of the original two integers. (See Algorithm 1.)

Algorithm 1: The Euclidean algorithm.

procedure $gcd(a, b$: positive integers)
$r_0 := a$
$r_1 := b$
$i := 1$
while $r_i \neq 0$
begin
$\quad r_{i+1} := r_{i-1} \bmod r_i$
$\quad i := i + 1$
end   {$gcd(a, b)$ is r_{i-1}}

---

2. *Lamé's theorem*: The number of divisions needed to find the greatest common divisor of two positive integers using the Euclidean algorithm does not exceed five times the number of decimal digits in the smaller of the two integers. (This was proved by Gabriel Lamé (1795–1870)). (See [BaSh96] or [Ro99] for a proof.)

3. The Euclidean algorithm finds the greatest common divisor of the Fibonacci numbers (§3.1.2) F_{n+1} and F_{n+2} (where n is a positive integer) using exactly n division steps. If the Euclidean algorithm uses exactly n division steps to find the greatest common divisor of the positive integers a and b (with $a < b$), then $a \geq F_{n+1}$ and $b \geq F_{n+2}$.

4. The Euclidean algorithm uses $O((\log b)^3)$ bit operations to find the greatest common divisor of two integers a and b with $a < b$.

5. The Euclidean algorithm uses $O(\text{Lg } a \cdot \text{Lg } b)$ bit operations to find the greatest common divisor of two integers a and b.

6. *Least remainder Euclidean algorithm*: The greatest common divisor of two integers a and b (with $a < b$) can be found by replacing a and b with a and the least remainder of b when divided by a. (The *least remainder* of b when divided by a is

Algorithm 2: The extended Euclidean algorithm.

procedure $gcdex(a, b$: positive integers)
$r_0 := a$
$r_1 := b$
$m_0 := 1$
$m_1 = 0$
$n_0 := 0$
$n_1 := 1$
$i := 1$
while $r_i \neq 0$
begin
$\quad r_{i+1} := r_{i-1} \bmod r_i$
$\quad m_{i+1} := m_{i-1} - \lfloor \frac{r_{i-1}}{r_i} \rfloor m_i$
$\quad n_{i+1} := n_{i-1} - \lfloor \frac{r_{i-1}}{r_i} \rfloor n_i$
$\quad i := i + 1$
end {gcd(a,b) is r_{i-1} and gcd$(a,b) = m_{i-1}a + n_{i-1}b$}

the integer of smallest absolute value congruent to b modulo a. It equals $b \bmod a$ if $b \bmod a \leq \frac{a}{2}$, and $(b \bmod a) - a$ if $b \bmod a > \frac{a}{2}$)). Repeating this procedure until a remainder of zero is reached produces the great common divisor of a and b as the last nonzero remainder.

7. The number of divisions used by the least remainder Euclidean algorithm to find the greatest common divisor of two integers is less than or equal the number of divisions used by the Euclidean algorithm to find this greatest common divisor.

8. *Binary greatest common divisor algorithm:* The greatest common divisor of two integers a and b can also be found using an algorithm known as the *binary greatest common divisor algorithm*. It is based on the following reductions: if a and b are both even, then gcd$(a,b) = 2\gcd(\frac{a}{2}, \frac{b}{2})$; if a is even and b is odd, then gcd$(a,b) = \gcd(\frac{a}{2}, b)$ (and if a is odd and b is even, switch them); and if a and b are both odd, then gcd$(a,b) = \gcd(\frac{|a-b|}{2}, b)$. To stop, the algorithm uses the rule that gcd$(a,a) = a$.

9. *Extended Euclidean algorithm:* The extended euclidean algorithm finds gcd(a,b) and expresses it in the form gcd$(a,b) = ma + nb$ for some integers m and n. The two-pass version proceeds by first working through the steps of the Euclidean algorithm to find gcd(a,b), and then working backwards through the steps to express gcd(a,b) as a linear combination of each pair of successive remainders until the original integers a and b are reached. The one-pass version of this algorithm keeps track of how each successive remainder can be expressed as a linear combination of successive remainders. When the last step is reached both gcd(a,b) and integers m and n with gcd$(a,b) = ma + nb$ are produced. The one-pass version is displayed as Algorithm 2.

Examples:

1. When the Euclidean algorithm is used to find gcd$(53, 77)$, the following steps result:
$\quad 77 = 1 \cdot 53 + 24,$
$\quad 53 = 2 \cdot 24 + 5,$
$\quad 24 = 4 \cdot 5 + 4,$
$\quad 5 = 1 \cdot 4 + 1,$
$\quad 4 = 4 \cdot 1.$

This shows that $\gcd(53, 77) = 1$. Working backwards through these steps to perform the two-pass version of the Euclidean algorithm gives

$$\begin{aligned} 1 &= 5 - 1 \cdot 4 \\ &= 5 - 1 \cdot (24 - 4 \cdot 5) = 5 \cdot 5 - 1 \cdot 24 \\ &= 5 \cdot (53 - 2 \cdot 24) - 1 \cdot 24 = 5 \cdot 53 - 11 \cdot 24 \\ &= 5 \cdot 53 - 11 \cdot (77 - 1 \cdot 53) = 16 \cdot 53 - 11 \cdot 77. \end{aligned}$$

2. The steps of the least-remainder algorithm when used to compute $\gcd(57, 93)$ are
$$\gcd(57, 93) = \gcd(57, 21) = \gcd(21, 6) = \gcd(6, 3) = 3.$$

3. The steps of the binary GCD algorithm when used to compute $\gcd(108, 194)$ are
$$\gcd(108, 194) = 2 \cdot \gcd(54, 97) = 2 \cdot \gcd(27, 97) = 2 \cdot \gcd(27, 35)$$
$$= 2 \cdot \gcd(4, 35) = 2 \cdot \gcd(2, 35) = 2 \cdot \gcd(1, 35) = 2.$$

4.3 CONGRUENCES

4.3.1 INTRODUCTION

Definitions:

If m is a positive integer and a and b are integers, then **a is congruent to b modulo m**, written $a \equiv b \pmod{m}$, if m divides $a - b$. If m does not divide $a - b$, a and b are **incongruent** modulo m, written $a \not\equiv b \pmod{m}$.

A **complete system of residues modulo m** is a set of integers such that every integer is congruent modulo m to exactly one of the integers in the set.

If m is a positive integer and a is an integer with $a = bm + r$, where $0 \leq r \leq m - 1$, then r is the **least nonnegative residue of a modulo m**. When a is not divisible by m, r is the **least positive residue of a modulo m**.

The **congruence class of a modulo m** is the set of integers congruent to a modulo m and is written $[a]_m$. Any integer in $[a]_m$ is called a **representative** of this class.

If m is a positive integer and a is an integer relatively prime to m, then $\bar{a}$ is an **inverse of a modulo m** if $a\bar{a} \equiv 1 \pmod{m}$. An inverse of a modulo m is also written $a^{-1} \bmod m$.

If m is a positive integer, then a **reduced residue system modulo m** is a set of integers such that every integer relatively prime to m is congruent modulo m to exactly one integer in the set.

If m is a positive integer, the set of congruence classes modulo m is written $\mathcal{Z}_m$. (See §5.2.1.)

If m is a positive integer greater than 1, the set of congruence classes of elements relatively prime to m is written $\mathcal{Z}_m^\star$; that is, $\mathcal{Z}_m^\star = \{\, [a]_m \in \mathcal{Z}_m \mid \gcd(a, n) = 1 \,\}$. (See §5.2.1.)

Facts:

1. If m is a positive integer and a, b, and c are integers, then:
   - $a \equiv a \pmod{m}$;
   - $a \equiv b \pmod{m}$ if and only if $b \equiv a \pmod{m}$;
   - if $a \equiv b \pmod{m}$ and $b \equiv c \pmod{m}$, then $a \equiv c \pmod{m}$.

   Consequently, congruence modulo m is an equivalence relation. (See §1.4.2 and §5.2.1.)

2. If m is a positive integer and a is an integer, then m divides a if and only if $a \equiv 0 \pmod{m}$.

3. If m is a positive integer and a and b are integers with $a \equiv b \pmod{m}$, then $\gcd(a, m) = \gcd(b, m)$.

4. If a, b, c, and m are integers with m positive and $a \equiv b \pmod{m}$, then $a + c \equiv b + c \pmod{m}$, $a - c \equiv b - c \pmod{m}$, and $ac \equiv bc \pmod{m}$.

5. If m is a positive integer and a, b, c, and d are integers with $a \equiv b \pmod{m}$ and $c \equiv d \pmod{m}$, then $ac \equiv bd \pmod{m}$.

6. If a, b, c, and m are integers, m is positive, $d = \gcd(c, m)$, and $ac \equiv bc \pmod{m}$, then $a \equiv b \pmod{\frac{m}{d}}$.

7. If a, b, c, and m are integers, m is positive, and c and m are relatively prime, and $ac \equiv bc \pmod{m}$, then $a \equiv b \pmod{m}$.

8. If a, b, k and m are integers with k and m positive and $a \equiv b \pmod{m}$, then $a^k \equiv b^k \pmod{m}$.

9. If a, b, and m are integers with $a \equiv b \pmod{m}$, then if c is an integer, it does not necessarily follow that $c^a \equiv c^b \pmod{m}$.

10. If $f(x_1, \ldots, x_n)$ is a polynomial with integer coefficients and $a_1 \ldots a_n$, $b_1, \ldots, b_n$ are integers with $a_i \equiv b_i \pmod{m}$ for all i, then $f(a_1, \ldots, a_n) \equiv f(b_1, \ldots, b_n) \pmod{m}$.

11. If a, b, and m_i are integers with m_i positive and $a \equiv b \pmod{m_i}$ for $i = 1, 2, \ldots, k$, then $a \equiv b \pmod{\text{lcm}(m_1, m_2, \ldots, m_k)}$.

12. If a and b are integers, m_i ($i = 1, 2, \ldots, k$) are pairwise relatively prime positive integers, and $a \equiv b \pmod{m_i}$ for $i = 1, 2, \ldots, k$, then $a \equiv b \pmod{m_1 m_2 \ldots m_k}$.

13. The congruence class $[a]_m$ is the set of integers $\{a, a \pm m, a \pm 2m, \ldots\}$. If $a \equiv b \pmod{m}$, then $[a]_m = [b]_m$. The congruence classes modulo m are the equivalence classes of the congruence modulo m equivalence relation. (See §5.2.1.)

14. Addition, subtraction, and multiplication of congruence classes modulo m, where m is a positive integer, are defined by $[a]_m + [b]_m = [a+b]_m$, $[a]_m - [b]_m = [a-b]_m$, and $[a]_m[b]_m = [ab]_m$. Each of these operations is well defined, in the sense that using representatives of the congruence classes other than a and b does not change the resulting congruence class.

15. If m is a positive integer, then $(\mathcal{Z}_n, +)$, where $+$ is the operation of addition of congruence classes defined in Fact 14 and in §5.2.1, is an abelian group. The identity element in this group is $[0]_m$ and the inverse of $[a]_m$ is $[-a]_m = [m-a]_m$.

16. If m is a positive integer greater than 1 and a is relatively prime to m, then a has an inverse modulo m.

17. An inverse of a modulo m, where m is a positive integer and $\gcd(a, m) = 1$, may be found by using the extended Euclidean algorithm to find integers x and y such that $ax + my = 1$, which implies that x is an inverse of a modulo m.

18. If m is a positive integer, then $(\mathcal{Z}_m^*, \cdot)$, where $\cdot$ is the multiplication operation on congruence classes, is an abelian group. (See §5.2.1.) The identity element of this group is $[1]_m$ and the inverse of the class $[a]_m$ is the class $[\bar{a}]_m$, where $\bar{a}$ is an inverse of a modulo m.

19. If a_i $(i = 1, \ldots, m)$ is a complete residue system modulo m, where m is a positive integer, and r and s are integers with $\gcd(m, r) = 1$, then $ra_i + s$ is a complete system of residues modulo m.

20. If a and b are integers and m is a positive integer with $0 \leq a < m$ and $0 \leq b < m$, then $(a+b) \bmod m = a+b$ if $a+b < m$, and $(a+b) \bmod m = a+b-m$ if $a+b \geq m$.

21. Computing the least positive residue modulo m of powers of integers is important in cryptology (see Chapter 14). An efficient algorithm for computing $b^n \bmod m$ where n is a positive integer with binary expansion $n = (a_{k-1} \ldots a_1 a_0)_2$ is to find the least positive residues of $b, b^2, b^4, \ldots, b^{2^{k-1}}$ modulo m by successively squaring and reducing modulo m, multiplying together the least positive residues modulo m of b^{2^j} for those j with $a_j = 1$, reducing modulo m after each multiplication.

22. *Wilson's theorem*: If p is prime, then $(p-1)! \equiv -1 \pmod{p}$.

23. If n is a positive integer greater than 1 such that $(n-1)! \equiv -1 \pmod{n}$ then n is prime.

24. *Fermat's little theorem*: If p is a prime and a is an integer not divisible by p then $a^{p-1} \equiv 1 \pmod{p}$.

25. *Euler's theorem*: If m is a positive integer and a is an integer relatively prime to m, then $a^{\phi(m)} \equiv 1 \pmod{m}$, where $\phi(m)$ is the number of positive integers not exceeding m that are relatively prime to m.

26. If a is an integer and p is a prime that does not divide a, then from Fermat's little theorem it follows that a^{p-2} is an inverse of a modulo p.

27. If a and m are relatively prime integers with $m > 1$, then $a^{\phi(m)-1}$ is an inverse of a modulo m. This follows directly from Euler's theorem.

28. *Linear congruential method*: One of the most common method used for generating pseudo-random numbers is the *linear congruential method*. It starts with integers m, a, c, and x_0 where $2 \leq a < m$, $0 \leq c < m$, and $0 \leq x_0 \leq m$. The sequence of pseudo-random numbers is defined recursively by

$$x_{n+1} = (ax_n + c) \bmod m, \quad n = 0, 1, 2, 3, \ldots.$$

Here m is the *modulus*, a is the *multiplier*, c is the *increment*, and x_0 is the *seed* of the generator.

29. Big-oh estimates for the number of bit operations required to do modular addition, modular subtraction, modular multiplication, modular inversion, and modular exponentiation is summarized in the following table.

name	operation	number of bit operations
modular addition	$(a+b) \bmod m$	$O(\log m)$
modular subtraction	$(a-b) \bmod m$	$O(\log m)$
modular multiplication	$(a \cdot b) \bmod m$	$O((\log m)^2)$
modular inversion	$(a^{-1}) \bmod m$	$O((\log m)^2)$
modular exponentiation	$a^k \bmod m, k < m$	$O((\log m)^3)$

Chapter 4 NUMBER THEORY

Examples:

1. $23 \equiv 5 \pmod{9}$, $-17 \equiv 13 \pmod{15}$, and $99 \equiv 0 \pmod{11}$, but $11 \not\equiv 3 \pmod 5$, $-3 \not\equiv 8 \pmod 6$, and $44 \not\equiv 0 \pmod 7$.

2. To find an inverse of 53 modulo 71, use the extended Euclidean algorithm to obtain $16 \cdot 53 - 11 \cdot 71 = 1$ (see Example 1 of 4.2.2). This implies that 16 is an inverse of 53 modulo 71.

3. Since 11 is prime, by Wilson's theorem it follows that $10! \equiv -1 \pmod{11}$.

4. $5! \equiv 0 \pmod 6$, which provides an impractical verification that 6 is not prime.

5. To find the least positive residue of 3^{201} modulo 11, note that by Fermat's little theorem $3^{10} \equiv 1 \pmod{11}$. Hence $3^{201} = (3^{10})^{20} \cdot 3 \equiv 3 \pmod{11}$.

6. *Zeller's congruence*: A congruence can be used to determine the day of the week of any date in the Gregorian calendar, the calendar used in most of the world. Let w represent the day of the week, with $w = 0, 1, 2, 3, 4, 5, 6$ for Sunday, Monday, Tuesday, Wednesday, Thursday, Friday, Saturday, respectively. Let k represent the day of the month. Let m represent the month with $m = 11, 12, 1, 2, 3, 4, 5, 6, 7, 8, 9, 10$ for January, February, March, April, May, June, July, August, September, October, November, December, respectively. Let N represent the previous year if the month is January or February or the current year otherwise, with C the century of N and Y the particular year of the century of N so that $N = 100Y + C$. Then the day of the week can be found using the congruence

$$w \equiv k + \lfloor 2.6m - 0.2 \rfloor - 2C + Y + \lfloor \tfrac{Y}{4} \rfloor + \lfloor \tfrac{C}{4} \rfloor \pmod 7.$$

7. January 1, 1900 was a Monday. This follows by Zeller's congruence with $C = 18$, $Y = 99$, $m = 11$, and $k = 1$, noting that to apply this congruence January is considered the eleventh month of the preceding year.

4.3.2 LINEAR AND POLYNOMIAL CONGRUENCES

Definitions:

A **linear congruence in one variable** is a congruence of the form $ax \equiv b \pmod m$, where a, b, and m are integers, m is positive, and x is an unknown.

If f is a polynomial with integer coefficients, an integer r is a **solution** of the congruence $f(x) \equiv 0 \pmod m$, or a **root** of $f(x)$ modulo m, if $f(r) \equiv 0 \pmod m$.

Facts:

1. If a, b, and m are integers, m is positive, and $\gcd(a, m) = d$, then the congruence $ax \equiv b \pmod m$ has exactly d incongruent solutions modulo m if $d | b$, and no solutions if $d \nmid b$.

2. If a, b, and m are integers, m is positive, and $\gcd(a, m) = 1$, then the solutions of $ax \equiv b \pmod m$ are all integers x with $x \equiv \bar{a}b \pmod m$.

3. If a and b are positive integers and p is a prime that does not divide a, then the solutions of $ax \equiv b \pmod p$ are the integers x with $x \equiv a^{p-2}b \pmod p$.

4. *Thue's lemma*: If p is a prime and a is an integer not divisible by p, then the congruence $ax \equiv y \pmod p$ has a solution x_0, y_0 with $0 < |x_0| < \sqrt{p}$, $0 < |y_0| < \sqrt{p}$.

5. *Chinese remainder theorem*: If m_i, $i = 1, 2, \ldots, r$, are pairwise relatively prime positive integers, then the system of simultaneous congruences $x \equiv a_i \pmod{m_i}$, $i = 1, 2, \ldots, r$, has a unique solution modulo $M = m_1 m_2 \ldots m_r$ which is given by $x \equiv a_1 M_1 y_1 + a_2 M_2 y_2 + \cdots + a_r M_r y_r$ where $M_k = \frac{M}{m_k}$ and y_k is an inverse of M_k modulo m_k, $k = 1, 2, \ldots, r$.

6. Problems involving the solution of a system of simultaneous congruences arose in the writing of ancient mathematicians, including the Chinese mathematician Sun-Tsu, and in other works by Indian and Greek mathematicians. (See [Di71] for details.)

7. The system of simultaneous congruences $x \equiv a_i \pmod{m_i}$, $i = 1, 2, \ldots, r$ has a solution if and only if $\gcd(m_i, m_j)$ divides $a_i - a_j$ for all pairs of integers (i, j) with $1 \leq i < j \leq r$. If a solution exists, it is unique modulo $\text{lcm}(m_1, m_2, \ldots, m_r)$.

8. If a, b, c, d, e, f, and m are integers with m positive such that $\gcd(ad - bc, m) = 1$, then the system of congruences $ax + by \equiv e \pmod{m}$, $cx + dy \equiv f \pmod{m}$ has a unique solution given by $x \equiv g(de - bf) \pmod{m}$, $y \equiv g(af - ce) \pmod{m}$ where g is an inverse of $ad - bc$ modulo m.

9. *Lagrange's theorem*: If p is prime, then the polynomial $f(x) = a_n x^n + \cdots + a_1 x + a_0$ where $a_n \not\equiv 0 \pmod{p}$ has at most n roots modulo p.

10. If $f(x) = a_n x^n + \cdots + a_1 x + a_0$, where a_i ($i = 1, \ldots, n$) is an integer and p is prime, has more than n roots modulo p, then p divides a_i for all $i = 1, \ldots, n$.

11. If $m_1, m_2, \ldots, m_r$ are pairwise relatively prime positive integers with product $m = m_1 m_2 \ldots m_r$, and f is a polynomial with integer coefficients, then $f(x)$ has a root modulo m if and only if $f(x)$ has a root modulo m_i, for all $i = 1, 2, \ldots, r$. Furthermore, if $f(x)$ has n_i incongruent roots modulo m_i and n incongruent roots modulo m, then $n = n_1 n_2 \ldots n_r$.

12. If p is prime, k is a positive integer, and s is a root of $f(x)$ modulo p^k, then:
    - if $p \nmid f'(s)$, then there is a unique root t of $f(x)$ modulo p^{k+1} with $t \equiv s \pmod{p^k}$, namely $t = s + p^k u$ where u is the unique solution of $f'(s) u \equiv -f(s)/p^k \pmod{p}$;
    - if $p | f'(s)$ and $p^{k+1} | f(s)$, then there are exactly p incongruent roots of $f(x)$ modulo p^{k+1} congruent to s modulo p, given by $s + p^k i$, $i = 0, 1, \ldots, p - 1$;
    - if $p | f'(s)$ and $p^{k+1} \nmid f(s)$, then there are no roots of $f(x)$ modulo p^{k+1} that are congruent to s modulo p^k.

13. *Finding roots of a polynomial modulo m, where m is a positive integer*: First find roots of the polynomial modulo p^r for each prime power in the prime-power factorization of m (Fact 14) and then use the Chinese remainder theorem (Fact 5) to find solutions modulo m.

14. Finding solutions modulo p^r reduces to first finding solutions modulo p. In particular, if there are no roots of $f(x)$ modulo p, there are no roots of $f(x)$ modulo p^r. If $f(x)$ has roots modulo p, choose one, say r with $0 \leq r < p$. By Fact 12, corresponding to r there are 0, 1, or p roots of $f(x)$ modulo p^2.

Examples:

1. There are 3 incongruent solutions of $6x \equiv 9 \pmod{15}$ since $\gcd(6, 15) = 3$ and $3 | 9$. The solutions are those integers x with $x \equiv 4, 9$, or $14 \pmod{15}$.

2. The linear congruence $2x \equiv 7 \pmod{6}$ has no solutions since $\gcd(2, 6) = 2$ and $2 \nmid 7$.

3. The solutions of the linear congruence $3x \equiv 5 \pmod{11}$ are those integers x with $x \equiv \overline{3} \cdot 5 \equiv 4 \cdot 5 \equiv 9 \pmod{11}$.

4. It follows from the Chinese remainder theorem (Fact 5) that the solutions of the systems of simultaneous congruences $x \equiv 1 \pmod 3$, $x \equiv 2 \pmod 4$, and $x \equiv 3 \pmod 5$ are all integers x with $x \equiv 1 \cdot 20 \cdot 2 + 2 \cdot 15 \cdot 3 + 3 \cdot 12 \cdot 3 \equiv 58 \pmod{60}$.

5. The simultaneous congruences $x \equiv 4 \pmod 9$ and $x \equiv 7 \pmod{15}$ can be solved by noting that the first congruence implies that $x - 4 = 9t$ for some integer t, so that $x = 9t + 4$. Inserting this expression for x into the second congruence gives $9t + 4 \equiv 7 \pmod{15}$. This implies that $3t \equiv 1 \pmod 5$, so that $t \equiv 2 \pmod 5$ and $t = 5u + 2$ for some integer u. Hence $x = 45u + 22$ for some integer u. The solutions of the two simultaneous congruences are those integers x with $x \equiv 22 \pmod{45}$.

4.4 PRIME NUMBERS

One of the most powerful tools in number theory is the fact that each composite integer can be decomposed into a product of primes. Primes may be thought of as the building blocks of the integers in the sense that they can be decomposed only in trivial ways, for example, $3 = 1 \times 3$. Prime numbers, once of only theoretical interest, now are important in many applications, especially in the area of cryptography were large primes play a crucial role in the area of public-key cryptosystems (see Chapter 14). From ancient to modern times, mathematicians have devoted long hours to the study of primes and their properties. Even so, many questions about primes have only partially been answered or remain complete mysteries, including questions that ask whether there are infinitely many primes of certain forms. There have been many recent discoveries concerning prime numbers, such as the discovery of new Mersenne primes. The current state of knowledge on some of these questions and the latest discoveries are described in this section.

Additional information about primes can be found in [CrPo99] and [Ri96] and on the Web. See the *Prime Pages* at the website
    http://www.utm.edu/research/primes/index.html#lists

4.4.1 BASIC CONCEPTS

Definitions:

A *prime* is a natural number greater than 1 that is exactly divisible only by 1 and itself.

A *composite* is a natural number greater than 1 that is not a prime. That is, a composite may be factored into the product of two natural numbers both smaller than itself.

Facts:

1. The number 1 is not considered to be prime.
2. Table 1 lists the primes up to 10,000.
3. *Fundamental theorem of arithmetic*: Every natural number greater than 1 is either prime or can be written as a product of prime factors in a unique way, up to the order of the prime factors. That is, every composite n can be expressed uniquely as $n = p_1 p_2 \ldots p_k$, where $p_1 \leq p_2 \leq \cdots \leq p_k$ are primes. This is sometimes also known as the *unique factorization theorem*.

4. The unique factorization of a positive integer n formed by grouping together equal prime factors produces the unique *prime-power factorization* $n = p_1^{a_1} p_2^{a_2} \ldots p_k^{a_k}$.

5. Table 2 lists the prime-power factorization of all positive integers below 2,500. Numbers appearing in boldface are prime.

Examples:
1. $6 = 2 \times 3$.
2. $245 = 5 \times 7^2$.
3. $10! = 2^8 \times 3^4 \times 5^2 \times 7$.
4. $68{,}718{,}821{,}377 = (2^{17} - 1) \cdot (2^{19} - 1)$ (both factors are Mersenne primes; see §4.4.3).
5. The largest prime known is $2^{3,021,377} - 1$. It has 909,526 decimal digits and was discovered in 1998. It is a Mersenne prime (see Table 3).

4.4.2 COUNTING PRIMES

Definitions:

The value of the **prime counting function** $\pi(x)$ at x where x is a positive real number equals the number of primes less than or equal to x.

The **li function** is defined by $\operatorname{li}(x) = \int_0^x \frac{dt}{\log t}$, for $x \geq 2$. (The principal value is taken for the integral at the singularity $t = 1$.)

Twin primes are primes that differ by exactly 2.

Facts:

1. Euclid (ca. 300 B.C.E.) proved that there are infinitely many primes. He observed that the product of a finite list of primes, plus one, must be divisible by a prime not on that list.

2. Leonhard Euler (1707–1783) showed that the sum of the reciprocals of the primes up to n tends toward infinity as n tends toward infinity, which also implies that there are infinitely many primes. (There are many other proofs as well.)

3. There is no useful, exact formula known which will produce the nth prime, given n. It is relatively easy to construct a useless (that is, impractical) one. For example, let $\alpha = \sum_{n=1}^{\infty} p_n / 2^{2^n}$, where p_n is the nth prime. Then the nth prime is $\lfloor 2^{2^n} \alpha \rfloor - 2^{2^{n-1}} \lfloor 2^{2^{n-1}} \alpha \rfloor$, where $\lfloor x \rfloor$ is the greatest integer less than or equal to x.

4. If $f(x)$ is a polynomial with integer coefficients that is not constant, then there are infinitely many integers n for which $|f(n)|$ is not prime.

5. There are polynomials with integer coefficients with the property that the set of positive values taken by each of these polynomials as the variables range over the set of nonnegative integers is the set of prime numbers. The existence of such polynomials has essentially no practical value for constructing primes. For example, there are polynomials in 26 variables of degree 25, in 42 variables of degree 5, and in 12 variables of degree 13697, with this property. (See [Ri96].)

6. $\frac{p_n}{n \log n} \to 1$ as $n \to \infty$. (This follows from the prime number theorem, Fact 10.)

7. An inexact and rough formula for the nth prime is $n \log n$.

8. $p_n > n \log n$ for all n. (J. B. Rosser)

Chapter 4 NUMBER THEORY

Table 1 Table of primes less than 10,000.

The prime number p_{10n+k} is found by looking at the row beginning with $n..$ and at the column beginning with $..k$.

	..0	..1	..2	..3	..4	..5	..6	..7	..8	..9
	2	3	5	7	11	13	17	19	23	
1..	29	31	37	41	43	47	53	59	61	67
2..	71	73	79	83	89	97	101	103	107	109
3..	113	127	131	137	139	149	151	157	163	167
4..	173	179	181	191	193	197	199	211	223	227
5..	229	233	239	241	251	257	263	269	271	277
6..	281	283	293	307	311	313	317	331	337	347
7..	349	353	359	367	373	379	383	389	397	401
8..	409	419	421	431	433	439	443	449	457	461
9..	463	467	479	487	491	499	503	509	521	523
10..	541	547	557	563	569	571	577	587	593	599
11..	601	607	613	617	619	631	641	643	647	653
12..	659	661	673	677	683	691	701	709	719	727
13..	733	739	743	751	757	761	769	773	787	797
14..	809	811	821	823	827	829	839	853	857	859
15..	863	877	881	883	887	907	911	919	929	937
16..	941	947	953	967	971	977	983	991	997	1009
17..	1013	1019	1021	1031	1033	1039	1049	1051	1061	1063
18..	1069	1087	1091	1093	1097	1103	1109	1117	1123	1129
19..	1151	1153	1163	1171	1181	1187	1193	1201	1213	1217
20..	1223	1229	1231	1237	1249	1259	1277	1279	1283	1289
21..	1291	1297	1301	1303	1307	1319	1321	1327	1361	1367
22..	1373	1381	1399	1409	1423	1427	1429	1433	1439	1447
23..	1451	1453	1459	1471	1481	1483	1487	1489	1493	1499
24..	1511	1523	1531	1543	1549	1553	1559	1567	1571	1579
25..	1583	1597	1601	1607	1609	1613	1619	1621	1627	1637
26..	1657	1663	1667	1669	1693	1697	1699	1709	1721	1723
27..	1733	1741	1747	1753	1759	1777	1783	1787	1789	1801
28..	1811	1823	1831	1847	1861	1867	1871	1873	1877	1879
29..	1889	1901	1907	1913	1931	1933	1949	1951	1973	1979
30..	1987	1993	1997	1999	2003	2011	2017	2027	2029	2039
31..	2053	2063	2069	2081	2083	2087	2089	2099	2111	2113
32..	2129	2131	2137	2141	2143	2153	2161	2179	2203	2207
33..	2213	2221	2237	2239	2243	2251	2267	2269	2273	2281
34..	2287	2293	2297	2309	2311	2333	2339	2341	2347	2351
35..	2357	2371	2377	2381	2383	2389	2393	2399	2411	2417
36..	2423	2437	2441	2447	2459	2467	2473	2477	2503	2521
37..	2531	2539	2543	2549	2551	2557	2579	2591	2593	2609
38..	2617	2621	2633	2647	2657	2659	2663	2671	2677	2683
39..	2687	2689	2693	2699	2707	2711	2713	2719	2729	2731
40..	2741	2749	2753	2767	2777	2789	2791	2797	2801	2803

Section 4.4 PRIME NUMBERS

	..0	..1	..2	..3	..4	..5	..6	..7	..8	..9
41..	2819	2833	2837	2843	2851	2857	2861	2879	2887	2897
42..	2903	2909	2917	2927	2939	2953	2957	2963	2969	2971
43..	2999	3001	3011	3019	3023	3037	3041	3049	3061	3067
44..	3079	3083	3089	3109	3119	3121	3137	3163	3167	3169
45..	3181	3187	3191	3203	3209	3217	3221	3229	3251	3253
46..	3257	3259	3271	3299	3301	3307	3313	3319	3323	3329
47..	3331	3343	3347	3359	3361	3371	3373	3389	3391	3407
48..	3413	3433	3449	3457	3461	3463	3467	3469	3491	3499
49..	3511	3517	3527	3529	3533	3539	3541	3547	3557	3559
50..	3571	3581	3583	3593	3607	3613	3617	3623	3631	3637
51..	3643	3659	3671	3673	3677	3691	3697	3701	3709	3719
52..	3727	3733	3739	3761	3767	3769	3779	3793	3797	3803
53..	3821	3823	3833	3847	3851	3853	3863	3877	3881	3889
54..	3907	3911	3917	3919	3923	3929	3931	3943	3947	3967
55..	3989	4001	4003	4007	4013	4019	4021	4027	4049	4051
56..	4057	4073	4079	4091	4093	4099	4111	4127	4129	4133
57..	4139	4153	4157	4159	4177	4201	4211	4217	4219	4229
58..	4231	4241	4243	4253	4259	4261	4271	4273	4283	4289
59..	4297	4327	4337	4339	4349	4357	4363	4373	4391	4397
60..	4409	4421	4423	4441	4447	4451	4457	4463	4481	4483
61..	4493	4507	4513	4517	4519	4523	4547	4549	4561	4567
62..	4583	4591	4597	4603	4621	4637	4639	4643	4649	4651
63..	4657	4663	4673	4679	4691	4703	4721	4723	4729	4733
64..	4751	4759	4783	4787	4789	4793	4799	4801	4813	4817
65..	4831	4861	4871	4877	4889	4903	4909	4919	4931	4933
66..	4937	4943	4951	4957	4967	4969	4973	4987	4993	4999
67..	5003	5009	5011	5021	5023	5039	5051	5059	5077	5081
68..	5087	5099	5101	5107	5113	5119	5147	5153	5167	5171
69..	5179	5189	5197	5209	5227	5231	5233	5237	5261	5273
70..	5279	5281	5297	5303	5309	5323	5333	5347	5351	5381
71..	5387	5393	5399	5407	5413	5417	5419	5431	5437	5441
72..	5443	5449	5471	5477	5479	5483	5501	5503	5507	5519
73..	5521	5527	5531	5557	5563	5569	5573	5581	5591	5623
74..	5639	5641	5647	5651	5653	5657	5659	5669	5683	5689
75..	5693	5701	5711	5717	5737	5741	5743	5749	5779	5783
76..	5791	5801	5807	5813	5821	5827	5839	5843	5849	5851
77..	5857	5861	5867	5869	5879	5881	5897	5903	5923	5927
78..	5939	5953	5981	5987	6007	6011	6029	6037	6043	6047
79..	6053	6067	6073	6079	6089	6091	6101	6113	6121	6131
80..	6133	6143	6151	6163	6173	6197	6199	6203	6211	6217

	..0	..1	..2	..3	..4	..5	..6	..7	..8	..9
81..	6221	6229	6247	6257	6263	6269	6271	6277	6287	6299
82..	6301	6311	6317	6323	6329	6337	6343	6353	6359	6361
83..	6367	6373	6379	6389	6397	6421	6427	6449	6451	6469
84..	6473	6481	6491	6521	6529	6547	6551	6553	6563	6569
85..	6571	6577	6581	6599	6607	6619	6637	6653	6659	6661
86..	6673	6679	6689	6691	6701	6703	6709	6719	6733	6737
87..	6761	6763	6779	6781	6791	6793	6803	6823	6827	6829
88..	6833	6841	6857	6863	6869	6871	6883	6899	6907	6911
89..	6917	6947	6949	6959	6961	6967	6971	6977	6983	6991
90..	6997	7001	7013	7019	7027	7039	7043	7057	7069	7079
91..	7103	7109	7121	7127	7129	7151	7159	7177	7187	7193
92..	7207	7211	7213	7219	7229	7237	7243	7247	7253	7283
93..	7297	7307	7309	7321	7331	7333	7349	7351	7369	7393
94..	7411	7417	7433	7451	7457	7459	7477	7481	7487	7489
95..	7499	7507	7517	7523	7529	7537	7541	7547	7549	7559
96..	7561	7573	7577	7583	7589	7591	7603	7607	7621	7639
97..	7643	7649	7669	7673	7681	7687	7691	7699	7703	7717
98..	7723	7727	7741	7753	7757	7759	7789	7793	7817	7823
99..	7829	7841	7853	7867	7873	7877	7879	7883	7901	7907
100..	7919	7927	7933	7937	7949	7951	7963	7993	8009	8011
101..	8017	8039	8053	8059	8069	8081	8087	8089	8093	8101
102..	8111	8117	8123	8147	8161	8167	8171	8179	8191	8209
103..	8219	8221	8231	8233	8237	8243	8263	8269	8273	8287
104..	8291	8293	8297	8311	8317	8329	8353	8363	8369	8377
105..	8387	8389	8419	8423	8429	8431	8443	8447	8461	8467
106..	8501	8513	8521	8527	8537	8539	8543	8563	8573	8581
107..	8597	8599	8609	8623	8627	8629	8641	8647	8663	8669
108..	8677	8681	8689	8693	8699	8707	8713	8719	8731	8737
109..	8741	8747	8753	8761	8779	8783	8803	8807	8819	8821
110..	8831	8837	8839	8849	8861	8863	8867	8887	8893	8923
111..	8929	8933	8941	8951	8963	8969	8971	8999	9001	9007
112..	9011	9013	9029	9041	9043	9049	9059	9067	9091	9103
113..	9109	9127	9133	9137	9151	9157	9161	9173	9181	9187
114..	9199	9203	9209	9221	9227	9239	9241	9257	9277	9281
115..	9283	9293	9311	9319	9323	9337	9341	9343	9349	9371
116..	9377	9391	9397	9403	9413	9419	9421	9431	9433	9437
117..	9439	9461	9463	9467	9473	9479	9491	9497	9511	9521
118..	9533	9539	9547	9551	9587	9601	9613	9619	9623	9629
119..	9631	9643	9649	9661	9677	9679	9689	9697	9719	9721
120..	9733	9739	9743	9749	9767	9769	9781	9787	9791	9803
121..	9811	9817	9829	9833	9839	9851	9857	9859	9871	9883
122..	9887	9901	9907	9923	9929	9931	9941	9949	9967	9973

Table 2 Prime power decompositions below 2500.

	0	1	2	3	4	5	6	7	8	9
0			**2**	**3**	2^2	**5**	$2 \cdot 3$	**7**	2^3	3^2
1	$2 \cdot 5$	**11**	$2^2 \cdot 3$	**13**	$2 \cdot 7$	$3 \cdot 5$	2^4	**17**	$2 \cdot 3^2$	**19**
2	$2^2 \cdot 5$	$3 \cdot 7$	$2 \cdot 11$	**23**	$2^3 \cdot 3$	5^2	$2 \cdot 13$	3^3	$2^2 \cdot 7$	**29**
3	$2 \cdot 3 \cdot 5$	**31**	2^5	$3 \cdot 11$	$2 \cdot 17$	$5 \cdot 7$	$2^2 \cdot 3^2$	**37**	$2 \cdot 19$	$3 \cdot 13$
4	$2^3 \cdot 5$	**41**	$2 \cdot 3 \cdot 7$	**43**	$2^2 \cdot 11$	$3^2 \cdot 5$	$2 \cdot 23$	**47**	$2^4 \cdot 3$	7^2
5	$2 \cdot 5^2$	$3 \cdot 17$	$2^2 \cdot 13$	**53**	$2 \cdot 3^3$	$5 \cdot 11$	$2^3 \cdot 7$	$3 \cdot 19$	$2 \cdot 29$	**59**
6	$2^2 \cdot 3 \cdot 5$	**61**	$2 \cdot 31$	$3^2 \cdot 7$	2^6	$5 \cdot 13$	$2 \cdot 3 \cdot 11$	**67**	$2^2 \cdot 17$	$3 \cdot 23$
7	$2 \cdot 5 \cdot 7$	**71**	$2^3 \cdot 3^2$	**73**	$2 \cdot 37$	$3 \cdot 5^2$	$2^2 \cdot 19$	$7 \cdot 11$	$2 \cdot 3 \cdot 13$	**79**
8	$2^4 \cdot 5$	3^4	$2 \cdot 41$	**83**	$2^2 \cdot 3 \cdot 7$	$5 \cdot 17$	$2 \cdot 43$	$3 \cdot 29$	$2^3 \cdot 11$	**89**
9	$2 \cdot 3^2 \cdot 5$	$7 \cdot 13$	$2^2 \cdot 23$	$3 \cdot 31$	$2 \cdot 47$	$5 \cdot 19$	$2^5 \cdot 3$	**97**	$2 \cdot 7^2$	$3^2 \cdot 11$
10	$2^2 \cdot 5^2$	**101**	$2 \cdot 3 \cdot 17$	**103**	$2^3 \cdot 13$	$3 \cdot 5 \cdot 7$	$2 \cdot 53$	**107**	$2^2 \cdot 3^3$	**109**
11	$2 \cdot 5 \cdot 11$	$3 \cdot 37$	$2^4 \cdot 7$	**113**	$2 \cdot 3 \cdot 19$	$5 \cdot 23$	$2^2 \cdot 29$	$3^2 \cdot 13$	$2 \cdot 59$	$7 \cdot 17$
12	$2^3 \cdot 3 \cdot 5$	11^2	$2 \cdot 61$	$3 \cdot 41$	$2^2 \cdot 31$	5^3	$2 \cdot 3^2 \cdot 7$	**127**	2^7	$3 \cdot 43$
13	$2 \cdot 5 \cdot 13$	**131**	$2^2 \cdot 3 \cdot 11$	$7 \cdot 19$	$2 \cdot 67$	$3^3 \cdot 5$	$2^3 \cdot 17$	**137**	$2 \cdot 3 \cdot 23$	**139**
14	$2^2 \cdot 5 \cdot 7$	$3 \cdot 47$	$2 \cdot 71$	$11 \cdot 13$	$2^4 \cdot 3^2$	$5 \cdot 29$	$2 \cdot 73$	$3 \cdot 7^2$	$2^2 \cdot 37$	**149**
15	$2 \cdot 3 \cdot 5^2$	**151**	$2^3 \cdot 19$	$3^2 \cdot 17$	$2 \cdot 7 \cdot 11$	$5 \cdot 31$	$2^2 \cdot 3 \cdot 13$	**157**	$2 \cdot 79$	$3 \cdot 53$
16	$2^5 \cdot 5$	$7 \cdot 23$	$2 \cdot 3^4$	**163**	$2^2 \cdot 41$	$3 \cdot 5 \cdot 11$	$2 \cdot 83$	**167**	$2^3 \cdot 3 \cdot 7$	13^2
17	$2 \cdot 5 \cdot 17$	$3^2 \cdot 19$	$2^2 \cdot 43$	**173**	$2 \cdot 3 \cdot 29$	$5^2 \cdot 7$	$2^4 \cdot 11$	$3 \cdot 59$	$2 \cdot 89$	**179**
18	$2^2 \cdot 3^2 \cdot 5$	**181**	$2 \cdot 7 \cdot 13$	$3 \cdot 61$	$2^3 \cdot 23$	$5 \cdot 37$	$2 \cdot 3 \cdot 31$	$11 \cdot 17$	$2^2 \cdot 47$	$3^3 \cdot 7$
19	$2 \cdot 5 \cdot 19$	**191**	$2^6 \cdot 3$	**193**	$2 \cdot 97$	$3 \cdot 5 \cdot 13$	$2^2 \cdot 7^2$	**197**	$2 \cdot 3^2 \cdot 11$	**199**
20	$2^3 \cdot 5^2$	$3 \cdot 67$	$2 \cdot 101$	$7 \cdot 29$	$2^2 \cdot 3 \cdot 17$	$5 \cdot 41$	$2 \cdot 103$	$3^2 \cdot 23$	$2^4 \cdot 13$	$11 \cdot 19$
21	$2 \cdot 3 \cdot 5 \cdot 7$	**211**	$2^2 \cdot 53$	$3 \cdot 71$	$2 \cdot 107$	$5 \cdot 43$	$2^3 \cdot 3^3$	$7 \cdot 31$	$2 \cdot 109$	$3 \cdot 73$
22	$2^2 \cdot 5 \cdot 11$	$13 \cdot 17$	$2 \cdot 3 \cdot 37$	**223**	$2^5 \cdot 7$	$3^2 \cdot 5^2$	$2 \cdot 113$	**227**	$2^2 \cdot 3 \cdot 19$	**229**
23	$2 \cdot 5 \cdot 23$	$3 \cdot 7 \cdot 11$	$2^3 \cdot 29$	**233**	$2 \cdot 3^2 \cdot 13$	$5 \cdot 47$	$2^2 \cdot 59$	$3 \cdot 79$	$2 \cdot 7 \cdot 17$	**239**
24	$2^4 \cdot 3 \cdot 5$	**241**	$2 \cdot 11^2$	3^5	$2^2 \cdot 61$	$5 \cdot 7^2$	$2 \cdot 3 \cdot 41$	$13 \cdot 19$	$2^3 \cdot 31$	$3 \cdot 83$
25	$2 \cdot 5^3$	**251**	$2^2 \cdot 3^2 \cdot 7$	$11 \cdot 23$	$2 \cdot 127$	$3 \cdot 5 \cdot 17$	2^8	**257**	$2 \cdot 3 \cdot 43$	$7 \cdot 37$
26	$2^2 \cdot 5 \cdot 13$	$3^2 \cdot 29$	$2 \cdot 131$	**263**	$2^3 \cdot 3 \cdot 11$	$5 \cdot 53$	$2 \cdot 7 \cdot 19$	$3 \cdot 89$	$2^2 \cdot 67$	**269**
27	$2 \cdot 3^3 \cdot 5$	**271**	$2^4 \cdot 17$	$3 \cdot 7 \cdot 13$	$2 \cdot 137$	$5^2 \cdot 11$	$2^2 \cdot 3 \cdot 23$	**277**	$2 \cdot 139$	$3^2 \cdot 31$
28	$2^3 \cdot 5 \cdot 7$	**281**	$2 \cdot 3 \cdot 47$	**283**	$2^2 \cdot 71$	$3 \cdot 5 \cdot 19$	$2 \cdot 11 \cdot 13$	$7 \cdot 41$	$2^5 \cdot 3^2$	17^2
29	$2 \cdot 5 \cdot 29$	$3 \cdot 97$	$2^2 \cdot 73$	**293**	$2 \cdot 3 \cdot 7^2$	$5 \cdot 59$	$2^3 \cdot 37$	$3^3 \cdot 11$	$2 \cdot 149$	$13 \cdot 23$
30	$2^2 \cdot 3 \cdot 5^2$	$7 \cdot 43$	$2 \cdot 151$	$3 \cdot 101$	$2^4 \cdot 19$	$5 \cdot 61$	$2 \cdot 3^2 \cdot 17$	**307**	$2^2 \cdot 7 \cdot 11$	$3 \cdot 103$
31	$2 \cdot 5 \cdot 31$	**311**	$2^3 \cdot 3 \cdot 13$	**313**	$2 \cdot 157$	$3^2 \cdot 5 \cdot 7$	$2^2 \cdot 79$	**317**	$2 \cdot 3 \cdot 53$	$11 \cdot 29$
32	$2^6 \cdot 5$	$3 \cdot 107$	$2 \cdot 7 \cdot 23$	$17 \cdot 19$	$2^2 \cdot 3^4$	$5^2 \cdot 13$	$2 \cdot 163$	$3 \cdot 109$	$2^3 \cdot 41$	$7 \cdot 47$
33	$2 \cdot 3 \cdot 5 \cdot 11$	**331**	$2^2 \cdot 83$	$3^2 \cdot 37$	$2 \cdot 167$	$5 \cdot 67$	$2^4 \cdot 3 \cdot 7$	**337**	$2 \cdot 13^2$	$3 \cdot 113$
34	$2^2 \cdot 5 \cdot 17$	$11 \cdot 31$	$2 \cdot 3^2 \cdot 19$	7^3	$2^3 \cdot 43$	$3 \cdot 5 \cdot 23$	$2 \cdot 173$	**347**	$2^2 \cdot 3 \cdot 29$	**349**
35	$2 \cdot 5^2 \cdot 7$	$3^3 \cdot 13$	$2^5 \cdot 11$	**353**	$2 \cdot 3 \cdot 59$	$5 \cdot 71$	$2^2 \cdot 89$	$3 \cdot 7 \cdot 17$	$2 \cdot 179$	**359**
36	$2^3 \cdot 3^2 \cdot 5$	19^2	$2 \cdot 181$	$3 \cdot 11^2$	$2^2 \cdot 7 \cdot 13$	$5 \cdot 73$	$2 \cdot 3 \cdot 61$	**367**	$2^4 \cdot 23$	$3^2 \cdot 41$
37	$2 \cdot 5 \cdot 37$	$7 \cdot 53$	$2^2 \cdot 3 \cdot 31$	**373**	$2 \cdot 11 \cdot 17$	$3 \cdot 5^3$	$2^3 \cdot 47$	$13 \cdot 29$	$2 \cdot 3^3 \cdot 7$	**379**
38	$2^2 \cdot 5 \cdot 19$	$3 \cdot 127$	$2 \cdot 191$	**383**	$2^7 \cdot 3$	$5 \cdot 7 \cdot 11$	$2 \cdot 193$	$3^2 \cdot 43$	$2^2 \cdot 97$	**389**
39	$2 \cdot 3 \cdot 5 \cdot 13$	$17 \cdot 23$	$2^3 \cdot 7^2$	$3 \cdot 131$	$2 \cdot 197$	$5 \cdot 79$	$2^2 \cdot 3^2 \cdot 11$	**397**	$2 \cdot 199$	$3 \cdot 7 \cdot 19$
40	$2^4 \cdot 5^2$	**401**	$2 \cdot 3 \cdot 67$	$13 \cdot 31$	$2^2 \cdot 101$	$3^4 \cdot 5$	$2 \cdot 7 \cdot 29$	$11 \cdot 37$	$2^3 \cdot 3 \cdot 17$	**409**
41	$2 \cdot 5 \cdot 41$	$3 \cdot 137$	$2^2 \cdot 103$	$7 \cdot 59$	$2 \cdot 3^2 \cdot 23$	$5 \cdot 83$	$2^5 \cdot 13$	$3 \cdot 139$	$2 \cdot 11 \cdot 19$	**419**
42	$2^2 \cdot 3 \cdot 5 \cdot 7$	**421**	$2 \cdot 211$	$3^2 \cdot 47$	$2 \cdot 3 \cdot 53$	$5^2 \cdot 17$	$2 \cdot 3 \cdot 71$	$7 \cdot 61$	$2^2 \cdot 107$	$3 \cdot 11 \cdot 13$
43	$2 \cdot 5 \cdot 43$	**431**	$2^4 \cdot 3^3$	**433**	$2 \cdot 7 \cdot 31$	$3 \cdot 5 \cdot 29$	$2^2 \cdot 109$	$19 \cdot 23$	$2 \cdot 3 \cdot 73$	**439**
44	$2^3 \cdot 5 \cdot 11$	$3^2 \cdot 7^2$	$2 \cdot 13 \cdot 17$	**443**	$2^2 \cdot 3 \cdot 37$	$5 \cdot 89$	$2 \cdot 223$	$3 \cdot 149$	$2^6 \cdot 7$	**449**
45	$2 \cdot 3^2 \cdot 5^2$	$11 \cdot 41$	$2^2 \cdot 113$	$3 \cdot 151$	$2 \cdot 227$	$5 \cdot 7 \cdot 13$	$2^3 \cdot 3 \cdot 19$	**457**	$2 \cdot 229$	$3^3 \cdot 17$

	0	1	2	3	4	5	6	7	8	9
46	$2^2 \cdot 5 \cdot 23$	**461**	$2 \cdot 3 \cdot 7 \cdot 11$	**463**	$2^4 \cdot 29$	$3 \cdot 5 \cdot 31$	$2 \cdot 233$	**467**	$2^2 \cdot 3^2 \cdot 13$	$7 \cdot 67$
47	$2 \cdot 5 \cdot 47$	$3 \cdot 157$	$2^3 \cdot 59$	$11 \cdot 43$	$2 \cdot 3 \cdot 79$	$5^2 \cdot 19$	$2^2 \cdot 7 \cdot 17$	$3^2 \cdot 53$	$2 \cdot 239$	**479**
48	$2^5 \cdot 3 \cdot 5$	$13 \cdot 37$	$2 \cdot 241$	$3 \cdot 7 \cdot 23$	$2^2 \cdot 11^2$	$5 \cdot 97$	$2 \cdot 3^5$	**487**	$2^3 \cdot 61$	$3 \cdot 163$
49	$2 \cdot 5 \cdot 7^2$	**491**	$2^2 \cdot 3 \cdot 41$	$17 \cdot 29$	$2 \cdot 13 \cdot 19$	$3^2 \cdot 5 \cdot 11$	$2^4 \cdot 31$	$7 \cdot 71$	$2 \cdot 3 \cdot 83$	**499**
50	$2^2 \cdot 5^3$	$3 \cdot 167$	$2 \cdot 251$	**503**	$2^3 \cdot 3^2 \cdot 7$	$5 \cdot 101$	$2 \cdot 11 \cdot 23$	$3 \cdot 13^2$	$2^2 \cdot 127$	**509**
51	$2 \cdot 3 \cdot 5 \cdot 17$	$7 \cdot 73$	2^9	$3^3 \cdot 19$	$2 \cdot 257$	$5 \cdot 103$	$2^2 \cdot 3 \cdot 43$	$11 \cdot 47$	$2 \cdot 7 \cdot 37$	$3 \cdot 173$
52	$2^3 \cdot 5 \cdot 13$	**521**	$2 \cdot 3^2 \cdot 29$	**523**	$2^2 \cdot 131$	$3 \cdot 5^2 \cdot 7$	$2 \cdot 263$	$17 \cdot 31$	$2^4 \cdot 3 \cdot 11$	23^2
53	$2 \cdot 5 \cdot 53$	$3^2 \cdot 59$	$2^2 \cdot 7 \cdot 19$	$13 \cdot 41$	$2 \cdot 3 \cdot 89$	$5 \cdot 107$	$2^3 \cdot 67$	$3 \cdot 179$	$2 \cdot 269$	$7^2 \cdot 11$
54	$2^2 \cdot 3^3 \cdot 5$	**541**	$2 \cdot 271$	$3 \cdot 181$	$2^5 \cdot 17$	$5 \cdot 109$	$2 \cdot 3 \cdot 7 \cdot 13$	**547**	$2^2 \cdot 137$	$3^2 \cdot 61$
55	$2 \cdot 5^2 \cdot 11$	$19 \cdot 29$	$2^3 \cdot 3 \cdot 23$	$7 \cdot 79$	$2 \cdot 277$	$3 \cdot 5 \cdot 37$	$2^2 \cdot 139$	**557**	$2 \cdot 3^2 \cdot 31$	$13 \cdot 43$
56	$2^4 \cdot 5 \cdot 7$	$3 \cdot 11 \cdot 17$	$2 \cdot 281$	**563**	$2^2 \cdot 3 \cdot 47$	$5 \cdot 113$	$2 \cdot 283$	$3^4 \cdot 7$	$2^3 \cdot 71$	**569**
57	$2 \cdot 3 \cdot 5 \cdot 19$	**571**	$2^2 \cdot 11 \cdot 13$	$3 \cdot 191$	$2 \cdot 7 \cdot 41$	$5^2 \cdot 23$	$2^6 \cdot 3^2$	**577**	$2 \cdot 17^2$	$3 \cdot 193$
58	$2^2 \cdot 5 \cdot 29$	$7 \cdot 83$	$2 \cdot 3 \cdot 97$	$11 \cdot 53$	$2^3 \cdot 73$	$3^2 \cdot 5 \cdot 13$	$2 \cdot 293$	**587**	$2^2 \cdot 3 \cdot 7^2$	$19 \cdot 31$
59	$2 \cdot 5 \cdot 59$	$3 \cdot 197$	$2^4 \cdot 37$	**593**	$2 \cdot 3^3 \cdot 11$	$5 \cdot 7 \cdot 17$	$2^2 \cdot 149$	$3 \cdot 199$	$2 \cdot 13 \cdot 23$	**599**
60	$2^3 \cdot 3 \cdot 5^2$	**601**	$2 \cdot 7 \cdot 43$	$3^2 \cdot 67$	$2^2 \cdot 151$	$5 \cdot 11^2$	$2 \cdot 3 \cdot 101$	**607**	$2^5 \cdot 19$	$3 \cdot 7 \cdot 29$
61	$2 \cdot 5 \cdot 61$	$13 \cdot 47$	$2^2 \cdot 3^2 \cdot 17$	**613**	$2 \cdot 307$	$3 \cdot 5 \cdot 41$	$2^3 \cdot 7 \cdot 11$	**617**	$2 \cdot 3 \cdot 103$	**619**
62	$2^2 \cdot 5 \cdot 31$	$3^3 \cdot 23$	$2 \cdot 311$	$7 \cdot 89$	$2^4 \cdot 3 \cdot 13$	5^4	$2 \cdot 313$	$3 \cdot 11 \cdot 19$	$2^2 \cdot 157$	$17 \cdot 37$
63	$2 \cdot 3^2 \cdot 5 \cdot 7$	**631**	$2^3 \cdot 79$	$3 \cdot 211$	$2 \cdot 317$	$5 \cdot 127$	$2^2 \cdot 3 \cdot 53$	$7^2 \cdot 13$	$2 \cdot 11 \cdot 29$	$3^2 \cdot 71$
64	$2^7 \cdot 5$	**641**	$2 \cdot 3 \cdot 107$	**643**	$2^2 \cdot 7 \cdot 23$	$3 \cdot 5 \cdot 43$	$2 \cdot 17 \cdot 19$	**647**	$2^3 \cdot 3^4$	$11 \cdot 59$
65	$2 \cdot 5^2 \cdot 13$	$3 \cdot 7 \cdot 31$	$2^2 \cdot 163$	**653**	$2 \cdot 3 \cdot 109$	$5 \cdot 131$	$2^4 \cdot 41$	$3^2 \cdot 73$	$2 \cdot 7 \cdot 47$	**659**
66	$2^2 \cdot 3 \cdot 5 \cdot 11$	**661**	$2 \cdot 331$	$3 \cdot 13 \cdot 17$	$2^3 \cdot 83$	$5 \cdot 7 \cdot 19$	$2 \cdot 3^2 \cdot 37$	$23 \cdot 29$	$2^2 \cdot 167$	$3 \cdot 223$
67	$2 \cdot 5 \cdot 67$	$11 \cdot 61$	$2^5 \cdot 3 \cdot 7$	**673**	$2 \cdot 337$	$3^3 \cdot 5^2$	$2^2 \cdot 13^2$	**677**	$2 \cdot 3 \cdot 113$	$7 \cdot 97$
68	$2^3 \cdot 5 \cdot 17$	$3 \cdot 227$	$2 \cdot 11 \cdot 31$	**683**	$2^2 \cdot 3^2 \cdot 19$	$5 \cdot 137$	$2 \cdot 7^3$	$3 \cdot 229$	$2^4 \cdot 43$	$13 \cdot 53$
69	$2 \cdot 3 \cdot 5 \cdot 23$	**691**	$2^2 \cdot 173$	$3^2 \cdot 7 \cdot 11$	$2 \cdot 347$	$5 \cdot 139$	$2^3 \cdot 3 \cdot 29$	$17 \cdot 41$	$2 \cdot 349$	$3 \cdot 233$
70	$2^2 \cdot 5^2 \cdot 7$	**701**	$2 \cdot 3^3 \cdot 13$	$19 \cdot 37$	$2^6 \cdot 11$	$3 \cdot 5 \cdot 47$	$2 \cdot 353$	$7 \cdot 101$	$2^2 \cdot 3 \cdot 59$	**709**
71	$2 \cdot 5 \cdot 71$	$3^2 \cdot 79$	$2^3 \cdot 89$	$23 \cdot 31$	$2 \cdot 3 \cdot 7 \cdot 17$	$5 \cdot 11 \cdot 13$	$2^2 \cdot 179$	$3 \cdot 239$	$2 \cdot 359$	**719**
72	$2^4 \cdot 3^2 \cdot 5$	$7 \cdot 103$	$2 \cdot 19^2$	$3 \cdot 241$	$2^2 \cdot 181$	$5^2 \cdot 29$	$2 \cdot 3 \cdot 11^2$	**727**	$2^3 \cdot 7 \cdot 13$	3^6
73	$2 \cdot 5 \cdot 73$	$17 \cdot 43$	$2^2 \cdot 3 \cdot 61$	**733**	$2 \cdot 367$	$3 \cdot 5 \cdot 7^2$	$2^5 \cdot 23$	$11 \cdot 67$	$2 \cdot 3^2 \cdot 41$	**739**
74	$2^2 \cdot 5 \cdot 37$	$3 \cdot 13 \cdot 19$	$2 \cdot 7 \cdot 53$	**743**	$2^3 \cdot 3 \cdot 31$	$5 \cdot 149$	$2 \cdot 373$	$3^2 \cdot 83$	$2^2 \cdot 11 \cdot 17$	$7 \cdot 107$
75	$2 \cdot 3 \cdot 5^3$	**751**	$2^4 \cdot 47$	$3 \cdot 251$	$2 \cdot 13 \cdot 29$	$5 \cdot 151$	$2^2 \cdot 3^3 \cdot 7$	**757**	$2 \cdot 379$	$3 \cdot 11 \cdot 23$
76	$2^3 \cdot 5 \cdot 19$	**761**	$2 \cdot 3 \cdot 127$	$7 \cdot 109$	$2^2 \cdot 191$	$3^2 \cdot 5 \cdot 17$	$2 \cdot 383$	$13 \cdot 59$	$2^8 \cdot 3$	**769**
77	$2 \cdot 5 \cdot 7 \cdot 11$	$3 \cdot 257$	$2^2 \cdot 193$	**773**	$2 \cdot 3^2 \cdot 43$	$5^2 \cdot 31$	$2^3 \cdot 97$	$3 \cdot 7 \cdot 37$	$2 \cdot 389$	$19 \cdot 41$
78	$2^2 \cdot 3 \cdot 5 \cdot 13$	$11 \cdot 71$	$2 \cdot 17 \cdot 23$	$3^3 \cdot 29$	$2^4 \cdot 7^2$	$5 \cdot 157$	$2 \cdot 3 \cdot 131$	**787**	$2^2 \cdot 197$	$3 \cdot 263$
79	$2 \cdot 5 \cdot 79$	$7 \cdot 113$	$2^3 \cdot 3^2 \cdot 11$	$13 \cdot 61$	$2 \cdot 397$	$3 \cdot 5 \cdot 53$	$2^2 \cdot 199$	**797**	$2 \cdot 3 \cdot 7 \cdot 19$	$17 \cdot 47$
80	$2^5 \cdot 5^2$	$3^2 \cdot 89$	$2 \cdot 401$	$11 \cdot 73$	$2^2 \cdot 3 \cdot 67$	$5 \cdot 7 \cdot 23$	$2 \cdot 13 \cdot 31$	$3 \cdot 269$	$2^3 \cdot 101$	**809**
81	$2 \cdot 3^4 \cdot 5$	**811**	$2^2 \cdot 7 \cdot 29$	$3 \cdot 271$	$2 \cdot 11 \cdot 37$	$5 \cdot 163$	$2^4 \cdot 3 \cdot 17$	$19 \cdot 43$	$2 \cdot 409$	$3^2 \cdot 7 \cdot 13$
82	$2^2 \cdot 5 \cdot 41$	**821**	$2 \cdot 3 \cdot 137$	**823**	$2^3 \cdot 103$	$3 \cdot 5^2 \cdot 11$	$2 \cdot 7 \cdot 59$	**827**	$2^2 \cdot 3^2 \cdot 23$	**829**
83	$2 \cdot 5 \cdot 83$	$3 \cdot 277$	$2^6 \cdot 13$	$7^2 \cdot 17$	$2 \cdot 3 \cdot 139$	$5 \cdot 167$	$2^2 \cdot 11 \cdot 19$	$3^3 \cdot 31$	$2 \cdot 419$	**839**
84	$2^3 \cdot 3 \cdot 5 \cdot 7$	29^2	$2 \cdot 421$	$3 \cdot 281$	$2^2 \cdot 211$	$5 \cdot 13^2$	$2 \cdot 3^2 \cdot 47$	$7 \cdot 11^2$	$2^4 \cdot 53$	$3 \cdot 283$
85	$2 \cdot 5^2 \cdot 17$	$23 \cdot 37$	$2^2 \cdot 3 \cdot 71$	**853**	$2 \cdot 7 \cdot 61$	$3^2 \cdot 5 \cdot 19$	$2^3 \cdot 107$	**857**	$2 \cdot 3 \cdot 11 \cdot 13$	**859**
86	$2^2 \cdot 5 \cdot 43$	$3 \cdot 7 \cdot 41$	$2 \cdot 431$	**863**	$2^5 \cdot 3^3$	$5 \cdot 173$	$2 \cdot 433$	$3 \cdot 17^2$	$2^2 \cdot 7 \cdot 31$	$11 \cdot 79$
87	$2 \cdot 3 \cdot 5 \cdot 29$	$13 \cdot 67$	$2^3 \cdot 109$	$3^2 \cdot 97$	$2 \cdot 19 \cdot 23$	$5^3 \cdot 7$	$2^2 \cdot 3 \cdot 73$	**877**	$2 \cdot 439$	$3 \cdot 293$
88	$2^4 \cdot 5 \cdot 11$	**881**	$2 \cdot 3^2 \cdot 7^2$	**883**	$2^2 \cdot 13 \cdot 17$	$3 \cdot 5 \cdot 59$	$2 \cdot 443$	**887**	$2^3 \cdot 3 \cdot 37$	$7 \cdot 127$
89	$2 \cdot 5 \cdot 89$	$3^4 \cdot 11$	$2^2 \cdot 223$	$19 \cdot 47$	$2 \cdot 3 \cdot 149$	$5 \cdot 179$	$2^7 \cdot 7$	$3 \cdot 13 \cdot 23$	$2 \cdot 449$	$29 \cdot 31$
90	$2^2 \cdot 3^2 \cdot 5^2$	$17 \cdot 53$	$2 \cdot 11 \cdot 41$	$3 \cdot 7 \cdot 43$	$2^3 \cdot 113$	$5 \cdot 181$	$2 \cdot 3 \cdot 151$	**907**	$2^2 \cdot 227$	$3^2 \cdot 101$

	0	1	2	3	4	5	6	7	8	9
91	$2 \cdot 5 \cdot 7 \cdot 13$	**911**	$2^4 \cdot 3 \cdot 19$	$11 \cdot 83$	$2 \cdot 457$	$3 \cdot 5 \cdot 61$	$2^2 \cdot 229$	$7 \cdot 131$	$2 \cdot 3^3 \cdot 17$	**919**
92	$2^3 \cdot 5 \cdot 23$	$3 \cdot 307$	$2 \cdot 461$	$13 \cdot 71$	$2^2 \cdot 3 \cdot 7 \cdot 11$	$5^2 \cdot 37$	$2 \cdot 463$	$3^2 \cdot 103$	$2^5 \cdot 29$	**929**
93	$2 \cdot 3 \cdot 5 \cdot 31$	$7^2 \cdot 19$	$2^2 \cdot 233$	$3 \cdot 311$	$2 \cdot 467$	$5 \cdot 11 \cdot 17$	$2^3 \cdot 3^2 \cdot 13$	**937**	$2 \cdot 7 \cdot 67$	$3 \cdot 313$
94	$2^2 \cdot 5 \cdot 47$	**941**	$2 \cdot 3 \cdot 157$	$23 \cdot 41$	$2^4 \cdot 59$	$3^3 \cdot 5 \cdot 7$	$2 \cdot 11 \cdot 43$	**947**	$2^2 \cdot 3 \cdot 79$	$13 \cdot 73$
95	$2 \cdot 5^2 \cdot 19$	$3 \cdot 317$	$2^3 \cdot 7 \cdot 17$	**953**	$2 \cdot 3^2 \cdot 53$	$5 \cdot 191$	$2^2 \cdot 239$	$3 \cdot 11 \cdot 29$	$2 \cdot 479$	$7 \cdot 137$
96	$2^6 \cdot 3 \cdot 5$	31^2	$2 \cdot 13 \cdot 37$	$3^2 \cdot 107$	$2^2 \cdot 241$	$5 \cdot 193$	$2 \cdot 3 \cdot 7 \cdot 23$	**967**	$2^3 \cdot 11^2$	$3 \cdot 17 \cdot 19$
97	$2 \cdot 5 \cdot 97$	**971**	$2^2 \cdot 3^5$	$7 \cdot 139$	$2 \cdot 487$	$3 \cdot 5^2 \cdot 13$	$2^4 \cdot 61$	**977**	$2 \cdot 3 \cdot 163$	$11 \cdot 89$
98	$2^2 \cdot 5 \cdot 7^2$	$3^2 \cdot 109$	$2 \cdot 491$	**983**	$2^3 \cdot 3 \cdot 41$	$5 \cdot 197$	$2 \cdot 17 \cdot 29$	$3 \cdot 7 \cdot 47$	$2^2 \cdot 13 \cdot 19$	$23 \cdot 43$
99	$2 \cdot 3^2 \cdot 5 \cdot 11$	**991**	$2^5 \cdot 31$	$3 \cdot 331$	$2 \cdot 7 \cdot 71$	$5 \cdot 199$	$2^2 \cdot 3 \cdot 83$	**997**	$2 \cdot 499$	$3^3 \cdot 37$
100	$2^3 \cdot 5^3$	$7 \cdot 11 \cdot 13$	$2 \cdot 3 \cdot 167$	$17 \cdot 59$	$2^2 \cdot 251$	$3 \cdot 5 \cdot 67$	$2 \cdot 503$	$19 \cdot 53$	$2^4 \cdot 3^2 \cdot 7$	**1009**
101	$2 \cdot 5 \cdot 101$	$3 \cdot 337$	$2^2 \cdot 11 \cdot 23$	**1013**	$2 \cdot 3 \cdot 13^2$	$5 \cdot 7 \cdot 29$	$2^3 \cdot 127$	$3^2 \cdot 113$	$2 \cdot 509$	**1019**
102	$2^2 \cdot 3 \cdot 5 \cdot 17$	**1021**	$2 \cdot 7 \cdot 73$	$3 \cdot 11 \cdot 31$	2^{10}	$5^2 \cdot 41$	$2 \cdot 3^3 \cdot 19$	$13 \cdot 79$	$2^2 \cdot 257$	$3 \cdot 7^3$
103	$2 \cdot 5 \cdot 103$	**1031**	$2^3 \cdot 3 \cdot 43$	**1033**	$2 \cdot 11 \cdot 47$	$3^2 \cdot 5 \cdot 23$	$2^2 \cdot 7 \cdot 37$	$17 \cdot 61$	$2 \cdot 3 \cdot 173$	**1039**
104	$2^4 \cdot 5 \cdot 13$	$3 \cdot 347$	$2 \cdot 521$	$7 \cdot 149$	$2^2 \cdot 3^2 \cdot 29$	$5 \cdot 11 \cdot 19$	$2 \cdot 523$	$3 \cdot 349$	$2^3 \cdot 131$	**1049**
105	$2 \cdot 3 \cdot 5^2 \cdot 7$	**1051**	$2^2 \cdot 263$	$3^4 \cdot 13$	$2 \cdot 17 \cdot 31$	$5 \cdot 211$	$2^5 \cdot 3 \cdot 11$	$7 \cdot 151$	$2 \cdot 23^2$	$3 \cdot 353$
106	$2^2 \cdot 5 \cdot 53$	**1061**	$2 \cdot 3^2 \cdot 59$	**1063**	$2^3 \cdot 7 \cdot 19$	$3 \cdot 5 \cdot 71$	$2 \cdot 13 \cdot 41$	$11 \cdot 97$	$2^2 \cdot 3 \cdot 89$	**1069**
107	$2 \cdot 5 \cdot 107$	$3^2 \cdot 7 \cdot 17$	$2^4 \cdot 67$	$29 \cdot 37$	$2 \cdot 3 \cdot 179$	$5^2 \cdot 43$	$2^2 \cdot 269$	$3 \cdot 359$	$2 \cdot 7^2 \cdot 11$	$13 \cdot 83$
108	$2^3 \cdot 3^3 \cdot 5$	$23 \cdot 47$	$2 \cdot 541$	$3 \cdot 19^2$	$2^2 \cdot 271$	$5 \cdot 7 \cdot 31$	$2 \cdot 3 \cdot 181$	**1087**	$2^6 \cdot 17$	$3^2 \cdot 11^2$
109	$2 \cdot 5 \cdot 109$	**1091**	$2^2 \cdot 3 \cdot 7 \cdot 13$	**1093**	$2 \cdot 547$	$3 \cdot 5 \cdot 73$	$2^3 \cdot 137$	**1097**	$2 \cdot 3^2 \cdot 61$	$7 \cdot 157$
110	$2^2 \cdot 5^2 \cdot 11$	$3 \cdot 367$	$2 \cdot 19 \cdot 29$	**1103**	$2^4 \cdot 3 \cdot 23$	$5 \cdot 13 \cdot 17$	$2 \cdot 7 \cdot 79$	$3^3 \cdot 41$	$2^2 \cdot 277$	**1109**
111	$2 \cdot 3 \cdot 5 \cdot 37$	$11 \cdot 101$	$2^3 \cdot 139$	$3 \cdot 7 \cdot 53$	$2 \cdot 557$	$5 \cdot 223$	$2^2 \cdot 3^2 \cdot 31$	**1117**	$2 \cdot 13 \cdot 43$	$3 \cdot 373$
112	$2^5 \cdot 5 \cdot 7$	$19 \cdot 59$	$2 \cdot 3 \cdot 11 \cdot 17$	**1123**	$2^2 \cdot 281$	$3^2 \cdot 5^3$	$2 \cdot 563$	$7^2 \cdot 23$	$2^3 \cdot 3 \cdot 47$	**1129**
113	$2 \cdot 5 \cdot 113$	$3 \cdot 13 \cdot 29$	$2^2 \cdot 283$	$11 \cdot 103$	$2 \cdot 3^4 \cdot 7$	$5 \cdot 227$	$2^4 \cdot 71$	$3 \cdot 379$	$2 \cdot 569$	$17 \cdot 67$
114	$2^2 \cdot 3 \cdot 5 \cdot 19$	$7 \cdot 163$	$2 \cdot 571$	$3^2 \cdot 127$	$2^3 \cdot 11 \cdot 13$	$5 \cdot 229$	$2 \cdot 3 \cdot 191$	$31 \cdot 37$	$2^2 \cdot 7 \cdot 41$	$3 \cdot 383$
115	$2 \cdot 5^2 \cdot 23$	**1151**	$2^7 \cdot 3^2$	**1153**	$2 \cdot 577$	$3 \cdot 5 \cdot 7 \cdot 11$	$2^2 \cdot 17^2$	$13 \cdot 89$	$2 \cdot 3 \cdot 193$	$19 \cdot 61$
116	$2^3 \cdot 5 \cdot 29$	$3^3 \cdot 43$	$2 \cdot 7 \cdot 83$	**1163**	$2^2 \cdot 3 \cdot 97$	$5 \cdot 233$	$2 \cdot 11 \cdot 53$	$3 \cdot 389$	$2^4 \cdot 73$	$7 \cdot 167$
117	$2 \cdot 3^2 \cdot 5 \cdot 13$	**1171**	$2^2 \cdot 293$	$3 \cdot 17 \cdot 23$	$2 \cdot 587$	$5^2 \cdot 47$	$2^3 \cdot 3 \cdot 7^2$	$11 \cdot 107$	$2 \cdot 19 \cdot 31$	$3^2 \cdot 131$
118	$2^2 \cdot 5 \cdot 59$	**1181**	$2 \cdot 3 \cdot 197$	$7 \cdot 13^2$	$2^5 \cdot 37$	$3 \cdot 5 \cdot 79$	$2 \cdot 593$	**1187**	$2^2 \cdot 3^3 \cdot 11$	$29 \cdot 41$
119	$2 \cdot 5 \cdot 7 \cdot 17$	$3 \cdot 397$	$2^3 \cdot 149$	**1193**	$2 \cdot 3 \cdot 199$	$5 \cdot 239$	$2^2 \cdot 13 \cdot 23$	$3^2 \cdot 7 \cdot 19$	$2 \cdot 599$	$11 \cdot 109$
120	$2^4 \cdot 3 \cdot 5^2$	**1201**	$2 \cdot 601$	$3 \cdot 401$	$2^2 \cdot 7 \cdot 43$	$5 \cdot 241$	$2 \cdot 3^2 \cdot 67$	$17 \cdot 71$	$2^3 \cdot 151$	$3 \cdot 13 \cdot 31$
121	$2 \cdot 5 \cdot 11^2$	$7 \cdot 173$	$2^2 \cdot 3 \cdot 101$	**1213**	$2 \cdot 607$	$3^5 \cdot 5$	$2^6 \cdot 19$	**1217**	$2 \cdot 3 \cdot 7 \cdot 29$	$23 \cdot 53$
122	$2^2 \cdot 5 \cdot 61$	$3 \cdot 11 \cdot 37$	$2 \cdot 13 \cdot 47$	**1223**	$2^3 \cdot 3^2 \cdot 17$	$5^2 \cdot 7^2$	$2 \cdot 613$	$3 \cdot 409$	$2^2 \cdot 307$	**1229**
123	$2 \cdot 3 \cdot 5 \cdot 41$	**1231**	$2^4 \cdot 7 \cdot 11$	$3^2 \cdot 137$	$2 \cdot 617$	$5 \cdot 13 \cdot 19$	$2^2 \cdot 3 \cdot 103$	**1237**	$2 \cdot 619$	$3 \cdot 7 \cdot 59$
124	$2^3 \cdot 5 \cdot 31$	$17 \cdot 73$	$2 \cdot 3^3 \cdot 23$	$11 \cdot 113$	$2^2 \cdot 311$	$3 \cdot 5 \cdot 83$	$2 \cdot 7 \cdot 89$	$29 \cdot 43$	$2^5 \cdot 3 \cdot 13$	**1249**
125	$2 \cdot 5^4$	$3^2 \cdot 139$	$2^2 \cdot 313$	$7 \cdot 179$	$2 \cdot 3 \cdot 11 \cdot 19$	$5 \cdot 251$	$2^3 \cdot 157$	$3 \cdot 419$	$2 \cdot 17 \cdot 37$	**1259**
126	$2^2 \cdot 3^2 \cdot 5 \cdot 7$	$13 \cdot 97$	$2 \cdot 631$	$3 \cdot 421$	$2^4 \cdot 79$	$5 \cdot 11 \cdot 23$	$2 \cdot 3 \cdot 211$	$7 \cdot 181$	$2^2 \cdot 317$	$3^3 \cdot 47$
127	$2 \cdot 5 \cdot 127$	$31 \cdot 41$	$2^3 \cdot 3 \cdot 53$	$19 \cdot 67$	$2 \cdot 7^2 \cdot 13$	$3 \cdot 5^2 \cdot 17$	$2^2 \cdot 11 \cdot 29$	**1277**	$2 \cdot 3^2 \cdot 71$	**1279**
128	$2^8 \cdot 5$	$3 \cdot 7 \cdot 61$	$2 \cdot 641$	**1283**	$2^2 \cdot 3 \cdot 107$	$5 \cdot 257$	$2 \cdot 643$	$3^2 \cdot 11 \cdot 13$	$2^3 \cdot 7 \cdot 23$	**1289**
129	$2 \cdot 3 \cdot 5 \cdot 43$	**1291**	$2^2 \cdot 17 \cdot 19$	$3 \cdot 431$	$2 \cdot 647$	$5 \cdot 7 \cdot 37$	$2^4 \cdot 3^4$	**1297**	$2 \cdot 11 \cdot 59$	$3 \cdot 433$
130	$2^2 \cdot 5^2 \cdot 13$	**1301**	$2 \cdot 3 \cdot 7 \cdot 31$	**1303**	$2^3 \cdot 163$	$3^2 \cdot 5 \cdot 29$	$2 \cdot 653$	**1307**	$2^2 \cdot 3 \cdot 109$	$7 \cdot 11 \cdot 17$
131	$2 \cdot 5 \cdot 131$	$3 \cdot 19 \cdot 23$	$2^5 \cdot 41$	$13 \cdot 101$	$2 \cdot 3^2 \cdot 73$	$5 \cdot 263$	$2^2 \cdot 7 \cdot 47$	$3 \cdot 439$	$2 \cdot 659$	**1319**
132	$2^3 \cdot 3 \cdot 5 \cdot 11$	**1321**	$2 \cdot 661$	$3^3 \cdot 7^2$	$2^2 \cdot 331$	$5^2 \cdot 53$	$2 \cdot 3 \cdot 13 \cdot 17$	**1327**	$2^4 \cdot 83$	$3 \cdot 443$
133	$2 \cdot 5 \cdot 7 \cdot 19$	11^3	$2^2 \cdot 3^2 \cdot 37$	$31 \cdot 43$	$2 \cdot 23 \cdot 29$	$3 \cdot 5 \cdot 89$	$2^3 \cdot 167$	$7 \cdot 191$	$2 \cdot 3 \cdot 223$	$13 \cdot 103$
134	$2^2 \cdot 5 \cdot 67$	$3^2 \cdot 149$	$2 \cdot 11 \cdot 61$	$17 \cdot 79$	$2^6 \cdot 3 \cdot 7$	$5 \cdot 269$	$2 \cdot 673$	$3 \cdot 449$	$2^2 \cdot 337$	$19 \cdot 71$
135	$2 \cdot 3^3 \cdot 5^2$	$7 \cdot 193$	$2^3 \cdot 13^2$	$3 \cdot 11 \cdot 41$	$2 \cdot 677$	$5 \cdot 271$	$2^2 \cdot 3 \cdot 113$	$23 \cdot 59$	$2 \cdot 7 \cdot 97$	$3^2 \cdot 151$

	0	1	2	3	4	5	6	7	8	9
136	$2^4 \cdot 5 \cdot 17$	**1361**	$2 \cdot 3 \cdot 227$	$29 \cdot 47$	$2^2 \cdot 11 \cdot 31$	$3 \cdot 5 \cdot 7 \cdot 13$	$2 \cdot 683$	**1367**	$2^3 \cdot 3^2 \cdot 19$	37^2
137	$2 \cdot 5 \cdot 137$	$3 \cdot 457$	$2^2 \cdot 7^3$	**1373**	$2 \cdot 3 \cdot 229$	$5^3 \cdot 11$	$2^5 \cdot 43$	$3^4 \cdot 17$	$2 \cdot 13 \cdot 53$	$7 \cdot 197$
138	$2^2 \cdot 3 \cdot 5 \cdot 23$	**1381**	$2 \cdot 691$	$3 \cdot 461$	$2^3 \cdot 173$	$5 \cdot 277$	$2 \cdot 3^2 \cdot 7 \cdot 11$	$19 \cdot 73$	$2^2 \cdot 347$	$3 \cdot 463$
139	$2 \cdot 5 \cdot 139$	$13 \cdot 107$	$2^4 \cdot 3 \cdot 29$	$7 \cdot 199$	$2 \cdot 17 \cdot 41$	$3^2 \cdot 5 \cdot 31$	$2^2 \cdot 349$	$11 \cdot 127$	$2 \cdot 3 \cdot 233$	**1399**
140	$2^3 \cdot 5^2 \cdot 7$	$3 \cdot 467$	$2 \cdot 701$	$23 \cdot 61$	$2^2 \cdot 3^3 \cdot 13$	$5 \cdot 281$	$2 \cdot 19 \cdot 37$	$3 \cdot 7 \cdot 67$	$2^7 \cdot 11$	**1409**
141	$2 \cdot 3 \cdot 5 \cdot 47$	$17 \cdot 83$	$2^2 \cdot 353$	$3^2 \cdot 157$	$2 \cdot 7 \cdot 101$	$5 \cdot 283$	$2^3 \cdot 3 \cdot 59$	$13 \cdot 109$	$2 \cdot 709$	$3 \cdot 11 \cdot 43$
142	$2^2 \cdot 5 \cdot 71$	$7^2 \cdot 29$	$2 \cdot 3^2 \cdot 79$	**1423**	$2^4 \cdot 89$	$3 \cdot 5^2 \cdot 19$	$2 \cdot 23 \cdot 31$	**1427**	$2^2 \cdot 3 \cdot 7 \cdot 17$	**1429**
143	$2 \cdot 5 \cdot 11 \cdot 13$	$3^3 \cdot 53$	$2^3 \cdot 179$	**1433**	$2 \cdot 3 \cdot 239$	$5 \cdot 7 \cdot 41$	$2^2 \cdot 359$	$3 \cdot 479$	$2 \cdot 719$	**1439**
144	$2^5 \cdot 3^2 \cdot 5$	$11 \cdot 131$	$2 \cdot 7 \cdot 103$	$3 \cdot 13 \cdot 37$	$2^2 \cdot 19^2$	$5 \cdot 17^2$	$2 \cdot 3 \cdot 241$	**1447**	$2^3 \cdot 181$	$3^2 \cdot 7 \cdot 23$
145	$2 \cdot 5^2 \cdot 29$	**1451**	$2^2 \cdot 3 \cdot 11^2$	**1453**	$2 \cdot 727$	$3 \cdot 5 \cdot 97$	$2^4 \cdot 7 \cdot 13$	$31 \cdot 47$	$2 \cdot 3^6$	**1459**
146	$2^2 \cdot 5 \cdot 73$	$3 \cdot 487$	$2 \cdot 17 \cdot 43$	$7 \cdot 11 \cdot 19$	$2^3 \cdot 3 \cdot 61$	$5 \cdot 293$	$2 \cdot 733$	$3^2 \cdot 163$	$2^2 \cdot 367$	$13 \cdot 113$
147	$2 \cdot 3 \cdot 5 \cdot 7^2$	**1471**	$2^6 \cdot 23$	$3 \cdot 491$	$2 \cdot 11 \cdot 67$	$5^2 \cdot 59$	$2^2 \cdot 3^2 \cdot 41$	$7 \cdot 211$	$2 \cdot 739$	$3 \cdot 17 \cdot 29$
148	$2^3 \cdot 5 \cdot 37$	**1481**	$2 \cdot 3 \cdot 13 \cdot 19$	**1483**	$2^2 \cdot 7 \cdot 53$	$3^3 \cdot 5 \cdot 11$	$2 \cdot 743$	**1487**	$2^4 \cdot 3 \cdot 31$	**1489**
149	$2 \cdot 5 \cdot 149$	$3 \cdot 7 \cdot 71$	$2^2 \cdot 373$	**1493**	$2 \cdot 3^2 \cdot 83$	$5 \cdot 13 \cdot 23$	$2^3 \cdot 11 \cdot 17$	$3 \cdot 499$	$2 \cdot 7 \cdot 107$	**1499**
150	$2^2 \cdot 3 \cdot 5^3$	$19 \cdot 79$	$2 \cdot 751$	$3^2 \cdot 167$	$2^5 \cdot 47$	$5 \cdot 7 \cdot 43$	$2 \cdot 3 \cdot 251$	$11 \cdot 137$	$2^2 \cdot 13 \cdot 29$	$3 \cdot 503$
151	$2 \cdot 5 \cdot 151$	**1511**	$2^3 \cdot 3^3 \cdot 7$	$17 \cdot 89$	$2 \cdot 757$	$3 \cdot 5 \cdot 101$	$2^2 \cdot 379$	$37 \cdot 41$	$2 \cdot 3 \cdot 11 \cdot 23$	$7^2 \cdot 31$
152	$2^4 \cdot 5 \cdot 19$	$3^2 \cdot 13^2$	$2 \cdot 761$	**1523**	$2^2 \cdot 3 \cdot 127$	$5^2 \cdot 61$	$2 \cdot 7 \cdot 109$	$3 \cdot 509$	$2^3 \cdot 191$	$11 \cdot 139$
153	$2 \cdot 3^2 \cdot 5 \cdot 17$	**1531**	$2^2 \cdot 383$	$3 \cdot 7 \cdot 73$	$2 \cdot 13 \cdot 59$	$5 \cdot 307$	$2^9 \cdot 3$	$29 \cdot 53$	$2 \cdot 769$	$3^4 \cdot 19$
154	$2^2 \cdot 5 \cdot 7 \cdot 11$	$23 \cdot 67$	$2 \cdot 3 \cdot 257$	**1543**	$2^3 \cdot 193$	$3 \cdot 5 \cdot 103$	$2 \cdot 773$	$7 \cdot 13 \cdot 17$	$2^2 \cdot 3^2 \cdot 43$	**1549**
155	$2 \cdot 5^2 \cdot 31$	$3 \cdot 11 \cdot 47$	$2^4 \cdot 97$	**1553**	$2 \cdot 3 \cdot 7 \cdot 37$	$5 \cdot 311$	$2^2 \cdot 389$	$3^2 \cdot 173$	$2 \cdot 19 \cdot 41$	**1559**
156	$2^3 \cdot 3 \cdot 5 \cdot 13$	$7 \cdot 223$	$2 \cdot 11 \cdot 71$	$3 \cdot 521$	$2^2 \cdot 17 \cdot 23$	$5 \cdot 313$	$2 \cdot 3^3 \cdot 29$	**1567**	$2^5 \cdot 7^2$	$3 \cdot 523$
157	$2 \cdot 5 \cdot 157$	**1571**	$2^2 \cdot 3 \cdot 131$	$11^2 \cdot 13$	$2 \cdot 787$	$3^2 \cdot 5^2 \cdot 7$	$2^3 \cdot 197$	$19 \cdot 83$	$2 \cdot 3 \cdot 263$	**1579**
158	$2^2 \cdot 5 \cdot 79$	$3 \cdot 17 \cdot 31$	$2 \cdot 7 \cdot 113$	**1583**	$2^4 \cdot 3^2 \cdot 11$	$5 \cdot 317$	$2 \cdot 13 \cdot 61$	$3 \cdot 23^2$	$2^2 \cdot 397$	$7 \cdot 227$
159	$2 \cdot 3 \cdot 5 \cdot 53$	$37 \cdot 43$	$2^3 \cdot 199$	$3^3 \cdot 59$	$2 \cdot 797$	$5 \cdot 11 \cdot 29$	$2^2 \cdot 3 \cdot 7 \cdot 19$	**1597**	$2 \cdot 17 \cdot 47$	$3 \cdot 13 \cdot 41$
160	$2^6 \cdot 5^2$	**1601**	$2 \cdot 3^2 \cdot 89$	$7 \cdot 229$	$2^2 \cdot 401$	$3 \cdot 5 \cdot 107$	$2 \cdot 11 \cdot 73$	**1607**	$2^3 \cdot 3 \cdot 67$	**1609**
161	$2 \cdot 5 \cdot 7 \cdot 23$	$3^2 \cdot 179$	$2^2 \cdot 13 \cdot 31$	**1613**	$2 \cdot 3 \cdot 269$	$5 \cdot 17 \cdot 19$	$2^4 \cdot 101$	$3 \cdot 7^2 \cdot 11$	$2 \cdot 809$	**1619**
162	$2^2 \cdot 3^4 \cdot 5$	**1621**	$2 \cdot 811$	$3 \cdot 541$	$2^3 \cdot 7 \cdot 29$	$5^3 \cdot 13$	$2 \cdot 3 \cdot 271$	**1627**	$2^2 \cdot 11 \cdot 37$	$3^2 \cdot 181$
163	$2 \cdot 5 \cdot 163$	$7 \cdot 233$	$2^5 \cdot 3 \cdot 17$	$23 \cdot 71$	$2 \cdot 19 \cdot 43$	$3 \cdot 5 \cdot 109$	$2^2 \cdot 409$	**1637**	$2 \cdot 3^2 \cdot 7 \cdot 13$	$11 \cdot 149$
164	$2^3 \cdot 5 \cdot 41$	$3 \cdot 547$	$2 \cdot 821$	$31 \cdot 53$	$2^2 \cdot 3 \cdot 137$	$5 \cdot 7 \cdot 47$	$2 \cdot 823$	$3^3 \cdot 61$	$2^4 \cdot 103$	$17 \cdot 97$
165	$2 \cdot 3 \cdot 5^2 \cdot 11$	$13 \cdot 127$	$2^2 \cdot 7 \cdot 59$	$3 \cdot 19 \cdot 29$	$2 \cdot 827$	$5 \cdot 331$	$2^3 \cdot 3^2 \cdot 23$	**1657**	$2 \cdot 829$	$3 \cdot 7 \cdot 79$
166	$2^2 \cdot 5 \cdot 83$	$11 \cdot 151$	$2 \cdot 3 \cdot 277$	**1663**	$2^7 \cdot 13$	$3^2 \cdot 5 \cdot 37$	$2 \cdot 7^2 \cdot 17$	**1667**	$2^2 \cdot 3 \cdot 139$	**1669**
167	$2 \cdot 5 \cdot 167$	$3 \cdot 557$	$2^3 \cdot 11 \cdot 19$	$7 \cdot 239$	$2 \cdot 3^3 \cdot 31$	$5^2 \cdot 67$	$2^2 \cdot 419$	$3 \cdot 13 \cdot 43$	$2 \cdot 839$	$23 \cdot 73$
168	$2^4 \cdot 3 \cdot 5 \cdot 7$	41^2	$2 \cdot 29^2$	$3^2 \cdot 11 \cdot 17$	$2^2 \cdot 421$	$5 \cdot 337$	$2 \cdot 3 \cdot 281$	$7 \cdot 241$	$2^3 \cdot 211$	$3 \cdot 563$
169	$2 \cdot 5 \cdot 13^2$	$19 \cdot 89$	$2^2 \cdot 3^2 \cdot 47$	**1693**	$2 \cdot 7 \cdot 11^2$	$3 \cdot 5 \cdot 113$	$2^5 \cdot 53$	**1697**	$2 \cdot 3 \cdot 283$	**1699**
170	$2^2 \cdot 5^2 \cdot 17$	$3^5 \cdot 7$	$2 \cdot 23 \cdot 37$	$13 \cdot 131$	$2^3 \cdot 3 \cdot 71$	$5 \cdot 11 \cdot 31$	$2 \cdot 853$	$3 \cdot 569$	$2^2 \cdot 7 \cdot 61$	**1709**
171	$2 \cdot 3^2 \cdot 5 \cdot 19$	$29 \cdot 59$	$2^4 \cdot 107$	$3 \cdot 571$	$2 \cdot 857$	$5 \cdot 7^3$	$2^2 \cdot 3 \cdot 11 \cdot 13$	$17 \cdot 101$	$2 \cdot 859$	$3^2 \cdot 191$
172	$2^3 \cdot 5 \cdot 43$	**1721**	$2 \cdot 3 \cdot 7 \cdot 41$	**1723**	$2^2 \cdot 431$	$3 \cdot 5^2 \cdot 23$	$2 \cdot 863$	$11 \cdot 157$	$2^6 \cdot 3^3$	$7 \cdot 13 \cdot 19$
173	$2 \cdot 5 \cdot 173$	$3 \cdot 577$	$2^2 \cdot 433$	**1733**	$2 \cdot 3 \cdot 17^2$	$5 \cdot 347$	$2^3 \cdot 7 \cdot 31$	$3^2 \cdot 193$	$2 \cdot 11 \cdot 79$	$37 \cdot 47$
174	$2^2 \cdot 3 \cdot 5 \cdot 29$	**1741**	$2 \cdot 13 \cdot 67$	$3 \cdot 7 \cdot 83$	$2^4 \cdot 109$	$5 \cdot 349$	$2 \cdot 3^2 \cdot 97$	**1747**	$2^2 \cdot 19 \cdot 23$	$3 \cdot 11 \cdot 53$
175	$2 \cdot 5^3 \cdot 7$	$17 \cdot 103$	$2^3 \cdot 3 \cdot 73$	**1753**	$2 \cdot 877$	$3^3 \cdot 5 \cdot 13$	$2^2 \cdot 439$	$7 \cdot 251$	$2 \cdot 3 \cdot 293$	**1759**
176	$2^5 \cdot 5 \cdot 11$	$3 \cdot 587$	$2 \cdot 881$	$41 \cdot 43$	$2^2 \cdot 3^2 \cdot 7^2$	$5 \cdot 353$	$2 \cdot 883$	$3 \cdot 19 \cdot 31$	$2^3 \cdot 13 \cdot 17$	$29 \cdot 61$
177	$2 \cdot 3 \cdot 5 \cdot 59$	$7 \cdot 11 \cdot 23$	$2^2 \cdot 443$	$3^2 \cdot 197$	$2 \cdot 887$	$5^2 \cdot 71$	$2^4 \cdot 3 \cdot 37$	**1777**	$2 \cdot 7 \cdot 127$	$3 \cdot 593$
178	$2^2 \cdot 5 \cdot 89$	$13 \cdot 137$	$2 \cdot 3^4 \cdot 11$	**1783**	$2^3 \cdot 223$	$3 \cdot 5 \cdot 7 \cdot 17$	$2 \cdot 19 \cdot 47$	**1787**	$2^2 \cdot 3 \cdot 149$	**1789**
179	$2 \cdot 5 \cdot 179$	$3^2 \cdot 199$	$2^8 \cdot 7$	$11 \cdot 163$	$2 \cdot 3 \cdot 13 \cdot 23$	$5 \cdot 359$	$2^2 \cdot 449$	$3 \cdot 599$	$2 \cdot 29 \cdot 31$	$7 \cdot 257$
180	$2^3 \cdot 3^2 \cdot 5^2$	**1801**	$2 \cdot 17 \cdot 53$	$3 \cdot 601$	$2^2 \cdot 11 \cdot 41$	$5 \cdot 19^2$	$2 \cdot 3 \cdot 7 \cdot 43$	$13 \cdot 139$	$2^4 \cdot 113$	$3^3 \cdot 67$

	0	1	2	3	4	5	6	7	8	9
181	$2 \cdot 5 \cdot 181$	**1811**	$2^2 \cdot 3 \cdot 151$	$7^2 \cdot 37$	$2 \cdot 907$	$3 \cdot 5 \cdot 11^2$	$2^3 \cdot 227$	$23 \cdot 79$	$2 \cdot 3^2 \cdot 101$	$17 \cdot 107$
182	$2^2 \cdot 5 \cdot 7 \cdot 13$	$3 \cdot 607$	$2 \cdot 911$	**1823**	$2^5 \cdot 3 \cdot 19$	$5^2 \cdot 73$	$2 \cdot 11 \cdot 83$	$3^2 \cdot 7 \cdot 29$	$2^2 \cdot 457$	$31 \cdot 59$
183	$2 \cdot 3 \cdot 5 \cdot 61$	**1831**	$2^3 \cdot 229$	$3 \cdot 13 \cdot 47$	$2 \cdot 7 \cdot 131$	$5 \cdot 367$	$2^2 \cdot 3^3 \cdot 17$	$11 \cdot 167$	$2 \cdot 919$	$3 \cdot 613$
184	$2^4 \cdot 5 \cdot 23$	$7 \cdot 263$	$2 \cdot 3 \cdot 307$	$19 \cdot 97$	$2^2 \cdot 461$	$3^2 \cdot 5 \cdot 41$	$2 \cdot 13 \cdot 71$	**1847**	$2^3 \cdot 3 \cdot 7 \cdot 11$	43^2
185	$2 \cdot 5^2 \cdot 37$	$3 \cdot 617$	$2^2 \cdot 463$	$17 \cdot 109$	$2 \cdot 3^2 \cdot 103$	$5 \cdot 7 \cdot 53$	$2^6 \cdot 29$	$3 \cdot 619$	$2 \cdot 929$	$11 \cdot 13^2$
186	$2^2 \cdot 3 \cdot 5 \cdot 31$	**1861**	$2 \cdot 7^2 \cdot 19$	$3^4 \cdot 23$	$2^3 \cdot 233$	$5 \cdot 373$	$2 \cdot 3 \cdot 311$	**1867**	$2^2 \cdot 467$	$3 \cdot 7 \cdot 89$
187	$2 \cdot 5 \cdot 11 \cdot 17$	**1871**	$2^4 \cdot 3^2 \cdot 13$	**1873**	$2 \cdot 937$	$3 \cdot 5^4$	$2^2 \cdot 7 \cdot 67$	**1877**	$2 \cdot 3 \cdot 313$	**1879**
188	$2^3 \cdot 5 \cdot 47$	$3^2 \cdot 11 \cdot 19$	$2 \cdot 941$	$7 \cdot 269$	$2^2 \cdot 3 \cdot 157$	$5 \cdot 13 \cdot 29$	$2 \cdot 23 \cdot 41$	$3 \cdot 17 \cdot 37$	$2^5 \cdot 59$	**1889**
189	$2 \cdot 3^3 \cdot 5 \cdot 7$	$31 \cdot 61$	$2^2 \cdot 11 \cdot 43$	$3 \cdot 631$	$2 \cdot 947$	$5 \cdot 379$	$2^3 \cdot 3 \cdot 79$	$7 \cdot 271$	$2 \cdot 13 \cdot 73$	$3^2 \cdot 211$
190	$2^2 \cdot 5^2 \cdot 19$	**1901**	$2 \cdot 3 \cdot 317$	$11 \cdot 173$	$2^4 \cdot 7 \cdot 17$	$3 \cdot 5 \cdot 127$	$2 \cdot 953$	**1907**	$2^2 \cdot 3^2 \cdot 53$	$23 \cdot 83$
191	$2 \cdot 5 \cdot 191$	$3 \cdot 7^2 \cdot 13$	$2^3 \cdot 239$	**1913**	$2 \cdot 3 \cdot 11 \cdot 29$	$5 \cdot 383$	$2^2 \cdot 479$	$3^3 \cdot 71$	$2 \cdot 7 \cdot 137$	$19 \cdot 101$
192	$2^7 \cdot 3 \cdot 5$	$17 \cdot 113$	$2 \cdot 31^2$	$3 \cdot 641$	$2^2 \cdot 13 \cdot 37$	$5^2 \cdot 7 \cdot 11$	$2 \cdot 3^2 \cdot 107$	$41 \cdot 47$	$2^3 \cdot 241$	$3 \cdot 643$
193	$2 \cdot 5 \cdot 193$	**1931**	$2^2 \cdot 3 \cdot 7 \cdot 23$	**1933**	$2 \cdot 967$	$3^2 \cdot 5 \cdot 43$	$2^4 \cdot 11^2$	$13 \cdot 149$	$2 \cdot 3 \cdot 17 \cdot 19$	$7 \cdot 277$
194	$2^2 \cdot 5 \cdot 97$	$3 \cdot 647$	$2 \cdot 971$	$29 \cdot 67$	$2^3 \cdot 3^5$	$5 \cdot 389$	$2 \cdot 7 \cdot 139$	$3 \cdot 11 \cdot 59$	$2^2 \cdot 487$	**1949**
195	$2 \cdot 3 \cdot 5^2 \cdot 13$	**1951**	$2^5 \cdot 61$	$3^2 \cdot 7 \cdot 31$	$2 \cdot 977$	$5 \cdot 17 \cdot 23$	$2^2 \cdot 3 \cdot 163$	$19 \cdot 103$	$2 \cdot 11 \cdot 89$	$3 \cdot 653$
196	$2^3 \cdot 5 \cdot 7^2$	$37 \cdot 53$	$2 \cdot 3^2 \cdot 109$	$13 \cdot 151$	$2^2 \cdot 491$	$3 \cdot 5 \cdot 131$	$2 \cdot 983$	$7 \cdot 281$	$2^4 \cdot 3 \cdot 41$	$11 \cdot 179$
197	$2 \cdot 5 \cdot 197$	$3^3 \cdot 73$	$2^2 \cdot 17 \cdot 29$	**1973**	$2 \cdot 3 \cdot 7 \cdot 47$	$5^2 \cdot 79$	$2^3 \cdot 13 \cdot 19$	$3 \cdot 659$	$2 \cdot 23 \cdot 43$	**1979**
198	$2^2 \cdot 3^2 \cdot 5 \cdot 11$	$7 \cdot 283$	$2 \cdot 991$	$3 \cdot 661$	$2^6 \cdot 31$	$5 \cdot 397$	$2 \cdot 3 \cdot 331$	**1987**	$2^2 \cdot 7 \cdot 71$	$3 \cdot 13 \cdot 17$
199	$2 \cdot 5 \cdot 199$	$11 \cdot 181$	$2^3 \cdot 3 \cdot 83$	**1993**	$2 \cdot 997$	$3 \cdot 5 \cdot 7 \cdot 19$	$2^2 \cdot 499$	**1997**	$2 \cdot 3^3 \cdot 37$	**1999**
200	$2^4 \cdot 5^3$	$3 \cdot 23 \cdot 29$	$2 \cdot 7 \cdot 11 \cdot 13$	**2003**	$2^2 \cdot 3 \cdot 167$	$5 \cdot 401$	$2 \cdot 17 \cdot 59$	$3^2 \cdot 223$	$2^3 \cdot 251$	$7^2 \cdot 41$
201	$2 \cdot 3 \cdot 5 \cdot 67$	**2011**	$2^2 \cdot 503$	$3 \cdot 11 \cdot 61$	$2 \cdot 19 \cdot 53$	$5 \cdot 13 \cdot 31$	$2^5 \cdot 3^2 \cdot 7$	**2017**	$2 \cdot 1009$	$3 \cdot 673$
202	$2^2 \cdot 5 \cdot 101$	$43 \cdot 47$	$2 \cdot 3 \cdot 337$	$7 \cdot 17^2$	$2^3 \cdot 11 \cdot 23$	$3^4 \cdot 5^2$	$2 \cdot 1013$	**2027**	$2^2 \cdot 3 \cdot 13^2$	**2029**
203	$2 \cdot 5 \cdot 7 \cdot 29$	$3 \cdot 677$	$2^4 \cdot 127$	$19 \cdot 107$	$2 \cdot 3^2 \cdot 113$	$5 \cdot 11 \cdot 37$	$2^2 \cdot 509$	$3 \cdot 7 \cdot 97$	$2 \cdot 1019$	**2039**
204	$2^3 \cdot 3 \cdot 5 \cdot 17$	$13 \cdot 157$	$2 \cdot 1021$	$3^2 \cdot 227$	$2^2 \cdot 7 \cdot 73$	$5 \cdot 409$	$2 \cdot 3 \cdot 11 \cdot 31$	$23 \cdot 89$	2^{11}	$3 \cdot 683$
205	$2 \cdot 5^2 \cdot 41$	$7 \cdot 293$	$2^2 \cdot 3^3 \cdot 19$	**2053**	$2 \cdot 13 \cdot 79$	$3 \cdot 5 \cdot 137$	$2^3 \cdot 257$	$11^2 \cdot 17$	$2 \cdot 3 \cdot 7^3$	$29 \cdot 71$
206	$2^2 \cdot 5 \cdot 103$	$3^2 \cdot 229$	$2 \cdot 1031$	**2063**	$2^4 \cdot 3 \cdot 43$	$5 \cdot 7 \cdot 59$	$2 \cdot 1033$	$3 \cdot 13 \cdot 53$	$2^2 \cdot 11 \cdot 47$	**2069**
207	$2 \cdot 3^2 \cdot 5 \cdot 23$	$19 \cdot 109$	$2^3 \cdot 7 \cdot 37$	$3 \cdot 691$	$2 \cdot 17 \cdot 61$	$5^2 \cdot 83$	$2^2 \cdot 3 \cdot 173$	$31 \cdot 67$	$2 \cdot 1039$	$3^3 \cdot 7 \cdot 11$
208	$2^5 \cdot 5 \cdot 13$	**2081**	$2 \cdot 3 \cdot 347$	**2083**	$2^2 \cdot 521$	$3 \cdot 5 \cdot 139$	$2 \cdot 7 \cdot 149$	**2087**	$2^3 \cdot 3^2 \cdot 29$	**2089**
209	$2 \cdot 5 \cdot 11 \cdot 19$	$3 \cdot 17 \cdot 41$	$2^2 \cdot 523$	$7 \cdot 13 \cdot 23$	$2 \cdot 3 \cdot 349$	$5 \cdot 419$	$2^4 \cdot 131$	$3^2 \cdot 233$	$2 \cdot 1049$	**2099**
210	$2^2 \cdot 3 \cdot 5^2 \cdot 7$	$11 \cdot 191$	$2 \cdot 1051$	$3 \cdot 701$	$2^3 \cdot 263$	$5 \cdot 421$	$2 \cdot 3^4 \cdot 13$	$7^2 \cdot 43$	$2^2 \cdot 17 \cdot 31$	$3 \cdot 19 \cdot 37$
211	$2 \cdot 5 \cdot 211$	**2111**	$2^6 \cdot 3 \cdot 11$	**2113**	$2 \cdot 7 \cdot 151$	$3^2 \cdot 5 \cdot 47$	$2^2 \cdot 23^2$	$29 \cdot 73$	$2 \cdot 3 \cdot 353$	$13 \cdot 163$
212	$2^3 \cdot 5 \cdot 53$	$3 \cdot 7 \cdot 101$	$2 \cdot 1061$	$11 \cdot 193$	$2^2 \cdot 3^2 \cdot 59$	$5^3 \cdot 17$	$2 \cdot 1063$	$3 \cdot 709$	$2^4 \cdot 7 \cdot 19$	**2129**
213	$2 \cdot 3 \cdot 5 \cdot 71$	**2131**	$2^2 \cdot 13 \cdot 41$	$3^3 \cdot 79$	$2 \cdot 11 \cdot 97$	$5 \cdot 7 \cdot 61$	$2^3 \cdot 3 \cdot 89$	**2137**	$2 \cdot 1069$	$3 \cdot 23 \cdot 31$
214	$2^2 \cdot 5 \cdot 107$	**2141**	$2 \cdot 3^2 \cdot 7 \cdot 17$	**2143**	$2^5 \cdot 67$	$3 \cdot 5 \cdot 11 \cdot 13$	$2 \cdot 29 \cdot 37$	$19 \cdot 113$	$2^2 \cdot 3 \cdot 179$	$7 \cdot 307$
215	$2 \cdot 5^2 \cdot 43$	$3^2 \cdot 239$	$2^3 \cdot 269$	**2153**	$2 \cdot 3 \cdot 359$	$5 \cdot 431$	$2^2 \cdot 7^2 \cdot 11$	$3 \cdot 719$	$2 \cdot 13 \cdot 83$	$17 \cdot 127$
216	$2^4 \cdot 3^3 \cdot 5$	**2161**	$2 \cdot 23 \cdot 47$	$3 \cdot 7 \cdot 103$	$2^2 \cdot 541$	$5 \cdot 433$	$2 \cdot 3 \cdot 19^2$	$11 \cdot 197$	$2^3 \cdot 271$	$3^2 \cdot 241$
217	$2 \cdot 5 \cdot 7 \cdot 31$	$13 \cdot 167$	$2^2 \cdot 3 \cdot 181$	$41 \cdot 53$	$2 \cdot 1087$	$3 \cdot 5^2 \cdot 29$	$2^7 \cdot 17$	$7 \cdot 311$	$2 \cdot 3^2 \cdot 11^2$	**2179**
218	$2^2 \cdot 5 \cdot 109$	$3 \cdot 727$	$2 \cdot 1091$	$37 \cdot 59$	$2^3 \cdot 3 \cdot 7 \cdot 13$	$5 \cdot 19 \cdot 23$	$2 \cdot 1093$	3^7	$2^2 \cdot 547$	$11 \cdot 199$
219	$2 \cdot 3 \cdot 5 \cdot 73$	$7 \cdot 313$	$2^4 \cdot 137$	$3 \cdot 17 \cdot 43$	$2 \cdot 1097$	$5 \cdot 439$	$2^2 \cdot 3^2 \cdot 61$	13^3	$2 \cdot 7 \cdot 157$	$3 \cdot 733$
220	$2^3 \cdot 5^2 \cdot 11$	$31 \cdot 71$	$2 \cdot 3 \cdot 367$	**2203**	$2^2 \cdot 19 \cdot 29$	$3^2 \cdot 5 \cdot 7^2$	$2 \cdot 1103$	**2207**	$2^5 \cdot 3 \cdot 23$	47^2
221	$2 \cdot 5 \cdot 13 \cdot 17$	$3 \cdot 11 \cdot 67$	$2^2 \cdot 7 \cdot 79$	**2213**	$2 \cdot 3^3 \cdot 41$	$5 \cdot 443$	$2^3 \cdot 277$	$3 \cdot 739$	$2 \cdot 1109$	$7 \cdot 317$
222	$2^2 \cdot 3 \cdot 5 \cdot 37$	**2221**	$2 \cdot 11 \cdot 101$	$3^2 \cdot 13 \cdot 19$	$2^4 \cdot 139$	$5^2 \cdot 89$	$2 \cdot 3 \cdot 7 \cdot 53$	$17 \cdot 131$	$2^2 \cdot 557$	$3 \cdot 743$
223	$2 \cdot 5 \cdot 223$	$23 \cdot 97$	$2^3 \cdot 3^2 \cdot 31$	$7 \cdot 11 \cdot 29$	$2 \cdot 1117$	$3 \cdot 5 \cdot 149$	$2^2 \cdot 13 \cdot 43$	**2237**	$2 \cdot 3 \cdot 373$	**2239**
224	$2^6 \cdot 5 \cdot 7$	$3^3 \cdot 83$	$2 \cdot 19 \cdot 59$	**2243**	$2^2 \cdot 3 \cdot 11 \cdot 17$	$5 \cdot 449$	$2 \cdot 1123$	$3 \cdot 7 \cdot 107$	$2^3 \cdot 281$	$13 \cdot 173$
225	$2 \cdot 3^2 \cdot 5^3$	**2251**	$2^2 \cdot 563$	$3 \cdot 751$	$2 \cdot 7^2 \cdot 23$	$5 \cdot 11 \cdot 41$	$2^4 \cdot 3 \cdot 47$	$37 \cdot 61$	$2 \cdot 1129$	$3^2 \cdot 251$

	0	1	2	3	4	5	6	7	8	9
226	$2^2 \cdot 5 \cdot 113$	$7 \cdot 17 \cdot 19$	$2 \cdot 3 \cdot 13 \cdot 29$	$31 \cdot 73$	$2^3 \cdot 283$	$3 \cdot 5 \cdot 151$	$2 \cdot 11 \cdot 103$	**2267**	$2^2 \cdot 3^4 \cdot 7$	**2269**
227	$2 \cdot 5 \cdot 227$	$3 \cdot 757$	$2^5 \cdot 71$	**2273**	$2 \cdot 3 \cdot 379$	$5^2 \cdot 7 \cdot 13$	$2^2 \cdot 569$	$3^2 \cdot 11 \cdot 23$	$2 \cdot 17 \cdot 67$	$43 \cdot 53$
228	$2^3 \cdot 3 \cdot 5 \cdot 19$	**2281**	$2 \cdot 7 \cdot 163$	$3 \cdot 761$	$2^2 \cdot 571$	$5 \cdot 457$	$2 \cdot 3^2 \cdot 127$	**2287**	$2^4 \cdot 11 \cdot 13$	$3 \cdot 7 \cdot 109$
229	$2 \cdot 5 \cdot 229$	$29 \cdot 79$	$2^2 \cdot 3 \cdot 191$	**2293**	$2 \cdot 31 \cdot 37$	$3^3 \cdot 5 \cdot 17$	$2^3 \cdot 7 \cdot 41$	**2297**	$2 \cdot 3 \cdot 383$	$11^2 \cdot 19$
230	$2^2 \cdot 5^2 \cdot 23$	$3 \cdot 13 \cdot 59$	$2 \cdot 1151$	$7^2 \cdot 47$	$2^8 \cdot 3^2$	$5 \cdot 461$	$2 \cdot 1153$	$3 \cdot 769$	$2^2 \cdot 577$	**2309**
231	$2 \cdot 3 \cdot 5 \cdot 7 \cdot 11$	**2311**	$2^3 \cdot 17^2$	$3^2 \cdot 257$	$2 \cdot 13 \cdot 89$	$5 \cdot 463$	$2^2 \cdot 3 \cdot 193$	$7 \cdot 331$	$2 \cdot 19 \cdot 61$	$3 \cdot 773$
232	$2^4 \cdot 5 \cdot 29$	$11 \cdot 211$	$2 \cdot 3^3 \cdot 43$	$23 \cdot 101$	$2^2 \cdot 7 \cdot 83$	$3 \cdot 5^2 \cdot 31$	$2 \cdot 1163$	$13 \cdot 179$	$2^3 \cdot 3 \cdot 97$	$17 \cdot 137$
233	$2 \cdot 5 \cdot 233$	$3^2 \cdot 7 \cdot 37$	$2^2 \cdot 11 \cdot 53$	**2333**	$2 \cdot 3 \cdot 389$	$5 \cdot 467$	$2^5 \cdot 73$	$3 \cdot 19 \cdot 41$	$2 \cdot 7 \cdot 167$	**2339**
234	$2^2 \cdot 3^2 \cdot 5 \cdot 13$	**2341**	$2 \cdot 1171$	$3 \cdot 11 \cdot 71$	$2^3 \cdot 293$	$5 \cdot 7 \cdot 67$	$2 \cdot 3 \cdot 17 \cdot 23$	**2347**	$2^2 \cdot 587$	$3^4 \cdot 29$
235	$2 \cdot 5^2 \cdot 47$	**2351**	$2^4 \cdot 3 \cdot 7^2$	$13 \cdot 181$	$2 \cdot 11 \cdot 107$	$3 \cdot 5 \cdot 157$	$2^2 \cdot 19 \cdot 31$	**2357**	$2 \cdot 3^2 \cdot 131$	$7 \cdot 337$
236	$2^3 \cdot 5 \cdot 59$	$3 \cdot 787$	$2 \cdot 1181$	$17 \cdot 139$	$2^2 \cdot 3 \cdot 197$	$5 \cdot 11 \cdot 43$	$2 \cdot 7 \cdot 13^2$	$3^2 \cdot 263$	$2^6 \cdot 37$	$23 \cdot 103$
237	$2 \cdot 3 \cdot 5 \cdot 79$	**2371**	$2^2 \cdot 593$	$3 \cdot 7 \cdot 113$	$2 \cdot 1187$	$5^3 \cdot 19$	$2^3 \cdot 3^3 \cdot 11$	**2377**	$2 \cdot 29 \cdot 41$	$3 \cdot 13 \cdot 61$
238	$2^2 \cdot 5 \cdot 7 \cdot 17$	**2381**	$2 \cdot 3 \cdot 397$	**2383**	$2^4 \cdot 149$	$3^2 \cdot 5 \cdot 53$	$2 \cdot 1193$	$7 \cdot 11 \cdot 31$	$2^2 \cdot 3 \cdot 199$	**2389**
239	$2 \cdot 5 \cdot 239$	$3 \cdot 797$	$2^3 \cdot 13 \cdot 23$	**2393**	$2 \cdot 3^2 \cdot 7 \cdot 19$	$5 \cdot 479$	$2^2 \cdot 599$	$3 \cdot 17 \cdot 47$	$2 \cdot 11 \cdot 109$	**2399**
240	$2^5 \cdot 3 \cdot 5^2$	7^4	$2 \cdot 1201$	$3^3 \cdot 89$	$2^2 \cdot 601$	$5 \cdot 13 \cdot 37$	$2 \cdot 3 \cdot 401$	$29 \cdot 83$	$2^3 \cdot 7 \cdot 43$	$3 \cdot 11 \cdot 73$
241	$2 \cdot 5 \cdot 241$	**2411**	$2^2 \cdot 3^2 \cdot 67$	$19 \cdot 127$	$2 \cdot 17 \cdot 71$	$3 \cdot 5 \cdot 7 \cdot 23$	$2^4 \cdot 151$	**2417**	$2 \cdot 3 \cdot 13 \cdot 31$	$41 \cdot 59$
242	$2^2 \cdot 5 \cdot 11^2$	$3^2 \cdot 269$	$2 \cdot 7 \cdot 173$	**2423**	$2^3 \cdot 3 \cdot 101$	$5^2 \cdot 97$	$2 \cdot 1213$	$3 \cdot 809$	$2^2 \cdot 607$	$7 \cdot 347$
243	$2 \cdot 3^5 \cdot 5$	$11 \cdot 13 \cdot 17$	$2^7 \cdot 19$	$3 \cdot 811$	$2 \cdot 1217$	$5 \cdot 487$	$2^2 \cdot 3 \cdot 7 \cdot 29$	**2437**	$2 \cdot 23 \cdot 53$	$3^2 \cdot 271$
244	$2^3 \cdot 5 \cdot 61$	**2441**	$2 \cdot 3 \cdot 11 \cdot 37$	$7 \cdot 349$	$2^2 \cdot 13 \cdot 47$	$3 \cdot 5 \cdot 163$	$2 \cdot 1223$	**2447**	$2^4 \cdot 3^2 \cdot 17$	$31 \cdot 79$
245	$2 \cdot 5^2 \cdot 7^2$	$3 \cdot 19 \cdot 43$	$2^2 \cdot 613$	$11 \cdot 223$	$2 \cdot 3 \cdot 409$	$5 \cdot 491$	$2^3 \cdot 307$	$3^3 \cdot 7 \cdot 13$	$2 \cdot 1229$	**2459**
246	$2^2 \cdot 3 \cdot 5 \cdot 41$	$23 \cdot 107$	$2 \cdot 1231$	$3 \cdot 821$	$2^5 \cdot 7 \cdot 11$	$5 \cdot 17 \cdot 29$	$2 \cdot 3^2 \cdot 137$	**2467**	$2^2 \cdot 617$	$3 \cdot 823$
247	$2 \cdot 5 \cdot 13 \cdot 19$	$7 \cdot 353$	$2^3 \cdot 3 \cdot 103$	**2473**	$2 \cdot 1237$	$3^2 \cdot 5^2 \cdot 11$	$2^2 \cdot 619$	**2477**	$2 \cdot 3 \cdot 7 \cdot 59$	$37 \cdot 67$
248	$2^4 \cdot 5 \cdot 31$	$3 \cdot 827$	$2 \cdot 17 \cdot 73$	$13 \cdot 191$	$2^2 \cdot 3^3 \cdot 23$	$5 \cdot 7 \cdot 71$	$2 \cdot 11 \cdot 113$	$3 \cdot 829$	$2^3 \cdot 311$	$19 \cdot 131$
249	$2 \cdot 3 \cdot 5 \cdot 83$	$47 \cdot 53$	$2^2 \cdot 7 \cdot 89$	$3^2 \cdot 277$	$2 \cdot 29 \cdot 43$	$5 \cdot 499$	$2^6 \cdot 3 \cdot 13$	$11 \cdot 227$	$2 \cdot 1249$	$3 \cdot 7^2 \cdot 17$

Algorithm 1: Sieve of Eratosthenes.

make a list of the numbers from 2 to N
$i := 1$
while $i \leq \sqrt{N}$
begin
  $i := i + 1$
  if i is not already crossed out **then** cross out all proper multiples of i that are less than or equal to N
end {The numbers *not* crossed out comprise the primes up to N}

---

9. *The sieve of Eratosthenes*: Eratosthenes (3rd century B.C.E.) developed Algorithm 1 for listing all prime numbers less than a fixed bound.

10. *Prime number theorem*: $\pi(x)$, when divided by $\frac{x}{\log x}$, tends to 1 as x tends to infinity. That is, $\pi(x)$ is asymptotic to $\frac{x}{\log x}$ as $x \to \infty$.

11. The prime number theorem was first conjectured by Carl Friedrich Gauss (1777–1855) in 1792, and was first proved in 1896 independently by Charles de la Vallée Poussin (1866–1962) and Jacques Hadamard (1865–1963). They proved it in the stronger form $|\pi(x) - \text{li}(x)| < c_1 x e^{-c_2 \sqrt{\log x}}$, where c_1 and c_2 are positive constants. Their proofs used functions of a complex variable. The first elementary proofs (not using complex variables) of the prime number theorem were supplied in 1949 by Paul Erdős (1913–1996) and Atle Seberg.

12. Integration by parts shows that $\operatorname{li}(x)$ is asymptotic to $\frac{x}{\log x}$ as $x \to \infty$.

13. $|\pi(x) - \operatorname{li}(x)| < c_3 x e^{-c_4 (\log x)^{3/5} (\log \log x)^{-1/5}}$ for certain positive constants c_3 and c_4. (I. M. Vinogradov and Nikolai Korobov, 1958.)

14. If the Riemann hypothesis (Open Problem 1) is true, $|\pi(x) - \operatorname{li}(x)|$ is bounded by $c\sqrt{x} \log x$ for some positive constant c.

15. J. E. Littlewood (1885–1977) showed that $\pi(x) - \operatorname{li}(x)$ changes sign infinitely often. However, no explicit number x with $\pi(x) - \operatorname{li}(x) > 0$ is known. Carter Bays and Richard H. Hudson have shown that such a number x exists below 1.4×10^{316}.

16. The largest exactly computed value of $\pi(x)$ is $\pi(10^{20})$. This value, computed by M. Deleglise in 1996, is about 2.23×10^8 below $\operatorname{li}(10^{20})$. (See the following table.)

n	$\pi(10^n)$	$\approx \pi(10^n) - \operatorname{li}(10^n)$
1	4	-2
2	25	-5
3	168	-10
4	1,229	-17
5	9,592	-38
6	78,498	-130
7	664,579	-339
8	5,761,455	-754
9	50,847,534	$-1,701$
10	455,052,511	$-3,104$
11	4,118,054,813	$-11,588$
12	37,607,912,018	$-38,263$
13	346,065,536,839	$-108,971$
14	3,204,941,750,802	$-314,890$
15	29,844,570,422,669	$-1,052,619$
16	279,238,341,033,925	$-3,214,632$
17	2,623,557,157,654,233	$-7,956,589$
18	24,739,954,287,740,860	$-21,949,555$
19	234,057,667,276,344,607	$-99,877,775$
20	2,220,819,602,560,918,840	$-223,744,644$

17. *Dirichlet's theorem on primes in arithmetic progressions*: Given coprime integers a, b with b positive, there are infinitely many primes $p \equiv a \pmod{b}$. G. L. Dirichlet proved this in 1837.

18. The number of primes p less than x such that $p \equiv a \pmod{b}$ is asymptotic to $\frac{1}{\phi(b)} \pi(x)$ as $x \to \infty$, if a and b are coprime and b is positive. (ϕ is the Euler phi-function; see §4.6.2.)

Open Problems:

1. *Riemann hypothesis*: The *Riemann hypothesis* (*RH*), posed in 1859 by Bernhard Riemann (1826–1866), is a conjecture about the location of zeros of the *Riemann zeta function*, the function of the complex variable s defined by the series $\zeta(s) = \sum_{n=1}^{\infty} n^{-s}$ when the real part of s is > 1, and defined by the formula

$$\zeta(s) = \frac{s}{s-1} - s \int_1^{\infty} (x - \lfloor x \rfloor) x^{-s-1} dx$$

in the larger region when the real part of s is > 0, except for the single point $s = 1$, where it remains undefined. The Riemann hypothesis asserts that all of the solutions to

$\zeta(s) = 0$ in this larger region lie on the vertical line in the complex number plane with imaginary part $\frac{1}{2}$. Its proof would imply a better error estimate for the prime number theorem. While believed to be true, it has not been proved.

2. *Extended Riemann hypothesis*: There is a generalized form of the Riemann hypothesis known as the *extended Riemann hypothesis (ERH)* or the *generalized Riemann hypothesis (GRH)*, which also has important consequences in number theory. (For example, see §4.4.4.)

3. *Hypothesis H*: The *hypothesis H* of Andrzej Schinzel and Waclaw Sierpinski (1882–1969) asserts that for every collection of irreducible nonconstant polynomials $f_1(x), \ldots, f_k(x)$ with integral coefficients and positive leading coefficients, if there is no fixed integer greater than 1 dividing the product $f_1(m) \ldots f_k(m)$ for all integers m, then there are infinitely many integers m such that each of the numbers $f_1(m), \ldots, f_k(m)$ is prime. The case when each of the polynomials is linear was previously conjectured by L. E. Dickson, and is known as the *prime k-tuples conjecture*. The only case of Hypothesis H that has been proved is the case of a single linear polynomial; this is Dirichlet's theorem (Fact 17). The case of the two linear polynomials x and $x + 2$ corresponds to the twin prime conjecture (Open Problem 4). Among many consequences of hypothesis H is the assertion that there are infinitely many primes of the form $m^2 + 1$.

4. *Twin primes*: It has been conjectured that there are infinitely many twin primes, that is, pairs of primes that differ by 2.

5. Let d_n denote the difference between the $(n+1)$st prime and the nth prime. The sequence d_n is unbounded. The prime number theorem implies that on average d_n is about $\log n$. The twin prime conjecture asks whether d_n is 2 infinitely often.

6. The best result known that shows that d_n has relatively small values infinitely often, proved by Helmut Maier in 1988, is that $d_n < c \log n$ infinitely often, where c is a constant slightly smaller than $\frac{1}{4}$.

7. It is conjectured that d_n can be as big as $\log^2 n$ infinitely often, but not much bigger. Roger Baker and Glyn Harman have recently shown that $d_n < n^{.535}$ for all large numbers n. In the other direction, Erdős and Robert Rankin have shown that $d_n > c \log n (\log \log n)(\log \log \log \log n)/(\log \log \log n)^2$ infinitely often. Several improvements have been made on the constant c, but this ungainly expression has stubbornly resisted improvement.

8. Christian Goldbach (1690–1764) conjectured that every integer greater than 5 is the sum of three primes.

9. *Goldbach conjecture*: Every even integer greater than 2 is a sum of two primes. (This is equivalent to the conjecture Goldbach made in Open Problem 8.)

- Matti Sinisalo, in 1993, verified the Goldbach conjecture up to 4×10^{11}. It has since been verified up to 1.615×10^{12} by J. M. Deshouillers, G. Effinger, H. J. J. te Riele, and D. Zinoviev.

- In 1937 Vinogradov proved that every sufficiently large odd number is the sum of three primes. In 1989 J. R. Chen and T. Z. Wang showed that this is true for every odd number greater than $10^{43,001}$. In 1998 Y. Saouter showed that this is true for every odd number below 10^{20}. Zinoviev showed in 1996 that it is true for the remaining odd numbers between 10^{20} and $10^{43,001}$ under the assumption of the ERH (Open Problem 2).

- In 1966 J. R. Chen proved that every sufficiently large even number is either the sum of two primes or the sum of a prime and a number that is the product of two primes.

Examples:

1. A method for showing that there are infinitely many primes is to note that the integer $n!+1$ must have a prime factor greater than n, so there is no largest prime. Note that $n! + 1$ is prime for $n = 1, 2, 3, 11, 27, 37, 41, 73, 77, 116, 154, 320, 340, 399,$ and 427, but is composite for all numbers less than 427 not listed.

2. Let $Q(p)$ (p a prime) equal one more than the product of the primes not exceeding p. For example $Q(5) = 2 \cdot 3 \cdot 5 + 1 = 31$. Then $Q(p)$ is prime for $p = 2, 3, 5, 7, 11, 31, 379, 1019, 1021, 2657, 3229, 4547, 4787, 11549, 13649$; it is composite for all $p < 11213$ not in this list. For example, $Q(13) = 2 \cdot 3 \cdot 5 \cdot 7 \cdot 11 \cdot 13 + 1$ is composite.

3. There are six primes not exceeding 16, namely $2, 3, 5, 7, 11,$ and 13. Hence $\pi(16) = 6$.

4. The expression $n^2 + 1$ is prime for $n = 1, 2, 4, 6, 10, \ldots$, but it is unknown whether there are infinitely many primes of this form when n is an integer. (See Open Problem 3.)

5. The polynomial $f(n) = n^2 + n + 41$ takes on prime values for $n = 0, 1, 2, \ldots, 39$, but $f(40) = 1681 = 41^2$.

6. Applying Dirichlet's theorem with $a = 123$ and $b = 1,000$, there are infinitely many primes that end in the digits 123. The first such prime is 1,123.

7. The pairs 17, 19 and 191, 193 are twin primes. The largest known twin primes have 11,755 decimal digits. They are $361,700,055 \times 2^{39,020} \pm 1$ and were found in 1999 by Henri Lifchitz.

4.4.3 NUMBERS OF SPECIAL FORM

Numbers of the form $b^n \pm 1$, for b a small number, are often easier to factor or test for primality than other numbers of the same size. They also have a colorful history.

Definitions:

A *Cunningham number* is a number of the form $b^n \pm 1$, where b and n are natural numbers, and b is "small" — 2, 3, 5, 6, 7, 10, 11, or 12. They are named after Allan Cunningham, who, along with H. J. Woodall, published in 1925 a table of factorizations of many of these numbers.

A *Fermat number* F_m is a Cunningham number of the form $2^{2^m} + 1$. (See Table 4.)

A *Fermat prime* is a Fermat number that is prime.

A *Mersenne number* M_n is a Cunningham number of the form $2^n - 1$.

A *Mersenne prime* is a Mersenne number that is prime. (See Table 3.)

The *cyclotomic polynomials* $\Phi_k(x)$ are defined recursively by the equation $x^n - 1 = \prod_{d|n} \Phi_d(x)$.

A *perfect number* is a positive integer that is equal to the sum of all its proper divisors.

Facts:

1. If M_n is prime, then n is prime, but the converse is not true.

2. If $b > 2$ or n is composite, then a nontrivial factorization of $b^n - 1$ is given by $b^n - 1 = \prod_{d|n} \Phi_d(b)$, though the factors $\Phi_d(b)$ are not necessarily primes.

3. The number $b^n + 1$ can be factored as the product of $\Phi_d(b)$, where d runs over the divisors of $2n$ that are not divisors of n. When n is not a power of 2 and $b \geq 2$, this factorization is nontrivial.

Algorithm 2: Lucas-Lehmer test.

$p :=$ an odd prime; $u := 4$; $i := 0$
while $i \leq p - 2$
begin
  $i := i + 1$
  $u := u^2 - 2 \bmod 2^p - 1$
end
{if $u = 0$ then $2^p - 1$ is prime, else $2^p - 1$ is composite}

4. Some numbers of the form $b^n \pm 1$ also have so-called *Aurifeuillian factorizations*, named after A. Aurifeuille. For more details, see [BrEtal88].

5. The only primes of the form $b^n - 1$ (with $n > 1$) are Mersenne primes.

6. The only primes of the form $2^n + 1$ are Fermat primes.

7. Fermat numbers are named after Pierre de Fermat (1601–1695), who observed that F_0, F_1, F_2, F_3 and F_4 are prime and stated (incorrectly) that all such numbers are prime. Euler proved this was false, by showing that $F_5 = 2^{32} + 1 = 641 \times 6{,}700{,}417$.

8. F_4 is the largest known Fermat prime. It is conjectured that all larger Fermat numbers are composite.

9. The smallest Fermat number that has not yet been completely factored is $F_{12} = 2^{2^{12}} + 1$, which has a 1187-digit composite factor.

10. In 1994 it was shown that F_{22} is composite. There are 141 values of $n > 22$ where a (relatively) small prime factor of F_n is known. In none of these cases do we know whether the remaining factor of F_n is prime or composite. Currently, F_{24} is the smallest Fermat number that has not been proved prime or shown to be composite. For up-to-date information about the factorization of Fermat numbers (maintained by Wilfrid Keller) consult http://vamri.xray.ufl.edu/proths/fermat.html#Prime.

11. *Pepin's criterion:* For $m \geq 1$, F_m is prime if and only if $3^{(F_m - 1)/2} \equiv -1 \pmod{F_m}$.

12. For $m \geq 2$, every factor of F_m is of the form $2^{m+2}k + 1$.

13. Mersenne numbers are named after Marin Mersenne (1588–1648), who made a list of what he thought were all the Mersenne primes M_p with $p \leq 257$. His list consisted of the primes $p = 2, 3, 5, 7, 13, 17, 19, 31, 67, 127$, and 257. However, it was later shown that M_{67} and M_{257} are composite, while M_{61}, M_{89}, and M_{107}, missing from the list, are prime.

14. It is not known whether there are infinitely many Mersenne primes, nor whether infinitely many Mersenne numbers with prime exponent are composite, though it is conjectured that both are true.

15. Euclid showed that the product of a Mersenne prime $2^p - 1$ with 2^{p-1} is perfect. Euler showed that every even perfect number is of this form. It is not known whether any odd perfect numbers exist. There are none below 10^{300}, a result of R. P. Brent, G. L. Cohen and H. J. J. teRiele in 1991.

16. The *Lucas-Lehmer test* can be used to determine whether a given Mersenne number is prime or composite. (See Algorithm 2.)

17. Table 3 lists all known Mersenne primes. The largest known Mersenne prime is $2^{6{,}972{,}593} - 1$. When a new Mersenne prime is found by computer, there may be other

numbers of the form M_p less than this prime not yet checked for primality. It can take months, or even years, to do this checking. A new Mersenne prime may even be found this way, as was the case for the 29th.

18. George Woltman launched the Great Internet Mersenne Prime Search (GIMPS) in 1996. GIMPS provides free software for PCs. GIMPS has played a role in discovering the last four Mersenne primes. Thousands of people participate in GIMPS over PrimeNet, a virtual supercomputer of distributed PCs, together running more than 0.7 Teraflops, the equivalent of more than a dozen of the fastest supercomputers, in the quest for Mersenne primes. Consult the GIMPS website at http://www.mersenne.org and the PrimeNet site at http://entropia.com/ips/ for more information about this quest and how to join it.

19. As of 1999, the two smallest composite Mersenne numbers not completely factored were $2^{617} - 1$, and $2^{619} - 1$.

20. The best reference for the history of the factorization of Cunningham numbers is [BrEtal88].

21. The current version of the Cunningham table, maintained by Sam Wagstaff, can be found at http://www.cs.purdue.edu/homes/ssw/cun/index.html

22. In Table 4, p_k indicates a k-digit prime, and c_k indicates a k-digit composite. All other numbers in the right column have been proved prime.

Examples:

1. The Mersenne number $M_{11} = 2^{11} - 1$ is not prime since $M_{11} = 23 \cdot 89$.
2. To factor $342 = 7^3 - 1$ note that $7^3 - 1 = (7-1)(7^2 + 7 + 1) = 6 \times 57$.
3. To factor $3^7 + 1$ note that $3^7 + 1 = \Phi_2(3)\Phi_{14}(3) = 4 \times 547$.
4. An example of an Aurifeuillian factorization is given by $2^{4k-2} + 1 = (2^{2k-1} - 2^k + 1) \cdot (2^{2k-1} + 2^k + 1)$.
5. $\Phi_1(x) = x - 1$ and $x^3 - 1 = \Phi_1(x)\Phi_3(x)$, so $\Phi_3(x) = (x^3 - 1)/\Phi_1(x) = x^2 + x + 1$.

4.4.4 PSEUDOPRIMES AND PRIMALITY TESTING

Definitions:

A *pseudoprime to the base b* is a composite number n such that $b^n \equiv b \pmod{n}$.

A *pseudoprime* is a pseudoprime to the base 2.

A *Carmichael number* is a pseudoprime to all bases.

A *strong pseudoprime to the base b* is an odd composite number $n = 2^s d + 1$, with d odd, and either $b^d \equiv 1 \pmod{n}$ or $b^{2^r d} \equiv -1 \pmod{n}$ for some integer r, $0 \leq r < s$.

A *witness* for an odd composite number n is a base b, with $1 < b < n$, to which n is not a strong pseudoprime. Thus, b is a "witness" to n being composite.

A *primality proof* is an irrefutable verification that an integer is prime.

Facts:

1. By Fermat's little theorem (§4.3.3), $b^{p-1} \equiv 1 \pmod{p}$ for all primes p and all integers b that are not multiples of p. Thus, the only numbers $n > 1$ with $b^{n-1} \equiv 1 \pmod{n}$ are primes and pseudoprimes to the base b (which are coprime to b). Similarly, the numbers n which satisfy the strong pseudoprime congruence conditions are the odd primes not dividing b and the strong pseudoprimes to the base b.
2. The smallest pseudoprime is 341.

Table 3 Mersenne primes.

n	exponent	decimal digits	year discovered	discoverer(s) (computer used)
1	2	1	ancient times	
2	3	1	ancient times	
3	5	2	ancient times	
4	7	3	ancient times	
5	13	4	1461	anonymous
6	17	6	1588	Cataldi
7	19	6	1588	Cataldi
8	31	10	1750	Euler
9	61	19	1883	Pervushin
10	89	27	1911	Powers
11	107	33	1913	Fauquembergue
12	127	39	1876	Lucas
13	521	157	1952	Robinson (SWAC)
14	607	183	1952	Robinson (SWAC)
15	1,279	386	1952	Robinson (SWAC)
16	2,203	664	1952	Robinson (SWAC)
17	2,281	687	1952	Robinson (SWAC)
18	3,217	969	1957	Riesel (BESK)
19	4,253	1,281	1961	Hurwitz (IBM 7090)
20	4,423	1,332	1961	Hurwitz (IBM 7090)
21	9,689	2,917	1963	Gillies (ILLIAC 2)
22	9,941	2,993	1963	Gillies (ILLIAC 2)
23	11,213	3,376	1963	Gillies (ILLIAC 2)
24	19,937	6,002	1971	Tuckerman (IBM 360/91)
25	21,701	6,533	1978	Noll and Nickel (Cyber 174)
26	23,209	6,987	1979	Noll (Cyber 174)
27	44,497	13,395	1979	Nelson and Slowinski (Cray 1)
28	86,243	25,962	1982	Slowinski (Cray 1)
29	110,503	33,265	1988	Colquitt and Welsh (NEC SX-W)
30	132,049	39,751	1983	Slowinski (Cray X-MP)
31	216,091	65,050	1985	Slowinski (Cray X-MP)
32	756,839	227,832	1992	Slowinski and Gage (Cray 2)
33	859,433	258,716	1994	Slowinski and Gage (Cray 2)
34	1,257,787	378,632	1996	Slowinski and Gage (Cray T94)
35	1,398,269	420,921	1996	Armengaud, Woltman, and team (90 MHz Pentium)
36	2,976,221	895,932	1997	Spence, Woltman, and others (100 MHz Pentium)
37	3,021,377	909,526	1998	Clarkson, Woltman, Kurowski, and others (200 MHz Pentium)
38	6,972,593	2,098,960	1999	Hajratwala, Woltman, and Kurowski (350 MHz Pentium)

3. There are infinitely many pseudoprimes; however, Paul Erdős has proved that pseudoprimes are rare compared to primes. The same results are true for pseudoprimes to any fixed base b. (See [Ri96] or [CrPo99] for details.)

Table 4 Fermat numbers.

m	known factorization of F_m
0	3
1	5
2	17
3	257
4	65,537
5	$641 \times p_7$
6	$274{,}177 \times p_{14}$
7	$59{,}649{,}589{,}127{,}497{,}217 \times p_{22}$
8	$1{,}238{,}926{,}361{,}552{,}897 \times p_{62}$
9	$2{,}424{,}833 \times 7{,}455{,}602{,}825{,}647{,}884{,}208{,}337{,}395{,}736{,}200{,}454{,}918{,}783{,}366{,}342{,}657$ $\times p_{99}$
10	$45{,}592{,}577 \times 6{,}487{,}031{,}809$ $\times 4{,}659{,}775{,}785{,}220{,}018{,}543{,}264{,}560{,}743{,}076{,}778{,}192{,}897 \times p_{252}$
11	$319{,}489 \times 974{,}849 \times 167{,}988{,}556{,}341{,}760{,}475{,}137$ $\times 3{,}560{,}841{,}906{,}445{,}833{,}920{,}513 \times p_{564}$
12	$114{,}689 \times 26{,}017{,}793 \times 63{,}766{,}529 \times 190{,}274{,}191{,}361$ $\times 1{,}256{,}132{,}134{,}125{,}569 \times c_{1{,}187}$
13	$2{,}710{,}954{,}639{,}361 \times 2{,}663{,}848{,}877{,}152{,}141{,}313$ $\times 3{,}603{,}109{,}844{,}542{,}291{,}969 \times 319{,}546{,}020{,}820{,}551{,}643{,}220{,}672{,}513 \times c_{2{,}391}$
14	c_{4933}
15	$1{,}214{,}251{,}009 \times 2{,}327{,}042{,}503{,}868{,}417 \times c_{9840}$
16	$825{,}753{,}601 \times c_{19{,}720}$
17	$31{,}065{,}037{,}602{,}817 \times c_{39{,}444}$
18	$13{,}631{,}489 \times c_{78{,}906}$
19	$70{,}525{,}124{,}609 \times 646{,}730{,}219{,}521 \times c_{157{,}804}$
20	$c_{315{,}653}$
21	$4{,}485{,}296{,}422{,}913 \times c_{631{,}294}$
22	$c_{1{,}262{,}611}$

4. In 1910, Robert D. Carmichael gave the first examples of Carmichael numbers. The first 16 Carmichael numbers are

$561 = 3 \cdot 11 \cdot 17$ $1{,}105 = 5 \cdot 13 \cdot 17$ $1{,}729 = 7 \cdot 13 \cdot 19$
$2{,}465 = 5 \cdot 17 \cdot 29$ $2{,}821 = 7 \cdot 13 \cdot 31$ $6{,}601 = 7 \cdot 23 \cdot 41$
$8{,}911 = 7 \cdot 19 \cdot 67$ $10{,}585 = 5 \cdot 29 \cdot 73$ $15{,}841 = 7 \cdot 31 \cdot 73$
$29{,}341 = 13 \cdot 37 \cdot 61$ $41{,}041 = 7 \cdot 11 \cdot 13 \cdot 41$ $46{,}657 = 13 \cdot 37 \cdot 97$
$52{,}633 = 7 \cdot 73 \cdot 103$ $62{,}745 = 3 \cdot 5 \cdot 47 \cdot 89$ $63{,}973 = 7 \cdot 13 \cdot 19 \cdot 37$
$75{,}361 = 11 \cdot 17 \cdot 31$

5. If n is a Carmichael number, then n is the product of at least three distinct odd primes with the property that if q is one of these primes, then $q-1$ divides $n-1$.

6. There are a finite number of Carmichael numbers that are the product of exactly r primes with the first $r-2$ primes specified.

7. If m is a positive integer such that $6m + 1$, $12m + 1$, and $18m + 1$ are all primes, then $(6m + 1)(12m + 1)(18m + 1)$ is a Carmichael number.

8. In 1994, W. R. Alford (born 1937), Andrew Granville (born 1962), and Carl Pomerance (born 1944) showed that there are infinitely many Carmichael numbers.

> **Algorithm 3:** Strong probable prime test (to a random base).
> input: positive numbers n, d, s, with d odd and $n = 2^s d + 1$.
> $b :=$ a random integer such that $1 < b < n$
> $c := b^d \bmod n$
> **if** $c = 1$ **or** $c = n - 1$, **then** declare n a *probable prime* and stop
> compute sequentially $c^2 \bmod n$, $c^4 \bmod n, \ldots, c^{2^{s-1}} \bmod n$
> **if** one of these is $n - 1$, **then** declare n a *probable prime* and stop
> **else** declare n composite and stop

9. There are infinitely many numbers that are simultaneously strong pseudoprimes to each base in any given finite set. Each odd composite n, however, can be a strong pseudoprime to at most one-fourth of the bases b with $1 \leq b \leq n - 1$.

10. J. L. Selfridge (born 1927) suggested Algorithm 3 (often referred to as the *Miller-Rabin test*).

11. A "probable prime" is not necessarily a prime, but the chances are good. The probability that an odd composite is not declared composite by Algorithm 3 is at most $\frac{1}{4}$, so the probability it passes k independent iterations is at most 4^{-k}. Suppose this test is applied to random odd inputs n with the hope of finding a prime. That is, random odd numbers n (chosen between two consecutive powers of 2) are tested until one is found that passes each of k independent iterations of the test. Ronald Burthe showed in 1995 that the probability that the output of this procedure is composite is less than 4^{-k}.

12. Gary Miller proved in 1976 that if the extended Riemann hypothesis (§4.4.2) is true, then every odd composite n has a witness less than $c \log^2 n$, for some constant c. Eric Bach showed in 1985 that one may take $c = 2$. Therefore, if an odd number $n > 1$ passes the strong probable prime test for every base b less than $2 \log^2 n$, and if the extended Riemann hypothesis is true, then n is prime.

13. In practice, one can test whether numbers under 2.5×10^{10} are prime by a small number of strong probable prime tests. Pomerance, Selfridge, and Samuel Wagstaff have verified (1980) that there are no numbers less than this bound that are simultaneously strong pseudoprimes to the bases 2, 3, 5, 7, and 11. Thus, any number less than 2.5×10^{10} that passes those strong pseudoprime tests is a prime.

14. Gerhard Jaeschke showed in 1993 that the test described in Fact 13 works almost 100 times beyond 2.5×10^{10}; the first number for which it fails is 2,152,302,898,747.

15. Only primes pass the strong pseudoprime tests to all the bases 2, 3, 5, 7, 11, 13, and 17 until the composite number 341,550,071,728,321 is reached.

16. While pseudoprimality tests are usually quite efficient at recognizing composites, the task of *proving* that a number is prime can be more difficult.

17. In 1983, Leonard Adleman, Carl Pomerance, and Robert Rumely developed the *APR algorithm*, which can prove that a number n is prime in time proportional to $(\log n)^{c \log \log \log n}$, where c is a positive constant. See [Co93] and [CrPo99] for details.

18. Recently, Oliver Atkin and François Morain developed an algorithm to prove primality. It is difficult to predict in advance how long it will take, but in practice it has been fast. One advantage of their algorithm is that, unlike APR, it produces a polynomial time primality proof, though the running time to find the proof may be a bit longer. An implementation called ECPP (*elliptic curve primality proving*) is available via ftp from

ftp.inria.fr

> **Algorithm 1: Trial division.**
>
> input: an integer $n \geq 2$
> output: j (smallest prime factor of n) or statement that n is prime
>
> $j := 2$
> while $j \leq \sqrt{n}$
> begin
>   if $j|n$ then print that j a prime factor of n and stop $\{n$ is not prime$\}$
> $j := j + 1$
> end
> if no factor is found then declare n prime

19. In 1986, Adleman and Ming-Deh A. Huang showed that there is a test for primality that can be executed in random polynomial time. The test, however, is not practical.

20. In 1987, Carl Pomerance showed that every prime p has a primality proof whose verification involves just $c \log p$ multiplications with integers the size of p. It may be difficult, however, to find such a short primality proof.

21. In 1995, Sergei Konyagin and Carl Pomerance gave a deterministic polynomial time algorithm which, for each fixed $\epsilon > 0$ and all sufficiently large x, succeeds in proving prime at least $x^{1-\epsilon}$ prime inputs below x. The degree of the polynomial in the time bound depends on the choice of ϵ.

4.5 FACTORIZATION

Determining the prime factorization of positive integers is a question that has been studied for many years. Furthermore, in the past two decades, this question has become relevant for an extremely important application, the security of public key cryptosystems. The question of exactly how to decompose a composite number into the product of its prime factors is a difficult one that continues to be the subject of much research.

4.5.1 FACTORIZATION ALGORITHMS

Definition:

A *smooth number* is an integer all of whose prime divisors are small.

Facts:

1. The simplest algorithm for factoring an integer is *trial division*, Algorithm 1. While simple, this algorithm is useful only for numbers that have a fairly small prime factor. It can be modified so that after $j = 3$, the number j is incremented by 2, and there are other improvements of this kind.

2. Currently, the fastest algorithm for numbers that are feasible to factor but do not have a small prime factor is the *quadratic sieve* (QS), Algorithm 2, invented by Carl Pomerance in 1981. (For numbers at the far range of feasibility, the *number field sieve* is faster; see Fact 9.)

> **Algorithm 2: Quadratic sieve.**
>
> input: n (an odd composite number that is not a power)
> output: g (a nontrivial factor of n)
>
> find $a_1, \ldots, a_k$ such that each ${a_i}^2 - n$ is smooth
> find a subset of the numbers $a_i^2 - n$ whose product is a square, say x^2
> reduce x modulo n
> $y :=$ the product of the a_i used to form the square
> reduce y modulo n
> {This gives a congruence $x^2 \equiv y^2 \pmod{n}$; equivalently $n|(x^2-y^2)$.}
> $g := \gcd(x-y, n)$
> **if** g is not a nontrivial factor **then** find new x and y (if necessary, find more a_i)

3. The greatest common divisor calculation may be quickly done via the Euclidean algorithm. If $x \not\equiv \pm y \pmod{n}$, then g will be a nontrivial factor of n. (Among all solutions to the congruence $x^2 \equiv y^2 \pmod{n}$ with xy coprime to n, at least half of them lead to a nontrivial factorization of n.) Finding the a_is is at the heart of the algorithm and is accomplished using a sieve not unlike the sieve of Eratosthenes, but applied to the consecutive values of the quadratic polynomial $a^2 - n$. If a is chosen near $\sqrt{n}$, then $a^2 - n$ will be relatively small, and thus more likely to be smooth. So one sieves the polynomial $a^2 - n$, where a runs over integers near $\sqrt{n}$, for values that are smooth. When enough smooth values are collected, the subset with product a square may be found via a linear algebra subroutine applied to a matrix formed out of the exponents in the prime factorizations of the smooth values. The linear algebra may be done modulo 2.

4. The current formulation of QS involves many improvements, the most notable of them the *multiple polynomial variation* of James Davis and Peter Montgomery.

5. In 1994, QS was used to factor a 129-digit composite that was the product of a 64-digit prime and a 65-digit prime. This number had been proposed as a challenge to those who would try to crack the famous RSA cryptosystem.

6. In 1985, Hendrik W. Lenstra, Jr. (born 1949) invented the *elliptic curve method* (ECM), which has the advantage that, like trial division, the running time is based on the size of the smallest prime factor. Thus, it can be used to find comparatively small factors of numbers whose size would be prohibitively large for the quadratic sieve. It can be best understood by first examining the $p-1$ *method* of John Pollard, Algorithm 3.

7. The Pollard algorithm (Algorithm 3) is successful and efficient if $p-1$ happens to be smooth for some prime $p|n$. If the prime factors p of n have the property that $p-1$ is not smooth, Algorithm 3 will eventually be successful if a high enough bound B is chosen, but in this case it will not be any more efficient than trial division, Algorithm 1. ECM gets around this restriction on the numbers that can be efficiently factored by randomly searching through various mathematical objects called *elliptic curve groups*, each of which has $p+1-a$ elements, where $|a| < 2\sqrt{p}$ and a depends on the curve. ECM is successful when a group is encountered such that $p+1-a$ is a smooth number.

8. As of 1998, prime factors as large as 49 digits have been found using ECM. (After such a factor is discovered it may turn out that the remaining part of the number is a prime and the factorization is now complete. This last prime may be very large, as with the tenth and eleventh Fermat numbers — see Table 4. In such cases the success of ECM is measured by the *second* largest prime factor in the prime factorization, though in some sense the method has discovered the largest prime factor as well.)

Algorithm 3: p-1 **factorization method.**

input: n (composite number), B (a bound)
output: a nontrivial factor of n

$b := 2$
{loop on b}
if $b \mid n$ then stop {b is a prime factor of n}
$M := 1$
while $M \leq B$
begin
    $g := \gcd(b^{\text{lcm}(1,2,\ldots,M)} - 1, n)$
    if $n > g > 1$ then output g and stop {g is a nontrivial factor of n}
    else if $g = n$ then choose first prime larger than b and go to beginning of
        the b-loop
    else $M := M + 1$
end

9. The *number field sieve* (NFS), originally suggested by Pollard for numbers of special form, and developed for general composite numbers by Joseph Buhler, Lenstra, and Pomerance, is currently the fastest factoring algorithm for very large numbers with no small prime factors.

10. The number field sieve is similar to QS in that one attempts to assemble two squares x^2 and y^2 whose difference is a multiple of n, and this is done via a sieve and linear algebra modulo 2. However, NFS is much more complicated than QS. Although faster for very large numbers, the complexity of the method makes it unsuitable for numbers much smaller than 100 digits. The exact crossover with QS depends a great deal on the implementations and the hardware employed. The two are roughly within an order of magnitude of each other for numbers between 100 and 150 digits, with QS having the edge at the lower end and NFS the edge at the upper end.

11. Part of the NFS algorithm requires expressing a small multiple of the number to be factored by a polynomial of moderate degree. The running time depends, in part, on the size of the coefficients of this polynomial. For Cunningham numbers, this polynomial can be easy to find. (For example, in the notation of §4.4.2, $8F_9 = 8(2^{2^9}+1) = f(2^{103})$, where $f(x) = x^5 + 8$.) This version is called the *special number field sieve* (SNFS). The version for general numbers, the *general number field sieve* (GNFS), has somewhat greater complexity. The greatest success of SNFS has been the factorization of a 180-digit Cunningham number, while the greatest success of GNFS has been the factorization of a 130-digit number of no special form and with no small prime factor.

12. See [Co93], [CrPo99], [Po90], and [Po94] for fuller descriptions of the factoring algorithms described here, as well as others, including the *continued fraction* (CFRAC) method. Until the advent of QS, this had been the fastest known practical algorithm.

13. The factorization algorithms QS, ECM, SNFS, and GNFS are fast in practice, but analyses of their running times depend on heuristic arguments and unproved hypotheses. The fastest algorithm whose running time has been rigorously analyzed is the *class group relations method* (CGRM). It, however, is not practical. It is a probabilistic algorithm whose expected running time is bounded by $e^{c\sqrt{\log n \log \log n}}$, where c tends to 1 as n tends to infinity through the odd composite numbers that are not powers. This result was proved in 1992 by Lenstra and Pomerance.

Table 1 Comparison of various factoring methods.

algorithm	year introduced	greatest success	running time	rigorously analyzed
trial division	antiquity	–	$\sqrt{n}$	yes
CFRAC	1970	63-digit number	$L\left(\frac{1}{2}, \sqrt{\frac{3}{2}}\right)$	no
$p-1$	1974	32-digit factor	–	yes
QS	1981	129-digit number	$L(\frac{1}{2}, 1)$	no
ECM	1985	47-digit factor	$L(\frac{1}{2}, 1)$	no
SNFS	1988	180-digit number	$L\left(\frac{1}{3}, \sqrt[3]{\frac{32}{9}}\right)$	no
CGRM	1992	–	$L(\frac{1}{2}, 1)$	yes
GNFS	1993	130-digit number	$L\left(\frac{1}{3}, \sqrt[3]{\frac{64}{9}}\right)$	no

14. These algorithms are summarized in Table 1. $L(a,b)$ means that the running time to factor n is bounded by $e^{c(\log n)^a (\log \log n)^{1-a}}$, where c tends to b as n tends to infinity through the odd composite non-powers. Running times are measured in the number of arithmetic steps with integers at most the size of n.

15. The running time for Trial Division in Table 1 is a worst case estimate, achieved when n is prime or the product of two primes of the same magnitude. When n is composite, Trial Division will discover the least prime factor p of n in roughly p steps. The record for the largest prime factor discovered via Trial Division is not known, nor is the largest number proved prime by this method, though the feat of Euler of proving that the Mersenne number $2^{31} - 1$ is prime, using only Trial Division and hand calculations, should certainly be noted. (Euler surely knew, though, that any prime factor of $2^{31} - 1$ is 1 mod 31, so only 1 out of every 31 trial divisors needed to be tested.)

16. The running time of the $p-1$ method is about B, where B is the least number such that for some prime factor p of n, $p-1$ divides lcm $(1, 2, \ldots, B)$.

17. There are variants of CFRAC and GNFS that have smaller heuristic complexity estimates, but the ones in the table above are for the fastest practical version.

18. The running time bound for ECM is a worst case estimate. It is more appropriate to measure ECM as a function of the least prime factor p of n. This heuristic complexity bound is $e^{c\sqrt{\log p \log \log p}}$, where c tends to $\sqrt{2}$ as p tends to infinity.

19. Table 2 was compiled with the assistance of Samuel Wagstaff. It should be remarked that there is no firm definition of a "hard number". What is meant here is that the number was factored by an algorithm that is not sensitive to any particular form the number may have, nor sensitive to the size of the prime factors.

20. It is unknown whether there is a polynomial time factorization algorithm. Whether there are any factorization algorithms that surpass the quadratic sieve, the elliptic curve method, and the number field sieve in their respective regions of superiority is an area of much current research.

21. A cooperative effort to factor large numbers called NFSNet has been set up. It can be found on the Internet at
    http://www.dataplex.net/NFSNet

Table 2 Largest hard number factored as a function of time.

year	method	digits
1970	CFRAC	39
1979	CFRAC	46
1982	CFRAC	54
1983	QS	67
1986	QS	87
1988	QS	102
1990	QS	116
1994	QS	129
1995	GNFS	130

22. A subjective measurement of progress in factorization can be gained by looking at the "ten most wanted numbers" to be factored. The list is maintained by Sam Wagstaff and can be found at http://www.cs.purdue.edu/homes/ssw/cun/index.html. As of May 1999, "number one" on this list is $2^{617} - 1$.

4.6 ARITHMETIC FUNCTIONS

Functions whose domains are the set of positive integers play an important role in number theory. Such functions are called arithmetic functions and are the subject of this section. The information presented here includes definitions and properties of many important arithmetic functions, asymptotic estimates on the growth of these functions, and algebraic properties of sets of certain arithmetic functions. For more information on the topics covered in this section see [Ap76].

4.6.1 MULTIPLICATIVE AND ADDITIVE FUNCTIONS

Definitions:

An **arithmetic function** is a function that is defined for all positive integers.

An arithmetic function is **multiplicative** if $f(mn) = f(m)f(n)$ whenever m and n are relatively prime positive integers.

An arithmetic function is **completely multiplicative** if $f(mn) = f(m)f(n)$ for all positive integers m and n.

If f is an arithmetic function, then $\sum_{d|n} f(d)$, the value of the **summatory function** of f at n, is the sum of $f(d)$ over all positive integers d that divide n.

An arithmetic function f is **additive** if $f(mn) = f(m) + f(n)$ whenever m and n are relatively prime positive integers.

An arithmetic function f is **completely additive** if $f(m,n) = f(m)+f(n)$ whenever m and n are positive integers.

Facts:

1. If f is a multiplicative function and $n = p_1^{a_1} p_2^{a_2} \ldots p_s^{a_s}$ is the prime-power factorization of n, then $f(n) = f(p_1^{a_1}) f(p_2^{a_2}) \ldots f(p_s^{a_s})$.
2. If f is multiplicative, then $f(1) = 1$.
3. If f is a completely multiplicative function and $n = p_1^{a_1} p_2^{a_2} \ldots p_s^{a_s}$, then $f(n) = f(p_1)^{a_1} f(p_2)^{a_2} \ldots f(p_s)^{a_s}$.
4. If f is multiplicative, then the arithmetic function $F(n) = \sum_{d|n} f(d)$ is multiplicative.
5. If f is an additive function, then $f(1) = 0$.
6. If f is an additive function and a is a positive real number, then $F(n) = a^{f(n)}$ is multiplicative.
7. If f is a completely additive function and a is a positive real number, then $F(n) = a^{f(n)}$ is completely multiplicative.

Examples:

1. The function $f(n) = n^2$ is multiplicative. Even more, it is completely multiplicative.
2. The function $I(n) = \lfloor \frac{1}{n} \rfloor$ (so that $I(1) = 1$ and $I(n) = 0$ if n is a positive integer greater than 1) is completely multiplicative.
3. The Euler phi-function, the number of divisors function, the sum of divisors function, and the Möbius function are all multiplicative. None of these functions is completely multiplicative.

4.6.2 EULER'S PHI-FUNCTION

Definition:

If n is a positive integer then $\phi(n)$, the value of the **Euler-phi function** at n, is the number of positive integers not exceeding n that are relatively prime to n. The Euler-phi function is also known as the **totient function**.

Facts:

1. The Euler ϕ function is multiplicative, but not completely multiplicative.
2. If p is a prime, then $\phi(p) = p - 1$.
3. If p is a positive integer with $\phi(p) = p - 1$, then p is prime.
4. If p is a prime and a is a positive integer, then $\phi(p^a) = p^a - p^{a-1}$.
5. If n is a positive integer with prime-power factorization $n = p_1^{a_1} p_2^{a_2} \ldots p_k^{a_k}$, then $\phi(n) = n \prod_{j=1}^{k} (1 - \frac{1}{p_j})$.
6. If n is a positive integer greater than 2, then $\phi(n)$ is even.
7. If n has r distinct odd prime factors, then 2^r divides $\phi(n)$.
8. If m and n are positive integers and $\gcd(m, n) = d$, then $\phi(mn) = \frac{\phi(m)\phi(n)d}{\phi(d)}$.
9. If m and n are positive integers and $m|n$, then $\phi(m)|\phi(n)$.
10. If n is a positive integer, then $\sum_{d|n} \phi(d) = \sum_{d|n} \phi(\frac{n}{d}) = n$.
11. If n is a positive integer with $n \geq 5$, then $\phi(n) > \frac{n}{6 \log \log n}$.
12. $\sum_{k=1}^{n} \phi(k) = \frac{3n^2}{\pi^2} + O(n \log n)$
13. $\sum_{k=1}^{n} \frac{\phi(k)}{k} = \frac{6n}{\pi^2} + O(n \log n)$

Examples:
1. Table 1 includes the value of $\phi(n)$ for $1 \leq n \leq 1000$.
2. To see that $\phi(10) = 4$, note that the positive integers not exceeding 10 relatively prime to 10 are 1, 3, 7, and 9.
3. To find $\phi(720)$, note that $\phi(720) = \phi(2^4 3^2 5) = 720(1 - \frac{1}{2})(1 - \frac{1}{3})(1 - \frac{1}{5}) = 192$.

4.6.3 SUM AND NUMBER OF DIVISORS FUNCTIONS

Definitions:

If n is a positive integer, then $\sigma(n)$, the value of the **sum of divisors function** at n, is the sum of the positive integer divisors of n.

A positive integer n is **perfect** if and only if it equals the sum of its proper divisors (or equivalently, if $\sigma(n) = 2n$).

A positive integer n is **abundant** if the sum of the proper divisors of n exceeds n (or equivalently, if $\sigma(n) > 2n$).

A positive integer n is **deficient** if the sum of the proper divisors of n is less than n (or equivalently, if $\sigma(n) < 2n$).

The positive integers m and n are **amicable** if $\sigma(m) = \sigma(n) = m + n$.

If n is a positive integer, then $\tau(n)$, the value of the **number of divisors function** at n, is the number of positive integer divisors of n.

Facts:
1. The number of divisors function is multiplicative, but not completely multiplicative.
2. The number of divisors function is the summatory function of $f(n) = 1$; that is, $\tau(n) = \sum_{d|n} 1$.
3. The sum of divisors function is multiplicative, but not completely multiplicative.
4. The sum of divisors function is the summatory function of $f(n) = n$; that is, $\sigma(n) = \sum_{d|n} d$.
5. If n is a positive integer with prime-power factorization $n = p_1^{a_1} p_2^{a_2} \ldots p_k^{a_k}$, then $\sigma(n) = \prod_{j=1}^{k} (p_j^{a_j+1} - 1)/(p_j - 1)$.
6. If n is a positive integer with prime-power factorization $n = p_1^{a_1} p_2^{a_2} \ldots p_k^{a_k}$, then $\tau(n) = \prod_{j=1}^{k} (a_j + 1)$.
7. If n is a positive integer, then $\tau(n)$ is odd if and only if n is a perfect square.
8. If k is an integer greater than 1, then the equation $\tau(n) = k$ has infinitely many solutions.
9. If n is a positive integer, then $(\sum_{d|n} \tau(d))^2 = \sum_{d|n} \tau(d)^3$.
10. A positive integer n is an even perfect number if and only if $n = 2^{m-1}(2^m - 1)$ where m is an integer, $m \geq 2$, and $2^m - 1$ is prime (so that it is a Mersenne prime (§4.4.3)). Hence, the number of known even perfect numbers equals the number of known Mersenne primes.
11. It is unknown whether there are any odd perfect numbers. However, it is known that there are no odd perfect numbers less than 10^{300} and that any odd perfect number must have at least eight different prime factors.

Table 1 Values of $\phi(n)$, $\sigma(n)$, $\tau(n)$, and $\mu(n)$ for $1 \leq n \leq 1000$.

Using Maple V, the numtheory package commands phi(n), sigma(n), tau(n), and mobius(n) can be used to calculate these functions.

n	ϕ	σ	τ	μ	n	ϕ	σ	τ	μ	n	ϕ	σ	τ	μ	n	ϕ	σ	τ	μ	n	ϕ	σ	τ	μ
1	1	1	1	1	2	1	3	2	-1	3	2	4	2	-1	4	2	7	3	0	5	4	6	2	-1
6	2	12	4	1	7	6	8	2	-1	8	4	15	4	0	9	6	13	3	0	10	4	18	4	1
11	10	12	2	-1	12	4	28	6	0	13	12	14	2	-1	14	6	24	4	1	15	8	24	4	1
16	8	31	5	0	17	16	18	2	-1	18	6	39	6	0	19	18	20	2	-1	20	8	42	6	0
21	12	32	4	1	22	10	36	4	1	23	22	24	2	-1	24	8	60	8	0	25	20	31	3	0
26	12	42	4	1	27	18	40	4	0	28	12	56	6	0	29	28	30	2	-1	30	8	72	8	-1
31	30	32	2	-1	32	16	63	6	0	33	20	48	4	1	34	16	54	4	1	35	24	48	4	1
36	12	91	9	0	37	36	38	2	-1	38	18	60	4	1	39	24	56	4	1	40	16	90	8	0
41	40	42	2	-1	42	12	96	8	-1	43	42	44	2	-1	44	20	84	6	0	45	24	78	6	0
46	22	72	4	1	47	46	48	2	-1	48	16	124	10	0	49	42	57	3	0	50	20	93	6	0
51	32	72	4	1	52	24	98	6	0	53	52	54	2	-1	54	18	120	8	0	55	40	72	4	1
56	24	120	8	0	57	36	80	4	1	58	28	90	4	1	59	58	60	2	-1	60	16	168	12	0
61	60	62	2	-1	62	30	96	4	1	63	36	104	6	0	64	32	127	7	0	65	48	84	4	1
66	20	144	8	-1	67	66	68	2	-1	68	32	126	6	0	69	44	96	4	1	70	24	144	8	-1
71	70	72	2	-1	72	24	195	12	0	73	72	74	2	-1	74	36	114	4	1	75	40	124	6	0
76	36	140	6	0	77	60	96	4	1	78	24	168	8	-1	79	78	80	2	-1	80	32	186	10	0
81	54	121	5	0	82	40	126	4	1	83	82	84	2	-1	84	24	224	12	0	85	64	108	4	1
86	42	132	4	1	87	56	120	4	1	88	40	180	8	0	89	88	90	2	-1	90	24	234	12	0
91	72	112	4	1	92	44	168	6	0	93	60	128	4	1	94	46	144	4	1	95	72	120	4	1
96	32	252	12	0	97	96	98	2	-1	98	42	171	6	0	99	60	156	6	0	100	40	217	9	0
101	100	102	2	-1	102	32	216	8	-1	103	102	104	2	-1	104	48	210	8	0	105	48	192	8	-1
106	52	162	4	1	107	106	108	2	-1	108	36	280	12	0	109	108	110	2	-1	110	40	216	8	-1
111	72	152	4	1	112	48	248	10	0	113	112	114	2	-1	114	36	240	8	-1	115	88	144	4	1
116	56	210	6	0	117	72	182	6	0	118	58	180	4	1	119	96	144	4	1	120	32	360	16	0
121	110	133	3	0	122	60	186	4	1	123	80	168	4	1	124	60	224	6	0	125	100	156	4	0
126	36	312	12	0	127	126	128	2	-1	128	64	255	8	0	129	84	176	4	1	130	48	252	8	-1
131	130	132	2	-1	132	40	336	12	0	133	108	160	4	1	134	66	204	4	1	135	72	240	8	0
136	64	270	8	0	137	136	138	2	-1	138	44	288	8	-1	139	138	140	2	-1	140	48	336	12	0
141	92	192	4	1	142	70	216	4	1	143	120	168	4	1	144	48	403	15	0	145	112	180	4	1
146	72	222	4	1	147	84	228	6	0	148	72	266	6	0	149	148	150	2	-1	150	40	372	12	0
151	150	152	2	-1	152	72	300	8	0	153	96	234	6	0	154	60	288	8	-1	155	120	192	4	1
156	48	392	12	0	157	156	158	2	-1	158	78	240	4	1	159	104	216	4	1	160	64	378	12	0
161	132	192	4	1	162	54	363	10	0	163	162	164	2	-1	164	80	294	6	0	165	80	288	8	-1
166	82	252	4	1	167	166	168	2	-1	168	48	480	16	0	169	156	183	3	0	170	64	324	8	-1
171	108	260	6	0	172	84	308	6	0	173	172	174	2	-1	174	56	360	8	-1	175	120	248	6	0
176	80	372	10	0	177	116	240	4	1	178	88	270	4	1	179	178	180	2	-1	180	48	546	18	0
181	180	182	2	-1	182	72	336	8	-1	183	120	248	4	1	184	88	360	8	0	185	144	228	4	1
186	60	384	8	-1	187	160	216	4	1	188	92	336	6	0	189	108	320	8	0	190	72	360	8	-1
191	190	192	2	-1	192	64	508	14	0	193	192	194	2	-1	194	96	294	4	1	195	96	336	8	-1
196	84	399	9	0	197	196	198	2	-1	198	60	468	12	0	199	198	200	2	-1	200	80	465	12	0
201	132	272	4	1	202	100	306	4	1	203	168	240	4	1	204	64	504	12	0	205	160	252	4	1
206	102	312	4	1	207	132	312	6	0	208	96	434	10	0	209	180	240	4	1	210	48	576	16	1
211	210	212	2	-1	212	104	378	6	0	213	140	288	4	1	214	106	324	4	1	215	168	264	4	1
216	72	600	16	0	217	180	256	4	1	218	108	330	4	1	219	144	296	4	1	220	80	504	12	0

n	φ	σ	τ	μ	n	φ	σ	τ	μ	n	φ	σ	τ	μ	n	φ	σ	τ	μ	n	φ	σ	τ	μ
221	192	252	4	1	222	72	456	8	-1	223	222	224	2	-1	224	96	504	12	0	225	120	403	9	0
226	112	342	4	1	227	226	228	2	-1	228	72	560	12	0	229	228	230	2	-1	230	88	432	8	-1
231	120	384	8	-1	232	112	450	8	0	233	232	234	2	-1	234	72	546	12	0	235	184	288	4	1
236	116	420	6	0	237	156	320	4	1	238	96	432	8	-1	239	238	240	2	-1	240	64	744	20	0
241	240	242	2	-1	242	110	399	6	0	243	162	364	6	0	244	120	434	6	0	245	168	342	6	0
246	80	504	8	-1	247	216	280	4	1	248	120	480	8	0	249	164	336	4	1	250	100	468	8	0
251	250	252	2	-1	252	72	728	18	0	253	220	288	4	1	254	126	384	4	1	255	128	432	8	-1
256	128	511	9	0	257	256	258	2	-1	258	84	528	8	-1	259	216	304	4	1	260	96	588	12	0
261	168	390	6	0	262	130	396	4	1	263	262	264	2	-1	264	80	720	16	0	265	208	324	4	1
266	108	480	8	-1	267	176	360	4	1	268	132	476	6	0	269	268	270	2	-1	270	72	720	16	0
271	270	272	2	-1	272	128	558	10	0	273	144	448	8	-1	274	136	414	4	1	275	200	372	6	0
276	88	672	12	0	277	276	278	2	-1	278	138	420	4	1	279	180	416	6	0	280	96	720	16	0
281	280	282	2	-1	282	92	576	8	-1	283	282	284	2	-1	284	140	504	6	0	285	144	480	8	-1
286	120	504	8	-1	287	240	336	4	1	288	96	819	18	0	289	272	307	3	0	290	112	540	8	-1
291	192	392	4	1	292	144	518	6	0	293	292	294	2	-1	294	84	684	12	0	295	232	360	4	1
296	144	570	8	0	297	180	480	8	0	298	148	450	4	1	299	264	336	4	1	300	80	868	18	0
301	252	352	4	1	302	150	456	4	1	303	200	408	4	1	304	144	620	10	0	305	240	372	4	1
306	96	702	12	0	307	306	308	2	-1	308	120	672	12	0	309	204	416	4	1	310	120	576	8	-1
311	310	312	2	-1	312	96	840	16	0	313	312	314	2	-1	314	156	474	4	1	315	144	624	12	0
316	156	560	6	0	317	316	318	2	-1	318	104	648	8	-1	319	280	360	4	1	320	128	762	14	0
321	212	432	4	1	322	132	576	8	-1	323	288	360	4	1	324	108	847	15	0	325	240	434	6	0
326	162	492	4	1	327	216	440	4	1	328	160	630	8	0	329	276	384	4	1	330	80	864	16	1
331	330	332	2	-1	332	164	588	6	0	333	216	494	6	0	334	166	504	4	1	335	264	408	4	1
336	96	992	20	0	337	336	338	2	-1	338	156	549	6	0	339	224	456	4	1	340	128	756	12	0
341	300	384	4	1	342	108	780	12	0	343	294	400	4	0	344	168	660	8	0	345	176	576	8	-1
346	172	522	4	1	347	346	348	2	-1	348	112	840	12	0	349	348	350	2	-1	350	120	744	12	0
351	216	560	8	0	352	160	756	12	0	353	352	354	2	-1	354	116	720	8	-1	355	280	432	4	1
356	176	630	6	0	357	192	576	8	-1	358	178	540	4	1	359	358	360	2	-1	360	96	1170	24	0
361	342	381	3	0	362	180	546	4	1	363	220	532	6	0	364	144	784	12	0	365	288	444	4	1
366	120	744	8	-1	367	366	368	2	-1	368	176	744	10	0	369	240	546	6	0	370	144	684	8	-1
371	312	432	4	1	372	120	896	12	0	373	372	374	2	-1	374	160	648	8	-1	375	200	624	8	0
376	184	720	8	0	377	336	420	4	1	378	108	960	16	0	379	378	380	2	-1	380	144	840	12	0
381	252	512	4	1	382	190	576	4	1	383	382	384	2	-1	384	128	1020	16	0	385	240	576	8	-1
386	192	582	4	1	387	252	572	6	0	388	192	686	6	0	389	388	390	2	-1	390	96	1008	16	1
391	352	432	4	1	392	168	855	12	0	393	260	528	4	1	394	196	594	4	1	395	312	480	4	1
396	120	1092	18	0	397	396	398	2	-1	398	198	600	4	1	399	216	640	8	-1	400	160	961	15	0
401	400	402	2	-1	402	132	816	8	-1	403	360	448	4	1	404	200	714	6	0	405	216	726	10	0
406	168	720	8	-1	407	360	456	4	1	408	128	1080	16	0	409	408	410	2	-1	410	160	756	8	0
411	272	552	4	1	412	204	728	6	0	413	348	480	4	1	414	132	936	12	0	415	328	504	4	1
416	192	882	12	0	417	276	560	4	1	418	180	720	8	-1	419	418	420	2	-1	420	96	1344	24	0
421	420	422	2	-1	422	210	636	4	1	423	276	624	6	0	424	208	810	8	0	425	320	558	6	0
426	140	864	8	-1	427	360	496	4	1	428	212	756	6	0	429	240	672	8	-1	430	168	792	8	-1
431	430	432	2	-1	432	144	1240	20	0	433	432	434	2	-1	434	180	768	8	-1	435	224	720	8	-1
436	216	770	6	0	437	396	480	4	1	438	144	888	8	-1	439	438	440	2	-1	440	160	1080	16	0
441	252	741	9	0	442	192	756	8	-1	443	442	444	2	-1	444	144	1064	12	0	445	352	540	4	1
446	222	672	4	1	447	296	600	4	1	448	192	1016	14	0	449	448	450	2	-1	450	120	1209	18	0
451	400	504	4	1	452	224	798	6	0	453	300	608	4	1	454	226	684	4	1	455	288	672	8	-1
456	144	1200	16	0	457	456	458	2	-1	458	228	690	4	1	459	288	720	8	0	460	176	1008	12	0

n	φ	σ	τ	μ	n	φ	σ	τ	μ	n	φ	σ	τ	μ	n	φ	σ	τ	μ	n	φ	σ	τ	μ
461	460	462	2	-1	462	120	1152	16	1	463	462	464	2	-1	464	224	930	10	0	465	240	768	8	-1
466	232	702	4	1	467	466	468	2	-1	468	144	1274	18	0	469	396	544	4	1	470	184	864	8	-1
471	312	632	4	1	472	232	900	8	0	473	420	528	4	1	474	156	960	8	-1	475	360	620	6	0
476	192	1008	12	0	477	312	702	6	0	478	238	720	4	1	479	478	480	2	-1	480	128	1512	24	0
481	432	532	4	1	482	240	726	4	1	483	264	768	8	-1	484	220	931	9	0	485	384	588	4	1
486	162	1092	12	0	487	486	488	2	-1	488	240	930	8	0	489	324	656	4	1	490	168	1026	12	0
491	490	492	2	-1	492	160	1176	12	0	493	448	540	4	1	494	216	840	8	-1	495	240	936	12	0
496	240	992	10	0	497	420	576	4	1	498	164	1008	8	-1	499	498	500	2	-1	500	200	1092	12	0
501	332	672	4	1	502	250	756	4	1	503	502	504	2	-1	504	144	1560	24	0	505	400	612	4	1
506	220	864	8	-1	507	312	732	6	0	508	252	896	6	0	509	508	510	2	-1	510	128	1296	16	1
511	432	592	4	1	512	256	1023	10	0	513	324	800	8	0	514	256	774	4	1	515	408	624	4	1
516	168	1232	12	0	517	460	576	4	1	518	216	912	8	-1	519	344	696	4	1	520	192	1260	16	0
521	520	522	2	-1	522	168	1170	12	0	523	522	524	2	-1	524	260	924	6	0	525	240	992	12	0
526	262	792	4	1	527	480	576	4	1	528	160	1488	20	0	529	506	553	3	0	530	208	972	8	-1
531	348	780	6	0	532	216	1120	12	0	533	480	588	4	1	534	176	1080	8	-1	535	424	648	4	1
536	264	1020	8	0	537	356	720	4	1	538	268	810	4	1	539	420	684	6	0	540	144	1680	24	0
541	540	542	2	-1	542	270	816	4	1	543	360	728	4	1	544	256	1134	12	0	545	432	660	4	1
546	144	1344	16	1	547	546	548	2	-1	548	272	966	6	0	549	360	806	6	0	550	200	1116	12	0
551	504	600	4	1	552	176	1440	16	0	553	468	640	4	1	554	276	834	4	1	555	288	912	8	-1
556	276	980	6	0	557	556	558	2	-1	558	180	1248	12	0	559	504	616	4	1	560	192	1488	20	0
561	320	864	8	-1	562	280	846	4	1	563	562	564	2	-1	564	184	1344	12	0	565	448	684	4	1
566	282	852	4	1	567	324	968	10	0	568	280	1080	8	0	569	568	570	2	-1	570	144	1440	16	1
571	570	572	2	-1	572	240	1176	12	0	573	380	768	4	1	574	240	1008	8	-1	575	440	744	6	0
576	192	1651	21	0	577	576	578	2	-1	578	272	921	6	0	579	384	776	4	1	580	224	1260	12	0
581	492	672	4	1	582	192	1176	8	-1	583	520	648	4	1	584	288	1110	8	0	585	288	1092	12	0
586	292	882	4	1	587	586	588	2	-1	588	168	1596	18	0	589	540	640	4	1	590	232	1080	8	-1
591	392	792	4	1	592	288	1178	10	0	593	592	594	2	-1	594	180	1440	16	0	595	384	864	8	-1
596	296	1050	6	0	597	396	800	4	1	598	264	1008	8	-1	599	598	600	2	-1	600	160	1860	24	0
601	600	602	2	-1	602	252	1056	8	-1	603	396	884	6	0	604	300	1064	6	0	605	440	798	6	0
606	200	1224	8	-1	607	606	608	2	-1	608	288	1260	12	0	609	336	960	8	-1	610	240	1116	8	-1
611	552	672	4	1	612	192	1638	18	0	613	612	614	2	-1	614	306	924	4	1	615	320	1008	8	-1
616	240	1440	16	0	617	616	618	2	-1	618	204	1248	8	-1	619	618	620	2	-1	620	240	1344	12	0
621	396	960	8	0	622	310	936	4	1	623	528	720	4	1	624	192	1736	20	0	625	500	781	5	0
626	312	942	4	1	627	360	960	8	-1	628	312	1106	6	0	629	576	684	4	1	630	144	1872	24	0
631	630	632	2	-1	632	312	1200	8	0	633	420	848	4	1	634	316	954	4	1	635	504	768	4	1
636	208	1512	12	0	637	504	798	6	0	638	280	1080	8	-1	639	420	936	6	0	640	256	1530	16	0
641	640	642	2	-1	642	212	1296	8	-1	643	642	644	2	-1	644	264	1344	12	0	645	336	1056	8	-1
646	288	1080	8	-1	647	646	648	2	-1	648	216	1815	20	0	649	580	720	4	1	650	240	1302	12	0
651	360	1024	8	-1	652	324	1148	6	0	653	652	654	2	-1	654	216	1320	8	-1	655	520	792	4	1
656	320	1302	10	0	657	432	962	6	0	658	276	1152	8	-1	659	658	660	2	-1	660	160	2016	24	0
661	660	662	2	-1	662	330	996	4	1	663	384	1008	8	-1	664	328	1260	8	0	665	432	960	8	-1
666	216	1482	12	0	667	616	720	4	1	668	332	1176	6	0	669	444	896	4	1	670	264	1224	8	-1
671	600	744	4	1	672	192	2016	24	0	673	672	674	2	-1	674	336	1014	4	1	675	360	1240	12	0
676	312	1281	9	0	677	676	678	2	-1	678	224	1368	8	-1	679	576	784	4	1	680	256	1620	16	0
681	452	912	4	1	682	300	1152	8	-1	683	682	684	2	-1	684	216	1820	18	0	685	544	828	4	1
686	294	1200	8	0	687	456	920	4	1	688	336	1364	10	0	689	624	756	4	1	690	176	1728	16	1
691	690	692	2	-1	692	344	1218	6	0	693	360	1248	12	0	694	346	1044	4	1	695	552	840	4	1
696	224	1800	16	0	697	640	756	4	1	698	348	1050	4	1	699	464	936	4	1	700	240	1736	18	0

Section 4.6 ARITHMETIC FUNCTIONS

n	φ	σ	τ	μ	n	φ	σ	τ	μ	n	φ	σ	τ	μ	n	φ	σ	τ	μ	n	φ	σ	τ	μ
701	700	702	2	-1	702	216	1680	16	0	703	648	760	4	1	704	320	1524	14	0	705	368	1152	8	-1
706	352	1062	4	1	707	600	816	4	1	708	232	1680	12	0	709	708	710	2	-1	710	280	1296	8	-1
711	468	1040	6	0	712	352	1350	8	0	713	660	768	4	1	714	192	1728	16	1	715	480	1008	8	-1
716	356	1260	6	0	717	476	960	4	1	718	358	1080	4	1	719	718	720	2	-1	720	192	2418	30	0
721	612	832	4	1	722	342	1143	6	0	723	480	968	4	1	724	360	1274	6	0	725	560	930	6	0
726	220	1596	12	0	727	726	728	2	-1	728	288	1680	16	0	729	486	1093	7	0	730	288	1332	8	-1
731	672	792	4	1	732	240	1736	12	0	733	732	734	2	-1	734	366	1104	4	1	735	336	1368	12	0
736	352	1512	12	0	737	660	816	4	1	738	240	1638	12	0	739	738	740	2	-1	740	288	1596	12	0
741	432	1120	8	-1	742	312	1296	8	-1	743	742	744	2	-1	744	240	1920	16	0	745	592	900	4	1
746	372	1122	4	1	747	492	1092	6	0	748	320	1512	12	0	749	636	864	4	1	750	200	1872	16	0
751	750	752	2	-1	752	368	1488	10	0	753	500	1008	4	1	754	336	1260	8	-1	755	600	912	4	1
756	216	2240	24	0	757	756	758	2	-1	758	378	1140	4	1	759	440	1152	8	-1	760	288	1800	16	0
761	760	762	2	-1	762	252	1536	8	-1	763	648	880	4	1	764	380	1344	6	0	765	384	1404	12	0
766	382	1152	4	1	767	696	840	4	1	768	256	2044	18	0	769	768	770	2	-1	770	240	1728	16	1
771	512	1032	4	1	772	384	1358	6	0	773	772	774	2	-1	774	252	1716	12	0	775	600	992	6	0
776	384	1470	8	0	777	432	1216	8	-1	778	388	1170	4	1	779	720	840	4	1	780	192	2352	24	0
781	700	864	4	1	782	352	1296	8	-1	783	504	1200	8	0	784	336	1767	15	0	785	624	948	4	1
786	260	1584	8	-1	787	786	788	2	-1	788	392	1386	6	0	789	524	1056	4	1	790	312	1440	8	-1
791	672	912	4	1	792	240	2340	24	0	793	720	868	4	1	794	396	1194	4	1	795	416	1296	8	-1
796	396	1400	6	0	797	796	798	2	-1	798	216	1920	16	1	799	736	864	4	1	800	320	1953	18	0
801	528	1170	6	0	802	400	1206	4	1	803	720	888	4	1	804	264	1904	12	0	805	528	1152	8	-1
806	360	1344	8	-1	807	536	1080	4	1	808	400	1530	8	0	809	808	810	2	-1	810	216	2178	20	0
811	810	812	2	-1	812	336	1680	12	0	813	540	1088	4	1	814	360	1368	8	-1	815	648	984	4	1
816	256	2232	20	0	817	756	880	4	1	818	408	1230	4	1	819	432	1456	12	0	820	320	1764	12	0
821	820	822	2	-1	822	272	1656	8	-1	823	822	824	2	-1	824	408	1560	8	0	825	400	1488	12	0
826	348	1440	8	-1	827	826	828	2	-1	828	264	2184	18	0	829	828	830	2	-1	830	328	1512	8	-1
831	552	1112	4	1	832	384	1778	14	0	833	672	1026	6	0	834	276	1680	8	-1	835	664	1008	4	1
836	360	1680	12	0	837	540	1280	8	0	838	418	1260	4	1	839	838	840	2	-1	840	192	2880	32	0
841	812	871	3	0	842	420	1266	4	1	843	560	1128	4	1	844	420	1484	6	0	845	624	1098	6	0
846	276	1872	12	0	847	660	1064	6	0	848	416	1674	10	0	849	564	1136	4	1	850	320	1674	12	0
851	792	912	4	1	852	280	2016	12	0	853	852	854	2	-1	854	360	1488	8	-1	855	432	1560	12	0
856	424	1620	8	0	857	856	858	2	-1	858	240	2016	16	1	859	858	860	2	-1	860	336	1848	12	0
861	480	1344	8	-1	862	430	1296	4	1	863	862	864	2	-1	864	288	2520	24	0	865	688	1044	4	1
866	432	1302	4	1	867	544	1228	6	0	868	360	1792	12	0	869	780	960	4	1	870	224	2160	16	1
871	792	952	4	1	872	432	1650	8	0	873	576	1274	6	0	874	396	1440	8	-1	875	600	1248	8	0
876	288	2072	12	0	877	876	878	2	-1	878	438	1320	4	1	879	584	1176	4	1	880	320	2232	20	0
881	880	882	2	-1	882	252	2223	18	0	883	882	884	2	-1	884	384	1764	12	0	885	464	1440	8	-1
886	442	1332	4	1	887	886	888	2	-1	888	288	2280	16	0	889	756	1024	4	1	890	352	1620	8	-1
891	540	1452	10	0	892	444	1568	6	0	893	828	960	4	1	894	296	1800	8	-1	895	712	1080	4	1
896	384	2040	16	0	897	528	1344	8	-1	898	448	1350	4	1	899	840	960	4	1	900	240	2821	27	0
901	832	972	4	1	902	400	1512	8	-1	903	504	1408	8	-1	904	448	1710	8	0	905	720	1092	4	1
906	300	1824	8	-1	907	906	908	2	-1	908	452	1596	6	0	909	600	1326	6	0	910	288	2016	16	1
911	910	912	2	-1	912	288	2480	20	0	913	820	1008	4	1	914	456	1374	4	1	915	480	1488	8	-1
916	456	1610	6	0	917	780	1056	4	1	918	288	2160	16	0	919	918	920	2	-1	920	352	2160	16	0
921	612	1232	4	1	922	460	1386	4	1	923	840	1008	4	1	924	240	2688	24	0	925	720	1178	6	0
926	462	1392	4	1	927	612	1352	6	0	928	448	1890	12	0	929	928	930	2	-1	930	240	2304	16	1
931	756	1140	6	0	932	464	1638	6	0	933	620	1248	4	1	934	466	1404	4	1	935	640	1296	8	-1
936	288	2730	24	0	937	936	938	2	-1	938	396	1632	8	-1	939	624	1256	4	1	940	368	2016	12	0
941	940	942	2	-1	942	312	1896	8	-1	943	880	1008	4	1	944	464	1860	10	0	945	432	1920	16	0
946	420	1584	8	-1	947	946	948	2	-1	948	312	2240	12	0	949	864	1036	4	1	950	360	1860	12	0

n	ϕ	σ	τ	μ	n	ϕ	σ	τ	μ	n	ϕ	σ	τ	μ	n	ϕ	σ	τ	μ	n	ϕ	σ	τ	μ
951	632	1272	4	1	952	384	2160	16	0	953	952	954	2	-1	954	312	2106	12	0	955	760	1152	4	1
956	476	1680	6	0	957	560	1440	8	-1	958	478	1440	4	1	959	816	1104	4	1	960	256	3048	28	0
961	930	993	3	0	962	432	1596	8	-1	963	636	1404	6	0	964	480	1694	6	0	965	768	1164	4	1
966	264	2304	16	1	967	966	968	2	-1	968	440	1995	12	0	969	576	1440	8	-1	970	384	1764	8	-1
971	970	972	2	-1	972	324	2548	18	0	973	828	1120	4	1	974	486	1464	4	1	975	480	1736	12	0
976	480	1922	10	0	977	976	978	2	-1	978	324	1968	8	-1	979	880	1080	4	1	980	336	2394	18	0
981	648	1430	6	0	982	490	1476	4	1	983	982	984	2	-1	984	320	2520	16	0	985	784	1188	4	1
986	448	1620	8	-1	987	552	1536	8	-1	988	432	1960	12	0	989	924	1056	4	1	990	240	2808	24	0
991	990	992	2	-1	992	480	2016	12	0	993	660	1328	4	1	994	420	1728	8	-1	995	792	1200	4	1
996	328	2352	12	0	997	996	998	2	-1	998	498	1500	4	1	999	648	1520	8	0	1000	400	2340	16	0

12. $\sum_{k=1}^{n} \sigma(k) = \frac{\pi^2 n^2}{12} + O(n \log n)$

13. $\sum_{k=1}^{n} \tau(k) = n \log n + (2\gamma - 1)n + O(\sqrt{n})$, where γ is Euler's constant.

14. If m and n are amicable, then m is the sum of the proper divisors of n, and vice versa.

Examples:

1. Table 1 lists the values of $\sigma(n)$ and $\tau(n)$ for $1 \leq n \leq 1000$.

2. To find $\tau(720)$, note that $\tau(720) = \tau(2^4 \cdot 3^2 \cdot 5) = (4+1)(2+1)(1+1) = 30$.

3. To find $\sigma(200)$ note that $\sigma(200) = \sigma(2^3 5^2) = \frac{2^4 - 1}{2 - 1} \cdot \frac{5^3 - 1}{5 - 1} = 15 \cdot 31 = 465$.

4. The integers 6 and 28 are perfect; the integers 9 and 16 are deficient; the integers 12 and 945 are abundant.

5. The integers 220 and 284 form the smallest pair of amicable numbers.

4.6.4 THE MÖBIUS FUNCTION AND OTHER IMPORTANT ARITHMETIC FUNCTIONS

Definitions:

If n is a positive integer, $\mu(n)$, the value of the **Möbius function**, is defined by:

$$\mu(n) = \begin{cases} 1, & \text{if } n = 1 \\ 0, & \text{if } n \text{ has a square factor larger than 1} \\ (-1)^s, & \text{if } n \text{ is squarefree and is the product of } s \text{ different primes.} \end{cases}$$

If $n > 1$ is a positive integer, with prime-power factorization $p_1^{a_1} p_2^{a_2} \ldots p_m^{a_m}$, then $\lambda(n)$, the value of **Liouville's function** at n, is given by $\lambda(n) = (-1)^{a_1 + a_2 + \cdots + a_m}$, with $\lambda(1) = 1$.

If n is a positive integer with prime-power factorization $n = p_1^{a_1} p_2^{a_2} \ldots p_m^{a_m}$, then the arithmetic functions Ω and ω are defined by $\Omega(1) = \omega(1) = 0$ and for $n > 1$, $\Omega(n) = \sum_{i=1}^{m} a_i$ and $\omega(n) = m$. That is, $\Omega(n)$ is the sum of the exponents in the prime-power factorization of n and $\omega(n)$ is the number of distinct primes in the prime-power factorization of n.

Facts:

1. The Möbius function is multiplicative, but not completely multiplicative.

2. *Möbius inversion formula*: If f is an arithmetic function and $F(n) = \sum_{d \mid n} f(d)$, then $f(n) = \sum_{d \mid n} \mu(d) F(\frac{n}{d})$.

3. If n is a positive integer, then $\phi(n) = \sum_{d|n} \mu(d)\frac{n}{d}$.
4. If f is multiplicative, then $\sum_{d|n} \mu(d)f(d) = \prod_{p|n}(1 - f(p))$.
5. If f is multiplicative, then $\sum_{d|n} \mu(d)^2 f(d) = \prod_{p|n}(1 + f(p))$.
6. If n is positive integer then $\sum_{d|n} \mu(d) = \begin{cases} 1 & \text{if } n = 1; \\ 0 & \text{if } n > 1. \end{cases}$
7. If n is a positive integer, then $\sum_{d|n} \lambda(d) = \begin{cases} 1 & \text{if } n \text{ is a perfect square}; \\ 0 & \text{if } n \text{ is not a perfect square}. \end{cases}$
8. In 1897 Mertens showed that $|\sum_{k=1}^{n} \mu(k)| < \sqrt{n}$ for all positive integers n not exceeding 10,000 and conjectured that this inequality holds for all positive integers n. However, in 1985 Odlyzko and teRiele disproved this conjecture, which went by the name *Mertens' conjecture* without giving an explicit integer n for which the conjecture fails. In 1987 Pintz showed that there is at least one counterexample n with $n \leq 10^{65}$, again without giving an explicit counterexample n. Finding such an integer n requires more computing power than is currently available.
9. Liouville's function is completely multiplicative.
10. The function ω is additive and the function Ω is completely additive.

Examples:

1. $\mu(12) = 0$ since $2^2 | 12$ and $\mu(105) = \mu(3 \cdot 5 \cdot 7) = (-1)^3 = -1$.
2. $\lambda(720) = \lambda(2^4 \cdot 3^2 \cdot 5) = (-1)^{4+2+1} = (-1)^7 = -1$.
3. $\Omega(720) = \Omega(2^4 \cdot 3^2 \cdot 5) = 4 + 2 + 1 = 7$ and $\omega(720) = \omega(2^4 \cdot 3^2 \cdot 5) = 3$.

4.6.5 DIRICHLET PRODUCTS

Definitions:

If f and g are arithmetic functions, then the **Dirichlet product** of f and g is the function $f \star g$ defined by $(f \star g)(n) = \sum_{d|n} f(d)g(\frac{n}{d})$.

If f and g are arithmetic functions such that $f \star g = g \star f = I$, where $I(n) = \lfloor \frac{1}{n} \rfloor$, then g is the **Dirichlet inverse** of f.

Facts:

1. If f and g are arithmetic functions, then $f \star g = g \star f$.
2. If f, g, and h are arithmetic functions, then $(f \star g) \star h = f \star (g \star h)$.
3. If f, g, and h are arithmetic functions, then $f \star (g + h) = (f \star g) + (f \star h)$.
4. Because of Facts 1–3, the set of arithmetic functions with the operations of Dirichlet product and ordinary addition of functions forms a ring. (See Chapter 5.)
5. If f is an arithmetic function with $f(1) \neq 0$, then there is a unique Dirichlet inverse of f, which is written as f^{-1}. Furthermore, f^{-1} is given by the recursive formulas $f^{-1}(1) = \frac{1}{f(1)}$ and $f^{-1} = -\frac{1}{f(1)} \sum_{\substack{d|n \\ d>n}} f(\frac{n}{d}) f^{-1}(d)$ for $n > 1$.
6. The set of all arithmetic functions f with $f(1) \neq 0$ forms an abelian group with respect to the operation $\star$, where the identity element is the function I.
7. If f and g are arithmetic functions with $f(1) \neq 0$ and $g(1) \neq 0$, then $(f \star g)^{-1} = f^{-1} \star g^{-1}$.

8. If u is the arithmetic function with $u(n) = 1$ for all positive integers n, then $\mu \star u = I$, so $u = \mu^{-1}$ and $\mu = u^{-1}$.
9. If f is a multiplicative function, then f is completely multiplicative if and only if $f^{-1}(n) = \mu(n)f(n)$ for all positive integers n.
10. If f and g are multiplicative functions, then $f \star g$ is also multiplicative.
11. If f and g are arithmetic functions and both f and $f \star g$ are multiplicative, then g is also multiplicative.
12. If f is multiplicative, then f^{-1} exists and is multiplicative.

Examples:
1. The identity $\phi(n) = \sum_{d|n} \mu(d)\frac{n}{d}$ (§4.6.4 Fact 3) implies that $\phi = \mu \star N$ where N is the multiplicative function $N(n) = n$.
2. Since the function N is completely multiplicative, $N^{-1} = \mu N$ by Fact 9.
3. From Example 1 and Facts 7 and 8, it follows that $\phi^{-1} = \mu^{-1} \star \mu N = \mu \star \mu N$. Hence $\phi^{-1}(n) = \sum_{d|n} d\mu(d)$.

4.7 PRIMITIVE ROOTS AND QUADRATIC RESIDUES

A primitive root of an integer, when it exists, is an integer whose powers run through a complete system of residues modulo this integer. When a primitive root exists, it is possible to use the theory of indices to solve certain congruences. This section provides the information needed to understand and employ primitive roots.

The question of which integers are perfect squares modulo a prime is one that has been studied extensively. An integer that is a perfect square modulo n is called a quadratic residue of n. The law of quadratic reciprocity provides a surprising link between the answer to the question of whether a prime p is a perfect square modulo a prime q and the answer to the question of whether q is a perfect square modulo p. This section provides information that helps determine whether an integer is a quadratic residue modulo a given integer n.

There are important applications of the topics covered in this section, including applications to public key cryptography and authentication schemes. (See Chapter 14.)

4.7.1 PRIMITIVE ROOTS

Definitions:

If a and m are relatively prime positive integers, then the **order of a modulo m**, denoted $\text{ord}_m a$, is the least positive integer x such that $a^x \equiv 1 \pmod{m}$.

If r and n are relatively prime integers and n is positive, then r is a **primitive root modulo m** if $\text{ord}_n r = \phi(n)$. A primitive root modulo m is also said to be a primitive root of m and m is said to have a primitive root.

If m is a positive integer, then the **minimum universal exponent modulo m** is the smallest positive integer $\lambda(m)$ for which $a^{\lambda(m)} \equiv 1 \pmod{m}$ for all integers a relatively prime to m.

Section 4.7 PRIMITIVE ROOTS AND QUADRATIC RESIDUES

Facts:

1. The positive integer n, with $n > 1$, has a primitive root if and only if $n = 2, 4, p^t$ or $2p^t$ where p is an odd prime and t is a positive integer.

2. There are $\phi(d)$ incongruent integers modulo p if p is prime and d is a positive divisor of $p - 1$.

3. There are $\phi(p - 1)$ primitive roots of p if p is a prime.

4. If the positive integer m has a primitive root, then it has a total of $\phi(\phi(m))$ incongruent primitive roots.

5. If r is a primitive root of the odd prime p, then either r or $r + p$ is a primitive root modulo p^2.

6. If r is a primitive root of p^2, where p is prime, then r is a primitive root of p^k for all positive integers k.

7. It is an unsettled conjecture (stated by E. Artin) whether 2 is a primitive root of infinitely many primes. More generally, given any prime p it is unknown whether p is a primitive root of infinitely many primes.

8. It is known that given any three primes, at least one of these primes is a primitive root of infinitely many primes. [GuMu84]

9. Given a set of n primes, $p_1, p_2, \ldots, p_n$, there are $\prod_{k=1}^{n} \phi(p_k - 1)$ integers x with $1 < x \leq \prod_{k=1}^{n} p_k$ such that x is a primitive root of p_k for $k = 1, 2, \ldots, n$. Such an integer x is a called a *common primitive root* of the primes $p_1, \ldots, p_n$.

10. Let g_p denote the smallest positive integer that is a primitive root modulo p where p is a prime. It is known that g_p is not always small; in particular it has been shown by Fridlender and Salié ([Ri96]) that there is a positive constant C such that $g_p > C \log p$ for infinitely many primes p.

11. Burgess has shown that g_p does not grow too rapidly; in particular he showed that $g_p \leq Cp^{\frac{1}{4}+\epsilon}$ for $\epsilon > 0$, C a constant, p sufficiently large. [Ri96]

12. The minimum universal exponent modulo the powers of 2 are: $\lambda(2) = 1$, $\lambda(2^2) = 2$, and $\lambda(2^k) = 2^{k-2}$ for $k = 3, 4, \ldots$.

13. If m is a positive integer with prime-power factorization $2^k q_1^{a_1} \ldots q_r^{a_r}$ where k is a nonnegative integer, then the least universal exponent of m is given by $\lambda(m) = \text{lcm}(\lambda(2^k), \phi(q_1^{a_1}), \ldots, \phi(q_r^{a_r}))$.

14. For every positive integer m, there is an integer a such that $\text{ord}_m a = \lambda(m)$.

15. There are six positive integers m with $\lambda(m) = 2$: $m = 3, 4, 6, 8, 12, 24$.

16. Table 1 displays the least primitive root of each prime less than 10,000.

Examples:

1. Since $2^1 \equiv 2, 2^2 \equiv 4$, and $2^3 \equiv 1 \pmod{7}$, it follows that $\text{ord}_7 2 = 3$.

2. The integers 2, 6, 7, and 8 form a complete set of incongruent primitive roots modulo 11.

3. The integer 10 is a primitive root of 487, but it is not a primitive root of 487^2.

4. There are $\phi(6)\phi(10) = 2 \cdot 4 = 8$ common primitive roots of 7 and 11 between 1 and $7 \cdot 11 = 77$. They are the integers 17, 19, 24, 40, 52, 61, 68, and 73.

5. From Facts 12 and 13 it follows that the minimum universal exponent of 1200 is $\lambda(7,200) = \lambda(2^5 \cdot 3^2 \cdot 5^2) = \text{lcm}(2^3, \phi(3^2), \phi(5^2)) = \text{lcm}(8, 6, 20) = 120$.

Table 1 Primes and primitive roots.

For each prime $p < 10{,}000$ the least primitive root ω is given.

p	ω	p	ω	p	ω	p	ω	p	ω	p	ω	p	ω	p	ω	p	ω
3	2	5	2	7	3	11	2	13	2	17	3	19	2	23	5	29	2
31	3	37	2	41	6	43	3	47	5	53	2	59	2	61	2	67	2
71	7	73	5	79	3	83	2	89	3	97	5	101	2	103	5	107	2
109	6	113	3	127	3	131	2	137	3	139	2	149	2	151	6	157	5
163	2	167	5	173	2	179	2	181	2	191	19	193	5	197	2	199	3
211	2	223	3	227	2	229	6	233	3	239	7	241	7	251	6	257	3
263	5	269	2	271	6	277	5	281	3	283	3	293	2	307	5	311	17
313	10	317	2	331	3	337	10	347	2	349	2	353	3	359	7	367	6
373	2	379	2	383	5	389	2	397	5	401	3	409	21	419	2	421	2
431	7	433	5	439	15	443	2	449	3	457	13	461	2	463	3	467	2
479	13	487	3	491	2	499	7	503	5	509	2	521	3	523	2	541	2
547	2	557	2	563	2	569	3	571	3	577	5	587	2	593	3	599	7
601	7	607	3	613	2	617	3	619	2	631	3	641	3	643	11	647	5
653	2	659	2	661	2	673	5	677	2	683	5	691	3	701	2	709	2
719	11	727	5	733	6	739	3	743	5	751	3	757	2	761	6	769	11
773	2	787	2	797	2	809	3	811	3	821	2	823	3	827	2	829	2
839	11	853	2	857	3	859	2	863	5	877	2	881	3	883	2	887	5
907	2	911	17	919	7	929	3	937	5	941	2	947	2	953	3	967	5
971	6	977	3	983	5	991	6	997	7	1009	11	1013	3	1019	2	1021	10
1031	14	1033	5	1039	3	1049	3	1051	7	1061	2	1063	3	1069	6	1087	3
1091	2	1093	5	1097	3	1103	5	1109	2	1117	2	1123	2	1129	11	1151	17
1153	5	1163	5	1171	2	1181	7	1187	2	1193	3	1201	11	1213	2	1217	3
1223	5	1229	2	1231	3	1237	2	1249	7	1259	2	1277	2	1279	3	1283	2
1289	6	1291	2	1297	10	1301	2	1303	6	1307	2	1319	13	1321	13	1327	3
1361	3	1367	5	1373	2	1381	2	1399	13	1409	3	1423	3	1427	2	1429	6
1433	3	1439	7	1447	3	1451	2	1453	2	1459	3	1471	6	1481	3	1483	2
1487	5	1489	14	1493	2	1499	2	1511	11	1523	2	1531	2	1543	5	1549	2
1553	3	1559	19	1567	3	1571	2	1579	3	1583	5	1597	11	1601	3	1607	5
1609	7	1613	3	1619	2	1621	2	1627	3	1637	2	1657	11	1663	3	1667	2
1669	2	1693	2	1697	3	1699	3	1709	3	1721	3	1723	3	1733	2	1741	2
1747	2	1753	7	1759	6	1777	5	1783	10	1787	2	1789	6	1801	11	1811	6
1823	5	1831	3	1847	5	1861	2	1867	2	1871	14	1873	10	1877	2	1879	6
1889	3	1901	2	1907	2	1913	3	1931	2	1933	5	1949	2	1951	3	1973	2
1979	2	1987	2	1993	5	1997	2	1999	3	2003	5	2011	3	2017	5	2027	2
2029	2	2039	7	2053	2	2063	5	2069	2	2081	3	2083	2	2087	5	2089	7
2099	2	2111	7	2113	5	2129	3	2131	2	2137	10	2141	2	2143	3	2153	3
2161	23	2179	7	2203	5	2207	5	2213	2	2221	2	2237	2	2239	3	2243	2
2251	7	2267	2	2269	2	2273	3	2281	7	2287	19	2293	2	2297	5	2309	2
2311	3	2333	2	2339	2	2341	7	2347	3	2351	13	2357	2	2371	2	2377	5
2381	3	2383	5	2389	2	2393	3	2399	11	2411	6	2417	3	2423	5	2437	2
2441	6	2447	5	2459	2	2467	2	2473	5	2477	2	2503	3	2521	17	2531	2
2539	2	2543	5	2549	2	2551	6	2557	2	2579	2	2591	7	2593	7	2609	3
2617	5	2621	2	2633	3	2647	3	2657	3	2659	2	2663	5	2671	7	2677	2
2683	2	2687	5	2689	19	2693	2	2699	2	2707	2	2711	7	2713	5	2719	3

p	ω	p	ω	p	ω	p	ω	p	ω	p	ω	p	ω	p	ω	p	ω
2729	3	2731	3	2741	2	2749	6	2753	3	2767	3	2777	3	2789	2	2791	6
2797	2	2801	3	2803	2	2819	2	2833	5	2837	2	2843	2	2851	2	2857	11
2861	2	2879	7	2887	5	2897	3	2903	5	2909	2	2917	5	2927	5	2939	2
2953	13	2957	2	2963	2	2969	3	2971	10	2999	17	3001	14	3011	2	3019	2
3023	5	3037	2	3041	3	3049	11	3061	6	3067	2	3079	6	3083	2	3089	3
3109	6	3119	7	3121	7	3137	3	3163	3	3167	5	3169	7	3181	7	3187	2
3191	11	3203	2	3209	3	3217	5	3221	10	3229	6	3251	6	3253	2	3257	3
3259	3	3271	3	3299	2	3301	6	3307	2	3313	10	3319	6	3323	2	3329	3
3331	3	3343	5	3347	2	3359	11	3361	22	3371	2	3373	5	3389	3	3391	3
3407	5	3413	2	3433	5	3449	3	3457	7	3461	2	3463	3	3467	2	3469	2
3491	2	3499	2	3511	7	3517	2	3527	5	3529	17	3533	2	3539	2	3541	7
3547	2	3557	2	3559	3	3571	2	3581	2	3583	3	3593	3	3607	5	3613	2
3617	3	3623	5	3631	15	3637	2	3643	2	3659	2	3671	13	3673	5	3677	2
3691	2	3697	5	3701	2	3709	2	3719	7	3727	3	3733	2	3739	7	3761	3
3767	5	3769	7	3779	2	3793	5	3797	2	3803	2	3821	3	3823	3	3833	3
3847	5	3851	2	3853	2	3863	5	3877	2	3881	13	3889	11	3907	2	3911	13
3917	2	3919	3	3923	2	3929	3	3931	2	3943	3	3947	2	3967	6	3989	2
4001	3	4003	2	4007	5	4013	2	4019	2	4021	2	4027	3	4049	3	4051	10
4057	5	4073	3	4079	11	4091	2	4093	2	4099	2	4111	12	4127	5	4129	13
4133	2	4139	2	4153	5	4157	2	4159	3	4177	5	4201	11	4211	6	4217	3
4219	2	4229	2	4231	3	4241	2	4243	2	4253	2	4259	2	4261	2	4271	7
4273	5	4283	2	4289	3	4297	5	4327	3	4337	3	4339	10	4349	2	4357	2
4363	2	4373	2	4391	14	4397	2	4409	3	4421	3	4423	3	4441	21	4447	3
4451	2	4457	3	4463	5	4481	3	4483	2	4493	2	4507	2	4513	7	4517	2
4519	3	4523	5	4547	2	4549	6	4561	11	4567	3	4583	5	4591	11	4597	5
4603	2	4621	2	4637	2	4639	3	4643	5	4649	3	4651	3	4657	15	4663	3
4673	3	4679	11	4691	2	4703	5	4721	6	4723	2	4729	17	4733	5	4751	19
4759	3	4783	6	4787	2	4789	2	4793	3	4799	7	4801	7	4813	2	4817	3
4831	3	4861	11	4871	11	4877	2	4889	3	4903	3	4909	6	4919	13	4931	6
4933	2	4937	3	4943	7	4951	6	4957	2	4967	5	4969	11	4973	2	4987	2
4993	5	4999	3	5003	2	5009	3	5011	2	5021	3	5023	3	5039	11	5051	2
5059	2	5077	2	5081	3	5087	5	5099	2	5101	6	5107	2	5113	19	5119	3
5147	2	5153	5	5167	6	5171	2	5179	2	5189	2	5197	7	5209	17	5227	2
5231	7	5233	10	5237	3	5261	2	5273	3	5279	7	5281	7	5297	3	5303	5
5309	2	5323	5	5333	2	5347	3	5351	11	5381	3	5387	2	5393	3	5399	7
5407	3	5413	5	5417	3	5419	3	5431	3	5437	5	5441	3	5443	2	5449	7
5471	7	5477	2	5479	3	5483	2	5501	2	5503	3	5507	2	5519	13	5521	11
5527	5	5531	10	5557	2	5563	2	5569	13	5573	2	5581	6	5591	11	5623	5
5639	7	5641	14	5647	3	5651	2	5653	5	5657	3	5659	2	5669	3	5683	2
5689	11	5693	2	5701	2	5711	19	5717	2	5737	5	5741	2	5743	10	5749	2
5779	2	5783	7	5791	6	5801	3	5807	5	5813	2	5821	6	5827	2	5839	6
5843	2	5849	3	5851	2	5857	7	5861	3	5867	5	5869	2	5879	11	5881	31
5897	3	5903	5	5923	2	5927	5	5939	2	5953	7	5981	3	5987	2	6007	3
6011	2	6029	2	6037	5	6043	5	6047	5	6053	2	6067	2	6073	10	6079	17
6089	3	6091	7	6101	2	6113	3	6121	7	6131	2	6133	5	6143	5	6151	3
6163	3	6173	2	6197	2	6199	3	6203	2	6211	2	6217	5	6221	3	6229	2
6247	5	6257	3	6263	5	6269	2	6271	11	6277	2	6287	7	6299	2	6301	10
6311	7	6317	2	6323	2	6329	3	6337	10	6343	3	6353	3	6359	13	6361	19
6367	3	6373	2	6379	2	6389	2	6397	2	6421	6	6427	3	6449	3	6451	3

p	ω	p	ω	p	ω	p	ω	p	ω	p	ω	p	ω	p	ω	p	ω
6469	2	6473	3	6481	7	6491	2	6521	6	6529	7	6547	2	6551	17	6553	10
6563	5	6569	3	6571	3	6577	5	6581	14	6599	13	6607	3	6619	2	6637	2
6653	2	6659	2	6661	6	6673	5	6679	7	6689	3	6691	2	6701	2	6703	5
6709	2	6719	11	6733	2	6737	3	6761	3	6763	2	6779	2	6781	2	6791	7
6793	10	6803	2	6823	3	6827	2	6829	2	6833	3	6841	22	6857	3	6863	5
6869	2	6871	3	6883	2	6899	2	6907	2	6911	7	6917	2	6947	2	6949	2
6959	7	6961	13	6967	5	6971	2	6977	3	6983	5	6991	6	6997	5	7001	3
7013	2	7019	2	7027	2	7039	3	7043	2	7057	5	7069	2	7079	7	7103	5
7109	2	7121	3	7127	5	7129	7	7151	7	7159	3	7177	10	7187	2	7193	3
7207	3	7211	2	7213	5	7219	2	7229	2	7237	2	7243	2	7247	5	7253	2
7283	2	7297	5	7307	2	7309	6	7321	7	7331	2	7333	6	7349	2	7351	6
7369	7	7393	5	7411	2	7417	5	7433	3	7451	2	7457	3	7459	2	7477	2
7481	6	7487	5	7489	7	7499	2	7507	2	7517	2	7523	2	7529	3	7537	7
7541	2	7547	2	7549	2	7559	13	7561	13	7573	2	7577	3	7583	5	7589	2
7591	6	7603	2	7607	5	7621	2	7639	7	7643	2	7649	3	7669	2	7673	3
7681	17	7687	6	7691	2	7699	3	7703	5	7717	2	7723	3	7727	5	7741	7
7753	10	7757	2	7759	3	7789	2	7793	3	7817	3	7823	5	7829	2	7841	12
7853	2	7867	3	7873	5	7877	2	7879	3	7883	2	7901	2	7907	2	7919	7
7927	3	7933	2	7937	3	7949	2	7951	6	7963	5	7993	5	8009	3	8011	14
8017	5	8039	11	8053	2	8059	3	8069	2	8081	3	8087	5	8089	17	8093	2
8101	6	8111	11	8117	2	8123	2	8147	2	8161	7	8167	3	8171	2	8179	2
8191	17	8209	7	8219	2	8221	2	8231	11	8233	10	8237	2	8243	2	8263	3
8269	2	8273	3	8287	3	8291	2	8293	2	8297	3	8311	3	8317	6	8329	7
8353	5	8363	2	8369	3	8377	5	8387	2	8389	6	8419	3	8423	5	8429	2
8431	3	8443	2	8447	5	8461	6	8467	2	8501	7	8513	5	8521	13	8527	5
8537	3	8539	2	8543	5	8563	2	8573	2	8581	6	8597	2	8599	3	8609	3
8623	3	8627	2	8629	6	8641	17	8647	3	8663	5	8669	2	8677	2	8681	15
8689	13	8693	2	8699	2	8707	5	8713	5	8719	3	8731	2	8737	5	8741	2
8747	2	8753	3	8761	23	8779	11	8783	5	8803	2	8807	5	8819	2	8821	2
8831	7	8837	2	8839	3	8849	3	8861	2	8863	3	8867	2	8887	3	8893	5
8923	2	8929	11	8933	2	8941	6	8951	13	8963	2	8969	3	8971	2	8999	7
9001	7	9007	3	9011	2	9013	5	9029	2	9041	3	9043	3	9049	7	9059	2
9067	3	9091	3	9103	6	9109	10	9127	3	9133	6	9137	3	9151	3	9157	6
9161	3	9173	2	9181	2	9187	3	9199	3	9203	2	9209	3	9221	2	9227	2
9239	19	9241	13	9257	3	9277	5	9281	3	9283	2	9293	2	9311	7	9319	3
9323	2	9337	5	9341	2	9343	5	9349	2	9371	2	9377	3	9391	3	9397	2
9403	3	9413	3	9419	2	9421	2	9431	7	9433	5	9437	2	9439	22	9461	3
9463	3	9467	2	9473	3	9479	7	9491	2	9497	3	9511	3	9521	3	9533	2
9539	2	9547	2	9551	11	9587	2	9601	13	9613	2	9619	2	9623	5	9629	2
9631	3	9643	2	9649	7	9661	2	9677	2	9679	3	9689	3	9697	10	9719	17
9721	7	9733	2	9739	3	9743	5	9749	2	9767	5	9769	13	9781	6	9787	3
9791	11	9803	2	9811	3	9817	5	9829	10	9833	3	9839	7	9851	2	9857	5
9859	2	9871	3	9883	2	9887	5	9901	2	9907	2	9923	2	9929	3	9931	10
9941	2	9949	2	9967	3	9973	11										

4.7.2 INDEX ARITHMETIC

Definition:

If m is a positive integer with primitive root r and a is an integer relatively prime to m, then the unique nonnegative integer x not exceeding $\phi(m)$ with $r^x \equiv a \pmod{m}$ is the **index of a to the base r modulo m**, or the **discrete logarithm of a to the base r modulo m**.

The index is denoted $\text{ind}_r a$ (where the modulus m is fixed).

Facts:

1. Table 2 displays, for each prime less than 100, the indices of all numbers not exceeding the prime using the least primitive root of the prime as the base.

Table 2 Indices for primes less than 100.

For each prime $p < 100$ two tables are given. Let g be least primitive element of the group F_p^* and assume $g^x = y$.

The table on the left has a y in position x, while the one on the right has an x in position y.

3:

N	0	1	2	3	4	5	6	7	8	9
0			2	1						

I	0	1	2	3	4	5	6	7	8	9
0		1	2	1						

5:

N	0	1	2	3	4	5	6	7	8	9
0			4	1	3	2				

I	0	1	2	3	4	5	6	7	8	9
0		1	2	4	3	1				

7:

N	0	1	2	3	4	5	6	7	8	9
0			6	2	1	4	5	3		

I	0	1	2	3	4	5	6	7	8	9
0		1	3	2	6	4	5	1		

11:

N	0	1	2	3	4	5	6	7	8	9
0			10	1	8	2	4	9	7	3
1	5									

I	0	1	2	3	4	5	6	7	8	9
0		1	2	4	8	5	10	9	7	3
1	1									

13:

N	0	1	2	3	4	5	6	7	8	9
0			12	1	4	2	9	5	11	3
1	8	10	7	6						

I	0	1	2	3	4	5	6	7	8	9
0		1	2	4	8	3	6	12	11	9
1	5	10	7	1						

17:

N	0	1	2	3	4	5	6	7	8	9
0			16	14	1	12	5	15	11	10
1	2	3	7	13	4	9	6	8		

I	0	1	2	3	4	5	6	7	8	9
0		1	3	9	10	13	5	15	11	16
1	14	8	7	4	12	2	6	1		

19:

N	0	1	2	3	4	5	6	7	8	9
0			18	1	13	2	16	14	6	3
1	8	17	12	15	5	7	11	4	10	9

I	0	1	2	3	4	5	6	7	8	9
0		1	2	4	8	16	13	7	14	9
1	18	17	15	11	3	6	12	5	10	1

23:

N	0	1	2	3	4	5	6	7	8	9
0		22	2	16	4	1	18	19	6	10
1	3	9	20	14	21	17	8	7	12	15
2	5	13	11							

I	0	1	2	3	4	5	6	7	8	9
0	1	5	2	10	4	20	8	17	16	11
1	9	22	18	21	13	19	3	15	6	7
2	12	14	1							

29:

N	0	1	2	3	4	5	6	7	8	9
0		28	1	5	2	22	6	12	3	10
1	23	25	7	18	13	27	4	21	11	9
2	24	17	26	20	8	16	19	15	14	

I	0	1	2	3	4	5	6	7	8	9
0	1	2	4	8	16	3	6	12	24	19
1	9	18	7	14	28	27	25	21	13	26
2	23	17	5	10	20	11	22	15	1	

31:

N	0	1	2	3	4	5	6	7	8	9
0		30	24	1	18	20	25	28	12	2
1	14	23	19	11	22	21	6	7	26	4
2	8	29	17	27	13	10	5	3	16	9
3	15									

I	0	1	2	3	4	5	6	7	8	9
0	1	3	9	27	19	26	16	17	20	29
1	25	13	8	24	10	30	28	22	4	12
2	5	15	14	11	2	6	18	23	7	21
3	1									

37:

N	0	1	2	3	4	5	6	7	8	9
0		36	1	26	2	23	27	32	3	16
1	24	30	28	11	33	13	4	7	17	35
2	25	22	31	15	29	10	12	6	34	21
3	14	9	5	20	8	19	18			

I	0	1	2	3	4	5	6	7	8	9
0	1	2	4	8	16	32	27	17	34	31
1	25	13	26	15	30	23	9	18	36	35
2	33	29	21	5	10	20	3	6	12	26
3	11	22	7	14	28	19	1			

41:

N	0	1	2	3	4	5	6	7	8	9
0		40	26	15	12	22	1	39	38	30
1	8	3	27	31	25	37	24	33	16	9
2	34	14	29	36	13	4	17	5	11	7
3	23	28	10	18	19	21	2	32	35	6
4	20									

I	0	1	2	3	4	5	6	7	8	9
0	1	6	36	11	25	27	39	29	10	19
1	32	28	4	24	21	3	18	26	33	34
2	40	35	5	30	16	14	2	12	31	22
3	9	13	37	17	20	38	23	15	8	7
4	1									

43:

N	0	1	2	3	4	5	6	7	8	9
0		42	27	1	12	25	28	35	39	2
1	10	30	13	32	20	26	24	38	29	19
2	37	36	15	16	40	8	17	3	5	41
3	11	34	9	31	23	18	14	7	4	33
4	22	6	21							

I	0	1	2	3	4	5	6	7	8	9
0	1	3	9	27	38	28	41	37	25	32
1	10	30	4	12	36	22	23	26	35	19
2	14	42	40	34	16	5	15	2	6	18
3	11	33	13	39	31	7	21	20	17	8
4	24	29	1							

47:

N	0	1	2	3	4	5	6	7	8	9
0		46	18	20	36	1	38	32	8	40
1	19	7	10	11	4	21	26	16	12	45
2	37	6	25	5	28	2	29	14	22	35
3	39	3	44	27	34	33	30	42	17	31
4	9	15	24	13	43	41	23			

I	0	1	2	3	4	5	6	7	8	9
0	1	5	25	31	14	23	21	11	8	40
1	12	13	18	43	27	41	17	38	2	10
2	3	15	28	46	42	22	16	33	24	26
3	36	39	7	35	34	29	4	20	6	30
4	9	45	37	44	32	19	1			

53:

N	0	1	2	3	4	5	6	7	8	9
0		52	1	17	2	47	18	14	3	34
1	48	6	19	24	15	12	4	10	35	37
2	49	31	7	39	20	42	25	51	16	46
3	13	33	5	23	11	9	36	30	38	41
4	50	45	32	22	8	29	40	44	21	28
5	43	27	26							

I	0	1	2	3	4	5	6	7	8	9
0	1	2	4	8	16	32	11	22	44	35
1	17	34	15	30	7	14	28	3	6	12
2	24	48	43	33	13	26	52	51	49	45
3	37	21	42	31	9	18	36	19	38	23
4	46	39	25	50	47	41	29	5	10	20
5	40	27	1							

59:

N	0	1	2	3	4	5	6	7	8	9
0		58	1	50	2	6	51	18	3	42
1	7	25	52	45	19	56	4	40	43	38
2	8	10	26	15	53	12	46	34	20	28
3	57	49	5	17	41	24	44	55	39	37
4	9	14	11	33	27	48	16	23	54	36
5	13	32	47	22	35	31	21	30	29	

I	0	1	2	3	4	5	6	7	8	9
0	1	2	4	8	16	32	5	10	20	40
1	21	42	25	50	41	23	46	33	7	14
2	28	56	53	47	35	11	22	44	29	58
3	57	55	51	43	27	54	49	39	19	38
4	17	34	9	18	36	13	26	52	45	31
5	3	6	12	24	48	37	15	30	1	

61:

N	0	1	2	3	4	5	6	7	8	9
0		60	1	6	2	22	7	49	3	12
1	23	15	8	40	50	28	4	47	13	26
2	24	55	16	57	9	44	41	18	51	35
3	29	59	5	21	48	11	14	39	27	46
4	25	54	56	43	17	34	58	20	10	38
5	45	53	42	33	19	37	52	32	36	31
6	30									

I	0	1	2	3	4	5	6	7	8	9
0	1	2	4	8	16	32	3	6	12	24
1	48	35	9	18	36	11	22	44	27	54
2	47	33	5	10	20	40	19	38	15	30
3	60	59	57	53	45	29	58	55	49	37
4	13	26	52	43	25	50	39	17	34	7
5	14	28	56	51	41	21	42	23	46	31
6	1									

67:

N	0	1	2	3	4	5	6	7	8	9
0		66	1	39	2	15	40	23	3	12
1	16	59	41	19	24	54	4	64	13	10
2	17	62	60	28	42	30	20	51	25	44
3	55	47	5	32	65	38	14	22	11	58
4	18	53	63	9	61	27	29	50	43	46
5	31	37	21	57	52	8	26	49	45	36
6	56	7	48	35	6	34	33			

I	0	1	2	3	4	5	6	7	8	9
0	1	2	4	8	16	32	64	61	55	43
1	19	38	9	18	36	5	10	20	40	13
2	26	52	37	7	14	28	56	45	23	46
3	25	50	33	66	65	63	59	51	35	3
4	6	12	24	48	29	58	49	31	62	57
5	47	27	54	41	15	30	60	53	39	11
6	22	44	21	42	17	34	1			

71:

N	0	1	2	3	4	5	6	7	8	9
0		70	6	26	12	28	32	1	18	52
1	34	31	38	39	7	54	24	49	58	16
2	40	27	37	15	44	56	45	8	13	68
3	60	11	30	57	55	29	64	20	22	65
4	46	25	33	48	43	10	21	9	50	2
5	62	5	51	23	14	59	19	42	4	3
6	66	69	17	53	36	67	63	47	61	41
7	35									

I	0	1	2	3	4	5	6	7	8	9
0	1	7	49	59	58	51	2	14	27	47
1	45	31	4	28	54	23	19	62	8	56
2	37	46	38	53	16	41	3	21	5	35
3	32	11	6	42	10	70	64	22	12	13
4	20	69	57	44	24	26	40	67	43	17
5	48	52	9	63	15	34	25	33	18	55
6	30	68	50	66	36	39	60	65	29	61
7	1									

73:

N	0	1	2	3	4	5	6	7	8	9
0		72	8	6	16	1	14	33	24	12
1	9	55	22	59	41	7	32	21	20	62
2	17	39	63	46	30	2	67	18	49	35
3	15	11	40	61	29	34	28	64	70	65
4	25	4	47	51	71	13	54	31	38	66
5	10	27	3	53	26	56	57	68	43	5
6	23	58	19	45	48	60	69	50	37	52
7	42	44	36							

I	0	1	2	3	4	5	6	7	8	9
0	1	5	25	52	41	59	3	15	2	10
1	50	31	9	45	6	30	4	20	27	62
2	18	17	12	60	8	40	54	51	36	34
3	24	47	16	7	35	29	72	68	48	21
4	32	14	70	58	71	63	23	42	64	28
5	67	43	69	53	46	11	55	56	61	13
6	65	33	19	22	37	39	49	26	57	66
7	38	44	1							

79:

N	0	1	2	3	4	5	6	7	8	9
0		78	4	1	8	62	5	53	12	2
1	66	68	9	34	57	63	16	21	6	32
2	70	54	72	26	13	46	38	3	61	11
3	67	56	20	69	25	37	10	19	36	35
4	74	75	58	49	76	64	30	59	17	28
5	50	22	42	77	7	52	65	33	15	31
6	71	45	60	55	24	18	73	48	29	27
7	41	51	14	44	23	47	40	43	39	

I	0	1	2	3	4	5	6	7	8	9
0	1	3	9	27	2	6	18	54	4	12
1	36	29	8	24	72	58	16	48	65	37
2	32	17	51	74	64	34	23	69	49	68
3	46	59	19	57	13	39	38	35	26	78
4	76	70	52	77	73	61	25	75	67	43
5	50	71	55	7	21	63	31	14	42	47
6	62	28	5	15	45	56	10	30	11	33
7	20	60	22	66	40	41	44	53	1	

83:

N	0	1	2	3	4	5	6	7	8	9
0		82	1	72	2	27	73	8	3	62
1	28	24	74	77	9	17	4	56	63	47
2	29	80	25	60	75	54	78	52	10	12
3	18	38	5	14	57	35	64	20	48	67
4	30	40	81	71	26	7	61	23	76	16
5	55	46	79	59	53	51	11	37	13	34
6	19	66	39	70	6	22	15	45	58	50
7	36	33	65	69	21	44	49	32	68	43
8	31	42	41							

I	0	1	2	3	4	5	6	7	8	9
0	1	2	4	8	16	32	64	45	7	14
1	28	56	29	58	33	66	49	15	30	60
2	37	74	65	47	11	22	44	5	10	20
3	40	80	77	71	59	35	70	57	31	62
4	41	82	81	79	75	67	51	19	38	76
5	69	55	27	54	25	50	17	34	68	53
6	23	46	9	18	36	72	61	39	78	73
7	63	43	3	6	12	24	48	13	26	52
8	21	42	1							

89:

N	0	1	2	3	4	5	6	7	8	9
0		88	16	1	32	70	17	81	48	2
1	86	84	33	23	9	71	64	6	18	35
2	14	82	12	57	49	52	39	3	25	59
3	87	31	80	85	22	63	34	11	51	24
4	30	21	10	29	28	72	73	54	65	74
5	68	7	55	78	19	66	41	36	75	43
6	15	69	47	83	8	5	13	56	38	58
7	79	62	50	20	27	53	67	77	40	42
8	46	4	37	61	26	76	45	60	44	

I	0	1	2	3	4	5	6	7	8	9
0	1	3	9	27	81	65	17	51	64	14
1	42	37	22	66	20	60	2	6	18	54
2	73	41	34	13	39	28	84	74	44	43
3	40	31	4	12	36	19	57	82	68	26
4	78	56	79	59	88	86	80	62	8	24
5	72	38	25	75	47	52	67	23	69	29
6	87	83	71	35	16	48	55	76	50	61
7	5	15	45	46	49	58	85	77	53	70
8	32	7	21	63	11	33	10	30	1	

97:

N	0	1	2	3	4	5	6	7	8	9
0		96	34	70	68	1	8	31	6	44
1	35	86	42	25	65	71	40	89	78	81
2	69	5	24	77	76	2	59	18	3	13
3	9	46	74	60	27	32	16	91	19	95
4	7	85	39	4	58	45	15	84	14	62
5	36	63	93	10	52	87	37	55	47	67
6	43	64	80	75	12	26	94	57	61	51
7	66	11	50	28	29	72	53	21	33	30
8	41	88	23	17	73	90	38	83	92	54
9	79	56	49	20	22	82	48			

I	0	1	2	3	4	5	6	7	8	9
0	1	5	25	28	43	21	8	40	6	30
1	53	71	64	29	48	46	36	83	27	38
2	93	77	94	82	22	13	65	34	73	74
3	79	7	35	78	2	10	50	56	86	42
4	16	80	12	60	9	45	31	58	96	92
5	72	69	54	76	89	57	91	67	44	26
6	33	68	49	51	61	14	70	59	4	20
7	3	15	75	84	32	63	24	23	18	90
8	62	19	95	87	47	41	11	55	81	17
9	85	37	88	52	66	39	1			

2. If m is a positive integer with primitive root r and a is a positive integer relatively prime to m, then $a \equiv r^{\text{ind}_r a} \pmod{m}$.

3. If m is a positive integer with primitive root r, then $\text{ind}_r 1 = 0$ and $\text{ind}_r r = 1$.

4. If $m > 2$ is an integer with primitive root r, then $\text{ind}_r(-1) = \frac{\phi(m)}{2}$.

5. If m is a positive integer with primitive root r, and a and b are integers relatively prime to m, then:
   - $\text{ind}_r 1 \equiv 0 \,(\text{mod } \phi(m))$;
   - $\text{ind}_r(ab) \equiv \text{ind}_r a + \text{ind}_r b \,(\text{mod } \phi(m))$;
   - $\text{ind}_r a^k \equiv k \cdot \text{ind}_r a \,(\text{mod } \phi(m))$ if k is a positive integer.

6. If m is a positive integer and r and s are both primitive roots modulo m, then $\text{ind}_r a \equiv \text{ind}_s a \cdot \text{ind}_r s \,(\text{mod } \phi(m))$.

7. If m is a positive integer with primitive root r, and a and b are integers both relatively prime to m, then the exponential congruence $a^x \equiv b \,(\text{mod } m)$ has a solution if and only if $d | \text{ind}_r b$. Furthermore, if there is a solution to this exponential congruence, then there are exactly $\gcd(\text{ind}_r a, \phi(m))$ incongruent solutions.

8. There is a wide variety of algorithms for computing discrete logarithms, including those known as the baby-step, giant-step algorithm, the Pollard rho algorithm, the Pollig-Hellman algorithm, and the index-calculus algorithm. (See [MevaVa96] for details.)

9. The fastest algorithms known for computing discrete logarithms, relative to a fixed primitive root, of a given prime p are index-calculus algorithms, which have subexponential computational complexity. In particular, there is an algorithm based on the number field sieve that runs using $L_p(\frac{1}{3}, 1.923) = O(\exp((1.923 + o(1))(\log p)^{\frac{1}{3}}(\log \log p)^{\frac{2}{3}}))$ bit operations. (See [MevaVa96].)

10. Many cryptographic methods rely on intractability of finding discrete logarithms of integers relative to a fixed primitive root r of a fixed prime p.

Examples:

1. To solve $3x^{30} \equiv 4 \,(\text{mod } 37)$ take indices to the base 2 (2 is the smallest primitive root of 37) to obtain $\text{ind}_2(3x^{30}) \equiv \text{ind}_2 4 = 2 \,(\text{mod } 36)$. Since $\text{ind}_2(3x^{30}) \equiv \text{ind}_2 3 + 30 \cdot \text{ind}_2 x = 26 + 30 \cdot \text{ind}_2 x \,(\text{mod } 36)$, it follows that $30 \cdot \text{ind}_2 x \equiv 12 (\text{mod } 36)$. The solutions to this congruence are those x such that $\text{ind}_2(x) \equiv 4, 10, 16, 22, 28, 34 (\text{mod } 36)$. From the Table of Indices (Table 2), the solutions are those x with $x \equiv 16, 25, 9, 21, 12, 28 \,(\text{mod } 37)$.

2. To solve $7^x \equiv 6 \,(\text{mod } 17)$ take indices to the base 3 (3 is the smallest primitive root of 17) to obtain $\text{ind}_3(7^x) \equiv \text{ind}_3 6 = 15 \,(\text{mod } 16)$. Since $\text{ind}_3(7^x) \equiv x \cdot \text{ind}_3 7 \equiv 11x \,(\text{mod } 16)$, it follows that $11x \equiv 15 \,(\text{mod } 16)$. Since all the steps in this computation are reversible, it follows that the solutions of the original congruence are the solutions of this linear congruence, namely those x with $x \equiv 13 \,(\text{mod } 16)$.

4.7.3 QUADRATIC RESIDUES

Definitions:

If m and k are positive integers and a is an integer relatively prime to m, then a is a **kth power residue** of m if the congruence $x^k \equiv a \,(\text{mod } m)$ has a solution.

If a and m are relatively prime integers and m is positive, then a is a **quadratic residue** of m if the congruence $x^2 \equiv a \,(\text{mod } m)$ has a solution. If $x^2 \equiv a \,(\text{mod } m)$ has no solution, then a is a **quadratic nonresidue** of m.

If p is an odd prime and p does not divide a, then the **Legendre symbol** $\left(\frac{a}{p}\right)$ is 1 if a is a quadratic residue of p and -1 if a is a quadratic nonresidue of p. This symbol is named after the French mathematician Adrien-Marie Legendre (1752–1833).

If n is an odd positive integer with prime-power factorization $n = p_1{}^{t_1} p_2{}^{t_2} \ldots p_m{}^{t_m}$ and a is an integer relatively prime to n, then the **Jacobi symbol** $\left(\frac{a}{n}\right)$ is defined by

$$\left(\frac{a}{n}\right) = \prod_{i=1}^{m} \left(\frac{a}{p_i}\right)^{t_i},$$

where the symbols on the right-hand side of the equality are Legendre symbols. This symbol is named after the German mathematician Karl Gustav Jacob Jacobi (1804–1851).

Let a be a positive integer that is not a perfect square and such that $a \equiv 0$ or $1 \pmod 4$. The **Kronecker symbol** (named after the German mathematician Leopold Kronecker (1823–1891)), which is a generalization of the Legendre symbol, is defined as:

- $\left(\frac{a}{2}\right) = \begin{cases} 1 & \text{if } a \equiv 1 \pmod 8 \\ -1 & \text{if } a \equiv 5 \pmod 8 \end{cases}$

- $\left(\frac{a}{p}\right) =$ the Legendre symbol $\left(\frac{a}{p}\right)$ if p is an odd prime such that p does not divide a

- $\left(\frac{a}{n}\right) = \prod_{j=1}^{r} \left(\frac{a}{p_j}\right)^{t_j}$ if $\gcd(a,n) = 1$ and $n = \prod_{j=1}^{r} p_j{}^{t_j}$ is the prime factorization of n.

Facts:

1. If p is an odd prime, then there are an equal number of quadratic residues modulo p and quadratic non-residues modulo p among the integers $1, 2, \ldots, p-1$. In particular, there are $\frac{p-1}{2}$ integers of each type in this set.

2. *Euler's criterion*: If p is an odd prime and a is a positive integer not divisible by p, then $\left(\frac{a}{p}\right) \equiv a^{(p-1)/2} \pmod p$.

3. If p is an odd prime and a and b are integers not divisible by p with $a \equiv b \pmod p$, then $\left(\frac{a}{p}\right) = \left(\frac{b}{p}\right)$.

4. If p is an odd prime and a and b are integers not divisible by p, then $\left(\frac{a}{p}\right)\left(\frac{b}{p}\right) = \left(\frac{ab}{p}\right)$.

5. If p is an odd prime and a and b are integers not divisible by p, then $\left(\frac{a^2}{p}\right) = 1$.

6. If p is an odd prime, then $\left(\frac{-1}{p}\right) = \begin{cases} 1 & \text{if } p \equiv 1 \pmod 4 \\ -1 & \text{if } p \equiv -1 \pmod 4 \end{cases}$

7. If p is an odd prime, then -1 is a quadratic residue of p if $p \equiv 1 \pmod 4$ and a quadratic nonresidue of p if $p \equiv -1 \pmod 4$. (This is a direct consequence of Fact 6.)

8. *Gauss' lemma*: If p is an odd prime, a is an integer with $\gcd(a, p) = 1$, and s is the number of least positive residues of $a, 2a, \ldots, \frac{p-1}{2}a$ greater than $\frac{p}{2}$, then $\left(\frac{a}{p}\right) = (-1)^s$.

9. If p is an odd prime, then $\left(\frac{2}{p}\right) = (-1)^{(p^2-1)/8}$.

10. The integer 2 is a quadratic residue of all primes p with $p \equiv \pm 1 \pmod 8$ and a quadratic nonresidue of all primes $p \equiv \pm 3 \pmod 8$. (This is a direct consequence of Fact 9.)

11. *Law of quadratic reciprocity*: If p and q are odd primes, then

$$\left(\frac{p}{q}\right)\left(\frac{q}{p}\right) = (-1)^{\frac{p-1}{2} \cdot \frac{q-1}{2}}.$$

This law was first proved by Carl Friedrich Gauss (1777–1855).

12. Many different proofs of the law of quadratic reciprocity have been discovered. By one count, there are more than 150 different proofs. Gauss published eight different proofs himself.

13. The law of quadratic reciprocity implies that if p and q are odd primes, then $\left(\frac{p}{q}\right) = \left(\frac{q}{p}\right)$ if either $p \equiv 1 \pmod 4$ or $q \equiv 1 \pmod 4$, and $\left(\frac{p}{q}\right) = -\left(\frac{q}{p}\right)$ if $p \equiv q \equiv 3 \pmod 4$.

14. If m is an odd positive integer and a and b are integers relatively prime to m with $a \equiv b \pmod{m}$, then $\left(\frac{a}{m}\right) = \left(\frac{b}{m}\right)$.

15. If m is an odd positive integer and a and b are integers relatively prime to m, then $\left(\frac{ab}{m}\right) = \left(\frac{a}{m}\right)\left(\frac{b}{m}\right)$.

16. If m is an odd positive integer and a is an integer relatively prime to m, then $\left(\frac{a^2}{m}\right) = 1$.

17. If m and n are relatively prime odd positive integers and a is an integer relatively prime to m and n, then $\left(\frac{a}{mn}\right) = \left(\frac{a}{m}\right)\left(\frac{a}{n}\right)$.

18. If m is an odd positive integer, then the value of the Jacobi symbol $\left(\frac{a}{m}\right)$ does not determine whether a is a perfect square modulo m.

19. If m is an odd positive integer, then $\left(\frac{-1}{m}\right) = (-1)^{\frac{m-1}{2}}$.

20. If m is an odd positive integer, then $\left(\frac{2}{m}\right) = (-1)^{\frac{m^2-1}{8}}$.

21. *Reciprocity law for Jacobi symbols*: If m and n are relatively prime odd positive integers, then
$$\left(\frac{m}{n}\right)\left(\frac{n}{m}\right) = (-1)^{\frac{m-1}{2}\frac{n-1}{2}}.$$

22. The number of integers in a reduced set of residues modulo n with $\left(\frac{k}{n}\right) = 1$ equals the number with $\left(\frac{k}{n}\right) = -1$.

23. The Legendre symbol $\left(\frac{a}{p}\right)$, where p is prime and $0 \leq a < p$, can be evaluated using $O((\log_2 p)^2)$ bit operations.

24. The Jacobi symbol $\left(\frac{a}{n}\right)$, where n is a positive integer and $0 \leq a < n$, can be evaluated using $O((\log_2 n)^2)$ bit operations.

25. Let p be an odd prime. Even though half the integers x with $1 \leq x < p$ are quadratic non-residues of p, there is no known polynomial-time deterministic algorithm for finding such an integer. However, picking integers at random produces a probabilistic algorithm that has 2 as the expected number of iterations done before a non-residue is found.

26. Let m be a positive integer with a primitive root. If k is a positive integer and a is an integer relatively prime to m, then a is a kth power residue of m if and only if $a^{\phi(m)/d} \equiv 1 \pmod{m}$ where $d = \gcd(k, \phi(m))$. Moreover, if a is a kth power residue of m, then there are exactly d incongruent solutions modulo m of the congruence $x^k \equiv a \pmod{m}$.

27. If p is a prime, k is a positive integer, and a is an integer with $\gcd(a,p) = 1$, then a is a kth power residue of p if and only if $a^{(p-1)/d} \equiv 1 \pmod{p}$, where $d = \gcd(k, p-1)$.

28. The kth roots of a kth power residue modulo p, where p is a prime, can be computed using a primitive root and indices to this primitive root. This is only practical for small primes p. (See §4.7.1.)

Examples:

1. The integers 1, 3, 4, 5, and 9 are quadratic residues of 11; the integers 2, 6, 7, 8, and 10 are quadratic nonresidues of 11. Hence $\left(\frac{1}{11}\right) = \left(\frac{3}{11}\right) = \left(\frac{4}{11}\right) = \left(\frac{5}{11}\right) = \left(\frac{9}{11}\right) = 1$ and $\left(\frac{2}{11}\right) = \left(\frac{6}{11}\right) = \left(\frac{7}{11}\right) = \left(\frac{8}{11}\right) = \left(\frac{10}{11}\right) = -1$.

2. To determine whether 11 is a quadratic residue of 19, note that using the law of quadratic reciprocity (Fact 12) and Facts 3, 4, and 10 it follows that $\left(\frac{11}{19}\right) = -\left(\frac{19}{11}\right) = -\left(\frac{8}{11}\right) = -\left(\frac{2}{11}\right)^3 = -(-1)^3 = 1$.

3. To evaluate the Jacobi symbol $\left(\frac{2}{45}\right)$ note that $\left(\frac{2}{45}\right) = \left(\frac{2}{3^2 \cdot 5}\right) = \left(\frac{2}{3}\right)^2 \cdot \left(\frac{2}{5}\right) = (-1)^2(-1) = -1$.

4. The Jacobi symbol $\left(\frac{5}{21}\right) = 1$, but 5 is not a quadratic residue of 21.

5. The integer 6 is a fifth power residue of 101 since $6^{(101-1)/5} = 6^{20} \equiv 1 \pmod{101}$.

6. From Example 5 it follows that 6 is a fifth power residue of 101. The solutions of the congruence $x^5 \equiv 6 \pmod{101}$, the fifth roots of 6, can be found by taking indices to the primitive root 2 modulo 101. Since $\text{ind}_2 6 = 70$, this gives $\text{ind}_2 x^5 = 5 \cdot \text{ind}_2 x \equiv 70 \pmod{100}$. The solutions of this congruence are the integers x with $\text{ind}_2 x \equiv 14 \pmod{20}$. This implies that the fifth roots of 6 are the integers with $\text{ind}_2 x = 14, 34, 54, 74$, and 94. These are the integers x with $x \equiv 22, 70, 85, 96$, and 30 (mod 101).

7. The integer 5 is not a sixth power residue of 17 since $5^{\overline{\gcd(6,16)}} = 5^8 \equiv -1 \pmod{17}$.

4.7.4 MODULAR SQUARE ROOTS

Definition:

If m is a positive integer and a is an integer, then r is a **square root of a modulo m** if $r^2 \equiv a \pmod{m}$.

Facts:

1. If p is a prime of the form $4n + 3$ and a is a perfect square modulo p, then the two square roots of a modulo p are $\pm a^{(p+1)/4}$.

2. If p is a prime of the form $8n + 5$ and a is a perfect square modulo p, then the two square roots of a modulo p are $x \equiv \pm a^{(p+3)/8} \pmod{p}$ if $a^{(p-1)/4} \equiv 1 \pmod{p}$ and $x \equiv \pm 2^{(p-1)/4} a^{(p+3)/8} \pmod{p}$ if $a^{(p-1)/4} \equiv -1 \pmod{p}$.

3. If n is a positive integer that is the product of two distinct primes p and q and a is a perfect square modulo n, then there are four distinct square roots of a modulo n. These square roots can be found by finding the two square roots of a modulo p and the two square roots of a modulo q and then using the Chinese remainder theorem to find the four square roots of a modulo n.

4. A square root of an integer a that is a square modulo p, where p is an odd prime, can be found by an algorithm that uses an average of $O((\log_2 p)^3)$ bit operations. (See [MevaVa96].)

5. If n is an odd integer with r distinct prime factors, a is a perfect square modulo n, and $\gcd(a, n) = 1$, then a has exactly 2^r incongruent square roots modulo n.

Examples:

1. Using Legendre symbols it can be shown that 11 is a perfect square modulo 19. Using Fact 1 it follows that the square roots of 11 modulo 19 are given by $x \equiv \pm 11^{(19+1)/4} = \pm 11^5 \equiv \pm 7 \pmod{19}$.

2. There are four incongruent square roots of 860 modulo $11021 = 103 \cdot 107$. To find these solutions, first note that $x^2 \equiv 860 \equiv 36 \pmod{103}$ so that $x \equiv \pm 6 \pmod{103}$ and $x^2 \equiv 860 \equiv 4 \pmod{107}$ so that $x \equiv \pm 2 \pmod{107}$. The Chinese remainder theorem can be used to find these square roots. They are $x \equiv -212, -109, 109, 212 \pmod{11021}$.

3. The square roots of 121 modulo 315 are 11, 74, 101, 151, 164, 214, 241, and 304.

4.8 DIOPHANTINE EQUATIONS

An important area of number theory is devoted to finding solutions of equations where the solutions are restricted to belong to the set of integers, or some other specified set, such as the set of rational numbers. An equation with the added proviso that the solutions must be integers (or must belong to some other specified countable set, such as the set of rational numbers) is called a *diophantine equation*. This name comes from the ancient Greek mathematician Diophantus (ca. 250 A.D.), who wrote extensively on such equations.

Diophantine equations have both practical and theoretical importance. Their practical importance arises when variables in an equation represent quantities of objects, for example. Fermat's last theorem, which states that there are no nontrivial solutions in integers $n > 2$, x, y, and z to the diophantine equation $x^n + y^n = z^n$ has long interested mathematicians and non-mathematicians alike. This theorem was proved only in the mid-1990s, even though many brilliant scholars sought a proof during the last three centuries.

More information about diophantine equations can be found in [Di71], [Gu94], and [Mo69].

4.8.1 LINEAR DIOPHANTINE EQUATIONS

Definition:

A *linear diophantine equation* is an equation of the form $a_1x_1 + a_2x_2 + \cdots + a_nx_n = c$, where $c, a_1, \ldots, a_n$ are integers and where integer solutions are sought for the unknowns $x_1, x_2, \ldots, x_n$.

Facts:

1. Let a and b be integers with $\gcd(a,b) = d$. The linear diophantine equation $ax + by = c$ has no solutions if $d \nmid c$. If $d | c$, then there are infinitely many solutions in integers. Moreover, if $x = x_0$, $y = y_0$ is a particular solution, then all solutions are given by $x = x_0 + \frac{b}{d}n$, $y = y_0 - \frac{a}{d}n$, where n is an integer.

2. A linear diophantine equation $a_1x_1 + a_2x_2 + \cdots + a_nx_n = c$ has solutions in integers if and only if $\gcd(a_1, a_2, \ldots, a_n) | c$. In that case, there are infinitely many solutions.

3. A solution (x_0, y_0) of the linear diophantine equation $ax + by = c$ where $\gcd(a,b) | c$ can be found by first expressing $\gcd(a,b)$ as a linear combination of a and b and then multiplying by $c/\gcd(a,b)$. (See §4.1.2.)

4. A linear diophantine equation $a_1x_1 + a_2x_2 + \cdots + a_nx_n = c$ in n variables can be solved by a reduction method. To find a particular solution, first let $b = \gcd(a_2, \ldots, a_n)$ and let (x_1, y) be a solution of the diophantine equation $a_1x_1 + by = c$. Iterate this procedure on the diophantine equation in $n-1$ variables, $a_2x_2 + a_3x_3 + \cdots + a_nx_n = y$ until an equation in two variables is obtained.

5. The solution to a system of r linear diophantine equations in n variables is obtained by using Gaussian elimination (§6.5.1) to reduce to a single diophantine equation in two or more variables.

6. If a and b are relatively prime positive integers and n is a positive integer, then the diophantine equation $ax + by = n$ has a *nonnegative* integer solution if $n \geq (a-1)(b-1)$.

7. If a and b are relatively prime positive integers, then there are exactly $(a-1)(b-1)/2$ nonnegative integers n less than $ab - a - b$ such that the equation $ax + by = n$ has a nonnegative solution.

8. If a and b are relatively prime positive integers, then there are no nonnegative solutions of $ax + by = ab - a - b$.

Examples:

1. To solve the linear diophantine equation $17x + 13y = 100$, express $\gcd(17, 13) = 1$ as a linear combination of 17 and 13. Using the steps of the Euclidean algorithm, it follows that $4 \cdot 13 - 3 \cdot 17 = 1$. Multiplying by 100 yields $100 = 400 \cdot 13 - 300 \cdot 417$. All solutions are given by $x = 400 + 17t$, $y = -300 - 13t$, where t ranges over the set of integers.

2. A traveller has exactly \$510 in travelers checks where each check is either a \$20 or a \$50 check. How many checks of each denomination can there be?

The solution to this question is given by the set of solutions in nonnegative integers to the linear diophantine equation $20x + 50y = 510$. There are infinitely many solutions in integers, which can be shown to be given by $x = -102 + 5n$, $y = 51 - 2n$. Since both x and y must be nonnegative, it follows that $n = 21, 22, 23, 24,$ or 25. Therefore there are 3 \$20 checks and 9 \$50 checks, 8 \$20 checks and 7 \$50 checks, 13 \$20 checks and 5 \$50 checks, 18 \$20 checks and 3 \$50 checks, or 23 \$20 checks and 1 \$50 check.

3. To find a particular solution of the linear diophantine equation $12x_1 + 21x_2 + 9x_3 + 15x_4 = 9$, which has infinitely many solutions since $\gcd(12, 21, 9, 15) = 3$, which divides 9, first divide both sides of the equation by 3 to get $4x_1 + 7x_2 + 3x_3 + 5x_4 = 3$. Now $1 = \gcd(7, 3, 5)$, so solve $4x_1 + 1y = 3$, as in Example 1, to get $x_1 = 1, y = -1$. Next solve $7x_2 + 3x_3 + 5x_4 = -1$. Since $1 = \gcd(3, 5)$, solve $7x_2 + 1z = -1$ to get $x_2 = 1, z = -8$. Finally, solve $3x_3 + 5x_4 = -8$ to get $x_3 = -1, x_4 = -1$.

4. To solve the following system of linear diophantine equations in integers:

$$x + y + z + w = 100$$
$$x + 2y + 3z + 4w = 300$$
$$x + 4y + 9z + 16w = 1000,$$

first reduce the system by elimination to:

$$x + y + z + w = 100$$
$$y + 2z + 3w = 200$$
$$2z + 6w = 300.$$

The solution to the last equation is $z = 150 + 3t$, $w = -t$, where t is an integer. Back-substitution gives

$$y = 200 - 2(150 + 3t) - 3(-t) = -100 - 3t$$
$$x = 100 - (-100 - 3t) - (150 + 3t) - (-t) = 50 + t.$$

4.8.2 PYTHAGOREAN TRIPLES

Definitions:

A **Pythagorean triple** is a solution (x, y, z) of the equation $x^2 + y^2 = z^2$ where $x, y,$ and z are positive integers.

A Pythagorean triple is **primitive** if $\gcd(x, y, z) = 1$.

Facts:
1. Pythagorean triples represent the lengths of sides of right triangles.
2. All primitive Pythagorean triples are given by
$$x = 2mn, \quad y = m^2 - n^2, \quad z = m^2 + n^2$$
where m and n are relatively prime positive integers of opposite parity with $m > n$.
3. All Pythagorean triples can be found by taking
$$x = 2mnt, \quad y = (m^2 - n^2)t, \quad z = (m^2 + n^2)t$$
where t is a positive integer and m and n are as in Fact 2.
4. Given a Pythagorean triple (x, y, z) with y odd, then m and n from Fact 2 can be found by taking $m = \sqrt{\frac{z+y}{2}}$ and $n = \sqrt{\frac{z-y}{2}}$.
5. The following table lists all Pythagorean triples with $z \leq 100$.

m	n	$x = 2mn$	$y = m^2 - n^2$	$z = m^2 + n^2$
2	1	4	3	5
3	1	6	8	10
3	2	12	5	13
4	1	8	15	17
4	2	16	12	20
4	3	24	7	25
5	1	10	24	26
5	2	20	21	29
5	3	30	16	34
5	4	40	9	41
6	1	12	35	37
6	2	24	32	40
6	3	36	27	45
6	4	48	20	52
6	5	60	11	61
7	1	14	48	50
7	2	28	45	53
7	3	42	40	58
7	4	56	33	65
7	5	70	24	74
7	6	84	13	85
8	1	16	63	65
8	2	32	60	68
8	3	48	55	73
8	4	64	48	80
8	5	80	39	89
8	6	96	28	100
9	1	18	80	82
9	2	36	77	85
9	3	54	72	90
9	4	72	65	97

6. The solutions of the diophantine equation $x^2 + y^2 = 2z^2$ can be obtained by transforming this equation into $\left(\frac{x+y}{2}\right)^2 + \left(\frac{x-y}{2}\right)^2 = z^2$, which shows that $(\frac{x+y}{2}, \frac{x-y}{2}, z)$ is a Pythagorean triple. All solutions are given by $x = (m^2 - n^2 + 2mn)t$, $y = (m^2 - n^2 - 2mn)t$, $z = (m^2 + n^2)t$ where m, n, and t are integers.

7. The solutions of the diophantine equation $x^2 + 2y^2 = z^2$ are given by $x = (m^2 - 2n^2)t$, $y = 2mnt$, $z = m^2 + 2n^2$ where m, n, and t are positive integers.

8. The solutions of the diophantine equation $x^2 + y^2 + z^2 = w^2$ where y and z are even are given by $x = \frac{m^2 + n^2 - r^2}{r}$, $y = 2m$, $z = 2n$, $w = \frac{m^2 + n^2 + r^2}{r}$, where m and n are positive integers and r runs through the divisors of $m^2 + n^2$ less than $(m^2 + n^2)^{1/2}$.

9. The solutions of the diophantine equation $x^2 + y^2 = z^2 + w^2$, with $x > z$, are given by $x = \frac{ms+nr}{2}$, $y = \frac{ns-mr}{2}$, $z = \frac{ms-nr}{2}$, $w = \frac{ns+mr}{2}$, where if m and n are both odd, then r and s are either both odd or both even.

4.8.3 FERMAT'S LAST THEOREM

Definitions:

The **Fermat equation** is the diophantine equation $x^n + y^n = z^n$ where x, y, z are integers and n is a positive integer greater than 2.

A **nontrivial solution** to the Fermat equation $x^n + y^n = z^n$ is a solution in integers x, y, and z where none of x, y, and z are zero.

Let p be an odd prime and let $\mathcal{K} = \mathcal{Q}(\omega)$ be the degree-p cyclotomic extension of the rational numbers (§5.6.2). If p does not divide the class number of $\mathcal{K}$ (see [Co93]), then p is said to be **regular**. Otherwise p is **irregular**.

Facts:

1. *Fermat's last theorem*: The statement that the diophantine equation $x^n + y^n = z^n$ has no nontrivial solutions in the positive integers for $n \geq 3$, is called *Fermat's last theorem*. The statement was made more than 300 years ago by Pierre de Fermat (1601–1665) and resisted proof until recently.

2. Fermat wrote in the margin of his copy of the works of Diophantus, next to the discussion of the equation $x^2 + y^2 = z^2$, the following: "However, it is impossible to write a cube as the sum of two cubes, a fourth power as the sum of two fourth powers and in general any power the sum of two similar powers. For this I have discovered a truly wonderful proof, but the margin is too small to contain it." In spite of this quotation, no proof was found of this statement until 1994, even though many mathematicians actively worked on finding such a proof. Most mathematicians would find it shocking if Fermat actually had found a proof.

3. Fermat's last theorem was finally proved in 1995 by Andrew Wiles [Wi95]. Wiles collected the Wolfskehl Prize, worth approximately $50,000 in 1997 for this proof.

4. That there are no nontrivial solutions of the Fermat equation for $n = 4$ was demonstrated by Fermat with an elementary proof using the *method of infinite descent*. This method proceeds by showing that for every solution in positive integers, there is a solution such that the values of each of the integers x, y, and z is smaller, contradicting the well-ordering property of the set of integers.

5. The method of infinite descent invented by Fermat can be used to show that the more general diophantine equation $x^4 + y^4 = z^2$ has no nontrivial solutions in integers x, y, and z.

6. The diophantine equation $x^4 - y^4 = z^2$ has no nontrivial solutions, as can be shown using the method of infinite descent.

7. The sum of two cubes may equal the sum of two other cubes. That is, there are nontrivial solution of the diophantine equation $x^3 + y^3 = z^3 + w^3$. The smallest solution is $x = 1$, $y = 12$, $z = 9$, $w = 10$.

8. The sum of three cubes may also be a cube. In fact, the solutions of $x^3 + y^3 + z^3 = w^3$ are given by $x = 3a^2 + 5b(a-b)$, $y = 4a(a-b) + 6b^2$, $z = 5a(a-b) - 3b^2$, $w = 6a^2 - 4b(a+b)$ where a and b are integers.

9. Euler conjectured that there were four fourth powers of positive integers whose sum is also the fourth power of an integer. In other words, he conjectured that there are nontrivial solutions to the diophantine equation $v^4 + w^4 + x^4 + y^4 = z^4$. The first such example was found in 1911 when it was discovered (by R. Norrie) that $30^4 + 120^4 + 272^4 + 315^4 = 353^4$.

10. Euler also conjectured that the sum of the fourth powers of three positive integers can never be the fourth power of an integer and that the sum of fifth powers of four positive integers can never be the fifth power of an integer, and so on. In other words, he conjectured that there were no nontrivial solutions to the Diophantine equations $w^4 + x^4 + y^4 = z^4$, $v^5 + w^5 + x^5 + y^5 = z^5$, and so on. He was mistaken. The smallest counterexamples known are $95{,}800^4 + 217{,}519^4 + 414{,}560^4 = 422{,}481^4$ and $27^5 + 84^5 + 110^5 + 133^5 = 144^5$.

11. If $n = mp$ for some integer m and p is prime, then the Fermat equation can be rewritten as $(x^m)^p + (y^m)^p = (z^m)^p$. Since the only positive integers greater than 2 without an odd prime factor are powers of 2 and $x^4 + y^4 = z^4$ has no nontrivial solutions in integers, Fermat's last theorem can be demonstrated by showing that $x^p + y^p = z^p$ has no nontrivial solutions in integers x, y, and z when p is an odd prime.

12. An odd prime p is regular if and only if it does not divide the numerator of any of the numbers $B_2, B_4, \ldots, B_{p-3}$, where B_k is the kth Bernoulli number. (See §3.1.4.)

13. There is a relatively simple proof of Fermat's last theorem for exponents that are regular primes.

14. The smallest irregular primes are 37, 59, 67, 101, 103, 149, and 157.

15. Wiles' proof of Fermat's last theorem is based on the theory of elliptic curves. The proof is based on relating to integers a, b, c, and n that supposedly satisfy the Fermat equation $a^n + b^n = c^n$ the elliptic curve $y^2 = x(x + a^n)(x - b^n)$ (called the *associated Frey curve*) and deriving a contradiction using sophisticated results from the theory of elliptic curves. (See Wiles' original proof [Wi95], the popular account [Si97], and http://www.best.com/ cgd/home/flt/flt01.htm (The Mathematics of Fermat's Last Theorem) and http://www.pbs.org/wgbh/nova/proof/ (NOVA Online | The Proof) for more details.)

4.8.4 PELL'S, BACHET'S, AND CATALAN'S EQUATIONS

Definitions:

Pell's equation is a diophantine equation of the form $x^2 - dy^2 = 1$, where d is a square-free positive integer. This diophantine equation is named after John Pell (1611–1685).

Bachet's equation is a diophantine equation of the form $y^2 = x^3 + k$. This diophantine equation is named after Claude Gaspar Bachet (1587–1638).

Catalan's equation is the diophantine equation $x^m - y^n = 1$, where a solution is sought with integers $x > 0$, $y > 0$, $m > 1$, and $n > 1$. This diophantine equation is named after Eugène Charles Catalan (1814–1894).

Facts:

1. If x, y is a solution to the diophantine equation $x^2 - dy^2 = n$ with d squarefree and $n^2 < d$, then the rational number $\frac{x}{y}$ is a convergent of the simple continued fraction for $\sqrt{d}$. (See §4.9.2.)
2. An equation of the form $ax'^2 + bx' + c = y'^2$ can be transformed by means of the relations $x = 2ax' + b$ and $y = 2y'$ into an equation of the form $x^2 - dy^2 = n$, where $n = b^2 - 4ac$ and $d = a$.
3. It is ironic that John Pell apparently had little to do with finding the solutions to the diophantine equation $x^2 - dy^2 = 1$. Euler gave this equation its name following a mistaken reference. Fermat conjectured an infinite number of solutions to this equation in 1657; this was eventually proved by Lagrange in 1768.
4. Let x, y be the least positive solution to $x^2 - dy^2 = 1$, with d squarefree. Then every positive solution is given by

$$x_k + y_k\sqrt{d} = (x + y\sqrt{d})^k$$

where k ranges over the positive integers.

5. Table 1 gives the smallest positive solutions to Pell's equation $x^2 - dy^2 = 1$ with d a squarefree positive integer less than 100.
6. If $k = 0$, then the formulae $x = t^2, y = t^3$ give an infinite number of solutions to the Bachet equation $y^2 = x^3 + k$.
7. There are no solutions to Bachet's equation for the following values of k: -144, -105, -78, -69, -42, -34, -33, -31, -24, -14, -5, 7, 11, 23, 34, 45, 58, 70.
8. The following table lists solutions to Bachet's equation for various values of k:

k	x
0	t^2 (t any integer)
1	$0, -1, 2$
17	$-1, -2, 2, 4, 8, 43, 52, 5334$
-2	3
-4	$2, 5$
-7	$2, 32$
-15	1

9. If $k < 0$, k is squarefree, $k \equiv 2$ or $3 \pmod 4$, and the class number of the field $\mathcal{Q}(\sqrt{-k})$ is not a multiple of 3, then the only solution of the Bachet equation $y^2 = x^3 + k$ for x is given by whichever of $-(4k \pm 1)/3$ is an integer. The first few values of such k are 1, 2, 5, 6, 10, 13, 14, 17, 21, and 22.
10. Solutions to the Catalan equation give consecutive integers that are powers of integers.
11. The Catalan equation has the solution $x = 3$, $y = 2$, $m = 2$, $n = 3$, so $8 = 2^3$ and $9 = 3^2$ are consecutive powers of integers. The *Catalan conjecture* is that this is the only solution.
12. Levi ben Gerson showed in the 14th century that 8 and 9 are the only consecutive powers of 2 and 3, so that the only solution in positive integers of $3^m - 2^n = \pm 1$ is $m = 2$ and $n = 3$.

Table 1 Smallest positive solutions to Pell's equation $x^2 - dy^2 = 1$ with d squarefree, $d < 100$.

d	x	y	d	x	y
2	3	2	51	50	7
3	2	1	53	66,249	9,100
5	9	4	55	89	12
6	5	2	57	151	20
7	8	3	58	19,603	2,574
10	19	6	59	530	69
11	10	3	61	1,766,319,049	226,153,980
13	649	180	62	63	8
14	15	4	65	129	16
15	4	1	66	65	8
17	33	8	67	48,842	5,967
19	170	39	69	7,775	936
21	55	12	70	251	30
22	197	42	71	3,480	413
23	24	5	73	2,281,249	267,000
26	51	10	74	3,699	430
29	9,801	1,820	77	351	40
30	11	2	78	53	6
31	1,520	273	79	80	9
33	23	4	82	163	18
34	35	6	83	82	9
35	6	1	85	285,769	30,996
37	73	12	86	10,405	1,122
38	37	6	87	28	3
39	25	4	89	500,001	53,000
41	2,049	320	91	1,574	165
42	13	2	93	12,151	1,260
43	3,482	531	94	2,143,295	221,064
46	24,335	3,588	95	39	4
47	48	7	97	62,809,633	6,377,352

13. Euler proved that the only solution in positive integers of $x^3 - y^2 = \pm 1$ is $x = 2$ and $y = 3$.

14. Lebesgue showed in 1850 that $x^m - y^2 = 1$ has no solutions in positive integers when m is an integer greater than 3.

15. The diophantine equations $x^3 - y^n = 1$ and $x^m - y^3 = 1$ with $m > 2$ were shown to have no solutions in positive integers in 1921, and in 1964 it was shown that $x^2 - y^n = 1$ has no solutions in positive integers.

16. R. Tijdeman showed in 1976 that there are only finitely many solutions in integers to the Catalan equation $x^m - y^n = 1$ by showing that there is a computable constant C such that for every solution, $x^m < C$ and $y^n < C$. However, the enormous size of the constant C makes it infeasible to establish the Catalan conjecture using computers.

Examples:

1. To solve the diophantine equation $x^2 - 13y^2 = 1$, note that the simple continued fraction for $\sqrt{13}$ is $[3; \overline{1,1,1,1,6}]$, with convergents $3, 4, \frac{7}{2}, \frac{11}{3}, \frac{18}{5}, \frac{119}{33}, \frac{137}{38}, \frac{256}{71}, \frac{393}{109}, \frac{649}{180}, \ldots$. The smallest positive solution to the equation is $x = 649$, $y = 180$. A second solution is given by $(649 + 180\sqrt{13})^2 = 842{,}401 + 233{,}640\sqrt{13}$, that is, $x = 842{,}401$, $y = 233{,}640$.

2. Congruence considerations can be used to show that there are no solutions of Bachet's equation for $k = 7$. Modulo 8, every square is congruent to 0, 1, or 4; therefore if x is even, then $y^2 \equiv 7 \pmod{8}$, a contradiction. Likewise if $x \equiv 3 \pmod 4$, then $y^2 \equiv 2 \pmod 8$, also impossible. So assume that $x \equiv 1 \pmod 4$. Add one to both sides and factor to get $y^2 + 1 = x^3 + 8 = (x+2)(x^2 - 2x + 4)$. Now $x^2 - 2x + 4 \equiv 3 \pmod 4$, so it must have a prime divisor $p \equiv 3 \pmod 4$. Then $y^2 \equiv -1 \pmod p$, which implies that -1 is a quadratic residue modulo p. (See §4.4.5.) But $p \equiv 3 \pmod 4$, so -1 cannot be a quadratic residue modulo p. Therefore, there are no solutions when $k = 7$.

4.8.5 SUMS OF SQUARES AND WARING'S PROBLEM

Definitions:

If k is a positive integer, then $g(k)$ is the smallest positive integer such that every positive integer can be written as a sum of $g(k)$ kth powers.

If k is a positive integer, then $G(k)$ is the smallest positive integer such that every *sufficiently large* positive integer can be written as a sum of $G(k)$ kth powers.

The determination of $g(k)$ is called **Waring's problem**. (Edward Waring, 1741–1793)

Facts:

1. A positive integer n is the sum of two squares if and only if each prime factor of n of the form $4k + 3$ appears to an even power in the prime factorization of n.

2. If $m = a^2 + b^2$ and $n = c^2 + d^2$, then the number mn can be expressed as the sum of two squares as follows: $mn = (ac + bd)^2 + (ad - bc)^2$.

3. If n is representable as the sum of two squares, then it is representable in $4(d_1 - d_2)$ ways (where the order of the squares and their signs matter), where d_1 is the number of divisors of n of the form $4k + 1$ and d_3 is the number of divisors of n of the form $4k + 3$.

4. An integer n is the sum of three squares if and only if n is not of the form $4^m(8k+7)$, where m is a nonnegative integer.

5. The positive integers less than 100 that are not the sum of three squares are $7, 15, 23, 28, 31, 39, 47, 55, 60, 63, 71, 79, 87, 92$, and 95.

6. *Lagrange's four-square theorem:* Every positive integer is the sum of 4 squares, some of which may be zero. (Joseph Lagrange, 1736–1813)

7. A useful lemma due to Lagrange is the following. If $m = a^2 + b^2 + c^2 + d^2$ and $n = e^2 + f^2 + g^2 + h^2$, then mn can be expressed as the sum of four squares as follows: $mn = (ae+bf+cg+dh)^2 + (af-be+ch-dg)^2 + (ag-ce+df-bh)^2 + (ah-de+bg-cf)^2$.

8. The number of ways n can be written as the sum of four squares is $8(s - s_4)$, where s is the sum of the divisors of n and s_4 is the sum of the divisors of n that are divisible by 4.

9. It is known that $g(k)$ always exists.

10. For $6 \leq k \leq 471{,}600{,}000$ the following formula holds except possibly for a finite number of positive integers k: $g(k) = \lfloor (\frac{3}{2})^k \rfloor + 2^k - 2$ where $\lfloor x \rfloor$ represents the floor (greatest integer) function.

11. The exact value of $G(k)$ is known only for two values of k, $G(2) = 4$ and $G(4) = 16$.

12. From Lagrange's results above it follows that $G(2) = g(2) = 4$.

13. If k is an integer with $k \geq 2$, then $G(k) \leq g(k)$.

14. If k is an integer with $k \geq 2$, then $G(k) \geq k + 1$.

15. Hardy and Littlewood showed that $G(k) \leq (k-2)2^{k-1} + 5$ and conjectured that $G(k) < 2k + 1$ when k is not a power of 2 and $G(k) < 4k$ when k is a power of 2.

16. The best upper bound known for $G(k)$ is $G(k) < ck \ln k$ for some constant c.

17. The known values and established estimates for $g(k)$ and $G(k)$ for $2 \leq k \leq 8$ are given in the following table.

$g(2) = 4$	$G(2) = 4$
$g(3) = 9$	$4 \leq G(3) \leq 7$
$g(4) = 19$	$G(4) = 16$
$g(5) = 37$	$6 \leq G(5) \leq 18$
$g(6) = 73$	$9 \leq G(6) \leq 27$
$143 \leq g(7) \leq 3{,}806$	$8 \leq G(7) \leq 36$
$279 \leq g(8) \leq 36{,}119$	$32 \leq G(8) \leq 42$

18. There are many related diophantine equations concerning sums and differences of powers. For instance $x = 1$, $y = 12$, $z = 9$, and $w = 10$ is the smallest solution to $x^3 + y^3 = z^3 + w^3$.

4.9 DIOPHANTINE APPROXIMATION

Diophantine approximation is the study of how closely a number θ can be approximated by numbers of some particular kind. Usually θ is an irrational (real) number, and the goal is to approximate θ using rational numbers $\frac{p}{q}$.

4.9.1 CONTINUED FRACTIONS

Definitions:

A **continued fraction** is a (finite or infinite) expression of the form

$$a_0 + \cfrac{1}{a_1 + \cfrac{1}{a_2 + \cfrac{1}{a_3 + \cfrac{1}{\ddots}}}}$$

The terms $a_0, a_1, \ldots$ are called the **partial quotients**. If the partial quotients are all integers, and $a_i \geq 1$ for $i \geq 1$, then the continued fraction is said to be **simple**. For convenience, the above expression is usually abbreviated as $[a_0, a_1, a_2, a_3, \ldots]$.

Algorithm 1: The continued fraction algorithm.

procedure $CFA(x$: real number)
$i := 0$
$x_0 := x$
$a_0 := \lfloor x_0 \rfloor$
output(a_0)
while $(x_i \neq a_i)$
begin
    $x_{i+1} := \frac{1}{x_i - a_i}$
    $i := i + 1$
    $a_i := \lfloor x_i \rfloor$
    output(a_i)
end
{returns finite or infinite sequence $(a_0, a_1, \ldots)$}

A continued fraction that has an expansion with a block that repeats after some point is called **ultimately periodic**. The ultimately periodic continued fraction expansion $[a_0, a_1, \ldots, a_N, a_{N+1}, \ldots, a_{N+k}, a_{N+1}, \ldots, a_{N+k}, a_{N+1}, \ldots]$ is often abbreviated as $[a_0, a_1, \ldots, a_N, \overline{a_{N+1}, \ldots, a_{N+k}}]$. The terms $a_0, a_1, \ldots, a_N$ are called the **pre-period** and the terms $a_{N+1}, a_{N+2}, \ldots, a_{N+k}$ are called the **period**.

Facts:

1. Every irrational number has a unique expansion as a simple continued fraction.

2. Every rational number has exactly two simple continued fraction expansion, one with an odd number of terms and one with an even number of terms. Of these, the one with the larger number of terms ends with 1.

3. The simple continued fraction for a real number r is finite if and only if r is rational.

4. The simple continued fraction for a real number r is infinite and ultimately periodic if and only if r is a quadratic irrational.

5. The simple continued fraction for $\sqrt{d}$, where d a positive integer that is not a square, is as follows: $\sqrt{d} = [a_0, \overline{a_1, a_2, \ldots, a_n, 2a_0}]$, where the sequence $(a_1, a_2, \ldots, a_n)$ is a palindrome.

6. The following table illustrates the three types of continued fractions.

type	kind of number	example
finite	rational	$\frac{355}{113} = [3, 7, 16]$
ultimately periodic	quadratic irrational	$\sqrt{2} = [1, 2, 2, 2, \ldots]$
infinite, but not ultimately periodic	neither rational nor quadratic irrational	$\pi = [3, 7, 15, 1, 292 \ldots]$

7. The continued fraction for a real number can be computed by Algorithm 1.
8. Continued fractions for $\sqrt{d}$, for $2 \leq d \leq 100$, are given in Table 1.
9. Continued fraction expansions for certain quadratic irrationals are given in Table 2.
10. Continued fraction expansions for some famous numbers are given in Table 3.

Table 1 Continued fractions for $\sqrt{d}$, $2 \leq d \leq 100$.

d	$\sqrt{d}$	d	$\sqrt{d}$
2	$[1,\overline{2}]$	53	$[7,\overline{3,1,1,3,14}]$
3	$[1,\overline{1,2}]$	54	$[7,\overline{2,1,6,1,2,14}]$
5	$[2,\overline{4}]$	55	$[7,\overline{2,2,2,14}]$
6	$[2,\overline{2,4}]$	56	$[7,\overline{2,14}]$
7	$[2,\overline{1,1,1,4}]$	57	$[7,\overline{1,1,4,1,1,14}]$
8	$[2,\overline{1,4}]$	58	$[7,\overline{1,1,1,1,1,14}]$
10	$[3,\overline{6}]$	59	$[7,\overline{1,2,7,2,1,14}]$
11	$[3,\overline{3,6}]$	60	$[7,\overline{1,2,1,14}]$
12	$[3,\overline{2,6}]$	61	$[7,\overline{1,4,3,1,2,2,1,3,4,1,14}]$
13	$[3,\overline{1,1,1,1,6}]$	62	$[7,\overline{1,6,1,14}]$
14	$[3,\overline{1,2,1,6}]$	63	$[7,\overline{1,14}]$
15	$[3,\overline{1,6}]$	65	$[8,\overline{16}]$
17	$[4,\overline{8}]$	66	$[8,\overline{8,16}]$
18	$[4,\overline{4,8}]$	67	$[8,\overline{5,2,1,1,7,1,1,2,5,16}]$
19	$[4,\overline{2,1,3,1,2,8}]$	68	$[8,\overline{4,16}]$
20	$[4,\overline{2,8}]$	69	$[8,\overline{3,3,1,4,1,3,3,16}]$
21	$[4,\overline{1,1,2,1,1,8}]$	70	$[8,\overline{2,1,2,1,2,16}]$
22	$[4,\overline{1,2,4,2,1,8}]$	71	$[8,\overline{2,2,1,7,1,2,2,16}]$
23	$[4,\overline{1,3,1,8}]$	72	$[8,\overline{2,16}]$
24	$[4,\overline{1,8}]$	73	$[8,\overline{1,1,5,5,1,1,16}]$
26	$[5,\overline{10}]$	74	$[8,\overline{1,1,1,1,16}]$
27	$[5,\overline{5,10}]$	75	$[8,\overline{1,1,1,16}]$
28	$[5,\overline{3,2,3,10}]$	76	$[8,\overline{1,2,1,1,5,4,5,1,1,2,1,16}]$
29	$[5,\overline{2,1,1,2,10}]$	77	$[8,\overline{1,3,2,3,1,16}]$
30	$[5,\overline{2,10}]$	78	$[8,\overline{1,4,1,16}]$
31	$[5,\overline{1,1,3,5,3,1,1,10}]$	79	$[8,\overline{1,7,1,16}]$
32	$[5,\overline{1,1,1,10}]$	80	$[8,\overline{1,16}]$
33	$[5,\overline{1,2,1,10}]$	82	$[9,\overline{18}]$
34	$[5,\overline{1,4,1,10}v]$	83	$[9,\overline{9,18}v]$
35	$[5,\overline{1,10}]$	84	$[9,\overline{6,18}]$
37	$[6,\overline{12}]$	85	$[9,\overline{4,1,1,4,18}]$
38	$[6,\overline{6,12}]$	86	$[9,\overline{3,1,1,1,8,1,1,1,3,18}]$
39	$[6,\overline{4,12}]$	87	$[9,\overline{3,18}]$
40	$[6,\overline{3,12}]$	88	$[9,\overline{2,1,1,1,2,18}]$
41	$[6,\overline{2,2,12}]$	89	$[9,\overline{2,3,3,2,18}]$
42	$[6,\overline{2,12}]$	90	$[9,\overline{2,18}]$
43	$[6,\overline{1,1,3,1,5,1,3,1,1,12}]$	91	$[9,\overline{1,1,5,1,5,1,1,18}]$
44	$[6,\overline{1,1,1,2,1,1,1,12}]$	92	$[9,\overline{1,1,2,4,2,1,1,18}]$
45	$[6,\overline{1,2,2,2,1,12}]$	93	$[9,\overline{1,1,1,4,6,4,1,1,1,18}]$
46	$[6,\overline{1,3,1,1,2,6,2,1,1,3,1,12}]$	94	$[9,\overline{1,2,3,1,1,5,1,8,1,5,1,1,3,2,1,18}]$
47	$[6,\overline{1,5,1,12}]$	95	$[9,\overline{1,2,1,18}]$
48	$[6,\overline{1,12}]$	96	$[9,\overline{1,3,1,18}]$
50	$[7,\overline{14}]$	97	$[9,\overline{1,5,1,1,1,1,1,1,5,1,18}]$
51	$[7,\overline{7,14}]$	98	$[9,\overline{1,8,1,18}]$
52	$[7,\overline{4,1,2,1,4,14}]$	99	$[9,\overline{1,18}]$

Table 2 Continued fractions for some special quadratic irrationals.

d	continued fraction expansion for $\sqrt{d}$
$\sqrt{n^2-1}$	$[n-1, \overline{1, 2n-2}]$
$\sqrt{n^2-2}$	$[n-1, \overline{1, n-2, 1, 2n-2}]$
$\sqrt{n^2+1}$	$[n, \overline{2n}]$
$\sqrt{n^2+2}$	$[n, \overline{n, 2n}]$
$\sqrt{n^2-n}$	$[n-1, \overline{2, 2n-2}]$
$\sqrt{n^2+n}$	$[n, \overline{2, 2n}]$
$\sqrt{4n^2+4}$	$[2n, \overline{n, 4n}]$
$\sqrt{4n^2-n}$	$[2n-1, \overline{1, 2, 1, 4n-2}]$
$\sqrt{4n^2+n}$	$[2n, \overline{4, 4n}]$
$\sqrt{9n^2+2n}$	$[3n, \overline{3, 6n}]$

Table 3 Continued fractions for some famous numbers. (See [Pe54].)

number	continued fraction expansion
π	$[3,7,15,1,292,1,1,1,2,1,3,1,14,2,1,1,2,2,2,2,1,84,2,1,1,15,3,\ldots]$
γ	$[0,1,1,2,1,2,1,4,3,13,5,1,1,8,1,2,4,1,1,40,1,11,3,7,1,7,1,1,5,\ldots]$
$\sqrt[3]{2}$	$[1,3,1,5,1,1,4,1,1,8,1,14,1,10,2,1,4,12,2,3,2,1,3,4,1,1,2,14,\ldots]$
$\log 2$	$[0,1,2,3,1,6,3,1,1,2,1,1,1,1,3,10,1,1,1,2,1,1,1,1,3,2,3,1,13,7,\ldots]$
e	$[2,1,2,1,1,4,1,1,6,1,1,8,1,1,10,1,1,12,\ldots]$
$e^{\frac{1}{n}}$	$[1, \ n-1, \ 1,1, \ 3n-1, \ 1,1, \ 5n-1, \ 1,1, \ 7n-1,\ldots]$
$e^{\frac{2}{2n+1}}$	$[1, \ \overline{(6n+3)k+n, \ (24n+12)k+12n+6, \ (6n+3)k+5n+2, \ 1, \ 1}_{k \geq 0}]$
$\tanh \frac{1}{n}$	$[0, \ n, \ 3n, \ 5n, \ 7n,\ldots]$
$\tan \frac{1}{n}$	$[0, n-1, 1, 3n-2, 1, 5n-2, 1, 7n-2, 1, 9n-2,\ldots]$
$\frac{1+\sqrt{5}}{2}$	$[1,1,1,1,\ldots]$

Examples:

1. To find the continued fraction representation of $\frac{62}{23}$, apply Algorithm 1 to obtain

$$\frac{62}{23} = 2 + \frac{1}{\frac{23}{16}}, \quad \frac{23}{16} = 1 + \frac{1}{\frac{16}{7}}, \quad \frac{16}{7} = 2 + \frac{1}{\frac{7}{2}}, \quad \frac{7}{2} = 3 + \frac{1}{2}.$$

Combining these equations shows that $\frac{62}{23} = [2,1,2,3,2]$. Since $2 = 1 + \frac{1}{1}$, it also follows that $\frac{62}{23} = [2,1,2,3,1,1]$.

2. Applying Algorithm 1 to find the continued fraction of $\sqrt{6}$, it follows that

$$a_0 = \lfloor \sqrt{6} \rfloor = 2, \quad a_1 = \lfloor \tfrac{\sqrt{6}+2}{2} \rfloor = 2, \quad a_2 = \lfloor \sqrt{6}+2 \rfloor = 4, \quad a_3 = a_1, \quad a_4 = a_2, \ldots.$$

Hence $\sqrt{6} = [2, \overline{2,4}]$.

3. The continued fraction expansion of e is $e = [2,1,2,1,1,4,1,1,6,\ldots]$. This expansion is often abbreviated as $[2, \overline{1, 2k, 1}_{k \geq 1}]$. (See [Pe54].)

4.9.2 CONVERGENTS

Definition:

Define $p_{-2} = 0$, $q_{-2} = 1$, $p_{-1} = 1$, $q_{-1} = 0$, and $p_n = a_n p_{n-1} + p_{n-2}$ and $q_n = a_n q_{n-1} + q_{n-2}$ for $n \geq 0$. Then $\frac{p_n}{q_n} = [a_0, a_1, \ldots, a_n]$. The fraction $\frac{p_n}{q_n}$ is called the nth *convergent*.

Facts:

1. $p_n q_{n-1} - p_{n-1} q_n = (-1)^{n+1}$ for $n \geq 0$.

2. Let $\theta = [a_0, a_1, a_2, \ldots]$ be an irrational number. Then $\left|\theta - \frac{p_n}{q_n}\right| < \frac{1}{a_{n+1} q_n^2}$.

3. If $n > 1$, $0 < q \leq q_n$, and $\frac{p}{q} \neq \frac{p_n}{q_n}$, then $\left|\theta - \frac{p}{q}\right| > \left|\theta - \frac{p_n}{q_n}\right|$.

4. $[\ldots, a, b, 0, c, d, \ldots] = [\ldots, a, b+c, d, \ldots]$.

5. Almost all real numbers have unbounded partial quotients.

6. For almost all real numbers, the frequency with which the partial quotient k occurs is $\log_2\left(1 + \frac{1}{k(k+2)}\right)$. Hence, the partial quotient 1 occurs about 41.5% of the time, the partial quotient 2 occurs about 17.0% of the time, etc.

7. For almost all real numbers,
$$\lim_{n \to \infty} (a_1 a_2 \ldots a_n)^{\frac{1}{n}} = K \approx 2.68545.$$
K is called *Khintchine's constant*.

8. *Lévy's law*: For almost all real numbers,
$$\lim_{n \to \infty} (p_n)^{\frac{1}{n}} = \lim_{n \to \infty} (q_n)^{\frac{1}{n}} = e^{\frac{\pi^2}{12 \log 2}}.$$

Examples:

1. Compute the first eight convergents to π:

$n =$	-2	-1	0	1	2	3	4	5	6	7	8
$a_n =$			3	7	15	1	292	1	1	1	2
$p_n =$	0	1	3	22	333	355	103,993	104,348	208,341	312,689	833,719
$q_n =$	1	0	1	7	106	113	33,102	33,215	66,317	99,532	265,381

2. Find a rational fraction $\frac{p}{q}$ in lowest terms that approximates e to within 10^{-6}. Compute the convergents q_n until $a_{n+1}(q_n)^2 < 10^{-6}$:

$n =$	-2	-1	0	1	2	3	4	5	6	7	8	9	10
$a_n =$			2	1	2	1	1	4	1	1	6	1	1
$p_n =$	0	1	2	3	8	11	19	87	106	193	1,264	1,457	2,721
$q_n =$	1	0	1	1	3	4	7	32	39	71	465	536	1,001

Hence, $\frac{2721}{1001} \approx 2.71828171$ is the desired fraction.

4.9.3 APPROXIMATION THEOREMS

Facts:

1. *Dirichlet's theorem*: If θ is irrational, then
$$\left|\theta - \frac{p}{q}\right| < \frac{1}{q^2}$$
for infinitely many p, q.

2. *Dirichlet's theorem in d dimensions*: If $\theta_1, \theta_2, \ldots, \theta_d$ are real numbers with at least one θ_i irrational, then
$$\left|\theta_i - \frac{p_i}{q}\right| < \frac{1}{q^{1+\frac{1}{d}}}$$
for infinitely many $p_1, p_2, \ldots, p_d, q$.

3. *Hurwitz's theorem*: If θ is an irrational number, then
$$\left|\theta - \frac{p}{q}\right| < \frac{1}{\sqrt{5}q^2}$$
for infinitely many p, q. The constant $\sqrt{5}$ is best possible.

4. *Liouville's theorem*: Let θ be an irrational algebraic number of degree n. Then there exists a constant c (depending on θ) such that
$$\left|\theta - \frac{p}{q}\right| > \frac{c}{q^n}$$
for all rationals $\frac{p}{q}$ with $q > 0$. The number θ is called a *Liouville number* if $\left|\theta - \frac{p}{q}\right| < q^{-n}$ has a solution for all $n \geq 0$. An example of a Liouville number is $\sum_{k \geq 1} 2^{-k!}$.

5. *Roth's theorem*: Let θ be an irrational algebraic number, and let ϵ be any positive number. Then
$$\left|\theta - \frac{p}{q}\right| > \frac{1}{q^{2+\epsilon}}$$
for all but finitely many rationals $\frac{p}{q}$ with $q > 0$.

4.9.4 IRRATIONALITY MEASURES

Definition:

Let θ be a real irrational number. Then the real number μ is said to be an **irrationality measure** for θ if for every $\epsilon > 0$ there exists a positive real $q_0 = q_0(\epsilon)$ such that $\left|\theta - \frac{p}{q}\right| > q^{-(\mu+\epsilon)}$ for all integers p, q with $q > q_0$.

Fact:

1. Here are the best irrationality measures known for some important numbers.

number θ	measure μ	discoverer
π	8.0161	Hata (1993)
π^2	5.4413	Rhin and Viola (1995)
$\zeta(3)$	8.8303	Hata (1990)
$\ln 2$	3.8914	Rukhadze (1987); Hata (1990)
$\frac{\pi}{\sqrt{3}}$	4.6016	Hata (1993)

4.10 QUADRATIC FIELDS

4.10.1 BASICS

Definitions:

A complex number α is an **algeraic number** if it is a root of a polynomial with integer coefficients.

An algebraic number α is an **algebraic integer** if it is a root of a monic polynomial with integer coefficients. (A **monic polynomial** is a polynomial with leading coefficient equal to 1.)

An algebraic number α is of **degree** n if it is a root of a polynomial with integer coefficients of degree n but is not a root of any polynomial with integer coefficients of degree less than n.

An **algebraic number field** is a subfield of the field of algebraic numbers.

If α is an algebraic number with minimal polynomial $f(x)$ of degree n, then the $n-1$ other roots of $f(x)$ are called the **conjugates** of α.

The **integers** of an algebraic number field are the algebraic integers that belong to this field.

If d is a squarefree integer, then $Q(\sqrt{d}) = \{a + b\sqrt{d} \mid a \text{ and } b \text{ are rational numbers}\}$ is called a **quadratic field**. If $d > 0$, then $Q(\sqrt{d})$ is called a **real quadratic field**; if $d < 0$, then $Q(\sqrt{d})$ is called an **imaginary quadratic field**.

A number α in $Q(\sqrt{d})$ is a **quadratic integer** (or an **integer** when the context is clear) if α is an algebraic integer.

If α and β are quadratic integers in $Q(\sqrt{d})$ and there is a quadratic integer γ in $Q(\sqrt{d})$ such that $\alpha\gamma = \beta$, then α divides β, written $\alpha|\beta$.

The integers of $Q(\sqrt{-1})$ are called the **Gaussian integers**. (These are the numbers in $\mathcal{Z}[i] = \{a + bi \mid a, b \text{ are integers}\}$. See §5.4.2.)

If $\alpha = a + b\sqrt{d}$ belongs to $Q(\sqrt{d})$, then its **conjugate**, denoted by $\overline{\alpha}$, is the number $a - b\sqrt{d}$.

If α belongs to $Q(\sqrt{d})$, then the **norm** of α is the number $N(\alpha) = \alpha\overline{\alpha}$.

An algebraic integer ϵ in $Q(\sqrt{d})$ is a **unit** if $\epsilon | 1$.

Facts:

1. The integers of the field $Q(\sqrt{d})$, where d is a squarefree integer, are the numbers $a + b\sqrt{d}$ when $d \equiv 2$ or $3 \pmod{4}$ and the numbers $\frac{a+b\sqrt{d}}{2}$, where a and b are integers which are either both even or both odd.

2. If $d < 0$, $d \neq -1$, $d \neq -3$, then there are exactly two units, ± 1, in $Q(\sqrt{d})$. There are exactly four units in $Q(\sqrt{-1})$, namely ± 1 and $\pm i$. There are exactly six units in $Q(\sqrt{-3})$: ± 1, $\pm\frac{-1+\sqrt{-3}}{2}$, $\pm\frac{-1-\sqrt{-3}}{2}$.

3. If $d > 0$, there are infinitely many units in $Q(\sqrt{d})$. Furthermore, there is a unit ϵ_0, called the *fundamental unit* of $Q(\sqrt{d})$ such that all units are of the form $\pm\epsilon_0^n$ where n is an integer.

Examples:

1. The conjugate of $-2+3i$ in the ring of Gaussian integers is $-2-3i$. Consequently, $N(-2+3i) = (-2-3i)(-2+3i) = 13$.

2. The number $1+\sqrt{2}$ is a fundamental unit of $Q(\sqrt{2})$. Therefore, all units are of the form $\pm(1+\sqrt{2})^n$ where $n = 0, \pm 1, \pm 2, \ldots$.

4.10.2 PRIMES AND UNIQUE FACTORIZATION

Definitions:

An integer π in $Q(\sqrt{d})$, not zero or a unit, is **prime** in $Q(\sqrt{d})$ if whenever $\pi = \alpha\beta$ where α and β are integers in $Q(\sqrt{d})$, either α or β is a unit.

If α and β are nonzero integers in $Q(\sqrt{d})$ and $\alpha = \beta\epsilon$ where ϵ is a unit, then β is called an **associate** of α.

A quadratic field $Q(\sqrt{d})$ is a **Euclidean field** if, given integers α and β in $Q(\sqrt{d})$ where β is not zero, there are integers δ and γ in $Q(\sqrt{d})$ such that $\alpha = \gamma\beta + \delta$ and $|N(\delta)| < |N(\beta)|$.

A quadratic field $Q(\sqrt{d})$ has the **unique factorization property** if whenever α is a nonzero, non-unit, integer in $Q(\sqrt{d})$ with two factorizations $\alpha = \epsilon\pi_1\pi_2\ldots\pi_r = \epsilon'\pi'_1\pi'_2\ldots\pi'_s$ where ϵ and ϵ' are units, then $r = s$ and the primes π_i and π'_j can be paired off into pairs of associates.

Facts:

1. If α is an integer in $Q(\sqrt{d})$ and $N(\alpha)$ is an integer that is prime, then α is a prime.
2. The integers of $Q(\sqrt{d})$ are a unique factorization domain if and only if whenever a prime $\pi | \alpha\beta$ where α and β are integers of $Q(\sqrt{d})$, then $\pi | \alpha$ or $\pi | \beta$.
3. A Euclidean quadratic field has the unique factorization property.
4. The quadratic field $Q(\sqrt{d})$ is Euclidean if and only if d is one of the following integers: $-11, -7, -3, -2, -1, 2, 3, 5, 6, 7, 11, 13, 17, 19, 21, 29, 33, 37, 41, 57, 73$.
5. If $d < 0$, then the imaginary quadratic field $Q(\sqrt{d})$ has the unique factorization property if and only if $d = -1, -2, -3, -7, -11, -19, -43, -67$, or -163. This theorem was stated as a conjecture by Gauss in the 10th century and proved in the 1960s by Harold Stark and Roger Baker independently.
6. It is unknown whether infinitely many real quadratic fields $Q(\sqrt{d})$ have the unique factorization property.
7. Of the 60 real quadratic fields $Q(\sqrt{d})$ with $2 \leq d \leq 100$, exactly 38 have the unique factorization property, namely those with $d = 2, 3, 5, 6, 7, 11, 13\ 14, 17, 19, 21, 22, 23, 29, 31, 33, 37, 38, 41, 43, 46, 47, 53, 57, 59, 61, 62, 67, 69, 71, 73, 77, 83, 86, 89, 93, 94$, and 97.

Examples:

1. The number $2+i$ is a prime Gaussian integer. This follows since its norm $N(2+i) = (2+i)(2-i) = 5$ is a prime integer. Its associates are itself and the three Gaussian integers $(-1)(2+i) = -2-i$, $i(2+i) = -1+2i$, and $-i(2+i) = 1-2i$.

2. The integers of $\mathcal{Q}(\sqrt{-5})$ are the numbers of the form $a + b\sqrt{-5}$ where a and b are integers. The field $\mathcal{Q}(\sqrt{-5})$ is not a unique factorization domain. To see this, note that $6 = 2 \cdot 3 = (1 + \sqrt{-5})(1 - \sqrt{-5})$ and each of 2, 3, $1 + \sqrt{-5}$, and $1 - \sqrt{-5}$ are primes in this quadratic field. For example, to see that $1 + \sqrt{-5}$ is prime, suppose that $1 + \sqrt{-5} = (a + b\sqrt{-5})(c + d\sqrt{-5})$. This implies that $6 = (a^2 + 5b^2)(c^2 + 5d^2)$, which is impossible unless $a = \pm 1$, $b = 0$ or $c = \pm 1$, $d = 0$. Consequently, one of the factors must be a unit.

REFERENCES

Printed Resources:

[An76] G. E. Andrews, *The Theory of Partitions*, Encyclopedia of Mathematics and Its Applications, vol. 2, Addison-Wesley, 1976. (Reissued: Cambridge University Press, 1984)

[Ap76] T. M. Apostol, *Introduction to Analytic Number Theory*, Springer-Verlag, 1976.

[BaSh96] E. Bach and J. Shallit, *Algorithmic Number Theory, Volume 1, Efficient Algorithms*, MIT Press, 1996.

[Ba90] A. Baker, *Transcendental Number Theory*, Cambridge University Press, 1990.

[Br89] D. M. Bressoud, *Factorization and Primality Testing*, Springer-Verlag, 1989.

[BrEtal88] J. Brillhart, D. H. Lehmer, J. L. Selfridge, B. Tuckerman, and S. S. Wagstaff, Jr., "Factorizations of $b^n \pm 1$, $b = 2, 3, 5, 6, 7, 10, 11, 12$ up to high powers", 2nd ed., *Contemporary Mathematics*, 22, American Mathematical Society, 1988.

[Ca57] J. W. S. Cassels, *An Introduction to Diophantine Approximation*, Cambridge University Press, 1957.

[Co93] H. Cohen, *A Course in Computational Algebraic Number Theory*, Springer-Verlag, 1993.

[CrPo99] R. E. Crandall and C. Pomerance, *Primes: A computational perspective*, Springer-Verlag, 1999.

[Di71] L. E. Dickson, *History of the Theory of Numbers*, Chelsea Publishing Company, 1971.

[GuMu84] R. Gupta and M. R. Murty, "A remark on Artin's Conjecture," *Inventiones Math.*, 78 (1984), 127–130.

[Gu94] R. K. Guy, *Unsolved Problems in Number Theory*, 2nd ed., Springer-Verlag, 1994.

[HaWr89] G. H. Hardy and E. M. Wright, *An Introduction to the Theory of Numbers*, 5th ed., Oxford University Press, 1980.

[Kn81] D. E. Knuth, *The Art of Computer Programming, Volume 2, Seminumerical Algorithms*, 2nd ed., Addison-Wesley, 1981.

[Ko93] I. Koren, *Computer Arithmetic Algorithms*, Prentice Hall, 1993.

[MevaVa96] A. J. Menezes, P. C. van Oorschot, S. A Vanstone, *Handbook of Applied Cryptography*, CRC Press, 1997.

[Mo69] L. J. Mordell, *Diophantine Equations*, Academic Press, 1969.

[NiZuMo91] I. Niven, H. S. Zuckerman, and H. L. Montgomery, *An Introduction to the Theory of Numbers*, 5th ed., Wiley, 1991.

[Pe54] O. Perron, *Die Lehre von den Kettenbrüchen*, 3rd ed., Teubner Verlagsgesellschaft, 1954.

[Po90] C. Pomerance, ed., *Cryptology and computational number theory*, Proceedings of Symposia in Applied Mathematics, 42, American Mathematical Society, 1990.

[Po94] C. Pomerance, "The number field sieve", in *Mathematics of computation 1943–1993: a half-century of computational mathematics*, W. Gautschi, ed., Proceedings of Symposia in Applied Mathematics, 48, American Mathematical Society, 1994, 465–480.

[Ri96] P. Ribenboim, *The New Book of Prime Number Records*, Springer-Verlag, 1996.

[Ro99] K. H. Rosen, *Elementary Number Theory and Its Applications*, 4th ed., Addison-Wesley, 1999.

[Sc80] W. M. Schmidt, *Diophantine Approximation*, Lecture Notes in Mathematics, 785, Springer-Verlag, 1980.

[Sc85] N. R. Scott, *Computer Number Systems and Arithmetic*, Prentice Hall, 1985.

[Si97] S. Singh, *The Quest to Solve the World's Greatest Mathematical Problem*, Walker & Co., 1997.

[St95] D. R. Stinson, *Cryptography: Theory and Practice*, CRC Press, 1995.

[Wi95] A. J. Wiles, "Modular Elliptic Curves and Fermat's Last Theorem", *Annals of Mathematics*, second series, vol. 141, no. 3, May 1995.

Web Resources:

http://www.best.com/~cgd/home/flt/flt01.htm   (The mathematics of Fermat's Last Theorem)

http://www.math.uga.edu/~ntheory/web.html (The Number Theory Web)

http://www.mersenne.org/ (The Great Internet Mersenne Prime Search)

http://www.utm.edu/research/primes (The Prime Pages)

http://www-groups.dcs.st-and.ac.uk/~history/HistTopics/Fermat's_last_theorem.html (The history of Fermat's Last Theorem)

http://www.cs.purdue.edu/homes/ssw/cun/index.html (The Cunningham Project)

http://www.pbs.org/wgbh/nova/proof/ (NOVA Online | The Proof)

5

ALGEBRAIC STRUCTURES

John G. Michaels

5.1 Algebraic Models
   5.1.1 Domains and Operations
   5.1.2 Semigroups and Monoids

5.2 Groups
   5.2.1 Basic Concepts
   5.2.2 Group Isomorphism and Homomorphism
   5.2.3 Subgroups
   5.2.4 Cosets and Quotient Groups
   5.2.5 Cyclic Groups and Order
   5.2.6 Sylow Theory
   5.2.7 Simple Groups
   5.2.8 Group Presentations

5.3 Permutation Groups
   5.3.1 Basic Concepts
   5.3.2 Examples of Permutation Groups

5.4 Rings
   5.4.1 Basic Concepts
   5.4.2 Subrings and Ideals
   5.4.3 Ring Homomorphism and Isomorphism
   5.4.4 Quotient Rings
   5.4.5 Rings with Additional Properties

5.5 Polynomial Rings
   5.5.1 Basic Concepts
   5.5.2 Polynomials over a Field

5.6 Fields
   5.6.1 Basic Concepts
   5.6.2 Extension Fields and Galois Theory
   5.6.3 Finite Fields

5.7 Lattices
   5.7.1 Basic Concepts
   5.7.2 Specialized Lattices

5.8 Boolean Algebras
   5.8.1 Basic Concepts
   5.8.2 Boolean Functions
   5.8.3 Logic Gates
   5.8.4 Minimization of Circuits

Chapter 5 ALGEBRAIC STRUCTURES

INTRODUCTION

Many of the most common mathematical systems, including the integers, the rational numbers, and the real numbers, have an underlying algebraic structure. This chapter examines the structure and properties of various types of algebraic objects. These objects arise in a variety of settings and occur in many different applications, including counting techniques, coding theory, information theory, engineering, and circuit design.

GLOSSARY

abelian group: a group in which $a \star b = b \star a$ for all a, b in the group.

absorption laws: in a lattice $a \vee (a \wedge b) = a$ and $a \wedge (a \vee b) = a$.

algebraic element (over a field): given a field F, an element $\alpha \in K$ (extension of F) such that there exists $p(x) \in F[x]$ ($p(x) \neq 0$) such that $p(\alpha) = 0$. Otherwise α is **transcendental** over F.

algebraic extension (of a field): given a field F, a field K such that F is a subfield of K and all elements of K are algebraic over F.

algebraic integer: an algebraic number that is a zero of a monic polynomial with coefficients in $\mathcal{Z}$.

algebraic number: a complex number that is algebraic over $\mathcal{Q}$.

algebraic structure: $(S, \star_1, \star_2, \ldots, \star_n)$ where S is a nonempty set and $\star_1, \ldots, \star_n$ are binary or monadic operations defined on S.

alternating group (on n elements): the subgroup A_n of all even permutations in S_n.

associative property: the property of a binary operator $\star$ that $(a \star b) \star c = a \star (b \star c)$.

atom: an element a in a bounded lattice such that $0 < a$ and there is no element b such that $0 < b < a$.

automorphism: an isomorphism of an algebraic structure onto itself.

automorphism φ fixes set S elementwise: $\varphi(a) = a$ for all $a \in S$.

binary operation (on a set S): a function $\star \colon S \times S \to S$.

Boolean algebra: a bounded, distributive, complemented lattice. Equivalent definition: $(B, +, \cdot, ', 0, 1)$ where B is a set with two binary operations, $+$ (addition) and $\cdot$ (multiplication), one monadic operation, $'$ (complement), and two distinct elements, 0 and 1, that satisfy the commutative laws ($a + b = b + a$, $ab = ba$), distributive laws ($a(b + c) = (ab) + (ac)$, $a + (bc) = (a + b)(a + c)$), identity laws ($a + 0 = a$, $a1 = a$), and complement laws ($a + a' = 1$, $aa' = 0$).

Boolean function of degree n: a function $f \colon \{0, 1\}^n = \{0, 1\} \times \cdots \times \{0, 1\} \to \{0, 1\}$.

bounded lattice: a lattice having elements 0 (*lower bound*) and 1 (*upper bound*) such that $0 \leq a$ and $a \leq 1$ for all a.

cancellation properties: if $ab = ac$ and $a \neq 0$, then $b = c$ (**left cancellation property**); if $ba = ca$ and $a \neq 0$, then $b = c$ (**right cancellation property**).

characteristic (of a field): the smallest positive integer n such that $1 + 1 + \cdots + 1 = 0$ (n summands). If no such n exists, the field has characteristic 0 (or characteristic ∞).

closure property: a set S is closed under an operation $\star$ if the range of $\star$ is a subset of S.

commutative property: the property of an operation $\star$ that $a \star b = b \star a$.

commutative ring: a ring in which multiplication is commutative.

complemented lattice: a bounded lattice such that for each element a there is an element b such that $a \vee b = 1$ and $a \wedge b = 0$.

conjunctive normal form (CNF) (of a Boolean function): a Boolean function written as a product of maxterms.

coset: For subgroup H of group G and $a \in G$, a **left coset** is $aH = \{ah \mid h \in H\}$; a **right coset** is $Ha = \{ha \mid h \in H\}$.

cycle of length n: a permutation on a set S that moves elements only in a single orbit of size n.

cyclic group: a group G with an element $a \in G$ such that $G = \{a^n \mid n \in \mathcal{Z}\}$.

cyclic subgroup (generated by a): $\{a^n \mid n \in \mathcal{Z}\} = \{\ldots, a^{-2}, a^{-1}, e, a, a^2, \ldots\}$, often written (a), $\langle a \rangle$, or $[a]$. The element a is a **generator** of the subgroup.

degree (of field K over field F): $[K:F]$ = the dimension of K as a vector space over F.

degree (of a permutation group): the size of the set on which the permutations are defined.

dihedral group: the group D_n of symmetries (rotations and reflections) of a regular n-gon.

disjunctive normal form (DNF) (of a Boolean function): a Boolean function written as a sum of minterms.

distributive lattice: a lattice that satisfies $a \wedge (b \vee c) = (a \wedge b) \vee (a \wedge c)$ and $a \vee (b \wedge c) = (a \vee b) \wedge (a \vee c)$ for all a, b, c in the lattice.

division ring: a nontrivial ring in which every nonzero element is a unit.

dual (of an expression in a Boolean algebra): the expression obtained by interchanging the operations $+$ and $\cdot$ and interchanging the elements 0 and 1 in the original expression.

duality principle: the principle stating that an identity between Boolean expressions remains valid when the duals of the expressions are taken.

Euclidean domain: an integral domain with a Euclidean norm defined on it.

Euclidean norm (on an integral domain): given an integral domain I, a function $\delta: I - \{0\} \to \mathcal{N}$ such that for all $a, b \in I$, $\delta(a) \leq \delta(ab)$; and for all $a, d \in I$ ($d \neq 0$) there are $q, r \in I$ such that $a = dq + r$, where either $r = 0$ or $\delta(r) < \delta(d)$.

even permutation: a permutation that can be written as a product of an even number of transpositions.

extension field (of field F): field K such that F is a subfield of K.

field: an algebraic structure $(F, +, \cdot)$ where F is a set closed under two binary operations $+$ and $\cdot$, $(F, +)$ is an abelian group, the nonzero elements form an abelian group under multiplication, and the distributive law $a \cdot (b + c) = a \cdot b + a \cdot c$ holds.

finite field: a field with a finite number of elements.

finitely generated group: a group with a finite set of generators.

fixed field (of a set of automorphisms of a field): given a set Φ of automorphisms of a field F, the set $\{a \in F \mid a\varphi = a \text{ for all } \varphi \in \Phi\}$.

free monoid (generated by a set): given a set S, the monoid consisting of all words on S under concatenation.

functionally complete: property of a set of operators in a Boolean algebra that every Boolean function can be written using only these operators.

Galois extension (of a field F): a field K that is a normal, separable extension of F.

Galois field: $GF(p^n)$ = the algebraic extension $Z_p[x]/(f(x))$ of the finite field Z_p where p is a prime and $f(x)$ is an irreducible polynomial over Z_p of degree n.

Galois group (of K over F): the group of automorphisms $G(K/F)$ of field K that fix field F elementwise.

group: an algebraic structure $(G, \star)$, where G is a set closed under the binary operation $\star$, the operation $\star$ is associative, G has an identity element, and every element of G has an inverse in G.

homomorphism of groups: a function $\varphi: S \to T$, where $(S, \star_1)$ and $(T, \star_2)$ are groups, such that $\varphi(a \star_1 b) = \varphi(a) \star_2 \varphi(b)$ for all $a, b \in S$.

homomorphism of rings: a function $\varphi: S \to T$, where $(S, +_1, \cdot_1)$ and $(T, +_2, \cdot_2)$ are rings such that $\varphi(a +_1 b) = \varphi(a) +_2 \varphi(b)$ and $\varphi(a \cdot_1 b) = \varphi(a) \cdot_2 \varphi(b)$ for all $a, b \in S$.

ideal: a subring of a ring that is closed under left and right multiplication by elements of the ring.

identity: an element e in an algebraic structure S such that $e \star a = a \star e = a$ for all $a \in S$.

improper subgroups (of G): the subgroups G and $\{e\}$.

index of H in G: the number of left (or tight) cosets of H in G.

integral domain: a commutative ring with unity that has no zero divisors.

inverse of an element a: an element a' such that $a \star a' = a' \star a = e$.

involution: a function that is the identity when it is composed with itself.

irreducible element in a ring: a noninvertible element that cannot be written as the product of noninvertible elements.

irreducible polynomial: a polynomial $p(x)$ of degree $n > 0$ over a field that cannot be written as $p_1(x) \cdot p_2(x)$ where $p_1(x)$ and $p_2(x)$ are polynomials of smaller degrees. Otherwise $p(x)$ is reducible.

isomorphic: property of algebraic structures of the same type, G and H, that there is an isomorphism from G onto H, written $G \cong H$.

isomorphism: a one-to-one and onto function between two algebraic structures that preserves the operations on the structures.

isomorphism of groups: for groups $(G_1, \star_1)$ and $(G_2, \star_2)$, a function $\varphi: G_1 \to G_2$ that is one-to-one, onto G_2, and satisfies the property $\varphi(a \star_1 b) = \varphi(a) \star_2 \varphi(b)$.

isomorphism of permutation groups: for permutation groups (G, X) and (H, Y), a pair of functions $(\alpha: G \to H, f: Y \to Y)$ such that α is a group isomorphism and f is a bijection.

isomorphism of rings: for rings $(R_1, +_1, \cdot_1)$ and $(R_2, +_2, \cdot_2)$, a function $\varphi: R_1 \to R_2$ that is one-to-one, onto R_2, and satisfies the properties $\varphi(a +_1 b) = \varphi(a) +_2 \varphi(b)$ and $\varphi(a \cdot_1 b) = \varphi(a) \cdot_2 \varphi(b)$.

kernel (of a group homomorphism): given a group homomorphism φ, the set $\varphi^{-1}(e) = \{x \mid \varphi(x) = e\}$, where e is the group identity.

kernel (of a ring homomorphism): given a ring homomorphism φ, the set $\varphi^{-1}(0) = \{x \mid \varphi(x) = 0\}$.

Klein four-group: the group under composition of the four rigid motions of a rectangle that leave the rectangle in its original location.

lattice: a nonempty partially ordered set in which $\inf\{a,b\}$ and $\sup\{a,b\}$ exist for all a,b. ($a \vee b = \sup\{a,b\}$, $a \wedge b = \inf\{a,b\}$.) Equivalently, a nonempty set closed under two binary operations $\vee$ and $\wedge$ that satisfy the associative laws, the commutative laws, and the absorption laws ($a \vee (a \wedge b) = a$, $a \wedge (a \vee b) = a$).

left divisor of zero: $a \neq 0$ with $b \neq 0$ such that $ab = 0$.

literal: a Boolean variable or its complement.

maximal ideal: an ideal in a ring R that is not properly contained in any ideal of R except R itself.

maxterm of the Boolean variables $x_1, \ldots, x_n$: a sum of the form $y_1 + \cdots + y_n$ where for each i, y_i is equal to x_i or x_i'.

minimal polynomial (of an element with respect to a field): given a field F and $\alpha \in F$, the monic irreducible polynomial $f(x) \in F[x]$ of smallest degree such that $f(\alpha) = 0$.

minterm of the Boolean variables $x_1, \ldots, x_n$: a product of the form $y_1 \cdots y_n$ where for each i, y_i is equal to x_i or x_i'.

monadic operation: a function from a set into itself.

monoid: an algebraic structure $(S, \star)$ such that $\star$ is associative and S has an identity.

normal extension of F: a field K such that K/F is algebraic and every irreducible polynomial in $F[x]$ with a root in K has all its roots in K (**splits** in K).

normal subgroup (of a group): given a group G, a subgroup $H \subseteq G$ such that $aH = Ha$ for all $a \in G$.

octic group: See *dihedral group*.

odd permutation: a permutation that can be written as a product of an odd number of transpositions.

orbit (of an object $a \in S$ under permutation σ): $\{\ldots, a\sigma^{-2}, a\sigma^{-1}, a, a\sigma, a\sigma^2, \ldots\}$.

order (of an algebraic structure): the number of elements in the underlying set.

order (of a group element): for an element $a \in G$, the smallest positive integer n such that $a^n = e$ ($na = 0$ if G is written additively). If there is no such integer, then a has **infinite** order.

p-group: for prime p, a group such that every element has a power of p as its order.

permutation: a one-to-one and onto function $\sigma: S \to S$, where S is any nonempty set.

permutation group: a collection of permutations on a set of objects that form a group under composition.

polynomial (in the variable x over a ring): an expression of the form $p(x) = a_n x^n + a_{n-1} x^{n-1} + \cdots + a_1 x^1 + a_0 x^0$ where $a_n, \ldots, a_0$ are elements of the ring. For a polynomial $p(x)$, the largest integer k such that $a_k \neq 0$ is the **degree** of $p(x)$. The **constant polynomial** $p(x) = a_0$ has degree 0, if $a_0 \neq 0$. If $p(x) = 0$ (**zero polynomial**), the degree of $p(x)$ is undefined (or $-\infty$).

polynomial ring (over a ring R): $R[x] = \{p(x) \mid p(x)$ is a polynomial in x over $R\}$ with the usual definitions of addition and multiplication.

prime ideal (of a ring R): an ideal $I \neq R$ with property that $ab \in I$ implies that $a \in I$ or $b \in I$.

proper subgroup (of a group G): any subgroup of G except G and $\{e\}$.

quotient group (**factor group**): for normal subgroup H of G, the group $G/H = \{aH \mid a \in G\}$, where $aH \cdot bH = (ab)H$.

quotient ring: for I an ideal in a ring R, the ring $R/I = \{a + I \mid a \in R\}$, where $(a + I) + (b + I) = (a + b) + I$ and $(a + I) \cdot (b + I) = (ab) + I$.

reducible (polynomial): a polynomial that is not irreducible.

right divisor of zero: $b \neq 0$ with $a \neq 0$ such that $ab = 0$.

ring: an algebraic structure $(R, +, \cdot)$ where R is a set closed under two binary operations $+$ and $\cdot$, $(R, +)$ is an abelian group, R satisfies the associative law for multiplication, and R satisfies the left and right distributive laws for multiplication over addition.

ring with unity: a ring with an identity for multiplication.

root field: a splitting field.

semigroup: an algebraic structure $(S, \star)$ where S is a nonempty set that is closed under the associative binary operation $\star$.

separable extension (of field F): a field K such that every element of K is the root of a separable polynomial in $F[x]$.

separable polynomial: a polynomial $p(x) \in F[x]$ of degree n that has n distinct roots in its splitting field.

sign (of a permutation): the value $+1$ if the permutation has an even number of transpositions when the permutation is written as a product of transpositions, and -1 otherwise.

simple group: a group whose only normal subgroups are $\{e\}$ and G.

skew field: a division ring.

splitting field (for nonconstant $p(x) \in F[x]$): the field $K = F(\alpha_1, \ldots, \alpha_n)$ where $p(x) = \alpha(x - \alpha_1) \ldots (x - \alpha_n)$, $\alpha \in F$.

subfield (of a field K): a subset $F \subseteq K$ that is a field using the same operations used in K.

subgroup (of a group G): a subset $H \subseteq G$ such that H is a group using the same group operation used in G.

subgroup generated by $\{a_i \mid i \in S\}$: for a given group G where $a_i \in G$ for all i in S, the smallest subgroup of G containing $\{a_i \mid i \in S\}$.

subring (of a ring R): a subset $S \subseteq R$ that is a ring using the same operations used in R.

Sylow p-subgroup (of G): a subgroup of G that is a p-group and is not properly contained in any p-group of G.

symmetric group: the group of all permutations on $\{1, 2, \ldots, n\}$ under the operation of composition.

transcendental element (over a field F): given a field F and an extension field K, an element of K that is not a root of any nonzero polynomial in $F[x]$.

transposition: a cycle of length 2.

unary operation: See *monadic operation*.

unit (in a ring): an element with a multiplicative inverse in the ring.

unity (in a ring): a multiplicative identity not equal to 0.
word (on a set): a finite sequence of elements of the set.
zero (of a polynomial f): an element a such that $f(a) = 0$.

5.1 ALGEBRAIC MODELS

5.1.1 DOMAINS AND OPERATIONS

Definitions:

An *n-ary operation* on a set S is a function $\star: S \times S \times \cdots \times S \to S$, where the domain is the product of n factors.

A *binary operation* on a set S is a function $\star: S \times S \to S$.

A *monadic operation* (or *unary operation*) on a set S is a function $\star: S \to S$.

An *algebraic structure* $(S, \star_1, \star_2, \ldots, \star_n)$ consists of a nonempty set S (the *domain*) with one or more n-ary operations $\star_i$ defined on S.

A binary operation can have some of the following properties:

- *associative property*: $a \star (b \star c) = (a \star b) \star c$ for all $a, b, c \in S$;
- *existence of an identity element*: there is an element $e \in S$ such that $e \star a = a \star e = a$ for all $a \in S$ (e is an *identity* for S);
- *existence of inverses*: for each element $a \in S$ there is an element $a' \in S$ such that $a' \star a = a \star a' = e$ (a' is an *inverse* of a);
- *commutative property*: $a \star b = b \star a$ for all $a, b \in S$.

Examples:

1. The most important types of algebraic structures with one binary operation are listed in the following table. A checkmark means that the property holds.

	closed	associative	commutative	existence of identity	existence of inverses
semigroup	✓	✓			
monoid	✓	✓		✓	
group	✓	✓		✓	✓
abelian group	✓	✓	✓	✓	✓

5.1.2 SEMIGROUPS AND MONOIDS

Definitions:

A *semigroup* $(S, \star)$ consists of a nonempty set S and an associative binary operation $\star$ on S.

A *monoid* $(S, \star)$ consists of a nonempty set S and an associative binary operation $\star$ on S such that S has an identity.

A nonempty subset T of a semigroup $(S, \star)$ is a **subsemigroup** of S if T is closed under $\star$.

A subset T of a monoid $(S, \star)$ with identity e is a **submonoid** of S if T is closed under $\star$ and $e \in T$.

Two semigroups [monoids] $(S_1, \star_1)$ and $(S_2, \star_2)$ are **isomorphic** if there is a function $\varphi \colon S_1 \to S_2$ that is one-to-one, onto S_2, and such that $\varphi(a \star_1 b) = \varphi(a) \star_2 \varphi(b)$ for all $a, b \in S_1$.

A **word** on a set S (the *alphabet*) is a finite sequence of elements of S.

The **free monoid [free semigroup] generated by** S is the monoid [semigroup] $(S^*, \star)$ where S^* is the set of all words on a set S and the operation $\star$ is defined on S^* by concatenation: $x_1 x_2 \ldots x_m \star y_1 y_2 \ldots y_n = x_1 x_2 \ldots x_m y_1 y_2 \ldots y_n$. $(S^*, \star)$ is also called the **free monoid [free semigroup] on** S^*.

Facts:

1. Every monoid is a semigroup.

2. Every semigroup $(S, \star)$ is isomorphic to a subsemigroup of some semigroup of transformations on some set. Hence, every semigroup can be regarded as a semigroup of transformations. An analogous result is true for monoids.

Examples:

1. *Free semigroups and monoids*: The free monoid generated by S is a monoid with the empty word $e = \lambda$ (the sequence consisting of zero elements) as the identity.

2. The possible input tapes to a computer form a free monoid on the set of symbols (such as the ASCII symbols) in the computer alphabet.

3. *Semigroup and monoid of transformations on a set S*: Let S be a nonempty set and let $\mathcal{F}$ be the set of all functions $f \colon S \to S$. With the operation $\star$ defined by composition, $(f \star g)(x) = f(g(x))$, $(\mathcal{F}, \star)$ is the semigroup [monoid] of transformations on S. The identity of $\mathcal{F}$ is the identity transformation $e \colon S \to S$ where $e(x) = x$ for all $x \in S$.

4. The set of closed walks based at a fixed vertex v in a graph forms a monoid under the operation of concatenation. The null walk is the identity. (§8.2.1.)

5. For a fixed positive integer n, the set of all $n \times n$ matrices with elements in any ring with unity (§5.4.1) where $\star$ is matrix multiplication (using the operations in the ring) is a semigroup and a monoid. The identity is the identity matrix.

6. The sets

$\mathcal{N} = \{0, 1, 2, 3, \ldots\}$ (natural numbers),
$\mathcal{Z} = \{\ldots, -2, -1, 0, 1, 2, \ldots\}$ (integers),
$\mathcal{Q}$ (the set of rational numbers),
$\mathcal{R}$ (the set of real numbers),
$\mathcal{C}$ (the set of complex numbers),

where $\star$ is either addition or multiplication, are all semigroups and monoids. Using either addition or multiplication, each semigroup is a subsemigroup of each of those following it in this list. Likewise, using either addition or multiplication, each monoid is a submonoid of each of those following it in this list. For example, $(\mathcal{Q}, +)$ is a subsemigroup and submonoid of $(\mathcal{R}, +)$ and $(\mathcal{C}, +)$. Under addition, $e = 0$; under multiplication, $e = 1$.

5.2 GROUPS

5.2.1 BASIC CONCEPTS

Definitions:

A *group* $(G, \star)$ consists of a set G with a binary operator $\star$ defined on G such that $\star$ has the following properties:

- *associative property*: $a \star (b \star c) = (a \star b) \star c$ for all $a, b, c \in G$;
- *identity property*: G has an element e (*identity* of G) that satisfies $e \star a = a \star e = a$ for all $a \in G$;
- *inverse property*: for each element $a \in G$ there is an element $a^{-1} \in G$ (*inverse* of a) such that $a^{-1} \star a = a \star a^{-1} = e$.

If $a \star b = b \star a$ for all $a, b \in G$, the group G is *commutative* or *abelian*. (Niels H. Abel, 1802-1829)

The *order* of a finite group G, denoted $|G|$, is the number of elements in the group.

The (**external**) *direct product* of groups $(G_1, \star_1)$ and $(G_2, \star_2)$ is the group $G_1 \times G_2 = \{(a_1, a_2) \mid a_1 \in G_1, a_2 \in G_2\}$ where multiplication $\star$ is defined by the rule $(a_1, a_2) \star (b_1, b_2) = (a_1 \star_1 b_1, a_2 \star_2 b_2)$. The direct product can be extended to n groups: $G_1 \times G_2 \times \cdots \times G_n$. The direct product is also called the *direct sum* and written $G_1 \oplus G_2 \oplus \cdots \oplus G_n$, especially if the groups are abelian. If $G_i = G$ for all i, the direct product can be written G^n.

The group G is *finitely generated* if there are $a_1, a_2, \ldots, a_n \in G$ such that every element of G can be written as $a_{k_1}^{\epsilon_1} a_{k_2}^{\epsilon_2} \ldots a_{k_j}^{\epsilon_j}$ where $k_i \in \{1, \ldots, n\}$ and $\epsilon_i \in \{1, -1\}$, for some $j \geq 0$; where the empty product is defined to be e.

Note: Frequently the operation $\star$ is multiplication or addition. If the operation is addition, the group $(G, +)$ is an *additive group*. If the operation is multiplication, the group $(G, \cdot)$ is a *multiplicative group*.

	operation $\star$	identity e	inverse a^{-1}
additive group	$a + b$	0	$-a$
multiplicative group	$a \cdot b$ or ab	1 or e	a^{-1}

Facts:

1. Every group has exactly one identity element.
2. In every group every element has exactly one inverse.
3. *Cancellation laws*: In all groups,
    - if $ab = ac$ then $b = c$ (left cancellation law);
    - if $ba = ca$, then $b = c$ (right cancellation law).
4. $(a^{-1})^{-1} = a$.
5. $(ab)^{-1} = b^{-1}a^{-1}$. More generally, $(a_1 a_2 \ldots a_k)^{-1} = a_k^{-1} a_{k-1}^{-1} \ldots a_1^{-1}$.
6. If a and b are elements of a group G, the equations $ax = b$ and $xa = b$ have unique solutions in G. The solutions are $x = a^{-1}b$ and $x = ba^{-1}$, respectively.
7. The direct product $G_1 \times \cdots \times G_n$ is abelian when each group G_i is abelian.

8. $|G_1 \times \cdots \times G_n| = |G_1| \cdot \cdots \cdot |G_n|$.

9. The identity for $G_1 \times \cdots \times G_n$ is $(e_1, \ldots, e_n)$ where e_i is the identity of G_i. The inverse of $(a_1, \ldots, a_n)$ is $(a_1, \ldots, a_n)^{-1} = (a_1^{-1}, \ldots, a_n^{-1})$.

10. The structure of a group can be determined by a single rule (see Example 2) or by a group table listing all products (see Examples 2 and 3).

Examples:

1. Table 1 displays information on several common groups. All groups listed have infinite order, except for the following: the group of complex nth roots of unity has order n, the group of all bijections $f\colon S \to S$ where $|S| = n$ has order $n!$, $\mathcal{Z}_n$ has order n, $\mathcal{Z}_n^*$ has order $\phi(n)$ (Euler phi-function), S_n has order $n!$, A_n has order $n!/2$, D_n has order $2n$, and the quaternion group has order 8. All groups listed in the table are abelian except for: the group of bijections, $GL(n, \mathcal{R})$, S_n, A_n, D_n, and $\mathbf{Q}$.

2. *The groups $\mathcal{Z}_n$ and $\mathcal{Z}_n^*$* (see Table 1): In the groups $\mathcal{Z}_n$ and $\mathcal{Z}_n^*$ an element a can be viewed as the equivalence class $\{\, b \in \mathcal{Z} \mid b \bmod n = a \bmod n \,\}$, which can be written $\overline{a}$ or $[a]$. To find the inverse a^{-1} of $a \in \mathcal{Z}_n^*$, use the extended Euclidean algorithm to find integers a^{-1} and k such that $aa^{-1} + nk = \gcd(a, n) = 1$. The following are the group tables for $\mathcal{Z}_2 = \{0, 1\}$ and $\mathcal{Z}_3 = \{0, 1, 2\}$:

+	0	1
0	0	1
1	1	0

+	0	1	2
0	0	1	2
1	1	2	0
2	2	0	1

3. Quaternion group: $\mathbf{Q} = \{1, -1, i, -i, j, -j, k, -k\}$ where multiplication is defined by the following relations:
$$i^2 = j^2 = k^2 = -1, \qquad ij = -ji = k, \qquad jk = -kj = i, \qquad ki = -ik = j$$
where 1 is the identity. These relations yield the following multiplication table:

$\cdot$	1	-1	i	$-i$	j	$-j$	k	$-k$
1	1	-1	i	$-i$	j	$-j$	k	$-k$
-1	-1	1	$-i$	i	$-j$	j	$-k$	k
i	i	$-i$	-1	1	k	$-k$	$-j$	j
$-i$	$-i$	i	1	-1	$-k$	k	j	$-j$
j	j	$-j$	$-k$	k	-1	1	i	$-i$
$-j$	$-j$	j	k	$-k$	1	-1	$-i$	i
k	k	$-k$	j	$-j$	$-i$	i	-1	1
$-k$	$-k$	k	$-j$	j	i	$-i$	1	-1

Inverses: $1^{-1} = 1$, $(-1)^{-1} = -1$, x and $-x$ are inverses for $x = i, j, k$. The group is nonabelian.

The quaternion group $\mathbf{Q}$ can also be defined as the following group of 8 matrices:

$$\begin{pmatrix} 1 & 0 \\ 0 & 1 \end{pmatrix}, \begin{pmatrix} -1 & 0 \\ 0 & -1 \end{pmatrix}, \begin{pmatrix} -i & 0 \\ 0 & i \end{pmatrix}, \begin{pmatrix} i & 0 \\ 0 & -i \end{pmatrix},$$

$$\begin{pmatrix} 0 & 1 \\ -1 & 0 \end{pmatrix}, \begin{pmatrix} 0 & -1 \\ 1 & 0 \end{pmatrix}, \begin{pmatrix} 0 & i \\ i & 0 \end{pmatrix}, \begin{pmatrix} 0 & -i \\ -i & 0 \end{pmatrix},$$

where i is the complex number such that $i^2 = -1$ and the group operation is matrix multiplication.

Table 1 Examples of groups.

set	operation	identity	inverses
$\mathcal{Z}, \mathcal{Q}, \mathcal{R}, \mathcal{C}$	addition	0	$-a$
$\mathcal{Z}^n$, n a positive integer (also $\mathcal{Q}^n, \mathcal{R}^n, \mathcal{C}^n$)	coordinatewise addition	$(0,\ldots,0)$	$-(a_1,\ldots,a_n) = (-a_1,\ldots,-a_n)$
the set of all complex numbers of modulus $1 = \{e^{i\theta} = \cos\theta + i\sin\theta \mid 0 \leq \theta < 2\pi\}$	multiplication	$e^{i0} = 1$	$(e^{i\theta})^{-1} = e^{-i\theta}$
the *complex nth roots of unity* (solutions to $z^n = 1$) $\{e^{2\pi i k/n} \mid k = 0,1,\ldots,n-1\}$	multiplication	1	$(e^{2\pi i k/n})^{-1} = e^{2\pi i(n-k)/n}$
$\mathcal{R}-\{0\}, \mathcal{Q}-\{0\}, \mathcal{C}-\{0\}$	multiplication	1	$1/a$
$\mathcal{R}^*$ (positive real numbers)	multiplication	1	$1/a$
all rotations of the plane around the origin; $r_\alpha =$ counterclockwise rotation through an angle of $\alpha°$: $r_\alpha(x,y) = (x\cos\alpha - y\sin\alpha, x\sin\alpha + y\sin\alpha)$	composition: $r_{\alpha_2} \circ r_{\alpha_1} = r_{\alpha_1+\alpha_2}$	r_0 (the 0° rotation)	$r_\alpha^{-1} = r_{-\alpha}$
all 1-1, onto functions (bijections) $f: S \to S$ where S is any nonempty set	composition of functions	$i: S \to S$ where $i(x) = x$ for all $x \in S$	$f^{-1}(y) = x$ if and only if $f(x) = y$
$\mathcal{M}_{m\times n} =$ all $m \times n$ matrices with entries in $\mathcal{R}$	matrix addition	$O_{m\times n}$ (zero matrix)	$-A$
$GL(n,\mathcal{R}) =$ all $n \times n$ invertible, or nonsingular, matrices with entries in $\mathcal{R}$; (the *general linear group*)	matrix multiplication	I_n (identity matrix)	A^{-1}
$\mathcal{Z}_n = \{0,1,\ldots,n-1\}$	$(a+b)$ mod n	0	$n-a$ ($a \neq 0$) $-0 = 0$
$\mathcal{Z}_n^* = \{k \mid k \in \mathcal{Z}_n, k$ relatively prime to $n\}$, $n > 1$	ab mod n	1	see Example 2
$S_n =$ all permutations of $\{1,2,\ldots,n\}$; (*symmetric group*) (See §5.3.)	composition of permutations	identity permutation	inverse permutation
$A_n =$ all even permutations of $\{1,2,\ldots,n\}$; (*alternating group*) (See §5.3.)	composition of permutations	identity permutation	inverse permutation
$D_n =$ symmetries (rotations and reflections) of a regular n-gon; (*dihedral group*)	composition of functions	rotation through 0°	$r_\alpha^{-1} = r_{-\alpha}$; reflections are their own inverses
$\mathcal{Q} =$ quaternion group (see Example 3)			

4. The set $\{a, b, c, d\}$ with either of the following multiplication tables is not a group. In the first case there is an identity, a, and each element has an inverse, but the associative law fails: $(bc)d \neq b(cd)$. In the second case there is no identity (hence inverses are not defined) and the associative law fails.

·	a	b	c	d
a	a	b	c	d
b	b	d	a	c
c	c	a	b	d
d	d	c	b	a

·	a	b	c	d
a	a	c	b	d
b	d	b	a	c
c	b	d	c	a
d	c	a	d	b

5.2.2 GROUP ISOMORPHISM AND HOMOMORPHISM

Definitions:

For groups G and H, a function $\varphi: G \to H$ such that $\varphi(ab) = \varphi(a)\varphi(b)$ for all $a, b \in G$ is a **homomorphism**. The notation $a\varphi$ is sometimes used instead of $\varphi(a)$.

For groups G and H, a function $\varphi: G \to H$ is an **isomorphism** from G to H if φ is a homomorphism that is 1–1 and onto H. In this case G is **isomorphic** to H, written $G \cong H$.

An isomorphism $\varphi: G \to G$ is an **automorphism**.

The **kernel** of φ is the set $\{g \in G \mid \varphi(g) = e\}$, where e is the identity of the group G.

Facts:

1. If φ is an isomorphism, φ^{-1} is an isomorphism.
2. Isomorphism is an equivalence relation: $G \cong G$ (reflexive); if $G \cong H$, then $H \cong G$ (symmetric); if $G \cong H$ and $H \cong K$, then $G \cong K$ (transitive).
3. If $\varphi: G \to H$ is a homomorphism, then $\varphi(G)$ is a group (a subgroup of H).
4. If $\varphi: G \to H$ is a homomorphism, then the kernel of φ is a group (a subgroup of G).
5. If p is prime there is only one group of order p (up to isomorphism), the group $(\mathcal{Z}_p, +)$.
6. *Cayley's theorem*: If G is a finite group of order n, then G is isomorphic to a subgroup of the group S_n of permutations on n objects. (Arthur Cayley, 1821–1895) The isomorphism is obtained by associating with each $a \in G$ the map $\pi_a: G \to G$ with the rule $\pi_a(g) = ga$ for all $g \in G$.
7. $\mathcal{Z}_m \times \mathcal{Z}_n$ is isomorphic to $\mathcal{Z}_{mn}$ if and only if m and n are relatively prime.
8. If $n = n_1 n_2 \ldots n_k$ where the n_i are powers of distinct primes, then $\mathcal{Z}_n$ is isomorphic to $\mathcal{Z}_{n_1} \times \mathcal{Z}_{n_2} \times \cdots \times \mathcal{Z}_{n_k}$.
9. *Fundamental theorem of finite abelian groups*: Every finite abelian group G (order ≥ 2) is isomorphic to a direct product of cyclic groups where each cyclic group has order a power of a prime. That is, G is isomorphic to $\mathcal{Z}_{n_1} \times \mathcal{Z}_{n_2} \times \cdots \times \mathcal{Z}_{n_k}$ where each cyclic order n_i is a power of some prime. In addition, the set $\{n_1, \ldots, n_k\}$ is unique.
10. Every finite abelian group is isomorphic to a subgroup of $\mathcal{Z}_n^*$ for some n.
11. *Fundamental theorem of finitely generated abelian groups*: If G is a finitely generated abelian group, then there are unique integers $n \geq 0$, $n_1, n_2, \ldots, n_k \geq 2$ where $n_{i+1} \mid n_i$ for $i = 1, 2, \ldots, k-1$ such that G is isomorphic to $\mathcal{Z}^n \times \mathcal{Z}_{n_1} \times \mathcal{Z}_{n_2} \times \cdots \times \mathcal{Z}_{n_k}$.

Table 2 Numbers of groups and abelian groups.

order	1	2	3	4	5	6	7	8	9	10	11	12	13	14	15	16	17	18	19	20
groups	1	1	1	2	1	2	1	5	2	2	1	5	1	2	1	14	1	5	1	5
abelian	1	1	1	2	1	1	1	3	2	1	1	2	1	1	1	5	1	2	1	2
order	21	22	23	24	25	26	27	28	29	30	31	32	33	34	35	36	37	38	39	40
groups	2	2	1	15	2	2	5	4	1	4	1	51	1	2	1	14	1	2	2	14
abelian	1	1	1	3	2	1	3	2	1	1	1	7	1	1	1	4	1	1	1	3
order	41	42	43	44	45	46	47	48	49	50	51	52	53	54	55	56	57	58	59	60
groups	1	6	1	4	2	2	1	52	2	5	1	5	1	15	2	13	2	2	1	13
abelian	1	1	1	2	2	1	1	5	2	2	1	2	1	3	1	3	1	1	1	2

Examples:
1. Table 2 lists the number of nonisomorphic groups and abelian groups of all orders from 1 to 60.
2. All groups of order 12 or less are listed by order in Table 3.

5.2.3 SUBGROUPS

Definitions:

A *subgroup* of a group $(G, \star)$ is a subset $H \subseteq G$ such that $(H, \star)$ is a group (with the same group operation as in G). Write $H \leq G$ if H is a subgroup of G.

If $a \in G$, the set $(a) = \{\ldots, a^{-2} = (a^{-1})^2, a^{-1}, a^0 = e, a, a^2, \ldots\} = \{a^n \mid n \in \mathcal{Z}\}$ is the *cyclic subgroup* generated by a. The element a is a *generator* of G.

G and $\{e\}$ are *improper subgroups* of G. All other subgroups of G are *proper subgroups* of G.

Facts:
1. If G is a group, then $\{e\}$ and G are subgroups of G.
2. If G is a group and $a \in G$, the set (a) is a subgroup of G.
3. Every subgroup of an abelian group is abelian.
4. If H is a subgroup of a group G, then the identity element of H is the identity element of G; the inverse (in the subgroup H) of an element a in H is the inverse (in the group G) of a.
5. Lagrange's theorem: Let G be a finite group. If H is any subgroup of G, then the order of H is a divisor of the order of G. (Joseph-Louis Lagrange, 1736–1813)
6. If d is a divisor of the order of a group G, there may be no subgroup of order d. (The group A_4, of order 12, has no subgroup of order 6. See §5.3.3.)
7. If G is a finite abelian group, then the converse of Lagrange's theorem is true for G.
8. If G is finite (not necessarily abelian) and p is a prime that divides the order of G, then G has a subgroup of order p.
9. If G has order $p^m n$ where p is prime and p does not divide n, then G has a subgroup of order p^m, called a *Sylow subgroup* or *Sylow p-subgroup*. See §5.2.6.

Table 3 All groups of order 12 or less.

order	groups
1	$\{e\}$
2	$\mathcal{Z}_2$
3	$\mathcal{Z}_3$
4	$\mathcal{Z}_4$, if there is an element of order 4 (group is cyclic) $\mathcal{Z}_2 \times \mathcal{Z}_2 \cong$ Klein four-group, if no element has order 4 (§5.3.2)
5	$\mathcal{Z}_5$
6	$\mathcal{Z}_6$, if there is an element of order 6 (group is cyclic) $S_3 \cong D_3$, if there is no element of order 6 (§5.3.1, 5.3.2)
7	$\mathcal{Z}_7$
8	$\mathcal{Z}_8$, if there is an element of order 8 (group is cyclic) $\mathcal{Z}_2 \times \mathcal{Z}_4$, if there is an element a of order 4, but none of order 8, and if there is an element $b \notin (a)$ such that $ab = ba$ and $b^2 = e$ $\mathcal{Z}_2 \times \mathcal{Z}_2 \times \mathcal{Z}_2$, if every element has order 1 or 2 D_4, if there is an element a of order 4, but none of order 8, and if there is an element $b \notin (a)$ such that $ba = a^3 b$ and $b^2 = e$ Quaternion group, if there is an element a of order 4, none of order 8, and an element $b \notin (a)$ such that $ba = a^3 b$ and $b^2 = a^2$ (§5.2.2)
9	$\mathcal{Z}_9$, if there is an element of order 9 (group is cyclic) $\mathcal{Z}_3 \times \mathcal{Z}_3$, if there is no element of order 9
10	$\mathcal{Z}_{10}$, if there is an element of order 10 (group is cyclic) D_5, if there is no element of order 10
11	$\mathcal{Z}_{11}$
12	$\mathcal{Z}_{12} \cong \mathcal{Z}_3 \times \mathcal{Z}_4$, if there is an element of order 12 (group is cyclic) $\mathcal{Z}_2 \times \mathcal{Z}_6 \cong \mathcal{Z}_2 \times \mathcal{Z}_2 \times \mathcal{Z}_3$, if group is abelian but noncyclic D_6, if group is nonabelian and has an element of order 6 but none of order 4 A_4, if group is nonabelian and has no element of order 6 The group generated by a and b, where a has order 4, b has order 3, and $ab = b^2 a$

10. A subset H of a group G is a subgroup of G if and only if the following are all true: $H \neq \emptyset$; $a, b \in H$ implies $ab \in H$; and $a \in H$ implies $a^{-1} \in H$.

11. A subset H of a group G is a subgroup of G if and only if $H \neq \emptyset$ and $a, b \in H$ implies that $ab^{-1} \in H$.

12. If H is a nonempty finite subset of a group G with the property that $a, b \in H$ implies that $ab \in H$, then H is a subgroup of G.

13. The intersection of any collection of subgroups of a group G is a subgroup of G.

14. The union of subgroups is not necessarily a subgroup. See Example 12.

Examples:

1. *Additive subgroups*: Each of the following can be viewed as a subgroup of all the groups listed after it: $(\mathcal{Z},+), (\mathcal{Q},+), (\mathcal{R},+), (\mathcal{C},+)$.
2. For n any positive integer, the set $n\mathcal{Z} = \{nz \mid z \in \mathcal{Z}\}$ is a subgroup of $\mathcal{Z}$.
3. $\mathcal{Z}_2$ is not a subgroup of $\mathcal{Z}_4$ (the group operations are not the same).
4. The set of odd integers under addition is not a subgroup of $(\mathcal{Z},+)$ (the set of odd integers is not closed under addition).
5. $(\mathcal{N},+)$ is not a subgroup of $(\mathcal{Z},+)$ ($\mathcal{N}$ does not contain its inverses).
6. The group $\mathcal{Z}_6$ has the following four subgroups: $\{0\}, \{0,3\}, \{0,2,4\}, \mathcal{Z}_6$.
7. *Multiplicative subgroups*: Each of the following can be viewed as a subgroup of all the groups listed after it: $(\mathcal{Q} - \{0\}, \cdot), (\mathcal{R} - \{0\}, \cdot), (\mathcal{C} - \{0\}, \cdot)$.
8. The set of n complex nth roots of unity can be viewed as a subgroup of the set of all complex numbers of modulus 1 under multiplication, which is a subgroup of $(\mathcal{C} - \{0\}, \cdot)$.
9. If $nd = 360$ (n and d positive integers) and r_k is the counterclockwise rotation of the plane about the origin through an angle of $k°$, then $\{r_k \mid k = 0, d, 2d, 3d, \ldots, (n-1)d\}$ is a subgroup of the group of all rotations of the plane around the origin.
10. The set of all $n \times n$ nonsingular diagonal matrices is a subgroup of the set of all $n \times n$ nonsingular matrices under multiplication.
11. If $n = mk$, then $\{0, m, 2m, \ldots, (k-1)m\}$ is a subgroup of $(\mathcal{Z}_n, +)$ isomorphic to $\mathcal{Z}_k$.
12. The union of subgroups need not be a subgroup: $\{2n \mid n \in \mathcal{Z}\}$ and $\{3n \mid n \in \mathcal{Z}\}$ are subgroups of $\mathcal{Z}$, but their union is not a subgroup of $\mathcal{Z}$ since $2 + 3 = 5 \notin \{2n \mid n \in \mathcal{Z}\} \cup \{3n \mid n \in \mathcal{Z}\}$.

5.2.4 COSETS AND QUOTIENT GROUPS

Definitions:

If H is a subgroup of a group G and $a \in G$, then the set $aH = \{ah \mid h \in H\}$ is a **left coset** of H in G. The set $Ha = \{ha \mid h \in H\}$ is a **right coset** of H in G. (If G is written additively, the cosets are written $a + H$ and $H + a$.)

The **index** of a subgroup H in a group G, written $(G{:}H)$ or $[G{:}H]$, is the number of left (or right) cosets of H in G.

A **normal** subgroup of a group G is a subgroup H of G such that $aH = Ha$ for all $a \in G$. The notation $H \triangleleft G$ means that H is a normal subgroup of G.

If H is a normal subgroup of G, the **quotient group** (or **factor group of G modulo H**) is the group $G/H = \{aH \mid a \in G\}$, where $aH \cdot bH = (ab)H$.

If G is a group and $a \in G$, an element $b \in G$ is a **conjugate** of a if $b = gag^{-1}$ for some $g \in G$.

If G is a group and $a \in G$, the set $\{x \mid x \in G, ax = xa\}$ is the **centralizer** (or **normalizer**) of a.

If G is a group, the set $\{x \mid x \in G, gx = xg \text{ for all } g \in G\}$ is the **center** of G.

If H is a subgroup of group G, the set $\{x \mid x \in G, xHx^{-1} = H\}$ is the **normalizer** of H.

Facts:
1. If H is a subgroup of a group G, then the following are equivalent:
    - H is a normal subgroup of G;
    - $aHa^{-1} = a^{-1}Ha = H$ for all $a \in G$;
    - $a^{-1}ha \in H$ for all $a \in G$, $h \in H$;
    - for all $a \in G$ and $h_1 \in H$, there is $h_2 \in H$ such that $ah_1 = h_2 a$.
2. If group G is abelian, then every subgroup H of G is normal. If G is not abelian, it may happen that H is not normal.
3. If group G is finite, then $(G:H) = |G|/|H|$.
4. $\{e\}$ and G are normal subgroups of group G.
5. In the group G/H, the identity is $eH = H$ and the inverse of aH is $a^{-1}H$.
6. **Fundamental homomorphism theorem**: If $\varphi: G \to H$ is a homomorphism and has kernel K, then K is a normal subgroup of G and G/K is isomorphic to $\varphi(G)$.
7. If H is a normal subgroup of a group G and $\varphi: G \to G/H$ is defined by $\varphi(g) = gH$, then φ is a homomorphism onto G/H with kernel H.
8. If H is a normal subgroup of a finite group G, then G/H has $|G|/|H|$ cosets.
9. If H and K are normal subgroups of a group G, then $H \cap K$ is a normal subgroup of G.
10. For all $a \in G$, the centralizer of a is a subgroup of G.
11. The center of a group is a subgroup of the group.
12. The normalizer of a subgroup of group G is a subgroup of G.
13. The index of the centralizer of $a \in G$ is equal to the number of distinct conjugates of a in G.
14. If a group G contains normal subgroups H and K such that $H \cap K = \{e\}$ and $\{hk \mid h \in H, k \in K\} = G$, then G is isomorphic to $H \times K$.
15. If G is a group such that $|G| = ab$ where a and b are relatively prime, and if G contains normal subgroups H of order a and K of order b, then G is isomorphic to $H \times K$.

Examples:
1. $\mathcal{Z}/n\mathcal{Z}$ is isomorphic to $\mathcal{Z}_n$, since $\varphi: \mathcal{Z} \to \mathcal{Z}_n$ defined by $\varphi(g) = g \bmod n$ has kernel $n\mathcal{Z}$.
2. The left cosets of the subgroup $H = \{0, 4\}$ in $\mathcal{Z}_8$ are $H + 0 = \{0, 4\}$, $H + 1 = \{1, 5\}$, $H + 2 = \{2, 6\}$, $H + 3 = \{3, 7\}$. The index of H in $\mathcal{Z}_8$ is $(\mathcal{Z}_8, H) = 4$.
3. $\{(1), (12)\}$ is not a normal subgroup of the symmetric group S_3 (§5.3.1).

5.2.5 CYCLIC GROUPS AND ORDER

Definitions:

A group $(G, \cdot)$ is **cyclic** if there is $a \in G$ such that $G = \{a^n \mid n \in \mathcal{Z}\}$, where $a^0 = e$ and $a^{-n} = (a^{-1})^n$ for all positive integers n. If G is written additively, $G = \{na \mid n \in \mathcal{Z}\}$, where $0a = 0$ and if $n > 0$, $na = a + a + a + \cdots + a$ (n terms) and $-na = (-a) + (-a) + \cdots + (-a)$ (n terms).

The element a is called a **generator** of G and the group $(G, \cdot)$ is written $((a), \cdot)$, (a), or $\langle a \rangle$.

The **order of an element** $a \in G$, written $|(a)|$ or $\text{ord}(a)$, is the smallest positive integer n such that $a^n = e$ ($na = 0$ if G is written additively). If there is no such integer, then a has **infinite** order.

A subgroup H of a group $(G, \cdot)$ is a **cyclic subgroup** if there is $a \in H$ such that $H = \{a^n \mid n \in \mathcal{Z}\}$.

Facts:

1. The order of an element a is equal to the number of elements in (a).
2. Every group of prime order is cyclic.
3. Every cyclic group is abelian. However, not every abelian group is cyclic; for example $(\mathcal{R}, +)$ and the Klein four-group.
4. If G is an infinite cyclic group, then $G \cong (\mathcal{Z}, +)$.
5. If G is a finite cyclic group of order n, then $G \cong (\mathcal{Z}_n, +)$.
6. If G is a group of order n, then the order of every element of G is a divisor of n.
7. *Cauchy's theorem*: If G is a group of order n and p is a prime that divides n, then G contains an element of order p. (Augustin-Louis Cauchy, 1789–1857)
8. If G is a cyclic group of order n generated by a, then $G = \{a, a^2, a^3, \ldots, a^n\}$ and $a^n = e$. If k and n are relatively prime, then a^k is also a generator of G, and conversely.
9. If G is a group and $a \in G$, then (a) is a cyclic subgroup of G.
10. Every subgroup of a cyclic group is cyclic.
11. If G is a group of order n and there is an element $a \in G$ of order n, then G is cyclic and $G = (a)$.

Examples:

1. $(\mathcal{Z}, +)$ is cyclic and is generated by each of 1 and -1.
2. $(\mathcal{Z}_n, +)$ is cyclic and is generated by each element of $\mathcal{Z}_n$ that is relatively prime to n. If $a \in \mathcal{Z}_n$, then a has order $n/\gcd(a, n)$.
3. $(\mathcal{Z}_p, +)$, p prime, is a cyclic group generated by each of the elements $1, 2, \ldots, p - 1$. If $a \neq 0$, a has order p.
4. $(\mathcal{Z}_n^*, \cdot)$ is cyclic if and only if $n = 2, 4, p^k$, or $2p^k$, where $k \geq 1$ and p is an odd prime.

5.2.6 SYLOW THEORY

The Sylow theorems are used to help classify the nonisomorphic groups of a given order by guaranteeing the existence of subgroups of certain orders. (Peter Ludvig Mejdell Sylow, 1832–1918)

Definitions:

For prime p, a group G is a **p-group** if every element of G has order p^n for some positive integer n.

For prime p, a **Sylow p-subgroup (Sylow subgroup)** of G is a subgroup of G that is a p-group and is not properly contained in any p-group in G.

Facts:

1. *Sylow theorem*: If G is a group of order $p^m \cdot q$ where p is a prime, $m \geq 1$, and $p \nmid q$, then:
   - G contains subgroups of orders $p, p^2, \ldots, p^m$ (hence, if prime p divides the order of a finite group G, then G contains an element of order p);
   - if H and K are Sylow p-subgroups of G, there is $g \in G$ such that $K = gHg^{-1}$ (K is *conjugate* to H);
   - the number of Sylow p-subgroups of G is $kp + 1$ for some integer k such that $(kp + 1) \mid q$.

2. If G is a group of order pq where p and q are primes and $p < q$, then G contains a normal subgroup of order q.

3. If G is a group of order pq where p and q are primes, $p < q$, and $p \nmid (q-1)$, then G is cyclic.

Examples:

1. Every group of order 15 is cyclic (by Fact 3).
2. Every group of order 21 contains a normal subgroup of order 7 (by Fact 2).

5.2.7 SIMPLE GROUPS

Simple groups arise as a fundamental part of the study of finite groups and the structure of their subgroups. An extensive, lengthy search by many mathematicians for all finite simple groups ended in 1980 when, as the result of hundreds of articles written by over one hundred mathematicians, the classification of all finite simple groups was completed. See [As86] and [Go82] for details.

Definitions:

A group $G \neq \{e\}$ is **simple** if its only normal subgroups are $\{e\}$ and G.

A **composition series** for a group G is a finite sequence of subgroups $H_1 = G, H_2, \ldots, H_{n-1}, H_n = \{e\}$ such that H_{i+1} is a normal subgroup of H_i and H_i/H_{i+1} is simple, for $i = 1, \ldots, n-1$.

A finite group G is **solvable** if it has a sequence of subgroups $H_1 = G, H_2, \ldots, H_{n-1}, H_n = \{e\}$ such that H_{i+1} is a normal subgroup of H_i and H_i/H_{i+1} is abelian, for $i = 1, \ldots, n-1$.

A **sporadic** group is one of 26 nonabelian finite simple groups that is not an alternating group or a group of Lie type [Go82].

Facts:

1. Every finite group has a composition series. Thus, simple groups (the quotient groups in the series) can be regarded as the building blocks of finite groups.
2. Some infinite groups, such as $(\mathcal{Z}, +)$, do not have composition series.
3. Every abelian group is solvable.
4. An abelian group G is simple if and only if $G \cong \mathcal{Z}_p$ where p is prime.
5. If G is a nonabelian solvable group, then G is not simple.
6. Every group of prime order is simple.
7. Every group of order p^n (p prime) is solvable.

8. Every group of order $p^n q^m$ (p, q primes) is solvable.
9. If G is a solvable, simple finite group, then G is either $\{e\}$ or $\mathcal{Z}_p$ (p prime).
10. If G is a simple group of odd order, then $G \cong \mathcal{Z}_p$ for some prime p.
11. There is no infinite simple, solvable group.
12. *Burnside conjecture/Feit-Thompson theorem*: In 1911 William Burnside conjectured that all groups of odd order are solvable. This conjecture was proved in 1963 by Walter Feit and John Thompson. (See Fact 13.)
13. Every nonabelian simple group has even order. (This follows from the Feit-Thompson theorem.)
14. The proof of the Burnside conjecture provided the impetus for a massive program to classify all finite simple groups. This program, organized by Daniel Gorenstein, led to hundreds of journal articles and concluded in 1980 when the classification problem was finally solved (Fact 15). [GoLySo94]
15. *Classification theorem for finite simple groups*: Every finite simple group is of one of the following types:
    - *abelian*: $\mathcal{Z}_p$ where p is prime (§5.2.1);
    - *nonabelian*:
        ◇ alternating groups A_n ($n \neq 4$) (§5.3.2);
        ◇ groups of *Lie* type, which fall into 6 classes of classical groups and 10 classes of exceptional simple groups [Ca72];
        ◇ *sporadic* groups. There are 26 sporadic groups, listed here from smallest to largest order. The letters in the names of the groups reflect the names of some of the people who conjectured the existence of the groups or proved the groups simple. M_{11} (order 7,920), M_{12}, M_{22}, M_{23}, M_{24}, J_1, J_2, J_3, J_4, HS, Mc, Suz, Ru, He, Ly, ON, .1, .2, .3, $M(22)$, $M(23)$, $M(24)'$, F_5, F_3, F_2, F_1 (the monster or Fischer-Griess group of order $\approx 10^{54}$).

5.2.8 GROUP PRESENTATIONS

Definitions:

The **balanced alphabet** on the set $X = \{x_1, \ldots, x_n\}$ is the set $\{x_1, x_1^{-1}, \ldots, x_n, x_n^{-1}\}$, whose elements are often called **symbols**.

Symbols x_j and x_j^{-1} of a balanced alphabet are **inverses** of each other. A double inverse $(x_j^{-1})^{-1}$ is understood as the identity operator.

A **word** in X is a string $s_1 s_2 \ldots s_n$ of symbols from the balanced alphabet on X.

The **inverse of a word** $s = s_1 s_2 \ldots s_n$ is the word $s^{-1} = s_n^{-1} \ldots s_2^{-1} s_1^{-1}$.

The **free semigroup** $W(X)$ has the set of words in X as its domain and string concatenation as its product operation.

A **trivial relator** in the set $X = \{x_1, \ldots, x_n\}$ is a word of the form $x_j x_j^{-1}$ or $x_j^{-1} x_j$.

A word u is **freely equivalent** to a word v, denoted $u \sim v$, if v can be obtained from u by iteratively inserting and deleting trivial relators, in the usual sense of those string operations. This is an equivalence relation, whose classes are called **free equivalence classes**.

A **reduced word** is a word containing no instances of a trivial relator as a substring.

The **free group** $F[X]$ has the set of free equivalence classes of words in X as its domain and class concatenation as its product operation.

A **group presentation** is a pair $(X:R)$, where X is an alphabet and R is a set of words in X called **relators**. A group presentation is **finite** if X and R are both finite.

A word u is **R-equivalent** to a word v under the group presentation $(X:R)$, denoted $u \sim_R v$, if v can be obtained from u by iteratively inserting and deleting relators from R or trivial relators. This is an equivalence relation, whose classes are called **R-equivalence classes**.

The group $\mathcal{G}(X:R)$ **presented** by the group presentation $(X:R)$ has the set of R-equivalence classes as its domain and class concatenation as its product operation. Moreover, any group G isomorphic to $\mathcal{G}(X:R)$ is said to be **presented** by the group presentation $(X:R)$.

The group G is **finitely presentable** if it has a presentation whose alphabet and relator set are both finite.

The **commutator** of the words u and v is the word $u^{-1}v^{-1}uv$. Any word of this form is called a commutator.

A **conjugate** of the word v is any word of the form $u^{-1}vu$.

Facts:

1. Max Dehn (1911) formulated three fundamental decision problems for finite presentations:

   - *word problem*: Given an arbitrary presentation $(X:R)$ and an arbitrary word w, decide whether w is equivalent to the empty word (i.e., the group identity).
   - *conjugacy problem*: Given an arbitrary presentation $(X:R)$ and two arbitrary words w_1 and w_2, decide whether w_1 is equivalent to a conjugate of w_2.
   - *isomorphism problem*: Given two arbitrary presentations $(X:R)$ and $(Y:S)$, decide whether they present isomorphic groups.

2. W. W. Boone (1955) and P. S. Ńovikov (1955) constructed presentations in which the word problem is recursively unsolvable. This implies that there is no single finite procedure that works for all finite presentations, thereby negatively solving Dehn's word problem and conjugacy problem.

3. M. O. Rabin (1958) proved that it is impossible to decide even whether a presentation presents the trivial group, which immediately implies that Dehn's isomorphism problem is recursively unsolvable.

4. The word problem is recursively solvable in various special classes of group presentations, including the following: presentations with no relators (i.e., free groups), presentations with only one relator, presentations in which the relator set includes the commutator of each pair of generators (i.e., abelian groups).

5. The group presentation $\mathcal{G}(X:R)$ is the quotient of the free group $F[X]$ by the normalizer of the relator set R.

6. More information on group presentations can be found in [CoMo72], [CrFo63], and [MaKaSo65].

Examples:

1. The cyclic group $\mathcal{Z}_k$ has the presentation $(x:x^k)$.
2. The direct sum $\mathcal{Z}_r \oplus \mathcal{Z}_s$ has the presentation $(x,y:x^r,y^s,x^{-1}y^{-1}xy)$.
3. The dihedral group $\mathcal{D}_q$ has the presentation $(x,y:x^q,y^2,y^{-1}xyx)$.

5.3 PERMUTATION GROUPS

Permutations, as arrangements, are important tools used extensively in combinatorics (§2.3 and §2.7). The set of permutations on a given set forms a group, and it is this algebraic structure that is examined in this section.

5.3.1 BASIC CONCEPTS

Definitions:

A **permutation** is a one-to-one and onto function $\sigma: S \to S$, where S is any nonempty set. If $S = \{a_1, a_2, \ldots, a_n\}$, a permutation σ is sometimes written as the $2 \times n$ matrix

$$\sigma = \begin{pmatrix} a_1 & a_2 & \cdots & a_n \\ a_1\sigma & a_2\sigma & \cdots & a_n\sigma \end{pmatrix}$$

where $a_i\sigma$ means $\sigma(a_i)$.

A permutation $\sigma: S \to S$ is a **cycle of length** n if there is a subset of S of size n, $\{a_1, a_2, \ldots, a_n\}$, such that $a_1\sigma = a_2, a_2\sigma = a_3, \ldots, a_n\sigma = a_1$, and $a\sigma = a$ for all other elements of S. Write $\sigma = (a_1\ a_2\ \cdots\ a_n)$. A **transposition** is a cycle of length 2.

A **permutation group** (G, X) is a collection G of permutations on a nonempty set X (whose elements are called *objects*) such that these permutations form a group under composition. That is, if σ and τ are permutations in G, $\sigma\tau$ is the permutation in G defined by the rule $a(\sigma\tau) = (a\sigma)\tau$. The **order** of the permutation group is $|G|$. The **degree** of the permutation group is $|X|$.

The **symmetric group on** n **elements** is the group S_n of all permutations on the set $\{1, 2, \ldots, n\}$ under composition. (See Fact 1.)

An **isomorphism** from a permutation group (G, X) to a permutation group (H, Y) is a pair of functions $(\alpha: G \to H, f: X \to Y)$ such that α is a group isomorphism and f is one-to-one and onto Y.

If $\sigma_1 = (a_{i_1}\ a_{i_2}\ \cdots\ a_{i_m})$ and $\sigma_2 = (a_{j_1}\ a_{j_2}\ \cdots\ a_{j_n})$ are cycles on S, then σ_1 and σ_2 are **disjoint cycles** if the sets $\{a_{i_1}, a_{i_2}, \ldots, a_{i_m}\}$ and $\{a_{j_1}, a_{j_2}, \ldots, a_{j_n}\}$ are disjoint.

An **even permutation** [**odd permutation**] is a permutation that can be written as a product of an even [odd] number of transpositions.

The **sign** of a permutation (where the permutation is written as a product of transpositions) is $+1$ if it has an even number of transpositions and -1 if it has an odd number of transpositions.

The **identity permutation** on S is the permutation $\iota: S \to S$ such that $x\iota = x$ for all $x \in S$.

An **involution** is a permutation σ such that $\sigma^2 = \iota$ (the identity permutation).

The **orbit** of $a \in S$ under σ is the set $\{\ldots, a\sigma^{-2}, a\sigma^{-1}, a, a\sigma, a\sigma^2, \ldots\}$.

Facts:

1. *Symmetric group of degree* n: The set of permutations on a nonempty set X is a group, where the group operation is composition of permutations: $\sigma_1\sigma_2$ is defined by $x(\sigma_1\sigma_2) = (x\sigma_1)\sigma_2$. The identity is the identity permutation ι. The inverse of σ is the permutation σ^{-1}, where $x\sigma^{-1} = y$ if and only if $y\sigma = x$. If $|X| = n$, the group of permutations is written S_n, the *symmetric group of degree* n.

2. Multiplication of permutations is not commutative. (See Examples 1 and 4.)
3. A permutation π is an involution if and only if $\pi = \pi^{-1}$.
4. The number of involutions in S_n, denoted inv(n), is equal to the number of Young tableaux that can be formed from the set $\{1, 2, \ldots, n\}$. (See §2.8.)
5. Permutations can be used to find determinants of matrices. (See §6.3.)
6. Every permutation on a finite set can be written as a product of disjoint cycles.
7. Cycle notation is not unique: for example, $(1\ 4\ 7\ 5) = (4\ 7\ 5\ 1) = (7\ 5\ 1\ 4) = (5\ 1\ 4\ 7)$.
8. Every permutation is either even or odd, and no permutation is both even and odd. Hence, every permutation has a unique sign.
9. Each cycle of length k can be written as a product of $k - 1$ transpositions:
$$(x_1\ x_2\ x_3\ \ldots\ x_k) = (x_1\ x_2)(x_1\ x_3)(x_1\ x_4)\ldots(x_1\ x_k).$$
10. S_n has order $n!$.
11. S_n is not abelian for $n \geq 3$. For example, $(1\ 2)(1\ 3) \neq (1\ 3)(1\ 2)$.
12. The order of a permutation that is a single cycle is the length of the cycle. For example, $(1\ 5\ 4)$ has order 3.
13. The order of a permutation that is written as a product of disjoint cycles is equal to the least common multiple of the lengths of the cycles.
14. *Cayley's theorem*: If G is a finite group of order n, then G is isomorphic to a subgroup of S_n. (See §5.2.2.)
15. Let G be a group of permutations on a set X (such a group is said to *act on* X). Then G induces an equivalence relation R on the set X by the following rule: for $a, b \in X$, aRb if and only if there is a permutation $\sigma \in G$ such that $a\sigma = b$.

Examples:
1. If $\sigma = \begin{pmatrix} 1 & 2 & 3 & 4 & 5 \\ 5 & 1 & 2 & 4 & 3 \end{pmatrix}$, $\tau = \begin{pmatrix} 1 & 2 & 3 & 4 & 5 \\ 4 & 5 & 1 & 3 & 2 \end{pmatrix}$, then $\sigma\tau = \begin{pmatrix} 1 & 2 & 3 & 4 & 5 \\ 2 & 4 & 5 & 3 & 1 \end{pmatrix}$ and $\tau\sigma = \begin{pmatrix} 1 & 2 & 3 & 4 & 5 \\ 4 & 3 & 5 & 2 & 1 \end{pmatrix}$. Note that $\sigma\tau \neq \tau\sigma$.

2. All elements of S_n can be written in *cycle notation*. For example,
$$\sigma = \begin{pmatrix} 1 & 2 & 3 & 4 & 5 & 6 & 7 \\ 4 & 6 & 3 & 7 & 1 & 2 & 5 \end{pmatrix} = (1\ 4\ 7\ 5)(2\ 6)(3).$$
Each cycle describes the orbit of the elements in that cycle. For example, $(1\ 4\ 7\ 5)$ is a cycle of length 4, and indicates that $1\sigma = 4$, $4\sigma = 7$, $7\sigma = 5$, and $5\sigma = 1$. The cycle (3) indicates that $3\sigma = 3$. If a cycle has length 1, that cycle can be omitted when a permutation is written as a product of cycles: $(1\ 4\ 7\ 5)(2\ 6)(3) = (1\ 4\ 7\ 5)(2\ 6)$.

3. Multiplication of permutations written in cycle notation can be performed easily. For example: if $\sigma = (1\ 5\ 3\ 2)$ and $\tau = (1\ 4\ 3)(2\ 5)$, then $\sigma\tau = (1\ 5\ 3\ 2)(1\ 4\ 3)(2\ 5) = (1\ 2\ 4\ 3\ 5)$. (Moving from left to right through the product of cycles, trace the orbit of each element. For example, $3\sigma = 2$ and $2\tau = 5$; therefore $3\sigma\tau = 5$.)

4. Multiplication of cycles need not be commutative. For example, $(1\ 2)(1\ 3) = (1\ 2\ 3)$, $(1\ 3)(1\ 2) = (1\ 3\ 2)$, but $(1\ 2\ 3) \neq (1\ 3\ 2)$. However, disjoint cycles commute.

5. If the group of permutations $G = \{\iota, (1\ 2), (3\ 5)\}$ acts on the set $S = \{1, 2, 3, 4, 5\}$, then the partition of S resulting from the equivalence relation induced by G is $\{\{1, 2\}, \{3, 5\}, \{4\}\}$. (See Fact 15.)

6. Let group $G = \{\iota, (1\ 2)\}$ act on $X = \{1, 2\}$ and group $H = \{\iota, (1\ 2)(3)\}$ act on $Y = \{1, 2, 3\}$. The permutation groups (G, X) and (H, Y) are not isomorphic since there is no bijection between X and Y (even though G and H are isomorphic groups).

5.3.2 EXAMPLES OF PERMUTATION GROUPS

Definitions:

The **alternating group** on n elements ($n \geq 2$) is the subgroup A_n of S_n consisting of all even permutations.

The **dihedral group** (**octic group**) D_n is the group of rigid motions (rotations and reflections) of a regular polygon with n sides under composition.

The **Klein four-group** (or **Viergruppe** or the **group of the rectangle**) is the group under composition of the four rigid motions of a rectangle that leave the rectangle in its original location. (Felix Klein, 1849–1925)

Given a permutation $\sigma: S \to S$, the **induced pair permutation** is the permutation $\sigma^{(2)}$ on unordered pairs of elements of S given by the rule $\sigma^{(2)}(\{x,y\}) = \{\sigma(x), \sigma(y)\}$.

Given a permutation group G acting on a set S, the **induced pair-action group** $G^{(2)}$ is the group of induced pair-permutations $\{\sigma^{(2)} \mid \sigma \in G\}$ under composition.

Given a permutation $\sigma: S \to S$, the **ordered pair-permutation** is the permutation $\sigma^{[2]}$ on the set $S \times S$ given by the rule $\sigma^{[2]}((x,y)) = (\sigma(x), \sigma(y))$.

Given a permutation group G acting on a set S, the **ordered pair-action group** $G^{[2]}$ is the group of ordered pair-permutations $\{\sigma^{[2]} \mid \sigma \in G\}$ under composition.

Facts:

1. Some common subgroups of S_n are listed in the following table.

subgroup	order	description
symmetric group S_n	$n!$	all permutations of $\{1, 2, \ldots, n\}$
alternating group A_n	$n!/2$	all even permutations of $\{1, 2, \ldots, n\}$
dihedral group D_n	$2n$	rigid motions of regular n-gon in 3-dimensional space (Example 2)
Klein 4-group (subgroup of S_4)	4	rigid motions of rectangle in 3-dimensional space (Example 3)
identity	1	consists only of identity permutation

2. The group A_n is abelian if $n = 2$ or 3, and is nonabelian if $n \geq 4$.

3. The group D_n has order $2n$. The elements consist of the n rotations and n reflections of a regular polygon with n sides. The n rotations are the counterclockwise rotations about the center through angles of $\frac{360k}{n}$ degrees ($k = 0, 1, \ldots, n-1$). (Clockwise rotations can be written in terms of counterclockwise rotations.) If n is odd, the n reflections are reflections in lines through a vertex and the center; if n is even, the reflections are reflections in lines joining opposite vertices and in lines joining midpoints of opposite sides.

322   Chapter 5   ALGEBRAIC STRUCTURES

4. The elements of D_n can be written as permutations of $\{1, 2, \ldots, n\}$. See the following figure for the rigid motions in D_4 (the rigid motions of the square) and the following table for the group multiplication table for D_4.

```
  2 1        1 4        4 3        3 2
  3 4        2 3        1 2        4 1

e = 0° CCW   90° CCW    180° CCW   270° CCW
 rotation    rotation   rotation   rotation
   (1)       (1 2 3 4)  (1 3)(2 4) (1 4 3 2)

  1 2        3 4        4 1        2 3
  3 4        2 1        3 2        1 4

reflection in  reflection in  reflection in  reflection in
vertical line  horizontal line  1-3 diagonal  2-4 diagonal
 (1 2)(3 4)   (1 4)(2 3)     (2 4)          (1 3)
```

·	(1)	(1234)	(13)(24)	(1432)	(12)(34)	(14)(23)	(24)	(13)
(1)	(1)	(1234)	(13)(24)	(1432)	(12)(34)	(14)(23)	(24)	(13)
(1234)	(1234)	(13)(24)	(1432)	(1)	(24)	(13)	(14)(23)	(12)(34)
(13)(24)	(13)(24)	(1432)	(1)	(1234)	(14)(23)	(12)(34)	(13)	(24)
(1432)	(1432)	(1)	(1234)	(13)(24)	(13)	(24)	(12)(34)	(14)(23)
(12)(34)	(12)(34)	(13)	(14)(23)	(24)	(1)	(13)(24)	(1432)	(1234)
(14)(23)	(14)(23)	(24)	(12)(34)	(13)	(13)(24)	(1)	(1234)	(1432)
(24)	(24)	(21)(34)	(13)	(14)(23)	(1234)	(1432)	(1)	(13)(24)
(13)	(13)	(14)(23)	(24)	(12)(34)	(1432)	(1234)	(13)(24)	(1)

5. The Klein four-group consists of the following four rigid motions of a rectangle: the rotations about the center through 0° or 180°, and reflections through the horizontal or vertical lines through its center, as illustrated in the following figure. The following table is the multiplication table for the Klein four-group.

```
  2 1        4 3        1 2        3 4
  3 4        1 2        4 3        2 1

e = 0° CCW   180° CCW    reflection in  reflection in
 rotation    rotation    vertical line  horizontal line
   (1)       (1 3)(2 4)   (1 2)(3 4)     (1 4)(2 3)
```

·	(1)	(13)(24)	(12)(34)	(14)(23)
(1)	(1)	(13)(24)	(12)(34)	(14)(23)
(13)(24)	(13)(24)	(1)	(14)(23)	(12)(34)
(12)(34)	(12)(34)	(14)(23)	(1)	(13)(24)
(14)(23)	(14)(23)	(12)(34)	(13)(24)	(1)

6. The Klein four-group is isomorphic to $\mathcal{Z}_8^*$.

7. The induced permutation group $S_n^{(2)}$ and the ordered-pair-action group $S_n^{[2]}$ are used in enumerative graph theory. (See §8.9.1.)

8. The induced permutation group $S_n^{(2)}$ has $\binom{n}{2}$ objects and $n!$ permutations.

9. The ordered-pair-action permutation group $S_n^{[2]}$ has n^2 objects and $n!$ permutations.

5.4 RINGS

5.4.1 BASIC CONCEPTS

Definitions:

A **ring** $(R, +, \cdot)$ consists of a set R closed under binary operations $+$ and $\cdot$ such that:
- $(R, +)$ is an abelian group; i.e., $(R, +)$ satisfies:
  - *associative property*: $a + (b + c) = (a + b) + c$ for all $a, b, c \in R$;
  - *identity property*: R has an *identity element*, 0, that satisfies $0 + a = a + 0 = a$ for all $a \in R$;
  - *inverse property*: for each $a \in R$ there is an *additive inverse element* $-a \in R$ (the *negative* of a) such that $-a + a = a + (-a) = 0$;
  - *commutative law*: $a + b = b + a$ for all $a, b \in R$;
- the operation $\cdot$ is *associative*: $a \cdot (b \cdot c) = (a \cdot b) \cdot c$ for all $a, b, c \in R$;
- the *distributive properties* for multiplication over addition hold for all $a, b, c \in R$:
  - *left distributive property*: $a \cdot (b + c) = a \cdot b + a \cdot c$;
  - *right distributive property*: $(a + b) \cdot c = a \cdot c + b \cdot c$.

A ring R is **commutative** if the multiplication operation is commutative: $a \cdot b = b \cdot a$ for all $a, b \in R$.

A ring R is a **ring with unity** if there is an identity, 1 ($\neq 0$), for multiplication; i.e., $1 \cdot a = a \cdot 1 = a$ for all $a \in R$. The multiplicative identity is the **unity** of R.

An element x in a ring R with unity is a **unit** if x has a multiplicative inverse; i.e., there is $x^{-1} \in R$ such that $x \cdot x^{-1} = x^{-1} \cdot x = 1$.

Subtraction in a ring is defined by the rule $a - b = a + (-b)$.

Facts:

1. Multiplication, $a \cdot b$, is often written ab or $a \times b$.

2. The order of precedence of operations in a ring follows that for real numbers: multiplication is to be done before addition. That is, $a + bc$ means $a + (bc)$ rather than $(a + b)c$.

3. In all rings, $a0 = 0a = 0$.

4. Properties of subtraction:

$$-(-a) = a \qquad (-a)(-b) = ab \qquad a(b-c) = ab - ac \qquad (a-b)c = ac - bc$$
$$a(-b) = (-a)b = -(ab) \qquad (-1)a = -a \text{ (if the ring has unity)}.$$

5. The set of all units of a ring is a group under the multiplication defined on the ring.

Examples:
1. Table 1 gives several examples of rings.
2. *Polynomial rings*: For a ring R, the set
$$R[x] = \{\, a_n x^n + \cdots + a_1 x + a_0 \mid a_0, a_1, \ldots, a_n \in R \,\}$$
forms a ring, where the elements are added and multiplied using the "usual" rules for addition and multiplication of polynomials. The additive identity, 0, is the constant polynomial $p(x) = 0$; the unity is the constant polynomial $p(x) = 1$ if R has a unity 1. (See §5.5.)
3. *Product rings*: For rings R and S, the set $R \times S = \{\, (r,s) \mid r \in R, s \in S \,\}$ forms a ring, where
$$(r_1, s_1) + (r_2, s_2) = (r_1 + r_2, s_1 + s_2);$$
$$(r_1, s_1) \cdot (r_2, s_2) = (r_1 r_2, s_1 s_2).$$
The additive identity is $(0,0)$. Unity is $(1,1)$ if R and S each have unity 1. Product rings can have more than two factors: $R_1 \times R_2 \times \cdots \times R_k$ or $R^n = R \times \cdots \times R$.

5.4.2 SUBRINGS AND IDEALS

Definitions:
A subset S of a ring $(R, +, \cdot)$ is a **subring** of R if $(S, +, \cdot)$ is a ring using the same operations $+$ and $\cdot$ that are used in R.

A subset I of a ring $(R, +, \cdot)$ is an **ideal** of R if:
- $(I, +, \cdot)$ is a subring of $(R, +, \cdot)$;
- I is closed under left and right multiplication by elements of R: if $x \in I$ and $r \in R$, then $rx \in I$ and $xr \in I$.

In a commutative ring R, an ideal I is **principal** if there is $r \in R$ such that $I = Rr = \{\, xr \mid x \in R \,\}$. I is the **principal ideal generated by** r, written $I = (r)$.

In a commutative ring R, an ideal $I \neq R$ is **maximal** if the only ideal properly containing I is R.

In a commutative ring R, an ideal $I \neq R$ is **prime** if $ab \in I$ implies that $a \in I$ or $b \in I$.

Facts:
1. If S is a nonempty subset of a ring $(R, +, \cdot)$, then S is a subring of R if and only if S is closed under subtraction and multiplication.
2. An ideal in a ring $(R, +, \cdot)$ is a subgroup of the group $(R, +)$, but not necessarily conversely.
3. The intersection of ideals in a ring is an ideal.
4. If R is any ring, R and $\{0\}$ are ideals, called *trivial* ideals.
5. In a commutative ring with unity, every maximal ideal is a prime ideal.
6. Every ideal I in the ring $\mathcal{Z}$ is a principal ideal. $I = (r)$ where r is the smallest positive integer in I.
7. If R is a commutative ring with unity, then R is a field (see §5.6) if and only if the only ideals of R are R and $\{0\}$.

Table 1 Examples of rings.

set and addition and multiplication operations	0	1
$\{0\}$, usual addition and multiplication; (*trivial ring*)	0	none
$\mathcal{Z}, \mathcal{Q}, \mathcal{R}, \mathcal{C}$, with usual $+$ and $\cdot$	0	1
$\mathcal{Z}_n = \{0, 1, \ldots, n-1\}$ (n a positive integer), $a+b = (a+b) \bmod n$, $a \cdot b = (ab) \bmod n$; (*modular ring*)	0	1
$\mathcal{Z}[\sqrt{2}] = \{a + b\sqrt{2} \mid a, b \in \mathcal{Z}\}$, $(a+b\sqrt{2})+(c+d\sqrt{2}) = (a+c)+(b+d)\sqrt{2}$, $(a+b\sqrt{2}) \cdot (c+d\sqrt{2}) = (ac+2bd) + (ad+bc)\sqrt{2}$ [Similar rings can be constructed using $\sqrt{n}$ (n an integer) if $\sqrt{n}$ not an integer.]	$0+0\sqrt{2}$	$1+0\sqrt{2}$
$\mathcal{Z}[i] = \{a + bi \mid a, b \in \mathcal{Z}\}$; (*Gaussian integers*; see §5.4.2, Example 2.)	$0+0i$	$1+0i$
$\mathcal{M}_{n \times n}(R) = $ all $n \times n$ matrices with entries in a ring R with unity, matrix addition and multiplication; (*matrix ring*)	O_n (zero matrix)	I_n (identity matrix)
$R = \{f \mid f: A \to B\}$ (A any nonempty set and B any ring), $(f+g)(x) = f(x)+g(x)$, $(f \cdot g)(x) = f(x) \cdot g(x)$; (*ring of functions*)	f such that $f(x)=0$ for all $x \in A$	f such that $f(x)=1$ for all $x \in A$ (if B has unity)
$\mathcal{P}(S) = $ all subsets of a set S, $A+B = A \triangle B = (A \cup B) - (A \cap B)$ (*symmetric difference*), $A \cdot B = A \cap B$; (*Boolean ring*)	$\emptyset$	S
$\{a+bi+cj+dk \mid a, b, c, d \in \mathcal{R}\}$, i, j, k in quaternion group, elements are added and multiplied like polynomials using $ij = k$, etc.; (*ring of real quaternions*, §5.2.2)	$0+0i+0j+0k$	$1+0i+0l+0k$

8. An ideal in a ring is the analogue of a normal subgroup in a group.

9. The second condition in the definition of ideal can be stated as $rI \subseteq I$ (I is a *left ideal*) and $Ir \subseteq I$ (I is a *right ideal*). (If A is a subset of a ring R and $r \in R$, then $rA = \{ra \mid a \in A\}$ and $Ar = \{ar \mid a \in A\}$.)

Examples:

1. With the usual definitions of $+$ and $\cdot$, each of the following rings can be viewed as a subring of all the rings listed after it: $\mathcal{Z}, \mathcal{Q}, \mathcal{R}, \mathcal{C}$.

2. *Gaussian integers*: $\mathcal{Z}[i] = \{a + bi \mid a, b \in \mathcal{Z}\}$ using the addition and multiplication of $\mathcal{C}$ is a subring of the ring of complex numbers.

3. The ring $\mathcal{Z}$ is a subring of $\mathcal{Z}[\sqrt{2}]$ and $\mathcal{Z}[\sqrt{2}]$ is a subring of $\mathcal{R}$.

4. Each set $n\mathcal{Z}$ (n an integer) is a principal ideal in the ring $\mathcal{Z}$.

5.4.3 RING HOMOMORPHISM AND ISOMORPHISM

Definitions:

If R and S are rings, a function $\varphi \colon R \to S$ is a *ring homomorphism* if for all $a, b \in R$:
- $\varphi(a + b) = \varphi(a) + \varphi(b)$ (φ preserves addition)
- $\varphi(ab) = \varphi(a)\varphi(b)$.   ($\varphi$ preserves multiplication)

Note: $\varphi(a)$ is sometimes written $a\varphi$.

If a ring homomorphism φ is also one-to-one and onto S, then φ is a *ring isomorphism* and R and S are *isomorphic*, written $R \cong S$.

A *ring endomorphism* is a ring homomorphism $\varphi \colon R \to R$.

A *ring automorphism* is a ring isomorphism $\varphi \colon R \to R$.

The *kernel* of a ring homomorphism $\varphi \colon R \to S$ is $\varphi^{-1}(0) = \{\, x \in R \mid \varphi(x) = 0 \,\}$.

Facts:

1. If φ is a ring isomorphism, then φ^{-1} is a ring isomorphism.
2. The kernel of a ring homomorphism from R to S is an ideal of the ring R.
3. If $\varphi \colon R \to S$ is a ring homomorphism, $\varphi(R)$ is a subring of S.
4. If $\varphi \colon R \to S$ is a ring homomorphism and R has unity, either $\varphi(1) = 0$ or $\varphi(1)$ is unity for $\varphi(R)$.
5. If φ is a ring homomorphism, then $\varphi(0) = 0$ and $\varphi(-a) = -\varphi(a)$.
6. A ring homomorphism is a ring isomorphism between R and $\varphi(R)$ if and only if the kernel of φ is $\{0\}$.
7. *Homomorphisms preserve subrings*: Let $\varphi \colon R \to S$ be a ring homomorphism. If A is a subring of R, then $\varphi(A)$ is a subring of S. If B is a subring of S, then $\varphi^{-1}(B)$ is a subring of R.
8. *Homomorphisms preserve ideals*: Let $\varphi \colon R \to S$ be a ring homomorphism. If A is an ideal of R, then $\varphi(A)$ is an ideal of S. If B is an ideal of S, then $\varphi^{-1}(B)$ is an ideal of R.

Examples:

1. The function $\varphi \colon Z \to Z_n$ defined by the rule $\varphi(a) = a \bmod n$ is a ring homomorphism.
2. If R and S are rings, then the function $\varphi \colon R \to S$ defined by the rule $\varphi(a) = 0$ for all $a \in R$ is a ring homomorphism.
3. The function $\varphi \colon Z \to R$ (R any ring with unity) defined by the rule $\varphi(x) = x \cdot 1$ is a ring homomorphism. The kernel of φ is the subring nZ for some nonnegative integer n, called the *characteristic* of R.
4. Let $\mathcal{P}(S)$ be the ring of all subsets of a set S (see Table 1). If $|S| = 1$, then $\mathcal{P}(S) \cong Z_2$ with the ring isomorphism φ where $\varphi(\emptyset) = 0$ and $\varphi(S) = 1$. More generally, if $|S| = n$, then $\mathcal{P}(S) \cong Z_2^n = Z_2 \times \cdots \times Z_2$.
5. $Z_n \cong Z/(n)$ for all positive integers n. (See §5.4.4.)
6. $Z_m \times Z_n \cong Z_{mn}$, if m and n are relatively prime.

5.4.4 QUOTIENT RINGS

Definitions:

If I is an ideal in a ring R and $a \in R$, then the set $a + I = \{a + x \mid x \in I\}$ is a *coset* of I in R.

The set of all cosets, $R/I = \{a + I \mid a \in R\}$, is a ring, called the **quotient ring**, where addition and multiplication are defined by the rules:
- $(a + I) + (b + I) = (a + b) + I$;
- $(a + I) \cdot (b + I) = (ab) + I$.

Facts:
1. If R is commutative, then R/I is commutative.
2. If R has unity 1, then R/I has the coset $1 + I$ as unity.
3. If I is an ideal in ring R, the function $\varphi \colon R \to R/I$ defined by the rule $\varphi(x) = x + I$ is a ring homomorphism, called the *natural map*. The kernel of φ is I.
4. *Fundamental homomorphism theorem for rings*: If φ is a ring homomorphism and K is the kernel of φ, then $\varphi(R) \cong R/K$.
5. If R is a commutative ring with unity and I is an ideal in R, then I is a maximal ideal if and only if R/I is a field (see §5.6).

Examples:
1. For each integer n, $\mathcal{Z}/n\mathcal{Z}$ is a quotient ring, isomorphic to $\mathcal{Z}_n$.
2. See §5.6.1 for Galois rings.

5.4.5 RINGS WITH ADDITIONAL PROPERTIES

Beginning with rings, as additional requirements are added, the following hierarchy of sets of algebraic structures is obtained:

$$\text{rings} \supset \text{commutative rings with unity} \supset \text{integral domains} \supset \text{Euclidean domains} \supset \text{principal ideal domains}$$

Definitions:

The **cancellation properties** in a ring R state that for all $a, b, c \in R$:

if $ab = ac$ and $a \neq 0$, then $b = c$ (left cancellation property)

if $ba = ca$ and $a \neq 0$, then $b = c$ (right cancellation property).

Let R be a ring and let $a, b \in R$ where $a \neq 0, b \neq 0$. If $ab = 0$, then a is a **left divisor of zero** and b is a **right divisor of zero**.

An **integral domain** is a commutative ring with unity that has no zero divisors.

A **principal ideal domain** (**PID**) is an integral domain in which every ideal is a principal ideal.

A **division ring** is a ring with unity in which every nonzero element is a unit (i.e., every nonzero element has a multiplicative inverse).

A **field** is a commutative ring with unity such that each nonzero element has a multiplicative inverse. (See §5.6.)

A **Euclidean norm** on an integral domain R is a function $\delta: R - \{0\} \to \{0, 1, 2, \ldots\}$ such that:
- $\delta(a) \leq \delta(ab)$ for all $a, b \in R - \{0\}$;
- the following generalization of the division algorithm for integers holds: for all $a, d \in R$ where $d \neq 0$, there are elements $q, r \in R$ such that $a = dq + r$, where either $r = 0$ or $\delta(r) < \delta(d)$.

A **Euclidean domain** is an integral domain with a Euclidean norm defined on it.

Facts:

1. The cancellation properties hold in an integral domain.
2. Every finite integral domain is a field.
3. Every integral domain can be imbedded in a field. Given an integral domain R, there is a field F and a ring homomorphism $\varphi: R \to F$ such that $\varphi(1) = 1$.
4. A ring with unity is a division ring if and only if the nonzero elements form a group under the multiplication defined on the ring.
5. *Wedderburn's theorem*: Every finite division ring is a field. (J. H. M. Wedderburn, 1882–1948)
6. Every commutative division ring is a field.
7. In a Euclidean domain, if $b \neq 0$ is not a unit, then $\delta(ab) > \delta(a)$ for all $a \neq 0$. For $b \neq 0$, b is a unit in R if and only if $\delta(b) = \delta(1)$.
8. In every Euclidean domain, a Euclidean algorithm for finding the gcd can be carried out.

Examples:

1. Some common Euclidean domains are given in the following table.

set	Euclidean norm
$\mathcal{Z}$	$\delta(a) = |a|$
$\mathcal{Z}[i]$ (Gaussian integers)	$\delta(a + bi) = a^2 + b^2$
F (any field)	$\delta(a) = 1$
polynomial ring $F[x]$ (F any field)	$\delta(p(x)) =$ degree of $p(x)$

2. The following table gives examples of rings with additional properties.

ring	commutative ring with unity	integral domain	principal ideal domain	Euclidean domain	division ring	field
$\mathcal{Z}$	yes	yes	yes	yes	no	no
$\mathcal{Q}, \mathcal{R}, \mathcal{C}$	yes	yes	yes	yes	yes	yes
$\mathcal{Z}_p$ (p prime)	yes	yes	yes	yes	yes	yes
$\mathcal{Z}_n$ (n composite)	yes	no	no	no	no	no
real quaternions	no	no	no	no	yes	no
$\mathcal{Z}[x]$	yes	no	no	no	no	no
$\mathcal{M}_{n \times n}$	no	no	no	no	no	no

5.5 POLYNOMIAL RINGS

5.5.1 BASIC CONCEPTS

Definitions:

A *polynomial in the variable x over a ring R* is an expression of the form
$$f(x) = a_n x^n + a_{n-1} x^{n-1} + \cdots + a_1 x^1 + a_0 x^0$$
where $a_n, \ldots, a_0 \in R$.

For a polynomial $f(x) \neq 0$, the largest integer k such that $a_k \neq 0$ is the **degree** of $f(x)$, written $\deg f(x)$.

A **constant polynomial** is a polynomial $f(x) = a_0$. If $a_0 \neq 0$, $f(x)$ has degree 0. If $f(x) = 0$ (the **zero polynomial**), the degree of $f(x)$ is undefined. (The degree of the zero polynomial is also said to be $-\infty$.)

The **polynomial ring** (in one variable x) over a ring R consists of the set
$$R[x] = \{ f(x) \mid f(x) \text{ is a polynomial over } R \text{ in the variable } x \}$$
with addition and multiplication defined by the rules:
$$(a_n x^n + \cdots + a_1 x^1 + a_0 x^0) + (b_m x^m + \cdots + b_1 x^1 + b_0 x^0)$$
$$= a_n x^n + \cdots + a_{m+1} x^{m+1} + (a_n + b_n) x^n + \cdots + (a_1 + b_1) x^1 + (a_0 + b_0) x^0$$
if $n \geq m$, and
$$(a_n x^n + \cdots + a_1 x^1 + a_0 x^0)(b_m x^m + \cdots + b_1 x^1 + b_0 x^0)$$
$$= c_{n+m} x^{n+m} + \cdots + c_1 x^1 + c_0 x^0$$
where $c_i = a_0 b_i + a_1 b_{i-1} + \cdots + a_i b_0$ for $i = 0, 1, \ldots, m+n$.

A polynomial $f(x) \in R[x]$ of degree n is **monic** if $a_n = 1$.

The **value** of a polynomial $f(x) = a_n x^n + a_{n-1} x^{n-1} + \cdots + a_1 x^1 + a_0 x^0$ at $c \in R$ is the element $f(c) = a_n c^n + a_{n-1} c^{n-1} + \cdots + a_1 c + a_0 \in R$.

An element $c \in R$ is a **zero** of the polynomial $f(x)$ if $f(c) = 0$.

If R is a subring of a commutative ring S, an element $a \in S$ is **algebraic** over R if there is a nonzero $f(x) \in R[x]$ such that $f(a) = 0$.

If $p(x)$ is not algebraic over R, then $p(x)$ is **transcendental** over R.

A polynomial $f(x) \in R[x]$ of degree n is **irreducible** over R if $f(x)$ cannot be written as $f_1(x) f_2(x)$ (**factors** of $f(x)$) where $f_1(x)$ and $f_2(x)$ are polynomials over R of degrees less than n. Otherwise $f(x)$ is **reducible** over R.

The **polynomial ring** (in the variables $x_1, x_2, \ldots, x_n$ with $n > 1$) over a ring R is defined by the rule $R[x_1, x_2, \ldots, x_n] = (R[x_1, x_2, \ldots, x_{n-1}])[x_n]$.

Facts:

1. Polynomials over an arbitrary ring R generalize polynomials with coefficients in $\mathcal{R}$ or $\mathcal{C}$. The notation and terminology follow the usual conventions for polynomials with real (or complex) coefficients:

- the elements $a_n, \ldots, a_0$ are *coefficients*;
- subtraction notation can be used: $a_i x^i + (-a_j) x^j = a_i x^i - a_j x^j$;
- the term $1x^i$ can be written as x^i;
- the term x^1 can be written x;
- the term x^0 can be written 1;
- terms $0x^i$ can be omitted.

2. There is a distinction between a polynomial $f(x) \in R[x]$ and the function it defines using the rule $f(c) = a_n c^n + a_{n-1} c^{n-1} + \cdots + a_1 c + a_0$ for $c \in R$. The same function might be defined by infinitely many polynomials. For example, the polynomials $f_1(x) = x \in \mathcal{Z}_2[x]$ and $f_2(x) = x^2 \in \mathcal{Z}_2[x]$ define the same function: $f_1(0) = f_2(0) = 0$ and $f_1(1) = f_2(1) = 1$.

3. If R is a ring, $R[x]$ is a ring.

4. If R is a commutative ring, then $R[x]$ is a commutative ring.

5. If R is a ring with unity, then $R[x]$ has the constant polynomial $f(x) = 1$ as unity.

6. If R is an integral domain, then $R[x]$ is an integral domain. If $f_1(x)$ has degree m and $f_2(x)$ has degree n, then the degree of $f_1(x) f_2(x)$ is $m + n$.

7. If ring R is not an integral domain, then $R[x]$ is not an integral domain. If $f_1(x)$ has degree m and $f_2(x)$ has degree n, then the degree of $f_1(x) f_2(x)$ can be smaller than $m + n$. (For example, in $\mathcal{Z}_6[x]$, $(3x^2)(2x^3) = 0$.)

8. *Factor theorem*: If R is a commutative ring with unity and $f(x) \in R[x]$ has degree ≥ 1, then $f(a) = 0$ if and only if $x - a$ is a factor of $f(x)$.

9. If R is an integral domain and $p(x) \in R[x]$ has degree n, then $p(x)$ has at most n zeros in R. If R is not an integral domain, then a polynomial may have more zeros than its degree; for example, $x^2 + x \in \mathcal{Z}_6[x]$ has four zeros — $0, 2, 3, 5$.

5.5.2 POLYNOMIALS OVER A FIELD

Facts:

1. Even though F is a field (§5.6.1), $F[x]$ is never a field. (The polynomial $f(x) = x$ has no multiplicative inverse in $F[x]$.)

2. If $f(x)$ has degree n, then $f(x)$ has at most n distinct zeros.

3. *Irreducibility over a finite field*: If F is a finite field and n is a positive integer, then there is an irreducible polynomial over F of degree n.

4. *Unique factorization theorem*: If $f(x)$ is a polynomial over a field F and is not the zero polynomial, then $f(x)$ can be uniquely factored (ignoring the order in which the factors are written) as $a f_1(x) \cdots f_k(x)$ where $a \in F$ and each $f_i(x)$ is a monic polynomial that is irreducible over F.

5. *Eisenstein's irreducibility criterion*: If $f(x) \in \mathcal{Z}[x]$ has degree $n > 0$, if there is a prime p such that p divides every coefficient of $f(x)$ except a_n, and if p^2 does not divide a_0, then $f(x)$ is irreducible over $\mathcal{Q}$. (F. G. M. Eisenstein, 1823–1852)

6. *Division algorithm for polynomials*: If F is a field with $a(x), d(x) \in F[x]$ and $d(x)$ is not the zero polynomial, then there are unique polynomials $q(x)$ (quotient) and $r(x)$ (remainder) in $F[x]$ such that $a(x) = d(x) q(x) + r(x)$ where $\deg r(x) < \deg d(x)$ or $r(x) = 0$. If $d(x)$ is monic, then the division algorithm for polynomials can be extended to all rings with unity.

7. Irreducibility over the real numbers $\mathcal{R}$: If $f(x) \in \mathcal{R}[x]$ has degree at least 3, then $f(x)$ is reducible. The only irreducible polynomials in $\mathcal{R}[x]$ are of degree 1 or 2; for example $x^2 + 1$ is irreducible over $\mathcal{R}$.

8. Fundamental theorem of algebra (irreducibility over the complex numbers $\mathcal{C}$): If $f(x) \in \mathcal{C}[x]$ has degree $n \geq 1$, then $f(x)$ can be completely factored:
$$f(x) = c(x - c_1)(x - c_2) \ldots (x - c_n)$$
where $c, c_1, \ldots, c_n \in \mathcal{C}$.

9. If F is a field and $f(x) \in F[x]$ has degree 1 (i.e., $f(x)$ is linear), then $f(x)$ is irreducible.

10. If F is a field and $f(x) \in F[x]$ has degree ≥ 2 and has a zero, then $f(x)$ is reducible. (If $f(x)$ has a as a zero, then $f(x)$ can be written as $(x - a)f_1(x)$ where $\deg f_1(x) = \deg f(x) - 1$. The converse is false: a polynomial may have no zeros, but still be reducible. (See Example 2.)

11. If F is a field and $f(x) \in F[x]$ has degree 2 or 3, then $f(x)$ is irreducible if and only if $f(x)$ has no zeros.

Examples:

1. In $\mathcal{Z}_5[x]$, if $a(x) = 3x^4 + 2x^3 + 2x + 1$ and $d(x) = x^2 + 2$, then $q(x) = 3x^2 + 2x + 4$ and $r(x) = 3x + 3$. To obtain $q(x)$ and $r(x)$, use the same format as for long division of natural numbers, with arithmetic operations carried out in $\mathcal{Z}_5$:

```
              3x² + 2x + 4
  x² + 2 ) 3x⁴ + 2x³ + 0x² + 2x + 1
           3x⁴       + x²
           ─────────────
                 2x³ + 4x²              [−x² = 4x² over Z₅]
                 2x³      + 4x
                 ─────────────
                       4x² + 3x         [2x − 4x = −2x = 3x over Z₅]
                       4x²      + 3
                       ─────────────
                             3x + 3
```

2. Polynomials can have no zeros, but be reducible. The polynomial $f(x) = x^4 + x^2 + 1 \in \mathcal{Z}_2[x]$ has no zeros (since $f(0) = f(1) = 1$), but $f(x)$ can be factored as $(x^2 + x + 1)^2$. Similarly, $x^4 + 2x^2 + 1 = (x^2 + 1)^2 \in \mathcal{R}[x]$.

5.6 FIELDS

5.6.1 BASIC CONCEPTS

Definitions:

A **field** $(F, +, \cdot)$ consists of a set F together with two binary operations, $+$ and $\cdot$, such that:
- $(F, +, \cdot)$ is a ring;
- $(F - \{0\}, \cdot)$ is a commutative group.

A **subfield** F of field $(K, +, \cdot)$ is a subset of K that is a field using the same operations as those in K.

If F is a subfield of K, then K is called an **extension field** of F. Write K/F to indicate that K is an extension field of F.

For K an extension field of F, the **degree of K over F** is $[K:F]$ = the dimension of K as a vector space over F. (See §6.1.3.)

A **field isomorphism** is a function $\varphi: F_1 \to F_2$, where F_1 and F_2 are fields, such that φ is one-to-one, onto F_2, and satisfies the following for all $a, b \in F_1$:
- $\varphi(a + b) = \varphi(a) + \varphi(b)$;
- $\varphi(ab) = \varphi(a)\varphi(b)$.

A **field automorphism** is an isomorphism $\varphi: F \to F$, where F is a field. The set of all automorphisms of F is denoted $\mathrm{Aut}(F)$.

The **characteristic** of a field F is the smallest positive integer n such that $1 + \cdots + 1 = 0$, where there are n summands. If there is no such integer, F has characteristic 0 (also called characteristic ∞).

Facts:

1. Every field is a commutative ring with unity. A field satisfies all properties of a commutative ring with unity, and has the additional property that every nonzero element has a multiplicative inverse.

2. Every finite integral domain is a field.

3. A field is a commutative division ring.

4. If F is a field and $a, b \in F$ where $a \neq 0$, then $ax + b = 0$ has a unique solution in F.

5. If F is a field, every ideal in $F[x]$ is a principal ideal.

6. If p is a prime and n is any positive integer, then there is exactly one field (up to isomorphism) with p^n elements, the *Galois field* $GF(p^n)$. (§5.6.2)

7. If $\varphi: F \to F$ is a field automorphism, then:
   - $-\varphi(a) = \varphi(-a)$
   - $\varphi(a^{-1}) = \varphi(a)^{-1}$

   for all $a \neq 0$.

8. The intersection of all subfields of a field F is a field, called the *prime field* of F.

9. If F is a field, $\mathrm{Aut}(F)$ is a group under composition of functions.

10. The characteristic of a field is either 0 or prime.

11. Every field of characteristic 0 is isomorphic to a field that is an extension of $\mathcal{Q}$ and has $\mathcal{Q}$ as its prime field.

12. Every field of characteristic $p > 0$ is isomorphic to a field that is an extension of $\mathcal{Z}_p$ and has $\mathcal{Z}_p$ as its prime field.

13. If field F has characteristic $p > 0$, then $(a + b)^p = a^p + b^p$ for all $a, b \in F$.

14. If field F has characteristic $p > 0$, $f(x) \in \mathcal{Z}_p[x]$, and $\alpha \in F$ is a zero of $f(x)$, then $\alpha^p, \alpha^{p^2}, \alpha^{p^3}, \ldots$ are also zeros of $f(x)$.

15. If p is not a prime, then $\mathcal{Z}_p$ is not a field since $\mathcal{Z}_p - \{0\}$ will fail to be closed under multiplication. For example, $\mathcal{Z}_6$ is not a field since $2 \in \mathcal{Z}_6 - \{0\}$ and $3 \in \mathcal{Z}_6 - \{0\}$, but $2 \cdot 3 = 0 \notin \mathcal{Z}_6 - \{0\}$.

Examples:

1. The following table gives several examples of fields.

set and operations	$-a$	a^{-1}	characteristic	order
$\mathcal{Q}, \mathcal{R}, \mathcal{C}$, with usual addition and multiplication	$-a$	$1/a$	0	infinite
$\mathcal{Z}_p = \{0, 1, \ldots, p-1\}$ (p prime) prime), addition and multiplication **mod** p	$p - a$ ($-0 = 0$)	$a^{-1} = b$, where $ab \bmod p = 1$	p	p
$F[x]/(f(x))$, $f(x)$ irreducible over field F, coset addition and multiplication (Example 2)	$-[a+(f(x))] =$ $-a + (f(x))$	$[a+(f(x))]^{-1} =$ $a^{-1} + (f(x))$	varies	varies
$GF(p^n) = \mathcal{Z}_p[x]/(f(x))$, $f(x)$ of degree n irreducible over $\mathcal{Z}_p$ (p prime), addition and multiplication of cosets (*Galois field*)	$-[a+(f(x))] =$ $-a + (f(x))$	$[a+(f(x))]^{-1} =$ $a^{-1} + (f(x))$	p	p^n

2. **The field $F[x]/(f(x))$**: If F is any field and $f(x) \in F[x]$ of degree n is irreducible over F, the quotient ring structure $F[x]/(f(x))$ is a field. The elements of $F[x]/(f(x))$ are cosets of polynomials in $F[x]$ modulo $f(x)$, where $(f(x))$ is the principal ideal generated by $f(x)$. Polynomials $f_1(x)$ and $f_2(x)$ lie in the same coset if and only if $f(x)$ is a factor of $f_1(x) - f_2(x)$.

Using the division algorithm for polynomials, any polynomial $g(x) \in F[x]$ can be written as $g(x) = f(x)q(x) + r(x)$ where $q(x)$ and $r(x)$ are unique polynomials in $F[x]$ and $r(x)$ has degree $< n$. The equivalence class $g(x) + (f(x))$ can be identified with the polynomial $r(x)$, and thus $F[x]/(f(x))$ can be regarded as the field of all polynomials in $F[x]$ of degree $< n$.

5.6.2 EXTENSION FIELDS AND GALOIS THEORY

Throughout this subsection assume that field K is an extension of field F.

Definitions:

For $\alpha \in K$, $F(\alpha)$ is the smallest field containing α and F, called the **field extension** of F by α.

For $\alpha_1, \ldots, \alpha_n \in K$, $F(\alpha_1, \ldots, \alpha_n)$ is the smallest field containing $\alpha_1, \ldots, \alpha_n$ and F, called the **field extension** of F by $\alpha_1, \ldots, \alpha_n$.

If K is an extension field of F and $\alpha \in K$, then α is **algebraic** over F if α is a root of a nonzero polynomial in $F[x]$. If α is not the root of any nonzero polynomial in $F[x]$, then α is **transcendental** over F.

A complex number is an **algebraic number** if it is algebraic over $\mathcal{Q}$.

An **algebraic integer** is an algebraic number α that is a zero of a polynomial of the form $x^n + a_{n-1}x^{n-1} + \cdots + a_1 x + a_0$ where each $a_i \in \mathcal{Z}$.

An extension field K of F is an **algebraic extension** of F if every element of K is algebraic over F. Otherwise K is a **transcendental extension** of F.

An extension field K of F is a **finite extension** of F if K is finite-dimensional as a vector space over F (see Fact 11). The dimension of K over F is written $[K:F]$.

Let α be algebraic over a field F. The **minimal polynomial of α with respect to F** is the monic irreducible polynomial $f(x) \in F[x]$ of smallest degree such that $f(\alpha) = 0$.

A polynomial $f(x) \in F[x]$ **splits** over K if $f(x) = \alpha(x - \alpha_1)\ldots(x - \alpha_n)$ where $\alpha, \alpha_1, \ldots, \alpha_n \in K$.

K is a **splitting field** (**root field**) of a nonconstant $f(x) \in F[x]$ if $f(x)$ splits over K and K is the smallest field with this property.

A polynomial $f(x) \in F[x]$ of degree n is **separable** if $f(x)$ has n distinct roots in its splitting field.

K is a **separable extension** of F if every element of K is the root of a separable polynomial in $F[x]$.

K is a **normal extension** of F if K/F is algebraic and every irreducible polynomial in $F[x]$ with a root in K has all its roots in K (i.e., splits in K).

K is a **Galois extension** of F if K is a normal, separable extension of F.

A field automorphism φ **fixes set S elementwise** if $\varphi(x) = x$ for all $x \in S$.

The **fixed field** of a subset $A \subseteq \text{Aut}(F)$ is $F_A = \{\, x \in F \mid \varphi(x) = x \text{ for all } \varphi \in A\,\}$.

The **Galois group** of K over F is the group of automorphisms $G(K/F)$ of K that fix F elementwise. If K is a splitting field of $f(x) \in F[x]$, $G(K/F)$ is also known as the **Galois group** of $f(x)$. (Évariste Galois, 1811–1832)

Facts:
1. The elements of K that are algebraic over F form a subfield of K.
2. The algebraic numbers in $\mathcal{C}$ form a field; the algebraic integers form a subring of $\mathcal{C}$, called the *ring of algebraic integers*.
3. Every nonconstant polynomial has a unique splitting field, up to isomorphism.
4. If $f(x) \in F[x]$ splits as $\alpha(x - \alpha_1)\ldots(x - \alpha_n)$, then the splitting field for $f(x)$ is $F(\alpha_1, \ldots, \alpha_n)$.
5. If F is a field and $p(x) \in F[x]$ is a nonconstant polynomial, then there is an extension field K of F and $\alpha \in K$ such that $p(\alpha) = 0$.
6. If $f(x)$ is irreducible over F, then the ring $F[x]/(f(x))$ is an algebraic extension of F and contains a root of $f(x)$.
7. The field F is isomorphic to a subfield of any algebraic extension $F[x]/(f(x))$. The element $0 \in F$ corresponds to the coset of the zero polynomial; all other elements of F appear in $F[x]/(f(x))$ as cosets of the constant polynomials.
8. Every minimal polynomial is irreducible.
9. If K is a field extension of F and $\alpha \in K$ is a root of an irreducible polynomial $f(x) \in F[x]$ of degree $n \geq 1$, then $F(\alpha) = \{c_{n-1}\alpha^{n-1} + \cdots + c_1\alpha + c_0 \mid c_i \in F \text{ for all } i\}$.
10. If K is an extension field of F and $\alpha \in K$ is algebraic over F, then:
    - there is a unique monic irreducible polynomial $f(x) \in F[x]$ of smallest degree (the *minimum polynomial*) such that $f(\alpha) = 0$;
    - $F(\alpha) \cong F[x]/(f(x))$;
    - if the degree of α over F is n, then $K = \{a_0 + a_1\alpha + a_2\alpha^2 + \cdots + a_{n-1}\alpha^{n-1} \mid a_0, a_1, \ldots, a_{n-1} \in F\}$; in fact, K is an n-dimensional vector space over F, with basis $1, \alpha, \alpha^2, \ldots, \alpha^{n-1}$.

11. If K is an extension field of F and $x \in K$ is transcendental over F, then $F(\alpha) \cong$ the field of all fractions $f(x)/g(x)$ where $f(x), g(x) \in F[x]$ and $g(x)$ is not the zero polynomial.

12. K is a splitting field of some polynomial $f(x) \in F[x]$ if and only if K is a Galois extension of F.

13. If K is a splitting field for separable $f(x) \in F[x]$ of degree n, then $G(K, F)$ is isomorphic to a subgroup of the symmetric group S_n.

14. If K is a splitting field of $f(x) \in F[x]$, then:
    - every element of $G(K/F)$ permutes the roots of $f(x)$ and is completely determined by its effect on the roots of $f(x)$;
    - $G(K/F)$ is isomorphic to a group of permutations of the roots of $f(x)$.

15. If K is a splitting field for separable $f(x) \in F[x]$, then $|G(K/F)| = [K:F]$.

16. For $[K:F]$ finite, K is a normal extension of F if and only if K is a splitting field of some polynomial in $F[x]$.

17. *The Fundamental theorem of Galois theory*: If K is a normal extension of F, where F is either finite or has characteristic 0, then there is a one-to-one correspondence Φ between the lattice of all fields K', where $F \subseteq K' \subseteq K$, and the lattice of all subgroups H of the Galois group $G(K/F)$:
$$\Phi(K') = G(K/K') \quad \text{and} \quad \Phi^{-1}(H) = K_H.$$
The correspondence Φ has the following properties:
    - for fields K' and K'' where $F \subseteq K' \subseteq K$ and $F \subseteq K'' \subseteq K$
$$K' \subseteq K'' \longleftrightarrow \Phi(K'') \subseteq \Phi(K');$$
    that is, $G(K/K'') \subseteq G(K/K')$.
    - Φ interchanges the operations meet and join for the lattice of subfields and the lattice of subgroups:
$$\Phi(K' \wedge K'') = G(K/K') \vee G(K/K'')$$
$$\Phi(K' \vee K'') = G(K/K') \wedge G(K/K'');$$
    (Note: In the lattice of fields [groups], $A \wedge B = A \cap B$ and $A \vee B$ is the smallest field [group] containing A and B.)
    - K' is a normal extension of F if and only if $G(K/K')$ is a normal subgroup of $G(K/F)$.

18. *Formulas for solving polynomial equations of degrees 2, 3, or 4*:
    - second-degree (quadratic) equation $ax^2 + bx + c = 0$: the quadratic formula gives the solutions $\dfrac{-b \pm \sqrt{b^2 - 4ac}}{2a}$;
    - third-degree (cubic) equation $a_3 x^3 + a_2 x^2 + a_1 x + a_0 = 0$:
       (1) divide by a_3 to obtain $x^3 + b_2 x^2 + b_1 x + b_0 = 0$,
       (2) make the substitution $x = y - \frac{b_2}{3}$ to obtain an equation of the form $y^3 + cy + d = 0$, with solutions $y = \sqrt[3]{\frac{-d}{2} + \sqrt{\frac{d^2}{4} + \frac{c^3}{27}}} + \sqrt[3]{\frac{-d}{2} - \sqrt{\frac{d^2}{4} + \frac{c^3}{27}}}$,
       (3) use the substitution $x = y - \frac{b_2}{3}$ to obtain the solutions to the original equation;

- fourth-degree (quartic) equation $a_4 x^4 + a_3 x^3 + a_2 x^2 + a_1 x + a_0 = 0$:
  (1) divide by a_4 to obtain $x^4 + ax^3 + bx^2 + cx + d = 0$,
  (2) solve the *resolvent* equation $y^3 - by^s + (ac - 4d)y + (-a^2 d + 4bd - c^2) = 0$ to obtain a root z,
  (3) solve the pair of quadratic equations:

$$x^2 + \tfrac{a}{2}x + \tfrac{z}{2} = \pm\sqrt{\left(\tfrac{a^2}{4} - b + z\right)x^2 + \left(\tfrac{a}{2}z - c\right)x + \left(\tfrac{z^2}{4} - d\right)}$$

to obtain the solutions to the original equation.

19. A general method for solving cubic equations algebraically was given by Nicolo Fontana (1500–1557), also called Tartaglia. The method is often referred to as Cardano's method because Girolamo Cardano (1501–1576) published the method. Ludovico Ferrari (1522–1565), a student of Cardano, discovered a general method for solving quartic equations algebraically.

20. *Equations of degree 5 or more*: In 1824 Abel proved that the general quintic polynomial equation $a_5 x^5 + \cdots + a_1 x + a_0 = 0$ (and those of higher degree) are not solvable by radicals; that is, there can be no formula for writing the roots of such equations using only the basic arithmetic operations and the taking of nth roots. Évariste Galois (1811–1832) demonstrated the existence of such equations that are not solvable by radicals and related solvability by radicals of polynomial equations to determining whether the associated permutation group (the Galois group) of roots is solvable. (See Application 1.)

Examples:

1. $\mathcal{C}$ *as an algebraic extension of* $\mathcal{R}$: Let $f(x) = x^2 + 1 \in \mathcal{R}[x]$ and $\alpha = x + (x^2 + 1) \in \mathcal{R}[x]/(x^2 + 1)$. Then $\alpha^2 = -1$. Thus, α behaves like i (since $i^2 = -1$). Hence $\mathcal{R}[x]/(x^2 + 1) = \{ c_1\alpha + c_0 \mid c_1, c_0 \in \mathcal{R} \} \cong \{ c_1 i + c_0 \mid c_0, c_1 \in \mathcal{R} \} = \mathcal{C}$.

2. *Algebraic extensions of* $\mathcal{Z}_p$: If $f(x) \in \mathcal{Z}_p$ is an irreducible polynomial of degree n, then the algebraic extension $\mathcal{Z}_p[x]/(f(x))$ is a *Galois field*.

3. If $f(x) = x^4 - 2x^2 - 3 \in \mathcal{Q}[x]$, its splitting field is

$$\mathcal{Q}(\sqrt{3}, i) = \{ a + b\sqrt{3} + ci + di\sqrt{3} \mid a, b, c, d \in \mathcal{Q} \}.$$

There are three intermediate fields: $\mathcal{Q}(\sqrt{3})$, $\mathcal{Q}(i)$, and $\mathcal{Q}(i\sqrt{3})$, as illustrated in Figure 1. The Galois group $G(\mathcal{Q}(\sqrt{3}, i)/\mathcal{Q}) = \{e, \phi_1, \phi_2, \phi_3\}$ where:

$$\phi_1(a + b\sqrt{3} + ci + di\sqrt{3}) = a + b\sqrt{3} - ci - di\sqrt{3},$$
$$\phi_2(a + b\sqrt{3} + ci + di\sqrt{3}) = a - b\sqrt{3} + ci - di\sqrt{3},$$
$$\phi_3(a + b\sqrt{3} + ci + di\sqrt{3}) = a - b\sqrt{3} - ci + di\sqrt{3} = \phi_2\phi_1 = \phi_1\phi_2,$$
$$e(a + b\sqrt{3} + ci + di\sqrt{3}) = a + b\sqrt{3} + ci + di\sqrt{3}.$$

$G(\mathcal{Q}(\sqrt{3}, i), \mathcal{Q})$ has the following subgroups:

$$G = G(\mathcal{Q}(\sqrt{3}, i), \mathcal{Q}) = \{e, \phi_1, \phi_2, \phi_3\},$$
$$H_1 = G(\mathcal{Q}(\sqrt{3}, i), \mathcal{Q}(\sqrt{3})) = \{e, \phi_1\},$$
$$H_2 = G(\mathcal{Q}(\sqrt{3}, i), \mathcal{Q}(i)) = \{e, \phi_2\},$$
$$H_3 = G(\mathcal{Q}(\sqrt{3}, i), \mathcal{Q}(i\sqrt{3})) = \{e, \phi_3\},$$
$$\{e\} = G(\mathcal{Q}(\sqrt{3}, i), \mathcal{Q}(\sqrt{3}, i)).$$

The correspondence between fields and Galois groups is shown in the following table and figure.

field	Galois group
$\mathcal{Q}(\sqrt{3}, i)$	$\{e\}$
$\mathcal{Q}(\sqrt{3})$	H_1
$\mathcal{Q}(i\sqrt{3})$	H_3
$\mathcal{Q}(i)$	H_2
$\mathcal{Q}$	G

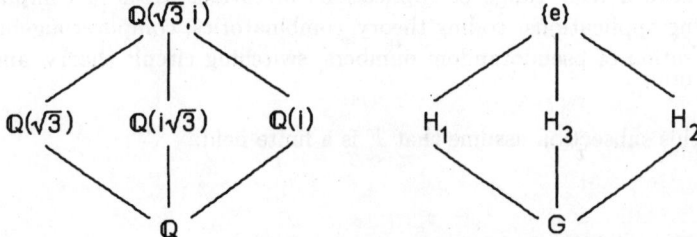

4. *Cyclotomic extensions*: The nth roots of unity are the solutions to $x^n - 1 = 0$: $1, \omega, \omega^2, \ldots, \omega^{n-1}$, where $\omega = e^{2\pi i/n}$. The extension field $\mathcal{Q}(\omega)$ is a *cyclotomic extension* of $\mathcal{Q}$. If $p > 2$ is prime, then $G(\mathcal{Q}(\omega), \mathcal{Q})$ is a cyclic group of order $p-1$ and is isomorphic to $\mathcal{Z}_p^*$ (the multiplicative group of nonzero elements of $\mathcal{Z}_p$).

Applications:

1. *Solvability by radicals*: A polynomial equation $f(x) = 0$ is *solvable by radicals* if each root can be expressed in terms of the coefficients of the polynomial, using only the operations of addition, subtraction, multiplication, division, and the taking of nth roots.

If F is a field of characteristic 0 and $f(x) \in F[x]$ has K as splitting field, then $f(x) = 0$ is solvable by radicals if and only if $G(K, F)$ is a solvable group. Since there are polynomials whose Galois groups are not solvable, there are polynomials whose roots cannot be found by elementary algebraic methods. For example, the polynomial $x^5 - 36x + 2$ has the symmetric group S_5 as its Galois group, which is not solvable. Hence, the roots of $x^5 - 36x + 2 = 0$ cannot be found by elementary algebraic methods. This example shows that there can be no algebraic formula for solving all fifth-degree equations.

2. *Straightedge and compass constructibility*: Using only a straightedge (unmarked ruler) and a compass, there is no general method for:
   - trisecting angles (given an angle whose measure is α, to construct an angle with measure $\frac{\alpha}{3}$);
   - duplicating the cube (given the side of a cube C_1, to construct the side of a cube C_2 that has double the volume of C_1);
   - squaring the circle (given a circle of area A, to construct a square with area A);
   - constructing a regular n-gon for all $n \geq 3$.

Straightedge and compass constructions yield only lengths that can be obtained by addition, subtraction, multiplication, division, and taking square roots. Beginning with lengths that are rational numbers, each of these operations yields field extensions $\mathcal{Q}(a)$ and $\mathcal{Q}(b)$ where a and b are coordinates of a point constructed from points in $\mathcal{Q} \times \mathcal{Q}$. These operations force $[\mathcal{Q}(a):\mathcal{Q}]$ and $[\mathcal{Q}(b):\mathcal{Q}]$ to be powers of 2. However, trisecting angles, duplicating cubes, and squaring circles all yield extensions of $\mathcal{Q}$ such that the degrees of the extensions are not powers of 2. Hence these three types of constructions are not possible with straightedge and compass.

5.6.3 FINITE FIELDS

Finite fields have a wide range of applications in various areas of computer science and engineering applications: coding theory, combinatorics, computer algebra, cryptology, the generation of pseudorandom numbers, switching circuit theory, and symbolic computation.

Throughout this subsection assume that F is a finite field.

Definitions:

A **finite field** is a field with a finite number of elements.

The **Galois field** $GF(p^n)$ is the algebraic extension $\mathcal{Z}_p[x]/(f(x))$ of the finite field $\mathcal{Z}_p$ where p is a prime and $f(x)$ is an irreducible polynomial over $\mathcal{Z}_p$ of degree n. (See Fact 1.)

A **primitive element** of $GF(p^n)$ is a generator of the cyclic group of nonzero elements of $GF(p^n)$ under multiplication.

Let α be a primitive element of $GF(p^n)$. The **discrete exponential function** (with base α) is the function $\exp_\alpha\colon \{0, 1, 2, \ldots, p^n-2\} \to GF(p^n)^*$ defined by the rule $\exp_\alpha k = \alpha^k$.

Let α be a primitive element of $GF(p^n)$. The **discrete logarithm** or **index function** (with base α) is the function $\operatorname{ind}_\alpha\colon GF(p^n)^* \to \{0, 1, 2, \ldots, p^n-2\}$ where $\operatorname{ind}_\alpha(x) = k$ if and only if $x = \alpha^k$.

Let α be a primitive element of $GF(p^n)$. The **Zech logarithm** (**Jacobi logarithm**) is the function $Z\colon \{1, \ldots, p^n-1\} \to \{0, \ldots, p^n-2\}$ such that $\alpha^{Z(k)} = 1 + \alpha^k$; if $1 + \alpha^k = 0$, then $Z(k) = 0$.

Facts:

1. *Existence of finite fields*: For each prime p and positive integer n there is exactly one field (up to isomorphism) with p^n elements — the field $GF(p^n)$, also written F_{p^n}.

2. *Construction of finite fields*: Given an irreducible polynomial $f(x) \in \mathcal{Z}_p[x]$ of degree n and a zero α of $f(x)$,
$$GF(p^n) \cong \mathcal{Z}_p[x]/(f(x)) \cong \{c_{n-1}\alpha^{n-1} + \cdots + c_1\alpha + c_0 \mid c_i \in \mathcal{Z}_p \text{ for all } i\}.$$

3. If F is a finite field, then:
   - F has p^n elements for some prime p and positive integer n;
   - F has characteristic p for some prime p;
   - F is an extension of $\mathcal{Z}_p$.

4. $[GF(p^n):\mathcal{Z}_p] = n$.

5. $GF(p^n) =$ the field of the p^n roots of $x^{p^n} - x \in \mathcal{Z}_p[x]$.

6. The minimal polynomial of $\alpha \in GF(p^n)$ with respect to Z_p is
$$f(x) = (x - \alpha)(x - \alpha^p)(x - \alpha^{p^2})\ldots(x - \alpha^{p^i})$$
where i is the smallest positive integer such that $\alpha^{p^{i+1}} = \alpha$.

7. If a field F has order p^n, then every subfield of F has order p^k for some k that divides n.

8. The multiplicative group of nonzero elements of a finite field F is a cyclic group.

9. If a field F has m elements, then the multiplicative order of each nonzero element of F is a divisor of $m - 1$.

10. If a field F has m elements and d is a divisor of $m - 1$, then there is an element of F of order d.

11. Each discrete logarithm function has the following properties:
$$\text{ind}_\alpha(xy) \equiv \text{ind}_\alpha x + \text{ind}_\alpha y \pmod{p^n - 1};$$
$$\text{ind}_\alpha(xy^{-1}) \equiv \text{ind}_\alpha x - \text{ind}_\alpha y \pmod{p^n - 1};$$
$$\text{ind}_\alpha(x^k) \equiv k \,\text{ind}_\alpha x \pmod{p^n - 1}.$$

12. The discrete logarithm function ind_α is the inverse of the discrete exponential function $\exp_\alpha$. That is, $\text{ind}_\alpha x = y$ if and only if $\exp_\alpha y = x$.

13. A discrete logarithm function can be used to facilitate multiplication and division of elements of $GF(p^n)$.

14. The Zech logarithm facilitates the addition of elements α^i and α^j $(i > j)$ in $GF(p^n)$, since $\alpha^i + \alpha^j = \alpha^j(\alpha^{i-j} + 1) = \alpha^j \cdot \alpha^{Z(i-j)} = \alpha^{j+Z(i-j)}$. (Note that the values of the Zech logarithm function depend on the primitive element used.)

15. There are $\frac{1}{k}\sum_{d|k}\mu(\frac{k}{d})p^{nd}$ irreducible polynomials of degree k over $GF(p^n)$, where μ is the Möbius function (§2.7).

Examples:

1. If p is prime, Z_p is a finite field and $Z_p \cong GF(p)$.
2. The field $Z_2 = F_2$:

+	0	1
0	0	1
1	1	0

·	0	1
0	0	0
1	0	1

3. The field $Z_3 = F_3$:

+	0	1	2
0	0	1	2
1	1	2	0
2	2	0	1

·	0	1	2
0	0	0	0
1	0	1	2
2	0	2	1

4. Construction of $GF(2^2) = F_4$:
$$GF(2^2) = Z_2[x]/(x^2 + x + 1) = \{c_1\alpha + c_0 \mid c_1, c_0 \in Z_2\} = \{0, 1, \alpha, \alpha + 1\}$$
where α is a zero of $x^2 + x + 1$; i.e., $\alpha^2 + \alpha + 1 = 0$. The nonzero elements of $GF(p^n)$ can also be written as powers of α as α, $\alpha^2 = -\alpha - 1 = \alpha + 1$, $\alpha^3 = \alpha \cdot \alpha^2 = \alpha(\alpha + 1) = \alpha^2 + \alpha = (\alpha + 1) + \alpha = 2\alpha + 1 = 1$.

Thus, $GF(2^2) = \{0, 1, \alpha, \alpha^2\}$ has the following addition and multiplication tables:

+	0	1	α	α^2
0	0	1	α	α^2
1	1	0	α^2	α
α	α	α^2	0	1
α^2	α^2	α	1	0

$\cdot$	0	1	α	α^2
0	0	0	0	0
1	0	1	α	α^2
α	0	α	α^2	1
α^2	0	α^2	1	α

5. *Construction of* $GF(2^3) = F_8$: Let $f(x) = x^3 + x + 1 \in \mathcal{Z}_2[x]$ and let α be a root of $f(x)$. Then $GF(2^3) = \{c_2\alpha^2 + c_1\alpha + c_0 \mid c_0, c_1, c_2 \in \mathcal{Z}_2\}$ where $\alpha^3 + \alpha + 1 = 0$. The elements of $GF(2^3)$ (using α as generator) are:

$$0, \quad \alpha, \quad \alpha^2, \quad \alpha^3 = \alpha + 1$$
$$\alpha^4 = \alpha^2 + \alpha, \quad \alpha^5 = \alpha^2 + \alpha + 1, \quad \alpha^6 = \alpha^2 + 1, \quad 1 \, (= \alpha^7).$$

Multiplication is carried out using the ordinary rules of exponents and the fact that $\alpha^7 = 1$. The following Zech logarithm values can be used to construct the table for addition: $Z(1) = 3, Z(2) = 6, Z(3) = 1, Z(4) = 5, Z(5) = 4, Z(6) = 2, Z(7) = 0$. For example $\alpha^3 + \alpha^5 = \alpha^3 \cdot \alpha^{Z(5-3)} = \alpha^3 \cdot \alpha^6 = \alpha^9 = \alpha^2$.

Using strings of 0s and 1s to represent the elements, $0 = 000$, $1 = 001$, $\alpha = 010$, $\alpha + 1 = 011$, $\alpha^2 = 100$, $\alpha^2 + \alpha = 110$, $\alpha^2 + 1 = 101$, $\alpha^2 + \alpha + 1 = 111$, yields the following tables for addition and multiplication:

+	000	001	010	011	100	101	110	111
000	000	001	010	011	100	101	110	111
001	001	000	011	010	101	100	111	110
010	010	011	000	001	110	111	100	101
011	011	010	001	000	111	110	101	100
100	100	101	110	111	000	001	010	011
101	101	100	111	110	001	000	011	010
110	110	111	100	101	010	011	000	001
111	111	110	101	100	011	010	001	000

$\cdot$	000	001	010	011	100	101	110	111
000	000	000	000	000	000	000	000	000
001	000	001	010	011	100	101	110	111
010	000	010	100	110	011	001	111	101
011	000	011	110	101	111	100	001	010
100	000	100	011	111	110	010	101	001
101	000	101	001	100	010	111	011	110
110	000	110	111	001	101	011	010	100
111	000	111	101	010	001	110	100	011

The same field can be constructed using $g(x) = x^3 + x^2 + 1$ instead of $f(x) = x^3 + x + 1$ and β as a root of $g(x)$ ($\beta^3 + \beta^2 + 1 = 0$). The elements (using β as generator) are: 0, β, β^2, $\beta^3 = \beta^2 + 1$, $\beta^4 = \beta^2 + \beta + 1$, $\beta^5 = \beta + 1$, $\beta^6 = \beta^2 + \beta$, $1 \, (= \beta^7)$.

The polynomial $g(x)$ yields the following Zech logarithm values, which can be used to construct the table for addition: $Z(1) = 5, Z(2) = 3, Z(3) = 2, Z(4) = 6, Z(5) = 1, Z(6) = 4, Z(\dot{7}) = 0$. This field is isomorphic to the field defined using $f(x) = x^3 + x + 1$.

6. Table 1 lists the irreducible polynomials of degree at most 8 in $\mathcal{Z}_2[x]$. For more extensive tables of irreducible polynomials over certain finite fields, see [LiNi94].

Table 1 Irreducible polynomials in $Z_2[x]$ of degree at most 8.

Each polynomial is represented by the string of its coefficients, beginning with the highest power. For example, $x^3 + x + 1$ is represented by 1011.

degree 1:	10	11				
degree 2:	111					
degree 3:	1011	1101				
degree 4:	10011	11001	11111			
degree 5:	100101	101001	101111	110111	111011	111101
degree 6:	1000011	1001001	1010111	1011011	1100001	1100111
	1101101	1110011	1110101			
degree 7:	10000011	10001001	10001111	10010001	10011101	10100111
	10101011	10111001	10111111	11000001	11001011	11010011
	11010101	11100101	11101111	11110001	11110111	11111101
degree 8:	100011011	100011101	100101011	100101101	100111001	100111111
	101001101	101011111	101100011	101100101	101101001	101110001
	101110111	101111011	110000111	110001011	110001101	110011111
	110100011	110101001	110110001	110111101	111000011	111001111
	111010111	111011101	111100111	111110011	111110101	111111001

5.7 LATTICES

5.7.1 BASIC CONCEPTS

Definitions:

A *lattice* $(L, \vee, \wedge)$ is a nonempty set L closed under two binary operations $\vee$ (*join*) and $\wedge$ (*meet*) such that the following laws are satisfied for all $a, b, c \in L$:

- *associative laws*: $\quad a \vee (b \vee c) = (a \vee b) \vee c \quad a \wedge (b \wedge c) = (a \wedge b) \wedge c$
- *commutative laws*: $\quad a \vee b = b \vee a \qquad\qquad\quad a \wedge b = b \wedge a$
- *absorption laws*: $\quad a \vee (a \wedge b) = a \qquad\quad\; a \wedge (a \vee b) = a.$

Lattices L_1 and L_2 are *isomorphic* (as lattices) if there is a function $\varphi: L_1 \to L_2$ that is one-to-one and onto L_2 and preserves $\vee$ and $\wedge$: $\varphi(a \vee b) = \varphi(a) \vee \varphi(b)$ and $\varphi(a \wedge b) = \varphi(a) \wedge \varphi(b)$ for all $a, b \in L_1$.

L_1 is a *sublattice* of lattice L if $L_1 \subseteq L$ and L_1 is a lattice using the same operations as those used in L.

The *dual* of a statement in a lattice is the statement obtained by interchanging the operations $\vee$ and $\wedge$ and interchanging the elements 0 (lower bound) and 1 (upper bound). (See §5.7.2.)

An *order relation* $\leq$ can be defined on a lattice so that $a \leq b$ means that $a \vee b = b$, or, equivalently, that $a \wedge b = a$. Write $a < b$ if $a \leq b$ and $a \neq b$. (See §2.7.1.)

Facts:
1. If L is a lattice and $a, b \in L$, then $a \wedge b$ and $a \vee b$ are unique.
2. *Lattices as partially ordered sets*: Every lattice is a partially ordered set using the order relation $\leq$. (See §1.4.3; also see Chapter 11 for extended coverage.)
3. Every partially ordered set L in which glb $\{a, b\}$ and lub $\{a, b\}$ exist for all $a, b \in L$ can be regarded as a lattice by defining $a \vee b = \text{lub}\{a, b\}$ and $a \wedge b = \text{glb}\{a, b\}$.
4. *The duality principle holds in all lattices*: If a theorem is the consequence of the definition of lattice, then the dual of the statement is also a theorem.
5. *Lattice diagrams*: Every finite lattice can be pictured in a poset diagram (Hasse diagram), called a *lattice diagram*.
6. *Idempotent laws*: $a \vee a = a$ and $a \wedge a = a$ for all $a \in L$.

Example:
1. The following table gives examples of lattices.

set	$\vee$ (join)	$\wedge$ (meet)
$\mathcal{N}$	$a \vee b = \text{lcm}\{a, b\}$	$a \wedge b = \gcd\{a, b\}$
$\mathcal{N}$	$a \vee b = \max\{a, b\}$	$a \wedge b = \min\{a, b\}$
$\mathcal{Z}_2^n$	$(a_1, \ldots, a_n) \vee (b_1, \ldots, b_n) =$ $(\max(a_1, b_1), \ldots, \max(a_n, b_n))$	$(a_1, \ldots, a_n) \vee (b_1, \ldots, b_n) =$ $(\min(a_1, b_1), \ldots, \min(a_n, b_n))$
all subgroups of a group G	$H_1 \vee H_2 =$ the intersection of all subgroups of G containing H_1 and H_2	$H_1 \wedge H_2 = H_1 \cap H_2$
all subsets of set S	$A_1 \vee A_2 = A_1 \cup A_2$	$A_1 \wedge A_2 = A_1 \cap A_2$

5.7.2 SPECIALIZED LATTICES

Definitions:

A lattice L is **distributive** if the following are true for all $a, b, c \in L$:
- $a \wedge (b \vee c) = (a \wedge b) \vee (a \wedge c)$;
- $a \vee (b \wedge c) = (a \vee b) \wedge (a \vee c)$.

A **lower bound** (**smallest element**, **least element**) in a lattice L is an element $0 \in L$ such that $0 \wedge a = 0$ (equivalently, $0 \leq a$) for all $a \in L$.

An **upper bound** (**largest element**, **greatest element**) in a lattice L is an element $1 \in L$ such that $1 \vee a = 1$ (equivalently, $a \leq 1$) for all $a \in L$.

A lattice L is **bounded** if L contains a lower bound 0 and an upper bound 1.

A lattice L is **complemented** if:
- L is bounded;
- for each $a \in L$ there is an element $b \in L$ (called a **complement** of a) such that $a \vee b = 1$ and $a \wedge b = 0$.

An element a in a bounded lattice L is an **atom** if $0 < a$ and there is no element $b \in L$ such that $0 < b < a$.

Facts:

1. Each of the distributive properties in a lattice implies the other.
2. Not every lattice is distributive. (See Example 1.)
3. If a lattice is not distributive, it must contain a sublattice isomorphic to one of the two lattices in the following figure.

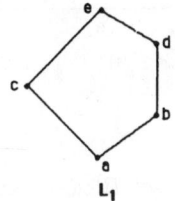
L_1

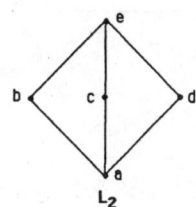
L_2

4. Every finite lattice is bounded: if $L = \{a_1, \ldots, a_n\}$, then
$$1 = a_1 \vee \cdots \vee a_n$$
and
$$0 = a_1 \wedge \cdots \wedge a_n.$$

5. Some infinite lattices are bounded, while others are not. (See Examples 2 and 3.)
6. In a complemented lattice, complements are not necessarily unique. See the lattice in Example 4.
7. If L is a finite, complemented, distributive lattice and $a \in L$, then there is exactly one set of atoms $\{a_1, \ldots, a_k\}$ such that $a = a_1 \vee \cdots \vee a_k$.

Examples:

1. Neither lattice in Fact 3 is distributive. For example, in lattice L_1,
$$d \vee (b \wedge c) = d, \text{ but } (d \vee b) \wedge (d \vee c) = b$$
and in L_2,
$$d \vee (b \wedge c) = d, \text{ but } (d \vee b) \wedge (d \vee c) = a.$$

2. The lattice $(\mathcal{N}, \vee, \wedge)$ where $a \vee b = \max(a, b)$ and $a \wedge b = \min(a, b)$ is not bounded; there is a lower bound (the integer 0), but there is no upper bound.

3. The following infinite lattice is bounded. The element 1 is an upper bound and the element 0 is a lower bound.

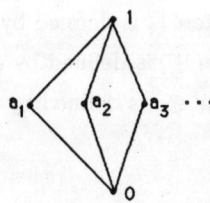

4. The lattice in Example 3 is complemented, but complements are not unique in that lattice. For example, the element a_1 has $a_2, a_3, \ldots$ as complements.

5. In lattice L_1 of Fact 3, b and c are atoms. In the lattice of all subsets of a set S (see Example 1), the atoms are the subsets of S of size 1.

5.8 BOOLEAN ALGEBRAS

Boolean algebra is a generalization of the algebra of sets and the algebra of logical propositions. It forms an abstract model of the design of circuits.

5.8.1 BASIC CONCEPTS

Definition:
A **Boolean algebra** $(B, +, \cdot, ', 0, 1)$ consists of a set B closed under two binary operations, $+$ (*addition*) and $\cdot$ (*multiplication*), and one monadic operation, $'$ (*complementation*), and having two distinct elements, 0 and 1, such that the following laws are true for all $a, b, c \in B$:
- *commutative* laws: $\quad a + b = b + a \qquad\qquad\qquad a \cdot b = b \cdot a$
- *distributive* laws: $\quad a \cdot (b + c) = (a \cdot b) + (a \cdot c) \quad a + (b \cdot c) = (a + b) \cdot (a + c)$
- *identity* laws: $\qquad a + 0 = a \qquad\qquad\qquad\quad a \cdot 1 = a$
- *complement* laws: $\;\; a + a' = 1 \qquad\qquad\qquad\;\; a \cdot a' = 0$.

(George Boole, 1813–1864)

Notes: It is common practice to omit the "$\cdot$" symbol in a Boolean algebra, writing ab instead of $a \cdot b$. The complement operation is also written using an overline: $x' = \overline{x}$. By convention, complementation is done first, then multiplication, and finally addition. For example, $a + bc'$ means $a + (b(c'))$.

The **dual** of a statement in a Boolean algebra is the statement obtained by interchanging the operations $+$ and $\cdot$ and interchanging the elements 0 and 1 in the original statement.

Boolean algebras B_1 and B_2 are **isomorphic** (as Boolean algebras) if there is a function $\varphi \colon B_1 \to B_2$ that is one-to-one and onto B_2 such that for all $a, b \in B_1$:
- $\varphi(a + b) = \varphi(a) + \varphi(b)$;
- $\varphi(ab) = \varphi(a)\varphi(b)$;
- $\varphi(a') = \varphi(a)'$.

An element $a \neq 0$ in a Boolean algebra is an **atom** if the following holds: if $xa = x$, then either $x = 0$ or $x = a$; that is, if $x \leq a$, then either $x = 0$ or $x = a$ (see Fact 1).

The binary operation NAND, written $|$, is defined by $a \,|\, b = (ab)'$.

The binary operation NOR, written $\downarrow$, is defined by $a \downarrow b = (a + b)'$.

The binary operation XOR, written $\oplus$, is defined by $a \oplus b = ab' + a'b$.

Facts:

1. Every Boolean algebra is a bounded, distributive, complemented lattice where $a \vee b = a + b$ and $a \wedge b = ab$. Hence, every Boolean algebra is a partially ordered set (where $a \leq b$ if and only if $a + b = b$, or, equivalently, $ab = a$ or $a' + b = 1$ or $ab' = 0$).

2. The *duality principle* holds in all Boolean algebras: if a theorem is the consequence of the definition of Boolean algebra, then the dual of the theorem is also a theorem.

3. *Structure of Boolean algebras*: Every finite Boolean algebra is isomorphic to $\{0,1\}^n$ for some positive integer n. Hence every finite Boolean algebra has 2^n elements. The atoms are the n n-tuples of 0s and 1s with a 1 in exactly one position.

4. If B is a finite Boolean algebra and $b \in B$ ($b \neq 0$), there is exactly one set of atoms $a_1, \ldots, a_k$ such that $b = a_1 + \cdots + a_k$.

5. If a Boolean algebra B has n atoms, then B has 2^n elements.

6. The following laws are true in all Boolean algebras B, for all $a, b, c \in B$:
   - *associative laws*: $\quad a + (b + c) = (a + b) + c,\ a(bc) = (ab)c$
     (Hence there is no ambiguity in writing $a + b + c$ and abc.)
   - *idempotent laws*: $\quad a + a = a,\ aa = a$
   - *absorption laws*: $\quad a(a + b) = a,\ a + ab = a$
   - *domination (boundedness) laws*: $\quad a + 1 = 1,\ a0 = 0$
   - *double complement (involution) law*: $(a')' = a$
   - *DeMorgan's laws*: $\quad (a + b)' = a'b',\ (ab)' = a' + b'$
   - *uniqueness of complement*: if $a + b = 1$ and $ab = 0$, then $b = a'$.

7. Since every Boolean algebra is a lattice, every finite Boolean algebra can be pictured using a partially ordered set diagram. (§11.1)

Examples:

1. $\{0, 1\}$ is a Boolean algebra, where addition, multiplication, and complementation are defined in the following tables:

+	0	1
0	0	1
1	1	1

·	0	1
0	0	0
1	0	1

x	x'
0	1
1	0

2. If S is any set, then $\mathcal{P}(S)$ (the set of all subsets of S) is a Boolean algebra where
$$A_1 + A_2 = A_1 \cup A_2, \quad A_1 \cdot A_2 = A_1 \cap A_2, \quad A' = \overline{A}$$
and $0 = \emptyset$ and $1 = S$.

3. Given n variables, the set of all compound propositions in these variables (identified with their truth tables) is a Boolean algebra where
$$p + q = p \vee q, \quad p \cdot q = p \wedge q, \quad \overline{p} = \neg p$$
and 0 is a contradiction (the truth table with only values F) and 1 is a tautology (the truth table with only values T).

4. If B is any Boolean algebra, then $B^n = \{(a_1, \ldots, a_n) \mid a_i \in B \text{ for all } i\}$ is a Boolean algebra, where the operations are performed coordinatewise:
$(a_1, \ldots, a_n) + (b_1, \ldots, b_n) = (a_1 + b_1, \ldots, a_n + b_n);$
$(a_1, \ldots, a_n) \cdot (b_1, \ldots, b_n) = (a_1 \cdot b_1, \ldots, a_n \cdot b_n);$
$(a_1, \ldots, a_n)' = (a_1', \ldots, a_n').$
In this Boolean algebra $0 = (0, \ldots, 0)$ and $1 = (1, \ldots, 1)$.

5. The statements in each of the following pairs are duals of each other:
$a + b = cd, \quad ab = c + d;$
$a + (b + c) = (a + b) + c, \quad a(bc) = (ab)c;$
$a + 1 = 1, \quad a0 = 0.$

5.8.2 BOOLEAN FUNCTIONS

Definitions:

A *Boolean expression* in the variables $x_1, \ldots, x_n$ is an expression defined recursively by:
- 0, 1, and all variables x_i are Boolean expressions in $x_1, \ldots, x_n$;
- if E and F are Boolean expressions in the variables $x_1, \ldots, x_n$, then (EF), $(E+F)$, and E' are Boolean expressions in the variables $x_1, \ldots, x_n$.

A *Boolean function* of degree n is a function $f \colon \{0,1\}^n \to \{0,1\}$.

A *literal* is a Boolean variable or its complement.

A *minterm* of the Boolean variables $x_1, \ldots, x_n$ is a product of the form $y_1 \ldots y_n$ where for each i, y_i is equal to x_i or x_i'.

A *maxterm* of the Boolean variables $x_1, \ldots, x_n$ is a sum of the form $y_1 + \cdots + y_n$ where for each i, y_i is equal to x_i or x_i'.

A Boolean function of degree n is in *disjunctive normal form* (**DNF**) (or **sum-of-products expansion**) if it is written as a sum of distinct minterms in the variables $x_1, \ldots, x_n$. (*Note*: disjunctive normal form is sometimes called *full* disjunctive normal form.)

A Boolean function is in *conjunctive normal form* (**CNF**) (or **product-of-sums expansion**) if it is written as a product of distinct maxterms.

A set of operators in a Boolean algebra is *functionally complete* if every Boolean function can be written using only these operators.

Facts:

1. Every Boolean function can be written as a Boolean expression.

2. There are 2^{2^n} Boolean functions of degree n. Examples of the 16 different Boolean functions with two variables, x and y, are given in the following table.

x	y	1	$x+y$	$x+y'$	$x'+y$	$x\|y$	x	y	$x \oplus y$	$(x \oplus y)'$	y'	x'	xy	xy'	$x'y$	$x \downarrow y$	0
1	1	1	1	1	1	0	1	1	0	1	0	0	1	0	0	0	0
1	0	1	1	1	0	1	1	0	1	0	1	0	0	1	0	0	0
0	1	1	1	0	1	1	0	1	1	0	0	1	0	0	1	0	0
0	0	1	0	1	1	1	0	0	0	1	1	1	0	0	0	1	0

3. Every Boolean function (not identically 0) can be written in disjunctive normal form. Either of the following two methods can be used:
   (a) Rewrite the expression for the function so that no parentheses remain. For each term that does not have a literal for a variable x_i, multiply that term by $x_i + x_i'$. Multiply out so that no parentheses remain. Use the idempotent law to remove any duplicate terms or duplicate factors.
   (b) Make a table of values for the function. For each row where the function has the value 1, form a minterm that yields 1 in only that row. Form the sum of these minterms.

4. Every Boolean function (not identically 1) can be written in conjunctive normal form. Any of the following three methods can be used:

(a) Write the negation of the expression in disjunctive normal form. Use DeMorgan's laws to take the negation of this expression.

(b) Make a table of values for the function. For each row where the function has the value 0, form a minterm that yields 1 in only that row. Form the sum of these minterms. Use DeMorgan's laws to take the complement of this sum.

(c) Make a table of values for the function. For each row where the function has the value 0, form a maxterm that yields 0 in only that row. Form the product of these maxterms.

5. The following are examples of functionally complete sets, with explanations showing how any Boolean function can be written using only these operations:

- $\{+, \cdot, '\}$ disjunctive normal form uses only the operators $+$, $\cdot$, and $'$
- $\{+, '\}$ DeMorgan's law $(a \cdot b)' = a' + b'$ allows the replacement of any occurrence of $a \cdot b$ with an expression that does not use $\cdot$
- $\{\cdot, '\}$ DeMorgan's law $a + b = (a' \cdot b')'$ allows the replacement of any occurrence of $a + b$ with an expression that does not use $+$
- $\{\,|\,\}$ write the expression for any function in DNF; use $a' = a\,|\,a$, $a + b = (a\,|\,a)\,|\,(b\,|\,b)$, and $a \cdot b = (a\,|\,b)\,|\,(a\,|\,b)$ to replace each occurrence of $'$, $+$, and $\cdot$ with $|$
- $\{\downarrow\}$ write the expression for any function in DNF; use $a' = a \downarrow a$, $a + b = (a \downarrow b) \downarrow (a \downarrow b)$, and $a \cdot b = (a \downarrow a) \downarrow (b \downarrow b)$ to replace each occurrence of $'$, $+$, and $\cdot$ with $\downarrow$.

6. The set $\{+, \cdot\}$ is not functionally complete.

Examples:

1. The function $f: \{0,1\}^3 \to \{0,1\}$ defined by $f(x,y,z) = x(z' + y'z) + x'$ is a Boolean function in the Boolean variables x, y, z. Multiplying out the expression for this function yields $f(x,y,z) = xz' + xy'z + x'$. In this form the second term, $xy'z$, is a minterm in the three variables x, y, z. The first and third terms are not minterms: the first term, xz', does not use a literal for y, and the third term, x', does not use literals for y and z.

2. *Writing a Boolean function in disjunctive normal form*: To write the function f from Example 1 in DNF using Fact 3(a), replace the terms xz' and x' with equivalent minterms by multiplying these terms by 1 ($= a + a'$) for each missing variable a:

$$xz' = xz' \cdot 1 = xz'(y + y') = xyz' + xy'z';$$
$$x' = x' \cdot 1 \cdot 1 = x'(y + y')(z + z') = x'yz + x'yz' + xy'z + xy'z'.$$

Therefore,
$$f(x,y,z) = x(z' + y'z) + x'$$
$$= xz' + xy'z + x'$$
$$= xyz' + xy'z' + xy'z + x'yz + x'yz' + x'y'z + xy'z'$$

Alternatively, using Fact 3(b), the table of values for f yields 1 in all rows except the row in which $x = y = z = 1$. Therefore minterms are obtained for the other rows, yielding the same sum of seven minterms.

3. *Writing a Boolean function in conjunctive normal form*: Using Fact 4(a) to write the function $f(x,y) = xy' + x'y$ in CNF, first rewrite the negation of f in DNF, obtaining $f'(x,y) = xy + x'y'$. The negation of f' is $f''(x,y) = f(x,y) = (x' + y')(x + y)$.

Alternatively, using Fact 4(c), the function f has value 0 only when $x = y = 1$ and $x = y = 0$. The maxterms that yield 0 in exactly one of these rows are $x' + y'$ and $x + y$. Therefore, in CNF $f(x,y) = (x' + y')(x + y)$.

5.8.3 LOGIC GATES

Boolean algebra can be used to model circuitry, with 0s and 1s as inputs and outputs. The elements of these circuits are *gates* that implement the Boolean operations.

Facts:

1. The following figure gives representations for the three standard Boolean operators, $+$, $\cdot$, and $'$, together with representations for three related operators. (For example, the AND gate takes two inputs, x and y, and produces one output, xy.)

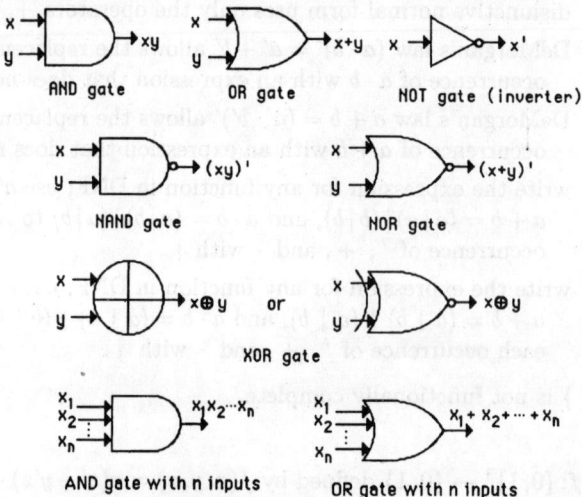

2. Gates can be extended to include cases where there are more than two inputs. The figure of Fact 1 also shows an AND gate and an OR gate with multiple inputs. These correspond to $x_1 x_2 \ldots x_n$ and $x_1 + x_2 + \cdots + x_n$. (Since both operations satisfy the associative laws, no parentheses are needed.)

Examples:

1. *The gate diagram for a half-adder*: A *half-adder* is a Boolean circuit that adds two bits, x and y, producing two outputs:

   a *sum bit* $s = (x+y)(xy)'$ ($s = 0$ if $x = y = 0$ or $x = y = 1$; $s = 1$ otherwise);
   a *carry bit* $c = xy$ ($c = 1$ if and only if $x = y = 1$).

The gate diagram for a half-adder is given in the following figure. This circuit is an example of a *multiple output circuit* since there is more than one output.

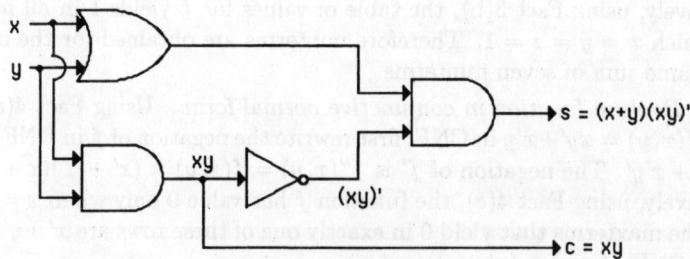

2. *The gate diagram for a full-adder*: A *full-adder* is a Boolean circuit that adds three bits (x, y, and a carry bit c) and produces two outputs (a sum bit s and a carry bit c'). The full-adder gate diagram is given in the following figure.

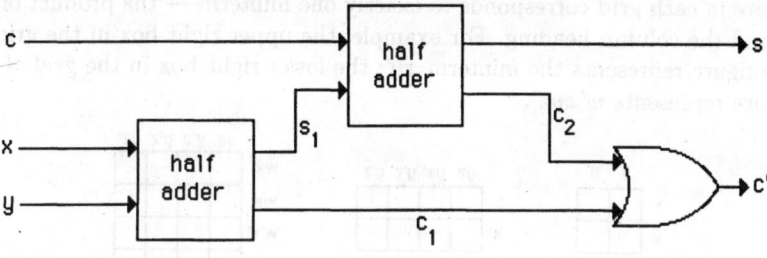

5.8.4 MINIMIZATION OF CIRCUITS

Boolean expressions that appear to be different can yield the same combinatorial circuit. For example, $xyz+xyz'+x'y$ and y (as functions of x and y) have the same table of values and hence yield the same circuit. (The first expression can be simplified to give the second: $xyz+xyz'+x'y = xy(z+z')+x'y = xy \cdot 1+x'y = xy+x'y = (x+x')y = 1y = y$.)

Definitions:

A Boolean expression is **minimal** (as a sum-of-products) if among all equivalent sum-of-products expressions it has the fewest number of summands, and among all sum-of-products expressions with that number of summands it uses the smallest number of literals in the products.

A **Karnaugh map** for a Boolean expression written in disjunctive normal form is a diagram (constructed using the following algorithm) that displays the minterms in the Boolean expression.

Facts:

1. Minimization of circuits is an NP-hard problem.

2. *Don't care conditions*: In some circuits, it may be known that some elements of the input set for the Boolean function will never be used. Consequently, the values of the expression for these elements is irrelevant. The values of the circuit function for these unused elements of the input set are called *don't care conditions*, and the values can be arbitrarily chosen to be 0 or 1. The blocks in the Karnaugh map where the function values are irrelevant are marked with d. In the simplification process of the Karnaugh map, 1s can be substituted for any or all of the ds in order to cover larger blocks of boxes and achieve a simpler equivalent expression.

Algorithm:

There is an algorithm for minimizing Boolean expressions by systematically grouping terms together. When carried out visually, the method uses a Karnaugh map (Maurice Karnaugh, born 1924). When carried out numerically using bit strings, the method is called the Quine-McCluskey method (Willard Quine, born 1908, Edward McCluskey, born 1929).

1. *Karnaugh map*: To minimize a Boolean expression:

(a) Write the Boolean expression in disjunctive normal form.

(b) Obtain the Karnaugh map for this Boolean expression. The layout of the table

350 Chapter 5 ALGEBRAIC STRUCTURES

depends on the number of variables under consideration.

The grids for Boolean expressions with two variables (x and y), three variables (x, y, and z), and four variables (w, x, y, and z) are shown in the following figure. Each square in each grid corresponds to exactly one minterm — the product of the row heading and the column heading. For example, the upper right box in the grid of part (a) of the figure represents the minterm xy'; the lower right box in the grid of part (c) of the figure represents $w'xyz'$.

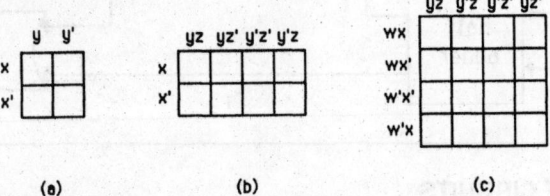

(a)   (b)   (c)

The headings are placed in a certain order — adjacent squares in any row (or column) differ in exactly one literal in their row headings (or column headings). The first and last squares in any row (or column) are to be regarded as adjacent. (The variable names can be permuted; for example, in part (b) of the figure, the row headings can be y and y' and the column headings can be xz, xz', $x'z'$, and $x'z$. The column headings could also have been written in order as yz, $y'z$, $y'z'$, yz' or $y'z$, $y'z'$, yz', yz.)

The Karnaugh map for the Boolean expression is obtained by placing a checkmark in each square corresponding to a minterm in the expression.

(c) Find the best covering. A geometric version of the distributive law is used to "cover" groups of the adjacent marked squares, with every marked square covered at least once and each group covered being as large as possible. The possible ways of covering squares depends on the number of variables.

(For example, working with two variables and using the distributive law, $x'y+x'y' = x'(y+y') = x'1 = x'$. This corresponds to covering the two boxes in the bottom row of the first 2×2 grid in the following figure and noting that the only literal common to both boxes is x'.)

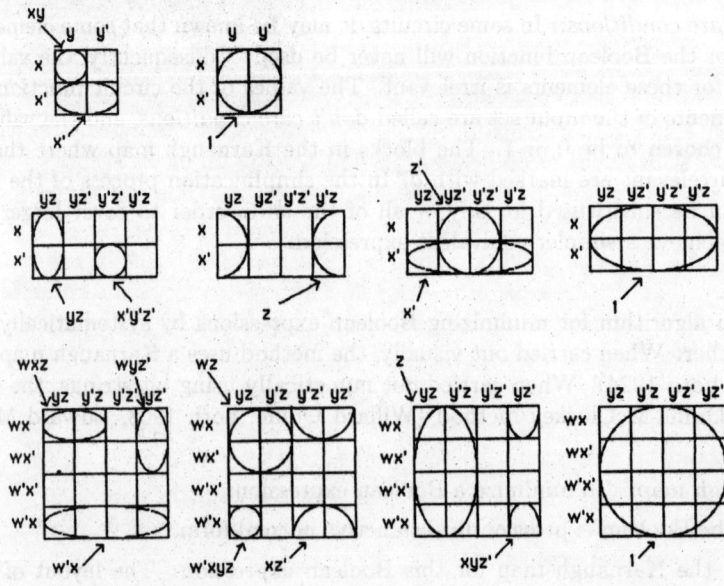

Similarly, working with three variables, $xyz' + xy'z' + x'yz' + x'y'z' = xz'(y+y') + x'z'(y+y') = xz' + x'z' = (x+x')z' = z'$. This corresponds to covering the four boxes in the second and third columns of the third 2×4 grid in the second row of the figure and noting that z' is the only common literal.)

The following table shows what groups of boxes can be covered, for expressions with 2, 3, and 4 variables. These are the combinations whose expressions can be simplified to a single minterm. Examples for 2, 3, and 4 variables are shown in the previous figure. (The method is awkward to use when there are more than 4 variables.)

# variables	groups of boxes that can be covered
2	1×1, 1×2, 2×1, 2×2
3	1×1, 1×2, 1×4, 2×1, 2×2, 2×4
4	1×1, 1×2, 1×4, 2×1, 2×2, 2×4, 4×1, 4×2, 4×4

To obtain the minimization, cover boxes according to the following rules:
- cover all marked boxes at least once
- cover the largest possible blocks of marked boxes
- do not cover any unmarked box
- use the fewest blocks possible.

(d) Find the product of common literals for each of the blocks and form the sum of these products to obtain the minimization.

2. *Quine-McCluskey method*:

(a) Write the Boolean expression in disjunctive normal form, and in each summand list the variables in alphabetical order. Identify with each term a bit string, using a 1 if the literal is not a complement and 0 if the literal is a complement. (For example, $v'wx'yz$ is represented by 01011.)

(b) Form a table with the following columns:

column 1: Make a numbered list of the terms and their bit strings, beginning with the terms with the largest number of uncomplemented variables. (For example, $wxy'z$ precedes $wx'yz'$.)

column 2: Make a list of pairs of terms from column 1 where the literals in the two terms differ in exactly one position. Use a distributive law to add and simplify the two terms and write the numbers of these terms and the sum of the terms in the second column, along with its bit string, using "−" in place of the variable that no longer appears in the sum. (For example, xyz' and $xy'z'$ can be combined to yield xz' with bit string $1-0$.)

columns 3, 4, etc.: To obtain column 3 combine the terms in column 2 in pairs according to the same procedure as that used to construct column 2. Repeat this process until no more terms can be combined.

(c) Form a table with a row for each of the terms that cannot be used to form terms with fewer variables and a column for each of the original terms in the disjunctive normal form of the original expression. Mark the square in the ij-position if the minterm in column j could be a summand for the term in row i.

(d) Find a set of rows, with as few rows as possible, such that every column has been marked at least once in at least one row. The sum of the products labeling these rows minimizes the original expression.

Examples:

1. Simplify $w'x'y + w'z(xy + x'y') + w'x'z' + w'xyz' + wx'y'z'$ (an expression in four variables) using a Karnaugh map.

First write the expression in disjunctive normal form:
$$w'x'y + w'z(xy + x'y') + w'x'z' + w'xyz' + wx'y'z' =$$
$$w'x'yz + w'x'yz' + w'xyz + w'x'y'z + w'x'y'z' + w'xyz' + wx'y'z'$$

Next, draw its Karnaugh map. See part (a) of the following figure. A covering is given in part (b) of the figure. Note that in order to use larger blocks, some squares have been covered more than once. Also note that $w'x'yz$, $w'xyz$, $w'x'yz'$, and $w'xyz'$ are covered with one 2×2 block rather than with two 1×2 blocks. In the three blocks the common literals are: $w'x'$, $w'y$, and $x'y'z'$.

Finally, form the sum of these products: $w'x' + w'y + x'y'z'$.

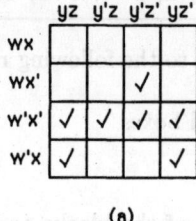 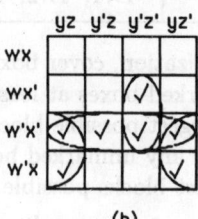

(a) (b)

2. Minimize $w'xy'z + wxyz' + wx'yz' + w'x'yz + wxyz + w'x'y'z + w'xyz$ (an expression in four variables) using the Quine-McCluskey method.

Step (b) of the Quine-McCluskey method yields the following table.

1	$wxyz$	1111	1,2	wxy	111−	3,5,6,7	$w'z$	0−−1
2	$wxyz'$	1110	1,3	xyz	−111			
3	$w'xyz$	0111	2,4	wyz'	1−10			
4	$wx'yz'$	1010	3,5	$w'yz$	0−11			
5	$w'x'yz$	0011	3,6	$w'xz$	01−1			
6	$w'xy'z$	0101	5,7	$w'x'z$	00−1			
7	$w'x'y'z$	0001	6,7	$w'y'z$	0−01			

The four terms $w'z$, wxy, xyz, wyz' were not used in combining terms, so they become the names of the rows in the following table.

	$wxyz$	$wxyz'$	$w'xyz$	$wx'yz'$	$w'x'yz$	$w'xy'z$	$w'x'y'z$
$w'x$			√		√	√	√
wxy	√	√					
xyz	√		√				
wyz'		√		√			

There are two ways to cover the seven minterms:
$$w'x, wxy, wyz' \quad \text{or} \quad w'x, xyz, wyz'.$$
This yields two ways to minimize the original expression:
$$w'x + wxy + wyz' \quad \text{and} \quad w'z + xyz + wyz'.$$

REFERENCES

Printed Resources:

[Ar91] M. Artin, *Algebra*, Prentice-Hall, 1991.

[As86] M. Aschbacher, *Finite Group Theory*, Cambridge University Press, 1986.

[BiBa70] G. Birkhoff and T. C. Bartee, *Modern Applied Algebra*, McGraw-Hill, 1970.

[BiMa77] G. Birkhoff and S. Mac Lane, *A Survey of Modern Algebra*, 4th ed., Macmillan, 1977.

[Bl87] N. J. Bloch, *Abstract Algebra with Applications*, Prentice-Hall, 1987.

[Ca72] R. W. Carter, *Simple Groups of Lie Type*, Wiley, 1972.

[Ch95] L. Childs, *A Concrete Introduction to Higher Algebra*, 2nd ed., Springer-Verlag, 1995.

[CoMo72] H. S. M. Coxeter and W. O. J. Moser, *Generators and Relators on Discrete Groups*, 3rd ed., Springer-Verlag, 1972.

[CrFo63] R. H. Crowell and R. H. Fox, *Introduction to Knot Theory*, Springer-Verlag, 1963.

[Fr89] J. B. Fraleigh, *A First Course in Abstract Algebra*, 4th ed., Addison-Wesley, 1989.

[Go82] D. Gorenstein, *Finite Simple Groups: An Introduction to Their Classification*, Plenum Press, 1982.

[GoLySo94] D. Gorenstein, R. Lyons, and R. Solomon, *The Classification of the Finite Simple Groups*, American Mathematical Society, 1994.

[Ha59] M. Hall, Jr., *The Theory of Groups*, Macmillan, 1959.

[He75] I. N. Herstein, *Topics in Algebra*, 2nd ed., Wiley, 1975.

[Hu74] T. W. Hungerford, *Algebra*, Holt, Rinehart and Winston, 1974.

[Ka72] I. Kaplansky, *Fields and Rings*, 2nd ed., University of Chicago Press, 1972.

[LiNi94] R. Lidl and H. Niederreiter, *Introduction to Finite Fields and Their Applications*, revised edition, Cambridge University Press, 1994.

[LiPi84] R. Lidl and G. Pilz, *Applied Abstract Algebra*, Springer-Verlag, 1984.

[MaBi79] S. Mac Lane and G. Birkhoff, *Algebra*, 2nd ed., Collier-Macmillan, 1979.

[MaKaSo65] W. Magnus, A. Karrass, and D. Solitar, *Combinatorial Group Theory*, Wiley-Interscience, 1965.

[Mc48] N. H. McCoy, *Rings and Ideals* (Carus Monograph No. 8), Mathematical Association of America, 1948.

[Mc64] N. H. McCoy, *The Theory of Rings*, Macmillan, 1964.

[McBe77] N. H. McCoy and T. Berger, *Algebra: Groups, Rings, and Other Topics*, Allyn and Bacon, 1977.

[Mc87] R. J. McEliece, *Finite Fields for Computer Scientists and Engineers*, Kluwer Academic Publishers, 1987.

[MeEtal93] A. Menezes, I. Blake, X. Gao, R. Mullin, S. Vanstone, and T. Yaghoobian, *Applications of Finite Fields*, Kluwer Academic Publishers, 1993.

[Ro84] J. J. Rotman, *An Introduction to the Theory of Groups*, 3rd ed., Allyn and Bacon, 1984.

[ThFe63] J. G. Thompson and W. Feit, "Solvability of Groups of Odd Order," *Pacific Journal of Mathematics*, 13 (1963), 775–1029.

[va49] B. L. van der Waerden, *Modern Algebra* (2 volumes), Ungar, 1949, 1950.

Web Resources:

http://www.maths.usyd.edu.au:8000/u/magma/ (The Magma Computer Algebra System, successor to CAYLEY, developed by the Computational Algebra Group at the University of Sydney.)

http://wwwmaths.anu.edu.au/research.groups/algebra/GAP/www/gap.html (Contains GAP — Groups, Algorithms and Programming — a system for computational discrete algebra.)

LINEAR ALGEBRA

6.1 Vector Spaces — Joel V. Brawley
 6.1.1 Basic Concepts
 6.1.2 Subspaces
 6.1.3 Linear Combinations, Independence, Basis, and Dimension
 6.1.4 Inner Products, Length, and Orthogonality

6.2 Linear Transformations — Joel V. Brawley
 6.2.1 Linear Transformations, Range, and Kernel
 6.2.2 Vector Spaces of Linear Transformations
 6.2.3 Matrices of Linear Transformations
 6.2.4 Change of Basis

6.3 Matrix Algebra — Peter R. Turner
 6.3.1 Basic Concepts and Special Matrices
 6.3.2 Operations of Matrix Algebra
 6.3.3 Fast Multiplication of Matrices
 6.3.4 Determinants
 6.3.5 Rank
 6.3.6 Identities of Matrix Algebra

6.4 Linear Systems — Barry Peyton and Esmond Ng
 6.4.1 Basic Concepts
 6.4.2 Gaussian Elimination
 6.4.3 LU Decomposition
 6.4.4 Cholesky Decomposition
 6.4.5 Conditioning of Linear Systems
 6.4.6 Pivoting for Stability
 6.4.7 Pivoting to Preserve Sparsity

6.5 Eigenanalysis — R. B. Bapat
 6.5.1 Eigenvalues and Characteristic Polynomial
 6.5.2 Eigenvectors and Diagonalization
 6.5.3 Localization
 6.5.4 Computation of Eigenvalues
 6.5.5 Special Classes

6.6 Combinatorial Matrix Theory — R. B. Bapat
 6.6.1 Matrices of Zeros and Ones
 6.6.2 Nonnegative Matrices
 6.6.3 Permanents

Chapter 6 LINEAR ALGEBRA

INTRODUCTION

Concepts from linear algebra play an important role in various applications of discrete mathematics, as in coding theory, computer graphics, generation of pseudo-random numbers, graph theory, and combinatorial designs. This chapter discusses fundamental concepts of linear algebra, computational aspects, and various applications.

GLOSSARY

access (of a class): The class C_i of vertices has access to class C_j if either $i = j$ or there is a path from a vertex in C_i to a vertex in C_j.

adjoint: See *Hermitian adjoint*.

algebraic multiplicity: given an eigenvalue, the multiplicity of the eigenvalue as a root of the characteristic equation.

augmented matrix (of a linear system): the matrix obtained by appending the right-hand side vector to the coefficient matrix as its rightmost column.

back substitution: a procedure for solving an upper triangular linear system.

basic class (of a matrix): a class such that the Perron root of the corresponding principal submatrix equals that of the entire matrix.

basis: an independent spanning set of vectors in a vector space.

characteristic equation: for a square matrix A, the equation $p_A(\lambda) = 0$, where $p_A(\lambda)$ is the characteristic polynomial of A.

characteristic polynomial: for a square matrix A, the polynomial (in the indefinite symbol λ) given by $p_A(\lambda) = \det(\lambda I - A)$.

Cholesky decomposition: expressing a matrix A as $A = LL^T$, where L is lower triangular and every entry on the main diagonal of L is positive.

circulant: a matrix in which every row is obtained by a single cyclic shift of the previous row.

class (of a matrix): a maximal set of row indices such that the corresponding vertices have mutual access in the directed graph of the matrix.

complete pivoting: an implementation of Gaussian elimination in which a pivot of largest magnitude is selected at each step.

condition number: given a matrix A, the number $\kappa(A) = \|A\| \, \|A^{-1}\|$.

conjugate sequence (of a sequence): the sequence whose nth term is the number of terms not less than n in the given sequence.

dependent set: a set of vectors in a vector space that are not independent.

determinant: given an $n \times n$ matrix A, $\det A = \sum_{\sigma \in S_n} \mathrm{sgn}(\sigma)\, a_{1\sigma(1)} a_{2\sigma(2)} \cdots a_{n\sigma(n)}$, where S_n is the symmetric group on n elements and the coefficient $\mathrm{sgn}(\sigma)$ is the sign of the permutation σ: 1 if σ is an even permutation and -1 if σ is an odd permutation.

diagonal matrix: a square matrix with nonzero elements only on the main diagonal.

diagonalizable matrix: a square matrix that is *similar* to a diagonal matrix.

difference (of matrices of the same dimensions): the matrix each of whose elements is the difference between corresponding elements of the original matrices.

dimension: for a vector space V, the number of vectors in any basis for V.

directed graph (of a matrix A): the graph $G(A)$ with vertices corresponding to the rows of A and an edge from i to j whenever a_{ij} is nonzero.

direct sum (of subspaces): given subspaces U and W, the sum of subspaces in which U and W have only the zero vector in common.

distance (between vectors): given vectors v and w, the length of the vector $v - w$.

dominant eigenvalue: given a matrix, an eigenvalue of the matrix of maximum modulus.

dot product (of real vectors): given real vectors $x = (x_1, \ldots, x_n)$ and $y = (y_1, \ldots, y_n)$, the number $x \cdot y = \sum_{i=1}^{n} x_i y_i$.

doubly stochastic matrix: a matrix with all entries nonnegative and with all row and column sums equal to 1.

eigenvalue: given a square matrix A, a scalar λ such that $Ax = \lambda x$ for some nonzero vector x.

eigenvector: given a square matrix A, a nonzero vector x such that the vector Ax is a scalar multiple of x.

eigenspace: given a square matrix A, the vector space $\{\, x \mid Ax = \lambda x \,\}$ for some scalar λ.

exponent (of a matrix): given a matrix A, the least positive integer m, if it exists, such that A^m has all positive entries.

fill: in Gaussian elimination, those nonzero entries created in the triangular factors of a matrix corresponding to zero entries in the original matrix.

final class: given a matrix, a class of the matrix with access to no other class.

flop: a multiply-add operation involving a single multiplication followed by a single addition.

forward substitution: a procedure for solving a lower triangular linear system.

fully indecomposable matrix: a matrix that is not partly decomposable.

Gaussian elimination: a solution procedure that at each step uses one equation to eliminate one variable from the system of equations.

geometric multiplicity: the dimension of the eigenspace.

Geršgorin discs: regions in the complex plane that collectively are guaranteed to contain all the eigenvalues of a given matrix.

growth factor: a ratio that measures how large the entries of a matrix become during Gaussian elimination.

Hermitian adjoint: given a matrix A, the matrix A^* obtained from the transpose A^T by replacing each entry by its complex conjugate.

Hermitian matrix: a complex matrix whose transpose is its (elementwise) complex conjugate.

idempotent matrix: a matrix A such that $A^2 = A$.

identity matrix: a diagonal matrix in which each diagonal element is 1.

ill-conditioned system: a linear system $Ax = b$ whose solution x is extremely sensitive to errors in the data A and b.

independent set: a set of vectors in a vector space that is not dependent.

index of cyclicity: for a matrix, the number of eigenvalues with maximum modulus.

inner product: a field-valued function of two vector variables used to define a notion of orthogonality (that is, perpendicularity). In real or complex vector spaces it is also used to introduce length, distance, and convergence.

inverse: given a square matrix A, the square matrix A^{-1} whose product with the original matrix is the identity matrix.

invertible matrix: a matrix that has an inverse.

irreducible matrix: a matrix that is not reducible.

isomorphic (vector spaces): vector spaces that are structurally identical.

kernel (of a linear transformation): the set of all vectors that are mapped to the zero vector by the linear transformation.

length (of a vector): the square root of the inner product of the vector with itself.

linear combination (of vectors): given vectors $v_1, v_2, \ldots, v_t$, a vector of the form $a_1v_1 + a_2v_2 + \cdots + a_tv_t$, where the a_i are scalars.

linear operator: a linear transformation from a vector space to itself.

linear system: a set of m linear equations in n variables x, represented by $Ax = b$; here A is the coefficient matrix and b is the right-hand side vector.

linear transformation: a function T from one vector space over F to another vector space over F satisfying $T(au+v) = aT(u)+T(v)$ for all vectors u, v and all scalars a.

lower triangular matrix: a matrix in which all nonzero elements occur either on or below the diagonal.

LU decomposition: expressing a matrix A as the product $A = LU$, where L is unit lower triangular and U is upper triangular.

Markowitz pivoting: a simple greedy strategy for reducing the number of nonzero entries introduced during the LU decomposition of a sparse matrix.

matrix (of a linear transformation): given a linear transformation T, a matrix associated with T that represents T with respect to a fixed basis.

minimal polynomial: for a matrix A, the monic polynomial $q(\cdot)$ of minimum degree such that $q(A) = 0$.

minimum degree algorithm: a version of the Markowitz pivoting strategy for symmetric coefficient matrices.

minor: the determinant of a square submatrix of a given matrix.

modulus: the absolute value of a complex number.

nilpotent matrix: a matrix A such that $A^k = 0$ for some positive integer k.

nonnegative matrix: a matrix with each entry nonnegative.

nonsingular matrix: a matrix that has an inverse.

normal matrix: a matrix A such that $AA^* = A^*A$ (A^* is the Hermitian adjoint of A).

nullity (of a linear transformation): the dimension of the kernel of the linear transformation.

nullity (of a matrix): the dimension of the null space of the matrix.

null space (of a matrix A): the set of all vectors x for which $Ax = 0$.

numerically stable algorithm: an algorithm whose accuracy is not greatly harmed by roundoff errors.

numerically unstable algorithm: an algorithm that can return an inaccurate solution even when the solution is relatively insensitive to errors in the data.

orthogonal matrix: a real square matrix whose inverse is its transpose.

orthogonal set (of vectors): a set of vectors in which any two distinct vectors have inner product zero.

orthonormal set (of vectors): a set of unit length orthogonal vectors.

partial pivoting: an implementation of Gaussian elimination which at step k selects the pivot of largest magnitude in column k.

partly decomposable (matrix): an $n \times n$ matrix containing a zero submatrix of size $k \times (n-k)$ for some $1 \leq k \leq n-1$.

permanent (of an $n \times n$ matrix A): $\text{per}(A) = \sum_{\sigma \in S_n} a_{1\sigma(1)} a_{2\sigma(2)} \cdots a_{n\sigma(n)}$, where S_n is the symmetric group on n elements.

permutation matrix: a square 0-1 matrix in which the entry 1 occurs exactly once in each row and exactly once in each column.

Perron root: the spectral radius of a nonnegative matrix.

pivot: the coefficient of the eliminated variable in the equation used to eliminate it.

positive definite matrix: a Hermitian matrix A such that $x^*Ax > 0$ for all $x \neq 0$.

positive matrix: a matrix with each entry positive.

positive semidefinite matrix: a Hermitian matrix A such that $x^*Ax \geq 0$ for all x.

power (of a square matrix): the square matrix obtained by multiplying the matrix by itself the required number of times.

primitive matrix: a matrix with a finite exponent.

principal minor (of a matrix): the determinant of a principal submatrix of the matrix.

principal submatrix (of a matrix A): the matrix obtained from A by deleting all but a specified set of rows and the same set of columns.

product (of matrices): for an $m \times n$ matrix A and an $n \times p$ matrix B, the $m \times p$ matrix AB whose ij-entry is the scalar product of row i of A and column j of B.

range (of a linear transformation T): the set of all vectors w for which $T(v) = w$ has a solution.

rank (of a linear transformation T): the dimension of the range of T.

rank (of a matrix): the maximum number of linearly independent rows (or columns) in the matrix.

reducible matrix: a matrix A with $a_{ij} = 0$ for all $i \in S$, $j \notin S$, for some set S.

roundoff errors: the errors associated with storing and computing numbers in finite precision arithmetic on a digital computer.

row stochastic matrix: a matrix with all entries nonnegative and row sums 1.

scalar: an element of a field.

scalar multiple (of a matrix): the matrix obtained by multiplying each element of the original matrix by the scalar.

scalar product: See *dot product*.

similar matrices: square matrices A and B satisfying the equation $P^{-1}BP = A$ for some invertible matrix P.

singular matrix: a matrix that has no inverse.

singular values (of a matrix A): the positive square roots of the eigenvalues of AA^*, where A^* is the Hermitian adjoint of A.

skew-Hermitian matrix: a matrix equal to the negative of its Hermitian adjoint.

skew-symmetric matrix: a matrix equal to the negative of its transpose.

span (of a set of vectors): all vectors obtainable as linear combinations of the given vectors.

spanning set: a set of vectors in a vector space V whose span equals V.

sparse matrix: a matrix that has relatively few nonzero entries.

spectral radius (of a matrix): the maximum modulus of an eigenvalue of the matrix.

square matrix: a matrix having the same number of rows and columns.

strictly diagonally dominant matrix: a square matrix each of whose diagonal elements exceeds in modulus the sum of the moduli of all other elements in that row.

strictly totally positive matrix: a matrix with all minors positive.

submatrix (of a matrix A): the matrix obtained from A by deleting all but a certain set of rows and a certain set of columns.

subspace: a vector space within a vector space.

sum (of matrices): for two matrices of the same dimensions, the matrix each of whose elements is the sum of the corresponding elements of the original matrices.

sum (of subspaces): given subspaces U and W, the subspace consisting of all possible sums $u + w$ where $u \in U$ and $w \in W$.

symmetric matrix: a matrix that equals its transpose.

term rank (of a 0-1 matrix): the maximum number of 1s such that no two are in the same row or column.

trace: given a square matrix, the sum of the diagonal elements of the matrix.

transpose (of a matrix): for a matrix A, the matrix A^T whose columns are the rows of the original matrix.

tridiagonal matrix: a matrix whose nonzero entries are either on the main diagonal or immediately above or below the main diagonal.

unitary matrix: a square matrix whose inverse is its Hermitian adjoint.

unit triangular matrix: a (lower or upper) triangular matrix having all diagonal entries 1.

upper triangular matrix: a matrix in which all nonzero elements occur either on or above the main diagonal.

vector: an individual object of a vector space.

vector space: a collection of objects that can be added and multiplied by scalars, always yielding another object in the collection.

well-conditioned system: a linear system $Ax = b$ whose solution x is relatively insensitive to errors in the data A and b.

0-1 matrix: a matrix with each entry either 0 or 1.

6.1 VECTOR SPACES

The concept of a "vector" comes initially from the physical world, where a vector is a quantity having both magnitude and direction (for example, force and velocity). The mathematical concept of a vector space generalizes these ideas, with applications in coding theory, finite geometry, cryptography, and other areas of discrete mathematics.

6.1.1 BASIC CONCEPTS

Definitions:

A *vector space* over a field F (§5.6.1) is a triple $(V, \oplus, \cdot)$ consisting of a set V and two operations, $\oplus$ (vector addition) and $\cdot$ (scalar multiplication), such that:

- $(V, \oplus)$ is an abelian group (§5.2.1); i.e., $\oplus$ is a function $(u, v) \to u \oplus v$ from $V \times V$ to V such that:

    $(u \oplus v) \oplus w = u \oplus (v \oplus w)$ for all $u, v, w \in V$;
    there is a vector 0 such that $v \oplus 0 = v$ for all $v \in V$;
    for each $v \in V$ there is $-v \in V$ such that $v \oplus (-v) = 0$;
    $u \oplus v = v \oplus u$ for all $u, v \in V$;

- the operation $\cdot$ is a function $(a, v) \to a \cdot v$ from $F \times V$ to V such that for all $a, b \in F$ and $u, v \in V$ the following properties hold:

    $a \cdot (b \cdot v) = (ab) \cdot v$;
    $(a + b) \cdot v = (a \cdot v) \oplus (b \cdot v)$;
    $a \cdot (u \oplus v) = (a \cdot u) \oplus (a \cdot v)$;
    $1 \cdot v = v$.

Here, ab and $a + b$ represent multiplication and addition of elements $a, b \in F$.

The **scalars** are the elements of F, the **vectors** are the elements of V, and the set V itself is often also called the **vector space**.

The **difference** of two vectors u and v is the vector $u - v = u \oplus (-v)$ where $-v$ is the negative of v in the abelian group $(V, \oplus)$.

Notation: While vector addition $\oplus$ and field addition $+$ can be quite different, it is customary to use the same notation $+$ for both. It is also customary to write av instead of $a \cdot v$, and to use the symbol 0 for the additive identities of the vector space V and the field F.

Facts:
Assume that V is a vector space over F.

1. $a0 = 0$ and $0v = 0$ for all $a \in F$ and $v \in V$.
2. $(-1)v = -v$ for all $v \in V$.
3. If $av = 0$ for $a \in F$ and $v \in V$, then either $a = 0$ or $v = 0$.
4. Cancellation property: For all $u, v, w \in V$, if $u + v = w + v$, then $u = w$.
5. $a(u - v) = au - av$ for all $a \in F$ and $u, v \in V$.

Examples:

1. *Force vectors*: Forces in the plane can be represented by geometric vectors such as F_1 and F_2 in part (a) of the following figure; addition of these vectors is carried out using the so-called parallelogram law. By introducing a coordinate system and locating the initial point of each directed line segment at the origin $(0,0)$, each geometric vector can be named by its terminal point. Thus, a vector in the plane becomes a pair $(x,y) \in \mathcal{R}^2$ of real numbers. The parallelogram law of addition translates into componentwise addition (part (c) of the figure), while stretching (respectively, shrinking, negating) translates to componentwise multiplication by a real number $r > 1$ (respectively, $0 < r < 1, r = -1$). Three-dimensional force vectors are similarly represented using triples $(x,y,z) \in \mathcal{R}^3$.

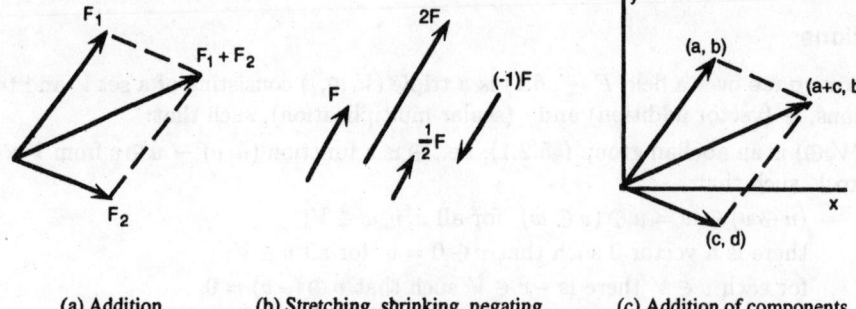

(a) Addition (b) Stretching, shrinking, negating (c) Addition of components

2. *Euclidean space*: Generalizing Example 1, n-dimensional Euclidean space consists of all n-tuples of real numbers $\mathcal{R}^n = \{(x_1, x_2, \ldots, x_n) \mid x_i \in \mathcal{R}\}$.

3. If F is any field, then $F^n = \{(x_1, x_2, \ldots, x_n) \mid x_i \in F\}$ is a vector space, where addition and scalar multiplication are componentwise:
$$(x_1, x_2, \ldots, x_n) + (y_1, y_2, \ldots, y_n) = (x_1 + y_1, x_2 + y_2, \ldots, x_n + y_n)$$
$$a(x_1, x_2, \ldots, x_n) = (ax_1, ax_2, \ldots, ax_n)$$
where $a \in F$. When $F = \mathcal{R}$, these are the vectors mentioned in Examples 1 and 2.

4. *A vector space over $\mathcal{Z}_2$*: V consists of the 128 subsets of the set $\{1, 2, \ldots, 7\}$ as represented by binary 7-tuples; for example, the subset $\{1, 4, 5, 7\}$ corresponds to $(1, 0, 0, 1, 1, 0, 1)$ and the subset $\{1, 2, 3, 4\}$ to $(1, 1, 1, 1, 0, 0, 0)$. The operations on V are componentwise addition and scalar multiplication mod 2. In this vector space, the sum of two members of V corresponds to the symmetric difference (§1.2.2) of the associated sets. (This example is a special case of Example 3.)

5. *A finite affine plane over $\mathcal{Z}_5$*: V consists of all pairs (x, y) where $x, y \in \mathcal{Z}_5$ and where addition and scalar multiplication are componentwise modulo 5. This special case of Example 3 arises in finite geometry where the 25 members of V are thought of as "points" and the sets of solutions to equations of the form $ax + by = c$ (where $a, b, c \in \mathcal{Z}_5$ with one of a or $b \neq 0$) are viewed as "lines".

6. *Infinite binary sequences*: V consists of all infinite binary sequences $\{(s_1, s_2, \ldots) \mid s_i \in \mathcal{Z}_2\}$ where addition and multiplication are componentwise mod 2. As in Example 4, each $s \in V$ may be viewed of as a subset of the positive integers, but each s may also be viewed as a potential "message" or "data" stream; for example, each group of 7 consecutive members of s could represent a letter in the 7-bit ASCII code.

7. $V = F^{m \times n}$, the set of all $m \times n$ matrices over F, is a vector space, where vector addition is the usual matrix addition and scalar multiplication is the usual scalar-by-matrix multiplication (§6.3.2). When $m = 1$, this reduces to Example 3.

8. Let $V = E$ be a field and F a subfield. Then V is a vector space over F where vector addition and scalar multiplication are the addition and multiplication of E. In particular, the finite field F_q of prime power order $q = p^n$ is a vector space over the subfield F_p.

9. Let $V = F[x]$, the set of all polynomials (§5.5.2) over F in an indeterminate x. Then V is a vector space over F, where addition is ordinary polynomial addition and scalar multiplication is the usual scalar-by-polynomial multiplication.

10. For a nonempty set X and a given vector space U over F, let V denote the set of all functions from X to U. The sum $f + g$ of two vectors (functions) $f, g \in V$ is defined by $(f + g)(x) = f(x) + g(x)$ for all $x \in X$ and the scalar multiplication af of $a \in F$ by $f \in V$ is defined by $(af)(x) = af(x)$. (For specific cases of this general vector space, see §6.1.2, Examples 13–15.)

6.1.2 SUBSPACES

Definitions:

A *subspace* of a vector space V is a nonempty subset W of V that is a vector space under the addition and scalar multiplication operations inherited from V.

The **sum** of two subspaces $U, W \subseteq V$ is the set $\{u + w \mid u \in U, w \in W\}$. If $U \cap W = \{0\}$, their sum is called the **direct sum**, denoted $U \oplus W$.

If A is an $m \times n$ matrix over F, the **null space** $NS(A)$ of A is $\{x \in F^{n \times 1} \mid Ax = 0\}$. The null space of A is also called the **right null space** when contrasted with the **left null space** $LNS(A)$ defined by $\{y \in F^{1 \times m} \mid yA = 0\}$.

Facts: Assume that V is a vector space over F.

1. $W \subseteq V$ is a subspace of V if and only if $W \neq \emptyset$ and for all $a, b \in F$ and $u, v \in W$, $au + bv \in W$.
2. $W \subseteq V$ is a subspace of V if and only if $W \neq \emptyset$ and for all $a \in F$ and $u, v \in W$, $u + v \in W$ and $au \in W$.
3. Every subspace of V contains 0, the zero vector.
4. The sets $\{0\}$ and V are subspaces of V.
5. The intersection of any collection of subspaces of V is a subspace of V.
6. The sum of any collection of subspaces of V is a subspace of V.
7. Each member of $U \oplus W$ can be expressed as a sum $u + w$ for a unique $u \in U$ and a unique $w \in W$.
8. The set of solutions to a homogeneous linear equation in the unknowns $x_1, x_2, \ldots, x_n$ is a subspace of F^n. Namely, for any fixed $(a_1, a_2, \ldots, a_n) \in F^n$, the set $W = \{x \in F^n \mid a_1 x_1 + a_2 x_2 + \cdots + a_n x_n = 0\}$ is a subspace of F^n.
9. The set of solutions to any collection of homogeneous linear equations in the unknowns $x_1, x_2, \ldots, x_n$ is a subspace of F^n. In particular, if W is a subspace of F^n then the set of all $x = (x_1, x_2, \ldots, x_n) \in F^n$ satisfying $a_1 x_1 + a_2 x_2 + \cdots + a_n x_n = 0$ for all $(a_1, a_2, \ldots, a_n) \in W$ is a subspace of V called the *orthogonal complement* of W and denoted $W^\perp$.
10. The null space $NS(A)$ of an $m \times n$ matrix A over F is a subspace of $F^{n \times 1}$.
11. The left null space $LNS(A)$ of an $m \times n$ matrix A is a subspace of $F^{1 \times m}$ and equals $(NS(A^T))^T$ where T denotes transpose.

Examples:

1. The set of all 3-tuples of real numbers of the form $(a, b, 2a + 3b)$ where $a, b \in \mathcal{R}$ is a subspace of $\mathcal{R}^3$. This subspace can also be described as the set of solutions (x, y, z) to the homogeneous linear equation $2x + 3y - z = 0$.

2. The set of all 4-tuples of real numbers of the form $(a, -a, 0, b)$ where $a, b \in \mathcal{R}$ is a subspace of $\mathcal{R}^4$. This subspace can also be described as the set of solutions (x_1, x_2, x_3, x_4) to the pair of equations $x_1 + x_2 = 0$ and $x_3 = 0$.

3. For $V = \mathcal{Z}_5^2$, the set of all solutions to the equation $x + 2y = 0$ forms a subspace. It consists of the finite set $\{(0,0), (3,1), (1,2), (4,3), (2,4)\}$ and can also be described as the set of all pairs in V of the form $(3a, a)$. The set S of solutions to $x + 2y = 1$, namely $\{(1,0), (4,1), (2,2), (0,3), (3,4)\}$, is not a subspace of V since for example $(1,0)+(4,1) = (0,1) \notin S$. However S is a "line" in the affine plane described in Example 5 of §6.1.1.

4. In the vector space $V = \mathcal{Z}_2^7$, the set of 7-tuples with an even number of 1s is a subspace. This subspace can also be described as the collection of all members of V whose components sum to 0.

5. *Coding theory*: In the vector space F^n over the finite field $F = GF(q)$, a linear code (§14.2) is simply any subspace of F^n. In particular, an (n, k) code is a k-dimensional subspace of F^n.

6. *Binary codes*: A linear binary code is any subspace of the vector space F^n where F is the finite field on two elements, $GF(2)$. Generalizing Example 4, the set of all binary n-tuples with an even number of 1s is a subspace of F^n and so is a linear binary code.

7. Consider the undirected graph (§8.1) in the following figure, where the edges have been labeled with the integers $\{1, 2, \ldots, 7\}$. Associate with this graph the vector space $V = \mathcal{Z}_2^7$ where, as in Example 4 (§6.1.1), each binary 7-tuple is identified with a subset of edges. One subspace W of V, called the *cycle space* of the graph, corresponds to the (edge-disjoint) union of cycles in the graph. For example, $(1, 1, 0, 1, 0, 1, 1) \in W$ as it corresponds to the cycle $1, 2, 6, 7, 4$, and so is $(1, 1, 1, 0, 1, 1, 1)$ which corresponds to the edge-disjoint union of cycles $1, 2, 3$ and $5, 6, 7$. The sum of these two members of W is $(0, 0, 1, 1, 1, 0, 0)$ which corresponds to the cycle $3, 4, 5$.

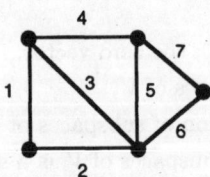

8. The set of $n \times n$ symmetric matrices (§6.3.1) over a field F is a subspace of $F^{n \times n}$, and so is the set of $n \times n$ upper triangular matrices (§6.3.1) over F.

9. For an $m \times m$ matrix A over F and $\lambda \in F$, the set $W = \{X \in F^{m \times n} \mid AX = \lambda X\}$ is a subspace of $F^{m \times n}$. (This space is related to the *eigenspaces* of A discussed in §6.5.2.)

10. For a given $n \times n$ matrix A over F, the set $W = \{X \in F^{n \times n} \mid XA = AX\}$ is a subspace of $F^{n \times n}$. (This is the space of matrices that commute with A.)

11. Let field E be a vector space over subfield F, and let K denote the set of all elements $\alpha \in E$ that satisfy a polynomial equation of the form $f(\alpha) = 0$ for some nonzero $f(x) \in F[x]$. Then K is a subfield of E containing F (the field of *algebraic elements* of E over F) and consequently is a subspace of E over F. (See §5.6.2.)

12. For each fixed $n \geq 1$, the set of all polynomials of degree $\leq n$ is a subspace of $F[x]$. (See §6.1.1 Example 9.)

13. In §6.1.1 Example 10, take $X = [a,b]$ where $a,b \in \mathcal{R}$ with $a < b$, and take $U = \mathcal{R}$ as a vector space over itself. The resulting V, the set of all real-valued functions on $[a,b]$, is a vector space. The set $C[a,b]$ of continuous real-valued functions on $[a,b]$ is a subspace of V.

14. In §6.1.1 Example 10, take $X = \{1,2,\ldots,7\}$ and take $U = \mathcal{Z}_2$ as a vector space over itself. The resulting V, the set of all functions from $\{1,2,\ldots,7\}$ to $\mathcal{Z}_2$, can be thought of as the vector space of binary 7-tuples $V = \mathcal{Z}_2^7$.

15. In §6.1.1 Example 10, take both X and U to be vector spaces over F. Then V is the vector space of all functions from X to U. The collection of those $T \in V$ satisfying $T(a\alpha + b\beta) = aT(\alpha) + bT(\beta)$ for all $a,b \in F$ and $\alpha,\beta \in X$ is a subspace of V. (This space is the space of *linear transformations* considered in §6.2.)

6.1.3 LINEAR COMBINATIONS, INDEPENDENCE, BASIS, AND DIMENSION

Definitions:

If $v_1, v_2, \ldots, v_t$ are vectors from a vector space V over F, then a vector $w \in V$ is a **linear combination** of $v_1, v_2, \ldots, v_t$ if $w = a_1 v_1 + a_2 v_2 + \cdots + a_t v_t$ for some scalars $a_i \in F$. The zero vector is considered a linear combination of $\emptyset$.

For $S \subseteq V$, the **span** of S, denoted $\mathrm{Span}(S)$, is the set of all (finite) linear combinations of members of S; that is, $\mathrm{Span}(S)$ consists of all finite sums $a_1 v_1 + a_2 v_2 + \cdots + a_t v_t$ where $v_i \in S$ and $a_i \in F$. (The span of the empty set is taken to be $\{0\}$.) $\mathrm{Span}(S)$ is also called the space **generated** or **spanned** by S. (See Fact 1.)

The **row space** $RS(A)$ of an $m \times n$ matrix A over F (§6.3.1) is $\mathrm{Span}(R_1, R_2, \ldots, R_m)$, where $R_1, R_2, \ldots, R_m$ are the rows of A viewed as vectors in $F^{1 \times n}$.

The **column space** $CS(A)$ of A is $\mathrm{Span}(C_1, C_2, \ldots, C_n)$, where $C_1, C_2, \ldots, C_n$ are the columns of A.

A subset $S \subseteq V$ is called a **spanning set** for V if $\mathrm{Span}(S) = V$.

A subset $S \subseteq V$ is **(linearly) independent** if every finite subset $\{v_1, v_2, \ldots, v_t\}$ of S has the property that the only scalars $a_1, a_2, \ldots, a_t$ satisfying $a_1 v_1 + a_2 v_2 + \cdots + a_t v_t = 0$ are $a_1 = a_2 = \cdots = a_t = 0$.

A subset $S \subseteq V$ is **(linearly) dependent** if it is not independent.

A **basis** for V is an independent spanning set.

A vector space V is **finite dimensional** if it has a finite basis; otherwise, V is **infinite dimensional**.

The **dimension**, $\dim V$, of a vector space V is the cardinality of any basis for V. (See Fact 8.)

If $\mathcal{B} = (v_1, v_2, \ldots, v_n)$ is an ordered basis for V, then the **coordinates** of v with respect to $\mathcal{B}$ are the scalars $a_1, a_2, \ldots, a_n$ such that $v = a_1 v_1 + a_2 v_2 + \cdots + a_n v_n$. (See Fact 14.) The **coordinate vector** $[v]_\mathcal{B}$ of v with respect to $\mathcal{B}$ (written as a column) is $[v]_\mathcal{B} = (a_1, a_2, \ldots, a_n)^T$ where T denotes transpose (§6.3.1).

Note: Some writers distinguish between the coordinates written as a row and as a column, calling the row $(a_1, a_2, \ldots, a_n)$ the *coordinate vector* of v with respect to $\mathcal{B}$ and the column $(a_1, a_2, \ldots, a_n)^T$ the *coordinate matrix* of v with respect to $\mathcal{B}$.

The **row rank** of a matrix A over F is $\dim RS(A)$, and the **column rank** of A is $\dim CS(A)$. The **rank** of A is the size of the largest square submatrix of A with nonzero determinant (§6.3.4); that is, rank $A = r$ if there exists an $r \times r$ submatrix of A whose determinant is nonzero, and every $t \times t$ submatrix of A with $t > r$ has zero determinant.

The **nullity** of a matrix A is $\dim NS(A)$.

Two vector spaces V and U over the same field F are **isomorphic** if there exists a bijective mapping $T: V \to U$ such that $T(v + w) = T(v) + T(w)$ and $T(av) = aT(v)$ for all $v, w \in V$ and $a \in F$. The mapping T is called an **isomorphism**.

Facts:

1. Span(S) is a subspace of V. In particular, $RS(A)$ is a subspace of $F^{1 \times n}$ and $CS(A)$ is a subspace of $F^{m \times 1}$.

2. Span(S) is the intersection of all subspaces of V that contain S; thus, Span(S) is the smallest subspace of V containing S in that it lies inside every subspace of V containing S.

3. A set $\{v\}$ consisting of a single vector from V is dependent if and only if $v = 0$.

4. A set of two or more vectors is dependent if and only if some vector in the set is a linear combination of the remaining vectors in the set.

5. Any superset of a dependent set is dependent, and any subset of an independent set is independent. (The empty set is independent.)

6. If V has a basis of n elements, then every subset of V with more than n elements is dependent.

7. If W is a subspace of V then $\dim W \leq \dim V$.

8. Every vector space V has a basis, and every two bases for V have the same number of elements (cardinality). For infinite-dimensional vector spaces, this fact relies on the *axiom of choice* (§1.2.4).

9. Every independent subset of V can be extended to a basis for V. More generally, if S is an independent set, then every maximal independent set containing S is a basis for V containing S. For infinite-dimensional vector spaces, this fact relies on the axiom of choice. (An independent set is *maximal* if every set properly containing it is dependent.)

10. Every spanning set contains a basis for V. More generally, if S is a spanning set, then every minimal spanning subset of S is a basis for V. For infinite-dimensional vector spaces, this fact relies on the axiom of choice. (A spanning set is *minimal* if it contains no proper subset that spans V.)

11. Rank-nullity theorem: If A is an $m \times n$ matrix over F, then:
    - $\dim RS(A) + \dim NS(A) = n$;
    - $\dim CS(A) + \dim NS(A) = n$;
    - $\dim RS(A) + \dim LNS(A) = m$;
    - $\dim CS(A) + \dim LNS(A) = m$.

12. For every matrix A, row rank A = column rank A = rank A. Thus, the (maximum) number of independent rows of A equals the (maximum) number of independent columns.

13. The set of solutions to the m homogeneous linear equations $\sum_{j=1}^{n} a_{ij} x_j = 0$ in n unknowns has dimension $n - r$, where r is the rank of the $m \times n$ coefficient matrix $A = (a_{ij})$.

14. If $\mathcal{B}$ is a basis for a vector space V (finite or infinite), then each $v \in V$ can be expressed as $v = a_1 v_1 + a_2 v_2 + \cdots + a_t v_t$, where $a_i \in F$ and $v_i \in \mathcal{B}$. If $v = b_1 v_1 + b_2 v_2 + \cdots + b_t v_t$ is another expression for v in terms of elements of $\mathcal{B}$ (where possibly some zero coefficients have been inserted to make the two expressions have equal length), then $a_i = b_i$ for $i = 1, 2, \ldots, t$. (If $\mathcal{B}$ is finite, this justifies the definition of the coordinate vector $[v]_\mathcal{B}$.)

15. If $\mathcal{B} = (v_1, v_2, \ldots, v_n)$ is an ordered basis for V, then the function $T: V \to F^{n \times 1}$ defined by $T(v) = [v]_\mathcal{B}$ is an isomorphism, so V is isomorphic to $F^{n \times 1}$.

16. Two vector spaces over F are isomorphic if and only if they have the same dimension.

Examples:

1. The vector space F^n has dimension n. The *standard basis* is the ordered basis $(e_1, e_2, \ldots, e_n)$ where e_i is the vector with 1 in position i and 0s elsewhere. (The spaces F^n, $F^{1 \times n}$, and $F^{n \times 1}$ are isomorphic and are often identified and used interchangeably.)

2. The vector space $F^{m \times n}$ of $m \times n$ matrices over F has dimension mn; the *standard basis* is $\{ E_{ij} \mid 1 \leq i \leq m, \ 1 \leq j \leq n \}$ where E_{ij} is the $m \times n$ matrix with a 1 in position (i, j) and 0s elsewhere. It is isomorphic to F^{mn}.

3. The subspace of $\mathcal{R}^3$ containing all 3-tuples of the form $(a, b, 2a + 3b)$ has dimension 2. One basis for this subspace is $\mathcal{B}_1 = ((1, 0, 2), (0, 1, 3))$ and another is $\mathcal{B}_2 = ((1, 1, 5), (1, -1, -1))$. The vector $w = (5, -1, 7)$ is in the subspace since $w = 5(1, 0, 2) + (-1)(0, 1, 3) = 2(1, 1, 5) + 3(1, -1, -1)$. The coordinate vector of w with respect to $\mathcal{B}_1$ is $(5, -1)^T$ and the coordinate vector of w with respect to $\mathcal{B}_2$ is $(2, 3)^T$.

4. If W is the subspace of $V = \mathcal{Z}_2^5$ containing all members of V whose components sum to 0, then W has dimension 4. In fact $W = \{ (a, b, c, d, a + b + c + d) \mid a, b, c, d \in \mathcal{Z}_2 \}$. One ordered basis for this space is $((1, 0, 0, 0, 1), (0, 1, 0, 0, 1), (0, 0, 1, 0, 1), (0, 0, 0, 1, 1))$.

5. *Binary codes*: More generally, consider the set of all binary n-tuples with an even number of 1s; this is the linear binary code mentioned in Example 6, §6.1.2. These vectors form a subspace W of $V = \mathcal{Z}_2^n$ of dimension $n - 1$. A basis for W consists of the following $n - 1$ vectors, each of which has exactly two 1s: $(1, 0, \ldots, 1), (0, 1, \ldots, 1), \ldots, (0, 0, \ldots, 1, 1)$. Consequently there are 2^{n-1} vectors in the code W.

6. The field $\mathcal{C}$ of complex numbers is two-dimensional as a vector space over $\mathcal{R}$; it has the ordered basis $(1, i)$, where $i = \sqrt{-1}$. Any two complex numbers, neither of which is a real multiple of the other, form a basis.

7. Both $\mathcal{C}$ and $\mathcal{R}$ are infinite-dimensional vector spaces over the rational field $\mathcal{Q}$.

8. The vector space $F[x]$ is an infinite-dimensional space over F; $(1, x, x^2, x^3, \ldots)$ is an ordered basis. The subspace of all polynomials of degree $\leq n$ has dimension $n + 1$; $(1, x, x^2, \ldots, x^n)$ is an ordered basis.

6.1.4 INNER PRODUCTS, LENGTH, AND ORTHOGONALITY

By imposing additional structure on real and complex vector spaces, the concepts of length, distance, and orthogonality can be introduced. These concepts are motivated by the corresponding geometric notions for physical vectors. Also, for real vector spaces the geometric idea of angle can be formulated analytically.

Definitions:

An **inner product** on a vector space V over $\mathcal{R}$ is a function $\langle \cdot, \cdot \rangle : V \times V \to \mathcal{R}$ such that for all $u, v, w \in V$ and $a, b \in \mathcal{R}$ the following hold:

- $\langle u, v \rangle = \langle v, u \rangle$;
- $\langle u, u \rangle \geq 0$ with equality if and only if $u = 0$;
- $\langle au + bv, w \rangle = a \langle u, w \rangle + b \langle v, w \rangle$.

An **inner product** on a vector space V over $\mathcal{C}$ is a function $\langle \cdot, \cdot \rangle : V \times V \to \mathcal{C}$ such that for all $u, v, w \in V$ and $a, b \in \mathcal{C}$ the following hold:

- $\langle u, v \rangle = \overline{\langle v, u \rangle}$ (where bar denotes complex conjugation);
- $\langle u, u \rangle \geq 0$ with equality if and only if $u = 0$;
- $\langle au + bv, w \rangle = a \langle u, w \rangle + b \langle v, w \rangle$.

Note: The first property implies that $\langle u, u \rangle$ is real, so the second property makes sense.

An **inner product space** is a vector space over $\mathcal{R}$ or $\mathcal{C}$ on which an inner product is defined. Such a space is called a real or complex inner product space, depending on its scalar field.

The **norm** (**length**) of a vector $v \in V$ is $\|v\| = \sqrt{\langle v, v \rangle}$.

A vector $v \in V$ is a **unit vector** if and only if $\|v\| = 1$.

The **distance** $d(v, w)$ from v to w is $d(v, w) = \|v - w\|$.

In a real inner product space, the **angle** between nonzero vectors v and w is the real number θ, $0 \leq \theta \leq \pi$, such that $\cos \theta = \dfrac{\langle v, w \rangle}{\|v\| \cdot \|w\|}$.

Two vectors v and w are **orthogonal** if and only if $\langle v, w \rangle = 0$.

A subset $S \subseteq V$ is an **orthogonal set** if $\langle v, w \rangle = 0$ for all $v, w \in S$ with $v \neq w$.

A subset $S \subseteq V$ is an **orthonormal set** if S is an orthogonal set and $\|v\| = 1$ for all $v \in S$.

If W is a subspace of an inner product space V, then the **orthogonal complement** $W^\perp = \{ v \in V \mid \langle v, w \rangle = 0 \text{ for all } w \in W \}$.

Facts:

1. *Standard inner product on $\mathcal{R}^n$*: The real-valued function defined by $\langle x, y \rangle = x_1 y_1 + x_2 y_2 + \cdots + x_n y_n$ is an inner product on $V = \mathcal{R}^n$.

2. *Standard inner product on $\mathcal{C}^n$*: The complex-valued function defined by $\langle x, y \rangle = x_1 \overline{y}_1 + x_2 \overline{y}_2 + \cdots + x_n \overline{y}_n$ is an inner product on $V = \mathcal{C}^n$.

3. If A is an $n \times n$ real positive definite matrix (§6.3.2), then the function defined by $\langle x, y \rangle = x^T A y$ is an inner product on $\mathcal{R}^n$. (Here x^T denotes the transpose of x.)

4. If H is an $n \times n$ complex positive definite matrix (§6.3.2), then the function defined by $\langle x, y \rangle = y^* H x$ is an inner product on $\mathcal{C}^n$. (y^* is the conjugate-transpose of y.)

5. The function $\langle f, g \rangle = \int_a^b f(x) g(x) dx$ is an inner product on the vector space $C[a, b]$ of continuous real-valued functions on the interval $[a, b]$.

6. The inner product $\langle \cdot, \cdot \rangle$ on an inner product space V is an inner product on any subspace W of V.

7. If W is a subspace of an inner product space V, then the orthogonal complement $W^\perp$ is a subspace of V and $V = W \oplus W^\perp$.

8. The norm function satisfies the following properties for all scalars a and all vectors $v, w \in V$:
   - $\|v\| \geq 0$ with equality if and only if $v = 0$;
   - $\|av\| = |a| \cdot \|v\|$, where $|a|$ denotes the absolute value of a;
   - $|\langle v, w \rangle| \leq \|v\| \cdot \|w\|$ (*Cauchy-Schwarz inequality*);
   - $\|v + w\| \leq \|v\| + \|w\|$ (*triangle inequality*);
   - if $v \neq 0$, then $\dfrac{1}{\|v\|} v$ is a unit vector (the *normalization* of v).

9. The distance function on a vector space V satisfies the following properties for all $v, w, z \in V$:
   - $d(v, w) \geq 0$ with equality if and only if $v = w$;
   - $d(v, w) = d(w, v)$;
   - $d(v, z) \leq d(v, w) + d(w, z)$ (*triangle inequality*).

10. For real inner product spaces, two nonzero vectors are orthogonal if and only if the angle between them is $\theta = \frac{\pi}{2}$.

11. An orthogonal set S of nonzero vectors can be converted to an orthonormal set by normalizing each vector in S.

12. An orthogonal set of nonzero vectors is independent. An orthonormal set is independent.

13. If V is an n-dimensional inner product space, any orthonormal set contains at most n vectors, and any orthonormal set of n vectors is a basis for V.

14. Every subspace W of an n-dimensional space V has an orthonormal (orthogonal) basis.

15. *Gram-Schmidt orthogonalization*: From any ordered basis $(w_1, w_2, \ldots, w_m)$ for a subspace W, an orthonormal basis $(u_1, u_2, \ldots, u_m)$ for W can be constructed using Algorithm 1. (Jörgen Gram, 1850–1916; Erhardt Schmidt, 1876–1959)

Algorithm 1: Gram-Schmidt orthogonalization process.

input: an ordered basis $(w_1, w_2, \ldots, w_m)$
output: an orthonormal basis $(u_1, u_2, \ldots, u_m)$

$u_1 := \dfrac{1}{a_1} w_1$, where $a_1 := \|w_1\|$

for $j := 2$ to m

$$a_j := \left\| w_j - \sum_{i=1}^{j-1} \langle w_j, u_i \rangle u_i \right\|$$

$$u_j := \dfrac{1}{a_j} \left(w_j - \sum_{i=1}^{j-1} \langle w_j, u_i \rangle u_i \right)$$

---

16. The standard basis is orthonormal with respect to the standard inner product.

17. If $(u_1, u_2, \ldots, u_m)$ is an orthonormal basis for a subspace W of V and $w \in W$, then $w = \langle w, u_1 \rangle u_1 + \langle w, u_2 \rangle u_2 + \cdots + \langle w, u_m \rangle u_m$.

18. *Projection vector*: Let W be a subspace of a vector space V and let v be a vector in V.
- There is a unique vector $p \in W$ nearest to v; that is, the vector p minimizes $\|v - w\|$ over all $w \in W$. This vector p is called the *projection* of v onto W, written $p = \text{proj}_W(v)$.
- If $(u_1, u_2, \ldots, u_m)$ is any orthonormal basis for W, then the projection of v onto W is given by $\text{proj}_W(v) = \langle v, u_1 \rangle u_1 + \langle v, u_2 \rangle u_2 + \cdots + \langle v, u_m \rangle u_m$.
- The vector $\text{proj}_W(v)$ is the unique vector $w \in W$ such that $v - w$ is orthogonal to every vector in W.

19. *Projection matrix*: If $V = \mathcal{R}^n$ is equipped with the standard inner product and $(u_1, u_2, \ldots, u_m)$ is an orthonormal basis for a subspace W, then the projection of each $x \in \mathcal{R}^n$ onto W is given by $\text{proj}_W(x) = Ax$, where $A = GG^T$ with $G = (u_1, u_2, \ldots, u_m)$ the $n \times m$ matrix with the u_i as columns.

20. The projection matrix A is symmetric and satisfies $A^2 = A$.

Examples:
Consider the vector space $\mathcal{R}^4$ with the standard inner product $\langle x, y \rangle = x^T y$, and let W be the subspace spanned by the three vectors $w_1 = (1, 1, 1, 1)^T$, $w_2 = (3, 1, 3, 1)^T$, $w_3 = (3, 1, 1, 1)^T$.

1. $\langle w_1, w_2 \rangle = 8$ and $\|w_1\| = 2$.
2. The angle θ between w_1 and w_2 satisfies $\cos\theta = \frac{8}{2\sqrt{20}} = \frac{2}{\sqrt{5}}$ (so $\theta \approx 0.4636$ radians).
3. The distance from w_1 to w_2 is $d(w_1, w_2) = \|w_1 - w_2\| = \|(-2, 0, -2, 0)^T\| = 2\sqrt{2}$.
4. The orthogonal complement $W^\perp$ of W is the set of vectors of the form $(0, a, 0, -a)$.
5. The Gram-Schmidt process applied to (w_1, w_2, w_3) yields:

$u_1 = \frac{1}{a_1} w_1 = (\frac{1}{2}, \frac{1}{2}, \frac{1}{2}, \frac{1}{2})^T$, where $a_1 = \|w_1\| = 2$;

$u_2 = \frac{1}{a_2}(w_2 - \langle w_2, u_1 \rangle u_1) = \frac{1}{a_2}((3, 1, 3, 1)^T - 4(\frac{1}{2}, \frac{1}{2}, \frac{1}{2}, \frac{1}{2})^T)$
$= \frac{1}{a_2}(1, -1, 1, -1)^T = (\frac{1}{2}, -\frac{1}{2}, \frac{1}{2}, -\frac{1}{2})^T$, where $a_2 = \|(1, -1, 1, -1)^T\| = 2$;

$u_3 = \frac{1}{a_3}(w_3 - \langle w_3, u_1 \rangle u_1 - \langle w_3, u_2 \rangle u_2)$
$= \frac{1}{a_3}((3, 1, 1, 1)^T - 3(\frac{1}{2}, \frac{1}{2}, \frac{1}{2}, \frac{1}{2})^T - 1(\frac{1}{2}, -\frac{1}{2}, \frac{1}{2}, -\frac{1}{2})^T)$
$= \frac{1}{a_3}(1, 0, -1, 0)^T = (\frac{1}{\sqrt{2}}, 0, -\frac{1}{\sqrt{2}}, 0)^T$, where $a_3 = \|(1, 0, -1, 0)^T\| = \sqrt{2}$.

6. The vector in W that is nearest to $v = (3, 6, 3, 4)^T$ is $p = \text{proj}_W(v) = \langle v, u_1 \rangle u_1 + \langle v, u_2 \rangle u_2 + \langle v, u_3 \rangle u_3 = 8u_1 + (-2)u_2 + 0u_3 = (3, 5, 3, 5)^T$. Further, $v - p = (0, 1, 0, -1)^T$ is orthogonal to every vector in W, and if $u_4 = (0, \frac{1}{\sqrt{2}}, 0, -\frac{1}{\sqrt{2}})^T$ is the normalization of $v - p$, then (u_1, u_2, u_3, u_4) is an orthonormal basis for $\mathcal{R}^4$.

7. The projection of any $x \in \mathcal{R}^4$ onto W is given by $\text{proj}_W(x) = Ax$, where

$$A = GG^T = (u_1, u_2, u_3)(u_1, u_2, u_3)^T = \begin{pmatrix} 1 & 0 & 0 & 0 \\ 0 & \frac{1}{2} & 0 & \frac{1}{2} \\ 0 & 0 & 1 & 0 \\ 0 & \frac{1}{2} & 0 & \frac{1}{2} \end{pmatrix}.$$

Thus, if $x = (3, 6, 3, 4)^T$, its projection onto W is computed as $Ax = (3, 5, 3, 5)^T$, consistent with the answer found in Example 6.

6.2 LINEAR TRANSFORMATIONS

Linear transformations are special types of functions that map one vector space to another. They are called "linear" because of their effect on the lines of a vector space, where by a "line" is meant a set of vectors w of the form $w = au + v$ where $u \neq 0$ and v are fixed vectors in the space and a varies over all values in the scalar field. Linear transformations carry lines in one vector space to lines or points in the other.

6.2.1 LINEAR TRANSFORMATIONS, RANGE, AND KERNEL

Definitions:

Let V and W be vector spaces over the same field F. A **linear transformation** is a function $T: V \to W$ satisfying $T(au + v) = aT(u) + T(v)$ for all $u, v \in V$ and $a \in F$.

The **range** R_T of a linear transformation T is $R_T = \{T(v) \mid v \in V\}$.

The **kernel** $\ker T$ of a linear transformation T is $\ker T = \{v \in V \mid T(v) = 0\}$.

The **rank** of T is the dimension of R_T. (R_T is a subspace of W by Fact 5.)

The **nullity** of T is the dimension of $\ker T$. ($\ker T$ is a subspace of V by Fact 5.)

A **linear operator** on V is a linear transformation from V to V.

Facts:

1. For any vector spaces V and W over F, the zero function $Z: V \to W$ defined by $Z(v) = 0$ for all $v \in V$ is a linear transformation from V to W.
2. For any vector space V over F, the identity function $I: V \to V$ defined by $I(v) = v$ for all $v \in V$ is a linear operator on V.
3. The following four statements are equivalent for a function $T: V \to W$:
   - T is a linear transformation;
   - $T(u + v) = T(u) + T(v)$ and $T(au) = aT(u)$ for all $u, v \in V$ and $a \in F$;
   - $T(au + bv) = aT(u) + bT(v)$ for all $u, v \in V$ and $a, b \in F$;
   - $T(\sum_{i=1}^{t} a_i v_i) = \sum_{i=1}^{t} a_i T(v_i)$ for all finite subsets $\{v_1, v_2, \ldots, v_t\} \subseteq V$ and scalars $a_i \in F$.
4. If $T: V \to W$ is a linear transformation, then:
   - $T(0) = 0$;
   - $T(-v) = -T(v)$ for all $v \in V$;
   - $T(u - v) = T(u) - T(v)$ for all $u, v \in V$.
5. If $T: V \to W$ is a linear transformation, then R_T is a subspace of W and $\ker T$ is a subspace of V.
6. If $T: V \to W$ is a linear transformation, then the rank of T plus the nullity of T equals the dimension of its domain: $\dim R_T + \dim(\ker T) = \dim V$.
7. If $T: V \to W$ is a linear transformation and if the vectors $\{v_1, v_2, \ldots, v_n\}$ span V, then $\{T(v_1), T(v_2), \ldots, T(v_n)\}$ span R_T.
8. If $T: V \to W$ is a linear transformation, then T is completely determined by its action on a basis for V. That is, if $\mathcal{B}$ is a basis for V and f is any function from $\mathcal{B}$ to W, then there exists a unique linear transformation T such that $T(v) = f(v)$ for all $v \in \mathcal{B}$.

9. A linear transformation $T: V \to W$ is one-to-one if and only if $\ker T = \{0\}$.

10. A linear transformation $T: V \to W$ is onto if and only if for *every* basis $\mathcal{B}$ of V, the set $\{T(v) \mid v \in \mathcal{B}\}$ spans W.

11. A linear transformation $T: V \to W$ is onto if and only if for *some* basis $\mathcal{B}$ of V, the set $\{T(v) \mid v \in \mathcal{B}\}$ spans W.

12. If $T: V \to W$ is a bijective linear transformation, then its inverse $T^{-1}: W \to V$ is also a bijective linear transformation.

13. For each fixed $m \times n$ matrix A over F, the function $T: F^{n \times 1} \to F^{m \times 1}$ defined by $T(x) = Ax$ is a linear transformation.

14. Every linear transformation $T: F^{n \times 1} \to F^{m \times 1}$ has the form $T(x) = Ax$ for some unique $m \times n$ matrix A over F.

15. The range R_T of the linear transformation $T(x) = Ax$ is equal to the column space of A, and $\ker T$ is equal to the null space of A. (See §6.1.2, §6.1.3.)

16. If T is a linear transformation from V to W and if $T(v_0) = w_0 \in R_T$, then the solution set S to the equation $T(v) = w_0$ is $S = \{v_0 + u \mid u \in \ker T\}$.

Examples:

1. The function $T: \mathcal{R}^{2 \times 1} \to \mathcal{R}^{2 \times 1}$ given by $T\begin{pmatrix} x_1 \\ x_2 \end{pmatrix} = \begin{pmatrix} x_1 - 3x_2 \\ -2x_1 + 6x_2 \end{pmatrix}$ is a linear transformation. It has the form $T(x) = Ax$, where $A = \begin{pmatrix} 1 & -3 \\ -2 & 6 \end{pmatrix}$. The kernel of T is $\{(3a, a)^T \mid a \in \mathcal{R}\}$ and the range of T is $\{(b, -2b)^T \mid b \in \mathcal{R}\}$.

2. For each fixed matrix $A \in F^{n \times n}$ the function $T: F^{n \times n} \to F^{n \times n}$ defined by $T(X) = AX - XA$ is a linear transformation whose kernel is the set of matrices commuting with A. Specifically, let $n = 2$, $F = \mathcal{R}$, and $A = \begin{pmatrix} 1 & -3 \\ -2 & 6 \end{pmatrix}$. Then $\dim \mathcal{R}^{2 \times 2} = 4$, and by computation $T\begin{bmatrix} \begin{pmatrix} 1 & 0 \\ 0 & 0 \end{pmatrix} \end{bmatrix} = \begin{pmatrix} 0 & 3 \\ -2 & 0 \end{pmatrix}$, $T\begin{bmatrix} \begin{pmatrix} 0 & 1 \\ 0 & 0 \end{pmatrix} \end{bmatrix} = \begin{pmatrix} 2 & -5 \\ 0 & -2 \end{pmatrix}$. Thus, $\dim R_T \geq 2$. Since both the identity matrix I and A itself are in $\ker T$, $\dim (\ker T) \geq 2$. By Fact 6, it follows that $\dim R_T = 2$ and $\dim (\ker T) = 2$. Therefore (I, A) forms a basis for $\ker T$, and the matrices $\begin{pmatrix} 0 & 3 \\ -2 & 0 \end{pmatrix}$ and $\begin{pmatrix} 2 & -5 \\ 0 & -2 \end{pmatrix}$ are a basis for R_T. From Fact 16, the solutions to $T(x) = \begin{pmatrix} 0 & 3 \\ -2 & 0 \end{pmatrix}$ are precisely the set of matrices of the form $\begin{pmatrix} 1 & 0 \\ 0 & 0 \end{pmatrix} + a\begin{pmatrix} 1 & 0 \\ 0 & 1 \end{pmatrix} + b\begin{pmatrix} 1 & -3 \\ -2 & 6 \end{pmatrix}$ with $a, b \in \mathcal{R}$.

3. The function $E(x_1, x_2, x_3, x_4) = (x_1, x_2, x_3, x_4, x_1 + x_3 + x_4, x_1 + x_2 + x_4, x_1 + x_2 + x_3)$, where $x_i \in \mathcal{Z}_2$, is a linear transformation important in coding theory. It represents an "encoding" of 4-bit binary vectors into 7-bit binary vectors ("codewords") before being sent over a "noisy" channel (§14.2). The kernel of the transformation consists of only the zero vector $0 = (0, 0, 0, 0)$, and so the transformation is one-to-one. The collection of codewords (that is, the range of E), is a 16-member, 4-dimensional subspace of $\mathcal{Z}_2^7$ having the special property that any two of its distinct members differ in at least three components. This means that if, during transmission of a codeword, an error is made in any single one of its components, then the error can be detected and corrected as there will be a unique codeword that differs from the received vector in a single component.

4. Continuing with Example 3, the linear transformation $D(z_1, z_2, z_3, z_4, z_5, z_6, z_7) = (z_1+z_3+z_4+z_5, z_1+z_2+z_4+z_6, z_1+z_2+z_3+z_7)$ is used in decoding the (binary) received vector z. This transformation has the special property that its kernel is precisely the set of codewords defined in Example 3. Thus, if $D(z) \neq 0$, then a transmission error has been made.

5. For $\mathcal{C}$ as a vector space over $\mathcal{R}$ and any $z_0 \in \mathcal{C}$, the function $T: \mathcal{C} \to \mathcal{C}$ defined by $T(z) = z_0 z$ is a linear operator; in particular, if $z_0 = \cos\theta + i\sin\theta$, then T is a *rotation* by the angle θ. ($T(z)$ is also a linear operator on $\mathcal{C}$ as a vector space over itself.)

6. For any fixed real-valued continuous function g on the interval $[a, b]$, the function T from the space $C[a,b]$ of continuous functions on $[a,b]$ to the space $D[a,b]$ of continuously differentiable functions on $[a,b]$ given by $T(f)(x) = \int_a^x g(t)f(t)dt$ is a linear transformation.

7. For the vector space V of functions $p: \mathcal{R} \to \mathcal{R}$ with continuous derivatives of all orders, the mapping $T: V \to V$ defined by $T(p) = p'' - 3p' + 2p$ (where p' and p'' are the first and second derivatives of p) is a linear transformation. Its kernel is the solution set to the homogeneous differential equation $p'' - 3p' + 2p = 0$: namely, $p(x) = Ae^x + Be^{2x}$, where $A, B \in \mathcal{R}$. Since $T(x^2) = 2 - 6x + 2x^2$, the set of all solutions to $T(p) = 2 - 6x + 2x^2$ is $x^2 + Ae^x + Be^{2x}$ (by Fact 16).

8. If v_0 is a fixed vector in a real inner product space V, then $T: V \to \mathcal{R}$ given by $T(v) = \langle v, v_0 \rangle$ is a linear transformation.

9. For W a subspace of the inner product space V, the projection proj_W of V onto W is a linear transformation. (See §6.1.4.)

6.2.2 VECTOR SPACES OF LINEAR TRANSFORMATIONS

Definitions:

If S and T are linear transformations from V to W, the **sum** (**addition**) of S and T is the function $S + T$ defined by $(S + T)(v) = S(v) + T(v)$ for all $v \in V$.

If T is a linear transformation from V to W, the **scalar product** (**scalar multiplication**) of $a \in F$ by T is the function aT defined by $(aT)(v) = aT(v)$ for all $v \in V$.

If $T: V \to W$ and $S: W \to U$ are linear transformations, then the **product** (**multiplication, composition**) of S and T is the function $S \circ T$ defined by $(S \circ T)(v) = S(T(v))$.

Note: Some writers use the notation vT to denote the image of v under the transformation T, in which case $T \circ S$ is used instead of $S \circ T$ to denote the product; that is, $v(T \circ S) = (vT)S$.

Facts:

1. The sum of two linear transformations from V to W is a linear transformation from V to W.

2. The product of a scalar and a linear transformation is a linear transformation.

3. If $T: V \to W$ and $S: W \to U$ are linear transformations, then their product $S \circ T$ is a linear transformation from V to U.

4. The set of linear transformations from V to W with the operations of addition and scalar multiplication forms a vector space over F. This vector space is denoted $L(V, W)$.

5. The set $L(V, V)$ of linear operators on V with the operations of addition, scalar multiplication, and multiplication forms an *algebra* with identity over F. Namely, $L(V, V)$ is a vector space over F and is a ring with identity under the addition and multiplication operations. In addition, $a(S \circ T) = (aS) \circ T = S \circ (aT)$ holds for all scalars $a \in F$ and all $S, T \in L(V, V)$. The identity mapping is the multiplicative identity of the algebra.

6. If $\dim V = n$ and $\dim W = m$, then $\dim L(V, W) = nm$.

Examples:

1. Consider $L(F^{n \times 1}, F^{m \times 1})$. If T and S are in $L(F^{n \times 1}, F^{m \times 1})$, then $T(x) = Ax$ and $S(x) = Bx$ for unique $m \times n$ matrices A and B over F. Then $(T + S)(x) = (A + B)x$, $(aT)(x) = aAx$, and in case $m = n$, $(T \circ S)(x) = ABx$.

2. Let $V = C[a, b]$ be the space of real-valued continuous functions on the interval $[a, b]$, and let T and S be linear operators defined by $T(f)(x) = \int_a^x e^{-t} f(t) dt$ and $S(f)(x) = \int_a^x e^t f(t) dt$. Then $(T + S)(f)(x) = \int_a^x (e^{-t} + e^t) f(t) dt$, $(cT)(f)(x) = \int_a^x ce^{-t} f(t) dt$, and $(T \circ S)(f)(x) = \int_a^x \int_a^t e^{s-t} f(s) ds dt$.

3. Let V be the real vector space of all functions $p: \mathcal{R} \to \mathcal{R}$ with continuous derivatives of all orders, and let D be the derivative function. Then $D: V \to V$ is a linear operator on V and so is a function such as $T = D^2 - 3D + 2I$ where $D^2 = D \circ D$ and I is the identity operator on V. The action of T on $p \in V$ is given by $T(p) = p'' - 3p' + 2p$.

6.2.3 MATRICES OF LINEAR TRANSFORMATIONS

Definitions:

If $T: V \to W$ is a linear transformation where $\dim V = n$, $\dim W = m$, and if $\mathcal{B} = (v_1, v_2, \ldots, v_n)$ and $\mathcal{B}' = (v_1', v_2', \ldots, v_m')$ are ordered bases for V and W, respectively, then the **matrix of T with respect to $\mathcal{B}$ and $\mathcal{B}'$** is the $m \times n$ matrix $[T]_{\mathcal{B}, \mathcal{B}'}$ whose jth column is $[T(v_j)]_{\mathcal{B}'}$, the coordinate vector (§6.1.3) of $T(v_j)$ with respect to $\mathcal{B}'$.

If $T: V \to V$ is a linear operator on V, then the **matrix of T with respect to $\mathcal{B}$** is the $n \times n$ matrix $[T]_{\mathcal{B}, \mathcal{B}}$ denoted simply as $[T]_{\mathcal{B}}$.

Facts:

Assume that T and S are linear transformations from V to W, $\mathcal{B}$ and $\mathcal{B}'$ are respective bases for V and W, and A and B are the matrices defined by $A = [T]_{\mathcal{B}, \mathcal{B}'}$ and $B = [S]_{\mathcal{B}, \mathcal{B}'}$.

1. $[T(v)]_{\mathcal{B}'} = [T]_{\mathcal{B}, \mathcal{B}'} [v]_{\mathcal{B}}$ for all $v \in V$; that is, if $y = [T(v)]_{\mathcal{B}'}$ and $x = [v]_{\mathcal{B}}$, then $y = Ax$.

2. $\ker T = \{x_1 v_1 + x_2 v_2 + \cdots + x_n v_n \mid (x_1, x_2, \ldots, x_n)^T \in NS(A)\}$, where $\mathcal{B} = (v_1, v_2, \ldots, v_n)$.

3. T is one-to-one if and only if $NS(A) = \{0\}$.

4. $R_T = \{y_1 v_1' + y_2 v_2' + \cdots + y_m v_m' \mid (y_1, y_2, \ldots, y_m)^T \in CS(A)\}$, where $\mathcal{B}' = (v_1', v_2', \ldots, v_m')$.

5. T is onto if and only if $CS(A) = F^{m \times 1}$.

6. T is bijective if and only if $m = n$ and A is invertible. In this case, $[T^{-1}]_{\mathcal{B}', \mathcal{B}} = A^{-1}$.

7. $[T+S]_{B,B'} = A+B$, $[aT]_{B,B'} = aA$ for all $a \in F$, and the mapping f from $L(V,W)$ to $F^{m \times n}$ defined by $f(T) = [T]_{B,B'}$ is an isomorphism.
8. If U is a vector space over F, B'' is a basis for U, and $R: W \to U$ is a linear transformation, then $[R \circ T]_{B,B''} = CA$ where $C = [R]_{B',B''}$; that is, $[R \circ T]_{B,B''} = [R]_{B',B''}[T]_{B,B'}$.
9. The algebra $L(V,V)$ is isomorphic to the matrix algebra $F^{n \times n}$.
10. If $I: V \to V$ is the identity mapping, then $[I]_{B,B} = [I]_B$ equals the identity matrix for any basis B.
11. If A is an $m \times n$ matrix over F with B and B' being arbitrary bases for V and W, respectively, then there exists a unique linear transformation $T: V \to W$ such that $A = [T]_{B,B'}$.
12. Linear transformations are used extensively in computer graphics. (See Example 5.) Further information can be found in [PoGe89].

Examples:

1. Consider $T: \mathcal{R}^{2 \times 1} \to \mathcal{R}^{2 \times 1}$ given by $T \begin{pmatrix} x_1 \\ x_2 \end{pmatrix} = \begin{pmatrix} x_1 - 3x_2 \\ -2x_1 + 6x_2 \end{pmatrix}$ and the bases $B = (v_1, v_2)$ and $B' = (v'_1, v'_2)$, where $v_1 = (1,0)^T$, $v_2 = (0,1)^T$ and $v'_1 = (1,1)^T$, $v'_2 = (2,1)^T$. Since
$$T(v_1) = (1,-2)^T = (-5)v'_1 + 3v'_2,$$
$$T(v_2) = (-3,6)^T = 15v'_1 + (-9)v'_2,$$
it follows that $[T(v_1)]_{B'} = (-5,3)^T$ and $[T(v_2)]_{B'} = (15,-9)^T$; hence, the matrix of T relative to B and B' is $[T]_{B,B'} = \begin{pmatrix} -5 & 15 \\ 3 & -9 \end{pmatrix}$. Similarly, $[T]_{B,B} = [T]_B = \begin{pmatrix} 1 & -3 \\ -2 & 6 \end{pmatrix}$, and $[T]_{B',B'} = [T]_{B'} = \begin{pmatrix} 10 & 5 \\ -6 & -3 \end{pmatrix}$.

2. Consider T of Example 1 where $A = [T]_{B,B'} = \begin{pmatrix} -5 & 15 \\ 3 & -9 \end{pmatrix}$. Since $NS(A) = \{(3a,a)^T \mid a \in \mathcal{R}\}$ and $CS(A) = \{(-5b, 3b)^T \mid b \in \mathcal{R}\}$, Fact 2 gives $\ker T = \{3av_1 + av_2 = (3a,a)^T \mid a \in \mathcal{R}\}$ and Fact 4 gives $R_T = \{(-5b)v'_1 + 3bv'_2 = (b,-2b)^T \mid b \in \mathcal{R}\}$. T is not one-to-one since $NS(A) \neq \{0\}$ and is not onto since $CS(A) \neq \mathcal{R}^{2 \times 1}$. (Any one of the three matrices found in Example 1 could have been used to determine $\ker T$ and R_T and to reach these same conclusions.)

3. Consider the linear operator on $\mathcal{R}^{2 \times 2}$ defined by $T(X) = AX - XA$ where $A = \begin{pmatrix} 1 & -3 \\ -2 & 6 \end{pmatrix}$, and let $B = (E_{11}, E_{12}, E_{21}, E_{22})$ be the standard basis. (Here, E_{ij} has a 1 in position (i,j) and 0s elsewhere.) Then
$$T(E_{11}) = AE_{11} - E_{11}A = \begin{pmatrix} 0 & 3 \\ -2 & 0 \end{pmatrix} = 0E_{11} + 3E_{12} + (-2)E_{21} + 0E_{22},$$
so $(0,3,-2,0)^T$ is the first column of $[T]_B$. Similar calculations yield
$$[T]_B = \begin{pmatrix} 0 & 2 & -3 & 0 \\ 3 & -5 & 0 & -3 \\ -2 & 0 & 5 & 2 \\ 0 & -2 & 3 & 0 \end{pmatrix}.$$
The null space of this 4×4 matrix is $\{(5a+b, 3a, 2a, b)^T \mid a,b \in \mathcal{R}\}$, so that those matrices X commuting with A (that is, in $\ker T$) have the form $X = \begin{pmatrix} 5a+b & 3a \\ 2a & b \end{pmatrix}$.

4. Consider $\mathcal{C}$ as a vector space over $\mathcal{R}$ and the rotation operator of §6.2.1 Example 5; namely, $T(z) = z_0 z$ where $z_0 = \cos\theta + i\sin\theta$. If $\mathcal{B}$ is the standard basis, $\mathcal{B} = (1, i)$, then the matrix of T relative to $\mathcal{B}$ is $[T]_\mathcal{B} = \begin{pmatrix} \cos\theta & -\sin\theta \\ \sin\theta & \cos\theta \end{pmatrix}$.

5. *Computer graphics*: The polygon in part (a) of the following figure can be rotated by applying the transformation T in Example 4 to its vertices $(-2, -2)$, $(1, -1)$, $(2, 1)$, $(-1, 3)$. The matrix of vertex coordinates is

$$X = \begin{pmatrix} -2 & 1 & 2 & -1 \\ -2 & -1 & 1 & 3 \end{pmatrix}.$$

For a rotation of $\frac{\pi}{3}$, the matrix of T is

$$A = \begin{pmatrix} \frac{1}{2} & -\frac{\sqrt{3}}{2} \\ \frac{\sqrt{3}}{2} & \frac{1}{2} \end{pmatrix}$$

and

$$AX \approx \begin{pmatrix} 0.732 & 1.366 & 0.134 & -3.098 \\ -2.732 & 0.366 & 2.232 & 0.634 \end{pmatrix},$$

giving the rotated polygon shown in part (b) of the following figure. To perform a "zoom in" operation, the original polygon can be rescaled by 50% by applying the transformation $S\begin{pmatrix} x \\ y \end{pmatrix} = \begin{pmatrix} 1.5x \\ 1.5y \end{pmatrix}$. Since the matrix for S relative to the standard basis is $D = \begin{pmatrix} 1.5 & 0 \\ 0 & 1.5 \end{pmatrix}$, the vertex coordinates X are transformed into $DX = \begin{pmatrix} -3 & 1.5 & 3 & -1.5 \\ -3 & -1.5 & 1.5 & 4.5 \end{pmatrix}$; see part (c) of the figure. Reflection through the x-axis would involve the transformation $R\begin{pmatrix} x \\ y \end{pmatrix} = \begin{pmatrix} x \\ -y \end{pmatrix}$, represented by the diagonal matrix $C = \begin{pmatrix} 1 & 0 \\ 0 & -1 \end{pmatrix}$. In computer graphics, the vertices of an object are actually given (x, y, z) coordinates and three-dimensional versions of the above transformations can be applied to move and reshape the object as well as render the scene when the user's viewpoint is changed.

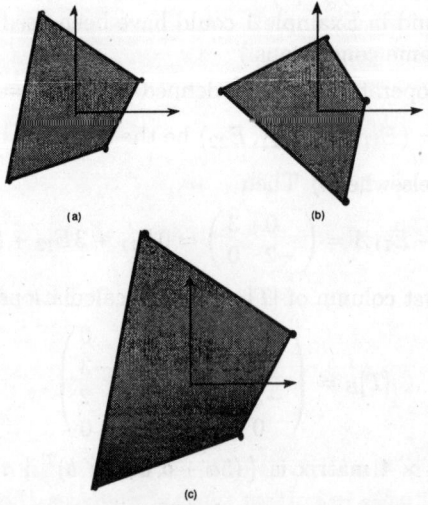

6.2.4 CHANGE OF BASIS

Definitions:

Let $B = (v_1, v_2, \ldots, v_n)$ and $B' = (v'_1, v'_2, \ldots, v'_n)$ be two ordered bases for V, and let I denote the identity mapping from V to V. The matrix $P = [I]_{B,B'}$ is the **transition matrix** from B to B'. It is also called the **change of basis matrix** from basis B to basis B'.

If A and B are two $n \times n$ matrices over a field F, then B is **similar** to A if there exists an invertible $n \times n$ matrix P over F such that $P^{-1}BP = A$.

Facts:

1. The transition matrix $P = [I]_{B,B'}$ is invertible; its inverse is $P^{-1} = [I]_{B',B}$.
2. If $x = [v]_B$ and $y = [v]_{B'}$, then $y = Px$ where $P = [I]_{B,B'}$.
3. When $B = B'$, the transition matrix $P = [I]_{B,B} = [I]_B$ is the $n \times n$ identity matrix.
4. If T is a linear operator on V with A and B the matrices of T relative to bases B and B', respectively, then B is similar to A. Specifically, $P^{-1}BP = A$ where $P = [I]_{B,B'}$.
5. If A and B are similar $n \times n$ matrices, then A and B represent the same linear operator T relative to suitably chosen bases. More specifically, suppose $P^{-1}BP = A$, $B = (v_1, v_2, \ldots, v_n)$ is any basis for V, and T is the unique linear transformation with $A = [T]_B$. Then $B = [T]_{B'}$ where $B' = (v'_1, v'_2, \ldots, v'_n)$ is the basis for V given by $v'_j = \sum_{i=1}^n p_{ij}^{-1} v_i$.

Examples:

1. Consider the $\mathcal{R}^{2 \times 1}$ bases $B = (v_1, v_2)$ and $B' = (v'_1, v'_2)$, where $v_1 = (1,0)^T$, $v_2 = (0,1)^T$ and $v'_1 = (1,1)^T$, $v'_2 = (2,1)^T$. Since $v_1 = (-1)v'_1 + v'_2$ and $v_2 = 2v'_1 + (-1)v'_2$, the transition matrix from B to B' is $P = [I]_{B,B'} = \begin{pmatrix} -1 & 2 \\ 1 & -1 \end{pmatrix}$, and its inverse $P^{-1} = \begin{pmatrix} 1 & 2 \\ 1 & 1 \end{pmatrix}$ is the transition matrix $[I]_{B',B}$. If $v = x_1 v_1 + x_2 v_2$ where $x_i \in \mathcal{R}$, then by Fact 2, $v = y_1 v'_1 + y_2 v'_2$ where $y_1 = (-1)x_1 + 2x_2$ and $y_2 = x_1 + (-1)x_2$.

2. Consider $T: \mathcal{R}^{2 \times 1} \to \mathcal{R}^{2 \times 1}$ given by $T \begin{pmatrix} x_1 \\ x_2 \end{pmatrix} = \begin{pmatrix} x_1 - 3x_2 \\ -2x_1 + 6x_2 \end{pmatrix}$, and the same bases B and B' specified in Example 1. The matrix of T with respect to B is $[T]_B = A = \begin{pmatrix} 1 & -3 \\ -2 & 6 \end{pmatrix}$ and the matrix of T with respect to B' is $[T]_{B'} = B = \begin{pmatrix} 10 & 5 \\ -6 & -3 \end{pmatrix}$. Moreover, A and B are similar; indeed, as Fact 4 shows, $A = P^{-1}BP$ where $P = \begin{pmatrix} -1 & 2 \\ 1 & -1 \end{pmatrix}$ is determined in Example 1.

6.3 MATRIX ALGEBRA

Matrices naturally arise in the analysis of linear systems and in representing discrete structures. This section studies important types of matrices, their properties, and methods for efficient matrix computation.

6.3.1 BASIC CONCEPTS AND SPECIAL MATRICES

Definitions:

The $m \times n$ matrix $A = (a_{ij})$ is a rectangular array of mn real or complex numbers a_{ij}, arranged into m **rows** and n **columns**.

The ith row of A, denoted $A(i,:)$, is the array $a_{i1}\ a_{i2}\ \ldots\ a_{in}$. The elements in the ith row can be regarded as a **row vector** $(a_{i1}, a_{i2}, \ldots, a_{in})$ in $\mathcal{R}^n$ or $\mathcal{C}^n$. The jth column of A, denoted $A(:,j)$, is the array

$$\begin{array}{c} a_{1j} \\ a_{2j} \\ \vdots \\ a_{mj} \end{array}$$

which can be identified with the **column vector** $(a_{1j}, a_{2j}, \ldots, a_{mj})^T$ (where the exponent T indicates the transpose).

A matrix is **sparse** if it has relatively few nonzero entries.

A **submatrix** of the matrix A contains the elements occurring in rows $i_1 < i_2 < \cdots < i_k$ and columns $j_1 < j_2 < \cdots < j_r$ of A. A **principal submatrix** of the matrix A contains the elements occurring in rows $i_1 < i_2 < \cdots < i_k$ and columns $i_1 < i_2 < \cdots < i_k$ of A. This principal submatrix has **order** k and is written $A[i_1, i_2, \ldots, i_k]$.

Two matrices A and B are **equal** if they are both $m \times n$ matrices with $a_{ij} = b_{ij}$ for all $i = 1, 2, \ldots, m$ and $j = 1, 2, \ldots, n$.

The **transpose** of the $m \times n$ matrix $A = (a_{ij})$ is the $n \times m$ matrix $A^T = (b_{ij})$ in which $b_{ij} = a_{ji}$.

The **Hermitian adjoint** of the $m \times n$ matrix $A = (a_{ij})$ is the $n \times m$ matrix $A^* = (b_{ij})$ in which b_{ij} is the complex conjugate of a_{ji}.

If $m = n$, the matrix $A = (a_{ij})$ is **square** with **diagonal** elements $a_{11}, a_{22}, \ldots, a_{nn}$. The **main diagonal** contains the diagonal elements of A. An **off-diagonal** element is any a_{ij} with $i \neq j$. The **trace** of A, tr A, is the sum of the diagonal elements of A.

Table 1 defines special types of square matrices.

Facts:

1. Triangular matrices arise in the solution of systems of linear equations (§6.4).

2. A tridiagonal matrix can be represented as follows, where the diagonal lines represent the (possibly) nonzero entries.

3. Tridiagonal matrices are particular types of sparse matrices. Such matrices arise in discretized versions of continuous problems, the solution of difference equations (§3.3, §3.4.4), and the solution of eigenvalue problems (§6.5).

Table 1 Special types of square matrices.

matrix	definition
identity	$I_n = (e_{ij})$ where $e_{ij} = \begin{cases} 1 & \text{if } i = j \\ 0 & \text{if } i \neq j \end{cases}$ ($n \times n$ matrix; each diagonal entry is 1; each off-diagonal entry is 0)
diagonal	$D = (d_{ij})$ where $d_{ij} = 0$ if $i \neq j$ (nonzero entries occur only on the main diagonal)
lower triangular	$L = (l_{ij})$ where $l_{ij} = 0$ if $j > i$ (nonzero entries occur only on or below the diagonal)
upper triangular	$U = (u_{ij})$ where $u_{ij} = 0$ if $j < i$ (nonzero entries occur only on or above the diagonal)
unit triangular	triangular matrix with all diagonal entries 1
tridiagonal	$A = (a_{ij})$ where $a_{ij} = 0$ if $|i - j| > 1$ (nonzero entries occur only on or immediately above or below the diagonal)
symmetric	real matrix A for which $A = A^T$
skew-symmetric	real matrix A for which $A = -A^T$
Hermitian	complex matrix A for which $A = A^*$
skew-Hermitian	complex matrix A for which $A = -A^*$

4. Sparse matrices frequently arise in the solution of large systems of linear equations (§6.4), since in many physical models a given variable typically interacts with relatively few others. Linear systems derived from sparse matrices require less storage space and can be solved more efficiently than those derived from a "dense" matrix.

5. Forming the transpose of a square matrix corresponds to "reflecting" the matrix elements with respect to the main diagonal.

6. Any skew-symmetric matrix A must have $a_{ii} = 0$ for all i.

7. Any Hermitian matrix A must have a_{ii} real for all i.

8. If A is real then $A^* = A^T$.

9. The columns of the identity matrix I_n are the standard basis vectors for $\mathcal{R}^n$ (§6.1.3).

10. Viewed as a linear transformation (§6.2), the identity matrix represents the identity transformation; that is, it leaves all vectors unchanged.

11. Viewed as linear transformations, diagonal matrices with positive diagonal entries leave the directions of the basis vectors unchanged, but alter the relative scale of the basis vectors.

Examples:

1. The 2×2 and 3×3 identity matrices are $I_2 = \begin{pmatrix} 1 & 0 \\ 0 & 1 \end{pmatrix}$ and $I_3 = \begin{pmatrix} 1 & 0 & 0 \\ 0 & 1 & 0 \\ 0 & 0 & 1 \end{pmatrix}$.

2. The matrix $A = \begin{pmatrix} 6 & 0 & 1 \\ 0 & 2 & 4 \\ 1 & 4 & 3 \end{pmatrix}$ is symmetric.

3. The matrix $A = \begin{pmatrix} 1 & 2 - 3i \\ 2 + 3i & -4 \end{pmatrix}$ is Hermitian.

4. A 2×2 diagonal matrix transforms the unit square in $\mathcal{R}^2$ into a rectangle with sides parallel to the coordinate axes. The following figure shows the effect of the diagonal matrix $\begin{pmatrix} 3 & 0 \\ 0 & 2 \end{pmatrix}$ on certain vectors and on the unit square in $\mathcal{R}^2$. The standard basis vectors $\{(1,0)^T, (0,1)^T\}$ have been transformed to $\{(3,0)^T, (0,2)^T\}$.

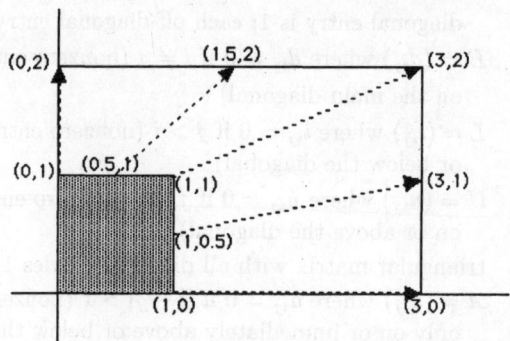

5. A 3×3 diagonal matrix transforms the unit cube into a rectangular parallelepiped.

6. The standard basis vectors are all eigenvectors of a diagonal matrix with the corresponding diagonal elements as their associated eigenvalues (§6.5).

6.3.2 OPERATIONS OF MATRIX ALGEBRA

Definitions:

The **scalar product** (**dot product**) of real vectors $x = (x_1, x_2, \ldots, x_n)$ and $y = (y_1, y_2, \ldots, y_n)$ is the number $x \cdot y = \sum_{i=1}^n x_i y_i$.

The $n \times n$ matrix A is **nonsingular** (**invertible**) if there exists an $n \times n$ matrix A^{-1} such that $AA^{-1} = A^{-1}A = I$. Any such matrix A^{-1} is an **inverse** of A.

An **orthogonal** matrix is a real square matrix A such that $A^T A = I$.

A **unitary** matrix is a complex square matrix A such that $A^* A = I$, where A^* is the Hermitian adjoint of A (§6.3.1).

A **positive definite** matrix is a real symmetric (or complex Hermitian) matrix A such that $x^* A x > 0$ for all $x \neq 0$.

The nonnegative **powers** of a square matrix A are given by $A^0 = I$, $A^n = AA^{n-1}$. If A is nonsingular then $A^{-n} = (A^{-1})^n$.

The following table defines various operations defined on matrices $A = (a_{ij})$ and $B = (b_{ij})$. (See Facts 1, 2, 5, 6 for restrictions on the sizes of the matrices.)

operation	definition
sum $A + B$	$A + B = (c_{ij})$ where $c_{ij} = a_{ij} + b_{ij}$
difference $A - B$	$A - B = (c_{ij})$ where $c_{ij} = a_{ij} - b_{ij}$
scalar multiple αA	$\alpha A = (c_{ij})$ where $c_{ij} = \alpha a_{ij}$
product AB	$AB = (c_{ij})$ where $c_{ij} = \sum_k a_{ik} b_{kj}$

Facts:

1. Matrices of different dimensions cannot be added or subtracted.
2. Square matrices of the same dimension can be multiplied.
3. Real or complex matrix addition satisfies the following properties:
   - commutative: $A + B = B + A$;
   - associative: $A + (B + C) = (A + B) + C$, $A(BC) = (AB)C$;
   - distributive: $A(B + C) = AB + AC$, $(A + B)C = AC + BC$;
   - $\alpha(A + B) = \alpha A + \alpha B$, $\alpha(AB) = (\alpha A)B = A(\alpha B)$ for all scalars α.
4. Matrix multiplication is not, in general, commutative — even when both products are defined. (See Example 3.)
5. The product AB is defined if and only if the number of columns of A equals the number of rows of B. That is, A must be an $m \times n$ matrix and B must be an $n \times p$ matrix.
6. The ijth element of the product $C = AB$ is the scalar product of row i of A and column j of B:

$$\begin{pmatrix} \overline{i \text{ th row}} \\ A \end{pmatrix} \begin{pmatrix} B \\ \end{pmatrix} = \begin{pmatrix} C \\ \end{pmatrix}$$

j th column $\qquad$ ij th element

7. Multiplication by identity matrices of the appropriate dimension leaves a matrix unchanged: if A is $m \times n$, then $I_m A = A I_n = A$.
8. Multiplication by diagonal matrices has the effect of scaling the rows or columns of a matrix. Pre-multiplication by a diagonal matrix scales the rows:

$$\begin{pmatrix} d_{11} & 0 & \cdots & 0 \\ 0 & d_{22} & \cdots & 0 \\ \vdots & \vdots & & \vdots \\ 0 & 0 & \cdots & d_{nn} \end{pmatrix} \begin{pmatrix} a_{11} & \cdots & a_{1p} \\ a_{21} & \cdots & a_{2p} \\ \vdots & & \vdots \\ a_{n1} & \cdots & a_{np} \end{pmatrix} = \begin{pmatrix} d_{11}a_{11} & \cdots & d_{11}a_{1p} \\ d_{22}a_{21} & \cdots & d_{22}a_{2p} \\ \vdots & & \vdots \\ d_{nn}a_{n1} & \cdots & d_{nn}a_{np} \end{pmatrix}.$$

Post-multiplication by a diagonal matrix scales the columns:

$$\begin{pmatrix} a_{11} & \cdots & a_{1n} \\ a_{21} & \cdots & a_{2n} \\ \vdots & & \vdots \\ a_{m1} & \cdots & a_{mn} \end{pmatrix} \begin{pmatrix} d_{11} & 0 & \cdots & 0 \\ 0 & d_{22} & \cdots & 0 \\ \vdots & \vdots & & \vdots \\ 0 & 0 & \cdots & d_{nn} \end{pmatrix} = \begin{pmatrix} d_{11}a_{11} & \cdots & d_{nn}a_{1n} \\ d_{11}a_{21} & \cdots & d_{nn}a_{2n} \\ \vdots & & \vdots \\ d_{11}a_{m1} & \cdots & d_{nn}a_{mn} \end{pmatrix}.$$

9. Any Hermitian matrix can be expressed as $A + iB$ where A is symmetric and B is skew-symmetric.
10. The inverse of a (nonsingular) matrix is unique.
11. If A is nonsingular, the solution of the system of linear equations (§6.4) $Ax = b$ is given by (but almost never computed by) $x = A^{-1}b$.

12. The product of nonsingular matrices A and B is nonsingular, with $(AB)^{-1} = B^{-1}A^{-1}$. Conversely, if A and B are square matrices with AB nonsingular, then A and B are nonsingular.

13. For a nonsingular matrix regarded as a linear transformation (§6.2), the inverse matrix represents the inverse transformation.

14. Sums of lower (upper) triangular matrices are lower (upper) triangular.

15. Products of lower (upper) triangular matrices are lower (upper) triangular.

16. A triangular matrix A is nonsingular if and only if $a_{ii} \neq 0$ for all i.

17. If a lower (upper) triangular matrix is nonsingular then its inverse is lower (upper) triangular.

18. *Properties of transpose:*
    - $(A^T)^T = A$;
    - $(A+B)^T = A^T + B^T$;
    - $(AB)^T = B^T A^T$;
    - AA^T and $A^T A$ are symmetric;
    - if A is nonsingular then so is A^T; moreover $(A^T)^{-1} = (A^{-1})^T$.

19. *Properties of Hermitian adjoint:*
    - $(A^*)^* = A$;
    - $(A+B)^* = A^* + B^*$;
    - $(AB)^* = B^* A^*$;
    - AA^* and A^*A are Hermitian;
    - if A is nonsingular, then so is A^*; moreover $(A^*)^{-1} = (A^{-1})^*$.

20. If A is orthogonal, then A is nonsingular and $A^{-1} = A^T$.

21. The rows (columns) of an orthogonal matrix are orthonormal with respect to the standard inner product on $\mathcal{R}^n$ (§6.1.4).

22. Products of orthogonal matrices are orthogonal.

23. If A is unitary, then A is nonsingular and $A^{-1} = A^*$.

24. The rows (columns) of a unitary matrix are orthonormal with respect to the standard inner product on $\mathcal{C}^n$ (§6.1.4).

25. Products of unitary matrices are unitary.

26. Positive definite matrices are nonsingular.

27. All eigenvalues (§6.5) of a positive definite matrix are positive.

28. Powers of a positive definite matrix are positive definite.

29. If A is skew-symmetric, then $I + A$ is positive definite.

30. If A is nonsingular, then $A^T A$ is positive definite.

Examples:

1. Let $A = \begin{pmatrix} 1 & 2 & 3 \\ 4 & 5 & 6 \end{pmatrix}$ and $B = \begin{pmatrix} 7 & 8 & 9 \\ 0 & 1 & 2 \end{pmatrix}$. Then $A + B = \begin{pmatrix} 8 & 10 & 12 \\ 4 & 6 & 8 \end{pmatrix}$ and $A - B = \begin{pmatrix} -6 & -6 & -6 \\ 4 & 4 & 4 \end{pmatrix}$.

2. The scalar product of the vectors $a = (1, 0, -1)$ and $b = (4, 3, 2)$ is $a \cdot b = (1)(4) + (0)(3) + (-1)(2) = 2$.

3. Let $A = \begin{pmatrix} 1 & 0 \\ 1 & 1 \end{pmatrix}$, $B = \begin{pmatrix} 2 & 3 \\ 4 & 1 \end{pmatrix}$, and $C = \begin{pmatrix} 1 & 1 & 2 \\ 0 & 2 & 3 \end{pmatrix}$. Then AB and BA are both defined with $AB = \begin{pmatrix} 1 & 0 \\ 1 & 1 \end{pmatrix} \begin{pmatrix} 2 & 3 \\ 4 & 1 \end{pmatrix} = \begin{pmatrix} 2 & 3 \\ 6 & 4 \end{pmatrix}$, whereas $BA = \begin{pmatrix} 2 & 3 \\ 4 & 1 \end{pmatrix} \begin{pmatrix} 1 & 0 \\ 1 & 1 \end{pmatrix} = \begin{pmatrix} 5 & 3 \\ 5 & 1 \end{pmatrix}$. Also, AC is defined but CA is not defined.

4. The matrices A, B of Example 1 cannot be multiplied since A has 3 columns and B has 2 rows; see Fact 5. However, all the products $A^T B$, AB^T, $B^T A$, BA^T exist:
$A^T B = \begin{pmatrix} 1 & 4 \\ 2 & 5 \\ 3 & 6 \end{pmatrix} \begin{pmatrix} 7 & 8 & 9 \\ 0 & 1 & 2 \end{pmatrix} = \begin{pmatrix} 7 & 12 & 17 \\ 14 & 21 & 28 \\ 21 & 30 & 39 \end{pmatrix}$, $AB^T = \begin{pmatrix} 1 & 2 & 3 \\ 4 & 5 & 6 \end{pmatrix} \begin{pmatrix} 7 & 0 \\ 8 & 1 \\ 9 & 2 \end{pmatrix} = \begin{pmatrix} 50 & 8 \\ 122 & 17 \end{pmatrix}$, $B^T A = \begin{pmatrix} 7 & 14 & 21 \\ 12 & 21 & 30 \\ 17 & 28 & 39 \end{pmatrix}$, $BA^T = \begin{pmatrix} 50 & 122 \\ 8 & 17 \end{pmatrix}$. Note that $(B^T A)^T = A^T B$, as guaranteed by Fact 18.

5. *Multiplication by a diagonal matrix:* $\begin{pmatrix} 3 & 0 \\ 0 & 2 \end{pmatrix} \begin{pmatrix} 1 & 2 & 3 \\ 4 & 5 & 6 \end{pmatrix} = \begin{pmatrix} 3 & 6 & 9 \\ 8 & 10 & 12 \end{pmatrix}$ and $\begin{pmatrix} 1 & 2 & 3 \\ 4 & 5 & 6 \end{pmatrix} \begin{pmatrix} 2 & 0 & 0 \\ 0 & 3 & 0 \\ 0 & 0 & 1 \end{pmatrix} = \begin{pmatrix} 2 & 6 & 3 \\ 8 & 15 & 6 \end{pmatrix}$.

6. The 2×2 matrix $A = \begin{pmatrix} a & b \\ c & d \end{pmatrix}$ is nonsingular if $\Delta = ad - bc \neq 0$; in this case $A^{-1} = \frac{1}{\Delta} \begin{pmatrix} d & -b \\ -c & a \end{pmatrix}$.

7. The matrix $A = \frac{1}{9} \begin{pmatrix} 4 & 8 & 1 \\ 7 & -4 & 4 \\ -4 & 1 & 8 \end{pmatrix}$ is orthogonal.

8. If $A = \frac{1}{2} \begin{pmatrix} 1 & -i & -1+i \\ i & 1 & 1+i \\ 1+i & -1+i & 0 \end{pmatrix}$ then $A^* = \frac{1}{2} \begin{pmatrix} 1 & -i & 1-i \\ i & 1 & -1-i \\ -1-i & 1-i & 0 \end{pmatrix}$.
Since $A^* A = I$ the matrix A is unitary.

9. Every 2×2 orthogonal matrix Q can be written as $Q = \begin{pmatrix} \cos\theta & -\sin\theta \\ \sin\theta & \cos\theta \end{pmatrix}$ for some real θ. Geometrically, the matrix Q effects a counterclockwise rotation by the angle θ.

10. For the matrix Q in Example 9, $Q^2 = \begin{pmatrix} \cos^2\theta - \sin^2\theta & -2\sin\theta\cos\theta \\ 2\sin\theta\cos\theta & \cos^2\theta - \sin^2\theta \end{pmatrix}$. Since this must be the same as a rotation by an angle of 2θ, then $Q^2 = \begin{pmatrix} \cos 2\theta & -\sin 2\theta \\ \sin 2\theta & \cos 2\theta \end{pmatrix}$. Equating these two expressions for Q^2 gives the double angle formulas of trigonometry.

11. The matrix $\begin{pmatrix} 4 & 2i & -3+i \\ -2i & -8 & 6+3i \\ -3-i & 6-3i & 5 \end{pmatrix}$ is Hermitian. It can be written as $A + Bi = \begin{pmatrix} 4 & 0 & -3 \\ 0 & -8 & 6 \\ -3 & 6 & 5 \end{pmatrix} + \begin{pmatrix} 0 & 2i & i \\ -2i & 0 & 3i \\ -i & -3i & 0 \end{pmatrix}$ where A is symmetric and B is skew-symmetric. (See Fact 9.)

Algorithm 1: Basic matrix multiplication.

input: $m \times n$ matrix A, $n \times p$ matrix B
output: $m \times p$ matrix $C = AB$

for $i := 1$ to m do
  for $j := 1$ to p do
    $C(i,j) := 0$
    for $k := 1$ to n do
      $C(i,j) := C(i,j) + A(i,k)B(k,j)$

6.3.3 FAST MULTIPLICATION OF MATRICES

A variety of methods have been devised to multiply matrices more efficiently than by simply using the definition in §6.3.2. This section presents alternative methods for carrying out matrix multiplication.

Definitions:

The **shift left operation** $\text{shL}(A(i,:),k)$ rotates elements of row i in matrix A exactly k places to the left, where data shifted off the left side of the matrix are wrapped around to the right side.

The **shift up operation** $\text{shU}(B(:,j),k)$ rotates elements of column j in matrix B exactly k places up, where data shifted off the top of the matrix are wrapped around to the bottom.

These operations can also be applied simultaneously to every row of A or every column of B, denoted $\text{shL}(A,k)$ and $\text{shU}(B,k)$ respectively.

Facts:

1. The basic definition given in §6.3.2 can be used to multiply the $m \times n$ matrix A and the $n \times p$ matrix B. The associated algorithm (Algorithm 1) requires $O(mnp)$ operations (additions and multiplications of individual elements).

2. *Matrix multiplication in scalar product form:* Algorithm 1 can be rewritten in terms of the scalar product operation, giving Algorithm 2.

3. Algorithm 2 is well-suited for fast multiplication on computers designed for efficient scalar product operations. It requires $O(mp)$ scalar products.

4. *Matrix multiplication in linear combination form:* Algorithm 3 carries out matrix multiplication by taking a linear combination of columns of A to obtain each column of the product.

5. The inner loop of Algorithm 3 performs a "vector + scalar × vector" operation, well-suited to a vector computer using efficiently pipelined arithmetic processing.

Algorithm 2: Scalar product form of matrix multiplication.

input: $m \times n$ matrix A, $n \times p$ matrix B
output: $m \times p$ matrix $C = AB$

for $i := 1$ to m do
  for $j := 1$ to p do
    $C(i,j) := A(i,:) \cdot B(:,j)$

Algorithm 3: Column linear combination form of matrix multiplication.

input: $m \times n$ matrix A, $n \times p$ matrix B
output: $m \times p$ matrix $C = AB$

for $j := 1$ to p do
  $C(:,j) := 0$
  for $k := 1$ to n do
    $C(:,j) := C(:,j) + B(k,j)A(:,k)$

---

6. Algorithm 3 is often used for fast general matrix multiplication on vector machines since it is based on a natural vector operation. If these vector operations can be performed on all elements simultaneously, then $O(np)$ vector operations are needed.

7. Access to matrix elements in Algorithm 3 is by column. There are other rearrangements of the algorithm which access matrix information by row.

8. *Fast multiplication on array processors:* Algorithm 4 multiplies two $n \times n$ (or smaller dimension) matrices on a computer with an $n \times n$ array of processors. It uses various shift operations on the arrays and the array-multiplication operation (∗) of elementwise multiplication.

9. At each step Algorithm 4 shifts A one place to the left and shifts B one place up so that components of the array product are correct new terms for the corresponding elements of $C = AB$. Each matrix is preshifted so the first step complies with this requirement.

10. Two $n \times n$ matrices can be multiplied in $O(n)$ time using Algorithm 4 on an array processor.

11. *The Strassen algorithm:* Algorithm 5 recursively carries out matrix multiplication for $n \times n$ matrices A and B where $n = 2^k$. The basis of Strassen's algorithm is partitioning the two factors into square blocks with dimension half that of the original matrices.

12. Strassen's algorithm ultimately requires the fast multiplication of 2×2 matrices (Algorithm 6).

13. Algorithm 6 multiplies two 2×2 matrices using only 7 multiplications and 18 additions instead of the normal 8 multiplications and 4 additions. For most modern computers saving one multiplication at the cost of 14 extra additions would not represent a gain.

14. Strassen's algorithm can be extended to $n \times n$ matrices where n is not a power of 2. The general algorithm requires $O(n^{\log_2 7}) \approx O(n^{2.807})$ multiplications. Details of this algorithm and its efficiency can be found in [GoVa96].

Algorithm 4: Array processor matrix multiplication.

input: $n \times n$ matrices A, B
output: $n \times n$ matrix $C = AB$
{Preshift the matrix arrays}
for $i := 1$ **to** n **do**
    shL$(A(i,:), i-1)$ {Shift ith row $i-1$ places left}
    shU$(B(:,i), i-1)$ {Shift ith column $i-1$ places up}

$C := 0$ {Initialize product array}
for $k := 1$ **to** n **do**
    $C := C + A * B$
    shL$(A, 1)$
    shU$(B, 1)$

Algorithm 5: Strassen's algorithm for $2^k \times 2^k$ matrices.

procedure $Strassen(A, B)$
input: $2^k \times 2^k$ matrices A, B
output: $2^k \times 2^k$ matrix $C = AB$
if $k = 1$ **then** use Algorithm 6
else
    partition A, B into 4 $2^{k-1} \times 2^{k-1}$ blocks $A = \begin{pmatrix} A_{11} & A_{12} \\ A_{21} & A_{22} \end{pmatrix}, B = \begin{pmatrix} B_{11} & B_{12} \\ B_{21} & B_{22} \end{pmatrix}$
    $P := Strassen((A_{11} + A_{22}), (B_{11} + B_{22}))$
    $Q := Strassen((A_{21} + A_{22}), B_{11}); \quad R := Strassen(A_{11}, (B_{12} - B_{22}))$
    $S := Strassen(A_{22}, (B_{21} - B_{11})); \quad T := Strassen((A_{11} + A_{12}), B_{22})$
    $U := Strassen((A_{21} - A_{11}), (B_{11} + B_{12}))$
    $V := Strassen((A_{12} - A_{22}), (B_{21} + B_{22}))$
    $C_{11} := P+S-T+V; \quad C_{12} := R+T; \quad C_{21} := Q+S; \quad C_{22} := P-Q+R+U$
end
$C := \begin{pmatrix} C_{11} & C_{12} \\ C_{21} & C_{22} \end{pmatrix}$

Algorithm 6: Strassen's algorithm for 2×2 matrices.

input: 2×2 matrices A, B
output: 2×2 matrix $C = AB$
$p := (a_{11} + a_{22})(b_{11} + b_{22}); \; q := (a_{21} + a_{22})b_{11}; \; r := a_{11}(b_{12} - b_{22})$
$s := a_{22}(b_{21} - b_{11}); \; t := (a_{11} + a_{12})b_{22}; \; u := (a_{21} - a_{11})(b_{11} + b_{12})$
$v := (a_{12} - a_{22})(b_{21} + b_{22})$
$c_{11} := p+s-t+v; \; c_{12} := r+t; \; c_{21} := q+s; \; c_{22} := p-q+r+u$

Examples:

1. This example illustrates Algorithm 4 for 4×4 array matrix multiplication. The preshift and the first array multiplication yield the arrays:

$$\begin{array}{cccc} a_{11} & a_{12} & a_{13} & a_{14} \\ a_{22} & a_{23} & a_{24} & a_{21} \\ a_{33} & a_{34} & a_{31} & a_{32} \\ a_{44} & a_{41} & a_{42} & a_{43} \end{array} \quad \begin{array}{cccc} b_{11} & b_{22} & b_{33} & b_{44} \\ b_{21} & b_{32} & b_{43} & b_{14} \\ b_{31} & b_{42} & b_{13} & b_{24} \\ b_{41} & b_{12} & b_{23} & b_{34} \end{array} \quad \begin{array}{cccc} a_{11}b_{11} & a_{12}b_{22} & a_{13}b_{33} & a_{14}b_{44} \\ a_{22}b_{21} & a_{23}b_{32} & a_{24}b_{43} & a_{21}b_{14} \\ a_{33}b_{31} & a_{34}b_{42} & a_{31}b_{13} & a_{32}b_{24} \\ a_{44}b_{41} & a_{41}b_{12} & a_{42}b_{23} & a_{43}b_{34} \end{array}$$

The next shifts and multiply-accumulate operation produce:

$$\begin{array}{cccc} a_{12} & a_{13} & a_{14} & a_{11} \\ a_{23} & a_{24} & a_{21} & a_{22} \\ a_{34} & a_{31} & a_{32} & a_{33} \\ a_{41} & a_{42} & a_{43} & a_{44} \end{array} \quad \begin{array}{cccc} b_{21} & b_{32} & b_{43} & b_{14} \\ b_{31} & b_{42} & b_{13} & b_{24} \\ b_{41} & b_{12} & b_{23} & b_{34} \\ b_{11} & b_{22} & b_{33} & b_{44} \end{array}$$

$$\begin{array}{cccc} a_{11}b_{11} + a_{12}b_{21} & a_{12}b_{22} + a_{13}b_{32} & a_{13}b_{33} + a_{14}b_{43} & a_{14}b_{44} + a_{11}b_{14} \\ a_{22}b_{21} + a_{23}b_{31} & a_{23}b_{32} + a_{24}b_{42} & a_{24}b_{43} + a_{21}b_{13} & a_{21}b_{14} + a_{22}b_{24} \\ a_{33}b_{31} + a_{34}b_{41} & a_{34}b_{42} + a_{31}b_{12} & a_{31}b_{13} + a_{32}b_{23} & a_{32}b_{24} + a_{33}b_{34} \\ a_{44}b_{41} + a_{41}b_{11} & a_{41}b_{12} + a_{42}b_{22} & a_{42}b_{23} + a_{43}b_{33} & a_{43}b_{34} + a_{44}b_{44} \end{array}$$

At subsequent stages the remaining terms get added in to the appropriate elements of the product matrix. The total cost of matrix multiplication is therefore reduced to n parallel multiply-accumulate operations plus some communication costs which for a typical distributed memory array processor are generally small.

2. Algorithm 6 is illustrated using the matrices $A = \begin{pmatrix} 3 & 4 \\ -1 & 2 \end{pmatrix}, B = \begin{pmatrix} 7 & 3 \\ 1 & -3 \end{pmatrix}$. Then

$p = 5 \cdot 4 = 20, \quad q = 1 \cdot 7 = 7, \quad r = 3 \cdot 6 = 18, \quad s = 3 \cdot (-6) = -12,$
$t = 7 \cdot (-3) = -21, \quad u = (-4) \cdot 10 = -40, \quad v = 2 \cdot (-2) = -4,$

giving the following elements of $C = AB$: $c_{11} = 20 - 12 + 21 - 4 = 25$, $c_{12} = 18 - 21 = -3$, $c_{21} = 7 - 12 = -5$, $c_{22} = 20 - 7 + 18 - 40 = -9$.

6.3.4 DETERMINANTS

Definitions:

For an $n \times n$ matrix A with $n > 1$, A_{ij} denotes the $(n-1) \times (n-1)$ matrix obtained by deleting row i and column j from A.

The **determinant** $\det A$ of an $n \times n$ matrix A can be defined recursively:
- if $A = (a)$ is a 1×1 matrix, then $\det A = a$;
- if $n > 1$, then $\det A = \sum_{j=1}^{n} (-1)^{j+1} a_{1j} \det A_{1j}$.

A **minor** of a matrix is the determinant of a square submatrix of the given matrix. A **principal minor** is the determinant of a principal submatrix.

Notation: The determinant of $A = (a_{ij})$ is commonly written using vertical bars:

$$\det A = |A| = \begin{vmatrix} a_{11} & a_{12} & \cdots & a_{1n} \\ a_{21} & a_{22} & \cdots & a_{2n} \\ \vdots & \vdots & & \vdots \\ a_{n1} & a_{n2} & \cdots & a_{nn} \end{vmatrix}$$

Facts:

1. *Laplace expansion*: For any r,
$$\det A = \sum_{j=1}^{n}(-1)^{r+j}a_{rj}\det A_{rj} = \sum_{i=1}^{n}(-1)^{i+r}a_{ir}\det A_{ir}.$$

2. If $A = (a_{ij})$ is $n \times n$, then $\det A = \sum_{\sigma \in S_n} \operatorname{sgn}(\sigma)\, a_{1\sigma(1)}a_{2\sigma(2)}\cdots a_{n\sigma(n)}$. Here S_n is the set of all permutations on $\{1, 2, \ldots, n\}$, and $\operatorname{sgn}(\sigma)$ equals 1 if σ is even and -1 if σ is odd (§5.3.1).

3. $\det \begin{pmatrix} a & b \\ c & d \end{pmatrix} = \begin{vmatrix} a & b \\ c & d \end{vmatrix} = ad - bc.$

4. $\det \begin{pmatrix} a & b & c \\ d & e & f \\ g & h & i \end{pmatrix} = \begin{vmatrix} a & b & c \\ d & e & f \\ g & h & i \end{vmatrix} = aei + bfg + cdh - afh - bdi - ceg.$

5. $\det AB = \det A \det B = \det BA$ for all $n \times n$ matrices A, B.

6. $\det A^T = \det A$ for all $n \times n$ matrices A.

7. $\det \alpha A = \alpha^n \det A$ for all $n \times n$ matrices A and all scalars α.

8. $\det I = 1$.

9. If A has two identical rows (or two identical columns), then $\det A = 0$.

10. Interchanging two rows (or two columns) of a matrix changes the sign of the determinant.

11. Multiplying one row (or column) of a matrix by a scalar multiplies its determinant by that same scalar.

12. Adding a multiple of one row (column) to another row (column) leaves the value of the determinant unchanged.

13. If $D = (d_{ij})$ is an $n \times n$ diagonal matrix, then $\det D = d_{11}d_{22}\cdots d_{nn}$.

14. If $T = (t_{ij})$ is an $n \times n$ triangular matrix, then $\det T = t_{11}t_{22}\cdots t_{nn}$.

15. If A and D are square matrices, then $\det \begin{pmatrix} A & B \\ 0 & D \end{pmatrix} = \det A \det D = \det \begin{pmatrix} A & 0 \\ C & D \end{pmatrix}$.

16. A is nonsingular if and only if $\det A \neq 0$.

17. If A is nonsingular then $\det(A^{-1}) = \dfrac{1}{\det A}$.

18. If A and D are nonsingular, then $\det \begin{pmatrix} A & B \\ C & D \end{pmatrix} = \det A \det(D - CA^{-1}B) = \det D \det(A - BD^{-1}C)$.

19. The determinant of a Hermitian matrix (§6.3.1) is real.

20. The determinant of a skew-symmetric matrix (§6.3.1) of odd size is zero.

21. The determinant of an orthogonal matrix (§6.3.1) is ± 1.

22. The $n \times n$ symmetric (or Hermitian) matrix A is positive definite if and only if all its leading principal submatrices $A[1], A[1,2], \ldots, A[1,2,\ldots,n]$ have positive determinant.

23. The $n \times n$ Vandermonde matrix
$$\begin{pmatrix} 1 & x_1 & \cdots & x_1^{n-1} \\ 1 & x_2 & \cdots & x_2^{n-1} \\ \vdots & \vdots & & \vdots \\ 1 & x_n & \cdots & x_n^{n-1} \end{pmatrix}$$
has determinant $\prod_{i<j}(x_j - x_i)$.

24. If the $n \times n$ matrix $A = (a_{ij})$ has diagonal elements $a_{ii} = x$ and off-diagonal elements $a_{ij} = y$, then $\det A = (x-y)^{n-1}(x-y+ny)$.

25. The equation of the straight line through points (a_1, b_1) and (a_2, b_2) is given by
$$\begin{vmatrix} x & y & 1 \\ a_1 & b_1 & 1 \\ a_2 & b_2 & 1 \end{vmatrix} = 0.$$

26. The equation of the circle through points (a_1, b_1), (a_2, b_2), (a_3, b_3) is given by
$$\begin{vmatrix} x^2+y^2 & x & y & 1 \\ a_1^2+b_1^2 & a_1 & b_1 & 1 \\ a_2^2+b_2^2 & a_2 & b_2 & 1 \\ a_3^2+b_3^2 & a_3 & b_3 & 1 \end{vmatrix} = 0.$$

27. If the three points (a_1, b_1), (a_2, b_2), (a_3, b_3) are listed in counterclockwise order, then the area of the triangle they form is given by
$$\frac{1}{2} \begin{vmatrix} a_1 & b_1 & 1 \\ a_2 & b_2 & 1 \\ a_3 & b_3 & 1 \end{vmatrix}.$$

28. The parallelepiped $P = \{\alpha_1 a_1 + \alpha_2 a_2 + \cdots + \alpha_n a_n \mid 0 \leq \alpha_i \leq 1\}$ spanned by the vectors $a_1, a_2, \ldots, a_n$ has volume $|\det A|$, where A has columns $a_1, a_2, \ldots, a_n$.

29. *Computation*: The determinant is (almost) never computed from the definition or from Fact 2. Instead it is calculated using Facts 12 and 14. (See Example 1.)

Examples:

1. Determinants can be calculated by using row operations to create a triangular matrix, and then applying Fact 14:
$$\det \begin{pmatrix} -1 & 2 & 1 \\ 0 & 5 & 2 \\ 3 & 4 & 3 \end{pmatrix} = \det \begin{pmatrix} -1 & 2 & 1 \\ 0 & 5 & 2 \\ 0 & 10 & 6 \end{pmatrix} = \det \begin{pmatrix} -1 & 2 & 1 \\ 0 & 5 & 2 \\ 0 & 0 & 2 \end{pmatrix} = -10.$$
Here the second matrix is obtained from the first by adding 3 times row 1 to row 3; the third matrix is obtained from the second by adding -2 times row 2 to row 3.

2. Determinants can be calculated by using row and column interchanges to obtain a form with exploitable zeros:
$$\det \begin{pmatrix} 4 & 5 & 1 & 6 \\ 0 & 6 & 0 & 3 \\ 0 & 5 & 0 & 2 \\ 3 & 3 & 2 & 4 \end{pmatrix} = -\det \begin{pmatrix} 4 & 1 & 5 & 6 \\ 0 & 0 & 6 & 3 \\ 0 & 0 & 5 & 2 \\ 3 & 2 & 3 & 4 \end{pmatrix} = \det \begin{pmatrix} 4 & 1 & 5 & 6 \\ 3 & 2 & 3 & 4 \\ 0 & 0 & 5 & 2 \\ 0 & 0 & 6 & 3 \end{pmatrix}.$$
Here the second matrix is obtained from the first by interchanging columns 2 and 3; the third matrix is obtained from the second by interchanging rows 2 and 4. The third matrix has block triangular form, with diagonal blocks $A = \begin{pmatrix} 4 & 1 \\ 3 & 2 \end{pmatrix}$ and $D = \begin{pmatrix} 5 & 2 \\ 6 & 3 \end{pmatrix}$. By Fact 15, the original determinant equals $\det A \det D = 5 \cdot 3 = 15$.

3. The symmetric matrix $A = \begin{pmatrix} 3 & 1 & 0 \\ 1 & 5 & 3 \\ 0 & 3 & 4 \end{pmatrix}$ is positive definite, since its leading principal minors (Fact 22) are positive: $\det(3) = 3 > 0$, $\det \begin{pmatrix} 3 & 1 \\ 1 & 5 \end{pmatrix} = 14 > 0$, and (by Fact 1) $\det A = 3 \det \begin{pmatrix} 5 & 3 \\ 3 & 4 \end{pmatrix} - \det \begin{pmatrix} 1 & 3 \\ 0 & 4 \end{pmatrix} = 3 \cdot 11 - 4 = 29 > 0$.

4. The equation of the line through points $(1,3)$ and $(4,5)$ can be found using Fact 25:
$$\begin{vmatrix} x & y & 1 \\ 1 & 3 & 1 \\ 4 & 5 & 1 \end{vmatrix} = \begin{vmatrix} x & y & 1 \\ 1 & 3 & 1 \\ 0 & -7 & -3 \end{vmatrix} = \begin{vmatrix} x & y-\frac{7}{3} & 1 \\ 1 & \frac{2}{3} & 1 \\ 0 & 0 & -3 \end{vmatrix} = (\tfrac{2}{3}x - y + \tfrac{7}{3})(-3) = 0,$$
giving $\tfrac{2}{3}x - y + \tfrac{7}{3} = 0$ or $y = \tfrac{2}{3}x + \tfrac{7}{3}$.

5. By Fact 27, the area of the triangle formed by the points $(0,0)$, $(1,3)$, and $(4,5)$ is
$$\tfrac{1}{2}\begin{vmatrix} 0 & 0 & 1 \\ 4 & 5 & 1 \\ 1 & 3 & 1 \end{vmatrix} = \tfrac{1}{2}\begin{vmatrix} 4 & 5 \\ 1 & 3 \end{vmatrix} = \tfrac{7}{2}.$$

6. *Cayley's formula*: The determinant of the $(n-1) \times (n-1)$ matrix
$$T_n = \begin{pmatrix} n-1 & -1 & \cdots & -1 \\ -1 & n-1 & \cdots & -1 \\ \vdots & \vdots & & \vdots \\ -1 & -1 & \cdots & n-1 \end{pmatrix}$$
counts the number of spanning trees of a complete graph. (See §9.2.2.) Using Fact 24, $\det T_n = n^{n-2}[n - (n-1)] = n^{n-2}$.

6.3.5 RANK

Definition:

The **rank** of an $m \times n$ matrix A, written rank A, is the size of the largest square nonsingular submatrix of A.

Facts:

1. rank A = rank A^T.
2. The rank of A equals the maximum number of linearly independent rows or linearly independent columns in A.
3. rank $(A+B) \leq$ rank A + rank B.
4. rank $AB \leq \min\{\text{rank } A, \text{rank } B\}$.
5. If A is nonsingular then rank AB = rank B and rank CA = rank C.
6. rank $A = \dim CS(A)$, where $CS(A)$ is the column space of A and $\dim V$ denotes the dimension of the vector space V. (See §6.1.3.)
7. rank $A = \dim RS(A)$, where $RS(A)$ is the row space of A. (See §6.1.3.)
8. An $n \times n$ matrix A is nonsingular if and only if rank $A = n$.
9. Every matrix of rank r can be written as a sum of r matrices of rank 1.
10. If a and b are nonzero $n \times 1$ vectors, then ab^T is an $n \times n$ matrix of rank 1.
11. The rank of a matrix is not always easy to compute. In the absence of severe roundoff errors, it can be obtained by counting the number of nonzero rows at the end of the Gaussian elimination procedure (§6.4.2).

12. The rank of a matrix is not always easy to compute. In the absence of severe roundoff errors, it can be obtained by counting the number of nonzero rows at the end of the Gaussian elimination procedure (§6.4.2).

13. *System of linear equations*: Consider the system $Ax = b$, where A is $m \times n$. Let $A_b = (A:b)$ denote the $m \times (n+1)$ matrix whose $(n+1)$st column is the vector b. Then the system $Ax = b$ has
   - a unique solution $\Leftrightarrow$ rank A = rank $A_b = n$;
   - infinitely many solutions $\Leftrightarrow$ rank A = rank $A_b < n$;
   - no solution $\Leftrightarrow$ rank A < rank A_b.

Examples:

1. The matrix $A = \begin{pmatrix} 1 & -1 & 2 \\ 3 & 4 & -1 \\ 5 & 2 & 3 \end{pmatrix}$ is singular since $\det A = 0$. However, the submatrix $A[1,2] = \begin{pmatrix} 1 & -1 \\ 3 & 4 \end{pmatrix}$ has determinant 7 and so is nonsingular, showing that rank $A = 2$. The matrix A has two linearly independent rows: row $3 = 2 \times$ (row 1) + (row 2). Likewise, it has two linearly independent columns: column 3 = (column 1) − (column 2). This again confirms (by Fact 2) that rank $A = 2$.

2. Consider the system of equations $Ax = b$, where A is the matrix in Example 1 and $b = (0, 7, 7)^T$. Since rank A = rank $A_b = 2 < 3$, this system has infinitely many solutions x. In fact, the set of solutions is given by $\{(1-\alpha, 1+\alpha, \alpha)^T \mid \alpha \in \mathcal{R}\}$.

3. The matrix $A = \begin{pmatrix} 1 & x & x^2 \\ x & x^2 & x^3 \\ x^2 & x^3 & x^4 \end{pmatrix}$ can be expressed as the product aa^T where a is the column vector $(1, x, x^2)^T$. By Fact 10, A has rank 1.

6.3.6 IDENTITIES OF MATRIX ALGEBRA

Facts:

1. *Cauchy-Binet formula*: If C is $m \times m$ and $C = AB$ where A is $m \times n$ and B is $n \times m$, then the determinant of C is given by the sum of all products of order m minors of A and the corresponding order m minors of B:

$$\det C = \sum_{1 \leq s_1 < s_2 < \cdots < s_m \leq n} \begin{vmatrix} a_{1s_1} & a_{1s_2} & \cdots & a_{1s_m} \\ a_{2s_1} & a_{2s_2} & \cdots & a_{2s_m} \\ \vdots & \vdots & & \vdots \\ a_{ms_1} & a_{ms_2} & \cdots & a_{ms_m} \end{vmatrix} \cdot \begin{vmatrix} b_{s_1 1} & b_{s_1 2} & \cdots & b_{s_1 m} \\ b_{s_2 1} & b_{s_2 2} & \cdots & b_{s_2 m} \\ \vdots & \vdots & & \vdots \\ b_{s_m 1} & b_{s_m 2} & \cdots & b_{s_m m} \end{vmatrix}$$

   - if $m = n$ there is only one possible selection; the Cauchy-Binet formula for this case reduces to $\det C = \det A \det B$; (see Fact 5, §6.3.4)
   - if $m > n$ no possible selections exist (the sum is empty), so $\det C = 0$.

2. *Courant-Fischer minimax identity*: If the eigenvalues (§6.5) of an $n \times n$ Hermitian matrix A are ordered so that $\lambda_1 \geq \lambda_2 \geq \cdots \geq \lambda_n$, then $\lambda_k = \max_{\dim V = k} \min_{0 \neq x \in V} \dfrac{x^T A x}{x^T x}$ where V is a linear subspace of $\mathcal{C}^n$.

3. *Hadamard's inequality*: This gives an upper bound for the determinant of an $n \times n$ matrix A in terms of the l_2 norms (§6.4.5) of its rows (or columns):

- in terms of rows: $\quad (\det A)^2 \leq \prod_{i=1}^{n} \left(\sum_{j=1}^{n} |a_{ij}|^2 \right) = \prod_{i=1}^{n} \|A(i,:)\|^2$

- in terms of columns: $\quad (\det A)^2 \leq \prod_{j=1}^{n} \left(\sum_{i=1}^{n} |a_{ij}|^2 \right) = \prod_{j=1}^{n} \|A(:,j)\|^2.$

4. *Sherman-Morrison identity*: If A is a nonsingular $n \times n$ matrix and $u, v \in \mathcal{R}^n$, then

$$(A - uv^T)^{-1} = A^{-1} + \frac{1}{1 - v^T A^{-1} u} (A^{-1} uv^T A^{-1}).$$

5. *Woodbury identity*: If A is nonsingular, then

$$(A - UV^T)^{-1} = A^{-1} + A^{-1} U (I - V^T A^{-1} U)^{-1} V^T A^{-1}.$$

6. Suppose A is a nonsingular $n \times n$ matrix, with S a set of k indices $i_1 < i_2 < \cdots < i_k$ and $\overline{S}$ the set of remaining indices in $\{1, 2, \ldots, n\}$. Then the principal minors of A^{-1} are related to the principal minors of A via

$$\det A^{-1}[S] = \frac{1}{\det A} \det A[\overline{S}].$$

7. *Jacobi's identity*: If the $n \times n$ system of linear differential equations $\dfrac{dx}{dt} = P(t)x$ has the linearly independent family of solutions $X(:,j)(t)$ for $j = 1, 2, \ldots, n$, then the determinant of the (variable) matrix $X(t)$ whose columns are the $X(:,j)(t)$ is given by

$$\det X(t_1) = c \exp \left(\int_{t_0}^{t_1} \operatorname{tr} P(t) dt \right),$$

where c is a constant and $\operatorname{tr} P(t) = p_{11}(t) + p_{22}(t) + \cdots + p_{nn}(t)$ is the trace of the matrix $P(t)$.

6.4 LINEAR SYSTEMS

The need to find solutions of linear systems arises in numerous branches of science and engineering (physics, biology, chemistry, structural engineering, electrical engineering, civil engineering) as well as statistics and applied mathematics. This section discusses various techniques for the efficient solution of such systems, especially important when these systems are large and sparse.

6.4.1 BASIC CONCEPTS

This subsection is concerned with representing and solving a system of m linear equations in n unknowns. Throughout, the focus will be on systems whose data are real numbers. The extension to linear systems with complex data is straightforward.

Algorithm 1: Forward substitution.

input: $n \times n$ nonsingular matrix L, $n \times 1$ vector b
output: $n \times 1$ vector $x = L^{-1}b$

$$x_1 := \frac{b_1}{l_{11}}$$

for $i := 2$ to n do

$$x_i := \frac{1}{l_{ii}}\left(b_i - \sum_{j=1}^{i-1} l_{ij}x_j\right)$$

Definitions:

A *linear equation* in unknowns $x_1, x_2, \ldots, x_n$ is an equation of the form $\sum_{j=1}^{n} a_j x_j = b$, where the *coefficients* $a_j \in \mathcal{R}$ and the *right-hand side* $b \in \mathcal{R}$. A *solution* of this equation is any set of values $x_1, x_2, \ldots, x_n$ satisfying the given equation.

A *system of linear equations* in unknowns $x_1, x_2, \ldots, x_n$ is a collection of m equations $\sum_{j=1}^{n} a_{ij}x_j = b_i$, $i = 1, 2, \ldots, m$ where all $a_{ij} \in \mathcal{R}$ and all $b_i \in \mathcal{R}$. A *solution* of this system is any set of values $x_1, x_2, \ldots, x_n$ satisfying (simultaneously) the m given equations. The *coefficient matrix* of this system is the $m \times n$ matrix $A = (a_{ij})$, and the *augmented matrix* is the $m \times (n+1)$ matrix $A_b = (A:b)$.

A *homogeneous* system has right-hand sides $b_i = 0$ for all $i = 1, 2, \ldots, m$; otherwise the system is *nonhomogeneous*.

Back substitution is a simple and efficient iterative procedure for solving an upper triangular linear system $Ux = b$, one unknown at a time.

Forward substitution is a simple and efficient iterative procedure for solving a lower triangular linear system $Lx = b$, one unknown at a time.

Facts:

1. A system of m linear equations in the unknowns $x_1, x_2, \ldots, x_n$ can be represented by the linear system $Ax = b$ where A is the $m \times n$ coefficient matrix, $x = (x_1, x_2, \ldots, x_n)^T$ is the column vector of unknowns, and $b = (b_1, b_2, \ldots, b_m)^T$ is the column vector of right-hand sides.

2. Given the linear system $Ax = b$, where A is $m \times n$,
   - the system has no solution when $\operatorname{rank} A < \operatorname{rank} A_b$;
   - the system has a unique solution when $\operatorname{rank} A = \operatorname{rank} A_b = n$;
   - the system has infinitely many solutions when $\operatorname{rank} A = \operatorname{rank} A_b < n$; in this case the set of solutions is an affine subspace of dimension $n - \operatorname{rank} A$ (§6.1.3).

3. If the square matrix A is nonsingular (§6.3.2), then $Ax = b$ has the unique solution vector $x = A^{-1}b$.

4. If the square matrix $L = (l_{ij})$ is lower triangular, then $Lx = b$ has a unique solution whenever $l_{ii} \neq 0$ for all i. In this case the solution can be found using *forward substitution* (Algorithm 1).

5. If the square matrix $U = (u_{ij})$ is upper triangular, then $Ux = b$ has a unique solution whenever $u_{ii} \neq 0$ for all i. In this case the solution can be found using *back substitution* (Algorithm 2).

Algorithm 2: Back substitution.

input: $n \times n$ nonsingular matrix U, $n \times 1$ vector b
output: $n \times 1$ vector $x = U^{-1}b$

$$x_n := \frac{b_n}{u_{nn}}$$
for $i := n-1$ down to 1 do
$$x_i := \frac{1}{u_{ii}}\left(b_i - \sum_{j=i+1}^{n} u_{ij}x_j\right)$$

Examples:

1. The system of linear equations
$$x_1 + 3x_2 + 4x_3 = 1$$
$$3x_1 + 5x_2 + 0x_3 = 7$$
corresponds to the linear system $Ax = b$, where $A = \begin{pmatrix} 1 & 3 & 4 \\ 3 & 5 & 0 \end{pmatrix}$ and $b = \begin{pmatrix} 1 \\ 7 \end{pmatrix}$. Since rank A = rank $A_b = 2 < 3$ the system has an infinite number of solutions. In fact the set of solutions can be expressed as $\{(4+5a, -1-3a, a)^T \mid a \in \mathcal{R}\}$. Equivalently, it can be expressed as the affine subspace $\{(4, -1, 0)^T + a(5, -3, 1)^T \mid a \in \mathcal{R}\}$ of dimension $n - \text{rank } A = 3 - 2 = 1$.

2. The system of linear equations
$$5x_1 - 3x_2 + 4x_3 = 4$$
$$-x_2 + 5x_3 = 7$$
$$3x_3 = 6$$
has the upper triangular coefficient matrix $U = \begin{pmatrix} 5 & -3 & 4 \\ 0 & -1 & 5 \\ 0 & 0 & 3 \end{pmatrix}$. Using Algorithm 2, the unique solution is $x_3 = \frac{6}{3} = 2$, $x_2 = \frac{1}{-1}(7 - 5 \cdot 2) = 3$, $x_1 = \frac{1}{5}(4 - (-3) \cdot 3 - 4 \cdot 2) = 1$.

6.4.2 GAUSSIAN ELIMINATION

Solving a system of linear equations via Gaussian elimination is one of the most common computations performed by scientists and engineers. Gaussian elimination successively eliminates variables from the original system, creating a triangular system that is easily solved (§6.4.1).

Note: This subsection deals only with linear systems $Ax = b$, where A is a nonsingular $n \times n$ real matrix and $b \in \mathcal{R}^n$.

Definitions:

Gaussian elimination is a method for solving systems of linear equations; at each step one equation is used to eliminate one variable from the rest of the equations. The coefficient of the eliminated variable in the eliminated equation is the ***pivot***.

Algorithm 3: Gaussian elimination and back substitution.

input: $n \times n$ nonsingular matrix A, $n \times 1$ vector b
output: $n \times 1$ vector $x = A^{-1}b$

{Gaussian elimination}
for $j := 1$ to $n-1$ do
   {a_{jj} is the pivot}
   for $i := j+1$ to n do
     {eliminate x_j from equation i}
     compute multiplier $m := -a_{ij}/a_{jj}$
     add $m \times$ row j to row i

{Back substitution}
use Algorithm 2 on resulting upper triangular matrix to obtain the values
   $x_n, x_{n-1}, \ldots, x_1$

A **flop** is a multiply-add operation of the form $t = s + ab$, especially when performed in floating point arithmetic on a digital computer.

Roundoff errors are the errors associated with storing and computing numbers in finite precision arithmetic on a digital computer.

A **numerically stable algorithm** is a method whose accuracy is not greatly harmed by roundoff errors.

A **numerically unstable algorithm** is a method that may return an inaccurate solution even when the solution is relatively insensitive to errors in the data.

Facts:

1. Gaussian elimination is easily extended to linear systems for which the data A and b are complex. Extension to rectangular $m \times n$ linear systems is more involved, but not difficult [GoVa96].

2. In Gaussian elimination, the coefficient of a variable in one of the equations can be used as a pivot if and only if its value is nonzero.

3. In practice, careful choice of pivots is needed to ensure accuracy or improve efficiency or both. (See Example 2.)

4. Assume freedom at each step to choose any available nonzero pivot. Then Gaussian elimination succeeds (using exact arithmetic) if and only if A is nonsingular.

5. Gaussian elimination transforms the initial linear system into a second linear system such that
   - the solutions of the two systems are identical;
   - the solution of the second system is easily obtained by back substitution.

6. $Ax = b$ can be solved by Gaussian elimination and back substitution (Algorithm 3), assuming that at each step there is a nonzero pivot on the main diagonal.

7. Algorithm 3, implemented to take advantage of created 0s, requires $\frac{1}{3}n^3 + O(n^2)$ flops.

8. To solve $Ax = b$ by computing the inverse and then forming the product $A^{-1}b$ requires $n^3 + O(n^2)$ flops.

9. *Cramer's rule*: This method for solving $Ax = b$ expresses each entry of the solution $x = (x_1, x_2, \ldots, x_n)^T$ as the ratio of two determinants:

$$x_1 = \frac{\det A_1}{\det A}, \; x_2 = \frac{\det A_2}{\det A}, \ldots, x_n = \frac{\det A_n}{\det A},$$

where A_i is obtained from A by substituting column vector b for the ith column of A. (Gabriel Cramer, 1704–1752)

10. Cramer's rule is of extremely limited use numerically because
    - it requires far more flops than Gaussian elimination and back substitution;
    - it is numerically unstable.

Examples:
1. The following system is solved by first applying Gaussian elimination:

$$\begin{aligned} x_1 + x_2 + 2x_3 + x_4 &= 1 \\ 2x_1 + 4x_2 + 5x_3 + 4x_4 &= 5 \quad (-\tfrac{2}{1} \times \text{equation 1}) \\ x_1 + 7x_2 + 7x_3 + 6x_4 &= 6 \quad (-\tfrac{1}{1} \times \text{equation 1}) \\ 2x_1 + 4x_2 + 9x_3 + 5x_4 &= 3 \quad (-\tfrac{2}{1} \times \text{equation 1}) \end{aligned}$$

$$\begin{aligned} x_1 + x_2 + 2x_3 + x_4 &= 1 \\ 2x_2 + x_3 + 2x_4 &= 3 \\ 6x_2 + 5x_3 + 5x_4 &= 5 \quad (-\tfrac{6}{2} \times \text{equation 2}) \\ 2x_2 + 5x_3 + 3x_4 &= 1 \quad (-\tfrac{2}{2} \times \text{equation 2}) \end{aligned}$$

$$\begin{aligned} x_1 + x_2 + 2x_3 + x_4 &= 1 \\ 2x_2 + x_3 + 2x_4 &= 3 \\ 2x_3 - x_4 &= -4 \\ 4x_3 + x_4 &= -2 \quad (-\tfrac{4}{2} \times \text{equation 3}) \end{aligned}$$

$$\begin{aligned} x_1 + x_2 + 2x_3 + x_4 &= 1 \\ 2x_2 + x_3 + 2x_4 &= 3 \\ 2x_3 - x_4 &= -4 \\ 3x_4 &= 6 \end{aligned}$$

The solution is then obtained by back substitution:

$$\begin{aligned} x_4 &= 6/3 = 2, \\ x_3 &= [-4 + 1 \cdot 2]/2 = -1, \\ x_2 &= [3 - 1 \cdot (-1) - 2 \cdot 2]/2 = 0, \\ x_1 &= [1 - 1 \cdot 0 - 2 \cdot (-1) - 1 \cdot 2]/1 = 1. \end{aligned}$$

2. Suppose the following system is solved, rounding all results to three significant digits:

$$\begin{aligned} 0.0001 x_1 + x_2 &= 1 \\ 0.5 x_1 + 0.5 x_2 &= 1 \quad (-\tfrac{0.5}{0.0001} \times \text{equation 1}) \end{aligned}$$

$$\begin{aligned} 0.0001 x_1 + x_2 &= 1 \\ -5000 x_2 &= -5000 \end{aligned}$$

Using back substitution produces $x_2 = 1$ and $x_1 = 0$. However, the correct solution to this simple linear system is $x_1 = \frac{10000}{9999}$ and $x_2 = \frac{9998}{9999}$, which to three significant digits becomes $x_1 = 1$ and $x_2 = 1$. Consequently, simply choosing any nonzero pivot can produce inaccurate results.

6.4.3 LU DECOMPOSITION

Gaussian elimination can be formulated as LU decomposition of the coefficient matrix.

Definitions:
An **LU decomposition** of a square matrix A expresses $A = LU$, where $L = (l_{ij})$ is unit lower triangular and $U = (u_{ij})$ is upper triangular.

A **permutation matrix** is a square matrix with entries 0 or 1, where the entry 1 occurs precisely once in each row and once in each column.

Facts:

1. A square matrix has an LU decomposition if and only if every principal submatrix (§6.3.1) is nonsingular.

2. If P is a permutation matrix, then the product PA rearranges the rows of A and the product AP rearranges the columns of A.

3. The matrix A is nonsingular if and only if there exists a permutation matrix P such that PA has an LU decomposition. The LU decomposition of PA is unique.

4. It may be necessary to rearrange the rows of A to avoid a zero pivot.

5. Assume A has an LU decomposition, and consider Gaussian elimination applied to $Ax = b$ with pivots on the main diagonal. The following statements express LU decomposition as a reformulation of Gaussian elimination:
    - the entry u_{ij} ($1 \leq i \leq j \leq n$) is the coefficient of x_j in equation i after Gaussian elimination has been completed;
    - to eliminate x_j from equation i, $i > j$, Gaussian elimination adds $-l_{ij} \times$ equation j to equation i.

6. If A has an LU decomposition, then the linear system $Ax = b$ can be solved as follows (see Algorithm 4):
    - compute the decomposition $A = LU$;
    - solve $Ly = b$; that is, perform forward substitution;
    - solve $Ux = y$; that is, perform back substitution.

7. It is inefficient to solve a nontrivial sequence of linear systems $Ax_1 = b_1$, $Ax_2 = b_2$, ..., $Ax_p = b_p$ by repeating Gaussian elimination for each system. Only one LU decomposition is needed, followed by p forward substitution steps and p back substitution steps.

8. An LU decomposition of an $n \times n$ matrix requires $n^2 + O(n)$ storage locations and $\frac{1}{3}n^3 + O(n^2)$ flops.

> **Algorithm 4:** *LU* **decomposition with forward and back substitution.**
>
> input: $n \times n$ nonsingular matrix A, $n \times 1$ vector b
> output: $n \times 1$ vector $x = A^{-1}b$
> {Compute $A = LU$}
> **for** $k := 1$ **to** $n - 1$ **do**
> $\quad u_{kk} := a_{kk}$
> $\quad$**for** $i := k + 1$ **to** n **do**
> $\quad\quad l_{ik} := a_{ik}/a_{kk};\; u_{ki} := a_{ki}$
> $\quad$**for** $j := k + 1$ **to** n **do**
> $\quad\quad$**for** $i := k + 1$ **to** n **do**
> $\quad\quad\quad a_{ij} := a_{ij} - l_{ik}u_{kj}$
> {Solve $Ly = b$; that is, perform forward substitution}
> **for** $i := 1$ **to** n **do**
> $\quad y_i := b_i - \sum_{j=1}^{i-1} l_{ij}y_j$
> {Solve $Ux = y$; that is, perform back substitution}
> **for** $i := n$ **down to** 1 **do**
> $\quad x_i := \dfrac{1}{u_{ii}}\left(y_i - \sum_{j=i+1}^{n} u_{ij}x_j\right)$

Examples:

1. The following matrix A has no LU decomposition because $a_{11} = 0$ (see Fact 1):
$$A = \begin{pmatrix} 0 & 1 \\ 2 & 3 \end{pmatrix}.$$
However, rearranging the rows of A (Fact 4) produces
$$PA = \begin{pmatrix} 0 & 1 \\ 1 & 0 \end{pmatrix}\begin{pmatrix} 0 & 1 \\ 2 & 3 \end{pmatrix} = \begin{pmatrix} 2 & 3 \\ 0 & 1 \end{pmatrix} = \begin{pmatrix} 1 & 0 \\ 0 & 1 \end{pmatrix}\begin{pmatrix} 2 & 3 \\ 0 & 1 \end{pmatrix} = LU.$$

2. The unique LU decomposition of the matrix A in Example 1, §6.4.2 is
$$\begin{pmatrix} 1 & 1 & 2 & 1 \\ 2 & 4 & 5 & 4 \\ 1 & 7 & 7 & 6 \\ 2 & 4 & 9 & 5 \end{pmatrix} = \begin{pmatrix} 1 & 0 & 0 & 0 \\ 2 & 1 & 0 & 0 \\ 1 & 3 & 1 & 0 \\ 2 & 1 & 2 & 1 \end{pmatrix}\begin{pmatrix} 1 & 1 & 2 & 1 \\ 0 & 2 & 1 & 2 \\ 0 & 0 & 2 & -1 \\ 0 & 0 & 0 & 3 \end{pmatrix}.$$

6.4.4 CHOLESKY DECOMPOSITION

For symmetric positive definite linear systems, Cholesky decomposition (which exploits symmetry in the coefficient matrix) is roughly twice as efficient as LU decomposition.

Definition:

A *Cholesky decomposition* of A expresses $A = LL^T$, where L is lower triangular and every entry on the main diagonal of L is positive.

Facts:

1. A matrix has a Cholesky decomposition if and only if it is symmetric and positive definite.

2. When A is symmetric and positive definite, the linear system $Ax = b$ can be solved as follows:
   - compute a Cholesky decomposition $A = LL^T$;
   - solve $Ly = b$; i.e., perform forward substitution;
   - solve $L^T x = y$; i.e., perform back substitution.

3. A simple symmetric variant of the standard LU decomposition algorithm is used to compute Cholesky decomposition [GoVa96, St88].

4. Cholesky decomposition requires $\frac{1}{2}n^2 + O(n)$ storage locations and $\frac{1}{6}n^3 + O(n^2)$ flops, in contrast to the $n^2 + O(n)$ storage locations and $\frac{1}{3}n^3 + O(n^2)$ flops required by LU decomposition.

Example:

1. The matrix $A = \begin{pmatrix} 1 & -1 & 3 \\ -1 & 2 & -1 \\ 3 & -1 & 14 \end{pmatrix}$ is clearly symmetric. It is positive definite since its principal submatrices (1), $\begin{pmatrix} 1 & -1 \\ -1 & 2 \end{pmatrix}$, and A have positive determinants. (See §6.3.4 Fact 22.) Matrix A can be written as $A = LL^T$, where L is the lower triangular matrix $\begin{pmatrix} 1 & 0 & 0 \\ -1 & 1 & 0 \\ 3 & 2 & 1 \end{pmatrix}$. To solve the linear system $Ax = b$, with $b = (1, 0, 6)^T$, first solve the lower triangular system $Ly = b$, yielding $y = (1, 1, 1)^T$. Then solve the upper triangular system $L^T x = y$, yielding $x = (-3, -1, 1)^T$.

6.4.5 CONDITIONING OF LINEAR SYSTEMS

Errors in the data A and b lead to errors in the solution x. The *condition number* of A can be used to bound relative error in the solution in terms of relative errors in the data.

Definitions:

A (*generalized*) **vector norm** on $\mathcal{R}^n$ is a real-valued function $\|\cdot\|$ satisfying the following properties for all real scalars a and all vectors $x, y \in \mathcal{R}^n$:
- $\|x\| \geq 0$ with equality if and only if $x = 0$;
- $\|ax\| = |a| \cdot \|x\|$, where $|a|$ denotes the absolute value of a;
- $\|x + y\| \leq \|x\| + \|y\|$.

The **matrix norm** induced by the vector norm $\|\cdot\|$ is defined by $\|A\| = \max_{\|x\|=1} \|Ax\|$.

The **condition number** of a nonsingular matrix A is the number $\kappa(A) = \|A\| \, \|A^{-1}\|$. The larger the condition number of a matrix, the more **ill conditioned** it is; the smaller the condition number of a matrix, the more **well conditioned** it is.

Facts:

1. The definition of a vector norm given here generalizes that of a vector norm derived from an inner product space (§6.1.4).
2. The matrix norm induced by a vector norm satisfies:
   - $\|X\| \geq 0$ with equality if and only if $X = 0$;
   - $\|aX\| = |a| \cdot \|X\|$, where $|a|$ denotes the absolute value of a;
   - $\|X + Y\| \leq \|X\| + \|Y\|$;
   - $\|XY\| \leq \|X\|\|Y\|$.
3. $\kappa(A) \geq 1$.
4. Consider the linear system $Ax = b \neq 0$, where A is nonsingular. Suppose that changing from A to $A + \Delta A$ and b to $b + \Delta b$ changes the solution from x to $x + \Delta x$. If $\|(\Delta A)\| \|A^{-1}\| < 1$, the relative error in x can be bounded in terms of relative errors in the data:
$$\frac{\|\Delta x\|}{\|x\|} \leq \left(\frac{\kappa(A)}{1 - \|(\Delta A)\| \|A^{-1}\|}\right)\left(\frac{\|\Delta b\|}{\|b\|} + \frac{\|\Delta A\|}{\|A\|}\right).$$
5. The following are consequences of Fact 4:
   - for an ill-conditioned linear system $Ax = b$, *some* small errors in A or b can potentially be amplified into large errors in x;
   - for a well-conditioned linear system $Ax = b$, where $1 - \|(\Delta A)\| \|A^{-1}\|$ is not approximately zero, *all* small errors in A or b result in no more than modest errors in x.
6. Assume A is nonsingular, let $Ax = b \neq 0$, and view $\hat{x}$ as an approximation to the solution x. Then the residual $r = A\hat{x} - b$ and the error $\hat{x} - x$ satisfy:
$$\frac{\|\hat{x} - x\|}{\|x\|} \leq \kappa(A)\frac{\|r\|}{\|b\|}.$$
7. Whenever A is ill conditioned, a small relative residual $\|r\|/\|b\|$ may not imply a small relative error $\|\hat{x} - x\|/\|x\|$.

Examples:

1. The standard Euclidean norm (§6.1.4) on $\mathcal{R}^n$ defined by $\|x\|_2 = (\sum_{i=1}^{n} x_i^2)^{1/2}$ is a (generalized) vector norm.
2. The l_1 norm on $\mathcal{R}^n$ defined by $\|x\|_1 = \sum_{i=1}^{n} |x_i|$ is a (generalized) vector norm.
3. In coding theory (§14.1), the *Hamming distance* between two codewords $x, y \in \mathcal{Z}_2^n$ is just $\|x - y\|_1$.
4. The l_∞ norm on $\mathcal{R}^n$ defined by $\|x\|_\infty = \max_{1 \leq i \leq n} |x_i|$ is a (generalized) vector norm.
5. The matrix norm induced by $\|x\|_1$ is given by $\|A\|_1 = \max_{1 \leq j \leq n} \sum_{i=1}^{n} |a_{ij}|$.
6. The matrix norm induced by $\|x\|_\infty$ is given by $\|A\|_\infty = \max_{1 \leq i \leq n} \sum_{j=1}^{n} |a_{ij}|$.
7. The matrix norm induced by $\|x\|_2$, also called the *spectral norm*, is given by $\|A\|_2 = \max\{\sqrt{\lambda} \mid \lambda$ an eigenvalue of $A^T A\}$.

8. Consider the linear system $Ax = b$, where $A = \begin{pmatrix} 1 & 2 \\ 2.001 & 4 \end{pmatrix}$ and $b = \begin{pmatrix} 2 \\ 4 \end{pmatrix}$. Then $\|A\|_\infty = 6.001$ and $\|A^{-1}\|_\infty = 3000$, so $\kappa(A) = 18{,}003$. The solution of $Ax = b$ is $x = (0,1)^T$ whereas the solution of the slightly perturbed system with $\hat{b} = \begin{pmatrix} 2 \\ 4.001 \end{pmatrix}$ is $x = (1, 0.5)^T$. Even though the change in the right-hand side is small, the large condition number allows for radical changes in the solution vector, as seen here.

6.4.6 PIVOTING FOR STABILITY

Gaussian elimination can be numerically unstable. Numerical stability can be vastly improved by the addition of pivoting strategies that select large pivots.

Definitions:

Let $a_{ij}^{(k)}$ denote the ij-entry of the current matrix after step k of Gaussian elimination (or LU decomposition). The **growth factor** is defined by $\dfrac{\max_{i,j,k} |a_{ij}^{(k)}|}{\max_{i,j} |a_{ij}|}$.

Partial pivoting is a solution strategy which at step k of Gaussian elimination exchanges row k with the row $i \geq k$ having the entry of largest magnitude in column k.

Complete pivoting is a solution strategy which at step k of Gaussian elimination exchanges row k and column k with, respectively, the row $i \geq k$ and the column $j \geq k$ containing the entry of largest magnitude.

Facts:

1. For general coefficient matrices, Gaussian elimination (that is, LU decomposition) without pivoting is numerically unstable.

2. To improve the numerical stability of Gaussian elimination, it suffices to introduce a pivoting strategy that keeps the growth factor small.

3. For Gaussian elimination with complete pivoting the growth factor is bounded above by
$$n^{1/2}\left(2^1 3^{1/2} 4^{1/3} \ldots n^{1/(n-1)}\right)^{1/2},$$
which is a relatively slow-growing function of n; hence, Gaussian elimination with complete pivoting is numerically stable.

4. For Gaussian elimination with partial pivoting, the growth factor is bounded above by 2^{n-1}, and moreover there are contrived examples for which the growth factor is 2^{n-1}. Hence, Gaussian elimination with partial pivoting can be numerically unstable.

5. In practice, partial pivoting is preferred over complete pivoting for the following two reasons:
   - despite contrived examples having an exponential growth factor, partial pivoting limits the growth factor in practice almost as well as complete pivoting;
   - partial pivoting is significantly more efficient than complete pivoting; it compares $\frac{1}{2}n^2 + O(n)$ pairs of potential pivots, while complete pivoting compares $\frac{1}{3}n^3 + O(n^2)$ pairs.

Example:

1. *LU* decomposition applied to the following matrix shows that partial pivoting can produce a growth factor of 2^{n-1} (see Fact 4). Observe that $\max_{i,j} |a_{ij}| = 1$ and $\max_{i,j,k} |a_{ij}^{(k)}| = u_{nn} = 2^{n-1}$; hence the growth factor is 2^{n-1}:

$$\begin{pmatrix} 1 & 0 & 0 & \cdots & 0 & 1 \\ -1 & 1 & 0 & \cdots & 0 & 1 \\ -1 & -1 & 1 & \cdots & 0 & 1 \\ \vdots & \vdots & \vdots & & \vdots & \vdots \\ -1 & -1 & -1 & \cdots & 0 & 1 \\ -1 & -1 & -1 & \cdots & 1 & 1 \\ -1 & -1 & -1 & \cdots & -1 & 1 \end{pmatrix}$$

6.4.7 PIVOTING TO PRESERVE SPARSITY

Many, if not most, linear systems that arise in practice have relatively few nonzero entries in the coefficient matrix. Some pivoting strategies aim to preserve many zero entries in the triangular factors; the *LU* decomposition algorithm can then save time and space by leaving zero entries out of the computation.

Definitions:

A matrix is **sparse** if it has relatively few nonzero entries. The number of nonzero entries of matrix A is denoted $|A|$. The ith row of A is denoted $A(i,:)$ and the jth column of A is denoted $A(:,j)$. (See §6.3.1.)

Fill refers to nonzero entries in the triangular factors whose corresponding positions in the coefficient matrix are occupied by zeros.

The **upper bandwidth** and **lower bandwidth** of a matrix A are given respectively by $\text{ub}(A) = \max\{(j - i) \mid a_{ij} \neq 0, i < j\}$, $\text{lb}(A) = \max\{(i - j) \mid a_{ij} \neq 0, i > j\}$.

A **banded** *LU* decomposition algorithm stores and computes all entries of L and U within the band defined by $\text{lb}(A)$ and $\text{ub}(A)$.

A **general sparse** *LU* decomposition algorithm stores and computes only the nonzero entries in the triangular factors, irrespective of the banded structure.

The **Markowitz pivoting strategy** for Gaussian elimination chooses at step k from among all available pivots one that minimizes the product $(|L(:,k)| - 1)(|U(k,:)| - 1)$.

The **minimum degree algorithm** is a restricted version of the Markowitz pivoting strategy; it assumes (and preserves) symmetry in the coefficient matrix. At step k of Gaussian elimination, this algorithm chooses from among the entries on the main diagonal a pivot that minimizes $|L(:,k)|$.

Note: The realistic "no-cancellation" assumption will be made throughout. Namely, once an entry becomes nonzero during a triangular decomposition, it will be nonzero upon termination.

Facts:

1. The amount of fill in triangular factors often varies greatly with the choice of pivots.

2. Under the no-cancellation assumption, bandwidth reduction and fill reduction become combinatorial optimization problems.

3. The following problems are provably intractable (i.e., NP-hard; see §16.5):
   - for a symmetric matrix A, find a permutation matrix P that minimizes the bandwidth $\text{lb}(PAP^T)$;
   - for a nonsingular matrix A, find permutation matrices P and Q such that the LU decomposition $PAQ = LU$ exists and $|L| + |U|$ is minimum;
   - for a symmetric positive definite matrix A, find a permutation matrix P that minimizes $|L|$, where L is the Cholesky factor of PAP^T.

4. In view of Fact 3, various heuristics are used to reduce bandwidth or to reduce fill.

5. Assume that A has an LU decomposition. Then $\text{lb}(L) = \text{lb}(A)$ and $\text{ub}(U) = \text{ub}(A)$.

6. The chief advantage of a banded LU decomposition algorithm over a general sparse LU decomposition algorithm is its simplicity. The same advantage holds for *profile* and *skyline* methods, both of which are generalizations of the banded approach [GeLi81].

7. For most problems encountered in practice, a banded LU decomposition algorithm, even if A has been permuted so that $\text{lb}(A)$ and $\text{ub}(A)$ are minimum, requires much more space and work than a general sparse LU decomposition algorithm coupled with the Markowitz pivoting strategy. The same comment applies to profile and skyline methods.

8. Let A be a symmetric positive definite matrix, and let P be a permutation matrix with the same number of rows and columns.
   - the Cholesky decomposition of PAP^T exists and is numerically stable;
   - the undirected graph (§8.1) G of the Cholesky factor of PAP^T is a *chordal graph* and P defines a *perfect elimination ordering* of G [GeLi81].

9. General sparse Cholesky decomposition can be handled in a clean, modular fashion:
   - using only the positions of nonzeros in A as input, compute a permutation P to reduce fill in the Cholesky factor of PAP^T (using, for example, the minimum degree algorithm);
   - construct data structures to contain the nonzeros of the Cholesky factor;
   - after putting the nonzero entries of PAP^T into the data structures, compute the Cholesky factor of PAP^T in the provided data structures;
   - perform forward and back substitutions to solve the linear system.

10. For symmetric positive definite matrices arising from two-dimensional and three-dimensional partial differential equations, the *nested dissection* algorithm often computes a more effective fill-reducing permutation than does the minimum degree algorithm [GeLi81].

11. The interplay between pivoting for stability and pivoting for sparsity complicates general sparse LU factorization. The best approach is not yet certain.

12. A number of robust and well-tested software packages are available for solving linear systems, including:
    - *LINPACK*: a collection of Fortran routines for relatively small dense systems; see http://www.netlib.org
    - *LAPACK/CLAPACK*: supersedes *LINPACK*, contains Fortran and C routines for dense and banded problems, ideal for shared-memory vector and parallel processors; see http://www.netlib.org
    - *NAG*: Fortran and C libraries for dense and sparse systems; see http://www.nag.com

- *IMSL*: Fortran and C libraries for dense and sparse systems; see
  http://www.vni.com/products/imsl
- *MATLAB*: high-level language for dense and sparse systems; see
  http://www.mathworks.com

Examples:

1. For any "arrowhead" matrix there is a pivot sequence that completely fills the matrix and another that creates no fill, making it the canonical example used to illustrate Fact 1. The following is a 4×4 arrowhead matrix that fills in completely. ($\star$ occupies a position that is nonzero in A, $\bullet$ is a fill entry in L or U, and a space is a zero.)

$$A = \begin{pmatrix} \star & \star & \star & \star \\ \star & \star & & \\ \star & & \star & \\ \star & & & \star \end{pmatrix}, \quad A = LU = \begin{pmatrix} \star & & & \\ \star & \star & & \\ \star & \bullet & \star & \\ \star & \bullet & \bullet & \star \end{pmatrix} \begin{pmatrix} \star & \star & \star & \star \\ & \star & \bullet & \bullet \\ & & \star & \bullet \\ & & & \star \end{pmatrix}.$$

Reversing the pivot sequence, however, results in no fill:

$$\widehat{A} = PAP^T = \begin{pmatrix} & & & 1 \\ & & 1 & \\ & 1 & & \\ 1 & & & \end{pmatrix} \begin{pmatrix} \star & \star & \star & \star \\ \star & \star & & \\ \star & & \star & \\ \star & & & \star \end{pmatrix} \begin{pmatrix} & & & 1 \\ & & 1 & \\ & 1 & & \\ 1 & & & \end{pmatrix} = \begin{pmatrix} \star & & & \star \\ & \star & & \star \\ & & \star & \star \\ \star & \star & \star & \star \end{pmatrix},$$

$$\widehat{A} = \widehat{L}\widehat{U} = \begin{pmatrix} \star & & & \\ & \star & & \\ & & \star & \\ \star & \star & \star & \star \end{pmatrix} \begin{pmatrix} \star & & & \star \\ & \star & & \star \\ & & \star & \star \\ & & & \star \end{pmatrix}.$$

2. The following table illustrates how Fact 7 typically manifests itself in practice. The four problems arise in finite element modeling of actual structures. The table records data for two distinct methods:

- a profile-reducing permutation from the reverse Cuthill-McKee algorithm [GeLi81] in tandem with a profile factorization algorithm;
- a fill-reducing permutation from the minimum degree algorithm [GeLi81] in tandem with a general sparse factorization algorithm.

Recorded for each method are the number of nonzero entries in the Cholesky factor (expressed in millions) and the number of flops needed to compute the factor (expressed in millions).

problem	n	$|A|$	$|L|(\times 10^{-6})$		No. flops $(\times 10^{-6})$	
			profile reduction	general sparse	profile reduction	general sparse
coliseum	1,806	63,454	0.190	0.112	11.803	4.952
winter sports arena	3,562	159,910	0.538	0.279	44.245	16.352
nuclear power station	11,948	149,090	5.908	0.663	2,135.163	70.779
76 story skyscraper	15,439	252,241	2.637	1.417	232.791	142.567

6.5 EIGENANALYSIS

Identifying the eigenvalues and eigenvectors of a matrix facilitates the study of complicated systems and the analysis of their behavior over time. A basis consisting of eigenvectors yields a particularly simple representation of a linear transformation (§6.2). Eigenvalues can also provide useful information about discrete structures (§8.10.1).

6.5.1 EIGENVALUES AND CHARACTERISTIC POLYNOMIAL

Definitions:

A complex number λ is an **eigenvalue** of the $n \times n$ complex matrix A if there exists a nonzero vector $x \in \mathcal{C}^n$ (an **eigenvector** of A corresponding to λ) such that $Ax = \lambda x$.

The **characteristic polynomial** of the square matrix A is the polynomial $p_A(\lambda) = \det(\lambda I - A)$.

The **characteristic equation** of A is the equation $p_A(\lambda) = 0$.

A **nilpotent** matrix is a square matrix A such that $A^k = 0$ for some positive integer k.

An **idempotent** matrix is a square matrix A such that $A^2 = A$.

Let $S_k(A)$ denote the sum of all order k principal minors of the matrix A.

Facts:

1. The characteristic polynomial $p_A(\lambda)$ of an $n \times n$ matrix A is a monic polynomial of degree n in λ.
2. The coefficient of λ^{n-1} in $p_A(\lambda)$ is $-\operatorname{tr} A$.
3. The constant term in $p_A(\lambda)$ is $(-1)^n \det A$.
4. $p_A(\lambda) = \sum_{k=0}^{n}(-1)^k S_k(A)\lambda^{n-k}$.
5. Similar matrices (§6.2.4) have the same characteristic polynomial.
6. The roots of the characteristic equation are the eigenvalues of A.
7. *Cayley-Hamilton theorem*: If $p_A(\cdot)$ is the characteristic polynomial of A then $p_A(A)$ is the zero matrix.
8. An $n \times n$ matrix has n (not necessarily distinct) eigenvalues.
9. The matrix A is singular if and only if 0 is an eigenvalue of A.
10. The characteristic equation of $A = \begin{pmatrix} a & b \\ c & d \end{pmatrix}$ is $p_A(\lambda) = \lambda^2 - (a+d)\lambda + (ad - bc)$.
11. The eigenvalues of $A = \begin{pmatrix} a & b \\ c & d \end{pmatrix}$ are given by $\dfrac{a + d \pm \sqrt{(a-d)^2 + 4bc}}{2}$.
12. If the $n \times n$ matrix A has eigenvalues $\lambda_1, \lambda_2, \ldots, \lambda_n$ then
    - $\sum_{i=1}^{n} \lambda_i = \operatorname{tr} A$;
    - $\prod_{i=1}^{n} \lambda_i = \det A$;
    - the k-th elementary symmetric function $\sum_{i_1 < \cdots < i_k} \lambda_{i_1} \ldots \lambda_{i_k}$ equals $S_k(A)$.

13. The eigenvalues are continuous functions of the entries of a matrix. More precisely, given an $n \times n$ matrix A with eigenvalues $\lambda_1, \lambda_2, \ldots, \lambda_n$ and $\epsilon > 0$, there exists $\delta > 0$ such that for any $n \times n$ matrix B, with eigenvalues $\mu_1, \mu_2, \ldots, \mu_n$ and satisfying $\max_{i,j} |a_{ij} - b_{ij}| < \delta$, there exists a permutation τ of $1, 2, \ldots, n$ such that $|\lambda_i - \mu_{\tau(i)}| < \epsilon$, $i = 1, 2, \ldots, n$.

14. The following table gives the eigenvalues of certain specialized matrices A, whose eigenvalues are $\lambda_1, \lambda_2, \ldots, \lambda_n$. In this table k is any positive integer.

matrix	eigenvalues
diagonal matrix	diagonal elements
upper (or lower) triangular matrix	diagonal elements
A^T	eigenvalues of A
A^*	complex conjugates of the eigenvalues of A
A^k	$\lambda_1^k, \ldots, \lambda_n^k$
A^{-k}, A nonsingular	$\lambda_1^{-k}, \ldots, \lambda_n^{-k}$
$q(A)$, where $q(\cdot)$ is a polynomial	$q(\lambda_1), \ldots, q(\lambda_n)$
SAS^{-1}, S nonsingular	eigenvalues of A
AB, where A is $m \times n$, B is $n \times m$, $m \geq n$	eigenvalues of BA; and 0 ($m - n$ times)
$(a - b)I_n + bJ_n$, where J_n is the $n \times n$ matrix of all 1s	$a + (n - 1)b$; and $a - b$ ($n - 1$ times)
A $n \times n$ nilpotent	0 (n times)
A $n \times n$ idempotent of rank r	1 (r times); and 0 ($n - r$ times)

Examples:

1. The characteristic polynomial for $A = \begin{pmatrix} 1 & 4 \\ 2 & 3 \end{pmatrix}$ is $p_A(\lambda) = \begin{vmatrix} \lambda - 1 & -4 \\ -2 & \lambda - 3 \end{vmatrix} = \lambda^2 - 4\lambda - 5 = (\lambda + 1)(\lambda - 5)$, so the eigenvalues are $\lambda = -1$ and $\lambda = 5$. The vector $x = (2, -1)^T$ is an eigenvector for $\lambda = -1$ since $Ax = (-2, 1)^T = -x$. The vector $x = (1, 1)^T$ is an eigenvector for $\lambda = 5$ since $Ax = (5, 5)^T = 5x$.

2. For the matrix in Example 1,

$$p_A(A) = A^2 - 4A - 5I = \begin{pmatrix} 9 & 16 \\ 8 & 17 \end{pmatrix} - \begin{pmatrix} 4 & 16 \\ 8 & 12 \end{pmatrix} - \begin{pmatrix} 5 & 0 \\ 0 & 5 \end{pmatrix} = \begin{pmatrix} 0 & 0 \\ 0 & 0 \end{pmatrix},$$

as required by Fact 7.

3. The characteristic polynomial of the matrix $A = \begin{pmatrix} 3 & 0 & 2 \\ 4 & 1 & 4 \\ 2 & 0 & 3 \end{pmatrix}$ can be calculated by using Facts 1-4. Since $\operatorname{tr} A = 7$, $\det A = 5$, and $S_2(A) = \begin{vmatrix} 1 & 4 \\ 0 & 3 \end{vmatrix} + \begin{vmatrix} 3 & 2 \\ 2 & 3 \end{vmatrix} + \begin{vmatrix} 3 & 0 \\ 4 & 1 \end{vmatrix} = 11$, it follows that $p_A(\lambda) = \lambda^3 - 7\lambda^2 + 11\lambda - 5$. Thus $p_A(5) = 0$, showing that $\lambda = 5$ is an eigenvalue of A. An eigenvector corresponding to $\lambda = 5$ is $x = (1, 2, 1)^T$ since $Ax = (5, 10, 5)^T = 5x$.

4. The matrix A in Example 3 is nonsingular since $p_A(0) \neq 0$, so 0 is not an eigenvalue of A (Fact 9). The inverse of A can be calculated using the Cayley-Hamilton theorem: $A^3 - 7A^2 + 11A - 5I = 0$, so $5I = A^3 - 7A^2 + 11A = A(A^2 - 7A + 11I)$ and $I = A[\frac{1}{5}(A^2 - 7A + 11I)]$. Consequently, $A^{-1} = \frac{1}{5}(A^2 - 7A + 11I) = \frac{1}{5}\begin{pmatrix} 3 & 0 & -2 \\ -4 & 5 & -4 \\ -2 & 0 & 3 \end{pmatrix}$.

6.5.2 EIGENVECTORS AND DIAGONALIZATION

Definitions:

Let λ be an eigenvalue of the $n \times n$ (complex) matrix A. The **algebraic multiplicity** of λ is its multiplicity as a root of the characteristic polynomial.

The **eigenspace** of A corresponding to λ is the vector space $\{x \in C^n \mid Ax = \lambda x\}$.

The **geometric multiplicity** of λ is the dimension of the eigenspace of A corresponding to λ.

The square matrix A is **diagonalizable** if there exists a nonsingular matrix P such that $P^{-1}AP$ is a diagonal matrix.

The **minimal polynomial** of the square matrix A is the monic polynomial $q(\cdot)$ of minimum degree such that $q(A) = 0$.

The square matrix A is **normal** if $AA^* = A^*A$.

The **singular values** of an $n \times n$ matrix A are the (positive) square roots of the eigenvalues of AA^*, written $\sigma_1(A) \leq \sigma_2(A) \leq \cdots \leq \sigma_n(A)$.

A **row stochastic** matrix is a matrix with all entries nonnegative and row sums 1.

Facts:

1. The eigenspace corresponding to λ is a subspace of the vector space C^n. Specifically, it is the null space (§6.1.2) of the matrix $A - \lambda I$.

2. Eigenvectors corresponding to distinct eigenvalues are linearly independent.

3. If λ, μ are distinct eigenvalues of A and if $Ax = \lambda x$ and $A^*y = \mu y$, then x, y are orthogonal.

4. The algebraic multiplicity is never less than the geometric multiplicity, but sometimes it is greater. (See Example 3.)

5. The minimal polynomial is unique.

6. If A can be diagonalized to a diagonal matrix D, then the eigenvalues of A appear along the diagonal of D.

7. The following conditions are equivalent for an $n \times n$ matrix A:
   - A is diagonalizable;
   - A has n linearly independent eigenvectors;
   - the minimal polynomial of A has distinct linear factors;
   - the algebraic multiplicity of each eigenvalue of A equals its geometric multiplicity.

8. If the $n \times n$ matrix A has n distinct eigenvalues then A is diagonalizable.

9. If the $n \times n$ matrix A has n linearly independent eigenvectors $v_1, v_2, \ldots, v_n$ then A is diagonalizable using the matrix P whose columns are the vectors $v_1, v_2, \ldots, v_n$.

10. Hermitian, skew-Hermitian and unitary matrices are normal matrices.

11. *Spectral theorem for normal matrices*: If A is an $n \times n$ normal matrix, then it can be diagonalized by a unitary matrix. That is, there exists an $n \times n$ unitary matrix U such that $U^*AU = \text{diag}(\lambda_1, \lambda_2, \ldots, \lambda_n)$, the diagonal matrix with the eigenvalues $\lambda_1, \lambda_2, \ldots, \lambda_n$ of A along its diagonal.

12. If A is normal, then it has a *spectral decomposition* $A = \sum_{i=1}^{n} \lambda_i u_i u_i^*$, where $\{u_1, u_2, \ldots, u_n\}$ is an orthonormal basis for $\mathcal{C}^n$.

13. Diagonalization results for special types of normal matrices are given in the following table:

matrix A	eigenvalues	diagonalization result
Hermitian	real	Fact 11
real symmetric	real	there exists a real orthogonal P such that $P^T A P$ is diagonal
skew-Hermitian	purely imaginary	Fact 11
real skew-symmetric	purely imaginary	there exists a real orthogonal Q such that $Q^T A Q$ is a direct sum of matrices, each of which is a 2×2 real skew-symmetric or null matrix
unitary	all with modulus 1	Fact 11

14. If A, B are normal and commute, they can be simultaneously diagonalized. Namely, there exists a unitary U such that U^*AU and U^*BU are both diagonal.

15. For any square matrix A, the rank of A is never less than the number of nonzero eigenvalues (counting multiplicities) of A.

16. If A is normal then its rank equals the number of nonzero eigenvalues.

17. *Schur's triangularization theorem*: If A is a square matrix, then there exists a unitary U such that U^*AU is upper triangular with the eigenvalues of A on its diagonal.

18. If A, B are square matrices that commute then there exists a unitary U such that U^*AU and U^*BU are both upper triangular.

19. *Jordan canonical form*: Let A be an $n \times n$ matrix with distinct eigenvalues $\lambda_1, \lambda_2, \ldots, \lambda_k$ having (algebraic) multiplicities $r_1, r_2, \ldots, r_k$ respectively. Then there exists a nonsingular matrix P such that $P^{-1}AP = \text{diag}(\Lambda_1, \Lambda_2, \ldots, \Lambda_k)$, where

$$\Lambda_i = \begin{pmatrix} \lambda_i & * & 0 & \cdots & 0 & 0 \\ 0 & \lambda_i & * & \cdots & 0 & 0 \\ \vdots & \vdots & & & \vdots & \vdots \\ 0 & 0 & 0 & \cdots & \lambda_i & * \\ 0 & 0 & 0 & \cdots & 0 & \lambda_i \end{pmatrix}$$

is an $r_i \times r_i$ matrix and each $*$ is either 0 or 1. Furthermore, the number of 1s is r_i minus the geometric multiplicity of λ_i.

20. The rank of a square matrix equals the number of nonzero singular values.

21. If A is a square matrix and if U and V are unitary, then A and UAV have the same singular values.

22. *Singular value decomposition*: If A is an $n \times n$ matrix then there exist $n \times n$ unitary matrices U, V such that UAV is diagonal with $\sigma_1(A), \sigma_2(A), \ldots, \sigma_n(A)$ on the diagonal.

23. *QR factorization*: If A is an $n \times n$ matrix then there exists a unitary matrix Q and an upper triangular matrix R such that $A = QR$.

24. The QR factorization of a matrix can be calculated using Gram-Schmidt orthogonalization (§6.1.4).

Examples:

1. Let x, y be vectors of size $n \times 1$ and let $A = xy^T$. Then the eigenvalues of A are given by (see §6.5.1, Table 1) $y^T x$ and 0, the latter with multiplicity $n - 1$.

2. The matrix of §6.5.1 Example 3 has the characteristic polynomial $p_A(\lambda) = \lambda^3 - 7\lambda^2 + 11\lambda - 5 = (\lambda - 1)^2(\lambda - 5)$. The eigenvalues are $\lambda = 1$ with algebraic multiplicity 2 and $\lambda = 5$ with algebraic multiplicity 1.

For $\lambda = 1$ the eigenspace is the null space of $A - \lambda I = \begin{pmatrix} 2 & 0 & 2 \\ 4 & 0 & 4 \\ 2 & 0 & 2 \end{pmatrix}$. It consists of all vectors of the form $(a, b, -a)^T$ and so is spanned by the linearly independent eigenvectors $(1, 0, -1)^T$ and $(0, 1, 0)^T$. Thus the geometric multiplicity of $\lambda = 1$ is 2, the same as its algebraic multiplicity 1.

The eigenvalue $\lambda = 5$ has the eigenvector $(1, 2, 1)^T$ (see §6.5.1 Example 3), linearly independent of the previous two eigenvectors (Fact 2). If $P = \begin{pmatrix} 1 & 0 & 1 \\ 0 & 1 & 2 \\ -1 & 0 & 1 \end{pmatrix}$ is the matrix containing these eigenvectors then $P^{-1}AP = \text{diag}(1, 1, 5)$, thereby diagonalizing A (Fact 9).

3. By using *Maple*, the characteristic polynomial of the matrix
$$A = \begin{pmatrix} 7 & 2 & 4 & 0 & 3 \\ 0 & 6 & 0 & 0 & 0 \\ 0 & -2 & 4 & 0 & 0 \\ 3 & 2 & 4 & 4 & 3 \\ 3 & 0 & 2 & 0 & 7 \end{pmatrix}$$
is found to be $\lambda^5 - 28\lambda^4 + 300\lambda^3 - 1552\lambda^2 + 3904\lambda - 3840 = (\lambda - 4)^3(\lambda - 6)(\lambda - 10)$, so the eigenvalue $\lambda = 4$ of A has algebraic multiplicity 3. The eigenspace for $\lambda = 4$ is the null space of
$$A - \lambda I = \begin{pmatrix} 3 & 2 & 4 & 0 & 3 \\ 0 & 2 & 0 & 0 & 0 \\ 0 & -2 & 0 & 0 & 0 \\ 3 & 2 & 4 & 0 & 3 \\ 3 & 0 & 2 & 0 & 3 \end{pmatrix},$$
which is spanned by $(1, 0, 0, 0, -1)^T$ and $(0, 0, 0, 1, 0)^T$. So $\lambda = 4$ has geometric multiplicity 2. By Fact 7, A is not diagonalizable. The minimal polynomial of A is $(\lambda - 4)^2(\lambda - 6)(\lambda - 10)$, which has the repeated linear factor $\lambda - 4$.

4. The conclusion of Fact 16 need not hold if A is not normal. For example, the matrix $\begin{pmatrix} 0 & 1 \\ 0 & 0 \end{pmatrix}$ has rank 1 but has no nonzero eigenvalues.

5. *Matrix powers*: The matrix $A = \begin{pmatrix} \frac{1}{2} & \frac{1}{2} \\ \frac{3}{4} & \frac{1}{4} \end{pmatrix}$ is a row stochastic matrix and the powers A^n of such matrices are important in the analysis of Markov chains (§7.7). The eigenvalues of A are $\lambda = 1$ and $\lambda = -\frac{1}{4}$, with corresponding eigenvectors $(1,1)^T$ and $(2,-3)^T$. Thus $P^{-1}AP = D = \text{diag}(1, -\frac{1}{4})$, where $P = \begin{pmatrix} 1 & 2 \\ 1 & -3 \end{pmatrix}$. Consequently $A = PDP^{-1}$, $A^2 = PDP^{-1}PDP^{-1} = PD^2P^{-1}$, and in general $A^n = PD^nP^{-1}$. Since $D^n = \text{diag}(1^n, (-\frac{1}{4})^n) = \text{diag}(1, \alpha^n)$, the nth power of A can be computed as $A^n = \frac{1}{5}\begin{pmatrix} 3 + 2\alpha^n & 2 - 2\alpha^n \\ 3 - 3\alpha^n & 2 + 3\alpha^n \end{pmatrix}$. Since $|\alpha| < 1$, $A^n \to \begin{pmatrix} \frac{3}{5} & \frac{2}{5} \\ \frac{3}{5} & \frac{2}{5} \end{pmatrix}$ as $n \to \infty$.

6.5.3 LOCALIZATION

Since analytic computation of eigenvalues can be complicated, there are several simple methods available for (geometrically) estimating the eigenvalues of a matrix. These methods can be informative in cases when only the approximate location of eigenvalues is needed.

Definitions:

The **spectral radius** of A, $\rho(A)$, is the maximum modulus of an eigenvalue of A.

Let A be an $n \times n$ matrix and let $\alpha_i = \sum_{j \neq i} |a_{ij}|$, $i = 1, 2, \ldots, n$.

The **Geršgorin discs** associated with A are the discs
$$\{ z \in \mathcal{C} \mid |z - a_{ii}| \leq \alpha_i \}, \quad i = 1, 2, \ldots, n.$$

The **ovals of Cassini** associated with A are the ellipses
$$\{ z \in \mathcal{C} \mid |z - a_{ii}||z - a_{jj}| \leq \alpha_i \alpha_j \}, \quad i \neq j.$$

A **strictly diagonally dominant** matrix is a square matrix A satisfying $|a_{ii}| > \alpha_i$ for $i = 1, 2, \ldots, n$.

Facts:

1. $\rho(A)$ is the radius of the smallest disc, centered at the origin of the complex plane, enclosing all of the eigenvalues of A.

2. $\rho(A) \leq \min \{\max_i \sum_j |a_{ij}|, \max_j \sum_i |a_{ij}|\}$.

3. The spectral radius of a row stochastic matrix is 1.

4. All the eigenvalues of A are contained in the union of the associated Geršgorin discs.

5. A connected region formed by precisely $k \leq n$ Geršgorin discs contains exactly k eigenvalues of A.

6. All the eigenvalues of A are contained in the union of the $\frac{n(n-1)}{2}$ ovals of Cassini associated with A.

Examples:
1. By Fact 2, the spectral radius of the symmetric matrix
$$A = \begin{pmatrix} 8 & -2 & 1 & 1 \\ -2 & -8 & -2 & -1 \\ 1 & -2 & 7 & 1 \\ 1 & -1 & 1 & 8 \end{pmatrix}$$
is bounded by the maximum absolute row (column) sum 13. Since the eigenvalues of a real symmetric matrix are real, the spectral radius bound gives the interval $[-13, 13]$ enclosing all eigenvalues. The Geršgorin discs are the intervals
$$8 \pm 4 = [4, 12], \ -8 \pm 5 = [-13, -3], \ 7 \pm 4 = [3, 11], \text{ and } 8 \pm 3 = [5, 11].$$
The second interval is disjoint from the others, so one eigenvalue is localized in the interval $[-13, -3]$ while the other three are in the interval $[3, 12]$. The actual eigenvalues of A are (approximately) $-8.51, 6.31, 7.03, 10.2$, consistent with the above intervals. Also, 0 is not in any of the four Geršgorin discs so 0 is not an eigenvalue and A is nonsingular. Since the eigenvalues of A^{-1} are the reciprocals of the eigenvalues of A (Table 1, §6.5.1), it follows that the eigenvalues of the symmetric matrix A^{-1} are localized to the intervals $[-\frac{1}{3}, -\frac{1}{13}]$ and $[\frac{1}{12}, \frac{1}{3}]$.

2. Using Fact 4, the eigenvalues of the matrix $A = \begin{pmatrix} 2 & 1 & -1 \\ 0 & 6 & 2 \\ 1 & -1 & 8 \end{pmatrix}$ are located in the union of the discs
$$D_1 = \{\, z \mid |z - 2| \leq 2 \,\}, \ D_2 = \{\, z \mid |z - 6| \leq 2 \,\}, \ D_3 = \{\, z \mid |z - 8| \leq 2 \,\}.$$
Since A and A^T have the same eigenvalues, an alternative set of disks can be formed based on the absolute *column* sums of A: namely
$$\widehat{D}_1 = \{\, z \mid |z - 2| \leq 1 \,\}, \ \widehat{D}_2 = \{\, z \mid |z - 6| \leq 2 \,\}, \ \widehat{D}_3 = \{\, z \mid |z - 8| \leq 3 \,\}.$$
Here $\widehat{D}_1$ is disjoint from both $\widehat{D}_2$ and $\widehat{D}_3$, and so one eigenvalue of A is localized to $\widehat{D}_1$, and the other two to $\widehat{D}_2 \cup \widehat{D}_3$. In fact, the eigenvalues of A are 2.24 and $6.88 \pm 0.91i$, approximately.

3. The row stochastic matrix $A = \begin{pmatrix} \frac{1}{2} & \frac{1}{2} \\ \frac{3}{4} & \frac{1}{4} \end{pmatrix}$ has Geršgorin discs
$$D_1 = \{\, z \mid |z - \tfrac{1}{2}| \leq \tfrac{1}{2} \,\} \text{ and } D_2 = \{\, z \mid |z - \tfrac{1}{4}| \leq \tfrac{3}{4} \,\}.$$
Since $D_1 \subseteq D_2$ all eigenvalues must lie in D_2. As seen in §6.5.2 Example 5, the eigenvalues of A are 1 and $-\frac{1}{4}$.

4. Suppose A is strictly diagonally dominant. Then all Geršgorin discs for A reside in the positive right-half plane so all the eigenvalues must have positive real part. In particular, 0 is not an eigenvalue and A must be nonsingular.

5. If the $n \times n$ matrix A satisfies $a_{ii} a_{jj} > \alpha_i \alpha_j$ for all $i \neq j$ then A must be nonsingular, since by Fact 6 zero is not an eigenvalue of A. The matrix of Example 2 satisfies this condition since $a_{ii} a_{jj} \geq 12 > 4 = \alpha_i \alpha_j$ for all $i \neq j$, and so it must be nonsingular.

6.5.4 COMPUTATION OF EIGENVALUES

The eigenvalues of a matrix can be obtained, in theory, by forming the characteristic equation and finding its roots. Since this is not a practical solution method for problems of realistic size, a variety of iterative techniques have been developed.

Algorithm 1: Power method.

input: $n \times n$ nonsingular matrix A
output: approximations x_k to an eigenvector of A

{Initialization}
choose any vector $x_0 \in \mathcal{C}^n$ with $\|x_0\| = 1$
{Iterative step}
for $k := 1$ to ... do
$$x_k := \frac{Ax_{k-1}}{\|Ax_{k-1}\|}$$

Algorithm 2: QR method.

input: $n \times n$ matrix A
output: $n \times n$ matrices A_k

{Initialization}
$A := Q_0 R_0$ (a QR factorization of A)
{Iterative step}
for $k := 1$ to ... do
  $A_k := R_{k-1} Q_{k-1}$
  obtain a QR factorization $A_k = Q_k R_k$

Definitions:

A *dominant* eigenvalue of a matrix is an eigenvalue with the maximum modulus.

Let $U(\theta; i, j)$ be the $n \times n$ matrix obtained by replacing the 2×2 principal submatrix of the identity matrix, corresponding to rows i and j, with the rotation matrix

$$\begin{pmatrix} \cos\theta & \sin\theta \\ -\sin\theta & \cos\theta \end{pmatrix}.$$

Facts:

1. *Power method*: The power method (Algorithm 1) is a simple technique for finding the dominant eigenvalue and an associated eigenvector of a nonsingular matrix A having a unique dominant eigenvalue.

2. In Algorithm 1, the kth estimate $x_k = \dfrac{A^k x_0}{\|A^k x_0\|}$.

3. The sequence x_k converges to an eigenvector of A.

4. The sequence $\|Ax_k\|$ approaches the dominant eigenvalue.

5. The power method is best suited for large sparse matrices.

6. The rate of convergence of the power method is dictated by the ratio of the largest to the second largest (in modulus) eigenvalue of A. The larger this ratio (the more separated these two eigenvalues in modulus), the faster the convergence of the method.

7. *QR method*: This method (Algorithm 2) calculates the eigenvalues of a given $n \times n$ matrix A.

Algorithm 3: Jacobi method.

input: $n \times n$ real symmetric matrix A
output: $n \times n$ matrices A_k

{Initialization}
$A_1 = (a_{ij}^{(1)}) := A$
{Iterative step}
for $k := 1$ **to** ... **do**
  choose r, s $(r < s)$ with $|a_{rs}^{(k)}|$ as large as possible
  define θ by $\cot 2\theta = \dfrac{a_{rr}^{(k)} - a_{ss}^{(k)}}{2a_{rs}^{(k)}}$
  $A_{k+1} = (a_{ij}^{(k+1)}) := U(\theta; r, s)^T A_k \, U(\theta; r, s)$

8. The QR factorization in Algorithm 2 produces a unitary matrix Q_k and an upper triangular matrix R_k. (See §6.5.2 Fact 23.)

9. Under certain conditions (for example, if the eigenvalues of A have distinct moduli) the sequence A_k in Algorithm 2 converges to an upper triangular matrix whose diagonal entries are the eigenvalues of A.

10. If A is real then its QR factors are real and can be calculated using real arithmetic. In this case, if A has nonreal eigenvalues then under certain conditions, the limiting matrix is block triangular with 1×1 and 2×2 diagonal blocks.

11. The QR method is not well suited for large sparse matrices since the factors Q, R can quickly fill with nonzeros.

12. Often as a preparatory step for the QR method the matrix is first reduced to *Hessenberg form* (upper triangular form in which there may be one nonzero diagonal below the main diagonal) by using Householder transformations [Da95].

13. The convergence of the QR method can be very slow if the matrix has two eigenvalues that are close in moduli.

14. More effective versions of the QR method are available which make use of certain *shift strategies* [GoVa96].

15. *Jacobi method*: This method (Algorithm 3) finds the eigenvalues of a real symmetric $n \times n$ matrix A having at least one nonzero off-diagonal entry.

16. The sequence A_k in Algorithm 3 converges to a real diagonal matrix with the eigenvalues of A on the diagonal.

17. The orthogonal matrix $U(\theta; r, s)$ represents a (clockwise) plane rotation by the angle θ.

18. The Jacobi method is particularly appropriate when A is nearly diagonal, although in general the QR method exhibits faster convergence.

19. A variant of the Jacobi method, the *serial Jacobi method*, uses plane rotation pairs cyclically — for example, $(1, 2), (1, 3), \ldots, (1, n), (2, 3), \ldots, (2, n), \ldots$.

20. For further information on numerical computation of eigenvalues, see [GoVa96, Da95].

414 Chapter 6 LINEAR ALGEBRA

21. A number of robust and well-tested software packages are available for carrying out eigensystem analysis, including:

- *EISPACK*: a collection of Fortran routines for analyzing eigenvalues and eigenvectors of several classes of matrices; see
  http://www.netlib.org
- *LAPACK/CLAPACK*: supersedes *EISPACK*, contains Fortran and C routines for dense and banded problems, ideal for shared-memory vector and parallel processors; see
  http://www.netlib.org
- *NAG*: Fortran and C libraries for eigenanalysis of dense and sparse matrices; see
  http://www.nag.com
- *IMSL*: contains Fortran and C libraries for eigenanalysis of dense and banded problems; see
  http://www.vni.com/products/imsl
- *MATLAB*: high-level language for eigenanalysis of dense and sparse matrices, calculation of characteristic polynomials; see
  http://www.mathworks.com

Examples:

1. The power method, when applied to the matrix in Example 3 of §6.5.1, produces the following sequence of vectors x_k and scalars $\|Ax_k\|$:

k	0	1	2	3	4	5
x_k	$\begin{pmatrix}1\\0\\0\end{pmatrix}$	$\begin{pmatrix}0.557\\0.743\\0.371\end{pmatrix}$	$\begin{pmatrix}0.436\\0.805\\0.403\end{pmatrix}$	$\begin{pmatrix}0.414\\0.814\\0.407\end{pmatrix}$	$\begin{pmatrix}0.409\\0.816\\0.408\end{pmatrix}$	$\begin{pmatrix}0.409\\0.816\\0.408\end{pmatrix}$
$\|Ax_k\|$	5.385	5.537	5.107	5.021	5.004	5.001

The scalars $\|Ax_k\|$ approach the dominant eigenvalue 5 and the vectors x_k approach a multiple of the eigenvector $(1,2,1)^T$. (See Example 3, §6.5.1.)

2. The eigenvalues of the matrix $A = \begin{pmatrix} 1 & 4 \\ 2 & 3 \end{pmatrix}$ can be approximated using the QR method. $A = Q_0 R_0$ with $Q_0 = \begin{pmatrix} -0.447 & -0.894 \\ -0.894 & 0.447 \end{pmatrix}$ and $R_0 = \begin{pmatrix} -2.236 & -4.472 \\ 0 & -2.236 \end{pmatrix}$. Then $A_1 = R_0 Q_0 = \begin{pmatrix} 5 & 0 \\ 2 & -1 \end{pmatrix}$. Continuing this process produces

$$A_2 = \begin{pmatrix} 4.862 & 2.345 \\ 0.345 & -0.862 \end{pmatrix}, \quad A_3 = \begin{pmatrix} 5.023 & -1.927 \\ 0.073 & -1.023 \end{pmatrix}, \quad A_4 = \begin{pmatrix} 4.995 & 2.014 \\ 0.014 & -0.995 \end{pmatrix}.$$

The sequence A_k approaches an upper triangular matrix with the eigenvalues 5 and -1 on its diagonal.

3. The eigenvalues of the matrix $A = \begin{pmatrix} 0 & 1 & 1 \\ 1 & 4 & -3 \\ 1 & -3 & 4 \end{pmatrix}$ can be approximated using the Jacobi method. The largest off-diagonal $|a_{rs}|$ of $A_1 = A$ occurs for $r = 2$, $s = 3$ giving $\theta = \frac{\pi}{4} = 0.7854$. Applying the matrix $U(\theta; 2, 3) = \begin{pmatrix} 1 & 0 & 0 \\ 0 & 0.7071 & 0.7071 \\ 0 & -0.7071 & 0.7071 \end{pmatrix}$ produces

$$A_2 = U(\theta;2,3)^T A_1 U(\theta;2,3) = \begin{pmatrix} 0 & 0 & 1.4142 \\ 0 & 7 & 0 \\ 1.4142 & 0 & 1 \end{pmatrix}.$$ The largest magnitude off-diagonal entry of A_2 is $|a_{13}|$, giving $\theta = 0.6155$, $U(\theta;1,3) = \begin{pmatrix} 0.8165 & 0 & 0.5774 \\ 0 & 1 & 0 \\ -0.5774 & 0 & 0.8165 \end{pmatrix}$ and $A_3 = U(\theta;1,3)^T A_2 U(\theta;1,3) = \begin{pmatrix} -1 & 0 & 0 \\ 0 & 7 & 0 \\ 0 & 0 & 2 \end{pmatrix}$. So the eigenvalues of A are $-1, 2, 7$.

6.5.5 SPECIAL CLASSES

This section discusses eigenvalues and eigenvectors of specially structured matrices, such as Hermitian, positive definite, nonnegative, totally positive, and circulant matrices.

Definitions:

If $x, y \in \mathcal{R}^n$, then x **majorizes** y if $\sum_{i=1}^{n} x_i = \sum_{i=1}^{n} y_i$ and for $k = 1, 2, \ldots, n-1$ the sum of the k largest components of x is at least as large as the sum of the k largest components of y. A similar definition holds for infinite sequences with finitely many nonzero terms.

A Hermitian $n \times n$ matrix A is **positive definite** if $x^* A x > 0$ for all nonzero $x \in \mathcal{C}^n$. It is **positive semidefinite** if $x^* A x \geq 0$ for all $x \in \mathcal{C}^n$.

If A and B are $n \times n$ Hermitian matrices then A **dominates** B in **Löwner order** if $A - B$ is positive semidefinite, written $A \succeq B$.

A matrix is **nonnegative** [**positive**] if each of its entries is nonnegative [positive].

The $n \times n$ matrix A is **reducible** if either it is the 1×1 zero matrix or there exists a permutation matrix P such that PAP^T is of the form $\begin{pmatrix} B & 0 \\ C & D \end{pmatrix}$, where B and D are square. A matrix is **irreducible** if it is not reducible.

A **strictly totally positive** matrix has all of its minors positive.

A **circulant** matrix has the form

$$\begin{pmatrix} a_0 & a_1 & a_2 & \cdots & a_n \\ a_n & a_0 & a_1 & \cdots & a_{n-1} \\ a_{n-1} & a_n & a_0 & \cdots & a_{n-2} \\ \vdots & \vdots & \vdots & & \vdots \\ a_1 & a_2 & a_3 & \cdots & a_0 \end{pmatrix}.$$

Notation: Let $\lambda_1(A) \leq \lambda_2(A) \leq \cdots \leq \lambda_n(A)$ be the eigenvalues of an $n \times n$ Hermitian matrix A.

Facts:

1. *Cauchy interlacing theorem*: Let A be an $n \times n$ Hermitian matrix and let B be a principal submatrix of A of order $n-1$. Then
$$\lambda_i(A) \leq \lambda_i(B) \leq \lambda_{i+1}(A), \quad i = 1, 2, \ldots, n-1.$$

2. *Weyl's theorem*: Let A, B be $n \times n$ Hermitian matrices and let j, k be integers satisfying $1 \leq j, k \leq n$.
   - If $j + k \geq n + 1$, then $\lambda_{j+k-n}(A + B) \leq \lambda_j(A) + \lambda_k(B)$;
   - If $j + k \leq n + 1$, then $\lambda_j(A) + \lambda_k(B) \leq \lambda_{j+k-1}(A + B)$.

3. Interpretations of the kth smallest eigenvalue of a Hermitian matrix are given in the following table:

eigenvalue	variational characterization
$\lambda_1(A)$	$\min(x^*Ax)$, minimum over all unit vectors x
$\lambda_n(A)$	$\max(x^*Ax)$, maximum over all unit vectors x
$\lambda_k(A)$, $k = 2, \ldots, n$	$\min(x^*Ax)$, minimum over all unit vectors x orthogonal to the eigenspaces of $\lambda_1, \ldots, \lambda_{k-1}$
$\lambda_{n-k}(A)$, $k = 1, \ldots, n-1$	$\max(x^*Ax)$, maximum over all unit vectors x orthogonal to the eigenspaces of $\lambda_{n-k+1}, \ldots, \lambda_n$

4. *Schur's majorization theorem*: If A is an $n \times n$ Hermitian matrix, then $(\lambda_1(A), \lambda_2(A), \ldots, \lambda_n(A))$ majorizes $(a_{11}, a_{22}, \ldots, a_{nn})$. Specifically, if $a_{11} \geq a_{22} \geq \cdots \geq a_{nn}$ then
$$\sum_{i=1}^{k} \lambda_{n-i+1}(A) \geq \sum_{i=1}^{k} a_{ii}, \quad k = 1, 2, \ldots, n.$$

5. *Hoffman-Wielandt theorem*: If A, B are $n \times n$ Hermitian matrices, then
$$\sum_{i=1}^{n} (\lambda_i(A + B) - \lambda_i(A))^2 \leq \sum_{i,j=1}^{n} |b_{ij}|^2.$$

6. *Sylvester's law of inertia*: If A is an $n \times n$ Hermitian matrix and if X is a nonsingular $n \times n$ matrix, then A and $X^T A X$ have the same number of positive eigenvalues as well as the same number of negative eigenvalues.

7. A Hermitian matrix is positive definite (positive semidefinite) if and only if all its eigenvalues are positive (nonnegative).

8. If A, B are $n \times n$ positive semidefinite matrices and $A \succeq B$, then $\lambda_i(A) \geq \lambda_i(B)$, $i = 1, 2, \ldots, n$.

9. If A, B are $n \times n$ positive semidefinite matrices, then $\lambda_{i+j-n}(AB) \leq \lambda_i(A)\lambda_j(B)$ holds for $1 \leq i, j \leq n$ and $i + j \geq n + 1$.

10. If A, B are $n \times n$ positive semidefinite matrices, then
$$\prod_{i=1}^{k} \lambda_i(AB) \geq \prod_{i=1}^{k} \lambda_i(A)\lambda_i(B), \quad k = 1, 2, \ldots, n.$$

11. *Kantorovich inequality*: If A is an $n \times n$ positive definite matrix and if $x \in \mathcal{C}^n$ is a unit vector, then
$$(x^*Ax)(x^*A^{-1}x) \leq \frac{(\lambda_1(A) + \lambda_n(A))^2}{4\lambda_1(A)\lambda_n(A)}.$$

12. *Perron-Frobenius theorem*: If A is an irreducible nonnegative square matrix, then the spectral radius of A (the *Perron root* of A) is an eigenvalue of A with algebraic multiplicity 1 and it has an associated positive eigenvector. If A is positive then the spectral radius exceeds the modulus of any other eigenvalue.

13. If A is a nonnegative square matrix, then the spectral radius of A is an eigenvalue of A and it has an associated nonnegative eigenvector.

14. Let A be an $n \times n$ strictly totally positive matrix. Then the eigenvalues of A are distinct and positive: $\lambda_1(A) < \lambda_2(A) < \cdots < \lambda_n(A)$. The real eigenvector corresponding to λ_{n-k} has exactly k variations in sign.

15. If A is an $n \times n$ strictly totally positive matrix, then $(\lambda_1(A), \lambda_2(A), \ldots, \lambda_n(A))$ majorizes $(a_{11}, a_{22}, \ldots, a_{nn})$.

16. An $(n+1) \times (n+1)$ circulant matrix has eigenvalues $\lambda_j = a_0 + a_1\omega^j + a_2\omega^{2j} + \cdots + a_n\omega^{nj}$, $j = 0, 1, \ldots, n$ with $(1, \omega^j, \omega^{2j}, \ldots, \omega^{nj})$, $j = 0, 1, \ldots, n$ the corresponding eigenvectors, where $\omega = e^{\frac{2\pi i}{n+1}}$.

Examples:

1. The matrix $A = \begin{pmatrix} 0 & 1 & 1 \\ 1 & 4 & -3 \\ 1 & -3 & 4 \end{pmatrix}$ has eigenvalues -1, 2, and 7 (§6.5.4, Example 3). The principal submatrix $A[1,2] = \begin{pmatrix} 0 & 1 \\ 1 & 4 \end{pmatrix}$ has eigenvalues $2 \pm \sqrt{5}$, which are approximately equal to -0.2361 and 4.2361. As required by Fact 1, these latter two eigenvalues interlace those of A: $-1 \leq -0.2361 \leq 2 \leq 4.2361 \leq 7$. Similarly, the principal submatrix $A[2,3] = \begin{pmatrix} 4 & -3 \\ -3 & 4 \end{pmatrix}$ has eigenvalues 1 and 7, which interlace those of A.

2. The matrix in Example 1 has the eigenvalue sequence $(-1, 2, 7)$. This sequence majorizes (see Fact 4) the sequence $(0, 4, 4)$ of diagonal elements: $7 \geq 4$, $7 + 2 \geq 4 + 4$, and $7 + 2 - 1 \geq 4 + 4 + 0$.

3. The irreducible matrix A in §6.5.3 Example 3 is positive with eigenvalues 1 and $-\frac{1}{4}$. Thus $\rho(A) = 1$ and it exceeds the modulus of any other eigenvalue. As required by Fact 12, there is a positive eigenvector associated with $\lambda = 1$, namely $(1, 1)^T$.

4. The matrix $A = \begin{pmatrix} 2 & 0 & 3 \\ 1 & 4 & 5 \\ 2 & 0 & 1 \end{pmatrix}$ is nonnegative with eigenvalues -1 (algebraic multiplicity 1) and 4 (algebraic multiplicity 2). So the spectral radius is 4 and (see Fact 13) $\lambda = 4$ must be an eigenvalue. In addition, there is a nonnegative eigenvector associated with $\lambda = 4$, namely $(0, 1, 0)^T$.

6.6 COMBINATORIAL MATRIX THEORY

Matrices and graphs represent two different ways of viewing certain discrete structures. At times a matrix perspective can lend insight into graphical or combinatorial structures. At other times the graph associated with a matrix can provide useful information about matrix properties.

6.6.1 MATRICES OF ZEROS AND ONES

Definitions:

A **0-1 matrix** is a matrix with each entry either 0 or 1.

The **term rank** of a 0-1 matrix is the maximum number of 1s such that no two are in the same row or column.

An $n \times n$ 0-1 matrix is **partly decomposable** if it has a $k \times (n-k)$ zero submatrix for some $1 \leq k \leq n-1$; otherwise A is **fully indecomposable**.

Let $\{x_n\}$ be a sequence of nonnegative integers with finitely many nonzero terms. The **conjugate sequence** of $\{x_n\}$ is the sequence $\{z_n\}$ in which z_n, $n \geq 1$, is the number of terms in $\{x_n\}$ that are not less than n.

Facts:

1. *König's theorem*: The term rank of a 0-1 matrix equals the minimum number of rows and columns required to cover all 1s in the matrix.

2. *Frobenius-König theorem*: Let A be an $n \times n$ 0-1 matrix. Then the term rank of A is less than n if and only if A has a zero submatrix of size $r \times s$ with $r + s = n + 1$.

3. Let A be an $n \times n$ 0-1 matrix each of whose row sums and column sums is k. Then A can be expressed as a sum of k permutation matrices (§6.4.3).

4. Let A be a square 0-1 matrix and let B be the matrix obtained from A by replacing each 0 entry on the main diagonal of A by 1. Then A is irreducible (§6.5.5) if and only if B is fully indecomposable.

5. Let A, B be $n \times n$ fully indecomposable matrices. Then the matrix obtained by replacing every nonzero entry in AB by 1 is fully indecomposable.

6. *Gale-Ryser theorem*: Let $x_1, x_2, \ldots, x_m; y_1, y_2, \ldots, y_n$ be nonnegative integers and let $\{z_n\}$ be the conjugate sequence of $x_1, x_2, \ldots, x_m, 0, 0, \ldots$. There exists an $m \times n$ 0-1 matrix with row sums $x_1, x_2, \ldots, x_m$ and column sums $y_1, y_2, \ldots, y_n$ if and only if $\{z_n\}$ majorizes $y_1, y_2, \ldots, y_n, 0, 0, \ldots$.

Examples:

1. The following matrix contains a 2×4 zero submatrix, occurring in rows 1, 3 and columns 1, 2, 4, 5. By Fact 2, this means that the matrix must have term rank less than 5. In fact, the matrix has term rank 3. Namely, the starred entries represent a set of 3 entries, no two of which are in the same row or column, and 3 is the largest number with this property. Rows 2, 4 and column 3 cover all the 1s in the matrix, and no smaller number suffices, as guaranteed by Fact 1.

$$\begin{pmatrix} 0 & 0 & 1^* & 0 & 0 \\ 1^* & 1 & 1 & 0 & 1 \\ 0 & 0 & 1 & 0 & 0 \\ 1 & 0 & 1 & 1 & 1^* \\ 0 & 0 & 1 & 0 & 0 \end{pmatrix}.$$

2. The matrix

$$A = \begin{pmatrix} 0 & 1 & 1 & 1 \\ 1 & 0 & 1 & 1 \\ 1 & 1 & 0 & 1 \\ 1 & 1 & 1 & 0 \end{pmatrix}$$

has all row and column sums equal to 3. By Fact 3, it can be expressed as the sum of three permutation matrices. For example

$$A = \begin{pmatrix} 0 & 0 & 1 & 0 \\ 0 & 0 & 0 & 1 \\ 1 & 0 & 0 & 0 \\ 0 & 1 & 0 & 0 \end{pmatrix} + \begin{pmatrix} 0 & 0 & 0 & 1 \\ 0 & 0 & 1 & 0 \\ 0 & 1 & 0 & 0 \\ 1 & 0 & 0 & 0 \end{pmatrix} + \begin{pmatrix} 0 & 1 & 0 & 0 \\ 1 & 0 & 0 & 0 \\ 0 & 0 & 0 & 1 \\ 0 & 0 & 1 & 0 \end{pmatrix}.$$

3. *Assignment problem*: There are n applicants for n vacant jobs. Each applicant is qualified for exactly $k \geq 1$ jobs and for each job there are exactly k qualified applicants. Is it possible to assign each applicant to a (distinct) job for which the applicant is qualified? To answer this question form the 0-1 matrix A where $a_{ij} = 1$ if applicant i is qualified for job j, otherwise $a_{ij} = 0$. All row and column sums of A equal k, so (by Fact 3) A can be expressed as the sum of $k \geq 1$ permutation matrices. Select any one of these permutation matrices and use it to define an assignment of applicants to jobs. Thus it *is* possible in this case to fill each job with a different qualified applicant.

4. In the matrix

$$\begin{pmatrix} 0 & 1 & 1 & 0 & 1 & 1 & 1 \\ 1 & 0 & 1 & 1 & 0 & 1 & 1 & 0 \\ 0 & 0 & 1 & 1 & 0 & 1 & 0 & 1 \\ 1 & 0 & 1 & 0 & 1 & 1 & 0 & 1 \\ 0 & 1 & 0 & 1 & 1 & 1 & 1 & 0 \end{pmatrix}$$

the conjugate sequence of the row sum sequence $6, 5, 4, 5, 5$ is $5, 5, 5, 5, 4, 1, 0, 0, \ldots$ and it majorizes the sequence $2, 2, 4, 3, 3, 5, 3, 3, 0, 0, \ldots$ obtained by appending zeros to the sequence of column sums (Fact 6).

6.6.2 NONNEGATIVE MATRICES

This subsection discusses nonnegative matrices and special classes of nonnegative matrices such as primitive and doubly stochastic matrices. Certain results highlight the relationship between the Perron root (§6.5.5, Fact 12) and the directed graph of a matrix.

Definitions:

The **directed graph** $G(A)$ of an $n \times n$ matrix A consists of n vertices $1, 2, \ldots, n$ with an edge from i to j if and only if $a_{ij} \neq 0$.

Vertices i and j of $G(A)$ are **equivalent** if $i = j$ or if there is a path in $G(A)$ from i to j and a path from j to i. The corresponding equivalence classes (§1.4.2) of this relation are the **classes** of A.

Class C_i has **access** to class C_j if $i = j$ or if there is a path in $G(A)$ from a vertex in C_i to a vertex in C_j. A class is **final** if it has access to no other class. Class C_i is **basic** if $\rho(A[C_i]) = \rho(A)$, where $\rho(\cdot)$ is the spectral radius (§6.5.3) and $A[C_i]$ is the principal submatrix of A defined by indices in class C_i.

Let A be an $n \times n$ nonnegative irreducible matrix. The number h of eigenvalues of A of modulus $\rho(A)$ is called the **index of cyclicity** of A. The matrix A is **primitive** if $h = 1$.

The **exponent** of A, written $\exp(A)$, is the least positive integer m with $A^m > 0$.

A square matrix is **doubly stochastic** if it is nonnegative and all row and column sums are 1.

If A is an $n \times n$ matrix and $\sigma \in S_n$, the symmetric group on n elements (§5.3.1), then the set $\{a_{1\sigma(1)}, a_{2\sigma(2)} \ldots, a_{n\sigma(n)}\}$ is the **diagonal** of A corresponding to σ.

A diagonal of A is **positive** if each entry in it is positive.

Matrices A and B of the same size have the same **pattern** if the following condition holds: $a_{ij} = 0$ if and only if $b_{ij} = 0$.

A matrix A has **doubly stochastic pattern** if there exists a doubly stochastic matrix B such that A and B have the same pattern.

Facts:

1. The matrix A is irreducible if and only if $G(A)$ is strongly connected (§8.3.2).

2. *Frobenius normal form*: If the $n \times n$ matrix A has k classes, then there exists a permutation matrix P such that
$$PAP^T = \begin{pmatrix} A_{11} & 0 & \cdots & 0 \\ A_{21} & A_{22} & \cdots & 0 \\ \vdots & \vdots & & \vdots \\ A_{k1} & A_{k2} & \cdots & A_{kk} \end{pmatrix}$$
where each A_{ii}, $1 \leq i \leq k$, is either irreducible or a 1×1 zero matrix.

3. The classes of a nonnegative $n \times n$ matrix A are in one-to-one correspondence with the strong components (§8.3.2) of $G(A)$ and hence can be found in linear time.

4. Let A be an $n \times n$ nonnegative matrix. There is a positive eigenvector corresponding to $\rho(A)$ if and only if the basic classes of A are the same as its final classes.

5. Let A be an $n \times n$ nonnegative matrix with eigenvalue λ. There exists a nonnegative eigenvector for λ if and only if there exists a class C_i satisfying both of the following:
   - $\rho(A[C_i]) = \lambda$;
   - if C_j ($j \neq i$) is any class that has access to C_i, then $\rho(A[C_j]) < \rho(A[C_i])$.

6. The $n \times n$ nonnegative matrix A is primitive if and only if $\exp(A) < \infty$.

7. A nonnegative irreducible matrix with positive trace is primitive.

8. Suppose A is an $n \times n$ nonnegative irreducible matrix. Let S_i be the set of all the lengths of cycles in $G(A)$ passing through vertex i, and let h_i be the greatest common divisor of all the elements of S_i. Then $h_1 = h_2 = \cdots = h_n$ and this common value equals the index of cyclicity of A.

9. Let A be a nonnegative irreducible $n \times n$ matrix with $p \geq 1$ nonzero elements on the main diagonal. Then A is primitive and $\exp(A) \leq 2n - p - 1$.

10. Let A be a primitive $n \times n$ matrix, and let s be the smallest length of a directed cycle in $G(A)$. Then $\exp(A) \leq n + s(n-2)$.

11. Let A be an $n \times n$ primitive 0-1 matrix, $n \geq 2$. Then $\exp(A) \leq (n-1)^2 + 1$. Equality holds if and only if there exists a permutation matrix P such that

$$PAP^T = \begin{pmatrix} 0 & 1 & 0 & \cdots & 0 \\ 0 & 0 & 1 & \cdots & 0 \\ \vdots & \vdots & \vdots & & \vdots \\ 1 & 0 & 0 & \cdots & 1 \\ 1 & 0 & 0 & \cdots & 0 \end{pmatrix}.$$

12. The set Ω_n of $n \times n$ doubly stochastic matrices is a compact convex set.

13. *Birkhoff-von Neumann theorem*: Every $A \in \Omega_n$ can be expressed as a convex combination of $n \times n$ permutation matrices: namely, $A = c_1 P_1 + c_2 P_2 + \cdots + c_t P_t$ for some permutation matrices $P_1, P_2, \ldots, P_t$ and some positive real numbers $c_1, c_2, \ldots, c_t$ with $c_1 + c_2 + \cdots + c_t = 1$.

14. The following conditions are equivalent for an $n \times n$ matrix A:
    - A has doubly stochastic pattern;
    - there exist permutation matrices P, Q such that PAQ is a direct sum of fully indecomposable matrices;
    - every nonzero entry of A is contained in a positive diagonal.

15. Let A be an $n \times n$ nonnegative idempotent matrix of rank k. Then there exists a permutation matrix P such that

$$PAP^T = \begin{pmatrix} J & JU & 0 & 0 \\ 0 & 0 & 0 & 0 \\ VJ & VJU & 0 & 0 \\ 0 & 0 & 0 & 0 \end{pmatrix},$$

where J is a direct sum of k positive idempotent matrices of rank 1.

16. A nonnegative symmetric matrix A of rank k is idempotent if and only if there exists a permutation matrix P such that

$$PAP^T = \begin{pmatrix} J & 0 \\ 0 & 0 \end{pmatrix},$$

where J is a direct sum of k positive symmetric rank one idempotent matrices.

Examples:
1. The following nonnegative matrix is in Frobenius normal form

$$\begin{pmatrix} 5 & 0 & 0 & 0 & 0 & 0 & 0 \\ 0 & 1 & 1 & 0 & 0 & 0 & 0 \\ 0 & 2 & 0 & 0 & 0 & 0 & 0 \\ 2 & 4 & 1 & 0 & 3 & 0 & 0 \\ 1 & 2 & 0 & 2 & 1 & 0 & 0 \\ 0 & 1 & 1 & 0 & 0 & 3 & 2 \\ 0 & 2 & 1 & 0 & 0 & 2 & 3 \end{pmatrix}$$

with four classes $C_1 = \{1\}$, $C_2 = \{2, 3\}$, $C_3 = \{4, 5\}$ and $C_4 = \{6, 7\}$. Class C_3 has access to C_1 and C_2 while class C_4 has access to C_2. Classes C_1 and C_2 are final since they have access to no other classes. The eigenvalues of A are $-2, -1; 1, 2, 3, 5, 5$ so $\rho(A) = 5$. Classes C_1 and C_4 are basic since $\rho(A[C_1]) = \rho(A[C_4]) = 5$. Since no class has access to C_3, Fact 5 shows there is a nonnegative eigenvector of A for the eigenvalue $\rho(A[C_3]) = 3$, namely $(0, 0, 0, 1, 1, 0, 0)^T$. However there is no nonnegative eigenvector of A for $\rho(A[C_2]) = 2$ since class C_3 has access to class C_2 and $\rho(A[C_3]) \geq \rho(A[C_2])$.

2. The directed graph of the matrix

$$A = \begin{pmatrix} 0 & 1 & 7 & 0 & 0 \\ 0 & 0 & 0 & 0 & 1 \\ 0 & 0 & 0 & 1 & 0 \\ 1 & 0 & 0 & 0 & 0 \\ 1 & 0 & 0 & 0 & 0 \end{pmatrix}$$

is the union of the cycles $1, 2, 5, 1$ and $1, 3, 4, 1$. The greatest common divisor of the lengths of all cycles passing through vertex 1 is 3, which by Fact 8 must be the index of cyclicity. In fact, A has eigenvalues $2, -1 \pm i\sqrt{3}, 0, 0$ and thus there are 3 eigenvalues with modulus $\rho(A) = 2$.

3. By Fact 13 every doubly stochastic matrix is a convex combination of permutation matrices. For example, the doubly stochastic matrix

$$\begin{pmatrix} .4 & .3 & .3 & 0 \\ .5 & 0 & .4 & .1 \\ .1 & .6 & 0 & .3 \\ 0 & .1 & .3 & .6 \end{pmatrix}$$

can be expressed as

$$.2 \begin{pmatrix} 0 & 0 & 1 & 0 \\ 1 & 0 & 0 & 0 \\ 0 & 1 & 0 & 0 \\ 0 & 0 & 0 & 1 \end{pmatrix} + .3 \begin{pmatrix} 0 & 1 & 0 & 0 \\ 1 & 0 & 0 & 0 \\ 0 & 0 & 0 & 1 \\ 0 & 0 & 1 & 0 \end{pmatrix} + .1 \begin{pmatrix} 0 & 0 & 1 & 0 \\ 0 & 0 & 0 & 1 \\ 1 & 0 & 0 & 0 \\ 0 & 1 & 0 & 0 \end{pmatrix} + .4 \begin{pmatrix} 1 & 0 & 0 & 0 \\ 0 & 0 & 1 & 0 \\ 0 & 1 & 0 & 0 \\ 0 & 0 & 0 & 1 \end{pmatrix}.$$

6.6.3 PERMANENTS

The permanent of a matrix is defined as a sum of terms, each corresponding to the product of elements along a diagonal of the matrix. Permanents arise in the study of systems of distinct representatives and in other combinatorial problems.

Definition:

The **permanent** of the $n \times n$ matrix A is $\text{per}(A) = \sum_{\sigma \in S_n} a_{1\sigma(1)} a_{2\sigma(2)} \cdots a_{n\sigma(n)}$, where S_n is the symmetric group on n elements. (See §5.3.1.)

Facts:

1. The permanent of A is an unsigned version of the determinant of A. (See §6.3.4 Fact 2.)

2. Computing the permanent of a square 0-1 matrix is #P-complete.

3. *Laplace expansion*: Suppose A is an $n \times n$ matrix and A_{ij} is the submatrix of A obtained by deleting the ith row and the jth column. Then for $i = 1, 2, \ldots, n$

$$\text{per}(A) = \sum_{j=1}^{n} a_{ij} \, \text{per}(A_{ij}).$$

A similar expansion holds with respect to any column.

4. $\text{per}(A^T) = \text{per}(A)$.

5. Interchanging two rows (or two columns) of A does not change $\text{per}(A)$.

6. Multiplying any row (or column) of A by the scalar α multiplies $\text{per}(A)$ by α.

7. Unlike the determinant, the permanent is not multiplicative with respect to matrix multiplication. (See Example 2.)

8. The permanent of a triangular matrix (§6.3.1) is equal to the product of its diagonal entries.

9. The permanent of a block diagonal matrix is equal to the product of the permanents of its diagonal blocks.

10. For each positive integer n the permanent of the $n \times n$ matrix

$$\begin{pmatrix} 0 & 1 & \cdots & 1 \\ 1 & 0 & \cdots & 1 \\ \vdots & \vdots & & \vdots \\ 1 & 1 & \cdots & 0 \end{pmatrix}$$

is $n! \sum_{r=0}^{n} (-1)^r \frac{1}{r!}$ and it represents the number of derangements (§2.4.2) of order n.

11. The permanent of an $n \times n$ 0-1 matrix A counts the number of assignments ($n \times n$ permutation submatrices) consistent with the 1 entries of A.

12. *Minc-Brègman inequality*: Let A be an $n \times n$ 0-1 matrix with row sums $r_1, r_2, \ldots, r_n$. Then

$$\text{per}(A) \leq \prod_{i=1}^{n} (r_i!)^{1/r_i}.$$

13. If A is a nonnegative $n \times n$ matrix with row sums $r_1, r_2, \ldots, r_n$ then $\text{per}(A) \leq r_1 r_2 \ldots r_n$.

14. Let A be a fully indecomposable nonnegative integral $n \times n$ matrix and let $s(A)$ denote the sum of the entries in A. Then

$$s(A) - 2n + 2 \leq \text{per}(A) \leq 2^{s(A)-2n} + 1.$$

15. *Alexandroff inequality*: Let A be a nonnegative $n \times n$ matrix and let A_i be the ith column of A, $i = 1, 2, \ldots, n$. Then

$$(\text{per}(A))^2 \geq \text{per}(A_1, \ldots, A_{n-2}, A_{n-1}, A_{n-1}) \text{per}(A_1, \ldots, A_{n-2}, A_n, A_n).$$

16. The definition of the permanent can be extended to $m \times n$ matrices with $m \leq n$ by summing over all permutations in S_m.

17. If A is an $m \times n$ 0-1 matrix, then $\text{per}(A) > 0$ if and only if A has term rank m.

18. *van der Waerden-Egorychev-Falikman inequality*: If A is a doubly stochastic $n \times n$ matrix then $\text{per}(A) \geq \frac{n!}{n^n}$, and equality holds if and only if $A = J_n$, the matrix with each entry $\frac{1}{n}$.

Note: This result was first conjectured by B. L. van der Waerden in 1926. Despite repeated attempts to prove it, the conjecture remained unresolved until finally established in 1980 by G. P. Egorychev. The conjecture was also proved independently by D. I. Falikman in 1981, apart from establishing the uniqueness of the minimizing matrix A. A self-contained exposition of Egorychev's proof is given in [Kn81].

19. Let A be the $m \times n$ incidence matrix of m subsets of a given n-set X: namely, $a_{ij} = 1$ if $j \in X_i$ and $a_{ij} = 0$ otherwise. Then $\text{per}(A)$ counts the number of SDRs (systems of distinct representatives, §1.2.2) selected from the sets $X_1, X_2, \ldots, X_m$.

Examples:

1. For the matrix
$$A = \begin{pmatrix} 1 & 0 & 2 \\ 3 & 1 & 1 \\ 1 & 5 & 2 \end{pmatrix},$$
evaluation of $\mathrm{per}(A)$ by the definition gives $\mathrm{per}(A) = 1\cdot 1\cdot 2 + 2\cdot 3\cdot 5 + 2\cdot 1\cdot 1 + 1\cdot 1\cdot 5 = 39$. Using the Laplace expansion on row 1 gives $\mathrm{per}(A) = 1\cdot \mathrm{per}\begin{pmatrix} 1 & 1 \\ 5 & 2 \end{pmatrix} + 2\cdot \mathrm{per}\begin{pmatrix} 3 & 1 \\ 1 & 5 \end{pmatrix} = 1\cdot 7 + 2\cdot 16 = 39$.

2. If $A = \begin{pmatrix} 1 & 1 \\ 0 & 1 \end{pmatrix}$ and $B = \begin{pmatrix} 1 & 0 \\ 1 & 1 \end{pmatrix}$, then $C = AB = \begin{pmatrix} 2 & 1 \\ 1 & 1 \end{pmatrix}$. Notice that $\mathrm{per}(AB) = 3 \neq 1\cdot 1 = \mathrm{per}(A)\,\mathrm{per}(B)$.

3. *Assignments*: Suppose there are 4 applicants for 4 jobs, where the qualifications of each applicant i for each job j is specified by the 0-1 matrix
$$A = \begin{pmatrix} 0 & 1 & 0 & 1 \\ 1 & 1 & 0 & 1 \\ 0 & 0 & 1 & 1 \\ 1 & 1 & 1 & 0 \end{pmatrix}.$$
Then the number of different assignments of jobs to qualified applicants (see §6.6.1, Example 3) equals $\mathrm{per}(A) = 4$. In fact, these are given by those permutations σ where $\{(\sigma(1), \sigma(2), \sigma(3), \sigma(4))\} = \{(2,1,4,3), (2,4,3,1), (4,1,3,2), (4,2,3,1)\}$.

4. *Ménage problem*: Suppose that 5 wives are seated around a circular table, leaving one vacant space between consecutive women. Find the number of ways to seat in these vacant spots their 5 husbands so that no man is seated next to his wife. Suppose that the wives occupy positions $W_1, W_2, \ldots, W_5$ listed in a clockwise fashion around the table and that X_i is the vacant position to the right of W_i. Let A be the 5×5 0-1 matrix where $a_{ij} = 1$ if and only if husband H_i can be assigned to position X_j without violating the requirements of the problem. Then
$$A = \begin{pmatrix} 0 & 0 & 1 & 1 & 1 \\ 1 & 0 & 0 & 1 & 1 \\ 1 & 1 & 0 & 0 & 1 \\ 1 & 1 & 1 & 0 & 0 \\ 0 & 1 & 1 & 1 & 0 \end{pmatrix}.$$
By Fact 11, the number of possible assignments for each fixed placement of wives is $\mathrm{per}(A) = 13$. (Also see §2.4.2, Example 7.)

5. Count the number of nontaking rooks on a chessboard with restricted positions (§2.4.2). Specifically, suppose that positions $(1,1), (2,3), (3,1), (4,2), (4,3)$ of a 4×4 chessboard cannot be occupied by rooks. In the remaining positions, 4 rooks are to be placed so they are nontaking: no two are in the same row or in the same column. This can be solved (see Fact 11) by finding all permutations consistent with the 1s in the matrix
$$A = \begin{pmatrix} 0 & 1 & 1 & 1 \\ 1 & 1 & 0 & 1 \\ 0 & 1 & 1 & 1 \\ 1 & 0 & 0 & 1 \end{pmatrix}.$$
Here $\mathrm{per}(A) = 6$ is easily found using the Laplace expansion on the first column of A, so there are 6 placements of nontaking rooks.

REFERENCES

Printed Resources:

[An87] T. Ando, "Totally positive matrices", *Linear Algebra and Its Applications* 90 (1987), 165–219. (Discusses basic properties of totally positive matrices.)

[An91] H. Anton, *Elementary Linear Algebra*, 6th ed., Wiley, 1991.

BaRa97] R. B. Bapat and T. E. S. Raghavan, *Nonnegative Matrices and Applications*, Encyclopedia of Mathematical Sciences, No. 64, Cambridge University Press, 1997.

[Be70] R. Bellman, *Introduction to Matrix Analysis*, 2nd ed., McGraw-Hill, 1970.

[BePl94] A. Berman and R. J. Plemmons, *Nonnegative Matrices in the Mathematical Sciences*, SIAM, 1994. (Second revised edition of an authentic modern introduction.)

[BlMi77] N. J. Bloch and J. G. Michaels, *Linear Algebra*, McGraw-Hill, 1977. Reprinted by Tech Books, 1994.

[BrRy91] R. A. Brualdi and H. J. Ryser, *Combinatorial Matrix Theory*, Cambridge University Press, 1991. (Discusses combinatorial results pertaining to matrices and graphs, as well as permanents.)

[Da95] B. N. Datta, *Numerical Linear Algebra and Applications*, Brooks/Cole, 1995.

[DuErRe89] I. S. Duff, A. M. Erisman, and J. K. Reid, *Direct Methods for Sparse Matrices*, 2nd ed., Oxford University Press, 1989.

[Ga59] F. R. Gantmacher, *The Theory of Matrices*, Volumes I and II, Chelsea, 1959. (The first volume contains an introduction to matrix theory; the second discusses special topics such as nonnegative matrices and totally positive matrices.)

[GeLi81] A. George and J. W-H. Liu, *Computer Solution of Large Sparse Positive Definite Systems*, Prentice-Hall, 1981.

[GoVa96] G. H. Golub and C. F. Van Loan, *Matrix Computations*, 3rd ed., Johns Hopkins University Press, 1996. (Discusses elementary and advanced matrix computation problems; a remarkable blend of theoretical aspects and numerical issues.)

[HoJo85] R. A. Horn and C. R. Johnson, *Matrix Analysis*, Cambridge University Press, 1985. (A comprehensive, modern introduction to matrix theory.)

[Kn81] D. E. Knuth, "A permanent inequality", *American Mathematical Monthly* 88 (1981), 731–740.

[Mi88] H. Minc, *Nonnegative Matrices*, Wiley, 1988. (An elementary exposition of basic results.)

[Mi78] H. Minc, *Permanents*, Addison-Wesley, 1978. (Contains an extensive annotated bibliography up to 1978.)

[PoGe89] C. K. Pokorny and C. F. Gerald, *Computer Graphics: The Principles Behind the Art and Science*, Franklin, Beedle & Associates, 1989.

[Sc86] H. Schneider, "The influence of the marked reduced graph of a nonnegative matrix on the Jordan form and on related properties: a survey", *Linear Algebra and Its Applications* 84 (1986), 161–189.

[St88] G. Strang, *Linear Algebra and Its Applications*, 3rd ed., Harcourt, Brace, Jovanovich, 1988.

[Wi65] J. H. Wilkinson, *The Algebraic Eigenvalue Problem*, Oxford University Press, 1965. (An excellent reference on eigenvalues and matrix computations in general.)

Web Resources:

http://www.mathworks.com (Contains *MATLAB*, high-level language for dense and sparse systems, eigenanalysis of dense and sparse matrices, and calculation of characteristic polynomials.)

http://www.nag.com (Contains *NAG*: Fortran and C libraries for dense and sparse systems and eigenanalysis of dense and sparse matrices.)

http://www.netlib.org (Contains *LINPACK*: a collection of Fortran routines for relatively small dense systems; also contains *EISPACK*: a collection of Fortran routines for analyzing eigenvalues and eigenvectors of several classes of matrices.)

http://www.netlib.org (Contains *LAPACK/CLAPACK* (supersedes *LINPACK*): a collection of Fortran and C routines for dense and banded problems, ideal for shared-memory vector and parallel processors.)

http://www.vni.com/products/imsl (Contains *IMSL*: Fortran and C libraries for dense and sparse systems and for eigenanalysis of dense and banded problems.)

7

DISCRETE PROBABILITY

7.1 Fundamental Concepts .... Joseph R. Barr

7.2 Independence and Dependence .... Joseph R. Barr
    7.2.1 Basic Concepts
    7.2.2 Urn Models

7.3 Random Variables .... Joseph R. Barr
    7.3.1 Distributions
    7.3.2 Mean, Variance, and Higher Moments
    7.3.3 Generating Functions

7.4 Discrete Probability Computations .... Peter R. Turner
    7.4.1 Integer Computations
    7.4.2 Floating-Point Computations

7.5 Random Walks .... Patrick Jaillet
    7.5.1 General Concepts
    7.5.2 One-Dimensional Simple Random Walks
    7.5.3 Generalized Random Walks
    7.5.4 Applications of Random Walks

7.6 System Reliability .... Douglas R. Shier
    7.6.1 General Concepts
    7.6.2 Coherent Systems
    7.6.3 Calculating System Reliability
    7.6.4 Improved Algorithms for Calculating Reliability

7.7 Discrete-Time Markov Chains .... Vidyadhar G. Kulkarni
    7.7.1 Markov Chains
    7.7.2 Transient Analysis
    7.7.3 Classification of States
    7.7.4 Limiting Behavior
    7.7.5 First Passage Times
    7.7.6 Branching Processes

7.8 Queueing Theory .... Vidyadhar G. Kulkarni
    7.8.1 Single-Station Queues
    7.8.2 General Systems
    7.8.3 Special Queueing Systems

7.9 Simulation .... Lawrence M. Leemis
    7.9.1 Input Modeling
    7.9.2 Output Analysis
    7.9.3 Simulation Languages

Chapter 7 DISCRETE PROBABILITY

INTRODUCTION

This chapter discusses aspects of discrete probability that are relevant to mathematics, computer science, engineering, and other disciplines. Topics covered include random variables, important discrete probability distributions, random walks, Markov chains, and queues. Various applications to genetics, telephone network performance and reliability, average-case algorithm analysis, and combinatorics are presented.

GLOSSARY

absorbing boundary: a boundary that stops the motion of a random walk whose trajectory comes into contact with it.

all-terminal reliability: the probability that a given network is connected.

antithetic variates: a variance reduction technique, based on negatively correlated variates, used in the simulation analysis of a given system.

aperiodic state: a state of a Markov chain that is not periodic.

arrival process: the statistical description of the time between successive arrivals to a queueing system.

average-case complexity (of an algorithm): the average number of operations required by the algorithm, taken over all problem instances of a given size.

Bernoulli random variable: the discrete random variable $X \in \{0, 1\}$ with probability distribution $Pr(X = 0) = 1 - p$ and $Pr(X = 1) = p$, for some $0 < p < 1$.

binomial random variable: the discrete random variable $X \in \{0, 1, \ldots, n\}$ with probability distribution $Pr(X = k) = \binom{n}{k} p^k (1-p)^{n-k}$, for some $0 < p < 1$.

Bose-Einstein model: a probability model in which k indistinguishable balls are randomly placed into n distinguishable urns; several balls are allowed to occupy the same urn.

boundary: a point or set of points restricting the trajectory of a random walk.

branching process: a special type of Markov chain used to model the growth, and possible extinction, of populations.

closed class: a communicating class of states of a Markov chain in which transitions from these states never lead to states outside the class.

coherent system: a system of components for which increasing the number of operating components will not degrade the performance of the system.

common random numbers: a variance reduction technique in which alternative system configurations are analyzed using the same set of random numbers.

communicating class: a maximal set of states in a Markov chain that are reachable from one another by a finite number of transitions.

conditional probability: the probability that one event, A, occurs, given that another event, B, has occurred, written $Pr(A|B)$.

cutset: a minimal set of edges in a graph the removal of which disconnects the graph.

density function: a nonnegative real-valued function $f(x)$ that determines the distribution of a continuous random variable X via $Pr(a < X < b) = \int_a^b f(x)\,dx$.

dependent (events): events that are not independent.

discrete-event simulation: a simulation of a time-evolving stochastic process in which changes to the state of the system can only occur at discrete instants.

discrete-time Markov chain: a probabilistic model of a randomly evolving system whose future is independent of the past if the present state is known.

distribution (of a random variable): a probability measure associated with the values attained by the random variable.

elastic boundary: a boundary that could be absorbing or reflecting, usually depending on some given probability.

event: a subset of the sample space.

expected value (of a random variable): the average value taken on by the random variable.

experiment: any physically or mentally conceivable action having a measurable result.

extinction probability: the probability in a branching process that the population eventually dies out.

Fermi-Dirac model: a probability model in which k indistinguishable balls are randomly placed into n distinguishable urns; at most one ball can occupy each urn.

first passage time: the time to first visit a given set of states in a Markov chain.

floating-point arithmetic: the "real number" arithmetic of computers.

flop: a unit for floating-point computations that is useful in assessing the complexity of an algorithm.

gambler's ruin: a one-dimensional random walk in which a gambler wins or loses one unit at each play of a game, with the game terminating whenever the gambler amasses a known amount or loses his entire initial stake.

geometric random variable: the discrete random variable $X \in \{1, 2, \ldots\}$ with probability distribution $Pr(X = k) = (1-p)^{k-1}p$, for some $0 < p < 1$.

hypergeometric random variable: the discrete random variable that counts the number of red balls obtained when randomly selecting a fixed number of balls from an urn containing a specified number of red and black balls.

independent events: events in which knowledge of whether one of the events did or did not occur does not alter the probability of occurrence of any of the other events.

independent random variables: random variables whose joint distribution is the product of their individual distributions.

irreducible chain: a Markov chain that can visit any state from any other state in a finite number of steps.

irrelevant edge: an edge of a two-terminal network not appearing on any simple path joining the two terminals of the network.

K-cutset: a minimal set of edges in a graph, the removal of which disconnects some pair of vertices in K.

K-tree: a minimal set of edges in a graph that connects all vertices in K.

machine unit: a measure of the precision of floating-point arithmetic.

Maxwell-Boltzmann model: a probability model in which k distinguishable balls are randomly placed into n distinguishable urns; several balls can occupy the same urn.

mincut: a minimal set of components in a coherent system such that the system fails whenever these specified components fail.

minpath: a minimal set of components in a coherent system such that the system operates whenever these specified components operate.

Monte Carlo simulation: a simulation used to study both deterministic and stochastic phenomenon in which the passage of time is not material.

overflow: the result of a floating-point arithmetic operation that exceeds the available range of numbers.

parallel system: a system of components that fails only when all components fail.

periodic state: a state of a Markov chain that can only be revisited at multiples of a certain number $d > 1$ (the ***period*** of the state).

Poisson random variable: the discrete random variable $X \in \{0, 1, \ldots\}$ with probability distribution $Pr(X = k) = \frac{e^{-\lambda}\lambda^k}{k!}$, for some $\lambda > 0$.

probability: a numerical value between 0 and 1 measuring the likelihood of occurrence of an event; the larger the number, the more likely the event.

pseudo-random numbers: numbers generated in a predictable fashion, but that appear to behave like independent and identically distributed random numbers.

purely multiplicative linear congruential generator: a widely used method of producing a stream of pseudo-random numbers.

queue capacity: the maximum number of customers allowed at any time in a queueing system, either waiting or being served.

queue discipline: the protocol according to which customers are selected for service from among those waiting for service.

queueing system: a stochastic process in which customers arrive, await service, and are served.

queue-length process: a stochastic process describing the number of customers in the queueing system.

random numbers: real numbers generated uniformly over the interval $(0, 1)$.

random variable: a function that assigns a real number to each outcome in the sample space.

random walk: a stochastic process based on the problem of determining the probable location of a point subject to random motions.

recurrent state: a state of a Markov chain from which the probability of return to itself is 1.

recurrent walk: a random walk that returns to its starting location with probability 1.

reflecting boundary: a boundary that redirects the motion of a random walk whose trajectory comes into contact with it.

relative error: the (percent) error in a computation relative to the true value.

reliability: the probability that a given system functions at a random instant of time.

roundoff error: the error resulting from abbreviating a number to the precision of the machine.

sample size: the number of possible outcomes of an experiment.

sample space: the set of all possible outcomes of an experiment.

series system: a system of components that operates only when all components operate.

service-time distribution: the statistical distribution of time required to serve a customer in a queueing system.

simple path: a path containing no repeated vertices.

simulation: a technique for studying numerically the behavior of complex stochastic systems and estimating their performance.

single-station queueing system: a system in which customers arrive, wait for service, and depart after service completion.

s-t cutset: a minimal set of edges in a graph the removal of which leaves no s-t path.

stability condition: the set of parameter values for which the queue-length process (or the waiting-time process) has a steady-state distribution.

steady-state distribution: in a queueing system, the limiting probability distribution of the number of customers in the system.

stochastic process: a collection of random variables, typically indexed by time (discrete or continuous).

structure function: a binary-valued function defined on all subsets of components; its value indicates whether or not the system operates when the specified components all operate.

traffic intensity: in a queueing system, the ratio of the maximum arrival rate to the maximum service rate.

trajectory: the successive positions traced out by a particle undergoing a random walk.

transient state: a state in a Markov chain from which the probability of return to itself is less than 1.

transient walk: a random walk that is not recurrent.

transition probability: the probability of reaching a specified state in a Markov chain by a single transition (step) from a given state.

transition probability matrix: the matrix of one-step transition probabilities for a Markov chain.

two-terminal network: a network in which two vertices (or terminals) are specified.

two-terminal reliability: the probability that the specified vertices of a two-terminal network are connected by a path of operating edges.

underflow: the result of a floating-point operation that is smaller than the smallest representable number.

uniform random variable: the continuous random variable $X \in (\alpha, \beta)$ with density function $f(x) = \frac{1}{\beta - \alpha}$.

variance (of a random variable): a measure of dispersion of the random variable, equal to the average square of the deviation of the random variable from its expected value.

variance reduction techniques: methods for obtaining greater precision for a fixed amount of sampling.

waiting-time process: a stochastic process describing the time spent in the system by the customers.

7.1 FUNDAMENTAL CONCEPTS

Definitions:

An **experiment** is any physically or mentally conceivable undertaking that results in a measurable outcome.

The **sample space** is the set Ω of all possible outcomes of an experiment.

The **sample size** of an experiment is the number of possible outcomes of the experiment.

An **event** in the sample space Ω is a subset of Ω.

For a family of events $\{\, A_j \mid j \in J \,\}$, the **union** $\bigcup_{j \in J} A_j$ is the set of outcomes belonging to at least one A_j; the **intersection** $\bigcap_{j \in J} A_j$ is the set of all outcomes belonging to every A_j.

The **complement** $\overline{A}$ of an event A is the set of outcomes in the sample space not belonging to A.

The events A and B are **disjoint** if $A \cap B = \emptyset$. The events $A_1, A_2, A_3, \ldots$ are **pairwise disjoint** if every pair A_i, A_j of distinct events are disjoint.

A **probability measure** on the sample space Ω is a function Pr from the set of subsets of Ω into the interval $[0, 1]$ satisfying:

- $Pr(\Omega) = 1$;
- $Pr(\bigcup_{k=1}^{\infty} A_k) = \sum_{k=1}^{\infty} Pr(A_k)$, if the events $\{A_k\}$ are pairwise disjoint.

A **fair (unbiased) coin** is a coin that is just as likely to land Heads (H) as it is to land Tails (T).

A **red/blue spinner** is a disk consisting of two sectors, one red with area r and one blue with area b.

Facts:

1. $Pr(\emptyset) = 0$.

2. $Pr(A)$ has the interpretation of the long-run proportion of time that the event A occurs in repeated trials of the experiment.

3. $Pr(\bigcup_{k=1}^{n} A_k) = \sum_{k=1}^{n} Pr(A_k)$, if the n events $\{A_k\}$ are pairwise disjoint.

4. If all outcomes are equally likely and the sample space has k elements, where k is a positive integer, then the probability of event A is the number of elements of A divided by the size of the sample space; that is, $Pr(A) = \dfrac{|A|}{k}$.

5. Principle of inclusion-exclusion (simple form): For events A and B,
$$Pr(A \cup B) = Pr(A) + Pr(B) - Pr(A \cap B).$$

6. Principle of inclusion-exclusion (general form): For any events $A_1, A_2, \ldots, A_n$,
$$Pr(\bigcup_{r=1}^{n} A_r) = \sum_{i} Pr(A_i) - \sum_{i<j} Pr(A_i \cap A_j) + \sum_{i<j<k} Pr(A_i \cap A_j \cap A_k) - \cdots + (-1)^{n+1} Pr(A_1 \cap A_2 \cap \cdots \cap A_n).$$

7. *Sieve principle*: If $A_1, A_2, \ldots, A_n$ are events, then
$$Pr(\text{exactly } k \text{ of the } A_j \text{ occur}) = \sum_{r=k}^{n} (-1)^{r+k} \binom{r}{k} \sum_{j_1 < j_2 < \cdots < j_r} Pr(A_{j_1} \cap A_{j_2} \cap \cdots \cap A_{j_r}).$$

8. *Boole's inequality*: If $A_1, A_2, \ldots, A_n$ are events, then
$$Pr(\bigcup_{k=1}^{n} A_k) \leq \sum_{k=1}^{n} Pr(A_k).$$
If $A_1, A_2, \ldots$ is an infinite sequence of events, then
$$Pr(\bigcup_{k=1}^{\infty} A_k) \leq \sum_{k=1}^{\infty} Pr(A_k).$$
(George Boole, 1815–1864.)

9. *Bonferroni's inequality*: If $A_1, A_2, \ldots, A_n$ are events,
$$Pr(\bigcup_{k=1}^{n} A_k) \geq \sum_{k=1}^{n} Pr(A_k) - \sum_{k<j} Pr(A_k \cap A_j).$$

10. $Pr(\overline{A}) = 1 - Pr(A)$.

11. *Monotonicity*: If $A \subseteq B$, then $Pr(A) \leq Pr(B)$.

12. If $A_1 \subseteq A_2 \subseteq A_3 \subseteq \cdots$ is an increasing sequence of events, then
$$\lim_{n \to \infty} Pr(A_n) = Pr(\bigcup_{n=1}^{\infty} A_n).$$

13. If $A_1 \supseteq A_2 \supseteq A_3 \supseteq \cdots$ is a decreasing sequence of events, then
$$\lim_{n \to \infty} Pr(A_n) = Pr(\bigcap_{n=1}^{\infty} A_n).$$

14. Web-based notes on basic probability concepts together with interactive experiments can be found at the site

    http://www.math.uah.edu/~stat/

Examples:

1. The following table gives examples of specific experiments, their sample spaces, and the corresponding sample size:

experiment	sample space	sample size
toss a coin	$\{H, T\}$	2
toss a coin n times	$\{(\omega_1, \ldots, \omega_n) \mid \omega_i \text{ is } H \text{ or } T\}$	2^n
roll a die	$\{1, 2, 3, 4, 5, 6\}$	6
roll a pair of dice	$\{(1,1), (1,2), \ldots, (6,5), (6,6)\}$	36
draw a card from a standard deck	$\{2\clubsuit, 2\diamondsuit, \ldots, A\heartsuit, A\spadesuit\}$	52
spin a red/blue spinner	$\{\text{red, blue}\}$	2

2. The following are various events defined for the experiment of rolling a pair of dice (see the table of Example 1):

   sum of dice is 9: $A = \{(3,6), (4,5), (5,4), (6,3)\}$
   both dice are multiples of 3: $B = \{(3,3), (3,6), (6,3), (6,6)\}$
   sum of dice ≤ 4: $C = \{(1,1), (1,2), (1,3), (2,1), (2,2), (3,1)\}$.

3. The events $A = \{\text{sum of dice} = 9\}$ and $C = \{\text{sum of dice} \leq 4\}$ are disjoint. The events $A_i = \{\text{sum of the dice is } i\}$, $2 \leq i \leq 12$, are pairwise disjoint.

4. *Random selection of an integer:* Let $\Omega = \{1, 2, 3, \ldots, n\}$ be the sample space corresponding to the experiment of randomly selecting an integer between 1 and n, and define $Pr(j) = Pr(\{j\}) = \frac{1}{n}$. By Fact 4, $Pr(3 \leq j \leq n) = Pr(\{3 \leq j \leq n\}) = \frac{n-2}{n}$.

5. For a red/blue spinner, the sample space is $\Omega = \{\text{red, blue}\}$. If the spinner is equally likely to land at any location, then $Pr(\text{red}) = \frac{r}{r+b}$ and $Pr(\text{blue}) = \frac{b}{r+b}$.

6. Toss a fair coin n times and interpret Heads as 1 and Tails as 0. The sample space $\Omega = \{(\omega_1, \omega_2, \ldots, \omega_n) \mid \omega_j \in \{0, 1\}\}$ consists of all possible 0-1 sequences of length n. Since $|\Omega| = 2^n$, each probability $Pr((\omega_1, \omega_2, \ldots, \omega_n))$ is assigned the value $\frac{1}{2^n}$. By Fact 4, $Pr(A) = \frac{|A|}{2^n}$ holds for all $A \subseteq \Omega$.

For example, the probability of no tails appearing in four coin tosses is the probability of event $A = \{(1,1,1,1)\}$, so $Pr(A) = \frac{1}{16}$. The probability of exactly one tail is the probability of event $B = \{(0,1,1,1), (1,0,1,1), (1,1,0,1), (1,1,1,0)\}$; hence $Pr(B) = \frac{4}{16} = \frac{1}{4}$. The probability of at least two tails is, using Fact 10, $1 - Pr(A) - Pr(B) = 1 - \frac{5}{16} = \frac{11}{16}$.

7. *Derangements:* Let D_n be the set of derangements (§2.4.2) on the n elements $\{1, 2, \ldots, n\}$ and define A_j to be the set of all permutations fixing j. For any permutation σ, $Pr(\sigma) = \frac{1}{n!}$. Also, for $j_1 < j_2 < \cdots < j_k$, $Pr(A_{j_1} \cap A_{j_2} \cap \cdots \cap A_{j_k}) = \frac{(n-k)!}{n!}$ and $\sum_{j_1 < j_2 < \cdots < j_k} Pr(A_{j_1} \cap A_{j_2} \cap \cdots \cap A_{j_k}) = \binom{n}{k} \frac{(n-k)!}{n!} = \frac{1}{k!}$. By Fact 6, $Pr(\bigcup_{r=1}^{n} A_r) = \sum_i Pr(A_i) - \sum_{i<j} Pr(A_i \cap A_j) + \sum_{i<j<k} Pr(A_i \cap A_j \cap A_k) - \cdots + (-1)^{n+1} Pr(A_1 \cap A_2 \cap \cdots \cap A_n) = 1 - \frac{1}{2!} + \frac{1}{3!} - \cdots + (-1)^{n+1} \frac{1}{n!}$. Hence, $Pr(D_n) = 1 - 1 + \frac{1}{2!} - \frac{1}{3!} - \cdots + \frac{(-1)^n}{n!} \approx e^{-1} \approx 0.36788$.

8. *5-card stud poker:* Five cards are drawn from a well-shuffled deck of 52 playing cards. The sample space consists of the $\binom{52}{5} = 2{,}598{,}960$ possible five-card hands. The approximate probabilities of various events are displayed in the following table. (See §2.3.2 Example 12 for further details.) As seen from the probabilities given in the table, obtaining a five-card hand containing three of a kind is approximately ten times more likely than obtaining a five-card hand containing a flush, which in turn is approximately ten times more likely than obtaining a five-card hand containing four of a kind.

type of hand	example	hand enumeration	probability
one pair	$7\heartsuit, 7\diamondsuit, K\clubsuit, J\spadesuit, 2\heartsuit$	$\binom{13}{1}\binom{4}{2}\binom{12}{3}4^3$	0.42
two pairs	$7\spadesuit, 7\heartsuit, K\diamondsuit, K\spadesuit, 3\clubsuit$	$\binom{13}{2}\binom{4}{2}\binom{4}{2}44$	0.048
three of a kind	$7\clubsuit, 7\heartsuit, 7\diamondsuit, 3\diamondsuit, 5\spadesuit$	$\binom{13}{1}\binom{4}{3}\binom{12}{2}4^2$	0.021
straight	$7\clubsuit, 8\spadesuit, 9\diamondsuit, 10\clubsuit, J\heartsuit$	$10(4^5 - 4)$	0.0039
flush	$3\diamondsuit, 6\diamondsuit, 7\diamondsuit, J\diamondsuit, K\diamondsuit$	$4(\binom{13}{5} - 10)$	0.0020
full house	$3\heartsuit, 3\diamondsuit, 3\spadesuit, 7\clubsuit, 7\heartsuit$	$13 \cdot 12 \cdot \binom{4}{3}\binom{4}{2}$	0.0014
four of a kind	$A\clubsuit, A\diamondsuit, A\heartsuit, A\spadesuit, 7\spadesuit$	$13 \cdot 48$	0.00024
straight flush	$7\diamondsuit, 8\diamondsuit, 9\diamondsuit, 10\diamondsuit, J\diamondsuit$	$\binom{4}{1}9$	0.000014
royal flush	$10\heartsuit, J\heartsuit, Q\heartsuit, K\heartsuit, A\heartsuit$	4	0.0000015

7.2 INDEPENDENCE AND DEPENDENCE

Sequences of independent events are often encountered when an experiment is repeated (without changes). Independent events correspond, intuitively, to events that do not affect the outcome of one another. The treatment of dependent events requires conditional probabilities.

7.2.1 BASIC CONCEPTS

Definitions:

Two events A and B are **independent** if $Pr(A \cap B) = Pr(A)Pr(B)$.

The n events $A_1, A_2, \ldots, A_n$ are **independent** if for all k ($2 \le k \le n$) and $j_1, j_2, \ldots, j_k$ ($1 \le j_1 < j_2 < \cdots < j_k \le n$),
$$Pr(A_{j_1} \cap A_{j_2} \cap \cdots \cap A_{j_k}) = Pr(A_{j_1})Pr(A_{j_2})\ldots Pr(A_{j_k}).$$

The infinite collection of events $\{A_n \mid n \ge 1\}$ is **independent** if for all finite $k \ge 2$ the events $A_1, A_2, \ldots, A_k$ are independent.

Let B be an event with $Pr(B) > 0$. The **conditional probability of A given B** is
$$Pr(A|B) = \frac{Pr(A \cap B)}{Pr(B)}.$$

Facts:

1. If events A and B are independent, then so are A and $\overline{B}$, $\overline{A}$ and B, and $\overline{A}$ and $\overline{B}$.
2. If $A_1, A_2, \ldots$ are independent events, then $Pr(\bigcap_{k=1}^{\infty} A_k) = \prod_{k=1}^{\infty} Pr(A_k)$.
3. The function $\phi_B : A \to Pr(A|B)$ is a probability measure (§7.1).
4. $Pr(A \cap B) = Pr(A|B)Pr(B)$.
5. If A and B are independent, then $Pr(A|B) = Pr(A)$. This equation captures the notion that for independent events A and B the knowledge that one of the events has occurred does not affect the probability of the other occurring.
6. Pairwise independence of a collection of events does not necessarily imply that all events are independent (see Example 3).
7. *Law of total probabilities*: For any event A and any partition of Ω into events $B_1, B_2, \ldots, B_n$,
$$Pr(A) = \sum_{i=1}^{n} Pr(A \cap B_i) = \sum_{i=1}^{n} Pr(A|B_i)Pr(B_i).$$
8. *Bayes' formula*: For any event A and any partition of Ω into events $B_1, B_2, \ldots, B_n$,
$$Pr(B_1|A) = \frac{Pr(B_1 \cap A)}{Pr(A)} = \frac{Pr(A|B_1)Pr(B_1)}{\sum_{i=1}^{n} Pr(A|B_i)Pr(B_i)}.$$

(Thomas Bayes, 1702–1761.)

9. *Chain rule*: For any events $A_1, A_2, \ldots, A_n$ satisfying $Pr(\bigcap_{k=1}^{n-1} A_k) > 0$,
$$Pr(A_1 \cap A_2 \cap \cdots \cap A_n) = Pr(A_1)\,Pr(A_2|A_1)\,Pr(A_3|A_1 \cap A_2)\ldots Pr(A_n | \bigcap_{k=1}^{n-1} A_k).$$

Examples:

1. *Tossing two fair coins*: The sample space for this experiment consists of the four outcomes HH, HT, TH, TT. For example, the outcome HT means that the first coin turns up Heads and the second Tails. Because both coins are fair, all four outcomes are equally likely and in particular $Pr(HT) = \frac{1}{4}$. Since $Pr(H) = Pr(T) = \frac{1}{2}$, $Pr(HT) = \frac{1}{4} = \frac{1}{2} \cdot \frac{1}{2} = Pr(H)Pr(T)$. Thus, the events "Heads on the first coin" and "Tails on the second coin" are independent.

2. *Tossing a fair coin n times*: As in Example 6 of §7.1, let 1 stand for Heads and 0 for Tails. For each $1 \leq i \leq n$ select $\epsilon_i \in \{0,1\}$ and define $A_i = \{\epsilon_i \text{ occurs on the } i\text{th toss}\}$. Since all outcomes are equally likely, $Pr(A_1 \cap A_2 \cap \cdots \cap A_n) = (\frac{1}{2})^n = \frac{1}{2} \times \frac{1}{2} \times \cdots \times \frac{1}{2} = Pr(A_1)Pr(A_2)\ldots Pr(A_n)$. Also, for all $j_1, j_2, \ldots, j_k$ $(2 \leq k \leq n)$, $Pr(A_{j_1} \cap A_{j_2} \cap \cdots \cap A_{j_k}) = (\frac{1}{2})^k = \frac{1}{2} \times \frac{1}{2} \times \cdots \times \frac{1}{2} = Pr(A_{j_1})Pr(A_{j_2})\ldots Pr(A_{j_k})$. Therefore the events $A_1, A_2, \ldots, A_n$ are independent.

3. Let $\Omega = \{a, b, c, d\}$ be a sample space with equiprobable outcomes. Let $A = \{a, b\}$, $B = \{a, c\}$, and $C = \{a, d\}$. Here $Pr(A) = Pr(B) = Pr(C) = \frac{1}{2}$. Also $Pr(A \cap B) = \frac{1}{4} = \frac{1}{2} \cdot \frac{1}{2} = Pr(A)Pr(B)$, $Pr(A \cap C) = \frac{1}{4} = \frac{1}{2} \cdot \frac{1}{2} = Pr(A)Pr(C)$, and $Pr(B \cap C) = \frac{1}{4} = \frac{1}{2} \cdot \frac{1}{2} = Pr(B)Pr(C)$. Yet $Pr(A \cap B \cap C) = \frac{1}{4} \neq \frac{1}{2} \cdot \frac{1}{2} \cdot \frac{1}{2} = Pr(A)Pr(B)Pr(C)$. In this example, any two of the events are independent, but all three are not.

4. *Gambler's fallacy*: Suppose that a fair coin is tossed five times, turning up Heads on all five tosses. What is the probability that the next (sixth) toss turns up Tails? A common fallacy is to believe that a Tail is more likely to turn up next, since in the long run 50% of the coins should turn up Tails (and 50% Heads).

The appropriate sample space consists of $2^6 = 64$ equiprobable outcomes, representing any sequence of six Heads and/or Tails. The required probability is $Pr(A|B)$, where $A = \{(H, H, H, H, H, T)\}$ and $B = \{(H, H, H, H, H, H), (H, H, H, H, H, T)\}$. Then $Pr(A|B) = \frac{Pr(A \cap B)}{Pr(B)} = \frac{Pr(A)}{Pr(B)} = \frac{1}{2}$. Consequently, a Tail turning up next is just as likely as a Head.

5. An urn contains 7 blue marbles and 5 red marbles. An experiment consists of drawing (without replacement) a marble at random, observing its color, and then drawing a second marble at random. Let B_i be the event "the ith marble drawn is blue" and let R_i be the event "the ith marble drawn is red", where $i \in \{1, 2\}$. Then $Pr(B_1) = \frac{7}{12}$, $Pr(R_2|B_1) = \frac{5}{11}$, $Pr(B_1 \cap R_2) = Pr(R_2|B_1)Pr(B_1) = \frac{5}{11} \cdot \frac{7}{12} = \frac{35}{132}$. By Fact 7:

$$Pr(R_2) = Pr(R_2|R_1)Pr(R_1) + Pr(R_2|B_1)Pr(B_1) = \frac{4}{11} \cdot \frac{5}{12} + \frac{5}{11} \cdot \frac{7}{12} = \frac{55}{132}.$$

By Fact 8:

$$Pr(B_1|R_2) = \frac{Pr(R_2|B_1)Pr(B_1)}{Pr(R_2|R_1)Pr(R_1) + Pr(R_2|B_1)Pr(B_1)} = \frac{\frac{5}{11} \cdot \frac{7}{12}}{\frac{4}{11} \cdot \frac{5}{12} + \frac{5}{11} \cdot \frac{7}{12}} = \frac{7}{11}.$$

6. A particular family is known to have two children (one 9 years old, the other 10 years old). When a census taker comes to the house, a girl answers the doorbell. What is the probability that the other child is also a girl?

To answer this question, construct the sample space $\Omega = \{(b, b), (b, g), (g, b), (g, g)\}$, where, for example, the ordered pair (b, g) means that the younger child is a boy and the older child is a girl. Assume that all four outcomes in the sample space are equiprobable. The required probability is $Pr(A|B)$, where $A = \{(g, g)\}$ and $B = \{(b, g), (g, b), (g, g)\}$. Then

$$Pr(A|B) = \frac{Pr(A \cap B)}{Pr(B)} = \frac{Pr(\{(g, g)\})}{Pr(\{(b, g), (g, b), (g, g)\})} = \frac{\frac{1}{4}}{\frac{3}{4}} = \frac{1}{3}.$$

7. *Genetics*: Genes are responsible for physical traits of all living things. Each gene is composed of two alleles. Dominant alleles are represented with capital letters and recessive alleles with lower case letters. The basic discoveries concerning genetics were made by Gregor Mendel (1822–1884). One of the genes that is responsible for eye color exhibits two alleles — a dominant one B, for brown eyes, and a recessive one b, for blue eyes. In a certain population the genotype probabilities are

Pr(an individual has genotype BB) $= 0.2$

Pr(an individual has genotype Bb) $= 0.5$

Pr(an individual has genotype bb) $= 0.3$.

Let E_b be the event that an offspring receives a b allele from its mother, and let F_b be the event that it receives a b allele from its father. Conditioning on the genotype (BB, Bb, bb) of the offspring produces

$$Pr(E_b) = Pr(E_b|BB)Pr(BB) + Pr(E_b|Bb)Pr(Bb) + Pr(E_b|bb)Pr(bb)$$
$$= 0 \times 0.2 + 0.5 \times 0.5 + 1 \times 0.3 = 0.55;$$

similarly $Pr(F_b) = 0.55$.

Let C be the event that the offspring has blue eyes (that is, has genotype bb). By independence,

$$Pr(C) = Pr(E_b \cap F_b) = Pr(E_b)Pr(F_b) = (0.55)^2 = 0.3025.$$

8. In Example 7, let A be the event that the father has blue eyes (i.e., has genotype bb). If the father has blue eyes, then the offspring will have blue eyes (event C) if and only if it receives a b allele from its mother, and so $Pr(C|A) = Pr(E_b) = 0.55$. Also, $Pr(C|$father is $Bb) = Pr(C$ and mother is $Bb|$father is $Bb) + Pr(C$ and mother is $bb|$father is $Bb) = 0.25 \times 0.5 + 0.5 \times 0.3 = 0.275$ and $Pr(C|$father is $BB) = 0$.

The conditional probability that the father has blue eyes if the offspring has blue eyes is obtained from Fact 8 (interchanging phenotype with genotype when convenient):

$$Pr(A|C) = \frac{Pr(C|\text{father has } bb)Pr(\text{father has } bb)}{Pr(C|bb)Pr(bb) + Pr(C|Bb)Pr(Bb) + Pr(C|BB)Pr(BB)}$$

$$= \frac{0.55 \times 0.3}{0.55 \times 0.3 + 0.275 \times 0.5 + 0} \approx 0.545.$$

9. *Let's Make a Deal*: A game show contestant is told there is a fabulous prize hidden behind one of three doors (A, B, or C). The contestant guesses that the prize is behind door A. At this point the game show host (who knows what is behind each door, and in particular knows that the prize is not behind door B) opens door B, revealing that the prize is not there. The contestant is then offered the opportunity to change her guess. Should she? Intuition might suggest that nothing is to be gained by changing the guess (the prize, it is argued, is now equally likely to be behind either door A or door C). Using conditional probabilities, however, shows that it is definitely worthwhile to now guess that the prize is behind door C, assuming that the host is known to always open a door with no prize and to choose randomly if both remaining doors do not hide the prize.

It is reasonable to assume that the prize is equally likely to be hidden behind each of the doors. Thus, if H_X denotes the event in which the prize is hidden behind door X, then $Pr(H_A) = Pr(H_B) = Pr(H_C) = \frac{1}{3}$. If O_X denotes the event that door X is opened, then $Pr(O_B|H_A) = Pr(O_C|H_A) = \frac{1}{2}$, whereas $Pr(O_B|H_B) = 0$ and $Pr(O_B|H_C) = 1$. By Fact 8,

$$Pr(H_A|O_B) = \frac{Pr(H_A \cap O_B)}{Pr(O_B)}$$

$$= \frac{Pr(H_A)Pr(O_B|H_A)}{Pr(H_A)Pr(O_B|H_A) + Pr(H_B)Pr(O_B|H_B) + Pr(H_C)Pr(O_B|H_C)}$$

$$= \frac{\frac{1}{3} \cdot \frac{1}{2}}{\frac{1}{3} \cdot \frac{1}{2} + 0 + \frac{1}{3} \cdot 1} = \frac{1}{3}.$$

Similarly $Pr(H_C|O_B) = \frac{2}{3}$, so it is twice as likely for the prize to be hidden behind door C as behind door A, given that door B is shown to contain no prize. A web-based simulation of this situation, in which prizes are randomly hidden behind doors, enables one to verify experimentally this conclusion; see the following World Wide Web site:
    http://www.intergalact.com/threedoor/threedoor.html.

7.2.2 URN MODELS

Several applications can be viewed as the result of placing balls into urns.

Definitions:

In the following models, k balls are randomly placed in n distinguishable urns labeled 1 through n.

- **Model 1** (**Maxwell-Boltzmann**): The balls are distinguishable and multiple occupancy is permitted.
- **Model 2**: The balls are distinguishable and multiple occupancy is not permitted.
- **Model 3** (**Fermi-Dirac**): The balls are indistinguishable and multiple occupancy is not permitted.
- **Model 4** (**Bose-Einstein**): The balls are indistinguishable and multiple occupancy is permitted.
- **Model 5**: The balls are distinguishable, no urn is allowed to remain empty, and multiple occupancy is permitted.

Facts:

1. The following table shows, for different urn models, the probability of the event $(k_1, k_2, \ldots, k_n)$, in which k_1 balls are in urn 1, k_2 balls are in urn 2, $\ldots$, k_n balls are in urn n, with the restrictions $\sum_{j=1}^{n} k_j = k$, $k_j \geq 0$. In model 2, $n^{\underline{k}}$ is a falling power (see §3.4.2). In models 2 and 3, every $k_j \in \{0, 1\}$ and the models are meaningful only if $k \leq n$. In model 5, every $k_j \geq 1$, $k \geq n$, and $\{{k \atop n}\}$ is a Stirling subset number (§2.5.2).

model	sample size	enumeration of $(k_1, \ldots, k_n)$	probability of $(k_1, \ldots, k_n)$
1	n^k	$\binom{k}{k_1 \ k_2 \ \ldots \ k_n}$	$\binom{k}{k_1 \ k_2 \ \ldots \ k_n} n^{-k}$
2	$n^{\underline{k}}$	$k!$	$\binom{n}{k}^{-1}$
3	$\binom{n}{k}$	1	$\binom{n}{k}^{-1}$
4	$\binom{n+k-1}{k}$	1	$\binom{n+k-1}{k}^{-1}$
5	$n!\{{k \atop n}\}$	$\binom{k}{k_1 \ k_2 \ \ldots \ k_n}$	$\binom{k}{k_1 \ k_2 \ \ldots \ k_n}/(n!\{{k \atop n}\})$

2. The Maxwell-Boltzmann model was originally proposed to explain the distribution of k subatomic particles into n different energy states. It has been replaced by the Bose-Einstein model (appropriate for particles with integer "spin", such as photons and pi mesons) and by the Fermi-Dirac model (appropriate for particles with half-integer "spin", such as protons and neutrons).

3. *Pólya's urn scheme* (George Pólya, 1887–1985): In this model, an urn contains b black and r red balls. At each step one ball is drawn and replaced, and c additional balls of the same color are placed in that urn. This scheme models the spread of a contagious disease where an infected person infects c other persons.

4. The case $c = 0$ in Pólya's urn scheme corresponds to sampling balls with replacement.

5. The case $c = -1$ in Pólya's urn scheme corresponds to sampling balls without replacement.

6. The following table shows how to calculate several types of probabilities using Pólya's urn scheme, where b is the number of black balls, r is the number of red balls, and c is the number of additional balls added each time:

event	probability
drawing a black	$\frac{b}{b+r}$
drawing a black then red	$\frac{br}{(b+r)(b+r+c)}$
drawing in order black, red, black	$\frac{br(b+c)}{(b+r)(b+r+c)(b+r+2c)}$
drawing k black and $n-k$ red balls in a prescribed order	$\frac{b(b+c)\ldots(b+(k-1)c)r(r+c)\ldots(r+(n-k-1)c)}{(b+r)(b+r+c)(b+r+2c)\ldots(b+r+(n-1)c)}$
drawing k black balls in n drawings; the order of drawing does not matter	$\frac{\binom{-b/c}{k}\binom{-r/c}{n-k}}{\binom{-(b+r)/c}{n}}$

Examples:

1. *Partial derivatives*: For analytic functions f, the order in which derivatives is taken does not matter. As an example, the mixed second partial derivatives f_{xy} and f_{yx} are equal, as are f_{xxy} and f_{xyx}. Consequently, the number of different third-order partial derivatives of a function of n variables is the number of ways to distribute $k = 3$ indistinguishable balls into n urns (variables). Each such distribution corresponds to selecting the number of times each variable occurs in forming the partial derivative. Using the entry for Model 4 in the table for Fact 1, there are $\binom{n+2}{3}$ third-order partial derivatives of f. When $n = 3$ this gives $\binom{5}{3} = 10$ different third-order partial derivatives of $f(x, y, z)$: namely, $f_{xxx}, f_{yyy}, f_{zzz}, f_{xxy}, f_{xxz}, f_{xyy}, f_{yyz}, f_{xzz}, f_{yzz}, f_{xyz}$. In general, there are $\binom{n+k-1}{k}$ different kth-order partial derivatives of f.

2. Model 3 provides a model for the occurrence of misprints on the pages of a book. Here the n urns correspond to the n symbols printed sequentially in the book and k is the number of misprints. Each symbol is either correct or a misprint, so multiple occupancy does not occur.

Also, assuming that the misprints are not generated in a systematic fashion, the k balls can be considered indistinguishable, with misprints equally likely to occur at any location on the page.

3. *Lottery odds*: A lottery is conducted by selecting five different numbers from $1, 2, \ldots, 9$. This can be viewed using urn model 3, in which the five selected numbers correspond to $k = 5$ identical balls placed into $n = 9$ distinguished urns. The number of such selections, by the table of Fact 1, is $\binom{9}{5} = 126$. Only one of these 126 selections matches all the five winning numbers, so $Pr(\text{match } 5) = \frac{1}{126}$.

To match exactly four of the five winning numbers, select the matching numbers in $\binom{5}{4} = 5$ ways and select the (single) nonmatching number in $\binom{4}{1} = 4$ ways, giving $Pr(\text{match } 4) = \frac{5 \cdot 4}{126} = \frac{20}{126}$.

To match exactly three of the winning numbers, select the matching numbers in $\binom{5}{3} = 10$ ways and select the two nonmatching numbers in $\binom{4}{2} = 6$ ways, giving $Pr(\text{match } 3) = \frac{10 \cdot 6}{126} = \frac{60}{126}$.

4. In a number of state lotteries, $k = 6$ numbers are drawn from $1, 2, \ldots, n$. The following table gives the probability of matching exactly six, exactly five, and exactly four of the six winning numbers, for various values of n.

n	match 6	match 5	match 4
35	1/1,623,160	87/811,580	87/23,188
36	1/1,947,792	15/162,316	2,175/649,264
37	1/2,324,784	31/387,464	2,325/774,928
38	1/2,760,681	64/920,227	2,480/920,227
39	1/3,262,623	66/1,087,541	2,640/1,087,541
40	1/3,838,380	17/319,865	561/255,892
41	1/4,496,388	35/749,398	2,975/1,498,796
42	1/5,245,786	108/2,622,893	675/374,699
43	1/6,096,454	111/3,048,227	4,995/3,048,227
44	1/7,059,052	57/1,764,763	10,545/7,059,052
45	1/8,145,060	39/1,357,510	741/543,004
46	1/9,366,819	80/3,122,273	3,900/3,122,273
47	1/10,737,573	82/3,579,191	4,100/3,579,191
48	1/12,271,512	21/1,022,626	4,305/4,090,504
49	1/13,983,816	43/2,330,636	645/665,896
50	1/15,890,700	22/1,324,225	473/529,690
51	1/18,009,460	27/1,800,946	1,485/1,800,946
52	1/20,358,520	69/5,089,630	3,105/4,071,704
53	1/22,957,480	141/11,478,740	3,243/4,591,496
54	1/25,827,165	32/2,869,685	376/573,937
55	1/28,989,675	98/9,663,225	392/644,215
56	1/32,468,436	25/2,705,703	875/1,546,116
57	1/36,288,252	17/2,016,014	2,125/4,032,028
58	1/40,475,358	52/6,745,893	1,105/2,248,631
59	1/45,057,474	53/7,509,579	3,445/7,509,579
60	1/50,063,860	81/12,515,965	4,293/10,012,772

5. Let an urn contain $c = 1$ red ball and $b = 9$ black balls. In Pólya's urn scheme with $c = 1$, the probability of obtaining the sequence RRB (two red balls and then a black ball) is found using conditional probabilities as

$$Pr(RRB) = Pr(B|RR)Pr(R|R)Pr(R) = \tfrac{9}{12} \cdot \tfrac{2}{11} \cdot \tfrac{1}{10}.$$

Likewise,
$$Pr(BRR) = Pr(R|BR)Pr(R|B)Pr(B) = \tfrac{2}{12} \cdot \tfrac{1}{11} \cdot \tfrac{9}{10},$$
and
$$Pr(RBR) = Pr(R|RB)Pr(B|R)Pr(R) = \tfrac{2}{12} \cdot \tfrac{9}{11} \cdot \tfrac{1}{10}.$$
Thus,
$$Pr(RRB) = Pr(BRR) = Pr(RBR) = \tfrac{3}{220},$$
agreeing with the value obtained using the table of Fact 6, with $k=1$ and $n=3$.

The probability of obtaining two red balls and one black ball in some order is then $Pr(RRB) + Pr(BRR) + Pr(RBR) = \tfrac{9}{220}$. Using the extended binomial coefficients (§2.3.2), the corresponding entry in the table of Fact 6 can be verified for this case, where $k=1$ and $n=3$:
$$Pr(\text{exactly one black ball}) = \frac{\binom{-9}{1}\binom{-1}{2}}{\binom{-10}{3}} = \frac{(-1)^1\binom{9}{1}(-1)^2\binom{2}{2}}{(-1)^3\binom{12}{3}} = \frac{9}{220},$$
agreeing with the value already found.

7.3 RANDOM VARIABLES

7.3.1 DISTRIBUTIONS

Definitions:

A **random variable** X is a real-valued function on a probability space Ω.

The random variable $X: \Omega \to \mathcal{R}$ is **discrete** if the range of X is finite or countable.

The real-valued function $f: \mathcal{R} \to \mathcal{R}$ is a **density function** if
- $f(x) \geq 0$ for all $x \in \mathcal{R}$;
- $\int_{-\infty}^{\infty} f(x)\,dx = 1$.

The random variable X is **(absolutely) continuous** if there exists a density function f such that $Pr(a < X < b) = \int_a^b f(x)\,dx$ for all $a < b$.

The **distribution** μ_X of the random variable X is given by $\mu_X(B) = Pr(X \in B)$ for every interval B.

The **cumulative distribution function** of a random variable X is given by $F(x) = Pr(X \leq x)$.

A **random vector** is a function $X = (X_1, \ldots, X_k): \Omega \to \mathcal{R}^k$.

The **joint distribution** $\mu_{X_1, \ldots, X_k}$ of the random vector $(X_1, \ldots, X_k)$ is defined by $\mu_{X_1, \ldots, X_k}(B_1, \ldots, B_k) = Pr(X_1 \in B_1, \ldots, X_k \in B_k)$ for any k intervals $B_1, \ldots, B_k$.

The random variables $X_1, \ldots, X_n$ are **independent** if for any intervals $B_1, \ldots, B_n$ $Pr(X_1 \in B_1, \ldots, X_n \in B_n) = Pr(X_1 \in B_1) \ldots Pr(X_n \in B_n)$.

Facts:

1. The cumulative distribution function $F(x)$ is a nondecreasing function of x.
2. $\lim_{x \to \infty} F(x) = 1$, $\lim_{x \to -\infty} F(x) = 0$.
3. $Pr(a < X \leq b) = F(b) - F(a)$ for $a < b$.

4. If X is a discrete random variable, then $\sum_k Pr(X = k) = 1$.
5. If X is a continuous random variable, then $\frac{d}{dx} F(x) = f(x)$.
6. If X_1 and X_2 are independent binomial random variables (see Table 1) with parameters n_1, p and n_2, p respectively, then $X_1 + X_2$ is also a binomial random variable with parameters $n_1 + n_2, p$.
7. If X_1 and X_2 are independent Poisson random variables (see Table 1) with parameters λ_1 and λ_2 respectively, then $X_1 + X_2$ is also a Poisson random variable with parameter $\lambda_1 + \lambda_2$.
8. If X_1 and X_2 are independent normal random variables (see Table 2) with parameters μ_1, σ_1^2 and μ_2, σ_2^2 respectively, then $X_1 + X_2$ is also a normal random variable with parameters $\mu_1 + \mu_2, \sigma_1^2 + \sigma_2^2$.

Examples:

1. A spinner has three sectors — red, white, and blue — with sector areas 0.2, 0.7, and 0.1, respectively. Define a random variable X according to the rule $X = 1$ if the spinner points on red, $X = 2$ if it points on white, and $X = 3$ if it points on blue. The distribution of the discrete random variable X is displayed in the following table.

event	i	$Pr(X = i)$
red	1	0.2
white	2	0.7
blue	3	0.1

2. *Bernoulli random variable*: Let $A \subset \Omega$ be a fixed subset, with $Pr(A) = p$ for some $0 < p < 1$. Define the random variable X by $X(\omega) = 1$ for $\omega \in A$ and $X(\omega) = 0$ for $\omega \notin A$. Often it is said that a *success* occurs whenever $\omega \in A$ and a *failure* occurs otherwise. Then X is a Bernoulli random variable with $Pr(X = 1) = p$ and $Pr(X = 0) = 1 - p$. (Jakob Bernoulli, 1654–1705)

3. *Binomial random variable*: Suppose that a die is thrown and that the occurrence of either a one or a six results in a "success". A single roll of the die constitutes a Bernoulli trial (Example 2) with probability of success $p = \frac{1}{3}$. The number of successes in 10 successive independent trials is a binomial random variable X with parameters $n = 10$ and $p = \frac{1}{3}$. In general, the number of successes X is a discrete random variable with possible values $0, 1, 2, \ldots, n$ and its distribution is given in Table 1.

4. A dart is thrown at a circular target of radius 1. Assume that the target is never missed and that any point on the target is as equally likely to be hit as any other point. Let X be the dart's distance from the center of the target. Since X can assume any value between 0 and 1, X is a continuous random variable. For $0 \le a < b \le 1$, $Pr(a < X < b) = Pr$(the dart lands in the annulus with radii a and b)$= \frac{1}{\pi} \times$(the area of the annulus with radii a and b)$= \frac{1}{\pi}(\pi b^2 - \pi a^2) = b^2 - a^2$.

5. *Hypergeometric random variable*: A total of n balls are selected from an urn containing N balls, of which m are red and $N - m$ are black. Let X be the number of red balls selected. Then X is a discrete random variable having the distribution $Pr(X = k) = \frac{\binom{m}{k}\binom{N-m}{n-k}}{\binom{N}{n}}$, for $0 \le k \le m$.

6. *Multinomial random variable*: Cast n identical balls into N labeled boxes in such a way that the probability that a ball ends up in box j is p_j, where $\sum_{j=1}^{N} p_j = 1$. Let X_j denote the number of balls in box j $(1 \le j \le N)$. For a vector $(x_1, x_2, \ldots, x_N)$ with

Table 1 Discrete random variables.

distribution	description of event $(X = k)$	range of X	$Pr(X = k)$
Bernoulli $B(1,p)$	$k=0$ indicates a failure, $k=1$ indicates a success	$0,1$	q p
binomial $B(n,p)$	k successes in n trials, each with probability p of success	$0,1,\ldots,n$	$\binom{n}{k}p^k q^{n-k}$
Poisson $P(\lambda)$	k arrivals to a counter over a unit period of time, at average rate λ	$0,1,2,\ldots$	$\frac{e^{-\lambda}\lambda^k}{k!}$
geometric $G(p)$	k trials before first success occurs	$1,2,\ldots$	$q^{k-1}p$
Pascal $NB(r,p)$	k trials before rth success occurs	$r, r+1, \ldots$	$\binom{k-1}{r-1}q^{k-r}p^r$
hypergeometric (N,m,n)	sample n items from N items, where m are defective and $N-m$ are not; $k =$ number of defectives selected	$0,1,\ldots,n$	$\frac{\binom{m}{k}\binom{N-m}{n-k}}{\binom{N}{n}}$

$\sum_{j=1}^{N} x_j = n$, the probability that box 1 contains x_1 balls, box 2 contains x_2 balls,..., box N contains x_N balls is given by

$$Pr(X_1 = x_1, X_2 = x_2, \ldots, X_N = x_N) = \frac{n!}{x_1! x_2! \ldots x_N!} p_1^{x_1} p_2^{x_2} \ldots p_N^{x_N}$$
$$= \binom{n}{x_1 \; x_2 \; \ldots \; x_N} p_1^{x_1} p_2^{x_2} \ldots p_N^{x_N},$$

expressed using the multinomial coefficients (§2.3.2).

7. *Joint distribution*: Two fair coins are tossed once, resulting in four equally likely outcomes $\{(T,T),(T,H),(H,T),(H,H)\}$. Let the random variable X be the total number of heads observed, and let the random variable Y be the number of heads on the first coin minus the number of heads on the second coin. The joint probability distribution $\mu_{X,Y}$ is given by $Pr(X=0, Y=0) = Pr(X=1, Y=-1) = Pr(X=1, Y=1) = Pr(X=2, Y=0) = \frac{1}{4}$. Thus $Pr(X=0) = Pr(X=2) = \frac{1}{4}$, $Pr(X=1) = \frac{1}{2}$ and $Pr(Y=1) = Pr(Y=-1) = \frac{1}{4}$, $Pr(Y=0) = \frac{1}{2}$. Since $Pr(X=0, Y=0) = \frac{1}{4} \neq \frac{1}{8} = Pr(X=0)Pr(Y=0)$, the variables X and Y are not independent.

8. Some important discrete random variables are described in Table 1. Here $q = 1-p$.
9. Some important continuous random variables are described in Table 2. Here it is understood that the density function $f(x) = 0$ outside the specified range.

7.3.2 MEAN, VARIANCE, AND HIGHER MOMENTS

Definitions:

The **mean** (*expected value*) EX of a discrete random variable X is given by $EX = \sum_k k Pr(X = k)$.

The **mean** of a continuous random variable X with density function f is given by $EX = \int_{-\infty}^{\infty} x f(x) \, dx$.

The **variance** $\text{Var}(X)$ of a random variable X is $\text{Var}(X) = E((X - EX)^2)$.

The **standard deviation** of X is $\sqrt{\text{Var}(X)}$.

Table 2 Continuous random variables.

distribution	range of X	density function $f(x)$
uniform (α, β)	(α, β)	$\frac{1}{\beta-\alpha}, \alpha < \beta$
exponential (λ)	$[0, \infty)$	$\lambda e^{-\lambda x}, \lambda > 0$
standard normal $(0,1)$	$(-\infty, \infty)$	$\frac{1}{\sqrt{2\pi}} e^{-x^2/2}$
normal (μ, σ^2)	$(-\infty, \infty)$	$\frac{1}{\sigma\sqrt{2\pi}} e^{-\frac{1}{2}(\frac{x-\mu}{\sigma})^2}, \sigma > 0$
gamma $\Gamma(n, \lambda)$	$[0, \infty)$	$\frac{\lambda^n x^{n-1} e^{-\lambda x}}{\Gamma(n)}, \Gamma(n) = \int_0^\infty t^{n-1} e^{-t}\, dt$
Cauchy (α)	$(-\infty, \infty)$	$\frac{\alpha}{\pi(\alpha^2+x^2)}, \alpha > 0$
beta (p, q)	$[0, 1]$	$\frac{\Gamma(p+q)}{\Gamma(p)\Gamma(q)} x^{p-1}(1-x)^{q-1}, p, q > 0$
chi square $\chi^2(r)$	$[0, \infty)$	$\frac{1}{2^{r/2}\Gamma(r/2)} x^{\frac{r}{2}-1} e^{-\frac{x}{2}}, r > 0$
F-distribution $F_{m,n}$	$[0, \infty)$	$\frac{\Gamma((m+n)/2)}{\Gamma(m/2)\Gamma(n/2)} \left(\frac{m}{n}\right)^{m/2} \frac{x^{(m-2)/2}}{(1+(m/n)x)^{(m+n)/2}}$
t-distribution t_k	$(-\infty, \infty)$	$\frac{\Gamma((k+1)/2)}{\Gamma(k/2)} \frac{1}{\sqrt{k\pi}} \frac{1}{(1+x^2/k)^{(k+1)/2}}$
Rayleigh $R(\sigma)$	$[0, \infty)$	$\frac{x e^{-x^2/2\sigma^2}}{\sigma^2}, \sigma > 0$

The **covariance** $\mathrm{Cov}(X, Y)$ of two random variables X and Y is given by $\mathrm{Cov}(X, Y) = E((X - EX)(Y - EY))$.

The **correlation** $\rho_{X,Y}$ of two random variables X and Y is $\rho_{X,Y} = \dfrac{\mathrm{Cov}(X,Y)}{\sqrt{\mathrm{Var}(X)\mathrm{Var}(Y)}}$.

The **kth moment** of a random variable X is $E(X^k)$.

Facts:

1. The expected value EX of a random variable X measures the "weighted average" of X or the "center of gravity" of its distribution.
2. $E(X + Y) = EX + EY$.
3. $E(cX) = cEX$ for all constants c.
4. $E(c) = c$ for all constants c.
5. If X is a nonnegative integer random variable, then $EX = \sum_{n=0}^{\infty} \Pr(X > n)$.
6. If X is a nonnegative continuous random variable, then $EX = \int_0^\infty \Pr(X > x)\, dx$.
7. If X and Y are independent, then $E(XY) = (EX)(EY)$.
8. If g is a real-valued function and X is a discrete random variable, then $E(g(X)) = \sum_k g(k) \Pr(X = k)$.
9. If g is an integrable real-valued function and X is continuous with density $f(x)$, then $E(g(X)) = \int g(t) f(t)\, dt$.
10. The variance $\mathrm{Var}(X)$ of a random variable X measures the "dispersion" of X about its expected value EX.
11. $\mathrm{Var}(X) \geq 0$; $\mathrm{Var}(X) = 0$ if and only if for some constant c, $\Pr(X = c) = 1$.

12. $\text{Var}(cX) = c^2 \text{Var}(X)$ for all constants c.
13. $\text{Var}(X+Y) = \text{Var}(X) + \text{Var}(Y) + 2\text{Cov}(X,Y)$.
14. $\text{Var}(X+Y) = \text{Var}(X) + \text{Var}(Y)$ if X and Y are independent.
15. $\text{Var}(X) = E(X^2) - (EX)^2$.
16. $\text{Cov}(X,Y) = 0$ if X and Y are independent. The converse is false. (Example 4.)
17. $\text{Cov}(X,Y) = E(XY) - (EX)(EY)$.
18. The correlation $\rho_{X,Y}$ is a scale-invariant measure of the degree of linear relationship between two random variables X and Y. Specifically, $\rho_{X,Y} = 1$ only when $Y = aX + b$ for some constants $a > 0$ and b. Similarly, $\rho_{X,Y} = -1$ only when $Y = aX + b$ for some constants $a < 0$ and b.
19. $|\rho_{X,Y}| \leq 1$.
20. *Bienaymé-Chebyshev's inequality*: $Pr(|X - EX| \geq t) \leq \frac{\text{Var}(X)}{t^2}$, for any value $t > 0$. (Irénée-Jules Bienaymé, 1796–1878 and Pafnuty Lvovich Chebyshev, 1821–1894.)
21. *Kolmogorov's inequality*: Suppose $X_1, X_2, \ldots, X_n$ are independent random variables, and let $S_k = X_1 + X_2 + \cdots + X_k$ for $1 \leq k \leq n$. Then for any value $t > 0$ the probability that $|S_k - ES_k| < t$ holds for all $k = 1, 2, \ldots, n$ is at least $1 - \frac{\text{Var}(S_n)}{t^2}$. (Andrey Nikolayevich Kolmogorov, 1903–1987.)

Examples:

1. The random variable X is the number of heads obtained in three tosses of a fair coin. It follows a binomial distribution (Table 1, §7.3.1), with $n = 3$ and $p = \frac{1}{2}$. Thus $Pr(X = 0) = \frac{1}{8}$, $Pr(X = 1) = \frac{3}{8}$, $Pr(X = 2) = \frac{3}{8}$, and $Pr(X = 3) = \frac{1}{8}$. Using the definition of expected value, $EX = \sum_k k Pr(X = k) = 0 \cdot \frac{1}{8} + 1 \cdot \frac{3}{8} + 2 \cdot \frac{3}{8} + 3 \cdot \frac{1}{8} = \frac{3}{2}$. In general, the mean of a binomial distribution with parameters n and p is np; see the corresponding entry in Table 3 of §7.3.3.

2. The variance of the discrete random variable X in Example 1 can be found using $\text{Var}(X) = E((X - EX)^2) = E((X - \frac{3}{2})^2) = (0 - \frac{3}{2})^2 \cdot \frac{1}{8} + (1 - \frac{3}{2})^2 \cdot \frac{3}{8} + (2 - \frac{3}{2})^2 \cdot \frac{3}{8} + (3 - \frac{3}{2})^2 \cdot \frac{1}{8} = \frac{3}{4}$. In general, the variance of a binomial distribution with parameters n and p is $np(1-p)$; see the corresponding entry in Table 3 of §7.3.3.

3. Suppose X is a Bernoulli random variable with parameter p, so $Pr(X = 0) = 1 - p$ and $Pr(X = 1) = p$. Then $EX = 0 \cdot (1-p) + 1 \cdot p = p$ and $\text{Var}(X) = E((X-p)^2) = (0-p)^2 \cdot (1-p) + (1-p)^2 \cdot p = p^2(1-p) + p(1-p)^2 = p(1-p)$. Also, $E(X^2) = 0^2 \cdot (1-p) + 1^2 \cdot p = p$ and using Fact 15 $\text{Var}(X) = E(X^2) - (EX)^2 = p - p^2 = p(1-p)$, as before.

4. *Covariance and independence*: In Example 7 of §7.3.1, $EX = 0 \cdot \frac{1}{4} + 1 \cdot \frac{1}{2} + 2 \cdot \frac{1}{4} = 1$ and $EY = -1 \cdot \frac{1}{4} + 0 \cdot \frac{1}{2} + 1 \cdot \frac{1}{4} = 0$. Also, $E(XY) = -1 \cdot \frac{1}{4} + 0 \cdot \frac{1}{2} + 1 \cdot \frac{1}{4} = 0$. By Fact 17, $\text{Cov}(X,Y) = E(XY) - (EX)(EY) = 0 - 1 \cdot 0 = 0$. In this example, variables X and Y have zero covariance (and zero correlation); however (see Example 7, §7.3.1) they are not independent random variables.

5. The moments of the normal random variable X with parameters $\mu = 0$ and $\sigma = 1$ are $E(X^{2k}) = 1 \cdot 3 \cdots (2k-1)$ and $E(x^{2k-1}) = 0$ for $k \geq 1$.

6. A manufacturing plant produces ball bearings with an average diameter of 50 mm and a variance of 11 mm^2. Without any further information about the shape of the distribution of the diameters X, Fact 20 shows that the probability $Pr(|X - 50| \geq 8)$ of exceeding the nominal diameter by more than 8 mm is no more than $\frac{\text{Var}(X)}{8^2} = \frac{11}{64} = 0.172$. Thus, no more than 17.2% of the ball bearings produced can exceed the stated tolerance.

7. Average-case algorithm analysis: A simple algorithm for locating an item in an (unordered) list $A = [a_1, a_2, \ldots, a_n]$ is called a *linear search*; it sequentially examines each entry of list A and compares the given item *key* with each a_k until a match is found, or until the entire list is searched, in which case *key* is known not to be in the list. To obtain the average case complexity of this algorithm, suppose that *key* is known to occur in A and that it is equally likely (with probability $\frac{1}{n}$) to be at each of the n positions of A. If *key* is in fact located at position k of A, then k comparisons are required by the algorithm. The expected number of comparisons needed is thus $EX = \sum_{k=1}^{n} k Pr(X = k) = \sum_{k=1}^{n} k \cdot \frac{1}{n} = \frac{1}{n} \sum_{k=1}^{n} k = \frac{1}{n} \frac{n(n+1)}{2} = \frac{(n+1)}{2}$. Consequently, the average-case complexity of linear search is $O(n)$; see §1.3.3.

7.3.3 GENERATING FUNCTIONS

Definitions:

The **probability generating function** of a discrete random variable X is the function $\phi(t) = E(t^X) = \sum_k Pr(X = k) t^k$, defined for $|t| \leq 1$.

The **moment generating function** of a discrete random variable X is the function $\psi(t) = E(e^{tX}) = \sum_k e^{tk} Pr(X = k)$, defined for all t such that $\psi(t)$ converges.

The **moment generating function** of a continuous random variable X with density f is the function $\psi(t) = E(e^{tX}) = \int e^{tx} f(x) \, dx$, defined for all t such that $\psi(t)$ converges.

The **characteristic function** (Fourier transform) of a discrete random variable X is $\chi(t) = E(e^{itX}) = \sum_k e^{itk} Pr(X = k)$, defined for all $t \in \mathcal{R}$.

The **characteristic function** of a continuous random variable X with density f is $\chi(t) = E(e^{itX}) = \int e^{itx} f(x) \, dx$, defined for all $t \in \mathcal{R}$.

Facts:

1. The expected value of a random variable X can be expressed in terms of the first derivative of its generating function: $EX = \phi'(1) = \psi'(0) = -i\chi'(0)$.

2. The variance of a random variable X can be expressed in terms of the first and second derivatives of its generating function: $\text{Var}(X) = \phi''(1) + \phi'(1) - [\phi'(1)]^2 = \psi''(0) - [\psi'(0)]^2 = [\chi'(0)]^2 - \chi''(0)$.

3. $\frac{d^k}{dt^k} \phi(1) = E(X(X-1)(X-2)\ldots(X-k+1))$.

4. $\frac{d^k}{dt^k} \psi(0) = E(X^k)$.

5. $\frac{d^k}{dt^k} \chi(0) = i^k E(X^k)$.

6. For any of the three types of generating functions defined, the generating function of the sum of independent random variables is the product of their respective generating functions.

Examples:

1. The binomial random variable with parameters n and p is the sum of n independent Bernoulli random variables with parameter p. The probability generating function for a Bernoulli random variable is $\phi(t) = E(t^X) = t^0(1-p) + t^1 p = q + pt$, where $q = 1-p$. By Fact 6 the probability generating function for a binomial random variable is $[\phi(t)]^n = (q + pt)^n$.

Table 3 Moments and generating functions for discrete distributions.

distribution	mean	variance	$\phi(t)$	$\psi(t)$	$\chi(t)$
Bernoulli $B(1,p)$	p	pq	$q+pt$	$q+pe^t$	$q+pe^{it}$
Binomial $B(n,p)$	np	npq	$(q+pt)^n$	$(q+pe^t)^n$	$(q+pe^{it})^n$
Poisson $P(\lambda)$	λ	λ	$e^{\lambda(t-1)}$	$e^{\lambda(e^t-1)}$	$e^{\lambda(e^{it}-1)}$
geometric $G(p)$	$\frac{1}{p}$	$\frac{q}{p^2}$	$\frac{pt}{1-qt}$	$\frac{pe^t}{1-qe^t}$	$\frac{pe^{it}}{1-qe^{it}}$
Pascal $NB(r,p)$	$\frac{r}{p}$	$\frac{rq}{p^2}$	$\left(\frac{pt}{1-qt}\right)^r$	$\left(\frac{pe^t}{1-qe^t}\right)^r$	$\left(\frac{pe^{it}}{1-qe^{it}}\right)^r$
hypergeometric (N,m,n)	$\frac{mn}{N}$	$\frac{m(N-m)n(N-n)}{N^2(N-1)}$	*	*	*

Table 4 Moments and generating functions for continuous distributions.

distribution	mean	variance	$\psi(t)$	$\chi(t)$
uniform (α,β)	$\frac{\alpha+\beta}{2}$	$\frac{(\beta-\alpha)^2}{12}$	$\frac{e^{\beta t}-e^{\alpha t}}{t(\beta-\alpha)}$	$\frac{e^{i\beta t}-e^{i\alpha t}}{it(\beta-\alpha)}$
exponential (λ)	$\frac{1}{\lambda}$	$\frac{1}{\lambda^2}$	$\frac{\lambda}{\lambda-t}$	$\frac{\lambda}{\lambda-it}$
standard normal $(0,1)$	0	1	$e^{t^2/2}$	$e^{-t^2/2}$
normal (μ,σ^2)	μ	σ^2	$e^{\mu t+\frac{\sigma^2 t^2}{2}}$	$e^{i\mu t-\frac{\sigma^2 t^2}{2}}$
gamma $\Gamma(n,\lambda)$	$\frac{n}{\lambda}$	$\frac{n}{\lambda^2}$	$\left(\frac{\lambda}{\lambda-t}\right)^n$	$\left(\frac{\lambda}{\lambda-it}\right)^n$
Cauchy (α)	∞	∞	∞	$\frac{1}{\alpha^2}e^{-\alpha|t|}$
beta (p,q)	$\frac{p}{p+q}$	$\frac{pq}{(p+q)^2(p+q+1)}$	*	*
chi square $\chi^2(r)$	r	$2r$	$(1-2t)^{-r/2}$	$(1-2it)^{-r/2}$
F-distribution $F_{m,n}$	$\frac{n}{n-2}$	$\frac{2n^2(m+n-2)}{m(n-2)^2(n-4)}$	*	*
t-distribution t_k	0	$\frac{k}{k-2}$	*	*
Rayleigh $R(\sigma)$	$\sqrt{\frac{\pi}{2}}\sigma$	$2(1-\frac{\pi}{4})\sigma^2$	*	*

2. Table 3 shows the mean, variance, probability generating function, moment generating function, and characteristic function of several important discrete distributions. Here $q = 1 - p$. An asterisk (*) signifies that the entry is not available in simple form.

3. Table 4 shows the mean, variance, moment generating function, and characteristic function of several important continuous distributions. An asterisk (*) indicates that the entry is not available.

4. The moments of a binomial random variable X can be found from its moment generating function $\psi(t) = (q + pe^t)^n$. For example, $\psi'(t) = n(q + pe^t)^{n-1}(pe^t)$ and by Fact 1 $EX = \psi'(0) = n(q+p)^{n-1}p = np$.

5. From Table 4 the moment generating function for the exponential distribution with parameter λ is $\psi(t) = \frac{\lambda}{\lambda-t}$. Then $\psi'(t) = \frac{\lambda}{(\lambda-t)^2}$ and $\psi''(t) = \frac{2\lambda}{(\lambda-t)^3}$, giving $\psi'(0) = \frac{\lambda}{\lambda^2} = \frac{1}{\lambda}$ and $\psi''(0) = \frac{2\lambda}{\lambda^3} = \frac{2}{\lambda^2}$. By Facts 1 and 2, $EX = \psi'(0) = \frac{1}{\lambda}$ and $\text{Var}(X) = \psi''(0) - [\psi'(0)]^2 = \frac{2}{\lambda^2} - (\frac{1}{\lambda})^2 = \frac{1}{\lambda^2}$.

7.4 DISCRETE PROBABILITY COMPUTATIONS

Many discrete probability computations are much less straightforward than may at first be imagined because of difficulties arising from the finiteness of computer arithmetic systems. Good algorithm design can usually avoid such problems.

7.4.1 INTEGER COMPUTATIONS

Enumeration of combinatorial objects, such as permutations and combinations, requires the computation of integer factorials. In practice, these factorials can only be computed as integers for small values. Since in most cases the factorials are not themselves the primary objective of the computation, potential numerical difficulties can be overcome by carefully designed recursive algorithms.

Definitions:

The N-bit **binary representation** of the positive integer n is $(b_{N-1}b_{N-2}\ldots b_1 b_0)_2$ where $b_i \in \{0,1\}$ and $n = \sum_{i=0}^{N-1} b_i 2^i$. (See §4.1.3.) Each b_i is a **binary digit** (**bit**).

The **two's complement** representation of the signed integer n is $(b_{N-1}b_{N-2}\ldots b_1 b_0)_{2'}$ where $n = -b_{N-1}2^{N-1} + \sum_{i=0}^{N-2} b_i 2^i$.

Integer **wraparound** is the phenomenon of adding 1 to the largest representable integer and obtaining the smallest representable integer.

Facts:

1. Signed integers are usually represented in a computer as two's complement binary words of a fixed wordlength; commonly 8, 16, or 32 bits are used.

2. A two's complement integer using N-bit words is interpreted by treating the most significant bit as a coefficient of 2^{N-1}.

3. The range of representable integers in N-bit two's complement is from $-2^{N-1} = (10\ldots 00)_{2'}$ to $2^{N-1} - 1 = (01\ldots 11)_{2'}$. Arithmetic operations can generate no carries beyond this range.

4. Integer wraparound is a consequence of Fact 3 since (in regular binary arithmetic) $(01\ldots 11)_2 + (00\ldots 01)_2 = (10\ldots 00)_2$. Some systems have integer range checking available to avoid the effect of wraparound.

5. Permutations and combinations (§2.3) are usually expressed in terms of integer factorials.

6. Binomial coefficients (§2.3.2) can be computed using integer arithmetic provided the result is within the range being used. Algorithm 1 breaks the computation of a binomial coefficient into a recursive loop using $\binom{n}{k} = \frac{n!}{k!(n-k)!} = \frac{n(n-1)\ldots(n-k+1)}{k(k-1)\ldots 1} = \left(\frac{n}{1}\right)\left(\frac{n-1}{2}\right)\ldots\left(\frac{n+1-k}{k}\right)$ and the result $\binom{n}{k} = \binom{n}{n-k}$. (See §2.3.2, Fact 7.)

7. By doing the multiplication before the integer division in Algorithm 1, the numerator necessarily has the appropriate factors to ensure an exact integer result.

> **Algorithm 1:** Integer computation of binomial coefficients.
>
> input: positive integers n, k
> output: $b = \binom{n}{k}$
> if $2k > n$ then $k := n - k$
> $b := 1$
> for $i := 1$ to k do
> $\quad b := [b \cdot (n+1-i)]$ div i

Examples:

1. For $N = 8, 16$, and 32 bits, the two's complement binary integer ranges are given in the following table:

N	minimum value	maximum value
8	-128	127
16	-32768	32767
32	-2147483648	2147483647

2. For $N = 8$, the integer 86 has the two's complement representation $(01010110)_{2'}$ and the integer -86 has the two's complement representation $-128 + 42 = (10101010)_{2'}$.

3. For 8-bit integers, the effect of integer wraparound is shown by $127 + 1 = -128$. Similarly, we would have $64 \times 2 = -128$.

4. For 8-bit two's complement integers, only $1!, 2!, 3!, 4!, 5!$ can be computed correctly. Subsequent factorials would generate integer answers—but wrong ones. In particular, $6!$ would evaluate to -48. Namely, the 8-bit two's complement representation of $5! = 120$ is 01111000 and $6 = 00000110$ so that, with no carries to the left of the 8th bit, $6! = 6 \times 5!$ is represented by the sum of 11100000 and 11110000 (which are respectively 01111000 shifted 2 and 1 places left.) This sum (again without carries to the left of the leading bit) is 11010000, which represents $(-128) + 64 + 16 = -48$.

5. Using 16-bit integers with wraparound, the binomial coefficient $\binom{12}{8}$ cannot be computed directly from its definition since neither $12!$ nor $8!$ can be computed correctly. Thus Algorithm 1 finds instead $\binom{12}{4} = \binom{12}{8}$ since $2 \times 8 > 12$. This is computed as $\left(\frac{12}{1}\right)\left(\frac{11}{2}\right)\left(\frac{10}{3}\right)\left(\frac{9}{4}\right)$, with each multiplication being performed before its associated division: $(12/1)$ is multiplied by 11, divided by 2, multiplied by 10, divided by 3, multiplied by 9, and divided by 4. This produces the intermediate results 12, 132, 66, 660, 220, 1980, 495, so that the correct final result is obtained without any intermediate computation exceeding the integer range.

7.4.2 FLOATING-POINT COMPUTATIONS

To compute discrete probabilities (e.g., binomial probabilities), careful attention must be given to the underlying floating-point computation model and its properties.

Definitions:

Let F be the set of numbers representable in a particular floating-point system, with Ω the largest positive number in F and ω the smallest positive number in F.

The floating-point arithmetic operations in F are denoted by $\oplus, \ominus, \otimes, \oslash$ when it is necessary to distinguish them from their real counterparts $+, -, \times, /$.

The number x is represented in the computer in binary **floating-point** form by the approximation $x \approx \pm f \times 2^E$ where the **fraction** or **mantissa** f is a binary fraction of fixed length and the **exponent** E is an integer within a fixed range. Usually the floating point representation is **normalized** so that $f \in [1, 2)$.

Floating-point arithmetic is subject to **roundoff error**, the error introduced by abbreviating the representation of a number or an arithmetic result to a finite wordlength.

The usual measure of error for floating-point computation is **relative error**, which is given for an approximation x^* to a quantity x by $\dfrac{|x^* - x|}{|x|} \approx \dfrac{|x^* - x|}{|x^*|}$.

A **floating-point operation** (or **flop**) is any arithmetic operation performed using floating-point arithmetic.

Overflow results from a floating-point operation where the magnitude of the result is too large for the available range of the floating-point system being used.

Underflow results from a floating-point operation where the magnitude of the result is too small for the available range of the floating-point system being used.

The **machine unit** μ of a floating-point system is the smallest positive number that can be added to 1 and produce a result recognized in the machine as greater than 1: namely, $\mu = \min\{\, x \in F \mid 1 \oplus x > 1 \,\}$.

Facts:

1. Roundoff errors are propagated in subsequent computations.
2. The two expressions given for relative error are often used interchangeably.
3. Overflow and underflow result from the finite range of available exponents. The limits of these ranges and the details of the implementation vary with both the hardware and software being used. See [IE85] for the most common implementations.
4. Usually an overflow condition terminates a program, while underflow results are normally replaced by 0.
5. Because of the finite mantissa length (and independent of the rounding rule), most axioms of the real number system fail for floating-point arithmetic [St74]. Table 1 summarizes similarities and differences between the real numbers $\mathcal{R}$ and the floating-point system F. The second column of the table describes the property, assuming $a, b, c \in \mathcal{R}$. If the property fails in F, a brief reason for the failure is also given.
6. In Table 1, most of the properties that fail in F hold approximately — at least for arguments of the same sign. These failures are not critical to most computations, but they can be important for computations such as summing sets of numbers and evaluating binomial probabilities.
7. The existence of the machine unit μ ensures that some of the order properties of $\mathcal{R}$ will not carry over to F.
8. The machine unit μ is not the same as the smallest representable positive number ω in F.
9. The relative error in subtraction is essentially unbounded due to cancellation.
10. IEEE arithmetic is required to deliver the same result as if rounding were performed on the infinite precision computation assuming that the data are exact.
11. A sum of terms of the same sign should generally be summed from smallest to largest.

Table 1 Properties of $\mathcal{R}$ and F.

property	description in $\mathcal{R}$	valid in F?
closure +	$a + b \in \mathcal{R}$	NO: overflow
closure ×	$a \times b \in \mathcal{R}$	NO: overflow
commutativity	$a + b = b + a, \ a \times b = b \times a$	YES
associativity +	$(a + b) + c = a + (b + c)$	NO: $a = 1, b = c = \frac{\mu}{2}$
special case	$(a + b) - a = b$	NO: $a = 1, b = \frac{\mu}{2}$
associativity ×	$(a \times b) \times c = a \times (b \times c)$	NO: roundoff, overflow, or underflow
distributive law	$a \times (b + c) = (a \times b) + (a \times c)$	NO: roundoff, overflow, or underflow
existence of zero	$(\exists\, 0) \ a + 0 = a$	YES
unique negative	$\exists !\, (-a) \ a + (-a) = 0$	NO: $[-(1 \oplus \mu) \otimes a] \oplus a = 0$ if $\mu \times a < \omega$
existence of one	$(\exists\, 1) \ a \times 1 = a$	YES
zero divisors	$a \times b = 0 \Rightarrow a = 0 \text{ or } b = 0$	NO: $a \otimes b = 0 \Rightarrow a < \sqrt{\omega}$ or $b < \sqrt{\omega}$
total ordering	$a < b \text{ or } a = b \text{ or } a > b$	YES
order-preservation	$a > b \Rightarrow a + c > b + c$	NO: roundoff
special case	$x > 0 \Rightarrow 1 + x > 1$	NO: $x < \mu$

Algorithm 2: Recursive computation of binomial probabilities.

input: positive integers N, k; real number p
output: $s = B(N, p; k)$

$q := 1 - p$
$t := q^N$
$s := t$
for $i := 1$ to k do
$\quad t := t * \left(\frac{p}{q}\right) * \frac{(N+1-i)}{i}$
$\quad s := s + t$

---

12. Improved accuracy in computing a summation is possible by regarding the partial sums as members of a (reduced) list of summands and always adding the two smallest terms in the current list. However, the overhead would be prohibitive in most cases.

13. Special care must be taken in computing a summation if its terms are computed recursively, since the smallest term can underflow.

14. There is no completely reliable method for summing terms of mixed sign.

15. For alternating series, special transformations such as Euler's method can be used ([BuTu92], Chapter 1).

16. Algorithm 2 computes the cumulative sum of binomial probabilities (§7.3.1) using the definition $B(N, p; k) = \sum_{i=0}^{k} \binom{N}{i} p^i (1-p)^{N-i}$.

17. Algorithm 2 will only work for small values of N.

18. If k is not too large, Algorithm 2 does in fact sum terms from smallest to largest.

> **Algorithm 3:** Logarithmic computation of binomial probability terms.
>
> input: positive integers N, r; real number p
> output: $b = \binom{N}{r} p^r (1-p)^{N-r}$
>
> $q := 1 - p$
> $t := r * \ln p + (N - r) \ln q$
> for $i := 1$ to r do
> $t := t + \ln(N + 1 - i) - \ln i$
> $b := e^t$

19. To compute $B(N, p; k)$ for large values of k, use the fact that $B(N, p; k) = 1 - B(N, 1 - p; N - k - 1)$ and then apply Algorithm 2.

20. If q^N underflows to 0, then Algorithm 2 returns 0 for all values of k.

21. Algorithm 3 gives an alternative way to calculate the individual binomial probability term $\binom{N}{r} p^r (1-p)^{N-r}$. It computes the logarithm of each factor recursively and then exponentiates this at the end.

22. Algorithm 3 must be safeguarded to ensure that e^t underflows to 0 for large negative arguments t.

23. Using logarithms is a frequently applied technique for computing products of many factors with widely varying magnitudes. It is one step along the way toward using the symmetric level-index scheme for number representation and arithmetic [ClOlTu89].

Examples:

1. *Summations*: If the first 2^{24} terms of the harmonic series are summed using IEEE single precision floating-point arithmetic, both forward and backward, then the sums differ by approximately 11%. Specifically, $\left(\cdots ((1 \oplus \frac{1}{2}) \oplus \frac{1}{3}) \oplus \cdots \oplus 2^{-24} \right) \approx 15.40$, while summing the same terms from right-to-left yields 17.23.

2. *Binomial probabilities*: The computation of binomial probabilities is thoroughly discussed in Section 2.6 of [St74] with reference to the specific case where $N = 2000$, $k = 200$, and $p = 0.1$. Using Algorithm 2 in this case gives the initial value $t = 0$ and therefore the final result is $s = 0$. The true value of the final probability is approximately 0.5.

3. If the final term in Example 2 is computed in IEEE single precision, the binomial coefficient itself would overflow. It is certainly greater than 10^{200}. Also, both $(0.1)^{200}$ and $(0.9)^{1800}$ would underflow. However the true value of this term is around 0.03.

7.5 RANDOM WALKS

Random walks are special stochastic processes whose applications include models for particle motion, crystallography, gambling, stock markets, biology, genetics, and astronomy. This section examines random walks whose trajectories are generated by the summation of independent and identically distributed discrete random variables.

7.5.1 GENERAL CONCEPTS

Definitions:

A *stochastic process* is a collection of random variables, typically indexed by time (discrete or continuous).

A *d-dimensional random walk* is a stochastic process on the integer lattice $\mathcal{Z}^d$ whose *trajectories* are defined by an initial position $S_0 = a$ and the sequence of sums $S_n = a + X_1 + X_2 + \cdots + X_n$, $n \geq 1$, where the displacements $X_1, X_2, \ldots$ are independent and identically distributed random variables on $\mathcal{Z}^d$.

A random walk is *simple* if the values X_i are restricted to the $2d$ points of $\mathcal{Z}^d$ of unit Euclidean distance from the origin. (That is, the random walk proceeds at each time step to a point one unit from the current point along some coordinate axis.) A *symmetric* random walk is a simple walk in which the $2d$ values of X_i have the same probability.

Random walks that return to the initial position with probability 1 are *recurrent*; otherwise they are *transient*.

An *absorbing boundary* is a point or a set of points on the lattice that stops the motion of a random walk whose trajectory comes into contact with it. A *reflecting boundary* is a point or a set of points that redirects the motion of a random walk. Both are special cases of an *elastic* boundary, which stops or redirects the motion depending on some given probability.

The *gambler's ruin problem* is a simple one-dimensional random walk with absorbing boundaries at values 0 and b. It colorfully illustrates the fortunes of a gambler, who starts with a dollars and who at each play of a game has a fixed probability of winning one dollar. The game ends once the gambler has either amassed the amount $S_n = b$ or goes broke $S_n = 0$.

For $k \in \mathcal{Z}^d$, the *first passage time* T_k into point k is the first time at which the random walk reaches the point k: namely, $T_k = \min\{i \geq 1 \mid S_i = k\}$. More generally, the *hitting time* T_A for entering set $A \subseteq \mathcal{Z}^d$ is the first time at which the random walk reaches some point in set A: namely, $T_A = \min\{i \geq 1 \mid S_i \in A\}$.

The following is the *basic initial problem* of random walks:
- For $k \in \mathcal{Z}^d$, find $Pr(S_n = k)$, the probability that a "particle", executing the random walk and starting at point a at time 0, will be at point k at time n.

The following are *first passage time problems*:
- Find the probability $Pr(T_k = n)$ that, starting at point a at time 0, the first visit to point k occurs at time n.
- Find the probability $Pr(T_A = n)$ that, starting at point a at time 0, the first visit to A occurs at time n; characterize S_{T_A}, the point at which A is first visited.

Other classical problems in random walks include:
- *Range problem*: Find or approximate the probability distribution and/or the mean of the number of distinct points visited by a random walk up to time n.
- *Occupancy problem*: Find or approximate the probability distribution and/or the mean of the number of times a given point or a set of points has been visited up to time n.
- *Boundary problem*: Address all previous problems under absorbing, reflecting, and/or elastic boundary conditions.

Examples:

1. *Coin tossing:* Tossing a coin n times can be viewed as a one-dimensional random walk ($d = 1$) on the integers $\mathcal{Z}$. This walk begins at the origin ($a = 0$) with $X_i = 1$ if the result of the ith toss is a Head and $X_i = -1$ if the result of the ith toss is a Tail. Since each step X_i is of unit length, this is a *simple* one-dimensional random walk. If the tosses are independent events, then $Pr(X_i = 1) = p$ and $Pr(X_i = -1) = 1-p$ holds for all i, where $0 < p < 1$. The random variable S_n is the cumulative number of Heads minus the cumulative number of Tails in n tosses. The walk is symmetric if $p = \frac{1}{2}$. A return to the origin means that $S_n = 0$: that is, the number of Heads and Tails have equalized after n tosses.

2. *Gambler's ruin:* A gambler repeatedly plays a game of chance, in which a dollar is won at each turn with probability p and a dollar is lost with probability $1 - p$. For example, suppose the gambler starts with 90 dollars, and stops whenever his current fortune is 0 (a ruin) or 100 (a positive net gain of 10 dollars). What is the gambler's ultimate probability of being ruined? Of success? On average how many expected plays does it take for the game to be over? What is the expected net gain for the gambler? If $p = 0.5$ the answers are 0.1, 0.9, 900, and 0 respectively. If $p = 0.45$ they are 0.866, 0.134, 765.6, and -76.6 respectively. (See §7.5.2, Example 4.)

7.5.2 ONE-DIMENSIONAL SIMPLE RANDOM WALKS

A number of results are known for random walks in one dimension that take a succession of unit steps (in either the positive or negative direction).

Definitions:

The **one-dimensional simple random walk** (see §7.5.1) corresponds to a particle moving randomly on the set $\mathcal{Z}$ of integers. It begins at the origin at time 0 and at each time $1, 2, \ldots$ thereafter, moves either one step up (right) with probability p, or one step down (left) with probability $1 - p$. This random walk is **symmetric** when $p = \frac{1}{2}$.

The **trajectory** of a one-dimensional simple random walk is described by $S_0 = 0$ and $S_n = X_1 + X_2 + \cdots + X_n$, $n \geq 1$, where the X_i are independent and have a Bernoulli distribution (§7.3.1), with $Pr(X_i = 1) = p$ and $Pr(X_i = -1) = q = 1 - p$ for $p \in (0, 1)$.

Suppose a trajectory is graphically represented by plotting S_n as a function of n, so that the point (n, k) corresponds to $S_n = k$. Linking successive points with straight lines produces a **path** between points. Define $N(n, k)$ to be the number of paths from $(0, 0)$ to (n, k).

In a random walk starting at $S_0 = a > 0$ with absorbing boundaries at 0 and $b > a$:

- q_a is the probability that the random walk will be absorbed at 0;
- p_a is the probability that the random walk will be absorbed at b;
- D_a is the time until absorption.

Facts:

1. *Reflection principle:* Let $n_2 > n_1 \geq 0$, $k_1 > 0$, $k_2 > 0$. The number of paths from (n_1, k_1) to (n_2, k_2) that touch or cross the x-axis equals the number of paths from $(n_1, -k_1)$ to (n_2, k_2).

2. $N(n, k) = \binom{n}{(n+k)/2}$, if $\frac{n+k}{2}$ is an integer in $\{0, 1, \ldots, n\}$; $N(n, k) = 0$ otherwise.

3. If $n \geq 1$ is fixed and $-n \leq k \leq n$, then $Pr(S_n = k) = N(n, k) p^{\frac{1}{2}(n+k)} q^{\frac{1}{2}(n-k)}$.

4. **Ballot theorem:** For $k > 0$, the number of paths from $(0,0)$ to (n,k) that do not return to or cross the x-axis is $\frac{k}{n}N(n,k)$.

5. For $n \geq 1$, the first return time T_0 to the origin satisfies:
   - $Pr(T_0 > n) = E(\frac{|S_n|}{n})$;
   - $Pr(T_0 = 2n) = \frac{1}{2n-1}\binom{2n}{n}p^n q^n$;
   - $Pr(T_0 > 2n) = Pr(S_{2n} = 0) = \binom{2n}{n}2^{-2n}$, if the walk is symmetric.

6. **Recurrent walks:** $Pr(T_0 < \infty) = 1$ (the walk is recurrent) if and only if $p = q = \frac{1}{2}$. In this case $E(T_0) = \infty$.

7. For $k \neq 0$ and $n > 0$, $Pr(T_k = n) = \frac{|k|}{n}Pr(S_n = k)$.

8. For $k > 0$ and $n > 0$, the maximum value $M_n = \max\{S_0, S_1, \ldots, S_n\}$ satisfies
   - $M_n \geq k$ if and only if $T_k \leq n$;
   - $Pr(M_n \geq k) = Pr(S_n = k) + \sum_{i \geq k+1}[1 + (\frac{q}{p})^{i-k}]Pr(S_n = i)$;
   - $Pr(M_n = k) = Pr(S_n = k) + Pr(S_n = k+1)$, if the walk is symmetric.

9. **Arc sine laws:** Let W_n be the number of times among among $\{0, 1, \ldots, n\}$ at which a random walk is positive and let L_n be the time of the last visit to 0 up to time n. For a symmetric random walk:
   - $Pr(W_{2n} = 2k) = Pr(L_{2n} = 2k) = Pr(S_{2k} = 0)Pr(S_{2n-2k} = 0)$;
   - as $n \to \infty$, $Pr(\frac{W_{2n}}{2n} \leq x) \approx \frac{2}{\pi}\arcsin\sqrt{x}$, for $x \in [0,1]$.

10. **Gambler's ruin problem:** In this random walk with absorbing boundaries (§7.5.1), q_a is the probability of the gambler (having an initial capital of a) being ruined and p_a is the probability of eventually winning (achieving a total of b). Facts 11–17 refer to the gambler's ruin problem.

11. If $p \neq q$, $q_a = \frac{(q/p)^b - (q/p)^a}{(q/p)^b - 1}$ and $p_a = 1 - q_a$.

12. If $p = q = \frac{1}{2}$, $q_a = 1 - \frac{a}{b}$ and $p_a = \frac{a}{b}$.

13. The expected gain in the gambler's ruin problem is $b(1 - q_a) - a$, which is 0 if and only if $p = q = \frac{1}{2}$.

14. $Pr(D_a = n) = b^{-1}2^n p^{(n-a)/2} q^{(n+a)/2} \sum_{k=1}^{b-1} \cos^{n-1}\frac{\pi k}{b} \sin\frac{\pi k}{b} \sin\frac{\pi ak}{b}$.

15. If $p \neq q$, $E(D_a) = \frac{a}{q-p} - \frac{b}{q-p}\frac{1-(q/p)^a}{1-(q/p)^b}$.

16. If $p = q = \frac{1}{2}$, $E(D_a) = a(b-a)$.

17. **Limiting case of the gambler's ruin problem:** When $b = \infty$:
    - $q_a = (\frac{q}{p})^a$ if $p > q$, and $q_a = 1$ otherwise;
    - $Pr(D_a = n) = 2^n p^{(n-a)/2} q^{(n+a)/2} \int_0^1 ; \cos^{n-1}\pi x \sin\pi x \sin\pi ax \, dx$
      $= \frac{a}{n}\binom{n}{(n+a)/2}p^{\frac{1}{2}(n-a)}q^{\frac{1}{2}(n+a)}$;
    - if $p < q$, $E(D_a) = \frac{a}{q-p}$;
    - if $p = q = \frac{1}{2}$, $E(D_a) = \infty$.

18. **Random walks with one reflecting boundary:** Consider a random walk starting at $S_0 = a \geq 0$ with a reflecting boundary at 0.
    - The position at time $n \geq 1$ is given by $S_n = \max\{0, S_{n-1} + X_n\}$.
    - When $p < q$ and as $n \to \infty$, there is a stationary distribution for the random walk, coinciding with the distribution of $M = \sup_{i \geq 0} S_i$, and given by $Pr(M = k) = (1 - \frac{p}{q})(\frac{p}{q})^k$ for all $k \geq 0$.

Examples:

1. A graphical representation of the trajectory for a one-dimensional simple random walk is shown in the following figure. Here $T_0 = 2$ and $T_2 = 4$; $M_3 = 1$ and $M_4 = 2$; $W_7 = 4$; and $L_7 = 6$.

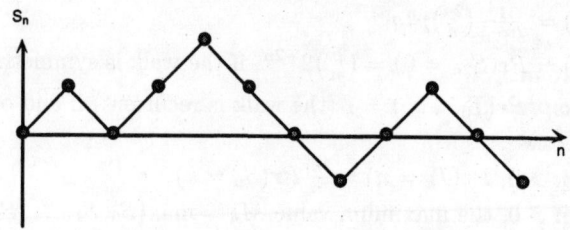

2. The ballot theorem takes its name from the following problem. Suppose that, in a ballot, candidate A scores x votes and candidate B scores y votes, $x > y$. What is the probability that, during the ballot, A is always ahead of B? By Fact 4, the answer is $\frac{x-y}{x+y}$. As an illustration, if $\frac{x}{x+y} = 0.52$, this probability is 0.04.

3. How much time does a symmetric random walk spend to the left of the origin? Contrary to intuition, with large probability, the fraction of time spent to the left (or to the right) of the origin is near 0 or 1, but not near $\frac{1}{2}$.

   For example, when n is large, Fact 9 shows that the probability a symmetric random walk spends at least 97.6% of the time to the left of the origin is approximately $0.1 = \frac{2}{\pi} \arcsin \sqrt{0.024}$. Symmetrically, there is a 0.1 probability that it spends at least 97.6% of the time to the right of the origin. Altogether, with probability 0.2 a symmetric random walk spends at least 97.6% of the time entirely on one side of the origin.

4. *Gambler's ruin*: Suppose that the probability of winning one dollar is $p = 0.45$, so $q = 0.55$. A gambler begins with an initial stake of $a = 90$ and will quit whenever the current winnings reach $b = 100$. Using Fact 11, the probability of ruin (i.e., losing the entire original stake) is

$$q_a = \frac{(11/9)^{100} - (11/9)^{90}}{(11/9)^{100} - 1} \approx 0.866.$$

The expected net gain is, by Fact 13, $b(1 - q_a) - a = 100(0.134) - 90 = -76.6$. The average duration (number of plays of the game) is found from Fact 15 to be 765.6 plays. Surprisingly, even though the probability p of winning is only slightly less than 0.5 and the gambler starts within close reach of the desired goal, the gambler can expect to be ruined with high probability and the average number of plays of the game is large.

5. The average duration of a fair game in the gambler's ruin problem is considerably longer than would be naively expected. When one player has one dollar and the adversary 1000 dollars, Fact 16 shows that the average duration is 1000 trials.

7.5.3 GENERALIZED RANDOM WALKS

Two generalizations of one-dimensional random walks are covered here. In the first case, a one-dimensional walk is now allowed to be based on an arbitrary (as opposed to Bernoulli) distribution. In the second case, a symmetric random walk is considered in higher dimensions.

Definitions:

In a one-dimensional random walk on $\mathcal{Z}$ with $S_0 = a$, let u_a be the probability that the particle arrives at a position ≤ 0 before reaching any position $\geq b$, where $b > 0$.

Let R_n be the number of distinct points visited by a random walk up to time n.

Facts:

1. If $X_1, X_2, \ldots$ are arbitrarily distributed independent random variables, many basic qualitative laws are preserved for one-dimensional walks.
   - In the case of two absorbing boundaries, the particle will reach one of them with probability 1.
   - In the case of a single absorbing boundary at 0, if $E(X_i) \leq 0$ then the particle will reach 0 with probability 1.
   - an unrestricted walk with $E(X_i) = 0$ and $\text{Var}(X_i) < \infty$ will return to its initial position with probability 1, and the expected return time is infinite.

2. *General ruin problem*: Assume that at each step the particle has probability p_k to move from any point i to $i + k$, where $k \in \mathcal{Z}$. The particle starts from position a.
   - $u_a = 1$ if $a \leq 0$, and $u_a = 0$ if $a \geq b$.
   - For $0 < a < b$, $u_a = \sum_{i \in \mathcal{Z}} u_i p_{i-a}$. This corresponds to a system of $b - 1$ linear equations in $b - 1$ unknowns that has a unique solution.

3. *Local central limit theorem*: For a d-dimensional symmetric (simple) random walk:
   - $|Pr(S_n = k) - 2(\frac{d}{2\pi n})^{d/2} e^{-\frac{d|k|^2}{2n}}| \leq O(n^{-(d+2)/2})$;
   - $|Pr(S_n = k) - 2(\frac{d}{2\pi n})^{d/2} e^{-\frac{d|k|^2}{2n}}| \leq |k|^{-2} O(n^{-d/2})$.

4. *Pólya's theorem*: For the symmetric random walks in 1 or 2 dimensions, there is probability 1 that the walk will eventually return to its initial position (recurrent random walk). In dimension $d \geq 3$ this probability is strictly less than 1. (George Pólya, 1887–1985.)

5. For symmetric random walks in $d = 3$ dimensions, the probability of an eventual return to the initial position is approximately 0.34 and the expected number of returns is approximately 0.53. The following table gives the approximate return probabilities $Pr(T_0 < \infty)$ for dimensions $d \leq 10$.

d	$Pr(T_0 < \infty)$
3	0.341
4	0.193
5	0.135
6	0.105
7	0.0858
8	0.0729
9	0.0634
10	0.0562

6. *Range problems*: As $n \to \infty$:
   - if $d = 1$, $E(R_n) \approx (\frac{8n}{\pi})^{1/2}$;
   - if $d = 2$, $E(R_n) \approx \frac{\pi n}{\log n}$;
   - if $d \geq 3$, $E(R_n) \approx c_d n$ for some constant c_d.

Example:

1. *Absorbing boundaries*: A particle starts at position $a \in \{0, 1, 2, 3, 4\}$ on a line and with equal probabilities moves one or two positions, either to the right or left. Upon reaching a position $x \leq 0$ or $x \geq 4$, the particle is stopped. This is a form of the general ruin problem with $p_{-2} = p_{-1} = p_1 = p_2 = \frac{1}{4}$. We are interested in the probability u_a of absorption at position $x \leq 0$, given that the particle starts at position a. Using Fact 2, $u_0 = 1$, $u_4 = 0$ and u_1, u_2, u_3 satisfy the following equations:

$$u_1 = \tfrac{1}{2} + \tfrac{1}{4}u_2 + \tfrac{1}{4}u_3$$
$$u_2 = \tfrac{1}{4} + \tfrac{1}{4}u_1 + \tfrac{1}{4}u_3$$
$$u_3 = \tfrac{1}{4}u_1 + \tfrac{1}{4}u_2.$$

Solving this linear system produces the unique solution $u_1 = \frac{7}{10}$, $u_2 = \frac{1}{2}$, $u_3 = \frac{3}{10}$. Intuitively, these are reasonable values since starting at the middle position $a = 2$ it should be equally likely for the particle to be absorbed at either boundary; starting at position $a = 1$ there should be a greater chance of absorption at the left boundary (probability $\frac{7}{10}$) than at the right boundary (probability $1 - \frac{7}{10} = \frac{3}{10}$).

7.5.4 APPLICATIONS OF RANDOM WALKS

Random walk methodology is central to a number of diverse problem settings. This section describes several important applications. Additional examples are found in [BaNi70], [Be93], and [We94].

Examples:

1. *Biological migration*: The name "random walk" first appears in a query sent by Karl Pearson (1857–1936) to the journal *Nature* in 1905. Pearson's problem refers to a walk in the plane, with successive steps of length $l_1, l_2, \ldots$ at angles $\Theta_1, \Theta_2, \ldots$ with respect to the x-axis, the Θ_i chosen randomly. The problem is to find, after some fixed time, the probability distribution of the distance from the initial position. The question was motivated by a theory of biological migration which Pearson developed at that time, but soon discarded. Nevertheless, Pearson's random walk was born and it has since been applied to many biological models.

2. *Biology*: Other, more recent, examples of random walk applications include DNA sequencing in genetics, bacterial migration in porous media, and molecular diffusion. In the latter example, diffusion of molecules occurs as a result of the thermal energy of the molecules. The motion of the molecules, perturbed through interactions with other molecules, is then modeled as a random walk. See [Be93] for further details.

3. *Physical sciences*: There are many applications in the physical sciences, including the classical Scher-Montroll model of electrical transport in amorphous semiconductors (a continuous-time random walk model), models of diffusion on tenuously connected structures such as percolation clusters, inference of molecular structure from data collected in x-ray scattering experiments, configurational statistics of polymers, and reaction kinetics in confined geometries. Details can be found in [BaNi70] and [We94].

4. *Sequential sampling*: A major application of random walks in statistics is in connection with Wald's theory of sequential sampling. In this context, the X_i represent certain characteristics of samples or observations. Measurements are taken as long as the random walk remains within a given region. Termination with acceptance or rejection of the appropriate hypothesis occurs depending on which part of the boundary is reached.

5. *Stock prices*: One of the early applications of computers in economics was to analyze economic time series and, in particular, the behavior of stock market prices over time. It first came as a surprise when Kendall found in 1953 that he could not identify any predictable patterns in stock prices. However, it soon became apparent that random price movements indicated a well-functioning or efficient market and that a random walk could be used as a model for the underlying market. See [Ma90]. In fact, at the beginning of this century, Bachelier had already developed a diffusion model for the stock market. Other macroeconomic time series have also been modeled using random walks.

6. *Astronomy*: The problem of the mean motion of a planet in the presence of perturbations due to other planets has a very long history (Lagrange). Statistical properties of perturbed orbits can be analyzed with the help of Pearson's random walk model. The escape of comets from the solar system has also been modeled as a random walk among energy states. Details are provided in [BaNi70].

7.6 SYSTEM RELIABILITY

System reliability involves the study of the overall performance of systems of interconnected components. Examples of such systems are communication, transportation, and electrical power distribution systems, as well as computer networks.

7.6.1 GENERAL CONCEPTS

Definitions:

Suppose a given system is composed of a set $N = \{1, 2, \ldots, n\}$ of failure-prone **components**. At any instant of time, each component is found in one of two **states**: either **operating** or **failed**.

The **reliability** of component i is the probability p_i that component i is operating at a given instant of time. The **unreliability** (or failure probability) of component i is $q_i = 1 - p_i$.

At any instant of time, the system is found in one of two **states**: **operating** or **failed**.

The **structure function** ϕ is a binary-valued function defined on all subsets $S \subseteq N$. Specifically, $\phi(S) = 1$ if the system operates when all components in S operate and all components of $N - S$ fail; otherwise $\phi(S) = 0$.

The structure function ϕ is **monotone** if $S \subseteq T \Rightarrow \phi(S) \leq \phi(T)$. In words, monotonicity means that the addition of more operating components to an already functioning system cannot result in system failure.

The structure function ϕ is **nontrivial** if $\phi(\emptyset) = 0$ and $\phi(N) = 1$.

A **coherent** system (N, ϕ) has a structure function ϕ that is monotone and nontrivial.

The **dual** of the system (N, ϕ) is the system (N, ϕ^D), defined by $\phi^D(S) = 1 - \phi(N - S)$.

The **reliability** $R_{N,\phi}$ of the system (N, ϕ) is the probability that the system functions at a random instant of time: $R_{N,\phi} = Pr(\phi(S) = 1)$.

The **unreliability** $U_{N,\phi}$ of the system is given by $U_{N,\phi} = Pr(\phi(S) = 0) = 1 - R_{N,\phi}$.

Facts:

1. If the state of any component is statistically independent of the state of any other component, then the probability that $S \subseteq N$ is precisely the set of operating components is given by $\text{prob}(S) = \prod_{i \in S} p_i \prod_{j \notin S} q_j$.

2. The dual of the dual of a system (N, ϕ) is the original system (N, ϕ).

Examples:

1. Consider a system built from the set of components $N = \{1, 2, 3, 4\}$. Associate with each component a known weight: $w_1 = 5$, $w_2 = 7$, $w_3 = 4$, $w_4 = 8$. If $S \subseteq N$ is the set of operating components, then the system is considered to operate if $\sum_{i \in S} w_i > 12$. Thus $\phi(S) = 1$ for precisely the following sets S:

$$\{1, 4\}, \{2, 4\}, \{1, 2, 3\}, \{1, 2, 4\}, \{1, 3, 4\}, \{2, 3, 4\}, \{1, 2, 3, 4\}.$$

In all other cases $\phi(S) = 0$. This structure function is nontrivial and monotone, so that (N, ϕ) is a coherent system. The reliability of the system is then

$$R_{N,\phi} = p_1 q_2 q_3 p_4 + q_1 p_2 q_3 p_4 + p_1 p_2 p_3 q_4 + p_1 p_2 q_3 p_4 + p_1 q_2 p_3 p_4 + q_1 p_2 p_3 p_4 + p_1 p_2 p_3 p_4.$$

The dual system (N, ϕ^D) has the structure function ϕ^D, where $\phi^D(T) = 0$ for precisely the following component sets T:

$$\emptyset, \{1\}, \{2\}, \{3\}, \{4\}, \{1, 3\}, \{2, 3\}.$$

In all other cases $\phi^D(T) = 1$. For example, $\phi^D(\{1, 3\}) = 1 - \phi(\{2, 4\}) = 1 - 1 = 0$.

2. For critical financial transactions, calculations are carried out simultaneously by three separate microprocessors. The three results are compared and the result is accepted if any two of the processors agree (or all three agree). Here the system has components $\{1, 2, 3\}$ corresponding to the three microprocessors. A component fails if it gives the wrong answer, and the system fails if this "majority rule" produces an incorrect (or inconclusive) answer. Thus, $\phi(S) = 1$ if and only if S is $\{1, 2\}$, $\{1, 3\}$, $\{2, 3\}$, or $\{1, 2, 3\}$.

This structure function is nontrivial and monotone, so the system is coherent. If the microprocessors are identical and each operates independently with probability p, then $Pr(\text{exactly two components work}) = 3p^2(1-p)$, $Pr(\text{all components work}) = p^3$, and

$$R_{N,\phi} = 3p^2(1-p) + p^3 = 3p^2 - 2p^3.$$

For example, if $p = 0.95$ then $R_{N,\phi} = 0.99275$. In this case, even though any single microprocessor has a 5% failure rate, the system as a whole has only a 0.7% failure rate.

3. *Telephone network*: The components of this system are individual communication links (or trunk lines) joining nearby locations. Any telephone call that is placed between two distant locations in this system needs to be routed along available communication links.

However, as a result of hardware or software malfunctions, or as a result of overloaded circuits, certain links may be unavailable at a given instant of time. Thus, a telephone network can be modeled as a system whose components are subject to failure at random times.

The reliability of the entire system is the probability that the system functions at a random instant of time; that is, that at any random point in time users can successfully complete their calls.

7.6.2 COHERENT SYSTEMS

It is assumed throughout this subsection that the system (N, ϕ) is coherent.

Definitions:

A **minpath** P is a minimal set of components such that $\phi(P) = 1$: i.e., $\phi(P) = 1$ and $\phi(S) = 0$ for all proper subsets $S \subset P$. The collection of all minpaths for (N, ϕ) is denoted $\mathcal{P}$.

A **mincut** C is a minimal set of components such that $\phi(N-C) = 0$: i.e., $\phi(N-C) = 0$ and $\phi(N-S) = 1$ for all proper subsets $S \subset C$. The collection of all mincuts for (N, ϕ) is denoted $\mathcal{C}$.

Let $G = (V, E)$ be an undirected graph with vertex set V and edge set E (§8.1.1).
- A **simple** path in G is a path that contains no repeated vertices.
- An **s-t cutset** of G is a minimal set of edges, the removal of which leaves no s-t path in G.
- A **cutset** of G is a minimal set of edges, the removal of which disconnects G.

If $K \subseteq V$, a **K-tree** is a minimal set F of edges in G such that every two vertices of K are joined by a path in F.

If $K \subseteq V$, a **K-cutset** is a minimal set of edges in G, the removal of which disconnects some pair of vertices in K.

Facts:

1. The dual of a coherent system is itself coherent.

2. The structure function of a coherent system (N, ϕ) can be completely described using its minpaths $\mathcal{P}$ or using its mincuts $\mathcal{C}$. Specifically:
   - $\phi(S) = 1$ if and only if S contains some minpath P;
   - $\phi(S) = 0$ if and only if $N - S$ contains some mincut C.

3. The minpaths of (N, ϕ) are the mincuts of the dual (N, ϕ^D), and conversely.

4. The mincuts of (N, ϕ) are the minpaths of the dual (N, ϕ^D), and conversely.

5. Every minpath of (N, ϕ) and every mincut of (N, ϕ) have nonempty intersection.

6. If P is a minimal set of components that has nonempty intersection with every mincut of (N, ϕ), then P is a minpath of (N, ϕ).

7. If C is a minimal set of components that has nonempty intersection with every minpath of (N, ϕ), then C is a mincut of (N, ϕ).

8. A K-tree has the topology of a tree (§9.1.1) whose leaf vertices are in K.

Examples:

1. *Series system*: This system (N, ϕ) operates only when all components of N operate. See the following figure. General characteristics are listed in Table 1.

—[1]—[2]—[3]— - - - —[n]—

Table 1 Characteristics of series and parallel systems.

	series	parallel
structure function	$\phi(S) = 0$, if $S \subset N$ $\phi(N) = 1$	$\phi(S) = 1$, if $S \neq \emptyset$ $\phi(\emptyset) = 0$
minpaths	$P_1 = N$	$P_1 = \{1\}, P_2 = \{2\}, \ldots, P_n = \{n\}$
mincuts	$C_1 = \{1\}, C_2 = \{2\}, \ldots, C_n = \{n\}$	$C_1 = N$
reliability	$p_1 p_2 \ldots p_n$	$1 - (1 - p_1)(1 - p_2) \ldots (1 - p_n)$
unreliability	$1 - p_1 p_2 \ldots p_n$	$(1 - p_1)(1 - p_2) \ldots (1 - p_n)$
dual	parallel, n components	series, n components

Table 2 Characteristics of k-out-of-n systems.

	k-out-of-n success	k-out-of-n failure
structure function	$\phi(S) = 1$ if $|S| \geq k$ $\phi(S) = 0$ if $|S| < k$	$\phi(S) = 0$ if $|S| \leq n - k$ $\phi(S) = 1$ if $|S| > n - k$
minpaths	$S \subseteq N$ with $|S| = k$	$S \subseteq N$ with $|S| = n - k + 1$
mincuts	$S \subseteq N$ with $|S| = n - k + 1$	$S \subseteq N$ with $|S| = k$
reliability	$\sum \{ \operatorname{prob}(S) \mid |S| \geq k \}$	$\sum \{ \operatorname{prob}(N - S) \mid |S| < k \}$
unreliability	$\sum \{ \operatorname{prob}(S) \mid |S| < k \}$	$\sum \{ \operatorname{prob}(N - S) \mid |S| \geq k \}$
dual	k-out-of-n failure	k-out-of-n success

2. *Parallel system*: This system (N, ϕ) fails only when all components of N fail. See the following figure. General characteristics are listed in Table 1.

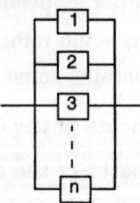

3. *k-out-of-n success system*: This system (N, ϕ) operates only when at least k out of the n components operate. The following figure illustrates a 2-out-of-3 success system. General characteristics are listed in Table 2. The special case $k = 1$ gives a parallel system; $k = n$ gives a series system.

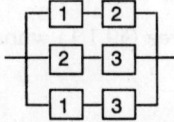

4. *k-out-of-n failure system*: This system (N, ϕ) fails only when at least k out of the n components fail. This is the same as an $(n - k + 1)$-out-of-n success system. General characteristics are listed in Table 2. The special case $k = 1$ gives a series system; $k = n$ gives a parallel system.

5. *Two-terminal network*: Two vertices s, t of an undirected graph $G = (V, E)$ are specified and a message is to be sent from vertex s to vertex t. Assume that only the edges are failure-prone, so $N = E$. The system operates when there exists some path of operating edges joining s to t in the graph.
   - *structure function*: $\phi(S) = 1$ if there exists a path from s to t in the subgraph defined by edges S; $\phi(S) = 0$ otherwise;
   - *minpaths*: all simple s-t paths of G;
   - *mincuts*: all s-t cutsets of G;
   - *reliability*: $R_{N,\phi} = $ the probability that a message sent from s will arrive at $t = \sum \{\operatorname{prob}(S) \mid S \text{ contains some simple } s\text{-}t \text{ path}\}$;
   - *unreliability*: $U_{N,\phi} = \sum \{\operatorname{prob}(N - S) \mid S \text{ contains some } s\text{-}t \text{ cutset}\}$.

6. *All-terminal network*: A message is to be disseminated among all vertices V in the undirected graph $G = (V, E)$. The system operates when the operating edges in the graph allow all vertices to mutually communicate.
   - *structure function*: $\phi(S) = 1$ if the subgraph defined by vertices V and edges S is connected; $\phi(S) = 0$ otherwise;
   - *minpaths*: all spanning trees (§9.2) of G;
   - *mincuts*: all cutsets of G;
   - *reliability*: $R_{N,\phi} = $ probability that G is connected $= \sum \{\operatorname{prob}(S) \mid S \text{ contains some spanning tree of } G\}$;
   - *unreliability*: $U_{N,\phi} = \sum \{\operatorname{prob}(N - S) \mid S \text{ contains some cutset of } G\}$.

7. *K-terminal network*: A message is to be disseminated among a fixed subset K of vertices in the undirected graph $G = (V, E)$. The system operates when the operating edges of the graph allow all vertices in K to mutually communicate.
   - *structure function*: $\phi(S) = 1$ if the subgraph defined by vertices K and edges S is connected; $\phi(S) = 0$ otherwise;
   - *minpaths*: all K-trees of G;
   - *mincuts*: all K-cutsets of G;
   - *reliability*: $R_{N,\phi} = $ probability that K is connected $= \sum \{\operatorname{prob}(S) \mid S \text{ contains some } K\text{-tree of } G\}$;
   - *unreliability*: $U_{N,\phi} = \sum \{\operatorname{prob}(N - S) \mid S \text{ contains some } K\text{-cutset of } G\}$;
   - *special cases*: $K = \{s, t\}$ gives the two-terminal network problem; $K = V$ gives the all-terminal network problem.

8. Examples 5-7 are defined in terms of undirected networks. The two-terminal, all-terminal, and K-terminal reliability problems described in these examples can also be defined for directed networks.

9. Consider the coherent system (N, ϕ) on components $N = \{1, 2, 3, 4, 5\}$ with minpaths $P_1 = \{1, 2\}, P_2 = \{1, 5\}, P_3 = \{3, 5\}, P_4 = \{2, 3, 4\}$. To illustrate Fact 2, notice that $\phi(\{3, 5\}) = 1$ and $\phi(\{2, 3, 5\}) = 1$ since P_3 is a minpath. Also, $\phi(\{2, 5\}) = 0$ since $\{2, 5\}$ contains no minpath. By Fact 7, $C_1 = \{1, 3\}$ is a mincut for this system since C_1 has nonempty intersection with each of $P_1, P_2, \ldots, P_4$ and since neither $\{1\}$ nor $\{3\}$ has this property. Likewise, $C_2 = \{2, 5\}$ and $C_3 = \{1, 4, 5\}$ are mincuts for this system. Fact 4 shows that the dual (N, ϕ^D) has as its minpaths the mincuts of (N, ϕ): namely, $\{1, 3\}, \{2, 5\}$, and $\{1, 4, 5\}$. This means $\phi^D(\{1, 3, 4\}) = 1$ since $\{1, 3, 4\}$ contains the minpath $\{1, 3\}$. Alternatively, from the definition $\phi^D(\{1, 3, 4\}) = 1 - \phi(\{2, 5\}) = 1 - 0 = 1$.

7.6.3 CALCULATING SYSTEM RELIABILITY

Four general approaches can be used to calculate the reliability $R_{N,\phi}$ of a coherent system (N,ϕ). These are state-space enumeration, inclusion-exclusion, disjoint products, and factoring.

Notation:

Let $\mathcal{P} = \{P_1, P_2, \ldots, P_m\}$ be the minpaths and $\mathcal{C} = \{C_1, C_2, \ldots, C_r\}$ be the mincuts of the coherent system (N, ϕ).

- E_i is the event that all components of minpath P_i operate (with no stipulation as to the states of the other components);
- F_i is the event that all components of C_i fail (with no stipulation as to the states of the other components);
- $E_i E_j$ denotes the event that both E_i and E_j occur;
- (N, ϕ_{+i}) is the system derived from (N, ϕ) in which component i always works;
- (N, ϕ_{-i}) is the system derived from (N, ϕ) in which component i always fails.

Facts:

1. Calculation of the reliability $R_{N,\phi}$ is in general quite difficult; it is a #P-complete problem.
2. $R_{N,\phi} = \sum \{\text{prob}(S) \mid \phi(S) = 1\}$.
3. $R_{N,\phi} = \sum \{\text{prob}(S) \mid S \text{ contains some } P \in \mathcal{P}\}$.
4. $R_{N,\phi} = 1 - U_{N,\phi} = 1 - \sum \{\text{prob}(N - S) \mid S \text{ contains some } C \in \mathcal{C}\}$.
5. *State-space enumeration*: System reliability can be found by enumerating all operating (or all failed) states of the system, using Facts 2–4.
6. $R_{N,\phi} = Pr(E_1 \cup E_2 \cup \cdots \cup E_m)$.
7. $U_{N,\phi} = Pr(F_1 \cup F_2 \cup \cdots \cup F_r)$.
8. Applying the inclusion-exclusion principle (§2.4) to Fact 6 produces
$$R_{N,\phi} = \sum_i Pr(E_i) - \sum_{i<j} Pr(E_i E_j) + \cdots + (-1)^{m+1} Pr(E_1 E_2 \ldots E_m).$$
9. Applying the inclusion-exclusion principle (§2.4) to Fact 7 produces
$$U_{N,\phi} = \sum_i Pr(F_i) - \sum_{i<j} Pr(F_i F_j) + \cdots + (-1)^{r+1} Pr(F_1 F_2 \ldots F_r).$$
10. *Inclusion-exclusion*: This approach calculates system reliability using Facts 8-9.
11. $R_{N,\phi} = Pr(E_1) + Pr(\overline{E_1}E_2) + \cdots + Pr(\overline{E_1}\overline{E_2}\ldots\overline{E_{m-1}}E_m)$.
12. $U_{N,\phi} = Pr(F_1) + Pr(\overline{F_1}F_2) + \cdots + Pr(\overline{F_1}\overline{F_2}\ldots\overline{F_{r-1}}F_r)$.
13. *Disjoint products*: This approach calculates system reliability using the law of total probabilities (§7.2.1). (See Facts 11-12.)
14. $R_{N,\phi} = p_i R_{N,\phi_{+i}} + (1 - p_i) R_{N,\phi_{-i}}$.
15. *Factoring*: Rather than requiring an enumeration of the minpaths or mincuts of the system, this method (based on Fact 14) concentrates on the state of an individual component i: it is either operating (with probability p_i) or failed (with probability $q_i = 1 - p_i$).
16. The factoring method is applied most productively when the system (N, ϕ) has additional structure. For example, this approach can be used to determine the reliability of k-out-of-n systems and two-terminal networks. (See §7.6.4.)

Examples:

1. *State-space enumeration*: Consider the coherent system (N, ϕ) with $N = \{1,2,3,4\}$ and minpaths $P_1 = \{2,3\}$, $P_2 = \{1,2,4\}$, $P_3 = \{1,3,4\}$. By Fact 2 of §7.6.2, the operating states of the system are $\{2,3\}, \{1,2,3\}, \{1,2,4\}, \{1,3,4\}, \{2,3,4\}$, and $\{1,2,3,4\}$. By Fact 3, $R_{N,\phi} = q_1 p_2 p_3 q_4 + p_1 p_2 p_3 q_4 + p_1 p_2 q_3 p_4 + p_1 q_2 p_3 p_4 + q_1 p_2 p_3 p_4 + p_1 p_2 p_3 p_4$.

2. *Inclusion-exclusion*: Consider the coherent system on $N = \{1,2,3,4\}$ with minpaths $P_1 = \{1,2\}$, $P_2 = \{2,4\}$, $P_3 = \{1,3,4\}$. Event E_1 has probability $p_1 p_2$, event $E_1 E_2$ has probability $p_1 p_2 p_4$, etc. Fact 8 gives

$$R_{N,\phi} = p_1 p_2 + p_2 p_4 + p_1 p_3 p_4 - p_1 p_2 p_4 - p_1 p_2 p_3 p_4 - p_1 p_2 p_3 p_4 + p_1 p_2 p_3 p_4$$
$$= p_1 p_2 + p_2 p_4 + p_1 p_3 p_4 - p_1 p_2 p_4 - p_1 p_2 p_3 p_4.$$

3. *Disjoint products*: Fact 11 is applied to the coherent system on $N = \{1,2,3,4,5,6\}$ with minpaths $P_1 = \{1,5\}$, $P_2 = \{1,3,6\}$, $P_3 = \{2,4,5\}$, $P_4 = \{2,6\}$. For simplicity of notation, let the event $\{e \text{ operates}\}$ be denoted by e, and let the event $\{e \text{ fails}\}$ be denoted by $\bar{e}$. Identities of set theory (Table 1, §1.2.2) can then be used to obtain

$$Pr(E_1) = p_1 p_5;$$
$$Pr(\overline{E}_1 E_2) = Pr((\bar{1} \cup \bar{5})136) = Pr(\bar{5}136) = p_1 p_3 q_5 p_6;$$
$$Pr(\overline{E}_1 \overline{E}_2 E_3) = Pr((\bar{1} \cup \bar{5})(\bar{1} \cup \bar{3} \cup \bar{6})245) = Pr(\bar{1}(\bar{1} \cup \bar{3} \cup \bar{6})245)$$
$$= Pr(\bar{1}245) = q_1 p_2 p_4 p_5;$$
$$Pr(\overline{E}_1 \overline{E}_2 \overline{E}_3 E_4) = Pr((\bar{1} \cup \bar{5})(\bar{1} \cup \bar{3} \cup \bar{6})(\bar{2} \cup \bar{4} \cup \bar{5})26)$$
$$= Pr((\bar{1} \cup \bar{5})(\bar{1} \cup \bar{3})(\bar{4} \cup \bar{5})26)$$
$$= Pr((\bar{1} \cup \bar{3}\bar{5})(\bar{4} \cup \bar{5})26)$$
$$= Pr((\bar{1}\bar{4} \cup \bar{1}\bar{5} \cup \bar{3}\bar{5})26).$$

Since the events $\bar{1}\bar{4}$, $\bar{1}\bar{5}$, and $\bar{3}\bar{5}$ above are not disjoint, Fact 11 can be reapplied to this new union of events, yielding

$$Pr(\overline{E}_1 \overline{E}_2 \overline{E}_3 E_4) = Pr((\bar{1}\bar{4} \cup 4\bar{1}\bar{5} \cup 1\bar{3}\bar{5})26)$$
$$= Pr(\bar{1}\bar{4}26 \cup 4\bar{1}\bar{5}26 \cup 1\bar{3}\bar{5}26)$$
$$= q_1 p_2 q_4 p_6 + q_1 p_2 p_4 q_5 p_6 + p_1 p_2 q_3 q_5 p_6.$$

The final expression for system reliability is then

$$R_{N,\phi} = p_1 p_5 + p_1 p_3 q_5 p_6 + q_1 p_2 p_4 p_5 + q_1 p_2 q_4 p_6 + q_1 p_2 p_4 q_5 p_6 + p_1 p_2 q_3 q_5 p_6.$$

7.6.4 SPECIALIZED ALGORITHMS FOR CALCULATING RELIABILITY

The general methods for calculating system reliability discussed in §7.6.3 can often be streamlined when the system has a special structure. This section describes algorithms for calculating the reliability of series-parallel, k-out-of-n, and certain network systems.

Definitions:

The **two-terminal reliability** of a network G is the probability $R_{st}(G)$ that vertices s and t are connected in G.

The **all-terminal reliability** of a network G is the probability $R_V(G)$ that G is connected.

Algorithm 1: Two-terminal reliability for undirected networks.

procedure $R_{st}(G)$

perform series and parallel reductions on G, producing network H
if H consists of the single edge (s,t) then return the reliability of (s,t)
else
    select an edge e from H
    let p_e be the reliability of e in H
    return $p_e\, R_{st}(H/e) + (1-p_e)\, R_{st}(H-e)$

If G is a two-terminal network with specified vertices s and t, then an **irrelevant** edge is one not appearing in any simple s-t path of G.

Let $G_S = (V_S, E_S)$ be the subgraph of a directed graph $G = (V, E)$ induced by edges $S \subseteq E$. If G_S is acyclic and has no irrelevant edges, the **domination** of G_S is $d_S = (-1)^{|E_S|-|V_S|+1}$; in all other cases, define $d_S = 0$.

Define $f_k(n)$ to be the reliability of a k-out-of-n success system having the components $N = \{1, 2, \ldots, n\}$ and corresponding reliabilities $p_1, p_2, \ldots, p_n$.

Facts:

1. A parallel system has reliability $f_1(n) = 1 - (1-p_1)(1-p_2)\ldots(1-p_n)$.
2. A series system has reliability $f_n(n) = p_1 p_2 \ldots p_n$.
3. *Series-parallel system*: If a system is constructed from series and parallel subsystems (with no component appearing in more than one subsystem), then the reliability of the overall system is calculated by successively applying Facts 1 and 2.
4. Applying the factoring approach of §7.6.3 to a k-out-of-n success system gives $f_k(n) = p_n f_{k-1}(n-1) + (1-p_n) f_k(n-1)$.
5. *k-out-of-n system*: Repeated application of Facts 1, 2, and 4 produces the reliability of any k-out-of-n success system. Since a k-out-of-n failure system is the same as an $(n-k+1)$-out-of-n success system, the reliability of any k-out-of-n failure system is found in a similar way.
6. For a two-terminal undirected network G, the system (N, ϕ_{-e}) corresponds to the two-terminal network $G - e$ with edge e deleted.
7. For a two-terminal undirected network G, the system (N, ϕ_{+e}) corresponds to the two terminal network G/e in which edge e is contracted.
8. *Two-terminal undirected network*: Algorithm 1 is a recursive procedure that calculates $R_{st}(G)$. Based on the factoring approach of §7.6.3, it splits the initial reliability calculation for G into calculations for the smaller networks $G - e$ and G/e.
9. For the sake of efficiency, Algorithm 1 carries out any applicable series and parallel reductions before selecting an edge on which to factor.
10. To avoid redundant calculations in Algorithm 1, edge e should be chosen from H so that H/e and $H - e$ do not contain irrelevant edges.
11. If G is a two-terminal directed network, then $R_{st}(G) = \sum_S d_S \prod_{i \in S} p_i$.
12. The expression in Fact 11 is obtained by using the inclusion-exclusion expansion (§7.6.3 Fact 8) applied to the simple s-t paths of G. Remarkably, a number of terms in this expansion cancel one another and the remaining coefficients are either $+1$ or -1.

Algorithm 2: All-terminal reliability for undirected networks.

input: undirected network G
output: $R_V(G)$
let $T_1, T_2, \ldots, T_m$ be the spanning trees of G, listed in lexicographic order
for $k := 1$ to m do
    $S_k := T_k$
    $F_k := \emptyset$
    for $j := 1$ to $k-1$ do
        $F_k := F_k \cup \min\{r \mid r \in T_j - T_k\}$
    $g_k := \prod_{i \in S_k} p_i \prod_{i \in F_k} q_i$
$R_V(G) := \sum_{k=1}^{m} g_k$

13. In an undirected network $G = (V, E)$, the minpaths for the all-terminal problem are the spanning trees of G.

14. *All-terminal reliability*: Algorithm 2 calculates $R_V(G)$ using the disjoint-products expansion (§7.6.3 Fact 11), applied to the spanning trees in lexicographic (dictionary) order. Here each term of the expansion reduces to a single product involving p_i and $q_i = 1 - p_i$.

Examples:

1. A system on four components is built up from series and parallel subsystems. Subsystem A has components 1 and 2 in series and subsystem B consists of component 3. Subsystem C has these two subsystems in parallel and its reliability is $1 - (1 - p_1 p_2)(1 - p_3) = p_1 p_2 + p_3 - p_1 p_2 p_3$. The entire system is constructed from subsystem D (component 4 alone) in series with subsystem C, so it has reliability $p_4(p_1 p_2 + p_3 - p_1 p_2 p_3) = p_1 p_2 p_4 + p_3 p_4 - p_1 p_2 p_3 p_4$.

2. To calculate the reliability of a 2-out-of-3 success system with components 1, 2, 3, Facts 1 and 2 are first used to obtain
$$f_1(2) = 1 - (1 - p_1)(1 - p_2) = p_1 + p_2 - p_1 p_2, \quad f_2(2) = p_1 p_2.$$
Fact 4 then gives the system reliability
$$R_{N,\phi} = f_2(3) = p_3 f_1(2) + (1 - p_3) f_2(2) = p_1 p_2 + p_1 p_3 + p_2 p_3 - 2 p_1 p_2 p_3.$$

3. The two-terminal *bridge* network G is shown in the following figure, with $s = a$ and $t = d$.

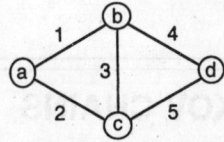

No series or parallel reductions can be performed on G, so factoring with respect to edge $e = 3$ produces the networks $G_1 = G/e$ and $G_2 = G - e$ shown in this figure:

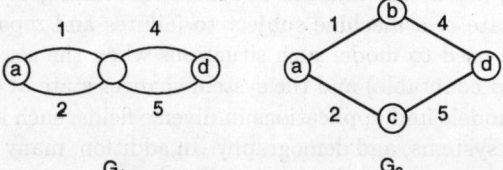

Since both G_1 and G_2 are series-parallel networks,
$$R_{st}(G_1) = [1 - (1-p_1)(1-p_2)][1 - (1-p_4)(1-p_5)]$$
$$= (p_1 + p_2 - p_1p_2)(p_4 + p_5 - p_4p_5),$$
$$R_{st}(G_2) = [1 - (1-p_1p_4)(1-p_2p_5)] = p_1p_4 + p_2p_5 - p_1p_2p_4p_5.$$
Algorithm 1 then produces
$$R_{st}(G) = p_3(p_1 + p_2 - p_1p_2)(p_4 + p_5 - p_4p_5) + (1-p_3)(p_1p_4 + p_2p_5 - p_1p_2p_4p_5)$$
$$= p_1p_4 + p_2p_5 + p_1p_3p_5 + p_2p_3p_4 - p_1p_3p_4p_5 - p_2p_3p_4p_5 - p_1p_2p_3p_4$$
$$- p_1p_2p_3p_5 - p_1p_2p_4p_5 + 2p_1p_2p_3p_4p_5.$$

4. Consider a directed version of the two-terminal network in the figure of Example 3, in which there are oppositely directed edges $3 = (b,c)$ and $6 = (c,b)$. Edges 1 and 2 are directed out of $s = a$, while edges 4 and 5 are directed into $t = d$. The cyclic subgraph defined by edges $S = \{1,2,3,4,5,6\}$ has $d_S = 0$. Also the subgraph defined by $S = \{1,2,3,4\}$ has the irrelevant edges 2 and 3, so that $d_S = 0$. On the other hand, $S = \{1,2,3,5\}$ defines an acyclic network without irrelevant edges, giving $d_S = (-1)^{4-4+1} = -1$ and the term $-p_1p_2p_3p_5$. Similarly, $S = \{1,4\}$ produces $d_S = (-1)^{2-3+1} = +1$ and the term $+p_1p_4$.

After generating all acyclic networks without irrelevant edges, Fact 11 is applied to obtain
$$R_{st}(G) = p_1p_4 + p_2p_5 + p_1p_3p_5 + p_2p_4p_6 - p_1p_2p_4p_5 - p_1p_3p_4p_5$$
$$- p_1p_2p_4p_6 - p_1p_2p_3p_5 - p_2p_4p_5p_6 + p_1p_2p_3p_4p_5 + p_1p_2p_4p_5p_6.$$

5. The bridge network G in Example 3 has eight spanning trees, given in lexicographic order by $T_1 = \{1,2,4\}$, $T_2 = \{1,2,5\}$, $T_3 = \{1,3,4\}$, $T_4 = \{1,3,5\}$, $T_5 = \{1,4,5\}$, $T_6 = \{2,3,4\}$, $T_7 = \{2,3,5\}$, $T_8 = \{2,4,5\}$. Applying Algorithm 2 gives:

$$S_1 = \{1,2,4\}, \quad F_1 = \emptyset, \quad g_1 = p_1p_2p_4$$
$$S_2 = \{1,2,5\}, \quad F_2 = \{4\}, \quad g_2 = p_1p_2q_4p_5$$
$$S_3 = \{1,3,4\}, \quad F_3 = \{2\}, \quad g_3 = p_1q_2p_3p_4$$
$$S_4 = \{1,3,5\}, \quad F_4 = \{2,4\}, \quad g_4 = p_1q_2p_3q_4p_5$$
$$\vdots \qquad \vdots \qquad \vdots$$
$$S_8 = \{2,4,5\}, \quad F_8 = \{1,3\}, \quad g_8 = q_1p_2q_3p_4p_5$$

Summing these eight terms then yields
$$R_V(G) = p_1p_2p_4 + p_1p_2q_4p_5 + p_1q_2p_3p_4 + p_1q_2p_3q_4p_5 + \cdots + q_1p_2q_3p_4p_5.$$

7.7 DISCRETE-TIME MARKOV CHAINS

Many physical systems evolve randomly in time, e.g., the population of a country, the value of a company's stock, the number of customers waiting at a checkout counter, and the functional state of a machine subject to failures and repairs. A discrete-time Markov chain can be used to model such situations when the set of possible states of the system is finite (or countable) and the system changes state at discrete time points. Such Markov chain models find applications in diverse fields, such as biology, inventory, production, queueing systems, and demography. In addition, many recursive algorithms can be viewed as a manifestation of an underlying discrete-time Markov chain.

7.7.1 MARKOV CHAINS

Definitions:

A sequence of random variables $\{X_n \mid n \geq 0\}$ is a (**discrete-time**) **Markov chain** (**DTMC**) on a (countable) state-space S if $X_n \in S$ for all $n \geq 0$ and X_{n+1} depends (probabilistically) on the previous states of the system only via X_n:

$$Pr(X_{n+1} = j \mid X_n = i, X_{n-1} = i_{n-1}, ..., X_0 = i_0) = Pr(X_{n+1} = j \mid X_n = i),$$

for all $i_0, i_1, ..., i_{n-1}, i, j \in S$.

A Markov chain $\{X_n \mid n \geq 0\}$ is **time-homogeneous** if

$$Pr(X_{n+1} = j \mid X_n = i) = p_{ij}, \text{ for all } n \geq 0.$$

Note: Only time-homogeneous discrete-time Markov chains will be considered in this and later sections.

The matrix $P = (p_{ij})$ is the (**one-step**) **transition probability matrix** of the discrete-time Markov chain.

The **initial distribution** for a DTMC is the vector $a = (a_i)$, where $a_i = Pr(X_0 = i)$ for $i \in S$.

The **transition diagram** of a DTMC is the directed graph (§8.3.1) $G = (V, E)$, where $V = S$ is the state-space and $E = \{(i,j) \in S \times S \mid p_{ij} > 0\}$.

A **stochastic** matrix $M = (m_{ij})$ has $m_{ij} \geq 0$ for all i, j and $\sum_j m_{ij} = 1$ for all i.

Facts:

1. The first systematic study of Markov chains was carried out by Andrei Andreevich Markov (1856–1922); this work initiated the study of stochastic processes (sequences of random variables).

2. A DTMC on state-space S is completely described by the initial distribution a and the transition probability matrix P.

3. The transition probability matrix P is a stochastic matrix.

Examples:

1. Consider a DTMC on the set $S = \{1, 2, 3, 4, 5, 6\}$ with the following transition probability matrix

$$P = \begin{pmatrix} 0.4 & 0.6 & 0 & 0 & 0 & 0 \\ 0.7 & 0.3 & 0 & 0 & 0 & 0 \\ 0 & 0 & 0 & 1 & 0 & 0 \\ 0 & 0 & 1 & 0 & 0 & 0 \\ 0 & 0 & 0 & 0 & 1 & 0 \\ 0.1 & 0.1 & 0.1 & 0.1 & 0.1 & 0.5 \end{pmatrix}.$$

To completely describe this DTMC, it is also necessary to specify the initial distribution. For example, $a = (0, 0, 0, 0, 0, 1)$ means that the system starts off in state 6.

2. *Simple random walk with absorbing states*: This is a DTMC on $S = \{0, 1, 2, ..., N\}$ with transition probabilities $p_{i,i+1} = p$, $p_{i,i-1} = 1 - p = q$, $1 \leq i \leq N - 1$, where $0 \leq p \leq 1$ is a given number. Also, $p_{0,0} = p_{N,N} = 1$, meaning that states 0 and N are

absorbing — once the DTMC visits these states it cannot leave them. The transition diagram of this DTMC is given in the following figure.

This Markov model is also (more colorfully) known as the *gambler's ruin problem* (§7.5.1, Example 2).

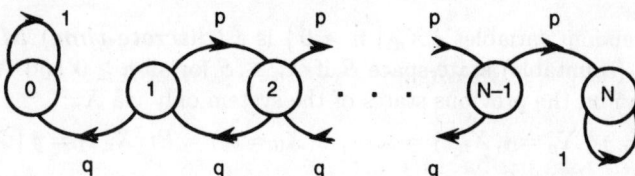

3. *Simple random walk with reflecting states*: This is a variant of Example 2, in which the boundary states 0 and N are reflecting: namely, $p_{0,1} = p_{N,N-1} = 1$. The transition diagram of this DTMC is given here.

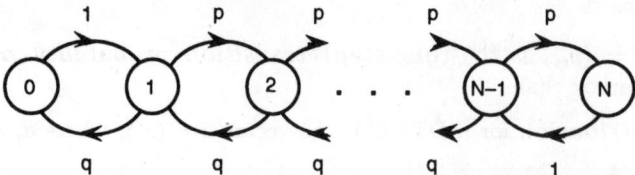

4. *Weather*: A simplified model of the daily weather results in a DTMC. Suppose that each day is either sunny (0) or rainy (1) and that tomorrow's weather depends only on today's weather. Specifically, suppose that a rainy day follows a sunny day with probability 0.3 and a sunny day follows a rainy day with probability 0.4. This is a DTMC with state-space $S = \{0, 1\}$ and transition probability matrix

$$P = \begin{pmatrix} 0.7 & 0.3 \\ 0.4 & 0.6 \end{pmatrix}.$$

5. *Urns*: Urn B contains 9 black and 1 white ball, while Urn R contains 6 red and 4 white balls. Balls are successively drawn with replacement from an urn. If the ball drawn is colored, the drawing continues from the same urn. If the ball drawn is white, the drawing continues from the other urn. Define the state of the system to be the urn being sampled, so $S = \{B, R\}$. This is a DTMC with transition probabilities $p_{BB} = 0.9$, $p_{BR} = 0.1$, $p_{RR} = 0.6$, $p_{RB} = 0.4$.

6. *Ehrenfest diffusion model*: Suppose that there are M molecules in a vessel, separated into two chambers by a membrane, across which molecules can pass. A state of the system at any instant is given by (k_1, k_2), where there are k_1 molecules in the first chamber and $k_2 = M - k_1$ in the second chamber. Transitions from the current state (k_1, k_2) occur by the movement of a single molecule from the first chamber to the second, resulting in state $(k_1 - 1, k_2 + 1)$, or from the second chamber to the first, resulting in state $(k_1 + 1, k_2 - 1)$.

In the Ehrenfest model of this process, the probability of transition from (k_1, k_2) to $(k_1 - 1, k_2 + 1)$ is given by $\frac{k_1}{M}$, whereas the probability of transition to $(k_1 + 1, k_2 - 1)$ is $\frac{k_2}{M} = 1 - \frac{k_1}{M}$. This quantifies the idea that if more molecules are present in (say) chamber 1, then it is more likely for some molecule to transfer next from chamber 1 to chamber 2. This is a DTMC with state-space $S = \{(0, M), (1, M-1), \ldots, (M, 0)\}$ and the transition probabilities specified.

7.7.2 TRANSIENT ANALYSIS

Transient analysis of a DTMC involves the computation of $Pr(X_n = j)$, the probability of the Markov chain being in state j after n steps.

Definitions:
For $i, j \in S$ the ***n*-step transition probability** $p_{ij}^{(n)}$ is the probability of being in state j after $n \geq 0$ steps, if the Markov chain starts in state i: $p_{ij}^{(n)} = Pr(X_n = j \mid X_0 = i)$.

The ***n*-step transition probability matrix** is given by $P^{(n)} = (p_{ij}^{(n)})$.

Facts:
1. $Pr(X_n = j) = \sum\limits_{i \in S} Pr(X_0 = i) Pr(X_n = j \mid X_0 = i) = \sum\limits_{i \in S} a_i p_{ij}^{(n)}$.
2. Chapman-Kolmogorov equations: $P^{(n+m)} = P^{(n)} P^{(m)} = P^{(m)} P^{(n)}$ for all $m, n \geq 0$.
3. If P^n denotes the nth power of P, then $P^{(n)} = P^n$, $n \geq 0$.
4. If a is the initial distribution of a DTMC, the (absolute) probabilities $Pr(X_n = j)$ are the entries of the vector aP^n.

Examples:
1. For §7.7.1 Example 4, the two-step transition probability matrix is
$$P^{(2)} = P^2 = \begin{pmatrix} 0.7 & 0.3 \\ 0.4 & 0.6 \end{pmatrix} \begin{pmatrix} 0.7 & 0.3 \\ 0.4 & 0.6 \end{pmatrix} = \begin{pmatrix} 0.61 & 0.39 \\ 0.52 & 0.48 \end{pmatrix}.$$
Note that $P^{(2)}$ is again a stochastic matrix. To illustrate, if Friday is sunny then the conditional probability that Sunday is sunny is given by $p_{00}^{(2)} = 0.61$.

2. A general two-state DTMC on $S = \{0, 1\}$ can be represented by the stochastic transition probability matrix
$$P = \begin{pmatrix} 1-p & p \\ q & 1-q \end{pmatrix}.$$
Direct calculation gives the two-step transition probability matrix
$$P^{(2)} = \begin{pmatrix} (1-p)^2 + pq & p(2-p-q) \\ q(2-p-q) & (1-q)^2 + pq \end{pmatrix},$$
which can be rewritten as
$$P^{(2)} = \tfrac{1}{p+q} \begin{pmatrix} q + p(1-p-q)^2 & p - p(1-p-q)^2 \\ q - q(1-p-q)^2 & p + q(1-p-q)^2 \end{pmatrix}.$$
In general,
$$P^{(n)} = \tfrac{1}{p+q} \begin{pmatrix} q + p(1-p-q)^n & p - p(1-p-q)^n \\ q - q(1-p-q)^n & p + q(1-p-q)^n \end{pmatrix}.$$

3. Limiting probabilities: Suppose in Example 2 that $0 < p < 1$ and $0 < q < 1$, so that $|1 - p - q| < 1$. From the final expression obtained in Example 2, it is seen that $P^{(n)}$ tends to the limiting matrix
$$\tfrac{1}{p+q} \begin{pmatrix} q & p \\ q & p \end{pmatrix}.$$
Consequently, if $a = (a_1, a_2)$ is any initial distribution, then the limiting probabilities aP^n approach
$$\tfrac{1}{p+q} (a_1 \ a_2) \begin{pmatrix} q & p \\ q & p \end{pmatrix} = \tfrac{1}{p+q} \begin{pmatrix} q & p \\ q & p \end{pmatrix}$$
since $a_1 + a_2 = 1$. For example, the limiting probability of being in state 0 is $\tfrac{q}{p+q}$, independent of the initial state of the DTMC. (See §7.7.4.)

7.7.3 CLASSIFICATION OF STATES

Definitions:

State $j \in S$ is **accessible** from state $i \in S$ (written $i \to j$) if it is possible to make a sequence of transitions leading from state i to state j: that is, $p_{ij}^{(n)} > 0$ for some $n \geq 0$.
States $i, j \in S$ **communicate** (written $i \leftrightarrow j$) if they are mutually accessible from one another: $i \to j$ and $j \to i$.
Set $C \subseteq S$ is a (maximal) **communicating class** if
- $i, j \in C \Rightarrow i \leftrightarrow j$;
- $i \in C, i \leftrightarrow j \Rightarrow j \in C$.

A communicating class C is **closed** if transitions from the states of C never lead to states outside C: $i \in C$, $j \notin C \Rightarrow j$ is not accessible from i.
A DTMC is **irreducible** if $i \leftrightarrow j$ for all $i, j \in S$; otherwise it is **reducible**.
For $j \in S$ define:
- $T_j = \min\{n > 0 \mid X_n = j\}$;
- $f_j = Pr(T_j < \infty \mid X_0 = j)$;
- $f_j(n) = Pr(T_j = n \mid X_0 = j)$;
- $m_j = E(T_j \mid X_0 = j)$.

State $j \in S$ is **recurrent** if return to that state is certain: $f_j = 1$; if $f_j < 1$ then state j is **transient**. A recurrent state $j \in S$ is **positive recurrent** if $m_j < \infty$ and **null recurrent** if $m_j = \infty$.

A recurrent state j has **period** d if d is the largest integer satisfying $\sum_{n=0}^{\infty} f_j(nd) = 1$. If $d = 1$ state j is **aperiodic**.

Facts:

1. Generally, all classes that are not closed can be lumped into a single set T of transient states. Thus, the state-space of a DTMC can be partitioned into closed classes $C_1, C_2, \ldots, C_K$ and the set T.

2. State j is accessible from state i if and only if there is a directed path from vertex i to vertex j in the transition diagram of the DTMC.

3. A set of states C is a communicating class if and only if the corresponding set of vertices forms a strongly connected component (§8.3.2) in the transition diagram.

4. Tarjan [Ta72] describes an algorithm to find the strongly connected components, which runs in time linear in the number of arcs in the transition diagram (i.e., the number of nonzero entries of P).

5. Transience, positive recurrence, and null recurrence are class properties:
   - if i is transient and $i \leftrightarrow j$ then j is transient;
   - if i is positive recurrent and $i \leftrightarrow j$ then j is positive recurrent;
   - if i is null recurrent and $i \leftrightarrow j$ then j is null recurrent.

In other words, states in a communicating class are all simultaneously transient or null recurrent or positive recurrent.

6. By Fact 5, a communicating class or an irreducible DTMC can be termed positive recurrent, null recurrent, or transient if all of its states are positive recurrent, null recurrent, or transient.

7. A finite communicating class is positive recurrent if it is closed, and transient otherwise.

8. A finite state irreducible DTMC is positive recurrent.

9. Null recurrent states do not occur in a finite state DTMC.

10. Establishing recurrence or transience in an infinite state DTMC is a difficult task and has to be done on a case-by-case basis.

11. We have not defined period for a transient state since for such a state the concept is not needed. Some references do however define period for all states.

12. The period of state j is the greatest common divisor of all integers $n \geq 0$ such that $p_{jj}^{(n)} > 0$.

13. The period of state j is the greatest common divisor of all the lengths of the directed cycles in the transition diagram that contain state j.

14. Periodicity is a class property: if i has period d and $i \leftrightarrow j$, then j has period d.

15. The period of a state in a finite irreducible DTMC is at most equal to the number of states in the DTMC.

16. By Fact 14, a recurrent communicating class or a recurrent irreducible DTMC can be termed periodic if all states in it are periodic with $d > 1$, else it is termed aperiodic.

Examples:

1. For the DTMC in §7.7.1 Example 1, it is seen that $1 \to 2$, $2 \to 1$, and $1 \leftrightarrow 2$. However, 3 is not accessible from 1. The communicating classes are $C_1 = \{1, 2\}$, $C_2 = \{3, 4\}$, $C_3 = \{5\}$, $C_4 = \{6\}$. This DTMC is reducible. Classes C_1, C_2, C_3 are closed, but C_4 is not. States $1, 2, 3, 4, 5$ are positive recurrent and state 6 is transient. Classes C_1, C_2, C_3 are positive recurrent.

2. Consider the random walk in §7.7.1 Example 2 with $0 < p < 1$. There are three communicating classes: $C_1 = \{0\}$, $C_2 = \{1, 2, ..., N-1\}$, and $C_3 = \{N\}$. This DTMC is reducible. Here C_1 and C_3 are closed, while C_2 is not. States 0 and N are positive recurrent. The rest are transient.

3. For the DTMC in §7.7.1 Example 1, states 3 and 4 have period 2; states 1, 2, and 5 are aperiodic. A period is not associated with state 6 since it is transient. Classes $\{1, 2\}$ and $\{5\}$ are aperiodic, while the class $\{3, 4\}$ is periodic with period 2.

4. For the DTMC in §7.7.1 Example 2, states 0 and N are aperiodic. Period is not defined for the rest of the states as they are transient.

5. §7.7.1 Example 3 is an irreducible chain. All states are positive recurrent and have period 2.

7.7.4 LIMITING BEHAVIOR

To establish possible equilibrium configurations of DTMCs, it is necessary to study the behavior of the n-step transition probabilities $Pr(X_n = j \mid X_0 = i)$ as $n \to \infty$.

Facts:

1. Let $\{X_n \mid n \geq 0\}$ be an irreducible DTMC with transition probability matrix P and finite state-space S. Then there exists a unique solution $\pi = (\pi_j)$ to the equations

$$\pi = \pi P, \quad \sum_{j \in S} \pi_j = 1.$$

2. The long run fraction of the visits to state j is given by π_j, regardless of the initial state. Also, m_j, the expected time between two consecutive visits to state j, is $\frac{1}{\pi_j}$.

3. If the DTMC is aperiodic, then $\lim_{n \to \infty} Pr(X_n = j \mid X_0 = i) = \pi_j$ for all $i \in S$.

4. Let $\{X_n \mid n \geq 0\}$ be a finite state reducible DTMC with K closed communicating classes $C_1, C_2, \ldots, C_K$ and the set of transient states T. Then $\lim_{n \to \infty} Pr(X_n = j \mid X_0 = i) = \pi_{ij}$, where $\{\pi_{ij}\}$ are the following:
   (a) If $j \in T$, then $\pi_{ij} = 0$.
   (b) If i and j belong to different closed classes, then $\pi_{ij} = 0$.
   (c) If i and j belong to the same closed class C_r, then $\pi_{ij} = \pi_j$, where $\{\pi_j\}$ are the limiting probabilities calculated by using Fact 1 for the irreducible DTMC formed by the states in C_r.
   (d) if $i \in T$ and $j \in C_r$, then $\pi_{ij} = \alpha_{ir} \pi_j$, where $\{\pi_j\}$ are as in (c) and α_{ir} is the probability that the DTMC eventually visits the class C_r starting from state i.

5. In Fact 4(c), if C_r is periodic then limiting probabilities do not exist and π_j is interpreted as the long run fraction of the time the DTMC spends in state j starting from state i.

6. The $\{\alpha_{ir}\}$ in Fact 4(d) are given by the unique solution to
$$\alpha_{ir} = \sum_{j \in C_r} p_{ij} + \sum_{j \in T} p_{ij} \alpha_{jr}.$$

Examples:

1. The DTMC in §7.7.1 Example 3 is irreducible and periodic with $d = 2$. Using Fact 1, its limiting behavior is described by the equations
$$\pi_0 = q\pi_1$$
$$\pi_1 = \pi_0 + q\pi_2$$
$$\pi_i = p\pi_{i-1} + q\pi_{i+1} \text{ for } 2 \leq i \leq N - 2$$
$$\pi_{N-1} = p\pi_{N-2} + \pi_N$$
$$\pi_N = p\pi_{N-1}$$
and $\sum_{j=0}^{N} \pi_j = 1$. Solving these equations gives $\pi_j = \frac{\rho_j}{c}$, where
$$\rho_0 = 1$$
$$\rho_j = \frac{(\frac{p}{q})^j}{p} \text{ for } 1 \leq j \leq N - 1$$
$$\rho_N = \left(\frac{p}{q}\right)^{N-1}$$
and the normalizing constant is $c = \sum_{j=0}^{N} \rho_j$. This DTMC is periodic and hence these π_j represent the long run fraction of the time the DTMC spends in state j. Here $\lim_{n \to \infty} Pr(X_n = j \mid X_0 = i)$ does not exist since the probabilities under question keep oscillating with period 2.

2. For the DTMC in §7.7.1 Example 1, $C_1 = \{1,2\}$, $C_2 = \{3,4\}$, $C_3 = \{5\}$, and $T = \{6\}$. Therefore, $\pi_1 = \frac{7}{13}$, $\pi_2 = \frac{6}{13}$, $\pi_3 = \frac{1}{2}$, $\pi_4 = \frac{1}{2}$, $\pi_5 = 1$, $\alpha_{61} = \frac{2}{5}$, $\alpha_{62} = \frac{2}{5}$, $\alpha_{63} = \frac{1}{5}$. By Fact 4 the limiting matrix (π_{ij}) is given by

$$\begin{pmatrix} \frac{7}{13} & \frac{6}{13} & 0 & 0 & 0 & 0 \\ \frac{7}{13} & \frac{6}{13} & 0 & 0 & 0 & 0 \\ 0 & 0 & \frac{1}{2} & \frac{1}{2} & 0 & 0 \\ 0 & 0 & \frac{1}{2} & \frac{1}{2} & 0 & 0 \\ 0 & 0 & 0 & 0 & 1 & 0 \\ \frac{14}{65} & \frac{12}{65} & \frac{1}{5} & \frac{1}{5} & \frac{1}{5} & 0 \end{pmatrix}$$

States 3 and 4 are periodic, and hence the third and fourth columns need to be interpreted as the long run fraction of the time the discrete-time Markov chain spends in those states.

3. The Ehrenfest diffusion model (§7.7.1 Example 6) is an irreducible DTMC with period 2. The solution π (Fact 1) is given by $\pi_j = \binom{M}{j} 2^{-M}$ ($0 \leq j \leq M$). The binomial distribution (§7.3.1) describes the long run fraction of time the system spends in each state.

7.7.5 FIRST PASSAGE TIMES

Definitions:

Let $\{X_n \mid n \geq 0\}$ be a DTMC on state-space S with transition probability matrix P. Let $A \subseteq S$ be a given subset of states.

The **first passage time** T_A into set A is the first time at which the Markov chain reaches some state in set A; i.e., $T_A = \min\{n \geq 0 \mid X_n \in A\}$.
For $i \in S$, let $\alpha_i = Pr(T_A < \infty \mid X_0 = i)$ and let $\tau_i = E(T_A \mid X_0 = i)$.

Facts:

1. The $\{\alpha_i\}$ are given by the unique solution to
$$\alpha_i = \sum_{j \in S} p_{ij} \alpha_j$$
with the boundary conditions $\alpha_i = 1$ if $i \in A$ and $\alpha_i = 0$ if no state in A is accessible from i.

2. If $\alpha_i = 1$ for all $i \in S$, then $\{\tau_i\}$ are given by the unique solution to
$$\tau_i = 1 + \sum_{j \in S} p_{ij} \tau_j$$
with the boundary condition $\tau_i = 0$ if $i \in A$.

Examples:

1. Consider the DTMC in §7.7.1 Example 2 with $A = \{0\}$. The equations of Fact 1 are $\alpha_i = q\alpha_{i-1} + p\alpha_{i+1}$, $1 \leq i \leq N-1$ with the boundary conditions $\alpha_0 = 1$ and $\alpha_N = 0$. If $q \neq p$, the solution is given by
$$\alpha_i = \frac{(\frac{q}{p})^i - (\frac{q}{p})^N}{1 - (\frac{q}{p})^N}, \quad 0 \leq i \leq N.$$
If $q = p$, the solution is $\alpha_i = 1 - \frac{i}{N}$.

2. Consider the DTMC in §7.7.1 Example 2 with $A = \{0, N\}$. In this case $\alpha_i = 1$ for all i. The equations of Fact 2 are $\tau_i = 1 + q\tau_{i-1} + p\tau_{i+1}$, $1 \leq i \leq N-1$, with the boundary conditions $\tau_0 = 0$ and $\tau_N = 0$. If $q \neq p$, the solution is given by
$$\tau_i = \frac{i}{q-p} - \frac{N}{q-p} \frac{1 - (\frac{q}{p})^i}{1 - (\frac{q}{p})^N}, \quad 0 \leq i \leq N.$$
If $q = p$, the solution is given by $\tau_i = i(N - i)$.

7.7.6 BRANCHING PROCESSES

Branching processes are a special type of Markov chain used to study the growth (and possible extinction) of populations in biology and sociology as well as particles in physics.

Definitions:

Suppose $\{Y_{ni} \mid n, i \geq 1\}$ are independent and identically distributed random variables having common probability distribution function $p_k = Pr(Y_{ni} = k)$, $k \geq 0$, with mean m and variance σ^2. Then the DTMC $\{X_n \mid n \geq 0\}$ is a **branching process** if $X_0 = 1$,
$$X_{n+1} = \sum_{i=1}^{X_n} Y_{ni}.$$

A branching process is **stable** if $m < 1$, **critical** if $m = 1$, and **unstable** if $m > 1$.

The **extinction probability** of a branching process is the probability that the population becomes extinct, where $X_0 = 1$.

Facts:

1. It is convenient to think of the random variable X_n as the number of individuals in the nth generation and the random variable Y_{ni} as the number of offspring of the ith individual in the nth generation.

2. The transient behavior of the branching process is given by:
$$E(X_n) = m^n,$$
$$\text{Var}(X_n) = \begin{cases} n\sigma^2, & \text{if } m = 1 \\ \sigma^2 m^{n-1} \dfrac{m^n - 1}{m - 1}, & \text{if } m \neq 1. \end{cases}$$

3. State 0 is absorbing for a branching process. Absorption in state 0 is certain if and only if $m \leq 1$, while the expected time until extinction (i.e., absorption in state 0) is finite if and only if $m < 1$.

4. The probability of extinction ρ in an unstable branching process is given by the unique solution in $(0, 1)$ to the equation
$$\rho = \sum_{n=0}^{\infty} p_n \rho^n.$$

5. The expected total number of individuals ever born in a stable branching process until it becomes extinct is given by
$$E(\sum_{n=0}^{\infty} X_n) = \frac{1}{1 - m}.$$

6. There is no simple expression for the expected time until absorption for a general stable branching process.

Examples:

1. The branching process with $p_0 = \frac{1}{2}$, $p_1 = \frac{1}{8}$, $p_2 = \frac{3}{8}$ has mean $m = \frac{7}{8} < 1$ and is stable. With probability 1, the population will die out.

2. The branching process with $p_0 = \frac{1}{4}$, $p_1 = \frac{1}{4}$, $p_2 = \frac{1}{2}$ has mean $m = \frac{5}{4} > 1$ and is unstable. The probability of extinction, ρ_0, is found as the smallest positive root of the

equation $\rho = \frac{1}{4} + \frac{1}{4}\rho + \frac{1}{2}\rho^2$. The roots of this equation are $\frac{1}{2}$ and 1, so the probability of extinction is $\rho_0 = \frac{1}{2}$.

If the initial population is $X_0 = 10$ instead of $X_0 = 1$, then the probability that the initial population eventually becomes extinct is $\rho_0^{10} = \frac{1}{1024}$.

7.8 QUEUEING THEORY

Queueing theory provides a set of tools for the analysis of systems in which customers arrive at a service facility. It has its origins in the works of A. K. Erlang (starting in 1908) in telephony. Since then it has found many applications in diverse areas such as manufacturing, inventory systems, computer science, analysis of algorithms, and telecommunications. Although queueing theory uses the terminology of servers providing service to customers, in actual applications the customers may be people, jobs, computational steps, or messages, and the servers may be human beings, machines, telephone circuits, communication channels, or computers.

7.8.1 SINGLE-STATION QUEUES

The simplest queueing system is a single-station queue in which customers arrive, wait for service, and depart after service completion. In this and subsequent sections we restrict ourselves to single-station queues.

Definitions:

A **queueing system** consists of a set of **customers**, who arrive at a service facility according to a specified **arrival process**. If a server is available then the customer is served immediately, with the length of time required to carry out the service determined by a **service-time distribution**. If a server is not free, the customer joins the **queue** and is later served according to a **service discipline**, which specifies the order in which customers are selected for service from the queue. Throughout, the service discipline is assumed to be First-Come-First-Served (FCFS). Alternative service disciplines include Last-Come-First-Served, randomly, or according to a tiered priority scheme.

The **queue capacity** is the maximum number of customers allowed in the system, either being served or awaiting service. Unless otherwise specified, the queue capacity is assumed to be infinite.

In a **single-station** queueing system, customers arrive, wait for service, and depart after service completion.

An **exponential distribution** with parameter λ is a density function (§7.3.1) having the form $f(x) = \lambda e^{-\lambda x}$ for $x \geq 0$.

An arrival process is **Poisson** if the interarrival times (times between successive arrivals) are independent and identically distributed exponential random variables.

A random variable has an **Erlang distribution** with phase parameter k if it is the sum of $k \geq 1$ independent and identically distributed exponential random variables.

Facts:

1. If the random variable X has an exponential distribution with parameter λ, then $E(X) = \frac{1}{\lambda}$ and $\text{Var}(X) = \frac{1}{\lambda^2}$.

2. If the arrival process is Poisson with parameter λ, then the number of customers arriving in an interval of time of length x is a Poisson random variable (§7.3.1) with parameter λx.

3. The Erlang distribution is a special type of gamma distribution (§7.3.1).

4. *Kendall's notation*: A single-station queueing system is described by the 5-tuple: interarrival-time distribution/service-time distribution/number of servers/waiting room capacity/service discipline.

5. The following symbols are standard in describing queueing systems according to the scheme in Fact 4:
   - M — exponential (M for Memoryless);
   - E_k — Erlang with phase parameter k;
   - D — deterministic (constant);
   - G — general.

6. More complicated queueing systems can consist of networks of queues, multiple types of customers, and priority schemes.

7. A World Wide Web site that provides a list of over 100 books on queueing theory can be found at the site:
   http://supernova.uwindsor.ca/people/hlynka/qbook.html

8. A compilation of queueing theory software can be found at the site:
   http://supernova.uwindsor.ca/people/hlynka/qsoft.html

Examples:

1. A single-station queueing system is depicted by the schematic diagram in the following figure. Here customers randomly join the system (according to the arrival process), wait for service in the waiting room, are served (which takes a random amount of time), and then depart from the system.

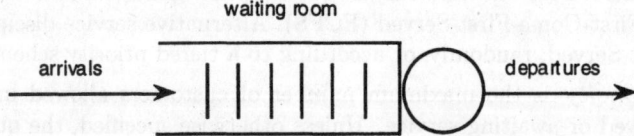

2. An $M/G/3/10/$LCFS system has Poisson arrivals (exponential interarrival times), general service times, three servers, room for ten customers (including those in service), and Last-Come-First-Served service discipline.

3. An $M/M/1$ queue has Poisson arrivals, exponential service times, a single server, infinite waiting room, and FCFS service discipline.

4. *Airplane landings*: The landing of aircraft at an airport can be viewed as a queueing system in which the aircraft are the customers and the runways are the servers. Aircraft arrive according to a certain stochastic arrival process, and the length of time to land follows a certain service-time distribution. Those aircraft that are unable to land must join the queue of circling aircraft, awaiting service (normally according to a FCFS discipline, except in the case of an emergency landing, which would be a type of priority scheme).

5. *Communication network*: Messages arrive according to a Poisson process at rate λ per second and are to be transmitted over a particular data link. The time required to transmit a single message is exponentially distributed, with average duration $\frac{1}{\mu}$ seconds. Messages waiting to be sent are stored in an input buffer. If the buffer has infinite capacity, then this system is an $M/M/1$ queue. If the input buffer has finite capacity c, then this system is an $M/M/1/c$ queue.

6. *Banking*: Customers arriving at a bank form a single common queue and are served by the s available tellers in FCFS order. If arrivals are Poisson and the length of time to service a customer is exponential, this system can be modeled as an $M/M/s$ queue.

7. *Remote computing*: A computer center has c dial-up telephone lines. Users can dial into the central computer from their remote devices using any of the lines. (Calls roll over to an available line if one is free.) If arrivals are Poisson and the length of time spent online follows an arbitrary distribution, then this is an $M/G/c/c$ queue. It is also known as a *loss system*, since any calls to the central computer receive a busy signal when all servers (lines) are occupied, and hence these calls are "lost" to the system.

7.8.2 GENERAL SYSTEMS

This section presents results applicable to single-station queues with general arrival patterns, service-time distributions, and queue disciplines.

Definitions:

For a single-station queueing system, define:
- A_n = the arrival time of the nth customer to the system;
- S_n = the service time of the nth customer;
- D_n = the departure time of the nth customer;
- $A(t)$ = the total number of arrivals up to and including time t;
- $D(t)$ = the total number of departures up to and including time t;
- $X(t)$ = the total number of customers waiting in the system at time t.

The stochastic process $\{X(t) \mid t \geq 0\}$ is the **queue-length process**.

The state distribution following an arbitrary departure is $\pi_j = \lim_{n \to \infty} Pr(X(D_n^+) = j)$, for $j \geq 0$.

The state distribution prior to an arbitrary arrival is $\pi_j^* = \lim_{n \to \infty} Pr(X(A_n^-) = j)$, for $j \geq 0$.

The state distribution at an arbitrary time point, or **steady-state distribution**, is $p_j = \lim_{t \to \infty} Pr(X(t) = j)$, for $j \geq 0$.

The queue-length process (or the queueing system) is **stable** if the steady-state distribution $\{p_j \mid j \geq 0\}$ exists and $\sum_{j=0}^{\infty} p_j = 1$.

The **waiting time** of the nth customer is $W_n = D_n - A_n$; it includes the service time.

The steady-state **expected waiting time** is $W = \lim_{n \to \infty} \frac{1}{n} \sum_{k=1}^{n} W_k$, if the limit exists.

The **long-run arrival rate** is $\lambda = \lim_{n \to \infty} \frac{n}{A_n}$, if the limit exists.

The steady-state **expected number** in the system is $L = \lim_{t \to \infty} \frac{1}{t} \int_0^t X(u)\, du$, if the limit exists.

Facts:

1. The number of customers in the system at any time equals the total number of arrivals up to that time minus the number of departures up to that time: that is, $X(t) = A(t) - D(t)$ for $t \geq 0$.

2. A sample path of the queue-length process $\{X(t) \mid t \geq 0\}$ is piecewise constant, with upward jumps at points of arrival and downward jumps at points of departure.

3. Suppose all the jumps in the sample paths of $\{X(t) \mid t \geq 0\}$ are of size ± 1, with probability 1. If either π_j or π_j^* exists, then $\pi_j = \pi_j^*$ for all $j \geq 0$.

4. *PASTA (Poisson Arrivals See Time Averages)*: If $\{A(t) \mid t \geq 0\}$ is Poisson and, for every $s \geq 0$, $\{A(t) \mid t \geq s\}$ is independent of $\{X(u) \mid 0 \leq u < s\}$, then $p_j = \pi_j^*$ for all $j \geq 0$.

5. *Little's Law* (J. D. C. Little, 1961): $L = \lambda W$.

Examples:

1. Suppose arrivals to a system occur deterministically, every 3 minutes, and service times are deterministic, each taking 2 minutes. Since every arriving customer is served immediately, either $X(t) = 0$ (no customers) or $X(t) = 1$ (a single customer). Every arrival finds an empty system and every departure leaves an empty system: $\pi_0 = 1 = \pi_0^*$, $\pi_1 = 0 = \pi_1^*$, as required by Fact 3. (The steady-state distribution does not exist.)

2. On average $\lambda = 24$ customers per hour arrive at a copy shop. Typically, there are $L = 9$ customers in the store at any time. Using Little's law, $W = \frac{L}{\lambda} = 0.375$ hour so that each customer spends on average 0.375 hours (or 22.5 minutes) in the shop.

3. The steady-state queue length or waiting time in a queueing system can be reduced by increasing the service rate. Suppose the long-run arrival rate doubles, but the service rate is increased so that the steady-state expected waiting time remains the same. Then by Little's law the steady-state expected queue length will also double.

4. *Machine repair*: A single machine is either working or being repaired. Suppose that the average time between breakdowns is exponentially distributed with mean $\frac{1}{\lambda}$ and the time to repair the machine is is exponentially distributed with mean $\frac{1}{\mu}$. This is then an $M/M/1/1$ queueing system with a single customer, corresponding to a broken down machine. $X(t) = 0$ signifies that the machine is working, and $X(t) = 1$ signifies that the machine is being repaired. A sample path of this system is shown in the following figure, with the machine initially working. Over a long period of time, after N breakdowns and subsequent repairs, the machine is working for $N(\frac{1}{\lambda})$ units of time and is being repaired for $N(\frac{1}{\mu})$ units of time. The long run proportion of time the machine is working is then

$$\frac{N(\frac{1}{\lambda})}{N(\frac{1}{\lambda}) + N(\frac{1}{\mu})} = \frac{\mu}{\lambda + \mu}.$$

This value also turns out to be the steady-state probability of finding the system in state 0, with the machine working.

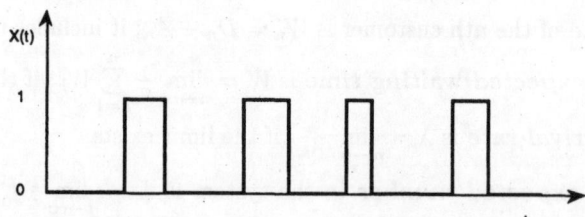

7.8.3 SPECIAL QUEUEING SYSTEMS

This section summarizes analytical results about special types of single-station queues.

Notation:
- $\frac{1}{\lambda}$ = expected interarrival time;
- $\tilde{A}(\cdot)$ = Laplace transform of the interarrival-time density;
- $\frac{1}{\mu}$ = expected service time;
- $\tilde{B}(\cdot)$ = Laplace transform of the service-time density;
- σ^2 = variance of the service time;
- s = number of servers;
- $\rho = \frac{\lambda}{s\mu}$ = traffic intensity.

The probability generating function (§7.3.3) for the steady-state distribution $\{p_j\}$ of a queueing system is $\phi(z) = \sum_{j=0}^{\infty} p_j z^j$, $|z| < 1$.

The Laplace transform for the waiting-time distribution $f(w)$ of a queueing system is $\psi(s) = \int_0^{\infty} e^{-sw} f(w)\, dw$, $Re(s) > 0$.

Facts:

1. The $M/M/1$ queue is stable if $\rho < 1$. The following results hold when the queue is stable:
$$p_j = (1-\rho)\rho^j = \pi_j = \pi_j^*$$
$$L = \frac{\rho}{1-\rho}$$
$$W = \frac{1}{\mu - \lambda}.$$

2. The $M/M/1/K$ queue is always stable. Assume $\rho \neq 1$.
$$p_j = \frac{1-\rho}{1-\rho^{K+1}}\rho^j, \quad 0 \le j \le K$$
$$\pi_j^* = \frac{p_j}{1-p_K} = \pi_j, \quad 0 \le j \le K-1$$
$$L = \frac{\rho}{1-\rho}\left(\frac{1-\rho^K}{1-\rho^{K+1}} - Kp_K\right)$$
$$W = \frac{1}{\mu - \lambda}\left(\frac{1-\rho^K}{1-\rho^{K+1}} - Kp_K\right).$$

If $\rho = 1$, the above formulas reduce to:
$$p_j = \frac{1}{K+1}, \quad 0 \le j \le K$$
$$\pi_j^* = \pi_j = \frac{1}{K}, \quad 0 \le j \le K-1$$
$$L = \frac{K}{2}$$
$$W = \frac{K}{2\mu}.$$

3. The $M/M/s$ queue is stable if $\rho < 1$. The following results hold when the queue is stable:

$$p_0 = \left(\sum_{n=0}^{s-1} \frac{(s\rho)^n}{n!} + \frac{s^s}{s!} \frac{\rho^s}{1-\rho} \right)^{-1}$$

$$p_j = \begin{cases} \frac{(s\rho)^j}{j!} p_0, & 0 \leq j < s \\ \frac{s^s}{s!} \rho^j p_0, & j \geq s \end{cases}$$

$$\pi_j = \pi_j^* = p_j, \quad j \geq 0$$

$$L = \rho \left(s + \frac{p_s}{(1-\rho)^2} \right)$$

$$W = \frac{1}{s\mu} \left(s + \frac{p_s}{(1-\rho)^2} \right)$$

E(number of busy servers) $= \rho$.

4. The $M/M/\infty$ queue is always stable:

$$p_j = e^{-(\lambda/\mu)} \frac{(\lambda/\mu)^j}{j!}, \quad j \geq 0$$

$$\pi_j = \pi_j^* = p_j, \quad j \geq 0$$

$$L = \frac{\lambda}{\mu}$$

$$W = \frac{1}{\mu}.$$

5. The $M/G/1$ queue is stable if $\rho < 1$. The following results hold when the queue is stable:

$$p_0 = 1 - \rho,$$

$$\pi_j = \pi_j^* = p_j$$

$$\phi(z) = (1-\rho) \frac{(1-z)\tilde{B}(\lambda(1-z))}{\tilde{B}(\lambda(1-z))-z}$$

$$\psi(s) = (1-\rho) \frac{s\tilde{B}(s)}{s - \lambda(1-\tilde{B}(s))}$$

$$L = \rho + \frac{\rho^2 + \lambda^2 \sigma^2}{2(1-\rho)}$$

$$W = \frac{1}{\mu} + \frac{\lambda((1/\mu)^2 + \sigma^2)}{2(1-\rho)}.$$

The last four equations are the various forms of the *Pollaczek-Khintchine formula*. No closed form results are available for $M/G/c$ queues for $2 \leq c < \infty$.

6. The $M/G/c/c$ queue, also called a *loss system*, is always stable. The main result is:

$$p_j = \frac{\rho^j/j!}{\sum_{n=0}^{c} \rho^n/n!}, \quad 0 \leq j \leq c.$$

7. The $M/G/\infty$ queue is always stable:

$$p_j = e^{-(\lambda/\mu)} \frac{(\lambda/\mu)^j}{j!}, \quad j \geq 0$$

$$\pi_j = \pi_j^* = p_j, \quad j \geq 0$$

$$L = \frac{\lambda}{\mu}$$

$$W = \frac{1}{\mu}.$$

8. The $G/M/1$ queue is stable if $\rho < 1$. When the queue is stable there is a unique solution $\alpha \in (0, 1)$ to $\alpha = \tilde{A}(\mu(1 - \alpha))$. The following results hold when the queue is stable:

$$\pi_j^* = (1 - \alpha)\alpha^j = \pi_j, \quad j \geq 0$$
$$p_0 = 1 - \rho$$
$$p_j = \rho \pi_{j-1}^*, \quad j \geq 1$$
$$L = \frac{\rho}{1-\alpha}$$
$$W = \frac{1}{\mu} \frac{1}{(1-\alpha)}.$$

The $G/M/c$ queue can be analytically solved, but the results are complicated.

9. The $G/M/\infty$ queue is always stable:

$$L = \frac{\lambda}{\mu}$$

variance of number in system $= \frac{\lambda}{\mu}(1 - \frac{\lambda}{\mu}) + \frac{\lambda/\mu}{2} \frac{\tilde{A}(\mu)}{1 - \tilde{A}(\mu)}.$

Examples:

1. At a drop-in legal clinic, the lawyer sees four clients during a typical (eight hour) day. Each client's case requires on average 1.5 hours of the lawyer's time. If arrivals are Poisson and service times are exponentially distributed, then this is an $M/M/1$ queue with $\lambda = \frac{4}{8} = \frac{1}{2}$ customers per hour and $\frac{1}{\mu} = \frac{3}{2}$. Here $\rho = \frac{\lambda}{\mu} = \frac{3}{4} < 1$, so the queue is stable. Using Fact 1, $p_0 = 1 - \rho = \frac{1}{4}$, so there is probability $\frac{1}{4}$ that the lawyer is idle. The expected number of clients in the clinic is $L = \frac{\rho}{1-\rho} = 3$ and the average wait of a client is $W = \frac{1}{\mu - \lambda} = 6$ hours.

2. Customers arrive at a service station according to a Poisson process with rate 10 per hour. The manager has two options: (a) employ a single fast server who can service a customer in 5 minutes on average, or (b) employ two slow servers each taking 10 minutes on average to serve a customer. Assume that the service times are exponential. Which option should the manager implement to minimize the expected waiting time in steady state?

Under (a) the system is an $M/M/1$ queue with $\lambda = 10$, $\mu = 12$. Since $\rho = \frac{10}{12} < 1$, the system is stable. By Fact 1, $W = \frac{1}{12-10} = 0.5$ hours. Under (b) the system is an $M/M/2$ queue with $\lambda = 10$ and $\mu = 6$. The system is stable since $\rho = \frac{10}{2 \cdot 6} < 1$. From Fact 3, $p_0 = \frac{1}{11}$, $p_2 = \frac{25}{198}$, and $W = \frac{6}{11} = 0.55$ hours. Thus option (a) is better. In general, it is better to employ a few fast servers than many slow servers with the same overall service capacity.

3. A system manager has the option of using one of three possible servers in a single-server system. The service times under the first server are exponential with mean of 6 minutes. Under the second server they are uniformly distributed over [4, 8] minutes. Under the third they are constant, equal to 6 minutes. The customers arrive according to a Poisson process with rate 8 per hour. Which server should be chosen to minimize the expected waiting time in steady state?

The mean service time is 6 minutes; i.e., $\mu = 10$ per hour, for all three servers. However, the variances σ^2 are different. This $M/G/1$ system is stable under all three servers since $\rho = \frac{8}{10} < 1$. For server one, $\sigma^2 = 0.01$ (hours)^2 and $W = 0.5$ hours. For the second server, $\sigma^2 = \frac{1}{2700} = 0.000370$ (hours)^2 and $W = \frac{83}{270} = 0.31$ hours. For the third server, $\sigma^2 = 0.0$ (hours)^2 and $W = \frac{3}{10} = 0.3$ hours. Thus, it is best to use the server with constant service times. In general, reducing the variance of the service times has a beneficial effect on the system.

4. A small business wants to install a telephone system with multiple lines, though without any capacity for call queueing. This is to be done to ensure that 95% of the calls made to the business get answered. Suppose that the arrival process is Poisson with $\lambda = 10$ calls per hour, and that the average call lasts 5 minutes. This is an $M/G/c/c$ loss system (Fact 6), and it is necessary to find the smallest value of c such that $p_c \leq 0.05$. Using Fact 6 with $\rho = \frac{10}{12} = \frac{5}{6}$ and $c = 1$ gives $p_1 = \frac{\rho}{1+\rho} = 0.45$. Similar calculations give $p_2 = 0.16$ for $c = 2$ and $p_3 = 0.042$ for $c = 3$. Consequently, three lines are needed to ensure the stipulated grade-of-service requirement.

7.9 SIMULATION

Simulation is a technique for numerically estimating the performance of a complex stochastic system when analytic solution is not feasible. This section discusses both *discrete-event* and *Monte Carlo* simulation. In discrete-event simulation models, the passage of time plays a key role, as changes to the state of the system occur only at certain points in simulated time. For example, queueing and inventory systems can be studied by discrete-event simulation models. Monte Carlo simulation models do not, however, require the passage of time. Such models are useful in estimating eigenvalues, estimating π, and estimating the quantiles of a mathematically intractable test statistic in hypothesis testing. Simulation has been described [BrEtal87] as "driving a model of a system with suitable inputs and observing the corresponding outputs." Accordingly, the following three subsections discuss input modeling, output analysis, and simulation programming languages.

7.9.1 INPUT MODELING

This section addresses three key issues in constructing a simulation model:
- determining a source of randomness to drive the probabilistic aspects of the model;
- input model selection to determine the appropriate probabilistic models to drive the simulation;
- random variate generation algorithms that transform random numbers to random variates.

Definitions:

Random numbers are real numbers generated uniformly over the interval $(0,1)$.

A **random number generator** is any mechanism or algorithm for generating random numbers.

Pseudo-random numbers are values generated deterministically, but that appear to behave like independent and identically distributed random numbers.

Let m be a large prime integer. A **purely multiplicative linear congruential generator** (§4.3.1) produces a stream of pseudo-random numbers $\{ \frac{x_i}{m} \mid i \geq 1 \}$ based on the recursive relationship $x_{i+1} = ax_i \bmod m$, where a is an integer **multiplier** between 1 and $m-1$, and x_0 is an integer **seed** between 1 and $m-1$.

An **input model** characterizes the stochastic elements of a discrete-event simulation model.

A **trace-driven** input model generates a process that is identical to the collected data values without relying on a parametric model.

A **random variate** is a realization of a random variable.

Facts:

1. Stochastic simulations typically derive their source of randomness from random numbers. That is, inputs to the simulated system need to be generated according to a specified probability model, a task that can be accomplished by suitably transforming (uniform) random numbers.

2. Desirable properties for random number generators include: uniformity, independence, speed, minimal memory requirements, ease of implementation, portability across various computer systems, reproducibility, and multiple stream capability.

3. Although numerous methods have been proposed for generating random numbers, multiplicative linear congruential generators are typically used to produce a stream of pseudo-random numbers.

4. Due to the prevalence of 32-bit computer architecture, m is often chosen to be $2^{31} - 1$, which is prime.

5. A *full period* generator, which cycles through all $m-1$ possible x_i values prior to repeating, is obtained by selecting a to be a primitive root modulo m. (See §4.7.1.)

6. Software for pseudo-random number generators can be found at the sites:
   - http://random.mat.sbg.ac.at/others/#MCSoftware
   - http://www.taygeta.com/random.html
   - http://www.isye.gatech.edu/informs-sim/#ware

7. Additional information on the theoretical and empirical performance of a variety of pseudo-random number generators is available at:
   - http://random.mat.sbg.ac.at/generators/

8. If the multiplier a is chosen so that it is "modulus-compatible" with m, then potential overflow can be averted for large values of m. Two values of a that are often used with $m = 2^{31} - 1$ are $a = 7^5 = 16807$ and $a = 48271$ [PaMi88].

9. Successful input modeling for a discrete-event simulation requires a close match between the input model and the true underlying probabilistic mechanism associated with the system.

10. One of the first steps in determining an appropriate input model for an element of a discrete-event simulation is to assess whether the observations are independent and identically distributed.

11. An input model can be specified in several ways: e.g., using a cumulative distribution function, joint probability density function, hazard function, intensity function, or variate-generation algorithm.

12. Many input models rely on parametric probabilistic models such as the binomial, normal, and Weibull distributions. Maximum likelihood is typically used to estimate parameters of these models.

> **Algorithm 1:** Inverse transformation method.
>
> input: cumulative distribution function F
> output: random variates X from this distribution
>
> generate U uniformly over $(0,1)$
> $X := F^{-1}(U)$

13. Bézier curves [FlWi93] offer a unique combination of the parametric and nonparametric approaches. After an initial distribution is fitted to the data set, the modeler decides whether differences between the empirical and fitted models represent sampling variability (chance variation) or an aspect of the distribution that should be included in the input model.

14. Multivariate distributions (e.g., the multivariate normal distribution with mean μ and variance-covariance matrix Σ) are considered by [Jo87].

15. Once an input model has been chosen, random variate generation algorithms are used to transform random numbers to variates from the input model.

16. Devroye [De86] gives algorithms for converting random numbers to random variates associated with input models chosen to drive the simulation.

17. Techniques commonly used for generating random variates from univariate probability distributions are: inverse transformation, composition, acceptance/rejection, and special properties.

18. Algorithm 1, which shows the inversion method, is based on the probability integral transformation. It is assumed that the cumulative distribution function $F(x)$ for the input model of interest has the inverse $F^{-1}(U)$.

19. Other topics in variate generation include table methods, generating from multivariate distributions, random sampling, estimating integrals, and generating processes correlated in time.

Examples:

1. Suppose that a sequence of arrival times (e.g., of customers at a bank) is collected over a 24-hour time period. A trace-driven input model for the arrival process is generated by having arrivals occur at the same times as the observed values.

2. Let $t_1, t_2, \ldots, t_n$ be the arrival times to a queue collected on the time interval $(0, c]$. If the times between arrivals are independent and identically distributed, a parametric or nonparametric model can be fitted to the data. In the former case, parameters are often estimated by maximizing the *likelihood function* [LaKe91]

$$L(\theta) = \prod_{i=1}^{n} f(x_i, \theta),$$

where $x_i = t_i - t_{i-1}$ for $i = 1, 2, \ldots, n$, $t_0 \equiv 0$, $\theta = (\theta_1, \theta_2, \ldots, \theta_p)$ is a vector of unknown parameters, and $f(x_i, \theta)$ is the probability density function of the interarrival times.

3. If the interarrival times to a queue (as in Example 2) are not independent and identically distributed, then a nonstationary point process might be considered, such as a nonhomogeneous Poisson process, where the arrival rate $\lambda(t; \theta)$ varies over time. One parametric model is the *power law* process, with intensity function $\lambda(t; \lambda, \kappa) = \lambda^\kappa \kappa t^{\kappa-1}$ for $t > 0$, where λ and κ are positive parameters. The likelihood function for the single realization on $(0, c]$ is $L(\lambda, \kappa) = \left(\prod_{i=1}^{n} \lambda(t_i; \lambda, \kappa)\right) \exp\left(-\int_0^c \lambda(t; \lambda, \kappa)\, dt\right)$. Maximum likelihood estimators can be determined by maximizing $L(\lambda, \kappa)$ or its logarithm with

respect to λ and κ. Confidence regions for the unknown parameters can be found by using asymptotic properties of the likelihood ratio statistic or the observed *information matrix* [LaKe91]. As with all statistical modeling, goodness-of-fit tests should be performed in order to assess the model adequacy.

4. *Weibull distribution*: The *Weibull distribution* has cumulative distribution function $F(x) = 1 - e^{-(\lambda x)^\kappa}$ for $x > 0$. The inverse cumulative distribution function is $F^{-1}(y) = \dfrac{[-\ln(1-y)]^{1/\kappa}}{\lambda}$ for $0 < y < 1$. Algorithm 1 can be used to generate a Weibull variate according to

$$X := \frac{[-\ln(1-U)]^{1/\kappa}}{\lambda}.$$

5. *$M/M/1$ queue*: To simulate the operation of a single-server queue (§7.8.1) with Poisson arrivals at rate λ and exponential service times with mean $\frac{1}{\mu}$, exponentially generated variates are needed for the interarrival times $\{I_n\}$ and the service times $\{S_n\}$. These are available as a special case of the Weibull distribution (Example 4) with $\kappa = 1$ and can be generated using $I_n = \dfrac{-\ln(1-U_n)}{\lambda}$ and $S_n = \dfrac{-\ln(1-V_n)}{\mu}$, with the $\{U_n\}$, $\{V_n\}$ generated uniformly over $(0,1)$.

A concrete example is provided in the following table, which shows one simulated run of an $M/M/1$ queue with $\lambda = 0.5$ and $\mu = 0.7$. The table shows, in successive columns, the following values for each customer n: the interarrival time I_n, the arrival time A_n, the service time S_n, the beginning time of service B_n, the departure time D_n, and the waiting time $W_n = D_n - A_n$. Notice that customers $1, 4, 7, 8, 9$ are served immediately and incur no waiting time in the queue.

customer	I_n	A_n	S_n	B_n	D_n	W_n
1	5.44	5.44	0.78	5.44	6.22	0.78
2	0.61	6.05	2.77	6.22	8.99	2.94
3	0.35	6.40	0.96	8.99	9.95	3.55
4	4.12	10.52	2.42	10.52	12.94	2.42
5	0.54	11.06	0.88	12.94	13.82	2.76
6	2.07	13.13	0.87	13.82	14.69	1.56
7	6.82	19.95	0.86	19.95	20.81	0.86
8	2.19	22.14	0.76	22.14	22.90	0.76
9	4.09	26.23	3.31	26.23	29.54	3.31
10	0.02	26.25	0.01	29.54	29.55	3.30

7.9.2 OUTPUT ANALYSIS

Once a verified and validated simulation model has been developed, a modeler typically wants to estimate measures of performance associated with outputs of the model. Although there are often several performance measures of interest, a single measure of performance θ (e.g., the mean waiting time in a queue) is studied here.

This section discusses using point estimation to compute an estimate for θ, determining a confidence interval for the point estimate, and using variance reduction techniques to obtain more precise point estimates.

Definitions:

Suppose $\{Y_i\}$ is the **output stochastic process**. If the output stochastic process consists of independent observations obtained from a population with cumulative distribution function F_Y, the pth **quantile** of F_Y is the value y_p such that $F_Y(y_p) = p$. The **median** of F_Y corresponds to $p = 0.5$.

The **sample mean** of the observations $Y_1, Y_2, \ldots, Y_n$ is given by $\overline{Y} = \frac{1}{n} \sum_{i=1}^{n} Y_i$.

If the values $Y_1, Y_2, \ldots, Y_n$ are rearranged so that $Y_{(1)} \leq Y_{(2)} \leq \cdots \leq Y_{(n)}$ then $Y_{(i)}$ is the ith **order statistic**.

The **mean** μ_Y of the process is the asymptotic mean of the output process $\{Y_i\}$.

The **variance** σ_Y^2 of the process is the asymptotic mean of the output process $\{(Y_i - \overline{Y})^2\}$.

The **probability** $Pr(A)$ of event A is the asymptotic mean of the output process $\{I(A)\}$, where I is the 0–1 indicator function for event A.

The output process $Y_1, Y_2, \ldots$ is **covariance stationary** if, for finite mean μ and finite variance $\sigma^2 > 0$, $E(Y_i) = \mu$, $i = 1, 2, \ldots$, $\text{Var}(Y_i) = \sigma^2$, $i = 1, 2, \ldots$, and $\text{Cov}(Y_i, Y_{i+j})$ is independent of i, for $j = 1, 2, \ldots$.

Variance reduction techniques are strategies for obtaining greater precision for a fixed amount of sampling.

Facts:

1. The two most common measures of performance to be estimated are means and quantiles.

2. Point estimates for μ_Y, σ_Y^2, $Pr(A)$ are typically given by the associated sample means.

3. A simple estimator of $y_p = F_Y^{-1}(p)$ is $Y_{(s)}$, where $s = \lfloor p(n+1) \rfloor$. This estimator can be improved (with respect to bias) by estimating $F_Y^{-1}(p)$ with the linear combination $(1 - \alpha)Y_{(s)} + \alpha Y_{(s+1)}$, where $\alpha = p(n+1) - \lfloor p(n+1) \rfloor$.

4. *Replication*: This is one of the simplest methods of interval estimation, in which several runs of a simulation model are used. Classical confidence intervals based on the central limit theorem for the measures of interest can then be applied to the output.

5. The presence of autocorrelation among observations (e.g., the waiting times of adjacent customers in a queue) significantly complicates the statistical analysis of simulation output from a single run.

6. To analyze a single simulation run with autocorrelation present, techniques have been developed for determining interval estimates whose actual coverage is close to the stated coverage. For many of these techniques, the output is assumed to be covariance stationary. These techniques include batch means, overlapping batch means, standardized time series, regeneration, spectral analysis, and autoregression.

7. *Common random numbers*: This is a variance reduction technique in which two or more alternative system configurations are analyzed using the same set of random numbers for particular purposes (e.g., generating service times). Using common random numbers insures that the output differences are due to the configurations rather than the sampling variability in the random numbers.

8. *Antithetic variates*: This is a second variance reduction technique, applicable to the analysis of a single system. If the random numbers $\{U_i\}$ are used for a particular purpose in one simulation run, then using $\{1 - U_i\}$ in a second run typically induces a negative correlation between the outputs of the two runs. Thus, the average of the output measures from the two runs will have a reduced variance.

9. There are a variety of variance reduction techniques. See Wilson [Wi84] for a detailed discussion.

10. Other topics in output analysis include initialization bias detection, ranking and selection, comparing alternative system designs, experimental design, and optimization.

Examples:

1. *Confidence intervals for expected waiting times*: Let $X_1, X_2, \ldots, X_n$ be the averages of the waiting times of customers in a single-server queue from n independent replications of a discrete-event simulation model. A $100(1 - \alpha)\%$ confidence interval for μ, the steady-state mean waiting time, is

$$\overline{X} - t_{\alpha/2, n-1} \frac{s}{\sqrt{n}} < \mu < \overline{X} + t_{\alpha/2, n-1} \frac{s}{\sqrt{n}},$$

where $\overline{X}$ is the sample mean, s is the sample standard deviation, and $t_{\alpha/2, n-1}$ is the $1 - \frac{\alpha}{2}$ fractile of the t distribution with $n-1$ degrees of freedom. The replications must be "warmed up" to avoid initialization bias. The asymptotic normality of $X_1, X_2, \ldots, X_n$ is assured by the central limit theorem and independence is based on the use of independent random number streams.

2. $M/M/1$ *queue*: The simulation of Example 5 of §7.9.1 was executed so that the first 200 customer wait times were collected. The measure of performance θ for the system is the steady-state expected customer wait time. The initial conditions for each replication are an empty system and an idle server. The stopping time for each replication is when the 200th customer departs. Running this simulation experiment for $n = 100$ replications gave $\overline{X} = 4.72$ and for $n = 500$ replications gave $\overline{X} = 4.76$. For this simple queueing system, the steady-state analytical solution is $W = \frac{1}{\mu - \lambda} = 5.0$ (§7.8.3, Fact 1). These averages are biased low since the early waiting times have a lower expected value than the subsequent waiting times as a result of the initial conditions. To improve these point estimates, the system was permitted to warm up for the first 100 customers and the average waiting time was then calculated for the last 100 customers. In this case, rerunning the simulation gave the improved estimates $\overline{X} = 5.20$ for $n = 100$ and $\overline{X} = 4.93$ for $n = 500$.

3. *Common random numbers*: Law and Kelton [LaKe91, pp. 620–621] compare the $M/M/1$ and $M/M/2$ queueing models with a utilization of $\rho = 0.9$ using the waiting times in the queue of the first 100 customers. With $n = 100$ independent replications of each system, they compare the two models in four ways:

- independent runs (I);
- arrival streams using common random numbers (A);
- service times using common random numbers (S);
- arrival streams and service times using common random numbers $(A \& S)$.

Common random numbers is a variance reduction technique that feeds identical interarrival and/or service times into the two different queueing models to increase the likelihood that observed differences in the waiting times are due to the system configurations ($M/M/1$ versus $M/M/2$) rather than sampling error. The mean half-widths of the confidence intervals ($\alpha = 0.10$) reported for their example are $0.70(I), 0.49(A), 0.49(S)$, and $0.04(A \& S)$.

7.9.3 SIMULATION LANGUAGES

This section considers the history and features of simulation programming languages developed over the years.

Facts:

1. The use of a general-purpose simulation programming language (SPL) expedites model development, input modeling, output analysis, and animation. In addition, SPLs have accelerated the use of simulation as an analysis tool by bringing down the cost of developing a simulation model.

2. In a history of the development of SPLs from 1955 to 1986, Nance [Na93] defines six requirements that a SPL must meet:
   - random number generation;
   - variate generation;
   - list processing capabilities so that objects can be created, altered, and deleted;
   - statistical analysis routines;
   - summary report generators;
   - a timing executive or event calendar to model the passage of time.

3. SPLs may take the form of:
   - a set of subprograms in a general purpose language (GPL) such as Fortran or C that can be called to meet these six requirements;
   - a preprocessor that converts statements or symbols to lines of code in a GPL;
   - a conventional programming language.

4. The following table shows a division of the historical record into five distinct periods, including the names of several languages that came into existence in each period.

period	characteristics	languages
1955–1960	period of search	GSP
1961–1965	the advent	CLP, CSL, DYNAMO, GASP, GPSS, MILITRAN, OPS, QUIKSCRIPT, SIMSCRIPT, SIMULA, SOL
1966–1970	formative period	AS, BOSS, Q-GERT, SLANG, SPL
1971–1978	expansion period	DRAFT, HOCUS, PBQ, SIMPL
1979–1986	consolidation and regeneration	INS, SIMAN, SLAM

5. The General Purpose System Simulator (GPSS) was first developed on various IBM computers in the early 1960s. Algol-based SIMULA was also developed in the 1960s and had features that were ahead of its time. These included abstract data types, inheritance, the co-routine concept, and quasi-parallel execution.

6. SIMSCRIPT was developed by the RAND Corporation with the purpose of decreasing model and program development times. SIMSCRIPT models are described in terms of entities, attributes, and sets. The syntax and program organization were influenced by Fortran.

7. The Control and Simulation Language (CSL) takes an "activity scanning" approach to language design, where the activity is the basic descriptive unit.

8. The General Activity Simulation Program (GASP), in common with several other languages, used flow-chart symbols to bridge the gap between personnel unfamiliar with programming and programmers unfamiliar with the application area. Although originally written in Algol, GASP provided Fortran subroutines for list-processing capabilities (e.g., queue insertion).

9. GASP was a forerunner to both the Simulation Language for Alternative Modeling (SLAM) and SIMulation ANalysis (SIMAN) languages.

10. SLAM was the first language to include three modeling perspectives in one language: network (process orientation), discrete-event, and continuous (state variables).

11. SIMAN was the first major SPL executable on an IBM PC.

12. Simulation software in the 1990s has mushroomed, with numerous packages and languages available both for general purpose and application-specific simulations. Special purpose and integrated packages are widespread and available on desktop computers. The 1997 survey [SW97] compares 46 products, having a wide range of features and capabilities, and the 1997 review [II97] compares 65 products.

13. A recent trend has been the addition of animation to intelligently view simulation output. Surveys of web-based simulations can be found at the sites:
    - http://ms.ie.org/websim/survey/survey.html
    - http://www.cise.ufl.edu/~fishwick/websim.html

14. Software for carrying out Monte Carlo simulation can be found at the site:
    - http://random.mat.sbg.ac.at/others/#MCSoftware

15. A number of commercial and freeware/shareware simulation packages are listed at the site:
    - http://www.isye.gatech.edu/informs-sim/#ware
    - http://www.isye.gatech.edu/informs-sim/comm.html
    - http://ws3.atv.tuwien.ac.at/eurosim/

REFERENCES

Printed Resources:

[BaNi70] M. N. Barber and B. W. Ninham, *Random and Restricted Walks*, Gordon and Breach, 1970.

[BaPr81] R. E. Barlow and F. Proschan, *Statistical Theory of Reliability and Life Testing*, To Begin With, 1981.

[Be93] H. C. Berg, *Random Walks in Biology*, Princeton University Press, 1993.

[BrFoSc87] P. Bratley, B. L. Fox, and L. E. Schrage, *A Guide to Simulation*, 2nd ed., Springer-Verlag, 1987.

[BuTu92] J. L. Buchanan and P. R. Turner, *Numerical Methods and Analysis*, McGraw-Hill, 1992.

[ClOlTu89] C. W. Clenshaw, F. W. J. Olver, and P. R. Turner, "Level-index arithmetic: an introductory survey", in *Numerical Analysis and Parallel Processing, Lecture Notes in Mathematics* 1397 (1989), Springer-Verlag, 95–168.

[Co87] C. J. Colbourn, *The Combinatorics of Network Reliability*, Oxford University Press, 1987.

[Co81] R. B. Cooper, *Introduction to Queueing Theory*, North-Holland, 1981.

[De86] L. Devroye, *Non-Uniform Random Variate Generation*, Springer-Verlag, 1986.

[Fe68] W. Feller, *An Introduction to Probability Theory and its Applications*, Vol. I, 3rd ed., Wiley, 1968. (A classic text with extensive coverage of probability theory, the combinatorics of simple random walks, and Markov chains.)

[Fe71] W. Feller, *An Introduction to Probability Theory and its Applications*, Vol. II, 2nd ed., Wiley, 1971. (A companion to the first volume with a treatment of random walks on continuous space and more advanced topics.)

[Fi96] G. S. Fishman, *Monte Carlo: Concepts, Algorithms, and Applications*, Springer-Verlag, 1996. (A comprehensive and integrated treatment of Monte Carlo methods and their applications.)

[FlWi93] M. Flanigan-Wagner and J. R. Wilson, "Using univariate Bézier distributions to model simulation input processes", in *Proceedings of the 1993 Winter Simulation Conference*, 1993, 365–373.

[GrHa85] D. Gross and C. M. Harris, *Fundamentals of Queueing Theory*, 2nd ed., Wiley, 1985.

[HeSo82] D. P. Heyman and M. J. Sobel, *Stochastic Models in Operations Research*, Vol. 1, McGraw-Hill, 1982.

[IE85] IEEE, *Binary Floating-Point Arithmetic*, IEEE Standard 754, IEEE, 1985.

[II97] IIE, "1997 simulation software buyer's guide", *IIE Solutions*, May 1997, 64–77.

[Jo87] M. E. Johnson, *Multivariate Statistical Simulation*, Wiley, 1987.

[KeSn60] J. G. Kemeny and L. J. Snell, *Finite Markov Chains*, Van Nostrand, 1960.

[Kl75] L. Kleinrock, *Queueing Systems, Vol. I: Theory*, Wiley, 1975.

[Ku95] V. C. Kulkarni, *Modeling and Analysis of Stochastic Systems*, Chapman-Hall, 1995.

[LaKe91] A. M. Law and W. D. Kelton, *Simulation Modeling and Analysis*, 2nd ed., McGraw-Hill, 1991.

[La91] G. Lawler, *Intersections of Random Walks*, Birkhäuser, 1991. (A recent monograph on the mathematical analysis of problems dealing with the non-intersection of paths of random walks. Sophisticated mathematical methods.)

[Ma90] B. G. Malkiel, *A Random Walk Down Wall Street*, W. W. Norton, 1990.

[Na93] R. E. Nance, "A history of discrete event simulation programming languages", *ACM SIGPLAN Notices* 28 (1993), 149–175.

[PaMi88] S. K. Park and K. W. Miller, "Random number generators: good ones are hard to find", *Communications of the ACM* 31 (1988), 1192–1201.

[Ro93] S. M. Ross, *Introduction to Probability Models*, 5th ed., Academic Press, 1993.

[Sc90] B. Schmeiser, "Simulation experiments", in *Handbooks in OR & MS*, D. P. Heyman and M. J. Sobel, eds., Elsevier, 1990, 296–330.

[Sh91] D. R. Shier, *Network Reliability and Algebraic Structures*, Clarendon Press, 1991.

[Sp76] F. Spitzer, *Principles of Random Walks*, 2nd ed., Springer-Verlag, 1976. (A classic monograph on mathematical properties of lattice random walks.)

[St74] P. H. Sterbenz, *Floating-Point Computation*, Prentice-Hall, 1974.

[Sw97] J. A. Swain, "Simulation goes mainstream", *ORMS Today*, October 1997, 35–46.

[Ta72] R. E. Tarjan, "Depth-first search and linear graph algorithms", *SIAM J. Computing* 1 (1972), 146–160.

[We94] G. H. Weiss, *Aspects and Applications of the Random Walk*, North Holland, 1994. (A recent treatment of random walks and their applications with a physical scientist's perspective.)

[Wi84] J. R. Wilson, "Variance reduction techniques for digital simulation", *American Journal of Mathematical and Management Sciences* 4 (1984), 277–312.

[Wo89] R. W. Wolff, *Stochastic Modeling and the Theory of Queues*, Prentice-Hall, 1989.

Web Resources:

http://ms.ie.org/websim/survey/survey.html (Surveys of web-based simulations.)

http://random.mat.sbg.ac.at/generators/ (Information on the theoretical and empirical performance of a variety of pseudo-random number generators.)

http://random.mat.sbg.ac.at/others/#MCSoftware (Software for Monte Carlo simulation and for pseudo-random number generators.)

http://supernova.uwindsor.ca/people/hlynka/qbook.html (A site that provides a list of over 100 books on queueing theory.)

http://supernova.uwindsor.ca/people/hlynka/qsoft.html (A compilation of software on queueing theory.)

http://ws3.atv.tuwien.ac.at/eurosim/ (Lists a number of commercial and freeware/shareware simulation packages.)

http://www.cise.ufl.edu/~fishwick/websim.html (Surveys of web-based simulations.)

http://www.intergalact.com/threedoor/threedoor.html (Simulation of the *Let's Make a Deal* game show, in which prizes are randomly hidden behind doors.)

http://www.isye.gatech.edu/informs-sim/#ware (Software for pseudo-random number generators; lists commercial and freeware/shareware simulation packages.)

http://www.isye.gatech.edu/informs-sim/comm.html (Lists a number of commercial and freeware/shareware simulation packages.)

http://www.math.uah.edu/~stat/ (Basic probability concepts together with interactive experiments.)

http://www.taygeta.com/random.html (Software for pseudo-random number generators.)

GRAPH THEORY

8.1 Introduction to Graphs — *Lowell W. Beineke*
   8.1.1 Varieties of graphs and graph models
   8.1.2 Graph operations
   8.1.3 Special graphs and graph families
   8.1.4 Graph representation and computation

8.2 Graph Models — *Jonathan L. Gross*
   8.2.1 Attributes of a graph model

8.3 Directed Graphs — *Stephen B. Maurer*
   8.3.1 Digraph models and representations
   8.3.2 Directed paths, cycles, and connectedness
   8.3.3 Orientation
   8.3.4 Directed acyclic graphs
   8.3.5 Tournaments

8.4 Distance, Connectivity, Traversability — *Edward R. Scheinerman*
   8.4.1 Walks, distance, and cycle rank
   8.4.2 Connectivity
   8.4.3 Euler trails and tours
   8.4.4 Hamilton cycles and paths

8.5 Graph Invariants and Isomorphism Type — *Bennet Manvel*
   8.5.1 Isomorphism invariants
   8.5.2 Isomorphism testing
   8.5.3 Graph reconstruction

8.6 Graph and Map Coloring — *Arthur T. White*
   8.6.1 Vertex colorings
   8.6.2 Edge colorings
   8.6.3 Cliques and independence
   8.6.4 Map colorings
   8.6.5 Graph multicoloring

8.7 Planar Drawings — *Jonathan L. Gross*
   8.7.1 Characterizing planar graphs
   8.7.2 Numerical planarity criteria
   8.7.3 Planarity algorithm
   8.7.4 Crossing numbers and thickness
   8.7.5 Stereographic projection
   8.7.6 Geometric drawings

8.8 Topological Graph Theory   Jonathan L. Gross
    8.8.1 Closed surfaces
    8.8.2 Drawing graphs on surfaces
    8.8.3 Combinatorial representation of graph imbeddings
    8.8.4 Formulas for genus and crosscap number

8.9 Enumerating Graphs   Paul K. Stockmeyer
    8.9.1 Counting graphs and multigraphs
    8.9.2 Counting digraphs and tournaments

8.10 Algebraic Graph Theory   Michael Doob
    8.10.1 Spectral graph theory
    8.10.2 Graph automorphisms

8.11 Analytic Graph Theory   Stefan A. Burr
    8.11.1 Extremal graph theory
    8.11.2 Ramsey theory for graphs
    8.11.3 Probabilistic graph theory

8.12 Hypergraphs   Andreas Gyarfas
    8.12.1 Hypergraphs as a generalization of graphs
    8.12.2 Hypergraphs as general combinatorial structures
    8.12.3 Numerical invariants of hypergraphs

INTRODUCTION

A graph is conceptually a spatial configuration with a finite set of points and a finite set of lines (possibly curved) joining one point to another (or to itself). Graph theory has its origins in many disciplines. Graphs are natural mathematical models of physical situations in which the points represent either objects or locations and the lines represent connections. Graphs are also used to model sociological and abstract situations in which each line represents a relationship between the entities represented by the points. Applications of graphs are wide-ranging — in areas such as circuit design, communications networks, ecology, engineering, operations research, counting, probability, set theory, information theory, and sociology.

This chapter contains an extensive treatment of the various properties of graphs. Further topics in graph theory are covered in Chapter 9 Trees and in Chapter 10 Networks and Flows.

GLOSSARY

acyclic digraph: a digraph containing no directed cycles.

acyclic graph: a graph containing no cycles.

adding a crosscap to a surface: an operation that increases the crosscap number of a nonorientable surface by 1.

adding a handle to a surface: an operation that increases the genus of an orientable surface by 2.

adjacency matrix (of a digraph): for a digraph D, the square matrix A_D with $A_D[i,j]$ = the number of edges from vertex v_i to vertex v_j.

adjacency matrix (of a graph): for a graph G, the square matrix A_G with $A_G[i,j]$ = the number of edges between vertices v_i and v_j.

adjacent edges: two edges with a common endpoint.

adjacent vertex (in a digraph) from [to] a vertex u: a vertex v such that there is an arc from u to v [to u from v].

adjacent vertices: two vertices that are endpoints of the same edge.

admittance matrix: given a graph G, the matrix $D - A$ where D is the diagonal matrix with the degree sequence of G on the diagonal and where A is the adjacency matrix; synonym for *Laplacian*.

algebraic specification (of a graph): a form of specification that uses group elements (see Chapter 5) in the vertex and edge names and uses the group operation in the incidence rule; a highly condensed form of specification because a single entry can specify the endpoints of all the edges in a class as large as the size of the group.

almost every (a. e.) graph has property P: the statement that the probability that a random n-vertex graph has property P approaches 1 as $n \to \infty$.

antichain: a hypergraph in which no edge contains any other edge.

arc: another name for a directed edge of a graph.

articulation point: synonym for cutpoint.

attachment of a bridge of a subgraph: given a bridge B of a subgraph H, a vertex of $B \cap H$.

attribute (of the edge-set or vertex-set): any additional feature, such as length, cost, or color, that enables a graph to model a real problem.

automorphism: given a graph or digraph, an isomorphism from the graph or digraph to itself.

automorphism group: the collection $Aut(G)$ of all automorphisms of a graph or digraph G under the operation of composition.

basis (for a digraph): a set of vertices V' of the digraph such that every vertex not in V' is reachable from V' and no proper subset of V' has this property.

bipartite: property of a graph that its vertices can be partitioned into two subsets, called the "parts", so that no two vertices within the same part are adjacent.

block: in a graph, a maximal nonseparable subgraph.

bond: a minimal disconnecting set of edges.

boundary (of a region of a graph imbedded in a surface): given a region R of a graph G imbedded in a surface, the subgraph of G containing all vertices and edges incident on R; it is denoted ∂R.

bouquet: a graph B_n with one vertex and n self-loops.

branch: synonym for arc (i.e., a directed edge).

bridge (*edge*): a cut-edge.

bridge (of a subgraph H): a maximal connected subgraph in which no vertex of H has degree greater than one.

cactus: a connected graph in which every block is either an edge or a cycle.

cartesian product: given graphs G and H, the graph $G \times H$ whose vertex set is the cartesian product $V_G \times V_H$ and whose edge set is $(V_G \times E_H) \cup (E_G \times V_H)$.

caterpillar: a tree that contains a path such that every edge has one or both endpoints in that path.

Cayley graph (or digraph): a graph that depicts a *group* with a prescribed set of generators; the vertices represent group elements, and the edges or arcs (often said to be "color-coded" for the generators) represent the product rule.

cellular imbedding: an imbedding such that every region is equivalent to the interior of the unit disk (the region $\{(x,y) \mid x^2 + y^2 < 1\}$ of the plane).

center: in a connected graph, the set of vertices of minimum eccentricity.

chain: a simple hypergraph in which, given any pair of edges, one edge contains the other.

characteristic polynomial (of a graph): the characteristic polynomial of its adjacency matrix.

characteristic value: See *eigenvalue*.

characteristic vector: See *eigenvector*.

chromatic index (of a graph or hypergraph): See *edge chromatic number*.

chromatic number (of a graph): the minimum number $\chi(G)$ of colors needed to color the vertices of a graph G so that no vertex is adjacent to a vertex of the same color; alternate notation $cr(G)$.

chromatic number (of a hypergraph): the smallest number $\chi(H)$ of independent sets required to partition the vertex set of H.

chromatic number (of a map): the minimum number $\chi(M)$ of colors needed to color the regions of the map M so that no color meets itself across an edge; alternate notation $cr(M)$.

chromatic number (of a surface): the largest map chromatic number $\chi(S)$ taken over all maps on the surface S; alternate notation $cr(S)$.

chromatically n-critical graph: an n-chromatic graph G such that $\chi(G-e) = n-1$ no matter what edge e is removed.

circuit: synonym for a closed walk, a closed trail, or a cycle, depending on the context.

clique (in a graph): in a graph G, a complete subgraph of G contained in no larger complete subgraph of G.

clique (in a hypergraph): a simple hypergraph such that every pair of edges has nonempty intersection.

clique number (of a graph): the number $\omega(G)$ of vertices of a largest clique in the graph G.

clique number (of a hypergraph): the largest number $\omega(H)$ of edges of any partial clique in the hypergraph H.

clique partition number: for a hypergraph H, the smallest number $cp(H)$ of cliques required to partition the edge set.

closed walk (trail or path): a walk, trail, or path whose origin and terminus are the same.

n-colorable graph: a graph having a vertex coloring using at most n colors.

n-colorable map: a map having a coloring using at most n colors.

comparability graph: a graph that admits a transitive orientation.

complement (of a graph): See *edge-complement*.

complete bipartite graph: a bipartite graph $K_{r,s}$ whose vertex set has two parts, of sizes r and s, respectively, such that every vertex in one part is adjacent to every vertex in the other part.

complete graph: the simple graph K_n with n vertices in which every pair of vertices is adjacent.

complete hypergraph: the simple n-vertex hypergraph K_n^* in which every subset of vertices is an edge.

complete multipartite (or *k-partite*) *graph*: a k-partite simple graph such that every pair of vertices from different parts is joined by an edge. Such a graph is denoted by $K_{n_1,\ldots,n_k}$, where $n_1,\ldots,n_k$ denote the sizes of the parts.

complete r-uniform hypergraph: the simple n-vertex hypergraph K_n^r in which every r-element subset is an edge.

complete set of invariants: a set of invariants that determine a graph or digraph up to isomorphism.

component: given a graph, a maximal connected subgraph; the number of components of a graph G is denoted $\beta_0(G)$.

connected: property of a graph that each pair of vertices is joined by a path.

connectivity: See *vertex connectivity*.

contraction, elementary (of a graph): the operation of shrinking an edge to a point, so that its endpoints are merged, without otherwise changing the graph.

contraction, elementary (of a simple graph): replacing two adjacent vertices u and v by one vertex adjacent to all other vertices to which u or v were adjacent.

contraction: for a graph, the composition of a sequence of elementary contractions.

converse: for a digraph, the digraph obtained by reversing the direction of every arc.

crosscap: a subportion of a surface that forms a Möbius band.

crosscap number (of a nonorientable surface): for a nonorientable surface S, the maximum number $\overline{\gamma}(S)$ of disjoint crosscaps one can find on the surface. The nonorientable surface of crosscap number k is denoted N_k.

crossing number: for a graph G, the minimum number $\nu(G)$ of edge-crossings taken over all normalized planar drawings of G.

cube graph: See *hypercube graph*.

cut-edge: given a graph G, an edge e such that $G - e$ has more components than G.

cut-vertex (or *cutpoint*): given a graph G, a vertex v such that $G - v$ has more components than G.

cycle: a closed path of positive length. See also *k-cycle*.

cycle, directed: a closed directed walk in which all the vertices except the first and last are distinct.

cycle graph: a graph C_n with n vertices and n edges that form a simple circuit.

cycle rank: given a connected graph G, the number $\beta_1(G)$ of edges in the complement of a spanning tree for G, that is, $|E_G| - |V_G| + 1$.

DAG: an acronym for directed acyclic graph.

degree (of a vertex in a graph): given a vertex v, the number $deg(v)$ of instances of v as an endpoint; that is, the number of proper edges incident on v plus twice the number of loops at v.

degree (of a hypergraph vertex): given a vertex x, the number $deg(x)$ of hypergraph edges containing x.

degree sequence of a graph: the sequence of the degrees of its vertices, most often sorted into size order, ascending or descending.

deleting an edge from a graph: given a graph G and an edge e of G, an operation that results in the subgraph $G - e$, which contains all the vertices of G and all edges except e.

deleting a vertex from a graph: given a graph G and a vertex v of G, an operation that results in the subgraph $G - v$, which contains all vertices of G except v and all the edges of G except those incident with v.

diameter: given a connected graph, the maximum distance between two of its vertices.

diconnected digraph: See *strongly connected digraph*.

digraph (or **directed graph**): a graph in which every edge is directed.

dipole: the graph D_n with two vertices and a multi-edge of multiplicity n joining them.

directed cycle, path, trail, walk: See *cycle, directed*, etc.

directed graph: See *digraph*.

direction (on an edge): a sense of forward progression from one end to the other, usually marked either by ordering its endpoints or by an arrowhead.

disconnected (digraph): a digraph whose underlying graph is disconnected.

disconnecting set of edges (in a connected graph): a set whose removal yields a nonconnected graph.

disconnecting set of vertices (in a connected graph): a set whose removal yields a nonconnected graph.

distance (between two vertices of a connected graph): given two vertices v and w, the length $d(v,w)$ of a shortest path between them.

distance (between two vertices of a connected digraph): given two vertices v and w, the length $d(v,w)$ of a shortest directed path between them.

dodecahedral graph: the 1-skeleton of the dodecahedron, which is a 3-dimensional polyhedron whose 12 faces are all pentagons; this graph has 20 vertices, each of degree 3, and 30 edges.

downset: a simple hypergraph in which every subset of every edge is also an edge of the hypergraph.

dual graph imbedding: a new graph imbedding obtained by placing a dual vertex in the interior of each existing ("primal") region and by drawing a dual edge through each existing ("primal") edge connecting the dual vertices on its opposite sides.

dual (of a hypergraph): given a hypergraph H, the hypergraph H^* whose incidence matrix is the transpose of the incidence matrix $M(H)$.

eccentricity (of a vertex): given a vertex v in a connected graph, the greatest distance from v to another vertex.

edge: a line, either joining one *vertex* to another or joining a *vertex* to itself; an element of the second constituent set of a *graph*.

edge chromatic number (of a graph): given a graph G, the smallest number n such that G is n-edge colorable, written $\chi_1(G)$ or $ecr(G)$.

edge chromatic number (of a hypergraph): given a hypergraph H, the smallest number $q(H)$ of matchings required to partition the edge set of H.

n-edge colorable: property of a graph that it has an edge coloring using at most n colors.

edge coloring: an assignment of colors to the edges of a graph so that adjacent edges receive different colors. See also *n-edge colorable*.

edge connectivity: the cardinality $\kappa'(G)$ of the smallest disconnecting set of edges in graph G. See also *k-edge connected*.

edge cut: See *disconnecting set*.

edge independence number: the cardinality $\alpha_1(G)$ of a largest independent set of edges in graph G.

edge-complement: given a graph G, the graph $\overline{G}$ with the same vertex set as G, but in which two vertices are adjacent if and only if they are not adjacent in G.

edge-deleted subgraph: any subgraph obtained from a graph by removing a single edge.

edge-reconstructible graph: a graph which is uniquely determined by its collection of edge-deleted subgraphs.

edge-reconstructible invariant: an invariant which is uniquely determined by the collection of edge-deleted subgraphs of a graph.

eigenvalue (of a matrix): given a matrix A, a number λ such that $Ax = \lambda x$ for some vector $x \neq 0$.

eigenvector (of a matrix): given a matrix A, a nonzero vector x such that $Ax = \lambda x$.

embedding: See *imbedding*.

empty graph: sometimes, a graph with no edges; other times, a graph with no vertices or edges. See *null graph*.

endpoints: the *vertices* that are joined by the edge.

Euler characteristic: given a surface S, an invariant $\chi(S)$ of the surface itself, given by the formula $\chi(S) = |V| - |E| + |F|$ where V, E, and F are the vertices, edges and faces of any cellular drawing of any graph on that surface; equivalently, $2 - 2g$ for the orientable surface S_g of genus g, and $2 - k$ for the nonorientable surface N_k of crosscap number k.

Euler tour: a closed Euler trail.

Euler trail: a trail that contains all the edges of the graph.

Eulerian graph: a graph that has an Euler tour.

exterior region: in a planar graph drawing, the region that extends to infinity.

extremal graph: given a set $\mathcal{G}$ of graphs and an integer n, an n-vertex graph with $ex(\mathcal{G};n)$ edges that contains no member of $\mathcal{G}$.

extremal number: given a set $\mathcal{G}$ of graphs, the greatest number $ex(\mathcal{G};n)$ of edges in any n-vertex simple graph that does *not* contain some member of $\mathcal{G}$ as a subgraph.

face: given an imbedding of a graph in a surface, a region plus its boundary.

forest: any graph without cycles.

four color theorem: the fact that every planar map can be properly colored with at most four colors, proved in 1976 by Haken and Appel after over a century of active investigation.

general graph: another name for a graph that might have loops.

generating set (for a group): a subset of group elements such that every group element is a product of generators. (*Note*: the identity of a group is the empty product).

genus (of an orientable surface): for a surface S, the maximum number $\gamma(S)$ of disjoint handles one can find on the surface; or equivalently, the maximum number of disjoint closed curves one can cut open without disconnecting the surface. The orientable surface of genus g is denoted S_g.

genus (of a graph): the minimum genus of a surface in which the graph can be cellularly imbedded.

girth: given a graph, the number of edges in a shortest cycle, if there is at least one cycle; undefined if the graph has no cycles.

graph: a set V of vertices and a set E of edges such that all the endpoints of edges in E are contained in V, written $G = (V, E)$, (V_G, E_G), or $(V(G), E(G))$.

graph model: any configuration with underlying graph structure, and possibly some additional attributes on its edges and/or vertices, such as length, direction, or cost.

graph sum: given graphs G and H, the graph $G + H$ whose vertex set and edge set are the disjoint unions, respectively, the disjoint union of the vertex sets and edge sets of G and H.

graphical sequence: a sequence of nonnegative integers such that there is a *simple graph* for which it is the *degree sequence*.

Gray code: a cyclic ordering of all 2^k bitstrings of length k, such that each bitstring differs from the next in exactly one bit entry.

Hamilton cycle: a spanning cycle, that is, a cycle including all the vertices of a graph.

Hamiltonian graph: a graph that contains a Hamiltonian cycle.

Hamilton path: a path that includes all the vertices of a graph.

head (of an arc): the vertex the arc goes to.

Hoffman polynomial (of a graph): a polynomial $p(x)$ of minimum degree such that $p(A) = J$, where A is the adjacency matrix and J is the matrix with every entry equal to 1.

homeomorphic graphs: two graphs that can both be obtained from the same graph by a sequence of edge subdivisions.

hypercube graph: a graph Q_d whose 2^d vertices could be labeled bijectively with the bitstrings of length d, so that two vertices are adjacent if and only if their labels differ in exactly one bit.

hypergraph: a finite set V of "vertices" together with a finite collection E of "edges" (sometimes, "hyperedges"), which are arbitrary subsets of V, written $H = (V, E)$.

icosahedral graph: the 1-skeleton of the icosahedron, which is a 3-dimensional polyhedron whose 20 faces are triangles; this graph has 12 vertices, each of degree 5, and 30 edges.

imbedding (of a graph in a surface): a drawing of the graph onto some surface so that there are no edge-crossings; also *embedding*.

incidence rule: any rule specifying the endpoints of every edge of a graph.

incidence matrix (of a digraph with no self-loops): given a digraph D with no self-loops, a matrix M_I (or $M_{I,D}$) with
$$M_I[i,j] = \begin{cases} 0, & \text{if vertex } v_i \text{ is not an endpoint of arc } e_j \\ +1, & \text{if } v_i \text{ is the head of arc } e_j \\ -1, & \text{if } v_i \text{ is the tail of arc } e_j. \end{cases}$$

incidence matrix (of a graph): given a graph G, a matrix M_I (or $M_{I,G}$) with
$$M_I[i,j] = \begin{cases} 0, & \text{if vertex } v_i \text{ is not an endpoint of edge } e_j \\ 1, & \text{if } e_j \text{ is a proper edge with endpoint } v_i \\ 2, & \text{if } e_j \text{ is a loop at } v_i. \end{cases}$$

incidence matrix (of a hypergraph): given a hypergraph $H = (V, E)$ with $E = \{e_1, e_2, \ldots, e_m\}$ and $V = \{x_1, x_2, \ldots x_n\}$, the matrix $[m_{i,j}]$ where
$$m_{i,j} = \begin{cases} 1, & \text{if } x_j \in e_i \\ 0, & \text{otherwise.} \end{cases}$$

incident edge (from [to] a digraph vertex): given a vertex u in a digraph, a directed edge e such that u is the tail [head] of e.

incident edge (in a graph): given a vertex u in a graph, an edge e such that u is an endpoint of e.

incident-edge table (for a graph): a table that lists, for each vertex, the edges having that vertex as an endpoint.

in-degree: given a vertex v, the number of arcs with head v.

independent subset (in a graph): given a graph G, a subset of either $V(G)$ or $E(G)$ such that no two elements are adjacent in G.

independent subset (of hypergraph vertices): a set of vertices which does not (completely) contain any edge of the hypergraph.

independence number (of a graph): the number $\alpha(G)$ of vertices in the largest independent subset in G.

independence number (of a hypergraph): the maximum number $\alpha(H)$ of vertices that form an independent set in H.

induced subgraph (on a vertex subset): the subgraph of a graph G containing every edge of G that joins two vertices of the prescribed vertex subset.

invariant: a parameter or property of graphs that is preserved by isomorphisms.

intersection graph (for a family of subsets): given a family $\mathcal{F} = \{S_j\}$ of subsets of a set S, the graph whose vertex set is $\mathcal{F}$ and such that there is an edge between each pair of subsets S_i and S_j whose intersection is nonempty.

intersection graph (of a hypergraph): given a hypergraph H, the simple graph $I(H)$ whose vertices are the edges of H, such that two vertices of $I(H)$ are adjacent if and only if the corresponding edges of H have nonempty intersection.

interval graph: the *intersection graph* of a family of subintervals of $[0, 1]$.

irreducible tournament: a tournament with no bipartition V_1, V_2 of the vertices such that all arcs between V_1 and V_2 go from V_1 to V_2.

isolated point: a vertex of a graph that is not the endpoint of any edge.

isomorphic (pair of graphs): a pair of graphs with identical mathematical structure; formally, a pair of graphs such that there is an isomorphism from one to the other.

isomorphism (of digraphs): an isomorphism of the underlying graphs of two digraphs such that the edge correspondence preserves direction.

isomorphism (of graphs): given graphs G and H, a pair of bijections $f_V: V_G \to V_H$ and $f_E: E_G \to E_H$ such that for every edge $e \in E_G$, the endpoints of e are mapped onto the endpoints of $f_E(e)$; f is usually used for both the vertex function f_V and the edge function f_E.

isomorphism (of simple graphs): a one-to-one correspondence between the vertices of two graphs such that a pair of vertices are adjacent in one graph if and only if the corresponding pair of vertices are adjacent in the other graph.

isomorphism type: given a graph [digraph] G, the class of all graphs [digraphs] isomorphic to G.

join: given graphs G and H, the graph $G * H$ obtained by adding to the disjoint union $G + H$ an edge from each vertex in G to each vertex in H.

k-connected: property of a graph G that the smallest size of a disconnecting set of vertices is at least k; that is, $\kappa(G) \geq k$.

k-cycle: a cycle of length k.

k-edge connected: property of a graph G that $\kappa'(G) \geq k$.

k-partite graph: a graph whose vertex set can be partitioned into at most k parts in such a way that each edge joins different parts, never the same part. Equivalent to a k-colorable graph.

k-regular: property of a graph or hypergraph that all its vertices have degree k.

king: a vertex in a digraph that can reach all other vertices by paths of length 1 or 2.

Klein bottle: the nonorientable surface N_2 with two crosscaps.

Kuratowski graphs: the complete graph K_5 and the complete bipartite graph $K_{3,3}$.

labeled graph: in applied graph theory, any graph in which the vertices and/or edges have been assigned labels; in pure graph theory, a graph in which standard labels $v_1, v_2, \ldots, v_n$ have been assigned to the vertices.

Laplacian (of a graph): given a graph G, the matrix $D - A$ where D is the diagonal matrix with the degree sequence of G on the diagonal and where A is the adjacency matrix.

length: given a walk, the number of edge-steps in the sequence that specifies the walk.

line: synonym for edge, or refers to what is modeled by an edge.

line graph: given a graph G, the graph $L(G)$ whose vertices correspond to the edges of G, with two vertices being adjacent in $L(G)$ whenever the corresponding edges have a common endpoint in G.

linear extension ordering: a consecutive labeling $v_1, v_2, \ldots, v_n$ of the vertices of a digraph such that, if there is an arc from v_i to v_j, then $i < j$.

link: See *proper edge*.

loop (or ***self-loop***): an edge joining a vertex to itself.

map: an imbedding of some graph on a surface.

map chromatic number: See *chromatic number of a map*.

map coloring: an assignment of colors to the regions of a map so that adjacent regions receive different colors.

mapping: given graphs G and H, a vertex function $f\colon V_G \to V_H$ and an edge function $f\colon E_G \to E_H$ that correspond together to a continuous function from a spatial representation of G in Euclidean space to a spatial representation of H.

matching: a set of pairwise disjoint edges in a graph or hypergraph.

matching number: in a graph, the maximum number of pairwise disjoint edges of the graph; in a hypergraph H, the maximum number $\nu(H)$ of pairwise disjoint edges of H, that is, the cardinality of the largest partial of H which forms a matching.

mesh (of trees): a graph obtained by construing each row and each column of a $2^d \times 2^d$ array of vertices as the leaves of a complete binary tree.

minor: given a graph G, any graph that can be obtained from G by a sequence of edge deletions and contractions.

Möbius band: the surface obtained from a rectangular sheet by pasting the left side to the right with a half-twist.

multi-arc: two or more arcs, all of which have the same head and all of which have the same tail.

multi-edge: a set of at least two edges, all of which have the same endpoints.

multi-graph: a graph with multi-edges.

neighbor: given a vertex v, any vertex adjacent to v.

node: a vertex, or refers to what is modeled by a vertex.

nonorientable surface: a surface such that some subportion forms a Möbius band.

nonorientable surface of crosscap number k: the surface N_k obtained by adding k crosscaps to a sphere.

nonplanar: property of a graph that it cannot be drawn in the plane without crossings.

nonseparable: property of a connected graph that it has no cut-vertices.

normal: property of a hypergraph H that $q(H) = \Delta(H)$.

normalized drawing: the usual way a graph is drawn, avoiding pathological contrivances such as overloaded crossings (i.e., more than two edges).

null graph: a graph with no vertices or edges.

obstruction to n-coloring: synonym for chromatically $(n+1)$-critical graph, since a chromatically $(n+1)$-critical subgraph prevents n-chromaticity.

octahedral graph: the 1-skeleton of the 3-dimensional octahedron, or sometimes, a generalization of this graph.

1-skeleton (of a polyhedron): the graph whose vertices and edges are, respectively, the vertices and edges of that polyhedron.

open: property of a walk, trail, or path that its final vertex is different from its initial vertex.

order (of a graph): given a graph G, the cardinality $|V_G|$ of the vertex set.

order (of a hypergraph edge): the number of vertices in the edge.

orientable surface: any surface obtainable from a sphere by adding handles, or (alternatively) any surface that does not contain a Möbius band.

orientable surface of genus g: the surface S_g obtained by attaching g handles to a sphere.

orientation: an assignment of a direction to every edge of a graph, making it a digraph.

origin (of a walk): the initial vertex of the walk.

out-degree: given a vertex v, the number of arcs with tail v.

partial: for a hypergraph $H = (V, E)$, a hypergraph $H' = (V, E')$ such that $E' \subseteq E$.

path: a trail in which all of its vertices are different, except that the initial and final vertices may be the same. See also u-v-path.

path, directed: a directed trail in which no vertex is repeated.

perfect graph: a graph such that every induced subgraph has vertex chromatic number equal to its clique number.

permutation graph: a graph whose vertices represent the objects of a permutation group and whose edges represent the action of a generating set of permutations.

Petersen graph: a 3-regular 10-vertex graph that looks like a 5-cycle joined by its vertices to the vertices of a 5-pointed star drawn in its interior.

planar: property of a graph that it can be drawn in the plane without crossings.

Platonic graph: the 1-skeleton of a Platonic solid.

Platonic solid: any of five 3-dimensional polyhedra whose sides are all identical regular polygons.

polyhedron: a generalization of a polygon to higher dimensions; usually a solid 3-dimensional figure subtended by planes.

product graph: See *cartesian product*.

projective plane: the nonorientable surface N_1 with one crosscap.

proper edge (or **link**): an edge with two distinct endpoints.

pseudo-graph: synonym for a graph with loops.

quotient: given a graph G, any graph H such that there exists a *graph mapping* of G onto H.

r-partite hypergraph: an r-uniform hypergraph whose vertex set can be partitioned into r blocks so that each edge intersects each block in exactly one vertex.

r-uniform: property of a uniform hypergraph that r is the common edge-order.

radius: for a connected graph G, the minimum eccentricity among the vertices of G.

Ramsey number, (classical): the number $r(m,n)$, which is the smallest positive integer k such that every simple graph with k vertices either contains K_m as a subgraph or has a set of n independent vertices.

Ramsey number: the number $R(G, H)$, which is the smallest positive integer k such that, if the edges of K_k are bipartitioned into red and blue classes, then either the red subgraph contains a copy of G or else the blue subgraph contains a copy of H.

random graph on n vertices: an n-vertex graph generated by a probability distribution, in which each edge is as likely to occur as any of the others.

reachable vertex (from vertex u): a vertex v such that there is a u,v-path in G.

reconstructible: property of a graph that it is uniquely determined by its collection of vertex-deleted subgraphs.

reconstructible invariant: an invariant which is uniquely determined by the collection of vertex-deleted subgraphs of a graph.

reducible: property of a digraph that its vertex set V can be partitioned into a disjoint union $V_1 \cup V_2$ so that all arcs joining V_1 and V_2 go from V_1 to V_2.

region: given a graph imbedded in a surface, a maximal expanse of surface containing no vertex and no part of any edge of the graph; i.e., any of the pieces of surface subtended by the graph.

regular: property of a graph or hypergraph that all its vertices have the same degree. See also *k-regular*.

representation (of a graph): a description of the graph, possibly without names for the vertices and edges.

rotation system (of an imbedding): a list of the cyclic orderings of the incidence of edges at each vertex.

Schreier graph: a graph that depicts the *cosets* of a prescribed subgroup of a group, with a prescribed set of *generators*; the vertices represent cosets, and the edges (often said to be "color-coded" for the generators) represent the product rule.

self-complementary: property of a graph that it is isomorphic to its edge-complement.

self-loop: an edge that joins a vertex to itself; see *loop*.

simple digraph: See *strict digraph*.

simple (or ***simplicial***) ***graph***: a graph with no loops or multi-edges.

simple hypergraph: a hypergraph with no repeated edges.

sink: a digraph vertex with out-degree zero.

source: a digraph vertex with in-degree zero.

spanning subgraph: a subgraph of a given graph G that includes all vertices of G.

specification (of a graph): a list of its vertices and a list of its edges, with an unambiguous incidence rule for determining the endpoints of every edge.

spectrum (of a graph): the spectrum of its adjacency matrix.

strict: property of a digraph that it has no self-loops and no pair of arcs with the same tail and head.

strong component: in a digraph, a maximal subdigraph that is strongly connected.

strong orientation: given a graph, an assignment of a direction to every edge, making it a strongly connected digraph.

strong tournament: a tournament in which there is a directed path from every vertex to every other vertex.

strongly connected: property of a digraph that every vertex is reachable from every other vertex.

strongly regular graph (with parameters (n, k, r, s)): an n-vertex, k-regular graph in which every adjacent pair of vertices is mutually adjacent to r other vertices, and in which every pair of nonadjacent vertices is mutually adjacent to s other vertices; by convention, strongly regular graphs are connected with at least one edge.

subdivision (of an edge): the operation of inserting a new vertex into the interior of the edge, thereby splitting it into two edges.

subdivision: given a graph, any new graph obtaining by subdividing one or more edges of the original graph one or more times.

subgraph: given a graph G, a graph H whose vertices and edges are all in G.

tail (of an arc): the vertex the arc goes from.

terminus (of a walk): the last vertex of the walk.

tetrahedral graph: another name for the complete graph K_4, resulting from the fact that it is equivalent to the 1-skeleton of the 4-sided Platonic solid called a tetrahedron.

thickness: given a graph G, the minimum number $\theta(G)$ of planar subgraphs whose union is G.

topological sort (or *topsort*): an algorithm that assigns a linear extension ordering to a DAG (not quite a sort, in the usual sense of sorting, and not used by topologists, despite the name).

torus: the surface of a doughnut; the orientable surface S_1 of genus 1.

tough graph: a connected graph G such that for every nonempty set S of vertices, the number of components of the graph $G - S$ does not exceed $|S|$.

tournament: a digraph with exactly one arc between each pair of distinct vertices.

trail: a walk in which no edge occurs more than once.

trail, directed: a directed walk in which no arc is repeated.

transitive: property of a digraph that whenever it contains an arc from u to v and an arc from v to w, it also contains an arc from u to w.

transitive orientation: given a graph, an assignment of a direction to every edge, making it a transitive digraph.

transmitter: in a digraph, a vertex that has an arc to every other vertex.

transversal: in a hypergraph, a set of vertices which has nonempty intersection with every edge of the hypergraph.

transversal number: the minimum number $\tau(H)$ of vertices taken over all transversals of H.

tree: a connected graph without any cycles as subgraphs.

trivial graph: the graph with one vertex and no edges.

Turán graph: the n-vertex k-partite simple graph $T_k(n)$ with the maximum number of edges.

u,v-path: a path whose origin is the vertex u and whose terminus is the vertex v.

underlying graph: given a digraph, the graph obtained from the digraph by stripping the directions off all the arcs.

uniform: property of a hypergraph that all edges have the same number of vertices. See also *r-uniform*.

unilaterally connected (or *unilateral*): property of a digraph that for every pair of vertices u, v, there is either a uv-path or a vu-path.

upset: a simple hypergraph in which every superset of every edge is also an edge of the hypergraph.

valence: a synonym for *degree*, adapted from molecular bonds in chemistry.

vertex: a point; an element of the first constituent set of a graph.

vertex coloring: an assignment of colors to the vertices of a graph so that adjacent vertices receive different colors.

(vertex) connectivity: the smallest number $\kappa(G)$ of vertices whose removal disconnects the graph; by convention, $\kappa(K_n) = n - 1$.

vertex cut: See *disconnecting set*.

vertex-deleted subgraph: any subgraph obtained from a graph by removing a single vertex and all of its incident edges.

vertex invariant: a property at a vertex which is preserved by every isomorphism.

walk: an alternating sequence $v_0, e_1, v_1, \ldots, e_r, v_r$ of vertices and edges where consecutive edges are adjacent, so that each edge e_i joins vertices v_{i-1} and v_i.

walk, directed: an alternating sequence of vertices and arcs $v_0, e_1, v_1, e_2, \ldots, e_n, v_n$ where the arcs align head to tail, so that each vertex is the head of the preceding arc and the tail of the subsequent arc.

weakly connected (or **weak**) **digraph**: a digraph whose underlying graph is connected.

weighted graph: a graph model in which each edge is assigned a number called the weight or the cost.

wheel graph: an $(n+1)$-vertex graph W_n that "looks like" a wheel whose rim is an n-cycle and whose hub vertex is joined by spokes to all the vertices on the rim.

8.1 INTRODUCTION TO GRAPHS

Graphs are highly adaptable mathematical structures, and they can be represented on a computer so that with each new application that arises, existing algorithms can be reused without rewriting. This section provides some of the basic terminology and operations needed for the study of graphs and lists several useful families of graphs.

8.1.1 VARIETIES OF GRAPHS AND GRAPH MODELS

Due to the vast breadth of the usefulness of graphs, the terminology varies widely, not only from one graph variety to another, but also from one application to another. The table in Fact 1 gives synonyms for several terms.

Definitions:

A *vertex* is usually conceptualized as a point. Abstractly, it is a member of the first of two sets that form a *graph*.

An *edge* is usually conceptualized as a line, either joining one vertex to another or joining a vertex to itself. Abstractly, it is a member of the second of two sets that form a *graph*.

A *proper edge* (or *link*) is an edge that joins one vertex to another.

A *loop* (or *self-loop*) is an edge joining a vertex to itself.

The *endpoints* of an edge are the vertices that it joins. A loop has only one endpoint.

An *incidence rule* specifies the endpoints of every edge.

An edge e is *incident with* a vertex v if v is an endpoint of e.

A *graph* is a set V of vertices and a set E of edges (both sets finite unless declared otherwise) such that all the endpoints of edges in E are contained in V. It is often denoted $G = (V, E)$, or (V_G, E_G), or $(V(G), E(G))$. Sometimes, each edge is regarded as a pair of vertices.

An *isolated point* of a graph $G = (V, E)$ is a vertex in V that is not the endpoint of any edge in E.

Vertices u and v are *adjacent* if there is an edge whose endpoints are u and v.

Two edges are *adjacent* if they have a common endpoint.

A *neighbor* of a vertex is any vertex to which it is adjacent.

An *attribute* of the edge-set or vertex-set is a feature such as length, cost, or color sometimes attached to graphs.

A *graph model* is a graph which (quite frequently, in applications) may have additional attributes on its edges and/or vertices. The vertices and edges of the model may represent arbitrary objects and relationships from the context of the application.

A *weighted graph* is a graph model in which each edge is assigned a number called the weight or the cost.

A *node* is sometimes a synonym for a vertex and sometimes refers to whatever is modeled by a vertex in a graph model.

A *line* is sometimes a synonym for an edge and sometimes refers to whatever is modeled by an edge in a graph model.

A *multi-edge* in an undirected graph is a set of more than one edge with the same endpoints, and in a digraph a set of more than one edge such that each edge in the set has the same head and each edge in the set has the same tail. A graph with a multi-edge is also said to have *multiple edges* or *parallel edges*.

A *multi-edge of multiplicity n* is a set of n edges with the same endpoints.

A set of *parallel edges* is a set of edges with the same endpoints, i.e., a multi-edge. A pair of *anti-parallel arcs* is a pair of oppositely directed arcs between the same two endpoints.

A graph is *simple* if it has no loops or multi-edges. Topologists often say *simplicial*, because such a graph is a special case of a "simplicial complex".

A *multi-graph* is another name for a graph with multi-edges but no self-loops, used for emphasis when the context is largely restricted to simple graphs.

A *pseudo-graph* (or *general graph*) is another name for a graph in which loops and multi-edges are permitted, used for emphasis when the context is largely restricted to loopless graphs.

A *direction* on an edge is an ordering for its endpoints so that the edge goes *from* one endpoint and *to* the other. Any edge, including a self-loop, can be directed by giving it a sense of forward progression, e.g., in a graph drawing, by placing an arrowhead to show which way is forward, or by ordering the endpoints.

An *arc* is another name for a directed edge.

The *tail* of an arc is the vertex at which the arc originates.

The *head* of an arc is the vertex at which the arc terminates.

A *directed graph* or *digraph* is a graph in which every edge is directed. (See §8.3.)

A *strict digraph* has no self-loops and no pair of arcs with the same tail and head.

The *degree* of vertex v, $deg(v)$, is the number of proper edges plus *twice* the number of loops incident with v. Thus, in a drawing, it is the number of edge-endings at v.

The *valence* of a vertex is a synonym for *degree* adapted from terminology in chemistry.

The *degree sequence* of a graph is the sequence of the degrees of its vertices, most often sorted into size order, ascending or descending.

A *regular* graph is a graph such that all vertices have the same degree. It is called *k-regular* if the vertices all have degree k.

A *graphical sequence* is a sequence of nonnegative integers that is the degree sequence of some simple graph.

The number of vertices of a graph is sometimes called the **order**.

The number of edges of a graph is sometimes called the **size**.

The *empty graph* is the graph whose vertex set and edge set are both empty. This is also called the *null graph* (and is sometimes not considered to be a graph).

The *trivial graph* is the graph with one vertex and no edges.

Facts:

1. The following table gives lists of some synonymous graph theory terms:

   > vertex: point, node
   > edge: line
   > loop: self-loop
   > neighbor: adjacent vertex
   > arc: directed edge
   > degree: valence
   > number of vertices: order
   > number of edges: size
   > nonsimple graph: pseudograph, general graph
   > loopless nonsimple graph: multi-graph
   > empty graph: null graph

2. The following table lists the varieties of graphs:

graph variety	loops allowed?	multi-edges allowed?
digraph	YES	YES
general graph	YES	YES
multi-graph	NO	YES
pseudo-graph	YES	YES
simple graph	NO	NO
strict digraph	NO	NO*

*at most one arc in each direction between two vertices

3. In a drawing of a graph, the degree of a vertex v equals the number of edge-ends at v. The degree of v need *not* equal the number of edges at v, since each loop contributes both its ends toward the degree.

4. *Euler's theorem*: In every graph, the sum of the degrees equals twice the number of edges. From this result it follows that in all graphs the sum of the degrees of all vertices is even.

5. In every graph the number of vertices of odd degree is even.

6. The name *handshaking lemma* is commonly applied to various elementary results about the degrees of simple graphs, especially Facts 4 and 5.

7. In every simple graph with at least two vertices, there is a pair of vertices with the same degree.

8. *Havel's theorem*: A sequence is a graphical sequence if and only if the sequence obtained by deleting the largest entry d and subtracting 1 from each of the d next largest entries is graphical. (V. Havel, 1955). This operation on a sequence is called *Havel's reduction*.

9. A nonincreasing sequence of nonnegative integers $d_1, d_2, \ldots, d_n$ is graphical if and only if its sum is even, and for $k = 1, 2, \ldots, n$,

$$\sum_{i=1}^{k} d_i \leq k(k-1) + \sum_{i=k+1}^{n} \min\{k, d_i\}.$$

(P. Erdős and T. Gallai, 1960)

10. In a computer, a graph is commonly represented as a structure with variable value. The empty graph is often used as the initial value of a graph variable, analogous to the way in which zero is used as the initial value of a numeric variable.

Examples:

1. The following figure gives examples of the various varieties of graphs. (See the table of Fact 2.)

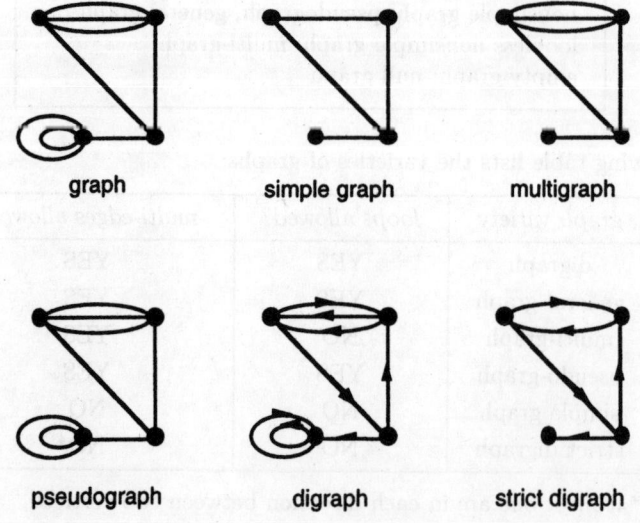

2. *Computer programming flowchart* (always a digraph): Each vertex represents some programmed operation or decision, and each arc represents the flow of control to the next operation or decision.

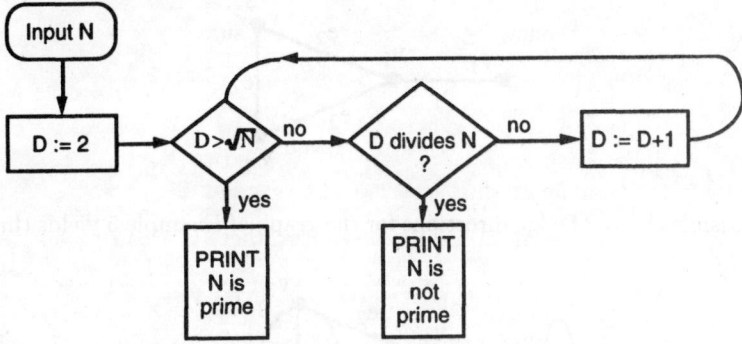

3. *Model for social networks* (usually undirected): Each vertex represents a person in the network, and each edge represents a form of interaction between the persons represented by its endpoints. This is illustrated by the following graph.

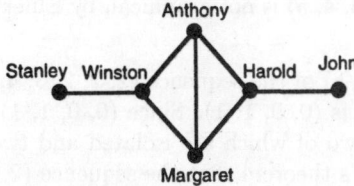

4. *Model for road networks* (most edges undirected): Each vertex represents either an intersection of two roads or the end of a dead-end street. The absence of an endpoint in the illustration indicates that the road continues beyond what is shown. Direction on an edge may be used to indicate a one-way road. Undirected edges are two-way roads.

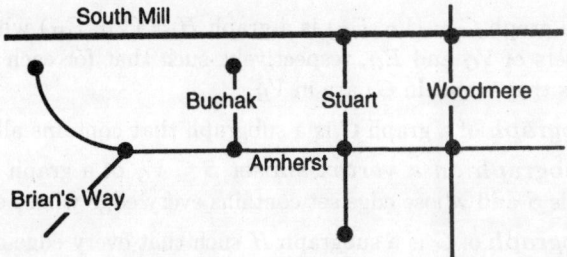

5. In the following graph with vertex set $V = \{v_1, v_2, v_3, v_4, v_5\}$ and edge set $E = \{e_1, e_2, e_3, e_4, e_5, e_6, e_7\}$, the vertex v_5 is an isolated point, and the degree sequence is $(0, 3, 3, 4, 4)$. The edge e_7 is a loop, and the three edges e_4, e_5, and e_6 form a multi-edge.

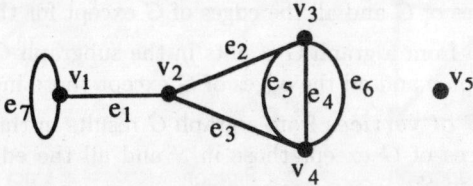

6. Deleting the isolated vertex and the self-loop in Example 5 and then reducing the multi-edge to a single edge yields the following simple graph, whose degree sequence is (1, 2, 2, 3).

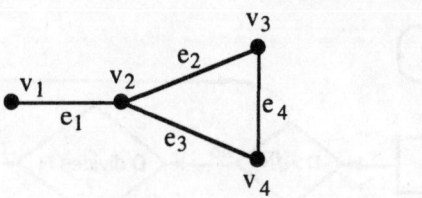

7. One possible choice of edge directions for the graph of Example 5 yields this digraph.

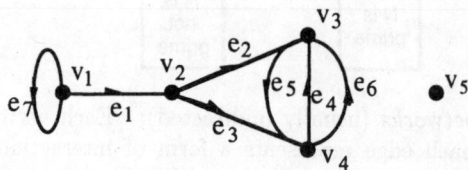

8. The sequence (1, 2, 2, 3, 4, 5) is not graphical, by Euler's theorem (Fact 4), because its sum is odd.

9. Havel's reduction (Fact 8) of the sequence (2, 2, 2, 3, 4, 5) is (1, 1, 1, 2, 3). Havel's reduction of that sequence is (0, 0, 1, 1). Since (0, 0, 1, 1) is the degree sequence of a graph with four vertices, two of which are isolated and two of which are joined by an edge, it follows from Havel's theorem that the sequence (2, 2, 2, 3, 4, 5) is graphical.

8.1.2 GRAPH OPERATIONS

Definitions:

A *subgraph* of a graph $G = (V_G, E_G)$ is a graph $H = (V_H, E_H)$ whose vertex set and edge set are subsets of V_G and E_G, respectively, such that for each edge e in E_H, the endpoints of e (as they occur in G) are in V_H.

A *spanning subgraph* of a graph G is a subgraph that contains all the vertices of G.

The *induced subgraph on a vertex subset* $S \subseteq V_G$ of a graph G is the subgraph whose vertex set is S and whose edge set contains every edge whose endpoints are in S.

An *induced subgraph* of G is a subgraph H such that every edge of G that joins two vertices of H is also an edge of H.

Deleting an edge e from a graph G results in the subgraph $G - e$ that contains all the vertices of G and all the edges of G except for e.

Deleting a subset Y of edges from a graph G results in the subgraph $G - Y$ that contains all the vertices of G and all the edges of G except for those in Y.

Deleting a vertex v from a graph G results in the subgraph $G - v$ that contains all the vertices of G except v and all the edges of G except those incident with v.

Deleting a subset S of vertices from a graph G results in the subgraph $G - S$ that contains all the vertices of G except those in S and all the edges of G except those incident with vertices in S.

Contracting an edge e in a graph G means shrinking the edge to a point, so that its endpoints are merged, without changing the rest of the graph. The resulting graph is denoted $G \downarrow e$ (or $G \cdot e$ or G/e). To construct $G \downarrow e$ from G, delete the edge e from the edge set and replace all instances of its endpoints in the vertex set and incidence rule by a new vertex.

A *minor* of a graph G is any graph that can be obtained from G by a sequence of edge deletions and contractions.

The *graph union* $G \cup H$ has as its vertices and edges those vertices and edges, respectively, that are either in G or in H.

The *graph intersection* $G \cap H$ has as its vertices and edges those vertices and edges, respectively, that are both in G and in H.

The *graph sum* (or *disjoint union*) $G + H$ has as its vertices and edges, respectively, the disjoint union of the vertex sets and the disjoint union of the edge sets of the graphs G and H.

The *iterated graph sum* nG is the union of n disjoint copies of G.

The *join* $G * H$ is obtained by adding to $G + H$ an edge from each vertex in G to each vertex in H.

The (*cartesian*) *product* $G \times H$ has as its vertices the cartesian product $V_G \times V_H$ and as its edges this union of two products: $(V_G \times E_H) \cup (E_G \times V_H)$. The endpoints of the edge (u, d) are the vertices (u, x) and (u, y), where x and y are the endpoints of d in H. The endpoints of the edge (e, w) are (u, w) and (v, w), where u and v are the endpoints of e.

An *isomorphism* $f: G \to H$ (of graphs) is a relationship between graphs that establishes their structural equivalence. It is given by a pair of set bijections $f_V: V_G \to V_H$ and $f_E: E_G \to E_H$ such that if u and v are the endpoints of edge e in graph G, then $f_V(u)$ and $f_V(v)$ are the endpoints of $f_E(e)$ in graph H. The vertex function and the edge function can both be denoted f without the subscript. (See §8.5.)

Two graphs are *isomorphic* if there is an isomorphism between them. This means that they are essentially the same graph except for the names of their vertices and edges.

A *graph mapping* $f: G \to H$ (of graphs) is a pair of set functions $f_V: V_G \to V_H$ and $f_E: E_G \to E_H$ such that if u and v are the endpoints of edge e in G, then $f(u)$ and $f(v)$ are the endpoints of $f(e)$ in H. Such a pair of functions is said to *preserve incidence*. The vertex function and the edge function can both be denoted f without the subscript.

A *quotient* of a graph G is a graph H such that there is a graph mapping from G onto H.

An *automorphism* of a graph G is an isomorphism of G to itself.

The *automorphism group* $\mathcal{A}ut(G)$ is the group of all automorphisms of graph G.

Subdivision of an edge e is the operation of inserting a new vertex in the interior of an edge. Combinatorially, this is achieved by joining a new vertex to the endpoints of edge e and then deleting e.

Two graphs are *homeomorphic* if there is a graph from which they can both be obtained by a sequence of edge subdivisions.

The *edge-complement* $\overline{G}$ of a simple graph G (often, *complement*) has the same vertex set as G, with every two distinct vertices being adjacent in $\overline{G}$ if and only if they are not adjacent in G.

516 Chapter 8 GRAPH THEORY

A **self-complementary** graph is a graph that is isomorphic to its complement.

The **line graph** $L(G)$ of a graph G has vertices corresponding to the edges of G, with two vertices being adjacent in $L(G)$ whenever the corresponding edges are adjacent in G.

Facts:

1. If a graph J is isomorphic to a subgraph of a graph G, then it is commonly said that J "is" a subgraph of G, even though V_J and E_J might not be subsets of V_G and E_G, respectively.
2. A graph is a subgraph of its union with any other graph.
3. The intersection of two graphs is a subgraph of both of them.
4. A graph mapping is the combinatorial counterpart of what is topologically a continuous function from one graph to the other.
5. A graph isomorphism is a graph mapping such that both the vertex function and the edge function are bijections.
6. There is a self-complementary graph of order n if and only if $n \equiv 0$ or $1 \pmod 4$.
7. The automorphism group $\mathcal{A}ut(G)$ of any simple graph is isomorphic to the automorphism group $\mathcal{A}ut(\overline{G})$ of its edge-complement.
8. A connected graph G is isomorphic to its line graph if and only if G is a cycle (§8.1.3).
9. If two connected graphs have isomorphic line-graphs, then they are either isomorphic to each other or they are K_3 and $K_{1,3}$ (§8.1.3).
10. $\mathcal{A}ut(K_n)$ is isomorphic to the symmetric group S_n (§5.3.1).
11. $\mathcal{A}ut(C_n)$ is isomorphic to the dihedral group D_n (§5.3.2).

Examples:

1. The dark subgraph spans the following graph, because it contains every vertex.

2. The cartesian product $C_4 \times K_2$ (§8.1.3) is illustrated as follows:

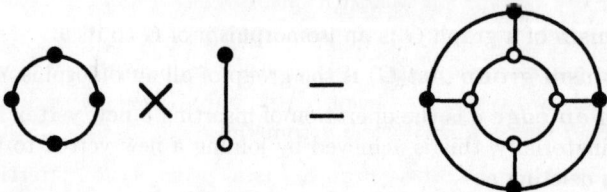

3. The join $\overline{K_2} * P_3$ (§8.1.3) is illustrated as follows:

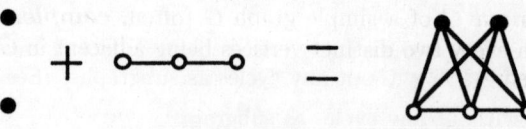

4. The following two graphs are homeomorphic, but not isomorphic.

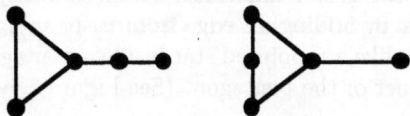

5. The graphs $K_{3,3}$ and $K_3 + K_3$ (§8.1.3) are edge-complements of each other.
6. The cycle graph C_5 (§8.1.3) is self-complementary.
7. The line graph $L(K_4)$ (§8.1.3) is isomorphic to the octahedral graph $\mathcal{O}_3$.

8.1.3 SPECIAL GRAPHS AND GRAPH FAMILIES

Definitions:

Note: Many of the following graphs are drawn in Figures 1 and 2.

A graph is **null** (sometimes, **empty**) if both its vertex set and edge set are empty.

The **bouquet** of n loops, denoted B_n, is a graph with one vertex and n self-loops.

The **dipole** D_n is the graph with two vertices and a multi-edge of multiplicity n joining them.

The **complete graph** K_n is a simple graph with n vertices in which every pair of vertices is adjacent.

The **n-path** P_n is a graph that "looks like" a path $n-1$ edges long. It consists of a sequence of n vertices $V = \{v_1, v_2 \ldots, v_n\}$ and a sequence of $n-1$ edges joining successive vertices in the sequence, that is, the $n-1$ edges are $\{v_1, v_2\}, \{v_2, v_3\}, \ldots, \{v_{n-1}, v_n\}$.

A **path** is a graph that is an n-path for some $n \geq 0$.

The **n-cycle** C_n is a graph that "looks like" a cycle. It has a "wraparound" sequence of vertices $V = \{v_1, v_2, \ldots, v_n\}$ and a sequence of edges joining successive vertices in the sequence, i.e., the n edges are $\{v_1, v_2\}, \{v_2, v_3\}, \ldots, \{v_{n-1}, v_n\}, \{v_n, v_1\}$.

A **cycle** is a graph which is an n-cycle, for any $n > 0$.

The **n-wheel** is the join of K_1 and the n-cycle.

A graph is **bipartite** if its vertices can be partitioned into two subsets (the **parts**, or **partite sets**) so that no two vertices in the same part are adjacent.

The **complete bipartite graph** $K_{r,s}$ is the simple bipartite graph in which the two parts have respective cardinalities r and s, such that every vertex in one part is adjacent to every vertex in the other part.

The **complete r-partite graph** $K_{n_1,n_2,\ldots,n_r}$ has r disjoint subsets of vertices of orders $n_1, n_2, \ldots, n_r$, with two vertices adjacent if and only if they lie in different subsets. If the r sets all have t vertices, this graph is sometimes denoted $K_{r(t)}$.

The **d-dimensional hypercube** graph Q_d is a graph with 2^d vertices that can be labeled with the 2^d bitstrings of length d so that two vertices are adjacent if and only if their labels differ in exactly one bit.

A graph G is **connected** if for each pair of vertices in G, there is a path in G that contains them both.

A **tree** is a connected graph without any cycles as subgraphs. (See Chapter 9.)

A **forest** is a graph without any cycles as subgraphs.

The **Kuratowski graphs** are the graphs K_5 and $K_{3,3}$.

The **Petersen graph** is the graph constructed from two disjoint 5-cycles u_0, u_1, u_2, u_3, u_4 and v_0, v_1, v_2, v_3, v_4 by adding an edge from u_j to $v_{2j \bmod 5}$, for $j = 0, 1, 2, 3, 4$. It is usually drawn to look like a 5-pointed star inside a pentagon, so that each point of the star is joined to a corner of the pentagon. (See Figure 3.)

A **polyhedron** is the generalization of a polygon to higher dimensions. Whereas a polygon is the intersection in $\mathcal{R}^2$ of several half-planes, an n-dimensional polyhedron is the intersection in $\mathcal{R}^n$ of several half-spaces of dimension n.

A **Platonic solid** is a regular 3-dimensional polyhedron.

The **1-skeleton** of a polyhedron is the graph that has as its vertices and edges the vertices and edges, respectively, of that polyhedron.

The **tetrahedral graph** is the 1-skeleton of the 4-sided Platonic solid called a tetrahedron. The faces of the tetrahedron are triangles.

The **cube graph** is the 1-skeleton of the 6-sided Platonic solid called a cube. The faces of the cube are squares.

The **octahedral graph** is the 1-skeleton of the 8-sided Platonic solid called an octahedron. The faces of the octahedron are triangles.

The **generalized octahedral graph** $\mathcal{O}_n$ is the graph that can be obtained from the complete graph K_{2n} by removing n mutually nonadjacent edges.

The **dodecahedral graph** is the 1-skeleton of the 12-sided Platonic solid called a dodecahedron. It has 20 vertices, each of degree 3, and 30 edges. The dodecahedron has 12 regular pentagons as its faces.

The **icosahedral graph** is the 1-skeleton of the 20-sided Platonic solid called an icosahedron. It has 12 vertices, each of degree 5, and 30 edges. The icosahedron has 20 equilateral triangles as its faces.

A **Platonic graph** is any graph isomorphic to the 1-skeleton of any Platonic solid.

The **intersection graph** for a collection $F = \{S_j\}$ of subsets of the same set has as its nodes the subsets themselves. There is an edge between each pair of subsets whose intersection is not empty.

An **interval graph** is any graph isomorphic to the intersection graph for a collection of intervals of the real line.

Facts:

1. In a computer program, the null graph is used as the initial value of a graph-valued variable, rather like the way that an integer-valued variable is initialized to zero. As the program runs, the graph-valued variable can be modified by adding vertices and edges.

2. Bouquets and dipoles are fundamental building blocks for graphs constructed by topological techniques.

3. The path graph P_n is a tree.

4. A graph is bipartite if and only if it has no cycles of odd length.

5. Trees are bipartite.

6. The hypercube graphs Q_n can be defined recursively as follows: $Q_0 = K_1$, $Q_n = K_2 \times Q_{n-1}$ for $n > 0$.

7. The hypercube graph Q_n is bipartite and is isomorphic to the lattice of subsets of a set of n elements. (See §13.2.)

8. The octahedral graphs $\mathcal{O}_n$ can be defined recursively as follows: $\mathcal{O}_0 = \overline{K_2}$, $\mathcal{O}_n = \overline{K_2} * \mathcal{O}_{n-1}$ for $n > 0$.

9. There are exactly five Platonic solids: the tetrahedron, the cube, the octahedron, the dodecahedron, and the icosahedron. Their 1-skeletons are $K_4, Q_3, \mathcal{O}_4$, the dodecahedral graph, and the icosahedral graph.

Examples:

1. Figure 1 shows some of the classes of graphs that occur most often in general constructions.

2. Figure 2 shows some of the graphs that occur most often as special examples.

8.1.4 GRAPH REPRESENTATION AND COMPUTATION

To apply a computer to graph-theoretic computations, it is necessary to specify the underlying graph completely and without ambiguity. Programming system designers use specifications that are efficient for practical computation and that can be reused in additional applications.

Definitions:

A **specification** of a graph is a list of its vertices, a list of its edges, and an unambiguous description of the incidence rule for determining the endpoints of every edge.

An **endpoint table** for a graph is a tabular description of the incidence rule, that gives the endpoints of every edge. In a digraph or partially directed graph, the tail and head of each directed edge are distinguished, for instance, by marking the head, or by always giving the tail first.

An **incident-edge table** for a graph is a tabular description of the incidence rule, that gives for each vertex v, a list of the edges having v as an endpoint. If the graph is directed, this list is partitioned into two sublists, according to whether v is tail or head.

A **representation** of a graph G is a graph description, such as a drawing, from which a formal specification could be constructed and labeled with the vertex names and edge names from G, so as to obtain a graph that conforms to the incidence rule for G.

The **incidence matrix** of a graph (without loops) G with vertices $v_1, v_2, \ldots, v_n$ and edges $e_1, e_2, \ldots, e_m$ is the $n \times m$ matrix M_I (or $M_{I,G}$, in the context of more than one graph) with

$$M_I[i,j] = \begin{cases} 0 & \text{if } v_i \text{ is not an endpoint of } e_j \\ 1 & \text{if } e_j \text{ is a proper edge and } v_i \text{ is an endpoint of } e_j \\ 2 & \text{if } e_j \text{ is a loop at } v_i. \end{cases}$$

(Sometimes an incidence matrix is written with $M_I[i,j] = 1$ for a loop, even though this violates the properties that every column-sum equals 2 and every row-sum equals the degree of the corresponding vertex.)

The **adjacency matrix** of a loopless graph G with vertices $v_1, v_2, \ldots, v_n$ and edges $e_1, e_2, \ldots, e_m$ is the $n \times n$ matrix A_G with

$A_G[i,j]$ = the number of edges between v_i and v_j if $i \neq j$.

If there are self-loops, then $A_G[i,i]$ is usually defined to be the number of loops at v_i.

Figure 1 Some fundamental infinite classes of graphs.

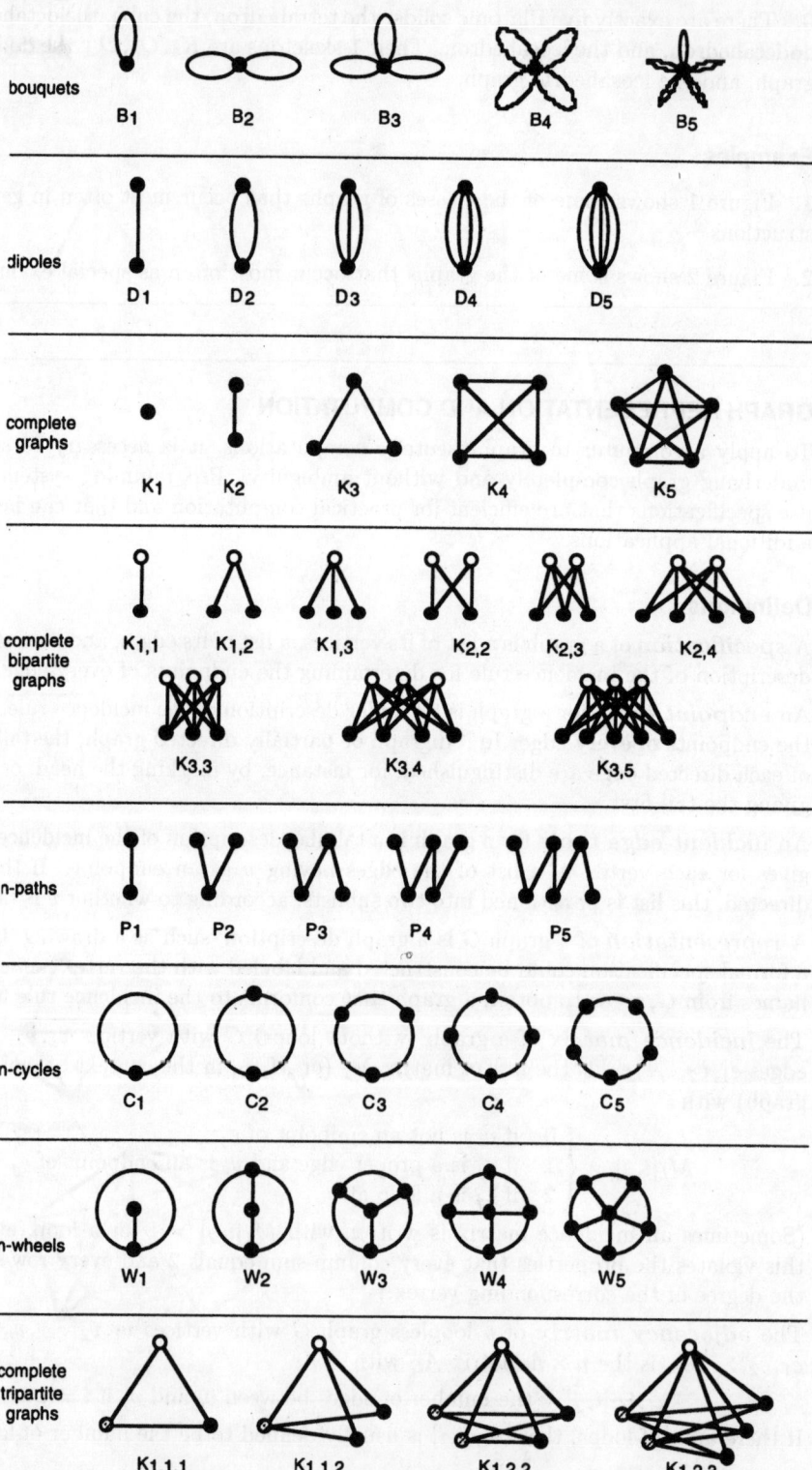

Section 8.1 INTRODUCTION TO GRAPHS 521

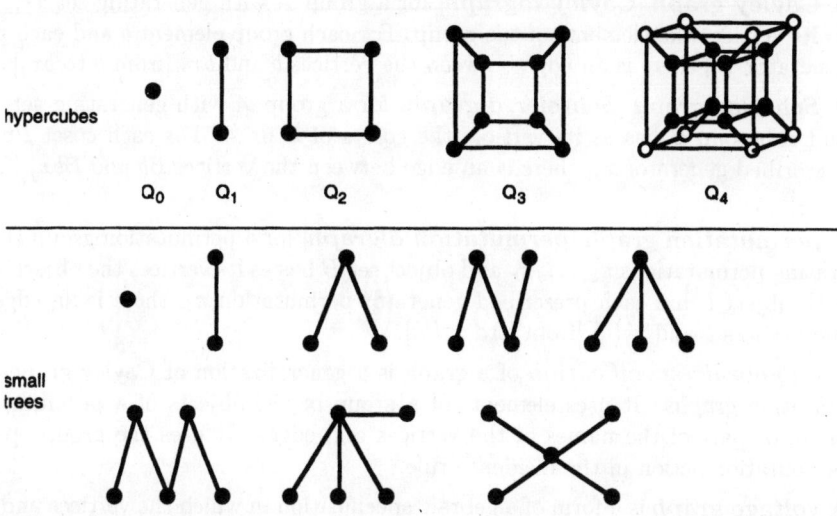

Figure 2 Some special graphs.

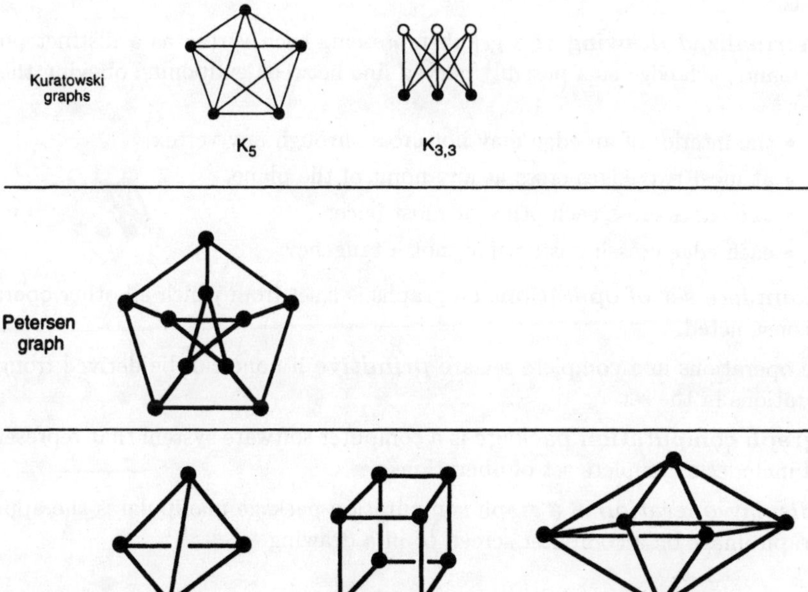

A **Cayley graph** [**Cayley digraph**] for a group $\mathcal{A}$ with generating set $x_1, \ldots, x_r$ has as its vertices the elements of the group. For each group element b and each prescribed generator x_j, there is an edge between the vertices b and bx_j [from b to bx_j].

A **Schreier graph** [**Schreier digraph**] for a group $\mathcal{A}$ with generating set $x_1, \ldots, x_r$ and subgroup $\mathcal{B}$ has as its vertices the cosets of $\mathcal{B}$ in $\mathcal{A}$. For each coset $\mathcal{B}b$ and each prescribed generator x_j, there is an edge between the vertices $\mathcal{B}b$ and $\mathcal{B}bx_j$ [from $\mathcal{B}b$ to $\mathcal{B}bx_j$].

A **permutation graph** [**permutation digraph**] for a permutation group Π with generating permutations $\pi_1, \ldots, \pi_r$ and object set B has as its vertices the object set B. For each object b and each prescribed generator permutation π_j, there is an edge between the vertices b and $\pi_j(b)$ [from b to $\pi_j(b)$].

An **algebraic specification** of a graph is a generalization of *Cayley graphs* and *permutation graphs*. It uses elements of a group or the objects of a permutation group as all or part of the names of the vertices and edges. It uses the group operation or permutation action in the incidence rule.

A **voltage graph** is a form of algebraic specification in which the vertices and edges are specified as a set of one or more symbols with subscripts, ranging over group elements or permuted objects. Its usual form is a digraph drawing with vertex labels and edge labels.

A **normalized drawing** of a graph represents each vertex as a distinct point in the plane and each edge as a possibly curved line between endpoints, obeying the following rules:

- the interior of an edge may not cross through any vertex;
- at most two edges cross at any point of the plane;
- two edges cross each other at most once;
- each edge crossing is normal, not a tangency.

A **complete set of operations** on graphs is a set from which all other operations can be constructed.

The operations in a complete set are **primitive** if none can be derived from the other operations in the set.

A **graph computation package** is a computer software system that represents graphs and includes a complete set of operations.

A **display operation** in a graph computation package manipulates the appearance of a graph image on a computer screen or in a drawing.

Facts:

1. Despite the redundancy of information, an incident-edge table is often used in combination with an endpoint table in computer software, because it facilitates the use of fast searching techniques at the cost of relatively little space.

2. If a graph is simple, then its edges can be represented as endpoint pairs uv. Thus, the graph can be specified as a list of endpoint pairs and a list of isolated vertices.

3. If a graph is simple, then its incident-edge table can be represented as a table that gives the list of neighbors of every vertex.

4. If $A = A_G$ is the adjacency matrix of graph G, then the (i,j)-entry of A^k is the number of walks (§8.4.1) of length k from v_i to v_j in G.

5. *Matrix-tree theorem* (Kirchoff, 1847): Let G be a graph, and let A be its adjacency matrix and D the diagonal matrix of the degrees of its vertices. Then the value of every cofactor of $D - A$ equals the number of spanning trees of G. (Gustav R. Kirchoff, 1824–1887)

6. Given the incidence matrix $M_{I,G}$, it is possible to obtain the incidence matrix $M_{I,H}$ for a subgraph H of G by deleting all rows and columns corresponding to vertices and edges, respectively, that are not in the subgraph.

7. Algebraic specification is useful when the graph is highly symmetric. It replaces an arbitrarily large table of endpoints by a concise algebraic rule.

8. Algebraic specification can be used to specify the graph model for a parallel architecture for a computer.

9. Every regular graph of even degree can be specified as a Schreier graph or as a permutation graph. (J. Gross, 1977)

10. The graph specified by a voltage graph drawing is topologically a covering space of the voltage graph. (J. Gross, 1974) Thus, its relationship to the voltage graph is exactly like the relationship of a Riemann surface to the complex plane.

11. The most commonly used complete set of operations is adding a vertex, deleting a vertex, adding an edge, and deleting an edge.

12. Graph computation packages are built into mathematical computation systems such as Maple and Mathematica.

13. Graph computation packages often include display operations.

Examples:

1. The following normalized drawing, endpoint table, incident-edge table, incidence matrix and adjacency matrix all specify the same graph G.

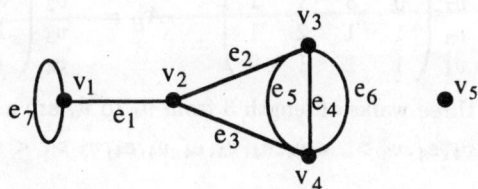

endpoint table

	e_1	e_2	e_3	e_4	e_5	e_6	e_7
	v_1	v_2	v_2	v_3	v_3	v_3	v_1
	v_2	v_3	v_4	v_4	v_4	v_4	

incident-edge table

v_1	e_1	e_7		
v_2	e_1	e_2	e_3	
v_3	e_2	e_4	e_5	e_6
v_4	e_3	e_4	e_5	e_6
v_5				

$$M_{I,G} = \begin{pmatrix} & e_1 & e_2 & e_3 & e_4 & e_5 & e_6 & e_7 \\ v_1 & 1 & 0 & 0 & 0 & 0 & 0 & 2 \\ v_2 & 1 & 1 & 1 & 0 & 0 & 0 & 0 \\ v_3 & 0 & 1 & 0 & 1 & 1 & 1 & 0 \\ v_4 & 0 & 0 & 1 & 1 & 1 & 1 & 0 \\ v_5 & 0 & 0 & 0 & 0 & 0 & 0 & 0 \end{pmatrix}$$

$$A_G = \begin{pmatrix} & v_1 & v_2 & v_3 & v_4 & v_5 \\ v_1 & 1 & 1 & 0 & 0 & 0 \\ v_2 & 1 & 0 & 1 & 1 & 0 \\ v_3 & 0 & 1 & 0 & 3 & 0 \\ v_4 & 0 & 1 & 3 & 0 & 0 \\ v_5 & 0 & 0 & 0 & 0 & 0 \end{pmatrix}$$

2. The following normalized drawing, list of endpoint pairs, lists-of-neighbors table, incidence matrix and adjacency matrix all specify the same simple graph H. It is a spanning subgraph of the graph G of Example 1, but it is *not* an induced subgraph. Compare the incidence matrix to Example 1 to see how the rows and columns are deleted.

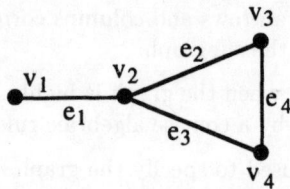

endpoint pairs:

$v_1v_2, v_2v_3, v_2v_4, v_3v_4$

lists-of-neighbors
$v_1 : v_2$
$v_2 : v_1 \quad v_3 \quad v_4$
$v_3 : v_2 \quad v_4$
$v_4 : v_2 \quad v_3$

$$M_{I,H} = \begin{pmatrix} & e_1 & e_2 & e_3 & e_4 \\ v_1 & 1 & 0 & 0 & 0 \\ v_2 & 1 & 1 & 1 & 0 \\ v_3 & 0 & 1 & 0 & 1 \\ v_4 & 0 & 0 & 1 & 1 \end{pmatrix} \qquad A_H = \begin{pmatrix} & v_1 & v_2 & v_3 & v_4 \\ v_1 & 0 & 1 & 0 & 0 \\ v_2 & 1 & 0 & 1 & 1 \\ v_3 & 0 & 1 & 0 & 1 \\ v_4 & 0 & 1 & 1 & 0 \end{pmatrix}$$

3. Squaring and cubing the adjacency matrix of Example 2 provides an illustration of Fact 4.

$$A_H^2 = \begin{pmatrix} & v_1 & v_2 & v_3 & v_4 \\ v_1 & 1 & 0 & 1 & 1 \\ v_2 & 0 & 3 & 1 & 1 \\ v_3 & 1 & 1 & 2 & 1 \\ v_4 & 1 & 1 & 1 & 2 \end{pmatrix} \qquad A_H^3 = \begin{pmatrix} & v_1 & v_2 & v_3 & v_4 \\ v_1 & 0 & 3 & 1 & 1 \\ v_2 & 3 & 2 & 4 & 4 \\ v_3 & 1 & 4 & 2 & 3 \\ v_4 & 1 & 4 & 3 & 2 \end{pmatrix}$$

For instance, the three walks of length 3 from v_4 to v_3 are as follows:

$< v_4, e_3, v_2, e_3, v_4, e_4, v_3 >, \quad < v_4, e_4, v_3, e_4, v_4, e_4, v_3 >, \quad < v_4, e_4, v_3, e_2, v_2, e_2, v_3 >$.

4. As an illustration of the Kirchoff matrix-tree theorem of Fact 5, observe that the graph of Example 2 has the following three spanning trees.

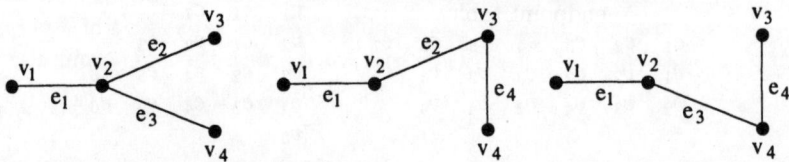

The value of the (2,2)-cofactor of the matrix $D - A$ is also equal to 3:

$$D - A = \begin{pmatrix} & v_1 & v_2 & v_3 & v_4 \\ v_1 & 1 & -1 & 0 & 0 \\ v_2 & -1 & 3 & -1 & -1 \\ v_3 & 0 & -1 & 2 & -1 \\ v_4 & 0 & -1 & -1 & 2 \end{pmatrix} \qquad \text{cofactor} = \begin{vmatrix} 1 & 0 & 0 \\ 0 & 2 & -1 \\ 0 & -1 & 2 \end{vmatrix} = 3.$$

5. (a) The following normalized graph drawing and algebraic specification using the group $\mathcal{Z}_5$ of integers mod 5 both specify the same graph:

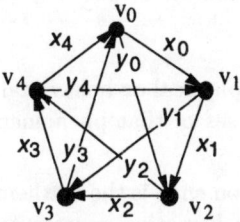

algebraic specification using group $\mathcal{Z}_5$ with generators $x = 1$ and $y = 2$:

vertex set $V = \{v_j \mid j \in \mathcal{Z}_5\}$;

edge set $E = \{x_j$ joining v_j and $v_{(j+1)\bmod 5}, y_j$ joining v_j and $v_{(j+2)\bmod 5} \mid j \in \mathcal{Z}_5\}$.

(b) The following voltage graph (see [GrTu87]) provides a highly compact visual form of algebraic specification of the same graph as in part (a).

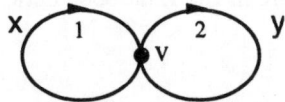

6. The Petersen graph has the following algebraic specification:

$V = \{u_j, v_j \mid j = 0, 1, 2, 3, 4\}$;

$E = \{x_j \ (u_j \rightarrow u_{(j+1)\bmod 5}), \ y_j \ (v_j \rightarrow v_{(j+2)\bmod 5}), \ z_j \ (u_j \rightarrow v_j) \mid j = 0, 1, 2, 3, 4\}$.

With appropriate labeling, the Petersen graph (left) corresponds to the following voltage graph specification (right).

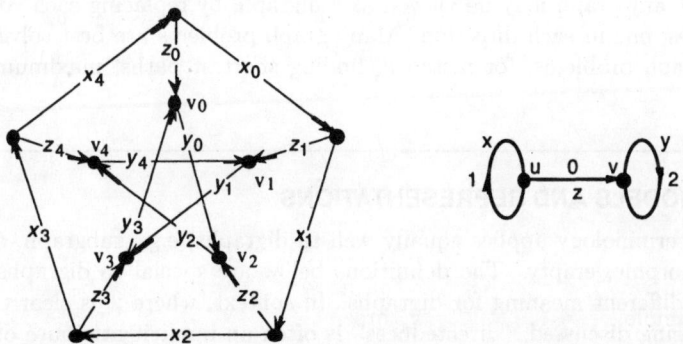

7. The Petersen graph as depicted in Example 6 is also an example of a permutation graph on the object set $V = \{u_j, v_j \mid j = 0, 1, 2, 3, 4\}$ with the two permutations:

$(u_0, u_1, u_2, u_3, u_4)(v_0, v_2, v_4, v_1, v_3)$ $(u_0, v_0)(u_1, v_1)(u_2, v_2)(u_3, v_3)(u_4, v_4)$.

8.2 GRAPH MODELS

Modeling with graphs is one of the main ways in which discrete mathematics has been applied to real world problems. This section gives a list of some of the ways in which graphs are used as mathematical models. Further information can be found in [HaNoCa65] and [Ro76].

8.2.1 ATTRIBUTES OF A GRAPH MODEL

Definitions:

A *mathematical representation* of a physical or behavioral phenomenon is a correspondence between the parts and processes of that phenomenon and a mathematical system of objects and functions.

A *model* of a physical or behavioral phenomenon is the mathematical object or function assigned to that phenomenon under a mathematical representation.

Modeling is the mathematical activity of designing models and comprehensive mathematical representations of physical and behavioral phenomena.

A *graph model* is a mathematical representation that involves a graph.

Examples:

1. Table 1 gives many examples of graph models. Each example states what the vertices and edges (or arcs) represent and where in the *Handbook* details on the application can be found.

8.3 DIRECTED GRAPHS

Assigning directions to the edges of a graph greatly enhances modeling capability, and is natural whenever order is important, e.g., in a hierarchical structure or a one-way road system. Also, any graph may be viewed as a digraph, by replacing each edge with two directed edges, one in each direction. Many graph problems are best solved as special cases of digraph problems, for instance, finding shortest paths, maximum flows, and connectivity.

8.3.1 DIGRAPH MODELS AND REPRESENTATIONS

Most graph terminology applies equally well to digraphs, e.g., subgraph, self-loop, bipartite, isomorphic, empty. The definitions below are special to digraphs or take on a somewhat different meaning for digraphs. In context, where it is clear that only digraphs are being discussed, "directedness" is often an implicit attribute of an "edge", "path", and other terms.

Definitions:

A *directed graph*, or *digraph*, consists of:
- a set V, whose elements are called **vertices**,
- a set E, whose elements are called **directed edges** or **arcs**, and
- an *incidence function* that assigns to each edge a **tail** and a **head**.

The **tail** of an arc is the vertex it leaves, and the **head** is the vertex it enters.

A *strict digraph* has no self-loops or multi-arcs.

The *underlying graph of a digraph* is the graph obtained from the digraph by replacing every directed edge by an undirected edge.

Table 1 Directory of graph models.

subject area and application	vertex attributes and meaning / edge/arc attributes and meaning	reference
computer programming / flowcharts	vertex labels are program steps / edge directions show flow	§8.1.1
social organization / social networks	vertices are persons / edges represent interactions	§8.1.1
civil engineering / road networks	vertices are road intersections / edges are roads	§8.1.1, §8.3.1
operations research / scheduling	vertices are activities / arcs show operational precedence	§8.3.1
sociology / hierarchical dominance	vertices are individuals / arcs show who reports to whom	§8.3.1
computer programming / subprogram calling diagram	vertices are subprograms / arcs show calling direction	§8.3.1
ecology / food webs	vertices are species / arcs show who eats whom	§8.3.1
operations research / scheduling	vertices are activities to be scheduled / edges are activity conflicts	§8.3.1, §8.6.1
genealogy / "family trees"	vertices are family members / arcs show parenthood	§8.3.1
set theory / binary relations	vertices are elements / arcs show relatedness	§8.3.1
probabilistic analysis / Markov models	vertices are process states / edges are state transitions	§8.3.2
traffic control / assigning one-way streets	vertices are intersection / edges are streets	§8.3.3
partially ordered sets / Hasse diagrams	vertices are elements / arcs show covering relation	§8.3.4
computer engineering / communications networks	vertices are computational nodes / arcs are communications links	§8.4.2
operations research / transportation networks	vertices are supply and demand nodes / arcs are supply lines	§8.4.2
walking tours / Seven Bridges of Königsberg	vertices are land masses / edges are bridges	§8.4.3
postal delivery routing / Chinese Postman Problem	vertices are street intersections / edges are streets	§8.4.3
information theory / Gray codes	vertices are binary strings / edges are single-bit changes	§8.4.4
radio broadcasting / assignment of frequencies	vertices are broadcast stations / edges are potential interference	§8.6.1
chemistry / preventing explosions	vertices are chemicals / edges are co-combustibility	§8.6.1

subject area and application	vertex attributes and meaning edge/arc attributes and meaning	reference
cartography map-coloring	regions are countries edges are borders	§8.6.4
highway construction avoiding overcrossings	vertices are road intersections edges are roads	§8.7.1
electrical network boards avoiding insulation	vertices are circuit components edges are wires	§8.7.1
VLSI computer chips minimizing layering	vertices are circuit components edges are wires	§8.7.4
information management binary search trees	vertices are data records edges are decisions	§17.1.4
computer operating systems priority trees	vertices are prioritized jobs edges are priority relations	§17.1.5
physical chemistry counting isomers	vertices are atoms edges are molecular bonds	§9.3.2
network optimization min-cost spanning trees	edges are connections edge-labels are costs	§10.1.1
bipartite matching personnel assignment	parts are people and jobs edges are job-capabilities	§10.2.2
network optimization shortest path	vertices are locations edge-labels are distances	§10.3.1
traveling salesman routing shortest complete tour	vertices are locations edge-labels are distances	§10.7.1

The **out-degree** of vertex v, denoted $\delta^+(v)$, is the number of arcs with tail at v.

The **in-degree** of vertex v, denoted $\delta^-(v)$, is the number of arcs with head at v.

A digraph D is **transitive** if whenever it contains an arc from u to v and an arc from v to w, it also contains an arc from u to w.

The **adjacency matrix** A_D of a digraph D is

$$A_D = [a_{ij}], \text{ where } a_{ij} = \text{number of arcs from } v_i \text{ to } v_j.$$

The **incidence matrix** M_D of a digraph D with no self-loops is $M_D = [b_{ij}]$, where

$$b_{ij} = \begin{cases} +1, & \text{if } v_i \text{ is the tail of } e_j \text{ but not the head} \\ -1, & \text{if } v_i \text{ is the head of } e_j \text{ but not the tail} \\ 0, & \text{otherwise.} \end{cases}$$

There is no standard convention for self-loops.

Facts:

1. *Strict-digraph terminology*: In a context focusing primarily on strict digraphs, there is often a different terminological convention:
   - "digraph" refers to a strict digraph;
   - a directed graph with multi-arcs is called a *multidigraph*;
   - a directed graph with self-loops is called a *pseudodigraph*;
   - an arc with tail u and head v is designated uv.

2. *Alternative "path" terminology*: There is an alternative convention in which a (directed) "path" may use vertices and arcs more than once, but an "elementary path" does not repeat arcs, and a "simple path" does not repeat vertices (and, hence, does not repeat arcs either). See §8.3.2.

3. The incidence structure of a digraph is frequently represented by an *arc list*, in which each arc is represented by an ordered pair uv, where u is its tail and v is its head. For each arc with tail u and head v, there is a separate entry, so that uv occurs as often as the number of such arcs. A list of the isolated vertices plus such an arc list completely specifies a digraph.

4. Another common specification of a digraph is the *lists-of-neighbors representation*. For each vertex u, there is a corresponding row, which has as an entry the head of each arc whose tail is u. Thus a vertex v occurs in that row as many times as there are arcs from u to v.

5. The incidence matrix is another common way to represent a digraph. Since all but one or two of the entries in every column are zero, the incidence matrix is a highly inefficient form of representation.

6. The adjacency matrix is also a common way to specify a digraph in some contexts when there is no reason to identify the arcs by name.

7. A digraph can be represented by a $2 \times |E|$ *incidence table* in which the tail and head of each arc e appear in column e. Direction on an arc can be indicated by a convention as to whether tail or head appears in the first row, which requires swapping the two column entries if the direction is changed. Alternatively, direction can be indicated by marking one of the two entries in each column as the head, and then moving the marker if the direction changes.

8. A row-sum in a directed adjacency matrix equals the out-degree of the corresponding vertex. A column-sum equals the in-degree.

9. In any digraph, the sum of the in-degrees, the sum of the out-degrees, and the number of edges are all equal to each other; i.e., $\sum_{v \in V} \delta^-(v) = \sum_{v \in V} \delta^+(v) = |E|$.

Examples:

1. The following arc list, incidence table, list-of-neighbors, and adjacency matrix all represent the digraph G.

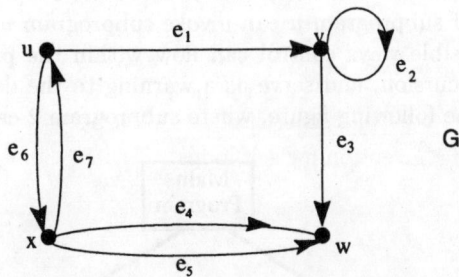

arc list:

uv, vv, vw, xw, xw, ux, xu

incidence table:

e_1	e_2	e_3	e_4	e_5	e_6	e_7
u	v	v	x	x	u	x
v	v	w	w	w	x	u

530   Chapter 8   GRAPH THEORY

$$\text{lists-of-neighbors:} \quad \begin{Bmatrix} u:v,x \\ v:v,w \\ w:\emptyset \\ x:w,w,u \end{Bmatrix} \qquad \text{adjacency matrix:} \quad \begin{matrix} & \begin{matrix} u & v & w & x \end{matrix} \\ \begin{matrix} u \\ v \\ w \\ x \end{matrix} & \begin{pmatrix} 0 & 1 & 0 & 1 \\ 0 & 1 & 1 & 0 \\ 0 & 0 & 0 & 0 \\ 1 & 0 & 2 & 0 \end{pmatrix} \end{matrix}$$

2. *Civil Engineering*: A road network in which at least some of the roads are one-way can be modeled by a digraph. The nodes are road junctures; each two-way road is represented by a pair of arcs, one in each direction. Loops are allowed, and they may represent "circles" that occur in housing developments and in industrial parks. Similarly, multiarcs may occur.

3. *Operations Research*: A large project consists of many smaller tasks with a *precedence relation* — some tasks must be completed before certain others can begin. The vertices represent tasks, and there is an arc from u to v if task u must be completed before v can begin. For instance, in the following figure it is necessary both that food is loaded and the cabin is cleaned before passengers are loaded.

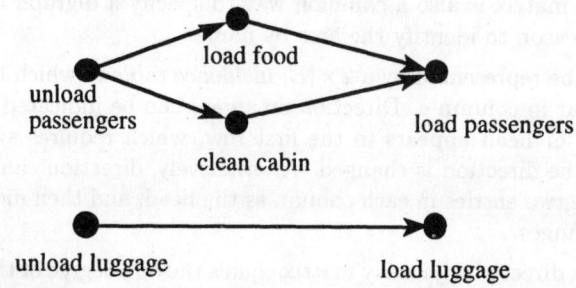

4. *Sociology and Sociobiology*: A business (or army, or society, or ant colony) has a *hierarchical dominance* structure. The nodes are the employees (soldiers, citizens, ants) and there is an arc from u to v if u dominates v. If the chain of command is unique, with a single leader, and if only arcs representing immediate authority are included, then the result is a *rooted tree*. (See §9.1.2.)

5. *Computer Software Design*: A large program consists of many subprograms, some of which can invoke others. Let the nodes of D be the subprograms, and let there be an arc from u to v if subprogram u can invoke subprogram v. Then the *call graph* D encapsulates all possible ways control can flow within the program. Directed cycles represent indirect recursion, and serve as a warning to the designer to ensure against infinite loops. See the following figure, where subprogram 2 can call itself indirectly.

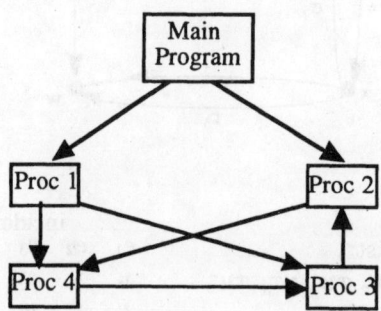

6. *Ecology*: A *food web* is a strict digraph in which nodes represent species and in which there is an arc from u to v if species u eats species v. The following figure shows a small food web.

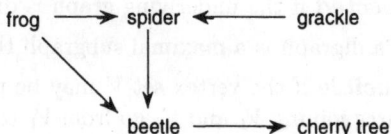

7. *Operations Research*: A sequence of books must be printed and bound, using one press and one binding machine. Suppose that book i requires time p_i for printing and time b_i for binding. It is desired to print the books in such an order that the binding machine is never idle: when it finishes one book, the next book should already be printed. The vertices of a digraph D can represent the books. There is an arc from book i to book j if $p_j \leq b_i$. Then any path through all the vertices corresponds to a permissible ordering.

8. *Genealogy*: A "family tree" is a digraph where the orientation is traditionally given not by arrows but by the direction down for later generations. Despite the name, a family tree is usually not a tree, since people commonly marry distant cousins, knowingly or unknowingly.

9. *Binary relations*: To any binary relation R on a set V (see §12.1) a digraph $D(V, R)$ can be associated: the vertices are the elements of V, and there is an arc from u to v if $(u, v) \in R$. Conversely, every digraph without multiple arcs defines a binary relation on its vertices. The relation R is transitive (see §12.1.2) if and only if the digraph $D(V, R)$ is transitive.

8.3.2 PATHS, CYCLES, AND CONNECTEDNESS

Definitions:

A **directed walk** is a sequence of arcs such that the head of one arc is the tail of the next arc.

The **length** of a directed walk is the number of arcs in the sequence.

A **closed directed walk** is a directed walk that begins and ends at the same vertex.

A **directed trail** is a directed walk in which no arc is repeated.

A **directed path** is a directed trail in which no vertex is repeated.

A **directed cycle** is a closed directed trail in which no vertices are repeated, except the starting and stopping vertex.

Vertex v is **reachable** from vertex u if there is a directed path from u to v.

A **basis for a digraph** is a set of vertices V' such that every vertex not in V' is reachable from V' and such that no proper subset of V' has this property.

The **distance** from a vertex u to a vertex v in a digraph D is the length of the shortest directed path from u to v.

A digraph is **strongly connected** (or **diconnected**, or **strong**) if every vertex is reachable from every other vertex.

A digraph is **unilaterally connected** (or **unilateral**) if for every pair of vertices u and v, there is either a uv-path or a vu-path.

A digraph D is **weakly connected** (or **weak**) if the underlying graph is connected.

The digraph D is **disconnected** if the underlying graph is disconnected.

A **strong component** of a digraph is a maximal subgraph that is strongly connected.

A digraph $D(V, E)$ is **reducible** if the vertex set V may be partitioned into a disjoint union $V_1 \cup V_2$ so that all arcs joining V_1 and V_2 go from V_1 to V_2.

The **condensation** D^* of a digraph D is the strict digraph whose nodes are the strong components $\{V_1, V_2, \ldots, V_k\}$ of D, with an arc $V_i V_j \in E_{D^*}$ if and only if there is an arc vv' in D such that $v \in V_i$ and $v' \in V_j$.

The **converse of a digraph** D is obtained by reversing the directions of all the arcs of D.

The **directional dual** of a theorem about digraphs is the statement obtained by replacing each property in the theorem statement by its converse.

Facts:

1. Using a pencil on a drawing of a digraph, a directed walk can be traversed by following the arrows without lifting the pencil from the graph.

2. Distance in digraphs need not be symmetric. That is, the distance from u to v might be different from the distance from v to u.

3. If A is the adjacency matrix of D, then the ij entry of A^n is the number of n-arc walks from v_i to v_j.

4. Let δ^+ be the smallest out-degree of a strict digraph D. If $\delta^+ > 0$, then D has a cycle of length at least $\delta^+ + 1$.

5. Let δ^- be the smallest in-degree of a strict digraph D. If $\delta^- > 0$, then D has a cycle of length at least $\delta^- + 1$.

6. The directional dual of a theorem about digraphs is a theorem about digraphs.

7. Fact 5 is the directional dual of Fact 4.

8. A digraph D is Eulerian (§8.4.3) if and only if the underlying graph is connected and in-degree equals out-degree at every vertex.

9. A digraph D has an Euler uv-trail (where $u \neq v$) if the following conditions hold:
   - the out-degree of vertex u exceeds the in-degree by one;
   - the in-degree of v exceeds the out-degree by one;
   - at every other vertex, the in-degree equals the out-degree.

That is,
   - $d^+(u) = d^-(u) + 1$;
   - $d^-(v) = d^+(v) + 1$;
   - $(\forall w \neq u, v) \, [d^-(w) = d^+(w)]$.

Other Euler-type results for graphs generalize to digraphs as well.

10. Let δ be the minimum of all in- and out-degrees of D. If D is strict and $\delta \geq \frac{|V|}{2} > 1$, then D contains a Hamilton cycle (§8.4.4).

11. Hamilton theory is much harder and less complete than Euler theory, for digraphs as for graphs.

12. The strong components of a digraph D partition the vertices of D, but not the arcs, since some arcs go from one component to another. However, the maximal unilateral subgraphs do partition the arcs. If V_1, V_2 are the vertex sets of two strong components of D, then all arcs between V_1 and V_2 face the same way — either all are from V_1 or all are to V_1. See the following figure.

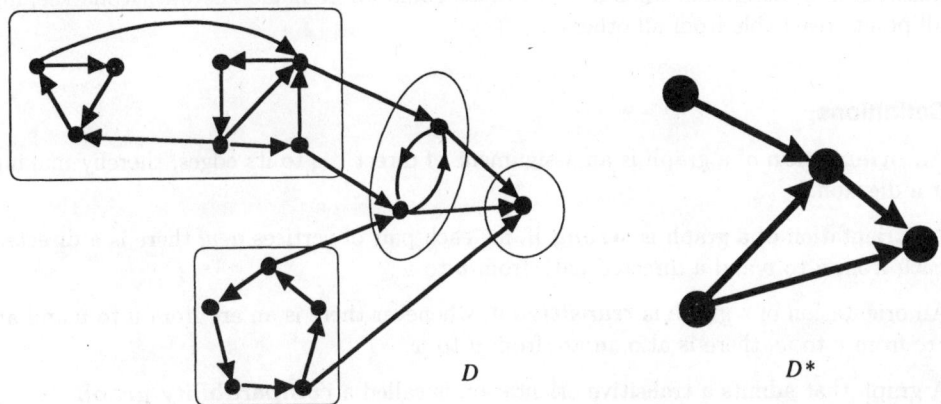

13. The condensation of any digraph is an acyclic digraph (§8.3.4). See the figure for Fact 12.

14. A digraph is reducible if and only if its condensation has at least two vertices.

15. A digraph is unilateral if and only if its condensation is a path.

16. A set V' is a basis of a digraph D if and only if V' consists of one vertex from each strong component of D that has in-degree 0 in D^*. Thus, every basis of a digraph has the same number of vertices.

17. The eigenvalues of a digraph D are the union (counting multiplicities) of the eigenvalues of its strong components. (See §8.9.3.)

Examples:

1. Let u, v be vertices of an n-vertex digraph D with adjacency matrix A. If v is reachable from u, then some uv-path has length $\leq n - 1$. Thus, D is strong if and only if every entry of $\sum_{k=0}^{n-1} A^k$ is positive. There are more computationally efficient tests for diconnectivity: Warshall's algorithm (§14.2) and directed depth first search (§13.3.2).

2. Let M be an arbitrary square matrix. Computation of the eigenvalues of M can sometimes be speeded up as follows. Create matrix A by replacing each nonzero entry of M by a '1', and then let D be the digraph with adjacency matrix A. The eigenvalues of M are the union of the eigenvalues of the minors of M indexed by the strong components of D. (If one component has vertices v_1, v_3, v_7, then one minor has rows and columns $1, 3, 7$ of M.) If M is *sparse* (few nonzeros), then digraph D will usually have many small components and this approach will be efficient.

3. *Markov models*: Let V represent a set of states and E the possible transitions of a Markov process (§7.7). Then walks through D represent "histories" that the process can follow.

8.3.3 ORIENTATION

There are many natural questions concerning when the edges of an undirected graph could be assigned directions so as to obtain a certain sort of digraph. For instance, when can a graph be oriented to obtain a strong digraph? An application of this last question is to determine when a set of roads could all be made one-way, while keeping all points reachable from all others.

Definitions:

An *orientation* of a graph is an assignment of directions to its edges, thereby making it a digraph.

An orientation of a graph is *strong* if, for each pair of vertices u, v, there is a directed path from u to v and a directed path from v to u.

An orientation of a graph is *transitive* if, whenever there is an arc from u to v and an arc from v to w, there is also an arc from u to w.

A graph that admits a transitive orientation is called a *comparability graph*.

A *cut-edge* (or *bridge*) of a graph is an edge whose removal would increase the number of components (§8.4.1).

A *2-edge-connected* graph G is connected and has no cut-edge.

A *generalized circuit* in a graph is a closed walk (§8.4.1) that uses each edge at most once in each direction.

A *triangular chord* for a closed walk (§8.4.1) $u_1, u_2, \ldots, u_k, u_1$ is a proper edge that joins two vertices exactly two apart on the walk.

Facts:

1. Let $\chi(G)$ be the chromatic number (§8.6.1) of graph G. Then every orientation of G has a path of length at least $\chi(G) - 1$.

2. A graph G has a strong orientation if and only if G is 2-edge-connected. (H. Robbins, 1939)

3. A graph G is a comparability graph if and only if every generalized circuit of G of odd length > 3 has a triangular chord.

4. Algorithms 1 and 2 give ways of creating a strong orientation in a 2-edge-connected graph.

Examples:

1. In the figure below the digraph D is a weak transitive orientation of the graph G and D' is a strong nontransitive orientation.

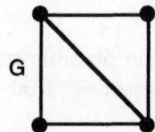
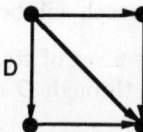
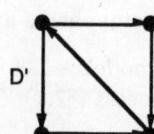

Section 8.3 DIRECTED GRAPHS   535

Algorithm 1:   Naive algorithm for creating a strong orientation.

{This algorithm is good to use by hand for small graphs.}

input: a 2-edge-connected graph G
output: a strong orientation of G

$H :=$ any cycle in G
direct H
while some vertex of G is not in directed subgraph H
  $v :=$ a vertex not in H
  find two edge-disjoint paths from v to H
    {Two such paths exist because G is 2-edge-connected}
  direct one path from v to H and the other from H to v
  $H := H$ with these two subgraph added
orient any remaining edges arbitrarily

Algorithm 2:   Better algorithm for creating a strong orientation.

{A good algorithm for large graphs or for computer implementation.}

input: a 2-edge-connected graph
output: a strong orientation

select an arbitrary vertex as root
construct the Depth-First-Search spanning tree from that root {See §9.2.2.}
orient the tree edges downward from the root
orient all back edges upward toward the root
orient all cross edges arbitrarily

---

2. The following graph is not transitively orientable, and $x, u, v, y, v, w, z, w, u, x$ are the vertices of a generalized circuit without a triangular chord.

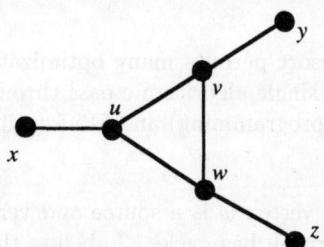

3. *Traffic control*: The flow of traffic on crowded city streets can sometimes be improved by making streets one-way. When this is done, it is necessary that a car can travel legally between any two locations. Assigning directions to the edges of the graph representing the street grid is an orientation of this graph, and cars can travel legally between any two points if and only this graph has a strong orientation. Consequently, by Robbins' theorem (Fact 2), to make all the streets one-way without losing mutual accessibility of locations, it is necessary and sufficient that the grid of streets be 2-edge-connected.

> **Algorithm 3:** Naive topological sort.
>
> {Construct a linear extension ordering for a DAG.}
>
> input: a DAG D
> output: a numbering of the vertices in a topsort order
>
> $H := D$; $k := 1$
> while $V_H \neq \emptyset$
> $v_k :=$ a vertex of H of in-degree 0 {This exists. See Fact 1.}
> $H := H - v_k$ {Remaining graph is still a DAG.}
> $k := k + 1$

8.3.4 DIRECTED ACYCLIC GRAPHS

Definitions:

A digraph is **acyclic** if it has no *directed* cycles. A directed acyclic graph is sometimes called a DAG.

A **source** of a digraph is a vertex of in-degree zero.

A **sink** of a digraph is a vertex of out-degree zero.

A **linear extension ordering** of the n vertices of a digraph is a consecutive labeling $v_1, v_2, \ldots, v_n$ so that, if there is an arc from v_i to v_j, then $i < j$. (See also §11.2.5.)

A **topological sort**, or **topsort**, is an algorithm that assigns a linear extension ordering to a DAG. This traditional name belies the facts that it is not quite a sort, in the usual sense of sorting, and that its relation to topology (in the sense understood by topologists) is obscure.

Facts:

1. Every DAG has at least one source, and by duality, at least one sink.

2. Every DAG has a unique basis (§8.3.2), namely, the set of all its sources.

3. Topsort yields a linear ordering for the vertices that makes the adjacency matrix of a DAG upper-triangular.

4. Doing a preliminary topsort permits many optimization problems about paths to be solved subsequently by a single algorithmic pass through the vertices in the topsort order; see §15.2.2 (dynamic programming) and §15.5 (critical paths). See Algorithm 3.

Examples:

1. In the following digraph vertex w is a source and vertex z is a sink. It is a DAG, even though the underling graph has cycles. Labeling the vertices either in the order w, x, y, z or w, y, x, z is a linear extension ordering.

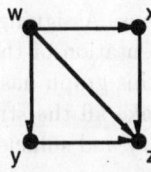

2. Consider any digraph whose vertices represent discrete events, and whose arcs go from earlier events to later events. Any such digraph is acyclic. Conversely, any digraph whose vertices represent procedural steps and whose arcs represent required precedence can be scheduled (using a topological sort) so that arcs do in fact go forward in time.

3. The Hasse diagram of a poset (§12.3.5) is a DAG, as is the entire graph of a poset (arc from u to v if and only if $u \geq v$).

8.3.5 TOURNAMENTS

Definitions:

A *tournament* is a digraph with exactly one arc between each pair of distinct vertices. An *n-tournament* has n vertices.

The *score vector* of a tournament is the sequence of out-degrees of the vertices (number of arcs leaving each vertex), usually in ascending order.

A tournament T is *regular* if every vertex has the same outdegree.

A tournament T is *strong* if there is a directed path between each pair of vertices in both directions.

A tournament T is *transitive* if, whenever there is an arc from u to v and from v to w, there is also an arc from u to w.

A tournament T is *irreducible* if there is no bipartition V_1, V_2 of the vertices such that all arcs between V_1 and V_2 go from V_1 to V_2.

Vertex u of a tournament *dominates* vertex v if there is an arc from u to v.

A *transmitter* in a digraph is a vertex that has an arc to every other vertex.

A *king* in a digraph is a vertex from which there is a path of length 1 or 2 to all other vertices.

A *single-elimination competition* is a contest from which a competitor is eliminated after the first loss.

Facts:

1. Every tournament has a Hamilton path (§8.4.4), in fact an odd number of them.

2. The following statements are equivalent for any n-tournament T:
    - T is strong;
    - T is irreducible;
    - T has a Hamilton cycle (§8.4.4);
    - T has cycles of all lengths $3, 4, \ldots, n$.;
    - Every vertex of T is on cycles of all lengths $3, 4, \ldots, n$.

3. Almost all tournaments are strong, in the sense that, as $n \to \infty$, the fraction of labeled n-tournaments that are strong approaches 1.

4. The following are equivalent for a tournament:
    - the tournament is transitive;
    - the tournament contains no cycles;
    - the tournament contains no 3-cycles;
    - the tournament is a total (i.e. linear) order;
    - the tournament has a unique Hamilton path.

5. Every tournament has a king.

6. The king of a tournament is unique if and only if it is a transmitter. Otherwise, there are at least three kings.

7. In a large tournament, almost every vertex is a king, for as $n \to \infty$, the fraction of n-tournaments in which every vertex is a king approaches 1.

8. *Score vector characterizations*: A nondecreasing sequence S of nonnegative integers $s_1, s_2, \ldots, s_n$ is the score vector of an n-tournament if and only if
$$\sum_{i=1}^{k} s_i \geq \binom{k}{2}, \text{ for } k = 1, 2, \ldots, n-1, \text{ and } \sum_{i=1}^{n} s_i = \binom{n}{2},$$
or equivalently, if and only if

the sequence S' obtained by deleting any one s_i and reducing the largest remaining $n - s_i - 1$ terms by 1 is a score vector of an $(n-1)$-tournament.

9. The second characterization of Fact 8 leads to a recursive algorithm to construct a tournament having a specified score vector. See Example 4.

10. A nonnegative integer sequence $s_1 \leq s_2 \leq \ldots \leq s_n$ is the score vector of a *strong* n-tournament if and only if
$$\sum_{i=1}^{k} s_i > \binom{k}{2}, \text{ for } k = 1, 2, \ldots, n-1, \text{ and } \sum_{i=1}^{n} s_i = \binom{n}{2}.$$

11. There are $2^{\binom{n}{2}}$ distinct *labeled* tournaments, because for each pair of vertices $\{u, v\}$, there are two choices which way to direct the edge. If c_n is the numbered of distinct *unlabeled* n-tournaments, then
$$c_n > \frac{2^{\binom{n}{2}}}{n!} \quad \text{and} \quad \lim_{n \to \infty} \frac{c_n}{2^{\binom{n}{2}}/n!} = 1.$$
The distinction between labeled and unlabeled tournaments is the same as between labeled and unlabeled graphs; see §8.9.1. The two tournaments in the following figure are isomorphic as unlabeled tournaments, but distinct as labeled tournaments.

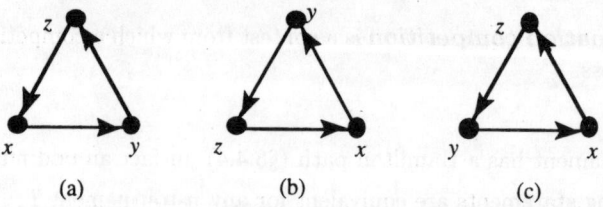

(a)   (b)   (c)

12. *Ranking real tournaments*: When a tournament models a competition, there is an obvious desire to rank the teams, or at least to pick a clear winner. Many ranking methods have been proposed, and continue to be proposed.

13. When a tournament is acyclic, it corresponds to a unique total ordering (Fact 4), so the ranking is unequivocal. However, almost all tournaments are strong (Fact 3). Moreover, in a large tournament, almost every vertex is a king (Fact 7). These are reasons why it is considered difficult to give a satisfactory general method to rank tournaments.

14. *Scheduling tournaments*: To speed up the play of an n-tournament, games can be scheduled in parallel. If n is even, then at most $\frac{n}{2}$ of the $\frac{n(n-1)}{2}$ games may be played at once, so at least $n - 1$ rounds are needed. However, if n is odd, then only $\frac{n-1}{2}$ games can be played at once, so at least n rounds are needed. In fact, this minimum number of rounds can be obtained, and several methods of scheduling tournaments, subject to various additional conditions, have been devised. See [Mo68].

Examples:

1. A round-robin sports tournament in which there are no ties is a tournament in the mathematical sense defined above. However, a single-elimination competition (e.g., most tennis tournaments) is not a tournament as defined above.

2. It has been observed that in every small flock of hens, *every* pair of hens establish a dominance relation — the weaker of the two allows the stronger to peck her. Thus, this pecking order is a tournament.

3. In a "paired comparison experiment", a subject is asked to state a preference in each pair chosen from n items. This amounts to a tournament, where there is an arc ij if item i is preferred to item j.

4. Is there a tournament on vertices (a, b, c, d, e) with respective scores $(1, 2, 2, 2, 3)$? Deleting e according to the second part of Fact 8 leaves vertices (a, b, c, d) with scores $(1, 2, 2, 1)$. Next deleting d leaves (a, b, c) with scores $(1, 1, 1)$. The obvious tournament with such a score vector is a 3-cycle. Next reinsert vertex d, making it dominate vertex a only. Then reinsert vertex e, making it dominate a, b, c. This 5-tournament has the specified score vector $(1, 2, 2, 2, 3)$.

5. *Ranking real tournaments*: Ranking teams by their order along a Hamilton path (see Fact 1) is rarely satisfactory, because that order is unique only for transitive tournaments (Fact 4); in most cases, there are a great many Hamilton paths. Ranking by score vector usually creates ties, and a team with few wins may deserve a better rank if those teams it beats have many wins. So one may consider the *second-order score vector*, where each team's score is the sum of the out-degrees of the teams it beats. This can be continued to nth-order score vectors. There is a limit ranking obtained this way (often quite satisfactory), related to the eigenvalues of the digraph. See [Mo68] for more detail and references.

8.4 DISTANCE, CONNECTIVITY, TRAVERSABILITY

Movement from one node to another in the network corresponds to the graph-theoretic notion of a walk. Graphs often serve as models for transportation and communication network problems. The capability for any two nodes in a network to communicate corresponds to connectedness. The connectivity of a graph is a measure of resistance to a communications cutoff.

8.4.1 WALKS, DISTANCE, AND CYCLE RANK

Definitions:

A **walk** in a graph is an alternating sequence $v_0, e_1, v_1, \ldots, e_r, v_r$ of vertices and edges in which each edge e_i joins vertices v_{i-1} and v_i. Such a walk is also called a v_0, v_r-*walk*.

The **length** of a walk is the number of occurrences of edges in it. An edge that occurs more than once is counted each time it occurs.

A **trail** is a walk in which all of the edges are different.

A **path** is a trail in which all the vertices are different, except that the initial and final vertices may be the same. A path from v_0 to v_r is called a v_0, v_r-*path*.

A walk, trail, or path is **open** if its final vertex is different from its initial vertex.

A walk, trail, or path is **closed** if its final vertex is the same as its initial vertex.

A graph is **connected** if each pair of vertices are joined by a path.

A **component** of a graph is a maximal connected subgraph of the graph.

The vertex v is **reachable** from vertex u in a graph if there is a u,v-path in the graph.

An **isolated vertex** of a graph is a vertex with no incident edges.

The **distance** $d(v,w)$ between two vertices v and w of a graph is the length of a shortest path between them, with $d(v,v) = 0$ and $d(v,w) = \infty$ if there is no path between v and w.

The **diameter** of a connected graph is the maximum distance between two of its vertices.

The **eccentricity** of a vertex v of a connected graph is the greatest distance from v to another vertex.

The **radius** of a connected graph is the minimum eccentricity among all the vertices of the graph.

The **center** of a connected graph is the set of vertices of minimum eccentricity.

A **cycle** is a closed path of positive length. (The word "cycle" also refers to a type of graph; see §8.1.3.)

The **cycle rank** (or **first Betti number**), denoted by $\beta_1(G)$, of a connected graph $G = (V, E)$ is $|E| - |V| + 1$.

Facts:

1. *Alternative terminology*: Sometimes "path" is used to mean what is here called a trail, in which case "simple path" is used to mean a path.

2. In a simple graph, a walk may be represented as a string of vertices $v_0 v_1 \ldots v_r$, without mentioning the edges.

3. The distance function on the vertex set of any connected graph G is a metric; i.e., the following rules hold for all vertices $u, v,$ and w in G:
    - $d(v,w) \geq 0$, with equality if and only if $v = w$;
    - $d(w,v) = d(v,w)$;
    - $d(u,w) \leq d(u,v) + d(v,w)$, with equality if and only if v is on a shortest path from u to w.

4. There are polynomial-time algorithms for finding a shortest path between vertices. (See §10.2.)

5. A graph is connected if and only if it has a spanning tree.

6. The graph G is nonconnected if and only if there is a partition of its vertex set into nonempty sets A and B so that no edge has one end in A and the other in B.

7. The relation "is reachable from" is an equivalence relation on the vertex set. The equivalence classes of this relation induce the components.

8. The graph G is connected if every vertex is reachable from every other vertex.

9. In a simple graph, the minimum length of a cycle is at least 3. In a general graph, a self-loop is a 1-cycle, and a 2-cycle is formed by a pair of vertices joined by a pair of parallel edges (§8.1.1).

10. The cycle rank $\beta_1(G)$ of a connected graph G is best conceptualized as the number of edges remaining in the complement of a spanning tree for G, and *not* as an abstract formula.

11. The cycle rank $\beta_1(G)$ of a connected graph G is equal to the rank of a vector space over Z_2 whose domain is the set of cycles of G.

12. The cycle rank of a connected planar graph G equals the number of regions in a plane drawing of G, minus the exterior region.

13. The following table gives the cycle rank of some infinite families of graphs.

graph	cycle rank
bouquet B_n	n
dipole D_n	$n-1$
complete graph K_n	$\frac{(n-2)(n-1)}{2}$
complete bipartite graph $K_{m,n}$	$(m-1)(n-1)$
cycle graph C_n	1
wheel W_n	n
hypercube Q_n	$(n-2)2^n + 1$
any tree	0

Example:

1. The following connected graph has diameter 3 and radius 2. The vertices in its center are indicated by solid dots.

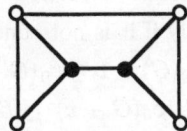

2. The cycle rank of the following connected graph is three. Observe that there are three edges in the complement of the indicated spanning tree.

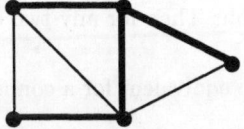

3. The following nonconnected graph has three components, one of which is an isolated vertex.

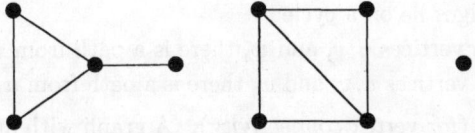

8.4.2 CONNECTIVITY

Definitions:

A **cut-vertex** of a graph G (or **cut-point** or **articulation point**) is a vertex v such that $G - v$ has more components than G. (In topological analysis of nonsimple graphs, sometimes a vertex attached to a self-loop is also considered to be a cut vertex.)

A **nonseparable** graph is a connected graph with no cut-vertices.

A **block** of a graph is a maximal nonseparable subgraph.

An **cut-edge** of a graph G is an edge e such that $G - e$ has more components than G (in which case there is just one more).

A **disconnecting set of vertices** in a connected graph is a set of vertices whose removal yields a nonconnected graph.

A **disconnecting set of edges** in a connected graph is a set of edges whose removal yields a nonconnected graph.

The **zeroth Betti number** $\beta_0(G)$ of a graph G is the number of components in G. Elsewhere this is sometimes is denoted by $c(G)$ or $\omega(G)$.

The (**vertex**) **connectivity** $\kappa(G)$ is the number of vertices in the smallest disconnecting set of vertices. By convention, $\kappa(K_n) = n - 1$.

The **edge connectivity** $\kappa'(G)$ is the number of edges in the smallest disconnecting set of edges.

A graph is **k-connected** if $\kappa(G) \geq k$.

A graph is **k-edge-connected** if $\kappa'(G) \geq k$.

Facts:

1. A vertex is a cut-vertex if and only if it lies on all paths between two other vertices.
2. Every nontrivial graph has at least two vertices that are not cut-vertices.
3. An edge is a cut-edge if and only if it is not contained in any cycle.
4. For any edge e of a graph G, $\beta_0(G) + 1 \geq \beta_0(G - e) \geq \beta_0(G)$.
5. For any vertex v of a graph G, $\beta_0(G - v) \geq \beta_0(G)$; however, $\beta_0(G - v)$ may be arbitrarily greater than $\beta_0(G)$.
6. Let G be a 2-connected graph. Then for any two vertices, there is a cycle containing those vertices.
7. Let G be a 2-connected graph. Then for any two edges, there is a cycle containing those edges.
8. The following statements are equivalent for a connected graph G with at least three vertices:
    - G is nonseparable;
    - every pair of vertices lie on a cycle;
    - every pair of edges lie on a cycle;
    - given any three vertices u, v, and w, there is a path from u to w containing v;
    - given any three vertices u, v, and w, there is a path from u to w not containing v.
9. *Menger's theorem (for vertex connectivity)*: A graph with at least $k + 1$ vertices is k-connected if and only if every pair of vertices is joined by k paths which are internally disjoint (i.e., disjoint except for their origin and terminus). (Menger, 1927)
10. *Menger's theorem (for edge connectivity)*: A graph is k-edge-connected if and only if every pair of vertices is joined by k edge-disjoint paths. (Ford and Fulkerson, 1956; also Elias, Feinstein, and Shannon, 1956)
11. For any graph G, the vertex connectivity is no more than the edge connectivity, and the edge connectivity is no more than the minimum degree. That is, $\kappa(G) \leq \kappa'(G) \leq \delta_{\min}(G)$, where $\delta_{\min}(G)$ denotes the minimum degree.
12. Furthermore, for any positive integers $a \geq b \geq c$, there exists a simple graph G for which $\delta_{\min}(G) = a$, $\kappa'(G) = b$, and $\kappa(G) = c$. (Chartrand and Harary, 1968)

13. The following table gives the vertex connectivity and edge connectivity of some infinite families of graphs.

graph	κ	κ'
complete graph K_n	$n-1$	$n-1$
complete bipartite graph $K_{m,n}$	$\min(m,n)$	$\min(m,n)$
cycle graph C_n	2	2
wheel W_n	3	3
hypercube Q_n	n	n
any nontrivial tree	1	1

Examples:

1. The following graph G has cut-vertices u and v. The blocks are illustrated at the right.

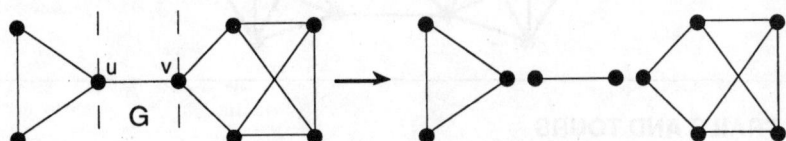

2. In the following graph, vertices u and v form a disconnecting set.

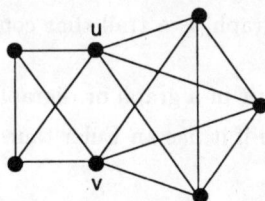

3. *Communication networks*: A communication network can be modeled as a graph with vertices representing the nodes and with undirected edges representing direct two-way communications links between nodes. In order that all pairs of nodes be in communication, the graph must be connected. Vertex connectivity and edge connectivity are measures of network reliability.

4. *Transportation networks*: Low connectivity in transportation networks results in "bottlenecks", in which many different shipments must all past through a small number of vertices. High connectivity implies (by Menger's theorem) several alternative routes between nodes.

5. Menger's theorem implies that a 2-connected graph has two disjoint paths between each pair of vertices. It does *not* imply that for any path between two vertices, there must be a second such path disjoint from the first, as indicated in the following graph. There are two disjoint paths from the leftmost vertex to the rightmost, but there is no such path disjoint from the one indicated by thick edges.

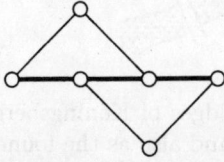

6. The following shows a graph G with $\kappa(G) = 2$ and $\kappa'(G) = 3$. On the left there are the two internally-disjoint paths between the upper-left vertex and the lower-right vertex, and on the right there are three edge-disjoint paths.

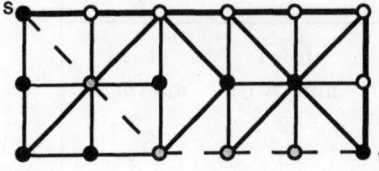 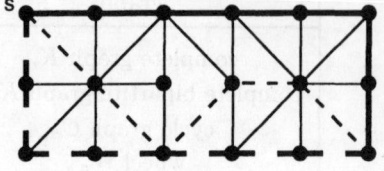

7. The following graph illustrates Fact 12, with $\kappa = 2, \kappa' = 3, \delta_{\min} = 4$.

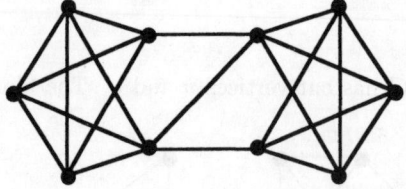

8.4.3 EULER TRAILS AND TOURS

Definitions:

An **Euler trail** in a graph [digraph] is a trail that contains all the edges [arcs] of the graph.

An **Euler tour** or **Euler circuit** in a graph or digraph is a closed Euler trail.

A graph or digraph is **Eulerian** if it has an Euler tour.

Facts:

1. *Seven bridges of Königsberg problem*: In Kaliningrad, Russia, two branches of the River Pregel meet and flow past an island into the Baltic Sea. In 1736, when this was the town of Königsberg in East Prussia, there were seven bridges joining the banks of the river, the headland, and the island, as illustrated below at the left. The celebrated Swiss mathematician Leonhard Euler (1707–1783) was invited by Emperor Frederick the Great to decide whether it was possible to cross all seven bridges without recrossing any of them. In the earliest known paper on graph theory, Euler proved it is impossible, because the graph at the right has no Euler trail.

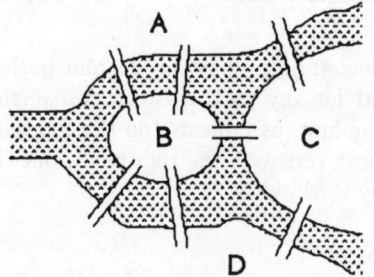 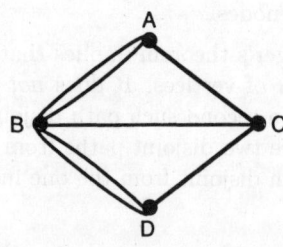

2. Euler's work on the seven bridges of Köningsberg problem is commonly described as the founding of graph theory and also as the founding of topology.

3. A connected graph is Eulerian if and only every vertex has even degree. The tour may begin/end at any vertex.

4. A connected digraph has a directed Euler tour if and only if the in-degree of every vertex v equals its out-degree.

5. A connected graph has an Euler trail between distinct vertices u and v if and only if u and v are the only vertices of odd degree.

6. A connected graph (digraph) is Eulerian if and only if there exists a collection of cycles (directed cycles) whose edges partition the edge set of the graph.

7. A connected planar (§8.7.1) graph is Eulerian if and only if its dual (§8.8.2) is bipartite.

8. A graph G can be oriented to have a directed Euler tour if and only if it is an Eulerian graph. (Traversing an Euler tour provides an orientation.)

9. The following table tells which members of several infinite families of graphs are Eulerian.

graph	Eulerian?
bouquet B_n	for all n
dipole D_n	for even n
complete graph K_n	for odd n
complete bipartite graph $K_{m,n}$	for m and n both even
cycle graph C_n	for all n
wheel W_n	never
hypercube Q_n	for even n
tree	only if trivial

10. Algorithm 1 gives a recursive method for finding an Eulerian tour on an Eulerian graph.

11. Fleury's algorithm for finding an Euler tour or trail is given in Algorithm 2.

Examples:

1. The following is an Eulerian graph and one of its Euler tours.

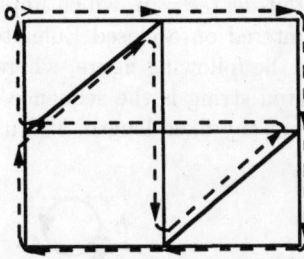

2. *Chinese postman problem* (due to Guan Meigu, 1962): A letter carrier begins at the post office, traverses every street in his territory at least once, and then returns to the post office. His objective is to walk as little as possible. Each edge of a graph representing the street configuration is labeled with the length of the corresponding block. If the graph is Eulerian, then an Euler tour gives an optimal solution. Otherwise, some edges must be retraced. Polynomial-time algorithms to solve this problem are known. See §10.2.3.

Algorithm 1: Recursive algorithm for finding an Eulerian tour.

input: a connected graph G, all of whose vertices have even degree
output: an Euler tour of G

$C :=$ a cycle in the graph G; place C on the cycle-queue $\mathcal{Q}$
partition the edge-complement $G - E(C)$ into components $H_1, H_2, \ldots, H_k$
recursively run this algorithm on each component H_i
{So far, E_G has been completely partitioned into the cycles on $\mathcal{Q}$}

merge the elements of $\mathcal{Q}$ into an Euler tour for G, by traversing the cycle C
   and splicing in the tours found for the components H_i whenever possible

Algorithm 2: Fleury's algorithm for finding an Euler tour/trail.

input: a connected graph G, an initial vertex v, and a final vertex w; if $v \neq w$,
   then every vertex except v and w must have even degree (if $v = w$, then all
   degrees must be even)
output: an Euler trail whose origin is v and whose terminus is w

{ find trail edge with origin v }
if $deg(v) > 1$ then $e :=$ any edge incident at v which is not a cut-edge
else { $deg(v) = 1$ }
   $e :=$ the unique edge incident at v
$u :=$ the other endpoint of e
recursively find an Euler trail from u to w in $G - e$
prepend the edge e to the trail found in the recursive step
{ This yields the required Euler trail of G. }

3. For every letter in an arbitrary n-letter alphabet $\mathcal{A}$, there is a string starting and ending with that letter, in which every possible substring of two letters appears consecutively exactly once. To see this, consider the digraph D with the letters of $\mathcal{A}$ as vertices and one arc for each ordered pair. The digraph D is connected and, at each of the n vertices, in-degree = out-degree = n, which implies that it is Eulerian. Thus, the sequence of vertices encountered on a closed Euler tour from any vertex to itself yields the specified string. See the following figure, where $e_1, e_2, \ldots, e_9$ are the arcs of an Euler cycle and the associated string is the sequence of vertices, $aabbccacba$. This result generalizes to substrings of any fixed length, also using Euler tours.

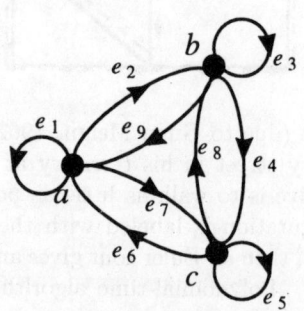

8.4.4 HAMILTON CYCLES AND PATHS

Definitions:

A **Hamilton cycle** in a graph [digraph] is a cycle [directed cycle] that includes all vertices of the graph.

A graph or digraph is **Hamiltonian** if it contains a Hamilton cycle.

A **Hamilton path** in a graph [digraph] is a path [directed path] that includes all vertices of the graph.

A **theta graph** is a subdivision of the complete bipartite graph $K_{2,3}$. Thus, the graph comprises three internally disjoint paths joining the two 3-valent vertices.

A **Gray code** is a cyclic ordering of all 2^k length-k bitstrings, such that each bitstring differs from the next in exactly one bit entry.

A **tough** graph is a connected graph G such that no matter what nonempty, proper vertex subset S is removed, the resulting number of components of $G - S$ is no more than $|S|$.

Facts:

1. The concept of a Hamilton cycle first arose in a puzzle within the Icosian Game, invented by Sir William Rowan Hamilton (1805–1865), an Irish mathematician. This puzzle involved a dodecahedron whose 20 vertices were labeled with world capitals. It required finding a complete tour of these 20 capitals.

2. The recognition problem for Hamiltonian graphs is NP-complete. Thus, unlike the case with Eulerian graphs, there is no easy test to decide whether a graph is Hamiltonian (unless P = NP). However, many of the following facts provide criteria that are often helpful in deciding.

3. A Hamiltonian graph has no cutpoint. (Thus, it is 2-connected.)

4. The previous fact has this generalization: Let G be a Hamiltonian graph and let $S \subseteq V_G$. Then the graph $G - S$ has at most $|S|$ components.

5. Bipartite Hamiltonian graphs have an equal number of vertices in the two parts of the bipartition.

6. If a simple graph has $n \geq 3$ vertices and minimum degree at least $\frac{n}{2}$, then it is Hamiltonian. (Dirac, 1952)

7. If a simple graph has $n \geq 3$ vertices, and if every pair of nonadjacent vertices u and v satisfies the inequality $deg(u) + deg(v) \geq n$, then it is Hamiltonian. (Ore, 1960)

8. Suppose that a simple graph with $n \geq 3$ vertices has degree sequence $d_1 \leq d_2 \leq \cdots \leq d_n$, and that for every i with $1 \leq i \leq \frac{n}{2}$ either $d_i > i$ or $d_{n-i} \geq n - i$. Then the graph is Hamiltonian.

9. Every simple graph with $n \geq 3$ vertices and at least $(n^2 - 3n + 6)/2$ edges is Hamiltonian.

10. Every graph with at least three vertices whose connectivity (κ) is at least as large as its independence number (α) is a Hamiltonian graph.

11. Every 4-connected planar graph is Hamiltonian.

12. If the edges of the complete graph K_n are assigned directions, then the resulting digraph always has a Hamilton directed path.

13. The edges of the complete graph K_{2n+1} can be partitioned into n Hamilton cycles.
14. A theta graph looks like a subdivided copy of the Greek letter theta (θ).
15. Theta graphs are non-Hamiltonian.
16. Every nonHamiltonian graph contains a theta subgraph.
17. A graph G is non-Hamiltonian if there is a subset of mutually nonadjacent vertices containing more than half the vertices of G.
18. Hamiltonian graphs are tough.
19. "Almost all" graphs are Hamiltonian. That is, of the exactly $2^{n(n-1)/2}$ simple graphs on n (labeled) vertices, the proportion that are Hamiltonian tends to 1 as $n \to \infty$.
20. Suppose that a simple graph is constructed by the following process: start with n vertices and no edges; until the minimum degree is 2, a possible edge is chosen uniformly at random from among the edges not already in the graph, and added to the graph. With probability tending to 1 as $n \to \infty$, the resulting graph is Hamiltonian.
21. The following table tells which members of several infinite families of graphs are Hamiltonian.

graph	Hamiltonian?
bouquet B_n	for all $n \geq 1$
dipole D_n	for all $n \geq 2$
complete graph K_n	for all $n \geq 3$
complete bipartite graph $K_{m,n}$	when $m = n$
cycle graph C_n	for all $n \geq 1$
wheel W_n	for all $n \geq 2$
hypercube Q_n	for all $n \geq 2$
any tree	if $|V| = 1$

Examples:

1. Finding a Hamilton cycle in the dodecahedral graph (see §8.2.3), as illustrated below, is equivalent to solving Hamilton's Icosian Game puzzle. An example of a Hamilton cycle in this graph is: $RSTVWXHJKLMNPCDFGBZQR$.

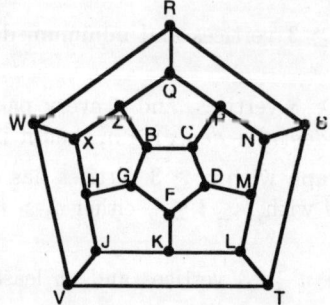

2. The following graph has the Hamilton cycle $acefdba$.

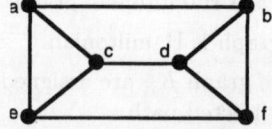

3. The following graph is non-Hamiltonian, by Fact 17, since the vertices u, v, w, and x are mutually nonadjacent.

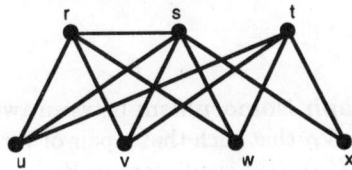

4. The 10-cycle C_{10} (§8.1.3 Figure 1) is an example of a graph that satisfies none of the sufficient conditions in Facts 7-9 above for Hamiltonicity, but is nonetheless Hamiltonian.

5. The traveling salesman problem (§10.7.1) is to find a minimum-cost Hamilton cycle in a complete graph whose edges are labeled with costs.

6. *Information theory — Gray codes:* In information theory, a cyclic ordering of the 2^n length-n bitstrings such that each bitstring differs from its predecessor in exactly one bit is called a *Gray code*. This corresponds to a Hamilton cycle in the k-dimensional hypercube.

The following figure shows a Hamilton cycle in the 3-cube giving the Gray code $000 \to 001 \to 011 \to 111 \to 101 \to 100 \to 110 \to 010 \to 000$.

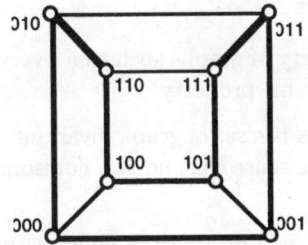

7. The Petersen graph (§8.1.3 Figure 2) is tough but not Hamiltonian.

8. The following graph is tough but not Hamiltonian.

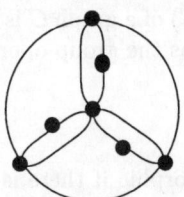

8.5 GRAPH INVARIANTS AND ISOMORPHISM

Deciding whether two graph descriptions actually specify structurally identical graphs is called isomorphism testing. Polynomial-time algorithms for isomorphism testing are known only for certain special classes of graphs. However, there are heuristic algorithms to test isomorphism of reasonable-sized graphs. The related problem of reconstructing a graph from its vertex-deleted subgraphs is also still unsettled.

8.5.1 ISOMORPHISM INVARIANTS

Definitions:

For simple graphs only, a **graph isomorphism** between two graphs G and H can be defined as a bijection $f: V_G \to V_H$ that such that a pair of vertices u, v is adjacent in V_G if and only if the image pair $f(u), f(v)$ is adjacent in V_H.

In full generality, a **graph isomorphism** $f: G \to H$ is a pair of bijections $f_V: V_G \to V_H$ and $f_E: E_G \to E_H$ such that for every edge $e \in E_G$, the endpoints of e are mapped onto the endpoints of $f_E(e)$.

Note: Except when confusion will result, the same notation f can be used for both the vertex function f_V and the edge function f_E.

A **digraph isomorphism** is an isomorphism of the underlying graphs such that the edge correspondence preserves all edge directions.

Two graphs are **isomorphic** if there is an isomorphism from one to the other, or informally, if their mathematical structures are identical.

The **isomorphism type** of a graph [digraph] G is the class of all graphs [digraphs] isomorphic to G.

A **graph invariant** is a property of graphs such that every two isomorphic graphs have the same value with regard to this property.

A **complete set of invariants** is a set of graph invariants that distinguishes any graph from any different graph, in the sense that no two nonisomorphic graphs have the same set of invariant values.

A **vertex invariant** is a property of a vertex which is preserved by isomorphism, in the following sense: if v is any vertex and f is any isomorphism, then the vertex $f(v)$ has the same value as v with regard to the property.

An **automorphism** is an isomorphism from a graph to itself.

The **automorphism group** $Aut(G)$ of a graph G is the collection of all automorphisms of G, with functional composition as the group operation.

Facts:

1. Two graphs G and H are isomorphic if there is a bijection $f: V_G \to V_H$ such that for every vertex pair $u, v \in V_G$ the number of edges joining u and v equals the number joining their images $f(u), f(v) \in V_H$.

2. Graph invariants are used to distinguish between nonisomorphic graphs.

3. Most graph invariants are either too tedious to compute or not strong enough at distinguishing similar but nonisomorphic graphs.

4. No good complete set of invariants is known, in the sense that each invariant value is easily computed and easily compared.

5. Vertex invariants are often used to organize the vertices of a graph into equivalence classes under graph automorphism, in order to discover the automorphism group of the graph.

6. Graphs have many different kinds of isomorphism invariants, including the following invariants:
- *elementary (ascertainable by counting)*:
    number of vertices,
    number of edges,
    sequence of vertex degrees;
- *structural invariants (concerning connectivity or cycles)*:
    cycle rank (§8.4.1),
    girth (§8.4.1),
    connectivity (§8.4.2),
    edge connectivity (§8.4.2);
- *topological invariants (concerning placement on surfaces)*:
    genus (§8.8.4),
    crosscap number (§8.8.4),
    crossing number (§8.7.4),
    thickness (§8.7.4);
- *chromatic invariants (concerning colorings)*:
    chromatic number (§8.6.1),
    edge-chromatic number (§8.6.2),
    chromatic polynomial (§8.6.1);
- *algebraic invariants (concerning groups or vector spaces)*:
    eigenvalues (§8.10.1),
    automorphism group (§8.10.2).

Examples:

1. The following figure illustrates an isomorphism f of simple graphs.

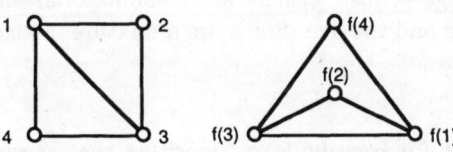

2. The following figure illustrates an isomorphism f of nonsimple graphs.

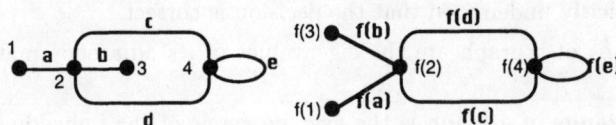

3. The following two digraphs are not isomorphic. Even though there are six different isomorphisms of their underlying graphs, none of them preserves the direction of all the edges.

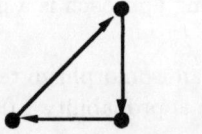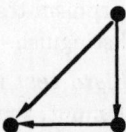

4. The following three graph drawings all look different. The table below shows some of their isomorphism invariants, from which it may be concluded that graph B cannot be isomorphic to either graph A or graph C, but that A and C might be isomorphic.

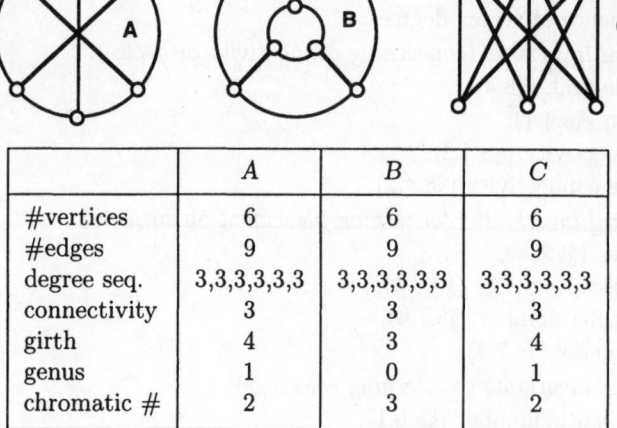

	A	B	C
#vertices	6	6	6
#edges	9	9	9
degree seq.	3,3,3,3,3,3	3,3,3,3,3,3	3,3,3,3,3,3
connectivity	3	3	3
girth	4	3	4
genus	1	0	1
chromatic #	2	3	2

To construct an isomorphism between graphs A and C, assign labels 0, 1, 2, 3, 4, 5 cyclically to the vertices of A. Then assign labels 0, 2, and 4 to the top three vertices of C, and 1, 3, and 5 to the bottom three.

8.5.2 ISOMORPHISM TESTING

Various concepts for isomorphism testing have been proposed. Some exploit or refine algebraic invariants, such as the automorphism group or the set of eigenvalues. Others exploit a decomposition into planar subgraphs or a refinement of a topological invariant, such as the average genus. Others are combinatorial and employ enumerative techniques, partitioning and the like. Some are a mixture of algebraic, topological, and combinatorial approaches.

Definitions:

An *isomorphism test for graphs* is an algorithm that accepts two graphs as input and outputs "yes" to indicate the decision that they are isomorphic or "no" to indicate that they are nonisomorphic. Unless the context explicitly mentions the possibility of error, it is implicitly understood that the decision is correct.

The *eigenvalues* of a graph are the eigenvalues of its adjacency matrix. (See §11.9 and §6.5.)

The *average genus* of a graph is the average genus of the imbedding surface, taken over all cellular imbeddings of that graph. (See §8.8.3.)

An *equitable partition* for a graph G is a partition $V_1, \ldots, V_n$ of its vertex set and a set of numbers $\{d_{i,j} \mid 1 \leq i, j \leq n\}$ such that every vertex in V_i is adjacent to exactly $d_{i,j}$ vertices in V_j.

A *devil's pair* for an isomorphism-testing approach is a pair of nonisomorphic graphs that the approach fails to distinguish.

A *probabilistic isomorphism test* is an isomorphism test such that no matter what pair of graphs is supplied as input, there is probability 1.0 of a correct decision.

Facts:

1. No polynomial-time isomorphism testing algorithm is known. Moreover, it is not known whether isomorphism testing is an NP-complete problem.

2. On an n-vertex simple graph, the naive graph isomorphism-testing algorithm below has worst-case time $O(n^2 \cdot n!)$.

3. It is easy to design an algorithm decide correctly with probability 1.0 whether two randomly selected graphs are isomorphic: they aren't. With probability 1.0, two randomly selected graphs will not have the same number of vertices or edges. This observation explains why the concept of probabilistic isomorphism testing is defined so that it must be able to decide correctly with probability 1.0 for all pairs, not just for randomly selected pairs.

4. If it were possible to quickly calculate the size of the automorphism group of a graph, such an algorithm could be a subprogram of a quick test for isomorphism of graph pairs, as follows:

   if $|Aut(G)| \neq |Aut(H)|$ **then** G and H are not isomorphic
   else
       if $|Aut(G \cup H)| = 2|Aut(G)|^2$ **then** G and H are isomorphic
       else G and H are not isomorphic

5. Another algebraic approach to isomorphism testing is based on eigenvalues. A devil's pair for simply comparing eigenvalues appears in Example 4.

6. One topological approach to isomorphism testing dissects each graph into planar components (§8.7) and combines known efficient tests for isomorphism of planar graphs with careful study of possible interconnections.

7. Another topological approach to isomorphism testing is based on the genus distribution (§8.8.3), taken over all cellular imbeddings. Although calculating the genus distribution by brute force would be tedious, one can estimate it by random sampling. Any pair of trees is a trivial devil's pair, but trees are easily tested by another isomorphism algorithm.

8. The best known practical isomorphism algorithm is "NAUTY" (an acronym for No AUTomorphisms, Yes?) by B. D. McKay. This backtrack algorithm repeatedly refines an initial vertex partition. At each stage of the refinement, a part of size greater than 1 is broken into two parts, one a single vertex, and the coarsest equitable partition is found. The discrete partitions generated in this way correspond to labelings of the graph, organized so as to determine the automorphism group. A complicated scheme is used to pick one of these labelings as the "canonical" one used for isomorphism testing. This algorithm can very quickly check isomorphism for most graphs, although it has no good theoretical bound.

9. To certify that two graphs are isomorphic, one can give a vertex bijection that realizes the isomorphism. Deciding whether a bijection between the vertex sets of two graphs is the vertex function of a graph isomorphism can be achieved in polynomial time.

10. Graph isomorphism is in NP, by the previous fact, but is not known to be in P. The computational complexity of the problem of deciding whether or not two graphs are isomorphic is unknown.

11. If graph isomorphism is NP-complete, then the complexity hierarchy between P and NP collapses, which is considered unlikely.

Algorithm 1: Naive graph isomorphism-testing algorithm.

input: simple graphs G, H

if $|V_G| \neq |V_H|$ **then** print "NO" and stop
if $|E_G| \neq |E_H|$ **then** print "NO" and stop
for each bijection $f: V_G \to V_H$
   for each pair $u, v \in V_G$
     if u, v adjacent and $f(u), f(v)$ not adjacent **then** print "NO" and stop
     if u, v not adjacent and $f(u), f(v)$ adjacent **then** print "NO" and stop
print "YES" and stop

12. The following table shows the best known time bounds for checking isomorphism in various classes of graphs. Almost all of these bounds have been achieved using an algebraic approach.

class of graphs (on n vertices)	time bound
graphs	$\exp\sqrt{cn \log n}$
trees	$O(n)$
planar graphs	$O(n)$
graphs of genus g	$n^{O(g)}$
cubic graphs	$O(n^3 \log n)$
graphs with max degree $\leq d$	$n^{O(d)}$
tournaments	$n^{O(\log n)}$

13. Algorithm 1 gives a naive method for testing whether or not two graphs are isomorphic.

Examples:

1. The labeling of these two isomorphic graphs indicates the correspondence between vertices. This is the famous Petersen graph.

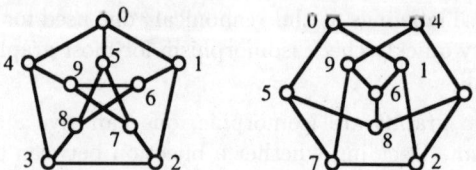

2. One devil's pair for isomorphism testing by degree sequence is

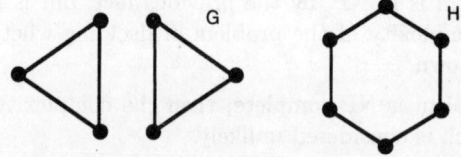

3. Another devil's pair for isomorphism testing by degree sequence is

4. The following graphs are a devil's pair for isomorphism testing by simple comparison of eigenvalues. They both have characteristic polynomial $\lambda^6 - 7\lambda^4 - 4\lambda^3 + 7\lambda^2 + 4\lambda - 1$. Yet they cannot be isomorphic because their degree sequences are different.

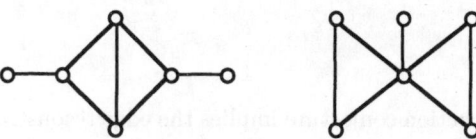

5. A 3-connected devil's pair for simply comparing average genus is shown below. Both graphs shown have the genus distribution $8, 536, 3416, 1224$, that is, 8 imbeddings of genus 0, 536 of genus 1, 3416 of genus 2 and 1224 of genus 3.

8.5.3 GRAPH RECONSTRUCTION

The question of whether a graph is reconstructible from its subgraphs is one of the most beguiling unsolved problems in graph theory.

Definitions:

A **vertex-deleted subgraph** of a graph G is a subgraph $G - v$ obtained by removing a single vertex v and all of its incident edges.

An **edge-deleted subgraph** of a graph G is a subgraph $G - e$ obtained by removing a single edge e.

The **vertex-deleted subgraph collection** of a graph G is the multi-set of all vertex-deleted subgraphs $G - v$. The number of times a graph appears in the collection equals the number of different vertices whose removal yields that graph. Thus, the cardinality of the collection equals the number of vertices of G.

The **edge-deleted subgraph collection** of a graph G is the multi-set of all edge-deleted subgraphs $G - e$. The number of times a graph appears in the collection equals the number of different edges whose removal yields that graph. Thus, the cardinality of the collection equals the number of edges of G.

A **reconstructible** graph is a graph G such that no other graph has the same vertex-deleted subgraph collection as G.

An **edge-reconstructible** graph is a graph G such that no other graph has the same edge-deleted subgraph collection as G.

A **reconstructible invariant** is a graph invariant such that all graphs with the same vertex-deleted subgraph collection have the same value with respect to this invariant.

An **edge-reconstructible invariant** is a graph invariant such that all graphs with the same edge-deleted subgraph collection have the same value with respect to this invariant.

Conjectures:

1. *The graph reconstruction conjecture* (P. Kelly and S. Ulam, 1941): Every graph with more than two vertices is reconstructible.

2. *The edge reconstruction conjecture*: Every graph with at least four edges is edge reconstructible.

3. *Halin's conjecture*: If two (possibly infinite) graphs with more than two vertices have the same vertex-deleted subgraph collection, then each graph is a subgraph of the other.

Facts:

1. The graph reconstruction conjecture implies the edge reconstruction conjecture, and both are implied by Halin's conjecture.

2. The graph reconstruction conjecture does not hold for graphs on two vertices, because K_2 and $\overline{K_2}$ have identical sets of deleted subgraphs.

3. The edge reconstruction conjecture does not hold for graphs on four edges, because $K_3 + K_1$ (disjoint union) and $K_{1,3}$ have identical collections of edge-deleted subgraphs.

4. Computer search has verified the reconstruction conjecture for graphs with nine or fewer vertices.

5. The following table lists some invariants and types of graphs which are known to be reconstructible.

both edge-reconstructible and reconstructible		other edge-reconstructible graphs
invariants	*graphs*	
number of vertices	regular	more edges than non-edges
number of edges	disconnected	only two vertex degrees
degree sequence	trees	no induced $K_{1,3}$ subgraph
connectivity	outerplanar	large with Hamilton path
characteristic polynomial	cacti	$2\log_2(2 \max \deg) \leq \text{avg deg}$

6. If graph F has fewer vertices than graph G then the number of subgraphs of G isomorphic to F is reconstructible.

7. The reconstruction conjecture is not true for directed graphs in general, because nonreconstructible tournaments of arbitrarily large size are known.

8. Infinite graphs are not reconstructible in general, but Halin's conjecture holds for all known nonreconstructible infinite pairs.

9. Almost every graph is uniquely determined by any three vertex-deleted subgraphs.

Example:

1. The following figure shows a graph (at the left) and its collection of vertex-deleted subgraphs.

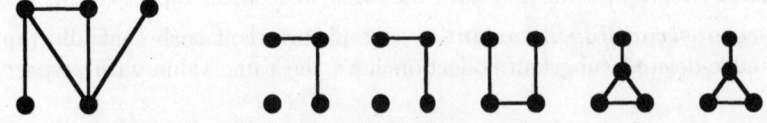

8.6 GRAPH AND MAP COLORING

The vertex set of a simple graph can be colored so that adjacent vertices are colored differently. Similarly, the edges of a graph without self-loops can be colored so that adjacent edges are colored differently. If a graph is imbedded in a surface so that there are no self-adjacent regions, then the regions can be colored so that adjacent regions receive different colors. These entertaining concepts have many important applications, including assignment and scheduling problems.

8.6.1 VERTEX COLORINGS

Definitions:

A (**proper**) **vertex k-coloring** (or **k-coloring**) of a simple graph G is a function $f: V_G \to \{1, \ldots, k\}$ such that adjacent vertices are assigned different numbers. Quite often, the set $\{1, \ldots, k\}$ is regarded as a set of colors.

A **coloring** of a graph is a k-coloring for some integer k.

An **improper coloring** of a graph permits two adjacent vertices to be colored the same.

A graph is **k-vertex colorable** (or **k-colorable**) if it has a vertex k-coloring.

The **vertex chromatic number** or (**chromatic number**) $\chi(G)$ (or $\chi_V(G)$) of a graph G is the minimum number k such that G is k-vertex colorable; that is, $\chi(G)$ is the smallest number of colors needed to color the vertices of G so that no adjacent vertices have the same color.

A graph G is **k-chromatic** if $\chi(G) = k$.

A graph G is **chromatically k-critical** if G is k-chromatic and if $\chi(G - e) = k - 1$ for each edge of G.

An **obstruction to** (or **for**) **k-coloring** is a chromatically $(k+1)$-critical graph, when that graph is regarded as a subgraph of other graphs, and thereby prevents them from having chromatic number k.

A (**complete**) **obstruction set for k-coloring** is a set of chromatically $(k+1)$-critical graphs such that every graph that is not k-colorable contains at least one of them as a subgraph.

An **elementary contraction** of a simple graph G on the edge e, denoted $G \downarrow e$ (or $G \cdot e$), is obtained by replacing the edge e and its two endpoints by one vertex adjacent to all the other vertices to which the endpoints were adjacent.

A graph G is (**combinatorially**) **contractible** to a subgraph H if H can be obtained from G by a sequence of elementary contractions. (The modifier "combinatorially" distinguishes this kind of contractibility used for graph colorings from topological contractibility.)

The **chromatic polynomial** of the graph G is the function $\pi_G(t)$ whose value at the integer t is the number of different functions $V_G \to \{1, \ldots, t\}$ that are proper colorings of G.

Algorithm 1:   Greedy coloring algorithm.

input: a graph G with vertex list $v_1, v_2, \ldots, v_n$
$c := 0$ {Initialize color at "color 0".}
while some vertex still has no color
$\quad c := c + 1$ {Get the next unused color.}
$\quad$ **for** $i := 1$ **to** n {Assign the new color to as many vertices as possible.}
$\quad\quad$ **if** v_i is uncolored and no neighbor of v_i has color c **then** assign color c to v_i

Facts:

1. A direct way to calculate the chromatic number of a reasonably small graph is in two steps. First derive an upper bound for the number of colors needed, either by finding a coloring by trial and error or by using the greedy coloring algorithm. Then prove that one fewer colors would be insufficient. This could be achieved by an exponential-time exhaustion algorithm, or by finding an insightful proof for the particular graph.

2. *Alternative notation*: In a topological context, where $\chi(S)$ means Euler characteristic (§8.8), $cr(G)$ can be used for the chromatic number of a graph.

3. Unlike a topological contraction along an edge, this operation of "elementary contraction" of two vertices of a simple graph always yields a simple graph.

4. $\chi(G) = 1$ if and only if the graph G is edgeless.

5. $\chi(G) = 2$ if and only if the graph is bipartite and its edgeset is nonempty.

6. *The four color theorem* (Appel and Haken, 1976): If G is planar, then $\chi(G) \leq 4$. That is, every planar graph has a proper coloring of its vertices with 4 or fewer colors.

7. $\{K_2\}$ is a complete obstruction set for 1-coloring.

8. The set of odd cycles is a complete obstruction set to 2-coloring.

9. The odd wheels $W_{2n+1}, n \geq 1,$ are obstructions to 3-coloring.

10. *Brooks' theorem*: If G is a connected graph which is neither an odd cycle nor a complete graph, then $\chi(G) \leq \Delta_{\max}(G)$, where $\Delta_{\max}$ denotes maximum degree. (The subscript "max" is often omitted.)

11. $\chi(G) \leq 1 + \Delta_{\max}(G)$.

12. $\chi(G) \leq 1 + \max \delta_{\min}(G')$, where $\delta_{\min}$ denotes minimum degree, and where the maximum is taken over all induced subgraphs G' of G.

13. $\chi(G) \leq diam(G)$, where the diameter $diam(G)$ is the length of a longest path in G.

14. *Hadwiger's conjecture.* If G is a connected graph with $\chi(G) = n$, then G is contractible to K_n; it is known to be true for $n \leq 5$.

15. *Nordhaus-Gaddum inequalities*: If G is a graph with $|V(G)| = p$ and $\overline{G}$ is its edge-complement, then
    - $2\sqrt{p} \leq \chi(G) + \chi(\overline{G}) \leq p + 1$;
    - $p \leq \chi(G) \cdot \chi(\overline{G}) \leq \left(\frac{p+1}{2}\right)^2$.

16. The greedy coloring algorithm (Algorithm 1) produces a vertex coloring of a graph G, whose vertices are ordered. (It is called "greedy" because once a color is assigned, it is never changed.) The number of colors it assigns depends on the vertex ordering, and it is not necessarily the minimum possible.

17. At least one ordering of the vertices of a graph G yields $\chi(G)$ under the greedy algorithm.

18. The number of colors used by the greedy coloring algorithm depends on the ordering in which the vertices of G are listed. At least one of the orderings of the vertices of G yields $\chi(G)$.

19. The number of colors used by the greedy coloring algorithm can exceed $\chi(G)$ by an arbitrarily large number.

20. There is no known polynomial-time algorithm for finding $\chi(G)$ exactly. Deciding whether a graph has a particular chromatic number is NP-complete, if that number is at least 3.

21. The following table gives the chromatic numbers and edge-chromatic numbers (§8.6.2) of the graphs in some common families.

graph G	$\chi(G)$	$\chi_1(G)$
path graph P_n, $n \geq 3$	2	2
cycle graph C_n, n even, $n \geq 2$	2	2
cycle graph C_n, n odd, $n \geq 3$	3	3
wheel W_n, n even, $n \geq 4$	3	n
wheel W_n, n odd, $n \geq 3$	4	n
complete graph K_n, n even, $n \geq 2$	n	$n-1$
complete graph K_n, n odd, $n \geq 3$	n	n
complete bipartite graph $K_{m,n}$, $m, n \geq 1$	2	$\max\{m, n\}$
bipartite G, at least one edge	2	$\Delta_{\max}(G)$
Petersen graph	3	4
complete k-partite $K_{m_1,\ldots,m_k}$, $m_i \geq 1$	k	$\max\{m_1, \ldots, m_k\}$

22. For every edge e of a simple graph G, $\pi_G(t) = \pi_{G-e}(t) - \pi_{G \cdot e}(t)$.

23. The chromatic polynomial $\pi_G(t)$ of a graph with $n \geq 1$ vertices and m edges is a polynomial in t of degree n, whose leading term is t^n, whose next term is $-mt^{n-1}$, and whose constant term is 0.

24. The following table gives the chromatic polynomials of some graphs.

graph	$\pi_G(t)$
n-vertex tree	$t(t-1)^{n-1}$
cycle graph C_n	$(t-1)^n + (-1)^n(t-1)$
wheel W_n	$t(t-2)^{n-1} + (-1)^{n-1}t(t-2)$
complete graph K_n	$t^{\underline{n}} = t(t-1)(t-2)\ldots(t-n+1)$

Examples:

1. *Time scheduling*: Let classes at a school be modeled by the vertices of a simple graph G, with two vertices adjacent if and only if there is at least one student in both of the corresponding classes. Then $\chi(G)$ gives the minimum number of time periods for scheduling the classes so as to accommodate all the students.

2. *Assignment of radio frequencies*: If the vertices of a graph G represent radio stations, with two stations adjacent precisely when their broadcast areas overlap, then $\chi(G)$ determines the minimum number of transmission frequencies required to avoid broadcast interference.

3. *Separating combustible chemical combinations*: Let the vertices of graph G represent different kinds of chemicals needed in some manufacturing process. An edge joins each pair of chemicals that might explode if they are combined. The chromatic number of this graph is the number of different storage areas required so that no two chemicals that mix explosively are stored together.

4. Proceeding in the direct way, as described in Fact 1, to color the graph in the following figure quickly yields its chromatic number. Applying the greedy coloring algorithm, with the vertices considered in cyclic order around the 8-cycle, yields a 3-coloring. Since this graph contains an odd cycle (a 5-cycle), it cannot be 2-colored. Thus, $\chi = 3$.

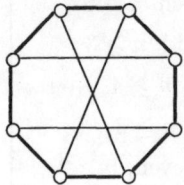

5. In the following figure vertex colorings are indicated for the cycle graphs C_3, C_4, and C_5; in each case three colors are used. Note that $\chi(C_3) = \chi(C_5) = 3$, whereas $\chi(C_4) = 2$ (since the vertex colored "3" could have been colored "1").

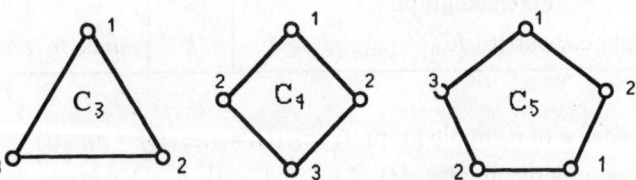

6. The following figure shows three chromatically 4-critical graphs.

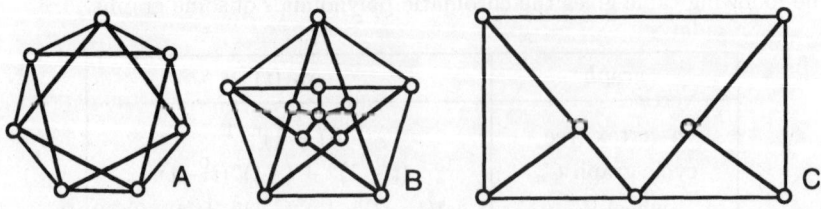

7. A 3-coloring of graph A in the figure of Example 6 would necessarily give some color to three different vertices. Two of these vertices would have to be adjacent (because the edge-complement contains no 3-cycle). Thus, a 3-coloring could not be proper, and hence $\chi = 4$.

8. A 3-coloring of graph B in the figure of Example 6 would need three different colors on the outer 5-cycle. These would force the use of three different colors on the points of the central 5-star. This would force the use of a fourth color on the central vertex. Thus, $\chi = 4$.

8.6.2 EDGE COLORINGS

Definitions:

An *edge coloring* of a graph is an assignment of colors to its edges such that adjacent edges receive different colors.

A graph G is *k-edge colorable* if there is an edge coloring of G using at most k colors.

The *edge chromatic number* $\chi_1(G)$ (or $\chi_E(G)$) of a graph G is the minimum k such that G is k-edge colorable. If $\chi_1(G) = k$, then G is *edge k-chromatic*.

Chromatic index is a synonym for edge chromatic number.

A graph is *edge-chromatically k-critical* if it is edge k-chromatic and $\chi_1(G-e) = \chi_1(G) - 1$ for every edge e of G.

For a graph G, the *line graph* $L(G)$ has as vertices the edges of G, with two vertices adjacent in $L(G)$ if and only if the corresponding edges are adjacent in G.

Facts:

1. *Alternative notation*: In a topological context where χ is used for Euler characteristic, the notation for edge-chromatic number is often $ecr(G)$.
2. Every edge coloring of a graph G can be interpreted as a vertex coloring of the associated line graph $L(G)$. Thus, $\chi_1(G) = \chi(L(G))$.
3. $\Delta_{\max}(G) \leq \chi_1(G)$.
4. *Vizing's theorem*: If G is a simple graph, then $\chi_1(G) \leq \Delta_{\max}(G) + 1$.
5. *Vizing's general theorem*: If G is a general graph whose maximum edge multiplicity is μ, then $\chi_1(G) \leq \Delta_{\max}(G) + \mu$.
6. Either $\chi_1(G) = \Delta_{\max}(G)$ (G is of *class one*) or $\chi_1(G) = \Delta_{\max}(G) + 1$ (G is of *class two*).
7. $\chi_1(K_{m,n}) = \chi(L(K_{m,n})) = \chi(K_m \times K_n) = \max\{m, n\}$, if $m, n \geq 1$.
8. If G is bipartite, then $\chi_1(G) = \Delta_{\max}(G)$.
9. $\chi_1(K_n) = n$ if n is odd ($n \neq 1$); $\chi_1(K_n) = n - 1$ if n is even.
10. If G is planar and $\Delta_{\max}(G) \geq 8$, then $\chi_1(G) = \Delta_{\max}(G)$.
11. If G is 3-regular and Hamiltonian, then $\chi_1(G) = \Delta_{\max}(G)$.
12. If G is regular with $|V_G|$ odd and $|E_G| > 0$, then $\chi_1(G) = \Delta_{\max}(G) + 1$.
13. The greedy edge-coloring algorithm (Algorithm 2) produces an edge-coloring of a graph G, whose vertices are ordered. The number of colors it assigns depends on the vertex ordering, and it is not necessarily the minimum possible. (It is equivalent to applying the greedy vertex-coloring algorithm to the line graph.)

Examples:

1. The following three graphs are all edge 3-chromatic. None of them is edge-chromatically 3-critical. Since each graph has a vertex of degree three, no 2-edge-coloring is possible.

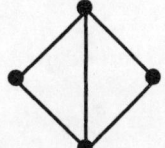  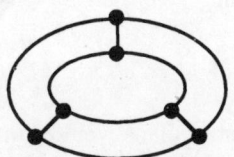

Algorithm 2: Greedy edge-coloring algorithm.

input: a graph G with edge list $e_1, e_2, \ldots, e_n$

$c := 0$ {Initialize color at "color 0".}
while some edge still has no color
   $c := c + 1$ {Get the next unused color.}
   for $i := 1$ **to** n
     {Assign the new color to as many edges as possible.}
     if e_i is uncolored and no neighbor of e_i has color c **then** assign color c to e_i

2. The following graph is 5-edge-chromatic. Since there are 14 edges, a 4-edge-coloring would have to give the same color to four of them. For this edge-coloring to be proper, these four edges would have to have no endpoints in common. That is impossible, because the graph has only seven vertices.

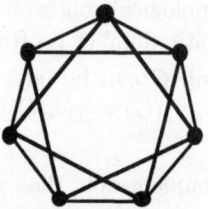

3. The Petersen graph is edge-chromatically 4-critical.

4. *Exam scheduling:* Suppose that each student at a university is to be examined orally by each of his or her professors at the end of the term. Then the minimum number of examination periods required is the edge chromatic number of the bipartite graph with vertices representing students and professors, and edges connecting students with their professors.

5. *Wiring electrical network boards:* A number of relays, switches, and other electronic devices $D_1, D_2, \ldots, D_n$ on a relay panel are to be connected into a network. The connecting wires are twisted into a cable, with those connected to D_1 emerging at one point, those connected to D_2 at another, and so forth. The wires emerging from the same point must be colored differently, so that they can be distinguished. The least number of colors required to color the wires is the edge chromatic number of the associated network.

6. The following nonsimple graph illustrates Vizing's general theorem. Its highest edge multiplicity is 3, its maximum degree is 6, and its edge chromatic number is 9.

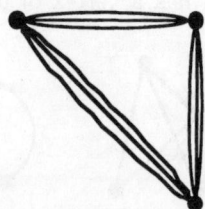

8.6.3 CLIQUES AND INDEPENDENCE

Definitions:

A *clique* of a graph G is a complete subgraph of G which is contained in no larger complete subgraph of G.

The *clique number* $\omega(G)$ of a graph G is the order (i.e., number of vertices) of a largest clique of G.

A subset W of $V(G)$ (or D of $E(G)$) is *independent* if no two elements of W (respectively D) are adjacent.

The *vertex independence number* $\alpha(G)$ of G is the order of a largest independent set of vertices in G.

The *edge independence number* $\alpha_1(G)$ of a graph G is the size of a largest independent set of edges in G.

A graph G is *perfect* if, for every induced subgraph H of G, the chromatic number equals the clique number, that is, $\chi(H) = \omega(H)$.

A graph G is *weakly γ-perfect* if $\chi(G) = \omega(G)$.

Facts:

1. The independence number of a graph is equal to the clique number of its edge-complement, and vice versa. That is, $\alpha(G) = \omega(\overline{G})$ and $\omega(G) = \alpha(\overline{G})$.

2. The chromatic number of a graph is at least as large as the clique number: $\chi(G) \geq \omega(G)$.

3. For each positive integer n, there is a graph G with chromatic number n and clique number equal to 2; that is, G contains no triangles.

4. If no induced subgraph of a graph is isomorphic to P_4, then its chromatic number equals its clique number and the greedy algorithm (§8.6.1 Algorithm 1) always produces a coloring with the minimum number of colors.

5. *Lovasz's perfect graph theorem*: A graph G is perfect if and only if its edge-complement $\overline{G}$ is perfect.

6. $\dfrac{|V(G)|}{\alpha(G)} \leq \chi(G) \leq |V(G)| + 1 - \alpha(G)$.

7. If $|E(G)| > \Delta_{\max}(G) \times \alpha_1(G)$, then $\chi_1(G) = \Delta_{\max}(G) + 1$.

Examples:

1. The following graph has three cliques — of sizes 2, 3, and 4. Thus, its clique number is 4.

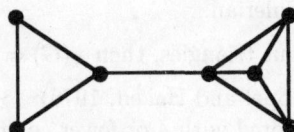

2. If $1 \leq m \leq n$, then $\omega(K_{m,n}) = 2, \alpha(K_{m,n}) = n$, and $\alpha_1(K_{m,n}) = m$.

3. Define $K_{n(m)}$ to be the graph whose edge complement is nK_m, the disjoint union of n copies of K_m. Then $\omega(K_{n(m)}) = n, \alpha(K_{n(m)}) = m$, and $\alpha_1(K_{n(m)}) = \left\lfloor \frac{mn}{2} \right\rfloor$.

8.6.4 MAP COLORINGS

Definitions:

An **orientable surface** S is a surface homeomorphic to a sphere with $g \geq 0$ handles attached and is denoted by S_g.

A **nonorientable surface** S is a surface homeomorphic to a sphere with $k \geq 1$ crosscaps attached and is denoted by N_k. (See §11.7.1.)

The **Euler characteristic** of a surface S is $\chi(S)$, where
$$\chi(S) = \begin{cases} 2 - 2g & \text{if } S \text{ is homeomorphic to } S_g \\ 2 - k & \text{if } S \text{ is homeomorphic to } N_k. \end{cases}$$
The most usual notation for Euler characteristic throughout mathematics is $\chi(S)$. However, ad hoc notation such as $eu(S)$ is sometimes used in chromatic graph theory.

A **map** on a surface is an imbedding of a graph on that surface. (See §8.7.)

A **map coloring** is an assignment of colors to the regions of a map so that adjacent regions (those sharing a one-dimensional boundary portion) receive different colors.

A map M is **n-colorable** if there is a map coloring of M using at most n colors.

The **chromatic number** $\chi(M)$ (or $cr(M)$ or $\chi_R(M)$) of a map M is the minimum n such that M is n-colorable.

The **chromatic number** $\chi(S)$ (or $cr(S)$) of a surface S is the largest chromatic number $\chi(M)$ for all maps M on S.

The (**empire**) **chromatic number** $\chi(S, c)$ for a surface S is the largest $\chi(M)$ for all maps M on S, where now a country has at most $c \geq 1$ components (regions) and all components of a fixed country are colored alike, but adjacent components of different countries must receive different colors. (Thus $\chi(S) = \chi(S, 1)$.)

Facts:

1. A region coloring can be regarded as a vertex coloring of the dual graph (§11.7). From this perspective, χ is the largest value of $\chi(G)$ for all graphs G imbeddable on S.

2. By stereographic projection (§11.6.5), $\chi(S_0)$ gives the chromatic number of the plane.

3. Let G be a planar cubic block; then $\chi_1(G) = 3$.

4. Let M be a plane map whose graph G is connected and bridgeless. Then $\chi(M) = 2$ if and only if G is Eulerian.

5. Let M be a plane map for a cubic connected bridgeless graph G; then $\chi(M) = 3$ if and only if the dual graph is Eulerian.

6. If G is a plane graph without triangles, then $\chi(G) = 3$. (Grötzsch, 1958)

7. *The four color theorem* (Appel and Haken, 1976): $\chi(S_0) = 4$. That is, every map on a sphere or plane can be colored with 4 or fewer colors.

8. *The Heawood map coloring theorem* (Ringel and Youngs, 1968): For $g > 0$,
$$\chi(S_g) = \left\lfloor \frac{7 + \sqrt{1 + 48g}}{2} \right\rfloor.$$

9. *The nonorientable Heawood map coloring theorem* (Ringel, 1954): For $k > 0$,
$$\chi(N_k) = \left\lfloor \frac{7 + \sqrt{1 + 24k}}{2} \right\rfloor.$$
except that $\chi(N_2) = 6$.

10. $\chi(S, c) \leq \left\lfloor \dfrac{6c + 1 + \sqrt{(6c + 1)^2 - 24eu(S)}}{2} \right\rfloor.$

11. $\chi(S_0, c) = 6c$ for $c \geq 2$.

12. $\chi(N_1, c) = 6c$ for $c \geq 1$.

13. $\chi(S_1, c) = 6c + 1$ for $c \geq 1$.

14. *History of the four color problem*: In 1852, Francis Guthrie first asked whether four colors suffice to color every planar map, and his brother Frederick communicated the question to Augustus De Morgan. Arthur Cayley in 1878 was first to mention the problem in print. In 1879, A. B. Kempe, a London barrister, published a "proof" of the four color conjecture: every planar map is 4-colorable. In 1890 Percy Heawood (1861–1955) found an error in Kempe's argument. A correct proof was established by Kenneth Appel and Wolfgang Haken in 1977.

15. *Concepts in the Haken-Appel proof of the four color theorem*: Appel and Haken found an "unavoidable" set with 1476 graphs, which means that at least one of these graphs must be a subgraph of any minimum counterexample to the four color conjecture. A method called "discharging", due to Heinrich Heesch, is used to find an unavoidable set. Using a computer, they proved that each of these graphs is "reducible", which means that it cannot be a subgraph of a minimum counterexample.

16. A simplified proof of the four color theorem can be found in [Th98] or at the following Web site:
    http://www.math.gatech.edu/~thomas/FC/ftpinfo.html

17. *History of the Heawood map coloring problem*: In 1890, Percy Heawood established the upper bound
$$\chi(M) \leq \left\lfloor \frac{7 + \sqrt{49 - 24eu(S)}}{2} \right\rfloor$$
for $\chi(M)$ for all maps M on all closed surfaces other than S_0. Heawood showed that his bound was exact for the torus, by the example of the dual of K_7 on S_1, and he asserted without proof that similar "verification figures" existed for all other cases. In 1934, Philip Franklin showed that Heawood's assertion was wrong for N_2 (the "Klein bottle"). For all other nonorientable surfaces, Gerhard Ringel provided the necessary figures in 1952. In 1968, Ringel and J. W. T. Youngs completed the verification for all orientable surfaces other than the sphere.

Examples:

1. Let M be the tetrahedral map, i.e., an imbedding of K_4 in S_0. By Fact 4, $\chi(M) \neq 2$, since K_4 is not Eulerian. By Fact 5, $\chi(M) \neq 3$, since the dual graph (isomorphic to K_4) is also not Eulerian. Thus, $\chi(M) = 4 = \chi(S_0)$.

2. *Cartography*: If countries on Earth are allowed two components, but no more, then by Fact 11 a map might require twelve colors, but no more.

3. By Fact 8, $\chi(S_1) = 7$. The dual map of the following figure imbeds K_7 in the torus. To obtain the torus, paste the left side of the rectangular sheet directly to the right side, and then paste the top to the bottom with a $\frac{2}{7}$ twist.

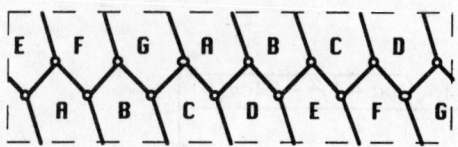

8.6.5 GRAPH MULTICOLORING

This subsection deals with the proper coloring of the vertices of a graph where each vertex is assigned more than one color (or label). Multicoloring of graphs has many applications to assignment and scheduling problems. See [MiRo91].

Definitions:

A (*proper*) **k-tuple coloring** (or **k-multicoloring**) of a graph is an assignment of a set of k distinct colors to each vertex of a graph so that whenever two vertices are adjacent, their sets of assigned colors are disjoint.

A (*proper*) **multicoloring** of a graph is a k-multicoloring for some k.

The **k-tuple chromatic number** $\chi_k(G)$ of a graph G is the smallest number of colors such that G has a k-tuple coloring.

A (*proper*) **set coloring** of a graph is an assignment of a set of colors to each vertex of G such that whenever two vertices are adjacent, the sets of colors assigned to the two vertices are disjoint. *Note*: The sets can have different sizes.

Facts:

1. $\chi_k(G) \leq k\chi(G)$.

2. If the clique number $\omega(G)$ (§8.6.3) of G is equal to $\chi(G)$ (i.e., G is weakly γ-perfect), then $\chi_k(G) = k\chi(G)$.

3. Set colorings generalize multicolorings since the sets of assigned colors in a set coloring can have different sizes.

4. Every set coloring of G where all sets assigned to the vertices are all k-sets is a k-tuple coloring of G.

5. $\chi_k(K_n) = nk$.

6. If G is bipartite (with at least one edge), then $\chi_k(G) = 2k$.

Examples:

1. *Multiple channel assignment*: Several cities each need to have four broadcast frequencies assigned to them (a generalization of §8.6.1, Example 7). The 4-tuple chromatic number $\chi_4(G)$ is the minimum number of frequencies needed so that there is no broadcast interference.

2. $\chi_2(C_5) = 5$, as illustrated in the following figure.

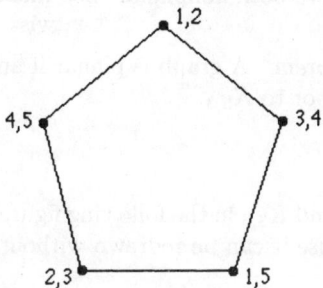

3. *Exam scheduling*: Each final exam at a school is given in two parts, with each part requiring one final exam period. If a graph G is constructed by using the courses as vertices, with an edge joining v and w if there is a student taking courses v and w, then $\chi_2(G)$ gives the minimum number of periods required to schedule all the exams so no student has a conflict.

4. Suppose that in the previous example the number of periods required for the final exams in courses varies. The problem of scheduling the final exams in the fewest number of periods is a set coloring problem.

8.7 PLANAR DRAWINGS

Planarity is an important consideration in physical networks of any kind, because it is usually less expensive to produce a planar network. For instance, overpasses are a costly feature in highway design. Also, it is less complicated to manufacture a planar electrical network than a nonplanar network.

8.7.1 CHARACTERIZING PLANAR GRAPHS

A graph cannot be drawn without edge-crossings in the plane if it "contains" either the graph K_5 or the graph $K_{3,3}$. Conversely, every graph that "contains" neither of those two graphs can be drawn without crossings.

Definitions:

A graph **imbedding** (or **embedding**) is a drawing with no crossings at all.

A graph is **planar** if it has an imbedding in the plane.

A graph is **nonplanar** if no imbedding in the plane is possible.

A drawing of a graph is **normalized** if there are no crossings, or if each crossing is a point where the interior of one edge crosses the interior of one other edge. (Edges may be drawn either straight or with curves.)

The graphs K_5 and $K_{3,3}$ are called the **Kuratowski graphs**, after the Polish mathematician Kazimierz Kuratowski (1896–1980).

Facts:

1. The graphs K_5 and $K_{3,3}$ are both nonplanar. See Examples 4 and 5 for proofs that they are not planar.

2. *Kuratowski planarity theorem*: A graph is planar if and only if it has no subgraph homeomorphic (§8.1.2) to K_5 or to $K_{3,3}$.

Examples:

1. The drawings of Q_3, K_5, and $K_{3,3}$ in the following figure all have crossings. However, the graph Q_3 is planar, because it can be redrawn without any crossings.

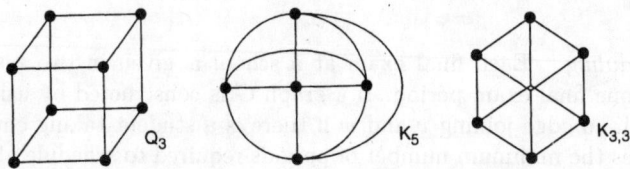

2. The drawings of Q_3 and K_5 in the figure of Example 1 are normalized, but the drawing of $K_{3,3}$ is not normalized, because three lines go through the same point.

3. The Petersen graph (§8.1) does not contain $K_{3,3}$ itself as a subgraph. However, if the two edges depicted by broken lines in the following figure are discarded, then the resulting graph is homeomorphic to $K_{3,3}$, so the Petersen graph is not planar.

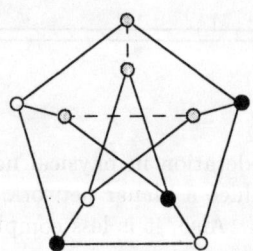

4. *Nonplanarity of K_5*: To draw the complete graph on the vertices v_1, v_2, v_3, v_4, v_5 in the plane, one might as well start by drawing the 4-cycle v_1, v_2, v_3, v_4, which separates the plane. Next draw the edges between v_1 and v_3 and between v_2 and v_4. To avoid crossing each other, one of these edges must go inside the 4-cycle and the other outside, as shown in the following figure. The net result so far is that there are four 3-sided regions, each with three vertices on its boundary. Thus, no matter which region is to contain the vertex v_5, that vertex cannot be joined to more than three other vertices without crossing the boundary.

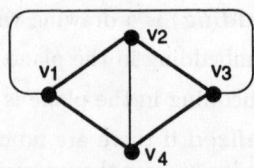

5. *Nonplanarity of $K_{3,3}$*: To form a planar drawing of the complete bipartite graph on the parts $\{v_1, v_3, v_5\}$ and $\{v_2, v_4, v_6\}$, one might as well start by drawing the 6-cycle $v_1, v_2, v_3, v_4, v_5, v_6$, which separates the plane. Next draw the edges between v_1 and v_4

and between v_2 and v_5. To avoid crossing each other, one of these edges must go inside the 6-cycle and the other outside. The net result so far is shown in the following figure. It is now clear that v_3 and v_6 cannot be joined without crossing some other edge.

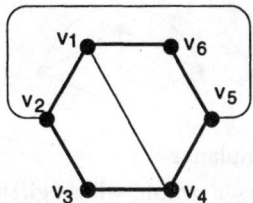

6. *Civil engineering*: Suppose that a number of towns are to be joined by a network of highways. If the network is planar, then the cost of bridges for underpasses and overpasses can be avoided.

7. *Electrical networks*: A planar electrical network with bare wires joining the nodes can be placed directly onto a flat board. Otherwise, insulation would be needed to prevent short circuits at wire crossings.

8.7.2 NUMERICAL PLANARITY CRITERIA

Certain numerical relationships are true of all planar graphs. One way to show that a graph is nonplanar is to show that it does not satisfy one of these relations.

Definitions:

A *region* of an imbedded graph is, informally, a piece of what results when the surface is cut open along all the edges. From a formal topological viewpoint, it is a maximal subsurface containing no vertex and no part of any edge of the graph.

The **boundary of a region** R of an imbedded graph is the subgraph containing all vertices and edges incident on R. It is denoted ∂R.

A *face* of an imbedded graph is a region plus its boundary.

The *exterior region* of a planar graph drawing is the region that extends to infinity.

The *girth* of a graph is the number of edges in a shortest cycle. The girth is undefined if the graph has no cycles.

Facts:

1. *Euler polyhedral equation*: Let $G = (V, E)$ be a connected graph imbedded in the plane with face set F. Then $|V| - |E| + |F| = 2$.

2. *Edge-face inequality*: Let $G = (V, E)$ be a simple, connected graph imbedded in a surface with face set F. Then $2|E| \geq 3|F|$.

3. *Edge-face inequality (strong version)*: Let $G = (V, E)$ be a connected graph, but not a tree, imbedded in a surface with face set F. Then $2|E| \geq \text{girth}(G) \cdot |F|$.

4. Let $G = (V, E)$ be a simple, connected graph. If G is planar then $3|V| - |E| \geq 6$.

5. Let $G = (V, E)$ be a connected graph that is not a tree. If G is planar then $(|V| - 2) \cdot \text{girth}(G) \geq |E| \cdot (\text{girth}(G) - 2)$.

6. Let $G = (V, E)$ be a simple, connected, bipartite graph that is not a tree. If G is planar then $|E| \leq 2 \cdot |V| - 4$.

Examples:

1. In the planar imbedding of the following figure, $|V| = 4$, $|E| = 6$, and $|F| = 4$. Thus, $|V| - |E| + |F| = 4 - 6 + 4 = 2$. (The "exterior" region counts as a face.)

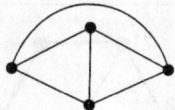

2. Fact 4 implies that K_5 is nonplanar.
3. Fact 5 implies that the Petersen graph, whose girth equals 5, is nonplanar.
4. Fact 6 implies that $K_{3,3}$ is nonplanar.

8.7.3 PLANARITY ALGORITHM

Definitions:

A **bridge of a subgraph** H in a graph G is a maximal connected subgraph of G in which no vertex of H has degree greater than 1.

An **attachment of a bridge** B of a subgraph H in a graph G is a vertex of $B \cap H$. (That is, an attachment is a vertex in which the bridge meets the rest of the graph.)

Facts:

1. Call two edges in the complement of a subgraph H of a graph G "related" if they are both contained in a path in G that has no vertices of H in its interior. Then the bridges of H are the induced subgraphs on the equivalence classes of edges under this relation.
2. Informally, a bridge is a subgraph obtained from one of the "pieces" that result by deleting H from G by reattaching the endpoints to the edges that attach to H. See Example 1.
3. The time needed to test planarity by searching directly for subdivided copies of K_5 and $K_{3,3}$ is an exponential function of the number of vertices.
4. J. Hopcroft and R. Tarjan [1974] have developed a planarity-testing algorithm that can be executed in time proportional to the number of vertices ("linear time").
5. Algorithm 1 can be implemented to run in time approximately proportional to the square of the number of vertices ("quadratic time").
6. None of the linear-time planarity algorithms is easy to describe and implement. However, Algorithm 1 is easily implemented, and its running time is satisfactory for reasonably large graphs.

Example:

1. The following figure shows a subgraph and its three bridges: B_1, B_2, and B_3. The subgraph H is the dark cycle. The attachments of the bridges are the vertices along the dark cycle.

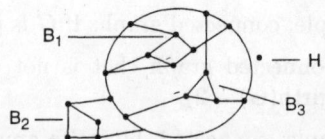

> **Algorithm 1:** Easy planarity-testing for graph G.
>
> input: a simple, connected graph G
>
> $G_0 :=$ an arbitrary cycle in G; draw G_0 in the plane; $j := 0$
> {Grow a sequence of nested subgraphs $G_0, G_1, \ldots$ until all of G has been drawn in the plane; if this does not happen, then G is nonplanar}
>
> **while** $G_j \neq G$ {this possible exit implies G is planar} **and** $(\forall B \in \text{bridges}\,(G_j))$
> $(\forall v \in \text{attachments}\,(B))(\exists \text{ region } R \text{ of } G_j \text{ in plane})\; v \in \partial R$ {this possible exit implies G is nonplanar} **do**
> {While-loop body says how to grow subgraph G_{j+1}}
> **if** $(\exists B \in \text{bridges}\,(G_j))(\forall v \in \text{attachments}\,(B))(\exists! \text{ region } R \text{ of } G_j)\, [v \in \partial R]$
> **then** {case 1 — a forced move exists}
>   select a path P between two attachments of B
>   obtain subgraph G_{j+1} by drawing path P in region R
> **else** {case 2 — no forced move exists}
>   select any bridge, and find two regions for its attachments
>   select any path between two attachments of that bridge
>   draw that path into either region to obtain G_{j+1}
> $j := j + 1$

2. Suppose that the figure in Example 1 occurred in the execution of Algorithm 1. At the next iteration of the while-loop body, suppose that bridge B_2 is selected, and suppose that a path in B_2 is drawn outside the dark cycle. Then, on the following iteration of the while-loop body, bridge B_3 would be a forced choice, and a path from B_3 would have to be drawn inside the dark cycle. Eventually, bridge B_1 would have to be drawn outside the dark cycle, thereby yielding a planar drawing of the entire graph.

8.7.4 CROSSING NUMBER AND THICKNESS

Definitions:

The **crossing number** of graph G, denoted $\nu(G)$, is the minimum number of edge-crossings possible in a normalized drawing of G in the plane.

The **thickness** of graph G, denoted $\theta(G)$, is the minimum number of planar graphs whose union is G.

Facts:

1. $\nu(K_n) \leq \frac{1}{4} \cdot \lfloor \frac{n}{2} \rfloor \cdot \lfloor \frac{n-1}{2} \rfloor \cdot \lfloor \frac{n-2}{2} \rfloor \cdot \lfloor \frac{n-3}{2} \rfloor$.

2. For all integers $n \leq 10$, $\nu(K_n) = \frac{1}{4} \cdot \lfloor \frac{n}{2} \rfloor \cdot \lfloor \frac{n-1}{2} \rfloor \cdot \lfloor \frac{n-2}{2} \rfloor \cdot \lfloor \frac{n-3}{2} \rfloor$.

3. R. Guy has conjectured that the equation of Fact 2 holds for all positive integers.

4. $\nu(K_{m,n}) \leq \lfloor \frac{m}{2} \rfloor \cdot \lfloor \frac{m-1}{2} \rfloor \cdot \lfloor \frac{n}{2} \rfloor \cdot \lfloor \frac{n-1}{2} \rfloor$.

5. For all integers m and n such that $\min(m,n) \leq 6$, D. Kleitman proved that $\nu(K_n) = \lfloor \frac{m}{2} \rfloor \cdot \lfloor \frac{m-1}{2} \rfloor \cdot \lfloor \frac{n}{2} \rfloor \cdot \lfloor \frac{n-1}{2} \rfloor$.

6. *Zarankiewicz's conjecture*: The equation of Fact 5 holds for all positive integers m and n.

7. $\theta(K_n) \geq \left\lfloor \frac{n+7}{6} \right\rfloor$.

8. $\theta(Q_n) = \left\lfloor \frac{n+1}{4} \right\rfloor$.

9. $\theta(G) \geq \left\lfloor \frac{|E|}{3|V|-6} \right\rfloor$ for all simple graphs.

Examples:

1. Fact 1 implies that $\nu(K_6) \leq 3$. Thus, it is possible to draw K_6 with at most three crossings.

2. *Computer engineering*: Facts 7 and 8 yield lower bounds for the minimum number of layers needed for a multi-layer layout of an electronic interconnection network whose architecture is a complete graph or a hypercube graph, respectively.

8.7.5 STEREOGRAPHIC PROJECTION

Definitions:

A continuous one-to-one function from one subset of Euclidean space onto another is a *topological equivalence* if its inverse is continuous. (Informally, this means that either subset could be reshaped into the other without tearing, but only by compressing, stretching, and twisting.)

The *stereographic projection* (due to Bernhard Riemann, 1826–1866) adds a single point to a plane and thereby closes the "hole at infinity" and converts it into a sphere, as follows:
- start with a sphere in 3-space, tangent at its south pole S to the plane $z = 0$ at the origin $(0,0,0)$, as shown in the following figure;
- through each point x of the sphere draw a ray from the north pole N, extending to the point $f(x)$ at which it meets the plane.

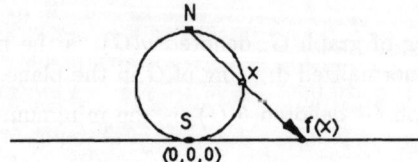

Facts:

1. The correspondence $x \rightarrow f(x)$ from the sphere minus its north pole onto the plane is a topological equivalence. In other words, the sphere minus a point could be stretched apart at the missing point and flattened out so that it covers the plane.

2. Any planar imbedding can be transformed into an imbedding in the sphere, which is a closed surface, by using the inverse of stereographic projection and closing up the pinhole. This eliminates the inelegant nuisance of having one "special" region with a hole.

8.7.6 GEOMETRIC DRAWINGS

Geometric drawing of graphs is a topic in computational geometry. Unlike ordinarily planarity and topological graph theory, its concerns include the exact coordinates in the plane of the images of the vertices and the edges.

Definitions:

A *straight-line drawing* of a graph is a drawing in which each edge is represented by a single straight line segment.

An *orthogonal drawing* of a graph is a drawing in which each edge is represented by a chain of horizontal and vertical line segments.

A *polyline drawing* of a graph is a drawing in which each edge is represented by a polygonal path, that is, by a chain of line segments with arbitrary slope.

A *bend* in a polyline drawing is a junction point of two line segments belonging to the same edge.

A *grid drawing* of a graph is a polyline drawing in which vertices, crossings, and bends have integer coordinates.

The *area of a graph drawing* is the area of the convex hull of the drawing.

A *distance-ranked partition* of a graph G with respect to a nonempty vertex subset S has cells C_j for $j = 0, 1, \ldots$. Vertex v is in cell C_j if and only if its shortest path to every vertex of S has length j.

A *distance-ranked drawing* of a graph G with respect to a nonempty vertex subset S has the cells of its distance-ranked partition organized into columns from left to right according to ascending distance from S.

Facts:

1. Straight-line and orthogonal drawings are special cases of polyline drawings.

2. Polyline drawings can approximate drawings with curved edges.

3. Computer programmed systems that support general polyline drawings are more complicated than systems that support only straight-line drawings.

4. Many graph drawing problems involve a trade-off between competing objectives, such as the desire to minimize both the area and the number of edge-crossings.

5. The area required for a planar polyline grid drawing of an n-vertex planar graph is $O(n^2)$.

6. The area required for a planar orthogonal grid drawing of an n-vertex planar graph is $O(n^2)$.

7. The area required for a planar straight line grid drawing of an n-vertex planar graph is $O(n^2)$.

8. Every planar graph of maximum degree 4 has an orthogonal planar drawing whose total number of bends is at most $2n + 2$.

9. Every planar graph of maximum degree 4 has an orthogonal planar drawing such that the maximum number of bends in an edge is at most 2.

Examples:

1. The following figure shows a nonplanar straight line drawing of the planar graph $K_6 - 3K_2$ and a planar polyline drawing of that same graph.

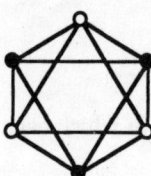 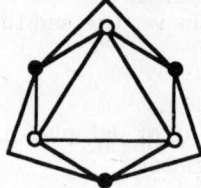

2. The following figure shows two orthogonal grid drawings of the graph $K_6 - 3K_2$. Whereas the lefthand drawing has two edges with three bends, the maximum number of bends in any edge of the middle drawing is two. The righthand drawing has the smallest total number of bends and the smallest area of the three drawings.

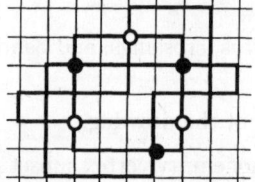 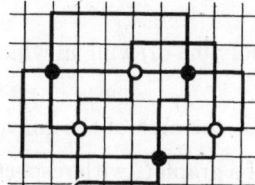 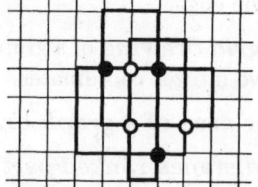

3. The following figure shows a distance-ranked drawing of the cube graph Q_3 with respect to the vertex 000.

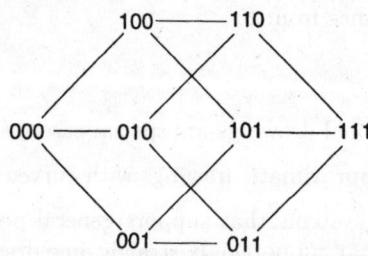

8.8 TOPOLOGICAL GRAPH THEORY

Topological graph theory mainly involves placing graphs on closed surfaces. Special emphasis is given to placements that are minimum with respect to some kind of cost or that are highly symmetric. Minimization helps to control the cost of manufacturing networks, and symmetry facilitates the task of routing information through a network.

8.8.1 CLOSED SURFACES

Holes in any surface can be closed off by operations like stereographic projection (§8.7.5). This enables topological graph theory to focus on drawings in closed surfaces.

Section 8.8 TOPOLOGICAL GRAPH THEORY

Definitions:

Adding a handle to a surface is accomplished in two steps (illustrated in Example 1):
- punch two disk-like holes into the surface;
- reclose the surface by installing a tube that runs from one hole to the other.

An **orientable surface** is defined recursively to be either the sphere S_0, or a surface that is obtained from an orientable surface by adding a handle. (See Example 2 for the construction.)

The **genus of an orientable surface** is the number of handles one must add to the sphere to obtain it. Thus, the surface obtained by adding g handles to S_0 has genus g. It is denoted S_g.

The **torus** is the surface S_1 of genus 1.

A **Möbius band** is the surface obtained by pasting the left side of a rectangular sheet to the right with a half-twist. A paper ring with a half-twist is a commonplace model of the Möbius band. (See Example 3.) (Augustus Ferdinand Möbius, 1790–1868)

Adding a crosscap to a surface is accomplished by the following two steps:
- punch one disk-like hole into the surface;
- reclose the hole by matching its boundary to the boundary of a Möbius band.

The **nonorientable surface** N_k is obtained by adding k crosscaps to the sphere. The sphere is sometimes regarded as the "surface with crosscap number 0" and denoted N_0, even though it is orientable. (See Example 4.)

The subscript k is called the **crosscap number** of the surface N_k.

The surfaces N_1 and N_2 are called the **projective plane** and the **Klein bottle**, respectively. (Felix Klein, 1849–1925)

Facts:

1. *Classification of closed surfaces*: Every closed surface is equivalent to exactly one of the surfaces S_g ($g \geq 0$) or N_k ($k \geq 1$).

2. Adding a handle to the nonorientable surface N_k is equivalent to adding two crosscaps. That is, the resulting surface is N_{k+2}.

3. If a loop is drawn around each handle of S_g and if these g loops are then cut open, the result is a (non-closed) surface that can be stretched and flattened out into a subset of the plane.

4. The subscript g equals the maximum number of closed curves on S_g that can be cut open without disconnecting that surface.

5. The subscript k equals the maximum number of closed curves on N_k that can be cut open without disconnecting that surface.

6. No nonorientable surface can be imbedded in $\mathcal{R}^3$.

7. *Network layouts*: The surfaces actually used for computer interconnection network layouts and other practical purposes rarely have graceful curved shapes, because among other reasons, that would obstruct miniaturization and ease of manufacture. Moreover, such surfaces usually have holes. However, the classification theorem and the closing of holes reduce the topology of the layout problems to placing graphs on closed surfaces.

Examples:

1. Adding a handle is achieved by punching two holes and connecting them with a tube, as illustrated.

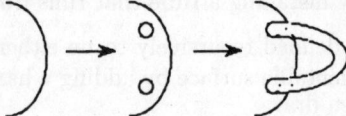

2. To construct the sequence of all orientable surfaces from the sphere S_0, each successive handle is added at the right of the previous surface.

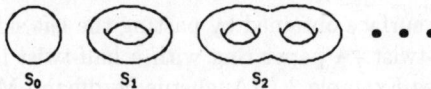

3. The Möbius band is constructed by giving a half-twist to a rectangular strip and then pasting the ends together.

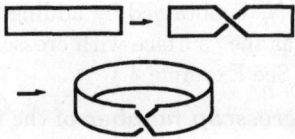

4. To construct the sequence of all nonorientable surfaces from the projective plane N_1, each successive crosscap is added at the right of the previous surface.

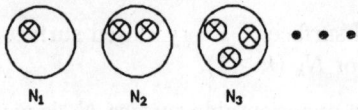

8.8.2 DRAWING GRAPHS ON SURFACES

Definitions:

A **flat polygon representation of a surface** S is a drawing of a flat polygon with markings to match the sides into pairs such that when the sides are pasted together as the markings indicate, the resulting surface S is obtained. (Certain special flat polygon representations are called **fundamental polygon representations**.) (See Example 1.)

An **imbedding** (or **embedding**) of a graph is a drawing with no edge-crossings.

A **face** of an imbedding means a region plus its boundary. The set of all faces is denoted F.

The **Euler characteristic of an imbedding** of a graph $G = (V, E)$ is the number $|V| - |E| + |F|$.

Section 8.8 TOPOLOGICAL GRAPH THEORY

A **flat-polygon drawing of a graph** on a surface has some graph edges drawn in two or more segments, so that one segment runs from one endpoint of an edge to a side of the flat polygon and another segment runs from the other endpoint to the corresponding position on the matched side of the flat polygon. Sometimes there are also some interior edge segments running between polygon sides. (Flat-polygon drawings are best used for small graphs.)

Imbedding modification (or "surgery") on a surface means adding handles and crosscaps to the surface and then drawing one or more edges that traverse the new handles and crosscaps.

Henri Poincarè (1854–1912) introduced a **duality construction** (see Example 3) as follows:

- insert into the interior of each (primal) face f a single dual vertex f^*;
- through each primal edge e draw a dual edge e^*; if edge e lies on the intersection of two primal faces f and f' (possibly $f = f'$), then the dual edge e^* joins the dual vertices f^* and f'^*;
- the **dual graph** is the graph $G^* = (\{f^* \mid f \in F\}, \{e^* \mid e \in E\})$;
- the **dual imbedding** is the resulting imbedding $G^* \to S$.

Facts:

1. Every closed surface has a flat polygon representation. This makes it possible to draw pictures of graph imbeddings in any surface.

2. *Euler polyhedral equation for orientable surfaces*: Let $G = (V, E)$ be a connected graph, cellularly imbedded (§8.8.3) into the surface S_g with face set F. Then
$$|V| - |E| + |F| = 2 - 2g = \chi(S_g).$$

3. *Euler polyhedral equation for nonorientable surfaces*: Let $G = (V, E)$ be a connected graph, cellularly imbedded into the surface N_k with face set F. Then
$$|V| - |E| + |F| = 2 - k = \chi(N_k).$$

4. *Edge-face inequality*: Let $G = (V, E)$ be a simple, connected graph imbedded in a surface with face set F. Then
$$2|E| \geq 3|F|.$$

5. *Edge-face inequality, strong version*: Let $G = (V, E)$ be a connected graph, but not a tree, imbedded in a surface with face set F. Then
$$2|E| \geq \text{girth}(G) \cdot |F|.$$

Examples:

1. Flat polygon representations of the double-torus and the torus are illustrated as follows:

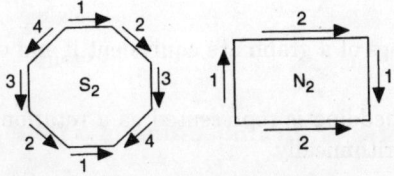

2. In the imbedding $K_5 \to S_1$ illustrated below, edges c and d cross through flat polygon sides 2 and 1, respectively. The "outer region" is actually 8-sided, with boundary circuit $\langle a, d, b, c, f, d, e, c \rangle$. Two pairs of sides of this region are pasted together. The appearance of this single region as four subregions at the corners of the flat polygon is a side-effect of the particular representation and not a true feature of the imbedding.

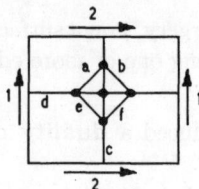

3. The Poincare duality construction is illustrated below.

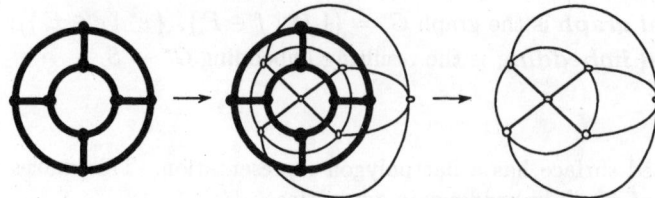

8.8.3 COMBINATORIAL REPRESENTATION OF GRAPH IMBEDDINGS

Definitions:

The *rotation* (*in "edge-format"*) at v is obtained from a flat-polygon drawing of a graph by the following sequence of steps:
- label one end of each edge $+$ and the other end $-$, or put an arrow on each edge so that the head faces the $+$ end;
- at each vertex, traverse a small circle centered at that vertex, and record the cyclically ordered list of edge-ends encountered; this list is the rotation.

The *vertex-format* of a rotation is obtained by replacing each edge-end in the edge-format by the vertex at the other end of that edge. The vertex format is used only for simple graphs.

A *rotation system* is a complete list of rotations, that is, one for every vertex. If the surface is orientable, it is assumed that the traversals of the small circles around the vertices are in a consistent direction, that is, all clockwise or else all counterclockwise.

An imbedding is *cellular* (or a "2-cell imbedding") if every region is planar and has connected boundary.

Facts:

1. Two cellular imbeddings of a graph are equivalent if and only if they have the same rotation system.

2. If a cellular graph imbedding is represented as a rotation system, then the regions can be reconstructed algorithmically.

Example:

1. An imbedding $K_4 \to S_1$ and both formats of its rotation system.

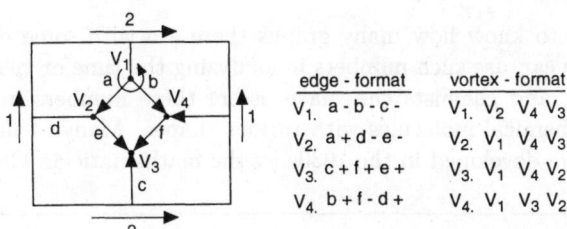

8.8.4 FORMULAS FOR GENUS AND CROSSCAP NUMBER

Definitions:

The **genus** $\gamma_{\min}(G)$ of a connected graph G is the minimum integer g such that there is an imbedding of G into the surface S_g.

The **crosscap number** $\overline{\gamma}_{\min}(G)$ is the minimum integer k such that there is an imbedding of G into N_k. Thus, a planar graph has crosscap number zero.

Facts:

1. The genus of any planar graph is 0.

2. $\gamma_{\min}(G) \geq \dfrac{|E| - 3|V| + 6}{6}$ if G is simple.

3. $\gamma_{\min}(G) \geq \dfrac{|E| - 2|V| + 4}{4}$ if G is simple and bipartite.

4. $\gamma_{\min}(K_n) = \left\lceil \dfrac{(n-3)(n-4)}{12} \right\rceil$. (Ringel and Youngs, 1968)

5. $\gamma_{\min}(K_{m,n}) = \left\lceil \dfrac{(m-2)(n-2)}{4} \right\rceil$. (Ringel, 1965)

6. $\gamma_{\min}(Q_n) = \left\lceil \dfrac{(m-2)(n-2)}{4} \right\rceil$. (Ringel, 1955)

7. $\overline{\gamma}_{\min}(G) \geq \left\lceil \dfrac{|E| - 3|V| + 6}{3} \right\rceil$ for every simple graph G.

8. $\overline{\gamma}_{\min}(G) \geq \left\lceil \dfrac{|E| - 2|V| + 4}{2} \right\rceil$ for every simple bipartite graph G.

9. $\overline{\gamma}_{\min}(K_n) = \left\lceil \dfrac{(n-3)(n-4)}{6} \right\rceil$, except that $\overline{\gamma}_{\min}(K_7) = 3$. (Ringel, 1959)

10. Many genus and crosscap number formulas can be derived by using *voltage graphs* or *current graphs* (§8.1.4). [GrTu87]

8.9 ENUMERATING GRAPHS

It is often valuable to know how many graphs there are with some desired property. Computer scientists can use such numbers in analyzing the time or space requirements of their algorithms, and chemists can make use of these numbers in organizing and cataloging lists of chemical molecules with various shapes. Many of the techniques for counting graphs were developed in the 1930s by the mathematician George Pólya.

8.9.1 COUNTING GRAPHS AND MULTIGRAPHS

Definitions:

A *labeled graph* is a graph with standard labels (commonly $v_1, v_2, \ldots, v_n$) assigned to the vertices. Two labeled graphs with the same set of labels are considered the same only if there is an isomorphism from one to the other that preserves the labels.

The *pair-permutation* $\gamma^{(2)}$ induced by a permutation γ on a set S is the permutation on the set of all subsets of S of size 2 defined by the rule $\gamma^{(2)}: \{x,y\} \mapsto \{\gamma(x), \gamma(y)\}$.

The *pair-action group* $\Gamma^{(2)}$ induced by a permutation group Γ on a set S is the group $\{\gamma^{(2)} \mid \gamma \in \Gamma\}$.

The *ordered-pair-permutation* $\gamma^{[2]}$ induced by a permutation γ on a set S is the permutation on the set $S \times S$ defined by the rule $\gamma^{[2]}: (x,y) \mapsto (\gamma(x), \gamma(y))$.

The *pair-action group* $\Gamma^{(2)}$ induced by a permutation group Γ on a set S is the group $\{\gamma^{[2]} \mid \gamma \in \Gamma\}$.

Facts:

1. The number of labeled simple graphs with n vertices and m edges is the binomial coefficient $\binom{\binom{n}{2}}{m}$.

2. For $m > \frac{\binom{n}{2}}{2}$, the number of labeled simple graphs with n vertices and m edges is the same as the number of labeled graphs with n vertices and $\binom{n}{2} - m$ edges.

3. The total number of labeled simple graphs with n vertices is $2^{\binom{n}{2}}$. See Table 1.

4. The number C_n of connected labeled simple graphs with n vertices can be determined from the following recurrence system. See Table 2.

$$C_1 = 1, \quad C_n = 2^{\binom{n}{2}} - \frac{1}{n}\sum_{i=1}^{n-1} i \binom{n}{i} 2^{\binom{n-i}{2}} C_i \quad \text{for } n > 1.$$

5. Most (unlabeled) graphical structures are counted with generating functions, using Burnside-Pólya enumeration (§2.6). In particular, the generating function for graphs with n vertices has the form

$$g_n(x) = \sum_{i=0}^{\binom{n}{2}} G_{n,i} \cdot x^i$$

where $G_{n,m}$ denotes the number of graphs with n vertices and m edges.

6. Pólya enumeration involves permutations (j) of the set $X_n = \{1, 2, \ldots, n\}$; j_k denotes the number of k-cycles in (j), for $k = 1, \ldots, n$. For example, if $(j) = (12)(34)(567)$, then $j_2 = 2, j_3 = 1$, and $j_1 = j_4 = j_5 = j_6 = j_7 = 0$.

Section 8.9 ENUMERATING GRAPHS 581

Table 1 Labeled graphs with n vertices and m edges.

$m \backslash n$	1	2	3	4	5	6	7	8
0	1	1	1	1	1	1	1	1
1		1	3	6	10	15	21	28
2			3	15	45	105	210	378
3			1	20	120	455	1,330	3,276
4				15	210	1,365	5,985	20,475
5				6	252	3,003	20,349	98,280
6				1	210	5,005	54,264	376,740
7					120	6,435	116,280	1,184,040
8					45	6,435	203,490	3,108,105
9					10	5,005	293,930	6,906,900
10					1	3,003	352,716	13,123,110
11						1,365	352,716	21,474,180
12						455	293,930	30,421,755
13						105	203,490	37,442,160
14						15	116,280	40,116,600
total	1	2	8	64	1,024	32,768	2,097,152	268,435,456

Table 2 Connected labeled graphs with n vertices.

n	1	2	3	4	5	6	7	8
C_n	1	1	4	38	728	26,704	1,866,256	251,548,592

7. The *cycle index polynomial* $Z(S_n^{(2)})$ used for counting simple graphs is

$$Z(S_n^{(2)}) = \frac{1}{n!} \sum_{(j) \in S_n} \frac{n!}{\prod_k k^{j_k} j_k!} \prod_k a_k^{k\binom{j_k}{2}} (a_k a_{2k}^{k-1})^{j_{2k}} a_{2k+1}^{kj_{2k+1}} \prod_{r<s} a_{\text{lcm}(r,s)}^{\gcd(r,s) j_r j_s}.$$

Here $\text{lcm}(r,s)$ and $\gcd(r,s)$ are the least common multiple and greatest common divisor of r and s, respectively. The following lists explicit formulas for $Z(S_n^{(2)})$ for small values of n:

$$Z(S_1^{(2)}) = 1$$

$$Z(S_2^{(2)}) = a_1$$

$$Z(S_3^{(2)}) = \frac{1}{3!}(a_1^3 + 3a_1 a_2 + 2a_3)$$

$$Z(S_4^{(2)}) = \frac{1}{4!}(a_1^6 + 9a_1^2 a_2^2 + 8a_3^2 + 6a_2 a_4)$$

$$Z(S_5^{(2)}) = \frac{1}{5!}(a_1^{10} + 10a_1^4 a_2^3 + 20a_1 a_3^3 + 15a_1^2 a_2^4 + 30a_2 a_4^2 + 20a_1 a_3 a_6 + 24a_5^2)$$

$$Z(S_6^{(2)}) = \frac{1}{6!}(a_1^{15} + 15a_1^7 a_2^4 + 40a_1^3 a_3^4 + 60a_1^3 a_2^6 + 180a_1 a_2 a_4^3 + 120a_1 a_2 a_3^2 a_6$$
$$+ 144a_5^3 + 40a_3^5 + 120a_3 a_6^2).$$

Table 3 Graphs with n vertices and m edges.

$m \backslash n$	1	2	3	4	5	6	7	8
0	1	1	1	1	1	1	1	1
1		1	1	1	1	1	1	1
2			1	2	2	2	2	2
3			1	3	4	5	5	5
4				2	6	9	10	11
5				1	6	15	21	24
6				1	6	21	41	56
7					4	24	65	115
8					2	24	97	221
9					1	21	131	402
10					1	15	148	663
11						9	148	980
12						5	131	1,312
13						2	97	1,557
14						1	65	1,646
total	1	2	4	11	34	156	1,044	12,346

Table 4 Multigraphs with n vertices and m edges.

$m \backslash n$	1	2	3	4	5	6
0	1	1	1	1	1	1
1		1	1	1	1	1
2		1	2	3	3	3
3		1	3	6	7	8
4		1	4	11	17	21
5		1	5	18	35	52
6		1	7	32	76	132
7		1	8	48	149	313
8		1	10	75	291	741
9		1	12	111	539	1,684
10		1	14	160	974	3,711

8. The generating function $g_n(x)$ for counting n-vertex graphs by number of edges is obtained from the cycle index $Z(S_n^{(2)})$ by replacing each variable a_i with $1 + x^i$. See Table 3.

9. The total number G_n of graphs with n vertices is obtained from the cycle index $Z(S_n^{(2)})$ by replacing each variable a_i with the number 2.

10. Asymptotically, the number G_n of n-vertex graphs satisfies $G_n \sim \frac{2^{\binom{n}{2}}}{n!}$.

11. The generating function $m_n(x) = \sum_i M_{n,i} x^i$ for counting n-vertex multigraphs according to their number of edges is obtained from the cycle index $Z(S_n^{(2)})$ by replacing each variable a_i with the infinite series $1 + x^i + x^{2i} + x^{3i} + \cdots$. See Table 4.

Examples:

1. The three different simple graphs with 4 vertices and 3 edges are given in the following figure. There are 4 essentially different ways to label each of the first two and 12 ways to label the third. Thus, there are 20 different labeled graphs with 4 vertices and 3 edges. The second and third graphs in this figure are connected.

2. There are six different multigraphs with 4 vertices and 3 edges, namely, the three graphs displayed in the previous figure plus the three additional multigraphs displayed in the following figure.

8.9.2 COUNTING DIGRAPHS AND TOURNAMENTS

Definitions:

A **digraph** (or **directed graph**) consists of a set V of vertices and a set A of arcs. When counting digraphs, two digraphs are considered the same if they are isomorphic.

A **labeled digraph** is a digraph in which standard labels such as $v_1, v_2, \ldots, v_n$ have been assigned to the vertices. Two labeled digraphs are considered the same only if there is an isomorphism from one to the other that preserves the labels.

A **tournament** is a digraph such that for each pair u, v of vertices, either there is an arc from u to v or an arc from v to u, but not both.

A tournament is **strong** (or **strongly connected**) if for each pair u, v of vertices, there exist directed paths from u to v and from v to u.

Facts:

1. The number of labeled digraphs with no loops that have n vertices and m arcs is $\binom{n(n-1)}{m}$.

2. For $m > n(n-1)$, the number of labeled digraphs with n vertices and m arcs is the same as the number of labeled digraphs with n vertices and $n(n-1) - m$ arcs.

3. The total number of labeled digraphs with n vertices is $2^{n(n-1)}$. See Table 5.

4. The number of labeled tournaments with n vertices is $2^{\binom{n}{2}}$, the same as the number of graphs with n vertices.

5. Like graphical structures, most (unlabeled) digraphical structures are counted with generating functions, using Burnside-Pólya enumeration. In particular, the generating function for digraphs with n vertices has the form

$$d_n(x) = \sum_{i=0}^{n(n-1)} D_{n,i} x^i$$

where $D_{n,m}$ denotes the number of digraphs with n vertices and m arcs.

Chapter 8 GRAPH THEORY

Table 5 Labeled digraphs with n vertices and m arcs.

$m \backslash n$	1	2	3	4	5
0	1	1	1	1	1
1		2	6	12	20
2		1	15	66	190
3			20	220	1,140
4			15	495	4,845
5			6	792	15,504
6			1	924	38,760
7				792	77,520
8				495	125,970
9				220	167,960
10				66	184,756
total	1	4	64	4,096	1,048,576

6. The *cycle index* $Z(S_n^{[2]})$ for counting digraphs is

$$Z(S_n^{[2]}) = \frac{1}{n!} \sum_{(j) \in S_n} \frac{n!}{\prod_k k^{j_k} j_k!} \prod_k a_k^{(k-1)j_k + 2k\binom{j_k}{2}} \prod_{r<s} a_{\text{lcm}(r,s)}^{2 \gcd(r,s) j_r j_s}.$$

The following lists explicit formulas for $Z(S_n^{[2]})$ for small values of n:

$Z(S_1^{[2]}) = 1$

$Z(S_2^{[2]}) = \frac{1}{2!}(a_1^2 + a_2)$

$Z(S_3^{[2]}) = \frac{1}{3!}(a_1^6 + 3a_2^3 + 2a_3^2)$

$Z(S_4^{[2]}) = \frac{1}{4!}(a_1^{12} + 6a_1^2 a_2^5 + 8a_3^4 + 3a_2^6 + 6a_4^3)$

$Z(S_5^{[2]}) = \frac{1}{5!}(a_1^{20} + 10a_1^6 a_2^7 + 20a_1^2 a_3^6 + 15a_2^{10} + 30a_4^5 + 20a_2 a_3^2 a_6^2 + 24a_5^4)$

$Z(S_6^{[2]}) = \frac{1}{6!}(a_1^{30} + 15a_1^{12} a_2^9 + 40a_1^6 a_3^8 + 45a_1^2 a_2^{14} + 90a_1^2 a_4^7 + 120a_2^3 a_3^4 a_6^2$
$+ 144a_5^6 + 15a_2^{15} + 90a_2 a_4^7 + 40a_3^{10} + 120a_6^5).$

7. The generating function $d_n(x)$ for counting n-vertex digraphs by number of arcs is obtained from the cycle index $Z(S_n^{[2]})$ by replacing each variable a_i with $1 + x^i$. See Table 6.

8. The total number D_n of digraphs with n vertices is obtained from the cycle index $Z(S_n^{[2]})$ by replacing each variable a_i with the number 2.

9. Asymptotically, D_n satisfies $D_n \sim \dfrac{2^{n(n-1)}}{n!}$.

Table 6 Digraphs with n vertices and m arcs.

$m \backslash n$	1	2	3	4	5
0	1	1	1	1	1
1		1	1	1	1
2		1	4	5	5
3			4	13	16
4			4	27	61
5			1	38	154
6			1	48	379
7				38	707
8				27	1,155
9				13	1,490
10				5	1,670
total	1	3	16	218	9,608

Table 7 Tournaments and strong tournaments with n vertices.

n	tournaments	strong tournaments
1	1	1
2	1	0
3	2	1
4	4	1
5	12	6
6	56	35
7	456	353
8	6,880	6,008
9	191,536	178,133
10	9,733,056	9,355,949
11	903,753,248	884,464,590
12	154,108,311,168	152,310,149,735

10. The number T_n of tournaments on n vertices is given by the formula

$$T_n = \frac{1}{n!} \sum_{(j)}' \frac{n!}{\prod_k k^{j_k} j_k!} 2^{D(j)},$$

where the sum is over all permutations (j) of X_n whose cycles are all of odd size, and where

$$D(j) = \frac{1}{2} \left(\sum_{r=1}^{n} \sum_{s=1}^{n} \gcd(r,s) j_r j_s - \sum_{k=1}^{n} j_k \right).$$

See Table 7.

11. Let $T(x) = x + x^2 + 2x^3 + 4x^4 + 12x^5 + 56x^6 + \cdots$ be the generating function for tournaments, from the formula of Fact 10. Then the generating function $S(x) = x + x^3 + x^4 + 6x^5 + 35x^6 + \cdots$ for strong tournaments can be computed from the relation $S(x) = \frac{T(x)}{1+T(x)}$. See Table 7. Note that there are no strong tournaments with exactly two vertices.

Examples:

1. The four digraphs with 3 vertices and 3 arcs are displayed in the following figure. The first three digraphs can each be labeled in six essentially different ways, while the fourth digraph can only be labeled in two essentially different ways. Thus there are 20 different labeled digraphs with 3 vertices and 3 arcs.

2. The four tournaments with 4 vertices are displayed in the following figure. Only the last tournament is strong.

 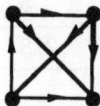  

8.10 ALGEBRAIC GRAPH THEORY

8.10.1 SPECTRAL GRAPH THEORY

Definitions:

The **characteristic polynomial of a graph** G is the characteristic polynomial $p(x)$ of its adjacency matrix A_G, that is, $p(x) = \det(xI - A_G)$.

An **eigenvector** (or **characteristic vector**) **of a matrix** A is a nonzero vector x such that $Ax = \lambda x$, for some value λ.

An **eigenvalue** (or **characteristic value**) **of a matrix** A is a number λ such that $Ax = \lambda x$, for some vector $x \neq 0$.

An **eigenvector of a graph** is an eigenvector of its adjacency matrix.

An **eigenvalue of a graph** is an eigenvalue of its adjacency matrix.

The **spectrum of a graph** is the spectrum of its adjacency matrix, i.e., the multiset of eigenvalues.

The **Laplacian** (or **admittance matrix**) **of a graph** G is the matrix $D_G - A_G$, where D_G is the diagonal matrix with the degree sequence of G on the diagonal and A_G is the adjacency matrix.

A graph G is **strongly regular with parameters** (n, k, r, s) if:
- $|V_G| = n$;
- G is k-regular;
- every adjacent pair of vertices is mutually adjacent to r other vertices;
- every pair of nonadjacent vertices is mutually adjacent to s other vertices.

By convention, strongly regular graphs are connected with at least one edge.

The **Hoffman polynomial of a graph** is a polynomial $p(x)$ of minimum degree such that $p(A_G) = J$, where A_G is the adjacency matrix and where J is the square matrix with every entry equal to 1.

A *cospectral pair of graphs* is a pair of nonisomorphic graphs that have the same spectrum.

Facts:
1. The eigenvalues of a graph are independent of the particular labeling of the vertices; thus, two isomorphic graphs have the same spectrum.
2. All the eigenvalues of a graph are real. This is a special case of the well-known linear algebra result that the eigenvalues of any Hermitian matrix are real.
3. From linear algebra, it follows that the characteristic polynomial of a graph G satisfies the equation $p(x) = \prod_{i=1}^{n}(x - \lambda_i)$, where $\lambda_1, \ldots, \lambda_n$ are the eigenvalues of G.
4. If a graph is connected, then its largest eigenvalue has multiplicity 1. This eigenvalue has a corresponding eigenvector with all positive entries, and it is the only such eigenvector.
5. If λ is the largest eigenvalue of a graph and μ is another eigenvalue, then $\lambda \geq |\mu|$; moreover, $-\lambda$ is an eigenvalue if and only if the graph is bipartite.
6. A graph is bipartite if and only if its spectrum is symmetric with respect to 0, that is, λ is an eigenvalue if and only if $-\lambda$ is also an eigenvalue.
7. The largest eigenvalue of a k-regular graph is k, and it has multiplicity equal to the number of connected components. The sum of the coordinates of an eigenvector corresponding to any other eigenvalue is 0.
8. The (i,j)th entry of the kth power A_G^k of the adjacency matrix of a graph G is the number of walks of length k starting at vertex v_i and terminating at v_j.
9. If $\lambda_1, \lambda_2, \ldots, \lambda_n$ are the eigenvalues of a graph G, then $\sum_{i=1}^{n} \lambda_i^2 = 2|E_G|$ where E_G is the edge-set of G. Also, $\sum_{i=1}^{n} \lambda_i^3 = 6T$ where T is the number of triangles in G.
10. If $p(x) = x^n + a_{n-1}x^{n-1} + a_{n-2}x^{n-2} + \cdots + a_1 x + a_0$ is the characteristic polynomial of a graph G, then $a_{n-1} = 0$, $-a_{n-2}$ is the number of edges, and $-a_{n-3}$ is the twice number of triangles.
11. The set of eigenvalues of the disjoint sum $G + H$ is the union of the sets of eigenvalues of G and H. The multiplicity of λ as an eigenvalue of $G + H$ is the sum of the multiplicity of λ as an eigenvalue of G and the multiplicity of λ as an eigenvalue of H.
12. The eigenvalues of the cartesian product $G \times H$ are $\{\lambda_i + \lambda_j \mid \lambda_i$ an eigenvalue of G and λ_j an eigenvalue of $H\}$. The multiplicity of $\lambda_i + \lambda_j$ as an eigenvalue of $G \times H$ is the product of the multiplicity of λ_i as an eigenvalue of G and λ_j as an eigenvalue of H.
13. If G is a k-regular graph and $\overline{G}$ is its complement, then $\lambda < k$ is an eigenvalue of G if and only if $-\lambda - 1$ is an eigenvalue of $\overline{G}$. In this case λ and $-\lambda - 1$ have the same multiplicities.
14. If λ is an eigenvalue of G with multiplicity m, then $-\lambda - 1$ is an eigenvalue of $\overline{G}$ with multiplicity $m - 1$, m, or $m + 1$.
15. If G has n vertices and $\lambda_1 \geq \lambda_2 \geq \cdots \geq \lambda_n$ as eigenvalues, and H is an induced subgraph with $n - 1$ vertices and eigenvalues $\mu_1 \geq \mu_2 \geq \cdots \geq \mu_{n-1}$, then $\lambda_1 \geq \mu_1 \geq \lambda_2 \geq \mu_2 \geq \cdots \geq \mu_{n-1} \geq \lambda_n$.
16. The eigenvalues of a line graph $L(G)$ are greater than or equal to -2. Equality is attained unless every connected component of G is a tree or has exactly one circuit, that circuit being odd.

17. If G is a k-regular graph with $\lambda > -k$ as an eigenvalue, then $\lambda + k - 2$ is an eigenvalue of $L(G)$.

18. A graph has a Hoffman polynomial if and only if it is regular and connected.

19. A regular connected graph has exactly three distinct eigenvalues if and only if it is strongly regular.

20. *Matrix-tree theorem*: If M_i is formed by deleting the i-th row and column from the Laplacian of G, then $\det(M_i)$ is independent of the choice of i and is equal to the number of spanning trees of G.

Examples:

1. The edgeless graph N_n on n vertices has one eigenvalue, namely 0 with multiplicity n.

2. The eigenvalues of the complete graph K_n are $n-1$ and -1 with respective multiplicities of 1 and $n-1$. For instance, the characteristic polynomial of K_4 is
$$\begin{vmatrix} x & -1 & -1 & -1 \\ -1 & x & -1 & -1 \\ -1 & -1 & x & -1 \\ -1 & -1 & -1 & x \end{vmatrix} = (x+1)^3(x-3).$$

3. The eigenvalues of the complete bipartite graph $K_{m,n}$ are $\sqrt{mn}$, 0, and $-\sqrt{mn}$, with respective multiplicities of 1, $mn-2$, and 1.

4. The eigenvalues of the Petersen graph are 3, 1, and -2, with respective multiplicities 1, 5, and 4.

5. The eigenvalues of the n-path P_n are $\{2\cos\frac{k\pi}{n+1} \mid k = 1, 2, \ldots, n\}$, each with multiplicity 1.

6. The eigenvalues of the n-cycle C_n, are $\{2\cos\frac{2k\pi}{n} \mid k = 1, 2, \ldots, n\}$. The eigenvalue 2, and the eigenvalue -2 when n is even, have multiplicity 1; all other eigenvalues have multiplicity 2.

7. The eigenvalues of the hypercube Q_d are $d, d-2, d-4, \ldots, -d+2, -d$, with respective multiplicities $\binom{d}{0}, \binom{d}{1}, \binom{d}{2}, \ldots, \binom{d}{d-1}, \binom{d}{d}$.

8. The eigenvalues of the line graph $L(K_n)$ are $2n-4$, $n-4$, and -2, with respective multiplicities 1, $n-1$, and $\frac{n(n-3)}{2}$.

9. The eigenvalues of the line graph $L(K_{m,n})$ are $m+n-2$, $m-2$, $n-2$, and -2, with respective multiplicities 1, $n-1$, $m-1$, and $(m-1)(n-1)$.

10. If G is strongly regular with parameters (n, k, r, s), then its eigenvalues are k and $\frac{1}{2}\left(r - s \pm \sqrt{(r-s)^2 - 4(s-k)}\right)$.

11. The smallest pair of cospectral graphs is $K_{1,4}$ and $C_4 + K_1$, each of which has spectrum $\{-2, 0, 0, 0, 2\}$. See the following figure. Observe that $K_{1,4}$ is connected and that $C_4 + K_1$ is not, and that the two graphs have different degree sequences. This implies that connectedness and degree sequences cannot be determined from spectral properties alone.

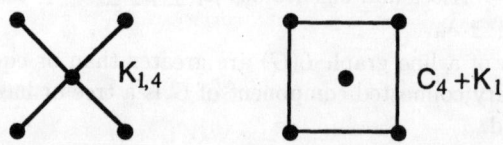

8.10.2 AUTOMORPHISMS OF GRAPHS

Definitions:

The *automorphism group $Aut(G)$ of a graph* G is the set of all automorphisms of graph G, under the operation of functional composition

A *generating subset for a group* Γ is a subset Σ of group elements such that every group element is a product of elements of Σ. (*Note*: The group identity is the empty product.)

The *Cayley digraph for group* Γ *and generating set* Σ has as vertices the elements of Γ, with an arc σ_γ from the vertex γ to the vertex γ' if and only if $\gamma\sigma = \gamma'$.

The *Cayley graph for group* Γ *and generating set* Σ is the graph obtained by removing all arc directions from the Cayley digraph.

The *Cayley graph for group* Γ *and generating set* Σ (alternative definition) is the graph obtained by removing all arc directions from the Cayley digraph, and by collapsing each pair of arcs corresponding to a generator of order two to a single edge.

Facts:

1. A simple graph G and its edge-complement $\overline{G}$ have the same automorphism group.

2. An automorphism φ of a graph G induces an automorphism $\overline{\varphi}$ on the line graph $L(G)$.

3. If G is a connected simple graph with at least 4 vertices, then G and its line graph $L(G)$ have isomorphic automorphism groups.

4. If the G graph has adjacency matrix A, and if the permutation φ of V_G has permutation matrix P, then φ is the vertex map of an automorphism of G if and only if $PA = AP$.

5. If all eigenvalues of a graph G have multiplicity 1, then every automorphism has order at most 2.

6. *Frucht's theorem*: Let Γ be any finite group. Then there exists a graph G whose automorphism group is isomorphic to Γ. It can be constructed by modifying a Cayley digraph for Γ.

Examples:

1. The n-vertex edgeless graph N_n and the complete graph K_n both have the symmetric group S_n as their automorphism group. They are the only n-vertex graphs with this automorphism group.

2. The automorphism group of the complete bipartite graph $K_{m,n}$ is $S_n \times S_m$ if $n \neq m$ and is the wreath product [Ro88] $S_n \wr S_2$ if $m = n$.

3. The automorphism group of the n-path graph P_n (with $n > 1$) is isomorphic to S_2.

4. The automorphism group of the n-cycle graph C_n is the dihedral group D_n of order $2n$. For instance, the 4-cycle C_4 with vertices a, b, c, and d (in cyclic order), has the following vertex automorphisms:

$$(a)(b)(c)(d) \quad (a\ b\ c\ d) \quad (a\ c)(b\ d) \quad (a\ d\ c\ b)$$
$$(a\ b)(c\ d) \quad (a\ d)(b\ c) \quad (a)(c)(b\ d) \quad (b)(d)(a\ c).$$

5. The Cayley digraph of the group S_3 with generating set $\{(1\ 2\ 3), (1\ 2)\}$ is illustrated in the following figure.

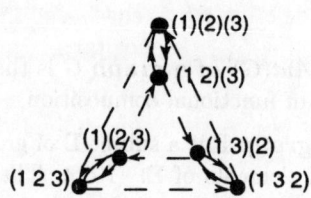

8.11 ANALYTIC GRAPH THEORY

Analytic graph theory involves three different perspectives on the properties of graphs that are sufficiently "dense". One analysis is what must happen in a simple n-vertex graph when the number of edges is sufficiently large. A second analysis is what must happen in at least one of the parts of a partition of the edges of a graph. The third analysis is what happens with a high probability when a graph is randomly chosen according to some distribution.

8.11.1 EXTREMAL GRAPH THEORY

Extremal graph theory is the analysis of the number of edges an n-vertex simple graph must have in order to guarantee that it contains a certain graph or type of graph. Elsewhere, it is sometimes taken to be the study of graph-theoretic inequalities in general.

Definitions:

The **extremal number** $ex(\mathcal{G}; n)$ for a set $\mathcal{G}$ of graphs is the greatest number of edges in any simple graph with n vertices that does not contain some member of $\mathcal{G}$ as a subgraph.

Notation: The notation $ex(G; n)$ is used when $\mathcal{G}$ consists of just one graph G.

An **extremal graph** for a set $\mathcal{G}$ of graphs and an integer n is a graph with n vertices and $ex(\mathcal{G}; n)$ edges that contains no member of $\mathcal{G}$.

The **Turán graph** $T_k(n)$ is the n-vertex k-partite simple graph with the maximum number of edges.

The **Turán number** $t_k(n)$ is the number of edges in the Turán graph $T_k(n)$.

Facts:

1. If $ex(\mathcal{G}; n) = \binom{n}{2}$, then no graph with n vertices contains any member of $\mathcal{G}$.
2. The Turán graph $T_k(n)$ is the unique complete k-partite graph with the property that the numbers of vertices in any two of its parts differ by at most 1. In the special case $k = 2$, $T_2(n) = K_{\lfloor n/2 \rfloor, \lceil n/2 \rceil}$. More generally, if $n = tk + r$, where $0 \leq r < k$, then there are r parts of size $t + 1$ and $k - r$ parts of size t.
3. The Turán number $t_k(n)$ equals $\binom{n}{2} + 1 - \frac{t(p-k+r)}{2}$, where $n = tk + r$, with $0 \leq r < k$. If $k = 2$, this greatly simplifies: $t_2(n) = \lfloor \frac{n}{2} \rfloor \lceil \frac{n}{2} \rceil = \lfloor \frac{n^2}{4} \rfloor$.
4. *Turán's theorem*: $ex(K_k; n) = t_{k-1}(n)$; furthermore, $T_{k-1}(n)$ is the only extremal graph for K_k and n.

5. Let $\chi = \chi(G)$ (chromatic number of G, §8.6), $p = |G|$, $c = 2 - \frac{1}{p}$. Then $ex(G;n) = \left(1 - \frac{1}{\chi-1}\right)\binom{n}{2} + O(n^c)$. Furthermore, all the extremal graphs differ from the Turán graph $T_{\chi-1}(n)$ by adding and deleting $O(n^c)$ edges, and the minimum degree of all such graphs is $(1 - \frac{1}{\chi-1})n + O(n^c)$. (Erdős, Simonovits)

6. Fact 5 is also true for $ex(\mathcal{G};n)$, where χ is the smallest chromatic number among the members of $\mathcal{G}$, p is the smallest order among these members, and $c = 2 - \frac{1}{p}$.

7. $ex(\mathcal{G};n) = O(n)$ if and only if $\mathcal{G}$ contains a (tree or) forest.

8. There exists a number t_0 such that, for $t > t_0$, every tree T of order t satisfies the inequality $ex(T;n) \leq \frac{n(t-2)}{2}$ for every $n \geq t+1$.

9. $ex(C_4;n) = \frac{1}{2}(n^{3/2}) + O(n^{4/3})$. (The exponent $\frac{4}{3}$ can be slightly improved.)

10. $ex(C_{2m};n) = O(n^{1+1/k})$. This is known to be sharp only for $2m = 4, 6, 10$, but is conjectured to be sharp for all m.

11. The ratio $\frac{ex(\mathcal{G};n)}{\binom{n}{2}}$ is monotone nonincreasing; that is, for every set $\mathcal{G}$ and for all $m \leq n$, $\frac{ex(\mathcal{G};m)}{\binom{m}{2}} \geq \frac{ex(\mathcal{G};n)}{\binom{n}{2}}$.

12. The following table summarizes many other facts that apply as the number of edges grows:

# edges	what must occur, but not for smaller # edges	what must occur if n is large enough
n	some cycle	
$\lfloor \frac{3n-1}{2} \rfloor$	some even cycle	
$3n - 5$	two disjoint cycles	
$t_2(n) + 1 = \lfloor \frac{n^2}{4} \rfloor + 1$	some odd cycle (i.e., $\chi \geq 3$), $C_3, \ldots, C_{\lfloor (n+3)/2 \rfloor}$	$K_{s,s} + e$ for fixed s
$t_2(n) + m$, m fixed		$m \lfloor \frac{n}{2} \rfloor$ copies of C_3, for fixed s $K_{s,s}$ plus m extra edges
$t_k(n) + 1$	K_k; also, $\chi \geq k$	
$t_k(n) + m$, m fixed		for fixed s, $K_{s;n}$ plus m extra edges
$\binom{n}{2} - n + 3$	a Hamilton cycle	

Examples:

1. $ex(K_2;n) = 0$. The extremal graph is the edgeless graph.

2. $ex(P_2;n) = \lfloor \frac{n}{2} \rfloor$. The extremal graph is the maximum matching.

3. $ex(K_{1,r};n) = \lfloor \frac{(r-1)n}{2} \rfloor$. If $(r-1)n$ is even, then any $(r-1)$-regular graph is an extremal graph. If $(r-1)n$ is odd, then any graph with one vertex of degree $r-2$ and all the others of degree $r-1$ is extremal.

4. $ex(K_3;n) = \lfloor \frac{n^2}{4} \rfloor$. The Turán graph $T_2(n)$ is the only extremal graph.

5. The Turán graph $T_3(10)$ is the 3-partite graph $K_{3,3,4}$. It has 33 edges, which is more than any other 3-partite graph on 10 vertices. Thus, $ex(K_4; 10) = 33$.

8.11.2 RAMSEY THEORY FOR GRAPHS

If the edges of a "dense" graph are partitioned into two parts, then at least one of the parts must still be fairly dense. Ramsey theory, which can also be studied in connection with many mathematical objects other than graphs, relies on this idea. (Also see §3.1.6.)

Definitions:

The (*classical*) **Ramsey number** $r(m,n)$ is the smallest positive integer k such that every k-vertex graph contains either the complete graph K_m or n mutually nonadjacent vertices.

The **Ramsey number** $r(G,H)$ is the smallest positive integer k such that, if the edges of K_k are bipartitioned into red and blue classes, then either the red subgraph contains a copy of G or else the blue subgraph contains a copy of H. Sometimes $r(G)$ denotes $r(G,G)$.

The **Ramsey number** $r(G_1,\ldots,G_s)$ is the smallest number k such that in any s-fold partition of the edgeset of K_k, there is an index j such that the jth part contains the graph G_j.

A ***k*-canonical coloring** of a complete graph is an edge-coloring in which the vertices can be partitioned into k or fewer parts, such that the color of each edge depends only on the two parts to which its endpoints belong.

The **arrows notation** $F \to (G,H)$ ("F arrows (G,H)") means that if the edges of the graph F are partitioned into two chromatic classes, e.g., into red edges and blue edges, then either the red subgraph contains a copy of G or else the blue subgraph contains a copy of H. When $G=H$, the notation $F \to G$ is often used. The notation $F \to (G_1,\ldots,G_k)$ means that k edge colors are involved.

Facts:

1. $r(K_m, K_n) = r(m,n)$ for all $m, n \geq 1$.
2. $r(G,H) = r(H,G)$. That is, Ramsey numbers are symmetric.
3. $r(K_n, K_1) = r(K_1, K_n) = 1$ for every $n \geq 1$.
4. $r(K_n, K_2) = r(K_2, K_n) = n$ for every $n \geq 1$.
5. $r(K_m, K_n) \leq r(K_m, K_{n-1}) + r(K_{m-1}, K_n)$ for all $m, n \geq 2$.
6. $r(K_m, K_n) \leq \binom{m+n-2}{m-1}$. (Erdős and Szekeres, 1935)
7. If $n \geq 3$, then $2^{n/2} \leq r(K_n, K_n) \leq \binom{2n+2}{n+1} < 4^{n+1}$.
8. $\frac{\sqrt{2}}{e}(1+o(1))n2^{n/2} \leq r(K_n, K_n) \leq \binom{2n+2}{n+1} \cdot O((\log n)^{-1})$.
9. There exist constants c_1 and c_2 such that $c_1 n \ln n \leq r(K_3, K_n) \leq c_2 n \ln n$.
10. A 1-canonical coloring assigns every edge the same color.
11. A 2-canonical coloring consists of two complete edge-monochromatic subgraphs, such that all edges joining them are of the same color.
12. If $\chi(G) = \chi$ and $|V_H| = n$, then $r(G,H) \geq (\chi-1)(n-1)+1$. This fact is based on a $(\chi-1)$-canonical coloring.
13. If T is an n-vertex tree, then $r(K_m, T) = (m-1)(n-1)+1$. In other words, the lower bound in the immediately preceding fact determines the Ramsey number.
14. Except for $r(C_3, C_3) = r(C_4, C_4) = 6$, $r(C_m, C_n)$ and $r(P_m, C_n)$ are determined by the best possible 2-canonical colorings, which are easy to find.

15. For every choice of graphs $G_1, G_2, \ldots, G_k$, there exists a graph F for which $F \to (G_1, \ldots, G_k)$. In particular, the Ramsey number $r(G_1, \ldots, G_k)$ is well-defined.

16. If $m, n \geq 3$, the values of only nine Ramsey numbers are known:

$$r(3,3) = 6 \qquad r(3,4) = 9 \qquad r(3,5) = 14$$
$$r(3,6) = 18 \qquad r(3,7) = 23 \qquad r(3,8) = 28$$
$$r(3,9) = 36 \qquad r(4,4) = 18 \qquad r(4,5) = 25.$$

Estimates on some other Ramsey numbers are given in §3.1.6. In addition to the nine exact results, only one other nontrivial Ramsey number for complete graphs is known:

$$r(K_3, K_3, K_3) = 17.$$

8.11.3 PROBABILISTIC GRAPH THEORY

Probabilistic graph theory takes two basic directions. It studies random graphs for themselves, and it uses random graphs in deriving graph-theoretical results that are not themselves probabilistic.

Definitions:

In **Model 1**, the **random graph** $G_{n,p}$ has n distinctly labeled vertices $v_1, \ldots, v_n$, and the probability of any pair of vertices being joined by an edge is p, where all these edge probabilities are mutually independent.

In **Model 2**, the **random graph** $G_{n,e}$ has n distinctly labeled vertices $v_1, \ldots, v_n$, and exactly e edges, and each such labeled graphs occurs with the same probability $1/\binom{N}{e}$, where $N = \binom{n}{2}$.

Almost every (a. e.) **graph** has a given property P under either Model 1 or Model 2, if the probability that a random graph has property P approaches 1 as $n \to \infty$, where the probability p stays constant under Model 1, but where one must specify how e varies with n under Model 2. If neither model is explicitly specified, then Model 1 with $p = \frac{1}{2}$ is implicit, so that all labeled graphs on n vertices have the same probability $2^{-\binom{n}{2}}$.

Facts:
1. The number of labeled graphs in the probability space for Model 1 is $2^{\binom{n}{2}}$.

2. While Model 2 is sometimes considered to be more natural and easier to define, it is, in practice, usually easier to work with Model 1. Fortunately, Model 1 with $p = \frac{e}{N}$ behaves very similarly to Model 2 in most cases, so that facts about Model 1 usually lead easily to facts about Model 2 as well.

3. In Model 1, a graph with e edges occurs with probability $p^e (1-p)^{\binom{n}{2} - e}$. If $p = \frac{1}{2}$, then every labeled graph on n vertices has the same probability $2^{-\binom{n}{2}}$.

4. Random graphs can be used to prove theorems about graphs, especially existence theorems. (See Example 1.)

5. Let $b = \frac{1}{p}$ and $d = 2 \log_b \frac{en}{2 \log_b n} = 1$, where $e = 2.718 \ldots$, not the number of edges. Then for every positive $\epsilon < \frac{1}{2}$ the clique number of a. e. graph is either $\lfloor d - \epsilon \rfloor$ or $\lfloor d + \epsilon \rfloor$, where these two values are usually the same when ϵ is small. This means that the clique number is determined for a. e. graph, unless d is close to an integer, in which case there are two possible values.

> **Algorithm 1:** Generate random graph $G_{n,p}$ (per Model 1).
>
> initialize graph G with vertex list $v_1, v_2, \ldots, v_n$
> for $i := 1$ to $n-1$
> for $j := i+1$ to n
>   join vertices v_i and v_j with probability p

> **Algorithm 2:** Generate random graph $G_{n,e}$ (per Model 2).
>
> initialize graph G with vertex list $v_1, v_2, \ldots, v_n$
> generate random integer $r \in \left\{ 1, \ldots, \binom{\binom{n}{2}}{e} \right\}$
> convert r to an e-combination C in $\left\{ 1, \ldots, \binom{n}{2} \right\}$
> convert e-combination C to e edges in $G_{n,e}$

6. Almost every graph satisfies $\chi \approx \frac{n}{2\log_2 n}$.

7. In connection with the fact immediately preceding, it can be shown that if $p = n^{-\alpha}$ for fixed $\alpha > \frac{5}{6}$, then in Model 1, there exists an $f(n,p)$ so that for almost every graph, $f(n,p) \leq \chi \leq f(n,p) + 3$. That is, the chromatic number χ takes on one of only four possible values.

8. Almost every graph in Model 1 has its connectivity and its edge connectivity equal to its minimum degree. Furthermore, the common value of these three parameters is $pn - (2p(1-p)n \log n)^{\frac{1}{2}} + o(n \log n)^{\frac{1}{2}}$.

9. Generating a random graph $G_{n,p}$ under Model 1 is straightforward, as indicated by Algorithm 1.

10. To generate a random graph $G_{n,e}$ under Model 2, the possible edges are placed in bijective correspondence with the integers $1, \ldots, \binom{n}{2}$ according to the rule $f(i,j) = \binom{n}{2} - \binom{n-i+1}{2} + j$. Also, the e-combinations of the integers $1, \ldots, \binom{n}{2}$ are placed in bijective correspondence with the integers $1, \ldots, \binom{\binom{n}{2}}{e}$ according to the lexicographic ordering of those e-combinations (§2.2.5). These bijections facilitate the formulation of Algorithm 2.

11. For every fixed s, almost every graph contains the complete graph K_s. Moreover, for every fixed graph H, almost every graph contains H.

Table 1 Properties of almost every n-vertex graph.

p under Model 1	e under Model 2	property of almost every graph
$o(\frac{1}{n})$	$o(n)$	no cycles
$\frac{2c}{n}$, $0 < c < \frac{1}{2}$	cn, $0 < c < \frac{1}{2}$	cycles are possible, and the largest component has order $\approx \ln n$
$\frac{1}{n}$	$\frac{n}{2}$	some cycle exists, and the largest component has order $\Theta(n^{2/3})$
$\frac{2c}{n}$, $c > \frac{1}{2}$	cn, $c > \frac{1}{2}$	the largest component has order $c'n$
$\frac{c \ln n}{n}$, $c < 1$	$\frac{c}{2} n \ln n$, $c < 1$	the graph is disconnected
$\frac{c \ln n}{n}$, $c > 1$	$\frac{c}{2} n \ln n$, $c > 1$	the graph is connected and Hamiltonian

Example:

1. *Using random graphs to prove theorems*: Here is a proof that the Ramsey number $r(K_n, K_n)$ is greater than $2^{n/2}$ for all $n \geq 3$. Consider a random red-blue edge-coloring of K_N for some $N > n$ with $p(red) = \frac{1}{2}$. The probability that any given K_n occurring within this 2-colored K_N is entirely red is $2^{-\binom{n}{2}}$. Of course, the probability that it is colored blue is the same. Thus, the probability that the given subgraph K_n is monochromatic in either color is $2^{1-\binom{n}{2}}$. Since there are $\binom{N}{n}$ different copies of K_n in the colored K_N, the expected number of monochromatic K_n is $\binom{N}{n} \cdot 2^{1-\binom{n}{2}}$.

With the choice of $N = \lfloor 2^{n/2} \rfloor$, this expectation is $\binom{N}{n} \cdot 2^{1-\binom{n}{2}} < \frac{2^{1+n/2}}{n!} \cdot \frac{N^n}{2^{n^2/2}} < 1$, i.e., less than 1. Therefore there must be some coloring with no monochromatic K_n at all. This completes the proof.

8.12 HYPERGRAPHS

In ordinary graph theory, an edge of a simple graph can be regarded as a pair of vertices. In hypergraph theory, an "edge" can be regarded as an arbitrary subset of vertices. In this sense, hypergraphs are a natural generalization of graphs. Their systematic study was initiated by C. Berge. They have evolved into a unifying combinatorial concept.

8.12.1 HYPERGRAPHS AS A GENERALIZATION OF GRAPHS

Definitions:

A **hypergraph** $H = (V, E)$ is a finite set V of "vertices" together with a finite multiset E of "edges" (sometimes, "hyperedges"), which are arbitrary subsets of V.

The **order** of a hypergraph edge is its cardinality.

A **partial hypergraph** (or simply a **partial**) of the hypergraph $H = (V, E)$ is a hypergraph $H' = (V, E')$ such that $E' \subseteq E$. This generalizes a spanning subgraph.

A hypergraph $H = (V, E)$ is **simple** if E has no repeated edges.

The **incidence matrix** of a hypergraph $H = (V, E)$ with $E = \{e_1, e_2, \ldots, e_m\}$ and $V = \{x_1, x_2, \ldots, x_n\}$ is the $m \times n$ matrix $M(H) = [m_{i,j}]$ with

$$m_{i,j} = \begin{cases} 1 & \text{if } x_j \in e_i \\ 0 & \text{otherwise.} \end{cases}$$

The **dual hypergraph** of the hypergraph H is the hypergraph H^* whose incidence matrix is the transpose of the incidence matrix $M(H)$. This concept of duality from block design theory differs from the Poincaré dual of graph theory.

The **degree** $deg(x)$ of a hypergraph vertex x is the number of hypergraph edges containing x.

A hypergraph is **regular** if all vertices have the same degree. If t is the common value of the degrees, then the hypergraph is **t-regular**.

A hypergraph is **uniform** if all edges have the same number of vertices. If r is the common value, then the hypergraph is r-**uniform**.

The **complete hypergraph** K_n^* has all subsets of n vertices as edges, so that it has 2^n edges.

The **complete r-uniform hypergraph** K_n^r is the simple hypergraph of order n with all r-element subsets as edges, so that it has $\binom{n}{r}$ edges.

The **intersection graph** $I(H)$ of the hypergraph H is a simple graph whose vertices are the edges of H. Two vertices of $I(H)$ are adjacent if and only if the corresponding edges of H have nonempty intersection.

An **independent set** of vertices in a hypergraph is a set of vertices that does not (completely) contain any edge of the hypergraph.

Facts:

1. *How to draw a hypergraph*: First draw the vertices and the hyperedges of order 2, as if they were vertices and edges, respectively, of a graph. Then shade triangular regions corresponding to hyperedges of order 3. Higher order hyperedges and hyperedges of order 1 can be indicated by drawing enclosures around their vertices.

2. Every hypergraph satisfies the generalized Euler equation for degree-sum:
$$\sum_{x \in V} \deg(x) = \sum_{e \in E} |e|.$$

3. Every simple graph is a 2-uniform simple hypergraph.

4. The intersection graph of a hypergraph generalizes the line graph $L(G)$ of a graph G. (See §11.1.)

5. Every graph is the intersection graph of some hypergraph.

6. Every graph of order n is isomorphic to the intersection graph of a hypergraph of order at most $\lfloor \frac{n^2}{4} \rfloor$.

7. When a graph G is regarded as a hypergraph, its dual is a hypergraph whose intersection graph is G.

Examples:

1. The hypergraph $H = (V, E)$ with $V = \{a, b, c, d\}$ and $E = \{ab, bc, bd, acd, c\}$ can be illustrated as follows:

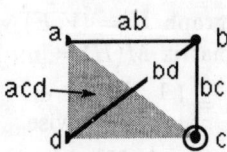

2. The hypergraph of Example 1 has the following incidence matrix:

	a	b	c	d
ab	1	1	0	0
acd	1	0	1	1
bc	0	1	1	0
bd	0	1	0	1
c	0	0	1	0

3. The dual of the hypergraph of Example 1 has the following incidence matrix:

	$v(ab)$	$v(acd)$	$v(bc)$	$v(bd)$	$v(c)$
$e(a)$	1	1	0	0	0
$e(b)$	1	0	1	1	0
$e(c)$	0	1	1	0	1
$e(d)$	0	1	0	1	0

This dual hypergraph may be illustrated as follows:

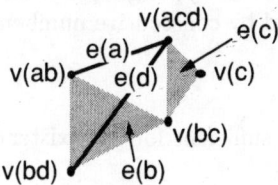

4. The hypergraph of Example 1 has the following intersection graph:

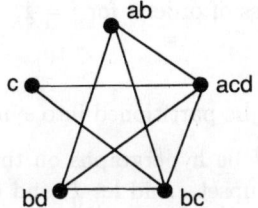

8.12.2 HYPERGRAPHS AS GENERAL COMBINATORIAL STRUCTURES

Definitions:

A *transversal* (or *cover* or *blocking set*) in a hypergraph is a set of vertices that has nonempty intersection with every edge of the hypergraph.

A *system of distinct representatives* (**SDR**) in a hypergraph $H = (V, E)$ with $E = \{e_1, e_2, \ldots, e_m\}$ is a transversal of m distinct vertices $x_1, x_2, \ldots, x_m$ such that $x_i \in e_i$ for $i = 1, \ldots, m$.

Hall's condition on a hypergraph is that, for each $t = 1, \ldots, m$ the union of every subset of t edges have at least t vertices. Thus, each partial must have at least as many vertices as edges.

A *matching* in a hypergraph is a set of pairwise disjoint edges.

An *antichain* is a hypergraph in which no edge contains any other edge.

A *chain* is a simple hypergraph in which, given any pair of edges, one edge contains the other.

A *symmetric chain* in an n-vertex hypergraph H is a chain with edges of order $\frac{n}{2} - t, \ldots, \frac{n}{2} + t$ for some $t \geq 0$.

A *downset* (or *ideal*) is a simple hypergraph in which every subset of every edge is also an edge of the hypergraph.

An *upset* (or *filter*) is a simple hypergraph in which every superset of every edge is also an edge of the hypergraph.

A **hypergraph clique** is a simple hypergraph such that every pair of edges has nonempty intersection.

An **r-partite** hypergraph is an r-uniform hypergraph whose vertex set can be partitioned into r blocks such that each edge intersects each block in exactly one vertex.

A hypergraph is **unimodular** if the determinant of every square submatrix of its incidence matrix is equal to 0, 1, or -1.

An n-vertex hypergraph is an **interval hypergraph** if its vertices can be labeled $1, 2, \ldots, n$ so that each edge is labeled by consecutive numbers.

Facts:

1. Hall's condition is necessary and sufficient for the existence of an SDR in a hypergraph.

2. *Sperner's lemma*: If the hypergraph H with n vertices and m edges is an antichain, then $m \leq \binom{n}{\lfloor \frac{n}{2} \rfloor}$.

3. If the hypergraph H with m_i edges of order i for $i = 1, \ldots, n$ is an antichain, then $\sum_{i=0}^{n} m_i \binom{n}{i}^{-1} \leq 1$.

4. The complete hypergraph K_n^* can be partitioned into symmetric chains.

5. *Kleitman's lemma*: Let D and U be hypergraphs on the same n vertices. Let D be a d-edge downset and U a u-edge upset. And let D and U have m common edges. Then $du \geq 2^n m$.

6. An n-vertex hypergraph clique has at most 2^{n-1} edges.

7. An r-uniform n-vertex hypergraph clique n has at most $\binom{n-1}{r-1}$ edges if $n \geq 2r$.

8. In any r-uniform hypergraph H, the maximum size r-partite partial hypergraph contains at least $\frac{r!}{r^r}$ of the edges of H.

9. Let H be an n-vertex, m-edge hypergraph clique, such that each pair of distinct edges intersect in exactly one vertex. Then $m \leq n$. (de Bruijn and Erdős)

10. *Fisher's inequality*: Let H be an n-vertex, m-edge hypergraph clique such that each pair of edges intersect in λ vertices. Then $m \leq n$.

11. *Modular intersection theorem*: Let L be a set of s integers, and let p be a prime number. Let H be an r-uniform hypergraph such that $r \notin L \bmod p$ and that the intersection size for each pair of distinct edges is in $L \bmod p$. Then $m \leq \binom{n}{s}$.

Examples:

1. The Fano plane (§12.1.1) is the hypergraph with (using mod 7 arithmetic):
$$V = \{1, 2, \ldots, 7\} \text{ and } E = \{\, \{1+i, 2+i, 4+i\} \mid 1 \leq i \leq 7 \,\}.$$

2. A block design is a regular, uniform hypergraph such that each pair of vertices is contained in precisely λ edges. Block designs often provide extremal examples in various extremal problems of hypergraph theory.

3. A matroid (§12.4.1) can be regarded as a hypergraph such that under every nonnegative weighting of the vertices, a greedy algorithm could find an edge of maximum weight.

8.12.3 NUMERICAL INVARIANTS OF HYPERGRAPHS

Calculating formulas for the values of some standard numerical invariants of hypergraphs tends to be quite difficult, even for complete hypergraphs. Two famous examples are Lovász's proof of the Kneser conjecture and Baranyai's proof of the factorization theorem.

Definitions:

The **maxdegree** $\Delta(H)$ is the largest degree of any vertex in the hypergraph H.

The **chromatic number** $\chi(H)$ is the smallest number of independent sets required to partition the vertex set of H. To ensure the existence of such partitions it is assumed that H does not contain any edges with just one vertex.

The **independence number** $\alpha(H)$ is the maximum number of vertices which form an independent set in H.

The **chromatic index** $q(H)$ is the smallest number of matchings required to partition the edges of H.

A hypergraph H is **normal** if $q(H) = \Delta(H)$.

The **transversal number** $\tau(H)$ is the minimum cardinality (i.e., number of vertices), taken over all transversals of H.

The **matching number** $\nu(H)$ is the maximum number of pairwise disjoint edges of H, i.e., the cardinality of the largest partial of H which forms a matching.

The **clique partition number** $cp(H)$ is the smallest number of cliques required to partition the edge set of H.

The **clique number** $\omega(H)$ is the largest number of edges of any partial clique in the hypergraph H.

Facts:

1. Many hypergraph invariants are representable as graph invariants. In particular,
$$\omega(H) = \omega(I(H)), \quad \nu(H) = \alpha(I(H)), \quad q(H) = \chi(I(H)), \quad cp(H) = \chi(\overline{I(H)})$$
where $\overline{G}$ denotes the edge-complement of a graph G.

2. Every hypergraph H satisfies the following two $min \geq max$ relations:
$$q(H) \geq cp(H) \geq \Delta(H) \qquad \tau(H) \geq cp(H) \geq \nu(H)$$

3. A hypergraph H is normal if and only if $\tau(H') = \nu(H')$ for all partials H' of H. (Lovász, 1972)

4. The following relations hold in every n-vertex hypergraph H:
$$\tau(H) = n - \alpha(H), \qquad \chi(H) \geq \tfrac{n}{\alpha(H)}, \qquad \chi(H) + \alpha(H) \leq n + 1.$$

5. The parameters $\chi(H)$ and $\tau(H)$ can be approximated by greedy algorithms.

6. The Kneser conjecture that $cp(K^r_{2r+k}) = k + 2$ was proved by topological methods. (Lovász and Bárány, 1978)

7. The *factorization theorem* that $q(K^r_{kr}) = \binom{kr-1}{r-1}$ was proved by using network flows. (Baranyai, 1975)

8. Hypergraphs in the following classes are known to be bicolorable (i.e., $\chi(H) = 2$): normal hypergraphs (including unimodular hypergraphs), r-uniform hypergraphs with size at most 2^{r-1}, r-uniform hypergraphs in which each edge intersects at most 2^{r-3} other edges (proved by probabilistic methods), finite planes of order at least three.

Examples:

1. Consider the hypergraph H of §8.11.1 Example 1 with $V = \{a, b, c, d\}$ and $E = \{ab, bc, bd, acd, c\}$. The maximum degree $\Delta(H)$ is 3, since vertex c has degree 3. The chromatic number $\chi(H)$ is 4, since every pair of vertices lies in some edge, so all four vertices must get different colors. The independence number $\alpha(H)$ is 1, since every pair of vertices lies in some edge. The chromatic index $q(H)$ is 4, using the matching c, ab.

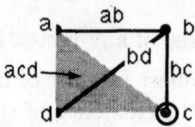

The hypergraph H is not normal, since $q(H) = 4$, but $\Delta(H) = 3$. The transversal number $\tau(H)$ is 2, using the transversal b, c. The matching number $\nu(H)$ is 2, using the matching ab, c.

2. The Fano plane (§8.12.2 Example 1) has the following parameters: $\omega = q = 7$, $\Delta = \tau = \chi = 3$, $\alpha = 4$, $\nu = cp = 1$.

REFERENCES

Printed Resources:

[BeWi83] L. W. Beineke and R. J. Wilson, *Selected Topics in Graph Theory 2*, Academic Press, 1983.

[Be89] C. Berge, *Hypergraphs*, North Holland 1989.

[Bi74] N. Biggs, *Algebraic Graph Theory*, Cambridge University Press, 1974.

[Bo85] B. Bollobás, *Random Graphs*, Academic Press, 1985.

[BoHe77] J. A. Bondy and R. L. Hemminger, Graph reconstruction — a survey, *J. Graph Theory* 1 (1977), 227–268.

[BoMu76] J. A. Bondy and U. S. R. Murty, *Graph Theory with Applications*, North-Holland, 1976.

[Br88] W. C. Brown, *A Second Course in Linear Algebra*, John Wiley & Sons, 1988.

[ChLe86] G. Chartrand and L. Lesniak, *Graphs & Digraphs*, Wadsworth & Brooks/Cole, 1986.

[CvEtAl88] D. M. Cvetković, M. Doob, I. Gutman, and A. Torgašev, *Recent Results in the Theory of Graph Spectra*, North-Holland, 1988.

[CvDoSa95] D. M. Cvetković, M. Doob, and H. Sachs, *Spectra of Graphs*, Johann Ambrosius Barth, 1995.

[Eu1736] L. Euler, "Solutio problematis ad geometriam situs pertinentis", *Comment. Academiae Sci. I. Petropolitanae* 8 (1736) 128–140.

[FiWi77] S. Fiorini and R. J. Wilson, *Edge Colorings of Graphs*, Pitman, 1977.

[Ga60] F. R. Gantmacher, *The Theory of Matrices, I, II*, Chelsea, 1960.

[GrTu87] J. L. Gross and T. W. Tucker, *Topological Graph Theory*, Wiley-Interscience, 1987.

[GrYe98] J. L. Gross and J. Yellen, *Graph Theory and Its Applications*, CRC Press, 1998.

[Ha69] F. Harary, *Graph Theory*, Addison-Wesley, 1969.

[HaNoCa65] F. Harary, R. Z. Norman, and D. Cartwright, *Structural Models: An Introduction to the Theory of Directed Graphs*, Wiley, 1965.

[HaPa73] F. Harary and E. M. Palmer, *Graphical Enumeration*, Academic Press, 1973.

[KöScTo93] J. Köbler, U. Schöning, and J. Torán, *The Graph Isomorphism Problem: Its Structural Complexity*, Birkhäuser, 1993.

[MaMi64] M. Marcus and H. Minc, *A survey of matrix theory and matrix inequalities*, Prindle, Weber & Schmidt, 1964.

[Mc81] B. D. McKay, Practical graph isomorphism, *Congressus Numerantium* 21 (1981), 45–87. (Describes the fast algorithm NAUTY.)

[MiRo91] J. G. Michaels and K. H. Rosen, eds., *Applications of Discrete Mathematics*, McGraw-Hill, 1991.

[Mo68] J. Moon, *Topics on Tournaments*, Holt, Rinehart & Winston, 1968.

[Na78] C. St.J. A. Nash-Williams, "The reconstruction problem", Chapter 8 of *Selected Topics in Graph Theory*, ed. L. W. Beineke and R. J. Wilson, Academic Press, 1978.

[PóRe87] G. Pólya and R. C. Read, *Combinatorial Enumeration of Groups, Graphs, and Chemical Compounds*, Springer-Verlag, 1987.

[Re98] R. C. Read, *Atlas of Graphs*, Oxford University Press (Clarendon Press), 1998.

[ReCo77] R. C. Read and D. G. Corneil, The graph isomorphism disease, *J. Graph Theory* 1 (1977), 339–363.

[Ri74] G. Ringel, *Map Color Theorem*, Springer-Verlag, 1974.

[Ro76] F. S. Roberts, *Discrete Mathematical Models with Applications to Social, Biological, and Environmental Problems*, Prentice-Hall, 1976. (Title tells it all; see especially Chapters 2 & 3.)

[Ro88] J. J. Rotman, *An Introduction to the Theory of Groups*, 3rd ed., Wm. C. Brown, 1988.

[Sk90] S. Skiena, *Implementing Discrete Mathematics*, Addison-Wesley, 1990. (Using Mathematica as an aid in graph theory.)

[Sl73] N. J. A. Sloane, *A Handbook of Integer Sequences*, Academic Press, 1973.

[Ta97] R. Tamassia, "Graph drawing", Article 44 in *Handbook of Discrete and Computational Geometry* J. E. Goodman and J. O'Rourke, eds., CRC Press, 1997.

[Th98] R. Thomas, "An Update of the Four-Color Theorem", *Notices Amer. Math Soc.* 45 (1998), 848–859.

[We96] D. B. West, *Introduction to Graph Theory*, Prentice-Hall, 1996.

[Wh84] A. T. White, *Graphs, Groups and Surfaces* (revised ed.), North-Holland, 1984.

[WiWa90] R. J. Wilson and J. J. Watkins, *Graphs: An Introductory Approach*, Wiley, 1990.

Web Resources:

http://dimacs.rutgers.edu/Projects/LINK.html (DIMACS Projects: LINK, a software system designed for experimentation with large graphs.)

http://msn.yahoo.com/Science/Mathematics/Graph_Theory/

http://www.astro.virginia.edu/~eww6n/math/Cros (Information on crossing numbers.)

http://www.cs.ualberta.ca/~joe/Coloring/index.html (Bibliographies and links on graph coloring.)

http://www.cs.sunysb.edu/~algorith/ (The Stony Brook Algorithm Repository; see Sections 1.4 and 1.5 on Graph Problems.)

http://www.cs.utexas.edu/users/kleyn/Kleyn-Graphs.html (A list of software for graph theory.)

http://www.ing.unlp.edu.ar/cetad/mos/Hamilton.html (Information on Hamiltonian problems.)

http://www.math.gatech.edu/~thomas/FC (Information on the Four Color Theorem.)

http://www.research.att.com/sw/tools/graphviz/ (AT&T Graphviz Package — graph theory software.)

http://wwwis.win.tue.nl/~percy/my/link-gd.html (A collection of links to sites relevant to graph drawing.)

TREES

9.1 Characterizations and Types of Trees — Lisa Carbone
    9.1.1 Properties of trees
    9.1.2 Roots and orderings
    9.1.3 Tree traversal
    9.1.4 Infinite trees

9.2 Spanning Trees — Uri Peled
    9.2.1 Depth-first and breadth-first spanning trees
    9.2.2 Enumeration of spanning trees

9.3 Enumerating Trees — Paul Stockmeyer
    9.3.1 Counting generic trees
    9.3.2 Counting trees in chemistry
    9.3.3 Counting trees in computer science

INTRODUCTION

A tree is a connected graph containing no cycles. Trees have applications in a wide variety of disciplines, particularly computer science. For example, they can be used to construct searching algorithms for finding a particular item in a list, to store data, to model decisions and their outcomes, or to design networks.

GLOSSARY

ancestor (of a vertex v in a rooted tree): any vertex on a path to v from the root.

m-ary tree: a rooted tree in which every internal vertex has at most m children.

backtrack: a pair of successive edges in a walk where the second edge is the same as the first, but traversed in the opposite direction.

balanced tree: a rooted m-ary tree of height h such that all leaves of the tree have height h or $h-1$.

bihomogeneous tree: a tree (usually infinite) in which there are exactly two values for the vertex degrees.

binary search tree: a type of binary tree used to represent a table of data, which is efficiently accessed by storage and retrieval algorithms, abbreviated BST.

binary tree: an ordered rooted tree in which each vertex has at most two children, that is, a possible "left child" and a possible "right child"; an only child must be designated either as a left child or a right child (this usage is normative for computer science); in pure graph theory, an m-ary tree in which $m = 2$.

bounded tree: a (possibly infinite) tree of finite diameter.

breadth-first search: a method for visiting all the vertices of a graph in a sequence, based on their proximity to a designated starting vertex.

caterpillar: a tree that contains a path such that every edge has one or both endpoints in that path.

center (of a tree): the set of vertices of minimum eccentricity.

child (of a vertex v in a rooted tree): a vertex such that v is its immediate ancestor.

chord: for a graph G with a spanning tree T, an edge e of G such that $e \notin T$.

complete binary tree: a binary tree where every parent has two children and all leaves are at the same depth.

decision tree: a rooted tree in which every internal vertex represents a decision and each path from the root to a leaf represents a cumulative choice.

dense graph: a graph in which the number of edges far exceeds the number of vertices.

depth (of a vertex in a rooted tree): the number of edges in the unique path from the root to that vertex.

depth-first search: a method for visiting every vertex of a graph by progressing as far as possible from the most recently visited vertex, before doing any backtracking.

descendant (of a vertex v in a rooted tree): a vertex that follows v on a path from the root.

diameter (of a tree): the maximum distance between two distinct vertices in the tree.

distance (between two vertices in a tree): the number of edges in the unique simple path between these vertices.

eccentricity (of a vertex in a connected graph): the length of the longest simple path beginning at that vertex.

finite tree: a tree with a finite number of vertices and edges.

forest: a graph with no cycles.

full m-ary tree: a rooted tree in which every internal vertex has exactly m children.

fundamental cycle of a connected graph G: the unique cycle created by adding the edge $e \in E_G$ not in T to a spanning tree T.

fundamental edge-cut of a connected graph G: the partition-cut $\langle X_1, X_2 \rangle$ where X_1 and X_2 are the vertex-sets of the two components of $T - e$, where e is an edge of a spanning tree T for G.

fundamental system of cycles of a connected graph G: the set of fundamental cycles corresponding to the various edges of $G - T$, where T is a spanning tree for G.

fundamental system of edge-cuts of a connected graph G: the set of fundamental edge-cuts that result from removal of an edge from a spanning tree T for G.

geodesic (between two vertices in a tree): the unique simple path between these vertices.

heap: a representation of a priority tree as an array.

height (of a rooted tree): the maximum of the levels of its vertices.

homogeneous: property of a tree (usually infinite) that every vertex has the same degree.

d-homogeneous: property of a tree (usually infinite) that every vertex has degree d.

infinite tree: a tree with an infinite number of vertices and edges.

inorder traversal (of an ordered rooted tree): a recursive listing of all vertices starting with the vertices of the first subtree of the root, next the root vertex itself, and then the vertices of the other subtrees as they occur from left to right.

internal vertex (of a rooted tree): a vertex with children.

isomorphism (of trees): for trees X and Y, a pair of bijections $f_V: V_X \to V_Y$ and $f_E: E_X \to E_Y$ such that if u and v are the endpoints of an edge e in the tree X, then $f_V(u)$ and $f_V(v)$ are the endpoints of the edge $f_E(e)$ in the tree Y (see §8.1).

isomorphism (of rooted trees): for rooted trees (T_1, r_1) and (T_2, r_2), a tree isomorphism $f: T_1 \to T_2$ that takes r_1 to r_2.

labeled tree: a tree with labels such as $v_1, v_2, \ldots, v_n$ assigned to its vertices.

leaf: in a rooted tree, a vertex that has no children.

left child (of a node in an ordered, rooted binary tree): the first child of that node.

left subtree (of an ordered, rooted binary tree): the tree rooted at a left child.

left-complete binary tree: a binary tree where each level except possibly the deepest is filled and the bottom level has no gaps as one traverses left to right.

left-right tree: a binary tree in which each vertex is a parent to either no children or to both a left and a right child.

level (of a vertex in a rooted tree): the length of the unique path from the root to this vertex.

locally finite tree: a tree in which the degree of every vertex is finite.

maximal tree (in a graph): a spanning tree.

mesh (of trees): a graph obtained by construing each row and each column of a $2^d \times 2^d$ array of vertices as the leaves of a complete binary tree.

minimum spanning tree (of a graph whose edges have weights assigned to them): a spanning tree with minimum total edge-weight.

nth level (of a rooted tree): the set of all vertices at depth n.

order (of a finite tree): the number of vertices in the tree.

ordered tree: a rooted tree in which the children of each internal vertex are linearly ordered.

parent (of a vertex v, other than the root, in a rooted tree): a vertex that is the immediate predecessor of v on the unique path from the root to v.

partition-cut of a graph: given a partition of the set of vertices of G into X_1 and X_2, the set $\langle X_1, X_2 \rangle$ of edges of G that have one endpoint in X_1 and the other in X_2.

postorder traversal: a recursive listing of the vertices in an ordered rooted tree starting with the vertices of subtrees as they occur from left to right, followed by the root.

preorder traversal: a recursive listing of the vertices in an ordered rooted tree starting with the root, then the vertices of the first subtree, followed by the vertices of other subtrees as they occur from left to right.

priority tree: a left-complete binary tree whose vertices have labels (from an ordered set) called "priorities", such that no vertex has higher priority than its parent.

reduced tree: a tree with no vertices of degree 2.

reduced walk: a walk in a graph without backtracking.

regular: Synonym for *homogeneous*.

d-regular: Synonym for *d-homogeneous*.

right child (of a node in an ordered rooted binary tree): the second child of that node.

right subtree (of an ordered, rooted binary tree): the tree rooted at the right child.

rooted tree: a tree in which one vertex is designated as the "root".

semi-homogeneous tree: a bihomogeneous tree (usually infinite) with a partition of the vertices into two sets, those of degree m and those of degree n, where each vertex of degree m is adjacent to one of degree n.

siblings (in a rooted tree): vertices with the same parent.

simplicial notation: notation for a tree or other simple graph in which each edge is specified by its endpoints and each path is specified by its vertex sequence.

spanning tree (of a connected graph): a tree that contains all the vertices of the graph.

subtree: a subgraph of a tree that is also a tree.

terminal vertex (of a tree): a vertex of degree 1.

tree: a connected graph with no cycles.

tree edge: for a graph G with a spanning tree T, an edge e of G such that $e \in T$.

tree traversal: a walk that visits all the vertices of a tree.

9.1 CHARACTERIZATIONS AND TYPES OF TREES

9.1.1 PROPERTIES OF TREES

For trees, as with other graphs, there is a wide variety of terminology in use from one application or specialty to another.

Definitions:

A graph is *acyclic* is it contains no subgraph isomorphic to a cycle C_n (§8.1.3).

A *forest* is an acyclic graph.

A *tree* is an acyclic connected graph. (*Note*: Unless stated otherwise, all trees are assumed to be finite, i.e., to have a finite number of vertices.)

The *eccentricity* of a vertex is the length of the longest simple path beginning at that vertex.

A *center* of a tree T is a vertex v with minimum eccentricity.

An *end vertex* of a tree is a vertex of degree 1.

A *caterpillar* is a tree that contains a path such that every edge has one or both endpoints in that path.

Facts:

1. A (finite) tree with at least two vertices has at least two end vertices.
2. A connected graph with n vertices is a tree if and only if has exactly $n-1$ edges.
3. A graph is a tree if and only if there is a unique simple path between any two vertices.
4. A graph is a forest if and only if every edge is a cut-edge (§8.3.3).
5. Trees are bipartite. Hence, every tree can be colored using two colors.
6. The center of a tree consists of either only one vertex or two adjacent vertices.

Examples:

1. A tree:

2. A forest:

3. A tree with two adjacent vertices a and b in its center:

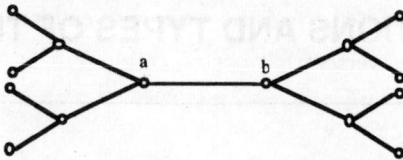

4. A caterpillar:

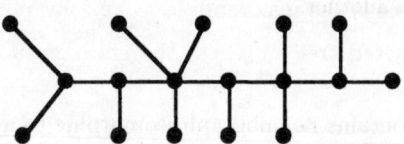

5. Neither of the graphs shown is a tree. One contains a 3-cycle, and the other contains a 1-cycle (i.e., a self-loop).

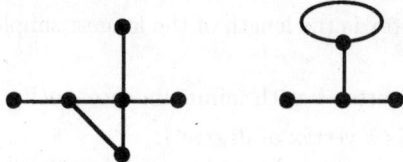

9.1.2 ROOTS AND ORDERINGS

Adding some extra structure to trees adapts them to applications in many disciplines, especially computer science.

Definitions:

A **rooted tree** (T, r) is a tree T with a distinguished vertex r (the **root**), in which all edges are implicitly directed away from the root.

Two rooted trees (T_1, r_1) and (T_2, r_2) are **isomorphic as rooted trees** if there is an isomorphism $f: T_1 \to T_2$ (§8.1.2) that takes r_1 to r_2.

A **child** of a vertex v in a rooted tree is a vertex that is the immediate successor of v on a path from the root.

A **descendant** of a vertex v in a rooted tree is v itself or any vertex that is a successor of v on a path from the root.

A **proper descendant** of a vertex v in a rooted tree is any descendant except v itself.

The **parent** of a vertex v in a rooted tree is a vertex that is the immediate predecessor of v on a path to v from the root.

The **parent function** of a rooted tree T maps the root of T to the empty set and maps every other vertex to its parent.

An **ancestor** of a vertex v in a rooted tree is v itself or any vertex that is the predecessor of v on a path to v from the root.

A **proper ancestor** of a vertex v in a rooted tree is any ancestor except v itself.

Siblings in a rooted tree are vertices with the same parent.

An *internal vertex* in a rooted tree is a vertex with children.

A *leaf* in a rooted tree is a vertex that has no children.

The *depth* of a vertex in a rooted tree is the number of edges in the unique path from the root to that vertex.

The *nth level* in a rooted tree is the set of all vertices at depth n.

The *height* of a rooted tree is the maximum depth of any vertex.

An *ordered tree* is a rooted tree in which the children of each internal vertex are linearly ordered.

A *left sibling* of a vertex v in an ordered tree is a sibling that precedes v in the ordering of v and its siblings.

A *right sibling* of a vertex v in an ordered tree is a sibling that follows v in the ordering of v and its siblings.

A *plane tree* is a drawing of an ordered tree such that the left-to-right order of the children of each node in the drawing is consistent with the linear ordering of the corresponding vertices in the tree.

In the *level ordering* of the vertices of an ordered tree, u precedes v under any of these circumstances:

- if the depth of u is less than the depth of v;
- if u is a left sibling of v;
- if the parent of u precedes the parent of v.

Two ordered trees (T_1, r_1) and (T_2, r_2) are *isomorphic as ordered trees* if there is a rooted tree isomorphism $f: T_1 \to T_2$ that preserves the ordering at every vertex.

An *m-ary tree* is a rooted tree such that every internal vertex has at most m children.

A *full m-ary tree* is a rooted tree such that every internal vertex has exactly m children.

A *(pure) binary tree* is a rooted tree such that every internal vertex has at most two children. This meaning of "binary tree" occurs commonly in pure graph theory.

A *binary tree* is a 2-ary tree such that every child, even an only child, is distinguished as *left child* or *right child*. This meaning of "binary tree" occurs commonly in computer science and in permutation groups.

The *principal subtree at a vertex* v of a rooted tree comprises all descendants of v and all edges incident to these descendants. It has v designated as its root.

The *left subtree* of a vertex v in a binary tree is the principal subtree at the left child. The *right subtree* of v is the principal subtree at the right child.

A *balanced tree* of height h is a rooted m-ary tree in which all leaves are of height h or $h-1$.

A *complete binary tree* is a binary tree in which every parent has two children and all leaves are at the same depth.

A *complete m-ary tree* is an m-ary tree in which every parent has two children and all leaves are at the same depth.

Algorithm 1: Find a Huffman code.

input: the probabilities $Pr(x_1), \ldots, Pr(x_n)$ on a set X
output: a Huffman code for (X, Pr)
initialize F to be a forest of isolated vertices, labeled $x_1, \ldots, x_n$, each to be
    regarded as a rooted tree
assign weight $Pr(x_j)$ to the rooted tree x_j, for $j = 1, \ldots, n$
repeat until forest F is a single tree
    choose two rooted trees, T and T', of smallest weights in forest F
    replace trees T and T' in forest F by a tree with a new root whose left subtree
        is T and whose right subtree is T'
    label the new edge to T with a 0 and the new edge to T' with a 1
    assign weight $w(T) + w(T')$ to the new tree
return tree F
{The Huffman code word for x_i is the concatenation of the labels on the unique
    path from the root to x_i.}

A *decision tree* a rooted tree in which every internal vertex represents a decision and each path from the root to a leaf represents a cumulative choice.

A *prefix code* for a finite set $X = \{x_1, \ldots, x_n\}$ is a set $\{c_1, \ldots, c_n\}$ of binary strings in X (called *codewords*) such that no codeword is a prefix of any other codeword.

A **Huffman code** for a set X with a probability measure Pr (see §7.1) is a prefix code $\{c_1, \ldots, c_n\}$ such that $\sum_{j=1}^{n} len(c_j) Pr(x_j)$ is minimum among all prefix codes, where $len(c_j)$ measures the length of c_j in bits.

A **Huffman tree** for a set X with a probability measure Pr is a tree constructed by Huffman's algorithm to produce a Huffman code for (X, Pr).

Facts:
1. Plane trees are usually drawn so that vertices of the same level in the corresponding ordered tree are represented at the same vertical position in the plane.
2. A rooted tree can be represented by its vertex set plus its parent function.
3. The concept of finite binary tree also has the following recursive definition: (basis clause) an ordered tree with only one vertex is a binary tree; (recursion clause) an ordered tree with more than one vertex is a binary tree if the root has two children and if both its principal subtrees are binary trees.
4. A full m-ary tree with k internal vertices has $mk + 1$ vertices and $(m-1)k + 1$ leaves.
5. A full m-ary tree with k vertices has $\dfrac{k-1}{m}$ internal vertices and $\dfrac{(m-1)k+1}{m}$ leaves.
6. There are at most m^h leaves in any m-ary tree of height h.
7. A *binary search tree* is a special kind of binary tree used to implement a *random access table* with $O(n)$ maintenance and retrieval algorithms. (See Chapter 17.)
8. A balanced binary tree can be used to implement a *priority queue* with $O(n)$ enqueue and dequeue algorithms. (See §17.2.4.)
9. Algorithm 1, due to D. Huffman in 1951, constructs a Huffman tree.

Examples:

1. A rooted tree (T, r):

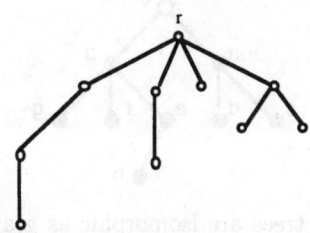

2. A rooted tree and its parent function:

vertex	a	b	c	d	e	f	g
parent	d	d	d	∅	c	b	c

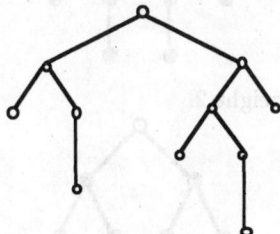

3. A 2-ary tree of height 4:

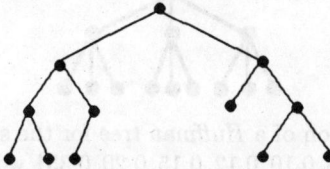

4. A balanced binary tree:

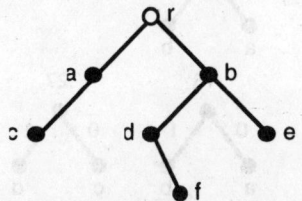

5. The following tree is rooted at vertex r. Vertices d and e are children of vertex b. Vertex f is a descendant of f, d, b, and r, but f is not a descendant of vertex a. Vertex a is the parent of c, which is the only proper descendant of vertex a. Vertices d and e are siblings, but c is not a sibling of d or of e.

6. The leaves of the following rooted tree are the vertices c, d, f, g, and h. The internal vertices are a, b, e, and s.

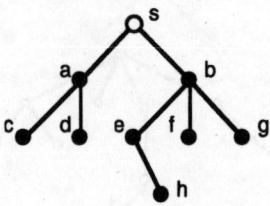

7. The following two rooted trees are isomorphic as graphs, but they are considered to be different as rooted trees, because there is no graph isomorphism from one to the other that maps root to root.

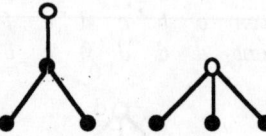

8. The following two plane trees are isomorphic as rooted trees, but they are not isomorphic as ordered rooted trees, because there is no rooted tree isomorphism from one to the other that preserves the child ordering at every vertex.

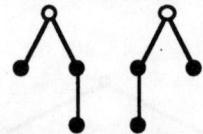

9. A complete binary tree of height 2.

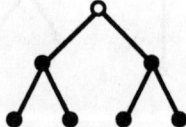

10. A complete 3-ary tree of height 2.

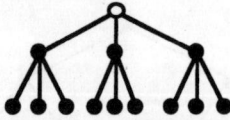

11. The iterative construction of a *Huffman tree* for the set $X = \{u, v, w, x, y, z\}$ with respective probabilities $\{0.08, 0.10, 0.12, 0.15, 0.20, 0.35\}$ would proceed as follows:

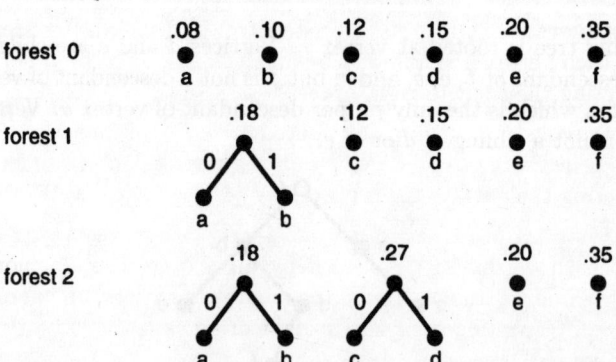

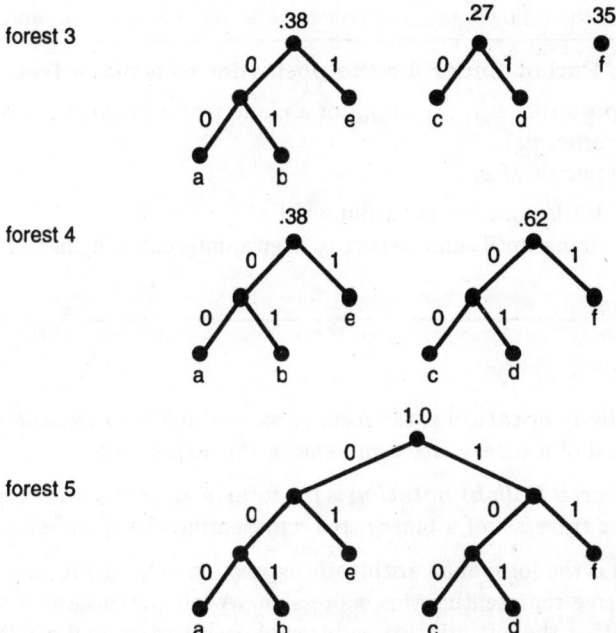

The codes are 000 for a, 001 for b, 100 for c, 101 for d, 01 for e, and 11 for f. Thus, the most frequently used objects in the set are represented by the shortest binary codes.

9.1.3 TREE TRAVERSAL

Ordered rooted trees can be used to store data or arithmetic expressions involving numbers, variables and operations. A tree traversal algorithm gives a systematic method for accessing the information stored in the tree.

Definitions:

A *boundary walk* of a plane tree is a walk around the boundary of the single region of the given plane imbedding of the tree, starting at the root.

A *backtrack* along a walk in a graph is an instance ..., $u, e, v, e, u, ...$ of two consecutive edge-steps in which an edge-step traverses the same edge as its predecessor, but in the opposite direction.

A *reduced walk* is a walk without backtracking.

A *preorder traversal* of an ordered rooted tree T lists the vertices of T (or their labels) so that each vertex v is followed by all the vertices, in preorder, in its principal subtrees, respecting their left-to-right order.

A *postorder traversal* of an ordered rooted tree T lists the vertices of T (or their labels) so that each vertex v is preceded by all the vertices, in postorder, in its principal subtrees, respecting their left-to-right order.

An *inorder traversal* of an ordered rooted tree T lists the vertices of T (or their labels) so that each vertex v is preceded by all the vertices, in inorder, in its first principal subtree and so that v is followed by the vertices, in inorder, of its other principal subtrees, respecting their left-to-right order.

Algorithm 2:   Parent-finder for the postorder of a plane tree.

input: the postorder $v_{p(1)}, \ldots, v_{p(n)}$ of a plane tree with sorted vertex labels and a vertex v_j

output: the parent of v_j

scan the postorder until v_j is encountered

continue scanning until some vertex v_i is encountered such that $i < j$

return (v_i)

Prefix (or Polish) notation is the form of an arithmetic expression obtained from a preorder traversal of a binary tree representing this expression.

Postfix (or reverse Polish) notation is the form of an arithmetic expression obtained from a postorder traversal of a binary tree representing this expression.

Infix notation is the form of an arithmetic expression obtained from an inorder traversal of a binary tree representing this expression. A left parenthesis is written immediately before writing the left principal subtree of each vertex, and a right parenthesis is written immediately after writing the right principal subtree.

The **universal address system** of an ordered rooted tree is a labeling in which the root is labeled 0 and in which for each vertex with label x, its m children are labeled $x.1, x.2, \ldots, x.m$, from left to right.

In the **level order** of the vertices of an ordered tree T, vertex u precedes vertex v if u is nearer the root, or if u and v are at the same level and u and v have ancestors u' and v' that are siblings and u' precedes v' in the ordering of T.

A bijective assignment of labels from an ordered set (such as alphabetic strings or the integers) to the vertices of an ordered tree is **sorted** if the level order of these labels is either ascending or descending.

Facts:

1. The preorder traversal of a plane tree is obtained by a counterclockwise traversal of the boundary walk of the plane region, that is, starting downward toward the left. As each vertex of the tree is encountered for the first time along this walk, it is recorded in the preorder.

2. The postorder traversal of a plane tree is obtained by a counterclockwise traversal of the boundary walk of the plane region, that is, starting downward toward the left. As each vertex of the tree is encountered for the last time along this walk, it is recorded in the postorder.

3. The inorder traversal of a plane tree is obtained by a counterclockwise traversal of the boundary walk of the plane region, that is, starting downward toward the left. As each interior vertex of the tree is encountered for the second time along this walk, it is recorded in the inorder. An end vertex is recorded whenever it is encountered for the only time.

4. Two nonisomorphic ordered trees with sorted vertex labels can have the same preorder but not the same postorder.

Examples:

1. A plane tree with pre-order $abefhcdgijk$, post-order $ehfbcijkgda$, and in-order $ebhfacigjkd$.

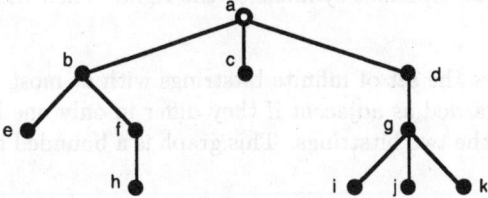

2. A binary tree representing the arithmetic expression $(x+y)/(x-2)$, with infix form $x+y\ /\ x-2$, prefix form $/+x\ y-x\ 2$, and postfix form $x\ y+x\ 2-/$.

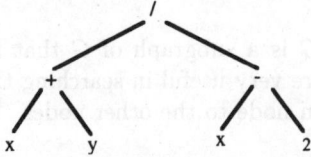

9.1.4 INFINITE TREES

Definitions:

An *infinite tree* is a tree with an infinite number of vertices or edges.

The *diameter* of a tree is the maximum distance between two distinct vertices in the tree.

A *bounded tree* is a tree of finite diameter.

A *locally finite tree* is a tree in which the degree of every vertex is finite.

A *homogeneous tree* is a tree in which every vertex has the same degree.

An *n-homogeneous tree* is a tree in which every vertex has degree n.

A *bihomogeneous tree* is a nonhomogeneous tree with a partition of the vertices into two subsets, such that all vertices in the same subset have the same degree.

A *semi-homogeneous tree* is a bihomogeneous tree such that each vertex of one of the two realized degrees is adjacent to a vertex of the other realized degree.

Examples:

1. Suppose that two finite bitstrings are considered adjacent if one bitstring can be obtained from the other by appending a 0 or a 1 at the right. The resulting graph is the infinite bihomogeneous tree, in which the empty string λ has degree 2 and all other finite bitstrings have degree 3.

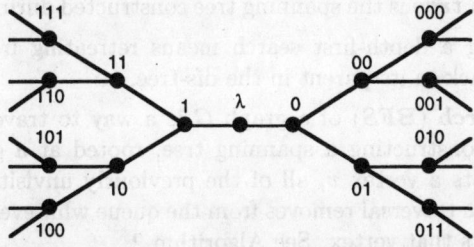

2. Consider the set of all finite strings on the alphabet $\{a, a^{-1}, b, b^{-1}\}$ containing no instances of the substrings aa^{-1} $a^{-1}a$, bb^{-1}, or $b^{-1}b$. Suppose that two such strings are considered to be adjacent if and only if one of them can be obtained from the other by appending one of the alphabet symbols at the right. Then the resulting graph is a 4-homogeneous tree.

3. Consider as vertices the set of infinite bitstrings with at most two 1s. Suppose two such bitstrings are regarded as adjacent if they differ in only one bit, and that bit is a rightmost 1 for one of the two bitstrings. This graph is a bounded tree of diameter four.

9.2 SPANNING TREES

A spanning tree of a graph G is a subgraph of G that is a tree and contains every vertex of G. Spanning trees are very useful in searching the vertices of a graph and in communicating from any given node to the other nodes. Minimum spanning trees are covered in §10.1.

9.2.1 DEPTH-FIRST AND BREADTH-FIRST SPANNING TREES

Definitions:

A *spanning tree* of a graph G is a tree that is a subgraph of G and that contains every vertex of G.

A *tree edge* of a graph G with a spanning tree T is an edge e such that $e \in T$.

A *chord* of a graph G with a spanning tree T is an edge e such that $e \notin T$.

A *back edge* of a digraph G with a spanning tree T is a chord e that joins one of its endpoints to an ancestor in T.

A *forward edge* of a digraph G with a spanning tree T is a chord e that joins one of its endpoints to a descendent in T.

A *cross edge* of a digraph G with a spanning tree T is a chord e that is neither a back edge nor a forward edge.

The *fundamental cycle* of a chord e with respect to a given spanning tree T of a graph G consists of the edge e and the unique path in T joining the endpoints of e.

A *depth-first search* (**DFS**) of a graph G is a way to traverse every vertex of a connected graph by constructing a spanning tree, rooted at a given vertex r. Each stage of the DFS traversal seeks to move to an unvisited neighbor of the most recently visited vertex, and backtracks only if there is none available. See Algorithm 1.

A *depth-first-search tree* is the spanning tree constructed during a depth-first search.

Backtracking during a depth-first search means retreating from a vertex with no unvisited neighbors back to its parent in the dfs-tree.

A *breadth-first search* (**BFS**) of a graph G is a way to traverse every vertex of a connected graph by constructing a spanning tree, rooted at a given vertex r. After the BFS traversal visits a vertex v, all of the previously unvisited neighbors of v are enqueued, and then the traversal removes from the queue whatever vertex is at the front of the queue, and visits that vertex. See Algorithm 2.

> **Algorithm 1: Depth-first search spanning tree.**
>
> input: a connected locally ordered n-vertex graph G and a starting vertex r
> output: the edgeset E_T of a spanning tree and an array $X[1..n]$ listing V_G in DFS-order
>
> initialize all vertices as *unvisited* and all edges as *unused*
> $E_T := \emptyset;\ loc := 1$
> $dfs(r)$
>
> **procedure** $dfs(u)$
> mark u as *visited*
> $X[loc] := u$
> $loc := loc + 1$
> **while** vertex u has any *unused* edges
> $e :=$ next *unused* edge at u
>   mark e as *used*
> $w :=$ the other endpoint of edge e
>   **if** w is *unvisited* **then**
>   add e to E_T
> $dfs(w)$

A **breadth-first-search tree** is the spanning tree constructed during a breadth-first search.

The **fundamental cycle** of a connected graph G associated with a spanning tree T and an edge $e \in E_G$ not in T is the unique cycle created by adding the edge e to the tree T.

The **fundamental system of cycles** of a connected graph G associated with a spanning tree T is the set of fundamental cycles corresponding to the various edges of $G - T$.

Given two vertex sets X_1 and X_2 that partition the vertex set of a graph G, the **partition-cut** $\langle X_1, X_2 \rangle$ is the set of edges of G that have one endpoint in X_1 and the other in X_2.

The **fundamental edge-cut** of a connected graph G associated with removal of an edge e from a spanning tree T is the partition-cut $\langle X_1, X_2 \rangle$ where X_1 and X_2 are the vertex-sets of the two components of $T - e$.

The **fundamental system of edge-cuts** of a connected graph G associated with a spanning tree T is the set of fundamental edge-cuts that result from removal of an edge from the tree T.

Facts:

1. Every connected graph has at least one spanning tree.

2. A connected graph G has k edge-disjoint spanning trees if and only if for every partition of V_G into m nonempty subsets, there are at least $k(m-1)$ edges connecting vertices in different subsets.

3. Let T and T' be spanning trees of a graph G and $e \in T - T'$. Then there exists an edge $e' \in T' - T$ such that both $T - e \cup \{e'\}$ and $T' - e' \cup \{e\}$ are spanning trees of G.

Algorithm 2: Breadth-first search spanning tree.

input: a connected locally n-vertex ordered graph G and a starting vertex r.
output: the edgeset E_T of a spanning tree and an array $X[1..n]$ listing V_G in BFS-order

initialize all vertices as *unvisited* and all edges as *unused*
$E_T := \emptyset$; $loc := 1$; $Q := r$ {Q is a queue}
while $Q \neq \emptyset$
  $x := front(Q)$
  remove x from Q
  $bfs(r)$

procedure: $bfs(u)$
mark u as *visited*
$X[loc] := u$
$loc := loc + 1$
while vertex u has any *unused* edges
  $e :=$ next *unused* edge at u
  mark e as *used*
  $w :=$ the other endpoint of edge e
  if w is *unvisited* **then**
    add e to E_T
    add w to the end of Q

---

4. In the column vector space of the incidence matrix of G over $GF(2)$, every edge set can be represented as a sum of column vectors. Let T be a spanning tree of G. Then each cycle C can be written in a unique way as a linear combination of the fundamental cycles of whatever chords of T occur in C.

5. Depth-first search on an n-vertex, m-edge graph runs in $O(m)$ time.

6. DFS-trees are used to find the components, cutpoints, blocks, and cut-edges of a graph.

7. The unique path in the BFS-tree T of a graph G from its root r to a vertex v is a shortest path in G from r to v.

8. Breadth-first search on an n-vertex, m-edge graph runs in $O(m)$ time.

9. A BFS-tree in a simple graph has no back edges.

10. Dijkstra's algorithm (§10.3.2) constructs a spanning tree T in an edge-weighted graph such that for each vertex v, the unique path in T from a specified root r to v is a minimum-cost path in the graph from r to v. When all edges have unit weights, Dijkstra's algorithm produces the BFS tree.

11. The level order of the vertices of an ordered tree is the order in which they would be traversed in a breadth-first search of the tree.

12. The fundamental cycle of an edge e with respect to a spanning tree T such that $e \notin T$ consists of edge e and those edges of T whose fundamental edge-cuts contain e.

13. The fundamental edge-cut with respect to removal of edge e from a spanning tree T consists of edge e and those edges of $E_G - E_T$ whose fundamental cycles contain e.

Examples:

1. Consider the following graph and spanning tree and a digraph on the same vertex and edge set. The tree edges are a, b, c, e, f, h, k, l, and the chords are d, g, i, j. Chord d is a forward edge, chord i is a back edge, and chords g and j are cross edges.

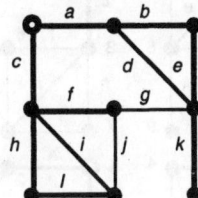 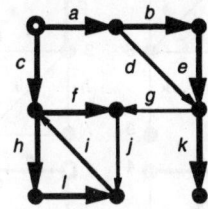

2. In the graph of Example 1, the fundamental cycles of the chords d, g, i, and j are $\{d, b, e\}$, $\{g, f, c, a, b, e\}$, $\{i, h, l\}$, and $\{j, f, h, l\}$, respectively. The non-fundamental cycle $\{a, d, g, c, f\}$ is the sum (mod 2) of the fundamental cycles of chords d and g.

3. A spanning tree and its fundamental system of cycles.

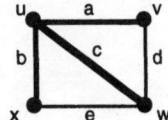 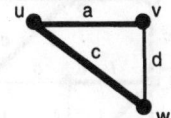 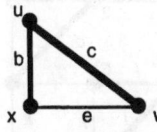

4. A spanning tree and its fundamental system of edge-cuts.

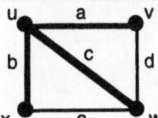 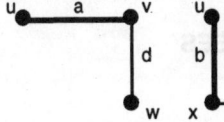 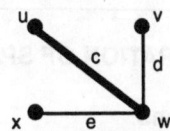

5. Suppose for the graph of Example 1, that the local order of adjacencies at each vertex is the alphabetic order of the edge labels. Then the construction of the DFS-tree is as follows:

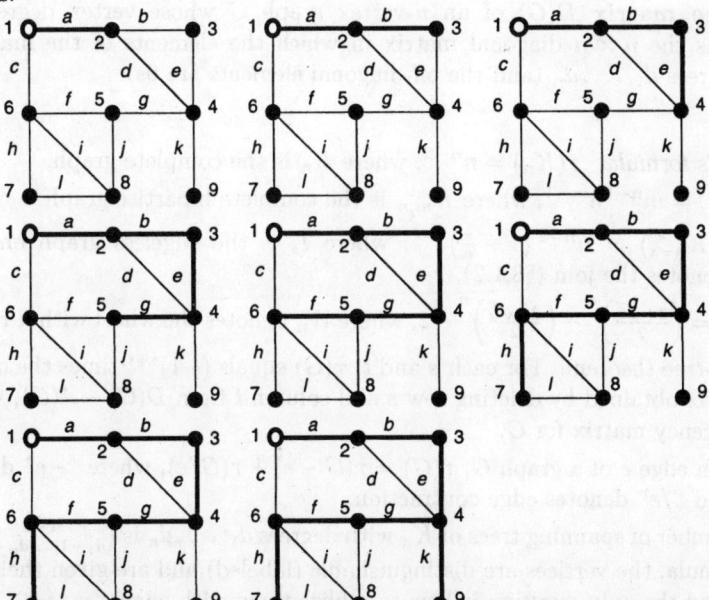

6. Suppose for the graph of Example 1, that the local order of adjacencies at each vertex is the alphabetic order of the edge labels. Then the construction of the BFS-tree is as follows:

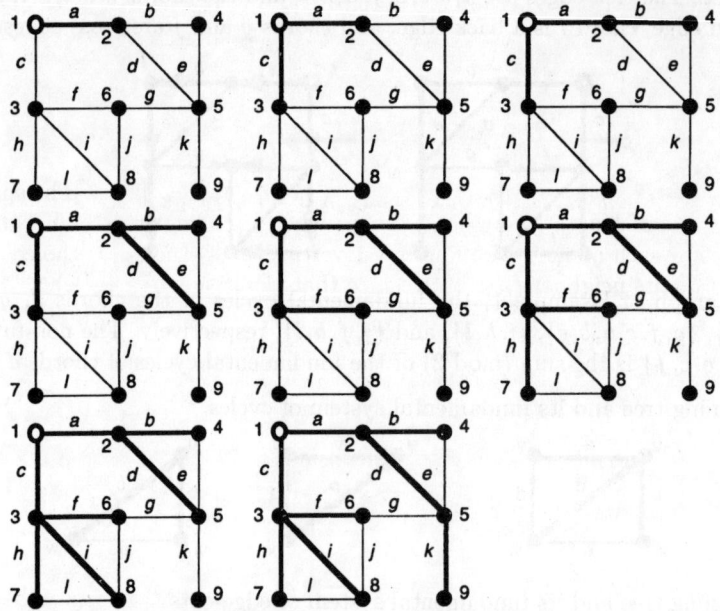

9.2.2 ENUMERATION OF SPANNING TREES

Definitions:

The **number of spanning trees** $\tau(G)$ of a graph G counts two spanning trees T_1 and T_2 as different if their edgesets are different, even if there is an automorphism of G mapping T_1 onto T_2.

The **degree matrix** $D(G)$ of an n-vertex graph G whose vertex degree sequence $d_1, \ldots, d_n$ is the $n \times n$ diagonal matrix in which the elements of the main diagonal are the degrees $d_1, \ldots, d_n$ (and the off-diagonal elements are 0s).

Facts:

1. Cayley's formula: $\tau(K_n) = n^{n-2}$, where K_n is the complete graph.
2. $\tau(K_{m,n}) = m^{n-1} n^{m-1}$, where $K_{m,n}$ is the complete bipartite graph.
3. $\tau(I_s + K_{n-s}) = n^{n-2} \left(1 - \frac{s}{n}\right)^{s-1}$, where I_s is the edgeless graph on n vertices and "+" denotes the join (§8.1.2).
4. $\tau(W_n) = \left(\frac{3+\sqrt{5}}{2}\right)^n + \left(\frac{3-\sqrt{5}}{2}\right)^n - 2$, where W_n denotes the wheel with n rim vertices.
5. Matrix-tree theorem: For each s and t, $\tau(G)$ equals $(-1)^{s+t}$ times the determinant of the matrix obtained by deleting row s and column t from $D(G) - A(G)$, where $A(G)$ is the adjacency matrix for G.
6. For each edge e of a graph G, $\tau(G) = \tau(G - e) + \tau(G/e)$, where "$-e$" denotes edge deletion and "$/e$" denotes edge contraction.
7. The number of spanning trees of K_n with degrees $d_1, \ldots, d_n$ is $\binom{n-2}{d_1-1, \ldots, d_n-1}$ (§2.3.2). In this formula, the vertices are distinguishable (labeled) and are given their degrees in advance, and the only question is how to realize them with edges.

Examples:

1. $\tau(K_3) = 3^{3-2} = 3$. Each of the three spanning trees is a path on two edges, as illustrated below. Also, $\tau(K_4) = 4^{4-2} = 16$.

2. $\tau(K_{2,n}) = n2^{n-1}$. To confirm this, let $X = \{x_1, x_2\}$ and $|Y| = n$. The spanning tree contains a path of length 2 joining x_1 to x_2, whose middle vertex in Y can be chosen in n ways. For each of the remaining $n - 1$ vertices of Y, there is a choice as to which of x_1 and x_2 is its neighbor (not both, since that would create a cycle).

3. $\tau(I_3 + K_2) = 5^3 \left(1 - \frac{3}{5}\right)^2 = 20$.

4. $\tau(W_4) = \left(\frac{3+\sqrt{5}}{2}\right)^4 + \left(\frac{3-\sqrt{5}}{2}\right)^4 - 2 = 45$.

5. To illustrate the matrix-tree theorem, consider the following graph G.

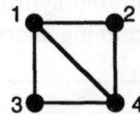

Then
$$D(G) - A(G) = \begin{pmatrix} 3 & -1 & -1 & -1 \\ -1 & 2 & -1 & 0 \\ -1 & -1 & 3 & -1 \\ -1 & 0 & -1 & 2 \end{pmatrix}.$$

Deleting row 2 and column 3, for example, yields
$$\tau(G) = (-1)^{2+3} \begin{vmatrix} 3 & -1 & -1 \\ -1 & -1 & -1 \\ -1 & 0 & 2 \end{vmatrix} = 8.$$

The 8 spanning trees of G are:

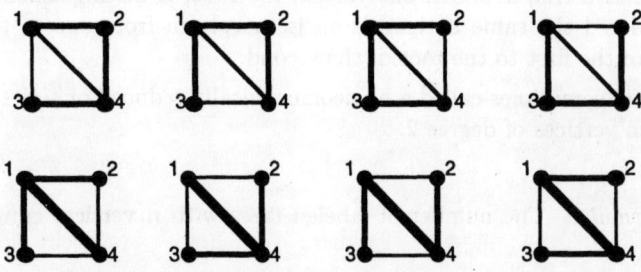

6. The recursive formula $\tau(G) = \tau(G-e) + \tau(G/e)$ is illustrated with the same graph G of the previous example and with $e = v_1 v_3$. In the computation G is drawn instead of writing $\tau(G)$, and similarly with the other graphs. This yields

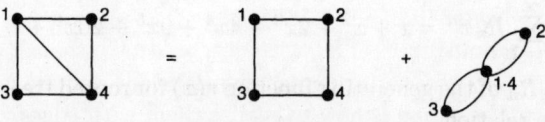

7. Let the vertices of K_5 be v_0, v_1, v_2, v_3, v_4. The number of spanning trees of K_5 in which the degrees of v_0, v_1, v_2, v_3, v_4 are $3, 2, 1, 1, 1$, respectively, is given by the multinomial coefficient $\binom{5-2}{3-1, 2-1, 1-1, 1-1, 1-1} = \frac{3!}{2! \cdot 1! \cdot 0! \cdot 0! \cdot 0!} = \frac{6}{2 \cdot 1 \cdot 1 \cdot 1 \cdot 1} = 3$. The three trees in question are:

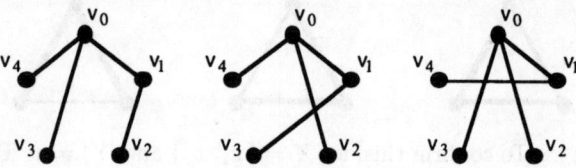

9.3 ENUMERATING TREES

Tree counting began with Arthur Cayley in 1889, who wanted to enumerate the saturated hydrocarbons. George Pólya developed an extensive theory in 1937 for counting families of organic chemicals, which was used by Richard Otter in 1948 in his solution of the specific problem of counting saturated hydrocarbons. Tree counting formulas are used in computer science to estimate running times in the design of algorithms.

9.3.1 COUNTING GENERIC TREES

Definitions:

A **tree** is a connected graph with no cycles. Two trees are considered the "same", for counting purposes, if they are isomorphic.

A **labeled tree** is a tree in which distinct labels such as $v_1, v_2, \ldots, v_n$ have been assigned to the vertices. Two labeled trees with the same set of labels are considered the same only if there is an isomorphism from one tree to the other such that each vertex is mapped to the vertex with the same label.

A **rooted tree** is a tree in which one vertex, the root, is distinguished. Two rooted trees are considered the same if there is an isomorphism from one to the other that maps the root of the first to the root of the second.

A **reduced tree**, sometimes called a homeomorphically reduced or series reduced tree, is a tree with no vertices of degree 2.

Facts:

1. *Cayley's formula*: The number of labeled trees with n vertices equals n^{n-2}. See Table 1.

2. The number of rooted labeled trees with n vertices equals n^{n-1}. See Table 1.

3. Rooted trees and most other tree structures can be counted by using generating functions.

4. The generating function $r(x)$ for the number R_n of rooted trees with n vertices (see Table 2) is $r(x) = \sum_{n=1}^{\infty} R_n x^n = x + x^2 + 2x^3 + 4x^4 + 9x^5 + 20x^6 + \cdots$.

5. The coefficients R_n of the generating function $r(x)$ for rooted trees can be determined from the recurrence relation

Table 1 Labeled trees and rooted labeled trees with n vertices.

n	labeled trees	rooted labeled trees
1	1	1
2	1	2
3	3	9
4	16	64
5	125	625
6	1,296	7,776
7	16,807	117,649
8	262,144	2,097,152
9	4,782,969	43,046,721
10	100,000,000	1,000,000,000
11	2,357,947,691	25,937,424,601
12	61,917,364,224	743,008,370,688
13	1,792,160,394,037	23,298,085,122,481
14	56,693,912,375,296	793,714,773,254,144
15	1,946,195,068,359,375	29,192,926,025,390,625
16	72,057,594,037,927,936	1,152,921,504,606,846,980

$$r(x) = x \prod_{i=1}^{\infty} (1 - x^i)^{-R_i}.$$

An alternative defining expression for this generating function is

$$r(x) = x \exp\left(\sum_{i=1}^{\infty} \frac{r(x^i)}{i} \right).$$

6. The generating function $t(x) = \sum_{n=1}^{\infty} T_n \cdot x^n = x + x^2 + x^3 + 2x^4 + 3x^5 + 6x^6 + \cdots$
for counting trees (see Table 2) is obtained from that for rooted trees by using Otter's formula

$$t(x) = r(x) - \tfrac{1}{2}\left(r(x)^2 - r(x^2) \right).$$

7. The generating function $h(x) = \sum_{n=1}^{\infty} H_n \cdot x^n = x + x^2 + x^4 + x^5 + 2x^6 + 2x^7 + 4x^8 + \cdots$
for counting reduced trees (see Table 2) is based on another function $f(x)$ determined by the equation

$$f(x) = \tfrac{x}{1+x} \prod_{i=1}^{\infty} (1 - x^i)^{-F_i} = \tfrac{x}{1+x} \exp\left(\sum_{i=1}^{\infty} \frac{f(x^i)}{i} \right).$$

Then

$$h(x) = (1+x)f(x) - \tfrac{1+x}{2} f(x)^2 + \tfrac{1-x}{2} f(x^2).$$

Note that there are no reduced trees with exactly 3 vertices.

Examples:

1. There are exactly three trees with five vertices:

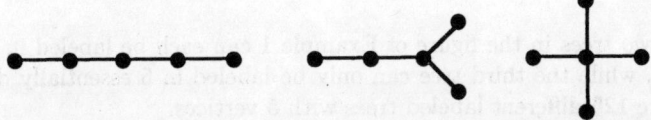

Table 2 Rooted trees, trees, and reduced trees with n vertices.

n	R_n (rooted trees)	T_n (trees)	H_n (reduced trees)
1	1	1	1
2	1	1	1
3	2	1	0
4	4	2	1
5	9	3	1
6	20	6	2
7	48	11	2
8	115	23	4
9	286	47	5
10	719	106	10
11	1,842	235	14
12	4,766	551	26
13	12,486	1,301	42
14	32,973	3,159	78
15	87,811	7,741	132
16	235,381	19,320	249
17	634,847	48,629	445
18	1,721,159	123,867	842
19	4,688,676	317,955	1,561
20	12,826,228	823,065	2,988
21	35,221,832	2,144,505	5,671
22	97,055,181	5,623,756	10,981
23	268,282,855	14,828,074	21,209
24	743,724,984	39,299,897	41,472
25	2,067,174,645	104,636,890	81,181
26	5,759,636,510	279,793,450	160,176
27	16,083,734,329	751,065,460	316,749
28	45,007,066,269	2,023,443,032	629,933
29	126,186,554,308	5,469,566,585	1,256,070
30	354,426,847,597	14,830,871,802	2,515,169
31	997,171,512,998	40,330,829,030	5,049,816
32	2,809,934,352,700	109,972,410,221	10,172,638
33	7,929,819,784,355	300,628,862,480	20,543,579
34	22,409,533,673,568	823,779,631,721	41,602,425
35	63,411,730,258,053	2,262,366,343,746	84,440,886
36	179,655,930,440,464	6,226,306,037,178	171,794,492
37	509,588,049,810,620	17,169,677,490,714	350,238,175
38	1,447,023,384,581,029	47,436,313,524,262	715,497,037
39	4,113,254,119,923,150	131,290,543,779,126	1,464,407,113
40	11,703,780,079,612,453	363,990,257,783,343	3,002,638,286

2. The first two trees in the figure of Example 1 can each be labeled in 60 essentially different ways, while the third tree can only be labeled in 5 essentially different ways. Thus, there are 125 different labeled trees with 5 vertices.

3. The first tree in the figure of Example 1 can be rooted in 3 essentially different ways, and thus corresponds to 3 different rooted trees. The second and third trees in that figure represent 4 and 2 different rooted trees, respectively. Thus there are 9 different rooted trees with 5 vertices.

4. The third tree in the figure of Example 1 is the only reduced tree with 5 vertices.

9.3.2 COUNTING TREES IN CHEMISTRY

Definitions:

A *1-4 tree* is a tree in which each vertex has degree 1 or 4.

A *1-rooted 1-4 tree* is a 1-4 tree rooted at a vertex of degree 1.

Facts:

1. Saturated hydrocarbons, also called alkanes, are compounds with the chemical formula $C_n H_{2n+2}$; they consist of n carbon atoms of valence 4 and $2n+2$ hydrogen atoms of valence 1. The molecular structure of alkanes is modeled by the 1-4 trees.

Note: It is convenient when counting alkanes to include the hydrogen molecule H_2, which has no carbon atoms and 2 hydrogen atoms, as an honorary alkane.

2. A monosubstituted hydrocarbon has n carbon atom, $2n+1$ hydrogen atoms, and an OH group. They have the chemical formula $C_n H_{2n+1} OH$; they include the familiar alcohols.

Note: It is convenient when counting alcohols to include the water molecule HOH as an honorary alcohol.

3. The number A_n (see Table 3) of 1-rooted 1-4 trees (alcohols) with n 4-valent vertices (carbon atoms), $2n+1$ non-root 1-valent vertices (hydrogen atoms), and a 1-valent root (the OH group) has the generating function

$$a(x) = \sum_{n=0}^{\infty} A_n x^n = 1 + x + x^2 + 2x^3 + 4x^4 + 8x^5 + 17x^6 + \cdots$$

whose coefficients can be determined from the recurrence relation

$$a(x) = 1 + \tfrac{x}{6}\left(a(x)^3 + 3a(x)a(x^2) + 2a(x^3)\right)$$

4. In counting unrooted 1-4 trees, a preliminary step is to count the number G_n of 1-4 trees rooted at a vertex of degree 4. The coefficients of the corresponding generating function

$$g(x) = \sum_{n=1}^{\infty} G^n x^n = x + x^2 + 2x^3 + 4x^4 + 9x^5 + 18x^6 + \cdots$$

are determined by the equation

$$g(x) = \tfrac{x}{24}\left(a(x)^4 + 6a(x)^2 a(x^2) + 8a(x)a(x^3) + 3a(x^2)^2 + 6a(x^4)\right).$$

5. The number B_n (see Table 3) of 1-4 trees (alkanes) with n 4-valent vertices (carbon atoms) and $2n+2$ 1-valent vertices (hydrogen atoms) has the generating function

$$b(x) = \sum_{n=0}^{\infty} B_n \cdot x^n = 1 + x + x^2 + x^3 + 2x^4 + 3x^5 + 5x^6 + \cdots$$

Table 3 1-Rooted 1-4 trees and 1-4 trees with n vertices of degree 4.

n	A_n: 1-rooted 1-4 trees (alcohols)	B_n: 1-4 trees (alkanes)
1	1	1
2	1	1
3	2	1
4	4	2
5	8	3
6	17	5
7	39	9
8	89	18
9	211	35
10	507	75
11	1,238	159
12	3,057	355
13	7,639	802
14	19,241	1,858
15	48,865	4,347
16	124,906	10,359
17	321,198	24,894
18	830,219	60,523
19	2,156,010	148,284
20	5,622,109	366,319
21	14,715,813	910,726
22	38,649,152	2,278,658
23	101,821,927	5,731,580
24	269,010,485	14,490,245
25	712,566,567	36,797,588
26	1,891,993,344	93,839,412
27	5,034,704,828	240,215,803
28	13,425,117,806	617,105,614
29	35,866,550,869	1,590,507,121
30	95,991,365,288	4,111,846,763
31	257,332,864,506	10,660,307,791
32	690,928,354,105	27,711,253,769
33	1,857,821,351,559	72,214,088,660
34	5,002,305,607,153	188,626,236,139
35	13,486,440,075,669	493,782,952,902
36	36,404,382,430,278	1,295,297,588,128
37	98,380,779,170,283	3,404,490,780,161
38	266,158,552,000,477	8,964,747,474,595
39	720,807,976,831,447	23,647,478,933,969
40	1,954,002,050,661,819	62,481,801,147,341

which can be determined from the equation

$$b(x) = g(s) + a(x) - \tfrac{1}{2}\left(a(x)^2 - a(x^2)\right).$$

Examples:

1. The three different 1-4 trees with 5 vertices of degree 4 are:

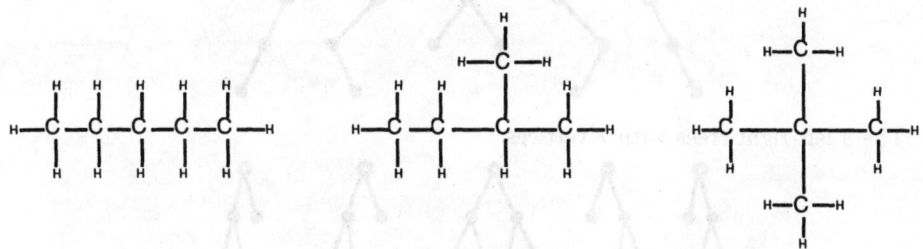

2. The first 1-4 tree in the figure of Example 1 can be rooted at a vertex of degree 1 in 3 essentially different way, the second in 4 essentially different ways, and the third in only 1 essential way. Thus there are 8 different 1-rooted 1-4 trees with 5 vertices.

9.3.3 COUNTING TREES IN COMPUTER SCIENCE

Definitions:

A **binary tree** is a rooted tree in which each vertex has at most two children, and such that each child is designated either a **left child** or a **right child**. An only child may be either a left child or a right child.

A **left-right tree** is a binary tree in which each vertex is a parent either to no children or to both a left child and a right child.

An **ordered tree** is a tree in which the children of every vertex are linearly ordered.

Facts:

1. Binary trees are counted by the Catalan numbers C_n (§3.1.3): the number of binary trees with n vertices is C_n.

2. Each principal subtree of a left-right tree is a left-right tree.

3. Left-right trees are frequently used to represent arithmetic expressions, in which the leaves of the tree correspond to numbers and the other vertices represent binary operations such as $+$, $-$, $\times$, or $\div$.

4. There is an obvious one-to-one correspondence between binary trees with n vertices and left-right trees with $2n+1$ vertices: delete all the leaves of a left-right tree to obtain a binary tree.

5. The number of left-right trees with n internal vertices and $n+1$ leaves is also C_n. This follows from Fact 4.

6. Ordered trees can represent structures such as family trees, showing all descendants of the person represented by the root. The children of each person in the tree would be represented as children of the corresponding vertex, ordered according to birth date.

7. The number of ordered trees with n vertices is C_{n-1}.

Examples:

1. The 5 binary trees with 3 vertices:

2. The 5 left-right trees with 7 vertices:

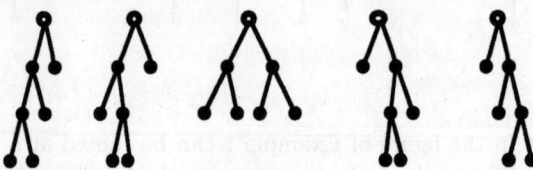

3. The 5 ordered trees with 4 vertices:

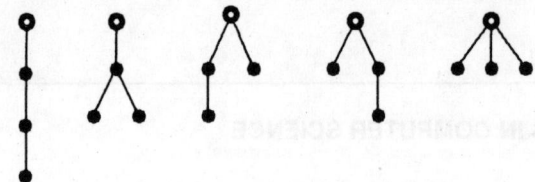

REFERENCES

Printed Resources:

[AhUl92] A. V. Aho and J. D. Ullman, *Foundations of Computer Science*, Computer Science Press, 1992.

[HaPa73] F. Harary and E. M. Palmer, *Graphical Enumeration*, Academic Press, 1973.

[PóRe87] G. Pólya and R. C. Read, *Combinatorial Enumeration of Groups, Graphs, and Chemical Compounds*, Springer-Verlag, 1987.

[Ro95] K. H. Rosen, *Discrete Mathematics and Its Applications*, 4th ed., McGraw-Hill, 1999.

[Se80] J. P. Serre, *Trees*, Springer-Verlag, 1980.

[SlPl95] N. J. A. Sloane and S. Plouffe, *The Encyclopedia of Integer Sequences*, Academic Press, 1995.

[Ta83] R. E. Tarjan, *Data Structures and Network Algorithms*, Soc. for Industrial and Applied Math., 1983.

[We96] D. B. West, *Introduction to Graph Theory*, Prentice-Hall, 1996.

Web Resources:

http://www.crl.com/~windbase/ (Software vendor, advertises software implementing data structures.)

http://www.cs.sunysb.edu/~algorith/ (The Stony Brook Algorithm Repository; see Sections 1.4 and 1.5 on Graph Problems.)

10

NETWORKS AND FLOWS

10.1 Minimum Spanning Trees	*J. B. Orlin and*
10.1.1 Basic Concepts	*Ravindra K. Ahuja*
10.1.2 Algorithms for Minimum Spanning Trees	
10.1.3 Parallel Algorithms	
10.1.4 Applications	
10.2 Matchings	*Douglas R. Shier*
10.2.1 Basic Concepts	
10.2.2 Matchings in Bipartite Networks	
10.2.3 Matchings in Nonbipartite Networks	
10.2.4 Applications	
10.3 Shortest Paths	*J. B. Orlin and*
10.3.1 Basic Concepts	*Ravindra K. Ahuja*
10.3.2 Algorithms for Single-Source Shortest Paths	
10.3.3 Algorithms for All-Pairs Shortest Paths	
10.3.4 Parallel Algorithms	
10.3.5 Applications	
10.4 Maximum Flows	*J. B. Orlin and*
10.4.1 Basic Concepts	*Ravindra K. Ahuja*
10.4.2 Algorithms for Maximum Flows	
10.4.3 Applications	
10.5 Minimum Cost Flows	*J. B. Orlin and*
10.5.1 Basic Concepts	*Ravindra K. Ahuja*
10.5.2 Algorithms for Minimum Cost Flows	
10.5.3 Applications	
10.6 Communication Networks	*David Simchi-Levi and*
10.6.1 Capacitated Minimum Spanning Tree Problem	*Sunil Chopra*
10.6.2 Capacitated Concentrator Location Problem	
10.6.3 Capacity Assignment Problem	
10.6.4 Models for Survivable Networks	
10.7 Difficult Routing and Assignment Problems	*Bruce L. Golden and*
10.7.1 Traveling Salesman Problem	*B. K. Kaku*
10.7.2 Vehicle Routing Problem	
10.7.3 Quadratic Assignment Problem	
10.8 Network Presentations and Data Structures	*Douglas R. Shier*
10.8.1 Network Representations	
10.8.2 Tree Data Structures	

INTRODUCTION

The vertices and edges of a graph often have quantitative information associated with them, such as supplies and demands (for vertices), and distance, length, capacity, and cost (for edges). Relative to such networks, a number of discrete optimization problems arise in a variety of disciplines: statistics, electrical engineering, operations research, combinatorics, and computer science. Typical applications include designing least cost telecommunication systems, maximizing throughput in a manufacturing system, finding a minimum cost route or set of routes for delivery vehicles, and distributing electricity from a set of supply points to meet customer demands at minimum cost. In this chapter, a number of classical network optimization problems are studied and algorithms are described for their exact or approximate solution.

GLOSSARY

adjacency matrix: a 0-1 matrix whose (i,j) entry indicates the absence or presence, respectively, of an arc joining vertex i to vertex j in a graph.

adjacency set: the set of arcs emanating from a specified vertex.

alternating path (in a matching): a path with edges that are alternately free and matched.

arc list: a list of the arcs of a graph, presented in no particular order.

assignment (from a set S to a set T): a bijective function from S onto T.

augmenting path (in a flow network): a directed path between two specified vertices in which each arc has a positive residual capacity.

augmenting path (in a matching): an alternating path between two free vertices.

backbone network: a collection of devices that interconnect vertices at which message exchanges occur in a communication network.

blossom: an odd length cycle formed by adding an edge joining two even vertices on an alternating path.

capacitated concentrator location problem: a network design problem in which a minimum cost configuration of concentrators and their connections to terminals is sought so that each concentrator's total capacity is not exceeded.

capacitated minimum spanning tree: a minimum cost collection of subtrees joined to a specified root vertex, in which the total amount of demand generated by each subtree is bounded above by a constant.

capacitated network: a network in which arc is assigned a capacity.

capacity (of an arc): the maximum amount of material that can flow along the arc.

capacity (of a cut): for a cut $[S,\overline{S}]$, the sum of the capacities of arcs (i,j) with $i \in S$ and $j \in \overline{S}$.

capacity (of a path): the smallest capacity of any arc on the path.

capacity assignment problem: a network design problem in which links of different capacities are to be installed at minimum cost to support a number of point-to-point communication demands.

complete matching: in a bipartite graph $G = (X \cup Y, E)$, a matching M in which each vertex of X is incident with an edge of M.

composite (*hybrid*) *method*: a heuristic algorithm that combines elements of both construction methods and improvement methods.

construction method: a heuristic algorithm that builds a feasible solution, starting with a trivial configuration.

cost (of a flow): $\sum_{(i,j)} c_{ij} x_{ij}$, where c_{ij} is the cost and x_{ij} is the flow on arc (i,j).

cut (in a graph): the set of edges $[S, \overline{S}]$ in the graph joining vertices in S to vertices in the complementary set $\overline{S}$.

directed network: a vertex set V and an arc set E, where each directed arc has an associated cost, length, weight, or capacity.

directed out-tree: a tree rooted at vertex s such that the unique path in the tree from vertex s to every other vertex is a directed path.

distance label: an estimate (in particular, an upper bound) on the shortest path length from the source vertex to each network vertex.

even vertex (in an alternating path): given an alternating path P, a vertex on P that is reached using an even number of edges of P, starting from the origin vertex of P.

exact algorithm: a procedure that produces a verifiable optimal solution to every problem instance.

flow: a feasible assignment of material that satisfies flow conservation and arc capacity restrictions.

forward star: a compact representation of a graph in which information about arcs leaving a vertex is stored using consecutive locations of an array.

free edge (in a matching): an edge that does not appear in the matching.

free vertex (in a matching): a vertex that is incident with no matched edges.

heuristic algorithm: a procedure that produces a feasible, though not necessarily optimal, solution to every problem instance.

improvement method: a heuristic algorithm that starts with a suboptimal solution (often randomly generated) and attempts to improve it.

length (of a path): the sum of all costs appearing on the arcs of the path.

linear assignment problem (*LAP*): an optimization problem in which an assignment is sought that minimizes an appropriate set-up cost.

link capacity: an upper bound on the amount of traffic that a communication link can carry at any one time.

linked adjacency list: a collection of singly-linked lists used to represent a graph.

local access network: a network used to transfer traffic between the backbone network and the end users.

matched edge (in a matching): an edge that appears in the matching.

matched vertex (in a matching): a vertex that is incident with a matched edge.

matching (in a graph): a set of pairwise nonadjacent edges in the graph.

mate (of a matched vertex): in a matching, the other endpoint of the matched edge incident with the given vertex.

maximum flow (in a network): a flow in the network having maximum value.

maximum size matching: a matching having the largest size.

maximum spanning tree (of a network): a spanning tree of the network with maximum cost.

maximum weight matching: a matching having the largest weight.

metaheuristic: a general-purpose heuristic procedure (such as tabu search, simulated annealing, genetic algorithms, or neural networks) for solving difficult optimization problems.

minimum cost flow (in a network): a flow in the network having minimum cost.

minimum cut (in a network): a cut in the network having minimum capacity.

minimum spanning tree (of a network): a spanning tree of the network with minimum cost.

negative cycle: a directed cycle of negative cost (or length).

odd vertex (in an alternating path): given an alternating path P, a vertex on P that is reached using an odd number of edges of the path P, starting from the origin vertex of P.

perfect matching: a matching in a graph in which each vertex of the graph is incident with exactly one edge of the matching.

predecessor: relative to a rooted tree, the vertex preceding a given vertex on the unique path from the root to the given vertex.

preflow: a relaxation of flow where inflow into a vertex can be greater than its outflow.

pseudoflow: a relaxation of flow where inflow into a vertex need not be equal to its outflow.

quadratic assignment problem (QAP): an optimization problem in which an assignment is sought that minimizes the sum of set-up and interaction costs.

reduced cost of arc (i, j): relative to given vertex potentials π, the quantity $c_{ij}^{\pi} = c_{ij} - \pi(i) + \pi(j)$.

residual capacity (of an arc): the maximum additional flow (with respect to a given flow) that can be sent on an arc.

residual network: a network consisting of arcs with positive residual capacity.

s-t cut: a cut $[S, \overline{S}]$ in which $s \in S$ and $t \in \overline{S}$.

savings: the reduction in cost from joining two vertices directly compared to joining both to a central vertex.

shortest path: a directed path between specified vertices having minimum total cost (or length).

size (of a matching): the number of edges in the matching.

survivable network: a network that can survive failures in some of its vertices or edges and still transfer a prespecified amount of traffic.

traveling salesman problem (TSP): an optimization problem in which a fixed set of cities must be visited in some order at minimum total cost.

two-phase method: a heuristic algorithm that implements a cluster first/route second philosophy.

undirected network: a vertex set V and an edge set E, where each undirected edge has an associated cost, length, weight, or capacity.

value of a flow: the total flow leaving the source vertex.

vehicle routing problem (VRP): an optimization problem in which a given set of customers must be serviced at minimum total cost, using a fleet of vehicles having fixed capacity.

vertex potential: a quantity $\pi(i)$ associated with each vertex i of a network.

weight (of a matching): the sum of the weights of edges in the matching.

10.1 MINIMUM SPANNING TREES

In an undirected network, the minimum spanning tree problem is the problem of identifying a spanning tree of the network that has the smallest possible sum of edge costs. This problem arises in a number of applications, both as a stand-alone problem and as a subproblem in more complex problem settings. It is assumed throughout this section that the network is connected.

10.1.1 BASIC CONCEPTS

Definitions:

An **undirected network** is a weighted graph (§8.1.1) $G = (V, E)$, where V is the set of vertices, E is the set of undirected edges, and each edge $(i,j) \in E$ has an associated **cost** (or **weight**, **length**) c_{ij}. Let $n = |V|$ and $m = |E|$.

If $T = (V, F)$ is a spanning tree (§9.2) of $G = (V, E)$, then every edge in $F \subseteq E$ is a **tree edge** and every edge in $E - F$ is a **nontree edge** (or **chord**).

A **minimum spanning tree (MST)** of G is a spanning tree of G for which the sum of the edge costs is minimum.

A **maximum spanning tree** of G is a spanning tree of G for which the sum of the edge costs is maximum.

A **cut** of $G = (V, E)$ is a partition of the vertex set V into two parts, S and $\overline{S} = V - S$. Each cut defines the set of edges $[S, \overline{S}] \subseteq E$ having one endpoint in S and the other endpoint in $\overline{S}$.

Facts:

1. Every spanning tree T of a network G with n vertices contains exactly $n - 1$ edges, and every two vertices of T are connected by a unique path.
2. Adding an edge to a spanning tree of G produces a unique cycle, called a *fundamental cycle* (§9.2.1).
3. Every cut $[S, \overline{S}]$ is a disconnecting set of edges (§8.4.2). However, not every disconnecting set of edges can be represented as a cut $[S, \overline{S}]$; see Example 2.
4. Removing an edge from a spanning tree of G produces two subtrees, on vertex sets S and $\overline{S}$, respectively. The associated cut $[S, \overline{S}]$ is called a *fundamental cut*.
5. *Path optimality conditions*: A spanning tree T^* is a minimum spanning tree of G if and only if for each nontree edge (k, l) of G, $c_{ij} \leq c_{kl}$ holds for all tree edges (i, j) in the fundamental cycle determined by edge (k, l).

6. *Cut optimality conditions*: A spanning tree T^* is a minimum spanning tree of G if and only if for each tree edge $(i,j) \in T^*$, $c_{ij} \leq c_{kl}$ holds for all nontree edges (k,l) in the fundamental cut determined by edge (i,j).

7. If all edge costs are different, then the minimum spanning tree is unique.

8. The minimum spanning tree can be unique even if some of the edge costs are equal; see Example 1.

9. Adding a constant to all edge costs of an undirected network does not change the minimum spanning tree(s) of the network. Thus, it is sufficient to have an algorithm that works when all edge costs are positive.

10. Multiplying each edge cost of an undirected network by -1 converts a maximum spanning tree into a minimum spanning tree, and vice versa. Thus, it is sufficient to have algorithms to find a minimum spanning tree.

Examples:

1. Part (a) of the following figure shows an undirected network G, with costs indicated on each edge. Part (b) shows a spanning tree T^* of G. Adding the nontree edge $(3,5)$ to T^* produces the fundamental cycle $[3,1,2,5,3]$; see part (c). Since each tree edge in this cycle has cost no more than that of the nontree edge $(3,5)$, the path optimality condition is satisfied by edge $(3,5)$. Similarly, it can be verified that the other nontree edges, namely $(2,3)$, $(4,5)$, and $(5,6)$, satisfy the path optimality conditions, establishing by Fact 5 that T^* is a minimum spanning tree. By Fact 7 this is the unique minimum spanning tree.

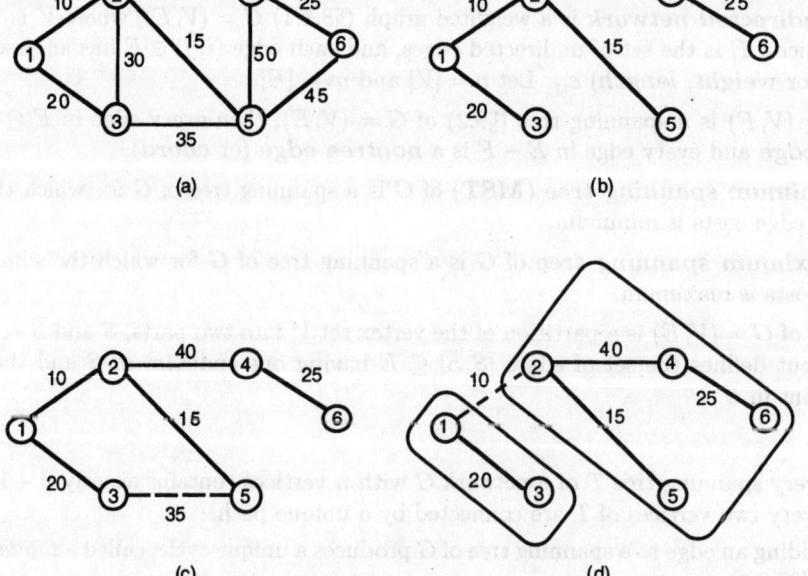

2. For the tree edge $(1,2)$ in part (b) of the figure of Example 1, the fundamental cut $[S, \overline{S}]$ formed by deleting edge $(1,2)$ has $S = \{1,3\}$ and $\overline{S} = \{2,4,5,6\}$; see part (d) of the figure. This cut contains two nontree edges, $(2,3)$ and $(3,5)$. Since each such nontree edge has cost greater than or equal to that of the tree edge $(1,2)$, the cut optimality condition is satisfied for edge $(1,2)$. Similarly, it can be verified that the other tree edges, namely $(1,3)$, $(2,4)$, $(2,5)$, and $(4,6)$, satisfy the cut optimality conditions, establishing by Fact 6 that T^* is a minimum spanning tree.

3. The undirected network of part (a) of the following figure has 4 vertices and 5 edges, with the edge cost shown beside each edge. This network contains 8 spanning trees, which are listed in the table in part (b) of the figure. The spanning tree T_5 achieves the minimum cost 7 among all the spanning trees and so is a minimum spanning tree. In fact, T_5 is the unique minimum spanning tree, even though the edge costs are not all distinct. See Fact 8.

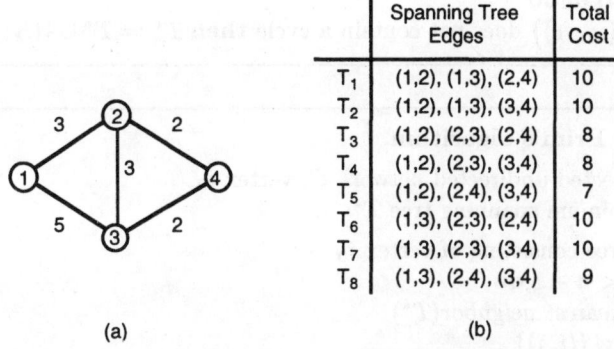

(a)   (b)

4. The set of edges $F = \{(2,3), (2,4), (3,4)\}$ is a disconnecting set in the network G of Example 3, since removal of these edges disconnects G. However, there is no partition of the vertex set of G into nonempty sets S and $\overline{S}$ for which $F = [S, \overline{S}]$.

10.1.2 ALGORITHMS FOR MINIMUM SPANNING TREES

There are several greedy algorithms for constructing minimum spanning trees, based on the optimality conditions in §10.1.1, Facts 5 and 6. Each of these algorithms myopically (greedily) adds an edge to the current configuration based on only local information; nonetheless, these procedures are guaranteed to produce a minimum spanning tree.

Definitions:

The **nearest neighbor** operation takes as input a tree T^* having vertex set S and produces a minimum cost edge (i, j) in the cut $[S, \overline{S}]$. That is, $c_{ij} = \min\{c_{ab} \mid a \in S, b \notin S\}$.

The **merge** operation takes as input an edge (i, j) whose two endpoints i and j belong to disjoint trees T_i and T_j and combines the trees into $T_i \cup T_j \cup \{(i, j)\}$.

The graph $G = (V, E)$ is assumed connected and has n vertices and m edges.

Facts:

1. *Kruskal's algorithm*: This greedy algorithm (Algorithm 1) is based on the path optimality conditions (§10.1.1) and builds a minimum spanning tree by examining edges of E one by one in nondecreasing order of their costs. The edge being examined is added to the current forest if its addition does not create a cycle. (J. B. Kruskal, born 1928)

2. Kruskal's algorithm can be terminated once $n - 1$ edges have been added to T^*.

3. Computer code (in Fortran) that implements Kruskal's algorithm can be found at the site:

    http://www.mat.uc.pt/~eqvm/cientificos/fortran/codigos.html

---

Algorithm 1: Kruskal's algorithm.

input: connected undirected network G
output: minimum spanning tree T^*

order the edges $(i_1, j_1), (i_2, j_2), \ldots, (i_m, j_m)$ so that $c_{i_1 j_1} \leq c_{i_2 j_2} \leq \cdots \leq c_{i_m j_m}$
$T^* := \emptyset$
for $k := 1$ to m do
   if $T^* \cup \{(i_k, j_k)\}$ does not contain a cycle then $T^* := T^* \cup \{(i_k, j_k)\}$

Algorithm 2: Prim's algorithm.

input: connected undirected network G, vertex i_0
output: minimum spanning tree T^*

$T^* :=$ the tree consisting of vertex i_0
while $|T^*| < n - 1$ do
   $(i, j) :=$ nearest_neighbor(T^*)
   $T^* := T^* \cup \{(i, j)\}$

---

4. Kruskal's algorithm can be implemented using several data structures yielding different time bounds:

   - [AhMaOr93] describes an implementation that runs in $O(m + n \log n)$ time plus the time needed for sorting the m edges;
   - [Ta83] describes an improved implementation that runs in $O(ma(n, m))$ time plus the time needed for sorting the m edges; here $a(n, m)$ is the inverse Ackermann function which for all practical purposes is less than 5.

5. Algorithm 1 was independently discovered by Kruskal (1956) and by H. Loberman and A. Weinberger (1957).

6. *Prim's algorithm*: This algorithm (Algorithm 2) is based on the cut optimality conditions (§10.1.1). It maintains a single tree T^*, which initially consists of an arbitrary vertex i_0. At each iteration, the algorithm adds the least cost edge emanating from T^* until a spanning tree is obtained. (R. C. Prim, born 1921)

7. Algorithm 2 was first proposed in 1930 by V. Jarník. Later it was independently discovered by Prim (1957) and by E. W. Dijkstra (1959).

8. Running times of several implementations of Prim's algorithm are shown in the following table. See [AhMaOr93] for a discussion of these implementations.

data structure	running time
binary heap	$O(m \log n)$
d-heap	$O(m \log_d n)$, with $d = \max\{2, \lceil \frac{m}{n} \rceil\}$
Fibonacci heap	$O(m + n \log n)$

9. A modification of Prim's algorithm has running time $O(m \log b(m, n))$, where the function $b(m, n)$ grows very slowly (for all practical purposes is less than 5) [GaEtal86]. This is currently the theoretically fastest algorithm for solving the minimum spanning tree problem.

Algorithm 3: Sollin's algorithm.

input: connected undirected network G
output: minimum spanning tree T^*

$T^* :=$ forest of all vertices of G, but no edges
while $|T^*| < n - 1$
    let $T_1, T_2, \ldots, T_p$ be the trees in the forest T^*
    for $k := 1$ to p
        $(i_k, j_k) :=$ nearest_neighbor(T_k)
    for $k := 1$ to p
        if i_k and j_k belong to different trees then
            merge(i_k, j_k)
            $T^* := T^* \cup \{(i_k, j_k)\}$

10. Computer codes (in Fortran) implementing Prim's algorithm can be found at the following three sites:

    http://www.netlib.org/toms/479

    http://www.netlib.org/toms/613

    http://www.mat.uc.pt/~eqvm/cientificos/fortran/codigos.html

11. *Sollin's algorithm:* This greedy algorithm (Algorithm 3) is also based on the cut optimality conditions (§10.1.1). It starts with a forest of n trees, each consisting of a single vertex, and builds a minimum spanning tree by repeatedly adding edges to the current forest. At each iteration a least cost edge emanating from each tree is added, leading to the merging of certain trees.

12. Each iteration of Algorithm 3 reduces the number of trees in the forest T^* by at least half.

13. Sollin's algorithm performs $O(\log n)$ iterations and can be implemented to run in $O(m \log n)$ time; see [AhMaOr93].

14. A variation of Sollin's algorithm that runs in time $O(m \log \log n)$ can be found in [Ya75].

15. The origins of Algorithm 3 can be traced to O. Boruvka (1926), who first formulated the minimum spanning tree problem in the context of electric power networks. This algorithm was independently proposed in 1938 by G. Choquet for points in a metric space and by G. Sollin in 1961 for arbitrary networks.

16. Sollin's algorithm lends itself to a parallel implementation (see §10.1.3), though care must be taken when edge costs are not distinct in order to ensure that no cycles are produced.

17. Computational studies have found that the Prim and Sollin algorithms consistently outperform Kruskal's algorithm. Prim's algorithm is faster when the network is dense, whereas Sollin's algorithm is faster when the network is sparse.

18. An excellent discussion of the history of the minimum spanning tree problem is provided in [GrHe85].

19. An important variant of the minimum spanning tree problem places constraints on the number of edges incident with a vertex in a candidate spanning tree. Such *degree-constrained* minimum spanning trees are investigated in [GlKl75] and [Vo89].

20. Another variant is the *capacitated* minimum spanning tree problem, which arises in the design of local access telecommunication networks. In this problem, a feasible spanning tree is one rooted at a specified central vertex such that the total traffic (number of calls) generated by each subtree connected to the central vertex does not exceed a known capacity. A feasible spanning tree having minimum total cost is then sought. (See §10.6.1 for further details.)

Examples:

1. For the network shown in part (a) of the following figure, ordering edges by nondecreasing cost produces the following sequence of edges: $(2,4)$, $(3,5)$, $(3,4)$, $(2,3)$, $(4,5)$, $(1,2)$, $(1,3)$. Kruskal's algorithm adds the edges $(2,4)$, $(3,5)$, $(3,4)$ to T^*; discards the edges $(2,3)$ and $(4,5)$; then adds the edge $(1,2)$ to T^* and terminates. Part (b) of the figure shows the resulting minimum spanning tree, having total cost 80.

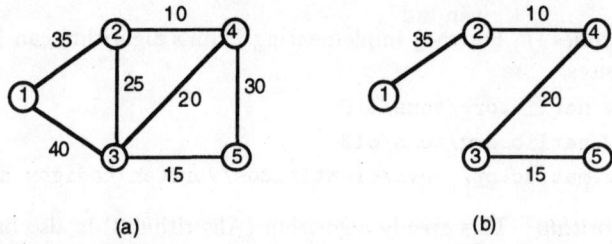

2. Prim's algorithm (Algorithm 2) is applied to the network of part (a) of the figure of Example 1, starting with the initial vertex $i_0 = 3$. The minimum cost edge out of vertex 3 is $(3,5)$, so $T^* = \{(3,5)\}$. Next, the minimum cost edge emanating from T^* is $(3,4)$, giving $T^* = \{(3,5),(3,4)\}$. Subsequent iterations add the edges $(2,4)$ and $(1,2)$, producing the minimum spanning tree $T^* = \{(3,5),(3,4),(2,4),(1,2)\}$. Starting from any other initial vertex i_0 would give the same result.

3. To apply Sollin's algorithm (Algorithm 3) to the network of part (a) of the figure of Example 1, begin with a forest containing five trees, each consisting of a single vertex. Part (a) of the following figure shows the least cost edge emanating from each of these trees. One iteration of Algorithm 3 produces the two trees shown in part (b) of the following figure. The least cost edge emanating from either of these two trees is $(3,4)$. Adding this edge completes the minimum spanning tree shown in part (c) of this figure.

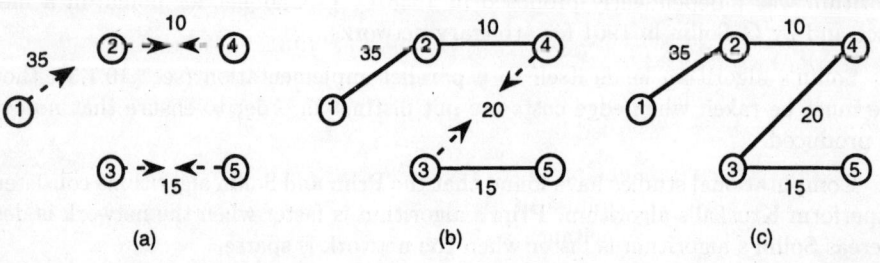

10.1.3 PARALLEL ALGORITHMS

Sollin's algorithm (§10.1.2) can be easily parallelized in EREW (exclusive-read, exclusive-write) PRAM (parallel random-access machine). (See §16.1.4.) This algorithm assigns a processor to each edge and each vertex of the network [KiLe88].

Facts:

1. Sollin's algorithm performs $O(\log n)$ iterations.

2. In each iteration, every component finds a least cost edge emanating from it in $O(\log n)$ time. To do this, each vertex finds a least cost edge emanating from it to a vertex in a different component. Next, a minimization is done over all vertices of the given component.

3. In each iteration, components that are connected by the newly found edges are merged using a procedure called recursive doubling. This operation can also be done in $O(\log n)$ time.

4. Overall the running time of the resulting algorithm is $O(\log^2 n)$ using $O(m)$ processors.

5. The most work-efficient parallel algorithm currently known for solving the minimum spanning tree problem is given in [CoKeTa94].

10.1.4 APPLICATIONS

Minimum spanning tree problems arise both directly and indirectly. For direct applications, the points in a given set are to be connected using the least cost collection of edges. For indirect applications, creative modeling of the original problem recasts it as a minimum spanning tree problem.

Applications:

1. *Designing physical systems*: A minimum cost network is to be designed to connect geographically dispersed system components. Each component is represented by a vertex, with potential network connections between vertices represented by edges. A cost is associated with each edge.

2. Examples of Application 1 occur in the following:
   - Connect terminals in cabling the panels of electrical equipment in order to use the least total cost of wire.
   - Construct a pipeline network to connect a number of towns using the smallest possible total cost of pipeline.
   - Link isolated villages in a remote region, which are connected by roads but not yet by telephone service. The problem is to determine along which stretches of roads to place telephone lines to link every pair of villages, using the minimum total miles of installed lines.
   - Construct a digital computer system, composed of high-frequency circuitry, when it is important to minimize the length of wires between different components to reduce both capacitance and delay line effects.
   - Connect a number of computer sites by high-speed lines. Each line is available for leasing at a certain monthly cost, and a configuration is required that connects all the sites at minimum overall cost.
   - Design a backbone network of high-capacity links that connect switching devices to support internet traffic. A minimum cost backbone network that maintains acceptable throughput is required.

3. *Clustering*: Objects having k measurable characteristics are to be clustered into groups of "similar" objects. First, construct an undirected network, where each object is represented by a vertex and every two distinct vertices are joined by an edge. The cost of edge (i,j) is the distance (in k-dimensional space) between the k-vectors for objects i and j. Applying Kruskal's algorithm to this network then yields a hierarchy of partitions of the vertex set; each partition is defined by the trees comprising the forest obtained at each iteration of Kruskal's algorithm. This hierarchy is then used to define clusters of the original objects.

4. Computation of minimum spanning trees sometimes arises as a subproblem in a larger optimization problem. For example, one heuristic approach to the *traveling salesman problem* (§10.7.1) involves the calculation of minimum spanning trees.

5. *Optimal message passing*: An intelligence service has agents operating in a non-friendly country. Each agent knows some of the other agents and has in place procedures for arranging a rendezvous with someone he knows. For each such possible rendezvous, say between agent i and agent j, any message passed between these agents will fall into hostile hands with a certain probability p_{ij}. The group leader wants to transmit a confidential message among all the agents while maximizing the probability that no message is intercepted.

If the agents are represented by vertices and each possible rendezvous by an edge, then in the resulting graph G a spanning tree T is required that maximizes the probability that no message is intercepted, given by $\Pi_{(i,j)\in T}(1 - p_{ij})$. Such a tree can be found by defining the cost of each edge (i,j) as $\log(1 - p_{ij})$ and solving a *maximum spanning tree problem*.

6. *All-pairs minimax path problem*: In this variant of the shortest path problem (see §10.3.1), the value of a path P is the maximum cost edge in P. The all-pairs minimax path problem is to determine a minimum value path between every pair of vertices in a network G. It can be shown that if T^* is a minimum spanning tree of G, then the unique path in T^* between any pair of vertices is also a minimax path between that pair of vertices.

7. Examples of Application 6 arise in the following contexts:
    - Determine the trajectory of a spacecraft that keeps the maximum temperature of the surface as small as possible.
    - When traveling through a desert, select a route that minimizes the length of the longest stretch between rest areas.
    - A person traveling in a wheelchair desires a route that minimizes the maximum ascent along the path segments of the route.

8. *Measuring homogeneity of bimetallic objects*: In this application minimum spanning trees are used to determine the degree to which a bimetallic object is homogenous in composition. First, the composition of the bimetallic object is measured at a set of sample points. A network is then constructed with vertices corresponding to the sample points and with an edge connecting physically adjacent sample points. The cost of edge (i,j) is the product of the physical (Euclidean) distance between sample points i and j, and a homogeneity factor between 0 and 1. The homogeneity factor is 0 if the composition of the corresponding samples is identical, and is 1 if the composition is very different. This cost structure gives greater weight to two points if they have different compositions and are far apart. Then the cost of the minimum spanning tree provides an overall measure of the homogeneity of the object.

9. Additional applications, with reference sources, are given in the following table.

application	references
two-dimensional storage schemes	[AhMaOr93], [AhEtal95]
chemical physics	[AhMaOr93]
manufacturing	[EvMi92]
network design	[AhMaOr93]
network reliability	[AhEtal95]
pattern classification	[GrHe85], [AhMaOr93]
picture processing	[GrHe85], [AhMaOr93]
automatic speech recognition	[GrHe85]
numerical taxonomy	[GrHe85]

10.2 MATCHINGS

In an undirected network, the maximum matching problem is to find a set of nonadjacent edges that has the largest total size or weight. This discrete optimization problem arises in a number of applications, often involving the optimal pairing of a set of objects.

10.2.1 BASIC CONCEPTS

Definitions:

Let $G = (V, E)$ be an undirected network with vertex set V and edge set E (see §10.1.1). Assume that G contains neither loops nor multiple edges. Each edge $e = (i, j) \in E$ has an associated **weight** $w_e = w_{ij}$. Let $n = |V|$ and $m = |E|$.

The **degree** of vertex $v \in V$ in G is the number of edges in G that are incident with v, written $deg(v)$. (See §8.1.1.)

A **matching** in $G = (V, E)$ is a set $M \subseteq E$ of pairwise nonadjacent edges (§8.1.1).

A **perfect matching** in $G = (V, E)$ is a matching M in which each vertex of V is incident with exactly one edge of M.

The **size (cardinality)** of a matching M is the number of edges in M, written $|M|$.

The **weight** of a matching M is $wt(M) = \sum_{e \in M} w_e$.

A **maximum size** matching of G is a matching M having the largest size $|M|$.

A **maximum weight** matching of G is a matching M having the largest weight $wt(M)$.

Relative to a matching M in $G = (V, E)$, edges $e \in M$ are **matched** edges, while edges $e \in E - M$ are **free** edges. Vertex v is **matched** if it is incident with a matched edge; otherwise vertex v is **free**.

Every matched vertex v has a **mate**, the other endpoint of the matched edge incident with v.

With respect to a matching M, the **weight** $wt(P)$ of path P is the sum of the weights of the free edges in P minus the sum of the weights of the matched edges in P.

An **alternating** path has edges that are alternately free and matched. An **augmenting** path is an alternating path that starts and ends at a free vertex.

Facts:

1. Matchings are useful in a wide variety of applications, such as assigning personnel to jobs, target tracking, crew scheduling, snowplowing streets, scheduling on parallel machines, among others (see §10.2.4).
2. In a matching M, each vertex of G has degree 0 or 1 relative to the edges in M. In a perfect matching M, each vertex of G has degree 1 relative to the edges in M.
3. If M is any matching in G, then $|M| \leq \lceil \frac{n}{2} \rceil$.
4. Every augmenting path has an odd number of edges.
5. If M is a matching and P is an augmenting path with respect to M, then the symmetric difference (§1.2.2) $M \triangle P$ is a matching of size $|M| + 1$.
6. If M is a matching and P is an augmenting path with respect to M, then $wt(M \triangle P) = wt(M) + wt(P)$.
7. *Augmenting path theorem*: M is a maximum size matching if and only if there is no augmenting path with respect to M.
8. Fact 7 was obtained independently by C. Berge (1957) and by R. Z. Norman and M. O. Rabin (1959). This result was also recognized in an 1891 paper of J. Petersen.
9. Suppose M is a matching having maximum weight among all matchings of a fixed size k. If P is an augmenting path of maximum weight, then $M \triangle P$ is a matching having maximum weight among all matchings of size $k + 1$.
10. Suppose paths $P_1, P_2, \ldots, P_k$ are obtained as in Fact 9 by augmenting along a maximum weight path. Then $wt(P_1) \geq wt(P_2) \geq \cdots \geq wt(P_k)$.
11. The number of perfect matchings of the complete graph (§8.1.3) K_{2n} on $2n$ vertices is $(2n-1)!! = 1 \cdot 3 \cdot 5 \ldots (2n-1)$.
12. An historical perspective on the theory of matchings is found in [Pl92].

Examples:

1. Part (a) of the following figure displays a network G with the weight w_e shown next to each edge e.

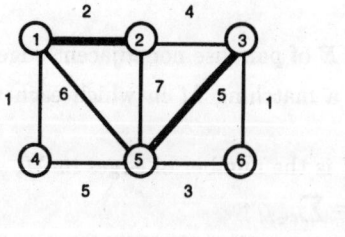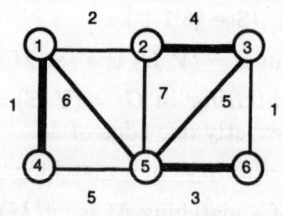

(a) (b)

The matching $M_1 = \{(1,2),(3,5)\}$ of size 2 is also shown, with the matched edges highlighted. The mate of vertex 1 is vertex 2, and the mate of vertex 5 is vertex 3. The weight of M_1 is $wt(M_1) = 7$. Relative to the matching M_1, vertices 4 and 6 are free vertices, and an augmenting path P from 4 to 6 is given by the set of edges $P = \{(1,4),(1,2),(2,3),(3,5),(5,6)\}$. Here $wt(P) = 1 + 4 + 3 - 2 - 5 = 1$ and (as guaranteed by Fact 4) path P has an odd number of edges. The matching $M_2 = M_1 \triangle P = \{(1,2),(3,5)\} \triangle \{(1,4),(1,2),(2,3),(3,5),(5,6)\} = \{(1,4),(2,3),(5,6)\}$ is a perfect matching and is highlighted in part (b) of the figure. There are no free vertices relative to matching M_2 and no augmenting paths, so M_2 is a maximum size matching of G. There are other maximum size matchings, such as $\{(1,4),(2,5),(3,6)\}$ and $\{(1,2),(4,5),(3,6)\}$.

2. Part (a) of the following figure shows a matching M_1 of size 1, with $wt(M_1) = 7$. Since edge $(2,5)$ has maximum weight among all edges, M_1 is a maximum weight matching of size 1. Relative to M_1 the augmenting path $P_1 = \{(1,5),(2,5),(2,3)\}$ has weight $6 + 4 - 7 = 3$, whereas the augmenting path $\{(3,6)\}$ has weight 1. It can be verified that P_1 is a maximum weight augmenting path relative to M_1. By Fact 9, $M_2 = M_1 \Delta P_1 = \{(1,5),(2,3)\}$ is a maximum weight matching of size 2 in the network, with $wt(M_2) = 10$; see part (b) of the figure.

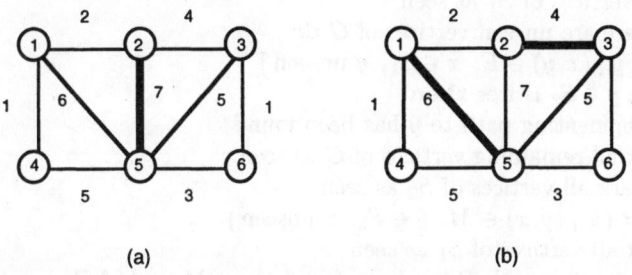

(a)   (b)

Relative to M_2 there are several augmenting paths between the free vertices 4 and 6:

$$Q_1 = \{(1,4),(1,5),(5,6)\}, \quad wt(Q_1) = 1 + 3 - 6 = -2,$$
$$Q_2 = \{(1,4),(1,5),(2,5),(2,3),(3,6)\}, \quad wt(Q_2) = 1 + 7 + 1 - 6 - 4 = -1,$$
$$Q_3 = \{(4,5),(1,5),(1,2),(2,3),(3,6)\}, \quad wt(Q_3) = 5 + 2 + 1 - 6 - 4 = -2.$$

The maximum weight augmenting path is Q_2 and so (by Fact 9) $M_3 = M_2 \Delta Q_2 = \{(1,4),(2,5),(3,6)\}$ is a maximum weight matching of size 3 in the network with $wt(M_3) = 9$. Overall, the maximum weight matching in G is M_2, as expected since all augmenting paths relative to M_2 have negative weight (see Fact 10).

10.2.2 MATCHINGS IN BIPARTITE NETWORKS

In this section, algorithms are described for finding maximum size and maximum weight matchings in bipartite networks (§8.1.3). Bipartite networks arise in a number of applications, such as in assigning personnel to jobs or tracking objects over time. Moreover, the algorithms developed for the case of bipartite networks are considerably simpler than those needed for the case of general networks (§10.2.3).

Definitions:

Let $G = (X \cup Y, E)$ be a bipartite network with n vertices and m edges, and edge weights w_{xy}.

If $S \subseteq X$ then $\Gamma(S) = \{y \in Y \mid (x,y) \in E \text{ for some } x \in X\}$ is the set of vertices in Y adjacent to some vertex of X.

A **complete matching** from X to Y in $G = (X \cup Y, E)$ is a matching M in which each vertex of X is incident with an edge of M.

The **directed two-terminal flow network** G' associated with $G = (X \cup Y, E)$ is defined by adding new vertices s and t, as well as arcs (s,x) for each $x \in X$ and arcs (y,t) for each $y \in Y$. All other arcs (x,y) of G' correspond to edges (x,y) of G where $x \in X$ and $y \in Y$. Every arc of G' has capacity 1.

Algorithm 1: Bipartite matching algorithm.

input: undirected bipartite network $G = (X \cup Y, E)$
output: maximum size matching M
$M := \emptyset$
while true **do**
  let S_1 consist of all free vertices of X
  mark all vertices of S_1 as seen
  while there are unseen vertices of G **do**
    $S_2 := \{ y \mid (x, y) \in E,\ x \in S_1,\ y \text{ unseen} \}$
    if some $y \in S_2$ is free **then**
      an augmenting path to y has been found
      mark all remaining vertices of G as seen
    else mark all vertices of S_2 as seen
      $S_1 := \{ x \mid (y, x) \in M,\ y \in S_2,\ x \text{ unseen} \}$
      mark all vertices of S_1 as seen
  if an augmenting path P has been found **then** $M := M \triangle P$
  else terminate with matching M

Facts:

1. *Hall's theorem:* $G = (X \cup Y, E)$ has a complete matching from X to Y if and only if $|\Gamma(S)| \geq |S|$ holds for every $S \subseteq X$. In words, a complete matching exists precisely when every set of vertices in X is adjacent to at least an equal number of vertices in Y. (Philip Hall, 1904–1982.)

2. *Sufficient condition for a complete matching:* Suppose there exists some k such that $deg(x) \geq k \geq deg(y)$ holds in $G = (X \cup Y, E)$ for all $x \in X$ and $y \in Y$. Then G has a complete matching from X to Y.

3. There is a one-to-one correspondence between matchings of size k in G and integral flows (§10.4.1) of value k in the associated two-terminal flow network G'.

4. A maximum flow in G', and thereby a maximum size matching of G, can be found in $O(m\sqrt{n})$ time.

5. Suppose that costs are added to the two-terminal flow network G', using $c_{ij} = 0$ if $i = s$ or $j = t$, and $c_{ij} = -w_{ij}$ otherwise. By starting with the flow (§10.5.1) $x = 0$, the successive shortest path algorithm (§10.5.2) can be repeatedly applied to G' until a shortest augmenting path has negative cost. The resulting minimum cost flow will yield (via Fact 3) a matching with maximum weight.

6. *Bipartite matching algorithm:* This method (Algorithm 1), based on §10.2.1 Fact 7, produces a maximum size matching of the bipartite network $G = (X \cup Y, E)$. Each iteration involves a modified breadth first search of G, starting with the free vertices in the set X. All vertices of G are structured into levels that alternate between free and matched edges.

7. Algorithm 1 can be implemented to run in $O(nm)$ time.

8. *Bipartite weighted matching algorithm:* This method (Algorithm 2), based on §10.2.1 Facts 9 and 10, produces a maximum weight matching of $G = (X \cup Y, E)$. Each iteration develops a longest path tree in G, rooted at the set of free vertices in X. The tentative largest weight of a path from a free vertex in X to vertex j is maintained in the label $d(j)$.

9. Algorithm 2 can be implemented to run in $O(nm)$ time.

Algorithm 2:   Bipartite weighted matching algorithm.

input: undirected bipartite network $G = (X \cup Y, E)$, weights w_e
output: maximum weight matching M

$M := \emptyset$
while true **do**
   let S_1 consist of all free vertices of X
   $d(j) := 0$ for $j \in S_1$, $d(j) := -\infty$ otherwise
   while $S_1 \neq \emptyset$ **do**
     $S_2 := \emptyset$
     for $(x, y) \in E - M$ with $x \in S_1$ **do**
       if $d(x) + w_{xy} > d(y)$ **then**
         $d(y) := d(x) + w_{xy}$, $S_2 := S_2 \cup \{y\}$
     $S_1 := \emptyset$
     for $(y, x) \in M$ with $y \in S_2$ **do**
       if $d(y) - w_{yx} > d(x)$ **then**
         $d(x) := d(y) - w_{yx}$, $S_1 := S_1 \cup \{x\}$
   $y :=$ a free vertex with maximum label $d(y)$
   $P :=$ the associated augmenting path
   if $d(y) > 0$ **then** $M := M \Delta P$
   else terminate with matching M

10. *Stable marriage problem*: A variation of the bipartite matching problem is the *stable marriage problem*, defined for a set X of n men and n women. Each person has a strict ranking of the n people of the opposite sex. A perfect matching is *stable* if it is impossible to find a man and a woman who are not matched to each other, yet each of these two prefers one another to their respective mates. For every set of rankings, a stable matching exists and can be found using a greedy algorithm [AhMaOr93].

Examples:
1. *Drug testing*: A drug company is testing n antibiotics on n volunteer patients in a hospital. Some of the patients have known allergic reactions to certain of these antibiotics. To determine whether there is a feasible assignment of the n different antibiotics to n different patients, construct the bipartite network $G = (X \cup Y, E)$, where X is the set of antibiotics and Y is the set of patients. An edge $(i, j) \in E$ exists when patient j is not allergic to antibiotic i. A complete matching of G is then sought.
2. Part (a) of the following figure shows a bipartite graph G with $X = \{1, 2, 3, 4\}$ and $Y = \{a, b, c, d\}$.

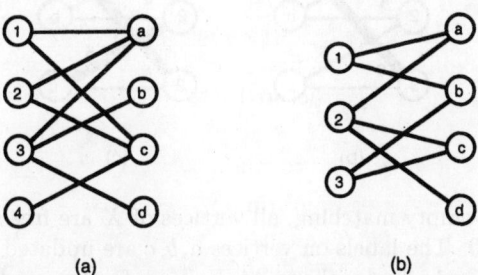

(a)   (b)

Using Fact 1, there cannot be a complete matching from X to Y: if $S = \{1, 2, 4\}$ then $\Gamma(S) = \{a, c\}$ and $|\Gamma(S)| < |S|$. There is, however, a (maximum) matching of size 3: for example, $\{(1, c), (2, a), (3, d)\}$.

3. Part (b) of the figure of Example 2 shows a bipartite graph G with $X = \{1, 2, 3\}$ and $Y = \{a, b, c, d\}$. Since $deg(x) \geq 2 \geq deg(y)$ holds for all $x \in X$ and $y \in Y$, there must exist a complete matching from X to Y. One such complete matching is given by $\{(1, a), (2, c), (3, b)\}$.

4. Algorithm 1 is used to find a maximum size matching in the bipartite graph of part (a) of the following figure.

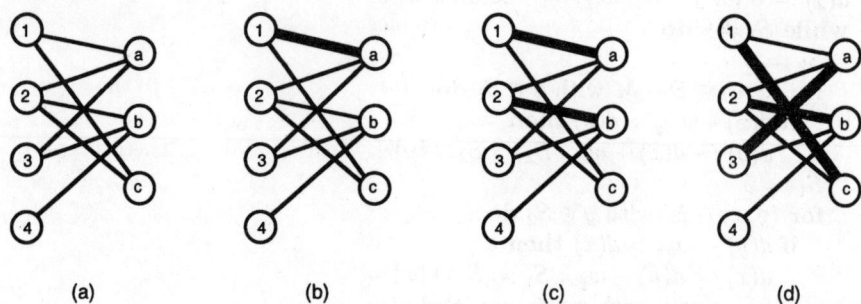

Relative to the initial empty matching, all vertices of X are free so $S_1 = \{1, 2, 3, 4\}$, giving $S_2 = \{a, b, c\}$. In particular, vertex $a \in S_2$ is free and an augmenting path to a is $P = \{(1, a)\}$. The resulting matching is $M = \{(1, a)\}$, shown in part (b) of the figure.

The second iteration of Algorithm 1 starts with $S_1 = \{2, 3, 4\}$, giving $S_2 = \{a, b, c\}$. An augmenting path to the free vertex b is $P = \{(2, b)\}$, resulting in $M = \{(1, a), (2, b)\}$; see part (c) of the figure.

At the next iteration, $S_1 = \{3, 4\}$ and $S_2 = \{a, b\}$. Since both vertices of S_2 are matched, the algorithm continues with $S_1 = \{1, 2\}$ and $S_2 = \{c\}$. Since $c \in S_2$ is free, with augmenting path $P = \{(3, a), (a, 1), (1, c)\}$, the new matching produced is $M = \{(1, c), (2, b), (3, a)\}$; see part (d) of the figure.

The fourth iteration produces $S_1 = \{4\}$, $S_2 = \{b\}$; $S_1 = \{2\}$, $S_2 = \{a, c\}$; and finally $S_1 = \{1, 3\}$, $S_2 = \emptyset$. No further augmenting paths are found, and the algorithm terminates with the maximum size matching $M = \{(1, c), (2, b), (3, a)\}$.

5. Algorithm 2 is used to find a maximum weight matching in the bipartite network of part (a) of the following figure.

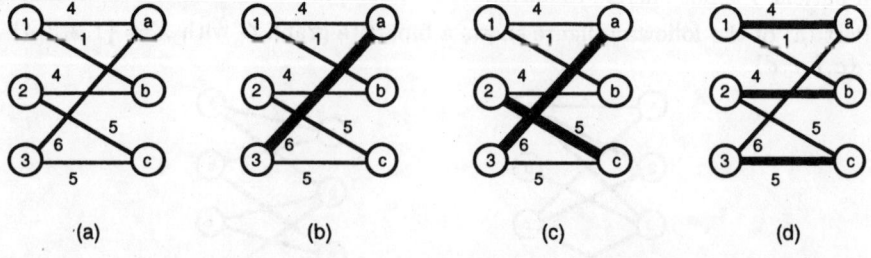

Relative to the initial empty matching, all vertices of X are free so $S_1 = \{1, 2, 3\}$, with $d(1) = d(2) = d(3) = 0$. The labels on vertices a, b, c are updated to $d(a) = 6$, $d(b) = 4$, $d(c) = 5$, giving $S_2 = \{a, b, c\}$. Since $M = \emptyset$ no further updates occur. The free vertex a has maximum label, and the associated path $P_1 = \{(3, a)\}$ has $wt(P_1) = 6$. The resulting matching $M = \{(3, a)\}$ is shown in part (b) of the following figure; it represents the largest weight matching of size 1, with $wt(M) = 6$.

The second iteration starts with $S_1 = \{1,2\}$. The labels on vertices a, b, c are then updated to $d(a) = 4$, $d(b) = 4$, $d(c) = 5$, so $S_2 = \{a, b, c\}$. Using the matched edge $(a, 3)$, vertex 3 has its label updated to $d(3) = -2$ and $S_1 = \{3\}$. No further updates occur, and free vertex c with maximum label $d(c) = 5$ is selected. This label corresponds to the augmenting path $P_2 = \{(2, c)\}$, with $wt(P_2) = 5$. The new matching is $M = \{(2, c), (3, a)\}$, with $wt(M) = 11$; see part (c).

At the third iteration, $S_1 = \{1\}$ and vertices a, b receive updated labels $d(a) = 4$, $d(b) = 1$.

Subsequently, updates are made to produce $d(3) = -2$, $d(c) = 3$, $d(2) = -2$, $d(b) = 2$.

Finally, the free vertex b is selected with $d(b) = 2$, corresponding to the augmenting path $P_3 = \{(1, a), (a, 3), (3, c), (c, 2), (2, b)\}$ with $wt(P_3) = 2$. This produces the maximum weight matching $M = \{(1, a), (2, b), (3, c)\}$, with $wt(M) = 13$; see part (d). As predicted by Fact 10 of §10.2.1, the weights of the augmenting paths are nonincreasing: $wt(P_1) \geq wt(P_2) \geq wt(P_3)$.

10.2.3 MATCHINGS IN NONBIPARTITE NETWORKS

This section covers matchings in more general (nonbipartite) networks. Algorithms for constructing maximum size and maximum weight matchings are considerably more intricate than for bipartite networks. The important new concept is that of a "blossom" in a network.

Definitions:

Suppose P is an alternating path from a free vertex s in network $G = (V, E)$. Then a vertex v on P is **even** (**outer**) if the subpath P_{sv} of P joining s to v has even length. Vertex v on P is **odd** (**inner**) if P_{sv} has odd length.

Suppose P is an alternating path from a free vertex s to an even vertex v and edge $(v, w) \in E$ joins v to another even vertex w on P. Then $P \cup \{(v, w)\}$ contains a unique cycle, called a **blossom**.

A **shrunken blossom** results when a blossom B is collapsed into a single vertex b, whereby every edge (x, y) with $x \notin B$ and $y \in B$ is transformed into the edge (x, b). The reverse of this process gives an **expanded blossom**.

Facts:

1. Every blossom B has odd length $2k + 1$ and contains k matched edges, for some $k \geq 1$.

2. A bipartite network contains no blossoms.

3. *Edmonds' theorem*: Suppose network G^B is formed from G by collapsing blossom B. Then G^B contains an augmenting path if and only if G does. (J. Edmonds, 1965.)

4. *General matching algorithm*: This method (Algorithm 3), based on Fact 7 of §10.2.1, produces a maximum size matching of G. At each iteration, a forest of trees is grown, rooted at the free vertices of G, to find an augmenting path. As encountered, blossoms B are shrunk, with the search continued in the resulting network G^B.

> **Algorithm 3:** General matching algorithm.
>
> input: undirected network $G = (V, E)$
> output: maximum size matching M
> $M := \emptyset$
> {Start iteration}
> mark all free vertices as even
> mark all matched vertices as unreached
> mark all free edges as unexamined
> **while** there are unexamined edges (v, w) and no augmenting path is found
>   mark (v, w) as examined
>   {Case 1}
>   **if** v is even and w is unreached **then**
>   mark w as odd and its mate z as even
>   extend the forest by (v, w) and the matched edge (w, z)
>   {Case 2}
>   **if** v, w are even and they belong to different subtrees **then**
>   an augmenting path has been found
>   {Case 3}
>   **if** v, w are even and they belong to the same subtree **then**
>   a blossom B has been found
>   shrink B to an even vertex b
> **if** an augmenting path P has been found **then**
> $M := M \Delta P$
>   go to {Start iteration}
> **else** terminate with matching M

5. Algorithm 3 was initially proposed by Edmonds [Ed65a] with a time bound of $O(n^4)$.

6. An improved implementation of Algorithm 3 runs in $O(nm)$ time.

7. There are other algorithms for maximum size matchings in nonbipartite networks:
   - an algorithm of Gabow [Ga76], which runs in time $O(n^3)$;
   - an algorithm of Micali and Vazirani [MiVa80], that runs in $O(m\sqrt{n})$ time.

Computer codes for these algorithms (in C, Pascal, and Fortran) can be found at these sites:

    ftp://dimacs.rutgers.edu/pub/netflow/matching/
    ftp://ftp.zib.de/pub/Packages/mathprog/matching/

8. *General weighted matching algorithms*: More complicated algorithms are required for solving weighted matching problems. The first such algorithm, also involving blossoms, was developed by Edmonds [Ed65b] and has a time bound of $O(n^4)$.

9. Improved algorithms exist for the weighted matching problem, with running times $O(n^3)$ and $O(nm \log n)$ respectively. Code (in C) for the first of these algorithms can be found at these sites:

    ftp://dimacs.rutgers.edu/pub/netflow/matching/
    ftp://ftp.zib.de/pub/Packages/mathprog/matching/

Examples:
1. In part (a) of the following figure, $P = \{(1,2),(2,3),(3,4),(4,5)\}$ is an alternating but not augmenting path, relative to the matching $M = \{(2,3),(4,5)\}$.

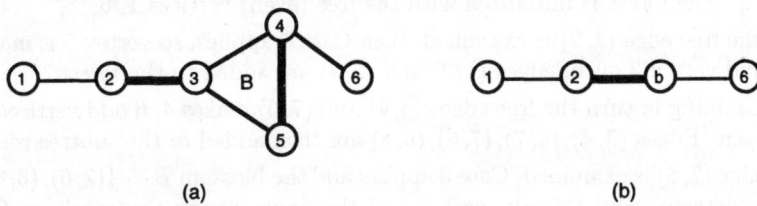

(a)   (b)

Relative to this path, vertices 1, 3, 5 are even while vertices 2, 4 are odd. Since $(3,5)$ is an edge joining two even vertices on P, the blossom $B = \{(3,4),(4,5),(5,3)\}$ is formed. On the other hand, $Q = \{(1,2),(2,3),(3,5),(5,4),(4,6)\}$ is an augmenting path relative to M so that $M \Delta P = \{(1,2),(3,5),(4,6)\}$ is a matching of larger size — in fact a matching of maximum size. Notice that relative to path Q, vertices 1, 3, 4 are even while vertices 2, 5, 6 are odd.

2. Shrinking the blossom B relative to path P in part (a) of the figure of Example 1 produces the network G^B shown in part (b) of that figure. The path $P^B = \{(1,2),(2,b),(b,6)\}$ is now augmenting in G^B. By expanding P^B so that $(2,3)$ remains matched and $(4,6)$ remains free, the augmenting path $Q = \{(1,2),(2,3),(3,5),(5,4),(4,6)\}$ in G is obtained.

3. Algorithm 3 is applied to the nonbipartite network shown in part (a) of the following figure. Suppose the matching $M = \{(3,4),(6,8)\}$ of size 2 is already available.

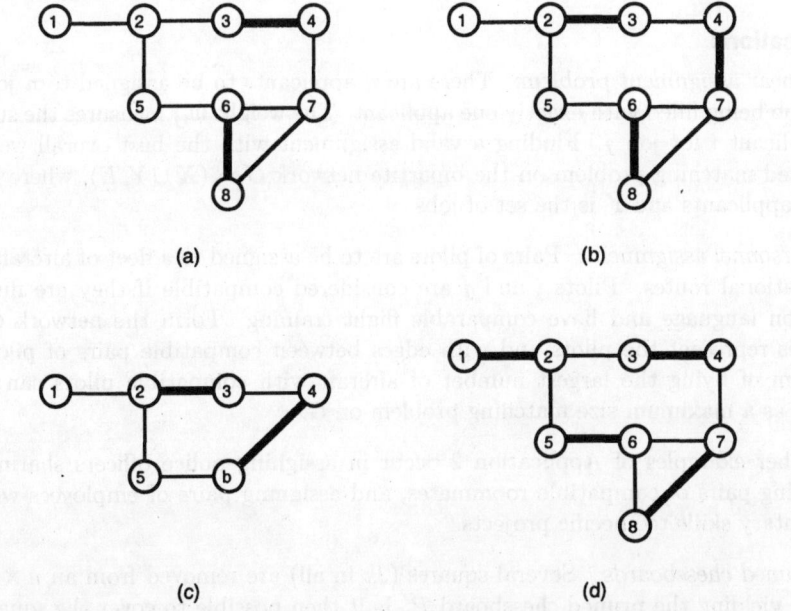

Iteration 1: The free vertices 1, 2, 5, 7 are marked as even, and the matched vertices 3, 4, 6, 8 are marked as unreached. The initial forest consists of the isolated vertices 1, 2, 5, 7.
- If the free edge $(2,3)$ is examined, then Case 1 applies, so vertex 3 is marked odd and vertex 4 even; the free edge $(2,3)$ and the matched edge $(3,4)$ are added to the forest.

- If the free edge $(4,7)$ is examined, then Case 2 applies, and the augmenting path $P = \{(2,3),(3,4),(4,7)\}$ is found. Using P the new matching $M = \{(2,3),(4,7),(6,8)\}$ of size 3 is obtained; see part (b) of the figure.

Iteration 2: The forest is initialized with the free (even) vertices 1, 5.
- If the free edge $(1,2)$ is examined, then Case 1 applies, so vertex 2 is marked odd and vertex 3 even; edges $(1,2)$ and $(2,3)$ are added to the forest.
- Examining in turn the free edges $(3,4)$ and $(7,6)$ makes 4, 6 odd vertices and 7, 8 even. Edges $(3,4),(4,7),(7,6),(6,8)$ are then added to the subtree rooted at 1.
- If edge $(7,8)$ is examined, Case 3 applies and the blossom $B = \{(7,6),(6,8),(8,7)\}$ is detected and shrunk; part (c) of the figure shows the resulting G^B. The current subtree rooted at 1 now becomes $\{(1,2),(2,3),(3,4),(4,b)\}$.
- If the free edge $(b,5)$ is examined, then Case 2 applies and the augmenting path $\{(1,2),(2,3),(3,4),(4,b),(b,5)\}$ is found in G^B. The corresponding augmenting path in G is $P = \{(1,2),(2,3),(3,4),(4,7),(7,8),(8,6),(6,5)\}$. Forming $M \triangle P$ produces the new matching $\{(1,2),(3,4),(7,8),(5,6)\}$, a maximum size matching; see part (d) of the figure.

10.2.4 APPLICATIONS

Matching problems, in both bipartite and nonbipartite networks, are useful models in a number of applied areas. This subsection presents some representative applications of matchings.

Applications:

1. *Linear assignment problem*: There are n applicants to be assigned to n jobs, with each job being filled with exactly one applicant. The weight w_{ij} measures the suitability of applicant i for job j. Finding a valid assignment with the best overall weight is a weighted matching problem on the bipartite network $G = (X \cup Y, E)$, where X is the set of applicants and Y is the set of jobs.

2. *Personnel assignment*: Pairs of pilots are to be assigned to a fleet of aircraft serving international routes. Pilots i and j are considered compatible if they are fluent in a common language and have comparable flight training. Form the network G whose vertices represent the pilots and with edges between compatible pairs of pilots. The problem of flying the largest number of aircraft with compatible pilots can then be solved as a maximum size matching problem on G.

3. Other examples of Application 2 occur in assigning police officers sharing beats, matching pairs of compatible roommates, and assigning pairs of employees with complementary skills to specific projects.

4. *Pruned chessboards*: Several squares ($2k$ in all) are removed from an $n \times n$ chessboard, yielding the pruned chessboard $\mathcal{P}$. Is it then possible to cover the squares of $\mathcal{P}$ using nonoverlapping dominoes, with no squares left uncovered? This can be formulated as a matching problem on the bipartite network $G = (R \cup B, E)$, where R is the set of red squares and B is the set of black squares in $\mathcal{P}$. An edge joins $r \in R$ to $b \in B$ if squares r and b share a common side. Each set of nonoverlapping dominoes on $\mathcal{P}$ corresponds to a matching in G.

All squares of $\mathcal{P}$ can be covered using nonoverlapping dominoes if and only if the maximum size matching in G has size $\frac{n^2}{2} - k$. More generally, the maximum size matching in G explicitly provides a way to cover the maximum number of squares of $\mathcal{P}$ using nonoverlapping dominoes.

5. *Target tracking*: The movements of n objects (such as submarines or missiles) are to be followed over time. The locations of the set of objects are known at two distinct times, though without identification of the individual objects. Suppose $X = \{x_1, x_2, \ldots, x_n\}$ and $Y = \{y_1, y_2, \ldots, y_n\}$ represent the spatial coordinates of the objects detected at times t and $t + \Delta t$. If Δt is sufficiently small, then the Euclidean distance between a given object's position at these two times should be relatively small. To aid in identifying the objects (as well as their velocities and directions of travel), a pairing between set X and set Y is desired that minimizes the sum of Euclidean distances.

This can be formulated as a maximum weight matching problem on the complete bipartite network $G = (X \cup Y, E)$, where the edge (i, j) indicates pairing position x_i with position y_j. The weight of this edge is the negative of the Euclidean distance between x_i and y_j. A maximum weight matching of size n in G then provides an optimal (minimum distance) pairing of observations at the two times t and $t + \Delta t$.

6. *Crew scheduling*: Bus drivers are hired to work two four-hour shifts each day. Union rules require a certain minimum amount of time between the shifts that a driver can work. There are also costs associated with getting the driver between the ending location of the first shift and the starting location of the second shift.

The problem of optimally combining pairs of shifts that satisfy union regulations and incur minimum total cost can be formulated as a maximum weight matching problem. Namely, define the network G with vertices representing each shift that must be covered and edges between pairs of compatible shifts (satisfying union regulations). The weight of edge (i, j) is the negative of the cost of assigning a single driver to shifts i and j. It is convenient also to add edges (i, i) to G to represent the possibility of needing a part-time driver to cover a single shift; edge (i, i) is given a sufficiently large negative weight to discourage single-shift assignments unless absolutely necessary.

A maximum weight matching in the network G then provides a minimum cost pairing of shifts for the bus drivers.

7. *Snowplowing streets*: The streets of an area of a city are to be plowed by a single snowplow. Let G be the network representing the street system of the city, with vertices representing street intersections and edges representing streets. Associated with each street (i, j) is its length c_{ij}.

If all vertices of G have even degree, then G is an Eulerian graph (§8.4.3) and a circuit that traverses each edge (street) exactly once can be found using the algorithms in §8.4.3.

Otherwise, a closed walk of G that covers each street at least once is needed, and one with minimum total length $\sum c_{ij}$ is desired. Let N be the set of vertices of G having odd degree; by Fact 4 of §8.1.1, $|N|$ is an even integer $2k$. Form the complete network $H = (N, E)$ in which the weight of edge (i, j) is the negative of the shortest path distance (§10.3.1) between vertices i and j in G. Determine a maximum weight (perfect) matching M of size k in H. For each (i, j) in M, add the edges of the shortest path between i and j to the network G, forming the network G'. Every vertex of G' now has even degree, and an Euler circuit of G' provides the required minimum cost traversal of the city streets.

This problem is known as the (undirected) *Chinese postman problem*. A directed version of the problem is discusses in §10.5.3, Application 4.

8. Additional applications, with reference sources, are given in the following table.

application	references
medical residents assignment	[AhMaOr93]
school bus driver assignment	[AhMaOr93]
oil well drilling	[LoPu86], [AhMaOr93], [Ge95]
chemical bonds	[AhMaOr93]
inventory depletion	[AhMaOr93]
scheduling on machines	[LoPu86], [AhMaOr93]
ranks of matrices	[AhMaOr93]
doubly stochastic matrices	[LoPu86]
nonnegative matrices	[LoPu86]
basketball conference scheduling	[EvMi92]
major league umpire scheduling	[EvMi92]
project scheduling	[Ge95]
plotting street maps	[Ge95]

10.3 SHORTEST PATHS

The shortest path problem requires finding paths of minimum cost (or length) from a specified source vertex to every other vertex in a directed network. Shortest path problems lie at the heart of network flows (§10.4–10.5). They are important both to researchers and to practitioners because:
- they arise frequently in application settings where material is to be sent between specified points as quickly, as cheaply, or as reliably as possible;
- they arise as subproblems when solving many combinatorial and network optimization problems;
- they can be solved very efficiently.

10.3.1 BASIC CONCEPTS

Definitions:

A *directed network* is a weighted graph $G = (V, E)$, where V is the set of vertices and E is the set of arcs (directed edges). Each arc $(i,j) \in E$ has an associated *cost* (or *weight, length*) c_{ij}. It is possible that certain of the c_{ij} are negative. Let $n = |V|$ and $m = |E|$.

The *adjacency set* $A(i)$ for vertex i is the set of all arcs incident from i, written $A(i) = \{ (i,j) \mid (i,j) \in E \}$.

A directed path (§8.3.2) P has **length** $\Sigma_{(i,j) \in P}\, c_{ij}$.

A directed cycle (§8.3.2) W for which $\Sigma_{(i,j) \in W}\, c_{ij} < 0$ is called a **negative cycle**.

Section 10.3 SHORTEST PATHS

A **shortest path** from vertex s to vertex j is a directed path from s to j having minimum length.

A **directed out-tree** is a tree rooted at vertex s in which all arcs are directed away from the root s.

A **shortest path tree** is an out-tree T^* rooted at vertex s with the property that the directed path in T^* from s to any other vertex j is a shortest s-j path.

A vector $d(\cdot)$ is called a vector of **distance labels** if for every vertex $j \in V$, $d(j)$ is the length of some directed path from the source vertex s to vertex j, with $d(s) = 0$. If these labels are the lengths of shortest s-j paths, they are called **shortest path distances**. The directed path $P = [i_0, i_1, \ldots, i_r]$ from vertex i_0 to vertex i_r can be represented using **predecessor indices**: $pred(i_1) = i_0, pred(i_2) = i_1, \ldots, pred(i_r) = i_{r-1}$.

Facts:

1. Shortest paths are useful in a wide variety of applications, such as in efficient routing of messages and distribution of goods, developing optimal investment strategies, scheduling personnel, and approximating piecewise linear functions (see §10.3.5).

2. If $P = [s, i_1, \ldots, i_r]$ is a shortest path from s to i_r then $Q = [s, i_1, \ldots, i_k]$ is a shortest path from s to i_k for each $1 \le k < r$.

3. **Shortest path optimality conditions**: The vector $d(\cdot)$ of distance labels represents shortest path distances if and only if $d(j) \le d(i) + c_{ij}$ for all $(i, j) \in E$.

4. If the network contains a negative cycle accessible from vertex s, then distance labels satisfying the conditions in Fact 3 do not exist.

5. If the network does not contain any negative cycle, then (unique) distance labels satisfying the conditions in Fact 3 always exist. Furthermore, there is a shortest path tree T^* realizing these shortest path distances.

Examples:

1. In the directed network of the following figure, arc costs are shown along each arc. Part (b) lists the nine paths from vertex 1 to vertex 6, together with their lengths. Path P_4, with length 10, is the (unique) shortest path joining these two vertices. This path can be represented using the predecessor indices: $pred(6) = 4$, $pred(4) = 3$, $pred(3) = 2$, $pred(2) = 1$. By Fact 2, the subpath $Q = [1, 2, 3, 4]$ of P_4 is a shortest path from vertex 1 to vertex 4.

Path		Length
P_1	[1, 2, 4, 6]	11
P_2	[1, 2, 5, 6]	14
P_3	[1, 2, 5, 4, 6]	13
P_4	[1, 2, 3, 4, 6]	10
P_5	[1, 2, 3, 5, 6]	13
P_6	[1, 2, 3, 5, 4, 6]	12
P_7	[1, 3, 4, 6]	12
P_8	[1, 3, 5, 4, 6]	14
P_9	[1, 3, 5, 6]	15

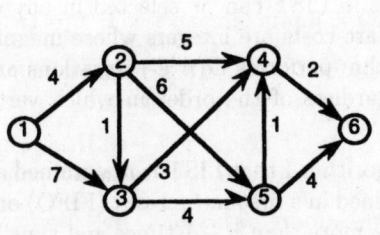

(a)   (b)

2. In the directed network of the following figure, arc costs are shown along each arc and a set of distance labels are shown at each vertex. Part (b) gives paths from vertex $s = 1$ whose lengths equal the corresponding distance labels. These distance labels do not satisfy the optimality conditions of Fact 3 because for the arc $(3,5)$, $d(5) > d(3)+c_{35}$. The out-tree T in this figure defined by predecessor indices $pred(2) = 5$, $pred(3) = 1$, $pred(4) = 2$ and $pred(5) = 3$ has distance labels $d = (0, 5, 5, 25, 0)$. It is a shortest path tree rooted at vertex 1 since the optimality conditions of Fact 3 are satisfied: namely

$$5 \leq 0 + 10 \text{ for arc } (1,2), \quad 5 \leq 5 + 10 \text{ for arc } (2,3), \quad 0 \leq 25 + 15 \text{ for arc } (4,5).$$

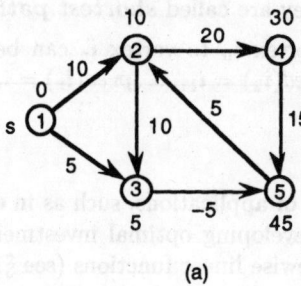

j	d(j)	path
2	10	[1,2]
3	5	[1,3]
4	30	[1,2,4]
5	45	[1,2,4,5]

(a)   (b)

10.3.2 ALGORITHMS FOR SINGLE-SOURCE SHORTEST PATHS

This subsection discusses algorithms for finding shortest path trees from a given source vertex s in a directed network G with n vertices and m arcs.

Facts:

1. *Label-correcting algorithm*: A general label-correcting algorithm (Algorithm 1) is based on the shortest path optimality conditions (§10.3.1 Fact 3) and is a very popular algorithm to solve shortest path problems with arbitrary arc costs (L. R. Ford, 1956 and R. E. Bellman, 1958).

2. Algorithm 1 maintains a list, LIST, of vertices with the property that if an arc (i, j) violates the optimality condition, then LIST must contain vertex i. If LIST is empty, then the current distance labels are optimal. Otherwise some vertex i is removed from LIST and the arcs of $A(i)$ are scanned. If an arc $(i, j) \in A(i)$ violates the optimality condition, then $d(j)$ is updated appropriately.

3. When Algorithm 1 terminates, the nonzero predecessor indices define a shortest path tree T^* rooted at the source vertex: namely, $T^* = \{ (pred(i), i) \mid i \in V - \{s\} \}$.

4. *Convergence*: In Algorithm 1, vertices in LIST can be selected in any order and the algorithm still converges finitely. If all arc costs are integers whose magnitudes are bounded by a constant C, then the algorithm performs $O(n^2 C)$ iterations and can be implemented to run in $O(nmC)$ time, regardless of the order in which vertices from LIST are selected.

5. *Queue implementation*: Suppose in Algorithm 1 that LIST is maintained as a queue (§17.1.2); that is, vertices in LIST are examined in a first-in-first-out (FIFO) order. This specific implementation examines no vertex more than $n - 1$ times and runs in $O(nm)$ time. This is the best strongly polynomial-time algorithm to solve the shortest path problem with arbitrary arc costs.

Algorithm 1: Label-correcting algorithm.

input: directed network G, source vertex s
output: shortest path tree T^* rooted at s
$d(s) := 0$
$pred(s) := 0$
$d(j) := \infty$ for all $j \in V - \{s\}$
LIST $:= \{s\}$
while LIST $\neq \emptyset$
  remove a vertex i from LIST
  for each $(i, j) \in A(i)$
    if $d(j) > d(i) + c_{ij}$ then
      $d(j) := d(i) + c_{ij}$
      $pred(j) := i$
      if $j \notin$ LIST then add j to LIST

6. *Dequeue implementation*: Suppose in Algorithm 1 that LIST is maintained as a dequeue (§17.1.2). Specifically, vertices are removed from the front of the dequeue, but vertices are added either at the front or at the rear. If the vertex has been in LIST earlier, the algorithm adds it to the front; otherwise, it adds the vertex to the rear. Empirical studies have found that the dequeue implementation is one of the most efficient algorithms to solve the shortest path problem in practice even though it is not a polynomial-time algorithm.

7. *Negative cycle detection*: The queue implementation (Fact 5) of the label-correcting algorithm can be used to detect the presence of a negative cycle. To do so, record the number of times that the algorithm examines each vertex. If the algorithm examines a vertex more than $n - 1$ times, there must exist a negative cycle. In this case, the subgraph formed by the arcs $(pred(i), i)$ will contain a negative cycle.

8. A variety of computer codes (in Fortran) that implement the label-correcting algorithm for shortest paths can be found at the following sites:

    http://www.netlib.org/toms/562
    ftp://ftp.zib.de/pub/Packages/mathprog/netopt-bertsekas/
    http://www.mat.uc.pt/~eqvm/cientificos/fortran/codigos.html
    http://www.neci.nj.nec.com/homepages/avg/soft/soft.html

9. *Dijkstra's algorithm* (1959): Dijkstra's algorithm (Algorithm 2) is a popular algorithm for solving shortest path problems with nonnegative arc costs (E. W. Dijkstra, born 1930).

10. Algorithm 2 performs two steps repeatedly: *vertex selection* and *distance update*. The vertex selection step chooses a vertex i with smallest distance label in LIST for examination. The distance update step scans each arc $(i, j) \in A(i)$ and updates the distance label $d(j)$, if necessary, to restore the optimality condition for arc (i, j).

11. Whenever a vertex is selected for examination in Algorithm 2, its distance label is the shortest path distance from s; consequently, each vertex is examined only once.

12. Using a simple array or linked list representation of LIST, vertex selections take a total of $O(n^2)$ time and distance updates take a total of $O(m)$ time. This implementation of Algorithm 2 runs in $O(n^2)$ time.

Algorithm 2:   Dijkstra's algorithm.

input: directed network G with $c_{ij} \geq 0$, source vertex s
output: shortest path tree T^* rooted at s

$d(s) := 0$
$pred(s) := 0$
$d(j) := \infty$ for all $j \in V - \{s\}$
LIST $:= V$
while LIST $\neq \emptyset$

{Vertex selection}
let $i \in$ LIST be a vertex for which $d(i) = \min\{ d(j) \mid j \in$ LIST $\}$
remove vertex i from LIST

{Distance update}
for each $(i,j) \in A(i)$
  if $d(j) > d(i) + c_{ij}$ then
    $d(j) := d(i) + c_{ij}$
    $pred(j) := i$

13. By using more sophisticated data structures, the efficiency of Dijkstra's algorithm can be improved. Currently, two of the best implementations:
    - use Fibonacci heaps, giving $O(m + n \log n)$ running time [FrTa84];
    - use radix heaps, giving $O(m + n(\log C)^{1/2})$ running time [AhEtal90].

14. Empirically, the fastest implementation of Dijkstra's algorithm is due to R. Dial [Di69], and it runs in $O(m + nC)$ time.

15. A comprehensive discussion of several implementations of Dijkstra's algorithm and the label-correcting algorithm is presented in [AhMaOr93].

16. A variety of computer codes (in C, Pascal, and Fortran) that implement Dijkstra's algorithm for shortest paths can be found at the following sites:

    ftp://ftp.zib.de/pub/Packages/mathprog/netopt-bertsekas/
    http://www.mat.uc.pt/~eqvm/cientificos/fortran/codigos.html
    http://orly1.snu.ac.kr/software/
    http://www.neci.nj.nec.com/homepages/avg/soft/soft.html

17. A useful extension of the shortest path problem involves finding the k *shortest paths* in a network. The case $k = 1$ corresponds to a shortest path. More generally, the kth shortest path is one having the kth smallest length among all paths from s to t. Several algorithms for solving the problem of finding the k shortest paths are discussed in [EvMi92].

Examples:

1. The following figure illustrates three iterations of the label-correcting algorithm applied to Example 2 of §10.3.1. LIST is maintained as a queue.
   In the first iteration, the source vertex $s = 1$ is examined, and the distance labels of vertices 2 and 3 are decreased to 10 and 5, respectively. At this point, LIST $= [2, 3]$.

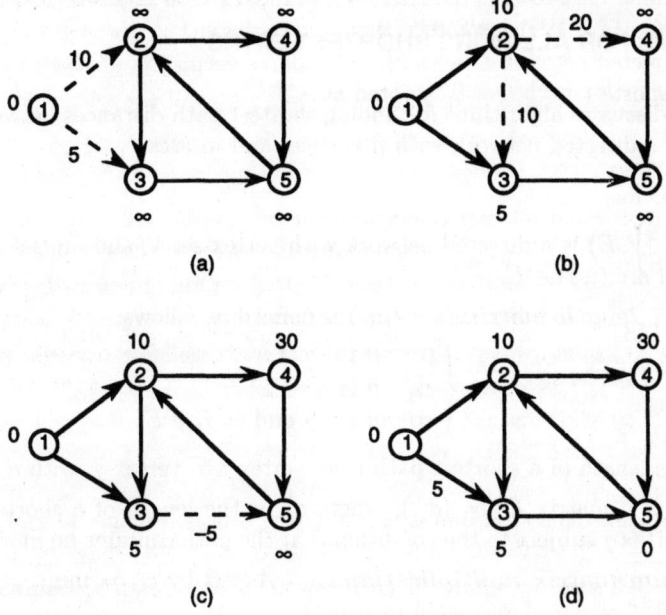

(a) (b)
(c) (d)

In the second iteration, vertex 2 is removed from LIST and examined. The distance label of vertex 4 decreases to 30, while the distance label of vertex 3 remains unchanged, giving LIST = [3, 4].

Next, vertex 3 is removed from LIST and examined, triggering a reduction of the distance label of vertex 5 to 0. At this point the current out-tree, defined by the predecessor indices $pred(\cdot)$, consists of arcs $(1,2)$, $(2,4)$, $(1,3)$, and $(3,5)$.

2. Part (b) of the following figure shows the application of Dijkstra's algorithm to the directed network drawn in part (a) with nonnegative arc costs and $s = 1$. Shown at each iteration are the current distance labels, the vertex selected, and the resulting distance updates. Upon termination, the shortest path lengths $d = (0, 6, 4, 9, 7)$ are realized by the optimal tree T^* having arcs $(1,3)$, $(3,2)$, $(2,4)$, and $(2,5)$.

iteration	labels	select	updates
1	$(0,\infty,\infty,\infty,\infty)$	1	$d(2) = 7$, $d(3) = 4$
2	$(0,7,4,\infty,\infty)$	3	$d(2) = \min\{7,6\} = 6$
			$d(5) = 9$
3	$(0,6,4,\infty,9)$	2	$d(4) = 9$
			$d(5) = \min\{9,7\} = 7$
4	$(0,6,4,9,7)$	5	$d(3) = \min\{4,9\} = 4$
5	$(0,6,4,9,7)$	4	$d(5) = \min\{7,13\} = 7$

(b)

10.3.3 ALGORITHMS FOR ALL-PAIRS SHORTEST PATHS

This section discusses algorithms for finding shortest path distances between every pair of vertices in a directed network with n vertices and m arcs.

Definitions:

Suppose $G = (V, E)$ is a directed network with vertex set V and arc set E, and let c_{ij} be the cost of arc $(i, j) \in E$.

The $n \times n$ **arc length matrix** $U = (u_{ij})$ is defined as follows:
$$u_{ij} = \begin{cases} 0 & \text{if } i = j \\ c_{ij} & \text{if } (i,j) \in E \\ \infty & \text{if } i \neq j \text{ and } (i,j) \notin E \end{cases}$$

Let d_{ij} be the length of a shortest path from vertex i to vertex j, with $d_{ii} = 0$.

Define the $n \times n$ matrix $D^k = (d_{ij}^k)$, where d_{ij}^k is the length of a shortest path from vertex i to vertex j subject to the condition that the path contains no more than k arcs.

Define **minsum matrix multiplication** $C = A \otimes B$ by $c_{ij} = \min_{1 \leq p \leq n}\{a_{ip} + b_{pj}\}$. Also, define $A^{\otimes k} = A \otimes A \otimes \cdots \otimes A$ (k times).

In the directed path $[i_0, i_1, \ldots, i_r]$ from i_0 to i_r, the vertices $i_1, i_2, \ldots, i_{r-1}$ are called **internal vertices**. Let $d^k[i, j]$ be the length of a shortest path from vertex i to vertex j subject to the condition that this path uses only $1, 2, \ldots, k-1$ as internal vertices. The $n \times n$ matrix $D^{[k]}$ contains the entries $d^k[i, j]$.

Facts:

1. The length of a shortest path containing at most k arcs can be expressed in terms of shortest path lengths involving at most $k-1$ arcs. Namely, for all vertices i and j

    - $d_{ij}^1 = u_{ij}$;
    - $d_{ij}^k = \min_{1 \leq p \leq n}\{d_{ip}^{k-1} + u_{pj}\}$ for $2 \leq k \leq n-1$;
    - if there is no negative cycle, then $d_{ij}^{n-1} = d_{ij}$.

2. $D^k = U^{\otimes k}$ for all $1 \leq k \leq n-1$.

3. For any pair of vertices i and j, the following conditions hold:

    - $d^1[i,j] = u_{ij}$;
    - $d^{k+1}[i,j] = \min\{d^k[i,j], d^k[i,k] + d^k[k,j]\}$, $1 \leq k \leq n$;
    - if there is no negative cycle then $d^{n+1}[i,j] = d_{ij}$.

4. The all-pairs shortest path problem can be solved by applying n times either Algorithm 1 or Algorithm 2 of §10.3.2, considering each vertex once as a source.

5. Specialized algorithms are available to solve the all-pairs shortest path problem: the matrix multiplication algorithm (Fact 6) and the Floyd-Warshall algorithm (Fact 8).

6. *Matrix multiplication algorithm*: This algorithm (Algorithm 3), based on Facts 1 and 2, computes the shortest path distances between all vertex pairs by multiplying two matrices repeatedly, using minsum matrix multiplication.

7. If there is no negative cycle, then Algorithm 3 finds all shortest path distances using $O(\log n)$ matrix multiplications, each of which takes $O(n^3)$ time. Hence this algorithm runs in $O(n^3 \log n)$ time and requires $O(n^2)$ space.

Algorithm 3: Matrix multiplication algorithm.

input: directed network G on n vertices
output: shortest distance matrix $D = (d_{ij})$

form the $n \times n$ arc length matrix U
compute $D := U^{\otimes(n-1)}$

Algorithm 4: Floyd-Warshall algorithm.

input: directed network G on n vertices
output: shortest distance matrix $D = (d[i,j])$

for all $(i,j) \in V \times V \quad d[i,j] := \infty$
for all $i \in V \quad d[i,i] := 0$
for all $(i,j) \in E \quad d[i,j] := c_{ij}$
for $k := 1$ to n
  for $(i,j) \in V \times V$
    if $d[i,j] > d[i,k] + d[k,j]$ then $d[i,j] := d[i,k] + d[k,j]$

---

8. *Floyd-Warshall algorithm*: This approach (Algorithm 4) to calculating all-pairs shortest path distances in a directed network G is based on computing conditional shortest path lengths $d[i,j]$.

9. If there is no negative cycle, the Floyd-Warshall algorithm correctly computes the matrix of shortest path distances. A single $n \times n$ array D is used to implement the algorithm.

10. Algorithm 4 can be used to detect (and identify) negative cycles by monitoring whenever $d[i,i] < 0$ occurs for some vertex i.

11. Algorithm 4 runs in $O(n^3)$ time and requires $O(n^2)$ space.

12. If the underlying network is dense, that is, $m = \Omega(n^2)$, then the $O(n^3)$ time bound for Algorithm 4 is as good as any other discussed in §10.3.2 or §10.3.3.

13. Algorithm 4 was first discovered by B. Roy in 1959 in the context of determining the transitive closure of a graph; this same algorithm was independently discovered by S. Warshall in 1962. The method was generalized to computing all shortest paths by R. W. Floyd, also in 1962.

14. Computer codes (in C, Pascal, and Fortran) that implement the Floyd-Warshall algorithm can be found at the site:

    http://orly1.snu.ac.kr/software/

Examples:

1. The matrix multiplication algorithm is applied to the directed network in the following figure.

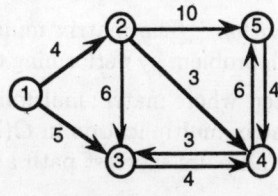

By Facts 1 and 2, the matrix D^4 is the matrix of shortest path distances.

$$D^1 = \begin{pmatrix} 0 & 4 & 5 & \infty & \infty \\ \infty & 0 & 6 & 3 & 10 \\ \infty & \infty & 0 & 4 & \infty \\ \infty & \infty & 3 & 0 & 6 \\ \infty & \infty & \infty & 4 & 0 \end{pmatrix} \quad D^2 = \begin{pmatrix} 0 & 4 & 5 & 7 & 14 \\ \infty & 0 & 6 & 3 & 9 \\ \infty & \infty & 0 & 4 & 10 \\ \infty & \infty & 3 & 0 & 6 \\ \infty & \infty & 7 & 4 & 0 \end{pmatrix}$$

$$D^4 = \begin{pmatrix} 0 & 4 & 5 & 7 & 13 \\ \infty & 0 & 6 & 3 & 9 \\ \infty & \infty & 0 & 4 & 10 \\ \infty & \infty & 3 & 0 & 6 \\ \infty & \infty & 7 & 4 & 0 \end{pmatrix}.$$

2. Algorithm 4 is illustrated with the network of the figure of Example 1.

$$D^{[1]} = \begin{pmatrix} 0 & 4 & 5 & \infty & \infty \\ \infty & 0 & 6 & 3 & 10 \\ \infty & \infty & 0 & 4 & \infty \\ \infty & \infty & 3 & 0 & 6 \\ \infty & \infty & \infty & 4 & 0 \end{pmatrix} \quad D^{[3]} = \begin{pmatrix} 0 & 4 & 5 & 7 & 14 \\ \infty & 0 & 6 & 3 & 10 \\ \infty & \infty & 0 & 4 & \infty \\ \infty & \infty & 3 & 0 & 6 \\ \infty & \infty & \infty & 4 & 0 \end{pmatrix}$$

$$D^{[5]} = \begin{pmatrix} 0 & 4 & 5 & 7 & 13 \\ \infty & 0 & 6 & 3 & 9 \\ \infty & \infty & 0 & 4 & 10 \\ \infty & \infty & 3 & 0 & 6 \\ \infty & \infty & 7 & 4 & 0 \end{pmatrix}.$$

It can be verified that $D^{[2]} = D^{[1]}$, $D^{[4]} = D^{[3]}$, and $D^{[6]} = D^{[5]}$. Consequently, the matrix $D^{[5]}$ above gives all shortest path distances.

10.3.4 PARALLEL ALGORITHMS

Parallel implementations of certain shortest path algorithms are described are described here relative to an EREW (exclusive-read, exclusive-write) PRAM (parallel random-access machine). For details of EREW PRAM, see §16.2.

Facts:

1. *Label-correcting algorithm*: The parallel implementation of the label-correcting algorithm (§10.3.2) associates a processor with each arc and with each vertex of the network. This algorithm maintains a distance label for each vertex, appropriately initialized. Suppose the distance labels are $d(i)$ at the beginning of an iteration. During the iteration, the processor attached to each arc (i,j) computes a temporary label $d'(i,j) = d(i) + c_{ij}$ in $O(1)$ time. Then the processor associated with vertex j examines incoming arcs at vertex j and sets $d(j) := \min\{d'(i,j) \mid 1 \leq i \leq n\}$.

2. Using a parallel prefix operation, the distance labels can be updated in $O(\log n)$ time. The label-correcting algorithm performs $O(n)$ iterations and so its running time is $O(n \log n)$ using $O(m)$ processors.

3. *Matrix multiplication algorithm*: The matrix multiplication algorithm of §10.3.3 solves the all-pairs shortest path problem by performing $O(\log n)$ matrix multiplications.

4. Unlike a sequential computer, where matrix multiplication takes $O(n^3)$ time, a parallel computer can perform matrix multiplication in $O(\log n)$ time using $O(n^3)$ processors [Le92]. Consequently, this all-pairs shortest path algorithm runs in $O(\log^2 n)$ time using $O(n^3)$ processors.

10.3.5 APPLICATIONS

Shortest path problems arise in a variety of applications, both as stand-alone models and as subproblems in more complex problem settings. Shortest path problems also arise in surprising ways that on the surface might not appear to involve networks at all. This subsection presents several models based on determining shortest paths.

Applications:

1. *Distribution*: Material needs to be shipped by truck from a central warehouse to various retailers at minimum cost. The underlying network is an undirected road network, with edges representing the roads joining various cities (vertices). The cost of an edge is the per unit shipping cost. Solving the single-source shortest path problem provides a least-cost shipping pattern for the material.

2. *Telephone routing*: A call is to be routed from a specified origin to a specified destination. Here the underlying network is the telephone system, with vertices representing individual users (or switching centers). Since a direct connection between the origin vertex s and the destination vertex t may not be available, one practice is to route the call along a path having the minimum number of arcs (i.e., involving the smallest number of switching centers). This means finding a shortest path with unit lengths on all arcs. Alternatively, each arc can be provided with a measure of delay, and routing can take place along a timewise shortest path from s to t.

3. *Salesperson routing*: A salesperson is to travel by air from city A to city B. The commission obtained by visiting each city along the way can be estimated. An optimal itinerary can be found by solving a shortest path problem on the underlying airline network, represented as a directed network of nonstop routes (arcs) connecting cities (vertices). Each arc (i,j) is given the net cost $c_{ij} = f_{ij} - r_j$, where f_{ij} is the cost of the flight from city i to city j and r_j is the commission obtained by visiting city j. A shortest path from A to B identifies an optimal itinerary.

4. *Investment strategy*: An investor has a fixed amount to invest at the beginning of the year. A variety of different financial opportunities are available for investing during the year, with each such opportunity assumed to be available only at the start of each month. Construct the directed network having a vertex for each month as well as a final vertex $t = 13$. The arc (i,j) corresponds to an investment opportunity beginning in month i and maturing at the start of month j, with its weight c_{ij} being the negative of the profit earned for the duration of the investment. An optimal investment strategy is identified by a shortest path from vertex 1 to vertex t.

5. *Equipment replacement*: A job shop must periodically replace its capital equipment because of machine wear. As the machine ages, it breaks down more frequently and so becomes more expensive to operate. Also, as a machine ages its salvage value decreases. Let c_{ij} denote the cost of buying a particularly important machine at the beginning of period i, plus the cost of operating the machine over the periods $i, i+1, \ldots, j-1$, minus the salvage cost of the machine at the beginning of period j. The problem is to design a replacement plan that minimizes the cost of buying, selling, and operating the machine over a planning horizon of n years, assuming that the job shop must have exactly one machine in service at all times.

This problem can be formulated as a shortest path problem on a network G with vertices $i = 1, 2, \ldots, n+1$; G contains an arc (i,j) with cost c_{ij} for all $i < j$. There is a one-to-one correspondence between directed paths in G from vertex 1 to vertex $n+1$ and equipment replacement plans. The following figure gives a sample network with $n = 5$.

The path $[1, 3, 6]$ corresponds to buying the equipment at the beginning of periods 1 and 3. A shortest path from vertex 1 to vertex $n + 1$ identifies an optimal replacement plan.

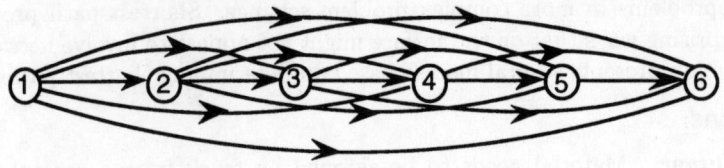

6. *Paragraph problem*: The document processing program TEX uses an optimization procedure to decompose a paragraph into several lines so that when lines are left- and right-justified, the appearance of the paragraph will be the most attractive. Suppose that a paragraph consists of words $i = 1, 2, \ldots, n$. Let c_{ij} denote the attractiveness of a line if it begins with the word i and ends with the word $j - 1$. The program TEX uses formulas to compute the value of each c_{ij}. Given the c_{ij}, the decision problem is to decompose the paragraph into several lines of text in order to maximize the total attractiveness (of all lines). This problem can be formulated as a shortest path problem in a manner similar to Application 5.

7. *Tramp steamer problem*: A ship travels from port to port carrying cargo and passengers. A voyage of the steamer from port i to port j earns p_{ij} units of profit and requires $t_{ij} \geq 0$ units of time. Here it is assumed that $\sum_{(i,j) \in W} t_{ij} > 0$ for every directed cycle W in G. The captain of the ship would like to know whether there exists a tour (directed cycle) W for which the daily profit is greater than a specified threshold μ_0; that is, $\sum_{(i,j) \in W} p_{ij} / \sum_{(i,j) \in W} t_{ij} > \mu_0$. By writing this inequality as $\sum_{(i,j) \in W} (\mu_0 t_{ij} - p_{ij}) < 0$, it is seen that there is a tour W with mean daily profit exceeding μ_0 if and only if G contains a negative cost directed cycle W. The shortest path label-correcting algorithm can be used to detect the presence (or absence) of negative cycles (see §10.3.2, Fact 7).

8. *System of difference constraints*: In some linear programming applications (§15.1) with constraints of the form $Ax \leq b$, the $m \times n$ constraint matrix A contains one $+1$ and one -1 in each row, with all other entries being zero. Suppose that the kth row has a $+1$ entry in column j_k and a -1 entry in column i_k; entries in the vector b have arbitrary signs. This linear program defines the following set of m difference constraints in n variables $x = (x(1), x(2), \ldots, x(n))$: $x(j_k) - x(i_k) \leq b(k)$ for each $k = 1, 2, \ldots, m$. The problem is to determine whether this system of difference constraints has a feasible solution, and if so, to obtain one.

Associate a graph G with this system of difference constraints; G has n vertices corresponding to the n variables, and the arc (i_k, j_k) of length $b(k)$ results from the constraint $x(j_k) - x(i_k) \leq b(k)$. These constraints are identical with the optimality conditions for the shortest path problem in G, and they can be satisfied if and only if G contains no negative cycle. In this case the shortest path distances give a solution x satisfying the constraints.

9. Examples of Application 8 occur in telephone operator scheduling, just-in-time scheduling, analyzing the consistency of measurements, and the scaling of data.

10. *Maximin paths*: In a network with capacities (that is, upper bounds on the amount of material that can be sent on each arc), the capacity of a path is the smallest capacity on any of its constituent arcs. A common problem in such networks is to find a path from vertex s to vertex t having the maximum capacity. This represents a path along which the maximum amount of material can flow. Such a *maximin path* can be found efficiently by adapting Dijkstra's shortest path algorithm.

11. Additional applications, with reference sources, are given in the following table.

application	references
approximating piecewise linear functions	[AhMaOr93], [AhEtal95]
DNA sequence alignment	[AhEtal95]
molecular confirmation	[AhMaOr93], [AhEtal95]
robot design	[AhMaOr93], [AhEtal95]
scaling of matrices	[AhMaOr93], [AhEtal95]
knapsack problems	[EvMi92], [AhMaOr93], [AhEtal95]
compact book storage	[AhMaOr93], [AhEtal95]
personnel planning	[AhMaOr93]
routing snow removal vehicles	[EvMi92]
production lot sizing	[EvMi92]
transportation planning	[EvMi92]
single-crew scheduling	[AhMaOr93]
dynamic facility location	[AhMaOr93]

10.4 MAXIMUM FLOWS

The maximum flow problem involves sending the maximum amount of material from a specified source vertex s to another specified sink vertex t, subject to capacity restrictions on the amount of material that can flow along each arc. A closely related problem is the minimum cut problem, which is to find a set of arcs with smallest total capacity whose removal separates s and t.

10.4.1 BASIC CONCEPTS

Definitions:

Let $G = (V, E)$ be a directed network with vertex set V and arc set E (see §10.3.1). Each arc $(i, j) \in E$ has an associated **capacity** $u_{ij} \geq 0$. Such a network is called a **capacitated** network. Let $n = |V|$ and $m = |E|$.

Suppose s is a specified **source** vertex and t is a specified **sink** vertex. A (**feasible**) **flow** is a function $x = (x_{ij})$ defined on arcs $(i, j) \in E$ satisfying:

- **mass balance constraints**: $\sum_{\{j|(i,j)\in E\}} x_{ij} = \sum_{\{j|(j,i)\in E\}} x_{ji}$ for all $i \in V - \{s, t\}$;

- **capacity constraints**: $0 \leq x_{ij} \leq u_{ij}$ for all $(i, j) \in E$.

The arc (i, j) is **saturated** in flow x if $x_{ij} = u_{ij}$.

The **value** of flow x is $v = \sum_{\{j|(s,j)\in E\}} x_{sj}$, the total flow leaving the source vertex.

A **maximum flow** is a flow having maximum value.

A **cut** $[S, \overline{S}]$ partitions the vertex set V into two subsets S and $\overline{S} = V - S$, and consists of all arcs with one endpoint in S and the other in $\overline{S}$. Arcs directed from S to $\overline{S}$ are **forward arcs**, and the set of forward arcs is denoted by $(S, \overline{S})$. Arcs directed from $\overline{S}$ to S are **backward arcs**, and the set of backward arcs is denoted by $(\overline{S}, S)$.

The cut $[S, \overline{S}]$ is an **s-t cut** if $s \in S$ and $t \in \overline{S}$. The **capacity** of the s-t cut $[S, \overline{S}]$ is
$$u[S, \overline{S}] = \sum_{(i,j) \in (S, \overline{S})} u_{ij}.$$

A **minimum cut** is an s-t cut having minimum capacity.

Facts:

1. The flow x_{ij} on arc (i, j) can represent the number of cars (per hour) traveling along a highway segment, the rate at which oil is pumped through a section of pipe in a distribution system, or the number of messages per unit time that can be sent along a data link in a communication system.

2. The mass balance constraints ensure that for all vertices i (other than the source or sink), the total flow out of i equals the total flow into i.

3. The capacity constraints ensure that the flow on an arc does not exceed its stated capacity.

4. Maximum flows arise in a variety of practical problems involving the flow of goods, vehicles, and messages in a network. Maximum flows can also be used to study the connectivity of graphs, the covering of chessboards, the selection of representatives, winning records in tournaments, matrix rounding, and staff scheduling (see §10.4.3).

5. For any s-t flow x, the flow out of s equals the flow into t; that is,
$$\sum_{\{j | (s,j) \in E\}} x_{sj} = v = \sum_{\{j | (j,t) \in E\}} x_{jt}.$$

6. Removal of the arcs in the s-t cut $Z = [S, \overline{S}]$ from G separates vertex s from vertex t: namely, there is no s-t path in $G - Z$.

7. Let $[S, \overline{S}]$ be any s-t cut in the network. Then the value of the flow x is given by
$$v = \sum_{(i,j) \in (S, \overline{S})} x_{ij} - \sum_{(j,i) \in (\overline{S}, S)} x_{ji}.$$
That is, the net flow across each s-t cut is the same and equals v.

8. **Weak duality theorem**: The value of every s-t flow is less than or equal to the capacity of every s-t cut in the network.

9. If x is some s-t flow whose value equals the capacity of some s-t cut $[S, \overline{S}]$, then x is a maximum flow and $[S, \overline{S}]$ is a minimum cut.

10. **Max-flow min-cut theorem**: The maximum value of the flow from vertex s to vertex t in a capacitated network equals the minimum capacity among all s-t cuts. (L. R. Ford and D. R. Fulkerson, 1956.)

11. A systematic study [FoFu62] of flows in networks was first carried out by Ford and Fulkerson, motivated by a simplified model of railway traffic flow.

Examples:

1. Part (a) of the following figure shows a flow network with $s = 1$ and $t = 6$; capacities are indicated along each arc. The function x given in part (b) satisfies the mass balance constraints and the capacity constraints, and hence is a feasible flow. Relative to this flow, arc $(1, 2)$ is not saturated since $x_{12} = 6 < 7 = u_{12}$; on the other hand, arc $(3, 5)$

is saturated since $x_{35} = 3 = u_{35}$. The flow has value $v = x_{12} + x_{13} = 6 + 2 = 8$. Here the flow into vertex 6 is $x_{46} + x_{56} = 5 + 3 = 8 = v$, as guaranteed by Fact 5. The flow value across the s-t cut $[S, \overline{S}]$ with $S = \{1, 3\}$ is $x_{12} + x_{34} + x_{35} - x_{23} = 6 + 3 + 3 - 4 = 8$. Similarly, the flow value across the s-t cut $[S, \overline{S}]$ with $S = \{1, 2, 3, 5\}$ is $x_{24} + x_{34} + x_{56} - x_{45} = 2 + 3 + 3 - 0 = 8$. (See Fact 7.) This flow is not, however, a maximum flow.

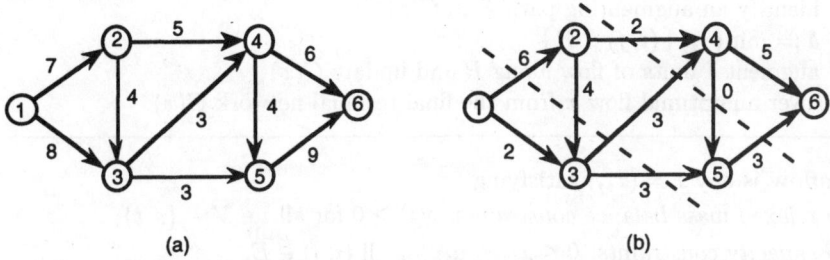

(a) (b)

2. In Example 1, the s-t cut $[S, \overline{S}]$ with $S = \{1, 2, 3, 5\}$ has capacity $u_{24} + u_{34} + u_{56} = 5 + 3 + 9 = 17$. Thus the value of any flow in the network is bounded above (see Fact 8) by 17. The s-t cut $[S, \overline{S}]$ with $S = \{1, 3\}$ has capacity $u_{12} + u_{34} + u_{35} = 7 + 3 + 3 = 13$. This cut capacity 13 provides an improved upper bound on the value of a flow. In particular, the flow defined in part (b) has value $v = 8 \leq 13$.

3. The following figure shows another feasible flow x' in the network of Example 1. For x', the flow value across the s-t cut $[S, \overline{S}]$ with $S = \{1, 2, 3\}$ is $v = x_{24} + x_{34} + x_{35} = 5 + 3 + 3 = 11$, which equals the s-t cut capacity $u[S, \overline{S}] = u_{24} + u_{34} + u_{35} = 5 + 3 + 3 = 11$. By Fact 9, x' is a maximum flow and $S = \{1, 2, 3\}$ defines a minimum cut $[S, \overline{S}]$.

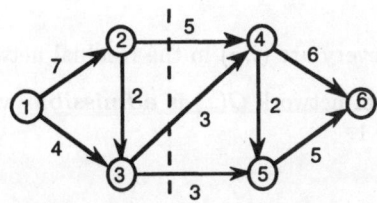

10.4.2 ALGORITHMS FOR MAXIMUM FLOWS

There are two main classes of maximum flow algorithms: augmenting path algorithms and preflow-push algorithms. Both types of algorithms work on an auxiliary network (called the residual network) associated with the current solution.

Definitions:

Let $G = (V, E)$ be a directed network with $n = |V|$ and $m = |E|$. Let s and t be the specified source and sink, and let U be the largest of the arc capacities u_{ij} in G.

Let $x = (x_{ij})$ be a function defined on the arcs (i, j) of G. Relative to x, the **outflow** from vertex i and **inflow** to vertex i are given, respectively, by

$$out(i) = \sum_{\{j|(i,j) \in E\}} x_{ij}, \quad in(i) = \sum_{\{j|(j,i) \in E\}} x_{ji}.$$

The **excess** of vertex i is $e(i) = in(i) - out(i)$.

Algorithm 1: Augmenting path algorithm.

input: directed network G, source vertex s, sink vertex t
output: maximum flow x

$x := 0$
while $G(x)$ contains a directed path from s to t
   identify an augmenting path P in $G(x)$
   $\delta := \min\{ r_{ij} \mid (i,j) \in P \}$
   augment δ units of flow along P and update $G(x)$
recover an optimal flow x from the final residual network $G(x)$

A **preflow** is any $x = (x_{ij})$ satisfying
- relaxed mass balance constraints: $e(i) \geq 0$ for all $i \in V - \{s,t\}$;
- capacity constraints: $0 \leq x_{ij} \leq u_{ij}$ for all $(i,j) \in E$.

Vertex i is **active** if $e(i) > 0$.

Given a flow (or a preflow) x, the **residual capacity** r_{ij} of the arc $(i,j) \in E$ is the maximum additional flow that can be sent from i to j using arcs (i,j) and (j,i).

The **residual network** $G(x)$ with respect to flow x consists of those arcs of G having positive residual capacity.

An **augmenting path** is a directed path from vertex s to vertex t in $G(x)$.

The **capacity** of a directed path is the minimum arc capacity appearing on the path.

A set of **distance labels** with respect to a preflow (or flow) x is a function $d: V \to \{0, 1, 2, \ldots\}$ satisfying
- $d(t) = 0$;
- $d(i) \leq d(j) + 1$ for every arc (i,j) in the residual network $G(x)$.

An arc (i,j) in the residual network $G(x)$ is **admissible** with respect to the distance labels $d(\cdot)$ if $d(i) = d(j) + 1$.

Facts:

1. The maximum flow problem on an undirected network can be converted to a maximum flow problem on a directed network. Namely, replace every undirected edge (i,j) of capacity u_{ij} by two oppositely directed arcs (i,j) and (j,i), each with capacity u_{ij}.
2. The residual capacity $r_{ij} = (u_{ij} - x_{ij}) + x_{ji}$. The first term $u_{ij} - x_{ij}$ represents the unused capacity of arc (i,j); the second term x_{ji} represents the amount of flow on arc (j,i) that can be canceled to increase flow from vertex i to vertex j.
3. The capacity of an augmenting path is always positive.
4. *Augmenting path property*: A flow x is a maximum flow if and only if the residual network $G(x)$ contains no augmenting path.
5. *Augmenting path algorithm*: A general augmenting path algorithm (Algorithm 1) is based on Fact 4. It identifies augmenting paths and sends flows on these paths until the residual network contains no such path.
6. *Integrality property*: For networks with integer capacities, Algorithm 1 starts with the zero flow and augments by an integral flow at each iteration. Hence the maximum flow problem with integral capacities always has an optimal integer flow.
7. An augmenting path in $G(x)$ can be identified by any search procedure that starts at vertex s and identifies all vertices reachable from s by directed paths (§9.2.1).

Algorithm 2: Preflow-push algorithm.

input: directed network G, source vertex s, sink vertex t
output: maximum flow x
compute the shortest path lengths $d(\cdot)$ to vertex t
$d(s) := n$; $x := 0$; $x_{sj} := u_{sj}$ for all arcs $(s,j) \in E$
while the network contains an active vertex
  select an active vertex i and $push\_relabel(i)$
recover an optimal flow x from the final residual network $G(x)$

procedure $push\_relabel(i)$
  if the network contains an admissible arc (i,j) **then**
    push $\delta := \min\{e(i), r_{ij}\}$ units of flow from i to j
  else $d(i) := \min\{d(j) + 1 \mid (i,j) \in E \text{ and } r_{ij} > 0\}$

8. Augmenting the flow along P by δ decreases the residual capacities of arcs in P by δ and increases the residual capacities of the reversals of arcs in P by δ.

9. At the last iteration of Algorithm 1, let S be the set of vertices reachable from s. Then $t \in \overline{S}$ and $[S, \overline{S}]$ is a minimum cut.

10. Upon termination of Algorithm 1, an optimal flow x can be reconstructed from the final $G(x)$ using Fact 2. Specifically, let $(i,j) \in E$. If $u_{ij} - r_{ij} \geq 0$ then set $x_{ij} = u_{ij} - r_{ij}$ and $x_{ji} = 0$; otherwise, set $x_{ji} = r_{ij} - u_{ij}$ and $x_{ij} = 0$.

11. Algorithm 1 was independently discovered by L. R. Ford and D. R. Fulkerson (1956) and by P. Elias, A. Feinstein, and C. E. Shannon (1956).

12. The distance label $d(i)$ is a lower bound on the length (number of arcs) of the shortest (directed) path from vertex i to vertex t in the residual network.

13. If some vertex j satisfies $d(j) \geq n$, then vertex j is separated from the sink vertex in the residual network.

14. Algorithm 1 runs in pseudopolynomial time $O(nmU)$ for networks with integer (or rational) arc capacities. The algorithm may not terminate finitely for networks with irrational capacities.

15. Two specific implementations of Algorithm 1 run in polynomial time:
    - by augmenting flow along a shortest path, the number of augmentations can be reduced to $O(nm)$, and using very sophisticated data structures this algorithm can be implemented to run in $O(nm \log n)$ time;
    - by augmenting flow along a path with maximum residual capacity, the number of augmentations is $O(m \log U)$ and this algorithm can be implemented to run in $O(nm \log U)$ time.

16. *Preflow-push algorithm:* The preflow-push algorithm (Algorithm 2) maintains a preflow at every step and pushes flow on individual arcs instead of along augmenting paths. The basic operation is to select an active vertex and try to remove its excess by pushing flow to neighbors that are "closer" to the sink.

17. The shortest path lengths calculated in Algorithm 2 represent the minimum number of arcs in a path to vertex t and can be efficiently found by carrying out a breadth-first search relative to t (§9.2.1).

18. In Algorithm 2, if the active vertex currently being examined has an admissible arc (i,j), then increasing the flow on (i,j) by δ decreases r_{ij} by δ and increases r_{ji} by δ. Also, $e(i)$ is decreased by δ and $e(j)$ is increased by δ.

19. In Algorithm 2, if the active vertex currently being examined has no admissible arc, then after its distance label is increased, at least one admissible arc is created.

20. The preflow-push algorithm can be implemented to run in $O(n^2 m)$ time. Variations of this algorithm with improved worst-case complexity are described in [AhOrTa89].

21. The highest-label preflow-push algorithm [GoTa86] is a specific implementation of Algorithm 2 that always examines vertices with the largest distance label. This $O(n^2 \sqrt{m})$ implementation is currently the fastest algorithm to solve the maximum flow problem in practice.

22. Algorithm 2 can be implemented to run in $O(nm \log(n^2/m))$ time using a dynamic tree data structure. This algorithm currently achieves the best strongly polynomial-time bound to solve the maximum flow problem, but is not as efficient in practice as its more straightforward implementation.

23. The books [AhMaOr93] and [CoLeRi90] discuss additional versions of augmenting and preflow-push algorithms, as well as specializations of these algorithms to unit capacity networks, bipartite networks, and planar networks.

24. Preflow-push algorithms are more general, more powerful, and more flexible than augmenting path algorithms for solving the maximum flow problem.

25. The best preflow-push algorithms currently outperform the best augmenting path algorithms in theory as well as in practice.

26. Computer codes (in C, Pascal, and Fortran) for solving maximum flow and minimum cut problems can be found at the sites:

    ftp://dimacs.rutgers.edu/pub/netflow/maxflow/
    ftp://ftp.zib.de/pub/Packages/mathprog/netopt-bertsekas/
    http://www.neci.nj.nec.com/homepages/avg/soft/soft.html
    http://orly1.snu.ac.kr/software/
    ftp://ftp.zib.de/pub/Packages/mathprog/mincut/

Examples:

1. Part (a) of the following figure illustrates a network G with capacities shown next to each arc.

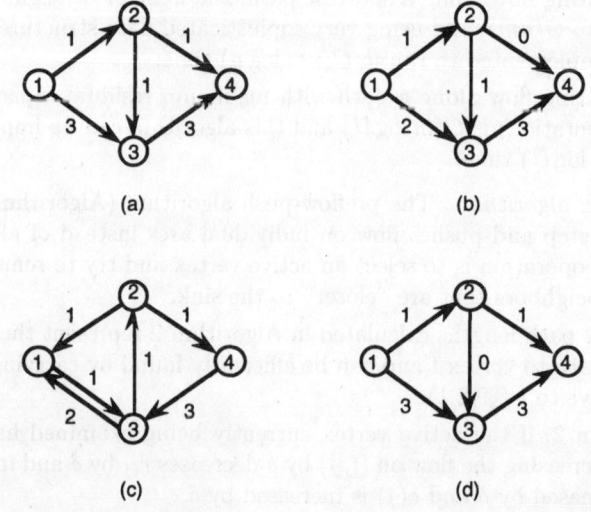

A feasible flow from vertex $s = 1$ to vertex $t = 4$ is displayed in part (b); this flow has value $v = x_{12} + x_{13} = 3$. Every path in G from s to t contains a saturated arc: paths $[1,2,4]$ and $[1,2,3,4]$ have the saturated arc $(1,2)$, and path $[1,3,4]$ has the saturated arc $(3,4)$. Consequently, no additional flow can be pushed in the "forward" direction from s to t. Yet, the current flow x is not a maximum flow.

To find additional flow from s to t, the residual network $G(x)$ is constructed; see part (c) of the figure. An augmenting path in part (c) is $P = [1,3,2,4]$ with (residual) capacity $\delta = 1$. Adding the flow on P to that in part (b) produces the new flow x' in part (d); notice that the flow on arc $(2,3)$ in x has been canceled in this process. The resulting flow x' has flow value $v = 4$. Since the s-t cut $[S, \overline{S}]$ with $S = \{1,2,3\}$ has capacity $u_{24} + u_{34} = 4 = v$, the flow x' is a maximum flow and $S = \{1,2,3\}$ defines a cut having minimum capacity.

2. The following figure illustrates three iterations of the augmenting path algorithm (Algorithm 1).

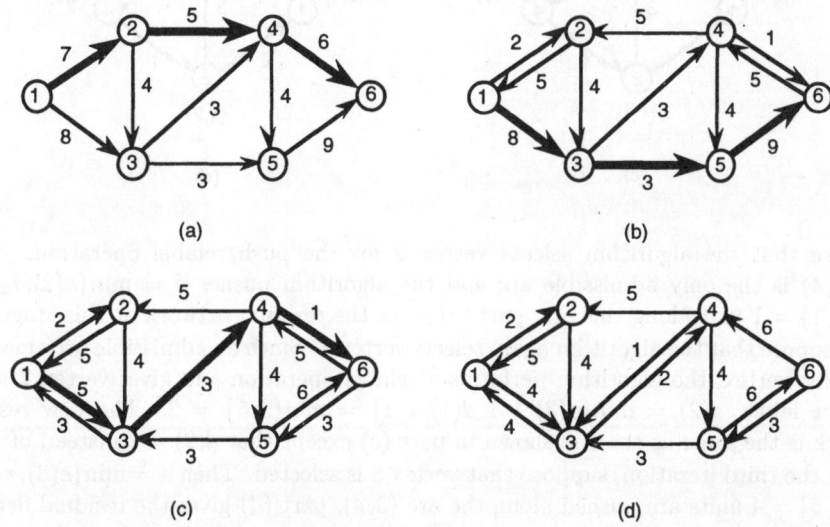

Part (a) of the figure shows a network with capacities indicated on each arc. Here $s = 1$ and $t = 6$. Initially the flow $x = 0$, so the residual network is identical to the original network with $r_{ij} = u_{ij}$ for every arc (i, j).

Suppose that the algorithm identifies path $P^1 = [1, 2, 4, 6]$ as the augmenting path. The algorithm augments $\delta = \min\{r_{12}, r_{24}, r_{46}\} = \min\{7, 5, 6\} = 5$ units of flow along P^1. This augmentation changes the residual capacities only of arcs in P^1 (or their reverse arcs), yielding the new residual network in part (b).

In the second iteration, suppose the algorithm identifies path $P^2 = [1, 3, 5, 6]$ as the next augmenting path. Then flow is increased by $\delta = \min\{8, 3, 9\} = 3$ units along P^2; part (c) shows the residual network after the second augmentation.

A third augmentation with $\delta = 1$ occurs along path $P^3 = [1, 3, 4, 6]$ in part (c), giving the residual network shown in part (d).

3. The following figure illustrates three iterations of the preflow-push algorithm on the flow network with capacities given in part (a). Here $s = 1$ and $t = 4$; in addition, the pair $(e(i), d(i))$ is shown beside each vertex i. Part (b) of the figure gives $G(x)$ corresponding to the initial preflow with $x_{12} = 2$ and $x_{13} = 4$.

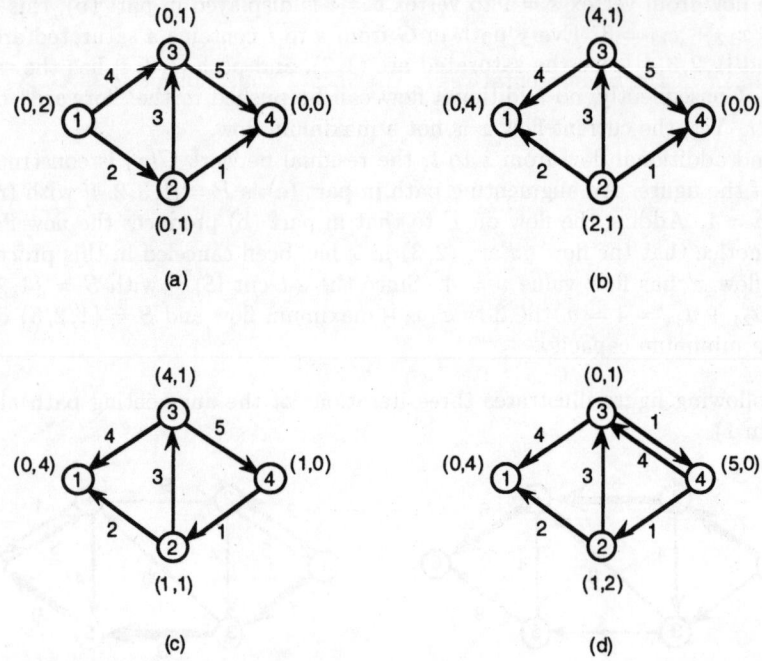

(a) (b) (c) (d)

Suppose that the algorithm selects vertex 2 for the push/relabel operation. Then arc $(2,4)$ is the only admissible arc and the algorithm pushes $\delta = \min\{e(2), r_{24}\} = \min\{2, 1\} = 1$ unit along this arc; part (c) gives the residual network at this stage.

Suppose that the algorithm again selects vertex 2. Since no admissible arc emanates from this vertex, the algorithm performs a relabel operation and gives vertex 2 a new distance label: $d(2) = \min\{d(3) + 1, d(1) + 1\} = \min\{2, 5\} = 2$. The new residual network is the same as the one shown in part (c) except that $d(2) = 2$ instead of 1.

In the third iteration, suppose that vertex 3 is selected. Then $\delta = \min\{e(3), r_{34}\} = \min\{4, 5\} = 4$ units are pushed along the arc $(3, 4)$; part (d) gives the residual network at the end of this iteration.

10.4.3 APPLICATIONS

A variety of applied problems can be modeled using maximum flows or minimum cuts. The max-flow min-cut theorem (§10.4.1, Fact 10) can also be used to deduce a number of min-max duality results in combinatorial theory. This section discusses a number of such applications.

Applications:

1. *Distribution network*: Oil needs to be shipped from a refinery to a storage facility using the pipelines of an underlying distribution network. Here the refinery corresponds to a particular vertex s in the distribution network, and the storage facility corresponds to another vertex t. The capacity of each arc is the maximum amount of oil per unit time that can flow along it. The maximum flow rate from the source vertex s to the sink vertex t is determined by the value of a maximum s-t flow.

2. Other examples of Application 1 occur in transportation networks, electrical power networks, and telecommunication networks.

3. *System of distinct representatives*: Given is a collection of sets $X_1, X_2, \ldots, X_m$ which are subsets of a given n-set X. A system of distinct representatives (§1.2.2) for the collection is sought, if one exists.

To solve this problem, set up the bipartite network $(V_1 \cup V_2, E)$ in which there is a vertex of V_1 for each set X_i and a vertex of V_2 for each element of X. An arc (i, j) of infinite capacity joins $i \in V_1$ to $j \in V_2$ if $j \in X_i$. Add a source vertex s joined by arcs of unit capacity to each $i \in V_1$, and a sink vertex t with arcs of unit capacity joining each $j \in V_2$ to t. Then a system of distinct representatives exists if and only if the maximum flow in this constructed network has value m. In this case, those arcs (i, j), with $i \in V_1$ and $j \in V_2$, having flow $x_{ij} = 1$ identify a system of distinct representatives selected from the m sets.

4. *Feasible flow problem*: This problem involves finding a flow x in $G = (V, E)$ so that the net flow at each vertex is a specified value $b(i)$, where $\sum_{i \in V} b(i) = 0$. That is, a flow x on the arcs of network G is required, satisfying:

- mass balance constraints: $\sum_{\{j|(i,j) \in E\}} x_{ij} - \sum_{\{j|(j,i) \in E\}} x_{ji} = b(i)$ for all $i \in V$;
- capacity constraints: $0 \leq x_{ij} \leq u_{ij}$ for all $(i, j) \in E$.

This can be modeled as a maximum flow problem. Construct the augmented network G' by adding a source vertex s and a sink vertex t to G. For each vertex i with $b(i) > 0$, an arc (s, i) is added to E with capacity $b(i)$; for each vertex i with $b(i) < 0$, an arc (i, t) is added to E with capacity $-b(i)$. Then solve a maximum flow problem from vertex s to vertex t in G'. It can be proved that the feasible flow problem for G has a solution if and only if the maximum flow in G' saturates all arcs emanating from vertex s in G'.

5. Application 4 frequently arises in distribution problems. For example, a known amount of merchandise is available at certain ports and is required at other ports in known quantities. Also the maximum quantity of merchandise that can be shipped on a particular sea route is specified. Determining whether it is possible to satisfy all of the demands by using the available supplies is a feasible flow problem.

6. *Graph connectivity*: In a directed graph G, the arc connectivity κ'_{ij} of vertices i and j is the minimum number of arcs whose removal from G leaves no directed path from i to j. The arc connectivity $\kappa'(G)$ is the minimum number of arcs whose removal from G separates some pair of vertices (see §8.4.2). The arc connectivity of a graph is an important measure of the graph's reliability or stability. Since $\kappa'(G) = \min\{\kappa'_{ij} \mid (i, j) \in V \times V, i \neq j\}$, the arc connectivity of a graph can be computed by determining the arc connectivity of $n(n-1)$ pairs of vertices. As a matter of fact, the arc connectivity of G can be found by determining only $n - 1$ arc connectivities.

The arc connectivity κ'_{ij} can be found by applying the max-flow min-cut theorem (§10.4.1) to the network obtained from G by setting the capacity of each arc (i, j) to 1. In such a unit capacity network, the maximum i-j flow value equals the maximum number of arc-disjoint paths from vertex i to vertex j, and the minimum i-j cut capacity equals the minimum number of arcs required to separate vertex i and vertex j. This shows that the maximum number of arc-disjoint paths from vertex i to vertex j equals the minimum number of arcs whose removal disconnects all paths from vertex i to vertex j. (This result is a variation of Menger's theorem in §8.4.2; it was independently discovered by Ford and Fulkerson and by Elias, Feinstein, and Shannon.) Consequently, κ'_{ij} equals the maximum i-j flow value in the network, and the arc connectivity $\kappa'(G)$ can be determined by solving $n-1$ maximum flow problems in a unit capacity network.

7. *Tournaments*: Consider a round-robin tournament between n teams, assuming each team plays against every other team c times and no game ends in a draw. It is claimed that α_i for $1 \leq i \leq n$ is the number of victories accrued by the ith team at the end of the tournament. Verifying whether the nonnegative integers $\alpha_1, \alpha_2, \ldots, \alpha_n$ are possible winning records for the n teams can be modeled as a feasible flow problem.

Define a directed network $G = (V, E)$ with vertex set $V = \{1, 2, \ldots, n\}$ and arc set $E = \{(i,j) \in V \times V \mid i < j\}$. Let x_{ij}, $i < j$, represent the number of times team i defeats team j. The total number of times team i defeats teams $i+1, i+2, \ldots, n$ is $\sum_{\{j \mid (i,j) \in E\}} x_{ij}$. Since the number of times team i defeats a team $j < i$ is $c - x_{ji}$, it follows that the total number of times that team i defeats teams $1, 2, \ldots, i-1$ is $(i-1)c - \sum_{\{j \mid (j,i) \in E\}} x_{ji}$. However, there are two constraints:

- the total number of wins α_i of team i must equal the total number of times it defeats teams $1, 2, \ldots, n$, giving

$$\sum_{\{j \mid (i,j) \in E\}} x_{ij} - \sum_{\{j \mid (j,i) \in E\}} x_{ji} = \alpha_i - (i-1)c \text{ for all } i \in V;$$

- a possible winning record must also satisfy

$$0 \leq x_{ij} \leq c \text{ for all } (i,j) \in E.$$

Consequently, $\{\alpha_i\}$ define a possible winning record if these two constraints have a feasible solution x. Let $b(i) = \alpha_i - (i-1)c$. Since $\sum_{i \in V} \alpha_i$ and $\sum_{i \in V} (i-1)c$ are both equal to $\frac{cn(n-1)}{2}$, the total number of games played, it follows that $\sum_{i \in V} b(i) = 0$. The problem of finding a feasible solution to the two constraints is then a feasible flow problem.

8. *Matchings and covers*: The max-flow min-cut theorem can also be used to prove a min-max result concerning matchings and covers in a directed bipartite graph $G = (V_1 \cup V_2, E)$. (See §8.1.3.) The subset $E' \subseteq E$ is a *matching* (§10.2.1) if no two arcs in E' are incident with the same vertex. The subset $V' \subseteq V_1 \cup V_2$ is a *vertex cover* if every arc in E is incident to at least one vertex in V'. Create the network G' from G by adding vertices s and t, as well as arcs (s,i) with capacity 1 for all $i \in V_1$ and arcs (j,t) with capacity 1 for all $j \in V_2$. All other arcs of G' correspond to arcs of G and have infinite capacity. Then each matching of cardinality v defines a flow of value v in G', and each s-t cut of capacity v induces a corresponding vertex cover with v vertices. Application of the max-flow min-cut theorem establishes the desired result: In a bipartite graph $G = (V_1 \cup V_2, E)$, the maximum cardinality of any matching equals the minimum cardinality of any vertex cover of G.

9. *0-1 matrices*: Suppose $A = (a_{ij})$ is a 0-1 matrix. Associate with A the directed bipartite graph $G = (V_1 \cup V_2, E)$, where V_1 is the set of row indices and V_2 is the set of column indices. There is an arc $(i,j) \in E$ whenever $a_{ij} = 1$. A matching in G now corresponds to a set of "independent" 1s in the matrix A: i.e., no two of these 1s are in the same row or the same column. Also, a vertex cover of G corresponds to a set of rows and columns in A that collectively cover all the 1s in the matrix. Applying the result in Application 8 shows that the maximum number of independent 1s in A equals the minimum number of lines (rows and/or columns) needed to cover all the 1s in A. This result is known as König's theorem (§6.6.1).

10. Additional applications, with reference sources, are given in the following table.

application	references
matrix rounding	[AhMaOr93], [AhEtal95]
distributed computing	[AhMaOr93], [AhEtal95]
network reliability	[AhMaOr93], [AhEtal95]
open pit mining	[AhMaOr93], [AhEtal95]
building evacuation	[AhMaOr93]
covering sports events	[AhEtal95]
nurse staff scheduling	[AhMaOr93], [AhEtal95]
bus scheduling	[AhEtal95]
machine scheduling	[AhMaOr93], [AhEtal95]
tanker scheduling	[AhMaOr93], [AhEtal95]
bottleneck assignment	[FoFu62]
selecting freight-handling terminals	[AhEtal95]
site selection	[EvMi92]
material-handling systems	[EvMi92]
decompositions of partial orders	[FoFu62]
matrices with prescribed row/column sums	[FoFu62]

10.5 MINIMUM COST FLOWS

The minimum cost flow problem involves determining the least cost shipment of a commodity through a capacitated network in order to satisfy demands at certain vertices using supplies available at other vertices. This problem generalizes both the shortest path problem (§10.3) and the maximum flow problem (§10.4).

10.5.1 BASIC CONCEPTS

Definitions:

Let $G = (V, E)$ be a directed network with vertex set V and arc set E (see §10.3.1). Each arc $(i, j) \in E$ has an associated *cost* c_{ij} and a *capacity* $u_{ij} \geq 0$. Let $n = |V|$ and $m = |E|$.

Each vertex $i \in V$ has an associated supply/demand $b(i)$. If $b(i) > 0$, then vertex i is a **supply vertex**; if $b(i) < 0$, then vertex i is a **demand vertex**.

A (**feasible**) **flow** is a function $x = (x_{ij})$ defined on arcs $(i, j) \in E$ satisfying:

- mass balance constraints: $\sum_{\{j|(i,j)\in E\}} x_{ij} - \sum_{\{j|(j,i)\in E\}} x_{ji} = b(i)$ for all $i \in V$,
- capacity constraints: $0 \leq x_{ij} \leq u_{ij}$ for all $(i, j) \in E$,

where $\sum_{i \in V} b(i) = 0$.

The **cost** of flow x is $\sum_{(i,j)\in E} c_{ij} x_{ij}$.

A **minimum cost flow** is a flow having minimum cost.

A **pseudoflow** is a function $x = (x_{ij})$ satisfying the arc capacity constraints; it may violate the mass balance constraints.

The **residual network** $G(x)$ corresponding to a flow (or pseudoflow) x is defined in the following manner. Replace each arc $(i,j) \in E$ by two arcs (i,j) and (j,i). Arc (i,j) has cost c_{ij} and **residual capacity** $r_{ij} = u_{ij} - x_{ij}$, and arc (j,i) has cost $-c_{ij}$ and **residual capacity** $r_{ji} = x_{ij}$. The residual network consists only of arcs with positive residual capacity.

The **potential** of vertex i is a quantity $\pi(i)$ associated with the mass balance constraint at vertex i. With respect to a given set of vertex potentials, the **reduced cost** of an arc (i,j) in the residual network $G(x)$ is $c_{ij}^\pi = c_{ij} - \pi(i) + \pi(j)$.

The **cost** of path P in $G(x)$ is $c(P) = \sum_{(i,j)\in P} c_{ij}$; its **reduced cost** is $c^\pi(P) = \sum_{(i,j)\in P} c_{ij}^\pi$.

A **negative cycle** is a directed cycle W in $G(x)$ for which $c(W) < 0$.

Facts:

1. The mass balance constraints ensure that the net flow out of each vertex i is equal to $b(i)$. Thus, if there is excess flow out of vertex i, then $b(i) > 0$ and i is a supply vertex. If $b(i) < 0$, then more flow enters i than leaves i, meaning that vertex i is a demand vertex.

2. Minimum cost flows arise in practical problems involving the least cost routing of goods, vehicles, and messages in a network. Minimum cost flows can also be used in models of warehouse layout, production and inventory problems, scheduling of personnel, automatic classification of chromosomes, and racial balancing of schools. (See §10.5.3.)

3. Let $\{\pi(i) \mid i \in V\}$ be any set of vertex potentials.
- If P is a path from i to j in $G(x)$, then $c^\pi(P) = c(P) - \pi(i) + \pi(j)$.
- If W is a cycle in $G(x)$, then $c^\pi(W) = c(W)$.

4. *Negative cycle optimality conditions*: A feasible flow x is a minimum cost flow if and only if the residual network $G(x)$ contains no negative cycle.

5. *Reduced cost optimality conditions*: A feasible flow x is a minimum cost flow if and only if some set of vertex potentials π satisfies $c_{ij}^\pi \geq 0$ for every arc (i,j) in $G(x)$.

6. *Complementary slackness optimality conditions*: A feasible flow x is a minimum cost flow if and only if there exist vertex potentials π such that for every arc $(i,j) \in E$:
- if $c_{ij}^\pi > 0$, then $x_{ij} = 0$;
- if $c_{ij}^\pi < 0$, then $x_{ij} = u_{ij}$;
- if $0 < x_{ij} < u_{ij}$, then $c_{ij}^\pi = 0$.

Examples:

1. In the flow network of part (a) of the following figure, $b(i)$ is shown next to each vertex i and (c_{ij}, u_{ij}) is shown next to each arc (i,j).

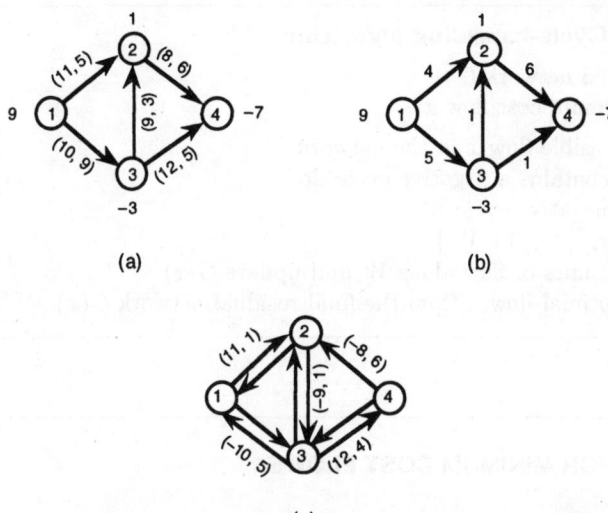

(a) (b)

(c)

The function $x = (x_{ij})$ given in part (b) satisfies the mass balance constraints for each vertex. For example, the flow out of vertex 2 is $x_{24} = 6$ and the flow into vertex 2 is $x_{12} + x_{32} = 5$, so that flow out minus flow in equals $6 - 5 = 1 = b(2)$. Also the capacity constraints for all arcs are satisfied: e.g., $x_{12} = 4 \leq 5 = u_{12}$. Thus x is a feasible flow, with cost 163. The residual network $G(x)$ corresponding to the flow x is shown in part (c). Selected arcs of $G(x)$ are labeled with their cost and residual capacity. The directed cycle $W = [1, 2, 3, 1]$ in $G(x)$ has cost $11 - 9 - 10 = -8$ and so W is a negative cycle. By Fact 4, this flow x is not a minimum cost flow.

2. Part (a) of the following figure shows another feasible flow x' for the network in Example 1, with cost 155.

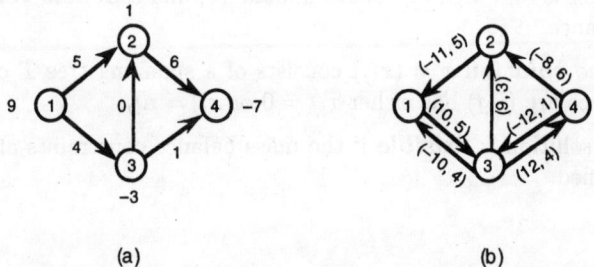

(a) (b)

The corresponding residual network $G(x')$ is given in part (b), in which each arc is labeled with its cost and its residual capacity. Using the vertex potentials $\pi = (0, -14, -10, -22)$, the reduced cost of arc $(2, 1)$ in the residual network is $c^\pi_{21} = -11 - (-14) + 0 = 3$; likewise $c^\pi_{32} = 9 - (-10) - 14 = 5$. The remaining reduced costs are found to be zero, so $c^\pi_{ij} \geq 0$ for all arcs (i, j) in $G(x)$. By Fact 5, x' is a minimum cost flow for the given network.

3. Alternatively, the optimality of the flow x' in part (a) of the figure of Example 2 can be verified using Fact 6. As in Example 2, let $\pi = (0, -14, -10, -22)$. Arc $(3, 2)$ of the original network G in part (a) of the figure of Example 1 has positive reduced cost $c^\pi_{32} = 9 - (-10) - 14 = 5$ and $x'_{32} = 0$. Arc $(1, 2)$ has $c^\pi_{12} = 11 - 0 - 14 = -3 < 0$ and $x'_{12} = u_{12}$. The remaining arcs $(1, 3)$, $(2, 4)$, and $(3, 4)$ have zero reduced cost. Consequently, the complementary slackness optimality conditions are satisfied and the flow x' achieves the minimum cost.

Algorithm 1:   Cycle-canceling algorithm.

input: directed network G
output: minimum cost flow x

establish a feasible flow x in the network
while $G(x)$ contains a negative cycle **do**
 identify a negative cycle W
 $\delta := \min\{\, r_{ij} \mid (i,j) \in W \,\}$
 augment δ units of flow along W and update $G(x)$
recover an optimal flow x from the final residual network $G(x)$

10.5.2   ALGORITHMS FOR MINIMUM COST FLOWS

A variety of algorithms are available to solve the minimum cost flow problem. Three algorithms are described in this section: the cycle-canceling algorithm, the successive shortest path algorithm, and the network simplex algorithm.

Definitions:

Let $G = (V, E)$ be a directed network with $n = |V|$ and $m = |E|$; let U denote the largest arc capacity and C denote the largest arc cost (in absolute value) in G.

For a given pseudoflow $x = (x_{ij})$, the **imbalance** of vertex $i \in V$ is $e(i) = b(i) + \sum_{\{j \mid (j,i) \in E\}} x_{ji} - \sum_{\{j \mid (i,j) \in E\}} x_{ij}$.

An **excess vertex** is one with a positive imbalance, and a **deficit vertex** is one with a negative imbalance.

A **spanning tree solution** $x = (x_{ij})$ consists of a spanning tree T of $G = (V, E)$ in which each nontree arc (i, j) has either $x_{ij} = 0$ or $x_{ij} = u_{ij}$.

A spanning tree solution is **feasible** if the mass balance constraints and capacity constraints are satisfied.

Facts:

1. *Cycle-canceling algorithm*: The cycle-canceling algorithm (Algorithm 1) is based on the negative cycle optimality conditions (§10.5.1, Fact 4). It starts with a feasible flow and successively augments flow along negative cycles in the residual network until there is no negative cycle.

2. As shown in §10.4.3, an initial feasible flow can be found by solving a maximum flow problem.

3. *Integrality property*: For problems with integer arc capacities and integer vertex supplies/demands, Algorithm 1 starts with an integer flow, at each iteration augments by an integral amount of flow, and thus produces an optimal flow that is integer. Thus any minimum cost flow problem with integer supplies, demands, and capacities always has an optimal solution that is integer.

4. A negative cycle W in the residual network can be identified in $O(nm)$ time by using a queue implementation of the label-correcting algorithm (§10.3.2, Fact 5).

Algorithm 2: Successive shortest path algorithm.

input: directed network G
output: minimum cost flow x

$x := 0$
$e(i) := b(i)$ for all $i \in V$
initialize $V^+ := \{\, i \mid e(i) > 0 \,\}$ and $V^- := \{\, i \mid e(i) < 0 \,\}$
while $V^+ \neq \emptyset$ do
  select a vertex $k \in V^+$ and a vertex $l \in V^-$
  identify a shortest path P in $G(x)$ from vertex k to vertex l
  $\delta := \min\{e(k), -e(l), \min\{\, r_{ij} \mid (i,j) \in P \,\}\}$
  augment δ units of flow along P
  update e, $G(x)$, V^+, and V^-
recover an optimal flow x from the final residual network $G(x)$

5. Augmenting the flow along W by δ decreases the residual capacities of arcs in W by δ and increases the residual capacities of the reversals of arcs in W by δ.

6. Upon termination of Algorithm 1, an optimal flow x can be reconstructed from the final $G(x)$; see §10.4.2 Fact 10.

7. For problems with integer supplies, demands, and arc capacities, the cycle-canceling algorithm runs in pseudopolynomial time $O(nm^2CU)$.

8. If flow is augmented along a negative cycle W in $G(x)$ that minimizes the ratio $\frac{1}{|W|} \sum_{(i,j) \in W} c_{ij}$ among all directed cycles in $G(x)$, then this implementation runs in polynomial time [GoTa88].

9. *Successive shortest path algorithm*: The successive shortest path algorithm (Algorithm 2) starts with the pseudoflow $x = 0$. It proceeds by selecting an excess vertex k and a deficit vertex l, and augmenting flow along a minimum cost path from vertex k to vertex l in $G(x)$.

10. If in Algorithm 2 reduced costs c_{ij}^π are used instead of arc costs c_{ij}, then Dijkstra's algorithm (§10.3.2, Algorithm 2) can be applied to determine a shortest path P in the residual network.

11. Augmenting the flow along P by δ decreases the residual capacities of arcs in P by δ and increases the residual capacities of the reversals of arcs in P by δ. It also decreases $e(k)$ by δ and increases $e(l)$ by δ.

12. The solution maintained by the successive shortest path algorithm always satisfies the reduced cost optimality conditions (§10.5.1, Fact 5). The final solution is in addition feasible, and so is an optimal solution of the minimum cost flow problem.

13. For problems with integer supplies, demands, and arc capacities, the shortest augmenting path algorithm runs in pseudopolynomial time.

14. Several implementations of the shortest augmenting path algorithm run in polynomial or even strongly polynomial time. [Or88] describes an implementation running in $O(m \log n(m + n \log n))$ time, currently the fastest strongly polynomial-time algorithm to solve the minimum cost flow problem.

15. If a minimum cost flow problem has an optimal solution, then it has an optimal spanning tree solution.

> **Algorithm 3:** Network simplex algorithm.
>
> input: directed network G
> output: minimum cost flow x
>
> determine an initial spanning tree solution with associated tree T
> let x be the flow and π the corresponding vertex potentials
> **while** some nontree arc violates the complementary slackness optimality conditions **do**
>   select an entering arc (k,l) violating its optimality condition
>   add arc (k,l) to T, augment the maximum possible flow in the cycle thus
>   created, and determine the leaving arc (p,q)
>   update the tree T, the flow x, and the vertex potentials π

16. Given a spanning tree solution x, with flows on nontree arcs (i,j) specified (at either 0 or u_{ij}), the flows on the tree arcs are uniquely determined by the mass balance constraints.

17. Given a spanning tree solution x, vertex potentials π can be determined such that:
    - $\pi(1) = 0$;
    - $c_{ij}^\pi = 0$ for all tree arcs (i,j).

18. *Complementary slackness optimality conditions*: Suppose x is a feasible spanning tree solution with vertex potentials determined as in Fact 17. Then x is a minimum cost flow if:
    - $c_{ij}^\pi \geq 0$ for all nontree arcs (i,j) with $x_{ij} = 0$;
    - $c_{ij}^\pi \leq 0$ for all nontree arcs (i,j) with $x_{ij} = u_{ij}$.

19. *Network simplex algorithm*: The network simplex algorithm (Algorithm 3) is a specialized version of the well-known linear programming simplex method (§15.1.3). It maintains a spanning tree solution and at each iteration transforms the current spanning tree solution into an improved spanning tree solution until optimality is reached.

20. Using appropriate data structures, the network simplex algorithm can be implemented very efficiently. The network simplex algorithm is one of the fastest algorithms to solve the minimum cost flow problem in practice.

21. The network simplex algorithm has a exponential worst-case time bound. [Or97] provides the first polynomial-time implementations of the (generic) network simplex algorithm.

22. Detailed descriptions of Algorithms 1-3, as well as several other algorithms for finding minimum cost flows, can be found in [AhMaOr93].

23. Computer codes (in C, Pascal, and Fortran) for solving the minimum cost flow problem can be found at the following sites:

    ftp://dimacs.rutgers.edu/pub/netflow/mincost/

    http://www.zib.de/Optimization/Software/Mcf/

    ftp://ftp.zib.de/pub/Packages/mathprog/netopt-bertsekas/

    http://www.neci.nj.nec.com/homepages/avg/soft/soft.html

    http://orly1.snu.ac.kr/software/

Examples:

1. The following figure illustrates the cycle-canceling algorithm.

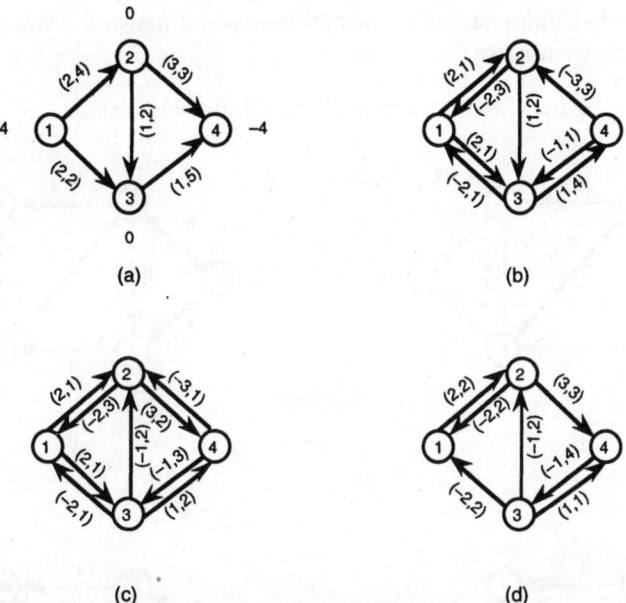

Part (a) of this figure depicts the given flow network, with $b(i)$ shown for each vertex i and (c_{ij}, u_{ij}) for each arc (i,j). Part (b) shows the residual network corresponding to the flow $x_{12} = x_{24} = 3$ and $x_{13} = x_{34} = 1$.

In the first iteration, suppose the algorithm selects the negative cycle $[2, 3, 4, 2]$ with cost -1. Then $\delta = \min\{r_{23}, r_{34}, r_{42}\} = \min\{2, 4, 3\} = 2$ units of flow are augmented along this cycle. Part (c) shows the modified residual network.

In the next iteration, the algorithm selects the cycle $[1, 3, 4, 2, 1]$ with cost -2 and augments $\delta = 1$ unit of flow. Part (d) depicts the updated residual network which contains no negative cycle, so the algorithm terminates. From part (d), an optimal flow pattern is deduced: $x_{12} = x_{13} = x_{23} = 2$ and $x_{34} = 4$.

2. The successive shortest path algorithm is illustrated using the flow network in part (a) of the figure for Example 1. The initial residual network $G(x)$ for $x = 0$ is the same as that of part (a). Initially, the imbalances are $e = (4, 0, 0, -4)$, so that $V^+ = \{1\}$ and $V^- = \{4\}$, giving $k = 1$ and $l = 4$. The shortest path from vertex 1 to 4 in $G(x)$ is $[1, 3, 4]$, and the algorithm augments $\delta = 2$ units of flow along this path.

The following figure shows the residual network after this augmentation, as well as the updated imbalance at each vertex.

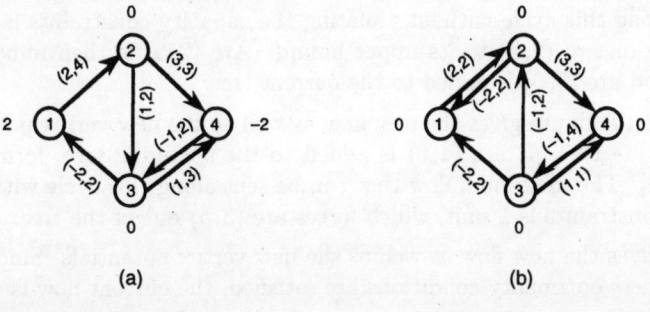

The sets V^+ and V^- do not change, so again $k = 1$ and $l = 4$. The shortest path from vertex 1 to vertex 4 is now $[1, 2, 3, 4]$, and the algorithm augments $\delta = 2$ units of flow along this path.

Part (b) of that figure shows the resulting residual network. Now $V^+ = V^- = \emptyset$ and the algorithm terminates.

3. The following figure illustrates the network simplex algorithm.

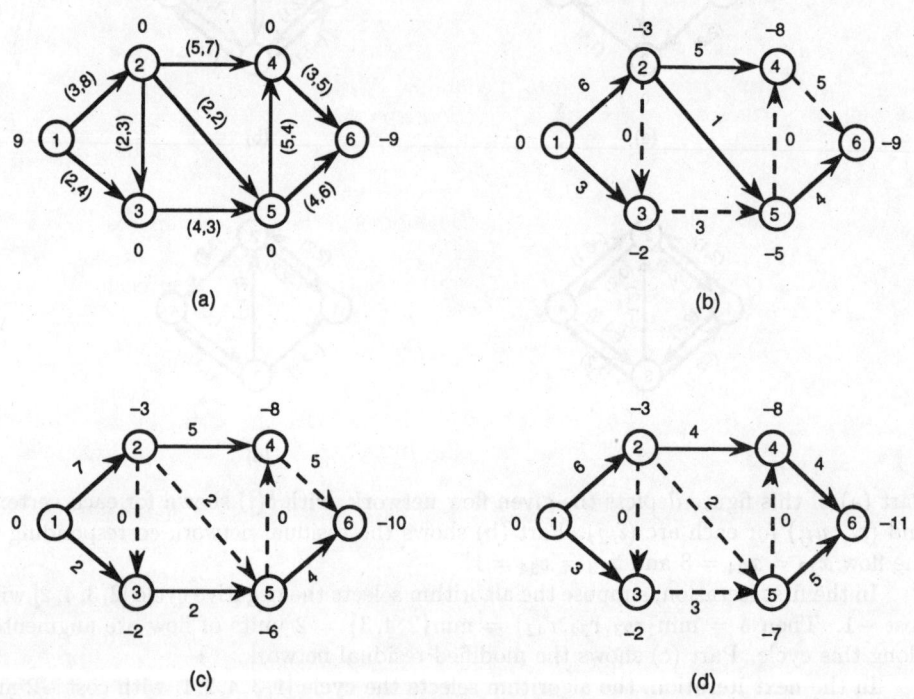

Part (a) of this figure depicts the given flow network, with $b(i)$ shown for each vertex i and (c_{ij}, u_{ij}) for each arc (i, j).

A feasible spanning tree solution is shown in part (b) of the figure; each nontree arc (dashed line) has flow at either its lower or upper bound. The unique flows on the tree arcs (solid lines) are determined by the mass balance constraints. A set of vertex potentials (obtained using Fact 17) are also shown in part (b). Relative to the these potentials π, the reduced costs for the nontree arcs are given by $c_{23}^\pi = 2 - (-3) - 2 = 3$, $c_{35}^\pi = 4 - (-2) - 5 = 1$, $c_{54}^\pi = 5 - (-5) - 8 = 2$, and $c_{46}^\pi = 3 - (-8) - 9 = 2$. Since arc $(3, 5)$, with flow at its upper bound, violates the optimality conditions of Fact 18, it is added to the current tree producing the cycle $[1, 2, 5, 3, 1]$. The maximum flow that can be sent along this cycle without violating the capacity constraints is 1 unit, which forces the flow on arc $(2, 5)$ to its upper bound. Arc $(2, 5)$ is then removed from the current tree and arc $(3, 5)$ is added to the current tree.

Part (c) of the figure gives the new flow as well as the new vertex potentials. Since $c_{46}^\pi = 3 - (-8) - 10 = 1$, arc $(4, 6)$ is added to the spanning tree, forming the cycle $[1, 3, 5, 6, 4, 2, 1]$. The maximum flow that can be sent along this cycle without violating the capacity constraints is 1 unit, which forces arc $(3, 5)$ out of the tree.

Part (d) gives the new flow as well as the new vertex potentials. Since the complementary slackness optimality conditions are satisfied, the current flow is optimal.

10.5.3 APPLICATIONS

Minimum cost flow problems arise in many industrial settings and scientific domains, often in the form of distribution or routing problems. The minimum cost flow problem also has less transparent applications, several of which are presented in this section.

Applications:

1. *Distribution*: A common application of the minimum cost flow problem involves the distribution at minimum cost of a product from manufacturing plants (with known supplies) to warehouses (with known demands). A similar scenario applies to the distribution of goods from warehouses to retailers as well as the flow of raw materials and intermediate goods through various machining stations in a production line.

2. *Routing*: The routing of cars through an urban street network and the routing of calls through a telephone system can be modeled using minimum cost flows. In either case, the items (cars, calls) must be sent from certain specified origins to other specified destinations, with capacity constraints on the total flow on each arc (road, communication link). This is done to minimize total (or average) delay in the system.

3. *Directed Chinese postman problem*: Leaving from the post office, a mail carrier needs to visit all houses on a postal route, delivering and collecting letters, and then return to the post office. The carrier would like to cover this route by traveling the minimum possible distance. (See also §8.4.3.) In this variation, known as the *directed Chinese postman problem*, each street is assumed to be directed, so the problem is defined on a directed network $G = (V, E)$ whose arcs (i, j) have an associated nonnegative length c_{ij}. It is desired to find a directed walk (§8.3.2) of minimum length that starts at some vertex (the post office), visits each arc of the network at least once, and returns to the starting vertex. In an optimal walk, some arcs may be traversed more than once. If x_{ij} represents the number of times arc (i, j) is traversed, then this problem can be formulated as:

$$\text{minimize:} \quad \sum_{(i,j) \in E} c_{ij} x_{ij},$$

$$\text{subject to:} \quad \sum_{\{j \mid (i,j) \in E\}} x_{ij} - \sum_{\{j \mid (j,i) \in E\}} x_{ji} = 0 \quad \text{for all } i \in V,$$

$$x_{ij} \geq 1 \quad \text{for all } (i,j) \in E.$$

This problem is a minor variant of the minimum cost flow problem where each arc has a lower bound of one unit of flow. From an optimal flow x^* for this problem, an optimal tour can be constructed in the following manner. First, replace each arc (i, j) with x_{ij}^* copies of the arc, each carrying a unit flow. Next, decompose the resulting network into a set of directed cycles. Finally, connect the directed cycles to form a closed walk.

4. *Optimal loading of a hopping airplane*: A small commuter airline uses a plane with capacity of at most p passengers on a "hopping flight", as shown in part (a) of the following figure. The flight visits the cities $1, 2, 3, \ldots, n$ in a fixed sequence. The plane can pick up passengers at any city and drop them off at any other city. Let b_{ij} denote the number of passengers available at city i who want to go to city j, and let f_{ij} denote the fare per passenger from city i to city j. The airline would like to determine the number of passengers that the plane should carry between various origins and destinations in order to maximize the total fare per trip while never exceeding the capacity of the plane.

Part (b) of the following figure shows a minimum cost flow formulation of this hopping plane flight problem. The network displays data only for those arcs with nonzero cost or finite capacity. Any arc without a displayed cost has zero cost; any arc without a displayed capacity has infinite capacity. For example, three types of passengers are available at vertex 1: those whose destination is vertex 2, vertex 3, or vertex 4. These three types of passengers are represented by the vertices 1-2, 1-3, and 1-4 with supplies b_{12}, b_{13}, and b_{14}. A passenger available at any such vertex, say 1-3, either boards the plane at its origin vertex by flowing through the arc (1-3, 1), and thus incurring a cost of $-f_{13}$ units, or never boards the plane, represented by flowing through the arc (1-3, 3).

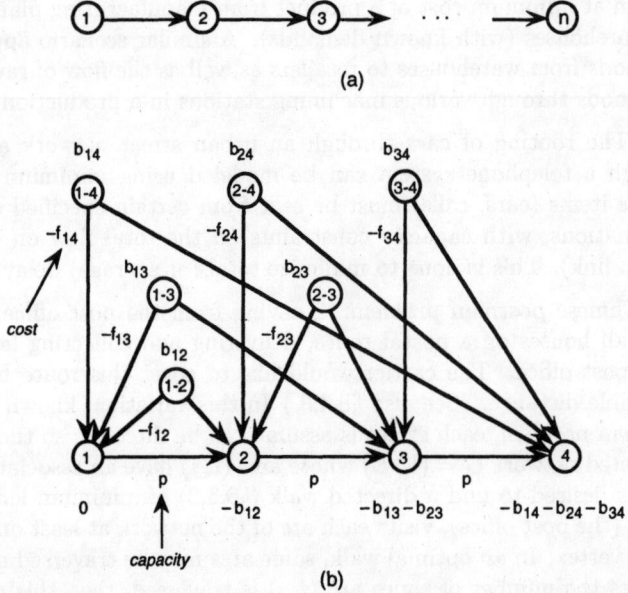

5. *Leveling mountainous terrain*: In building road networks through hilly or mountainous terrain, civil engineers must determine how to distribute earth from high points to low points of the terrain to produce a leveled roadbed. To model this, construct a *terrain graph*, an undirected graph G whose vertices represent locations with a demand for earth (low points) or locations with a supply of earth (high points). An edge of G indicates an available route for distributing the earth, and the cost of this edge is the cost per truckload of moving earth between the corresponding two locations. The following figure shows a portion of a sample terrain graph. A leveling plan for a terrain graph is a flow (set of truckloads) that meets the demands at vertices (levels the low points) by the available supplies (earth obtained from high points) at minimum trucking cost. This can be solved as a minimum cost flow problem on the terrain graph.

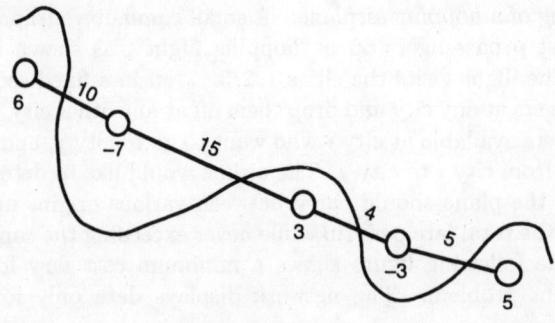

6. Additional applications, with reference sources, are given in the following table.

application	references
medical tomography	[AhMaOr93], [AhEtal95]
automatic chromosome classification	[AhMaOr93], [AhEtal95]
racial balancing of schools	[AhMaOr93], [AhEtal95]
controlled matrix rounding	[AhMaOr93], [AhEtal95]
building evacuation	[AhMaOr93], [AhEtal95]
just-in-time scheduling	[AhMaOr93], [AhEtal95]
telephone operator scheduling	[AhMaOr93], [AhEtal95]
nurse staff scheduling	[AhMaOr93]
machine scheduling	[AhMaOr93]
production scheduling	[EvMi92]
equipment replacement	[AhMaOr93]
microdata file merging	[EvMi92], [AhEtal95]
warehouse layout	[AhMaOr93], [AhEtal95]
facility location	[AhMaOr93], [AhEtal95]
determining service districts	[EvMi92], [AhMaOr93], [AhEtal95]
capacity expansion	[AhMaOr93]
vehicle fleet planning	[AhMaOr93]

10.6 COMMUNICATION NETWORKS

Modern communication networks consist of two main components. Using high-capacity links, the *backbone network* interconnects switching centers and gateway vertices that carry and direct traffic through the communication system. *Local access networks* transfer traffic between the backbone network and the end users. This section presents several optimization models used in the design of communication networks.

10.6.1 CAPACITATED MINIMUM SPANNING TREE PROBLEM

The capacitated minimum spanning tree problem arises in the design of local access tree networks in which end users generate and retrieve data from other sources, always through a specified control center (e.g., a communication switch of the backbone network). In this problem, user sites are to be interconnected at minimum cost by means of subtrees, which are in turn connected to the control center. The total traffic in each subtree is limited by a capacity constraint.

Definitions:

Let $N = \{1, 2, \ldots, n\}$ be a set of **terminals** and let 0 denote a specified **control center**. The complete undirected graph $G = (V, E)$ has vertex set $V = N \cup \{0\}$ and contains all possible edges between distinct vertices of V (§8.1.3).

The *cost* of connecting distinct vertices $i, j \in V$ is $c_{ij} = c_e$, where $e = (i, j) \in E$.

The **demand** w_i at vertex $i \in N$ is the amount of traffic to be transmitted to the control center.

Relative to a spanning tree T (§9.2.1) of G, vertex j is a **root vertex** if it is adjacent to vertex 0. Vertex i is **assigned** to root vertex j if j is on the unique path in T joining i to the control center. The set of all vertices assigned to j defines the **subtree** T_j of T. This subtree has **demand** $D(T_j) = \sum_{i \in T_j} w_i$.

A **capacitated minimum spanning tree (CMST)** is a spanning tree T of G composed of subtrees $T_{j_1}, T_{j_2}, \ldots, T_{j_r}$ such that:

- $\sum_{e \in T} c_e$ is minimum;
- the demand in each T_j is at most Q, a specified capacity.

Let $i, j \in N$ and define

$$y_j = \begin{cases} 1 & \text{if vertex } j \text{ is a root vertex} \\ 0 & \text{otherwise} \end{cases}$$

$$x_{ij} = \begin{cases} 1 & \text{if vertex } i \text{ is assigned to root vertex } j \\ 0 & \text{otherwise} \end{cases}$$

and for $i \neq j$

$$z_{ij} = \begin{cases} 1 & \text{if } (i,j) \in T \\ 0 & \text{otherwise.} \end{cases}$$

Given a vector z, the subgraph $G(N, E_z)$ of G **induced by** z has vertex set N and edges $e \in E_z$ if $z_e > 0$. Similarly, given a vector (z, y), the subgraph **induced by** (z, y), written $G(V, E_{zy})$, has vertex set V; it contains every edge of E_z plus each edge from j to 0 where $y_j > 0$.

Relative to a given vector z, $C(i,j)$ denotes the set of all **i-j cuts** $[S, \overline{S}]$ in the graph $G(N, E_z)$. (See §10.4.1.)

If $S \subseteq V$, then $E(S) = \{ (i,j) \in E \mid i, j \in S \}$ contains all edges between vertices of S.

For $I \subset N$, let $b(I)$ be the minimum number of subtrees needed to pack all terminals in I. That is, $b(I)$ is the optimal solution to the **bin packing problem** (§15.3.2) with bins of capacity Q and items of size w_i for every $i \in I$.

A set $S \subset N$ is a **cover** if $\sum_{i \in S} w_i > Q$. If also $(\sum_{i \in S} w_i) - w_k \leq Q$ for all $k \in S$, then S is a **minimal cover**.

Facts:

1. The CMST problem has the following 0-1 integer linear programming formulation (§15.1.8):

$$\min \left\{ \sum_{i=1}^{n-1} \sum_{j=i+1}^{n} c_{ij} z_{ij} + \sum_{j=1}^{n} c_{0j} y_j \right\}$$

subject to:

$$\sum_{j=1}^{n} x_{ij} = 1, \text{ for all } i \in \{1, 2, \ldots, n\}$$

$$\sum_{i=1}^{n} w_i x_{ij} \leq Q y_j, \text{ for all } j \in \{1, 2, \ldots, n\}$$

$$x_{ij} \leq y_j, \text{ for all } i, j \in \{1, 2, \ldots, n\}$$

Algorithm 1: Savings heuristic.

input: undirected network G, control center 0, capacity limit Q
output: an approximate capacitated minimum spanning tree T^*

$U := \{1, 2, \ldots, n\}$
$T_u := \{u\}$ for $u \in U$
while true
  for $u \in U$
    compute f_u, the minimum cost of connecting the control center 0 to component T_u
  $S := \emptyset$
  for $u, v \in U$ $(u \neq v)$
    if $D(T_u \cup T_v) \leq Q$ then
      compute s_{uv}, the difference between $\max\{f_u, f_v\}$ and the minimum cost of connecting T_u to T_v
      if $s_{uv} > 0$ then $S := S \cup \{(u, v)\}$
  if $S = \emptyset$ then return
  else
    choose u_0, v_0 such that $s_{u_0 v_0} = \max\{s_{uv} \mid (u, v) \in S\}$
    merge T_{u_0} and T_{v_0}, creating a new subtree indexed by $\min\{u_0, v_0\}$, and update U appropriately

$$x_{ij} \leq \sum_{e \in K} z_e, \text{ for all } i, j \in \{1, 2, \ldots, n\} \ (i \neq j) \text{ and for all } K \in C(i, j)$$

$$\sum_e z_e + \sum_j y_j = n$$

$$x_{ij} \in \{0, 1\}, \text{ for all } i, j \in \{1, 2, \ldots, n\}$$

$$y_j \in \{0, 1\}, \text{ for all } j \in \{1, 2, \ldots, n\}$$

$$z_{ij} \in \{0, 1\}, \text{ for all } i, j \in \{1, 2, \ldots, n\} \ (i \neq j).$$

2. In Fact 1:
   - the first set of constraints ensures that each vertex is assigned to a root vertex;
   - the second set of constraints ensures that the flow through any root vertex is no more than the capacity Q;
   - third set of constraints ensures that vertex i can be assigned to vertex j only if j is a root vertex;
   - the fourth set of constraints ensures that if vertex i is assigned to root vertex j, then there must be a path between i and j;
   - the fifth set of constraints guarantees that $G(V, E_{zy})$ is a tree.

3. *Savings heuristic*: This greedy heuristic (Algorithm 1) begins with n components, each a single vertex, and successively merges pairs of components to reduce the total cost by the largest amount.

4. The quantity s_{uv} computed in Algorithm 1 represents the *savings* in joining subtrees T_u and T_v to one another, compared to joining both to vertex 0.

5. The savings heuristic, developed by Esau and Williams [EsWi66], was one of the first heuristics developed for the CMST problem. It is surprisingly effective in practice.

> **Algorithm 2:** Optimal tour partitioning heuristic.
>
> input: undirected network G, control center 0, capacity limit Q
> output: an approximate capacitated minimum spanning tree T^*
>
> find a traveling salesman tour on the vertex set $V = N \cup \{0\}$
> let $0 = x^{(0)}, x^{(1)}, \ldots, x^{(n)}$ be an ordering of the vertices on the tour
> construct the directed graph H with vertex set V and arc costs C_{jk}:
>   if $j < k$ and $\sum_{i=j+1}^{k} w_{x^{(i)}} \leq Q$ then $C_{jk} := c_{x^{(0)}, x^{(j+1)}} + \sum_{i=j+1}^{k-1} c_{x^{(i)}, x^{(i+1)}}$
>   else $C_{jk} := \infty$
> find a shortest path P from $x^{(0)}$ to $x^{(n)}$ in H
> use $P = [x^{(0)}, x^{(u)}, x^{(v)}, \ldots, x^{(t)}, x^{(n)}]$ to define T^* via the subtrees
> $\{x^{(0)}, x^{(1)}, x^{(2)}, \ldots, x^{(u)}\}, \{x^{(0)}, x^{(u+1)}, x^{(u+2)}, \ldots, x^{(v)}\}, \ldots,$
> $\{x^{(0)}, x^{(t+1)}, x^{(t+2)}, \ldots, x^{(n)}\}$

6. *Optimal tour partitioning heuristic*: This heuristic (Algorithm 2), developed by Altinkemer and Gavish [AlGa88], is based on finding a traveling salesman tour (§10.7.1) in a certain derived graph.

7. In Algorithm 2, every path from $x^{(0)}$ to $x^{(n)}$ in the directed graph H generates a collection of subtrees satisfying the capacity restriction.

8. The performance of Algorithm 2 depends on the initial traveling salesman tour chosen. If an optimal traveling salesman tour is used, then the worst-case relative error bound of the algorithm is $4 - \frac{4}{Q}$. That is, $\widehat{Z}/Z^* \leq 4 - \frac{4}{Q}$, where $\widehat{Z}$ is the cost of the heuristic solution generated and Z^* is the cost of the optimal design.

9. *Exact algorithms*: A number of exact algorithms are based on mathematical programming approaches:

   - Gavish [Ga85] develops a Lagrangian relaxation based algorithm and uses it to solve problems with homogeneous (unit) demands;
   - Araque, Hall, and Magnanti [ArHaMa90] derive valid inequalities and facets for the CMST problem;
   - Hall [Ha96] and Bienstock, Deng, and Simchi-Levi [BiDeSi94] develop valid inequalities and facets and use them in a branch-and-cut algorithm.

10. The CMST formulation given in Fact 1 can be improved by adding the various inequalities listed in Facts 11-14.

11. *Knapsack inequalities*: Let S be a minimal cover. For every $l \in N$, the inequality
$$\sum_{i \in S} x_{il} \leq (|S| - 1) y_l$$
is valid for the CMST problem.

12. *Subtour elimination inequalities*: For any $I \subset N$, let $\mathcal{P} = \{S_1, S_2, \ldots, S_{|I|}\}$ be a partition of $N - I$ into $|I|$ subsets, some of which may be empty. For every $i \in I$, let S_i be the unique subset from $\mathcal{P}$ associated with it. Then
$$\sum_{e \in E(I)} z_e + \sum_{j \in I} y_j + \sum_{i \in I} \sum_{j \in S_i} x_{ij} \leq \sum_{i \in I} \sum_{j \in N} x_{ij}$$
is valid for the CMST problem.

13. *Generalized subtour elimination inequalities*: For any $I \subset N$, the inequality
$$\sum_{e \in E(I)} z_e \leq |I| - b(I)$$
is valid for the CMST problem.

14. *Cluster inequalities:* Consider p sets of vertices $S_1, S_2, \ldots, S_p \subset N$ with $p \geq 3$. If the conditions:

- $S_0 = \bigcap_{i=1}^{p} S_i \neq \emptyset$, and $S_i - S_0 \neq \emptyset$ for $i = 1, 2, \ldots, p$
- $\sum_{i \in S_k \cup S_l} w_i > Q$ for all $1 \leq k < l \leq p$

are satisfied, then

$$\sum_{i=1}^{p} \sum_{e \in E(S_i)} z_e \leq \sum_{i=1}^{p} |S_i| - 2p + 1$$

is valid for the CMST problem.

Examples:

1. The following figure presents data for a problem involving $n = 5$ terminals and a control center 0. Part (a) gives the cost c_{ij} of constructing each edge (i, j) as well as the demand w_i at each vertex i. The objective is to construct a minimum cost set of subtrees connected to vertex 0, in which the demand generated by any subtree is at most $Q = 150$. Part (b) shows a feasible capacitated spanning tree T, which contains two root vertices (at 2 and 3). The total demand in subtree T_2 is $w_2 + w_4 + w_5 = 150 \leq Q$ and the total demand in subtree T_3 is $w_1 + w_3 = 95 \leq Q$. The spanning tree T has total cost 21, the sum of the displayed edge costs c_{ij}.

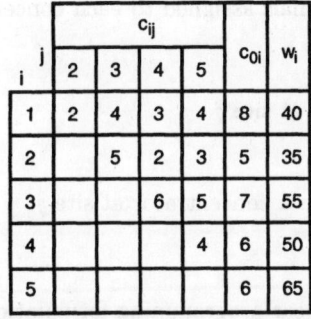
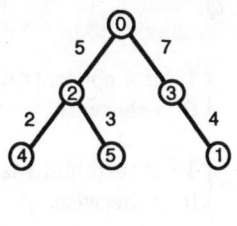

(a)   (b)

2. Algorithm 1 is applied to the problem data of the figure of Example 1. To begin, five subtrees are selected, each a single vertex and each joined to the control center 0. Thus, $f_1 = 8$, $f_2 = 5$, $f_3 = 7$, $f_4 = 6$, and $f_5 = 6$. Then $s_{12} = 8 - 2 = 6$, $s_{13} = 8 - 4 = 4$, $\ldots$, $s_{35} = 7 - 5 = 2$, and $s_{45} = 6 - 4 = 2$. The largest savings occurs for $(1, 2)$ so T_1 and T_2 are merged, giving the new tree T_1 with root vertex 2 and the single edge $(1, 2)$. Next, the f_u and s_{uv} are updated. For example, $f_1 = \min\{8, 5\} = 5$, $f_3 = 7$, and $s_{13} = 7 - \min\{c_{13}, c_{23}\} = 7 - 4 = 3$. The largest savings is found to be s_{14}, so T_1 and T_4 are merged, giving the new tree T_1 with root vertex 2 and edges $(1, 2)$ and $(2, 4)$. At the next stage T_3 and T_5 are merged, giving the new tree T_3 with root vertex 5 and the single edge $(3, 5)$. Since no further merging can take place (without violating the capacity constraint), the savings heuristic terminates with the spanning tree shown in the following figure, having total cost 20.

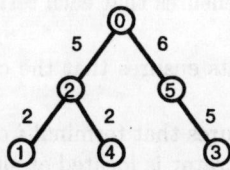

10.6.2 CAPACITATED CONCENTRATOR LOCATION PROBLEM

The capacitated concentrator location problem is frequently used to locate concentrators in local access networks and switching centers in the backbone network. In either case, concentrators of fixed capacity are to be located at a subset of possible sites. Each given terminal of the network is to be connected to exactly one concentrator, so that the concentrator's capacity is not exceeded. A feasible configuration having minimum total cost is then sought.

Definitions:

$N = \{1, 2, \ldots, n\}$ is a specified set of **terminals**, where terminal i uses w_i units of **capacity**. $M = \{1, 2, \ldots, m\}$ is a given set of possible **sites** for concentrators, each of fixed capacity Q.

If a concentrator is located at site j, the **set-up cost** is v_j, for $j \in M$. The **connection cost** of connecting terminal i to a concentrator at site j is c_{ij}, for $i \in N$ and $j \in M$.

The **capacitated concentrator location problem** (**CCLP**) involves finding locations for concentrators and an assignment of terminals to concentrators such that:

- the sum of set-up and connection costs is minimum;
- the total capacity required by the terminals assigned to each concentrator is at most Q.

Define
$$y_j = \begin{cases} 1 & \text{if a concentrator is located at site } j \\ 0 & \text{otherwise} \end{cases}$$
and
$$x_{ij} = \begin{cases} 1 & \text{if terminal } i \text{ is connected to a concentrator at site } j \\ 0 & \text{otherwise.} \end{cases}$$

Facts:

1. The CCLP has the following 0-1 integer linear programming formulation (§15.1.8):

$$\min\left\{ \sum_{i=1}^{n} \sum_{j=1}^{m} c_{ij} x_{ij} + \sum_{j=1}^{m} v_j y_j \right\}$$

subject to

$$\sum_{j=1}^{m} x_{ij} = 1 \quad \text{for all } i \in N$$

$$\sum_{i=1}^{n} w_i x_{ij} \leq Q y_j \quad \text{for all } j \in M$$

$$x_{ij} \leq y_j \quad \text{for all } i \in N, \; j \in M$$

$$x_{ij} \in \{0, 1\} \quad \text{for all } i \in N, \; j \in M$$

$$y_j \in \{0, 1\} \quad \text{for all } j \in M.$$

2. In Fact 1:
    - the first set of constraints ensures that each terminal is connected to exactly one concentrator;
    - the second set of constraints ensures that the concentrator's capacity is not exceeded;
    - third set of constraints ensures that terminal i can be assigned to a concentrator at site j only if a concentrator is located at site j.

3. A number of algorithms have been proposed for this problem, most of which are based on a Lagrangian relaxation approach, while some are based on polyhedral analysis [NeWo88].

Example:

1. The following figure shows the data for a problem with four terminals i and three possible sites j for locating concentrators. Let the capacity of any concentrator be $Q = 30$. One feasible configuration is to connect terminals 2, 3, and 4 to a concentrator at site 2, and to connect terminal 1 to a concentrator at site 3. The concentrator at site 2 has total capacity $w_2 + w_3 + w_4 = 29 \leq 30$ and the concentrator at site 3 has total capacity $w_1 = 11 \leq 30$. The connection cost is $c_{22} + c_{32} + c_{42} + c_{13} = 15$ and the set-up cost is $v_2 + v_3 = 15$, giving a total cost of 30. Another feasible configuration is to connect terminals 2 and 4 to a concentrator at site 1, and to connect terminals 1 and 3 to a concentrator at site 3. The connection cost is 12 and the set-up cost is 17, giving the smaller total cost 29.

c_{ij}

$i \backslash j$	1	2	3	w_i
1	8	6	4	11
2	2	4	3	12
3	5	5	3	9
4	3	2	6	8
v_j	9	7	8	

10.6.3 CAPACITY ASSIGNMENT PROBLEM

Fiber-optic and opto-electronic cable technologies, together with traditional copper cables, provide many possible choices for link capacities and offer economies of scale. In the capacity assignment problem, a point-to-point communication demand is given between various pairs of vertices of the (typically, backbone) network. The objective is to install links of several types (capacities) to transfer all communication demand without violating link capacities and to do so at minimum total cost. The special case involving two types of transmission media is discussed here.

Definitions:

Let $G = (V, E)$ be an undirected graph with vertex set V and edge set E.

Each communication demand is represented by a ***commodity*** $k \in K$, where K is the set of commodities. Commodity $k \in K$ has a required flow in G of d_k units between its ***origin vertex*** $O(k)$ and its ***destination vertex*** $D(k)$.

Two types of cables can be installed: ***low capacity*** cables have capacity L, and ***high capacity*** cables have capacity H. Let a_e (b_e) be the ***installation cost*** for each low capacity (high capacity) cable on edge $e \in E$.

The ***capacity assignment problem*** (***CAP***) involves finding a mix of low and high capacity cables for each edge of G such that:

- the total installation cost is minimum;
- all communication demands d_k are met;
- the flow on each edge does not exceed its installed capacity.

Let $x_e = x_{ij}$ ($y_e = y_{ij}$) be the number of low capacity (high capacity) cables installed on edge $e = (i, j)$.

Let f_{ij}^k be the amount of commodity k that flows from i to j on edge (i, j).

Facts:

1. The CAP has the following mixed-integer linear programming formulation (§15.1.8):

$$\min\left\{\sum_{e \in E}(a_e x_e + b_e y_e)\right\}$$

subject to

$$\sum_{j \in V} f_{j,O(k)}^k - \sum_{j \in V} f_{O(k),j}^k = -d_k \text{ for all } k \in K$$

$$\sum_{j \in V} f_{j,D(k)}^k - \sum_{j \in V} f_{D(k),j}^k = d_k \text{ for all } k \in K$$

$$\sum_{j \in V} f_{ji}^k - \sum_{j \in V} f_{ij}^k = 0 \text{ for all } k \in K \text{ and for all } i \in V - \{O(k), D(k)\}$$

$$\sum_{k \in K} (f_{ij}^k + f_{ji}^k) \leq L x_{ij} + H y_{ij} \text{ for all } (i,j) \in E$$

$$x_e, y_e \geq 0 \text{ integer } \text{ for all } e \in E$$

$$f_{ij}^k, f_{ji}^k \geq 0 \text{ for all } (i,j) \in E \text{ and for all } k \in K.$$

2. In Fact 1:
   - the first three sets of constraints are the standard mass balance constraints (§10.4.1);
   - the next set of constraints enforces the capacity constraint on the total flow through edge $e = (i,j)$.

3. Various models and algorithms for capacity assignment problems are discussed in [MaMiVa95] and [BiGu95].

Example:

1. In the network G of the following figure, the costs (a_e, b_e) are shown for each edge e; here $L = 2$ and $H = 5$. There are $k = 3$ communication demands (commodities): $d_1 = 12$ between vertices 1 and 4, $d_2 = 10$ between vertices 2 and 5, and $d_3 = 9$ between vertices 1 and 5. A feasible assignment of flows and capacities to edges is displayed in part (b) of the following figure. For instance, edge $(1,3)$ carries 7 units of commodity 1 and 9 units of commodity 3, for a total flow of 16 units. There are 3 high capacity cables and 1 low capacity cable installed on this edge giving a total capacity of $3H + L = 17$, at a cost of $3 \cdot 5 + 1 \cdot 3 = 18$. The total installation cost for this assignment is 114.

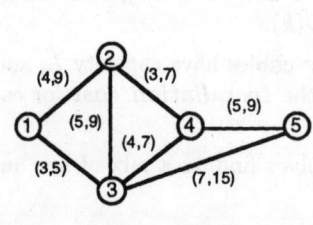

e	1	2	3	x_e	y_e
(1,2)	5			0	1
(1,3)	7		9	1	3
(2,3)		4		0	1
(2,4)	5	6		3	1
(3,4)	7		3	0	2
(3,5)		4	6	0	2
(4,5)		6	3	0	2

(a)   (b)

10.6.4 MODELS FOR SURVIVABLE NETWORKS

The introduction of fiber-optic technology has provided high capacity links and makes it possible to design communication networks with low-cost sparse topologies. Unfortunately, sparse networks are very vulnerable; a failure in one edge or vertex can disconnect many users from the rest of the network. This is the prime motivation for studying the design of survivable networks.

Definitions:

Let $G = (V, E)$ be an undirected graph with vertex set V and edge set E.

The **cost** of establishing edge $e \in E$ is given by c_e. The cost of a subnetwork $H = (V, F)$ of G is $\sum_{e \in F} c_e$.

Associated with every vertex $s \in V$ is a corresponding number r_s, indicating a desired level of **redundancy**.

A spanning subnetwork $H = (V, F)$ of G is said to satisfy the **edge (vertex) connectivity requirement** if for every distinct pair $s, t \in V$ there are at least $r_{st} = \min\{r_s, r_t\}$ edge-disjoint (vertex-disjoint) paths between s and t in H.

Define x_e, for $e = (i, j)$, to be the number of edges connecting vertex i to vertex j.

Facts:

1. The problem of designing a minimum cost subnetwork that satisfies all edge connectivity requirements has the following integer linear programming formulation (§15.1.8):

$$\min \left\{ \sum_{e \in E} c_e x_e \right\}$$

subject to

$$\sum_{e \in [S, \overline{S}]} x_e \geq \max_{(i,j) \in [S, \overline{S}]} r_{ij}, \text{ for all } S \subset V, \ S \neq \emptyset$$

$$x_e \geq 0 \text{ integer, for all } e \in E.$$

2. The model in Fact 1, analyzed by Goemans and Bertsimas [GoBe93], allows multiple edges connecting the same two vertices.

3. Grötschel, Monma, and Stoer [GrMoSt92] analyze a related survivability model in which multiple edges are forbidden. In this case, x_e is restricted to be 0 or 1 in the formulation of Fact 1.

Examples:

1. Part (a) of the following figure shows a network G having four vertices and six edges; the cost c_e of each edge e is also displayed.

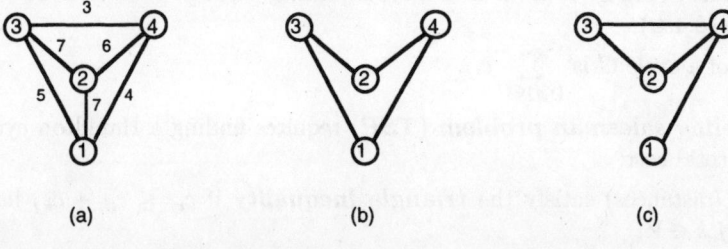

(a)   (b)   (c)

Suppose that the specified redundancies are $r_1 = 1$ and $r_2 = r_3 = r_4 = 2$. The spanning subnetwork H shown in part (b), with cost 22, satisfies the vertex connectivity requirement. For example, there are $\min\{r_3, r_4\} = 2$ vertex-disjoint paths joining vertices 3 and 4 in H: namely, $[3, 2, 4]$ and $[3, 1, 4]$. Also, there are $\min\{r_2, r_4\} = 2$ vertex-disjoint paths joining vertices 2 and 4: $[2, 4]$ and $[2, 3, 1, 4]$.

2. Part (c) of the figure for Example 1 shows another spanning subnetwork H' of cost 20 that satisfies the stated vertex connectivity requirement. For example, there are $\min\{r_2, r_4\} = 2$ vertex-disjoint paths joining vertices 2 and 4 in H': $[2, 4]$ and $[2, 3, 4]$. Notice that there is $\min\{r_1, r_3\} = 1$ path joining 1 and 3 in H', but not two such vertex-disjoint paths.

10.7 DIFFICULT ROUTING AND ASSIGNMENT PROBLEMS

An *exact* algorithm for a combinatorial optimization problem is a procedure that produces a verifiable optimal solution to every instance of this problem. A *heuristic* algorithm produces a feasible (although not necessarily optimal) solution to each problem instance. This section discusses exact and heuristic approaches to three classical combinatorial optimization problems: the traveling salesman problem, the vehicle routing problem, and the quadratic assignment problem. These three problems have in common the goal of minimizing the cost of movement or travel, generally of people or of materials.

10.7.1 TRAVELING SALESMAN PROBLEM

In the traveling salesman problem, a salesman starts out from a home city and is to visit in some order a specified set of cities, returning home at the end. This journey is to be designed to incur the minimum total cost (or distance). While the traveling salesman problem has attracted the attention of many mathematicians and computer scientists, it has resisted attempts to develop an efficient solution algorithm.

Definitions:

Let $G = (V, E)$ be a complete graph (§8.1.3) with $V = \{1, 2, \ldots, n\}$ the set of vertices and E the set of all edges joining pairs of distinct vertices.

Each edge $(i, j) \in E$ has an associated **cost** or **distance** c_{ij}.

The **distance** between set $S \subseteq V$ and vertex $j \notin S$ is $d_S(j) = \min\{c_{ij} \mid i \in S\}$.

A **Hamilton cycle** or **tour** in G is a cycle passing through each vertex $i \in V$ exactly once. (See §8.4.4.)

The **cost** of a cycle C is $\sum_{(i,j) \in C} c_{ij}$.

The **traveling salesman problem** (**TSP**) requires finding a Hamilton cycle in G of minimum total cost.

The costs (distances) satisfy the **triangle inequality** if $c_{ij} \leq c_{ik} + c_{kj}$ holds for all distinct $i, j, k \in V$.

In a **Euclidean** TSP, each vertex i corresponds to a point x_i in $\mathcal{R}^2$ and c_{ij} is the distance between x_i and x_j, relative to the standard real inner product (§6.1.4).

Tour construction procedures generate an approximately optimal TSP tour from the costs c_{ij}.

Tour improvement procedures attempt to find a smaller cost tour given an initial (often random) tour.

Composite (hybrid) procedures construct a starting tour from one of the tour construction procedures and then attempt to find a smaller cost tour using one or more of the tour improvement procedures.

A **k-change** or **k-opt exchange** of a given tour is obtained by deleting k edges from the tour and adding k other edges to form a new tour. A tour is **k-optimal (k-opt)** if it is not possible to improve the tour via a k-change.

Metaheuristics are general-purpose procedures (such as tabu search, simulated annealing, genetic algorithms, or neural networks) for heuristically solving difficult optimization problems; these general methodologies for searching complex solution spaces can be specialized to handle specific types of optimization problems.

Facts:

1. The TSP is possibly the most well-known network optimization problem, and it serves as a prototype for difficult combinatorial optimization problems in the theory of algorithmic complexity (§16.5.2).

2. The first use of the term "traveling salesman problem" in a mathematical context appears to have occurred in 1931–1932.

3. There are numerous applications of the TSP: drilling of printed circuit boards, cluster analysis, sequencing of jobs, x-ray crystallography, archaeology, cutting stock problems, robotics, and order-picking in a warehouse (see Examples 7-11).

4. There are $\frac{(n-1)!}{2}$ different Hamilton cycles in the complete graph G. This means that brute force enumeration of all Hamilton cycles to solve the TSP is not practical. (See Example 1.)

5. The TSP is an NP-hard optimization problem (§16.5.2). This remains true even when the distances satisfy the triangle inequality or represent Euclidean distances.

6. If certain edges (i, j) of G are missing, then c_{ij} can be assigned a sufficiently large value M — for example, M greater than the sum of the n largest edge costs. The TSP can then be solved on the complete graph G. If the (exact) solution obtained has any edges with cost M, then there is no Hamilton cycle in the original graph.

7. *Asymmetric traveling salesman problem*: Certain applications require finding a minimum cost directed Hamilton cycle in a directed network H; here it is not required that $c_{ij} = c_{ji}$ holds for all arcs (i, j) of H. This asymmetric (directed) TSP can be transformed into a TSP problem on an undirected network; see [JüReRi95].

8. A seminal paper of G. B. Dantzig, D. R. Fulkerson, and S. M. Johnson (1954) solved a 49-city TSP to optimality by adding cutting planes (§15.1.8) to a linear programming relaxation of the problem.

9. Although ingenious exact algorithms for the TSP have been proposed by numerous authors, most encounter problems with storage and/or running time for cases with more than five hundred vertices.

Algorithm 1: Nearest neighbor heuristic.

input: undirected network $G = (V, E)$
output: a traveling salesman tour

$i_0 :=$ any vertex of G {the starting vertex}
$W := V - \{i_0\}$
$P := \emptyset$
$v := i_0$
while $W \neq \emptyset$
 let $k \in W$ be such that $c_{vk} = \min\{c_{vj} \mid j \in W\}$
 add (v, k) to P
 $W := W - \{k\}$
 $v := k$
add (k, i_0) to the path P to produce a tour

10. Exact approaches to the TSP are computationally intensive, especially for large networks. Thus a large number of heuristic approaches have been developed to produce useful, but not necessarily optimal, solutions to the TSP.

11. The wealth of TSP heuristics can be categorized into four broad classes — tour construction procedures, tour improvement procedures, composite procedures, and metaheuristics.

12. *Nearest neighbor heuristic*: This construction method (Algorithm 1) builds up a tour by successively adding new vertices that are closest to a growing path.

13. Using appropriate data structures, Algorithm 1 can be implemented to run in $O(n^2)$ time.

14. Suppose z_{NN} is the cost of a tour constructed by the nearest neighbor heuristic and z_{OPT} is the cost of an optimal TSP tour. Then there are examples for which $\frac{z_{\text{NN}}}{z_{\text{OPT}}}$ is $\Theta(\log n)$. This means that the cost of the tour produced by Algorithm 1 cannot be bounded above by a constant times the cost of an optimal TSP tour.

15. *Nearest insertion heuristic*: This construction method (Algorithm 2) builds up a tour from smaller cycles by successively adding a vertex that is closest to the current cycle C. The new vertex is inserted between two successive vertices in the cycle, in the best possible way.

16. Using appropriate data structures, Algorithm 2 can be implemented to run in $O(n^2)$ time.

17. Suppose z_{NI} is the cost of a tour constructed by the nearest insertion heuristic and that z_{OPT} is the cost of an optimal TSP tour. If the values c_{ij} satisfy the triangle inequality, then $\frac{z_{\text{NI}}}{z_{\text{OPT}}} \leq 2$ holds for all TSP instances.

18. *Clarke and Wright savings heuristic*: This construction method (Algorithm 3) builds up a tour by successively adding an edge (i, j) having the largest savings s_{ij}, the benefit from directly connecting vertices i and j compared with joining each directly to a central vertex.

19. Using appropriate data structures, Algorithm 3 can be implemented to run in $O(n^2 \log n)$ time.

Algorithm 2: Nearest insertion heuristic.

input: undirected network $G = (V, E)$
output: a traveling salesman tour

$i :=$ any vertex of G {the starting vertex}
$j :=$ subscript such that $c_{ij} = \min\{ c_{ir} \mid r \in V - \{i\} \}$
$S := \{i, j\}$
$C := \{(i, j), (j, i)\}$
while $S \neq V$
  let k be such that $d_S(k) = \min\{ d_S(r) \mid r \in V - S \}$
  $S := S \cup \{k\}$
  find an edge $(u, v) \in C$ so $c_{uk} + c_{kv} - c_{uv} = \min\{ c_{xk} + c_{ky} - c_{xy} \mid (x, y) \in C \}$
  add (u, k) and (k, v) to C, and remove (u, v) from C

Algorithm 3: Clarke and Wright savings heuristic.

input: undirected network G
output: a traveling salesman tour

select any vertex (for example, 1) as the starting vertex
compute $s_{ij} = c_{1i} + c_{1j} - c_{ij}$ for distinct $i, j \in V - \{1\}$
order the savings $s_{i_1 j_1} \geq s_{i_2 j_2} \geq \cdots \geq s_{i_t j_t}$
$P := \emptyset$
$k := 0$
while $|P| < n - 2$
  $k := k + 1$
  if $P \cup \{(i_k, j_k)\}$ is a vertex-disjoint union of paths then add (i_k, j_k) to P
connect the endpoints of P to vertex 1, forming a tour

Algorithm 4: Christofides' heuristic.

input: undirected network G
output: a traveling salesman tour

$T :=$ minimum spanning tree of G (see §10.1)
let S contain all odd-degree vertices in T
find a minimum cost perfect matching M (§10.2) relative to vertices S of G and
  using the costs c_{ij}
obtain a closed trail C by adding M to the edges of T
remove all edges but two incident with vertices of degree greater than 2 by exploiting the triangle inequality, transforming C into a tour

20. *Christofides' heuristic:* This construction method (Algorithm 4) builds up a tour from a minimum spanning tree to which are added certain other small cost edges. It is assumed that the costs satisfy the triangle inequality.

21. Using appropriate data structures, Algorithm 4 can be implemented to run in $O(n^3)$ time.

> **Algorithm 5: General edge-exchange heuristic.**
>
> input: undirected network G, initial tour
> output: a traveling salesman tour
>
> **repeat** improve the tour using an allowable edge exchange
> **until** no additional improvement can be made

22. Suppose z_C is the cost of a tour constructed by Christofides' heuristic and z_{OPT} is the cost of an optimal TSP tour. If the c_{ij} satisfy the triangle inequality, then $\frac{z_C}{z_{OPT}} \leq \frac{3}{2}$ holds for all TSP instances.

23. The following table [JüReRi95] compares several of the most popular tour construction procedures on a set of 30 Euclidean TSPs from the literature with known optimal solutions. These problems range in size from 105 to 2392 vertices. Surprisingly, the savings heuristic is the best tour construction heuristic of those tested. These results are consistent with those of other studies.

heuristic	average percent above optimality
nearest neighbor	24.2
nearest insertion	20.0
Christofides	19.5
modified nearest neighbor	18.6
cheapest insertion	16.8
random insertion	11.1
farthest insertion	9.9
savings	9.8
modified savings	9.6

24. The best known tour improvement heuristics for the TSP involve edge exchanges (Algorithm 5). Often the initial tour is chosen randomly from the set of all possible tours.

25. Specialized versions of Algorithm 5 typically use 2-opt exchanges, 3-opt exchanges, and more complicated *Lin-Kernighan* [JüReRi95] edge exchanges. Such exchange techniques have been used to generate excellent solutions to large-scale TSPs in a reasonable amount of time.

26. Edge-exchange procedures are typically more expensive computationally than tour construction procedures.

27. Tour improvement procedures typically require a "downhill move" (i.e., a strict reduction in cost) in order for edge exchanges to be made. As a result, they terminate with a local minimum solution.

28. Since the 2-opt exchange procedure is weaker than the 3-opt procedure, Algorithm 5 will generally terminate at an inferior local optimum using 2-opt exchanges instead of 3-opt exchanges. The Lin-Kernighan procedure will generally terminate with a better local optimum than will a 3-opt exchange procedure.

29. In practice, it often makes sense to apply a composite procedure. The strategy is to get a good initial solution rapidly (by tour construction), which is then improved by an edge-exchange procedure.

30. The following table [JüReRi95] compares several composite procedures on the same sample problems described in Fact 23. In each case, the initial tour is constructed using the nearest neighbor heuristic (Algorithm 1). The improvement procedures include 2-opt, 3-opt, two variants of Lin-Kernighan, and iterated Lin-Kernighan. Iterated Lin-Kernighan is the most computationally burdensome of the edge-exchange procedures, but it consistently obtains results that are within 1% of optimality.

heuristic	average percent above optimality
2-opt	8.3
3-opt	3.8
Lin-Kernighan (variant 1)	1.9
Lin-Kernighan (variant 2)	1.5
Iterated Lin-Kernighan	0.6

31. *Metaheuristics*: Unlike Algorithm 5 (which permits only downhill moves), metaheuristics [OsKe95] allow the possibility of nonimproving moves. For example, uphill moves can be accepted either randomly (simulated annealing) or based upon deterministic rules (threshold accepting). Memory can be incorporated in order to prevent revisiting local minima already evaluated and to encourage discovering new ones (tabu search).

Other metaheuristics such as evolutionary strategies, genetic algorithms, and neural networks have also been applied to the TSP. To date, neural networks and tabu search have been less successful than the other approaches.

32. For a detailed history of the traveling salesman problem see the first chapter of [LaEtal85].

33. Software, research papers, and other heuristic approaches for the traveling salesman and related problems are described on the web pages:

   http://www.ing.unlp.edu.ar/cetad/mos/TSPBIB_home.html

   http://www.netlib.org/toms/750

34. A library of sample problems, with their best known solutions, is available at:

   http://www.iwr.uni-heidelberg.de/iwr/comopt/soft
         /TSPLIB95/TSPLIB.html

Examples:

1. *Brute force enumeration*: Suppose that a TSP solution is required for the complete graph G on $n = 25$ cities. By Fact 4, there are $\frac{24!}{2} \approx 3.1 \times 10^{23}$ Hamilton tours in the graph G. Even with a supercomputer that is capable of finding and evaluating each such tour in one nanosecond (10^{-9} seconds), it would take over 9.8 million years of uninterrupted computations to determine an optimal TSP tour.

This example illustrates how quickly brute force enumeration of Hamilton tours becomes impractical.

2. Part (a) of the following figure shows the costs c_{ij} for a five city TSP. An initial tour can be constructed using the nearest neighbor heuristic (Algorithm 1). Let the initial vertex be $i_0 = 1$, so $W = \{2, 3, 4, 5\}$. The closest vertex of W to 1 is 2, with $c_{12} = 1$, so edge $(1, 2)$ is added to the current path. A closest vertex of $W = \{3, 4, 5\}$ to 2 is 5, so edge $(2, 5)$ is added to the path. Continuing in this way, edges $(5, 3)$ and edge $(3, 4)$ are added, giving the path $P = [1, 2, 5, 3, 4]$ and the tour $[1, 2, 5, 3, 4, 1]$ with total cost $1 + 3 + 2 + 3 + 5 = 14$. This tour is displayed in part (b).

c_{ij} i \ j	2	3	4	5
1	1	2	5	4
2		3	4	3
3			3	2
4				3

(a)

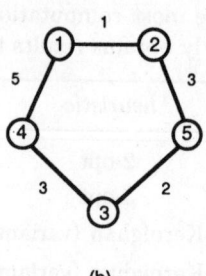

(b)

3. Suppose that the nearest insertion heuristic (Algorithm 2) is applied to the problem data in part (a) of the figure for Example 2, starting with the initial vertex $i = 1$. The nearest vertex to i is $j = 2$, giving the initial cycle $C = \{(1, 2), (2, 1)\}$. The closest vertex to this cycle is $k = 3$, producing the new cycle $C = \{(1, 2), (2, 3), (3, 1)\}$. Relative to $S = \{1, 2, 3\}$, $d_S(4) = 3$ and $d_S(5) = 2$, so vertex 5 will next be added to the cycle. Since $c_{15} + c_{52} - c_{12} = 6$, $c_{25} + c_{53} - c_{23} = 2$, and $c_{15} + c_{53} - c_{13} = 4$, vertex 5 is inserted between vertices 2 and 3 in the current cycle, giving $C = \{(1, 2), (2, 5), (5, 3), (3, 1)\}$. Finally, vertex 4 is added between vertices 2 and 5, producing the tour $C = \{(1, 2), (2, 4), (4, 5), (5, 3), (3, 1)\}$ with total cost 12.

4. The savings heuristic (Algorithm 3) can alternatively be applied to the problem specified in part (a) of the figure of Example 2. The savings $s_{23} = c_{12} + c_{13} - c_{23} = 1 + 2 - 3 = 0$. Similarly, $s_{24} = 2$, $s_{25} = 2$, $s_{34} = 4$, $s_{35} = 4$, and $s_{45} = 6$. This produces the ordered list of edges $[(4, 5), (3, 4), (3, 5), (2, 4), (2, 5), (2, 3)]$. Considering edges in turn from this list gives the path $P = [3, 4, 5, 2]$. Adding edges from the endpoints of P to vertex 1 produces the tour $[1, 3, 4, 5, 2, 1]$ with total cost 12.

5. Christofides' heuristic (Algorithm 4) is now applied to the problem given in part (a) of the figure of Example 1. A minimum spanning tree T consists of the following edges: $(1, 2), (1, 3), (3, 5), (3, 4)$; see part (a) of the following figure. Vertices $2, 3, 4, 5$ have odd degree and $\{(2, 1), (3, 5)\}$ constitutes a minimum cost perfect matching on these vertices. Adding these edges to those of T produces the multi-graph in part (b) of the following figure. Replacing edges $(4, 3)$ and $(3, 5)$ having aggregate cost 5 by the single edge $(4, 5)$ of cost 3 produces the tour in part (c), having total cost 12.

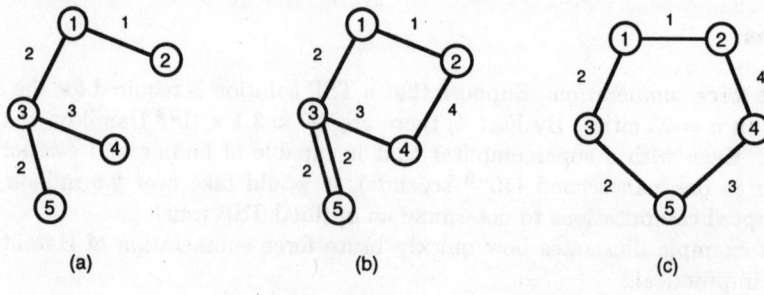

6. To illustrate edge exchanges, consider the tour of cost 14 in part (b) of the figure of Example 2. Removal of edges (1, 4) and (3, 5) disconnects the cycle into two disjoint paths. Join the endpoints of one path to the endpoints of the other with edges (1, 3) and (4, 5) to create a new tour [1, 2, 5, 4, 3, 1] of smaller cost 12. No further pairwise exchanges reduce the cost of this tour, so this tour is a 2-opt local minimum solution.

7. *Delivery routes*: A delivery truck must visit a set of customers in a city and then return to the central garage after completing the route. Determining an optimal (i.e., shortest time) delivery route can be modeled as a traveling salesman problem on a city street network. Here the vertices represent the customer locations and the cost c_{ij} of edge (i, j) is the driving time between locations i and j.

8. *Printed circuit boards*: One application of the TSP occurs in fabricating printed circuit boards. Holes at a number of fixed locations have to be drilled through the board. The objective is to minimize the total time needed to move the drilling head from position to position. Here the vertices i correspond to the locations of the holes as well as the starting position of the drill. The cost c_{ij} represents the time required to move the drilling head from i and reposition it at j. A minimum cost traveling salesman tour gives an optimal way of sequencing the drilling of the holes.

9. *Order-picking*: In a warehouse, a customer order requires a certain subset of the items stored there. A vehicle must be sent to pick up these items and then return to the central dispatch location. Here the vertices are the locations of the items as well as the central dispatch location. The costs are the times needed to move the vehicle from one location to the other. A minimum cost traveling salesman tour then gives an optimal order in which to retrieve items from the warehouse.

10. *Job sequencing*: In a factory, materials must be processed by a series of operations on a machine. The set-up time between operations varies depending on the order in which the operations are scheduled. Determining an optimal ordering that minimizes the total set-up time can be formulated as a traveling salesman problem.

11. Additional applications, with reference sources, are given in the following table.

application	references
dating archaeological finds	[AhEtal95]
DNA mapping	[AhEtal95]
x-ray crystallography	[JüReRi95]
engine design	[AhEtal95]
robotics	[JüReRi95]
clustering	[LaEtal85], [AhEtal95]
cutting stock problems	[HoPa96]
aircraft route assignment	[HoPa96]
computer wiring	[LaEtal85], [EvMi92], [JüReRi95]

10.7.2 VEHICLE ROUTING PROBLEM

Private firms and public organizations that distribute goods or provide services to customer locations rely on a fleet of vehicles. Given demands for service at numerous points in a transportation network, the vehicle routing problem requires determining which customers are to be serviced by each vehicle and the order in which customers on a route are to be visited.

Definitions:

Let $G = (V, E)$ be a complete graph with $V = \{1, 2, \ldots, n\}$ the set of vertices and E the set of all edges joining pairs of distinct vertices (§8.1.3).

Vertex 1 is the **central depot**, whereas the other vertices represent **customer locations**. Customer location i has a known **demand** w_i.

Each edge $(i, j) \in E$ has an associated **distance** or **cost** c_{ij}.

There are also available a number of **vehicles**, each having the same **capacity** Q.

A **route** is sequence of customers visited by a vehicle that starts and ends at the central depot.

The **vehicle routing problem** (**VRP**) requires partitioning the set of customers into a set of delivery routes such that:
- the total distance traveled by all vehicles is minimum;
- the total demand generated by the customers assigned to each route is $\leq Q$.

In a **construction** heuristic for the VRP, subtours are joined as long as the resulting subtour does not violate the vehicle capacity.

An **improvement** heuristic employs successive edge exchanges that reduce the total distance without violating any vehicle capacity constraint.

A **two-phase** heuristic implements a cluster first-route second philosophy, in which customers are first partitioned into groups G_k with $\sum_{i \in G_k} w_i \leq Q$, after which a minimum distance sequencing of customers is found within each group.

Facts:

1. The TSP (§10.7.1) is a special case of the VRP in which there is a single vehicle with unlimited capacity.

2. The VRP is an NP-hard optimization problem (§16.5.2).

3. VRPs with more than 50 vertices are difficult to solve to optimality.

4. Most solution strategies for large VRPs are heuristic in nature, involving construction, improvement, and two-phase methods as well as metaheuristics.

5. In 1959 G. B. Dantzig and J. H. Ramser first formulated the general vehicle routing problem and developed a heuristic solution procedure. This solution technique was applied to a problem involving the delivery of gasoline to service stations.

6. The Clarke and Wright savings heuristic (§10.7.1) is a construction approach, originally proposed for the VRP. Algorithm 6 outlines this heuristic, which begins with each customer served by a different vehicle and successively combines routes in order of nonincreasing savings $s_{ij} = c_{1i} + c_{1j} - c_{ij}$ to form a smaller set of feasible routes.

7. In two-phase methods, a minimum distance ordering of customers within each specified cluster of vertices can be found by solving a TSP (§10.7.1).

8. In recent years, metaheuristics such as simulated annealing and tabu search have been applied quite successfully to VRPs. In particular, on the twelve benchmark problems in the literature, which range in size from 50 to 199 vertices, tabu search heuristics currently outperform the competition [GeHeLa94].

> **Algorithm 6:** Clarke and Wright savings heuristic.
> input: undirected network G, capacity limit Q
> output: a set of delivery routes
> $R_i :=$ route consisting of edges $(1,i)$ and $(i,1)$ for $i \in V - \{1\}$
> compute $s_{ij} = c_{1i} + c_{1j} - c_{ij}$ for distinct $i,j \in V - \{1\}$
> order the savings $s_{i_1 j_1} \geq s_{i_2 j_2} \geq \cdots \geq s_{i_t j_t}$
> for $k := 1$ to t
>   if R_{i_k} and R_{j_k} have combined demand at most Q then merge R_{i_k} and R_{j_k}

9. Extensions to the basic VRP include modifications for asymmetric distances (c_{ij} need not equal c_{ji}), differing vehicle capacities, constraints on the total distance traveled, multiple depots, and constraints on the time intervals for visiting customers.

10. A survey of 20 commercial software products for vehicle routing problems is available in [HaPa97]. This survey discusses interfaces with geographic information systems, computer platforms supported, extensions to the basic VRP that are incorporated, and significant installations of the product for industrial customers.

11. Data sets, software, and research papers on vehicle routing problems are available on the web page:

    http://www.geocities.com/ResearchTriangle/7279/vrp.html

12. Data sets and software for vehicle routing problems with time windows are available on the web page:

    http://dmawww.epfl.ch/~rochat/rochat_data/solomon.html

Examples:

1. The following table gives the data for a VRP involving six customers in which vehicle capacity is 820. The route $[1, 2, 4, 6, 1]$ is not feasible since the total demand of customers on this route is $w_2 + w_4 + w_6 = 486 + 326 + 24 = 836 > 820$. Since $\sum_{i=2}^{7} w_i = 1967$, at least $\lceil 1967/820 \rceil = 3$ routes will be needed to service all demands. The routes $[1, 5, 2, 6, 1]$, $[1, 3, 1]$, and $[1, 4, 7, 1]$ constitute a feasible set of routes with (respective) demands 800, 541, and 626. In this feasible solution, the total distance traveled by the first vehicle is $c_{15} + c_{52} + c_{26} + c_{61} = 131$, by the second is $c_{13} + c_{31} = 114$, and by the third is $c_{14} + c_{47} + c_{71} = 181$, for a total distance of 426.

customer	2	3	4	5	6	7
demand	486	541	326	290	24	300

c_{ij}	2	3	4	5	6	7
1	19	57	51	49	4	92
2		51	10	53	25	53
3			49	18	30	47
4				50	11	38
5					68	9
6						94

2. The Clarke and Wright heuristic is applied to the problem specified in the table of Example 1. For instance, $s_{35} = c_{13} + c_{15} - c_{35} = 57 + 49 - 18 = 88$. The largest savings occurs for $s_{57} = 132$ and $w_5 + w_7 = 590 \leq 820$, so the initial routes $[1, 5, 1]$ and $[1, 7, 1]$ are merged to produce the feasible route $[1, 5, 7, 1]$ with distance 150. The next largest savings occur for $s_{47} = 105$, $s_{37} = 102$, and $s_{35} = 88$; however, neither customer 3 nor customer 4 can be inserted into the route $[1, 5, 7, 1]$ without exceeding the vehicle capacity. The next largest savings is $s_{24} = 60$, giving the new feasible route $[1, 2, 4, 1]$ with demand 812 and distance 80. Continuing in this fashion eventually finds $s_{36} = 31$ and constructs the route $[1, 3, 6, 1]$ with demand 565 and distance 91. This feasible set of three routes has total distance $150 + 80 + 91 = 321$, smaller than that for the feasible solution given in Example 1.

10.7.3 QUADRATIC ASSIGNMENT PROBLEM

The quadratic assignment problem deals with the relative location of facilities that interact with one another in some manner. The objective is to minimize the total cost of interactions between facilities, with distance often used as a surrogate for measures such as dollar cost, fatigue, or inconvenience.

Definitions:

There are n **facilities** to be assigned to n predefined **locations**, where each location can accommodate any one facility.

The **fixed cost** c_{ij} is the cost of assigning facility i to location j.

The **flow** f_{ij} is the level of interaction between facilities i and j.

The **distance** d_{ij} between locations i and j is the per unit cost of interaction between the two locations. Typically, it is measured using the rectilinear or Euclidean distance between the locations.

An **assignment** is a bijection ρ from the set of facilities onto the set of locations.

The **linear assignment problem** (**LAP**) is the problem of finding an assignment ρ that minimizes $\sum_i c_{i,\rho(i)}$.

The **quadratic assignment problem** (**QAP**) is the problem of finding an assignment ρ that gives the minimum value z_{QAP} of $\sum_i c_{i,\rho(i)} + \sum_{i,p} f_{ip} d_{\rho(i),\rho(p)}$.

In some partial assignment for the QAP, let $\mathcal{F}$ be the set of facilities (possibly empty) that have already been assigned and $\mathcal{L}$ be the set of locations having assigned facilities.

Facts:

1. The following table gives a variety of situations that can be formulated using the QAP model.

facilities	interaction
departments in a manufacturing plant	flow of materials
departments in an office building	flow of information, movement of people
departments in a hospital	movement of patients and medical staff
buildings on a campus	movement of students and staff
electronic component boards	connections
typewriter/computer keyboard keys	movement of fingers

2. The interdependence of facilities due to interactions between them leads to the quadratic nature of the objective function in the QAP.

3. If the facilities are independent of each other (there are no interactions between them), the QAP reduces to the LAP, which can be solved in polynomial time (§10.2.2).

4. The TSP is a special case of the QAP (see Example 4).

5. The QAP is an NP-hard optimization problem (§16.5.2).

6. Exact solution of the QAP is limited to fairly small problems, generally of size 16 or smaller.

7. A lower bound on completions of a partial assignment for the QAP is given by

$$\min \sum_{i \in \mathcal{F}} c_{i,\rho(i)} + \sum_{i \in \mathcal{F}} \sum_{p \in \mathcal{F}} f_{ip} d_{\rho(i),\rho(p)}$$
$$+ \sum_{i \in \mathcal{F}} \sum_{p \notin \mathcal{F}} \left(f_{ip} d_{\rho(i),\rho(p)} + f_{pi} d_{\rho(p),\rho(i)} \right)$$
$$+ \sum_{i \notin \mathcal{F}} c_{i,\rho(i)} + \sum_{i \notin \mathcal{F}} \sum_{p \notin \mathcal{F}} f_{ip} d_{\rho(i),\rho(p)}.$$

The first two terms above are the known fixed and interaction costs of assignments already made; the third term captures the interaction costs between assigned facilities and those yet to be assigned; and the last two terms represent the fixed and interaction costs of assignments not yet made.

8. A minimum value z^* can be calculated for the last three terms in the lower bound expression of Fact 7 by solving a LAP such that each cost term is a lower bound on the incremental costs that would be incurred if facility $i \notin \mathcal{F}$ is assigned to location $j \notin \mathcal{L}$.

9. *Gilmore-Lawler lower bound*: This lower bound for z_{QAP} is given by

$$\sum_{i \in \mathcal{F}} c_{i,\rho(i)} + \sum_{i \in \mathcal{F}} \sum_{p \in \mathcal{F}} f_{ip} d_{\rho(i),\rho(p)} + z^*,$$

where z^* is found as in Fact 8.

10. The Gilmore-Lawler lower bound allows the QAP to be solved using a branch-and-bound (implicit enumeration) technique (§15.1.8).

11. Alternative tighter bounds are available. However, considering the quality of these bounds and the effort involved in computing them, the Gilmore-Lawler lower bound still seems to be the most effective bound to use within a branch-and-bound scheme.

12. There are several ways to linearize the QAP by defining additional variables and constraints. However, none of the linearizations proposed so far has proved to be computationally effective.

13. Heuristic methods for solving the QAP can be classified as limited enumeration, construction methods, improvement methods, hybrid methods, and metaheuristics. A survey of exact and heuristic solution methods for the QAP is found in [KuHe87]; experimental comparisons of heuristic approaches appear in [BuSt78] and [Li81].

14. *Limited enumeration*: There are two distinct approaches for limiting the search for an optimal QAP solution using a branch-and-bound approach:
   - The search can be curtailed by placing a limit on the computation time or the number of subproblems examined. Since an optimal solution is often found fairly early in a branch-and-bound procedure, especially if a good branching rule is available, this approach may find an optimal (or a near-optimal) solution while saving on the significant cost of proving optimality.

- The gap between the lower and upper bound is largest at higher levels of a branch-and-bound tree. Thus a relatively large gap can be used to fathom subproblems at higher levels, and this gap can be decreased gradually as the search reaches lower levels of the tree.

15. *Construction methods*: These heuristics start with an empty assignment and add assignments one at a time until a complete solution is obtained. The rule used to choose the next assignment can employ:
    - a *local view*: select a facility having the maximum interaction with a facility already assigned; locate it to minimize the cost of interaction between facilities;
    - a *global view*: take into account assignments already made as well as future assignments to be made.

16. Suppose that k assignments have already been made. Using statistical properties, the expected value for the completion of the partial assignment is given by the following expression, whose terms are analogous to those in Fact 7:

$$EV = \sum_{i \in \mathcal{F}} c_{i,\rho(i)} + \sum_{i \in \mathcal{F}} \sum_{p \in \mathcal{F}} f_{ip} d_{\rho(i),\rho(p)}$$
$$+ \frac{\sum_{i \in \mathcal{F}} \sum_{p \notin \mathcal{F}} \sum_{j \notin \mathcal{L}} \left(f_{ip} d_{\rho(i),j} + f_{pi} d_{j,\rho(i)} \right)}{n - k}$$
$$+ \frac{\sum_{i \notin \mathcal{F}} \sum_{j \notin \mathcal{L}} c_{ij}}{n - k} + \frac{\sum_{i,p \notin \mathcal{F}} f_{ip} \left(\sum_{j,q \notin \mathcal{L}} d_{jq} \right)}{(n - k)(n - k - 1)}.$$

The low computational requirements of computing EV make this a good choice to guide a construction heuristic [GrWh70].

17. *Improvement methods*: These heuristics start with a suboptimal solution (often randomly generated) and attempt to improve it through partial changes in the assignments. Several important issues arise in designing an improvement heuristic:
    - *type of exchange*: The choices are pairwise, triple, or higher-order exchanges. The use of pairwise exchanges has been found to be the most effective in terms of solution quality and computational burden. Higher-order exchanges can be beneficial but are generally used in a limited way because of the significant increase in computation time.
    - *scope of exchange*: The procedure can use a local approach that considers only the exchange of adjacent facilities, or a global approach that considers all possible exchanges. Current computing capabilities allow the use of a global approach, which has been found to be more effective.
    - *choice of exchange*: The procedure can effect an exchange as soon as an improving move is found, or can evaluate all possible exchanges and choose the best. The first improvement option is more common.
    - *order of evaluation*: The possible exchanges can be evaluated in a random or some predetermined order. This is relevant only if the "first improvement" approach is used, as is often the case. One simple but effective solution is to consider facilities in the fixed order of decreasing total interactions, so that exchanges with potentially large savings are evaluated first.

18. *Hybrid methods*: Unlike improvement procedures, which tend to get trapped at local minima, hybrid methods use multiple restarts from a set of diversified solutions. Hybrid procedures combine the power of improvement routines with diversified solutions obtained through construction methods.

19. *Metaheuristics*: In recent years, metaheuristics such as simulated annealing, tabu search, and genetic algorithms have been developed to help improvement procedures avoid the trap of local minima and have been applied with success to the QAP. Metaheuristics have been able to find the best known solutions for the commonly used benchmark problems in the literature and remain an active area of research on the QAP.

20. Computer codes (in Fortran and C) for heuristically solving the QAP can be found at the sites:

http://www.netlib.org/toms/608
http://www.netlib.org/toms/754
http://www.netlib.org/toms/769
http://rtm.science.unitn.it/~battiti/archive/code/rts_qap/

Examples:

1. The following table gives the data c_{ij}, f_{ij} for a QAP with four facilities and four locations.

c_{ij}	1	2	3	4
1	1	3	2	1
2	2	1	4	3
3	4	2	4	4
4	3	1	2	2

f_{ij}	1	2	3	4
1	0	1	3	4
2	1	0	2	1
3	3	2	0	3
4	4	1	3	0

The fixed locations $1, 2, 3, 4$ occur at equally spaced points along a line, with unit distances between successive points, so that $d_{ij} = |i - j|$. For the assignment ρ specified by $\rho(1) = 1, \rho(2) = 4, \rho(3) = 2, \rho(4) = 3$ the fixed cost is $c_{11} + c_{24} + c_{32} + c_{43} = 8$. Because the flows and distances are symmetric, the interaction cost is $2(f_{12}d_{14} + f_{13}d_{12} + f_{14}d_{13} + f_{23}d_{42} + f_{24}d_{43} + f_{34}d_{23}) = 44$. The total cost of assignment ρ is then $8 + 44 = 52$.

2. The assignment in Example 1 can be improved by a pairwise exchange. Namely, instead of assigning facilities 1 and 2 (respectively) to locations 1 and 4, they are assigned to the interchanged locations 4 and 1. This gives $\sigma(1) = 4, \sigma(2) = 1, \sigma(3) = 2, \sigma(4) = 3$. Then the fixed cost incurred is $c_{14} + c_{21} + c_{32} + c_{43} = 7$ and the interaction cost is $2(f_{12}d_{41} + f_{13}d_{42} + f_{14}d_{43} + f_{23}d_{12} + f_{24}d_{13} + f_{34}d_{23}) = 40$. The total cost 47 is lower than that for the assignment ρ in Example 1. In fact σ is an optimal QAP assignment.

3. The QAP arises in designing the layout of a manufacturing facility. A number of products are to be made in this facility and different products require different operations in given sequences for completion. These operations are performed by n departments: e.g., turning, milling, drilling, heat treatment, and assembly. Knowing the sequence of operations and the volume of each product to be produced, it is possible to calculate the flow f_{ij} from any department i to another department j. There are n physical locations, with distance d_{ij} between locations i and j. The fixed cost of assigning department i to location j is c_{ij}, representing the cost of building foundations and installing support equipment (cables, pipes) for the machines. Then the objective is to assign departments to locations in order to minimize the sum of fixed and interaction costs.

4. The TSP (§10.7.1) can be formulated as a special case of the QAP, where the n cities correspond to locations and a position number (facility) in the tour is to be associated with each city. Let $f_{12} = f_{23} = \cdots = f_{n1} = 1$ and $f_{ij} = 0$ otherwise. The distance d_{ij} represents the cost of traveling between cities i and j, and let all fixed costs c_{ij} be zero. Then a solution to this QAP gives an optimal labeling of cities with their positions in an optimal TSP tour.

10.8 NETWORK REPRESENTATIONS AND DATA STRUCTURES

To carry out network optimization algorithms efficiently, careful attention needs to paid to the design of the data structures supporting these algorithms. There are alternative ways to represent a network — differing in their storage requirements and their efficacy in executing certain fundamental operations. These representations need to incorporate both the topology of the underlying graph and also any quantitative information present in the network (such as cost, length, capacity, demand, or supply). Standard representations of networks, and trees in particular, are discussed in this section.

10.8.1 NETWORK REPRESENTATIONS

There are various ways to represent networks, just as there are various ways to represent graphs (§8.1.4, §8.3.1). In addition it is necessary to incorporate quantitative information associated with the vertices and edges (or arcs) of the network. While the description here concentrates on directed networks, extensions to undirected networks are also indicated.

Definitions:

Let $G = (V, E)$ be a directed graph (§8.3.1) with vertex set $V = \{1, 2, \ldots, n\}$ and arc set E. Define $m = |E|$ to be the number of arcs in G.

The **adjacency set** $A(i) = \{(i, j) \mid (i, j) \in E\}$ for vertex i is the set of arcs emanating from i. (See §10.3.1.)

The **adjacency matrix** for G is the 0-1 matrix $A_G = (a_{ij})$ having $a_{ij} = 1$ if $(i, j) \in E$ and $a_{ij} = 0$ if $(i, j) \notin E$. (See also §8.3.1.)

The **arc list** for G (see §8.3.1) can be implemented using two arc-length arrays FROM and TO:

- For each arc $(i, j) \in E$ there is a unique $1 \leq k \leq m$ satisfying $\text{FROM}(k) = i$ and $\text{TO}(k) = j$.
- Arcs are listed sequentially in the FROM and TO arrays in no particular order.

The **linked adjacency list** for G is given by a vertex-length array START and a singly-linked list ARCLIST of arc records:

- START(i) points to the first record for vertex i in this list, corresponding to a specified first element of $A(i)$.
- Each arc $(i, j) \in A(i)$ has an associated arc record, which contains the fields TO and NEXT. Specifically, ARCLIST.TO gives the adjacent vertex j, and ARCLIST.NEXT points to the next arc record in $A(i)$. If there is no such following record, ARCLIST.NEXT = **null**.

The **forward star** for G is given by a vertex-length array START and an arc-length array TO, with the latter in one-to-one correspondence with arcs $(i, j) \in E$:

- START(i) gives the position in array TO of the first arc leaving vertex i.
- The arcs of $A(i)$ are found in the consecutive positions START(i), START(i) + 1, $\ldots$, START($i + 1$) − 1 of array TO. If arc (i, j) corresponds to position k of TO, then TO(k) = j.
- By convention, an additional dummy vertex $n + 1$ is added, with START($n + 1$) = $m + 1$.

Section 10.8 NETWORK REPRESENTATIONS AND DATA STRUCTURES

Facts:

1. An undirected graph can be represented by replacing each undirected edge (i,j) by two oppositely directed arcs (i,j) and (j,i).

2. The adjacency matrix, the arc list, the linked adjacency list, and the forward star are four standard representations of a directed (or undirected) graph.

3. The linked adjacency list and forward star structures are commonly used implementations of the *lists-of-neighbors* representation (§8.3.1).

4. The following table shows the (worst-case) computational effort required to carry out certain fundamental operations on G: finding an arc, deleting an arc (once found), adding an arc, and scanning the adjacency set of an arbitrary vertex i. Here $\alpha_i = |A(i)| \leq n$.

representation	find arc	delete arc	add arc	scan $A(i)$
adjacency matrix	$O(1)$	$O(1)$	$O(1)$	$O(n)$
arc list	$O(m)$	$O(1)$	$O(1)$	$O(m)$
linked adjacency list	$O(\alpha_i)$	$O(1)$	$O(1)$	$O(\alpha_i)$
forward star	$O(\alpha_i)$	$O(n+m)$	$O(n+m)$	$O(\alpha_i)$

5. The storage requirements of the four representations are given in the following table for both directed and undirected graphs. For the last two representations, each undirected edge appears twice: once in each direction.

representation	storage (directed)	storage (undirected)	exploit sparsity?
adjacency matrix	n^2	$\frac{n^2}{2}$	no
arc list	$2m$	$2m$	yes
linked adjacency list	$n+2m$	$n+4m$	yes
forward star	$n+m$	$n+2m$	yes

6. As seen in the table of Example 5, all representations other than the adjacency matrix representation can exploit sparsity in the graph G. That is, the storage requirements are sensitive to the actual number of arcs and the computations will generally proceed more rapidly when G has relatively few arcs.

7. Quantitative data for network vertices (such as supply and demand) can be stored in an associated vertex-length array, thus supplementing the standard graph representations.

8. Quantitative data for network arcs (such as cost, length, capacity, and flow) can be accommodated as follows:
    - For the adjacency matrix representation, costs (or lengths) c_{ij} can be imbedded in the matrix A_G itself. Namely, redefine $A_G = (a_{ij})$ so that $a_{ij} = c_{ij}$ if $(i,j) \in E$, whereas a_{ij} is an appropriate special value if $(i,j) \notin E$. For instance, in the shortest path problem (§10.3.1), $a_{ij} = \infty$ can be used to signify that $(i,j) \notin E$. Additional $n \times n$ arrays would be needed however to represent more than one type of arc data.

- For the arc list representation, additional arrays parallel to the arrays FROM and TO can be used to store quantitative arc data.
- For the linked adjacency list representation, additional fields within the arc record can be used to store quantitative arc data.
- For the forward star representation, additional arrays parallel to the array TO can be used to store quantitative arc data.

9. The arc list representation is best suited for arc-based processing of a network, such as occurs in Kruskal's minimum spanning tree algorithm (§10.1.2).

10. The arc list representation is a convenient form for the input of a network to an optimization algorithm. Often this external representation is converted within the algorithm to a more suitable internal representation (linked adjacency list or forward star) before executing the steps of the optimization algorithm.

11. The linked adjacency list and forward star representations are best suited to carrying out vertex-based explorations of a graph, such as a breadth-first search or a depth-first search (§9.2.1). It is also ideal for carrying out Prim's minimum spanning tree algorithm (§10.1.2) as well as most shortest path algorithms (§10.3.2).

12. Especially in the case of undirected graphs, the linked adjacency list and forward star representations can be enhanced by use of an additional arc-length array MIRROR. The array MIRROR allows one to move from the location of arc (i, j) to the location of arc (j, i) in constant time.

13. The linked adjacency list is typically used when the structure of the graph can dynamically change (as by addition/deletion of arcs or vertices). On the other hand, the forward star representation is appropriate for static graphs, in which the graph structure does not change.

Examples:

1. A directed graph G with 5 vertices and 8 arcs is shown in the following figure.

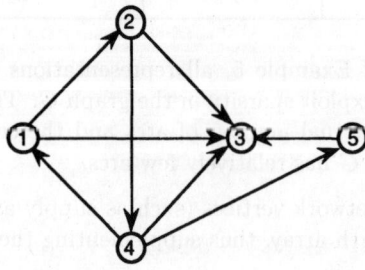

The 5×5 adjacency matrix for G is given by

$$A_G = \begin{pmatrix} & 1 & 2 & 3 & 4 & 5 \\ 1 & 0 & 1 & 1 & 0 & 0 \\ 2 & 0 & 0 & 1 & 1 & 0 \\ 3 & 0 & 0 & 0 & 0 & 0 \\ 4 & 1 & 0 & 1 & 0 & 1 \\ 5 & 0 & 0 & 1 & 0 & 0 \end{pmatrix}$$

Section 10.8   NETWORK REPRESENTATIONS AND DATA STRUCTURES

2. An arc list representation of the directed graph in the figure of Example 1 is given in the following table:

FROM	1	2	1	4	2	4	4	5
TO	2	4	3	1	3	5	3	3

3. The following figure shows a linked adjacency list representation of the directed graph of Example 1. The symbol ⊙ is used to indicate a **null** pointer.

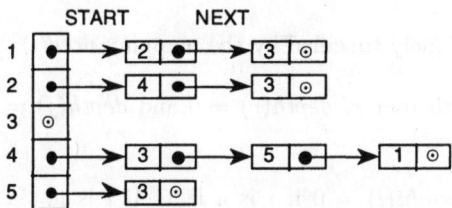

4. The following figure shows a forward star representation of the directed graph in Example 1. Since $A(3) = \emptyset$, it is necessary to set $\text{START}(3) = \text{START}(4) = 5$. For example, the arcs in $A(4)$ are associated with positions $[\text{START}(4), \ldots, \text{START}(5) - 1] = [5, 6, 7]$ of the TO array. Similarly, the single arc emanating from vertex 5 is associated with position $[\text{START}(5), \ldots, \text{START}(6) - 1] = [8]$ of the TO array.

START		TO	
1	1	1	2
2	3	2	3
3	5	3	4
4	5	4	3
5	8	5	3
6	9	6	5
		7	1
		8	3

10.8.2   TREE DATA STRUCTURES

Since trees are important objects in optimization problems, as well as useful data structures in their own right (see §9.1), additional representations and features of trees are given here.

Definitions:

If T is a rooted tree with root r (§9.1.2), then the **predecessor function** $pred: V \to V$ is defined by $pred(r) = 0$, and $pred(j) = i$ if vertex i is the parent of j in T. (See §10.3.1.)

The **principal subtree** T_j rooted at vertex j is the subgraph of T induced by all descendants of j (including j). (See §9.1.2.)

The **cardinality** $card(j)$ of vertex j in a rooted tree T is the number of vertices in its principal subtree T_j.

The **least common ancestor** $least(i, j)$ of vertices i and j in a rooted tree is the vertex of largest depth (§9.1.2) that is an ancestor of both i and j.

Facts:

1. A rooted tree is uniquely specified by the mapping $pred(\cdot)$.

2. In a rooted tree with root r, $depth(r) = 0$ and $depth(j) = 1 + depth(pred(j))$ for $j \neq r$.

3. In a rooted tree, $height(i) = 0$ if i is a leaf. If i is not a leaf, then $height(i) = 1 + \max\{\, height(j) \mid j$ a child of $i\,\}$.

4. In a rooted tree, $card(i) = 1$ if i is a leaf. If i is not a leaf, $card(i) = 1 + \sum\{\, card(j) \mid j$ a child of $i\,\}$.

5. The predecessor, depth, height, and cardinality of a rooted tree T can all be calculated while carrying out a preorder or postorder traversal (§9.1.3) of T:

   - The predecessor and depth can be calculated while advancing from the current vertex to an unvisited vertex.
   - The height and cardinality can be updated when retreating from a vertex all of whose children have been visited.

6. The depth of a vertex is a monotone increasing function on each path from the root.

7. The height and cardinality are monotone decreasing functions on each path from the root.

Examples:

1. The following figure shows a tree T rooted at vertex 1. The vertices have been numbered according to a preorder traversal of T. The following table gives the predecessor, depth, height, and cardinality of each vertex of T.

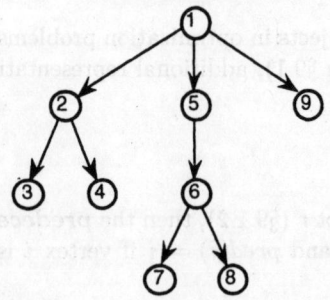

Algorithm 1: Least common ancestor.

input: rooted tree T, vertices i and j
output: least common ancestor of i and j

procedure $least(i, j)$

    while $i \neq j$ **do**
        if $depth(i) > depth(j)$ **then** $i := pred(i)$
        else if $depth(i) < depth(j)$ **then** $j := pred(j)$
        else $i := pred(i), j := pred(j)$
    return i

vertex	1	2	3	4	5	6	7	8	9
pred	0	1	2	2	1	5	6	6	1
depth	0	1	2	2	1	2	3	3	1
height	3	1	0	0	2	1	0	0	0
card	9	3	1	1	4	3	1	1	1

2. Certain applications (such as cycle detection in the network simplex algorithm, §10.5.2) require finding the least common ancestor $least(i, j)$ of vertices i and j in a rooted tree.

3. The calculation of $least(i, j)$ can be carried out efficiently, in $O(n)$ time, by using Algorithm 1.

4. Algorithm 1 is based on Fact 6 and employs two auxiliary data structures, the predecessor and depth functions. It repeatedly backs up from a vertex of larger depth until the least common ancestor is found.

REFERENCES

Printed Resources:

[AhMaOr93] R. K. Ahuja, T. L. Magnanti, and J. B. Orlin, *Network Flows: Theory, Algorithms, and Applications*, Prentice-Hall, 1993.

[AhEtal95] R. K. Ahuja, T. L. Magnanti, J. B. Orlin, and M. R. Reddy, "Applications of network optimization" in *Network Models*, M. Ball, T. Magnanti, C. Monma, and G. Nemhauser (eds.), North-Holland, 1–83, 1995.

[AhEtal90] R. K. Ahuja, K. Mehlhorn, J. B. Orlin, and R. E. Tarjan, "Faster algorithms for the shortest path problem", *Journal of ACM* 37 (1990), 213–223.

[AhOrTa89] R. K. Ahuja, J. B. Orlin, and R. E. Tarjan, "Improved time bounds for the maximum flow problem", *SIAM Journal on Computing* 18 (1989), 939–954.

[AlGa88] K. Altinkemer and B. Gavish, "Heuristics with constant error guarantees for topological design of local access tree networks", *Management Science* 34 (1988), 331–341.

[ArHaMa90] G. J. R. Araque, L. Hall, and T. Magnanti, "Capacitated trees, capacitated routing and associated polyhedra", C.O.R.E. discussion paper, 1990.

[BiDeSi94] D. Bienstock, Q. Deng, and D. Simchi-Levi, "A branch-and-cut algorithm for the capacitated minimum spanning tree problem", working paper, Columbia University, 1994.

[BiGü95] D. Bienstock and O. Günlük, "Computational experience with a difficult mixed-integer multicommodity flow problem", *Mathematical Programming* 68 (1995), 213–237.

[BuSt78] R. Burkard and K. Stratmann, "Numerical investigations on quadratic assignment problems", *Naval Research Logistics Quarterly* 25 (1978), 129–148.

[CoKeTa94] R. Cole, P. Kelvin, and R. E. Tarjan, "A linear-work parallel algorithm for finding minimum spanning trees", *Proceedings of the 6th ACM Symposium on Parallel Algorithms and Architectures*, 1994, 11–15.

[CoLeRi90] T. H. Cormen, C. L. Leiserson, and R. L. Rivest, *Introduction to Algorithms*, MIT Press and McGraw-Hill, 1990.

[Di69] R. Dial, "Algorithm 360: Shortest path forest with topological ordering", *Communications of the ACM* 12 (1969), 632–633.

[Ed65a] J. Edmonds, "Paths, trees, and flowers", *Canadian Journal of Mathematics* 17 (1965), 449–467.

[Ed65b] J. Edmonds, "Maximum matching and a polyhedron with 0, 1-vertices", *Journal of Research of the National Bureau of Standards* B-69 (1965), 125–130.

[EsWi66] L. R. Esau and K. C. Williams, "On teleprocessing system design", *IBM Systems Journal* 5 (1966), 142–147.

[EvMi92] J. R. Evans and E. Minieka, *Optimization Algorithms for Networks and Graphs*, Marcel Dekker, 1992.

[FoFu62] L. R. Ford and D. R. Fulkerson, *Flows in Networks*, Princeton University Press, 1962.

[FrTa84] M. L. Fredman and R. E. Tarjan, "Fibonacci heaps and their uses in improved network optimization algorithms", *Proceedings of the 25th Annual IEEE Symposium on Foundations of Computer Science*, 1984, 338–346. Full paper in *Journal of the ACM* 34 (1987), 596–615.

[Ga76] H. N. Gabow, "An efficient implementation of Edmonds' algorithm for maximum matching on graphs", *Journal of the ACM* 23 (1976), 221–234.

[GaEtal86] H. N. Gabow, Z. Galil, T. Spencer, and R. E. Tarjan, "Efficient algorithms for finding minimum spanning trees in undirected and directed graphs", *Combinatorica* 6 (1986), 109–122.

[Ga85] B. Gavish, "Augmented Lagrangian based algorithms for centralized network design", *IEEE Transactions on Communications* COM-33 (1985), 1247–1257.

[Ga91] B. Gavish, "Topological design of telecommunication networks—local access design methods", *Annals of Operations Research* 33 (1991), 17–71.

[GeHeLa94] M. Gendreau, A. Hertz, and G. Laporte, "A tabu search heuristic for the vehicle routing problem", *Management Science* 40 (1994), 1276–1290.

[Ge95] A. M. H. Gerards, "Matching" in *Network Models*, M. Ball, T. Magnanti, C. Monma, and G. Nemhauser (eds.), North-Holland, 135–224, 1995.

[GlKl75] F. Glover and D. Klingman, "Finding minimum spanning trees with a fixed number of links at a node", in *Combinatorial Programming: Methods and Applications*, B. Roy, et al. (eds.), 191–201, 1975.

[GoBe93] M. X. Goemans and D. J. Bertsimas, "Survivable networks, linear programming relaxations and the parsimonious property", *Mathematical Programming* 60 (1993), 145–166.

[GoTa86] A. V. Goldberg and R. E. Tarjan, "A new approach to the maximum flow problem", *Proceedings of the 18th ACM Symposium on the Theory of Computing*, 1986, 136–146. Full paper appears in *Journal of the ACM* 35 (1988), 921–940.

[GoTa88] A. V. Goldberg and R. E. Tarjan, "Finding minimum-cost circulations by canceling negative cycles", *Proceedings of the 20th ACM Symposium on the Theory of Computing*, 1988, 7–18. Full paper appears in *Journal of the ACM* 36 (1989), 873–886.

[GrHe85] R. L. Graham and P. Hell, "On the history of the minimum spanning tree problem", *Annals of the History of Computing* 7 (1985), 43–57.

[GrWh70] G. Graves and A. Whinston, "An algorithm for the quadratic assignment problem", *Management Science* 17 (1970), 453–471.

[GrMoSt92] M. Grötschel, C. L. Monma, and M. Stoer, "Computational results with a cutting plane algorithm for designing communication networks with low-connectivity constraints", *Operations Research* 40 (1992), 309–330.

[Ha96] L. Hall, "Experience with a cutting plane algorithm for the capacitated spanning tree problem", *INFORMS Journal on Computing* 8 (1996), 219–234.

[HaPa97] R. W. Hall and J. G. Partyka, "On the road to efficiency", *OR/MS Today* 24 (3) (1997), 38–47.

[HoPa96] K. L. Hoffman and M. Padberg, "Traveling salesman problem" in *Encyclopedia of Operations Research*, S. I. Gass and C. M. Harris (eds.), Kluwer, 697–700, 1996.

[JüReRi95] M. Jünger, G. Reinelt, and G. Rinaldi, "The traveling salesman problem" in *Network Models*, M. Ball, T. Magnanti, C. Monma, and G. Nemhauser (eds.), North-Holland, 225–330, 1995.

[KiLe88] G. A. P. Kindervator and J. K. Lenstra, "Parallel computing in combinatorial optimization", *Annals of Operations Research* 14 (1988), 245–289.

[KuHe87] A. Kusiak and S. Heragu, "The facility layout problem", *European Journal of Operational Research* 29 (1987), 229–251.

[LaEtal85] E. Lawler, J. Lenstra, A. Rinnooy Kan, and D. Shmoys, eds., *The Traveling Salesman Problem*, Wiley, 1985.

[Le92] F. T. Leighton, *Introduction to Parallel Algorithms and Architectures*, Morgan Kaufmann, 1992.

[Li81] R. Liggett, "The quadratic assignment problem: an experimental evaluation of solution strategies", *Management Science* 27 (1981), 442–458.

[LoPu86] L. Lovász and M. D. Plummer, *Matching Theory*, North-Holland, 1986.

[MaMiVa95] T. L. Magnanti, P. Mirchandani, and R. Vachani, "Modeling and solving the two-facility capacitated network loading problem", *Operations Research* 43 (1995), 142–157.

[MiVa80] S. Micali and V. V. Vazirani, "An $O(\sqrt{|V|} \cdot |E|)$ algorithm for finding maximum matching in general graphs", *Proceedings of the 21st Annual Symposium on Foundations of Computer Science*, 1980, 17–27.

[NeWo88] G. L. Nemhauser and L. A. Wolsey, *Integer and Combinatorial Optimization*, Wiley, 1988.

[Or88] J. B. Orlin, "A faster strongly polynomial minimum cost flow algorithm", *Proceedings of the 20th ACM Symposium on the Theory of Computing*, 1988, 377–387. Full paper in *Operations Research* 41 (1993), 338–350.

[Or97] J. B. Orlin, "A polynomial time primal network simplex algorithm for minimum cost flows", *Mathematical Programming* 78 (1997), 109–129.

[OsKe95] I. Osman and J. Kelly, eds., *Meta-Heuristics: Theory and Applications*, Kluwer, 1995.

[Pl92] M. D. Plummer, "Matching theory—a sampler: from Dénes König to the present", *Discrete Mathematics* 100 (1992), 177–219.

[Re93] C. Reeves, ed., *Modern Heuristic Techniques for Combinatorial Problems*, Wiley, 1993.

[Ta83] R. E. Tarjan, *Data Structures and Network Algorithms*, SIAM, 1983.

[Vo89] A. Volgenant, "A Lagrangean approach to the degree-constrained minimum spanning tree problem", *European Journal of Operational Research* 39 (1989), 325–331.

[Ya75] A. Yao, "An $O(|E| \log \log |V|)$ algorithm for finding minimum spanning trees", *Information Processing Letters* 4 (1975), 21–23.

Web Resources:

ftp://dimacs.rutgers.edu/pub/netflow/matching/ (Computer code in C for an algorithm for the weighted matching problem; computer code in C, Pascal, and Fortran for algorithms for maximum size matchings in nonbipartite networks.)

ftp://dimacs.rutgers.edu/pub/netflow/maxflow/ (Computer code in Fortran for solving maximum flow and minimum cut problems.)

ftp://dimacs.rutgers.edu/pub/netflow/mincost/ (Computer code in Fortran for solving the minimum cost flow problem.)

ftp://ftp.zib.de/pub/Packages/mathprog/matching/ (Computer code in C for an algorithm for the weighted matching problem; computer code in C and Fortran for algorithms for maximum size matchings in nonbipartite networks.)

ftp://ftp.zib.de/pub/Packages/mathprog/mincut/ (Computer code in C for solving minimum cut problems.)

ftp://ftp.zib.de/pub/Packages/mathprog/netopt-bertsekas/ (Computer code in Fortran for implementing the label-correcting algorithm and Dijkstra's algorithm for shortest paths; computer code in Fortran for solving maximum flow, minimum cut, and minimum cost flow problems.)

http://dmawww.epfl.ch/~rochat/rochat_data/solomon.html (Data sets and software for vehicle routing problems with time windows.)

http://orly1.snu.ac.kr/software/ (Computer code in C, Pascal, and Fortran for implementing Dijkstra's algorithm for shortest paths and the Floyd-Warshall algorithm; computer code in C, Pascal, and Fortran for solving maximum flow, minimum cut, and minimum cost flow problems.)

http://rtm.science.unitn.it/~battiti/archive/code/rts_qap/ (Computer code in Fortran and C for heuristically solving the QAP.)

http://www.cs.sunysb.edu/~algorith/ (The Stony Brook Algorithm Repository; see Sections 1.4 and 1.5 on Graph Problems.)

http://www.geocities.com/ResearchTriangle/7279/vrp.html (Data sets, software, and research papers on vehicle routing problems.)

http://www.ing.unlp.edu.ar/cetad/mos/TSPBIB_home.html (Software, research papers, and other heuristic approaches for the traveling salesman and related problems.)

http://www.iwr.uni-heidelberg.de/iwr/comopt/soft/TSPLIB95/TSPLIB.html (A library of sample problems related to the traveling salesman problem, with their best known solutions.)

http://www.mat.uc.pt/~eqvm/cientificos/fortran/codigos.html (Fortran code for implementing Kruskal's algorithm and Prim's algorithm for minimum spanning trees; Fortran code for implementing the label-correcting algorithm and Dijkstra's algorithm for shortest paths.)

http://www.neci.nj.nec.com/homepages/avg/soft/soft.html (Computer code for implementing the label-correcting algorithm and Dijkstra's algorithm for shortest paths; computer code for solving maximum flow, minimum cut, and minimum cost flow problems.)

http://www.netlib.org/toms/479 (Fortran code for implementing Prim's algorithm.)

http://www.netlib.org/toms/562 (Fortran code for implementing the label-correcting algorithm for shortest paths.)

http://www.netlib.org/toms/608 (Fortran code for heuristically solving the QAP.)

http://www.netlib.org/toms/613 (Fortran code for implementing Prim's algorithm.)

http://www.netlib.org/toms/750 (Fortran code for solving the TSP.)

http://www.netlib.org/toms/754 (Fortran code for heuristically solving the QAP.)

http://www.netlib.org/toms/769 (Fortran code for heuristically solving the QAP.)

http://www.zib.de/Optimization/Software/Mcf/ (Computer code for solving the minimum cost flow problem.)

11

PARTIALLY ORDERED SETS

11.1 Basic Poset Concepts
    11.1.1 Comparability
    11.1.2 Chains, antichains, poset operations
    11.1.3 Rank, ideals and filters
    11.1.4 Lattices
    11.1.5 Distributive and modular lattices

11.2 Poset Properties
    11.2.1 Poset partitions
    11.2.2 LYM property
    11.2.3 Rankings, semiorders, and interval orders
    11.2.4 Application to social choice
    11.2.5 Linear extensions and dimension
    11.2.6 Posets and graphs

Graham Brightwell and
Douglas B. West

Graham Brightwell and
Douglas B. West

INTRODUCTION

Partially ordered sets play important roles in a wide variety of applications, including the design of sorting and searching methods, the scheduling of tasks, the study of social choice, and the study of lattices. This chapter covers the basic concepts involving partially ordered sets, the various types of partially ordered sets, the fundamental properties of these sets, and their important applications.

A table of notation used in the study of posets is given following the glossary.

GLOSSARY

antichain: a subset of a poset in which no two distinct elements are comparable.

atom: in a poset, an element of height 1.

atomic lattice: a lattice such that every element is a join of atoms (or equivalently, such that the atoms are the only join-irreducible elements).

auxiliary graph (of a simple graph G): the graph G' whose vertices are the edges of G, with vertex e_1 adjacent to vertex e_2 in G' if and only if e_1 and e_2 are adjacent edges in G, but do not lie on a 3-cycle in G.

biorder representation (on a digraph D): a pair of real-valued functions f, g on the vertex set V_D such that $u \to v$ is an arc if and only if $f(u) > g(v)$.

bipartite poset: a poset of height at most 2.

Boolean algebra: the poset whose domain is all subsets of a given set, partially ordered by inclusion.

Borda consensus function (on a set of social choice profiles): the consensus function that ranks the alternatives by their Borda count.

Borda count (of an alternative social choice x): the sum, over all individual rankings, of the number of alternatives x "beats".

bounded poset: a poset with both a unique minimal element and a unique maximal element.

u,v-bypass (in a directed graph): a u,v-path of length at least two such that there is also an arc from u to v.

Cartesian product (of two posets $P = (X, R)$ and $P' = (X', R')$): the poset $P \times P = (X \times X', S)$, such that $(x, x')S(y, y')$ if and only if xRy and $x'R'y'$.

chain: a subset of a poset in which every two elements are comparable.

k-chain: a chain of size k, i.e., a chain on k elements.

chain-product: the Cartesian product of a collection of chains.

comparability digraph (of a poset (X, R)): the simple digraph whose vertex set is the domain X and which has an arc from x to y if and only if $x \leq y$.

comparability graph (of a poset (X, R)): the simple graph whose vertex set is the domain X and which has an edge joining distinct vertices x and y if $x \leq y$.

comparability invariant (for posets): an invariant f such that $f(P) = f(Q)$ whenever posets P and Q have the same comparability graph.

comparable elements (in a poset (X, R)): elements x and y such that either $(x, y) \in R$ or $(y, x) \in R$.

consecutive chain (in a ranked poset): a chain whose elements belong to consecutive ranks.

consensus function (on a set of social choice profiles): a function that assigns to each possible profile $P = \{ P_i \mid i \in I \}$ on a set of alternatives a linear ordering (ties allowed) of those alternatives.

consensus ranking (on a set of social choice profiles): the linear ordering of the alternatives assigned by the consensus function.

cover graph (of a poset (X, R)): the graph with vertex set X and edge set consisting of the pairs satisfying the cover relation.

cover relation (of a poset (X, R)): the relation on X consisting of the pairs (x, y) such that $x > y$ in R and such that there is no "intermediate" element z with $x > z > y$.

cover diagram: a synonym for the Hasse diagram.

critical pair (in a poset): an ordered incomparable pair that cannot be made comparable by adding any other single incomparable pair as a relation.

dependent edge (in an acyclic directed graph): an arc from u to v such that the graph contains a u,v-bypass.

dimension (of a poset): the minimum number of chains in a realizer of the poset.

distributive lattice: a lattice in which the meet operator distributes over the join operator, so that $x \wedge (y \vee z) = (x \wedge y) \vee (x \wedge z)$ for all x, y, z.

divisor lattice: the poset $D(n)$ of divisors of n, in which $x \leq y$ means that y is an integer multiple of x.

down-set (in a poset): a subposet I such that if $x \in I$ and if $y < x$, then $y \in I$, also called an ideal.

dual (of a poset $P = (X, R)$): the poset $P^* = (X, S)$ such that $x \leq y$ in S if and only if $y \leq x$ in R.

extension (of a poset $P = (X, R)$): a poset $Q = (X, S)$ such that $R \subseteq S$; meaning that xRy implies xSy.

k-family (in a poset): a subposet containing no chain of size $k+1$.

Ferrers digraph: a digraph having a biorder representation.

filter (generated by an element x in a poset P): the up-set $U[x] = \{ y \in P \mid y \geq x \}$.

filter (generated by a subset in a poset P): given a subset A of P, the up-set $U[A] = \bigcup_{x \in A} U[x]$.

filter (in a poset): a subposet whose domain is the set-theoretic complement of the domain of an ideal.

forbidden subposet description (of a class of posets): a characterization of that class as the class of all posets that does not contain any of the posets in a specified collection.

geometric lattice: an atomic, upper semimodular lattice of finite height.

greatest lower bound (of elements x and y in a poset): a common lower bound z such that every other common lower bound z' satisfies the inequality $z \geq z'$. Such an element, if it exists, is denoted $x \wedge y$.

graded poset: a poset in which all maximal chains have the same length.

Hasse diagram (of a poset): a straight-line drawing of the cover graph in the plane so that the lesser element of each adjacent pair is below the greater.

height (of a poset): the maximum size of a chain in that poset.

height (of an element x of a poset): the maximum length $h(x)$ of a chain that has x as its maximal element.

ideal (generated by an element x in a poset P): the down-set $D[x] = \{\, y \in P \mid y \leq x \,\}$.

ideal (generated by a subset A in a poset P): the down-set $D[A] = \bigcup_{x \in A} D[x]$.

ideal (in a poset): a subposet I such that if $x \in I$ and if $y < x$, then $y \in I$.

incomparability graph (of a poset P): the edge-complement of the comparability graph $G(P)$.

incomparable pair (in a poset (X, R)): a pair $x, y \in X$ such that neither $x \leq y$ nor $y \leq x$ in R.

integer partition: a nonincreasing nonnegative integer sequence having finitely many nonzero terms, with trailing zeros added as needed for comparison.

intersecting family: a collection of subsets of a set such that every pair of members has nonempty intersection.

intersection (of partial orderings $P = (X, R)$ and $Q = (X, S)$ on the set X): the poset $(X, R \cap S)$ that includes the comparisons in both.

intersection (of posets (X, R) and (X, S)): the poset $(X, R \cap S)$.

interval (in a poset): the subposet which contains all elements z such that $x \leq z \leq y$.

interval order: a poset in which there is an assignment to its members of real intervals, such that $x < y$ if and only if the interval for y is totally to the right of the interval for x.

interval representation (of a poset P): a collection of real intervals corresponding to an interval order for P.

isomorphic (posets): posets $P = (X, R)$ and $Q = (Y, S)$ such that there is a poset isomorphism $P \to Q$.

isomorphism (of lattices): an order-preserving bijection from one lattice to another that also preserves greatest lower bounds and least upper bounds of pairs.

isomorphism (of posets): a bijection from one poset to another that preserves the order relation.

join: given $\{x, y\}$, another name for the least upper bound $x \vee y$.

join-irreducible element (in a lattice): a nonzero element that cannot be expressed as the join of two other elements.

Jordan-Dedekind chain condition: the condition for a poset that every interval has finite length.

lattice: a poset in which every pair of elements has both a greatest lower bound and a least upper bound.

lattice (of bounded sequences): the set $L(m, n)$ of length-m real sequences $a_1, \ldots, a_n$ such that $0 \leq a_1 \leq \ldots \leq a_n \leq n$.

lattice (of order ideals in a poset $P = (X, R)$): the set $J(P)$ of order ideals of P, ordered by inclusion.

least upper bound (of elements x and y in a poset): a common upper bound z such that every other common upper bound z' satisfies the inequality $z' \geq z$. Such an element, if it exists, is denoted $x \vee y$.

length (of a chain): the number of cover relations in the chain; in other words, one less than the number of elements in the chain.

length (of a poset): the length of a longest chain, which is one less than the height of that poset. (Sometimes *height* is used synonymously with length.)

lexicographic ordering (of the Cartesian product of posets): the ordering for the Cartesian product of the domains in which $(x_1, x_2) \leq (y_1, y_2)$ if and only if $x_1 < y_1$ or $x_1 = y_1$ and $x_2 \leq y_2$; this is not the usual ordering of the Cartesian product of posets.

linear extension (of a poset): an extension of the poset that is a chain.

linear order: See *total order*.

linearly ordered set: a poset in which every pair of elements is comparable.

linear sum (of two disjoint posets P and P'): the poset in which all the elements of poset P lie "below" all those of poset P'.

locally finite poset: a poset in which every interval is finite.

lower bound (of elements x and y in a poset): an element z such that $x \geq z$ and $y \geq z$.

lower semimodular lattice: a lattice whose dual is upper semimodular.

majority rule property (for a consensus function): the property that it prefers x to y if and only if a majority of the individuals prefer x to y.

maximal element (in a poset): an element such that no other element is greater.

meet (of elements x and y): another name for the greatest lower bound $x \wedge y$.

meet-irreducible element (of a lattice): a nonzero element that cannot be expressed as the meet of two other elements.

minimal element (in a poset): an element such that no other element is less.

minimum realizer encoding (of a poset): a poset that lists for each element its position on each extension in a minimum realizer.

modular lattice: a lattice in which $x \wedge (y \vee z) = (x \wedge y) \vee z$ for all x, y, z such that $z \leq x$.

module (in a graph G): a vertex subset $U \subseteq V_G$ such that each vertex outside U is adjacent to all or none of the vertices in U.

k-norm (of a sequence $a = \{a_i\}$): the sum $\sum \min\{k, a_i\}$, whose value is commonly denoted $m_k(a)$.

k-norm of a chain partition: the k-norm of its sequence of chain sizes.

normalized matching property (for a graded poset): the property that for every rank k and every subset A of rank P_k, the set A^* of elements in the rank P_{k+1} that are comparable to at least one element of A satisfies the inequality $\frac{|A^*|}{N_{k+1}} \geq \frac{|A|}{N_k}$.

order module in a poset: a set S of elements such that every element outside S is above all of S, below all of S, or incomparable to all of S.

order-preserving mapping (from poset $P = (X, R)$ to poset $Q = (Y, S)$): a function $f: X \to Y$ such that $f(x) \leq f(y)$ whenever $x \leq y$ in P.

order relation (on a set X): a relation R such that (X, R) is a partially ordered set.

partially ordered set: a pair $P = (X, R)$ consisting of a set X and a relation R that is reflexive, antisymmetric, and transitive.

partition lattice: the poset Π_n of partitions of the set $[n] = \{1, \ldots, n\}$, where $\pi < \sigma$ if π is a refinement of σ.

permutation graph: a graph whose vertices can be placed in 1-1 correspondence with the elements of a permutation of $[n] = \{1, \ldots, n\}$, such that v_i is adjacent to v_j if and only if the larger of $\{i, j\}$ comes first in the permutation.

planar poset: a poset with a Hasse diagram that has no edge-crossings.

plurality consensus function (on a set of social choice profiles): the consensus function in which the winner(s) is(are) the alternative(s) appearing in the greatest number of top ranks, after which the winner(s) is(are) deleted and the procedure is repeated to select the next rank of the consensus ranking, etc.

poset: a partially ordered set.

profile (on a set of alternative social choices): a set $P = \{\, P_i \mid i \in I \,\}$ of linear rankings (ties allowed) of the alternatives, one for each member of a set I of "individuals" participating in the decision process.

quasi-transitive orientation (on a simple graph G): an assignment of directions to the edges of G so that whenever there is an xy-arc and a yz-arc, there is also an arc between x and z.

rank (of a graded poset): the length of any maximal chain in the poset.

rank function (on a poset): an integer-valued function r on the elements of the poset so that "y covers x" implies that $r(y) = r(x) + 1$.

ranked poset: a poset having a rank function.

kth rank of a ranked poset: the subset P_k of elements for which $r(x) = k$.

rank parameters (of a subset F of elements in a ranked poset P): the numbers $f_k = |F \cap P_k|$.

ranking: a poset P whose elements are partitioned into ranks $P_1, \ldots, P_k$ such that two elements are incomparable in the poset if and only if they belong to the same rank.

realizer (of a poset P): a set of linear extensions of P whose intersection is P.

refinement (of a set partition σ): replacement of each block $B \in \sigma$ by some partition of B.

regular covering (of a poset by chains): a multiset of maximal chains such that for each element x the fraction of the chains containing x is $\frac{1}{N_{r(x)}}$, where $N_{r(x)}$ is a Whitney number.

self-dual poset: a poset isomorphic to its dual.

semimodular lattice: an upper semimodular lattice.

semiorder: a poset on which there is a real-valued function f and a real number $\delta > 0$ such that $x < y$ if and only if $f(y) - f(x) > \delta$.

shadow (of a family of sets $\mathbf{F}$): the collection of sets containing every set that is obtainable by selecting a set in $\mathbf{F}$ and deleting one of its elements.

size (of a finite poset): the number of elements.

Sperner property (for a graded poset): the property that some single rank is a maximum antichain.

k-Sperner property (for a graded poset): the property that the poset has a maximum k-family consisting of k ranks.

standard k-chain: the poset $\{1,\ldots,n\}$, under the usual ordering of the integers, written $\underline{k}$.

standard example of an n-dimensional poset: the subposet S_n of the Boolean algebra $\underline{2}^n$ induced by the singletons and their complements.

strict Sperner property: the property of a graded poset that all maximum antichains are single ranks.

strong Sperner property: the property that a graded poset is k-Sperner for all $k \leq r(P)$.

Steinitz exchange axiom: for a closure operator $\sigma: 2^E \to 2^E$, the rule that $p \notin \sigma(A)$ and $p \in \sigma(A \cup q)$ imply $q \in \sigma(A \cup p)$.

sublattice (of a lattice): a subposet that contains the meet and join of every pair of its elements.

submodular height function (in a lattice): a height function h such that $h(x \wedge y) + h(x \vee y) \leq h(x) + h(y)$ for all x, y.

subposet (of a poset (X, R)): a poset (Y, S) such that $Y \subseteq X$ and $S = R \cap (Y \times Y)$.

subset lattice: the Boolean algebra $\underline{2}^n$, that is, the Cartesian product of n copies of the standard 2-chain.

subspace lattice: the set $L_n(q)$ of subspaces of an n-dimensional vector space over a q-element field, partially ordered by set inclusion.

symmetric chain (in a ranked poset P): a chain that has an element of rank $r(P) - k$ whenever it has an element of rank k.

symmetric chain decomposition (of a ranked poset): a partition of that poset into symmetric consecutive chains.

symmetric chain order: a poset with a symmetric chain decomposition.

topological ordering (of an acyclic digraph): a linear extension of the poset it represents.

topological sort: an algorithm that arranges the elements of a partially ordered set into a total ordering that is compatible with the original partial ordering.

total order (of a set): an order relation in which each pair of distinct elements is comparable.

transitive orientation (on a simple graph): an assignment of directions to the edges of a simple graph G so that whenever there is an xy-arc and a yz-arc, there is also a xz-arc.

triangular chord (for a walk $x_1, \ldots, x_k$ in an undirected graph): an edge between vertices x_{i-1} and x_{i+1}, two apart on the walk.

upper bound (of elements x and y in a poset): an element z such that $x \leq z$ and $y \leq z$.

upper semimodular lattice: a lattice in which whenever x covers $x \wedge y$, it is also true that $x \vee y$ covers y.

up-set (in a poset): a filter.

weak order: a ranking, i.e., a poset P whose elements are partitioned into ranks $P_1, \ldots, P_k$ such that two elements are incomparable if and only if they belong to the same rank.

kth Whitney number (of a ranked poset P): the cardinality $|P_k|$ of the kth rank; written $N_k(P)$.

width (of a poset): the maximum size of an antichain in the poset.

Young lattice: the lattice of integer partitions under component-wise ordering.

poset notation	
notation	meaning
$y \geq x$	$x \leq y$
$x < y$	$x \leq y$ and $x \neq y$
$x \| y$	$x \not\leq y$ and $y \not\leq x$
0	minimal element in a bounded poset
1	maximal element in a bounded poset
$[x, y]$	the interval $\{ z \mid x \leq z \leq y \}$
$\underline{k}$	standard k-chain
$P_1 + P_2$	disjoint union of posets
$P_1 \oplus P_2$	linear sum of two posets
$P_1 \times P_2$	Cartesian product of two posets
P^n	iterated Cartesian product of copies of P
P^*	dual of poset P
$D(n)$	divisibility poset of the integer n
$r(P)$	rank of a graded poset P
$N_k(P)$	kth Whitney number (= cardinality of kth rank) of P
$w(P)$	width of P (= maximum size of an antichain)
$D[x]$	down-set (ideal) $\{ y \mid y \leq x \}$
$D(x)$	down-set (ideal) $\{ y \mid y < x \}$
$U[x]$	up-set (filter) $\{ y \mid y \geq x \}$
$U(x)$	up-set (filter) $\{ y \mid y > x \}$
$x \vee y$	lub of x and y
$x \wedge y$	glb of x and y

11.1 BASIC POSET CONCEPTS

11.1.1 COMPARABILITY

The integers and real numbers are totally ordered sets, since every pair of distinct elements can be compared. In a partially ordered set, some pairs of elements may be incomparable. For example, under the containment relation, the sets $\{1, 2\}$ and $\{1, 3\}$ are incomparable.

Definitions:

A **partial ordering** (or **order relation**) R on a set X is a binary relation that is:
- reflexive: for all $x \in S$, xRx;
- antisymmetric: for all $x, y \in S$, if xRy and yRx, then $x = y$;
- transitive: for all $x, y, z \in S$, if xRy and yRz, then xRz.

Note: $x \leq y$ or $x \leq_P y$ are often written in place of xRy or $(x, y) \in R$. Also, $y \geq x$ means $x \leq y$. The notation $\preceq$ is sometimes used in place of $\leq$. See the table following the Glossary for further poset notation.

A **partially ordered set** (or **poset**) $P = (X, R)$ is a pair consisting of a set X, called the **domain**, and a partial ordering R on X. Writing $x \in P$ means that $x \in X$. The notation $(X, \leq)$ is often used instead of (X, R) to designate a poset.

The **size** of a finite poset P is the number of elements in the domain.

A **totally ordered set** (or **linearly ordered set**) is a poset in which every element is comparable to every other element.

The elements x and y are **comparable** (**related**) in P if either $x \leq y$ or $y \leq x$ (or both, in which case $x = y$).

The elements x and y are **incomparable** (**unrelated**) if they are not comparable. Writing $x \parallel y$ indicates incomparability.

Element x is **less than** element y, written $x < y$, if $x \leq y$ and $x \neq y$. (The notation $\prec$ is sometimes used in place of $<$.)

Element x is **greater than** element y, written $x > y$, if $x \geq y$ and $x \neq y$.

An element x of a poset is **minimal** if the poset has no element less than x.

An element x of a poset is **maximal** if the poset has no element greater than x.

A poset is **bounded** if it has both a unique minimal element (denoted "0") and a unique maximal element (denoted "1").

The **comparability digraph** $D(P)$ of a poset $P = (X, R)$ is the digraph with vertex set X, such that there is an arc from x to y if and only if $x \leq y$.

The **comparability graph** $G(P)$ of a poset $P = (X, R)$ is the simple graph with vertex set X, such that $xy \in E_G$ if and only if x and y are comparable in P, where $x \neq y$.

The **incomparability graph** of a poset P is the edge-complement of the comparability graph $G(P)$.

The **induced poset** of an acyclic digraph D is the poset whose elements are the vertices of D and such that $x \leq y$ if and only if there is a directed path from x to y.

The element y **covers** the element x in a poset if $x < y$ and there is no intermediate element z such that $x < z < y$.

The **cover graph** of poset P is the graph with vertex set X such that x and y are adjacent if and only if one of them covers the other in P.

A **Hasse diagram** (or **cover diagram** or **diagram**) of poset P is a straight-line drawing of the cover graph in the plane such that the lesser element of each pair satisfying the cover relation is lower in the drawing.

A poset is **planar** if it has a Hasse diagram without edge-crossings.

A *subposet* of $P = (X, \leq)$ is a subset $Y \subseteq X$ with the relation $x \leq y$ in Y if and only if $x \leq y$ in X.

The *interval* $[x, y]$ in poset P is the subposet that contains all elements z such that $x \leq z \leq y$.

A poset P is **locally finite** if every interval in P has finitely many elements.

An **order-preserving mapping** from poset $P = (X, \leq_P)$ to poset $Q = (Y, \leq_Q)$ is a function $f: X \to Y$ such that $x \leq_P x'$ implies $f(x) \leq_Q f(x')$.

An *isomorphism of posets* $P = (X, R)$ and $Q = (Y, S)$ is a bijection $f: X \to Y$ that preserves the order relation: whenever $x_1 \leq_P x_2$, then $f(x_1) \leq_Q f(x_2)$.

Isomorphic posets are posets $P = (X, R)$ and $Q = (Y, S)$ such that there is a poset isomorphism $P \to Q$. This is sometimes indicated informally by writing $P = Q$.

Poset $Q = (Y, S)$ *is contained in* (or *imbeds in*) poset $P = (X, R)$ if Q is isomorphic to a subposet of P.

A poset P is Q-*free* if P does not contain a poset isomorphic to Q.

Facts:

1. Every finite nonempty poset has a minimal element and a maximal element.

2. The comparability digraph $D(P)$ of a poset P is an acyclic digraph.

3. The minimal elements of a poset P induced by a digraph D are the sources of D; that is, they are the vertices at which every arc points outward.

4. The maximal elements of a poset P induced by a digraph D are the sinks of D; that is, they are the vertices at which every arc points inward.

5. The element y covers the element x in a poset P induced by a digraph D if and only if there is an arc in digraph D from x to y and there is no other directed path from x to y.

6. Suppose that the poset P is induced from an acyclic digraph D. Then the comparability digraph of P is the transitive closure of D.

7. Two different posets cannot have the same Hasse diagram, but they may have the same cover graph or the same comparability graph.

8. There is a polynomial-time algorithm to check whether a graph G is a comparability graph, but the problem of deciding whether there exists a poset for which G is the cover graph is NP-complete.

Examples:

1. Any collection of subsets of the same set forms a poset when the subsets are partially ordered by the usual inclusion relation $X \subseteq Y$.

2. *Boolean algebra*: The Boolean algebra on a set X is the poset consisting of all the subsets of X, ordered by inclusion.

3. The Boolean algebra on the set $\{a, b, c\}$ has the following Hasse diagram. The only maximal element is $\{a, b, c\}$. The only minimal element is $\emptyset$.

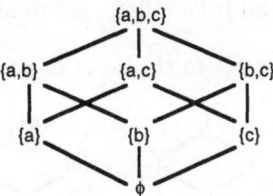

4. There are five different isomorphism types of posets of size three, whose Hasse diagrams are as follows.

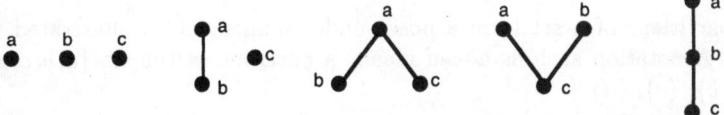

5. There are 16 different isomorphism types of posets of size four, whose Hasse diagrams are as follows.

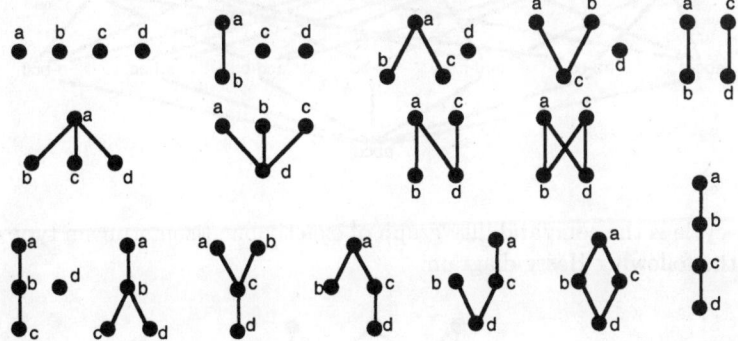

6. *Divisibility poset:* The divisibility poset on the set I of positive integers, denoted $D(I)$, has the relation $x \leq y$ if y is an integer multiple of x. A number y covers a number x if and only if the quotient $\frac{y}{x}$ is prime.

7. The set $D(n)$ of divisors of n forms a subposet of $D(I)$, for any positive integer n. The set $D(n)$ is identical to the interval $[1, n]$ in $D(I)$. For instance, the following figure is the Hasse diagram of $D(24) = [1, 24]$.

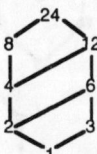

8. The interval $[3, 30]$ in $D(I)$ has domain $\{3, 6, 12, 15, 30\}$. The interval $[2, 24]$ has domain $\{2, 4, 6, 8, 12, 24\}$.

9. The poset $D(I)$ is infinite, but locally finite.

10. The Boolean algebra of all subsets of an infinite set is not a finite poset. Nor is it locally finite, since each interval from a finite set to an infinite superset is infinite.

11. The poset $D(6)$ is isomorphic to the poset of subsets of $\{a,b\}$:

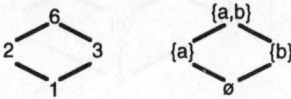

12. To generalize Example 11, if $p_1, \ldots, p_n$ are distinct primes, then the divisibility poset $D(p_1 \ldots p_n)$ is isomorphic to the poset of subsets of a set of n objects.

13. The partitions of a set form a poset under refinement, as illustrated for the set $\{a, b, c, d\}$. A notation such as $ab\text{-}c\text{-}d$ means a partition of the set $\{a, b, c, d\}$ into the subsets $\{a, b\}, \{c\}, \{d\}$.

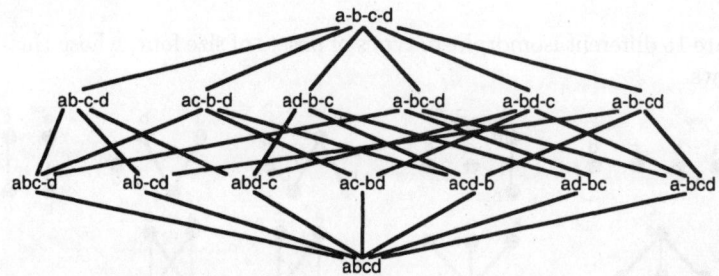

14. The 6-cycle is the comparability graph of exactly one (isomorphism type of) poset, which has the following Hasse diagram:

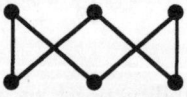

15. The 6-cycle is the cover graph of seven posets, all of which are planar. They have the following Hasse diagrams:

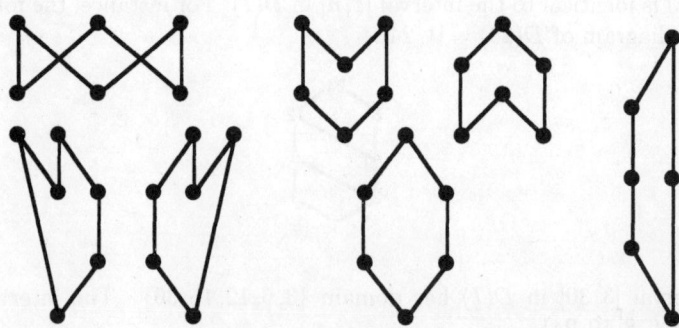

11.1.2 CHAINS, ANTICHAINS, POSET OPERATIONS

Definitions:

A **chain** is a subset S of mutually comparable elements of a poset P, or sometimes the subposet of P formed by such a subset.

The **length** of a finite chain C is $|C|-1$, i.e., the number of edges in the Hasse diagram of that chain, regarded as a poset.

A **k-chain** is a chain of size k, i.e., a chain on k elements.

The **standard k-chain** $\underline{k}$ is a fixed k-chain, presumed to be disjoint from other objects in the universe of discourse.

The **height of a poset** P is the maximum size of a chain in P.

The **height of an element** x in a poset P is the maximum length $h(x)$ of a chain in P that has x as its maximal element.

A **bipartite poset** is a poset of height at most 2.

A **chain-product** (or **grid**) is the Cartesian product of a collection of chains.

An **antichain** (or **clutter** or **Sperner family**) is a subset S of pairwise incomparable elements of a poset P, or sometimes the subposet of P formed by such a subset.

A chain or antichain is **maximal** if it is contained in no other chain or antichain.

A chain or antichain in a finite poset is a **maximum** chain or antichain if it is one of maximum size.

The **disjoint union** of two posets $P = (X, R)$ and $P' = (X', R')$ with $X \cap X' = \emptyset$ is the poset $(X \cup X', R \cup R')$, denoted $P + P'$.

The **linear sum** of two posets $P = (X, R)$ and $P' = (X', R')$ with $X \cap X' = \emptyset$ is the poset $(X \cup X', R \cup R' \cup (X \times X'))$, denoted $P \oplus P'$. (This puts all of poset P "below" poset P').

The **Cartesian product** $P \times P'$ (or **direct product** or **product**) of two posets $P = (X, R)$ and $P' = (X', R')$ is the poset $(X \times X', S)$, such that $(x, x')S(y, y')$ if and only if xRy and $x'R'y'$.

The **iterated Cartesian product** of n copies of a poset $P = (X, \leq)$, written P^n, is the set of n-tuples in P, such that $(x_1, \ldots, x_n) \leq (y_1, \ldots, y_n)$ if and only if either $x_j \leq y_j$, for $j = 1, \ldots, n$.

The **lexicographic ordering** of the Cartesian product $P_1 \times P_2$ of the domains of two posets is the partial ordering in which $(x_1, x_2) \leq (y_1, y_2)$ if and only if $x_1 < y_1$, or $x_1 = y_1$ and $x_2 \leq y_2$.

The **dual** of a poset P, denoted P^*, is the poset on the elements of P defined by the relation $y \leq_{P^*} x$ if and only if $x \leq_P y$.

A **self-dual poset** is a poset that is isomorphic to its dual.

Facts:

1. Every k-chain is isomorphic to the linear sum $\underline{1} \oplus \underline{1} \oplus \cdots \oplus \underline{1}$ of k copies of $\underline{1}$.

2. Every antichain of size k is isomorphic to the disjoint union $\underline{1} + \underline{1} + \cdots + \underline{1}$ of k copies of $\underline{1}$.

3. The chains are characterizable as the class of $(\underline{1} + \underline{1})$-free posets.

4. The cover graph of a chain is a path.

5. The comparability graph of a chain of size n is the complete graph K_n.

6. The antichains are the class of $\underline{2}$-free posets;

7. The comparability graph of an antichain has no edges.

8. The maximum size of a chain in a finite poset P equals the minimum number of antichains needed to cover the elements of P, that is, the minimum number of antichains whose union equals the domain of poset P.

9. The bipartite posets are precisely the $\underline{3}$-free posets.

10. The bipartite posets are the posets whose comparability graph and cover graph are the same.

11. Every maximal chain of a finite poset P extends from a minimal element of P to a maximal element of P, and successive pairs on a maximal chain satisfy the cover relation of P.

12. The Cartesian product of two posets is a poset.

13. A poset and its dual have the same comparability graph and the same cover graph.

14. The Hasse diagram of the dual of a poset P can be obtained from the Hasse diagram of P either by reflecting through the horizontal axis or by rotating 180 degrees.

15. The set of order-preserving maps from a poset P to a poset Q forms a poset, denoted by Q^P, under "coordinate-wise ordering": $f \leq g$ in Q^P if and only if $f(x) \leq_Q g(x)$ for all $x \in P$.

Examples:

1. The following figure shows: (A) a 3-chain, (B) an antichain of width 4, and (C) a bipartite poset.

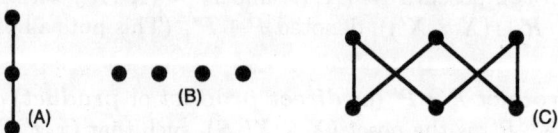

2. The poset $\underline{2}^3$ is not planar, even though it has a planar cover graph. However, deleting its minimal element or maximal element leaves a planar subposet.

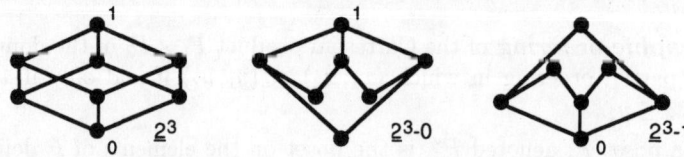

3. The cover graph of the poset $\underline{2}^n$ is isomorphic to the n-dimensional cube, whose vertices are the bitstrings of length n, with bitstrings adjacent if they differ in one position. Each bit encodes the possible presence of an element of the set of which B_n is the Boolean algebra (§5.8.1).

4. The interval in the Boolean algebra $\underline{2}^n$ between an element of rank k and an element of rank $l \geq k$ is isomorphic to the poset $\underline{2}^{l-k}$.

5. Every maximal chain in the Boolean algebra $\underline{2}^n$ has size $n+1$ and length n, and there are $n!$ such chains. There are maximal antichains as small as 1 element.

6. In general, the poset $D(n)$ is isomorphic to a chain product, one factor for each prime divisor of n. The elements of $D(n)$ can be encoded as integer vectors $\{a_1,\ldots,a_n \mid 0 \leq a_i < e_i\}$, where n is a product of distinct primes with powers $e_1,\ldots,e_n$, and $a \leq b$ if and only if $a_i \leq b_i$ for all i.

7. The Hasse diagrams for two possible partial orderings on the Cartesian product of the domains of two posets is shown in the following figure:

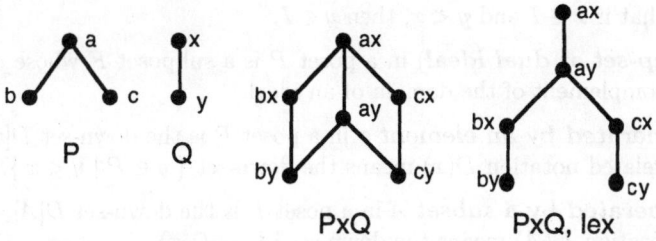

8. The two posets $M_5 = \underline{1} \oplus (\underline{1}+\underline{1}+\underline{1}) \oplus \underline{1}$ and $N_5 = \underline{1} \oplus (\underline{2}+\underline{1}) \oplus \underline{1}$ are used in §11.1.4 in a forbidden subposet description.

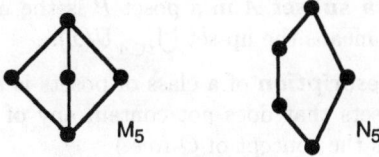

9. The following posets are self-dual:

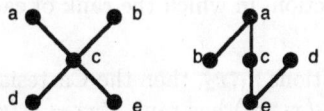

11.1.3 RANK, IDEALS, AND FILTERS

Definitions:

A **graded poset** is a poset in which all maximal chains have the same length.

The **rank** $r(P)$ of a graded poset P is the length of any maximal chain.

A **rank function** r on a poset P is an assignment of integers to the elements so that the relation y covers x implies that $r(y) = r(x) + 1$.

A **ranked poset** is a poset having a rank function.

The **kth rank** of a ranked poset P is the subset P_k of elements for which $r(x) = k$.

The **kth rank parameter** of a subset of elements F in a ranked poset P is the cardinality $|F \cap P_k|$ of the number of elements of F in the kth rank of P.

The **kth Whitney number** $N_k(P)$ of a ranked poset P is the cardinality $|P_k|$ of the kth rank.

The **length of a poset** P is the length of a longest chain in P, which is one less than the height of P. *Note*: Sometimes "height" is used synonymously with length.

The **Jordan-Dedekind chain condition** for a poset is that every interval has finite length.

The **width of a poset** P, denoted $w(P)$, is the maximum size of an antichain in P.

An **ideal** (or **down-set**, **order ideal**, or **hereditary family**) in a poset P is a subposet I such that if $x \in I$ and $y < x$, then $y \in I$.

A **filter** (or **up-set** or **dual ideal**) in a poset P is a subposet F whose domain is the set-theoretic complement of the domain of an ideal.

The **ideal generated by an element** x in a poset P is the down-set $D[x] = \{\, y \in P \mid y \leq x \,\}$. The related notation $D(x)$ means the down-set $\{\, y \in P \mid y < x \,\}$.

The **ideal generated by a subset** A in a poset P is the down-set $D[A] = \bigcup_{x \in A} D[x]$. The related notation $D(A)$ means the down-set $\bigcup_{x \in A} D(x)$.

The **filter generated by an element** x in a poset P is the up-set $U[x] = \{\, y \in P \mid y \geq x \,\}$. The related notation $U(x)$ means the up-set $\{\, y \in P \mid y > x \,\}$.

The **filter generated by a subset** A in a poset P is the up-set $U[A] = \bigcup_{x \in A} U[x]$. The related notation $U(A)$ means the up-set $\bigcup_{x \in A} U(x)$.

A **forbidden subposet description** of a class of posets is a characterization of that class as the class of all posets that does not contain any of the posets in a specified collection. (This generalizes the concept of Q-free.)

Facts:

1. A graded poset has a rank function, in which the rank of each element is defined to be its height.

2. If posets P_1, P_2 have rank functions r_1, r_2, then the Cartesian product $P = P_1 \times P_2$ is ranked, so that the element $x = (x_1, x_2)$ has rank $r(x) = r_1(x_1) + r_2(x_2)$.

3. In a Cartesian product of finite ranked posets P_1 and P_2, the Whitney numbers for the Cartesian product $P = P_1 \times P_2$ satisfy the equation $N_k(P) = \sum_i N_i(P_1) N_{k-i}(P_2)$.

4. The Boolean algebra on a set X of cardinality n is isomorphic to $\underline{2}^n$, the Cartesian product of n copies of $\underline{2}$. This poset isomorphism type is often denoted B_n.

5. The Boolean algebra on a set X of cardinality n is a graded poset, with rank function $r(S) = |S|$, and with Whitney numbers $N_k(\underline{2}^n) = \binom{n}{k}$.

6. The sequence of Whitney numbers on the Boolean algebra on a set X of cardinality n is symmetric, since $\binom{n}{k} = \binom{n}{n-k}$. It is also unimodal, since the sequence rises monotonically to the maximum and then falls monotonically.

7. **Sperner's theorem**: The only maximum antichains in the Boolean algebra $\underline{2}^n$ are the middle ranks (one such rank if n is even, two if n is odd). Thus the width of $\underline{2}^n$ is $\binom{n}{\lfloor n/2 \rfloor}$.

8. The maximal elements of an ideal form an antichain, as do the minimal elements of a dual ideal; these yield natural bijections between the set of antichains in a poset P and the sets of ideals or dual ideals of P.

9. The divisibility poset $D(I)$ on the integers satisfies the Jordan-Dedekind chain condition.

Examples:

1. In the poset P of partitions of $\{a,b,c,d\}$ under inverse refinement, illustrated below, the Whitney numbers are $N_1(P) = 1$, $N_2(P) = 6$, $N_3(P) = 7$, and $N_4(P) = 1$.

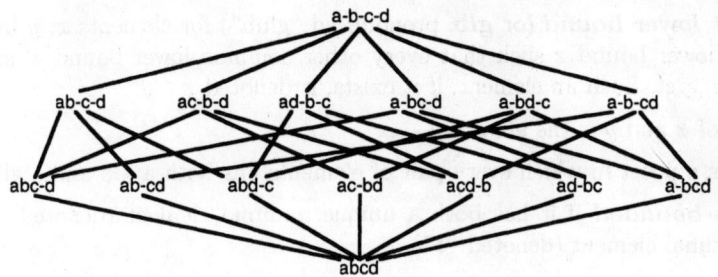

2. In the poset of partitions of $\{a,b,c,d\}$ under inverse refinement, the ideal $D[ac\text{-}bd]$ is the set $\{ac\text{-}bd,\, a\text{-}c\text{-}bd,\, ac\text{-}b\text{-}d,\, a\text{-}b\text{-}c\text{-}d\}$, and the ideal $D(ac\text{-}bd)$ is the set $\{a\text{-}c\text{-}bd,\, ac\text{-}b\text{-}d,\, a\text{-}b\text{-}c\text{-}d\}$.

3. In the poset of partitions of $\{a,b,c,d\}$ under inverse refinement, the filter $U[a\text{-}c\text{-}bd]$ is the set $\{ac\text{-}bd,\, abd\text{-}c,\, a\text{-}bcd,\, abcd\}$, and the filter $U(a\text{-}c\text{-}bd)$ is the set $\{abd\text{-}c,\, a\text{-}bcd,\, abcd\}$.

4. In the graded poset $D(n)$ of divisors of n, the rank $r(x)$ is the sum of the exponents in the prime power factorization of x. The Whitney numbers of $D(n)$ are symmetric because the divisors x and $\frac{n}{x}$ have "complementary" ranks. If n is a product of k distinct primes, then $D(n) \cong \underline{2}^k$.

5. The following Hasse diagram corresponds to an ungraded poset, because the lengths of its maximum chains differ, i.e., they are length 2 and length 3.

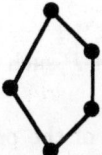

6. In the divisibility poset $D(I)$, the subposet $D(n)$ of divisors of n is a finite ideal, and the non-multiples of n form an infinite ideal, whose complement is the infinite filter $U(n)$ of numbers that are divisible by n.

11.1.4 LATTICES

Lattices are posets with additional properties that capture some aspects of the intersection and the union of sets and (more generally) of the greatest common divisor and least common multiple of positive integers. (See also §5.7.)

Definitions:

A (*common*) **upper bound** for elements x, y in a poset is an element z such that $x \leq z$ and $y \leq z$.

A **least upper bound** (or **lub**, pronounced "lub") for elements x, y in a poset is a common upper bound z such that every other common upper bound z' satisfies the inequality $z \leq z'$. Such an element, if it exists, is denoted $x \vee y$.

The **join** of x and y is the lub $x \vee y$.

A **(common) lower bound** for elements x, y in a poset is an element z such that $x \geq z$ and $y \geq z$.

A **greatest lower bound** (or **glb**, pronounced "glub") for elements x, y in a poset is a common lower bound z such that every other common lower bound z' satisfies the inequality $z \geq z'$. Such an element, if it exists, is denoted $x \wedge y$.

The **meet** of x and y is the glb $x \wedge y$.

A **lattice** is a poset in which every pair of elements has both a lub and a glb.

A lattice is **bounded** if it has both a unique minimal element (denoted "0") and a unique maximal element (denoted "1").

A nonzero element of a lattice L is **join-irreducible** (or simply **irreducible**) if it cannot be expressed as the lub of two other elements. The subposet formed by the join-irreducible elements of L is denoted by $P(L)$.

A nonzero element of a lattice L is **meet-irreducible** if it cannot be expressed as a glb of two other elements. The subposet formed by the meet-irreducible elements of L is denoted by $Q(L)$.

A **complement** of an element x of a lattice is an element $\bar{x}$ such that $x \vee \bar{x} = 1$ and $x \wedge \bar{x} = 0$.

A **complemented lattice** is a lattice in which every element has a complement.

A **lattice isomorphism** is an order-preserving bijection from one lattice to another that also preserves glbs and lubs.

An **atom** of a poset is an element of height 1.

A lattice is **atomic** if every element is a lub of atoms (or equivalently, if the atoms are the only join-irreducible elements).

A **sublattice** of a lattice L is a subposet P such that $x \wedge y$ and $x \vee y$ are in P for all x and $y \in P$.

The **divisor lattice** is the poset $D(n)$ of the positive integer divisors of n, in which $x \leq y$ means that y is an integer multiple of x.

The **subset lattice** is the Boolean algebra $\underline{2}^n$, that is, the Cartesian product of n copies of the standard 2-chain.

The **subspace lattice** $L_n(q)$ is the set of subspaces of an n-dimensional vector space over a q-element field, partially ordered by set inclusion.

The **lattice of (order) ideals** $J(P)$, for any poset $P = (X, R)$, is the set of order ideals of P, ordered by inclusion.

The **lattice of bounded sequences** $L(m, n)$ has as members the length-m real sequences $a_1, \ldots, a_n$ such that $0 \leq a_1 \leq \cdots \leq a_m \leq n$.

An **integer partition** is a nonincreasing nonnegative integer sequence having finitely many nonzero terms, with trailing zeros added as needed for comparison.

The **Young lattice** is the lattice of integer partitions under component-wise ordering.

A **refinement** of a set partition σ replaces each block $B \in \sigma$ by some partition of B.

The **partition lattice** is the poset Π_n of partitions of the set $[n] = \{1, \ldots, n\}$, where $\pi < \sigma$ if π is a refinement of σ.

Facts:

1. If $z \leq x$ in a lattice, then $x \wedge (y \vee z) \geq (x \wedge y) \vee z$.
2. *4-point lemma*: If each of the elements z, w is less than or equal to each of the elements x, y in a lattice, then $z \vee w \leq x \wedge y$.
3. An element z is a least upper bound for x and y if and only if it is a unique minimal element among their common upper bounds.
4. Every finite lattice is bounded.
5. Every chain-product is a lattice.
6. If a locally finite poset P with a unique maximal element 1 also has a well-defined glb operation, then P is a lattice.
7. Not all lattices are ranked. In particular, the lattice of integer partitions under dominance ordering is unranked.
8. Every interval in a lattice is a sublattice, but not every sublattice is an interval.
9. In the subspace lattice $L_n(q)$, the Whitney numbers (§11.1.3) satisfy the equation $N_k(L_n(q)) = \frac{(q^n-1)(q^{n-1}-1)...(q^{n-k+1}-1)}{(q^k-1)(q^{k-1}-1)...(q^1-1)}$.
10. In the subspace lattice $L_n(q)$, the Whitney number $N_k(L_n(q))$ equals the *Gaussian coefficient* $\begin{bmatrix} n \\ k \end{bmatrix}_q$ (§2.3.2), which appears in algebraic identities and in analogues of results on subsets.
11. The Whitney number $N_k(\Pi_n)$ of partitions of the set $[n]$ into $n - k$ blocks is the *Stirling subset number* $\left\{ {n \atop n-k} \right\}$ (§2.5.2). This has no closed formula, but the inclusion-exclusion principle yields $\left\{ {n \atop t} \right\} = \sum_{i=0}^{t}(-1)^i \frac{(t-i)^n}{t!}$.

Examples:

1. The poset specified by the following Hasse diagram is a lattice.

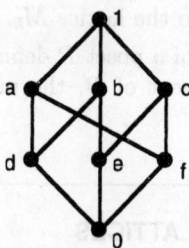

2. The poset specified by the following Hasse diagram is *not* a lattice. Although every pair of elements has a common upper bound, none of the three common upper bounds for the elements c and d is a least upper bound.

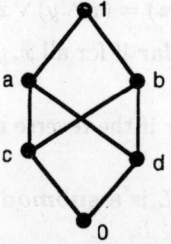

3. Two 5-element lattices that occur as subposets of $\underline{2}^3$ but not as sublattices of $\underline{2}^3$ are $M_5 = \underline{1} \oplus (\underline{1} + \underline{1} + \underline{1}) \oplus \underline{1}$ and $N_5 = \underline{1} \oplus (\underline{2} + \underline{1}) \oplus \underline{1}$.

4. In the divisor lattice $D(n)$, $a \wedge b = \gcd(a,b)$ and $a \vee b = \operatorname{lcm}(a,b)$.

5. The join-irreducible elements of the divisibility lattice $D(I)$ are the powers of primes.

6. In the subset lattice $\underline{2}^n$, $a \wedge b = a \cap b$ and $a \vee b = a \cup b$.

7. The join-irreducible elements of the subset lattice $\underline{2}^n$ are the singleton sets.

8. The subspace lattice $L_n(q)$ is a graded lattice, with rank $r(U) = \dim U$.

9. In the subspace lattice $L_n(q)$, the meet of subspaces U and V is their intersection $U \cap V$, and the join is the unique minimal subspace containing their union.

10. In the lattice of order ideals $J(P)$, glb and lub are given by intersection and union. Hence $J(P)$ is a sublattice of the Boolean lattice $\underline{2}^{|P|}$; equality holds if and only if P is an antichain.

11. The lattice $J(P)$ of ideals of a poset $P = (X, R)$ is finite, with $1_{J(P)} = X$ and $0_{J(P)} = \emptyset$. It is graded, with rank function $r(I) = |I|$.

12. By the correspondence between ideals of a poset $P = (X, R)$ and their antichains of maximal elements, the lattice $J(P)$ of ideals is also a lattice on the antichains of P. The corresponding ordering on antichains is $A \leq B$ if every element of A is less than or equal to some element of B.

13. The lattice $L(m,n)$ of bounded sequences is a sublattice of $\underline{n+1}^m$, and $L(m,n) = J(\underline{m} \times \underline{n}) = L(n,m)$. The natural isomorphism maps a sequence $a \in L(m,n)$ to the order ideal of $\underline{m} \times \underline{n}$ generated by $\{ (m+1-i, a_i) \mid a_i > 0 \}$.

14. The lattice $L(m,n)$ is a sublattice of the Young lattice.

15. In the partition lattice Π_n, $13|4|2|5 < 123|45$; the order of the blocks and the order of elements within each block are irrelevant.

16. The partition lattice Π_n is a graded poset, with $1_{\Pi_n} = [n]$ and $0_{\Pi_n} = 1|2|\cdots|n$. The common refinement of π and σ with the fewest blocks is the greatest lower bound (meet) of π and σ.

17. The lattice Π_3 is isomorphic to the lattice M_5.

18. In the ordering on antichains of a poset P defined by $A \leq B$ if every element of A is less than or equal to some element of B, the maximum antichains of P induce a sublattice.

11.1.5 DISTRIBUTIVE AND MODULAR LATTICES

Definitions:

A lattice L is **distributive** if glb distributes over lub in L, that is, if $x \wedge (y \vee z) = (x \wedge y) \vee (x \wedge z)$ for all $x, y, z \in L$.

A lattice L is **modular** if $x \wedge (y \vee z) = (x \wedge y) \vee z$ for all $x, y, z \in L$ such that $z \leq x$.

A lattice L is (**upper**) **semimodular** if for all $x, y \in L$, "x covers $x \wedge y$" implies "$x \vee y$ covers y".

A lattice L is **lower semimodular** if the reverse implication holds (equivalently, if the dual lattice L^* is semimodular).

The height function h of a lattice L is a **submodular height function** if $h(x \wedge y) + h(x \vee y) \leq h(x) + h(y)$ for all $x, y \in L$.

A lattice is **geometric** if it is atomic and semimodular has finite height.

A **closure operator** on the subsets of a set E is a function $\sigma: 2^E \to 2^E$ that maps each set to a superset of itself, is order-preserving with respect to set inclusion, and is "idempotent": $\sigma(\sigma(A)) = \sigma(A)$.

The **closed subsets** of a set E, with respect to a closure operator $\sigma: 2^E \to 2^E$, are the sets with $\sigma(A) = A$.

The **Steinitz exchange axiom** for a closure operator $\sigma: 2^E \to 2^E$ is the rule that $p \notin \sigma(A)$ and $p \in \sigma(A \cup q)$ imply $q \in \sigma(A \cup p)$.

Facts:

1. The smallest nondistributive lattices are $M_5 = \underline{1} \oplus (\underline{1} + \underline{1} + \underline{1}) \oplus \underline{1}$ and $N_5 = \underline{1} \oplus (\underline{2} + \underline{1}) \oplus \underline{1}$, which are illustrated in §11.1.1, Example 18.
2. A lattice is distributive if and only if it occurs as a sublattice of $\underline{2}^n$ for some n.
3. Every sublattice of a distributive lattice is a distributive lattice.
4. The product of distributive lattices L_1 and L_2 is a distributive lattice, with $(x_1, x_2) \wedge (y_1, y_2) = (x_1 \wedge y_1, x_2 \wedge y_2)$ and $(x_1, x_2) \vee (y_1, y_2) = (x_1 \vee y_1, x_2 \vee y_2)$.
5. In a lattice L, distributivity and the dual property that $x \vee (y \wedge z) = (x \vee y) \wedge (x \vee z)$ for all $x, y, z \in L$ are equivalent. Hence the dual of a distributive lattice is a distributive lattice.
6. A lattice L is modular if and only if $c \in [a \wedge b, a]$ implies $a \wedge (b \vee c) = c$ for all $a, b \in L$ (equivalently, if $c \in [b, b \vee d]$ implies $b \vee (c \wedge d) = d$ for all $b, d \in L$).
7. Let $\mu_a: L \to L$ be the operation "take the glb with a", and let $\nu_b: L \to L$ be the operation "take the lub with b". A lattice L is modular if and only if for all $a, b \in L$, the intervals $[a \wedge b, a]$ and $[b, a \vee b]$ are isomorphic sublattices of L, with lattice isomorphisms given by ν_b and μ_a.
8. If y covers x in a semimodular lattice L, then for all $z \in L$, $x \vee z = y \vee z$ or $x \vee z$ is covered by $y \vee z$.
9. A lattice L with a lower bound is semimodular if and only if the following is true: the height function of L is submodular and in each interval the maximal chains all have the same length.
10. A lattice is modular if and only if it does not have N_5 as a sublattice.
11. Every distributive lattice is modular, because in a distributive lattice $x \wedge (y \vee z) = (x \wedge y) \vee (x \wedge z) = (x \wedge y) \vee z$ if $z \leq x$.
12. A modular lattice is distributive if and only if it does not have M_5 as a sublattice.
13. Given a closure operator, the closed sets form a lattice under inclusion with meet and lub given by intersection and closure of the union, respectively.
14. If a closure operator σ satisfies the Steinitz exchange axiom, then the lattice of closed sets is semimodular.
15. The lattice $L_n(q)$ is semimodular. (This follows from the previous fact.)
16. A poset is a geometric lattice if and only if it is the lattice of closed sets of a matroid, ordered by inclusion (§12.4). (The span operator in a matroid, which adds to X every element whose addition to X does not increase the rank, is a closure operator that satisfies the Steinitz exchange axiom.)

17. A geometric lattice is distributive if and only if it has the form $\underline{2}^n$, and the corresponding matroid is the free matroid, in which all subsets of the elements are independent.

18. A complemented distributive lattice is a Boolean algebra.

Examples:

1. Among nondistributive lattices, the lattice M_5 is modular, and the lattice N_5 is not (which explains the notation).

2. The subspace lattices $L_n(q)$ are not distributive.

3. The partition lattice Π_n is semimodular but not modular for $n > 3$. The lattice $L_n(q)$ is semimodular.

4. The partition lattice Π_n is geometric, and it is the lattice of closed sets of the cycle matroid of the complete graph K_n.

5. For $n \geq 3$, the lattice $\Pi(n)$ is not distributive.

6. The Boolean lattice $\underline{2}^n$, the divisor lattice $D(N)$, the lattice $J(P)$ of order ideals of a poset, and the bounded sequence lattice $L(m,n)$ are distributive.

11.2 POSET PROPERTIES

11.2.1 POSET PARTITIONS

Definitions:

A **chain partition** of a poset is a partition of the domain of that poset into chains.

The **k-norm of a sequence** $x = \{x_i\}$ of real numbers is the sum $\sum_i \min\{k, x_i\}$, whose value is commonly denoted $m_k(x)$.

The **k-norm of a chain partition** $\mathbf{C}$ of a poset, denoted $m_k(\mathbf{C})$, means the k-norm of the sequence of sizes of the chains in the partition.

A **k-family** in a poset P is a subposet containing no chain of size $k+1$. The size of a maximum k-family in P is denoted by $d_k(P)$.

A partition of a poset P into chains is **k-saturated** if $m_k(\mathbf{C}) = d_k(P)$.

A chain in a ranked poset is **symmetric** if it has an element of rank $r(P) - k$ whenever it has an element of rank k.

A chain is **consecutive** if its elements belong to consecutive ranks.

A **symmetric chain decomposition** of P is a partition of P into symmetric consecutive chains.

A **symmetric chain order** is a poset with a symmetric chain decomposition.

A graded poset has the **Sperner property** if some single rank is a maximum antichain.

A graded poset has the **strict Sperner property** if all maximum antichains are single ranks.

A poset P has the **k-Sperner property** if it has a maximum k-family consisting of k ranks.

A poset has the ***strong Sperner property*** if it is k-Sperner for all $k \leq r(P)$.

A graded poset P satisfies the ***normalized matching property*** if for every k and every subset A of P_k, the set A^* of elements in P_{k+1} that are comparable to at least one element of A satisfies the inequality $\frac{|A^*|}{N_{k+1}} \geq \frac{|A|}{N_k}$, where N_k and N_{k+1} are Whitney numbers.

A ***regular covering by chains*** is a multiset of maximal chains such that for each $x \in P$ the fraction of the chains containing x is $\frac{1}{N_{r(x)}}$.

To obtain the ***bracket representation*** of a subset S of $[n] = \{1, \ldots, n\}$, first represent the subset S as a length-n "parenthesis-vector", in which the jth bit is a right parenthesis if $j \in S$ and a left parenthesis if $j \notin S$. Then wherever possible, recursively, match a left parenthesis to the nearest unmatched right parenthesis that is separated from it only by previously matched entries.

An ***on-line partitioning algorithm*** processes the elements of a poset as they are "revealed". Once an element is assigned to a cell, it remains there; there is no backtracking to change earlier decisions.

Facts:

1. *Dilworth's theorem*: If P is a finite poset, then the width of P equals the minimum number of chains needed to cover the elements of P.
2. Dilworth's theorem also holds for infinite posets of finite width.
3. The 1-families are the antichains.
4. Every k-family is a union of k antichains.
5. A k-family in P can be transformed into an antichain in $P \times \underline{k}$ of the same size, and vice versa, and hence $d_k(P) = w(P \times \underline{k})$.
6. The discussion of saturated partitions is generally restricted to finite posets.
7. If α_k is the number of chains of size at least k in a k-saturated chain partition of P, then $\Delta_k(P) \geq \alpha_k \geq \Delta_{k+1}(P)$, where $\Delta_k(P) = d_k(P) - d_{k-1}(P)$ for $k \geq 1$.
8. *Littlewood-Offord problem*: Let $A = \{a_1, \ldots, a_n\}$ be a set of vectors in $\mathcal{R}^d$, with each vector having length at least 1. Let $R_1, \ldots, R_k$ be regions in $\mathcal{R}^d$ of diameter at most 1. Then of the 2^n subsets of $A = \{a_i\}$, the number whose sum lies in $\bigcup_i R_i$ is at most $d_k(2^n)$.
9. *Greene-Kleitman (GK) theorem*: For every finite poset P and every $k \geq 0$, there is a chain partition of P that is both k-saturated and $(k+1)$-saturated.
10. The GK theorem is best possible, since there are infinitely many posets for which no chain partition is both k-saturated and l-saturated for any nonconsecutive nontrivial values for k, l; the smallest has 6 elements (illustration). The GK theorem extends in various ways to directed graphs.
11. Dilworth's theorem is the special case of the GK theorem for $k = 0$ (every chain partition is 0-saturated).
12. Every product of symmetric chain orders is a symmetric chain order.
13. The lattice of bounded sequences $L(m, n)$ (§11.1.4) has a symmetric chain decomposition if $\min\{m, n\} \leq 4$. It is not known whether $L(m, n)$ in general has a symmetric chain decomposition.

14. The lattice $L(m,n)$ has the Sperner property.

15. The partition lattice $\Pi(n)$ fails to satisfy the Sperner property if n is sufficiently large.

16. The Boolean lattice $\underline{2}^n$ and the subspace lattice $L_n(q)$ satisfy the strict Sperner property.

17. Every symmetric chain order has the strong Sperner property, and a symmetric chain decomposition is k-saturated for all k.

18. The class of graded posets that have the strong Sperner property and a symmetric unimodal sequence of Whitney numbers is closed under Cartesian product.

19. When $N_k \leq N_{k+1}$, the normalized matching property implies Hall's condition for the existence of a matching saturating P_k in the bipartite graph of the relations between the two levels.

20. Two subsets of the Boolean lattice in $[n]$ are on the same chain of the "bracketing decomposition" if and only if they have the same bracketing representation. This provides an explicit symmetric chain decomposition of $\underline{2}^n$. This generalizes for multisets $(D(N))$.

21. *Dedekind's problem*: This is the problem of computing the total number of antichains in the Boolean algebra $\underline{2}^n$. By using the bracketing decomposition, this number is calculated to be at most $3^{\binom{n}{\lfloor n/2 \rfloor}}$. Asymptotically, for even n, the number is

$$2^{\binom{n}{n/2}} e^{\binom{n}{n/2-1}[2^{-n/2} + n^2 2^{-n-5} - n 2^{-n-4}(1+o(1))]}.$$

The exact values for $n \leq 7$ are 3, 6, 20, 168, 7581, 7828354, and 2,414,682,040,998, with the estimate giving 7996118 for $n = 6$.

22. *Universal set sequences*: A universal set sequence on a set S is a sequence that contains every subset of S as a consecutive subsequence. The bracketing decomposition yields a universal set sequence on $[n]$ of length asymptotic to $\frac{2}{\pi} 2^n$.

23. If two sets, x and y, are chosen independently according to a probability distribution on the Boolean lattice $\underline{2}^n$, then the probability that x is contained in y is at least $\binom{n}{\lfloor n/2 \rfloor}^{-1}$.

24. There is an on-line algorithm that partitions posets of height k into $\binom{k+1}{2}$ antichains. This is best possible, even for 2-dimensional posets.

25. There is an on-line algorithm that partitions posets of width k into $\frac{5^k-1}{4}$ chains.

26. It is impossible to design an on-line algorithm partitions every poset of width k into fewer than $\binom{k+1}{2}$ chains.

27. There is an on-line algorithm to partition every poset of width 2 into 5 chains, and this is best possible.

Examples:

1. This is a symmetric chain decomposition of the Boolean lattice $\underline{2}^3$:

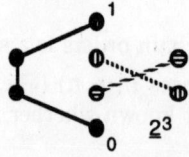

2. This is a regular chain covering of the Boolean lattice $\underline{2}^3$:

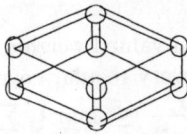

3. The poset $\underline{3} \times \underline{4}$ has a regular covering by six chains, using two chains twice and two other chains once each:

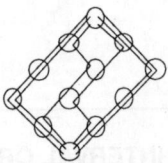

4. The poset $\underline{3} \times \underline{4}$ satisfies the Sperner property but not the strict Sperner property, since it has a maximum antichain (size three) that is not confined to a single rank.

11.2.2 LYM PROPERTY

Definitions:

The **LYM inequality** for a family F in a ranked poset P is the inequality $\sum_{x \in F} \frac{1}{N_{r(x)}} \leq 1$, where $N_{r(x)}$ is a Whitney number.

A poset P is an **LYM order** (or satisfies the **LYM property**) if every antichain $F \subseteq P$ satisfies the LYM inequality.

Facts:

1. The LYM property was discovered independently for $\underline{2}^n$ by Lubell, Yamamoto, and Meshalkin.

2. The LYM property, the normalized matching property, and the existence of a regular covering by chains are equivalent.

3. The LYM property implies the Sperner property and also implies the strong Sperner property (but not the strict Sperner property).

4. Every LYM order that has symmetric unimodal Whitney numbers has a symmetric chain decomposition. In particular, $L_n(q)$ is a symmetric chain order.

5. It is not known whether every LYM poset has a chain decomposition that is k-saturated for all k.

6. A product of LYM orders may fail the LYM property.

7. A product of LYM orders whose sequence of Whitney numbers is log-concave is an LYM order with a log-concave sequence of Whitney numbers. (A sequence $\{a_n\}$ is log-concave if $a_n^2 \geq a_{n-1} a_{n+1}$ for all n.)

8. The divisor lattice $D(N)$ is an LYM order, which follows from the previous fact.

9. The partition lattice $\Pi(n)$ if an LYM order if and only if $n < 20$.

10. The Boolean lattice $\underline{2}^n$ and the subspace lattice $L_n(q)$ have regular coverings by chains and hence are LYM orders.

11. If $\{\lambda_x\}$ is an assignment of real-valued weights to the elements of an LYM poset P, then for every subset $G \subset P$ and every regular covering $\mathbf{C}$ of P,

$$\sum_{x \in G} \frac{\lambda_x}{N_{r(x)}} \leq \max_{C \in \mathbf{C}} \{ \sum_{y \in C \cap G} \lambda_y \}.$$

Example:

1. The lattice $L(m,n)$ of bounded sequences is an LYM order if and only if $\min\{m,n\} \leq 2$ or $(m,n) = (3,3)$.

11.2.3 RANKINGS, SEMIORDERS, AND INTERVAL ORDERS

A chain names the "better" of any pair according to a single scale. Realistically, some comparisons may yield indifference. Several families of "chain-like" partial orders successively relax the requirements on indifference.

Definitions:

A poset P is a **ranking** or **weak order** if its elements are partitioned into ranks $P_1, \ldots, P_k$ such that two elements are incomparable if and only if they belong to the same rank.

A poset P is a **semiorder** if there is a real-valued function f and a fixed real number $\delta > 0$ (δ may be taken to be 1) such that $x < y$ if and only if $f(y) - f(x) > \delta$. The pair (f, δ) is a **semiorder representation** of the poset P.

A poset P is an **interval order** if there is an assignment of real intervals to its members such that $x < y$ if and only if the interval for y is totally to the right of the interval for x. The collection of intervals is called an **interval representation** of the poset P.

A **biorder representation** on a digraph D is a pair of real-valued functions f, g on the vertex set V_D such that $u \to v$ is an arc if and only if $f(u) > g(v)$.

A **Ferrers digraph** (or **Ferrers relation** or **biorder**) is a digraph having a biorder representation. (Also see §2.5.1.)

Facts:

1. Rankings model a single criterion of comparison with "ties" allowed, as in voting.

2. A poset is a ranking if and only if its comparability graph is a complete multipartite graph.

3. A ranking assigns a score $f(z)$ to each element z such that $x < y$ if and only if $f(x) < f(y)$.

4. The forbidden subposet characterization of a ranking is $\underline{1} + \underline{2}$.

5. Semiorders were introduced to model intransitivity of indifference; a difference of a few grains of sugar in a coffee cup or a few dollars in the price of a house is not likely to affect one's attitude, but pounds of sugar or thousands of dollars will. The threshold δ in a semiorder representation indicates a "just-noticeable difference".

6. A poset is a semiorder if and only if its incomparability graph is a unit interval graph, that is, an interval graph (§8.1.3) such that all intervals are of unit length.

7. An interval representation of a semiorder P with semiorder representation (f, δ) is obtained by setting $I_x = [f(x) - \frac{\delta}{2} + \epsilon, f(x) + \frac{\delta}{2} - \epsilon]$.

8. *Scott-Suppes theorem*: The forbidden subposet characterization of a semiorder is $\{\underline{1} + \underline{3}, \underline{2} + \underline{2}\}$.

9. The number of nonisomorphic semiorders on an n-element set is the Catalan number $C_n = \frac{1}{n+1}\binom{2n}{n}$ (§3.1.3).

10. Interval orders model a situation where the value assigned to an element is imprecise.

11. The incomparability graph of an interval order is an interval graph.

12. Every poset whose incomparability graph is an interval graph is an interval order. This follows from the forbidden subposet characterization of interval orders.

13. The forbidden subposet characterization of an interval order is $\underline{2} + \underline{2}$.

14. A poset P is an interval order if and only if both the collections of "upper holdings" $U(x) = \{y \in P \mid y > x\}$ and "lower holdings" $D(x) = \{y \in P \mid y < x\}$ form chains under inclusion, in which case the number of distinct nonempty upper holding sets and distinct nonempty lower holding sets is the same. Construction of these chains yields a fast algorithm to compute a representation for an interval order or semiorder.

15. The strict comparability digraph of an interval order is a Ferrers digraph, with $f(x)$ and $g(x)$ denoting the left and right endpoints of the interval assigned to x. This is the strict comparability digraph of a poset because $f(x) \leq g(x)$ for all x. The "upper holdings" and "lower holdings" for an interval order become predecessor and successor sets for a Ferrers digraph.

16. For a digraph D with adjacency matrix $A(D)$, the following are equivalent:
    - D has a biorder representation (and is a Ferrers digraph);
    - $A(D)$ has no 2 by 2 submatrix that is a permutation matrix;
    - the successor sets of D are ordered by inclusion;
    - the predecessor sets of D are ordered by inclusion;
    - the rows and columns of $A(D)$ can be permuted independently so that to the left of a 1 is a 1.

17. The greedy algorithm is an optimal on-line algorithm for partitioning an interval order into the minimum number of antichains. It uses at most $2h - 1$ antichains to recursively partition an interval order of height h, and this is best possible.

18. There is an on-line algorithm to partition every interval order of width k into $3k - 2$ chains, and this is best possible. Equivalently, the maximum number of colors needed for on-line coloring of interval graphs with clique size k is $3k - 2$.

19. No on-line partitioning algorithm colors all trees with a bounded number of colors.

20. *"Universal" interval orders*: Since the ordering of the interval endpoints is all that matters, interval representations may be restricted to have integer endpoints. The poset $I[0, n]$ or $\mathbf{I}_n$ denotes the interval order whose interval representation consists of all intervals with integer endpoints in $\{0, \ldots, n\}$.

21. Every finite interval order is a subposet of some $I[0, n]$.

Examples:

1. This Hasse diagram represents a poset that is a ranking. Its three ranks are indicated by the levels in the diagram.

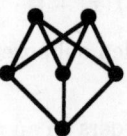

2. This Hasse diagram represents a poset that is a semiorder: for instance, with $\delta = 1$, define $f(a) = 2$, $f(b) = 1.3$, $f(c) = 0.8$, and $f(d) = 0$. It is not a ranking, by Fact 4, because $\underline{1} + \underline{2}$ is a subposet. The interval representation of its incomparability graph is at the right.

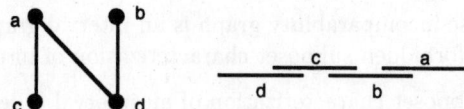

3. This Hasse diagram represents a poset that is an interval order. The interval representation of its incomparability graph is at the right. By Fact 8, it is not a semi-order.

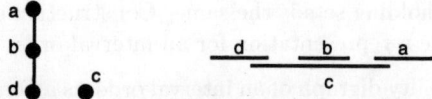

4. The skill of a tennis player may vary from day to day, leading to use of an interval $[a_x, b_x]$ to represent player x. In this case player x always beats player y if $a_x > b_y$.

5. The interval order $I[0,3]$ is not a semiorder.

11.2.4 APPLICATION TO SOCIAL CHOICE

When there are more than two candidates for a public office, it is not obvious what is the "best" way to select a winner. Any rule has its pluses and minuses, from the standpoint of public policy. Social choice theory analyzes the effect of various rules for deciding the outcomes of preferential rankings.

Definitions:

A **profile** on a set A of "alternatives" (e.g., candidates for a public office) is a set $P = \{P_i \mid i \in I\}$ of linear rankings (ties allowed) of A, one for each member of a set I of "individuals" (e.g., voters).

A **consensus function** (or **social choice function**) is a function ϕ that assigns to each possible profile $P = \{P_i \mid i \in I\}$ on a set A of alternatives a linear order (ties allowed) of A called the **consensus ranking** for P.

A consensus function upholds **majority rule** provided that it prefers x to y if and only if a majority of the individuals prefer x to y.

Plurality is the consensus function in which the winner(s) is(are) the alternative(s) appearing in the greatest number of top ranks, after which the winner(s) is(are) deleted and the procedure is repeated to select the next rank of the consensus ranking, etc.

The **Borda count** of an alternative x is the sum, over individual rankings, of the number of alternatives x "beats". The resulting **Borda consensus function** ranks the alternatives by their Borda count.

Facts:

1. Plurality can elect some ranked last by a majority.

2. *Condorcet's paradox*: Some profiles have no decisive consensus (i.e., producing a single winner) that upholds majority rule.

3. The Borda count is subject to abuse.

4. *Arrow's impossibility theorem*: No consensus function exists that satisfies the following four axioms, which were formulated in an attempt to develop a consensus function ϕ that avoids the difficulties cited in the facts above:

  - *monotonicity*: If $a > b$ in $\phi(P)$ and if profile P' agrees with profile P except for moving alternative a upward in some or all rankings, then $a > b$ in $\phi(P')$.
  - *independence of irrelevant alternatives*: If profiles P and P' agree within a set $A' \subseteq A$, then $\phi(P)$ and $\phi(P')$ have the same restriction to A'. This axiom implies that votes for extraneous alternatives do not affect the determination of the consensus ranking among the alternatives within the subset A'.
  - *nondegeneracy*: Given $a, b \in A$, there is a profile P such that $a > b$ in $\phi(P)$. This axiom implies that the structure of the outcome is independent of renaming the alternatives.
  - *nondictatorship*: There is no $i \in I$ such that $a > b$ in P_i implies $a > b$ in $\phi(P)$.

Examples:

1. Suppose that $A = \{a, b, c\}$ is the set of alternatives, and suppose that the profile consists of the three rankings $a > b > c$, $c > a > b$, and $b > c > a$. Then for each alternative, there is another alternative that is preferred by $\frac{2}{3}$ of the population.

2. The U.S. presidential election of 1912 had three candidates: Wilson (W), Roosevelt (R), and Taft (T). It is estimated that 45% of the voters ranked $W > R > T$, that 30% ranked $R > T > W$, and that 25% ranked $T > R > W$. Wilson won the election, garnering a plurality of the popular vote, but a majority of the population preferred Roosevelt to Wilson. Moreover, 55% regarded Wilson as the least desirable candidate.

3. Consider a close election, with four individuals preferring x to y to all other alternatives. A fifth individual prefers y to x. If there are enough other alternatives, the fifth individual can throw a Borda-count election to y by placing x at the bottom.

11.2.5 LINEAR EXTENSIONS AND DIMENSION

By adding additional comparison pairs to a partial ordering on a set, ultimately a total ordering is obtained. Each of the many ways to do this is called an extension of the original partial ordering.

Definitions:

An **extension** of a poset $P = (X, R)$ is a poset $Q = (X, S)$ such that $R \subseteq S$ (i.e., xRy implies xSy).

A **linear extension** of a poset P is an extension of P that is a chain.

A **topological sort** is an algorithm that accepts a finite poset as input and produces a linear extension of that poset as output.

A **topological ordering** of an acyclic digraph is a linear extension of the poset arising from it.

The **intersection of two partial orderings** $P = (X, R)$ and $Q = (X, S)$ on the same set X is the poset $(X, R \cap S)$ that includes the relations common to both.

A **realizer** of a poset P is a set of linear extensions of P whose intersection is P.

The **order dimension** (or **dimension**) of P, written $\dim(P)$, is the minimum cardinality of a realizer of P.

The **standard example** S_n of an n-dimensional poset is the subposet of $\underline{2}^n$ induced by the singletons and their complements.

An **alternating k-cycle** in a poset P is a sequence of ordered incomparable pairs $\{(x_i, y_i)\}_{i=1}^{k}$ such that $y_i \leq x_{i+1}$, where subscripts are taken modulo k.

A **critical pair** (or **unforced pair**) in a poset P is an ordered incomparable pair that cannot be made comparable by adding any other single incomparable pair as a relation.

A linear extension L of a poset P **puts Y over X**, where X and Y are disjoint subposets, if y is above x in L whenever (x, y) is an incomparable pair with $x \in X$, $y \in Y$.

Given a subposet $Q \subseteq P$, an **upper extension** of Q is a linear extension of P that puts $P - Q$ over Q.

Given a subposet $Q \subseteq P$, a **lower extension** of Q is a linear extension of P that puts $P - Q$ below Q.

The **minimum realizer encoding** of a poset lists for each element its position on each extension in a minimum realizer.

The **probability space** on the set of all linear extensions of a (finite) poset P is obtained by taking each linear extension to be equally likely. The notation $Pr(x < y)$ denotes the proportion of linear extensions in which element x comes below element y.

Facts:

1. Every poset is the intersection of all its linear extensions, from which it follows that the concept of dimension is well-defined.

2. Given incomparable elements x and y in a poset P, there is a linear extension of P in which x appears above y.

3. The chains are the only posets of dimension 1.

4. Every antichain has dimension 2, because the intersection of a linear order and its dual is an antichain.

5. Topological sort is used to organize activities with a precedent ordering into a sequential schedule.

6. The list of minimal forbidden subposets for dimension 2 consists of 10 isolated examples and 7 one-parameter families.

7. If Q is a subposet of P, then $\dim(Q) \leq \dim(P)$.

8. The dimension of a product of k chains (each of size at least 2) is k.

9. A poset has dimension at most k if and only if it imbeds in a product of k chains.

10. The dimension of a poset P equals the minimum integer n such that P is a subposet of $\mathcal{R}^n$.

Algorithm 1: Topological sort.

input: a finite poset $(X = \{x_1, \ldots, x_n\}, \leq)$
output: a compatible total ordering $A = x_1 \leq x_2 \leq \cdots \leq x_n$ of the elements of X
for $j := 1$ to n
  $x_j :=$ a minimal element of X
  $X := X - \{x_j\}$

11. The standard example S_n is a bipartite poset whose comparability graph is obtained from the complete bipartite graph $K_{n,n}$ by deleting a complete matching.

12. The minimum realizer encoding of an n-element poset of dimension k takes only $O(kn \log n)$ bits, instead of the $O(n^2)$ bits of the order relation. Thus, posets of small dimension have concise representations.

13. In the sense of Fact 12, the dimension of a poset may be regarded measure of its "space complexity".

14. The dimension of a poset P equals the minimum number of linear extensions containing all the critical pairs of P.

15. The dimension of a poset P is equal to the chromatic number of the hypergraph whose vertex set is the set of critical pairs and whose edges are the sets of critical pairs forming minimal alternating cycles.

16. The cover graph of the standard example S_n (of an n-dimensional poset) is $K_{n,n}$-(1-factor).

17. If X and Y are disjoint subposets of a poset P, then P has a linear extension L putting Y over X if and only if P contains no $\underline{2} + \underline{2}$ with minimal elements in Y and maximal elements in X.

18. $\dim(P) \leq w(P)$. A realizer of size $w(P)$ can be formed by taking upper extensions of the chains in a partition of P into $w(P)$ chains.

19. $\dim(P) \leq \frac{|P|}{2}$. The standard example S_n shows that this bound is the best possible.

20. *One-point removal theorem:* For every $x \in P$, $\dim(P) \leq 1 + \dim(P - x)$.

21. For every poset P, there exist four elements $\{x, y, z, w\}$ such that $\dim(P) \leq 2 + \dim(P - \{x, y, z, w\})$. It is conjectured that, for every poset P, there exist two elements $\{x, y\}$ such that $\dim(P) \leq 1 + \dim(P - \{x, y\})$.

22. A poset has dimension 2 if and only if the complement of its comparability graph is also a comparability graph; thus there is a polynomial time algorithm to decide whether a poset has dimension 2. However, recognizing posets of dimension k is NP-complete for every fixed k at least 3.

23. If P is a finite poset that is not a chain, then P has a pair of elements x, y such that
$$0.2764 \simeq \tfrac{1}{2} - \tfrac{1}{2\sqrt{5}} \leq Pr(x < y) \leq \tfrac{1}{2} + \tfrac{1}{2\sqrt{5}} \simeq 0.7236.$$

24. *The $\tfrac{1}{3}$-$\tfrac{2}{3}$ conjecture:* This conjecture states that there is always a pair of elements, x and y, such that $\tfrac{1}{3} \leq Pr(x < y) \leq \tfrac{2}{3}$.

25. The traditional name *topological sort* (Algorithm 1) is commonly used in applications. However, a topological sort is *not* a sort in the standard meaning of that word. Nor is it directly related to what mathematicians call *topology*.

Examples:

1. The following poset has the linear extensions abc and acb and it is the intersection of these extensions. Thus its dimension is 2.

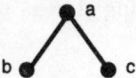

2. The following poset has six linear extensions: $abcd$, $acbd$, $acdb$, cad, and $cdab$. Since it is the intersection of $abcd$ and $cdab$, its dimension is 2.

3. The bipartite poset S_3 whose comparability graph and cover graph is the 6-cycle $1, \bar{2}, 3, \bar{1}, 2, \bar{3}$ has dimension 3. The realizer $\{23\bar{1}1\bar{2}\bar{3}, 13\bar{2}2\bar{1}\bar{3}, 12\bar{3}3\bar{1}\bar{2}\}$ establishes the upper bound. Every realizer must have an extension with $\bar{1}$ below 1, one with $\bar{2}$ below 2, and one with $\bar{3}$ below 3. No two of these can occur in the same linear extension, so the dimension is at least three.

4. More generally, for the elements $i \in [n]$ of the standard example S_n, a realizer must include distinct linear extensions in which the singleton $\{i\}$ appears above its complement, and any n such extensions suffice.

5. For the standard example S_n of dimension n, the critical pairs are $\{\bar{i}, i\}$; this reflects the fact that, in a realizer, the extensions need to put i above $\bar{i}$, for each i. Each pair of critical pairs forms a minimal alternating cycle. Viewing the minimal alternating cycles as edges creates a hypergraph, namely the complete graph K_n, with chromatic number n.

6. Let N be the bipartite poset with minimal elements a and b and maximal elements c and d, in which a lies below c, and b lies below c and d. This poset has five linear extensions, namely $a < b < c < d$, $a < b < d < c$, $b < a < c < d$, $b < a < d < c$, and $b < d < a < c$. Thus $Pr(a < b) = \frac{2}{5}$.

7. *Application of posets to sorting*: The objective of a sort is to arrange the elements of a set X into a sequence by posing sequential queries of the form: "is $x < y$ true?". At any time, the state of cumulative knowledge is representable by a poset $P = (X, R)$, such that the linear extensions of P are remaining candidates for the final sequence order. A desirable query substantially reduces the number of candidates for extensions no matter whether the answer is yes or no, most especially finding a pair, x and y, such that $Pr(x < y)$ is close to $\frac{1}{2}$. Thus, Fact 22 shows that the worst case time to sort, in the presence of partial information given by a poset P, is $\Omega(\log P)$.

8. *Application of posets to searching*: The objective of searching a poset P in which item $s(x)$ is stored at location x is to determine whether a target item α is present in P. Each step of the search probes a location and compares its value against the target item. The worst case requires determining for each $x \in P$ whether the item at location x is greater or less than α, so the searching problem is the problem of identifying the downset $D_\alpha = \{x \in P \mid s(x) < \alpha\}$. A probe of location x splits the remaining possible downsets into those that contain x and those that do not. The former remain as candidates if $s(x) < \alpha$; the latter remain if $s(x) > \alpha$. A hypothetical adversary would arrange the value $s(x)$ so that the response would leave the larger portion of the ideals. Thus, the number $c(P)$ of probes required in the worst case is at least $\lceil \log_2 i(P) \rceil$, where $i(P)$ denotes the number of ideals in poset P.

11.2.6 POSETS AND GRAPHS

From the graph-theoretic viewpoint, a comparability graph is by definition a simple graph (§8.6.3) that has a transitive orientation. Comparability graphs are perfect graphs, which motivates most study of comparability graphs.

Definitions:

A *transitive orientation* on a simple graph G is an assignment of directions to the edges so that whenever there is an xy-arc and a yz-arc, there is also an xz-arc.

A *quasi-transitive orientation* on a simple graph G is an assignment of directions to the edges so that whenever there is an xy-arc and a yz-arc, there is also an arc between x and z.

A *triangular chord* for a walk $x_1, \ldots, x_k$ in an undirected graph G is an edge between vertices x_{i-1} and x_{i+1}, two apart on the walk.

The *auxiliary graph* for a simple graph G is the graph G' whose vertices are the edges of G, with vertex e_1 adjacent to vertex e_2 in G' if and only if edges e_1 and e_2 are adjacent in graph G but do not lie on a cycle.

A *module* in a graph G is a vertex subset U such that each vertex outside U is adjacent to all or none of the vertices in U.

An *order module* in a poset P (or *autonomous set*) is a set S of elements such that every element outside S is above all of S, below all of S, or incomparable to all of S.

A *comparability invariant* for posets is an invariant f such that $f(P) = f(Q)$ whenever posets P and Q have the same comparability graph.

A *permutation graph* is a graph whose vertices can be placed in 1-1 correspondence with the elements of a permutation of $[n] = \{1, \ldots, n\}$ such that v_i is adjacent to v_j if and only if the larger of i and j comes first in the permutation.

A *u,v-bypass* in a directed graph is a u,v-path of length at least two such that there is also an arc from u to v. v

A *dependent edge* in an acyclic directed graph is an arc from u to v such that the graph contains a u,v-bypass.

Facts:

1. For a simple graph G, the following are equivalent:
    - G has a transitive orientation;
    - G has a quasi-transitive orientation;
    - every closed odd walk of G has a triangular chord;
    - the auxiliary graph G' is bipartite.

The implications from top to bottom are straightforward, as is the proof that if the auxiliary graph G' is bipartite then G has a quasi-transitive orientation. The proof that if G has a quasi-transitive orientation then G has a transitive orientation takes more work. The last characterization gives an algorithm to decide whether G is a comparability graph in $O(n^3)$ time, where n is the number of vertices. The proof is constructive, so a transitive orientation can also be obtained.

2. In any graph, the set of all vertices, the singleton sets of vertices, and the empty set are always modules.

3. Modules yield a forbidden subgraph characterization of comparability graphs. The minimal forbidden induced subgraphs consist of eight infinite families and ten special examples.

4. If two partial orders have the same comparability graph, then one can be transformed into the other by a sequence of moves involving reversing all the relations inside an order module S, i.e., by replacing the partial order induced on S by its dual, and preserving all relations between S and its complement.

5. Let f be a poset invariant such that $f(P) = f(P^*)$ for all posets P, and such that, if poset Q is obtained from P by replacing a module in P with another module having the same value of f, then $f(Q) = f(P)$. Then the invariant f is a comparability invariant.

6. Height, width, dimension, and number of linear extensions are all comparability invariants.

7. A graph is the complement of a comparability graph if and only if it is the intersection graph of the curves representing a collection of continuous real-valued functions on $[0, 1]$.

8. The following conditions are equivalent for a graph G:
   - G is a permutation graph (adjacency representing the inversions of a permutation);
   - $\overline{G}$ is the comparability graph of a 2-dimensional partial order;
   - G and $\overline{G}$ are comparability graphs.

9. Isomorphism of permutation graphs be tested in $O(n^2)$ time. Some NP-complete scheduling problems become polynomial when the poset of precedence constraints is 2-dimensional.

10. A directed graph corresponds to the diagram of some partial order if and only if it contains no cycles or bypasses.

11. Every graph that is the cover graph of some poset is triangle-free.

12. If a graph has chromatic number less than its girth, then it is the cover graph of some poset. In particular, a 3-chromatic graph is a cover graph if and only if it is triangle-free.

13. It is NP-complete to decide whether a 4-chromatic graph is a covering graph.

14. The smallest triangle-free graph that is not a cover graph is the 4-chromatic Grötzsch graph with 11 vertices.

15. The maximum number of dependent edges among the orientations of a graph G is equal to the cycle rank $\beta(G) = |E| - |V| + 1$.

16. If a graph G has chromatic number less than its girth (§8.7.2), then for all i such that $0 \leq i \leq \beta(G)$, the graph has an acyclic orientation with exactly i dependent edges.

17. The cover graph of a modular lattice is bipartite.

18. A modular lattice is distributive if and only if its cover graph does not contain the complete bipartite graph $K_{2,3}$.

19. The subgraph of the cover graph of $\underline{2}^{2k+1}$ induced by the k-sets and $k + 1$-sets is a vertex-transitive $k + 1$-regular bipartite graph. The graph is known to contain cycles using more than 80% of its vertices. The *Erdős revolving door conjecture* asserts that this graph is Hamiltonian.

20. *Gallai-Milgram theorem*: The vertices of a digraph D can be covered using at most $\alpha(D)$ disjoint paths, where $\alpha(D)$ is the maximum size of an independent set in D.

21. Dilworth's theorem (§11.2.1) is the special case of the Gallai-Milgram theorem for comparability digraphs.

Examples:

1. A transitive orientation for a bipartite graph can be obtained by assigning all the edge directions from one part to the other, as shown here for $K_{3,3}$:

2. An odd cycle of length ≥ 5 has no quasi-transitive orientation (see Fact 1).

3. Inserting a triangular chord into a 5-cycle permits the resulting graph to have a transitive orientation, as shown in the following figure:

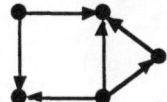

4. The following figure shows a graph and its auxiliary graph:

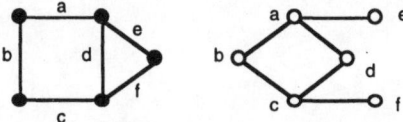

5. Any subset of either part of a complete bipartite graph is a module, since the other vertices in its part are not adjacent to any vertex in the module, and the vertices in the other part of the bipartition are each adjacent to all the vertices in the module.

6. Deleting a 1-factor from $K_{n,n}$, for $n \geq 3$, yields a graph with no module other than the complete set of vertices, the singletons, and the empty set.

REFERENCES

Printed Resources:

[An87] I. Anderson, *Combinatorics of Finite Sets*, Oxford University Press, 1987.

[Bi67] G. Birkhoff, *Lattice Theory*, 3rd ed, American Mathematical Society, 1967.

[Bo86] B. Bollobás, *Combinatorics: Set Systems, Hypergraphs, Families of Vectors, and Combinatorial Probability*, Cambridge University Press, 1986.

[BoMo94] V. Bouchitté and M. Morvan eds., *Orders, Algorithms and Applications*, Springer, 1994.

[DaPr90] B. A. Davey and H. A. Priestley, *Introduction to Lattices and Order*, Cambridge University Press, 1990.

[Fi85] P. C. Fishburn, *Interval Orders and Interval Graphs: A Study of Partially Ordered Sets*, Wiley, 1985.

[Ri89] I. Rival, ed., *Algorithms and Order*, Kluwer Academic Press, 1989.

[Ri86] I. Rival, ed., *Combinatorics and Ordered Sets*, Contemporary Mathematics vol. 57, 1986.

[Ri85] I. Rival, ed., *Graphs and Order*, Reidel Publishing, 1985.

[Ri82] I. Rival, ed., *Proceedings of the Symposium on Ordered Sets*, Reidel Publishing, 1982. (Contains a fairly complete bibliography of publications on partial orders up to 1981.)

[Ro99] K. H. Rosen, *Discrete Mathematics and Its Applications*, 4th ed., WCB/McGraw-Hill, 1999.

[Tr92] W. T. Trotter, *Combinatorics and Partially Ordered Sets: Dimension Theory*, The Johns Hopkins University Press, 1992.

12

COMBINATORIAL DESIGNS

12.1 Block Designs
    12.1.1 Balanced Incomplete Block Designs   Charles J. Colbourn
    12.1.2 Isomorphism and Automorphism   and Jeffrey H. Dinitz
    12.1.3 Subdesigns
    12.1.4 Resolvable Designs
    12.1.5 t-designs and Steiner Systems
    12.1.6 Pairwise Balanced Designs
    12.1.7 Group Divisible Designs and Transversal Designs

12.2 Symmetric Designs and Finite Geometries   Charles J. Colbourn
    12.2.1 Finite Geometries   and Jeffrey H. Dinitz
    12.2.2 Symmetric Designs
    12.2.3 Projective and Affine Planes
    12.2.4 Hadamard Designs and Matrices
    12.2.5 Difference Sets

12.3 Latin Squares and Orthogonal Arrays   Charles J. Colbourn
    12.3.1 Latin Squares   and Jeffrey H. Dinitz
    12.3.2 Mutually Orthogonal Latin Squares
    12.3.3 Orthogonal Arrays

12.4 Matroids   James G. Oxley
    12.4.1 Basic Definitions and Examples
    12.4.2 Alternative Axiom Systems
    12.4.3 Duality
    12.4.4 Fundamental Operations
    12.4.5 Characterizations
    12.4.6 The Greedy Algorithm

INTRODUCTION

In broad terms, the study of combinatorial designs is the study of the structure of collections of subsets of a finite set when these collections of subsets satisfy certain prescribed properties. In particular, a block design has the property that every one of these subsets has the same size k and every *pair* of points in the set is in exactly the same number of these subsets. Latin squares are also fundamental in this area and can be thought of in this context, but they are commonly thought of as $n \times n$ arrays with the property that each cell contains one element from an n-set and each row and each column contain each element exactly once. Some of the questions of general interest include: existence of designs, enumeration of nonisomorphic designs, and the study of subdesigns of designs.

Matroids generalize a variety of combinatorial objects, such as matrices and graphs. These structures arise naturally in a variety of combinatorial contexts and provide a framework for the study of many problems in combinatorial optimization and graph theory.

Much of the information in §12.1–12.3 is condensed from [CoDi96], which provides a comprehensive treatment of combinatorial designs. The main source for material in §12.4 is [Ox92].

GLOSSARY

affine plane: a set of points and a set of subsets of points (called lines) such that every two points lie on exactly one line, if a point does not lie on a line L there is exactly one line through the point that does not intersect L, and there are three points that are not collinear.

affine space (of dimension n): the set $AG(n,q)$ of all cosets of subspaces of an n-dimensional vector space over a field of order q.

automorphism (a design D): an isomorphism from D onto D.

balanced incomplete block design (BIBD): given a finite set X (of *points*), a collection of subsets (called *blocks*) of X of the same size such that every point belongs to the same number of blocks, and each pair of points belongs to the same number of blocks. The BIBD is described by five parameters: size of X, number of blocks, number of blocks to which every element of X belongs, size of each block, and number of blocks to which each pair of distinct points belongs.

basis (for a matroid): a maximal independent set in the matroid.

basis axioms: a set of axioms that specifies the set of bases of a matroid.

binary matroid: a matroid that is isomorphic to a vector matroid of a matrix over the field $GF(2)$.

biplane: symmetric design in which every pair of distinct points belongs to exactly two blocks.

block: each of the subsets in a design.

circuit: a minimal dependent set in a matroid.

circuit axioms: a set of axioms that specifies the set of circuits of a matroid.

closed set: in a matroid, a subset of its ground set that is equal to its closure.

closed under duality: property of a class of matroids that the dual of a matroid in the class is also in the class.

closure (of a subset of the ground set in a matroid): given a subset X of the ground set E in a matroid, the set of all points $x \in E$ such that the rank of $X \cup \{x\}$ is equal to the rank of X.

closure axioms: a set of axioms that specifies the properties that a closure operator of a matroid must have.

closure operation: the mapping $K \to \mathbf{B}(K)$, where K is a set of positive integers and $\mathbf{B}(K)$ the set of positive integers v for which there exists a (v, K)-PBD.

cobasis (of a matroid): a basis of the dual of a matroid.

cocircuit (of a matroid): a circuit of the dual of a matroid.

cographic matroid: a matroid isomorphic to the cocyle matroid of a graph.

coindependent set (of a matroid): an independent set of the dual of a matroid.

coloop (of a matroid): a loop of the dual of a matroid.

combinatorial geometry: a simple matroid.

complete set of mutually orthogonal latin squares: a set of $n - 1$ mutually orthogonal latin squares of side n.

conjugate: Let L be an $n \times n$ latin square on symbol set E_3, with rows indexed by the elements of the n-set E_1 and columns indexed by the elements of the n-set E_2. Let $\mathcal{T} = \{(x_1, x_2, x_3) \mid L(x_1, x_2) = x_3\}$. Let $\{a, b, c\} = \{1, 2, 3\}$. The (a, b, c)-conjugate of L, $L_{(a,b,c)}$, has rows indexed by E_a, columns by E_b, and symbols by E_c, and is defined by $L_{(a,b,c)}(x_a, x_b) = x_c$ for each $(x_1, x_2, x_3) \in \mathcal{T}$.

connected: property of a matroid that it cannot be written as the direct sum of two nonempty matroids.

cycle matroid (of a graph): the matroid on the edge-set of the graph whose circuits are the cycles of the graph.

t-design: a t-(v, k, λ) design.

t-(v, k, λ) design: a pair $(X, \mathcal{A})$ where X is a set of v elements (points), $\mathcal{A}$ is a family of k-subsets (blocks) of X, and every t-subset of X occurs in exactly λ blocks.

development (of a difference set D): the incidence structure $\text{dev}(D)$ whose points are the elements of G and whose blocks are the translates $D + g = \{d + g \mid d \in D\}$.

dual (of an incidence structure): the incidence structure obtained by interchanging the roles of points and lines.

dual (of a matroid): given a matroid M, the matroid on the same set as M whose bases are the complements of the bases of M.

equivalent (latin squares): Two latin squares L and L' of side n are equivalent if there are three bijections, from the rows, columns, and symbols of L to the rows, columns, and symbols, respectively, of L', that map L to L'.

Fano plane (or ***projective plane of order 2***): the $(7, 7, 3, 3, 1)$ design with point set $X = \{0, \ldots, 6\}$ and the block set $\mathcal{A} = \{013, 124, 235, 346, 450, 561, 602\}$.

flat: closed set.

t-flat: a subspace of projective dimension t of a projective space; a coset of a subspace of affine dimension t of an affine space.

k-GDD: a group divisible design with $\lambda = 1$ and $K = \{k\}$.

graphic matroid: a matroid that is isomorphic to the cycle matroid of some graph.

ground set: the set of points of a matroid.

group divisible design (or (K, λ)-**GDD**): given an integer λ and a set of positive integers K, a triple $(X, \mathcal{G}, \mathcal{A})$ where X is a set (of *points*), $\mathcal{G}$ is a partition of X into at least two subsets (called *groups*), $\mathcal{A}$ is a family of subsets of X (called *blocks*) such that: if A in $\mathcal{A}$, then $|A| \in K$, a group and a block contain at most one common point, and every pair of points from distinct groups occurs in exactly λ blocks.

group-type (or **type**): for a group divisible design, the multiset $\{|G| : G \in \mathcal{G}\}$.

Hadamard design: a symmetric $(4n - 1, 2n - 1, n - 1)$ design.

Hadamard matrix: an $n \times n$ matrix H with all entries ± 1 that satisfies $H^{\mathrm{T}} H = nI$.

hyperplane: a subspace of projective dimension $n-1$ of projective space of projective dimension n; a coset of a subspace of affine dimension $n-1$ of an affine space of affine dimension n; a maximal nonspanning set of a matroid.

idempotent: property of a latin square (or partial latin square) that for all i, cell (i, i) is occupied by i.

imbedded latin square: An $n \times n$ partial latin square P is imbedded in a latin square L if the upper $n \times n$ left corner of L agrees with P.

incidence matrix (of a (v, b, r, k, λ) design): the $b \times v$ matrix with (i, j)-entry equal to 1 if the ith block contains the jth element, and 0 otherwise.

incidence structure: the structure $(V, B, \mathcal{I})$ consisting of a finite set V of *points*, a finite set B of *lines*, and an incidence relation $\mathcal{I}$ between them.

independent set: any set in a special collection of subsets of the ground set in a matroid.

index: the number of blocks to which each pair of distinct points in a design belongs.

isomorphism (of block designs $(V, \mathcal{B})$ and $(W, \mathcal{C})$): a bijection $\psi : (V, \mathcal{B}) \to (W, \mathcal{C})$ under which $\psi(B)$ occurs as a block in $\mathcal{C}$ the same number of times that B occurs as a block in $\mathcal{B}$.

isomorphism (of matroids): a bijection between the ground sets of two matroids that preserves independence.

isotopic: equivalent.

Kirkman schoolgirl problem: the problem of arranging 15 schoolgirls in 5 subsets of size 3 for a walk on each of 7 days so that every pair of girls walk together exactly once.

Kirkman triple system: a $(v, 3, 1)$ resolvable BIBD, together with a resolution of it.

Kronecker product: for $m \times p$ matrix $M = (m_{ij})$ and $n \times q$ matrix $N = (n_{ij})$, the $mn \times pq$ matrix given by

$$M \times N = \begin{pmatrix} m_{11}N & m_{12}N & \cdots & m_{1p}N \\ \vdots & \vdots & & \vdots \\ m_{m1}N & m_{m2}N & \cdots & m_{mp}N \end{pmatrix}.$$

latin rectangle: a $k \times n$ ($k < n$) array in which each cell contains a single element from an n-set such that each element occurs exactly once in each row and at most once in each column.

latin square: A latin square of side n is an $n \times n$ array in which each entry contains a single element from a set S of size n such that each element occurs exactly once in each row and exactly once in each column.

line: a subspace of projective dimension 1 of a projective space; a coset of a subspace of affine dimension 1 of an affine space.

loop: in a matroid, element e of the matroid such that $\{e\}$ is a circuit.

matroid: an ordered pair $M = (E(M), \mathcal{I}(M))$ where E (the ground set) is a finite set and $\mathcal{I}$ is a collection of subsets (independent sets) of E such that: the empty set is independent; every subset of an independent set is independent; and if X and Y are independent and $X < Y$, then there is an element e in $Y - X$ such that $X \cup \{e\}$ is independent.

matroid representable over a field: a matroid that is isomorphic to the vector matroid of some matrix over the field F.

multiplier (of a difference set D in a group G): an automorphism φ of G such that $\varphi(D) = D + g$ for some $g \in G$.

mutually orthogonal: property of a set of latin squares that every two are orthogonal.

orthogonal: property of two $n \times n$ latin squares $A = (a_{ij})$ and $B = (b_{ij})$ that all n^2 ordered pairs (a_{ij}, b_{ij}) are distinct.

orthogonal array (of size N with k constraints, s levels, and strength t): a $k \times N$ array with entries from a set of $s \geq 2$ symbols, having the property that in every $t \times N$ submatrix every $t \times 1$ column vector appears the same number of times.

pairwise balanced design (PBD): for a set K of positive integers, a design (v, K, λ) consisting of an ordered pair $(X, \mathcal{A})$ where X is a set of size v and $\mathcal{A}$ is a collection of subsets of X with the property that every pair of elements of X occurs in exactly λ blocks, and for every block $A \in \mathcal{A}$, $|A| \in K$; a pairwise balanced design is called a (v, K)-PBD when $\lambda = 1$.

parallel class: a collection of blocks that partition the point set of a design.

parallel elements: in a matroid, two elements that form a circuit.

partial latin square: an $n \times n$ array with cells, each of which is either empty or else contains exactly one symbol, such that no symbol occurs more than once in any row or column.

partial transversal (of length k): in a latin square, a set of k cells, each from a different row and each from a different column, such that no two contain the same symbol.

paving matroid: a matroid such that the number of elements in every circuit is at least as large as the rank of the matroid.

PBD-closure: for a set K of positive integers, the set $\mathbf{B}(K) = \{ v \mid$ there exists a (v, K)-PBD $\}$.

planar: property of a matroid that it is isomorphic to the cycle matroid of a planar graph.

plane: a subspace of projective dimension 2 of a projective space; a coset of a subspace of affine dimension 2 of an affine space.

projective plane: a finite set (of points) and a set of subsets of points (called lines) such that every two points lie on exactly one line, every two lines intersect in exactly one point, and there are four points with no three collinear; equivalently, a symmetric $(n^2 + n + 1, n + 1, 1)$ design.

projective space (of dimension n): for a field F of order q and an $(n+1)$-dimensional vector space S over F, the set $PG(n, q)$ of all subspaces of S.

rank (of a matroid): the rank of the ground set of the matroid.

rank (of a set in a matroid): the cardinality of every maximal independent subset of the set.

rank axioms: a set of axioms that specifies the properties that a rank function on a matroid must have.

reduced latin square: a latin square such that the elements in the first row and the elements in the first column occur in natural order.

regular matroid: a matroid that is representable over all fields.

replication number: the number of blocks to which each point in a design belongs.

representable over a field: property of a matroid that it is isomorphic to a vector matroid of some matrix over the field.

resolution: a partition of the family of blocks of a balanced incomplete block design into parallel classes.

resolvable: the property of a balanced incomplete block design that it has at least one resolution.

simple matroid: a matroid that has no loops or parallel elements.

simple (t-design): a t-design that contains no repeated blocks.

spanning set (of a matroid): for a matroid M, a subset of the ground set E of rank $r(M)$.

Steiner triple system: a balanced incomplete block design in which each block has 3 elements and each pair of points occurs in exactly 1 block; that is, a $(v, 3, 1)$ design.

subdesign: a collection of points and blocks in a block design that is itself a block design.

subsquare: for $k < n$, a latin square of side k whose rows and columns are chosen from a latin square of side n.

symmetric block design: a (v, b, r, k, λ) design where the number of points (v) equals the number of blocks (b).

ternary matroid: a matroid that is isomorphic to a vector matroid of a matrix over $GF(3)$.

transversal design: a k-GDD having k groups of size n and uniform block size k.

transversal matroid: given a family of sets, the matroid whose independent sets are partial transversals of this family.

traversal: in a latin square of side n, a set of n cells, one from each row and column, containing each of the n symbols exactly once.

type: See group-type.

uniform matroid: the matroid with $1, 2, \ldots, n$ as ground set, and all subsets of size less that a specified number as independent sets.

(v, k, λ) **design**: a BIBD with parameters (v, b, r, k, λ).

$(v, k, \lambda; n)$ **difference set** (of order $n = k - \lambda$): a k-subset D of a group G (of order v) where every nonzero element of G has exactly λ differences $d - d'$ with elements from D.

vector matroid: the matroid on the columns of a matrix whose independent sets are the linearly independent sets of columns.

void design: a BIBD with at most one element.

12.1 BLOCK DESIGNS

12.1.1 BALANCED INCOMPLETE BLOCK DESIGNS

Definitions:

A *balanced incomplete block design* (**BIBD**) with parameters (v, b, r, k, λ) is a pair $(X, \mathcal{A})$, where X is a set, $\mathcal{A}$ is a collection of subsets of X, the five parameters are nonnegative integers, either $v \in \{0, 1\}$ (the *void* designs) or $v > k > 0$, and the parameters represent the following:

- v (**order**): the size of X (elements of X are **points**, **varieties**, or **treatments**);
- b (**block number**): the number of elements of $\mathcal{A}$ (elements of $\mathcal{A}$ are **blocks**);
- r (**replication number**): the number of blocks to which every point belongs;
- k (**block size**): the common size of each block;
- λ (**index**): the number of blocks to which every pair of distinct points belongs.

Note: A BIBD is often referred to as a *design*. Different notations are used for balanced incomplete block designs: (v, b, r, k, λ) BIBD, (v, k, λ) BIBD and $S_\lambda(2, k, v)$. In this chapter (v, k, λ) design will be used. See Fact 6.

A **Steiner triple system** is a $(v, \frac{v(v-1)}{6}, \frac{v-1}{2}, 3, 1)$ design, i.e., a BIBD in which each block has size 3 and each pair of points occurs in exactly one block. A Steiner triple system is denoted STS(v) or $S(2, 3, v)$. (Jakob Steiner, 1796–1863)

The **incidence matrix** of a (v, b, r, k, λ) design is the $b \times v$ matrix $A = (a_{ij})$ defined by
$$a_{ij} = \begin{cases} 1 & \text{if the } i\text{th block contains the } j\text{th point} \\ 0 & \text{otherwise}. \end{cases}$$

Facts:

1. Balanced incomplete block designs are used in the design of experiments when the total number (v) of objects to be tested is greater than the number (k) that can be tested at any one time. They are used to design experiments where the subjects must be divided into subsets (blocks) of the same size to receive different treatments, such that each subject is tested the same number of times and every pair of subjects appears in the same number of subsets.

2. Designs are useful in many areas, such as coding theory, cryptography, group testing, and tournament scheduling. Detailed coverage of these and other applications of designs can be found in Chapter V of [CoDi96].

3. The word "balanced" refers to the fact that λ remains constant. If λ changes depending on the pair of points chosen, the design is not balanced.

4. The word "incomplete" refers to the fact that $k < v$, that is, the size of each block is less than the number of varieties.

5. *Necessary conditions for existence*: If there is a (v, b, r, k, λ) design for particular v, b, r, k, and λ, then the parameters must satisfy:
   - $vr = bk$;
   - $\lambda(v-1) = r(k-1)$;
   - $b \geq v$. (Fisher's inequality, 1940) (Ronald A. Fisher, 1890–1962)

6. If a (v, b, r, k, λ) design exists, $r = \frac{\lambda(v-1)}{k-1}$ and $b = \frac{\lambda v(v-1)}{k(k-1)}$. In view of these two relationships, (v, b, r, k, λ) designs are commonly referred to simply by the three parameters — v, k, λ — as a (v, k, λ) design.

7. *Necessary conditions for existence*: If there is a (v, k, λ) design for particular v, k, and λ, then:
   - $\lambda(v-1) \equiv 0 \pmod{k-1}$;
   - $\lambda v(v-1) \equiv 0 \pmod{k(k-1)}$.

8. *Existence of (v, k, λ) designs*:
   - $(v, 3, \lambda)$ design: exists for all v satisfying the necessary conditions given in Fact 5, namely:
     ◇ if $\lambda \equiv 2$ or $4 \pmod 6$ and $v \equiv 0$ or $1 \pmod 3$
     ◇ if $\lambda \equiv 1$ or $5 \pmod 6$ and $v \equiv 1$ or $3 \pmod 6$
     ◇ if $\lambda \equiv 3 \pmod 6$ and $v \equiv 1 \pmod 2$
     ◇ if $\lambda \equiv 0 \pmod 6$ and $v \neq 2$;
   - $(v, 4, \lambda)$ design: exists for all v and λ satisfying the necessary conditions given in Fact 5;
   - $(v, 5, \lambda)$ design: exists for all v satisfying the necessary conditions given in Fact 5 except for the case $v = 15, \lambda = 2$;
   - $(v, 6, \lambda)$ design: exists for all v satisfying the necessary conditions given in Fact 5, if $\lambda > 1$;
   - $(v, 6, 1)$ design: exists for all $v \equiv 1$ or $6 \pmod{15}$, $v \geq 31$, $v \neq 36$, with 56 possible exceptions, the largest being 2241. (The first few open cases are 46, 51, 61, 81, and 141.)

9. *Existence of Steiner triple systems*: A Steiner triple system with v points exists if and only if $v \equiv 1$ or $3 \pmod 6$. (Kirkman)

10. *Wilson's asymptotic existence theorem*: Given k and λ, there exists a $v_0(k, \lambda)$ such that a (v, k, λ) design exists for all $v \geq v_0(k, \lambda)$ that satisfy the necessary conditions given in Fact 5 and make b and r integral. It is known that $v_0(k, \lambda) < \exp(\exp(k^{k^2}))$.

11. Assume that $(V, \mathcal{B})$ is a (v, k, λ) design. Let $\overline{\mathcal{B}} = \{V - B \mid B \in \mathcal{B}\}$. Then $(V, \overline{\mathcal{B}})$ is a $\left(v, v-k, \lambda\frac{(v-k)(v-k-1)}{k(k-1)}\right)$ design, the *complement* of $(V, \mathcal{B})$.

12. Given two Steiner triple systems with v_1 and v_2 points, respectively, a Steiner triple system with v_1v_2 points can be constructed as follows: Let an STS(v_1) be defined on the point set $\{x_1, \ldots, x_{v_1}\}$ and an STS(v_2) be defined on the point set $\{y_1, \ldots, y_{v_2}\}$. Define an STS(v_1v_2) on the point set $\{z_{ij} \mid 1 \leq i \leq v_1, 1 \leq j \leq v_2\}$ where $z_{mn}z_{pq}z_{rs}$ is a block in STS(v_1v_2) if and only if one of the following holds:
- $m = p = r$ and $y_ny_qy_s$ is a block in STS(v_2);
- $n = q = s$ and $x_mx_px_r$ is a block in STS(v_1);
- $x_mx_px_r$ is a block in STS(v_1) and $y_ny_qy_s$ is a block in STS(v_2).

13. The following table lists different types of block designs and their features.

name	block size	size of subset covered	# times covered	other properties
balanced incomplete block design BIBD §12.1.1	k	2	λ	
pairwise balanced design PBD §12.1.6	various	2	λ	also called a linear space if $\lambda = 1$
Steiner triple system STS §12.1.1, §12.1.5	3	2	1	
Kirkman triple system KTS §12.1.4	3	2	1	resolvable
resolvable balanced incomplete block design RBIBD §12.1.4	k	2	λ	resolvable
projective plane $PG(2, q)$ §12.2.3	$q + 1$	2	1	#points = #blocks $= q^2 + q + 1$
affine plane $AG(2, q)$ §12.2.3	q	2	1	resolvable
symmetric design SBIBD §12.2.2	k	2	λ	#points = #blocks
t-design t-(v, k, λ) §12.1.5	k	$t \geq 2$	λ	
Steiner system $S(t, v, k)$ §12.1.5	k	$t \geq 2$	1	

Examples:

1. The following is a $(4, 4, 3, 3, 2)$ design: $X = \{a, b, c, d\}$, blocks $\{abc, abd, acd, bcd\}$.

2. *Affine plane of order 3* (a $(9, 3, 1)$ *design*): The point set is $X = \{0, \ldots, 8\}$ and the block set is $\mathcal{A} = \{012, 345, 678, 036, 147, 258, 048, 156, 237, 057, 138, 246\}$. Also see §12.2.3. This design is known as AG(2,3). This is a Steiner triple system.

3. Each of the following is a Steiner triple system (a $(v, 3, 1)$ design). In each of the following a set of *base blocks* $B_i = \{b_{i1}, b_{i2}, b_{i3}\}$ in the group $\mathcal{Z}_v$ is given. To get all the blocks of the design, take all distinct translates $B_i + g = \{b_{i1} + g, b_{i2} + g, b_{i3} + g\}$, for all $g \in \mathcal{Z}_v$, for each of the base blocks B_i.

$v = 7$: $\{0, 1, 3\}$ (mod 7) [Fano plane]

$v = 15$: $\{0, 1, 4\}$ $\{0, 2, 8\}$ $\{0, 5, 10\}$ (mod 15) [The last base block has only 5 ($=\frac{v}{3}$) distinct translates. This is a *short orbit* and occurs for all orders $v \equiv 3$ (mod 6).]

$v = 19$: $\{0,1,4\}$ $\{0,2,9\}$ $\{0,5,11\}$ (mod 19)

$v = 21$: $\{0,1,3\}$ $\{0,4,12\}$ $\{0,5,11\}$ $\{0,7,14\}$ (mod 21)

$v = 25$: $\{0,1,3\}$ $\{0,4,11\}$ $\{0,5,13\}$ $\{0,6,15\}$ (mod 25)

$v = 27$: $\{0,1,3\}$ $\{0,4,11\}$ $\{0,5,15\}$ $\{0,6,14\}$ $\{0,9,18\}$ (mod 27)

$v = 31$: $\{0,1,3\}$ $\{0,4,11\}$ $\{0,5,15\}$ $\{0,6,18\}$ $\{0,8,17\}$ (mod 31)

$v = 33$: $\{0,1,3\}$ $\{0,4,10\}$ $\{0,5,18\}$ $\{0,7,19\}$ $\{0,8,17\}$ $\{0,11,22\}$ (mod 33)

$v = 37$: $\{0,1,3\}$ $\{0,4,9\}$ $\{0,6,21\}$ $\{0,7,18\}$ $\{0,8,25\}$ $\{0,10,24\}$ (mod 37)

$v = 39$: $\{0,1,3\}$ $\{0,4,9\}$ $\{0,6,20\}$ $\{0,7,18\}$ $\{0,8,23\}$ $\{0,10,22\}$ $\{0,13,26\}$ (mod 39).

4. *Fano plane* or *projective plane of order 2*, PG(2,2): A (7,7,3,3,1) design with point set $X = \{0, \ldots, 6\}$ and block set $\mathcal{A} = \{013, 124, 235, 346, 450, 561, 602\}$, shown in the following figure. (Often, as here, a block $\{a,b,c\}$ is written as abc.) Also see §12.2.3. (Gino Fano, 1871–1952)

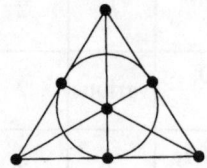

The incidence matrix of the Fano plane is

$$\begin{pmatrix} 1 & 1 & 0 & 1 & 0 & 0 & 0 \\ 0 & 1 & 1 & 0 & 1 & 0 & 0 \\ 0 & 0 & 1 & 1 & 0 & 1 & 0 \\ 0 & 0 & 0 & 1 & 1 & 0 & 1 \\ 1 & 0 & 0 & 0 & 1 & 1 & 0 \\ 0 & 1 & 0 & 0 & 0 & 1 & 1 \\ 1 & 0 & 1 & 0 & 0 & 0 & 1 \end{pmatrix}$$

12.1.2 ISOMORPHISM AND AUTOMORPHISM

Definitions:

Two designs $(V, \mathcal{B})$ and $(W, \mathcal{C})$ are **isomorphic** if there is a bijection $\psi: V \to W$ under which $\psi(B) = \{\psi(x) \mid x \in B\}$ occurs as a block in $\mathcal{C}$ the same number of times that B occurs as a block in $\mathcal{B}$. Such a bijection is an **isomorphism**.

An **automorphism** of a design D is an isomorphism from D onto D.

The **automorphism group** of a design D is the set of all automorphisms for D with composition as the group operation.

Facts:

1. Nonisomorphic Steiner triple systems of order v have been enumerated for $v \leq 15$. Up to isomorphism, there are unique designs of order 3, 7, and 9; there are precisely two nonisomorphic designs of order 13, and 80 of order 15. At that point, an explosion occurs: the number of nonisomorphic STS(19) exceeds 2,000,000.

2. The number of nonisomorphic STS(v) is at least $(e^{-5}v)^{v^2/6}$ for large v. (Wilson).

3. Table 1 lists the parameter sets (v, k, λ) that satisfy the necessary conditions for the existence of a block design, with $r \leq 15$ and $3 \leq k \leq \frac{v}{2}$. The parameter sets are ordered lexicographically across the rows of the table by r, k and λ (in this order). The column N contains the number of pairwise nonisomorphic (v, k, λ) designs or the best known lower bound for this number. A "?" indicates that no design with these parameters is known to exist, but that existence has not been ruled out.

12.1.3 SUBDESIGNS

Definition:

Let Y be a subset of w points in a (v, k, λ) design. If every block of the BIBD contains 0, 1, or k of the points in Y, then a (w, k, λ) design is obtained by taking those blocks that contain k points from Y. This BIBD on w points is a **subdesign**, called a (w, k, λ) subdesign.

Facts:

1. If there is a $(v, k, 1)$ design containing a $(w, k, 1)$ subdesign, then $v \geq (k-1)w + 1$. (The parameter lists $(v, k, 1)$ and $(w, k, 1)$ must satisfy the necessary conditions of §12.1.1 Fact 6.)

2. In the cases $k = 3$ and $k = 4$, the necessary conditions of §12.1.1 Fact 5 for the presence of a subdesign are sufficient. That is, in the case of $k = 3$, for all $v \geq 2w + 1$, with both $v, w \equiv 1$ or $3 \pmod{6}$, there exists a $(v, 3, 1)$ design that contains a $(w, 3, 1)$ subdesign. In the case $k = 4$, for all $v \geq 3w + 1$, with both $v, w \equiv 1$ or $4 \pmod{12}$ there exists a $(v, 4, 1)$ design that contains a $(w, 4, 1)$ subdesign.

Example:

1. *A construction for a Steiner triple system of order $2v+1$ given a Steiner triple system of order v*: A variant of this construction dates back at least to Thomas P. Kirkman in 1847. The original STS(v) is a subdesign of the resulting STS($2v + 1$).

Let $(X, \mathcal{A})$ be an STS(v) with $X = \{x_0, x_1, \ldots, x_{v-1}\}$. For each $i = 0, 1, \ldots, v - 1$, let $F_i = \{ \{x + i, -x + i\} \mid x \in \mathcal{Z}_v, x \neq 0 \} \cup \{i, \infty\}$. Then for each $i = 0, \ldots, v - 1$, construct the triples $\{a, b, x_i\}$ where $\{a, b\} \in F_i$. The set of all such triples in addition to the original triples in $\mathcal{A}$ is the desired STS($2v+1$) on the point set $X \cup \{0, 1, \ldots, v, \infty\}$.

For $v = 7$, the following STS(15) is obtained. The last row of triples is an STS(7).

$\{0, \infty, x_0\}$ $\{1, 6, x_0\}$ $\{2, 5, x_0\}$ $\{3, 4, x_0\}$

$\{1, \infty, x_1\}$ $\{2, 0, x_1\}$ $\{3, 6, x_1\}$ $\{4, 5, x_1\}$

$\{2, \infty, x_2\}$ $\{3, 1, x_2\}$ $\{4, 0, x_2\}$ $\{5, 6, x_2\}$

$\{3, \infty, x_3\}$ $\{4, 2, x_3\}$ $\{5, 1, x_3\}$ $\{6, 0, x_3\}$

$\{4, \infty, x_4\}$ $\{5, 3, x_4\}$ $\{6, 2, x_4\}$ $\{0, 1, x_4\}$

$\{5, \infty, x_5\}$ $\{6, 4, x_5\}$ $\{0, 3, x_5\}$ $\{1, 2, x_5\}$

$\{6, \infty, x_6\}$ $\{0, 5, x_6\}$ $\{1, 4, x_6\}$ $\{2, 3, x_6\}$

$\{x_0, x_1, x_3\}$ $\{x_1, x_2, x_4\}$ $\{x_2, x_3, x_5\}$ $\{x_3, x_4, x_6\}$ $\{x_4, x_5, x_0\}$ $\{x_5, x_6, x_1\}$ $\{x_6, x_0, x_2\}$.

Table 1 (v,b,r,k,λ) designs with $r \leq 15$.

v	k	λ	N	v	k	λ	N	v	k	λ	N
7	3	1	1	9	3	1	1	13	4	1	1
6	3	2	1	16	4	1	1	21	5	1	1
11	5	2	1	13	3	1	2	7	3	2	4
10	4	2	3	25	5	1	1	31	6	1	1
16	6	2	3	15	3	1	80	8	4	3	4
15	5	2	0	36	6	1	0	43	7	1	0
22	7	2	0	15	7	3	5	9	3	2	36
25	4	1	18	13	4	2	2,461	9	4	3	11
21	6	2	0	49	7	1	1	57	8	1	1
29	8	2	0	19	3	1	$\geq 1.1 \times 10^9$	10	3	2	960
7	3	3	10	28	4	1	≥ 145	10	5	4	21
46	6	1	?	16	6	3	18,920	28	7	2	7
64	8	1	1	73	9	1	1	37	9	2	4
25	9	3	78	19	9	4	6	21	3	1	$\geq 2 \times 10^6$
6	3	4	4	16	4	2	22,859	41	5	1	≥ 5
21	5	2	≥ 35	11	5	4	4,393	51	6	1	?
21	7	3	3,809	36	8	2	0	81	9	1	7
91	10	1	4	46	10	2	0	31	10	3	151
12	3	2	$\geq 10^6$	12	4	3	$\geq 17,172,470$	45	5	1	≥ 16
12	6	5	116,034	45	9	2	≥ 11	100	10	1	0
111	11	1	0	56	11	2	≥ 5	23	11	5	1,102
25	3	1	$\geq 10^{14}$	13	3	2	$\geq 92,714$	9	3	3	22,521
7	3	4	35	37	4	1	≥ 3	19	4	2	≥ 423
13	4	3	$\geq 3,702$	10	4	4	$\geq 1,759,613$	25	5	2	≥ 28
61	6	1	?	31	6	2	≥ 72	21	6	3	≥ 1
16	6	4	≥ 111	13	6	5	$\geq 2,572,156$	22	8	4	?
33	9	3	$\geq 3,375$	55	10	2	0	121	11	1	≥ 1
133	12	1	≥ 1	67	12	2	0	45	12	3	$\geq 3,752$
34	12	4	0	27	3	1	$\geq 10^{11}$	40	4	1	$\geq 10^6$
66	6	1	≥ 1	14	7	6	$\geq 17,896$	27	9	4	$\geq 8,071$
40	10	3	?	66	11	2	≥ 2	144	12	1	?
157	13	1	?	79	13	2	≥ 2	53	13	3	0
40	13	4	≥ 389	27	13	6	208,310	15	3	2	$\geq 685,521$
22	4	2	$\geq 7,921$	8	4	6	2,310	15	5	4	≥ 103
36	6	2	≥ 5	15	6	5	≥ 117	85	7	1	?
43	7	2	≥ 4	29	7	3	≥ 1	22	7	4	≥ 34
15	7	6	$\geq 57,810$	78	12	2	0	169	13	1	≥ 1
183	14	1	≥ 1	92	14	2	0	31	3	1	$\geq 6 \times 10^{16}$
16	3	2	$\geq 10^{13}$	11	3	3	$\geq 436,800$	7	3	5	109
6	3	6	6	16	4	3	$\geq 6 \times 10^{13}$	61	5	1	≥ 10
31	5	2	≥ 1	21	5	3	$\geq 10^9$	16	5	4	≥ 11
13	5	5	≥ 30	11	5	6	≥ 127	76	6	1	≥ 1
26	6	3	≥ 1	16	6	5	≥ 15	91	7	1	≥ 2
16	8	7	$\geq 9 \times 10^7$	21	9	6	$\geq 10^4$	136	10	1	?
46	10	3	?	28	10	5	≥ 3	56	12	3	≥ 4
91	13	2	0	196	14	1	0	211	15	1	0
106	15	2	0	71	15	3	≥ 8	43	15	5	0
36	15	6	$\geq 25,634$								

12.1.4 RESOLVABLE DESIGNS

Definitions:

A *parallel class* is a collection of blocks that partition the point set.

A *resolution* of a BIBD is a partition of the family of blocks into parallel classes. A resolution contains exactly r parallel classes.

A BIBD is *resolvable*, denoted RBIBD, if it has at least one resolution.

A $(v, 3, 1)$ RBIBD, together with a resolution of it, is a **Kirkman triple system**, written KTS(v).

Facts:

1. Necessary conditions for existence of a (v, k, λ) RBIBD are
   - $\lambda(v-1) \equiv 0 \pmod{(k-1)}$;
   - $v \equiv 0 \pmod{k}$.

2. If a (v, k, λ) RBIBD exists, then $b \geq v + r - 1$ where b is the number of blocks. When $b = v + r - 1$ (or equivalently, $r = k + \lambda$) the RBIBD has the property that two nonparallel lines intersect in exactly $\frac{k^2}{v}$ points. (R. C. Bose, 1901–1987)

3. A KTS(v) exists if and only if $v \equiv 3 \pmod 6$.

4. The following table summarizes the current state of knowledge concerning the existence of resolvable designs.

    For the values of k and λ given, the number of parameter sets (v, k, λ) satisfying all necessary conditions for the existence of a resolvable (v, k, λ) design for which the existence of a resolvable (v, k, λ) design is not known is given under the column headed "exceptions". The column headed "largest possible exception" gives the largest v satisfying the necessary conditions for the existence of a resolvable (v, k, λ) design for which a resolvable (v, k, λ) design is not known.

k	λ	exceptions	largest possible exception
3	1	none	
3	2	6	
4	1	none	
4	3	none	
5	1		4,965
5	2	15	50,722,390
5	4	10	195
6	5		3,042
6	10	none	
7	6	14	33,936
8	1		24,480
8	7		2,928

Example:

1. *Kirkman's schoolgirl problem*: In 1850, Kirkman posed the following: fifteen young ladies in a school walk out three abreast for seven days in succession; it is required to arrange them daily, so that no two walk twice abreast. (Thomas P. Kirkman, 1806–1895)

This is equivalent to finding a resolution of some $(15, 3, 1)$ design (or a KTS(15)). The following is a solution to Kirkman's schoolgirl problem:

Monday	Tuesday	Wednesday	Thursday	Friday	Saturday	Sunday
9, 10, 12	10, 11, 13	11, 12, 14	12, 13, 15	13, 14, 9	14, 15, 10	15, 9, 11
15, 8, 1	9, 8, 2	10, 8, 3	11, 8, 4	12, 8, 5	13, 8, 6	14, 8, 7
13, 2, 7	14, 3, 1	15, 4, 2	9, 5, 3	10, 6, 4	11, 7, 5	12, 1, 6
11, 3, 6	12, 4, 7	13, 5, 1	14, 6, 2	15, 7, 3	9, 1, 4	10, 2, 5
14, 4, 5	15, 5, 6	9, 6, 7	10, 7, 1	11, 1, 2	12, 2, 3	13, 3, 4

12.1.5 *t*-DESIGNS AND STEINER SYSTEMS

Definitions:

A ***t*-(v, k, λ) design** (also denoted $S_\lambda(t, k, v)$ and written ***t*-design**) is a pair $(X, \mathcal{A})$ that satisfies the properties:

- X is a set of v elements (called *points*);
- $\mathcal{A}$ is a family of subsets (*blocks*) of X, each of cardinality k;
- every t-subset of distinct points occurs in exactly λ blocks.

A t-design is **simple** if it contains no repeated blocks.

A **Steiner system** is a t-$(v, k, 1)$ design.

A **Steiner triple system**, denoted STS(v), is a $(v, 3, 1)$ design. (See §12.1.1.)

A **Steiner quadruple system**, denoted SQS(v), is a 3-$(v, 4, 1)$ design.

Facts:

1. A (v, k, λ) design (a BIBD) is a 2-(v, k, λ) design.

2. If $s < t$, then a t-(v, k, λ) design is also an s-(v, k, μ) design, where $\mu = \lambda \frac{\binom{v-s}{t-s}}{\binom{k-s}{t-s}}$.

3. t-(v, k, λ) designs exist for all t. A t-$(v, t+1, ((t+1)!)^{2t+1})$ design exists if $v \geq t+1$ and $v \equiv t \pmod{[(t+1)!]^{2t+1}}$. (Teirlinck)

4. If a t-(v, k, λ) design exists, where $t = 2s$ is even, then the number of blocks $b \geq \binom{v}{s}$. (This generalizes Fisher's inequality, §12.1.1, Fact 5.)

5. When $\lambda = 1$, t-designs are known only for $t \leq 5$. Construction of a 6-$(v, k, 1)$ design remains one of the outstanding open problems in the study of t-designs.

6. Much less is known about the existence of t-(v, k, λ) designs with $t \geq 3$ compared to BIBDs:

- For $t = 3$, several infinite families are known.
- For every prime power q and $d \geq 2$, there exists a 3-$(q^d + 1, q + 1, 1)$ design, known as an *inversive geometry*. When $d = 2$, these designs are known as *inversive planes*.
- A 3-$(v, 4, 1)$ design (Steiner quadruple system) exists if and only if $v \equiv 2$ or $4 \pmod 6$.

Examples:

1. The following is a 3-$(8, 4, 1)$ design:

    $X = \{\infty, 0, 1, 2, 3, 4, 5, 6\}$

    $\mathcal{A} = \{\ \{0, 1, 3, \infty\},\ \{1, 2, 4, \infty\},\ \{2, 3, 5, \infty\},\ \{3, 4, 6, \infty\},\ \{4, 5, 0, \infty\},$
    $\{5, 6, 1, \infty\},\ \{6, 0, 2, \infty\},\ \{2, 4, 5, 6\},\ \{3, 5, 6, 0\},\ \{4, 6, 0, 1\},$
    $\{5, 0, 1, 2\},\ \{6, 1, 2, 3\},\ \{0, 2, 3, 4\},\ \{1, 3, 4, 5\}\ \}.$

2. *Simple t-designs* ($t = 4, 5$): For $t = 4$ or 5 and $v \leq 30$, the only t-$(v, k, 1)$ designs known to exist are those having the following parameters:

4-$(11, 5, 1)$	5-$(12, 6, 1)$	4-$(23, 7, 1)$
5-$(24, 8, 1)$	4-$(27, 6, 1)$	5-$(28, 7, 1)$.

3. *Simple t-designs* ($t = 6$): For $t = 6$ and $v \leq 30$, the only t-(v, k, λ) designs known to exist are those having the following parameters:

    6-$(14, 7, 4)$ 6-$(20, 9, 112)$ 6-$(22, 7, 8)$ 6-$(30, 7, 12)$.

12.1.6 PAIRWISE BALANCED DESIGNS

Definitions:

Given a set K of positive integers and a positive integer λ, a **pairwise balanced design**, written (v, K, λ)-PBD, is an ordered pair $(X, \mathcal{A})$ where X is a set (of *points*) of size v and $\mathcal{A}$ is a collection of subsets (*blocks*) of X such that:

- every pair of elements of X occurs together in exactly λ blocks;
- for every block $A \in \mathcal{A}$, $|A| \in K$.

When $\lambda = 1$, λ can be omitted from the notation and the design is called a (v, K)-PBD or a **finite linear space**.

Given a set K of positive integers, let $\mathbf{B}(K)$ denote the set of positive integers v for which there exists a (v, K)-PBD. The mapping $K \to \mathbf{B}(K)$ is a **closure operation** on the set of subsets of the positive integers, as it satisfies the properties:

- $K \subseteq \mathbf{B}(K)$;
- $K_1 \subseteq K_2 \Rightarrow \mathbf{B}(K_1) \subseteq \mathbf{B}(K_2)$;
- $\mathbf{B}(\mathbf{B}(K)) = \mathbf{B}(K)$.

The set $\mathbf{B}(K)$ is the **closure** of the set K.

If K is any set of positive integers, then K is **PBD-closed** (or **closed**) if $\mathbf{B}(K) = K$.

If K is a closed set, then there exists a finite subset $J \subseteq K$ such that $K = \mathbf{B}(J)$. This set J is a **generating set** for the PBD-closed set K.

If J is a generating set for K and if $s \in J$ is such that $J - \{s\}$ is also a generating set for K, then s is **inessential** in K; otherwise s is **essential**.

A **basis** is a generating set consisting of essential elements.

Facts:

1. A (v, k, λ) design is a special case of a PBD in which the blocks are only permitted to be of one size, k.

2. *Necessary conditions for existence*: The existence of a (v, K)-PBD (with $v > 0$) implies:
   - $v \equiv 1 \pmod{\alpha(K)}$
   - $v(v-1) \equiv 0 \pmod{\beta(K)}$

   where $\alpha(K)$ is the greatest common divisor of the integers $\{\, k - 1 \mid k \in K \,\}$ and $\beta(K)$ is the greatest common divisor of the integers $\{\, k(k-1) \mid k \in K \,\}$.

3. *Asymptotic existence*: Given K, there exists a constant c_k such that a (v, K)-PBD exists for all $v \geq c_k$ that satisfy the necessary conditions of Fact 2. The constant c_k is, in general, unspecified. In practice, considerable further work is usually required to obtain a concrete upper bound on c_k.

Examples:

1. The following is a $(10, \{3, 4\})$-PBD:

$$\{1,2,3,4\}, \{1,5,6,7\}, \{1,8,9,10\}, \{2,5,8\}, \{2,6,9\}, \{2,7,10\}$$
$$\{3,5,10\}, \{3,6,8\}, \{3,7,9\}, \{4,5,9\}, \{4,6,10\}, \{4,7,8\}$$

2. Table 2 lists closures of some subsets of $\{3, 4, \ldots, 8\}$. From Fact 3, for a given set K there are only a finite number of values of v (satisfying the necessary conditions) for which there does not exist a (v, K)-PBD. These exceptional cases are listed in this table for some small sets K. Since $7 \in B(3)$, it is not necessary to include 7 in the list of sets whose closures are given, when 3 is present. Genuine exceptions (values of v satisfying the necessary conditions for the existence of a (v, K)-PBD for which it has been *proven* that no such design can exist) are shown in boldface, while possible exceptions (neither existence or nonexistence of a (v, K)-PBD is known) are shown in normal type.

12.1.7 GROUP DIVISIBLE DESIGNS AND TRANSVERSAL DESIGNS

Definitions:

A *group divisible design* (or (K, λ)-GDD) is a triple $(X, \mathcal{G}, \mathcal{A})$ where X is a set (of *points*), $\mathcal{G}$ is a partition of X into at least two subsets (called *groups*), $\mathcal{A}$ is a family of subsets of X (called *blocks*) such that:
- if A in $\mathcal{A}$, then $|A| \in K$;
- a group and a block contain at most one common point;
- every pair of points from distinct groups occurs in exactly λ blocks.

If $\lambda = 1$, a (K, λ)-GDD is often denoted by K-GDD. If $K = \{k\}$, a K-GDD is written k-GDD.

The *group-type* (or *type*) of a GDD is the multiset $\{\, |G| \mid G \in \mathcal{G} \,\}$. Usually an "exponential notation" is used to describe the type of a GDD: a GDD of type $t_1^{u_1} t_2^{u_2} \ldots t_k^{u_k}$ is a GDD where there are u_i groups of size t_i for $1 \leq i \leq k$.

A *transversal design* $\text{TD}(k, n)$ is a k-GDD of type n^k (that is, one having k groups of size n and uniform block size k).

Fact:

1. The existence of a $\text{TD}(k, n)$ is equivalent to the existence of $k-2$ mutually orthogonal latin squares of side n. (§12.3.2.)

Table 2 Closures of some subsets of $\{3, 4, \ldots, 8\}$.

subset K	necessary conditions	exceptions
3	1, 3 mod 6	–
3, 4	0, 1 mod 3	–
3, 5	1 mod 2	–
3, 6	0, 1 mod 3	4, 10, 12, 22
3, 8	$\mathcal{N}$ (natural numbers)	4, 5, 6, 10, 11, 12, 14, 16, 17, 18, 20, 23, 26, 28, 29, 30, 34, 35, 36, 38
3, 4, 5	$\mathcal{N}$	6, 8
3, 4, 6	0, 1 mod 3	–
3, 4, 8	$\mathcal{N}$	5, 6, 11, 14, 17
3, 5, 6	$\mathcal{N}$	4, 8, 10, 12, 14, 20, 22
3, 5, 8	$\mathcal{N}$	4, 6, 10, 12, 14, 16, 18, 20, 26, 28, 30, 34
3, 6, 8	$\mathcal{N}$	4, 5, 10, 11, 12, 14, 17, 20, 23
3, 4, 5, 6	$\mathcal{N}$	8
3, 4, 5, 8	$\mathcal{N}$	6
3, 4, 6, 8	$\mathcal{N}$	5, 11, 14, 17
3, 5, 6, 8	$\mathcal{N}$	4, 10, 14, 20
3, 4, 5, 6, 8	$\mathcal{N}$	–
4	1, 4 mod 12	–
4, 5	0, 1 mod 4	8, 9, 12
4, 6	0, 1 mod 3	7, 9, 10, 12, 15, 18, 19, 22, 24, 27, 33, 34, 39, 45, 46, 51, 55, 75, 87
4, 7	1 mod 3	10, 19
4, 8	0, 1 mod 4	5, 9, 12, 17, 20, 21, 24, 33, 41, 44, 45, 48, 53, 60, 65, 69, 77, 89, 101, 161, 164, 173
4, 5, 6	$\mathcal{N}$	7, 8, 9, 10, 11, 12, 14, 15, 18, 19, 23, 47
4, 5, 7	$\mathcal{N}$	6, 8, 9, 10, 11, 12, 14, 15, 18, 19, 23, 26, 27, 30, 39, 42, 50, 51, 54, 62, 63, 66, 74, 78
4, 5, 8	0, 1 mod 4	9, 12
4, 6, 7	0, 1 mod 3	5, 9, 10, 12, 15, 19, 24, 27, 33, 45, 75, 87
4, 6, 8	$\mathcal{N}$	5, 7, 9, 10, 11, 12, 14, 15, 17, 18, 19, 20, 22, 23, 24, 26, 27, 33, 34, 35, 39, 41, 47, 50, 51, 53, 55, 59, 62, 65, 71, 74, 75, 77, 87, 89, 95, 98, 101, 110, 122, 131, 161, 170, 182, 194, 242, 266, 290
4, 7, 8	$\mathcal{N}$	5, 6, 9, 10, 11, 12, 14, 15, 17, 18, 19, 20, 21, 23, 24, 26, 27, 30, 33, 35, 38, 39, 41, 42, 44, 45, 47, 48, 51, 54, 59, 62, 65, 66, 69, 74, 75, 77, 78, 83, 86, 87, 89, 90, 93, 101, 102, 107, 110, 111, 114, 122, 123, 126, 129, 131, 135, 138, 143, 146, 150, 158, 159, 161, 162, 164, 165, 167, 170, 171, 173, 174, 186, 194, 195, 198

subset K	necessary conditions	exceptions
4, 5, 6, 7	$\mathcal{N}$	8, 9, 10, 11, 12, 14, 15, 18, 19, 23
4, 5, 6, 8	$\mathcal{N}$	7, 9, 10, 11, 12, 14, 15, 18, 19, 23, 47
4, 5, 7, 8	$\mathcal{N}$	6, 9, 10, 11, 12, 14, 15, 18, 19, 23, 26, 27, 30, 38, 42, 51, 62, 66, 74, 78
4, 6, 7, 8	$\mathcal{N}$	5, 9, 10, 11, 12, 14, 15, 17, 18, 19, 20, 23, 24, 26, 27, 33, 35, 41, 65, 74, 75, 77, 123, 131, 143
4, 5, 6, 7, 8	$\mathcal{N}$	9, 10, 11, 12, 14, 15, 18, 19, 23
5, 6	0, 1 mod 5	10, 11, 15, 16, 20, 35, 40, 50, 51, 80

12.2 SYMMETRIC DESIGNS AND FINITE GEOMETRIES

12.2.1 FINITE GEOMETRIES

Definitions:

A finite *incidence structure* $(V, \mathcal{B}, \mathcal{I})$ consists of a finite set V of *points*, a finite set $\mathcal{B}$ of *lines*, and an *incidence relation* $\mathcal{I}$ between them. (Equivalently, a finite incidence structure is a pair $(V, \mathcal{B})$, where $\mathcal{B} = \{\{v \mid (v, b) \in \mathcal{I}\} \mid b \in \mathcal{B}\}$. In this case, lines are sets of points.)

The *dual* incidence structure is obtained by interchanging the roles of points and lines.

Let F be a finite field, and let S be an $(n+1)$-dimensional vector space over F. The set of all subspaces of S is the **projective space of projective dimension** n over F. When F is the Galois field $GF(q)$ (see §5.6.3), the projective space of projective dimension n is denoted $\text{PG}(n, q)$.

Subspaces of projective dimensions 0, 1, 2, and n are **points**, **lines**, **planes**, and **hyperplanes**, respectively; in general, subspaces of projective dimension t are **t-flats**. $\text{PG}_t(n, q)$ denotes the incidence structure of points and t-flats in $\text{PG}(n, q)$ (incidence is just containment as subspaces). Often, $\text{PG}_1(n, q)$ is denoted $\text{PG}(n, q)$ (taking the structure of points and lines as the natural geometry of the underlying space).

Let S be an n-dimensional vector space over a finite field F. The set of all cosets of subspaces of S is the **affine space of affine dimension** n over F. When F is the Galois field $GF(q)$, the affine space of affine dimension n is denoted $\text{AG}(n, q)$.

Cosets of subspaces of (affine) dimension 0, 1, 2, and $n-1$ of an affine space of affine dimension n are **points**, **lines**, **planes**, and **hyperplanes**, respectively. In general, cosets of subspaces of affine dimension t are **t-flats**. $\text{AG}_t(n, q)$ denotes the incidence structure of points and t-flats in $\text{AG}(n, q)$ (incidence is containment). Often, $\text{AG}_1(n, q)$ is denoted $\text{AG}(n, q)$ (taking the structure of points and lines as the natural geometry of the underlying space).

Note: The term *finite geometry* often just means finite incidence structure. However, incidence structures are often too unstructured to be of much (geometric) interest. Hence the term is sometimes reserved to cover only incidence structures satisfying additional axioms such as those given in §12.2.3 Fact 3 for projective planes.

Facts:

1. *Projective geometries*: For q a prime power and $1 \leq t < n$, $\mathrm{PG}_t(n,q)$ is a
   $\left(\frac{q^{n+1}-1}{q-1}, \frac{q^{t+1}-1}{q-1}, \frac{(q^{n-1}-1)(q^{n-2}-1)\cdots(q^{n-t-1}-1)}{(q^{t-1}-1)(q^{t-2}-1)\cdots(q-1)} \right)$ design.

2. *Affine geometries*: For q a prime power and $1 \leq t < n$, $\mathrm{AG}_t(n,q)$ is a
   $\left(q^n, q^t, \frac{(q^{n-1}-1)(q^{n-2}-1)\cdots(q^{n-t-1}-1)}{(q^{t-1}-1)(q^{t-2}-1)\cdots(q-1)} \right)$ design.

12.2.2 SYMMETRIC DESIGNS

Definitions:

A (v, b, r, k, λ) block design is **symmetric** if the number of points equals the number of blocks, that is, $v = b$.

A symmetric design with $\lambda = 2$ is a **biplane**. The parameters of a biplane are $v = \binom{k}{2} + 1$, $k = k$, $\lambda = 2$.

Facts:

1. In a symmetric (v, k, λ) design, $r = k$.
2. If a symmetric (v, k, λ) design exists, then:
   - if v is even, then $k - \lambda$ is a perfect square;
   - if v is odd, then the Diophantine equation $x^2 = (k-\lambda)y^2 + (-1)^{(v-1)/2}\lambda z^2$ has a solution in integers, not all of which are zero. (Bruck-Ryser-Chowla. The theorem is often referred to as BRC.)
3. For every positive integer k there is a symmetric $(2^{k+2}-1, 2^{k+1}-1, 2^k-1)$ block design.
4. If p is prime and k is a positive integer, there is a symmetric $(p^{2k}+p^k+1, p^k+1, 1)$ block design.
5. In a symmetric design any two blocks intersect in exactly λ points.
6. The *dual* incidence structure obtained by interchanging the roles of points and blocks is also a BIBD with the same parameters (hence the term symmetric).
7. The dual of a symmetric design need not be isomorphic to the original design.
8. Given a symmetric (v, k, λ) design, and a block A of this design, if the points not in A are deleted from all blocks which intersect A, the design obtained is the *derived design*. Its parameters are $(k, v-1, k-1, \lambda, \lambda-1)$.
9. Given a symmetric (v, k, λ) design, and given a block A of this design, delete the block A, and delete all points in A from all other blocks. The resulting design is the *residual design*, and has parameters $(v-k, v-1, k, k-\lambda, \lambda)$. (See Example 1.)
10. Any $(v-k, v-1, k, k-\lambda, \lambda)$ design is a *quasi-residual design*.
11. Any quasi-residual BIBD with $\lambda = 1$ or 2 is residual (Hall-Connor); but for $\lambda = 3$, there are examples of quasi-residual BIBDs that are not residual.
12. If there is a symmetric (v, k, λ) design with $n = k - \lambda$, then $4n-1 \leq v \leq n^2+n+1$. When $v = 4n - 1$ it is a *Hadamard design*; when $v = n^2+n+1$ it is a *projective plane*.
13. The only known biplanes have parameters $(7, 4, 2)$, $(11, 5, 2)$, $(16, 6, 2)$, $(37, 9, 2)$, $(56, 11, 2)$, and $(79, 13, 2)$.

14. The only known symmetric designs with $\lambda = 3$ have parameters $(11, 6, 3)$, $(15, 7, 3)$, $(25, 9, 3)$, $(31, 10, 3)$, $(45, 12, 3)$, and $(71, 15, 3)$.

15. Although infinitely many symmetric designs with $\lambda = 1$ are known, there is no other value of λ for which this is known to be true.

Example:

1. In the symmetric $(15, 7, 3)$ design in the following table, if the block b_0 is removed and if all the points in b_0 are removed from the blocks $b_1, \ldots, b_{14}$, the resulting design is the residual design. It has parameters $(8,4,3)$ and its blocks are given on the right.

b_0:	0	1	2	3	4	5	6
b_1:	0	1	2	7	8	9	10
b_2:	0	1	2	11	12	13	14
b_3:	0	3	4	7	8	11	12
b_4:	0	3	4	9	10	13	14
b_5:	0	5	6	7	8	13	14
b_6:	0	5	6	9	10	11	12
b_7:	1	3	5	7	9	11	13
b_8:	1	3	6	7	10	12	14
b_9:	1	4	5	8	10	11	14
b_{10}:	1	4	6	8	9	12	13
b_{11}:	2	3	5	8	10	12	13
b_{12}:	2	3	6	8	9	11	14
b_{13}:	2	4	5	7	9	12	14
b_{14}:	2	4	6	7	10	11	13

12.2.3 PROJECTIVE AND AFFINE PLANES

Definitions:

A **projective plane** is a finite set of *points* and a set of subsets of points (called *lines*) such that:
- every two points lie on exactly one line;
- every two lines intersect in exactly one point;
- there are four points with no three collinear.

An **affine plane** is a set of *points* and a set of subsets of points (called *lines*) such that:
- every two points lie on exactly one line;
- if a point does not lie on a line L, there is exactly one line through the point that does not intersect L;
- there are three points that are not collinear.

Facts:

1. A finite projective plane is a symmetric $(n^2 + n + 1, n + 1, 1)$ design, for some positive integer n, called the *order* of the projective plane. The projective plane has $n^2 + n + 1$ points and $n^2 + n + 1$ lines. Each point lies on $n + 1$ lines and every line contains exactly $n + 1$ points.

2. *Principle of duality*: Given any statement about finite projective planes that is a theorem, the dual statement (obtained by interchanging "point" and "line" and interchanging "point lying on a line" with "line passing through a point") is a theorem.

3. Any symmetric design with $\lambda = 1$ is a projective plane.

4. The existence of a projective plane of order n is equivalent to the existence of a set of $n-1$ mutually orthogonal latin squares (MOLS) of side n.

5. Existence of projective planes: Very little is known about the existence of projective planes:
   - There exists a projective plane of order p^k whenever p is prime and k is a positive integer. (See Fact 10.)
   - There is no projective plane known for any order n that is not a power of a prime. The smallest open order is 12.
   - There is no projective plane of order 10 or any $n \equiv 6 \pmod{8}$.
   - There are nondesarguesian planes (Fact 9) known for every order q^2 and q^3 when q is a prime power. (See Fact 11.)
   - There are four nonisomorphic projective planes of order 9, three of which are nondesarguesian.
   - The following table summarizes the known facts about the existence and number of projective planes of order n, for $1 \leq n \leq 12$:

order	2	3	4	5	6	7	8	9	10	11	12
number of projective planes	1	1	1	1	0	1	1	4	0	≥ 1	?

6. The proof by Lam, Thiel, and Swiercz in 1989 that there is no projective plane of order 10 involved great amounts of computer power and time. For details, see [CoDi96].

7. The existence of a projective plane of order n is equivalent to the existence of an affine plane of order n.

8. A finite affine plane is a $(n^2, n, 1)$ design, for some positive integer n. The affine plane has n^2 points and $n^2 + n$ lines. Each point lies on $n + 1$ lines and every line contains exactly n points. The integer n is the *order* of the affine plane.

9. Any affine plane of order n has the property that the lines can be partitioned into $n + 1$ parallel classes each containing n lines and hence is a resolvable block design.

10. *A direct construction of a projective plane of every order $q = p^k$, when p is prime and k a positive integer*: Consider the three-dimensional vector space $\mathcal{F}_q^3$ over GF(q). This vector space contains $\frac{q^3-1}{q-1} = q^2 + q + 1$ 1-dimensional subspaces (lines through the origin $(0,0,0)$) and an equal number of 2-dimensional subspaces (planes through the origin). Now construct an incidence structure where the points are the 1-dimensional subspaces, the lines are the 2-dimensional subspaces and a point is on a line if the 1-dimensional subspace (associated with the point) is contained in the 2-dimensional subspace (associated with the line). This structure satisfies the axioms and thus is a projective plane (of order q). Projective planes, such as this, coming from finite fields via the construction in this example are *desarguesian planes*.

11. *A construction of a projective plane of order n from an affine plane of order n*: To construct the projective plane of order n from the affine plane of order n, use the fact that the lines of the affine plane can be partitioned into $n + 1$ parallel classes each containing n lines. To each line in the ith parallel class adjoin the new symbol ∞_i. Add one new line, namely $\{\infty_1, \infty_2, \ldots, \infty_{n+1}\}$. Now each line contains $n + 1$ points, there are $n^2 + n + 1$ total points, and each pair of points is on a unique line — this is the projective plane of order n.

12. *A construction of an affine plane of order n from a projective plane of order n*: The affine plane of order n is the residual design of the projective plane of order n. See §12.2.2 Fact 9 for the construction.

Examples:

1. The Fano plane (§12.1.1 Example 1) is the projective plane of order 2.

2. The affine plane of order 2 is given in part (a) of the following figure. The set of points is $\{1, 2, 3, 4\}$. The six lines are:

$$\{1,2\} \quad \{3,4\} \quad \{1,3\} \quad \{2,4\} \quad \{1,4\} \quad \{2,3\}.$$

The three parallel classes are:

$$\{\{1,2\},\{3,4\}\} \quad \{\{1,3\},\{2,4\}\} \quad \{\{1,4\},\{2,3\}\}.$$

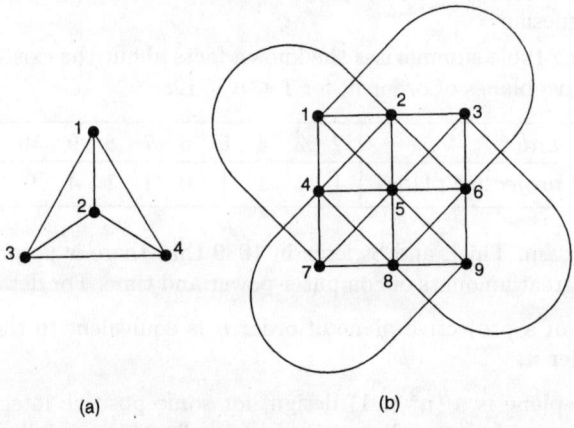

(a)   (b)

3. The affine plane of order 3 is given in part (b) of the figure of Example 2. The set of points is $\{1, 2, 3, \ldots, 9\}$. The twelve lines (listed in order in four parallel classes of three lines each) are:

$$\{1,2,3\}, \{4,5,6\}, \{7,8,9\} \quad \{1,4,7\}, \{2,5,8\}, \{3,6,9\}$$

$$\{1,5,9\}, \{6,2,7\}, \{4,8,3\} \quad \{3,5,7\}, \{2,4,9\}, \{8,6,1\}.$$

4. Using the construction in Fact 10 on the affine plane of order 3 (Example 3) yields the projective plane of order 3 with thirteen points $\{1, 2, 3, 4, 5, 6, 7, 8, 9, \infty_1, \infty_2, \infty_3, \infty_4\}$ and thirteen lines:

$$\{1,2,3,\infty_1\} \quad \{4,5,6,\infty_1\} \quad \{7,8,9,\infty_1\}$$

$$\{1,4,7,\infty_2\} \quad \{2,5,8,\infty_2\} \quad \{3,6,9,\infty_2\}$$

$$\{1,5,9,\infty_3\} \quad \{6,2,7,\infty_3\} \quad \{4,8,3,\infty_3\}$$

$$\{3,5,7,\infty_4\} \quad \{2,4,9,\infty_4\} \quad \{8,6,1,\infty_4\}$$

$$\{\infty_1,,\infty_2,\infty_3,\infty_4\}.$$

12.2.4 HADAMARD DESIGNS AND MATRICES

Definitions:

A **Hadamard matrix** H of order n is a square $n \times n$ matrix all of whose entries are ± 1 that satisfies the property that $H^t H = nI$ where I is the $n \times n$ identity matrix and H^t is the transpose of H.

If $M = (m_{ij})$ is a $m \times p$ matrix and $N = (n_{ij})$ is an $n \times q$ matrix, the **Kronecker product** is the $mn \times pq$ matrix $M \times N$ given by

$$M \times N = \begin{pmatrix} m_{11}N & m_{12}N & \cdots & m_{1p}N \\ m_{21}N & m_{22}N & \cdots & m_{2p}N \\ \vdots & \vdots & & \vdots \\ m_{m1}N & m_{m2}N & \cdots & m_{mp}N \end{pmatrix}.$$

A **Hadamard design** of order $4n$ is a symmetric $(4n-1, 2n-1, n-1)$ design. The **dimension** of the Hadamard design is n.

Facts:

1. A necessary condition for the existence of a Hadamard matrix of order n is that $n = 1$, $n = 2$, or $n \equiv 0 \pmod 4$.

2. $HH^t = nI = H^t H$.

3. The rows [columns] of a Hadamard matrix are pairwise orthogonal [when considered as vectors of length n].

4. If a row of a Hadamard matrix is multiplied by -1, the result is a Hadamard matrix. Similarly, if a column of a Hadamard matrix is multiplied by -1, the result is a Hadamard matrix.

5. By multiplying rows and columns of a Hadamard matrix by -1, a Hadamard matrix can be obtained where the first row and column consist entirely of $+1$'s. A Hadamard matrix of this type is *normalized*.

6. In a normalized Hadamard matrix of order $4n$, every row and column (except the first) contains $+1$ and -1 exactly $2n$ times each.

7. Of all $n \times n$ matrices with entries from $\{-1, +1\}$, a Hadamard matrix has the maximal determinant. $|\det H| = n^{n/2}$.

8. The Kronecker product of two Hadamard matrices is a Hadamard matrix. Thus, if there are Hadamard matrices of order m and n, then there is a Hadamard matrix of order mn.

9. If there exist Hadamard matrices of orders $4m, 4n, 4p, 4q$, then there exists a Hadamard matrix of order $16mnpq$. (Craigen, Seberry, and Zhang)

10. Let q be a positive integer. Then there exists a Hadamard matrix of order $2^s q$ for every $s \geq \lfloor 2\log_2(q-3) \rfloor$. (Seberry)

11. A Hadamard design of order $4n$ exists if and only if a Hadamard matrix of order $4n$ exists. See the construction in Fact 14.

12. *Hadamard conjecture*: The fundamental question concerning Hadamard matrices remains the *existence question*. The *Hadamard conjecture* is that there exist Hadamard matrices of order $4n$ for all $n \geq 1$. This remains unproved.

13. Currently the two smallest orders for which the existence of a Hadamard matrix is open are 428 and 668. Because of Fact 7 and the existence of a Hadamard matrix of order 2, if q is an odd number and there exists a Hadamard matrix of order $2^s q$, then there exists a Hadamard matrix of order $2^t q$ for all $t \geq s$.

Thus, to tabulate known existence results, it is only necessary to give the odd numbers q and the smallest power s of 2 such that a Hadamard matrix of order $2^s q$ exists.

The following table is given in this manner. The number q is obtained by adding the indices at the top, left, and bottom of the entry t.

	000	100	200	300	400	500	600	700	800	900
00	22222	22232	22222	22222	22222	22222	22222	22222	22222	22232
10	22222	22222	22222	32222	22223	22322	22222	22224	22222	22233
20	22222	22222	23222	22222	22222	23222	22222	32222	23222	22222
30	22222	22222	22224	22222	22222	22232	32222	22225	22224	24222
40	22222	22222	22222	22232	23222	22222	23232	22224	22222	22232
50	22222	22222	32222	22224	22222	22222	23224	32222	23223	22322
60	22222	22232	22222	22222	23222	22222	22223	22222	23222	22222
70	22222	22223	22222	22222	22224	33222	22222	22222	22222	52222
80	22222	22222	23222	22222	22232	22222	22222	22233	23222	22222
90	22222	32222	22222	22222	52222	22225	22222	22222	22222	32222
	13579	13579	13579	13579	13579	13579	13579	13579	13579	13579

14. *A general construction for Hadamard matrices of order $q + 1$ when q is an odd prime power and $q \equiv 3 \pmod 4$*:

- construct a $q \times q$ matrix $C = (c_{ij})$ indexed by the elements of the field $GF(q)$ by letting
$$c_{ij} = \begin{cases} 1, & \text{if } i - j \text{ is a square in } GF(q) \\ -1, & \text{if } i - j \text{ is not a square in } GF(q); \end{cases}$$

- construct a Hadamard matrix H of order $q + 1$ from C by adding a first column of all -1s and then a top row of all 1s.

This method is used in Example 1 to construct the Hadamard matrix of order 12.

15. *Constructing Hadamard designs from Hadamard matrices, and vice versa*: Assume that there exists a Hadamard matrix of order $4n$. Let H be a normalized Hadamard matrix of this order. Remove the first row and column of H and replace every -1 in the resulting matrix by a 0. The final $(4n - 1) \times (4n - 1)$ matrix can be shown to be the incidence matrix of a $(4n - 1, 2n - 1, n - 1)$ Hadamard design.

This process can be reversed to construct a Hadamard matrix from a Hadamard design.

Example:

1. The smallest examples of Hadamard matrices are the following:

$$(1), \begin{pmatrix} 1 & 1 \\ 1 & -1 \end{pmatrix},$$

$$\begin{pmatrix} -1 & 1 & 1 & 1 \\ 1 & -1 & 1 & 1 \\ 1 & 1 & -1 & 1 \\ 1 & 1 & 1 & -1 \end{pmatrix}, \begin{pmatrix} 1 & 1 & 1 & 1 & 1 & 1 & 1 & 1 \\ 1 & -1 & 1 & -1 & 1 & -1 & 1 & -1 \\ 1 & 1 & -1 & -1 & 1 & 1 & -1 & -1 \\ 1 & -1 & -1 & 1 & 1 & -1 & -1 & 1 \\ 1 & 1 & 1 & 1 & -1 & -1 & -1 & -1 \\ 1 & -1 & 1 & -1 & -1 & 1 & -1 & 1 \\ 1 & 1 & -1 & -1 & -1 & -1 & 1 & 1 \\ 1 & -1 & -1 & 1 & -1 & 1 & 1 & -1 \end{pmatrix},$$

$$\begin{pmatrix} 1 & 1 & 1 & 1 & 1 & 1 & 1 & 1 & 1 & 1 & 1 & 1 \\ -1 & 1 & 1 & -1 & 1 & 1 & 1 & -1 & -1 & -1 & 1 & -1 \\ -1 & -1 & 1 & 1 & -1 & 1 & 1 & 1 & -1 & -1 & -1 & 1 \\ -1 & 1 & -1 & 1 & 1 & -1 & 1 & 1 & 1 & -1 & -1 & -1 \\ -1 & -1 & 1 & -1 & 1 & 1 & -1 & 1 & 1 & 1 & -1 & -1 \\ -1 & -1 & -1 & 1 & -1 & 1 & 1 & -1 & 1 & 1 & 1 & -1 \\ -1 & -1 & -1 & -1 & 1 & -1 & 1 & 1 & -1 & 1 & 1 & 1 \\ -1 & 1 & -1 & -1 & -1 & 1 & -1 & 1 & 1 & -1 & 1 & 1 \\ -1 & 1 & 1 & -1 & -1 & -1 & 1 & -1 & 1 & 1 & -1 & 1 \\ -1 & 1 & 1 & 1 & -1 & -1 & -1 & 1 & -1 & 1 & 1 & -1 \\ -1 & -1 & 1 & 1 & 1 & -1 & -1 & -1 & 1 & -1 & 1 & 1 \\ -1 & 1 & -1 & 1 & 1 & 1 & -1 & -1 & -1 & 1 & -1 & 1 \end{pmatrix}.$$

12.2.5 DIFFERENCE SETS

Note: In this section only difference sets in abelian groups are considered.

Definitions:

Let G be an additively written group of order v. A k-subset D of G is a $(v, k, \lambda; n)$ **difference set of order** $n = k - \lambda$ if every nonzero element of G has exactly λ representations as a difference $d - d'$ $(d, d' \in D)$. The difference set is **abelian**, **cyclic**, etc., if the group G has the respective property.

The **development** of a difference set D is the incidence structure dev(D) whose points are the elements of the group G and whose blocks are the *translates* $D + g = \{d + g \mid d \in D\}$, $g \in G$.

A **multiplier** of a difference set D in a group G is an automorphism φ of G such that $\varphi(D) = D + g$ for some $g \in G$. If φ is a multiplier and $\varphi(h) = th$ for all $h \in G$, then t is a **numerical** multiplier.

Facts:

1. Both the group G itself and $G - \{g\}$ (for an arbitrary $g \in G$) are $(v, v, v; 0)$ and $(v, v - 1, v - 2; 1)$ difference sets. In Table 3 these *trivial* difference sets are excluded.

2. The complement of a $(v, k, \lambda; n)$ difference set is again a difference set with parameters $(v, v - k, v - 2k + \lambda; n)$. Therefore only $k \leq \frac{v}{2}$ (the case $k = \frac{v}{2}$ is actually impossible) is considered.

3. The existence of a $(v, k, \lambda; n)$ difference set is equivalent to the existence of a symmetric (v, k, λ) design $\mathcal{D}$ admitting G as a point regular automorphism group; that is, for any two points p and q, there is a unique group element g which maps p to q. The design $\mathcal{D}$ is isomorphic with dev(D).

4. There are many symmetric designs which do not have difference set representations.

5. Since difference sets can yield symmetric designs, the parameters v, k, and λ must satisfy the trivial necessary conditions for the existence of a symmetric design ($\lambda(v-1) = k(k-1)$) and must also satisfy the Bruck-Ryser-Chowla condition (§12.2.2, part 2 of Fact 2).

6. If φ is a multiplier of the difference set D, then there is at least one translate $D + g$ of D which is fixed by φ.

7. *The multiplier theorem*: Let D be an abelian $(v, k, \lambda; n)$ difference set. If p is a prime that satisfies $(p, v) = 1$, $p|n$, and $p > \lambda$, then p is a numerical multiplier.

8. *The multiplier conjecture*: Every prime divisor p of n that is relatively prime to v is a multiplier of a $(v, k, \lambda; n)$ difference set; that is, the condition $p > \lambda$ in Fact 7 is unnecessary.

Examples:

1. A $(11, 5, 2; 3)$ difference set in the group $\mathcal{Z}_{11}$ is $\{1, 3, 4, 5, 9\}$.

2. Table 1 lists abelian difference sets of order $n \leq 15$. (See Fact 1.) One difference set for each abelian group is listed. In general, there will be many more examples. There are no other groups or parameters with $n \leq 15$ for which the existence of a difference set is undecided.

   In the column "group" the decomposition of the group as a product of cyclic subgroups is given. If the group is cyclic, the integers modulo the group order are used to describe the difference set.

12.3 LATIN SQUARES AND ORTHOGONAL ARRAYS

12.3.1 LATIN SQUARES

Definitions:

A *latin square* of *side n* is an $n \times n$ array in which each entry contains a single element from an n-set S, such that each element of S occurs exactly once in each row and exactly once in each column.

A latin square of side n (on the set $\{1, 2, \ldots, n\}$ or on the set $\{0, 1, \ldots, n-1\}$) is **reduced** or in **standard form** if in the first row and column the elements occur in increasing order.

Let L be an $n \times n$ latin square on symbol set E_3, with rows indexed by the elements of the n-set E_1 and columns indexed by the elements of the n-set E_2. Let $\mathcal{T} = \{(x_1, x_2, x_3) \mid L(x_1, x_2) = x_3\}$ and $\{a, b, c\} = \{1, 2, 3\}$. The (a, b, c)-**conjugate** of L, $L_{(a,b,c)}$, has rows indexed by E_a, columns by E_b, and symbols by E_c, and is defined by $L_{(a,b,c)}(x_a, x_b) = x_c$ for each $(x_1, x_2, x_3) \in \mathcal{T}$.

Table 1 Abelian difference sets of order $n \leq 15$.

n	v	k	λ	group	difference set
2	7	3	1	(7)	1 2 4
3	13	4	1	(13)	0 1 3 9
3	11	5	2	(11)	1 3 4 5 9
4	21	5	1	(21)	3 6 7 12 14
4	16	6	2	(8)(2)	(00) (10) (11) (20) (40) (61)
				$(4)^2$	(00) (01) (10) (12) (20) (23)
				$(4)(2)^2$	(000) (010) (100) (101) (200) (211)
				$(2)^4$	(0000) (0010) (1000) (1001) (1100) (1111)
4	15	7	3	(3)(5)	0 1 2 4 5 8 10
1	2	3	4	5	6
5	31	6	1	(31)	1 5 11 24 25 27
5	19	9	4	(19)	1 4 5 6 7 9 11 16 17
6	23	11	5	(23)	1 2 3 4 6 8 9 12 13 16 18
7	57	8	1	(57)	1 6 7 9 19 38 42 49
7	27	13	6	$(3)^3$	(001) (011) (021) (111) (020) (100) (112) (120) (121) (122) (201) (202) (220)
8	73	9	1	(73)	1 2 4 8 16 32 37 55 64
8	31	15	7	(31)	1 2 3 4 6 8 12 15 16 17 23 24 27 29 30
9	91	10	1	(91)	0 1 3 9 27 49 56 61 77 81
9	45	12	3	$(3)^2(5)$	(000) (001) (002) (003) (010) (020) (101) (112) (123) (201) (213) (222)
9	40	13	4	(40)	1 2 3 5 6 9 14 15 18 20 25 27 35
9	36	15	6	$(4)(3)^2$	(010) (011) (012) (020) (021) (022) (100) (110) (120) (200) (211) (222) (300) (312) (321)
				$(2)^2(3)^2$	(0010) (0011) (0012) (0020) (0021) (0022) (0100) (0110) (0120) (1000) (1011) (1022) (1100) (1112) (1121)
9	35	17	8	(35)	0 1 3 4 7 9 11 12 13 14 16 17 21 27 28 29 33
11	133	12	1	(133)	1 11 16 40 41 43 52 60 74 78 121 128
11	43	21	10	(43)	1 4 6 9 10 11 13 14 15 16 17 21 23 24 25 31 35 36 38 40 41
12	47	23	11	(47)	1 2 3 4 6 7 8 9 12 14 16 17 18 21 24 25 27 28 32 34 36 37 42
13	183	14	1	(183)	0 2 3 10 26 39 43 61 109 121 130 136 141 155
15	59	29	14	(59)	1 3 4 5 7 9 12 15 16 17 19 20 21 22 25 26 27 28 29 35 36 41 45 46 48 49 51 53 57

The **transpose** of a latin square L, denoted L^T, is the latin square which results from L when the role of rows and columns are exchanged; that is, $L^T(i,j) = L(j,i)$.

A latin square L of side n is **symmetric** if $L(i,j) = L(j,i)$ for all $1 \leq i \leq n$, $1 \leq j \leq n$.

A latin square L of side n is **idempotent** if $L(i,i) = i$ for all $1 \leq i \leq n$.

A **transversal** in a latin square of side n is a set of n cells, one from each row and column, containing each of the n symbols exactly once.

A **partial transversal of length** k in a latin square of side n is a set of k cells, each from a different row and each from a different column, such that no two contain the same symbol.

Two latin squares L and L' of side n are **equivalent** (or **isotopic**) if there are three bijections, from the rows, columns, and symbols of L to the rows, columns, and symbols, respectively, of L', that map L to L'.

Two latin squares L and L' of side n are **main class isotopic** if L is isotopic to some conjugate of L'.

Let $k < n$. If in a latin square L of side n the k^2 cells defined by k rows and k columns form a latin square of side k, then the cells are a latin **subsquare** of L.

An n by n array L with cells that are either empty or contain exactly one symbol is a **partial** latin square if no symbol occurs more than once in any row or column.

A partial latin square is **symmetric** (or **commutative**) if whenever cell (i,j) is occupied by x, cell (j,i) is also occupied by x, for every $1 \le i \le n$, $1 \le j \le n$.

A partial latin square is **idempotent** if cell (i,i) is occupied by i, for all i.

A **latin rectangle** is a $k \times n$ ($k < n$) array in which each cell contains a single element from an n-set such that each element occurs exactly once in each row and at most once in each column.

An $n \times n$ partial latin square P is said to be **imbedded** in a latin square L if the upper $n \times n$ left corner of L agrees with P.

Facts:

1. The multiplication table of any (multiplicative) group is a latin square.

2. For each positive integer k a reduced latin square can be constructed using the following format:

$$\begin{array}{ccccc} 1 & 2 & 3 & \cdots & k-1 & k \\ 2 & 3 & 4 & \cdots & k & 1 \\ 3 & 4 & 5 & \cdots & 1 & 2 \\ \vdots & \vdots & \vdots & \ddots & \vdots & \vdots \\ k-1 & k & 1 & \cdots & k-3 & k-2 \\ k & 1 & 2 & \cdots & k-2 & k-1 \end{array}$$

3. Every latin square has 1, 2, 3, or 6 distinct conjugates.

4. A symmetric latin square of even side can never be idempotent.

5. A symmetric idempotent latin square of side n is equivalent to a 1-factorization of the complete graph on $n+1$ points, K_{n+1}. A latin square of side n is equivalent to a 1-factorization of the complete graph $K_{n,n}$.

6. Every idempotent latin square has a transversal (the main diagonal).

7. Some latin squares have no transversals. One such class of latin squares is composed of the addition tables of $\mathcal{Z}_{2n}$ for every $n \ge 1$, or in general the addition table of any group that has a unique element of order 2.

8. Every latin square of side n has a partial transversal of length k where
$$k \ge \max\{n - \sqrt{n}, n - 15(\log n)^2\}. \quad \text{(P. W. Shor)}$$

9. A latin square of side n with a proper subsquare of side k exists if and only if $k \le \lfloor \frac{n}{2} \rfloor$.

10. There exists a latin square of side n with *no* proper subsquares if $n \ne 2^a 3^b$ or if $n = 3, 9, 12, 16, 18, 27, 81, 243$.

Section 12.3 LATIN SQUARES AND ORTHOGONAL ARRAYS   781

11. A partial latin square of side n with at most $n-1$ filled cells can always be completed to a latin square of side n.

12. A $k \times n$ ($k < n$) latin rectangle can always be completed to a latin square of side n.

13. Let L be a partial latin square of order n in which cell (i,j) is filled if and only if $i \leq r$ and $j \leq s$. Then L can be completed to a latin square of order n if and only if $N(i) \geq r + s - n$ for $i = 1, 2, \ldots, n$, where $N(i)$ denotes the number of elements in L that are equal to i. (H. Ryser)

14. A partial $n \times n$ latin square can be imbedded in a $t \times t$ latin square for every $t \geq 2n$.

15. An $n \times n$ partial symmetric latin square can be imbedded in a $t \times t$ symmetric latin square for every even $t \geq 2n$.

16. The number of distinct latin squares, the number of main classes, and the number of equivalence classes of latin squares of side n go to infinity as $n \to \infty$. The number of main classes and equivalence classes of latin squares of side $n \leq 8$ is given in the following table:

n	1	2	3	4	5	6	7	8
main classes	1	1	1	2	2	12	147	283,657
equivalence classes	1	1	1	2	2	22	563	1,676,257

Examples:

1. A 4×4 latin square on $\{1, 2, 3, 4\}$, where $1, 2, 3, 4$ represent four brands of tires, gives a way to test each brand of tire on each of the four wheel positions on each of four cars: the i,j-entry of the latin square is the brand of tire to be tested on wheel position i of car j.

2. A latin square of side 8 on the symbols $0, 1, \ldots, 7$:

$$\begin{array}{cccccccc}
0 & 1 & 2 & 3 & 4 & 5 & 6 & 7 \\
1 & 0 & 3 & 4 & 5 & 6 & 7 & 2 \\
2 & 3 & 5 & 0 & 6 & 7 & 4 & 1 \\
3 & 4 & 0 & 7 & 1 & 2 & 5 & 6 \\
4 & 5 & 6 & 1 & 7 & 0 & 2 & 3 \\
5 & 6 & 7 & 2 & 0 & 3 & 1 & 4 \\
6 & 7 & 4 & 5 & 2 & 1 & 3 & 0 \\
7 & 2 & 1 & 6 & 3 & 4 & 0 & 5
\end{array}$$

3. A latin square of side 4 and its six conjugates:

$$\begin{array}{cccc}
1 & 4 & 2 & 3 \\
2 & 3 & 1 & 4 \\
4 & 1 & 3 & 2 \\
3 & 2 & 4 & 1
\end{array} \qquad
\begin{array}{cccc}
1 & 2 & 4 & 3 \\
4 & 3 & 1 & 2 \\
2 & 1 & 3 & 4 \\
3 & 4 & 2 & 1
\end{array} \qquad
\begin{array}{cccc}
1 & 3 & 2 & 4 \\
2 & 4 & 1 & 3 \\
4 & 2 & 3 & 1 \\
3 & 1 & 4 & 2
\end{array}$$
$(1,2,3)$-conjugate $(2,1,3)$-conjugate $(3,2,1)$-conjugate

$$\begin{array}{cccc}
1 & 2 & 4 & 3 \\
3 & 4 & 2 & 1 \\
2 & 1 & 3 & 4 \\
4 & 3 & 1 & 2
\end{array} \qquad
\begin{array}{cccc}
1 & 3 & 4 & 2 \\
3 & 1 & 2 & 4 \\
2 & 4 & 3 & 1 \\
4 & 2 & 1 & 3
\end{array} \qquad
\begin{array}{cccc}
1 & 3 & 2 & 4 \\
3 & 1 & 4 & 2 \\
4 & 2 & 3 & 1 \\
2 & 4 & 1 & 3
\end{array}$$
$(2,3,1)$-conjugate $(1,3,2)$-conjugate $(3,1,2)$-conjugate

4. The following gives the main classes of latin squares of sides 4, 5, and 6. No two latin squares listed are main class isotopic.

$n = 4:$
```
0 1 2 3      0 1 2 3
1 0 3 2      1 0 3 2
2 3 0 1      2 3 1 0
3 2 1 0      3 2 0 1
```

$n = 5:$
```
0 1 2 3 4      0 1 2 3 4
1 2 3 4 0      1 0 3 4 2
2 3 4 0 1      2 3 4 0 1
3 4 0 1 2      3 4 1 2 0
4 0 1 2 3      4 2 0 1 3
```

$n = 6:$
```
0 1 2 3 4 5      0 1 2 3 4 5      0 1 2 3 4 5
1 0 3 2 5 4      1 0 3 2 5 4      1 0 3 4 5 2
2 3 4 5 0 1      2 3 4 5 0 1      2 3 0 5 1 4
3 2 5 4 1 0      3 2 5 4 1 0      3 4 5 0 2 1
4 5 0 1 2 3      4 5 0 1 3 2      4 5 1 2 0 3
5 4 1 0 3 2      5 4 1 0 2 3      5 2 4 1 3 0

0 1 2 3 4 5      0 1 2 3 4 5      0 1 2 3 4 5
1 0 3 4 5 2      1 0 3 4 5 2      1 0 3 4 5 2
2 3 0 5 1 4      2 3 1 5 0 4      2 3 4 5 0 1
3 4 5 0 2 1      3 4 5 1 2 0      3 4 5 2 1 0
4 5 1 2 3 0      4 5 0 2 3 1      4 5 0 1 2 3
5 2 4 1 0 3      5 2 4 0 1 3      5 2 1 0 3 4

0 1 2 3 4 5      0 1 2 3 4 5      0 1 2 3 4 5
1 0 3 2 5 4      1 0 3 2 5 4      1 0 3 2 5 4
2 4 0 5 1 3      2 4 0 5 1 3      2 4 0 5 1 3
3 5 1 4 0 2      3 5 1 4 0 2      3 5 1 4 2 0
4 2 5 0 3 1      4 2 5 1 3 0      4 3 5 1 0 2
5 3 4 1 2 0      5 3 4 0 2 1      5 2 4 0 3 1

0 1 2 3 4 5      0 1 2 3 4 5      0 1 2 3 4 5
1 0 3 2 5 4      1 0 3 4 5 2      1 2 0 4 5 3
2 4 0 5 3 1      2 3 1 5 0 4      2 0 1 5 3 4
3 5 4 0 1 2      3 5 4 1 2 0      3 5 4 1 0 2
4 2 5 1 0 3      4 2 5 0 1 3      4 3 5 2 1 0
5 3 1 4 2 0      5 4 0 2 3 1      5 4 3 0 2 1
```

5. The following are a latin square of side 7 with a subsquare of side 3 (3 × 3 square in upper left corner) and a latin square of order 12 with no proper subsquares.

```
1 2 3 4 5 6 7
2 3 1 6 4 7 5
3 1 2 7 6 5 4
4 7 5 1 3 2 6
7 5 6 3 2 4 1
6 4 7 5 1 3 2
5 6 4 2 7 1 3
```

```
1 2 3 4 5 6 7 8 9 a b c
2 3 4 5 6 1 8 9 a b c 7
3 1 5 2 7 8 4 a 6 c 9 b
4 5 6 7 1 9 b c 8 3 2 a
5 6 2 8 a 7 9 b c 4 1 3
6 c 8 1 3 a 2 7 b 9 4 5
7 8 1 a c b 5 4 2 6 3 9
8 9 b 3 4 c a 6 5 1 7 2
9 b 7 c 2 5 1 3 4 8 a 6
a 7 c b 9 4 6 1 3 2 5 8
b 4 a 9 8 3 c 2 7 5 6 1
c a 9 6 b 2 3 5 1 7 8 4
```

6. The partial latin square $\begin{array}{ccc} 1 & \cdot & 2 \\ \cdot & 3 & 1 \\ 4 & 1 & \cdot \end{array}$ is imbedded in the latin square $\begin{array}{cccc} 1 & 4 & 2 & 3 \\ 2 & 3 & 1 & 4 \\ 4 & 1 & 3 & 2 \\ 3 & 2 & 4 & 1 \end{array}$.

12.3.2 MUTUALLY ORTHOGONAL LATIN SQUARES

Definitions:

Two latin squares $A = (a_{ij})$ and $B = (b_{ij})$ of order n are **orthogonal** if the n^2 ordered pairs (a_{ij}, b_{ij}) $(1 \le i, j \le n)$ are distinct. (The relation of orthogonality is symmetric.)
A set of latin squares $\{A_1, \ldots, A_k\}$ is a set of **mutually orthogonal latin squares** (MOLS) if A_i and A_j are orthogonal for all $i, j \in \{1, \ldots, k\}$ $(i \ne j)$. The maximum number of MOLS of order n is written $N(n)$. It is customary to define $N(0) = N(1) = \infty$.
A set of $n-1$ MOLS of side n is a **complete set** of MOLS.

Facts:
1. If $n \ge 2$, then $N(n) \le n-1$.
2. If n is a prime power, then $N(n) = n-1$.
3. $N(n) \ge 2$ for all $n \ge 3$, except $n = 6$. (Bose-Parker-Shrikhande)
4. $N(n) \ge 3$ for all $n \ge 4$ except for $n = 6$ and possibly $n = 10$. The following table gives the best known lower bounds for $N(n)$ for $0 \le n \le 499$. Add the row and column indices to obtain the order.

	0	1	2	3	4	5	6	7	8	9	10	11	12	13	14	15	16	17	18	19
0	∞	∞	1	2	3	4	1	6	7	8	2	10	5	12	3	4	15	16	3	18
20	4	5	3	22	5	24	4	26	5	28	4	30	31	5	4	5	5	36	4	4
40	7	40	5	42	5	6	4	46	6	48	6	5	5	52	5	5	7	7	5	58
60	4	60	4	6	63	7	5	66	5	6	6	70	7	72	5	5	6	6	6	78
80	9	80	8	82	6	6	6	6	7	88	6	7	6	6	6	6	7	96	6	8
100	8	100	6	102	7	7	6	106	6	108	6	6	13	112	6	7	6	8	6	6
120	7	120	6	6	6	124	6	126	127	7	6	130	6	7	6	7	7	136	6	138
140	6	7	6	10	10	7	6	7	6	148	6	150	7	8	8	7	6	156	7	6
160	9	7	6	162	6	7	6	166	7	168	6	8	6	172	6	6	14	9	6	178
180	6	180	6	6	7	8	6	10	6	8	6	190	7	192	6	7	6	196	6	198
200	7	8	6	7	6	8	6	8	14	11	10	210	6	7	6	7	7	8	6	10
220	6	12	6	222	13	8	6	226	6	228	6	7	7	232	6	7	6	7	6	238
240	7	240	6	242	6	7	6	12	7	7	6	250	6	12	9	7	255	256	6	12
260	6	8	8	262	7	8	6	10	6	268	6	270	15	16	6	13	10	276	6	9
280	7	280	6	282	6	12	6	7	15	288	6	6	6	292	6	6	7	10	10	12
300	6	7	6	6	15	15	6	306	6	7	6	310	7	312	6	10	7	316	6	10
320	15	15	6	16	6	12	6	7	7	9	6	330	6	8	6	6	8	336	6	7
340	6	10	10	342	7	7	6	346	6	348	8	12	18	352	6	9	6	9	6	358
360	8	360	6	7	6	10	6	366	15	15	6	15	6	372	6	15	7	13	6	378
380	6	12	6	382	15	15	6	15	6	388	6	16	7	8	6	7	6	396	6	7
400	15	400	7	15	11	7	6	15	8	408	6	13	8	12	10	9	18	15	6	418
420	6	420	6	15	7	16	6	7	6	10	6	430	15	432	6	15	6	18	6	438
440	7	15	6	442	6	13	6	11	15	448	6	15	6	7	6	15	7	456	6	16
460	6	460	6	462	15	15	6	466	6	7	6	15	7	15	10	18	6	15	6	478
480	15	15	6	15	6	7	6	486	7	15	6	490	6	16	6	7	15	15	6	498

5. $N(n) \to \infty$ as $n \to \infty$. (Chowla, Erdős, Straus)
6. $N(n \times m) \geq \min\{N(n), N(m)\}$. (MacNeish)
7. The existence of $n-1$ MOLS of order n is equivalent to the existence of a projective plane of order n (an $(n^2 + n + 1, n + 1, 1)$ design) and an affine plane of order n (an $(n^2, n, 1)$ design). (§12.2.3)
8. The existence of a set of $k-2$ mutually orthogonal latin squares of order n is equivalent to the existence of a transversal design $\mathrm{TD}(k, n)$. (§12.1.7)
9. A set of $k-2$ MOLS of order n is equivalent to an $OA(n, k)$ (§12.3.3).
10. *Constructing a complete set of MOLS of order q for q a prime power*: A complete set of MOLS of order q for q a prime power can be constructed as follows:
    - for each $\alpha \in GF(q) - \{0\}$, define the latin square $L_\alpha(i, j) = i + \alpha j$, where $i, j \in GF(q)$ and the algebra is performed in $GF(q)$.

    The set of latin squares $\{L_\alpha \mid \alpha \in GF(q) - \{0\}\}$ is a set of $q-1$ MOLS of side q.
11. Let n_k be the largest order for which the existence of k MOLS is unknown. So if $n > n_k$, then there exist at least k MOLS of order n. See the following table:

k	n_k	k	n_k	k	n_k	k	n_k	k	n_k
2	6	5	62	8	2,774	11	7,222	14	7,874
3	10	6	75	9	3,678	12	7,286	15	8,360
4	42	7	780	10	5,804	13	7,288		

12. *Constructing a set of r MOLS of size $mn \times mn$ from a set of r MOLS of size $m \times m$ and a set of r MOLS of size $n \times n$*: Let $A_1, \ldots, A_r$ and $B_1, \ldots, B_r$ be two sets of MOLS, where each $A_i = (a_{xy}^{(i)})$ is of size $m \times m$ and each $B_i = (b_{xy}^{(i)})$ is of size $n \times n$. Construct a set $C_1, \ldots, C_r$ of $mn \times mn$ MOLS as follows: for each $k = 1, \ldots, r$, let

$$C_k = \begin{pmatrix} D_{11}^{(k)} & D_{12}^{(k)} & \cdots & D_{1m}^{(k)} \\ D_{21}^{(k)} & D_{22}^{(k)} & \cdots & D_{2m}^{(k)} \\ \vdots & \vdots & & \vdots \\ D_{m1}^{(k)} & D_{m2}^{(k)} & \cdots & D_{mm}^{(k)} \end{pmatrix}$$

where

$$D_{ij}^{(k)} = \begin{pmatrix} (a_{ij}^{(k)}, b_{11}^{(k)}) & (a_{ij}^{(k)}, b_{12}^{(k)}) & \cdots & (a_{ij}^{(k)}, b_{1n}^{(k)}) \\ (a_{ij}^{(k)}, b_{21}^{(k)}) & (a_{ij}^{(k)}, b_{22}^{(k)}) & \cdots & (a_{ij}^{(k)}, b_{2n}^{(k)}) \\ \vdots & \vdots & & \vdots \\ (a_{ij}^{(k)}, b_{n1}^{(k)}) & (a_{ij}^{(k)}, b_{n2}^{(k)}) & \cdots & (a_{ij}^{(k)}, b_{nn}^{(k)}) \end{pmatrix}.$$

Note:
1. In 1782 Leonhard Euler considered the following problem:

    A very curious question, which has exercised for some time the ingenuity of many people, has involved me in the following studies, which seem to open a new field of analysis, in particular in the study of combinations. The question revolves around arranging 36 officers to be drawn from 6 different ranks and at the same time from 6 different regiments so that they are also arranged in a square so that in each line (both horizontal and vertical) there are 6 officers of different ranks and different regiments.

A solution to Euler's problem would be equivalent to a pair of orthogonal latin squares of order 6, the symbol set of the first consisting of the 6 ranks and the symbol set of the second consisting of the 6 regiments. Euler convinced himself that his problem was incapable of solution and goes even further:

> I have examined a very great number of tables ... and I do not hesitate to conclude that one cannot produce an orthogonal pair of order 6 and that the same impossibility extends to 10, 14, ... and in general to all the orders which are unevenly even.

Euler was proven correct in his claim that an orthogonal pair of order 6 does not exist [G. Tarry, 1900]; however in 1960 Euler was shown to be wrong for all orders greater than 6. (See the Bose-Parker-Shrikhande theorem, Fact 3.)

Examples:

1. Two mutually orthogonal latin squares of side 3:

    ```
    1 2 3     1 2 3
    2 3 1     3 1 2
    3 1 2     2 3 1
    ```

2. Three mutually orthogonal latin squares of side 4:

    ```
    1 2 3 4     1 2 3 4     1 2 3 4
    4 3 2 1     3 4 1 2     2 1 4 3
    2 1 4 3     4 3 2 1     3 4 1 2
    3 4 1 2     2 1 4 3     4 3 2 1
    ```

3. Two MOLS of order 10:

    ```
    0 4 1 7 2 9 8 3 6 5     0 7 8 6 9 3 5 4 1 2
    8 1 5 2 7 3 9 4 0 6     6 1 7 8 0 9 4 5 2 3
    9 8 2 6 3 7 4 5 1 0     5 0 2 7 8 1 9 6 3 4
    5 9 8 3 0 4 7 6 2 1     9 6 1 3 7 8 2 0 4 5
    7 6 9 8 4 1 5 0 3 2     3 9 0 2 4 7 8 1 5 6
    6 7 0 9 8 5 2 1 4 3     8 4 9 1 3 5 7 2 6 0
    3 0 7 1 9 8 6 2 5 4     7 8 5 9 2 4 6 3 0 1
    1 2 3 4 5 6 0 7 8 9     4 5 6 0 1 2 3 7 8 9
    2 3 4 5 6 0 1 8 9 7     1 2 3 4 5 6 0 9 7 8
    4 5 6 0 1 2 3 9 7 8     2 3 4 5 6 0 1 8 9 7
    ```

12.3.3 ORTHOGONAL ARRAYS

Definition:

An *orthogonal array of size* N, with k *constraints* (or of *degree* k), s *levels* (or of *order* s), and *strength* t, denoted $OA(N, k, s, t)$, is a $k \times N$ array with entries from a set of $s \geq 2$ symbols, having the property that in every $t \times N$ submatrix, every $t \times 1$ column vector appears the same number $\lambda = \frac{N}{s^t}$ of times. The parameter λ is the *index* of the orthogonal array.

Note: An $OA(N, k, s, t)$ is also denoted by $OA_\lambda(t, k, s)$; in this notation, if t is omitted it is understood to be 2, and if λ is omitted it is understood to be 1.

Chapter 12 COMBINATORIAL DESIGNS

Facts:

1. An $OA_\lambda(k,v)$ is equivalent to a transversal design $TD_\lambda(k,v)$.

2. $OA_1(t,k,s)$ are known as *MDS codes* in coding theory.

3. An $OA_\lambda(k,n)$ exists only if $k \le \left\lfloor \frac{\lambda v^2 - 1}{v-1} \right\rfloor$ (Bose-Bush bound). Generally one is interested in finding the largest k for which there exists an $OA_\lambda(k,n)$ (for a given λ and n).

The following table gives the best known upper bounds and lower bounds for the largest k for which there exists a $OA_\lambda(k,n)$. Entries for which the upper and lower bounds match are shown in boldface.

$\lambda \backslash n$	1	2	3	4	5	6	7	8	9	10	11	12	13	14	15	16	17	18
2	**3**	**7**	**11**	**15**	**19**	**23**	**27**	**31**	**35**	**39**	**43**	**47**	**51**	**55**	**59**	**63**	**67**	**71**
3	**4**	**7**	**13**	16	22	**25**	31	34	**40**	43	49	52	58	61	67	70	76	**79**
				13	10		13	25		31	13	49	13	25	37	49	25	
4	5	9	14	**21**	25	30	37	**41**	46	53	57	62	69	73	78	**85**	89	94
			13		10	13	13		37	21	13	61	21	57	21		37	37
5	**6**	11	17	23	**31**	36	42	48	56	**61**	67	73	81	86	92	98	106	111
			8	21		16	18	21	21		18	26	21	21	43	81	36	91
6	**3**	13	20	27	34	43	49	56	63	70	79	85	92	99	106	115	121	128
		7	8	8	8	12	9	12	11	12	8	19	9	17	13	23	11	18
7	**8**	15	23	30	38	46	**57**	64	72	79	87	95	106	**113**	121	128	136	144
			9	29	12	16		29	29	16	29	37	29		36	38	29	64
8	9	17	26	34	43	52	61	**73**	81	90	98	107	116	125	137	**145**	154	162
			9	33	9	17	57		22	41	33	33	22	57	57		41	73
9	**10**	19	29	38	48	58	68	78	**91**	100	110	119	129	139	149	159	172	**181**
			28	37	19	55	28	73		37	28	109	37	55	55	73	73	
10	9	21	32	42	53	64	75	86	97	111	121	132	142	153	164	175	186	197
	4	10	12	10	10	20	12	11	12	12	11	28	12	12	12	19	12	30

12.4 MATROIDS

Linearly independent sets of columns in a matrix and acyclic sets of edges in a graph share many similar properties. Hassler Whitney (1907–1989) aimed to capture these similarities when he defined matroids in 1935. These structures arise naturally in a variety of combinatorial contexts. Moreover, they are precisely the hereditary families of sets for which a greedy strategy produces an optimal set.

12.4.1 BASIC DEFINITIONS AND EXAMPLES

Definitions:

A *matroid* M (also written $(E, \mathcal{I})$ or $(E(M), \mathcal{I}(M))$) is a finite set E (the **ground set** of M) and a collection $\mathcal{I}$ of subsets of E (**independent sets**) such that:
- the empty set is independent;
- every subset of an independent set is independent ($\mathcal{I}$ is *hereditary*);
- if X and Y are independent and $|X| < |Y|$, then there is $e \in Y - X$ such that $X \cup \{e\}$ is independent.

Subsets of E that are not in $\mathcal{I}$ are **dependent**.

A *basis* of a matroid is a maximal independent set. The collection of bases of M is denoted $\mathcal{B}(M)$.

A *circuit* of a matroid is a minimal dependent set. The collection of circuits of M is denoted $\mathcal{C}(M)$.

Matroids M_1 and M_2 are **isomorphic** ($M_1 \cong M_2$) if there is a one-to-one function φ from $E(M_1)$ onto $E(M_2)$ that preserves independence; that is, a subset X of $E(M_1)$ is in $\mathcal{I}(M_1)$ if and only if $\varphi(X)$ is in $\mathcal{I}(M_2)$.

For a matroid M with ground set E and $A \subseteq E$, all maximal independent subsets of A have the same cardinality, called the **rank** of A, written $r(A)$ or $r_M(A)$. The rank $r(M)$ of M is $r(E)$.

A *spanning set* of a matroid M is a subset of the ground set E of rank $r(M)$.

A *hyperplane* of a matroid M is a maximal nonspanning set.

The *closure* $\mathrm{cl}(X)$ (or $\sigma(X)$) of X is $\{\, x \in E \mid r(X \cup \{x\}) = r(X)\,\}$.

A set X is a *closed set* or **flat** if $\mathrm{cl}(X) = X$.

A *loop* of M is an element e such that $\{e\}$ is a circuit.

If $\{f, g\}$ is a circuit, then f and g are **parallel elements**.

Matroid M is a **simple matroid** (or **combinatorial geometry**) if it has no loops and no parallel elements.

A *paving matroid* is a matroid M in which all circuits have at least $r(M)$ elements.

Various classes of matroids are defined in Table 1:

Matroid M is in the specified class if M satisfies the indicated condition:
- **graphic**: $M \cong M(G)$ for some graph G;
- **planar**: $M \cong M(G)$ for some planar graph G;
- **representable over** F: $M \cong M[A]$ for some matrix A over the field F;
- **binary**: representable over $GF(2)$, the 2-element field;
- **ternary**: representable over $GF(3)$;
- **regular**: representable over all fields.

Table 1 Classes of matroids.

matroid M	ground set $E(M)$	independent sets $\mathcal{I}(M)$	bases $\mathcal{B}(M)$	circuits $\mathcal{C}(M)$
uniform matroid, $U_{m,n}$ $(0 \leq m \leq n)$	$\{1, 2, \ldots, n\}$	$\{I \subseteq E :$ $|I| \leq m\}$	$\{B \subseteq E :$ $|B| = m\}$	$\{C \subseteq E :$ $|C| = m+1\}$
$M(G)$, **cycle matroid** of graph G	$E(G)$, edge-set of G	$\{I \subseteq E(G) \mid$ I contains no cycle$\}$	For connected G: edge-sets of spanning trees	edge-sets of cycles
$M[A]$, **vector matroid** of matrix A over field F	column labels of A	$\{I \subseteq E \mid I$ labels a linearly independent multiset of columns$\}$	labels of maximal linearly independent sets of columns	labels of minimal linearly dependent multisets of columns
transversal matroid, $M(\mathcal{A})$, of family $\mathcal{A} = (A_1, A_2, \ldots, A_m)$ where $A_j \subseteq E$	E	partial transversals of $\mathcal{A}$: sets $\{x_{i_1}, \ldots, x_{i_k}\}$, $i_1 < \ldots < i_k$ and $x_{i_j} \in A_{i_j}$	maximal partial transversals of $\mathcal{A}$	minimal sets that are not partial transversals

Facts:

1. If a matroid M is graphic, then $M \cong M(G)$ for some connected graph G.

2. *Whitney's 2-isomorphism theorem*: Two graphs have isomorphic cycle matroids if and only if one can be obtained from the other by performing a sequence of the following operations:
   - choose one vertex from each of two components and identify the chosen vertices;
   - produce a new graph from which the original can be recovered by applying the previous operation;
   - in a graph that can be obtained from the disjoint union of two graphs G_1 and G_2 by identifying vertices u_1 and v_1 of G_1 with vertices u_2 and v_2 of G_2, *twist* the graph by identifying, instead, u_1 with v_2 and u_2 with v_1.

3. If A' is obtained from the matrix A over the field F by elementary row operations, deleting or adjoining zero rows, permuting columns, and multiplying columns by nonzero scalars, then $M[A'] \cong M[A]$. The converse of this holds if and only if F is $GF(2)$ or $GF(3)$.

4. If a matroid M is representable over F and $r(M) \geq 1$, then $M \cong M[I_{r(M)}|D]$, where $I_{r(M)}|D$ consists of an $r(M) \times r(M)$ identity matrix followed by some other matrix D over F.

5. A matroid M is regular if and only if M can be represented over the real numbers by a *totally unimodular matrix* (a matrix for which all subdeterminants are 0, 1, or -1).

6. A matroid M is regular if and only if M is both binary and ternary.

7. The smallest matroids not representable over any field have 8 elements.

8. *Conjecture*: For all n, more than half of all matroids on $\{1, 2, \ldots, n\}$ are paving.

9. The following table lists the numbers of nonisomorphic matroids, simple matroids, and binary matroids with up to 8 elements:

$|E(M)|$	0	1	2	3	4	5	6	7	8
matroids	1	2	4	8	17	38	98	306	1,724
simple	1	1	1	2	4	9	26	101	950
binary	1	2	4	8	16	32	68	148	342

Examples:

1. Let M be the matroid with $E(M) = \{1, 2, \ldots, 6\}$ and $\mathcal{C}(M) = \{\{1\}, \{5,6\}, \{3,4,5\}, \{3,4,6\}\}$. Then $\mathcal{B} = \{\{2,3,4\}, \{2,3,5\}, \{2,3,6\}, \{2,4,5\}, \{2,4,6\}\}$. The following figure shows that M is graphic and binary since $M = M(G_1) = M(G_2)$ and $M = M[A]$ with A being interpreted over $GF(2)$. M is regular since $M = M[A]$ when A is interpreted over any field F. Also M is transversal since $M = M(\mathcal{A})$ where $\mathcal{A} = (\{2\}, \{3,4\}, \{4,5,6\})$.

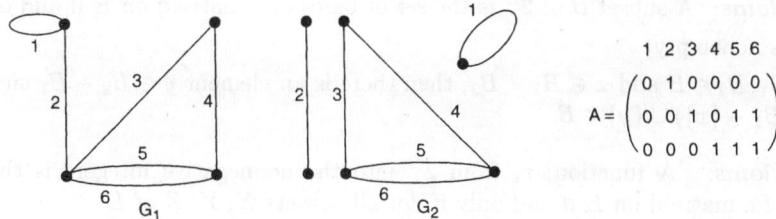

2. *Fano and non-Fano matroids*: Given a finite set E of points in the plane and a collection of lines (subsets of E with at least three elements), no two of which share more than one common point, there is a matroid with ground set E whose circuits are all sets of three collinear points and all sets of four points no three of which are collinear. Two such matroids are shown in the following figure. Each has ground set $\{1, 2, \ldots, 7\}$. On the right is the non-Fano matroid, F_7^-. It differs from the Fano matroid, F_7, on the left by the collinearity of 4, 5, and 6 in the latter.

The matrix in this figure represents F_7 over all fields of characteristic 2, and represents F_7^- over all other fields.

F_7 is binary but non-ternary; F_7^- is ternary but non-binary. Both are non-uniform, non-regular, non-graphic, and non-transversal.

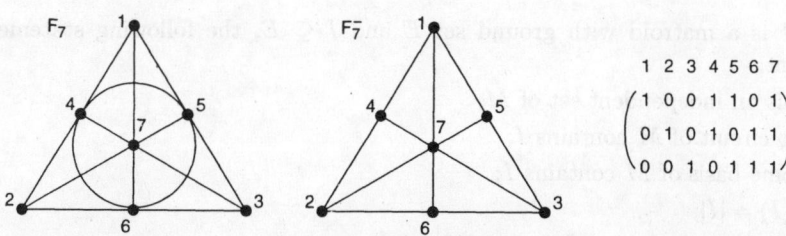

12.4.2 ALTERNATIVE AXIOM SYSTEMS

Matroids can be characterized by many different axiom systems. Some examples of these systems follow. Throughout, E is assumed to be a finite set and 2^E stands for the set of subsets of E.

Definitions:

Circuit axioms: A subset $\mathcal{C}$ of 2^E is the set of circuits of a matroid on E if and only if $\mathcal{C}$ satisfies:

- $\emptyset \notin \mathcal{C}$;
- no member of $\mathcal{C}$ is a proper subset of another;
- *circuit elimination*: if C_1, C_2 are distinct members of $\mathcal{C}$ and $e \in C_1 \cap C_2$, then $\mathcal{C}$ has a member C_3 such that $C_3 \subseteq (C_1 \cup C_2) - \{e\}$.

Note: The circuit elimination axiom can be strengthened to the following:

- *strong circuit elimination*: if $C_1, C_2 \in \mathcal{C}$, $f \in C_1 - C_2$, and $e \in C_1 \cap C_2$, then $\mathcal{C}$ has a member C_3 such that $f \in C_3 \subseteq (C_1 \cup C_2) - \{e\}$.

Basis axioms: A subset $\mathcal{B}$ of 2^E is the set of bases of a matroid on E if and only if:

- $\mathcal{B}$ is nonempty;
- if $B_1, B_2 \in \mathcal{B}$ and $x \in B_1 - B_2$, then there is an element $y \in B_2 - B_1$ such that $(B_1 - \{x\}) \cup \{y\} \in \mathcal{B}$.

Rank axioms: A function, r, from 2^E into the nonnegative integers is the rank function of a matroid on E if and only if, for all subsets X, Y, Z of E:

- $0 \leq r(X) \leq |X|$;
- if $Y \subseteq Z$, then $r(Y) \leq r(Z)$;
- *submodularity*: $r(X \cup Y) + r(X \cap Y) \leq r(X) + r(Y)$.

Closure axioms: A function, cl, from 2^E into 2^E is the closure operator of a matroid on E if and only if, for all subsets X and Y of E:

- $X \subseteq \text{cl}(X)$;
- if $X \subseteq Y$, then $\text{cl}(X) \subseteq \text{cl}(Y)$;
- $\text{cl}(\text{cl}(X)) = \text{cl}(X)$;
- *MacLane-Steinitz exchange*: if $x \in E$ and $y \in \text{cl}(X \cup \{x\}) - \text{cl}(X)$, then $x \in \text{cl}(X \cup \{y\})$.

Fact:

1. If M is a matroid with ground set E and $I \subseteq E$, the following statements are equivalent:
   - I is an independent set of M;
   - no circuit of M contains I;
   - some basis of M contains I;
   - $r(I) = |I|$;
   - for every element e of I, $e \notin \text{cl}(I - \{e\})$.

12.4.3 DUALITY

Definitions:

For a matroid M, let $\mathcal{B}^*(M) = \{ E(M) - B \mid B \in \mathcal{B}(M) \}$. Then $\mathcal{B}^*(M)$ is the set of bases of a matroid M^*, called the **dual** of M, whose ground set is also $E(M)$.

Bases, circuits, loops, and independent sets of M^* are called **cobases**, **cocircuits**, **coloops**, and **coindependent sets** of M.

For a graph G, the **cocycle matroid** (or **bond matroid**) of G is the dual of $M(G)$ and is denoted by $M^*(G)$.

A matroid M is **cographic** if $M \cong M^*(G)$ for some graph G.

A class of matroids is **closed under duality** if the dual of every member of the class is also in the class.

Facts:

1. For all matroids M, $(M^*)^* = M$.

2. For all matroids M, the rank function of M^* is given by $r^*(X) = |X| - r(M) + r(E - X)$.

3. The cocircuits of every matroid M are the minimal sets having nonempty intersection with every basis of M.

4. The cocircuits of every matroid M are the minimal nonempty sets C^* such that $|C^* \cap C| \neq 1$ for every circuit C of M.

5. For every graph G, the circuits of $M^*(G)$ are the minimal edge cuts of G.

6. A graphic matroid is cographic if and only if it is planar.

7. The following classes of matroids are closed under duality: uniform matroids, matroids representable over a fixed field F, planar matroids, and regular matroids. The classes of graphic and transversal matroids are not closed under duality.

8. The following are special sets and their complements in a matroid M and M^*:

| X | basis of M | independent set of M | circuit of M |
| $E - X$ | basis of M^* | spanning set of M^* | hyperplane of M^* |

Example:

1. The following are duals of some basic examples:

matroid	dual
$U_{m,n}$	$U_{n-m,n}$
$M(G)$ (G plane)	$M(G^*)$, where G^* is the dual of G
$M[I_r \mid D]$ ($[I_r \mid D]$ an $r \times n$ matrix)	$M[-D^T \mid I_{n-r}]$, same order of column labels as $[I_r \mid D]$

12.4.4 FUNDAMENTAL OPERATIONS

Definitions:

Three basic constructions for matroids M, M_1, and M_2 are defined in the following table. $M \backslash T$ and M/T are also written as $M|(E-T)$ and $M.(E-T)$ and are called the **restriction** and **contraction** of M to $E-T$. $M \backslash \{e\}$ and $M/\{e\}$ are written as $M \backslash e$ and M/e.

matroid	$\mathcal{I}$	$\mathcal{C}$	rank
$M \backslash T$ (**deletion** of T from M)	$\{I \subseteq E(M) - T \mid I \in \mathcal{I}(M)\}$	$\{C \subseteq E(M) - T \mid C \in \mathcal{C}(M)\}$	$r_{M \backslash T}(X) = r_M(X)$
M/T (**contraction** of T from M)	$\{I \subseteq E(M) - T \mid I \cup B_T \in \mathcal{I}(M)$ for some B_T in $\mathcal{B}(M|T)\}$	minimal nonempty empty members of $\{C-T \mid C \in \mathcal{C}(M)\}$	$r_{M/T}(X) = r_M(X \cup T) - r_M(T)$
$M_1 \oplus M_2$ (**direct sum** of M_1 and M_2)	$\{I_1 \cup I_2 \mid I_j \in \mathcal{I}(M_j)\}$	$\mathcal{C}(M_1) \cup \mathcal{C}(M_2)$	$r_{M_1 \oplus M_2}(X) = r_1(X \cap E(M_1)) + r_2(X \cap E(M_2))$

Matroid N is a **minor** of M if N can be obtained from M by a sequence of deletions and contractions. The minor N is **proper** if $N \neq M$.

A matroid is **connected** if it cannot be written as the direct sum of two *nonempty matroids* (matroids with nonempty ground sets).

Facts:

In each of the following, M, M_1, and M_2 are matroids.

1. $M \backslash X \backslash Y = M \backslash (X \cup Y) = M \backslash Y \backslash X$; $M/X/Y = M/(X \cup Y) = M/Y/X$; and $M \backslash X / Y = M/Y \backslash X$.
2. $M_1 \oplus M_2 = M_2 \oplus M_1$.
3. $(M/T)^* = M^* \backslash T$; and $(M \backslash T)^* = M^*/T$. (Deletion and contraction are dual operations.)
4. *The scum theorem*: Every minor of M can be written as $M \backslash X / Y$ for some independent set Y and coindependent set X. (The name derives from the fact that an isomorphic copy of every simple minor of a matroid occurs at (that is, floats to) the top of the lattice.) (D. A. Higgs) [CrRo70]
5. The following are equivalent:
    - M is connected;
    - M^* is connected;
    - every two distinct elements of M are in a circuit;
    - there is no proper nonempty subset T of $E(M)$ such that $M \backslash T = M/T$;
    - there is no proper nonempty subset T of $E(M)$ such that $r(T) + r(E(M) - T) = r(M)$;
    - there is no proper nonempty subset T of $E(M)$ such that $r(T) + r^*(T) = |T|$.

6. If M is connected, then M is uniquely determined by the set of circuits containing some fixed element of $E(M)$.
7. If M is connected and $e \in E(M)$, then $M\backslash e$ or M/e is connected.
8. $F_7 \oplus F_7^-$ is not representable over any field.

Examples:
1. $U_{m,n}\backslash e = U_{m,n-1}$ unless $m = n$ when $U_{m,n}\backslash e = U_{m-1,n-1}$.
2. $U_{m,n}/e = U_{m-1,n-1}$ unless $m = 0$ when $U_{m,n}/e = U_{m,n-1}$.
3. $M(G)\backslash e = M(G\backslash e)$ where $G\backslash e$ is obtained from G by deleting the edge e.
4. $M(G)/e = M(G/e)$ where G/e is obtained from G by contracting the edge e.
5. $M[A]\backslash e$ is the vector matroid of the matrix obtained by deleting column e from A.
6. If e corresponds to a standard basis vector in A, then $M[A]/e$ is the vector matroid of the matrix obtained by deleting both the column e and the row containing the one of e.

12.4.5 CHARACTERIZATIONS

Many matroid results characterize various classes of matroids. Some examples of such results appear below. The Venn diagram in the following figure indicates the relationship between certain matroid classes.

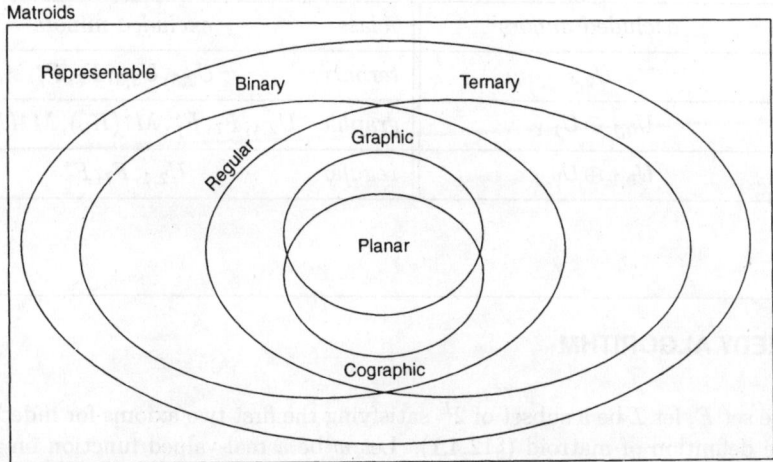

Definition:
Let M_1 and M_2 be two binary matroids such that $E(M_1) \cap E(M_2) = T$, $M_1|T = M_2|T$, and no cocircuit of M_1 or M_2 is contained in T. The **2-sum** and **3-sum** of M_1 and M_2 are matroids on $(E(M_1) \cup E(M_2)) - T$ whose flats are those sets $F - T$ such that $F \cap E(M_i)$ is a flat of M_i for $i = 1, 2$. The 2-sum occurs when $|T| = 1$, $|E(M_i)| \geq 3$, and T is not a loop of M_i, and the 3-sum occurs when $|E(M_i)| \geq 7$ and T is a 3-element circuit of M_i.

Facts:

1. The following are equivalent for a matroid M:
   - M is uniform;
   - every circuit of M has $r(M) + 1$ elements;
   - every circuit of M meets every cocircuit of M.

2. The following are equivalent for a matroid M:
   - M is binary;
   - for every circuit C and every cocircuit C^*, $|C \cap C^*|$ is even;
   - for every circuit C and every cocircuit C^*, $|C \cap C^*| \neq 3$;
   - for all $C_1, C_2 \in \mathcal{C}$, $(C_1 - C_2) \cup (C_2 - C_1)$ is a disjoint union of circuits.

3. The class of regular matroids is the class of matroids that can be constructed by direct sums, 2-sums, and 3-sums from graphic matroids, cographic matroids, and copies of R_{10} (the matroid that is represented over $GF(2)$ by the ten 5-tuples with exactly three ones). (This last fact is the basis of a polynomial-time algorithm to determine whether a real matrix is totally unimodular.)

4. *Excluded-minor theorems*: Many classes of matroids are *minor-closed*; that is, every minor of a member of the class is also in the class. Such classes can be characterized by listing their *excluded minors* (those matroids that are not in the class but have all their proper minors in the class). Some important examples of such results are given in the following table. The class of transversal matroids is not minor-closed since a contraction of a transversal matroid need not be transversal.

class	excluded minors	class	excluded minors
binary	$U_{2,4}$	ternary	$U_{2,5}, U_{3,5}, F_7, F_7^*$
uniform	$U_{0,1} \oplus U_{1,1}$	graphic	$U_{2,4}, F_7, F_7^*, M^*(K_5), M^*(K_{3,3})$
paving	$U_{0,1} \oplus U_{2,2}$	regular	$U_{2,4}, F_7, F_7^*$

12.4.6 THE GREEDY ALGORITHM

For a finite set E, let $\mathcal{I}$ be a subset of 2^E satisfying the first two axioms for independent sets in the definition of matroid (§12.4.1). Let w be a real-valued function on E. For $X \subseteq E$, let $w(X)$, the *weight* of X, be $\sum_{x \in X} w(x)$, and let $w(\emptyset) = 0$.

Facts:

1. Matroids have an important relationship to the greedy algorithm, Algorithm 1, that makes them important in optimization problems.

2. $\mathcal{I}$ (a subset of 2^E satisfying the first two axioms for independent sets in the definition of matroid) is the set of independent sets of a matroid on E if and only if, for all real-valued weight functions w on E, the set B_G produced by the greedy algorithm is a maximal member of $\mathcal{I}$ of maximum weight.

> **Algorithm 1:** The greedy algorithm for $(\mathcal{I}, w)$.
> $X_0 := \emptyset; \ j := 0$
> while $E - X_j$ contains an element e such that $X_j \cup \{e\} \in \mathcal{I}$
> $\quad e_{j+1} :=$ an element e of maximum weight such that $X_j \cup \{e\} \in \mathcal{I}$
> $\quad X_{j+1} := X_j \cup \{e_{j+1}\}$
> $\quad j := j + 1$
> $B_G := X_j$

Example:

1. Let G be a connected graph with each edge e having a cost $c(e)$. Define $w(e) = -c(e)$. Then the greedy algorithm is just Kruskal's algorithm (§10.1.2) and B_G is the edge-set of a spanning tree of minimum cost.

REFERENCES

Printed Resources:

[An93] I. Anderson, *Combinatorial Designs: Construction Methods*, Ellis Horwood, 1993.

[BeJuLe86] T. Beth, D. Jungnickel, and H. Lenz, *Design Theory*, Cambridge University Press, 1986.

[CoDi96] C. J. Colbourn and J. H. Dinitz, eds., *The CRC Handbook of Combinatorial Designs*, CRC Press, 1996. (A comprehensive source of information on combinatorial designs.)

[CrRo70] H. H. Crapo and G.-C. Rota, *On the Foundations of Combinatorial Theory: Combinatorial Geometries*, MIT Press, 1970.

[DéKe74] J. Dénes and A. D. Keedwell, *Latin Squares and Their Applications*, Academic Press, 1974.

[DiSt92] J. H. Dinitz and D. R. Stinson, eds., *Contemporary Design Theory: A Collection of Surveys*, John Wiley & Sons, 1992.

[Ox92] J. G. Oxley, *Matroid Theory*, Oxford University Press, 1992. (The main source for §12.4; contains references for all stated results.)

[Re89] A. Recski, *Matroid Theory and Applications in Electric Network Theory and Statics*, Springer-Verlag, 1989.

[StSt87] A. P. Street and D. J. Street, *Combinatorics of Experimental Design*, Clarendon Press, 1987.

[Wa88] W. D. Wallis, *Combinatorial Designs*, Marcel Dekker, 1988.

[We76] D. J. A. Welsh, *Matroid Theory*, Academic Press, 1976.

[Wh86] N. White, ed., *Theory of Matroids*, Cambridge University Press, 1986.

Web Resources:

http://cacr.math.uwaterloo.ca/ dstinson/papers/designnotes.ps (Contains a complete set of notes (approximately 100 pages, 615K, in postscript) from Doug Stinson for a "Combinatorial Designs with Selected Applications" course.)

http://gams.nist.gov/ (Guide To Available Mathematical Software at the National Institute of Science and Technology: a cross-index and virtual repository of mathematical and statistical software components of use in computational science and engineering; contains a program for finding t-designs.)

http://kapis.www.wkap.nl/kapis/CGI-BIN/WORLD/journalhome.htm?0925-1022 (Designs, Codes and Cryptography has a site with its table of contents.)

http://lib.stat.cmu.edu/designs/ (The Designs Archive at Statlib: has some very useful programs for making orthogonal arrays.)

http://members.aol.com/matroids (Contains information on matroids, has a software package for binary matroids, and has links to several other sites on matroid theory.)

http://sdcc12.ucsd.edu/ xm3dg/cover.html (La Jolla Covering Repository: contains coverings $C(v, k, t)$ with $v \leq 32$, $k \leq 16$, $t \leq 8$, and less than 5,000 blocks.)

http://winnie.math.tu-berlin.de/~ziegler (Source of information on oriented matroids.)

http://www.cecm.sfu.ca/organics/papers/lam/paper/html/paper.html (Information on the search for a finite projective plane of order 10.)

http://www.cse.cuhk.edu.hk/ luk036/mdc/ (Contains tables of minimal difference covers.)

http://www.emba.uvm.edu/ jcd/ (Journal of Combinatorial Designs: has information including its contents back to volume 1.)

http://www.emba.uvm.edu/ dinitz/hcd.html (The Handbook of Combinatorial Designs has a website that lists new results in design theory that have been discovered since its 1996 publication.)

http://www.netlib.org/ (Netlib Repository: a collection of mathematical software, papers, and databases.)

http://www.research.att.com/ njas/gosset/index.html (GOSSET: A general purpose program for designing experiments.)

http://www.research.att.com/ njas/hadamard/ (Contains an extensive collection of Hadamard matrices.)

http://www.research.att.com/ njas/oadir/index.html (Contains an extensive collection of orthogonal arrays.)

http://www.utu.fi/ honkala/designs.ps (Notes from Ian Anderson and Iiro Honkala for a "Short Course in Combinatorial Designs"; about 40 pages, 272K, in postscript.)

13

DISCRETE AND COMPUTATIONAL GEOMETRY

13.1 Arrangements of Geometric Objects   *Ileana Streinu*
   13.1.1 Point Configurations
   13.1.2 Line and Hyperplane Arrangements
   13.1.3 Pseudoline Arrangements
   13.1.4 Oriented Matroids

13.2 Space Filling   *Karoly Bezdek*
   13.2.1 Packing
   13.2.2 Covering
   13.2.3 Tiling

13.3 Combinatorial Geometry   *János Pach*
   13.3.1 Convexity
   13.3.2 Incidences
   13.3.3 Distances
   13.3.4 Coloring

13.4 Polyhedra   *Tamal K. Dey*
   13.4.1 Geometric Properties of Polyhedra
   13.4.2 Triangulations
   13.4.3 Face Numbers

13.5 Algorithms and Complexity in Computational Geometry   *Jianer Chen*
   13.5.1 Convex Hulls
   13.5.2 Triangulation Algorithms
   13.5.3 Voronoi Diagrams and Delaunay Triangulations
   13.5.4 Arrangements
   13.5.5 Visibility

13.6 Geometric Data Structures and Searching   *Dina Kravets*
   13.6.1 Point Location
   13.6.2 Range Searching
   13.6.3 Ray Shooting and Lines in Space

13.7 Computational Techniques   *Nancy M. Amato*
   13.7.1 Parallel Algorithms
   13.7.2 Randomized Algorithms
   13.7.3 Parametric Search
   13.7.4 Finite Precision
   13.7.5 Degeneracy Avoidance

13.8 Applications of Geometry   W. Randolph Franklin
    13.8.1 Mathematical Programming
    13.8.2 Polyhedral Combinatorics
    13.8.3 Computational Convexity
    13.8.4 Motion Planning in Robotics
    13.8.5 Convex Hull Applications
    13.8.6 Nearest Neighbor
    13.8.7 Computer Graphics
    13.8.8 Mechanical Engineering Design and Manufacturing
    13.8.9 Layout Problems
    13.8.10 Graph Drawing
    13.8.11 Geographic Information Systems
    13.8.12 Geometric Constraint Solving
    13.8.13 Implementations
    13.8.14 Geometric Visualization

INTRODUCTION

This chapter outlines the theory and applications of various concepts arising in two rapidly growing, interrelated areas of geometry: discrete geometry (which deals with topics such as space filling, arrangements of geometric objects, and related combinatorial problems) and computational geometry (which deals with the many aspects of the design and analysis of geometric algorithms). A more extensive treatment of discrete and computational geometry can be found in [GoO'R97].

GLOSSARY

anti-aliasing: the filtering out of high-frequency spatial components of a signal, to prevent artifacts, or aliases, from appearing in the output image.

aperiodic (prototile): a prototile in d-dimensional Euclidean space such that the prototile admits a tiling of the space, yet all such tilings are nonperiodic.

arrangement (of lines in the plane): the planar straight-line graph whose vertices are the intersection points of the lines and whose edges connect consecutive intersection points on each line (it is assumed that all lines intersect at a common point at infinity).

arrangement graph: a graph associated with a Euclidean or projective line arrangement, or a big circle arrangement.

aspect ratio (of a simplex): the ratio of the radius of the circumscribing sphere to the radius of the inscribing sphere of the simplex; for a triangulation, the largest aspect ratio of a simplex in the triangulation.

basis (of a point configuration in $\mathcal{R}^d$): a subset of the point configuration that is a simplex of the ambient space $\mathcal{R}^d$.

basis (of a vector configuration in $\mathcal{R}^d$): a subset of the vector configuration that is a basis of the ambient space $\mathcal{R}^d$.

big-circle arrangement: the intersection of a central plane arrangement with the unit sphere in $\mathcal{R}^3$.

boundary (of a polyhedron): the vertices, edges, and higher dimensional facets of the polyhedron.

cell (of a line arrangement): a connected component of the complement in $\mathcal{R}^2$ of the union of the points on the lines.

centerpoint (of a point configuration P of size n): a point q, not necessarily in P, such that for any hyperplane containing q there are at least $\lceil \frac{n}{d+1} \rceil$ points in each semi-space induced by the hyperplane.

central hyperplane arrangement (in $\mathcal{R}^d$): a finite set of central hyperplanes, not all of them going through the same point.

central plane arrangement (in $\mathcal{R}^3$): a finite set of central planes.

chain: a planar straight-line graph with vertices $v_1, \ldots, v_n$ and edges $\{v_1, v_2\}, \{v_2, v_3\}, \ldots, \{v_{n-1}, v_n\}$.

chirotope: for an ordered point configuration, the set of all signed bases of the configuration; for an ordered vector configuration, the set of all signed bases of the configuration.

circuit (of a set of labeled vectors $V = \{v_1, \ldots, v_n\}$): the signed set $C = (C^+, C^-)$, where $C^+ = \{j \mid \alpha_j > 0\}$ and $C^- = \{j \mid \alpha_j < 0\}$, of indices of the non-null coefficients α_j in a minimal linear dependency $V' = \{v_{i_1}, \ldots, v_{i_k}\}$ of V with $\sum_{j=1}^{k} \alpha_j v_{i_j} = \underline{0}$.

class library: in an object-oriented computer language, a set of new data types and operations on them, activated by sending a data item a message.

closed halfspace: the set of all points on a hyperplane and the points on one side of the same hyperplane.

cluster of rank 3 hyperline sequences (associated with a vector configuration): the ordered set of stars, one for each point in the configuration.

cluster of stars (associated with a point configuration): the ordered set of stars, one for each point in the configuration.

cocircuit (of a labeled vector configuration V): a signed set $C = (C^+, C^-)$ of the set $\{1, 2, \ldots, n\}$, induced by a subset of $d - 1$ vectors spanning a central hyperplane h. For an arbitrary orientation of h, C^+ is the set of indices of elements in V lying in h^+ and C^- is the set of indices of elements in V lying in h^-.

computational convexity: the study of high-dimensional convex bodies.

contraction (on element i in a rank d central plane arrangement): the arrangement obtained by identifying h_i with $\mathcal{R}^{d-1}$ and intersecting it with all the other hyperplanes to obtain a rank $d-1$ arrangement with one element less.

convex: property of a subset of a Euclidean space that for every pair of points in the set the linear segment joining them is contained in the set.

convex body: a closed and bounded convex set with nonempty interior.

convex d-polyhedron: the intersection of a finite number of closed halfspaces in $\mathcal{R}^d$.

convex decomposition (of a polyhedron): its partition into interior disjoint convex pieces.

convex hull (of a set of points): the smallest convex set containing the given set of points.

convex polygon: a polytope in the plane.

convex polyhedron: the intersection of a finite number of half-spaces.

convex polytope: a bounded convex polyhedron.

convex position: property of a set of points that it is the vertex set of a polytope.

convex set: a subset of d-dimensional Euclidean space such that for every pair of distinct points in the set, the segment with these two points as endpoints is contained in the set.

covering: a family of convex bodies in d-dimensional Euclidean space such that each point belongs to at least one of the convex bodies.

cyclic d-polytope: the convex hull of a set of $n \geq d+1$ points on the moment curve in $\mathcal{R}^d$. The moment curve in $\mathcal{R}^d$ is defined parametrically by $x(t) = (t, t^2, ..., t^d)$.

Davenport-Schinzel sequence (of order s): a sequence of characters over an alphabet of size n such that no two consecutive characters are the same, and for any pair of characters, a and b, there is no alternating subsequence of length $s+2$ of the form $...a...b...a...b....$

deletion: the removal of a point (vector, line, etc.) from a configuration and recording the oriented matroid (chirotope, circuits, cluster of stars, etc.) only for the remaining points.

density (of a covering): the common value (if it exists) of the lower density and upper density of the covering.

density (of a packing): the common value (if it exists) of the lower density and upper density of the packing.

dual polytopes: two polytopes P and Q such that there exists a one-to-one correspondence δ between the set of faces of P and Q where two faces $f_1, f_2 \in P$ satisfy $f_1 \subset f_2$ if and only if $\delta(f_1) \supset \delta(f_2)$ in Q.

duality transformation: a mapping of points to lines and lines to points that preserves incidences.

Euclidean hyperplane arrangement (in $\mathcal{R}^d$): a finite set of affine hyperplanes, not all of them going through the same point.

Euclidean line arrangement: a finite set of planar lines, not all of them going through the same point.

Euclidean pseudoconfiguration of points: a pair consisting of a planar set of points and a pseudoline arrangement, such that for every pair of distinct points there exists a unique pseudoline incident with them.

k-face: an open set of dimension k that is part of the boundary of a polyhedron. (0-faces, 1-faces and $(d-1)$-faces of a d-polyhedron are called *vertices*, *edges*, and *facets*.

face vector (of a d-polyhedron): the d-dimensional vector $(f_0, f_1, ..., f_{d-1})$, where f_i is the number of i-dimensional faces of the d-polyhedron.

face-to-face (tiling): a tiling of d-dimensional Euclidean space by convex d-polytopes such that the intersection of any two tiles is a face of each tile, possibly the (improper) empty face.

general position: property of a set of vectors that every subset of d elements is a basis; property of a set of points that every subset of d elements is a simplex.

genus (of a manifold 3-polyhedron): the genus number of its boundary, if the boundary is a 2-manifold.

geographic information system (GIS): an information system designed to capture, store, manipulate, analyze, and display spatial or geographically-referenced data.

geometric constraint solving: the problem of locating a set of geometric elements given a set of constraints between them.

graphical user interface (GUI): a mechanism that allows a user to interactively control a computer program with a bitmapped display by using a mouse or pointer to select menu items, move sliders or valuators, etc.

Grassmann-Plücker relations (rank 3): the identities $[123][145] - [124][135] + [125][134] = 0$ satisfied by the determinants $[ijk] = \det(v_i, v_j, v_k)$, for any five vectors $v_i, 1 \leq i \leq 5$.

half-space: one of the two connected components of the complement of a hyperplane.

ham-sandwich cut: a hyperplane that simultaneously bisects d point configurations in d-dimensional Euclidean space.

hyperplane: in d dimensions the set of all points on a $(d-1)$-dimensional plane.

hyperplane arrangement: the partitioning of the Euclidean space $\mathcal{R}^d$ into connected regions of different dimensions (vertices, edges, etc.) by a finite set of hyperplanes.

isogonal (tiling): a tiling such that the group of symmetries acts transitively on the vertices of the tiles.

isomorphic (vector or point configurations): configurations having the same order type, after possibly relabeling their elements.

isotoxal (tiling): a tiling such that the group of symmetries acts transitively on the edges of the tiles.

lattice (tiling): a tiling of d-dimensional Euclidean space by translates of a tile such that the corresponding translation vectors form a d-dimensional lattice.

k-level (in a nonvertical arrangement of n lines): the lower boundary of the set of points in $\mathcal{R}^2$ having exactly k lines above and $n-k$ below.

line arrangement: the partitioning of the plane into connected regions (cells, edges and vertices) induced by a finite set of lines.

lower density (of a covering $\mathcal{C}$): $\underline{\nu}(\mathcal{C}) = \liminf_{R \to +\infty} \frac{\sum_{K_i \cap B_R \neq \emptyset} \mathrm{Vol}(K_i)}{\mathrm{Vol}(B_R)}$, where each K_i is a convex body in the covering $\mathcal{C}$ of d-dimensional Euclidean space and B_R is the closed ball of radius R centered at the origin.

lower density (of a packing $\mathcal{P}$): $\underline{\delta}(\mathcal{P}) = \liminf_{R \to +\infty} \frac{\sum_{K_i \subset B_R} \mathrm{Vol}(K_i)}{\mathrm{Vol}(B_R)}$ where each K_i is a convex body in the packing $\mathcal{P}$ of d-dimensional Euclidean space and B_R is the closed ball of radius R centered at the origin.

lower envelope (of a nonvertical line arrangement): the half-plane intersection of the half-planes below the lines of the arrangement.

manifold d-polyhedron: a polyhedron whose boundary is topologically the same as a $(d-1)$-manifold; i.e., every point on the boundary of a manifold d-polyhedron has a small neighborhood that looks like an open d-ball.

mathematical programming: the large-scale-optimization of an objective function of many variables subject to constraints.

minor (of an oriented matroid given by hyperline sequences): an oriented matroid obtained by a sequence of deletions and/or contractions.

monohedral (tiling): a tiling $\mathcal{T}$ of d-dimensional Euclidean space in which all tiles are congruent to one fixed set T, the (metrical) prototile of $\mathcal{T}$.

nonconvex polyhedron: the union of a set of convex polyhedra such that the underlying space is connected and nonconvex.

non-manifold d-polyhedron: a d-polyhedron that does not have manifold boundary.

nonperiodic (tiling): a tiling such that its group of symmetries contains no translation other than the identity.

normal (tiling): a tiling of d-dimensional Euclidean space by convex polytopes such that there exist positive real numbers r and R such that each tile contains a Euclidean ball of radius r and is contained in a Euclidean ball of radius R.

oracle: an algorithm that gives information about a convex body

order type (of a vector or point configuration): the collection of all semi-spaces of the configuration.

oriented matroid: a pair $\mathcal{M} = (n, \mathcal{L})$, where $\mathcal{L}$, the set of covectors of $\mathcal{M}$, is a subset of $\{+, -, 0\}^n$ and satisfies the properties: $0 \in \mathcal{L}$; if $X \in \mathcal{L}$, then $-X \in \mathcal{L}$; if $X, Y \in \mathcal{L}$, then $X \circ Y \in \mathcal{L}$; if $X, Y \in \mathcal{L}$ and $i \in S(X,Y) = \{i \mid X_i = -Y_i \neq 0\}$, then there is $Z \in \mathcal{L}$ such that $Z_i = 0$; for each $j \notin S(X,Y)$, $Z_j = (X \circ Y)_j = (Y \circ X)_j$.

oriented matroid given by a chirotope: an abstract set of points labeled $\{1, \ldots, n\}$, together with a function satisfying the chirotope axioms.

packing: a family of convex bodies in d-dimensional Euclidean space such that no two have an interior point in common.

parallel algorithm: an algorithm that concurrently uses more than one processing element during its execution.

parallel random access machine (PRAM): a synchronous machine in which each processor is a sequential RAM, and processors communicate using a shared memory.

parametric search: an algorithmic technique for solving optimization problems.

periodic (tiling): a tiling of d-dimensional Euclidean space such that the group of all symmetries of the tiling contains translations in d linearly independent directions.

planar straight-line graph: a planar graph such that each edge is a straight line.

point configuration (of dimension d): a finite set of points affinely spanning $\mathcal{R}^d$.

point location problem: the problem of determining which region of a given subdivision of $\mathcal{R}^d$ contains a given point.

polar-duality of vectors and central planes (in $\mathcal{R}^3$): a mapping associating with a vector v in $\mathcal{R}^3$ an oriented central plane h having v as its normal vector, and vice versa.

polyhedron: the intersection of a finite number of closed half-spaces in d-dimensional Euclidean space.

polytope: a bounded polyhedron.

d-polytope: a convex d-polyhedron for which there exists a d-dimensional cube containing it inside; that is, a bounded convex d-polyhedron.

$\mathcal{H}$-polytope: a polytope defined as the intersection of d half-spaces in d-dimensional Euclidean space.

$\mathcal{V}$-polytope: a polytope defined as the convex hull of d points in d-dimensional Euclidean space.

projective line arrangement: a finite set of projective lines in the projective plane.

prototile: the single tile used repeatedly in a monohedral tiling.

pseudoline arrangement: a finite collection of simple planar curves that intersect pairwise in exactly one point, where they cross.

Radon partition (of a set of labeled points P): a signed set $C = (C^+, C^-)$ of points of P such that the convex hull of the points in C^+ intersects the convex hull of the points in C^-.

randomized algorithm: an algorithm that makes random choices during its execution.

range counting problem: the problem of counting the number of points of a given set of points that lie in a query range.

range emptiness problem: the problem of determining if a query range contains any points of a given set of points.

range reporting problem: the problem of determining all the points of a given set of points that lie in a query range.

rank 3 hyperline sequence (associated with a vector $v \in V \subseteq \mathcal{R}^3$): an alternating circular sequence of subsets of indices in $\overline{E}_n$ obtained by rotating an oriented central plane in counterclockwise order around the line through v.

ray: a half-line that is directed away from its endpoint.

ray shooting problem: the problem of determining the first object in a set of geometric objects that is hit by a query ray.

real random access machine (***real RAM***): a model of computation in which values can be arbitrarily long real numbers, and all standard operations such as $+, -, \times$, and $\div$ can be performed in unit time regardless of operand length.

realizable (pseudoline arrangement): a pseudoline arrangement isomorphic to a line arrangement.

reflex edges: edges of a nonconvex 3-polyhedron that subtend an inner dihedral angle greater than 180°.

regular (polygon): a polygon with all sides congruent and all interior angles equal.

regular (polytope): a d-polytope ($d > 0$) with all its facets regular $(d-1)$-polytopes that are combinatorially equivalent; a regular 0-polytope is a vertex.

regular (tiling): a monohedral tiling of the plane with a regular polygon as prototile.

reorientation (of a vector configuration $V = \{v_1, \ldots, v_n\}$): a vector configuration $V' = \{v'_1, \ldots, v'_n\}$ such that each v'_i is equal to v_i or $-v_i$.

semiregular (polyhedron): a convex polyhedron with each face a regular polygon, but where more than one regular polygon can be used as a face.

semiregular (tiling): a tiling of the plane using n prototiles with the same numbers of polygons around each vertex.

semi-space (of a configuration induced by a hyperplane): the set of indices of the configuration lying on one side of the hyperplane.

semi-space (of a vector or point configuration): a semi-space induced by some hyperplane.

k-set (of a point configuration): a semi-space of the configuration of size k.

d-dimensional simplex (or **d-simplex**): a d-polytope with $d+1$ vertices.

simplicial complex: a triangulation of a polyhedron such that for any two simplices in the triangulation, either the intersection of the simplices is empty or is a face of both simplices.

simplicial polytope: a polytope in which all faces are simplices.

(standard affine) pseudo polar-duality: the association between an x-monotone pseudoline arrangement $L = \{l_1, \ldots, l_n\}$ given in slope order and a pseudo configuration of points (P, L'), $P = \{p_1, \ldots, p_n\}$, given in increasing order of the x-coordinates and with L' being x-monotone, satisfying the property that the cluster of stars associated with L and to P are the same.

straight line dual: given the Voronoi diagram of a set $\{p_1, \ldots, p_n\}$ of points in the plane, the planar straight-line graph whose vertices are the points in the set, with two vertices p_i and p_j adjacent if and only if the regions $V(p_i)$ and $V(p_j)$ share a common edge.

strictly convex: the property of a convex set that its boundary contains no line segment.

symmetry (of a tiling): a Euclidean motion that maps each tile of the tiling onto a tile of the tiling.

tile: an element of a tiling.

tiling (of Euclidean d-space): a countable family $\mathcal{T}$ of closed topological d-cells of $\mathcal{R}^d$ that cover $\mathcal{R}^d$ without gaps and overlaps.

triangulation (of a d-polyhedron): a convex decomposition where each convex piece of the decomposition is a d-simplex.

triangulation (of a simple polygon): an augmentation of the polygon with non-intersecting diagonal edges connecting vertices of the polygon such that in the resulting planar straight-line graph every bounded face is a triangle.

uniform chirotope: a chirotope function that takes nonzero values on all d-tuples

upper density (of a covering $\mathcal{C}$): $\overline{\nu}(\mathcal{C}) = \limsup\limits_{R \to +\infty} \frac{\sum_{K_i \cap B_R \neq \emptyset} \text{Vol}(K_i)}{\text{Vol}(B_R)}$, where each K_i is a convex body in the covering $\mathcal{C}$ of d-dimensional Euclidean space and B_R is the closed ball of radius R centered at the origin.

upper density (of a packing $\mathcal{P}$): $\overline{\delta}(\mathcal{P}) = \limsup\limits_{R \to +\infty} \frac{\sum_{K_i \subset B_R} \text{Vol}(K_i)}{\text{Vol}(B_R)}$ where each K_i is a convex body in the packing $\mathcal{P}$ of d-dimensional Euclidean space and B_R is the closed ball of radius R centered at the origin.

upper envelope (of a nonvertical line arrangement): the half-plane intersection of the half-planes above the lines of the arrangement.

vector configuration (of dimension d): a finite set of vectors spanning $\mathcal{R}^d$.

visibility graph: given n nonintersecting line segments in the plane, the graph whose vertices are the endpoints of the line segments, with two vertices adjacent if and only if they are visible from each other.

visibility problem: the problem of finding what is visible, given a configuration of objects and a viewpoint.

Voronoi cell (with center c_i): the convex polyhedral set $V_i = \{x \in \mathcal{R}^d : |x - c_i| = \min_j |x - c_j|\}$, where $c_1, c_2, \ldots$ are centers of unit balls in a packing of d-dimensional Euclidean space.

Voronoi diagram (of a set of points $\{p_1, \ldots, p_n\}$ in d-dimensional Euclidean space): the partition of d-dimensional Euclidean space into convex polytopes $V(p_i)$ such that $V(p_i)$ is the locus of points that are closer to p_i than to any other point in p_j.

zone (of a line in an arrangement): the set of cells of the arrangement intersected by the line.

zonotope: the vector (Minkowski) sum of a finite number of line segments.

13.1 ARRANGEMENTS OF GEOMETRIC OBJECTS

A wide range of applied fields (statistics, computer graphics, robotics, geographical databases) depend on solutions to geometric problems: polygon intersection, visibility computations, range searching, shortest paths among obstacles, just to name a few.

These problems typically start with "consider a finite set of points (or lines, segments, curves, hyperplanes, polygons, polyhedra, etc.)". The *combinatorial* properties of these sets, or arrangements, of objects (incidence, order, partitioning, separation, convexity) set the foundations for the *algorithms* developed in the field of computational geometry.

In this chapter attention is focused on the most studied and best understood arrangements of geometric objects: points, lines and hyperplanes. Introducing the concepts relies on linear algebra. The combinatorial properties studied belong however to a relatively new field, the theory of oriented matroids, which has sometimes been described as *linear algebra without coordinates*.

Several fundamental types of questions are asked about these arrangements. The most basic is the classification problem, whose goal is to find combinatorial parameters allowing the partitioning of the (uncountable) set of all possible arrangements of n objects into a finite number of equivalence classes. Examples of such structures for point and line arrangements include semi-spaces, Radon partitions, chirotopes, hyperline sequences, etc. They satisfy simple properties known as axiomatic systems for oriented matroids, which lead to the definition of an abstract class of objects generalizing finite point and vector sets. In dimension 2 oriented matroids can be visualized topologically as pseudoline arrangements. The numerous definitions needed to introduce arrangements and oriented matroids will be complemented in this section by the most important facts, such as counting the number of finite point, line and pseudoline arrangements, deciding when a pseudoline arrangement is equivalent to a line arrangement, and basic algorithmic results.

13.1.1 POINT CONFIGURATIONS

The simplest geometric objects are points in some d-dimensional space. Most of the other objects of interest for applications of computational geometry (sets of segments, polygons, polyhedra) are built on top of, and inherit, geometric structure from sets of points.

The setting for computational geometry problems is the Euclidean (affine) space $\mathcal{R}^d$ and most of its fundamental concepts (convexity, proximity) belong here in a natural way. However, some standard techniques, such as polarity and duality, as well as the abstraction to oriented matroids, are better explained in the context of vector spaces.

Several categories of concepts are introduced in this section, and developed and used in the subsequent subsections: vector and point configurations, hyperplanes and half-spaces, convexity, and some combinatorial parameters associated with vector or point configurations, relevant to applications (in statistics, pattern recognition or computational geometry): signed bases, semi-spaces, k-sets, centerpoints.

Definitions:

The (**standard**) **real vector space** of dimension d is the vector space $\mathcal{R}^d = \{\underline{x} \mid \underline{x} = (x_1, \ldots, x_d), x_i \in \mathcal{R}\}$, with vector addition $\underline{x} + \underline{y} = \{x_1 + y_1, \ldots, x_d + y_d\}$ and scalar multiplication $\alpha \underline{x} = \{\alpha x_1, \ldots, \alpha x_d\}$. A vector in $\mathcal{R}^d$ is a **d-dimensional vector**.

A **linear combination** of a set of vectors $\{v_1, \ldots, v_n\}$ is a vector of the form $\sum_{i=1}^{n} \alpha_i v_i$, for coefficients $\alpha_1, \ldots, \alpha_n \in \mathcal{R}$.

A **linearly independent set** of vectors is a set of vectors $\{v_1, \ldots, v_k\}$ such that a linear combination of them equals the zero vector ($\sum_{i=1}^{k} \alpha_i v_i = \underline{0}$) if and only if $\alpha_i = 0$ for all $i = 1, \ldots, k$.

A **basis** of $\mathcal{R}^d$ is a maximal set of linearly independent vectors, i.e., one that is no longer independent if a new element is added.

A basis is an **ordered basis** if it is given as an ordered set.

The **sign** of an ordered basis is the sign of the determinant of the $d \times d$ matrix with columns given in order by the vectors of the ordered basis.

A **linearly dependent set** of vectors $V = \{v_1, \ldots, v_k\}$ is a set of vectors for which there exists a linear combination with at least one nonzero coefficient yielding the $\underline{0}$ vector; i.e., $\sum_{i=1}^{k} \alpha_i v_i = \underline{0}$ with some $\alpha_i \neq 0$.

The **linear space** spanned by a set of vectors $V = \{v_1, \ldots, v_k\}, v_i \in \mathcal{R}^d$, is the set of all linear combinations of vectors of V.

A **linear k-dimensional subspace** of $\mathcal{R}^d$ ($k \leq d$) is the set of all linear combinations of k linearly independent vectors $v_1, \ldots, v_k$ in $\mathcal{R}^d$.

A **line** through $v \in \mathcal{R}^d$ is the 1-dimensional linear subspace of $\mathcal{R}^d$ induced by $v \neq \underline{0}$.

Euclidean space of dimension d is $\mathcal{R}^d$ seen as an affine space. It is sometimes identified with the d-dimensional affine hyperplane $x_{d+1} = 1$ in $\mathcal{R}^{d+1}$.

A (**d-dimensional**) **point** is an element of $\mathcal{R}^d$ seen as a Euclidean space.

An **affine combination** of a set of points $\{p_1, \ldots, p_n\}$ is a point of the form $\sum_{i=1}^{n} \alpha_i p_i$, with $\alpha_i \in \mathcal{R}$ and $\sum_{i=1}^{n} \alpha_i = 0$.

An **affinely independent set of points** is a set of points $\{p_1, \ldots, p_k\}$ such that no point is an affine combination of the others.

A **simplex** of $\mathcal{R}^d$ is a maximal set of affinely independent vectors. It is an **ordered simplex** if it is given as an ordered set.

The **extended matrix** of an ordered simplex $\{p_1, \ldots, p_{d+1}\}$ is the $(d+1) \times (d+1)$ matrix with its ith column equal to $(p_i, 1)$.

The **sign** of an ordered simplex is the sign of the determinant of the extended matrix of the simplex.

An **affinely dependent** set of points $P = \{p_1, \ldots, p_k\}$ is a set of points such that one of the points is an affine combination of the others.

The **affine space** spanned by a set of points $P = \{p_1, \ldots, p_k\}$, with $p_i \in \mathcal{R}^d$, is the set of all affine combinations of points of P. It is an affine subspace of $\mathcal{R}^d$.

The **affine k-dimensional subspace** of $\mathcal{R}^d$ ($k \leq d$) is the set of all affine combinations of k affinely independent points $p_1, \ldots, p_k$ in $\mathcal{R}^d$.

A **linear function** is a function $h: \mathcal{R}^d \to \mathcal{R}$, such that $h(x_1, \ldots, x_d) = \sum_{i=1}^{d} a_i x_i + a_{d+1}$.

A linear function is **homogeneous** if $a_{d+1} = 0$.

The **affine hyperplane** induced by a linear function h is the set $h^0 = \{x \in \mathcal{R}^d \mid h(x) = 0\}$.

An affine hyperplane is a **central hyperplane** if h is a homogeneous linear function.

An **oriented hyperplane** is a hyperplane, together with a choice of a positive side for the hyperplane. This amounts to choosing a (homogeneous or affine) linear function h to generate it, together with all those of the form αh, $\alpha > 0$.

An **reorientation** of an oriented hyperplane is a swapping of the negative and positive sides of the hyperplane (or, changing the generating linear function h to $-h$).

The **open half-spaces** induced by an oriented hyperplane h are $h^+ = \{\underline{x} \mid h(x) > 0\}$ (the **positive side** of h^0) and $h^- = \{x \mid h(x) < 0\}$ (the **negative side** of h^0). The sets h^+, h^0, and h^- form a partition of $\mathcal{R}^d$: $\mathcal{R}^d = h^+ \cup h^- \cup h^0$, and h^+, h^-, and h^0 are pairwise disjoint.

The **closed half-spaces** induced by h are $h^+ \cup h^0$ and $h^- \cup h^0$.

A **convex combination** of a set of points $\{p_1, \ldots, p_n\}$ is a point of the form $\sum_{i=1}^{n} \alpha_i p_i$ with $\alpha_i \in \mathcal{R}$, $\alpha_i > 0$, and $\sum_{i=1}^{n} \alpha_i = 1$.

The **segment** with endpoints $p_1 \neq p_2$ is the set of all convex combinations of p_1 and p_2.

A set of points $\{p_1, \ldots, p_k\}$ is **convexly independent** if no point is a convex combination of the others. The points are also said to be in **convex position**.

A **convex set** in $\mathcal{R}^d$ is a set $S \subseteq \mathcal{R}^d$ such that if p_1 and p_2 are distinct points in S, then the segment with endpoints p_1 and p_2 is contained in S.

The **convex hull** of a finite set of points P is the set of all convex combinations of points of P.

A **convex polytope** is the convex hull of a finite set of points. Its boundary consists of faces of dimension 0 (**vertices**), 1 (**edges**), $\ldots$, $d-1$ (**facets**).

The **face description** of a convex polytope is a data structure storing information about all the faces and their incidences.

A **convex polygon** is the convex hull of a finite set of points in $\mathcal{R}^2$.

A **vector configuration** [**point configuration**] of dimension d is a finite set of n vectors $\{v_1, \ldots, v_n\}$ ($v_i \in \mathcal{R}^d$) spanning $\mathcal{R}^d$ [points $\{p_1, \ldots, p_n\}$ ($p_i \in \mathcal{R}^d$) affinely spanning $\mathcal{R}^d$].

A configuration is **labeled** if its elements are given as an ordered set. (It may be given as the set of columns of a $d \times n$ matrix with real entries.)

The **rank** of a vector configuration [point configuration] in $\mathcal{R}^d$ is the number d [$d+1$].

A set of vectors [points] in $\mathcal{R}^d$ is in **general position** if every subset of d elements is a basis [simplex].

An **affine configuration** (or **acyclic vector configuration**) is a configuration with a central hyperplane containing all the vectors of the configuration on one side.

A **reorientation** of a vector configuration $V = \{v_1, \ldots, v_n\}$ is a vector configuration $V' = \{v'_1, \ldots, v'_n\}$ with each v'_i equal to either v_i or $-v_i$.

A **reorientation class** is the set of all labeled vector configurations which are reorientation equivalent.

A **point configuration** $P \subset \mathcal{R}^d$ **induced by an acyclic vector configuration** $V \subset \mathcal{R}^{d+1}$ contained in a half-space h^+ is the set of all points p obtained as follows: take the affine plane h' in h^+ parallel to h and tangent to the unit sphere S^d in $\mathcal{R}^{d+1}$; the intersection of the line through vector $v \in V$ with the plane h' is a point $p \in h'$. $\mathcal{R}^d$ is identified with the affine plane h'.

A **semi-space** of a vector configuration [point configuration] V induced by an oriented central hyperplane [affine hyperplane] h is the set of indices of the elements in V lying on one side of the hyperplane.

A **semi-space of a vector configuration [point configuration]** V is a semi-space induced by some hyperplane h.

The **order type** of a vector or point configuration V is the collection of all semi-spaces of V.

Isomorphic vector or point configurations are configurations having the same order type, after possibly relabeling their elements.

A **k-set** of a point configuration P is a semi-space of P of size k ($0 \leq k \leq n$).

A **centerpoint** of a point configuration P of size n is a point q, not necessarily in P, such that if h is a hyperplane containing q there are at least $\lceil \frac{n}{d+1} \rceil$ points in each semi-space induced by h.

A **ham-sandwich cut** is a hyperplane that simultaneously bisects d point configurations $P_1, P_2, \ldots, P_d$ in $\mathcal{R}^d$.

Facts:

1. A basis of the vector space $\mathcal{R}^d$ has d elements and a simplex of the affine space $\mathcal{R}^d$ has $d+1$ elements.

2. The rank of the $d \times n$ matrix associated with an n-vector configuration in $\mathcal{R}^d$ is d; the rank of the matrix associated with a point configuration in $\mathcal{R}^d$, extended with a row of 1s, is $d+1$.

3. The determinant of a $d \times d$ matrix whose columns are a basis of $\mathcal{R}^d$ and the determinant of the $(d+1) \times (d+1)$ extended matrix of a simplex in $\mathcal{R}^d$ are nonnull.

4. If $\{v_1, \ldots, v_d\}$ is a basis, then for any vector v_{d+1}, $\{v_1, \ldots, v_d, v_{d+1}\}$ is linearly dependent.

5. The intersection of linear subspaces [affine subspaces, convex subspaces] of $\mathcal{R}^d$ is linear [affine, convex].

6. The intersection of k central hyperplanes [affine hyperplanes] in $\mathcal{R}^d$ ($k \le d$) is a linear subspace [affine subspace]. Its dimension is at least $d - k$.
7. Every affine subspace of $\mathcal{R}^d$ is a convex set.
8. *Carathéodory's theorem*: Each point in the convex hull of a set of points $P \subset \mathcal{R}^d$ lies in the convex hull of a subset of P with at most $d + 1$ points.
9. *Radon's theorem*: Each set of at least $d + 2$ points in $\mathcal{R}^d$ can be partitioned into two disjoint sets whose convex hulls intersect in a nonempty set.
10. *Helly's theorem*: Let $\mathcal{S}$ be a finite family of n convex sets in $\mathcal{R}^d$ ($n \ge d + 1$). If every subfamily of $d + 1$ sets in $\mathcal{S}$ has a nonempty intersection, then there is a point common to all the sets in $\mathcal{S}$.
11. Every point configuration admits a centerpoint.
12. For every d configurations of points in $\mathcal{R}^d$, there exists a ham-sandwich cut.
13. *Upper bound theorem*: The number of facets of the convex hull of an n-point configuration in $\mathcal{R}^d$ is $O(n^{\lfloor d/2 \rfloor})$. This bound is obtained for configurations of points on the moment curve $P = \{p(t) \mid t \in \{t_1, \ldots, t_n\} \subseteq \mathcal{R}\}$, where $p(t) = (t, t^2, \ldots, t^d) \in \mathcal{R}^d$.
14. The number of semi-spaces of a rank $d + 1$ vector or point configuration of n elements is $O(n^d)$. The maximum is attained for points in general position.
15. For $d = 2$, let $e_k(n)$ be the maximum number of k-sets of any planar n-point configuration. Then $\Omega(n \log k) \le e_k(n) \le O(nk^{\frac{1}{3}})$.
16. *Erdős-Szekeres problem*: If $c(k)$ is the maximum number of planar points in general position such that no k are in convex position, then $2^{k-2} \le c(k) \le \binom{2n-4}{n-2}$.
17. The face description of the convex hull of a point set of size n in $\mathcal{R}^d$ can be computed optimally in $O(n \log n)$ time if $d = 2$ or $d = 3$, and $O(n^{\lfloor \frac{d}{2} \rfloor})$ for $d > 3$.
18. A ham-sandwich cut in dimension 2 can be found in linear time.

Examples:

1. The configuration of points in the following figure is given by the columns of the matrix $\begin{pmatrix} 0 & -1 & 3 & 1 & 2 \\ 0 & 1 & 0 & 1 & 2 \end{pmatrix}$. It is not in general position: the three points 1, 4, and 5 are collinear. The extended matrix is $\begin{pmatrix} 0 & -1 & 3 & 1 & 2 \\ 0 & 1 & 0 & 1 & 2 \\ 1 & 1 & 1 & 1 & 1 \end{pmatrix}$. Because $\det \begin{pmatrix} 0 & -1 & 3 \\ 0 & 1 & 0 \\ 1 & 1 & 1 \end{pmatrix} < 0$, the simplex 123 is negative. Some semi-sets are: $\{3\}$ (1-set), $\{2, 5\}$ (2-set), $\{1, 3, 4\}$ (3-set), etc. The convex hull is $\{1, 3, 5, 2\}$.

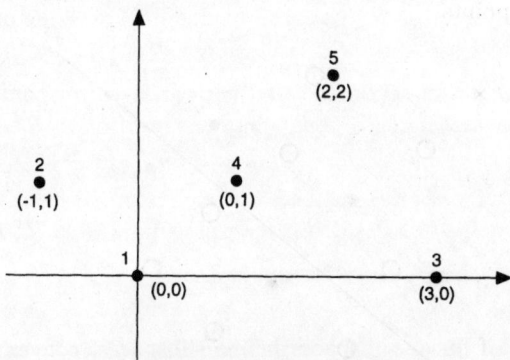

2. The two configurations of points from the figure of Example 1 and part (a) of the following figure are isomorphic, but those in parts (a) and (b) of the following figure are not. This can be seen because, for example, they have different numbers of points on their convex hulls.

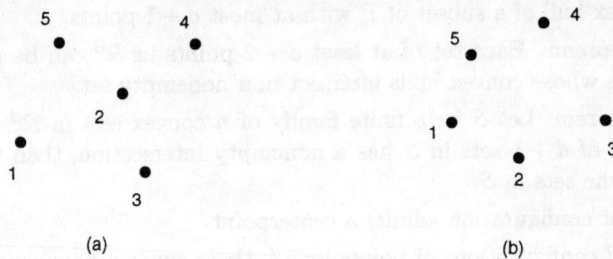

(a)   (b)

3. The grey point in the following figure is a centerpoint of the point configuration of black points. Some of the separating lines have been shown: they have at least $\frac{1}{3}$ of the points on each side.

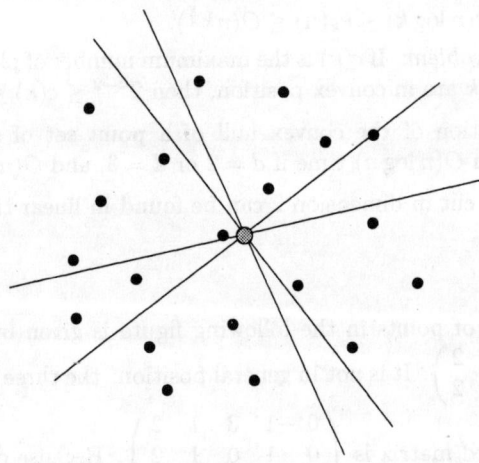

4. The line in the following figure is a ham-sandwich cut: it simultaneously bisects the black and the white points.

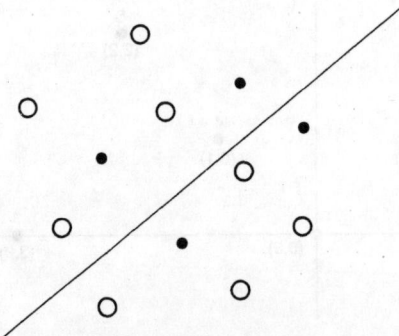

13.1.2 LINE AND HYPERPLANE ARRANGEMENTS

Line arrangements and affine point configurations in the plane are related via polar-duality, a transformation which is better understood in terms of 3-dimensional vectors and central planes or using the projective and spherical models. Several types of combinatorial data can be directly translated from the primal setting to the polar. As a consequence, theorems and algorithms on line arrangements follow directly from their counterparts on point configurations, and vice versa. This powerful tool has been used successfully in computational geometry for the design of efficient algorithms. It also generalizes to higher dimensions, where hyperplane arrangements are polar-dual to point configurations in $\mathcal{R}^d$.

Definitions:

A (*Euclidean*) *line in* $\mathcal{R}^2$ is an affine subspace of dimension 1. A line is induced by a linear function $l(x,y) = ax + by + c$ and any of its multiples of the form αl. A line is *oriented* if a direction has been chosen for it. Its induced half-spaces are called half-planes and denoted by l^+ and l^-.

A *nonvertical line* is a line given by an equation of the form $y = ax + b$, where a is called the *slope* and b the y-intercept of the line. It is oriented in increasing order of the x-coordinates of its points and its induced half-planes are above/below it.

A (*Euclidean*) *line arrangement* is a set $\mathcal{L} = \{l_1, \ldots, l_n\}$ of planar lines, not all of them going through the same point; that is, $\bigcap_{i=1}^n l_i = \emptyset$. If the lines are oriented, this is an *arrangement of oriented lines*. It is *labeled* if it is given as an ordered set.

A line arrangement is in *general position* if no three lines have a point in common.

An *x-monotone curve* is a curve intersecting each vertical line in exactly one point.

A *half-plane intersection* is the planar region lying in the intersection of a finite set of half-planes. It is described by the (circular) list of the lines incident to its boundary.

An *upper envelope* [*lower envelope*] of a nonvertical line arrangement is the half-plane intersection of the half-planes above [below] the lines of the arrangement.

The *k-level* in a nonvertical arrangement of n lines ($1 \leq k \leq n$) is the lower boundary of the set of points in $\mathcal{R}^2$ having exactly k lines above and $n - k$ below.

The *cell of a line arrangement* is a connected component of the complement in $\mathcal{R}^2$ of the union of the points on the lines.

The *zone of a line* l *in an arrangement* $\mathcal{L}$ ($l \notin \mathcal{L}$) is the set of cells of the arrangement $\mathcal{L}$ intersected by l.

A *central plane arrangement* in $\mathcal{R}^3$ is a finite set of central planes. The arrangement is *oriented* if the planes are oriented.

An *acyclic* (or *affine*) *central plane arrangement* in $\mathcal{R}^3$ is a central plane arrangement such that there is a point in $\mathcal{R}^3$ that lies on the positive side of all these planes.

The (*standard*) *line arrangement* induced by a central plane arrangement is the arrangement of the lines of intersection of the central planes with an affine plane h in $\mathcal{R}^3$ that is not parallel with any plane of the arrangement. If the central planes are oriented, an orientation is induced on the lines by keeping the positive side of a line within the positive side of the corresponding plane.

A **big circle** is the intersection of the unit sphere S^2 in $\mathcal{R}^3$ with a central plane. If the plane is oriented, the circle is given an orientation so that the positive side of the plane lies on the left of the circle.

A **big-circle arrangement** is the intersection of a central plane arrangement with the unit sphere S^2 in $\mathcal{R}^3$. It is oriented if the planes are oriented.

The **big-circle arrangement induced by a central plane arrangement** is the arrangement of the big circles of intersection of the central planes with the sphere S^2 in $\mathcal{R}^3$. It is oriented if the planes are oriented.

The **projective plane** P^2 is the sphere S^2 in $\mathcal{R}^3$ with the antipodal points identified.

A **projective line** is the projective curve induced by identifying the antipodal points of a big circle on S^2.

A **projective line arrangement** is a finite set of projective lines in the projective plane P^2.

The **projective line arrangement induced by a central plane arrangement** is the projective arrangement obtained by the antipodal point identification of the big circle arrangement on S^2 induced by the central plane arrangement.

An **arrangement graph** is a graph associated with a Euclidean or projective line arrangement, or a big circle arrangement. Its **vertices** correspond to intersection points of lines (or circles) and its **edges** correspond to line (or arc) segments between two intersection points. [Note: For the Euclidean case, typically only the bounded line segments are considered as edges (but by adding extra dummy vertices "at infinity", the infinite extremities of each line among the edges can be included). For the Euclidean or spherical case, if the lines are oriented, the arrangement graph is directed, with the edges oriented to be compatible with the orientation of the lines or circles.]

Isomorphic arrangements are arrangements having isomorphic arrangement graphs. (This applies to Euclidean lines, big-circles (oriented or not), and projective lines.)

A **polar-duality of vectors and central planes** in $\mathcal{R}^3$ is a mapping $\mathcal{D}$ associating a vector $v \in \mathcal{R}^3$ with an oriented central plane having v as its normal vector, and vice versa.

A **polar-duality of points and lines** in the affine space $\mathcal{R}^2$ is any mapping $\mathcal{D}$ associating a point $p \in \mathcal{R}^2$ with an oriented line l in $\mathcal{R}^2$ and vice versa, by the following general procedure: map the points to vectors via some imbedding of $\mathcal{R}^2$ as an affine plane in $\mathcal{R}^3$, apply the polar-duality of vectors and central planes, and then intersect the polar central planes with some affine plane (identified with $\mathcal{R}^2$) to get lines.

A (**standard**) **affine polar-duality** is a mapping $\mathcal{D}$ between nonvertical lines and points in $\mathcal{R}^2$, associating the point $(a, -b)$ with the line $y = ax + b$, and vice versa.

A **Euclidean hyperplane** [**central hyperplane**] **arrangement** in $\mathcal{R}^d$ is a finite set $\mathcal{H} = \{h_1, \ldots, h_n\}$ of affine hyperplanes [central hyperplanes], not all of them going through the same point. If the hyperplanes are oriented, the arrangement is oriented.

The following are generalizations to an arbitrary affine space $\mathcal{R}^d$ [vector space $\mathcal{R}^{d+1}$] of previously defined concepts in affine dimension 2 [vector space $\mathcal{R}^3$]:

- **arrangements of big $(d-1)$-spheres** on S^d generalize big-circle arrangements on S^2;
- **projective arrangements** of hyperplanes in P^d generalize projective arrangements of lines in P^2;

- the **polar-duality** between vectors and central hyperplanes in $\mathcal{R}^{d+1}$ associates with a vector the hyperplane normal to it;
- the **face lattice** of a hyperplane (central, affine, projective) or sphere arrangement, a data structure storing information on faces and their incidences, generalizes the arrangement graph, and is used to define isomorphism of arrangements.
- the **k-level** in an affine arrangement of nonvertical hyperplanes is the lower boundary of the set of points having exactly k hyperplanes above them.

Facts:

1. A bounded cell in a Euclidean line arrangement is a convex polygon.

2. The k-level of a nonvertical line arrangement is an x-monotone piecewise linear curve incident with vertices and lines of the arrangement.

3. The upper envelope is the 0-level of an arrangement of nonvertical lines.

4. In a simple big-circle arrangement, every pair of big circles intersect in exactly two points, which are antipodal on the sphere.

5. The arrangement graphs of planar line arrangements or spherical big-circle arrangements are planar imbedded graphs. The arrangement graph of a projective line arrangement is projective-planar. The *faces* or *cells* of these graphs are the connected components of the complement of the union of lines or circles.

6. The association among central plane arrangements, big-circle arrangements, and projective arrangements preserves isomorphisms. The standard association of an affine line arrangement and big-circle arrangement to an acyclic plane arrangement preserves isomorphisms.

7. In a simple Euclidean arrangement of n lines, the number of vertices is $\binom{n}{2}$, the number of segments (bounded or unbounded) is n^2, and the number of cells (bounded or unbounded) is $\binom{n}{2} + n + 1$. No nonsimple arrangement exceeds these values.

8. *Zone theorem*: The total number of edges (bounded or unbounded) in a zone of an arrangement of n lines is at most $6n$.

9. $\mathcal{D}(\mathcal{D}(v)) = p$ and $\mathcal{D}(\mathcal{D}(h)) = h$, for every vector v and hyperplane h.

10. *Incidence preserving*: If $v \in h$, then $\mathcal{D}(h) \in \mathcal{D}(v)$, for every vector v and hyperplane h.

11. *Orientation preserving*: If $v \in h^+$, then $\mathcal{D}(h) \in \mathcal{D}(v)^+$, for every vector v and hyperplane h.

12. *Basic properties of the standard affine polar-dual transformation*:
    - polar-duality preserves above/below properties: if a point p is above line l, then the polar line $\mathcal{D}(p)$ is below the polar point $\mathcal{D}(l)$;
    - the polar-dual of a configuration of points P is an arrangement of nonvertical lines $\mathcal{L} = \mathcal{D}(P)$, and vice versa;
    - the polar-dual of a set of points given in increasing order of their x-coordinates is a set of lines given in increasing order of their slopes;
    - the polar-dual of the set of points on the convex hull of P is the set of lines on the upper and lower envelopes of the polar arrangement $\mathcal{D}(P)$; convex hull computation dualizes to half-plane intersection;

- semi-spaces of P dualize to vertices, edges, and cells of the polar arrangement $\mathcal{D}(P)$;
- isomorphic arrangements of lines dualize to isomorphic configurations of points;
- the polar-duals of lines $p_i p_j$ inducing $(k-1)$-sets and k-sets in a point configuration $P = \{p_1, \ldots, p_n\}$ are the vertices $l_i \cap l_j$ on levels k and $n-k$ of the polar-dual arrangement $L = \{l_1, \ldots, l_n\}$.

13. The polar-dual of an acyclic vector configuration in $\mathcal{R}^3$ is an acyclic central plane arrangement.

14. The upper envelope of a line arrangement can be computed optimally in $O(n \log n)$ time.

15. The arrangement graph of a line arrangement can be computed in $O(n^2)$ time and space.

16. The face incidence lattice of a hyperplane arrangement in $\mathcal{R}^d$ can be computed in $O(n^d)$ time and space.

17. The standard polar-duality is ubiquitously used in computational geometry. For example, it is used to derive algorithms for half-plane intersection from convex hull algorithms, to translate between line slope and point x-coordinate selection, and to compute the visibility graph in $O(n^2)$ time using the polar-dual arrangement graph. See [Ed87] (Chapter 12) for a collection of such problems.

18. In computational geometry, k-levels are related to the furthest k-neighbors Voronoi diagrams via a lifting transformation that reduces the computation of dimension d Voronoi diagrams to dimension $d + 1$ arrangements of hyperplanes.

19. k-levels in arrangements, as polar-duals to k-sets, have an abundance of applications in statistics.

Examples:

1. The following figure shows a line arrangement in general position. The arrangement graph has 10 vertices corresponding to the black points (which could be labeled with pairs of indices of lines such as $12, 13$, etc.), 15 edges corresponding to the bounded line segments such as $(12, 13)$, and $2 \times 5 = 10$ unbounded edges. The upper envelope is bounded from below by the 0-level, whose list of lines is $\{1, 2, 3, 5\}$. The dashed piecewise linear curve is the 2-level.

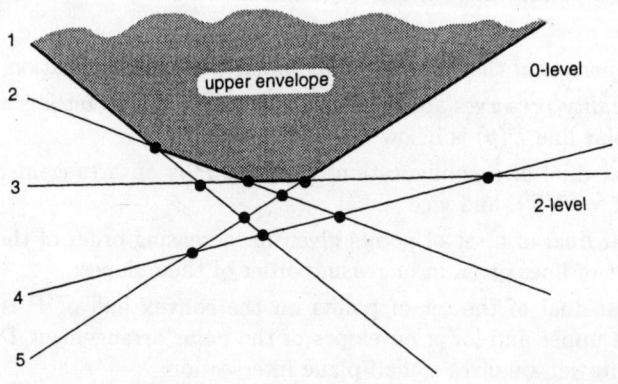

2. The zone of line 3 in the line arrangement $\{1, 2, 4, 5\}$ from the figure of Example 1 is depicted in the following figure. It has 5 cells (1 bounded, 4 unbounded), whose boundaries sum up to 12 segments.

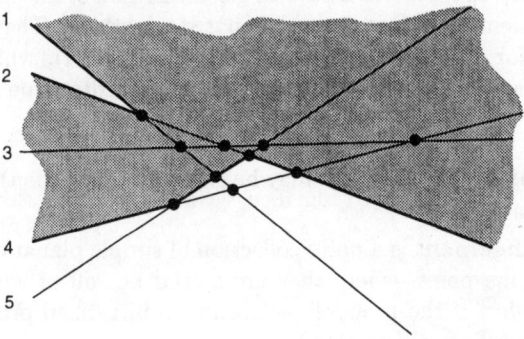

3. The line arrangements in the figure for Example 1 and part (a) of the following figure are isomorphic. Those in parts (a) and (b) of the following figure are not isomorphic.

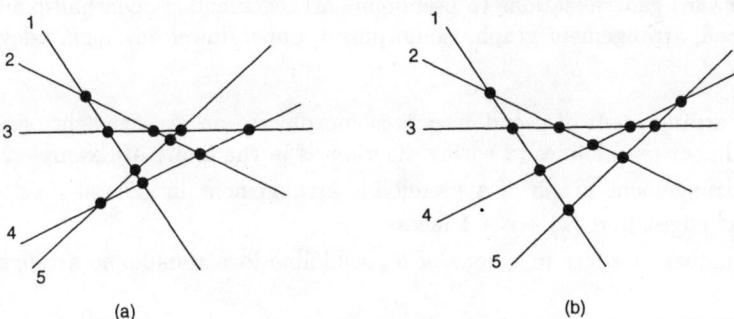

4. The following figure illustrates the standard polar-duality. The arrangement is polar-dual to the configuration of points in the figure of §13.1.1 Example 1; hence the lines are given by the equations $1: y = 0$, $2: y = -x - 1$, $3: y = 3x$, $4: y = x - 1$ and $5: y = 2x - 2$. In the primal configuration, point 2 is above line 13. In the polar-dual, line 2 is below the intersection point of lines 1 and 3.

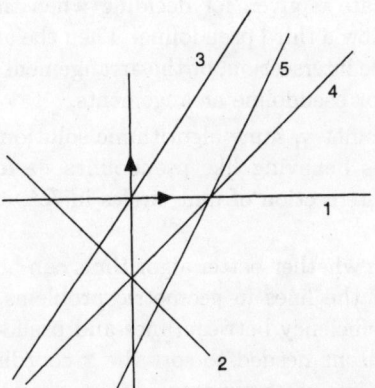

13.1.3 PSEUDOLINE ARRANGEMENTS

Pseudoline arrangements represent a natural generalization of line arrangements, retaining incidence and orientation properties, but not straightness. They provide a topological representation for rank 3 oriented matroids (see §13.1.4), which in turn abstract combinatorial properties of vector configurations and oriented line arrangements.

Definitions:

A *pseudoline* is a planar curve (which may be given an orientation). It is *x-monotone* if the curve is x-monotone.

A *pseudoline arrangement* is a finite collection of simple planar curves that intersect pairwise in exactly one point, where they cross, and not all of which have a point in common. It is **labeled** if the pseudolines are given in a fixed order $\{l_1, \ldots, l_n\}$, and **oriented** if the pseudolines are oriented.

A pseudoline arrangement is **realizable** or **stretchable** if it is isomorphic to a line arrangement.

Note: The following terms, defined in §13.1.1 and §13.1.2 for line arrangements, have straightforward generalizations to pseudoline arrangements: open half-planes, general position, cell, arrangement graph, isomorphism, upper/lower envelope, k-level, zone.

Facts:

1. Every arrangement of pseudolines is isomorphic to an arrangement of x-monotone piecewise linear pseudolines (a wiring diagram as in the figure of Example 2).
2. The arrangement graph of a pseudoline arrangement in general position has $\binom{n}{2}$ vertices, n^2 edges, and $\binom{n}{2} + n + 1$ faces.
3. The number of edges in a zone of a pseudoline in a pseudoline arrangement is at most $6n$.
4. Let $e_k(n)$ be the number of edges on the k-level of a pseudoline arrangement. Then $\Omega(n \log k) \leq e_k(n) \leq O(nk^{\frac{1}{3}})$.
5. The logarithm of the number of isomorphism classes of pseudoline arrangements is $\Theta(n^2)$. The same number for line arrangements is $\Theta(n \log n)$.
6. There exist nonstretchable pseudoline arrangements.
7. It is NP-hard to decide whether a pseudoline arrangement is stretchable.
8. Pseudoline stretchability is decidable in PSPACE.
9. Assume that a predicate is given for deciding when the intersection point of two pseudolines is above or below a third pseudoline. Then the algorithms for computing the upper envelopes, half-space intersection, or the arrangement graph of a line arrangement can be adapted to work for pseudoline arrangements.
10. In computational geometry, some algorithmic solutions can be found by reducing the problem to structures behaving like pseudolines — for example, computing the boundary of a union or intersection of unit circles in $\mathcal{R}^2$, all having at least one point in common.
11. It is an open problem whether better algorithms can be devised by making explicit use of the straightness of the lines in geometric problems. So far, the only problem where an explicit gap in efficiency between lines and pseudolines has been displayed is in the number of comparisons needed to sort the x-coordinates of the vertices of line versus x-monotone pseudoline arrangements.

Section 13.1 ARRANGEMENTS OF GEOMETRIC OBJECTS   817

Examples:

1. The pseudoline arrangement in this figure is stretchable, because it is isomorphic to the line arrangement in the figure of §13.1.2, Example 1.

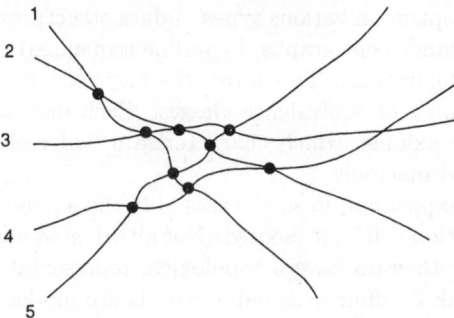

2. The following figure shows a standard way of representing a pseudoline arrangement as an x-monotone piecewise linear curve arrangement called a *wiring diagram*. The arrangement is the same as in the figure of Example 1.

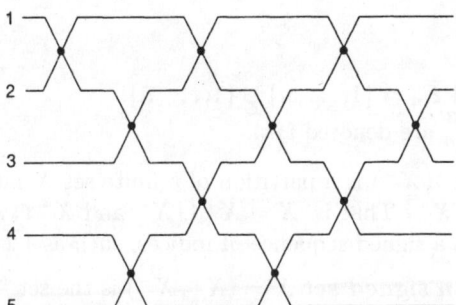

3. *A nonstretchable pseudoline arrangement*: The theorem of Pappus in plane geometry states that if the points 1, 2, 3 and 4, 5, 6 are collinear and in this order on two lines, then the three intersection points of the pairs of lines $7 = (15, 24)$, $8 = (16, 34)$, and $9 = (26, 35)$ are also collinear. See the following figure. The perturbed arrangement obtained by replacing the line through 7, 8, 9 with the dashed pseudoline is not stretchable, since it violates the theorem of Pappus.

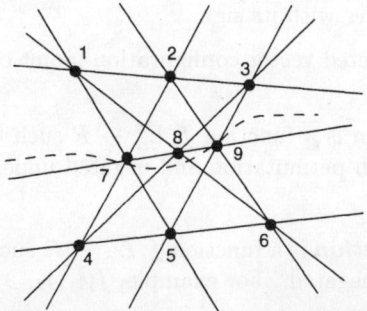

13.1.4 ORIENTED MATROIDS

General oriented matroids are abstractions of vector configurations and oriented hyperplane arrangements. Affine oriented matroids model the corresponding situation in affine spaces. They capture in various types of data structures (semi-spaces, chirotopes, Radon partitions, arrangement graphs, hyperline sequences) combinatorial information about n-element configurations. This forms the basis of the classification of all n-point sets into a finite number of equivalence classes. Each data structure satisfies a set of simple properties, or axioms, which characterize a wider class of objects collectively referred to as oriented matroids.

To simplify the exposition, in some cases only the axiomatization corresponding to points in general position will be presented. Not all oriented matroids arise geometrically from vector sets, but they do have a topological representation via pseudohyperplane arrangements. In rank 3, affine oriented matroids are modeled by pseudoline arrangements. Many geometric algorithms working with line arrangements or, by polarity, point configurations, make use of no more than oriented matroid properties and can be extended to pseudolines.

As potential applications, oriented matroids lay the foundations for a rigorous theory of geometric program verification and testing.

Definitions:

Notation:
$E_n = \{1, \ldots, n\}$ and $\overline{E}_n = \{1, \ldots, n\} \cup \{\overline{1}, \ldots, \overline{n}\}$.
Triplets $(i, j, k) \in \overline{E}_n^3$ are denoted ijk.

A **signed set** $X = (X^+, X^-)$ is a partition of a finite set X into a **positive part** X^+ and a **negative part** X^-. That is, $X = X^+ \cup X^-$ and $X^+ \cap X^- = \emptyset$. In E_n a signed set may be denoted as a signed sequence of indices, such as $1\overline{2}34$ for $(\{1, 3, 4\}, \{2\})$.

The **complement of a signed set** $X = (X^+, X^-)$ is the set $-X = (X^-, X^+)$.

The **support of a signed set** $X = (X^+, X^-)$ is the unsigned set X.

The **size of a signed set** $X = (X^+, X^-)$ is the size of the support of X.

The **signed double covering** of a finite set X is the set $\overline{X} = X^+ \cup X^-$, where $X^+ = X$ and $X^- = \{\overline{x} \mid x \in X\}$ is a signed distinct copy of X (its elements called *negated elements*), $X^+ \cap X^- = \emptyset$. If $x \in X^-$, then $\overline{x}$ is the corresponding nonnegated element in X^+.

A **basis** of a vector configuration [point configuration] $V \subset \mathcal{R}^d$ is a subset of V, identified by a d-set of indices, which is a basis [simplex] of the ambient space $\mathcal{R}^d$. A **signed basis** is an ordered basis together with its sign.

The **chirotope** of an ordered vector configuration [point configuration] is the set of all signed bases of V.

An **alternating function** is a function $f : E_n^d \to R$ such that the sign of $f(i_1, \ldots, i_d)$ is preserved under an even permutation and negated under an odd permutation of the d-tuple $(i_1, \ldots, i_d)$.

An **antisymmetric function** is a function $f : \overline{E}_n^d \to \mathcal{R}$ such that its sign changes when one of the parameters is negated. [For example, $f(\overline{i}_1, i_2, \ldots, i_d) = -f(i_1, i_2, \ldots, i_d)$.]

The (**rank 3**) **Grassmann-Plücker relations** are the identities
$$[1\,2\,3][1\,4\,5] - [1\,2\,4][1\,3\,5] + [1\,2\,5][1\,3\,4] = 0$$
satisfied by the determinants $[i\,j\,k] = \det(v_i, v_j, v_k)$, for any five vectors $v_i, 1 \leq i \leq 5$.

The (**rank d**) **Grassmann-Plücker relations** are the identities
$$[i_1 \ldots i_{d-2}1\,2][i_1 \ldots i_{d-2}3\,4] - [i_1 \ldots i_{d-2}1\,3][i_1 \ldots i_{d-2}2\,4] +$$
$$[i_1 \ldots i_{d-2}1\,4][i_1 \ldots i_{d-2}2\,3] = 0$$
satisfied by the determinants $[i_1 \ldots i_{d-2}j\,k] = \det(v_{i_1}, \ldots, v_{i_{d-2}}, v_j, v_k)$, for any $d+2$ vectors v_{i_j} ($1 \leq j \leq d-2$) and v_i ($1 \leq i \leq 4$).

Chirotope axioms (rank d): A function $\chi: \overline{E}_n^d \to \{-1, 0, +1\}$ is a chirotope of rank d if it satisfies the following conditions:
- χ is alternating and antisymmetric;
- for any $d+2$ generic points $i_1 \ldots i_{d-2}1\,2\,3\,4$, the signs $\chi(i_1 \ldots i_{d-2}j\,k)$ of the six triplets involved in the Grassmann-Plücker relations are such that equality is possible.

A **uniform chirotope** is a chirotope function that takes nonzero values on all d-tuples.

Chirotope axioms (uniform, rank 3): A function $\chi: \overline{E}_n^3 \to \{-1, +1\}$ is a uniform chirotope of rank 3 if it satisfies the following conditions:
- χ is alternating and antisymmetric;
- for any 5 generic points a, b, i, j, k, if $\chi(a\,b\,i) = \chi(a\,b\,j) = \chi(a\,b\,k) = +1$ and $\chi(a\,i\,j) = \chi(a\,j\,k) = +1$, then $\chi(a\,i\,k) = +1$.

The chirotope χ is **affine** if, in addition, it satisfies the axiom:
- for any four points i, j, k, l, if $\chi(i\,j\,k) = \chi(i\,k\,l) = \chi(i\,l\,j) = +1$, then $\chi(j\,k\,l) = +1$.

An **oriented matroid given by a chirotope** is an abstract set of points labeled $\{1, \ldots, n\}$, together with a function χ satisfying the chirotope axioms.

The **circuit** of a set of labeled vectors $V = \{v_1, \ldots, v_n\}$ is the signed set $C = (C^+, C^-)$, where $C^+ = \{j \mid \alpha_j > 0\}$ and $C^- = \{j \mid \alpha_j < 0\}$, of indices of the nonnull coefficients α_j in a *minimal* linear dependency $V' = \{v_{i_1}, \ldots, v_{i_k}\}$ of V with $\sum_{j=1}^k \alpha_j v_{i_j} = \underline{0}$. If C is a circuit, its complement $-C$ is also a circuit.

A **Radon partition** of a set of labeled points P is a signed set $C = (C^+, C^-)$ of points of P such that the convex hull of the points in C^+ intersects the convex hull of the points in C^-.

A **minimal Radon partition** (or **circuit**) is a Radon partition whose support is minimal with respect to set inclusion.

An **oriented matroid** of an ordered vector [point] configuration given by its circuits is the set of all circuits of V.

A set $\mathcal{C}$ of signed subsets of E_n satisfies the **circuit axioms** if:
- $\emptyset \notin \mathcal{C}$;
- if $C \in \mathcal{C}$ then $-C \in \mathcal{C}$;
- (minimality): if $C = (C^+, C^-)$ is a circuit, then no subset of the support of C is the support of another circuit;
- (exchange): if C_1 and C_2 are two circuits such that $C_1 \neq -C_2$ and $e \in C_1^+ \cap C_2^-$, then there exists another circuit D such that $e \notin D$, $D^+ \subset C_1^+ \cup C_2^+$, and $D^- \subset C_1^- \cup C_2^-$.

The **oriented matroid given by its circuits** is an abstract set E_n together with a set of signed sets $\mathcal{C}$ satisfying the circuit axioms.

A **cocircuit** of a labeled vector configuration V is a signed set $C = (C^+, C^-)$ of E_n, induced by a subset of $d-1$ vectors spanning a central hyperplane h. For an arbitrary orientation of h, C^+ is the set of indices of elements in V lying in h^+ and C^- is the set of indices of elements in V lying in h^-.

The **cocircuit axioms** are the conditions obtained from the circuit axioms by replacing "circuit" with "cocircuit".

An **oriented matroid given by its cocircuits** is an abstract set E_n, together with a set of signed sets $\mathcal{C}$ satisfying the cocircuit axioms.

A **circular sequence** of period k is a doubly infinite sequence $(q_i)_{i \in \mathcal{Z}}$ with $q_i = q_{i+k}$ for all $i \in \mathcal{Z}$.

A **signed permutation** of a set S is a permutation of S whose elements are also assigned a sign; for example, $1\overline{3}42$, where 3 is negative and 1, 2, and 4 are positive.

An **alternating circular sequence** is a circular sequence $(q_i)_{i \in \mathcal{Z}}$ with half-period k, defined with elements from a signed double covering $q_i \in \overline{X}$ and satisfying the property $q_i = \overline{q}_{i+k}$ for all $i \in \mathcal{Z}$.

A **representation** of an alternating circular sequence can be obtained by any of its subsequences of k consecutive elements (half period) $\{q_1, \ldots, q_k\}$.

A **star** (or **rank 3 hyperline sequence**) associated with a point $p_i \in P \subseteq \mathcal{R}^2$ [vector $v_i \in V \subseteq \mathcal{R}^3$] is an alternating circular sequence of subsets of indices in $\overline{E}_n$ obtained by rotating an oriented line [oriented central plane] in counterclockwise order around p_i [the line through vector v_i] and recording the successive positions where it coincides with lines [central planes] defined by pairs of points (p_i, p_j) with $p_j \in P \setminus \{p_i\}$ [vectors (v_i, v_j), with $v_j \in V \setminus \{v_i\}$]. If a point p_j is encountered by the rotating line in the positive direction from p_i, it will be recorded as a positive index, otherwise it will be recorded as a negative index. When the points are not in general position, several may become simultaneously collinear with the rotating line, and they are recorded as one subset L^i_j. The sequence is denoted by a half-period $s_I = (L^i_1, L^i_2, \ldots, L^i_{k_i})$, where $L^i_j \subset \overline{E}_n \setminus \{i, \overline{i}\}$.

A **cluster of stars** (or **rank 3 hyperline sequences**) **associated with a point** (or **vector**) **configuration** P is the ordered set of n stars $s_1, \ldots, s_n$, one for each point $p_i \in P$.

A **uniform cluster of stars** is a cluster of stars corresponding to a set of points in general position. (Each star is a sequence of individual indices.)

An **oriented matroid of a vector** (or **point**) **set** V **given by its cluster of stars** is the cluster of stars associated with V.

A **star** (or **rank 3 hyperline sequence**) **associated with an element** c_i **of a big-circle arrangement** $C = \{c_1, \ldots, c_n\}$ on S^2 is an alternating circular sequence of subsets of indices in $\overline{E}_n$ obtained by traversing the oriented big-circle in its given direction and recording in order the intersections of c_i with the other big-circles c_j ($j \neq i$). Each intersection is recorded as a signed index j: positive if c_j crosses c_i from left to right, negative otherwise.

The **cluster of stars** (or **rank 3 hyperline sequences**) **associated with a big-circle arrangement** is the set of n stars $s_1, \ldots, s_n$, one for each circle $c_i \in C$.

The *cluster of stars associated with an oriented central plane arrangement* in $\mathcal{R}^3$ [*line arrangement in $\mathcal{R}^2$*] is the cluster of stars of the big-circle arrangement associated with the central plane arrangement [to the central plane arrangement induced by the line arrangement via the imbedding of $\mathcal{R}^2$ as the plane $z = 1$ in $\mathcal{R}^3$].

The *cluster of stars associated with a pseudoline arrangement* [*a pseudoconfiguration of points*] is the generalization from straight lines to pseudolines obtained by recording the order of the vertices of the arrangement along a pseudoline (positive or negative according to whether the line crossing at that vertex comes from right or left) [the circular counterclockwise order of the pseudolines incident with a point].

A *cluster of star permutations* is an ordered set of alternating circular sequences $s_1, \ldots, s_n$ with the property that the representative half-period of sequence s_i is a signed permutation of the set $E_n \setminus \{i\}$.

A *chirotope function associated with a set of cluster of stars permutations* $s_1, \ldots, s_n$ is a function $\chi : E_n^3 \to \{-1, +1\}$ defined by $\chi(i\,j\,k) = +1$ if, in the ith sequence s_i and in a half period of it where both j and k occur positively, j occurs before k. Otherwise $\chi(i\,j\,k) = -1$.

A set E_n, together with an ordered set of alternating circular sequences $s_1, \ldots, s_n$, satisfies the *cluster of stars axioms* (*uniform, rank 3*) if the set of sequences are cluster of star permutations whose associated chirotope function is alternating.

A (*uniform, rank 3*) *oriented matroid given by its cluster of stars* is a set E_n together with n alternating sequences satisfying the cluster of stars axioms.

An abstract set $\overline{E}_n$, together with a set of n^{d-2} (uniform) alternating sequences (indexed by $(d-2)$-tuples $(i_1, \ldots, i_{d-2})$), is an *oriented matroid given by its hyperline sequences* (*uniform, rank d*) if the chirotope function $\chi : E_n^d \to \{-1, +1\}$ associated with it is alternating. [The function $\chi : E_n^d \to \{-1, +1\}$ is defined by $\chi(i_1 \ldots i_{d-2}j\,k) = +1$ if in the star indexed by $i_1, \ldots, i_{d-2}$ and in a half period where both j and k occur positively, j occurs before k. Otherwise $\chi(i_1 \ldots i_{d-2}j\,k) = -1$.]

Deletion is the removal of a point (vector, line, etc.) from a configuration and recording the oriented matroid (chirotope, circuits, cluster of stars, etc.) only for the remaining points.

In a rank d central plane arrangement, the *contraction* on element i is obtained by identifying h_i with $\mathcal{R}^{d-1}$ and intersecting it with all the other hyperplanes to obtain a rank $d-1$ arrangement with one element less.

The *oriented matroid obtained by a one-element deletion* in the hypersequence representation is the matroid obtained by removing the element from all the hyperline sequences of the original oriented matroid, and discarding all hyperline sequences whose labels contain that element.

The *oriented matroid obtained by a one-element contraction* in the hypersequence representation is the matroid obtained by retaining only the hyperline sequences whose labels contain the element, and dropping it from the labels.

A *rank 2 contraction* of a cluster of stars is one of the stars.

A *minor* of an oriented matroid given by hyperline sequences is an oriented matroid obtained by a sequence of deletions and/or contractions.

A (**Euclidean**) **pseudoconfiguration of points** is a pair (P, L) with $P = \{p_1, \ldots, p_n\}$ a planar set of points and $L = \{l_1, \ldots, l_m\}$ a pseudoline arrangement, such that for every pair of distinct points (p_i, p_j) there exists a unique pseudoline $l_{ij} \in L$ incident with them.

If a pseudoline arrangement is intersected with a vertical line l_v ($x = -M$), for some very large constant M and all the vertices of the arrangement lie to the right of l_v, then the order in which the pseudolines in L cross v_h (decreasing by the y-coordinates of the crossings) is the (increasing) **slope order of the pseudolines**.

The (**standard affine**) **pseudo polar-duality** is the association between an x-monotone pseudoline arrangement $L = \{l_1, \ldots, l_n\}$ given in slope order and a pseudo configuration of points (P, L'), $P = \{p_1, \ldots, p_n\}$, given in increasing order of the x-coordinates and with L' being x-monotone, satisfying the property that the cluster of stars associated with L and to P are the same.

Facts:
1. Cocircuits correspond to semi-spaces, when the defining hyperplane is incident with $d-1$ independent elements of the configuration.
2. In the rank d uniform oriented matroid associated with a vector configuration in general position in $\mathcal{R}^d$, all the d-tuples are bases, all the $(d+1)$-tuples are supports of circuits, and all the $(n-d+1)$-tuples are supports of cocircuits.
3. The oriented matroid associated with an affine vector configuration V and the affine oriented matroid associated with the affine point configuration induced by V are the same.
4. The chirotope function χ associated with the cluster of stars of a vector or point configuration is an alternating and antisymmetric function.
5. The two given systems of chirotope axioms are equivalent (for the uniform case).
6. The hyperline sequences of a contraction by one element of a central plane arrangement are the induced rank $(d-1)$ contraction by that element of the set of rank d hyperline sequences of the original arrangement.
7. The induced rank $(d-1)$ contraction of a set of rank d hyperline sequences is a rank $(d-1)$ set of hyperline sequences.
8. A minor of an oriented matroid (given by its hyperline sequences) is an oriented matroid.
9. For two labeled vector (or point) configurations V_1 and V_2, the following statements are equivalent:
    - V_1 and V_2 have the same chirotope;
    - V_1 and V_2 have the same order type;
    - V_1 and V_2 have the same hyperline sequences;
    - V_1 and V_2 have the same minors.

Moreover, for any reorientation of V_1 or V_2:
    - V_1 and V_2 have the same oriented matroid given by circuits;
    - V_1 and V_2 have the same oriented matroid given by cocircuits.

This justifies the unique name *oriented matroid* for the equivalence class of vector configurations with the same chirotope (or clusters, order type, etc.), and for a reorientation class of an oriented matroid.

10. For a labeled vector configuration V in $\mathcal{R}^{d+1}$, the following statements are equivalent:
   - V is acyclic (affine);
   - there is a labeled point configuration P in $\mathcal{R}^d$ whose oriented matroid (or order type or chirotope) is the same as the oriented matroid of V.

11. Multiplying the elements of a vector configuration by a positive scalar yields a vector configuration with the same oriented matroid. In particular, for any vector configuration V in $\mathcal{R}^{d+1}$, there exists an equivalent vector configuration on the d-sphere.

12. For any vector configuration, there exists a reorientation of some of its vectors which makes it affine.

13. The set of circuits of an affine (acyclic) vector configuration does not contain the *positive* cycle $(E_n, \emptyset)$.

14. Two projective line arrangements with the same arrangement graph have polar-dual configurations of points in the same reorientation class (and this can be generalized to arbitrary dimension d).

15. If a labeled vector configuration in $\mathcal{R}^3$ and a labeled oriented arrangement of central planes are polar-dual, then they have the same hyperline sequences (and this can be generalized to arbitrary dimension d).

16. The number of oriented matroids of rank d and size n is $2^{O(n^{d-1})}$.

17. The number of realizable oriented matroids of rank d and size n is $2^{O(n \log n)}$.

18. *Folkman-Lawrence topological representation theorem*: Every oriented matroid of rank d can be represented as a $(d-1)$-pseudosphere arrangement on S^d.

19. Every affine oriented matroid of rank 3 can be represented as a pseudoline arrangement and as a (polar-dual) pseudoconfiguration of points.

20. There exist nonrealizable oriented matroids of any rank.

21. Realizable oriented matroids cannot be characterized by a finite set of excluded minors.

Examples:

1. The function $\chi(i,j,k) = sign \; \det(v_i, v_j, v_k)$, for $v_l \in V = \{v_1, \cdots, v_n\} \subset \mathcal{R}^3$ is alternating antisymmetric.

2. The following figure shows an example of a point configuration in general position, together with the connecting lines.

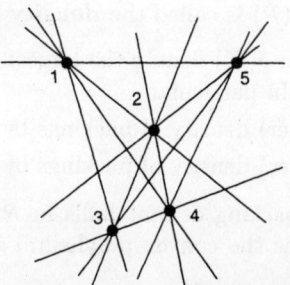

Its oriented matroid is given by:
- chirotope: $123-$, $124-$, $125+$, $134+$, $135+$, $234+$, $235+$, $345+$. The other signed triplets are computed by antisymmetry and alternation.
- minimal Radon partitions: $1\overline{2}\overline{3}4$, $1\overline{2}35$, $1\overline{2}45$, $\overline{1}3\overline{4}5$, $\overline{2}3\overline{4}5$ and their complements.
- cocircuits: $34\overline{5}$, 245, $2\overline{3}5$, 234, $\overline{1}45$, 135, $\overline{1}34$, 125, $12\overline{4}$, 123 and their complements.
- cluster of stars: $1\colon 3\,4\,2\,5$, $2\colon 1\,\overline{5}\,3\,4$, $3\colon 1\,\overline{4}\,\overline{5}\,2$, $4\colon 5\,2\,1\,3$, and $5\colon 1\,2\,3\,4$. (An arbitrary half-period was chosen for the circular sequences.)

3. The oriented matroid given as a cluster of stars for the line arrangement in the figure of §13.1.2 Example 1 (or for the pseudoline arrangement in the figure of §13.1.3 Example 1) is:

$$1\colon 2\,3\,5\,4,\quad 2\colon \overline{1}\,3\,5\,4,\quad 3\colon \overline{1}\,\overline{2}\,5\,4,\quad 4\colon 5\,\overline{1}\,\overline{2}\,3,\quad 5\colon \overline{4}\,\overline{1}\,\overline{2}\,3.$$

The orientation of the lines is assumed to be in increasing order of the x-coordinates. Its minor by the deletion of element 3 is: $1\colon 2\,5\,4$, $2\colon \overline{1}\,5\,4$, $4\colon 5\,\overline{1}\,\overline{2}$, and $5\colon \overline{4}\,\overline{1}\,\overline{2}$. Its contraction on point 3 is $\overline{1}\,\overline{2}\,5\,4$.

13.2 SPACE FILLING

13.2.1 PACKING

The central notion in the theory of packing is the density of a packing.

Definitions:

A **convex body** in d-dimensional Euclidean space $\mathcal{R}^d$ is a compact convex subset of $\mathcal{R}^d$ with nonempty interior.

A family $\mathcal{P} = \{K_1, K_2, \ldots\}$ of the convex bodies $K_1, K_2, \ldots$ in $\mathcal{R}^d$ forms a **packing** of $\mathcal{R}^d$ if no two of the convex bodies $K_1, K_2, \ldots$ have an interior point in common.

Let $\mathcal{P} = \{K_1, K_2, \ldots\}$ be a packing of $\mathcal{R}^d$ and B_R the closed ball of radius R centered at the origin in $\mathcal{R}^d$. The **lower density** and **upper density** of $\mathcal{P}$ are defined by

$$\underline{\delta}(\mathcal{P}) = \liminf_{R \to +\infty} \frac{\sum_{K_i \subset B_R} \mathrm{Vol}(K_i)}{\mathrm{Vol}(B_R)} \quad \text{and} \quad \overline{\delta}(\mathcal{P}) = \limsup_{R \to +\infty} \frac{\sum_{K_i \subset B_R} \mathrm{Vol}(K_i)}{\mathrm{Vol}(B_R)}.$$

If $\underline{\delta}(\mathcal{P}) = \overline{\delta}(\mathcal{P}) = \delta(\mathcal{P})$, then $\delta(\mathcal{P})$ is called the **density** of $\mathcal{P}$.

For a convex body $K \subset \mathcal{R}^d$, let $\delta(K)$ denote the largest (upper) density of packings by congruent copies of K in $\mathcal{R}^d$. In particular:

$\delta_T(K)$ = the largest (upper) density of packings by translates of K in $\mathcal{R}^d$;

$\delta_L(K)$ = the largest (upper) density of packings by lattice translates of K in $\mathcal{R}^d$.

Let $\mathcal{P} = \{B_1^d, B_2^d, \ldots\}$ be a packing of unit balls in $\mathcal{R}^d$ with centers $c_1, c_2, \ldots$. The **Voronoi cell** with center c_i is the convex polyhedral set $V_i = \{\, x \in \mathcal{R}^d \mid |x - c_i| = \min_j |x - c_j|\,\}$.

Facts:
1. There are two major problems concerning density:
   - Given a convex body $K \subseteq \mathcal{R}^d$, find efficient packings with congruent copies of K; i.e., find packings with congruent copies of K in $\mathcal{R}^d$ having "relatively high" density.
   - Find a "good" upper bound for $\delta(K)$.
2. Sphere packing in n-dimensional space, for n large, is important in designing codes that are efficient and unlikely to contain errors when data is transmitted.
3. If K is a convex body in $\mathcal{R}^d$, $\dfrac{2\zeta(d)}{\binom{2d}{d}} \leq \delta_L(K) \leq \delta_T(K) \leq \delta(K)$, where $\zeta(d) = 1 + \frac{1}{2^d} + \frac{1}{3^d} + \cdots$ denotes Riemann's zeta function.
4. If K is a centrally symmetric convex body in $\mathcal{R}^d$, $\dfrac{\zeta(d)}{2^{d-1}} \leq \delta_L(K) \leq \delta_T(K) \leq \delta(K)$.
5. Facts 3 and 4 have been improved slightly for different classes of convex bodies and subclasses of centrally symmetric convex bodies. (See [DaRo47], [ElOdRu91], and [Sc63].)
6. For each d, $(d-1)\dfrac{\zeta(d)}{2^{d-1}} \leq \delta_L(B^d) \leq \delta_T(B^d) = \delta(B^d)$.
7. For every convex domain D, $\delta_L(D) = \delta_T(D)$.
8. For every centrally symmetric convex domain D, $\delta_L(D) = \delta_T(D) = \delta(D)$.
9. There exists an ellipsoid C in $\mathcal{R}^3$ for which $\delta_L(C) < \delta(C)$.
10. The class of convex bodies $C \subseteq \mathcal{R}^d$ for which $\delta_L(C)$, $\delta_T(C)$, and $\delta(C)$ can be determined (for a given d) is very small.
11. It is possible to extend some of the above theorems to spherical as well as hyperbolic spaces. (In short, a space of constant curvature means the corresponding Euclidean space or spherical space or hyperbolic space.) For example, see Fact 15.
12. Let $\mathcal{P} = \{B_1^d, B_2^d, \ldots\}$ be a packing of $\mathcal{R}^d$ by unit balls with centers $c_1, c_2, \ldots$. For a regular simplex of edge length 2 in $\mathcal{R}^d$ with a unit ball drawn around each vertex, let σ_d be the ratio of the volume of the portion of the simplex covered by balls to the volume of the whole simplex. For each Voronoi cell V_i ($i = 1, 2, \ldots$) with center c_i, let $\widehat{V}_i = \left\{x \in V_i \mid |x - c_i| \leq \sqrt{\frac{2d}{d+1}}\right\}$. Then $\dfrac{\text{Vol}(B_i^d)}{\text{Vol}(\widehat{V}_i)} \leq \sigma_d$ for all $i = 1, 2, \ldots$, and hence $\overline{\delta}(\mathcal{P}) \leq \sigma_d$.

Note: This result is sharp for $d = 1$ and 2, but has been improved for sufficiently large d and for $d = 3$ in Facts 13 and 14.

13. $\delta(B^d) \leq \dfrac{1}{2^{(0.599 + o(1))d}}$ as $d \to \infty$.
14. $\delta(B^3) \leq 0.7731$.
15. The densest lattice packing of unit balls is determined up to dimension 8. The following table lists the optimal lattices. (See [CoSl93].)

dimension	1	2	3	4	5	6	7	8
densest lattice packing	$\mathcal{Z}$	A_2	A_3	D_4	D_5	E_6	E_7	E_8

16. Given a set of n ($n \geq 3$) nonoverlapping circles of radius r in a plane of constant curvature, the density of the circles with respect to the outer parallel domain of the convex hull of their centers at distance r is less than $\dfrac{\pi}{\sqrt{12}} = 0.90689\ldots$.

17. The densest packing of circles of radius r in the plane is a hexagonal arrangement, with each circle tangent to its six neighbors.

18. No packing of spheres in $\mathcal{R}^3$ can fill more than 78 percent of the possible volume.

19. A face-centering packing of spheres in $\mathcal{R}^3$ fills slightly more than 74 percent of the total possible volume. (A *face-centering* packing of spheres consists of first arranging a layer of spheres so each is tangent to six neighbors. A second layer is added by placing spheres in the depressions that occur in any triangle formed by three adjacent spheres in the first layer. This pattern continues, giving a configuration much like a pyramid of oranges seen in a supermarket produce display.)

20. For numerous other important concepts and results in the field, see [FeKu93].

Open Questions and Conjectures:

1. The major outstanding question is to estimate $\delta_L(B^d)$ and $\delta_T(B^d) = \delta(B^d)$ for the d-dimensional unit ball B^d in $\mathcal{R}^d$.

2. *Kepler's conjecture*: $\delta(B^3) \leq \dfrac{\pi}{\sqrt{18}} = 0.74048\ldots$.

3. *The dodecahedral conjecture*: The volume of any Voronoi cell in a packing of unit balls in E^3 is at least as large as the volume of a regular dodecahedron with inradius 1.

4. It is widely believed, but not proved, that $\delta_L(B^3) = \delta(B^3)$ and $\delta_L(B^d) < \delta(B^d)$ for sufficiently large d.

13.2.2 COVERING

Definitions:

A family $\mathcal{C} = \{K_i \mid i \in I\}$ of convex bodies in $\mathcal{R}^d$ forms a **covering** of $\mathcal{R}^d$ (that is, **covers** $\mathcal{R}^d$) if each point of $\mathcal{R}^d$ belongs to at least one convex body of $\mathcal{C}$.

The **lower density** and **upper density** of a covering $\mathcal{C}$ are

$$\underline{\nu}(\mathcal{C}) = \liminf_{R \to +\infty} \frac{\sum_{K_i \cap B_R \neq \emptyset} \text{Vol}(K_i)}{\text{Vol}(B_R)} \quad \text{and} \quad \overline{\nu}(\mathcal{C}) = \limsup_{R \to +\infty} \frac{\sum_{K_i \cap B_R \neq \emptyset} \text{Vol}(K_i)}{\text{Vol}(B_R)}$$

where B_R denotes the closed ball of radius R centered at the origin in $\mathcal{R}^d$.

If $\underline{\nu}(\mathcal{C}) = \overline{\nu}(\mathcal{C}) = \nu(\mathcal{C})$, then $\nu(\mathcal{C})$ is called the **density** of $\mathcal{C}$.

For a convex body $K \subseteq \mathcal{R}^d$, let

$\nu(K) =$ the smallest (lower) density of coverings of $\mathcal{R}^d$ by congruent copies of K;
$\nu_T(K) =$ the smallest (lower) density of coverings of $\mathcal{R}^d$ by translates of K;
$\nu_L(K) =$ the smallest (lower) density of coverings of $\mathcal{R}^d$ by lattice translates of K.

Facts:

1. There are two major problems concerning covering:
- Given a convex body $K \subseteq \mathcal{R}^d$, find efficient coverings of $\mathcal{R}^d$ with congruent copies of K; that is, find coverings of $\mathcal{R}^d$ by congruent copies of K having "relatively small" density.
- Find a "good" lower bound for $\nu(K)$. (This is a highly nontrivial task for most of the convex bodies $K \subseteq \mathcal{R}^d$.)

2. If K is a convex body in $\mathcal{R}^d$, then
$$\nu(K) \leq \nu_T(K) \leq d(\ln d) + d(\ln \ln d) + 4d$$
and
$$\nu_T(K) \leq \nu_L(K) \leq d^{(\log_2 \log_2 d) + c} \quad \text{for some constant } c.$$

3. If B^d denotes the d-dimensional closed unit ball in $\mathcal{R}^d$, then
$$\nu(B^d) = \nu_T(B^d) \leq \nu_L(B^d) \leq cd(\ln d)^{\frac{1}{2}\log_2(2\pi e)}$$
for some constant c.

4. Take a regular simplex inscribed in a unit ball in $\mathcal{R}^d$ and draw unit balls around each vertex. Let τ_d be the ratio of the sum of the volumes of the portions of these balls lying in the regular simplex to the volume of the regular simplex. Then $\tau_d \leq \nu(B^d)$.

5. $\tau_d \sim \frac{d}{e^{3/2}}$. (Thus, Facts 2 and 3 give strong estimates for $\nu(B^d)$. Moreover, for $d = 1$ and 2, the lower bound τ_d is sharp.)

6. The thinnest lattice covering of $\mathcal{R}^d$ by unit balls has been determined up to dimension 5 only. The following table lists the optimal lattices. (See [CoSl93].)

dimension	1	2	3	4	5
thinnest lattice covering	$\mathcal{Z}$	A_2	A_3^*	A_4^*	A_5^*

13.2.3 TILING

Only a "diagonal" view of the basic definitions and theorems of this area will be given. See [GrSh86] for additional material.

Definitions:

A *tiling* $\mathcal{T}$ of Euclidean d-space $\mathcal{R}^d$ is a countable family of closed topological d-cells of $\mathcal{R}^d$, the *tiles* of $\mathcal{T}$, which cover $\mathcal{R}^d$ without gaps and overlaps.

A *monohedral tiling* is a tiling $\mathcal{T}$ of $\mathcal{R}^d$ in which all tiles are congruent to one fixed tile T, the (metrical) *prototile* of $\mathcal{T}$. In this case, T *admits* the tiling $\mathcal{T}$.

A *regular polygon* is a polygon with all sides congruent and all interior angles equal.

A *regular tiling* is a monohedral tiling of the plane ($\mathcal{R}^2$) with a regular polygon as prototile.

A *semiregular tiling* is a tiling of the plane using n prototiles with the same numbers of polygons around each vertex.

A *semiregular polyhedron* is a convex polyhedron with each face a regular polygon, but where more than one regular polygon can be used as a face.

A tiling $\mathcal{T}$ of $\mathcal{R}^d$ by convex polytopes is *normal* if there exist positive real numbers r and R such that each tile contains a Euclidean ball of radius r and is contained in a Euclidean ball of radius R.

A *face-to-face* tiling is a tiling $\mathcal{T}$ by convex d-polytopes such that the intersection of any two tiles is a face of each tile, possibly the (improper) empty face. When $d = 2$, such a tiling is an *edge-to-edge* tiling.

A **lattice tiling**, with lattice L, is a tiling $\mathcal{T}$ by translates of a single tile T such that the corresponding translation vectors form a d-dimensional lattice L in $\mathcal{R}^d$.

A Euclidean motion σ of $\mathcal{R}^d$ is a **symmetry** of a tiling $\mathcal{T}$ if σ maps (each tile of) $\mathcal{T}$ onto (a tile of) $\mathcal{T}$. The set of all symmetries of $\mathcal{T}$ (a group under composition) is the **symmetry group** $S(\mathcal{T})$ of $\mathcal{T}$.

A **periodic tiling** is a tiling $\mathcal{T}$ of $\mathcal{R}^d$ such that $S(\mathcal{T})$ contains translations in d linearly independent directions. A tiling $\mathcal{T}$ is **nonperiodic** if $S(\mathcal{T})$ contains no translation other than the identity.

An **isohedral tiling** is a tiling $\mathcal{T}$ such that $S(\mathcal{T})$ acts transitively on the tiles of $\mathcal{T}$.

An **isogonal tiling** is a tiling $\mathcal{T}$ such that $S(\mathcal{T})$ acts transitively on the vertices of $\mathcal{T}$.

An **isotoxal tiling** is a tiling $\mathcal{T}$ such that $S(\mathcal{T})$ acts transitively on the edges of $\mathcal{T}$.

Let $\mathcal{T}$ and $\mathcal{T}'$ be tilings of $\mathcal{R}^d$ with symmetry groups $S(\mathcal{T})$ and $S(\mathcal{T}')$. Let $\Phi: \mathcal{R}^d \to \mathcal{R}^d$ be a homeomorphism that maps $\mathcal{T}$ onto $\mathcal{T}'$. Φ **compatible with a symmetry** σ of $\mathcal{T}$ if there exists a symmetry σ' of $\mathcal{T}'$ such that $\sigma'\Phi = \Phi\sigma$. Φ is **compatible** with $S(\mathcal{T})$ if Φ is compatible with each σ in $S(\mathcal{T})$. The tilings $\mathcal{T}$ and $\mathcal{T}'$ of $\mathcal{R}^d$ are **homeomeric**, or of the **same homeomeric type**, if there exists a homeomorphism $\Phi: \mathcal{R}^d \to \mathcal{R}^d$ that maps $\mathcal{T}$ onto $\mathcal{T}'$ such that Φ is compatible with $S(\mathcal{T})$ and Φ^{-1} is compatible with $S(\mathcal{T}')$.

A prototile T in $\mathcal{R}^d$ is **aperiodic** if T admits a tiling of $\mathcal{R}^d$, yet all such tilings are nonperiodic. In general, a set S of prototiles in $\mathcal{R}^d$ is said to be **aperiodic** if S admits a tiling of $\mathcal{R}^d$, yet all such tilings are nonperiodic.

Facts:

1. There are three monohedral edge-to-edge tilings of $\mathcal{R}^2$ with regular polygons; the prototile must be a triangle, a square, or a hexagon. See the following figure.

2. There are eight semiregular tilings of $\mathcal{R}^2$. These tilings use two or three prototiles.

3. Shapes that are not regular polygons [polyhedra] can be used in monohedral tilings of $\mathcal{R}^2$ [$\mathcal{R}^3$].

4. Any triangle can be used in a monohedral tiling of the plane. (Join two to form a parallelogram and tile a strip using these parallelograms. Repeat this process with parallel strips to tile the plane.) See the following figure.

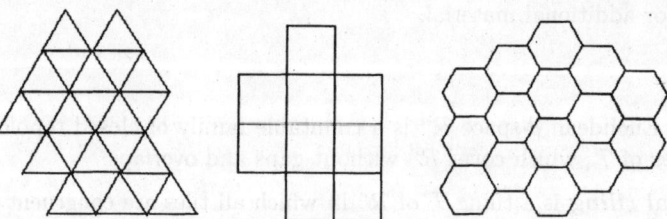

5. Any quadrilateral can be used in a monohedral tiling of the plane. See the following figure. (Take a second copy of the quadrilateral and rotate it 180 degrees. Join the two to form a hexagon. Use the hexagons to tile the plane.)

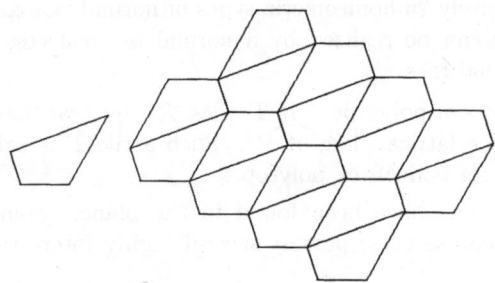

6. Any pentagon with a pair of parallel sides can be used in a monohedral tiling of the plane.

7. There are at least fourteen types of convex pentagons that can be used in a monohedral tiling of the plane. It is not known if there are more.

8. There are three types of convex hexagons that can be used in a monohedral tiling of the plane. Assume that the hexagon has vertices a, b, c, d, e, f in clockwise order. See the following figure. The prototile must be of one of the following forms:
   - sum of angles at a, b, c is 360°; length of $\{a, f\}$ = length of $\{c, d\}$;
   - sum of angles at a, b, e is 360°; length of $\{a, f\}$ = length of $\{d, e\}$ and length of $\{b, c\}$ = length of $\{e, f\}$;
   - angles at a, b, and c are each equal to 120°; length of $\{a, b\}$ = length of $\{a, f\}$, length of $\{c, b\}$ = length of $\{c, d\}$, and length of $\{e, d\}$ = length of $\{e, f\}$.

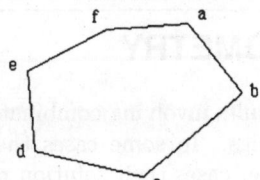

9. No convex polygon with more than six sides can be used as prototile in a monohedral tiling of $\mathcal{R}^2$.

10. Of the five regular polyhedra (tetrahedron, hexahedron (cube), octahedron, dodecahedron, icosahedron), only the tetrahedron and cube can be used as a prototile in a regular tiling of $\mathcal{R}^3$.

11. If $\mathcal{T}$ is a tiling of $\mathcal{R}^d$ with convex tiles, then each tile in $\mathcal{T}$ is a convex d-polyhedron.

12. If $\mathcal{T}$ is a tiling of $\mathcal{R}^d$ with compact convex tiles, then each tile in $\mathcal{T}$ is a convex d-polytope.

13. The following classification results have a long history. (See [GrSh86].)
   - There exist precisely 11 distance edge-to-edge isogonal plane tilings, the tiles of which are convex regular polygons (called Archimedean tilings).
   - There exist precisely 81 homeomeric types of normal isohedral plane tilings. Precisely 47 of these can be realized by a normal isohedral edge-to-edge tiling with convex polygonal tiles.

- There exist precisely 91 homeomeric types of normal isogonal plane tilings. Precisely 63 types can be realized by normal isogonal edge-to-edge tilings with convex polygonal tiles.
- There exist precisely 26 homeomeric types of normal isotoxal plane tilings. Precisely 6 types can be realized by a normal isotoxal edge-to-edge tiling with convex polygonal tiles.

14. Let T be a convex d-polytope. If T tiles $\mathcal{R}^d$ by translation, then T admits (uniquely) a face-to-face lattice tiling of $\mathcal{R}^d$. Such a tile T is called a *parallelotope*. This result is not true for nonconvex polytopes.

15. Several aperiodic sets have been found in the plane. Some of them, such as the Wang tiles and Penrose tiles, possess several highly interesting properties. (See [GrSh86].)

16. Very recently, considerable progress has been achieved for aperiodic tilings in higher dimensions via dynamical systems. (See [Ra95].)

Open Questions:

1. Extend the classification problems to higher dimensions. (At present, this looks hopeless.)

2. Classify all convex d-polytopes which are prototiles of monohedral tilings of $\mathcal{R}^d$. (This problem is not even solved for the plane.) However, under suitable restrictions the complexity of the problem changes. (See Fact 4.)

3. For $d \geq 5$, determine whether each d-parallelotope is a Voronoi cell (see §9.2) for some d-lattice. (This is known to be true for $1 \leq d \leq 4$.)

13.3 COMBINATORIAL GEOMETRY

This section studies geometric results involving combinatorics in the areas of convexity, incidences, distances, and colorings. In some cases the problems themselves have a combinatorial flavor, while in other cases their solution requires combinatorial tools.

13.3.1 CONVEXITY

In this subsection, questions of two different kinds are studied. Most of them belong to geometric transversal theory, a subject originating in Helly's theorem. Another group of problems grew out of the Erdős-Szekeres theorem, which turned out to be a starting point of Ramsey theory.

Definitions:

A subset C of d-dimensional Euclidean space (d-space) $\mathcal{R}^d$ is **convex** if the following is true: for any pair of points in C, the straight-line segment connecting them is entirely contained in C.

A convex set is **strictly convex** if its boundary contains no line segment.

A **convex body** is a compact (i.e., bounded and closed) convex set with nonempty interior.

A **polytope** is a bounded convex body that can be obtained as the intersection of finitely many closed half-spaces. (§13.1.1.)

A **convex polygon** is a polytope in the plane.

A **vertex** of a polytope P is a point $q \in P$, for which there is a hyperplane (§13.1.1) H such that $H \cap P = \{q\}$.

A point set is in **convex position** if it is the vertex set of a polytope.

The **convex hull** of a set $S \subseteq \mathcal{R}^d$ is the smallest convex set containing S.

A family $\mathcal{C} = \{C_1, C_2, \ldots\}$ of sets in d-space is said to be **intersecting** if all members of $\mathcal{C}$ have a point in common.

A set $T \subseteq \mathcal{R}^d$ is a **transversal** of a family $\mathcal{C}$ of sets if $T \cap C_i$ is nonempty for every i. If $\mathcal{C}$ has a k-element transversal ($|T| = k$), its members can be **pierced** by k points.

Two sequences $P = \{p_1, \ldots, p_n\}$ and $Q = \{q_1, \ldots, q_n\}$ of points in $\mathcal{R}^k$ have the same **order type** if, for all $1 \leq i_1 < i_2 < \cdots < i_{k+1} \leq n$, the orientations of the simplices induced by $\{p_{i_1}, \ldots, p_{i_{k+1}}\}$ and $\{q_{i_1}, \ldots, q_{i_{k+1}}\}$ are the same. This order type is **nontrivial** if P and Q are not contained in any hyperplane of $\mathcal{R}^k$.

A k-flat (an oriented k-dimensional plane) F intersects a sequence of d-dimensional convex bodies $\mathcal{C} = \{C_1, \ldots, C_n\}$ **consistently** with the above order type if there are $x_i \in F \cap C_i$ such that the sequences $X = \{x_1, \ldots, x_n\}$ and P have the same order type.

Facts:

1. The convex hull of a set S is the intersection of all convex sets containing S.

2. For any set $S \subset \mathcal{R}^d$ of finitely many points, not all of which lie in the same hyperplane, the convex hull of S is a polytope. In particular, if S has $d+1$ points, then its convex hull is a simplex whose vertices are the elements of S.

3. The convex hull of the vertex set of any convex polytope P is identical with P.

4. **Helly's theorem**: If a family $\mathcal{C}$ of at least $d+1$ convex bodies in $\mathcal{R}^d$ has the property that every $d+1$ of its members have a point in common, then $\mathcal{C}$ is intersecting (i.e., all its members have a point in common). [He23]

5. **Carathéodory's theorem**: If the convex hull of a set $S \subseteq \mathcal{R}^d$ contains a point p, then there exists a subset of S with at most $d+1$ elements whose convex hull contains p.

6. Let S be a compact set in $\mathcal{R}^d$ with the property that for every $(d+1)$-element subset $T \subset S$, there is a point $s \in S$ such that each segment connecting s to an element of T lies in S. Then S has a point such that every segment connecting it to an element of S is entirely contained in S. [Kr46]

7. Any set of $(k-1)(d+1)+1$ points in $\mathcal{R}^d$ can be partitioned into k parts whose convex hulls have a point in common. [Ra21], [Tv66]

8. Let $\mathcal{C}$ be any family of convex bodies in $\mathcal{R}^d$ with the property that the volume of the intersection of any $2d$ of them is at least 1. Then the volume of the intersection of all members of $\mathcal{C}$ is at least a positive constant depending only on d. [BáKaPa82]

9. For any $\epsilon > 0$ and for any d there is a $\delta > 0$ satisfying the following condition: if $\mathcal{C}$ is a family of n ($> d+1$) convex bodies in $\mathcal{R}^d$ having at least $\epsilon \binom{n}{d+1}$ intersecting $(d+1)$-tuples, then $\mathcal{C}$ has at least δn members with a point in common. [Ka84]

10. For any $d < q \leq p$, there exists $k = k(p,q,d)$ satisfying the following condition: if $\mathcal{C}$ is a family of convex bodies in $\mathcal{R}^d$ such that every subfamily of $\mathcal{C}$ of size p contains q members with a point in common, then $\mathcal{C}$ can be pierced by k points. [AlKl92]

11. A sequence $\mathcal{C} = \{C_1, \ldots, C_n\}$ of convex bodies in $\mathcal{R}^d$ has a hyperplane transversal if and only if for some $0 \leq k \leq d-1$, there is a nontrivial k-dimensional order type of n points such that every $(k+2)$-member subfamily of $\mathcal{C}$ can be met by a suitable k-flat consistently with that order type. [PoWe90]

12. If $\mathcal{S}_k$ is any family of k-dimensional linear subspaces of $\mathcal{R}^d$ with the property that any $\binom{k+l}{l}$ of them can be intersected by an l-dimensional subspace, then all members of $\mathcal{S}_k$ can be intersected by an l-dimensional subspace. (Two subspaces intersect each other if they have at least one point in common, different from the origin.)

13. Any set of five points in the plane, no three of which are on a line, has four elements in convex position.

14. *Erdős-Szekeres theorem*: For every $k > 2$, there exists a smallest integer $n(k)$ with the property that every set of at least $n(k)$ points in the plane, no three of which are on a line, contains k points in convex position. If $k = 3, 4, 5, 6$, then $n(k) = 2^{n-2} + 1$. (These are the only known values.) Furthermore, $2^{k-2} + 1 \leq n(k) \leq \binom{2k-5}{k-2}$. [ErSz35], [ToVa98]

Examples:

1. Let $S = \{(1,0,0), (0,1,0), (0,0,1)\}$. The convex hull of $S \subset \mathcal{R}^3$ is a triangular region, which is *not* a convex body in 3-space because its interior is empty.

2. Let $S = \{(0,0), (1,0), (0,1)\}$. The convex hull of $S \subset \mathcal{R}^2$ is a triangular region, which *is* a convex body (polygon) in the plane.

3. Let $S = \{(x, y) \mid 0 \leq x \leq 2, 0 \leq y \leq 2\}$ and $S' = \{(x, y) \mid 0 \leq x \leq 3, 0 \leq y \leq 3\}$. The family of all axis-parallel unit squares lying in S is intersecting because each of them contains the point $(1, 1)$. The family of axis parallel unit squares in S' can be pierced by four points: $(1,1), (1,2), (2,1), (2,2)$.

4. In the line, $\{1, 3, 4, 2\}$ and $\{0, 4, 25, 3\}$ have the same (1-dimensional) order type. The 3-dimensional closed unit balls centered at $(0, 1, 5), (0, 0, 9.6), (0, 0, 9.4), (1, 0, 7)$ are met by the z-axis consistently with the above order type, because these balls contain the points $(0, 0, 5), (0, 0, 9), (0, 0, 10)$, and $(0, 0, 7)$, respectively, and the order type of this sequence along the z-axis is the same as the 1-dimensional order type of $\{1, 3, 4, 2\}$.

13.3.2 INCIDENCES

This subsection studies the structure (and number) of incidences between a set of points and a set of lines (or planes, spheres, etc.). The starting point of many investigations in this field was the Sylvester-Gallai theorem.

Definitions:

Given a point set P and a set L of lines (or k-flats, spheres, etc.) in Euclidean d-space $\mathcal{R}^d$, a point $p \in P$ and a line $l \in L$ are **incident** with each other, if $p \in l$.

Given a set L of lines in the plane, a point incident with precisely two elements of L is called an **ordinary crossing**. Given a set of points $P \subseteq \mathcal{R}^d$, a hyperplane passing through precisely d elements of P is called an **ordinary hyperplane** (for $d = 2$, an **ordinary line**).

Given a set of points P, a **Motzkin hyperplane** is a hyperplane h such that all but one element of $h \cap P$ lie in a $(d-2)$-flat.

A family Γ of curves in the plane has d **degrees of freedom** if there exists an integer s such that:
- no two curves in Γ have more than s points in common;
- for any d points, there are at most d curves in Γ passing through all the points.

A family of **pseudolines** is a family of simple curves in the plane with the property that every two of them meet precisely once.

A family of **pseudocircles** is a family of simple closed curves in the plane with the property that every two of them meet in at most two points.

Facts:

1. *Sylvester-Gallai theorem*: Every finite set of points in the plane, not all of which are on a line, determines an ordinary line. In dual version: every finite set of straight lines in the plane, not all of which pass through the same point, determines an ordinary crossing.

2. For every finite set of points in Euclidean d-space, not all of which lie on a hyperplane, there exists a Motzkin hyperplane. [Ha65], [Ha80]

3. Every set of n points in d-space, not all of which lie on a hyperplane, determine at least n distinct hyperplanes.

4. In 3-space, every set of n non-coplanar points determines at least $\frac{2n}{5}$ Motzkin hyperplanes.

5. If n is sufficiently large, then every set of n non-cocircular points in the plane determines at least $\binom{n-1}{2}$ distinct circles, and this bound is best possible. [El67]

6. Every set of n (>7) noncollinear points in the plane determines at least $\frac{6n}{13}$ ordinary lines. This bound is sharp for $n = 13$ and false for $n = 7$. [CsSa93]

7. There is a positive constant c such that every set of n points in the plane, not all on a line, has an element incident with at least cn connecting lines. Moreover, any set of n points in the plane, no more than $n - k$ of which are on the same line, determines at least $c'kn$ distinct connecting lines, for a suitable constant $c' > 0$. According to the $d = 2$ special case of Fact 4, due to de Bruijn-Erdős, for $k = 1$ the number of distinct connecting lines is at least n. For $k = 2$, the corresponding bound is $2n - 4$ (for $n \geq 10$). [Be83], [SzTr83]

8. Every set of n noncollinear points in the plane always determines at least $2\lfloor \frac{n}{2} \rfloor$ lines of different slopes. Furthermore, every set of n points in the plane, not all on a line, permits a spanning tree, all of whose $n - 1$ edges have different slopes. [Un82], [Ja87]

9. The number of incidences between a set P of points and a set L of lines can be obtained by summing over all $l \in L$ the number of points in l belonging to P, or, equivalently, by summing over all $p \in P$ the number of lines in L passing through p.

10. Let Γ be a family of curves in the plane with d degrees of freedom. Then the maximum number of incidences between n points in the plane and m elements of Γ is
$$O(n^{d/(2d-1)} m^{(2d-2)/(2d-1)} + n + m). \quad [\text{PaSh98}]$$

From the most important special case, when Γ is the family of all straight lines ($d = 2$), it follows that for any set P of n points in the plane, the number of distinct straight lines containing at least k elements of P is $O(\frac{n^2}{k^3} + \frac{n}{k})$ [SzTr83]. This bound is asymptotically tight. The same result holds for pseudolines.

11. The maximum number of incidences between n points and m spheres in $\mathcal{R}^3$ is

$$O(n^{\frac{4}{7}}m^{\frac{9}{7}}\beta(n,m) + n^2),$$

where $\beta(n,m) = o(\log(nm))$ is an extremely slowly growing function.

If no three spheres contain the same circle, then the following better bound is obtained:

$$O(n^{\frac{3}{4}}m^{\frac{3}{4}} + n + m).$$

Neither of these estimates is known to be asymptotically tight. [ClEtal90]

12. The maximum number of collinear triples determined by n points in the plane, no four of which are on a line, is at least $\frac{n^2}{6} - O(n)$. This bound is asymptotically tight. [BuGrSl79]

13. If $M(n)$ denotes the minimum number of different midpoints of the $\binom{n}{2}$ line segments determined by n points in convex position in the plane, then

$$\binom{n}{2} - \left\lfloor \frac{n(n+1)(1-e^{-1/2})}{4} \right\rfloor \leq M(n) \leq \binom{n}{2} - \left\lfloor \frac{n^2-2n+12}{20} \right\rfloor.$$

[ErFiFü91]

Examples:

1. Let P be a set of 7 points in the plane, consisting of the vertices, the centroid (the point of intersection of the medians), and the midpoints of all sides of an equilateral triangle. Then P determines 3 ordinary lines (the lines connecting the midpoints of two sides).

2. Let P be a $4k$-element set in the plane that can be obtained from the vertex set $\{v_1, v_2, \ldots, v_{2k}\}$ of a regular $2k$-gon by adding the intersection of the line at infinity with every line $v_i v_j$. Then the set P determines precisely $2k$ ordinary lines: every line connecting some v_i to the intersection point of $v_{i-1}v_{i+1}$ and the line at infinity ($1 \leq i \leq 2k$, the indices are taken modulo $2k$). (It can be achieved by a suitable projective transformation that no point of P is at infinity, and the number of ordinary lines remains $\frac{|P|}{2} = 2k$.)

3. Let P be a set of $n \geq 4$ points lying on two noncoplanar lines in 3-space so that there are at least two points on each line. Not all points of P are coplanar, but P does not determine any ordinary plane.

4. The family of all straight lines in the plane and the family of all unit circles both have 2 degrees of freedom. The family of all circles with arbitrary radii has 3 degrees of freedom. The family of the graphs of all polynomials of one variable and degree d has d degrees of freedom.

5. Let P be an $n^{\frac{1}{2}} \times n^{\frac{1}{2}}$ part of the integer grid; i.e.,

$$P = \{(i,j) \mid 1 \leq i \leq n^{\frac{1}{2}}, 1 \leq j \leq n^{\frac{1}{2}}\}.$$

Let $k = \left(\frac{cm}{n^{1/2}}\right)^{1/3} > 2$, where $c > 0$ is a sufficiently small constant. For every $1 \leq s < r \leq k$ and for every $1 \leq i \leq r, 1 \leq j \leq \frac{n^{1/2}}{2}$, consider the line passing through (i,j) and $(i+r, j+s)$. If c is sufficiently small, then the number of these lines is at most m. There is a constant $c' > 0$ such that the total number of incidences between these lines and the elements of P is at least $c' n^{\frac{2}{3}} m^{\frac{2}{3}}$. (See the case $d = 2$ of Fact 10.)

13.3.3 DISTANCES

The systematic study of the distribution of the $\binom{n}{2}$ distances determined by n points was initiated by Erdős. Given a set of n points $P = \{p_1, p_2, \ldots, p_n\}$, let $g(P)$ denote the number of distinct distances determined by P, and let $f(P)$ denote the number of times that the unit distance occurs between two elements of P. That is, $f(P)$ is the number of pairs $p_i p_j, i < j$, such that $|p_i - p_j| = 1$. In [Er46], Erdős raised the following general questions: What is the minimum of $g(P)$ and what is the maximum of $f(P)$ over all n-element subsets of Euclidean d-space or of any other fixed metric space?

Definitions:

For any point set P in a metric space, the **unit distance graph** of P is the graph $G(P)$ whose vertex set is P and two points (vertices) are connected by an edge if and only if their distance is 1.

Let P be a finite set of points in a metric space. If the distance between two points $p, q \in P$ is minimum, then p and q form a **closest pair**.

A point $q \in P$ is a **nearest neighbor** of $p \in P$, if no point of P is closer to p than q.

A set P in a metric space is a **separated set** if the minimum distance between the points of P is at least 1.

The **diameter** of a finite set of points in a metric space is the maximum distance between two points of the set.

A point $q \in P$ is a **farthest neighbor** of $p \in P$, if no point of P is farther from p than q.

A set of points in the plane is said to be in **general position** if no three are on a line and no four on a circle.

Facts:

1. $f(P)$ is equal to the number of edges of $G(P)$.

2. If p and q form a closest pair in P, then q is a nearest neighbor of p and p is a nearest neighbor of q.

3. If the distance between p and q is equal to the diameter of P, then q is a farthest neighbor of p and p is a farthest neighbor of q.

4. The maximum number of times that the unit distance can occur among n points in the plane is $O(n^{4/3})$. Conjecture: the asymptotically best bound is $O(n^{1+c/\log\log n})$. [SpSzTr84]

5. The maximum number of times that the unit distance can occur in a separated set of $n \leq 3$ points is $\lfloor 3n - \sqrt{12n - 3} \rfloor$. [Ha74]

6. The maximum number of times that the unit distance can occur in a set of n points in the plane with unit diameter is n. [HoPa34]

7. For any set of $n > 3$ points in the plane, the total number of farthest neighbors of all elements is at most $3n - 3$ if n is even, and at most $3n - 4$ if n is odd. These bounds cannot be improved. [EdSk89]

8. The maximum number of times that the unit distance can occur among n points in convex position in the plane is $O(n \log n)$. For $n > 15$, the best known lower bound is $2n - 7$. [Fü90], [EdHa91].

9. The minimum number of distinct distances determined by n points in the plane is $\Omega(n^{\frac{4}{5}})$. It is conjectured that the best bound is $\Omega(\frac{n}{\sqrt{\log n}})$. [Sz97]

10. The minimum number of distinct distances determined by $n > 3$ points in convex position in the plane is $\lfloor \frac{n}{2} \rfloor$. [Al63]

11. The minimum number of distinct distances determined by $n > 3$ points in the plane, no three of which are on a line, is at least $\lceil \frac{n-1}{3} \rceil$. Conjecture: the best possible bound is $\lfloor \frac{n}{2} \rfloor$.

12. The minimum number of distinct distances determined by n points in general position in the plane is $O(n^{1+c/\sqrt{\log n}})$, for some positive constant c. However, it is not known whether this function is superlinear in n. [ErEtal93]

13. There are arbitrarily large noncollinear finite point sets in the plane such that all distances determined by them are integers, but there exists no infinite set with this property.

14. In an n-element planar point set, the maximum number of noncollinear triples that determine the same angle is $O(n^2 \log n)$, and this bound is asymptotically tight. [PaSh90]

15. Let $f_3(n)$ denote the maximum number of times that the unit distance can occur among n points in $\mathcal{R}^3$. Then

$$\Omega(n^{\frac{4}{3}} \log \log n) \leq f_3(n) \leq n^{\frac{3}{2}} \beta(n),$$

where $\beta(n) = o(\log \log n)$ is an extremely slowly growing function. [ClEtal90]

16. The maximum number of times that the unit distance can occur in a set of $n \geq 4$ points in $\mathcal{R}^3$ with unit diameter is $2n - 2$. [Gr56]

17. If n is sufficiently large, then for any set of n points in $\mathcal{R}^3$, the total number of farthest neighbors of all elements is at most $\frac{n^2}{4} + \frac{3n}{2} + 3$ if n is even, at most $\frac{n^2}{4} + \frac{3n}{2} + \frac{9}{4}$ if $n \equiv 1 \pmod{4}$, and at most $\frac{n^2}{4} + \frac{3n}{2} + \frac{13}{4}$ if $n \equiv 3 \pmod{4}$. These bounds cannot be improved. [Cs96]

18. Let $f_d(n)$ denote the maximum number of times that the unit distance can occur among n points in $\mathcal{R}^d$. If $d \geq 4$ is even, then

$$f_d(n) = \frac{n^2}{2}\left(1 - \frac{1}{\lfloor \frac{d}{2} \rfloor}\right) + n - O(d).$$

If $d \geq 5$ is odd, then

$$f_d(n) = \frac{n^2}{2}\left(1 - \frac{1}{\lfloor \frac{d}{2} \rfloor}\right) + \Theta(n^{\frac{4}{3}}). \quad [\text{Er60}], [\text{ErPa90}]$$

19. Let $\Phi_d(n)$ denote the maximum of the total number of farthest neighbors of all points over all n-element sets in $\mathcal{R}^d$. For every $d \geq 4$,

$$\Phi_d(n) = n^2\left(1 - \frac{1}{\lfloor \frac{d}{2} \rfloor} + o(1)\right). \quad [\text{ErPa90}]$$

Examples:

1. Let P be the vertex set of a regular n-gon ($n > 3$) in the plane. Then $g(P)$, the number of distinct distances determined by P, is equal to $\lfloor \frac{n}{2} \rfloor$. The number of times that the diameter of P is realized is equal to n if n is odd, and $\frac{n}{2}$ if n is even.

2. Take a regular hexagon of side length k and partition it into $6k^2$ equilateral triangles with unit sides. Let P denote the union of the vertex sets of these triangles. Then P is a separated set, and $|P| = n = 3k^2 + 3k + 1$. The number of times that the minimum (unit) distance occurs between two elements of P is $9k^2 + 3k = 3n - \sqrt{12n - 3}$.

3. Let P denote an $n^{\frac{1}{2}} \times n^{\frac{1}{2}}$ part of the integer grid; i.e., let $P = \{(x,y) \mid 1 \leq x, y \leq n^{\frac{1}{2}}\}$. It follows from classical number theoretic results that there exists an integer k ($\frac{n}{16} \leq k \leq \frac{n}{8}$) that can be written as the sum of two squares in $2n^{\frac{c}{\log \log n}}$ different ways, for a constant $c > 0$. Thus, for every $(x,y) \in P$, the number of points $(x', y') \in P$ satisfying $(x - x')^2 + (y - y')^2 = k$ is at least $2n^{\frac{c}{\log \log n}}$. In other words, the distance $k^{\frac{1}{2}}$ occurs $n^{1 + \frac{c}{\log \log n}}$ times among the elements. By proper scaling, an n-element point set P' is obtained in which the unit distance occurs $n^{1 + \frac{c}{\log \log n}}$ times. That is, $f(P') = n^{1 + \frac{c}{\log \log n}}$. It can also be shown that the number of distinct distances determined by P' satisfies $g(P') = g(P) = \frac{c'n}{\sqrt{\log n}}$ for a suitable positive constant c'.

4. **Lenz' construction**: Let $C_1, \ldots, C_{\lfloor \frac{d}{2} \rfloor}$ be circles of radius $\frac{1}{\sqrt{2}}$ centered at the origin of $\mathcal{R}^d$, and assume that the supporting planes of these circles are mutually orthogonal. Choose n_i points on C_i, where $n_i = \lfloor n/\lfloor \frac{d}{2} \rfloor \rfloor$ or $n_i = \lceil n/\lfloor \frac{d}{2} \rfloor \rceil$, so that $\sum_i n_i = n$. It is clear that any pair of points belonging to different circles C_i are at unit distance from each other. Hence, this point system determines at least

$$\frac{n^2}{2}\left(1 - \frac{1}{\lfloor \frac{d}{2} \rfloor}\right) + n - O(d)$$

unit distances.

5. Let p_1, p_2, p_3, p_4 be the vertices of a regular tetrahedron with side length 1 in $\mathcal{R}^3$. The locus of points in 3-space lying at unit distance from both p_1 and p_2 is a circle passing through p_3 and p_4. Choose distinct points $p_5, p_6, \ldots, p_n$ on the shorter arc of this circle between p_3 and p_4. An n-element point set in $\mathcal{R}^3$ is obtained with diameter 1 and in which the diameter occurs $2n - 2$ times.

13.3.4 COLORING

One of the oldest problems in graph theory is the Four Color Problem (§8.6.4). This problem has attracted much interest among professional and amateur mathematicians, and inspired a lot of research about colorings, including Ramsey theory [GrRoSp90] and the study of chromatic numbers, polynomials, etc. In this section, some coloring problems are discussed in a geometric setting.

Definitions:

A *coloring* of a set with k colors is a partition of the set into k parts. Two points that belong to the same part are said to have the same color.

The **chromatic number of a graph** G is the minimum number of colors, $\chi(G)$, needed to color the vertices of G so that no two adjacent vertices have the same color.

The **chromatic number of a metric space** is the chromatic number of the unit distance graph of the space; that is, the minimum number of colors needed to color all points of the space so that no two points of the same color are at unit distance.

The **polychromatic number of a metric space** is the minimum number of colors, χ, needed to color all points of the space so that for each color class C_i ($1 \leq i \leq \chi$) there is a distance d_i with the property that no two points of this color are at distance d_i from each other.

A point set P in $\mathcal{R}^d$ is **k-Ramsey** if for any coloring of $\mathcal{R}^d$ with k colors, at least one of the color classes has a subset congruent to P. If for every k, there exists $d(k)$ such that P is k-Ramsey in $\mathcal{R}^{d(k)}$, then P is called **Ramsey**.

A point set P' is called a **homothetic copy** (or a **homothet**) of P, if P and P' are similar to each other and they are in parallel position.

Facts:

1. The minimum number of colors needed for coloring the plane so that no two points at unit distance receive the same color is at least 4 and at most 7. That is, the chromatic number of the plane is between 4 and 7. [JeTo95]

2. The following table contains the best known upper and lower bounds on the chromatic numbers of various metric spaces. ($S^{d-1}(r)$ denotes the sphere of radius r in d-space, where the distance between two points is the length of the chord connecting them.)

space	lower bound	upper bound
line	2	2
plane	4	7
rational points of plane	2	2
3-space	5	21
rational points of $\mathcal{R}^3$	2	2
$S^2(r)$, $\frac{1}{2} \leq r \leq \frac{\sqrt{3-\sqrt{3}}}{2}$	3	4
$S^2(r)$, $\frac{\sqrt{3-\sqrt{3}}}{2} \leq r \leq \frac{1}{\sqrt{3}}$	3	5
$S^2(r)$, $r \geq \frac{1}{\sqrt{3}}$	4	7
$S^2(\frac{1}{\sqrt{2}})$	4	4
rational points of $\mathcal{R}^4$	4	4
rational points of $\mathcal{R}^5$	6	$< \infty$
$\mathcal{R}^d$	$(1+o(1))(1.2)^d$	$(3+o(1))^d$
$S^{d-1}(r)$, $r \geq \frac{1}{2}$	d	$< \infty$

3. The polychromatic number of the plane is at least 4 and at most 6. [So94]

4. For any finite d-dimensional point configuration P and for any coloring of d-space with finitely many colors, at least one of the color classes will contain a homothetic copy of P. The corresponding statement is false if "homothetic copy of P" is replaced by "translate of P".

5. A necessary condition for a finite set P to be Ramsey is that it be spherical; i.e., all its points lie on a sphere. [GrRoSp90]

6. The following conditions are sufficient for a finite set P to be Ramsey:
   - P is the vertex set of a right parallelepiped;
   - P is the set of points in d-space with exactly k ($k < d$) nonzero coordinates having values $x_1, \ldots, x_k$ in this order, where $x_1, \ldots, x_k$ is an arbitrary sequence of nonzero reals;
   - P is the vertex set of a regular n-gon;
   - P is a subset of a Ramsey set;
   - P is the cartesian product of two Ramsey sets. [FrRö86], [FrRö90]

7. It follows from the first two and the last two conditions of Fact 6 that all "triangles" are Ramsey. Moreover, given any nondegenerate point configuration ("simplex") S, there is a constant $c(S) > 1$ such that for every $k < c^d(S)$, S is k-Ramsey in d-space.

Examples:
1. Let G be a graph on the vertex set $\{v_1, \ldots, v_7\}$, whose edges are v_1v_2, v_1v_3, v_1v_4, v_1v_5, v_2v_3, v_2v_6, v_3v_6, v_4v_5, v_4v_7, v_5v_7, and v_6v_7. The chromatic number of G is 4.
2. The graph G of Example 1 can be embedded in the plane so that if two of its vertices are connected by an edge, then the corresponding points in the plane are at unit distance. In other words, G is a *subgraph* of the *unit distance graph* of the plane. (In every such imbedding, the points corresponding to $\{v_1, v_2, v_3, v_6\}$ and $\{v_1, v_4, v_5, v_7\}$ form two rhombi of side length 1 that share a vertex.) Hence, the chromatic number of the plane is at least 4.
3. Let P be a 2-element point set in Euclidean space. For every positive integer k, P is k-Ramsey in k-space. (To see this, consider a regular simplex in k-space, whose side length is equal to the distance between the elements of P. Any coloring of $\mathcal{R}^k$ induces a coloring of the vertices of this simplex, and, by the pigeonhole principle, one can always find two vertices that get the same color. They form a 2-element set congruent to P. Thus, P is Ramsey.)

13.4 POLYHEDRA

This section presents basic properties of polyhedra, commonly known as (planar) solids. Any application such as geometric modeling that models the three-dimensional world of objects must deal with polyhedra. Basic geometric and combinatorial properties of polyhedra as well as their convex decompositions and triangulations are discussed.

13.4.1 GEOMETRIC PROPERTIES OF POLYHEDRA

Definitions:

A **$(d\text{-}1)$-dimensional plane** is the solution set of the linear equation $a_1x_1 + a_2x_2 + \cdots + a_dx_d = a_{d+1}$, where $a_1, a_2, \ldots, a_{d+1}$ are constants and $x_1, \ldots, x_d$ are d variables.

A **hyperplane** in d-dimensional Euclidean space $\mathcal{R}^d$ is the set of all points on a $(d-1)$-dimensional plane.

A **closed halfspace** in $\mathcal{R}^d$ is the set of all points on the hyperplane together with the points on one side of the same hyperplane.

A **convex d-polyhedron** is the intersection of a finite number of closed halfspaces in $\mathcal{R}^d$.

A **nonconvex polyhedron** is the union of a set of convex polyhedra such that the underlying space is connected and nonconvex.

A **k-face**, part of the boundary of a polyhedron, lies on at least $d-k$ hyperplanes forming the boundary. In particular, 0-faces, 1-faces and $(d-1)$-faces of a d-polyhedron are **vertices**, **edges**, and **facets**, respectively.

A **polytope** (**d-polytope**) is a convex d-polyhedron that is contained in the interior of some d-dimensional cube; that is, a *bounded* convex d-polyhedron.

A d-polytope is **regular** if all its facets are regular $(d-1)$-polytopes that are combinatorially equivalent. A vertex is a regular 0-polytope.

Two polytopes P and Q are **dual polytopes** if there exists a one-to-one correspondence δ between the set of faces of P and Q such that two faces $f_1, f_2 \in P$ satisfy $f_1 \subset f_2$ if and only if $\delta(f_1) \supset \delta(f_2)$ in Q.

A **manifold** (**d-manifold**) is a topological space that is locally homeomorphic to $\mathcal{R}^d$ everywhere.

A **manifold d-polyhedron** is a polyhedron whose boundary is topologically the same as a $(d-1)$-manifold. That is, every point on the boundary of a manifold d-polyhedron has a small neighborhood that looks like $\mathcal{R}^d$.

A **non-manifold d-polyhedron** is a d-polyhedron whose boundary is not a manifold.

A manifold 3-polyhedron has **genus** g if its boundary is a 2-manifold with genus g. A 2-manifold surface has genus g if every set of $g+1$ circular cuts separate the surface, but not all sets of g circular cuts do.

Edges in a 3-polyhedron are **reflex edges** if the inner angle subtended by two faces meeting at that edge is greater than 180°.

Facts:

1. Every polytope has a dual polytope.
2. Every polytope is the convex hull of its vertices.
3. A k-face is an open set of dimension k.
4. *Curvature*: The curvature κ_v of a manifold 3-polyhedron at a vertex v is
$$\kappa_v = \frac{2\pi - \sum_i \theta_i}{2\pi},$$
where θ_i is the angle between two consecutive edges incident with v. Intuitively, curvature at a vertex measures its "sharpness".
5. *Gauss-Bonnet theorem*: $\sum_v \kappa_v = 2 - 2g$.
6. *Angle sums*: Let f be a face of a polytope P and p an interior point of f. The angle at f is measured as the fraction of P covered by a sufficiently small $(d-1)$-dimensional sphere centered at p. If α_k is the sum of angles at all k-dimensional faces, then
$$\sum_{k=0}^{d-1} (-1)^k \alpha_k = (-1)^{d-1}. \quad (Gram's\ formula)$$
[Gr67].
7. A 3-polyhedron is convex if and only if it does not have any reflex edges.

Examples:

1. Tetrahedra and cubes are manifold 3-polyhedra with genus 0. See the following figure.

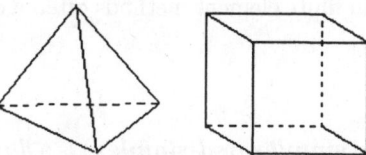

2. Two cubes meeting at a single edge, or two tetrahedra meeting at a single vertex, form non-manifold 3-polyhedra. See the following figure.

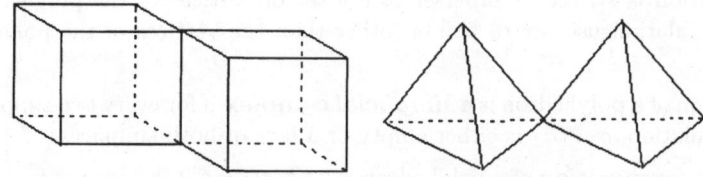

3. A cube has genus 0, but a cube with a cubical through-hole is a manifold 3-polyhedron with genus 1. See the following figure.

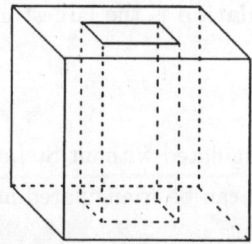

4. A cube and a octahedron (bipyramid) are dual to each other; a tetrahedron is dual to itself.

5. There are five regular polytopes in three dimensions: tetrahedron, cube (hexahedron), octahedron, dodecahedron, icosahedron. They are also called *Platonic solids*. See the following figure.

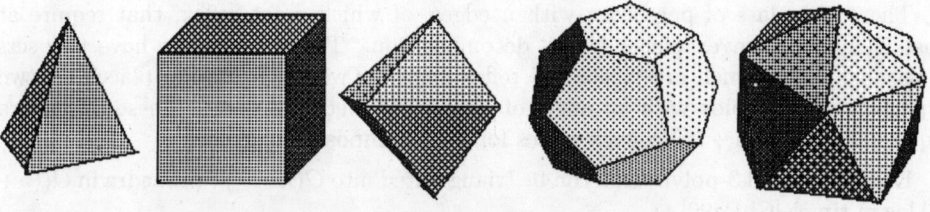

6. There is a circular cut for a toroidal surface that does not separate it, though any two circular cuts always separate it.

13.4.2 TRIANGULATIONS

A complex domain is decomposed into simple parts for computational simplicity in many applications. For example, in finite element methods often a domain is triangulated into simplices.

Definitions:

A *simplex* (*d-dimensional simplex* or *d-simplex*) is a d-polytope with $d+1$ vertices.

A *triangulation* of a d-polyhedron is a convex decomposition where each convex piece of the decomposition is a d-simplex.

A polyhedron is *triangulated with Steiner points* if the vertex set of simplices in the triangulation is strictly a superset of the set of vertices of the polyhedron. This type of triangulation uses extra points (other than the vertices of the polyhedron) as vertices.

A triangulation of a polyhedron is a *simplicial complex* if for every two simplices σ_1, σ_2 in the triangulation, $\sigma_1 \cap \sigma_2$ is either empty or a face of both simplices.

A *convex decomposition* of a polyhedron is its partition into convex pieces that have disjoint interiors.

The *aspect ratio of a simplex* is the ratio of the radius of the circumscribing sphere to the radius of the inscribing sphere of the simplex.

The *aspect ratio of a triangulation* is the largest aspect ratio of a simplex in the triangulation.

Facts:

1. Every d-polytope can be triangulated without Steiner points.

2. Every d-polytope with n faces can be triangulated into $O(n)$ simplices in $O(n)$ time and space.

3. There are nonconvex 3-polyhedra that can't be triangulated without Steiner points.

4. The problem of deciding if a nonconvex 3-polyhedron can be triangulated without Steiner points or not is NP-complete. [RuSe92]

5. The problem of decomposing a polyhedron into the minimum number of convex pieces is NP-hard.

6. Every polyhedron can be decomposed into disjoint convex pieces by repeatedly slicing the polyhedron through reflex edges. [BaDe92] and [Ch84]

7. There is a class of polyhedra with n edges, of which r are reflex, that require at least $\Omega(n + r^2)$ convex pieces for its decomposition. These polyhedra have two sets of parallel edges which are created as reflex edges. Two such sets are placed on two hyperbolic paraboloids with an angle of almost 90° between them. These polyhedra require at least $\Omega(n + r^2)$ convex pieces for its decomposition. [Ch84]

8. Every manifold 3-polyhedron can be triangulated into $O(n+r^2)$ tetrahedra in $O((n+r^2)\log r)$ time. [ChPa90]

9. There exists a polynomial time algorithm that produces a triangulation of any 3-polyhedron with an aspect ratio and size that are within a constant factor of the optimal. [MiVa92]

Examples:

1. A triangle is a 2-simplex; a tetrahedron is a 3-simplex.
2. Part (a) of the following figure shows a tetrahedron with a bad aspect ratio; part (b) shows a tetrahedron with a good aspect ratio.

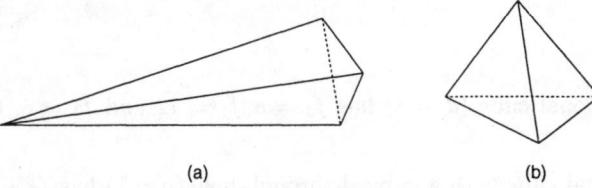

(a)   (b)

3. The Schönhardt polyhedron is a nonconvex 3-polyhedron that cannot be triangulated without Steiner points. This polyhedron can be constructed out of a prism whose base and top facets are equilateral triangles. Twist the top triangle, keeping the base fixed. This destroys the planarity of vertical facets. To maintain the planarity, triangulate these facets appropriately. [RuSe92]

13.4.3 FACE NUMBERS

In many cases complexity of algorithms dealing with polyhedra depend on the number of their faces. Therefore, combinatorial bounds on these numbers play a significant role in analyzing these algorithms.

Definitions:

A *cyclic d-polytope* is the convex hull of a set of n ($n \geq d+1$) points on the moment curve in $\mathcal{R}^d$, $x(t) = (t, t^2, \ldots, t^d)$.

A *face vector* of a d-polyhedron P is the d-dimensional vector $(f_0, f_1, \ldots, f_{d-1})$, where $f_i = f_i(P)$ is the number of i-dimensional faces of P.

A *simplicial polytope* is a polytope in which all faces are simplices.

Facts:

1. For $2k \leq d$, every k vertices of a cyclic polytope define a $(k-1)$-face.
2. *Euler's relation*: For any d-polytope, $\sum_{i=0}^{d-1}(-1)^i f_i = 1 - (-1)^d$.
3. For a manifold 3-polyhedron with genus g, $\sum_{i=0}^{2}(-1)^i f_i = 2 - 2g$.
4. The edges on the boundary of a manifold 3-polyhedron with genus 0 form a planar graph. By the property of planarity, the number of vertices, edges, and facets of such polyhedra are within a constant factor of each other.
5. *Dehn-Sommerville's equations*: The face-vectors of simplicial polytopes satisfy the following equations for $-1 \leq k \leq d-2$ with $f_{-1} = 1$:

$$E_d^k : \sum_{j=k}^{d-1}(-1)^j \binom{j+1}{k+1} f_j = (-1)^{d-1} f_k.$$

In particular, E_d^{-1} is Euler's relation.

6. *Upper bound theorem*: For any d-polytope P with n vertices, $f_i(P) = O(n^{\lfloor d/2 \rfloor})$ for $1 \leq i \leq d-1$.

7. *Optimality of cyclic polytopes*: Cyclic polytopes achieve the upper bound since they have $\binom{n}{k} = \Omega(n^k)$ $(k-1)$-faces for $2k \leq d$. This implies that they have $\Omega(n^{\lfloor d/2 \rfloor})$ $\lfloor \frac{d}{2} \rfloor$-faces.

Examples:

1. The 3-dimensional cube ($d = 3$) has $f_0 = 8, f_1 = 12$ and $f_2 = 6$; thus, by Fact 2, $f_0 - f_1 + f_2 = 2$.

2. A 3-dimensional cube with a cubical through-hole ($g = 1$) has $f_0 = 16, f_1 = 24$ and $f_2 = 10$; thus, by Fact 3, $f_0 - f_1 + f_2 = 0$.

13.5 ALGORITHMS AND COMPLEXITY IN COMPUTATIONAL GEOMETRY

Computational geometry studies efficient algorithms for solving geometric problems and has applications in computer graphics, robotics, VLSI design, computer-aided design, pattern recognition, statistics, and other fields. The study of computational geometry uses concepts and results from classical geometry, topology, combinatorics, as well as standard techniques from design and analysis of computer algorithms. See [PrSh85] and [GoO'R97].

13.5.1 CONVEX HULLS

Finding efficient algorithms for the construction of convex hulls has been a central topic in computational geometry. Several efficient algorithms for constructing boundaries of convex hulls of sets of points in the plane have been developed.

Definition:

The **convex hull** of a set of points in $\mathcal{R}^d$ is the smallest convex set containing the points.

Algorithms:

1. *Finding boundaries of convex hulls by rotational sweeping*:

   - *GrahamScan*: Given a set S of n points in the plane, Algorithm 1 scans the points rotationally around a fixed point and eliminates those that are not hull vertices. The remaining points are the vertices of the boundary of the convex hull of S. The running time of GrahamScan is $O(n \log n)$, which is dominated by the sorting of the points. The remaining steps take only linear time.

   - *Jarvis' March*: Given a set S of n points in the plane, Jarvis' March algorithm constructs the boundary of the convex hull by "marching around" the outer perimeter of S. This method is also called "gift-wrapping". Jarvis' March runs in time $O(hn)$, where h is the number of vertices of the convex hull, which, in the worst case is n.

> **Algorithm 1: Graham scan.**
>
> input: a finite set S of points in the plane
> output: the vertices of the boundary of the convex hull of S
>
> $p_0 :=$ the point in S with the minimum y-coordinate
> sort remaining points by polar angle around p_0; append the point p_0 to the
>   end of the sorted list; let the resulting list be $(p_1, p_2, \ldots, p_n)$, where $p_n = p_0$.
> $H[1] := p_n$; $H[2] := p_1$; $j := 2$
> for $i := 2$ to n
>   while the path $\{H[j-1], H[j], p_i\}$ does not form a left turn
> $j := j - 1$
> $j := j + 1$
> $H[j] := p_i$
> $\{H[1], H[2], \ldots, H[j]$ is the boundary of the convex hull.$\}$

2. *Divide-and-conquer algorithms*:

   • *QuickHull*: This algorithm recursively constructs a chain on the boundary of the convex hull, connecting two hull vertices u and v. It first finds a hull vertex w on the chain (for example, w is the farthest point from the line $\overline{uv}$). Then the subchains connecting u and w, w and v, respectively, are constructed recursively and are concatenated. [PrSh85]

   QuickHull runs practically fast, but in the worst case the running time of QuickHull is $O(n^2)$.

   • *MergeHull*: This algorithm first partitions the set S of points into two subsets S_1 and S_2 of equal size and then recursively constructs the boundaries of the convex hulls $CH(S_1)$ and $CH(S_2)$. Finally, $CH(S_1)$ and $CH(S_2)$ are "merged" into the convex hull of the set S. [PrSh85]

   The boundary of the convex hull for S is the same as the boundary of the convex hull for the hull vertices of $CH(S_1)$ and $CH(S_2)$. Thus, to construct the boundary of $CH(S)$, first sort the hull vertices of $CH(S_1)$ and $CH(S_2)$ (this sorting can be done in linear time), then apply the linear scan of GrahamScan to construct $CH(S)$. Therefore, the boundary of the convex hull $CH(S)$ can be constructed from $CH(S_1)$ and $CH(S_2)$ in linear time. The running time of MergeHull is $O(n \log n)$.

3. *Other methods*:

   • *incremental method*: The *incremental method* for constructing the boundary of the convex hull of a set of points in the plane adds one point at a time to an already constructed boundary of a convex hull. This method has time complexity $O(n \log n)$ [PrSh85]. An advantage of this method is that it can be generalized to construct boundaries of convex hulls in higher dimensions [Ed87].

   • *An algorithm by Kirkpatrick and Siedel based on the prune-and-search method*: This algorithm partitions a given set of points in the plane into two linearly separable subsets of equal size, finds the two edges of the boundary of the convex hull that "bridge" these two subsets, and recursively constructs the subchains on the boundary of the convex hull between these two bridges. This method has time complexity $O(n \log h)$, where h is the number of vertices on the boundary of the convex hull [Ya90].

Facts:

1. The problem of finding convex hulls is at least as hard as sorting. The lower bound $\Omega(n \log n)$ of sorting on comparison decision trees also applies to the convex hull problem. This lower bound $\Omega(n \log n)$ can be extended to a more general computation model, the bounded-degree algebraic decision trees.

2. An $O(n \log n)$ time algorithm for constructing the boundary of the convex hull of a set of points in $\mathcal{R}^3$ has been developed, which is a generalization of the MergeHull algorithm.

3. For dimension $d > 3$, the convex hull of n points in $\mathcal{R}^d$ can have up to $O(n^{\lfloor \frac{d}{2} \rfloor})$ faces. An algorithm based on the incremental method has been proposed to construct the convex hull for a set of n points in $\mathcal{R}^d$ in time $O(n^{\lfloor \frac{d+1}{2} \rfloor})$ [Ya90]. An optimal algorithm of time $O(n^{\lfloor \frac{d}{2} \rfloor})$ has been developed recently by Chazelle. (See the bibliography of [Mu94] for a reference.)

Examples:

1. The following figure shows a set S of 7 points in the plane and the convex hull of S.

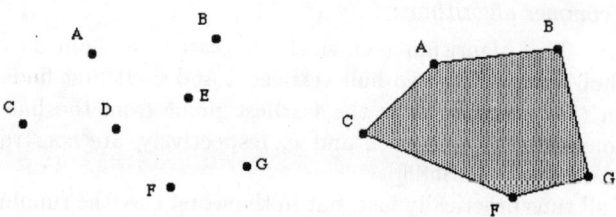

2. The convex hull of the set $\{(0,0,0), (0,0,1), (0,1,0), (0,1,1), (1,0,0), (1,0,1), (1,1,0), (1,1,1)\}$ in $\mathcal{R}^3$ is the surface of the unit cube together with its interior.

13.5.2 TRIANGULATION ALGORITHMS

Triangulation plays an important role in many applications. On a triangulated planar straight-line graph, many problems can be solved more efficiently. Triangulation of a set of points arises in numerical interpolation of bivariate data and in the finite element method.

Definitions:

A **planar straight-line graph** (**PSLG**) is a planar graph such that each edge is a straight line.

A **triangulation of a simple polygon** P is an augmentation of P with nonintersecting diagonal edges connecting vertices of P such that in the resulting PSLG, every bounded face is a triangle.

A **triangulation of a PSLG** G is an augmentation of G with nonintersecting edges connecting vertices of G so that every point in the interior of the convex hull of G is contained in a face that is a triangle. In particular, the PSLG to be triangulated can be simply n discrete points in the plane.

A *chain* is a PSLG with vertices $v_1, \ldots, v_n$ and edges $\{v_1, v_2\}, \{v_2, v_3\}, \ldots, \{v_{n-1}, v_n\}$.

A chain is *monotone* if there is a straight line L such that every line perpendicular to L intersects the chain in at most one point.

A simple polygon is *monotone* if its boundary can be decomposed into two monotone chains.

Two vertices v and u in a polygon are *visible* from each other if the open line segment $\overline{uv}$ is entirely in the interior of the polygon.

Facts:

1. Every simple polygon can be triangulated.

2. Every triangulation of a simple polygon with n vertices has $n-2$ triangles and $n-3$ diagonals.

3. Given a simple polygon with n vertices, there is a diagonal that divides the polygon into two polygons that have at most $\lceil \frac{2n}{3} \rceil + 1$ vertices.

4. For a history of triangulations, see [O'R87].

5. Simple polygons can be triangulated in $O(n)$ time using an algorithm developed by Chazelle. [Ch91].

6. PSLGs can be triangulated in $O(n \log n)$ time. (See the *triangulation of a general PSLG* algorithm — item 3 in the following list of algorithms.) This is optimal because a lower bound $\Omega(n \log n)$ has been derived for the time complexity of triangulation of a PSLG.

Algorithms:

1. *Triangulation of a monotone polygon*: A monotone polygon P can be triangulated in linear time based on the following greedy method. Observe that the monotone polygon P is triangulated if nonintersecting edges are added so that no two vertices are visible from each other.

   If necessary, rotate the polygon so that it is monotone with respect to the y-axis. Sort the vertices of P by decreasing y-coordinate. (This sorting can be done in linear time by merging the two monotone chains of P.) Move through the sorted list, and for a vertex v, examine each vertex u lower than v, in the sorted order, and add an edge between vertices v and u as long as u is visible from v. The edge addition process for the vertex v stops at a lower vertex that is not visible from v. Then move to the next vertex and perform the edge addition process. Note that once an edge is added between vertices v and u, then no vertices between v and u in the sorted list are visible from a vertex that is lower than u. Therefore, such vertices can be ignored in the later edge addition process. The edge addition process for all vertices can be performed in linear time if a stack is used to hold the sorted list.

2. *Triangulation of a simple polygon*: Given a general simple polygon P, partition P in time $O(n \log n)$ into monotone polygons, then apply the previous linear time algorithm to triangulate each monotone polygon. This gives a triangulation of a simple polygon in time $O(n \log n)$.

3. *Triangulation of a general PSLG*: To triangulate a general PSLG G, first add edges to G so that each face is a simple polygon (no nonconsecutive edges intersect), then apply Chazelle's linear time algorithm (Fact 5) to triangulate each face.

The complexity of Chazelle's algorithm can be avoided since there is an efficient algorithm that adds edges to a PSLG so that each face is a monotone polygon. To do this, observe that in a PSLG G every face is a monotone polygon if and only if each vertex (except the highest one) has a higher neighbor and each vertex (except the lowest one) has a lower neighbor. Thus, to make each face of G a monotone polygon, check each vertex of G and for those that do not have desired neighbors, add proper edges to them. This process can be accomplished in time $O(n \log n)$ using the plane sweeping method [PrSh85]. Now, the simpler linear time algorithm for triangulating a monotone polygon (see item 1) is applied to triangulate each face.

Examples:

1. The following figure illustrates a simple polygon and two of its triangulations.

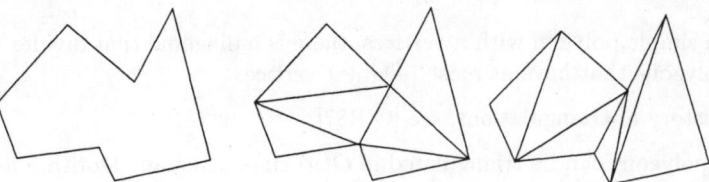

2. In part (a) of the following figure the chain is monotone (with respect to any horizontal line); the chain in part (b) is not monotone.

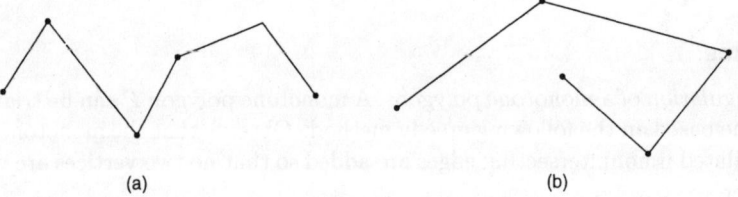

13.5.3 VORONOI DIAGRAMS AND DELAUNAY TRIANGULATIONS

Definitions:

Given a set $S = \{p_1, \ldots, p_n\}$ in $\mathcal{R}^d$, the **Voronoi diagram** Vor(S) of S is a partition of $\mathcal{R}^d$ into n convex polytopes (**Voronoi cells** or **Dirichlet cells**) $V(p_1), \ldots, V(p_n)$ such that the region $V(p_i)$ is the locus of points that are closer to p_i than to any other point in S.

Given the Voronoi diagram Vor(S) of a set $S = \{p_1, \ldots, p_n\}$ of points in the plane, the **straight line dual** $D(S)$ of Vor(S) is a PSLG whose vertices are the points in S and two vertices p_i and p_j in $D(S)$ are connected if and only if the regions $V(p_i)$ and $V(p_j)$ share a common edge.

The PSLG $D(S)$ is a triangulation of the set S, called the **Delaunay triangulation** of S.

> **Algorithm 2: Construction of Voronoi diagrams.**
>
> input: a set S of points in the plane
> output: the Voronoi diagram of S
>
> **if** $|S| < 4$ **then** construct Vor(S) directly and stop
> **else**
>   partition S into two equal size subsets S_L (left subset) and S_R (right subset)
>   separated by a vertical line
>   construct Vor(S_L) and Vor(S_R) recursively;
>   merge Vor(S_L) and Vor(S_R) into Vor(S);

Facts:

1. The Voronoi diagram of a set of n points in the plane can be constructed in time $O(n \log n)$.

2. The Delaunay triangulation of a set S of points in the plane has the property that the circuit with the three vertices of a triangle of the triangulation on its boundary contains no other point of the set S. This property makes the Delaunay triangulation useful in interpolation applications.

3. The convex hull problem in the plane can be reduced in linear time to the Voronoi diagram problem in the plane: a point p in a set S is a hull vertex if and only if $V(p)$ is unbounded, and two hull vertices p_i and p_j are adjacent if and only if the two unbounded regions $V(p_i)$ and $V(p_j)$ share a common edge. Thus, the $O(n \log n)$ time algorithm for constructing Voronoi diagrams (Algorithm 2) is optimal.

4. The Voronoi diagram problem for n points in $\mathcal{R}^d$ can be reduced in linear time to the convex hull problem for n points in $\mathcal{R}^{d+1}$ [Ed87]. Thus, the Voronoi diagram of a set of n points in $\mathcal{R}^d$ can be constructed in time $O(n^{\lfloor (d+1)/2 \rfloor})$ based on the optimal algorithm for constructing the convex hull of a set of n points in $\mathcal{R}^{d+1}$.

Algorithms:

1. *Construction of Voronoi diagrams in the plane*: The Voronoi diagram of a set of points in $\mathcal{R}^2$ can be constructed using the divide-and-conquer method of Algorithm 2.

To efficiently partition a set S into a left subset and a right subset of equal size in each recursive construction, pre-sort the set S by x-coordinate. To merge the Voronoi diagrams Vor(S_L) and Vor(S_R) into the Voronoi diagram Vor(S), add the part of Vor(S) that is missing in Vor(S_L) and Vor(S_R) and to delete the part of Vor(S_L) and Vor(S_R) that does not appear in Vor(S).

2. *Voronoi diagrams and geometric optimization problems*: An $O(n \log n)$ time optimal algorithm can be derived via the Voronoi diagram for the problem of finding for each point in a set S of n points in the plane the nearest point in S. This is so because each point p in S and its nearest neighbor correspond to two regions in Vor(S) that share a common edge. This also implies an $O(n \log n)$ time optimal algorithm for the problem of finding the closest pair in a set of n points in the plane.

The Voronoi diagram can be used to design an $O(n \log n)$ time optimal algorithm for constructing a Euclidean minimum spanning tree for a set S of n points in the plane because edges of any Euclidean minimum spanning tree must be contained in the Delaunay triangulation $D(S)$ of S. This algorithm implies an $O(n \log n)$ time approximation algorithm for the Euclidean traveling salesman problem which produces a traveling salesman tour of length at worst twice the optimum.

Example:

1. The left half of the following figure illustrates a set of 6 points in the plane and the Voronoi diagram for the set. The right half shows the Delaunay triangulation of the set.

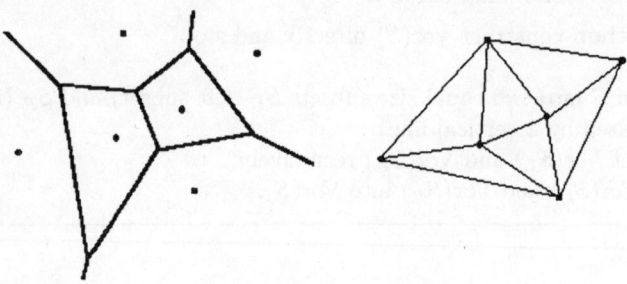

13.5.4 ARRANGEMENTS

Definition:

Given n lines in the plane, the **arrangement** of the lines in the plane is the PSLG whose vertices are the intersection points of the lines and whose edges connect consecutive intersection points on each line (it is assumed that all lines intersect at a common point at infinity).

Facts:

1. The arrangement of n lines in the plane partitions the plane into a collection of $O(n^2)$ faces, edges, and vertices.
2. The arrangement of n lines can be constructed in $O(n^2)$ time (Algorithm 4), which is optimal.
3. An arrangement can be represented by a *doubly-connected-edge-list* in which the edges incident with a vertex can be traversed in clockwise order in constant time per edge. [PrSh85]
4. Applications of arrangements include finding the smallest-area triangle among n points, constructing Voronoi diagrams, and half-plane range query.
5. The arrangement of n hyperplanes in $\mathcal{R}^d$ can be defined similarly, which partitions $\mathcal{R}^d$ into $O(n^d)$ faces of dimension at most d.
6. Algorithm 3 can be generalized to construct the arrangement of n hyperplanes in $\mathcal{R}^d$ in $O(n^d)$ time, which is optimal. [Ed87]

Algorithm:

1. *Constructing the arrangement of a set of lines*: Algorithm 3 constructs the arrangement $\mathcal{A}$ of a set H of n lines $L_1, \ldots, L_n$ in the plane by the incremental method.

    To traverse the faces of $\mathcal{A}$ that intersect the line L_i, start from a face F that has the point p_i on its boundary, and traverse the boundary of F until an edge e is encountered such that e intersects L_i at a point q. A new vertex q is introduced in $\mathcal{A}$ and the adjacencies of the two ends of the edge e are updated. Then reverse the traversing direction on the edge e and start traversing the face that shares the edge e with F, and so on. The total number of edges traversed in this process in order to insert the line L_i is bounded by $O(i)$. [Mu94]

Algorithm 3: Incremental method for constructing the arrangement of a set H of n lines.
input: a set H of n lines $L_1, L_2, \ldots, L_n$ in the plane
output: the arrangement $\mathcal{A}$ of the set H
$\mathcal{A} := L_1$;
for $i := 2$ to n
    find the intersection point p_i of L_i and L_1
    starting from p_i, traverse the faces of $\mathcal{A}$ that intersect L_i and update the vertex set and edge set of $\mathcal{A}$

Example:

1. The following figure shows an arrangement of four lines in the plane. The graph has 7 vertices (including the vertex at infinity), 16 edges (of which 8 are unbounded), and 11 regions (of which 8 are unbounded).

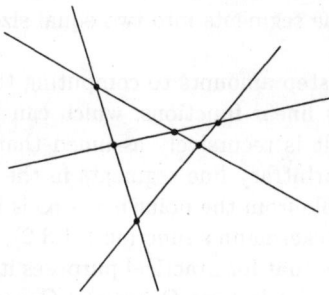

13.5.5 VISIBILITY

Visibility problems are concerned with determining what can be viewed from a given point (or points) in the plane or three-dimensional space. [O'R87], [O'R93], [Sh92].

Definitions:

The **visibility problem** is the problem of finding what is visible, given a configuration of objects and a viewpoint.

Given n nonintersecting line segments in the plane, the **visibility graph** is the graph whose vertices are the endpoints of the line segments, with two vertices adjacent if and only if they are visible from each other; i.e., there is an edge joining a and b if and only if the open line segment ab does not intersect any other line segments.

A **star polygon** is a polygon with an interior point p such that each point on the polygon is visible from p.

Facts:

1. Visibility problems have important applications in computer graphics and robotics and have served as motivation for research in computational geometry.

2. Constructing the visibility graph for a set of n nonintersecting line segments is a critical component of the shortest path problem in the plane. The visibility graph problem can be solved in optimal time $O(n^2)$ [Ya90].

3. *Art gallery theorems*: Given a simple polygon with n vertices,
   - there is a set S of $\lfloor \frac{n}{3} \rfloor$ vertices of the polygon such that each point on or inside the polygon is visible from a point in S;
   - there is a set S of $\lceil \frac{n}{3} \rceil$ points on the polygon such that each point on or outside the polygon is visible from a point in S;
   - there is a set S of $\lceil \frac{n}{2} \rceil$ vertices of the polygon such that each point on, inside, or outside the polygon is visible from a point in S.

In each case the number given is the best possible.

Algorithms:

1. Given n line segments in the plane, compute the sequence of subsegments that are visible from the point $y = -\infty$, that is, by using parallel rays. The problem can be solved by a modified version of the plane sweeping algorithm for computing all intersection points of the line segments. The algorithm has worst case time complexity $O(n^2 \log n)$.

2. An alternative algorithm is based on the divide-and-conquer approach: arbitrarily partition the set of the n line segments into two equal size halves, solve both subproblems, and merge the results.

Note that the merging step amounts to computing the minimum of two piecewise (not necessarily continuous) linear functions, which can be easily done in time linear to the number of pieces if it is recursively assumed that the pieces are sorted by x-coordinate. For a set of n arbitrary line segments in the plane, in the worst case, the number of subsegments visible from the point $y = -\infty$ is bounded by $O(n\alpha(n))$ [Ya90], where $\alpha(n)$, the inverse of Ackermann's function (§1.3.2), is a monotonically increasing function that grows so slowly that for practical purposes it can be treated as a constant. Therefore, the merging step runs in time $O(n\alpha(n))$. Consequently, the time complexity of the algorithm based on the divide-and-conquer method is $O(n\alpha(n) \log n)$.

3. *Three-dimensional visibility*: Given a set of disjoint opaque polyhedra in $\mathcal{R}^3$, find the part of the set that is visible from the viewpoint $z = -\infty$ (that is, with parallel rays). The problem can be solved in time $O(n^2 \log n)$ by a modified plane sweeping algorithm for computing the intersection points of the line segments that are the projections of the edges of the polyhedra on the xy-plane. Optimal algorithms of time complexity $O(n^2)$ have been developed based on line arrangements. [Do94], [Ya90]

Examples:

1. The following figure shows three line segments and the visibility graph with six vertices determined by the line segments. The edges of the visibility graph are shown as dotted lines.

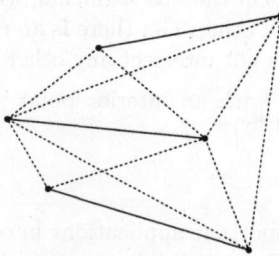

2. *Surveillance problems*: A variety of problems require that "guards" be posted at points of a polygon so that corners and/or edges are visible. (See the art gallery theorems of Fact 3.)

3. *Hidden surface removal*: An important problem in computer graphics is the problem of finding (and removing) the portions of shapes in three-dimensional space that are hidden from view, when the object is viewed from a given point. (See the Three-dimensional visibility algorithm.)

13.6 GEOMETRIC DATA STRUCTURES AND SEARCHING

This section describes the use of data structures for searching, or querying, among a set S of geometric objects. For each of the following problems, there are algorithms that perform a single search in time proportional to n, the total complexity of all the geometric objects in S. These single-search algorithms use minimal data structures and minimal preprocessing time. When the application searches multiple times among the elements of the same set S, it becomes more efficient to preprocess the objects in S into a data structure that would allow a faster searching procedure.

This section presents four fundamental searching problems in computational geometry for which clever data structures reduce the search time to $O(\log_2^k n)$, where k is a small constant (often equal to 1 for problems in the plane and three-dimensional space). This section covers only the static versions of these problems; that is, S never changes. The dynamic versions allow deletions from S and/or insertions into S in between queries. In the dynamic versions of the problem, in addition to polylogarithmic query time, the goal is to keep the update time polylogarithmic. The dynamic versions are, as a rule, much more difficult.

13.6.1 POINT LOCATION

Definition:

Let p be a point and S a subdivision of $\mathcal{R}^d$. (S can be a single geometric object, such as a polytope, or can be a general subdivision of $\mathcal{R}^d$.) The **point location problem** is the problem of determining which region of S contains p.

Examples:

1. Locate a point in the subdivision of the space induced by an *arrangement* of a set of hyperplanes. (See §13.5.4.)

2. Locate a point in a subdivision all of whose regions are convex (a *convex subdivision*). For example, an arrangement of hyperplanes is a convex subdivision.

3. Search for nearest neighbors in Voronoi diagrams. (See §13.5.3.)

4. Range searching. (See §13.6.2.)

5. Ray shooting. (See §13.6.3.)

Algorithms:

1. *Point location in straight-line subdivision using triangulation hierarchy*: Given an n-vertex triangulation $R' = (V', T')$, where V' is the set of vertices and T' is the set of triangles, an $(n+3)$-vertex *enclosed* triangulation $R = (V, T)$ is the triangulation R' together with the triangulation of the region between U and the convex hull of R', where $U = (V_U, T_U)$ is a triangle that contains R' in its interior. The *triangulation-hierarchy* of [Ki83] consists of a sequence of triangulations $\mathcal{R} = \langle R_1, R_2, \ldots, R_{c \log_2 n} \rangle$, where $R_1 = R$, $R_{c \log_2 n} = U$, and R_i is created from R_{i-1} as follows (illustrated in the following figure):

- remove from $V_{i-1} - V_U$ a set X of independent (that is, nonadjacent) vertices and remove from T_{i-1} the set Z of all triangles incident with any vertex in X: $V_i = V_{i-1} - X$, $T_i = T_{i-1} - Z$;
- retriangulate any polygons in $R_i = (V_i, T_i)$.

Part (a) of the following figure shows triangulation R_1 (the vertices that are removed from R_1 are circled). Part (b) shows triangulation R_2 (dotted lines are edges in re-triangulation). Part (c) gives a list of the pointers from triangles in R_2 to triangles in R_1.

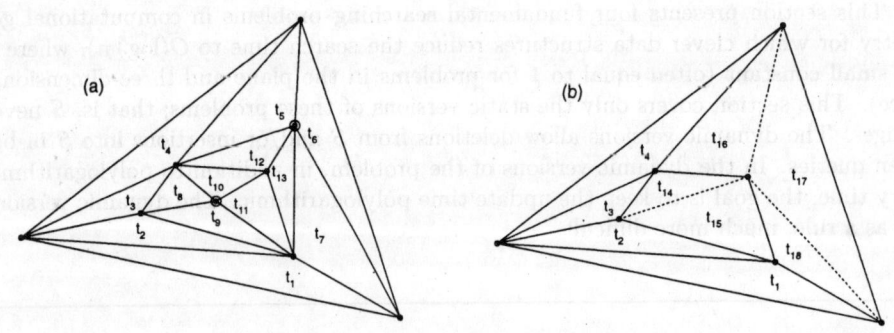

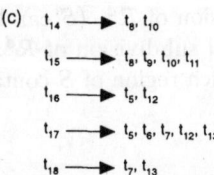

Algorithm 1 produces a hierarchy for planar subdivision. With minor modifications (for example, "triangles" become tetrahedrons), it can be used for subdivisions in $\mathcal{R}^3$. It can be proven that $|T_{c \log_2 n}| = 1$ for some constant c, and that $|\tau(t)|$ is a constant for every t.

This algorithm runs in $O(n)$ time and produces a triangulation hierarchy that takes $O(n)$ space. Algorithm 2 takes $O(\log_2 n)$ time.

Algorithm 1:   Computing the triangulation hierarchy.

input: planar straight-line subdivision S
output: triangulation hierarchy of S

compute triangulation R' of S; compute enclosed triangulation R of R'; choose a
  small constant k; $R_1 := R$; $i := 1$
while $|T_i| > 1$
  $i := i + 1$; $R_i := R_{i-1}$; mark all vertices in V_i having degree $< k$
  while there exists some marked vertex v
    $P := (V_P, E_P)$ (polygon consisting of vertices adjacent to v; that is,
      $V_P := \{ v_j \mid (v, v_j, v_k) \in T_i \}$; $E_P := \{ (v_j, v_k) \mid (v, v_j, v_k) \in T_i \})$
    remove v from V_i; remove all the triangles incident with v from T_i; that is
      $T_{rem} := \{ (v, v_j, v_k) \mid v_j \in V_P \}$; $T_i := T_i - T_{rem}$
    compute the triangulation R_P of P
    for each triangle t in R_P
      $\tau(t) :=$ the set of triangles in T_{rem} that overlap with t
      create a pointer from t to every triangle in $\tau(t)$ {See part (c) of figure.}
    unmark v and any marked vertices in V_P.

Algorithm 2:   Performing point location.

input: a point q and a triangulation hierarchy $\mathcal{R}$
output: triangle that contains q

check if $R_{c \log_2 n}$ contains q
$i := c \log_2 n - 1$; $t := U$
while $i \geq 1$
  determine the triangle t' in $\tau(t)$ that contain q using pointers from t to $\tau(t)$
  $t := t'$; $i := i - 1$.

---

2. The following table shows the complexity of various point location algorithms. The number m denotes the number of regions (or cells) in the subdivision S; n denotes the total combinatorial complexity of S.

dimension	subdivision type	query time	space	preprocessing time
2	convex subdivision	$O(\log_2 n)$	$O(n)$	$O(n)$
3	simple polytope	$O(\log_2 n)$	$O(n)$	$O(n)$
3	convex subdivision	$O(\log_2^2 n)$	$O(n \log_2^2 n)$	$O(n \log_2^2 n)$
d	arrangement of n hyperplanes	$O(\log_2 n)$	$O(n^d)$	$O(n^d)$
d	subdivision of m $(d-1)$-simplices with a total of n faces, $\epsilon > 0$	$O(\log_2^3 m)$	$O(m^{d-1+\epsilon} + n)$	$O(m^{d-1+\epsilon} + n \log_2 m)$

13.6.2 RANGE SEARCHING

Definitions:

The **range counting problem** is the problem of counting the number of points in a given set $S \subseteq \mathcal{R}^d$ that lie in a given query range q.

The **range reporting problem** is the problem of determining all points in a given set $S \subseteq \mathcal{R}^d$ that lie in a given query range q.

The **range emptiness problem** is the problem of determining if a given query range q contains any points from a given set $S \subseteq \mathcal{R}^d$.

Facts:

1. The following table gives information on various range searching algorithms. The integer n is the number of points in S; ϵ is an arbitrarily small positive constant. When the query is reporting, the query time has an additive factor of k, which is the size of the output.

dim	range type	query time	space	preprocessing time
2	orthogonal	$O(\log_2 n)$	$O(n \log_2^{2+\epsilon} n)$	$O(n \log_2 n)$
2	convex polygon	$O(\sqrt{n} \log_2 n)$	$O(n)$	$O(n^{1+\epsilon})$
2	convex polytope	$O(n^{2/3} \log_2^2 n)$	$O(n \log_2 n)$	$O(n^{1+\epsilon})$
d	convex polytope for $n \leq m \leq n^d$	$O(\log_2^{d+1} n)$ $O(n^{1-1/d})$ $O((n/m^{1/d}) \log_2^{d+1} n)$	$O(n^d)$ $O(n)$ $O(m^{1+\epsilon})$	$O(n^d (\log_2 n)^\epsilon)$ $O(n^{1+\epsilon})$ $O(m^{1+\epsilon})$
d	half-space for $n \leq m \leq n^d$	$O(\log_2 n)$ $O(n/m^{1/d})$	$O(n^d / \log_2^d n)$ $O(m)$	$O(n^d / \log_2^{d-\epsilon} n)$ $O(n^{1+\epsilon} + m (\log_2 n)^\epsilon)$

2. The following table gives information on various range reporting algorithms.

dim	range type	query time	space	preprocessing time
2	half-plane fixed-radius circle	$O(\log_2 n + k)$	$O(n)$	$O(n \log_2 n)$
2	orthogonal	$O(\log_2 n + k)$	$O(n \log_2^\epsilon n)$	$O(n \log_2 n)$
3	half-space	$O(\log_2 n + k)$	$O(n \log_2 n)$	$O(n \log_2^3 n \log_2 \log_2 n)$
d	half-space $n \leq m \leq n^{\lfloor d/2 \rfloor}$	$O(\log_2 n + k)$ $O(n^{1-1/\lfloor d/2 \rfloor + \epsilon} + k)$ $O(\frac{n}{m^{1/d}} \log_2 n + k)$	$O(n^{\lfloor d/2 \rfloor + \epsilon})$ $O(n)$ $O(m^{1+\epsilon})$	$O(n^{\lfloor d/2 \rfloor + \epsilon})$ $O(n)$ $O(m^{1+\epsilon})$
d	orthogonal	$O(\log_2^{d-1} n + k)$ $O(dn^{1-\frac{1}{d}} + k)$	$O(\frac{n \log_2^{d-1} n}{\log_2 \log_2 n})$ $O(dn)$	$O(n \log_2^{d-1} n)$ $O(dn \log_2 n)$

Examples:

1. *Orthogonal range search:* The query range q is a cartesian product of intervals on different coordinates axes.
2. *Bounded distance search:* The query range q is a sphere in $\mathcal{R}^d$.
3. Other typical search domains are half-spaces and simplices.
4. *Machine learning:* Points are labelled as positive or negative examples of a concept, and range query determines the relative number of positive and negative examples in the range (thus enabling the range to be classified as either positive or negative example of the concept).
5. *Multikey searching in databases:* Records identified by a d-tuple of keys can be viewed as a point in $\mathcal{R}^d$, and the range query on records corresponds to orthogonal range query.

Algorithm:

1. *Orthogonal range searching in $\mathcal{R}^2$ using range trees:* The *range tree* is defined recursively by Algorithm 3. Each node stores a subset of point organized into a threaded binary search tree by the y-coordinates of the points. The left child contains half the parent's points, in particular those with lesser x-coordinates; the right child contains the other half of the parents' points with greater x-coordinates. See the following figure.

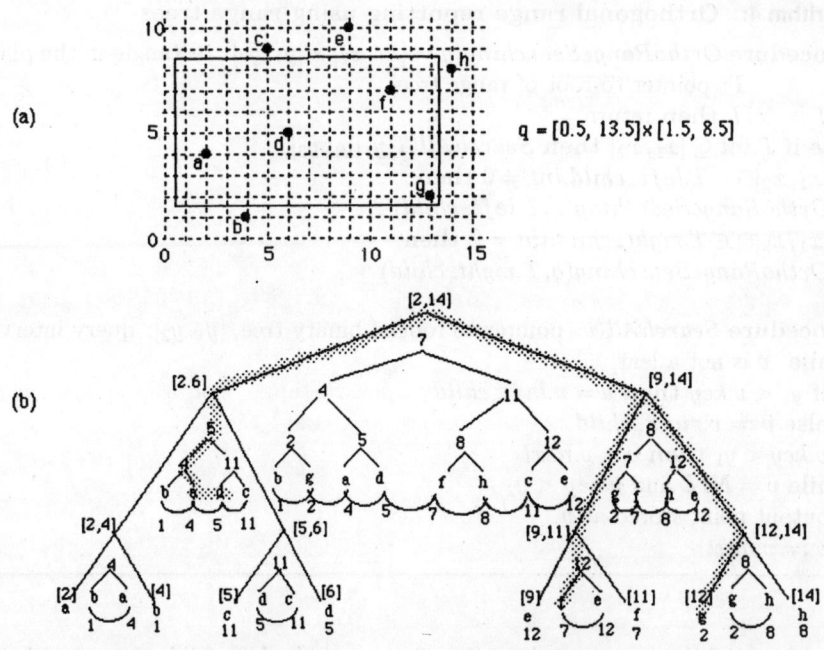

Each node also stores the range of x-coordinates of its points. For simplicity, all coordinates of all points are assumed to be distinct. It is also assumed that all points of $S = \{(x_1, y_1), (x_2, y_2), \ldots, (x_n, y_n)\}$ have been presorted by their x-coordinate so that $x_1 < x_2 < \cdots < x_n$.

Orthogonal range reporting proceeds as follows down the range tree. If the range of the current node x is a subset of the x range of the query, then all the points in the node's binary search tree with y-coordinate in the y range of the query are output.

> **Algorithm 3:** Computing the range tree.
>
> **procedure** $RangeTree(S = \{(x_1, y_1), (x_2, y_2), \ldots, (x_n, y_n)\}$: set of points,
> T: pointer to root of a range tree)
> **if** $S = \emptyset$ **then** return
> **else** store the interval $[x_1, x_n]$ in $T.int$
>   store $BinarySearchTree(S)$ in $T.y$
> $RangeTree(\{(x_1, y_1), \ldots, (x_{\frac{n}{2}}, y_{\frac{n}{2}})\}, T.left\_child)$
> $RangeTree(\{(x_{\frac{n}{2}+1}, y_{\frac{n}{2}+1}), \ldots, (x_n, y_n)\}, T.right\_child)$
>
> **procedure** $BinarySearchTree(S' = \{(x_1, y_1), \ldots, (x_n, y_n)\}$: set of points)
> sort the points of S' by y-coordinate so that $y_1 < y_2 < \cdots < y_n$
> create a threaded balanced binary search tree B for S':
>   store point (x_i, y_i) in the ith leftmost leaf ℓ_i
> $\ell_i.next := \ell_{i+1}$ {connect the leaves into a linked list}
> $\ell_i.key := y_i$
>   for each node v, $v.key := \min\{\ell_i.key \mid \ell_i \in subtree(v.right\_child)\}$

> **Algorithm 4:** Orthogonal range reporting using range trees
>
> **procedure** $OrthoRangeSearching(q = [x_1, x_2] \times [y_1, y_2]$: rectangle in the plane,
> T: pointer to root of range tree)
> **if** $T = NIL$ **then** return
> **else if** $T.int \subseteq [x_1, x_2]$ **then** $SearchAll(T.y, [y_1, y_2])$
> **if** $[x_1, x_2] \cap \in T.left\_child.int \neq \emptyset$ **then**
> $OrthoRangeSearching(q, T.left\_child)$
> **if** $[x_1, x_2] \cap \in T.right\_child.int \neq \emptyset$ **then**
> $OrthoRangeSearching(q, T.right\_child)$
>
> **procedure** $SearchAll(v$: pointer to root of binary tree, $[y_1, y_2]$: query interval)
> **while** v is not a leaf
>   **if** $y_1 < v.key$ **then** $v = v.left\_child$
>   **else** $v := v.right\_child$
> **if** $v.key < y_1$ **then** $v = v.next$
> **while** $v \neq NIL$ and $v.key < y_2$
>   output point stored at v
> $v := v.next$

If the x range of the query overlaps the x range of the left child, then the algorithm proceeds recursively to the left child. If the x range of the query overlaps the x range of the right child, then the algorithm proceeds recursively to the right child.

The running time of Algorithm 3 is $O(n \log_2^2 n)$ and the space taken by the range tree is $O(n \log_2 n)$. The running time can be improved by a $\log_2 n$ factor. Essentially the same procedure can be used to build range trees in any dimension. Algorithm 4 takes $O(\log_2^2 n + k)$ time, where k is the number of reported points. This running time can be improved to $O(\log_2 n + k)$.

13.6.3 RAY SHOOTING AND LINES IN SPACE

Definitions:

A **ray** r is a half-line that is directed away from its endpoint; that is, it satisfies the equation $r = p + \lambda \vec{v}$, $\lambda \geq 0$, where p is the **starting point** of r and $\vec{v}$ is the **direction** of r.

Given a set S of geometric objects in $\mathcal{R}^d$ and a query ray r, the **ray shooting problem** is the problem of determining the first object in S that is hit by r, that is, the object $s \in S$ whose intersection with r is closer to p than the intersection between r and any other object in S.

A polyhedron is **axis-parallel** if each of its edges is parallel to a coordinate axis.

Facts:

1. The following table gives information on various ray shooting algorithms.

dim	subdivision type	query time	space	preprocessing time
2	simple polygon	$O(\log_2 n)$	$O(n)$	$O(n)$
2	line segments	$O(\log_2 n)$ $O(\sqrt{n} \log_2 n)$	$O(n^2 \alpha^2(n))$ $O(n \log_2^2 n)$	$O(n^2 \alpha^2(n))$ $O(n \log_2^2 n)$
3, fix p	axis-parallel polyhedra	$O(\log_2 n)$	$O(n \log_2 n)$	$O(n \log_2 n)$
3, fix p	polyhedra for any $\epsilon > 0$, $n \leq m \leq n^2$	$O(\log_2 n)$ $O(n^{1+\epsilon}/\sqrt{m})$	$O(n^2 \alpha(n))$ $O(m^{1+\epsilon})$	$O(n^2 \alpha(n))$ $O(m^{1+\epsilon})$
3, fix $\vec{v}$	axis-parallel polyhedra for any $\epsilon > 0$	$O(\log_2 n \times (\log_2 \log_2 n)^2)$ $O(\log_2 n)$	$O(n \log_2 n)$ $O(n^{1+\epsilon})$	$\tilde{O}(n \log_2^2 n)$ $O(n^{1+\epsilon})$
3, fix $\vec{v}$	polyhedra for any $\epsilon > 0$, $n \leq m \leq n^3$	$O(\log_2 n)$ $O(n^{1+\epsilon}/m^{1/3})$	$O(n^{3+\epsilon})$ $O(m^{1+\epsilon})$	$O(n^{3+\epsilon})$ $O(m^{1+\epsilon})$
3	axis-parallel polyhedra	$O(\log_2 n)$	$O(n^{2+\epsilon})$	$O(n^{2+\epsilon})$
3	polyhedra	$O(\log_2 n)$	$O(n^{4+\epsilon})$	$\tilde{O}(n^{4+\epsilon})$

2. Applications of ray shooting include hidden surface removal, visibility questions and ray tracing in computer graphics, and computing shortest paths in presence of obstacles in robotics.

Examples:

1. S can be a single object, such as a simple polygon in the plane.

2. S can be a collection of objects, such as a set of polyhedra in three-dimensional space.

Algorithm:

1. *Ray shooting from a fixed point among planar nonintersecting segments*: For simplicity, assume that the fixed point p in the plane is at the origin. Define two relations using the same notation: for two points q_j and q_k, $q_j \prec q_k$ if q_j makes a smaller polar angle with respect to the origin than does q_k; for two nonintersecting segments s_j and s_k, $s_j \prec s_k$ if for every ray r that starts at the origin and crosses both s_j and s_k, r crosses s_j before crossing s_k. Segment (q_j, q_k) starts at q_j and ends at q_k if $q_j \prec q_k$. A null segment is denoted s_∞; that is, a query ray hitting s_∞ does not intersect any of the given segments.

Algorithm 5, *VisibilityMap*, creates an array $\mathcal{I}$ of nonoverlapping angle intervals, sorted by their polar angle, with the property that consecutive entries in $\mathcal{I}$ have different "smallest" segments according to "$\prec$" relation.

This algorithm uses a technique called sweep-plane: the algorithm sweeps the polar coordinates originating at p with a ray, stopping the sweep-ray at all the angles where the sweep-ray intersects a segment endpoint. The set S' stores all the segments intersected by the current sweep-ray; S' is organized as a binary search tree ordered by the "$\prec$" relation on the segments. When the sweep-ray encounters a segment endpoint that starts a segment, the segment is added to S'; when the sweep-ray encounters a segment endpoint that ends a segment, the segment is removed from S'. At every stop of the sweep-ray, if the smallest (under the "$\prec$" relation) segment of S' is different from the sweep-ray's last stop, a new interval is added to $\mathcal{I}$. See the following figure.

The thick lines in this figure are the segments in S and the thin lines are the boundaries between intervals in the visibility map for S. The intervals are labeled by their names.

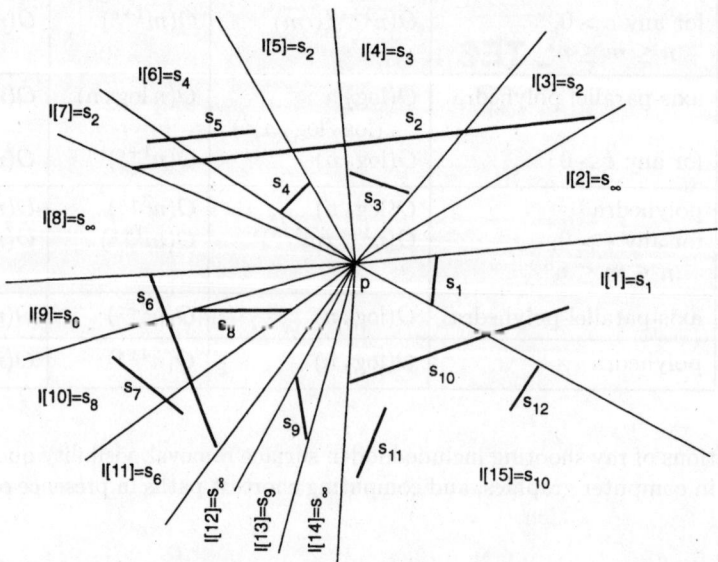

Algorithm 5, *VisibilityMap*, takes $O(n \log_2 n)$ time and can be used for ray shooting among simple polygons. The visibility map consists of array $\mathcal{I}$ and takes $O(n)$ space. The problem is harder if the segments are allowed to intersect. Algorithm 6, *RayShoot*, takes $O(\log_2 n)$ time.

> **Algorithm 5:** Computing visibility map.
>
> **procedure** $VisibilityMap(p$: fixed origin point, S: set of n segments)
> sort endpoints of all segments by their polar angles so that $q_1 \prec q_2 \prec \cdots \prec q_{2n}$
> $no\_of\_intervals := 0;\ S' := \langle s_\infty \rangle$
> {S' is a binary search tree containing segments ordered by "$\prec$" relation}
> **for** $i = 1$ **to** $2n$
> $first :=$ the "smallest" (under the "$\prec$" relation) segment in S'
>   **if** q_i starts a segment s_j **then** insert s_j into S'
>   **if** q_i ends a segment s_j **then** remove s_j from S'
>   **if** $first \neq$ smallest segment in S' **then**
> $no\_of\_intervals := no\_of\_intervals + 1$
> $\mathcal{I}[no\_of\_intervals].angle :=$ polar angle of q_i
> $\mathcal{I}[no\_of\_intervals].name :=$ the smallest segment in S'

> **Algorithm 6:** Ray shooting using the visibility map.
>
> **procedure** $RayShoot(r = (p, \vec{v}))$: query ray, $\mathcal{I}$: visibility map)
> consider $\vec{v}$ as a polar angle
> do a binary search among $\mathcal{I}[*].angle$ to find the segment $\mathcal{I}[k].name$ such that
> $\mathcal{I}[k].angle \leq \vec{v}$ and $\mathcal{I}[k+1].angle > \vec{v}$

13.7 COMPUTATIONAL TECHNIQUES

This section describes some techniques used in the design of geometric algorithms.

13.7.1 PARALLEL ALGORITHMS

The goal of parallel computing is to solve problems faster than would be possible on a sequential machine — through the use of parallel algorithms. The complexity of a parallel algorithm is given in terms of its *time* and the number of *processors* used [Já92]. The *work* of a parallel algorithm is bounded above by the *processor-time product*, the product of the number of processors and the time.

Definitions:

A *parallel algorithm* is an algorithm that concurrently uses more than one processing element during its execution.

A *parallel machine* is a computer that can execute multiple operations concurrently.

A *parallel random access machine* (**PRAM**) is a synchronous machine in which each processor is a sequential RAM and processors communicate using a shared memory. Depending upon whether concurrent accesses are allowed to the shared memory cells, a PRAM is either **exclusive read** (**ER**) or **concurrent read** (**CR**), and either **exclusive write** (**EW**) or **concurrent write** (**CW**).

> **Algorithm 1:** ConvexHull(S: a set presorted by x-coordinate).
>
> if $|S| \leq 3$ then construct $CH(S)$ directly
> else
>   partition S into two subsets of equal size, S_L (the left subset) and S_R (the right subset), which are separated by a vertical line
>   in parallel, recursively construct $CH(S_L)$ and $CH(S_R)$
>   construct $CH(S)$ from $CH(S_L)$, $CH(S_R)$, and common tangents between them

Facts:

1. *Parallel divide-and-conquer*: Parallel algorithms can be obtained using the *divide-and-conquer* paradigm. The subproblems resulting from the divide phase are solved in parallel, and the combining phase of the algorithm is parallelized. In traditional divide-and-conquer algorithms, the problem is partitioned into two subproblems.

2. *Many-way divide-and-conquer*: Sometimes faster parallel algorithms can be obtained by partitioning the original problem into multiple, smaller subproblems, often referred to as *many-way divide-and-conquer* [Já92]. The solution to the original problem is obtained from the solutions to the subproblems.

3. *Cascading divide-and-conquer*: Some divide-and-conquer algorithms can be speeded up by pipelining (cascading) the work performed in the recursive applications as follows. Consider a binary tree representing the solutions to the recursive computations of the original divide-and-conquer algorithm, where leaves represent terminal subproblems. To obtain a faster algorithm, information is passed from a child to its parent before the solution to the child's subproblem is completely known. The parent then does some precomputation so that the solution to its problem can be computed as soon as the solutions of both of its children's problems are available. Typically, the information passed from a child to its parent is a constant sized sample of the child's current state. Often, such algorithms run in time proportional to the height of the recursion tree.

4. Cole first used pipelining to design a work-optimal $O(\log n)$ time parallel version of merge sort. Atallah, Cole, and Goodrich used this strategy to solve several geometric problems including the three-dimensional maxima problem and computing the visibility of a polygon from a point [AtGo93].

Algorithms:

1. *Using parallel divide-and-conquer to compute the convex hull $CH(S)$ of a set S of n points in the plane*: In Algorithm 1 the points in S are presorted by x-coordinate so that the division in the second step can be accomplished in constant time. The sorting takes $O(\log n)$ time using $O(n \log n)$ work [Já92]. The tangents needed in the third step can be computed in constant time using $|S_L| + |S_R|$ processors on a CREW PRAM [AtGo93]. Thus, as there are $O(\log n)$ recursive calls, Algorithm 1 runs in $O(\log n)$ time using $O(n \log n)$ work, which is both worst-case work-optimal and time-optimal for the CREW PRAM.

2. *Computing the convex hull of a three-dimensional point set*: The convex hull of a three-dimensional point set can be computed using a two-way divide-and-conquer algorithm similar to Algorithm 1. The running time of this algorithm is $O(\log^2 n)$ because the combining in the last step of Algorithm 1 takes $O(\log n)$ time since it is

more complex than in the planar case. However, a faster algorithm can be designed using many-way divide-and-conquer. The point set is partitioned into $O(n^{\frac{1}{2}})$ groups, each of size $O(n^{\frac{1}{2}})$, and then, even though the combining still takes $O(\log n)$ time, a total time of $O(\log n)$ can be obtained.

13.7.2 RANDOMIZED ALGORITHMS

Randomization is a powerful technique that has been used in algorithms for many geometric problems.

Definition:

A *randomized algorithm* is an algorithm that makes random choices during its execution.

Facts:

1. Randomized algorithms are often faster, simpler, and easier to generalize to higher dimensions than deterministic algorithms.

2. For many problems, efficient algorithms can be obtained by processing the input objects in a particular order or by grouping them into equal sized subsets. Although significant computation may be required to exactly determine an appropriate order or a good partition into subsets, in many cases simple random choices can be used instead.

3. *Randomized incremental methods*: One of the simplest ways to construct a geometric structure is incrementally. In an incremental construction algorithm, the input objects are inserted one at a time and the current structure is updated after each addition. The desired structure is obtained after all input objects have been inserted. Although the cost of some updates could be very large, in many cases it can be shown that if the objects are inserted in *random order*, then the amount of work for each update is expected to be small and the expected running time of the algorithm will be small as well. Thus, the expectation of the running time depends not on the input distribution but on the ratio of good to bad insertion sequences.

4. The power of randomization in incremental algorithms was first noted by Clarkson and Shor, and by Mulmuley. Randomized incremental algorithms have been proposed for constructing many geometric structures including convex hulls, Delaunay triangulations, trapezoidal decompositions, and Voronoi diagrams [Mu94].

5. *Randomized divide-and-conquer*: In randomized divide-and-conquer a random subset of the input is used to partition the original problem into subproblems. Then, as in any divide-and-conquer algorithm, the subproblems are solved, perhaps recursively, and their solutions are combined to obtain the solution to the original problem. Ideally, the partition produces subproblems of nearly equal size and the sum of the subproblem sizes roughly equals the input size. This ideal can almost be achieved for many geometric problems. For example, with probability at least $\frac{1}{2}$, $O(r)$ subproblems of size $O(\frac{n \log r}{r})$ can often be obtained, where n is the number of input objects and r is the size of the random subset.

6. The fact that a random sample of the input items can often be used to produce subproblems of almost equal size was first shown by Clarkson, and by Haussler and Welzl. Randomized divide-and-conquer algorithms are known for many geometric problems including answering closest-point queries, constructing arrangements, triangulating point sets, and constructing convex hulls [Mu94].

> **Algorithm 2:** Intersect(S: a set of n halfplanes).
>
> $R :=$ a random sample of size r chosen from S.
>
> compute the intersection $I(R)$ and triangulate it by connecting each vertex of $I(R)$ to the origin to obtain $T(R)$
>
> for each triangle $t \in T(R)$, determine its *conflict list* $c(t)$ (the set of halfplanes in S whose bounding lines intersect t)
>
> for each triangle $t \in T(R)$, compute the intersection of the halfplanes in $c(t)$ restricted to t {This may be done recursively}

Examples:

1. *Incrementally constructing the intersection of a set H of n halfplanes in the plane*: First, a random permutation $(h_1, h_2, \ldots, h_n)$ of H is formed. Let I_i denote the intersection of halfplanes h_1 through h_i. During the ith iteration of the algorithm, I_i is computed from I_{i-1} by removing the portion of I_{i-1} that is not contained in h_i.

   To make this update easier, a vertex of the current intersection I_{i-1} that is not contained in h_j is maintained, for all $j \geq i$; such a vertex is said to *conflict* with h_j. Given a conflicting vertex v for h_i, the portion of I_{i-1} that must be removed is determined by traversing its boundary in both directions from v until reaching vertices that are contained in h_i. Since I_n has size $O(n)$, the amortized cost of each update is $O(1)$ and the algorithm spends a total of $O(n)$ time updating the intersection. After I_i is computed, the conflicting vertex for h_j is updated, for all $j > i$. It can be shown that the total cost of maintaining the conflicting vertices in the algorithm is $O(n \log n)$.

2. *Using randomized divide-and-conquer to construct the intersection $I(S)$ of a set S of n halfplanes, each of which contains the origin*: See Algorithm 2. In Step 2, the intersection of the halfplanes in the sample is used to create $O(r)$ triangles, each of which corresponds to a subproblem. For technical reasons, the region corresponding to each subproblem should have constant descriptive complexity.

   Often, this condition is achieved by triangulating the resulting structure. Next, the input is distributed to the subproblems. Usually, this is done by finding the objects that intersect, or *conflict* with, a subproblem's region.

   Then, the subproblems are solved, recursively if desired, and their solutions are combined to form the solution to the original problem. In the intersection algorithm the final solution is obtained by "gluing" the subproblem solutions together.

   If r is some suitably chosen constant, then Algorithm 2 runs in expected time $O(n \log n)$.

13.7.3 PARAMETRIC SEARCH

Definition:

Parametric search is an algorithmic technique for solving optimization problems by using an algorithm for solving an associated decision problem as a subroutine.

Algorithm:

1. Parametric search is a powerful algorithmic technique that can be applied to a diverse range of optimization problems. It is best explained in terms of an abstract optimization problem. Consider two versions of this problem:
   - a *search problem* P, which can be parametrized by some real, nonnegative value t and whose solution is a value t^* that satisfies the optimization criteria;
   - a *decision problem* $D(t)$, whose input includes a real, nonnegative value t and whose solution is either YES or NO.

   In parametric search, the search problem P is solved using an algorithm A_s for the decision problem $D(t)$. In order to apply parametric search, the points corresponding to YES answers for $D(t)$ must form a connected, possibly unbounded, interval in $[0, \infty)$. Then, assuming that $D(0) = $ YES, the search problem P is to find the largest value of t for which $D(t) = $ YES.

   The basic idea of parametric search is to use A_s to find t^*. This is done by simulating A_s, but using a variable (parameter) instead of a value for t. Assume that A_s can detect if $t = t^*$, and that the computation in A_s consists of comparisons, each of which tests the sign of a bounded degree polynomial in t and in the original input. During each comparison in the simulation:
   - the roots of the appropriate polynomial are computed,
   - A_s is run on each root,
   - t^* is located between two consecutive roots.

   Thus, each comparison in the simulation reduces the interval known to contain t^*, which is originally $[0, \infty)$.

Facts:

1. In most geometric applications it can be shown that at some point the simulation will test a root that is equal to t^*.

2. If T_s denotes the worst-case running time of A_s, the total cost of the parametric search is $O(T_s^2)$ since there are $O(T_s)$ operations in the simulation of A_s, and each comparison operation in this simulation takes $O(T_s)$ time.

3. With a parallel algorithm A_p to solve the decision problem $D(t)$, the (sequential) parametric search can be done faster. Suppose that A_p runs in T_p parallel time steps using p processors. Then, A_p performs at most p *independent* comparison operations during each parallel time step. The parametric search simulates A_p sequentially as follows. To simulate a parallel time step of A_p:
   - compute the $O(p)$ roots of the at most p bounded degree polynomials corresponding to independent comparisons in the parallel time step;
   - sort these roots;
   - use a binary search to locate t^* between two consecutive roots. (The sequential algorithm A_s is used to evaluate the roots during the binary search.)

   The cost of simulating each parallel time step is $O(p \log p + T_s \log p)$, and the total cost of the parametric search is $O(T_p(p \log p + T_s \log p))$.

4. Parametric search was originally proposed by Megiddo. It has been used for many geometric problems including ray shooting, slope selection, and computing the diameter of a three-dimensional point set [Mu94].

Example:

1. *Answering ray shooting queries in a convex polytope Q*: In the search problem P, a ray r is given originating at a point p contained in Q, and the face of Q hit by r needs to be found. The decision problem $D(t)$ is the *polytope membership problem*: given a point t, determine if t lies in Q. The value t^* sought by the parametric search is the point where r intersects Q. The ray shooting problem can be solved in $O(\log^2 n)$ time by parametric search [Mu94] (since, given an appropriate data structure, the polytope membership problem can be solved sequentially in $O(\log n)$ time).

13.7.4 FINITE PRECISION

Geometric algorithms are usually designed assuming the *real random access machine* model of computation. In this model, values can be arbitrarily long real numbers, and all standard operations such as $+$, $-$, $\times$, and $\div$ can be performed in unit time regardless of operand length. In reality, however, computers have finite precision and can only approximate real numbers. Several techniques have been suggested for dealing with this problem [Fo93].

Facts:

1. *Algorithms directly designed for a discrete domain*: This approach can dramatically increase the complexity of the algorithm and has not gained wide use.

2. *Floating point numbers*: Floating point numbers provide a convenient and efficient way to approximate real numbers. Unfortunately, naively rounding numbers in geometric algorithms can create serious problems such as topological inversions. There are cases when floating point arithmetic can safely be used. For example, for certain inputs the result of the floating point operation will be unambiguous. Some algorithms have been shown to be sufficiently stable using floating point arithmetic. However, no general method is known for designing stable algorithms.

3. *Exact arithmetic*: In exact arithmetic, numbers are represented by vectors of integers and all primitive operations are guaranteed to give correct answers. Integer arithmetic is sufficient for many geometric algorithms since symbolic or algebraic numbers are rarely needed, and in many cases homogeneous coordinates can remove the need for rational numbers. However, since exact representations can have large bit complexity, exact arithmetic can be expensive — typically increasing the cost of arithmetic operations by an order of magnitude. This cost can be decreased somewhat by optimizing the expressions and computations involving exact arithmetic.

4. *A combination of floating point and exact arithmetic*: First, the operation is performed using floating point arithmetic. Then, if the result is ambiguous, an exact computation is performed.

5. *Adaptive-precision arithmetic*: In adaptive-precision arithmetic each number is approximated by an interval whose endpoints require lower precision. If the exact computation using the approximation is ambiguous, the method iterates using smaller and smaller intervals with higher precision endpoints.

6. Although no clear consensus has been reached, a combination of the above strategies may yield the best results.

13.7.5 DEGENERACY AVOIDANCE

To simplify exposition by reducing the number of cases that must be considered, many geometric algorithms assume that the input is in *general position*. The general position assumption depends on the problem. For example, in problems involving planar point sets, the assumption might be that no two points have the same x-coordinate, that no three points lie on the same line, or that no four points lie on the same circle.

Definitions:

A set of objects is in **general position** if the objects satisfy certain specified conditions.

A set of input objects that violates the general position assumption is **degenerate**.

Facts:

1. *Perturbation*: Several schemes have been proposed that apply small perturbations to transform the input so that it does not contain degeneracies [Fo93]. The object of perturbation schemes is to allow the design of simpler algorithms which may validly assume the input is in general position.

    Algorithms using perturbation schemes may not always produce correct output. For example, in a convex hull algorithm, points on the boundary could potentially be perturbed into the interior of the polytope and vice versa. Also, perturbation schemes can affect, perhaps adversely, *output-sensitive* algorithms whose running times depend on the size of the output they produce.

2. *Deal with degeneracies directly*: Algorithms that deal with degeneracies in a problem specific manner have been designed for problems such as triangulating point sets. Although this approach is not as general and usually leads to more complex algorithms than those employing perturbation schemes, it can provide superior performance.

3. *Symbolically perturbing the coordinates of each input point* (Edelsbrunner and Mücke): This is done by adding to each coordinate a suitable power of a small, positive real number represented by a symbolic parameter ϵ. Then, since values are now polynomials in ϵ, the arithmetic operations in the algorithm are replaced by polynomial arithmetic operations. Geometric primitives are implemented symbolically, typically by evaluating a sequence of determinants. Assuming that the dimension of the problem is fixed, this scheme increases the running time of the algorithm by at most a constant factor. However, the overhead incurred can in fact be quite large.

13.8 APPLICATIONS OF GEOMETRY

Geometry overlays the entire computing spectrum, having applications in almost every area of science and engineering, including astrophysics, molecular biology, mechanical design, fluid mechanics, computer graphics, computer vision, geographic information systems, robotics, multimedia, and mechanical engineering.

Some excellent general references to geometry applications include a report on application challenges to computational geometry at the website

    http://www.cs.princeton.edu/~chazelle/taskforce/CGreport.ps

a status report on theoretical computer science, including applications of computational geometry, at the website

    http://hercule.csci.unt.edu/sigact/longrange/contributions.html

Eppstein's list of general geometric references at the website

    http://www.ics.uci.edu/~eppstein/geom.html

and Amenta's directory of software at the website

    http://www.geom.umn.edu/software/cglist/

including planar algorithms, arbitrary dimensional convex hull, Voronoi diagram, Delaunay triangulation, polygon decomposition, point location, intersection, linear programming, smallest enclosing ball and center point, visualization, mesh generation, shape reconstruction and collision in robotics.

13.8.1 MATHEMATICAL PROGRAMMING

Definition:

Mathematical programming is the large-scale optimization of an objective function (such as cost) of many variables subject to constraints, such as supplies and capacities.

Facts:

1. Mathematical programming includes both *linear programming* (continuous, integer, and network) and *nonlinear programming* (quadratic, convex, general continuous, and general integer).

2. Applications include transportation planning and transshipment, factory production scheduling, and even determining a least-cost but adequate diet.

3. A *modeling language*, such as AMPL, is often used in applications. [FoGaKe93]

Example:

1. *Linear programming in low dimensions*: This is a special case of linear programming since there are algorithms whose time is linear in the number of constraints but exponential in the number of dimensions.

Application:

1. Find the *smallest enclosing ball* of a set of points or a set of balls in arbitrary dimension, [We91]. This uses a randomized incremental algorithm employing the move to front heuristic.

13.8.2 POLYHEDRAL COMBINATORICS

Polyhedra have been important to geometry since the classification of regular polyhedra in classical times.

Fact:

1. The following are some of the many possible operations that can be performed on polygons and polyhedra:
   - *Boolean operations*, such as intersection, union, and difference;
   - *point location* of new query points in a preprocessed set of polygons, which may partition a larger region;
   - *range search* of a preprocessed set of points to find those inside a new query polygon;
   - *decomposition* (or *triangulation*) of a polygon into triangles or a polyhedron into tetrahedra (§13.4.2). A simple n-gon is always decomposable into $n-2$ triangles, in linear time, and all triangulations have exactly $n-2$ triangles. However, in $\mathcal{R}^3$, some polyhedra cannot be partitioned intro tetrahedra without additional Steiner points. Also, different triangulations of the same polyhedron may have different numbers of tetrahedra.

Applications:

1. *Aperiodic tilings and quasicrystals*: Tilings (in the plane) and crystallography (in three-dimensional space) are classic applications of polygons and polyhedra. A recent development is the study of aperiodic (Penrose) tilings [Ga77] and quasicrystals [Ap94]. These are locally but not globally symmetric under 5-fold rotations, quasi-periodic with respect to translations, and self-similar. See the website

   http://www.geom.umn.edu/apps/quasitiler/

   Large-scale symmetries, such as 5-fold, that are impossible in traditional crystallography, can be visible with X-ray diffraction. Tilings can be constructed by projecting simple objects from, say, $\mathcal{R}^5$. One application is the surface reinforcement of soft metals.

2. *Error-correcting codes*: Some error-correcting codes can be visualized with polytopes as follows. Assume that the goal is k-bit symbols, where any error of up to b bits can be detected. The possible symbols are some of the 2^k vertices of the hypercube in k-dimensional space. The set of symbols must contain no two symbols less than $b+1$ distance apart, where the metric is the number of different bits. If errors of up to c bits are to be correctable, then no two symbols can be closer than $2c+1$.

   Similarly, in quantum computing, a *quantum error-correcting code* can be designed using Clifford groups and binary orthogonal geometry. [CaEtal95]

13.8.3 COMPUTATIONAL CONVEXITY

Definitions:

An *$\mathcal{H}$-polytope* is a polytope defined as the intersection of m half-spaces in $\mathcal{R}^n$.

A *$\mathcal{V}$-polytope* is a polytope defined as the convex hull of m points in $\mathcal{R}^n$.

A *zonotope* is the vector (Minkowski) sum of a finite number of line segments.

Computational convexity is the study of high-dimensional convex bodies.

An *oracle* is an algorithm that gives information about a convex body.

Facts:

1. Computational complexity is related to linear programming, polyhedral combinatorics, and the algorithmetic theory of polytopes and convex bodies.

2. In contrast to computational geometry, computational complexity considers convex structures in normed vector spaces of finite but not restricted dimension. If the body under consideration is more complex than a polytope or zonotope, it may be represented as an oracle. Here the body is a black-box, and all information about it, such as membership, is supplied by calls to the oracle function. Typical algorithms involve *volume computation*, either deterministically, or by *Monte-Carlo* methods, perhaps after decomposition into simpler bodies, such as simplices.

3. When the dimension n is fixed, the volume of $\mathcal{V}$-polytopes and $\mathcal{H}$-polytopes can be computed in polynomial time.

4. There does not exist a polynomial-space algorithm for the exact computation of the volume of $\mathcal{H}$-polytopes (where n is part of the input).

5. Additional information on computational convexity can be found at the following website:

   http://dimacs.rutgers.edu/techps/1994/94-31.ps

13.8.4 MOTION PLANNING IN ROBOTICS

In Computer Assisted Manufacturing, both the tools and the parts being assembled must often be moved around each other in a cluttered environment. Their motion should be planned to avoid collisions, and then to minimize cost.

Definitions:

A **Davenport-Schinzel sequence** of order s over an alphabet of size n, or $DS(n, s)$, is a sequence of characters such that:

- no two consecutive characters are the same;
- for any pair of characters, a and b, there is no alternating subsequence of length $s + 2$ of the form $\ldots a \ldots b \ldots a \ldots b \ldots$.

Facts:

1. Practical general motion, *path planning*, is solvable with Davenport-Schinzel sequences [Sh95]. Upper bounds on $\lambda_s(n)$, the length of the longest $DS(n, s)$, determine upper bounds on the complexity of the lower envelopes of certain functions.

   For example, given n points in the plane that are moving with positions that are polynomials of degree s in time, the number of times that the closest pair of points can change is $\lambda_{2s}(C(n, 2))$.

2. Visibility graphs (§13.5.5) are useful in finding shortest path between two points in the plane, in the presence of obstacles.

3. The problem of moving a finite object in the presence of obstacles may also be mapped into a *configuration space* (or *C-space*) problem of moving a corresponding point in a higher dimension. If translational and rotational motion in 2-dimensional (respectively 3-dimensional) is allowed, then the C-space is 3-dimensional (respectively 6-dimensional).

4. Articulated objects, multiple simultaneous motion, and robot hands also increase the number of degrees of freedom.

5. Current problems:
   - *representation* of objects, since, although planar, faceted, models are simpler, the objects should be algebraic surfaces, and, even if they are planar, in C-space their corresponding versions will be curved;
   - *grasping*, or placing a minimal number of fingers to constrain the object's motion;
   - *sequence planning* of the assembly of a collection of parts;
   - *autonomous navigation* of robots in unstructured environments.

13.8.5 CONVEX HULL APPLICATIONS

Facts:

1. The convex hull (§13.5.1) is related to the Voronoi diagram (§13.5.3) since a convex hull problem in $\mathcal{R}^k$ is trivially reducible to a Voronoi diagram problem in $\mathcal{R}^k$, and a Voronoi diagram problem in $\mathcal{R}^k$ is reducible to a convex hull problem in $\mathcal{R}^{k+1}$.

2. The definition of convex hull is not constructive, in that it does not lead to a method for finding the convex hull. Nevertheless, there are many constructive algorithms and implementations. One common implementation is *QuickHull* (§13.5.1), a general dimension code for computing convex hulls, Delaunay triangulations, Voronoi vertices, furthest-site Voronoi vertices, and halfspace intersections. [BaDoHu95]

Applications:
See http://www.geom.umn.edu/~bradb/qhull-news.html

1. *Mathematics*:
   - determining the principal components of spectral data;
   - studying circuits of matroids that form a Hilbert base;
   - studying the neighbors of the origin in the $\mathcal{R}^8$ lattice.

2. *Biology and medicine*:
   - classifying molecules by their biological activity;
   - determining the shapes of left ventricles for electrical analysis of the heart.

3. *Engineering*:
   - computing support structures for objects in layered manufacturing in rapid prototyping, [StBrEa95]. By supporting overhanging material, these prevent the object from toppling while partially built.
   - designing nonlinear controllers for controlling vibration;
   - finding invariant sets for delta-sigma modulators;
   - classifying handwritten digits;
   - analyzing the training sets for a multilayer perceptron model;
   - determining the operating characteristics of process equipment;
   - navigating robots;
   - creating 6-dimensional wrench spaces to measure the stability of robot grasps;
   - building micromagnetic models with irregular grain structures;
   - building geographical information systems;
   - simulating a spatial database system to evaluate spatial tesselations for indexing;
   - producing virtual reality systems;
   - performing discrete simulations of incompressible viscous fluids using vortex methods;
   - modeling subduction zones of tectonic plates and studying fluid flow and crystal deformation;
   - computing 3-dimensional unstructured meshes for computational fluid dynamics.

13.8.6 NEAREST NEIGHBOR

Variants of the problem of finding the nearest pair of a set of points have applications in fields from handwriting recognition to astrophysics.

Applications:

1. *Fixed search set, varying query point*: A fixed set, $\mathcal{P}$, of n points in $\mathcal{R}^d$ is preprocessed so that the closest point $p \in \mathcal{P}$ can be found for each query point, q. The search time per query can range from $\log n$ (if $d = 2$) to $n^{\frac{d}{2}}$ (for large d). The Voronoi diagram is commonly used in low dimension. However, because of the Voronoi diagram's complexity in higher dimension, hierarchical search structures, bucketing, and probabilistic methods perhaps returning approximate answers are common.
2. *Moving points*: The points in $\mathcal{P}$ may be moving and the close pairs of points over time is of interest.

Examples:

1. *Examples of fixed search sets with varying query point*:
   - *Character recognition in document processing*: Each representative character is defined by a vector of features. Each new, unknown character must be mapped to the closest representative character in feature space.
   - *Color map optimization in computer graphics*: Many frame buffers allow only the 256 colors in the current color map to be displayed simultaneously, from a palette of 2^{24} possible colors. Thus, each color in a new image must be mapped to the closest color in the color map. A related problem is the problem of determining what colors to use in the color map.
   - *Clustering algorithms for speech and image compression in multimedia systems*: As in the color map problem, a large number of points must be quantized down to a smaller set.
2. *Examples of moving points*:
   - *Simulation of star motion in astrophysics*: Calculating the gravitational attraction between every pair of stars is too costly, so only close pairs are individually calculated. Otherwise the stars are grouped, and the attraction between close groups is calculated. The groups may themselves be grouped hierarchically.
   - *Molecular modeling*: In molecular modeling, close pairs of atoms will be subject to van der Waals forces.
   - *Air traffic control*: Air traffic controllers wish to know about pairs of aircraft closer than a minimum safe distance. Here the metric is nonuniform; small vertical separations are more tolerable than small horizontal separations.
   - During path planning in robotics and numerically controlled machining, unintended close pairs of objects must also be avoided.

13.8.7 COMPUTER GRAPHICS

Computer graphics may be divided into *modeling* of surfaces, and *simulation* of the models. The latter includes *rendering* a scene and its light sources to generate synthetic imagery with respect to some viewpoint. Rendering involves *visibility*, or determining which parts of the surfaces are visible, and *shading* them according to some lighting model.

Definitions:

Anti-aliasing refers to filtering out high-frequency spatial components of a signal, to prevent artifacts, or aliases, from appearing in the output image. In graphics, a high frequency may be an object whose image is smaller than one pixel or a sharp edge of an object.

A *GUI (graphical user interface)* is a mechanism that allows a user to interactively control a computer program with a bitmapped display by using a mouse or pointer to select menu items, move sliders or valuators, and so on. The keyboard is only occasionally used. A GUI contrasts with typing the program name followed by options on a command line, or by preparing a text file of commands for the program. A GUI is easier and more intuitive to use, but can slow down an expert user.

Examples:

1. *Visibility*: Visibility algorithms may be *object-space*, where the visible parts of each object are determined, or *image-space*, where the color of each pixel in the frame buffer is determined. The latter is often simpler, but the output has less meaning, since it is not referred back to the original objects. Techniques include *ray tracing* and *radiosity*.

    - *Ray tracing*: Ray tracing extends a line from viewpoint through each pixel of the frame buffer until the first intersecting object. If that surface is a mirror, then the line is reflected from the surface and continues in a different direction until it hits another object (or leaves the scene). If the object is glass, then both a reflecting and a refracting line are continued, with their colors to be combined according to Fresnel's law.

        One geometry problem here is that of *sampling for subpixel averaging*. The goal is to color a square pixel of a frame buffer according to the fraction of its area occupied by each visible object. Given a line diagonally crossing a pixel, the fraction of the pixel covered by that face must be obtained for anti-aliasing. If the edges of two faces intersect in this pixel, each face cannot be handled independently, for example with an anti-aliased Bresehnam algorithm. If this is done badly, then it is very obvious in the final image as a possible fringe of a different color around the border of the object, [Mi96].

        The solution is to pick a small set of points in the pixel (typically 9, 16, or 64 points), determine which visible object projects to each point, and combine those colors. The problem is then to select a set of sampling points in the pixel, such that given a subset region, the number of points in it approximates its area. Four possible methods, from worst to best, are:

        ⋄ pick the points independently and uniform randomly;
        ⋄ use a nonrandom uniform distribution;
        ⋄ start with the above distribution, then jitter the points, or perturb each one slightly;
        ⋄ use simulated annealing to improve the point distribution.

    - *Radiosity*: Radiosity partitions the scene into facets, computes a *form factor* of how much light from each facet will impinge on each other, and solves a system of linear equations to determine each facet's brightness. This models diffuse lighting particularly well.

    - *Windowing systems*: Another visibility problem is designing the appropriate data structure for representing the windows in a GUI, so that the window that is in front at any particular pixel location can be determined, in order to receive the input focus.

874   Chapter 13 DISCRETE AND COMPUTATIONAL GEOMETRY

- *Radio wave propagation*: The transmission of radio waves, as from cellular telephones, which are reflected and absorbed by building contents, is another application of visibility. [Fo96]

2. *Computer vision*: Applications of geometry to vision include *model-based recognition* (or *pattern matching*), and *reconstruction or recovery of 3-D structure* from 2-D images, such as stereopsis, and structure from motion. See the website
   http://www.cs.princeton.edu/~chazelle/taskforce/CGreport.ps

In recognition, a model of an object is transformed into a sensor-based coordinate system and the transformation must be recovered. In reconstruction, the object must be determined from multiple projections.

3. *Medical image shape reconstruction*: Various medical imaging methods, such as computer tomography, produce data in the form of successive parallel slices through the body. The basic step in reconstructing the 3-dimensional object from these slices in order to view it involves joining the corresponding vertices and edges of two polygons in parallel planes by triangles to form a simple polyhedron. However, there exists a pair of polygons that cannot be so joined. [GiO'RSu96]

13.8.8 MECHANICAL ENGINEERING DESIGN AND MANUFACTURING

Geometry is very applicable in CAD/CAM, such as in the design and manufacture of automobile bodies and parts, aircraft fuselages and parts such as turbine blades, and ship hulls and propellers.

Examples:

1. *Representations*: How should mechanical parts be represented? One problem is that geometric descriptions are verbose compared to 2-dimensional descriptions, such as draftings, since those assume certain things that the users will fill in as needed, but which must be explicit in the geometric description. Possible methods include the following:
   - *constructive solid geometry*: Primitive objects, such as cylinders and blocks, are combined with the regularized Boolean operators union, intersection, and difference.
   - *faceted boundary representation*: The object is a polyhedron with a boundary of planar faces.
   - *exact boundary representation*: The object is defined by boundary "faces", but now each face can be curved, such as a NURBS (Non-Uniform Rational B-Spline), or an implicit piecewise quadric, Dupin cyclide (a quartic surface that is good for blending two quadric surfaces), or supercyclide.

The possible methods can be evaluated with the following criteria:
   - *robustness* against numerical errors;
   - *elegance*;
   - *accuracy* in representing complex, curved, shapes, especially blends between the two surfaces at the intersection of two components;
   - *ease* of explicitly obtaining geometry such as the boundary.

2. *Mesh generation*: A *mesh* is the partition of a polyhedron into, typically, tetrahedra or hexahedra to facilitate finite element modeling. A good mesher conforms to constraints, can change scale over a short distance, has no unnecessary long thin elements, and has fewer elements when possible. In some applications, periodic *remeshing* is required.

If the elements are tetrahedra, then a Delaunay criterion that the circumsphere of each tetrahedron contains no other vertices may be used. However, this is inappropriate in certain cases, such as just exterior to an airfoil, where a (hexahedral) element may have an aspect ratio of 100,000:1. This raises numerical computation issues.

Applications of meshing outside mechanical design include *computational fluid dynamics*, *contouring* in GIS, *terrain databases* for real time simulations, and Delaunay applications in general. See

   http://www.cs.cmu.edu/~quake/triangle.html

3. *Minimizing workpiece setup in NC machining*: In 4- and 5-axis numerically controlled machining, in order to machine all the faces, the workpiece must be repeatedly dismounted, recalibrated, and remounted. This setup can take much more time than the actual machining. Minimizing the number of setups by maximizing the number of faces that can be machined in one setup is a visibility problem harder than finding an optimal set of observers to cover some geographic terrain. Exact solutions are NP-hard; approximate solutions use geometric duality, topological sweeping, and efficient construction and searching of polygon arrangements on a sphere.

4. *Dimensional tolerancing*: *Tolerancing* refers to formally modeling the relationships between mechanical function and geometric form while assigning and analyzing dimensional tolerances to ensure that parts assemble interchangeably. [SrVo93]

A tolerance may be specified *parametrically*, as a variation in a parameter, such as the width of a rectangle, or as a *zone* that the object's boundary must remain in. The latter is more general but must be restricted to prohibit pathologies, such as the object's boundary being not connected.

Tolerance synthesis attempts to optimize the tolerances so as to minimize the manufacturing cost of an object, considering that, while large tolerances are cheaper to manufacture, the resulting product may function poorly. [Sk96]

Unsolved Problems:

The following lists some of the many remaining unsolved problems in applying geometry:

1. *Blending* between two surfaces in mechanical design, especially at the ends of the blend, where these surfaces meet others. (A *blending surface* smooths the intersection of two surfaces by being tangent to them, each along a curve.)

2. *Variational design* of a class of objects subject to constraints. Well-designed constraint systems may have multiple solutions; the space must be searched for the correct one. Labeling derivative entities, such as the edge resulting from the intersection of two inputs is an issue, partly because this edge may not exist for some parameter values.

3. Generally *formalizing the semantics* of solid modeling. [Ho96]

4. Updating *simplifying assumptions*, such as the linearity of random access memory, and points being in general position, which were useful in the past, but which cause problems now.

5. *Accounting for dependencies* between geometric primitives, and *maintaining topological consistency*.

6. Designing *robust algorithms* when not only is there numerical roundoff during the computation, but also the input data are imprecise, for example, with faces not meeting properly.

7. Better *3-dimensional anti-aliasing* to remove crevices and similar database errors before *rapid prototyping*.

8. There still remains a need for many features in geometry implementations, such as more *geometric primitives* at all levels, *default visualization or animation* easily callable for each data structure, more *rapid prototyping* with visualization, a *visual debugger* for geometric software, including changing objects online, and generally *more interactivity*, not just data-driven programs.

13.8.9 LAYOUT PROBLEMS

Efficiently laying out objects has wide-ranging applications in geometry.

Examples:

1. *Textile part layout:* The clothing industry cuts parts from stock material after performing a tight, nonoverlapping, layout of the parts, in order to minimize the costs of expensive material. Often, because the cloth is not rotationally symmetric, the parts may be translated, but not rotated. Therefore, geometric algorithms for minimizing the overlap of translating polygons are necessary. Since this problem is PSPACE-hard, heuristics must be used. [Da95] [LiMi95].

2. *VSLI layout:* Both laying out circuits and analyzing the layouts are important problems. The masks and materials of a VLSI integrated circuit design are typically represented as rectangles, mostly isothetic, although 45 degrees or more general angles of inclination for the edges are becoming common. The rectangles of different layers may overlap. One integrated circuit may be 50MB of data before its hierarchical data structure is flattened, or 2GB after. See the website:

    `http://ams.sunysb.edu/~held/proc_usb_comp_geo-95.html`

Geometry problems include the following.

- *design rule verification:* It is necessary to check that objects are separated by the proper distances and that average metal densities are appropriate for the fabrication process.
- *polygon simplification*: A design described by a complex set of polygons may perhaps be optimized into a smaller set of isothetic polygons (with only horizontal and vertical sides), such that the symmetric difference from the original design is as small as possible.
- *logic verification*: The electrical *circuit* is determined by the graph extracted from the adjacency information of the rectangles, and whether it matches the original logic design is determined. A subproblem is determining *devices* (*transistors*), which occur when rectangles of two particular different layers overlap.
- *capacitance*: This depends on the closeness of the component rectangles, which might be overlapping or separated, representing two conductors.
- *PPC* (*process proximity correction*): This means to correct the effect that, when *etching* a circuit, a rectangle's edges are displaced outward, possibly causing it to come too close to another rectangle, and change the circuit.

13.8.10 GRAPH DRAWING

The classic field of *graph drawing* [TaTo94] aims automatically to display a graph, emphasizing fundamental properties such as symmetry while minimizing the ratio between longest and shortest edges, number of edge crossings, etc. Applications include advanced GUIs, visualization systems, databases, showing the interactions of individuals and groups in sociology, and illustrating connections between components in software engineering. Recent 3-dimensional visualization hardware now permits 3-dimensional graph drawing. See

> file://ftp.cs.brown.edu/pub/papers/compgeo/gdbiblio.ps.gz

Facts:

1. Graph G can be drawn as the 1-skeleton of a convex polytope in $\mathcal{R}^3$ if and only if G is planar and 3-connected. (Steinitz) [Gr67].

2. Given a 3-connected planar graph, the graph can be drawn as a convex polyhedron in $\mathcal{R}^3$ using $O(n)$ volume while requiring the vertices to be at least unit distance apart, which allows them to be visually distinguished. This can be done in time $O(n^{1.5})$. (Chrobak, Goodrich, Tamassia) See

> http://ams.sunysb.edu/~held/proc_usb_comp_geo-95.html

13.8.11 GEOGRAPHIC INFORMATION SYSTEMS

A *map* (§8.6.4) is a planar graph. Minimally, it contains *vertices*, *edges*, and *polygons*. However, a sequence of consecutive edges and 2-vertices is often called a *chain* (or *polyline*), and its interior vertices *points*. For example, if each polygon is one nation, then the southern border of Canada with the USA, is one chain.

Definition:

A **geographic information system** (**GIS**) is an information system designed to capture, store, manipulate, analyze, and display spatial or geographically-referenced data.

Facts:

Typical simple geometric operations are given in Facts 1–6. More complex ones are given in Facts 7–9.

1. *Projecting data from one map projection to another, and determining the appropriate projection*: Since the earth is not a developable surface, no projection meets all the following criteria simultaneously: *equal-area*, *equidistant* (preserving distances from one central point to every other point), *conformal* (preserving all angles), and *azimuthal* (correctly showing the compass angle from one central point to every other point). Since a projection that meets any one criterion exactly is quite bad in the others, the most useful projections tend to be compromises, such as the recent *Robinson projection*. [Da95]

2. *Rubber-sheeting*, or nonlinear stretching, to align a map with calibration points, and for *edge joining* of adjacent map sheets or databases, which may have slightly different coordinate systems.

3. *Generalizing* or reducing the number of points in a chain while preserving certain error properties.

4. *Topological cleanup* so that edges that are supposed to meet at one vertex do so, the boundary of each polygon is a closed sequence of vertices and polylines, adjacency information is correct, and so on.

5. Choice of the correct *data structure*. Should elevation data be represented in a *gridded* form (as an array of elevations) or should a *triangulated irregular network* (TIN) be used (the surface is partitioned into triangles)?

6. *Zone of influence calculation*: For example, find all the national monuments within ten miles of the highway.

7. *Overlaying*: Overlaying two maps to produce a third, where one polygon of the overlay map will be those points that are all from the same two polygons of the two input maps is one of the most complex operations in a GIS. If only the area or other mass property of the overlay polygons is desired, then it is not necessary completely to find the overlay polygons first; it is sufficient to find the set of vertices and their neighborhoods of each overlay polygon. [FrEtal94]

8. *Name placement*: Consider a cartographic map containing point features such as cities, line features such as rivers, and area features such as states. The *name placement* problem involves locating the features' names so as to maximize readability and aesthetics [FrAh84]. Efficient solutions become more important as various mapping packages now produce maps on demand. The techniques also extend to *labelling CAD drawings*, such as piping layouts and wiring diagrams.

9. *Viewsheds and visibility indices*: Consider a terrain database, and an observer and target, both of which may be some distance above the terrain. The observer can see the target if and only if a line between them does not intersect the terrain. Note that if they are at different heights above the terrain, then this relation is not necessarily commutative.

The (not necessarily connected) polygon of possible targets visible by a particular observer is his *viewshed*. The viewshed's area is the observer's *visibility index*. In order to site observers optimally, the visibility index for each possible observer in the database may be required. Calculating this exactly for an $n \times n$ gridded database takes time $O(n^5)$ so sampling techniques are used. [FrRa94]

13.8.12 GEOMETRIC CONSTRAINT SOLVING

Applications of geometric constraint solving include mechanical engineering, molecular modeling, geometric theorem proving, and surveying.

Definition:

Geometric constraint solving is the problem of locating a set of geometric elements given a set of constraints among them.

Fact:

1. The problem may be under-constrained, with an infinite number of solutions, or over-constrained, with no solutions without some relaxation.

Examples:

1. A *receptor* is a rigid cavity in a protein, which is the center of activity for some reaction. A *ligand* is a small molecule that may bind at a receptor. The activity level of a drug may depend on how the ligand fits the receptor, which is made more complicated by the protein molecule's bending.

2. In CAD/CAM, there may be constraints such as that opposite sides of a feature shall be parallel. For example, commercial systems like Pro/Engineer allow the user to freehand-sketch a part, and then apply constraints such as right angles, to snap the drawing to fit. Then the user is required to add more constraints until the part is well-constrained.

3. *Molecular modeling*: There is often a lock-and-key relationship between a flexible protein molecule's receptor and the ligand that it binds. In addition to geometrically matching the fitted shapes, the surface potentials of the molecules is also important. This fitting problem, called *molecular docking*, is important in *computer-aided drug design*. Generally, a heuristic strategy is used to move the molecules to achieve no overlap between the two molecules while maximizing their contact area. (Ierardi and Park) See

    http://ams.sunysb.edu/~held/proc_usb_comp_geo-95.html

3.8.13 IMPLEMENTATIONS

One major application of geometry is in implementations of geometric software packages, either as *subroutine packages* callable from user programs, or as *standalone systems*, which the user prepares input data files for and directs with either input command files or a GUI.

Definition:

In an object-oriented computer language, a **class library** is a set of new data types and operations on them, activated by sending an **object**, or data item, a message. (For example, a plane object may respond to a message to rotate itself. The internal representation of an object is private, and it may be accessed only by sending it a message.)

Examples:

1. *Leda*, started in 1989, is a major $C++$ class library, whose design goals are correctness, ease of use and elegance, and efficiency [MeNä95]. Its geometry has been moved to CGAL, which often uses exact computation and aims for efficiency in a general-purpose professional-quality library of geometric algorithms written in $C++$ for Unix first, and then PCs. See the website

    http://www.cs.ruu.nl/people/geert/CGAL/

2. *Stand-alone systems*:
   - *Geomview* is an interactive program for viewing and manipulating geometric objects, from the University of Minnesota Geometry Center. Examples like Penrose quasi-tiling, Pascal's theorem in projective conics, Teichmuller space, and families of Riemann surfaces with a specified group of symmetries. See the website

    http://www.geom.umn.edu/apps/gallery.html

   - *Geomamos*, a "Geometric Object Manipulation/Monitoring System" is an X-based visualization system with a 2-D display, GeomSheet, based on X-fig, which allows mouse input. It includes a library based on LEDA. See the website

    http://web.eecs.nwu.edu/~theory/geomamos.html

- *XYZ Geobench* assists the implementation of geometric algorithms. [Sc91]

13.8.14 GEOMETRIC VISUALIZATION

There are many packages to display geometric objects. D. Banks lists many examples, by himself and others, such as D. Cox, G. Francis, and R. Idaszak, including a torus rotating in $\mathcal{R}^4$, a Steiner surface showing the triple point, crosscaps, Klein bottles, a Sudanese surface, a complex reciprocal, a knotted sphere, the Etruscan Venus, a stable mapping of a Klein bottle into 3-dimensional Euclidean space. See the website

    http://www.icase.edu/~banks/math.html

There are also libraries of minimal and other surfaces, and knots, at the Center for Geometry Analysis Numerics and Graphics, U Mass Amherst.

REFERENCES

Printed Resources:

[AgEtal96] P. K. Agarwal, B. Aronov, J. Pach, R. Pollack, and M. Sharir, "Quasi-planar graphs have a linear number of edges", in *Graph Drawing '95*, Lecture Notes in Computer Science 1027, Springer-Verlag, 1996, 1–7.

[AjEtal82] M. Ajtai, V. Chvátal, M. M. Newborn, and E. Szemerédi, "Crossing-free subgraphs", *Ann. Discrete Math.* 12 (1982), 9–12.

[AkAl89] J. Akiyama and N. Alon, "Disjoint simplices and geometric hypergraphs", *Combinatorial Mathematics*, Annals of the New York Academy of Sciences 555 (1989), 1–3.

[AlKl92] N. Alon and D. Kleitman, "Piercing convex sets and the Hadwiger-Debrunner (p,q)-problem", *Advances Math.* 96 (1992), 103–112.

[Al63] E. Altman, "On a problem of Erdős", *Amer. Math. Monthly* (70) 1963, 148–157.

[Ap94] *Aperiodic '94, International Conference on Aperiodic Crystals*, Les Diablerets, Switzerland 18–22 September 1994.

[ArEtal91] B. Aronov, B. Chazelle, H. Edelsbrunner, L. Guibas, M. Sharir, and R. Wenger, "Points and triangles in the plane and halving planes in space", *Discrete Comput. Geom.* 6 (1991), 435–442.

[AtGo93] M. J. Atallah and M. T. Goodrich, "Deterministic Parallel Computational Geometry", *Synthesis of Parallel Algorithms*, J. Reif, ed., Morgan Kaufmann, 1993.

[BaKe92] A. Bachem and W. Kerns, *Linear Programming Duality, An Introduction to Oriented Matroids*, Springer-Verlag, 1992.

[BaDe92] C. Bajaj and T. K. Dey, "Convex decomposition of polyhedra and robustness", *SIAM J. Comput.* 21 (1992), 339-364.

[Bá82] I. Bárány, "A generalization of Carathéodory's theorem", *Discrete Mathematics* 40 (1982), 141–152.

[BáFüLo90] I. Bárány, Z. Füredi, and L. Lovász, "On the number of halving planes", *Combinatorica* 10 (1990), 175–183.

[BáKaPa82] I. Bárány, M. Katchalski, and J. Pach, "Quantitative Helly-type theorems", *Proc. Amer. Math. Soc.* 86 (1982), 109–114.

[BaDoHu95] C. B. Barber, D. P. Dobkin, and H. T. Huhdanpaa, "The Quickhull algorithm for convex hulls", May, 1995, (submitted for publication).

[Be83] J. Beck, "On the lattice property of the plane and some problems of Dirac, Motzkin and Erdős in combinatorial geometry", *Combinatorica* 3 (1983), 281–297.

[BeKu90] A. Bezdek and W. Kuperberg, "Examples of space-tiling polyhedra related to Hilbert's Problem 18, Question 2", in *Topics in Combinatorics and Graph Theory*, R. Bodendiek and R. Henn, eds., Physica-Verlag, 1990, 87–92.

[BjEtal93] A. Björner, M. Las Vergnas, B. Sturmfels, N. White and G. Ziegler, *Oriented Matroids*, Cambridge University Press, 1993. (A comprehensive monograph on the theory of oriented matroids.)

[Bo93] J. Bokowski, *Oriented Matroids*, in [GrWi93].

[BoSt89] J. Bokowski and B. Sturmfels, *Computational Synthetic Geometry*, LNM 1355, Springer-Verlag, 1989.

[BoFü84] E. Boros and Z. Füredi, "The number of triangles covering the center of an n-set", *Geometria Dedicata* 17 (1984), 69–77.

[BuGrSl79] S. Burr, B. Grünbaum, and N. Sloane, "The orchard problem", *Geometria Dedicata* 2 (1979), 397–424.

[CaEtal95] A. R. Calderbank, E. Rains, P. W. Shor, and N. J. A. Sloane, *Quantum Error Correction and Orthogonal Geometry*, 1995, (submitted for publication).

[Ch84] B. Chazelle, "Convex partitions of polyhedra: a lower bound and worst-case optimal algorithm", *SIAM J. Comput.* 13 (1984), 488–507.

[Ch91] B. Chazelle, "Triangulating a simple polygon in linear time", *Discrete Comput. Geom.* 6 (1991), 485–524.

[ChPa90] B. Chazelle and L. Palios, "Triangulating a nonconvex polytope", *Discrete Comput. Geom.* 5 (1990), 505-526.

[ChGoTa95] M. Chrobak, M. Goodrich, and R. Tamassia, "On the volume and resolution of 3-dimensional convex graph drawing", *Electronic Proceedings of the Fifth MSI-Stony Brook Workshop on Computational Geometry*, 20–21 Oct, 1995, http://ams.sunysb.edu/~held/proc_usb_comp_geo-95.html

[ClEtal90] K. Clarkson, H. Edelsbrunner, L. Guibas, M. Sharir, and E. Welzl, "Combinatorial complexity bounds for arrangements of curves and spheres", *Discrete Comput. Geom.* 5 (1990), 99–160.

[CoSl93] J. H. Conway and N. J. A. Sloane, *Sphere packings, lattices and groups*, Springer-Verlag, 1993.

[CsSa93] J. Csima and E. Sawyer, "There exist $6n/13$ ordinary points", *Discrete Comput. Geom.* 9 (1993), 187–202.

[Cs96] G. Csizmadia, "Furthest neighbors in space", *Discrete Mathematics*, 150 (1996), 81–88.

[Da95] K. Daniels, *Containment Algorithms for Nonconvex Polygons with Applications to Layout*, PhD thesis, Harvard University, 1995.

[de93] M. de Berg, *Ray Shooting, Depth Orders and Hidden Surface Removal*, Springer-Verlag, 1993.

[DeEd93] T. K. Dey and H. Edelsbrunner, "Counting triangle crossings and halving planes", *9th Sympos. Comput. Geom.* 1993, 270–273. Also: *Discrete Comput. Geom.* 12 (1994), 281–289.

[Do94] S. E. Dorward, "A survey of object-space hidden surface removal", *International Journal of Computational Geometry and Its Applications* 4 (1994), 325–362.

[Ed87] H. Edelsbrunner, *Algorithms in Combinatorial Geometry*, Springer-Verlag, 1987. (Monograph on combinatorial and algorithmic aspects of point configurations and hyperplane arrangements.)

[EdHa91] H. Edelsbrunner and P. Hajnal, "A lower bound on the number of unit distances between the points of a convex polygon", *J. Combinat. Theory, Ser. A* 56 (1991), 312–316.

[EdSk89] H. Edelsbrunner and S. Skiena, "On the number of furthest neighbor pairs in a point set", *Amer. Math. Monthly* 96 (1989), 614–618.

[El67] P. D. T. A. Elliott, "On the number of circles determined by n points", *Acta Math. Acad. Sci. Hung.* 18 (1967), 181–188.

[Er46] P. Erdős, "On sets of distances of n points", *Amer. Math. Monthly* 53 (1946), 248–250.

[Er60] P. Erdős, "On sets of distances of n points in Euclidean space", *Magyar Tudományos Akadmía Közleményei* 5 (1960), 165–169.

[ErFiFü91] P. Erdős, P. Fishburn, and Z. Füredi, "Midpoints of diagonals of convex n-gons", *SIAM Journal of Discrete Mathematics* 4 (1991), 329–341.

[ErEtal93] P. Erdős, Z. Füredi, J. Pach, and Z. Ruzsa, "The grid revisited", *Discrete Math.* 111 (1993), 189–196.

[ErPa90] P. Erdős and J. Pach, "Variations on the theme of repeated distances", *Combinatorica* 10 (1990), 261–269.

[ErSz35] P. Erdős and G. Szekeres, "A combinatorial problem in geometry", *Comp. Math.* 2 (1935), 463–470.

[FeKu93] G. Fejes Tóth and W. Kuperberg, "Packing and covering with convex sets", *Handbook of Convex Geometry*, P. Gruber and J. Wills, eds., North-Holland, 1993, 799–860.

[Fo93] S. Fortune, "Computational Geometry", *Geometric Computing*, R. Martin, ed., Information Geometers Ltd., 1993.

[Fo96] S. Fortune, "A beam tracing algorithm for prediction of indoor radio propagation", *Proceedings First ACM Workshop on Applied Computational Geometry*, M. C. Lin and D. Manocha, eds., Philadelphia, 27–28 May 1996, 76–81.

[FoGaKe93] R. Fourer, D. M. Gay, and B. W. Kernighan, *AMPL: A Modeling Language for Mathematical Programming*, Duxbury Press/Wadsworth Publishing Co., 1993.

[FrRö86] P. Frankl and V. Rödl, "All triangles are Ramsey", *Trans. Amer. Math. Soc.* 297 (1986), 777–779.

[FrRö90] P. Frankl and V. Rödl, "A partition property of simplices in Euclidean space", *J. Amer. Math. Soc.* 3 (1990), 1–7.

[FrRa94] W. R. Franklin and C. Ray, "Higher isn't necessarily better: visibility algorithms and experiments", *Advances in GIS Research: Sixth International Symposium on Spatial Data Handling*, T. C. Waugh and R. G. Healey, eds., Edinburgh, 5–9 Sept, 1994, 751–770.

[FrEtal94] W. R. Franklin, V. Sivaswami, D. Sun, M. Kankanhalli, and C. Narayanaswami, "Calculating the area of overlaid polygons without constructing the overlay", *Cartography and Geographic Information Systems*, April 1994, 81–89.

[FrAh84] H. Freeman and J. Ahn, "A System for automatic name placement", *4th Jerusalem Conference on Information Technology (JCIT); Next Decade in Information Technology*, 1984, IEEE Computer Society Press, 134-143.

[Fü90] Z. Füredi, "The maximum number of unit distances in a convex n-gon", *J. Combinat. Theory, Ser. A* 55 (1990), 316–320.

[Ga77] M. Gardner, "Mathematical recreations", *Scientific American* 236 (Jan. 1977), 110–121.

[GiO'RSu96] C. Gitlin, J. O'Rourke, and V. Subramanian, "On reconstruction of polyhedra from slices", *Int. J. Comput. Geom. & Appl.* 6(1) (1996), 103–112.

[Go97] J. E. Goodman, *Pseudoline arrangements*, in [GoO'R97].

[GoO'R97] J. E. Goodman and J. O'Rourke, eds., *Handbook of Discrete and Computational Geometry*, CRC Press, 1997. (An extensive, comprehensive reference source in discrete and computational geometry.)

[GrRoSp90] R. Graham, B. Rothschild, and J. Spencer, *Ramsey Theory*, 2nd ed., Wiley, 1990.

[GrWi93] P. M. Gruber and J. M. Wills, eds., *Handbook of Convex Geometry*, Vol. A and B, 555-602, North Holland, 1993.

[Gr56] B. Grünbaum, "A proof of Vázsonyi's conjecture", *Bull. Research Council Israel, Section A* 6 (1956), 77–78.

[Gr67] B. Grünbaum, *Convex Polytopes*, Wiley, 1967.

[Gr72] B. Grünbaum, *Arrangements and Spreads*, CBMS Regional Conf. Ser. in Math., Amer. Math. Soc., Vol. 10, 1972.

[GrSh85] B. Grünbaum and G. C. Shephard, *Tilings and Patterns*, Freeman, 1986.

[GuEtal95] P. Gupta, R. Janardan, J. Majhi, and T. Woo, "Efficient geometric algorithms for workpiece orientation in 4- and 5-axis NC-machining", *Electronic Proceedings of the Fifth MSI-Stony Brook Workshop on Computational Geometry*, 20-21 Oct, 1995,
http://ams.sunysb.edu/~held/proc_usb_comp_geo-95.html.

[Ha94] T. C. Hales, "The status of the Kepler conjecture", *Math. Int.*, 16(3) (1994), 47-58.

[Ha65] S. Hansen, "A generalization of a theorem of Sylvester on lines determined by a finite set", *Math. Scand.* 16 (1965), 175-180.

[Ha80] S. Hansen, "On configurations in 3-space without elementary planes and on the number of ordinary planes", *Math. Scand.* 47 (1980), 181-194.

[Ha74] H. Harborth, "Solution to Problem 664A", *Elemente der Mathematik* 29 (1974), 14-15.

[He23] E. Helly, "Über Mengen konvexer Körper mit gemeinschaftlichen Punkten", *Jber. Deutsch. Math. Verein.* 32 (1923), 175-176.

[Ho96] C. M. Hoffmann, "How solid is solid modeling?", *Proceedings First ACM Workshop on Applied Computational Geometry*, M. C. Lin and D. Manocha, eds., Philadelphia, 27-28 May 1996, 19.

[HoPa34] H. Hopf and E. Pannwitz, "Aufgabe Nr. 167", *Jber. Deutsch. Math. Verein.* 43 (1934), 114.

[IePa95] D. Ierardi and S. Park, "Rigid molecular docking by surface registration at multiple resolutions", *Electronic Proceedings of the Fifth MSI-Stony Brook Workshop on Computational Geometry*, 20-21 Oct, 1995,
http://ams.sunysb.edu/~held/proc_usb_comp_geo-95.html.

[Já92] J. JáJá, *An Introduction to Parallel Algorithms*, Addison-Wesley, 1992.

[Ja87] R. Jamison, "Direction trees", *Discrete Comput. Geom.* 2 (1987), 249-254.

[JeTo95] T. R. Jensen and B. Toft, *Graph Coloring Problems*, Wiley-Interscience, 1995.

[Ka84] G. Kalai, "Intersection patterns of convex sets", *Israel J. Math.* 48 (1984), 161-174.

[KáPaTó96] G. Károlyi, J. Pach, and G. Tóth, "Ramsey-type results for geometric graphs", *12th Sympos. Comput. Geom.*, 1996.

[Ki83] D. Kirkpatrick, "Optimal Search in Planar Subdivisions", *SIAM Journal of Computing* 12 (1983), 28-35.

[Kn92] D. E. Knuth, *Axioms and Hulls*, Lecture Notes in Computer Science 606, Springer-Verlag, 1992.

[Ko93] P. Komjáth, "Set theoretic constructions in Euclidean spaces", in *New Trends in Discrete and Computational Geometry*, J. Pach, ed., Springer-Verlag, 1993.

[Kr46] M. A. Krasnoselskiĭ, "Sur un critère pour qu'un domain soit étoilé", (Russian, with French summary), *Matematicheskiĭ Sbornik, N. S.* 19 (1946), 309–310.

[Ku79] Y. Kupitz, "Extremal Problems in Combinatorial Geometry", *Aarhus University Lecture Notes Series*, 53, Aarhus University, 1979.

[Le83] F. T. Leighton, *Complexity Issues in VLSI, Foundations of Computing Series*, MIT Press, 1983.

[LiMi95] Z. Li and V. Milenkovic, "Compaction and separation algorithms for non-convex polygons and their applications", *European J. Operations Res.* 84 (1995), 539–561.

[LiMa96] M. C. Lin and D. Manocha, eds., *Proceedings First ACM Workshop on Applied Computational Geometry*, Philadelphia, 27–28 May 1996.

[Lo71] L. Lovász, "On the number of halving lines", *Annales Universitatis Scientiarum Budapest, Eötvös, Sectio Mathematica* 14 (1971), 107–108.

[Ma93] J. Matoušek, "Geometric Range Searching", Technical Report, *FB Mathematik und Informatik*, Freie Universität Berlin, 1993.

[Mc80] P. McMullen, "Convex bodies which tile space by translation", *Mathematika* 27 (1980), 113–121.

[MeNä95] K. Mehlhorn and S. Näher, "LEDA: a platform for combinatorial and geometric computing", *Commun. ACM*, 38(1) (1995), 96–102.

[Mi96] J. S. B. Mitchell, "On some application of computational geometry in manufacturing and virtual environments", *Proceedings First ACM Workshop on Applied Computational Geometry*, M. C. Lin and D. Manocha, eds., Philadelphia, 27–28 May 1996, 73–74.

[MiVa92] S. A. Mitchell and S. A. Vavasis, "Quality mesh generation in three dimensions", *Proc. of the ACM Computational Geometry*, 1992, 212–221.

[Mu94] K. Mulmuley, *Computational Geometry: An Introduction Through Randomized Algorithms*, Prentice Hall, 1994.

[O'R87] J. O'Rourke, *Art Gallery Theorems and Algorithms*, Oxford University Press, 1987.

[O'R93] J. O'Rourke, "Computational Geometry Column 18", *International Journal of Computational Geometry and Its Applications* 3 (1993), 107–113.

[O'R94] J. O'Rourke, *Computational Geometry in C*, Cambridge University Press, 1994.

[Pa93] J. Pach, ed., *New Trends in Discrete and Computational Geometry*, Springer-Verlag, 1993.

[PaAg95] J. Pach and P. K. Agarwal, *Combinatorial Geometry*, Wiley, 1995.

[PaShSz94] J. Pach, F. Shahrokhi, and M. Szegedy, "Applications of crossing numbers", *10th ACM Sympos. Comput. Geom.*, 1994, 198–202.

[PaSh98] J. Pach and M. Sharir, "On the number of incidences between points and curves", *Combinatorics, Probability and Computing* 7 (1998), 121–127.

[PaSh90] J. Pach and M. Sharir, "Repeated angles in the plane and related problems", *J. Combinat. Theory, Ser. A* 59 (1990), 12–22.

[PaStSz92] J. Pach, W. Steiger, and M. Szemerédi, "An upper bound on the number of planar k-sets", *Discrete Comput. Geom.* 7 (1992), 109–123.

[PaTö93] J. Pach and J. Törőcsik, "Some gemetric applications of Dilworth's theorem", *9th Sympos. Comput. Geom.*, 1993, 264–269. Also in *Discrete Comput. Geom.* 12 (1994), 1–7.

[PoWe90] R. Pollack and R. Wenger, "Necessary and sufficient conditions for hyperplane transversals", *Combinatorica* 10 (1990), 307–311.

[PrSh85] F. P. Preparata and M. I. Shamos, *Computational Geometry: An Introduction*, Springer-Verlag, 1985.

[Ra95] C. Radin, "Aperiodic tilings in higher dimensions", *Proc. Amer. Math. Soc.* 123/11 (1995), 3543–3548.

[Ra21] J. Radon, "Mengen konvexer Körper, die einen gemeinsamen Punkt enthalten", *Math. Ann.* 83 (1921), 113–115.

[Ra95] V.T. Rajan, "Computational geometry problems in an integrated circuit design and layout tool", *Electronic Proceedings of the Fifth MSI-Stony Brook Workshop on Computational Geometry*, 20–21 Oct, 1995, http://ams.sunysb.edu/~held/proc_usb_comp_geo-95.html.

[Ri96] J. Richter-Gebert, *Realization Spaces of Polytopes*, Lecture Notes in Mathematics, Vol. 1643, Springer-Verlag, 1996.

[RiZi97] J. Richter-Gebert and G. Ziegler, *Oriented Matroids*, in [GoO'R97].

[RuSe92] J. Ruppert and R. Seidel, "On the difficulty of triangulating three-dimensional nonconvex polyhedra", *Discrete Comput. Geom.* 7 (1992), 227-253.

[Sc91] P. Schorn, "Implementing the XYZ GeoBench: A programming environment for geometric algorithms", in *Computational Geometry — Methods, Algorithms and Applications: Proc. Internat. Workshop Comput. Geom. CG '91*, Lecture Notes in Computer Science, 553, Springer-Verlag, 1991, 187–202.

[ShAg95] M. Sharir and P. K. Agarwal, *Davenport-Schinzel Sequences and Their Geometric Applications*, Cambridge University Press, 1995.

[Sh92] T. C. Shermer, "Recent results in art galleries", *Proc. IEEE* 80 (1992), 1384–1399.

[Sk96] V. Skowronski, *Synthesizing Tolerances for Optimal Design Using the Taguchi Quality Loss Function*, PhD thesis, Rensselaer Polytechnic Institute, Feb 1996.

[So94] A. Soifer, "Six-realizable set x_6", *Geombinatorics* 111 (1994), 140–145.

[SpSzTr84] J. Spencer, E. Szemerédi, and W. T. Trotter, "Unit distances in the Euclidean plane", in *Graph Theory and Combinatorics*, B. Bollobás, ed., Academic Press, 1984, 293–303.

[SrVo93] V. Srinivasan and H. B. Voelcker, eds., *Proceedings of the 1993 International Forum on Dimensional Tolerancing and Metrology*, American Society of Mechanical Engineers, Center for Research and Technology Development, and Council on Codes and Standards, 17–19 June, 1993.

[StBrEa95] P. Stucki, J. Bresenham, and R. Earnshaw, eds., "Rapid prototyping technology", special issue of *IEEE Comp. Graphics and Applications*, 15 (1995), 17–55.

[Sz97] L. A. Székely, "Crossing numbers and hard Erdős problems in discrete geometry", *Combinatorics, Probability & Computing*, 7 (1997), 353–358.

[SzTr83] E. Szemerédi and W. T. Trotter, "Extremal problems in discrete geometry", *Combinatorica*, 3 (1983), 381–392.

[TaTo94] R. Tamassia and I. G. Tollis, eds., *Graph Drawing (Proceedings of GD '94)*, Lecture Notes in Computer Science, 894, Springer-Verlag.

[TóVa98] G. Tóth and P. Valtr, "Note on the Erdős-Szekeres Theorem", *Discrete Comput. Geom.* 19 (1998), 457–459.

[Tv66] H. Tverberg, "A generalization of Radon's theorem", *J. London Mathematical Society* 41 (1966), 123–128.

[Un82] P. Ungar, "$2N$ noncollinear points determine at least $2N$ directions", *J. Combinat. Theory, Ser. A* 33 (1982), 343–347.

[We91] Emo Weltz, "Smallest enclosing disks (balls and ellipsoids)", *New Results and New Trends in Computer Science, Lecture Notes in Computer Science*, 555, 1991.

[Ya90] F. F. Yao, "Computational Geometry", in *Handbook of Theoretical Computer Science*, Vol. A, J. van Leeuwen, ed., The MIT Press, Elsevier, 1990.

[Zi94] G. M. Ziegler, *Lectures on polytopes*, Springer-Verlag, 1994.

[Zi96] G. M. Ziegler, *Oriented matroids today: Dynamic survey and updated bibliography*, Electron. J. Combin., 3:DS#4, 1996+.

[ŽiVr91] R. Živaljević and S. Vrećica, "The colored Tverberg's problem and complexes of injective functions", *J. Combin. Theory Ser. A* 61 (1992), 309–318.

Web Resources:

file://ftp.cs.brown.edu/pub/papers/compgeo/gdbiblio.ps.gz. (Algorithms for the drawing of graphs: an annotated bibliography.)

http://ams.sunysb.edu/~held/proc_usb_comp_geo-95.html (*Electronic Proceedings of the Fifth MSI-Stony Brook Workshop on Computational Geometry*, Oct. 20–21, 1995. Contains: M. Chrobak, M. Goodrich, and R. Tamassia, "On the volume and resolution of 3-dimensional convex graph drawing"; P. Gupta, R. Janardan, J. Majhi, and T. Woo, "Efficient geometric algorithms for workpiece orientation in 4- and 5-axis NC-machining"; D. Ierardi and S. Park, "Rigid molecular docking by surface registration at multiple resolutions"; V. T. Rajan, "Computational geometry problems in an integrated circuit design and layout tool".)

http://dimacs.rutgers.edu/techps/1994/94-31.ps (On the Complexity of Some Basic Problems in Computational Convexity: II. Volume and mixed volumes, DIMACS Technical Report 94-31.)

http://hercule.csci.unt.edu/sigact/longrange/contributions.html (SIGACT Long Range Planning Committee, Contributions of Theoretical Computer Science.)

http://web.eecs.nwu.edu/~theory/geomamos.html (GeoMANOS.)

http://www.cs.cmu.edu/~quake/triangle.html (Triangle: A Two-dimensional Quality Mesh Generator and Delaunay Triangulator.)

http://www.cs.duke.edu/~jeffe/compgeom/ (Jeff Erickson's Computational Geometry pages; extensive computational geometry site.)

http://www.cs.hope.edu/~alganim/ccaa/algo.html (The Complete Collection of Animated Algorithms.)

http://www.cs.princeton.edu/~chazelle/taskforce/CGreport.ps (Computational Geometry Impact Task Force, Application Challenges to Computational Geometry.)

http://www.cs.ruu.nl/people/geert/CGAL/ (The CGAL Kernel Manual.)

http://www.cs.sunysb.edu/~algorith/ (The Stony Brook Algorithm Repository; see Section 1.6 on Computational Geometry.)

http://www.geom.umn.edu/~bradb/qhull-news.html (News about QuickHull.)

http://www.geom.umn.edu/apps/gallery.html (Gallery of Interactive Geometry, U. Minnesota Geometry Center.)

http://www.geom.umn.edu/apps/quasitiler/ (QuasiTiler 3.0, U. Minnesota Geometry Center.)

http://www.geom.umn.edu/software.cglist/ (Software repository of The Geometry Center at the University of Minnesota Geometry Center, collected by Nina Amenta.)

http://www.icase.edu/~banks/math.html (Mathematical Visualization.)

http://www.ics.uci.edu/~eppstein/geom.html (David Eppstein's Geometry in Action pages, a collection of many applications of discrete and computational geometry.)

http://www.inria.fr/prisme/personnel/bronnimann/cgt/ (Hervé Brönnimann's CG Tribune, a journal of articles in computational geometry; contains extensive list of software and Web sites in computational geometry.)

http://www.math.tu-berlin.de/diskregeom/polymake/doc/ (Polymake, a tool for the algorithmic treatment of polytopes and polyhedra.)

http://www.research.ibm.com/xw-p4205-description00.html (IBM 3D Interaction Accelerator.)

http://www.utexas.edu/depts/grg/gcraft/notes/mapproj/mapproj.html (Map Projections.)

Chapter 14

CODING THEORY AND CRYPTOLOGY

Alfred J. Menezes and
Paul C. van Oorschot

14.1 Communication Systems and Information Theory
  14.1.1 Basic Concepts
  14.1.2 Entropy
  14.1.3 The Noiseless Coding Theorem
  14.1.4 Channels and Channel Capacity

14.2 Basics of Coding Theory
  14.2.1 Fundamental Concepts
  14.2.2 Maximum Likelihood Decoding
  14.2.3 The Noisy Channel Coding Theorem

14.3 Linear Codes
  14.3.1 Introduction
  14.3.2 Syndrome Decoding
  14.3.3 Constructing New Codes from Old
  14.3.4 Hamming Codes
  14.3.5 Cyclic Codes
  14.3.6 BCH Codes
  14.3.7 Reed-Solomon Codes
  14.3.8 Weight Enumerators

14.4 Bounds for Codes
  14.4.1 Constraints on Code Parameters
  14.4.2 Perfect Codes

14.5 Nonlinear Codes
  14.5.1 Nordstrom-Robinson Code
  14.5.2 Preparata Codes

14.6 Convolutional Codes
  14.6.1 Background
  14.6.2 Shift Registers
  14.6.3 Encoding

14.7 Basics of Cryptography
  14.7.1 Basic Concepts
  14.7.2 Security of Cryptosystems

14.8 Symmetric-Key Systems
 14.8.1 Redundancy
 14.8.2 Substitution and Transposition Ciphers
 14.8.3 Block Ciphers
 14.8.4 Stream Ciphers
 14.8.5 Key Distribution Problem

14.9 Public-Key Systems
 14.9.1 Basic Information
 14.9.2 Knapsack Encryption Scheme
 14.9.3 RSA Cryptosystem
 14.9.4 El Gamal Cryptosystem
 14.9.5 McEliece Encryption Scheme
 14.9.6 Digital Signature Algorithm
 14.9.7 Fiat-Shamir Signature Scheme
 14.9.8 Key Distribution
 14.9.9 Public-Key Certificates
 14.9.10 Authentication
 14.9.11 Secret Sharing
 14.9.12 Hash Functions

INTRODUCTION

This chapter deals with techniques for the efficient, reliable, and secure transmission of data over communications channels that may be subject to non-malicious errors and adversarial intrusion. The general topic areas related to these techniques are information theory, coding theory, and cryptology.

Information theory is concerned with the mathematical theory of communication, and includes the study of redundancy and the underlying limits of communications channels.

Coding theory, in its broadest sense, deals with the translation between source data representations and the corresponding representative symbols used to transmit source data over a communications channels, or store this data. Error-correcting coding is the part of coding theory that adds systematic redundancy to messages to allow transmission errors to not only be detected, but also to be corrected.

Cryptology is the field which includes both cryptography, which deals with the protection of data from malicious or unauthorized actions, and cryptanalysis, which attempts to defeat cryptographic mechanisms.

GLOSSARY

affine cipher: a cipher that replaces the plaintext letter x (represented as the appropriate integer in the set $\{0, 1, \ldots, 25\}$) by $ax + b \bmod 26$, where a and b are integers relatively prime to 26.

analog channel: a channel that is continuous in amplitude and time.

authentication: corroboration that a party, or the origin of a message, is as claimed.

BCH code: a code from a special family of cyclic codes.

binary symmetric channel (BSC): a memoryless channel with binary input and output alphabets, and fixed probability p that a symbol is transmitted incorrectly.

block cipher: a cipher that processes the plaintext after grouping it into pieces called blocks.

burst error: a vector whose only nonzero entries are among a string of successive components, the first and last of which are nonzero.

Caesar cipher: the cipher that shifts each letter forward three positions in the alphabet, wrapping the letters at the end of the alphabet around to the beginning letters.

capacity of a channel: a measure of the ability of a channel to transmit information reliably.

certification authority: a trusted authority who verifies the identity and public key of a party, and signs this data.

chosen-plaintext attack: an attack when the adversary has some chosen plaintext and its corresponding ciphertext.

chosen-ciphertext attack: an attack when the adversary has some chosen ciphertext, and its corresponding plaintext.

cipher: an encryption scheme.

cipher-block chaining (CBC) mode: a mode of operation of an n-bit block cipher in which plaintext is processed n bits at a time, an initialization block is used, and to encrypt each successive n-bit block the bitwise XOR of the block with the encrypted version of the previous block is formed and the resulting n-bit block is encrypted by the block cipher.

cipher feedback (CFB) mode: a mode of operation of an n-bit block cipher in which plaintext may be processed r bits at a time where $1 \leq r \leq n$ and in which ciphertext depends on the current block and previous blocks.

ciphertext: transformed plaintext that is supposed to be unintelligible to all but an authorized recipient.

ciphertext-only attack: an attack when the adversary has possession of some ciphertext and nothing else.

code: a map from the set of words to the set of all finite strings of elements of a designated alphabet.

codeword: a string produced when a code is applied to a word.

coding theory: the subject concerned with the translation between source data representations and the corresponding representative symbols used to transmit source data over a communications channel.

complete maximum likelihood decoding (CMLD): the decoding scheme that decodes a received n-tuple to the unique codeword of minimum distance from this n-tuple, if such a codeword exists. Otherwise, the scheme arbitrarily decodes the n-tuple to one of the codewords closest to this n-tuple.

computational security: the amount of computational effort required by the best currently-known attacks to defeat a system.

convolutional code: a code in which the encoder has memory, so that an n-tuple produced by the encoder not only depends on the message k-tuple u, but also on some message k-tuples produced prior to u.

coset: the set $C + x = \{\, c + x \mid c \in C \,\}$ determined by a word x, given a code C.

coset leader: a coset member of smallest Hamming weight.

cryptanalysis: the science devoted to the defeat of cryptographic protection mechanisms.

cryptography: the science of protecting data from malicious or unauthorized actions.

cryptology: the field that includes both cryptography and cryptanalysis.

cryptosystem (or ***cryptographic system***): a system comprised of a space of plaintext messages, a space of ciphertext messages, a space of keys, and families of enciphering and deciphering functions.

cyclic code: a linear code in which every cyclic shift of a codeword is also a codeword.

data compression: the transformation of data into a representation which is more compact yet maintains the information content of the original data.

data encryption standard (***DES***): a block cipher adopted as a standard in the United States and which is widely used for commercial applications.

data integrity: the ability to detect data manipulation by unauthorized parties.

data origin authentication: corroboration that the origin of data is as claimed.

decryption: the process of recovering plaintext from ciphertext.

digital signature: a number dependent on some secret known only to the signer, and on the message being signed.

dual code (of a code): the orthogonal complement of the code.

electronic codebook (***ECB***) ***mode***: a mode of operation of a n-bit block cipher in which long messages are partitioned into n-bit blocks and encrypted separately.

El Gamal cryptosystem: a public-key cryptosystem based on the discrete logarithm problem.

encryption: the process of mapping plaintext to ciphertext designed to render data unintelligible to all but the intended recipient.

entity authentication: corroboration that a party's identity is as claimed.

entropy: a measure of the amount of information provided by an observation of a random variable.

equivalent codes: codes for which there is a fixed permutation of the coordinate positions which transform one code to the other.

error-correction coding: coding that adds systematic redundancy to messages to allow transmission errors to be detected and corrected.

error-detection coding: coding that adds systematic redundancy to messages to allow transmission errors to be detected (but not necessarily corrected).

extended code: the code obtained by adding a parity check symbol to each codeword of a code.

generator matrix for a code: a matrix whose rows form a basis for that code.

generator polynomial: a monic polynomial of least degree in a cyclic code.

Golay code: a particular perfect code.

Hamming code: a perfect single-error correcting code.

Hamming distance between two n-tuples: the number of coordinate positions in which they differ.

Hamming distance of a code: the smallest Hamming distance over all pairs of distinct codewords in that code.

Hamming weight of an n-tuple: the number of nonzero coordinates.

hash function: a function that maps arbitrary length bit strings to small fixed-length outputs that is easy to compute and in addition may have preimage-resistance, weak collision-resistance, and/or strong collision-resistance.

Hill cipher: a cipher that has an $m \times m$ matrix K as its key and which encrypts plaintext by splitting it into blocks of size m and sending the plaintext block $x = (x_1, x_2, \ldots, x_m)$ to the m-tuple xK.

homophonic substitution: a cipher where plaintext characters in the source language are associated with disjoint sets of ciphertext characters, and each time a character is to be encrypted, one element of the associated set of ciphertext characters is randomly chosen.

incomplete maximum likelihood decoding (IMLD): the decoding scheme that decodes a received n-tuple to a unique codeword such that the distance between the n-tuple and the codeword is a minimum if such a codeword exists. If no such codeword exists, then the scheme reports that errors have been detected, but no correction is possible.

information theory: the mathematical theory of communication concerned with both the study of redundancy and the underlying limits of communication channels.

Kerberos protocol: an authenticated key distribution protocol developed as part of Project Athena at M.I.T. based on symmetric cryptographic techniques and the use of a key distribution center.

key agreement: a key establishment mechanism in which two parties jointly establish a shared secret key which is a function of information contributed by each.

key distribution center (KDC): a trusted third party who distributes short-term secret keys for secure communications from a particular party to another.

key distribution problem: the problem of how to securely distribute secret keys between two or more parties.

key establishment: a mechanism with the specific objective of making a symmetric key secretly available to two authorized parties for subsequent cryptographic use.

key transfer: a key establishment mechanism in which a key created by one party is securely transmitted to another.

knapsack cryptosystem: a cryptosystem in which encryption is carried out using a super-increasing sequence of integers.

known-plaintext attack: an attack when the adversary has some plaintext and its corresponding ciphertext.

linear code: a subspace of the set of n-tuples with entries from a finite field.

McEliece cryptosystem: a public-key cryptosystem based on linear codes from the theory of error-correcting codes.

message: a finite string of source words.

minimum error probability decoding (**MED**): the decoding scheme that decodes a received n-tuple r to a codeword c for which the conditional probability $P(c$ is sent $|$ r is received$)$, $c \in C$, is largest.

memoryless source: a source for which the probability of a particular word being emitted at any point in time is fixed.

modem: a device that transforms between analog channel data and discrete encoder-decoder data; modem is short for modulator/demodulator.

non-repudiation: a provision for the resolution of disputes arising related to digital signatures where the purported sender of a message denies having sent it.

Nordstrom-Robinson code: a special nonlinear code.

one-time pad: a stream cipher where each bit of plaintext is encrypted by XOR-ing it to the next bit of a truly random key, which is never reused for encryption and is of bit length equal to that of the plaintext.

output feedback (**OFB**) ***mode***: a mode of operation of an n-bit block cipher in which a message may be split into blocks of r bits where $1 \leq r \leq n$ for processing and in which error propagation is avoided.

parity check bit: a bit added to a bit string so that the total number of 1s in the extended string is even.

parity check matrix (for a code): a generator matrix for the dual code of the code.

perfect code: a code of distance d for which every word is within distance $t = \lfloor \frac{d-1}{2} \rfloor$ of some codeword.

plaintext: a message in some source language.

polyalphabetic cipher: a cipher that uses multiple substitutions for mapping plaintext letters to ciphertext letters.

Preparata code: a code from an infinite family of nonlinear codes that have efficient encoding and decoding algorithms.

privacy: preventing confidential data from being available in an intelligible form to unauthorized parties.

prefix code: a code in which no codeword is a prefix of another codeword.

provable security (of a cryptographic method): security where the difficulty of defeating the method is essentially as difficult as solving a well-known and supposedly difficult problem.

public-key certificate: data that binds together a party's identification and public key.

public-key cryptosystem: a cryptosystem in which each user has his/her own pair of encryption (public) and decryption (private) keys.

punctured code: the code obtained by removing any column of a generator matrix of a linear code.

Rabin cryptosystem: a public-key cryptosystem whose security depends on the difficulty of finding square roots modulo the product of two large primes.

Reed-Muller code: a code from a particular family of linear codes.

Reed-Solomon code: a code from a special family of BCH codes.

RSA cryptosystem: a public-key cryptosystem in which encryption is based on modular exponentiation with a modulus that is the product of two primes.

secret sharing scheme: a scheme where the contents of a secret can be recovered if and only if particular groups of people sharing information relating to the secret collaborate.

self-dual code: a linear code that is equal to its dual code.

self-orthogonal code: a linear code that is contained in its dual code.

self-synchronizing stream cipher: a stream cipher capable of reestablishing proper decryption automatically after loss of synchronization, with only a fixed number of plaintext characters unrecoverable.

shift cipher: a cipher that replaces each plaintext letter by the letter shifted a fixed number of positions in the alphabet, with letters at the end of the alphabet shifted to the beginning of the alphabet.

shortened code: the set of all codewords in a linear code which are 0 in a fixed coordinate position with that position deleted.

syndrome (of a word x): the vector xH^T, where H a parity check matrix for a linear code C.

stream cipher: a cipher which encrypts individual characters of a plaintext message.

substitution cipher: a cipher that replaces each plaintext character by a fixed substitute according to a permutation on the source alphabet.

super-increasing sequence: a set $\{a_1, a_2, \ldots, a_n\}$ of positive integers with the property that $a_i > \sum_{j=1}^{i-1} a_j$ for each $i = 2, \ldots, n$.

symmetric-key system: a cryptosystem where each pair of users share a secret key.

synchronous stream cipher: a stream cipher in which the keystream is generated independently of the message.

systematic code: a linear code that has a generator matrix of the form $[I_k \mid A]$.

(n,k)-threshold scheme: a scheme whereby a secret datum S can be divided up into n pieces, in such a way that knowledge of any k or more pieces allows S to be easily recovered, but knowledge of $k-1$ or fewer pieces provides no information about S.

transposition cipher: a cipher that divides plaintext into blocks of a fixed size and rearranges the characters in each block according to a fixed permutation.

turbo code: a special type of code built using convolutional codes and an interleaver which permutes the original bits before sending them to the second encoder.

unconditional security (for encryption schemes): the security condition where observation of the ciphertext provides no information to an adversary.

uniquely decodable code: a code for which every string of symbols is the image of at most one message.

Vernam cipher: a one-time pad.

Vigenère cipher: a cipher with a d-tuple $(k_1, \ldots, k_d)$ as its key that encrypts plaintext messages in blocks of size d so that the ith letter in a block is shifted k_i positions in the alphabet, modulo 26.

14.1 COMMUNICATION SYSTEMS AND INFORMATION THEORY

14.1.1 BASIC CONCEPTS

Definitions:

A **communication system**, as illustrated in the following figure, is modeled as a data source providing either continuous or discrete output, a **source encoder** transforming source data into binary digits (**bits**), a **channel encoder**, and a channel.

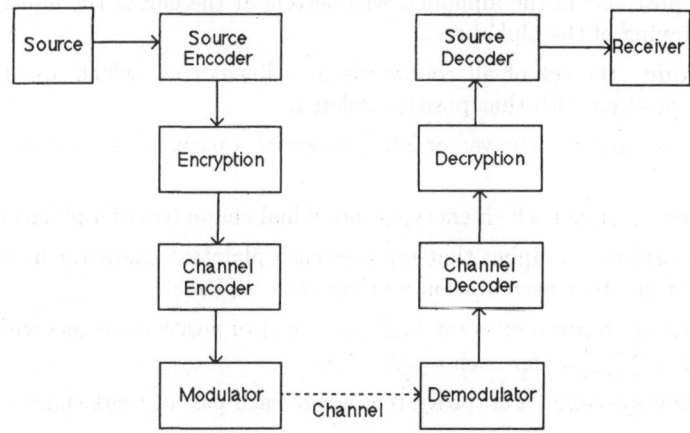

In many communication systems the channel is **analog**, that is, continuous in amplitude and time, in which case a modulator/demodulator (**modem**) is required to transform between analog channel data and discrete encoder/decoder data.

The source encoder, the aim of which is to minimize the number of bits required to represent source data while still allowing subsequent reconstruction, typically includes **data compression** to remove unnecessary redundancy.

The objective of the **channel encoder** is to maximize the rate at which information can be reliably conveyed by the channel, in the presence of disruptive channel noise.

Coding theory is the study of the translation between source data representations and the corresponding representative symbols (coded data) used to transmit source data over a communication channel.

Error-correction coding, located in the channel encoder, adds systematic redundancy to messages to allow transmission errors not only to be detected but also to be corrected.

Encryption, located after the source encoder but not always before the channel encoder, is designed to render data unintelligible to all but the intended recipient, and thereby preserve the **secrecy** of messages in the presence of unfriendly monitoring of the channel.

Cryptography is the science of maintaining secrecy of data, that is, protecting data from malicious or unauthorized actions, including **passive intrusion** (eavesdropping) and **active intrusion** (injection or modification of messages).

Authentication is corroboration that a party, or the origin of a message, is as claimed.

Data integrity is the property that data have not been modified in an unauthorized manner.

Non-repudiation is the preclusion of parties from making undetectable false denials.

14.1.2 ENTROPY

Definitions:

Let X be a random variable that takes on a finite set of values $x_1, x_2, \ldots, x_n$ with probability $P(X = x_i) = p_i$, where $0 \leq p_i \leq 1$ for each i, $1 \leq i \leq n$, and $\sum_{i=1}^{n} p_i = 1$. Also, let Y be a random variable that takes on a finite set of values.

Information theory is concerned with a mathematical theory of communication and includes the study of redundancy and the underlying limits of communication channels.

The ***entropy*** (or ***uncertainty***) of X is defined to be $H(X) = -\sum_{i=1}^{n} p_i \log_2 p_i$, where $p_i \log_2 p_i = 0$ if $p_i = 0$.

The ***joint entropy*** of X and Y is defined to be
$$H(X, Y) = -\sum_{x,y} P(X=x, Y=y) \log_2 P(X=x, Y=y).$$

If X and Y are random variables, the ***conditional entropy of X given $Y = y$*** is
$$H(X \mid Y=y) = -\sum_{x} P(X=x \mid Y=y) \log_2 P(X=x \mid Y=y).$$

The ***conditional entropy of X given Y*** (or ***equivocation of Y about X***), is $H(X \mid Y) = -\sum_{y} P(Y=y) H(X \mid Y=y)$. (The summation indices x and y range over all values of X and Y, respectively.)

Facts:

1. Useful books that cover information theory include [Ha80], [HaHaJo97], [Mc77], [Re94], and [We88].

2. Information theory provides a theoretical basis for many results in error-correcting codes and cryptography, and provides theoretical bounds useful as metrics for evaluating conjectures in both areas.

3. The entropy of X is a measure of the amount of information provided by an observation of X.

4. The entropy of X is also useful for approximating the number of bits required to encode the elements of X.

5. If X and Y are random variables, then:
    - $0 \leq H(X) \leq \log_2 n$;
    - $H(X) = 0$ if and only if $p_i = 1$ for some i, and $p_j = 0$ for all $j \neq i$ (that is, there is no uncertainty of the result);
    - $H(X) = \log_2 n$ if and only if $p_i = \frac{1}{n}$ for each i, $1 \leq i \leq n$ (that is, all outcomes are equally likely);
    - $H(X, Y) \leq H(X) + H(Y)$;
    - $H(X, Y) = H(X) + H(Y)$ if and only if X and Y are independent.

6. The quantity $H(X \mid Y)$ measures the amount of uncertainty remaining about X after Y has been observed.

7. If X and Y are random variables, then:
   - $H(X\,|\,Y) \geq 0$;
   - $H(X\,|\,X) = 0$;
   - $H(X,Y) = H(Y) + H(X\,|\,Y)$;
   - $H(X\,|\,Y) \leq H(X)$;
   - $H(X\,|\,Y) = H(X)$ if and only if X and Y are independent. .

Example:

1. If X is the random variable on the set $\{x_1, x_2, x_3, x_4\}$ with $X(x_1) = 0.4$, $X(x_2) = 0.3$, $X(x_3) = 0.2$, and $X(x_4) = 0.1$, then the entropy of X is
$$H(X) = -(0.4 \log_2 0.4 + 0.3 \log_2 0.3 + 0.2 \log_2 0.2 + 0.1 \log_2 0.1) \approx -1.84644.$$

14.1.3 THE NOISELESS CODING THEOREM

Definitions:

A **source** is a stream of **words** from a set $W = \{w_1, w_2, \ldots, w_M\}$.

Let X_i denote the ith word produced by a source. The source is said to be **memoryless** if for each word $w_j \in W$, the probability $P(X_i = w_j) = p_j$ is independent of i, that is, the X_i are independent and identically distributed random variables.

The **entropy** of a memoryless source is $H = -\sum_{j=1}^{M} p_j \log_2 p_j$.

A **code** is a map f from W to A^*, the set of all finite strings of elements of A where A is a finite set called the **alphabet**.

For each source word $w_j \in W$, the string $f(w_j)$ is a **codeword**.

The **length** of the codeword $f(w_j)$, denoted $|f(w_j)|$, is the number of symbols in the string.

A **message** is any finite string of source words. If $m = v_1 v_2 \ldots v_r$ is a message, then its encoding is obtained by concatenation: $f(m) = f(v_1)f(v_2)\ldots f(v_r)$.

The **average length** of a code f is $\sum_{j=1}^{M} p_j |f(w_j)|$.

A code is **uniquely decodable** if every string from A^* is the image of at most one message.

A **prefix code** is a code such that there do not exist distinct words w_i and w_j such that $f(w_i)$ is an initial segment, or prefix, of $f(w_j)$.

Facts:

1. Prefix codes are uniquely decodable.
2. Prefix codes have the advantage of being *instantaneous*. That is, they can be decoded online without looking at future codewords.
3. *Kraft's inequality*: A prefix code $f: W \to A^*$ with codeword length $|f(w_i)| = l_i$ for $i = 1, 2, \ldots, M$ exists if and only if $\sum_{j=1}^{M} n^{-l_j} \leq 1$, where n is the size of the alphabet A.
4. *Macmillan's inequality*: If a uniquely decodable code $f: W \to A^*$ with codeword lengths $l_1, l_2, \ldots, l_M$ exists, then $\sum_{j=1}^{M} n^{-l_j} \leq 1$, where n is the size of the alphabet A.
5. A uniquely decodable code with prescribed word lengths exists if and only if a prefix code with the same word lengths exists. As a result, attention can be restricted to prefix codes.

6. *Shannon's noiseless coding theorem*: For a memoryless source of entropy H, any uniquely decodable code for the source into an alphabet of size n must have average length at least $\frac{H}{\log_2 n}$. Moreover, there exists such a code having average length less than $1 + \frac{H}{\log_2 n}$.

7. For a memoryless source, a prefix code with smallest possible average length can be constructed by the Huffman coding algorithm. (See §9.1.2.)

Examples:

1. The code that maps the letters A, B, C, D to 1, 01, 001, 0001, respectively, is a prefix code on this set of four letters.

2. The code that maps the letters A, B, C, D to 11, 111, 11111, 111111, respectively, is not a prefix code since the code for A forms the first part of the code for B (and for the codes for C and D as well). It is also not uniquely decodable since a bit string can correspond to more than one string of the letters A, B, C, D. For example, 11111 corresponds to AB, BA, and C.

14.1.4 CHANNELS AND CHANNEL CAPACITY

Definitions:

A *channel* is a medium that accepts strings of symbols from a finite alphabet $A = \{a_1, \ldots, a_n\}$ and produces strings of symbols from a finite alphabet $B = \{b_1, \ldots, b_m\}$.

Let X_i denote the ith input symbol and let Y_i denote the ith output symbol. The channel is said to be *memoryless* if the probability $P(Y_i = b_j \mid X_i = a_k) = p_{jk}$ (for $1 \leq j \leq m$ and $1 \leq k \leq n$) is independent of i.

A *binary symmetric channel* (*BSC*) is a memoryless channel with input and output alphabets $\{0, 1\}$, and probability p that a symbol is transmitted incorrectly. The probability p is called the *symbol error probability* of the channel. See the following figure.

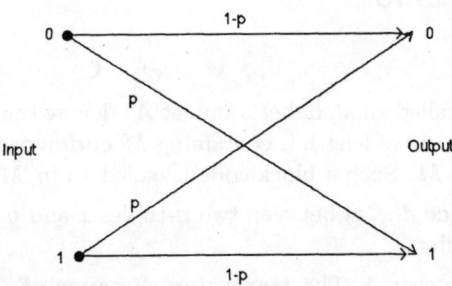

A *q-ary symmetric channel* is a memoryless channel with input and output alphabets each of size q and such that the probability that an error occurs on symbol transmission is a constant p. Furthermore, if an error does occur then each of the $q-1$ symbols different from the correct symbol is equally likely to be received.

The *capacity* of a binary symmetric channel with symbol error probability p is $C(p) = 1 + p \log_2 p + (1 - p) \log_2 (1 - p)$.

Facts:

1. The capacity of a communications channel is a (unitless) measure of its ability to transmit information reliably.

2. The capacity of a BSC with symbol error probability p is a monotone decreasing function of p for $0 \leq p \leq \frac{1}{2}$, with $1 \geq C(p) \geq 0$. Moreover, $C(0) = 1$ and $C(\frac{1}{2}) = 0$.

Example:

1. The capacity of a BSC with symbol error probability 0.01 is given by
$$C(0.01) = 1 + 0.01 \log_2(0.01) + 0.99 \log_2(0.99) \approx 0.92.$$

14.2 BASICS OF CODING THEORY

Coding theory is the subject devoted to the theory of error-correcting codes. Error-correcting codes were invented to correct errors over unreliable transmission links. With digital communications and digital storage media ubiquitous in the modern world, error-correcting codes have grown in importance. Advances in error-correcting codes have made it possible to transmit information across the solar system using weak transmitters and to store data robustly on storage media so that it is resistant to damage, such as scratches on a compact disk.

Error-correcting codes work by encoding data as strings of symbols, such as bit strings, that contain redundant information that helps identify which codeword may have been sent when a string of symbols, potentially different than the string sent, is received. Coding theory is an active area, with new and better codes being devised at a steady pace.

14.2.1 FUNDAMENTAL CONCEPTS

Definitions:

Let A be any finite set (called an *alphabet*), and let A^n denote the set of all n-tuples with entries in A. A **block code** of length n containing M **codewords** over the alphabet A is a subset of A^n of size M. Such a block code is called an $[n, M]$-**code** over A.

The **Hamming distance** $d(x, y)$ between two n-tuples x and $y \in A^n$ is the number of entries in which they differ.

Let C be an $[n, M]$-code over A. The **Hamming distance of** C is the smallest Hamming distance over all pairs of distinct codewords in C. If C has Hamming distance d, then C is sometimes referred to as an $[n, M, d]$-code.

The **information rate** (or **rate**) of an $[n, M]$-code over an alphabet of size q is $R = \frac{\log_q M}{n}$.

Suppose that a codeword c from a block code is transmitted and r is received. The **error-vector** is $e = r - c$ (formed by subtracting componentwise). The **number of errors** is the number of nonzero components in e.

A code is said to **detect** t errors if the decoder is capable of detecting any pattern of t or fewer errors per codeword that may be introduced by the channel.

A code is said to **correct** t errors if the decoder is capable of correcting any pattern of t or fewer errors per codeword that may be introduced by the channel.

If C_1 and C_2 are two $[n, M]$-codes over an alphabet A, then C_1 and C_2 are said to be **equivalent codes** if there is a fixed permutation of the coordinate positions which transform one code to the other.

The **parity check bit** of a bit string is 0 if there are an even number of bits in the string and is 1 if there are an odd number of bits in the string.

Facts:

1. Some of the many introductory-level books in coding theory are [Ba97], [Hi86], [HoEtal92], [Pl89], [Pr92], [Ro96], [Vava89], and [We98]. For more extensive treatments, see [Be84], [Bl83], [LiCo83], [PlHuBr98], [va90], [va99], [PeWe72], and especially [MaSl77], which contains a bibliography of 1478 entries.

2. The Error Correcting Codes (ECC) home page provides free software implementing several important error-correcting codes:
   http://imailab.iis.u-tokyo.ac.jp/~robert/codes.html

3. The main objective of coding theory is the design of codes such that:
   - an efficient algorithm is known for encoding messages;
   - an efficient algorithm is known for decoding;
   - the error-correcting capability of the code is high;
   - the information rate of the code is high.

4. For applications in which a two-way communications channel is available (for example, a telephone circuit), it is sometimes economical to use error detection and retransmission upon error, in a so-called *automatic repeat request* (ARQ) strategy, rather than so-called *forward error correction* (FEC) techniques capable of actually correcting errors at the cost of more complex decoding equipment. This is not an option when the communications channel is effectively one-way or unperturbed source data is not available for retransmission (for example in CD-ROM storage and deep-space communications systems).

5. For any n-tuples $x, y, z \in A^n$, the Hamming distance satisfies the following:
   - $d(x, y) \geq 0$ with equality if and only if $x = y$;
   - $d(x, y) = d(y, x)$;
   - $d(x, y) + d(y, z) \geq d(x, z)$.

6. The information rate R of a block code measures the fraction of information of the code which is non-redundant; the information rate R satisfies the inequality $0 < R \leq 1$.

7. When a word r is received, the decoder must make some decision. This decision may be one of the following:
   - no errors have occurred; accept r as a codeword;
   - errors have occurred; correct r to a codeword c;
   - errors have occurred; no correction is possible.

8. Let C be an $[n, M, d]$-code.
   - If used only for error detection, C can detect $d-1$ errors.
   - If used for error correction, C can correct $\lfloor \frac{d-1}{2} \rfloor$ errors.

9. Equivalent codes have the same distance, and hence the same error-correcting capabilities.

10. Adding a parity check bit to a bit string of length n produces a bit string of length $n+1$ with an even number of 0s.

11. Different families of error-correcting codes have been, and continue to be, designed to meet the particular requirements of applications. One type of requirement is the ability to correct specific types of errors. For example, when signals are sent over radio channels, including those from deep space, interference can produce errors in a run of bits. Similarly, damage to storage media, such as a compact disk, can produce errors that come in clusters. Some of the codes designed to correct errors of these types, known as *burst errors*, are Reed Solomon codes (§14.3.7), interleaved Reed-Solomon codes (see [Vava89] for more information) and fire codes (see [Bl83]).

Examples:

1. The code produced by adding a parity check bit to each bit string of length n can detect a single error. (It detects an odd number of errors, but not an even number of errors; no error correction is possible using this code.) For example, suppose the bit string 0111 is received where the code word sent is a bit string of length three with a parity check bit added. Since 0111 contains three 1s, it cannot be a codeword. Hence, an error was made in transmission. This error cannot be corrected. To see this, note that if exactly one bit error was made in the transmission, any of the codewords 0110, 0101, 0011, and 1111 could have been sent.

2. $C = \{0100011, 1010101, 1101111\}$ is a $[7,3,3]$-code over the binary alphabet. The information rate of C is $R = \frac{\log_2 3}{7} \approx 0.226$.

3. The *binary repetition code* of length n is the code $C = \{00\ldots0, 11\ldots1\}$. The code has distance n, and so can correct $\lfloor \frac{n-1}{2} \rfloor$ errors. If used only for error detection, then C can detect $n-1$ errors. Although the error-correcting capabilities of C are very good, its information rate $R = \frac{1}{n}$ is very poor.

14.2.2 MAXIMUM LIKELIHOOD DECODING

Definitions:

Suppose C is an $[n, M, d]$-code. Different decoding schemes can be used to recover a codeword from a transmitted bit string received with possible errors. These schemes include the following:

- **Minimum Error Probability Decoding (MED)**: If an n-tuple r is received, then correct r to a codeword c for which the conditional probability $P(c$ is sent $|\ r$ is received$)$, $c \in C$, is largest.
- **Incomplete Maximum Likelihood Decoding (IMLD)**: If an n-tuple r is received, and there is a unique codeword $c \in C$ such that $d(r, c)$ is a minimum, then correct r to c. If no such c exists, then report that errors have been detected, but no correction is possible.
- **Complete Maximum Likelihood Decoding (CMLD)**: If an n-tuple r is received, and there is a unique codeword $c \in C$ such that $d(r, c)$ is a minimum, then correct r to c. Otherwise, arbitrarily select one of the codewords $c \in C$ that is the closest to r, and correct r to c.

Facts:

1. For any fixed probability distribution of the source messages, the probability of a decoding error, given that an n-tuple r is received, is minimized by MED among all decoding schemes.

2. MED has the disadvantage that the decoding algorithm depends on the probability distribution of the source messages. The decoding strategy that is used in practice is CMLD.

3. Suppose that the probability that a symbol is transmitted incorrectly in a q-ary symmetric channel is p, where $0 < p < \frac{q-1}{q}$. Let r be a received word and $c_1, c_2 \in C$ with $d(c_1, r) = d_1$ and $d(c_2, r) = d_2$. Let $P(r \mid c)$ denote the probability that r is received, given that c was sent. Then $P(r \mid c_1) \leq P(r \mid c_2)$ if and only if $d_1 \geq d_2$.

4. CMLD chooses a codeword c for which the conditional probability $P(r$ is received $\mid c$ is sent$)$, $c \in C$, is largest.

5. If all source messages are equally likely, then CMLD performs in exactly the same way as MED.

14.2.3 THE NOISY CHANNEL CODING THEOREM

Definitions:

Let C be an $[n, M]$-code, each word occurring with equal probability. Let r_i be the probability of making an incorrect decision using complete maximum likelihood decoding given that the ith codeword was transmitted. The **error probability** of the code C is $P_C = \frac{1}{M} \sum_{j=1}^{M} r_j$.

Let parameters n and M be fixed. Define $P^*(n, M, p)$ to be the smallest error probability P_C of any $[n, M]$-code using a BSC with symbol error probability p.

Facts:

1. *Shannon's noisy channel coding theorem*: Let $C(p)$ denote the capacity of a BSC with symbol error probability p, and define the quantity $M_n = 2^{\lfloor Rn \rfloor}$. If $0 < R < C(p)$, then $P^*(n, M_n, p) \to 0$ as $n \to \infty$.

2. By Shannon's noisy channel coding theorem, arbitrarily reliable communication with a fixed information rate is possible on a channel provided that the information rate is less than the channel capacity. Unfortunately, all proofs of the theorem are non-constructive and hence does specify how to construct such codes. Moreover, the good codes promised by the theorem may have very large word lengths.

14.3 LINEAR CODES

Linear codes are an important type of codes with a particular type of structure. In particular, a linear code is a code that is a subspace of a finite-dimensional vector space over a finite field. The main advantages of using linear codes arise from the efficient procedures for correcting errors. These procedures are based on matrix computations that can be carried out easily and rapidly.

14.3.1 INTRODUCTION

Definitions:

Let F_q^n denote the vector space of all n-tuples having components from the finite field F_q (§5.6.3). The elements of F_q^n are called *vectors* or *words*.

An (n, k)-**linear code** C over F_q is a k-dimensional subspace of F_q^n over F_q. More precisely, C is a **linear block code**, but the qualification "block" is generally omitted. The code C is referred to as an (n, k, d)-code, where n is the **length** of the code, k is the **dimension** of the subspace, and d is the **distance**.

The **Hamming weight** of a word $v \in F_q^N$ is the number of nonzero coordinates in v.

Let C be an (n, k)-code over F_q. A **generator matrix** G for C is a $k \times n$ matrix with entries from F_q whose rows form a basis for C.

If an (n, k)-code C has a generator matrix of the form $G = [I_k \,|\, A]$, then C is called a **systematic code**, and the generator matrix G is said to be in **standard form**.

Let $x = (x_1, x_2, \ldots, x_n)$ and $y = (y_1, y_2, \ldots, y_n)$ be two vectors in F_q^n. The **inner product** of x and y is the field element $x \circ y = \sum_{i=1}^n x_i y_i$. If $x \circ y = 0$, x and y are **orthogonal**.

Let C be an (n, k)-code over F_q. The **orthogonal complement** of C, denoted $C^\perp$ (read "C perp"), is the set of vectors orthogonal to every vector in C:

$$C^\perp = \{\, x \in F_q^n \mid x \circ y = 0 \text{ for all } y \in C \,\}.$$

$C^\perp$ is usually called the **dual code** of C.

A **parity check matrix** for an (n, k)-code C is a generator matrix for $C^\perp$.

A linear code C is **self-orthogonal** if $C \subseteq C^\perp$. It is **self-dual** if $C = C^\perp$.

Facts:

1. Round parentheses (used to denote an (n, k)-code or an (n, k, d)-code) denote that a code is linear, while square brackets (used to denote an $[n, M]$-code or an $[n, M, d]$-code as defined in §14.2.1) are used for all codes, linear or not.

2. An (n, k)-code over F_q, the finite field of q elements, is an $[n, q^k]$-block code.

3. The information rate of an (n, k)-code is $R = \frac{k}{n}$.

4. The distance of a linear code C is the minimum Hamming weight of a nonzero vector in C.

5. A linear code is often described by its generator matrix.

6. A linear code can have many different generator matrices.

7. If G is a generator matrix for a code, then any matrix obtained from G by applying a sequence of elementary row operations is also a generator matrix for that code.

8. Let C be an (n, k)-code over F_q. Then there exists an equivalent code C' with generator matrix $[I_k \,|\, A]$, where I_k is the $k \times k$ identity matrix, and A is a $k \times (n-k)$ matrix with entries from F_q.

9. If G is a generator matrix for an (n, k)-code C, then $C = \{\, mG \mid m \in F_q^k \,\}$. The source messages can be taken to be the elements of F_q^k, and hence encoding is simply multiplication by G. Systematic codes are advantageous because if G is in standard form and $c = mG$ is the codeword corresponding to a message m, then the first k components of c are identically m.

> **Algorithm 1:** Constructing a parity check matrix H from a generator matrix G.
>
> $G' :=$ the reduced row echelon form of G {use elementary row operations}
>
> $A :=$ the $k \times (n-k)$ matrix obtained from G' by deleting the leading columns of G'
>
> $H :=$ the $(n-k) \times n$ matrix H obtained by placing, in order, the rows of $-A$ in the columns of H which correspond to the leading columns of G', and placing in the remaining $n-k$ columns of H, in order, the columns of the $(n-k) \times (n-k)$ identity matrix I_{n-k}

10. If C is an (n,k)-code over F_q, then $C^\perp$ is an $(n, n-k)$-code over F_q.

11. If C is an (n,k)-code over F_q, then the dual code of $C^\perp$ is C itself.

12. An interesting and useful way to describe an (n,k)-code is in terms of $C^\perp$.

13. There are many important special types and families of linear codes, including Hamming codes (§14.3.4), Golay codes (§14.4.2), Reed-Muller codes (see Chapter 4 in [Vava89] for details) and cyclic codes (§14.3.5). Among cyclic codes, BCH codes form an important class (§14.3.6) and among BCH codes there is an important class of codes known as Reed-Solomon codes (§14.3.7).

14. Reed-Muller codes were used by the Mariner 9 spacecraft on its mission to Mars. A Golay code was used by the Voyager 2 on its mission to Jupiter and Saturn. A Reed-Solomon code was used by the Voyager 2 on its mission to Uranus. (See [Vava89] for more details on these applications.)

15. Algorithm 1 uses linear algebra to construct a parity check matrix for a linear code from a generator matrix.

16. *Parity check matrices*: Let C be an (n,k)-code over F_q with a generator matrix G, and let H be a parity check matrix for C.

    - A vector $x \in F_q^n$ belongs to C if and only if $xH^T = 0$; it follows that $GH^T = 0$.
    - If $G = [I_k \,|\, A]$ is a generator matrix for C, then $H = [-A^T \,|\, I_{n-k}]$ is a parity check matrix for C.
    - C has distance at least s if and only if every set of $s-1$ columns of H are linearly independent over F_q; in other words, the distance of C is equal to the smallest number of columns of H that are linearly dependent over F_q.

17. Let C be an (n,k)-code with generator matrix G. C is self-orthogonal if and only if $GG^T = 0$.

18. Let C be an (n,k)-code with generator matrix G. C is self-dual if and only if it is self-orthogonal and $k = \frac{n}{2}$ (and hence n is even).

Examples:

1. Let C be a binary $(7,4)$-code with generator matrix

$$G = \begin{pmatrix} 0 & 0 & 1 & 0 & 1 & 0 & 1 \\ 1 & 1 & 0 & 0 & 1 & 0 & 1 \\ 0 & 0 & 1 & 0 & 0 & 1 & 1 \\ 1 & 1 & 1 & 0 & 1 & 1 & 1 \end{pmatrix}.$$

Elementary row operations yields the reduced row echelon form of G:
$$G' = \begin{pmatrix} 1 & 1 & 0 & 0 & 0 & 1 & 0 \\ 0 & 0 & 1 & 0 & 0 & 1 & 0 \\ 0 & 0 & 0 & 0 & 1 & 1 & 0 \\ 0 & 0 & 0 & 0 & 0 & 0 & 1 \end{pmatrix}.$$

The leading columns of G' are columns 1, 3, 5 and 7, and
$$A = \begin{pmatrix} 1 & 0 & 1 \\ 0 & 0 & 1 \\ 0 & 0 & 1 \\ 0 & 0 & 0 \end{pmatrix}.$$

Hence, the following parity check matrix is obtained
$$H = \begin{pmatrix} 1 & 1 & 0 & 0 & 0 & 0 & 0 \\ 0 & 0 & 0 & 1 & 0 & 0 & 0 \\ 1 & 0 & 1 & 0 & 1 & 1 & 0 \end{pmatrix}.$$

2. The *extended Hamming code* of order 3 is a binary $(8,4,4)$-code with generator matrix
$$G = \begin{pmatrix} 1 & 0 & 0 & 0 & 1 & 1 & 0 & 1 \\ 0 & 1 & 0 & 0 & 1 & 0 & 1 & 1 \\ 0 & 0 & 1 & 0 & 0 & 1 & 1 & 1 \\ 0 & 0 & 0 & 1 & 1 & 1 & 1 & 0 \end{pmatrix}.$$

The code is self-dual since $GG^T = 0$.

14.3.2 SYNDROME DECODING

Syndrome decoding is a general decoding technique for linear codes that is useful if the information rate of the code is high. Let C be an (n,k,d)-code over F_q with parity check matrix H.

Definitions:

For any $x \in F_q^n$, the *coset* of C determined by x is the set $C + x = \{c + x \mid c \in C\}$.

For any $x \in F_q^n$, the *syndrome* of x is the vector xH^T.

A *coset leader* of a coset of C is one of the coset members of smallest weight.

Facts:

1. The coset determined by 0 is C.
2. For all $x \in F_q^n$, $x \in C + x$.
3. For all $x, y \in F_q^n$, if $y \in C + x$, then $C + y = C + x$, that is, each word in a coset determines that coset.
4. The cosets of C partition F_q^n into q^{n-k} cosets, each of size q^k.
5. A syndrome is a vector of length $n - k$.
6. Two vectors x_1 and $x_2 \in F_q^n$ are in the same coset of C if and only if they have the same syndrome, that is, $x_1 H^T = x_2 H^T$.
7. A vector $x \in F_q^n$ is a codeword if and only if its syndrome is 0.

> **Algorithm 2: Syndrome decoding for linear codes.**
>
> *precomputation*: set up a one-to-one correspondence between coset leaders and syndromes; let r be a received word and H the parity check matrix.
>
> compute the syndrome $s = rH^T$ of r
> find the coset leader e associated with s
> correct r to $r - e$

8. Suppose that a codeword c is transmitted and r is received. If $e = r - c$, then $rH^T = eH^T$, which means that the error-vector is in the same coset as the received word. By maximum likelihood decoding, the decoder should choose a vector of smallest weight in this coset as the error vector.

9. The fact that there is a one-to-one correspondence between syndromes and coset leaders leads to *syndrome decoding*, a decoding algorithm for linear codes, which is described as Algorithm 2.

Example:

1. Consider the binary $(5,2)$-code C with generator matrix
$$G = \begin{pmatrix} 1 & 0 & 0 & 0 & 1 \\ 0 & 1 & 1 & 1 & 1 \end{pmatrix}$$
and parity check matrix
$$H = \begin{pmatrix} 0 & 1 & 1 & 0 & 0 \\ 0 & 1 & 0 & 1 & 0 \\ 1 & 1 & 0 & 0 & 1 \end{pmatrix}.$$
The 8 cosets of C are

$\{00000, 10001, 01111, 11110\}$ $\{10000, 00001, 11111, 01110\}$
$\{01000, 11001, 00111, 10110\}$ $\{00100, 10101, 01011, 11010\}$
$\{00010, 10011, 01101, 11100\}$ $\{11000, 01001, 10111, 00110\}$
$\{10100, 00101, 11011, 01010\}$ $\{01100, 11101, 00011, 10010\}$

The following is a list of coset leaders and their syndromes:

coset leader	00000	10000	01000	00100	00010	11000	10100	01100
syndrome	000	001	111	100	010	110	101	011

If the word $r = 01101$ is received, compute the syndrome $01101 \cdot H^T = 010$, which corresponds to a coset leader $e = 00010$. Hence, r is corrected to $r - e = 01111$.

14.3.3 CONSTRUCTING NEW CODES FROM OLD

There are several methods for modifying a linear code to produce a new linear code. Some of these methods are extending a code, puncturing a code, and shortening a code.

Definitions:

If C is a linear code of length n over the field F_q, then the **extended code** $\overline{C}$ of C is $\overline{C} = \{ (c_1, c_2, \ldots, c_n, c_{n+1}) \mid (c_1, c_2, \ldots, c_n) \in C, \sum_{i=1}^{n+1} c_i = 0 \}$. The symbol c_{n+1} is called the **overall parity check symbol**.

If C is a linear code over F_q, the code obtained by removing any column of a generator matrix of C is called a **punctured** C, denoted C^*.

If C is a linear code of length n, a **shortened code** C' of C is a linear code of length $n-1$ which equals the set of all codewords in C having 0 in a fixed coordinate position, with that position deleted.

Facts:
1. If C is an (n, k, d)-code over F_q with generator matrix G and parity check matrix H, then:
    - $\overline{C}$ is an $(n+1, k, \overline{d})$-code over F_q;
    - if C is a binary code, then $\overline{d} = \begin{cases} d, & \text{if } d \text{ is even} \\ d+1, & \text{if } d \text{ is odd}; \end{cases}$
    - a generator matrix for $\overline{C}$ is $\overline{G}$, which is obtained by adding a column to G in such a way that the sum of the elements of each row of $\overline{G}$ is 0;
    - a parity check matrix for $\overline{C}$ is $\overline{H}$, where $\overline{H} = \begin{pmatrix} 1 & 1 & 1 & 1 & \cdots & 1 \\ & & & & & 0 \\ & & H & & & 0 \\ & & & & & \vdots \\ & & & & & 0 \end{pmatrix}$.

2. Puncturing a code is the reverse process to extending a code.
3. If C is an (n, k, d)-code over F_q, then C^* is a linear code over F_q of length $n-1$, dimension k or $k-1$, and distance d or $d-1$.
4. If C is an (n, k, d)-code over F_q, $k \geq 2$, and C has at least one codeword for which the deleted position has a nonzero entry, then C' is an $(n-1, k-1, d')$-code over F_q, with $d' \geq d$.

14.3.4 HAMMING CODES

Definition:

A **Hamming code** of order r over F_q, denoted $H_r(q)$, is an (n, k)-code where $n = \frac{q^r - 1}{q - 1}$ and $k = n - r$, with a parity check matrix whose columns are nonzero and such that no two columns are scalar multiples of each other.

Facts:
1. A decoding algorithm for Hamming codes is shown in Algorithm 3.
2. In the binary case ($q = 2$), the Hamming code $H_r(2)$ has a parity check matrix whose columns consist of all nonzero binary vectors of length r, each used exactly once.
3. $H_r(q)$ has distance 3, and so is a 1-error correcting code.
4. Any two binary Hamming codes of order r are equivalent.
5. $H_r(q)$ is a perfect code (§14.4.2).

Example:
1. Consider $H_3(2)$, the binary Hamming code of order 3. The code has length $n = 7$ and dimension $k = 4$, and a parity check matrix is

$$H = \begin{pmatrix} 1 & 0 & 0 & 1 & 1 & 0 & 1 \\ 0 & 1 & 0 & 1 & 0 & 1 & 1 \\ 0 & 0 & 1 & 0 & 1 & 1 & 1 \end{pmatrix}.$$

If the received word is $r = 1011101$, compute the syndrome $s = 1011101 \cdot H^T = 001$, which is the third column of H. Hence $e = 0010000$, and correct r to 1001101.

Algorithm 3: Decoding algorithm for Hamming codes.

$H :=$ a parity check matrix for a Hamming code $H_r(q)$
$r :=$ a received word
compute the syndrome $s = rH^T$ of r
if $s = 0$ **then** accept r as the transmitted word
else
    compare s^T with the columns of H
    if $s^T = \alpha h_i$ (where h_i is the ith column of H) **and** $\alpha \in F_q$ **then**
        the error vector e is the vector with α in position i and 0s elsewhere
        correct r to $c = r - e$

14.3.5 CYCLIC CODES

Definitions:

A linear code C of length n is **cyclic** if whenever $(a_0, a_1, a_2, \ldots, a_{n-1})$ is a codeword in C, then the cyclic shift $(a_{n-1}, a_0, a_1, \ldots, a_{n-2})$ is also a codeword in C.

Let $g(x)$ be a polynomial in $F_q[x]/(x^n - 1)$. The ideal generated by $g(x)$, namely $\{ a(x)g(x) \mid a(x) \in F_q[x]/(x^n - 1) \}$, is called the **code generated by** $g(x)$, and denoted $\langle g(x) \rangle$.

Let C be a nonzero cyclic code in $F_q[x]/(x^n - 1)$. A monic polynomial $g(x)$ of least degree in C is called a **generator polynomial** of C.

The polynomial $h(x) = \dfrac{x^n - 1}{g(x)}$ is called the **check polynomial** of C.

Let H be a parity check matrix for a cyclic code. If r is a received word, the **syndrome polynomial** of r is the polynomial $s(x)$ corresponding to the syndrome $s = rH^T$.

Facts:

1. The study of cyclic codes is facilitated by the attachment of some additional algebraic structure to the vector space F_q^n.

2. If the vector $(a_0, a_1, a_2, \ldots, a_{n-1})$ in F_q^n is identified with the polynomial $a_0 + a_1 x + a_2 x^2 + \cdots + a_{n-1} x^{n-1}$, then:

   - the ring $F_q[x]/(x^n - 1)$ can be viewed as a vector space over F_q;
   - the vector spaces F_q^n and $F_q[x]/(x^n - 1)$ are isomorphic;
   - multiplication of a polynomial in $F_q[x]/(x^n - 1)$ by x corresponds to a cyclic shift of the corresponding vector;
   - a linear code C in the vector space F_q^n is cyclic if and only if C is an ideal in the ring $F_q[x]/(x^n - 1)$.

3. An ideal may contain many elements which will generate the ideal. One of these generators is singled out as *the* generator.

4. If $g(x)$ is a generator polynomial of a cyclic code C, then $g(x)$ generates C; that is, $\langle g(x) \rangle = C$.

5. The following are consequences of the fact that the ring $F_q[x]/(x^n - 1)$ is a principal ideal domain (§5.4.5). Here C is a nonzero cyclic code in $F_q[x]/(x^n - 1)$ with generator polynomial $g(x)$.
   - the generator polynomial of C is unique;
   - $g(x)$ divides $x^n - 1$ in $F_q[x]$;
   - if the degree of $g(x)$ is $n-k$, that is, $g(x) = g_0 + g_1 x + g_2 x^2 + \cdots + g_{n-k} x^{n-k}$ (and $g_{n-k} = 1$), then a basis for C is $\{g(x), xg(x), x^2 g(x), \ldots, x^{k-1} g(x)\}$; hence C has dimension k and a generator matrix for C is

$$\begin{pmatrix} g_0 & g_1 & g_2 & \cdots & \cdots & g_{n-k} & 0 & 0 & \cdots & 0 \\ 0 & g_0 & g_1 & \cdots & \cdots & g_{n-k-1} & g_{n-k} & 0 & \cdots & 0 \\ 0 & 0 & g_0 & \cdots & \cdots & g_{n-k-2} & g_{n-k-1} & g_{n-k} & \cdots & 0 \\ \vdots & \vdots & & \ddots & & & \ddots & \ddots & \ddots & \vdots \\ 0 & 0 & \cdots & 0 & g_0 & & \cdots & & & g_{n-k} \end{pmatrix}.$$

6. Any $c(x) \in C$ can be written uniquely as $c(x) = f(x)g(x)$ in the ring $F_q[x]$, where $f(x) \in F_q[x]$ has degree less than k. Hence, encoding a message polynomial $f(x)$ consists simply of polynomial multiplication by $g(x)$.

7. The dual code $C^\perp$ is also cyclic.

8. Let $h(x) = h_0 + h_1 x + h_2 x^2 + \cdots + h_k x^k = \dfrac{x^n - 1}{g(x)}$ in $F_q[x]$. Then the *reciprocal polynomial* $h^*(x) = x^k h(\frac{1}{x})$ of $h(x)$ is a generator of $C^\perp$. (In fact, $(\frac{1}{h_0})h^*(x)$ is the generator polynomial of $C^\perp$.) Hence, a parity check matrix for C is

$$\begin{pmatrix} h_k & h_{k-1} & h_{k-2} & \cdots & \cdots & h_0 & 0 & 0 & \cdots & 0 \\ 0 & h_k & h_{k-1} & \cdots & \cdots & h_1 & h_0 & 0 & \cdots & 0 \\ 0 & 0 & h_k & \cdots & \cdots & h_2 & h_1 & h_0 & \cdots & 0 \\ \vdots & \vdots & & \ddots & & & \ddots & \ddots & \ddots & \\ 0 & 0 & \cdots & 0 & h_k & & \cdots & & & h_0 \end{pmatrix}.$$

9. A cyclic code of length n over F_q is characterized by its generator polynomial.

10. There is a one-to-one correspondence between cyclic codes in F_q^n and monic polynomials in $F_q[x]$ which divide $x^n - 1$.

11. Table 1 gives the complete factorization of $x^n - 1$ over F_2 for some small values of odd n.

12. If C is an (n, k)-cyclic code generated by $g(x)$, then another parity check matrix for C is the matrix H whose ith column is $x^i \bmod g(x)$, for $i = 0, 1, \ldots, n - 1$.

13. If $r(x)$ is the polynomial corresponding to the received word r, then the syndrome polynomial of r is simply $s(x) = r(x) \bmod g(x)$.

Example:

1. Over F_2, the factorization of $x^7 - 1$ is $x^7 - 1 = (1 + x)(1 + x + x^3)(1 + x^2 + x^3)$. The monic divisors of $x^7 - 1$ are:

Table 1 Factorization of $x^n - 1$ over F_2, n odd, $1 \leq n \leq 31$.

n	factorization of $x^n - 1$ over F_2
1	$1 + x$
3	$(1+x)(1+x+x^2)$
5	$(1+x)(1+x+x^2+x^3+x^4)$
7	$(1+x)(1+x+x^3)(1+x^2+x^3)$
9	$(1+x)(1+x+x^2)(1+x^3+x^6)$
11	$(1+x)(1+x+x^2+\cdots+x^{10})$
13	$(1+x)(1+x+x^2+\cdots+x^{12})$
15	$(1+x)(1+x+x^2)(1+x+x^2+x^3+x^4)(1+x+x^4)(1+x^3+x^4)$
17	$(1+x)(1+x+x^2+x^4+x^6+x^7+x^8)(1+x^3+x^4+x^5+x^8)$
19	$(1+x)(1+x+x^2+\cdots+x^{18})$
21	$(1+x)(1+x+x^2)(1+x^2+x^3)(1+x+x^3)(1+x^2+x^4+x^5+x^6)$ $(1+x+x^2+x^4+x^6)$
23	$(1+x)(1+x+x^5+x^6+x^7+x^9+x^{11})(1+x^2+x^4+x^5+x^6+x^{10}+x^{11})$
25	$(1+x)(1+x+x^2+x^3+x^4)(1+x^5+x^{10}+x^{15}+x^{20})$
27	$(1+x)(1+x+x^2)(1+x^3+x^6)(1+x^9+x^{18})$
29	$(1+x)(1+x+x^2+\cdots+x^{28})$
31	$(1+x)(1+x^2+x^5)(1+x^3+x^5)(1+x+x^2+x^3+x^5)(1+x+x^2+x^4+x^5)$ $(1+x+x^3+x^4+x^5)(1+x^2+x^3+x^4+x^5)$

$$g_1(x) = 1$$
$$g_2(x) = 1 + x$$
$$g_3(x) = 1 + x + x^3$$
$$g_4(x) = 1 + x^2 + x^3$$
$$g_5(x) = (1+x)(1+x+x^3) = 1 + x^2 + x^3 + x^4$$
$$g_6(x) = (1+x)(1+x^2+x^3) = 1 + x + x^2 + x^4$$
$$g_7(x) = (1+x+x^3)(1+x^2+x^3) = 1 + x + x^2 + x^3 + x^4 + x^5 + x^6$$
$$g_8(x) = 1 + x^7$$

The polynomial $g_5(x)$ generates the binary $(7, 3)$-cyclic code
$$C = \{0000000, 1011100, 0101110, 0010111, 1001011, 1100101, 1110010, 0111001\}.$$

A generator matrix for C is
$$G = \begin{pmatrix} 1 & 0 & 1 & 1 & 1 & 0 & 0 \\ 0 & 1 & 0 & 1 & 1 & 1 & 0 \\ 0 & 0 & 1 & 0 & 1 & 1 & 1 \end{pmatrix}.$$

A parity check matrix for C is
$$H = \begin{pmatrix} 1 & 1 & 0 & 1 & 0 & 0 & 0 \\ 0 & 1 & 1 & 0 & 1 & 0 & 0 \\ 0 & 0 & 1 & 1 & 0 & 1 & 0 \\ 0 & 0 & 0 & 1 & 1 & 0 & 1 \end{pmatrix}.$$

Algorithm 4: Decoding algorithm for BCH codes.

Suppose a codeword c is transmitted and r is received.

Compute $S_j = r(\beta^{a+j})$ for $j = 0, 1, \ldots, \delta - 2$, and form the polynomial $S(z) = \sum_{j=0}^{\delta-2} S_j z^j$.

Use the extended Euclidean algorithm to calculate the greatest common divisor of $S(z)$ and $z^{\delta-1}$ in the ring $F_{q^m}[z]$; stop as soon as the remainder $r_i(z)$ has degree $< \frac{\delta-1}{2}$; this yields polynomials $s_i(z)$ and $t_i(z)$ such that $s_i(z)z^{\delta-1} + t_i(z)S(z) = r_i(z)$; $\sigma(z) := t_i(z)$; $w(z) := r_i(z)$

Find B, the set of roots of $\sigma(z)$ in F_{q^m} {the roots will actually lie in the subgroup of $F_{q^m}^*$ generated by β}

For each $\gamma \in B$, set $E_\gamma = \frac{-\gamma^{-1} w(\gamma)}{\sigma'(\gamma)}$, where $\sigma'(z)$ denotes the formal derivative of $\sigma(z)$.

The error vector is $e = (e_0, e_1, \ldots, e_{n-1})$, where $e_i = \begin{cases} 0, & \text{if } \beta^{-i} \notin B, \\ E_\gamma, & \text{if } \beta^{-i} = \gamma \in B \end{cases}$

decode r to $r - e$

{it is assumed that the number of errors is $l \leq \lfloor \frac{\delta-1}{2} \rfloor$; if the number of errors is such, then the decoding is correct}

{there are more efficient ways of obtaining $\sigma(z)$ and $w(z)$ than by using the Euclidean algorithm, for example by using the Berlekamp-Massey algorithm (see [MevaVa96])}

14.3.6 BCH CODES

Definitions:

Let β be a primitive nth root of unity in an extension field of F_q. Let $g(x)$ be the least common multiple of the minimal polynomials over F_q of $\beta^a, \beta^{a+1}, \ldots, \beta^{a+\delta-2}$ where a is an integer. The cyclic code of length n over F_q with generator polynomial $g(x)$ is called a **BCH code** (after its discoverers: R. C. Bose, D. Ray-Chaudhuri, and A. Hocquenghem) with **designed distance** δ.

If $a = 1$ in the definition of a BCH code, the code is called **narrow-sense**. If $n = q^m - 1$ for some positive integer m (that is, β is primitive in F_{q^m}), the code is **primitive**.

Facts:

1. BCH codes are special types of cyclic codes, discovered by A. Hocquenghem in 1959 and independently by R. C. Bose and D. K. Ray-Chaudhuri in 1960.

2. **BCH bound:** Let C be a BCH code over F_q with designed distance δ. Then C has distance at least δ.

3. Algorithm 4 is one method for decoding BCH codes. In the algorithm, $g(x)$ be a generator polynomial for a BCH code over F_q of designed distance δ and length n. Hence $g(x) = \text{lcm}\{ m_i(x) \mid a \leq i \leq a + \delta - 2 \}$, where $m_i(x)$ is the minimal polynomial of β^i over F_q, and β is a primitive nth root of unity in an extension field F_{q^m}.

Table 2 Elements of F_{3^3} as powers of α, where α is a root of $f(x) = 1 + 2x^2 + x^3$.

i	α^i	i	α^i	i	α^i
0	1	9	$2 + 2\alpha + 2\alpha^2$	18	$1 + \alpha$
1	α	10	$1 + 2\alpha + \alpha^2$	19	$\alpha + \alpha^2$
2	α^2	11	$2 + \alpha$	20	$2 + 2\alpha^2$
3	$2 + \alpha^2$	12	$2\alpha + \alpha^2$	21	$1 + 2\alpha + 2\alpha^2$
4	$2 + 2\alpha + \alpha^2$	13	2	22	$1 + \alpha + \alpha^2$
5	$2 + 2\alpha$	14	2α	23	$2 + \alpha + 2\alpha^2$
6	$2\alpha + 2\alpha^2$	15	$2\alpha^2$	24	$1 + 2\alpha$
7	$1 + \alpha^2$	16	$1 + 2\alpha^2$	25	$\alpha + 2\alpha^2$
8	$2 + \alpha + \alpha^2$	17	$1 + \alpha + 2\alpha^2$		

Examples:

1. Consider the finite field F_{3^3} generated by a root α of the primitive polynomial $f(x) = 1 + 2x^2 + x^3 \in F_3[x]$. A table of powers of α is given in Table 2.

   The element $\beta = \alpha^2$ is a primitive 13th root of unity in F_{3^3}. If $m_i(x)$ denotes the minimal polynomial of β^i over F_3, then

   $m_0(x) = 2 + x$
   $m_1(x) = 2 + 2x + 2x^2 + x^3$
   $m_2(x) = 2 + 2x + x^3$
   $m_4(x) = 2 + x + 2x^2 + x^3$
   $m_7(x) = 2 + x^2 + x^3$.

   Since $m_1(x) = m_3(x)$, the polynomial
   $$g(x) = \text{lcm}(m_0(x), m_1(x), m_2(x), m_3(x)) = m_0(x)m_1(x)m_2(x)$$
   has among its roots the elements β^0, β^1, β^2, and β^3. Hence $g(x)$ is a generator polynomial for a BCH code over F_3 of designed distance $\delta = 5$ and length $n = 13$.

2. Using the BCH code in Example 1, suppose that the decoder received the word $r = (220\ 021\ 110\ 2110)$. The following steps follow Algorithm 4 to decode r:

   - Compute $S_0 = r(\beta^0) = 1$, $S_1 = r(\beta^1) = \alpha^{14}$, $S_2 = r(\beta^2) = \alpha^{23}$, and $S_3 = r(\beta^3) = \alpha^{16}$. This gives $S(z) = 1 + \alpha^{14}z + \alpha^{23}z^2 + \alpha^{16}z^3$.
   - Applying the extended Euclidean algorithm in $F_{3^3}[z]$ to $S(z)$ and z^4 yields:

i	$s_i(z)$	$t_i(z)$	$r_i(z)$	$\deg r_i(z)$
-1	1	0	z^4	4
0	0	1	$1 + \alpha^{14}z + \alpha^{23}z^2 + \alpha^{16}z^3$	3
1	1	$\alpha^{17} + \alpha^{23}z$	$\alpha^{17} + \alpha^{16}z + \alpha^{13}z^2$	2
2	$\alpha^3 + \alpha^{16}z$	$\alpha^{15} + \alpha^3 z + \alpha^{13}z^2$	$\alpha^{15} + \alpha^{16}z$	1

   Stop, since $\deg(r_2(z)) < \frac{\delta-1}{2} = 2$. Hence $\sigma(z) = \alpha^{15} + \alpha^3 z + \alpha^{13}z^2$ and $w(z) = \alpha^{15} + \alpha^{16}z$.
   - By trying all possibilities, find that the set of roots of $\sigma(z)$ is $B = \{\beta^5, \beta^9\}$.
   - Compute $E_{\beta^5} = -\beta^{-5}\frac{w(\beta^5)}{\sigma'(\beta^5)} = 2$, and $E_{\beta^9} = -\beta^{-9}\frac{w(\beta^9)}{\sigma'(\beta^9)} = 2$.
   - Hence, the error vector is $e = (000\ 020\ 002\ 0000)$ and the word r is decoded to $(220\ 001\ 111\ 2110)$.

14.3.7 REED-SOLOMON CODES

Definition:

A **Reed-Solomon (RS)** code is a primitive BCH code of length $n = q - 1$ over F_q.

Facts:

1. Reed-Solomon codes are special types of BCH codes, and hence they have the same encoding and decoding algorithms.

2. Reed-Solomon codes are important because, for a fixed n and k, no linear code can have greater distance.

3. Reed-Solomon codes are useful for correcting burst errors. (A *binary burst* of length b is a bit string whose only nonzero entries are among b successive components, the first and last of which are nonzero.)

4. A Reed-Solomon code was used to encode the data transmissions from the Voyager 2 spacecraft during its encounter with Uranus in January, 1986.

5. If C is an (n, k)-RS code over F_q with designed distance δ, then the generator polynomial for C has the form $g(x) = (x - \beta^a)(x - \beta^{a+1})\ldots(x - \beta^{a+\delta-2})$, where β is a primitive element of F_q.

6. If C is an (n, k)-RS code over F_q with designed distance δ, then the distance of C is exactly δ.

7. Error correction in compact disks (developed by Philips and Sony) uses a code known as the Cross-Interleaved Reed-Solomon Code (CIRC). The CIRC code is obtained by cross-interleaving two Reed-Solomon codes, one a (28,24)-RS code and the other a (32,28)-RS code. See [Vava89] for more information and further references.

Example:

1. Consider the finite field F_5 generated by $\beta = 2$. Then $g(x) = (x - \beta)(x - \beta^2) = (x - 2)(x - 4) = x^2 + 4x + 3$ generates a $(4, 2)$-RS code over F_5 with distance $\delta = 4$.

14.3.8 WEIGHT ENUMERATORS

Definitions:

Let C be an $[n, M]$-code and let A_i be the number of codewords of weight i in C, for $i = 0, 1, \ldots, n$. The vector $(A_0, A_1, \ldots, A_n)$ is called the **weight distribution** of C.

Let C be an (n, k)-code over F_q with weight distribution $(A_0, A_1, \ldots, A_n)$. The **weight enumerator** of C is defined to be the polynomial $W_C(z) = \sum_{i=0}^{n} A_i z^i$.

Facts:

1. Let C be an (n, k)-code over F_q, and let the symbol error probability on the q-ary symmetric channel be p. If C is used only for error detection, then the probability of an error going undetected is $\sum_{i=0}^{n} A_i \left(\frac{p}{q-1}\right)^i (1-p)^{n-i}$.

2. *MacWilliams identity*: Let C be an (n, k)-code over F_q with dual code $C^\perp$. Then
$$W_{C^\perp}(z) = \frac{1}{q^k}\left[1 + (q-1)z\right]^n W_C\left(\frac{1-z}{1+(q-1)z}\right).$$

Examples:
1. The weight distribution of a binary Hamming code of length n satisfies the recurrence $A_0=1$, $A_1 = 0$,
$$(i+1)A_{i+1} + A_i + (n-i+1)A_{i-1} = \binom{n}{i}, \quad i \geq 1.$$
2. The weight enumerator of the Golay code (§14.4.2) is $1 + 253z^7 + 506z^8 + 1288z^{11} + 1288z^{12} + 506z^{15} + 253z^{16} + z^{23}$.

4.4 BOUNDS FOR CODES

How many codewords can a code have if its codewords are n-tuples of elements of F_q and it has distance d? Although this is a difficult question for all but special sets of values of n, q, and d, there are several different useful bounds on M, the number of codewords in the code. There are also special types of codes, called perfect codes, that achieve the maximum number of codewords possible, given values of n, q, and d.

4.4.1 CONSTRAINTS ON CODE PARAMETERS

Definitions:

Let $A_q(n,d)$ be the maximum M for which there exists an $[n, M, d]$-code over F_q. A code that achieves this bound is called **optimal**.

Let $V_q(n,d)$ be the number of words in F_q^n that have distance at most d from a fixed word.

An (n, k, d)-code for which $k = n - d + 1$ is called a **maximum distance separable (MDS)** code.

Facts:
1. Little is known about $A_q(n, d)$ except for some specific values of q, n, and d.
2. For all $n \geq 1$, $A_q(n, 1) = q^n$ and $A_q(n, n) = q$.
3. For all $n \geq 2$, $A_q(n, d) \leq qA_q(n - 1, d)$.
4. If d is even, then $A_2(n, d) = A_2(n - 1, d - 1)$.
5. $V_q(n, d) = \sum_{i=0}^{d} \binom{n}{i}(q-1)^i$.
6. *Hamming bound* (or *sphere-packing bound*): If $t = \lfloor \frac{d-1}{2} \rfloor$, $A_q(n,d) \leq \frac{q^n}{V_q(n,t)}$.
7. *Singleton bound*: $A_q(n,d) \leq q^{n-d+1}$. Hence, for any (n, k, d)-code over F_q, $k \leq n - d + 1$.
8. *Gilbert-Varshamov bound*:
   - $A_q(n,d) \geq \frac{q^n}{V_q(n,d-1)}$;
   - If $V_q(n-1, d-2) < q^{n-k}$, then there exists an (n, k, d)-linear code over F_q; hence, if k is the largest integer for which this inequality holds, then $A_q(n, d) \geq q^k$.

9. For thirty years, the asymptotic version of the Gilbert-Varshamov bound (not discussed here) was believed to be the best possible lower bound for good codes. In 1982, using some sophisticated ideas from algebraic geometry, it was proved that the Gilbert-Varshamov bound can be bettered. A good survey of these results appears in [Va92].

10. Let C be an (n, k)-MDS code. If G is a generator matrix for C, then any k columns of G are linearly independent.

11. If C is an (n, k)-MDS code, then $C^\perp$ is also an MDS code.

12. Johnson bound: If $d = 2t + 1$, then

$$A_2(n,d) \le \frac{2^n}{\sum_{i=0}^{t} \binom{n}{i} + \frac{1}{\lfloor \frac{n}{t+1} \rfloor} \binom{n}{t}\left(\frac{n-t}{t+1} - \lfloor \frac{n-t}{t+1} \rfloor\right)}.$$

This is an improvement of the Hamming bound (Fact 6) for binary codes.

Example:

1. The Reed-Solomon codes (§14.3.7) are MDS codes.

14.4.2 PERFECT CODES

Definitions:

An $[n, M, d]$-code over F_q is said to be **perfect** if it meets the Hamming bound, that is, $M = \frac{q^n}{V_q(n,t)}$, where $t = \lfloor \frac{d-1}{2} \rfloor$.

The **binary Golay code** is a $(23, 12, 7)$-code over F_2 with generator matrix $G = [I_{12} \mid A]$, where

$$A = \begin{pmatrix} 1 & 1 & 0 & 1 & 1 & 1 & 0 & 0 & 0 & 1 & 0 \\ 1 & 0 & 1 & 1 & 1 & 0 & 0 & 0 & 1 & 0 & 1 \\ 0 & 1 & 1 & 1 & 0 & 0 & 0 & 1 & 0 & 1 & 1 \\ 1 & 1 & 1 & 0 & 0 & 0 & 1 & 0 & 1 & 1 & 0 \\ 1 & 1 & 0 & 0 & 0 & 1 & 0 & 1 & 1 & 0 & 1 \\ 1 & 0 & 0 & 0 & 1 & 0 & 1 & 1 & 0 & 1 & 1 \\ 0 & 0 & 0 & 1 & 0 & 1 & 1 & 0 & 1 & 1 & 1 \\ 0 & 0 & 1 & 0 & 1 & 1 & 0 & 1 & 1 & 1 & 0 \\ 0 & 1 & 0 & 1 & 1 & 0 & 1 & 1 & 1 & 0 & 0 \\ 1 & 0 & 1 & 1 & 0 & 1 & 1 & 1 & 0 & 0 & 0 \\ 0 & 1 & 1 & 0 & 1 & 1 & 1 & 0 & 0 & 0 & 1 \\ 1 & 1 & 1 & 1 & 1 & 1 & 1 & 1 & 1 & 1 & 1 \end{pmatrix}.$$

The **ternary Golay code** is an $(11, 6, 5)$-code over $F_3 = \{0, 1, 2\}$ with generator matrix $G = [I_6 \mid B]$, where

$$B = \begin{pmatrix} 1 & 1 & 1 & 1 & 1 \\ 0 & 1 & 2 & 2 & 1 \\ 1 & 0 & 1 & 2 & 2 \\ 2 & 1 & 0 & 1 & 2 \\ 2 & 2 & 1 & 0 & 1 \\ 1 & 2 & 2 & 1 & 0 \end{pmatrix}.$$

Facts:

1. A necessary condition for a code to be perfect is that d be odd.
2. The binary Golay code is a perfect code.
3. The extended binary Golay code is a $(24, 12, 8)$-code that is self-dual.
4. The ternary Golay code is a perfect code.
5. The set of all perfect codes over F_q, determined in 1973 by Aimo Tietäväinen, consists of the following:
   - the linear code consisting of all words in F_q^n;
   - the binary repetition codes of odd lengths;
   - the Hamming codes and all codes of the same parameters as them;
   - the binary Golay code and all codes equivalent to it;
   - the ternary Golay code and all codes equivalent to it.
6. There do exist perfect codes with the same parameters as the Hamming codes, but which are not equivalent to them.

14.5 NONLINEAR CODES

Although linear codes are studied and used extensively, there are several important types of nonlinear codes. In particular, there are nonlinear codes with efficient encoding and decoding algorithms, as well as nonlinear codes that are important for theoretical reasons.

14.5.1 NORDSTROM-ROBINSON CODE

Definitions:

Permute the coordinates of the extended binary Golay code so that one of the weight 8 codewords is $1111111100\ldots 0$, and call this new code C'. For each of the 8-bit words 00000000, 10000001, 01000001, 00100001, 00010001, 00001001, 00000101, 00000011, there are exactly 32 codewords in C' that begin with that word. The **extended Nordstrom-Robinson code** is the code whose codewords are obtained from these 256 words by deleting the first 8 coordinate positions. The **Nordstrom-Robinson code** is obtained by puncturing the last digit of the extended Nordstrom-Robinson code.

Facts:

1. The extended Nordstrom-Robinson code is a binary $[16, 256, 6]$-nonlinear code.
2. The Nordstrom-Robinson code is a binary $[15, 256, 5]$-nonlinear code.
3. The Johnson bound (§14.4.1, Fact 12) yields $A_2(15, 5) \leq 256$, and hence it follows that $A_2(15, 5) = 256$. On the other hand, it has been proved that no linear code of length 15 and distance 5 has more codewords than the binary 2-error correcting BCH code, which has 128.

14.5.2 PREPARATA CODES

Definitions:

The **Preparata codes** are an infinite family of nonlinear codes that have efficient encoding and decoding algorithms. Let β be a primitive element of F_{2^m}, and label the elements of F_{2^m} as $\alpha_i = \beta^i$, $0 \leq i \leq 2^m - 2$, and $\alpha_{2^m-1} = 0$. For a subset $X \subseteq F_{2^m}$, let $\chi(X)$ denote the characteristic vector of X; that is, $\chi(X)$ is a binary vector of length 2^m whose ith coordinate is 1 if $\alpha_i \in X$ and 0 otherwise, for each $0 \leq i \leq 2^m - 1$.

If $m \geq 3$ is odd, the **extended Preparata code** $\overline{P}(m)$ is the set of words of the form $(\chi(X), \chi(Y))$, where X and Y are subsets of F_{2^m} such that:

- $|X|$ and $|Y|$ are even;
- $\sum_{x \in X} x = \sum_{y \in Y} y$;
- $\sum_{x \in X} x^3 + \left(\sum_{x \in X} x\right)^3 = \sum_{y \in Y} y^3$.

The **Preparata code** $P(m)$ is obtained from $\overline{P}(m)$ by puncturing the coordinate corresponding to the field element 0 in the first half of each codeword.

Facts:

1. If $m \geq 3$ is odd, then $\overline{P}(m)$ is a binary nonlinear code with parameters $n = 2^{m+1}$, $M = 2^{2^{m+1}-2m-2}$, $d = 6$.
2. If $m \geq 3$ is odd, then $P(m)$ is a binary nonlinear code with parameters $n = 2^{m+1} - 1$, $M = 2^{2^{m+1}-2m-2}$, $d = 5$.
3. $P(3)$ is the same as the Nordstrom-Robinson code.
4. The Preparata codes can be viewed as linear codes over $\mathcal{Z}_4$.

14.6 CONVOLUTIONAL CODES

Convolutional codes are a powerful class of error-correcting codes. They work differently than block codes do. Instead of grouping message symbols into blocks for encoding, check digits are interwoven within streams of information symbols. Convolutional codes can be considered to have memory, since n symbols of information are encoded using these n symbols and previous information symbols.

14.6.1 BACKGROUND

Definitions:

The figure in §14.1.1 can be used to distinguish two approaches to decoding. For a **hard decision** decoder, the demodulator maps received coded data symbols into the set of transmitted data symbols (for example, 0 and 1). In contrast, the demodulator of **soft decision** decoders may pass extra information to the decoder (for example, 3 bits of information for each received channel data symbol, indicating the degree of confidence in its being a 0 or a 1).

Facts:

1. Convolutional codes were introduced by P. Elias in 1955, and are widely used in practice today.

2. Convolutional codes are used extensively in radio and satellite links and have been used by NASA for deep-space missions since the late 1970s.

3. There are linear codes that differ from block codes in that the codewords do not have constant length.

4. Convolutional codes also differ from block codes in that the n-tuple produced by an encoder depends not only on the message k-tuple u, but also on some message k-tuples produced prior to u; that is, the encoder has memory.

5. Soft decision decoding typically allows performance improvements.

6. Hard and soft decision techniques can be used in both block and convolutional codes, although soft decision techniques can typically be used to greater advantage in convolutional codes.

7. Theoretical results, particularly with respect to BCH codes, position block codes as superior to convolutional codes.

8. The minimum distances of BCH codes are typically much larger than the corresponding free distances (§14.6.3) of comparable convolutional codes.

9. Decoding techniques for block codes are generally applicable only to q-ary (or binary) symmetric channels, which are an appropriate model for only a relatively small fraction of channels that arise in practice.

10. Efficient decoding of BCH codes requires hard-decision decoding, which suffers information loss relative to soft-decision strategies, precipitating a performance penalty. The resulting performance of the BCH decoder is significantly inferior to that for a comparable convolutional code, despite the BCH codes being inherently more powerful. Consequently, convolutional codes are used in a majority of practical applications, due to their relative simplicity and performance, and the large number of communication channels which benefit from soft decoding techniques.

11. A recently developed classes of codes, known as *turbo codes*, are built using convolutional codes. The basic idea behind a turbo encoder is to combine two simple convolutional encoders. Input to the encoder is a block of bits. The two constituent encoders generate parity bits and the information bits are sent unchanged. The key innovation is an interleaver, which permutes the original information bits before they are provided as input to the second encoder. The permutation causes input sequences which produce low-weight codewords for one encoder to generally produce high-weight codewords for the other encoder. See [HeWi99] for information on turbo codes.

12. A good starting point for information on turbo codes is the JPL Turbo Codes Web page:
    http://www331.jpl.nasa.gov/public/JPLtcodes.html

Example:

1. In the simplest version of soft decision decoding, known as the *binary erasure channel* (and usually classified as a hard-decision technique), the demodulator output is one of three values: 0, 1, or "erasure" (indicating that neither a 0 nor a 1 was clearly recognized).

14.6.2 SHIFT REGISTERS

Definitions:

An *m-stage shift register* is a hardware device that consists of m **delay elements** (or **flip-flops**), each having one input and one output, and a clock which controls the movement of data. During each unit of time, the following operations are performed:
- a new input bit and the contents of some of the delay elements are added modulo 2 to form the output bit;
- the content of each delay element (with the exception of the last delay element) is shifted one position to the right;
- the new input bit is fed into the first delay element.

The **generator** of an m-stage shift register is a polynomial $g(x) = 1 + g_1 x + g_2 x^2 + \cdots + g_m x^m \in F_2[x]$, where $g_i = 1$ if the contents of the ith delay element is involved in the modulo 2 sum that produces the output, and 0 otherwise.

Fact:

1. Assume that the initial contents of a shift register are all 0s. Suppose that a shift register has generator $g(x)$. Let the input stream $u_0, u_1, u_2, \ldots$ be described by the formal power series $u(x) = u_0 + u_1 x + u_2 x^2 + \cdots$ over F_2. (If the input stream is finite of length t, let $u_i = 0$ for $i \geq t$.) Similarly, let the output stream $c_0, c_1, c_2, \ldots$ be described by the formal power series $c(x) = c_0 + c_1 x + c_2 x^2 + \cdots$ over F_2. Then $c(x) = u(x)g(x)$.

Examples:

1. *Shift-example*: Suppose that the delay elements of the 4-stage shift register in the following figure initially contain all 0s:

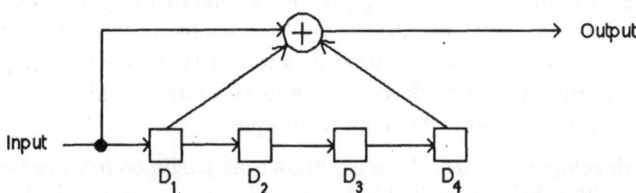

If the input stream to the register is 11011010 (from left to right), the updated contents of the delay elements and the output bits are summarized in the following table:

time	input	D_1	D_2	D_3	D_4	output
0	–	0	0	0	0	–
1	1	1	0	0	0	1
2	1	1	1	0	0	0
3	0	0	1	1	0	1
4	1	1	0	1	1	1
5	1	1	1	0	1	1
6	0	0	1	1	0	0
7	1	1	0	1	1	1
8	0	0	1	0	1	0

2. The generator of the shift register in Example 1 is $g(x) = 1 + x + x^4$.

14.6.3 ENCODING

Note: Throughout this subsection assume that the initial contents of a shift register are all 0s.

Definitions:

An $(n, 1, m)$-*convolutional code* with generators $g_1(x), g_2(x), \ldots, g_n(x) \in F_2[x]$ ($m = \max(\deg g_i(x))$) consists of all codewords of the form $c(x) = (c_1(x), c_2(x), \ldots, c_n(x))$, where $c_i(x) = u(x)g_i(x)$, and $u(x) = u_0 + u_1 x + u_2 x^2 + \cdots$ represents the input stream. The *system memory* of the code is m.

A convolutional code is *catastrophic* if a finite number of channel errors can cause an infinite number of decoding errors.

The *rate* of an (n, k, m)-convolutional code is $\frac{k}{n}$.

The *free distance* d_{free} of a convolutional code is the minimum weight of all nonzero output streams.

Facts:

1. A convolutional code is linear.

2. Convolutional codes are not block codes since the codewords have infinite length. They are, however, similar to block codes, and in fact can be viewed as block codes over certain infinite fields.

3. An $(n, 1, m)$-convolutional code can be described by a single shift register with n outputs, where $c_i(x)$ is the output of the single-output shift register with generator $g_i(x)$ when $u(x)$ is the input. In practice, $c_1(x), c_2(x), \ldots, c_n(x)$ are interleaved to produce one output stream.

4. Let C be an $(n, 1, m)$-convolutional code with generators $g_1(x), g_2(x), \ldots, g_n(x)$. Let $G(x) = \sum_{i=1}^{n} x^{i-1} g_i(x^n)$. If the message is $u(x)$, then the corresponding interleaved codeword is $\overline{c}(x) = G(x)u(x^n)$.

5. The *Viterbi algorithm* is a maximum likelihood decoding algorithm for convolutional codes. See [LiCo83]. For an algebraic treatment of convolutional codes, see [Pi88].

6. If $\gcd(g_1(x), g_2(x), \ldots, g_n(x)) = 1$ in $F_2[x]$ then C is not catastrophic.

7. An (n, k, m)-convolutional code can be described by k multi-output shift registers, each of maximum length m. The message is divided into k streams, each stream being the input to one of the k shift registers. There are n output streams, each formed using some or all of the shift registers.

8. The free distance of a convolutional code is a measure of the error-correcting capability of the code, and is a concept analogous to the distance of a block code.

9. In contrast to block codes, there are few algebraic constructions known for convolutional codes.

10. The convolutional codes used in practice are usually those found by a computer search designed to maximize the free distance among all encoders with fixed parameters n, k, and m. The following table lists the best codes with a rate of $\frac{1}{2}$ ($n = 2$, $k = 1$). The polynomials $g_1(x)$ and $g_2(x)$ are represented by their coefficients, from low order to high order.

m	$g_1(x)$	$g_2(x)$	d_{free}
2	101	111	5
3	1101	1111	6
4	10011	11101	7
5	110101	101111	8
6	1011011	1111001	10
7	11100101	10011111	10
8	101110001	111101011	12
9	1001110111	1101100101	12
10	10011011101	11110110001	14
11	100011011101	101111010011	15
12	1000101011011	1111110110001	16

Examples:

1. Consider the $(2,1,3)$-convolutional code with generators $g_1(x) = 1+x^3$ and $g_2(x) = 1+x+x^3$. The code can be described by the shift register of the following figure. The message $u(x) = 1+x^3+x^4$, corresponding to the bit string 10011, gets encoded to $c(x) = (u(x)g_1(x), u(x)g_2(x)) = (1+x^4+x^6+x^7, 1+x+x^5+x^6+x^7)$, or in interleaved form to $\bar{c} = 11\ 01\ 00\ 00\ 10\ 01\ 11\ 11\ 00\ 00\ 00\ \ldots$.

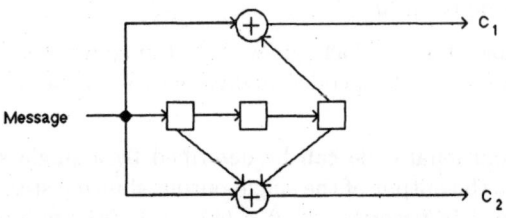

2. Suppose that the input stream contains an infinite number of 1s, and the output stream has only finitely many 1s. If the channel introduces errors precisely in the positions of these 1s, then the resulting all-zero output stream will be decoded by the receiver to $m(x) = 0$.

3. The following figure is a shift register encoder for a $(3,2,3)$-convolutional code C.

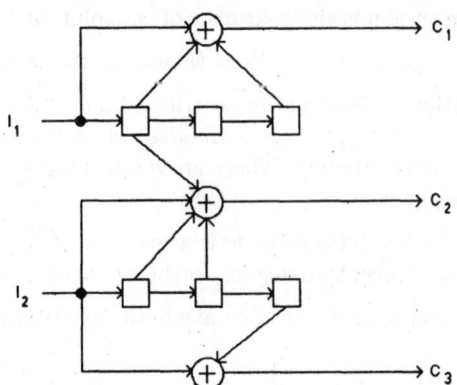

If the input stream is $u = 1011101101$, it is first divided using alternating bits into 2 streams $I_1 = 11110$ and $I_2 = 01011$. The 3 output streams are $c_1 = 10010$, $c_2 = 00011$ and $c_3 = 01010$, and the interleaved output is $\bar{c} = 100\ 001\ 000\ 111\ 010$.

4.7 BASICS OF CRYPTOGRAPHY

Protecting the secrecy of information goes back to ancient times. For example, the ancient Romans used a secret code to send messages so the messages could not be read by their enemies. In modern times, there is a constant need to protect information from unauthorized access and from malicious actions. The science of cryptography is devoted to methods that offer such protection. Sending secret messages, authenticating messages, distributing secret keys, and sharing secrets are only some of the applications addressed by modern cryptography.

4.7.1 BASIC CONCEPTS

Definitions:

Cryptography is the science and study of protecting data from malicious or unauthorized actions, including access to, manipulation, impersonation, and forgery.

Cryptanalysis is the use of mathematical, statistical, and other techniques to defeat cryptographic protection mechanisms.

Cryptology is the study of both cryptography and cryptanalysis, although "cryptography" is often used in place of "cryptology".

A *cipher* is a method whereby a message in some source language (the *plaintext*) is transformed by a mapping, called an *encryption* algorithm to yield an output, called the *ciphertext*, which is unintelligible to all but an authorized recipient.

A recipient of an encrypted message is able to recover the plaintext from the ciphertext by use of a corresponding *decryption* algorithm.

A *key* is a secret number or other significant information which parametrizes an encryption or decryption algorithm.

The *message space* $\mathcal{M}$ is the set of all possible plaintexts, the *ciphertext space* $\mathcal{C}$ consists of all possible ciphertexts, and the *key space* $\mathcal{K}$ consists of all possible keys.

An *encryption algorithm* E is a family of mappings parametrized by a key $k \in \mathcal{K}$, such that each value k defines a mapping $E_k \in \mathcal{E}$, where $\mathcal{E}$ is the set of all invertible mappings from $\mathcal{M}$ to $\mathcal{C}$. A specific plaintext message m is mapped by E_k to a ciphertext $c = E_k(m)$.

The set $\mathcal{D}$ of *decryption algorithms* consists of all invertible mappings from $\mathcal{C}$ back to $\mathcal{M}$, such that for each encryption key $k \in \mathcal{K}$, there is some mapping $D \in \mathcal{D}$ such that $D_{f(k)}(E_k(m)) = m$ for all $m \in \mathcal{M}$, where $f(k)$ is some key dependent on k and $D_{f(k)}$ is the decryption algorithm corresponding to the decryption key $f(k)$. For so-called *symmetric-key* systems, this decryption key $f(k)$ is equal to k itself.

Facts:

1. Useful books that cover cryptography include [BePi82], [Br88], [DaPr89], [De83], [Ko94], [Sc96], [SePi89], [Si93], [St95], and [We88]. A comprehensive treatment can be found in [MevaVa96].

2. Useful Internet sites on cryptography include:
   - A-Z Cryptology:
     `http://www.achiever.com/freehmpg/cryptology/crypto.html`
   - Cryptographic Software Archive:
     `ftp://ftp.funet.fi/pub/crypt`
   - Introduction to Cryptography:
     `http://www.cs.hut.fi/ssh/crypto/intro.html`
   - RSA Laboratories:
     `http://www.rsa.com/rsalabs`
   - Some Classic Ciphers:
     `http://rschp2.anu.edu.au:8080/cipher.html`

3. Separating cryptography from cryptanalysis is, in fact, difficult, as the design of secure cryptographic systems requires that all possible cryptanalytic attacks be taken into account.

4. Cryptography differs from *steganography* in that while the former involves use of techniques to secure data (for example, codes and ciphers), the latter involves the use of techniques which obscure the existence of data itself (for example, invisible ink, secret compartments, use of subliminal channels).

5. While falling under the broader category of communications security, cryptography is generally concerned with the more mathematical details, rather than system-level aspects such as traffic-flow analysis and electronic security aspects such as monitoring electromagnetic emanations.

6. Cryptographic mechanisms can be used to support a number of fundamental security services, including:
   - *privacy*: preventing confidential data from being available in an intelligible form to unauthorized parties;
   - *data integrity*: detection of data manipulation by unauthorized parties (including alteration, insertion, deletion, substitution, delay and replay); it should be noted that encryption alone does not guarantee data integrity;
   - *authentication*: corroboration that a party's identity is as claimed (*entity authentication*), or that the origin of data is as claimed (*data origin authentication*); related to this is the assurance that data has not been subjected to unauthorized manipulation (cf. data integrity), possibly including assurances regarding uniqueness and timeliness;
   - *non-repudiation*: provision for the resolution of disputes related to digital signatures; digital signatures can be used as the basis of authorization of certain actions; disputes may occasionally arise subsequently due to either false denials (*repudiated* signatures) or fraudulent claims (*forged* signatures).

In addition, entity authentication and/or data authentication may be the basis for granting access to certain controlled resources. *Access control* mechanisms often rely upon cryptographic support to restrict access (of information or other resources) to authorized parties; access is generally granted upon proof of authorization, which may be based on an entity's identity or possession of anonymous *tokens*, either physical or digital.

7. The traditional objectives of privacy and authentication (although not both required in all cases) lead to the following requirements for a cryptosystem:
   - *fundamental requirement*: (to maintain secrecy of key) it should be infeasible for an adversary to deduce the key k given one or more plaintext-ciphertext pairs (m, c);
   - *privacy requirement*: (to maintain confidentiality) it should be infeasible for an adversary to deduce the plaintext m corresponding to any given ciphertext c;
   - *authentication requirement*: (to prevent forgery or substitution) it should be infeasible for an adversary to deduce a ciphertext c' corresponding to any message m' of his choosing, or corresponding to any other (meaningful) message m.

In the encryption model above, the decryption mapping was such that $D_{f(k)}(E_k(m)) = m$, and it was stated that in symmetric-key systems the decryption key $f(k)$ is the same as (or easily computed from) the encryption key k. In their landmark 1976 paper, W. Diffie and M. Hellman introduced the concept of *public-key cryptosystems*. Here each user has his own pair of encryption and decryption keys $(k, f(k))$ where $k \neq f(k)$, with the property that it is infeasible for anyone to deduce $f(k)$ from k. If the so-called *public key*, k, of a user A is published, then anyone looking up that key can encrypt a message for A, such that A alone, having knowledge of $f(k)$, the *private key*, can decrypt it. User A is able to compute both k and $f(k)$ from another key k'.

8. A *digital signature* is intended to be the digital analogue of a handwritten signature; it should be a number dependent on some secret known only to the signer, and, additionally, on the content of the message being signed.

9. Signatures must be verifiable in the sense that, should a dispute arise as to whether a party signed a document (caused by either a lying signer trying to *repudiate* a signature it did create, or a fraudulent claimant), an unbiased third party can resolve the matter equitably, without requiring access to the signer's secret information (private key).

10. Signatures must be easy for the signer to computer.

11. Signatures must be easy to verify by anyone.

12. The following describes the general method of constructing digital signatures: A has a message m which it wishes to send to B. A sends to B the quantity $D_{f(k)}(m)$ obtained by applying the *decryption* function. Then, upon reception, B can use A's public-key algorithm E_k to recover the message $m = E_k(D_{f(k)}(m))$. (Here it is required that for each $k \in \mathcal{K}$, $D_{f(k)}$ is a mapping from $\mathcal{M}$ to $\mathcal{C}$, E_k is a mapping from $\mathcal{C}$ to $\mathcal{M}$, and E_k is the inverse of $D_{f(k)}$.) Provided the message m recovered by B is meaningful (e.g. contains a sufficient degree of redundancy — to ensure it is not simply the result of applying E_k to a random quantity anyone might have generated), B has assurance that the message is authentic and originated from A, since by assumption no one aside from A knows or can feasibly compute A's secret key $f(k)$. Moreover, B can keep the signature $D_{f(k)}(m)$ to prove to any third party, at a later point in time, that A actually did send the message m; such a party would similarly uses A's public key to recover m as verification. This provides a digital analogue to handwritten signatures.

13. Digital signatures are also possible using symmetric-key techniques, but this generally requires use of an on-line trusted third party or new keying material for each signature (*one-time signature schemes*). For these reasons, digital signatures based on public-key cryptography are used in practice.

14.7.2 SECURITY OF CRYPTOSYSTEMS

Definitions:

Adversaries are either **passive** or **active**. Passive adversaries are a threat to confidentiality; they do not interrupt, alter or insert any transmitted or stored data. Active adversaries additionally threaten integrity and authentication.

There are many models under which one can assume a cryptanalyst is able to attack a cryptographic system. The following types of attack can be hypothesized for increasingly powerful adversaries:

- **ciphertext-only**: the adversary has possession only of some ciphertext;
- **known-plaintext**: the adversary has some plaintext and its corresponding ciphertext;
- **chosen-plaintext**: the adversary has some plaintext of his choosing, and its corresponding ciphertext;
- **chosen-ciphertext**: the adversary has some ciphertext of his choosing, and its corresponding plaintext.

The most stringent measure of the security of a cryptographic algorithm is **unconditional security** where an adversary is assumed to have unlimited computational resources, and the question is whether there is enough information available to defeat the system. Unconditional security for encryption systems is called *perfect secrecy*.

To measure an adversary's uncertainty in the key after observing n ciphertext characters, C. Shannon defined the **key equivocation function** $Q(n) = H(K|C_1 C_2 \ldots C_n)$ and defined the **unicity distance** of the cipher to be the first value $n = n_0$ such that $Q(n) \approx 0$.

A cryptographic method is said to be **provably secure** if the difficulty of defeating it can be shown to be essentially as difficult as (that is, polynomially equivalent to) solving a well-known and *supposedly* difficult (typically number-theoretic) problem, such as integer factorization or the computation of discrete logarithms. (Thus, "provable" here means provable subject to as yet unproved assumptions.)

A proposed technique is said to be **computationally secure** if the (perceived) level of computation required to defeat it exceeds, by a comfortable margin, the computational resources of the hypothesized adversary.

Facts:

1. It is a standard cryptographic assumption that an adversary will have access to ciphertext.

2. *Kerckhoff's assumption*: The security of a system should rest entirely in the secret key — the adversary is assumed to have complete knowledge of the rest of the cryptographic mechanism(s).

3. In determining whether the security of a particular cryptosystem is adequate for a particular application, the powers and resources of the anticipated adversary must be taken into account. Potential adversaries may have powers ranging from minimal to unlimited.

4. The security of a cryptographic algorithm can be measured according to several different metrics, including unconditional security, provable security, and computational security.

5. Let M, C, and K be random variables ranging over the message space $\mathcal{M}$, ciphertext space $\mathcal{C}$, and key space $\mathcal{K}$. Unconditional security for encryption systems can be specified by the condition $H(M|C) = H(M)$; that is, the uncertainty in the plaintext, after observing the ciphertext, is equal to the a priori uncertainty about the plaintext — observation of the ciphertext provides no information (whatsoever) to an adversary.

6. A necessary condition for an encryption scheme to be unconditionally secure is that the key should be at least as long as the message. The one-time pad (§14.8.4) is an example of an unconditionally secure encryption algorithm.

7. In general, encryption schemes do not offer perfect secrecy, and each ciphertext character observed decreases the uncertainty in the encryption key k used.

8. Let n_0 be the unicity distance of a cipher. After observing n_0 characters, the key uncertainty is zero, meaning an information-theoretic adversary can narrow the set of possible keys down to a single candidate, thus defeating the cipher.

9. The computational security measures the amount of computational effort required, by the best currently-known attacks, to defeat a system; it must be assumed here that the system has been well-studied to determine which attacks are relevant.

14.8 SYMMETRIC-KEY SYSTEMS

Classical cryptosystems such as the Caesar cipher have the property that the decryption key can easily be found from the corresponding encryption key. Until recently, all cryptosystems had this general property. Such systems are known as symmetric-key systems, to distinguish them from cryptosystems that do not have this property, that is, where knowing an encryption key does not provide adequate information for finding the corresponding decryption key.

14.8.1 REDUNDANCY

Definitions:
Let plaintext messages be composed from an alphabet $\mathcal{A}$ of L characters. Let H_n denote the entropy (§14.1.2) of n-character messages, or equivalently, the expected bit length of n-character messages under an optimal encoding.

The **absolute rate** R of a language is the maximum number of bits of information that each character could encode, assuming all combinations of characters are equiprobable in the language. $R = \log_2 L$.

The **rate** r_n **for** n**-character messages** is $r_n = \frac{H_n}{n}$.

The **rate** r_∞ **of a language** is $r_\infty = \lim_{n \to \infty} r_n$.

The **redundancy** D_n for n-character messages is $D_n = n(R - r_n)$.

The **redundancy** D **of a language** (measured in bits per plaintext character) is $D = \lim_{n \to \infty} \frac{D_n}{n} = R - r_\infty$.

Let D be a family of parametrized decryption algorithms, K be a random variable ranging over keyspace $\mathcal{K}$, and C be a random variable ranging over ciphertext space $\mathcal{C}$. A **random cipher** is a cipher such that the decipherment $D_K(C)$ is a random variable uniformly distributed over all preimages of C.

Facts:

1. Knowing the rate and redundancy of a language allows estimation of the unicity distance (§14.7.2 Fact 8) for a certain class of "statistically perfect" ciphers known as random ciphers.

2. A reasonable estimate of unicity distance for a random cipher is $n_0 = \frac{H(K)}{D}$ ciphertext characters, where $D = R - r_\infty$ is the redundancy of the language and $H(K)$ is the key entropy.

3. The redundancy in the denominator of this estimate indicates that data compression prior to encryption increases the unicity distance of a cipher (increasing security).

Examples:

1. Estimates for the English language with 26-character alphabet indicate $r_1 \approx 4.2$, $r_2 \approx 3.6$, and $r_3 \approx 3.2$ bits/character; these estimate the number of bits of information per character in messages of lengths one, two and three characters.

2. The rate r_1 differs from $R = \log_2 26 \approx 4.7$ due to the fact that characters in English messages are not equiprobable; r_n decreases as n grows due to the decreasing likelihood that random character strings are meaningful messages — effectively due to redundancy in the language.

3. Estimates suggest that for English, $1 \leq r_\infty \leq 1.5$, yielding a redundancy of between 3.2 and 3.7 bits per character in long messages, or between 68% and 79%.

14.8.2 SUBSTITUTION AND TRANSPOSITION CIPHERS

Definitions:

Let $P = p_1 p_2 \ldots p_n$ represent a plaintext message of n characters; often p_i are interpreted as integers $0, 1, \ldots, 25$ corresponding to the characters $a, b, \ldots, z$.

A **simple substitution cipher** S replaces each plaintext character by a fixed substitute according to a permutation π on the source alphabet. This means that S replaces the string $p_1 p_2 \ldots p_n$ with $\pi(p_1)\pi(p_2)\ldots\pi(p_n)$. This can be written as $S(p_1 p_2 \ldots p_n) = \pi(p_1)\pi(p_2)\ldots\pi(p_n)$.

An **affine cipher** replaces the plaintext character x (represented as an integer) by the ciphertext character $(ax + b) \bmod 26$, where a and b are integers with a relatively prime to 26. When $a = 1$, an affine cipher is called a **shift cipher** since each letter is shifted a fixed number of positions, with wrap around, in the alphabet. The shift cipher where each character is shifted three positions, that is, where x is mapped to $(x + 3) \bmod 26$, is known as the *Caesar cipher*.

A **simple transposition cipher** T (with fixed period d) divides plaintext into d-character blocks and rearranges these characters by a permutation π on the numbers $1, 2, \ldots, d$. This can be written as $T(p_1 p_2 \ldots p_n) = p_{\pi(1)} p_{\pi(2)} \cdots p_{\pi(d)} p_{d+\pi(1)} \cdots$.

A **full Vigenère cipher** V of period d consists of d simple substitutions defined by permutations $\pi_0, \ldots, \pi_{d-1}$ used in sequence: $V(p_i) = \pi_{i \bmod d}(p_i)$. Ciphers such as the full Vigenère are called **polyalphabetic** — d different alphabetic substitutions are used.

The **simple Vigenère cipher** of period d restricts each permutation π_i to a simple shift, so that the key may be represented by a d-letter sequence $k_0, \ldots, k_{d-1}$; encryption then consists of simply adding key characters to plaintext: $V(p_i) = (p_i + k_{i \bmod d}) \bmod 26$. Such ciphers are also called **periodic substitution ciphers**.

A **Hill cipher** is a cipher with an $m \times m$ matrix K as its key that encrypts plaintext by splitting it into blocks of size m and mapping the plaintext block $x = (x_1, x_2, \ldots, x_m)$ to the ciphertext block $(y_1, y_2, \ldots, y_m) = xK$.

A **homophonic substitution** is a cipher in which each plaintext character x in the source language is associated with a set S_x of ciphertext characters, and each time x is to be encrypted, one element of S_x is randomly chosen and where the sets S_x are pairwise disjoint. The cardinality of S_x is chosen to be proportional to the frequency of x in the source language, to flatten out frequency distributions.

Facts:

1. Simple transposition and substitution ciphers, and related symmetric-key ciphers, are often called *classical ciphers* since they were designed and used in ancient times.

2. The permutation that serves as key of a substitution cipher can often be represented more compactly than by specifying the permutation in full.

3. The size of the full key space of substitution ciphers provides an upper bound on security, but is often a poor indication. For example, for a simple substitution, there are 26! possible keys providing key entropy of 88 bits, or approximately 2^{88} keys to search through if one resorts to exhaustive cryptanalysis.

4. The unicity distance of a simple substitution can be estimated to be 28 characters; all simple substitutions applied to English messages can be trivially cryptanalyzed given about this many characters.

5. Periodic substitution ciphers may be cryptanalyzed by first deducing the period of the cipher (by one of several known techniques, for example, the *index-of-coincidence* introduced by W. Friedman, c. 1920, or the *Kasiski method*), and then solving d simple substitution ciphers. (See [St95] for details.)

6. The simple Vigenère cipher of period $d = 1$ is a shift cipher.

7. Decryption of the affine cipher with encryption function $e(x) = (ax + b) \bmod 26$ is carried out using the decryption function $d(y) = \bar{a}(y - b) \bmod 26$ where $\bar{a}$ is an inverse of a modulo 26 and y is the ciphertext character associated with the plaintext character x.

8. Decryption of the Hill cipher with encryption function $e(x) = xK$ where $x = (x_1, x_2, \ldots, x_m)$ and K is an $m \times m$ matrix is carried out using the decryption function $d(y) = yK^{-1}$ where K^{-1} is the inverse of K modulo 26 and y is the ciphertext block $(y_1, y_2, \ldots, y_m)$ associated with the plaintext block x. Note for K^{-1} to exist, it must be the case that $\gcd(\det K, 26) = 1$.

9. The ideas of simple transposition and substitution can be combined and compounded in countless ways. While not secure individually, they can be combined to construct the powerful class of *product ciphers* which include DES (§14.8.3).

10. Data expansion is inherent in homophonic substitutions as the ciphertext character set must be larger than the plaintext set.

11. *Codes* differ from ciphers in that codes employ a *codebook* or dictionary which specifies words or phrases that are used to substitute for plain text words or phrases

requiring encryption. Decryption is accomplished by a reverse-codebook, indexed by the entries (rather than index terms) of the encryption codebook. The use of such codes is not easily automated, and they no longer see much use.

Examples:

1. The Caesar cipher (shifting every letter by three modulo 26) sends the plaintext message ZERO to the ciphertext message CHUR.

2. The affine cipher with encryption function $e(x) = (7x + 10) \bmod 26$ sends the plaintext message PLEASESENDMONEY to the ciphertext message LJMKGMGM-FQEXMW. For example, the letter P, which corresponds to the number 15, is sent to $e(15) = (7 \cdot 15 + 10) = 11 \bmod 26$, which corresponds to the letter L. Decryption is done using the function $d(y) = (15y + 6) \bmod 26$.

3. The Vigenère cipher of period 4 with key $(1, 0, 13, 3)$ sends the plaintext message RENAISSANCE to the ciphertext message SEADJSFDOCR.

4. The Hill cipher that has $\begin{pmatrix} 11 & 3 \\ 8 & 7 \end{pmatrix}$ as its key encrypts the plaintext JULY (which corresponds to the string 9 20 11 24) to the string DELW (which corresponds to the string 3 4 11 22). Decryption is carried out using the matrix $\begin{pmatrix} 7 & 23 \\ 18 & 11 \end{pmatrix}$, which is an inverse of the encryption matrix.

14.8.3 BLOCK CIPHERS

Definitions:

A *block cipher* derives its name from the property that it processes the plaintext stream after grouping it into pieces or *blocks* consisting of a fixed number of characters, thereafter operating on the block as a whole. Each block of plaintext is enciphered independently of preceding and succeeding plaintext input.

The *U. S. Data Encryption Standard* (*DES*) is a block cipher widely used in commerical applications. It has been adopted as a standard by the United States government.

The *mode of operation* of an n-bit block cipher describes how the cipher processes messages with more than n bits.

In *Electronic Codebook* (*ECB*) *mode* the plaintext message m is split into n-bit blocks $m = m_1 m_2 \ldots m_l$. Each message block is encrypted independently using the same secret key k:

$$c_i = E_k(m_i), \quad 1 \leq i \leq l.$$

In *Cipher Block Chaining* (*CBC*) *mode* each ciphertext block is dependent on all previous plaintext blocks. Encryption is performed as follows, given an n-bit initialization vector IV:

$$c_1 = E_k(m_1 \oplus IV); \quad c_i = E_k(m_i \oplus c_{i-1}), \quad 2 \leq i \leq l.$$

Decryption is performed as follows:

$$m_1 = D_k(c_1) \oplus IV; \quad m_i = D_k(c_i) \oplus c_{i-1}, \quad 2 \leq i \leq l.$$

In **Cipher Feedback (CFB) mode** the plaintext message is split into t-bit blocks $m = m_1 m_2 \ldots m_l$, where $1 \leq t \leq n$. An n-bit shift register is initialized to the value $s_0 = IV$. Encryption is then performed as follows:

$$c_i = m_i \oplus \mathrm{MSB}_t(E_k(s_{i-1})), \quad 1 \leq i \leq l,$$

where $\mathrm{MSB}_t(v)$ denotes the t most significant bits of v, and where s_i is obtained from s_{i-1} by shifting the contents of the register t positions to the left and moving c_i into the rightmost t positions of the register.

In **Output Feedback (OFB) mode**: the plaintext message is split into t-bit blocks $m = m_1 m_2 \ldots m_l$, where $1 \leq t \leq n$. Encryption is performed as follows:

$$c_i = m_i \oplus \mathrm{MSB}_t(s_{i-1}), \quad 1 \leq i \leq l,$$

where $s_0 = E_k(IV)$ and $s_i = E_k(s_{i-1})$ for $1 \leq i \leq l-1$ and MSB has the same meaning as in the definition of CFB mode.

Facts:

1. Transposition ciphers are examples of block ciphers.

2. DES is the most widely used block cipher.

3. DES was published as a U.S. Federal Information Processing Standard in 1977. It resulted from an IBM submission to a 1974 request by the U.S. National Bureau of Standards (NBS) (which has now become NIST) soliciting encryption algorithms for the protection of computer data.

4. DES processes plaintext blocks of $n = 64$ bits, producing ciphertext blocks of 64 bits.

5. The encryption mapping E_k is parametrized by a secret 56-bit key k. Since decryption requires that the mapping be invertible, E_k is a bijection.

6. The total number of distinct permutations on an n-bit space is $(2^n)!$; DES implements only a tiny fraction of these — at most 2^{56}, corresponding to the number of distinct DES keys.

7. DES, and in fact all block ciphers, can be viewed as large substitution ciphers. For a fixed key k, each 64-bit plaintext "character" is substituted by a fixed 64-bit ciphertext "character". The same techniques that make simple substitution ciphers trivial to cryptanalyze do not directly threaten the security of DES or similar ciphers, however, due to the large block size.

8. Encryption of each 64-bit block proceeds in sixteen stages or *rounds*.

9. The 56-bit key k is used to create sixteen 48-bit subkeys k_i, one for each round.

10. Within each round, eight fixed, carefully selected 6-to-4 bit substitution mappings (S-boxes) S_i, collectively denoted S, are used.

11. The initial 64-bit plaintext is divided into two 32-bit halves, L_0 and R_0.

12. Each round is functionally equivalent, taking 32-bit inputs L_{i-1} and R_{i-1} from the previous round and producing outputs L_i and R_i for $1 \leq i \leq 16$, as follows:

$$L_i = R_{i-1}; \quad R_i = L_{i-1} \oplus F(R_{i-1}, K_i); \quad F(R_{i-1}, K_i) = P(S(E(R_{i-1} \oplus K_i))).$$

Here E is a fixed expansion permutation mapping 32 bits to 48 bits (all bits are used once, some are taken twice), and P is another fixed permutation on 32 bits.

13. An initial permutation (IP) precedes the first round, and its inverse is applied following the last round.

14. Decryption makes use of the same algorithm and the same key, except that subkeys are applied to the internal rounds in the reverse order.

15. Each round involves both (bitwise) substitution and transposition.

16. A complete description of the U.S. Data Encryption Standard (DES) algorithm can be found in the U.S. Federal Information Processing Standards Publication 46 (FIPS 46: Data Encryption Standard).

17. Given a plaintext-ciphertext pair, exhaustive cryptanalysis of DES is possible — all 2^{56} keys could be checked to determine which maps the given plaintext to the given ciphertext. With current technologies, such an attack is now feasible in practice for strongly-motivated adversaries. See [Gi98] for example.

18. Results of experiments indicate that all bits of the ciphertext depend on all bits of the key and all bits of the plaintext; changing any single bit of the plaintext or key causes each ciphertext bit to change with probability about 0.5.

19. There are several ways in which a block cipher can be employed. Let E be a block cipher parametrized by a key k. Suppose that E processes plaintext blocks of n bits, producing ciphertext blocks of n bits. The initial value IV is a randomly chosen n-bit block known to the sender and receiver; IV may be exchanged in the clear (except in output feedback mode (OFB)) or it may be transmitted by the sender to the receiver by encrypting it in ECB mode.

20. The weakness of ECB mode is that two identical plaintext blocks are always encrypted to the same ciphertext block. An advantage of ECB mode is that transmission errors are not propagated from block to block.

21. If a different IV is selected for each message encrypted in CBC mode, then two identical plaintexts will, in general, be encrypted to different ciphertexts. If the integrity of the IV is not protected, then an opponent can selectively manipulate the bits of the first message block by manipulating the bits of the IV. This situation may be avoided by encrypting the IV.

22. In CBC mode there is no propagation of transmission errors since a message block m_i depends on only two ciphertext blocks, c_{i-1} and c_i.

23. Decryption in CFB mode is performed by initializing the shift register to the value $s_0 = IV$, and then computing
$$m_i = c_i \oplus \text{MSB}_t(E_k(s_{i-1})), \quad 1 \leq i \leq l.$$

24. In CFB mode a transmission error may affect several message blocks. Note that the block cipher E is operated in encryption mode at both the sending and receiving ends.

25. In OFB mode decryption is performed by computing
$$m_i = c_i \oplus \text{MSB}_t(s_{i-1}), \quad 1 \leq i \leq l.$$

26. In OFB there is no error propagation. A single bit error in the ciphertext causes a single bit error in the recovered plaintext. As with CBC mode, the block cipher E is operated in encryption mode at both the sending and receiving ends.

14.8.4 STREAM CIPHERS

Definitions:

A *stream cipher* is a symmetric-key cipher that encrypts individual characters of a plaintext message, or small units. (In contrast, a *block cipher* tends to encrypt groups of characters, or larger units.)

A *synchronous stream cipher* is a stream cipher in which the keystream is generated independently of the message.

A *self-synchronizing stream cipher* is a stream cipher capable of reestablishing proper decryption automatically after loss of synchronization, with only a fixed number of plaintext characters unrecoverable.

The *one-time pad* is a stream cipher with the following encryption function: each bit of plaintext is XORed to the next bit of a truly random key, which is never reused for encryption and is of bit length equal to that of the plaintext. Decryption is accomplished by applying the same process, with the same key, to the ciphertext string.

Facts:

1. Stream ciphers are more appropriate, and in some cases mandatory (for example, in some telecommunications applications), when buffering is limited and characters must be individually processed as they are received.

2. A stream cipher typically consists of a generator which produces a pseudorandom bit sequence (the key) which is then XORed (added modulo 2) with the plaintext bits.

3. In a synchronous stream cipher, both the sender and receiver must be *synchronized* — using the same key and operating at the same position (state) within that key — in order for proper decryption. If synchronism is lost then decryption fails and can only be restored through additional techniques for resynchronization (for example, reinitialization, or the receiver trying possible offsets).

4. The OFB mode of a block cipher is an example of a synchronous stream cipher.

5. The CFB mode of a block cipher is an example of a self-synchronizing stream cipher.

6. For such ciphers, self-synchronization is possible because the encryption/decryption mappings depend only on a fixed number of preceding ciphertext characters.

7. The one-time pad is the most well-known stream cipher. It is also referred to as the *Vernam cipher*, originating from work of G. Vernam in 1917.

8. The one-time pad offers unconditional security, at the price of a key of length equal to that of the plaintext, which can be used only once. It is an example of a synchronous stream cipher.

9. An extensive study of stream ciphers can be found in [Ru86].

14.8.5 KEY DISTRIBUTION PROBLEM

Definitions:

The problem of producing secure keys that can be used by each of a group of users to be able to communicate in secret with every other user is called the *key distribution problem*.

A *key distribution center* (*KDC*) is a trusted third party that distributes short-term secret keys for secure communications from a particular party to another.

Facts:

1. The security of all cryptographic mechanisms depends on the secrecy and/or authenticity of keying material.

2. Consider a system that uses only symmetric cryptographic algorithms such as the U.S. Data Encryption Standard (DES), with n users. If each pair of users is to be able to communicate privately, then each user must acquire and maintain $n-1$ secrets keys, one for each other party; overall, this requires $\frac{n(n-1)}{2}$ keys in the system. These keys must be distributed by secure means, for example by each pair meeting in person or by trusted couriers, prior to the commencement of a secure communication. Such distribution is typically both inconvenient and costly, and increasingly unmanageable as n grows.

3. A solution to the key distribution problem is obtained by using public-key cryptography (§14.9.1).

4. Another solution of the key distribution problem is to make use of a trusted third party T, as follows. Each party A shares a unique long-term secret key K_{AT} with T. Any party A may acquire a short-term secret key or *session key* to communicate securely with any other party B, using the third party T as a key distribution center in the following way:

   - using K_{AT} to establish a secure channel with T, A requests from T a new random secret key to use with B;
   - T creates such a key K_{AB}, transfers one copy of it to A using the secure channel facilitated by K_{AT}, and makes another copy of it available either directly to B using the secure channel facilitated by K_{BT}, or sends a copy of K_{AB} encrypted under K_{BT} to A over the secure channel to A; A then transfers this encrypted key to B.

5. The *Kerberos protocol*, originating from Project Athena at M.I.T. in 1987 and based on a 1978 protocol of R. Needham and M. Schroeder, is a particular example of an authenticated key distribution protocol based on symmetric cryptographic techniques and the use of a KDC.

6. An alternative to a KDC is to use a trusted third party as a *key translation center* (KTC); in this case, party A itself creates the key K_{AB} intended for use with a party B, transfers it securely to the trusted party under the channel secured by K_{AT}, and relies on the trusted party to decrypt the key intended for B, secure it specifically for B by reencrypting it under the key K_{BT}, and then make this encrypted version available to B either directly or via A.

7. The use of both KDCs and KTCs for cryptographic key establishment was popularized within the U.S. financial community by the ANSI X9.17 standard.

8. The use of a KDC or KTC to solve the key distribution problem can be pictured as a spoked-wheel, with each user on the perimeter at the end of a spoke, and the trusted party at the center. A secure channel between any two users A and B on the perimeter can be established by using the secure channels provided by the spokes (keys K_{AT} and K_{BT}), set up during system initialization, to establish a secure channel directly (by key K_{AB}). Each party now needs initially to acquire only a single secret key (corresponding to a single spoke), rather than $n-1$ keys as before.

14.9 PUBLIC-KEY SYSTEMS

The invention of public-key cryptosystems in the mid-1970s has had a profound impact on cryptography. In a public-key cryptosystem, there are public encryption keys that can be publicly shared and secret decryption keys that cannot be found, using a practical amount of computation, from the public encryption keys.

This fundamental difference between public-key cryptosystems and symmetric-key cryptosystems, where knowledge of an encryption key brings knowledge of the corresponding decryption key, makes public-key cryptography useful for many different types of practical applications.

14.9.1 BASIC INFORMATION

Definitions:
Public-key cryptography is the study of codes where each party has a pair of keys (a *private key* and a *public key*) and only needs to keep the private key secret.

Facts:

1. Public-key cryptography was first proposed by W. Diffie and M. Hellman in 1976 [DiHe76].

2. The fundamental concept is that the public key, which can be made known to everyone, allows anyone to encrypt messages for A, but the decryption of these messages can be carried out only with knowledge of the corresponding private key, which only party A knows.

3. A necessary condition for a public-key cryptosystem to be secure is that it be infeasible to derive a private key from the corresponding public key.

4. Three advantages offered by public-key systems over private-key systems are the following:
    - They provide a solution to the key distribution problem without using a KDC or a KTC. Each party requires only one public key and one private key in order to communicate securely with all other parties, as opposed to a separate private key to communicate with each. A new prerequisite however is that authentic copies of other parties' public keys be available by some means.
    - They provide an elegant solution to the problem of digital signatures.
    - They allow public key distribution systems, whereby, surprisingly, secret symmetric keys can be derived jointly by two remote parties through communications over unsecured public channels.

5. Diffie and Hellman proposed solutions to the public key distribution problem, and shortly after they conceived the notion, practical instantiations of both public-key encryption and public-key signature systems were proposed: knapsack encryption schemes and both RSA signature and encryptions schemes.

6. Many different public-key cryptosystems have been proposed. Among the most important of these are:
   - the *RSA cryptosystem* (security based on the difficulty of factoring large integers and the problem of finding eth roots modulo a composite integer where e is an integer);
   - the *Rabin cryptosystem* (security based on the difficulty of factoring large integers and the problem of finding square roots modulo a composite integer);
   - the *El Gamal cryptosystem* (security based on the difficulty of finding discrete logarithms);
   - the *McEliece cryptosystem* (security based on the difficulty of decoding certain linear codes);
   - the *Merkle-Hellman cryptosystem* (security based on the difficulty of the subset sum problem);
   - the *elliptic curve cryptosystem* (security based on the theory of elliptic curves); see the Certicom ECC Tutorials and Whitepapers page at the website:
     http://www.certicom.com/ecc/index.htm
   - the *NTRU cryptosystem* (security based on the difficulty of lattice problems); see the NTRU Public Key Cryptosystem Overview at the website:
     http://www.ntru.com/tutorials/techsummary.htm

7. The current status of public-key cryptosystems is that they offer many advantages for key establishment and digital signatures, but so far have generally been too computationally expensive for bulk encryption.

8. Typically, in practice symmetric ciphers like DES continue to be used for encryption, public-key systems like RSA are used for digital signatures, and a combination of symmetric and public-key techniques are used for key establishment.

14.9.2 KNAPSACK ENCRYPTION SCHEME

Definitions:

Given a set of positive integers $\{a_1, a_2, \ldots, a_n\}$ and a specified sum s, the **knapsack problem** is the problem of finding a 0-1 vector $(x_1, x_2, \ldots, x_n)$ such that $\sum_{i=1}^{n} a_i x_i = s$, or determining that such a vector does not exist.

A **super-increasing sequence** is a set $\{a_1, a_2, \ldots, a_n\}$ of positive integers with the property that $a_i > \sum_{j=1}^{i-1} a_j$ for each $i = 2, \ldots, n$.

The **knapsack cryptosystem** is a cryptosystem in which encryption is carried out using a super-increasing sequence of integers.

Facts:

1. The knapsack encryption scheme, due to R. Merkle and M. Hellman, was the first concrete realization of a public-key encryption scheme. Its security is based on the knapsack problem.

2. Although the knapsack problem, also known as the *subset sum* problem, is known to be NP-hard, there are special instances of the problem which are easy to solve.

3. For a super-increasing sequence the knapsack problem is very easy to solve. The Merkle-Hellman knapsack encryption scheme disguises a super-increasing sequence by modular multiplication and a permutation.

4. Key generation for the knapsack encryption scheme can be carried out as follows. An integer n is fixed. Each user does the following:
   - choose a super-increasing sequence of positive integers $a_1, a_2, \ldots, a_n$, and choose a modulus M with $M > a_1 + a_2 + \cdots + a_n$;
   - choose an integer W, $1 \leq W \leq M - 1$, with $\gcd(W, M) = 1$;
   - choose a random permutation π of the integers $\{1, 2, \ldots, n\}$;
   - compute $b_i = W a_{\pi(i)} \bmod M$, for $i = 1, 2, \ldots, n$;
   - publish the public key $(b_1, b_2, \ldots, b_n)$; the private key is $(\pi, M, W, a_1, a_2, \ldots, a_n)$.

5. The *basic Merkle-Hellman knapsack encryption scheme* operates as follows, where person B sends a message to person A:
   - *encryption:* B does the following:
     ◇ look up A's public key $(b_1, b_2, \ldots, b_n)$;
     ◇ represent the message m as a binary string of length n, $m = m_1 m_2 \ldots m_n$; if the message is too big, break it into blocks;
     ◇ compute $c = m_1 b_1 + m_2 b_2 + \cdots + m_n b_n$;
     ◇ send c to A.
   - *decryption:* A does the following:
     ◇ compute $d = W^{-1} c \bmod M$;
     ◇ solve the super-increasing knapsack by finding integers $r_1, r_2, \ldots, r_n$, $r_i \in \{0, 1\}$, such that $d = r_1 a_1 + r_2 a_2 + \cdots + r_n a_n$;
     ◇ conclude that the message bits are $m_i = r_{\pi(i)}$, $i = 1, 2, \ldots, n$.

6. The Merkle-Hellman scheme, as well as sundry variations of it, have all been shown to be insecure. Essentially, this is because the underlying easy knapsack can be recovered from the public knapsack with minimal effort.

Examples:

1. The sequence $1, 2, 5, 10, 20, 40$ is super-increasing, but $1, 2, 6, 10, 18, 30$ is not.

2. The solution to the super-increasing subset problem with super-increasing sequence $1, 2, 5, 10, 20, 40$ and subset sum 27 is $27 = 20 + 5 + 2$.

14.9.3 RSA CRYPTOSYSTEM

Definition:

The *RSA cryptosystem* is a public-key cryptosystem that encrypts messages using modular exponentiation, where the modulus is the product of two very large primes.

Facts:

1. The RSA cryptosystem was invented by R. Rivest, A. Shamir, and L. Adleman in 1978, and is the most widely used public-key cryptosystem today.

2. The RSA cryptosystem supports both secrecy and digital signatures, and its security is based on the difficulty of factoring integers.

3. Keys are generated in RSA when each user does the following:
   - pick two large primes p and q, each roughly the same size. Compute $n = pq$ and $\phi(n) = (p-1)(q-1)$;
   - select a random integer e, $1 < e < \phi(n)$, such that $\gcd(e, \phi(n)) = 1$;
   - using the extended Euclidean algorithm (§4.2.2), compute the unique integer d, $1 < d < \phi(n)$, such that $ed \equiv 1 \pmod{\phi(n)}$;
   - publish the public key (n, e); the private key is d.

4. The RSA system is used as follows for B to send a message to A:
   - *encryption*: B does the following:
     ◇ look up A's public key (n, e);
     ◇ represent the message as an integer m in the interval $[0, n-1]$; if the message is too big, break it into blocks;
     ◇ compute $c = m^e \bmod n$;
     ◇ send c to A.
   - *decryption*: A does the following:
     ◇ use the private key d to recover $m = c^d \bmod n$.

5. The RSA system can be used to send signed messages as follows, where A signs a message for B:
   - *signature generation*: A does the following:
     ◇ represent the message m as an integer in the interval $[0, n-1]$; if the message is too big, break it into blocks;
     ◇ use his/her private key d to compute $s = m^d \bmod n$;
     ◇ send the signature s to B (the message m will be recovered from s itself).
   - *signature verification*: B does the following:
     ◇ look up A's public key (n, e);
     ◇ recover the message $m = s^e \bmod n$;
     ◇ accept A's signature, provided that m is "meaningful".

6. In practice, one does not select an exponent e at random, but instead chooses some small value such as 3, 17, or $2^{16} + 1$.

7. One technique for rendering a message meaningful is to add some prearranged redundancy to the message, for example by requiring that m begin with a predetermined 64-bit pattern. Another is to use a suitable hash function (§14.9.2) before signing, even when the message m is short enough to fit in a single block.

8. A common technique used to avoid having to sign each block of a long message m is to first compute $m^* = H(m)$, where H is a public one-way hash function that outputs integers in the interval $[0, n-1]$, and then send the message m along with the signature of the hash value, $s = (m^*)^d \bmod n$ to B. Person B can verify the signature by computing $s^e \bmod n$ and $H(m)$, and checking that these two quantities are the same.

9. Breaking the RSA encryption or signature schemes is widely believed to be as difficult as factoring the modulus n in these schemes, although such an equivalence has never been proven.

10. Given the latest progress on the factorization of large integers, a 512 bit modulus n will provide only marginal security from concerted attack; as of 1999, a modulus n of at least 768 bits is recommended.

11. More information about the RSA cryptosystem can be obtained on the Internet at the RSA Laboratories site

    www.rsa.com/rsalabs

Example:

1. Take the modulus in the public key encryption scheme to be $2537 = 43 \cdot 59$ and the exponent to be 13. The plaintext message PUBLICKEYCRYPTOGRAPHYX, corresponding to 1520 0111 0802 1004 2402 1724 1519 1406 1700 1507 2423 when letters are replaced with the corresponding integers in the set $\{0, 1, \ldots, 25\}$, is mapped to the ciphertext message 0095 1648 1410 1299 0811 2333 2132 0370 1185 1457 1084. For example, the first block 1520 is mapped to 0095 since $1520^{13} \bmod 2537 = 95$.

14.9.4 EL GAMAL CRYPTOSYSTEM

Definition:

The *El Gamal cryptosystem* is a public-key cryptosystem based on the discrete logarithm problem (see Chapter 4).

Facts:
1. The El Gamal cryptosystem was proposed in 1985 by T. El Gamal.
2. The El Gamal cryptosystem supports both secrecy and digital signatures and its security is based on the difficulty of the discrete logarithm problem.
3. Keys for the El Gamal cryptosystem are generated when each user does the following:
    - pick a large prime p and a generator α of the multiplicative group Z_p^* of the integers modulo p;
    - select a random integer a, $1 < a < p-1$, and compute $\alpha^a \bmod p$;
    - publish the public key (p, α, α^a); the private key is a.
4. The *El Gamal encryption scheme* works as follows where B sends a message to A:
    - *encryption*: B does the following:
        - look up A's public key (p, α, α^a);
        - represent the message as an integer m in the interval $[0, p-1]$; if the message is too big, break it into blocks;
        - select a random integer k, $1 \leq k \leq p-2$;
        - compute $\alpha^k \bmod p$ and $m \cdot (\alpha^a)^k \bmod p$
        - send $(\alpha^k, m\alpha^{ak})$ to A;
    - *decryption*: A does the following:
        - use the private key a to compute $\alpha^{ak} = (\alpha^k)^a \bmod p$ and then compute $\alpha^{-ak} \bmod p$;
        - recover m by computing $(\alpha^{-ak})(m\alpha^{ak}) \bmod p$.
5. The *El Gamal signature scheme* operates as follows where A signs a message for B:
    - *signature generation*: to sign a message m of arbitrary length, A does the following:
        - select a random integer k, $1 \leq k \leq p-2$, with $\gcd(k, p-1) = 1$;
        - use the extended Euclidean algorithm to compute an integer I, $1 \leq I \leq p-2$, such that $kI \equiv 1 \pmod{p-1}$;
        - compute $r = \alpha^k \bmod p$;
        - compute $H(m)$, the hash of m, using a one-way hash function H;
        - compute $s = I \cdot (H(m) - ra) \bmod (p-1)$
        - send the signature (r, s) along with the message m to B.
    - *signature verification*: B does the following:
        - look up A's public key (p, α, α^a);
        - compute $H(m)$;
        - compute $u_1 = (\alpha^a)^r \cdot (r^s) \bmod p$;
        - compute $u_2 = \alpha^{H(m)} \bmod p$;
        - accept the signature only if $u_1 = u_2$.
6. Breaking the El Gamal encryption or signature scheme is widely believed to be as difficult as computing logarithms in Z_p^*, although such an equivalence has never been proven.

7. Given the latest progress on the discrete logarithm problem, a 512-bit modulus p will provide only marginal security from concerted attack; as of 1999, a modulus p of at least 768 bits is recommended.

8. The parameters p and α can be common to a group of users, in which case user A's public key is just α^a.

9. The ElGamal cryptosystem can be generalized to work in any finite cyclic group G instead of the multiplicative group $\mathcal{Z}_p^*$. Groups that have been proposed for this purpose for reasons of practical efficiencies include the following: the multiplicative group $F_{2^m}^*$ of finite fields of characteristic two and the group of points on an elliptic curve over a finite field.

14.9.5 MC ELIECE ENCRYPTION SCHEME

Definition:

The **McEliece encryption scheme** is the encryption method that is the foundation of a public-key cryptosystem based on linear codes from the theory of error-correcting codes.

Facts:

1. R. McEliece introduced the McEliece encryption scheme in 1978 as the basis of a public-key cryptosystem.

2. The security of the McEliece encryption scheme is based on the fact that the general decoding problem for linear codes is NP-hard.

3. Keys are generated in the McEliece encryption scheme in the following way: Integers k and n are first fixed. Each user does the following:
   - choose a $k \times n$ generator matrix G for a binary (n, k)-code that can correct t errors, and for which there is an efficient decoding algorithm;
   - choose a random $k \times k$ binary nonsingular matrix S;
   - choose a random $n \times n$ permutation matrix P;
   - compute the $k \times n$ matrix $\widehat{G} = SGP$;
   - publish the public key $(\widehat{G}, t)$; the private key is (S, G, P).

4. A party B sends a message to a party A in the McEliece encryption scheme as follows:
   - B does the following to encrypt the message:
     ◇ look up A's public key $(\widehat{G}, t)$;
     ◇ represent the message m as a binary string of length k; if the message is too big, break it into blocks;
     ◇ choose a random error vector z of length n and Hamming weight $\leq t$;
     ◇ compute $c = m\widehat{G} + z$;
     ◇ send c to A.
   - A does the following to decrypt the received message:
     ◇ compute $\widehat{c} = cP^{-1}$;
     ◇ use the decoding algorithm for the code generated by G to decode $\widehat{c}$ to $\widehat{m}$;
     ◇ compute $m = \widehat{m}S^{-1}$.

5. McEliece suggested using a special type of error-correcting code called a *Goppa code* with parameters $n = 1024$, $k = 524$, $t = 50$, in step 1 of the key generation procedure. (For each irreducible polynomial $g(x)$ of degree t over the finite field of 2^m elements, there exists a binary Goppa code of length $n = 2^m$ and dimension $k \geq n - mt$ capable of correcting any pattern of t or fewer errors. Furthermore, efficient decoding algorithms are known for Goppa codes. For further information, see [MaSl77].) With these parameters, the McEliece encryption scheme is believed to form the basis of a secure public-key cryptosystem.

6. Two disadvantages of the scheme are the large size of public keys and the message expansion.

14.9.6 DIGITAL SIGNATURE ALGORITHM

The *digital signature algorithm* (DSA) was adopted in 1994 as a signature standard by the U.S. Government. Its security is based on the difficulty of the discrete logarithm problem in a large subgroup of the multiplicative group Z_p^*. The scheme can be viewed as a variant of the El Gamal signature scheme.

Facts:
1. To generate keys, each user does the following:
   - pick a prime p such that $p - 1$ has a prime factor q, where $2^{159} < q < 2^{160}$;
   - select a random integer h, $1 < h < p-1$, and such that $h^{\frac{p-1}{q}} \bmod p > 1$; let $g = h^{\frac{p-1}{q}} \bmod p$;
   - select a random integer x, $0 < x < q$, and compute $y = g^x \bmod p$;
   - the user's public key is (p, q, g, y); the user's private key is x.

2. The following steps make up a digital signature algorithm where A signs a message that is to be sent to B:
   - *signature generation*: to sign a message m of arbitrary length, A does the following:
     ◇ choose a random integer k, $0 < k < q$;
     ◇ compute $k^{-1} \bmod q$;
     ◇ compute $r = (g^k \bmod p) \bmod q$;
     ◇ compute $H(m)$, the hash of the message, using a one-way hash function $H()$;
     ◇ compute $s = k^{-1} \cdot (H(m) + xr) \bmod q$;
     ◇ send the signature (r, s) along with message m to B.
   - *signature verification*: B does the following:
     ◇ look up A's public key (p, q, g, y);
     ◇ compute $w = s^{-1} \bmod q$;
     ◇ compute $H(m)$;
     ◇ compute $u_1 = H(m)w \bmod q$ and $u_2 = rw \bmod q$;
     ◇ compute $v = ((g^{u_1} y^{u_2}) \bmod p) \bmod q$;
     ◇ accept the signature only if $v = r$.

3. The U.S. Government standard specifies that the prime p must be between 512 and 1024 bits in length, however it is generally recommended that p be at least 768 bits in length.

4. The parameters p, q, and g can be common to a group of users, in which case the public key is just y.

5. The specific hash algorithm specified for use within the DSA standard is the *secure hash algorithm* (SHA-1) as specified in the U.S. Federal Information Processing Standards Publication 180-1 (FIPS 180-1: Secure Hash Standard).

14.9.7 FIAT-SHAMIR SIGNATURE SCHEME

Definition:

The **Fiat-Shamir signature scheme** is a signature scheme based on the difficulty of extracting square roots modulo a composite number n.

Facts:

1. The Fiat-Shamir signature scheme was introduced in 1986.

2. The security of the Fiat-Shamir signature scheme is based on the difficulty of extracting square roots modulo a composite number n, a problem that is equivalent in difficulty to the problem of factoring n.

3. Key generation is done using the Fiat-Shamir scheme as follows. Integers k and t are fixed; each user does the following:
   - pick two primes p and q, and compute $n = pq$;
   - select k random integers $s_1, s_2, \ldots, s_k$ in the interval $[1, n-1]$, such that for each i, $\gcd(s_i, n) = 1$;
   - compute $v_i = s_i^{-2} \bmod n$, for $1 \leq i \leq k$.
   - the user's public key is $(v_1, v_2, \ldots, v_k, n)$; the user's private key is $(s_1, s_2, \ldots, s_k)$.

4. The *Fiat-Shamir signature scheme* operates as follows, where A signs a message for B:
   - *signature generation*: to sign a message m of arbitrary length, A does the following:
     ◇ choose random integers $r_1, r_2, \ldots, r_t$ in the range $[0, n-1]$, and compute $x_i = r_i^2 \bmod n$ for each i, $1 \leq i \leq t$;
     ◇ compute $H(m, x_1, x_2, \ldots, x_t)$, where H is a one-way hash function, and use its first kt bits as entries e_{ij} of a $t \times k$ binary matrix E;
     ◇ compute $y_i = (r_i \prod_{e_{ij}=1} s_j) \bmod n$ for $i = 1, 2, \ldots, t$;
     ◇ send the signature $(y_1, y_2, \ldots, y_t, E)$ along with message m to B.
   - *signature verification*: B does the following:
     ◇ look up A's public key $(v_1, v_2, \ldots, v_k, n)$;
     ◇ compute $z_i = (y_i^2 \prod_{e_{ij}=1} v_j) \bmod n$ for $i = 1, 2, \ldots, t$;
     ◇ compute $h = H(m, z_1, z_2, \ldots, z_t)$;
     ◇ accept the signature only if the first kt bits of h are the same as the entries e_{ij} of E.

5. The Fiat-Shamir scheme is provably secure, provided that factoring is difficult and H is a truly random function.

6. The modulus n should be large enough to withstand the best algorithms known for factoring integers; as of 1999, a size of at least 768 bits is recommended.

7. To avoid forgeries, the parameters k and t should be chosen so that kt is at least 72.

14.9.8 KEY DISTRIBUTION

Definitions:

Entity (identity) authentication is some form of positive corroboration of the identity of another party in a protocol.

A key establishment mechanism provides **explicit key authentication** (key confirmation) if one party receives some indication that a second party whose identity has been corroborated actually knows the secret key established; it provides (only) **implicit key authentication** if no such indication is received, but nonetheless the identified second party is the only other party who could feasibly derive that key.

Facts:

1. The solution to the key distribution problem (§14.8.5) by symmetric techniques and a trusted third party has several disadvantages, two of which are the requirement and involvement of an on-line trusted third party, and that compromise of long-term keys shared between a user and the trusted party will compromise all other keys established for that user based on that key.

2. If public-key cryptographic techniques are used, even in the case where each party computes its own private keys of (public, private) key pairs, two types of keys generally need to be distributed between parties: *public keys* (for use in algorithms such as RSA), and *symmetric keys* (for use in symmetric algorithms such as DES).

3. Public keys can be delivered in person, but this is costly; other appropriate means are generally used, such as public-key certificates (§14.9.9).

4. *Key establishment* mechanisms are generally used to make a symmetric key secretly available to two authorized parties for subsequent cryptographic use.

5. Key establishment mechanisms may be divided into *key transfer* mechanisms, in which a key created by one party is securely transmitted to another; and *key agreement* mechanisms, whereby two parties jointly establish a shared secret key which is a function of information contributed by each.

6. Two basic requirements in key establishment are the secrecy of the established key, and that each party learn the true identity of the other party sharing the key.

7. Authentication may be either *unilateral* or *mutual*.

8. The number of *passes* refers to the number of messages exchanged between the parties.

9. In 1976, W. Diffie and M. Hellman provided the first practical solution to the key distribution problem by presenting a key agreement protocol with the following properties: two parties who may possibly have never met before nor shared any information related to keys are able to establish a shared secret key by exchanging two messages over an unsecured public channel. To set up the *Diffie-Hellman key agreement* where two parties establish a secret key over a public channel, carry out the following steps:
   - fix an appropriate prime p and generator α of $\mathcal{Z}_p^*$;
   - A chooses a random secret $r_A \in \{1, 2, \ldots, p-2\}$, and sends α^{r_A} **mod** p to B;
   - B chooses a random secret $r_B \in \{1, 2, \ldots, p-2\}$, and sends α^{r_B} **mod** p to A;
   - B computes the shared key as $k = (\alpha^{r_A})^{r_B}$ **mod** p;
   - A receives α^{r_B} and computes $k = (\alpha^{r_B})^{r_A}$ **mod** p.

10. As of 1999, it is recommended that p be at least 768 bits in length.

11. The basic mechanism provides secrecy against passive intruders but not authentication — neither party is assured about the other's identity, and neither obtains entity authentication or implicit key authentication.

12. *El Gamal key agreement* (one-pass; B sends to A information allowing key agreement): El Gamal's encryption scheme (§14.9.4) is a variation of the Diffie-Hellman key agreement protocol, and can be used for a one-pass key transfer protocol with implicit key authentication of the intended recipient A to the originator B, as follows (assume the same setup as in §14.9.4):
   - B chooses a random integer l, $1 \leq l \leq p-2$, and sends A the quantity $\alpha^l \bmod p$;
   - B looks up A's public key α^a and computes for himself the key $k = (\alpha^a)^l \bmod p$;
   - A computes the same quantity upon receipt of B's message, as $k = (\alpha^l)^a \bmod p$.

13. The recipient A has no corroboration of whom it shares the secret key with; the protocol does not provide entity authentication to either party.

14. If A independently initiates an analogous protocol simultaneously with B, resulting in the key k', and each party then computes $K = kk' \bmod p$, then the combined two-pass scheme provides key agreement with mutual implicit key authentication (but still provides neither entity authentication nor explicit key authentication).

15. Both the one-pass and two-pass schemes have the advantage that public keys could be exchanged (for example, by including certificates) within the protocol itself without additional passes.

16. The following three-pass variation of the basic Diffie-Hellman protocol allows the establishment of a shared secret key between two parties with mutual entity authentication and mutual explicit key authentication. This technique makes use of digital signatures. Set up is the same as in basic Diffie-Hellman key agreement, plus:
   - A has RSA public signature key (e_A, n_A), and private key d_A; B has analogous keys;
   - RSA signature generation is done using an appropriate one-way hash function H prior to exponentiation; A's signature on m is $S_A(m) = H(m)^{d_A} \bmod n_A$;
   - A and B have access to authentic copies of the other's public signature keys.

17. *Diffie-Hellman with explicit authentication* (three-pass):
   - A generates a secret random number $r_A \in \{1, 2, \ldots, p-2\}$ and sends to B: $\alpha^{r_A} \bmod p$;
   - B generates a secret random number $r_B \in \{1, 2, \ldots, p-2\}$, and computes the shared key $k = (\alpha^{r_A})^{r_B} \bmod p$. B signs the concatenation of both exponentials and the computed key, and sends to A: α^{r_B}, $S_B(\alpha^{r_B}, \alpha^{r_A}, k)$;
   - A computes the shared key $k = (\alpha^{r_B})^{r_A} \bmod p$, and verifies with B's public key that the message recovered on signature verification of the received message is the hash of the following three quantities: the cleartext exponential received, the exponential sent in the first message, and the computed key k;
   - if signature verification fails, A terminates with failure; otherwise, A accepts that k is actually shared with B, and sends to B the message: $S_A(\alpha^{r_A}, \alpha^{r_B}, k)$.
   - B analogously verifies A's signature on the received message;
   - if signature verification fails, B terminates with failure; otherwise B accepts that k is actually shared with A.

18. Inclusion of the key k within the hashed, signed portion of the second and third messages provides explicit key authentication.

19. RSA can also be used in a one-pass protocol for key establishment by key transfer. The basic protocol consists of A using B's public encryption key, encrypting a randomly generated key k, and sending it to B. This provides B with no authentication regarding the source of the key, but can be modified by having the sender RSA-sign the message using its own private signature key, before RSA-encrypting it with the intended recipient's public encryption key.

14.9.9 PUBLIC-KEY CERTIFICATES

Definitions:

A public-key certificate consists of a *data part* and a *signature part*.

The data part consists of the name of an entity, the public key corresponding to that entity (for example, RSA public key), and possibly additional relevant information (for example, entity's street or network address, validity period for public key, etc.).

The signature part consists of the signature of a trusted authority, called a *central authority* or *certification authority* (*CA*), over the data part.

Facts:

1. The distribution of public keys is generally easier than that of symmetric keys, since secrecy is not required. However, the integrity (authenticity) of public keys is critical.

2. In 1979 L. Kohnfelder suggested the idea of using *public-key certificates* to facilitate the distribution of public keys over unsecured channels, such that their authenticity can be verified.

3. For any party B to verify the authenticity of the public key of any party A, B must have an authentic copy of the public (signature verification) key of the CA. (For simplicity, assume that the authenticity of this public key is provided to party B by non-cryptographic means, for example, by having party B obtain it from the CA in person.)

4. Given the Fact 3, B can then carry out the following steps:
    - acquire the public-key certificate of A over some unsecured channel, either from a central database of certificates, from A directly, or otherwise;
    - using the CA's public key, verify the CA's signature on A's certificate;
    - if this signature verifies correctly, accept the public key in the certificate as A's authentic public key; otherwise, assume the public key is invalid.

5. Before creating a public-key certificate for a A, the CA must take appropriate measures to verify the identity of A and the fact that the public key to be certified actually belongs to that party.

6. One method might be to require that A appear before the CA with a conventional government passport as proof of identity, and obtain A's public key from A in person.

7. Once the CA creates a certificate for a party, the trust that all other entities have in the authenticity of the CA's public key can be used transitively to gain trust in the authenticity of that party's public key, through acquisition and verification of the certificate.

14.9.10 AUTHENTICATION

Definitions:

Two important types of authentication are **entity authentication** (also known as **identification**), which authenticates the identity of a party, and **message authentication**, which authenticates the validity of a message.

Facts:

1. One authentication method, allowing identification of one party using a 2-pass challenge-response protocol based on a shared secret key, is called an *IFF scheme* (*identification, friend or foe*). This terminology originates from the original use of this technique for identifying aircraft during times of war — the challenger is a military radar station, and the challenged entity is an aircraft, either friendly or foreign.

    An entity B (the *challenger*), which wishes to be able to identify a second entity A (the *responder*), distributes a shared secret key to that entity ahead of time.
    - the challenger B sends to an unidentified entity X a time-varying number (the *challenge*);
    - entity X receives the challenge, and replies with an answer (the *response*) expected by B to be a one-way function of the challenge and the shared secret key;
    - if the response is that which was expected from entity A, then the challenger accepts X to be A; otherwise X remains unidentified.

2. Any one-way function of the shared secret key can be used, including encryption with a block cipher such as DES, or an appropriate keyed one-way hash function.

3. The protocol must be modified in environments where the roles of challenger and responder can be reversed, otherwise a challenged party, upon being challenged, can initiate a new protocol by reflecting the challenge back to the challenger, extracting the correct response from that entity, and then using that response to respond to the original challenge.

4. An authentication scheme, proposed by L. Guillou and J.-J. Quisquater in 1988 known as the *GQ scheme*, allows identification of one party and is based on public-key techniques. It is an optimization, with respect to number of messages and memory requirements, of an earlier scheme of A. Fiat and A. Shamir; the Fiat-Shamir Signature Scheme discussed in §14.9.7. The GQ Scheme involves three messages between entities A and B, where A is the prover (entity whose identity is to be corroborated) and B is the verifier (or challenger). It was designed with the specific application in mind where the prover is a processor such as a "smart card" (integrated circuit mounted on a credit card) with limited processing power and memory. This scheme is set up as follows:
    - a trusted authority C randomly selects two appropriate primes p and q as in RSA, and computes $n = pq$;
    - C defines as the public exponent an integer v coprime to $\phi(n)$;
    - the values n and v are made public; C keeps p and q secret;
    - each entity X has a unique identity I_X from which an integer $J_X < n$ is derived using publicly known redundancy rules;
    - for each integer $J = J_X$, C computes $U = (J)^{v^{-1}} \bmod n$, and gives to entity X the secret $W = U^{-1} = (J)^{-(v^{-1})} \bmod n$. (Note: $J \cdot W^v \equiv 1 \pmod{n}$.)

The GQ identification scheme operates as follows to provide unilateral identification of prover A to verifier B:
- entity A with identifier I_A selects a random integer r, $1 < r < n-1$, and computes the *initial witness* $T = r^v \bmod n$;
- A sends to B the pair of integers (T, I_A);
- B selects a random *challenge* d, $0 \leq d \leq v-1$, and sends d to A;
- A computes the *response* $t = r \cdot W^d \bmod n$, and sends t to B;
- B receives t, computes $J^d \cdot t^v \bmod n$, and accepts A's identity as authentic if this quantity is equal to T (which it will be if A carried out the protocol properly).

5. For security reasons: a new random value r must be chosen each time the GQ identification scheme is run and the prover must respond to only one challenge d for each initial witness T.

6. A fraudulent prover can defeat the GQ identification scheme with a 1 in v chance by guessing d correctly *a priori* (and then forming $T = J^d \cdot t^v \bmod n$ as the verifier would). Thus the recommended bit length of v depends on the environment under which attacks could be mounted. For a fraudulent prover who must participate locally and is subject to being apprehended in person upon failure, 8 to 16 bits may suffice; if remote attacks are possible, for example, by telecommunications linkups, 30 or more bits may be required.

7. Extracting vth roots modulo n appears necessary to defeat the GQ identification scheme, and this is believed to be intractable in the absence of knowledge of the factorization of n.

8. The security of the GQ identification scheme relies on the fact that a fraudulent prover has only a 1 in v chance of guessing d correctly *a priori* (and then forming $T = J^d \cdot t^v \bmod n$ as the verifier would); and it can be shown that if a fraudulent prover is able to correctly respond to two different challenges for the same initial witness, then that prover can recover W, i.e. compute a vth root modulo n, which is believed to be infeasible unless the factorization of n is known.

9. The following algorithm computes a short quantity called a *message authentication code* (MAC), which can be appended to a message as a data integrity mechanism to allow the receiver to verify that the message has not been altered by an unauthorized party. In addition, this provides a type of symmetric data origin authentication — the identity of the party which originated the message can be implicitly verified. MAC algorithms such as this have been used in the financial services industry for over 15 years. The algorithm is set up as follows:
- let w be the required bit length of the MAC;
- select a fixed n-bit block cipher algorithm E (for example, DES, yielding $n = 64$), such that $w \leq n$;
- the originator of the message and the intended recipient must share a secret key k for the block cipher E.

The CBC-based MAC scheme operates as follows to append a keyed checksum to message m for data integrity. The originator generates the MAC as follows:
- a single 1-bit is appended to the message m (this allows unambiguous recovery of the original message even after padding as outlined below);
- the augmented message is broken into n-bit blocks m_i; the last block is padded by zero or more 0-bits as required to fill it completely; label the resulting blocks $x_1, \ldots, x_t$;

- the MAC is defined to be the leftmost w bits of the value c_t computed by the following sequence of computations:
$$c_1 = E_k(x_1); \qquad c_i = E_k(x_i \oplus c_{i-1}), \quad 2 \leq i \leq t;$$
- the message m, along with the MAC, are sent to the recipient in such a format that the recipient can separate the MAC from the message; the padding bits may or may not be sent along with the message (as agreed by the two parties).

The MAC verification by recipient is carried out as follows:
- the recipient receives the message m, adds padding bits as the sender did (if these were not transmitted along with m), and computes a MAC on the message using the shared key k, as outlined above;
- the recipient compares the computed MAC to the received MAC; if they agree, the recipient is satisfied that the message was not altered during transit (and that it originated from the party with whom the key k is shared).

10. The strength of the MAC algorithm depends on the secrecy and bit length of the key k, the strength of the block cipher E, and the bit length w of the MAC.

11. As an option in the MAC construction, the last block c_t can be subjected to additional processing to make the algorithm more resistant to certain types of attacks.

12. Any digital signature scheme, and, in particular, public key schemes such as RSA, can also be used to provide message authentication.

14.9.11 SECRET SHARING

Definitions:

A scheme whereby a secret datum S can be divided up into n pieces, in such a way that knowledge of any k or more of the n pieces allows S to be easily recovered, but knowledge of $k-1$ or fewer pieces provides no information about S whatsoever (that is, no more information than 0 pieces) is called a (k, n) **threshold scheme**, and is an instance of a more general class of techniques known as **secret sharing** schemes.

Facts:

1. Threshold schemes were introduced by A. Shamir in 1979.

2. *Shamir's secret sharing scheme*: To set up Shamir's Secret Sharing Scheme, first do the following:
   - define an upper bound $S_{\max}$ on any secret number S to be shared;
   - define an upper bound $n_{\max}$ on the number of participants;
   - select a prime number p which exceeds both $n_{\max}$ and $S_{\max}$.

To use the scheme to split a secret S so that any k of n users can recover it, the following steps are used:
   - *splitting up the secret*: a trusted party does the following:
     ◊ obtain a secret number S to be shared, $S < S_{\max}$, and define $a_0 = S$;
     ◊ define the number of active participants to be $n < n_{\max}$ (additional active participants can easily be added later);

⋄ define a recovery threshold $k \leq n$;
⋄ select $k-1$ random integers a_i, $1 \leq i \leq k-1$, from $[0, p-1]$, and consider the polynomial $f(x) = \sum_{i=0}^{k-1} a_i x^i$ of degree at most $k-1$;
⋄ pick n distinct values x_j, $1 \leq j \leq n$ (for example, $x_j = j$), and compute $S_j = f(x_j) \bmod p$ (S_j is a share);
⋄ give the point (x_j, S_j) to participant j; x_j can be made public, but the share S_j should be a secret revealed only to participant j;

- *to recover the secret*: any k of n active participants do the following (without loss of generality, label these as participants 1 through k):
    ⋄ pool their k shares (x_j, S_j), $1 \leq j \leq k$, allowing the recovery of the coefficients of $f(x)$ by polynomial interpolation;
    ⋄ from $f(x)$, compute the secret S by evaluating $f(0)$.

3. Distributing one piece or share to each of n parties yields a method of distributing trust in a secret (such as a cryptographic key) jointly among any k-subset of them. The built-in redundancy of the secret sharing scheme also provides reliability — loss of any number of shares that leaves k or more shares remaining does not result in overall loss of the secret.

4. Shamir's Secret Sharing scheme is based on polynomial interpolation and the fact that a polynomial $y = f(x)$ of degree $k-1$ is uniquely defined by any k points (x_i, y_i) for distinct x_i, where $f(x_i) = y_i$. By construction, $f(0) = S$, that is, S is the y-intercept of the graph $y = f(x)$. No partial information regarding S is obtained from any $k-1$ (or fewer) shares because given $k-1$ shares, a kth point is needed to uniquely determine $f(x)$, and each of the p candidate points $(0, S)$ for S in $\{0, 1, \ldots, p-1\}$ defines a different (equally probable) polynomial. Polynomial evaluation can be done by Horner's rule in k multiplications and k additions. Polynomial interpolation can be done using either Lagrange's formula or Newton's formula, with the greatest computational cost being $O(k^2)$ multiplications or divisions.

9.12 HASH FUNCTIONS

Definitions:

A **hash function** h maps arbitrary length bit strings to small fixed-length (for example, 64 or 128 bit) outputs called **hash-values**. (See §17.4.1 for a discussion of hash functions in a more general setting.)

A **collision** is a pair of bit strings mapped to the same output by a hash function.

The following are common properties a hash function h may have:

- **preimage-resistance**: given any y in the range of h (for which a corresponding input is not known), it should be computationally infeasible to find any preimage x^* such that $h(x^*) = y$;
- **weak collision-resistance**: given any one input x, it should be computationally infeasible to find a second preimage $x^* \neq x$ such that $h(x) = h(x^*)$;
- **strong collision-resistance**: it should be computationally infeasible to find any two distinct inputs, x and x^*, such that $h(x) = h(x^*)$.

A ***one-way hash function*** (**OWHF**) is a function h that maps arbitrary length inputs to fixed length outputs, and has the properties of preimage-resistance and weak collision-resistance.

A ***collision-resistant hash function*** (**CRHF**) is a function h that maps arbitrary length inputs to fixed length outputs, and has the property of strong collision-resistance.

An n-bit hash function is said to have ***ideal security*** if the following properties hold:
- given a hash output, producing both a preimage and a second preimage given a first, requires approximately 2^n operations;
- producing a collision requires approximately $2^{n/2}$ operations.

Facts:

1. The basic idea is that a hash-value serves as a compact representative image (sometimes called a *digital fingerprint*, or *message digest*) of the input string, and can be used as if it were uniquely identifiable with that string.

2. The problem of checking the integrity of the potentially large original input is reduced to verifying that of a small, fixed-size hash-value.

3. A hash-value should be uniquely identifiable with a single input *in practice*, and collisions should be *computationally* difficult to find.

4. While the utility of hash functions is widespread, the most common cryptographic uses are with digital signatures and for data integrity.

5. Regarding digital signatures, long messages are typically hashed first, and then the hash-value is signed rather than signing individual blocks of the original message. Advantages of this over signing the individual blocks of the original message directly include efficiency with respect to both time and space.

6. Regarding data integrity, hash functions together with appropriate additional techniques can be used to verify the integrity of data. Specific integrity applications include virus protection and software distribution.

7. MACs (§14.9.10) are a special class of hash functions, which take in addition to message input a secret key as a second input, allowing for the verification of both data integrity and data origin authentication.

8. Given a hash function h and an input x, $h(x)$ should be easy to compute.

9. The complete specification of h is usually assumed to be publicly available.

10. Collision-resistance is required for applications such as digital signatures and data integrity, otherwise an adversary might find two messages, x and x', that have the same hash-value, obtain a signature on x, and claim it as a signature on x'.

11. Depending on the intended application and the susceptibility of the environment to certain attacks, weak or strong collision-resistance may be required.

12. There are no known instances of functions that have been proven to be one-way, that is, for which it can be *proven* (without assumptions) that finding a preimage is difficult. However, it would be most surprising if such functions indeed did not exist. All instances of "one-way functions" given to date should thus properly be qualified as "conjectured" or "candidate" one-way functions.

13. Most hash functions process fixed-size blocks of the input iteratively as follows:
    - A prespecified starting value or *initializing value* (IV) is defined.
    - The hash input $x = x_1 x_2 \ldots x_t$ of arbitrary finite length is divided into fixed-length n-bit blocks x_i. This preprocessing typically involves appending extra bits (*padding*) as necessary to extend the input to an overall bit length that is a multiple of the block length n. The padding also often includes a partial block indicating the bit length of the unpadded input.
    - Each block x_i is then used as input to a simpler hash function f called an *m-bit compression function*, which computes a new intermediate result of some fixed bit length m as a function of the previous m-bit intermediate result (initially the IV) and the block x_i. Letting H_i denote the partial result after the ith stage, the hash $h(x)$ of an input $x = x_1 x_2 \ldots x_t$ is defined as follows:
    $$H_0 = IV; \qquad H_i = f(H_{i-1}, x_i), \ 1 \le i \le t; \qquad h(x) = H_t.$$
    - H_{i-1} serves as the *chaining variable* between stages $i - 1$ and i.

14. Particular hash functions differ in the nature of the compression function and preprocessing of the input.

Examples:

1. A typical usage for data integrity is as follows:
    - the hash-value corresponding to a particular input is computed at some point in time;
    - the integrity of this hash-value is then protected in some manner;
    - at a subsequent point in time, to verify the input data has not been altered, the hash-value is recomputed, using purportedly the same input, and compared for equality with the original hash-value.

2. *Matyas-Meyer-Oseas hash function*: Let E be an n-bit block cipher, such as DES, parametrized by a symmetric key k, and let g be a function that maps an n-bit string to a key k suitable for E. Fix an initial value IV. The following algorithm is then an n-bit hash function which, given any input string x, outputs an n-bit hash $h(x)$:
    - divide x into n-bit blocks and pad if necessary by some method such that all blocks are complete, yielding a padded message of t n-bit blocks $x_1 x_2 \ldots x_t$;
    - define $h(x) = H_t$ where:
    $$H_0 = IV; \qquad H_i = E_{g(H_{i-1})}(x_i) \oplus x_i, \ 1 \le i \le t.$$

This is believed to be a one-way hash function requiring 2^n operations to find a preimage, and $2^{n/2}$ operations to find a collision. For underlying ciphers, such as DES, which have relatively small blocklength (for example, with blocks of no more than 64 bits), this is not a collision-resistant hash function since 2^{32} operations is well within current computational capability.

REFERENCES

Printed Resources:

[Ba97] J. Baylis, *Error-Correcting Codes: A Mathematical Introduction*, Chapman & Hall, 1997.

[BePi82] H. Beker and F. Piper, *Cipher Systems: The Protection of Communications*, Wiley, 1982.

[Be84] E. R. Berlekamp, *Algebraic Coding Theory* (Revised 1984 ed.), Aegean Press, 1984.

[Bl83] R. E. Blahut, *Theory and Practice of Error Control Codes*, Addison-Wesley, 1983.

[Br88] G. Brassard, *Modern Cryptology: a Tutorial*, Springer-Verlag, 1988.

[DaPr89] D. W. Davies and W. L. Price, *Security for Computer Networks*, Wiley, 1989.

[De83] D. E. Denning, *Cryptography and Data Security*, Addison-Wesley, 1983.

[DiHe76] W. Diffie and M. E. Hellman, "Multiuser cryptographic techniques", *Proceedings of AFIPS National Computer Conference*, 109-112, 1976.

[Gi98] J. Gilmore, editor, *Cracking DES: Secrets of Encryption Research, Wiretap Politics & Chip Design*, O'Reilly, 1998.

[GoPi91] C. M. Goldie and R. G. E. Pinch, *Communication Theory*, Cambridge University Press, 1991.

[Ha80] R. W. Hamming, *Coding and Information Theory*, Prentice-Hall, 1980.

[HaHaJo97] D. Hankerson, G. A. Harris, and P. D. Johnson, Jr., *Introduction to Information Theory and Data Compression*, CRC Press, 1997.

[HeWi99] C. Heegard and S. B. Wicker, *Turbo Coding*, Kluwer, 1999.

[Hi86] R. Hill, *A First Course in Coding Theory*, Oxford, 1986.

[HoEtal92] D. G. Hoffman, D. A. Leonard, C. C. Lindner, K. T. Phelps, C. A. Rodger and J. R. Wall, *Coding Theory: The Essentials*, Marcel Dekker, 1992.

[Ko94] N. Koblitz, *A Course in Number Theory and Cryptography*, second edition, Springer-Verlag, 1987.

[LiCo83] S. Lin and D. J. Costello, Jr., *Error Control Coding: Fundamentals and Applications*, Prentice-Hall, 1983.

[MaSl77] F. J. MacWilliams and N. J. A. Sloane, *The Theory of Error-Correcting Codes*, North-Holland, 1977.

[Mc77] R. J. McEliece, *The Theory of Information and Coding*, Addison-Wesley, 1977.

[MevaVa96] A. J. Menezes, P. C. van Oorschot and S. A. Vanstone, *Handbook of Applied Cryptography*, CRC Press, 1996.

[PeWe72] W. W. Peterson and E. J. Weldon, Jr., *Error-Correcting Codes*, MIT Press, 1972.

[Pi88] P. Piret, *Convolutional Codes*, MIT Press, 1988.

[Pl89] V. Pless, *Introduction to the Theory of Error-Correcting Codes*, Wiley, 1989.

[PlHuBr98] V. Pless, W. C. Huffman, and R. A. Brualdi, *Handbook of Coding Theory*, Elsevier, 1998.

[Pr92] O. Pretzel, *Error-Correcting Codes and Finite Fields*, Clarendon Press, 1992.

[Re94] F. M. Reza *An Introduction to Information Theory*, Dover, 1994.

[Ro96] S. Roman, *Introduction to Coding and Information Theory*, Springer-Verlag, 1996.

[Ru86] R. A. Rueppel, *Analysis and Design of Stream Ciphers*, Springer-Verlag, 1986.

[Sa90] A. Salomaa, *Public-Key Cryptography*, Springer-Verlag, 1990.

[Sc96] B. Schneier, *Applied Cryptography: Protocols, Algorithms, and Source Code in C*, 2nd ed., Wiley, 1996.

[SePi89] J. Seberry and J. Pieprzyk, *Cryptography: An Introduction to Data Security*, Prentice-Hall, 1989.

[Si93] G. J. Simmons, ed., *Contemporary Cryptology: the Science of Information Integrity*, IEEE Press, 1992.

[St95] D. Stinson, *Cryptography: Theory and Practice*, CRC Press, 1995.

[va90] J. H. van Lint, "Algebraic geometric codes", *Coding Theory and Design Theory*, Part 1, Springer-Verlag, 1990, 137-162.

[va99] J. H. van Lint, *Introduction to Coding Theory*, third edition, Springer-Verlag, 1999.

[Vava89] S. A. Vanstone and P. C. van Oorschot, *An Introduction to Error Correcting Codes with Applications*, Kluwer Academic Publishers, 1989.

[We98] R. B. Wells, *Applied Coding and Information Theory for Engineers*, Prentice-Hall, 1998.

[We88] D. Welsh, *Codes and Cryptography*, Clarendon Press, 1988.

Web Resources:

ftp://ftp.funet.fi/pub/crypt (Cryptographic Software Archive.)

http://imailab.iis.u-tokyo.ac.jp/~robert/codes.html (The Error Correcting Codes (ECC) Home Page.)

http://rschp2.anu.edu.au:8080/cipher.html (Some Classic Ciphers and Their Weaknesses.)

http://www.achiever.com/freehmpg/cryptology/crypto.html (A-Z Cryptology.)

http://www.cacr.math.uwaterloo.ca/hac/ (Handbook of Applied Cryptography Site.)

http://www.certicom.com (Certicom Corporation.)

http://www.cs.hut.fi/ssh/crypto/intro.html (Introduction to Cryptography.)

http://www.entrust.com (Entrust Technologies.)

http://www.ntru.com (NTRU Cryptosystems.)

http://www.rsa.com/rsalabs/ (RSA Laboratories.)

http://www331.jpl.nasa.gov/public/JPLtcodes.html (JPL Turbo Codes Page.)

15

DISCRETE OPTIMIZATION

15.1 Linear Programming
    15.1.1 Basic Concepts   Beth Novick
    15.1.2 Tableaus
    15.1.3 Simplex Method
    15.1.4 Interior Point Methods
    15.1.5 Duality
    15.1.6 Sensitivity Analysis
    15.1.7 Goal Programming
    15.1.8 Integer Programming

15.2 Location Theory   S. Louis Hakimi
    15.2.1 p-Median and p-Center Problems
    15.2.2 p-Medians and p-Centers on Networks
    15.2.3 Algorithms for Location on Networks
    15.2.4 Capacitated Location Problems
    15.2.5 Facilities in the Plane
    15.2.6 Obnoxious Facilities
    15.2.7 Equitable Locations

15.3 Packing and Covering   Sunil Chopra and
    15.3.1 Knapsacks   David Simchi-Levi
    15.3.2 Bin Packing
    15.3.3 Set Covering and Partitioning

15.4 Activity Nets   S. E. Elmaghraby
    15.4.1 Deterministic Activity Nets
    15.4.2 Probabilistic Activity Nets
    15.4.3 Complexity Issues

15.5 Game Theory   Michael Mesterton-Gibbons
    15.5.1 Noncooperative Games
    15.5.2 Matrix and Bimatrix Games
    15.5.3 Characteristic-Function Games
    15.5.4 Applications

15.6 Sperner's Lemma and Fixed Points   Joseph R. Barr
    15.6.1 Sperner's Lemma
    15.6.2 Fixed-Point Theorems

INTRODUCTION

This chapter discusses various topics in discrete optimization, especially those that arise in applying operations research techniques to applied problems. Linear programming provides a fundamental operations research tool for studying, formulating, and solving a number of combinatorial optimization problems — either exactly or approximately. For example, linear programming is an important tool in solving packing and covering problems, in which a given resource must be optimally utilized subject to constraints. Location theory studies the optimal placement of facilities in order to service a finite number of customers on a network or in the plane. Activity networks are commonly used in the planning and scheduling of interrelated activities to complete a project; in this case the completion time, resources used, and total cost are important considerations. Game theory is a discipline with applications to many areas, in which several agents compete or cooperate to maximize their respective gains. Fixed-point theorems have applications to economics, nonlinear optimization, and game theory.

GLOSSARY

active constraint: an inequality satisfied with equality by a given vector.

balanced matrix: a 0-1 matrix having no square submatrix of odd order with exactly two 1s in each row and column.

basic feasible solution (of an LP): a basic solution that is also a feasible solution.

basic solution (of an LP): a solution obtained by setting certain **nonbasic** variables to zero and solving for the remaining **basic** variables.

bin packing problem: an optimization problem in which a given set of items are to be packed using the fewest number of bins.

bounded LP: a linear programming problem having a finite optimal solution.

capacitated location problem: a location problem in which bounds are placed on the amount of demand that can be handled by individual facilities.

p-center: a set of p locations for facilities that minimizes the maximum distance from any demand point to its closest facility.

characteristic function: a mapping from the set of all coalitions to the nonnegative real numbers.

characteristic-function game: a model for distributing a cooperative benefit fairly among players when the concept of fairness is based on the bargaining strengths of coalitions that could form if the players had not already agreed to cooperate.

coalition: any subset of the players in a game.

complete information: a situation arising when a game's structure is known to all players.

convex hull: the smallest convex set containing a given set of points.

convex set: a set containing the line segment joining any two of its points.

CPM model: a deterministic activity net with strict precedence among the activities.

critical path: a sequence of activities that determines the completion time of a project.

criticality index: the probability that a given path (activity) is (lies on) a critical path of a project.

cutting plane: a constraint that can be added to an existing set of constraints without excluding any feasible integer solution.

decision variables: the unknowns in an optimization problem.

demand point: a point in a metric space that is a source of demand for the service provided by the facilities.

deterministic activity net: a directed network in which all the parameters (such as duration, resource requirements, precedence) are known deterministically.

dual LP: a minimization LP problem associated with a given maximization LP problem.

equilibrium: a strategy combination from which no player has a unilateral incentive to depart.

facility: a place where a service (or product) is provided.

facility location: a point in a metric space where a facility is located.

feasible direction: a direction that preserves feasibility in a sufficiently small neighborhood of a given feasible solution.

feasible LP: an LP with a nonempty feasible region.

feasible region: the set of all feasible solutions to a given LP.

feasible solution: a vector that satisfies the given set of constraints.

fixed point (of a function): given a function f, a point x such that $f(x) = x$.

float: in a deterministic activity net, a measure of the flexibility available in scheduling an activity without delaying the project completion time.

c-game: a characteristic-function game.

GAN model: a probabilistic activity net with conditional progress and probabilistic realization of activities.

general position: a set of points $x_1, x_2, \ldots, x_{p+1} \in \mathcal{R}^n$ such that the vectors $x_2 - x_1$, $x_3 - x_1, \ldots, x_{p+1} - x_1$ are linearly independent.

GERT model: a probabilistic activity net with exclusive-or branching.

goal programming (GP) problem: an LP having multiple objective functions.

improving direction: a feasible direction that improves the objective function value.

imputation: a distribution among players of the cooperative benefit in a c-game.

infeasible LP: an LP with an empty feasible region.

integer programming (IP) problem: a linear programming problem in which some of the decision variables are required to be integers.

interior point method: a technique for solving an LP that iteratively moves through the interior of the feasible region.

knapsack problem: an optimization problem in which items are to be selected to maximize the total benefit without exceeding the capacity of the knapsack.

linear programming (LP) problem: an optimization problem involving the selection of decision variables that optimize a given linear function and that satisfy linear inequality constraints.

location problem: an optimization problem in which p facilities are to be established to minimize the cost of meeting known demands arising from n locations.

LP relaxation: the linear programming problem obtained by dropping the integrality requirements of an IP.

p-median: a set of p locations for facilities that minimizes the total (transportation) cost of satisfying all demands.

metric space: a set of points on which a distance function has been defined.

mixed strategy: a probability distribution over a set of pure strategies.

noncooperative game: a mathematical model of strategic behavior in the absence of binding agreements.

normalized characteristic function: a mapping from the set of all coalitions to $[0, 1]$.

nucleolus: a c-game solution concept based on minimizing the dissatisfaction of the most dissatisfied coalitions.

objective function: the function associated with a given optimization problem that is to be maximized or minimized.

optimal solution: a feasible solution to an optimization problem achieving the largest (or perhaps smallest) value of the objective function.

packing: a subset of items from a given list that can be placed in a bin of specified capacity.

payoff function: a mapping from the set of feasible strategy combinations to $\mathcal{R}^n$, where n is the number of players.

perfect information: a situation arising when the history of a game is known to all players.

PERT model: a probabilistic activity net with strict precedence and activity durations that are known only in probability.

pivot: a move from a given basic solution of an LP to one differing in only one active constraint.

players: a collection of interacting decisionmakers.

polyhedron: the set of points satisfying a given finite set of linear inequalities.

probabilistic activity net: a directed network in which some or all of the parameters, including the realization of the activities, are probabilistically known.

pure strategy: a plan of action available to a player.

reduced cost: the unit change in the objective function incurred by increasing the value of a given decision variable.

redundant constraint: a constraint that can be removed from a given set of constraints without changing the set of feasible solutions.

set cover: a family of subsets such that each of a specified list of elements is contained in at least one subset.

set covering problem: an optimization problem in which a minimum cost set cover is needed.

set partition: a family of subsets such that each of a specified list of elements is contained in exactly one subset.

set partitioning problem: an optimization problem in which a minimum cost set partition is needed.

Shapley value: a c-game solution concept based on players' marginal worths to coalitions on joining, assuming all orders of formation are equally likely.

p-simplex: the convex hull of a collection of $p+1$ points in general position.

simplex method: a technique for solving an LP that moves from vertex to neighboring vertex along the boundary of the feasible region.

simplicial subdivision (of a simplex): a decomposition of the simplex into a collection of simplices that intersect only along entire common faces.

slack variables: the components of $b - Ax^*$ where x^* is a feasible solution to an LP with constraints $Ax \leq b$, $x \geq 0$.

solution: an equilibrium or set of equilibria in a noncooperative game, or an imputation or set of imputations in a c-game.

strategic behavior: behavior such that the outcome of an individual's actions depends on actions yet to be taken by others.

strategy combination: a vector of strategies, one for each player.

tableau: a table storing all information pertinent to a given basic solution for an LP.

totally unimodular matrix: a 0-1 matrix such that every square submatrix has determinant 0, +1, or −1.

unbounded LP: a linear programming problem that is not bounded.

vertex (of a feasible region): given a feasible region S, a point $x \in S \subseteq \mathcal{R}^n$ defined by the intersection of exactly n linearly independent constraints.

5.1 LINEAR PROGRAMMING

Linear programming involves the optimization of a linear function under linear inequality constraints. Applications of this model are widespread, including problems arising in marketing, finance, inventory, capital budgeting, computer science, transportation, and production. Algorithms are available that, in practice, solve LP problems efficiently.

5.1.1 BASIC CONCEPTS

Definitions:

A **linear programming (LP) problem** is an optimization problem that can be written

$$\text{maximize:} \quad cx$$
$$\text{subject to:} \quad \widetilde{A}x \leq \widetilde{b} \qquad (1)$$

where $\widetilde{A}$ is a given $q \times n$ matrix, c is a given row vector of length n, and $\widetilde{b}$ is a given column vector of length q. The **decision variables** of problem (1) are represented by the column vector x of length n.

A **feasible solution** is a vector x satisfying $\widetilde{A}x \leq \widetilde{b}$. The **feasible region** is the subset of all feasible solutions in $\mathcal{R}^n$. If no feasible solution exists (so that the feasible region is empty), the LP problem is **infeasible**; otherwise it is **feasible**.

Each of the q inequalities in $\widetilde{A}x \leq \widetilde{b}$ is a **constraint**. A constraint is **redundant** if removing it from (1) doesn't change the feasible region.

For a feasible solution x, the function $z = cx$ is the **objective function**, with cx the **objective value** of x. When the objective value $z^* = cx^*$ is also maximum, then the feasible x^* is an **optimal** solution. If the objective value can be made arbitrarily large over the feasible region, the LP problem is **unbounded**. Otherwise it is **bounded**.

A vector y is a **feasible direction** at x^* if there is some $\tau > 0$ such that $\widetilde{A}(x^* + \lambda y) \leq \widetilde{b}$ for all $0 \leq \lambda \leq \tau$. If $cy > 0$ also holds, then y is an **improving direction**.

A constraint of the system $\widetilde{A}x \leq \widetilde{b}$ that is satisfied with equality by a feasible solution x^* is **active** at x^*.

A set of constraints $\{a_i x \leq b_i \mid i = 1, 2, \ldots, k\}$ is **linearly independent** if the vectors $\{a_1, a_2, \ldots, a_k\}$ are linearly independent (see §6.1.3).

A **vertex** is a feasible solution with n linearly independent active constraints. A vertex with more than n active constraints is **degenerate**. An LP problem with a degenerate vertex is **degenerate**.

A set S is **convex** if the line segment joining any two of its points is contained in S: i.e., for all $x, y \in S$ and $0 \leq \lambda \leq 1$, then $\lambda x + (1 - \lambda)y \in S$.

Let L be the line segment connecting the two vertices x^1 and x^2. Then x^1 and x^2 are **adjacent** if for all points $y \neq x^1, x^2$ on L and all feasible y^1 and y^2, the only way y can equal $\frac{1}{2}y^1 + \frac{1}{2}y^2$ is if y^1 and y^2 are also on L. In this case, L is an **edge**.

Facts:

1. Linear programming models arise in a wide variety of applications, which typically involve the allocation of scarce resources in the best possible way. A sample of such application areas, with reference sources, is given in the following table.

application	references
production scheduling and inventory control	[Ch83], [Ga85]
tanker scheduling	[BaJaSh90]
airline scheduling	[Ga85]
cutting stock problems	[BaJaSh90], [Ch83]
workforce planning	[Ga85]
approximation of data	[Ch83]
matrix games	[Ch83]
blending problems	[BaJaSh90]
petroleum refining	[Ga85]
capital budgeting	[BaJaSh90]
military operations	[Ga85]
land use planning	[Ga85]
agriculture	[Ga85]
banking and finance	[Ga85]
environmental economics	[Ga85]
health care	[Ga85]
marketing	[Ga85]
public policy	[Ga85]

2. The general concepts of linear programming were first developed by G. B. Dantzig in 1947 in connection with military planning problems for the U. S. Air Force. Earlier, in 1939, L. V. Kantorovich formulated and solved a particular type of LP problem in production planning.

3. The term "linear programming" conveys its historical origins and purpose: it is a mathematical model involving *linear* constraints and a *linear* objective function, used for the optimal planning (*programming*) of operations.

4. Form (1) of an LP naturally occurs in the selection of levels for production activities that maximize profit subject to constraints on the utilization of the given resources.

5. These transformations on an LP do not change feasible (or optimal) solutions:
   - *constraints*: change the sense of an inequality by multiplying both sides by -1; or replace $a_i x = b_i$ with $a_i x \leq b_i$ and $-a_i x \leq -b_i$; or replace $a_i x \leq b_i$ with $a_i x + s_i = b_i$ and $s_i \geq 0$;
   - *variables*: for x_j unrestricted, set $x_j := x_j' - x_j''$ with $x_j', x_j'' \geq 0$; or for $x_j \leq 0$, set $x_j := -x_j'$ with $x_j' \geq 0$;
   - *objective function*: change a minimization (maximization) problem to a maximization (minimization) problem by setting $c := -c$.

6. *Farkas' lemma*: Suppose $\widetilde{A}$ is a $q \times n$ matrix and c is an n-row vector. Then the following are equivalent:
   - $cy \geq 0$ for all $y \in \mathcal{R}^n$ such that $\widetilde{A}y \geq 0$;
   - there exists some $u \in \mathcal{R}^q$ such that $u \geq 0$, $c = u\widetilde{A}$.

This result is important in establishing the optimality conditions for linear programming problems; it can also be applied to show the existence (and uniqueness) of solutions to linear models of economic exchange and stationary distributions in finite Markov chains (§7.7). (J. Farkas, born 1902)

7. A feasible solution with an improving direction can not be optimal for (1).

8. A feasible solution with no improving direction is always optimal for (1).

9. If a feasible solution to (1) has an improving direction y and if $\widetilde{A}y \leq 0$ then the LP problem is unbounded.

10. Each LP problem is either infeasible, unbounded, or has an optimal solution. This need not be the case for nonlinear optimization problems.

11. Form (1) of an LP is helpful for understanding the geometric properties of an LP.

12. For algorithmic purposes the following form, form (2), of an LP is preferred:

$$\text{maximize:} \quad cx$$
$$\text{subject to:} \quad Ax \leq b \qquad (2)$$
$$x \geq 0$$

Here A is an $m \times n$ matrix.

13. The most general form of an LP problem is:

$$\text{maximize (or minimize):} \quad dx_1 + ex_2 + fx_3$$
$$\text{subject to:} \quad Ax_1 + Bx_2 + Cx_3 \leq a$$
$$Dx_1 + Ex_2 + Fx_3 \geq b$$
$$Gx_1 + Hx_2 + Kx_3 = c$$
$$x_1 \geq 0, \ x_2 \leq 0, \ x_3 \text{ unrestricted.}$$

In this formulation A, B, C, D, E, F, G, H, and K are matrices; a, b, and c are column vectors; and d, e, and f are row vectors.

14. The feasible region of an LP problem is a convex set.

15. *Equivalence of forms*: The general form in Fact 13 is equivalent to both form (1) and form (2) in the following sense: any of these three forms can be transformed into another using the operations of Fact 5. Each form possesses the same set of feasible (or optimal) solutions.

16. An excellent glossary of linear programming terms, as well as concepts in general mathematical optimization, can be found at the site:

- http://www-math.cudenver.edu/~hgreenbe/glossary/glossary.html

Examples:

1. *Feed mix*: A manufacturer produces a special feed for farm animals. To ensure that the feed is nutritionally balanced, each bag of feed must supply at least 1250 mg of Vitamin A, 250 mg of Vitamin B, 900 mg of Vitamin C, and 232.5 mg of Vitamin D. Three different grains (1, 2, 3) are blended to create the final product. Each ounce of Grain 1 supplies 2, 1, 5, 0.6 mg of Vitamins A, B, C, D, respectively. Each ounce of Grain 2 provides 3, 1, 3, 0.25 mg of Vitamins A, B, C, D, while each ounce of Grain 3 provides 7, 1 mg of Vitamins A, D. The costs (per ounce) of the constituent grains are 41, 35, and 96 cents for Grains 1, 2, and 3, respectively.

The manufacturer wants to determine the minimum cost mix of grains that satisfies all four nutritional requirements. If x_i is the number of ounces of Grain i that are blended in the final product, then the manufacturer's problem is modeled by the following LP:

$$\begin{aligned}
\text{minimize:} \quad & 0.41x_1 + 0.35x_2 + 0.96x_3 \\
\text{subject to:} \quad & 2x_1 + 3x_2 + 7x_3 \geq 1250 \\
& x_1 + x_2 \geq 250 \\
& 5x_1 + 3x_2 \geq 900 \\
& 0.6x_1 + 0.25x_2 + x_3 \geq 232.5 \\
& x_1, x_2, x_3 \geq 0.
\end{aligned}$$

Each constraint in this LP corresponds to a nutritional requirement. It turns out that the optimal solution to the LP is $x_1^* = 200.1$, $x_2^* = 49.9$, $x_3^* = 100.01$ with $z^* = 195.5$. Note that the amount of Vitamin C supplied by this solution is in excess of 900 mg, while the other vitamins are supplied in exactly the minimum amounts.

2. The LP in Example 1 is not in either form (1) or form (2). However, using Fact 5 it can be transformed into form (2), giving the equivalent representation:

$$\begin{aligned}
\text{maximize:} \quad & -0.41x_1 - 0.35x_2 - 0.96x_3 \\
\text{subject to:} \quad & -2x_1 - 3x_2 - 7x_3 \leq -1250 \\
& -x_1 - x_2 \leq -250 \\
& -5x_1 - 3x_2 \leq -900 \\
& -0.6x_1 - 0.25x_2 - x_3 \leq -232.5 \\
& x_1, x_2, x_3 \geq 0.
\end{aligned}$$

3. The following figure shows the feasible region of the LP problem:

$$\text{maximize: } -x_1$$

$$\begin{array}{rl} \text{subject to:} & -x_2 \leq 0 \quad \text{(A)} \\ & -x_1 - x_2 \leq -4 \quad \text{(B)} \\ & -x_1 + x_2 \leq 4 \quad \text{(C)} \\ & -3x_1 + 5x_2 \leq 30 \quad \text{(D)} \\ & -x_1 + 3x_2 \leq 22 \quad \text{(E)} \end{array}$$

This LP has $n = 2$ decision variables x_1, x_2 and it has the vertices $(4, 0)$, $(0, 4)$, and $(5, 9)$. Vertex $(x_1, x_2) = (0, 4)$ is the optimal solution, achieving the maximum objective value $z = 0$. Thus, the LP is bounded, even though its feasible region is not bounded. Constraint (D) is redundant, since dropping it doesn't change the feasible region. Vertex $(5, 9)$ is degenerate, since $3 > n$ constraints are active at this vertex. All vectors are feasible directions at $(6, 5)$. At vertex $(5, 9)$, the direction $(1, -1)$ is feasible, but the direction $(1, 1)$ is not. Vertices $(0, 4)$ and $(5, 9)$ are adjacent, as are $(4, 0)$ and $(0, 4)$.

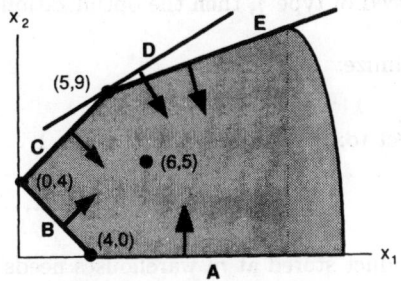

4. *Farkas' lemma*: The row vectors a^1, a^2 of $\widetilde{A} = \begin{pmatrix} 2 & 4 \\ 5 & 2 \end{pmatrix}$ are shown in the following figure.

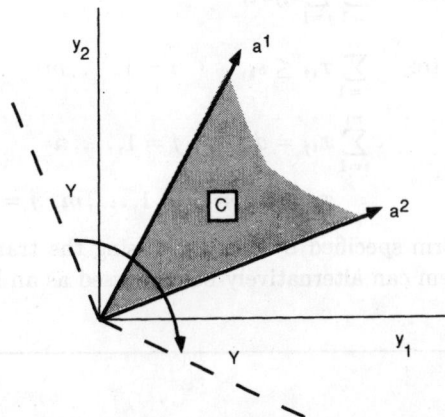

The set $Y = \{ y \mid \widetilde{A}y \geq 0 \}$ is the region bounded by the two dashed lines. Notice that if $c = u\widetilde{A}$ for some $u \geq 0$ then c must lie in the cone C bounded by the vectors a^1 and a^2. Geometrically, any $c \in C$ makes an acute angle with every $y \in Y$, hence $cy \geq 0$. Conversely, any c making an acute angle with every $y \in Y$ must be in C.

5. Fact 10 is illustrated using the following LP problem:

$$\text{maximize:} \quad -x_1 - x_2$$
$$\text{subject to:} \quad -2x_1 + x_2 \leq -1$$
$$-x_1 - 2x_2 \leq -2$$
$$x_1, x_2 \geq 0.$$

This LP has the optimal solution $(x_1^*, x_2^*) = (\frac{4}{5}, \frac{3}{5})$. Suppose the objective function is changed to $z = x_1 - x_2$. Then $(x_1, x_2) = (a, 0)$ is feasible for $a \geq 2$ with objective value a. Thus z can be made arbitrarily large and the LP is unbounded. On the other hand, if the second constraint is changed to $4x_1 - x_2 \leq -1$, the feasible region is empty and the LP is infeasible.

6. *Product mix*: A company manufactures n types of a product, using m shops. Type j requires a_{ij} machine-hours in shop i. There is a limitation of b_i machine-hours for shop i and the sale of each type j unit brings the company a profit c_j. The optimization problem facing the company is given by an LP problem in form (2). Namely, if x_j is the number of units produced of type j, then the optimization problem is

$$\text{maximize:} \quad \sum_{j=1}^{n} c_j x_j$$
$$\text{subject to:} \quad \sum_{j=1}^{n} a_{ij} x_j \leq b_i, \quad i = 1, \ldots, m$$
$$x_j \geq 0, \quad j = 1, \ldots, n.$$

7. *Transportation*: A product stored at m warehouses needs to be shipped to satisfy demands at n markets. Warehouse i has a supply of s_i units of the product, and market j has a demand of d_j units. The cost of shipping a unit of product from warehouse i to market j is c_{ij}. The problem is to determine the number of units x_{ij} to ship from warehouse i to market j in order to satisfy all demands while minimizing cost:

$$\text{minimize:} \quad \sum_{i=1}^{m} \sum_{j=1}^{n} c_{ij} x_{ij}$$
$$\text{subject to:} \quad \sum_{j=1}^{n} x_{ij} \leq s_i, \quad i = 1, \ldots, m$$
$$\sum_{i=1}^{m} x_{ij} = d_j, \quad j = 1, \ldots, n,$$
$$x_{ij} \geq 0, \quad i = 1, \ldots, m, \; j = 1, \ldots, n.$$

This is an LP in the form specified by Fact 13. Using the transformations in Fact 5, this optimization problem can alternatively be expressed as an LP in the form (1).

15.1.2 TABLEAUS

Definitions:
Suppose that an LP is expressed in form (2) of §15.1.1, with A an $m \times n$ matrix. A *tableau* is any table $\begin{array}{c|c} u & z \\ \hline D & f \end{array}$ with the following properties:

- D is an $m \times (m+n)$ matrix with entries d_{tj}; z is a real number; u is an $(m+n)$-row vector; and f is an m-column vector.
- Associated with the tableau is a partition Σ_B, Σ_N of the integers $1, \ldots, m + n$. The set Σ_B, with cardinality m, is the **basic set** and Σ_N is the **nonbasic set**.
- For every row index $t = 1, \ldots, m$, there is a column of D equal to zero in all coordinates except for the tth coordinate, which equals 1. The index of this column is $\varphi(t)$ where φ is a function from $\{1, \ldots, m\}$ to Σ_B associated with the tableau.
- $\begin{array}{c|c} u & z \\ \hline D & f \end{array}$ can be obtained from $\begin{array}{c|c|c} -c & 0 & 0 \\ \hline A & I & b \end{array}$ (where $\Sigma_B = \{n + 1, \ldots, m + n\}$ and $\varphi(t) = n + t$, $t = 1, \ldots, m$) by performing the following **pivot** operation a finite number of times:

    P1. choose a row index $t^* \in \{1, \ldots, m\}$ and a column index $j^* \in \Sigma_N$ with $d_{t^* j^*} \neq 0$;

    P2. multiply row t^* by $1/(d_{t^* j^*})$;

    P3. add appropriate multiples of row t^* to all other rows to make $u_{j^*} = 0$ and to make $d_{tj^*} = 0$ for all $t \neq t^*$;

    P4. remove j^* from Σ_N and place it in Σ_B; remove $\varphi(t^*)$ from Σ_B and place it in Σ_N; set $\varphi(t^*) = j^*$.

In the pivot operation, $\varphi(t^*)$, before replacement, is the index of the **leaving variable** and j^* is the index of the **entering variable**.

The set of variables $\{x_i \mid i \in \Sigma_B\}$ are the **basic variables** and the remaining variables are the **nonbasic variables**.

A **basic solution** is a vector x^* with its basic variables defined by $x_i^* = f_t$ where $t = \varphi^{-1}(i)$; its nonbasic variables have $x_i^* = 0$. If $f \geq 0$ then x^* is a **basic feasible solution** (BFS).

The **basis matrix** B is the $m \times m$ matrix consisting of the columns of $[A \ I]$ corresponding to the basic variables; the **nonbasis matrix** N is the $m \times n$ matrix corresponding to the nonbasic variables. Let c_B $[c_N]$ denote the vector of basic [nonbasic] components of c. Let x_B $[x_N]$ denote the vector of basic [nonbasic] components of x.

The **reduced cost** of nonbasic variable x_j is the negative of u_j in the associated tableau.

The **slack variables** are given by $(x_{n+1}, x_{n+2}, \ldots, x_{n+m}) = b - Ax$.

Facts:

1. Every BFS of (2) corresponds to a vertex of (1), where $q = m + n$, $\widetilde{A} = \begin{bmatrix} A \\ -I \end{bmatrix}$, and $\widetilde{b} = \begin{bmatrix} b \\ 0 \end{bmatrix}$.

2. In the absence of degeneracy the correspondence in Fact 1 is one-to-one; otherwise it is many-to-one.

3. Every LP problem (2) with an optimal solution has an optimal solution that is a vertex. Since the number of vertices is finite, LP problems are combinatorial in nature; that is, an LP can be solved in theory by enumerating its vertices and then selecting one with maximum objective function value.

4. Let x^* be a BFS of (2). All information pertinent to x^* is contained in its tableau which (after possibly permuting the first $m+n$ columns) is

$$\begin{array}{c|c} 0 \quad c_B B^{-1} N - c_N & c_B B^{-1} b \\ \hline I \quad B^{-1} N & B^{-1} b \end{array}.$$

Here $c_B B^{-1} b$ is the objective value z of x^*. The value of the basic variable x_i^* is the tth component of $B^{-1} b$, where $i = \varphi(t)$. Every nonbasic variable has value 0.

5. A tableau expresses the set of equations below, called a *dictionary* [Ch83]:

$$x_B = B^{-1} b - B^{-1} N x_N$$
$$z = c_B B^{-1} b + (c_N - c_B B^{-1} N) x_N.$$

6. The reduced costs of the nonbasic variables are given by the vector $c_N - c_B B^{-1} N$. Basic variables have zero reduced cost.

7. Column i of B^{-1} is identical to d^{m+i}, the column of D associated with the slack variable x_{n+i}.

Examples:

1. When slack variables x_4, x_5, x_6 are added to the LP

$$\begin{aligned}
\text{maximize:} \quad & 3x_1 + 4x_2 + 4x_3 \\
\text{subject to:} \quad & 3x_1 \quad\quad\quad - x_3 \le 5 \\
& -9x_1 + 4x_2 + 3x_3 \le 12 \\
& -6x_1 + 2x_2 + 4x_3 \le 2 \\
& x_1, x_2, x_3 \ge 0
\end{aligned}$$

the following equivalent LP is formed:

$$\begin{aligned}
\text{maximize:} \quad & 3x_1 + 4x_2 + 4x_3 + 0x_4 + 0x_5 + 0x_6 \\
\text{subject to:} \quad & 3x_1 \quad\quad\quad - x_3 + x_4 = 5 \\
& -9x_1 + 4x_2 + 3x_3 + x_5 = 12 \\
& -6x_1 + 2x_2 + 4x_3 + x_6 = 2 \\
& x_1, x_2, \ldots, x_6 \ge 0.
\end{aligned}$$

The associated tableau, with $\Sigma_B = \{4, 5, 6\}$ and $\Sigma_N = \{1, 2, 3\}$, is then

-3	-4	-4	0	0	0	0
3	0	-1	1	0	0	5
-9	4	3	0	1	0	12
-6	2	4	0	0	1	2

Here $\varphi(1) = 4$, $\varphi(2) = 5$, $\varphi(3) = 6$. The basic variables are $x_4 = 5$, $x_5 = 12$, $x_6 = 2$ and the nonbasic variables are $x_1 = 0$, $x_2 = 0$, $x_3 = 0$. The basic feasible solution associated with this tableau is $x = (0, 0, 0, 5, 12, 2)^T$ with objective value $z = 0$. The nonbasic variables x_1, x_2, x_3 have reduced costs $3, 4, 4$, respectively.

2. A pivot is now performed on the tableau in Example 1 using $t^* = 3$ and $j^* = 2$, so the entering variable is x_2 and the leaving variable is x_6. The resulting tableau (a) follows, where $\Sigma_B = \{2, 4, 5\}$ and $\varphi(1) = 4$, $\varphi(2) = 5$, $\varphi(3) = 2$. The corresponding BFS is $x = (0, 1, 0, 5, 8, 0)^T$ with objective value $z = 4$. If a pivot is performed on (a) using $t^* = 1$ and $j^* = 1$, then tableau (b) results. Here $\varphi(1) = 1$, $\varphi(2) = 5$, $\varphi(3) = 2$ and the new BFS is $x = (\frac{5}{3}, 6, 0, 0, 3, 0)^T$ with objective value $z = 29$.

	tableau (a)					
-15	0	4	0	0	2	4
3	0	-1	1	0	0	75
3	0	-5	0	1	-2	8
-3	1	2	0	0	$\frac{1}{2}$	1

	tableau (b)					
0	0	-1	5	0	2	29
1	0	$-\frac{1}{3}$	$\frac{1}{3}$	0	0	$\frac{5}{3}$
0	0	-4	-1	1	-2	3
0	1	1	1	0	$\frac{1}{2}$	6

For tableau (b), the basis matrix B corresponds to columns $1, 5, 2$ of $[A\ I]$, namely
$B = \begin{pmatrix} 3 & 0 & 0 \\ -9 & 1 & 4 \\ -6 & 0 & 2 \end{pmatrix}$. From Fact 7, the inverse matrix $B^{-1} = \begin{pmatrix} \frac{1}{3} & 0 & 0 \\ -1 & 1 & -2 \\ 1 & 0 & \frac{1}{2} \end{pmatrix}$ consists of columns $4, 5, 6$ in tableau (b).

15.1.3 SIMPLEX METHOD

The simplex method is in practice remarkably efficient and it is widely used for solving LP problems. The solution idea dates back to J. B. J. Fourier (1768–1830); it was developed and popularized in 1947 by G. B. Dantzig (born 1914). This section presents two descriptions of the same algorithm — the first geometrically intuitive, the second closer to its actual implementation.

Facts:

1. *Simplex algorithm I*: This method (Algorithm 1) solves a linear programming problem in form (1) of §15.1.1. Assuming that an initial vertex is known, this algorithm travels from vertex to vertex along improving edges until an optimal vertex is reached or an unboundedness condition is detected. In Algorithm 1, the rows of $\widetilde{A}$ are denoted $a_1, a_2, \ldots, a_q$ and $\widetilde{b}$ has the corresponding components $b_1, b_2, \ldots, b_q$.

2. *Simplex algorithm II*: This method (Algorithm 2) solves a linear programming problem in form (2) of §15.1.1, assuming that $b \geq 0$. It proceeds by successively identifying nonbasic variables having positive reduced cost and pivoting them into the current basis in a way that maintains a basic feasible solution (BFS).

3. There are examples for which Algorithm 1 requires exponential running time, and similarly for Algorithm 2.

4. In practice the number of iterations of Algorithms 1 and 2 is proportional to the number of constraints m and grows slowly with the number of variables n.

5. There is a one-to-one correspondence between the vertex x_k of Algorithm 1 and the BFS x_k of Algorithm 2, when $\widetilde{A}$ is set to $\begin{bmatrix} A \\ -I \end{bmatrix}$ and $\widetilde{b}$ is set to $\begin{bmatrix} b \\ 0 \end{bmatrix}$.

6. Interchanging a basic and a nonbasic variable in Algorithm 2 corresponds to interchanging a nonactive and an active constraint in Algorithm 1.

7. In the absence of degeneracy, the objective value strictly increases at each step (in both algorithms). The method of breaking ties by choosing the smallest index prevents *cycling* and ensures termination in finite time. In practice, though, cycling is rare and other rules are used.

8. When a vertex is not known in Algorithm 1 (when $b \not\geq 0$ in Algorithm 2) a preliminary LP problem, *Phase I*, can be solved to get an initial vertex (a starting tableau).

Algorithm 1: Simplex algorithm — form (1).

input: LP in form (1), initial vertex x_0
output: an optimal vertex or an indication of unboundedness
$k := 0$
find a subsystem $Bx \leq r$ of (1) consisting of n linearly independent constraints active at x_k
$S :=$ list containing the indices of these active constraints
{Main loop}
if $u \equiv cB^{-1} \geq 0$ then x_k is an optimal solution — stop
else {an improving direction}
  $i^* :=$ the smallest index such that $u_{i^*} < 0$
  $y :=$ column i^* of $-B^{-1}$
  if $\tilde{A}y \leq 0$ then the LP problem is unbounded — stop
  else {move to next vertex (possibly same as last)}
    $j^* :=$ smallest index j attaining minimum $\lambda \equiv \min\{\frac{b_j - a_j x_k}{a_j y} \mid j \notin S, a_j y > 0\}$
    $x_{k+1} := x_k + \lambda y$
    $S[i^*] := j^*$; update B
    $k := k + 1$
{Continue with next iteration of main loop}

Algorithm 2: Simplex algorithm — form (2).

input: LP in form (2), with $b \geq 0$
output: an optimal BFS or an indication of unboundedness

begin with the initial tableau: $\frac{-c \ \ 0 \ | \ 0}{A \ \ I \ | \ b}$, where $\Sigma_B = \{n+1, \ldots, m+n\}$
and $\varphi(t) = n + t$, $t = 1, \ldots, m$
$x_0 := (x_B, x_N)$ where $x_B = b \geq 0$ and $x_N = 0$
$k := 0$
{Main loop}
if $u_j \geq 0$ for all $j \in \Sigma_N$ then x_k is an optimal solution — stop
else {select entering variable}
  $j^* :=$ the smallest index with $u_{j^*} < 0$
  if $d_{tj^*} \leq 0$ for $t = 1, \ldots, m$ then the LP is unbounded — stop
  else
    $t^* :=$ an index t achieving the minimum
       $\min\{\frac{f_t}{d_{tj^*}} \mid t = 1, \ldots, m; \ d_{tj^*} > 0\}$
    (if there are several such t^*, make $\varphi(t^*)$ as small as possible)
    do a pivot with entering index j^*, leaving index $\varphi(t^*)$
    set component $\varphi(t)$ of x_{k+1} to f_t for $t = 1, \ldots, m$ and the remaining components of x_{k+1} to zero
    $k := k + 1$
{Continue with next iteration of main loop}

9. The *revised simplex method* is a variation of Algorithm 2. Instead of maintaining the entire tableau at each step only B^{-1} is kept. Columns of $[\,A\ I\,]$ are brought in from storage as needed to find j^* and t^*. This method is good for sparse matrices A with many columns.

10. A survey of 39 software packages for solving linear programming problems is described in [Fo97]. Virtually all of these products run on PCs. In many cases, the LP solvers are linked to more general modeling packages that provide a single environment for carrying out the formulation, solution, and analysis of LP problems.

11. Many software packages are available to solve LP problems on mainframes and personal computers. Commercial packages include LINDO, CPLEX, OSL, C-WHIZ, and MINOS. Most use a version of the revised simplex method:

- http://www.lindo.com/
- http://www.cplex.com/
- http://www.research.ibm.com/osl/
- http://www.ketronms.com/products.html
- http://www-leland.stanford.edu/~saunders/brochure/brochure.html

12. An extensive tabulation of software packages to solve LP problems is found at the site:

- http://www-c.mcs.anl.gov/home/otc/Guide/SoftwareGuide
  /Categories/linearprog.html

13. Computer codes (in C, Pascal, and Fortran) that implement the simplex method are catalogued at the sites:

- http://plato.la.asu.edu/guide.html#LP
- http://ucsu.colorado.edu/~xu/software/lp/
- http://www.wior.uni-karlsruhe.de/Bibliothek/Title_Page1.html

Examples:

1. The LP in Example 1 of §15.1.2 can be placed in the form (1) with

$$\widetilde{A} = \begin{pmatrix} 3 & 0 & -1 \\ -9 & 4 & 3 \\ -6 & 2 & 4 \\ -1 & 0 & 0 \\ 0 & -1 & 0 \\ 0 & 0 & -1 \end{pmatrix},\ \widetilde{b} = \begin{pmatrix} 5 \\ 12 \\ 2 \\ 0 \\ 0 \\ 0 \end{pmatrix},\ c = (3\ 4\ 4).$$

If $x_0 = (0, 1, 0)^T$ then constraints $3, 4, 6$ are active at x_0 and $S = [3, 4, 6]$. Thus

$$B = \begin{pmatrix} -6 & 2 & 4 \\ -1 & 0 & 0 \\ 0 & 0 & -1 \end{pmatrix},\ B^{-1} = \begin{pmatrix} 0 & -1 & 0 \\ \frac{1}{2} & -3 & 2 \\ 0 & 0 & -1 \end{pmatrix},\ u = cB^{-1} = (2\ -15\ 4).$$

Here $i^* = 2$, $y = (1,3,0)^T$, $\widetilde{A}y = (3,3,0,-1,-3,0)^T$, and $\widetilde{b} - \widetilde{A}x_0 = (5,8,0,0,1,0)$. Then $\lambda = \min\{\frac{5}{3}, \frac{8}{3}\} = \frac{5}{3}$ and $j^* = 1$. The new vertex is $x_1 = x_0 + \lambda y = (\frac{5}{3}, 6, 0)^T$ and S is updated to $S = [3, 1, 6]$, so B now contains rows $3, 1, 6$ of $\widetilde{A}$. Additional iterations of Algorithm 1 can then be carried out using the updated S and B.

2. The same LP can alternatively be solved using Algorithm 2. For illustration, suppose that the tableau (a) from Example 2 (§15.1.2) is given, corresponding to the BFS $x_1 = (0, 1, 0, 5, 8, 0)^T$ and $\Sigma_B = \{2, 4, 5\}$. Here $u = (-15, 0, 4, 0, 0, 2)$ and $j^* = 1$ is chosen. The minimum ratio test gives $\min\{\frac{5}{3}, \frac{8}{3}\} = \frac{5}{3}$ and $t^* = 1$. The next pivot produces tableau (b) in Example 2 (§15.1.2), with $\Sigma_B = \{1, 2, 5\}$ and $x_2 = (\frac{5}{3}, 6, 0, 0, 3, 0)^T$. Here $u = (0, 0, -1, 5, 0, 2)$ so a further pivot is performed using $j^* = 3$ and $t^* = 3$, giving the tableau below. Since $u \geq 0$ the BFS $x_3 = (\frac{11}{3}, 0, 6, 0, 27, 0)^T$ is an optimal solution to the LP, with optimal objective value $z^* = 35$.

0	1	0	6	0	$\frac{5}{2}$	35
1	$\frac{1}{3}$	0	$\frac{2}{3}$	0	$\frac{1}{6}$	$\frac{11}{3}$
0	4	0	3	1	0	27
0	1	1	1	0	$\frac{1}{2}$	6

15.1.4 INTERIOR POINT METHODS

There are numerous interior point methods for solving LP problems. In contrast to the simplex method, which proceeds from vertex to vertex along edges of the feasible region, these methods move through the interior of the feasible region. In particular this section discusses N. Karmarkar's "projective scaling" algorithm (1984).

Definitions:

The **norm** of $x \in \mathcal{R}^n$ is given by $\|x\| = \sqrt{x_1^2 + x_2^2 + \cdots + x_n^2}$. (See §6.1.4.)

Let e denote the row vector of n 1s.

The LP problem

$$\begin{aligned} \text{minimize:} \quad & z = cx \\ \text{subject to:} \quad & Ax = 0 \\ & ex = 1 \\ & x \geq 0 \end{aligned} \quad (3)$$

is in **standard form** for Karmarkar's method if $\frac{1}{n} e$ is a feasible vector and if the optimal objective value is $z^* = 0$.

The $n \times n$ diagonal matrix $\mathbf{diag}(x_1, x_2, \ldots, x_n)$ has diagonal entries $x_1, x_2, \ldots, x_n$. (See §6.3.1.)

The **unit simplex** in n dimensions is $S_n = \{x \in \mathcal{R}^n \mid ex = 1, x \geq 0\}$.

If $\overline{x}$ is feasible to (3), Karmarkar's **centering transformation** $T_{\overline{x}} : S_n \to S_n$ is

$$T_{\overline{x}}(x) = \frac{\text{diag}(\overline{x})^{-1} x}{e \, \text{diag}(\overline{x})^{-1} x} .$$

The **projection** of a vector v onto the subspace $X \equiv \{x \in \mathcal{R}^n \mid \widetilde{A}x = 0\}$ is the unique vector $p \in X$ for which $(v - p)^T x = 0$ for all $x \in X$. (See §6.1.4.)

Karmarkar's **potential function** for (3) is $f(x) = \sum_{j=1}^{n} \ln\left(\frac{cx}{x_j}\right)$.

Algorithm 3: Karmarkar's method.

input: LP in form (3)
output: an optimal solution to (3)
$x_0 := \frac{e}{n}$
$k := 0$
{Main loop}
{test for optimality within ϵ}
if $cx_k < \epsilon$ then stop
else {find new point y in transformed unit simplex}
$$P := \begin{pmatrix} A \operatorname{diag}(x_k) \\ 1\,1\,\cdots\,1 \end{pmatrix}$$
$c_P := [I - P^T(PP^T)^{-1}P]\operatorname{diag}(x_k)\,c^T$
$y_k := \frac{e}{n} - \left(\frac{\theta}{\sqrt{n(n-1)}}\right)\frac{c_P}{\|c_P\|}$
{find new feasible point in the original space}
$x_{k+1} := T_{x_k}^{-1}(y_k)$
$k := k + 1$
{Continue with next iteration of main loop}

Facts:

1. Any LP problem can be transformed into form (3); see [Sc86], [BaJaSh90] for details.
2. The centering transformation $T_{\bar{x}}$ is 1-1 and onto.
3. The inverse of the centering transformation is
$$T_{\bar{x}}^{-1}(y) = \frac{\operatorname{diag}(\bar{x})\,y}{e\operatorname{diag}(\bar{x})\,y}.$$
4. The transformation $T_{\bar{x}}$ places $\bar{x}$ at the center of the transformed unit simplex: $T_{\bar{x}}(\bar{x}) = \frac{1}{n}e$.
5. The transformation $T_{\bar{x}}$ maps the feasible region of (3) to
$$Y = \{y \in \mathcal{R}^n \mid A\operatorname{diag}(\bar{x})\,y = 0,\ ey = 1,\ y \geq 0\}.$$
6. $W = \{w \in \mathcal{R}^n \mid A\operatorname{diag}(\bar{x})\,w = 0,\ ew = 0,\ w \geq 0\}$ is the set of all feasible directions for Y.
7. The projection of v onto W is $[I - P^T(PP^T)^{-1}P]v$, where $P = \begin{pmatrix} A\operatorname{diag}(\bar{x}) \\ 1\,1\,\cdots\,1 \end{pmatrix}$.
8. *Karmarkar's algorithm*: This method (Algorithm 3) moves through the interior of the feasible region of (3), transforming the problem at each iteration to place the current point at the "center" of the transformed region.
9. In Algorithm 3, $\epsilon > 0$ is a fixed tolerance chosen arbitrarily small. The parameter θ is a constant, $0 < \theta < 1$, associated with convergence of the algorithm. The value $\theta = \frac{1}{4}$ ensures the convergence of Algorithm 3.
10. There is a positive constant δ with $f(x_k) - f(x_{k+1}) \geq \delta$ for all iterations k of Karmarkar's method. To ensure this inequality $\operatorname{diag}(x_k)\,c$, rather than c, is projected onto the space of feasible directions W.

11. For large problems, Karmarkar's method requires many fewer iterations than does the simplex method.

12. Letting L be the maximum number of bits needed to represent any number associated with the LP problem, the running time of Karmarkar's algorithm is *polynomial*, namely $O(n^{3.5}L^2)$.

13. The earliest polynomial-time algorithm for LP problems is the ellipsoid method, proposed by L. G. Khachian in 1979. (See [Ch83] or [Sc86].)

14. The ellipsoid method has worst-case complexity $O(n^6 L^2)$, where L is defined in Fact 12. Because its calculations require high precision, this method is very inefficient in practice.

15. Karmarkar's polynomial-time algorithm was announced in 1984 and it has proven to be seriously competitive with the simplex method. Typically, Karmarkar's algorithm reduces the objective function by fairly significant amounts at the early iterations, often converging within 50 iterations regardless of the problem size.

16. Other versions of Karmarkar's algorithm are faster than the one described here, but are more complicated to explain. Efficient implementations of these faster versions solve some classes of large LP problems over 50 times faster than the simplex method.

17. Computer codes (in C, Pascal, and Fortran) that implement interior point methods are catalogued at the sites:

    - http://plato.la.asu.edu/guide.html#LP
    - http://ucsu.colorado.edu/~xu/software/lp/
    - http://www.wior.uni-karlsruhe.de/Bibliothek/Title_Page1.html

18. LP problems can be submitted online for solution by different interior point algorithms using the NEOS home page:

    - http://www-c.mcs.anl.gov/home/otc/Server/

19. An archive of technical papers and other information on interior point algorithms is available at the site:

    - http://www-c.mcs.anl.gov/home/otc/InteriorPoint/index.html

Example:

1. In the following LP the vector $x = (\frac{1}{3}, \frac{1}{3}, \frac{1}{3})^T$ is feasible and the problem has the optimal objective value $z^* = 0$, achieved for $x^* = (0, \frac{2}{3}, \frac{1}{3})^T$.

$$\begin{aligned}\text{minimize:} \quad & x_1 \\ \text{subject to:} \quad & x_1 + x_2 - 2x_3 = 0 \\ & x_1 + x_2 + x_3 = 1 \\ & x_1, x_2, x_3 \geq 0\end{aligned}$$

Karmarkar's algorithm is started with $x_0 = (\frac{1}{3}, \frac{1}{3}, \frac{1}{3})^T$, yielding $cx_0 = \frac{1}{3}$. For illustrative purposes the value $\theta = 0.9$ is used throughout. Since $A = (1\ 1\ -2)$ the matrix $P = \begin{pmatrix} \frac{1}{3} & \frac{1}{3} & -\frac{2}{3} \\ 1 & 1 & 1 \end{pmatrix}$, giving $c_P = (\frac{1}{6}, -\frac{1}{6}, 0)$ and $y_0 = (0.0735, 0.5931, 0.3333)^T = x_1$. The new objective value is $cx_1 = 0.0735$.

Additional iterations of Algorithm 3 are tabulated in the following table, showing convergence to the optimal $x^* = (0, \frac{2}{3}, \frac{1}{3})^T$ after just a few iterations.

k	0	1	2	3	4
x_k	$\begin{pmatrix}0.3333\\0.3333\\0.3333\end{pmatrix}$	$\begin{pmatrix}0.0735\\0.5931\\0.3333\end{pmatrix}$	$\begin{pmatrix}0.0056\\0.6611\\0.3333\end{pmatrix}$	$\begin{pmatrix}0.0004\\0.6663\\0.3333\end{pmatrix}$	$\begin{pmatrix}0.0000\\0.6666\\0.3333\end{pmatrix}$
y_k	$\begin{pmatrix}0.0735\\0.5931\\0.3333\end{pmatrix}$	$\begin{pmatrix}0.0349\\0.5087\\0.4564\end{pmatrix}$	$\begin{pmatrix}0.0333\\0.4852\\0.4814\end{pmatrix}$	$\begin{pmatrix}0.0333\\0.4835\\0.4832\end{pmatrix}$	$\begin{pmatrix}0.0333\\0.4833\\0.4833\end{pmatrix}$
cx_k	0.3333	0.0735	0.0056	0.0004	0.0000

15.1.5 DUALITY

Associated with every LP problem is its dual problem, which is important in devising alternative solution procedures for the original LP. The dual also provides useful information for conducting postoptimality analyses on the given LP.

Definitions:

Associated with every LP problem is another LP problem, its **dual**. The original problem is called the **primal**.

The dual of an LP in form (2)

$$\text{maximize:} \quad cx$$
$$\text{subject to:} \quad Ax \leq b \quad\quad (2)$$
$$x \geq 0$$

is defined to be the LP

$$\text{minimize:} \quad ub$$
$$\text{subject to:} \quad uA \geq c \quad\quad (4)$$
$$u \geq 0.$$

The components $u_1, u_2, \ldots, u_m$ of u are the **dual variables**.

Facts:

1. To find the dual of an arbitrary LP problem either transform it (§15.1.1 Fact 5) into form (2) or use the following table:

primal	dual
maximization problem	minimization problem
unrestricted variable	equality constraint
nonnegative variable	$\geq$ constraint
nonpositive variable	$\leq$ constraint
equality constraint	unrestricted variable
$\leq$ constraint	nonnegative variable
$\geq$ constraint	nonpositive variable

2. The dual of the dual LP is the primal LP.

3. *Weak duality theorem*: For any feasible solution x to the primal and any feasible solution u to the dual $cx \leq ub$.

4. *Strong duality theorem*: If x^* is an optimal solution to (2) then there exists an optimal solution u^* for (4) and $cx^* = u^*b$.

5. A given primal LP and its associated dual LP can only produce certain combinations of outcomes, as specified in the following table. For example, if one problem is unbounded then the other must be infeasible.

primal	dual
optimal	optimal
infeasible	unbounded
unbounded	infeasible
infeasible	infeasible

6. Let x^* be an optimal BFS of the primal LP (2), with the corresponding tableau $\dfrac{u \mid z}{D \mid f}$. Then u is an optimal BFS of the dual LP (4).

7. *Complementary slackness*: An optimal dual (primal) variable u_i^* (x_j^*) can be nonzero only if it corresponds to a primal (dual) constraint active at x^* (u^*).

8. *Economic interpretation*: Suppose in the LP (2) that b_i is the amount of resource i available to a firm maximizing its profit. Then the optimal dual variable u_i^* is the price the firm should be willing to pay (over and above its market price) for an extra unit of resource i.

9. *Dual simplex algorithm*: This approach (Algorithm 4) can be used when a basic solution for (2) is known that is not necessarily feasible but which has nonnegative reduced costs (i.e., it is a *dual feasible* basic solution). The main idea of the algorithm is to start with the dual feasible basic solution and to maintain dual feasibility at each pivot. An optimal BFS is found once primal feasibility is achieved.

10. The dual simplex method was devised in 1954 by C. E. Lemke.

11. Computer code (in C) that implements the dual simplex algorithm can be found at the site:
    - http://ucsu.colorado.edu/~xu/software/lp/minit.html

Examples:

1. Using the table of Fact 1, the dual of

$$\begin{aligned}
\text{maximize:} \quad & 5x_1 - 7x_2 \\
\text{subject to:} \quad & x_1 + 3x_2 - x_3 + x_4 \leq -1 \\
& 2x_1 + x_2 - 4x_3 - x_4 \geq 3 \\
& x_1 + x_2 - 3x_3 + 2x_4 = 2 \\
& x_2 \geq 0, \ x_4 \leq 0, \ x_1, x_3 \text{ unrestricted}
\end{aligned}$$

is

Algorithm 4: Dual simplex algorithm.

input: LP in form (2), dual feasible basic solution x_0
output: an optimal BFS or an indication of infeasibility

associate with $x_0 = (x_B, x_N) = (f, 0)$ the tableau $\begin{array}{c|c} u & z \\ \hline D & f \end{array}$, where $u \geq 0$

$k := 0$
{Main loop}
{optimality test}
if $f \geq 0$ then x_k is an optimal solution — stop
else
$\quad t^* :=$ the smallest index with $f_{t^*} < 0$
$\quad$ if $d_{t^* j} \geq 0$ for all j then the LP is infeasible — stop
$\quad$ else
$\qquad j^* :=$ the smallest index attaining the maximum
$\qquad \max \{ \frac{u_j}{d_{t^* j}} \mid j = 1, \ldots, m+n; d_{t^* j} < 0 \}$
$\qquad$ do a pivot with entering index j^*, leaving index $\varphi(t^*)$
$\qquad$ set component $\varphi(t)$ of x_{k+1} to f_t for $t = 1, \ldots, m$ and the remaining components of x_{k+1} to zero
$\qquad k := k + 1$
{Continue with next iteration of main loop}

$$\begin{aligned} \text{minimize:} \quad & -u_1 + 3u_2 + 2u_3 \\ \text{subject to:} \quad & u_1 + 2u_2 + u_3 = 5 \\ & 3u_1 + u_2 + u_3 \geq -7 \\ & -u_1 - 4u_2 - 3u_3 = 0 \\ & u_1 - u_2 + 2u_3 \leq 0 \\ & u_1 \geq 0, \; u_2 \leq 0, \; u_3 \text{ unrestricted.} \end{aligned}$$

2. The LP of §15.1.2 Example 1 has the dual

$$\begin{aligned} \text{minimize:} \quad & 5u_1 + 12u_2 + 2u_3 \\ \text{subject to:} \quad & 3u_1 - 9u_2 - 6u_3 \geq 3 \\ & 4u_2 + 2u_3 \geq 4 \\ & -u_1 + 3u_2 + 4u_3 \geq 4 \\ & u_1, u_2, u_3 \geq 0. \end{aligned}$$

The optimal solution to the primal LP (see §15.1.3 Example 2) is $x^* = (\frac{11}{3}, 0, 6)^T$ with optimal objective value $z^* = 35$. The associated tableau has $u = (0, 1, 0, 6, 0, \frac{5}{2})$. The optimal dual variables for (4) are recovered from the reduced costs of the slack variables x_4, x_5, and x_6, so that $u^* = (6, 0, \frac{5}{2})$. As guaranteed by Fact 4, the optimal dual objective value $5u_1^* + 12u_2^* + 2u_3^* = 30 + 5 = 35 = z^*$. The complementary slackness conditions in Fact 7 hold here: the second primal constraint holds with strict inequality ($x_5^* = 27 > 0$), so the second dual variable $u_2^* = 0$; also, the second dual constraint holds with strict inequality ($u_5^* = 1 > 0$), so the second primal variable $x_2^* = 0$.

3. Using the transformations of §15.1.1 Fact 5, the LP problem

$$\text{minimize:} \quad 2x_1 + 3x_2 + 4x_3$$
$$\text{subject to:} \quad 2x_1 - x_2 + 3x_3 \geq 4$$
$$x_1 + 2x_2 + x_3 \geq 3$$
$$x_1, x_2, x_3 \geq 0$$

can be written in form (2), with the corresponding tableau (a) below. Since $u \geq 0$ the current solution $x_4 = -4$, $x_5 = -3$ is dual feasible but not primal feasible. Algorithm 4 can then be applied, giving $t^* = 1$ and $j^* = 1$. The variable x_4 leaves the basis and the variable x_1 enters, giving tableau (b) and the new basic but not feasible solution $x = (2, 0, 0, 0, -1)^T$ with $z = 4$. One additional dual simplex pivot achieves primal feasibility and produces the optimal solution $x^* = (\frac{11}{5}, \frac{2}{5}, 0, 0, 0)^T$ with $z^* = \frac{28}{5}$.

tableau (a)

2	3	4	0	0	0
-2	1	-3	1	0	-4
-1	-2	-1	0	1	-3

tableau (b)

0	4	1	1	0	-4
1	$-\frac{1}{2}$	$\frac{3}{2}$	$-\frac{1}{2}$	0	2
0	$-\frac{5}{2}$	$\frac{1}{2}$	$-\frac{1}{2}$	1	-1

15.1.6 SENSITIVITY ANALYSIS

Since the data to an LP are often estimates or can vary over time, the analysis of many problems requires studying the behavior of the optimal LP solution to changes in the input data. This form of sensitivity analysis typically uses the solution of the original LP as a starting point for solving the altered LP.

Definitions:

The *original tableau* for the LP problem (2) is $\dfrac{-c \quad 0 \;\big|\; 0}{A \quad I \;\big|\; b}$.

The *final tableau* for the optimal basic solution x^* (possibly after a permutation of the columns $1, \ldots, m+n$) is

$$\frac{u \;\big|\; z}{D \;\big|\; f} = \frac{0 \quad c_B B^{-1} N - c_N \;\big|\; c_B B^{-1} b}{I \quad B^{-1} N \;\big|\; B^{-1} b}.$$

Row 0 of a tableau refers to the row u of associated dual variables.

A tableau is **suboptimal** if some entries of row 0 are negative. A tableau is **infeasible** if some entries of column f are negative.

Let a^j be the column of $[A \; I]$ associated with variable x_j and let d^j be the column of D associated with variable x_j.

Facts:

1. The formulas in Table 1 show how to construct an updated tableau T' from the final tableau T of an LP problem:
   - if T' is suboptimal, reoptimize using the simplex method starting with T';
   - if T' is infeasible, reoptimize using the dual simplex method starting with T';
   - otherwise, T' corresponds to an optimal BFS for the altered problem.

Table 1 Formulas for constructing the updated tableau T'.

change in LP data	possible changes in tableau	tableau updates
change in c_s, s nonbasic: $c_s := c_s + \Delta$	only entry s of row 0 can change	$u_s := u_s - \Delta$
change in c_s, s basic: $c_s := c_s + \Delta$	row 0 and z can change	update c_B $u_j := c_B B^{-1} a^j - c_j$, j nonbasic $z := c_B B^{-1} b$
change in b_r: $b_r := b_r + \Delta$	decision variables and z can change	$f := f + \Delta(d^{n+r})$ update b $z := c_B B^{-1} b$
change nonbasic column s: $a^s := \tilde{a}^s$ $c_s := \tilde{c}_s$	tableau column s and u_s can change	update a^s and c_s $u_s := c_B B^{-1} a^s - c_s$ $d^s := B^{-1} a^s$
add a new column a^ℓ with cost c_ℓ	new tableau column ℓ and new u_ℓ	$u_\ell := c_B B^{-1} a^\ell - c_\ell$ $d^\ell := B^{-1} a^\ell$

2. *Ranging*: Table 2 shows how to calculate the (maximal) ranges over which the current basis B remains optimal. In the "range" column of Table 2, b and c_B refer to entries of T, rather than T'.

3. When b_i is changed within the allowable range (Table 2), the change in the objective value is $-\Delta$ times the reduced cost of the slack variable associated with row i.

4. To add a new constraint $a_\ell x \leq b_\ell$ to the original LP do the following:
   - add a new (identity) column to the tableau corresponding to the slack variable of the new constraint;
   - add a new row ℓ to the tableau corresponding to the new constraint;
   - for each basic j with $d_{\ell j} \neq 0$, multiply row $i = \varphi^{-1}(j)$ by $-d_{\ell j}$ and add to row ℓ;
   - if the updated $f_\ell < 0$ use the dual simplex method to reoptimize.

5. For changes in more than one component of c, or in more than one right-hand side b, use the "100% rule":
   - *objective function changes*: If all changes occur in variables j with $u_j > 0$, the current solution remains optimal as long as each c_j is within its allowable range (Table 2). Otherwise, let Δc_j be the change to c_j. If $\Delta c_j \geq 0$ set $r_j := \frac{\Delta c_j}{\Delta_U}$, else set $r_j := -\frac{\Delta c_j}{\Delta_L}$, where Δ_U, Δ_L are computed from Table 2. If $\sum r_j \leq 1$, the current solution remains optimal (if not, the rule tells nothing).
   - *right-hand side changes*: If all changes are in constraints not active at x^*, the current basis remains optimal as long as each b_i is within its allowable range (Table 2). Otherwise, let Δb_i be the change to b_i. If $\Delta b_i \geq 0$ set $r_i := \frac{\Delta b_i}{\Delta_U}$, else set $r_i := -\frac{\Delta b_i}{\Delta_L}$, where Δ_U, Δ_L are computed from Table 2. If $\sum r_i \leq 1$, the current solution remains optimal (if not, the rule tells nothing).

Table 2 Ranges over which current basis is optimal.

change in LP data	range
change in c_s, s nonbasic: $c_s := c_s + \Delta$	$\Delta \leq u_s$
change in c_s, s basic: $c_s := c_s + \Delta$	$\Delta_L \leq \Delta \leq \Delta_U$, where $p = $ sth row of B^{-1} $\Delta_L = \max \{ \frac{c_j - c_B B^{-1} a^j}{pa^j} \mid pa^j > 0, j \text{ nonbasic} \}$ $\Delta_U = \min \{ \frac{c_j - c_B B^{-1} a^j}{pa^j} \mid pa^j < 0, j \text{ nonbasic} \}$
change in b_r: $b_r := b_r + \Delta$	$\Delta_L \leq \Delta \leq \Delta_U$, where $q = $ rth column of B^{-1} $\Delta_L = \max \{ \frac{-(B^{-1}b)_i}{q_i} \mid q_i > 0 \}$ $\Delta_U = \min \{ \frac{-(B^{-1}b)_i}{q_i} \mid q_i < 0 \}$

Examples:

1. The LP problem

$$\text{maximize:} \quad 3x_1 + 4x_2 + 4x_3$$
$$\text{subject to:} \quad 3x_1 \quad\quad - x_3 \leq 5$$
$$-9x_1 + 4x_2 + 3x_3 \leq 12$$
$$-6x_1 + 2x_2 + 4x_3 \leq 2$$
$$x_1, x_2, x_3 \geq 0$$

has the final tableau T

0	1	0	6	0	$\frac{5}{2}$	35
1	$\frac{1}{3}$	0	$\frac{2}{3}$	0	$\frac{1}{6}$	$\frac{11}{3}$
0	4	0	3	1	0	27
0	1	1	1	0	$\frac{1}{2}$	6

corresponding to the optimal BFS $x^* = (\frac{11}{3}, 0, 6, 0, 27, 0)^T$ with $z^* = 35$. The associated basis matrix B contains columns $1, 5, 3$ and the inverse basis matrix is B^{-1}, where

$$B = \begin{pmatrix} 3 & 0 & -1 \\ -9 & 1 & 3 \\ -6 & 0 & 4 \end{pmatrix}, \quad B^{-1} = \begin{pmatrix} \frac{2}{3} & 0 & \frac{1}{6} \\ 3 & 1 & 0 \\ 1 & 0 & \frac{1}{2} \end{pmatrix}.$$

If the nonbasic objective coefficient c_2 is changed to $4 + \Delta$, the current BFS remains optimal for $\Delta \leq u_2 = 1$, that is for $c_2 \leq 5$. If the basic objective coefficient c_1 is changed to $3 + \Delta$, then $p = (\frac{2}{3}, 0, \frac{1}{6})$ and $\Delta_L = \max \{ \frac{-1}{1/3}, \frac{-6}{2/3}, \frac{-5/2}{1/6} \} = -3$. This gives $-3 \leq \Delta$, so the current BFS remains optimal over the range $c_1 \geq 0$. If however c_3 is changed to the value 2, meaning $\Delta = -2$, the current basis with $\Sigma_B = \{1, 5, 3\}$ will no longer be optimal. Using Table 2 the vector c_B is updated to $c_B = (3, 0, 2)$ and the nonbasic u_j are computed as $u_2 = -1$, $u_4 = 4$, $u_6 = \frac{3}{2}$. The updated $u = (0, -1, 0, 4, 0, \frac{3}{2})$ and $z = 23$ are inserted in tableau T. Since $u_2 < 0$ a simplex pivot with $j^* = 2$ and $t^* = 3$ is performed, leading to the new optimal solution $x^* = (\frac{5}{3}, 6, 0, 0, 3, 0)^T$ with $z^* = 29$.

2. Suppose that the right-hand side b_1 in the original LP of Example 1 is changed to $b_1 = 3$, corresponding to the change $\Delta = -2$. From Table 2 f is updated to $(\frac{11}{3}, 27, 6)^T - 2(\frac{2}{3}, 3, 1)^T = (\frac{7}{3}, 21, 4)^T$, giving the optimal BFS $x^* = (\frac{7}{3}, 0, 4, 0, 21, 0)^T$. Since $b = (3, 12, 2)^T$ the objective value found from Table 2 is $z = 23$. Notice that the change in objective value is $\Delta z = 23 - 35 = -12$, which is the same as $-\Delta$ times the reduced cost $-u_4$ of x_4: namely, $-12 = 2 \cdot (-6)$. To determine the range of variation of b_1 so that the basis defined by $\Sigma_B = \{1, 5, 3\}$ remains unchanged, Table 3 is used. Here $q = (\frac{2}{3}, 3, 1)^T$ and $\Delta_L = \max \{-\frac{11/3}{2/3}, -\frac{27}{3}, -\frac{6}{1}\} = -\frac{11}{2}$. Thus $-\frac{11}{2} \leq \Delta$ so that the current basis is optimal for $b_1 \geq -\frac{1}{2}$.

3. If the new constraint $3x_1 + 2x_2 - x_3 \leq 4$ is added to the LP in Example 1, the (previous) optimal solution $x^* = (\frac{11}{3}, 0, 6, 0, 27, 0)^T$ is no longer feasible. Using Fact 4, a new row and column are added to the tableau $\mathcal{T}$, giving tableau (a) below. By adding (-3) times row 1 and $+1$ times row 3 to the last row, a new tableau (b) is produced corresponding to the basic set $\Sigma_B = \{1, 5, 3, 7\}$. Since $b_4 < 0$ the dual simplex algorithm is then used with $t^* = 4$ and $j^* = 4$, producing a new tableau that is primal feasible, with the new optimal BFS $(3, 0, 5, 1, 24, 0, 0)^T$ and objective value 29.

tableau (a)

0	1	0	6	0	$\frac{5}{2}$	0	35
1	$\frac{1}{3}$	0	$\frac{2}{3}$	0	$\frac{1}{6}$	0	$\frac{11}{3}$
0	4	0	3	1	0	0	27
0	1	1	1	0	$\frac{1}{2}$	0	6
3	2	-1	0	0	0	1	4

tableau (b)

0	1	0	6	0	$\frac{5}{2}$	0	35
1	$\frac{1}{3}$	0	$\frac{2}{3}$	0	$\frac{1}{6}$	0	$\frac{11}{3}$
0	4	0	3	1	0	0	27
0	1	1	1	0	$\frac{1}{2}$	0	6
0	2	0	-1	0	0	1	-1

4. One example of the practical use of sensitivity analysis occurred in the airline industry. When the price of aviation fuel was relatively high, and varied by airport location, a linear programming model was successfully used to determine an optimal strategy for refueling aircraft. The key idea is that it might be more economical to take on extra fuel at an enroute stop if the fuel cost savings for the remainder of the flight are greater than the extra fuel burned because of the excess weight of additional fuel. A linear programming model of this situation ended up saving millions of dollars annually. An important feature was providing pilots with *ranges* of fuel prices for each airport location, with associated optimal policies for taking on extra fuel based on the cost range.

5. Another example of the beneficial use of sensitivity analysis occurred in a 1997 study to assess the effectiveness of mandatory minimum-length sentences for reducing drug use. One finding of the study was that the longer sentences become more effective than conventional enforcement only when it costs more than $30,000 to arrest a drug dealer. Thus, rather than producing a single optimal policy, this study identified conditions (parameter ranges) under which each alternative policy is to be preferred.

5.1.7 GOAL PROGRAMMING

Goal programming refers to a multicriteria decision-making problem in which a given LP problem can have multiple objectives or goals. This technique is useful when it is impossible to satisfy all goals simultaneously. For example, a model for optimizing the operation of an oil refinery might seek not only to minimize production cost, but also to reduce the amount of imported crude oil and the amount of oil having a high sulfur content. In another instance, the routing of hazardous waste might consider minimizing not only the total distance traveled but also the number of residents living within ten miles of the selected route.

Definitions:

A **goal programming (GP) problem** has linear constraints that can be written

$$Ax \leq b$$
$$Hx + \tilde{x} - \bar{x} = h$$
$$x \geq 0,\ \tilde{x} \geq 0,\ \bar{x} \geq 0$$

and objective functions

$$G_1:\quad \text{minimize } z_1 = c_1\tilde{x}_1 + d_1\bar{x}_1$$
$$G_2:\quad \text{minimize } z_2 = c_2\tilde{x}_2 + d_2\bar{x}_2$$
$$\vdots$$
$$G_\ell:\quad \text{minimize } z_\ell = c_\ell\tilde{x}_\ell + d_\ell\bar{x}_\ell$$

where A is an $m \times n$ matrix and H is an $\ell \times n$ matrix.

The value h_k is the **target value** of the kth goal. Goal k is **satisfied** if $(Hx)_k = h_k$ holds for a given vector x of **decision variables**.

The variables $\tilde{x}$ are the **underachievement variables** while the variables $\bar{x}$ are the **overachievement variables**.

Facts:

1. In a GP problem, the aim is to find decision variables that approximately satisfy the given goals, which is achieved by jointly minimizing the magnitudes of the underachievement and overachievement variables.
2. Assuming $c_k > 0$ and $d_k > 0$, then goal k is satisfied by making $z_k = 0$.
3. If all c_t and d_t are positive then for each $k = 1, \ldots, \ell$ at most one of $\tilde{x}_k, \bar{x}_k$ will be positive in an optimal solution.
4. One important case of a GP problem has $c_k = d_k = 1$ for $k = 1, \ldots, \ell$, making the objective to (approximately) satisfy all constraints $Hx = h$.
5. When the relative importance of $G_1, \ldots, G_\ell$ is known precisely, an ordinary LP can be used with the objective function being a weighted sum of $z_1, \ldots, z_\ell$.
6. *Preemptive goal programming*: Here the goals are prioritized $G_1 \gg G_2 \gg \cdots \gg G_\ell$, meaning that goal G_1 is the most important and goal G_ℓ is the least important. Solutions are sought that satisfy the most important goal. Among all such solutions, those are retained that best satisfy the second highest goal, and so forth.
7. *Goal programming simplex method*: The simplex method (§15.1.3, Algorithm 2) can be extended to preemptive GP (minimization) problems, with the following modifications:
   - ℓ "objective rows" are maintained in the tableau instead of just one.
   - Let i^* be the highest-priority index with $z_{i^*} > 0$ for which there exists a nonbasic j^* with $u^{i^*}_{j^*} > 0$ and with $u^i_{j^*} \geq 0$ for any (higher-priority) objective row $i < i^*$. If there is no such i^*, stop; else pick j^* corresponding to the most positive $u^{i^*}_{j^*}$.
   - All ℓ objective rows are updated when a pivot is performed.
   - At completion, if the solution fails to satisfy all goals, then every nonbasic variable that would decrease the objective value z_i if it entered the basis, would increase $z_{i'}$ for some higher-priority goal $G_{i'}$, $i' < i$.

8. Computer codes (in C, Pascal, and Fortran) that implement goal programming are available at the sites:
   - ftp://garbo.uwasa.fi/pc/ts/tslin35c.zip
   - http://www.iiasa.ac.at/~marek/soft/descr.html#MCMA

5.1.8 INTEGER PROGRAMMING

Integer programming problems are LPs in which some of the variables are constrained to be integers. Such problems more accurately model a wide range of application areas, including capital budgeting, facility location, manufacturing, scheduling, logical inference, physics, engineering design, environmental economics, and VLSI circuit design. However, integer programming problems are much more difficult to solve than LPs.

Definitions:

Let $\mathcal{Z}^n$ [$\mathcal{Z}_+^n$] denote the set of all n-vectors with all components integers [nonnegative integers], and let $\mathcal{R}^n$ [$\mathcal{R}_+^n$] denote the set of all n-vectors with all components real numbers [nonnegative real numbers].

A **pure integer programming (IP) problem** is an optimization problem of the form

$$\begin{aligned}
\text{maximize:} \quad & z_{IP} = cx \\
\text{subject to:} \quad & Ax \leq b \\
& x \in \mathcal{Z}_+^n
\end{aligned} \quad (5)$$

where A is an $m \times n$ matrix, b is an m-column vector, and c is an n-row vector.

A **0-1 IP problem** is an IP with each $x_j \in \{0, 1\}$.

A **mixed integer programming (MIP) problem** is of the form

$$\begin{aligned}
\text{maximize:} \quad & z_{MIP} = cx + hy \\
\text{subject to:} \quad & Ax + Gy \leq b \\
& x \in \mathcal{Z}_+^n, \, y \in \mathcal{R}_+^p
\end{aligned}$$

where A is an $m \times n$ matrix, G is an $m \times p$ matrix, b is an m-column vector, c is an n-row vector, and h is a p-row vector.

For IP problem (5), the **feasible region** is $S \equiv \{x \in \mathcal{Z}_+^n \mid Ax \leq b\}$.

A **polyhedron** is a set of points in $\mathcal{R}^n$ satisfying a finite set of linear inequalities.

If X is a finite set of points in $\mathcal{R}^n$, the **convex hull** of X is

$$\text{conv}(X) \equiv \{\sum \lambda_i x_i \mid x_i \in X, \sum \lambda_i = 1, \lambda_i \geq 0\}.$$

The **LP relaxation** of (5) is the linear programming problem

$$\begin{aligned}
\text{maximize:} \quad & z_{LP} = cx \\
\text{subject to:} \quad & Ax \leq b \\
& x \geq 0
\end{aligned} \quad (6)$$

More generally, a **relaxation** of (5) is any problem $\max\{cx \mid x \in T\}$, where $S \subset T$.

The problem $\max\{cx \mid \bar{A}x \leq \bar{b}, x \in \mathcal{Z}_+^n\}$ is a **formulation** for (5) if it contains exactly the same set of feasible integral points as (5). If the feasible region of the LP relaxation of one formulation is strictly contained in another, the first is a **tighter** formulation.

Algorithm 5: Cutting plane algorithm for (5).

input: IP in form (5)
output: an optimal solution x^* with objective value z^*
let R be the LP relaxation: $\max\{cx \mid Ax \leq b, x \geq 0\}$
{Main loop}
optimally solve problem R, obtaining $\bar{x}$
if $\bar{x} \in \mathcal{Z}_+^n$ **then** stop with $x^* := \bar{x}$ and $z^* := c\bar{x}$
else
    find a cutting plane $\pi x \leq \pi_0$ with $\pi\bar{x} > \pi_0$ and $\pi x \leq \pi_0$ for all feasible solutions of (5)
    modify R by adding the constraint $\pi x \leq \pi_0$
{Continue with next iteration of main loop}

Suppose $\tilde{x}$ is a feasible solution to (6) but not to (5). A ***cutting plane*** is any inequality $\pi x \leq \pi_0$ satisfied by all points in $conv(S)$ but not by $\tilde{x}$.

A family $\mathcal{S}$ of subsets of S is a **separation** of S if $\bigcup_{S_k \in \mathcal{S}} S_k = S$; a separation is usually a partition of the set S.

A ***lower bound*** $\underline{z}$ for z_{IP} is an underestimate of z_{IP}.

Facts:

1. $z_{IP} \leq z_{LP}$. More generally, any relaxation of (5) has an optimal objective value at least as large as z_{IP}.

2. If x' is feasible to (5) then $z' = cx'$ satisfies $z' \leq z_{IP}$.

3. The feasible region of an LP problem is a polyhedron, and every polyhedron is the feasible region of some LP problem.

4. The set $conv(S)$ is a polyhedron, so there is an LP problem $\max\{cx \mid \tilde{A}x \leq \tilde{b}, x \geq 0\}$ with the feasible region $conv(S)$.

5. An optimal solution to the LP in Fact 4 is an optimal solution to (5). However, finding all necessary constraints, called *facets*, of this LP is extremely difficult.

6. IP is an NP-hard optimization problem (§16.5.2). Consequently, such problems are harder to solve in practice than LPs. The inherent complexity of solving IPs stems from the nonconvexity of their feasible region, which makes it difficult to verify the optimality of a proposed optimal solution in an efficient manner.

7. Formulation of an IP is critical: achieving problem tightness is more important than reducing the number of constraints or variables appearing in the formulation.

8. Solution techniques for (5) usually involve some preliminary operations that improve the formulation, called *preprocessing*, followed by an iterative use of heuristics (§10.7) to quickly find feasible solutions.

9. Popular solution techniques for solving (5) include cutting plane methods (Fact 10), branch and bound techniques (Fact 12), and (hybrid) branch and cut methods.

10. *Cutting plane method*: This approach (Algorithm 5) proceeds by first finding an optimal solution $\bar{x}$ to a relaxation R of the original problem (5). If $\bar{x}$ is not optimal, a cutting plane is added to the constraints of the current relaxation and the new LP is then solved. This process is repeated until an optimal solution is found.

11. General methods for finding cutting planes for IP or MIP problems are relatively slow. Cutting plane algorithms using facets for specific classes of IP problems are better, since facets make the "deepest" cuts.

12. *Branch and bound method*: This approach (Algorithm 6) decomposes the original problem P into subproblems or *nodes* by breaking S into subsets. Each subproblem P_j is implicitly investigated (and possibly discarded) until an optimal one is found. In this algorithm z_j^* is the optimal value of problem P_j, $\overline{z_j}$ is the optimal value of the relaxation R_j of P_j, and $\underline{z_j}$ is the best known lower bound for z_j^*.

13. In Algorithm 6 the optimal value $\overline{z_j}$ of relaxation R_j is an upper bound for z_j^*. Also $\underline{z_0}$ is the objective function value for the best known feasible solution to (5).

14. LP relaxations are often used in the bounding portion of Algorithm 6.

15. There are specializations of Algorithm 6 for 0-1 IP problems and for MIP problems.

16. Branch and bound tends to be a computationally expensive solution method. Usually it is applied only when other methods appear to be stalling.

17. In the survey [Fo97] of linear programming software, several of the packages listed will handle IP problems as well. When available, these extensions to binary and/or integer-valued variables are indicated by the survey.

18. There are several commercial software packages that solve IP and MIP problems, such as CPLEX, OSL, MIPIII, XPRESS-MP, XA, and LINDO:

   - http://www.cplex.com/
   - http://www.research.ibm.com/osl/
   - http://www.ketronms.com/mipiii.html
   - http://www.dash.co.uk/
   - http://www.sunsetsoft.com/
   - http://www.lindo.com/

19. An extensive tabulation of software packages to solve IP and MIP problems is found at the site:

   - http://www-c.mcs.anl.gov/home/otc/Guide/SoftwareGuide/Categories/intprog.html

20. Computer codes (in C and Pascal) for solving IP and MIP problems are available at the sites:

   - http://www.mpi-sb.mpg.de/~barth/opbdp/opbdp.html
   - http://www.iiasa.ac.at/~marek/soft/descr.html#MOMIP
   - http://www.netcologne.de/~nc-weidenma/readme.htm

Examples:

1. The following figure shows the convex hull of feasible solutions to the IP

$$\begin{aligned}
\text{maximize:} \quad & -x_1 + \tfrac{1}{2}x_2 \\
\text{subject to:} \quad & x_2 \leq 4 \\
& -x_1 - x_2 \leq -\tfrac{5}{2} \\
& 8x_1 + x_2 \leq 24 \\
& -3x_1 + 4x_2 \leq 10 \\
& x_1, x_2 \geq 0 \\
& x_1, x_2 \text{ integers.}
\end{aligned}$$

Algorithm 6: Branch and bound algorithm for (5).

input: IP in form (5)
output: an optimal solution x_0 with objective value $\underline{z_0}$
let P be the problem: $\max\{cx \mid x \in S\}$
$P_0 := P; S_0 := S; \underline{z_0} := -\infty; \overline{z_0} := +\infty$
put P_0 on the list of *live* nodes
{branching}
   if no live node exists **then** go to {termination}
   else select a live node P_j
{bounding}
   solve a relaxation R_j of P_j
   if R_j is infeasible **then** discard P_j and go to {branching}
   if $\overline{z_j} = +\infty$ **then** go to {separation}
   {$\overline{z_j}$ is finite}
     if $\overline{z_j} \leq \underline{z_0}$ **then** discard node P_j and go to {branching}
     if $\overline{z_j} = \underline{z_j}$ **then** update $\underline{z_0} := \max\{\underline{z_0}, \underline{z_j}\}$ and discard any node P_i for
        which $\overline{z_i} \leq \underline{z_0}$
{separation}
   choose a separation S_j^* of S_j forming new live nodes and go to {branching}
{termination}
   if $\underline{z_0} = -\infty$ **then** problem (5) is infeasible
   if $\underline{z_0}$ is finite **then** $\underline{z_0}$ is the optimal objective value and the associated x_0 is
       an optimal solution

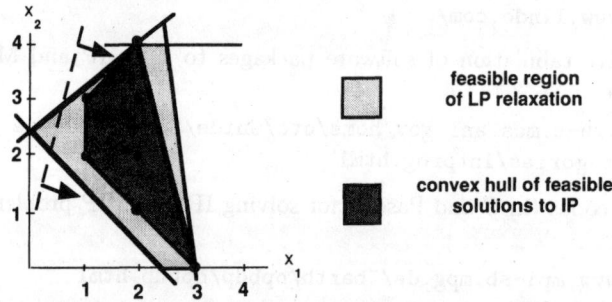

Here $S = \{(1,2), (1,3), (2,1), (2,2), (2,3), (2,4), (3,0)\}$. The optimal solution occurs at $(1,3)$, with $z_{IP} = \frac{1}{2}$. The feasible region of the LP relaxation is also shown in Figure 3, with the optimal LP value $z_{LP} = \frac{5}{4}$ attained at $(0, \frac{5}{2})$; a cutting plane is shown as the dashed line. For this problem, $conv(S)$ is defined by the following constraints:

$$-x_1 - x_2 \leq -3$$
$$-x_1 + x_2 \leq 2$$
$$-x_1 + \leq -1$$
$$4x_1 + x_2 \leq 12$$
$$x_2 \leq 4$$
$$x_1, x_2 \geq 0.$$

All of these constraints are facets except for $x_2 \leq 4$ and the nonnegativity constraints, which are redundant.

2. The following IP has the feasible region $S = \{(0,0), (1,1), (2,2), (3,3), (4,4)\}$ and the optimal solution occurs at $(4,4)$ with $z_{IP} = 8$:

$$\begin{aligned} \text{maximize:} \quad & x_1 + x_2 \\ \text{subject to:} \quad & 2x_1 - 2x_2 \leq 1 \\ & -7x_1 + 8x_2 \leq 4 \\ & x_1, x_2 \geq 0 \\ & x_1, x_2 \text{ integers.} \end{aligned}$$

The LP relaxation has a feasible region defined by vertices $(0,0)$, $(\frac{1}{2}, 0)$, $(0, \frac{1}{2})$, $(8, \frac{15}{2})$, so its optimal solution occurs at $(8, \frac{15}{2})$ with $z_{LP} = \frac{31}{2}$. Consequently, the LP solution is a poor approximation to the optimal IP solution. Moreover, simply rounding the LP solution gives either $(8,7)$ or $(8,8)$, both of which are infeasible to the given IP problem.

3. The following IP can be solved using Algorithm 6:

$$\begin{aligned} \text{maximize:} \quad & 3x_1 + 3x_2 - 8x_3 \\ \text{subject to:} \quad & -3x_1 + 6x_2 + 7x_3 \leq 8 \\ & 6x_1 - 3x_2 + 7x_3 \leq 8 \\ & x_1, x_2, x_3 \geq 0 \\ & x_1, x_2, x_3 \text{ integers.} \end{aligned}$$

The initial problem P_0 has an LP relaxation R_0 that is obtained by removing the integer restrictions; solving this LP gives $x = (2.667, 2.667, 0)$ with $z = 16$. A separation is achieved by creating the two subproblems P_1 and P_2; the constraint $x_1 \leq 2$ is appended to P_0 creating P_1 while the constraint $x_1 \geq 3$ is appended to P_0 creating P_2. Now the live nodes are P_1 and P_2. Solving the LP relaxation R_1 gives $x = (2, 2.333, 0)$ with $z = 13$. New subproblems P_3 and P_4 are obtained from P_1 by appending the constraints $x_2 \leq 2$ and $x_2 \geq 3$, respectively. Now the live nodes are subproblems P_2, P_3, P_4. The LP relaxation R_2 of P_2 is infeasible, as is the LP relaxation R_4 of P_4. Solving the LP relaxation R_3 gives the feasible integer solution $x = (2, 2, 0)$ with $z = 12$. Since there are no more live nodes, this represents the optimal solution to the stated problem.

4. *Fixed-charge problems*: Find optimal levels of n activities to satisfy m constraints while minimizing total cost. Each activity j has per unit cost c_j. In addition, there is a startup cost d_j for certain undertaken activities j.

This problem can be modeled as a MIP problem, with a real variable x_j for the level of each activity j. If activity j has a startup cost, introduce the additional 0-1 variable y_j, equal to 1 when $x_j > 0$ and 0 otherwise. For example, this condition can be enforced by imposing the constraints $M_j y_j \geq x_j$, $y_j \in \{0,1\}$, where M_j is a known upper bound on the value of x_j. The objective is then to minimize $z = cx + dy$.

5. *Queens problem*: On an $n \times n$ chessboard, the task is to place as many nontaking queens as possible.

This problem can be formulated as a 0-1 IP problem, having binary variables x_{ij}. Here $x_{ij} = 1$ if and only if a queen is placed in row i and column j of the chessboard. The objective function is to maximize $z = \sum_i \sum_j x_{ij}$ and there is a constraint for each row, column, and diagonal of the chessboard. Such a constraint has the form $\sum_{(i,j) \in S} x_{ij} \leq 1$, where S is the set of entries in the row, column, or diagonal. For example, one optimal solution of this IP for the 7×7 chessboard is the assignment $x_{16} = x_{24} = x_{37} = x_{41} = x_{53} = x_{65} = x_{72} = 1$, with all other $x_{ij} = 0$.

15.2 LOCATION THEORY

Location theory is concerned with locating a fixed number of facilities at points in some space. The facilities provide a service (or product) to the customers whose locations and levels of demand (for the service) are known. The object is to find locations for the facilities to optimize some specified criterion, e.g., the cost of providing the service. Interest in location theory has grown very rapidly because of its variety of applications to such fields as operations research, city planning, geography, economics, electrical engineering, and computer science.

15.2.1 p-MEDIAN AND p-CENTER PROBLEMS

Definitions:

A **metric space** is a space S consisting of a set of points with a real-valued function $d(x, y)$ defined on all pairs of points $x, y \in S$ with the following properties (§6.1.4):
- $d(x, y) = d(y, x) \geq 0$ for all $x, y \in S$
- $d(x, y) = 0$ if and only if $x = y$
- $d(x, z) \leq d(x, y) + d(y, z)$ for all $x, y, z \in S$.

The value $d(x, y)$ is called the **distance** between points $x, y \in S$.

There are p facilities that are to be located at some set $X_p = \{x_1, x_2, \ldots, x_p\}$ of p points in the (metric) space S. The elements of X_p are the **facility locations**.

The facilities are to provide a service to the customers whose positions are given by a set $V = \{v_1, v_2, \ldots, v_n\}$ of n points in S. The points in V are the **demand points** and the level of **demand** at $v_i \in V$ is given by $w(v_i) \geq 0$.

For $x \in S$ and $X_p \subseteq S$, let $d(x, X_p)$ be the minimum distance from x to a point of X_p: $d(x, X_p) = \min_{x_i \in X_p} \{d(x, x_i)\}$.

Suppose X_p is a candidate set of points in S for locating the p facilities. The following two objective functions are defined on $X_p \subseteq S$:
- $F(X_p) = \sum_{i=1}^{n} w(v_i) d(v_i, X_p)$;
- $H(X_p) = \max_{1 \leq i \leq n} \{w(v_i) d(v_i, X_p)\}$.

$X_p^m \subseteq S$ is a **p-median** if $F(X_p^m) \leq F(X_p)$ for all possible $X_p \subseteq S$.
$X_p^c \subseteq S$ is a **p-center** if $H(X_p^c) \leq H(X_p)$ for all possible $X_p \subseteq S$.

Facts:

1. It is customary to assume that the demand at v_i is satisfied by its closest facility. Then $w(v_i) d(v_i, X_p)$ indicates the total (transportation) cost associated with having the demand at $v_i \in S$ satisfied by its closest facility in the candidate set X_p.

2. $F(X_p)$ represents the total transportation cost of satisfying the demands if the facilities are located at X_p.

3. $H(X_p)$ represents the cost (or unfairness) associated with a farthest demand point not being in close proximity to any facility.

4. A p-median formulation is designed for locating p facilities in a manner than minimizes the average cost of serving the customers.

5. A p-center formulation is designed for locating p emergency facilities (police, fire, and ambulance services), in which the maximum time to respond to an emergency is to be made as small as possible.

6. The p-median and p-center problems are only interesting if $p < n$; otherwise, it is possible to locate at least one facility at each demand point, thereby reducing $F(X_p)$ or $H(X_p)$ to 0.

Examples:

1. Suppose that a single warehouse is to be located in a way to service n retail outlets at minimum cost. Here $w(v_i)$ is the number of shipments made per week to the outlet at location v_i. This can be modeled as a 1-median problem, since the objective is to locate the single warehouse to minimize the total distance traveled by delivery vehicles.

2. A new police station is to be located within a portion of a city to serve residents of that area. Neighborhoods in that area can be taken as the demand points, and locating the police station can be formulated as a 1-center problem. Here, the maximum distance from the source of an emergency is critical so the police station should be located to minimize the maximum distance from a neighborhood. The weights at each demand point might be taken to be equal, or in some situations differing weights could signify conversion factors that translate distance into some other measure such as the value of residents' time.

3. *Statistics*: Suppose that n given data values $x_1, x_2, \ldots, x_n$ are viewed as points placed along the real line. If the distance between points x_i and x_j is their absolute difference $|x_i - x_j|$, then a 1-median of this set of points (unweighted customer locations) is a point (facility) $\hat{x}$ that minimizes $\sum_{i=1}^{n} |x_i - \hat{x}|$. In fact, $\hat{x}$ corresponds to a *median* of the n data values. If the distance between points x_i and x_j is their squared difference $(x_i - x_j)^2$, then a 1-median is a point $\bar{x}$ minimizing $\sum_{i=1}^{n} (x_i - \bar{x})^2$, which is precisely the *mean* of the n data values. Alternatively, for either distance measure the 1-center of this set of points turns out to correspond to the *midrange* of the data set: namely, the 1-center is the point located halfway between the largest and the smallest data values.

15.2.2 p-MEDIANS AND p-CENTERS ON NETWORKS

Definitions:

A *network* is a weighted graph $G = (V, E)$ with vertex set $V = \{v_1, v_2, \ldots, v_n\}$ and edge set E, where $m = |E|$; see §8.1.1.

The *weight* of vertex $v \in V$ represents the demand at v and is denoted by $w(v) \geq 0$.

The *length* of edge $e \in E$ represents the cost of travel (or distance) across e and is denoted by $\ell(e) > 0$. Each edge is assumed to be a line segment joining its end vertices.

A *point on a network* G is any point along any edge of G. The precise location of the point x on edge $e = (u, v)$ is indicated by the distance of x from u or v.

If x and y are any two points on G, the *distance* $d(x, y)$ is the length of a shortest path between x and y in G, where the length of a path is the sum of the lengths of the edges (or partial edges) in the path.

Facts:

1. Network G with the above definition of distance constitutes a metric space (§15.2.1).

2. *p-median theorem*: Given a positive integer p and a network $G = (V, E)$, there exists a set of p vertices $V_p \subseteq V$ such that $F(V_p) \leq F(X_p)$ for all possible sets X_p of p points on G. That is, a p-median can always be found that consists entirely of vertices.

3. A p-median of a network G can be found by a finite search through all possible $\binom{n}{p}$ choices of p vertices out of n. This is still a formidable task, but if p is a small number (say $p < 5$) it is certainly manageable.

4. The p-median theorem also holds if the cost of satisfying the demand at $v_i \in V$ is $f_i(d(v_i, X_p))$, instead of $w(v_i) \cdot d(v_i, X_p)$, provided that $f_i : \mathcal{R}^+ \to \mathcal{R}^+$ is a concave nondecreasing function for all $i = 1, 2, \ldots, n$. ($\mathcal{R}^+$ denotes the set of nonnegative real numbers.) In this case, the objective function for the p-median problem becomes $F(X_p) = \sum_{i=1}^n f_i(d(v_i, X_p))$.

5. Each point x of a p-center X_p^c of network $G = (V, E)$ is a point on some edge e such that for some pair of distinct vertices u and $v \in V$, $w(u)d(u, x) = w(v)d(v, x)$; i.e., the point x is the "center" of a shortest path from u to v in G that passes through edge e.

6. There are at most n^2 predetermined choices of points on each edge of G that could be potential points in X_p^c; thus there are $n^2 m$ predetermined choices for points in X_p^c.

7. A p-center X_p^c of network G can be found by examining all possible $\binom{n^2 m}{p}$ choices of p points out of $n^2 m$. Even for small values of p, this is a formidable task.

Examples:

1. In the following network the levels of demand are given at the vertices and the lengths of the edges are shown on the edges. The 1-median of this network is at the vertex labeled x_1. The total transportation cost is $F(\{x_1\}) = 35$.

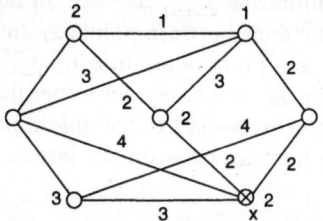

2. A tree network T is shown in the following figure, with the vertex demands and edge lengths displayed.

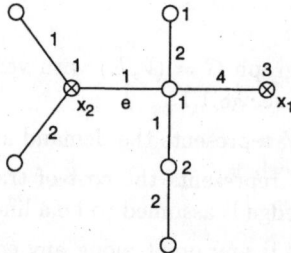

A 2-median of T is the set of vertices $X_2 = \{x_1, x_2\}$, with total transportation cost $F(X_2) = 25$. If x_1 is kept fixed and t is an arbitrary point along edge e, then $\{x_1, t\}$ also constitutes a 2-median of T. This is consistent with Fact 2, which only states that there is a p-median that is a subset of V, not that every p-median occurs in this way.

3. A 1-center X_1^c is found for the network in part (a) of the following figure. For illustration, suppose that the 1-center is along the edge (u_1, u_2), thereby limiting the search to candidate points on this edge. Let $X(x)$ be an arbitrary point along edge (u_1, u_2), parametrized by the scalar $x = d(u_1, X(x))$. Note that $x \in [0, 3]$ with $X(0) = u_1$ and $X(3) = u_2$.

Part (b) of the figure shows plots of $w(u_i)d(u_i, X(x))$ as a function of x for $i = 1, 2, \ldots, 4$. The plot of $D(x) \equiv \max_{1 \leq i \leq 4} w(u_i)d(u_i, X(x))$ is indicated in bold and $D(x)$ assumes its minimum value when $x = \frac{8}{7}$. The 1-center of the network in part (a) is then located along edge (u_1, u_2) a distance of $\frac{8}{7}$ from u_1. Note that for this value of x, $w(u_4)d(u_4, X(x)) = w(u_3)d(u_3, X(x))$, consistent with Fact 5. In general, X_1^c of a network is not necessarily a unique point; however, here X_1^c is unique and $H(X_1^c) = \frac{30}{7}$.

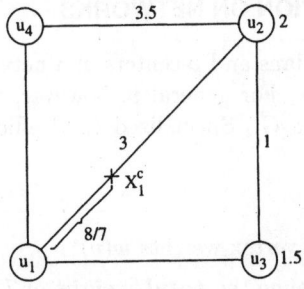

(a) A simple network.

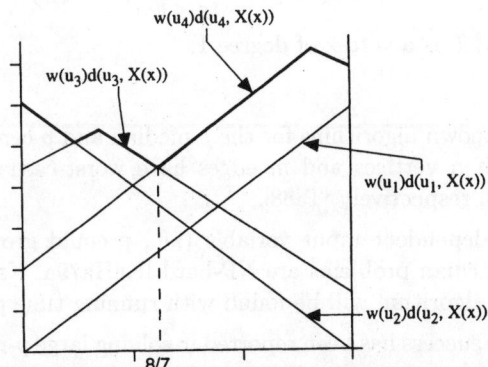

(b) Plots of $w(u_i)d(u_i, X(x))$ for $1 \leq i \leq 4$.

4. Transportation planners are trying to decide where to locate a single school bus stop along a major highway. Situated along a one-mile stretch of the highway are 8 communities. The table below gives the number of school-age students in each community who ride the bus on a daily basis. The distance of each community from the westernmost edge of the one-mile stretch of highway is also shown.

community	1	2	3	4	5	6	7	8
# students	9	4	8	11	5	3	5	11
distance (mi)	0.0	0.2	0.3	0.4	0.6	0.7	0.9	1.0

The data of this problem can be represented by an undirected path (§8.1.3) with each vertex v corresponding to a community and $w(v)$ being the number of students from that community riding the bus. Edges join adjacent communities and have a length given by the difference in distance entries from the table. To minimize the total (weighted) distance traveled by the students, a 1-median is sought. By Fact 2, only vertex locations need to be considered. For example, situating the bus stop at vertex 3 incurs a cost of $F(\{3\}) = 9(0.3) + 4(0.1) + 8(0) + 11(0.1) + 5(0.3) + 3(0.4) + 5(0.6) + 11(0.7) = 17.6$. The minimum cost is incurred for vertex 4, with $F(\{4\}) = 16.3$, so that the bus stop should be located at community 4.

15.2.3 ALGORITHMS FOR LOCATION ON NETWORKS

Algorithms for finding p-medians and p-centers of a network G can be devised that are feasible for small values of p. For general p, however, there are no efficient methods known for arbitrary networks G. Specialized (and efficient) algorithms are available when G is a tree.

Definitions:

Let T be a tree (§9.1.1) with vertex weights $w(v)$.

If T' is a subnetwork of T, define the **total weight** of T' by $W(T') = \sum_{v \in V(T')} w(v)$.

For $v \in V(T)$, let $T_{v1}, T_{v2}, \ldots, T_{vd(v)}$ be the components of $T - v$, where $d(v)$ is the degree (§8.1.1) of vertex v. Define $M_v = \max_{1 \leq i \leq d(v)} \{W(T_{vi})\}$.

A *leaf vertex* of T is a vertex of degree 1.

Facts:

1. The fastest known algorithms for the p-median and p-center problems on a network $G = (V, E)$ with n vertices and m edges have worst-case complexities $O(n^{p+1})$ and $O(m^p n^p (\log n)^2)$, respectively [Ta88].

2. If p is an independent input variable (i.e., p could grow with n), then both the p-center and p-median problems are NP-hard [KaHa79a, KaHa79b]. Thus it is highly unlikely that an algorithm will be found with running time polynomial in n, m, and p.

3. Considerable success has been reported in solving large p-median problems by heuristic methods that do not necessarily guarantee optimal solutions. The best known such procedure is a dual-based integer programming approach due to Erlenkotter [Er78].

4. Algorithms of complexities $O(n^2 p)$ and $O(n \log n)$ for the p-median and p-center problems on tree networks have been reported by Tamir [Ta96] and by Frederickson and Johnson [MiFr90, Chapter 7].

5. Vertex u is a 1-median of a tree network T if and only if $M_u \leq \frac{1}{2} W(T)$.

6. *1-median of a tree:* This algorithm (Algorithm 1) is based on Fact 5. The main idea is to repeatedly remove a leaf vertex, confining the problem to a smaller tree T'.

7. Algorithm 1 can be implemented to run in $O(n)$ time.

8. Let T be a tree network with $w(v) = c$ for all $v \in V(T)$. Then the 1-center of T is the unique middle point of a longest path in T.

9. Select any vertex v_0 in a tree network T. Let v_1 be a farthest vertex from v_0, and let v_2 be a farthest vertex from v_1. Then the path from v_1 to v_2 is a longest path in T.

> **Algorithm 1:** 1-median of a tree.
>
> input: tree T
> output: 1-median $\widetilde{v}$
>
> $T' := T$; $W_0 := \sum_{v \in V(T')} w(v)$; $W(v) := w(v)$ for each $v \in V(T')$
>
> {Main loop}
> if T' consists of a single vertex $\widetilde{v}$ then stop
> else
> $\quad \widetilde{v} := $ a leaf vertex of T'
> $\quad$ if $W(\widetilde{v}) \geq \frac{1}{2} W_0$ then stop
> $\quad$ else
> $\quad\quad u := $ the vertex adjacent to $\widetilde{v}$ in T'
> $\quad\quad W(u) := W(u) + W(\widetilde{v})$
> $\quad\quad T' := T' - \widetilde{v}$
> {Continue with next iteration of main loop}

> **Algorithm 2:** 1-center of an unweighted tree.
>
> input: tree T with $w(v) = c$ for all $v \in V(T)$
> output: 1-center x
>
> find a longest path P in T (using Fact 9)
> let u_1 and u_2 be the end vertices of P
> find the middle point of this path: i.e., the point x such that $d(x, u_1) = d(x, u_2)$

10. *1-center of a tree*: This algorithm (Algorithm 2) applies to "unweighted" trees, in which there are identical weights at each vertex. It is based on Facts 8 and 9.

11. Algorithm 2 can be implemented to run in O(n) time.

Examples:

1. Suppose that the vertices of the tree T in Figure 2 (§15.2.2) are labeled $v_1, v_2, \ldots, v_8$ in order from top to bottom and left to right at each height. Algorithm 1 can be applied to find the 1-median of T. First, the leaf vertex v_1 is selected and since $W(v_1) = 1$ is less than $\frac{1}{2} W_0 = \frac{15}{2}$, its weight is added to vertex v_3. The following table shows the progress of the algorithm, which eventually identifies vertex v_4 as the 1-median of T. As guaranteed by Fact 5, $M_{v_4} = \max\{6, 1, 3, 4\} \leq \frac{15}{2}$.

iteration	$\widetilde{v}$	1	2	3	$W(v_i)$ 4	5	6	7	8
0		1	1	1	1	3	4	2	2
1	v_1	—	1	2	1	3	4	2	2
2	v_2	—	—	2	2	3	4	2	2
3	v_5	—	—	2	5	—	4	2	2
4	v_6	—	—	6	5	—	—	2	2
5	v_3	—	—	—	11	—	—	2	2

2. Let the vertices of the tree T in the figure of §15.2.2 Example 2 be labeled as in Example 1 of this section. Algorithm 2 can be applied to find the 1-center of T, with all vertex weights being 1. First, select v_1 and find a farthest vertex from it, namely v_5. A farthest vertex from v_5 is then v_8, giving a longest path $P = [v_5, v_4, v_7, v_8]$ in T. The midpoint x of P, located $\frac{1}{2}$ unit from v_4 along edge (v_4, v_5), is then the 1-center of T. If instead the longest path $Q = [v_6, v_3, v_4, v_5]$ in T had been identified, then the same midpoint x would be found.

15.2.4 CAPACITATED LOCATION PROBLEMS

Definitions:

Let $X_p = \{x_1, \ldots, x_p\}$ be a set of locations for p facilities in the metric space S with n demand points $V = \{v_1, \ldots, v_n\} \subseteq S$ where $w(v_i) \geq 0$ is the demand at $v_i \in V$.

For each $v_i \in V$ and $x_j \in X_p$, let $w(v_i, x_j)$ be the portion of the demand at v_i satisfied by the facility at x_j.

Let $W(x_j)$ be the sum of the demands satisfied by (or allocated to) the facility at x_j. In a *capacitated location problem*, upper (and/or lower) bounds are placed on $W(x_j)$.

Given the positive integer p and positive constant α, two versions of the **capacitated p-median (CPM) problem** in network $G = (V, E)$ are defined:

(a) Find a set of locations X_p such that $F(X_p) = \sum_{v_i \in V} w(v_i) d(v_i, X_p)$ is minimized subject to $W(x_j) \leq \alpha$ for all $x_j \in X_p$. Here it is assumed that the demands are satisfied by their closest facility, and in the case of ties, a demand, say at v, may be allocated in an arbitrary way among the closest facilities to v.

(b) Find X_p and $\{w(v_i, x_j) \mid v_i \in V \text{ and } x_j \in X_p\}$ to minimize

$$\sum_{j=1}^{p} \sum_{i=1}^{n} w(v_i, x_j) d(v_i, x_j)$$

subject to

$$\sum_{i=1}^{n} w(v_i, x_j) \leq \alpha, \quad j = 1, 2, \ldots, p$$

$$\sum_{j=1}^{p} w(v_i, x_j) = w(v_i), \quad i = 1, 2, \ldots, n.$$

Facts:

1. Capacitated facility location problems occur in several applied settings, including:
   - the location of manufacturing plants (with limited output) to serve customers;
   - the location of landfills (with limited capacity), which receive solid waste from the members of a community;
   - the location of concentrators in a telecommunication network, where each concentrator bundles messages received from individual users and can handle only a certain amount of total message traffic.

2. Version (a) of CPM may not have a solution if α is too small.

3. If α is sufficiently large in version (a) and CPM has a solution, there may not exist a solution consisting entirely of vertices of G. See Example 1.

4. Version (b) of the CPM has a solution consisting entirely of vertices of G. This was shown by J. Levy; see [HaMi79].

Examples:

1. Suppose $p = 2$ and $\alpha = 3$ in the following network G. A solution to version (a) of the CPM problem consists of the points $X_2 = \{x_1, x_2\}$ with $W(x_1) = W(x_2) = 3$ and $F(X_2) = 9$. It is easy to see that the choice of any two vertices for X_2 would violate the allocation constraint to one facility.

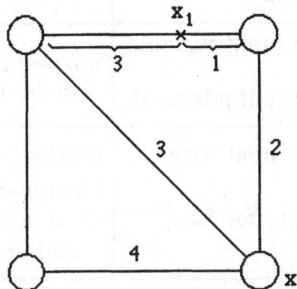

2. If $p = 2$ and $\alpha = 3$ for the network G of the figure of Example 1, then version (b) of the CPM problem has a solution containing only vertices of G. Suppose that the top two vertices in the figure are v_1 and v_2 (from left to right) and the bottom two vertices are v_3 and v_4. Then $X_2 = \{v_1, v_2\}$ is an optimal solution, where all the demand $w(v_3) = 1$ is allocated to v_1 whereas the demand $w(v_4) = 2$ is equally split between v_1 and v_2. (Here, not all demand from v_4 is sent to its closest facility v_2.) In this solution, $W(v_1) = W(v_2) = 3$ and $F(X_2) = 7$.

15.2.5 FACILITIES IN THE PLANE

The p-median and p-center problems can be defined in the plane $\mathcal{R}^2$. Several measures of distance are commonly considered for these location problems in the plane.

Definitions:

Let $x = (x_1, x_2)$ and $y = (y_1, y_2)$ be points of $S = \mathcal{R}^2$.

The **Euclidean** (ℓ_2) **distance** between x and y is $d(x, y) = [(x_1 - y_1)^2 + (x_2 - y_2)^2]^{1/2}$.

The **rectilinear** (ℓ_1) **distance** between x and y is $d(x, y) = |x_1 - y_1| + |x_2 - y_2|$.

The (**generalized**) **Weber problem** is the p-median problem in $\mathcal{R}^2$ with ℓ_2 as the measure of distance.

The **unweighted Euclidean 1-center problem** is the 1-center problem in $\mathcal{R}^2$ with the ℓ_2 measure of distance and with $w(v_i) = c$ for all $v_i \in V$.

Facts:

1. No polynomial-time algorithm for the Weber problem, even when $p = 1$, has been discovered.

2. In practice, an iterative method due to Weiszfeld [FrMcWh92] has been shown to be highly successful for the Weber problem with $p = 1$.

3. The p-median and p-center problems in $\mathcal{R}^2$ with either ℓ_1 or ℓ_2 as the measure of distance have been proven to be NP-hard if p is an input variable [MeSu84].

4. The unweighted Euclidean 1-center problem is equivalent to finding the center of the smallest (radius) circle that encloses all points in V.

5. The following table provides a summary of time complexity results of the best known algorithms for location problems in the plane [Me83, MeSu84].

	p arbitrary	$p = 1$
p-median	NP-hard if p is an input variable under ℓ_1 or ℓ_2 unknown complexity if p is fixed	unknown complexity even when $p = 1$ under ℓ_1 or ℓ_2
p-center	NP-hard if p is an input variable under ℓ_1 or ℓ_2 unknown complexity for fixed $p > 1$	$O(n \log^2 n)$ under ℓ_2 $O(n)$ under ℓ_2 in unweighted case $O(n)$ under ℓ_1 for both the weighted and unweighted cases

Examples:

1. The floor plan of a factory contains existing machines A, B, C at the coordinate locations $a = (0, 4)$, $b = (2, 0)$, $c = (5, 2)$. A new central storeroom, to house materials needed by the machines, is to be placed at some location $x = (x_1, x_2)$ on the factory floor. Because the aisles of the factory floor run north-south and east-west, transportation between the storeroom and the machines must take place along these perpendicular directions. For example, the distance between the storeroom and the machine C is $|x_1 - 5| + |x_2 - 2|$. Management wants to locate the storeroom so that the weighted sum of distances between the new storeroom and each machine is minimized, taking into account that the demand for material by machine A is twice the demand by machine B, and demand for material by machine C is three times that by machine B. This is a weighted 1-median problem in the plane with the ℓ_1 measure of distance. The point $(3, 2)$ is an optimal location for the storeroom. In fact, for any $2 \leq u \leq 5$ the point $(u, 2)$ is also an optimal location.

2. Suppose that the ℓ_2 distance measure is used instead in Example 1. Then a weighted 1-median is a location $x = (x_1, x_2)$ that minimizes

$$2\sqrt{x_1^2 + (x_2 - 4)^2} + \sqrt{(x_1 - 2)^2 + x_2^2} + 3\sqrt{(x_1 - 5)^2 + (x_2 - 2)^2}.$$

The minimizing point in this case is $(5, 2)$, which is the *unique* optimal location for the storeroom. On the other hand, if the demands for material are the same for all three machines, then the unweighted 1-median occurs at the unique location $(2.427, 1.403)$.

15.2.6 OBNOXIOUS FACILITIES

In the preceding subsections, it has been assumed that the consumers at v_i wish to be as close as possible to a facility. That is, the facilities are desirable. In contrast, this subsection discusses location problems where the facilities are undesirable or obnoxious.

Definitions:

For $v_i \in V$, $w(v_i)d(v_i, X_p)$ represents the **utility** (in contrast to cost) associated with having an obnoxious facility located at distance $d(v_i, X_p)$ from v_i.

The following two **obnoxious facility location problems** are defined:

(a) find $X_p \subseteq S$ to maximize $F(X_p) = \sum_{i=1}^{n} w(v_i)d(v_i, X_p)$;

(b) find $X_p \subseteq S$ to maximize $G(X_p) = \min_{1 \leq i \leq n} w(v_i)d(v_i, X_p)$.

If space S is a network $G = (V, E)$, for each edge $e = (u, v) \in E$, let $x(e) = x$ be the point on e such that $w(u)d(u, x) = w(v)d(v, x) \equiv w(e)$.

Facts:

1. If S is a network, problem (a) may not have a solution that is a subset of vertices (see Example 2).

2. Suppose S is a network G with $w(v_i) > 0$ for all $v_i \in V$; further assume that at most one point of X_p can be on any particular edge. Renumber the m edges of G so that $w(e_1) \geq w(e_2) \geq \cdots \geq w(e_m)$. Then $x(e_1), x(e_2), \ldots, x(e_p)$ is a solution to problem (b).

3. Additional results on this subject can be found in [BrCh89].

Examples:

1. In the location of obnoxious facilities, the distance to a closest facility is to be made as large as possible. This type of problem arises in siting nuclear power plants, sewage treatment facilities, and landfills, for example.

2. In the following network, a solution X_1 to problem (a) when $p = 1$ is the midpoint of any edge, and $F(X_1) = 5$. If the facility is located at any vertex v then $F(\{v\}) = 4$.

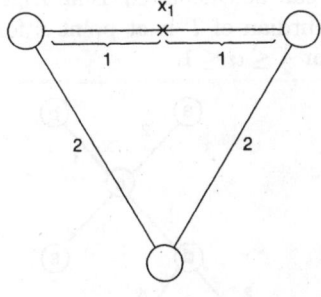

15.2.7 EQUITABLE LOCATIONS

The p-median problem is a widely used model for locating public or private facilities. However, it may leave some demand points (communities) too far from their closest facility and thus be perceived as inequitable. To remedy this situation, the p-median problem can be modified in several ways.

Definitions:

Suppose S is a metric space, $V \subseteq S$, $w(v_i) \geq 0$ for all $v_i \in V$, and p is a positive integer. Let $w'(v_i) = w(v_i)/\sum_{j=1}^{n} w(v_j)$ and define $F'(X_p) = \sum_{i=1}^{n} w'(v_i)d(v_i, X_p)$, $Z(X_p) = \sum_{i=1}^{n} w'(v_i)[d(v_i, X_p) - F'(X_p)]^2$.

The following three **equitable facility location problems** are defined:

(a) given a constant β, find a set of p points $X_p \subseteq S$ to minimize
$$F(X_p) = \sum_{i=1}^{n} w(v_i)d(v_i, X_p)$$
subject to
$$d(v_i, X_p) \leq \beta, \quad \text{for all } v_i \in V;$$

(b) given a constant α, $0 < \alpha < 1$, find a set of p points $X_p \subseteq S$ to minimize $\alpha F(X_p) + (1-\alpha)H(X_p)$;

(c) find X_p to minimize $Z(X_p)$.

Facts:

1. Since the objective function in (b) is a linear combination of the objective functions for the p-median and p-center problems, the solution X_p^* is called a *centdian*.

2. $F'(X_p) = F(X_p)/\sum_{j=1}^{n} w(v_j)$ is the mean distance to the consumers given that the facilities are located at X_p.

3. $Z(X_p)$ is the variance of the distance to the consumers given that the facilities are located at X_p.

4. Additional results are discussed in [HaMi79, Ma86].

Example:

1. The following figure shows a tree network T on 7 vertices, with edge lengths displayed. Suppose that all vertex weights are 1. Then the 1-median of T is located at vertex c, while the 1-center of T is located at the point x one unit from vertex d along (d, g). These locations can be calculated using Algorithms 1 and 2 from §15.2.3. It can be verified that the centdian of T is at point x for $0 \leq \alpha \leq \frac{1}{6}$, at vertex d for $\frac{1}{6} \leq \alpha \leq \frac{1}{2}$, and at vertex c for $\frac{1}{2} \leq \alpha \leq 1$.

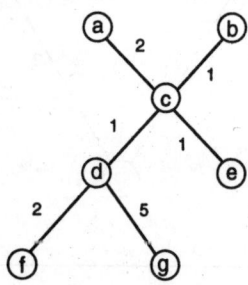

15.3 PACKING AND COVERING

Many practical problems can be formulated as either packing or covering problems. In *packing problems*, known activities are given each of which requires several resources for its completion. The problem is to select a most valuable set of activities to undertake without using any common resources. Such a problem arises for example in scheduling as many computational activities as possible on a set of machines (resources) that cannot be used simultaneously for more than one activity.

In *covering problems*, a specified set of tasks must be performed, and the objective is to minimize the resources required to perform the tasks. For example, a number of delivery trucks (resources) operating on overlapping geographical routes need to be dispatched to pick up items at customer locations (tasks). The fewest number of trucks are to be sent so that each customer location is "covered" by at least one of the dispatched trucks.

Both exact and heuristic solution algorithms for packing and covering problems are discussed in this section.

15.3.1 KNAPSACKS

The *knapsack problem* arises when a single critical resource is to be optimally allocated among a variety of options. Specifically, there are available a number of items, each of which consumes a known amount of the resource and contributes a known benefit. Items are to be selected to maximize the total benefit without exceeding the given amount of the resource. Knapsack problems arise in many practical situations involving cutting stock, cargo loading, and capital budgeting.

Definitions:

Let $N = \{1, 2, \ldots, n\}$ be a given set of n *items*. Utilizing item j consumes (requires) $a_j > 0$ units of the given resource and confers the benefit $c_j > 0$.

The **knapsack problem** (**KP**) is the following 0-1 integer linear programming problem:

$$\begin{aligned}
\text{maximize:} \quad & \sum_{j \in N} c_j x_j \\
\text{subject to:} \quad & \sum_{j \in N} a_j x_j \leq b \\
& x_j \in \{0, 1\}
\end{aligned} \quad (1)$$

It is assumed that $a_j \leq b$ for all $j \in N$. Let $z(b)$ denote the optimal objective value in (1) for a given integer b.

The **LP relaxation** (§15.1.8) of (1) is the linear programming problem:

$$\begin{aligned}
\text{maximize:} \quad & \sum_{j \in N} c_j x_j \\
\text{subject to:} \quad & \sum_{j \in N} a_j x_j \leq b \\
& 0 \leq x_j \leq 1
\end{aligned} \quad (2)$$

Let z_{LP} denote the optimal objective value to the LP relaxation (2).

Let N^1 and N^0 be the set of variables taking values 1 and 0, respectively, in the optimal solution to (2). Let λ^* be the dual variable (§15.1.5) associated with the knapsack inequality in the optimal solution to (2).

A **cover** is a set $S \subseteq N$ such that $\sum_{j \in S} a_j > b$. The cover S is **minimal** if no proper subset of S is a cover.

A **branch and bound tree** for KP is a tree T whose nodes correspond to subproblems obtained by fixing certain variables of (1) to either 0 or 1. The **root** of T corresponds to the original problem (1).

Algorithm 1: Greedy heuristic for the KP.
input: KP with $\frac{c_1}{a_1} \geq \frac{c_2}{a_2} \geq \cdots \geq \frac{c_n}{a_n}$
output: feasible solution $x = (x_1, x_2, \ldots, x_n)$
for $k := 1$ to n
  if $\sum_{j=1}^{k} a_j x_j \leq b$ then $x_k := 1$
  else $x_k := 0$

A node t of T is specified by its **level** $k = 0, 1, \ldots, n$ and the index set $N_t \subseteq \{1, \ldots, k\}$ of variables currently fixed to 1.

Associated with the set N_t is the current benefit $z_t = \sum_{j \in N_t} c_j$ and the available amount of resource $b_t = b - \sum_{j \in N_t} a_j$.

Facts:

1. Formulation (1) is a 0-1 integer linear programming problem with a single (knapsack) constraint. It expresses the optimization problem in which a subset of the n items is to be selected to maximize the total benefit without exceeding the available amount of the given resource (the capacity of the knapsack).

2. KP is an NP-hard optimization problem (§16.5.2).

3. KP can be solved in polynomial time for fixed b.

4. Given a rational $\epsilon > 0$, a $\{0,1\}$-vector x^* can be found satisfying $\sum_{j \in N} a_j x_j^* \leq b$ and $\sum_{j \in N} c_j x_j^* \geq (1 - \epsilon) z(b)$ in time polynomially bounded by $\frac{1}{\epsilon}$ and by the sizes of $a = (a_1, \ldots, a_n)$, $c = (c_1, \ldots, c_n)$, and b.

5. If the coefficients a_j can be ordered such that each coefficient is an integer multiple of the previous one, then KP can be solved in polynomial time.

6. If $a_{j-1} \geq a_j + \cdots + a_n$ holds for $j = 2, \ldots, n$ then KP can be solved in polynomial time.

7. *Greedy heuristic*: This heuristic (Algorithm 1) for the KP processes the variables x_j in nonincreasing order of $\frac{c_j}{a_j}$, making each variable equal to 1 if possible.

8. Suppose z_H is the objective value for the solution x produced by Algorithm 1. Then $z(b) \geq z_H \geq \frac{1}{2} z(b)$.

9. Algorithm 1 is most effective if the coefficients a_j are small relative to b.

10. The LP relaxation (2) can be solved explicitly by filling the knapsack in turn with items j in order of nonincreasing $\frac{c_j}{a_j}$, ignoring the integer restriction. The solution x obtained has at most one fractional component.

11. *Core heuristic*: This heuristic (Algorithm 2) for the KP first solves the LP relaxation (2) as in Fact 10, in which at most one variable x_k can be fractional. A smaller knapsack problem is then solved, by setting to 0 any variable x_j with index j sufficiently greater than k and by setting to 1 any variable x_j with index j sufficiently smaller than k.

Algorithm 2: Core heuristic for the KP.
input: KP with $\frac{c_1}{a_1} \geq \frac{c_2}{a_2} \geq \cdots \geq \frac{c_n}{a_n}$
output: feasible solution $x = (x_1, x_2, \ldots, x_n)$
{solve LP relaxation}
  find the smallest value k such that $\sum_{j=1}^{k} a_j \geq b$
{solve restricted KP}
  select any $r > 0$
  $x_j := 1$ for $j \leq k - r$
  $x_j := 0$ for $j \geq k + r$
  solve to optimality the smaller knapsack problem:
    maximize $\sum_{j=k-r+1}^{k+r-1} c_j x_j$
    subject to $\sum_{j=k-r+1}^{k+r-1} a_j x_j \leq b - \sum_{j=1}^{k-r} a_j$
    $x_j \in \{0, 1\}$

12. Algorithm 2 is effective if the number of variables n is large since values of r between 10 and 25 give very good approximations in most cases. For further details see [BaZe80].

13. $z(b) \leq z_{LP}$.

14. If z^l is the objective value of a feasible solution to (1), then $z^l \leq z(b)$.

15. Node t of the branch and bound tree T corresponds to a subproblem having a nonempty set of feasible solutions if and only if $b_t \geq 0$. When this holds, z_t is a lower bound for $z(b)$.

16. An upper bound on the objective value over the subproblem corresponding to node t is $z_t^u = \lfloor z_t^* \rfloor$, where $z_t^* = z_t + \max\{\sum_{j=k+1}^{n} c_j x_j \mid \sum_{j=k+1}^{n} a_j x_j \leq b_t, 0 \leq x_j \leq 1\}$.

17. *Implicit enumeration*: This is an exact technique based on the branch and bound method (§15.1.8). It is implemented using a branch and bound tree T, with the following specifications:

- The initial tree T consists of the root t, with lower bound z^l on $z(b)$ obtained using the greedy heuristic. An upper bound for node t is $z_t^u = z_{LP}$.
- If $z_t^u \leq z^l$, then node t is discarded since it cannot provide a better solution.
- If $z_t^u > z^l$, there are three cases (where node t is at level k of T):
  ◇ $a_{k+1} < b_t$: If $k + 1 < n$, create a new node with $x_{k+1} = 1$. If $k + 1 = n$ an optimal solution for node t has $x_n = 1$. Since this solution is feasible for KP, set $z^l = z_t^u$ and discard node t.
  ◇ $a_{k+1} = b_t$: An optimal solution for node t (and a feasible solution for KP) is obtained by setting $x_{k+1} = 1$ and $x_j = 0$ for $j > k + 1$. Set $z^l = z_t^u$ and discard node t.
  ◇ $a_{k+1} > b_t$: Discard the (infeasible) node with $x_{k+1} = 1$ and create a new node with $x_{k+1} = 0$.
- To backtrack from node t let $N_t = \{j_1, \ldots, j_r\} \subseteq \{1, \ldots, k\}$ with $j_1 < \cdots < j_r$. If $k \notin N_t$ retreat to level j_r and set $x_{j_r} = 0$. If $k = j_r$ retreat to level j_{r-1} and set $x_{j_{r-1}} = 0$.

18. *Variable fixing*: Given z_{LP}, z^l, N^1, N^0, and λ^*, the following tests can be used to fix variables and reduce the size of the knapsack problem:
   - If $k \in N^1$ and $z_{LP} - (c_k - \lambda^* a_k) \leq z^l$, then fix $x_k = 1$.
   - If $k \in N^0$ and $z_{LP} + (c_k - \lambda^* a_k) \leq z^l$, then fix $x_k = 0$.
   - Given $k \in N^1$ define
   $$z_{LP}^k = \sum_{j \in N^1 - \{k\}} c_j + \max\left\{\sum_{j \in N - N^1} c_j x_j \,\bigg|\, \sum_{j \in N - N^1} a_j x_j \leq b,\ 0 \leq x_j \leq 1\right\}.$$
   If $z_{LP}^k \leq z^l$, then x_k can be fixed to 1.
   - Given $k \in N^0$ define
   $$z_{LP}^k = c_k + \max\left\{\sum_{j \in N - N^0} c_j x_j \,\bigg|\, \sum_{j \in N - N^0} a_j x_j \leq b - a_k,\ 0 \leq x_j \leq 1\right\}.$$
   If $z_{LP}^k \leq z^l$, then x_k can be fixed to 0.

19. *Minimal cover inequality*: If S is a minimal cover then each feasible solution x to KP satisfies $\sum_{j \in S} x_j \leq |S| - 1$.

20. *Lifted minimal cover inequality*: The minimal cover inequality can be further strengthened. Without loss of generality, assume that $a_1 \geq a_2 \geq \cdots \geq a_n$ and $S = \{j_1 < j_2 < \cdots < j_r\}$. Let $\mu_h = \sum_{k=1}^h a_{j_k}$ for $h = 1, \ldots, r$ and define $\lambda = \mu_r - b \geq 1$. Then each feasible solution x to KP satisfies $\sum_{j \in N-S} \alpha_j x_j + \sum_{j \in S} x_j \leq |S| - 1$, where:
   - if $\mu_h \leq a_j \leq \mu_{h+1} - \lambda$ then $\alpha_j = h$;
   - if $\mu_{h+1} - \lambda + 1 \leq a_j \leq \mu_{h+1} - 1$ then: (a) $\alpha_j \in \{h, h+1\}$, and (b) there is at least one lifted minimal cover inequality with $\alpha_j = h + 1$.

21. Algorithms and computer codes to solve knapsack problems are given in [MaTo90].

22. Fortran code for solving knapsack problems can be found at the site:
   - http://www.netlib.org/toms/632

23. Further details on the material in this section are available in [NeWo88], [Sc86].

Examples:

1. *Investment problem*: An investor has $50,000 to place in any combination of five available investments $(1, 2, 3, 4, 5)$. All investments have the same maturity but are issued in different denominations and have different (one-year) yields, as shown here:

investment	1	2	3	4	5
denomination ($)	10,000	20,000	30,000	10,000	20,000
yield (%)	20	14	18	9	13

Let variable $x_j = 1$ if Investment j is selected and $x_j = 0$ if it is not. The interest earned for Investment 1 is $(0.20)10{,}000 = 2{,}000$; the values of the other investments are found similarly. Then the investor's problem is the knapsack problem

maximize: $\quad 2{,}000 x_1 + 2{,}800 x_2 + 5{,}400 x_3 + 900 x_4 + 2{,}600 x_5$

subject to: $\quad 10{,}000 x_1 + 20{,}000 x_2 + 30{,}000 x_3 + 10{,}000 x_4 + 20{,}000 x_5 \leq 50{,}000$

which has the optimal solution $x_1 = x_3 = x_4 = 1$, $x_2 = x_5 = 0$ with maximum interest of $8,300.

2. Consider the knapsack problem in 0-1 variables x

$$\text{maximize:} \quad 30x_1 + 8x_2 + 16x_3 + 20x_4 + 12x_5 + 9x_6 + 5x_7 + 3x_8$$
$$\text{subject to:} \quad 10x_1 + 3x_2 + 7x_3 + 9x_4 + 6x_5 + 5x_6 + 3x_7 + 2x_8 \leq 27.$$

Here the variables x_j are indexed in nonincreasing order of $\frac{c_j}{a_j}$. The optimal solution to the LP relaxation (2) is $x_1 = x_2 = x_3 = 1$, $x_4 = \frac{7}{9}$, $x_j = 0$ otherwise, with $z_{LP} = 69\frac{5}{9}$. The greedy heuristic (Algorithm 1) gives the feasible solution $x_1 = x_2 = x_3 = x_5 = 1$, $x_j = 0$ otherwise, with $z_H = 66$. Using $r = 3$, the core heuristic gives $x_1 = x_2 = x_4 = x_6 = 1$, $x_j = 0$ otherwise. This solution is optimal, with objective value 67.

3. Consider the knapsack problem in 0-1 variables x

$$\text{maximize:} \quad x_1 + x_2 + x_3 + x_4 + x_5 + x_6$$
$$\text{subject to:} \quad 10x_1 + 8x_2 + 4x_3 + 3x_4 + 3x_5 + 2x_6 \leq 11.$$

The set $S = \{3, 4, 5, 6\}$ is a minimal cover which gives the lifted minimal cover inequality $3x_1 + 2x_2 + x_3 + x_4 + x_5 + x_6 \leq 3$. Adding this inequality and solving the resulting linear program gives $x_4 = x_5 = x_6 = 1$, $x_j = 0$ otherwise. This solution is optimal.

4. *General knapsack problem:* The general (or unbounded) knapsack problem allows the decision variables x_j to be any nonnegative integers, not just 0 and 1. The following site provides an interactive algorithm for solving such knapsack problems (having up to 10 integer variables):

- http://www.maths.mu.oz.au/~moshe/recor/knapsack/knapsack.html

15.3.2 BIN PACKING

Minimizing the number of copies of a resource required to perform a specified set of tasks can be formulated as a *bin packing problem*. It is assumed that no such task can be split between two different units of the resource.

For example, this type of problem arises in allocating a set of customer loads to (identical) trucks, with no load being split between two trucks. Also, the scheduling of heterogeneous tasks on identical machines can be viewed as a bin packing problem. Namely, find the fewest number of machines of capacity C such that each task is executed on one of the machines and the total capacity of jobs assigned to any machine does not exceed C.

Definitions:

The positive integer C denotes the **bin capacity**.

Let $L = (p_1, p_2, \ldots, p_n)$ be a list of n *items*, where item p_i has an integer *size* $s(p_i) \leq C$.

A subset $P \subseteq L$ is a **packing** if $\sum_{p_i \in P} s(p_i) \leq C$.

The **gap** of a packing P is given by the quantity $C - \sum_{p_i \in P} s(p_i)$.

The **bin packing problem** is the problem of finding the minimum number of bins (each of capacity C) needed to pack all items so that the gap in each bin is nonnegative. The minimum number of bins needed for the list L is denoted $b^*(L)$.

> **Algorithm 3:** MFFD algorithm for bin packing.
>
> input: list L, bin capacity C
> output: a packing of L
> partition L into the three sublists $L_A = \{p_i \mid s(p_i) \in (\frac{1}{3}C, C]\}$,
> $L_D = \{p_i \mid s(p_i) \in (\frac{11}{71}C, \frac{1}{3}C]\}$, $L_X = \{p_i \mid s(p_i) \in (0, \frac{11}{71}C]\}$
> pack the sublist L_A using the FFD algorithm.
> {pack as much of L_D into A-bins as possible}
>   1. let bin B_j be the A-bin with the currently largest gap; if the two smallest unpacked items in L_D will not fit together in B_j, go to 4
>   2. place the smallest unpacked item p_i from L_D in B_j
>   3. let p_k be the largest unpacked item in L_D that will now fit in B_j; place p_k in B_j and go to 1
>   4. combine the unpacked portion of L_D with L_X and add these items to the packing using FFD

Facts:

1. The bin packing problem is an NP-hard optimization problem (§16.5.2).

2. *First fit (FF) method*: In this heuristic algorithm, item p_i ($i = 1, 2, \ldots, n$) is placed in the first bin into which it fits. A new bin is started only when p_i will not fit into any nonempty bin.

3. Let $b^{FF}(L)$ denote the number of bins produced by the FF algorithm for a list L. Then $b^{FF}(L) \leq \min\{\lceil \frac{17}{10}b^*(L)\rceil, 1.75b^*(L)\}$.

4. *First fit decreasing (FFD) method*: In this heuristic algorithm, the items are first ordered by decreasing size so that $s(p_1) \geq s(p_2) \geq \cdots \geq s(p_n)$. Then the FF algorithm is applied to the reordered list.

5. Let $b^{FFD}(L)$ denote the number of bins produced by the FFD algorithm for a list L. Then $b^{FFD}(L) \leq \min\{\frac{11}{9}b^*(L) + 3, 1.5b^*(L)\}$.

6. If all item sizes are of the form $C(\frac{1}{k})^j$, $j \geq 0$, for some fixed positive integer k, then $b^{FFD}(L) = b^*(L)$.

7. If the item sizes are uniformly distributed on $[0, a]$ with $0 < a \leq \frac{C}{2}$, then asymptotically $\frac{b^{FFD}(L)}{b^*(L)} \to 1$.

8. *Modified first fit decreasing (MFFD) method*: This heuristic method (Algorithm 3) produces a packing using relatively few bins. After the initial phase of packing the largest size items L_A, let an "A-bin" denote one containing only a single item from L_A.

9. Let $b^{MFFD}(L)$ denote the number of bins produced by the MFFD algorithm for a list L. Then asymptotically, as $b^*(L)$ gets large, $b^{MFFD}(L) \leq 1.183b^*(L)$.

10. *Best fit (BF) method*: In this heuristic algorithm, item p_i is placed in the bin into which it will fit with the smallest gap left over. Ties are broken in favor of the lowest indexed bin.

11. *Best fit decreasing (BFD) method*: In this heuristic algorithm, the items are first ordered so that $s(p_1) \geq s(p_2) \geq \cdots \geq s(p_n)$. Then the BF algorithm is applied to the reordered list.

12. Asymptotic worst-case bounds for BF [BFD] are the same as those for FF [FFD]. In practice the BF version performs somewhat better.

13. Further details on the material in this section are provided in [CoGaJo84].

Examples:

1. Television commercials are to be assigned to station breaks. This is a bin packing problem where the duration of each station break is C and the duration of each commercial is $s(p_i)$.

2. Material such as cable, lumber, or pipe is supplied in a standard length C. Demands for pieces of the material are for arbitrary lengths $s(p_i)$ not exceeding C. The objective is to use the minimum number of standard lengths to supply a given list of required pieces. This is also a bin packing problem.

3. A set of independent tasks with known execution times $s(p_i)$ are to be executed on a collection of identical processors. Determining the minimum number of processors needed to complete all tasks by the deadline C is a bin packing problem.

4. Consider the list $L = (4, \ldots, 4, 7, \ldots, 7, 8, \ldots, 8, 13, \ldots, 13)$, in which there are twelve 4s and six each of 7s, 8s, and 13s in the list. Each bin has capacity $C = 24$. Either FF (or BF) when applied to L result in a packing with twelve bins: two bins are packed as $(4, 4, 4, 4, 4, 4)$, two as $(7, 7, 7)$, two as $(8, 8, 8)$, and six as (13).

5. If FFD (or BFD) is applied to the list in Example 4, a packing with ten bins results: six bins are packed as $(13, 8)$, two as $(7, 7, 7)$, and two as $(4, 4, 4, 4, 4, 4)$.

6. If MFFD is applied to the list in Example 4, then L_A contains the six 13s and L_D contains the remaining items. Packing L_A using FFD results in six A-bins, each containing a single 13 and having gap 11. Steps 1–3 of Algorithm 3 result in six bins packed as $(13, 7, 4)$, and Step 4 yields two bins packed as $(8, 8, 8)$ and one bin packed as $(4, 4, 4, 4, 4, 4)$. This is an optimal solution since all nine bins are completely packed.

15.3.3 SET COVERING AND PARTITIONING

Set covering or set partitioning problems arise when a specified set of tasks must be performed while minimizing the cost of resources used. Such problems arise in scheduling fleets of vehicles or aircraft, locating fire stations in an urban area, political redistricting, and fault testing of electronic circuits.

Definitions:

Let e denote the column vector of all 1s.

Let $A = (a_{ij})$ be a 0-1 *incidence matrix* and let $c = (c_j)$ be a row vector of **costs**.

The set $A_j = \{i \mid a_{ij} = 1\}$ contains all rows **covered** by column j.

The **set covering (SC) problem** is the 0-1 integer linear programming problem:

$$\begin{aligned}\text{minimize:} \quad & cx \\ \text{subject to:} \quad & Ax \geq e \\ & x_j \in \{0, 1\}.\end{aligned}$$

Let v^* be the optimal objective value to this problem.

The **set partitioning (SP) problem** has the same form as the set covering problem except the constraints are $Ax = e$.

The **LP relaxation** of SC or SP is obtained by replacing the constraints $x_j \in \{0,1\}$ by $0 \leq x_j \leq 1$. Let v_{LP} be the optimal objective value to the LP relaxation.

The matrix A is **totally unimodular** if the determinant of every square submatrix of A is 0, $+1$, or -1.

The matrix A is **balanced** if A has no square submatrix of odd order, containing exactly two 1s in each row and column.

The matrix A is in **canonical block form** if, by reordering, its columns can be partitioned into t nonempty subsets $B_1, \ldots, B_t$ such that for each block B_j there is some row i of A with $a_{ik} = 1$ for all $k \in B_j$ and $a_{ik} = 0$ for $k \in \cup_{l=j+1}^{t} B_l$. The rows of A are then ordered so that the row defining B_j becomes the jth row for $j = 1, \ldots, t$.

Facts:

1. Formulation SC expresses the problem of selecting a set of columns (sets) that together cover all rows (elements) at minimum cost. In Formulation SP, the covering sets are required to be disjoint.

2. Both SC and SP are NP-hard optimization problems (§16.5.2).

3. Checking whether a set partitioning problem is feasible is NP-hard.

4. In many instances (including bin packing, graph partitioning, and vehicle routing) the LP relaxation of the set covering (partitioning) formulation of the problem is known to give solutions very close to optimality.

5. For the bin packing and vehicle routing problems (see Examples 2, 3) $v^* \leq \frac{4}{3}\lceil v_{LP} \rceil$.

6. If A is totally unimodular or balanced, then the polyhedra $\{x \mid Ax \geq e, 0 \leq x_j \leq 1\}$ and $\{x \mid Ax = e, 0 \leq x_j \leq 1\}$ have only integer extreme points (vertices). In this case, SC and SP can be solved in polynomial time using linear programming.

7. Checking whether a given matrix A is totally unimodular or balanced can be done in polynomial time.

8. Every 0-1 matrix that is totally unimodular is also balanced. The converse however is not true (see Example 4).

9. The matrix A is totally unimodular if and only if each collection of columns of A can be split into two parts so that the sum of the columns in one part minus the sum of the columns in the other part is a vector with entries 0, $+1$, -1.

10. *Greedy heuristic*: This heuristic (Algorithm 4) for the set covering problem successively chooses columns that have smallest cost per covered row.

11. *Randomized greedy heuristic*: This heuristic for the set covering problem is similar to Algorithm 4 except that at iteration k the column $j^k \in N^k$ is selected at random from among those columns j satisfying $\frac{c_j}{|A_j \cap M^k|} \leq (1+\alpha) \min\{ \frac{c_r}{|A_r \cap M^k|} \mid r \in N^k \}$, where $\alpha \geq 0$.

12. Whereas Algorithm 4 is run only once, the randomized greedy heuristic is repeated several times and the best solution is selected.

13. *Implicit enumeration*: This exact approach (Algorithm 5) for SP works well for dense matrices. In this algorithm, S is the index set of the variables fixed at 1, z is the associated objective value, and R is the set of rows satisfied by S. Also z^* denotes the objective value of the best feasible solution found so far.

14. Other implicit enumeration approaches to set partitioning and set covering are discussed in [BaPa76].

Algorithm 4:   Greedy heuristic for the set covering problem.

input: 0-1 $m \times n$ matrix A, costs c
output: feasible set cover x
$M^1 := \{1, 2, \ldots, m\}; N^1 := \{1, 2, \ldots, n\}; k := 1$
{Main loop}
select $j^k \in N^k$ to minimize $\frac{c_j}{|A_j \cap M^k|}$
$N^{k+1} := N^k - \{j^k\}$
obtain M^{k+1} from M^k by deleting all rows containing a 1 in column j^k
if $M^{k+1} = \emptyset$ then $x_j := 1$ for $j \notin N^{k+1}$ and $x_j := 0$ otherwise
else $k := k + 1$
{Continue with next iteration of main loop}

Algorithm 5:   Implicit enumeration method for SP.

input: 0-1 matrix A, costs c
output: optimal set of columns S (if any)

place A in canonical block form with blocks B_j
order the columns within B_j by nondecreasing $c_t / \sum_i a_{it}$
$S := \emptyset; R := \emptyset; z := 0, z^* := \infty$

1. $r := \min\{i \mid i \notin R\}$; set a marker in the first column of B_r
2. examine all columns of B_r in order starting from the marked column
   if column j is found with $a_{ij} = 0$ for all $i \in R$ and $z + c_j < z^*$ then go to 3
   if B_r is exhausted then go to 4
3. $S := S \cup \{j\}; R := R \cup \{i \mid a_{ij} = 1\}; z := z + c_j$
   if all rows are included in R then $z^* := z$ and go to 4 else go to 1
4. if $S = \emptyset$ then terminate with the best solution found (if any)
   else let $k := $ the last index included in S
   $S := S - \{k\}$; update z and R
   $B_r :=$ the block to which column k belongs
   move the marker in B_r forward by one column and go to 2

---

15. *Cutting plane methods*: Cutting plane methods (§15.1.8) have been used successfully to solve large set partitioning and set covering problems. For details regarding an implementation used to solve crew scheduling problems see [HoPa93].

16. Further details on the material in this section are in [GaNe72], [NeWo88], [Sc86].

Examples:

1. *Crew scheduling problem*: An airline must cover a given set of flight segments with crews. There are specified work rules that restrict the assignment of crews to flights. The objective is to cover all flights at minimum total cost. The rows of the matrix A correspond to the flights that an airline has to cover. The columns of A are the incidence vectors of flight "rotations": sequences of flight segments for each flight that begin and end at individual base locations and that conform to all applicable work rules. The objective is to minimize crew costs. This problem can be formulated as either a set covering or set partitioning problem.

2. *Bin packing*: The bin packing problem (§15.3.2) can be formulated as a set partitioning problem. The rows of the matrix A correspond to the items and the columns are incidence vectors of any feasible packing of items to a bin. The cost of each variable is 1 if the number of bins is to be minimized. In general, a weighted version can also be formulated where different bins have different costs.

3. *Vehicle routing*: Given are a set of customers and the quantity that is to be supplied to each from a warehouse. A fleet of trucks of a specified capacity are available. The objective is to service all the customers at minimum cost. The rows of the matrix A correspond to the customers and the columns are incidence vectors of feasible assignments of customers to trucks (a bin packing problem). The cost of each variable is the cost of the corresponding assignment of customers to the truck. This problem can be formulated as either a set covering or set partitioning problem.

4. The following matrix A is not totally unimodular, since $\det(A) = -2$. This can also be seen using Fact 9. If A has columns C_j then $(C_1 + C_2) - (C_3 + C_4) = (0, 2, 0, 0)^T$ has an entry greater than one in absolute value. However, A is a balanced matrix.

$$A = \begin{pmatrix} 1 & 1 & 1 & 1 \\ 1 & 1 & 0 & 0 \\ 1 & 0 & 1 & 0 \\ 1 & 0 & 0 & 1 \end{pmatrix}$$

5. There are four requests R_1, R_2, R_3, R_4 for information stored in a database, which is comprised of five large files $\{1, 2, 3, 4, 5\}$. Request R_1 can be fulfilled by retrieving files 1, 3, or 4; request R_2 by retrieving files 2 or 3; request R_3 by retrieving files 1 or 5; and request R_4 by retrieving files 4 or 5. The lengths of the files are $7, 3, 12, 7, 6$ (gigabytes) respectively, and the time to retrieve each file is proportional to its length. Filling all requests in the minimum amount of time is then a set covering problem, with costs $c = (7, 3, 12, 7, 6)$ and incidence matrix

$$A = \begin{pmatrix} 1 & 0 & 1 & 1 & 0 \\ 0 & 1 & 1 & 0 & 0 \\ 1 & 0 & 0 & 0 & 1 \\ 0 & 0 & 0 & 1 & 1 \end{pmatrix}$$

Applying the greedy heuristic (Algorithm 4) produces $j^1 = 2$, $j^2 = 5$, and $j^3 = 1$, giving $x = (1, 1, 0, 0, 1)$ with total cost 16. This is an optimal solution to the SC problem.

15.4 ACTIVITY NETS

Activity nets are important tools in the planning, scheduling, and control of projects. In particular, the CPM (Critical Path Method) and PERT (Program Evaluation and Review Technique) models are widely used in the management of large projects, such as those occurring in construction, shipbuilding, aerospace, computer system design, urban planning, marketing, and accounting.

15.4.1 DETERMINISTIC ACTIVITY NETS

The scheduling of large complex projects can be aided by modeling as a directed network of activities having known durations and resource requirements, with the network structure defining the activity precedences. The commonly used critical path method is described as well as extensions that address constrained resources, financial considerations, and project compression.

Definitions:

A **project** is defined by a set of **activities** that are related by **precedence** relations. An **activity** consumes time and resources to accomplish, whereas a **dummy activity** consumes neither.

Activity u (strictly) **precedes** activity v, written $u \prec v$, if activity u must be completed before activity v can be initiated.

A project can be represented using a directed acyclic network G (§8.3.4).

In the **activity-on-node** (**AoN**) representation of a project, the network G contains a **node** for each activity and the **arcs** of G represent the precedence relations between nodes (activities).

In the **activity-on-arc** (**AoA**) representation of a project, the network G contains an **arc** for each activity and the **nodes** of G represent certain **events**. Precedence relations are described by the network arcs, possibly requiring the use of **dummy arcs** (dummy activities). In the AoA representation, the network is assumed to have no multiple arcs joining the same pair of nodes, so an activity can be unambiguously referred to by (i, j) for some nodes i and j, with the corresponding activity **duration** being a_{ij}.

Network G is a **deterministic activity net** if the precedence relations and the parameters associated with the activities are known deterministically. Such a network is also referred to as a **Critical Path Method** (**CPM**) **model**.

An **initial node** of G has no entering directed arcs; a **terminal node** has no exiting directed arcs.

Generalized precedence relations (**GPRs**) relax the necessity of a strict precedence between activities. They can be specified in the form of certain lead or lags between a pair of activities, commonly by **start-to-start**, **finish-to-finish**, **start-to-finish**, and **finish-to-start** relations.

The **optimal project compression problem** is that of achieving a target project completion time with least cost, or alternatively minimizing the duration of the project subject to a specified budget constraint.

The complex interaction between the required resources and the duration of an activity is assumed to be given by the functional relationship $c_a = \phi(y_a)$, where y_a is the duration of activity a, $\ell_a \leq y_a \leq u_a$, and c_a is its cost. The upper limit u_a is the **normal** duration and the lower limit ℓ_a is the **crash** duration of activity a.

Facts:

1. The CPM model arose out of the need to solve industrial scheduling problems; the original work was jointly sponsored by Dupont and Sperry-Rand in the late 1950s.

2. In the AoA representation, the network can be assumed to have a single initial node and a single terminal node. These conditions can in general be guaranteed, possibly through the introduction of dummy arcs.

3. Suppose that the AoA representation of a network has n nodes, with initial node 1 and terminal node n. Then the nodes can always be renumbered (*topologically sorted*) such that each arc leads from a smaller numbered node to a larger numbered one. (See §8.3.4.)

4. In the AoA representation, the *earliest time* of realization of node j, written $t_j(E)$, is determined recursively from $t_j(E) = \max_{i \in B(j)}\{t_i(E) + a_{ij}\}$ and $t_1(E) = 0$, where $B(j)$ is the set of nodes immediately preceding node j.

5. Suppose the time of realization of node n is specified as $t_n(L) \geq t_n(E)$. The *latest time of realization* of node i, written $t_i(L)$, is determined recursively from $t_i(L) = \min_{j \in A(i)} \{t_j(L) - a_{ij}\}$, where $A(i)$ is the set of nodes immediately succeeding node i.

6. $t_j(L) \geq t_j(E)$ holds for any node j. The difference $t_j(L) - t_j(E) \geq 0$ is called the *node slack* for j.

7. For each activity (i,j) there are four *activity floats* corresponding to the differences $t_j(X) - t_i(Y) - a_{ij}$, where $X, Y \in \{E, L\}$:
    - total float: $TF(i,j) = t_j(L) - t_i(E) - a_{ij}$
    - safety float: $SF(i,j) = t_j(L) - t_i(L) - a_{ij}$
    - free float: $FF(i,j) = t_j(E) - t_i(E) - a_{ij}$
    - interference float: $IF(i,j) = t_j(E) - t_i(L) - a_{ij}$.

8. $TF(i,j)$, $SF(i,j)$, and $FF(i,j)$ are always nonnegative, whereas $IF(i,j)$ can be negative, indicating infeasibility of realization under the specified conditions (all activities succeeding node j are accomplished as early as possible and all activities preceding node i are accomplished as late as possible).

9. A *critical* activity (i,j) has total float $TF(i,j) = 0$. If $t_n(L) = t_n(E)$, then the set of critical activities contains at least one path from node 1 to node n, which represents a longest path in the network from node 1 to node n. Such a path is called a *critical path* (CP).

10. Floats play an important role in both resource allocation and activity scheduling, since floats give a measure of the flexibility in scheduling activities during project execution without delaying the project completion time.

11. The problems of optimal resource allocation and activity scheduling subject to the known precedence constraints are NP-hard optimization problems (§16.5.2).

12. Practical solutions to optimal resource allocation and activity scheduling problems are based on heuristics. Virtually all heuristics used in practice rely on ranking the activities according to their float (TF, SF, FF, IF).

13. The measure TF is the only float in the AoA mode that is representation-invariant; this measure is the same in both modes of representation and in all AoA models of the same project.

14. The SF, FF, IF measures are representation-dependent: they do indeed depend on the structure of the AoA and they may also vary from their AoN values.

15. A simple redefinition of $t_i(E)$ for nodes i with all outgoing arcs dummy and $t_j(L)$ for nodes j with all incoming arcs dummy reestablishes the invariance of the activity floats to the mode of representation [ElKa90].

16. A plethora of "off-the-shelf" project planning and control software packages for PCs are currently available [Ho85], [DeHe90]. The review [DeHe90] also outlines criteria against which a software package should be judged. To a varying degree of sophistication, all these software packages satisfy the basic requirements of analysis and reporting. However, these packages are typically incapable of correctly carrying out optimization procedures.

17. A listing of commercial and noncommercial software for project planning can be found at the site:
    - http://www.wior.uni-karlsruhe.de/Bibliothek/Title_Page1.html

18. GPRs afford the flexibility of modeling relations that are present among activities in many practical situations, gained at a price in computational effort and interpretation of results. The concepts of criticality and float of an activity take on a new meaning, since activities may be "compressed" (speeded up) or "expanded" (slowed down) from their "normal" durations [ElKa92].

19. Considerations of resource availabilities are important in project planning and its dynamic control. The most common planning criteria are:
    - minimization of the project duration
    - smoothing of resource usage
    - minimization of the maximum resource utilization
    - minimization of the cost of resource usage
    - maximization of the present value of the project.

20. In the presence of limited resources, the "critical path" may no longer be a "path", in the sense of a connected chain of activities. What emerges is the concept of a *critical sequence* of activities [El77], which need not form a connected chain in the network. (See Example 4.)

21. The scheduling of activities related by arbitrary precedence relations subject to resource availabilities is an NP-hard problem. Consequently, such problems are typically approached by integer programming techniques (§15.1.8) or heuristics (e.g., simulated annealing, tabu search, genetic algorithms, neural nets).

22. The book [SlWe89] discusses project scheduling under constrained resources; in particular, Chapter 5 of Part I evaluates various heuristics that have been proposed. Also [HeDeDe98] gives a review of recent contributions to this area.

23. Typically, resources are available in one or several units or may be acquired at a cost. Mathematical models (large-scale integer linear programs) abound for the minimization of the project duration [El77]. Various branch and bound approaches have been proposed for these models [DeHe92, Sp94].

24. Heuristic procedures, let alone optimization algorithms, for activity scheduling under the other criteria mentioned in Fact 19 are not generally available.

25. In the CPM model of activity networks there is little problem in defining the cost of an activity, and subsequently the cost of the project.

26. Generally, there are two streams of cash flow (from the contractor's point of view): an *in-stream* representing payments to the contractor by the owner, and an *out-stream* representing payments by the contractor in the execution of the activities.

27. From the owner's point of view there is only one stream of cash flows: namely, payments to the contractor for work accomplished. Given a particular schedule for the activities, the two streams of cash flow can be easily obtained. The problem is then scheduling activities to maximize the net present value (NPV) of the project.

28. Issues concerning the NPV of a project are equally important to those interested in bidding on a proposed project and those who are committed to carry out an already agreed-upon project. Succinctly stated, the problem is to determine the dates of the deliverables in order to maximize the NPV.

29. Suppose that the function $\phi(y_a)$ is nonincreasing over the interval $[\ell_a, u_a]$. Reference [El77] gives a treatment of linear, convex, concave, and discrete functions, while [ElKa92] discusses the case in which $\phi(y_a)$ is piecewise linear and convex over the interval $[\ell_a, u_a]$.

Examples:

1. *Construction planning*: The construction of a house involves carrying out the nine activities listed in the following table. Their durations and (immediate) predecessor activities are also indicated.

activity	duration (days)	predecessors
foundation/frame	12	—
wiring/plumbing	4	foundation/frame
sheetrock	7	wiring/plumbing
interior paint	2	sheetrock, windows
carpet	3	interior paint
roof	3	foundation/frame
siding	7	roof
windows	2	siding
exterior paint	2	windows

An AoA representation of this project is shown in the following figure. It is necessary to use a dummy activity to ensure that the given precedences are faithfully depicted. The nodes have been numbered in topological order, with node 1 the initial node and node 9 the terminal node. The longest path from node 1 to node 9 is $[1, 2, 4, 5, 6, 7, 8, 9]$ with length 29, corresponding to a project completion time of 29 days.

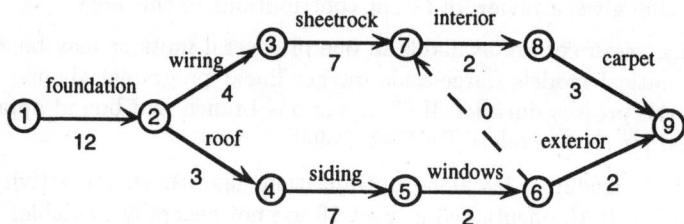

2. A project is composed of the four activities a, b, c, d with precedence relations $a \prec c$, $a \prec d$, and $b \prec d$. The AoN representation of this project is shown in part (a) of the following figure. The AoA representation is shown in part (b) of the figure, where the nodes have been numbered in topological order (Fact 3). The dummy activity joining nodes 2 and 3 is needed to maintain the integrity of the precedence relations. Activity durations are indicated on the arcs of part (b) of the figure.

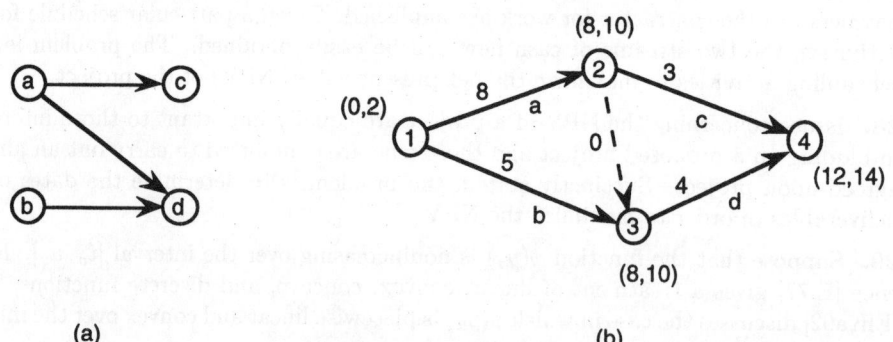

3. The earliest and latest event times $t_j(E), t_j(L)$ are shown next to each node j in part (b) of the figure of Example 2, where $t_4(L) = 14$, which is 2 units more than $t_4(E) = 12$. Here all nodes have the same slack of 2, which provides no information on the various activity floats, given in the following table. Since $TF(i,j) > 0$ for all activities (i,j), there are no critical activities and no critical path, since a small delay in any activity will not delay the completion time of the project. The critical path can be determined if instead $t_4(L) = t_4(E) = 12$. Then the critical path is given by $[1,2,3,4]$.

activity	TF	SF	FF	IF
$(1,2)$	2	0	0	-2
$(1,3)$	5	3	3	1
$(2,4)$	3	1	1	-1
$(3,4)$	2	0	0	-2

4. The following figure gives a project with six activities in AoN representation. There is a single resource, with availability of 6 units. The duration of each activity and the required quantity of the resource are indicated next to each activity (node). The CP (based solely on durations) is $[1,3,5,6]$ of duration 5. If the integrity of the CP is maintained as long as possible, then activity 4 must be inserted before activity 6 (thus breaking the continuity of the CP), which is then followed by activity 2, as shown in part (b) of the figure. The total duration of the project under this schedule is 11 time units.

Now consider the schedule shown in part (c) of the figure, in which the CP is split after activity 1; the total duration of the project is thereby reduced to only 8 time units.

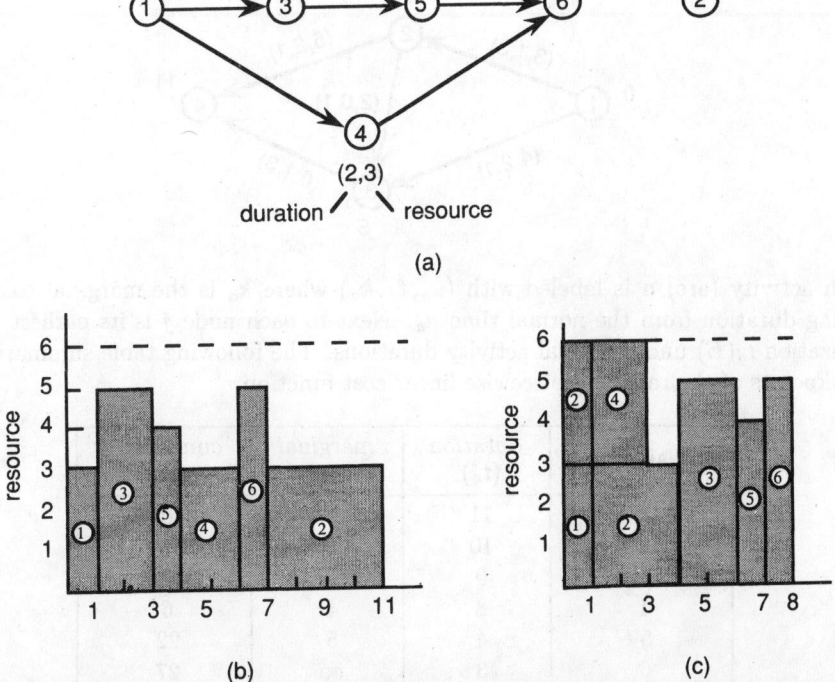

5. The project of the following figure is shown in AoA mode, with the duration of each activity written beside each arc. The payment shown next to a node is the income accrued (if positive) or expense incurred (if negative) at the time of realization of that event (node). The CP is $[1, 3, 4]$ with duration 11. Ignoring the time value of money (i.e., assuming a discount factor $\beta = 0$) gives 1000 as the estimate of project profit. Assuming a discount factor $\beta = 0.99$ and that activities are done as early as possible to maintain the CP, the estimate of project profitability shrinks to $-5000(.99)^2 + 3000(.99)^8 + 3000(.99)^{11} = 553.75$.

Now suppose that the schedule of activities is modified as follows: delay activity $(1, 2)$ to complete at time $t_2 = 4$ (instead of 2); do activity $(1, 3)$ as early as possible to complete at time $t_3 = 8$; and do activity $(2, 4)$ as early as possible (after the realization of node 2) to complete at time $t_4 = 12$. Then the project profitability increases to $-5000(.99)^4 + 3000(.99)^8 + 3000(.99)^{12} = 624.41$. Note that the increase in project profitability comes as a consequence of ignoring the CP, and in fact delaying the project beyond its normal duration.

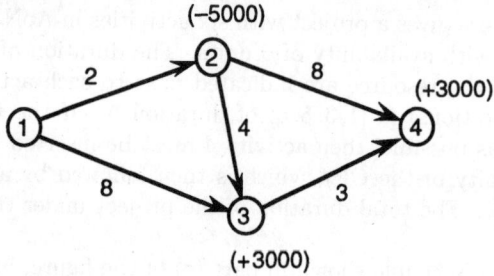

6. A project involving five activities is shown in the following figure in AoA mode.

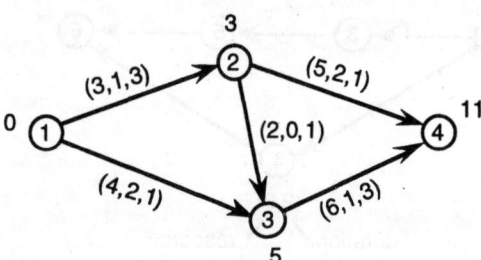

Each activity (arc) a is labeled with (u_a, ℓ_a, k_a) where k_a is the marginal cost of reducing duration from the normal time u_a. Next to each node j is its earliest time of realization $t_j(E)$ under normal activity durations. The following table summarizes the breakpoints of the resulting piecewise linear cost function.

breakpoint	duration (t_4)	marginal cost	cumulative cost
1	11	1	0
2	10	2	1
3	9	3	3
4	8	4	6
5	4	5	22
6	3	∞	27

The function itself is shown in the following figure. With the complete cost function in hand it is easy to answer various questions. For example, the least additional cost required to reduce the project duration from its normal value 11 to 7 is seen to be 10. Alternatively, if 6 additional units of money are available, then the maximum reduction achievable in the project duration is 3 units of time (from 11 to 8).

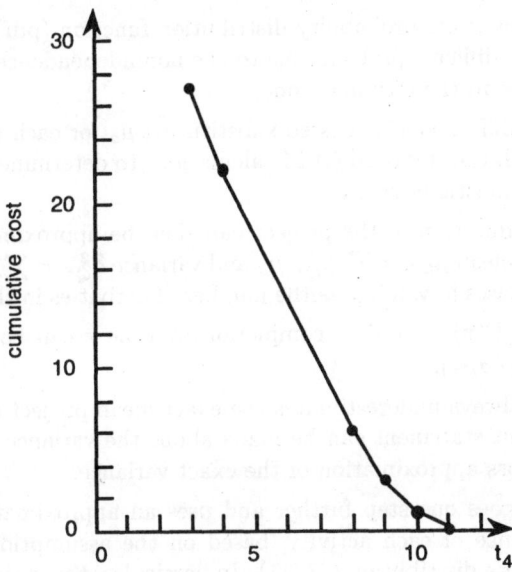

15.4.2 PROBABILISTIC ACTIVITY NETS

The CPM model can be extended to incorporate uncertainty or randomness. If the durations of activities are random variables, then the network is a PERT (Program Evaluation and Review Technique) model. Alternatively, the very undertaking of an activity may be determined by chance and this consideration has led to the development of GAN (Generalized Activity Network) models.

Definitions:

A *probabilistic activity net* is a directed network in which some or all of the parameters, including the realization of the activities, are probabilistically known.

In a **PERT model**, activity durations are random variables. The duration of activity a has expected value μ_a and variance σ_a^2.

Let $P(\tau)$ be the probability that the project is completed by time τ.

The *criticality index* of a path Q in the network is the probability that Q is a critical path in any realization of the project.

The *criticality index* of an activity a is the probability that a lies on a critical path in any realization of the project.

A **GAN model** is a probabilistic activity net with conditional progress and probabilistic realization of activities.

If X is a standard normal deviate (§7.3.1), then its (cumulative) distribution function is denoted by $\Phi(x) = Pr(X \leq x)$.

Facts:

1. The original PERT model evolved in the late 1950s from the U.S. Navy's efforts to plan and accelerate the Polaris submarine missile project.

2. A detailed account of the original PERT model, its analysis, and the criticisms levied against it is found in [El77, Chapter 3].

3. Estimation of the exact probability distribution function (pdf) of the project duration is an extremely difficult problem due to the nonindependence of the paths leading from the initial node to the terminal node.

4. The original PERT model suggested substituting μ_a for each activity duration and then proceeding with the standard CPM calculations to determine a critical path Q^* in the resulting deterministic network.

5. The pdf of the duration of the project can then be approximated using a normal distribution having mean $\hat{\mu}_{Q^*} = \sum_{a \in Q^*} \mu_a$ and variance $\hat{\sigma}^2_{Q^*} = \sum_{a \in Q^*} \sigma^2_a$. The normal approximation increases in validity as the number of activities in the path Q^* increases.

6. The probability $P(\tau)$ of project completion by time τ can be approximated using $\hat{P}(\tau) = \Phi((\tau - \hat{\mu}_{Q^*})/\hat{\sigma}_{Q^*})$.

7. The value $\hat{\mu}_{Q^*}$ always underestimates the exact mean project duration (often, seriously). No equivalent statement can be made about the variance estimate $\hat{\sigma}^2_{Q^*}$ except that it is often a gross approximation of the exact variance.

8. PERT analysis goes one step further and uses an approximation to the expected value and the variance of each activity, based on the assumption that each activity duration follows a beta distribution (§7.3.1). In particular, the variance is approximated by $\frac{1}{36}(\text{range})^2$. These additional assumptions render the procedure even more suspect.

9. An immediate consequence of randomness in the activity durations is that (virtually) any path can be the CP in some realization of the project. Thus, the criticality index of a path and the criticality index of an activity are more meaningful concepts. See [Wi92] for a critique of the latter.

10. In general, it is extremely difficult to determine the exact values of the criticality indices analytically. Monte Carlo sampling is typically used to estimate these values.

11. Since the early days of PERT, significant strides have been made in estimating the various parameters in the PERT model. The approaches can be classified into the categories of *exact*, *approximating*, and *bounding* procedures. See [El89, Ka92].

12. The concept of a *uniformly directed cutset* has been used to evaluate some common network performance criteria under the assumption of exponentially distributed activity durations [KuAd86]. Attempts to extend the concept to applications in optimal resource allocation have had limited success thus far.

13. The restriction of GANs to "exclusive-or" type nodes renders the network a graphical representation of a *semi-Markov process*. The resulting *GERT* (Graphical Evaluation and Review Technique) model has been expanded into SLAM II, an extremely powerful discrete event simulation language.

14. The analysis of stochastic activity nets with exclusive-or type nodes (*STEOR-nets*) is thoroughly discussed in [Ne90].

Examples:

1. The following figure shows a project with six activities whose durations are random variables that assume discrete values with equal probabilities. For example, activity $(1,2)$ has duration 1, 2, or 5 with probability $\frac{1}{3}$ each. The exact distribution of project

completion time (secured by complete enumeration of the 324 realizations) is shown in the following table, from which it is seen that the true mean project duration is $\mu = 12.315$ and the true standard deviation is $\sigma = 2.5735$. The probability that the project duration is no more than 12 time units is $P(12) = 0.4815$. The PERT estimates of these same parameters, based on the deterministic critical path $[1, 2, 3, 4]$, are $\widehat{\mu}_{Q^*} = 10$, $\widehat{\sigma}_{Q^*} = 1$, and $\widehat{P}(12) = 0.9772$.

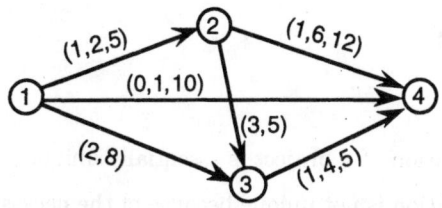

duration (t_4)	frequency	relative frequency
17	36	0.1111
15	12	0.0370
14	48	0.1481
13	72	0.2222
12	42	0.1296
11	30	0.0926
10	36	0.1111
9	28	0.0864
8	10	0.0309
7	6	0.0185
6	2	0.0062
5	2	0.0062
	324	1.0000

2. The paths from initial node 1 to terminal node 4 for the project in the figure of Example 1 are: $Q_1 = [1, 2, 4]$, $Q_2 = [1, 2, 3, 4]$, $Q_3 = [1, 4]$, $Q_4 = [1, 3, 4]$. The following table lists the frequency and relative frequency that each path Q_i, or combination of paths, is a critical path.

t_4	Q_1	Q_2	Q_3	Q_4	Q_1,Q_2	Q_1,Q_4	Q_2,Q_3	Q_2,Q_4
17	36							
15		12						
14	36	12						
13	30	6		24		6		6
12		12		24				6
11	6	18			6			
10		8	24			4		
9		10		16				2
8	2	6			2			
7	2	2			2			
6		2						
5		2						
freq.	112	90	24	64	10	6	4	14
rel. freq.	.3457	.2778	.0741	.1975	.0309	.0185	.0123	.0432

The path criticality indices are then easily determined from this table. For example, the criticality index of Q_1 is 0.3951 ($= 0.3457 + 0.0309 + 0.0185$), and of Q_4 is 0.2592 ($= 0.1975 + 0.0185 + 0.0432$). The criticality index of each activity can be easily determined from the criticality indices of the paths. For instance, the criticality index of activity $(1,2)$, which lies on paths Q_1 and Q_2, is 0.7284 ($= 1 - 0.0741 - 0.1975$).

15.4.3 COMPLEXITY ISSUES

Facts:

1. The AoN representation of a project is essentially unique.

2. The AoA representation is not unique because of the necessity to introduce dummy activities (e.g., to maintain the integrity of the precedence relations).

3. Construction of the AoA representation can be carried out with different objectives in mind: to minimize the number of nodes, to minimize the number of dummy activities, or to minimize the *complexity index* of the resulting AoA network [MiKaSt93].

4. Analytical solutions to optimization problems for project networks often proceed by conditioning upon certain activities, and then removing the conditioning through either enumeration or multiple integration. Minimizing the computing effort then involves minimizing the number of activities on which such conditioning takes place.

5. If the network is series-parallel then no conditioning is required and its analysis is straightforward, though it may be computationally demanding.

6. If the network is not series-parallel, then the minimum number of activities for conditioning can be secured by the optimal node reduction procedure of [BeKaSt92], which has polynomial complexity.

7. Patterson [Pa83] collected a set of 110 standard test problems, useful for comparing alternative solution procedures. These problems have been supplanted by a more recent set of test problems [KaSpDr92].

8. Several measures of the complexity of a project network were proposed in the 1960s, with questionable validity. The significance of the complexity index [BeKaSt92] in accounting for the difficulty in analysis is discussed in [DeHe96].

15.5 GAME THEORY

Games, mathematical models of conflict or bargaining, can be classified in three ways: by mood of play (noncooperative or cooperative), by field of application (e.g., biology or economics), and by mathematical structure (e.g., discrete, continuous, or differential). Correspondingly, game theory is a vast and diverse subject with different traditions in each of many specialties.

This section discusses discrete games, in which finitely many strategies are available to finitely many players. Combinatorial and other games form largely separate disciplines to which appropriate references appear in §15.5.4.

15.5.1 NONCOOPERATIVE GAMES

This section discusses noncooperative games involving a finite number of players. Collusion among the players is not allowed in these types of games. Such games can model a wide variety of situations, as indicated in §15.5.4.

Definitions:

An n-player game Γ in **extensive form** consists of:

- a set $\{1, \ldots, n\} \cup \{0\}$ of n decisionmakers (or **players**) augmented by a fictitious player, called 0 (or **chance**), whose actions are random
- a tree, in which each nonterminal vertex represents a decision point for some player, whose possible actions correspond to arcs emanating from the vertex
- a **payoff function** that assigns an n-vector to each terminal vertex
- a partition of the nonterminal vertices into $n+1$ vertex sets, one for each player and for chance
- a subpartition of each player's vertex set into subsets (**information sets**), such that no vertex follows another in the same subset and all vertices in a subset are followed by the same number of arcs
- a probability distribution on arcs emanating from any chance vertex.

A **subgame** of Γ is a game whose tree is a subtree of the tree for Γ. A subgame is **proper** if the information set that contains its root contains no other vertices.

A game is **finite** if its tree is finite.

A game has **perfect** information if all information sets contain a single vertex; otherwise, it has **imperfect** information.

A game has **complete** information if all players know the entire extensive form including all terminal payoffs; otherwise it has **incomplete** information.

A **pure strategy** is a function that maps each of a player's information sets to an emanating arc.

An n-person game in **normal** (or **strategic**) **form** consists of a set $N = \{1, 2, \ldots, n\}$ of players, a set S_k of possible pure strategies for each $k \in N$, and a payoff function $f = (f_1, f_2, \ldots, f_n)$ that assigns $f_k(w)$ to Player k for every pure strategy combination $w = (w^1, w^2, \ldots, w^n)$, where $w^k \in S_k$. Payoffs are computed by taking expected values over distributions associated with chance vertices in the corresponding extensive form.

Let $D \subseteq S_1 \times S_2 \times \cdots \times S_n$ be the set of all possible pure strategy combinations w.

Let $w \,\|\, \overline{w}^k$ denote the joint pure strategy combination that is identical to w except for the strategy of Player k:

$$w \,\|\, \overline{w}^k = (w^1, \ldots, w^{k-1}, \overline{w}^k, w^{k+1}, \ldots, w^n).$$

$w^* \in D$ is a **Nash equilibrium pure strategy combination** (or simply **equilibrium**) if, for every $k \in N$, $f_k(w^*) \geq f_k(w^* \,\|\, \overline{w}^k)$ holds for all $\overline{w}^k \in S_k$. (J. F. Nash, born 1928) Let E denote the set of all such equilibria.

For $k \in N$ define the function m_k that minimizes $f_k(w)$ over components of w that k does not control:

$$m_k(w^k) = \min_{\{\overline{w} \,|\, \overline{w}^k = w^k\}} f_k(\overline{w}).$$

If $\widetilde{w}^k$ maximizes $m_k(w^k)$, then $\widetilde{w}^k$ is a **max-min strategy** for k and $\widetilde{f}_k = m_k(\widetilde{w}^k)$ is the corresponding **max-min payoff**.

Let $D^* = \{\, w \in D \mid f_k(w) \geq \widetilde{f}_k \text{ for all } k \in N \,\}$.

The strategy combination w is **individually rational** for all players if $w \in D^*$.

The combination $w \in D$ is **group rational** (or **Pareto-optimal**) if no $\overline{w} \in D$ exists such that $f_k(\overline{w}) \geq f_k(w)$ for all $k \in N$ and $f_i(\overline{w}) > f_i(w)$ for some $i \in N$.

Let P denote the set of all Pareto-optimal w. The set $P^* = P \cap D^*$ is the **bargaining set** and each $w \in P^*$ is a **cooperative** strategy combination.

An equilibrium is **subgame perfect** if its restriction to any proper subgame is also an equilibrium. Let E_S denote the set of subgame perfect equilibria.

Facts:

1. Information sets are constructed so that in making a decision a player knows the identity of the information set, but not the particular vertex of the set at which the decision is being made.

2. At an equilibrium $w^* \in D$, no $k \in N$ has a unilateral incentive to depart from $(w^*)^k$ if each $j \in N$, $j \neq k$, holds fast to $(w^*)^j$.

3. $w \parallel w^k = w$.

4. Different equilibria can yield identical outcomes.

5. The bargaining set can also be defined with "threat" strategies in lieu of max-min (or "security") strategies as criteria of individual rationality. Context determines which definition is apt.

6. If E is a singleton, or if all elements of E yield the same outcome (see Example 7), then the game is usually regarded as solved.

7. In general, however, E may either be empty or yield a multiplicity of outcomes (see Example 8).

8. A sufficient condition for $E \neq \emptyset$ in a finite game is that information be perfect (although E need not be computable by all players unless information is also complete). The above condition is not necessary; see Examples 7 and 8.

9. If E yields a multiplicity of outcomes, then an equilibrium selection criterion is necessary. One criterion is to reduce E to $E \cap P^*$, thus preferring cooperative equilibria (of a noncooperative game) to noncooperative equilibria. Another criterion is to reduce E to $E \cap E_S$.

10. Rationales for the above criteria are discussed in [Me93]. Other equilibrium selection criteria are discussed in [Fr90] and [My91].

11. The equilibrium selection problem is one of the important unsolved problems of game theory; see [BiKiTa93].

Examples:

1. A university (Player 3) must offer a faculty position to either or both of two individuals, a distinguished researcher (Player 1) and a younger colleague in the same area (Player 2), each of whom can say either YES or NO to an offer but cannot communicate with the other. The payoff to Player $i = 1, 2$ (in well-being) is σ_i (> 0) for an offer, b_i ($> \sigma_i$) for an appointment, and B_i ($> b_i$) if both are appointed. To the university, hiring Player 1 alone is worth 4 (in prestige); but hiring both merits 3, hiring neither is worth 2, and hiring Player 2 alone merits zero, because appointing Player 2 prevents the appointment of another distinguished researcher. The university hides from each candidate whether it has made an offer to the other. The extensive form of this game is shown in the following figure. Each player has a single information set (denoted by a rectangle). There are no chance vertices and no proper subgames. The payoffs to Players 1, 2, and 3 are indicated by the 3-vector at each terminal vertex of the tree.

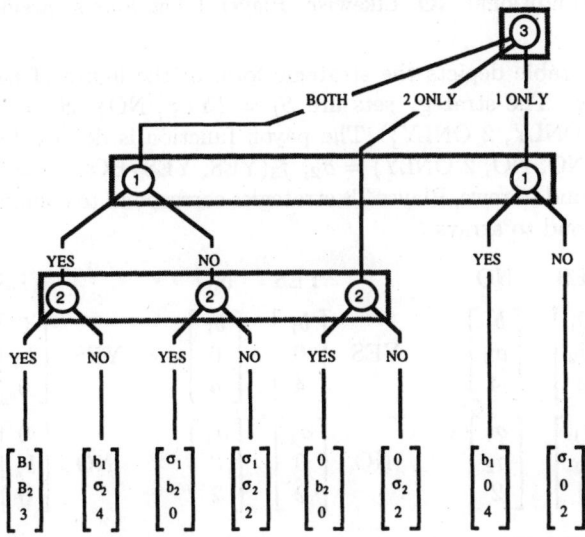

2. Suppose in Example 1 that the university now reveals to whom it has made an offer. Also, the university need not offer the position to either candidate this year, in which case a single individual is appointed next year and chance decides with equal probability which current candidate the appointee matches in caliber, giving the university a payoff of $0.5 \times 4 + 0.5 \times 0 = 2$. The extensive form of this game is shown in the following figure. Player 1 has information sets I, J whereas Player 2 has information sets K, L. There is a single chance vertex. Information sets I, J, K each contain the root of a proper subgame.

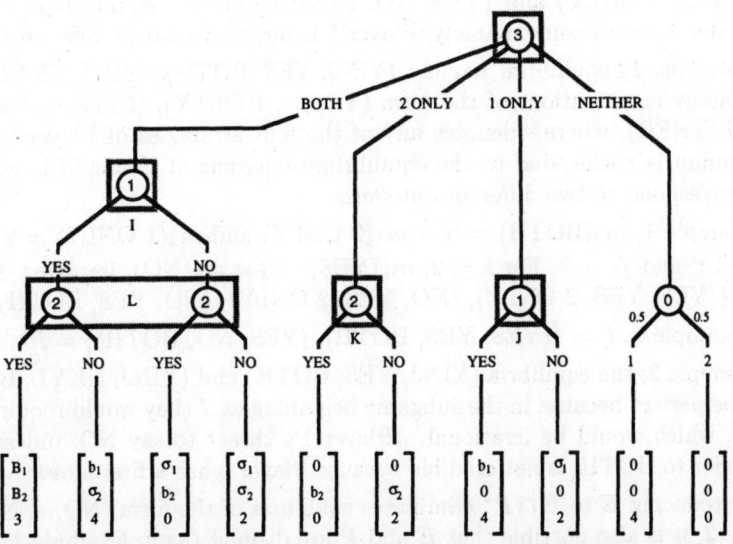

3. The figures of Examples 1 and 2 are finite games of imperfect information since in both cases Player 2 has an information set with more than one vertex. Each game has incomplete information if players know only their own terminal payoffs.

4. In the figure of Example 2, Player 2 can say YES or NO at each of K or L. Hence Player 2 has four possible strategies: YKNL (yes if K, no if L), NKYL, an unconditional YES, and an unconditional NO. Likewise, Player 1 has four strategies: YINJ, NIYJ, YES, and NO.

5. The following table depicts the strategic form of the figure of Example 1 as a 3-dimensional array. The strategy sets are $S_1 = \{\text{YES, NO}\}$, $S_2 = \{\text{YES, NO}\}$, and $S_3 = \{\text{BOTH, 1 ONLY, 2 ONLY}\}$. The payoff function is defined by $f_1(\text{YES, NO, 1 ONLY}) = b_1$, $f_2(\text{NO, NO, 2 ONLY}) = \sigma_2$, $f_3(\text{YES, YES, BOTH}) = 3$, etc. Player 1's strategies correspond to rows, Player 2's strategies correspond to columns, and Player 3's strategies correspond to arrays.

$$
\begin{array}{c}
 \\
\text{YES} \\
\\
\text{NO}
\end{array}
\begin{array}{cc}
\text{YES} & \text{NO} \\
\begin{bmatrix} B_1 \\ B_2 \\ 3 \end{bmatrix} & \begin{bmatrix} b_1 \\ \sigma_2 \\ 4 \end{bmatrix} \\
\begin{bmatrix} \sigma_1 \\ b_2 \\ 0 \end{bmatrix} & \begin{bmatrix} \sigma_1 \\ \sigma_2 \\ 2 \end{bmatrix}
\end{array}
\quad
\begin{array}{c}
 \\
\text{YES} \\
\\
\text{NO}
\end{array}
\begin{array}{cc}
\text{YES} & \text{NO} \\
\begin{bmatrix} b_1 \\ 0 \\ 4 \end{bmatrix} & \begin{bmatrix} b_1 \\ 0 \\ 4 \end{bmatrix} \\
\begin{bmatrix} \sigma_1 \\ 0 \\ 2 \end{bmatrix} & \begin{bmatrix} \sigma_1 \\ 0 \\ 2 \end{bmatrix}
\end{array}
\quad
\begin{array}{c}
 \\
\text{YES} \\
\\
\text{NO}
\end{array}
\begin{array}{cc}
\text{YES} & \text{NO} \\
\begin{bmatrix} 0 \\ b_2 \\ 0 \end{bmatrix} & \begin{bmatrix} 0 \\ \sigma_2 \\ 2 \end{bmatrix} \\
\begin{bmatrix} 0 \\ b_2 \\ 0 \end{bmatrix} & \begin{bmatrix} 0 \\ \sigma_2 \\ 2 \end{bmatrix}
\end{array}
$$

$$\text{BOTH} \qquad\qquad\qquad \text{1 ONLY} \qquad\qquad\qquad \text{2 ONLY}$$

6. The following table depicts the strategic form of the figure of Example 2 as a 3-dimensional array. Player 1's strategies correspond to rows, Player 2's strategies correspond to columns, and Player 3's strategies correspond to arrays.

The strategy sets now are $S_1 = \{\text{YES, YINJ, NIYJ, NO}\}$, $S_2 = \{\text{YES, YKNL, NKYL, NO}\}$, and $S_3 = \{\text{BOTH, 1 ONLY, 2 ONLY, NEITHER}\}$. The sets S_1, S_2 contain more strategies than in Example 5 because Players 1 and 2 have better information: the game is less imperfect. The payoff to Player 3 from NEITHER is an expectation over arcs emanating from the game's single chance vertex.

7. $\{\text{YES, YES, 1 ONLY}\}$ and $\{\text{YES, NO, 1 ONLY}\}$ are the equilibria of Example 1; both yield the same outcome, namely, Player 1 is hired without an offer to Player 2.

8. Example 2 has 14 equilibria: namely, (YINJ, YES, BOTH), (YINJ, NKYL, BOTH), and all strategy combinations of the form (YES, ·, 1 ONLY), (NIYJ, ·, 1 ONLY), or (NO, ·, NEITHER), where · denotes any of the four strategies of Player 2. Eight of these 14 equilibria correspond to the equilibrium outcome of Example 1, whereas the other six correspond to two different outcomes.

9. For Example 1, $m_3(\text{BOTH}) = 0 = m_3(\text{2 ONLY})$ and $m_3(\text{1 ONLY}) = 2$, implying $\widetilde{w}^3 = 1$ ONLY and $\widetilde{f}_3 = 2$. For $k \leq 2$, $m_k(\text{YES}) = 0 = m_k(\text{NO})$, implying $\widetilde{f}_k = 0$. So $D^* = D - \{(\text{YES, YES, 2 ONLY}), (\text{NO, YES, 2 ONLY}), (\text{NO, YES, BOTH})\}$.

10. For Example 1, $P = \{(\text{YES, YES, BOTH}), (\text{YES, NO, BOTH})\} = P^*$.

11. In Example 2, the equilibria (YINJ, YES, BOTH) and (YINJ, NKYL, BOTH) are not subgame perfect because in the subgame beginning at J they would require Player 1 to say NO, which would be irrational. (Player 1's threat to say NO, unless Player 3 makes an offer to BOTH, is not credible because Player 3 has a first mover advantage.)

12. While reducing E to $E \cap P^*$ eliminates equilibria of the form (NO, ·, NEITHER) in Example 2, it is also possible that E and P are disjoint (as in Example 1).

	YES	YKNL	NKYL	NO
YES	$\begin{bmatrix}B_1\\B_2\\3\end{bmatrix}$	$\begin{bmatrix}b_1\\\sigma_2\\4\end{bmatrix}$	$\begin{bmatrix}B_1\\B_2\\3\end{bmatrix}$	$\begin{bmatrix}b_1\\\sigma_2\\4\end{bmatrix}$
YINJ	$\begin{bmatrix}B_1\\B_2\\3\end{bmatrix}$	$\begin{bmatrix}b_1\\\sigma_2\\4\end{bmatrix}$	$\begin{bmatrix}B_1\\B_2\\3\end{bmatrix}$	$\begin{bmatrix}b_1\\\sigma_2\\4\end{bmatrix}$
NIYJ	$\begin{bmatrix}\sigma_1\\b_2\\0\end{bmatrix}$	$\begin{bmatrix}\sigma_1\\\sigma_2\\2\end{bmatrix}$	$\begin{bmatrix}\sigma_1\\b_2\\0\end{bmatrix}$	$\begin{bmatrix}\sigma_1\\\sigma_2\\2\end{bmatrix}$
NO	$\begin{bmatrix}\sigma_1\\b_2\\0\end{bmatrix}$	$\begin{bmatrix}\sigma_1\\\sigma_2\\2\end{bmatrix}$	$\begin{bmatrix}\sigma_1\\b_2\\0\end{bmatrix}$	$\begin{bmatrix}\sigma_1\\\sigma_2\\2\end{bmatrix}$

BOTH

	YES	YKNL	NKYL	NO
YES	$\begin{bmatrix}b_1\\0\\4\end{bmatrix}$	$\begin{bmatrix}b_1\\0\\4\end{bmatrix}$	$\begin{bmatrix}b_1\\0\\4\end{bmatrix}$	$\begin{bmatrix}b_1\\0\\4\end{bmatrix}$
YINJ	$\begin{bmatrix}\sigma_1\\0\\2\end{bmatrix}$	$\begin{bmatrix}\sigma_1\\0\\2\end{bmatrix}$	$\begin{bmatrix}\sigma_1\\0\\2\end{bmatrix}$	$\begin{bmatrix}\sigma_1\\0\\2\end{bmatrix}$
NIYJ	$\begin{bmatrix}b_1\\0\\4\end{bmatrix}$	$\begin{bmatrix}b_1\\0\\4\end{bmatrix}$	$\begin{bmatrix}b_1\\0\\4\end{bmatrix}$	$\begin{bmatrix}b_1\\0\\4\end{bmatrix}$
NO	$\begin{bmatrix}\sigma_1\\0\\2\end{bmatrix}$	$\begin{bmatrix}\sigma_1\\0\\2\end{bmatrix}$	$\begin{bmatrix}\sigma_1\\0\\2\end{bmatrix}$	$\begin{bmatrix}\sigma_1\\0\\2\end{bmatrix}$

1 ONLY

	YES	YKNL	NKYL	NO
YES	$\begin{bmatrix}0\\b_2\\0\end{bmatrix}$	$\begin{bmatrix}0\\b_2\\0\end{bmatrix}$	$\begin{bmatrix}0\\\sigma_2\\2\end{bmatrix}$	$\begin{bmatrix}0\\\sigma_2\\2\end{bmatrix}$
YINJ	$\begin{bmatrix}0\\b_2\\0\end{bmatrix}$	$\begin{bmatrix}0\\b_2\\0\end{bmatrix}$	$\begin{bmatrix}0\\\sigma_2\\2\end{bmatrix}$	$\begin{bmatrix}0\\\sigma_2\\2\end{bmatrix}$
NIYJ	$\begin{bmatrix}0\\b_2\\0\end{bmatrix}$	$\begin{bmatrix}0\\b_2\\0\end{bmatrix}$	$\begin{bmatrix}0\\\sigma_2\\2\end{bmatrix}$	$\begin{bmatrix}0\\\sigma_2\\2\end{bmatrix}$
NO	$\begin{bmatrix}0\\b_2\\0\end{bmatrix}$	$\begin{bmatrix}0\\b_2\\0\end{bmatrix}$	$\begin{bmatrix}0\\\sigma_2\\2\end{bmatrix}$	$\begin{bmatrix}0\\\sigma_2\\2\end{bmatrix}$

2 ONLY

	YES	YKNL	NKYL	NO
YES	$\begin{bmatrix}0\\0\\2\end{bmatrix}$	$\begin{bmatrix}0\\0\\2\end{bmatrix}$	$\begin{bmatrix}0\\0\\2\end{bmatrix}$	$\begin{bmatrix}0\\0\\2\end{bmatrix}$
YINJ	$\begin{bmatrix}0\\0\\2\end{bmatrix}$	$\begin{bmatrix}0\\0\\2\end{bmatrix}$	$\begin{bmatrix}0\\0\\2\end{bmatrix}$	$\begin{bmatrix}0\\0\\2\end{bmatrix}$
NIYJ	$\begin{bmatrix}0\\0\\2\end{bmatrix}$	$\begin{bmatrix}0\\0\\2\end{bmatrix}$	$\begin{bmatrix}0\\0\\2\end{bmatrix}$	$\begin{bmatrix}0\\0\\2\end{bmatrix}$
NO	$\begin{bmatrix}0\\0\\2\end{bmatrix}$	$\begin{bmatrix}0\\0\\2\end{bmatrix}$	$\begin{bmatrix}0\\0\\2\end{bmatrix}$	$\begin{bmatrix}0\\0\\2\end{bmatrix}$

NEITHER

13. While reducing E to $E \cap E_S$ eliminates equilibria of the form (YINJ, YES, BOTH) and (YINJ, NKYL, BOTH) in Example 2, it is also possible that $E = E_S$ (as in Example 1, where there are no proper subgames).

15.5.2 MATRIX AND BIMATRIX GAMES

This subsection discusses two-player noncooperative games. Such games can be represented in normal form by a pair of matrices.

Definitions:

Suppose $S_1 = \{1, \ldots, r\}$ and $S_2 = \{1, \ldots, s\}$.

The $r \times s$ **payoff matrices** $A = (a_{ij})$ and $B = (b_{ij})$, with $a_{ij} = f_1(i,j)$ and $b_{ij} = f_2(i,j)$, define a **bimatrix game**.

The game is **zero-sum** if $a_{ij} + b_{ij} = 0$ for all $i \in S_1$, $j \in S_2$. The game is **symmetric** if $r = s$ and $B = A^T$. In either case, the game is completely determined by A and is called a **matrix game**.

For Player 1, $i \in S_1$ is **dominated** by $i' \in S_1$ if $a_{i'j} \geq a_{ij}$ for all $j \in S_2$, with strict inequality for at least one j. For Player 2, $j \in S_2$ is dominated by $j' \in S_2$ if $b_{ij'} \geq b_{ij}$ for all $i \in S_1$ with strict inequality for at least one i.

Let 1_k denote the k-dimensional vector in which every entry is 1, and let X_k denote the $(k-1)$-dimensional unit simplex: $X_k = \{(x_1,\ldots,x_k) \mid x1_k = 1,\ x \geq 0\}$.

A **mixed strategy** for Player 1 is a vector $p = (p_1,\ldots,p_r) \in X_r$, where p_i is the probability that Player 1 selects $i \in S_1$. Similarly, a mixed strategy for Player 2 is $q = (q_1,\ldots,q_s) \in X_s$, where q_j is the probability that Player 2 selects $j \in S_2$.

In a **mixed strategy combination** $(p,q) \in X_r \times X_s$, the expected payoffs to Players 1 and 2, respectively, are given by $\phi_1(p,q) = pAq^T$ and $\phi_2(p,q) = pBq^T$.

The pair $(p^*, q^*) \in X_r \times X_s$ is a **Nash equilibrium mixed strategy combination**, or simply an **equilibrium in mixed strategies**, if $\phi_1(p^*,q^*) \geq \phi_1(p,q^*)$ for all $p \in X_r$ and $\phi_2(p^*,q^*) \geq \phi_2(p^*,q)$ for all $q \in X_s$. If the game is zero-sum, then p^* is called an **optimal strategy** for Player 1 and q^* is called an optimal strategy for Player 2.

Facts:

1. Every bimatrix game has at least one equilibrium in mixed strategies.

2. All equilibria in mixed strategies of a zero-sum game yield the same expected payoffs, v to Player 1 and $-v$ to Player 2; v is known as the *value* of the game.

3. The value v of a zero-sum game and a pair (p^*, q^*) of optimal strategies can always be computed by solving a dual pair of linear programming (LP) problems (§15.1). The primal LP problem finds p to maximize v subject to $A^T p \geq v1_s$, $p \in X_r$, whereas the dual LP problem finds q to minimize v subject to $Aq \leq v1_r$, $q \in X_s$.

4. Player 1 can achieve the value v of a zero-sum game with a mixed strategy that attaches zero probability to any dominated pure strategy. Likewise, Player 2 can achieve $-v$ by playing dominated pure strategies with zero probability.

5. Graphical methods can be used to compute efficiently all equilibria of zero-sum games where $r = 2$ or $s = 2$, or of matrix games (of either type) where $r = s = 3$; see [Dr81], [Ow95], and [Me92]. There is no general method for computing all equilibria.

6. The definition of mixed strategy and the existence of equilibria are readily extended to n-player games. This result was one of the fundamental contributions to game theory for which John Nash was awarded the 1994 Nobel Prize in Economic Science.

Examples:

1. Two advertising agencies are involved in a campaign to promote competing beverages. The payoffs of various promotional strategies are shown in this table:

	j	1	2
i		old	new
1	old	0	-2
2	new	-2	-1
3	diet	3	-3

The promotional strategies for the first agency are to: stress the old formula, advertise a new formula, or advertise a diet drink. The second agency has the possible strategies:

stress the old formula, or advertise a new formula. The payoffs in this case indicate the net change in millions of sales gained (by Advertiser 1). For example, if the first agency promotes a diet drink while the other agency promotes the old formula, three million more drinks will be sold. On the other hand, if the other agency happens to promote the new formula, then the first agency will end up losing three million unit sales to the second agency.

2. An investor has just taken possession of jewels worth \$45,000 and must store them for the night in one of two locations (A, B). The safe in location A is relatively secure, with a probability $\frac{1}{15}$ of being opened by a thief. The safe at location B is not as secure as the safe at location A, and has a probability $\frac{1}{5}$ of being opened. A notorious thief is aware of the jewels, but doesn't know where they will be stored. Nor is it possible for the thief to visit both locations in one evening. This is a (symmetric) zero-sum game between the investor (Player 1), who selects where to keep the jewels and the thief (Player 2), who decides which safe to try. If the investor puts the jewels in the most secure location (A) and the jewel thief goes to this location, the expected loss in this case is $\frac{1}{15}(-45,000) + \frac{14}{15}(0) = -3,000$. The other entries of the payoff matrix in the following table are computed similarly, and are expressed in thousands of dollars (to the investor).

	j	1	2
i		A	B
1	A	-3	0
2	B	0	-9

No pure strategy combination is a Nash equilibrium, since it is always tempting for one player to defect from the current strategy. However, there is a Nash equilibrium mixed strategy combination: $p^* = (\frac{3}{4}, \frac{1}{4}) = q^*$, with value $v = -\$2,250$ to the investor. The mixed strategy p^* is found by solving the following linear program, in which Player 1 wants to find the largest value of v so that he is guaranteed of receiving at least v (regardless of what Player 2 does). The associated optimal dual LP solution gives q^*.

$$\text{maximize:} \quad v$$
$$\text{subject to:} \quad -3p_1 + 0p_2 \geq v$$
$$0p_1 - 9p_2 \geq v$$
$$p_1 + p_2 = 1$$
$$p_1, p_2 \geq 0$$

3. The zero-sum game of *chump* is played between two camels, a dromedary (Player 1) and a bactrian (Player 2). Player k must simultaneously flash F_k humps and guess that its opponent will flash G_k. Possible strategies (F_k, G_k) satisfy $0 \leq F_1, G_2 \leq 1$ and $0 \leq F_2, G_1 \leq 2$. If both players are right or wrong, then the game is a draw; if one is wrong and the other is right, then the first pays $F_1 + F_2$ piasters to the second. The following table shows the strategy sets and corresponding payoffs a_{ij} to Player 1.

	j	1	2	3	4	5
i		(0,0)	(0,1)	(1,0)	(1,1)	(2,0)	(2,1)
1	(0,0)	0	0	-1	0	-2	0
2	(0,1)	0	0	0	1	-2	0
3	(0,2)	0	0	-1	0	0	2
4	(1,0)	1	0	0	-1	0	-3
5	(1,1)	0	-1	2	0	0	-3
6	(1,2)	0	-1	0	-2	3	0

The first row and column can be deleted from the full payoff matrix because $(0,0)$ is dominated by $(0,1)$ for both players (Fact 4). Thus it suffices to analyze the reduced payoff matrix in which $r = s = 5$. The value of the game is $-\frac{6}{35}$ for Player 1 ($\frac{6}{35}$ for Player 2). Optimal strategies $p^* = (\frac{3}{35}, \frac{18}{35}, \frac{8}{35}, \frac{6}{35}, 0)$ and $q^* = (\frac{4}{7}, \frac{2}{7}, 0, \frac{3}{35}, \frac{2}{35})$ are found by linear programming (Fact 3). Note that strategies $i = 5$ and $j = 3$ have zero probability at this equilibrium, despite being undominated.

4. The symmetric game of *four ways* [Me92] is played by two left-turning motorists who arrive simultaneously from opposite directions at a 4-way junction. Each has three pure strategies: the first is to go, the second to wait, and the third a conditional strategy of going only if the other appears to be waiting. It takes 2 seconds for one motorist to cross the junction while the other waits. If initially both either go or wait, then both motorists incur an extra "posturing" delay of either 3 or 2 seconds, respectively. Also, the one who ultimately waits is equally likely to be either player. For example, $a_{11} = 0.5 \times (-3 - 2) + 0.5 \times (-3) = -4$ and $a_{22} = 0.5 \times (-2 - 2) + 0.5 \times (-2) = -3$. This game has the payoff matrix

$$\begin{pmatrix} -4 & 0 & 0 \\ -2 & -3 & -2 \\ -2 & 0 & -4 \end{pmatrix}.$$

There are infinitely many equilibria in mixed strategies; these are described in the following table, where $0 \leq a \leq 1$ and $\frac{1}{2} \leq b \leq 1$.

p^*	q^*
$(1, 0, 0)$	$(0, a, 1-a)$
$(0, a, 1-a)$	$(1, 0, 0)$
$(0, 1, 0)$	$(b, 0, 1-b)$
$(b, 0, 1-b)$	$(0, 1, 0)$
$\frac{1}{11}(6, 2, 3)$	$\frac{1}{11}(6, 2, 3)$

15.5.3 CHARACTERISTIC-FUNCTION GAMES

When there exists a binding agreement among all players to cooperate, attention shifts from strategies to the bargaining strengths of coalitions. These strengths are assumed to be measured in terms of a freely transferable benefit (e.g., money or time) and players are assumed to seek a fair distribution of the total benefit available. Also, without loss of generality, the benefit of cooperation will be taken as the savings in costs.

Definitions:

A *coalition* is a subset S of $N = \{1, \ldots, n\}$; equivalently, $S \in 2^N$.

The *cost* associated with coalition S is denoted $c(S)$.

Let $\mathcal{R}^+$ denote the set of nonnegative reals. The **characteristic function** $\overline{\nu}\colon 2^N \to \mathcal{R}^+$ assigns to each S its cooperative benefit, using $\overline{\nu}(S) = \max\{0, \sum_{i \in S} c(\{i\}) - c(S)\}$.

A *characteristic-function game*, or *c-game*, is the pair $\Gamma = (N, \overline{\nu})$.

The game Γ is **inessential** if $\overline{\nu}(N) = 0$. If $\overline{\nu}(N) > 0$ then the game is **essential**, with **normalized characteristic function** $\nu\colon 2^N \to [0, 1]$ defined by $\nu(S) = \frac{\overline{\nu}(S)}{\overline{\nu}(N)}$.

The game Γ is **convex** if $\nu(S \cup T) \geq \nu(S) + \nu(T) - \nu(S \cap T)$ for all $S, T \in 2^N$.

Let $X = X_n$ be the $(n-1)$-dimensional unit simplex (§15.5.2). Any $x \in X$ is called an **imputation**; it allocates x_i of the total normalized benefit $\nu(N) = 1$ to Player i. An imputation is **unreasonable** if it allocates more to some $i \in N$ than the maximum that i could contribute to any coalition $T - \{i\}$ by joining it.

The **reasonable set** is $X_{RS} = \{\, x \in X \mid x_i \leq \max_T [\nu(T) - \nu(T - \{i\})] \text{ for all } i \in N \,\}$.

For any $x \in X$ and $S \in 2^N$, the **excess** of coalition S at x is $e(S, x) = \nu(S) - \sum_{i \in S} x_i$.

The **core** of Γ is $C = \{\, x \in X \mid e(S, x) \leq 0 \text{ for all } S \in 2^N \,\}$.

The **marginal worth** of Player i to the coalition $T - \{i\}$ is $\nu(T) - \nu(T - \{i\})$.

The **Shapley value** of a c-game is the imputation $x^S = (x_1^S, x_2^S, \ldots, x_n^S)$ defined by $x_i^S = \frac{1}{n!} \sum_{T \in \Pi^i} (|T| - 1)!\, (n - |T|)!\, (\nu(T) - \nu(T - \{i\}))$, where $\Pi^i = \{\, T \in 2^N \mid T \supseteq \{i\} \,\}$.

Facts:

1. An imputation is both individually rational and group rational (see §15.5.1).

2. Convexity is a sufficient (but not necessary) condition for the core to exist.

3. If $C \neq \emptyset$ then $C \subseteq X_{RS}$.

4. If C contains a single imputation, then the c-game is usually regarded as solved.

5. In general, C may either be empty (see Example 1) or contain infinitely many imputations (see Example 2).

6. If C contains infinitely many imputations, then there are several ways to single one out as the solution to the c-game. One approach is to define a "center" of C, which leads to the important concept of the *nucleolus* [Me92].

7. Every c-game solution concept assumes that players have agreed to enter coalition N. If its order of formation were known, players could be allocated their marginal worths; in general, however, this order of formation (and hence marginal worth) is a random variable.

8. If all orders of formation of N are equally likely, then the probability that Player i enters N by joining the coalition $T - \{i\}$ is $\frac{(|T|-1)!(n-|T|)!}{n!}$.

9. The Shapley value distinguishes a single imputation as the solution of a c-game by allocating to players the expected values of their marginal worths, based on the assumption that all orders of formation of N are equally likely.

10. $x^S \in X_{RS}$.

11. $x^S \in C$ if Γ is convex.

Examples:

1. In the c-game *log-hauling* [Me92], three lone drivers of pickup trucks discover a pile of 150 logs too heavy for any one to lift. Players 1, 2, and 3 can haul up to 45, 60, and 75 logs, respectively. Thus $\overline{\nu}(\{1,2\}) = 105$, $\overline{\nu}(\{1,3\}) = 120$, $\overline{\nu}(\{2,3\}) = 135$, and $\overline{\nu}(\{1,2,3\}) = 150$ so that $\nu(\{1,2\}) = \frac{7}{10}$, $\nu(\{1,3\}) = \frac{4}{5}$, and $\nu(\{2,3\}) = \frac{9}{10}$. This c-game is not convex; for example, if $S = \{1,2\}$ and $T = \{2,3\}$, then $1 = \nu(S \cup T) < \nu(S) + \nu(T) - \nu(S \cap T) = \frac{8}{5}$. Also, $C = \emptyset$.

2. The c-game *car pool* [Me92] is played by three co-workers whose office is d miles from their residential neighborhood, shown in the following figure.

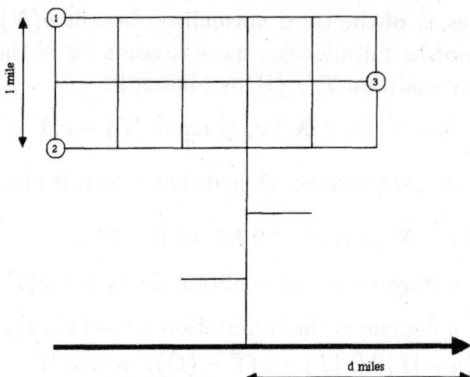

Driving to work costs $\$k$ per mile, and the shortest route is always used. The benefit of cooperation is car pool savings, leading to the characteristic function in this table.

S	$c(S)$	$\bar{\nu}(S)$	$\nu(S)$	$e(S,x)$ for $d=1$
$\emptyset$	0	0	0	0
$\{1\}$	$(4+d)k$	0	0	$-x_1$
$\{2\}$	$(3+d)k$	0	0	$-x_2$
$\{3\}$	$(3+d)k$	0	0	$x_1 + x_2 - 1$
$\{1,2\}$	$(4+d)k$	$(3+d)k$	$\frac{3+d}{3+2d}$	$\frac{4}{5} - x_1 - x_2$
$\{1,3\}$	$(6+d)k$	$(1+d)k$	$\frac{1+d}{3+2d}$	$x_2 - \frac{3}{5}$
$\{2,3\}$	$(6+d)k$	dk	$\frac{d}{3+2d}$	$x_1 - \frac{4}{5}$
$\{1,2,3\}$	$(7+d)k$	$(3+2d)k$	1	0

Because $x_3 = 1 - x_1 - x_2 \ (\geq 0)$, a set of imputations is determined by its projection onto $x_3 = 0$. In these terms, for $d = 1$, X is the largest triangle in the following figure, X_{RS} is the shaded hexagon, and C is the shaded quadrilateral. Here $C \subset X_{RS} \subset X$ because the c-game is convex (for all $d \geq 0$).

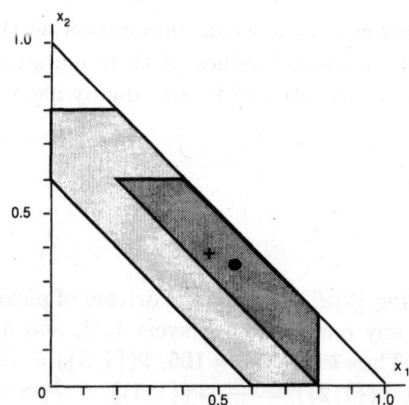

3. For the c-game in Example 2, it is easy enough to locate a center for C; see the figure for Example 2, where the nucleolus is marked by a dot.

4. In Example 1, the six possible orders of formation of N are 123, 132, 213, 231, 312, 321. Thus, the Shapley value is the imputation $x^S = (\frac{17}{60}, \frac{1}{3}, \frac{23}{60})$; see the following table.

i	$T \in \Pi^i$	$\nu(T) - \nu(T - \{i\})$	probability i enters N by joining $T - \{i\}$	x_i^S
1	$\{1\}$	0	$\frac{1}{3}$	$\frac{17}{60}$
	$\{1,2\}$	$\frac{7}{10}$	$\frac{1}{6}$	
	$\{1,3\}$	$\frac{4}{5}$	$\frac{1}{6}$	
	$\{1,2,3\}$	$\frac{1}{10}$	$\frac{1}{3}$	
2	$\{2\}$	0	$\frac{1}{3}$	$\frac{1}{3}$
	$\{1,2\}$	$\frac{7}{10}$	$\frac{1}{6}$	
	$\{2,3\}$	$\frac{9}{10}$	$\frac{1}{6}$	
	$\{1,2,3\}$	$\frac{1}{5}$	$\frac{1}{3}$	
3	$\{3\}$	0	$\frac{1}{3}$	$\frac{23}{60}$
	$\{1,3\}$	$\frac{4}{5}$	$\frac{1}{6}$	
	$\{2,3\}$	$\frac{9}{10}$	$\frac{1}{6}$	
	$\{1,2,3\}$	$\frac{3}{10}$	$\frac{1}{3}$	

5. By a calculation very similar to that laid out in the table of Example 4, the Shapley value for Example 2 is the imputation $x^S = (\frac{7}{15}, \frac{11}{30}, \frac{1}{6})$. Because the c-game is convex, $x^S \in C$. This is illustrated in the previous figure, where x^S is marked by a cross.

15.5.4 APPLICATIONS

Discrete (noncooperative or characteristic-function) games have numerous applications and merge with other categories of games not examined here. The references in the following table provide sources for the definitions, concepts, and applications of such games. This table also lists some representative areas of application of game theory.

15.6 SPERNER'S LEMMA AND FIXED POINTS

A fixed point of a function from a set X to itself is a point of X that is mapped into itself. Brouwer (1912) proved that every continuous mapping f on the unit ball has a fixed point. Sperner (1928) gave an elegant proof of Brouwer's fixed-point theorem using a combinatorial lemma known today as *Sperner's lemma*. This lemma has a number of applications to economics, nonlinear programming, and game theory.

category and references	selected applications	remarks
characteristic function games [Me92, 93], [Ow95], [Wa88]	airport landing fees, voting, water resources	utility is usually assumed to be transferable: in essence, players value benefits identically
classical game theory [LuRa57], [voMo53]	microeconomics, parlor games	economic (as opposed to evolutionary) game theory
combinatorial games [Gu91]	chess, go, nim, other parlor games	two players; complete, perfect information; no chance moves; zero-sum
continuous games [Dr81], [Fr90]	duels, military combat, oligopoly theory	a discrete game with mixed strategies is a special case of a continuous game
cooperative games in strategic form (as opposed to c-games) [Fr90], [Me92]	wage bargaining, motoring behavior	agreements among players are binding
differential games [BaHa94], [Me93]	fishery and forest management	extension of optimal control theory
economic game theory [Fr90], [My91]	microeconomics	equilibria are the result of rational thought processes
evolutionary game theory [Cr92], [Ma82], [Me92]	animal behavior	equilibria are the result of natural selection or equivalent populational processes
iterated games [Fr90], [Me92]	rationality of cooperation	often infinitely many iterations
resource games [Me93]	fisheries, forestry, water resources	discrete, continuous, and differential games all used
symmetric matrix games [Cr92], [Ma82], [Me92]	evolutionary game theory	dynamical systems theory provides a rationale for strategic equilibrium
zero-sum matrix games [Dr81]	military science	

15.6.1 SPERNER'S LEMMA

Sperner's lemma is a combinatorial result applicable to certain triangulations of a p-dimensional convex set, in which the vertices of the triangulation are given labels from $\{1, 2, \ldots, p+1\}$.

Definitions:

The $p+1$ points $x_1, x_2, \ldots, x_{p+1} \in \mathcal{R}^n$ are said to be in **general position** if the vectors $x_2-x_1, x_3-x_1, \ldots, x_{p+1}-x_1$ are linearly independent (§6.1.3).

The set $C \subseteq \mathcal{R}^n$ is **convex** if for all $x, y \in C$ and $0 \leq \lambda \leq 1$, $\lambda x + (1-\lambda)y \in C$.

The **convex hull** of a finite set of points $v_1, \ldots, v_{p+1} \in \mathcal{R}^n$ is the set $\langle v_1, \ldots, v_{p+1} \rangle = \{ \sum_{i=1}^{p+1} \lambda_i v_i \mid \sum_{i=1}^{p+1} \lambda_i = 1, \lambda_i \geq 0 \}$.

A **p-simplex** σ is the convex hull of $p+1$ points $x_1, \ldots, x_{p+1} \in \mathcal{R}^n$ in general position.

The **vertices** of the p-simplex $\sigma = \langle x_1, \ldots, x_{p+1} \rangle$ are the points $x_1, \ldots, x_{p+1}$. The **face** $\tau = \langle x_{j_1}, \ldots, x_{j_k} \rangle$ of σ is the simplex spanned by the subset $\{x_{j_1}, \ldots, x_{j_k}\}$ of $\{x_1, \ldots, x_{p+1}\}$. Write $\tau \prec \sigma$ when τ is a face of σ.

A **simplicial complex** K is a collection of simplices satisfying:
- if $\sigma \in K$ and $\tau \prec \sigma$ then $\tau \in K$;
- if $\sigma, \tau \in K$ intersect, their intersection is a face of each.

The **p-skeleton** of a simplicial complex K is the set of all simplices of dimension p or less. The 0-skeleton is the **vertex set**, denoted $V(K)$.

A **simplicial subdivision** $\mathcal{F}$ of a simplex σ is a collection of simplices $\{ \tau_j \mid 1 \leq j \leq m \}$ satisfying:
- $\sigma = \bigcup_{j=1}^m \tau_j$;
- the intersection of any two τ_j is either empty or a face of each.

A simplicial subdivision $\mathcal{F}'$ of a simplicial complex $\mathcal{K}$ is a **refinement** of the simplicial subdivision $\mathcal{F}$ of $\mathcal{K}$ if every simplex of $\mathcal{F}$ is a union of simplices of $\mathcal{F}'$.

Given a simplicial subdivision $\mathcal{F}$ of the p-simplex $\sigma = \langle x_1, \ldots, x_{p+1} \rangle$, a **proper labeling** of $\mathcal{F}$ is a mapping $\ell : V(\mathcal{F}) \to \{1, 2, \ldots, p+1\}$ satisfying:
- $\ell(x_m) = m$ for $m = 1, \ldots, p+1$;
- if vertex v lies on a face $\langle x_{k_1}, \ldots, x_{k_q} \rangle$ of σ, then $\ell(v) \in \{k_1, \ldots, k_q\}$.

Here $\{1, 2, \ldots, p+1\}$ is the **label set**, and if $\ell(v) = k$ then v **receives the label** k.

A **distinguished simplex** is a p-simplex that receives all $p+1$ labels 1 through $p+1$.

Facts:

1. The convex hull $\langle v_1, \ldots, v_{p+1} \rangle$ is the intersection of all convex sets containing the points $v_1, \ldots, v_{p+1}$.

2. The dimension of any p-simplex is p.

3. A p-simplex contains $2^{p+1} - 1$ simplices of dimension p or less.

4. Sperner's lemma (1928): Every properly labeled subdivision of a simplex σ has an odd number of distinguished simplices. (E. Sperner, 1906–1980)

5. Algorithm 1 gives a method for finding a distinguished triangle in a properly labeled subdivision of a triangle T. Each iteration of the outer loop starts at a distinguished 1-simplex and traces out a path, terminating either at a distinguished 2-simplex or at an outer edge of T.

6. Since there are an odd number of distinguished 1-simplices along the bottom of T (Fact 4) and since each "failed" outer loop iteration produces a path joining two such distinguished 1-simplices, Algorithm 1 must eventually produce a path terminating at a distinguished 2-simplex.

Algorithm 1: Distinguished simplex of a 2-simplex.

input: properly labeled subdivision of triangle T
output: a distinguished triangle of T

{Outer loop}
   find a distinguished 1-simplex τ along the bottom of T
   {Inner loop}
   repeat
      if the unique triangle containing τ is distinguished **then** stop
      else proceed to a neighboring triangle whose common edge is distinguished
   until either a distinguished triangle is found or the search leads to the bottom edge of T
continue outer loop with a new distinguished 1-simplex τ

Examples:

1. A 0-simplex is a point, a 1-simplex is a line segment, and a 2-simplex is a triangle (interior included). A 3-simplex includes the vertices, edges, faces, and interior of a tetrahedron. See the following figure.

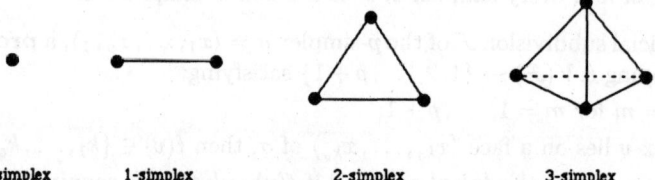

0-simplex   1-simplex   2-simplex   3-simplex

2. The 0-skeleton of a simplex σ is its vertex set; the 1-skeleton of σ is the edge set of σ including their endpoints; if σ is a 3-simplex, the 2-skeleton is the union of the faces of σ.

3. Part (a) of the following figure shows a simplicial subdivision of a 2-simplex. The subdivision in part (b) of the figure is not simplicial because $\tau_1 \cap \tau_3$ is not a face of the simplex τ_3.

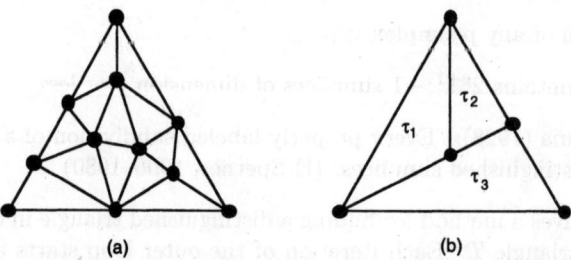

(a)   (b)

4. The following figure shows a proper labeling of a simplicial subdivision of a 1-simplex. A distinguished 1-simplex is a subinterval that receives both the labels 1 and 2. In this example, there are five such 1-simplices, an odd number (as guaranteed by Fact 4).

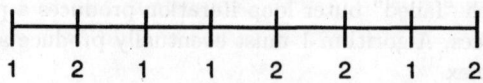

1   2   1   1   2   2   1   2

5. The following figure shows a proper labeling of a simplicial subdivision of a 2-simplex. There is one distinguished 2-simplex, receiving all three labels, which is shown shaded in the figure. If the vertex in the interior of the triangle is instead labeled 3, then there will be three distinguished 2-simplices, still an odd number.

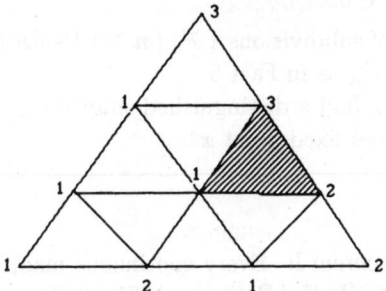

6. Several possible paths from executing Algorithm 1 are displayed in the following figure. The rightmost path terminates in a bottom edge, while the leftmost path leads to a distinguished triangle. Note that there are three distinguished triangles in this example, an odd number as required by Sperner's lemma.

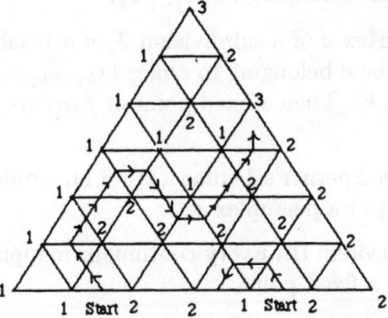

15.6.2 FIXED-POINT THEOREMS

Fixed-point theorems have applicability to a number of problems in economics, as well as to game theory and optimization.

Definitions:

The point $x \in B$ is a **fixed point** of the mapping $f: B \to B$ if $f(x) = x$.

The mapping f defined on a subset X of a normed space B is a **contraction** if there is some $0 \leq \beta < 1$ such that $\|f(x) - f(y)\| \leq \beta \|x - y\|$ for all $x, y \in X$.

The function F is a **set mapping** on X if $F(x)$ is a nonempty subset of X for all $x \in X$.

The set mapping F is **convex** if $F(x)$ is a convex subset of X for all $x \in X$.

Facts:

1. Fixed-point theorems can be used to demonstrate the existence of economic equilibria, solutions to a system of nonlinear equations, and Nash equilibria in two-person nonzero-sum games.

> **Algorithm 2: Fixed point of a p-simplex.**
>
> input: function f defined on a p-simplex σ
> output: fixed point $x^* \in \sigma$
>
> construct a sequence of subdivisions $\{\mathcal{F}_n \mid n \geq 1\}$ such that $\mathcal{F}_{n+1}$ refines $\mathcal{F}_n$
> label the vertex set of $\mathcal{F}_n$ as in Fact 5
> for each subdivision $\mathcal{F}_n$ find a distinguished simplex τ_n
> $\bigcap \tau_n$ contains the desired fixed point x^*

2. *Brouwer fixed-point theorem I*: Every continuous mapping $f: \sigma \to \sigma$ where σ is a p-simplex has a fixed point. (L. E. J Brouwer, 1881–1966)

3. For a simplex $\sigma = \langle x_1, x_2, \ldots, x_{p+1}\rangle$ and a continuous mapping $f: \sigma \to \sigma$, let $f(\sum_{k=1}^{p+1} \lambda_k x_k) = \sum_{k=1}^{p+1} \mu_k x_k$. Then $\sum_{k=1}^{p+1} \mu_k = \sum_{k=1}^{p+1} \lambda_k = 1$.

4. Relative to the mapping $f: \sigma \to \sigma$, define $T_j = \{\sum_{k=1}^{p+1} \lambda_k x_k \mid \mu_j \leq \lambda_j\}$. Then a fixed point of f is any point belonging to $\bigcap_{j=1}^{p+1} T_j$.

5. Suppose an interior vertex v of a subdivision $\mathcal{F}$ of σ is labeled with j provided that $v \in T_j$, and suppose a vertex v belonging to a face $\langle x_{k_1}, x_{k_2}, \ldots, x_{k_t}\rangle$ is labeled with any one of the labels $k_1, k_2, \ldots, k_t$. Then a fixed point of f occurs in a distinguished simplex of σ.

6. Algorithm 2, based on Sperner's lemma (§15.6.1), produces a sequence of points converging to a fixed point of a p-simplex σ.

7. *Brouwer fixed-point theorem II*: Every continuous mapping from a convex compact set $B \subseteq \mathcal{R}^n$ into itself has a fixed point.

8. *Contraction mapping theorem*: Every contraction $f: X \to X$ has a fixed point. The fixed point is the limit of the sequence $\{f(x_n) \mid n \geq 0\}$, where x_0 is an arbitrary element of X and $x_{n+1} = f(x_n)$.

9. *Kakutani fixed-point theorem*: Let $X \subseteq \mathcal{R}^n$ be a convex and compact set and suppose that F is a convex mapping on X. If the graph $\{(x, y) \mid y \in F(x)\} \subseteq \mathcal{R}^{2n}$ is closed, then there exists a point $x^* \in X$ such that $x^* \in F(x^*)$.

10. *Schauder fixed point theorem*: Every continuous mapping f on a convex compact subset X in a normed space B has a fixed point.

11. Reference [Bo85] gives applications of fixed-point theorems to determining market equilibria, maximal elements of binary relations, solutions to complementarity problems, as well as solutions to various types of games (cooperative and noncooperative).

Examples:

1. The real-valued function $f(x) = 1 - x$ is a mapping from the 1-simplex $\sigma = [0, 1]$ to itself. It is not a contraction since $|f(x) - f(y)| = |(1 - x) - (1 - y)| = 1 \cdot |x - y|$ holds for all $x, y \in \sigma$ so $\beta \geq 1$. The function f has a fixed point at $x = \frac{1}{2}$. However, the iterative procedure in Fact 8 will not generally locate this fixed point. For example, using $x_0 = \frac{1}{4}$ produces the sequence $x_1 = f(x_0) = \frac{3}{4}$, $x_2 = f(x_1) = \frac{1}{4}$, $x_3 = f(x_2) = \frac{3}{4}$, and so forth, with no limiting value.

2. The following figure shows a real-valued function $f : \sigma \to \sigma$ defined over the 1-simplex $\sigma = [0, 1]$. This function has three fixed points, identified by the intersection of the graph of f with the dashed line $y = x$. The sets T_j of Fact 4 relative to $x_1 = 0$, $x_2 = 1$ are also indicated in the figure, and it is verified that $T_1 \cap T_2$ contains the three fixed points of f. A subdivision of σ into five subintervals is shown in the figure. Using Fact 5, the associated vertices (at $x = 0.0, 0.2, 0.4, 0.6, 0.8, 1.0$) receive the labels $1, 2, 1, 1, 2, 2$ respectively, and so there are three distinguished simplices (each containing a fixed point).

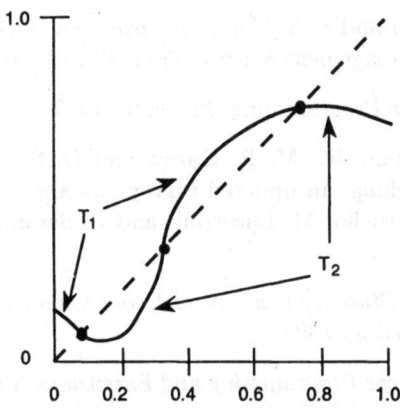

3. The real-valued function $f(x) = \frac{1}{4+x^2}$ is a mapping from $\mathcal{R}$ to itself. It can also be shown to be a contraction mapping with $\beta = \frac{1}{16} < 1$. If the iterative procedure in Fact 8 is applied using $x_0 = 1$, then $x_1 = 0.2$, $x_2 = 0.24752$, $x_3 = 0.24623$, $x_4 = 0.24627$, and $x_5 = 0.24627$, yielding the (approximate) fixed point $x^* = 0.24627$.

4. *Perron's theorem*: This theorem (§6.5.5), which assures that every positive matrix has a positive eigenvalue-eigenvector pair, can be proved using the fixed-point theorem in Fact 2. Let $A = (a_{ij})$ be an $n \times n$ matrix, with all $a_{ij} > 0$. The set $\sigma = \{\, x \in \mathcal{R}^n \mid \sum_{k=1}^n x_k = 1,\ x_k \geq 0 \,\}$ is an $(n{-}1)$-simplex, and the continuous function defined by $f(x) = \frac{Ax}{\|Ax\|_1}$ maps σ into itself. (Here $\|w\|_1$ is the l_1 norm of vector w; see §6.4.5.) By Fact 2, f has a fixed point $\bar{x}$, so that $A\bar{x} = \|A\bar{x}\|_1 \bar{x}$. Since at least one component of $\bar{x}$ is positive and A is positive, the vector $A\bar{x}$ has positive components. It then follows that the eigenvalue $\|A\bar{x}\|_1$ is positive and that the corresponding eigenvector $\bar{x}$ has all positive components.

REFERENCES

Printed Resources:

[AdKu89] V. G. Adlakha and V. G. Kulkarni, "A classified bibliography of research on stochastic PERT networks: 1966–1987", *INFOR* 27 (1989), 272–296.

[BaPa76] E. Balas and M. W. Padberg, "Set partitioning: a survey", *SIAM Review* 18 (1976), 710–760.

[BaZe80] E. Balas and E. Zemel, "An algorithm for large zero-one knapsack problems", *Operations Research* 28 (1980), 1130–1154.

[BaHa94] T. Basar and A. Haurie (eds.), *Advances in Dynamic Games and Applications*, Birkhäuser, 1994.

[BaJaSh90] M. S. Bazaraa, J. J. Jarvis, and H. D. Sherali, *Linear Programming and Network Flows*, 2nd ed., Wiley, 1990.

[BeKaSt92] W. W. Bein, J. Kamburowski, and M. F. M. Stallmann, "Optimal reduction of two-terminal directed acyclic graphs", *SIAM Journal on Computing* 21 (1992), 1112–1129.

[BiKiTa93] K. Binmore, A. Kirman, and P. Tani (eds.), *Frontiers of Game Theory*, MIT Press, 1993.

[BrCh89] M. L. Brandeau and S. S. Chiu, "An overview of representative problems in location research", *Management Science* 35 (1989), 645–674.

[Ch83] V. Chvátal, *Linear Programming*, Freeman, 1983.

[CoGaJo84] E. G. Coffman, Jr., M. R. Garey, and D. S. Johnson, "Approximation algorithms for bin-packing: an updated survey" in *Algorithm Design for Computer System Design*, G. Ausiello, M. Lucertini, and P. Serafini (eds.), Springer-Verlag, 1984, 49–106.

[Cr92] R. Cressman, *The Stability Concept of Evolutionary Game Theory: A Dynamic Approach*, Springer-Verlag, 1992.

[Da63] G. B. Dantzig, *Linear Programming and Extensions*, Princeton University Press, 1963.

[DeHe92] E. Demeulemeester and W. S. Herroelen, "A branch-and-bound procedure for the multiple resource-constrained project scheduling problem", *Management Science* 38 (1992), 1803–1818.

[DeHe96] B. De Reyck and W. S. Herroelen, "On the use of the complexity index as a measure of complexity in activity networks", *European Journal of Operational Research* 91 (1996), 347–366.

[DeHe90] J. De Wit and W. S. Herroelen, "An evaluation of microcomputer-based software packages for project management", *European Journal of Operational Research* 49 (1990), 102–139.

[Dr81] M. Dresher, *The Mathematics of Games of Strategy: Theory and Applications*, Dover, 1981. (First published in 1961.)

[El77] S. E. Elmaghraby, *Activity Networks: Project Planning and Control by Network Models*, Wiley, 1977.

[El89] S. E. Elmaghraby, "The estimation of some network parameters in the PERT model of activity networks: review and critique", in R. Słowinski and J. Weglarz (eds.), *Advances in Project Scheduling*, Elsevier, 1989, 371–432.

[ElKa90] S. E. Elmaghraby and J. Kamburowski, "On project representation and activity floats", *Arabian Journal of Science and Engineering* 15 (1990), 627–637.

[ElKa92] S. E. Elmaghraby and J. Kamburowski, "The analysis of activity networks under generalized precedence relations (GPRs)", *Management Science* 38 (1992), 1245–1263.

[Er78] D. Erlenkotter, "A dual-based procedure for uncapacitated facility location", *Operations Research* 26 (1978), 992–1009.

[Fo97] R. Fourer, "Software survey: linear programming", *OR/MS Today* 24(2) (1997), 54–63.

[FrMcWh92] R. L. Francis, L. F. McGinnis, Jr., and J. A. White, *Facility Layout and Location: An Analytical Approach*, Prentice-Hall, 1992.

[Fr90] J. W. Friedman, *Game Theory with Applications to Economics*, 2nd ed., Oxford University Press, 1990.

[GaNe72] R. S. Garfinkel and G. L. Nemhauser, *Integer Programming*, Wiley, 1972.

[Ga85] S. I. Gass, *Linear Programming: Methods and Applications*, 5th ed., Boyd & Fraser, 1985.

[Gu91] R. Guy (ed.), *Combinatorial Games*, Proceedings of Symposia in Applied Mathematics, American Mathematical Society, Volume 43, 1991.

[HaMi79] G. Y. Handler and P. B. Mirchandani, *Location on Networks: Theory and Algorithms*, MIT Press, 1979.

[HeDeDe98] W. S. Herroelen, B. De Reyck, and E. Demeulemeester, "Resource-constrained project scheduling: a survey of recent developments", *Computers & Operations Research* 25 (1998), 279–302.

[HoPa93] K. Hoffman and M. W. Padberg, "Solving airline crew scheduling problems by branch-and-cut", *Management Science* 39 (1993), 657–682.

[Ho85] T. Hogan, et al., "Project planning programs put to the test", *Business Software* 3(3) (1985), 21–56.

[Ka92] J. Kamburowski, "Bounding the distribution of project duration in PERT networks", *Operations Research Letters* 12 (1992), 17–22.

[KaSpDr92] J. Kamburowski, A. Sprecher, and A. Drexl, "Characterization and generation of a general class of resource-constrained project scheduling problems: easy and hard instances", *Research Report N0301*, Institut für Betriebswirtschaftslehre, Christian-Albrechts Universität zu Kiel, 1992.

[KaHa79a] O. Kariv and S. L. Hakimi, "An algorithmic approach to network location problems. I: the p-centers", *SIAM Journal on Applied Mathematics* 37 (1979), 513–538.

[KaHa79b] O. Kariv and S. L. Hakimi, "An algorithmic approach to network location problems. II: the p-medians", *SIAM Journal on Applied Mathematics* 37 (1979), 539–560.

[KuAd86] V. G. Kulkarni and V. G. Adlakha, "Markov and Markov-regenerative PERT networks", *Operations Research* 34 (1986), 769–781.

[LuRa57] R. D. Luce and H. Raiffa, *Games and Decisions*, Wiley, 1957. (Covers much the same ground as [voMo53] but at a more elementary level.)

[Ma86] O. Maimon, "The variance equity measures in location decisions on trees", *Annals of Operations Research* 6 (1986), 147–160.

[MaTo90] S. Martello and P. Toth, *Knapsack Problems: Algorithms and Computer Implementations*, Wiley, 1990.

[Ma82] J. Maynard Smith, *Evolution and the Theory of Games*, Cambridge University Press, 1982. (The founding treatise on evolutionary game theory.)

[Me83] N. Megiddo, "Linear-time algorithms for linear programming in $\mathcal{R}^3$ and related problems", *SIAM Journal on Computing* 12 (1983), 759–776.

[MeSu84] N. Megiddo and K. J. Supowit, "On the complexity of some common geometric location problems", *SIAM Journal on Computing* 13 (1984), 182–196.

[Me92] M. Mesterton-Gibbons, *An Introduction to Game-Theoretic Modelling*, Addison-Wesley, 1992. (A unified introduction to classical and evolutionary game theory.)

[Me93] M. Mesterton-Gibbons, "Game-theoretic resource modeling", *Natural Resource Modeling* 7 (1993), 93–147. (An up-to-date survey of game theory and applications accessible to the general reader.)

[MiKaSt93] D. J. Michael, J. Kamburowski, and M. Stallmann, "On the minimum dummy-arc problem", *RAIRO Recherche Opérationnelle* 27 (1993), 153–168.

[MiFr90] P. B. Mirchandani and R. L. Francis (eds.), *Discrete Location Theory*, Wiley, 1990.

[My91] R. B. Myerson, *Game Theory: Analysis of Conflict*, Harvard University Press, 1991. (General introduction to economic game theory.)

[NeWo88] G. L. Nemhauser and L. A. Wolsey, *Integer and Combinatorial Optimization*, Wiley, 1988.

[Ne90] K. Neumann, *Stochastic Project Networks: Temporal Analysis, Scheduling, and Cost Minimization*, Lecture Notes in Economics and Mathematical Systems 344, Springer-Verlag, 1990.

[Ow95] G. Owen, *Game Theory*, 3rd ed., Academic Press, 1995. (An overview of the mathematical theory of games.)

[Pa83] J. H. Patterson, "Exact and heuristic solution procedures for the constrained-resource, project scheduling problem: Volumes I, II, and III", *Research Monograph*, privately circulated, 1983.

[Sc86] A. Schrijver, *Theory of Linear and Integer Programming*, Wiley, 1986.

[SlWe89] R. Słowinski and J. Weglarz (eds.), *Advances in Project Scheduling*, Elsevier, 1989.

[Sp94] A. Sprecher, *Resource-Constrained Project Scheduling: Exact Methods for the Multi-Mode Case*, Lecture Notes in Economics and Mathematical Systems 409, Springer-Verlag, 1994.

[Ta88] A. Tamir, "Improved complexity bounds for center location problems on networks using dynamic data structures", *SIAM Journal on Discrete Mathematics* 1 (1988), 377–396.

[Ta96] A. Tamir, "An $O(pn^2)$ algorithm for the p-median and related problems on tree graphs", *Operations Research Letters* 19 (1996), 59–64.

[TaFrLo83] B. C. Tansel, R. L. Francis, and T. J. Lowe, "Location on networks: a survey, Parts I and II", *Management Science* 29 (1983), 482–511.

[voMo53] J. von Neumann and O. Morgenstern, *Theory of Games and Economic Behavior*, 3rd ed., Princeton University Press, 1953. (The founding treatise on economic game theory.)

[Wa88] J. Wang, *The Theory of Games*, Oxford University Press, 1988. (An overview of the mathematical theory of games emphasizing bimatrix games and c-games.)

[Wi92] T. M. Williams, "Criticality of stochastic networks", *Operations Research* 43 (1992), 353–357.

[Wi94] W. L. Winston, *Operations Research: Applications and Algorithms*, 3rd ed., Duxbury, 1994.

Web Resources:

ftp://garbo.uwasa.fi/pc/ts/tslin35c.zip (Computer code that implements goal programming.)

http://plato.la.asu.edu/guide.html#LP (Computer codes in C, Pascal, Fortran that implement the simplex method and interior point algorithms.)

http://ucsu.colorado.edu/~xu/software/lp/ (Computer code that implements the simplex method.)

http://ucsu.colorado.edu/~xu/software/lp/minit.html (Computer code in C that implements the dual simplex algorithm.)

http://www.cplex.com/ (Software package CPLEX to solve LP and IP/MIP problems.)

http://www.dash.co.uk/ (Software package XPRESS-MP to solves IP/MIP problems.)

http://www.iiasa.ac.at/~marek/soft/descr.html#MCMA (Computer code that implements goal programming.)

http://www.iiasa.ac.at/~marek/soft/descr.html#MOMIP (Computer code to solve IP/MIP problems.)

http://www.ketronms.com/mipiii.html (Software package MIPII to solve IP/MIP problems.)

http://www.ketronms.com/products.html (Software package C-WHIZ to solve LP problems.)

http://www.lindo.com/ (Software packages LINDO to solve LP problems and LINGO to solve IP/MIP problems.)

http://www.maths.mu.oz.au/~worms/javaorjava/knapsack/knapsack.html/ (Interactive algorithm for solving general knapsack problems.)

http://www.mpi-sb.mpg.de/~barth/opbdp/opbdp.html (Computer code in C for solving binary IP problems.)

http://www.netcologne.de/~nc-weidenma/readme.htm (Computer code in Pascal for solving IP/MIP problems.)

http://www.netlib.org/toms/632 (Fortran code for solving knapsack problems.)

http://www.research.ibm.com/osl/ (Software package OSL to solve LP and IP/MIP problems.)

http://www.sunsetsoft.com/ (Software package to solve IP/MIP problems.)

http://www.wior.uni-karlsruhe.de/Bibliothek/Title_Page1.html (Listing of software to solve LP problems; listing of software for project planning.)

http://www-c.mcs.anl.gov/home/otc/Guide/SoftwareGuide/Categories/linearprog.html (An extensive tabulation of software packages to solve LP problems.)

http://www-c.mcs.anl.gov/home/otc/Guide/SoftwareGuide/Categories/intprog.html (An extensive tabulation of software packages to solve IP/MIP problems.)

http://www-c.mcs.anl.gov/home/otc/InteriorPoint/index.html (An archive of technical papers and other information on interior point algorithms.)

http://www-c.mcs.anl.gov/home/otc/Server/ (LP problems can be submitted on-line for solution by different interior point algorithms using the NEOS home page.)

http://www-leland.stanford.edu/~saunders/brochure/brochure.html (Software package MINOS to solve LP problems.)

http://www-math.cudenver.edu/~hgreenbe/glossary/glossary.html (Glossary of linear programming terms and concepts in general mathematical optimization.)

16

THEORETICAL COMPUTER SCIENCE

16.1 Computational Models — *Jonathan L. Gross*
  16.1.1 Finite state machines
  16.1.2 Pushdown automata
  16.1.3 Turing machines
  16.1.4 Parallel computational models

16.2 Computability — *William Gasarch*
  16.2.1 Recursive functions and the Church's thesis
  16.2.2 Recursive sets and solvable problems

16.3 Languages and Grammars — *Aarto Salomaa*
  16.3.1 Alphabets and languages
  16.3.2 Operations on strings and languages
  16.3.3 Grammars and the Chomsky hierarchy
  16.3.4 Regular and context-free languages
  16.3.5 Combinatorics on words

16.4 Algorithmic Complexity — *Thomas Cormen*
  16.4.1 Overview of complexity
  16.4.2 Worst-case analysis
  16.4.3 Average-case analysis

16.5 Complexity Classes — *Lane Hemaspaandra*
  16.5.1 Oracles and the polynomial hierarchy
  16.5.2 Reducibility and NP-completeness
  16.5.3 Probabilistic Turing machines

16.6 Randomized Algorithms — *Milena Mihail*
  16.6.1 Overview and general paradigms
  16.6.2 Las Vegas and Monte Carlo algorithms

INTRODUCTION

Theoretical computer science is concerned with modeling computational problems and solving them algorithmically. It strives to distinguish what can be computed from what cannot. If a problem can be solved by an algorithm, it is important to know the amount of space and time needed.

GLOSSARY

abelian square: a word having the pattern xx^p, where x^p is any permutation of the word x.

acceptance probability (of an input word by a probabilistic TM): the sum of the probabilities over all acceptance paths of computation.

Ackermann function: a very rapidly growing function that is recursive, but not primitive recursive.

algorithm: a finite list of instructions that is supposed to accomplish a specified computation or other task.

alphabet: a finite nonempty set whose elements are called *symbols*.

ambiguous context-free grammar: a grammar whose language has a string having two different leftmost derivations.

analysis of an algorithm: an estimation of its cost of execution, especially of its running time.

antecedent of a production $\alpha \to \beta$: the string α that precedes the arrow.

average-case running time: the expected running time of an algorithm, usually expressed asymptotically in terms of the input size.

Backus-Naur (or **Backus normal**) **form** (**BNF**): a metalanguage for specifying computer language syntax.

busy beaver function: the function $BB(n)$ whose value is the maximum number of 1s that an n-state Turing machine can print and still halt.

busy beaver machine, **n-state**: an n-state Turing machine on the alphabet $\Sigma = \{\#, 1\}$ that accepts an input tape filled with blanks (#s) and halts after placing a maximum number of 1s on the tape.

cellular automaton, (n-dimensional): an interconnection network in which there is a processor at each integer lattice point of n-dimensional Euclidean space, and each processor communicates with its immediate neighbors.

characteristic function (of a language): the function on strings in the alphabet for that language that has value *yes* for elements in the language, and *no* otherwise.

characteristic function (of a set): the function whose value is 1 for elements of the set, and 0 otherwise.

Chomsky hierarchy: four classes of grammars, with gradually increasing restrictions.

Chomsky normal form (for a production rule): the form $A \to BC$ where B and C are nonterminals or the form $A \to a$ where a is a terminal.

Church's thesis (or the ***Church-Turing thesis***): the premise that the intuitive notion of what is computable or partially computable should be formally defined as computable by a Turing machine.

code (for an alphabet V): a nonempty language $C \subseteq V^+$, such that whenever a word w in V can be written as a catenation of words in C, the write-up is always unique. That is, if $w = x_1 \ldots x_m = y_1 \ldots y_n$, where $m, n \geq 1$, and $x_i, y_j \in C$, then $m = n$ and $x_i = y_i$ for $i = 1, \ldots, m$.

code indicator (of a language): the sum of the code indicators of all words in the language.

code indicator (of a word $w \in V^*$): the number $ci(w) = |V|^{-|w|}$.

collapse (of the polynomial hierarchy to the ith rank): the circumstance that PH $= \Sigma_i^p$, for some $i \geq 0$.

common PRAM (or $CRCW^{com}$): a CRCW PRAM model in which concurrent writes to the same location are permitted if all processors are trying to write the same data.

comparison sort: a sorting algorithm that uses only comparisons between record keys to determine the sorted order.

complement (of a language L over an alphabet V): the language $\overline{L}$, where complementation is taken with respect to V^*.

C-complete language (where C is a class of languages): a language A such that A is C-hard and $A \in C$.

complexity (of an algorithm): an asymptotic measure of the number of operations or the running time needed for a complete execution; sometimes, a measure of the total amount of computational space needed.

complexity (of a function): usually, the minimum complexity of any algorithm representing the function; sometimes, the length or complicatedness of the list of instructions.

complexity (of a function), ***Kolmogorov-Chaitin type***: a measure of the minimum complicatedness of any algorithm representing the function, usually according to number of instructions in the algorithm (and *not* related to its running time).

***complexity class* coNP**: the class Π_1^p, which contains every language A such that $\overline{A} \in \Sigma_1^p$.

***complexity class* NP**: the minimal class that contains every language that is nondeterministically TM-decidable in polynomial time.

***complexity class* P**: the class comprising every language that is deterministically TM-decidable in polynomial time.

***complexity class* PSPACE**: the minimal class that contains every language that is TM-decidable in polynomial space.

concatenation (of two languages L_1 and L_2): the set $\{xy \mid x \in L_1, y \in L_2\}$, denoted $L_1 L_2$.

concatenation (of two strings): the result of appending the second string to the right end of the first.

consequent (of a production $\alpha \to \beta$): the string β that follows the arrow.

context-free (or ***type 2***) ***grammar***: a grammar in which the antecedent α of each production $\alpha \to \beta$ is a nonterminal.

context-sensitive (or **type 1**) **grammar** $G = (N, T, S, P)$: a grammar such that every production $\alpha \to \beta$ (except possibly $S \to \lambda$) has the form $\alpha = uAv$ and $\beta = uxv$, for $u, v \in (N \cup T)^*, A \in N, x \in (N \cup T)^+$.

CRCW concurrent read concurrent write: a PRAM model in which concurrent reads from and concurrent writes to the same location are both allowed.

CREW concurrent read exclusive write: a PRAM model in which concurrent reads are allowed, but not concurrent writes to the same location.

cube (over an alphabet): a word having the pattern xxx.

derivation (of the string y from the string x): a sequence of substitutions, according to the production rules, that transforms string y into string z. The notation $x \Longrightarrow^* y$ means that such a derivation exists.

emptiness problem for grammars: deciding whether the language generated by a grammar is empty.

empty string: the string of length zero, that is, the string with no symbols; often written λ.

equivalence problem for grammars: deciding whether two grammars are equivalent.

equivalent automata: two automata that accept the same language.

equivalent grammars: grammars that generate the same language.

EREW exclusive read exclusive write: a PRAM model in which concurrent reads from and concurrent writes to the same location are not allowed.

existential lower bound (for an algorithm): a lower bound for its number of execution steps that holds for at least one input.

existential lower bound (for a problem): a lower bound for every algorithm that could solve that problem.

finite automaton: either a finite state recognizer or a nondeterministic finite state recognizer.

finite-state recognizer (**FSR**): a model of a computer for deciding membership in a set.

finite-state machine: a finite automaton or a finite transducer.

finite-state machine with output: is another name for a finite transducer.

finiteness problem (for a grammar): deciding whether the language generated by that grammar is finite.

finite transducer: a model of a computer for calculating a function, like an FSR, except that it also produces an output string each time it reads an input symbol.

free monoid (generated by an alphabet): the set of all strings composable from symbols in the alphabet, with the semigroup operation of string concatenation.

frequency (of a symbol in a string): the number of occurrences of the symbol in the string.

Game of Life: a 2-dimensional cellular automaton designed by John H. Conway.

Gödel numbering (of a set): a method for encoding Turing machines as products of prime powers; more generally, a similar one-to-one recursive function on an arbitrary set whose image in $\mathcal{N}$ is a recursive set.

grammar: a quadruple $G = (N, T, S, P)$, where N is a finite nonempty alphabet of *nonterminals*, T is a finite nonempty alphabet of *terminals*, with $N \cap T = \emptyset$, S is a nonterminal called the *start symbol*, and P is a finite set of *production* rules of the form $\alpha \to \beta$.

halting problem: the problem of designing an algorithm capable of deciding which computations $P(x)$ halt and which do not, where P is a computer program (or a Turing machine), and x is a possible input.

C-hard language (C a class of languages): a language A such that every language in class C is polynomial-time reducible to A.

Hilbert's tenth problem: the (recursively unsolvable) problem of deciding for an arbitrary multivariate polynomial equation $p(x_1, \ldots, x_n) = 0$ whether there exists a solution consisting of integers.

inclusion problem for grammars: deciding whether one language is included in another.

inherently ambiguous context-free language: a context-free language such that every context-free grammar for the language is ambiguous.

input size: the quantity of data supplied as input to a computation.

interconnection network model: a parallel computation model as a digraph in which each vertex represents a processor, and in each phase of the computational process, each processor communicates with its neighbors and makes a computation.

inverse (of a morphism $h: V^* \longrightarrow U^*$): the mapping $h^{-1}: U^* \longrightarrow 2^{V^*}$ defined by $h^{-1}(x) = \{\, y \in V^* \mid h(y) = x \,\}, x \in U^*$.

Kleene closure (or *Kleene star*) *of a language* L: the set of all iterated concatenations of zero or more words in L, denoted L^*.

language (accepted by a machine, such as an FSR, a pushdown automaton, or a Turing machine): the set of all accepted strings.

language (generated by the grammar G): the language $L(G) = \{\, x \in T^* \mid S \Longrightarrow^* x \,\}$ of words consisting of terminal symbols derivable from the starting symbol.

language (over an alphabet V): a subset of the free monoid V^*.

Las Vegas algorithm: an algorithm that always produces correct output, whose running time is a random variable.

Las Vegas to Monte Carlo transformation: the Monte Carlo algorithm obtained by running the Las Vegas scheme for $kE[T]$ steps and halting, where $E[T]$ is the expected Las Vegas running time.

leftmost derivation $x \Longrightarrow_{left} y$: a derivation $x \Longrightarrow y$ in which at each step the leftmost nonterminal is replaced.

leftmost language (generated by the grammar G): the language $L_{left}(G)$ of strings of terminals with leftmost derivations from the start symbol S.

length set (of a language L): the set $\{\, |x| \mid x \in L \,\}$.

length-increasing (or *type 1*) *grammar*: a grammar in which the consequent β of each production $\alpha \to \beta$ (except $S \to \lambda$, if present) is at least as long as its antecedent α.

linear grammar: a context-free grammar in which each production $\alpha \to \beta$ has $\alpha \in N$ and $\beta \in T^* \cup T^* N T^*$.

Mealy machine: a finite transducer whose output function always produces a single symbol.

membership problem (for a grammar G): given an arbitrary string x, deciding whether $x \in L(G)$.

mirror image (of a language L): the language $mi(L) = \{\, x^R \mid x \in L \,\}$ obtained by reversing every string in L.

Monte Carlo algorithm: an algorithm that has a bounded number of computational steps and might produce incorrect output with some low probability.

Monte Carlo to Las Vegas transformation: the Las Vegas algorithm of repeatedly running that Monte Carlo algorithm until correct output occurs.

Moore machine: a Mealy machine such that for every state k and every pair of input symbols s_1 and s_2, the outputs $\tau(k, s_1)$ and $\tau(k, s_2)$ are the same.

morphism (from the alphabet V to the alphabet U): a function $s{:}V \longrightarrow 2^{U^*}$ with $s(a)$ a singleton set for all symbols $a \in V$.

nondeterministic finite-state recognizer (**NDFSR**): a model like a finite state recognizer, but there may be several different states to which a transition is possible, instead of only one.

nondeterministic polynomial-time computation on a TM: a computation for which there exists a polynomial function $p(n)$ such that for any input of size n there is a computational path on the TM whose length is at most $p(n)$ steps.

nondeterministic Turing machine: a 5-tuple $M = (K, s, h, \Sigma, \Delta)$ otherwise like a deterministic Turing machine, except that the transition function Δ maps each state-symbol pair (q, b) to a set of state-symbol-direction triples.

nonterminal (in a grammar): a symbol that may be replaced when a production is applied.

nontrivial family of languages: a family that contains at least one language different from $\emptyset$ and $\{\lambda\}$.

NP-complete language: a language A such that A is NP-hard and $A \in NP$.

NP-complete problem: a decision problem equivalent to deciding membership in an NP-complete language.

NP-hard language: a language A such that every language in complexity class NP is polynomial-time reducible to A.

oracle (for a language): a machine state that decides whether or not a given string is in the language.

oracle Turing machine: a 6-tuple $M = (K, s, h, \Sigma, \delta \text{ or } \Delta, L)$, equipped with a special second tape on which it can write a string in the alphabet of an oracle for language L (which might be different from Σ). (Aside from oracle steps, it is a Turing machine.)

palindrome: a string that is identical to its reverse.

parallel computation model: a computational model that permits more than one instruction to be executed simultaneously, instead of requiring that instructions be executed sequentially.

parsing a string: in theoretical computer science, a *derivation*.

partial function: an incomplete rule that assigns values to some elements in its domain but not necessarily to all of them.

partial function ϕ_M induced by a TM M: the rule that associates to each input v for which the M-computation halts the output $\phi_M(v)$, and is otherwise undefined.

partial recursive function: a partial function derivable from the constant zero functions $\zeta_n(x_1, \ldots, x_n) = 0$, the successor function $\sigma(n) = n + 1$, and the projection functions $\pi_i^n(x_1, \ldots, x_n) = x_i$, using multivariate composition, multivariate primitive recursion, and unbounded minimalization.

pattern (over an alphabet V): a string of variables over that alphabet; regarded as present in a particular word $w \in V^*$ if there exists an assignment of strings from V^+ to the variables in that pattern such that the word formed thereby is a substring of w.

***polynomial hierarchy* PH**: the union of the complexity classes Σ_n^p, for $n \geq 0$.

polynomial-space computation (of a function by a TM M): a computation by M of that function such that there exists a polynomial function $p(n)$ such that for every input of size n, the calculation workspace takes at most $p(n)$ positions on the tape.

polynomial-time computation (of a function by a TM M): a computation by M of that function such that there exists a polynomial function $p(n)$ such that for every input of size n, the calculation takes at most $p(n)$ steps.

positive closure* (or *Kleene plus*) *of a language L: the set of all iterated concatenations of words in L excluding the empty word, denoted L^+.

nth power of a language: the set of all iterated concatenations $w_1 w_2 \ldots w_n$ where each w_i is a word in the language.

PRAM memory conflict: the conflict that occurs when more than one processor attempts concurrently to write into or read from the same global memory register.

PRAM parallel random access machine: a model of parallel computation as a set of global memory registers and a set of processors, each with access to an infinite sequence of its own local registers.

primitive recursion: a restricted way of defining $f(n + 1)$ in terms of $f(n)$.

primitive recursive function: any function derivable from the constant zero functions $\zeta_k(x_1, \ldots, x_k) = 0$, the successor function $\sigma(n) = n + 1$, and the projection functions $\pi_i^n(x_1, \ldots, x_n) = x_i$, using multivariate composition and multivariate primitive recursion.

probabilistic Turing machine: a nondeterministic Turing machine M with exactly two choices of a next state at each step, both with probability $\frac{1}{2}$ and independent of all previous choices.

production rule* (in a grammar) *of the form $\alpha \to \beta$: a rule for making a substitution in a string; iterative application of the production rules generates all the words of the language of the grammar.

projection function, n-place: a function $\pi_i^n(x_1, \ldots, x_n) = x_i$ that maps an n-tuple to its ith coordinate.

pumping lemma: any one of several results in formal language theory concerned with rewriting strings.

***pushdown automaton* (PDA)**: a (possibly non-deterministic) finite-state automaton equipped with an auxiliary stack.

random access machine (RAM): a computation model with several arithmetic registers and an infinite number of memory registers.

randomized algorithm: an algorithm that makes random choices during its execution, guided by the output of a random (or pseudo-random) number generator.

recursive language: a language with a decidable membership question.

recursive set: a set whose characteristic function is recursive.

recursively enumerable set: a set that is either empty or the image of a recursive function.

reducibility in polynomial-time (of language A to language B): the existence of a polynomial-time computable function f such that $x \in A$ if and only if $f(x) \in B$, for each string x in the alphabet of language A; denoted by $A \leq_m^P B$.

reduction: a strategy for solving a problem by transforming its natural form of input into the input form for another problem, solving that other problem on the transformed input, and transforming the answer back into the original problem domain.

regular expression (over an alphabet V): a string w in the symbols of V and the special set $\{\epsilon,), (, +, *\}$ such that $w \in V$ or $w = \epsilon$, or (continuing recursively) $w = (\alpha\beta), (\alpha + \beta)$, or α^*, where α and β are regular expressions.

regular (or type 3) grammar: a grammar such that every production $\alpha \to \beta$ has antecedent $\alpha \in N$ and consequent $\beta \in T \cup TN \cup \{\lambda\}$.

regular language: a language that can be obtained from elements of its alphabet V using finitely many times the operations of union, concatenation and Kleene star.

regularity problem (for grammars): deciding whether $L(G)$ is a regular language.

reverse (of the string x): the string x^R obtained by writing x backwards.

running time: the number of primitive operation steps executed by an algorithm, usually expressed in big-O asymptotic notation (or sometimes Θ-notation) as a formula based on the input size variables.

solvable problem: a problem that can be decided by a recursive function.

space complexity (of an algorithm): a measure of the amount of computational space needed in the execution, relative to the size of the input.

sparse language: a language A for which there is a polynomial function $p(n)$ such that for every $n \in \mathcal{N}$, there are at most $p(n)$ elements of length n in A.

square (over an alphabet): the pattern xx, or any word having that pattern.

square-free word: a word having no subwords with the pattern xx.

start symbol (in a grammar): a designated nonterminal from which every word of the language is generated.

state diagram (for an FSR): a labeled digraph whose vertices represent the states and whose arcs represent the transitions.

string (accepted by an FSR or NDFSR): a string such that the automaton ends up in an accepting state, immediately after the last transition.

string (accepted by a PDA): a string that, when supplied as input, ultimately can lead to the stack being empty and the PDA being in an acceptance state after the last transition.

string (accepted by a TM M): a string w such that M halts on input w.

string (over an alphabet): a finite sequence of symbols from that alphabet.

substitution (for the alphabets V in the alphabet U): a mapping $s: V \longrightarrow 2^{U^*}$, which means that each symbol $b \in V$ may be replaced by any of the strings in the set $s(b)$; extends to strings of V^*.

terminal (in a grammar): a symbol that cannot be replaced by other symbols.

time complexity (of an algorithm): a function representing the number of operations or the running time needed, using the size of the input as its argument.

total function: a partial function defined on all of its domain, i.e., a function.

tractable problem: a problem that can be solved by an algorithm with polynomial-time complexity.

λ-transition (in a NDFSR): a transition that could occur without reading any symbols of the input string.

transition table (for an FSR): a table whose rows are indexed by the states and whose columns are indexed by the symbols, such that the entry in row r and column c is the state to which the FSR moves if it reads symbol c while in state r.

trapping state (of a finite automaton): a non-accepting state q from which every outward arc is a self-loop back into q.

trio: a nontrivial family of languages closed under λ-free morphisms, inverse morphisms, and intersection with regular languages.

Turing-acceptable language: a language such that has a TM M that accepts it.

Turing-computable function: a function such that there is a TM M with $f = \phi_M$.

Turing-decidable language: a language whose characteristic function is Turing-computable.

Turing machine (TM): an automaton whose tape can move one character in either direction and that can replace the symbol it reads by a different symbol.

Turing-p-reducibility (of language A to language B): the existence of a deterministic oracle TM M^B that decides language A in polynomial time. Notation: $A \leq^P_T B$.

Turing's test (of whether a given computer can think): are its responses to written questions distinguishable from human responses by a person who does not know whether a computer or a person gave the response?

type 0 grammar: a grammar with no restrictions.

type 1 grammar: a length-increasing grammar, or equivalently, a context-sensitive grammar.

type 2 grammar: a context-free grammar.

type 3 grammar: a regular grammar.

unambiguous context-free language: a context-free language L that has a context free grammar that is not ambiguous.

unbounded minimalization: a way of using a function or partial function to define a new function or partial function.

uncomputable function: a function whose values cannot be calculated by a Turing machine (or by a computer program).

undecidable problem: a decision problem whose answers cannot be given by a Turing machine (or by a computer program).

universal Turing machine: a TM that can simulate every other TM.

variable (over an alphabet V): a symbol not in V whose values range over V^*.

word (over an alphabet): usually a finite sequence of symbols (same as *string*), sometimes a countably infinite sequence.

word equation (over an alphabet V): an expression $\alpha = \beta$, such that α and β are words containing letters from V and some variables over V.

word inequality: the negation of a word equation, commonly written as $\alpha \neq \beta$.

worst-case running time: the maximum number of execution steps of an algorithm, usually expressed in big-O asymptotic notation (or sometimes Θ-notation) as a formula based on the input size variables.

16.1 COMPUTATIONAL MODELS

The objectives of a computer, no matter what special input/output or memory devices are attached, are ultimately to make logical decisions and to calculate the values of a function. A decision problem can be represented as recognizing whether an input string is in a specified subset. Calculating a function amounts to accepting an input string and producing an output string. At this fundamental level, the fundamental models in Table 1 can serve as the theoretical basis for all sequential computers.

16.1.1 FINITE STATE MACHINES

Definitions:

A (*deterministic*) *finite-state recognizer* (often abbreviated **FSR**) models a computer for decision-making as a 5-tuple $M = (K, s, F, \Sigma, \delta)$ such that:
- K is a finite set whose members are called the **states**;
- $s \in K$ (s is called the **starting state**);
- $F \subseteq K$ (each member of F is called an **acceptance state**);
- Σ is a finite set called the **alphabet of symbols**
- $\delta : K \times \Sigma \to K$ (δ is called the **transition function**).

The **computer model** for a finite-state recognizer $M = (K, s, F, \Sigma, \delta)$ consists of a logic box, programmed by the transition function δ. It is equipped with a read-only head that examines an input tape that moves in only one direction. Whenever it reads symbol c on the input tape while in state q, the computer switches into state $\delta(q, c)$ and moves on to read the next symbol. The string is considered to be accepted if the automaton is in an acceptance state after the last transition.

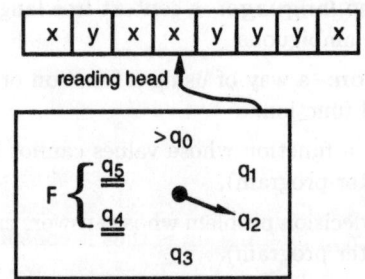

Table 1 Fundamental computational models.

model	description	comment
$FSR = (K, s, F, \Sigma, \delta)$ K = set of states, start at $s \in K$ accepting states $F \subseteq K$ input alphabet Σ transition fn $\delta: K \times \Sigma \to K$	Finite state recognizer: scans a tape once; decides whether to accept.	recognizes regular languages
$NDFSR = (K, s, F, \Sigma, \Delta)$ K, s, F, Σ like FSR trans. relation $\Delta: K \times \Sigma^* \to 2^K$	nondeterministic FSR.	equivalent to FSR
$FSM = (K, s, \Sigma_I, \Sigma_O, \delta, \tau)$ K = set of states, start at $s \in K$ in-alphabet Σ_I, out-alphabet Σ_O transition fn $\delta: K \times \Sigma_I \to K$ output fn $\delta: K \times \Sigma_I \to \Sigma_O^*$	Finite state transducer: (also called "finite state machine with output")	
Mealy = $(K, s, \Sigma_I, \Sigma_O, \delta, \tau)$ $K, s, \Sigma_I, \Sigma_O, \delta$ like FSM output fn $\delta: K \times \Sigma_I \to \Sigma_O$	Mealy machine: writes a single output symbol for each input symbol.	equivalent to FSM
Moore = $(K, s, \Sigma_I, \Sigma_O, \delta, \tau)$ $K, s, \Sigma_I, \Sigma_O, \delta$ like FSM output function $\delta: K \to \Sigma_O$	Moore machine: output symbol depends only on state prior to transition.	equivalent to FSM
$PDA = (K, s, F, \Sigma, \Gamma, \Delta)$ with K, s, F, Σ like FSR, stack alphabet Γ $\Delta \subseteq (K \times \Sigma^* \times \Gamma^*) \times (K \times \Gamma^*)$ transition relation Δ is finite set	Pushdown automaton: uses a stack as a computational resource, nondeterministic.	recognizes context-free languages
$TM = (K, s, h, \Sigma, \delta)$ K = set of states, start at $s \in K$ halting state $h \notin K$, alphabet Σ $\delta: K \times \Sigma \to K \times \Sigma \times \{L, R\} \cup \{h\}$	Turing machine: has two-way tape with rewritable symbols.	decides membership in recursive sets

A **state diagram** for an FSR $M = (K, s, F, \Sigma, \delta)$ is a labeled digraph whose vertex set is K, and such that for each state $q \in K$ and each symbol $c \in \Sigma$ there is an arc from vertex q to vertex $\delta(q, c)$, labeled with the symbol c. Sometimes a single arc is labeled with more than one symbol, instead of drawing two arcs from the same state to the same state. The starting state is designated by an entering arrow "$\longrightarrow$" and the accepting states are indicated by a double circle.

A **transition table** for an FSR $M = (K, s, F, \Sigma, \delta)$ is a table whose rows are indexed by the states in K and whose columns are indexed by the symbols in Σ, such that the entry in row q and column c is $\delta(q, c)$. The starting-state row label is marked with a ">" and the acceptance state row labels are underscored.

A **configuration** for an FSR $M = (K, s, F, \Sigma, \delta)$ is a pair (q, w) such that $q \in K$ and $w \in \Sigma^*$. The pair (q, w) signifies that the automaton is in state q with the read-only head positioned at the initial character of the string w. (Since the read-only head moves in only one direction, a common assumption is that it consumes each character that it reads.)

The FSR configuration (q, w) **yields the configuration** (q', w') **in one step** if deleting the initial symbol, call it c, of the string w yields the string w' and if $\delta(q, c) = q'$. This relationship between configurations is denoted by $(q, w) \vdash_M (q', w')$.

A *nondeterministic finite-state recognizer* (often abbreviated **NDFSR**) is a 5-tuple $M = (K, s, F, \Sigma, \Delta)$ just like an FSR, except that Δ is a finite subset of $K \times \Sigma^* \times K$ and is called the **transition relation**.

The **computer model** for an NDFSR $M = (K, s, F, \Sigma, \Delta)$ is like the computer model for an FSR. However, whenever it reads string u on the input string while in state q, the computer switches into any of the states in the set $\Delta(q, u)$ and moves on to read the next symbol.

A **state diagram** for an NDFSR $M = (K, s, F, \Sigma, \Delta)$ is a labeled digraph whose vertex set is K, and such that for each triple $(q, u, p) \in \Delta$ there is an arc from vertex q to vertex p, labeled with the string u, which may be the empty string λ.

A **transition** of an NDFSR $M = (K, s, F, \Sigma, \Delta)$ is a triple $(q, u, p) \in \Delta$. The idea is that from state q, the NDFSR M may read the substring u, and then transfer into state p.

A **λ-transition** in a NDFSR $M = (K, s, F, \Sigma, \Delta)$ is a transition $(q, \lambda, q') \in \Delta$ that could occur without reading any symbols off the input string. That is, it reads the empty string λ.

A **configuration for an NDFSR** $M = (K, s, F, \Sigma, \Delta)$ is a pair (q, w) such that $q \in K$ and $w \in \Sigma^*$.

The NDFSR configuration (q, w) **yields the configuration** (q', w') **in one step** if there is an initial prefix u on the string w whose deletion yields the string w', and if $(q, u, q') \in \Delta$. Notation: $(q, w) \vdash_M (q', w')$.

A **finite automaton** is either an FSR or an NDFSR.

A **computation** for a finite automaton M is a sequence of configurations (q_0, w_0), $(q_1, w_1), \ldots, (q_n, w_n)$ such that $(q_{i-1}, w_{i-1}) \vdash_M (q_i, w_i)$, for $i = 1, \ldots, n$. This is called a **computation of** (q_n, w_n) **from** (q_0, w_0).

For any finite automaton, the configuration (q, w) **yields** the configuration (q', w'), denoted $(q, w) \vdash_M^* (q', w')$, if there is a computation of (q', w') from (q, w).

A string $w \in \Sigma^*$ is **accepted by an FSR or NDFSR** $M = (K, s, F, \Sigma, \delta \text{ or } \Delta)$ if there is an accepting state $q \in F$ such that $(s, w) \vdash_M^* (q, \lambda)$. That is, machine M accepts string w if, starting in state s at the first symbol, its transition sequence ultimately leads to an accepting state, immediately after its last transition.

The **language accepted by a finite automaton** M is the set of all strings accepted by M. It is denoted $L(M)$.

Finite automata M_1 and M_2 are **equivalent** if $L(M_1) = L(M_2)$, that is, if they accept the exact same language.

A **trapping state** of a finite automaton M is a non-accepting state q from which every outward arc is a self-loop back into q.

An **implicit trapping state** is a convention used to simplify state diagrams. If from some state there is no exiting arc labeled with a particular symbol, then that combination is deemed to lead to the implicit trapping state.

A **trapping group** of a finite automaton M is set of non-accepting states from which there is no directed path in the state diagram to an accepting state.

A (*deterministic*) **finite transducer** models a function-calculating computer as a 6-tuple $M = (K, s, \Sigma_I, \Sigma_O, \delta, \tau)$ such that:
- K is a finite set (its members are called **states**);
- $s \in K$ (s is the **starting state**);
- Σ_I is a finite alphabet of **input symbols**;
- Σ_O is a finite alphabet of **output symbols**;
- $\delta: K \times \Sigma_I \to K$ (δ is called the **transition function**);
- $\tau: K \times \Sigma_I \to \Sigma_O^*$ (τ is called the **output function**).

A **finite-state machine with output** is another name for a finite transducer.

A **Mealy machine** is a finite transducer whose output function always produces a single symbol.

A **Moore machine** is a Mealy machine such that for every state k and every pair of input symbols s_1 and s_2, the outputs $\tau(k, s_1)$ and $\tau(k, s_2)$ are the same.

A **finite-state machine** is a finite automaton or a finite transducer.

Facts:

1. Finite state machines are the design plan of many practical types of electronic control devices, for instance in wristwatches or automobiles.

2. Terminological usage has evolved over several decades. The following table provides a quick guide to current usage regarding output capacity:

terminology	output capacity
recognizer	none
Mealy machine	one output symbol for each input symbol
transducer	arbitrary output string for each input symbol

The phrase "finite state machine" refers to a finite state model that may or may not have output capacity and that may or may not be nondeterministic.

3. The nondeterminism of an NDFSR is that possibly $u = \lambda$ or that there might also be a transition (q, u, p'), so that from the same state q the NDFSR M might also read substring u and transfer either into state p or into state p'.

4. For every NDFSR, there is an equivalent FSR. (M. Rabin and D. Scott, 1959)

5. In software design, NDFSRs are commonly used in preference to deterministic FSRs because they often achieve the same task with fewer states.

6. NDFSRs are often defined so that λ-transitions are the only possible instances of non-determinism. In this seemingly more restrictive kind of NDFSR, the second component of a transition (q, u, p) is either a single symbol or the empty string.

7. The class of languages accepted by finite automata is closed under all of the following operations:
- union;
- concatenation;
- Kleene star (see §16.3.2);
- complementation;
- intersection.

8. *Kleene's theorem*: A language is regular (§16.3.4) if and only if it is the language accepted by some finite automaton. (S. Kleene, 1956)

9. Some lexical scanning processes of compilers are modeled after finite automata.

10. The relation "yields" for finite automata is the reflexive, transitive closure of the relation $\vdash_M$.

11. Both Moore machines and Mealy machines have the computational capability of an unrestricted finite transducer.

12. More comprehensive coverage of finite state machines is provided by many textbooks, including [Gr97] and [LePa81].

Examples:

1. The FSR specified by the following transition table and state diagram decides whether a binary string has evenly many 1s. Formally, $M = (K, s, F, \Sigma, \delta)$ with $K = \{\text{Even}, \text{Odd}\}$, $s = \text{Even}$, $F = \{\text{Even}\}$, and $\Sigma = \{0, 1\}$.

	0	1
> Even	Even	Odd
Odd	Odd	Even

2. In one early form of the language BASIC, an identifier could be a letter or a letter followed by a digit. The following state diagram specifies an FSR that accepts this restricted form of BASIC identifier.

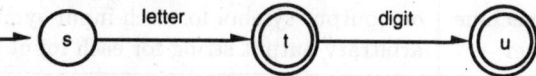

3. A "proper fixed-point numeral" is a nonempty string of decimal digits (the "whole part"), followed by a decimal point, and then another non-empty string (the "fractional part") of digits. For instance, the number zero would be represented as "0.0". The following FSR decides whether the input string is a fixed-point numeral.

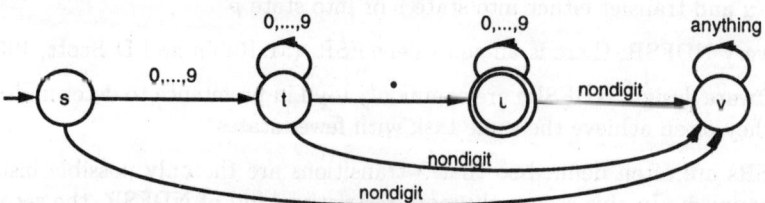

4. An "integer" in some programming languages is a nonempty string of decimal digits, possibly preceded by a sign + or −. The following NDFSR decides whether the input string is an integer.

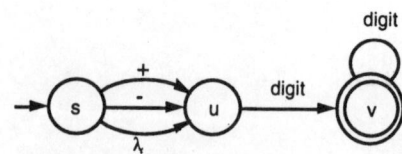

5. The following finite-state transducer has $\{0, 1, \$\}$ and $\{0, 1\}$ for its input and output alphabets, respectively, where "$\$$" serves as an end-of-string marker. It reads a binary numeral, starting at the units digit, and prints a binary numeral whose value is double the input numeral.

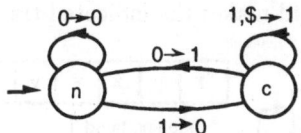

6. The following finite state machine models a vending machine for a 20-cent local newspaper. The possible inputs are a nickel, a dime, and a push of a button that releases the newspaper if enough change has been deposited. The states indicate the amount of money that has been deposited. This machine may be regarded as a transducer that produces symbol N (newspaper) if it receives input B while in state 20.

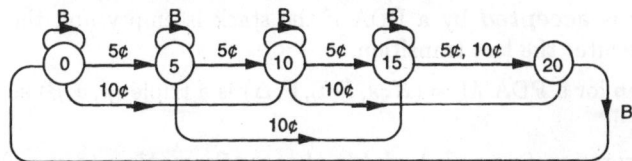

7. NDFSRs can be used to model various kinds of solitaire games and puzzles. For instance, making a complete knight's tour of a chessboard is such a puzzle. At each stage, there may be some moves that ultimately permit a complete tour and some other moves that are traps.

16.1.2 PUSHDOWN AUTOMATA

Definitions:

A *pushdown automaton* (**PDA**) is essentially a (possibly non-deterministic) finite-state machine equipped with an auxiliary stack. A pushdown automaton is given by a 6-tuple $M = (K, s, F, \Sigma, \Gamma, \Delta)$ such that:
- K is a finite set (its members are called *states*);
- $s \in K$ (s is called the *starting state*);
- $F \subseteq K$ (each member of F is called an *acceptance state*);
- Σ is a finite set called the **alphabet of input symbols**;
- Γ is a finite set called the **alphabet of stack symbols**;
- Δ is a finite subset of $(K \times \Sigma^* \times \Gamma^*) \times (K \times \Gamma^*)$ (Δ is the *transition relation*).

A *transition* of a PDA $M = (K, s, F, \Sigma, \Gamma, \Delta)$ is a pair $((p, u, \beta), (q, \gamma)) \in \Delta$. The idea is that from state p, the PDA M may read the substring u and the stack substring β, and transfer into state q while popping β and pushing γ, thereby replacing β by γ.

Note: A PDA is frequently defined so that the only strings that can be read or written or pushed or popped are single characters and the empty string. This has no effect on the computational generality, but it can lead to the need for more states to accomplish a given task.

The **computer model** for a PDA $M = (K, s, F, \Sigma, \Gamma, \Delta)$ consists of a logic box, programmed by Δ, equipped with a read-only head that examines an input tape that moves in only one direction, and also equipped with a stack. When it reads substring u from input while in state p with substring β at the top of the stack, the computer selects a corresponding entry from Δ and makes the indicated transition.

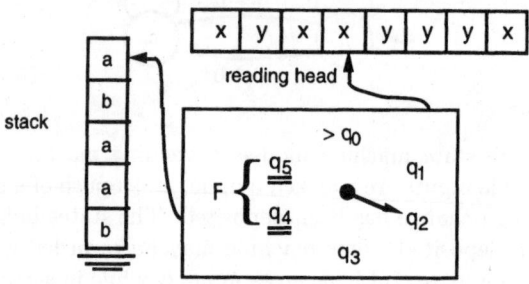

An input string is **accepted** by a PDA if the stack is empty and the PDA is in an acceptance state after the last transition.

A **configuration** for a PDA $M = (K, s, F, \Sigma, \Gamma, \Delta)$ is a triple (p, u, β) such that $p \in K$, $u \in \Sigma^*$, and $\beta \in \Gamma^*$.

The PDA configuration $(p, ux, \beta\alpha)$ **yields the configuration** $(q, x, \gamma\alpha)$ **in one step** if and only if there is a transition $((p, u, \beta), (q, \gamma))$. This relationship between configurations is denoted by $(p, ux, \beta\alpha) \vdash_M (q, x, \gamma\alpha)$.

A **computation** for a PDA M is a sequence of configurations $C_0, C_1, \ldots, C_n$ such that $C_{i-1} \vdash_M C_i$, for $i = 1, \ldots, n$.

The PDA configuration C **yields** the configuration C', denoted $C \vdash_M^* C'$, if there is a computation of C' from C.

A string $w \in \Sigma^*$ is **accepted** by a PDA $M = (K, s, F, \Sigma, \Gamma, \Delta)$ if there is an accepting state $q \in F$ such that $(s, w) \vdash_M^* (q, \lambda, \lambda)$. That is, machine M accepts string w if, starting in state s at the first symbol, its transition sequence can ultimately lead to an accepting state and an empty stack after it has read the last symbol.

The **language accepted** by a PDA M is the set of all strings accepted by M. It is denoted $L(M)$.

A **state diagram** for a PDA $M = (K, s, F, \Sigma, \Gamma, \Delta)$ is a labeled digraph whose vertex set is K, and such that for each transition $((p, u, \beta), (q, \gamma)) \in \Delta$ there is an arc from vertex p to vertex q, labeled $(u, \beta) \mapsto \gamma$. Sometimes a single arc is labeled with more than one symbol, instead of drawing two arcs from the same state to the same state. The starting state is usually designated by an entering arrow, and the accepting states are usually indicated by a double circle.

Facts:

1. The PDA model was invented by A. G. Oettinger in 1961.

2. A language L is context-free (see §16.3.3) if and only if there is a pushdown automaton M such that L is the language accepted by M. (M. Schutzenberger 1963, and independently by N. Chomsky and by J. Evey)

3. A PDA can test whether a string is a palindrome or whether all the left and right parentheses are matched, but an FSR cannot.

4. The class of languages accepted by deterministic PDAs is smaller than the class accepted by non-deterministic PDAs.

5. More comprehensive coverage of pushdown automata is provided by many textbooks, including [Gr97] and [LePa81].

Examples:

1. The following PDA decides whether a sequence of left and right parentheses is well-nested, in the sense that every left parenthesis is uniquely matched to a right parenthesis, and vice versa. It is necessary and sufficient that in counting left and right parentheses while reading from left to right, the number of right parentheses never exceeds the number of left parentheses and that the total counts are the same.

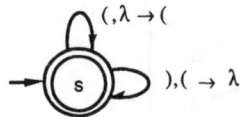

2. The following PDA decides whether a string in the alphabet $\{0, 1, m\}$ has the form bmb^r where b is a bitstring and b^r its reverse, i.e., the same string written backwards. The m in the middle signals when to switch from pushing symbols onto the stack to popping them off.

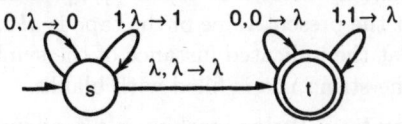

3. The following non-deterministic PDA decides whether a binary string is of the form bb^r, that is, a bitstring followed by its reverse. In effect, it considers every character interspace in the string as the possible middle.

16.1.3 TURING MACHINES

Definitions:

A **Turing machine** (**TM**) models a computer as a 5-tuple $M = (K, s, h, \Sigma, \delta)$ such that:
- K is a finite set not containing h (elements of K are called **states**; h is called the **halting state**);
- $s \in K$ (s is called the **starting state**);
- Σ is a finite set of symbols, including the **blank symbol** #; (Σ is called the **alphabet**);
- $\delta: K \times \Sigma \to (K \times \Sigma \times \{L, R\}) \cup \{h\}$ (δ is called a **transition function**).

An $m \times n$ **Turing machine** is a Turing machine with m states and n symbols.

The **computer model** for a Turing machine $M = (K, s, h, \Sigma, \delta)$ consists of a logic box, programmed by δ, equipped with a read-write head that examines an input tape with a left end, but no right end. To start a computation, the input string is written at the left end of the tape, and the rest of the tape is filled with blanks. The read-write head starts at the leftmost symbol. Whenever the Turing machine reads symbol b on the input string while in state q, its internal logic produces the triple $\delta(q, b) = (p, c, D)$ the computer switches into state p, replaces b by c, and moves one space in direction D, that is, either to the left (L) or to the right (R), whereupon it is ready to read the next symbol.

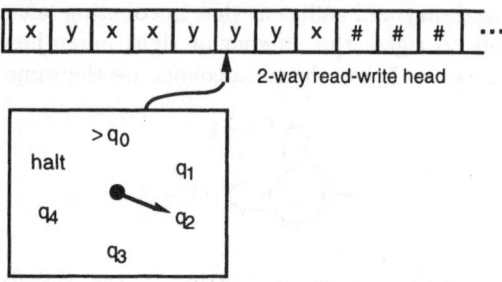

A **transition table** for an $m \times n$ Turing machine is an $m \times n$ table whose rows are labeled with the states, whose columns are labeled with the symbols, such that the entry in row q and column b is $\delta(q, b)$. Thus, a typical table entry is a triple indicating to which state to switch from q, the new symbol to replace b, and whether to move one square to the right or one to the left. However, another possibility is that a table entry could be the halt state h.

In an **alternative definition of a Turing machine**, each transition is a change of state and either a change of symbol or a one-symbol move to the right or left. In this case the transition table entries are pairs, and it tends to take more states and symbols to achieve a given objective.

A **configuration** for a Turing machine is a quadruple (q, u, b, v), such that $q \in K \cup \{h\}$, $u, v \in \Sigma^*$, and $b \in \Sigma$, commonly written as a pair $(q, u\underline{b}v)$. This means that the Turing machine is in state q, that the present value of the tape is ubv, that the present location of the read-write head is at the indicated instance of the symbol b, and that the rest of the tape to the right of the string ubv is filled with blanks.

A **starting configuration** for a Turing machine is a configuration of the form $(s, \lambda \underline{b} v)$. This means that the string bv is supplied to the given Turing machine as input, in the starting state s.

A **halting configuration** for a Turing machine is a configuration of the form $(h, u\underline{b}v)$. This means that the Turing machine has entered the halting state h, and that whatever is on the tape is to be interpreted as the output.

A **hanging configuration** for a Turing machine is a configuration of the form $(q, \lambda \underline{b} v)$, such that the transition value $\delta(q, b)$ tells the Turing machine to move left (L), i.e., off the left end of the tape.

The Turing machine configuration $(p, u\underline{b}v)$ **yields the configuration** $(q, x\underline{c}y)$ **in one step** if and only if the transition $\delta(p, b)$ would change configuration $(p, u\underline{b}v)$ to configuration $(q, x\underline{c}y)$. This relationship between configurations is denoted by $(p, u\underline{b}v) \vdash_M (q, x\underline{c}y)$.

An **infinite loop** for a Turing machine is an infinite sequence of configurations $C_0, C_1, C_2, \ldots$ such that $C_{i-1} \vdash_M C_i$, for $i = 1, 2, \ldots$.

The **M-computation** for input v to a Turing machine M is one of three possibilities: (1) the finite sequence of configurations $C_0 = (s, \lambda \underline{\#} v), C_1, \ldots, C_n$ such that $C_{i-1} \vdash_M C_i$, for $i = 1, \ldots, n$, in which C_n is a halting configuration; (2) the finite sequence of configurations $C_0 = (s, \lambda \underline{\#} v), C_1, \ldots, C_n$ such that $C_{i-1} \vdash_M C_i$, for $i = 1, \ldots, n$, in which C_n is a hanging configuration; (3) the infinite sequence $C_0 = (s, \lambda \underline{\#} v), C_1, C_2, \ldots$ such that $C_{i-1} \vdash_M C_i$, for $i = 1, \ldots, n$.

The **output** of an M-computation for input v to a Turing machine M is the string $\phi_M(v)$ from the left end of the tape up to the last non-blank character if the M-computation halts, and undefined otherwise.

The **partial function ϕ_M induced by a Turing machine M** is the rule that associates to each input v for which the M-computation halts the output $\phi_M(v)$, and is otherwise undefined.

A function $f: \Sigma^* \to \Sigma^*$ is **M-computable** by a Turing machine $M = (K, s, h, \Sigma, \delta)$ if the machine M halts for all inputs $v \in \Sigma^*$ and if the machine M computes the function f, that is, $f(v) = \phi_M(v)$ for all $v \in \Sigma^*$.

A function $f: \Sigma^* \to \Sigma^*$ is **Turing-computable** if there is a Turing machine $M = (K, s, h, \Sigma, \delta)$ such that f is M-computable.

A Turing machine $M = (K, s, h, \Sigma, \delta)$ **simulates** another Turing machine $M' = (K', s', h, \Sigma', \delta')$ if there exists a Turing-computable function $\beta: \Sigma'^* \to \Sigma^*$ such that $\phi_M(\beta(w)) = \phi_{M'}(w)$ for all $w \in domain(\phi_{M'})$ and $\phi_M(\beta(w))$ is undefined for all $w \notin domain(\phi_{M'})$.

A **universal Turing machine** is a Turing machine $U = (K_U, s_U, h_U, \Sigma_U, \delta_U)$ that can simulate every other Turing machine, in the following sense. There is a rule α_U for encoding any given Turing machine M and a rule β_U for encoding any given input w to M, such that $\phi_U(\alpha_U(M) \# \beta_U(w))$ is defined and equals $\phi_M(w)$ whenever $\phi_M(w)$ is defined, and is undefined otherwise.

A string $w \in \Sigma^*$ is **Turing machine M accepted by the Turing machine** $M = (K, s, h, \Sigma, \delta)$ if M halts on input w.

A language $L \subseteq \Sigma^*$ is **Turing machine M accepted by the Turing machine** $M = (K, s, h, \Sigma, \delta)$ if $L = \{ w \in \Sigma^* \mid M \text{ accepts } w \}$.

A language $L \subseteq \Sigma^*$ is **Turing-acceptable** if there exists a Turing machine M that accepts it.

The **characteristic function** $\chi_L: \Sigma^* \to \{yes, no\}$ of a language $L \subseteq \Sigma^*$ is given by the rule

$$\chi_L(w) = \begin{cases} yes, & \text{if } w \in L; \\ no, & \text{if } w \notin L. \end{cases}$$

A language $L \subseteq \Sigma^*$ is a **Turing-decidable language** if its characteristic function is Turing-computable.

A subset-membership decision problem is **unsolvable** if it does not correspond to a Turing-decidable language.

An **n-state busy beaver machine** is an n-state Turing machine on the alphabet $\Sigma = \{\#, 1\}$ that accepts a two-way infinite input tape filled with #s and halts after placing a maximum number of 1s on the tape. (The name *busy beaver* derives from an analogy between the machine piling up 1s and a beaver piling up logs.)

The **busy beaver function** $BB(n)$ has as its value the number of 1s on the output tape of an n-state busy beaver machine.

A **linear bounded automaton** (or **LBA**) is representable as a Turing machine that is fed only a finite stretch of tape containing the input word, rather than an infinite tape.

A **nondeterministic Turing machine** is defined like a Turing machine, except that instead of a transition function that assigns a unique change of symbol and direction of motion for the read-write head, there is a transition relation that may permit more than one possibility.

Facts:

1. A Turing machine is commonly regarded as a program to compute the partial function ϕ_M.
2. Every Turing-decidable language is Turing-acceptable.
3. If a language $L \subseteq \Sigma^*$ is Turing-decidable, then its complement $\overline{L}$ is also Turing-decidable.
4. Every Turing-acceptable language $L \subseteq \Sigma^*$ whose complement $\overline{L}$ is also Turing-acceptable is a Turing-decidable language.
5. The following problems about Turing machines are unsolvable:
    (a) Given a TM M and an input string w, does M halt on input w?
    (b) Given a TM M, does M halt on the empty tape?
    (c) Given a TM M, does there exist an input w for which M halts?
    (d) Given a TM M, does M halt on every input string.
    (e) Given two TMs M_1 and M_2, do they accept the same input?
    (f) Given two numbers n and k, is $BB(n) > k$?
6. In view of part (d) of the preceding fact, there is no way to tell whether an arbitrary computer program in a general language always halts, much less whether it calculates what it is supposed to calculate.
7. It is possible to construct a universal Turing machine.
8. A universal Turing machine with six states and four symbols was constructed in 1982 by Y. Rogozhin.
9. The busy beaver problem was invented by Tibor Rado in 1962.
10. Turing machines have been extended in several ways, including the following: infiniteness in two directions, more work tapes, two or more tapes, two- or more dimensional tapes, nondeterminism.
11. Some of the extensions of Turing machine can perform computations more quickly and are easier to program.
12. Any function that can be computed by a Turing machine with a two-way infinite tape can also be computed by some standard Turing machine.
13. Any function that can be computed by a Turing machine with k tapes can also be computed by some standard Turing machine.
14. Any function that can be computed by a Turing machine with a two-dimensional tape can also be computed by some standard Turing machine.
15. Any function that can be computed by a nondeterministic Turing machine can also be computed by some standard Turing machine.

16. The following table gives known values and lower bounds for the busy beaver function:

n	1	2	3	4	5	6	8
$BB(n)$	1	4	6	13	≥ 4098	$\geq 95{,}524{,}079$	$\geq 10^{44}$

17. The finite amount of workspace to which a linear bounded automaton is restricted causes it to be less powerful than a Turing machine with infinite tape. However, it is more powerful than a pushdown automaton.

18. Alan M. Turing (1912-1954) was a British mathematician whose cryptanalytic work during World War II lead to the decryption of ciphertext from the German cipher machine called the Enigma.

19. Turing proposed that a machine be regarded as "thinking" if its responses to written questions could not be distinguished from those of a person. This criterion is called *Turing's test*.

20. More comprehensive coverage of Turing machines is provided by many textbooks, including [Gr97] and [LePa81].

Examples:

1. This is a 1-state Turing machine with alphabet $\Sigma = \{0, 1, \#\}$ that changes every character preceding the first blank into a blank. It accepts any string over its alphabet.

$$\begin{array}{cccc} & 0 & 1 & \# \\ \rightarrow a & \#aR & \#aR & h \end{array}$$

2. This 3-state Turing machine with alphabet $\Sigma = \{0, 1, \#\}$ doesn't change its input tape at all. It halts whenever it encounters the third '1'. Thus, it accepts any tape with at least three 1's but accepts no other strings.

$$\begin{array}{cccc} & 0 & 1 & \# \\ \rightarrow a & 0aR & 1bR & \#aR \\ b & 0bR & 1cR & \#bR \\ c & 0cR & h & \#cR \end{array}$$

3. This 3-state Turing machine with alphabet $\Sigma = \{1, \#\}$ adds two positive integers, each represented as a string of 1s. For instance, the tape $111\#11\#\#\#\cdots$ becomes $11111\#\#\#\cdots$.

$$\begin{array}{ccc} & 1 & \# \\ \rightarrow a & 1aR & 1bR \\ b & 1bR & \#cL \\ c & \#cR & h \end{array}$$

4. This 2-state Turing machine shows that $BB(2)$ is at least 4.

$$\begin{array}{ccc} & \# & 1 \\ \rightarrow a & 1bL & 1bR \\ b & 1aR & h \end{array}$$

5. This 3-state Turing machine shows that $BB(3)$ is at least 6.

$$\begin{array}{ccc} & \# & 1 \\ \rightarrow a & 1bR & 1cL \\ b & 1cR & 1h \\ c & 1aL & \#bL \end{array}$$

16.1.4 PARALLEL COMPUTATIONAL MODELS

Definitions:

A *parallel computation model* permits more than one instruction to be executed simultaneously, instead of requiring that they be executed sequentially.

An *interconnection network* models parallel computation as a digraph in which each vertex represents a processor. In each phase of the computational process, a processor communicates with its neighbors and makes a computation.

An n-dimensional *cellular automaton* is an interconnection network in which there is a processor at each integer lattice point of n-dimensional Euclidean space, and each processor communicates with its immediate neighbors.

A *random access machine* (**RAM**) has several arithmetic registers and an infinite number of memory registers (often modeled as an infinite array), and of which can be accessed immediately via its address (the index in the array).

A *parallel random access machine* (**PRAM**) models parallel computation as a set of *global memory registers* $\{\, M_j \mid j = 1, 2, \ldots \,\}$ and a set of *processors* $\{\, P_j \mid j = 1, 2, \ldots \,\}$. Each processor P_j has access to an infinite sequence of *local registers* $\{\, R_{j,k} \mid k = 1, 2, \ldots \,\}$.

In a PRAM, a *register* (global or local) may contain a single integer. It is **local** if it can be accessed only by a single processor and **global** if it can be accessed by all processors.

In a PRAM, a *processor* performs *read* and *write* instructions involving global memory and other instructions involving only its local memory. All processors of a PRAM perform the same program in perfect synchrony, so that at any given time all processors that are not idle are all performing their task under the same instruction of the program.

In a PRAM, the concurrent construct $\mathbf{par}[a \leq j \leq b] P_j : S_j$ means that each of the processors P_j for $a \leq j \leq b$ is performing the operation S_j.

In a PRAM, the *read instruction* $READ(j)$ tells processor P_i to read the content of global register M_j into local register $R_{i,0}$.

In a PRAM, the *write instruction* $WRITE(j)$ tells processor P_i to write the content of local register $R_{i,0}$ into global register M_j.

In a PRAM, a *computation* starts when all the processors execute the first instruction. It stops when processor P_1 halts. The contents of the global memory are regarded as the output.

In a PRAM, a *memory conflict* occurs when more than one processor attempts concurrently to write into or read from the same global memory register.

In an *exclusive read exclusive write* (**EREW**) PRAM model, concurrent reads from and concurrent writes to the same location are not allowed.

In a *concurrent read exclusive write* (**CREW**) PRAM model, concurrent reads are allowed, but not concurrent writes to the same location.

In a *concurrent read concurrent write* (**CRCW**) PRAM model, concurrent reads from and concurrent writes to the same location are both allowed.

In a *common PRAM* ($CRCW^{com}$ **PRAM**) model, concurrent writes to the same location are permitted if all processors are trying to write the same data.

Facts:

1. Commercially available parallel computers often have an array of elements in which a single broadcast instruction applying to every element is executed simultaneously for all the elements.

2. Random access machines are commonly thought to be close theoretical models of commercially available sequential computers.

3. The indexing of registers and processors of a PRAM may be over any finite or countably infinite set.

4. PRAM programs are often described using high-level programming language constructs for array-processing that are similar to sequential array processing, except that the PRAM array locations are processed in parallel.

5. Whereas linear time is regarded as fast for sequential processing, a parallel algorithm tends to be regarded as fast if it runs in $O(\lg n)$ time or less.

Examples:

1. A parallel computer for sorting up to n items can be modeled as a row of processors $P_1, P_1, \ldots, P_n$. Joining each processor P_j such that $2 \leq j \leq n-1$ to and from its immediate predecessor P_{j-1} and its immediate successor P_{j+1} are arcs in both directions, as shown here.

On the first phase and on all subsequent odd-numbered phases, each processor pair (P_{2j+1}, P_{2j+2}) compares items and swaps, if necessary, so that the smaller item ends up in the lower-indexed processor. On the second phase and on all subsequent even numbered phases, each processor pair (P_{2j}, P_{2j+1}) compares items and swaps, if necessary, so that the smaller item ends up in the lower-indexed processor. After n phases, the items are completely sorted into ascending order.

2. *EREW PRAM: finding the maximum:* Given n numbers, with $n = 2^r$, store the numbers in global registers $M_n, \ldots, M_{2n-1}$. Then execute the following program:

> **for** $i = r - 1$ **downto** 0
>   **par**$[2^i \leq j \leq 2^{i+1}] M_j := \max\{M_{2j}, M_{2j+1}\}$
> **next** i

After r iterations of the loop body, the maximum appears in global register M_1.

3. $CRCW^{com}$ *PRAM: finding the maximum:* Given n numbers, store the numbers in global registers $M_1, \ldots, M_n$. Use processors $P_{i,j}, 1 \leq i, j \leq n$. Then execute the following program:

> **par**$[1 \leq i, j \leq n] P_{i,j} : M_{i+n} := 0$
> **par**$[1 \leq i, j \leq n] P_{i,j} :$ **if** $M_i < M_j$ **then** $M_{i+n} := 1$
> $\{M_{i+n} = 0$ if and only if $M_i = \max\{M_1, \ldots, M_n\}\}$
> **par**$[1 \leq i, j \leq n] P_{i,j} :$ **if** $M_{n+i} = 0$ **then** $M_0 := M_i$

This program is much faster than the EREW PRAM program, because all pairs are compared simultaneously in a single parallel step.

1062   Chapter 16   THEORETICAL COMPUTER SCIENCE

4. *Game of Life*: The Game of Life, invented at Cambridge by John H. Conway, a mathematician now at Princeton University, is played on an infinite checkerboard. The *neighbors* of a square are the eight squares that touch it, including those four at its corners. In the initial configuration c_0 of the game, some squares are regarded as *alive* and all others *dead*. Each configuration c_k gives birth to a new configuration c_{k+1}, according to the following rules:

- a live cell in configuration c_k remains alive if its has either two or three live neighbors, but no more;
- a dead cell in configuration c_k becomes alive if and only if it has exactly three live neighbors.

The following sequence of configurations illustrates these rules:

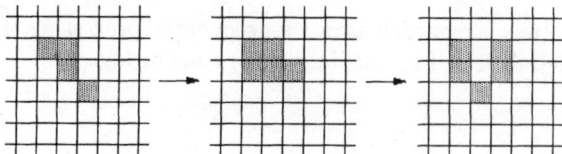

The Game of Life can be regarded as a cellular automaton in which the squares are the processors and each processor is joined to its eight neighbors.

5. A configuration in the Game of Life has *periodicity* n if the sequence of configurations to which its gives birth repeats every n configurations, and if n is the smallest such number. Here are three periodic configurations.

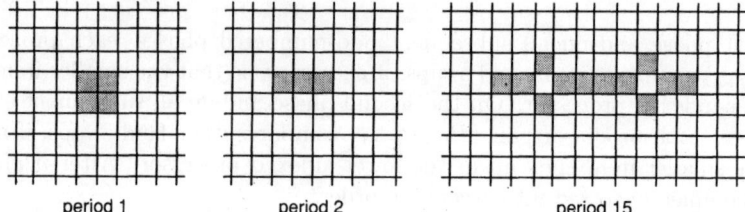

period 1   period 2   period 15

16.2   COMPUTABILITY

The theory of computability is concerned with distinguishing what can be computed from what cannot. This is not a question of skill at performing calculations. The remarkable truth is that the impossibility of computing certain functions can be proved from the definition of what it means to compute a function.

16.2.1   RECURSIVE FUNCTIONS AND CHURCH'S THESIS

The implicit domain for the theory of computability is the set $\mathcal{N}$ of natural numbers. The encoding of problems concerned with arbitrary objects into terms of natural numbers permits general application of this theory.

Definitions:

The n-place **constant zero function** is the function $\zeta_n(x_1,\ldots,x_n) = 0$.

The **successor function** is the function $\sigma(n) = n+1$.

The ith n-place **projection function** is the function $\pi_i^n(x_1,\ldots,x_n) = x_i$.

The (**multivariate**) **composition** of the n-place function $f(x_1,\ldots,x_n)$ and the n m-place functions $g_1(x_1,\ldots,x_m),\ldots,g_n(x_1,\ldots,x_m)$ is the m-place function
$$h(x_1,\ldots,x_m) = f(g_1(x_1,\ldots,x_m),\ldots,g_n(x_1,\ldots,x_m)).$$

(**Multivariate**) **primitive recursion** uses a previously defined $(n+2)$-place function $f(x_1,\ldots,x_{n+2})$ and a previously defined n-place function $g(x_1,\ldots,x_n)$ to define the following new $(n+1)$-place function:
$$h(x_1,\ldots,x_{n+1}) = \begin{cases} g(x_1,\ldots,x_n) & \text{if } x_{n+1} = 0; \\ f(x_1,\ldots,x_{n+1},h(x_1,\ldots,x_n,x_{n+1}-1)) & \text{otherwise.} \end{cases}$$

Unbounded minimalization uses an $(n+1)$-place function $f(x_1,\ldots,x_{n+1})$ to define the following new n-place function, which is denoted $\mu_m[f(x_1,\ldots,x_n,m)=0]$:
$$g(x_1,\ldots,x_{n+1}) = \begin{cases} \text{the least } y \text{ such that } f(x_1,\ldots,x_n,y)=0 & \text{if it exists;} \\ 0 & \text{otherwise.} \end{cases}$$

An $(n+1)$-place function $f(x_1,\ldots,x_{n+1})$ is **regular** if for every n-tuple $(x_1,\ldots,x_n)$ there is a $y \in \mathcal{N}$ such that $f(x_1,\ldots,x_n,y) = 0$.

The class $\mathcal{P}$ of **primitive recursive functions** is the smallest class of functions that contains:
- the constant zero functions $\zeta_n(x_1,\ldots,x_n) = 0$, for all $n \in \mathcal{N}$;
- the successor function $\sigma(n) = n+1$;
- the projection functions $\pi_i^n(x_1,\ldots,x_n) = x_i$, for all $n \in \mathcal{N}$ and $1 \leq i \leq n$;

and is closed under multivariate composition and multivariate primitive recursion.

The class $\mathcal{RF}$ of **recursive functions** is the smallest class of functions that contains:
- the constant zero functions $\zeta_n(x_1,\ldots,x_n) = 0$, for all $n \in \mathcal{N}$;
- the successor function $\sigma(n) = n+1$;
- the projection functions $\pi_i^n(x_1,\ldots,x_n) = x_i$, for all $n \in \mathcal{N}$ and $1 \leq i \leq n$;

and is closed under multivariate composition, multivariate primitive recursion, and the application of unbounded minimalization to regular functions.

A **recursive function** is a function in $\mathcal{RF}$.

Church's thesis, or the **Church-Turing thesis**, is the premise that recursive functions and Turing machines are capable of representing every function that is computable or partially computable.

A **partial function** on $\mathcal{N}$ is a function whose values are possibly undefined for certain natural numbers.

A partial function on $\mathcal{N}$ is called **total** if it is defined on every natural number.

The class $\mathcal{PR}$ of **partial recursive functions** is the smallest class of partial functions that contains the constant zero functions ζ, the successor function σ, and the projection functions π_i^n, and is closed under multivariate composition, multivariate primitive recursion, and the arbitrary application of unbounded minimalization.

A **partial recursive function** is a function in $\mathcal{PR}$.

A partial recursive function f is **represented by the Turing machine** M if machine M calculates the value $f(n)$ for every number n on which f is defined and fails to halt for every number on which f is undefined.

The **Ackermann function** $A \colon \mathcal{N} \times \mathcal{N} \to \mathcal{N}$ is defined as follows:
$$A(0, j) = j + 1;$$
$$A(i + 1, 0) = A(i, 1);$$
$$A(i + 1, j + 1) = A(i, A(i + 1, j)).$$

Facts:

1. The standard integer functions of arithmetic, including addition, subtraction, multiplication, division, and exponentiation, are all primitive recursive functions.

2. A function is a partial recursive function if and only if it can be represented by a Turing machine.

3. There are several other models of computation that are equivalent to partial recursive functions and to Turing machines, including labeled Markov algorithms and Post production systems (see [BrLa74]).

4. Church's thesis identifies formal concepts (recursive functions and Turing machines) with the intuitive concept of what is computable, so it is not something that is subject to proof.

5. Church's thesis is often invoked in the proof of theorems about computable functions to avoid dealing with low-level details of the model of computation.

6. The Ackermann function is recursive but not primitive recursive.

7. The Ackermann function grows faster than any primitive recursive function, in the following sense. For every primitive recursive function $f(n)$, there is an integer n_0 such that $f(n) < A(n, n)$ for all $n > n_0$.

Examples:

1. Addition is primitive recursive.
$$a(x, 0) = \pi_1^1(x);$$
$$a(x, y + 1) = \sigma(\pi_3^3(x, y + 1, a(x, y)).$$
Then $a(x, y) = x + y$.

2. Multiplication is primitive recursive.
$$m(x, 0) = 0;$$
$$m(x, y + 1) = a(m(x, y), \pi_2^2(x, y)), \text{ where } a(x, y) \text{ is addition.}$$
Then $m(x, y) = x \cdot y$.

3. Predecessor is primitive recursive.
$$p(0) = 0.$$
$$p(x + 1) = \pi_1^1(x).$$
Then $p(x) = x \dot{-} 1$.

4. Nonnegative subtraction is primitive recursive.
$$s(x, 0) = \pi_1^1(x).$$
$$s(x, y + 1) = p(x \dot{-} y).$$
Then $s(x, y) = x \dot{-} y$.

5. The function $p(n)$ = the nth prime number is a primitive recursive function.
6. The Ackermann function is recursive but not primitive recursive.

16.2.2 RECURSIVE SETS AND SOLVABLE PROBLEMS

Definitions:

The **characteristic function** of a set A is the function
$$f(x) = \begin{cases} 1 & \text{if } x \in A; \\ 0 & \text{if } x \notin A. \end{cases}$$

A set A is **recursive** if its characteristic function is recursive.

A problem is (**computationally**) **solvable** if it can be represented as a membership problem that can be decided by a recursive function.

A set $A \subseteq \mathcal{N}$ is **recursively enumerable** (**r.e.**) if $A = \emptyset$ or A is the image of a recursive function.

A **Gödel numbering** of a set S is a one-to-one recursive function $g \colon S \to \mathcal{N}$ whose image in $\mathcal{N}$ is a recursive set.

Facts:

1. If a set A and its complement $\overline{A}$ are both recursively enumerable, then A is recursive.
2. If an recursively enumerable set A is the image of a non-decreasing function, then A is recursive.
3. A set is recursively enumerable if and only if it is the image of a partial recursive function.
4. A set is recursively enumerable if and only if it is the domain of a partial recursive function.
5. The set of Turing machines has a Gödel numbering.

Examples:

1. Every finite set of numbers is recursive.
2. The prime numbers are a recursive set.
3. The problem of deciding which Turing machines halt on all inputs is unsolvable. The set of Gödel numbers for these Turing machines is neither recursive not recursively enumerable
4. For any fixed $c \in \mathcal{N}$, the problem of deciding which Turing machines halt when the number c is supplied as input is unsolvable. The set of Gödel numbers for these Turing machines is recursively enumerable, but not recursive.
5. The problem of deciding which Turing machines halt when their own Gödel number is supplied as input is unsolvable. The set of Gödel numbers for these Turing machines is recursively enumerable, but not recursive.
6. *Hilbert's tenth problem*: Hilbert's tenth problem (posed in 1900) was the problem of devising an algorithm to determine, given a polynomial $p(x_1, \ldots, x_n)$ with integer coefficients, whether there exists an integer root. Y. Matiyasevich proved in 1970 that no such algorithm exists. That is, the set of polynomials with integer coefficients that have an integer solution is not recursive. Hilbert's tenth problem is called a "natural example" of an unsolvable problem, since the concepts used to define it are not from within computability theory (i.e., unlike problems concerned with the behavior of Turing machines). [Ma93]

16.3 LANGUAGES AND GRAMMARS

Strings of symbols are a general way to represent information, both in written text and in a computer. A language is a set of strings that are used within some domain of discourse, and a grammar is a system for generating a language. A grammar is what enables a compiler to determine whether a body of code is syntactically correct in a given computer language. Formal language theory is concerned with languages, grammars, and rudiments of combinatorics on strings. The range of applications of formal language theory extends from natural and programming languages, developmental biology, and computer graphics to semiotics, artificial intelligence, and artificial life.

16.3.1 ALPHABETS

Definitions:

An *alphabet* is a finite nonempty set.

A *symbol* is an element of an alphabet.

A *string* in an alphabet is a finite sequence of symbols over that alphabet.

A *word* in an alphabet is a finite or countably infinite sequence of symbols over that alphabet.

The *empty string* λ is the string of length zero, that is, the string with no symbols.

The *length of a string* w is the number of symbols in w, denoted $|w|$.

The *frequency* $|w|_a$ of a symbol a in a string w is the number of occurrences of a in string w.

A *substring* of a string w is a sequence of consecutive symbols that occurs in w.

A *subword* of a word w is a sequence of consecutive symbols that occurs in w.

A *prefix* of a string w is a substring that starts at the leftmost symbol.

A *suffix* of a string w is a substring that ends at the rightmost symbol.

The *reverse* or *mirror image* x^R of the string $x = a_1 a_2 \ldots a_n$, is the string $a_n \ldots a_2 a_1$.

A *palindrome* is a string that is identical to its reverse.

A *pseudopalindrome* in an alphabet (such as English) that includes punctuation symbols (such as comma, hyphen, or blank) is a word that becomes a palindrome when all of its punctuation symbols are deleted.

The *concatenation* xy of two strings $x = a_1 a_2 \ldots a_m$ and $y = b_1 b_2 \ldots b_n$ is the string $a_1 a_2 \ldots a_m b_1 b_2 \ldots b_n$ obtained by appending string y to the right of string x.

The *nth power of a string* w, denoted w^n, is the concatenation of n copies of w.

The *shuffle* $x ⧢ y$ of two strings $x = x_1 \ldots x_n$ and $y = y_1 \ldots y_n$ is the string $x_1 y_1 \ldots x_n y_n$.

Facts:

1. A symbol of an alphabet is usually conceptualized as something that it can be represented by a single byte or by a written character

2. A finite word is a string.

3. The sum of the frequencies $|w|_a$ taken over all the symbols a in the alphabet equals the length of the string w.

4. The length $|xy|$ of a concatenation equals the sum $|x|+|y|$ of the lengths of the strings x and y from which it is formed.

5. The length of the nth power of a string w is $n \cdot |w|$.

6. When a pseudopalindrome occurs in a natural language, it is commonly called a palindrome.

Examples:

1. The English alphabet includes lower and upper case English letters, the blank symbol, the digits $0, 1, \ldots, 9$, and various punctuation symbols.

2. ASCII (*American Standard Code for Information Interchange*) is an alphabet of size 128 for many common computer languages. See Table 1.

3. ABLE WAS I ERE I SAW ELBA is a palindrome.

4. The names EVE, HANNAH, and OTTO are palindromes.

5. MADAM I'M ADAM and SIX AT-NOON TAXIS are pseudopalindromes.

6. A list of palindromes can be found on the website
   http://freenet.buffalo.edu/~cd431/palindromes.html

7. The third power of the string 011 is 011011011.

8. The concatenation of BOOK and KEEPER is BOOKKEEPER.

9. The shuffle of FLOOD and RIVER is FRLIOVOEDR.

16.3.2 LANGUAGES

Definitions:

The **free monoid** V^* generated by the alphabet V is the structure whose domain is the set of all strings composable from symbols over V, with the semigroup operation of string concatenation.

A **(formal) language** on the alphabet V is a subset L of the free monoid V^*.

The **λ-free semigroup** V^+ on an alphabet V is the set $V^* - \{\lambda\}$, with the concatenation operation.

A **λ-free language** on the alphabet V is a subset of the λ-free semigroup V^+.

The **length set** of a language L is the set $length(L) = \{ |x| \mid x \in L \}$.

The **concatenation** $L_1 L_2$ of two languages is the set $\{ xy \mid x \in L_1, y \in L_2 \}$.

The **ith power** of a language L is the language L^i defined recursively by the rule $L^0 = \{\lambda\}$ and $L^{i+1} = L^i L$, $i \geq 0$.

The **Kleene closure** (or **Kleene star**) L^* of a language L is the union $\bigcup_{i \geq 0} L^i$ of all its powers.

The **positive closure** (or **Kleene plus**) L^+ of a language L is the union $\bigcup_{i \geq 1} L^i$ of all its powers excluding the zeroth power.

The **union** of two languages L_1 and L_2 is $L_1 \cup L_2$, using the usual set operation.

The **intersection** of two languages L_1 and L_2 is $L_1 \cap L_2$, using the usual set operation.

Table 1 ASCII codes.

000 0000	NUL	010 0000	SP	100 0000	@	110 0000	`
000 0001	SOH	010 0001	!	100 0001	A	110 0001	a
000 0010	STX	010 0010	"	100 0010	B	110 0010	b
000 0011	ETX	010 0011	#	100 0011	C	110 0011	c
000 0100	EOT	010 0100	$	100 0100	D	110 0100	d
000 0101	ENQ	010 0101	%	100 0101	E	110 0101	e
000 0110	ACK	010 0110	&	100 0110	F	110 0110	f
000 0111	BEL	010 0111	'	100 0111	G	110 0111	g
000 1000	BS	010 1000	(	100 1000	H	110 1000	h
000 1001	HT	010 1001	)	100 1001	I	110 1001	i
000 1010	LF	010 1010	*	100 1010	J	110 1010	j
000 1011	VT	010 1011	+	100 1011	K	110 1011	k
000 1100	FF	010 1100	,	100 1100	L	110 1100	l
000 1101	CR	010 1101	-	100 1101	M	110 1101	m
000 1110	SO	010 1110	.	100 1110	N	110 1110	n
000 1111	SI	010 1111	/	100 1111	O	110 1111	o
001 0000	DLE	011 0000	0	101 0000	P	111 0000	p
001 0001	DC1	011 0001	1	101 0001	Q	111 0001	q
001 0010	DC2	011 0010	2	101 0010	R	111 0010	r
001 0011	DC3	011 0011	3	101 0011	S	111 0011	s
001 0100	DC4	011 0100	4	101 0100	T	111 0100	t
001 0101	NAK	011 0101	5	101 0101	U	111 0101	u
001 0110	SYN	011 0110	6	101 0110	V	111 0110	v
001 0111	ETB	011 0111	7	101 0111	W	111 0111	w
001 1000	CAN	011 1000	8	101 1000	X	111 1000	x
001 1001	EM	011 1001	9	101 1001	Y	111 1001	y
001 1010	SUB	011 1010	:	101 1010	Z	111 1010	z
001 1011	ESC	011 1011	;	101 1011	[	111 1011	}
001 1100	FS	011 1100	<	101 1100	\	111 1100	|
001 1101	GS	011 1101	=	101 1101	]	111 1101	}
001 1110	RS	011 1110	>	101 1110	^	111 1110	~
001 1111	US	011 1111	?	101 1111	_	111 1111	DEL

Control codes: ACK: acknowledge, BEL: bell, BS: backspace, CAN: cancel, CR: carriage return, DC1-4: device controls, DEL: delete, DLE: data link escape, EM: end of medium, ENQ: enquiry, EOT: end of transmission, ESC: escape, ETB: end of transmission block, ETX: end of text, FF: form feed, FS: file separator, GS: group separator, HT: horizontal tab, LF: line feed, NAK: negative acknowledgment, NUL: null, RS: record separator, SI: shift in, SO: shift out, SOH: start of heading, SP: space, STX: start of text, SUB: substitute, SYN: synchronous/idle, US: united separator, VT: vertical tab

The **complement** of a language L over an alphabet V is the language $\overline{L}$, where complementation is taken with respect to the free monoid V^* as the universe of discourse.

A language is **regular** if it is any of the languages $\emptyset$, $\{\lambda\}$, or $\{b\}$, where b is a symbol of its alphabet, or if it can be obtained by applying the operations of union, concatenation and Kleene star finitely many times to one of those languages.

The **shuffle** $L_1 ⧢ L_2$ of two languages L_1, L_2 is the language $\{\, w \in V^* \mid w = x ⧢ y$, for some $x \in L_1, y \in L_2 \,\}$.

The **mirror image** $mi(L)$ of the language L is the language $\{\, x^R \mid x \in L \,\}$. It is also called the **reverse** of the language L.

The **left quotient** of the language L_1 with respect to the language L_2 on the same alphabet V, is the language $L_2 \backslash L_1$ containing every string of V^* that can be obtained from a string in L_1 by erasing a prefix from L_2. That is, $L_2 \backslash L_1 = \{\, w \in V^* \mid$ there is $x \in L_2$ such that $xw \in L_1 \,\}$.

The **left derivative** of language L with respect to the string x over the same alphabet V is the language $\partial_x(L) = \{x\} \backslash L$.

The **right quotient** is the notion symmetric to left quotient.

The **right derivative** is the notion symmetric to left derivative.

A **substitution** for the alphabet V in the alphabet U is a mapping $s: V \longrightarrow 2^{U^*}$. This means that each symbol $b \in V$ may be replaced by any of the strings in the set $s(b)$.

A **finite substitution** is a substitution such that the replacement set $s(a)$ for each symbol $a \in V$ is finite.

The **extension of a substitution** $s: V \longrightarrow 2^{U^*}$ from its domain alphabet V to the set V^* of strings over V is given by the rules $s(\lambda) = \{\lambda\}$ and $s(ax) = s(a)s(x)$, for $a \in V, x \in V^*$.

A **morphism from the alphabet** V to the alphabet U is a substitution $s: V \longrightarrow 2^{U^*}$ such that the replacement set $s(a)$ for every symbol $a \in V$ is a singleton set.

A **λ-free substitution** is a substitution such that λ is never substituted for a symbol. That is, $\lambda \notin s(a)$, for every symbol $a \in V$.

A **λ-free morphism** is a morphism such that $s(a) \neq \{\lambda\}$, for every symbol $a \in V$.

The **extension of a substitution** $s: V \longrightarrow 2^{U^*}$ to the language $L \subseteq V^*$ is the language $s(L) = \bigcup_{x \in L} s(x)$ that contains every string in U^* obtainable from a string in L by making replacements permissible under substitution s.

The **inverse of a morphism** $h: V^* \longrightarrow U^*$ is the mapping $h^{-1}: U^* \longrightarrow 2^{V^*}$ defined by $h^{-1}(x) = \{\, y \in V^* \mid h(y) = x \,\}$, $x \in U^*$.

A family of languages is **nontrivial** if it contains at least one language different from $\emptyset$ and $\{\lambda\}$.

Facts:

1. The set of all binary strings with at least as many 1s as 0s is a language.

2. The set of all binary strings in which no two occurrences of 1 are consecutive is a language.

3. Some strings of a natural language such as English are categorized as nouns, verbs, and adjectives. Other more complicated strings are categorized as sentences.

4. Some strings of common computer languages are categorized as identifiers and arithmetic expressions. Other more complicated strings are categorized as statements, with subcategories such as assignment statements, if-statements, and while-statements.

Examples:

1. Natural languages and computer languages are formal languages.

2. The Kleene closure of the language $\{00, 01, 10, 11\}$ is the language of all strings of even length.

3. The left derivative {bee}\English includes the following strings: f, n, p, r, s, t, tle, ts, keeper, swax, feater, ping.

4. The substitution $0 \mapsto \{00, 01\}, 1 \mapsto \{10, 11\}$ over the free monoid $\{0, 1\}^*$ is the language of all strings of even length.

5. Given the alphabet $\{a, b\}$, define the morphism $\phi: \{a, b\} \longrightarrow \{a, b\}^*$ by the replacements $\phi(a) = ab$ and $\phi(b) = ba$, and define the string w_n by the recursion $w_0 = a$ and $w_{n+1} = \phi(w_n)$. Then,
$$w_1 = ab, \quad w_2 = abba, \quad w_3 = abbabaab, \quad w_4 = abbabaabbaababba, \ldots$$

6. In Example 5 each word w_n is a prefix of the next word w_{n+1}. The *Thue ω-word* is the infinite word $\lim_{n \to \infty} w_n$.

7. Given the alphabet $\{a, b\}$, define the morphism $\rho: \{a, b\} \longrightarrow \{a, b\}^*$ by the replacements $\rho(a) = ab$ and $\rho(b) = a$, and define the string w_n by the recursion $w_0 = a$ and $w_{n+1} = \phi(w_n)$. Then,
$$w_1 = ab, \quad w_2 = aba, \quad w_3 = abaab, \quad w_4 = abaababa, \ldots$$

8. In Example 7 each word w_n is a prefix of the next word w_{n+1}. The *Fibonacci ω-word* is the infinite word $\lim_{n \to \infty} w_n$.

9. A language is regular if and only if it is the language of strings accepted by some finite state recognizer.

16.3.3 GRAMMARS AND THE CHOMSKY HIERARCHY

Definitions:

A **phrase-structure grammar** (or **unrestricted grammar** or **type 0 grammar**) is a quadruple $G = (N, T, S, P)$ such that:
- N is a finite nonempty alphabet of symbols called **nonterminals**;
- T is a finite nonempty alphabet, disjoint from N, of symbols called **terminals**;
- S is a nonterminal called the **start symbol**;
- P is a finite set of **production rules** of the form $\alpha \to \beta$, where α is a string in $N \cup T$ that contains at least one nonterminal and β is a string in $N \cup T$.

The **antecedent** of a production $\alpha \to \beta$ is α.

The **consequent** of a production $\alpha \to \beta$ is β.

The string y is **directly derivable** from the string x with respect to the grammar G if there is a production rule $u \to v \in P$ and if there are strings $w_1, w_2 \in (N \cup T)^*$ such that $x = w_1 u w_2$ and $y = w_1 v w_2$.

The **direct derivability relation** $x \Longrightarrow_G y$ (or $x \Longrightarrow y$, when the grammar G is implicitly understood) means that y is directly derivable from string x.

A **derivation** of the string y from the string x is a sequence of direct derivations $x \Longrightarrow z_1, z_1 \Longrightarrow z_2, \ldots, z_n \Longrightarrow y$. This is sometimes called **parsing**.

The string y is **derivable** from the string x with respect to the grammar G if there is a derivation of y from x. Notation: $x \Longrightarrow^* y$.

The **Chomsky normal form for a production rule** is $A \to BC$, where B and C are nonterminals or the form $A \to a$ where a is a terminal.

The **language generated by the grammar** G is the language $L(G) = \{x \in T^* \mid S \Longrightarrow^* x\}$.

Grammars G_1 and G_2 are **equivalent** if $L(G_1) = L(G_2)$.

A **leftmost derivation** $x \Longrightarrow_{left} y$ is a derivation $x \Longrightarrow y$ in which at each step the leftmost nonterminal is replaced.

The **leftmost language generated by the grammar** G is the language $L_{left}(G)$ of strings of terminals with leftmost derivations from the start symbol S.

A grammar $G = (N, T, S, P)$ is **length-increasing** (or of **type 1**) if $|u| \leq |v|$ for all $u \to v \in P$. (However, the production $S \to \lambda$ is allowed, provided that S does not appear in the consequents of rules in P.)

A grammar $G = (N, T, S, P)$ is **context-sensitive** if for each production $u \to v \in P$, the antecedent and consequent have the form $u = u_1 A u_2$ and $v = u_1 x u_2$, for $u_1, u_2 \in (N \cup T)^*, A \in N, x \in (N \cup T)^+$. (The production $S \to \lambda$ is allowed, provided that S does not appear in the right-hand members of rules in P.)

A grammar $G = (N, T, S, P)$ is **context-free** (or of **type 2**) if the antecedent of each production $u \to v \in P$ is a nonterminal.

An **L-system** is a production-based model for growth and life development.

A grammar $G = (N, T, S, P)$ is **monotonic** if the consequent of each production (except possibly $S \to \lambda$) has at least as many symbols as the antecedent, and S does not occur in any consequent.

A grammar $G = (N, T, S, P)$ is **linear** if each production $u \to v \in P$ has its antecedent $u \in N$ and its consequent $v \in T^* \cup T^*NT^*$.

A grammar $G = (N, T, S, P)$ is **right-linear** if each production $u \to v \in P$ has $u \in N$ and $v \in T^* \cup T^*N$.

A grammar $G = (N, T, S, P)$ is **left-linear** if each production $u \to v \in P$ has $u \in N$ and $v \in T^* \cup NT^*$.

A grammar $G = (N, T, S, P)$ is **regular** (or **type 3**) if each rule $u \to v \in P$ has $u \in N$ and $v \in T \cup TN \cup \{\lambda\}$.

Given a class of grammars, there are some basic decision problems about arbitrary grammars G_1, G_2 in the class:

equivalence: are the grammars G_1 and G_2 equivalent?
inclusion: is the language $L(G_1)$ included in the language $L(G_2)$?
membership: given an arbitrary string x, is x an element of $L(G_1)$?
emptiness: is the language $L(G_1)$ empty?
finiteness: is the language $L(G_1)$ finite?
regularity: is $L(G_1)$ a regular language? (see §16.3.2)

The **recursive** languages are the languages with a decidable membership question.

The various classes of languages are denoted as follows:

RE (type 0): the class of all unrestricted languages;
CS (type 1): the class of all context-sensitive languages;
CF (type 2): the class of all context-free languages;
LIN: the class of all linear languages;
REG (type 3): the class of all regular languages.

Table 2 Closure properties for Chomsky hierarchy classes.

	RE	CS	CF	LIN	REG
union	yes	yes	yes	yes	yes
intersection	yes	yes	no	no	yes
complement	no	yes	no	no	yes
concatenation	yes	yes	yes	no	yes
Kleene star	yes	yes	yes	no	yes
intersection with regular languages	yes	yes	yes	yes	yes
substitution	yes	no	yes	no	yes
λ-free substitution	yes	yes	yes	no	yes
morphisms	yes	no	yes	yes	yes
λ-free morphisms	yes	yes	yes	yes	yes
inverse morphisms	yes	yes	yes	yes	yes
left/right quotient	yes	no	no	no	yes
left/right quotients with regular languages	yes	no	yes	yes	yes
left/right derivative	yes	yes	yes	yes	yes
shuffle	yes	yes	no	no	yes
mirror image	yes	yes	yes	yes	yes

Facts:

1. *Chomsky hierarchy*: The following strict inclusions hold:
$$\mathbf{REG} \subset \mathbf{LIN} \subset \mathbf{CF} \subset \mathbf{CS} \subset \mathbf{RE}.$$

2. The language of an unrestricted grammar is recursively enumerable (RE).

3. **CS** (context sensitive) $\subset$ {recursive languages} $\subset$ **RE** (unrestricted).

4. The class of languages generated by monotonic grammars is identical to the class of languages generated by context-sensitive grammars.

5. L-systems were introduced in 1968 by Aristid Lindenmayer (1922–1990), a Dutch biologist, to model the development of some plant systems. (See [Gr97].)

6. The classes of languages generated by right-linear or by left-linear grammars coincide. This class is identical to the family of languages generated by regular grammars, as well as to the class of regular languages (§16.3.2).

7. $L_{left}(G) \in \mathbf{CF}$ (context-free) for each type-0 grammar G.

8. If G is a context-free grammar, then $L_{left}(G) = L(G)$.

9. Let G be a context-free grammar. Then there is a grammar $G' = (N, T, S, P)$, with every rule in Chomsky normal form. Moreover, there is constructive method for transforming grammar G into the grammar G'.

10. *Rice's theorem*: Let P be a nontrivial property of recursively enumerable languages (i.e., a property such that there exists at least one grammar having property P and at least one grammars not having property P). Then property P is undecidable.

11. A language is context-free if and only if it is the language accepted by some (possibly nondeterministic) pushdown automaton.

12. The following table summarizes the decidability properties of the grammar classes in the Chomsky hierarchy. In this table U stands for *undecidable*, D for *decidable*, and T for *trivial*.

	RE (type 0)	CS (type 1)	CF (type 2)	LIN	REG (type 3)
equivalence	U	U	U	U	D
inclusion	U	U	U	U	D
membership	U	D	D	D	D
emptiness	U	U	D	D	D
finiteness	U	U	D	D	D
regularity	U	U	U	U	T
intersection	yes	yes	no	no	yes
complement	no	yes	no	no	yes

Examples:

1. In the grammar $G = (N, T, S, P)$, where $N = \{S, x, y\}$, $T = \{0, 1\}$, and $P = \{S \to 0S1, S \to \lambda\}$, a derivation of the string 0011 is $S \Longrightarrow 0S1 \Longrightarrow 00S11 \Longrightarrow 0011$.

2. The following are examples of languages generated by grammar $G = (N, T, S, P)$ with $N = \{S, x, y, z\}$, $T = \{0, 1, 2\}$, and the following sets P of productions:

production set P	language $L(G)$	class
$S \to 0x, x \to 1y, y \to 0x, x \to 1, y \to \lambda$	$\{01, 0101, 010101, \ldots\}$	regular
$S \to \lambda, S \to 0x, S \to 01, x \to S1$	$\{0^n 1^n \mid n \geq 0\}$	linear
$S \to \lambda, S \to 0Sx2, 2x \to x2, 0x \to 01, 1x \to 11$	$\{0^n 1^n 2^n \mid n \geq 0\}$	unrestricted

16.3.4 REGULAR AND CONTEXT-FREE LANGUAGES

Definitions:

Given an alphabet V, a **regular expression** over V is a string w over the alphabet $V \cup \{\epsilon,), (, +, *\}$ that has one of the following forms:
- $w \in V$ or $w = \epsilon$;
- $w = (\alpha\beta)$, where α and β are regular expressions;
- $w = (\alpha + \beta)$, where α and β are regular expressions;
- $w = \alpha^*$, where α is a regular expression.

The set of all regular expressions over alphabet V is denoted $\mathbf{rex}_V$.

The function L maps $\mathbf{rex}_V$ to the set of all languages over the alphabet V, using the following rules:
- $L(\epsilon) = \emptyset$, and $L(a) = \{a\}$ for all $a \in V$;
- $L((\alpha\beta)) = L(\alpha)L(\beta)$, $L((\alpha + \beta)) = L(\alpha) \cup L(\beta)$, and $L(\alpha^*) = (L(\alpha))^*$.

A context-free grammar G is **ambiguous** if there is string $x \in L(G)$ having two different leftmost derivations in G.

A context-free language L is **inherently ambiguous** if every context-free grammar of L is ambiguous; otherwise, language L is called **unambiguous**.

Facts:

1. *Kleene theorem*: A language L is regular if and only if there is a regular expression e such that $L = L(e)$.

2. Every context-free language over a one-letter alphabet is regular.

3. Every regular language L can be represented in the form $L = h_4(h_3^{-1}(h_2(h_1^{-1}(a^*b))))$, where h_1, h_2, h_3, h_4 are morphisms.

4. Each regular language is unambiguous.

5. There are inherently ambiguous linear languages.

6. The ambiguity problem for context-free grammars is undecidable.

7. The length set of a context-free language is a finite union of arithmetical progressions.

8. Every language L can be represented in the form $L = h(L_1 \cap L_2)$, as well as in the form $L = L_3 \backslash L_4$, where h is a morphism and L_1, L_2, L_3, L_4 are linear languages.

9. *Pumping lemma for regular languages*: If L is a regular language over the alphabet V, then there are numbers p and q such that every string $z \in L$ with length $|z| > p$ can be written in the form $z = uvw$, with $u, v, w \in V^*$, where $|uv| \leq q, v \neq \lambda$, so that $uv^iw \in L$ for all $i \geq 0$.

10. *Pumping lemma for linear languages*: If L is a linear language on the alphabet V, then there are numbers p and q such that every string $z \in L$ with length $|z| > p$ can be written in the form $z = uvwxy$, with $u, v, w, x, y \in V^*$, where $|uvxy| \leq q$ and $vx \neq \lambda$, so that $uv^iwx^iy \in L$ for all $i \geq 0$.

11. *Bar-Hillel ($uvwxy$, pumping) lemma for context-free languages*: If L is a context-free language over the alphabet V, then there are numbers p and q such that every string $z \in L$ with length $|z| > p$ can be written in the form $z = uvwxy$, with $u, v, w, x, y \in V^*$, where $|vwx| \leq q$ and $vx \neq \lambda$, so that $uv^iwx^iy \in L$ for all $i \geq 0$.

12. *Ogden pumping lemma (pumping with marked positions)*: If L is a context-free language on the alphabet V, then there is a number p such that for every string $z \in L$ and for every set of at least p marked occurrences of symbols in z, we can write $z = uvwxy$, where:

    - either each of u, v, w or each of w, x, y contains at least one marked symbol;
    - vwx contains at most p marked symbols;
    - $uv^iwx^iy \in L$ for all $i \geq 0$.

13. Let G be a context-free grammar G. Then there is a grammar $G' = (N, T, S, P)$, with every rule in P of the form $A \to a\alpha$, for $A \in N, a \in T, \alpha \in (N \cup T)^*$, such that $L(G') = L(G) - \{\lambda\}$. Moreover, there is constructive method for transforming grammar G into the grammar G', which is said to be in the *Greibach normal form*.

14. Let G be a context-free grammar G and (k,l,m) a triple of nonnegative integers. Then an equivalent grammar $G' = (N,T,S,P)$ can be effectively constructed whose every rule is in one of the following two forms:
   - $A \to xByCz$, with $A,B,C \in N, x,y,z \in T^*$, and $|x| = k, |y| = l, |z| = m$;
   - $A \to x$, with $A \in N, x \in T^*, |x| \in length(L(G))$.

Such a grammar G' is said to be in *super normal form*.

15. Variants of the Chomsky and Greibach normal forms can be obtained by particularizing the parameters k, l, m in the super normal form.

Examples:

1. The following are some regular expressions over $\{0,1\}$ and the languages they represent:

1^*	all strings with no 0s
1^*01^*	all strings with exactly one 0
$1^*(0+\epsilon)1^*$	all strings with one or no 0s
$(0+1)(0+1)$	all strings of length 2
$(0+1)(0+1+\epsilon)$	all strings of length 1 or 2.

2. *Backus-Naur form* (*BNF*) (or *Backus normal form*) for specifying computer language syntax uses context-free production rules. Nonterminals are enclosed in brackets; the symbol ::= is used in place of →; and all the consequents of the same antecedent are written in the same statement, with the alternative consequents separated by vertical bars. For instance, in some programming languages, this might be the BNF for the lexical token called an identifier.

    ⟨identifier⟩ ::== ⟨letter⟩|⟨letter⟩⟨alphameric string⟩
    ⟨letter⟩ ::= a|b|c|d|e|f|g|h|i|j|k|l|m|n|o|p|q|r|s|t|u|v|w|x|y|z
    ⟨alphameric string⟩ ::= ⟨alphameric⟩|⟨alphameric string⟩⟨alphameric⟩
    ⟨alphameric⟩ ::= ⟨letter⟩|⟨digit⟩
    ⟨digit⟩ ::= 0|1|2|3|4|5|6|7|8|9

16.3.5 COMBINATORICS ON WORDS

Note: In this subsection, a *word* is taken to be finite.

Definitions:

A (*word*) *variable* over an alphabet V is a symbol (such as x or y) not in V whose values range over V^*.

A *pattern* in a word is a string of word variables.

A pattern is *present* in a word $w \in V^*$ if there exists an assignment of strings from V^+ to the variables in that pattern such that the word formed thereby is a substring of w.

A *square* is a word of the pattern "xx".

A *square-free* word is a word with no subwords of the pattern "xx".

A *cube* is a word of the pattern "xxx".

An *Abelian square* is a word of the form xx^p, where x^p is any permutation of the word x.

A *word equation* over an alphabet V is an expression of the form $\alpha = \beta$ such that α and β are words containing letters of an alphabet V and some variables over V.

A *word inequality* is the negation of a word equation, which is commonly written in the form $\alpha \neq \beta$.

A *solution* to a *system* S of (finitely many) word equations and word inequalities is a list of words whose substitutions for their respective variables converts every word equation and word inequality in the system into a true proposition.

A *code* is a nonempty language $C \subseteq V^+$ such that whenever a word w in V can be written as a catenation of words in C, the write-up is always unique. That is, if $w = x_1 \ldots x_m = y_1 \ldots y_n$, where $m, n \geq 1$, and $x_i, y_j \in C$, then $m = n$ and $x_i = y_i$ for $i = 1, \ldots, m$. This property is called *unique decodability*.

The *code indicator of a word* $w \in V^*$ is the number $ci(w) = |V|^{-|w|}$.

The *code indicator of a language* is the sum of the code indicators of all words in the language.

Facts:

1. Certain patterns are *unavoidable* in sufficiently long words.
2. Squares are avoidable in alphabets with three or more letters; that is, there are arbitrarily long square-free words.
3. Cubes are avoidable over two letter alphabets.
4. Although squares are avoidable in three letter alphabets, Abelian squares are unavoidable. Every word of length ≥ 8 over $V = \{a, b, c\}$ contains a subword of the form $xx^p, x \in V^+$, where x^p is a permutation of x.
5. Abelian squares are avoidable in alphabets with four or more letters.
6. It is *decidable* (by the so-called *Makanin's algorithm*) whether or not a system S of word equations and inequalities has a solution.
7. It is decidable whether or not a given finite language is a code.
8. Every code C satisfies the inequality $ci(C) \leq 1$.
9. If a language $C = \{w_1, \ldots, w_n\}$ over V is not a code then, according to the so-called *defect theorem*, the algebraic structure of C^* can be simulated by an alphabet with at most $n-1$ letters: the smallest free submonoid of V^* containing C is generated by at most $n-1$ words.
10. The following three conditions are equivalent for any two words u and v:
    - $\{u, v\}$ is not a code;
    - u and v are powers of the same word;
    - $uv = vu$.

(This is a corollary to Fact 9.)

11. For every word $w \in V^+$, there are a unique shortest word $\rho(w)$ and an integer $n \geq 1$ such that $w = (\rho(w))^n$. (The word $\rho(w)$ is called the *primitive root* of w.)
12. Lyndon's theorem: If $uv = vw$ with $u, v, w \in V^*$, then there exist words $x, y \in V^*$ and a number $n \geq 0$ such that $u = xy, w = yx$ and $v = (xy)^n x = x(yx)^n$.

13. If $uv = vu$ with $u, v \in V^+$, then $\rho(u) = \rho(v)$ and, consequently, u and v are powers of the same word. This is a corollary to Lyndon's theorem.

14. Assume that words u^m and v^n have a common prefix or suffix of length $|u|+|v|-d$, where $u, v \in V^+, m, n \geq 1$ and $d = \gcd(|u|, |v|)$. Then $\rho(u) = \rho(v)$ and $|\rho(u)| \leq d$. Thus, if $d = 1$ then u and v are powers of the same letter.

15. If $u^m = v^n$, where $m, n \geq 1$, then u and v are powers of the same word. (This is a corollary to Fact 14.)

16. If $u^m v^n = w^p$, where $m, n, p \geq 2$, then $\rho(u) = \rho(v) = \rho(w)$.

Examples:

1. In the alphabet $V = \{a, b\}$, the only square-free three-letter words are *aba* and *bab*. The two possible extensions of *aba* by one letter are *abaa*, which contains the square *aa*, and *abab*, which is a square. Similarly, both extensions of *bab* by one letter contain a square. Thus, squares are unavoidable in words of length ≥ 4 over two-letter alphabets.

2. All solutions for the system $xaba = abax, xx \neq x, x \neq aba$, over the alphabet $V = \{a, b\}$ are (by the corollary to Lyndon's theorem) of the form $x = (aba)^n, n \geq 2$.

16.4 ALGORITHMIC COMPLEXITY

The "complexity of an algorithm" has come to mean, most often, a measure of the computational effort or cost of execution, relative to the "size" of the problem. Other factors that may affect this kind of complexity are the characteristics of the particular input and the values returned by random number generators. The most common complexity measure is *running time*, but other measures, such as *space utilized* and *number of comparisons* are sometimes used. Another view of complexity focuses on the complicatedness of the algorithm, rather than on the effort needed to execute it.

16.4.1 OVERVIEW OF COMPLEXITY

To simplify discussion, it is assumed that every function and algorithm under consideration here has one argument. (Everything is easily generalized to multivariate functions by regarding the list of arguments as an n-tuple.)

Definitions:

A function $f: \mathcal{N} \to \mathcal{N}$ is **asymptotic** to a function $g: \mathcal{N} \to \mathcal{N}$ if $\lim_{n \to \infty} \frac{f(n)}{g(n)} = 1$. Notation: $f(n) \sim g(n)$. (See §1.3.3.)

The **input size** of the argument of an algorithm is either its numeric value or the number of bits required to specify a value of that argument.

A (**cost-based**) **complexity measure** for an algorithm is any of several different asymptotic measures of cost or difficulty in running that algorithm, relative to the input size. It is given in big-O notation (or sometimes in Θ-notation). (See §1.3.3.)

A ***time-complexity*** measure of an algorithm is a big-O expression for the number of operations or the running time needed for a complete execution of that algorithm, represented as a function of the size of the input.

A ***space-complexity*** measure of an algorithm is a big-O expression for the amount of computational space needed in the execution of that algorithm, represented as a function of the size of the input.

An algorithm runs in ***polynomial time*** if its time-complexity is dominated by a polynomial.

An algorithm runs in ***polynomial space*** if its space-complexity is dominated by a polynomial.

A ***Kolmogorov-Chaitin complexity measure*** of an algorithm is a measure based on the number of instructions of the algorithm, which is taken as an estimate of the logical complicatedness.

The ***time-complexity*** of a computable function is the minimum time-complexity taken over all algorithms that compute the function.

The ***parallel time-complexity*** of a computable function is the minimum time-complexity taken over all parallel algorithms that compute the function.

The ***space-complexity*** of a computable function is the minimum space-complexity taken over all algorithms that compute the function.

A ***decision function*** is a function on a countably infinite domain that decides whether an object is in some specified subset of that domain.

A computable decision function is in ***class P*** (*polynomial*) if its time-complexity is polynomial.

A computable decision function is in ***class NP*** (*nondeterministic polynomial*) if its parallel time-complexity is polynomial.

A function g ***reduces*** a decision function h to a decision function f if $h = f \circ g$.

A computable decision function f is ***NP-hard*** if every decision function in class NP can be reduced to f by a polynomial-time function.

A computable decision function is **NP-complete** if it is NP-hard and in class NP.

A ***tractable problem*** is a set membership problem with a decision function in class P.

Facts:

1. The previous definitions can be rephrased in terms of problems and algorithms:
    - a problem is in *class P* (or *tractable*) if it can be solved by an algorithm that runs in polynomial time;
    - a problem is in *class NP* if, given a tentative solution (obtained by any means), it is possible to check that the solution is correct in polynomial time;
    - a problem is *NP-complete* if it is in class NP and NP-hard.

2. When considering whether a given problem belongs to P or NP, and whether it might be NP-complete, it is helpful to rewrite the problem, or an associated problem, as a decision problem (which has a yes/no answer) because decision problems have been easier to characterize and classify than general problems. For example, see the description of the traveling salesman problem in Example 3 in this section.

3. Time-complexity of sorting algorithms is typically measured according to the number of comparisons needed.

4. The words *good*, *efficient*, and *feasible* are commonly used interchangeably to mean polynomial-time.

5. Additive and multiplicative constants that are ignored in big-O analysis of an algorithm can sometimes be too large for practical application.

6. That a problem belongs to P does not necessarily imply that it can be solved in a practical amount of time, since the polynomial bound of its complexity can be of high degree. Fortunately, however, for most problems in P arising in practical applications, the polynomial bound is of relatively small degree.

7. Belonging to class NP means that a solution can be *checked* in polynomial time, but not necessarily *found* in polynomial time.

8. When a problem is in class NP, it may be possible to solve the problem for cases arising in practical applications in a reasonable amount of time, even though there are other cases for which this is not true. Also, such problems can often be attacked using *approximation algorithms* which do not produce the exact solution, but instead produce a solution guaranteed to be close in some precise sense to the actual solution sought.

9. Every problem in class P is in class NP.

10. It often requires only a small change to transform a problem in class P to one in class NP. For example, the first four problems in Example 2 (Euler graph, edge cover, linear Diophantine equation, 2-satisfiability) are in class P, but the similar first four problems in Example 3 (Hamilton graph, vertex cover, quadratic Diophantine equation, 3-satisfiability), each of which results from seemingly small changes in the respective problem from class P, are in class NP.

11. To show a problem is NP-complete, the problem can be transformed (in a specific way) to a problem already known to be NP-complete. This is often much easier than showing directly that the problem is NP-complete. See [GaJo78] for details.

12. If there is an NP-hard problem that belongs to P, then P = NP.

13. Not all NP problems are NP complete. (See Example 4 for such a problem.)

14. Deciding whether P = NP is the outstanding problem in the theory of computational complexity. It is the common belief that P $\neq$ NP, based on an extensive search for polynomial-time solutions to various NP problems.

15. The first problem to be shown to be NP-complete was the *satisfiability problem* (Example 3). That the satisfiability problem is NP-complete is called *Cook's theorem*, after Steven A. Cook, who discovered it in 1971. [Co71]

16. In 1972 Richard Karp proved that the traveling salesman problem (TSP) (and many others) were NP-complete. [Ka72]

17. Currently, over 2500 problems (in many areas, including mathematics, computer science, operations research, physics, biology) are known to be NP-complete. Further information on NP-complete problems can be found in the "NP-completeness column: an ongoing guide", authored by David S. Johnson, in the *Journal of Algorithms*. See [Jo81] for the first such column.

18. Extensive information on NP-completeness (methods of proof, examples, etc.) can be found in [At99], [GaJo78], [Tu97], and [va90].

19. A more formal approach to complexity, given in terms of Turing machines, appears in §16.5.

Examples:

1. The following table gives some different input size variable for different problem problem types:

problem type	typical input size parameters
database sorting	number of records
graph algorithms	number of vertices and/or number of edges
arithmetic computation	numbers of digits in the numerals
convex hull construction	number of points

2. The following problems are in class P:

- *Euler graph*: given a graph, determine whether the graph has an Euler circuit;
- *edge cover*: given a graph G and positive integer n, determine whether there is a subset E of edges of G with $|E| \leq n$ and every vertex of G an endpoint of an edge in E;
- *linear Diophantine equation*: given positive integers a, b, c, determine whether $ax + by = c$ has a solution in positive integers x and y;
- *2-satisfiability*: given a Boolean expression in conjunctive normal form in which each sum contains only two variables, determine whether the expression is "satisfiable" (i.e., there is an assignment of 0 and 1 to the variables such that the expression has value 1);
- *circuits*: given a graph G and positive integer n, determine whether there is a subset E of edges of G with $|E| \leq n$ such that each circuit in G contains an edge in E;
- *linear programming*: maximize cx subject to $\widetilde{A}x \leq \widetilde{b}$ where $\widetilde{A}$ is a given $q \times n$ matrix, c is a given row vector of length n, and $\widetilde{b}$ is a given column vector of length q (see §15.1.1).

3. The following problems are NP-complete:

- *Hamilton graph*: given a graph, determine whether the graph has a Hamilton circuit;
- *vertex cover*: given a graph G and positive integer n, determine whether there is a subset V of vertices of G with $|V| \leq n$ with every edge of G having an endpoint in V;
- *quadratic Diophantine equation*: given positive integers a, b, c, determine whether the equation $ax^2 + by = c$ has a solution in positive integers x, y;
- *3-satisfiability*: given a Boolean expression in conjunctive normal form in which each sum contains only three variables, determine whether the expression is "satisfiable" (i.e., there is an assignment of 0 and 1 to the variables such that the expression has value 1);

- *satisfiability*: given a Boolean expression in conjunctive normal form, determine whether the expression is "satisfiable" (i.e., there is an assignment of 0 and 1 to the variables such that the expression has value 1) (see Fact 15);
- *traveling salesman problem* given a weighted graph and positive number k, determine whether there is a Hamilton circuit of weight at most k (see §10.7.1);
- *independent vertex set*: given a graph G and a positive integer n, determine whether G contains an independent vertex set of size at least n;
- *knapsack problem*: given a set S, values a_i and b_i for each $i \in S$, and numbers a and b, determine whether there is a subset $T \subseteq S$ such that $\sum_{i \in T} a_i \leq a$ and $\sum_{i \in T} b_i \geq b$ (see §15.3.1);
- *bin packing problem*: given k bins (each of capacity c) and a collection of weights, determine whether the weights can be placed in the bins so that no bin has its capacity exceeded (§15.3.2);
- *3-coloring*: given a graph G, determine whether its vertices can be colored with 3 colors;
- *clique problem*: given a graph G and positive integer n, determine whether G has a clique of size at least n;
- *dominating set*: given a graph G and positive integer n, determine whether G has a dominating set of size at most n;
- *graph isomorphism*: given two graphs, determine whether they are isomorphic.

4. The following problem is an NP problem, but not NP complete: given vertices v, w in graph G, determine whether v and w are joined by a path in G.

16.4.2 WORST-CASE AND AVERAGE-CASE ANALYSIS

Definitions:

A **worst-case complexity measure** of an algorithm is based on the maximum computational cost for any input of that size. It is usually expressed in big-O asymptotic notation (or sometimes Θ-notation) as a formula based on the input size variables.

An **average-case complexity measure** of an algorithm is based on the expected computational cost over a random distribution of its inputs of a given size.

Facts:

1. Algorithmic analysis of deterministic algorithms often assumes a uniform random distribution of the possible inputs, when the actual distribution is unknown.

2. For sorting algorithms, an average-case analysis may assume that all input permutations of the keys to be sorted are equally likely. In practice, however, some permutations may be far more likely than others, e.g., already sorted, almost sorted, or reverse sorted.

3. The input size measures for average-case analysis are usually the same as for worst-case analysis.

Examples:

1. The following table gives the worst-case running times of some sorting algorithms [CoLeRi90], where the input size parameter n is the number of records:

sorting method	worst-case complexity
insertion sort	$\Theta(n^2)$
selection sort	$\Theta(n^2)$
bubble sort	$\Theta(n^2)$
heapsort	$\Theta(n \log n)$
quicksort	$\Theta(n^2)$
mergesort	$\Theta(n \log n)$

2. The following table gives the worst-case running times of some graph algorithms, based on input size parameters $|V|$ and $|E|$, which are the numbers of vertices and edges:

graph algorithm	worst-case complexity
Kruskal's MST algorithm	$\Theta(|E| \log |V|)$
Dijkstra's shortest-path algorithm with linked-list priority queue	$O(|V|^2)$
Dijkstra's shortest-path algorithm with heap-based priority queue	$O(|E| \log |V|)$ [CoEtal90]
Dijkstra's shortest-path algorithm with Fibonacci-heap priority queue	$O(|E| + |V| \log |V|)$ [CoEtal90]
Edmonds-Karp max-flow algorithm	$O(|V| \cdot |E|^2)$

3. The following table gives the worst-case running times of some plane convex hull algorithms (§13.5.1), based on the number n of points supplied as input:

convex hull algorithm	worst-case complexity
Graham scan	$\Theta(n \log n)$
Jarvis march ("gift-wrapping")	$\Theta(nh)$, $h = \#$ corners (convex hull)
QuickHull	$O(n^2)$
MergeHull	$O(n \log n)$

4. The following table gives the average-case running times of some sorting algorithms, where the input size parameter n is the number of records:

sorting method	average-case complexity
insertion sort	$O(n^2)$
selection sort	$O(n^2)$
bubble sort	$O(n^2)$
heapsort	$O(n \log n)$
quicksort	$O(n \log n)$
mergesort	$O(n \log n)$

> **Algorithm 1: Randomized quicksort.**
>
> procedure *randomized-quicksort*(A, p, r)
> if $p < r$ then
> begin
> $i := random(p, r)$
>   exchange $A[p]$ and $A[i]$
> $q := partition(A, p, r)$
>   *randomized-quicksort*(A, p, q)
>   *randomized-quicksort*$(A, q + 1, r)$
> end {subarray $A[p..r]$ is now sorted}

5. *Randomized quicksort* (Algorithm 1) [CoLeRi90]: A subarray from index p to index r of an array A is sorted, using an external subroutine $random(p, r)$ that generates a number in the set $\{p, \ldots, r\}$ within $O(1)$ worst-case running time. Another external subroutine $partition(A, p, r)$ rearranges the subarray $A[p..r]$ and returns an index q, $p \leq q < r$, such that for $i = p, \ldots, q$, $A[i] \leq A[q]$ and such that for $i = q + 1, \ldots, r$, $A[i] > A[q]$; this subroutine runs in $\Theta(r - p)$ worst-case time.

To sort n keys, randomized quicksort takes $\Theta(n^2)$ time in the worst case (when unlucky enough to have partition sizes always unbalanced), but only $\Theta(n \log n)$ time in the average case (partition sizes are usually at least a constant fraction of the total).

6. *Convex hull*: For certain distributions of n points in the plane, the expected value $E[h]$ of the number of vertices on the convex hull, is known. This bound implies that the average-case running time of Jarvis's march is an additional factor of n greater:

distribution	$E[h]$	average-case running time
uniform in convex polygon	$O(\log n)$	$O(n \log n)$
uniform in circle	$O(n^{\frac{1}{3}})$	$O(n^{\frac{4}{3}})$
normal in plane	$O(\sqrt{\log n})$	$O(n\sqrt{\log n})$

16.4.3 LOWER BOUNDS

Lower bounds on running times of algorithms are typically given as functions of input size using Ω-notation (§1.3.3).

Definitions:

An *existential lower bound* for an algorithm is a lower bound for its running time that holds for at least one input.

An *existential lower bound* for a problem is a lower bound for every algorithm that could solve that problem.

A *comparison sort* is a sorting method that rearranges records based only on comparisons between keys.

The **Euclidean minimum spanning tree** (or **Euclidean MST**) **problem** has as input vertices a set of n points in the plane and as output a spanning tree of minimum total edge-length.

A **reduction of a problem** A to another problem B is the following sequence of steps:
- the input to problem A is transformed into an input to problem B;
- problem B is solved on the transformed input;
- the output of problem B is transformed back into a solution to problem A for the original input.

An $f(n)$ **time reduction** of problem A to problem B is a reduction such that the time for the three steps together is $f(n)$.

Facts:

1. For a given model of computation, if problem A has a lower bound of $T(n)$ and it reduces in $f(n)$ time to problem B, then problem B has a lower bound of $T(n) - f(n)$.
2. Every comparison sort on n records requires $\Omega(n \log n)$ comparisons in the worst case.
3. Computing the Euclidean minimum spanning tree on n points takes $\Omega(n \log n)$ time in the worst case.
4. Unlike the Euclidean MST problem, most graph *problems* have no known nontrivial lower bound. Some graph *algorithms*, however, have lower bounds on their implementation.
5. Running Dijkstra's algorithm (§10.3.2) on a directed graph with $|V|$ vertices takes $\Omega(|V| \log |V|)$ time in the worst case.
6. Finding the vertices for the convex hull of n points in the plane, in any order, takes $\Omega(n \log n)$ time in the worst case.
7. Constructing the Voronoi diagram (§13.5.2) on n points in the plane takes in the worst case $\Omega(n \log n)$ time.

Examples:

1. *An $O(n)$-time reduction of sorting to a gift-wrap of a convex hull*: Given a set of n positive numbers $\{x_1, \ldots, x_n\}$, first produce in $\Theta(n)$ time their respective squares $\{x_1^2, \ldots, x_n^2\}$. Since each point (x_j, x_j^2) lies on the parabola given by $y = x^2$, the Jarvis march on the convex hull of the points (x_j, x_j^2) is a list of points, ordered by abscissa. Sequentially read off the first coordinate of every point of the convex hull in $\Theta(n)$ time, thereby producing the sorted list of numbers. This implies that finding the gift-wrapped convex hull of n points requires at least $\Omega(n \log n) - \Theta(n) = \Omega(n \log n)$ time.
2. *An $O(n)$-time reduction of sorting numbers to Dijkstra's algorithm*: To sort a list of n nonnegative numbers $\{x_1, \ldots, x_n\}$, first create a *star graph*, with vertices $\{v_0, \ldots, v_n\}$, and with an edge (v_0, v_i) weighted x_i, for $1 \leq i \leq n$. Next designate v_0 as the root vertex, and apply Dijkstra's algorithm. Dijkstra's algorithm proceeds according to increasing order of the edge weights, which yields the sorted order. This implies that Dijkstra's algorithm requires at least $\Omega(n \log n) - \Theta(n) = \Omega(n \log n)$ time.
3. *An $O(n)$-time reduction of sorting numbers to a Voronoi diagram*: To sort n numbers $\{x_1, \ldots, x_n\}$, create n points $\{(x_i, 0) \mid 1 \leq i \leq n\}$ in the Euclidean plane. The Voronoi diagram consists of the $n-1$ bisectors separating adjacent points $(x_i, 0)$ on the line $y = 0$. Since the Voronoi diagram description includes ordering of Voronoi edges around each Voronoi vertex, the Voronoi diagram gives the ordering of the bisectors and hence the n numbers. This implies that the Voronoi diagram requires at least $\Omega(n \log n) - \Theta(n) = \Omega(n \log n)$ time.

4. An $O(n)$-time reduction of sorting numbers to Euclidean MST: To sort n numbers $\{x_1,\ldots,x_n\}$, create n points $\{\,(x_i,0) \mid 1 \leq i \leq n\,\}$ in the Euclidean plane. The Euclidean MST of this set contains an edge between points $(x_i,0)$ and $(x_j,0)$ if and only if the numbers x_i and x_j are consecutive in the sorted list of numbers. The Euclidean MST is easily converted back to a sorted list of numbers in $O(n)$ time. This implies that Euclidean MST requires at least $\Omega(n \log n) - \Theta(n) = \Omega(n \log n)$ time.

16.5 COMPLEXITY CLASSES

From a formal viewpoint, complexity theory is concerned with classifying the difficulty of testing for membership in various languages. This means deciding whether any given string is in the language. The general application of complexity theory is achieved by encoding decision problems on natural topics such as graph coloring and finding integer solutions to equations as set membership problems.

16.5.1 ORACLES AND THE POLYNOMIAL HIERARCHY

Throughout this section, whenever the *alphabet* is unspecified, it may be assumed to be the binary set $\{0,1\}$. Also, throughout this section a *Turing machine* is assumed to have among its states a unique *acceptance state* q_A and a unique *rejection state* q_R. All other states continue the computation.

Definitions:

A *language over an alphabet* is a set of strings on that alphabet (see §16.1.1).

A *nondeterministic Turing machine* is a 5-tuple $M = (K, s, h, \Sigma, \Delta)$ otherwise like a deterministic Turing machine, except that the transition function Δ maps each state-symbol pair (q, b) to a set of state-symbol-direction triples.

An *oracle for a language* L is a special computational state to which a machine presents a string w, which switches to special state Y ("yes") if $w \in L$ and to special state N ("no") if $w \notin L$.

An *oracle Turing machine* M is a 6-tuple $M = (K, s, h, \Sigma, \delta \text{ or } \Delta, L)$, equipped with an oracle for language L and with a special second tape on which it can write a string over the alphabet of language L (which might be different from Σ). Aside from oracle steps, it is a Turing machine.

A Turing machine M *accepts* string w if there exists a computational path from the starting configuration with input w to the acceptance state q_A.

A Turing machine M *rejects* string w if it does not accept w. (Either M halts in rejection state q_R or does not halt.)

The *language accepted by a Turing machine* M is the set of all the strings it accepts. It is denoted $\mathcal{L}(M)$.

The Turing machine M **decides** the language $\mathcal{L}(M)$ if it always halts, even for input strings not in $\mathcal{L}(M)$.

The **time $\mathrm{Time}_M(\mathbf{w})$** taken by Turing machine M on input word w is the number of steps on the shortest accepting path if M accepts w, the number of steps on the longest rejecting path if M rejects w but always halts, and $+\infty$ otherwise.

The **space $\mathrm{Space}_M(\mathbf{w})$** taken by Turing machine M on input word w is the maximum number of different tape cells on which M writes during the computation, possibly $+\infty$.

The **time complexity** of a Turing machine M is the function t defined by $t(n) = \max\{\mathit{Time}_M(x) \mid |x| = n\}$.

A Turing machine M has **polynomial time complexity** if there exists a polynomial function $p(n)$ such that $\{\mathit{Time}_M(x) \leq p(n) \mid n = 0, 1, \ldots\}$.

A Turing machine M has **polynomial space complexity** if there exists a polynomial function $p(n)$ such that $\{\mathit{Space}_M(x) \leq p(n) \mid n = 0, 1, \ldots\}$.

The **complexity class P** contains every language that can be decided by a deterministic TM with polynomial time complexity.

The **complexity class PSPACE** contains every language that can be decided by a deterministic TM with polynomial space complexity.

The **complexity class NP** contains every language that can be decided by a nondeterministic TM with polynomial time complexity.

For any language L, the **complexity class $\mathbf{P}^L$** contains every language that is decided in polynomial time by a deterministic TM with oracle L.

For any language L, the **complexity class $\mathbf{NP}^L$** contains every language that is decided in polynomial time by a nondeterministic TM with oracle L.

For any class $\mathcal{C}$ of languages, the **complexity class $\mathbf{P}^\mathcal{C}$** contains every language that is decidable in polynomial time by a deterministic TM with oracle $L \in \mathcal{C}$.

For any class $\mathcal{C}$ of languages, the **complexity class $\mathbf{NP}^\mathcal{C}$** contains every language that is decidable in polynomial time by a nondeterministic TM with oracle $L \in \mathcal{C}$.

The **complexity class Σ_n^P** is defined recursively:

$$\Sigma_k^P = \begin{cases} P & \text{if } k = 0 \\ NP^{\Sigma_{k-1}^P} & \text{if } k \geq 1 \end{cases}$$

The **polynomial hierarchy PH** is the collection comprising every language A for which there exists an n such that $A \in \Sigma_n^p$.

The polynomial hierarchy is said to **collapse** (to the ith rank) if $PH = \Sigma_i^p$, for some $i \geq 0$.

Complexity class $\mathrm{coNP} = \Pi_1^P$.

For $n \geq 0$, the **complexity class Π_n^P** contains every language A such that $\overline{A} \in \Sigma_n^p$.

For any class $\mathcal{C}$, the **complexity class $\mathbf{P}^\mathcal{C} = \{L \mid \text{there is } B \in \mathcal{C} \text{ such that } L \leq_T^p B\}$**.

Complexity class $\Delta_0^P = P$. For $n \in \mathcal{Z}^+$, the **class $\Delta_n^P = P^{\Sigma_{n-1}^P}$**.

The language A is **sparse** if there exists a polynomial $p(n)$ such that for every $n \in \mathcal{N}$, there are at most $p(n)$ elements of length n in A.

Facts:

1. The following identity for complexity classes holds:
$$\Sigma_0^p = \Pi_0^p = \Delta_0^p = \Delta_1^p = P.$$

2. For $n \geq 0$, the following relationships hold:
$$\Delta_n^p \subseteq \Sigma_n^p \cap \Pi_n^p \begin{subarray}{c}\subseteq \Sigma_n^p \subseteq \\ \subseteq \Pi_n^p \subseteq\end{subarray} \Sigma_n^p \cup \Pi_n^p \subseteq \Delta_{n+1}^p.$$

3. The polynomial hierarchy PH is a subset of the complexity class PSPACE.

4. If PH = PSPACE, then the polynomial hierarchy collapses.

5. *Downward separation*: If $\Sigma_n^p = \Sigma_{n+1}^p$, then PH = Σ_n^p. In particular, P = NP if and only if P = PH.

6. *Downward separation*: If $\Sigma_n^p = \Pi_n^p$, then PH = Σ_n^p.

7. Complexity class NP = Σ_1^p.

8. The complexity class Σ_n^p is closed under union and intersection, for all $n \geq 0$.

9. $P^{NP \cap coNP} = NP \cap coNP$. More generally, $P^{\Sigma_n^p \cap \Pi_n^p} = \Sigma_n^p \cap \Pi_n^p$ and $P^{\Delta_n^p} = \Delta_n^p$, for all $n \geq 0$.

10. *Upward separation*: Nondeterministic exponential time $\left(\bigcup_{c>0} \text{NTIME}[2^{cn}]\right)$ is equal to deterministic exponential time $\left(\bigcup_{c>0} \text{DTIME}[2^{cn}]\right)$ if and only if NP − P contains no sparse sets.

11. *Succinct certificates*: For every language in NP there is a proof scheme in which each member (and only members) has a polynomial-size "proof" of membership that can be checked in deterministic polynomial time. Such a short membership proof is sometimes called a *succinct certificate*.

Examples:

1. *Logical proposition problems*: The problem of deciding whether a particular assignment of TRUE-FALSE values to the variables satisfies a logical proposition is in P. Deciding whether a proposition has an assignment that satisfies it is in NP. Deciding whether all assignments satisfy it (i.e., whether the proposition is a tautology) is in coNP.

2. *Graph isomorphism problems*: Deciding whether a given vertex bijection between two graphs realizes a graph isomorphism is in class P. Deciding whether two graphs are isomorphic is in class NP.

3. *Graph coloring problems*: Deciding whether an assignment of colors from a set of three colors to the vertices of a graph is a proper coloring is in P. Deciding whether a graph has a proper 3-coloring is in NP.

4. *Unique maximum clique problem*: Define UMC to be the set of graphs G with a clique $U \subseteq V_G$ such that every other clique is strictly smaller than U. Then UMC is in the class $\Delta_2^p = P^{NP}$.

5. To prove by succinct certificate that a given graph has some clique of size at least k, one can provide a list of k adjacent vertices that are mutually adjacent. (The mutual adjacency condition for the k vertices can be verified in polynomial time.)

16.5.2 REDUCIBILITY AND NP-COMPLETENESS

Definitions:

The language A over alphabet Σ is **polynomial-time reducible** (or **m-p-reducible**) to the language B, denoted $A \leq_m^P B$, if there exists a polynomial-time computable function f such that $x \in A$ if and only if $f(x) \in B$, for each $x \in \Sigma^*$.

The language A is **NP-hard** if every language in class NP is polynomial-time reducible to A.

The language A is **NP-complete** if A is NP-hard and $A \in$ NP.

The language A is **C-hard** if C is a class of languages that represent computational problems and every language in class C is polynomial-time reducible to A.

The language A is **C-complete** if A is C-hard and $A \in C$.

The language A is **Turing-p-reducible** to the language B, denoted $A \leq_T^P B$ if there is a deterministic oracle TM M^B that decides language A in polynomial time.

The language A is **C-Turing-p-hard** if C is a class of languages that represent computational problems and every language in class C is Turing-p-reducible to A.

The language A is **C-Turing-p-complete** if A is C-Turing-p-hard and $A \in C$.

Facts:

1. For most NP-complete problems, showing membership in NP is easy.

2. For integer linear programming, however, it is easy to show NP-hardness, but showing membership in NP is nontrivial.

3. Polynomial-time reducibility is also called *Karp reducibility* after R. M. Karp.

4. Turing-p-reducibility is also called *Cook reducibility* after S. A. Cook.

5. The complement of any NP-complete problem is coNP-complete.

6. If A is polynomial-time reducible to B, then A is Turing-p-reducible to B.

7. If $A \leq_m^P B$ and $B \leq_m^P C$, then $A \leq_m^P C$.

8. If $A \leq_T^P B$ and $B \leq_T^P C$, then $A \leq_T^P C$.

9. *Downward closure*: If $A \in \Sigma_n^p$ and $B \leq_m^p A$, then $B \in \Sigma_n^p$, for every $n \geq 1$. In particular, if any NP-complete set is in P, then P = NP.

10. *Karp-Lipton theorem*: If there is a sparse NP-$\leq_T^p$-hard set, then PH = Σ_2^p.

11. If there is a sparse NP-$\leq_T^p$-complete set, then PH = Δ_2^p.

12. *Mahaney's theorem*: If there is a sparse NP-hard (or NP-complete) set, then P = NP.

13. *Ladner's theorem*: If P $\neq$ NP, then there exists a set in NP $-$ P that is not NP-complete.

14. A large catalog of NP-complete problems appears in [GaJo79]. A few of the most commonly cited appear in §16.4.1.

Examples:

1. For examples of NP-complete problems, see §16.4.1 Example 3.

2. *Quantified Boolean formulas*: Let QBF be the class of true statements of the form
$$(\exists x_1)(\forall x_2)(\exists x_3)(\forall x_4) \ldots (Q_z x_z)[F(x_1, x_2, \ldots, x_z)],$$
where F is a quantifier-free formula over the Boolean variables $x_1, \ldots, x_z$ and where Q_i is $\exists$ if z is odd and $\forall$ if z is even. Then QBF is PSPACE-complete.

3. *Tautologies problem*: The classic coNP-complete language is the set TAUTOLOGY of all logical propositions that are satisfied by every assignment of logical values to its variables.

4. *Graph isomorphism problem*: It is not known whether the set GI of isomorphic graph pairs is in coNP or whether GI is NP-complete, though it is known that GI is NP-complete only if the polynomial hierarchy collapses.

16.5.3 PROBABILISTIC TURING MACHINES

Definitions:

A *probabilistic Turing machine* is a nondeterministic Turing machine M with exactly two choices at each step. Each such choice occurs with probability $\frac{1}{2}$, and is independent of all previous choices.

The *acceptance probability* $p_M(w)$ that a probabilistic Turing machine accepts input word w is the sum of the probabilities over all acceptance paths of computation.

A probabilistic Turing machine M **accepts language L with one-sided error** if
$$p_M(w) > \tfrac{1}{2} \text{ if } w \in L$$
$$p_M(w) = 0 \text{ if } w \notin L.$$

A probabilistic Turing machine M **accepts language L with two-sided error** if
$$p_M(w) > \tfrac{1}{2} \text{ if } w \in L$$
$$p_M(w) \leq \tfrac{1}{2} \text{ if } w \notin L.$$

A probabilistic Turing machine M **accepts language L with bounded two-sided error** if for some $\epsilon > 0$
$$p_M(w) > \tfrac{1}{2} + \epsilon \text{ if } w \in L$$
$$p_M(w) < \tfrac{1}{2} - \epsilon \text{ if } w \notin L.$$

The *complexity class* **RP** *of random polynomial-time languages* is the class of languages that are decided by Turing machines with one-sided error in polynomial time.

The *complexity class* **coRP** contains the language A if $\overline{A} \in$ RP.

The *complexity class* **ZPP** is the intersection RP $\cap$ coRP.

The *complexity class* **PP** *of probabilistic polynomial-time languages* is decided by Turing machines with two-sided error in polynomial time.

The *complexity class* **BPP** *of bounded-error probabilistic polynomial-time languages* is decided by Turing machines with bounded two-sided error in polynomial time.

Facts:

1. ZPP is exactly the class of languages accepted by error-free probabilistic Turing machines running in expected polynomial time.

2. ZPP = RP ∩ coRP $\genfrac{}{}{0pt}{}{\subseteq\text{ RP }\subseteq}{\subseteq\text{coRP}\subseteq}$ RP ∪ coRP ⊆ BPP ⊆ PP ⊆ PSPACE.

3. RP ⊆ NP ⊆ PP.

4. P^{ZPP} = ZPP; P^{BPP} = BPP.

5. BPP ⊆ $\Sigma_2^p \cap \Pi_2^p$.

6. PH ⊆ P^{PP}.

7. PP is closed under all Boolean operations.

8. If NP ⊆ BPP then BPP = PH and RP = NP.

9. PP is the class of languages L that for some Turing machine M running in polynomial time and given access to a fair two-sided coin has the property that, for each x, $x \in L$ if and only if M accepts x with probability greater than $\frac{1}{2}$.

10. It remains an open question whether BPP, RP, coRP, or ZPP have complete languages.

Examples:

1. *SAT ∈ PP*: Consider a probabilistic polynomial-time Turing machine M that, given a proposition F, immediately flips its coin. If the result is "heads", then proposition F is accepted and machine M halts. If "tails", then the machine, via a series of coin flips, randomly assigns each variable to be either true or false, and ultimately accepts F if the resulting assignment satisfies the proposition. Thus, F is accepted with probability exactly $\frac{1}{2}$ if F is unsatisfiable, but is accepted with probability at least $\frac{1}{2} + \frac{1}{2^k}$ if F is satisfiable, where k is the number of logical variables in F. Thus, SAT ∈ PP. This implies that NP ⊆ PP, since the language SAT is NP-complete.

2. *MAJORITY-SAT is PP-complete*: The language *MAJORITY-SAT* is the set of (quantifier-free) Boolean formulas F such that F is satisfied by more than half of the possible variable assignments.

3. *PRIMES ∈ ZPP*: The language *PRIMES* consists of the bitstrings that represent prime numbers when interpreted as binary numerals. If the Extended Riemann Hypothesis holds, then *PRIMES* ∈ P.

4. *Equality of polynomial products*: Given two lists of rational-coefficient polynomials, where each polynomial in the lists has been specified by a list of (coefficient,degree) pairs, the problem of deciding whether the product of the polynomials in the first list yields the same polynomial as the product of the polynomials in the second list is in the class coRP.

Intuitively, this is because if the two products are equal, then they will evaluate to the same value on any argument, yet it can be argued that if an argument is chosen in an appropriate "random" fashion the products evaluated at that argument will probably differ if the product polynomials are not identical.

16.6 RANDOMIZED ALGORITHMS

Some general randomization principles for algorithms have many specific applications. In particular, random algorithms from number theory have applications in cryptography and fingerprinting, Also, randomized algorithms for partitioning, for searching and sorting, and for graph problems such as mincut and matching, including some heuristics for NP-complete problems, have applications in testing and applications for parallel or distributed environments.

16.6.1 OVERVIEW AND GENERAL PARADIGMS

Most randomized algorithms follow a few general paradigms that guide the effective use of probabilistic strategies. For many further topics not covered here, see the excellent survey papers [Ka91] and [We83] and also the textbook [MoRa95].

Definition:

A *randomized algorithm* is an algorithm that makes random choices during its execution. Such random choices can be guided by the output of a random (or, in practice, pseudo-random) number generator.

Facts:

1. Intuitively, the power of randomization is analogous to the standard game-theoretic fact that probabilistic game strategies are substantially more effective than deterministic ones.

2. In the game-theoretic analogy, an algorithm can be regarded as a player, and the problem to be solved can be regarded as an adversary trying to present the player with input instances on which the algorithm exhibits worst case performance.

3. If an algorithm is deterministic, then the game-theoretic adversary knows in advance the entire strategy of the player. Thus, the worst case instances are well defined and can be presented as input to the algorithm.

4. If an algorithm is probabilistic, then the game-theoretic adversary does not know in advance the output of the random number generator that guides part of the algorithm's choices. In particular, worst case instances under deterministic strategies may be smoothed out by randomization.

5. Worst case instances of a randomized algorithm occur when the algorithm performs badly for the overwhelming majority of its probabilistic choices.

6. Many problems have no known deterministic algorithms to match the efficiency of randomized algorithms. Even for problems for which efficient deterministic algorithms are known, randomized algorithms are often remarkably easier to understand and implement.

7. *Abundance of witnesses paradigm*: Deciding whether a given input has a certain property sometimes reduces to finding a combinatorial object "witnessing" the property. When the space of all potential witnesses is too large to be searched exhaustively, it

sometimes suffices to inspect a small random sample, selected so that one of the elements of the sample will be a suitable witness with very high probability.

8. *Random sampling*: Sometimes a small random sample is indicative of the population as a whole.

Examples:

1. *Cryptography*: Most of public-key cryptography is based on the sharp dichotomy between the efficiency of deciding whether a number is prime or composite and the apparent hardness of actually factoring composite numbers.

2. *Fingerprinting*: A large data object is represented by a much smaller "fingerprint" such that, with very high probability, distinct objects map to distinct fingerprints. A similar strategy is used for "hashing" (§17.4) where large objects are mapped to much smaller keys with very low probabilities of collisions.

3. *Testing identities*: It is often possible to check if an algebraic expression is identically equal to zero by substituting random values for the variables and checking whether the expression evaluates to zero.

4. *Symmetry breaking*: It is often necessary for a set of distributed or parallel processes to come collectively to an arbitrary but consistent decision among a set of indistinguishable possibilities. There is a method to break such symmetries using randomization and for an indication that gives an efficient parallel perfect matching algorithm, as well as applications to many protocols for distributed environments, to computation in the presence of errors, and to Byzantine agreements.

5. *Load balancing*: For problems involving choice between a number of resources, such as processors or communication links in parallel or distributed networks, randomization can be useful in spreading out the load.

6. *The probabilistic method*: The probabilistic method is to demonstrate that a combinatorial object of interest occurs with non-zero probability in a suitably defined probability space. Sometimes the probabilistic method yields efficient algorithmic constructions rather than mere existential arguments.

16.6.2 LAS VEGAS AND MONTE CARLO ALGORITHMS

Randomized algorithms are classified into two types — Monte Carlo algorithms and Las Vegas algorithms.

Definitions:

A *Monte Carlo algorithm* has bounded running time and produces correct output with probability bounded away from zero.

The *success amplification method for a Monte Carlo algorithm* is to perform k independent runs of the algorithm.

A *Las Vegas algorithm* always produces correct output. However, its running time is a random variable, whose expectation and variance must be quantified in the analysis of the algorithm.

The *success amplification method for a Las Vegas algorithm* is to perform $\frac{k}{2}$ independent Las Vegas runs of $2E[T]$ steps each.

The **Monte Carlo to Las Vegas transformation**, starting from a Monte Carlo algorithm, is the Las Vegas algorithm of repeatedly running that Monte Carlo algorithm until a success occurs.

The **Las Vegas to Monte Carlo transformation**, starting from a Las Vegas algorithm, is the Monte Carlo algorithm obtained by running the Las Vegas scheme for $kE[T]$ steps and halting, where $E[T]$ is the expected Las Vegas running time.

Facts:

1. If the probability of success of a single run is p, then the probability under the success amplification method that k independent runs fail is $(1-p)^k$. Thus, the probability of success becomes $1-(1-p)^k$.

2. If p is the probability of success of a Monte Carlo algorithm, then the expected number of Las Vegas trials before a success occurs is
$$p + 2(1-p)p + 3(1-p)^2 p + 4(1-p)^3 p + \cdots = \frac{1}{p}.$$

3. *Markov's inequality*: The probability that a positive random variable exceeds k times its expectation is at most $\frac{1}{k}$.

4. Markov's inequality yields a general method to bound variances of Las Vegas algorithms. If T is the running time of a Las Vegas algorithm, then
$$\Pr[T > kE[T]] < \tfrac{1}{k}.$$

5. The probability that a transformed Las Vegas to Monte Carlo algorithm is successful is at least $1 - \frac{1}{k}$.

6. If the expected running time of a Las Vegas algorithm is $E(T)$, then the running time of the amplified algorithm is $kE[T]$. However, the probability of success becomes $1-(\frac{1}{2})^{\frac{k}{2}}$, since each $2E[T]$ run has probability of failure at most $\frac{1}{2}$. Thus, the probability that $\frac{k}{2}$ independent runs fail is at most $(\frac{1}{2})^{\frac{k}{2}}$.

Examples:

1. *A database problem*: In a large database whose keys are stored in no particular order, find a key that is not contained in that database, within time $O(N)$, where N is the size of the database. (Assume $N = 2^{30}$ and that the keys are 32 digits long.) This would match the natural lower bound of $\Omega(N)$, the time required just to read the entire database. The deterministic strategy of sorting and checking for the first missing key would take time $O(N \log N)$, where $N = 2^{30}$ is the size of the database.

   - a Monte Carlo randomized strategy: Pick a random 32-digit key and then scan the database! There are 2^{32} potential 32-digit keys and only $\frac{N}{32} = 2^{25}$ keys in the database, a fraction of $\frac{2^{25}}{2^{32}} = 2^{-7}$. Thus, the probability that a randomly chosen key is not in the database is at least $1 - 2^{-7}$, which is greater than 99%. The running time is dominated by a single scan of the database to check whether the randomly chosen key is suitable. Thus, it completes in $O(N)$.
   - *success amplification*: The probability that among k independently chosen random keys none are found suitable is less than $0.01^k = 10^{-2k}$; this quantity becomes neg-

> **Algorithm 1:** In a set A of n distinct keys, find the mth smallest.
>
> input: a set A with $n = |A|$, and an integer m with $1 \leq m \leq n$
> FIND (A, m)
> if $A = \{s\}$ then return s
> else
>   pick s uniformly at random from A
>   compute $X = \{a \in A \mid a < s\}$
>   compute $Y = \{a \in A \mid a > s\}$
>   if $|X| \geq m$ then call FIND(X, m)
>   if $n - |Y| \leq m$ then call FIND$(Y, m - (n - |Y|))$
>   if $|X| < m < n - |Y|$ then return s
> end

ligible, even for very small values of k. The running time of this amplified algorithm is $O(kN)$.

- *from Monte Carlo to Las Vegas:* Repeatedly pick random keys until a suitable key is found. The expected number of trials before a suitable key is found is $\frac{1}{p} < \frac{100}{99}$. Thus, the expected running time is $O(\frac{100N}{99}) = O(N)$.

2. *A modified database problem:* Among the $n = \frac{N}{32}$ keys of the database of Example 1, find the mth in increasing lexicographic order, within time $O(n)$. (Algorithm 1 does this.) The deterministic strategy of sorting would take time $O(n \log n)$.

- *a Las Vegas randomized strategy:* Pick a random key s from the database and consider the sets X and Y of keys in the database that are smaller and larger, respectively, than s. If $|X| \geq m$, then the problem reduces to finding the mth key in X. If $n - |Y| \leq m$, then the problem reduces to finding the $(m-(n-|Y|))$th key in Y. Finally, if $|X| < m < n - |Y|$, then s is the mth key.

- *expected running time:* The randomly chosen key s splits the database into pieces X and Y which are, on average, of size $\frac{n}{2}$, and in most cases substantially smaller than n. Thus, the problem of looking for a key in a set of size n reduces to a problem of looking for a key in a set of size "approximately" $\frac{n}{2}$ and a running time of the type

$$T(n) \approx T(\tfrac{n}{2}) + O(n) = O(n)$$

can intuitively be expected. More precisely, let $T(n, m)$ denote the running time to find the mth key. Since any of the keys could equally likely be picked as the splitter s, the expectation $E[T(n, m)]$ satisfies this recurrence:

$$E[T(n,m)] =$$
$$\tfrac{1}{n}E[T(n-1, m-1)] + \tfrac{1}{n}E[T(n-2, m-2)] + \cdots + \tfrac{1}{n}E[T(n-(m-1), 1)]$$
$$+ \tfrac{1}{n}E[T(m+1, m)] + \tfrac{1}{n}E[T(m+2, m)] + \cdots + \tfrac{1}{n}E[T(n-1, m)] + cn,$$

for some constant c. The solution is $E[T(n, m)] = O(n)$, for all m.

- *variance:* Markov's inequality bounds the variance of the running time by

$$\Pr[T(n,m) > kE[T(n,m)]] < \tfrac{1}{k}.$$

Algorithm 2: Test primality of n with k witnesses.

PRIMALITY TEST(n, k)
input: positive integers n and k with n odd and $k > 2$
if n is odd **then**
  pick $a_1, \ldots, a_k$, each a_i independently and uniformly at random from $[1, n-1]$
  compute $\gcd(n, a_i)$ for all $1 \leq i \leq k$
    $\{\gcd(n, a_i)$ can be computed efficiently using Euclid's algorithm$\}$
  if there exists an a_i with $\gcd(n, a_i) \neq 1$ **then** output "*composite*" and halt
  compute $a_i^{\frac{n-1}{2}}$ (**mod** n) for all a_i with $1 \leq i \leq k$
    $\{a_i^{\frac{n-1}{2}}$ (**mod** n) can be computed efficiently by repeated squaring$\}$
  if for some a_i, $a_i^{\frac{n-1}{2}} \not\equiv \pm 1$ (**mod** n) **then** output "*composite*" and halt
  if for some a_i, $a_i^{\frac{n-1}{2}} \equiv -1$ (**mod** n) **then** output "*prime, with high confidence*"
    and halt
  if for all a_i, $a_i^{\frac{n-1}{2}} \equiv 1$ (**mod** n) **then** output "*composite, with high confidence*"
    and halt

3. Algorithm 2, *Primality Test*, produces correct output with probability at least $1 - (\frac{1}{2})^k$. After $\log n$ trials of selecting a random integer less than n and testing, the likelihood is very high for reasonably large n, that a prime number will be obtained. This follows from the prime number theorem and Markov's inequality.

REFERENCES

Printed Resources:

[AhUl77] A. V. Aho and J. D. Ullman, *Principles of Compiler Design*, Addison-Wesley, 1977.

[AhUl73] A. V. Aho and J. D. Ullman, *The Theory of Parsing, Translation, and Compiling*, Vols. I and II, Prentice Hall, Vol. I: 1971, Vol. II: 1973.

[AhMaOr93] R. K. Ahuja, T. L. Magnanti, and J. B. Orlin, *Network Flows: Theory, Algorithms, and Applications*, Prentice Hall, 1993. (Excellent reference for shortest paths and network flows.)

[At99] M. J. Atallah, *Algorithms and Theory of Computation Handbook*, CRC Press, 1999.

[BoCr93] D. Bovet and P. Crescenzi, *Introduction to the Theory of Complexity*, Prentice Hall, 1993.

[BrLa74] W. S. Brainerd and L. H. Landweber, *Theory of Computation*, Wiley, 1974.

[Co71] S. A. Cook, "The complexity of theorem-proving procedures", *Proceedings of the Third Annual ACM Symposium on the Theory of Computing*, 1971, 151–158.

[CoLeRi90] T. H. Cormen, C. E. Leiserson, and R. L. Rivest, *Introduction to Algorithms*, The MIT Press, 1990. (Algorithms and techniques, with full analysis, across many problem domains.)

[DaWe83] M. D. Davis, E. J. Weyuker, *Computability, Complexity, and Languages*, Academic Press, 1983.

[DeDeQu78] P. J. Denning, J. B. Dennis, and J. E. Qualitz, *Machines, Languages, and Computation*, Prentice-Hall, 1978.

[Ed65] J. Edmonds, "Paths, Trees and Flowers", *Canadian Journal of Mathematics* (17) 1965, 449–467.

[Ei7476] S. Eilenberg, *Automata, Languages, and Machines*, Academic Press, Vol. A: 1974, Vol. B: 1976.

[FlBe94] R. W. Floyd and R. Beigel, *The Language of Machines: An Introduction to Computability and Formal Languages*, Computer Science Press, 1994.

[Fr77] R. Frievalds, "Probabilistic machines can use less running time", B. Gilchrist, ed., *Information Processing 77, Proceedings of IFIP 77*, North-Holland, 1977, 839–842.

[GaJo78] M. R. Garey and D. S. Johnson, *Computers and Intractability, A Guide to the Theory of NP-Completeness*, W. H. Freeman, 1979.

[Gi75] S. Ginsburg, *Algebraic and Automata-Theoretic Properties of Formal Languages*, North-Holland, 1975.

[Gi66] S. Ginsburg, *The Mathematical Theory of Context-Free Languages*, McGraw-Hill, 1966.

[GrHoRu95] R. Greenlaw, H. Hoover, and W. Ruzzo, *Limits to Parallel Computation: P-Completeness Theory*, Oxford University Press, 1995.

[Gr97] J. Gruska, *Foundations of Computing*, International Thomson, 1997.

[Ha78a] M. Harrison, *Introduction to Formal Language Theory*, Addison-Wesley, 1978.

[Ha78b] J. Hartmanis, *Feasible Computations and Provable Complexity Properties*, SIAM, 1978.

[HeRo75] G. T. Herman and G. Rozenberg, *Developmental Systems and Languages*, North-Holland, 1975.

[HoUl69] J. E. Hopcroft and J. D. Ullman, *Formal Languages and Their Relations to Automata*, Addison-Wesley, 1969.

[HoUl79] J. E. Hopcroft and J. D. Ullman, *Introduction to Automata Theory, Languages, and Computing*, Addison-Wesley, 1979.

[Jo81] D. S. Johnson, "The NP-completeness column: an ongoing guide (1st edition)", *Journal of Algorithms* 2 (1981), 393–405.

[Jo90] D. S. Johnson, "A Catalog of Complexity Classes", *Handbook of Theoretical Computer Science*, J. van Leeuwen ed., MIT Press/Elsevier, 1990, 67–161.

[Ka93] D. R. Karger, "Global mincuts in RNC, and other ramifications of a simple mincut algorithm", *Proceedings of the 4th Annual ACM-SIAM Symposium on Discrete Algorithms* 1993, 21–30.

[Ka91] R. M. Karp, "An Introduction to Randomized Algorithms", *Discrete Applied Mathematics* (34) 1991, 165–201.

[Ka72] R. M. Karp, "Reducibility among combinatorial problems", *Complexity of Computer Computations*, R. E. Miller and J. W. Thatcher, eds., Plenum Press, 1972, 85-103.

[Kn71] D. E. Knuth, *Seminumerical Algorithms*, Volume 2 of *The Art of Computer Programming*, Addison Wesley, 1971.

[KuSa86] W. Kuich and A. Salomaa, *Semirings, Automata, Languages*, Springer-Verlag, 1986.

[LePa81] H. R. Lewis and C, H, Papadimitriou, *Elements of the Theory of Computation*, Prentice-Hall, 1981.

[Li90] P. Linz, *An Introduction to Formal Languages and Automata*, D. C. Heath, 1990.

[Lo83] M. Lothaire, *Combinatorics on Words*, Addison-Wesley, 1983.

[Lo79] L. Lovasz, "On Determinants, Matchings, and Random Algorithms", L. Budach, ed., *Fundamentals of Computing Theory*, Akademia-Verlag, 1979.

[Ma93] Y. Matijasevic, *Hilbert's Tenth Problem*, MIT Press, 1993. (A book for non-logicians on the proof and ramifications of the unsolvability of Hilbert's 10th problem.)

[MoRa95] R. Motwani and P. Ranghavan, *Randomized Algorithms*, Cambridge University Press, 1995.

[MuVaVa87] K. Mulmeley, U. V. Vazirani and V. V. Vazirani, "Matching is as Easy as Matrix Inversion", *Combinatorica* (7) 1987, 105–113.

[Od89] P. Odifreddi, *Classical Recursion Theory* Vol. I, North-Holland, 1989. (Contains much on models of computation and the Church-Turing thesis.)

[Pa94] C. Papadimitriou, *Computational Complexity*, Addison-Wesley, 1994.

[PaYa91] C. H. Papadimitriou and M. Yannakakis, "Optimization, Approximation, and Complexity Classes", *Journal of Computer and Systems Sciences* (43) 1991, 425–440.

[Par94] I. Parberry, *Circuit Complexity and Neural Networks*, MIT Press, 1994.

[Pi86] J. E. Pin, *Varieties of Formal Languages*, Plenum Press, 1986.

[PrSh85] F. P. Preparata and M. I. Shamos, *Computational Geometry: An Introduction*, Springer-Verlag, 1985. (Upper and lower bounds for computational-geometry problems.)

[Ra80] M. O. Rabin, "Probabilistic Algorithms for Testing Primality", *Journal of Number Theory* (12) 1980, 128–138.

[Ré83] G. E. Révész, *Introduction to Formal Languages*, McGraw-Hill, 1983.

[RiShAd78] R. L. Rivest, A. Shamir, and L. Adleman, "A method for obtaining digital signatures and public key cryptosystems", *Communications of the ACM* (21) 1978, 120–126.

[Ro67] H. Rogers, Jr., *Theory of Recursive Functions and Effective Computability*, McGraw-Hill, 1967.

[Sa85] A. Salomaa, *Computation and Automata*, Cambridge University Press, 1985.

[Sa73] A. Salomaa, *Formal Languages*, Academic Press, 1973.

[Sa81] A. Salomaa, *Jewels of Formal Language Theory*, Computer Science Press, 1981.

[Sa69] A. Salomaa, *Theory of Automata*, Pergamon, 1969.

[SaSo78] A. Salomaa and M. Soittola, *Automata-Theoretic Aspects of Formal Power Series*, Springer-Verlag, 1978.

[Se88] R. Sedgewick, *Algorithms*, 2nd ed., Addison-Wesley, 1988. (A compendium of many algorithms, light on analysis.)

[SiSo88] S. Sippu and E. Soisalon-Soininen, *Parsing Theory. Vol. I: Languages and Parsing*, Springer-Verlag, 1988.

[So87] R. I. Soare, *Recursively Enumerable Sets and Degrees*, Perspectives in Mathematical Logic, Springer-Verlag, 1987.

[SoSt77] R. Solovay and V. Strassen, "A fast Monte Carlo Test for Primality", *SIAM Journal of Computing* 6(1) 1977, 84–85.

[Ta83] R. E. Tarjan, *Data Structures and Network Algorithms*, SIAM, 1983. (Covers minimum spanning trees, shortest paths, network flows, and matchings.)

[Tu97] A. B. Tucker, Jr., ed., *The Computer Science and Engineering Handbook*, CRC Press, 1997.

[va90] J. van Leeuwen, ed., *Handbook of Theoretical Computer Science, Vol. A: Algorithms and Complexity*, Elsevier, 1990.

[Va82] L. G. Valiant, "A scheme for Fast Parallel Communication", *SIAM Journal of Computing* (11) 1982, 350–361.

[We83] D. J. A. Welsh, "Randomized Algorithms", *Discrete Applied Mathematics* (5) 1983, 133-145.

[Wo80] D. Wood, *Grammar and L Forms. An Introduction*, Lecture Notes in Computer Science (91), Springer-Verlag, 1980.

[Ya] C. Yap, *Introduction to Complexity Classes*, Oxford University Press, to appear.

Web Resources:

http://alife.santafe.edu/alife/topics/cas/ca-faq/ca-faq.html   (Frequently Asked Questions About Cellular Automata.)

http://grail.cba.csuohio.edu/~somos/bb.html (Busy Beaver Turing Machines.)

http://robotics.stanford.edu/~suresh/theory/ (Theoretical Computer Science on the Web.)

http://www.bvu.edu/~daveh/dhturing.html (The Turing Link page.)

http://www.cs.jhu.edu/~callahan/lifepage.html (Paul's Page of the Game of Life Miscellany.)

http://www.cs.sunysb.edu/~algorith/ (The Stony Brook Algorithm Repository; see Section 1.7.7 Finite State Machine Minimization.)

http://www.drb.insel.de/~heiner/BB/ (Busy Beaver Turing Machines.)

http://www.halcyon.com/hkoenig/LifeInfo/LifeInfo.html (Game of Life information.)

http://www.stanford.edu/~mariaf/ra/knowledgeUnits/AL.html (A compendium of sites on complexity analysis, computability and undecidability, and parallel algorithms.)

http://www.uni-bielefeld.de/~achim/gol.html (Achim's Game of Life page.)

17

INFORMATION STRUCTURES

17.1 Abstract Datatypes — *Charles H. Goldberg*
   17.1.1 Abstract specification of records and databases
   17.1.2 Stacks and queues
   17.1.3 Two-way sequential lists
   17.1.4 Dictionaries and random access lists
   17.1.5 Priority queues

17.2 Concrete Data Structures — *Jonathan L. Gross*
   17.2.1 Modeling computer storage and retrieval
   17.2.2 Arrays and linked lists
   17.2.3 Binary search trees
   17.2.4 Priority trees and heaps
   17.2.5 Network incidence structures

17.3 Sorting and Searching — *Jianer Chen*
   17.3.1 Generic concepts for sorting and searching
   17.3.2 Sorting by expanding a sorted subsequence
   17.3.3 Sorting by exchanging out-of-order pairs
   17.3.4 Sorting by divide-and-conquer
   17.3.5 Sorting by distribution
   17.3.6 Searching

17.4 Hashing — *Viera Krnanova Proulx*
   17.4.1 Hash functions
   17.4.2 Collision resolution

17.5 Dynamic Graph Algorithms — *Joan Feigenbaum and Sampath Kannan*
   17.5.1 Dynamically maintainable properties
   17.5.2 Techniques
   17.5.3 Applications
   17.5.4 Recent results and open questions

INTRODUCTION

Information structures are groupings of related information into records and organization of the records into databases. The mathematical structure of a record is specified as an abstract datatype and represented concretely as a linkage of segments of computer memory. General chapter references are [AhHoUl83], [Kn68], and [Kn73].

GLOSSARY

abstract datatype (ADT): a mathematically specified datatype equipped with operations that can be performed on its data objects.

adaptive bubblesort: a bubblesort that stops the first time a scan produces no transpositions.

ADT-constructor: any of the three operations *string of, set of,* or *tuple of* used to build more complex ADTs from simpler ADTs.

alphabetic datatype: an elementary datatype whose domain is a finite set of *symbols*, and whose only primary operation is a total ordering query.

ambivalent data structure: a structure that keeps track of several alternatives at many of its vertices, even though a global examination of the structure would determine which of these alternatives is optimal.

array data structure: an indexed sequence of cells $\langle a_j \mid j = d, \ldots, u \rangle$ of fixed size, with consecutive indices.

AVL tree: a binary search tree with the property that the two subtrees of each node differ by at most one in height.

binary search: a recursive search method that proceeds by comparing the target key to the key in the middle of the list, in order to determine which half of the list could contain the target item, if it is present.

binary search tree: a binary tree in which the key at each node is larger than all the keys in its left subtree, but smaller than all the keys in its right subtree.

binary-search-tree structure: a binary-tree structure in which for every cell, all cells accessible through the left child have lower keys, and all cells accessible through the right child have higher keys.

binary-tree structure: a tree structure such that each cell has two pointers.

bubblesort: a sort that repeatedly scans an array from the highest index to the lowest, on each iteration swapping every out-of-order pair of consecutive items that is encountered.

cell (in a concrete data structure): a storage unit within the data structure that may contain data and pointers to other cells.

certificate (for a property of a graph G): another graph that has the specified property if and only if the graph G has the property.

chaining method (for hash tables): a hashing method that resolves collisions by placing all the records whose keys map to the same location in the main array into a linked list (chain), which is rooted at that location, but stored in the secondary array.

circular linked list: a set of cells, each with two pointers, one designated as its *forward pointer* and the other as its *backward pointer*, plus a header with one or more pointers to *current cells*, such that these conditions hold:
- the sequence of cells formed by following the forward pointers, starting from any cell, traverses the entire set and returns to the starting cell;
- the sequence of cells formed by following the backward pointers, starting from any cell, traverses the entire set and returns to the starting cell.

closed hash table: a hash table in which collisions are resolved without the use of secondary storage space, that is, by probing in the main array to find available locations.

cluster (in a spanning tree): a set of vertices whose induced subgraph is connected.

clustering property (of a probe function): the undesirable possibility that parts of the probe sequences generated for two different keys are identical.

collision instance (of a hash function): a pair of different keys for which the value of the hash function is the same.

collision resolution (of a hashing process): a procedure within the hashing process used to define a sequence of alternative locations for storage of a record whose key collides with the key of an existing record in the table.

comparison sort: a sorting method in which the final sorted order is based solely on comparisons between elements in the input sequence.

concrete data structure: a mathematical model for storing the current value of a structured variable in computer memory.

database: a set of *records*, stored in a computer.

datatype: a set of objects, called the *domain*, and a set of mappings, called *primary operations*, from the domain to itself or to the domain of some other datatype.

deheaping: removing the highest priority entry from a heap and patching the result so that the heap properties are restored.

dictionary: an abstract datatype whose domain is a set of keyed pairs, in which arbitrary pairs may be accessed directly.

domain (of a datatype): the set of objects within that datatype.

dyadic graph property: a property defined with respect to pairs of vertices.

dynamic structure (for a database): an information structure for the database whose configuration may be changed, for instance, by the insertion or deletion of elements.

dynamic update operation: (on a graph) an operation that changes the graph and keeps track of whether the graph has some designated property.

edge-incidence table (for a graph): a *dictionary* whose keys are the vertices of a graph or digraph. The data component for each key vertex is a list of all the edges that are incident on that vertex. Each self-loop occurs twice in the list.

elementary datatype: an alphabetic datatype or a numeric datatype, usually intended for direct representation in the hardware of a computer.

endpoint table (for a graph): a *dictionary* whose keys are the edges. The data component for each key edge is the set of endpoints for that edge. If an edge is directed, then its endpoints are marked as *head* and *tail*.

enheaping: placing a new entry into its correctly prioritized position in a heap.

entry (in a database): a 2-tuple, whose first component is a *key*, and whose second component is some data; also called a *record*.

external sorting method: a method that uses external storage, such as hard disk or tape, outside the main memory during the sorting process.

far end (of a one-way linked list): the cell that contains a null pointer.

Fibonacci heap: a modification of a heap, using the Fibonacci sequence, that permits more efficient implementation of a priority queue than a heap based on a left-complete binary tree.

FIFO property (of a database): the property that the item retrieved is always the item inserted the longest ago. FIFO means "first-in-first-out".

flat notation (in a postcondition of a primary operation specification): the value X^b of the variable X before the specified operation is executed.

fullness (of a closed hash table): the ratio of the number of records presently in the table to the size of the table.

generic datatype: a specification in an ADT-template that means that there are no restrictions whatsoever on that datatype.

hash function (for storing records in a table): a function that maps each *key* to a location in the table.

hash table: an array of locations for records (entries) in which each record is identified by a unique key, and in which a *hash function* is used to perform the table-access operations (of insertion, deletion, and search), possibly involving the use of a secondary array to resolve competition for locations.

hashing: storage-retrieval in a large table in which the table location is computed from the key of each data entry.

header (of a concrete data structure): a special memory unit (but not a cell) that contains current information about the entire configuration and pointers to some critical cells in the structure.

heap: a concrete data structure that represents a priority tree as an array.

heapsort: sorting a set of entries by first enheaping them all and then deheaping them all.

incidence matrix (for a graph): a 0-1 matrix that specifies the incidence relation. The rows are indexed by the vertices and the columns by the edges. The entry in the row corresponding to vertex v and edge e is 1 if v is an endpoint of e, and 0 otherwise.

in-place realization (of a sorting method): a method that uses, beyond the space needed for one copy of each data entry, only a constant amount of additional space, regardless of the size of the list to be sorted.

insertion sort: a sort that transforms an unsorted list into a sorted list by iteratively transferring the next item from the remaining items in the unsorted input list and inserting it into correct position in the sorted output list.

internal sorting method: any method that keeps all the entries in the primary memory of the computer during the process of rearrangement.

key (in a database entry): a value from an ordered set, used to store and retrieve data.

key domain: the ordered set from which values of keys are drawn.

key entry (of a record in a table): a value from an ordered set (e.g., integer identification codes or alphabetic strings) used to store records in the table.

key randomization: a "preliminary" procedure within the hashing process for mapping non-numeric keys (or keys with poor distribution) into (more uniformly) random distributed integers in the domain of the hash function.

keyed pair: a 2-tuple whose first component, called a *key*, is used to locate the data in the second component.

left-child (of a cell in a binary tree structure): the cell to which the first pointer points.

left-complete binary tree: either a binary tree that is complete, or a balanced binary tree (§9.1.2) such that at depth one less than the maximum, the following hold:
- all nodes with two children are to the left of all nodes with one or no children;
- all nodes with no children are to the right of all nodes with one or two children;
- there is at most one node with only one child, which must be a left-child.

LIFO property (of a database): the property that the item retrieved is always the item most recently inserted. LIFO means "last-in-first-out".

linear search: the technique of scanning the entries of a list in sequence, until either some stopping condition occurs or the entire list has been scanned.

mergesort: a sort that partitions an unsorted list into lists of length one and then iteratively merges them until a single sorted list is obtained.

near end (of a one-way linked list): the cell that is pointed to by the header and by no other cell.

nearly complete (property of a binary tree): the possible property that the binary tree is complete at every level except possibly at the bottom level. At the bottom level, all the missing leaves are to the right of all the present leaves.

null pointer: a pointer that points to an artificial location, which serves as a signal to an algorithm to react somewhat differently than to a pointer to an actual location.

numeric datatype: an elementary datatype whose domain is a set of *numbers* and whose primary operations are a total ordering query and the arithmetic operators + (addition), × (multiplication), and − (change of sign).

one-way linked list: a set of cells, each with one pointer, such that:
- exactly one of these cells is pointed to by the header but by no cell;
- exactly one cell contains a null pointer;
- the sequence of cells formed by following the pointers, starting from the header, traverses the entire set, ending with the cell containing the null pointer.

open hash table: a hash table that uses a secondary array to resolve collisions.

ordered datatype: a datatype with an order relation such that any two elements can be compared.

pivot (in a quicksort): an entry at which the sequence is split.

plane graph: a planar graph, together with a particular imbedding in the plane.

pointer (to a cell): a representation of that cell's location in computer memory.

postcondition (of a primary operation): a list of conditions that must hold after the operation is executed, if the precondition is satisfied when the operation commences.

precondition (of a primary operation): a list of conditions that must hold immediately before the operation is executed, for the operation to execute as specified.

primary key: the key component of highest precedence, when the key has more than one component.

primary operation (for a datatype): a basic operation that retrieves information from an object in the domain or modifies the object.

priority queue: an abstract datatype whose domain is a set of records, in which only the entry with the largest key is immediately accessible.

priority tree: a nearly complete binary tree whose nodes are assigned data entries from an ordered set of "priorities", such that there is no node whose priority supersedes the priority of its parent node.

probe function: a function used iteratively to calculate an alternative location in a closed hash table when the initial location calculated from the key or the previous probe location is already occupied.

probe sequence (for a hash table location): the sequence of locations calculated by the probe function in its effort to find an unoccupied place in the table.

query (to a datatype): a primary or secondary operation that changes nothing and returns a logical value, i.e., *true* or *false*.

queue: an abstract datatype that organizes the records into a sequence, such that records are inserted at one end (called the *back* of the queue) and extractions are made from the other end (called the *front*).

quicksort: sorting by recursively partitioning a list around an entry called the *pivot* so that all smaller items precede the pivot and all larger items follow it.

radix sort: a sort using iterative partitioning into queues and recombining by concatenation, in which the partitioning is based on a digit in a numeral.

random access list: an abstract datatype whose domain is a set of records such that the values of the *key* field range within an interval of integers $a \leq k \leq b$; this permits implementations that execute primary operations faster than a general table.

rank (of an element of a finite ordered set): the number of elements that it exceeds or equals.

rank-counting sort: sorting by calculating the "rank" for each element, and then assigning each element to its correct position according to its rank.

record: a 2-tuple, whose first component is a *key*, and whose second component is some data; also called an *entry*.

record in a table: a table entry containing a *key* and some data.

right-child (of a cell in a binary-tree structure): the cell to which the second pointer points.

root cell (of a tree structure): the cell to which the header points.

scanning a database (or a portion of a database): examining every record in that database or portion.

searching (a database): seeking either a target entry with a specific key or a target entry whose key has some specified property.

secondary key: the key component of next highest precedence, when the key has more than one component.

secondary operation (for a datatype): an operation constructed from primary operations and from previously defined secondary operations.

selection sort: a sort that transforms an unsorted list into a sorted list by iteratively finding the item with smallest key from the remaining items in the unsorted input list and appending it to the end of the sorted output list.

sequence of: an ADT-constructor that converts a datatype with domain D into a new datatype, whose domain is the set of all finite sequences of objects from domain D, and whose primary operations are some sequence operations.

set of: an ADT-constructor that converts a datatype with domain D into a new datatype, whose domain is the set of all subsets of objects from domain D, and whose primary operations are set operations.

shakersort: a bubblesort variation that alternates between bubbling upward and sinking downward on alternate scans.

Shellsort: a sorting method that involves partitioning a list into sublists and insertion sorting each of the sublists.

sinking sort: a "reverse bubblesort" that scans an array repeatedly from the lowest index to the highest, each time swapping every out-of-order pair of consecutive items that is encountered.

size (of a cell in a data structure): the number of bytes of computer memory that the cell occupies.

size (of a hash table): the number of locations in the main array in which the records are stored. (If chaining is used to resolve collision, the total number of records stored may exceed the size of the main array.)

sorting algorithm: a method for arranging the entries of a database into a sequence that conforms to the order of their keys.

sparse certificate: a strong certificate (for a property of a graph G) in which the number of edges is $O(|V_G|)$.

sparse sequence: a sequence in which nearly all the entries are zeros.

stable certificate: a certificate produced by a stable function.

stable (**certificate**) **function**: a function that maps graphs to strong certificates such that:
- $A(G \cup H) = A(A(G) \cup H)$;
- $A(G - e)$ differs from $A(G)$ by $O(1)$ edges, where e is an edge in G.

stack: an abstract datatype that organizes the records into a sequence, in which insertion and extraction are made at the same end (called the *top* of the stack).

static structure (for a database): an information structure for the database whose configuration does not change during an algorithmic process.

strong certificate (for a property of a graph G): a certificate graph G' for G with the same vertex set as G such that, for every graph H, the graph $G \cup H$ has property $\mathcal{P}$ if and only if $G' \cup H$ has property $\mathcal{P}$.

table: a set of keyed pairs, in which arbitrary pairs may be accessed directly; used as the domain of a *dictionary*.

target (of a database search): an entry whose key has been designated as the objective of the search.

2-3 tree: a tree in which each non-leaf node has 2 or 3 children, and in which every path from the root to a leaf is of the same length.

tree structure: a concrete data structure such that the header points to a single cell, and such that from that cell to each other cell, there is a single chain of pointers.

k-tuple of: an ADT-constructor that converts a list of k datatypes into a new datatype, whose domain is the cartesian product of the domains of the datatypes in that list, and whose primary operations are the projection functions from a k-tuple to each of its coordinates.

two-way incidence structure (for a graph): a pair consisting of an edge-incidence table and an endpoint table.

two-way linked list: a set of cells, each with two pointers, one designated as its *forward pointer* and the other as its *backward pointer*, plus a header with a forward pointer and a backward pointer, such that these conditions hold:
- considering only the forward pointers, it is a one-way linked list;
- following the sequence of backward pointers yields the reverse of the sequence obtained by following the forward pointers.

two-way sequential list: an ADT-template whose domain is strings, in which an entry is reached by applying the access operations *forward* and *backward*. Insertions are made before or after the *current* location.

union-find datatype: an abstract datatype whose records are mutually disjoint sets, in which there is a primary operation to locate the set containing a specified target element and a primary operation of merging two sets.

17.1 ABSTRACT DATATYPES

Organizing numbers and symbols into various kinds of records is a principal activity of information engineering. The organizational structure of a record is called a datatype. Abstractly, a datatype is characterized by a formal description of its domain and of the intrinsic operations by which information is entered, modified, and retrieved. Providing the specification at this abstract level ensures that the datatype is independent of the underlying types of information elements stored within the structure, and independent also of the hardware and software used to implement this organization. [AhHoUl83], [Kn68]

17.1.1 ABSTRACT SPECIFICATION OF RECORDS AND DATABASES

Information engineering uses discrete mathematics as a source of models for various kinds of records and databases. The language of abstract mathematics is used to specify a complex structure in terms of its elements. Constructors and templates are used to create new kinds of data from old kinds.

Definitions:

A **datatype** consists of a set of objects, called the *domain*, and a set of mappings, called *primary operations*, from the domain to itself or to the domain of some other datatype.

The **domain** of a datatype is its set of objects.

A **primary operation** for a datatype is a basic operation that retrieves information from an object in the domain or modifies the object.

A *secondary operation* on the domain of a datatype is an operation constructed from primary operations and previously defined secondary operations.

A *query* is a primary or secondary operation on a datatype domain that preserves the values of all its arguments and returns a logical value, i.e., *true* or *false*.

An *alphabetic datatype* is a datatype whose domain is a finite set of *symbols*. Its only primary operation is a total ordering query.

A *numeric datatype* is a datatype whose domain is a set of *numbers* and whose primary operations are a total ordering query and the arithmetic operators $+$ (addition), $\times$ (multiplication), and $-$ (change of sign).

An *elementary datatype* is an alphabetic datatype or a numeric datatype, usually intended for direct representation in the hardware of a computer.

An *abstract datatype* (**ADT**) is a mathematically specified datatype equipped with operations that can be performed on its data objects.

An **ADT-constructor** is a template that converts a datatype into a new ADT.

The constructor *sequence of* transforms a datatype X-type with domain D into a new datatype "sequence of X-type" whose domain is the set Seq_D of all finite sequences of elements of D. The primary operations of the resulting datatype are:

- *header(s)*, which yields a singleton sequence whose only element is the first object in the sequence s (or the empty sequence, if the sequence s is empty);
- *trailer(s)*, which deletes the first entry of sequence s (or yields the empty sequence, if the sequence s is empty);
- *concat(s, t)*, which concatenates the two sequences;
- *first(s)*, which gives the value of the first entry of a non-empty sequence s;
- *append(s, d)*, which appends to sequence $s \in Seq_D$ an entry $d \in D$;
- *nullseq( )*, whose value is the null sequence λ.

The constructor *set of* converts a datatype X-type with domain D into a new datatype "set of X-type" whose domain is the set of all subsets of D. The primary operations are:

- *inclusion(S, T)*, a query whose value is true if $S \subseteq T$;
- *union(S, T)*, whose value is $S \cup T$;
- *intersection(S, T)*, whose value is $S \cap T$;
- *difference(S, T)*, whose value is $S - T$;
- *choose(S)*, whose value is an arbitrary element of a nonempty subset S;
- *singleton(d)*, which transforms an element of D into the singleton set whose only entry is d;
- *emptyset( )*, whose value is the emptyset $\emptyset$;
- *universe( )*, whose value is the underlying domain D.

The constructor *k-tuple of* converts a list of k datatypes
$$X_1\text{-type}, X_2\text{-type}, \ldots, X_k\text{-type}$$
into a new datatype "k-tuple $(X_1, \ldots, X_k)$" whose domain is the cartesian product $D_1 \times D_2 \times \cdots \times D_k$ of the domains of the respective datatypes in that list. The primary operations are the projection functions:

- $coord_j(s)$, which gives the value of the jth coordinate of the k-tuple s;
- *entuple* $(d_1, \ldots, d_k)$, whose value is the k-tuple whose jth coordinate is the element d_j of domain D_j.

An **elementary ADT-constructor** is any of the three operations *sequence of, set of,* or *tuple of* used to build more complex ADTs from simpler ADTs.

The **Iverson truth function** assigns to a proposition p the integer value (p) such that
$$(p) = \begin{cases} 1 & \text{if } p \text{ is true;} \\ 0 & \text{otherwise.} \end{cases}$$

A **datatype specification** uses a combination of elementary datatypes and ADT-constructors to specify the domain and the primary operations. It may also use the following mathematical notation:

- $\emptyset$ denotes the empty set;
- λ denotes the empty sequence;
- · denotes the operation of appending one element to a sequence;
- ∘ denotes the sequence concatenation operation;

Moreover, every primary operation is either a query or a procedure.

Specifying a datatype as **generic** in an ADT-template means that any datatype can be used in that part of the template as a block in the construction of the new datatype.

Specifying a datatype as **ordered** in an ADT-template means that any ordered datatype can be used in that part of the template as a block in the construction of the new datatype.

The **precondition** of a primary operation is a list of conditions that must hold immediately before the operation is executed, for it to execute as described.

The **postcondition** of a primary operation is a specification of conditions that must hold after the operation is executed, if the precondition is satisfied when the operation commences.

The **flat notation** $X^\flat$ in a postcondition of a primary operation specification means the value of the variable X before that operation is executed. Unadorned X (without the $\flat$) means the value of X after the operation.

Facts:

1. The domain of an ADT is specified as a mathematical model, without saying how its elements are to be represented.

2. Sometimes the domain of a datatype is specified by roster. Other times it is specified with the use of set-theoretic operations.

3. A primary operation of an ADT is specified functionally. That is, its value on every element of the domain is declared, but the choice of an algorithm to be used in its implementation is omitted.

4. A primary operation may be implemented so that it has direct access to the data representing the value of the computational variable to which it is applied.

5. A primary operation can modify the information within a variable in its datatype or retrieve information from a variable.

6. A secondary function is implemented through calls to the primary operations from which it is ultimately composed.

7. There is no set of standard conventions for writing ADTs.

8. Software designers frequently specify a particular concrete information structure (see §17.2), instead of writing an ADT.

9. The advantage of writing an ADT, rather than a concrete datatype, is that it leaves the implementer room to find a new (and possibly improved) way to meet the requirements of the task.

10. In a datatype specification, a functional subprogram is represented by a procedure that produces a non-boolean value, and a variable to receive that value is specified as its last parameter.

Examples:

1. Elementary numeric datatypes include the integers and the reals.

2. Elementary alphabetic datatypes include the ASCII set and the decimal digits.

3. *Complex_number* is a datatype that represents complex numbers and their addition and multiplication.

ADT **complex_number**:
Domain
    2-tuple (*re*: real, *im*: real)

Primary Operations
    sum (*w*: complex_number, *z*: complex_number)
    Comment: add two complex numbers.
    {pre: *none*}
    {post: $sum(w,z) = entuple(re(w) + re(z),\ im(w) + im(z))$}
    prod (*w*: complex_number, *z*: complex_number)
    Comment: multiply two complex numbers.
    {pre: *none*}
    {post: $prod(w,z) = entuple$
    $(re(w) \cdot re(z) - im(w) \cdot im(z),\ re(w) \cdot im(z) + im(w) \cdot re(z))$}

4. *Baseten_digit* is a datatype that might be used in the construction of base-ten numerals representing arbitrarily large integers and their addition.

ADT **baseten_digit**:
Domain
    $\{0,1,2,3,4,5,6,7,8,9\}$: integers

Primary Operations
    add_digits (*x*: baseten_digit, *y*: baseten_digit)
    {pre: *none*}
    {post: $add\_digits(x,y) = x + y \bmod 10$}
    addcarry: (*x*: baseten_digit, *y*: baseten_digit)
    {pre: *none*}
    {post : $addcarry(x,y) = (x + y \geq 10)$}

5. The datatype *alphastring* represents sequences of lowercase English letters.

ADT **alphastring**:
Domain
    sequence of $\{a,b,c,d,e,f,g,h,i,j,k,l,m,n,o,p,q,r,s,t,u,v,w,x,y,z\}$

Primary Operations
    none except from the *constructor* sequence of

6. The **union-find** constructor transforms a datatype on domain set S into a datatype whose objects are of three kinds: elements of S, subsets of S, and partitions of S. There is a primary operation to merge two cells of a partition, and a primary operation to locate the cell of a partition that contains a specified target element.

17.1.2 STACKS AND QUEUES

Access to entries in the interior of a list is unnecessary much of the time. Restricting access to the first and last entries is a precaution to prevent mistakes.

Definitions:

A *stack* is an ADT whose domain is a sequence, one end of which is called the *top*, and the other the *bottom*. One primary operation, called *pushing* appends a new entry to the top, and the other, called *popping* removes the entry at the top and returns its value. No entry may be examined, added to the stack, or deleted from the stack except by an iterated composition of these operations.

The *top of a stack* is the end of that *stack* that can be accessed directly.

Pushing an entry onto a stack means appending it to the top of the stack.

Popping an entry from a stack means deleting it from the top of the stack and possibly examining the data it contains.

The **LIFO property** of a database is that the item retrieved is always the item most recently inserted. LIFO means "last-in-first-out".

A *queue* is an ADT whose domain is a sequence, one end of which is called the *front*, and the other the *back*. One primary operation, called *enqueueing* appends a new entry to the back, and the other, called *dequeueing* removes the entry at the front and returns its value. No entry may be examined, added to the queue, or deleted from the queue except by an iterated composition of these operations.

The *back of a queue* is the end to which entries may be appended.

The *front of a queue* is the end from which entries may be deleted and possibly examined.

Enqueueing an entry into a queue means appending it to the back of the queue.

Dequeueing an entry from a queue means deleting it from the front of the queue, and possibly examining the data it contains.

The **FIFO property** of a database is that the item retrieved is always the item inserted the longest ago. FIFO means "first-in-first-out".

Facts:

1. Abstract specification of stacks and queues mention only the behavior of those datatypes, and totally avoid all details of implementation. This permits a skillful implementer to innovate with efficient concrete structures (see §17.2) that meet the behavioral specification.

2. Abstract specification of stacks and queues is consistent with the principles of object-oriented programming, in which details of implementation are hidden inside the data objects, so that the rest of the program perceives only the specified functional behavior.

3. All stacks have the LIFO property. For any stack S and for any element b, after executing the sequence of instructions
$$\text{push}(S, b), \text{pop}(S, x)$$
the resulting value of the stack S is whatever it was before the operations.

4. A stack is most commonly implemented as a linked list (see §17.2.2).

5. After changing the value of the variable y to that of the top entry on the stack S, the sequence of instructions
$$\text{pop}(S, x), y := x, \text{push}(x)$$
restores S to its previous state.

6. All queues have the FIFO property. Given an empty queue Q and two elements b_1 and b_2, the sequence of operations
$$\text{enqueue}(Q, b_1), \text{enqueue}(Q, b_2), \text{dequeue}(Q, x_1), \text{dequeue}(Q, x_2),$$
yields $x_1 = b_1$, $x_2 = b_2$, and $Q = \lambda$.

7. A queue is most commonly implemented as a linked list (see §17.2.2).

Examples:

1. The following pseudocode specifies the ADT **stack of D**, where D is an arbitrary datatype.

Domain
    sequence of D: generic

Primary Operations
    create_stack (S: stack)
    Comment: Initialize variable S as an empty stack.
        {pre: *none*}
        {post: $S = \lambda$}
    push (S: stack, x: element of D)
    Comment: Put value of x at **top** of stack S
        {pre: *none*}
        {post: $S = x \cdot S^b$}
    pop (S: stack, x: element of D)
    Comment: Remove top item of stack S; return it as value of variable x.
        {pre: $S \neq \lambda$}
        {post: $x \cdot S = S^b$}
    query_empty_stack (S: stack)
    Comment: Decide whether stack S is empty.
        {pre: *none*}
        {post: query_empty_stack $= (S = \lambda)$}

2. The following pseudocode specifies the ADT **queue of D**, where D is an arbitrary datatype.

Domain
    sequence of D: generic

Primary Operations
　create_queue (Q: queue)
　Comment: Initialize Q as an empty queue.
　　{pre: *none*}
　　{post: $Q = \lambda$ }
　enqueue (Q: queue, x: element of D)
　Comment: Put x at the **back** of queue Q.
　　{pre: *none*}
　　{post: $Q = Q^b \cdot x$}
　dequeue (Q: queue, x: element of D)
　Comment: Delete **front** of Q; return as x.
　　{pre: $Q \neq \lambda$}
　　{post: $Q^b = x \cdot Q$}
　query_empty_queue (Q: queue)
　Comment: Decide whether queue Q is empty.
　　{pre: *none*}
　　{post: query_empty_queue = $(Q = \lambda)$}

3. The following figure illustrates the difference between stacking (last-in-first-out) and queueing (first-in-first out).

empty stack		empty queue	
push A	A	enqueue A	A
push B	B A	enqueue B	A B
push C	C B A	enqueue C	A B C
pop ⇒ C	B A	dequeue ⇒ A	B C
pop ⇒ B	A	dequeue ⇒ B	C
push D	D A	enqueue D	C D

17.1.3 TWO-WAY SEQUENTIAL LISTS

A two-way sequential list conceptualizes a linear list as having a current location, so that entries may be inserted or deleted only at the current location.

Definitions:

A **two-way sequential list** is a list with a designated location at which access is permitted.

The **current location** of a two-way sequential list is the location at which access is permitted.

The **forepart** of a two-way sequential list is the part preceding the current location, which is empty when the current location is at the start of the list.

The **aftpart** of a two-way sequential list is the part following the current location, which is empty when the current location is at the finish of the list.

Facts:

1. A two-way sequential list does not maintain place-in-list numbers for the entries. The result of such an additional requirement would force the insert operation to renumber the part of the list following a newly inserted entry. This would slow the performance.

2. A two-way sequential list is easily implemented as a pair of stacks.

Example:

1. ADT *seq_list of D*
Domain
    2-tuple (*fore*: sequence of D, *aft*: sequence of D)
    type D: generic
Primary Operations
    create_list (L: seq_list)
    Comment: Initialize an empty list L.
        {pre: *none*}
        {post: $fore(L) = \lambda \wedge aft(L) = \lambda$}
    reset_to_start (L: seq_list)
    Comment: Reset to start of list.
        {pre: *none*}
        {post: $fore(L) = \lambda \wedge aft(L) = fore(L^b) \circ aft(L^b)$}
    advance (L: seq_list)
    Comment: Advance current position by one element.
        {pre: $aft(L) \neq \lambda$}
        {post: $(\exists x : D)[fore(L) = fore(L^b) \cdot x \wedge aft(L^b) = x \cdot aft(L)]$}
    query_atstart (L: seq_list)
        {pre: *none*}
        {post: $query\_atstart = (fore(L) = \lambda)$}
    query_atfinish (L: seq_list)
        {pre: *none*}
        {post: $query\_atfinish = (aft(L) = \lambda)$}
    insert (L: seq_list, x: element of D)
        {pre: *none*}
        {post: $aft(L) = x \cdot aft(L^b) \wedge fore(L) = fore(L^b)$}
    remove (L: seq_list, x: element of D)
        {pre: $aft(L) \neq \lambda$}
        {post: $aft(L^b) = x \cdot aft(L) \wedge fore(L) = fore(L^b)$}
    swap_right (L: seq_list, M: seq_list)
        {pre: *none*}
        {post: $fore(L) = fore(L^b) \wedge fore(M) = fore(M^b)$
            $\wedge aft(L) = aft(M^b) \wedge aft(M) = aft(L^b)$}

17.1.4 DICTIONARIES AND RANDOM ACCESS LISTS

Definitions:

A **keyed pair** is a 2-tuple whose first entry, which is called a *key*, is from an ordered datatype and is used to access data in the second entry.

A **table** is a set of keyed pairs such that no two keys are identical.

A *random access list* is a table whose keys are consecutive integers.

A *dictionary* is another name for a table.

Facts:

1. A static table (whose size does not change) can be implemented as an array.

2. A dynamic table (which permits inserts and deletes) is often implemented as a binary search tree (see §17.2.4).

3. Specifying a datatype as a dictionary means that its primary retrieval operation can execute in $\Theta(n)$ time.

4. Specifying a datatype as a random access list means that its primary retrieval operation can execute in $\Theta(1)$ time.

Examples:

1. ADT *table*
Domain
    set of table_entry
    type table_entry: 2-tuple (*key*: ordered, *data*: generic)

Primary Operations
    create_table (*T*: table)
      {pre: *none*}
      {post: $T = \lambda$}
    insert_entry (*T*: table, *e*: table_entry)
      {pre: $(\forall e' \in T)[\text{key}(e') \neq \text{key}(e)]$}
      {post: $T = T^b \cup \{e\}$}
    remove_entry (*T*: table, *e*: table_entry)
      {pre: $e \in T$}
      {post: $T = T^b - \{e\}$}
    find_entry (*T*: table, *k*: key, *found*: boolean, *e*: table_entry)
      {pre: *none*}
      {post: $((\exists e' \in T)[e'.key = k]) \wedge (found = true) \wedge (e = e'))$
        $\vee (\neg(\exists e' \in T)[e'.key = k] \wedge found = false)\}$

2. ADT **Random access list**
Domain
    set of table_entry
    type table_entry: 2-tuple (*key*: subrange of integers, *data*: generic)
Primary Operations
    Exactly the same as for the ADT table.

17.1.5 PRIORITY QUEUES

A priority queue is an "unfair queue", in which entries are not dequeued on a first-enqueued basis. Instead, each entry has a priority, and is dequeued on a highest priority basis.

Definition:

A *priority queue* is a set of keyed pairs, such that the key of the entry returned by a dequeue operation is not exceeded by the key of any other entry currently in the queue.

Facts:

1. A priority queue is usually implemented as a heap (see §17.2.5).
2. Two different entries in a priority queue may have the same key.
3. The operating system for a multi-user programming environment places computational tasks into a priority queue.

Example:

1. ADT *P_queue*
Domain
    set of Pq_entry
    type Pq_entry: 2-tuple (key: ordered, data: generic)

Primary Operations
    create_Pq (PQ: P_queue)
        { pre: *none* }
        { post: $PQ = \lambda$ }
    enPqueue(PQ: P_queue, e: Pq_entry)
        { pre: *none* }
        { post: $PQ = PQ^b \cup \{e\}$ }
    dePqueue (PQ: P_queue, e: Pq_entry)
        { pre: $PQ \neq \emptyset$ }
        { post: $(e \in PQ^b) \wedge (\forall e' \in PQ)[key(e) \leq key(e')] \wedge PQ = PQ^b - \{e\}$ }
    query_empty_Pqueue(PQ: P_queue)
        { pre: *none* }
        { post: query_empty_Pqueue $= (PQ = \lambda)$ }

7.2 CONCRETE DATA STRUCTURES

Concrete data structures configure computer memory into containers of related information. They are used to implement abstract datatypes. Contiguous stretches of memory are regarded as arrays, and noncontiguous portions are linked with pointers.

7.2.1 MODELING COMPUTER STORAGE AND RETRIEVAL

There are a few generic concepts common to nearly all concrete data structures.

Definitions:

A *concrete data structure* is a mathematical model for storing the current value of a structured variable in computer memory.

A *cell* in a concrete data structure S is a unit within the data structure that may contain data and pointers to other cells.

The *header* of a concrete data structure is a special unit that contains current information about the entire configuration and pointers to some critical cells in the structure. It is *not* a cell.

An **insert operation** insert(S: structure, c: cell, loc: location) inserts a new cell c into structure S at location loc.

A **delete operation** delete(S: structure, loc: location) deletes from a structure S the cell at location loc.

A **target predicate** for a concrete data structure is a predicate that applies to the cells.

A **find operation** find(S: structure, t: target, loc: location) searches a structure S for a cell that satisfies target predicate t. It returns $false$ if there is no such cell. In addition to returning the boolean value $true$ if there is such a cell, it also assigns to its location parameter loc the location of such a cell.

A **next operation** next(S: structure, loc: location) returns the boolean value $true$ if the structure S is nonempty, in which case it also assigns to its location parameter loc the location of whatever cell it regards as next; it returns $false$ if S is empty.

The **size of a cell** is the number of bytes of computer memory it occupies.

A **pointer to a cell** is a representation of its location in computer memory.

A **null pointer** is a pointer that points to an artificial location. Detecting a null pointer is a signal to an algorithm to react somewhat differently than to a pointer to an actual location.

Facts:

1. There may be several alternative suitable concrete data structures that can be used to implement a given abstract datatype.

2. If the records of a database are all of the same fixed size, then the records themselves may be in the cells of a concrete data structure.

3. If the size of records is variable, then the cells of the concrete data structure often contain pointers to the actual data, rather than the data itself. This permits faster execution of operations.

4. The most common form of target predicate for a concrete data structure is an assertion that a key component of the cell matches some designated value.

17.2.2 ARRAYS AND LINKED LISTS

Definitions:

An **array** is an indexed sequence of identically structured cells $\langle a_j \mid j = d, \ldots, u \rangle$, with consecutive indices.

An array is **zero-based** if its lowest index is zero.

A **one-way linked list** is a set of cells, each with one pointer, such that:
- exactly one of these cells is pointed to by the header but by no cell;
- exactly one cell contains a null pointer;
- the sequence of cells formed by following the pointers, starting from the header, traverses the entire set, ending with the cell containing the null pointer.

The **far end of a one-way linked list** is the cell that contains a null pointer.

The **near end of a one-way linked list** is the cell that is pointed to by the header and by no other cell.

A *two-way linked list* is a set of cells, each with two pointers, one designated as its *forward pointer* and the other as its *backward pointer*, plus a header with a forward pointer and a backward pointer, such that:
- considering only the forward pointers, it is a one-way linked list;
- following the sequence of backward pointers yields the reverse of the sequence obtained by following the forward pointers.

A *sparse sequence* is a sequence in which nearly all the entries are zeros.

A *circular linked list* is a set of cells, each with two pointers, one designated as its *forward pointer* and the other as its *backward pointer*, plus a header with one or more pointers to *current cells*, such that:
- the sequence of cells formed by following the forward pointers, starting from any cell, traverses the entire set and returns to the starting cell;
- the sequence of cells formed by following the backward pointers, starting from any cell, traverses the entire set and returns to the starting cell.

Facts:

1. A random-access list (§17.1.4) can be implemented as an array so that a *find* operation executes in $O(1)$ time.

2. A stack (§17.1.2) can be implemented as a one-way linked list with its top at the near end, so that *push* and *pop* both execute in $O(1)$ time.

3. A queue (§17.1.2) can be implemented as a two-way linked list with its back at the near end of the forward list and its front at the far end, so that *enqueue* and *dequeue* both execute in $O(1)$ time.

4. A two-way sequential list (§17.1.3) can be implemented as a two-way linked list, or as a pair of one-way linked lists.

Examples:

1. The following figure illustrates an array with cells $a_d, \ldots, a_u$.

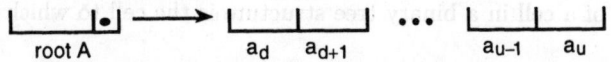

2. The following figure illustrates a one-way linked list, with cell a_d at the near end and cell a_u at the far end.

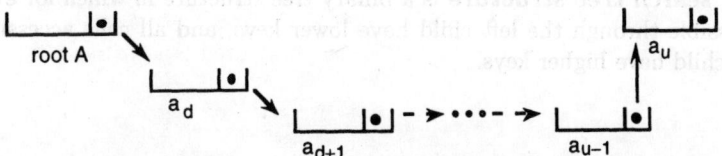

3. The following figure illustrates a two-way linked list.

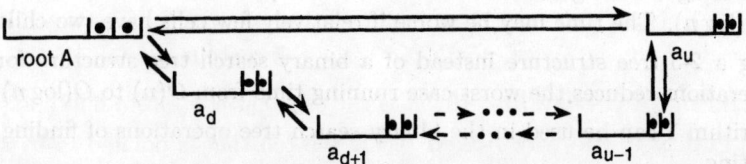

> **Algorithm 1:** BSTsearch(T, t).
>
> input: a binary-search tree T and a target key t
> output: if $t \in T$, the address of the vertex with key t, else the address where t could be inserted
> if $root(T) =$ NULL then return address of $root$
> else if $t < key(root)$ then BSTsearch$(leftsubtree(T), t)$
> else if $t = key(root)$ then return address of $root$
> else BSTsearch$(rightsubtree(T), t)$

4. Representing a sparse finite sequence by a linked list can save space. The cell given to each nonzero entry includes its position in the sequence and points to the cell with the next nonzero entry.

5. A queue whose maximum length is bounded can be represented by a circular linked list with two header pointers, one to the back and one to the front of the queue. The number of cells equals the maximum queue length. This eliminates the need for "garbage collection".

17.2.3 BINARY SEARCH TREES

Definitions:

A **tree structure** is a concrete data structure such that the header points to a single cell, and such that from that cell to each other cell, there is a single chain of pointers.

The **root cell** of a tree structure is the cell to which the header points.

A **binary tree structure** is a tree structure such that each cell has two pointers.

The **left child** of a cell in a binary tree structure is the cell to which the first pointer points.

The **right child** of a cell in a binary tree structure is the cell to which the second pointer points.

A **binary search tree structure** is a binary tree structure in which for every cell, all cells accessible through the left child have lower keys, and all cells accessible through the right child have higher keys.

Facts:

1. The *ADT table* is commonly implemented as a binary search tree structure.

2. The average running time for the *ADT table* operations of *insertion*, *deletion*, and *find* is $O(\log n)$. The time may be worse if relatively few cells have two children.

3. Using a *2-3 tree structure* instead of a binary search tree structure for the *ADT table* operations reduces the worst case running time from $O(n)$ to $O(\log n)$.

4. Algorithm 1 can be used in the binary search tree operations of finding, inserting, and deleting.

5. To find a target key t in the binary search tree T, first apply BSTsearch(T, t). If the address of a null pointer is returned, then there is no node with key t. Otherwise, the address returned is a node with key t.

6. To insert a node with target key t into the binary search tree T, first apply the algorithm BSTsearch(T, t). Then install the node at the location returned.

7. To delete node t from the binary search tree T, first apply BSTsearch(T, t). Then replace node t either by the node with the largest key in the left subtree or by the node with the smallest key in the right subtree.

Examples:

1. The following figure illustrates a binary search tree.

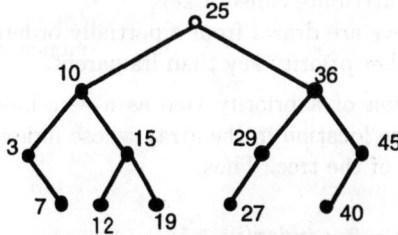

2. Inserting 32 into the BST of Example 1 yields the following BST.

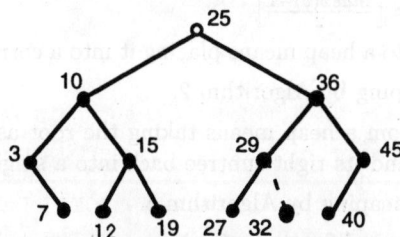

3. Deleting node 10 from the BST of Example 1 would yield one of the following two BSTs.

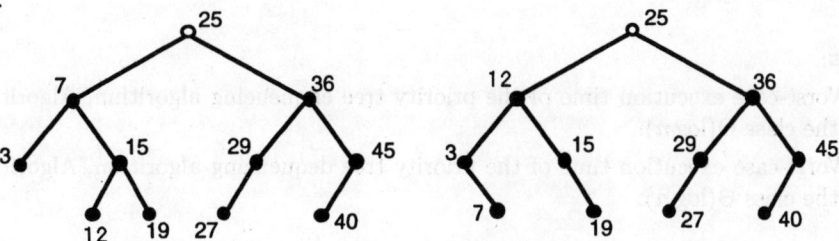

17.2.4 PRIORITY TREES AND HEAPS

Definitions:

A binary tree is **left-complete** if it is complete or if it is a balanced binary tree (§9.1.2) such that at depth one less than the maximum, the following conditions hold:
- all nodes with two children are to the left of all nodes with one or no children;
- all nodes with no children are to the right of all nodes with one or two children;
- there is at most one node with only one child, which must be a left-child.

Algorithm 2: PriorityTreeEnqueue (T, x).

input: a priority tree T and a new entry x
output: tree T with the new vertex x inserted so that it remains a priority tree
install entry x into the first vacant spot in the left-complete tree T
while $x \neq root(T)$ and $priority(x) > priority(parent(x))$
    swap x with $parent(x)$

A *priority tree* is a left-complete binary tree, with the following additional structure:
- each vertex has an attribute called a *key*;
- the values of the keys are drawn from a partially ordered set;
- no vertex has a higher priority key than its parent.

A *heap* is a representation of a priority tree as a zero-based array, such that each vertex is represented at the location in the array whose index equals its location in the breadth-first-search order of the tree. Thus:
- $index(root) = 0$;
- $index(leftchild(v)) = 2 \times index(v) + 1$;
- $index(rightchild(v)) = 2 \times index(v) + 2$;
- $index(parent(v)) = \lfloor \frac{index(v)-1}{2} \rfloor$.

Enheaping an entry into a heap means placing it into a correctly prioritized position.

Trickle-up means enheaping by Algorithm 2.

Deheaping an entry from a heap means taking the root as the deheaped entry and patching its left subtree and its right subtree back into a single tree.

Trickle-down means deheaping by Algorithm 3.

A *Fibonacci heap* is a modification of a heap, using the Fibonacci sequence, that permits more efficient implementation of a priority queue than a heap based on a left-complete binary tree.

Facts:

1. Worst-case execution time of the priority tree enqueueing algorithm, Algorithm 2, is in the class $\Theta(\log n)$.

2. Worst-case execution time of the priority tree dequeueing algorithm, Algorithm 3, is in the class $\Theta(\log n)$.

Examples:

1. This is a left-complete binary tree.

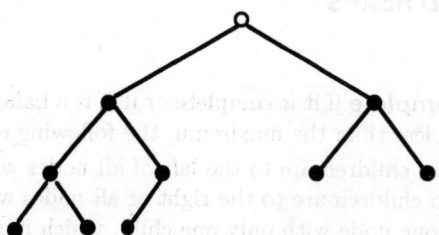

Algorithm 3: PriorityTreeDequeue (T).

input: a priority tree T
output: tree $T - root(T)$ with priority-tree shape restored

replace $root(T)$ by rightmost entry y at bottom level of T
while y is not a leaf and $[priority(y) \leq priority(leftchild(y))$ or
$\quad priority(y) \leq priority(rightchild(y))]$
$\quad$ if $priority(leftchild(y)) > priority(rightchild(y))$
$\quad\quad$ then swap y with $leftchild(y)$
$\quad\quad$ else swap y with $rightchild(y)$

2. The following is a priority tree of height 3.

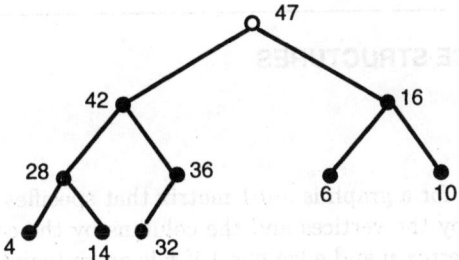

3. The following figure illustrates a priority tree insertion. It shows how 45 is inserted into the priority tree of Example 2 in the correct location to maintain the left-compete binary tree shape and then rises until the priority property is restored.

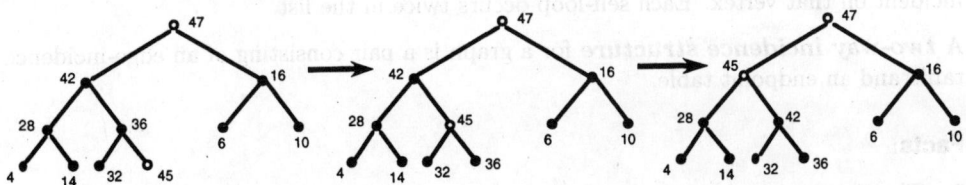

4. The following figure illustrates a priority tree deletion. It shows how the left-complete binary tree shape and priority property are restored after the root is removed from the priority tree of Example 2.

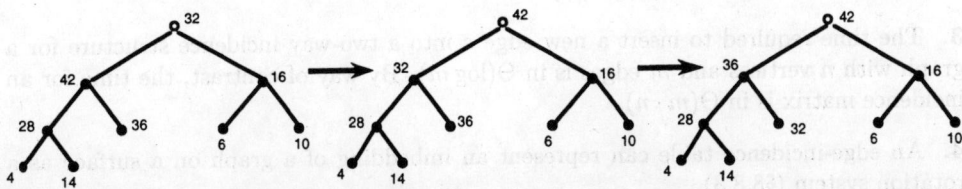

5. The following heap corresponds to the priority tree of Example 2. Observe that the keys occur in the array according to the breadth-first-search order of their vertices.

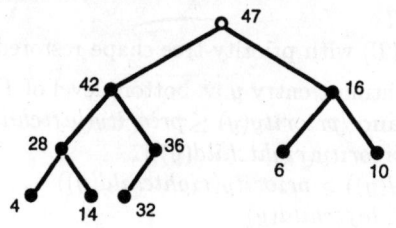

index	0	1	2	3	4	5	6	7	8	9
key	47	42	16	28	36	6	10	4	14	32

17.2.5 NETWORK INCIDENCE STRUCTURES

Definitions:

An *incidence matrix* for a graph is a 0-1 matrix that specifies the incidence relation. The rows are indexed by the vertices and the columns by the edges. The entry in the row corresponding to vertex v and edge e is 1 if v is an endpoint of e, and 0 otherwise.

An *endpoint table* for a graph (§8.1) is a *dictionary* whose keys are the edges. The data component for each key edge is the set of endpoints for that edge. If an edge is directed, then its endpoints are marked as *head* and *tail*.

An *edge-incidence table* is a *dictionary* whose keys are the vertices of a graph or digraph. The data component for each key vertex is a list of all the edges that are incident on that vertex. Each self-loop occurs twice in the list.

A *two-way incidence structure* for a graph is a pair consisting of an edge-incidence table and an endpoint table.

Facts:

1. The time required to insert a new vertex v into a two-way incidence structure for a graph with n vertices and m edges is in $\Theta(\log n)$. By way of contrast, the time for an incidence matrix is in $\Theta(n \cdot m)$.

2. The time required to delete a vertex v from a two-way incidence structure for a graph with n vertices and m edges is in $\Theta(\log n + \deg(v))$. By way of contrast, the time for an incidence matrix is in $\Theta(n \cdot m)$.

3. The time required to insert a new edge e into a two-way incidence structure for a graph with n vertices and m edges is in $\Theta(\log m)$. By way of contrast, the time for an incidence matrix is in $\Theta(m \cdot n)$.

4. An edge-incidence table can represent an imbedding of a graph on a surface as a rotation system (§8.8.3).

Example:

1. The following graph corresponds to the network incidence structure given below.

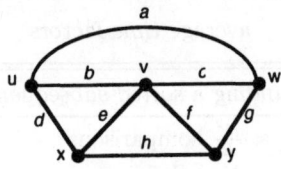

EDGE-INCIDENCE TABLE

$u.$	a	b	d
$v.$	b	c	e	f
$w.$	a	c	g
$x.$	d	e	h
$y.$	f	g	h

ENDPOINT TABLE

$a.$	u	w
$b.$	u	v
$c.$	v	w
$d.$	u	x
$e.$	v	x
$f.$	v	y
$g.$	w	y
$h.$	x	y

17.3 SORTING AND SEARCHING

Since commercial data processing involves frequent sorting and searching of large quantities of data, efficient sorting and searching algorithms are of great practical importance. Sorting and searching strategies are also of fundamental theoretical importance, since sorting and searching steps occur in many algorithms. Table 1 compares the performance of some of the most common sorting methods.

17.3.1 GENERIC CONCEPTS FOR SORTING AND SEARCHING

Definitions:

A *database* is a set of *entries*, stored in a computer as an *information structure*.

An *entry* in a database is a 2-tuple whose first component is a *key* and whose second component is some data.

A *key* in a database entry is a value from an ordered set, used to store and retrieve data.

The *key domain* is the ordered set from which keys are drawn.

The *primary key* is the component of highest precedence, when the key for database records has more than one component.

The *secondary key* is the component of next highest precedence after the primary key.

A *record* is another name for a database entry.

Table 1 Comparison of sorting methods.

sorting method (grouped by type)	average time factors	comments
expanding a sorted subsequence		
selection sort	$\approx \frac{N^2}{2}$ comparisons $\approx N$ exchanges	
insertion sort	$\approx \frac{N^2}{4}$ comparisons $\approx \frac{N^2}{8}$ exchanges	linear if input file is "almost sorted"
Shellsort	$< N^{3/2}$ comparisons	for "good" increments $1, 4, 13, 40, 131, \ldots$
exchanging out-of-order pairs		
bubblesort	$\approx \frac{N^2}{2}$ comparisons $\approx \frac{N^2}{4}$ exchanges	one pass if input file is already sorted
sinking sort	$\approx \frac{N^2}{2}$ comparisons $\approx \frac{N^2}{4}$ exchanges	one pass if input file is already sorted
shakersort	$\approx \frac{N^2}{2}$ comparisons $\approx \frac{N^2}{4}$ exchanges	
heapsort	$< 2N \lg N$ comparisons	always $\Theta(N \log N)$
divide-and-conquer		
mergesort	$\approx N \lg N$ comparisons	always $\Theta(N \log N)$
quicksort	$\approx 2N \lg N$ comparisons	worst-case $\frac{N^2}{2}$
sorting by distribution		
rank counting	$\Theta(N)$	
radix sort on k-digit key	$\approx N \lg N$ comparisons	

Sorting is the process of arranging a collection of database entries into a sequence that conforms to the order of their keys.

Searching a database means using a systematic procedure to find an entry with a key designated as the objective of the search.

Scanning a database (or a portion of a database) means examining every record in that database (or portion).

The **target** of a database search is an entry whose key that has been designated as the objective of the search.

A **comparison sort** is any sorting method that uses only comparisons of keys.

An **internal sorting method** keeps all the entries in the primary memory of the computer during the process of rearrangement.

An **external sorting method** uses external storage outside the main memory during the sorting process.

An **in-place realization** of a sorting method uses beyond the space needed for one copy of each data entry, only a constant amount of additional space, regardless of the size of the list to be sorted.

A **dynamic structure** for a database is an information structure whose configuration may change during an algorithmic process, for instance, by the insertion or deletion of elements.

A **static structure** for a database is a data structure whose configuration does not change during an algorithmic process.

Facts:

1. Several different general strategies for sorting are given in the following subsections. Each leads to more than one method for sorting.
2. Some elementary sorting methods take $O(n^2)$ time. Most practical comparison sorting methods require $O(n \log n)$ time.
3. The worst case running time of any comparison sort is at least $\Omega(n \log n)$.

Examples:

1. Selection sort (§17.3.2), insertion sort (§17.3.2), Shellsort (§17.3.2), bubblesort (§17.3.3), heapsort (§17.3.3), and quicksort (§17.3.4) are all internal comparison sorts.
2. Mergesort (§17.3.4) is a comparison sort that may be either internal or external.
3. *Database model for a telephone directory*: Each entry has as key the name of a person and as data that persons's telephone number. The target of a search is the entry for a person whose number one wishes to call. Names of persons form an ordered key domain under a modified lexicographic ("alphabetic") ordering, in which it is understood that a family name (a "last name" in European-based cultures) has higher precedence than a given name.
4. *Database model for a reverse telephone directory*: In a reverse telephone directory entry the key is a telephone number and the data is the name of the person with that number. This permits the telephone company to retrieve the name of the person who has a particular phone number, for instance, if someone inquires why some particular telephone number occurs on a long-distance phone bill.
5. *Database model for credit-card information*: In a credit-card database, the key to each entry is a credit-card number, and the data include the name of the cardholder, the maximum credit limit, and the present balance.

17.3.2 SORTING BY EXPANDING A SORTED SUBSEQUENCE

One general strategy for sorting iteratively expands a sorted subsequence, most often implemented as an array or a linked list, until the expanded subsequence ultimately contains all the entries of the database.

Definitions:

A **selection sort** iteratively transforms an unsorted input sequence into a sorted output sequence. At each iteration, it *selects* the item with smallest key from the remaining items in the unsorted input sequence and appends that item at the end of the sorted output sequence.

An **insertion sort** iteratively transforms an unsorted input sequence into a sorted output sequence. At each iteration, it takes the first remaining item from the unsorted input subsequence and *inserts* it into its proper position in the sorted output sequence.

A **Shellsort** of an unsorted sequence $a_1, \ldots, a_n$ is based on a list of *increments* of decreasing size: $h_1 > h_2 > \cdots > h_t = 1$. On the kth iteration, the sequence is partitioned into h_k subsequences, such that for $j = 1, \ldots, h_k$, the jth subsequence is

$$\langle a_{j+rh_k} \mid 0 \leq r \leq \tfrac{n-j}{h_k} \rangle$$

and each of these j subsequences is sorted by an insertion sort.

Facts:

1. Selection sorts and insertion sorts both have time-complexity $O(n^2)$ in the worst case.

2. The time-complexity of a selection sort is independent of the order of the input sequence, since finding the smallest remaining item requires scanning all the remaining items.

3. The running time of an insertion sort can be significantly reduced for "almost sorted" sequences, with time $O(n)$ as the limiting case.

4. Optimizing the running time of a Shellsort involves some very difficult mathematical problems, many of which have not yet been solved. In particular, it is not known which choice of increments yields the best result.

5. It is known that Shellsort increments should not be multiples of each other, if the objective is to achieve fast execution.

6. Evidence supporting the efficiency of the Shellsort increment list $\ldots, 40, 13, 4, 1$ according to the rule $h_{i-1} = 3h_i + 1$ is given by Knuth [Kn73]. The increment list $\ldots, 15, 7, 3, 1$ satisfying the rule $h_{i-1} = 2h_i + 1$ is also recommended.

7. Shellsort is a refinement of a straight insertion sort. The motivation for its design in 1959 by D. L. Shell is based on the observation that an insertion sort works very fast for "almost sorted" sequences.

8. Shellsort is guaranteed to produce a sorted list, because on the last pass, it applies an insertion sort to the whole sequence.

9. An in-place realization of the strategy of expanding a sorted subsequence conceptually partitions the array into a sorted subsequence at the front of the array $A[1..n]$ and an unsorted subsequence of remaining items at the back. Initially, the sorted subsequence is the empty sequence and the unsorted subsequence is the whole list. At each step of the iteration, the sorted front part expands by one item and the unsorted back part contracts by one item.

Algorithms:

1. Algorithm 1 is an in-place realization of a selection sort.
2. Algorithm 2 is an in-place realization of an insertion sort.

Examples:

In the following examples of single-list implementations of SelectionSort and InsertionSort, the symbol " | " separates the sorted subsequence at the front from the remaining unsorted subsequence at the back. The arrows "←" and "→" indicate how far the index j moves during an iteration.

> **Algorithm 1:** SelectionSort of array $A[1..n]$.
>
> for $i := 1$ to $n - 1$ do
>   minindex $:= i$; minkey $:= A[i]$
>   for $j := i + 1$ to n do
>   if $A[j] <$ minkey then minindex $:= j$; minkey $:= A[j]$ end-if
>   swap $A[i]$ with $A[\text{minindex}]$

> **Algorithm 2:** InsertionSort of array $A[1..n]$.
>
> for $i := 2$ to n do
>   nextkey $:= A[i]$; $j := i - 1$
>   while $j > 0$ and $A[j] >$ nextkey do $A[j+1] := A[j]$; $j := j - 1$ end-while
> $A[j+1] :=$ nextkey

1. On the sequence 15, 8, 10, 6, 13, 17, SelectionSort would progress as follows:

	minkey							minkey							minkey			
15	8	10	$\hat{6}$	13	17		6	$\hat{8}$	10	15	13	17		6	8	$\hat{10}$	15	13	17
i	j $\xrightarrow{\text{search for minkey}}$						i	j $\longrightarrow$						i	j $\longrightarrow$			

			minkey							minkey							
6	8	10	15	$\hat{13}$	17		6	8	10	13	$\hat{15}$	17		6	8	10	13	15	17
		i	j $\longrightarrow$							i	j		

2. On the sequence $\{15, 8, 10, 6, 13, 17\}$, InsertionSort would progress as follows:

   shift: 15 ↑ 8 10 6 13 17   shift: 8 15 ↑ 10 6 13 17   shift shift shift: 8 10 15 ↑ 6 13 17
   ← j i ← j i ←——— j i

   shift: 6 8 10 15 ↑ 13 17   6 8 10 13 15 | 17   6 8 10 13 15 17 |
   ← j i j i

3. If $n = 13$ and $h_3 = 4$, then on the third iteration, ShellSort would insertion sort the following subsequences:

$$a_1, a_5, a_9, a_{13}$$
$$a_2, a_6, a_{10}$$
$$a_3, a_7, a_{11}$$
$$a_4, a_8, a_{12}$$

17.3.3 SORTING BY EXCHANGING OUT-OF-ORDER PAIRS

A standard measure of the totality of disorder of a sequence of n items is the number of pairs (a_i, a_j) such that $i < j$ but $a_i > a_j$. Thus, the disorder ranges from 0 (i.e., totally ordered) to $\binom{n}{2}$ (i.e., in reverse order). The strategy of exchange sorts is to swap out-of-order pairs until all pairs are in order.

Definitions:

A *bubblesort* scans an array $A[1..n]$ repeatedly from the highest index to lower indices, each time swapping every out-of-order pair of consecutive items that is encountered.

A *sinking sort* scans an array $A[1..n]$ repeatedly from the lowest index to higher indices, each time swapping every out-of-order pair of consecutive items that is encountered.

A *shakersort* scans an array $A[1..n]$ repeatedly, and alternates between bubbling upward and sinking downward on alternate scans.

A bubblesort, sinking sort, or shakersort is *adaptive* if it stops the first time a scan produces no transpositions.

Heapsort sorts a sequence of entries by iteratively enheaping them all into a heap (§17.2.4) and then iteratively deheaping them all. The order in which they deheap is sorted.

Facts:

1. The name "bubblesort" suggests imagery in which lighter items (i.e., earlier in the prescribed order of the key domain) bubble to the top of the list.

2. The name "sinking sort" suggests that heavier items sink to the bottom.

3. The name "shakersort" suggests a salt shaker that is turned upside down.

4. Since each swap during an exchange sort reduces the total disorder, it follows that each scan brings the list closer to perfect order. By transitivity of the order relation, it follows that if every consecutive pair in a sequence is in the correct order, then the entire sequence is in order.

5. After the first pass of a bubblesort from bottom to top, the smallest element is certain to be in its correct final position at the beginning of the list. After the second pass, the second largest element must be in its correct position, and so on.

6. Bubblesort has worst-case time complexity $O(n^2)$.

7. For "almost sorted" sequences, an adaptive bubblesort can run much faster than $O(n^2)$ time.

8. The priority property implies that the root of a priority tree is assigned the data entry with first precedence.

9. Whereas a sequence of length n has $\binom{n}{2}$ pairs that might be out of order, a binary tree of n elements has at most $n \log n$ pairs that could be out of order, if one compares only those pairs such that one node is an ancestor of the other.

10. Heapsort improves upon the idea of bubblesort because it bubbles only along tree paths between a bottom node and the root, instead of along the much longer path in a linear sequence from a last item to the first.

11. Heapsort runs in $O(n \log n)$ time.

12. Heapsort was invented by J. W. J. Williams in 1964.

Algorithms:

1. Algorithm 3 is an adaptive version of bubblesort.
2. Algorithm 4 is a heapsort algorithm.

> **Algorithm 3:** BubbleSort of array A[1..n].
>
> $first := 1;\ last := n;\ exchange := true$
> **while** $exchange$ **do**
> $\quad first := first + 1;\ exchange := false;$
> $\quad$ **for** $i := last$ **to** $first$ **with step** -1 **do**
> $\quad\quad$ **if** $A[i] < A[i-1]$ **then** {swap $A[i]$ and $A[i-1]$; $exchange := true$}

> **Algorithm 4:** HeapSort of array A[0..n] into array B[0..n].
>
> **procedure** heapify(i)
> **if** ($A[i]$ is not a leaf) **and** (a child of $A[i]$ is larger than $A[i]$) **then**
> $\quad$ let $A[k]$ be the larger child of $A[i]$
> $\quad$ swap $A[i]$ and $A[k]$
> $\quad$ heapify(k)
>
> **procedure** buildheap
> **for** $i := \lfloor \frac{n}{2} \rfloor$ **downto** 0 **do**
> $\quad$ heapify(i)
>
> **main program** heapsort
> buildheap
> **for** $i = 0$ **to** n **do**
> $\quad$ deheap root of A and transfer its value to $B[n-i]$

Examples:

1. When canceled checks are returned to the payer by a bank, they may be in nearly sorted order, since the payees are likely to deposit checks quite soon after they arrive. Thus, they arrive for collection in an order rather close to the order in which they are written. A shakersort might work quite quickly on such a distribution.

2. Starting with the unsorted list $L = 15, 8, 10, 6, 17, 13$, bubblesort would produce the following sequence of lists.

$$
\begin{array}{rcccccc}
\text{initial list}: & 15 & 8 & 10 & 6 & 17 & 13 \\
\text{after one pass}: & 6 & 15 & 8 & 10 & 13 & 17 \\
\text{after two passes}: & 6 & 8 & 15 & 10 & 13 & 17 \\
\text{after three passes}: & 6 & 8 & 10 & 15 & 13 & 17 \\
\text{after four passes}: & 6 & 8 & 10 & 13 & 15 & 17 \\
\end{array}
$$

3. Starting with the unsorted list $L = 15, 8, 10, 6, 17, 13$, sinking sort would produce the following sequence of lists.

$$
\begin{array}{rcccccc}
\text{initial list}: & 15 & 8 & 10 & 6 & 17 & 13 \\
\text{after one pass}: & 8 & 10 & 6 & 15 & 13 & 17 \\
\text{after two passes}: & 8 & 6 & 10 & 13 & 15 & 17 \\
\text{after three passes}: & 6 & 8 & 10 & 13 & 15 & 17 \\
\end{array}
$$

4. Starting with the unsorted list $L = 15, 8, 10, 6, 17, 13$, shakersort would produce the following sequence of lists:

$$\begin{array}{rcccccc} \text{initial list}: & 15 & 8 & 10 & 6 & 17 & 13 \\ \text{after one pass}: & 6 & 15 & 8 & 10 & 13 & 17 \\ \text{after two passes}: & 6 & 8 & 10 & 13 & 15 & 17 \end{array}$$

17.3.4 SORTING BY DIVIDE-AND-CONQUER

The strategy of a divide-and-conquer sort is to partition the given sequence into smaller subsequences, to sort the subsequences recursively, and finally to merge the sorted subsequences into a single sorted sequence.

Definitions:

A **top-down mergesort** splits the input sequence into two equal (or nearly equal) sized subsequences, recursively mergesorts the two subsequences, and finally merges the two sorted subsequences into a single sorted sequence.

A **bottom-up mergesort** initially regards each entry in its input sequence as a list of length one. It merges two consecutive pairs at a time into lists of length two. Then it merges the lists of length two into lists of length four. Ultimately, all the initial items are merged into a single list.

A **quicksort** selects an element x (called the *pivot*) in the input list and splits the input list into two subsequences S_1 and S_2 such that every element in S_1 is no larger than x and every element in S_2 is no smaller than x. Next it recursively sorts S_1 and S_2. Then it concatenates the two sorted subsequences into a single sorted sequence.

The **pivot** in a quicksort iteration is the element x at which the sequence is split.

Facts:

1. A top-down mergesort is usually implemented as an internal sort.
2. A bottom-up mergesort is a common form of external sort.
3. An outstanding merit of quicksort is that it can be performed quickly within a single array.
4. Quicksort was first described by C. A. R. Hoare in 1962.
5. The running time of a mergesort is $O(n \log n)$.
6. In the worst case, a quicksort takes time $\Omega(n^2)$.
7. Choosing the quicksort pivot at random tends to avoid worst case behavior.
8. The average running time for a quicksort is $O(n \log n)$.
9. External sorting is used to process very large files, much too large to fit into the primary memory of any computer.
10. The emphasis in devising good external sorting algorithms is on decreasing the number of times the data are accessed because the time required to transfer data back and forth between the the primary memory and the tape usually outweighs far the time required to perform comparisons on data in the primary memory.

Algorithm 5:   Merge two sequences.

procedure merge($A[1..m], B[1..h], C[\ ]$)
{merge two sorted sequences A and B into a single sorted sequence C}
$i_A := 1;\ \ i_B := 1;\ \ i_C := 1$
while $i_A \leq m$ and $i_B \leq h$ do
  if $A[i_A] \leq B[i_B]$ then $\{C[i_C] := A[i_A];\ i_A := i_A + 1\}$
  else $\{C[i_C] := B[i_B];\ i_B := i_B + 1\}$
  $i_C := i_C + 1$
if $i_A > m$ then move the remaining elements in B to C
else move the remaining elements in A to C

Algorithm 6:   MergeSort S.

procedure mergesort(S)
if $length(S) \leq 1$ then return else
  split S into two (equal or nearly equal)-sized subsequences S_1 and S_2
  mergesort S_1
  mergesort S_2
  merge(S_1, S_2)

Algorithm 7:   External MergeSort sequence S of length n.

for $i := 1$ to $\lceil \log n \rceil$
  for $j := 1$ to $\lceil \frac{\log n}{4i} \rceil$
    merge next sublist from input A with next sublist from input B,
      writing merged sublist onto output tape C
    merge next sublist from input A with next sublist from input B,
      writing merged sublist onto output tape D
  reset output tape C as input tape A and vice versa
  reset output tape D as input tape B and vice versa

---

11. Formal algorithms and more detailed discussions of external can be found in [Kn73].

12. In an external sort, the number of times each element is read from or written to the external memory is $\log(\frac{n}{m}) + 1$, where m is the available internal memory size. Improvements on the construction of runs as well as on the merging process are possible (see [Kn73]).

Algorithms:

1. Algorithm 5 merges two sorted sequences into a single sorted sequence.

2. Algorithm 6 mergesorts a sequence internally.

3. In a typical external mergesort such as Algorithm 7, there are two input tapes and two output tapes. The entries are initially arranged onto the two input tapes, with half the entries on each tape, and regarded as sublists of length one. A sublist from the first

Algorithm 8: QuickSort.

procedure split(x, S)
 for each element y in S **do**
  if $x \geq y$ **then** put y in S_1 **else** put y in S_2

main program
 if $length(S) \leq 1$ **then** return **else**
  choose an arbitrary element x in sequence S
  split(x, S) into S_1 and S_2
  recursively sort S_1 and S_2
  concatenate the two sorted subsequences

input tape is merged with a sublist from the second input tape and written as a sublist of doubled length onto the first output tape. Then the next sublist from the first input tape is merged with the next sublist from the second input tape and written as a sublist of doubled length onto the second output tape. The alternating process is iterated until the sublists from the input tapes have all been merged into sublists of doubled length onto the two output tapes. Then the two output tapes become input tapes to another iteration of the merging process. This continues until all the original entries are in a single list.

4. The generic quicksort algorithm QuickSort (Algorithm 8) does not specify how to select a pivot.

Example:

1. The following illustrates MergeSort on the sequence $S = 21, 6, 8, 11; 10, 17, 15, 13$.

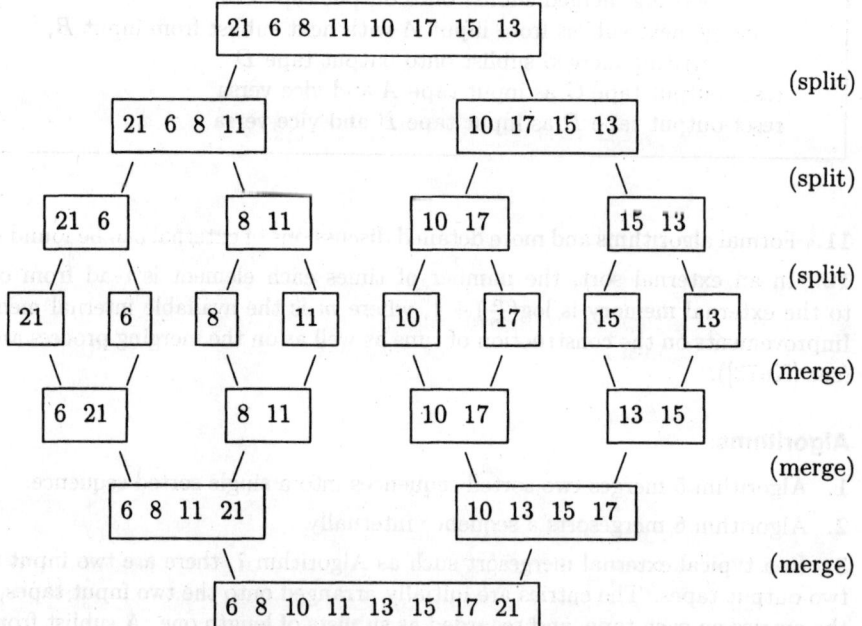

Algorithm 9: RankCountingSort of array $A[1..]$ into array $B[1..n]$.

{pre : $max(A[i]) \leq cn$}
for $i := 1$ to cn do $C[i] := 0$
for $j := 1$ to n do $C[A[j]] := 1$
for $i := 2$ to cn do $C[i] := C[i] + C[i-1]$
for $j := 1$ to n do $B[C[A[j]]] := A[j]$

17.3.5 SORTING BY DISTRIBUTION

Prior knowledge of the distribution of the elements of the input sequence sometimes permits sorting algorithms to break the lower bound of $\Omega(n \log n)$ for running time of comparison sorts

Definitions:

The **rank** of an element of a finite ordered set is the number of elements that it exceeds or equals.

A **rank counting sort** calculates the "rank" for each element, and then assigns the elements directly to their correct position according to their rank.

In a **base-ten radix sort**, the keys are base-ten integer numerals with at most k digits. Each entry is appended to one of ten queues $Q_0, \ldots, Q_9$, according to the value of its least significant digit, after which the list $Q_0 \circ \cdots \circ Q_9$ is formed by concatenation. The concatenated list is then similarly separated into ten queues, according to the values of the next least significant digit. This process is iterated up to the most significant digit.

A **radix sort** is a sort like the base-ten radix sort, using an arbitrary radix, not necessarily ten.

Facts:

1. A rank counting sort gives favorable results when the input keys are n different positive integers, all bounded by cn for some constant c.

2. The running time of a rank counting sort is in $O(n)$. The RankCountingSort (Algorithm 9) can be modified so that it sorts in linear time even when the input elements are not all distinct. [CoLeRi90]

3. It can be proved that a radix sort correctly sorts the input. [Kn73]

4. The running time of RadixSort is bounded by $O(kn)$, where k is the maximum number of digits in a key. When k is a constant independent of n, RadixSort sorts in linear time. Note, however, that if the input consists of n distinct numbers and the base of the numbers is fixed, then k is of order $\Omega(\log n)$.

Algorithms:

1. In the rank counting sort, Algorithm 9, the array A contains n input elements, the array B is the output array, and the array C is an auxiliary array of size cn used for counting. Step 3 causes count $C[A[j]]$ to be the rank of entry $A[j]$.

2. The base-ten radix sort, Algorithm 10, starts with an input list A whose keys have at most k digits.

> **Algorithm 10:** RadixSort of array $A[1..n]$.
>
> for $d := 1$ to k do
>   for $i := 0$ to 9 do make Q_i an empty queue
>   for $j := 1$ to n do
>   let h be the jth digit of $A[j]$
>   append $A[j]$ to queue Q_h
> $A := Q_0 \circ \cdots \circ Q_9$ (concatenation)

17.3.6 SEARCHING

Definitions:

Searching a database means seeking either a target entry with a specific key or a target entry whose key has some specified property.

Linear search is the technique of scanning the entries of a list in sequence, either until some stopping condition occurs or the entire list has been scanned.

Binary search is a recursive technique for seeking a specific target entry in a list. The target key is compared to the key in the middle of the list, in order to determine which half of the list could contain the target item, if it is present.

Hashing is storage-retrieval in a large table in which the table location is computed from the key of each data entry. (§17.4)

A ***binary search tree*** is a binary tree in which each note has an attribute called its *key*, and the keys are elements of an ordered datatype (e.g., the integers or alphabetic strings). Moreover, at each node v the key is larger that all the keys in its left subtree, but smaller than all the keys in its right subtree.

A ***2-3 tree*** is a tree in which each non-leaf node has 2 or 3 children, and in which every path from the root to a leaf is of the same length.

An ***AVL tree*** is a binary search tree with the property that the two subtrees of each node differ by at most 1 in height.

Facts:

1. Some common database search objectives are for a specified target entry, for the maximum entry, for the minimum entry, or for the kth smallest entry.

2. The performance of a dynamic database structure that permits insertions and deletions is measured by the time needed for insertions and deletions, as well as the time needed for searching.

3. A binary search runs in average time $O(\log n)$ to search for a specified element x in a sorted list of n elements.

4. In the worst case of searching by comparison-based algorithms for a specified target element in a sorted list of length n, $\Omega(\log n)$ comparisons are necessary.

5. A randomly constructed n-node binary search tree has expected height of at most $2 \log n$.

6. An AVL tree of n nodes has depth $O(\log n)$.

Algorithm 11: BinarySearch (A, L, U, x).

{Look for x in $A[L..U]$. Report its position if found, else report 0.}
if $L = U$ then
   if $x = A[L]$ then return L else return 0
else $M := \lfloor \frac{L+U}{2} \rfloor$
   if $x > A[M]$ then return BinarySearch $(A, M + 1, U, x)$
   else return BinarySearch (A, L, M, x)

Algorithm 12: 23TSearch(x, r).

case 1: *r is a leaf*
   if r is labeled with x then return "yes" else return "no"
case 2: *r is not a leaf*
   if $x \leq L[r]$ then return 23TSearch$(x, \text{leftchild}(r))$
   else if $x \leq M[r]$ then return 23TSearch$(x, \text{midchild}(r))$
   else if x has a right child then return 23TSearch$(x, \text{rightchild}(r))$
   else return "no"

---

7. Insertion and deletion on an AVL tree, with patching if needed so that the result is an AVL tree, can be performed in $O(\log n)$ worst-case running time. [Kn73]

8. AVL trees are named for their inventors, G. M. Adelson-Velskii and Y. M. Landis.

9. A 2-3 tree for a set S of entries can be constructed by assigning the entries to the leaves of the tree in order of increasing key from left to right. Each non-leaf node v is labeled with two elements $L[v]$ and $M[v]$, which are the largest keys in the subtrees rooted at its left child and middle child, respectively.

10. The operations of searching, finding a maximum or minimum, inserting a new entry, and deleting an entry all execute within $O(\log n)$ time.

Algorithms:

1. To search for a specified target key x in a sorted list $A[1..n]$, a call to the recursive algorithm BinarySearch $(A, 1, n, x)$ in Algorithm 11 can be used. Its technique is to compare the target to the middle entry of the list and to decide thereby in which half the target might occur; then that half remains as the active portion of the list for the next iteration of the search step, while the other half becomes inactive.

2. *Find the maximum [minimum] in an unsorted list:* Scan the list from start to finish and keep track of the largest [smallest] seen so far.

3. *Finding the maximum in a binary search tree:* Start at the root and follow the right child pointers until some node has no right child. That node must contain the maximum.

4. *Finding the minimum in a binary search tree:* Start at the root and follow the left child pointers until some node has no left child. That node must contain the minimum.

5. *Searching for a target entry x in a 2-3 tree:* Start at the root, and use the keys at non-leaf nodes to locate the correct leaf, as described by Algorithm 12.

Algorithm 13: Finding-The-kth-Smallest.

divide the n input elements into $\lceil \frac{n}{5} \rceil$ groups of five elements
find the median for each of the $\lceil \frac{n}{5} \rceil$ groups
recursively find the median m^* of these group medians
partition the input into two sets S_1 and S_2 such that each element in S_1 is no
    larger than m^* and each element in S_2 is no smaller than m^*
if S_1 has $\geq k$ elements **then** recursively find the kth smallest element in S_1
else recursively find the $(k - |S_1|)$th smallest element in S_2

6. *Finding the maximum in a 2-3 tree*: Starting from the root, follow the right-child pointers to the rightmost leaf, which contains the maximum entry.

7. *Finding the minimum in a 2-3 tree*: Starting from the root, follow the left-child pointers to the leftmost leaf, which contains the minimum entry.

8. *To insert a new entry x into a 2-3 tree*: First locate the non-leaf node v whose child x "should" be. If v is a 2-child node v, then simply install x as a third child of v. If v already has three children, then let v keep as children the two smallest of the set comprising its three children and x. A new non-leaf node u becomes the parent of the largest member of this set. Now recursively insert node u as a new child to the parent of node v. If the process eventually makes the root of the tree a 4-child node, then the last step is to create a new root with two new children, each of which has two of the four children of the former root. Note that the labels of some non-leaf nodes may be updated in this process.

9. *To delete an entry x from a 2-3 tree*: Essentially, reverse the manner by which an element is inserted. First find the leaf v containing x. If the parent p of v has three children, then the leaf v is simply deleted. If p has only two children v and v', then select an adjacent sibling p' of p. If p' has only two children, then make v' a child of p', and recursively delete the node p from the tree. If p' has three children, then make an appropriate child of p' into a new child of p and delete the node v (note that now both p and p' have two children). Again the process may progress recursively to the root of the tree, such that it is necessary to delete one of the only two children of the root. In this case, delete the root and make the remaining child of the root into a new root of the tree. Labels of some non-leaf nodes may need to be updated in this process.

10. *Searching in a random list*: Finding the kth smallest element, for an arbitrary k, in a random list can also be done in linear time. The algorithm, Algorithm 13, is based on the method of "Prune and Search". That is, the process first prunes away in linear time a constant factor of the elements in the input, then recursively searches the rest. A careful analysis shows that each of the two sets S_1 and S_2 contains at most $\frac{7n}{10}$ elements. Therefore, if $T(n)$ is the running time for Algorithm 13 on input of n elements, it follows that $T(n) \leq T(\frac{n}{5}) + T(\frac{7n}{10}) + O(n)$. This relation gives $T(n) = O(n)$.

Example:

1. To search for the target 64 in the following 16-element list
    5, 8, 9, 13, 16, 22, 25, 36, 47, 49, 64, 81, 100, 121, 144, 169
first split it into these two 8-element sublists,
    5, 8, 9, 13, 16, 22, 25, 36   47, 49, 64, 81, 100, 121, 144, 169
and then compare 64 to the largest item in the first list. Since $64 > 36$, it follows that

64, if present in the original list, would have to be in the second sublist. Next split the active sublist further into these two 4-element sublists

    47, 49, 64, 81   100, 121, 144, 169

and then compare 64 to the largest item in the first new sublist. Since $64 \leq 81$, it follows that 64, if present, would have to be in the second sublist. Therefore, resplit the active sublist further into these two 2-element sublists

    47, 49   64, 81

and then compare target 64 to the largest item in the first new sublist. Since $64 > 49$, it follows that 64, if present, would have to be in the second sublist. Therefore, resplit the active sublist further into these two 1-element sublists

    64   81

and then compare 64 to the largest item in the first new sublist. Since $64 \leq 64$, it follows that 64, if present, would have to be in the first sublist. Since that sublist has only one element, namely 64, the target 64 is compared to that one element. Since they are a match, the target has been located as the 11th item in the original list.

7.4 HASHING

Hashing, also known as "address calculation" and as "scatter storage", is a mathematical approach to organizing records within a table. The objective is to reduce the amount of time needed to find a record with a given key. Hashing is best suited for "dynamic" tables, that is, for databases whose use involves interspersed lookups and insertions. Dynamic dictionaries (such as spelling checkers) and compiler-generated symbol tables are examples of applications where hashing may be useful.

7.4.1 HASH FUNCTIONS

Hashing is an approach to placing records into a table and retrieving them when needed, in which the location for a record is calculated by an algorithm called a hash function.

Definitions:

A *record* is a pair of the form (k:key, d:data), in which the second component is data and the first component is a *key* used to store it in a table and to retrieve it subsequently from that table.

A *key domain* is an ordered set, usually the integers, whose members serve as keys for the *records* of a table. No two different records have the same key.

A *hash table* is an array, in which the location for storing and retrieving a *record* are calculated from the *key* of that record.

A *hash function* h is a function that maps a *key domain* to an interval of integers $[0..m-1]$. The intent is that a record with key k is to be stored in or retrieved from the location $h(k)$ in the table.

A *collision* occurs when a hash function h assigns the same table location to two different keys $k_1 \neq k_2$, i.e., when $h(k_1) = h(k_2)$.

Collision resolution is the process of finding an alternative location for a new record whose key collides with that of a record already in the table.

The *fullness* of a hash table T is the ratio $\alpha(T) = \frac{n}{m}$ of the number n of records in the table to the capacity m of the table.

Facts:

1. Hashing is often used when the set of keys (of the records in the database) is not a consecutive sequence of integers or easily convertible to a consecutive sequence of integers.

2. Keys that are non-numeric (or do not have a good random distribution) can be transformed into integers by using or modifying their binary representation. The resulting integers are called *virtual keys*.

3. It is desirable for a hash function to have the *simplicity property*, i.e., that it takes only a few simple operations to compute the hash function value.

4. It is desirable for a hash function to have the *uniformity property*, i.e., that each possible location in the range $0, \ldots, m-1$ of the hash function $h: K \to [0..m-1]$ is generated with equal likelihood, that is, with probability $\frac{1}{m}$.

Examples:

1. The *division method* $h(k) = k \bmod m$ is a simple hash function that can be used when the keys of the records are integers that extend far beyond a feasible table size. The table size m must be chosen carefully to avoid high instance of collision, without wasting too much storage space. Selecting m to be a prime not close to a power of 2 is typically a good choice.

2. The *multiplication method* is another simple hashing rule. First the key (in base 2) is multiplied by a constant value A such that $0 < A < 1$, and then $p = \log m$ bits are selected for the hash function value from somewhere in the fractional part of the resulting base-2 numeral, often required to be away from its low end. (This is similar to some methods for generating random numbers.)

3. As a simplified example of a multiplicative hash function, consider table size $m = 16$, address size $p = \log(16) = 4$ bits, and keysize $w = 6$ bits. With fractional constant $A = 0.101011_2$, use first four bits of the fractional part. For instance, given key $k = 011011_2$, first calculate $R = A \cdot k = 010010.001001_2$. Then take $k = 0010_2$, the low-end four bits of the fractional part. Knuth [Kn73] suggests $A = 0.6180339887\ldots$ as a good choice of a multiplier.

17.4.2 COLLISION RESOLUTION

Definitions:

Collision resolution is the process of computing an alternative location for a colliding record. The two basic methods are *chaining* and *rehashing*.

Chaining is a *collision resolution* method that involves auxiliary storage space for records outside the confines of the main array. Each slot in the main table can be used as the root of a linked list that contains all the records assigned to that location by the hash function. Each additional colliding record is inserted at the front of the linked list. When searching for a record, the list is traversed until the record is found, or until the end of the list is reached.

The *size of a chained hash table* is the number of linked list headers (i.e., the size of the main array). Thus, a chained hash table said to be of size m may be used to store a database with more than m records.

Rehashing is a *collision resolution* method in which there is no auxiliary storage outside the main table, so that a colliding record must be stored elsewhere in the main table, that is, at a location other than that assigned by the hash function to its key k. A *collision resolution function* finds the substitute location.

A **collision resolution function** under rehashing generates a *probe sequence* $\langle h_0(k) = h(k), h_1(k), h_2(k), \ldots, h_{m-1}(k)\rangle$. The new record is inserted into the first unoccupied probe location. When searching for a record, the successive probes are tried until the record is found or the probe finds an unoccupied location (i.e. unsuccessful search).

A **probe sequence** for key k is a sequence $\langle h_0(k), h_1(k), h_2(k), \ldots, h_{m-1}(k)\rangle$ of possible storage locations in the table T that runs without repetition through the entire set $\langle 0, 1, 2, \ldots, m-1\rangle$ of locations in the table, as possible places to store the record with key k.

Clustering is a hashing phenomenon in which after two keys collide, their probe sequences continue to collide.

Linear probing means trying to resolve a collision at location $h(k)$ with a probe sequence of the form $h_i(k) = (h_0(k) + ci) \bmod m$.

Quadratic probing means trying to resolve a collision at location $h(k)$ by using a probe sequence of the form $h_i(k) = (h_0(k) + c_1 i + c_2 i^2) \bmod m$.

A **secondary hash function** is a hash function used to generate a probe sequence, once a collision has occurred.

Double hashing means using a primary hash function h and a secondary hash function h' to resolve collisions, so that $h_i(k) = (h(k) + ih'(k)) \bmod m$.

Facts:

1. In designing a hash function, an objective is to keep the length of the probe sequences short, so that records are stored and retrieved quickly.

2. Under chain hashing, inserting a record always requires $O(1)$ time.

3. Under chain hashing, the time to find a record is proportional to the length of the chain from its key, and the average length of a chain in the table equals the fullness α of the table.

4. Under chain hashing, if the number of records in the table is proportional to the table capacity, then a *find* operation needs $O(1)$ time, on average.

5. Under chain hashing, a *delete* operation consists of a *find* with time proportional to table fullness, followed by removal from a linked list, which requires only $O(1)$ time.

6. Analysis of rehashing performance is based on the following assumptions:
    - uniform distribution of keys: each possible key in the key domain is equally likely to occur as the key of some record;
    - uniform distribution of initial hash locations (see §17.4.1);
    - uniform distribution of probe sequences: each possible probe sequence
    $$\langle h_0(k), h_1(k), h_2(k), \ldots, h_{m-1}(k)\rangle,$$
    regarded as a permutation on the set of all table locations is equally likely.

7. Under rehashing, the expected time to perform an *insert* operation is the same as the expected time for unsuccessful search, and is at most $\frac{1}{(1-\alpha)}$.

8. Under rehashing, the expected time $E(\alpha)$ to perform a successful find operation is at most $\frac{1}{\alpha} \ln \frac{1}{1-\alpha} + \frac{1}{\alpha}$. For instance, $E(0.5) = 3.386$, and $E(0.9) = 3.670$. That means that if a table is 90% full, a record will be found, on average, with 3.67 probes.

9. Under rehashing, location of the deleted record needs to be marked as *deleted* so that subsequent probe sequences do not terminate prematurely. Moreover, the running time of a *delete* operation is the same as for a successful find operation. (It also causes the measure of fullness for searches to be different from that used for insertions, because a new record can be inserted into the location marked as *deleted*). However, in most applications that use hashing, records are never deleted.

Examples:

1. The following example of linear probing uses prime modulus $m = 1013$ and prime multiplier $c = 367$. The keys are taken to be social security numbers.

key k	$h_0(k)$	$h_1(k)$
113548956	773	
146834522	172	
207639313	651	
359487245	896	
378545523	592	
435112760	896	250
670149788	651	
721666437	172	539
762456748	12	

2. Linear probing suffers from clustering. That is, if $h_i(k_1) = h_j(k_2)$, then $h_{i+p}(k_1) = h_{j+p}(k_2)$ for all $p = 1, 2, \ldots$. All probe sequences follow the same (linear) pattern, from which it follows that long chains of filled locations will cause a large number of probes needed to insert a new record (and to locate it later).

3. Quadratic probing suffers from clustering. That is, if $h_0(k_1) = h_0(k_2)$, then $h_i(k_1) = h_i(k_2)$ for $i = 1, 2, \ldots$.

4. Pairing the primary hash function $h(k) = k \bmod p$ with the secondary hash function $h'(k) = k \operatorname{div} p$, where p is a prime, yields the double hash function $h_i(k) = (h_0(k) + ih'(k)) \bmod p$.

17.5 DYNAMIC GRAPH ALGORITHMS

Dynamic graph algorithms are algorithms that maintain information in regard to properties of a (possibly edge-weighted) graph while the graph is changing. These algorithms are useful in a number of application areas, including communication networks, VLSI design, distributed computing, and graphics, where the underlying graphs are subject to dynamic changes. Efficient dynamic graph algorithms are also used as subroutines in algorithms that build and modify graphs as part of larger tasks, e.g., the algorithm for constructing Voronoi diagrams by building planar subdivisions.

Notation: Throughout this section, n and m denote the number of vertices and the number of edges, respectively, of the graph that is being maintained and queried dynamically.

7.5.1 DYNAMICALLY MAINTAINABLE PROPERTIES

Definitions:

A (*dynamic*) *update operation* is an operation on a graph that keeps track whether the graph has a designated property.

A *query* is a request for information about the designated property.

Facts:

1. The primitive update operations for most dynamic graph algorithms are edge insertions and deletions and, in the case of edge-weighted graphs, changes in edge weights.
2. For most dynamic graph algorithms, insertion or deletion of an isolated vertex can be accomplished by an easy modification of a non-dynamic algorithm.
3. The insertion or deletion of vertices together with their incident edges is usually harder and has to be done by iterating the associated edge update operation.
4. There is a trade-off between the time required for update operations and the time required to respond to queries about the property being maintained. Thus, running times of the update operations depend strongly on the property being maintained.
5. Nontrivial dynamic algorithms corresponding to several graph properties are known (see Examples).

Examples:

1. *Connectivity*: The permitted query is whether two vertices x and y are in the same component. Permitted updates are edge insertions, edge deletions, and isolated vertex insertions. Frederickson [Fr85] provides an algorithm for maintaining minimum spanning forests that can easily be adapted to this problem. Improvements in running times have been achieved by [EpEtal92] and [EpGaIt93].
2. *Bipartiteness*: Update operations are the same as for Connectivity (Example 1). A query simply asks whether a graph is bipartite. An algorithm is presented in [EpEtal92], with an improvement in [EpGaIt93].
3. *Minimum spanning forests*: The query is whether an edge is in a minimum spanning forest. The graph is weighted, and the update operations are increments and decrements of weights. (Edge insertion is accomplished by lowering the edge weight from ∞ and edge deletion by incrementing the edge weight to ∞.) [Fr85] contains the early result, with improvements by [EpGaIt93]. The plane and planar graph cases have been considered by [EpEtal92] and [EpEtal93].
4. *Biconnectivity and 2-Edge Connectivity*: Update operations are the same as for Connectivity (Example 1). Queries ask whether two given vertices lie in the same biconnected (resp., 2-edge connected) component. Efficient algorithms for maintaining biconnectivity are found in [EpEtal92], [EpGaIt93], [Ra93], and [HeRaSu94]. Efficient algorithms for maintaining 2-edge connectivity are found in [EpEtal92], [Fr91], and [Fr85]. Any algorithm for dynamically maintaining biconnectivity translates to an algorithm with the same time bounds for 2-edge connectivity [GaIt91].
5. *Planarity*: Update operations include edge insertions and deletions. Queries ask whether the graph is currently planar. Variants include queries that would test whether the addition of a particular edge would destroy the current imbedding. Algorithms are described in [EpEtal93] and [Ra93].

17.5.2 TECHNIQUES

Definitions:

A *partially dynamic algorithm* is usually an algorithm that handles only edge insertions and, for edge-weighted graphs, decrements in edge weights. Less commonly, this term can refer to an algorithm that handles only edge deletions or weight increments.

A *cluster* in a spanning tree T for a graph G is a set of vertices such that the subgraph of T induced on these vertices is connected.

An *ambivalent data structure* is a structure that, at many of its vertices, keeps track of several alternatives, despite the fact that a global examination of the structure would determine which of these alternatives is optimal.

A *certificate* for property $\mathcal{P}$ and graph G is a graph G' such that G has property $\mathcal{P}$ if and only if G' has property $\mathcal{P}$.

A *strong certificate* for property $\mathcal{P}$ and graph G is a graph G', on the same vertex set as G, such that, for every graph H, $G \cup H$ has property $\mathcal{P}$ if and only if $G' \cup H$ has property $\mathcal{P}$.

A *sparse certificate* is a strong certificate in which the number of edges is $O(n)$.

A function A that maps graphs to strong certificates is *stable* if it satisfies:
- $A(G \cup H) = A(A(G) \cup H)$;
- $A(G - e)$ differs from $A(G)$ by $O(1)$ edges, where e is an edge in G.

A *stable certificate* is one produced by a stable mapping.

A *plane graph* is a planar graph, together with a particular imbedding in the plane.

A *compressed certificate* for a property $\mathcal{P}$ of G, where $G = (V, E)$ is a subgraph of a larger graph F and $X \subset V$ separates G from $F - G$, is a small certificate $G' = (V', E')$ with $X \subset V'$ such that, for any graph H that is attached to G only at the vertices of X, $H \cup G$ has property $\mathcal{P}$ if and only if $H \cup G'$ does, and $|V'| = O(|X|)$.

A graph property $\mathcal{P}$ is *dyadic* if it is defined with respect to a pair of vertices (x, y).

A graph C is a *certificate of a dyadic property* $\mathcal{P}$ for X in G if and only if, for any H with $V(H) \cap V(G) \subset X$ and every x and y in $V(H)$, $\mathcal{P}$ is true for (x, y) in the graph $G \cup H$ if and only if it is true for (x, y) in the graph $C \cup H$.

Facts:

1. Using the union-find data structure [Ta75], it is possible to maintain connectivity information in $O(\alpha(m, n))$ amortized time per update or query.

2. For other graph properties, such as 2-edge connectivity and biconnectivity, a data structure called the *link/condense tree* [WeTa92] maintains information in $O(\alpha(m, n))$ amortized time per update or query.

3. The link/condense tree supports the operation of condensing an entire path in the tree into a single vertex. This is important in the applications considered, because the insertion of an edge may cause several biconnected components or 2-edge connected components to be combined into one.

4. Link/condense trees are based on *dynamic trees*. [SlTa83]

> **Algorithm 1:** Frederickson: to maintain a minimum spanning tree.
>
> *Preprocessing*:
>   find a minimum spanning tree T of the initial graph G
>   maintain a dynamic tree of T, using [SlTa83]
>   for $z := n^{2/3}$ group the vertices of T into *clusters* whose sizes are between z
>   and $3z - 2$ {There will be $\Theta(n^{1/3})$ clusters.}
>   for each pair of clusters i, j maintain the set of edges E_{ij} as a min-heap
>
> *Updates*: Decreases in tree edge weights do not change anything, and increases in non-tree weights can be handled by a suitable update of the appropriate min-heap. Handle decreases in non-tree edge weights by using the dynamic tree appropriately. If tree edge e increases in weight, remove it, thus partitioning the clusters into two sets. Find an edge of minimum cost between clusters on opposite sides of the partition.

5. The dynamic tree data structure maintains a set of rooted trees. It supports the operations of *linking* the root of one tree as the child of a vertex in another tree, *cutting* a tree at a specified edge, and *everting* a tree to make a specified vertex the root in worst-case $O(\log n)$ time per operation. It also supports other operations based on keys stored at vertices, such as finding the minimum key on the path from a given vertex to the root in $O(\log n)$ time.

6. To maintain minimum spanning trees in an edge-weighted, connected graph subject to changes in edge weights, Frederickson [Fr85] uses *clustering* and *topology trees* in Algorithm 1.

7. Ambivalence may permit faster updates, possibly at the cost of slower queries.

8. Frederickson [Fr91] presents an ambivalent data structure for spanning forests that builds upon the ideas of multilevel partitions and (2-dimensional) topology trees that he developed in [Fr85].

9. Let $\mathcal{P}$ be a property for which sparse certificates can be found in time $f(n, m)$. Suppose that there is a data structure that can be built in time $g(n, m)$ and permits static testing of property $\mathcal{P}$ in time $q(n, m)$. Then there is a fully dynamic data structure for testing whether a graph has property $\mathcal{P}$; update time for this structure is $f(n, O(n))O(\log(\frac{m}{n})) + g(n, O(n))$, and query time is $q(n, O(n))$. This "basic sparsification technique" is used to dynamize static algorithms. To use it, one need only be able to compute sparse certificates efficiently.

10. The *sparsification* method of [EpEtal92] is to partition the input graph into sparse subgraphs (with $O(n)$ edges) and summarize the relevant information about each subgraph in an even sparser "certificate". Certificates are merged in pairs, producing larger subgraphs that are themselves sparsified using the certificate technique. The result is a balanced binary tree in which each vertex is a sparse certificate. Each insertion or deletion of an edge in the input graph causes changes in $\log(\frac{m}{n})$ tree vertices. Because these changes occur in graphs with $O(n)$ edges, instead of the m edges in the input graph, time bounds for updates are reduced in most natural problems.

11. Let $\mathcal{P}$ be a property for which stable sparse certificates can be maintained in time $f(n, m)$ per update. Suppose that there is a fully dynamic algorithm for $\mathcal{P}$ with update time $g(n, m)$ and query time $q(n, m)$. Then this algorithm can be sped up; specifically, $\mathcal{P}$ can be maintained fully dynamically in time $f(n, O(n))O(\log(\frac{m}{n})) + g(n, O(n))$

per update, with query time $q(n, O(n))$. Because this "stable sparsification technique" is used to speed up existing dynamic algorithms, it often yields better results than the basic sparsification technique described above. However, to use it, one must be able to maintain stable sparse certificates efficiently; this is a more stringent requirement than what is needed to apply basic sparsification.

12. Eppstein, Galil, and Italiano [EpGaIt93] improve the sparsification technique to get rid of the $\log(\frac{m}{n})$ factor in these bounds. They achieve this improvement by partitioning the edge set of the original graph more carefully.

13. Dynamic algorithms restricted to plane graphs have been considered by several authors. Eppstein *et al.* [EpEtal93] introduce a variant of sparsification that permits the design of efficient dynamic algorithms for *planar* graphs in an imbedding-independent way, as long as the updates to the graph preserve planarity. Because these graphs are already sparse, Eppstein *et al.* design a *separator-based* sparsification technique.

14. The fact that separator sizes are sublinear ($O(\sqrt{n})$ for planar graphs) allows the possibility of maintaining sublinear certificates. Eppstein *et al.* [EpEtal93] use a separator-based decomposition tree as the sparsification tree and show how to compute it in linear time and maintain it dynamically. They use it to show the following: For a property $\mathcal{P}$ for which compressed certificates can be built in time $T(n)$, a data structure for testing $\mathcal{P}$ built in time $P(n)$, and queries answered in time $Q(n)$, a fully dynamic algorithm for maintaining $\mathcal{P}$ under planarity-preserving insertions and deletions takes amortized time $P(O(n^{1/2})) + T(O(n^{1/2}))$ per update and $Q(O(n^{1/2}))$ per query.

15. A dyadic property $\mathcal{P}$, for which compressed certificates can be built in time $T(n)$, a data structure for testing $\mathcal{P}$ built in time $P(n)$, and queries answered in time $Q(n)$, can be maintained with updates taking $T(O(n^{1/2}))$ amortized time and queries taking $P(O(n^{1/2})) + Q(O(n^{1/2})) + T(O(n^{1/2}))$ worst-case time.

16. In dealing with plane (as opposed to planar) graphs and allowing only updates that can be performed in a planarity-preserving manner on the existing imbedding, simpler techniques that rely on planar duality can be used [EpEtal92].

17. When maintaining minimum spanning trees under updates that change only edge weights, the most difficult operation to handle is an increase in the weight of an MST edge. However, in the dual graph this can be viewed as a decrease in the weight of a non-MST edge. This idea and the handling of edge insertions and deletions are addressed by the data structures of [GuSt85] and the *edge-ordered tree* data structure of [EpEtal92]. These data structures help maintain the subdivision and its dual in the face of general updates and also help perform required access operations efficiently. Edge-ordered trees are an adaptation of the dynamic trees of [SlTa83].

18. Knowledge of the imbedding allows one to use topology trees in more efficient ways. Specifically, Rauch [Ra94] partitions the non-tree edges into equivalence classes called *bundles*. In the cyclical ordering of edges emanating from a cluster, bundles are carefully chosen, consecutive subsets of edges.

17.5.3 APPLICATIONS

Examples:

1. *Bipartiteness* [EpGaItNi92], [EpGaIt93]: A graph that is not bipartite contains an odd cycle. The graph formed by adding the shortest edge inducing an odd cycle (if any) to the minimum spanning forest is a stable certificate of (non-)bipartiteness. Using the clustering techniques of [Fr85] and the improved sparsification techniques of [EpGaIt93],

this certificate can be maintained in $O(n^{1/2})$ time per update. The query time in this example is $O(1)$; one bit is used to indicate whether the operation is maintaining a certificate of bipartiteness or of non-bipartiteness.

2. *Minimum spanning forests* [EpGaItNi92], [EpGaIt93], [Fr85]: In this example, the goal is not to maintain a data structure that supports efficient testing of a property, but rather to maintain the minimum spanning forest itself as edges are added to and deleted from the input graph. It is shown in [EpGaItNi92] how to define a canonical minimum spanning forest that serves as the analogue of a stable sparse certificate. Frederickson [Fr85] uses the topological approach to obtain a fully dynamic algorithm that maintains minimum spanning forests in time $O(m^{1/2})$ per update. Applying the improved stable sparsification technique with $f(n,m) = g(n,m) = O(m^{1/2})$ yields a fully dynamic minimum spanning forest algorithm with update time $O(n^{1/2})$. For plane graphs, [EpEtal92] show that both updates and queries can be performed in $O(\log n)$ time per operation; in planar graphs, [EpEtal93] show that $O(\log^2 n)$ per deletion and $O(\log n)$ per insertion are sufficient.

3. *Connectivity* [EpGaItNi92], [EpGaIt93], [Fr85]: Simple enhancements to the minimum spanning forest algorithms in [Fr85] yield fully dynamic algorithms for the connectivity problem in which the update times are the same as they are for minimum spanning forests, and the query times are $O(1)$. Thus, as in the previous example, applying improved stable sparsification with $f(n,m) = g(n,m) = O(m^{1/2})$ yields a fully dynamic connectivity algorithm with update time $O(n^{1/2})$ and query time $O(1)$. Similarly, the planar and plane graph algorithms for minimum spanning trees can be generalized to work for minimum spanning forests and adapted to maintain connected components.

4. *Biconnectivity* [EpGaItNi92], [EpGaIt93], [Ra93], [HeRaSu94]: Cheriyan, Kao, and Thurimella [ChKaTh93] show that $C_2 = C_1 \cup B_2$ is a sparse certificate for biconnectivity, where C_1 is a breadth-first spanning forest of the input graph G, and B_2 is a breadth-first spanning forest of $G - C_1$. Eppstein et al. [EpGaItNi92] show that C_2 is in fact a strong certificate of biconnectivity. These strong certificates can be found in time $O(m)$, using classical breadth-first search algorithms. Applying improved sparsification with $f(n,m) = g(n,m) = O(m)$ yields a fully dynamic algorithm for maintaining the biconnected components of a graph that has update time $O(n)$.

The approach to biconnectivity in [Ra94] is to partition the graph G into clusters and decompose a query that asks whether vertices u and v lie in the same biconnected component into a query in the cluster of u, a query in the cluster of v, and a query between clusters. The 2-dimensional topology tree is adapted in a novel way, and the ambivalent data structures previously defined for connectivity and 2-edge connectivity are extended to test biconnectivity between clusters. To test biconnectivity within a cluster C, the entire subgraph induced by C and a compressed certificate of $G - C$ are maintained. Using all these ingredients, [Ra94] obtains amortized $O(m^{1/2})$ time per update and $O(1)$ worst-case time per query.

Using clever data structures based on topology trees, bundles, and the idea of *recipes*, first introduced in this context in [HeRaSu94], the problem of fully dynamic biconnectivity for plane graphs can be solved in $O(\log^2 n)$ time per update and $O(\log n)$ time per query.

5. *2-edge connectivity* [EpGaItNi92], [Fr91], [Fr85]: Thurimella [Th89] and Nagamochi and Ibaraki [NaIb92] show that the following structure U_2 is a certificate for 2-edge connectivity: $U_2 = U_1 \cup F_2$, where U_1 is a spanning forest of G, and F_2 is a spanning forest of $G - U_1$. Eppstein et al. [EpGaItNi92] show that U_2 is in fact a stable,

sparse certificate. Frederickson's minimum spanning forest algorithm [Fr85] can be adapted to maintain U_2 in time $f(n,m) = O(m^{1/2})$. Frederickson's ambivalent data structure technique [Fr91] can be used to test 2-edge connectivity with update time $g(n,m) = O(m^{1/2})$ and query time $q(n,m) = O(\log n)$. Here a "query" is a pair of vertices, and the answer is "yes" if they are in the same 2-edge connected component and "no" otherwise. Applying improved stable sparsification yields a fully dynamic algorithm with update time $O(n^{1/2})$ and query time $O(\log n)$.

6. *Planarity* [EpEtal93], [Ra93]: Eppstein et al. [EpEtal93] use the separator-based sparsification technique described above to obtain a fully dynamic planarity-testing algorithm for general graphs that answers queries of the form "is the graph currently planar?" and "would the insertion of this edge preserve planarity?". Their algorithm requires amortized running time $O(n^{1/2})$ per update or query. Italiano, La Poutré, and Rauch [ItLaRa93] use topology trees, bundles, and recipes to obtain a fully dynamic algorithm on plane graphs that tests whether the insertion of a particular edge would destroy the given imbedding. Their algorithm requires time $O(\log^2 n)$ for updates and queries.

17.5.4 RECENT RESULTS AND OPEN QUESTIONS

Examples:

1. Alberts and Henzinger [AlHe95] investigate dynamic algorithms on random graphs with n vertices and m_0 edges on which a sequence of k arbitrary update operations is performed. They obtain expected update times of $O(k \log n + \sum_{i=1}^{k} \frac{n}{\sqrt{m_i}})$ for minimum spanning forest, connectivity, and bipartiteness and $O(k \log n + \sqrt{\log n} \sum_{i=1}^{k} \frac{n}{\sqrt{m_i}})$ for 2-edge connectivity. The data structures required for these algorithms use linear space, and the preprocessing times match those of the best algorithms for finding a minimum spanning forest.

2. Fredman and Rauch [FrRa94] investigate lower bounds in the cell probe model of computation and obtain good results for k-edge connectivity, k-vertex connectivity, and planarity-testing of imbedded planar graphs. Both average-case analysis and lower bounds are important topics for future research on dynamic graph algorithms.

3. Klein et al. [KlEtal94] give a fully dynamic algorithm for the all-pairs shortest path problem on planar graphs. If the sum of the absolute values of the edge-lengths is D, then the time per operation is $O(n^{9/7} \log D)$ (worst case for queries, edge deletion, and length changes, and amortized for edge insertion); the space requirement is $O(n)$. Several types of partially dynamic algorithms for shortest paths appear in [AuEtal90], [EvGa85], [FrMaNa94], and [Ro85].

Although it is one of the most important dynamic graph algorithms problems, there is less known about shortest paths than about many other problems, and this is an important topic for future study.

4. In a recent breakthrough, Henzinger and King [HeKi95] obtained fully dynamic, randomized algorithms for connectivity, 2-edge connectivity, bipartiteness, cycle equivalence, and constant-weight minimum spanning trees that have polylogarithmic expected time per operation.

REFERENCES

Printed Resources:

[AhHoUl83] A. V. Aho, J. E. Hopcroft, and J. D. Ullman, *Data Structures and Algorithms*, Addison-Wesley, 1983.

[AhHoUl74] A. V. Aho, J. E. Hopcroft, and J. D. Ullman, *The Design and Analysis of Computer Algorithms*, Addison-Wesley, 1974.

[AlHe95] D. Alberts and M. Rauch Henzinger, "Average Case Analysis of Dynamic Graph Algorithms", *Proceedings of the 6th Symposium on Discrete Algorithms*, ACM/SIAM (1995), 312–321.

[AuEtal90] G. Ausiello, G. Italiano, A. Marchetti-Spaccamela, and U. Nanni, "Incremental Algorithms for Minimal Length Paths", *Journal of Algorithms* 12 (1991), 615–638.

[ChKaTh93] J. Cheriyan, M.-Y. Kao, and R. Thurimella, "Scan-First Search and Sparse Certificates: An Improved Parallel Algorithm for k-vertex Connectivity", *SIAM Journal on Computing* 22 (1993), 157–174.

[CoLeRi90] T. H. Cormen, C. E. Leiserson, R. L. Rivest, *Introduction to Algorithms*, McGraw-Hill and MIT Press, 1990.

(A comprehensive study of algorithms and their analysis.)

[EpGaIt93] D. Eppstein, Z. Galil, and G. Italiano, *Improved Sparsification*, Technical Report 93-20, Department of Information and Computer Science, University of California at Irvine, 1993.

[EpGaItNi92] D. Eppstein, Z. Galil, G. Italiano, and A. Nissenzweig, "Sparsification – A Technique for Speeding up Dynamic Graph Algorithms", *Proceedings of the 33rd Symposium on Foundations of Computer Science*, IEEE Computer Society Press, 1992, 60–69.

[EpEtal93] D. Eppstein, Z. Galil, G. Italiano, and T. Spencer, "Separator Based Sparsification for Dynamic Planar Graph Algorithms", *Proceedings of the 25th Symposium on Theory of Computing*, ACM, 1993, 208–217.

[EpEtal92] D. Eppstein, G. Italiano, R. Tamassia, R. Tarjan, J. Westbrook, and M. Yung, "Maintenance of a Minimum Spanning Forest in a Dynamic Plane Graph", *Journal of Algorithms* 13 (1992), 33–54.

[EvGa85] S. Even and H. Gazit, "Updating Distances in Dynamic Graphs", *Methods of Operations Research* 49 (1985), 371–387.

[FrRa94] M. Fredman and M. Rauch, "Lower Bounds for Dynamic Connectivity Problems in Graphs", *Algorithmica*, to appear.

[Fr91] G. Frederickson, "Ambivalent Data Structures for Dynamic 2-Edge-Connectivity and k Smallest Spanning Trees", *Proceedings of the 32nd Symposium on Foundations of Computer Science*, IEEE Computer Society Press, 1991, 632–641.

[Fr85] G. N. Frederickson, "Data Structures for On-Line Updating of Minimum Spanning Trees, with Applications", *SIAM Journal on Computing* 14 (1985), 781–798.

[FrMaNa94] D. Frigioni, A. Marchetti-Spaccamela, and U. Nanni, "Incremental Algorithms for the Single-Source Shortest Path Problem", in *Proceedings of 1994 Symposium on Foundations of Software Technology and Theoretical Computer Science*, 113–124.

[GaIt91] Z. Galil and G. Italiano, "Reducing Edge Connectivity to Vertex Connectivity", *SIGACT News* 22 (1991), 57–61.

[GoBa91] G. H. Gonnett, R. Baeze-Yates, *Handbook of Algorithms and Data Structures in Pascal and C*, 2nd ed., Addison Wesley, 1991. (An extensive collection of algorithms with analysis and implementation; includes mathematical formulas used in analysis.)

[GuSt85] L.J. Guibas and J. Stolfi, "Primitives for the Manipulation of General Subdivisions and the Computation of Voronoi Diagrams", *ACM Transactions on Graphics* 4 (1985), 74–123.

[HeKi95] M. Rauch Henzinger and V. King, "Randomized Dynamic Algorithms with Polylogarithmic Time Per Operation", *Proceedings of the 27th Symposium on the Theory of Computing*, ACM, 1995, to appear.

[HeRaSu94] J. Hershberger, M. Rauch, and S. Suri, "Fully Dynamic 2-Edge-Connectivity in Planar Graphs", *Theoretical Computer Science* 130 (1994), 139–161.

[ItLaRa93] G. Italiano, H. La Poutré, and M. Rauch, "Fully Dynamic Planarity Testing in Embedded Graphs", *Algorithms – ESA '93, Lecture Notes in Computer Science* 726 (1993) Springer, 212–223.

[KlEtal94] P. Klein, S. Rao, M. Rauch, and S. Subramanian, "Faster Shortest-Path Algorithms for Planar Graphs", *Proceedings of the 26th Symposium on Theory of Computing*, ACM, 1994, 27–37.

[Kn68] D. E. Knuth, *The Art of Computer Programming. Vol. 1: Fundamental Algorithms*, Addison-Wesley, 1968. (A seminal reference on the subject of algorithms.)

[Kn73] D. E. Knuth, *The Art of Computer Programming. Vol. 3: Sorting and Searching*, Addison-Wesley, 1973. (A seminal reference on the subject of algorithms.)

[NaIb92] H. Nagamochi and T. Ibaraki, "Linear Time Algorithms for Finding a Sparse k-connected Spanning Subgraph of a k-Connected Graph", *Algorithmica* 7 (1992), 583–596.

[Ra93] M. Rauch, "Fully Dynamic Graph Algorithms and Their Data Structures", PhD Thesis, Computer Science Department, Princeton University, 1993.

[Ra94] M. Rauch, "Improved Data Structures for Fully Dynamic Biconnectivity", *Proceedings of the 26th Symposium on Theory of Computing*, ACM, 1994, 686–695.

[Ro85] H. Rohnert, "A Dynamization of the All-Pairs Least-Cost Path Problem", *Proceedings of the 2nd Symposium on Theoretical Aspects of Computer Science, Lecture Notes in Computer Science* 182 (1985) Springer, 279–286.

[SlTa83] D. Sleator and R. Tarjan, "A Data Structure for Dynamic Trees", *Journal of Computer and System Sciences* 26 (1983), 362–391.

[Ta75] R. Tarjan, "Efficiency of a Good but not Linear Set Union Algorithm", *Journal of the ACM* 22 (1975), 215–225.

[Th89] R. Thurimella, "Techniques for the Design of Parallel Graph Algorithms", PhD Thesis, Computer Science Department, University of Texas at Austin, 1989.

[WeTa92] J. Westbrook and R. Tarjan, "Maintaining Bridge-Connected and Biconnected Components On-Line", *Algorithmica* 7 (1992), 433–464.

Web Resources:

http://members.xoom.com/killough/heaps.html (Priority queues.)

http://ciips.ee.uwa.edu.au/~morris/Year2/PLDS210/ds_ToC.html (Data Structures amd Algorithms: Table of Contents.)

http://www.cs.brockport.edu/cs/javasort.html (Collection of applets illustrating operation of and complexity incurred by various sorting algorithms.)

http://www.cs.hope.edu/~alganim/ccaa/site.html (The complete collection of algorithm animations on sorting, binary search trees, and other well-known mathematical structures.)

http://www.cs.oswego.edu/~birgit/html/diplom/links.html (Algorithm animation links, including animation of graph algorithms and data structures.)

http://www.cs.sunysb.edu/~algorith/ (The Stony Brook Algorithm Repository; see Sections 1.1 Data Structures.)

http://www.ctc.dcu.ie/ctcweb/courses/algorithms/links.html (A collection of links to various algorithm animation applets.)

http://www.geocities.com/Area51/Vault/3150/book.htm (Sorting and Searching Algorithms: A Cookbook.)

http://www.stanford.edu/~mariaf/ra/knowledgeUnits/AL.html (A compendium of sites on algorithms and basic data structures.)

BIOGRAPHIES

Victor J. Katz

Niels Henrik Abel (1802–1829), born in Norway, was self-taught and studied the works of many mathematicians. When he was nineteen years old, he proved that there is no closed formula for solving the general fifth degree equation. He also worked in the areas of infinite series and elliptic functions and integrals. The term *abelian* group was coined in Abel's honor in 1870 by Camille Jordan.

Abraham ibn Ezra (1089–1164) was a Spanish-Jewish poet, philosopher, astrologer, and biblical commentator who was born in Tudela, but spent the latter part of his life as a wandering scholar in Italy, France, England, and Palestine. It was in an astrological text that ibn Ezra developed a method for calculating numbers of combinations, in connection with determining the number of possible conjunctions of the seven "planets" (including the sun and the moon). He gave a detailed argument for the cases $n = 7$, $k = 2$ to 7, of a rule which can easily be generalize to the modern formula $C(n, k) = \sum_{i=k-1}^{n-1} C(i, k - 1)$. Ibn Ezra also wrote a work on arithmetic in which he introduced the Hebrew-speaking community to the decimal place-value system. He used the first nine letters of the Hebrew alphabet to represent the first nine numbers, used a circle to represent zero, and demonstrated various algorithms for calculation in this system.

Aristotle (384–322 B.C.E.) was the most famous student at Plato's academy in Athens. After Plato's death in 347 B.C.E., he was invited to the court of Philip II of Macedon to educate Philip's son Alexander, who soon thereafter began his successful conquest of the Mediterranean world. Aristotle himself returned to Athens, where he founded his own school, the Lyceum, and spent the remainder of his life writing and lecturing. He wrote on numerous subjects, but is perhaps best known for his works on logic, including the *Prior Analytics* and the *Posterior Analytics*. In these works, Aristotle developed the notion of logical argument, based on several explicit principles. In particular, he built his arguments out of syllogisms and concluded that demonstrations using his procedures were the only certain way of attaining scientific knowledge.

Emil Artin (1898–1962) was born in Vienna and in 1921 received a Ph.D. from the University of Leipzig. He held a professorship at the University of Hamburg until 1937, when he came to the United States. In the U.S. he taught at the University of Notre Dame, Indiana University, and Princeton. In 1958 he returned to the University of Hamburg. Artin's mathematical contributions were in number theory, algebraic topology, linear algebra, and especially in many areas of abstract algebra.

Charles Babbage (1792–1871) was an English mathematician best known for his invention of two of the earliest computing machines, the *Difference Engine*, designed to calculate polynomial functions, and the *Analytical Engine*, a general purpose calculating machine. The Difference Engine was designed to use the idea that the nth order differences in nth degree polynomials were always constant and then to work backwards from those differences to the original polynomial values. Although Babbage received a grant from the British government to help in building the Engine, he never was able to complete one because of various difficulties in developing machine parts of sufficient accuracy. In addition, Babbage became interested in his more advanced Analytical Engine. This latter device was to consist of a *store*, in which the numerical variables were kept, and a *mill*, in which the operations were performed. The entire machine was to be controlled by instructions on punched cards. Unfortunately, although Babbage made numerous engineering drawings of sections of the Analytical Engine and gave a series of seminars in 1840 on its workings, he was never able to build a working model.

Paul Gustav Heinrich Bachmann (1837–1920) studied mathematics at the University of Berlin and at Göttingen. In 1862 he received a doctorate in group theory and held positions at the universities at Breslau and Münster. He wrote several volumes on number theory, introducing the big-*O* notation in his 1892 book.

John Backus (born 1924) received bachelor's and master's degrees in mathematics from Columbia University. He led the group at IBM that developed FORTRAN. He was a developer of ALGOL, using the Backus-Naur form for the syntax of the language. He received the National Medal of Science in 1974 and the Turing Award in 1977.

Abu-l-'Abbas Ahmad ibn Muhammad ibn al-Banna al-Marrakushi (1256–1321) was an Islamic mathematician who lived in Marrakech in what is now Morocco. Ibn al-Banna developed the first known proof of the basic combinatorial formulas, beginning by showing that the number of permutations of a set of n elements was $n!$ and then developing in a careful manner the multiplicative formula to compute the values for the number of combinations of k objects in a set of n. Using these two results, he also showed how to calculate the number of permutations of k objects from a set of n. The formulas themselves had been known in the Islamic world for many years, in connection with specific problems like calculating the number of words of a given length which could be formed from the letters of the Arabic alphabet. Ibn al-Banna's main contribution, then, was to abstract the general idea of permutations and combinations out of the various specific problem situations considered earlier.

Thomas Bayes (1702–1761) an English Nonconformist, wrote an *Introduction to the Doctrine of Fluxions* in 1736 as a response to Berkeley's *Analyst* with its severe criticism of the foundations of the calculus. He is best known, however, for attempting to answer the basic question of statistical inference in his *An Essay Towards Solving a Problem in the Doctrine of Chances*, published three years after his death. That basic question is to determine the probability of an event, given empirical evidence that it has occurred a certain number of times in a certain number of trials. To do this, Bayes gave a straightforward definition of probability and then proved that for two events E and F, the probability of E given that F has happened is the quotient of the probability of both E and F happening divided by the probability of F alone. By using areas to model probability, he was then able to show that, if x is the probability of an event happening in a single trial, if the event has happened p times in n trials, and if $0 < r < s < 1$, then the probability that x is between r and s is given by the quotient of two integrals. Although in principle these integrals can be calculated, there has been a great debate since Bayes' time about the circumstances under which his formula gives an appropriate answer.

James Bernoulli (Jakob I) (1654–1705) was one of eight mathematicians in three generations of his family. He was born in Basel, Switzerland, studied theology in addition to mathematics and astronomy, and entered the ministry. In 1682 be began to lecture at the University of Basil in natural philosophy and mechanics. He became professor at the University of Basel in 1687, and remained there until his death. His research included the areas of the calculus of variations, probability, and analytic geometry. His most well-known work is *Ars Conjectandi*, in which he described results in combinatorics and probability, including applications to gambling and the law of large numbers; this work also contained a reprint of the first formal treatise in probability, written in 1657 by Christiaan Huygens.

Bhaskara (1114–1185), the most famous of medieval Indian mathematicians, gave a complete algorithmic solution to the *Pell equation* $Dx^2 \pm 1 = y^2$. That equation had been studied by several earlier Indian mathematicians as well. Bhaskara served much of his adult life as the head of the astronomical observatory at Ujjain, some 300 miles northeast of Bombay, and became widely respected for his skills in astronomy and the mechanical arts, as well as mathematics. Bhaskara's mathematical contributions are chiefly found in two chapters, the *Lilavati* and the *Bijaganita*, of a major astronomical work, the *Siddhāntasiromani*. These include techniques of solving systems of linear equations with more unknowns than equations as well as the basic combinatorial formulas, although without any proofs.

George Boole (1815–1864) was an English mathematician most famous for his work in logic. Born the son of a cobbler, he had to struggle to educate himself while supporting his family. But he was so successful in his self-education that he was able to set up his own school before he was 20 and was asked to give lectures on the work of Isaac Newton. In 1849 he applied for and was appointed to the professorship in mathematics at Queen's College, Cork, despite having no university degree. In 1847, Boole published a small book, *The Mathematical Analysis of Logic*, and seven years later expanded it into *An Investigation of the Laws of Thought*. In these books, Boole introduced what is now called Boolean algebra as part of his aim to "investigate the fundamental laws of those operations of the mind by which reasoning is performed; to give expression to them in the symbolical language of a Calculus, and upon this foundation to establish the science of Logic and construct its method." In addition to his work on logic, Boole wrote texts on differential equations and on difference equations that were used in Great Britain until the end of the nineteenth century.

William Burnside (1852–1927), born in London, graduated from Cambridge in 1875, and remained there as lecturer until 1885. He then went to the Royal Naval College at Greenwich, where he stayed until he retired. Although he published much in applied mathematics, probability, and elliptic functions, he is best known for his extensive work in group theory (including the classic book *Theory of Groups*). His conjecture that groups of odd order are solvable was proved by Walter Feit and John Thompson and published in 1963.

Georg Ferdinand Ludwig Philip Cantor (1845–1918) was born in Russia to Danish parents, received a Ph.D. in number theory in 1867 at the University of Berlin, and in 1869 took a position at Halle University, where he remained until his retirement. He is regarded as a founder of set theory. He was interested in theology and the nature of the infinite. His work on the convergence of Fourier series led to his study of certain types of infinite sets of real numbers, and ultimately to an investigation of transfinite numbers.

Augustin-Louis Cauchy (1789–1857) the most prolific mathematician of the nineteenth century, is most famous for his textbooks in analysis written in the 1820s for use at the École Polytechnique, textbooks which became the model for calculus texts for the next hundred years. Although born in the year the French Revolution began,

Cauchy was a staunch conservative. When the July Revolution of 1830 led to the overthrow of the last Bourbon king, Cauchy refused to take the oath of allegiance to the new king and went into a self-imposed exile in Italy and then in Prague. He did not return to his teaching posts until the Revolution of 1848 led to the removal of the requirement of an oath of allegiance. Among the many mathematical subjects to which he contributed besides calculus were the theory of matrices, in which he demonstrated that every symmetric matrix can be diagonalized by use of an orthogonal substitution, and the theory of permutations, in which he was the earliest to consider these from a functional point of view. In fact, he used a single letter, say S, to denote a permutation and S^{-1} to denote its inverse and then noted that the powers S, S^2, S^3, ... of a given permutation on a finite set must ultimately result in the identity. He also introduced the current notation $(a_1 a_2 \ldots a_n)$ to denote the cyclic permutation on the letters $a_1, a_2, \ldots, a_n$.

Arthur Cayley (1821–1895), although graduating from Trinity College, Cambridge as Senior Wrangler, became a lawyer because there were no suitable mathematics positions available at that time in England. He produced nearly 300 mathematical papers during his fourteen years as a lawyer, and in 1863 was named Sadlerian professor of mathematics at Cambridge. Among his numerous mathematical achievements are the earliest abstract definition of a group in 1854, out of which he was able to calculate all possible groups of order up to eight, and the basic rules for operating with matrices, including a statement (without proof) of the Cayley-Hamilton theorem that every matrix satisfies its characteristic equation. Cayley also developed the mathematical theory of *trees* in an article in 1857. In particular, he dealt with the notion of a *rooted tree*, a tree with a designated vertex called a root, and developed a recursive formula for determining the number of different rooted trees in terms of its *branches* (edges). In 1874, Cayley applied his results on trees to the study of chemical isomers.

Pafnuty Lvovich Chebyshev (1821–1894) was a Russian who received his master's degree in 1846 from Moscow University. From 1860 until 1882 he was a professor at the University of St. Petersburg. His mathematical research in number theory dealt with congruences and the distribution of primes; he also studied the approximation of functions by polynomials.

Avram Noam Chomsky (born 1928) received a Ph.D. in linguistics at the University of Pennsylvania. For many years he has been a professor of foreign languages and linguistics at M.I.T. He has made many contributions to the study of linguistics and the study of grammars.

Chrysippus (280–206 B.C.E.) was a Stoic philosopher who developed some of the basic principles of the propositional logic, which ultimately replaced Aristotle's logic of syllogisms. He was born in Cilicia, in what is now Turkey, but spent most of his life in Athens, and is said to have authored more than 700 treatises. Among his other achievements, Chrysippus analyzed the rules of inference in the propositional calculus, including the rules of *modus ponens*, *modus tollens*, the hypothetical syllogism, and the alternative syllogism.

Alonzo Church (1903–1995) studied under Hilbert at Göttingen, was on the faculty at Princeton from 1927 until 1967, and then held a faculty position at UCLA. He is a founding member of the Association for Symbolic Logic. He made many contributions in various areas of logic and the theory of algorithms, and stated the Church-Turing thesis (if a problem can be solved with an effective algorithm, then the problem can be solved by a Turing machine).

George Dantzig (born 1914) is an American mathematician who formulated the general linear programming problem of maximizing a linear objective function subject to several linear constraints and developed the simplex method of solution in 1947. His study of linear programming grew out of his World War II service as a member of Air Force Project SCOOP (Scientific Computation of Optimum Programs), a project chiefly concerned with resource allocation problems. After the war, linear programming was applied to numerous problems, especially military and economic ones, but it was not until such problems could be solved on a computer that the real impact of their solution could be felt. The first successful solution of a major linear programming problem on a computer took place in 1952 at the National Bureau of Standards. After he left the Air Force, Dantzig worked for the Rand Corporation and then served as a professor of operations research at Stanford University.

Richard Dedekind (1831–1916) was born in Brunswick, in northern Germany, and received a doctorate in mathematics at Göttingen under Gauss. He held positions at Göttingen and in Zurich before returning to the Polytechnikum in Brunswick. Although at various times he could have received an appointment to a major German university, he chose to remain in his home town where he felt he had sufficient freedom to pursue his mathematical research. Among his many contributions was his invention of the concept of ideals to resolve the problem of the lack of unique factorization in rings of algebraic integers. Even though the rings of integers themselves did not possess unique factorization, Dedekind showed that every ideal is either prime or uniquely expressible as the product of prime ideals. Dedekind published this theory as a supplement to the second edition (1871) of Dirichlet's *Vorlesungen über Zahlentheorie*, of which he was the editor. In the supplement, he also gave one of the first definitions of a field, confining this concept to subsets of the complex numbers.

Abraham deMoivre (1667–1754) was born into a Protestant family in Vitry, France, a town about 100 miles east of Paris, and studied in Protestant schools up to the age of 14. Soon after the revocation of the Edict of Nantes in 1685 made life very difficult for Protestants in France, however, he was imprisoned for two years. He then left France for England, never to return. Although he was elected to the Royal Society in 1697, in recognition of a paper on "A method of raising an infinite Multinomial to any given Power or extracting any given Root of the same", he never achieved a university position. He made his living by tutoring and by solving problems arising from games of chance and annuities for gamblers and speculators. DeMoivre's major mathematical work was *The Doctrine of Chances* (1718, 1736, 1756), in which he devised methods for calculating probabilities by use of binomial coefficients. In particular, he derived the normal approximation to the binomial distribution and, in essence, invented the notion of the standard deviation.

Augustus DeMorgan (1806–1871) graduated from Trinity College, Cambridge in 1827. He was the first mathematics professor at University College in London, where he remained on the faculty for 30 years. He founded the London Mathematical Society. He wrote over 1000 articles and textbooks in probability, calculus, algebra, set theory, and logic (including DeMorgan's laws, an abstraction of the duality principle for sets). He gave a precise definition of limit, developed tests for convergence of infinite series, and gave a clear explanation of the Principle of Mathematical Induction.

René Descartes (1596–1650) left school at 16 and went to Paris, where he studied mathematics for two years. In 1616 he earned a law degree at the University of Poitiers. In 1617 he enlisted in the army and traveled through Europe until 1629, when he settled in Holland for the next 20 years. During this productive period of his life he wrote on mathematics and philosophy, attempting to reduce the sciences to mathematics. In 1637 his *Discours* was published; this book contained the development of analytic geometry. In 1649 he has invited to tutor the Queen Christina of Sweden in philosophy. There he soon died of pneumonia.

Leonard Eugene Dickson (1874–1954) was born in Iowa and in 1896 received the first Ph.D. in mathematics given by the University of Chicago, where he spent much of his faculty career. His research interests included abstract algebra (including the study of matrix groups and finite fields) and number theory.

Diophantus (c. 250) was an Alexandrian mathematician about whose life little is known except what is reported in an epigram of the *Greek Anthology* (c. 500), from which it can calculated that he lived to the age of 84. His major work, however, the *Arithmetica*, has been extremely influential. Despite its title, this is a book on algebra, consisting mostly of an organized collection of problems translatable into what are today called indeterminate equations, all to be solved in rational numbers. Diophantus introduced the use of symbolism into algebra and outlined the basic rules for operating with algebraic expressions, including those involving subtraction. It was in a note appended to Problem II-8 of the 1621 Latin edition of the *Arithmetica* — to divide a given square number into two squares — that Pierre de Fermat first asserted the impossibility of dividing an nth power ($n > 2$) into the sum of two nth powers. This result, now known as Fermat's Last Theorem, was finally proved in 1994 by Andrew Wiles.

Charles Lutwidge Dodgson (1832–1898) is more familiarly known as Lewis Carroll, the pseudonym he used in writing his famous children's works *Alice in Wonderland* and *Through the Looking Glass*. Dodgson graduated from Oxford University in 1854 and the next year was appointed a lecturer in mathematics at Christ Church College, Oxford. Although he was not successful as a lecturer, he did contribute to four areas of mathematics: determinants, geometry, the mathematics of tournaments and elections, and recreational logic. In geometry, he wrote a five-act comedy, "Euclid and His Modern Rivals", about a mathematics lecturer Minos in whose dreams Euclid debates his *Elements* with various modernizers but always manages to demolish the opposition. He is better known, however, for his two books on logic, *Symbolic Logic* and *The Game of Logic*. In the first, he developed a symbolical calculus for analyzing logical arguments and wrote many humorous exercises designed to teach his methods, while in the second, he demonstrated a game which featured various forms of the syllogism.

Eratosthenes (276–194 B.C.E) was born in Cyrene (North Africa) and studied at Plato's Academy in Athens. He was tutor of the son of King Ptolemy III Euergetes in Alexandria and became chief librarian at Alexandria. He is recognized as the foremost scholar of his time and wrote in many areas, including number theory (his sieve for obtaining primes) and geometry. He introduced the concepts of meridians of longitude and parallels of latitude and used these to measure distances, including an estimation of the circumference of the earth.

Paul Erdős (1913–1996) was born in Budapest. At 21 he received a Ph.D. in mathematics from Eötvös University. After leaving Hungary in 1934, he traveled extensively throughout the world, with very few possessions and no permanent home, working with other mathematicians in combinatorics, graph theory, number theory, and many other areas. He was author or coauthor of approximately 1500 papers with 500 coauthors.

Euclid (c. 300 B.C.E.) is responsible for the most famous mathematics text of all time, the *Elements*. Not only does this work deal with the standard results of plane geometry, but it also contains three chapters on number theory, one long chapter on irrational quantities, and three chapters on solid geometry, culminating with the construction of the five regular solids. The axiom-definition-theorem-proof style of Euclid's work has become the standard for formal mathematical writing up to the present day. But about Euclid's life virtually nothing is known. It is, however, generally assumed that he was among the first mathematicians at the Museum and Library of Alexandria, which was founded around 300 B.C.E by Ptolemy I Soter, the Macedonian general of Alexander the Great who became ruler of Egypt after Alexander's death in 323 B.C.E.

Leonhard Euler (1707–1783) was born in Basel, Switzerland and became one of the earliest members of the St. Petersburg Academy of Sciences. He was the most prolific mathematician of all time, making contributions to virtually every area of the subject. His series of analysis texts established many of the notations and methods still in use today. He created the calculus of variations and established the theory of surfaces in differential geometry. His study of the Königsberg bridge problem led to the formulation and solution of one of the first problems in graph theory. He made numerous discoveries in number theory, including a detailed study of the properties of residues of powers and the first statement of the quadratic reciprocity theorem. He developed an algebraic formula for determining the number of partitions of an integer n into m distinct parts, each of which is in a given set A of distinct positive integers. And in a paper of 1782, he even posed the problem of the existence of a pair of orthogonal latin squares: If there are 36 officers, one of each of six ranks from each of six different regiments, can they be arranged in a square in such a way that each row and column contains exactly one officer of each rank and one from each regiment?

Kamāl al-Dīn al-Fārisī (died 1320) was a Persian mathematician most famous for his work in optics. In fact, he wrote a detailed commentary on the great optical work of Ibn al-Haytham. But al-Farisi also made major contributions to number theory. He produced a detailed study of the properties of *amicable numbers* (pairs of numbers in which the sum of the proper divisors of each is equal to the other). As part of this study, al-Fārisī developed and applied various combinatorial principles. He showed that the classical figurate numbers (triangular, pyramidal, etc.) could be interpreted as numbers of combinations and thus helped to found the theory of combinatorics on a more abstract basis.

Pierre de Fermat (1601–1665) was a lawyer and magistrate for whom mathematics was a pastime that led to contributions in many areas: calculus, number theory, analytic geometry, and probability theory. He received a bachelor's degree in civil law in 1631, and from 1648 until 1665 was King's Counsellor. He suffered an attack of the plague in 1652, and from then on he began to devote time to the study of mathematics. He helped give a mathematical basis to probability theory when, together with Blaise Pascal, he solved Méré's paradox: why is it less likely to roll a 6 at least once in four tosses of one die than to roll a double 6 in 24 tosses of two dice. He was a discoverer of analytic geometry and used infinitesimals to find tangent lines and determine maximum and minimum values of curves. In 1657 he published a series of mathematical challenges, including the conjecture that $x^n + y^n = z^n$ has no solution in positive integers if n is an integer greater than 2. He wrote in the margin of a book that he had a proof, but the proof would not fit in the margin. His conjecture was finally proved by Andrew Wiles in 1994.

Fibonacci (Leonardo of Pisa) (c. 1175–c. 1250) was the son of a Mediterranean merchant and government worker named Bonaccio (hence his name *filius Bonaccio*, "son of Bonaccio"). Fibonacci, born in Pisa and educated in Bougie (on the north coast of Africa where his father was administrator of Pisa's trading post), traveled extensively around the Mediterranean. He is regarded as the greatest mathematician of the Middle Ages. In 1202 he wrote the book *Liber Abaci*, an extensive treatment of topics in arithmetic and algebra, and emphasized the benefits of Arabic numerals (which he knew about as a result of his travels around the Mediterranean). In this book he also discussed the rabbit problem that led to the sequence that bears his name: $1, 1, 2, 3, 5, 8, 13, \ldots$. In 1225 he wrote the book *Liber Quadratorum*, studying second degree diophantine equations.

Joseph Fourier (1768–1830), orphaned at the age of 9, was educated in the military school of his home town of Auxerre, 90 miles southeast of Paris. Although he hoped to become an army engineer, such a career was not available to him at the time because he was not of noble birth. He therefore took up a teaching position. During the Revolution, he was outspoken in defense of victims of the Terror of 1794. Although he was arrested, he was released after the death of Robespierre and was appointed in 1795 to a position at the École Polytechnique. After serving in various administrative posts under Napoleon, he was elected to the Académie des Sciences and from 1822 until his death served as its perpetual secretary. It was in connection with his work on heat diffusion, detailed in his *Analytic Theory of Heat* of 1822, and, in particular, with his solution of the *heat equation* $\frac{\partial v}{\partial t} = \frac{\partial^2 v}{\partial x^2} + \frac{\partial^2 v}{\partial y^2}$, that he developed the concept of a Fourier series. Fourier also analyzed the relationship between the series solution of a partial differential equation and an appropriate integral representation and thereby initiated the study of Fourier integrals and Fourier transforms.

Georg Frobenius (1849–1917) organized and analyzed the central ideas of the theory of matrices in his 1878 memoir "On linear substitutions and bilinear forms". Frobenius there defined the general notion of *equivalent* matrices. He also dealt with the special cases of *congruent* and *similar* matrices. Frobenius showed that when two symmetric matrices were similar, the transforming matrix could be taken to be *orthogonal*, one whose inverse equaled its transpose. He then made a detailed study of orthogonal matrices and showed that their eigenvalues were complex numbers of absolute value 1. He also gave the first complete proof of the Cayley-Hamilton theorem that a matrix satisfies its characteristic equation. Frobenius, a full professor in Zurich and later in Berlin, made his major mathematical contribution in the area of group theory. He was instrumental in developing the concept of an abstract group, as well as in investigating the theory of finite matrix groups and group characters.

Evariste Galois (1811–1832) led a brief, tragic life which ended in a duel fought under mysterious circumstances. He was born in Bourg-la-Reine, a town near Paris. He developed his mathematical talents early and submitted a memoir on the solvability of equations of prime degree to the French Academy in 1829. Unfortunately, the referees were never able to understand this memoir nor his revised version submitted in 1831. Meanwhile, Galois became involved in the revolutionary activities surrounding the July revolution of 1830 and was arrested for threatening the life of King Louis-Phillipe and then for wearing the uniform of a National Guard division which had been dissolved because of its perceived threat to the throne. His mathematics was not fully understood until fifteen years after his death when his manuscripts were finally published by Liouville in the *Journal des mathématique*. But Galois had in fact shown the relationship between subgroups of the group of permutations of the roots of a polynomial equation and the various extension fields generated by these roots, the relationship at the basis of what is now known as *Galois theory*. Galois also developed the notion of a finite field in connection with solving the problem of finding solutions to congruences $F(x) \equiv 0 \pmod{p}$, where $F(x)$ is a polynomial of degree n and no residue modulo the prime p is itself a solution.

Carl Friedrich Gauss (1777–1855), often referred to as the greatest mathematician who ever lived, was born in Brunswick, Germany. He received a Ph.D. from the University of Helmstedt in 1799, proving the Fundamental Theorem of Algebra as part of his dissertation. At age 24 Gauss published his important work on number theory, the *Disquisitiones Arithmeticae*, a work containing not only an extensive discussion of the theory of congruences, culminating in the quadratic reciprocity theorem, but also a detailed treatment of cyclotomic equations in which he showed how to construct regular n-gons by Euclidean techniques whenever n is prime and $n-1$ is a power of 2. Gauss also made fundamental contributions to the differential geometry of surfaces as well as to complex analysis, astronomy, geodesy, and statistics during his long tenure as a professor at the University of Göttingen. It was in connection with using the method of least squares to solve an astronomical problem that Gauss devised the systematic procedure for solving a system of linear equations today known as Gaussian elimination. (Unknown to Gauss, the method appeared in Chinese mathematics texts 1800 years earlier.) Gauss' notebooks, discovered after his death, contained investigations in numerous areas of mathematics in which he did not publish, including the basics of non-Euclidean geometry.

Sophie Germain (1776–1831) was forced to study in private due to the turmoil of the French Revolution and the opposition of her parents. She nevertheless mastered mathematics through calculus and wanted to continue her study in the École Polytechnique when it opened in 1794. But because women were not admitted as students, she diligently collected and studied the lecture notes from various mathematics classes and, a few years later, began a correspondence with Gauss (under the pseudonym Monsieur LeBlanc, fearing that Gauss would not be willing to recognize the work of a woman) on ideas in number theory. She was, in fact, responsible for suggesting to the French general leading the army occupying Brunswick in 1807 that he insure Gauss' safety. Germain's chief mathematical contribution was in connection with Fermat's Last Theorem. She showed that $x^n + y^n = z^n$ has no positive integer solution where xyz is not divisible by n for any odd prime n less than 100. She also made contributions in the theory of elasticity and won a prize from the French Academy in 1815 for an essay in this field.

Kurt Gödel (1906–1978) was an Austrian mathematician who spent most of his life at the Institute for Advanced Study in Princeton. He made several surprising contributions to set theory, demonstrating that Hilbert's goal of showing that a reasonable axiomatic system for set theory could be proven to be complete and consistent was in fact impossible. In several seminal papers published in the 1930s, Gödel proved that it was impossible to prove internally the consistency of the axioms of any reasonable system of set theory containing the axioms for the natural numbers. Furthermore, he showed that any such system was inherently incomplete, that is, that there are propositions expressible in the system for which neither they nor their negations are provable. Gödel's investigations were stimulated by the problems surrounding the axiom of choice, the axiom that for any set S of nonempty disjoint sets, there is a subset T of the union of S that has exactly one element in common with each member of S. Since that axiom led to many counterintuitive results, it was important to show that the axiom could not lead to contradictions. But given his initial results, the best Gödel could do was to show that the axiom of choice was relatively consistent, that its addition to the Zermelo-Fraenkel axiom set did not lead to any contradictions that would not already have been implied without it.

William Rowan Hamilton (1805–1865), born in Dublin, was a child prodigy who became the Astronomer Royal of Ireland in 1827 in recognition of original work in optics accomplished during his undergraduate years at Trinity College, Dublin. In 1837, he showed how to introduce complex numbers into algebra axiomatically by considering $a + ib$ as a pair (a, b) of real numbers with appropriate computational rules. After many years of seeking an appropriate definition for multiplication rules for triples of numbers which could be applied to vector analysis in 3-dimensional space, he discovered that it was in fact necessary to consider quadruplets of numbers, which Hamilton named quaternions. Although quaternions never had the influence Hamilton forecast for them in physics, their noncommutative multiplication provided the first significant example of a mathematical system which did not obey one of the standard arithmetical laws of operation and thus opened the way for more "freedom" in the creation of mathematical systems. Among Hamilton's other contributions was the development of the *Icosian* game, a graph with 20 vertices on which pieces were to be placed in accordance with various conditions, the overriding one being that a piece was always placed at the second vertex of an edge on which the previous piece had been placed. One of the problems Hamilton set for the game was, in essence, to discover a cyclic path on his game board which passed through each vertex exactly once. Such a path in a more general setting is today called a Hamilton circuit.

Richard W. Hamming (1915–1998) was born in Chicago and received a Ph.D. in mathematics from the University of Illinois in 1942. He was the author of the first major paper on error correcting and detecting codes (1950). His work on this problem had been stimulated in 1947 when he was using an early Bell System relay computer on weekends only. During the weekends the machine was unattended and would dump any work in which it discovered an error and proceed to the next problem. Hamming realized that it would be worthwhile for the machine to be able not only to detect an error but also to correct it, so that his jobs would in fact be completed. In his paper, Hamming used a geometric model by considering an n-digit code word to be a vertex in the unit cube in the n-dimensional vector space over the field of two elements. He was then able to show that the relationship between the word length n and the number m of digits which carry the information was $2^m \leq \frac{2^n}{n+1}$. (The remaining $k = n - m$ digits are check digits which enable errors to be detected and corrected.) In particular, Hamming presented a particular type of code, today known as a Hamming code, with $n = 7$ and $m = 4$. In this code, the set of actual code words of 4 digits was a 4-dimensional vector subspace of the 7-dimensional space of all 7-digit binary strings.

Godfrey Harold Hardy (1877–1947) graduated from Trinity College, Cambridge in 1899. From 1906 until 1919 he was lecturer at Trinity College, and, recognizing the genius of Ramanujan, invited Ramanujan to Cambridge in 1914. Hardy held the Sullivan chair of geometry at Oxford from 1919 until 1931, when he returned to Cambridge, where he was Sadlerian professor of pure mathematics until 1942. He developed the Hardy-Weinberg law which predicts patterns of inheritance. His main areas of mathematical research were analysis and number theory, and he published over 100 joint papers with Cambridge colleague John Littlewood. Hardy's book *A Course in Pure Mathematics* revolutionized mathematics teaching, and his book *A Mathematician's Apology* gives his view of what mathematics is and the value of its study.

Abū 'Alī al-Hasan ibn al-Haytham (**Alhazen**) (965–1039) was one of the most influential of Islamic scientists. He was born in Basra (now in Iraq) but spent most of his life in Egypt, after he was invited to work on a Nile control project. Although the project, an early version of the Aswan dam project, never came to fruition, ibn al-Haytham did produce in Egypt his most important scientific work, the *Optics*. This work was translated into Latin in the early thirteenth century and was studied and commented on in Europe for several centuries thereafter. Although there was much mathematics in the *Optics*, ibn al-Haytham's most interesting mathematical work was the development of a recursive procedure for producing formulas for the sum of any integral powers of the integers. Formulas for the sums of the integers, squares, and cubes had long been known, but ibn al-Haytham gave a consistent method for deriving these and used this to develop the formula for the sum of fourth powers. Although his method was easily generalizable to the discovery of formulas for fifth and higher powers, he gave none, probably because he only needed the fourth power rule in his computation of the volume of a paraboloid of revolution.

Hypatia (c. 370–415), the first woman mathematician on record, lived in Alexandria. She was given a very thorough education in mathematics and philosophy by her father Theon and became a popular and respected teacher. She was responsible for detailed commentaries on several important Greek works, including Ptolemy's *Almagest*, Apollonius' *Conics*, and Diophantus' *Arithmetica*. Unfortunately, Hypatia was caught up in the pagan-Christian turmoil of her times and was murdered by an enraged mob.

Leonid Kantorovich (1912–1986) was a Soviet economist responsible for the development of linear optimization techniques in relation to planning in the Soviet economy. The starting point of this development was a set of problems posed by the Leningrad timber trust at the beginning of 1938 to the Mathematics Faculty at the University of Leningrad. Kantorovich explored these problems in his 1939 book *Mathematical Methods in the Organization and Planning of Production*. He believed that one way to increase productivity in a factory or an entire industrial organization was to improve the distribution of the work among individual machines, the orders to various suppliers, the different kinds of raw materials, the different types of fuels, and so on. He was the first to recognize that these problems could all be put into the same mathematical language and that the resulting mathematical problems could be solved numerically, but for various reasons his work was not pursued by Soviet economists or mathematicians.

Abū Bakr al-Karajī (died 1019) was an Islamic mathematician who worked in Baghdad. In the first decade of the eleventh century he composed a major work on algebra entitled *al-Fakhrī* (*The Marvelous*), in which he developed many algebraic techniques, including the laws of exponents and the algebra of polynomials, with the aim of systematizing methods for solving equations. He was also one of the early originators of a form of mathematical induction, which was best expressed in his proof of the formula for the sum of integral cubes.

Stephen Cole Kleene (1909–1994) studied under Alonzo Church and received his Ph.D. from Princeton in 1934. His research has included the study of recursive functions, computability, decidability, and automata theory. In 1956 he proved Kleene's Theorem, in which he characterized the sets that can be recognized by finite-state automata.

Felix Klein (1849–1925) received his doctorate at the University of Bonn in 1868. In 1872 he was appointed to a position at the University of Erlanger, and in his opening address laid out the *Erlanger Programm* for the study of geometry based on the structure of groups. He described different geometries in terms of the properties of a set that are invariant under a group of transformations on the set and gave a program of study using this definition. From 1875 until 1880 he taught at the Technische Hochschule in Munich, and from 1880 until 1886 in Leipzig. In 1886 Klein became head of the mathematics department at Göttingen and during his tenure raised the prestige of the institution greatly.

Donald E. Knuth (born 1938) received a Ph.D. in 1963 from the California Institute of Technology and held faculty positions at the California Institute of Technology (1963–1968) and Stanford (1968–1992). He has made contributions in many areas, including the study of compilers and computational complexity. He is the designer of the mathematical typesetting system TEX. He received the Turing Award in 1974 and the National Medal of Technology in 1979.

Kazimierz Kuratowski (1896–1980) was the son of a famous Warsaw lawyer who became an active member of the Warsaw School of Mathematics after World War I. He taught both at Lwów Polytechnical University and at Warsaw University until the outbreak of World War II. During that war, because of the persecution of educated Poles, he went into hiding under an assumed name and taught at the clandestine Warsaw University. After the war, he helped to revive Polish mathematics, serving as director of the Polish National Mathematics Institute. His major mathematical contributions were in topology; he formulated a version of a maximal principle equivalent to the axiom of choice. This principle is today known as Zorn's lemma. Kuratowski also contributed to the theory of graphs by proving in 1930 that any non-planar graph must contain a copy of one of two particularly simple non-planar graphs.

Joseph Louis Lagrange (1736–1813) was born in Turin into a family of French descent. He was attracted to mathematics in school and at the age of 19 became a mathematics professor at the Royal Artillery School in Turin. At about the same time, having read a paper of Euler's on the calculus of variations, he wrote to Euler explaining a better method he had recently discovered. Euler praised Lagrange and arranged to present his paper to the Berlin Academy, to which he was later appointed when Euler returned to Russia. Although most famous for his *Analytical Mechanics*, a work which demonstrated how problems in mechanics can generally be reduced to solutions of ordinary or partial differential equations, and for his *Theory of Analytic Functions*, which attempted to reduce the ideas of calculus to those of algebraic analysis, he also made contributions in other areas. For example, he undertook a detailed review of solutions to quadratic, cubic, and quartic polynomials to see how these methods might generalize to higher degree polynomials. He was led to consider permutations on the roots of the equations and functions on the roots left unchanged by such permutations. As part of this work, he discovered a version of Lagrange's theorem to the effect that the order of any subgroup of a group divides the order of the group. Although he did not complete his program and produce a method of solving higher degree polynomial equations, his methods were applied by others early in the nineteenth century to show that such solutions were impossible.

Gabriel Lamé (1795–1870) was educated at the École Polytechnique and the École des Mines before going to Russia to direct the School of Highways and Transportation in St. Petersburg. After his return to France in 1832, he taught at the École Polytechnique while also working as an engineering consultant. Lamé contributed original work to number theory applied mathematics, and thermodynamics. His best-known work is his proof of the case $n = 5$ of Fermat's Last Theorem in 1839.

Eight years later, he announced that he had found a general proof of the theorem, which began with the factorization of the expression $x^n + y^n$ over the complex numbers as $(x + y)(x + \alpha y)(x + \alpha^2 y) \ldots (x + \alpha^{n-1} y)$, where α is a primitive root of $x^n - 1 = 0$. He planned to show that the factors in this expression are all relatively prime and therefore that if $x^n + y^n = z^n$, then each of the factors would itself be an nth power. He would then use the technique of infinite descent to find a solution in smaller numbers. Unfortunately Lamé's idea required that the ring of integers in the cyclotomic field of the nth roots of unity be a unique factorization domain. And, as Kummer had already proved three years earlier, unique factorization in fact fails in many such domains.

Edmund Landau (1877–1938) received a doctorate under Frobenius and taught at the University of Berlin and at Göttingen. His research areas were analysis and analytic number theory, including the distribution of primes. He used the big-O notation (also called a Landau symbol) in his work to estimate the growth of various functions.

Pierre-Simon de Laplace (1749–1827) entered the University of Caen in 1766 to begin preparation for a career in the church. He soon discovered his mathematical talents, however, and in 1768 left for Paris to continue his studies. He later taught mathematics at the École Militaire to aspiring cadets. Legend has it that he examined, and passed, Napoleon there in 1785. He was later honored by both Napoleon and King Louis XVIII. Laplace is best known for his contributions to celestial mechanics, but he was also one of the founders of probability theory and made many contributions to mathematical statistics. In fact, he was one of the first to apply his theoretical results in statistics to a genuine problem in statistical inference, when he showed from the surplus of male to female births in Paris over a 25-year period that it was "morally certain" that the probability of a male birth was in fact greater than $\frac{1}{2}$.

Gottfried Wilhelm Leibniz (1646–1716), born in Leipzig, developed his version of the calculus some ten years after Isaac Newton, but published it much earlier. He based his calculus on the inverse relationship of sums and differences, generalized to infinitesimal quantities called differentials. Leibniz hoped that his most original contribution to philosophy would be the development of an alphabet of human thought, a way of representing all fundamental concepts symbolically and a method of combining these symbols to represent more complex thoughts. Although he never completed this project, his interest in finding appropriate symbols ultimately led him to the d and $\int$ symbols for the calculus that are used today. Leibniz spent much of his life in the diplomatic service of the Elector of Mainz and later was a Counsellor to the Duke of Hanover. But he always found time to pursue his mathematical ideas and to carry on a lively correspondence on the subject with colleagues all over Europe.

Levi ben Gerson (1288–1344) was a rabbi as well as an astronomer, philosopher, biblical commentator, and mathematician. He lived in Orange, in southern France, but little is known of his life. His most famous mathematical work is the *Maasei Hoshev* (The Art of the Calculator) (1321), which contains detailed proofs of the standard combinatorial formulas, some of which use the principle of mathematical induction. About a dozen copies of this medieval manuscript are extant, but it is not known whether the work had any direct influence elsewhere in Europe.

Augusta Ada Byron King Lovelace (1815–1852) was the child of the famous poet George Gordon, the sixth Lord Byron, who left England five weeks after his daughter's birth and never saw her again. She was raised by her mother, Anna Isabella Millbanke, a student of mathematics herself, so she received considerably more mathematics education than was usual for girls of her time. She was tutored privately by well-known mathematicians, including William Frend and Augustus DeMorgan. Her husband, the Earl of Lovelace, was made a Fellow of the Royal Society in 1840, and through this connection, Ada was able to gain access to the books and papers she needed to continue her mathematical studies and, in particular, to understand the workings of Babbage's Analytical Engine. Her major mathematical work is a heavily annotated translation of a paper by the Italian mathematician L. F. Menabrea dealing with the Engine, in which she gave explicit descriptions of how it would solve specific problems and described, for the first time in print, what would today be called a computer program, in this case a program for computing the Bernoulli numbers. Interestingly, only her initials, A.A.L., were used in the published version of the paper. It was evidently not considered proper in mid-nineteenth century England for a woman of her class to publish a mathematical work.

Jan Łukasiewicz (1878–1956) studied at the University of Lwów and taught at the University of Lwów, the University of Warsaw, and the Royal Irish Academy. A logician, he worked in the area of many-valued logic, writing papers on three-valued and m-valued logics, He is best known for the parenthesis-free notation he developed for propositions, called Polish notation.

Percy Alexander MacMahon (1854–1929) was born into a British army family and joined the army himself in 1871, reaching the rank of major in 1889. Much of his army service was spent as an instructor at the Royal Military Academy. His early mathematical work dealt with invariants, following on the work of Cayley and Sylvester, but a study of symmetric functions eventually led to his interest in partitions and to his extension of the idea of a partition to higher dimensions. MacMahon's two volume treatise *Combinatorial Analysis* (1915–16) is a classic in the field. It identified and clarified the basic results of combinatorics and showed the way toward numerous applications.

Mahāvīra (ninth century) was an Indian mathematician of the medieval period whose major work, the *Ganitasārasaṅgraha*, was a compilation of problems solvable by various algebraic techniques. For example, the work included a version of the hundred fowls problem: "Doves are sold at the rate of 5 for 3 coins, cranes at the rate of 7 for 5, swans at the rate of 9 for 7, and peacocks at the rate of 3 for 9. A certain man was told to bring at these rates 100 birds for 100 coins for the amusement of the king's son and was sent to do so. What amount does he give for each?" Mahāvīra also presented, without proof and in words, the rule for calculating the number of combinations of r objects out of a set of n. His algorithm can be easily translated into the standard formula. Mahavira then applied the rule to two problems, one about combinations of tastes and another about combinations of jewels on a necklace.

Andrei Markov (1856–1922) was a Russian mathematician who first defined what are now called Markov chains in a paper of 1906 dealing with the Law of Large Numbers and subsequently proved many of the standard results about them. His interest in these chains stemmed from the needs of probability theory. Markov never dealt with their application to the sciences, only considering examples from literary texts, where the two possible states in the chain were vowels and consonants. Markov taught at St. Petersburg University from 1880 to 1905 and contributed to such fields as number theory, continued fractions, and approximation theory. He was an active participant in the liberal movement in pre-World War I Russia and often criticized publicly the actions of state authorities. In 1913, when as a member of the Academy of Sciences he was asked to participate in the pompous ceremonies celebrating the 300th anniversary of the Romanov dynasty, he instead organized a celebration of the 200th anniversary of Jacob Bernoulli's publication of the Law of Large Numbers.

Marin Mersenne (1588–1648) was educated in Jesuit schools and in 1611 joined the Order of Minims. From 1619 he lived in the Minim Convent de l'Annonciade near the Place Royale in Paris and there held regular meetings of a group of mathematicians and scientists to discuss the latest ideas. Mersenne also served as the unofficial "secretary" of the republic of scientific letters in Europe. As such, he received material from various sources, copied it, and distributed it widely, thus serving as a "walking scientific journal". His own contributions were primarily in the area of music theory as detailed in his two great works on the subject, the *Harmonie universelle* and the *Harmonicorum libri*, both of which appeared in 1636. As part of his study of music, he developed the basic combinatorial formulas by considering the possible tunes one could create out of a given number of notes. Mersenne was also greatly interested in the relationship of theology to science. He was quite concerned when he learned that Galileo could not publish one of his works because of the Inquisition and, in fact, offered his assistance in this matter.

Hermann Minkowski (1864–1909) was a German Jewish mathematician who received his doctorate at the University of Königsberg. He became a lifelong friend of David Hilbert and, on Hilbert's suggestion, was called to Göttingen in 1902. In 1883, he shared the prize of the Paris Academy of Sciences for his essay on the topic of the representations of an integer as a sum of squares. In his essay, he reconstructed the entire theory of quadratic forms in n variables with integral coefficients. In further work on number theory, he brought to bear geometric ideas beginning with the realization that a symmetric convex body in n-space defines a notion of distance and hence a geometry in that space. The connection with number theory depends on the representation of forms by lattice points in space.

Muhammad ibn Muhammad al-Fullāni al-Kishnāwī (died 1741) was a native of northern Nigeria and one of the few African black scholars known to have made contributions to "pure" mathematics before the modern era. Muhammad's most important work, available in an incomplete manuscript in the library of the School of Oriental and African Studies in London, deals with the theory of magic squares. He gave a clear treatment of the "standard" construction of magic squares and also studied several other constructions — using knight's moves, borders added to a magic square of lower order, and the formation of a square from a square number of smaller magic squares.

Peter Naur (born 1928) was originally an astronomer, using computers to calculate planetary motion. In 1959 he became a full-time computer scientist; he was a developer of the programming language ALGOL and worked on compilers for ALGOL and COBOL. In 1969 he took a computer science faculty position at the University of Copenhagen.

Amalie Emmy Noether (1882–1935) received her doctorate from the University of Erlangen in 1908 and a few years later moved to Göttingen to assist Hilbert in the study of general relativity. During her eighteen years there, she was extremely influential in stimulating a new style of thinking in algebra by always emphasizing its structural rather than computational aspects. In 1934 she became a professor at Bryn Mawr College and a member for the Institute for Advanced Study. She is most famous for her work on Noetherian rings, and her influence is still evident in today's textbooks in abstract algebra.

Blaise Pascal (1623–1662) showed his mathematical precocity with his *Essay on Conics* of 1640, in which he stated his theorem that the opposite sides of a hexagon inscribed in a conic section always intersect in three collinear points. Pascal is better known, however, for his detailed study of what is now called Pascal's triangle of binomial coefficients. In that study Pascal gave an explicit description of mathematical induction and used that method, although not quite in the modern sense, to prove various properties of the numbers in the triangle, including a method of determining the appropriate division of stakes in a game interrupted before its conclusion. Pascal had earlier discussed this matter, along with various other ideas in the theory of probability, in correspondence with Fermat in the 1650s. These letters, in fact, can be considered the beginning of the mathematization of probability.

Giuseppe Peano (1858–1932) studied at the University of Turin and then spent the remainder of his life there as a professor of mathematics. He was originally known as an inspiring teacher, but as his studies turned to symbolic logic and the foundations of mathematics and he attempted to introduce some of these notions in his elementary classes, his teaching reputation changed for the worse. Peano is best known for his axioms for the natural numbers, first proposed in the *Arithmetices principia, nova methodo exposita* of 1889. One of these axioms describes the principle of mathematical induction. Peano was also among the first to present an axiomatic description of a (finite-dimensional) vector space. In his *Calcolo geometrico* of 1888, Peano described what he called a *linear system*, a set of quantities provided with the operations of addition and scalar multiplication which satisfy the standard properties. He was then able to give a coherent definition of the *dimension* of a linear system as the maximum number of linearly independent quantities in the system.

Charles Sanders Peirce (1839–1914) was born in Massachusetts, the son of a Harvard mathematics professor. He received a master's degree from Harvard in 1862 and an advanced degree in chemistry from the Lawrence Scientific School in 1863. He made contributions to many areas of the foundations and philosophy of mathematics. He was a prolific writer, leaving over 100,000 pages of unpublished manuscript at his death.

George Pólya (1887–1985) was a Hungarian mathematician who received his doctorate at Budapest in 1912. From 1914 to 1940 he taught in Zurich, then emigrated to the United States where he spent most of the rest of his professional life at Stanford University. Pólya developed some influential enumeration ideas in several papers in the 1930s, in particular dealing with the counting of certain configurations that are not equivalent under the action of a particular permutation group. For example, there are 16 ways in which one can color the vertices of a square using two colors, but only six are non-equivalent under the various symmetries of the square. In 1937, Pólya published a major article in the field, "Combinatorial Enumeration of Groups, Graphs and Chemical Compounds", in which he discussed many mathematical aspects of the theory of enumeration and applied it to various problems. Pólya's work on problem solving and heuristics, summarized in his two volume work *Mathematics and Plausible Reasoning*, insured his fame as a mathematics educator; his ideas are at the forefront of recent reforms in mathematics education at all levels.

Qin Jiushao (1202–1261), born in Sichuan, published a general procedure for solving systems of linear congruences — the Chinese remainder theorem — in his *Shushu jiuzhang* (*Mathematical Treatise in Nine Sections*) in 1247, a procedure which makes essential use of the Euclidean algorithm. He also gave a complete description of a method for numerically solving polynomial equations of any degree. Qin's method had been developed in China over a period of more than a thousand years; it is similar to a method used in the Islamic world and is closely related to what is now called the Horner method of solution, published by William Horner in 1819. Qin studied mathematics at the Board of Astronomy, the Chinese agency responsible for calendrical computations. He later served the government in several offices, but because he was "extravagant and boastful", he was several times relieved of his duties because of corruption. These firings notwithstanding, Qin became a wealthy man and developed an impressive reputation in love affairs.

Srinivasa Ramanujan (1887–1920) was born near Madras into the family of a bookkeeper. He studied mathematics on his own and soon began producing results in combinatorial analysis, some already known and others previously unknown. At the urging of friends, he sent some of his results to G. H. Hardy in England, who quickly recognized Ramanujan's genius and invited him to England to develop his untrained mathematical talent. During the war years from 1914 to 1917, Hardy and Ramanujan collaborated on a number of papers, including several dealing with the theory of partitions. Unfortunately, Ramanujan fell ill during his years in the unfamiliar climate of England and died at age 32 soon after returning to India. Ramanujan left behind several notebooks containing statements of thousands of results, enough work to keep many mathematicians occupied for years in understanding and proving them.

Frank Ramsey (1903–1930), son of the president of Magdalene College, Cambridge, was educated at Winchester and Trinity Colleges. He was then elected a fellow of King's College, where he spent the remainder of his life. Ramsey made important contributions to mathematical logic. What is now called Ramsey theory began with his clever combinatorial arguments to prove a generalization of the pigeonhole principle, published in the paper "On a Problem of Formal Logic". The problem of that paper was the *Entscheidungsproblem* (the decision problem), the problem of searching for a general method of determining the consistency of a logical formula. Ramsey also made contributions to the mathematical theory of economics and introduced the subjective interpretation to probability. In that interpretation, Ramsey argues that different people when presented with the same evidence, will have different degrees of belief. And the way to measure a person's belief is to propose a bet and see what are the lowest odds the person will accept. Ramsey's death at the age of 26 deprived the mathematical community of a brilliant young scholar.

Bertrand Arthur William Russell (1872–1970) was born in Wales and studied at Trinity College, Cambridge. A philosopher/mathematician, he is one of the founders of modern logic and wrote over 40 books in different areas. In his most famous work, *Principia Mathematica*, published in 1910–13 with Alfred North Whitehead, he attempted to deduce the entire body of mathematics from a single set of primitive axioms. A pacifist, he fought for progressive causes, including women's suffrage in Great Britain and nuclear disarmament. In 1950 he won a Nobel Prize for literature.

al-Samaw'al ibn Yahyā ibn Yahūda al-Maghribī (1125–1180) was born in Baghdad to well-educated Jewish parents. Besides giving him a religious education, they encouraged him to study medicine and mathematics. He wrote his major mathematical work, *Al-Bāhir* (*The Shining*), an algebra text that dealt extensively with the algebra of polynomials. In it, al-Samaw'al worked out the laws of exponents, both positive and negative, and showed how to divide polynomials even when the division was not exact. He also used a form of mathematical induction to prove the binomial theorem, that $(a+b)^n = \sum_{k=0}^{n} C(n,k) a^{n-k} b^k$, where the $C(n,k)$ are the entries in the Pascal triangle, for $n \leq 12$. In fact, he showed why each entry in the triangle can be formed by adding two numbers in the previous row. When al-Samaw'al was about 40, he decided to convert to Islam. To justify his conversion to the world, he wrote an autobiography in 1167 stating his arguments against Judaism, a work which became famous as a source of Islamic polemics against the Jews.

Claude Elwood Shannon (born 1916) applied Boolean algebra to switching circuits in his master's thesis at M.I.T in 1938. Shannon realized that a circuit can be represented by a set of equations and that the calculus necessary for manipulating these equations is precisely the Boolean algebra of logic. Simplifying these equations for a circuit would yield a simpler, equivalent circuit. Switches in Shannon's calculus were either open (represented by 1) or closed (represented by 0); placing switches in parallel was represented by the Boolean operation "+", while placing them in parallel was represented by " · ". Using the basic rules of Boolean algebra, Shannon was, for example, able to construct a circuit which would add two numbers given in binary representation. He received his Ph.D. in mathematics from M.I.T. in 1940 and spent much of his professional life at Bell Laboratories, where he worked on methods of transmitting data efficiently and made many fundamental contributions to information theory.

James Stirling (1692–1770) studied at Glasgow University and at Balliol College, Oxford and spent much of his life as a successful administrator of a mining company in Scotland. His mathematical work included an exposition of Newton's theory of cubic curves and a 1730 book entitled *Methodus Differentialis* which dealt with summation and interpolation formulas. In dealing with the convergence of series, Stirling found it useful to convert factorials into powers. By considering tables of factorials, he was able to derive the formula for $\log n!$, which leads to what is now known as Stirling's approximation: $n! \approx (\frac{n}{e})^n \sqrt{2\pi n}$. Stirling also developed the Stirling numbers of the first and second kinds, sequences of numbers important in enumeration.

Sun Zi (4th century) is the author of *Sunzi suanjing* (*Master Sun's Mathematical Manual*), a manual on arithmetical operations which eventually became part of the required course of study for Chinese civil servants. The most famous problem in the work is one of the first examples of what is today called the Chinese remainder problem: "We have things of which we do not know the number; if we count them by threes, the remainder is 2; if we count them by fives, the remainder is 3; if we count them by sevens, the remainder is 2. How many things are there?" Sun Zi gives the answer, 23, along with some explanation of how the problem should be solved. But since this is the only problem of its type in the book, it is not known whether Sun Zi had developed a general method of solving simultaneous linear congruences.

James Joseph Sylvester (1814–1897), who was born into a Jewish family in London and studied for several years at Cambridge, was not permitted to take his degree there for religious reasons. Therefore, he received his degree from Trinity College, Dublin and soon thereafter accepted a professorship at the University of Virginia. His horror of slavery, however, and an altercation with a student who did not show him the respect he felt he deserved led to his resignation after only a brief tenure. After his return to England, he spent 10 years as an attorney and 15 years as professor of mathematics at the Royal Military Academy at Woolwich. Sylvester returned to the United States in 1871 to accept the chair of mathematics at the newly opened Johns Hopkins University in Baltimore, where he founded the *American Journal of Mathematics* and helped initiate a tradition of graduate education in mathematics in the United States. Sylvester's primary mathematical contributions are in the fields of invariant theory and the theory of partitions.

John Wilder Tukey (born 1915) received a Ph.D. in topology from Princeton in 1939. After World War II he returned to Princeton as professor of statistics, where he founded the Department of Statistics in 1966. His work in statistics included the areas of spectra of time series and analysis of variance. He invented (with J. W Cooley) the fast Fourier transform. He was awarded the National Medal of Science and served on the President's Science Advisory Committee. He also coined the word "bit" for a binary digit.

Alan Turing (1912–1954) studied mathematics at King's College, Cambridge and in 1936 invented the concept of a Turing machine to answer the questions of what a computation is and whether a given computation can in fact be carried out. This notion today lies at the basis of the modern all-purpose computer, a machine which can be programmed to do any desired computation. At the outbreak of World War II, Turing was called to serve at the Government Code and Cypher School in Bletchley Park in Buckinghamshire. It was there, during the next few years, that he led the successful effort to crack the German "Enigma" code, an effort which turned out to be central to the defeat of Nazi Germany. After the war, Turing continued his interest in automatic computing machines and so joined the National Physical Laboratory to work on the design of a computer, continuing this work after 1948 at the University of Manchester. Turing's promising career came to a grinding halt, however, when he was arrested in 1952 for homosexual acts. The penalty for this "crime" was submission to psychoanalysis and hormone treatments to "cure" the disease. Unfortunately, the cure proved worse than the disease, and, in a fit of depression, Turing committed suicide in June, 1954.

Alexandre-Théophile Vandermonde (1735–1796) was directed by his physician father to a career in music. However, he later developed a brief but intense interest in mathematics and wrote four important papers published in 1771 and 1772. These papers include fundamental contributions to the theory of the roots of equations, the theory of determinants, and the knight's tour problem. In the first paper, he showed that any symmetric function of the roots of a polynomial equation can be expressed in terms of the coefficients of the equation. His paper on determinants was the first logical, connected exposition of the subject, so he can be thought of as the founder of the theory. Toward the end of his life, he joined the cause of the French revolution and held several different positions in government.

François Viète (1540–1603), a lawyer and advisor to two kings of France, was one of the earliest cryptanalysts and successfully decoded intercepted messages for his patrons. In fact, he was so successful in this endeavor that he was denounced by some who thought that the decipherment could only have been made by sorcery. Although a mathematician only by avocation, he made important contributions to the development of algebra. In particular, he introduced letters to stand for numerical constants, thus enabling him to break away from the style of verbal algorithms of his predecessors and treat general examples by formulas rather than by giving rules for specific problems.

Edward Waring (1734–1798) graduated from Magdalen College, Cambridge in 1757 with highest honors and shortly thereafter was named a Fellow of the University. In 1760, despite opposition because of his youth, he was named Lucasian Professor of Mathematics at Cambridge, a position he held until his death. To help solidify his position, then, he published the first chapter of his major work, *Miscellanea analytica*, which in later editions was renamed *Meditationes algebraicae*. Waring is best remembered for his conjecture that every integer is the sum of at most four squares, at most nine cubes, at most 19 fourth powers, and, in general, at most r

kth powers, where r depends on k. The general theorem that there is a finite r for each k was proved by Hilbert in 1909. Although the result for squares was proved by Lagrange, the specific results for cubes and fourth powers were not proved until the twentieth century.

Hassler Whitney (1907–1989) received bachelor's degrees in both physics and music from Yale; in 1932 he received a doctorate in mathematics from Harvard. After a brief stay in Princeton, he returned to Harvard, where he taught until 1952, when he moved to the Institute for Advanced Study. Whitney produced more than a dozen papers on graph theory in the 1930s, after his interest was aroused by the four color problem. In particular, he defined the notion of the *dual graph* of a map. It was then possible to apply many of the results of the theory of graphs to gain insight into the four color problem. During the last twenty years of his life, Whitney devoted his energy to improving mathematical education, particularly at the elementary school level. He emphasized that young children should be encouraged to solve problems using their intuition, rather than only be taught techniques and results which have no connection to their experience.

REFERENCES

Printed Resources:

Dictionary of Scientific Biography, Macmillan, 1998.

D. M. Burton, *The History of Mathematics, An Introduction*, 3rd ed., McGraw-Hill, 1996.

H. Eves, *An Introduction to the History of Mathematics*, 6th ed., Saunders, 1990.

H. Eves, *Great Moments in Mathematics (After 1650)*, Dolciani Mathematical Expositions, No. 7, Mathematical Association of America, 1983.

H. Eves, *Great Moments in Mathematics (Before 1650)*, Dolciani Mathematical Expositions, No. 5, Mathematical Association of America, 1983.

V. J. Katz, *History of Mathematics, an Introduction*, 2nd ed., Addison-Wesley, 1998.

Web Resource:

http://www-groups.dcs.st-and.ac.uk/~history (The MacTutor History of Mathematics archive.)

INDEX

A

Abel, Niels Henrik, 1153
Abel's transformation, 198
abelian, 300, 307
abelian difference set, 777
abelian group, 305, 361
abelian square, 1040, 1076
Abraham ibn Ezra, 1153
absolute rate, 927
absolutely continuous, 441
absorbing boundary, 428, 453
absorption laws, 17, 25, 300, 341
abstract datatype, 1102, 1108, 1109
abundance of witnesses, 1092
abundant, 261
acceptance probability, 1040, 1089
acceptance state, 1048, 1053
accepted, 1054
accepts language, 1085, 1089
access, 356, 419
access control, 924
accessible, 472
Ackermann, Wilhelm, 36
Ackermann function, 36, 1040, 1064
action, 2, 75
active, 960
active adversary, 926
active constraint, 956
active intrusion, 896
active vertex, 666
activity, 1007
activity nets, deterministic, 1006
activity-on-arc (AoA), 1007
activity-on-node (AoN), 1007
acyclic central plane arrangement, 811
acyclic digraph, 496, 536
acyclic graph, 496, 607
acyclic vector configuration, 808
adaptive bubblesort, 1102
adaptive sort, 1130
adaptive-precision arithmetic, 866
adding a crosscap to a surface, 496, 575

adding a handle to a surface, 496, 575
addition of linear transformations, 373
additive arithmetic function, 259
additive group, 307
adjacency matrix, 497, 519, 528, 630, 706
adjacency set, 630, 652, 706
adjacent edges, 497
adjacent vertices, 510, 960
  in a digraph, 497
adjoint, 357
admissible, 666
admittance matrix, 497, 586
ADT, 1102, 1109
ADT-constructor, 1102, 1109
adversary,
  active, 926
  passive, 926
affine central plane arrangement, 811
affine chirotope, 819
affine cipher, 890, 928
affine combination, 806
affine configuration, 808
affine geometry, 771
affine hyperplane, 807
affine plane, 362, 754, 761, 772
affine polar-duality, standard, 812
affine space, 754, 770, 807
affine subspace, k-dimensional, 807
affinely dependent, 807
affinely independent, 806
aftpart, 1114
aleph-null, 2
Alexandroff inequality, 423
algebra of linear transformations, 374
algebraic, 300
algebraic element, 329, 333
algebraic extension, 300, 333
algebraic integer, 295, 300, 333
algebraic multiplicity, 356, 407
algebraic number, 214, 219, 295, 300, 333
  degree of, 295
algebraic number field, 295
algebraic specification, 497, 522

algebraic structure, 300, 305
algorithm, 1040
   all-terminal reliability for undirected networks, 467
   analysis of, 1040
   arrangement of lines construction, 851
   array processor matrix multiplication, 386
   augmenting path, 666
   back substitution, 394
   base b expansion, 224
   BCH code decoding, 912
   binary greatest common divisor, 230
   binary search, 1137
   binary search tree, 1120
   binomial probabilities, logarithmic computation, 452
   binomial probabilities, recursive computation, 451
   branch and bound, 984
   bubblesort, 1131
   Christofides' heuristic, 695
   Clarke and Wright heuristic, 701
   Clarke and Wright savings heuristic, 695
   clustering, 872
   k-combinations, generating, 106
   k-combinations, generating random, 106
   complexity of, 1041
   computing range tree, 858
   computing triangulation hierarchy, 855
   constructing parity check matrix, 905
   continued fraction, 290
   convex hull, 862
   core heuristic for KP, 999
   creating a strong orientation, 535
   cutting plane, 982
   cycle-canceling, 676
   de Bruijn sequence, generating, 156
   decryption, 923
   Dijkstra, 656
   distinguished simplex of a 2-simplex, 1030
   dual simplex, 975
   dynamic graph, 1142
   eigenvalue, power method for computing dominant, 412
   eigenvalue, QR method for computing, 412
   encryption, 923
   Euclidean, 229
   Euclidean least remainder, 230
   Eulerian tour, 546
   exact, 631
   extended Euclidean, 230
   find kth smallest element in a list, 1138
   find mth smallest key, 1094
   fixed point of a p-simplex, 1032
   Fleury, 546
   Floyd-Warshall, 659
   forward substitution, 393
   Frederickson, 1145
   Gaussian elimination and back substitution, 395
   general edge-exchange heuristic, 696
   generating permutations, 105
   Gosper's, 207
   Graham scan, 845
   Gram-Schmidt, 369
   graph, generate random, 594
   greedy, 795
   greedy coloring, 558
   greedy edge-coloring, 562
   greedy heuristic for KP, 998
   greedy heuristic for set covering problem, 1005
   Greene-Nijenhuis-Wilf, 131
   Hamming code decoding, 909
   heapsort, 1131
   heuristic, 631
   Huffman code, 610
   implicit enumeration method for SP, 1005
   insertion sort, 1129
   integer computation of binomial coefficients, 449
   intersection of halfplanes, 864
   inverse transformation method, 486
   isomorphism testing, 554
   isomorphism quick test, 553
   Jacobi method, 413
   Karmarkar's method, 971
   Kruskal, 636
   label-correcting, 655
   Las Vegas, 1043, 1092
   least common ancestor, 711
   LU decomposition, 398
   Lucas-Lehmer test, 250

INDEX   1175

maintaining a minimum spanning tree, 1145
Makanin, 1076
matching, bipartite weighted, 645
matching, general, 648
matrix multiplication, 384, 659
matrix multiplication using linear combinations of columns, 385
mergesort, 1133
Miller-Rabin test, 254
modified first fit decreasing method, 1002
Monte Carlo, 1044, 1092
naive topological sort, 536
nearest insertion heuristic, 695
nearest neighbor heuristic, 694
network simplex, 678
numerically stable, 359
numerically unstable, 359
1-center of an unweighted tree, 991
1-median of a tree, 991
optimal tour partitioning, 686
orthogonal range reporting using range trees, 858
parallel, 802, 861
parent finder, 614
partially dynamic, 1144
k-permutations, generating 106
permutation, generating random, 106
k-permutations, generating random, 106
planarity-testing, 571
point location, 855
Pollard, 256
preflow-push, 667
Prim, 636
primality test, 1095
priority tree dequeueing, 1123
priority tree enqueueing, 1122
quadratic sieve, 256
quicksort, 1134
radix sort, 1136
randomized, 803, 863, 1046, 1091
randomized quicksort, 1083
range reporting, 856
range searching, 856
rank counting sort, 1135
ray shooting, 859
Robinson-Schensted, 132
savings heuristic, 685

scalar product form of matrix multiplication, 385
searching a 2-3 tree, 1137
selection sort, 1129
sieve of Eratosthenes, 246
simplex, 968
Sister Celine's, 207
Sollin, 637
sorting, 1107
spanning tree, breadth-first search, 618
spanning tree, depth-first search, 617
standard tableau generating, 131
Strassen, 189
Strassen for 2×2 matrices, 386
Strassen for $2^k \times 2^k$ matrices, 386
strong probable prime test, 254
successive shortest path, 677
syndrome decoding for linear codes, 907
topological sort, 747
trial division, 255
two-terminal reliability for undirected networks, 466
visibility map, 861
Viterbi, 921
Voronoi diagram construction, 849
all-pairs minimax path problem, 640
all-terminal network, 463
all-terminal reliability, 428, 465
almost every, 593
almost every graph has property P, 497
alphabet, 306, 898, 900, 1040, 1053, 1055, 1066
   balanced, 317
   of symbols, 1048
alphabetic datatype, 1102, 1109
alphastring, 1111
alternating, 641
alternating circular sequence, 820
alternating k-cycle, 746
alternating function, 818
alternating group, 300, 321
alternating path, 630
alternative axiom systems, 790
ambiguous, 1074
   context-free grammar, 1040
ambivalent data structure, 1102, 1144
American Standard Code for Information Interchange (ASCII), 1067
amicable, 261

analog channel, 890, 896
analytic graph theory, 590
analysis of an algorithm, 1040
ancestor, 604, 608
    least commmon, 710
    proper, 608
AND, 2
AND gate, 348
angle, 368
answer to a goal, 68
antecedent, 2, 13, 75, 1070
    of a production, 1040
anti-aliasing, 798, 873
anti-parallel arcs, 510
antichain, 2, 46, 497, 597, 718, 729
antidifference, 136, 194, 197
antisymmetric, 2, 41, 725
antisymmetric function, 818
antithetic variates, 428, 489
AoA representation, 1007
AoN representation, 1007
aperiodic prototile, 798, 828
aperiodic state, 428, 472
aperiodic tiling, 869
Appel, Kenneth, 565
arc(s), 497, 510
    anti-parallel, 510
arc connectivity, 671
arc length matrix, 658
arc list, 630, 706
arc sine laws, 455
area of a graph drawing, 573
argument form, 2, 50
Aristotle, 1153
arithmetic function, 214, 259
    additive, 259
    completely additive, 259
    completely multiplicative, 259
    multiplicative, 259
arrangement, 798
    of big $(d-1)$-spheres, 812
    of geometric objects, 805
    of lines, 850
    of oriented lines, 811
arrangement graph, 798, 812
array, 1118
array data structure, 1102
arrival process, 428, 477
Arrow's impossibility theorem, 745
arrows notation, 592

art gallery theorems, 852
articulation point, 497, 541
Artin, Emil, 1153
ascent, 136, 148
ASCII codes, 1068
aspect ratio, 798
    of a simplex, 842
    of a triangulation, 842
assertion, 2, 62
assigned, 684
assignment, 630, 702
assignment instruction, 63
assignment problem, 419, 424
    quadratic, 632
associate, 296
associated Stirling number,
    of the first kind, 157, 163
    of the second kind, 157, 163
associative law, 17, 25, 300, 305, 307, 341, 381
asymmetry, 2, 41
asymptotic, 2, 1077
    behavior, 37
    equality, 38, 136
    estimates
        analytic methods for deriving, 202
        of multiply-indexed sequences, 204
asymptotics of sequences, 201
atom, 2, 68, 300, 342, 344, 718, 734
atomic,
    formula, 2, 19, 68
    lattice, 718
    proposition, 2, 13
attachment, 497
    of a bridge, 570
attack,
    chosen-ciphertext, 891
    chosen-plaintext, 891
    ciphertext-only, 891
attribute, 497, 510
augmented matrix, 356, 393
augmenting path, 630, 641, 666
augmenting path algorithm, 666
augmenting path theorem, 642
Aurifeuillian factorization, 250
authentication, 891, 897, 924, 946
    data origin, 892, 924
    entity, 892, 924, 943, 946
    explicit key, 943
    identity, 943

implicit key, 943
message, 946
mutual, 943
unilateral, 943
automated reasoning, 2, 76, 77
automatic repeat request, 901
automata, equivalent, 1042
automaton,
cellular, 1040, 1060
finite, 1042, 1050
linear bounded, 1058
pushdown, 1045
automorphism(s), 300, 310, 326, 332, 497, 515, 550, 752, 754, 762
group of, 497, 515, 550, 589, 762
autonomous set, 749
auxiliary graph, 718, 749
average genus, 552
average length, 898
average-case complexity, 428, 1081
average-case running time, 1040
AVL tree, 1102, 1136
avoidability, 1076
avoidance of degeneracy, 867
axiom, 2, 22
of assignment, 63
of choice, 2, 29, 366
of equality, 30
of if-then, 64
of if-then-else, 65
of NOP, 63
of procedure, 67
of sequence, 63
of while, 66
axiomatic program verification, 61
axioms,
chirotope, 819
circuit, 819
cluster of stars, 821
cocircuit, 820
for set theory, 29
axis-parallel, 859

B

Babbage, Charles, 1153
Bachet's equation, 214, 285
Bachmann, Paul Gustav Heinrich, 38, 1154

back, of a queue, 1112
back edge, 616
back substitution, 356, 393
backbone network, 630, 683
backtrack, 604, 613, 616
Backus, John, 1154
Backus normal form (BNF), 1040, 1075
Backus-Naur form (BNF), 1040, 1075
backward arcs, 664
backward pointer, 1103
Baker's linear forms in logarithms, 220
balanced
alphabet, 317
incomplete block design (BIBD), 754, 759
matrix, 956, 1004
tree, 604, 609
ballot theorem, 455
banded, 402
Abu-l-'Abbas Ahmad ibn Muhammad ibn al-Banna al-Marrakushi, 1154
barber paradox, 23
Bar-Hillel ($uvwxy$, pumping) lemma, 1074
bargaining set, 1018
base b expansion, 223, 224
algorithm, 224
base-ten radix sort, 1135
baseten_digit, 1111
base, 214, 223
bases, converting between, 224
basic class, 356, 419
basic feasible solution, 956, 965
basic initial problem of random walks, 453
basic Merkle-Hellman knapsack encryption scheme, 937
basic set, 965
basic solution, 956
basic variables, 965
basis, 356, 365, 754, 767, 787, 806
for a digraph, 497, 531
of a point configuration, 799, 818
of a vector configuration, 799, 818
signed, 818
basis axioms, 754, 790
basis matrix, 965
basis step, 2, 57
basis, signed, 818
Bayes, Thomas, 1154

Bayes' formula, 435
BCH code, 891, 912
Bell number, 117, 164
bend, 573
Bernoulli, James (Jakob I), 442, 1154
Bernoulli distribution, 443
Bernoulli numbers, 136, 147, 164
    exponential generating function, 140
    recurrence relation, 141
Bernoulli polynomials, 136, 147
Bernoulli random variable, 428, 442
best fit (BF) method, 1002
best fit decreasing (BFD) method, 1002
beta distribution, 444
Betti number,
    first, 540
    zeroth, 542
BF method, 1002
BFD method, 1002
BFS, 616, 965
Bhaskara, 1155
BIBD, 754, 759
biconditional, 13
    as implication, 17
biconnectivity, 1147
Bienaymé-Chebyshev's inequality, 445
big circle, 812
big omega, 38, 136
big theta, 136
big-circle arrangement, 799, 812
big-oh, 2, 38, 136
bihomogeneous, 615
bihomogeneous tree, 604
bijection, 2, 31
bimatrix game, 1021
bin capacity, 1001
bin packing, 956, 1001, 1006, 1081
binary burst, 914
binary, 32
binary code, 364
binary coded decimal, 224
    expansion, 214
binary digit, 223, 448
binary erasure channel, 919
binary expansion, 223
binary Golay code, 916
binary greatest common divisor algorithm, 230
binary logarithm, 33
binary matroid, 754, 787

binary operation, 300, 305
binary relation, 3, 41
binary repetition code, 902
binary representation, 214, 448
binary search, 189, 1102, 1136
binary search algorithm, 1137
binary search tree, 25, 604, 610, 1102, 1120, 1136
binary search tree algorithm, 1120
binary symmetric channel (BSC), 891, 899
binary tree, 604, 609, 627, 1102, 1120
    left-complete, 1105, 1121
    complete, 609
    left-complete, 605
Binet form, 142
binomial coefficient, 82, 97, 161
    identities, 103
    recurrence relation, 141
binomial convolution, 136, 176
binomial distribution, 443
binomial random variable, 428, 442
binomial theorem, 104, 191
Biographies, 1153
biological migration, 458
biorder representation, 718, 742
biorder, 742
bipartite, 497, 517, 718, 729, 1147
bipartite matching algorithm, 644
biplane, 754, 771
Birkhoff-Von Neumann theorem, 421
bit, 223, 448, 896
    parity check, 894, 901
bit string, 25, 179
blending, 875
block, 497, 542, 754, 759, 766, 767, 768, 930
block cipher, 891, 930
block code, 900
block designs, properties of, 761
block number, 759
block size, 759
blocking set, 597
blossom, 630, 647
BNF, 1040, 1075
body, of a clause, 3, 68
bond, 497
bond matroid, 791
Bonferroni's inequality, 433
Boole, George, 433, 1155

Boole's inequality, 433
Boolean algebra, 300, 344, 718
Boolean expression, 346
Boolean function, 300, 346
Boolean lattice, 734
Borda consensus function, 718, 745
Borda count, 718, 745
Bose-Einstein, 428, 438
bottom-up mergesort, 1132
bound,
    Gilbert-Varshamov, 915
    Hamming, 915
    Johnson, 916
    singleton, 915
    sphere-packing, 915
bound variable, 19
boundary, 428, 497, 799
    absorbing, 428, 453
    elastic, 453
    of a region, 569
    reflecting, 430, 453
boundary problem, 453
boundary walk, 613
bounded, 38, 725, 960
bounded distance search, 857
bounded lattice, 300, 342, 734
bounded LP, 956
bounded poset, 718
bounded tree, 604, 615
bounded two-sided error, 1089
bouquet, 497, 517
bracket representation, 739
branch, 91, 497
branch and bound algorithm, 984
branch and bound method, 983
branch and bound tree, 997
branching process, 428, 476
breadth-first search, 604, 616
breadth-first search tree, 617, 618
bridge, 497, 534, 570
Brooks' theorem, 558
Brouwer fixed-point theorem, 1032, 1032
Bruck-Ryser-Chowla Theorem, 771
BSC, 899
bubblesort, 1102, 1130
    adaptive, 1102
    algorithm, 1131
bundle, 1146
Burnside, William, 124, 1155
Burnside conjecture, 317

Burnside's lemma, 124
burst error, 891
burst vector, 902
busy beaver function, 1040, 1058
busy beaver machine, 1040, 1057

C

c-game, 1024
CA, 945
cactus, 497
CAD/CAM, 874, 879
Caesar cipher, 891, 928
calculus of differences, 191
call instruction, 67
cancellation laws, 300, 307, 327, 361
canonical block form, 1004
Cantor, Georg Ferdinand Ludwig Philip, 22, 28, 1155
Cantor-Schröder-Bernstein Theorem, 28
CAP, 689
capacitated concentrator location problem, 630, 688
capacitated location problem, 956, 992
capacitated minimum spanning tree, 630, 683, 684
capacitated minimum spanning tree problem, 683
capacitated network, 630, 663
capacitated p-median (CPM) problem, 992
capacity assignment problem, 630, 689
capacity, 663, 664, 666, 673, 688, 700, 899
    of an arc, 630
    of a channel, 891
    of a cut, 630
    of a path, 630
    residual, 674
car pool, 1026
Carathéodory's Theorem, 809, 831
cardinal number, 3, 27
cardinality, 3, 22, 641, 710
Carmichael number, 167, 214, 251
carry bit, 348
Cartesian product, 3, 24, 47, 498, 515, 718, 729
Cascading divide-and-conquer, 862
Cassini's identity, 142

Catalan conjecture, 286
Catalan number, 136, 145, 164, 176, 180, 627
  generating function, 140
  recurrence relation, 141
Catalan's equation, 215, 285
catatrophic code, 921
caterpillar, 498, 604, 607
Cauchy, Augustin-Louis, 1155
Cauchy distribution, 444
Cauchy interlacing theorem, 415
Cauchy's theorem, 315
Cauchy-Binet formula, 391
Cayley, Arthur, 1156
Cayley digraph, 498, 522, 589
Cayley graph, 498, 522, 589
Cayley's formula, 620, 622, 390
Cayley's theorem, 310, 320
Cayley-Hamilton theorem, 405
CBC, 930
CCLP, 688
C-complete, 1088
ceiling, 3, 34
cell, 799, 811, 1102, 1117
  Voronoi, 824
cellular, 578
cellular automaton, 1040, 1060
cellular imbedding, 498
centdian, 996
center, 313, 498, 540, 604, 607
p-center, 956, 986
centering transformation, 970
centerpoint, 799, 808
central authority, 945
central depot, 700
central hyperplane, 807
central hyperplane arrangement, 799, 812
central plane arrangement, 799, 811
centralizer, 313
certificate, 209, 1102, 1144
  compressed, 1144
  of a dyadic property, 1144
  public-key, 945
  sparse, 1107, 1144
  stable, 1107, 1144
  strong, 1107, 1144
certificate function, stable, 1107
certification authority (CA), 891, 945
certifying the truth of combinatorial identities, 209

CFB, 931
CFRAC, 257
CGRM, 258
chain, 3, 46, 498, 597, 718, 729, 799, 847, 877
  irreducible, 429
  k-, 729
  monotone, 847
chain partition, 738
  k-norm of, 721
chain rule, 435
chaining, 1140
chaining method, 1102
chaining variable, 951
chain-product, 718, 729
challenge, 946
challenger, 946
chance, 1017
k-change, 693
change of basis matrix, 377
channel, 899
  analog, 890, 896
  binary erasure, 919
  binary symmetric (BSC), 899
  capacity of, 891
  q-ary symmetric, 899
channel encoder, 896
Chapman-Kolmogorov equations, 471
character recognition, 872
characteristic, 300, 332
characteristic equation, 136, 180, 356, 405
characteristic function, 3, 34, 446, 956, 1024, 1040, 1057, 1065
characteristic polynomial, 356, 405, 498, 586
characteristic root, 180
characteristic value, 498, 586
characteristic vector, 498, 586
characteristic-function game, 956, 1024
characterization of matroid, 793
Chazelle's algorithm, 848
Chebyshev, Pafnuty Lvovich, 445, 1156
check polynomial, 909
chemistry, and trees, 625
chi square distribution, 444
child, 604, 608, 609, 627
Chinese postman problem, 545
Chinese remainder theorem, 215, 235
chirotope, 799, 818

affine, 819
uniform, 804, 819
chirotope axioms, 819
chirotope function, 821
Cholesky decomposition, 356, 398
Chomsky, Avram Noam, 1156
Chomsky hierarchy, 1040, 1072
Chomsky normal form, 1040, 1070
Chor-Rivest Cryptosystem, 936
chord, 604, 616, 633
    triangular, 534
chosen-ciphertext attack, 891
chosen-plaintext attack, 891
Christofides' heuristic algorithm, 695
k-chromatic, 557
chromatic index, 561, 599
chromatic number, 170, 498, 557, 564, 566, 599, 837
    edge, 561
    empire, 564
    k-tuple-, 566
chromatic polynomial, 557
chromatically k-critical, 498, 557
Chrysippus, 1156
Church, Alonzo, 1156
Church's thesis, 1041, 1063
Church-Turing thesis, 1041, 1063
chump, 1023
cipher, 891, 923
    affine, 928
    block, 891, 930
    Caesar, 891, 928
    full Vigenère, 928
    Hill, 893, 929
    homophonic substitution, 893
    mode of operation, 930
    polyalphabetic, 894, 928
    random, 927
    self-synchronizing stream, 895, 933
    shift, 895
    simple substitution, 928
    simple transposition, 928
    simple Vigenère, 929
    stream, 895, 933
    substitution, 895
    synchronous stream, 895, 933
    transposition, 895
    U.S. Data Encryption Standard (DES), 930
    Vernam, 895, 933
    Vigenère, 895
cipher block chaining (CBC) mode 891, 930
cipher feedback (CFB) mode, 891, 931
cipher shift, 928
ciphertext, 891, 923
ciphertext space, 923
ciphertext-only attack, 891
circuit, 498, 755, 787, 799, 819, 1080
    Euler, 544
    generalized, 534
circuit axioms, 755, 790, 819
circuit elimination, 790
circulant, 356, 415
circular linked list, 1103, 1119
circular permutation, 91
circular sequence, 820
Clarke and Wright heuristic algorithm, 701
Clarke and Wright savings heuristic, 695
class, 356, 419
class group relations method, 258
class library, 799, 879
class NP, 1078
class P, 1078
classification,
    of closed surfaces, 575
    of states, 472
classification theorem for finite simple groups, 317
clause, 3, 68
Clausen's $_4F_3$ identity, 206
clique, 498, 563, 598
clique number, 498, 563, 599
clique partition number, 498, 599
clique problem, 1081
closed, 767
closed class, 428
closed directed walk, 531
closed form, 136, 139, 195
closed formula, 3, 35, 68
closed halfspace, 799, 807, 840
closed hash table, 1103
closed path, 498, 540
closed set, 755, 787
closed subset, 737
closed surfaces, classification of, 575
closed trail, 498, 540
closed under duality, 755, 791
closed walk, 498, 540

closed wff, 19
closest pair, 835
closure, 3, 41, 300, 755, 767, 787
   Kleene, 1043, 1067
   positive, 1067
closure axioms, 755, 790
closure operation, 755, 767
closure operator, 737
cluster, 1103, 1144
   of rank 3 hyperline sequences, 799
   of stars, 799, 820
   of star permutations, 821
   of stars associated with a pseudoline arrangement, 821
cluster of stars axioms, 821
clustering, 640, 1141
clustering algorithms, 872
clustering property, 1103
clutter, 729
CMLD, 891, 902
CMST, 684
CNF, 16
CNF, 301, 346
coalition, 956, 1024
cobasis, 755, 791
cocircuit, 755, 791, 799, 820
cocircuit axioms, 820
cocycle matroid, 791
code(s), 891, 898, 1076
   $[n, M]$-, 900
   ASCII, 1068
   average length of, 898
   BCH, 891, 912
   binary Golay, 916
   binary representation, 902
   block, 900
   catastrophic, 921
   convolutional, 892
   $(n, 1, m)$-convolutional, 921
   cross-interleaved Reed-Solomon, 914
   cyclic, 892, 909
   dual, 892, 904
   equivalent, 892, 901
   error-correcting, 869, 892
   error-detecting, 892
   extended, 892, 907
   extended Hamming, 906
   extended Preparata, 918
   for an alphabet, 1041
   generated by a polynomial, 909
   Golay, 893
   Goppa, 941
   Hamming, 893, 908
   Huffman, 610
   linear, 893, 904
   (n, k)-linear, 904
   linear block, 904
   maximum distance separable (MDS), 915
   message authentication (MAC), 947
   narrow-sense, 912
   $[n, M]$-, 900
   Nordstrom-Robinson, 894, 917
   optimal, 915
   perfect, 894, 916
   prefix, 610, 894, 898
   Preparata, 894, 918
   primitive, 912
   punctured, 894, 907
   quantum error-correcting, 869
   Reed-Muller, 894
   Reed-Solomon (RS), 894, 914
   self-dual, 895, 904
   self-orthogonal, 895, 904
   shortened, 895, 908
   systematic, 895, 904
   ternary Golay, 916
   turbo, 895, 919
   uniquely decodable, 895
code indicator, 1041
   of a language, 1076
   of a word, 1076
codebook, 929
codeword, 610, 891, 898
coding theory, 891, 896, 900
codomain, 3, 31
coefficient, 393
coefficient matrix, 393
cographic matroid, 755, 791
Cohen, Paul, 28
coherent, 459
coherent system, 428, 461
coin tossing, 454
coindependent set, 755, 791
collapse, 1041, 1086
Collatz conjecture, 35
collision, 949, 1103, 1139
collision-resistant hash function (CRHF), 950
collision resolution, 1103, 1140

INDEX   1183

chaining, 1140
rehashing, 1141
collision resolution function, 1141
coloop, 755, 791
color map optimization, 872
colorable,
   k-edge, 501, 561
   k-vertex, 557
coloring, 122, 837
   k-canonical, 592
   of edges, 561
   of edges, algorithm for, 562
   k-, 557
   of a map, 564
   of vertices, 557
   of vertices, algorithm for, 558
coloring pattern, 82
column, of a matrix, 378
column rank, 366
column space, 365
column vector, 378
combination, 82
   k-, 97
combination coefficient, 82, 97
combination-with-replacement, 82
   coefficient, 82, 101
combinatorial geometry, 755, 787
combinatorial matrix theory, 417
combinatorially contractible, 557
combinatorics, polyhedral, 868
commodity, 689
common logarithm, 33
common PRAM, 1041, 1060
common random numbers, 428, 488, 489
communicate, 472
communicating class, 428, 472
commutative, 307, 780
commutative law, 17, 25, 300, 305, 341, 344, 381
commutative ring, 301
commutator, 318
comparability, 724
comparability digraph, 718, 725
comparability graph, 499, 534, 718, 725
comparability invariant, 718, 749
comparable, 3, 725
comparable elements, 46, 719
comparison sort, 1041, 1083, 1103, 1126
compatible with a symmetry, 828
complement, 23, 432, 734, 1041, 1068

   in a lattice, 342
   of a relation, 3, 41
   of a set, 3
   of a signed set, 818
complement laws, 25, 344
complement operator, 3, 71
complementary slackness, 974
complemented lattice, 301, 342, 734
complete, 3, 22
C-complete, 1041, 1088
complete maximum likelihood decoding (CMLD), 902
complete k-partite graph, 499, 517
complete m-ary tree, 609
complete r-uniform hypergraph, 499, 596
complete binary tree, 604, 609
complete bipartite graph, 499, 517
complete graph, 499, 517
complete hypergraph, 499, 596
complete information, 956, 1017
complete matching, 630, 643
complete maximum likelihood decoding (CMLD), 891
complete multipartite graph, 499
complete pivoting, 356, 401
complete set
   of invariants, 499, 550
   of mutually orthogonal latin squares, 755, 783
   of operations, 522
complete system of residues, 215, 231
completely additive, 259
completely multiplicative, 259
complex number, 3, 21, 306
complex_number, 1111
complexity,
   of an algorithm, 1041
   average-case, 1081
   cost-based, 1077
   of a function, 1041
   space, 1046, 1078
   time, 1047, 1078
   worst-case, 1081
complexity class,
   Δ_0^p, 1086
   Π_n^p, 1086
   Σ_n^p, 1086
   coNP, 1041, 1086
   coRP, 1089
   NP, 1041, 1086

NP^C, 1086
NP^L, 1086
P, 1041, 1086
P^C, 1086
P^L, 1086
PP, 1089
PSPACE, 1041, 1086
RP, 1089
ZPP, 1089
complexity issues, 1016
component, 459, 499, 540
   strong, 532
composite, 215, 236
composite hybrid method, 631
composite hybrid procedure, 693
composite key, 3, 49
composition, 3, 32, 42, 114, 144, 373
composition series, 316
compound formula, 19
compound interest, 179, 181
compound proposition, 3, 13
compressed certificate, 1144
computable, M-, 1057
computation, 1050, 1054, 1060
   M-, 1057
   polynomial-space, 1045
   polynomial-time, 1045
computational convexity, 799, 869
computational fluid dynamics, 875
computational geometry, applications, 867
computational models, 1049
computational representation, 35
computational security, 891
computationally secure, 926
computationally solvable, 1065
computer assisted manufacturing, 870
computer graphics, 872
   linear transformations in, 375
computer model, 1048, 1050, 1054, 1056
computer vision, 874
computer science, counting trees in, 627
computer-assisted proof, 3, 77
computing range tree algorithm, 858
computing triangulation hierarchy algorithm, 855
concatenation, 306, 1066, 1067
   of languages, 1041
   of strings, 1041
conclusion,
   of an argument, 50
   of an argument form, 4
   of a proof, 4
concrete data structure, 1103, 1117
concurrent read, 861
concurrent read concurrent write (CRCW) PRAM, 1042, 1060
concurrent read exclusive write (CREW) PRAM, 1042, 1060
concurrent write, 861
condensation, 532
condition, 4, 75
condition number, 356, 399
conditional, 13
   as disjunction, 17
conditional assignment construct, 64
conditional entropy, 897
conditional probability, 428, 435
conditional statement, 4
Condorcet's paradox, 745
configuration, 1049, 1054, 1056
   for an NDFSR, 1050
congruence,
   of integers, 231
   root of, 234
   solution, 234
congruence class, 215, 231
congruent integers, 215
conjugacy problem, 318
conjugate, 295, 313, 318, 755, 778
   of an algebraic number, 295
   of a partition, 114
conjugate sequence, 356, 418
conjunction, 4, 13
conjunctive addition, 51
conjunctive normal expression, 16
conjunctive normal form, 4, 16, 301, 346
   algorithm, 16
conjunctive simplification, 51
connected, 41, 499, 517, 531, 540
   k-edge, 504, 542
   k-, 504, 542
   2-, 534
connected matroid, 755, 792
connection cost, 688
connectivity, 499, 1147
   2-edge, 1148
   vertex, 509
connectivity relation, 42
coNP, 1041, 1086

INDEX   1185

consecutive chain, 719, 738
consensus function, 719, 744
consensus ranking, 719, 744
consequent, 4, 13, 75, 1070
    of a production, 1041
consistent, 4, 22
consistently intersect, 831
constant coefficients, linear homogeneous recurrence relation with, 179
constant polynomial, 303, 329
constant zero function, 1063
constraint, 785, 959
construct, 4
constructing parity check matrix algorithm, 905
construction method, 631
construction, 700
    of arrangement of lines algorithm, 851
    of matroids, 792
    of Voronoi diagrams, 849
constructive dilemma, 51
constructive direct proof, 54
constructive solid geometry, 874
constructor, 71, 1109
    ADT, 1110
contained in, 726
context-free grammar, 1041, 1071
context-free language, unambiguous, 1047
context-sensitive grammar, 1042, 1071
continued fraction, 215, 289
    ultimately periodic, 290
continued fraction algorithm, 290
continued fraction method, 257
continuous, absolutely, 441
continuum hypothesis, 4, 28
continuum problem, 28
contouring, 875
contractible, combinatorially, 557
contracting an edge, 515
contraction, 499, 792, 799, 821, 1031
    elementary, 499, 557
contraction mapping theorem, 1032
contradiction, 4, 15, 17
contrapositive, 4, 16
contrapositive law, 17
control and simulation language, 491
control center, 683
convergent, 215, 293
converse, 4, 16, 499, 532

of a relation, 4, 41
converting between bases, 224
convex, 799, 830, 960
convex d-polyhedron, 800, 840
convex body, 799, 824, 830
convex combination, 807
convex decomposition, 800, 842
convex game, 1025
convex hull, 800, 807, 831, 844, 956, 981, 1029, 1083
    algorithm, 862
    applications, 871
convex polygon, 800, 807, 831
convex polyhedron, 800
convex polytope, 800, 807
convex position, 800, 831
convex set, 152, 807, 956
convex set mapping, 1031
convex subdivision, 853
convexity, computational, 869
convexly independent, 807
convolution, 136, 171
convolutional code, 892
$(n, 1, m)$-convolutional code, 921
Conway, John H., 1062
Cook reducibility, 1088
Cook's theorem, 1079
cooperative strategy combination, 1018
coordinate, 365
coordinate matrix, 365
coordinate vector, 365
coprime, 215, 227
core, 1025
core heuristic for KP algorithm, 999
corner coloring, 122
corollary, 4, 50
coRP, 1089
correct, 901
correct conclusion, 4
correlation, 444
coset, 301, 313, 327, 892, 906
coset leader, 892, 906
cospectral pair of graphs, 587
cost, 631, 652, 673, 674, 683, 691, 692, 700
    reduced, 632
cost-based complexity, 1077
countable, 27
countable set, 4
countably infinite, 27

counterexample, 4, 56
counting methods, 89
counting problems, summary, 84
counting subsets, 104
Courant-Fischer minimax identity, 391
covariance, 444
covariance stationary, 488
cover, 47, 597, 725, 826, 997
  minimal, 684, 997
cover diagram, 47, 719, 725
cover graph, 719, 725
cover relation, 719
covered, 1003
covering, 800, 826
  lower density of, 826
  upper density of, 826
covering problems, 997
CP, 1008
CPM model, 956, 1007
CPM problem, 992
Cramer's rule, 396
crash duration, 1007
CRCW, 1042, 1060
CREW PRAM, 1042, 1060
CRHF, 950
k-critical, 557
critical pair, 719, 746
critical path, 956, 1008
critical path method (CPM) model, 1007
criticality index, 956, 1013
cross edge, 616
cross-interleaved Reed-Solomon code (CIRC), 914
crosscap, 499
crosscap number, 499, 575, 579
crossing number, 400, 571
cryptanalysis, 892, 923
cryptographic system, 892
cryptography, 892, 896, 923, 1092
  public-key, 935
cryptology, 892, 923
cryptosystem, 892
  Chor-Rivest, 936
  El Gamal, 892, 936, 939
  elliptic curve, 936
  knapsack, 893, 936
  McEliece, 893, 936
  Merkle-Hellman, 936
  public-key, 894, 925
  Rabin, 894, 936

RSA, 895, 936, 937
CSL, 491
cube, 1042, 1076
cube graph, 499, 518
cumulative distribution function, 441
Cunningham number, 249
current cell, 1103
current location, 1114
curvature, 840
curve, x-monotone, 811
customer, 477
customer locations, 700
cut, 631, 632, 633, 664, 684
  minimum, 632, 664
  s-t, 664, 632
cut optimality conditions, 634
cut-edge, 499, 534, 542
cut-point, 499, 541
cutset, 428, 461
  K-, 429, 461
  s-t, 431, 461
cut-vertex, 499, 541
cutting plane, 957, 982
cutting plane algorithm, 982
cutting plane method, 982, 1005
cycle(s), 301, 319, 499, 517, 531, 540
  alternating k-, 746
  directed, 499, 500, 531
  fundamental, 605, 616, 617
  fundamental system of, 617
  Hamilton, 547
  k-, 504
  negative, 632, 652, 674
cycle decomposition, 120
cycle graph, 499
cycle index, 82, 121
cycle matroid, 755
cycle rank, 499, 540
cycle space, 364
cycle structure, 82, 120
cycle-canceling algorithm, 676
cyclic code, 892, 909
cyclic difference set, 777
cyclic group, 301, 314
cyclic partition, 117
cyclic permutation, 120
cyclic d-polytope, 800, 843
cyclic subgroup, 301, 311, 315
cyclotomic extension, 337
cyclotomic polynomial, 249

D

DAG (directed acyclic graph), 500, 536
Dantzig, George, 1157
Data Encryption Standard (DES), 892
data compression, 892, 896
data integrity, 892, 897, 924
data origin authentication, 892, 924
data part, 945
data structure,
   ambivalent, 1102, 1144
   array, 1102
   concrete, 1103, concrete, 1117
database, 1103, 1125
datatype, 1103, 1108
   abstract, 1102, 1108, 1109
   alphabetic, 1102, 1109
   elementary, 1103, 1109
   generic, 1104, 1110
   numeric, 1105, 1109
   ordered, 1110
   union-find, 1108
datatype specification, 1110
Davenport-Schinzel sequence, 800, 870
day of the week, determining, 234
De Morgan, Augustus, 565
de Bruijn diagram, 155
de Bruijn sequence, 136, 155
   algorithm for generating, 156
de Laplace, Pierre-Simon, 1165
decidability properties of grammar
   classes, 1073
decides, 1086
decimal digit, 223
decimal expansion, 223
decision function, 1078
decision tree, 604, 610
decision variables, 957, 959
decodable, uniquely, 898
decoding, 902
   BCH codes algorithm, 912
   complete maximum likelihood
     (CMLD), 891, 902
   Hamming codes algorithm, 909
   incomplete maximum likelihood
     (IMLD), 893, 902
   minimum error probability (MED),
     894, 902
   syndrome, 907
decoder,

hard decision, 918
soft decision, 918
decryption, 892, 923
   algorithm, 923
Dedekind, Richard, 1157
Dedekind sum, 215
Dedekind's problem, 740
defect theorem, 1076
deficient, 261
deficit vertex, 676
defining predicate, 21
definite answer property, 69
definite answer to a goal, 68
definite clause, 4, 68
degeneracy avoidance, 867
degenerate, 867
degenerate LP, 960
degenerate vertex, 960
degree, 303, 500, 511, 641
   of an algebraic number, 295
   of an extension field, 332
   of a field extension, 301
   of a permutation group, 301, 319
   of a polynomial, 329
   of a relation, 48
degree matrix, 620
degree sequence, 500, 511
degrees of freedom, 833
deheaping, 1103, 1122
Dehn-Sommerville's equations, 843
Delaunay criterion, 875
Delaunay triangulation, 848
delay element, 920
delete operation, 1118
deleting,
   a subset of edges, 514
   a subset of vertices, 514
   a vertex, 500, 514
   an edge, 500, 514
deletion, 792, 800, 821
demand, 684, 700, 986
demand point, 957, 986
demand vertex, 673
deMoivre, Abraham, 1157
deMoivre's theorem, 181
DeMorgan, Augustus, 1157
DeMorgan's laws, 17, 25
dense graph, 604
density,
   of a covering, 800

of a packing, 800, 824
density function, 428, 441
denumerable, 27
denumerable set, 4
dependent, 365, 787
dependent edge, 719, 749
dependent events, 428
dependent set, 356
depth, 604, 609
depth-first search, 604, 616
depth-first search spanning tree algorithm, 617
depth-first search tree, 616
dequeue, 1112
dequeue implementation, 655
derangement(s), 82, 96, 109
   exponential generating function, 140
   number of, 166, 434
   recurrence relation, 141
derivable, 1070
   directly, 1070
derivation, 1070
   leftmost, 1043, 1071
   of a string, 1042
derivative, 1069
derived design, 771
DES, 892, 930
Descartes, René, 1157
descendant, 604, 608
   proper, 608
descent, 148
design, 759
   (v, k, λ), 759
   t-, 755, 766
   t-(v, k, λ), 755, 766
designing physical systems, 639
destination vertex, 689
detect, 901
determinant, 356, 387
deterministic activity net, 957, 1006, 1007
deterministic finite transducer, 1051
development, 777
   of a difference set, 755
devil's pair, 552
DFS, 616
diagonal, 60, 378, 420
diagonal matrix, 35, 970
diagonal relation, 44
diagonalizable, 407

diagonalizable matrix, 356
diagonalization proof, 4, 60, 61
diagram, 725
diameter, 500, 540, 604, 615, 835
   of a tree, 169
Dickson, Leonard Eugene, 1158
diconnected, 531
diconnected digraph, 500
dictionary, 1103, 1116
difference, 4
   of matrices, 356
   of relations, 41
   of sets, 4
   of vectors, 361
difference equation, 194
difference operator, 136, 189
difference sequence, 136, 192
difference set, 777
   trivial, 777
   $(v, k, \lambda; n)$ difference set, 759
difference table, 137, 192
Diffie-Hellman key agreement, 943
digit, 223
digital fingerprint, 950
digital signature, 892, 925
digital signature algorithm (DSA), 941
digraph(s), 170, 419, 500, 511, 526, 583
   acyclic, 496
   Cayley, 522
   comparability, 725
   condensation of, 532
   converse of, 532
   disconnected, 500, 532
   isomorphism of, 550
   labeled, 583
   of a matrix, 357
   number of, 170
   permutation, 522
   reducible, 507, 532
   Schreier, 522
   simple, 507
   strict, 511, 526
   strongly connected, 507, 531
   transitive, 528
   unilateral, 532
   weak, 532
   weakly connected, 532
dihedral group, 301, 321
Dijkstra's algorithm, 618, 656
nth dilutor, 71

Dilworth's theorem, 739, 751
dimension, 357, 365, 775
    of a poset, 719, 746
d-dimensional point, 806
dimensional tolerancing, 875
Diophantine approximation, 215, 289
Diophantine equation, 215, 281
Diophantus, 1158
dipole, 500, 517
direct derivability relation, 1070
direct product, 307
direct proof, 4, 52, 54
direct sum, 307, 357, 792
directed acyclic graph, 536
directed Chinese postman problem, 681
directed cycle, 500, 531
directed graph(s), 170, 419, 500, 511, 526, 583
    of a matrix, 357
    number of, 170
directed network, 63, 652
directed ordering, 46
directed out-tree, 631, 653
directed path, 500, 531
directed trail, 500, 531
directed two-terminal flow network, 643
directed walk, 500, 531
    closed, 531
direction, 500, 510, 859
directional dual, 532
directly derivable, 1070
Dirichlet, Gustav Lejeune, 92
Dirichlet drawer principle, 92
Dirichlet inverse, 267
Dirichlet product, 267
Dirichlet's theorem in d dimensions, 215, 247, 294
disconnected, 500, 532
disconnecting set of edges, 500, 542
disconnecting set of vertices, 500, 542
discordant permutation, 137, 158
discrete calculus, fundamental theorem of, 197
discrete exponential function, 338
discrete logarithm, 215, 273, 338
discrete random variable, 441
discrete-event simulation, 429
discrete-time Markov chain, 429, 469
disjoint, 4, 23, 432
    mutually, 23
    pairwise, 23
disjoint cycles, 319
disjoint union, 515, 729
disjunction, 4, 13
disjunctive addition, 51
disjunctive normal expression, 15
disjunctive normal form, 5, 15, 301, 346
    algorithm, 16
display operation, 522
disproof, 5, 56
dissimilar hypergeometric terms, 137
distance, 357, 368, 500, 531, 540, 604, 692, 700, 702, 904, 986, 987
    Euclidean, 993
    Hamming, 893, 900
    rectilinear, 993
    unicity, 926
distance label, 631, 653, 666
distance-ranked drawing, 573
distance-ranked partition, 573
distinguished simplex, 1029, 1030
distribution, 429, 441
    Bernoulli, 443
    beta, 444
    binomial, 443
    Cauchy, 444
    chi square, 444
    exponential, 444
    F-, 444
    gamma, 444
    geometric, 443
    hypergeometric, 443
    joint, 443
    normal, 444
    Pascal, 443
    Poisson, 443
    Rayleigh, 444
    standard normal, 444
    t-, 444
    uniform, 444
distributive lattice, 301, 719, 736
distributive law, 17, 25, 342, 344, 381
divide-and-conquer algorithm, 137
    cascading, 862
    many-way, 862
    parallel, 862
    randomized, 863
divide-and-conquer recurrence relations, 188
divide-and-conquer sort, 1132

dividend, 221
divides, 215, 221
    exactly, 221
divisibility, tests for, 222
divisibility lattice, 5
divisibility poset, 727
division algorithm, 222, 320
division ring, 301, 327
divisor, 215, 221
divisor lattice, 719, 734
divisor of zero, 327
Dixon's identity, 206
DNF, 15, 301, 346
dodecahedral conjecture, 826
dodecahedral graph, 500, 518
Dodgson, Charles Lutwidge, 1158
domain, 5, 18, 31, 48, 305, 725, 1103, 1108
dominance laws, 17, 25
dominant, 412
dominant eigenvalue, 357
dominate, 537
    in Löwner order, 415
dominated, 1022
dominating set, 1081
domination, 188, 466
don't care conditions, 349
dot product, 357, 380
double complement law, 25
double factorial, 157
double hashing, 1141
double negation law, 17
doubly hypergeometric, 137, 206
doubly stochastic, 357, 420
downset, 500, 597, 719, 732
downward separation, 1087
DSA, 941
dual, 301, 341, 344, 459, 500, 719, 729, 755, 770, 791
dual code, 892, 904
dual graph, 577
dual graph imbedding, 500
dual hypergraph, 595
dual ideal, 732
dual imbedding, 577
dual LP problem, 957, 973
dual polytopes, 800, 840
dual simplex algorithm, 975
dual variables, 973
duality, 772

duality construction, 577
duality principle, 301, 342, 344
duality transformation, 800
dummy activity, 1007
dyadic graph property, 1103, 1144
dynamic graph algorithms, 1142
dynamic structure, 1103, 1127
dynamic update operation, 1103, 1143

E

ECB, 892, 930
eccentricity, 500, 540, 605, 607
ECM, 215, 256
edge(s), 501, 509, 840, 960
    dependent, 749
    matched, 631
    multiple, 510
    parallel, 510
edge k-chromatic, 561
edge chromatic number, 501, 561
edge chromatically k-critical, 561
edge coloring, 122, 501
    algorithm, 562
edge connectivity, 501, 542, 691
edge cover, 1080
edge cut, 501
edge independence number, 501, 563
edge reconstruction conjecture, 556
edge-complement, 501, 515
edge-cut, fundamental, 605, 617
edge-deleted subgraph, 501, 555
edge-deleted subgraph collection, 555
edge-face inequality, 569, 577
edge-format, 578
edge-incidence table, 1103, 1124
edge-ordered tree, 1146
edge-reconstructible graph, 501
edge-reconstructible invariant, 501, 555
edge-to-edge tiling, 827
Edmonds' theorem, 647
Ehrenfest diffusion model, 470, 475
eigenanalysis, 405
eigenspace, 357, 407
eigenvalue, 357, 405, 501, 552, 586
    computing, 411
eigenvector, 357, 405, 501, 552, 586
Eisenstein's irreducibility criterion, 330
EISPACK, 414

El Gamal cryptosystem, 892, 936, 939
elastic boundary, 429, 453
electronic codebook (ECB), 892, 930
element, 5, 21
elementary ADT-constructor, 1110
elementary contraction, 557
elementary datatype, 1103, 1109
elementary projection function, 5
ellipsoid method, 972
elliptic curve, 215
elliptic curve cryptosystem, 936
elliptic curve method, 215, 256
embedding, see imbedding
empire chromatic number, 564
emptiness, 1071
emptiness problem, for grammars, 1042
empty graph, 501, 511, 517
empty relation, 44
empty set, 5, 21, 30, 1152
empty string, 1042, 1066
encryption, 892, 896, 923
    algorithm, 923
end vertex, 607
endomorphism, 326
endpoint, 501, 510
endpoint table, 519, 1103, 1124
enheaping, 1103, 1122
enqueue, 1112
entity authentication, 892, 943, 946
entropy, 892, 897, 898
    conditional, 897
    joint, 897
entry, 36, 1104, 1106, 1125
enumeration
    of graphs 580
    of spanning trees, 620
    of trees, 622
Epimenides, paradox of, 22
epimorphism, 5, 31
EQP, 78
equality,
    of sets, 5, 21
    of matrices, 378
equation,
    diophantine, 281
    linear diophantine, 281
    word, 1048
equilibrium, 957, 1017
    in mixed strategies, 1022
equinumerous, 27

equitable facility location problems, 996
equitable locations, 995
equitable partition, 552
equivalence, 1050, 1071
    topological, 572
equivalence class, 5, 45, 419
equivalence problem for grammars, 1042
equivalence relation, 5, 4, 310
equivalent automata, 1042
equivalent codes, 892, 901
equivalent colorings, 123
equivalent grammars, 1042, 1071
equivalent latin squares, 755, 780
equivalent propositions, 5
equivalent vertices, 419
equivocation, 897
Eratosthenes, 1158
    sieve of, 246
Erdős, Paul, 1158
Erdős-Szekeres problem, 809
Erdős-Szekeres theorem, 832
EREW (exclusive read exclusive write), 1042
EREW PRAM, 1060
Erlang distribution, 477
error,
    relative, 430
    roundoff, 430
error probability, 903
error-correction code, 869, 892, 896
error-detection code, 892
error-vector, 900
essential, 767
essential game, 1024
Euclid, 1159
Euclidean algorithm, 229
Euclidean distance, 993
Euclidean domain, 301, 328
Euclidean field, 296
Euclidean hyperplane arrangement, 800, 812
Euclidean line arrangement, 800
Euclidean minimum spanning tree problem (MST), 1084
Euclidean norm, 301, 328
Euclidean pseudoconfiguration of points, 800
Euclidean space, 362, 806
Euclidean TSP, 693
Euler, Leonhard, 1159

Euler characteristic, 501, 564, 576
Euler circuit, 544
Euler constant, 221
Euler criterion, 278
Euler formula, 569
Euler numbers, 137, 148, 153, 164
    exponential generating function, 140
    recurrence relation, 141
    sums of powers, 200
Euler phi-function, 215, 247, 260
Euler polyhedral equation, 569, 577
Euler polynomials, 152
Euler relation, 843
Euler theorem, 215, 233, 512
Euler totient function, 167
Euler tour, 501, 544
Euler trail, 501, 544
Eulerian graph(s), 501
    number of, 171
even, 647
even permutation, 301, 319
even vertex, 631
event(s), 429, 432
    complement of, 432
    dependent, 428
    disjoint, 432
    independent, 429
exact algorithm, 631
exact arithmetic, 866
exact boundary representation, 874
exactly divides, 215, 221
excedance, 137, 148
excess, 665, 1025
excess vertex, 676
exchange, 819
    k-opt, 693
excluded middle, 17
excluded-minor theorems, 794
exclusive or, 13
exclusive read, 861
exclusive read exclusive write (EREW), 1042, 1060
exclusive write, 861
existence proof, 55
existential generalization, 51
existential lower bound, 1042, 1083
existential quantification, 18
existential quantifier, 5
existential specification, 51
existentially quantified predicate, 5

expanded blossom, 647
expected number, 479
expected value, 429, 443
expected waiting time, 479
experiment, 429, 432
explicit key authentication, 943
exponent, 357, 420, 450
exponential distribution, 444, 477
exponential function, 5, 33
exponential generating function, 82, 137, 139, 176
exponentiation of cardinals, 27
exportation law, 17
extended binomial coefficient, 98
extended code, 892, 907
extended Euclidean algorithm, 230
extended Hamming code, 906
extended matrix, 807
extended Preparata code, 918
extended Riemann hypothesis, 248
extension, 719, 745
    cyclotomic, 337
    of a function, 31
    Galois, 334
    normal, 334
    separable, 334
    of a substitution, 1069
extension field, 301, 333
extensionality, 30
extensive form, 1017
exterior region, 501, 569
external direct product, 307
external sorting method, 1104, 1126
extinction probability, 429, 476
extremal graph, 501, 590
extremal graph theory, 590
extremal number, 501, 590

F

face, 502, 569, 576, 840
    k-, 800
face description, 807
face lattice, 813
face vector, 843
face-to-face tiling, 801, 827
facet, 807, 840, 892
faceted boundary representation, 874
facility, 702, 957

INDEX   1193

facility location, 957, 986
fact set, 5, 75
factor, 216, 221
   of a polynomial, 329
factor group, 313
factor theorem, 330
factorial, 5, 96
factorial expansion, 104
factorial function, 37
factorial numbers, 161
factorization, QR, 409
factorization algorithm, 216
failed, 459
failure system, k-out-of-n, 462
fair coin, 432
falling power, 82, 96, 137, 191
families of graphs, 517
Fano matroid, 789
Fano plane, 598, 755, 762
far end, 1104, 1118
Farey series, 216, 227
Kamāl al-Dīn al-Fārisī, 1159
Farkas' lemma, 961
farthest neighbor, 835
fast multiplication of matrices, 384
feasible direction, 957, 960
feasible flow, 663, 671, 673
feasible LP, 957
feasible region, 957, 959, 981
feasible solution, 676, 957, 959
Feit-Thompson theorem, 317
de Fermat, Pierre, 1159
Fermat equation, 216, 284
Fermat number(s), 216, 249
   table of, 253
Fermat prime, 216, 249
Fermat's last theorem, 216, 284
Fermat's little theorem, 216, 233
Fermi-Dirac model, 429, 438
Ferrers diagram, 82, 114
Ferrers digraph, 719, 742
Ferrers relation, 742
FF method, 1002
FFD method, 1002
Fiat-Shamir Signature Scheme, 942
Fibonacci (Leonardo of Pisa), 141, 1160
Fibonacci heap, 1104, 1122
Fibonacci numbers, 137, 139, 140, 164,
    179, 182, 216, 229
   Binet form, 142

   generalized, 140
   generating function, 140
   recurrence relation, 141
field, 301, 327, 331
   extension, 333
   finite, 338
   Galois, 338
field automorphism, 332
field extension, 333
field isomorphism, 332
FIFO property, 1104, 1112
figurate number, 137, 158, 168
fill, 357, 402
filter, 597, 719, 732, 732
final class, 357, 419
final tableau, 976
find operation, 1118
finding kth smallest element algorithm,
   1138
finding mth smallest key algorithm, 1094
fingerprint, 950, 1092
finite automaton, 1042, 1050
finite differences, 189
   properties of, 190
finite dimensional, 365
finite extension, 334
finite field, 301, 338
finite game, 1017
finite geometry, 770
finite linear space, 767
finite precision, 866
finite sequence, 36
finite set, 5, 22
finite substitution, 1069
finite sum, 195
finite transducer, 1042
finite tree, 605
finite-state machine, 1042, 1051
finite-state machine with output, 1042,
   1051
finite-state recognizer (FSR), 1042
   nondeterministic, 1044
finitely generated, 307
finitely generated group, 301
finitely presentable, 318
finiteness, 1071
finiteness problem, 1042
first fit (FF) method, 1002
first fit decreasing (FFD) method, 1002
first Betti number, 540

first passage time, 429, 453, 475
first-come-first-served, 477
first-order linear recurrence relation with constant coefficients, 137
first-order logic, 5, 18
first-order recurrence relation, solution of, 181
Fisher's inequality, 598, 760
  generalization of, 766
fix a set, 334
fix element, 300
fixed cost, 702
fixed field, 301, 334
fixed point, 123, 957, 1031
  of a p-simplex algorithm, 1032
fixed-charge problems, 985
flat, 756, 787
  t-, 756, 770
flat notation, 1104, 1110
flat polygon representation, 576
flat-polygon drawing of a graph, 577
flip-flop, 920
float, 957
floating-point arithmetic, 429
floating-point computations, 449
floating-point form, 450
floating-point numbers, 866
floating-point operation, 450
floor, 5, 34
flop, 357, 395, 429, 450
flow, 631, 702
  feasible, 663, 673
  maximum, 663
  minimum cost, 632
flowchart, 513
Floyd-Warshall algorithm, 659
$f(n)$ time reduction, 1084
Folkman-Lawrence theorem, 823
forbidden subposet description, 719, 732
force vectors, 362
forepart, 1114
forest, 502, 517, 605, 607
formal language, 1067
formula, 5, 68
formulation, 981
forward arcs, 664
forward edge, 616
forward error correction, 901
forward pointer, 1103
forward star, 631, 706

forward substitution, 357, 393
foundation, 30
four color theorem, 502, 558, 564
4-point lemma, 735
four ways, 1024
Fourier, Joseph, 1160
Fourier transform, 446
fraction, 450
Frame-Robinson-Thrall hook formula, 130
Frame-Stewart algorithm, 94
Frederickson algorithm, 1145
free distance, 921
free edge, 631, 641
free equivalence class, 317
free group, 318
free monoid, 301, 306, 1042, 1067
Q-free poset, 726
free semigroup, 306, 317
free vertex, 631
freely equivalent, 317
frequency, 1042, 1066
Frobenius, Georg, 124, 1160
Frobenius determinantal formula, 130
Frobenius normal form, 420
Frobenius-König theorem, 418
front, of a queue, 1112
Frucht's theorem, 589
FSR, 1042, 1048
full m-ary tree, 605, 609
full disjunctive normal form, 5, 346
full Vigenère cipher, 928
full-adder, 349
fullness, 1104, 1140
fully indecomposable, 357, 418
fully parenthesized proposition, 5, 13
function, 5, 31
  Ackermann, 1040, 1064
  alternating, 818
  Borda consensus, 745
  busy beaver, 1040, 1058
  characteristic, 446, 1024, 1040, 1065
  collision resolution, 1141
  collision-resistant hash, 950
  complexity of, 1041
    Kolmogorov-Chaitin type, 1041
  consensus, 744
  constant zero, 1063
  contraction, 1031
  density, 428

doubly hypergeometric, 206
generating, 446
hash, 893, 949, 1104
induced by a TM, partial, 1045
Iverson truth, 1110
key equivocation, 926
linear homogeneous, 807
Matyas-Meyer-Oseas hash, 951
moment generating, 446
normalized characteristic function, 1024
number of, 33, 91
objective, 960
ω-, 266
Ω-, 266
one-way hash, 950
onto, 180
output, 1051
partial, 1045, 1063
partial recursive, 1064
predecessor, 709
primitive recursive, 1045
probe, 1106
projection, 1063
secondary hash, 1141
social choice, 744
stable, 1107, 1144
successor, 1063
total, 1047, 1063
transition, 1051
Turing-computable, 1047
uncomputable, 1047
functionally complete, 15, 302, 346
set, 5
fundamental cycle, 605, 616, 617
fundamental edge-cut, 605, 617
fundamental homomorphism theorem, 314, 327
fundamental polygon representations, 576
fundamental system of cycles, 605, 617
fundamental system of edge-cuts, 605, 617
fundamental theorem,
 of algebra, 331
 of arithmetic, 216, 222, 236
 of discrete calculus, 197
 of finite abelian groups, 310
 of finitely generated abelian groups, 310

 of Galois theory, 335
fuzzy complement, 71
fuzzy intersection, 71, 71
fuzzy logic, 6, 70, 72
fuzzy set, 6, 70, 71
fuzzy system, 71
fuzzy union, 71

G

Gale-Ryser theorem, 418
Gallai-Milgram Theorem, 751
Galois extension, 302, 334
Galois field, 302, 332, 338
Galois group, 302, 334
Galois, Evariste, 1160
gambler's fallacy, 436
gambler's ruin, 429, 454, 455, 456
game, 1016
 c-, 957
 characteristic-function, 1024
 convex, 1025
 essential, 1024
 extensive form, 1017
 finite, 1017
 imperfect, 1017
 inessential, 1024
 in normal form, 1017
 in strategic form, 1017
 symmetric, 1022
 zero-sum, 1022
Game of Life, 1042, 1062
game perfect, 1017
game theory, 1016
gamma distribution, 444
gamma function, 202
GAN model, 957, 1013
gap, 1001
GASP, 491
Gauss, Carl Friedrich, 1161
Gauss' $_2F_1$ identity, 205
Gauss' lemma, 278
Gauss-Bonnet theorem, 840
Gaussian binomial coefficient, 82, 98
Gaussian elimination, 357, 394
Gaussian integer, 216, 295, 325
gcd, 216
k-GDD, 756
(K, λ)-GDD, 756

Gelfond-Schneider theorem, 220
general activity simulation program, 491
general knapsack problem, 1001
general purpose system simulator, 490
general edge-exchange heuristic algorithm, 696
general graph, 502, 510
general matching algorithm, 648
general number field sieve, 257
general position, 801, 808, 811, 835, 867, 957
general sparse, 402
generalized activity network (GAN) model, 1013
generalized axiom of choice, 30
generalized binomial coefficient, 98
generalized circuit, 534
generalized continuum hypothesis, 6, 28
generalized octahedral graph, 518
generalized pigeonhole principle, 92
generalized precedence relation (GPR), 1007
generalized random walk, 456
generalized vector norm, 399
generalizing from the generic particular, 51
generating function(s), 82, 137, 139, 171, 446
  binomial-type, 174
  for counting reduced trees, 623
  for counting trees, 623
  exponential, 176
  ordinary, 171
  for particular sequences, 140
  used to solve counting problems, 95
  used to solve recurrence relations, 186
generating set, 502, 767
generating subset, 589
generator, 301, 311, 314
  of a shift register, 920
generator matrix, 892, 904
generator polynomial, 892, 909
generic datatype, 1104, 1110
genetics, 437
genus, 170, 502, 575, 579, 801, 840
geodesic, 605
geographic information system, 801, 877
Geomamos, 879
geometric constraint solving, 801, 878
geometric distribution, 443

geometric lattice, 719, 737
geometric multiplicity, 357, 407
geometric progression, 180
geometric random variable, 429
geometric series, 137, 204
geometric visualization, 880
Geomview, 879
Germain, Sophie, 1161
Geršgorin discs, 357, 410
GERT model, 957, 1014
gift-wrapping, 844
Gilbert-Varshamov bound, 915
Gilmore-Lawler lower bound, 703
GIMPS, 251
girth, 502, 569
GIS, 801, 877
glb, 734
GNFS, 257
goal, 6, 68
goal programming (GP), 957, 980
  preemptive, 980
goal programming simplex method, 980
Gödel, Kurt, 28, 1161
Gödel numbering, 1042, 1065
Golay code, 893
Goldbach conjecture, 248
Golomb's self-generating sequence, 157
Goppa code, 941
Gosper's algorithm, 207
GP problem, 980
GPR, 1007
GPSS, 490
graded poset, 720, 731
Graham scan algorithm, 845
Gram-Schmidt orthogonalization, 369
grammar(s), 1043
  ambiguous context-free, 1040, 1074
  context-free, 1041, 1071
  context-sensitive, 1042, 1071
  decision problems for, 1071
  equivalent, 1042, 1071
  inherently ambiguous context-free, 1074
  language generated by, 1070
  left-linear, 1071
  leftmost language generated by, 1071
  length-increasing, 1043, 1071
  linear, 1043, 1071
  monotonic, 1071
  phrase-structure, 1070

regular, 1046, 1071
right-linear, 1071
type 0, 1047, 1070
type 1, 1042, 1043, 1047, 1071
type 2, 1041, 1047, 1071
type 3, 1046, 1047, 1071
unambiguous context-free, 1074
unrestricted, 1070
grammar classes, decidability properties of, 1073
graph(s), 6, 31, 502, 510
   a.e., 593
   acyclic, 496, 536, 607
   arrangement, 812
   bipartite, 497, 517
   bouquet, 517
   Cayley, 522
   k-chromatic, 557
   n-colorable, 498
   coloring, 557
   comparability, 499, 534, 725
   complete, 499, 517
   complete bipartite, 499, 517
   complete multipartite, 499, 517
   computation package for, 522
   connected, 517, 540
   connectivity of, 671
   cover, 725
   cube, 518
   cycle, 499, 517
   n-cycle, 517
   dense, 604
   d-dimensional hypercube, 517
   dipole, 517
   directed, 500, 511, 526, 583
   dodecahedral, 500, 518
   dual, 577
   edge-reconstructible, 555
   empty, 501, 511, 517
   extremal, 501
   general, 510
   generalized octahedral, 518
   Hamiltonian, 547
   icosahedral, 502, 518
   incomparability, 725
   intersection, 503, 515, 518
   interval, 503, 518
   invariant for, 550
   isomorphism of, 550, 1089
   Kuratowski, 518, 567
   labeled, 504, 580
   line, 504, 516
   nonplanar, 567
   nonseparable, 541
   normalized drawing of, 522
   null, 505, 511, 517, 518
   number of, 170
   number of connected, 170
   k-partite, 504
   n-path, 517
   perfect, 506, 563
   permutation, 506, 522, 749
   Petersen, 506, 518
   planar, 506, 567
   planar straight-line, 802, 846
   plane, 1105, 1144
   Platonic, 506, 518
   product, 506
   quotient, 506, 515
   random, 593
   reconstructible, 555
   reconstruction conjecture for, 556
   regular, 507, 511
   k-regular, 511
   Schreier, 507, 522
   self-complementary, 507, 516
   simple, 507, 510
   simplicial, 507, 510
   strongly regular, 50, 586
   sum, 502, 515
      iterated, 515
   tetrahedral, 50, 518
   tough, 508, 547
   tree, 517
   trivial, 50, 511
   Turán, 508, 590
   2-edge connected, 534
   underlying, 508, 526
   union, 515
   unit distance, 835
   voltage, 522
   weakly γ-perfect, 563
   weighted, 509, 510
   wheel, 509
   n-wheel, 517
graph drawing, 877
graph mapping, 515
graph model, 502, 510, 526, 525
graph representation, 519
graph specification, 519

graphic matroid, 756, 787
graphical evaluation and review technique (GERT) model, 1014
graphical sequence, 502, 511
graphical user interface (GUI), 801, 873
graphics, 872
Grassmann-Plücker relations, 801, 819
Gray code, 137, 155, 502, 547
great internet Mersenne prime search (GIMPS), 251
greater than, 725
greatest common divisor, 216, 227
greatest element, 46, 342
greatest integer function, 34
greatest lower bound, 6, 47, 719, 734
greedy algorithm, 795
greedy coloring algorithm, 558
greedy edge-coloring algorithm, 562
greedy heuristic for KP algorithm, 998
greedy heuristic for set covering problem algorithm, 1005
Greene-Kleitman (GK) theorem, 739
Greene-Nijenhuis-Wilf algorithm, 131
Greibach normal form, 1074
grid, 729
grid drawing, 573
ground formula, 6, 68
ground set, 756, 787
group(s), 302, 305, 307
  abelian, 305
  automorphism, 515, 550, 589
  classification theorem for finite simple, 317
  Galois, 334
  number of, 165
  p-, 303, 315
  presented by a group presentation, 318
  of the rectangle, 321
  simple, 316
  solvable, 316
  trapping, 1050
group divisible design (GDD), 756, 768
  k-, 756
  (K, λ)-, 756
group presentation, 318
group rational, 1018
group-type, 756, 768
growth factor, 357, 401
GUI, 801, 873
Guthrie, Francis, 565

H

Hadamard conjecture, 776
Hadamard design, 756, 771, 775
Hadamard inequality, 392
Hadamard matrix, 756, 775
Hadwiger's conjecture, 558
Haken, Wolfgang, 565
half-adder, 348
half-plane intersection, 811
half-space, 801
Halin's conjecture, 556
Hall's condition, 597
Hall's theorem, 26, 644
halting configuration, 1056
halting function, 6, 37
halting problem, 1043
halting state, 1055
ham-sandwich cut, 801, 808
Hamilton, William Rowan, 1162
Hamilton cycle, 502, 547, 692
Hamilton graph(s), 502, 1080
  number of, 171
Hamilton path, 502, 547
Hamilton tour, 692
Hamiltonian, 547
Hamming, Richard W., 1162
Hamming bound, 915
Hamming code, 893, 908
  extended, 906
Hamming distance, 400, 893, 900
Hamming weight, 893, 904
handshaking lemma, 512
hanging configuration, 1056
hard decision decoder, 918
C-hard language, 1043, 1088
Hardy, Godfrey Harold, 1162
harmonic number(s), 137, 154, 164
  generating function, 140
  recurrence relation, 141
harmonic sum, 189
Abū 'Alī al-Hasan ibn al-Haytham (Alhazen), 1163
hash function, 893, 949, 1104, 1139
  collision-resistant, 950
  Matyas-Meyer-Oseas, 951
  one-way, 950
  secondary, 1141
hash table, 1104, 1139
  closed, 1103

hash-value, 949
hashing, 1104, 1136, 1139
  double, 1141
hashing rule,
  division method, 1140
  multiplication method, 1140
Hasse diagram, 6, 47, 720, 725
hatcheck problem, 112
Hausdorff maximal principle, 30
Havel's reduction, 512
Havel's theorem, 512
head, 6, 68, 502, 511, 1103
header, 1104, 1117
heap, 605, 1104, 1122
  Fibonacci, 1104, 1122
heapsort, 1104, 1130
heapsort algorithm, 1131
Heawood map coloring theorem, 564
  nonorientable, 565
Heawood, Percy, 565
hedge, 71
height, 169, 605, 609, 720, 729, 732
Helly's theorem, 809, 831
heptagonal numbers, 168
hereditary family, 732
Hermitian adjoint, 357, 378
  properties of, 382
Hermitian matrix, 357, 379
Hessenberg form, 413
heuristic algorithm, 631
hexadecimal expansion, 223
hexadecimal representation, 216
hexagonal numbers, 168
high capacity, 689
higher-order logic, 19
Hilbert, David, 28
Hilbert's tenth problem, 1043, 1065
Hill cipher, 893, 929
hitting time, 453
Hoare, C.A.R., 1132
Hoffman polynomial, 502, 587
Hoffman-Wielandt theorem, 416
Hofstadter G-sequence, 157
homeomeric, 828
homeomorphic, 502, 515
homogeneous, 393, 605, 615, 807
d-homogeneous, 605
homogeneous recurrence relation, 137, 179, 180
homogeneous tree, 605

homomorphism, 302, 310, 326
homophonic substitution, 893, 929
homothet, 838
homothetic copy, 838
hook, 82, 129
hooklength, 82, 129
Huffman code, 610
Huffman code algorithm, 610
Huffman tree, 610
Hypatia, 1163
hypercube, d-dimensional, 517
hypercube graph, 502
hypergeometric distribution, 443
hypergeometric random variable, 429, 442
hypergeometric series, 137, 204
hypergraph, 502, 595
  complete, 499, 596
  complete r-uniform, 499, 596
  normal, 599
  partial, 595
  r-partite, 506, 598
  regular, 595
  simple, 507, 595
  uniform, 596
  unimodular, 598
hypergraph clique, 598
hyperplane, 756, 770, 787, 801, 839
  affine, 807
  central, 807
  Motzkin, 832
  oriented, 807
hypothesis H, 248
hypothetical syllogism, 51

I

icosahedral graph, 502, 518
icosian game, 547
ideal, 302, 324, 597, 720, 732
  generated by an element, 732
  generated by a subset, 732
  maximal, 324
  prime, 324
  principal, 324
ideal security, 950
idempotent, 32, 405, 779, 780
idempotent latin square, 756
idempotent laws, 17, 25, 342

idempotent matrix, 357
identification, 946
identity, 302, 305, 307, 379
identity function, 33
identity laws, 17, 25, 344
identity matrix, 357
identity permutation, 319
identity property, 307
if-then instruction, 64
if-then-else instruction, 64
ill-conditioned system, 357
image set, 6, 31
imaginary quadratic field, 295
imbalance, of a vertex, 676
imbedded, 780
imbedded latin square, 756
imbedding, 501, 503, 567
   cellular, 578
   of a graph, 576
imbedding modification, 577
imbeds in, 726
IMLD, 893, 902
imperfect game, 1017
implication, 6
implicit enumeration method for SP algorithm, 1005
implicit key authentication, 943
implicit trapping state, 1050
improper coloring, 557
improper subgroup, 302, 311
improvement, 700
improvement method, 631
improving direction, 957, 960
imputation, 957, 1025
   unreasonable, 1025
IMSL, 404, 414
in-degree, 503, 528
in-place realization, 1104, 1127
incidence matrix, 503, 519, 528, 595, 756, 759, 1003, 1104, 1124
incidence relation, 770
incidence rule, 503, 510
incidence structure, 756, 770
   network, 1124
incident, 832
incident edge, 503
incident with, 510
incident-edge table, 503, 519
inclusion, 1071
inclusion problem for grammars, 1043

inclusion-exclusion principle, 107, 432
incomparability graph, 720, 725
incomparable, 6, 725
incomparable elements, 46
incomparable pair, 720
incomplete maximum likelihood decoding (IMLD), 893, 902
incomplete information, 1017
incremental method, 845
indefinite clause, 68
indefinite sum, 137, 197
independence,
   of continuum hypothesis, 28
   of events, 429
   of irrelevant alternatives, 745
independence number, 503, 563, 599
independent, 6, 22, 365, 435
   linearly, 960
independent events, 429
independent random variables, 429, 441
independent set, 358, 59, 756, 787
independent subset, 503, 563
independent vertex set, 1081
index, 216, 273, 302, 313, 756, 759
index arithmetic, 273
index function, 338
index of cyclicity, 358, 419
index of summation, 195
index-of-coincidence, 929
indices for primes, table of, 273
indirect proof, 6, 53, 54
individually rational, 1018
induced by, 684
induced pair permutation, 321
induced pair-action group, 321
induced partition, 6, 45
induced poset, 725
induced subgraph, 503, 514
induction, 6
   principle of mathematical, 57
induction hypothesis, 6, 57
induction step, 6, 57
inductive proof, 6
inequality, word, 1048
inessential, 767
inessential game, 1024
infeasible, 959
infeasible LP, 957
infeasible tableau, 976
infinite, 6, 22, 30

infinite dimensional, 365
infinite loop, 1057
infinite sequence, 36
infinite sets, 27
infinite tree, 605, 615
infix notation, 13, 614
inflow, 665
information rate, 900
information sets, 1017
information theory, 893, 897
inherently ambiguous, 1043, 1074
initial conditions, 179
initial distribution, 469
injection, 6, 31
injective, 31
inner, 647
inner corner, 131
inner product, 358, 368, 904
inner product space, 368
inorder traversal, 605, 613
input model, 485
input size, 1043, 1077
input symbols, 1051
insert operation, 1118
insertion sort, 1104, 1128
insertion sort algorithm, 1129
installation cost, 689
instance,, 68
instantiation, 6, 18, 50, 75
integer(s), 6, 21, 219, 295, 306, 1052
integer partition, 720, 734
integer programming (IP), 957, 981
integral domain, 302, 327
integrity, data, 892
interchanging variables in a double sum, 196
interconnection network model, 1043, 1060
interior point method, 957, 970
internal sorting method, 1104, 1126
internal vertex, 605, 609, 658
intersecting, 831
intersecting family, 720
intersection, 7, 23, 41, 720, 746, 1067
intersection graph, 503, 515, 518, 596
intersection of halfplanes, algorithm for finding, 864
intersection relation, 7
interval, 7, 46, 720, 726
interval graph, 503, 518

interval hypergraph, 598
interval order, 720, 742
interval representation, 720, 742
intransitivity, 41
intrusion, 896
invariant(s), 503, 550
  complete set of, 499
  edge-reconstructible, 556
  reconstructible, 555
inverse, 16, 231, 302, 305, 307, 317, 380, 1043, 1069
  of an integer a modulo m, 216
  of a matrix, 358
  of a word, 317
inverse function, 7, 32
inverse image, 7, 31
inverse property, 307
inverse relation, 7
inverter, 348
invertible function, 7
invertible matrix, 358, 380
involution, 32, 130, 302, 319
IP problem, 981
irrational number, 7, 216, 219
irrationality measure, 294
irreducibility, 330, 331
irreducible, 329, 472, 734
irreducible chain, 429
irreducible element, 302
irreducible matrix, 358
irreducible polynomial, 302, 340
irreducible tournament, 503, 537
irreflexive, 7, 41
irregular, 284
irrelevant, 429, 466
isogonal tiling, 801, 828
isohedral tiling, 828
isolated, 540
isolated point, 503, 510
isomorphic, 550
isomorphism, 302, 310, 319, 1081
  algorithm for testing, 553, 554
  of arrangements, 812
  of block designs, 756
  of Boolean algebras, 344
  of configurations, 801, 808
  of designs, 762
  of digraphs, 504, 550
  of fields, 332
  of graphs, 504, 515, 550

of groups, 302
of lattices, 341, 720
of matroids, 756, 787
of monoids, 306
of ordered trees, 609
of permutation groups, 302
of posets, 720, 726
of rings, 302, 326
of semigroups, 306
of simple graphs, 504
of rooted trees, 605, 608
of trees, 605
of vector spaces, 358, 366
isomorphism problem, 318
isomorphism test, 552
isomorphism type, 504, 550
isotopic, 756, 780
isotoxal tiling, 801, 828
item, 36, 997
iterated Cartesian product, 729
iterated function, 32
iterated graph sum, 515
iterated logarithm, 33
iterated sum, 195
Iverson truth function, 1110

J

Jacobi logarithm, 338
Jacobi method, 413
Jacobi symbol(s), 216, 278
  reciprocity law for, 279
Jacobi's identity, 392
Jarvis' march, 844
Johnson bound, 916
join, 49, 341, 504, 515, 720, 734
join-irreducible, 720, 734
joint distribution, 441, 443
joint entropy, 897
Jordan canonical form, 408
Jordan-Dedekind chain condition, 720, 732
JPL turbo codes, 919

K

k-tuple of, 1108, 1109
Kakutani fixed-point theorem, 1032

Kantorovich, Leonid, 1163
Abū Bakr al-Karajī, 1163
Karmarkar's method, 970
  algorithm, 971
Karnaugh map, 349
Karp reducibility, 1088
Karp-Lipton theorem, 1088
Kasiski method, 929
KDC, 893, 933
Kempe, A. B., 565
Kendall's notation, 478
Kepler's conjecture, 826
Kerberos protocol, 893, 934
Kerckhoff's assumption, 926
kernel, 310, 326
  of a group homomorphism, 302
  of a linear transformation, 358, 371
  of a ring homomorphism, 302, 326
key, 923, 1104, 1106, 1115, 1125
  primary, 1106
  private, 925, 935
  public, 92, 935
  session, 934
  virtual, 1140
key agreement, 893, 943
key agreement mechanism, 943
key authentication,
  explicit, 943
  implicit, 943
key distribution, 943
key distribution center (KDC), 893, 933
key distribution problem, 893, 933
key domain, 110, 1125, 1139
key entry, 1105
key equivocation function, 926
key establishment, 893, 943
key randomization, 1105
key space, 923
key transfer, 893
key translation center (KTC), 934
keyed pair, 1105, 1115
king, 504, 537
Kirkman schoolgirl problem, 756, 766
Kirkman triple system, 756, 765
Kleene, Stephen Cole, 1163
Kleene closure, 1043, 1067
Kleene plus, 1045
Kleene star, 1043, 1067
Kleene theorem, 1052, 1074
Klein, Felix, 1163

Klein bottle, 504, 575
Klein four-group, 303, 321
Kleitman's lemma, 598
knapsack cryptosystem, 893, 936
knapsack problem (KP), 936, 957, 997, 1081
    general, 1001
Kneser conjecture, 599
k-norm,
    of a chain partition, 721
    of a sequence, 721, 738
known-plaintext attack, 893
Knuth, Donald E., 1164
Kolmogorov's inequality, 445
Kolmogorov-Chaitin complexity, 1041, 1078
König's theorem, 418, 672
Königsberg bridges, 544
KP, 997
k-Ramsey, 838
Kraft's inequality, 898
Kronecker delta function, 82, 127
Kronecker product, 756, 775
Kronecker symbol, 216, 278
Kruskal's algorithm, 635
KTC, 934
Kummer's $_2F_1$ identity, 205
Kuratowski, Kazimierz, 1164
Kuratowski graphs, 504, 518, 567
Kuratowski planarity theorem, 568

L

L-system, 1071
label-correcting algorithm, 655, 660
labeled arrangement of oriented lines, 811
labeled configuration, 808
labeled digraph, 583
labeled graph, 504, 580
labeled tree, 605, 622
Ladner's theorem, 1088
Lagrange, Joseph Louis, 1164
Lagrange's four-square theorem, 288
Lagrange's theorem, 235, 311
Lah coefficient, 137, 157, 164
λ-free language, 1067
λ-free morphism, 1069
λ-free semigroup, 1067

λ-free substitution, 1069
λ-transition, 1047, 1050
Lamé, Gabriel, 1164
Lamé's theorem, 229
Landau symbol, 38
Landau, Edmund, 1165
Landis, Y. M., 1137
language(s), 1943
    accepted, 1054
    accepted by a finite automaton, 1050
    accepted by a machine, 1043
    accepted by a Turing machine, 1085
    code indicator of, 1076
    C-complete, 1041, 1088
    concatenation of, 1041
    formal, 1067
    generated by a grammar, 1043, 1070
    C-hard, 1043, 1088
    inherently ambiguous context-free, 1043
    λ-free, 1067
    leftmost, 1043, 1071
    nontrivial family of, 1044
    NP-complete, 1044
    over an alphabet, 1085
    powers of, 1045, 1067
    rate, 927
    recursive, 1046, 1071
    redundancy, 927
    regular, 1046, 1068
    sparse, 1046, 1086
    Turing-acceptable, 1047
    Turing-decidable, 1047, 1057
    C-Turing-p-complete, 1088
    C-Turing-p-hard, 1088
    Turing-p-reducible, 1088
    unambiguous context-free, 1047
LAP, 702
LAPACK/CLAPACK, 403, 414
Laplace expansion, 388, 422
Laplacian, 497, 504
Laplacian matrix, 586
largest element, 342
largest prime known, 237
Las Vegas algorithm, 1043, 1092
Las Vegas to Monte Carlo transformation, 1043, 1093
last-come-first-served, 477
latin rectangle, 757, 780
latin square, 757, 778

lattice(s), 7, 47, 303, 341, 720, 734
   as a partially ordered set, 342
   bounded, 342
   complemented, 342
   distributive, 342, 736
   geometric, 737
   isomorphism of, 734
   lower semimodular, 736
   modular, 736
   upper semimodular, 736
lattice diagram, 342
lattice tiling, 801, 828
law of quadratic reciprocity, 278
law of total probabilities, 435
layout problems, 876
LBA, 1058
lcm, 216
leaf, 605, 609, 990
least common ancestor, 710
   algorithm, 711
least common multiple, 216, 227
least element, 46, 342
least positive residue, 216
least remainder Euclidean algorithm, 230
least upper bound, 7, 47, 721, 733
leaving variable, 965
Leda, 879
left cancellation law, 300, 307, 327
left child, 605, 609, 627, 1105, 1120
left coset, 301, 313
left derivative, 1069
left divisor of zero, 30, 327
left null space, 363
left quotient, 1069
left sibling, 609
left subtree, 605, 609
left-complete, 605
left-complete binary tree, 1105, 1121
left-linear grammar, 1071
left-right tree, 605, 627
leftmost derivation, 1043, 1071
leftmost language, 1043
  generated by the grammar, 1071
Legendre symbol, 216, 277
Leibniz, Gottfried Wilhelm, 1165
Leibniz's theorem, 190
lemma, 7, 50
length, 358, 368, 504, 531, 539, 631, 652,
     721, 729, 732, 898, 904, 987, 1066
length set, 1043, 1067

length-increasing, 1043, 1071
Lenz' construction, 837
Leonardo of Pisa (Fibonacci), 141, 1160
less than, 27, 725
Let's Make a Deal, 437
level, 605, 606, 609, 785
k-level, 609, 801, 811, 813
level order, 614, 609
Levi ben Gerson, 1165
Lévy's law, 293
lexical scanning, 1052
lexicographic order, 83, 105, 721, 729
liar's paradox, 22
LIFO property, 1105, 1112
lifted minimal cover inequality, 1000
limit ordinal, 27
limiting behavior, 473
line, 504, 510, 757, 770, 806, 811
   nonvertical, 811
   oriented, 811
line arrangement, 801, 811
line graph, 504, 516, 561
linear, 1071
linear assignment problem, 631, 650, 702
linear binary code, 367
linear block code, 904
linear bounded automaton (LBA), 1058
linear code, 364, 893, 904
linear combination, 358, 365, 806
linear congruence in one variable, 234
linear congruential method, 216, 233
linear diophantine equation, 281, 1080
linear equation, 393
linear extension, 721, 745
linear extension ordering, 504, 536
linear function, 807
linear grammar, 1043, 1071
linear operator, 358, 371
linear order, 721
linear probing, 1141
linear programming (LP) problem, 95,
     959, 1080
linear recurrence relation, 179
  with constant coefficients, 137
linear search, 1105, 1136
linear space, 806
linear subspace, k-dimensional, 806
linear sum, 721, 729
linear system, 358
linear transformation, 358, 371

linearly dependent, 365
linearly dependent sequences, 180
linearly dependent set, 806
linearly independent, 365, 960
linearly independent sequences, 180
linearly independent set, 806
linearly ordered, 7, 11
linearly ordered set, 46, 721, 725
link, 504, 506, 509
link capacity, 631
linked adjacency list, 631, 706
linked list,
    circular, 1119
    one-way, 1118
    two-way, 1119
LINPACK, 403
Liouville's function, 266
Liouville's theorem, 294
list,
    two-way linked, 1108
    two-way sequential, 1108, 1114
literal, 7, 303, 346
little omega, 38, 138
Little's Law, 480
little-oh, 7, 38, 137
Littlewood-Offord problem, 739
load balancing, 1092
local access network, 631
local central limit theorem, 457
localization, 410
locally finite, 615, 726
locally finite poset, 721
locally finite tree, 605
location, 702
location problem, 957
location theory, 986
log-concave, 741
log-hauling, 1025
logarithm, 7, 33
    discrete, 215, 273, 338
    Jacobi, 338
    Zech, 338
logic gate, 348
logic program, 7, 68
logical connective, 12
logical dual, 16
logical identity, 15
logical implication, 62
logical proof, 52
logically equivalent, 7, 15

logically implies, 7, 15
long-run arrival rate, 479
loop, 504, 510, 757, 787
loop invariant, 7, 66
loop-body, 65
lottery odds, 440
Lovasz's perfect graph theorem, 563
Lovelace, Augusta Ada Byron King, 1165
low capacity, 689
lower bandwidth, 402
lower bound, 7, 47, 300, 342, 721, 734, 982
    existential, 1042, 1083
lower density, 801, 824, 826
lower envelope, 801, 811
lower extension, 746
lower semimodular lattice, 721, 736
lower triangular, 379
lower triangular matrix, 358
Löwner order, 415
LP problem, 959
LP relaxation, 958, 981, 997, 1004
LU decomposition, 358, 397
lub, 733
Lucas numbers, 138, 140, 164, 182
    generating function, 140
    recurrence relation, 141
Lucas-Lehmer test, 250
Łukasiewicz, Jan, 13, 1166
LYM inequality, 741
LYM order, 741
LYM property, 741
Lyndon's theorem, 1076

M

m-p-reducible, 1088
machine,
    busy beaver, 1040
    n-state busy beaver, 1057
    finite-state, 1042, 1051
    finite-state with output, 1042, 1051
    Mealy, 1044, 1051
    Moore, 1044, 1051
    parallel, 861
    parallel random access, 861, 1060
    random access (RAM), 1046, 1060
    Turing, 1047, 1055
machine unit, 429, 450

MacMahon, Percy Alexander, 1166
Macmillan's inequality, 898
MacWilliams identity, 914
Mahaney's theorem, 1088
Mahāvīra, 1166
main class isotopic, 780
main diagonal, 378
maintaining a minimum spanning tree algorithm, 1145
majority rule, 744
majority rule property, 721
majorize, 415
Makanin's algorithm, 1076
manifold, 840
d-manifold, 840
manifold d-polyhedron, 802, 840
mantissa, 450
many-way divide-and-conquer, 862
map, 504, 564, 877
  chromatic number of, 564
  n-colorable, 499, 564
map chromatic number, 504
map coloring, 505, 564
mapping, 505
marginal worth, 1025
Markov, Andrei Andreevich, 469, 1166
Markov chain, 961
  discrete-time, 469
  irreducible, 429
  time-homogeneous, 469
Markov's inequality, 1093
Markowitz pivoting, 358, 402
matched, 641
matched edge, 631
matched vertex, 631
matching, 505, 597, 631, 641, 672
  complete, 630, 643
  maximum size, 631
  maximum weight, 632
  perfect, 632, 641
  stable, 645
matching number, 505, 599
mate, 631, 641
mathematical induction, 7, 57
mathematical programming, 802, 868
mathematical proof, 52
mathematical representation, 526
MATLAB, 404, 414
matrix,
  adjacency, 497, 519, 528, 630

admittance, 497, 586
arc length, 658
balanced, 956, 1004
circulant, 415
code generator, 892
degree, 620
diagonal, 970
doubly stochastic, 420
extended, 807
fully decomposable, 418
generator, 904
Hadamard, 775
incidence, 519, 528, 1104, 1124
irreducible, 415
Laplacian, 586
of a linear transformation, 358, 374
nonnegative (positive), 415
parity check, 894, 904
partly decomposable, 418
positive definite, 415
positive semidefinite, 415
primitive, 419
row stochastic, 407
sparse, 402
stochastic, 469
strictly diagonally dominant, 410
strictly totally positive, 415
transition probability, 431
  n-step, 471
0-1, 360, 418, 672
matrix game, 1022
matrix multiplication algorithm, 659, 660
matrix norm, 399
matrix theory, combinatorial, 417
matrix-tree theorem, 523, 588, 620
matroid(s), 598, 757, 787
  binary, 787
  characterization of, 793
  cographic, 791
  connected, 792
  construction of, 792
  Fano, 789
  graphic, 787
  minor-closed, 794
  non-Fano, 789
  oriented, 802, 819
  paving, 787
  planar, 787
  regular, 787
  representable, 757, 787

INDEX 1207

simple, 758, 787
ternary, 787
transversal, 758
uniform, 758
vector, 759
Matyas-Meyer-Oseas hash function, 951
Maurolico, Francesco, 58
max-flow min-cut theorem, 664
max-min payoff, 1018
max-min strategy, 1018
maxdegree, 599
maximal, 725
maximal antichain, 729
maximal chain, 729
maximal element, 7, 46, 721
maximal ideal, 303, 324
maximal independent set, 366
maximal tree, 606
maximum antichain, 729
maximum chain, 729
maximum distance separable (MDS) code, 915
maximum element, 7, 46
maximum flow, 631, 663
maximum size, 641
maximum size matching, 631
maximum spanning tree, 632, 633
maximum weight, 641
maximum weight matching, 632
maxterm, 303, 346
Maxwell-Boltzmann model, 429, 438
McEliece cryptosystem, 936, 893
McEliece encryption scheme, 940
M-computable, 1057
M-computation, 1057
MDS code, 786
Mealy machine, 1044, 1051
mean, 443, 488
mechanical summation procedure, 204
MED, 894, 902
median, 488
   p-, 958, 986, 988
medical image shape reconstruction, 874
meet, 341, 721, 734
meet-irreducible, 721, 734
member, 21
membership, 1071
membership function, 7, 71
membership problem, 1044
membership table, 8, 24

memory conflict, 1060
memoryless channel, 898, 899
   q-ary symmetric, 899
memoryless source, 894
ménage problem, 424
ménage numbers, 166
Mendel, Gregor, 437
Menger's theorem, 542
merge, 635
merge sort (mergesort), 189, 1105, 1132
   algorithm, 1133
MergeHull, 845
Merkle-Hellman cryptosystem, 936
Mersenne, Marin, 1167
Mersenne prime, 167, 216, 249
   table of, 252
mesh, 505, 606, 875
mesh generation, 875
message, 894, 898
   redundancy, 927
   rate, 927
message authentication, 946
message authentication code (MAC), 947
message digest, 950
message space, 923
metaheuristic, 632, 693
method, QR, 412
method of affirming, 51
method of denying, 51
metric space, 958, 986
MFFD method, 1002
Miller-Rabin test, 254
mimimal element, 721
Minc-Bregman inequality, 423
mincut, 430, 461
miniguide of sequences, 158
minimal, 725
minimal cover, 684
minimal cover inequality, 1000
minimal Boolean expression, 349
minimal element, 8, 46
minimal independent set, 366
minimal polynomial, 303, 334, 358, 407
minimal Radon partition, 819
minimality, 819
minimization, unbounded, 1047
minimum cost flow, 632
minimum cut, 632, 664
minimum degree algorithm, 358, 402
minimum element, 8, 46

minimum error probability decoding (MED), 902
minimum realizer encoding, 721, 746
minimum spanning forest, 1147
minimum spanning tree, 606, 632, 633
   capacitated, 630, 684
minimum universal exponent modulo m, 268
Minkowski, Hermann, 1167
minor, 358, 387, 505, 515, 792, 802, 821
minor-closed matroids, 794
minpath, 430, 461
minsum matrix multiplication, 658
minterm, 303, 346
MIP problem, 981
mirror image, 1044, 1066, 1069
mixed integer programming (MIP) problem, 981
mixed strategy, 958, 1022
mixed strategy combination, 1022
Möbius, Augustus Ferdinand, 575
Möbius band, 505, 575
Möbius function, 83, 128, 217, 266
Möbius inversion formula, 128, 266
mod, 33, 221
mode of operation, 930
model, 526
   computational, 1048, 1049
model-based recognition, 874
modeling, 526
modem, 89, 896
modified first fit decreasing (MFFD) method, 1002
   algorithm, 1002
modular intersection theorem, 598
modular lattice, 721, 736
modular square roots, 280
module, 721, 749
modulus, 217, 358
modus ponens, 51
molecular modeling, 879
MOLS, 783
moment, 444
moment generating function, 446
monadic operation, 32, 303, 305
monic polynomial, 329
monohedral tiling, 802, 827
monoid, 303, 305
   free, 1042, 1067
   of transformations, 306

monomorphism, 8, 31
monotone chain, 847
monotone simple polygon, 847
monotone structure function, 459
monotonic grammar, 1071
monotonicity, 104, 433, 745
Monte Carlo algorithm, 1044, 1092
Monte Carlo simulation, 430
Monte Carlo to Las Vegas transformation, 1044, 1093
Monte-Carlo method, 870
Moore machine, 1044, 1051
morphism, 1044, 1069
   inverse of, 1069
   λ-free, 1069
Motzkin hyperplane, 832
MST, 633, 1084
Muhammad ibn Muhammad al-Fullāni al-Kishnāwī, 1167
multi-arc, 505
multi-edge, 505, 510
multi-graph, 505, 510
multi-valued logic, 8, 72
multicoloring, 566
   k-tuple, 566
multicombination coefficient, 98
multilevel security policy, 48
multinomial coefficient, 83, 98
multinomial random variable, 442
multiple, 217
multiple edges, 510
multiplication of linear transformations, 373
multiplicative, 259
multiplicative function, 217
multiplicative group, 307
multiplicity, 22
multiplier, 485, 757, 777
multiplier conjecture, 778
multiplier theorem, 778
multiset, 8, 22
multivariate composition, 1063
multivariate function, 32
multivariate primitive recursion, 1063
mutual authentication, 943
mutually disjoint, 8, 23
mutually orthogonal, 757
mutually orthogonal latin squares (MOLS), 783
mutually relatively prime, 217, 227

N

NAG, 403, 414
naive set theory, 8, 22
NAND, 8, 344
narrow-sense code, 912
Nash equilibrium mixed strategy combination, 1022
Nash equilibrium pure strategy combination, 1017
natural logarithm, 33
natural number, 8, 21, 219, 306
Naur, Peter, 1167
NAUTY, 553
NDFSR, 1050
near end, 1105, 1118
nearest insertion heuristic algorithm, 695
nearest neighbor, 635, 835, 872
nearest neighbor heuristic algorithm, 694
nearly complete, 1105
necklace, 158, 166
negated elements, 818
negation, 8, 13
  of conditional, 17
  rules for, 19
negative cycle, 632, 652, 674
negative cycle detection, 655
negative part, 818
negative side, 807
neighbor, 505, 510
  nearest, 635
network(s), 987
  backbone, 630, 683
  capacitated, 630, 663
  directed, 631, 652
  directed two-terminal flow, 643
  incidence structure of, 1124
  interconnection, 1060
  K-terminal, 463
  layout of, 575
  representation of, 706
  residual, 632, 674
  simplex algorithm for, 678
  survivable, 632
  undirected, 632, 633
Newton's theorem, 191
next operation, 1118
NFS, 257
nilpotent, 405
nilpotent matrix, 358

node, 505, 510
Noether, Amalie Emmy, 1167
noisy channel coding theorem, 903
non-Fano matroid, 789
non-manifold d-polyhedron, 802, 840
non-repudiation, 894, 897, 924
nonbasic set, 965
nonbasic variables, 965
nonconstructive direct proof, 54
nonconvex polyhedron, 802, 840
noncooperative game, 958
nondegeneracy, 745
nondeterministic finite-state recognizer (NDFSR), 1044, 1048, 1050
nondeterministic polynomial-time computation, 1044
nondeterministic Turing machine, 1044, 1058, 1085
nondictatorship, 745
nonhomogeneous recurrence relation, 138, 179, 184
  general solution of, 184
nonlinear recurrence relation, 179
nonnegative (positive), 415
nonnegative matrix, 358
nonorientable surface, 505, 564, 575
  of crosscap number k, 505
nonperiodic tiling, 802, 828
nonplanar, 505, 567
nonseparable, 505, 541
nonsingular, 358, 380
nonterminal, 1044, 1070
nontree edge, 633
nontrivial, 1069
nontrivial family of languages, 1044
nontrivial structure function, 459
nonvertical line, 811
NOP, 8, 63
NOR, 8, 344
Nordhaus-Gaddum inequalities, 558
Nordstrom-Robinson code, 894, 917
norm, 295, 368, 970
  of a chain partition, 721, 738
  l_1, 400
  l_∞, 400
  of a sequence, 721, 738
normal distribution, 444
normal duration, 1007
normal extension, 303, 334
normal form game, 1017

normal hypergraph, 505, 599
normal matrix, 358, 407
normal subgroup, 303, 313
normal tiling, 802, 827
normalization, 369
normalized, 450, 567, 775
normalized characteristic function, 958, 1024
normalized drawing, 505, 522
normalized matching property, 721, 739
normalizer,
   of an element, 313
   of a subgroup, 313
NOT, 8
NOT gate, 348
not and, 13
not or, 13
NP, 1041, 1078, 1086
NP^C, 1086
NP^L, 1086
NP-complete, 1078, 1088
NP-complete language, 1044
NP-complete problem, 1044
NP-hard, 1078, 1088
NP-hard language, 1044
nucleolus, 958
null graph, 505, 511, 517
null pointer, 1105, 1118
null set, 8, 21
null space, 358, 363
nullity, 366, 371
   of a linear transformation, 358
   of a matrix, 358
number(s),
   abundant, 261
   algebraic, 214, 219, 295, 300, 333
   amicable, 261
   associated Stirling,
      of the first kind, 157, 163
      of the second kind, 157, 163
   Bell, 117, 164
   Bernoulli, 136, 147, 164
   Betti, 540, 542
   binomial coefficient, 82, 97, 103, 141, 161
   block, 759
   cardinal, 3, 27
   Carmichael 167, 214, 251
   Catalan, 136, 145, 164, 176, 180, 627
   chromatic, 170, 498, 557, 564, 566, 599, 837
   clique, 498, 563, 599
   clique partition, 498, 599
   complex, 3, 21, 306
   composite, 215, 236
   condition, 356, 399
   crosscap, 499, 575, 579
   crossing, 499, 571
   Cunningham, 249
   deficient, 261
   edge chromatic, 501, 561
   edge independence, 501, 563
   empire chromatic, 564
   Euler, 137, 140, 141, 148, 153, 164, 200
   expected, 479
   extremal, 501, 590
   factorial, 161
   Fermat, 216, 249
   Fibonacci, 137, 139, 140, 164, 179, 182, 216, 229
   figurate, 137, 158, 168
   floating-point, 866
   harmonic, 137, 154, 164
   heptagonal, 168
   hexagonal, 168
   independence, 503, 563, 599
   irrational, 7, 216, 219
   Lucas, 138, 140, 164, 182
   map chromatic, 504
   matching, 505, 599
   ménage, 166
   Mersenne, 249
   natural, 8, 21, 219, 306
   octagonal, 168
   ordinal, 27
   pentagonal, 168
   perfect, 249, 261
   prime, 167, 196, 217, 221, 236, 238, 241
      Fermat, 249
      irregular, 284
      largest known, 237
      Mersenne, 249
      regular, 284
   pseudo-random, 430, 484
   pyramidal, 168
   Ramsey, 138, 150, 506, 592
   k-Ramsey, 838
   random, 428, 488, 489
   rencontre, 109

replication, 758, 759
Schur, 150
smooth, 218, 255
Stirling, of first kind, 83, 117, 157,
Stirling, of second kind, 83, 117, 157, 162
Stirling cycle, 83, 117, 140, 141,192
Stirling subset, 83, 117, 140, 162, 192, 200, 735,
surface chromatic, 564
tangent, 138, 140, 153, 164
"ten most wanted", 218, 259
transcendental, 218, 219
transversal, 508, 599
triangular, 168
Turán, 590
vertex chromatic, 557
Whitney, 735
kth, 724, 731
number of divisors function, 261
number field sieve, 217, 257
numbering, Gödel, 1042
numeric datatype, 1105, 1109
numerical multiplier, 777
numerically stable algorithm, 359, 395
numerically unstable algorithm, 359, 395

O

object, 879
objective function, 958, 960
objective value, 960
obnoxious facility, 994
  location problems, 995
obstruction, 557
  to coloring, 557
  to n-coloring, 505
occupancy problem, 453
octagonal numbers, 168
octahedral graph, 505, 518
  generalized, 518
octal expansion, 223
octal representation, 217
octic group, 303
odd, 647
odd permutation, 303, 319
odd vertex, 632
OFB, 894, 931
off-diagonal, 378

Ogden pumping lemma, 1074
omega notation, 8
On-Line Encyclopedia of Integer Sequences, 158
on-line partitioning algorithm, 739
one's complement expansion, 217, 224
one-dimensional simple random walk, 454
one-point removal theorem, 747
one-sided error, 1089
one-time pad, 894, 933
one-to-one correspondence, 8, 31
one-way hash function (OWHF), 950
one-way linked list, 1105, 1118
onto, 8, 31, 180
open half-spaces, 807
open hash table, 1105
open path, 539
open trail, 539
open walk, 539
operating, 459
operation,
  binary, 305
  monadic, 305
  n-ary, 32, 305
  unary, 305
operator, S-norm, 71
k-opt exchange, 693
optimal, 960
optimal code, 915
optimal message passing, 640
optimal project compression problem, 1007
optimal solution, 958
optimal strategy, 1022
optimal tour partitioning algorithm, 686
optimization, color map, 872
OR, 8
OR gate, 348
oracle, 802, 869, 1044, 1085
oracle Turing machine, 1044, 1085
orbit, 121, 303, 319
order, 307, 511, 606, 773, 785
  of an algebraic structure, 303
  of an element, 315
  of a graph, 505
  of a group element, 303
  of a hypergraph, 595
  of a hypergraph edge, 505
  of an integer, 217
  of a modulo m, 268

of operations, 14
of a permutation group, 319
of a recurrence relation, 179
order dimension, 746
order ideal, 732
order module, 721, 749
order relation, 27, 341, 722, 725
order statistic, 488
order type, 802, 808, 831
order-preserving mapping, 721, 726
ordered basis, 806
ordered datatype, 1105, 1110
ordered pair-action group, 321
ordered pair-permutation, 321
ordered selection, 83, 96
   with replacement, 83, 100
   with specified replacement, 100
ordered simplex, 807
ordered tree(s), 606, 609, 627
   isomorphic, 609
ordered-pair-permutation, 580
ordinal number, 27
ordinary crossing, 832
ordinary generating function, 82, 83, 171
ordinary hyperplane, 832
ordinary line, 832
orientable surface, 505, 564, 575
   of genus g, 506
orientation, 506, 534
   strong, 534
   transitive, 534
oriented hyperplane, 807
oriented line, 811
oriented matroid, 802, 819, 820, 821
   given by a chirotope, 802, 819
   given by its circuits, 820
   given by its cocircuits, 820
   obtained by a one-element contraction, 821
   obtained by a one-element deletion, 821
oriented plane arrangement, 811
oriented pseudoline arrangement, 816
origin of a walk, 506
origin vertex, 689
original tableau, 976
orthogonal, 368, 380, 757, 783, 904
orthogonal array, 757, 785
orthogonal complement, 363, 368, 904
orthogonal drawing, 573

orthogonal latin squares of order 6, 785
orthogonal matrix, 359
orthogonal range reporting using range trees algorithm, 858
orthogonal range search, 857
orthogonal set, 359, 368
orthonormal set, 359, 368
OTTER system, 78
Otter's formula, 623
out-degree, 506, 528
   directed, 631, 653
outer, 647
outer corner, 131
outflow, 665
output, 1057
output analysis, 487
output feedback (OFB) mode, 894, 931
output function, 1051
output stochastic process, 488
output symbols, 1051
ovals of Cassini, 410
overall parity check symbol, 907
overflow, 430, 450
OWHF, 950

P

P, 1041, 1078, 1086
P^C, 1086
P^L, 1086
packing, 802, 824, 958, 1001
   density of, 824
   lower density of, 824
   upper density of, 824
packing problems, 996
padding, 951
pair-action group, 580
pair-permutation, 580
pairing axiom, 30
pairwise balanced design, 757, 767
pairwise disjoint, 8, 23
pairwise relatively prime, 217, 227
palindrome, 91, 217, 1044, 1066
paradox, 8
parallel algorithm, 638, 802, 861
parallel class, 757, 765
parallel computation model, 1044, 1060
parallel divide-and-conquer, 862
parallel edges, 510

parallel elements, 757, 787
parallel machine, 861
parallel random access machine (PRAM), 802, 861, 1060
parallel system, 430, 462
parallel time-complexity, 1078
parametric search, 802, 864
parent, 606, 608
parent function, 608
parent-finder algorithm, 614
parentheses, well-formed (balanced) sequences of, 145
Pareto-optimal, 1018
parity check bit, 894, 901
parity check matrix, 894, 904
parity check symbol, overall, 907
parsing, 1070
parsing a string, 1044
partial, 506
partial function(s), 8, 31, 1045, 1063
   induced by a Turing machine, 1045, 1057
   number of, 100
partial hypergraph, 595
partial latin square, 757, 780
partial order, 8, 46, 725
partial pivoting, 359, 401
partial quotient, 217
partial recursive function, 1045, 1064
partial sum, 195
partial transversal, 757, 779
partially dynamic algorithm, 1144
partially ordered set(s), 8, 46, 83, 722, 725
   number of, 165
partite set, 517
partition(s), 8, 23, 83, 114, 175
   number of, 167
partition lattice, 722, 734
partition-cut, 606, 617
partly decomposable, 359, 418
Pascal, Blaise, 1168
Pascal distribution, 443
Pascal's identity, 104
Pascal's recursion, 98
Pascal's triangle, 83, 99
passage time, first, 429
passive adversary, 926
passive intrusion, 896
PASTA, 480

path, 454, 506, 517, 531, 539
   alternating, 630
   augmenting, 666
   closed, 498
   directed, 500, 506, 531
   in a flow network, augmenting, 630
   Hamilton, 547
   in a matching, augmenting, 630
   shortest, 632, 653
   simple, 431, 461
   u,v-, 508
path optimality conditions, 633
pattern, 420, 1045, 1075
pattern inventory, 83, 125
pattern matching, 874
paving matroid, 757, 787
payoff function, 958, 1017
payoff matrix, 1021
PBD, 757
PBD-closed, 767
PBD-closure, 757
PDA, 1053
Peano, Giuseppe, 27, 1168
Peano definition for the natural numbers, 8, 27
Peirce, Charles Sanders, 1168
Peirce arrow, 13
Pell's equation, 217, 285
pentagonal numbers, 168
Pepin's criterion, 250
perfect, 261
perfect code, 894, 916
perfect game, 1017
perfect graph, 506, 563
perfect information, 958
perfect matching, 632, 641
perfect number, 167, 217, 249
perfect secrecy, 926
period, 224, 290, 472
periodic, 224
periodic base b expansion, 217, 224
periodic state, 430
periodic tiling, 802, 828
permanent, 359, 422
permutation(s), 34, 83, 96, 120, 180, 303, 319
   discordant, 158
   even, 301, 319
   k-, 96
   number of, 91

discordant, 166
odd, 303, 319
without certain types of cycles, number of, 165
permutation coefficient, 83, 96
permutation digraph, 522
permutation graph, 506, 522, 722, 749
permutation group, 83, 120, 303, 319
permutation matrix, 359, 397
permutation-with-replacement coefficient, 83, 101
Perron root, 359, 416
Perron's theorem, 1033
Perron-Frobenius theorem, 416
PERT model, 958, 1013
perturbation, 196, 867
Petersen graph, 506, 518
phi-function, 247
phrase-structure grammar, 1070
PID, 327
pierced, 831
pigeonhole principle, 92
pivot, 359, 394, 958, 1105, 1132
pivoting, to preserve sparsity, 402
pivoting, for stability, 401
placing balls into bins, 97, 99, 102, 103, 117
plaintext, 894, 923
planar, 506, 567, 722, 725, 757, 787, 802, 846
planar graphs, number of, 171
planarity, 1148
  algorithm for testing, 571
plane, 757, 770, 839
plane graph, 1105, 1144
plane tree, 609
Platonic graph, 506, 518
Platonic solid, 506, 518
players, 958, 1017
plurality, 744
plurality consensus function, 722
Poincaré, Henri, 577
point, 759, 766, 767, 768, 770, 987
  d-dimensional, 806
point configuration, 802, 807, 808
point location algorithm, 855
point location problem, 802, 853
pointer, 1105, 1118
  null, 1105, 1118
Poisson arrival process, 477

Poisson distribution, 443
Poisson random variable, 430
poker hands, 100
  probabilities of, 434
polar-duality, 802, 813
  standard affine, 812
Polish notation, 8, 13, 614
Pollard algorithm, 256
Pólya, George, 439, 457, 1168
Pólya's enumeration formula, 125
Pólya's theorem, 457
Pólya's urn scheme, 439
polyalphabetic cipher, 928, 894
polychromatic number of a metric space, 838
polygon,
  convex, 831
  monotone simple, 847
  regular, 827
polyhedral combinatorics, 868
polyhedron, 506, 518, 802, 827, 840, 958, 981
  1-skeleton of, 505
  d-polyhedron, convex, 840
polyline, 877
polyline drawing, 573
polynomial, 303
  check, 909
  chromatic, 557
  generator, 892, 909
  irreducible, 329, 340
  over a field, 330
  over a ring, 329
  reducible, 329
  separable, 334
  syndrome, 909
polynomial hierarchy, 1045, 1086
polynomial ring, 303, 324, 329
polynomial space, 1078
polynomial space complexity, 1086
polynomial space computation, 1045
polynomial time, 1078
polynomial time complexity, 1086
polynomial time computation, 1045
  nondeterministic, 1044
polynomial-time reducible, 1046, 1088
polyomino, 138, 158, 168
polytope, 803, 831, 840
  d-, 803
  dual, 840

$\mathcal{H}$-, 803, 869
  regular, 840
  $\mathcal{V}$-, 803, 869
popping, 1112
poset(s), 9, 46, 83, 722, 725
  bipartite, 729
  divisibility, 727
  graded, 731
  and graphs, 749
  induced, 725
  ranked, 731
  self-dual, 729
positive closure, 1045, 1067
positive definite, 359, 380, 415
positive matrix, 359
positive part, 818
positive semidefinite, 359, 415
positive side, 807
postcondition, 9, 62, 1105, 1110
postfix notation, 9, 13, 614
postorder traversal, 606, 613
postulate, 22
potential, 674, 970
power(s),
  of integers, 159
  of a language, 1045, 1067
  of a matrix, 359, 380
  of a relation, 9, 42
  of a string, 1066
power residue, 277
power set, 9, 24, 30
power sum, 138, 157, 199
powerful, 217, 221
PP, 1089
PRAM, 802, 861, 1041, 1045, 1060
PRAM memory conflict, 1045
pre-image, 31
pre-period, 224, 290
precedence, 1007
precision, finite, 866
precondition, 9, 62, 1105, 1110
predecessor, 632
predecessor function, 709
predecessor indices, 653
predicate, 9, 18
  n-ary, 8, 68
predicate calculus, 9, 18
predicate logic, 12, 18
preemptive goal programming, 980
prefix, 1066

prefix code, 610, 894, 898
prefix notation, 9, 13, 614
preflow, 632, 666
preflow-push algorithm, 667
preimage-resistance, 949
premise, 9, 50
prenex normal form, 9, 19
preorder, 9, 46
preorder traversal, 606, 613
Preparata code, 894, 918
preprocessing, 982
present, 1075
presented, 318
Prim's algorithm, 636
primal LP problem, 973
primality proof, 251
primality test, 217
  algorithm, 1095
primary key, 9, 48, 1106, 1125
primary operation, 1103, 1106, 1108
prime(s), 167, 217, 221, 236, 296
  Fermat, 249
  irregular, 284
  largest known, 237
  Mersenne, 249
  regular, 284
  table of, 238
prime counting function, 237
prime factorization, 217
prime ideal, 303, 324
prime number theorem, 217, 246
prime power decompositions, 241
prime-power factorization, 217, 237
primitive, 419, 522
primitive code, 912
primitive element, 338
primitive matrix, 359
primitive Pythagorean triple, 282
primitive recursion, 1045
primitive recursive function, 1045, 1063
primitive root(s), 217, 1076
  table of, 269
  modulo m, 268
principal ideal, 324
  generated by an element, 324
principal ideal domain, 327
principal minor, 359, 387
principal submatrix, 359, 378
principal subtree, 609, 710
principle of inclusion-exclusion, 107, 432

principle of mathematical induction, 57
print command, 75
priority queue, 610, 1106, 1116
priority tree(s), 606, 1106, 1122
   dequeueing algorithm for, 1123
   enqueueing algorithm for, 1122
privacy, 894, 924
private key, 925, 935
probabilistic activity net, 958, 1013
probabilistic graph theory, 593
probabilistic isomorphism test, 552
probabilistic method, 1092
probabilistic primality test, 217
probabilistic Turing machine, 1045, 1089
probability, 430, 488
   acceptance, 1040
   conditional, 428
   transition, 431
probability generating function, 446
probability measure, 432
probability space, 746
probe function, 1106
probe sequence, 1106, 1141
probing,
   linear, 1141
   quadratic, 1141
problem,
   NP-complete, 1044
   tractable, 1047
   undecidable, 1047
problème des ménages, 83, 112
problème des rencontres, 83, 112
procedure, 67
   body, 67
   name, 67
processor, 1060
product,
   of cardinals, 27
   of linear transformations, 373
   of matrices, 359
   of ordinals, 27
product graph, 506
product ring, 324
product rule, 90
product-of-sums expansion, 346
production,
   antecedent of, 1040
   consequent, 1041
production rule, 9, 75, 1045, 1070
   Chomsky normal form for, 1070

production system, 9, 75
products, well-parenthesized, 145
profile, 744
program assertion, 2
program construct, 4
program evaluation and review technique (PERT) model, 1013
program fragment, 9
program semantics, 9
programming,
   goal, 980
   linear, 868
   mathematical, 868
   nonlinear, 868
project, 1007
projection function, 9, 34, 49, 370, 970, 1045, 1063
projective arrangements, 812
projective geometry, 771
projective line, 812
projective line arrangement, 803, 812
projective plane, 506, 575, 758, 771, 772, 812
   of order 2, 755, 762
projective scaling, 970
projective space, 758, 770
proof, 9, 52
   by cases, 54
   by contradiction, 9, 53, 54
   by contraposition, 54
   by cycle of implications, 54, 56
   direct, 54
   done by hand, 9, 77
   existence, 55
   indirect, 54
   techniques of, 50, 54
proper k-tuple coloring, 566
proper k-tuple multicoloring, 566
proper ancestor, 608
proper coloring, 557
proper descendant, 608
proper edge, 506, 509
proper fixed-point numeral, 1052
proper hypergeometric term, 206
proper labeling, 1029
proper minor, 792
proper multicoloring, 566
proper set coloring, 566
proper subgame, 1017
proper subgroup, 304, 311

proper subset, 9, 22
proposition, 10, 12, 50
propositional calculus, 10, 12
propositional logic, 12
propositional variable, 12
protocol, Kerberos, 893
prototile, 803
provable security, 894
provably secure, 926
prune and search, 1138
pseudo polar-duality, 822
    standard affine, 804
pseudo-graph, 506, 510
pseudo-random number(s), 430, 484
    generator for, 218
pseudocircles, 833
pseudoconfiguration of points, 822
pseudoflow, 632
pseudoline, 816, 833
pseudoline arrangement, 803, 816
    x-monotone, 816
    oriented, 816
    realizable, 816
    stretchable, 816
pseudopalindrome, 1066
pseudoprime, 218, 251
    to base b, 251
PSLG, 846
PSPACE, 1041, 1086
public key, 935
public-key certificate, 894, 945
public-key cryptography, 935
public-key cryptosystem, 894, 925
pumping, with marked positions, 1074
pumping lemma, 1045
    for linear languages, 1074
    for regular languages, 1074
punctured code, 894, 907
pure binary tree, 609
pure integer programming (IP) problem, 981
pure strategy, 958, 1017
purely multiplicative linear congruential generator, 430, 485
pushdown automaton (PDA), 1045, 1053
pushing, 1112
puts over, 746
pyramidal numbers, 168
Pythagorean triple, 218, 282
    primitive, 282

Q

QAP, 632, 702
QED Project, 78
Qin Jiushao, 1168
QS, 256
quadratic assignment problem, 632, 702
quadratic Diophantine equation, 1080
quadratic field, 218
quadratic integer, 295
quadratic irrational, 218
quadratic nonresidue, 218, 277
quadratic probing, 1141
quadratic reciprocity, 218
quadratic residue, 218, 277
quadratic sieve, 218, 256
quantified Boolean formula, 1089
quantile, 488
quantum error-correcting code, 869
quasi-residual design, 771
quasi-transitive orientation, 722, 749
quaternion, 309, 325
queens problem, 985
query, 1106, 1109, 1143
queue, 477, 1106, 1112
    priority, 1106
queue capacity, 430, 477
queue discipline, 430
queue implementation, 654
queue-length process, 430, 479
queueing system, 430, 477
queueing theory, 477
QuickHull, 845
quicksort, 1106, 1132
    algorithm, 1134
Quine-McCluskey method, 351
quotient, 221, 506, 515
    of a language, 1069
quotient group, 304, 313
quotient ring, 304, 327
quotient rule, 90

R

Rabin cryptosystem, 894, 936
radio wave propagation, 874
radiosity, 873
radius, 506, 540

of convergence, 138, 202
radix, 223
radix sort, 1106, 1135
    algorithm, 1136
Radon partition, 803, 819
    minimal, 819
Radon's theorem, 809
RAM, 1060
Ramanujan, Srinivasa, 1169
Ramsey, Frank, 1169
Ramsey number, 138, 150, 506, 592
k-Ramsey number, 838
random access list, 1106, 1116
random access machine (RAM), 1046, 1060
    parallel (PRAM), 861
random cipher, 927
random graph, 506, 593
    algorithm, 594
random number, 430, 484
    generator, 484
random selection of an integer, 434
random variable, 430, 441
    Bernoulli, 428, 442
    binomial, 428, 442
    geometric, 429
    hypergeometric, 429, 442
    independent, 429, 441
    multinomial, 442
    Poisson, 430
random variate, 485
random vector, 441
random walk, 182, 430
    d-dimensional, 453
    recurrent, 453
    simple, 453
    transient, 453
randomized algorithm, 803, 863, 1046, 1091
randomized divide-and-conquer, 863
randomized incremental methods, 863
randomized quicksort algorithm, 1083
range, 10, 31, 359, 371
range problem, 453, 803, 856
range reporting algorithms, 856
range searching, 856
    algorithms, 856
ranging, 977
rank, 359, 366, 371, 390, 722, 731, 758, 787, 808, 1106, 1135

axioms, 758, 790
rank counting sort, 1106, 1135
rank counting sort algorithm, 1135
rank function, 722, 731
rank parameters, 722, 731
rank 3 hyperline sequence, 803, 820
rank 2 contraction, 821
ranked poset, 722, 731
ranking, 538, 722, 742
rank-nullity theorem, 366
rate, 900, 921
    absolute, 927
    of a language, 927
    for messages, 927
rational cuboid problem, 218
rational number, 10, 21, 218, 219, 306
ray, 803, 859
ray shooting algorithms, 859
ray shooting problem, 803, 859
ray tracing, 873
Rayleigh distribution, 444
RBIBD, 765
reachable, 531, 540
reachable vertex, 506
read instruction, 1060
real number, 10, 21, 306
real quadratic field, 295
real RAM, 803
real random access machine, 803
realizable, 803, 816
realizer, 722, 746
reasonable set, 1025
recipe, 1147
reciprocity law for Jacobi symbols, 279
recognizer, finite-state, 1042
reconstructible, 506, 555
reconstructible invariant, 507, 555
reconstruction or recovery of 3-D structure, 874
record, 1103, 1104, 1106, 1125, 1139
rectilinear distance, 993
recurrence relation, 138, 139, 178
    approximate solutions, 201
    divide-and-conquer, 188
    homogeneous, 179, 180
    linear, 179
    nonhomogeneous, 179, 184
    nonlinear, 179
    solution of, 181
    used to solve counting problems, 93

recurrent, 472
recurrent random walk, 453
recurrent state, 430
recurrent walk, 430, 455
recursion, primitive, 1045
recursive, 1065
recursive definition, 36
   of a set, 10
   of a function, 10
recursive description, 21
recursive divide-and-conquer algorithm, 188
recursive function, 1045
recursive language, 1046, 1071
recursive set, 1046
recursively enumerable, 1065
recursively enumerable set, 1046
reduced, 778
reduced cost, 632, 674, 958, 965
reduced latin square, 758
reduced residue system modulo m, 231
reduced system of residues, 218
reduced tree(s), 606, 622
   generating function for counting, 623
reduced walk, 606, 613
reduced word, 317
reduces, 1078
reducibility,
   Cook, 1088
   Karp, 1088
   in polynomial-time, 1046
reducible, 24, 302, 304, 329, 415, 472, 507, 532
   polynomial-time, 1046, 1088
reducible matrix, 359
reduction, 1046, 1084
   $f(n)$ time, 1084
redundancy, 691
   of a language, 927
   for messages, 927
redundant, 959
redundant constraint, 958
Reed-Muller code, 894
Reed-Solomon (RS) code, 894, 914
refinement, 10, 23, 722, 734, 1029
reflecting boundary, 430, 453
reflection principle, 454
reflex edges, 803, 840
reflexive, 10, 41, 310, 725
reflexive closure, 43

region, 507, 569
register, 1060
   global, 1060
   local, 1060
regular, 284, 507, 511, 606, 803, 840, 1063
d-regular, 606
k-regular, 504, 511
regular covering, 722, 739
regular expression, 1046, 1073
regular grammar, 1046, 1071
regular graphs, number of, 171
regular hypergraph, 595
regular language, 1046, 1068
regular matroid, 758, 787
regular polygon, 803, 827
regular polytope, 80
regular tiling, 803, 827
regular tournament, 537
regularity, 30, 1071
regularity problem, 1046
rehashing, 1141
rejects, 1085
related, 725
relation, 10, 40
   direct derivability, 1070
   n-ary, 8, 48
   number of, 165
relative error, 430, 450
relatively prime, 218, 227
relator, trivial, 317
relaxation, 981
reliability, 430, 459
remainder, 218, 221
rencontre number, 109
reorientation, 803, 807, 808
reorientation class, 808
replacement, 30
replication, 488
replication number, 758, 759
representable, over a field, 758
representable matroid, 787
representation, 507, 519, 820
represented by the Turing machine, 1064
repudiation, 925
residual capacity, 632, 666, 674
residual network, 632, 666, 674
resolution, 758, 765
   of a collision, 1140
resolvable, 758

resolvable balanced incomplete block design, 765
responder, 946
response, 946
restriction, of a function, 10, 31
result of replacing each instance, 63
return instruction, 67
Reve's puzzle, 94
reverse, 1046, 1066
reverse Polish notation, 10, 13, 614
Rice's theorem, 1072
Riemann hypothesis, 247
Riemann zeta function, 247
right cancellation law, 300, 307, 327
right child, 606, 609, 627, 1106, 1120
right coset, 301, 313
right derivative, 1069
right divisor of zero, 304, 327
right null space, 363
right quotient, 1069
right sibling, 609
right subtree, 606, 609
right-linear grammar, 1071
ring, 304, 323
   commutative, 323
   polynomial, 324
   product, 324
   with unity, 304, 323
   of real quaternions, 325
ring automorphism, 326
ring endomorphism, 326
ring homomorphism, 326
ring isomorphism, 326
rising power, 138, 191
road network, 513
Robbins problem, 78
Robinson-Schensted algorithm, 132
robotics, 870
rook polynomial, 113
root, 91, 218, 234, 608, 1076
   of unity, 309
root cell, 1106, 1120
root field, 304, 334
root vertex, 684
rooted tree(s), 606, 608, 622
   isomorphism, 605, 608
   number of, 169
roster, 21
rotation, 578
rotation system, 507, 578
rotational sweeping, 844
Roth's theorem, 294
roundoff error, 359, 395, 430, 450
route, 700
row, of a matrix, 378
row rank, 366
row space, 365
row stochastic, 359, 407
row vector, 378
Roy-Warshall algorithm, 43
RP, 1089
RS code, 914
RSA cryptosystem, 895, 936, 937
rule,
   of inference, 10, 50
   100%, 977
   of product, 90
   of quotient, 90
   of sum, 90
ruler and compass construction, 337
running time, 1046
   average-case, 1040
   worst-case, 1048
Russell, Bertrand, 23
Russell's paradox, 23

S

Saalschütz's $_3F_2$ identity, 206
al-Samaw'al ibn Yahyā ibn Yahūda al-Maghribī, 1169
sample mean, 488
sample size, 430, 432
sample space, 430, 432
satisfiability, 1079, 1081
   3-, 1080
   2-, 1080
saturated, k-, 738
savings, 632
savings heuristic algorithm, 685
scalar, 359, 361
scalar multiple, 359, 380
scalar multiplication, 361, 373
scalar product, 359, 373, 380
scanning a database, 1106, 1126
Schauder fixed-point theorem, 1032
scheduling, 538
scheme,
   secret sharing, 895, 948

threshold, 948
(n,k)-threshold, 895, 948
Schönhardt polyhedron, 843
Schreier digraph, 522
Schreier graph, 507, 522
Schröder-Bernstein Theorem, 28
Schur number, 150
Schur's majorization theorem, 152, 408, 416
scope, 10, 18
score vector, 537
    characterization, 538
Scott-Suppes theorem, 743
scum theorem, 792
SDR, 24, 597
search,
    binary, 1136
    breadth-first, 604, 616
    depth-first, 604, 616
    linear, 1105, 1136
search tree, 1120
searching, 748, 1106, 1126, 1136
searching a 2-3 tree algorithm, 1137
second-order linear recurrence relation with constant coefficients, 138
second-order logic, 19
second-order recurrence relation, solution of, 181
secondary hash function, 1141
secondary key, 1106, 1125
secondary operation, 1106, 1109
secrecy, 896
    perfect, 926
secret sharing scheme, 895, 948
secure,
    computationally, 926
    provably, 926
security,
    ideal, 950
security, provable, 894
    unconditional, 895, 926
seed, 485
segment, 807
selection,
    ordered, with repetition, 100
    ordered, without repetition, 96
    unordered, with repetition, 100
    unordered, without repetition, 97
selection sort, 1107, 1127
    algorithm, 1129

self-complementary, 507, 516
self-contradiction, 15
self-dual code, 895
self-dual linear code, 904
self-dual poset, 722, 729
self-generating sequences, 156
self-loop, 504, 507, 510
self-orthogonal code, 895, 904
self-synchronizing stream cipher, 895, 933
semantic axiom, 2, 10, 62
semantics, 9, 10
semi-homogeneous, 606, 615
semi-Markov process, 1014
semi-space, 804, 808
semigroup, 304, 305, 306
    free, 317
    λ-free, 1067
semimodular lattice, 722
semiorder, 722, 742
semiorder representation, 742
semiregular polyhedron, 803, 827
semiregular tiling, 804, 827
semistandard Young tableau (SSYT), 129
sensitivity analysis, 976
sentence, 10, 19
separable extension, 304, 334
separable polynomial, 304, 334
separated set, 835
separation, 982
sequence(s), 10, 36, 138, 139
    see also *number*
    k-norm of, 721, 738
    list of, 158
    miniguide to, 157
    probe, 1106
    representations of, 138
    sparse, 1107, 1119
    super-increasing, 895, 936
sequence of, 1107, 1109
sequential sampling, 458
series system, 431, 461
service discipline, 477
service-time distribution, 431, 477
session key, 934
set(s), 10, 21, 27
    adjacency, 630
    coloring of, 566
    difference of, 24
    k-, 804, 808

obstruction, 557
operations on, 23
partition of, 958
recursive, 1046
recursively enumerable, 1046
set cover, 958, 1003
set covering problem, 958
set equation, 24
set expression, 24
set identity, 24
set mapping, 1031
set of, 1107, 1109
set paradox, 22
set partitioning (SP) problem, 958
set theory, 21
  axioms, 29
set-up cost, 688
seven bridges of Königsberg, 544
shadow, 722
shakersort, 1107, 1130
Shamir's secret sharing scheme, 948
Shannon, Claude Elwood, 1170
Shannon's noiseless coding theorem, 899
Shannon's noisy channel coding theorem, 903
shape, of a Young tableau, 129
Shapley value, 959, 1025
Sheffer stroke, 13
Shell, D.L., 1128
Shellsort, 1107, 1128
Sherman-Morrison identity, 392
shift cipher, 895, 928
shift operator, 138, 189
shift left operation, 384
shift register, m-stage, 920
shift strategies, 413
shift up operation, 384
shortened code, 895, 908
shortest path, 632, 653
shortest path distances, 653
shortest path tree, 653
shrunken blossom, 647
shuffle, 1066, 1069
sibling, 606, 609
  left, 609
  right, 609
side, of a latin square, 778
sieve, of Eratosthenes, 111, 218, 246
sieve principle, 433
Σ_n^p, 1086

sigma expression, 195, 195
sign, 304, 319, 806, 807
signature, digital, 892, 925
signature part, 945
signed basis, 818
signed double covering, 818
signed permutation, 820
signed set, 818
SIMAN, 491
similar hypergeometric series, 205
similar hypergeometric terms, 138
similar, 377, 360
simple digraph, 507
simple graph, 507, 510
simple group, 304, 316
  order of, 165
simple hypergraph, 507, 595
simple matroid, 758, 787
simple path, 431, 461
simple proposition, 13
simple random walk, 453
simple substitution cipher, 928
simple t-design, 758, 766
simple transposition cipher, 928
simple Venn diagram 24
simple Vigenère cipher, 929
simplex, 804, 807, 842, 1029
  d-, 842
  d-dimensional, 804, 842
  distinguished, 1029
  p-, 959
  unit, 970
simplex algorithm, 968
simplex method, 959, 967
  goal programming, 980
simplicial complex, 804, 842, 1029
simplicial graph, 507, 510
simplicial notation, 606
simplicial polytope, 804, 843
simplicial subdivision, 959, 1029
simplicity property, 1140
SIMSCRIPT, 490
SIMULA, 490
simulation, 431, 484, 1057
SIMulation ANalysis (SIMAN), 491
simulation language, 490
Simulation Language for Alternative Modeling (SLAM), 491
single-elimination competition, 537
single-station, 431, 477

singleton, 10, 21
singleton bound, 915
singular matrix, 360
singular value, 360, 407
singular value decomposition, 409
sink, 507, 536
sinking sort, 1107, 1130
Sister Celine's algorithm, 207
size, 511, 632, 641, 722, 725, 818, 1107, 1118, 1141
skeleton,
   1-, 518
   p-, 1029
skew field, 304
skew-Hermitian, 360, 379
skew-symmetric, 360, 379
slack variables, 959, 965
SLAM, 491
SLAM II, 1014
Sloane's On-line Encyclopedia of Integer Sequences, 158
slope, 811
slope order of the pseudolines, 822
smallest element, 342
smooth number, 218, 255
S-norm operator, 71
social choice function, 744
social network, 513
soft decision decoder, 918
Sollin's algorithm, 637
solution,
   of a congruence, 234
   to Fermat equation, 284
   in a game, 959
      optimal, 960
   of a recurrence relation, 139
   of a system of linear equations, 393
   of a system of word equations, 1076
solvable, 316, 1046
   computationally, 1065
solvable by radicals, 337
solving polynomial equations, 335
sort,
   adaptive, 1130
   base-ten radix, 1135
   bubblesort, 1130
   comparison, 1041, 1083, 1103, 1126
   divide-and-conquer, 1132
   insertion, 1104, 1128
   mergesort, 1105, 1132

   quicksort, 1106, 1132
   radix, 1106, 1135
   rank counting, 1106, 1135
   selection, 1107, 1127
   shakersort, 1107, 1130
   Shellsort, 1107
   sinking, 1107, 1130
   topological, 536
sorted assignment, 614
sorting, 748, 1126
sorting algorithm(s), 1107
   external, 1104, 1126
   internal, 1104, 1126
   table of, 1126
source, 507, 536, 663, 898
   memoryless, 894
source encoder, 896
SP problem, 1004
space,
   ciphertext, 923
   key, 923
   message, 923
   taken by a Turing machine, 1086
space complexity, 1046, 1078
   polynomial, 1086
span, 360, 365
spanning set,
   of a matroid, 758, 787
   of a subset of a vector space, 365
   of a vector space, 360
spanning subgraph, 507, 514
spanning tree(s), 606, 616
   capacitated minimum, 684
   enumeration of, 620
   Euclidean minimum, 1084
   maximum, 632, 633
   minimum, 606, 632, 633
spanning tree solution, 676
sparse, 360, 378, 402, 1086
sparse certificate, 1107, 1144
sparse language, 1046
sparse sequence, 1107, 1119
sparsification, 1145
special Burnside's lemma, 124
specification, 10, 30, 507, 519
spectral radius, 360, 410
spectral theorem for normal matrices, 408
spectrum, 507, 586
Sperner family, 729

Sperner property, 723, 738
  k-, 723, 739
Sperner's lemma, 598, 1028, 1029
Sperner's theorem, 732
sphere-packing bound, 915
spinner, 432
split, 334
splitting field, 304, 334
sporadic group, 316
square, 1046, 1075
  abelian, 1040, 1076
square matrix, 360
square root, 218
  of a modulo m, 280
square-free, 1075
square-free integer, 218, 221
square-free word, 1046
stability condition, 431
stable, 476, 479, 645
stable certificate, 1107, 1144
stable function, 1107, 1144
stable marriage problem, 645
stack, 1107, 1112
stand-alone systems, 879
standard affine polar-duality, 812
standard affine pseudo polar-duality, 804
standard basis, 367
standard k-chain, 723, 729
standard deviation, 444
standard Euclidean norm, 400
standard form, 778, 904, 970
standard inner product, 368
standard line arrangement, 811
standard normal distribution, 444
standard poset (n-dimensional) example, 723, 736
standard real vector space, 806
standard Young tableau (SYT), 129
standardized form, 138, 195
star, 820
  Kleene, 1043, 1067
star polygon, 851
start symbol, 1046, 1070
starting configuration, 1056
starting point, 859
starting state, 1048, 1051, 1053, 1055
state, 459, 1048, 1053, 1055
  acceptance, 1048
  implicit trapping, 1050
  recurrent, 430

  starting, 1048, 1051
  transient, 431
  trapping, 1047, 1050
state diagram, 1046, 1049, 1050, 1054
statement, 19
statement form, 10, 50
static structure, 1107, 1127
steady-state distribution, 431, 479
steganography, 924
Steiner quadruple system, 766
Steiner system, 766
Steiner triple system, 758, 759, 766
Steinitz exchange axiom, 723, 737
stereographic projection, 572
Stirling, James, 1170
Stirling approximation, 138, 202
Stirling cycle number, 83, 117, 192
  exponential generating function, 140
  recurrence relation, 141
Stirling number of the first kind, 83, 117
  associated, 157
Stirling number of the second kind, 83, 117, 162
  associated, 157
Stirling subset number, 83, 117, 162, 192, 735
  exponential generating function, 140
  recurrence relation, 141
  sums of powers, 200
stochastic matrix, 469
stochastic process, 431, 453
stock market prices, 459
straight-line drawing, 573
straight-line dual, 804, 848
straight-line graph, 802
straightedge and compass construction, 337
Strassen algorithm, 189, 386
strategic behavior, 959
strategic form game, 1017
strategy,
  mixed, 1022
  optimal, 1022
  pure, 1017
strategy combination, 959
stream cipher, 895, 933
  self-synchronizing, 895, 933
  synchronous, 895, 933
strength, 785
stretchable pseudoline arrangement, 816

strict digraph, 507, 511, 526
strict Sperner property, 723, 738
strictly convex, 804, 830
strictly diagonally dominant matrix, 360, 410
strictly totally positive matrix, 360, 415
string(s), 10, 36, 1046, 1066
   accepted by a finite-state recognizer, 1050
   accepted by a PDA, 1046
   accepted by a TM, 1046
   over an alphabet, 1047
   concatenation of, 1041
   empty, 1042, 1066
   length of, 1066
   nth power of, 1066
strong certificate, 1107, 1144
strong circuit elimination, 790
strong collision-resistance, 949
strong component, 507, 532
strong duality theorem, 974
strong orientation, 507, 534
strong probable prime test, 254
strong pseudoprime to the base b, 251
strong Sperner property, 723, 739
strong tournament, 507, 537, 583
strongly connected, 507, 531, 583
strongly correct code, 10, 66
strongly regular graph, 507, 586
structure function, 431, 459
subdesign, 758, 763
subdivision,
   of an edge, 507
   of a graph, 508, 515
   simplicial, 1029
subfield, 304, 332
subgame, 1017
   proper, 1017
subgame perfect, 1018
subgraph, 508, 514
   edge-deleted, 555
   induced, 514
   spanning, 514
   vertex-deleted, 555
subgroup, 304, 311
   generated by elements, 304
sublattice, 341, 723, 734
submatrix, 360, 378
submodular height function, 723, 736
submonoid, 306

suboptimal tableau, 976
subposet, 723, 726
subring, 304, 324
subsemigroup, 306
subset, 10, 21
subset axiom, 30
subset lattice, 723, 734
subset sum problem, 936
subspace, 360, 363
   spanned by vectors, 365
subspace lattice, 723, 734
subsquare, 758, 780
substitution, 10, 68, 1047, 1069
   extension of, 1069
   finite, 1069
   homophonic, 929
   λ-free, 1069
substitution cipher, 895
substitution rule, 50
substring, 1066
subtraction,
   in a ring, 323
   of vectors, 361
subtree, 606, 684
   left, 609
   principal, 609, 710
   right, 606, 609
subword, 1066
success amplification method, 1092
successive shortest path algorithm, 677
success system, k-out-of-n, 462
successor function, 27, 1063
succinct certificate, 1087
suffix, 1066
sum bit, 348
sum,
   of cardinals, 27
   of divisors function, 261
   of linear transformations, 373
   of matrices, 360
   of ordinals, 27
   of powers, 199
   of subspaces, 360, 363
   3-, 793
   2-, 793
sum rule, 90
sum-of-products expansion, 346
summation, 195
summation formulas, 175
summation variable, 195

summatory function, 259
Sun Zi, 1170
super-increasing sequence, 895, 936
supply vertex, 673
support of a signed set, 818
surface,
   chromatic number of, 564
   nonorientable, 505, 564, 575
   nonorientable with crosscap number k, 505
   orientable, 505, 564, 575
   orientable of genus g, 506
surjection, 10, 31
surjective, 31
survivable network, 632
Sylow subgroup, 315
Sylow p-subgroup, 304, 315
Sylow theorem, 316
Sylvester, James Joseph, 1170
Sylvester's law of inertia, 416
Sylvester-Gallai theorem, 833
symbol, 1066
symmetric block design, 758
symmetric chain, 597, 723, 738
symmetric chain decomposition, 723, 738
symmetric chain order, 723, 738
symmetric channel, q-ary, 899
symmetric closure, 43
symmetric design, 771
symmetric difference,
   of relations, 11, 41
   of sets, 11, 24
symmetric game, 1022
symmetric group, 304, 310, 319
symmetric latin square, 779
symmetric matrix, 360, 379
symmetric partial latin square, 780
symmetric random walk, 454
symmetric relation, 11
symmetric-key, 923
symmetric-key systems, 895
symmetry,
   of a binary relation 41
   Euclidean motion 828
   of a figure, 83, 121
   of a tiling, 804
symmetry breaking, 1092
symmetry group, 828
synchronous stream cipher, 895, 933
syndrome, 895, 906

syndrome decoding, 907
   algorithm, 907
syndrome polynomial, 909
system,
   of cycles, fundamental, 605, 617
   of distinct representatives, 11, 24, 596, 671
   of edge-cuts, fundamental, 605, 617
   of linear equations, 393
   symmetric-key, 895
system reliability, 459
   calculating, 464
systematic code, 895, 904

T

table, 1107, 1115
   transition, 1047
   of values, 35
tableau, 959, 964
   final, 976
   infeasible, 976
   original, 976
   suboptimal, 976
tableaux algorithms, 131
tabular form, 120
tail, 508, 511, 526, 1103
tangent numbers, 138, 153, 164
   exponential generating function, 140
target, 1107
target predicate, 1118
tautologies problem, 1089
tautology, 11, 15
techniques of proof, 54
telescoping, 196
"ten most wanted numbers", 218, 259
term, 11, 36, 68, 139
term rank, 360, 418
terminal, 683, 688, 1047, 1070
terminal vertex, 606
terminating expansion, 218, 224
terminus, 508
ternary Golay code, 916
ternary matroid, 758, 787
terrain databases, 875
tetrahedral graph, 508, 518
textile part layout, 876
theorem, 11, 50
theta, 38

theta graph, 547
theta notation, 11
thickness, 508, 571
third-order logic, 19
threshold scheme, 948
   (n, k)-, 895
Thue's lemma, 234
tighter formulation, 981
tile, 804
tiling, 804, 827
   aperiodic, 869
   edge-to-edge, 827
   face-to-face, 801, 827
   homeomeric, 828
   isogonal, 801, 828
   isohedral, 828
   isotoxal, 801, 828
   lattice, 801, 828
   monohedral, 802, 827
   nonperiodic, 802, 828
   normal, 802, 827
   periodic, 802, 828
   regular, 803, 827
   semiregular, 804, 827
time complexity, 1047, 1078, 1086
time complexity function, 188
time reduction, $f(n)$, 1084
time taken by a Turing machine, 1086
time-homogeneous, 469
TM, 1047, 1055
T-norm operator, 71
tolerance synthesis, 875
tomography, 874
top of a stack, 1112
top-down mergesort, 1132
topological equivalence, 572
topological graph theory, 574
topological ordering, 723, 746
topological sort algorithm, 747
topological sort, 508, 536, 723, 746
topsort, 508, 536
torus, 508, 575
total function, 1047, 1063
total order, 723
total weight, 990
totally ordered, 7, 11 ,46, 725
totally unimodular, 959, 1004
totient function, 218, 260
tough, 508, 547
tour,

Euler, 544
   k-optimal, 693
tour construction, 693
tour improvement, 693
tournament, 508, 537, 583, 672
   irreducible, 503, 537
   n-, 537
   regular, 537
   scheduling, 538
   strong, 507, 537, 583
   strongly connected, 583
   transitive, 537
tower of Hanoi, 94, 185
trace, 360, 378
trace-driven, 485
tractable, 1047, 1078
traffic intensity, 431
trail, 508, 539
   closed, 498
   directed, 500, 508, 531
   Euler, 544
trajectory, 431, 454
transcendental element, 300, 304, 329, 333
transcendental extension, 333
transcendental number, 218, 219
transducer, deterministic finite, 1051
transformation rules for sums, 196
transformation,
   Las Vegas to Monte Carlo, 1043
   Monte Carlo to Las Vegas, 1044
transient, 472
transient analysis, 471
transient random walk, 453
transient state, 431
transient walk, 431
transition, 1050, 1053
   λ-, 1047, 1050
transition diagram, 469
transition function, 1048, 1051, 1055
transition matrix, 377, 431
transition probability, 431
   n-step, 471
transition relation, 1050, 1053
transition table, 1047, 1049, 1056
transitive, 11, 41, 508, 725
transitive closure, 11, 43
transitive digraph, 528
transitive orientation, 508, 534, 723, 749
transitive reduction, 11, 42

transitive relations, number of, 165
transitive tournament, 537
translate, 777
transmitter, 508, 537
transpose, 360, 378, 779
transpose, properties of, 382
transposition, 304, 319
transposition cipher, 895
transversal, 508, 597, 779, 831
transversal design, 758, 768
transversal matroid, 758
transversal number, 508, 599
trapping group, 1050
trapping state, 1047, 1050
traveling salesman problem (TSP), 632, 692, 1081
 asymmetric, 693
traversal, 613, 758
 inorder, 605, 613
 postorder, 606, 613
treatment, 759
tree(s), 508, 517, 606, 607, 622
 m-ary, 609
 AVL, 1102, 1136
 balanced, 609
 bihomogeneous, 604, 615
 binary, 604, 627
 binary search, 604, 610, 1102, 1136
 bounded, 604, 615
 branch and bound, 997
 breadth-first-search, 617
 capacitated minimum spanning, 684
 and chemistry, 625
 complete m-ary, 609
 complete binary, 604
 and computer science, 627
 decision, 604, 610
 depth-first-search, 616
 edge-ordered, 1146
 enumeration of, 622
 Euclidean minimum spanning, 1084
 finite, 605
 full m-ary, 605, 609
 generating function for counting, 623
 homogeneous, 605, 615
 Huffman, 610
 infinite, 605, 615
 isomorphism, 605
 K-, 429, 461
 labeled, 605, 622
 left-complete binary, 1105, 1121
 left-right, 605, 627
 locally finite, 605, 615
 m-ary, 604, 609
  complete, 609
 maximal, 606
 maximum spanning, 633
 minimal spanning, 606, 633
 number of, 169
 number of, by diameter, 169
 number of, by height, 169
 1-4, 625
 1-rooted 1-4, 625
 ordered, 606, 609, 627
 plane, 609
 priority, 606, 1106, 1122
 pure binary, 609
 reduced, 606, 622
 rooted, 606, 608, 622
 semi-homogeneous, 606, 615
 shortest path, 653
 spanning, 606, 616
 2-3, 1107, 1136
tree diagram, 84, 91
tree edge, 606, 616
tree structure, 1108, 1120
 binary, 1120
tree traversal, 606
trial division, 218, 255
triangle inequality, 692
triangular chord, 534, 723, 749
triangular numbers, 168
triangulated with Steiner points, 842
triangulation, 146, 804, 842, 846
trichotomy, 30
trickle-down, 1122
trickle-up, 1122
tridiagonal, 360, 379
trio, 1047
trivial difference set, 777
trivial graph, 508, 511
trivial relator, 317
truth table, 11, 12, 13, 72
TSP, 632, 692
Tukey, John Wilder, 1171
tuple of, k- 1108, 1109
Turán graph, 508, 590
Turán number, 590
Turán's theorem, 590
turbo code, 895, 919

INDEX   1229

Turing, Alan M., 1059, 1171
Turing halting problem, 60
Turing machine accepted language, 1057
Turing machine accepted string, 1057
Turing machine (TM), 1047, 1055
    alternate definition, 1056
    $m \times n$, 1055
    nondeterministic, 1044, 1058, 1085
    oracle, 1044
    probabilistic, 1045, 1089
    universal, 1048, 1057
Turing's test, 1047, 1059
Turing-acceptable, 1057
Turing-acceptable language, 1047
Turing-computable, 1057
Turing-computable function, 1047
Turing-decidable language, 1047, 1057
$\mathcal{C}$-Turing-p-complete, 1088
$\mathcal{C}$-Turing-p-hard, 1088
Turing-p-reducible, 1047, 1088
twin primes, 219, 237, 248
two's complement, 219, 224, 448
two-phase, 632, 700
two-sided error, 1089
two-terminal network, 431, 463
two-terminal reliability, 431, 465
two-valued logic, 11, 71
two-way incidence structure, 1108, 1124
two-way linked list, 1108, 1119
two-way sequential list, 1108, 1114
type, 756, 758, 768
type 0 grammar, 1047, 1070
type 1 grammar, 1042, 1043, 1047, 1071
type 2 grammar, 1041, 1047, 1071
type 3 grammar, 1046, 1047, 1071

U

ultimately periodic, 219, 290
unambiguous context-free language, 1047, 1074
unary, 32, 304, 305
unavoidability, 1076
unbiased coin, 432
unbounded, 960
unbounded LP, 959
unbounded minimalization, 1047, 1063
uncertainty, 897
uncomputable function, 1047

unconditional security, 895, 926
uncountable, 27
undecidable problem, 1047
underflow, 431, 450
underlying graph, 508, 526
undirected network, 632, 633
unforced pair, 746
unicity distance, 926
uniform, r-, 506
uniform chirotope, 804, 819
uniform cluster of stars, 820
unform distribution, 444
uniform hypergraph, 508, 596
uniform matroid, 758
uniform random variable, 431
uniform, rank 3 oriented matroid, 821
uniformity property, 1140
unilateral, 508, 532
unilateral authentication, 943
unilaterally connected, 508, 532
unimodular, 598
union, 11, 23, 30, 41, 515, 1067
union relation, 11
union-find, 1108, 1112
unique existential quantification, 18
unique existential quantifier, 18
unique factorization property, 296
unique factorization theorem, 236, 330
uniquely decodable, 895, 898
unit, 219, 295, 304, 323
unit distance graph, 835
unit simplex, 970
unit triangular matrix, 360, 379
unit vector, 368
unitary matrix, 360, 380
unity, 305, 323
universal address system, 614
universal domain, 11
universal instantiation, 51
universal quantifier, 11, 18
universal relation, 44
universal set sequence, 740
universal Turing machine, 1048, 1057
universally quantified predicate, 11
universe of discourse, 11, 18
unordered selection,
    with replacement, 84, 100, 175
    without replacement, 97
unreasonable imputation, 1025
unrelated, 725

unreliability, 459
unrestricted grammar, 1070
unsolvable, 1057
unstable, 476
unweighted Euclidean 1-center problem, 993
up-set, 723, 732
update operation, 1143
upper bound, 11, 47, 300, 342, 723, 733
upper bound theorem, 809
upper bandwidth, 402
upper density,
  of a covering, 804, 826
  of a packing, 804, 824
upper envelope, 804, 811
upper extension, 746
upper semimodular lattice, 723, 736
upper triangular matrix, 360, 379
upset, 508, 597
upward separation, 1087
urn models, 438
U. S. Data Encryption Standard (DES), 930
utility, 994

V

valence, 508, 511
valid argument form, 11, 50
value, of a function, 31
  of a flow, 632, 663
  objective, 960
  of a polynomial, 329
van der Waerden-Egorychev-Falikman inequality, 423
Vandermonde, Alexandre-Théophile, 1171
Vandermonde convolution, 104
  identity, 174
Vandermonde matrix, 388
variable, 1048, 1075
variance, 431, 443, 488
variance reduction technique, 431, 488
variety, 759
vector, 360, 361, 904
vector composition, 114
vector configuration, 804, 807
vector matroid, 759
vector partition, 114

vector space, 360, 361
  standard, 806
vehicle, 700
vehicle routing problem, 633, 700
Venn, John, 24
Venn diagram, 11, 24
verification, 11
Vernam cipher, 895, 933
vertex, 509, 831, 840, 960
  active, 666
  destination, 689
  end, 607
  equivalent, 419
  even, 631, 647
  of a feasible region, 959
  inner, 647
  internal, 605, 609
  isolated, 540
  matched, 631
  odd, 632, 647
  origin, 689
  outer, 647
  reachable, 540
  supply, 673
  terminal, 606
vertex chromatic number, 557
vertex coloring, 509, 557
vertex connectivity, 509, 542
vertex connectivity requirement, 691
vertex cover, 672, 1080
vertex cut, 509
vertex invariant, 509, 550
vertex potential, 633
vertex-deleted subgraph, 509, 555
vertex-format, 578
Viergruppe, 321
Viète, François, 1171
Vigenère cipher, 895
  simple, 929
virtual key, 1140
visibility, 873
visibility graph, 805, 851
visibility map algorithm, 861
visibility problem, 805, 851
visible vertices, 847
visualization, geometric, 880
Viterbi algorithm, 921
Vizing's general theorem, 561
Vizing's theorem, 561
void design, 759

voltage graph, 522
Voronoi cell, 805, 824
Voronoi diagram, 805, 848, 871
VRP, 633, 700
VSLI layout, 876

W

waiting time, 479
waiting-time process, 431
walk, 509, 539
   boundary, 613
   closed, 498
   directed, 500, 509, 531
   recurrent, 430
   reduced, 606, 613
   transient, 431
Waring's problem, 219, 288
Waring, Edward, 1171
Warshall algorithm, 43
weak, 509, 532
weak collision-resistance, 949
weak duality theorem, 664, 974
weak excedance, 148
weak order, 724, 742
weakly connected, 509, 532
weakly correct, 66
weakly correct code, 12
weakly γ-perfect, 563
Weber problem, 993
Wedderburn's theorem, 328
Weibull distribution, 487
weight, 641, 652, 987
   Hamming, 893, 904
   of a matching, 633
   total, 990
weight distribution, 914
weight enumerator, 914
weighted graph, 509, 510
well-conditioned system, 360
well-defined, 31
well-formed formula, 12, 19
well-ordered set, 12, 47
well-ordering principle, 12, 30, 58
Weyl's theorem, 416
wff, 12, 19
wheel graph, 509
while-condition, 65
while-loop instruction, 65

Whitehead, Alfred North, 23
Whitney, Hassler, 1172
Whitney number, 735
   kth, 724, 731
Whitney's 2-isomorphism theorem, 788
width, 724, 732
Wiles, Andrew, 1158
Wilson's asymptotic existence theorem, 760
Wilson's theorem, 233
windowing systems, 873
witness, 251
Woodbury identity, 392
word(s), 305, 317, 898, 904, 1048, 1066
   code indicator of, 1076
   R-equivalent, 318
   square-free, 1046
word equation, 1048, 1076
word inequality, 1048, 1076
word problem, 318
word variable, 1075
Worpitzky's identity, 149
worst-case complexity, 1081
worst-case running time, 1048
wraparound, 448
write instruction, 1060
WZ certificate, 209
WZ mate, 209
WZ pair, 209
WZ proof, 209

X

XOR, 12, 344
XYZ Geobench, 879

Y

yields the configuration, 1050, 1054, 1056
Young lattice, 724, 734
Young tableau, 84, 129

Z

Zech logarithm, 338
Zeller's congruence, 234
Zermelo-Fraenkel axioms, 12, 30

zero, 305
  of a polynomial, 329
zero divisor, 327
zero polynomial, 303, 329
zero-based array, 1118
zero-order logic, 12
zero-sum game, 1022

zeroth Betti number, 542
zeta function, 248
zone, 805, 811
zone theorem, 813
zonotope, 805, 869
Zorn's lemma, 30
ZPP, 1089